LIVERPOOL JMU LIBRARY

3 1111 01513 1673

Supramolecular Chemistry:
From Molecules to Nanomaterials

EDITORIAL BOARD

Editors-in-Chief

Philip A. Gale
University of Southampton, Southampton, UK

Jonathan W. Steed
Durham University, Durham, UK

Section Editors

David B. Amabilino
Institut de Ciència de Materials de Barcelona (CSIC), Catalonia, Spain

Eric V. Anslyn
The University of Texas at Austin, Austin, TX, USA

Pavel Anzenbacher, Jr
Bowling Green State University, Bowling Green, OH, USA

Bradley Smith
Department of Chemistry and Biochemistry, University of Notre Dame, Notre Dame, IN, USA

Leonard J. Barbour
University of Stellenbosch, Stellenbosch, South Africa

David K. Smith
University of York, York, UK

Enrique García-España
Universidad de Valencia, Valencia, Spain

Douglas Philp
University of St Andrews, St Andrews, UK

Jon A. Preece
School of Chemistry, The University of Birmingham, Birmingham, UK

Paula M. Mendes
School of Chemical Engineering, The University of Birmingham, Birmingham, UK

Marcey Waters
Department of Chemistry, The University of North Carolina at Chapel Hill, Chapel Hill, NC, USA

Edwin C. Constable
University of Basel, Basel, Switzerland

International Advisory Board

Jerry L. Atwood
University of Missouri, Columbia, MO, USA

Jonathan L. Sessler
The University of Texas at Austin, Austin, TX, USA
Yonsei University, Seoul, South Korea

Makoto Fujita
The University of Tokyo, Tokyo, Japan

Andrew D. Hamilton
Department of Chemistry, Yale University, New Haven, CT, USA

Ivan Huc
Université de Bordeaux, Bordeaux, France

Paul D. Beer
University of Oxford, Oxford, UK

Michaele J. Hardie
University of Leeds, Leeds, UK

Kate Jolliffe
School of Chemistry, The University of Sydney, Sydney, NSW, Australia

David A. Leigh
School of Chemistry, University of Edinburgh, Edinburgh, Scotland, UK

Tom Fyles
Department of Chemistry, University of Victoria, Victoria, BC, Canada

George Shimizu
Department of Chemistry, University of Calgary, Alberta, Canada

Kimoon Kim
Center for Smart Supermolecules (CSS), Pohang University of Science and Technology (POSTECH), Pohang, South Korea

Edwin C. Constable
The University of Basel, Basel, Switzerland

Yun-Bao Jiang
Stake Key Laboratory of Marine Environmental Science, Xiamen University, Xiamen, China

Wiley is grateful to the following sources for permission to use the images shown on the cover of this publication (clockwise from top):

An [18]crown-6 host-guest complex of calcium
Image supplied with kind permission of Professor Jonathan W. Steed and Professor Philip A. Gale

Hyperpolarised Xe exchange NMR spectrum showing exchange of guests within the cavity of 'gallium wheels'
Image taken from Volume 2, NMR Spectroscopy in Solution
Reprinted with permission from C-Y. Cheng, T. C. Stamatatos, G. Christou and C. R. Bowers, *J. Am. Chem. Soc.*, 2010, **132**, 5387. Copyright 2010 American Chemical Society

Sulfate binding by a diindolylurea anion receptor
Image adapted from Volume 3, Podands, with kind permission of the authors.

DNA
Image reproduced from Volume 4, Nucleic Acid Mimetics, with kind permission of the authors.

An octahedral hexammine coordination complex
Image adapted from Volume 5, Self-Assembly of Coordination Compounds: Design Principles, with kind permission of the authors.

Interactions to an amide probed using the Cambridge Structural Database
Image adapted from Volume 6, The Cambridge Structural Database System and Its Applications in Supramolecular Chemistry and Materials Design, with kind permission of the authors.

SEM image of a chiral supramolecular gel
Image supplied with kind permission of Professor Jonathan W. Steed and Professor Philip A. Gale

Adsorption of a molecular guest onto graphene
Image taken from Volume 8, Physisorption for Self-Assembly of Supramolecular Systems: A Scanning Tunneling Microscopy Perspective
Reprinted in part with permission from M. Linares, P.Iavicoli, K. Psychogyiopoulou, D. Beljonne, S. De Feyter, D. B. Amabilino, R. Lazzaroni, *Langmuir*, 2008, **24**, 9566. Copyright 2008 American Chemical Society

Supramolecular Chemistry:

From Molecules to Nanomaterials

Volume 3: Molecular Recognition

Editors-in-Chief

Philip A. Gale
University of Southampton, Southampton, UK

Jonathan W. Steed
Durham University, Durham, UK

WILEY

This edition first published 2012
© 2012 John Wiley & Sons, Ltd

Registered office
John Wiley & Sons Ltd, The Atrium, Southern Gate, Chichester, West Sussex, PO19 8SQ, United Kingdom

For details of our global editorial offices, for customer services and for information about how to apply for permission to reuse the copyright material in this book please see our website at www.wiley.com.

The right of the author to be identified as the author of this work has been asserted in accordance with the Copyright, Designs and Patents Act 1988.

All rights reserved. No part of this publication may be reproduced, stored in a retrieval system, or transmitted, in any form or by any means, electronic, mechanical, photocopying, recording or otherwise, except as permitted by the UK Copyright, Designs and Patents Act 1988, without the prior permission of the publisher.

Wiley also publishes its books in a variety of electronic formats. Some content that appears in print may not be available in electronic books.

Designations used by companies to distinguish their products are often claimed as trademarks. All brand names and product names used in this book are trade names, service marks, trademarks or registered trademarks of their respective owners. The publisher is not associated with any product or vendor mentioned in this book. This publication is designed to provide accurate and authoritative information in regard to the subject matter covered. It is sold on the understanding that the publisher is not engaged in rendering professional services. If professional advice or other expert assistance is required, the services of a competent professional should be sought.

The Publisher and the Author make no representations or warranties with respect to the accuracy or completeness of the contents of this work and specifically disclaim all warranties, including without limitation any implied warranties of fitness for a particular purpose. The advice and strategies contained herein may not be suitable for every situation. In view of ongoing research, equipment modifications, changes in governmental regulations, and the constant flow of information relating to the use of experimental reagents, equipment, and devices, the reader is urged to review and evaluate the information provided in the package insert or instructions for each chemical, piece of equipment, reagent, or device for, among other things, any changes in the instructions or indication of usage and for added warnings and precautions. The fact that an organization or Website is referred to in this work as a citation and/or a potential source of further information does not mean that the author or the publisher endorses the information the organization or Website may provide or recommendations it may make. Further, readers should be aware that Internet Websites listed in this work may have changed or disappeared between when this work was written and when it is read. No warranty may be created or extended by any promotional statements for this work. Neither the Publisher nor the Author shall be liable for any damages arising herefrom.

Library of Congress Cataloging-in-Publication Data

Steed, Jonathan W., 1969-
Supramolecular chemistry : from molecules to nanomaterials / Jonathan W. Steed, Philip A. Gale. – 1st ed.
p. cm.
Includes bibliographical references and index.
ISBN 978-0-470-74640-0 (hardback)
1. Supramolecular chemistry. I. Gale, Philip A. II. Title.
QD878.S75 2012
547'.1226–dc23

2011039272

British Library Cataloguing in Publication Data

A catalogue record for this book is available from the British Library

ISBN: 978-0-470-74640-0

Typeset in 10 / 12.5 pt Times by Laserwords (Private) Limited, Chennai, India.
Printed and bound in Singapore by Markono Print Media Pte Ltd.

Contents

VOLUME 1: CONCEPTS

VOLUME 2: TECHNIQUES

VOLUME 3: MOLECULAR RECOGNITION

VOLUME 4: SUPRAMOLECULAR CATALYSIS, REACTIVITY AND CHEMICAL BIOLOGY

VOLUME 5: SELF-ASSEMBLY AND SUPRAMOLECULAR DEVICES

VOLUME 6: SUPRAMOLECULAR MATERIALS CHEMISTRY

VOLUME 7: SOFT MATTER

VOLUME 8: NANOTECHNOLOGY

Contributors to Volume 3

Ignacio Alfonso
IQAC-CSIC, Barcelona, Spain

Laura Baldini
Università di Parma, Parma, Italy

Rowshan Ara Begum
University of Kansas, Lawrence, KS, USA

Raquel Belda
Universidad de Valencia, Valencia, Spain

Antonio Bianchi
University of Florence, Florence, Italy

Kristin Bowman-James
University of Kansas, Lawrence, KS, USA

Alessandro Casnati
Università di Parma, Parma, Italy

Christian G. Claessens
Universidad Autónoma de Madrid, Madrid, Spain

J. Grant Collins
UNSW@ADFA, Canberra, ACT, Australia

Matthew P. Conley
Institute of Chemical Research of Catalonia (ICIQ), Tarragona, Spain

Antonella Dalla Cort
Universitá "La Sapienza," Roma, Italy

Peter J. Cragg
University of Brighton, Brighton, UK

Anthony Ivan Day
UNSW@ADFA, Canberra, ACT, Australia

Victor W. Day
University of Kansas, Lawrence, KS, USA

Javier de Mendoza
Institute of Chemical Research of Catalonia (ICIQ), Tarragona, Spain

John S. Fossey
University of Birmingham at Birmingham, Birmingham, UK

Philip A. Gale
University of Southampton, Southampton, UK

Francisco Galindo
University Jaume I, Castellón, Spain

Enrique García-España
Universidad de Valencia, Valencia, Spain

Carolina Godoy-Alcántar
Universidad Autónoma del Estado de Morelos, Cuernavaca, México

Jorge González
Universidad de Valencia, Valencia, Spain

Michaele J. Hardie
University of Leeds, Leeds, UK

Cally J. E. Haynes
University of Southampton, Southampton, UK

Md. Alamgir Hossain
Jackson State University, Jackson, MS, USA

Tony D. James
University of Bath, Bath, UK

Elizabeth Karnas
The University of Texas at Austin, Austin, TX, USA

Stefan Kubik
Technische Universität Kaiserslautern, Kaiserslautern, Germany

William Levason
University of Southampton, Southampton, UK

Shim Sung Lee
Gyeongsang National University, Jinju, South Korea

Stephen F. Lincoln
University of Adelaide, Adelaide, South Australia, Australia

Leonard F. Lindoy
Gyeongsang National University, Jinju, South Korea

Stephen J. Loeb
University of Windsor, Windsor, Ontario, Canada

Santiago V. Luis
University Jaume I, Castellón, Spain

M. Victoria Martínez-Díaz
Universidad Autónoma de Madrid, Madrid, Spain

Ki-Min Park
Gyeongsang National University, Jinju, South Korea

Duc-Truc Pham
University of Adelaide, Adelaide, South Australia, Australia

Javier Pitarch
Universidad de Valencia, Valencia, Spain

Gillian Reid
University of Southampton, Southampton, UK

Francesco Sansone
Università di Parma, Parma, Italy

Elaine Sedenberg
The University of Texas at Austin, Austin, TX, USA

Jonathan L. Sessler
The University of Texas at Austin, Austin, TX, USA • Yonsei University, Seoul, South Korea

Bellam Sreenivasulu
National University of Singapore, Singapore, Singapore

Jonathan W. Steed
Durham University, Durham, UK

Adam N. Swinburne
Durham University, Durham, UK

Tomás Torres
Universidad Autónoma de Madrid, Madrid, Spain • IMDEA-Nanociencia, Madrid, Spain

Rocco Ungaro
Università di Parma, Parma, Italy

Rafik Vahora
University of Brighton, Brighton, UK

Julián Valero
Institute of Chemical Research of Catalonia (ICIQ), Tarragona, Spain

Ralf Warmuth
Rutgers, The State University of New Jersey, Piscataway, NJ, USA

Andrew J. Wilson
University of Leeds, Leeds, UK

Anatoly K. Yatsimirsky
Universidad Nacional Autónoma de México, México D.F., México

Foreword

Supramolecular chemistry has been defined as "chemistry beyond the molecule". It aims at constructing highly complex, functional chemical systems from components held together by intermolecular noncovalent forces. It has relied on the development of preorganized molecular receptors for effecting molecular recognition, on the basis of the molecular information stored in the covalent framework of the components and read out at the supramolecular level through specific interactional algorithms. Suitably functionalized receptors may display supramolecular reactivity and catalysis and selective transport processes.

A most basic and far-reaching contribution of supramolecular chemistry to chemical sciences has been the implementation of the concept of molecular information. It involved the storage of information at the molecular level; in the structural features; and its retrieval, transfer, and processing at the supramolecular level, through molecular recognition processes operating via specific spatial relationships and interaction patterns. Supramolecular chemistry has thus paved the way toward apprehending chemistry also as an information science.

The control provided by recognition processes allowed the development of functional molecular and supramolecular devices, defined as structurally organized and functionally integrated systems built from suitably designed molecular components performing a given action (e.g., photoactive, electroactive, and ionoactive) and endowed with the structural features required for assembly into an organized supramolecular architecture. Thus emerged the areas of supramolecular photonics, electronics, and ionics.

Beyond mastering preorganization and taking advantage of it, supramolecular chemistry has been actively exploring the design of systems undergoing self-organization, that is systems capable of spontaneously generating well-defined, organized supramolecular architectures by self-assembly from their components, under the control of molecular information processes. They operate as programed chemical systems and are of major interest for supramolecular science and engineering. They give access to advanced functional supramolecular materials, such as supramolecular polymers, liquid crystals and lipid vesicles as well as solid-state assemblies.

The implementation of 'programed' self-organizing systems amounts to performing self-organization by design.

It also provides an original approach to nanoscience and nanotechnology. In particular, the generation of well-defined, functional supramolecular architectures of nanometric size through self-organization represents a means of performing programed engineering and processing of nanomaterials. Technologies resorting to self-organization processes are, in principle, able to provide a powerful complement and/or alternative to nanofabrication and nanomanipulation procedures by making use of the spontaneous but controlled generation of the desired superstructures and devices from suitably instructed and functional building blocks. The long-range goal is to shift from entities that need to be made to entities that make themselves, that is, from fabrication to self-fabrication.

From another point of view, self-organization is, in principle, able to select the correct molecular components for the generation of a given supramolecular entity from a diverse collection of building blocks. It may thus take place with selection, by virtue of a basic feature inherent in supramolecular chemistry, that is, its dynamic character.

Indeed, supramolecular chemistry is intrinsically a dynamic chemistry in view of the lability of the interactions connecting the molecular components of a supramolecular entity and the resulting ability of supramolecular species to exchange their constituents. Such a dynamic character is also conferred to molecular chemistry when the molecular entity contains covalent bonds that may form and break reversibility, so as to allow a continuous change in constitution by reorganization and exchange of building blocks. Thus, supramolecular chemistry has also fertilized molecular chemistry, leading to the definition of a Constitutional Dynamic Chemistry on both the molecular and supramolecular levels. It takes advantage of dynamic diversity to allow variation and selection. It operates on dynamic constitutional diversity in response to either internal or external factors to achieve adaptation.

Supramolecular chemistry has progressed over the years along three overlapping phases. The first is that of molecular recognition and its corollaries, supramolecular reactivity, catalysis, and transport; it relies on design and preorganization and implements information storage and processing.

The second concerns self-assembly and self-organization, that is self-processes in general; it relies on design and implements programing and programed systems for controlling the generation of specific entities in complex mixtures.

The third concerns constitutional dynamics of both molecular and supramolecular entities, defining a constitutional dynamic chemistry as a unifying concept. It relies on self-organization with selection in addition to design, and leads to the emergence of adaptive and evolutive chemistry.

Since it has been named in 1978 by the undersigned, about 10 years after the seed had been planted, the field of supramolecular chemistry has experienced a spectacular growth at the triple meeting point of chemistry with biology and physics. Its concepts and the perspectives it opens have been delineated, attracting scientists with a wide range of expertise. It has given rise to numerous review articles and special issues of journals and books. The present monumental work comes very timely. It provides thorough reviews and discussions, covering a broad range of topics, authored by many of the major players in the field. It takes stake and opens perspectives to the creative imagination of all participants in our common adventure.

I would like to very warmly congratulate and thank the editors and the contributors alike for this precious gift to the science of chemistry.

Jean-Marie Lehn
July 2011

Preface

Over the past decade, there have been tremendous advances in our understanding of the way in which chemical concepts at the molecular level build up into materials and systems with fascinating, emergent properties on the nanoscale. Creating that link between the chemist's understanding of the way in which molecules interact with one another and the understanding a materials scientist, engineer, or biologist has of the resulting properties of a material or system composed of those molecules is one of the huge, grand challenges facing modern molecular science. This vision of a molecular-level approach to complex systems and materials is the underlying drive for this project.

In 1996, the impressive *Comprehensive Supramolecular Chemistry* was published. This substantial 11-volume work summarized all of the major systems studied in fields based in supramolecular chemistry since its inception in clathrate chemistry in the early nineteenth century and cation receptor chemistry in the mid-1960s. In the 15 years since, the field has blossomed enormously and supramolecular concepts have become much more integrated into modern science underlying many areas that are based, fundamentally, on molecules. In attempting to capture and catalyze that continuing development, we have adopted a very different vision for this project. We aim to produce an enmeshed overview of the concepts and techniques of modern supramolecular chemistry and show, based on fluent chapters by the leading international experts, how these paradigms evolve seamlessly into nanoscale systems chemistry and materials science, and of course beyond. The scope and coverage has been carefully designed by the Editorial Advisory Board and the Editors of the 10 sections to avoid mere summative descriptions and instead to produce an interlocking series of tutorial-style articles that guide advanced students and veteran practitioners swiftly to the key science and techniques used in addressing modern supramolecular and nanoscale chemistry. We and the Board have taken particular care to try to break down the barriers between synthetic chemistry and materials science and show how modern techniques allow access increasingly far along the "synthesising up" pathway. We hope that this conceptual basis and forward-looking narrative is useful and complements the fascinating descriptions of earlier work published in 1996.

The origins of this work lie in a very successful "fun day" of science organised by Thorri Gunnlaugsson at the Trinity College, Dublin, in 2008, and we thank Thorri for being the initial catalyst and such a tremendous host. A large subsection of the editorial and advisory boards met at the *International Symposium for Supramolecular and Macrocyclic Chemistry* at Maastricht, Netherlands, in 2009, and between them defined the scope and structure of the project. The 10 editors then translated these concepts into a detailed vision for their own sections. We are hugely grateful to everyone on the advisory and editorial boards who gave their time, energy, and reputations to this project. Their belief has been invaluable, and as editors-in-chief, we feel that the result has vindicated their commitment. Our greatest debt goes, of course, to the authors themselves who have had the hugely challenging task of translating this concept-based vision into reality in the 169 individual chapters, each one a significant scientific product in its own right. We feel they have done an excellent job and salute their fidelity to the project's values.

We would also like to express our tremendous gratitude to Paul Deards at Wiley who believed in us all through this idea and brought it to reality. The project would also have come to nothing without the mountain-moving organizational skills of Stacey Woods and Anne Hunt at Wiley who have worked tirelessly to keep the momentum moving forward and "herd cats" as well as bring the book to the standard and accessibility it needs to have. J. W. S. is very grateful to the Durham University for providing two terms of research leave, which made this project and the travel it needed much easier to achieve, and we are both as ever indebted to the many fine coworkers who have passed through our laboratories over the years who make chemistry such an enjoyable subject to work in. P. A. G. thanks Nittaya for her love and support. J. W. S. would like to offer an ongoing thanks to his partner Kirsty, an ever-present source of wisdom and voice of sense.

Philip A. Gale
Southampton, UK

Jonathan W. Steed
Durham, UK

March 2011

Abbreviations and Acronyms

17-AAG	17-Allylamino-17-demethoxygeldanamycin
2-AP	2-Aminopyrimidine
9-ap	Anthracene-9-propionic Acid
4-bn-res	4-Benzylresorcinol
4,4′-bpe	*Trans*-1,2-bis(4-Pyridyl)Ethylene
bpea	1,4-bis(4-Pyridyl)Ethane
1,4-bpeb	1,4-bis[2-(4-Pyridyl)Ethenyl] benzene
bpee	1,4-bis(4-Pyridyl)Ethene
1,4-bpef	*p*-Di-[2-(4-Pyridyl)Ethenyl]-2-fluorobenzene
bpp	1,3-bis(4-Pyridyl)Propane
1,5-bppo	Bis(4-Pyridyl)-1,4-pentadiene-3-one
2-CH_3THF	2-Methyltetrahydrofuran
4-Cl dpcb	*Rctt*-1,2-bis(4-Pyridyl)-3,4-bis (*p*-Chlorophenyl)Cyclobutane
4-Cl stilbz	4-Chlorostilbazole
1D	One-Dimensional
2D	Two-Dimensional
3D	Three-Dimensional
9EA	9-Ethyladenine
6HB	Six-Helix Bundle
8HB	Eight-Helix Bundle
2,3-nap	2,3-bis(4-Methylenethiopyridyl)-Naphthalene
1,8-nda	1,8-naphthalenedicarboxylic Acid
4-pa	(*E*)-3-(4-Pyridyl)Acrylic Acid
6PE	Sixfold Phenyl Embraces
6PGL	6-Phosphogluconoylation
4-py-but	*Trans*-1,4-(4-Pyridyl)-1,3-butadiene
4-py-hex	*Trans*-1,6-(4-Pyridyl)-1,3,5-hexatriene
4-vp	4-Vinylpyridine
AAO	Anodized Aluminum Oxide
ABC	Adenosine-5′-triphosphate Binding Cassette
ABZ	Albendazole
AC	Alternating Current
ACA	Acetoxychavicol Acetate
acac	Acetylacetonate
ACE	Affinity Capillary Electrophoresis or Angiotensin Converting Enzyme
ACh	Acetylcholine
ACHC	Aminocyclohexanecarboxylic Acid
AChE	Acetylcholine Esterase
AcOH	Acetic Acid
ACR	Aza-Crown Resorcinarene
aCTG	Triamino Cyclotriguaiacylene
ACU	Undecyl-Aza-18-crown-6
AD	Acceptor Dendrimers or Activating Domain
Ad	Adamantane
Ad-PEG	Ad-Modified Polyethylene Glycol
ADA	Acceptor Donor Acceptor
ADMET	Acyclic Diene Metathesis
ADP	Adenosine 5′-Diphosphate
ADR	Adriamycin
ADV	Adenovirus
aeg	*N*-(2-Aminoethyl)-Glycine
AEM	Arylene Ethynylene Macrocycle
AFM	Atomic Force Microscopy
AFP	Alpha-Fetoprotein
AgNP	Silver Nanoparticle
AHX	ε-aminohexanoic Acid
AIBN	2,2′-Azobis(Isobutyronitrile)
AIEE	Aggregation-Induced Enhanced Emission
AK	Attenuated *K*
ALD	Atomic Layer Deposition
ALK	Ala-Leu-Lys-Arg-Gln-Gly-Arg-Thr-Leu-Tyr-Gly-Phe
ALP	Alkaline Phosphatase
ALP	Amphiphilic Lipopeptide
AM1	Austin Model 1
AM1.5	Air Mass 1.5
AMF	Alternating Magnetic Field
AMFE	Anomalous Mole Fraction Effect
amm-stilb	Bis(Dialkylammonium)-Substituted Stilbene
AMP	Adenosine 5′-Monophosphate
AMT	Amitriptyline Hydrochloride
ANB	5-Azido-2-nitrobenzoic Acid Chloride
ANN	Artificial Neural Network

ANTS — 8-Aminonaphthalene-1,3,6-trisulfonate
AO — Atomic Orbital
AP — Aptamer–Photosensitizer
APC — 2,4-bis(4-Dialkylaminophenyl)-3-Hydroxy-4-alkylsulfanylcyclobut-2-enone
APED — Alternating Polyelectrolyte Deposition
API — Active Pharmaceutical Ingredient
aPP — Avian Pancreatic Polypeptide
APTES — Aminopropyltriethoxysilane
AQ — Anthraquinone
AR — Aromatic Resin
AR — Aviram-Ratner
ASGPR — Asialoglycoprotein Receptor
Atb — *S*-2-Amino-4-trifluorobutyric Acid
ATCh — Acetylthiocholine
ATP — Adenosine 5′-Triphosphate
ATR-FTIRS — Attenuated Total Reflection-Fourier Transform Infrared Spectroscopy
ATR-IR — Attenuated Total Reflectance Infrared
ATRP — Atom Transfer Radical Polymerization
AU — Analytical Ultracentrifugation
Au-SNP — Au Supramolecular Nanoparticle
AUC — Analytical Ultracentrifugation
AuNP — Gold Nanoparticle
AV — Ala-Val
AZTDP — Azido-3′-deoxythymidene 5′-diphosphate
AZTMP — 3′-azido-3′-deoxythymidene 5′-monophosphate

BAM — Brewster Angle Microscopy
BAMP-ligand — Bis(aminomethyl)pyridine Ligand
BAPTA — 1,2-bis(o-aminophenoxy)ethane-*N,N,N′,N′*-tetraacetic acid
BAR — Barbituric Acid
BASE — Boron Affinity Saccharide Electrophoresis
BASF — Baden Aniline and Soda Factory
BBV — Boronic Acid-Substituted Benzylviologen
BCA — Bio-bar-code amplification
BCB — Benzocyclobutane
BCC — Body Centered Cubic
BCD — β-Cyclodextrin
BCP — Block Copolymer
bdc — 1,4-benzenedicarboxylate
BDC — Benzenedicarboxylic Acid
BDE — Bond-Dissociation Energies
BDG — Benzodiguanamine
bdta — 1,2,4,5-benzenetetracarboxylic Acid
BET — Brunauer Emmett Teller or Back Electron Transfer
BFDMA — Bis-(11-Ferrocenylundecyl)-Dimethylammonium Bromide
bFGF — Basic Fibroblast Growth Factor
BH — Bcl-2 Homology
BI — Bovine Insulin
BIA — Biomolecular Interaction Analysis
BIC — 5-(Benzyloxy)-Isophthalic Acid
BINAM — 2,2′-diamino-1,1′-binaphthalene
BINAP — 2,2′-bis(Diphenylphosphino)-1,1′-binaphthyl
bipy — Bipyridine
BiTE — Bispecific T-Cell Engager Molecules
bix — 1,4-bis(Imidazol-1-ylmethyl)-Benzene
BLM — “Black” Lipid Membrane
BM — Ball Milling
BMP — Bone Morphogenetic Protein
BNCT — Boron Neutron Capture Therapy
BNP — Binaphthyl Phosphate
BNPP — Bis(4-Nitrophenyl)Phosphate
BO — Butylene Oxide
BODIPY — Boron Dipyrromethane
BoNT — *Botulinum Neurotoxin*
BP — Biphenol
BPA — Bipyridine Amine
BPB — Bromophenol Blue
BPEA — Bis-(Phenylethynyl)Anthracene
bPP — Bovine Pancreatic Polypeptide
BPP34C10 — Bis(*p*-Phenylene)-34-Crown-10
bpp-34-crown-10 — Bisparaphenylene-34-crown-10
BPPM — *N*-tert-butoxycarbonyl-4-diphenylphosphino-2-diphenylphosphinomehtyl-pyrrolidine
bpy — Bipyridine
BRGD-PA — Aspartate-Arg-Gly-Asp
BSA — Bovine Serum Albumin
BSM — Bovine Submaxillary Mucin
BSP — Bone Sialoprotein
BTA — Benzene-1,3,5-tricarboxamide
BTB — 1,3,5-benzenetribenzoate
BTC — 1,3,5-benzenetricarboxylic Acid
BTC — Benzene-1,3,5-tricarboxylate
BTE — Backbone Thioester Exchange
BTF6 — 1,2-bis(2-Methylbenzo[b]thiophen-3-yl)Hexafluorocyclopentene
BTM — Benzotetramisole
BTMA — *n*-Butyltrimethylammonium
BTX — Bent Triple-Crossover
BZ — Belousov–Zhabotinsky
BZD — Benzidine

CA	Carbonic Anhydrase or Cyanuric Acid
CAC	Critical Aggregation Concentration
CAHBs	Charge-assisted H-bonds
cAMP	Cyclic Adenosine Monophosphate
CAP	Chloramphenicol
CAP-MR	Chloramphenicol-methyl Red
CAS	Chrome Azurol S
CB or CB[*n*]	Curcurbit[*n*]uril
CB[6]	Cucurbit[6]uril
CB[7]	Cucurbit[7]uril
CB[8]	Cucurbit[8]uril
CBA	4-Carboxyphenylboronic Acid
CBED	Convergent-Beam Electron Diffraction
$CBPQT^{4+}$	Cyclobis(Paraquat-*p*-phenylene)
cbta	Cyclobutanetetracarboxylic Acid
CC	Coupled Cluster
CCA	Colloidal Crystalline Array
CCD	Charge Coupled Device
CCDC	Cambridge Crystallographic Data Centre
ccdc	Cobaltocenium-1,1′-dicarboxylate
CCK8	Cholecystokinin Octapeptide
CCMV	Cowpea Chlorotic Mottle Virus
ccnm	Carbamoylcyanonitrosomethanide
ccp	Cubic Close-packed
CCW	Counterclockwise
CD	Circular Dichroism or Cyclodextrin
α-CD	α-Cyclodextrin
β-CD	β-Cyclodextrin
γ-CD	γ-Cyclodextrin
CD-PEI	Cyclodextrin-Modified Polyethylenimine
CD/Ad	Cyclodextrin/adamantine
CDCs	Cholesterol-Dependent Cytolysins
CDI	1-(3-Dimethylaminopropyl)-3-Ethylcarbodiimide
CDI	Coherent Diffraction Imaging
cdo	Diolefin Chelidonic Acid
CDP	Cyclodextrin-Based Polymer
CdSe	Cadmium Selenide
CDV	Cyclodextrin Vesicle
CE	Capillary Electrophoresis
CEC	Capillary Electrochromatography
CEST	Chemical Exchange Saturation Transfer
CF	5(6)-Carboxyfluorescein
CFET	Chemical-Field-Effect Transistor
CFSE	Crystal Field Stabilization Energy
CGOM	Crystal Growth of Organic Materials
CHAPS	3-[(3-Cholamidopropyl)Dimethylammonio]-1-propanesulfonate
CHEF	Chelation-Enhanced Fluorescence
CHEMFET	Chemically Modified Field-Effect Transistor
CHO	Chinese Hamster Ovarian
CH_3OH	Methanol
CHO-K1	Chinese Hamster Ovary
ChS	Chondroitin 4-Sulfate
CHTE	Cyclohepta-1,2,4,6-tetraene
CI	Configuration Interaction
CID	Collision-Induced Dissociation
C-IDA	Colorimetric Indicator Displacement Assay
CIF	Crystallographic Information File
CIGS	$Cuin_xGa_{(1-x)}Se_2$
CK II	Casein Kinase II
CL	Chemiluminescence
CLC	Cholesteric or Columnar Liquid-Crystalline
CLIO	Crosslinked Iron Oxide
CLs	Chemical Leitmotifs
CLSM	Confocal Laser Scanning Microscopy
ClSubPc	Chlorosubphthalocyanine
CMC	Critical Micellar Concentration
CME	Chemically Modified Electrode
CMOS	Complementary Metal Oxide Semiconductor
CMP	Cytosine Monophosphate
CMT	Critical Micellization Temperature
CMV	Cytomegalovirus
CN	Coordination Number
cNRG	Cyclic Asparagine-Glycine-Arginine
CNT	Carbon Nanotube
cod	1,5-cyclooctadiene
Col	Collagen
Col_h	Columnar Hexagonal
Col_r	Columnar Rectangular
COM	Center of Mass
CoMoCat	Cobalt Molybdenum Catalyzed
Con A	Concanavalin A
CONTIN	Continuous Distributions of Exponentials
COR	Coronene
CORE	Component Resolved
CP	Cross-Polarization
Cp	Cyclopentadienyl
μCP	Microcontact Printing
m-CPBA	*meta*-Chloro-Perbenzoic Acid
CP-MAS	Cross-Polarized Magic-Angle Spinning
CPD	Cyclophanediene
CPK	Corey–Pauling–Koltun
CPL	Circularly Polarized Luminescence

CPMAS	Cross Polarization Magic Angle Spinning
CPMV	Cowpea Mosaic Virus
CPP	Cell-Penetrating Peptide
CPs	Conjugated Polymers
CPs	Cyclic Peptides
CPT	Camptothecin
CRAMPS	Combined Rotation and Multipulse Sequence
CREST	Core Research for Evolutional Science and Technology
CRP	Controlled Radical Polymerization
Cryo	Cryogenic
cryo EM	Cryo Electron Microscopy
cryo-TEM	Cryogenic Transmission Electron Microscopy
CS	Circumsporozoite
CS	Citrate Synthase
CSA	Chemical Shift Anisotropy
CSD	Cambridge Structural Database
CSI	Coldspray Ionization
CSNP	Core–Shell Nanoparticle
CSP	Coiled-Coil Switch Peptide
CSP	Crystal Structure Prediction
CT	Charge Transfer
CT-AFM	Conducting-Tip Atomic Force Microscope
CTA	Cellulose Triacetate
CTAB	Cetyltrimethylammonium Bromide
CTAHS	Cetyltrimethylammonium Hydrogensulfate
CTAOH	Cetyltrimethylammonium Hydroxide
CTB	Cyclotribenzylene
CTC	Cyclotricatechylene
CTG	Cyclotriguaiacylene
CTP	Cytidine Triphosphate
CTTV	Cyclotetraveratrylene
CTV	Cyclotriveratrylene
CuAAC	Copper-Catalyzed Azide–Alkyne Cycloaddition
CV	Cyclic Voltammetry
CVD	Chemical Vapor Deposition
CW	Clockwise
CWA	Chemical Warfare Agent
CyD	Cyclodextrin
cyt *c*	Cytochrome c
D–A	Donor–Acceptor
D–EZ	Dendrimer–Ez
d.e.	Diastereoisomeric Excess
DA	Diels–Alder
DAB	Diaminobutane
DABCO	1,4-diazabicyclo[2.2.2]octane
DABCYL	(Dimethylamino)phenyl)azo)benzoic Acid
DAD	Donor–Acceptor–Donor
DAMA	*N*-(*N*′,*N*′-Dicarboxymethylamino-propyl)Methacrylamide
DAN	2,7-diamido-1,8-naphthyridine
DAP	Diamidopyridine
DASP	Dimethylaminostyryl Pyridinium
DB-24-C8	Dibenzo-[24]-crown-8
DB-CTCDI	1,7-di(Butyl)-Coronene-3,4 : 9,10-tetracarboxylic Acid Bisimide
DBD	Dna-Binding Domain
DBO	1,8-Dibromooctane
DBPT	Dibutylphthalate
DBSA	Dodecylbenzene Sulfonic Acid
dbsf	4,4′-sulfonyldibenzoate
DC	Direct Current
dc	Double-Chained
DCA	Deoxycholic Acid
dca	Dicyanamide
DCC	*N*,*N*′-Dicyclohexyl Carbodiimide
DCC	Dynamic Combinatorial Chemistry
DCH	4,4-diphenyl-2,5-cyclohexadienone
DCL	Dynamic Combinatorial Library
DCM	Dual-Core Microreactor
DCTX	Docetaxel
DD	Donor Dendrimer
DDAB	Didodecyldimethylammonium Bromide
ddn	1,12-dodecanedinitrile
DDQ	2,3-dichloro-5,6-dicyanobenzo-quinone
DDS	Drug Delivery Systems
DDSCs	Dye Sensitized Solar Cells
DDSNPs	Dye-Doped Silica Nanoparticles
DeAp	Deazapterin
DECRA	Direct Exponential Curve Resolution Algorithm
DEFRET	DElayed Fluorescence Resonance Energy Transfer
DELFIA	Dissociation Enhanced Lanthanide Fluoroimmunoassay
Den–CD–NTs	Dendron–Cd–Nanotubes
dex	Dexamethasone
DFBZ	Difluorinated Benzidine
Dfeg	4,4-difluoroethylglycine
Dfp	4,4-difluoroproline
DFT	Density Functional Theory
DGR	Asp-Gly-Arg-Gly-Asp-Ser-Val-Ala-Tyr-Gly
DGU	Density Gradient Ultracentrifugation
DβH	Dopamine β-Hydroxylase
DHA	Dicyclohexylammonium

DIC	Differential Interference Contrast
DiFMU	Difluoromethylumbelliferone
DIO	1,8-Diiodooctane
DIOP	4,5-bis(diphenylphosphinomethyl)-2,2-dimethyl-1,3-dioxolane
α,ω-DIPFA	α,ω-Diiodoperfluoroalkane
DITFB	Diiodotetrafluorobenzene
DITFE	1,2-diiodotetrafluoroethane
DLS	Dynamic Light Scattering
DβM	Dopamine β-Monooxygenase
DMA	Dynamic Mechanical Analysis
dmaa	9-(N,N-Dimethylamino)Anthracene
DMAB	Dimethylamine Borane
DMAc	Dimethylacetamide
DMB	Dimethylbenzil
DMDG	Digital Microfluidic Droplet Generator
dme	1,2-dimethoxyethane
DME	Danish Micro Engineering
DMEM	Dulbecco'S Modified Eagle'S Medium
DMF	N,N-Dimethylformamide
DMG	Dimethylglyoxime
DMP	Double Minimum Potential
DMPC	1,2-dimyristoyl-sn-glycero-3-phosphocholine
DMPC	DL-α-dimyristoylphosphatidylcholine
DMPG	Dimyristoylphosphatidylglycerol
dMSC	Dog Mesenchymal Stem Cell
DMSO	Dimethylsulfoxide
DMT-MM	4-(4,6-dimethoxy-1,2,5-triazin-2-yl)-4-methylmorpholinium
DMTA	Dynamic Mechanical Thermal Analysis
DNA	Deoxyribonucleic Acid
DNNS	Dinonylnaphtalenesulfonic Acid
DNP	1,5-dioxynaphthalene
DNPA	2,4-dinitrophenylacetate
DNS	Dansyl
DO3A	1,4,7,10-tetraazacyclododecane-1,4,7-triacetic Acid
DODAC	Dioctadecyldimethylammonium Chloride
DOE	Department of Energy
DOF	Depth of Focus
DOGSDSO	1,2-dioleoyl-sn-glycerol-3-succinyl-2-hydroxyethyl Disulfide Ornithine
DON	Dioxynaphthalene
DOPA	Dihydroxyphenylalanine
DOPC	Dioctadecylphosphatidylcholine
DOPC	Dioleoylphosphatidylcholine
DOPE	1,2-dioleoyl-*Sn*-glycero-3-phosphatidylethanolamine
DOR	δ Type Opioid Receptors
DOS	Density of States
DOSY	Diffusion Ordered Spectroscopy
DOT	Diffuse Optical Tomography
DOTA	1,4,7,10-tetra(carboxymethyl)-1,4,7,10-tetraazacyclododecane
DOTAP	N-[1-(2,3-dioleoyloxy)Propyl]-N,N,N-trimethylammonium Chloride
DOX	Doxorubicin
DP	Degree of Polymerization
DP-PTCDI	1,7-dipropylthio-perylene-3,4 : 9,10-tetracarboxydiimide
DPB	Diphenylbutadiene
DPC	Diphenylcarbazide
DPDI	4,9-diaminoperylene-quinone-3,10-diimine
DPhPC	Diphytanoyl Phosphatidylcholine
DPK	2,2′-dipyridylketone
dpn	1,8-bis(4-Pyridyl)Naphthalene
DPN	Dip-Pen Nanolithography
DPNTs	Dipeptide Nanotubes
dpp	1,3-di(4-Pyridyl)Propane
DPP	Differential Pulse Polarographic
DPPA	Dipalmitoylphosphatidic Acid
DPPC	1,2-dipalmitoyl-*Sn*-glycero-3-phosphocholine
DPPE	1,2-dipalmitoyl-*Sn*-glycero-3-phophoethanolamine
DPPS	Dipalmitoylphosphatidylserine
DPQ	2,3-di-(1H-2-Pyrrolyl)Quinoxaline
dps	4,4′-dipyridylsulfide
DPSC	Dental Pulp Stem Cell
DPV	Differential Pulse Voltammetry
DQ	Double-Quantum
DR	Design Rule
ds	Double-Stranded
DSC	Differential Scanning Calorimetry
DSCG	Disodium Cromoglycate
DSIDA	Copper(II) Distearylglycerotriethyleneglycyl Iminodiacetic Acid
DSNP	Dye inside Silica Nanoparticle
DSPC	1,2-dimyristoyl-*Sn*-glycero-3-phosphoethanol-amine
Dspp	Dentin Sialoprotein
DSSC	Dye Sensitized Solar Cell
DTA	Differential Thermal Analysis
DTAB	Dodecyl Trimethyl Ammonium Bromide
DTAR	4-n-Dodecyl-6-(2-Thiazolylazo)-Resorcinol
DTBC	3,5-di-t-Bu-catechol
DTBQ	3,5-di-t-Bu-quinone
DTE	Dithienylethene
DTPA	Diethylene Triamine Pentaacetic Acid

DTPO	Di-(*Tert*-butyl)Phenoxy
DTT	Dithiothreitol
DTTA	Diethylenediamine-Tetraacetate
DUMBO	Decoupling Using Mind-Boggling Optimization
DVB	Divinylbenzene
DWNTs	Double-Wall Nanotubes
DX	Double Crossover
EA	Electron-Acceptor
EB	Ethyl Benzoate
EB^+	Ethidium Bromide
EBHTL	Electron Blocking Hole Transport Layer
EBL	Electron Beam Lithography
EC-STM	Electrochemical Scanning Tunneling Microscope
ECBJ	Electrically Controlled Break Junction
ECD	Electron Capture Dissociation
ECD	Electron-Emitting Cathode or Electronic Circular Dichroism
ECHBM	Electrostatic-Covalent H-Bond Model
ECL	Electrogenerated Chemiluminescence
ECM	Extracellular Matrix
ED	Electron-Donor
EDA	Electron Donor-Acceptor
EDC	*N*-Ethyl-*N*′-(3-Dimethylaminopropyl)-Carbodiimide Hydrochloride
EDCI	1-Ethyl-3-(3′-Dimethylaminopropyl)-Carbodiimide
EDGA	*N*,*N*-Eicosanedioyl-Di-L-glutamic Acid
EDMA	Ethylene Dimethacrylate
EDOT	3,4-ethyldioxythiophene
EDS	Energy Dispersive X-Ray Spectroscopy
EDTA	Ethylenediamine Tetraacetic Acid
EDX	Energy Dispersive X-Ray
ee	Enantiomeric Excess
EEB	2-Ethoxyethyl Ester Benzoic Acid
EELS	Electron Energy-Loss Spectroscopy
EET	Electronic Energy Transfer
EF	Edge-Face
EFJC	Extended Freely Jointed Chain
EFM	Electrical Force Microscopy
EG	Ethylene Glycol
EGDMA	Ethylene Glycol Dimethacrylate
EGF	Epidermal Growth Factor
EGFP	Enhanced Green Fluorescent Protein
EGTA	ethylene glycol tetraacetic acid
eIDA	Enantioselective Indicator Displacement Assay
EK	Equal *K*
ELISA	Enzyme-Linked Immunosorbant Assay
ELM	Emulsion Liquid Membrane
EM	Effective Molarity
EMBJ	Electromigrated Break Junction
EMF	Electromotive Force
EMSA	Electrophoretic Mobility Shift Assay
en	1,2-diaminoethane
EnFETs	Enzyme Field Effect Transistor
ENT	Energy Transfer
EO	Ethylene Oxide
EOE	Enamel Organ Epithelial
EOF	Electroosmotic Flow
EPA	Environmental Protection Agency
EPEG	Ethyl Phthalyl Ethyl Glycolate
EPR	Electron Paramagnetic Resonance
ePTFE	Extended Polytetrafluoroethylene
EQE	External Quantum Efficiency
ES	Excited-State
ESA	Esterase-Sensitive Amphiphile
ESAC	Encoded Self-Assembling Chemical
esd	Estimated Standard Deviation
ESEM	Environmental Scanning Electron Microscope
ESI	Electrospray Ionization
ESI-FTICR	Electrospray Ionization–Fourier-Transform Ion-Cyclotron-Resonance
ESI-MS	Electrospray Ionization Mass Spectrometry
ESP	Equilibrium Spreading Pressure
ESPL	Esterase-Sensitive Phospholipid
ESR	Electron Spin Resonance
ET	Electron Transfer
$E-T$	Energy–Temperature
Et_2-en	*N*,*N*-Diethyl-Ethylenediamine
ETD	Electron-Transfer Dissociation
ETO	Etoposide
EU	European Union
EURYI	European Young Investigator
EUV	Extreme Ultraviolet
EUVL	Extreme UV Lithography
EWC	Equilibrium Water Content
EXAFS	Extended X-Ray Absorption Fine Structure
EXSY	Exchange Spectroscopy
EYFP	Enhanced Yellow Fluorescent Protein
EYPC	Egg Yolk Phosphatidylcholine
FA	Frontal Analysis
FAB	Fast Atom Bombardment
FAB-MS	Fast Atom Bombardment Mass Spectrometry
FACCE	Frontal Analysis Continuous Capillary Electrophoresis

FACE	Fluorophore Assisted Carbohydrate Electrophoresis
FACS	Fluorescent Activated Cell Sorting
FAD	Flavin Adenine Dinucleotide
Fc	Ferrocene
FCCP	4-(Trifluoromethoxy)Phenylhydrazone
fcdc	1,1′-ferrocenedicarboxylate
FCI	Fonds der Chemischen Industrie
$FcN^+(CH_3)_3$	Ferrocenylmethyltrimethylammonium
FCP	Flexible Coordination Polymer
FCS	Fluorescence Correlation Spectroscopy
FDA	Food and Drug Administration
[^{18}F]FDG	2-[^{18}F]-fluoro-deoxy-D-glucose
FE	Field Emission
FED	Field Emission Device
FEL	Free Electron Laser
FEM	Finite Element Modeling
FESEM	Field Emission Scanning Electron Microscope
FET	Field-Effect Transistor
FF	Force Field
[^{18}F]6-Fluoro-L-DOPA	6-[^{18}F]-fluoro-3,4-dihydroxyphenylalanine
α-FFP	α-Helical Fiber-Forming Peptide
FFT	Fast Fourier Transformation
FGF	Fibroblast Growth Factor
FIA	Flow Injection Analysis
FID	Free Induction Decay
FIM	Fixed Interference Method
FIT probe	Forced Intercalation Probe
FITC	Fluorescein Isothiocyanate
FJC	Freely Jointed Chain
FLAB	Fluorophore Linker Boronic Acid Biotin
Flp	4(*R*)-Fluoroproline
FLuc	Firefly Luciferase
FM	Fluorescence Microscopy
FMN	Flavin Mononucleotide
FMNP	Ferromagnetic Nanoparticle
Fmoc	*N*-Fluorenyl-9-methoxycarbonyl
Fmoc-FF	Fmoc-Diphenylalanine
Fmoc-PEG-COOH	Poly(Ethylene Glycol) *O*-(*N*-Fmoc-2-aminoethyl)-*O*′-(2-Carboxyethyl) Undeca(Ethylene Glycol)
FMT	Fluorescence-Mediated Tomography
FP	Fluorescence Polarization
FPIM	Fixed Primary Ion Method
FRAP	Fluorescence Recovery after Photobleaching
FRET	Fluorescence or Förster Resonance Energy Transfer
fs	Femtosecond
FSLG	Frequency-Switched Lee–Goldburg
f-SPR	Fluorescence Surface Plasmon Resonance
FTICR	Fourier-Transform Ion-Cyclotron-Resonance
FTIR	Fourier Transform Infrared
FTO	Fluorine-Doped Tin Oxide
fum	Fumaric Acid
GA	Gum Arabic
GABA	γ-Aminobutyric Acid
GAG	Glycosaminoglycan
β-gal	β-Galactosidase
GalCer	Galactosylceramides
GalNAc	*N*-Acetylgalactosamine
GAO	Galactose Oxidase
GAPDH	Glyceraldehyde-3-phosphate Dehydrogenase
GC	Gas Chromatography
GC–MS	Gas Chromatography–Mass Spectrometry
GCMC	Grand Canonical Monte Carlo
Gd-DTPA	Gd-Diethylenetriaminepentaacetate
GdL	Glucono-δ-lactone
GFE	Gibbs Free Energy
GFP	Green Fluorescent Protein
GHK	Goldman–Hodgkin–Katz
GIAO	Gauge-Invariant Atomic Orbital
GID	Grazing Incidence Diffraction
GISANS	Grazing Incidence Small Angle Neutron Scattering
GISAXS	Grazing Incidence Small Angle X-Ray Scattering
GIXD	Grazing Incidence X-Ray Diffraction
GlcNAc	*N*-Acetylglucosamine
GMP	Guanosine-5′-monophosphate
GMR	Giant Magnetoresistance
Gn	Generation
GNA	Glycol Nucleic Acid
GNP	Gold Nanoparticle
GOA	Geconcerteerde Onderzoeksacties
GOx	Glucose Oxidase
βGP	β-Glycerophosphate
GPC	Gel Permeation Chromatography
GPCR	G-Protein Coupled Receptor
GPI	Glycosylphosphatidylinositol
GRAS	Generally Regarded As Safe
GRID	Gd(DTPA)-tetramethylrhodamine-aminedextran
GS	Ground-State
GSH	Glutathione
GSK	Glaxo Smithkline
GSNPs	Grafted Silica Nanoparticles
GTP	Guanosine Triphosphate

Gua	Guanidinocarbonyl Pyrrole
GUV	Giant Unilamellar Vesicle
^{1}H NMR	Proton Nuclear Magnetic Resonance
HA-PHP	Hydroxyapatite-Polyhipe Polymer
HAA	Haloacetic Acid
HABA	2-(4′-Hydroxybenzeneazo)Benzoic Acid
HAP	Hydroxyapatite
HAT	1,4,5,8,9,12-Hexaazatriphenylene
HB	Hydrogen Bond
HBC	Hexabenzocoronene
H_2bdc	1,4-benzenedicarboxylic Acid
HBETL	Hole-Blocking Electron Transport Layers
hBMSC	Human Bone Mesenchymal Stem Cell
HBPA	Heparin-Binding Peptide Amphiphile
HBR	Hydrogen-Bonded Rotaxane
HBS	Hepes Buffered Saline
HC	Hydrocarbon
HCA	Hierarchical Cluster Analysis
hcp	Hexagonal Close-Packed
HD	Hummel–Dreyer
HDA	9-Hexadecyladenine
HDC	Hydrodynamic Chromatography
HDD	High Dye Density
HDL	High-Density Lipoprotein
HDOR	Heteronuclear Dipolar-Order Rotor
HEEDTA	2-Hydroxyethylenediaminetriacetic Acid
HEK	Human Embryonic Kidney
HEMA	Hydroxyethyl Methacrylate
HEMT	High Electron Mobility Transistor
HEP	Heparin
HEPES	4-(2-Hydroxyethyl)-1-Piperazineethanesulfonic Acid
hER-α	Human Estrogen Receptor-α
HETCOR	Heteronuclear Correlation
HF	Hartree-Fock
hfac	Hexafluoroacetylacetonate
Hfl	5,5,5,5′,5′,5′-hexafluoroleucine
hGH	Human Growth Hormone
hGHbp	Human Growth Hormone Receptor
Hhfpd	Hexafluoro-2,4-pentanedione
hIAPP	Human Islet Amyloid Polypeptide
HIF-1α	Hypoxia-Inducible Factor 1α
HiPco	High-Pressure Carbon Monoxide
His-GFP	His-Tagged Green Fluorescent Protein
HIV-1	Human Immunodeficiency Virus-1
HLB	Hydrophilic–Lipophilic Balance
HMPA	*N*-(2-Hydroxypropyl)Methacrylamide
hMSC	Human Mesenchymal Stem Cell
HMT	Histone Lysine Methyltransferase
HMTA	Hexamethylenetetramine
HOM	Ordered Monolithic Silica
HOMO	Higher Occupied Molecular Orbital
HOPG	Highly Ordered Pyrolytic Graphite
HP1	Heterochromatin Protein 1
HPCE	High-Performance Capillary Electrophoresis
HPL	Hexagonally Perforated Lamellae
HPLC	High Perfomance Liquid Chromotography
HPMA	*N*-(2-Hydroxypropyl)-Methacrylamide
HPNP	2-Hydroxypropyl *p*-Nitrophenyl Phosphate
HPNP	Hydroxypropyl-*Para*-nitrophenyl Phosphate
HPTS	8-Hydroxy-1,3,6-pyrenetrisulfonate
HRP	Horseradish Peroxidase
HRTEM	High-Resolution Transmission Electron Microscopy
HS	High Spin
HSA	Human Serum Albumin
HSAB	Hard and Soft Acids and Bases
HSM	Hot-Stage Microscopy
HSQ	Hydrogen Silsesquioxane
HSV	Herpes Simplex Virus
HTS	High-Throughput Screening
IAP	Interuniversity Attraction Pole
IAPP	Islet Amyloid Polypeptide
ic	Internal Conversion
IC	Ion Chromatography
ICD	Induced Circular Dichroism
ICG	Indocyanine Green
ICP-MS	Inductively Coupled Plasma Spectrometry-Mass Spectrometry
ICR-EMP	Internal Charge Repulsion-External Membrane Pressure
ICS	Ion Channel Sensor
ICT	Internal Charge Transfer
ID	Injected Dose
IDA	Indicator Displacement Assay
IEC	International Electrotechnical Commission
IETS	Inelastic Electron Tunneling Spectroscopy
IgG	Immunoglobulin G
IgM	Immunoglobulin M
IK	Incremental *K*
IL	Intraligand
IL-2	Interleukin-2
imc	4-Imidazole-Carboxylate
imdtc	4-Imidazole-Dithiocarboxylate
IP	Intellectual Property

IPCE	Incident Photon to Collected Electron Efficiency
IPER	Irradiation-Promoted Exchange Reaction
IQEs	Internal Quantum Efficiencies
IR	Infrared
IRE-BP	Iron-Responsive Element-Binding Protein
IRMOF	Isoreticular Metal–Organic Framework
IRMPD	Infrared Multiphoton Dissociation
IRRAS	Infrared Reflection Absorption Spectroscopy
ISA	Isophthalic Acid
ISA-DIA	5-(10,12-tricosadiinyloxy) Isophthalic Acid
ISE	Ion Selective Electrode
ISFET	Ion-Selective Field-Effect Transistor
ITC	Isothermal Titration Calorimetry
ITO	Indium Tin Oxide
IUCr	International Union of Crystallography
IUPAC	International Union of Pure and Applied Chemistry
IVCT	Intervalence Charge-Transfer
IVT	Intervalence Transfer
JST	Japan Science and Technology Agency
KER	Kinetic Energy Release
LADH	Liver Alcohol Dehydrogenase
LAG	Liquid-Assisted Grinding
LAM	Lamellae
LAOS	Large Amplitude Oscillatory Shear
LASC	Lewis-Acid—Surfactant-Combined Catalyst
LB	Langmuir–Blodgett
LBF	Langmuir–Blodgett Film
LBHB	Low Barrier Hydrogen Bond
LBL	Layer-by-Layer
LC	Ligand-Centered
LC	Liquid Crystal
LCAO	Linear Combination of Atomic Orbital
LCM	Large Compound Micelle
LCSM	Laser Confocal Scanning Microscopy
LD	Linear Dichroism
LDA	Linear Discriminant Analysis
LDD	Low Dye Density
LDH	Lactate Dehydrogenase
L-DOPA	L-3,4-dihydroxyphenylalanine
LDOS	Local Density of States
LE	Locally Excited
LEC	Light-Emitting Electrochemical Cell
LED	Light Emitting Diode
LEED	Low Energy Electron Diffraction
LEP	Lysyl-Endopeptide
LF	Ligand-Field
LFER	Linear-Free Energy Relationships
LGP	Lipoglycopolymer
LH	Light-Harvesting
LHA	Light Harvesting Antennas
LHP	Logit Hydrogen-Bonding Propensity
LIF	Laser-Induced Fluorescence
LILBID	Laser-Induced Liquid Bead Ion Desorption
LiSuNS	Liquid Supramolecular Nanostamping
LLC	Lyotropic Liquid Crystal
LM	Ligand-to-Metal
LMCT	Ligand-to-Metal Charge Transfer
LMWG	Low Molecular Weight Gelator
LMWH	Low Molecular Weight Heparin
LNA	Locked Nucleic Acid
LNT	Lipid Nanotube
LOD	Limit of Detection
LPS	Lipopolysaccharides
LR	Lawesson's Reagent
LS	Langmuir–Schaefer
LSM	Light-Scattering Microscopy
LSPR	Localized Surface Plasmon Resonance
LUMO	Lowest Unoccupied Molecular Orbital
LUV	Large Unilamellar Vesicle
LV	Liquid–Vapor
MA	Magic Angle
MAA	Methacrylic Acid
mAbs	Monoclonal Antibodies
MACE	Micellar Affinity Capillary Electrophoresis
MALDI	Matrix-Assisted Laser Desorption Ionization
MALDI-TOF	Matrix-Assisted Laser Desorption Ionization Time-of-Flight
MALS	Multiangle Light Scattering
MAMS	Mesh Adjustable Molecular Sieve
MAPL	Molecular Assembly Patterning by Lift-off
MARS	Magnetic Activating Release System
MAS NMR	Magic Angle Spinning Nuclear Magnetic Resonance
MBA	Monobromoacetic
MBB	Molecular Building Block
MBD	Molecular Basis of Disease
MBDCA	Monobromodichloroacetic
MBE	Molecular Beam Epitaxy
MC	Merocyanine
M-C	Metal-Centred

MCA	Monochlorocacetic
MCBJ	Mechanically Controlled Break Junction
MCDBA	Monochlorodibromoacetic
MCH	Methylcyclohexane
MCID	Molecular Computational Identification
MCM	Mobil Composition of Matter
MCN	Molecular Coordination Number
MCR	Matlab Component Runtime or Multivariate Curve Resolution
MD	Molecular Dynamic
M-DNA	Metal DNA
MeCHTE	5-Methyl-Cyclohepta-1,2,4,6-tetraene
MEF	Metal Enhanced Fluorescence
MEH-PPV	(Poly[2-methoxy-5-(2′-Ethyl-Hexyloxy)Phenylene Vinylene])
MEKC	Micellar Electrokinetic Chromatography
MEM	Maximum Entropy Method
MEP	Molecular Electrostatic Potential
metalloCD	Metallocyclodextrin
MF	Mole Fraction
MFT	Mean-Field Theory
MHA	Mercaptohexadecanoic Acid
mhsa	*N*-(*m*-Hydroxyphenylene)Salicylideneaminate
MI	Myocardial Infarction
MIB	Methylisoborneol
MIC	Minimal Inhibitory Concentration
micro-PET	Micro-Positron Emission Tomography
MIDA	Multicomponent Indicator Displacement Assay
MIM	Mechanically Interlocked Molecule
MIMIC	Micromolding in Capillaries
MIP	Molecularly Imprinted Polymer
MISPE	Molecularly Imprinted Solid Phase Extraction
MJF	Multiple Juxtapositional Fixedness
ML	Metal-Ligand
MLAM	Modulated Lamellae
MLCT	Metal to Ligand Charge Transfer
MLM	Monolayer Lipid Membrane
MLP-ANN	Multilayer Perceptron Artificial Neural Networks
MLV	Multilamellar Vesicle
MM	Molecular Mechanics
MMA	Methyl Methacrylate
MMFF	Merck Molecular Force Field
MMO	Methane Monooxygenase
MMP	Matrix Metalloprotease
MMPC	Mixed Monolayer Protected (Gold) Cluster
MMPs	Matrix Metalloproteinases
MMR	Mannose-Specific Macrophage Receptor
MNP	Magnetic Nanoparticle
MO	Molecular Orbital
MOC	Metal-Organic Cube
MOCVD	Metal Organic Chemical Vapor Deposition
MOF	Metal Organic Framework
MOMs	Metal-Organic Materials
MONTs	Metal–Organic Nanotubes
MOPs	Metal-Organic Polyhedra
MOs	Molecular Orbitals
MOSFET	Metal Oxide Semiconductor Field Effect Transistor
MP	Møller–Plesset
MPC	2-(Methacryloyloxy)Ethylphosphorylcholine
MPc	Metallophthalocyanine
mPE	*Meta*-phenylene Ethynylene
MPEG	Methoxy Polyethylene Glycol
MPO	Myeloperoxidase
mppe	1-Methyl-1′-(4-Pyridyl)-2-(4-Pyrimidyl)Ethylene
MPS	Mononuclear Phagocyte
MQ	Menaquinone
MQMAS	Multiple Quantum Magic Angle Spinning
MR	Magnetoresistive
MR-CD	Methyl-Red-Modified β-cyclodextrin
MRI	Magnetic Resonance Imaging
M-βRib	Methyl β-D-ribofuranoside
MRS	Magnetic Relaxation Switches
MRSA	*Meticilline-Resistant Staphylococcus Aureus*
MS	Mass Spectrometry
MSG	Monosodium Glutamate
MSH	*o*-Mesitylsulfonylhydroxylamine
MSTJs	Molecular Switch Tunnel Junction
m-SWNT	Metallic Single-Walled Carbon Nanotube
mtDNA	Mitochondrial DNA
M-TPP	Metal Tetraphenyl Porphyrin
MUA	Mercaptoundecanoic Acid
muco	Muconic Acid
MV	Methyl Viologen
MVV	Multivesicular Vesicle
MW	Molecular Weight
MWCNTs	Multiwall Carbon Nanotube
MX	Mitoxantrone
NA	Numerical Aperture
NAD	Nicotinamide Adenine Dinucleotide

NADPH	Nicotinamide Adenine Dinucleotide Phosphate-Oxidase
NAG	*p*-Nitro-Phenyl-*N*-acetyl-β-Chitooligoside
NaLS	Sodium Lauryl Sulfate
NBED	Nanobeam Electron Diffraction
NBS	*N*-Bromosuccinimide
NBT	4′-Nitro-1,1′-biphenyl-4-thiol
NC	Nanocrystal
NCA	*N*-Carboxyanhydride
NCCM	Noncovalently Connected Polymer Micelle
NCS	*N*-chlorosuccinimide
ND	Nanodiamond
NDDO	Neglect of Diatomic Differential Overlap
NDI	Naphthalene Diimide
NDR	Negative Differential Resistance
NDT	1,9-nonanedithiol
NF	Nanofiltration
NF	Nuclear Transcription Factor
NGH	Natural Gas Hydrates
NHC	*N*-Heterocyclic Carbene
NHE	Normal Hydrogen Electrode
NHS	*N*-Hydroxylsuccinimide
NI	1,8-Naphthalimide
NIL	Nanometer-Scale Imprint Lithography
NIM	Negative Index of Refraction Material
NIPAM	*N*-Isopropylacrylamide
NIR	Near Infrared
NK	Natural Killer
NLO	Nonlinear Optical
NLS	Nuclear Localization Signal
NMPX	*N*-Methylpyridinium Salts
NMR	Nuclear Magnetic Resonance
NMRD	Nuclear Magnetic Relaxation Dispersion
NN	Neural Network
NNR	Nearest Neighbor Recognition
NO	Nitric Oxide
NOE	Nuclear Overhauser Effect
NOESY	Nuclear Overhauser Effect Spectroscopy
NP	Nanoparticle
NPHE	*o*-Nitrophenyl Hexyl Ether
NPLIN	Nonphotochemical Laser-Induced Nucleation
NPOE	2-Nitrophenyl Octyl Ether
NPPOC	Nitrophenylpropyloxycarbonyl
NR	Nile Red
NS	Nanoshell
NSC	Neural Stem Cell
NSET	Nonradiative Surface Energy Transfer
NSF	Nephrogenic Systemic Fibrosis
NSOM	Near-Field Scanning Optical Microscopy
NT-3	Neurotrophin-3
NTA	Nitrilotriacetic Acid
NTCDI	Naphthalene-1,4 : 5,8-tetracarboxylic Diimide
NTFET	Nanotube Field-Effect Transistor
Nvoc	6-Nitroveratryloxycarbonyl
NWs	Nanowires
oba	Oxybis-4,4′-benzenecarboxylare
OCN	Osteocalcin
ODS	Octadecylsilyl-Silica
ODT	1,8-Octanedithiol
ODT	Order Disorder Transition
ODTNB	5-(Octyldithio)-2-Nitrobenzoic Acid
OEG	Oligo(Ethylene Glycol)
OF	Oligofluorene
OFET	Organic Field Effect Transistor
OG	Octylglucoside
OGP	Osteogenic Growth Peptide
OHB	Ordinary H-bonds
OLED	Organic Light Emitting Diode
OLS	Ostwald's Law of Stages
OM	Optical Microscopy
OMBD	Organic Molecular Beam Deposition
OMT	Orbital-mediated Tunneling
OPD	Osteopromotive Domain
OPDA	Osteopromotive Domain Peptide Amphiphile
OPE	One-Photon Excitation
OPs	Oxoporphyrinogens
OPV	Oligo(*p*-Phenylenevinylene)
OPV3-CHO	Aldehyde-Substituted Oligo-(*p*-Phenylene Vinylene)
ORD	Optical Rotary Dispersion
ORTEP	Oak Ridge Thermal Ellipsoid Plot
OTAB	Octadecyltrimethylammonium Bromide
OTS	Octadecyltrichlorosilane
OVA	Ovalbumin
PA	Peptide Amphiphile
PAA-*b*-PS	Polyacrylic Acid and Polystyrene
PAGE	Polyacrylamide Gel Electrophoresis
PAH	Polycyclic Aromatic Hydrocarbons
PAHB	Polarization-Assisted H-bonds
Pal	4-Pyridyl Alanine
PALM	Photoactivated Localization Microscopy
PAMAM	Poly(Amido Amine)
PAN	Polyacrylonitrile

PANI-CSA	Polyaniline-Camphorsulfonic Acid
PARACEST	Paramagnetic Chemical Exchange Saturation Transfer
PazoMA-*b*-PAA	Polymethacrylate with Side Chains Containing Azobenzenes and Poly(Acrylic Acid)
PB	Polybutadiene
PbAE	Poly(β-Aminoester)
PBC	Porous Block Copolymer
PBD	2-(4-Biphenyl)-5-(4-Tert-Butyphenyl)-1,3,4-oxidiazole
PBI	Perylene Bisimide
PBL	Peptide-Based Boronic Acid Lectin
PBLG	Poly(γ-Benzyl L-glutamate)
PBMA	Poly(Butyl Methacrylate)
PBS	Phosphate Buffered Saline
PCA	Principal Component Analysis
PCBM	Phenyl-C_{61}-Butyric Acid Methyl Ester
PCCAs	Polymerized Colloidal Crystalline Arrays
PCL	Poly(ε-Caprolactone)
PCM	Polarizable Continuum Model
PCP	Porous Coordination Polymer
PCR	Polymerase Chain Reaction
Pcs	Phthalocyanines
PCT	Photoinduced Charge Transfer
pcu	Primitive Cubic
PDA	Polydiacetylene
PDB	Protein Databank
PDC	Pyridinedicarboxylic Acid
PDEA	Poly[2-(*N*,*N*-diethylamino)ethyl Methacrylate]
PDGF	Platelet-Derived Growth Factor
PDH	Peptide–Dendron Hybrid
PDI	Polydispersity Index
PDMS	Poly(Dimethylsiloxane)
PDPA	Poly(2-(Diisopropylamino)Ethyl Methacrylate)
P3DT	Poly(3-Decylthiophene)
*m*PE	*Meta*-phenylene Ethynylene
*o*PE	*Ortho*-substituted Phenylene Ethynylene
pE-*co*-B	Polyethylene-*Co*-butylene
PEA	2-Phenylethylamonium
PEAAc	Poly(2-Ethyl Acrylic Acid)
PEB	Poly(Ethylene Butylene)
PECs	Polyelectrolyte β-Sheet Complexes
PEDOT	poly(3,4-ethyldioxythiophene)
PEDOT : PSS	Poly(3,4-ethylenedioxythiophene) : Poly(Styrenesulfonate)
PEG	Poly(Ethylene Glycol)
PEG–PS	Poly(Ethylene Glycol)–Polystyrene
PEHO-star-PPO	3-Ethyl-3-(Hydroxymethyl)Oxetane-Propylene Oxide
PEI	Polyethyleneimine
PEK	poly(etherketone)
PEO	Poly(Ethylene Oxide)
PEO_{113}-*b*-PS_{410}	Poly(Ethylene Oxide)-*b*-Poly(Styrene)
PEO-*b*-PB	Poly(Ethylene Oxide)-*b*-Poly (Butadiene)
PEO-*b*-PBO	Poly(Ethylene Oxide)-*b*-Poly(Butylene Oxide)
PEO-*b*-PCL	Poly(Ethylene Oxide)-*b*-Poly (Caprolactone)
PEO-*b*-PDEAMA	Peo-*b*-poly(*N*,*N*-Diethylaminoethyl Methacrylate)
PEO-*b*-PPO-*b*-PEO	Poly(Ethylene Oxide)-*b*-Poly (Propylene Oxide)-*b*-Poly(Ethylene Oxide)
PEO-*b*-PS	Poly(Ethylene Oxide)-*b*-Poly (Butadiene)
PEO–PBD	Poly(Ethylene Oxide)–Poly(Butadiene)
PEO–PBD	Poly(Ethylene Oxide)–Polybutadiene
PEO–PEE	Polyethylene Oxide–Polyethylethylene
pery	Pentaerythritol
PET	Photo-Induced Electron Transfer
PET	Positron Emission Tomography
PFC	Perfluorocarbon
PFDMS-*b*-PDMS	Poly(Ferrocenyldimethylsilane)-*b*-Poly(Dimethylsiloxane)
Pff	*S*-Pentafluorophenylalanine
PFG	Pulsed-Field Gradient
PFIB	pentafluoroiodobenzene
PFO	Perfringolysin O
PFS	Polyferrocenyldimethylsilane
pg	Phloroglucinol
PG	Phosphatidylglycerol
PGA	Polyglycolic Acid
PGE_2	Prostaglandine E_2
PGSE	Pulsed-Field Gradient Spin Echo
PHD	π-Halogen Dimer
PHEMA	Poly(Hydroxyethyl Methacrylate)
phen	1,10-phenanthroline
PHM	Peptidylglycine α-Hydroxylating Monooxygenase
P3HT	Poly(3-Hexylthiophene)
PHTP	Perhydrotriphenylene
Pi	Inorganic Phosphate
PI	Proportional-integral
PIB	poly(isobutylene)
PIC	Polyion Complex
PIE	Pseudophase Ion Exchange
PIHn	Polymer-Induced Heteronucleation
PIT	Phase Inversion Temperature

PK	Pharmacokinetic
PL	Phenyllactic Acid
PL	Photoluminescence
PL-PEG	Pegylated Phospholipids
PLA	Poly(L-lactic Acid)
PLAM	Perforated Lamellae
PLL	Poly(L-lysine)
PLL-*g*-PEG	Poly(L-lysine)-*Graft*-poly(Ethylene Glycol)
PLM	Polymer Inclusion Membrane
PLT	Poly-Lysine-Tyrosine
PLV	Pulsed-Laser Vaporization
PM	Puramatrix
PM	*Plasmepsin*
PM-IRRAS	Polarization-Modulation Infrared Reflection Absorption Spectroscopy
PM3	Parameterized Model 3
PMAA	Poly(Methacrylic Acid)
PMF	Potential Mean Force
PMI	Perylene Monoimide
PMLG	Phase-Modulated Lee–Goldburg
PMMA	Poly(Methyl Methacrylate)
p(MMA-GMA)	Poly (methacrylate-ran-glycidyl methacrylate)
PMPC	Poly(2-Methacryloyloxyethyl Phosphorylcholine)
PmPV	Poly(Metaphenylenevinylene)
PMSA	Prostate-Specific Membrane Antigen
PN	Phenylnitrene
PN	Phosphoramidate
PNA	Peptide Nucleic Acid
PNIPAAm	Poly(*N*-Isopropyl Acrylamide)
PNIPAM	Poly(*N*-Isopropylacrylamide)
P-*n*-PIMB	1,3-Phenylene Bis[4-(4-*n*-alkyl-phenyliminomethyl) Benzoates]
POM	Polarized Optical Microscopy or Poly-Oxo-Metalate
POP	Oligophenylene
POPAM	Polypropylenamine
POPC	1-Palmitoyl-2-eleoyl-*Sn*-glycero-3-phosphocholine
POPS	1-Palmitoyl-2-oleoyl-*Sn*-glycero-3-phospho-L-serine
PPBR	Polyproline Helix-Basic Region
PPE	Poly(*p*-Phenylene Ethynylene)
PPi	Inorganic Pyrophosphate
PPI	Poly(Propylenimine)
PPII	Polyproline II
PPn	Polyphenol
PPO	Polypropylene Oxide
PPP	4-(3-Phenylpropyl)-Pyridine
PPS	Poly(Propylene Sulfide)
PPV	Poly(p-phenylene Vinylene)
PPY	Polypyrrole
Ppy-BDSA	Polypyrrole-Dodecylbenzene Sulfonic Acid
PQ	Phenanthrenequinone
PQQ	Pyrroloquinoline Quinine
PR	Pyrogallol Red
PRC	Photosynthetic Reaction Center
PRE	Proton Relaxation Enhancement
PRG	Pro-Arg-Gly-Asp-Ser-Gly-Tyr-Arg-Gly-Asp-Ser
Pro-BASE	Protein-Boron Assisted Saccharide Electrophoresis
PRODAN	6-Propionyl-2-dimethylamino-naphthalene
PRP	Platelet-Rich Plasma
PrP^C	Cellular Prion Protein
PrP^{Sc}	Abbrrantly Folded (scrapie) Prion Protein
PS	Photosystem
PSA	Prostate Specific Antigen
PSar	Polysarcosine
PS-*b*-P4VP	Polystyrene-*Block*-poly(4-Vinyl Pyridine)
PS-*b*-PAA	Polystyrene-*b*-poly(Acrylic Acid)
PS-*b*-PBD	Polystyrene-*b*-polybutadiene
PS-*b*-PEO	Polystyrene-*Block*-poly(Ethylene Oxide)
PS-*b*-PFS	Polystyrene-*Block*-polyferrocenyldi-methylsilane
PS-*b*-PI	Polystyrene-*Block*-polyisoprene
PS-*b*-PMMA	Polystyrene-*b*-poly(Methyl Methacrylate)
PS-*b*-PVP	Polystyrene-*b*-poly(2-Vinylpyridine)
PS-*b*-PVP	Polystyrene-*b*-Polyvinylpyridine
P(S-*r*-4VP)-*b*-PMMA)	Poly(Styrene-*r*-4-vinylpyridine)-*b*-Poly(Methyl Methacrylate)
PS-DVB	Polysterene-Divinylbenzene
PS-PEP	Polystyrene-block-poly(ethylene-alt-propylene)
PS-PIAT	$Polystyrene_{40}$-b-poly(L-isocyanoalanine(2-Thiophen-3-yl-ethyl)$Amide)_{50}$
PS-PMMA	Polystyrene-block-poly(methyl Methacrylate)
PSD	Position-Sensitive Detector
PSM	Porcine Stomach Mucin
PSS	Polystyrene Sulfonic Acid
PSTM	Photon Scanning Tunneling Microscopy
PT	Polythiophene
PTA	1,3,5-triaza-7-phosphaadamantane
PTC	Phase Transfer Catalysis
PTCDA	Perylene Tetracarboxylic Acid-Dianhydride

PTCDI Perylene-3,4 : 9,10-tetracarboxylic-diimide
p53TD P53 Tetramerization Domain
PTFE Polytetrafluoroethylene
pTHF Poly(Tetrahydrofuran)
pTIRF Polarized Total Internal Reflection Fluorescence
PTK Protein Tyrosine Kinase
PTMBPEC Poly(2,4,6-trimethoxybenzyli Denepentaerythritolcarbonate)
p-TRIM Polytrimethyloltrimethacrylate
PTX Paclitaxel
PTZ Phenothiazine
PU Polyurethane
PV Pyrocatechol Violet
PVA Polyvinyl Alcohol
PVC Polyvinyl Chloride
PVD Physical Vapor Deposition
PVK Poly(*N*-Vinyl Carbazole)
PVL Poly(Vinyl Laurate)
PVP Poly(4- or 2-Vinylpyridine)
P2VP-*b*-PEO Poly(2-Vinylpyridine)-*Block*-poly (Ethylene Oxide)
P4VPy Poly(4-Vinyl Pyridine)
PXRD Powder X-Ray Diffraction
Py Pyridine
PyFs Pyrrolidinofullerenes
PyM 1-Pyrenemethanol
Pymo Hydroxypyrimidinolate
pyta 4-Pyridylthioacetate

Qa Quinic Acid
QB-TTP 8-Quinolynylboronic Acid-Modified Thymidine-5′-triphosphate
QCM Quartz Crystal Microbalance
QD Quantum Dot
QELS Quasielastic Laser Light Scattering
QIOEt *N*-(Quinoline-8-yl)-2-(3-Triethoxy-silyl-Propylamino)-Acetamide
QM Quantum Mechanical
QTL Quencher-tether-ligand
Quats Quaternary Ammonium Cation

RAFT Reversible Addition Fragmentation Chain Transfer
RAHB Resonance-assisted H-bonds
RAIRS Reflection–Absorption Infrared Spectroscopy
RASMC Rat Aortic Smooth Muscle Cells
RB Rose Bengal
RBDCC Resin-Bound Dynamic Combinatorial Chemistry
RBM Radial Breathing Mode
RC Reaction Center
RCM Ring-Closing Metathesis
RCSR Reticular Chemistry Structural Resource
rctt-4,4′-tpcb *Rctt*-tetrakis-(4-Pyridyl)Cyclobutane
RDC Residual Dipolar Couplings
RE Recognition Element
REAPDOR Rotational-Echo Adiabatic-Passage Double-Resonance
Reb-im Rebek's Imide
REMPI Resonance-Enhanced Multiphoton Ionization
REREDOR Rotor-Encoded Redor
res Resorcinol
RES Reticuloendothelial System
RET Resonance Energy Transfer
ReTOF Reflectron Time-of-Flight
RF Radio Frequency
RFR Resorcinol-Formaldehyde Resin
RGB Red-Green-Blue
RGDS Arg-Gly-Asp-Ser
RHCl Ranitidine Hydrochloride
RIE Reactive Ion Etching
RIME Receptor-Induced Magnetization Enhancement
RIT Radioimmunotherapy
RLS Resonance Light Scattering
RM Reverse Micelle
RM1 Recife Model 1
rms Root Mean Square
RMSD Root Mean Square Deviation
RNA Ribonucleic Acid
ROESY Rotating Frame Overhauser Effect Spectroscopy
ROMP Ring Opening Metathesis Polymerization
RR Rectification Ratio
RSA Rabbit Serum Albumin
RT-PCR Reverse Transcription Polymerase Chain Reaction
RTP Room Temperature Phosphorescence
($Rubb_n$) Dinuclear Ruthenium

SAA Solvent Assisted Annealing
SAC Symmetry Adapted Cluster
SAED Selected Area Electron Diffraction
SAF Self-Assembling Fiber
SAHB Stabilized Alpha-Helix of Bcl-2 Domain
SAHSA Solvent-Accessible Hydrophobic Surface Area
SAM Self-Assembled Monolayer
SAMIM Solvent-Assisted Micromolding

SAN Styrene Acrylonitrile
SANS Small Angle Neutron Scattering
SAP Self-Assembling Peptide
SAPSA Solvent-Accessible Polar Surface Area
SAR Structure–Activity Relationship
SASA Solvent-Accessible Surface Area
SAW Surface Acoustic Wave
SAXS Small Angle X-Ray Scattering
SB Schottky Barrier
SBBs Supramolecular Building Blocks
SBPs Square-Based Pyramids
SBU Secondary Building Unit
sc Single-Chained
SC-XRD Single-Crystal X-Ray Diffraction
SCAM Substituted-Cysteine Accessibility Method
SCC-7 Squamous Carcinoma Cells
SCE Saturated Calomel Electrode
SCF Supercritical Fluid
SCFT Self-Consistent Field Theory
scFv Single-Chain Variable Fragment
SCLCPs Side-Chain Liquid Crystal Polymers
scr Sacrificial
SCSC Single Crystal to Single Crystal
SCSS Silica-Core/Surfactant-Shell
SD Supradendrimer
SDAs Structure Directing Agents
SDBS Sodium Dodecyl Benzene Sulfonate
sdp 4,4′-sulfonyldiphenolate
SDS Sodium Dodecyl Sulfate
SDS-PAGE Sodium Dodecyl Sulfate Polyacrylamide Gel Electrophoresis
sebn Sebaconitrile
SEC Size Exclusion Chromatography
SEC-MALS Size-Exclusion Chromatography with Multi-angle Light Scattering
SECM Scanning Electrochemical Microscopy
SED Selective Electrochemical Deposition
SEDOR Spin-Echo Double Resonance
SELEX Systematic Evolution of Ligands by Exponential Enrichment
SEM Scanning Electron Microscope
SERS Surface-Enhanced Raman Scattering
SET Single-Electron Transistor or Surface Energy Transfer
SFG Sum-Frequency Generation
SFM Scanning Force Microscopy
SH2 Src Homology 2
SHED Stem Cells from Human Exfoliated Deciduous Teeth
SHG Second-Harmonic Generation
SIA Semiconductor Industry Association
SIAM Scanning Interferometric Apertureless Microscopy
SIMS Secondary Ion Mass Spectroscopy
SIPP Surface Initiated Polymerization
SiRNA Small Interfering RNA
SKIE Steric Kinetic Isotope Effect
SL Solid—Liquid
SLBs Supported Lipid Bilayers
sLe^X Sialyl $Lewis^X$
SLF Separated Local Field Spectroscopy
SLM Supported Liquid Membrane
SLS Static Light Scattering
SLT Shiga-like Toxin
SmA Smectic A
SmC Smectic C
SMFS Single-Molecule Force Spectroscopy
SNARE Soluble *N*-Ethylmaleimide-Sensitive Factor Attachment Protein Receptor
SNase Staphylococcal Nuclease
SNOM Scanning Near-Field Optical Microscope
SNP Single-Nucleotide Polymorphism
SNP Supramolecular Nanoparticle
soc Square-Octahedral
SOF Site Occupancy Factor
SOG Singlet Oxygen Generation
SOPY Spirooxazine-Containing Pyridine
SOSG Singlet Oxygen Sensor Green
SP Spiropyran
SPB Spherical Polyelectrolyte Brushes
SPBJ Scanned Probe Microscopy Break Junction
SPDS2 Spermidine Synthase 2
SPE Solid-Phase Extraction
SPECT Single Photon Emission Computed Tomography
SPEF Single-Photon Excited Fluorescence
SPIO Superparamagnetic Iron Oxide
SPL Scanning Probe Lithography
SPLMOD Spline Model
SPM Scanned Probe Microscopy
SPMNP Superparamagnetic Nanoparticle
SPNs Self-Assembling Peptide Nanotubes
SPPS Solid-Phase Peptide Synthesis
SPR Surface Plasmon Resonance
SPS Poly(Sodium *p*-Sulfonate Styrene)
SPW Surface Plasmon Wave
SQ Squaraine
SQ–DQ Single Quantum–Double Quantum
SQ–SQ Single Quantum–Single Quantum
SQUID Superconducting Quantum Interference Device
SR Supramolecular Reagent

ss	Single-Stranded
SS	Supramolecular Synthon
SS-NMR	Solid-State NMR Spectroscopy
SSB	Solid-Supported Bilayer
ssDNA	Single-Strand DNA
SSL	Strong Segregation Limit
SSM	Separate Solution Method
SSME	*N*-Stearoylserine Methyl Ester
SSP	Strand-Swapping Peptide
s-SWNT	Semiconducting Single-Walled Carbon Nanotube
STD	Saturation Transfer Difference
STD-NMR	Saturation Transfer Difference NMR
STED	Stimulated Emission Depletion
STEM	Scanning Transmission Electron Microscopy
stilbz	Stilbazole
STM	Scanning Tunnel Microscopy
STORM	Stochastic Optical Reconstruction Microscopy
STP	Standard Temperature and Pressure
STS	Scanning Tunneling Spectroscopy
Stx-1	Shiga Toxin Type 1
STXM	Scanning Transmission X-Ray Microscopy
SubPc	Subphthalocyanine
SUV	Small Unilamellar Vesicle
SV	Solid–Vapor
SVLP	Synthetic Virus-like Particle
SWCNH	Single Wall Carbon Nanohorn
SWCNT	Single-Walled Carbon Nanotube
SWV	Square Wave Voltammetry
TβH	Tyramine β-Monooxygenase
TA	Thermally Annealed
TAA	Tetraamide Acycle
tae	Dibasic Tetraacetylethane
TAEC	Total Antioxidant Equivalent Capacity
TAM	Tetraamide Macrocycle
TAMRA	5-(and-6)-Carboxytetramethylrhodamine
TAP	2,4,6-triaminopyrimidine
TAPP	1,3,8,10-tetraazaperopyrene
TATA	Tripropyl-4,8,12-triazatriangulenium
TB	Toluidine Blue
tba	2-Thiobarbituric Acid
TBA	Tetrabutyl Ammonium
TBAB	Tetra-*n*-butyl Ammonium Bromide
TBACl	Tetrabutylammonium Chloride
TBAF	Tetrabutylammonium Fluoride
TBAI	Tetrabutylammonium Iodide
TBBA	*N*,*N*′-Terephthalylidene-Bis-4-*n*-butylaniline
TBC	*P-Tert*-butylcalix[4]arene
TBEP	Tris(2-Butoxyethyl) Phosphate
TBHP	*t*-Butyl Hydroperoxide
TBME	Tert-Butylmethylether
TBP	Tetraethylphosphonium
tca	Tricarballylic Acid
TCA	Trichloroacetic
TCDB	1,3,5-tris(10-Carboxydecyloxy)-Benzene
TCh	Thiocholine
TCID	Threshold Collision-Induced Dissociation
TClPP	Tetrakis(4-Chlorophenyl)Porphyrin
tcm	Tricyanomethanide
TCNQ	7,7,8,8-tetracyano-1,4-quinodimethane
TCPO	Bis(Trichlorophenyl)Oxalate
TD-DFT	Time-Dependent Density Functional Theory
TDA	Taylor Dispersion Analysis
TDCA	Taurodeoxycholic Acid
TDDFT	Time-Dependent Density Functional Theory
TDP	Thymidine Diphosphate
TDPPP	Time-Dependent Pariser–Parr–Pople
TE	Transverse Electric
TEA	Tetraethylammonium
TEAP	Tetraethylammonium Perchlorate
TEB	1,3,5-tris(4-Ethynylbenzonitrile) Benzene
TEG	Triethylene Glycol
TEM	Transmission Electron Microscopy
TEOA	Triethanolamine
TEOS	Tetraethyl Orthosilicate
TEP	Tetrabutylammonium
TF	Thomsen–Friedenreich
Tf	Transferrin
Tfa	3,3,3-Trifluoroalanine
TFA	Trifluoroacetate
TFA	Trifluoroacetic Acid
TFACA	*o*-Trifluoroacetylcarboxanilide
TFDBB	1,4-Dibromotetrafluorobenzene
Tfl	5,5,5-Trifluoroleucine
TFO	Triplex-Forming Oligonucleotide
TfOH	Trifluoromethanesulfonic Acid
TfR	Transferrin Receptor
Tfv	4,4,4-Trifluorovaline
TG	Thermogravimetric
TGA	Thermogravimetric Analysis
TGF	Transforming Growth Factor
THF	Tetrahydrofuran
THP	Tetrahydro-2-pyrimidinone
THPP	5,10,15,20-tetrakis(Hexadecyloxyphenyl)-21*H*,23*H*-Porphine

TIM	Triose Phosphate Isomerase
TIRF	Total Internal Reflection Fluorescence
TIRFM	Total Internal Reflection Fluorescence Microscopy
TLC	Thin Layer Chromotography
TM	Transition Metal
TM	Transverse Magnetic
TMA	Tetramethylammonium
TMACl	Tetramethylammonium Chloride
TMC	Twisted Molecular Chromophore
TMDN	*N,N,N′,N′*-tetramethyl-1,5-diamino-naphthalene
tmdta	Trimethylenediamine Tetraacetate
tmeda	*N,N,N′,N′*-Tetramethylethylenediamine
TMIO	1,1,3,3-tetramethylisoindolin-2-yloxyl
TMO	Trimethylene Oxide
TMP	Thymidine Monophosphate
TMPA	Tris[(2-pyridyl)methyl]amine
TMPyP	tetra(1-methylpyridynium-4-yl) porphyrin p-toluenesulfonate
TMS	Tetramethylsilane
TMSS	Tetrakis(Trimethylsilyl)Silane
TMV	Tobacco Mosaic Virus
TMVCP	Tobacco Mosaic Virus Coat Protein
TNF	2,4,7-Trinitro-9-fluorenone
TNF	Tumor Necrosis Factor
TNFα	Tumor Necrosis Factor-α
TNS	6-(*p*-Toluidino)Naphthalene-2-sulfonate
TNT	2,4,6-trinitrotoluene
TOCSY	Total-Correlation Spectroscopy
TOF	Time-of-Flight or Turn-over Frequency
ToF-SIMS	Time-of-Flight Secondary Ion Mass Spectrometry
TON	Turnover Number
TOT	Tri-*o*-thymotide
TPE	Two-Photon Excitation
TPEF	Two-Photon Excited Fluorescence
TPEN	*N,N,N′,N′*-Tetrakis(2-Pyridylmethyl) Ethylenediamine
TPhP	Tetrabutylphosphonium
TPhP	Tetraphenylphosphonium
tPNAs	Thioester Peptide Nucleic Acids
TPP	Tetraarylporphyrin
TPP	Tris-*o*-phenylenedioxycyclotriphos-phazene
TPPC	Meso-Tetrakis(4-Carboxyphenyl) Porphyrin
TPPS	Tetraphenylporphyrin Tetrasulfonic Acid
TPrA	Tripropylamine
TPSLM	Two-Photon Scanning Laser Microscopy
tpt	2,4,6-tri(4-Pyridyl)-1,3,5-triazine
TPTC	*p*-Terphenyl-3,5,3′,5′-tetracarboxylic Acid
tpy	2,2′ : 6′,2″-terpyridine
trans-ACHC	*Trans*-2-amino-cyclohexanecarboxylic Acid
TRAP	Telomeric Repeat Amplification Protocol
tren	Tris(2-Aminoethyl)Amine
TRI	Triphenylene
TRIS	Tris-(Hydroxymethyl)Aminomethane
Trk	Tropomyosin Receptor Kinase
TRMC	Time-Resolved Microwave Conductivity
TS	Transition-State
TSA	Transition-State Analog
TSB35	1,3,5-tris[(E)-2-(3,5-didecyloxy-phenyl)-Ethenyl]-benzene
tsc	Tetrasulfonatocalix[4]arene
TsCl	Tosylchloride
TsOH	Tosylic Acid
TSP	Travelling Salesman Problem
TSPP	Tetrakis(4-Sulfonatophenyl)Porphyrin
TSQ	6-Methoxy-8-*p*-toluenesulfonami-doquinoline
TTA	Triplet–Triplet Annihilation
TTA-DIA	2,5(10,12-heneicosadiinyloxy) Terephthalic Acid
TTF	Tetrathiafulvalene
TTF–TCNQ	Tetrathiafulvalene—Tetracyanoquino-dimethane
TTHA	triethylenetetramine-*N,N,N′,N″,N‴,N‴*-hexaacetic acid
TTP	Thymidine Triphosphate
UCNC	Upconversion Nanocrystals
UCST	Upper Critical Solution Temperature
UDP	Underpotential Deposition or Uridine Diphosphate
UE	Unimolecular Electronics
UHV	Ultra-High Vacuum
UMP	Uridine Monophosphate
UNI-FF	UNI Force Field
UPD	Underpotential Deposition
UPy	Ureido Pyrimidinone
UQ	Ubiquinone
USANS	Ultra-Small Angle Neutron Scattering
USAXS	Ultra-Small Angle X-Ray Scattering
USFDA	United States Food and Drug Administration
UTP	Uridine triphosphate
5′-UTR	5′-Untranslated Region

UV	Ultraviolet
UV/vis	Ultraviolet/visible
V-UV	Vacuum Ultraviolet
VA	Val-Ala
VACE	Vacancy Affinity Capillary Electrophoresis
VAPG	Val-Ala-Pro-Gly
VBZ	4-Vinylbenzoic Acid
VCAM-1	Vascular Cell Adhesion Molecule 1
VCD	Vibrational Circular Dichroism
VDW	Van der Waals
VEGF	Vascular Endothelial Growth Factor
VEGFR	Vascular Endothelial Growth Factor Receptor
vis	Visible
VP	Vacancy Peak
VSFS	Vibrational Sum-Frequency Spectroscopy
VSM	Vibrating Sample Magnetometer
VT	Variable Temperature
VT-CD	Variable Temperature Circular Dichroism
VUV-CD	Vacuum–Ultraviolet Circular Dichroism
WAXS	Wide Angle X-Ray Scattering
WGA	Wheat Germ Agglutinin
WLC	Wormlike Chain
WPI	World Premier International Research Center Initiative
WS	Water–Surfactant
WT	Wild-Type
XB	Halogen Bonding
XFEL	X-Ray Radiation from Free Electron Lasers
XOR	EXclusive-OR
XPCS	X-Ray Photon Correlation Spectroscopy
XPS	X-Ray Photoelectron Spectroscopy
XRD	X-Ray Diffraction
XRF	X-Ray Fluorescence
XRM	X-Ray Microscopy
XRPD	X-Ray Powder Diffraction
YFP	Yellow Fluorescent Protein
YIGSR	Tyr-Ile-Gly-Ser-Arg
ZFS	Zero-Field Splitting
ZIF	Zeolitic Imidazole Framework
ZMOF	Zeolitic Metal-Organic Framework
ZnNCs	Zinc-Doped Iron Oxide Nanocrystals
ZnPP	Zinc Protoporphyrin IX
ZPE	Zero-Point Energy

Molecular Recognition

Crown and Lariat Ethers

Peter J. Cragg and Rafik Vahora
University of Brighton, Brighton, UK

1 INTRODUCTION

The origin of supramolecular chemistry is inextricably linked to the recognition of cations. Perhaps the primary reason that cation recognition was an early target for supramolecular chemists was the level of success they could achieve by following that path. Flexible crown ethers are able to encapsulate alkali, alkaline earth, and lanthanide metal cations, as all the target cations exhibit essentially spherical electron densities and are of similar sizes, and the more rigid tetraamide calix[4]arene derivatives bind both Na^+ and Tb^{3+} in a well-defined cavity. In many cases, complexation results in crystalline products that are amenable to X-ray diffraction and other studies. Transition metal cations also have well-known binding preferences so that macrocycles or podands can be designed to complement specific metal cations. These factors allowed rapid advances in complex characterization to occur. Organic cations, for example, protonated amines and guanidinium derivatives, have also been relatively easy to target through complementary hydrogen bonding motifs. By way of contrast, the design of hosts for neutral or anionic guests is a far harder task and less likely to yield informative results.

Pioneering supramolecular chemists were inspired by nature where small biological molecules function as cation receptors (see **Biological Small Molecules as Receptors**, Volume 3). For example, calixarenes (see **Calixarenes in Molecular Recognition**, Volume 3), though discovered as a result of phenol–formaldehyde polymer research, have strong biological connections. Research that had begun in the 1950s, and which continues to this day, has demonstrated the antitubercular effects of the larger members of the class. The rigid structure of the calix[4]arenes inspired a renaissance in their chemistry during the 1970s and 1980s when their potential to act as the framework for synthetic enzymes was explored. With two areas for functionalization available, cations could be bound at the lower rim and small reactants attracted to the upper rim. Recently, this goal has been realized and has had a major impact on supramolecular bioinorganic chemistry (see **Supramolecular Bioinorganic Chemistry**, Volume 4).

Synthetic macrocycles such as the cucurbiturils (see **Cucurbituril Receptors and Drug Delivery**, Volume 3), which are cyclic polymers of glycoluril, have been shown to bind K^+ through multiple weak interactions with convergent ketone groups reminiscent of K^+–valinomycin complexes. Even stronger cation complexation is seen with the encapsulating cryptands and spherands (see **Cryptands and Spherands**, Volume 3). More complicated biological pore-forming molecules form cation-specific transmembrane channels and others transport ion pairs (see **Anion Receptors Containing Heterocyclic Rings**, Volume 3; **Membrane Transport**, Volume 4).

The cyclic tetrapyrrole structural motif (see **Porphyrins and Expanded Porphyrins as Receptors**, Volume 3)

Supramolecular Chemistry: From Molecules to Nanomaterials.
Edited by Philip A. Gale and Jonathan W. Steed.
© 2012 John Wiley & Sons, Ltd. ISBN: 978-0-470-74640-0.

found at the active sites of hemoglobin, chlorophyll, and some hydrogenase enzymes has been employed by supramolecular chemists as a means of creating a metal-binding planar macrocycle. Phthalocyanine, a structurally similar but entirely artificial tetraaza planar macrocycle, was first identified as its iron complex, which formed as an unintended by-product in the manufacture of phthalimide from phthalic anhydride and ammonia in iron vats (see **Supramolecular Phthalocyanine-Based Systems**, Volume 3). Since then it has been used in highly stable dyes, as its iron or copper complex and as a sensitizing agent in photodynamic chemotherapy. Salicylaldehyde-derived Schiff bases are also able to bind transition metals, thereby providing analogs of vitamin B_{12} and manganese superoxide dismutatse.

When sequestering transition metal cations, Nature uses siderophores such as enterobactin to provide convergent donor atoms that match the metals' preferred octahedral coordination environment. The threefold symmetry of enterobactin and its phenolic termini can be mimicked by podands (see **Podands**, Volume 3), particularly tripodal amines (see **Schiff Base and Reduced Schiff Base Ligands**, Volume 3). Ionophores such as monensin, nonactin, and valinomycin undertake highly selective alkali metal cation transport across cell membranes; their supramolecular analogs can be found in host molecules with convergent donor atoms, notably the crown and lariat ethers, which can function as phase transfer catalysts. It is these compounds that will be the focus of this chapter.

Crown ethers are recognized throughout chemistry and chemistry-related disciplines: their discovery merited a Nobel Prize and their academic and industrial applications are legion. The simple cyclic polyether structure of the original all-oxygen-containing crowns had a pleasing symmetry and fascinating cation-binding properties, but the field rapidly moved on leading to the syntheses of crowns in which oxygen and other functional groups that were originally present in the macrocyclic framework were substituted with other donor atoms. These developments allowed functionality to be increased through the incorporation of side arms, attached either to a carbon atom or to a nitrogen atom (substituting for an ether oxygen) to give the lariat ethers.

Lariat ethers have shown us how much supramolecular chemistry can be advanced by linking one receptor site, the crown ether, to another such as an electroresponsive group, a fluorescent antenna, an aromatic ring system, or a complementary binding motif for a carboxylic acid. The lariat ethers have helped us understand cation-π interactions, have been incorporated into chromatographic columns for chiral separations, have been used to develop logic gates, and have found applications in critical analyte detectors that are used in clinical settings worldwide.

The following account will attempt to illustrate how crown ethers rose from their obscure origin to such prominence in the world of supramolecular chemistry, from being regarded as impurities in a reaction performed in an industrial laboratory to the iconic status these molecules have today. The development of the lariat ethers is charted, and important derivatives are discussed with a focus on their uses as molecular sensors.

2 CROWN ETHERS

2.1 Origins

In 1957, a patent was published on the reaction of ethylene oxide with a range of bases that yielded an unexpected cyclic product containing the $-CH_2CH_2O-$polyether repeat pattern.[1] The synthesis of the cyclic tetramer was described but no applications, beyond its proposed use as a high boiling solvent, were described. Ten years later a research chemist approaching retirement was credited with the discovery of macrocycles, now known as crown ethers, that include this cyclotetrameric ether.

The origins of crown ethers are usually traced back to Charles Pedersen's groundbreaking 1967 paper[2] in which he describes the synthesis and metal binding properties of almost 50 macrocycles containing the $-CH_2CH_2O-$ repeating structural motif. Famously, he was working for Du Pont in Wilmington, Delaware and attempting to prepare bis[2-(hydroxyphenoxy)ethyl]ether from 2-(hydroxyphenoxy)tetrahydropyran and diethylene glycol dichloride as shown in Figure 1. His target was a compound designed to bind the vanadyl cation, VO^+, through reaction with $VOCl_4$ or VCl_4, in order to produce an olefin polymerization catalyst. On 5 July 1962, to his surprise, he isolated an unusual material that had unexpected solubilizing effects on alkali metal cations. Given that the material was isolated in less than 1% yield, most chemists would have ignored it and turned their attention to the major products. Fortunately, Pedersen was very inquisitive and, in his own words:

> *Crown ethers were in no danger, because of my natural curiosity.*[3]

Upon careful analysis, the product was found to be a macrocycle containing two benzene rings, six oxygen atoms, and ethylene spacers between the oxygens. It had formed through the reaction of diethylene glycol dichloride and catechol. Pedersen had known that the 2-(hydroxyphenoxy)tetrahydropyran was contaminated with about 10% of the unprotected catechol but had decided to proceed with the reaction in any case. Following his chance result Pedersen carefully repeated the experiments

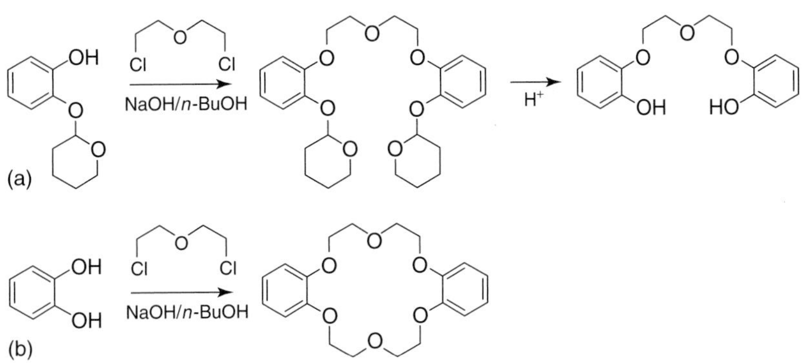

Figure 1 Pedersen's fortuitous route to dibenzo-18-crown-6: (a) target compound; (b) minor product.[2]

with a variety of starting materials and was able to prepare many similar macrocycles in yields of up to 80%. In the introduction to his 1967 paper, which reflected five years of his research, he noted that the structures of his heterocyclic compounds bore similarities to cyclic ethers obtained from resorcinol by Ziegler and Luttringhaus,[4] the cyclic furan "tetroxaquaterenes" prepared by Ackman *et al.*,[5] and the macrocycle 1,1′,4,4′-bis(trimethenedioxy)dibenzene reported by Adams and Whitehill.[6] Reference was made to the synthesis of macrocycles by the polymerization of propylene oxide using $BF_3 \cdot Et_2O$ by the Wilkinson group.[7] Their product was later observed to give a blue solution in the presence of metallic potassium and a sodium–potassium alloy, though these observations were not investigated further. Pedersen was also aware of the 1957 patent granted to Stewart *et al.* for the cyclotetramer of ethylene oxide[1] as it was also referenced. Similarities between the compounds that pre-date the crown ethers and the crowns themselves can be seen in Figure 2. Significantly, Pedersen's compounds were synthesized in "one-pot" reactions and could be carried out on large scales: the first experiment described in the 1967 paper, a reaction between catechol and 1,11-dichloro-3,6,9-trioxaundecane, yielded 174 g of the target macrocycle.

After Pedersen's work became more widely known, Dale and colleagues reinvestigated the macrocyclization of ethylene oxide and demonstrated that not only could the cyclic tetramer be prepared but that it also was possible to make a series of larger macrocycles.[8] The relative yields of these macrocycles were highly dependent on the nature of the alkali metal used as a base in the reaction, leading to the observation that the compounds required a template in order to cyclize. When no metal was used, only small amounts of macrocycles were obtained as indicated in Table 1.

2.2 Nomenclature

Pedersen realized that the complex systematic chemical names for his compounds did not reveal much about

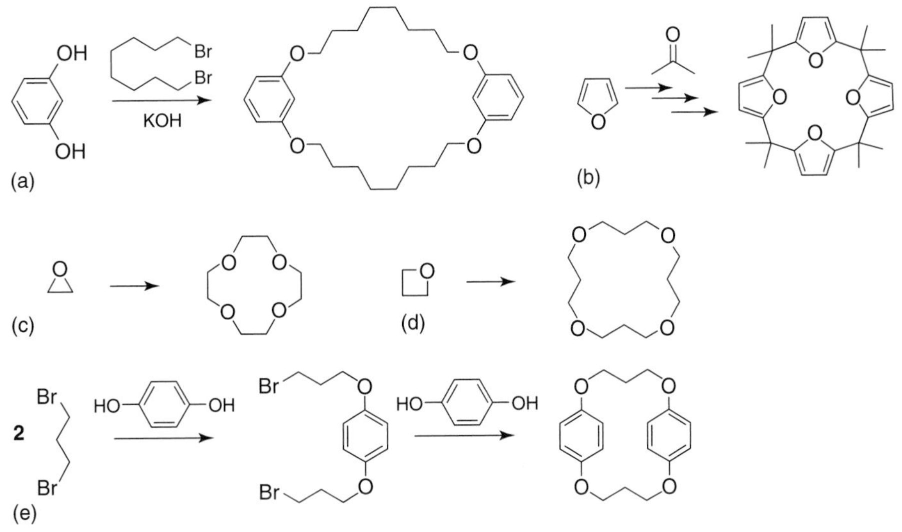

Figure 2 Crown ethers and related compounds reported prior to 1960: (a) Lüttringhaus and Ziegler[4]; (b) Ackman *et al.*[5]; (c) Adams and Whitehill[6]; (d) Down *et al.*[7]; (e) Stewart *et al.*[1]

Table 1 Templated synthesis of crown ethers from ethylene oxide.

	12-crown-4	15-crown-5	18-crown-6
ethylene oxide + BF_3	15%	5%	4%
ethylene oxide + $BF_3/LiBF_4$	30%	70%	–
ethylene oxide + $BF_3/NaBF_4$	25%	50%	25%
ethylene oxide + BF_3/KBF_4	–	50%	50%
ethylene oxide + $BF_3/RbBF_4$	–	–	100%
ethylene oxide + $BF_3/CsBF_4$	–	–	100%

their interesting structures and proposed a nomenclature on the basis of the number of atoms in the macrocycle, which is still in use today. The convention is to start with the number of atoms, carbon or otherwise, in the macrocyclic backbone. This is followed by the word "crown," in recognition of the cyclic nature of the compounds, before giving the number of oxygen atoms in the structure. Thus, an 18-membered macrocycle containing six oxygen atoms is named 18-crown-6 rather than 1,4,7,10,13,16-hexaoxacyclooctadecane. When the macrocycle contains other groups such as a benzene ring, or different heteroatoms such as sulfur or nitrogen, this information is included as a prefix. Consequently, the original 18-crown-6 derivative discovered by Pedersen is named dibenzo-18-crown-6, rather than 2,3,11,12-dibenzo-1,4,7,10,13,16-hexaoxacyclooctadeca-2,11-diene. Using the same approach, 15-crown-5 containing one nitrogen atom in place of an oxygen is called aza-15-crown-5, rather than the more unwieldy 1,4,7,10-tetraoxa-13-azacyclopentadecane.

A vast majority of crown ethers have the $-CH_2CH_2X-$ repeating structure (where X is oxygen or another heteroatom) so that there is one heteroatom for every two carbons. This leads to the sequence of 9-crown-3, 12-crown-4, 15-crown-5, and so on. Other examples have been prepared where extra carbon atoms have been introduced as in 16-crown-5 or the compound known as cyclam (1,4,8,11-tetraazacyclotetradecane) both of which are in the crown class of macrocycles.

2.3 Synthesis

2.3.1 *Aromatic crowns*

Following his accidental discovery of dibenzo-18-crown-6, Pedersen optimized its synthesis from catechol and diethylene glycol dichloride in *n*-butanol. A solution of catechol in *n*-butanol was refluxed with solid sodium hydroxide and a solution of diethylene glycol dichloride added in stages with the latter in ~3% excess. Following a 16 h reflux, the mixture was treated with concentrated hydrochloric acid and the *n*-butanol removed by distillation. Water was added to keep the overall volume constant until the vapor temperature reached 100 °C. Upon cooling, the mixture was filtered and washed with water before being dispersed in acetone. Further filtration, washing with acetone, and oven drying at 100 °C gave the product in yields of 44–48%. Pedersen worked on an appropriately industrial scale, producing about 250 g in a single reaction, and in a typically off-hand manner noted that:

> *the acetone filtrate and wash contained about 16 g of the desired product, but its recovery from a single run is not worth the trouble.*[2]

When an aromatic group is required in a crown ether, it is usually incorporated through the reaction of catechol because it gives an ideal coordination environment for a deprotonating metal to remain bound to the ligand. The same metal then forms a template that preorganizes the difunctionalized polyether, introduced to form the remainder of the macrocycle, such that it can react with the phenolic nucleophile.

2.3.2 *Alicyclic crowns*

Shortly after Pedersen's initial work, which led to a share of the 1987 Nobel Prize for Chemistry,[9] Greene reported a much improved synthesis of 18-crown-6 along with its higher homologs.[10] The problem encountered by Pedersen was that while crowns containing aromatic groups could be formed in yields of 20–80%, simple crowns such as 18-crown-6 were only isolated in less than a 2% yield. He had already noted that the size of the metal cation appeared to influence the success of the reaction:

> *The conclusion that the relative sizes of the holes and the cation control the stoichiometry seems inescapable.*[11]

This observation led to the notion that particular alkali metal cations appear to have specific affinities for different crown ethers. Greene's insight was to use the appropriate metal cation to template the formation of the crown in much the same way as transition metals had been used as templates for other, nitrogen containing, ligands and to experiment with different solvents. In every example of crown ether synthesis reported by Pedersen, the only metal cation present was Na^+ (except in one case where K^+ was used) as no link had been made between the size of the metal cation used and the target crown at that time. The purpose of a template in the synthesis of 18-crown-6 can be seen in Figure 3. The relationship between metal

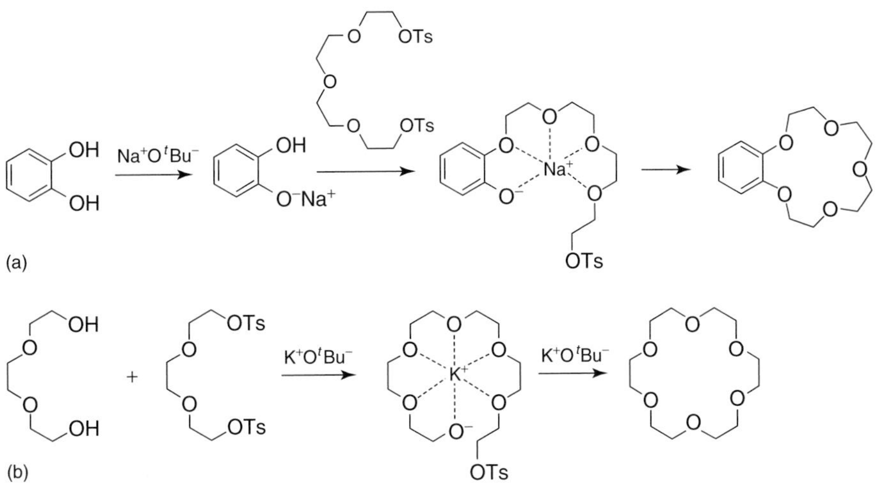

Figure 3 Template synthesis: (a) benzo-15-crown-5; (b) 18-crown-6.

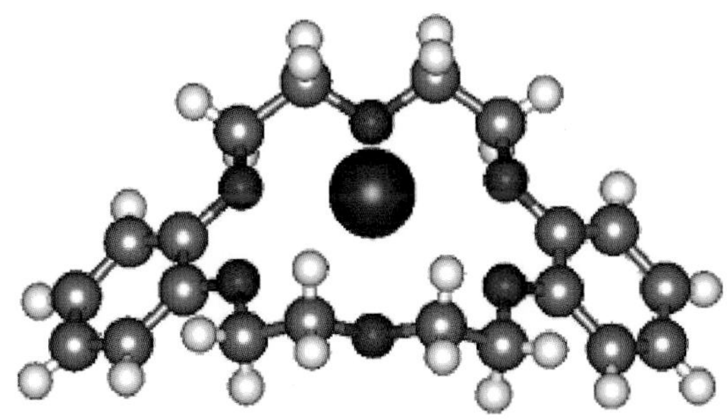

Figure 4 The crystal structure of the [Rb·dibenzo-18-crown-6]$^+$ cation.[12]

Table 2 Crown ether cavity sizes and cation diameters.

	Cavity size (Å)	Best fit cation/ diameter (Å)
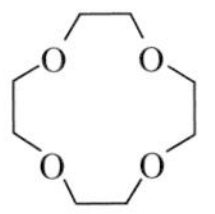	1.2 – 1.5	Li^+/1.36
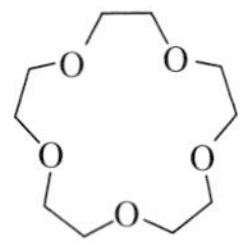	1.7 – 2.2	Na^+/1.90
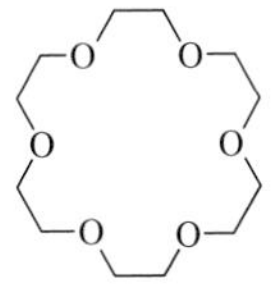	2.6 – 3.2	K^+/2.66

size and fit to a specific crown ether was later supported by crystallographic evidence from Bright and Truter who reported the solid state structure of [(Rb·dibenzo-18-crown-6)$^+$ NCS^-] in 1970.[12]

The size mismatch between crown cavity and metal size is clear in Figure 4 as the cation is forced to "perch" on top of the crown ether. Consideration of data derived from crystallography, computer modeling, and other sources has led to the notion that crowns' affinities for alkali metal cations relates to the match between the cavity size of the former and the ionic diameters of the latter, as indicated in Table 2. This relationship is broadly true; however, certain exceptions occur. For example, many cations form sandwich structures with crown ethers in ratios of 2 : 1 or 3 : 2, and 18-crown-6 appears to be the best crown ligand not only for K^+ but also for NH_4^+, H_3O^+, Na^+, and Ca^{2+}. The attraction to cations is greatly influenced by the interaction between the metal guest and the heteroatoms in the host crown: trithia-9-crown-3 binds transition metals more effectively than its all-oxygen analog, 9-crown-3.

The template effect has a very practical consequence. Ordinarily, when a macrocycle is to be prepared from a bifunctional precursor, linear polymers are the major products. To promote cyclization over linear polymerization, the so-called high dilution conditions were often used. The principle underlying this approach is that if the species are in very dilute solutions, the probability of their reactive ends meeting up, which is a first-order phenomenon, will be greater than the probability of two molecules meeting and reacting with each other, which is a second-order phenomenon. Reactions of this type are generally carried out using motorized syringes that simultaneously but slowly add the reactive compound and the reagent required to initiate its activation. Rapid stirring is essential as is a vast volume of solvent necessary to deliver the compounds at concentrations of 10^{-3}–10^{-4} M. With moves to limit the amount of solvents used in chemical synthesis, both within industry and academia, processes that reduce solvent volumes are to be welcomed. The importance of the template

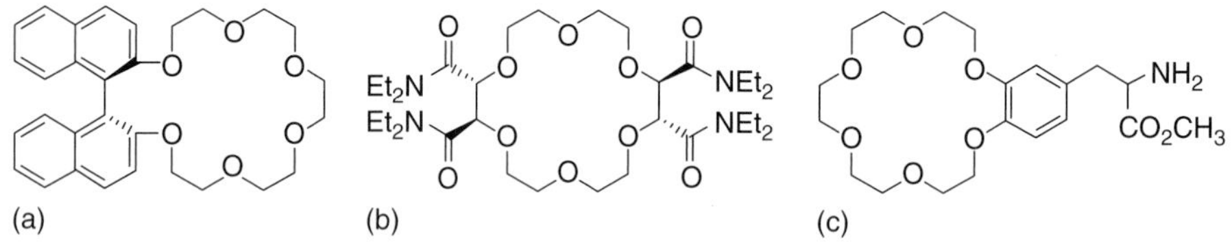

Figure 5 Chiral crowns: (a) a binaphthyl 18-crown-6 derivative[13]; (b) an 18-crown-6 derivative prepared from maleic acid[14]; (c) an 18-crown-6 DOTA derivative.[15]

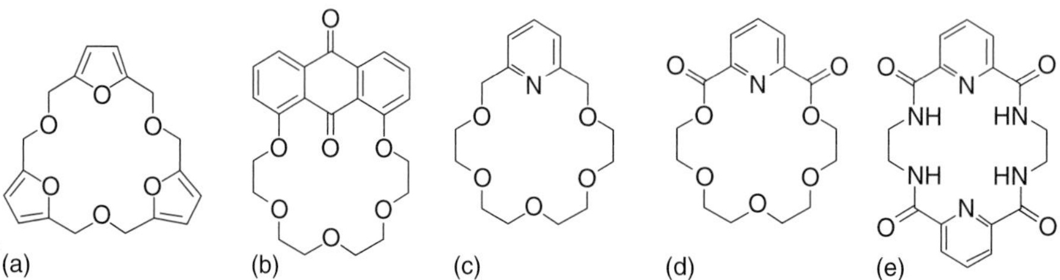

Figure 6 Heterocrowns: (a) a furanyl derivative[16]; (b) an anthraquinone crown[17]; (c) a pyridylcrown[18]; (d) a macrocyclic polyester[19]; (e) a cyclotetralactam.[20]

method is that it organizes the reactive molecules such that the major product will be the desired macrocycle. Consequently, many macrocyclization reactions can be performed in very concentrated solutions: the reagents in Pedersen's dibenzo-18-crown-6 reaction are present at approximately 0.8 M. Template synthesis therefore promotes the formation of macrocycles over linear polymers to give very high yields while greatly reducing the amount of solvent needed.

2.3.3 *Optically active crowns*

The crown ethers described so far are able to bind simple metal cations and, in the case of 18-crown-6, the ammonium cation. The latter observation led one of Pedersen's fellow Nobel Laureates, Donald Cram, to consider if crown ethers could be used to resolve racemic mixtures of amino acids. None of the crown ethers prepared by Pedersen was chiral; however, the Cram group was able to synthesize 18-crown-6 derivatives that incorporated one, two, or three binaphthyl groups.[13] These crown ethers could be separated into their enantiomers and attached to a silica support. The modified silica was subsequently used to separate racemic mixtures of amino acids by chromatography. Early experiments gave separation factors for racemic mixtures of simple amino acid methyl esters in the region of 1.5–3.5.

Using a different approach, the group of Jean-Marie Lehn, the third of Pedersen's fellow Nobel Laureates in 1987, prepared chiral derivatives of 18-crown-6 from protected tartaric acid to give crowns with four acid groups.[14] As an alternative strategy, the Voyer group later used the catechol functional group of L-3,4-dihydroxyphenylalanine (L-DOPA) as a way to incorporate a chiral center into a crown ether.[15] The approach was remarkable in its simplicity: the methyl ester of L-DOPA was *N*-Boc protected and reacted with pentaethylene glycol ditosylate in the presence of Cs^+ in dimethylformamide at 60 C to yield the protected chiral benzo-18-crown-6 derivative. Examples of these chiral crown ethers are shown in Figure 5.

2.3.4 *Heterocyclic crowns*

Incorporation of heterocycles, exemplified by the compound prepared by Ackman and colleagues,[5] has been investigated by many groups. Variations on the 18-crown-6 motif were reported by Timko and Cram who prepared bifunctional furan derivatives and formed macrocycles in the presence of alkali metal templates as shown in Figure 6.[16] Incorporation of anthraquinone was achieved by the Gokel group in order to investigate the effect that quinine reduction would have on metal binding.[17] Pyridylcrowns and macrocyclic polyesters containing pyridine were reported by the groups of Cram[18] and Vögtle,[19] respectively. Using similar methods, Szumna and Jurczak prepared a tetralactam as an anion binding macrocycle.[20] This cyclization method has recently been applied in the synthesis of rotaxane "beads" by the groups of Leigh[21] and Smith,[22] in which the macrocycle is formed around the rotaxane "thread" that acts as the template.

2.3.5 *Azacrowns*

Several methods have been developed for azacrown ether synthesis. The most obvious approach was to allow a bifunctional polyether, such as a diiodide, react with an amine. Calverly and Dale used this method to prepare

N-benzylaza-12-crown-4 by the sodium templated reaction of benzylamine with tetraethylene glycol diiodide.[23] Subsequent hydrogenolysis with 10% Pd/C in aqueous acetic acid gave the parent aza-12-crown-4 in a 40% yield from the diiodide. Aza-15-crown-5 can also be prepared this way; however, aza-9-crown-3 cannot: when triethylene glycol diiodide reacts with benzylamine, a 2 + 2 cyclization occurs to give N,N'-dibenzyldiaza-18-crown-6, as the Gokel group discovered.[24]

Lockhart and coworkers incorporated nitrogen into a crown ether through the reaction of *ortho*-aminophenol with triethylene glycol dichloride[25] and obtained two products, azabenzo-15-crown-5 and N-(2-phenol)aza-12-crown-4. The latter was an early example of a lariat ether, discussed below.

Greene's paper on the improved synthesis of crown ethers also included the first example of a crown with mixed heteroatoms.[10] He prepared N-tritylaza-18-crown-6 from tetraethylene glycol ditosylate and N-trityldiethanolamine. The trityl group was then removed by hydrochloric acid to give a 28% yield of aza-18-crown-6 from the glycol. A variation of Greene's synthesis was reported by Gokel and Garcia[26] who reacted N-benzyldiethanolamine with tetraethylene glycol ditosylate to give the N-protected derivative in a 25% yield. This was debenzylated to form aza-18-crown-6. A similar method was used by the Sutherland group in 1979 where N-tosyldiethanolamine was treated with tri- or tetraethylene glycol ditosylate in dry tetrahydrofuran with sodium hydride acting as the base.[27] Detosylation was achieved with $LiAlH_4$. Okahara and coworkers[28] subsequently undertook a thorough study of azacrown ether synthesis. From diethanolamine and the appropriate polyethylene glycol ditosylate, the group was able to harness the template effects of alkali metals and produce a family of azacrowns, from aza-12-crown-4 through to aza-21-crown-7, in yields ranging from 3% (aza-12-crown-4) to 78% (aza-18-crown-6). Examples of these routes are shown in Figure 7.

One of the driving forces behind the interest in azacrowns was the increased functionality that they offered. As will be seen later, nitrogen can act as a bridgehead from which a variety of side arms are easily attached. As well as the monosubstituted compounds, routes to diazacrowns, such as 1,9-diaza-18-crown-6, were being developed. This work had been pioneered by the Lehn group as 1,9-diaza-18-crown-6 was required as a starting point for the synthesis of their encapsulating cryptands.[29] The synthesis of 1,9-diaza-18-crown-6 was similar to that used to prepare macrocyclic polyesters: a triethylene glycol diamine reacted with triethylene glycol diacid chloride in high dilution to give the diamide that was reduced to the diazacrown as shown in Figure 8. Many other strategies have been adopted to generate different regioisomers of azacrowns containing two or more nitrogen atoms as detailed in early reviews.[30,31] In addition to their potential as points for side arm attachment, nitrogen donor atoms also have different cation binding preferences to oxygens as formalized in Pearson's "hard soft acid base" principle. So, while 18-crown-6 has no affinity for transition metals, 1,9-diaza-18-crown-6 forms 1 : 1 complexes with Cu^{2+} and other transition metals.[32] The smaller azacrowns, aza-15-crown-5 and 1,7-diaza-15-crown-5, bind La^{3+} and Hg^{2+} respectively, among many metals.[33,34] Increasing the number of nitrogen donors, as in 1,4,7-triaza-15-crown-5, leads to a Cu^{2+} complexing ligand.[35] The small size of these cations compared to those bound by all-oxygen crowns often leads to 2 : 1 complexes with the larger crowns or 1 : 1 complexes with the smallest crowns such as the 9-crown-3 derivatives.

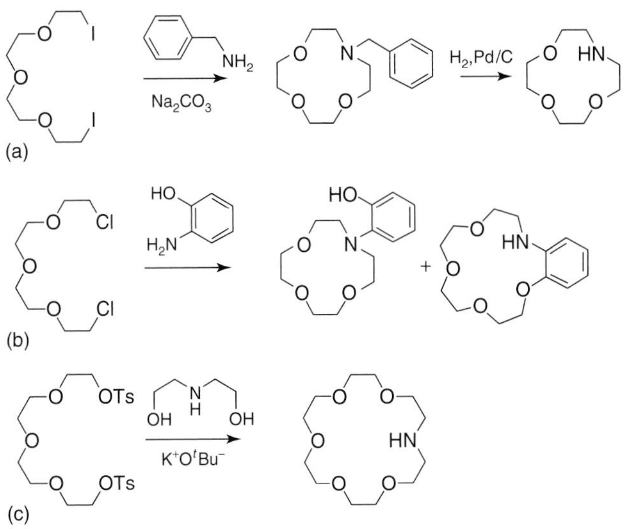

Figure 7 Azacrown ethers: (a) aza-12-crown-4 synthesis[23]; (b) an early azacrown lariat ether[25]; (c) a convenient route to azacrowns.[28]

2.3.6 *Thiacrowns*

Thiacrown ethers may well have first been prepared by the remarkable Bengali chemist Sir Prafulla Chandra Rây who reported the synthesis of cyclic triethylene tri- and tetrasulfides in 1920,[36] though this was later disputed by Ochrymowycz.[37] In 1934, Meadow and Reid reported the synthesis of numerous cyclic compounds from dimercaptans and dihalides including tetrathia-18-crown-6 and hexathia-18-crown-6.[38] In 1961, the synthesis of 1,10-dithia-18-crown-6 was described by Dann and coworkers who reacted Na_2S with 1,2-bis(2-chloroethoxy)ethane in the presence of Na_2CO_3.[39] In addition to producing linear polymers, the reaction yielded a crystalline solid with a molecular mass of 295, an empirical formula of $C_6H_{12}SO_2$ and no reactive thiol groups. Other data led the authors to note that:

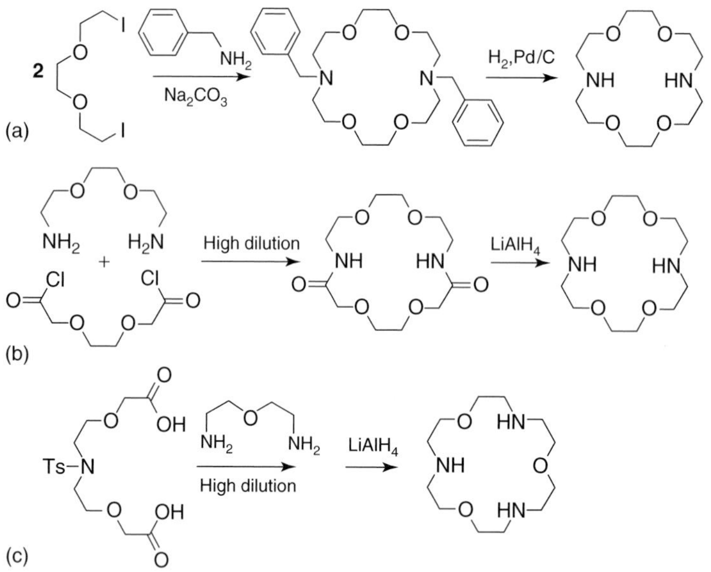

Figure 8 Di- and triazacrown ethers: (a) diazacrown ether synthesis by a 2 + 2 reaction[24]; (b) diazacrown ether synthesis via the diester[29]; (c) triazacrown ether synthesis via the diester.

> *The best structure which can be written to conform to the facts just given. . . is that of an eighteen-membered ring compound. . .*[39]

The authors proceeded to make the disulfone and diethioiodide salt to confirm their suspicions. In the same paper, they reported the syntheses of thia-12-crown-4, 1,7-dithia-12-crown-4, and 1,13-dithia-24-crown-8. The work was not referenced in Pedersen's 1967 paper but he went on to make a systematic investigation of the synthetic routes to thiacrowns in 1971.[40]

Bradshaw and colleagues prepared numerous thiaoxacrown ethers by the reaction of polyethylene glycol dihalides with dithiols and sulfur-containing polyethers in ethanol with sodium or sodium hydroxide acting as the base.[41,42] The yields were quite variable and most compounds required chromatographic separation.

In 1981, Buter and Kellogg noted that the problem facing macrocyclic chemists wishing to use thiacrowns was one of synthetic difficulty and devised a general method that could be used to synthesize a range of sulfur-containing macrocycles from dithiols and dibromides.[43] Dimethylformamide was found to be the preferred solvent but the important factors in the success of the reactions were that the reagents had to be of high purity and, crucially, that the base used to deprotonate the dithiol had to be cesium carbonate. Unlike the synthesis of other crown ethers, where the choice of metal is dictated by the size of crown that is to be made, there appeared to be a special cesium effect in the synthesis of thiacrown ethers. The reaction between decane-1,10-dithiol and 1,5-dibromopentane with five metal carbonates produced 1,7-dithiacycloheptadecane in yields varying from 0% (lithium) up to 90% (cesium). Mixed thiaoxocrowns could be prepared by this method; for example, 1,10-dithia-18-crown-6 was isolated in a 90% yield. A summary of these methods is illustrated in Figure 9.

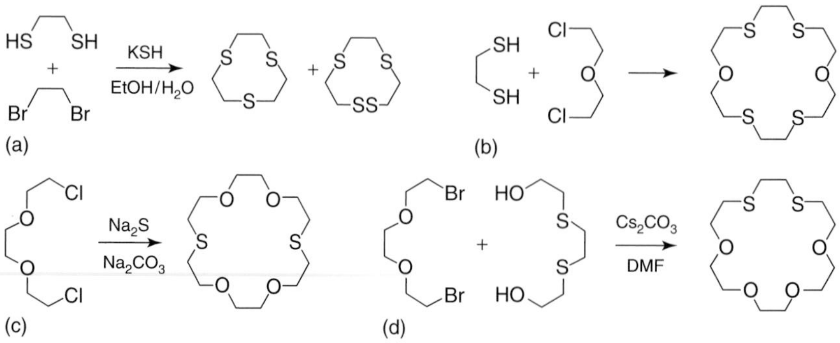

Figure 9 Thiacrown ethers: (a) trithia-9-crown-3[36]; (b) 1,4,10,13-tetrathia-18-crown-6[38]; (c) 1,10-dithia-18-crown-6[39]; (d) 1,3-dithia-18-crown-6.[43]

2.3.7 *Azathiacrowns*

Nitrogen, oxygen, and sulfur all have slightly different affinities for guest metals so, to take advantage of these subtle effects when targeting specific metals, it would be useful to have synthetic routes to crown ethers with mixtures of donor atoms. As all three elements have been incorporated into crown ethers, the strategy is to determine which bonds are to be formed in the macrocyclization. It is possible to form either nitrogen–carbon or sulfur–carbon bonds at the ring closing stage. For nitrogen–carbon bond formation, terminal amine groups may be either unprotected, to react with acid chlorides or other similar functional groups, or protected as the monotosylates, to react with polyethylene glycol ditosylates. For sulfur–carbon bond formation, the amines also need to be protected, usually as the tosylates, so that terminal thiols can react with alkyl halides or polyethylene glycol dihalides rather than the amines. Following cyclization, the conventional method to remove the tosyl groups is treatment with an alkali metal in liquid ammonia because the use of concentrated sulfuric acid, which is effective with other nitrogen-containing macrocycles such as the cyclams, is too harsh for the ether or thioether links. These approaches are illustrated in Figure 10.

2.3.8 *Anticrowns*

All the crown ethers discussed so far have electronegative heteroatoms incorporated into their frameworks. What if this was reversed and the heteroatoms were electropositive with respect to the carbon backbone of the macrocycle? Rather than attract cations, such macrocycles could act as anion hosts as in the example of the mercury-linked tetrafluorobenzenediol system reported by Shur and colleagues.[44] It was with this in mind that the Newcomb group designed and synthesized neutral macrocycles in which the heteroatoms were tin.[45] The di- and tetrastannyl macrocycles, shown in Figure 11, bound fluoride[46] and would later be classed as "anticrowns," a term coined by the Hawthorne group for their trimeric mercury-linked carborane compound.[47]

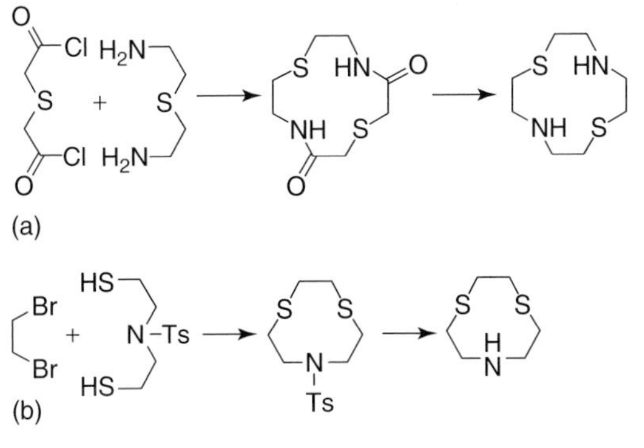

Figure 10 Azathiacrown synthesis; (a) N–C bond formation; (b) S–C bond formation.

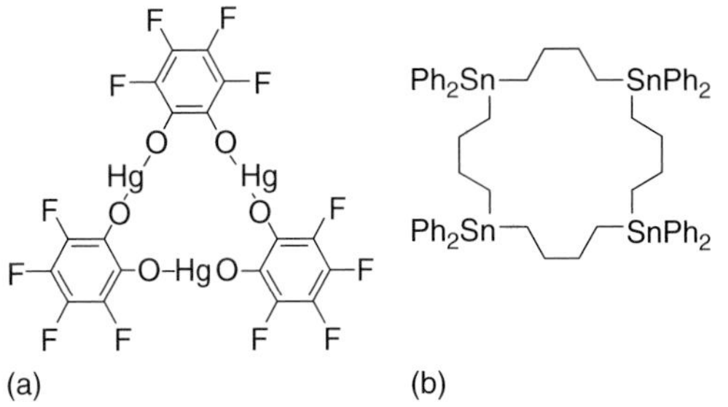

Figure 11 Anticrowns: (a) a mercurial anticrown[44]; (b) a stannyl anticrown.[45]

2.4 Crown ether crystal structures

Initial assertions made by Pedersen concerning the selectivity and size specificity of crown ethers were based on his observations that certain combinations of crowns and salts gave precipitates or crystalline products. It was assumed that the crowns adopted a conformation in which the heteroatoms' lone pairs were directed inwards toward the guest ion. This size specificity hypothesis was supported by Corey–Pauling–Koltun (CPK) molecular models and, later, in the form of single crystal X-ray studies, as noted above.[12] Subsequent structures showed that the relationship between crown size and preferred guest applied broadly to the basic crown ethers.[48] Three binding modes have been found to predominate. Where the crown cavity and size of cation are a good fit, then the guest lies in the plane of the macrocycle as defined by the positions of the heteroatoms. In the absence of other factors, the heteroatoms of 18-crown-6 complexes are often observed to alternate above and below the macrocyclic plane such that the cation appears to be held between two triangles of donor atoms. Where the guest is slightly too large for the macrocyclic cavity, as in $[K\cdot 15\text{-crown-}5]^+$ complexes, a "perching" mode is seen. The cation still interacts with all the donor atoms but now lies above the macrocyclic plane. In an extension of this type of behavior, the guest may be large enough to bind two macrocycles in a 2 : 1 sandwich complex. Finally, if the guest is much smaller than the macrocyclic cavity, then asymmetric binding may be seen where only some of the donor atoms are involved in guest binding. Large, flexible macrocycles may fold over to involve as many donor atoms as possible or they may bind multiple guests. Examples of all these binding modes can be seen in Figure 12.

Crystallographic evidence is of great importance in supramolecular chemistry as it is sometimes the only way to prove the formation of the target supermolecule. This is particularly true of weakly interacting host–guest systems,

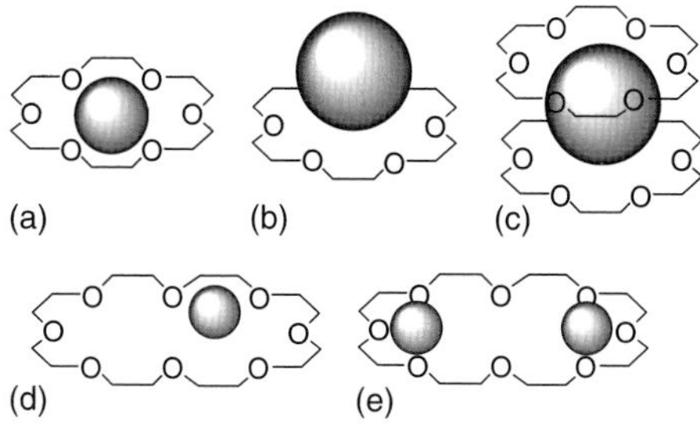

Figure 12 Crown ether-cation binding modes; (a) good fit; (b) perching; (c) sandwich; (d) small cation; (e) multiple cations.

such as those that form when the crown acts as part of a molecular sensor and the guest is an analyte, and is the best evidence for selectivity. After crystallographic evidence emerged for host–guest complementarity, computational methods were quickly employed to probe the generality of this apparently simple relationship.[49] In addition to alkali metals and alkaline earth metals, lanthanides also formed stable crystalline complexes with crowns.[50] Crystallographic evidence showed that 18-crown-6 was an ideal size and geometry to bind both NH_4^+ and H_3O^+ cations[51,52] in addition to the lanthanides and K^+.[53] These structures are illustrated in Figure 13. The complementarity between crown and cation sizes is emphasized when the relative stability constants of the complexes[54] are plotted as in Figure 14.

3 LARIAT ETHERS

3.1 *C*-Pivot lariat ethers

Many of the first crown ethers to be prepared were benzo or dibenzo derivatives. These were often amenable to arene substitution reactions that made the products lariat ethers in the current sense but, unlike later examples, the intention was not to enhance guest binding. The first report of a true lariat ether was from Gokel in 1980 whose group reacted epichlorohydrin with several phenols and alcohols resulting in the formation of diols. These

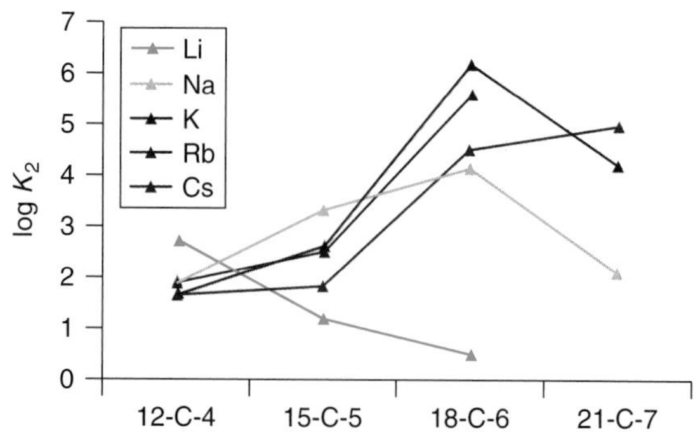

Figure 14 Alkali metal complexation by 21-crown-7, 18-crown-6, 15-crown-5, and 12-crown-4 in methanol.[53]

subsequently formed *C*-pivot lariat ethers upon reaction with tetraethylene glycol ditosylate (or dimesylate) in the presence of NaH.[55] The authors acknowledge a similarity to hydroxymethyl-18-crown-6 designed by Montanari and Tundo[56] as a polymerizable extracting agent; however, in the latter's paper, the purpose of the side arm was not to bind guest cations but as a point of attachment to a polymer. Despite being in the title of Gokel's 1980 communication, the use of the term lariat ether was only explained in later papers from the group.[57] According to the authors:

> *The physical resemblance of CPK molecular models of these compounds to rope lassoes coupled with the concept of "roping and tying" a cation suggested the name "lariat ether."*[57]

In 1983, Gandour and coworkers reported the preparation of a family of 18-crown-6 alkoxymethyl derivatives.[58] Cleavage of the side chain gave the methylol derivative that could be further derivatized. In the same year, Inoue, Hakushi, and coworkers reported the syntheses of three 16-crown-5 compounds functionalized at C15.[59] Later, the group expanded the range of these compounds, introducing polyether side chains to both 16-crown-5 and 19-crown-6 derivatives.[60] It was noted that while additional donor ether groups in the side arms can enhance binding, the unfavorable steric effects of groups (other than hydrogen)

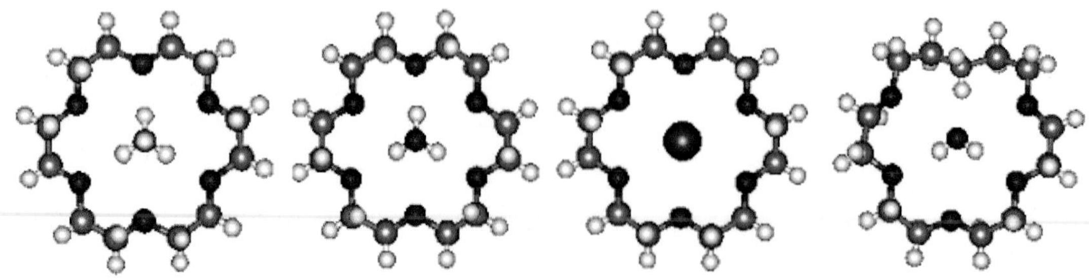

Figure 13 X-Ray crystal structures of some simple 18-crown-6 ether complexes: (left to right) with NH_4^+ [51]; with H_3O^+ [52]; with K^+ [53]; aza-18-crown-6 with H_3O^+.[26]

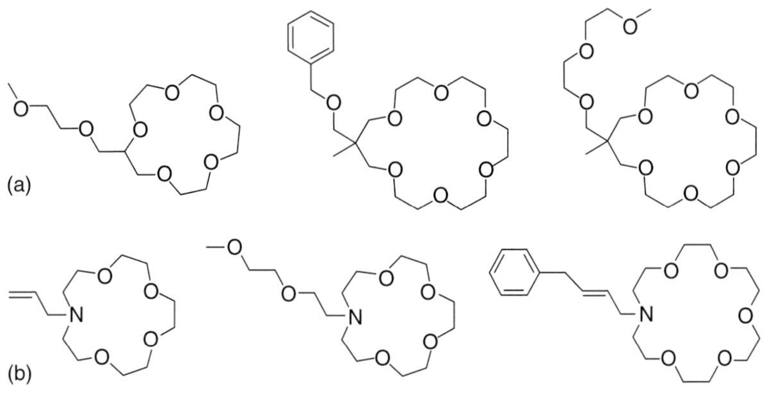

Figure 15 Lariat ethers: (a) *C*-pivot lariat ethers[55,60]; (b) azacrown lariat ethers.[57,65]

attached to the pivotal carbon atom negate this enhancement in several cases. Despite quite a wide interest in these compounds, particularly as metal extracting modifiers for chromatographic separations,[61–63] the ease with which azacrowns can be *N*-functionalized makes the latter far more popular compounds from which to prepare lariat ethers.

3.2 *N*-Pivot lariat ethers

Once routes to the monazacrowns had been rationalized by the groups of Gokel and Maeda, the next step was to incorporate further functional groups as side arms. The first *N*-functionalized azacrowns were prepared by Gokel and coworkers who reported a series of compounds with polyether side arms.[64] The aim was to determine the effect of increasing the number of donor atoms available to bind guests held within the central cavity of the crown ether. Figure 15 reinforces how the structures of these compounds resemble the lariats, or lassos, used by cattle herders.

Two routes are available to prepare *N*-functionalized azacrowns: addition of the side arm to the nitrogen of an appropriately sized azacrown, or reaction of the side arm with diethanolamine followed by azacrown ether formation through templated synthesis with the appropriate ditosyl- or diiodopolyether. The former route is preferred when researchers want to attach a range of side arms to the same sized crown and the latter is preferred when the same side arm is to be introduced to crowns of different sizes. These routes are shown in Figure 16.

The effect of increasing the number of available donor atoms was at first glance quite surprising. As shown in Figure 17, selectivity for sodium was greatest when the cation was surrounded by one nitrogen and six oxygen atoms regardless of crown size, but the selectivity for ammonium showed no such trend. It was proposed that the alkali metal had no directional preference in its interactions

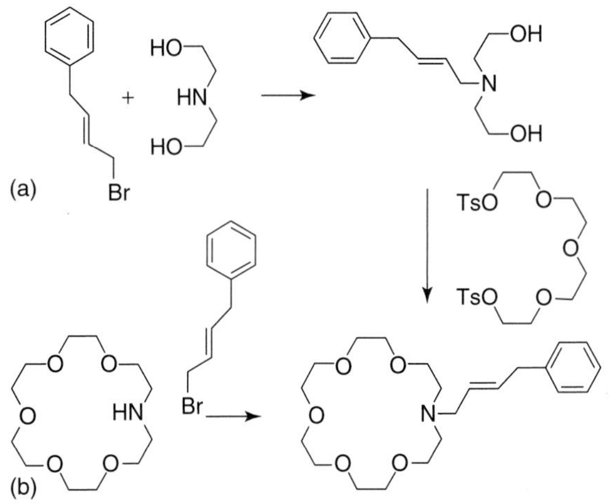

Figure 16 Routes to azacrown lariat ethers: (a) via diethanolamine functionalization; (b) direct addition of a side arm.

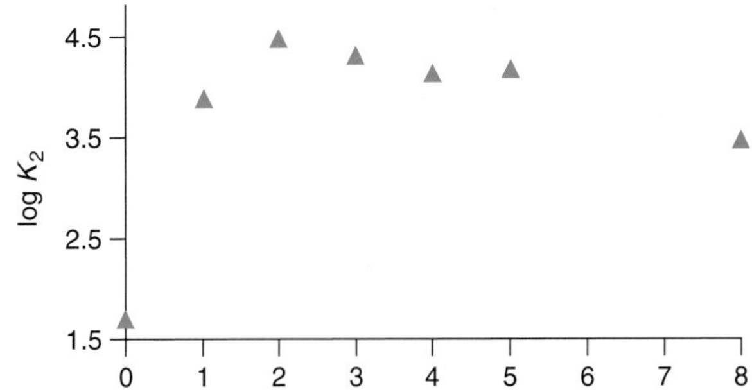

Figure 17 Na^+ complexation by aza-15-crown-5 lariat ethers with n oxygen donor atoms in the side arm ($n = 0$–8).[57]

with the donor atoms whereas the ammonium cation relied on hydrogen bonds organized in a tetrahedral disposition.

Since these original reports, many *N*-pivot lariat ethers have been prepared. In addition to the Gokel group's series of polyether homologs, which probed the effect of donor group numbers on complex stability, they also prepared a cholesterol derivative, which self-assemble into simple cell-like vesicles, and other derivatives with side arms designed to bind metal cations.[66] The last endeavor illustrates how simple model systems can shed light on poorly understood biological phenomena. In another example of crown ethers mimicking natural systems, azacrown lariat ethers have been used to investigate cation–π interactions.

Cation–π interactions have been widely discussed in the literature and there has been much speculation about their importance in biology, specifically in the nature of the interactions between ubiquitous alkali and alkaline earth cations Na^+, K^+, and Ca^{2+} and amino acids with aromatic side chains. In early work, the Gokel group focussed on single and double armed lariat ethers with benzyl and allyl side chains. In these compounds, the π-donor group was linked to the nitrogen atom by a single carbon and X-ray crystal structures showed no cation–π interactions, although metals were bound within the macrocyclic cavities.[67] Independently, the groups of Steed and Cragg prepared azacrown ethers with allyl and butenyl side arms. The KPF_6 complex of *N*-allylaza-15-crown-5 crystallized as a C_2 symmetric dimer comprising two crowns bound to two potassium cations that were in turn weakly bound to both anions.[65] Significantly, no cation–π interactions were observed. In the $AgSbF_6$ complex of the same lariat ether, the silver cations were held within the macrocycle and interacted strongly with the allyl group of another lariat ether to produce an infinite chain structure.[68] In the *N*-butenylaza-18-crown-6 derivative, the π-donor group was two carbon atoms from the nitrogen and was able to reach over the top of the macrocyclic ring to bind Ag^+ held within the cavity in a scorpion-like manner.[65] The importance of a two carbon separation between a donor group and azacrown nitrogen atom was noticed by Gokel, leading his group to revise its design strategy. In a key paper on the new compounds, this change was rationalized succinctly:

> *The recognition that Nature separates the aromatic residues of histidine, phenylalanine, tryptophan, and tyrosine from nitrogen by two carbons, rather than one, led to an alteration of the receptor system.*[69]

Needless to say, this alteration was successful: incorporation of these π-donor groups into azacrowns and diazacrowns with a two carbon spacer resulted in several examples of crystalline complexes in which the π-donor group bound Na^+, K^+, and Ca^{2+} held within the macrocyclic cavity. X-ray crystal structures, illustrated in Figure 18, showed that the cation–π interactions were of clear importance in the cases of Na^+ and K^+ but not in the case of Ca^{2+} even when the macrocycle was sufficiently large enough to accommodate the latter.[69–71]

Some examples of azacrown lariat ethers acting as ion-specific sensors are described below but it is important to consider the other developments in the field before moving on. Once research into the azacrown lariat ethers had been initiated, it did not take long for the Gokel group to realize that di- and triazacrowns may also hold some promise as encapsulating ligands. Lariat ethers based on these azacrowns were dubbed bibracchial and tribracchial lariat ethers, or BiBLEs and TriBLEs for short, after the Latin *brachium* for arm. Although a number of derivatives were synthesized, perhaps the most interesting were those 1,10-diaza-18-crown-6 compounds with adenine and thymine side arms. Three derivatives were prepared in which the nucleobases were separated from the

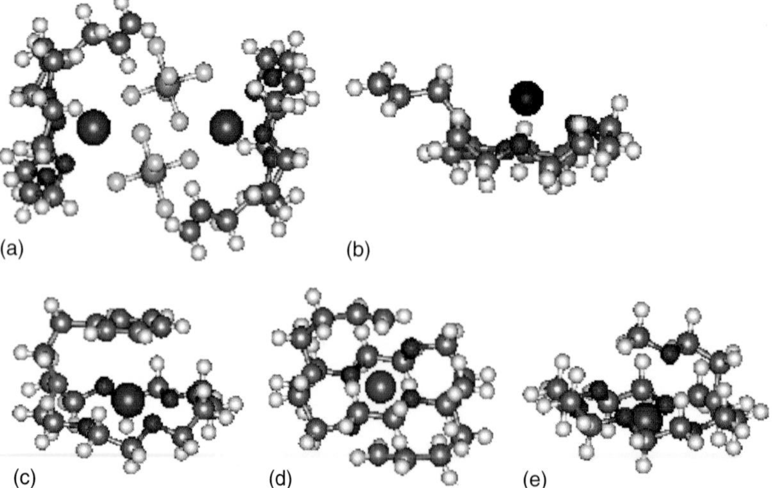

Figure 18 X-Ray crystal structures of some lariat ether complexes: (a) $[(K^+ \cdot N\text{-(allyl)aza-15-crown-5)}PF6^-]_2$[65]; (b) $Pb^{2+} \cdot N$-(allyl)aza-15-crown-5[65]; (c) $K^+ \cdot N$-(ethylbenzene)aza-18-crown-6[69]; (d) $K^+ \cdot N,N'$-di(but-1-enyl)diaza-18-crown-6[72]; (e) $K^+ \cdot N$-(methoxyethyl)aza-18-crown-6.[57]

crown by a propyl spacer to give compounds with adenine–crown–adenine, thymine–crown–thymine, and adenine–crown–thymine structural motifs. As expected these compounds interacted through both Crick–Watson and Hoogsteen hydrogen bonding modes as determined by nuclear magnetic resonance (NMR) methods.[73]

3.2.1 Industrial and clinical applications of crown and lariat ethers

Phase transfer catalysts

One of the first applications found for the crown ethers related to their amphiphilic properties and concomitant ability to bind metal cations in both aqueous and nonaqueous solvents. The ligand facilitates the movement of cations across the interface between the aqueous phase and organic solvent where they can catalyze reactions of hydrophobic substrates, as shown in Figure 19. The effect is very similar to that seen in more conventional phase transfer catalysts such as the quaternary ammonium salts. Classic examples include the use of 18-crown-6 to solubilize the oxidant $KMnO_4$ in organic solvents through the formation of the complex salt $[K{\cdot}18\text{-crown-}6]^+$ MnO_4^-.[74] The reagent is able to oxidize organic molecules with much greater efficacy than $KMnO_4$ alone. Similarly, NaCN can be solubilized in organic solvents, through the formation of $[Na{\cdot}18\text{-crown-}6]^+$ CN^-, to enable the transfer of cyanide.[75]

Other phase transfer examples include chiral crowns that have been linked to polystyrene beads and have shown promise as heterogeneous phase transfer catalysts for halide exchange between 1-bromooctane and KI.[76] Dibenzocrown ethers and azacrown ethers with perfluorinated side arms have started to see use as phase transfer catalysts between fluorous solvents and organic or aqueous phases in which the reactants are soluble. Perfluorocrown ethers are not themselves particularly effective cation binding agents as the presence of the electron withdrawing $-CF_2-$groups

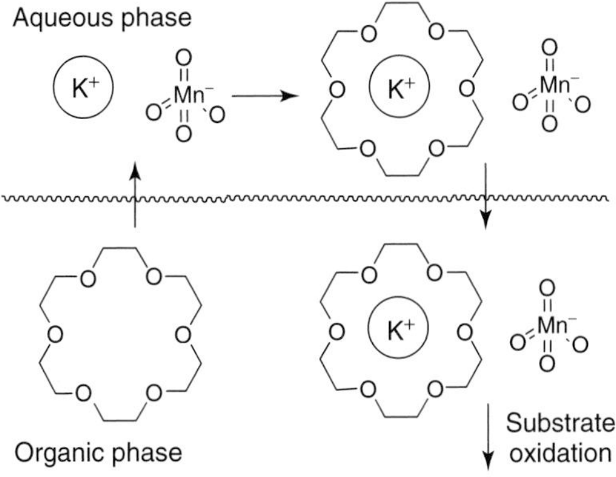

Figure 19 Phase transfer catalysis by crown ethers.

detract from the electron donor character of the crown oxygen atoms.[77] Also, in the area of fluorine chemistry, the enantioselective trifluoromethylation of alkyl aryl ketones, aryl aldehydes, and alkyl aldehydes has been shown to be moderately enhanced by the addition of catalytic levels of chiral crown ethers based on Cram's binapthyl derivatives.[78]

Separation science

Crown ethers' variable affinities for cations have led to their attachment to chromatographic media and as extractants, notably of alkali and alkaline earth metal cations. Izatt, Bradshaw, and colleagues at Brigham Young University demonstrated that crown ethers could be bound to silica to produce a metal sequestering material for column chromatography.[79]

The threefold symmetry of quaternarized amines makes them ideal guests for 18-crown-6 and its derivatives. Consequently, reverse-phase chromatographic column materials modified with chiral 18-crown-6 derivatives have been used to successfully separate racemic mixtures of the common amino acids. The separation coefficients were enhanced at lower temperatures and at higher loading levels of the crown ether.[80]

The nuclear industry has been a major focus for research involving crown ethers. In addition to the well-known radioactive actinide byproducts, such as uranium and plutonium oxides, other radioactive species are associated with the industry. ^{90}Sr and ^{137}Cs are found in low-level acidified nuclear waste and can be targeted by crown ethers and derivatives, chief among them being the calixcrowns.[81]

Sensor applications

Numerous sensors based on crown and lariat ethers have been reported. Many are simply based on the size selectivity of the crown coupled to a functional side arm that responds to variations in electrostatics that occur when a cation is bound. This has led to crown ethers being incorporated into sensors that function through a classic receptor–linker–reporter relay motif. The molecular architecture of these sensors requires that one end of the molecule contains a group that binds to a specific target while the other has a functional group that responds to the binding event. The two groups are linked by a variety of spacers. These may be simple alkyl chains, conjugated links, or more complex structures, depending on the requirements of the sensor. If the electrostatic change that occurs upon guest binding has to be transmitted to the reporter group, then a short conjugated link is needed. Conversely, if the reporting molecule has to be kept remote from the receptor, as is often the case when labeling proteins, then a long flexible polyether may be more appropriate.

Two pioneering approaches to the use of crown ether-based sensors can be found in the work of the Beer and de Silva groups. In 1988, Beer reported the interesting finding that the electrochemistry of ferrocene was affected when it was linked to a crown ether and exposed to a solution of alkali–metal cations.[82] Where benzo-15-crown-5 was attached to ferrocene by an alkene spacer, the Na^+-binding event was signaled by an anodic shift of 60 mV in the ferrocene oxidation wave. This allowed direct detection of Na^+ and determination of its concentration by comparing the response to a series of standards. A similar electroresponsive lariat ether was prepared from an aza-15-crown-5-ferrocene conjugate.

The de Silva group incorporated crown ethers as recognition elements in its logic gate molecules. These molecules generally included a fluorophore, often anthracene, linked to a crown ether by a single carbon spacer. If a metal is bound to the macrocycle, then a change in fluorescence intensity resulted to give a binary "on–off" signal, or a TRUE/FALSE logic gate. The principle can be extended to include another function to the sensor such as an amine group to give a positive response only if the appropriate metal and a proton are present: an AND gate.[83]

The combination of sulfur and nitrogen donor atoms in crown ethers appears to provide an ideal binding site for Hg^{2+} as several sensors, mainly electrochemical, which use azathiacrown ethers to target the metal have been reported. The selectivity for mercury over other metals is often astonishingly high.[84] A 15-crown-5 *C*-pivot lariat ether incorporating a thioether side chain was found to extract 46% of Hg^{2+} from aqueous solution into dichloroethane, although it was also able to remove almost 100% of Ag^+ under the same conditions.[85]

An interesting example of a luminescent acid sensor with an integral reporting group was reported by the Sykes group.[86] Gokel's anthraquinone[17] incorporating 18-crown-6 derivative, illustrated in Figure 6, was treated with several oxoacids and found to yield luminescent solutions. Notably, the perchloric acid complex gave the most intense response, and crystal structures revealed that this complex included H_3O^+ in the macrocycle's central cavity. Nitric and sulfuric acid responses were much weaker, proportional to their relative pK_as, and their crystal structures revealed that they did not dissociate to provide hydronium ions to bind in the macrocyclic cavity.

Lariat ethers can also be used as detectors for toxins. Gawley, Leblanc, and coworkers reported a coumaryl derivative of aza-18-crown-6 that responded to saxitoxin, a neurotoxin originating in marine bacteria and puffer fish.[87] The fluorescence of the lariat ether increased with increasing saxitoxin concentration but was unaffected by either Na^+ or K^+, the two main chemical species that affect the response of crown ether-based sensors.

Pharmaceutical activity enhancement

The crown ethers have generally low toxicities; however, they may be able to enhance the activities of drug molecules. In an interesting report from the Kagabu group, a series of benzocrown lariat ethers that incorporated the insecticide imidacloprid as a side arm was synthesized.[88] The parent benzocrowns had very low toxicities, with flies exhibiting no ill effects at 1100 ng per fly, yet the lariat ether derivatives were three to five times more potent than the insecticide itself with LD_{50} values of 4.3 ng per fly. It appears that the crown ethers increased the drug's amphiphilic properties and allowed it to enter the fly's central nervous system with greater ease. The crowns may also have acted in their more familiar ionophore role, transporting alkali–metal cations as well as the insecticide thereby enhancing the drug's efficacy.

Clinical diagnostics

Alongside the crown ethers' affinities for metal cations, Pedersen also showed that 18-crown-6 provides a binding site for NH_4^+. This has led to the incorporation of crown ethers in molecular sensors that required a hydrogen bond acceptor. Foremost among these are derivatives of 18-crown-6 as their threefold symmetry is complementary to that of the H_3N^+R terminus of protonated primary amines. By extension it can be used to bind the *N*-terminus of a peptide or a biologically important amine such as the neurotransmitter γ-aminobutyric acid (GABA).[89] Using this approach, a GABA sensor has been prepared from aza-18-crown-6 linked to a terminal guanidinium group via a fluorescent anthracene spacer. Here, both ends of the target analyte were bound by the crown ether derivative: the amine by the crown and the carboxylic acid by the guanidinium group. A similar example, terminating in boronic acid, shows promise as a D-glucosamine sensor. Again, both ends of the analyte are bound to functional groups in the sensor.[90]

These examples highlight the importance of the fit of the crown incorporated into the sensor and the size of the analyte. Not only must the individual binding sites be incorporated into the sensor but their relative positions must also be at the correct angle and distance apart if they are to be selective for the analyte. Thus a sensor for γ-aminobutyric acid will need a different structure than that for α-aminobutyric acid. Although the two receptor sites could be the same, the linker between them would have to be shorter and the angle between them much more acute.

The de Silva group laid the groundwork using this approach to sensing that has since found application in the clinic. Critical care analysis kits are small portable detectors that determine the concentrations of key inorganic species (Na^+, K^+, and Ca^{2+}) in blood samples and are often used at the scene of accidents as the first diagnostic technique. The

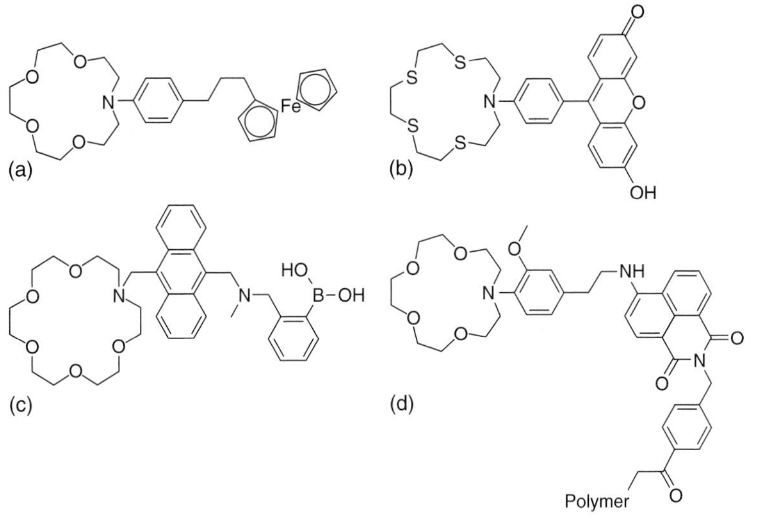

Figure 20 Sensor and diagnostic crown ethers: (a) a redox active sensor[83]; (b) a Hg^{2+} sensor[85]; (c) a D-glucosamine sensor[90]; (d) Na^+ sensor.[91]

kits contain cassettes with sensors for each target analyte that comprise a recognition site and fluorescent linker to a polymer matrix. The presence of target analytes is signaled by fluorescent responses. The Na^+ sensor, illustrated in Figure 20, uses aza-15-crown-5 as its cation recognition element whereas the K^+ and Ca^{2+} sensors are based on cryptand and bis(carboxylic acid) motifs, respectively.[91]

4 BIOLOGICAL RELEVANCE

4.1 Ion transport

In their 1970 paper, Bright and Truter picked up on the similarity between the crown ethers and naturally occurring antibiotics:

> *Cyclic polyethers have been shown to increase the permeability of reconstituted biological membranes to alkali metals in a similar way to naturally occurring antibiotics, such as enniatin, nonactin and valinomycin, the similarity extending to the order of preference for metal transfer across the membrane.*[12]

A year later Smid noted that:

> . . . crown compounds can exert specific effects on cation transport across biological membranes. This property makes them useful as model compounds of neutral ion carriers and resembles the behavior of the more complex macrocyclic antibiotics monactin and valinomycin which are known to make phospholipid bilayer membranes selectively permeable to cations.[92]

Similarities between cyclic peptides such as valinomycin and the crown ethers was clearly not lost on the pioneering researchers in the field, as is obvious from these quotations taken from early papers, so it is not surprising that the crown structure has been incorporated into molecules designed to mimic cation selective biomolecules. This structural similarity between natural ionophores and crown ethers, as illustrated in Figure 21, has resulted in the crown ether motif being incorporated as an amphiphilic size selective filter in compounds designed to mimic the behavior of transmembrane channel-forming proteins.

4.2 Transmembrane channel models

Lehn had shown that 18-crown-6 derivatives with substituents emanating from their carbon backbones could function as rudimentary ion channels but size selectivity was not discussed in his paper.[95] The Fyles group decorated crown ethers with thioglucose-terminated cyclic alkyl esters to mimic amphotericin-like compounds that acted as transmembrane channels through a "barrel-stave" mechanism.[93] These compounds transported K^+ with reasonable selectivity over other monovalent cations at rates similar to the natural ionophore valinomycin. The Voyer group made a biologically inspired channel-forming protein mimic by synthesizing an α-helical polypeptide with an artificial, tetraazacyclododecanetetraacetic acid (DOTA)-derived benzo-18-crown-6 inserted every three or four residues to give helices comprised of 7, 14, or 21 residues.[15] The crown ethers stacked one over another down one side of the α-helix to form a channel for Na^+. Later work by the group showed that some 14-mer, 2-nm long derivatives inhibited prostate cancer proliferation in two cell lines but did not induce hemolysis in normal human cells and so had low

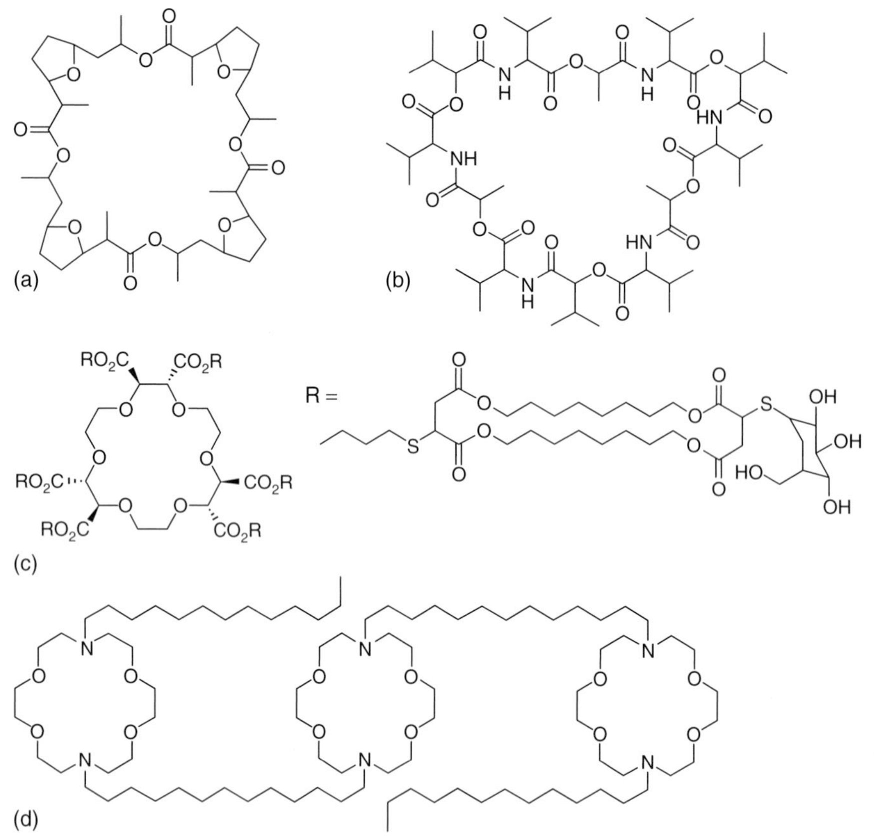

Figure 21 Natural crown ethers and biomimetic crown ethers: (a) nonactin; (b) valinomycin; (c) a barrel-stave model[93]; (d) a hydraphile.[94]

general toxicity.[96] The active 14-mers were terminated by glutamate or aspartate residues, which are susceptible to cleavage by a prostate-specific membrane antigen. Presumably, once this occurred, the helices could self-assemble into membrane-spanning tubes, a behavior that has been seen in the gramicidin class of α-helical polypeptides. This would then allow cations to flood through the crown ether side chains and into, or out of, the affected cells leading to necrotic cell death.

In a final example, the binding affinity that each size of crown has for guest cations was cleverly exploited by Gokel and coworkers when they prepared a group of compounds later named as hydraphiles.[94] Their diaza-18-crown-6 molecules were linked together by long alkyl chains; the terminal crown ethers also had alkyl chains appended to them that ended in various hydrophilic groups attached as shown in Figure 21. The entire assembly was able to span a lipid bilayer, a mimic for a cell membrane, and facilitated the transmembrane transport of Na^+ but not K^+. The reason for the selectivity was easier to understand: while 18-crown-6 derivatives bind both metals, the binding constants of the K^+ complexes were far higher than the Na^+ analogs and as a consequence any K^+ present would be bound whereas any Na^+ would be allowed to pass through the crown ethers relatively unhindered.

4.3 Enzyme models

Much of the initial work involving macrocycles resulted from their similarity to known biological catalysts. These enzymes, such as cytochromes and hydrolases, had rigid binding sites, often for transition metals, where the catalytic activity was focused. The crown ethers, by way of contrast, were far more flexible and better suited to bind simple inorganic cations and amino acids. The groups of Lehn and Mertes synthesized a number of polyazacrown ethers designed to bind the phosphate region of ATP.[97–99] When protonated, these crowns were able to cleave the terminal ATP phosphate group thus catalyzing one of nature's most important reactions, the release of energy through ATP hydrolysis. Consequently, the authors considered their compounds to be functional ATPase mimics.

Other examples of crown ethers being employed in model enzymes include transacylation,[100] peptide synthases,[101] urease,[102] and cyclotransferase mimics.[103] Taken together, they demonstrate the breadth of activity that has been undertaken to use simple crown-containing model compounds to further our understanding of complex natural processes.

5 CONCLUSIONS

In this chapter, we have endeavored to show how crown ethers first came into prominence in the late 1960s through the pioneering work of Charles Pedersen. The discovery that they bind cations based on complementarity between that of the crown ether cavity and the cation has been a driving force in macrocyclic and supramolecular chemistry ever since. Many derivatives have been synthesized, including the *C*- and *N*-pivot lariat ethers, and their cation selectivity harnessed for use in molecular sensors. A consideration of the history of crown and lariat ethers together with their syntheses, structural and binding evidence for size selectivity, and their industrial and clinical applications indicates the broad scope of this iconic class of macrocyclic compounds. As discussed above, crown and lariat ethers have found use as phase transfer catalysts, modifiers for chromatographic media, sensors for numerous chemical species, critical analyte detectors, and can even enhance insecticide activity. Their derivatives have been used to probe important biological phenomena, including enzyme specificity, the nature of *in vivo* cation–π interactions in proteins and transmembrane ion transport, through the synthesis of model compounds.

It is impossible, in a chapter as short as this, to do justice to the work done by the vast number of researchers who have worked on crown and lariat ethers. A survey of the literature indicates that almost 15 000 papers have been written about these compounds since their inception to the present day. Numerous excellent reviews and several monographs as well as chapters in other works are available. Some of these key sources are given in the bibliography, which covers general aspects of crown and lariat ether chemistry, their syntheses, and data relating to the complexes that they form.

ACKNOWLEDGMENTS

The authors wish to acknowledge the use of the Chemical Database Service at Daresbury.

REFERENCES

1. D. G. Stewart, D. Y. Waddan, and E. T. Borrows, GB Patent 785229, Oct. 23 1957.
2. C. J. Pedersen, *J. Am. Chem. Soc.*, 1967, **89**, 7017.
3. Quoted in G. W. Gokel, *Crown Ethers and Cryptands*, The Royal Society of Chemistry, Cambridge, 1991, p. 5.
4. A. Lüttringhaus and K. Ziegler, *Liebigs Ann. Chem.*, 1937, **528**, 155.
5. R. G. Ackman, W. H. Brown, and G. F. Wright, *J. Org. Chem.*, 1955, **20**, 1147.
6. R. Adams and L. N. Whitehill, *J. Am. Chem. Soc.*, 1941, **63**, 2073.
7. J. L. Down, J. Lewis, B. Moore, and G. W. Wilkinson, *J. Chem. Soc.*, 1959, 3767.
8. J. Dale and P. O. Kristiansen, *Chem. Commun.*, 1971, 670.
9. C. J. Pedersen, *J. Inclusion Phenom.*, 1988, **6**, 337.
10. R. N. Greene, *Tetrahedron Lett.*, 1972, **18**, 1793.
11. C. J. Pedersen, *J. Am. Chem. Soc.*, 1970, **92**, 386.
12. D. Bright and M. R. Truter, *J. Chem. Soc. B*, 1970, 1544.
13. E. P. Kyba, M. G. Siegel, L. R. Sousa, *et al.*, *J. Am. Chem. Soc.*, 1973, **95**, 2691.
14. J.-M. Girodeau, J.-M. Lehn, and J.-P. Sauvage, *Angew. Chem. Int. Ed. Engl.*, 1975, **14**, 764.
15. N. Voyer and B. Guérin, *Chem. Commun.*, 1997, 2329.
16. J. M. Timko and D. J. Cram, *J. Am. Chem. Soc.*, 1974, **96**, 7159.
17. M. Delgado, D. A. Gustowski, H. K. Yoo, *et al.*, *J. Am. Chem. Soc.*, 1988, **110**, 119.
18. M. Newcomb, G. W. Gokel, and D. J. Cram, *J. Am. Chem. Soc.*, 1974, **96**, 6810.
19. K. Frensch and F. Vögtle, *Tetrahedron Lett.*, 1977, **18**, 2573.
20. A. Szumna and J. Jurczak, *Eur. J. Org. Chem.*, 2001, 4031.
21. A. G. Johnson, D. A. Leigh, R. J. Prichard, and M. D. Deegan, *Angew. Chem. Int. Ed. Engl.*, 1995, **34**, 1209.
22. J.-J. Lee, A. G. White, J. M. Baumes, and B. D. Smith, *Chem. Commun.*, 2010, **46**, 1068.
23. M. J. Calverley and J. Dale, *Acta Chem. Scand. B*, 1982, **36**, 241.
24. V. J. Gatto and G. W. Gokel, *J. Am. Chem. Soc.*, 1984, **106**, 8240.
25. J. C. Lockhart, A. C. Robson, M. E. Thompson, *et al.*, *J. Chem. Soc., Perkin Trans. 1*, 1973, 577.
26. G. W. Gokel and B. J. Garcia, *Tetrahedron Lett.*, 1977, **4**, 317.
27. M. R. Johnson. I. O Sutherland, and R. F. Newton, *J. Am. Chem. Soc., Perkin 1*, 1979, 357.
28. H. Maeda, S. Furuyoshi, Y. Nakatsuji, and M. Okahara, *Bull. Chem. Soc. Jpn.*, 1983, **56**, 212.
29. B. Dietrich, J.-M. Lehn, and J.-P. Sauvage, *Tetrahedron Lett.*, 1969, **4**, 2885.
30. G. W. Gokel, D. M. Dishong, R. A. Schultz, and V. J. Gatto, *Synthesis*, 1982, 997.
31. K. E. Krakowiak, J. R. Bradshaw, and R. M. Izatt, *Synlett*, 1993, 611.
32. J. Muhle and W. S. Sheldrick, *Z. Anorg. Allg. Chem.*, 2003, **629**, 2097.

33. P. J. Cragg, S. G. Bott, and J. L. Atwood, *Lanth. Act. Res.*, 1988, **2**, 265.
34. J. Pickardt and S. Wiese, *Z. Naturforsch. B: Chem. Sci.*, 1997, **52**, 847.
35. C. Bazzicalupi, A. Bencini, A. Bencini, *et al.*, *Inorg. Chem.*, 1996, **35**, 5540.
36. P. C. Rây, *J. Chem. Soc., Trans.*, 1920, **117**, 1090.
37. D. Gerber, P. Chongsawangvirod, A. K. Leung, and L. A. Ochrymowycz, *J. Org. Chem.*, 1977, **42**, 2644.
38. J. R. Meadow and E. E. Reid, *J. Am. Chem. Soc.*, 1934, **56**, 2177.
39. J. R. Dann, P. P. Chiesa, and J. W. Gates Jr., *J. Org. Chem.*, 1961, **26**, 1991.
40. C. J. Pedersen, *J. Org. Chem.*, 1971, **36**, 254.
41. J. S. Bradshaw, J. Y. Hui, B. L. Haymore, *et al.*, *J. Heterocycl. Chem.*, 1973, **10**, 1.
42. J. S. Bradshaw, J. Y. Hui, Y. Chan, *et al.*, *J. Heterocycl. Chem.*, 1974, **11**, 45.
43. J. Buter and R. M. Kellogg, *J. Org. Chem.*, 1981, **46**, 4481.
44. V. B. Shur, I. A. Tikhonova, A. I. Vanovsky *et al.*, *J. Organomet. Chem.*, 1991, **418**, C29.
45. M. Newcomb, Y. Azuma, and A. R. Courtney, *Organometallics*, 1983, **2**, 175.
46. M. Newcomb, J. H. Horner, M. T. Blanda, and P. J. Squattrito, *J. Am. Chem. Soc.*, 1989, **111**, 6294.
47. X. Yang, Z. Zheng, C. B. Knobler, and M. F. Hawthorne, *J. Am. Chem. Soc.*, 1993, **115**, 193.
48. G. W. Gokel, W. M. Leevy, and M. E. Weber, *Chem. Rev.*, 2004, **104**, 2723.
49. R. D. Hancock, *J. Chem. Ed.*, 1992, **69**, 615.
50. R. D. Rogers, A. N. Rollins, R. D. Etzenhouser, *et al.*, *Inorg. Chem.*, 1993, **32**, 3451.
51. A. D. Bokare and A. Patnaik, *Cryst. Res. Technol.*, 2004, **39**, 465.
52. J. L. Atwood, S. G. Bott, A. W. Coleman, *et al.*, *J. Am. Chem. Soc.*, 1987, **109**, 8100.
53. S. V. Rosokha, J. Lu, T. Y. Rosokha, and J. K. Kochi, *Phys. Chem. Chem. Phys.*, 2009, **11**, 324.
54. R. M. Izatt, K. Pawlak, J. S. Bradshaw, *et al.*, *Chem. Rev.*, 1992, **92**, 1261–1354.
55. G. W. Gokel, D. M. Dishong, and C. J. Diamond, *J. Chem. Soc., Chem. Commun.*, 1980, 1053.
56. F. Montanari and P. Tundo, *Tetrahedron Lett.*, 1979, **20**, 5055.
57. R. A. Schultz, B. D. White, D. M. Dishong, *et al.*, *J. Am. Chem. Soc.*, 1985, **107**, 6659.
58. S. J. Jungk, J. A. Moore, and R. D. Gandour, *J. Org. Chem.*, 1983, **48**, 1116.
59. M. Ouchi, Y. Inoue, H. Sakamoto, *et al.*, *J. Org. Chem.*, 1983, **48**, 3168.
60. M. Ouchi, Y. Inoue, K. Wada, *et al.*, *J. Org. Chem.*, 1987, **52**, 2420.
61. L. R. Sousa, G. D. Y. Sogah, D. H. Hoffman, and D. J. Cram, *J. Am. Chem. Soc.*, 1978, **100**, 4569.
62. J. Rivello, M. Ray, J. Jagodzinski, and C. Pohl, US Patent 5968363, Oct. 19 1999.
63. M. H. Hyun, *Bull. Korean Chem. Soc.*, 2005, **26**, 1153.
64. R. A. Schultz, D. M. Dishong, and G. W. Gokel, *Tetrahedron Lett.*, 1981, **22**, 2623.
65. P. Arya, A. Channa, P. J. Cragg, *et al.*, *New J. Chem.*, 2002, **26**, 440.
66. G. W. Gokel, J. C. Hernandez, A. M. Viscariello, *et al.*, *J. Org. Chem.*, 1987, **52**, 2963.
67. K. A. Arnold, A. M. Viscariello, M. Kim, *et al.*, *Tetrahedron Lett.*, 1988, **29**, 3025.
68. P. D. Prince, P. J. Cragg, and J. W. Steed, *Chem. Commun.*, 1999, 1179.
69. J. Hu, L. J. Barbour, and G. W. Gokel, *Chem. Commun.*, 2002, 1808.
70. J. Hu, L. J. Barbour, R. Ferdani, and G. W. Gokel, *Chem. Commun.*, 2002, 1806.
71. J. Hu, L. J. Barbour, R. Ferdani, and G. W. Gokel, *Chem. Commun.*, 2002, 1810.
72. J. Hu, L. J. Barbour, and G. W. Gokel, *Collect. Czech Chem. Commun.*, 2004, **69**, 1050.
73. O. F. Schall and G. W. Gokel, *J. Am. Chem. Soc.*, 1994, **116**, 6089.
74. D. G. Lee and V. S. Chang, *J. Org. Chem.*, 1978, **43**, 1532.
75. F. L. Cook, C. W. Bowers, and C. L. Liotta, *J. Org. Chem.*, 1974, **39**, 3416.
76. H. Watanabe, T. Iijima, W. Fukuda, and M. Tomoi, *React. Funct. Polym.*, 1998, **37**, 101.
77. G. Pozzi, S. Quici, and R. H. Fish, *J. Fluorine Chem.*, 2008, **129**, 920.
78. H. Kawai, A. Kusuda, S. Mizuta, *et al.*, *J. Fluorine Chem.*, 2009, **130**, 762.
79. J. S. Bradshaw, R. M. Izatt, V. B. Christensen, and R. L. Bruening, US Patent 5179213, Jan. 12 1993.
80. T. Shinbo, T. Yamaguchi, K. Nishimura, and M. Sugiura, *J. Chromatogr., Biomed. Appl.*, 1987, **405**, 145.
81. B. A. Moyer, R. A. Sachleben, P. V. Bonnesen, and D. J. Presley, US Patent 6174503, Jan. 16 2001.
82. P. D. Beer, H. Sikanyika, C. Blackburn, *et al.*, *J. Organomet. Chem.*, 1988, **356**, C19.
83. A. P. de Silva, T. P. Vance, M. E. S. West, and G. D. Wright, *Org. Biomol. Chem.*, 2008, **6**, 2468.
84. E. W. Miller, A. E. Albers, A. Pralle, *et al.*, *J. Am. Chem. Soc.*, 2005, **127**, 16030.
85. T. Nabeshima, K. Nishijima, N. Tsukada, *et al.*, *J. Chem. Soc., Chem. Commun.*, 1992, 1092.
86. V. G. Young Jr., H. L. Quiring, and A. G. Sykes, *J. Am. Chem. Soc.*, 1997, **119**, 12477.
87. P. Kele, J. Orbulescu, T. L. Calhoun, *et al.*, *Tetrahedron Lett.*, 2002, **43**, 4413.
88. S. Kagabu, M. Takagi, I. Ohno, *et al.*, *Bioorg. Med. Chem. Lett.*, 2009, **19**, 2947.
89. A. P. de Silva, H. Q. N. Gunaratne, C. McVeigh, *Chem. Commun.*, 1996, 2191.

90. C. R. Cooper and T. D. James, *Chem. Commun.*, 1997, 1419.
91. J. K. Tusa and H. He, *J. Mater. Chem.*, 2005, **15**, 2640.
92. U. Takaki, T. E. H. Esch, and J. Smid, *J. Am. Chem. Soc.*, 1971, **93**, 6760.
93. T. M. Fyles, T. D. James, and K. C. Kaye, *J. Am. Chem. Soc.*, 1993, **115**, 12315.
94. G. W. Gokel, *Chem. Commun.*, 2001, 1.
95. L. Jullien and J.-M. Lehn, *Tetrahedron Lett.*, 1988, **29**, 3803.
96. P.-L. Boudreault, M. Arseneault, F. Otis, and N. Voyer, *Chem. Commun.*, 2008, 2118.
97. M. W. Hosseini, J.-M. Lehn, and M. P. Mertes, *Helv. Chim. Acta*, 1983, **66**, 2454.
98. M. W. Hosseini, J.-M. Lehn, L. Maggiora, *et al.*, *J. Am. Chem. Soc.*, 1987, **109**, 537.
99. M. W. Hosseini, A. J. Blacker, and J.-M. Lehn, *J. Chem. Soc., Chem. Commun.*, 1988, 596.
100. T. Matsui and K. Koga, *Tetrahedron Lett.*, 1978, 1115.
101. S. Sasaki, M. Shionoya, and K. Koga, *J. Am. Chem. Soc.*, 1985, **107**, 3371.
102. A. M. Reichwein, W Verboom, and D. N. Reinhoudt, *Rec. Trav. Chim. Pays-Bas*, 1993, **112**, 595.
103. M. Kunishima, K. Hioki, T. Moriya, *et al.*, *Angew. Chem. Int. Ed.*, 2006, **45**, 1252.

FURTHER READING

General

G. W. Gokel, *Crown Ethers and Cryptands*, The Royal Society of Chemistry, Cambridge, 1991.

G. W. Gokel, *Chem. Soc. Rev.*, 1992, **21**, 39–47.

M. Hiraoka, *Crown Compounds: Their Characteristics and Applications, Studies in Organic Chemistry, Vol. 12*, Kodansha Ltd., Tokyo, 1982.

Synthesis

A. A. Abbas and A. H. M. Elwahy, *J. Heterocyclic Chem.*, 2010, **46**, 1035.

P. J. Cragg, *A Practical Guide to Supramolecular Chemistry*, John Wiley & Sons, Ltd, Chichester, 2005.

D. Parker, ed., *Macrocycle Synthesis: A Practical Approach*, Oxford University Press, Oxford, 1996.

Complex formation

J. J. Christensen, D. J. Eatough, and R. M. Izatt, *Chem. Rev.*, 1974, **74**, 351–384.

R. M. Izatt, J. S. Bradshaw, S. A. Nielsen, *et al.*, *Chem. Rev.*, 1985, **85**, 271–339.

R. M. Izatt, J. S. Bradshaw, K. Pawlak, and R. L. Bruening, *Chem. Rev.*, 1991, **91**, 1721–2085.

R. M. Izatt, K. Pawlak, J. S. Bradshaw, *et al.*, *Chem. Rev.*, 1992, **92**, 1261–1354.

R. M. Izatt, K. Pawlak, J. S. Bradshaw, and R. L. Bruening, *Chem. Rev.*, 1995, **95**, 2529–2586.

J. W. Steed, *Coord. Chem. Rev.*, 2001, **215**, 171–221.

Azacycloalkanes and Azacyclophanes

Antonio Bianchi[1] and Enrique García-España[2]

[1] *University of Florence, Florence, Italy*
[2] *Universidad de Valencia, Valencia, Spain*

1 INTRODUCTION

Azamacrocycles, namely, macrocyclic ligands containing nitrogen donor atoms, are associated with a number of fundamental biological systems, where they are present as metal complexes. The iron–phorphyrin complex in heme proteins, the chlorin–magnesium complex of chlorophyll, and the corrin–cobalt complex of vitamin B_{12} (Figure 1) are well-known examples of azamacrocycles implicated in life processes. Nature has good reasons for adopting similar ligands, since the high thermodynamic and kinetic stabilities of these complexes ensure the preservation of their biological functions against all kinds of processes that could determine the loss of the active metal center.

Inspired by these natural complexes, chemists were interested, since the early 1960s, in obtaining synthetic azamacrocycles that could be employed as model systems for understanding their particular properties, and later to mimic their functions. Before 1960, only the phthalocyanine family (Figure 1), a class of highly conjugated synthetic ligands with strong resemblance to natural azamacrocycles, was well established. In 1960, Curtis described the reaction of the blue-violet triethylenediaminenickel(II) perchlorate complex with acetone yielding a yellow complex[1] that was later[2] identified to contain the tetraazadiene macrocyclic ligand shown in Figure 1. At approximately the same time, Busch and Thomson,[3,4] Schrauzer,[5] Thierig and Umland,[6] and Busch and Melson[7] described the synthesis of other azamacrocycles.

In 1937, van Alphen described the synthesis of the azacycloalkane (saturated azamacrocycle) 1,4,8,11-tetraazacyclotetradecane (**6**), more commonly known as cyclam, obtained by reaction of 1,3-dibromopropane and 1,4,8,11-tetraazaundecane in the presence of an alkali.[8] Twenty-four years later, Stetter and Mayer synthesized the ligand by a different route and showed that the material described by van Alphen contained only a small amount of the claimed ligand,[9] but it took another five years for the first metal complex with cyclam to be reported.[10] After these early works, a huge number of saturated azamacrocycles were synthesized, making the azacycloalkanes one of the most studied class of macrocyclic ligands.[11–15]

In 1955, Stetter and Roos described the synthesis of three tetraazacycloalkanes (**7**–**9**) containing two biphenyl groups in the cyclic ring and reported that both **8** and **9** form stable solid-state 1 : 1 adducts with benzene or dioxane, unlike the smaller **7** which is unable to form similar compounds.[16] Until a crystal structure showing that the 1 : 1 adduct between **9** and benzene is a clathrate in which benzene is accommodated between ligand molecules in the crystal lattice was reported by Hilgenfeld and Saenger in 1982,[17] these compounds were thought to be intramolecular cavity inclusion complexes. Another azacyclophane

Supramolecular Chemistry: From Molecules to Nanomaterials.
Edited by Philip A. Gale and Jonathan W. Steed.

© 2012 John Wiley & Sons, Ltd.

ISBN: 978-0-470-74640-0.

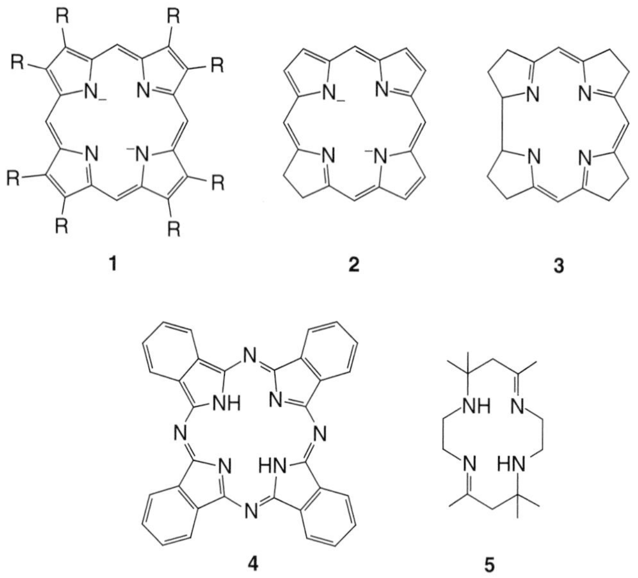

Figure 1 Porphyrin (**1**), chlorin (**2**), and corrin (**3**) rings, and synthetic phthalocyanine (**4**) and tetraazadiene (**5**) macrocycles.

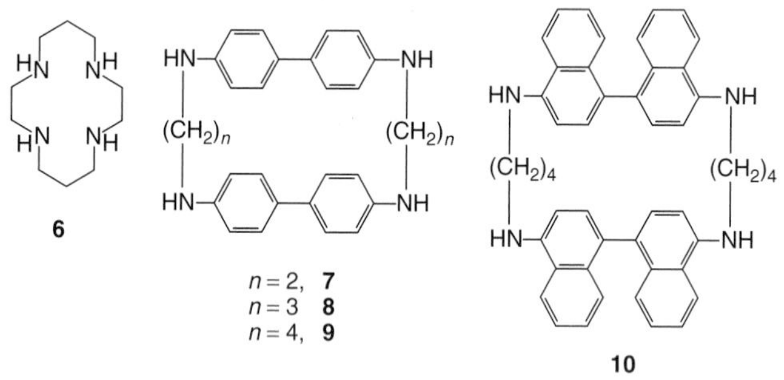

(azacycloalkane with aromatic groups in the cyclic structure) containing two binaphthyl groups (**10**) was reported by Faust and Pallas in 1960.[18] Also, this ligand forms a solid-state 1 : 1 adduct with dioxane which must be reasonably regarded as a crystal lattice inclusion compound. The first unambiguous evidence of inclusion complex formation by azacyclophanes was reported in 1980 by Koga and coworkers, who demonstrated that the tetraprotonated form of the water-soluble ligand **11** binds apolar molecules into the ligand cavity both in aqueous solution and in the solid state.[19] This work was a milestone in the development of azacyclophanes, since the introduction of polyamine chains into the hydrophobic cyclophane rings gives rise to water-soluble ligands that can benefit from the hydrophobic effect in the binding of apolar substrates in aqueous media.[20,21] In addition to water solubility, several other advantages, such as straightforward synthesis, well-defined and preorganized molecular architectures that are promoted by the rigid aromatic groups, and easy-to-get information on binding events reported by variation of the physicochemical properties of the aromatic groups, fostered the development of azacyclophanes. Furthermore, the presence of several amine groups in these ligands makes them interesting not only for the binding of neutral molecules but also for cation and anion coordination.[11–15,22,23] To this purpose, the binding ability of azacyclophanes was enriched by the insertion of heteroaromatic groups, such as pyridine, bipyridine, terpyridine, and phenanthroline, whose donor atoms can be involved in substrate binding, and may even be the preferential binding site in the case of metal ions. The last ligands can be regarded as belonging to a subcategory of azacyclophanes, the azaheterocyclophanes, and frequently the suffixes pyridilophanes, phenathrolinophanes, and so on, are used to refer to azacyclophanes containing the specified heteroaromatic groups.

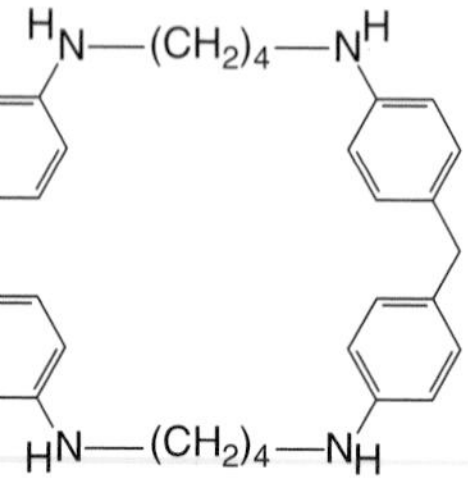

The nomenclature of macrocyclic ligands is difficult. For this reason, many shorthand nomenclatures have been proposed for these molecules, but in several cases these can be ambiguous. For instance, the IUPAC (International Union of Pure and Applied Chemistry) name of cyclam is 1,4,8,11-tetraazacyclotetradecane (Figure 2) but it is also referred to as [14]aneN_4, a term which makes reference to the number of atoms constituting the macrocyclic ring ([14]), to the fact that this molecule is saturated (ane), and to the type and number of donor atoms (N_4). Nevertheless, the name [14]aneN_4 does not discriminate between isomeric ligands in which ethylenic and propylenic chains have different positions within the macrocycle. In the IUPAC rules to name cyclophanes, each aromatic group is collapsed to a single atom ("superatom") so that the cyclophane is reduced to an aliphatic ring (Figure 2) which is named by replacing the ending *-ne* with *-phane*. The superatoms are then designated with the name of the corresponding arene by changing the terminal *-e* with *-a*. For instance, "benzene" becomes "benzena." The arene position is denoted according to the ring position of the superatom. Examples of nomenclature and notations are shown in Figure 2.

2 SYNTHETIC PROCEDURES

Cyclization reactions involve intermolecular condensation of at least two molecular moieties, each of which contains two reactive groups, or intramolecular reaction of an appropriately bifunctionalized molecule. For this reason, the

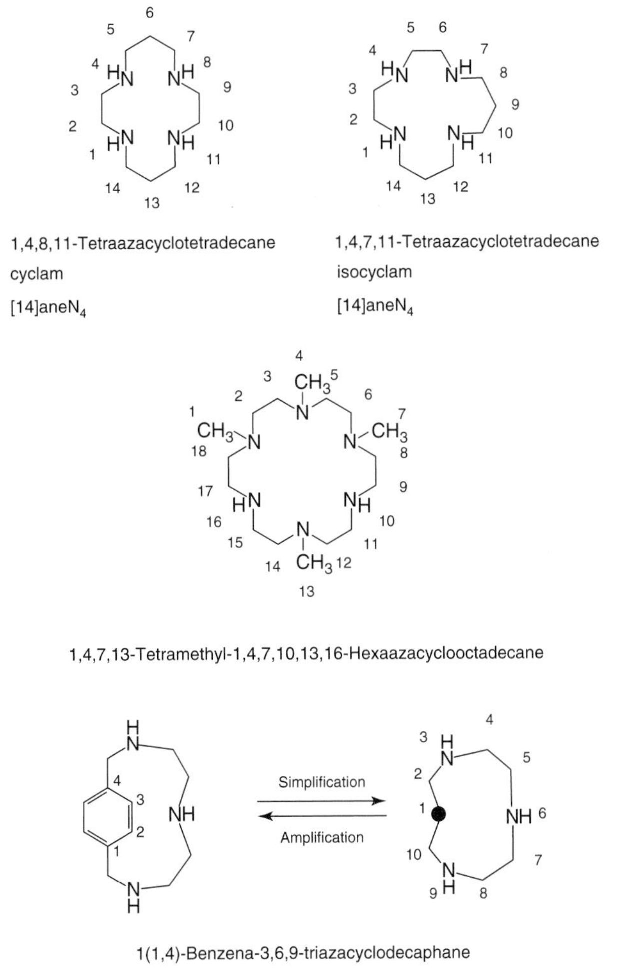

1,4,8,11-Tetraazacyclotetradecane
cyclam
[14]aneN_4

1,4,7,11-Tetraazacyclotetradecane
isocyclam
[14]aneN_4

1,4,7,13-Tetramethyl-1,4,7,10,13,16-Hexaazacyclooctadecane

Simplification
Amplification

1(1,4)-Benzena-3,6,9-triazacyclodecaphane

Figure 2 Examples of nomenclature and notations for polyazacycloalkanes and polyazacyclophanes.

formation of 1 : 1 condensation or intramolecular reaction products can be accompanied by formation of higher order oligomers and/or polymeric species. Formation of such by-products can be prevented, or reduced, by adopting high-dilution conditions. A typical way of carrying out this procedure consists in the dropwise addition, at a very slow rate, of reactants to a large volume of solvent kept under the required experimental conditions.[24] If the reaction rate is faster than the rate of addition of the reactants, then the concentration of the latter will always be small during the entire course of the reaction. When the cyclization reaction is slow or the ring closure is favored by the conformation of reactants or by the presence of a template (see below), high-dilution conditions do not offer any advantage, but may lead to longer reaction times.

Although many synthetic routes have been adopted for the preparation of azacycloalkanes and -cyclophanes, the vast majority of cyclization procedures can be grouped into two major categories: (i) direct synthesis by conventional organic reactions and (ii) metal ion-promoted (template) reactions.

2.1 Conventional organic synthesis

An important method that has been successfully employed for the preparation of many macrocycles of these types was proposed by Richman and Atkins in 1974.[25] This method, shown in Scheme 1 for the synthesis of 1,5,9,13-tetraazacyclohexadecane, involves the conversion of a linear polyamine in its *N*-tosylated (tosyl = *p*-toluenesulfonyl; Ts) derivative. Tosyl groups increase the acidity of primary amines, which can be easily deprotonated to form more reactive species, and protect secondary ones from further reactions. Cyclization performed in dipolar aprotic solvents, such as dimethyl formamide (DMF), involves the displacement of a leaving group from a complementary reactant by the deprotonated tosylamide to produce a tosylated macrocycle. The removal of protecting tosyl groups requires fairly drastic conditions. The most used detosylation process is acid hydrolysis, which can be carried out with hot concentrated (98%) sulfuric acid. However, acidic hydrolysis of tosylate does not always give good yields. Reductive cleavage with HBr/acetic acid mixture in the presence of phenol is another widely used procedure to remove tosyl groups, although other reductive agents, including sodium in liquid ammonia, and lithium aluminum hydride have

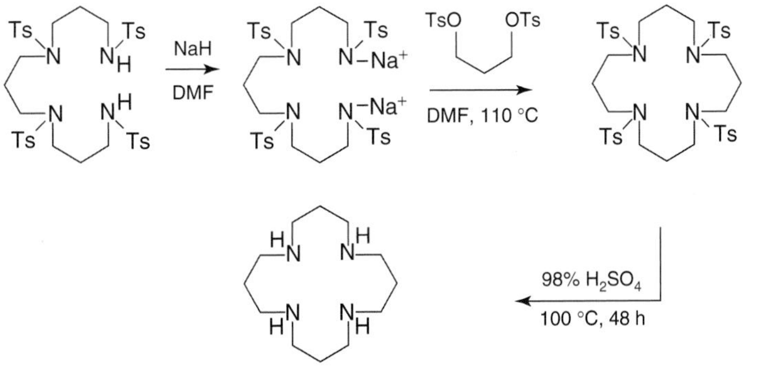

NaH
DMF
TsO OTs
DMF, 110 °C
98% H_2SO_4
100 °C, 48 h

Scheme 1 Synthesis of 1,5,9,13-tetraazacyclohexadecane according to the Richman and Atkins' method.

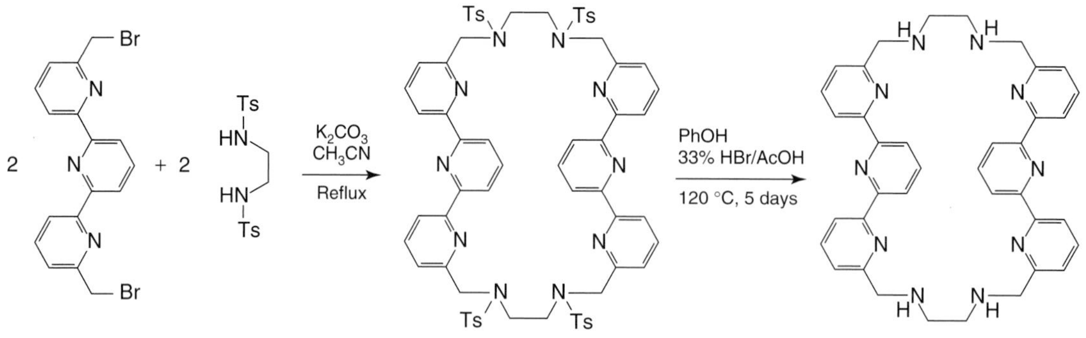

Scheme 2 Synthesis of the terpyridinophane **12**.

been successfully employed. The Richman and Atkins procedure does not require high-dilution conditions to achieve high yields of the cyclic products. The bulky tosyl groups reduce the conformational freedom of reactants, and this is believed to facilitate cyclization relative to oligomerization and polymerization reactions.

In some cases, oligomeric species are formed in enough amount to be isolated, and may even be the target compounds of the synthetic process as in the case of the 2 : 2 condensation reaction shown in Scheme 2 for the synthesis of the terpyridinophane **12**, obtained in 56% yield.[26]

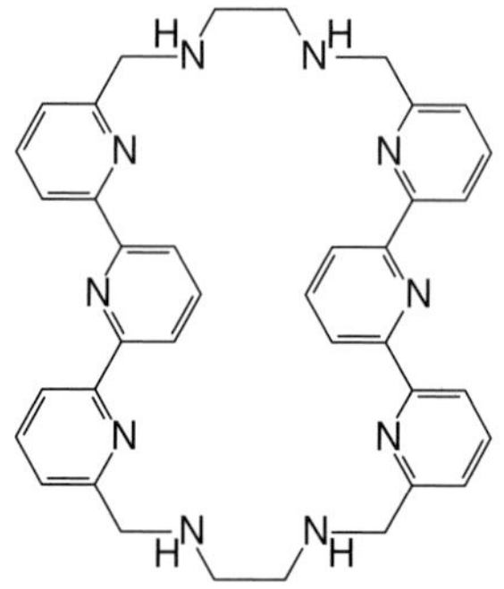

Azacyclophanes can also be synthesized by means of conventional organic reactions involving the halide–amine cyclization, under high-dilution conditions, or N-alkylation of diethyl phosphoramidates using bis(bromomethyl)arenes in the presence of base.[20] These reactions give rise to the formation of cyclic amides and diethoxyphosphorylazacyclophanes, respectively, which can be readily converted into the corresponding polyazacyclophanes by amide reduction with $LiAlH_4$ and acid cleavage of the phosphoryl groups. An example of N-alkylation of diethyl phosphoramidates affording 1 : 1, 2 : 2, and 3 : 3 condensation products is shown in Scheme 3.[27]

2.2 Metal ion template reactions

As already said, Curtis in 1960 discovered a template reaction between the triethylenediaminenickel(II) complex and acetone to yield the tetraazadiene macrocycle shown in Figure 1.[1] After this work, template reactions involving different metal ions were routinely used for the preparation of azacycloalkanes and -cyclophanes.[28] A typical template cyclization reaction performed in the presence of Ni(II) is depicted in Scheme 4. Ni(II) forms a 1 : 1 complex with the triamine 1,5,9-triazanonane according to a square-planar coordination geometry involving a water molecule. This labile water molecule is replaced by the nitrogen atom of a 2,6-diacetylpyridine molecule that brings the two carbonyl groups within reaction distance of the two primary amine groups of the coordinated triamine, and a Schiff base condensation reaction takes place with ring closure. The C=N imine bond is successively hydrogenated. Similar condensation reactions involving Ni(II) complexes of tetraamine and glyoxal give rise, after reduction of the Schiff base, to tetraazacycloalkane complexes (Scheme 5). The conversion of the imine groups into amine ones can be performed with the use of conventional reductants, including H_2 in the presence of a catalyst, $NaBH_4$, as well as by electrochemical means. The removal of the metal ion from the complex, to yield the free macrocycle, is usually carried out by treatment with alkaline cyanide or sulfide solutions, followed by ligand extraction. In the case of labile complexes, addition of excess acid may be sufficient to demetallate the ligand which is obtained in its protonated form, while for inert complexes, such as those of Co(III) and Cr(III), metal ion reduction can be necessary before it may be removed from the macrocycle.

Another route to the synthesis of azacycloalkanes was inspired by Sargeson's template synthesis of Co(III) sepulcrate and sarcophagine complexes.[29,30] Examples of this procedure are shown in Scheme 6. The bis-ethylenediamine complexes of Cu(II) or Ni(II) first undergo Schiff base condensation with four molecules of formaldehyde and then two molecules of a diprotic acid, such as $CH_3CH_2NO_2$

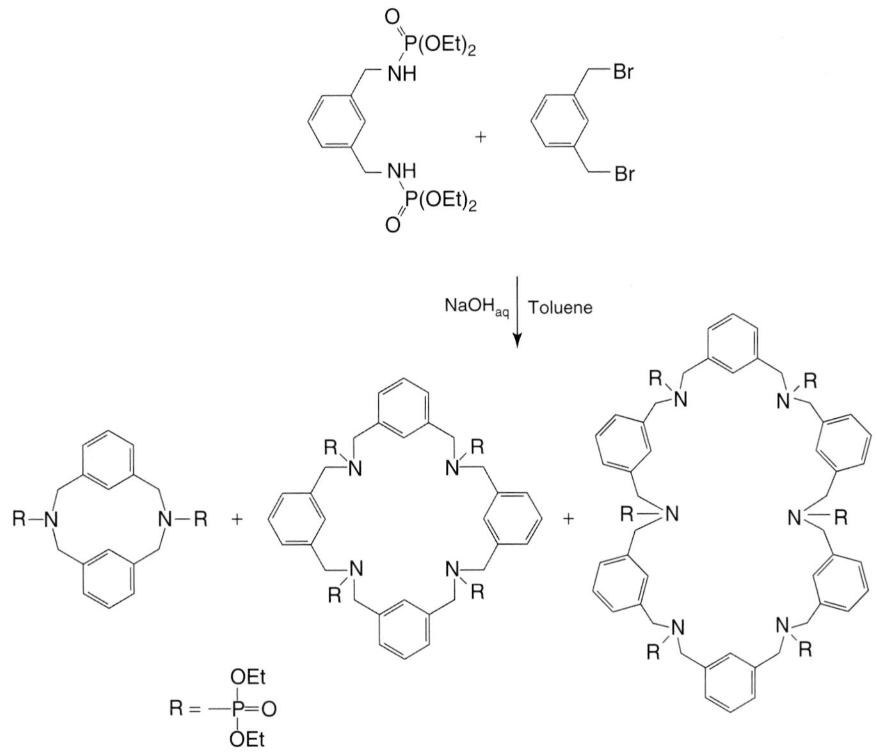

Scheme 3 Synthesis of polyazacyclophanes via N-alkylation of diethyl phosphoramidates.

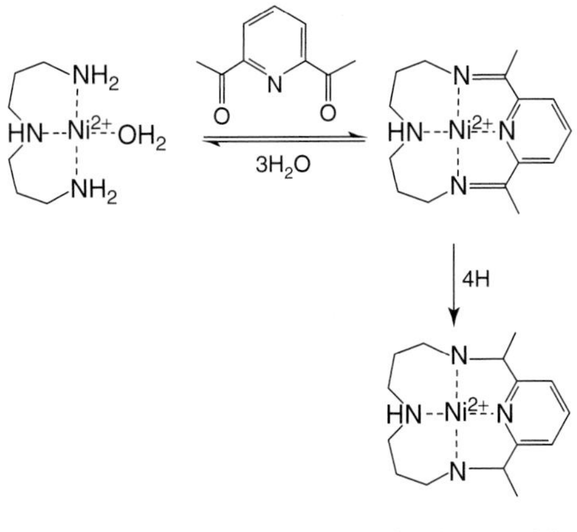

Scheme 4 Ni(II) template synthesis of 1(2,6)-pyridina-3,7,11-triaza-2,12-dimethyldodecaphane.

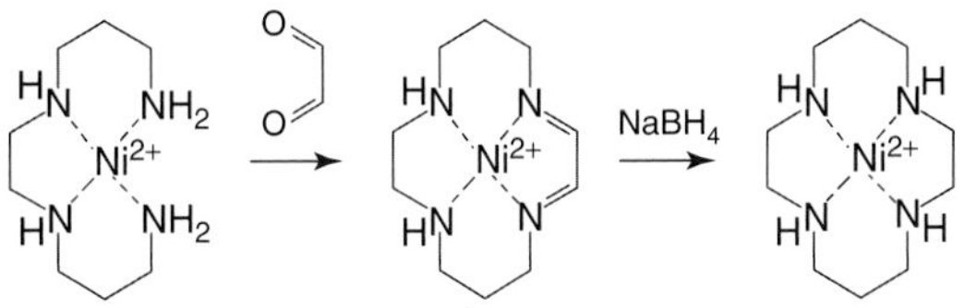

Scheme 5 Ni(II) template synthesis of 1,4,8,11-tetraazacyclotetradecane (cyclam).

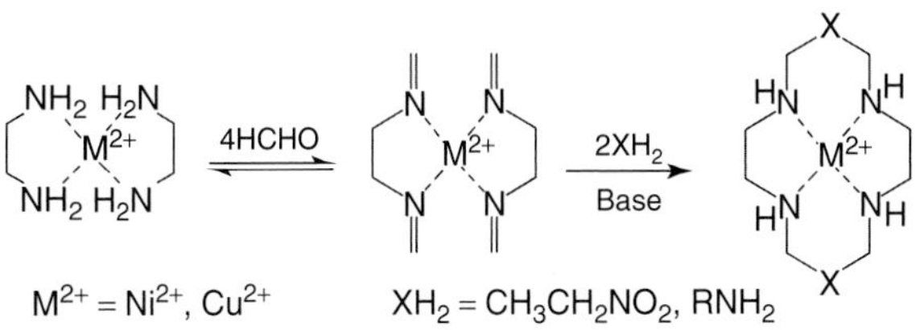

Scheme 6 Template synthesis of cyclam-like macrocyclic complexes according to the Sargeson's method.

or R–NH_2, deprotonate in the presence of a base and the resulting anion reacts with the imine bonds to form a macrocycle.[31,32] Similar reactions can be performed by using complexes with tetraamine ligands as starting compounds.[33,34] Removal of the metal ion can be performed according to the methods indicated above.

Different roles for the metal ion in a template reaction have been delineated and termed *thermodynamic template effect* and *kinetic template effect*.[4] In the first case, the metal ion picks out the macrocyclic ligand from an equilibrating mixture of products, thus driving the reaction equilibrium over to the product of higher stability. On the other hand, in the kinetic template effect, the metal ion influences the steric course of the condensation, thus increasing the rate of formation of the cyclic product. Very often, the role of the metal ion is quite complex and may involve aspects of both effects.

2.3 Further synthetic procedures

In addition to the above syntheses involving metal ions as external templates (also called *exo-templates*), certain azamacrocycles can also be prepared by the use of *endo-templates*. This type of template effect involves a temporary covalent linkage maintaining the reactant molecule in appropriate conformation for ring closure. The *endo*-templated synthesis of cyclam is shown in Scheme 7.[35] Condensation of 1,5,8,12-tetraazadodecane with butanedione gives rise to a tricyclic aminal intermediate, which is successively reacted with 1,2-dibromoethane for ring closure, after which the *endo*-template group is removed by acid cleavage. Similar procedures have been employed for the synthesis of various azacycloalkanes.[36]

Synthetic methods for the preparation of azacycloalkanes and -cyclophanes also include modification of a preformed macrocycle. This is commonly performed by means of conventional functionalization reactions which do not follow particular strategies and are adequately described in the cited literature.[11–13,20,21] Conversely, a large strategic effort has been dedicated to the preparation of topologically constrained ligands[36] and linked-ring[37] systems. The first case mostly consists of bridging superstructures that are added to small macrocycles to obtain macropolycyclic ligands, which are extensively described in **Cryptands and Spherands**, Volume 3. Several procedures, including polycyclization reactions and linkage of preformed macrocycles, have been reported for linking together two or more macrocycles. The latter route has proved to be the most common and practical form of joining macrocycles, in particular, via nitrogen–nitrogen (N–N) linking, owing to the reactivity of the amine functionalities, but carbon–carbon (C–C) linkages are also not uncommon. Different strategies to synthesize linked azamacrocyclic ligands were described in a comprehensive review dedicated to this topic.[37] Examples of linked polyazacycloalkanes and -cyclophanes are shown in Figure 3.

3 BASICITY PROPERTIES

Polyamines are bases and, accordingly, they can be involved in protonation equilibria. In solution, the proton does not exist as an elementary particle. In water, for instance, it forms a covalently bound hydronium ion, H_3O^+,

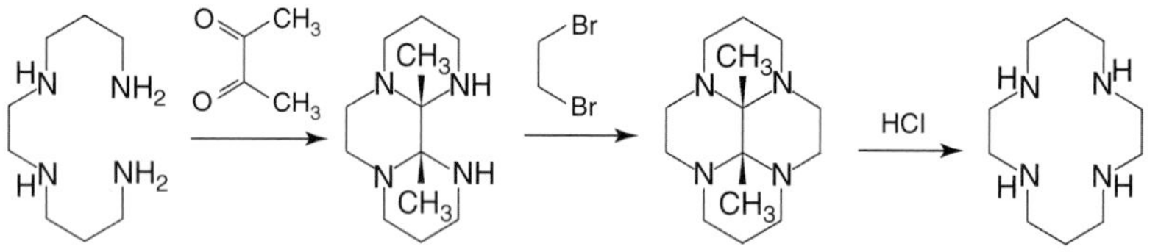

Scheme 7 Synthesis of cyclam promoted by an *endo*-template obtained via tetraamine-diketone condensation.

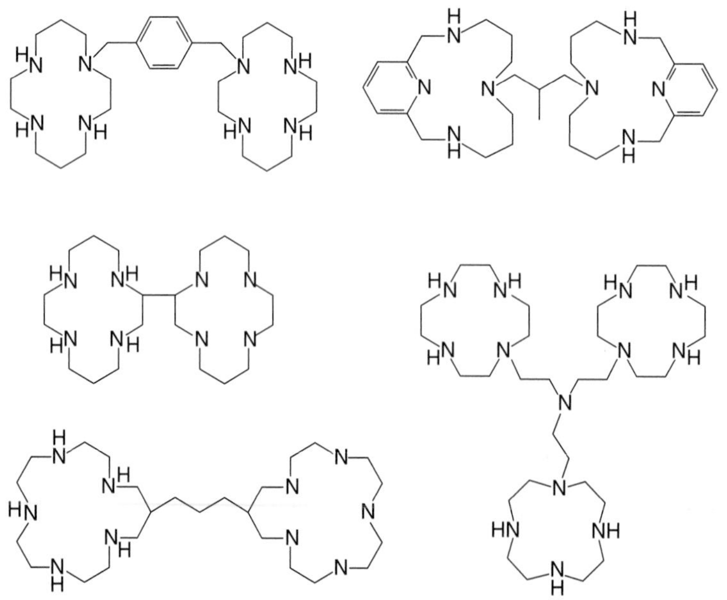

Figure 3 Nitrogen–nitrogen and carbon–carbon linked azacycloalkanes and -cyclophanes.

which through hydrogen bonding agglomerates other water molecules, forming more hydrated species, such as $H_9O_4^+$. Despite the complexity of the proton–solvent systems, the name "proton" is used here with the meaning "proton in solution" and "protonation" is "the transfer of protons from the solvent to the species being protonated."

Protonation of azacycloalkanes and -cyclophanes is a process that competes with metal ion coordination and gives rise to the formation of positively charged ammonium species that may function as anion receptors. For this reason, the knowledge of the protonation properties of these ligands is paramount for both metal cation and anion coordination studies, in particular, when the variation of a ligand property with pH is followed to determine the complex species formed in solution and the relevant stability constants (see **Binding Constants and Their Measurement**, Volume 2).

3.1 Basicity properties of azacycloalkanes

The protonation behavior of azacycloalkanes is similar to that shown by the acyclic counterparts, being mostly determined by the basicity of the amine nitrogen atoms and the electrostatic repulsion between the ammonium groups generated upon increasing protonation.[38] An example of increasing protonation is offered by the [3k]aneN_k (k = 2–12) series of azacycloalkanes constituted by secondary amine groups connected by ethylenic chains and spanning from piperazine ([6]aneN_2) to the large [36]aneN_{12}. The protonation constants of these molecules, determined in aqueous solution by means of potentiometric measurements, are listed in Table 1. As can be seen from this table, the basicity of these ligands decreases in the successive protonation steps, in agreement with the general behavior of polyamines, giving rise to some grouping of the protonation constants. Such behavior shows a general correlation with the number of nitrogen atoms in the molecules. Actually, [3k]aneN_k molecules with an even k number of nitrogen atoms form two separated groups of protonation constants, while for [3k]aneN_k molecules with odd k numbers the two groups are separated by an intermediate value. In the first case (even k number), $k/2$ ammonium groups can be generated without great electrostatic repulsion (Figure 4) while further protons will be necessarily located between charged nitrogen atoms with increasing repulsion. As a consequence, a gap is produced between the protonation constants of the first and the second half of nitrogen atoms (Table 1). On the other hand, when the k number of nitrogen atoms in the macrocycle is odd, only $(k-1)/2$ protons can be bound without great electrostatic repulsion (Figure 4). The successive proton will bind to a nitrogen atom next to one ammonium group and the following $(k-1)/2$ protons will be placed next to two ammonium groups. Accordingly, two groups of $(k-1)/2$ protonation constants corresponding to low and high electrostatic repulsion, respectively, are generated with a further protonation constant falling in the gap between them. This produces attenuation or disappearance of the grouping effect previously described.

The overall basicities of polyazacycloalkanes are lower than those of their acyclic counterparts, because of greater gathering of positive charge occurring for the cyclic molecules upon successive protonation (Figure 5). Nevertheless, the parallelism between the variation of the overall basicities of macrocyclic and open-chain polyamines and the number of amine groups is consistent with similar protonation patterns.[38]

Table 1 Logarithms of protonation constants of [3k]aneN_k (k = 2–12) polyazacycloalkanes determined at 298 K, I = 0.15 M (0.1 M for **13**, 0.2 M for **16**).

	13	**14**	**15**	**16**	**17**	**18**	**19**	**20**	**21**	**22**	**23**
Log K_1	9.71	12.6	10.38	10.85	10.15	9.76	9.65	9.59	9.85	9.79	9.75
Log K_2	5.59	7.55	9.71	9.65	9.48	9.28	9.33	9.40	9.44	9.48	9.65
Log K_3	—	2.53	2.05	6.00	8.89	8.63	8.76	8.77	8.95	9.02	8.88
Log K_4	—	—	<1	1.74	4.27	6.42	7.87	8.27	8.56	8.64	8.96
Log K_5	—	—	—	1.16	2.21	3.73	4.55	6.37	7.79	8.06	8.12
Log K_6	—	—	—	—	1.0	2.13	3.42	4.22	5.24	6.44	7.82
Log K_7	—	—	—	—	—	2.0	2.71	3.24	3.84	4.49	5.66
Log K_8	—	—	—	—	—	—	1.95	2.31	3.02	3.58	4.27
Log K_9	—	—	—	—	—	—	—	1.8	1.97	2.76	3.58
Log K_{10}	—	—	—	—	—	—	—	—	1.8	2.26	2.62
Log K_{11}	—	—	—	—	—	—	—	—	—	1.7	2.3
Log K_{12}	—	—	—	—	—	—	—	—	—	—	1.0
Log β^a	15.3	22.68	22.1	29.4	36.0	42.0	48.24	54.0	60.5	66.2	72.6

^{a}Log $\beta = \Sigma \log K_i$.

$n = 0$, **13**
$n = 1$, **14**
$n = 2$, **15**
$n = 3$, **16**
$n = 4$, **17**
$n = 5$, **18**
$n = 6$, **19**
$n = 7$, **20**
$n = 8$, **21**
$n = 9$, **22**
$n = 10$, **23**

An excellent linear correlation is observed between the logarithm of overall basicities of $[3k]aneN_k$ polyamines and the number of amine groups (Figure 5), evidencing a mean contribution of 6.2 logarithmic units per nitrogen atom to the overall basicity. A very high basicity is shown by the triazacycloalkane $[9]aneN_3$ in the first protonation step (Table 1), which shows a higher affinity for protons than the acyclic analogs at this stage, while the acyclic triamines are more basic in the following steps. This is a general behavior of triazacycloalkanes, which is representative of the effect of ring closure on proton binding properties.[38] Such characteristics can be ascribed to an inside orientation of the lone pairs of nitrogen atoms, promoted by the cyclic structure, which stabilize the monoprotonated forms of these molecules by the formation of intramolecular hydrogen bond networks involving all amino groups as shown by the crystal structure[39] of the $H(\mathbf{14})ClO_4$ (**14** = 1,4,7-trimethyl-1,4,7-triazacyclononane, $(CH_3)_3[9]aneN_3$) salt (Figure 6).

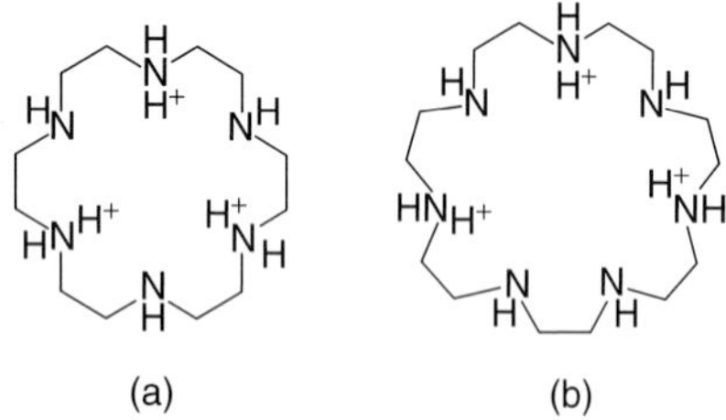

Figure 4 Localizations of H^+ ions in $H_3(\mathbf{17})^{3+}$ (a) and $H_3(\mathbf{18})^{3+}$ (b), leading to the minimum electrostatic repulsion.

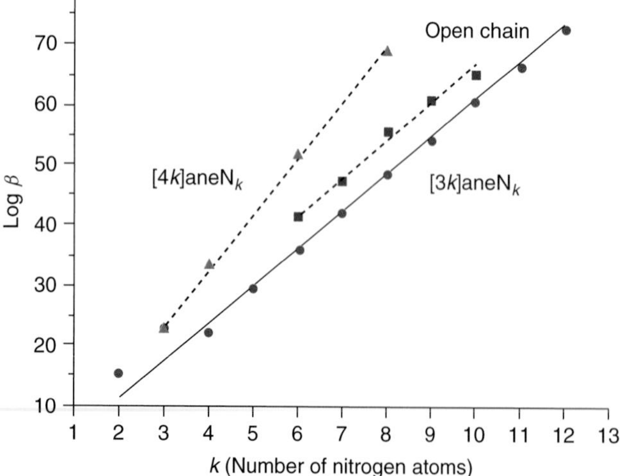

Figure 5 Overall basicity of $[3k]aneN_k$, $[4k]aneN_k$, and open-chain analogs of $[3k]aneN_k$ polyamines.

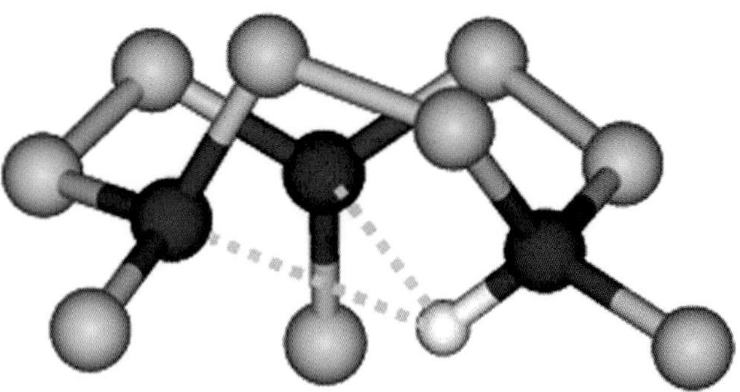

Figure 6 View of the monoprotonated cation $H(\mathbf{14})ClO_4$ showing the proton binding. Carbon (Grey), Nitrogen (Blue), and Hydrogen (White).

Also, for the series of $[4k]aneN_k$ azacycloalkanes (**24**–**27**), containing secondary amine groups connected by propylenic chains, there is an excellent correlation between the overall basicity and the number of amine groups in the molecules (Figure 5). For these macrocyclic polyamines, the mean contribution to the overall basicity is 9.1 logarithmic units per amine group. The greater basicity of these polyamines relative to $[3k]aneN_k$ homologs is due to a greater inductive effect of propylenic versus ethylenic nitrogen substituents and to the longer separation between nitrogen atoms in $[4k]aneN_k$ macrocycles giving rise to a lower electrostatic repulsion between the ammonium groups in protonated species.[38]

$n = 1$, **24**
$n = 2$, **25**
$n = 4$, **26**
$n = 6$, **27**

Interesting protonation patterns are shown by the ditopic hexaazacycloalkanes, **28**–**32**, containing two polyamine units separated by long aliphatic chains (Table 2). Such molecules represent for protons what ditopic ligands are for metal ions. As shown in Table 2, the two triaminic subunits behave as almost independent entities toward protonation, since protons can alternatively occupy the two subunits, which are far apart from each other, giving rise to similar values of the equilibrium constants for the same protonation step in each subunit.[38]

$n = 4$, **28**
$n = 6$, **29**
$n = 9$, **30**

$n = 7$, **31**
$n = 10$, **32**

Table 2 Logarithms of protonation constants of hexaazacycloalkanes **28–32** determined at 298 K, $I = 0.1$ M (0.5 M for **28** and **29**).

	28	**29**	**30**	**31**	**32**
Log K_1	10.64	10.73	*a*	10.70	*a*
Log K_2	10.12	10.31	*a*	10.70	*a*
Log K_3	9.37	9.93	9.60	9.85	10.10
Log K_4	8.86	9.47	9.25	9.60	9.60
Log K_5	3.44	3.82	4.15	7.90	7.95
Log K_6	3.42	3.57	3.55	7.30	7.30

[a]Values not determined because of low solubility of the compound.

Protonation constants for many triaza-, tetraaza-, and pentaazacycloakanes, which are characterized by different connections between the amine groups, have been reported.[40] Also the protonation behavior of these polyamines is mostly determined by the basicity of the amine nitrogen atoms and the electrostatic repulsion between the ammonium groups generated upon increasing protonation. As already shown for the smaller triazacycloalkanes, the formation of intraring hydrogen bond networks can afford particular stability to protonated species, giving rise, in some cases, to special protonation features. Another interesting example of this type is furnished by cyclam (**6**). The successive protonation constants of cyclam are log $K = 11.58$, 10.62, 1.61, and 2.41. The high values of the first two protonation constants have been associated with the involvement of two amine groups in the binding of each proton, as shown in the solid state by the molecular structure[41] of the $H_2(\mathbf{6})^{2+}$ cation (Figure 7). Breaking of such intramolecular hydrogen bond network, caused by further protonation, would determine the considerable drop in stability found for the successive protonation steps and the interconversion of the molecule from the *in* nitrogen configuration observed in the structure of $H_2(\mathbf{6})^{2+}$ to the *out* configuration found[42] for $H_4(\mathbf{6})^{4+}$ (Figure 7), which allows a better minimization of the electrostatic repulsion between positive charges. Such *in–out* interconversion is thought to occur at the third protonation step. For this reason, the fourth proton-binding $H_3(\mathbf{6})^{3+}$ would be favored by such expanded ligand conformation, leading to a fourth protonation constant being greater than the third one. Similar

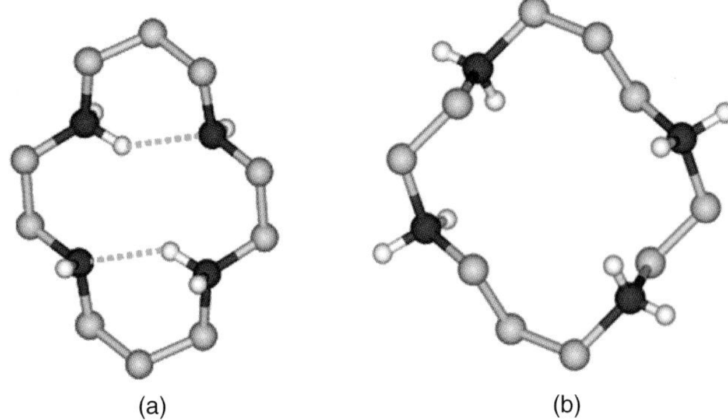

Figure 7 Molecular structures of the diprotonated (a) and tetraprotonated (b) forms of cyclam. Carbon (Grey), Nitrogen (Blue), and Hydrogen (White).

in–out interconversions are expected to be a general feature of all azacycloalkanes arising not only from intramolecular forces generated by protonation but also from medium effects and, in particular, by the presence of anionic species that may form stable adducts with polyammonium cations. This phenomenon, which is known as "anion coordination," (see Section 5 of this Chapter) can influence the proton binding properties of all kinds of amines, although it is particularly evident for macrocyclic polyamines and highly charged anions.[38]

N-functionalization modifies the basicity of amine groups. For instance, methylation in the case of azacycloalkanes gives rise to the conversion of secondary into tertiary amino groups. Since tertiary amines are less basic than secondary ones in aqueous solution, one could predict that in the stepwise protonation of azacycloalkanes containing both secondary and tertiary amino groups, the former will be protonated before the latter, although a statistical effect favoring protonation of the largest set of identical amino groups also has to be taken into account. In the successive protonation steps, however, the effect of electrostatic repulsion between positive charges becomes crucial in determining the protonation sites. Nevertheless, the macroscopic effect of N-methylation in polyazacycloalkanes is a general trend of decreasing basicity in aqueous solution. Furthermore, N-methylation also modifies the hydrogen bond ability of amino groups. As a matter of fact, the singular behavior of cyclam (**6**), having the third protonation constant smaller than the fourth one, is no longer observed for its mono-, di-, and tetramethylated forms, indicating that N-methylation prevents the formation of the particularly stable hydrogen-bonded structure found for $H_2(\mathbf{6})^{2+}$.[38]

As noted before, the knowledge of ligand protonation behaviors is important in the study of polyamine interaction with substrate species. In many cases, the knowledge of proton location in the ligands would also be important, since the specificity of the interaction depends on the matching between substrate and ligand binding groups. In the case of [3*k*]aneN$_k$ and [4*k*]aneN$_k$ azacycloalkanes, as well as for any other highly symmetrical ligand, the determination of protonation sites may be very problematic. For instance, ^{1}H and ^{13}C NMR (nuclear magnetic resonance) spectroscopies, which are powerful instruments for structure analysis in solution, are almost insensitive in the analysis of protonation patterns of such azacycloalkanes, since all carbon and hydrogen atoms remain magnetically equivalent independent of pH. The insertion of methyl groups removes the magnetic equivalence, and the pH dependence of ^{1}H and ^{13}C signals becomes a probe to recognize protonation sites.[38]

3.2 Basicity properties of azacyclophanes

The protonation behavior of azacyclophanes depends not only on the general features delineated above for azacycloalkanes but also on interactions with the π-clouds of the aromatic groups and the hydrophobic environments these units may generate. As an example, let us consider the azacyclophanes **33**–**36** containing *p*-xylyl spacers.[38] The protonation constants for this group of ligands are included in Table 3. As already observed for azacycloalkanes, the overall basicity of **33**–**36** increases with increasing number of amine groups in the ligand. The trends in the stepwise basicity constants follow the general criteria of minimum electrostatic repulsion between positively charged groups. In this sense, the tetraazacyclophane **35** with only ethylenic chains shows two relatively large protonation constants, one intermediate and one much lower constant for the last protonation step, due to the necessary entry of this proton on a nitrogen atom between already protonated amine groups. Polyazacyclophane **36**, however, with a symmetrical array of one ethylenic and two propylenic chains, shows much larger constants in the third and fourth protonation steps due to the presence of larger propylenic chains. Similarly, the triazacyclophane **34** displays higher basicity than **33** in all protonation steps. In contrast to the triazacycloalkane **14**, no special stabilization of protonated species is observed

Table 3 Logarithms of protonation constants of polyazacyclophanes **33**–**36** determined at 298 K, $I = 0.15$ M.

	33	34	35	36
Log K_1	9.42	10.13	9.39	9.93
Log K_2	7.31	8.34	8.45	9.09
Log K_3	3.26	6.82	5.38	7.44
Log K_4	—	—	2.51	3.61
Log β^a	20.0	25.3	25.7	30.1

^{a}Log $\beta = \Sigma \log K_i$.

for **33**, since the long xylyl spacer enlarges the macrocyclic cavity, thus preventing the formation of intramolecular hydrogen bond networks. The presence of the aromatic unit also affects the protonation pattern of these molecules. ^{1}H and ^{13}C NMR spectroscopies are useful tools to assess protonation sites in these types of compounds. For example, the variation of ^{1}H and ^{13}C NMR signals with pH for the azacyclophane **35** is shown in Figure 8. The very significant shift of the ^{13}C resonance assigned to the quaternary carbon atom CB1 indicates that the first protonation of **35** involves the benzylic amine groups. In agreement with this result, thermodynamic data for the first protonation step of **35** ($\Delta H^\circ = -34.4\,\text{kJ mol}^{-1}$, $T\Delta S^\circ = 19.7\,\text{kJ mol}^{-1}$ for **35** + H$^+$) showed a rather low enthalpic contribution and a significant entropic term, which can be attributed to protonation occurring on rather hydrophobic protonation sites such as the benzylic amine groups.[38]

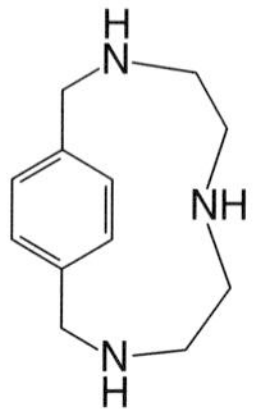

33 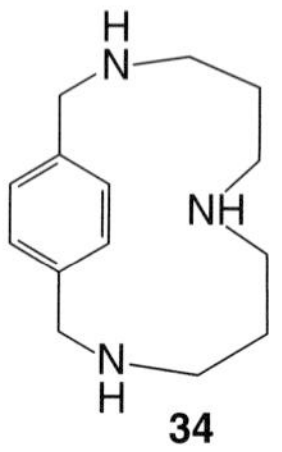 34 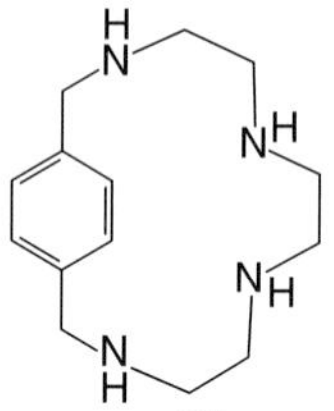 35

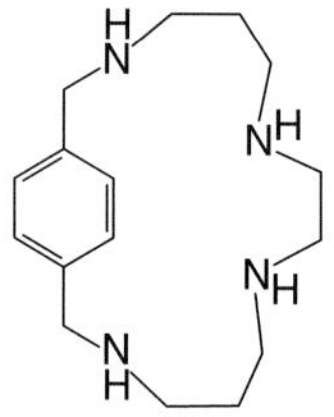

36

The effect of the aromatic group on the protonation behavior of these azacyclophanes becomes less effective on increasing the macrocycle size by increasing the number of amine groups and the length of the hydrocarbon fragments of the polyamine chain. Also, the type of substitution at the aromatic ring affects the basicity properties of such molecules, as shown, for instance, by the differently substituted ligands **37**–**39** and **40**–**42** containing a common tetraamine chain and *o*-, *m*-, and *p*-xylyl groups, respectively.[43,44] The protonation constants of these molecules are included in Table 4. The first two protonation processes involving the *o*-substituted **37** and **40** occur with constants significantly larger than those found for the *m*- and *p*-substituted analogs, while the latter is considerably more basic in the two successive protonation steps. The increased basicity of **37** and **40** in the early protonation steps was attributed to the formation of intramolecular hydrogen bonds between the protonated and unprotonated nitrogen

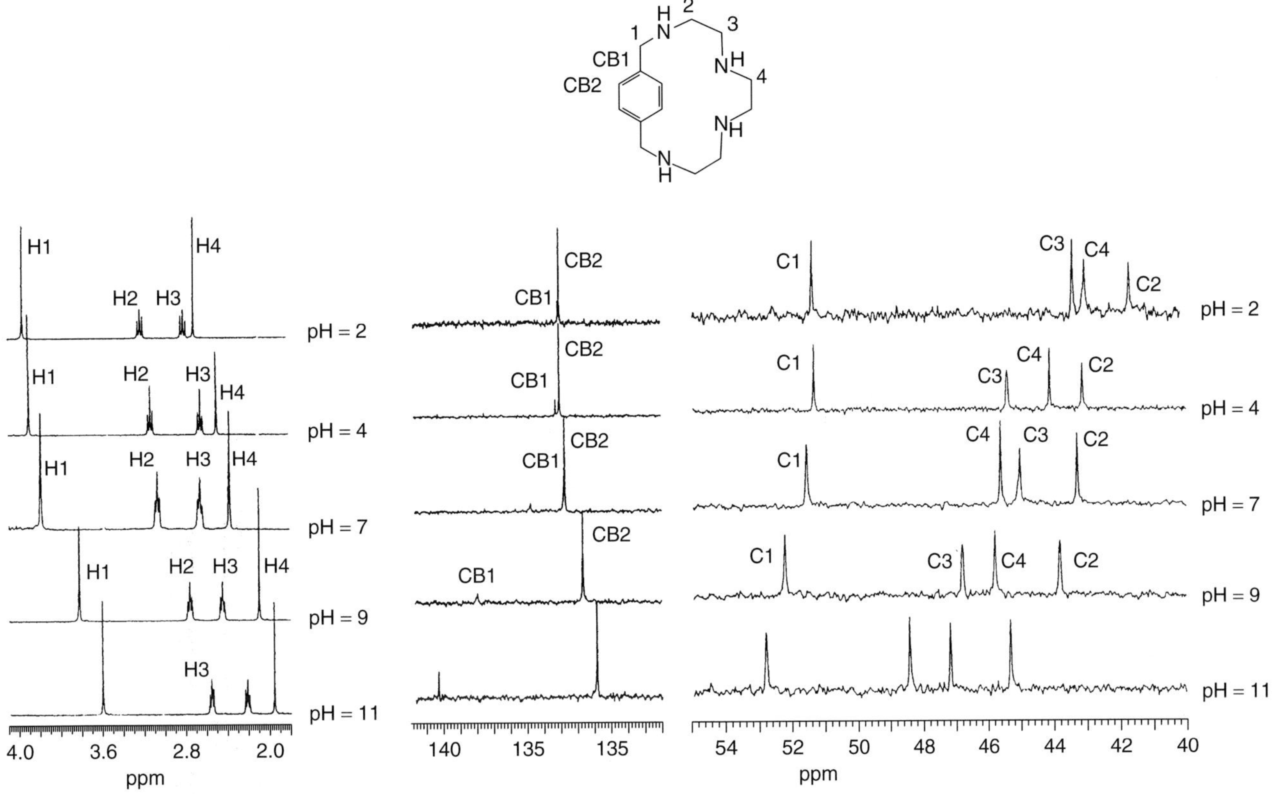

Figure 8 Variation of ^1H and ^{13}C NMR signals of **35** with pH. (Reproduced from Ref. 38. © with permission from Elsevier, 1999.)

Table 4 Logarithms of protonation constants of polyazacyclophanes **37–42** determined at 298 K, $I = 0.15$ M.

	37	**38**	**39**	**40**	**41**	**42**
Log K_1	10.38	9.53	9.39	11.01	9.83	9.55
Log K_2	9.30	8.74	8.45	10.88	8.60	8.59
Log K_3	2.96	5.61	5.38	2.20	5.92	5.81
Log K_4	<2	2.77	2.51	<2	3.88	4.50

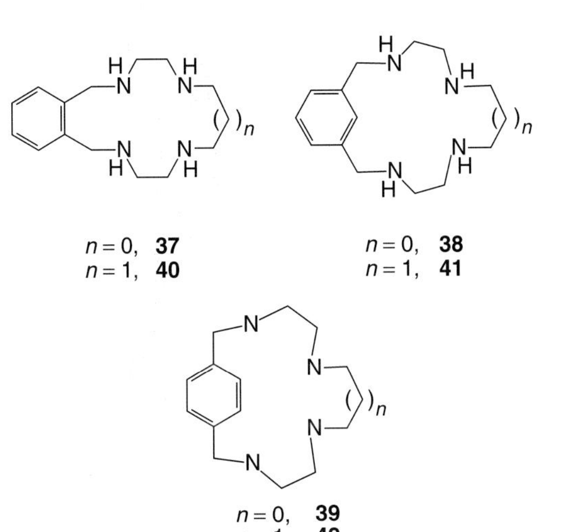

atoms, in agreement with NMR information and the crystal structure of the $H_2(\mathbf{40})^{2+}$ cation (Figure 9).[43,44] As the following protonation steps involve the breaking of such hydrogen bond networks, the values of the corresponding equilibrium constants are smaller than those for the other ligands. The formation of intramolecular hydrogen bonds is strongly favored by the presence of the *o*-xylyl spacer,

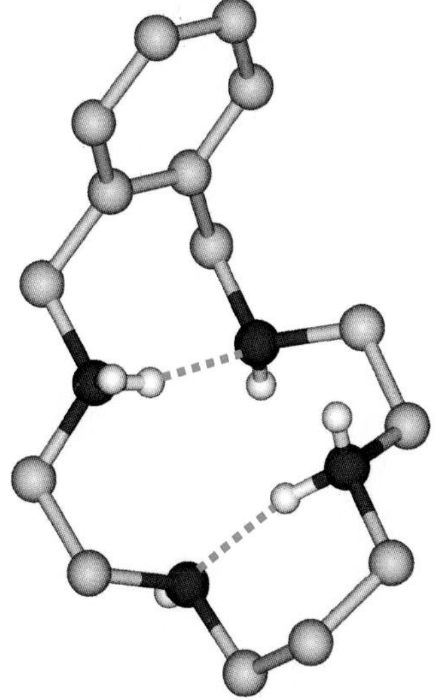

Figure 9 Crystal structure of the diprotonated $H_2(\mathbf{40})^{2+}$ cation. Carbon (Grey), Nitrogen (Blue), and Hydrogen (White).

but they do not appear to be relevant in the protonation of the related macrocycles containing *m*- and *p*-xylyl spacers which bring the amine groups to much longer distances.

Also 2 : 2 polyazacyclophanes, such as **43–45**, have been studied with regard to their basicity in solution. Their protonation behaviors are consistent with those previously presented for ditopic hexaazacycloalkanes, in which the two triaminic subunits behave as almost independent entities toward protonation, giving rise to similar values of the protonation constants for the same protonation step in each subunit. Also for these compounds, the criterion of minimum electrostatic repulsion operates since all of them present a group of four relatively high constants that are well separated from the much lower values corresponding to the last protonation step.[38]

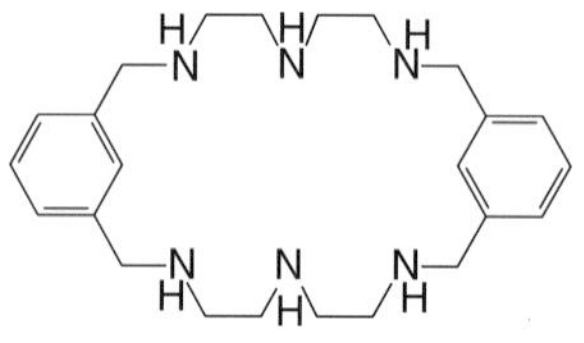

43

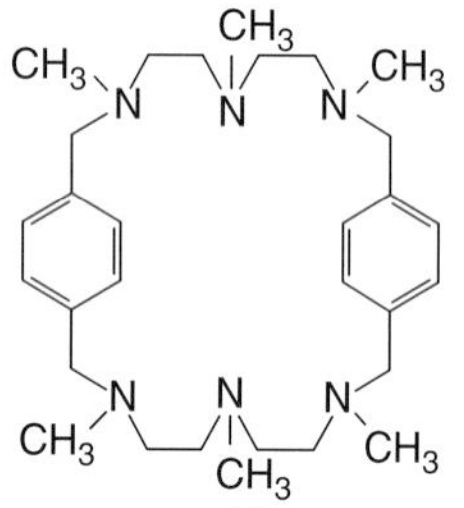

44

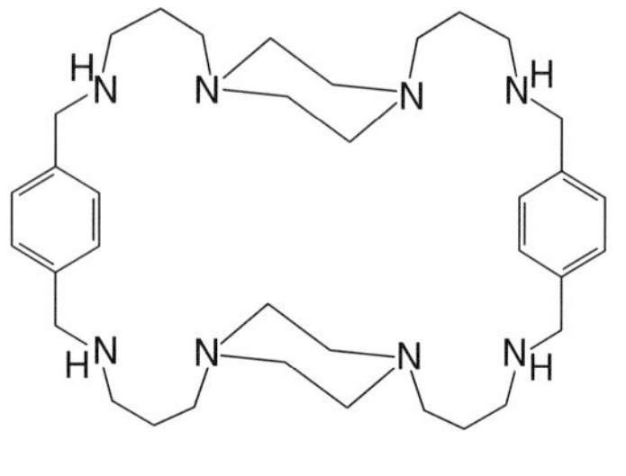

45

Protonation properties of azaheterocyclophanes, namely, azacyclophane containing heteroaromatic groups, are consistent with the behaviors displayed by azacycloalkanes and -cyclophanes. Compounds **46–49** are representative examples of these types of ligands. Their protonation constants are collected in Table 5.[45–48] Unfortunately, only the first three protonation constants were reported for **46**.[45]

Table 5 Logarithms of protonation constants of polyazacyclophanes **46–49** determined at 298 K, $I = 0.15$ M.

	46	47	48	49
Log K_1	9.63	9.42	9.31	9.38
Log K_2	9.05	8.76	8.63	8.74
Log K_3	7.56	7.37	7.22	7.18
Log K_4	—	4.16	3.92	3.85
Log K_5	—	2.23	2.80	1.79
Log K_6	—	1.5	2.10	—

Nevertheless, the first three protonation processes display very similar constants, corresponding, as shown by spectroscopic data for **47–49**,[46–48] to protonation of the common aliphatic pentaamine chain. Three protons can be located in this chain on alternate amine groups, with low electrostatic repulsion. For this reason, the first three protonation constants are relatively high and grouped, and are considerably higher than those corresponding to the successive protonation steps. Pyridine, bipyridine, terpyridine, and phenanthroline have lower protonation constants than aliphatic secondary amines, and accordingly, these etheroaromatic groups undergo protonation only in very acidic solutions. For instance, clear participation of bipyridine and terpyridine groups of **47** and **48** in proton binding occurs only in the fifth and sixth protonation steps, as shown by the pH dependence of ^{1}H NMR and UV–vis spectra.[46,47] Smaller azaheterocyclophanes resemble small azacycloalkanes and -cyclophanes; electrostatic repulsion due to accumulation of positive charge upon protonation becomes of great importance, determining, for instance, protonation of the pyridine group of **50** in the second protonation step, and a higher constant for the first protonation step (log $K = 10.33$ for $\mathbf{50} + H^+$) is observed concomitant with the participation of two nitrogen atoms in proton binding.[49]

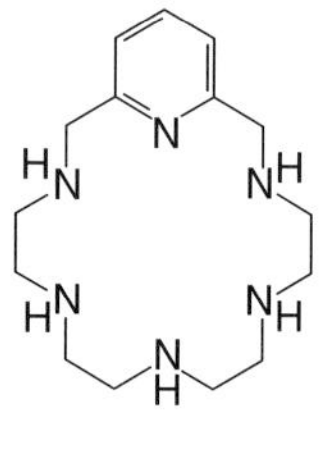

46 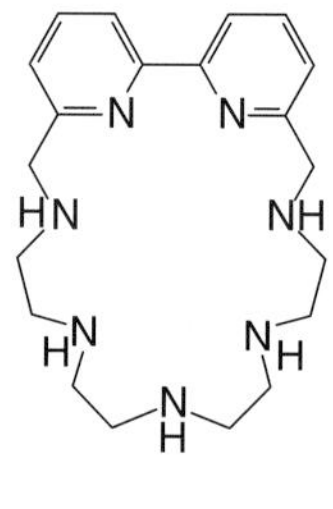 **47** 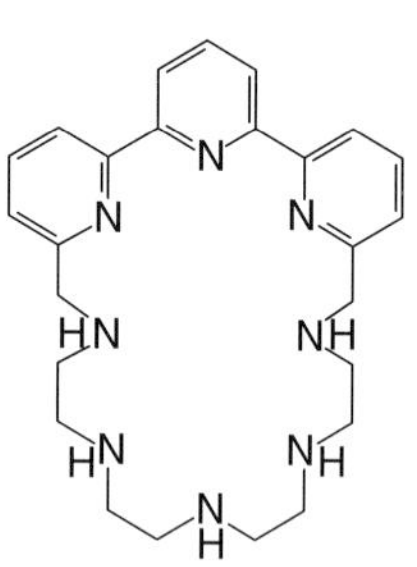 **48**

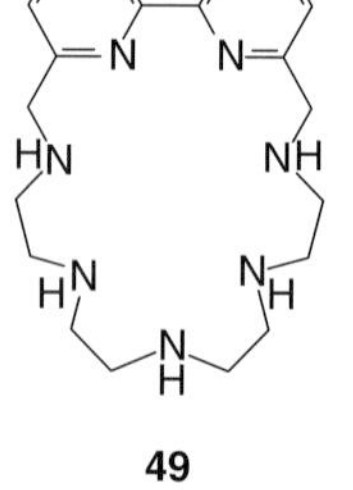

49 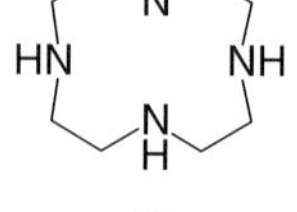 **50**

4 METAL ION COMPLEXES

Metal ion complexes with azamacrocycles, and, in particular, complexes with transition metal ions, have turned out to have special properties since the synthesis of the first similar ligands performed by template reactions in the presence of Ni(II). The strong resistance to acid dissociation of complexes, such as $Ni(\mathbf{6})^{2+}$ (**6** = cyclam), which was estimated to have a half-life of approximately 30 years in very acidic solutions, required the use of strong scavengers, such as cyanides, to remove metal ions from the complexes. The greater, and sometimes exceptionally higher, thermodynamic stability of these complexes relative to their analogs with acyclic ligands—a phenomenon known as macrocyclic effect (see **Cooperativity and the Chelate, Macrocyclic and Cryptate Effects**, Volume 1)—contributed to generation of large interest toward complexes with these kinds of ligands. The reduced flexibility of macrocyclic polyamines makes complexation reactions more complex than those for acyclic ones, and the role played by the ligand is more important.

4.1 Metal complexes with azacycloalkanes

The properties and the structure of metal ion complexes with azacycloalkanes critically depend on the ring size, number of donor atoms, as well as on the size and number of chelate rings formed upon coordination. The $[3k]aneN_k$ ($k = 2-12$) series of homologous ligands enables us to analyze a part of these parameters.[40,50]

The triazacycloalkane **14** ($[9]aneN_3$), constituted by nine atoms in the macrocyclic ring, is too small to encircle a metal ion and is consequently constrained to occupy three facial positions in an octahedral coordination geometry, in contrast to its acyclic analogs that may adopt both facial and meridional coordination modes (Figure 10). Most likely, this is the reason why the $Cu(\mathbf{14})^{2+}$ complex is less stable than the analogous open-chain complex, in contrast to Co(II), Ni(II), Zn(II), and Cd(II) complexes displaying a macrocyclic effect, since the meridional coordination of the acyclic triamine is favored in the case of Cu(II) by the Jahn–Teller distortion. The second molecule of **14** can occupy the remaining coordination positions of the metal ion in the 1 : 1 complexes, forming 1 : 2 metal-to-ligand species in which the metal ion is sandwiched between two triamine rings (Figure 10b). On increasing the ring size of the cyclic triamine ligand, however, the ability to form 1 : 2 complexes rapidly extinguishes.

Also, the tetraazacycloalkane **15** (cyclen, $[12]aneN_4$) is too small to accommodate the metal ion in the plane of the four nitrogen atoms. The coordinated metal ions remain above this plane, giving rise to square-pyramidal or trigonal-bipyramidal environments by the coordination of solvent molecules or counterions. Alternatively, the ligand can fold along the axis passing through opposite nitrogen atoms to form octahedral complexes in which two cis positions are occupied by the solvent molecules or counterions (Figure 10). Nevertheless, for $Ni(\mathbf{15})^{2+}$, a yellow, diamagnetic species has also been observed, which is consistent with a square-planar geometry. In aqueous solution, such yellow species is in equilibrium with a blue, octahedral paramagnetic form. Increasing the temperature and ionic strength displaces the equilibrium toward the formation of the yellow complex.

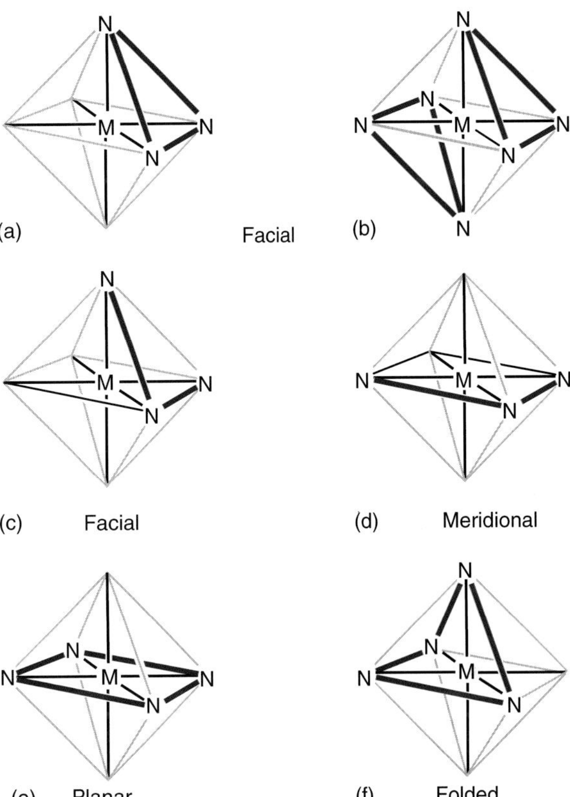

Figure 10 Coordination of triamine and tetraamine in octahedral complexes: facial coordination mode of triazacycloalkanes (a and b); facial and meridional coordination modes of acyclic triamines (c and d); and planar and folded coordination modes of tetraazacycloalkanes (e and f).

The pentaazacycloalkane **16** ($[15]aneN_5$) is large enough to include metal ions in its macrocyclic cavity. The ligand can arrange around the metal ion to form octahedral complexes, as in the case of $Ni(\mathbf{16})(H_2O)_2{}^{2+}$, or dispose on the equatorial plane of a pentagonal bipyramid, as found for $M(\mathbf{16})Cl_2$ (M = Mn, Cd) complexes,[51] while a pentacoordinated square-planar geometry is preferred by $Cu(\mathbf{16})^{2+}$.[52]

The hexaaza- and heptaazacycloalkanes **17** ($[18]aneN_6$) and **18** ($[21]aneN_7$) are two intermediate elements between the smaller $[3k]aneN_k$ ligands already described, which are able to accommodate a single metal ion within the macrocyclic cavity, and the larger ones that can host more than one metal ion. Indeed, **17** forms mononuclear (1 : 1) complexes with all the studied metal ions in aqueous solution, with the exception of Pd(II), which gives rise to the binuclear (1 : 2) $Pd_2(\mathbf{17})_2Cl_2{}^{2+}$ complex, where each Pd(II) ion is coordinated by three ligand nitrogen donors and one

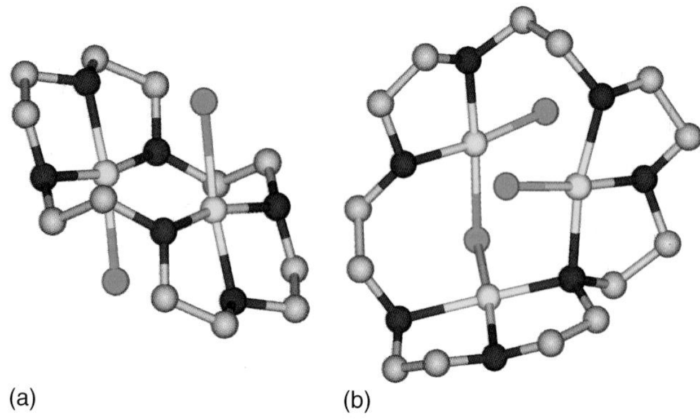

Figure 11 Crystal structures of the complexes $Pd_2(\mathbf{17})Cl_2^{2+}$ and $Pd_3(\mathbf{18})Cl_3^{2+}$. Carbon (Grey), Nitrogen (Blue), Chlorine (Green), and Palladium (Yellow).

chloride anion in a square-planar geometry (Figure 11).[53] On the other hand, **18**, which is able to bind up to three Pd(II) ions forming the $Pd_3(\mathbf{18}\text{-H})Cl_3^{2+}$ complex in which deprotonation of secondary amine groups occurs upon bridging coordination to two Pd(II) ions (Figure 11),[53] also forms binuclear complexes with Cu(II). From **17** on, the larger $[3k]$aneN_k polyazacycloalkanes display an increasing tendency to form polynuclear complexes, while their ability to bind a single metal ion rapidly extinguishes. As far as the formation of complexes with Cu(II), Ni(II), Co(II), Zn(II), and Cd(II) is considered, **19** ([24]aneN_8) forms mononuclear complexes with all these metal ions but Cu(II), while it forms dinuclear complexes with all of them but Co(II); **20** ([27]aneN_9) forms binuclear complexes with all these cations, but it is able to form mononuclear species only with Co(II); the larger ligands form only binuclear species, but **23** ([36]aneN_{12}) can also bind three Cu(II) ions. Some crystal structures of binuclear complexes with $[3k]$aneN_k ligands are shown in Figure 12.

Regarding the stability of these complexes, a survey of thermodynamic data showed some discrepancy in the values of complex stability constants obtained by different research groups, in particular, for the smaller azacycloalkanes.[40] A critical evaluation of these data made it possible to define some general trends.[54] As an example, the stability constants (log K) of the mononuclear complexes formed by $[3k]$aneN_k molecules with Mn(II), Co(II), Ni(II), Cu(II), and Zn(II) versus the number of ligand amine groups are plotted in Figure 13. All these metal ions show a clear trend of increasing stability of the complex from the smaller tridentate ligand [9]aneN_3 (**14**) to the larger pentadentate [15]aneN_5 (**16**), in the case of Mn(II), Cu(II), and Zn(II), and to the hexadentate [18]aneN_6 (**17**) for Co(II) and Ni(II). Then, a general decrease in stability is observed with larger macrocycles. The maximum of stability for the complexes with Cu(II), Co(II), and Ni(II) coincides with the preferred coordination number of these metal ions in their high-spin state. In the case of Mn(II), an almost invariable stability is observed for its complexes with the penta-, hexa-, and heptadentate ligands **16–18**, even though all donor atoms of these ligands are involved in metal coordination. Accumulation of intramolecular strain with increasing size of the macrocycles, occurring upon full ligand coordination to the large Mn(II) ion, is thought to be at the origin of this behavior. In the case of $Zn(\mathbf{17})^{2+}$, only five amine donors are coordinated to the metal ion and, consequently, the formation of a large chelate ring containing the uncoordinated nitrogen atoms reduces the stability of the complex relative to $Zn(\mathbf{16})^{2+}$, in which all donor atoms are coordinated.[54]

As far as the binuclear complexes with $[3k]$aneN_k ligands are concerned, a general trend of increasing stability with increasing number of donor atoms in the macrocycle is observed.[50]

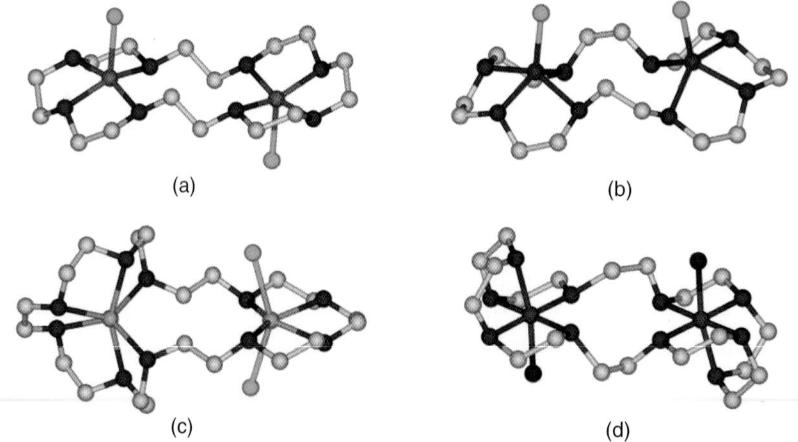

Figure 12 Crystal structures of the binuclear complexes $Cu_2(\mathbf{19})Cl_2^{2+}$ (a), $Zn_2(\mathbf{19})Cl_2^{2+}$ (b), $Cd_2(\mathbf{21})Cl_2^{2+}$ (c), and $Ni_2(\mathbf{21})(H_2O)_2^{4+}$ (d). Carbon (Grey), Nitrogen (Blue), Copper (Orange), Zinc (Black), Cadmium (Yellow), Nickel (Violet), Chlorine (Green), and Oxygen (Red).

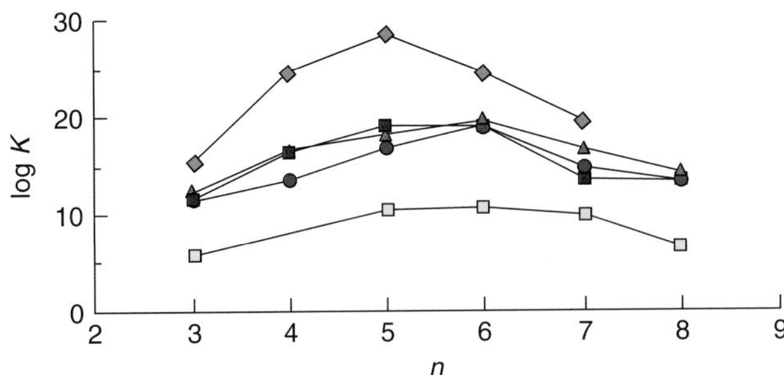

Figure 13 Logarithms of the stability constants for the formation of $[ML]^{2+}$ complexes with $[3k]aneN_k$ $(k = 3-8)$ azacycloalkanes as a function of the number of nitrogen donor atoms, n, in each macrocycle. M = Mn (yellow), Co (blue), Ni (pink), Cu (green), and Zn (red). (Reproduced from Ref. 54. © Royal Society of Chemistry, 1991.)

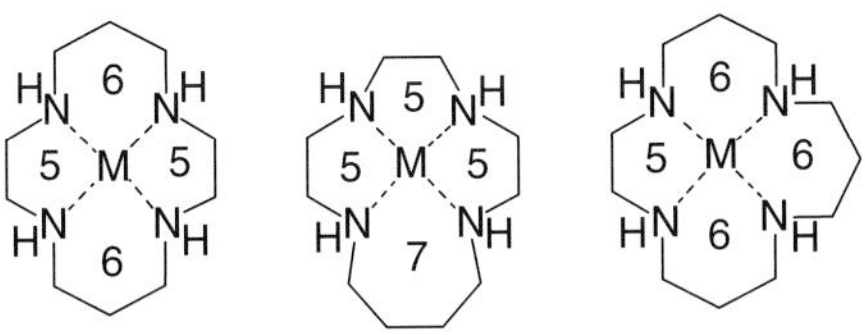

Figure 14 Chelate rings in metal complexes. Figures indicate the number of atoms defining the chelate ring. Examples: 5 = five-membered chelate ring; 6 = six-membered chelate ring.

It is a frequent observation in coordination chemistry that an increase in chelate ring size determines a decrease in the stability of the complex, even though a reverse trend can be observed for small metal ions. Examples of chelate rings in macrocyclic complexes are shown in Figure 14. Both entropic and enthalpic contributions to this effect can be devised. A longer spacer connecting two donor atoms lowers the probability that the second donor atom attaches to the metal ion after the first one had already attached. This leads to an adverse entropic contribution which is expected to be more evident the longer the spacer. But there are many evidences that chelate ring size effects are driven by enthalpic contributions, which have been explained by considering that larger chelate rings produce steric strain, weakening the metal-to-ligand interaction. These observations are especially relevant to individual chelate rings. When more chelate rings are joined together, different effects can be found. For example, insertion of a six-membered chelate ring between two five-membered ones may lead to increased complex stability, while the union of two six-membered chelate rings generally reduces the stability. Such effects can be correlated with the reduction or the increase of steric strain brought about by joined chelate rings, which are dependent on metal ion size, in particular, in the case of macrocyclic complexes.

In this respect, an instructive example is given by the variation of complex stability through the series of

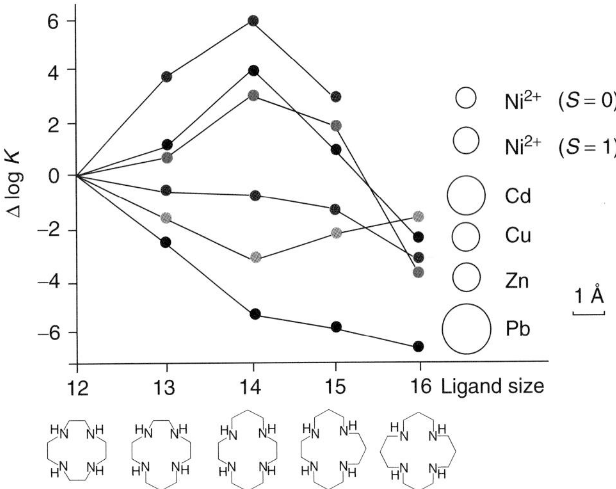

Figure 15 Change in complex stability (Δlog K) relative to cyclen complexes (**15**) with increasing size of tetraazacycloalkanes. The metal ions are indicated as circle of size proportional to their ionic radii. (Reproduced from Ref. 55. © Springer, 1996.)

tetraazacycloalkanes **15, 51, 6, 52**, and **25**. In Figure 15, such variation has been shown relative to the stability of the complexes with **15** (cyclen) as a function of Δlog K. As can be seen from this figure, there are different behaviors for smaller and larger metal ions. Ni(II), both high and low spins, and Cu(II) display increased stability as one ethylenic chain of cyclen is converted into a propylenic one, and further increase is observed when a second ethylenic chain is converted into a propylenic one to generate the ligand cyclam. On the contrary, the presence of joined six-membered rings in larger macrocycles, **52** and **25**, reduces the stability of these complexes. For large metal ions, such as Pb(II), the six-membered chelate rings are less appropriate, and a steady decrease in stability occurs for complexes with increasing number of such chelate rings.[55]

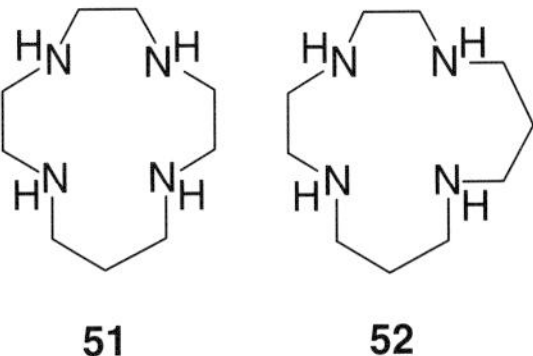

For metal complexes with azacycloalkanes containing more than one secondary amine group, different configurational isomers deriving from the combination of the coordinated chiral nitrogens are possible. As an example, the five most stable conformers of cyclam are shown in Figure 16. The energy of these structures is much dependent on the conformation assumed by the individual five- and six-membered chelate rings. The most stable isomer, the *trans*-(III), having the five-membered rings in gauche conformation and the six-membered ones in

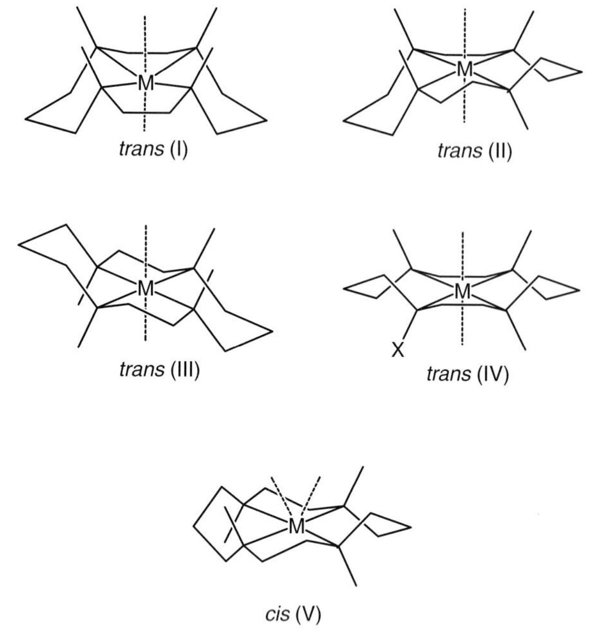

Figure 16 The five most stable conformers of cyclam.

chair conformation, is the most common form, especially for octahedral complexes, while the *trans*-(I) form is predominant in square-planar and five-coordinated complexes. Cyclam can also assume a folded conformation (Figure 10) in its complexes. Such conformation can be forced by the action of chelating ligands, but it also forms spontaneously as a minor species in equilibrium with other complex species. Ni(**6**)$^{2+}$, for instance, in aqueous solution (298 K, $I = 0.1$ M) is a mixture of 69% square-planar, 29% *trans*-diaquo, and 2% *cis*-diaquo octahedral species.[40] The cis (V) form gives the most favorable chelate ring conformations for the folded ligand complex.

4.2 Metal complexes with azacyclophanes

Relative to azacycloalkanes, azacyclophanes show additional properties due to the presence of aromatic groups in the macrocyclic ring. These groups act as rigid spacers, stiffening the ligand structure; impose a large separation to some donor atoms of the aliphatic polyamine chain; and provide polarizable π-clouds that can interact with guest species. Furthermore, when these groups are heteroaromatic, they can provide preferential binding sites for metal ion coordination.

As an effect of their restricted conformational flexibility, azacyclophanes commonly give rise to slower reactions for the formation of metal complexes than their cycloalkane counterparts. For example, dibenzocyclam (**53**) reacts with Cu(II) approximately 25 times slower than cyclam (**6**).[56] Also, dissociation reactions of metal complexes are more sluggish for azacyclophanes. The dissociation of the Cu(**40**)$^{2+}$ complex in acidic media, for instance, is 4–5 orders of magnitude slower than the analogous reaction for the corresponding acyclic Cu(II) complex. Nevertheless, when the substitution at the aromatic ring changes from ortho in **40** to meta or para in **41** and **42**, respectively, the ligand flexibility increases and the dissociation reactions become faster.[44]

53

The coordination properties of ligands **40–42**, along with those of the structurally related **37–39**, are representative of this class of compounds.[43,44] For example, the stability constants of their Cu(II) complexes in aqueous solution are reported in Table 6. The smallest ligands of this series, the *o*-substituted **37** and **40**, display the appropriate structure for tight coordination to a single metal ion using all four nitrogen donors, and very stable complexes are formed. When the substitution at the aromatic ring changes from ortho to meta or para, the separation between the benzylic amine groups becomes too large and their simultaneous coordination to the same metal center is no longer possible.

Table 6 Logarithms of stability constants for the formation of Cu(II) complexes with tetraazacyclophanes **37–42** determined at 298 K, $I = 0.15$ M.

	37	**38**	**39**	**40**	**41**	**42**
$Cu^{2+} + L = CuL^{2+}$	19.58	13.52	10.41	17.73	13.29	12.71
$CuL^{2+} + H^+ = Cu(HL)^{3+}$	—	5.90	6.51	3.89	7.19	6.09
$Cu(HL)^{3+} + H^+ = Cu(H_2L)^{4+}$	—	3.58	—	—	3.63	4.89
$Cu(H_2L)^{3+} + H^+ = Cu(H_3L)^{5+}$	—	—	—	—	—	4.14
$CuL + H_2O = CuL(OH)^+ + H^+$	−11.42	−8.82	−8.14	—	−9.14	—
$2Cu^{2+} + L = Cu_2L^{4+}$	—	—	—	—	—	16.81
$2Cu^{2+} + L + H_2O = Cu_2L(OH)^{3+} + H^+$	—	10.67	—	—	9.31	—
$2Cu^{2+} + L + 2H_2O = Cu_2L(OH)_2{}^{2+} + 2H^+$	—	3.60	3.44	—	3.28	5.96

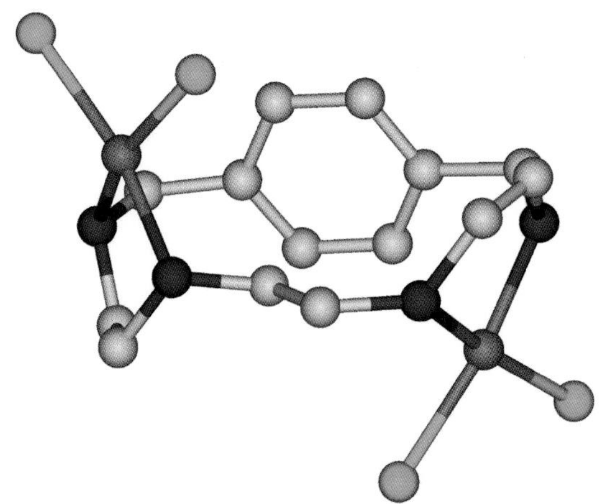

Figure 17 Crystal structure of the binuclear Cu_2(**39**)Cl_4 complex. Carbon (Grey), Nitrogen (Blue), Chlorine (Green), and Copper (Salmon pink).

Accordingly, the *m*- and *p*-substituted ligands involve only three donors in metal ion coordination, with an evident loss of complex stability, and the formation of binuclear complexes becomes possible (Table 6). As shown by the crystal structure of the binuclear Cu_2(**39**)Cl_4 complex (Figure 17), the two Cu(II) ions lie on opposite sides of the macrocyclic plane, the rather long distance between them (6.077 Å) leading to a moderate electrostatic repulsion between the two positively charged metal centers.[57] This crystal structure clearly shows that p-substitution at the aromatic ring is a key structural element in determining the bis(chelating) properties of the ligands. In this regard, it is worth remembering that the smallest polyazacycloalkane that is able to form binuclear Cu(II) complexes in aqueous solution is [21]aneN_7 (**18**) containing seven amine groups separated by ethylenic chains. Accordingly, small azacyclophanes are good tools for generating metal complexes with low coordination numbers—a structural feature of great interest for efficient binding and activation of exogenous species.

The binucleating ability of azacyclophanes considerably increases when ligands with ditopic structures are constructed. The hexaazametacyclophane **43**, for instance, contains two independent triamine binding sites and, accordingly, can form both mono- and binuclear Cu(II) complexes in which each metal ion is coordinated by only three ligand donor atoms. Nevertheless, the Cu(II) complexes are characterized by relatively high thermodynamic stability in aqueous solution (log $K = 13.79$ for Cu(II) + **43** = Cu(**43**)$^{2+}$, log $K = 9.68$ for Cu(**43**)$^{2+}$ + Cu(II) = Cu_2(**43**)$^{4+}$). The presence in Cu_2(**43**)$^{4+}$ of two metal ions with low coordination numbers at relatively short distance between them provides the complex with strong hydrolytic ability to form mono- and dihydroxo complexes in which the hydroxide ions likely bridge the two metal centers.[58] Increasing the distance between the two binding units, as is evident in the case of **44** upon p-substitution at the aromatic ring, reduces the hydrolytic properties of the binuclear Cu(II) complex. The Cu_2(**44**)$^{4+}$ complex, however, is still able to form mono- and dihydroxo species, although of lower stability relative to Cu_2(**43**)$^{4+}$, in which the hydroxide ions are located on separate metal centers. As with **43, 44** also forms both mono- and binuclear Cu(II) complexes, even though their stability (log $K = 10.03$ for Cu(II) + **44** = Cu(**44**)$^{2+}$, log $K = 8.32$ for Cu(**44**)$^{2+}$ + Cu(II) = Cu_2(**44**)$^{4+}$) is somewhat lower than that shown by the analogous species formed by **43**, as expected as a consequence of N-methylation in **44**. Nevertheless, the loss of stability is less evident for the binding of the second metal ion, since the p-substitution at the aromatic ring brings the two metal centers at longer distance, reducing the electrostatic repulsion between the two coordinated metal ions. A crystal structure of the Cu_2(**44**)$Cl_2{}^{2+}$ complex cation showed the two metal ions in a square-planar coordination environment, located 6.945 Å apart from each other.[59]

An early work on the complexation properties of azacyclophanes containing single heteroaromatic groups was undertaken by using the tetrazapyridilophanes **50, 54**, and **55** and several divalent transition metal ions, such as Cd(II) and Pb(II).[49] The stability of these complexes in aqueous solution was shown to be lower than that of the analogous tetraazacycloalkanes, except for the Ni(II) complex with **50**. Despite the lower σ-donating properties of pyridine nitrogen atoms, relative to aliphatic amines, which can be compensated by their greater π-acceptor ability, the main reason for the observed lower stability seems to reside in the different conformations adopted by the pyridine-containing ligands upon complexation. For example, as shown by spectroscopic data, Ni(II) seems to form a pentacoordinated complex with **55**, with a high-spin, square-pyramidal arrangement, in contrast to the planar arrangement assumed by **52**, in both square-planar and diaquo-octahedral species, in its Ni(II) complex. Accordingly, the lower stability of Ni(**55**)$^{2+}$ has to be ascribed to the greater strain and the loss in crystal field stabilization energy brought about in the complex by the square-pyramidal arrangement of the complex.[49]

54 **55** **56**

At almost the same time,[60,61] the coordination properties of the hexaaza-ligand **56** toward some transition metal ions (Mn(II), Cu(II), Zn(II), Cd(II)), alkaline earth cations (Ca(II), Mg(II)), and lanthanide ions (La(III), Gd(III)) were reported. The study was later extended to Ba(II), Ni(II), Pb(II), and Sm(III).[62] This ligand containing two pyridine rings is closely related to the hexazacycloalkane **17** ([18]aneN_6). The main coordination characteristic of **56** is its ability to form 1 : 1 complexes in which the ligand wraps around the metal ion in octahedral geometry with only meridional disposition of two sets of three nitrogen donors, including a pyridine in the central position, as shown by the crystal structure of Zn(**56**)$^{2+}$ (Figure 18).[61] Such ligand arrangement has to be ascribed to the presence of the pyridine groups, since the structurally related hexaazacycloalkane **17** ([18]aneN_6) may adopt both facial and meridional dispositions, and it is thought to be close to the conformation of the metal-free ligand. Such ligand preorganization would represent a favorable contribution to the high stability of the complexes formed by **56**, which binds metal ions at least as well as **17** and in some cases, mostly involving filled-shell cations, it binds stronger by up to three orders of magnitude.[60,62]

As already observed for the octaazacycloalkanes, the formation of binuclear complexes is possible with metal ions such as Cu(II), Ni(II), Cd(II), and Pb(II) for the octaaza-pyridilophane **57** and its *N*-methylated derivative **58** also.[63]

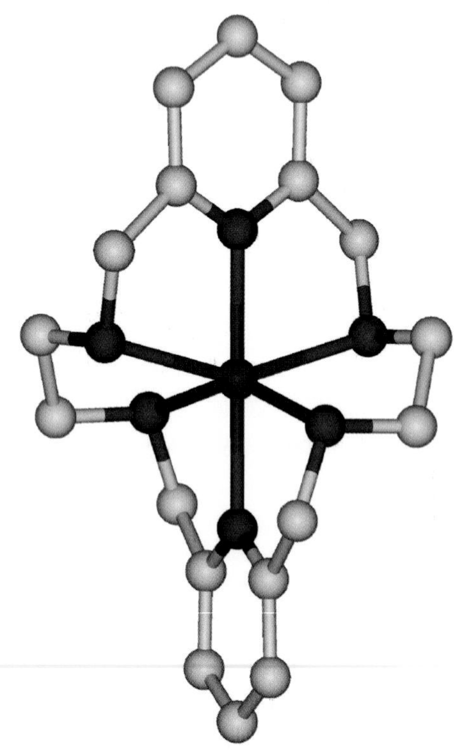

Figure 18 Crystal structure of the complex Zn(**56**)$^{2+}$. Carbon (Grey), Nitrogen (Blue), and Zinc (Black).

Mononuclear species are also formed with all these metal ions and Zn(II).[63] With the exception of Pb(II), which favors the formation of binuclear complexes even for metal-to-ligand molar ratios of 1 : 1, mononuclear complexes are the main species in solution containing ligands and metal ions in equimolar concentrations. On the basis of spectroscopic data and modeling calculation, it seems that the most likely coordination geometry in solution is octahedral for all complexes. The crystal structure of the binuclear complex Ni_2(**57**)$(H_2O)_4{}^{4+}$ showed each symmetric Ni(II) center in a distorted octahedral coordination environment with four ligand donors occupying meridional positions. The complexes formed by these ligands are thermodynamically less stable than the analogous species with the octaazacycloalkane **19**, mostly due to the formation of several six-membered chelate rings by **57** and **58**. However, the difference in stability is attenuated when the dinuclear complexes are compared and even inverted for the Pb(II) complexes.[63]

R = H, **57**
R = CH_3, **58**

n = 1, **59**
n = 2, **60**
n = 3, **49**

The presence of chelating heteroaromatic groups in azacyclophanes may afford special ligand properties. The phenanthroline group in ligands **49, 59**, and **60**, for instance, was shown to be a preferential binding site for metal ions (Cu(II), Zn(II), Cd(II), and Pb(II)) within the macrocyclic ring, although other amine groups of the ligands participate in the coordination. The rigidity of this heteroaromatic group, however, stiffens the cyclic molecules, precluding the simultaneous coordination of all donor atoms to a single metal center. In the case of the Zn(II) complexes with these ligands, the benzylic nitrogen atoms in close proximity of the phenanthroline unit are not involved in the coordination to the metal ion, independent of the number of donor atoms and ligand size.[64,65] The crystal structure of the Zn(**59**)H_2O^{2+} complex, shown in Figure 19(a), evidences the low coordination environment provided by the ligand to Zn(II). As a consequence of this structural feature, large chelate rings are present in the complexes, lowering their thermodynamic stability relative to the analogous azacycloalkanes.

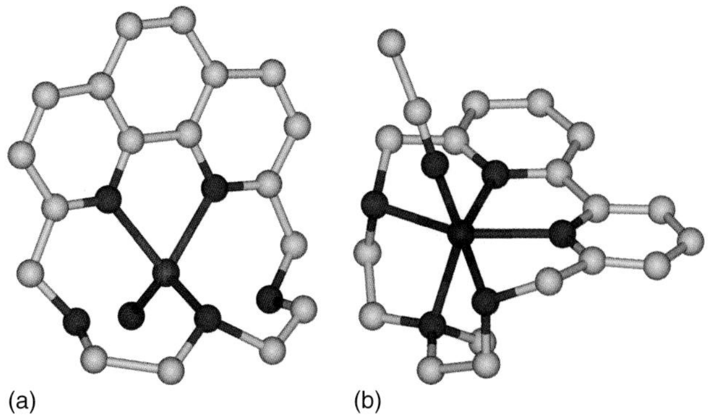

Figure 19 Crystal structures of the complexes $Zn(\mathbf{59})H_2O^{2+}$ (a) and $Zn(\mathbf{61})^{2+}$ (b). Carbon (Grey), Nitrogen (Blue), Zinc (Black), Oxygen (Red), and Sulphur (Yellow).

On the other hand, the presence of nitrogen atoms that are not involved in the coordination offers the opportunity of exploiting the fluorescence emission properties of phenanthroline for metal ion sensing. The phenanthroline group is emissive, but its emission can be efficiently quenched by electron transfer processes involving the lone pairs of the nitrogen atoms in the aliphatic ligand chains. Coordination to metal ions, such as Zn(II), or protonation of these nitrogen atoms prevents such quenching effect, allowing the emission to be active. Nevertheless, a peculiarity of these ligands is to form nonemissive Zn(II) complexes, because of their inability to accommodate all donor atoms in the coordination sphere of the metal ion. As shown in Figure 20, where the emission of the Zn(II)/**59** system is superimposed on the distribution curves of the complex species present at equilibrium, the ligand emission is completely quenched upon metal ion binding below pH 4, while the free ligand would be emissive up to pH 7.

The stability of the $Zn(L)^{2+}$ (L = **49**, **59**, **60**) complexes decreases with increasing ligand size (log $K = 16.15$, 14.29, and 12.38 for $Zn(II) + L = Zn(L)^{2+}$L = **59**, **60**, and **49**, respectively), indicating that the presence of larger

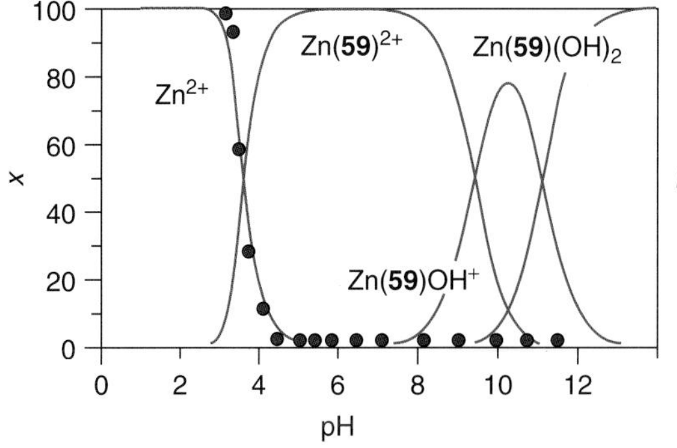

Figure 20 Fluorescence emission intensity (I, dots) of **59** in the presence of an equimolar amount of Zn(II) and molar fractions of its Zn(II) complexes as a function of pH ([**59**] = 2×10^{-5} M).

number of donor atoms does not enhance the binding ability of these ligands toward Zn(II) and suggesting that the number of donor atoms involved in the coordination does not parallel the increasing number of donors in the ligands. The formation of larger chelate rings accounts for the observed trend of decreasing complex stability.[65] On the other hand, **49** containing seven amine groups can bind a second Zn(II) even if this ligand does not exhibit a marked tendency to form binuclear complexes (log $K = 4.39$ for $Zn(\mathbf{49})^{2+} + Zn(II) = Zn_2(\mathbf{49})^{4+}$).

Analogous ligands containing bipyridine units instead of the phenanthroline unit display some different coordination features. Bipyridine has more conformational freedom than phenanthroline, leading to more flexibility of the macrocyclic ligands. The main consequence of this is the ability of azabipyridilophanes containing the bipyridine unit linked via the 6 and 6′ positions to bind metal ions by using these heteroaromatic groups, which remain as the preferential binding units, and involve both nitrogen donors in benzylic positions. As shown in the crystal structure of the Zn(II) complex with ligand **61**, which is closely related to **59**, the bipyridine-based macrocycle uses all its five donor atoms to bind the metal ion (Figure 19b) although some metal-to-ligand bond distances are relatively long.[66] This corresponds to a greater stability of the $Zn(\mathbf{61})^{2+}$ complex (log $K = 17.5$) relative to $Zn(\mathbf{59})^{2+}$ (log $K = 16.15$) in aqueous solution. Nevertheless, when the ligand size and the number of donor atoms available for metal ion coordination increase, the difference between the stabilities of metal complexes with analogous azabipyridilophanes and azaphenanthrolinophanes extinguishes as observed, for instance, for both mono- and binuclear complexes with ligands **47** and **49**.[64–66] Also, **61** and similar azabipyridilophane form complexes with many other transition metal ions.

61 **62**

A marked tendency to form binuclear complexes is shown by the ligand **62**, containing two bipyridine units connected by triamine chains. The main coordination feature of this ligand is related to its ditopic nature defined by the presence of the two bipyridine groups acting as preferential binding sites, located at relatively long distance from each other. Ligand **62** forms both mononuclear and

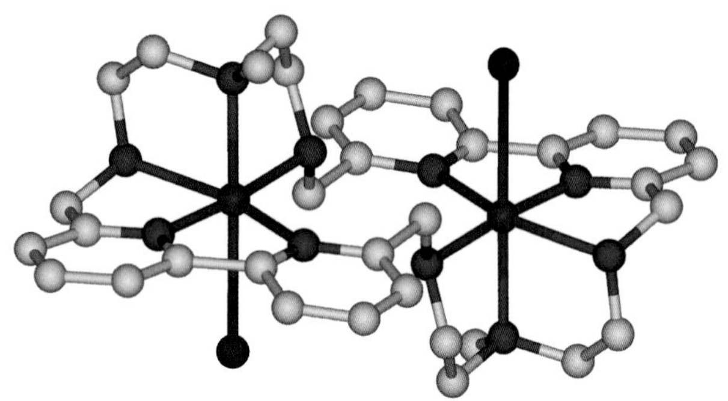

Figure 21 Crystal structure of $Zn_2(\mathbf{62})^{4+}$. Carbon (Grey), Nitrogen (Blue), Zinc (Black), and Iodine (Violet).

binuclear complexes with Zn(II). Despite the large number of donor atoms available for coordination and the greater flexibility of this ligand, the $Zn(\mathbf{62})^{2+}$ complex displays a stability ($\log K = 12.69$ for $Zn(II) + \mathbf{62} = Zn(\mathbf{62})^{2+}$) similar to that of $Zn(\mathbf{59})^{2+}$, because of the formation of at least one very large, destabilizing chelate ring in the former. On the contrary, the second Zn(II) ion is coordinated with rather high stability constant ($\log K = 8.79$ for $Zn(\mathbf{62})^{2+} + Zn(II) = Zn_2(\mathbf{62})^{4+}$).[66] The crystal structure of $Zn_2(\mathbf{62})^{4+}$ is shown in Figure 21.

Also, terpyridine has been considered as a chelating unit for inclusion in azacyclophanes.[67] The three terpyridinophanes **48, 63**, and **64** are representative of this class of compounds.

63 **64**

The two octa-aza ligands **48** and **63** differ in that **63** has two propylene instead of two ethylene spacers, and in that the polyamine chain is attached at 5,5″ for **63** and 6,6″ for **48**. These structural differences give rise to quite different coordination properties and complex reactivity. The 5,5″ attachment of the aliphatic chain to terpyridine in **63** means that metal ions coordinated to the terpyridine site, which is the preferential binding site, are not able to interact with the amine groups in benzylic positions. Hence, the propylenic chains separate two distinct binding units where two metal ions can be accommodated. Accordingly, **63** forms both mono- and binuclear complexes with metal ions, such as Cu(II) and Zn(II). In the binuclear complexes, the distance between the two metal centers is large enough to favor the coordination of bridging substrates, as shown

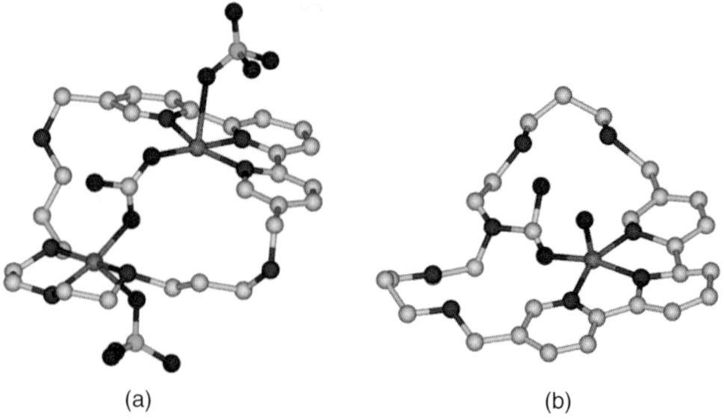

Figure 22 Crystal structures of the complexes $Cu_2(H_2\mathbf{63})CO_3(ClO_4)_2^{2+}$ (a) and $Cu_2(H\mathbf{63}\text{-cabamate})(H_2O)(ClO_4)_3$. Carbon (Grey), Nitrogen (Blue), Oxygen (Red), Chlorine (Green), and Copper (salmon pink).

by the crystal structure of the $Cu_2(H_2\mathbf{63})CO_3(ClO_4)_2{}^{2+}$ complex cation (Figure 22a) showing a bridging carbonate anion. Interestingly, this carbonate complex is formed by spontaneous absorption of atmospheric CO_2 in slightly alkaline solutions. The avidity of the Cu(II) complexes with **63** toward CO_2 was also manifested by the mononuclear complex, which readily absorbs atmospheric CO_2 to form a carbamate complex via metal-ion-assisted reaction of CO_2 with the central aliphatic amine groups of the ligand (Figure 22b).[68]

On the other hand, the 6,6″ attachment of the aliphatic chain to terpyridine in **48** gives rise to a more flexible ligand that is able to use both terpyridine nitrogen atoms and aliphatic amine groups, including those in benzylic positions, for metal ion coordination. Both mono- and binuclear complexes are formed by **48** with Cu(II), Zn(II), Cd(II), and Pb(II). The main coordinative characteristic of this ligand, however, is the formation of dimeric forms of the mononuclear complexes with Cu(II) and Zn(II). The crystal structure of the $[Zn(H\mathbf{48})]_2(\mu\text{-OH})^{5+}$ complex cation showed that the two mononuclear units of this complex are held together by a OH^- anion bridging the two metal centers and by π-stacking interactions between the facing terpyridine groups (Figure 23). Similar assembly, characterized by high thermodynamic stability, is also formed in aqueous solution by Cu(II) and Zn(II). Nevertheless, protonation of the dimeric complexes or the addition of a second metal ion leads to the disruption of the dimer due to the increased electrostatic repulsion between the monomeric units.[47]

Similarly, the Cu(II) complex with the heptaaza ligand **64** also forms stable dimeric species while only monomeric complexes are produced with Zn(II), Cd(II), and Pb(II). In the case of Cu(II) and Zn(II), the ligand can bind a second metal ion. The crystal structures of the Cd(II) and Pb(II) complexes showed the metal ions surrounded by the three terpyridine nitrogen atoms and the two benzylic

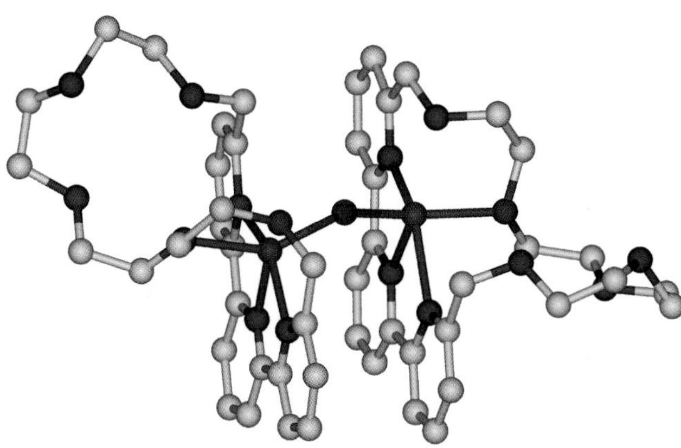

Figure 23 Crystal structure of the complex $[Zn(H\mathbf{48})]_2(\mu\text{-}OH)^{5+}$. Carbon (Grey), Nitrogen (Blue), Zinc (Black), and Oxygen (Red).

amine groups forming the pentagonal basis of a pentagonal bipyramid (Cd(II)) or a pentagonal pyramid (Pb(II)) whose apical positions are occupied by water molecules (Cd(II)) or a counterion (Pb(II)).[69]

5 ANION COMPLEXES

Anions are guest species of great interest in supramolecular chemistry, both for their ubiquitous presence in the natural world and for their introduction into the environment as pollutants deriving from human activities. For this reason, the design and synthesis of anion receptors has become a field of intense research in the last years.[14,15] The most effective way to bind anions consists in taking advantage of their negative charge, and, accordingly, polyammonium (positively charged) ligands, in particular, protonated polyamines, have been the principal receptors of choice since they ensure an adequate electrostatic attraction reinforced by hydrogen bond contacts with the coordinated anions. Polyammonium ligands deriving from azacycloalkanes were the first synthetic receptors used for anion binding. They attracted large interest due to their ability to form species of high positive charge density, relative to their acyclic analogs, and to provide preformed molecular cavities for anion inclusion. Later, polyammonium receptors deriving from azacyclophanes were also considered, because of their ability to provide, thanks to their aromatic groups, additional binding sites for anionic species containing aromatic residues. Noncovalent forces that are relatively strong, namely, coulomb attraction and hydrogen bonds, can combine with weaker ones such as dispersive, π-stacking, and anion–π interactions to achieve tight anion coordination. Accordingly, many other receptors containing a variety of functionalities, such as guanidinium (see **Guanidinium-Based Receptors for Oxoanions**, Volume 3), pyrrole, indole, imidazolium, 1,2,3-triazole (see **Anion Receptors Containing Heterocyclic Rings**, Volume 3), amide, urea (see **Amide and Urea-Based Receptors**, Volume 3), imine (see **Schiff Base and Reduced Schiff Base Ligands**, Volume 3), peptide (see **Synthetic Peptide-Based Receptors**, Volume 3), and boronic (see **Hydrogen-Bonding Receptors for Molecular Guests**, Volume 3) groups, or containing metal centers (see **Metal Complexes as Receptors**, Volume 3), or based on calixarenes (see **Calixarenes in Molecular Recognition**, Volume 3), porphyrins and expanded porphyrins (see **Porphyrins and Expanded Porphyrins as Receptors**, Volume 3) rapidly appeared on the stage of supramolecular chemistry to target the largest variety of anionic species including those constituting ion pairs (see **Ion-Pair Receptors**, Volume 3) and zwitterionic forms (see **Receptors for Zwitterionic Species**, Volume 3).

5.1 Anion coordination by polyammonium receptors deriving from azacycloalkanes

Anion coordination chemistry took its first steps in the second half of the 1970s with the study of cryptand polyammonium receptors.[14,15] In 1981, while working on the identification and separation of several azacycloalkanes, Kimura noted a special behavior of a few compounds, including [18]aneN_6 (**17**), in the presence of citrate anions. In particular, in electrophoretic experiments, these compounds migrated toward the anode, as expected for negatively charged species, while the polyamines were present in solution in protonated cationic forms. Such special behavior was ascribed to the ability of the carboxylate anion to form stable complexes with macrocyclic polyammonium receptors having proper symmetry.[70] A successive study, performed by means of a polarographic method, revealed the formation of 1 : 1 complexes between the azacycloalkanes **16, 17, 65**, and **66** in their triprotonated forms and several polycarboxylate anions, at neutral pH, showing stability constants up to 1000 for citrate (trinegative anion) binding to $H_3\mathbf{66}^{3+}$.[71] When the stability constants for carboxylate complexes with the cyclic receptors and relevant acyclic ligands were considered, it was observed that electrostatic forces play a major role in the association process, the more charged species interacting stronger, that macrocyclic ligands form more stable complexes than their acyclic analogs, and that ring size, as well as other structural features of the interacting partners, is important in controlling the interaction.

In the successive 10 years, many other azacycloalkanes were used for the coordination of a wide range of anionic species, including simple inorganic anions (halides, ClO_4^-, IO_3^-, NO_3^-, SO_4^{2-}, CO_3^{2-}, PO_4^{3-}, etc.), anionic complexes ($Co(CN)_6{}^{3-}$, $Fe(CN)_6{}^{4-}$, $Fe(CN)_6{}^{3-}$), nucleotide phosphate anions (ATP^{4-}, ADP^{3-}, AMP^{2-}), and carboxylate and polyacarboxylate anions, which led to the definition of the main features of anion coordination processes and showed that the receptors may promote modifications of the coordinated anions.[40]

The interaction with anionic complexes represented an instructive field for analyzing the characteristics of anion complexes and the new properties generated in the anions upon binding to polyammonium azacycloalkanes. The ability of such ligands to interact with anionic complexes, such as $Co(CN)_6{}^{3-}$ and $Fe(CN)_6{}^{4-}$, was firstly reported by Lehn in a paper dealing with the hexa- and octa-aza ligands **26** and **27**, where it was shown that the structural complementarity between the anionic substrates and the macrocyclic receptors is of extreme importance in maximizing electrostatic and hydrogen bond interactions. In the same paper, it was also anticipated that the formation of such *complexes of complexes*, also referred to as *supercomplexes*, permits regulation of the physical properties of the anion, as revealed by the strong shift toward positive values of the redox potential of the $Fe(CN)_6{}^{4-}/Fe(CN)_6{}^{3-}$ couple of coordinated relative to free anions.[72]

The study on metallocyanide interaction with protonated forms of azacycloalkanes was successively extended by Bianchi and Garcia-España to the series of cyclic polyamines from [21]aneN_7 to [36]aneN_{12} (**18**–**23**).[23] As an example of the thermodynamic properties of the relevant complexes, the logarithms of stepwise stability constants for the equilibria $M(CN)_6^{(n-6)} + H_pL^{p+} = [M(CN)_6](H_pL)^{(n+p-6)}$ (n = metal ion charge) for the interaction of $Fe(CN)_6{}^{4-}$ and $Co(CN)_6{}^{3-}$ with **18, 19**, and their acyclic counterparts 1,19-dimethyl-1,4,7,10,13,16,19-heptaazanonadecane (**67**) and 1,22-dimethyl-1,4,7,10,13,16,19,22-octaazadocosane (**68**) are plotted in Figure 24 versus the number of acidic protons (p) in the ligands.[73] This plot evidences that (i) the stepwise constants for a given receptor increase with increasing protonation degree (positive charge) of the receptor, (ii) for a given protonation degree of the receptor, the highest stability is shown by the receptor with the highest charge density: smaller ligands display higher constants than larger ones and cyclic ligands show higher constants than the acyclic counterparts (**18** > **19** > **67** > **68**), and (iii) complexes with $Fe(CN)_6{}^{4-}$ are more stable than complexes with $Co(CN)_6{}^{3-}$ for a given protonation state of the ligands.

These data show that electrostatic attraction is the main driving force in metallocyanide–polyammonium azacycloalkanes' interaction. Although charge–charge interactions do not provide much control in the modulation of selective discrimination of a guest over another, the variation in ring size along the series of ligands **18**–**23** leads to some binding selectivity toward metallocyanide complexation. As shown in Figure 25, where the logarithms of the stability constants for the equilibria $Co(CN)_6{}^{3-} + H_pL^{p+} = [Co(CN)_6](H_pL)^{(3-p)}$ (L = **18**–**23**) are plotted for each ligand, for a given protonation degree (p) the complex stability decreases when going from one macrocycle to the next one in the order of increasing size until [30]aneN_{10} (**21**) is reached, to successively increase with larger ligands. This change in the pattern was ascribed to the inclusion of the anion into the macrocyclic cavity, favoring stronger charge–charge and hydrogen bond interactions.[74]

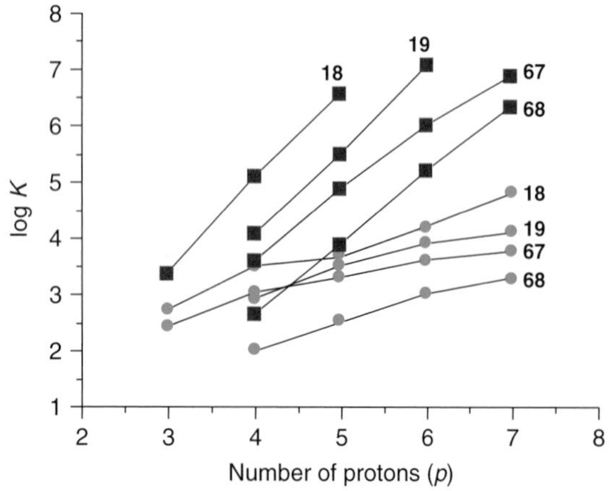

Figure 24 Plot of the logarithms of the stability constants for the equilibria $M(CN)_6{}^{(n-6)} + H_pL^{p+} = [M(CN)_6](H_pL)^{(n+p-6)}$ (n = metal ion charge) versus the number of positive charges (p) in the receptor for the interaction of $Fe(CN)_6{}^{4-}$ (squares) and $Co(CN)_6{}^{3-}$ (circles) with receptors **18, 19, 67**, and **68**. (Reproduced from Ref. 73. © Royal Society of Chemistry, 1992.)

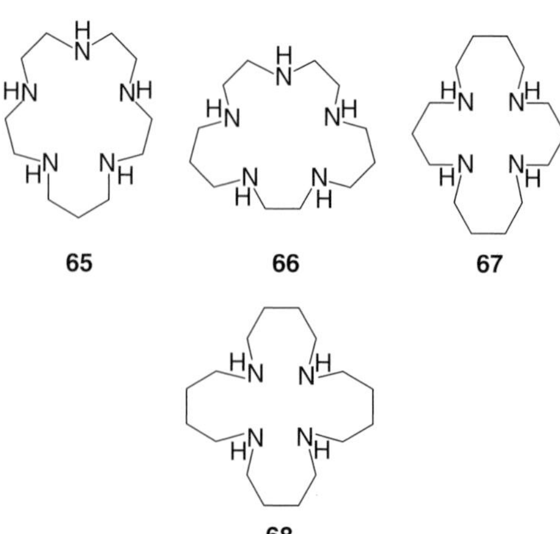

Inclusion of $Co(CN)_6{}^{3-}$ inside the cavity of protonated polyazacycloalkanes was also postulated in early work of Lehn and Balzani to explain the variation of the quantum yield in the reaction of photoaquation of this metallocyanide anion in the presence of polyammonium forms of **26** and

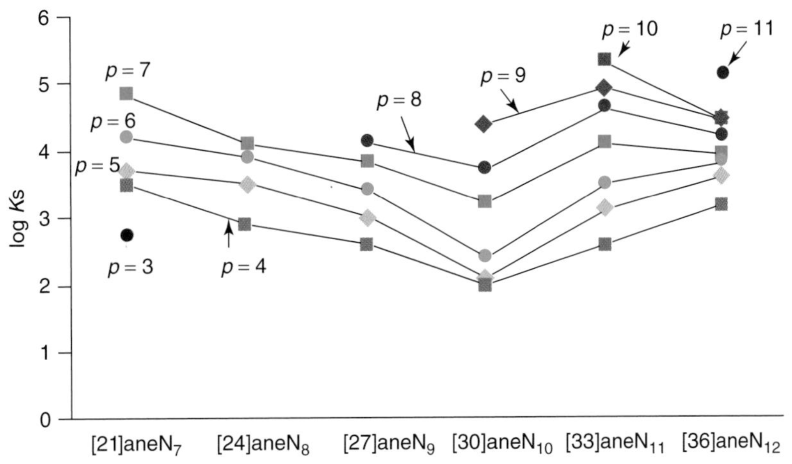

Figure 25 Plot of the logarithms of the stability constants for the equilibria $[Co(CN)_6]^{3-} + H_p([3k]aneN_k)^{p+} = [Co(CN)_6](H_p([3k]aneN_k))^{(3-p)}$. (Modified from Ref. 74. © American Chemical Society, 1992.)

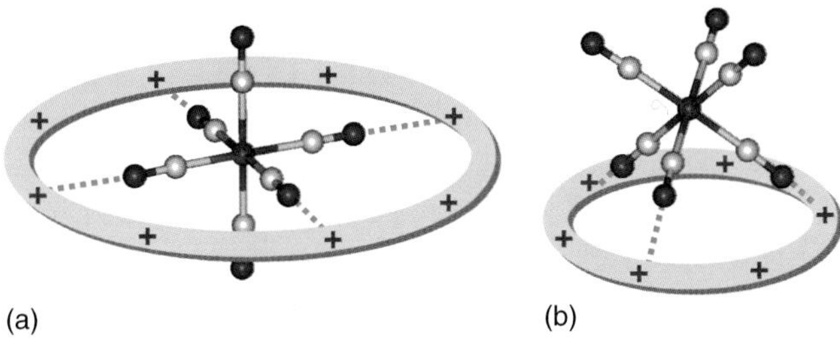

Figure 26 Inclusive (a) and external (b) complexation of $Co(CN)_6{}^{3-}$ by polyammonium receptors. Carbon (Grey), Nitrogen (Blue), and Cobalt (Deep pink).

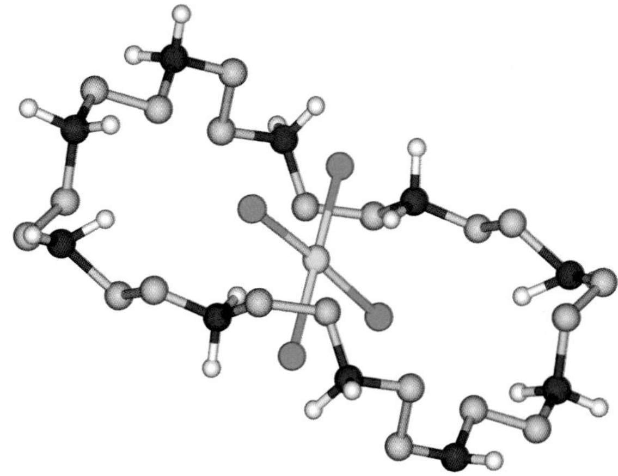

Figure 27 Crystal structure of the inclusion complex between $PdCl_4{}^{2-}$ and $H_{10}[30]aneN_{10}{}^{10+}$ ($H_{10}\mathbf{21}^{10+}$). Carbon (Grey), Nitrogen (Blue), Hydrogen (White), Chlorine (Green), and Palladium (Yellow).

27.[75] These authors noted that the quantum yield for the light-induced reaction:

$$Co(CN)_6{}^{3-} + H_3O^+ \rightarrow [Co(CN)_5(H_2O)]^{2-} + HCN$$

was reduced to one-third upon interaction with $H_8\mathbf{27}^{7+}$, suggesting that inclusion of the anion into the macrocycle, in such a way that four cyanide anion were hydrogen bonded to ammonium groups of the ligand, had occurred and only two cyanide groups of $Co(CN)_6{}^{3-}$ out of six (one-third) were available for the photoaquation reaction (Figure 26a). Similarly, reduction of the quantum yield by half, observed in the presence of the smaller $H_6\mathbf{26}^{6+}$, was ascribed to the external interaction with the polyammonium receptor through a triangular face of $Co(CN)_6{}^{3-}$ (Figure 26b).

Inclusion of an anionic complex within the macrocyclic cavity of a polyammonium azacycloalkanes was verified in the solid state for the $PdCl_4{}^{2-}$ complex with $H_{10}[30]aneN_{10}{}^{10+}$ ($H_{10}\mathbf{21}^{10+}$), as shown by the crystal structure displayed in Figure 27.[76]

As anticipated, modification of the physical properties of metallocyanides can also be achieved by coordination of polyammonium receptors as second-sphere ligands to $Fe(CN)_6{}^{4-}$. Indeed, addition of protonated azacycloalkanes to aqueous solution of $Fe(CN)_6{}^{4-}$ yields an anodic shift of the redox potential associated with this anion. The Fe(II)/Fe(III) redox couple in this species is reversible and highly sensitive to the second coordination sphere of the iron ion. As shown in Figure 28, where the shift in redox potential as a function of pH is superimposed on the distribution curves of the complex species formed by $Fe(CN)_6{}^{4-}$ with protonated forms of $[27]aneN_9$ (**20**), the anion becomes progressively more resistant to oxidation as the protonation state (positive charge) of the receptor increases, in agreement with increasing difficulty in extracting an electron from more positively charged complexes.[77]

Size matching between polyprotonated azacycloalkanes and anions offers the possibility of escaping the control of

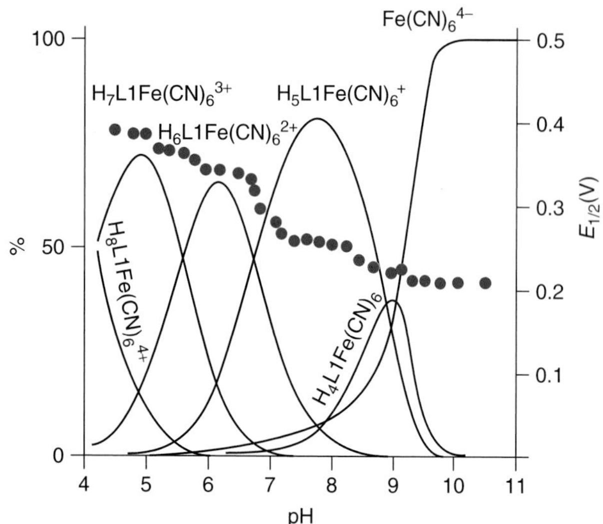

Figure 28 Distribution diagram for the system $H^+/Fe(CN)_6{}^{4-}$/**20** and $E_{1/2}$ in cyclic voltammetry (•) for the couple (H_n**20**$^{n+}$) $[Fe(CN)_6]^{(n-4)+}$/(H_n**20**$^{n+}$) $[Fe(CN)_6]^{(n-3)+}$ versus pH. (Reproduced from Ref. 77. © American Chemical Society, 1987.)

electrostatic forces on anion complex formation between charged partners to achieve binding selectivity. Although this is particularly relevant to aza-cryptands, azacycloalkanes also display interesting cases. The stability profile of $Co(CN)_6{}^{3-}$ complexes with the size of ligands **18–23**, shown in Figure 25, represents one of these cases. Another example of this type was shown in an early work by Suet and Handel on the interaction of F^- with the tetraprotonated forms of the tetraaza macrocycles **25, 67**, and **68**. It was shown that the stability of these anion complexes ($\log K = 1.9$, 2.0, and 2.8, respectively) increases with increasing macrocycle ring size, as the F^- anion (ionic radius 1.36 Å) is better accommodated within the receptor cavity, whose radius was estimated to be 0.7 Å for H_4**25**$^{4+}$, 1.0 Å for H_4**67**$^{4+}$, and 1.4 Å for H_4**68**$^{4+}$, even though the density of charge on the ligand decreases with increasing ring size.[78]

Linear recognition of ditopic anions in aqueous solution was achieved by spatial matching of complementary functionalities between the hexaprotonated ditopic azacycloalkanes **31** and **32**, containing two 1,5,9-triazanonane subunits linked by hydrocarbon bridges, and α,ω-dicarboxylate anions $^-O_2C-(CH_2)_n-CO_2{}^-$ of the series oxalate^{2-}–sebacate^{2-} ($n = 0$–8).[79] The fully protonated species H_6**31**$^{6+}$ and H_6**32**$^{6+}$ form strong complexes with such dicarboxylate anions showing structural dependence of binding selectivity. H_6**31**$^{6+}$ associates more strongly with succinate^{2-} and glutarate^{2-} ($n = 2$, 3) than with shorter and longer dicarboxylates, while the selectivity peak shifts to pimelate^{2-} and suberate^{2-} ($n = 5$, 6) for H_6**32**$^{6+}$ corresponding to the same increase in length, by three CH_2 groups, both for the most strongly bound dicarboxylates and for the hydrocarbon chains linking the triammonium subunits of the protonated receptors. A similar linear recognition of α,ω-dicarboxylate anions was reported for the tetra- and pentaprotonated forms of the ditopic ligand **30** showing selective binding of glutarate^{2-} ($n = 3$) within the series of homologous dianions.[80]

Anion-receptor complementarity was also shown to lead to selective binding of *cis,cis*-1,3,5-trimethyl-1,3,5-cyclohexanetricarboxylate over the corresponding *cis,trans* isomer performed by protonated forms of **18**. Greater stability, by up to 4 orders of magnitude, was found for the complexes with the *cis,cis* isomer in aqueous solution relative to the complexes of equal stoichiometry with the *cis,trans* isomer.[81] This behavior can be rationalized considering that the *cis,cis* isomer, having all three carboxylate groups oriented in the same direction, provides a negatively charged surface that is well matched by the flat, positively charged surface of protonated **18** (Figure 29). Such a favorable situation is not possible for the *cis,trans* isomer having divergent orientations of the carboxylate groups (Figure 29).

Coordination of $SO_4{}^{2-}$, $PO_4{}^{3-}$, and $P_2O_7{}^{4-}$ with polyammonium cations, formed by several azamacrocycles, including, among others, **17, 18**, and some of their methylated derivatives, leading to the formation of 1 : 1 complexes, was studied by both potentiometric titrations and isothermal titration calorimetry to obtain full sets of thermodynamic parameters of complexation ($\log K$, ΔH°, $T\Delta S^\circ$) in water.[82,83] The stability of $SO_4{}^{2-}$ complexes was shown to be determined by electrostatic forces, increasing with the receptor charge, and to be driven by invariably favorable entropic contributions ($T\Delta S^\circ > 0$), with the enthalpic terms being endothermic ($\Delta H^\circ > 0$) or almost athermic ($\Delta H^\circ \approx 0$), as expected for association processes mostly controlled by desolvation of the interacting species.[82] When two solvated species of opposite charge come in contact, the associated neutralization of charge produces an important

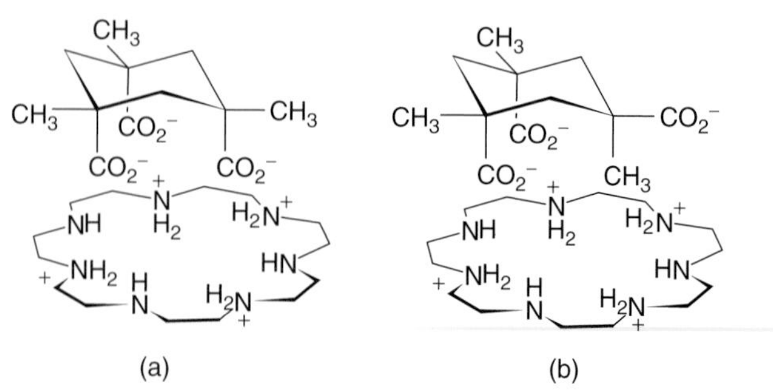

Figure 29 Schematic representation of the 1 : 1 complexes of *cis,cis*- (a) and *cis,trans*-1,3,5-trimethyl-1,3,5-cyclohexanetricarboxylate (b) with $H_4[21]aneN_7{}^{7+}$ (H_4**18**$^{4+}$).

release of solvent molecules from the interacting species, determining an entropic gain opposed by the enthalpic cost of desolvation. Conversely, the stability trends observed for the PO_4^{3-} and $P_2O_7^{4-}$ complexes are not strictly determined by electrostatic contributions. In some cases, for instance, the stability of the complexes formed by HPO_4^{2-} decreases with increasing charge on the ligand, while the stability of pyrophosphate complexes increases with decreasing negative charge from $HP_2O_7^{3-}$ to $H_3P_2O_7^{-}$.[83] Also, the stability of the complexes formed by phosphate and pyrophosphate, in a given protonation degree, with the same receptor, does not show strict trends, and, in contrast to electrostatic expectations, the less charged anion can form more stable complexes, although, in a general sense, pyrophosphate displays a greater propensity to form complexes.

Many of these complexation processes are almost athermic, or endothermic, and are promoted by favorable entropic contributions ($T\Delta S^\circ > 0$), in agreement with the ideal electrostatic model, although there are also a considerable number of reactions promoted by large favorable enthalpy changes ($\Delta H^\circ < 0$) and accompanied by evident entropy loss. Phosphate and pyrophosphate anions, and polyammonium receptors can be involved in the formation of many hydrogen bonds in which both the anions and the receptors can act as acceptors or donors. There are five possible modes (1–5) of hydrogen bonding involving amine or ammonium groups, and phosphate or protonated phosphate anions:

$$-\text{N}-\text{H}^+\cdots{}^-\text{O}- \quad \Delta H^\circ > 0, \quad T\Delta S^\circ > 0 \qquad (1)$$

$$-\text{N}-\text{H}^+\cdots\text{OH}- \quad \Delta H^\circ > 0, \quad T\Delta S^\circ \approx 0 \qquad (2)$$

$$-\text{N}-\text{H}\cdots{}^-\text{O}- \quad \Delta H^\circ > 0, \quad T\Delta S^\circ \approx 0 \qquad (3)$$

$$-\text{N}-\text{H}\cdots\text{OH}- \quad \Delta H^\circ > 0, \quad T\Delta S^\circ < 0 \qquad (4)$$

$$-\text{NI}\cdots\text{H}-\text{O}- \quad \Delta H^\circ < 0, \quad T\Delta S^\circ < 0 \qquad (5)$$

Taking into account that deprotonation of an amine group is a strongly endothermic reaction while protonation of phosphate anions is almost athermic, the partial amine-to-anion proton transfer processes involved in the four hydrogen bonding modes 1–4 are expected to give unfavorable enthalpic contributions ($\Delta H^\circ > 0$), while the partial proton transfer process of the bonding mode 5, occurring from the anion to the amine group, is the unique mode furnishing favorable enthalpy changes ($\Delta H^\circ < 0$).

Accordingly, the unusual stability trends observed for the formation of phosphate and pyrophosphate complexes can be explained by assuming the formation of different hydrogen bonds whose contribution is decisive even in a competitive solvent such as water which is a good donor and acceptor of hydrogen bonds. Actually, the stability decrease previously highlighted for some complexes formed by HPO_4^{2-} when the positive charge of the receptor increases can be interpreted in terms of increasing hydrogen bond donor properties (type (1) bonds) of the receptors, leading to unfavourable enthalpic contributions, while the stability increase of the complexes observed as the charge on the anion decreases from $HP_2O_7^{3-}$ to $H_3P_2O_7^{-}$ can be attributed to the greater donor ability of more protonated anions (type (5) bonds) determining more favorable enthalpic and less favorable entropic contributions.[83]

Protonated forms of azacycloalkanes were also shown to bind nucleotide phosphate anions (ATP^{4-}, ADP^{3-}, AMP^{2-}).[14,15,40] To this interesting issue featuring, among other aspects, the ability of azacycloalkanes to behave as synthetic models for a variety of phosphoryl transfer processes is dedicated an entire chapter of this volume (see **Receptors for Nucleotides**, Volume 3). The reader is referred to this chapter for a comprehensive presentation of the topic.

5.2 Anion coordination by polyammonium receptors deriving from azacyclophanes

Relative to azacycloalkanes, azacyclophanes implement the multifunctional approach to anion recognition with additional weak forces involving their aromatic moieties and with the stiffening of the ligand structure, brought about by the insertion of aromatic groups, which enhances the ligand preorganization. Azacyclophanes, in their protonated forms, have been employed for the binding of many inorganic and aliphatic anions, but, of course, these types of receptors show their best performance with anionic species containing aromatic groups. For instance, the azacyclophane **69**, containing two naphthalene groups connected by two triamine chains, was reported by Lehn, in an early work, to bind dicarboxylate anions in water. In an acidic solution, the protonated amine chains furnish electrostatic and hydrogen bond anchorages to the carboxylate groups, while the two aromatic walls of the macrocycle can provide additional binding sites via π-stacking interactions. Indeed, the tetraprotonated $H_4\mathbf{69}^{4+}$ receptor shows a clear preference for terephthalate ($\log K = 5.2$) and isophthalate ($\log K = 5.0$) over fumarate ($\log K = 4.4$) and maleate ($\log K = 3.5$).[84]

More recently, Delgado reported the recognition of dicarboxylate anions by **70**, a ligand showing some structural similarity to **69**.[85] Ligand **70** contains propylenic chains, instead of ethylenic ones, which increases the basicity of the amine groups and favors full ligand protonation at relatively high pH (pH > 6), when carboxylate anions are completely deprotonated, but contains less extended aromatic groups, relative to **69**, which means that a more strict positioning of the aromatic anions in the complexes is necessary for efficient π-stacking interactions between the interacting partners to take place. The stability constants of the complexes formed in aqueous solution with the aromatic dicarboxylates phthalate, isophthalate, terephthalate, and benzoate, as well as with the aliphatic oxalate, malonate, succinate, glutarate, and adipate, were determined by means of potentiometric methods and ^{1}H NMR spectroscopy. Both methods afforded comparable stability constants for 1 : 1 complexes, although the NMR measurements, which were performed spanning the concentration of anion up to a relatively large excess, evidenced the ability of $H_6\mathbf{70}^{6+}$ to also form 1 : 2 ligand-to-anion species. The constants for the formation of 1 : 1 complexes with the aromatic anions are significantly higher than those for aliphatic anions, and, in particular, for terephthalate that forms a more stable complex with $H_6\mathbf{70}^{6+}$ ($\log K = 4.3$ for $H_6\mathbf{70}^{6+} + tph^{2-} = (H_6\mathbf{70})tph^{4-}$), by at least one order of magnitude, than the other aromatic anions. NMR spectroscopy allowed for the conclusion that the interaction with the aromatic guests might involve hydrogen bonding, and electrostatic and π-stacking interactions, the importance of such forces depending on the substrate. The best cooperativity of these binding interactions occurs with terephthalate. In agreement with NMR data and molecular dynamic calculations, the terephthalate anion is included in the ligand cavity where it forms all these types of interactions, while phthalate and isophthalate remain outside the cavity where significant π-stacking interactions with the ligand are not possible.[85]

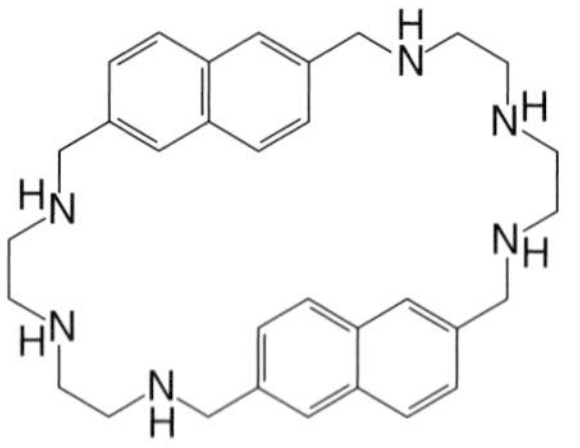

69 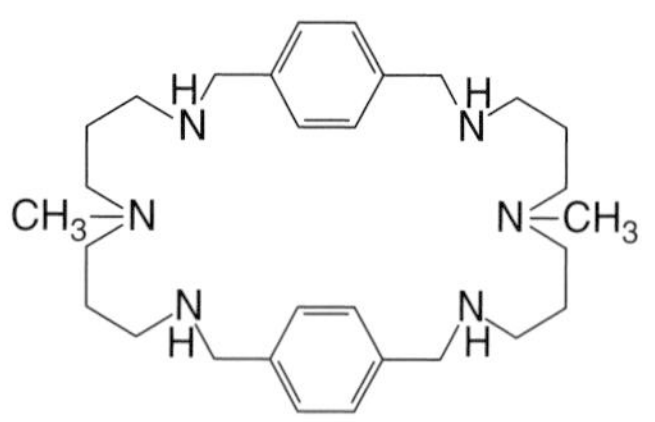 **70**

π-Stacking interactions are not the only forces that may arise in anion coordination with azacyclophanes in addition to electrostatic attraction and hydrogen bonding. In an early work, dealing with the binding of benzoate (C_6H_6–COO^-) and phenylacetate (C_6H_6–CH_2–COO^-) and their cyclohexane analogs by the quaternary tetraazaparacyclophane **71**, performed in D_2O/CD_3OD (80 : 20, v/v) solutions, Schneider showed that the aromatic anions are more strongly retained inside the host cavity than their aliphatic analogs, with 40–60% enhancement in the free energy change of association.[86] Comparison with cyclophanes that do not bear positive charge in the cavity showed that, in the case of **71**, cation–π interactions occurring between the ligand ammonium groups and the π benzene system of the aromatic anions are the major driving force in the formation of these complexes.

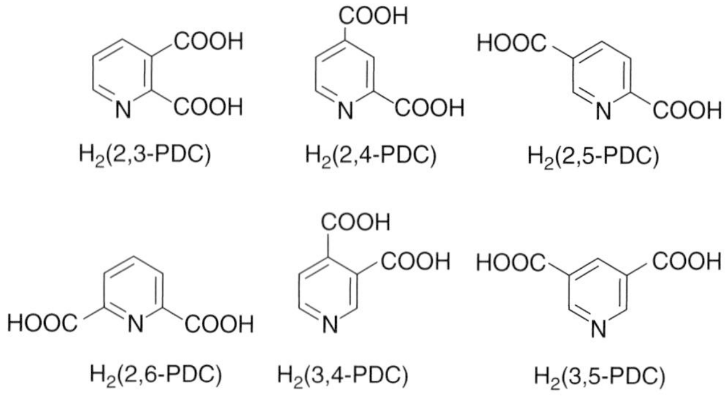

H_2(2,3-PDC) H_2(2,4-PDC) H_2(2,5-PDC)

H_2(2,6-PDC) H_2(3,4-PDC) H_2(3,5-PDC)

Figure 30 Pyridinedicarboxylic acids (H_2PDC).

Azacyclophanes containing heteroaromatic amine groups are also of interest for anion binding and recognition. The electronegativity of their nitrogen atoms polarizes the π clouds of the aromatic rings, affecting the geometry and the strength of π-stacking interaction with substrates, while the negative charge density on these nitrogens may provide additional anchorage for positively charged groups. The heteroaromatic azacyclophanes **47** and **49** were analyzed by Bianchi and Garcia-España for the recognition of the anionic forms of the six isomers of pyridinedicarboxylic acid (H_2PDC) in aqueous solution, as shown in Figure 30.[87] The structurally analogous ligand [21]aneN_7 (**18**) was also considered as a fully aliphatic counterpart of **47** and **49**. The occurrence of π-stacking interaction between **47, 49**, and pyridinedicarboxylate substrates was denoted by the upfield shift of ^{1}H NMR signals of all aromatic protons of the interacting partners, indicating that in the complex they are arranged in a face-to-face geometry.

The ability of **47** and **49** to give rise to π-stacking interactions characterizes, to a large extent, the binding properties of these ligands. As a matter of fact, potentiometric measurements showed that, for equally charged forms of receptors and substrates, **47** and **49** give rise to more stable 1 : 1 complexes than **18**, which binds these anions only via charge–charge attraction and hydrogen bonding. Among pyridinedicarboxylate isomers, 2,6-PDC^{2-} and H(2,6-$PDC)^-$ are efficiently recognized by **47** and **49**, since in these species the two carboxylate/carboxylic groups and the pyridine nitrogen are conveniently arranged for a synergistic binding action (Figure 31). The heteroaromatic azacyclophane **47**, however, binds these anions more strongly than **49** over a large pH range, being the best receptor among the group. Molecular modeling calculations suggested that such recognition ability of **47** is due to a superior structural and electrostatic complementarity with the substrate, relative to **49**, promoted, at least in part,

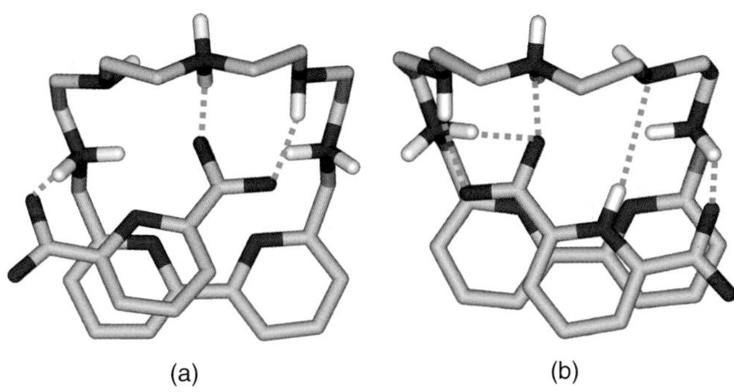

Figure 31 Calculated structure for the complexes H_3**47**(2,6-PDC)$^+$ (a) and H_3**47**[H(2,6-PDC)]$^{2+}$ (b). Carbon (Grey), Nitrogen (Blue), Hydrogen (White), and Oxygen (Red).

by the greater flexibility of the bipyridine moiety in **47**, compared to phenanthroline in **49**. Interestingly, protonation of 2,6-PDC^{2-} to give H(2,6-PDC)$^-$, taking place on the pyridine nitrogen, produces an enhancement of complex stability, in contrast to the loss of charge experienced by the protonated anion. According to modeling calculations (Figure 31), anion protonation promotes a repositioning of the anion within the complex, giving rise to the formation of a greater number of hydrogen bonds and salt bridges and to an additional electrostatic attraction between the acidic proton of the pyridinium group and one heteroaromatic nitrogen of the receptor (Figure 31b).[87]

71 **72**

73

Interesting azacyclophanes, such as **72** and **73**, containing two phenanthroline groups were synthesized by Félix and coworkers.[88] The interaction of these ligands with a variety of aromatic carboxylates, including benzoate, 1-naphthalate, 9-anthracenate, pyrene-1-carboxylate, phthalate, isophthalate, terephthalate, and 1,3,5-benzenetricarboxylate, was studied in aqueous solution by means of potentiometric and ^{1}H NMR measurements. The ligand **72**, in its protonated forms, was found to be the most suitable for binding these anions in 1 : 1 stoichiometry, showing a remarkable selectivity toward the extended pyrene-1-carboxylate anion and a decreasing affinity for anions with smaller aromatic surface, indicating that π-stacking interactions play a major role. Molecular dynamic simulations carried out for anion binding by H_4**72**$^{4+}$ and H_6**73**$^{6+}$ in water showed that the receptors may adopt a folded conformation with the anion sandwiched between the two phenanthroline groups and forming salt bridges with ligand ammonium groups, as shown in Figure 32 for H_6**73**$^{6+}$. While for H_6**73**$^{6+}$ an open conformation also seems to be possible, the folded conformation is favored for H_4**72**$^{4+}$ in the presence of similar anionic substrates.[88]

Among anionic species bearing aromatic residues, nucleotides are of special interest according to their biological relevance and, obviously, have been extensively considered for interaction with azacyclophanes. This topic, however, is not considered here since it is thoroughly treated in another chapter of this volume (see **Receptors for Nucleotides**, Volume 3).

As said above, azacyclophanes have also been used for the coordination of many inorganic and aliphatic anions. Being deprived of the possibility to take advantage of the interaction forces provided by the aromatic groups, azacyclophanes show binding properties toward these anions that are very similar to those of azacycloalkanes. Nevertheless, the introduction of rigid aromatic groups produces stiffening of these cyclic molecules, reducing their possible conformations, and may give rise to preformed ligand cavities. For instance, the diphenylmethane moieties of ligands **11** and **71** impart curvature to the molecules and

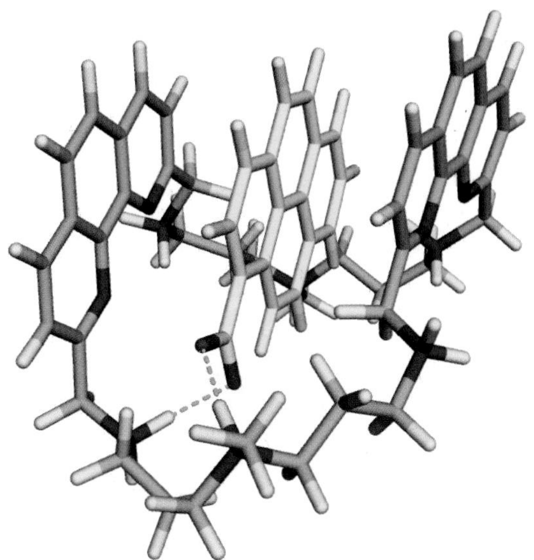

Figure 32 Calculated structure for the complex of pyrene-1-carboxylate with H_6**73**$^{6+}$. Carbon (Both grey and yellow), Nitrogen (Blue), Hydrogen (White), and Oxygen (Red).

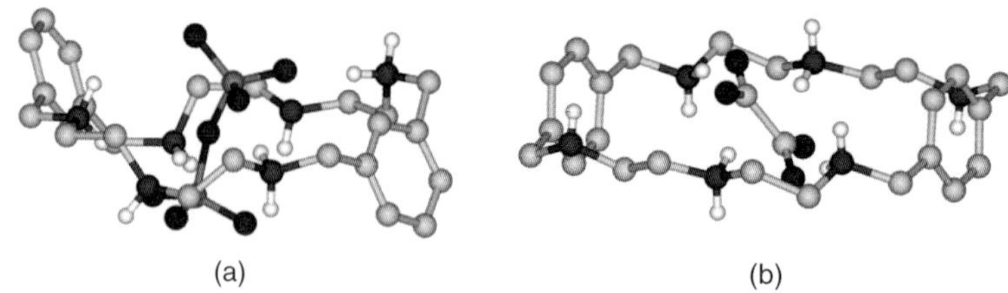

Figure 33 Crystal structures of the inclusion complexes (a) $(H_4\mathbf{43})(H_2P_2O_7{}^{2-})^{2+}$ and (b) $(H_6\mathbf{76})(ox)^{4+}$. Carbon (Grey), Nitrogen (Blue), Hydrogen (White), Oxygen (Red), and Phosphorus (Orange).

define their cavities. Accordingly, inclusive complexation of simple nonaromatic anions is more frequent with azacyclophanes than with azacycloalkanes.

Apart from the extreme cases, such as the very rigid quaternarized ligands **74** and **75**, which were shown to include two Br^- anions in the solid state,[89,90] azacycloalkanes endowed with more ligand flexibility also display inclusive complexation of anions. Crystal structures of protonated forms of the azacyclophane **43**, for instance, showed $H_6\mathbf{43}^{6+}$ and $H_4\mathbf{43}^{4+}$ hosting, respectively, two Br^- anions and a diprotonated pyrophosphate $(H_2P_2O_7{}^{2-})$ anions into the ligand cavity.[91] In particular, in the case of pyrophosphate, the anion is located across the macrocyclic ring, almost axially to it, with each phosphate group being hydrogen bonded to the amine/ammonium ligand chain, as shown in Figure 33(a). The results of the analysis of the stability constants of the complexes formed by differently protonated forms of **43** and $P_2O_7{}^{4-}$ agreed with inclusive complexation occurring also in aqueous solution. Complexes with 1 : 1 stoichiometry are also formed with phosphate and triphosphate, with the strength of the interaction increasing in the order mono- < di- < triphosphate.[91]

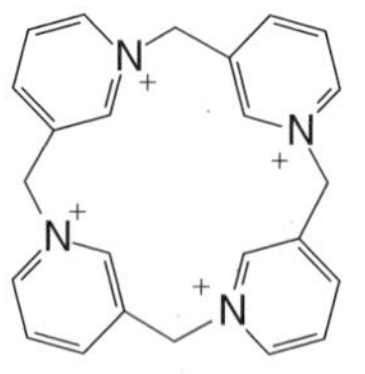

74 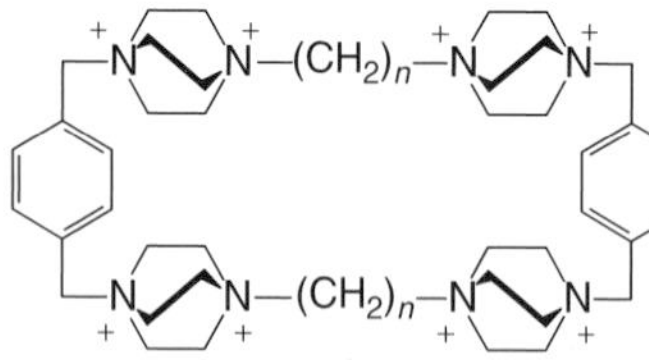 **75**

Another nice example of inclusive coordination was shown by the azacycloalkane **76**.[92] Protonated forms of **76** interact with oxalate (ox^{2-}) and oxydiacetate (od^{2-}) anions, both in solution and in the solid state. In the solid state, $H_6\mathbf{76}^{6+}$ was found to host oxalate inside the cavity, with the inversion center of the centrosymmetric $H_6\mathbf{76}^{6+}$ cation residing in the middle of the oxalate C–C bond (Figure 33b). The oxalate anion, forming an angle of 50.29° with the plane defined by the ligand nitrogen atoms, is held into the cavity by a network of eight salt bridges involving all its nitrogen atoms. In solution, various protonated species from $H_2\mathbf{76}^{2+}$ to $H_6\mathbf{76}^{6+}$ are able to bind both oxalate and oxydiacetate, with stability constants increasing as the charge of the ligand increases. A very high stability constant ($\log K = 6.08$) was determined for the formation of the $H_6\mathbf{76}(ox)^{4+}$ complex, which is thought to retain in solution the inclusive nature observed in the solid state, while a lower stability constant ($\log K = 4.28$) was found for the complex $(H_6\mathbf{76}(od)^{4+})$ with the more flexible oxydiacetate anion.[92]

6 BINDING OF NEUTRAL MOLECULES

The complexation of apolar solutes in water, rather than in organic solvents, has always attracted special interest, because it may be used for modeling recognition processes occurring in biological systems.[20,21] Furthermore, the interaction between apolar species is stronger in water, where van der Waals and dispersive forces between the interacting partners do not suffer the competitive effect of solvation by organic solvents and the hydrophobic effect (see **Supramolecular Interactions**, Volume 1) becomes a significant, frequently decisive, driving force for the association process.

Azacyclophanes are an efficient category of receptors for the binding of apolar substrates in water, since the aromatic walls constituting their host cavities provide the appropriate binding sites for these types of substrates, and the amine groups ensure water solubility, especially in acidic solution where they are converted into protonated, positively charged ammonium groups. This is the main reason that fostered the synthesis of the first azacyclophanes. Alternatively, pH-independent water solubility of azacyclophanes can be achieved by quaternarization of the amine groups, as is the case with ligands **71, 74**, and **75**. These solvated positively charged groups in the periphery of the binding sites, however, while ensuring water solubility of these compounds and favoring anion complexation, may reduce the driving forces of apolar association, since they mostly result from the desolvation of apolar surfaces.

As already noted in the introduction to this chapter, earlier studies[16,18] performed with receptors such as **7–10** afforded the first information about the formation of adducts

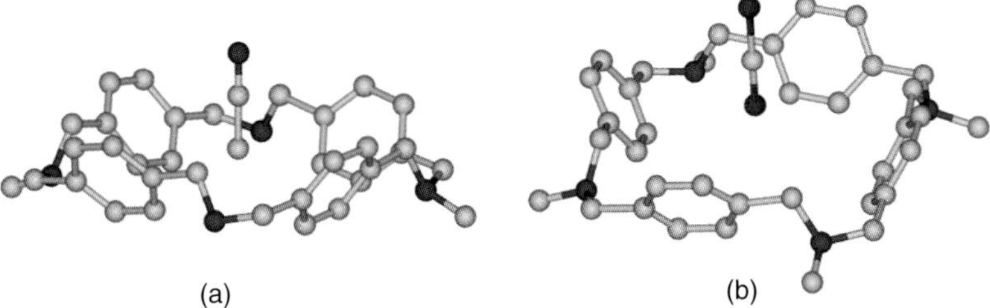

Figure 34 Crystal structures of the inclusion complexes of CH_3CN (a) and CO_2 with **78** (b). Carbon (Grey), Nitrogen (Blue), and Oxygen (Red).

between azacyclophanes and neutral molecules which were later shown to be of external type.[17]

In 1981, Tabushi reported the ability of the water-soluble azacyclophane **77** to bind apolar solutes in aqueous solution.[93] Despite the observation of significant association with organic molecules, such as hydroxynaphthoic acid (log K values of about 3), CPK models showed that inclusion of molecules of this size would be only possible at the expense of considerable receptor strain. For this reason, surface interaction or, at most, a "nesting complexation" mode involving partial inclusion of the substrate was suggested to occur. Nevertheless, three years later, the same author reported the crystal structures of 1 : 1 complexes formed by the tetraazacyclophane **78** with $CHCl_3$, CH_2Cl_2, CH_3CN, and CO_2, showing the inclusive complexation of the substrates inside the receptor cavity, whose shape resembles a square box with walls formed by benzene rings and the floor defined by the plane containing the four nitrogen atoms, and with the width being 4.6–6.4 Å and the depth being about 6 Å (Figure 34).[94,95]

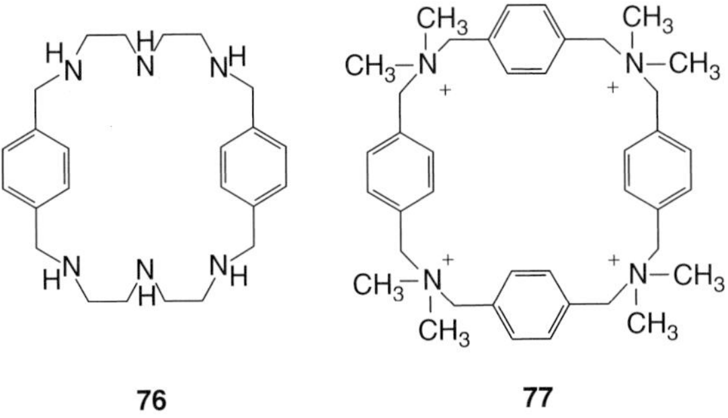

The first clear evidence of inclusion compounds formed by azacyclophanes, both in solution and in the solid state, with aromatic guests was reported in 1980 for the tetraprotonated form of **11**, existing in aqueous solution below pH 2.[19] Solid 1 : 1 complexes were obtained between $H_4\mathbf{11}^{4+}$ and 1,3-dihydroxynaphthalene, 2,7-dihydroxynaphthalene, naphthalene, *p*-xylene, and durene from acidic aqueous solution or upon extraction into acidic aqueous solution of substrates dissolved in hexane. Complexation in solution, as evidenced by ^{1}H NMR spectra, revealed an intimate contact

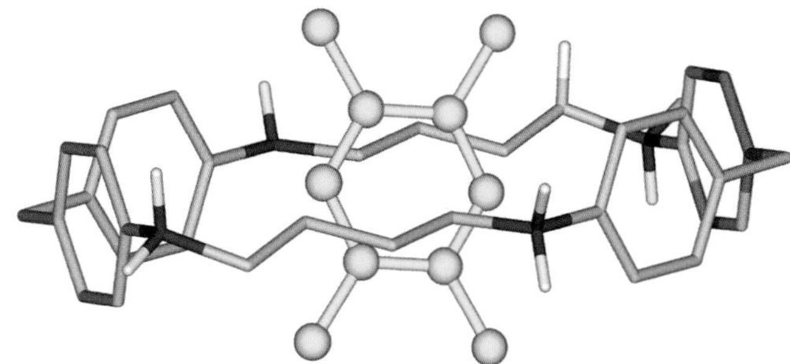

Figure 35 Crystal structure of the inclusion complex of durene with $H_4\mathbf{11}^{4+}$. Carbon (Both grey and yellow), Nitrogen (Blue), and Hydrogen (White).

between the interacting partners, which was not observed for a reference acyclic compound, and showed interacting patterns indicative of inclusive complexation. In the case of 2,7-dihydroxynaphthalene, for instance, ^{1}H NMR spectra were consistent with the formation of a complex with significant stability (log $K = 3.4$), in which the substrate was axially included in the receptor cavity.[96] The crystal structure of the durene complex with $H_4\mathbf{11}^{4+}$ (Figure 35) showed the guest molecule fully included within the rectangular host cavity, defined by benzene rings perpendicular to the mean plane of the macrocycle, and located on a center of symmetry, exactly in the middle of the cavity having a width of about 3.5 × 7.9 Å and a depth of 6.5 Å. The included durene molecule fits well with the cavity, being nearly parallel to the inner walls and with the methyl groups partly protruding outside, and forms several close contacts (3.6–3.8 Å) with the receptor.[19]

78 **79**

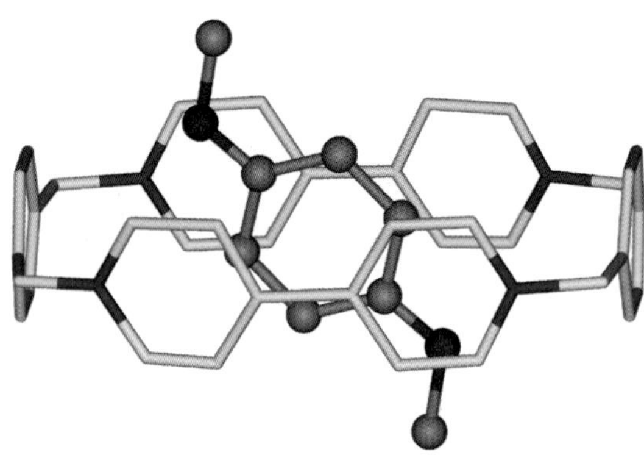

Figure 36 Crystal structure of the inclusion complex of *p*-dimethoxybenzene with **79**. Carbon (Grey), Nitrogen (Blue), and Oxygen (Red).

A very appealing application of azacyclophanes has been the formation of inclusion complexes with neutral substrates bound via charge-transfer interactions. In this respect, the ligand **79** is one of the most famous cyclophane in supramolecular chemistry. This tetracationic receptor, independently synthesized by Hünig[97] and Stoddart[98] in 1988, is constituted by two 4,4′-bipyridinium groups (paraquat) joined by *p*-xylylene units to form a boxlike structure in which two almost parallel π-electron-deficient walls (paraquat) are held at 6.8 Å, as shown by the crystal structure[98] of its **77**$(PF_6)_4$ salt. The cavity of **77** provides a tight space for inclusion of neutral π-electron-rich benzene derivatives forming intensively colored charge-transfer complexes.[98,99] For instance, **79** forms an intense-red complex with *p*-dimethoxybenzene, both in solution and in the solid state, in which the benzene ring of the guest molecule is sandwiched between the paraquat ligand walls (Figure 36), forming strong π–π interactions, while the methoxy residues protrude out of the cavity.[99] Such inclusion complexes showed a variety of new electrochemical and photochemical properties, but the most important outcome of their discovery was to point the way to the formation of molecular threads. The possibility to thread molecular rings such as **79** with linear components, to form stable complexes in solution, and the successive functionalization of the included molecules opened the horizons to new fascinating classes of supramolecular assemblies, such as rotaxanes and catenanes (see **Self-Assembled Links: Catenanes, Rotaxanes—Self-Assembled Links**, Volume 5).[100]

7 CONCLUSIONS

Azacoronands and azacyclophanes have played a central role in the birth and development of supramolecular chemistry. Along with the crown ethers, they have made a decisive contribution to the definition of the basic structures and the main properties of macrocyclic molecules, acting as a stepping stone for a great variety of macrocycles that characterize the supramolecular chemistry of today. They have contributed to the formation of new concepts, such as the macrocyclic effect in the formation of coordination compounds with metal ions, which are now part of the culture of all chemists and can be found both on specialized literature and on textbooks. Azacoronands and azacyclophanes represent two of the most eclectic groups of macrocyclic molecules, due to their ability to form complexes both with charged species (cations and anions) and with neutral molecules. Azacyclophanes receptors provide also the possibility to evaluate the relevance of π-stacking interactions like cation–π, anion–π, and π–π with appropriate guest species. Despite their long life, in the era of supramolecular chemistry, azacycloalkanes and azacyclophanes are still the subject of intense study and continue to produce important results even in the youngest branches of supramolecular chemistry.

REFERENCES

1. N. F. Curtis, *J. Chem. Soc.*, 1960, 4409.
2. N. F. Curtis and D. A. House, *Chem. Ind.*, 1961, **42**, 1708.
3. M. C. Thomson and D. H. Busch, *Chem. Eng. News*, 1962, **40**(Sept. 17), 57.
4. M. C. Thomson and D. H. Busch, *J. Am. Chem. Soc.*, 1964, **86**, 3651.
5. G. N. Schrauzer, *Chem. Ber.*, 1962, **95**, 1438.
6. F. Umland and D. Thierig, *Angew. Chem.*, 1962, **74**, 388.
7. G. A. Melson and D. H. Busch, *Proc. Chem. Soc.*, 1963, 223.
8. J. Van Alphen, *Rec. Trav. Chim.*, 1937, **56**, 343.
9. H. Stetter and K. H. Mayer, *Chem. Ber.*, 1961, **94**, 1410.
10. B. Bosnich, C. K. Poon, and M. L. Tobe, *Inorg. Chem.*, 1965, **4**, 1102.
11. L. F. Lindoy, *The Chemistry of Macrocyclic Ligand Complexes*, Cambridge University Press, Cambridge, 1989.
12. J. S. Bradshaw, K. E. Krakowiak, and R. M. Izatt, *Aza-Crown Macrocycles*, John Wiley & Sons, New York, 1993.
13. K. Gloe, *Macrocyclic Chemistry: Current Trends and Future*, Springer, Dordrecht, 2005.
14. J. L. Sessler, P. A. Gale, and W.-S. Cho, *Anion Receptor Chemistry*, RSC Publishing, Cambridge, 2006.
15. A. Bianchi, K. Bowman-James, and E. Garcia-España, eds, *Supramolecular Chemistry of Anions*, Wiley-VCH, New York, 1997.
16. H. Stetter and E. E. Roos, *Chem. Ber.*, 1955, **88**, 1390.
17. R. Hilgenfeld and W. Saenger, *Angew. Chem. Int. Ed. Engl.*, 1982, **21**, 787.
18. G. Faust and M. Pallas, *J. Prakt. Chem.*, 1960, **11**, 146.

19. K. Odashima, A. Itai, Y. Iitaka, and K. Koga, *J Am. Chem. Soc.*, 1980, **102**, 2504.
20. I Cyclophanes and F. Vögtle, *Top. Cur. Chem.*, 1983, **113**, 1–182.
21. F. Diederich, *Cyclophanes*, The Royal Society of Chemistry, Cambridge, 1991.
22. E. Garcia-España and S. L. Luis, *Supramol. Chem.*, 1996, **6**, 257.
23. E. Garcia-España, P. Diaz, J. M. Llinares, and A. Bianchi, *Coord. Chem. Rev.*, 2006, **250**, 2952.
24. S. Karbach, W. Löhr, and F. Vögtle, *J. Chem. Res.*, 1981, 314.
25. J. E. Richman and T. J. Atkins, *J. Am. Chem. Soc.*, 1974, **96**, 2268.
26. C. Bazzicalupi, A. Bencini, A. Bianchi, *et al.*, *J. Am. Chem. Soc.*, 2008, **130**, 2440.
27. H. Takemura, G. Wen, and T. Shinmyozu, *Synthesis*, 2005, **17**, 2845.
28. N. V. Gerbeleu, V. B. Arion, and J. Burgess, *Template Synthesis in Macrocyclic Chemistry*, John Wiley and Sons, Chichester, 1999.
29. I. I. Creaser, J. M. Harrowfield, A. J. Herit, *et al.*, *J. Am. Chem. Soc.*, 1977, **99**, 3181.
30. R. J. Geue, T. W. Hambley, J. M. Harrowfield, *et al.*, *J. Am. Chem. Soc.*, 1984, **106**, 5478.
31. P. Comba, N. F. Curtis, G. A. Lawrance, *et al.*, *Inorg. Chem.*, 1986, **25**, 4260.
32. M. Paik Suh and S.-G. Kangl, *Inorg. Chem.*, 1988, **27**, 2544.
33. A. De Blas, G. De Santis, L. Fabbrizzi, *et al.*, *J. Chem. Soc., Dalton. Trans.*, 1993, 1411.
34. F. Abbà, G. De Santis, L. Fabbrizzi, *et al.*, *Inorg. Chem.*, 1994, **33**, 1366.
35. G. Hervé, H. Bernard, N. Le Bris, *et al.*, *Tetrahedron Lett.*, 1998, **39**, 6861.
36. T. J. Hubin, *Coord. Chem. Rev.*, 2003, **241**, 27.
37. J. C. Timmons and T. J. Hubin, *Coord. Chem. Rev.*, 2010, **254**, 1661.
38. A. Bencini, A. Bianchi, E. Garcia-España, *et al.*, *Coord. Chem. Rev.*, 1999, **188**, 97.
39. K. Wieghardt, S. Brodka, E. M. Peters, and A. Simon, *Z. Naturforsch. Teil B*, 1987, **42**, 279.
40. A. Bianchi, M. Micheloni, and P. Paoletti, *Coord. Chem. Rev.*, 1991, **110**, 17.
41. C. Nave and M. R. Truter, *J. Chem. Soc.*, 1974, 2351.
42. M. Studer, A. Riesen, and T. A. Kaden, *Helv. Chim. Acta*, 1989, **72**, 1253.
43. M. Chadim, P. Diaz, E. Gracia-España, *et al.*, *New. J. Chem.*, 2003, 1132.
44. B. Verdejo, M. Garcia Basallote, A. Ferrer, *et al.*, *Eur. J. Inorg. Chem.*, 2008, 1497.
45. H. Fujioka, E. Kimura, and M. Kodama, *Chem. Lett.*, 1982, 737.
46. C. Bazzicalupi, A. Bencini, V. Fusi, *et al.*, *Inorg. Chem.*, 1998, **37**, 941.
47. C. Bazzicalupi, A. Bencini, E. Berni, *et al.*, *Inorg. Chem.*, 2004, **43**, 5134.
48. C. Lodeiro, A. J. Parola, F. Pina, *et al.*, *Inorg. Chem.*, 2001, **40**, 2968.
49. J. Costa and R. Delgado, *Inorg. Chem.*, 1993, **32**, 5257.
50. A. Bencini, A. Bianchi, P. Paoletti, and P. Paoli, *Coord. Chem. Rev.*, 1992, **120**, 51.
51. G. W. Franklin, D. P. Riley, and W. L. Neumann, *Coord. Chem. Rev.*, 1998, **174**, 133.
52. J. C. A. Boeyens and E. L. Oosthuizen, *J. Crystallogr. Spectrosc. Res.*, 1992, **22**, 3.
53. A. Bencini, A. Bianchi, P. Dapporto, *et al.*, *J. Chem. Soc., Chem. Commun.*, 1990, 1382.
54. A. Bencini, A. Bianchi, M. Micheloni, *et al.*, *J. Chem. Soc. Dalton Trans.*, 1991, 1171.
55. A. E. Martell and R. D. Hancock, *Metal Complexes in Aqueous Solutions*, Plenum Press, New York, 1996.
56. D.-D. Klaehn, H. Paulus, R. Grewe, and H. Elias, *Inorg. Chem.*, 1984, **23**, 483.
57. A. Andrés, C. Bazzicalupi, A. Bianchi, *et al.*, *J. Chem. Soc. Dalton Trans.*, 1994, 2995.
58. R. Menif, A. E. Martell, P. J. Squattrito, and A. Clearfield, *Inorg. Chem.*, 1990, **29**, 4723.
59. C. Bazzicalupi, A. Bencini, A. Bianchi, *et al.*, *Inorg. Chem.*, 1995, **34**, 552.
60. G. L. Rothermel, L. Miao, A. L. Hill, and S. C. Jackels Jr., *Inorg. Chem.*, 1992, **31**, 4854.
61. L. H. Bryant, A. Lachgar, K. S. Coates, and S. C. Jackels Jr. *Inorg. Chem.*, 1994, **33**, 2219.
62. L. Branco, J. Costa, R. Delgado, *et al.*, *J. Chem. Soc. Dalton Trans.*, 2002, 3539.
63. C. Cruz, S. Carvalho, R. Delgado, *et al.*, *Dalton Trans.*, 2003, 3172.
64. C. Bazzicalupi, A. Bencini, A. Bianchi, *et al.*, *Inorg. Chem.*, 1999, **38**, 3806.
65. A. Bencini, M. A. Bernardo, A. Bianchi, *et al.*, *Eur. J. Inorg. Chem.*, 1999, 1911.
66. C. Bazzicalupi, A. Bencini, E. Berni, *et al.*, *Inorg. Chem.*, 2004, **43**, 6255.
67. C. Bazzicalupi, A. Bencini, A. Bianchi, *et al.*, *Coord. Chem. Rev.*, 2008, **252**, 1052.
68. E. Garcia-España, P. Gaviña, J. Latorre, *et al.*, *J. Am. Chem. Soc.*, 2004, **126**, 5082.
69. C. Bazzicalupi, A. Bencini, A. Bianchi, *et al.*, *Dalton Trans.*, 2006, 5743.
70. T. Yatsunami, A. Sakonaka, and E. Kimura, *Anal. Chem.*, 1981, **53**, 477.
71. E. Kimura, A. Sakonaka, T. Yatsunami, and M. Kodama, *J. Am. Chem. Soc.*, 1981, **103**, 3041.
72. B. Dietrich, M. W. Hosseini, and J.-M. Lehn, *J. Am. Chem. Soc.*, 1981, **103**, 1282.

73. J. Aragó, A. Bencini, A. Bianchi, *et al.*, *J. Chem. Soc. Dalton Trans.*, 1992, 319.
74. A. Bencini, A. Bianchi, P. Dapporto, *et al.*, *Inorg. Chem.*, 1992, **31**, 1902.
75. M. F. Manfrin, L. Moggi, V. Castelvetro, *et al.*, *J. Am. Chem. Soc.*, 1985, **107**, 6888.
76. A. Bencini, A. Bianchi, P. Dapporto, *et al.*, *J. Chem. Soc. Chem. Commun.*, 1990, 753.
77. A. Bencini, A. Bianchi, E. Garcia-España, *et al.*, *Inorg. Chem.*, 1987, **26**, 3902.
78. E. Suet and H. Handel, *Tetrahedron Lett.*, 1984, **25**, 645.
79. M. W. Hosseini and J.-M. Lehn, *J. Am. Chem. Soc.*, 1982, **104**, 3525.
80. M. W. Hosseini and J.-M. Lehn, *Helv. Chim. Acta*, 1986, **69**, 587.
81. A. Bencini, A. Bianchi, M. I. Burguete, *et al.*, *J. Am. Chem. Soc.*, 1992, **114**, 1919.
82. P. Arranz, A. Bencini, A. Bianchi, *et al.*, *J. Chem. Soc. Perkin Trans. 2*, 2001, 1765.
83. C. Bazzicalupi, A. Bencini, A. Bianchi, *et al.*, *J. Am. Chem. Soc.*, 1999, **121**, 6807.
84. M. Dhaenens, J.-M. Lehn, and J.-P. Vigneron, *J. Chem. Soc Perkin Trans. 2*, 1993, 1379.
85. S. Carvalho, R. Delgado, N. Fonseca, and V. Félix, *New. J. Chem.*, 2006, **30**, 247.
86. H.-J. Schneider, T. Blatter, S. Simova, and I. Theis, *J. Chem. Soc. Chem. Commun.*, 1989, 580.
87. C. Bazzicalupi, A. Bencini, A. Bianchi, *et al.*, *J. Org. Chem.*, 2008, **73**, 8286.
88. C. Cruz, V. Calisto, R. Delgado, and V. Félix, *Chem. Eur. J.*, 2009, **15**, 3277.
89. S. Shinoda, M. Tadokoro, H. Tsukube, and R. Arakawa, *Chem. Commun.*, 1998, 181.
90. F. M. Menger and K. K. Catlin, *Angew. Chem. Int. Ed. Eng.*, 1995, **34**, 2147.
91. D. A. Nation, J. Reibenspies, and A. E. Martell, *Inorg. Chem.*, 1996, **35**, 4597.
92. C. Anda, A. Llobet, A. E. Martell, *et al.*, *Inorg. Chem.*, 2004, **43**, 2793.
93. I. Tabushi, Y. Kimura, and K. Yamamura, *J. Am. Chem. Soc.*, 1981, **103**, 6486.
94. I. Tabushi, K. Yamamura, H. Nonoguchi, *et al.*, *J. Am. Chem.Soc.*, 1984, **106**, 2621.
95. K. Hirotsu, S. Kamitori, T. Higuchi, *et al.*, *J. Incl. Phenom.*, 1984, **2**, 215.
96. K. Odashima, A. Itai, Y. Iitaka, *et al.*, *Tetrahedron Lett.*, 1980, **21**, 4347.
97. M. Bühner, W. Geuder, W.-K. Gries, *et al.*, *Angew. Chem. Int. Ed. Engl.*, 1988, **27**, 1553.
98. B. Odell, M. V. Reddington, A. M. Z. Slawin, *et al.*, *Angew. Chem. Int. Ed. Engl.*, 1988, **27**, 1547.
99. P. R. Ashton, B. Odell, M. V. Reddington, *et al.*, *Angew. Chem. Int. Ed. Engl.*, 1988, **27**, 1550.
100. P. L. Anelli, P. R. Ashton, R. Ballardini, *et al.*, *J. Am. Chem. Soc.*, 1992, **114**, 193.

Hetero-Crown Ethers—Synthesis and Metal-Binding Properties of Macrocyclic Ligands Bearing Group 16 (S, Se, Te) Donor Atoms

William Levason and Gillian Reid

University of Southampton, Southampton, UK

1 INTRODUCTION AND SCOPE

The preparations and properties of chalcogenoether macrocycles have been the subject of several previous reviews—thioethers,[1–7] selenoethers,[4–11] and telluroethers.[4,5,8–10,12] The aim of this review therefore is to provide a survey of the most significant advances in the preparations and metal-binding properties (especially toward d-block and p-block acceptors) of chalcogenoether macrocyclic ligands (hetero-crown ethers) over the past 10 years or so. In the space available it is not possible to be comprehensive, and so we have chosen to focus on developments in the preparations of new macrocyclic ligands containing S, Se, or Te donor atoms, and mixed donor derivatives where the donor types are both chalcogens. Therefore, while crown ethers themselves are excluded, mixed donor S/O, Se/O, and other macrocycles are included, whereas those containing, for example, S/N or S/P donor atoms are excluded. In addition, we have tried to present an overview of the coordination chemistry of the hetero-crown ethers toward a range of acceptors grouped under (i) d- and f-block metal ions, and (ii) p-block elements (groups 13–17), emphasizing in particular the variety of coordination modes observed. It is not our intention here to revisit work described prior to 2000 unless it is particularly relevant to more recent work.

Supramolecular Chemistry: From Molecules to Nanomaterials.
Edited by Philip A. Gale and Jonathan W. Steed.
© 2012 John Wiley & Sons, Ltd. ISBN: 978-0-470-74640-0.

2 NEW CHALCOGENOETHER MACROCYCLE PREPARATIONS

2.1 Thioether macrocycles and mixed S/O crowns

The preparations of the most widely studied saturated thioether crowns are well established and have been reviewed previously.[1–5] New examples include two hexathiacrowns that incorporate pendant alcohol functions on the C backbone (**L**^{1} and **L**^{2}), prepared by high-dilution cyclizations. These compounds were of interest for Ag(I) extraction.[13] Also, the thioalkyne $(CH_3C{\equiv}CSCH_2CH_2)_2S$ reacted with the benzyne-nickel(0) complex $[Ni(\eta^2\text{-}C_6H_4)(PEt_3)_2]$ in a double-insertion reaction to give the trithioether macrocycle 2,3-naphthotrithiacyclononane (**L**^{3}), which was structurally characterized as its complex with Ru(II)—Section 3.7.[14]

$\mathbf{L^1}$ $\mathbf{L^2}$

$\mathbf{L^3}$

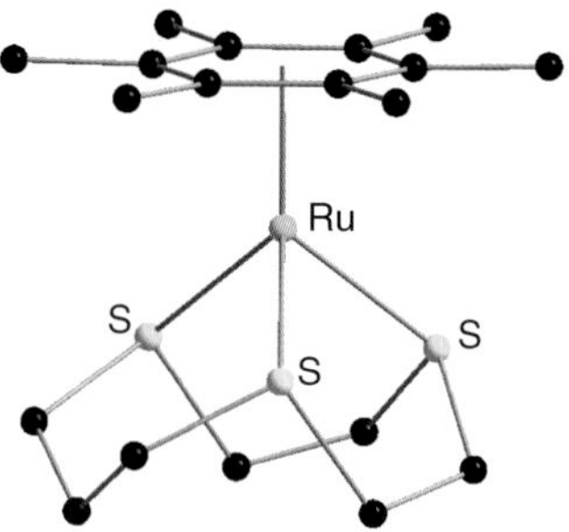

Figure 1 View of the structure of $[(\eta^6\text{-}(CH_3)_6C_6)Ru([10]aneS_3)]^{2+}$.[17]

Unsaturated thia-crowns with ring sizes between 6 and 27 (**I**) have been prepared by reaction of Na_2S with *cis*-ClCH=CHCl at 50 °C in CH_3CN in the presence of 15-crown-5. Using 0.4 mol equivalent of 15-crown-5 gave 40% total yield of the cyclic oligomers. The 6-, 15-, 18-, 21-, 24-, and 27-membered rings were separated chromatographically and crystal structures of the five larger rings were determined and their conformations compared with those of the saturated analogs.[15]

$n = 0–7$

(**I**)

Tetrathia-crowns containing maleonitrile units, $\mathbf{L^4}$ and $\mathbf{L^5}$, were obtained in modest yield (<20%) by high-dilution cyclization of disodium 1,2-dicyanoethene-1,2-dithiolate with 1,8-dichloro-3,6-dithiaoctane and 1,9-dichloro-3,7-dithianonane, respectively, in ethanol/water solution. Crystal structure determinations established the integrity of the rings and showed that while the S atoms closest to the rigid maleonitrile fragment adopt an *endocyclic* arrangement, with the two other S atoms in the more flexible parts of the rings tending toward an *exocyclic* orientation. The solid-state structures were consistent with the lowest energy conformation from molecular modeling calculations.[16]

$\mathbf{L^4}$ $\mathbf{L^5}$

Several trithioether macrocyclic ligands have been produced in high yields as their complexes of the $(\eta^6\text{-arene})Ru^{2+}$ fragment via template cyclization reactions based on $[(\eta^6\text{-arene})Ru\{S(CH_2CH_2S)_2\}]$ and $Br(CH_2)_nBr$ ($n = 1–5$). The crystal structures of three examples confirm the *pseudo*-octahedral coordination geometry at Ru, with the thia-crowns *facially* coordinated (Figure 1).[17]

Several series of styryl-based dyes[18,19] incorporating thia-crown ether moieties (e.g., **II**) have been prepared and studied as selective optical sensors for heavy metal ions in aqueous solutions.

(**II**)

R = CH_3, NO_2, CH_2OH, $CH_2OSiR'R''_2$

(**III**)

Kaden and coworkers have prepared several hexathia-cryptands (**III**) with variable cavity sizes and capping groups. The preparations used Cs_2CO_3/dmf to promote the key cyclization step.[20]

Derivatives of dithia-13-crown-4 and dithia-16-crown-5 have been prepared (**IV**), and the alcohol functions may be modified further to give thia-oxa-crowns with a range of pendant functional groups on the C backbone.[21] Functionalized benzo-thia-oxa crown ethers (e.g., **V**) have also been prepared.[22,23]

A series of cage-annulated thia-crown ethers and cryptands[24] and an adamantylidene-derived hexathioether macrocycle have been prepared; the crystal structure of the latter is shown in Figure 2.[25]

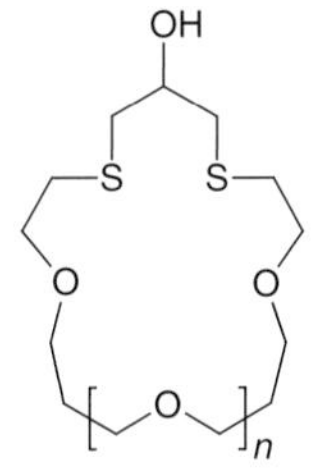

$n = 1, 2$

(IV)

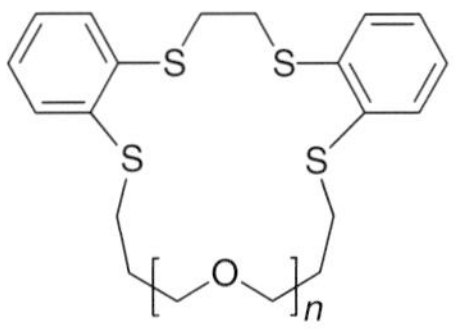

$n = 1, 2$

(V)

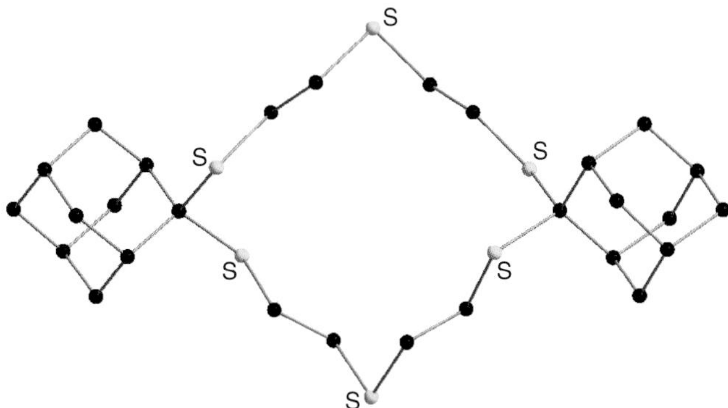

Figure 2 View of the structure of the adamantylidene-based hexathia-crown.[25]

Thioethers are widely regarded as very poor donors to H^+ cations and well-characterized protonated thioethers are rare, with protonation requiring the use of a superacid. However, it has been shown recently that reaction of [9]aneS_3 with NbF_5 in CH_2Cl_2 solution unexpectedly produced some crystals of the highly reactive monoprotonated sulfonium salt, [([9]aneS_3)H][NbF_6].[26] The presence of the proton was clearly evident from the X-ray crystal structure (Figure 3) and the proton formed H bonds to the two remaining thioether S atoms, d(S–H) = 1.19(4) Å; S···H = 2.50(4), 2.51(4) Å. The formation of this sulfonium salt may be promoted by the "preorganized" geometry of the [9]aneS_3 ring, allowing stabilization via the S–H···S interactions, as well as the presence of the large, weakly coordinating $[NbF_6]^-$ anion. The conformation of the protonated [9]aneS_3 macrocycle was different from that of the neutral form. This was the first evidence for the sulfonium salt of [9]aneS_3 despite the extensive studies of the coordination chemistry of this macrocycle with metal ions in a variety of oxidation states from across the periodic table.

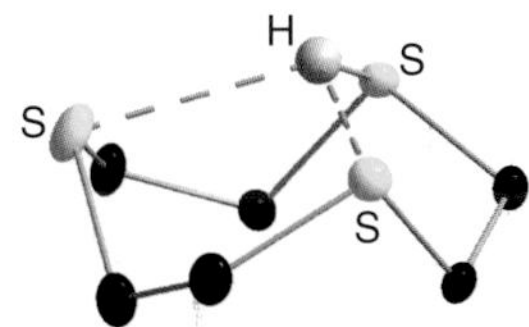

Figure 3 View of the structure of $[([9]aneS_3)H]^+$.[26]

A notable development associated with the chemistry of thioether and mixed thia-oxa-crowns over the last decade has been the reaction of these rings with sulfimidating reagents leading to a general method for modifying and functionalizing the crowns post-cyclization—a process that is much more challenging for group 16 donor ligands compared to group 15 donor systems. Kelly and coworkers have shown that reaction of [9]aneS_3 with 1 or 2 mol equivalents of *o*-mesitylsulfonylhydroxylamine (MSH) in CH_2Cl_2 solution gave the mono- or disulfimidinium cation, with terminal (protonated) sulfimide functions (Scheme 1).[27] These compounds have been isolated as salts of the $[(CH_3)_3C_6H_2SO_3]^-$ anion derived from the MSH, and their crystal structures showed significant H-bonding interactions between the protons on the sulfimidinium cations with the O atoms of the anion, as shown for example in Figure 4. The S–N bond distances are typically ~1.6 Å, and IR spectroscopy shows ν(NH) ~3180 cm^{-1}. Excess MSH led to the introduction of a sulfimide nitrogen bridge between two S atoms, with one terminal SNH_2^+ function (Scheme 2), possibly resulting from intramolecular rearrangement of the tricationic $\{[9]aneS(NH_2)S(NH_2)S(NH_2)\}^{3+}$. The structure of the N-bridged species has been confirmed crystallographically (Figure 5).[27] Using the tetrathioether crown, [14]aneS_4, with excess MSH under similar conditions gave the tetrasulfimidinium cation $\{14]ane(SNH_2)_4\}^{4+}$,

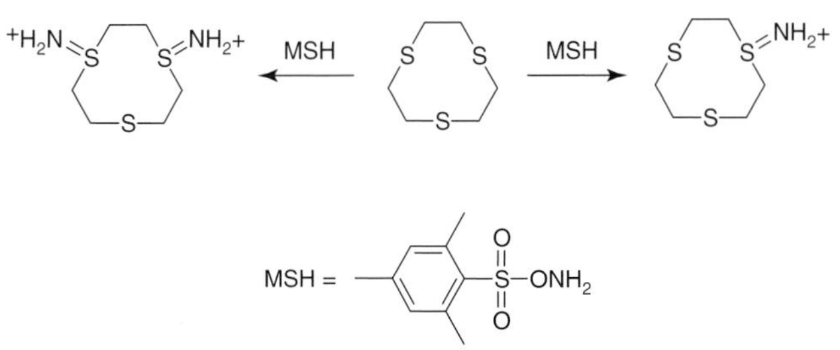

MSH = $C_6H_2(CH_3)_3$–SO_2–ONH_2

Scheme 1

>3 MSH

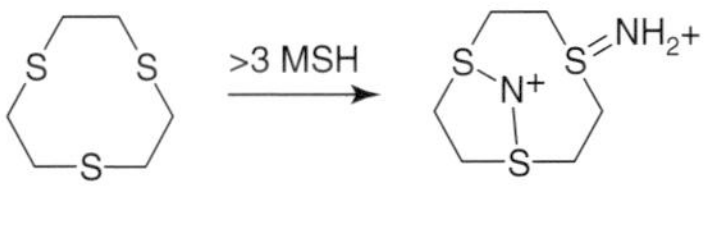

Scheme 2

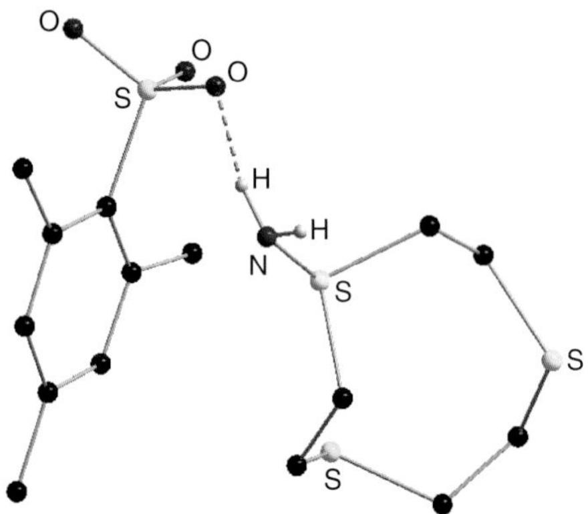

Figure 4 View of the structure of $\{[9]aneS_2S(NH_2)\}[(CH_3)_3C_6H_2SO_3]$.[27]

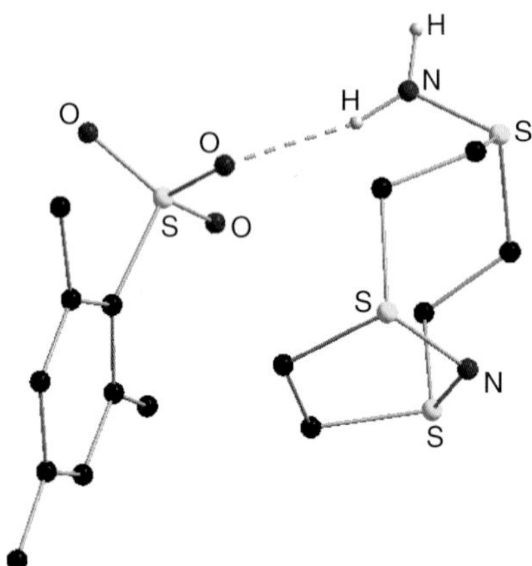

Figure 5 View of the structure of $\{[9]aneS(NH_2)S_2(N)\}^{2+}$ showing the H-bonding to one of the $[(CH_3)_3C_6H_2SO_3]^-$ anions.[27]

whose structure (as its $[(CH_3)_3C_6H_2SO_3]^-$ salt) revealed a centrosymmetric tetracation and an extensive H-bonding network through N–H···O interactions which formed a one-dimensional chain.[28] This reaction also showed that multiple sulfimidation reactions can be achieved in polythioether crowns.

Reaction of $[9]aneS_3$ with the bromosulfimide Ph_2SNBr in CH_2Cl_2 solution gave the sulfimide-bridged cation $\{[9]aneS_2S(NSPh_2)\}^+$ whose structure (as its $[BPh_4]^-$ salt) has been authenticated (Figure 6), $\angle S\text{–}N\text{–}S = 108.55(10)^\circ$, demonstrating the potential of this method as a general approach for introduction of a range of functionalities at the S atom.[29] Addition of further Ph_2SNBr did not lead to further reaction at the remaining thioether donor groups. However, $1,4\text{-}(PhSNBr)_2C_6H_4$ reacted with $[9]aneS_3$ in a 1 : 2 molar ratio to form $[1,4\text{-}\{[9]aneS_2S(NSPh)\}_2C_6H_4]Br_2$, in which a thia-crown unit was added to each sulfimide unit of the starting material.[29] These sulfimidation reactions formally involve oxidation of the S(II) to S(IV), reducing the number of available lone pairs on

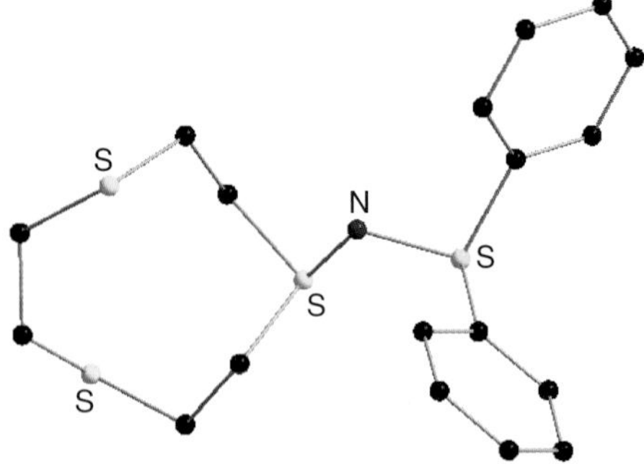

Figure 6 View of the structure of $\{[9]aneS_2S(NSPh_2)\}^+$.[29]

S from two to one and the introduction of the possibility of N-donation to metal ion guests through the sulfimide groups.

This approach has been developed further to allow mixed thia-oxa-crowns to be linked together, with sulfimidation occurring selectively at the S-donor atoms. Thus, reaction of $[18]aneSO_5$ with 1 mol equivalent of MSH in diethyl ether gave the $\{[18]aneS(NH_2)O_5\}^+$ cation subsequent and addition of excess $Na[BPh_4]$ in methanol solution gave $[\{[18]aneS(NH_2)O_5\}Na(CH_3OH)_2][BPh_4]_2$.[30] The crystal structure of the latter confirmed (Figure 7) that the amination reaction occurred selectively at the sulfur atom and that the cationic ligand could still coordinate to a sodium cation via four ether functions and two CH_3OH molecules. The hinge S–N–S angle in this dication = $112.30(19)^\circ$.[30]

The sulfimidium cation $\{[18]aneS(NH_2)O_5\}^+$ was readily deprotonated by reaction with LDA (lithium diisopropylamide) at $-78\,^\circ C$, followed by bromination with *N*-bromosuccinimide. Treatment with a further equivalent of $[18]aneSO_5$ in a "one-pot" reaction formed the linked-crown system $[(\{[18]aneSO_5\}_2N)Na][BPh_4]_2$. The crystal structure of the latter (Figure 8) revealed that the two ring systems sandwich the sodium cation, which is positioned away from the $S\text{–}N\text{–}S^+$ moiety,

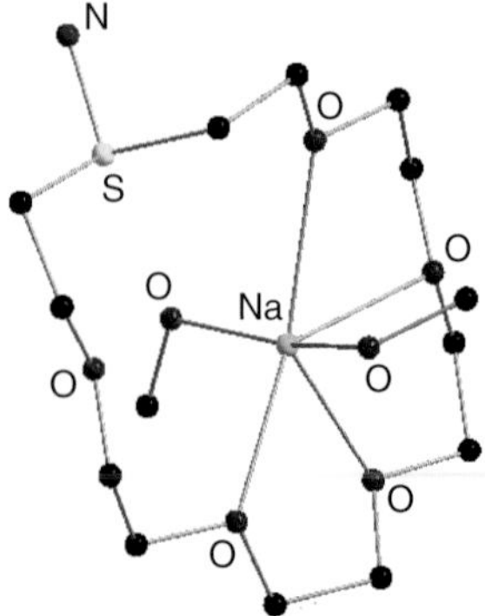

Figure 7 View of the structure of $[\{[18]aneS(NH_2)O_5\}Na(CH_3OH)_2]^{2+}$.[30]

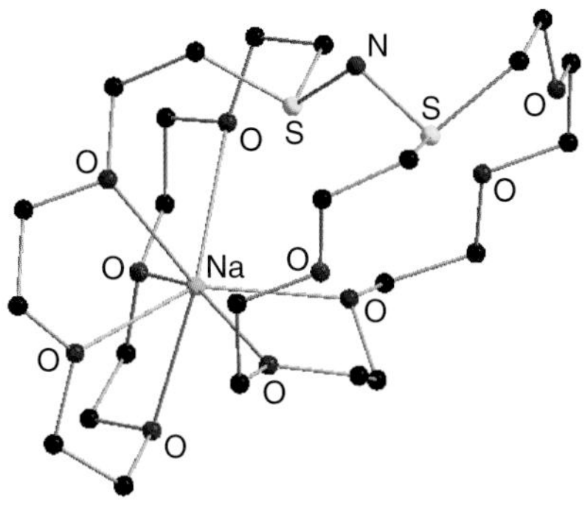

Figure 8 View of the structure of $[(\{[18]aneSO_5\}_2N)Na]^{2+}$.[30]

presumably minimizing repulsions between the positive charges, forming short Na–O bonds to five oxygen atoms, two from one ring, and three from the other.[30] This methodology has been applied with a range of other mixed S/O crowns including $[9]aneSO_2$, $[12]aneSO_3$ and $[18]aneS_2O_4$ which form the corresponding mono-sulfimidated cations in good yield, while using 2 mol equivalents of MSH with $[18]aneS_2O_4$ gave $\{[18]ane(SNH_2)_2O_4\}^{2+}$. Treatment of $\{[9]aneS(NH_2)O_2\}^+$ or $\{[12]aneS(NH_2)O_3\}^+$ with LDA and NBS (*N*-bromosuccinimide) at −78 °C, followed by addition of a further equivalent of the parent thia-oxa-crown, produced the monocationic N-bridged sulfimide bicyclic compounds, containing S–N–S linkages. In contrast, reaction of $\{[18]aneSS(NH_2)O_4\}^+$ with LDA and NBS gave the $\{[18]aneS_2(N)O_4\}^+$ cation containing an intramolecular S–N–S bridge.[31]

2.2 Selenoether macrocycles and mixed Se/O crowns

Several selenoether macrocycles have been prepared by Pinto and coworkers over the last two decades and the details described in previous reviews.[4,5,8–11] The preparation of the 12-membered triselena-crown $[12]aneSe_3$ was also originally reported by the same group[32] via high-dilution cyclization from $Se\{(CH_2)_3OTs\}_2$ and $NCSe(CH_2)_3SeCN$ under reducing conditions ($NaBH_4$ in thf/EtOH); the crystal structure (Figure 9) showed that the ring has approximate local mirror symmetry, with the Se atoms pointing out of the ring. The C–Se–C–C units adopt the preferred *gauche* torsions.[33] Recently, this method has been scaled up (with relatively lower volumes of solvent) to give multigram quantities of $[12]aneSe_3$ without compromising the yield.[34]

Modifying the route using $Se\{(CH_2)_3OTs\}_2$ and o-$C_6H_4(CH_2SeCN)_2$ also formed $\mathbf{L^6}$ in high yield (Scheme 3).[34] The 11-membered Se_3-donor ring incorporating the rigid *o*-phenylene unit, $\mathbf{L^7}$ was also obtained by a modification of this method using simultaneous dropwise

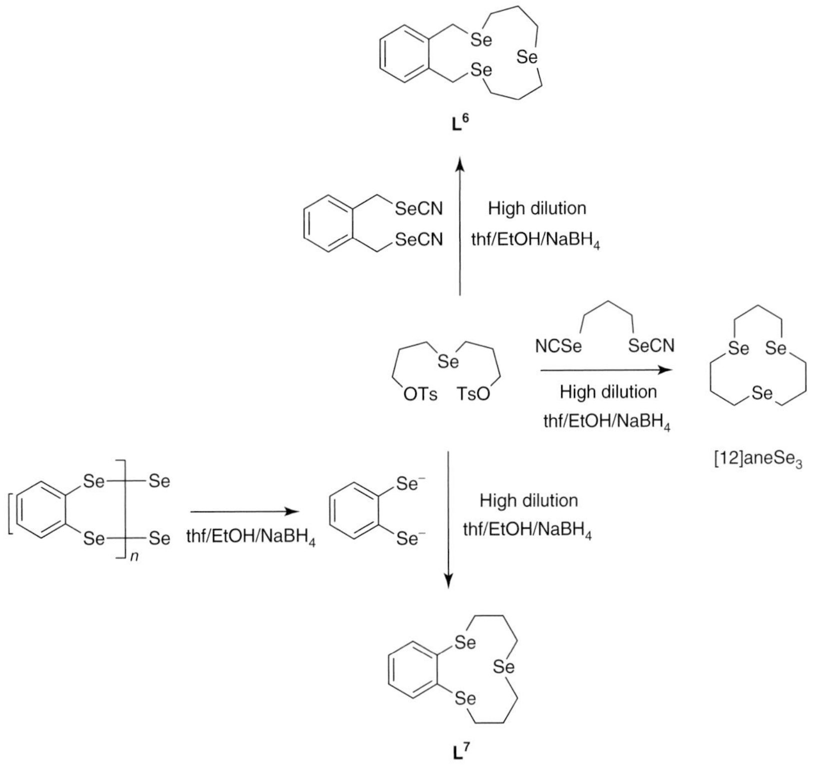

L^6

SeCN
SeCN
High dilution
thf/EtOH/$NaBH_4$

Se
OTs TsO
NCSe SeCN
High dilution
thf/EtOH/$NaBH_4$
$[12]aneSe_3$

thf/EtOH/$NaBH_4$
High dilution
thf/EtOH/$NaBH_4$

L^7

Scheme 3

LIVERPOOL JOHN MOORES UNIVERSITY
LEARNING SERVICES

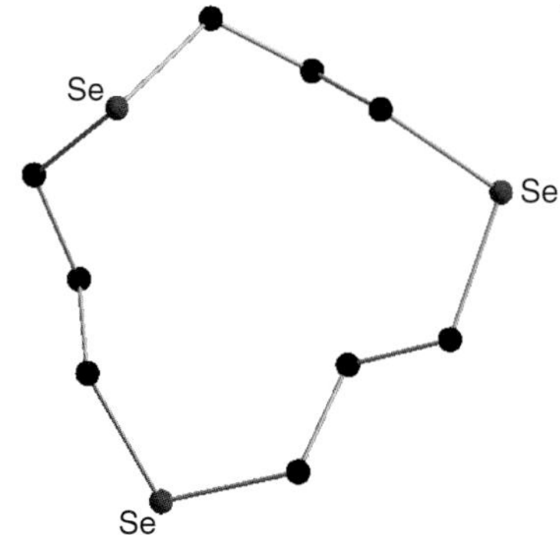

Figure 9 View of the structure of [12]aneSe_3.[33]

addition of thf/EtOH solutions of $Na_2[o\text{-}C_6H_4Se_2]$ (prepared freshly by $NaBH_4$ reduction of the $[o\text{-}C_6H_4Se_2]_n$ polymer) and $Se\{(CH_2)_3OTs\}_2$ to a suspension of $NaBH_4$ in thf/EtOH at room temperature. Selenium-77 NMR data showed significant dependence of the chemical shifts on the ring sizes and the nature of the groups linking the Se atoms. The crystal structure of $\mathbf{L^6}$ showed that the two Se atoms connected through the *o*-xylyl ring adopt *anti* positions, lying above and below the aromatic ring, while the macrocycle ring conformation led to the third Se atom pointing *exo* to the ring, minimizing lone-pair interactions (Figure 10).[34] The versatility of this cyclization method has also been demonstrated by its application to give Se_2N(pyridyl)-donor macrocycles $\mathbf{L^8}$ and $\mathbf{L^9}$ in essentially quantitative yield from 2,6-bis(bromomethyl)pyridine and either $o\text{-}C_6H_4(CH_2SeCN)_2$ or $NCSe(CH_2)_3SeCN$.[34]

$\mathbf{L^8}$ $\mathbf{L^9}$

The preparations of the cyclic ferrocenophane-based Se-donor macrocycles, 1,5,9-triselena[9]ferrocenophane ($\mathbf{L^{10}}$) and 1,5,9,21,25,29-hexaselena[9.9]ferrocenophane ($\mathbf{L^{11}}$) have been reported recently according to Scheme 4, and without the need for a templating agent, giving the rings in modest yields. Structure determinations confirmed

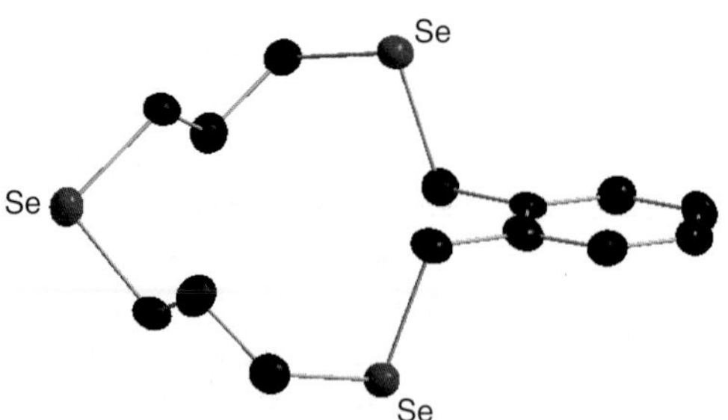

Figure 10 View of the structure of $\mathbf{L^6}$.[34]

(i) 8 $NaBH_4$/EtOH
(ii) Excess $Br(CH_2)_3Br$
Na_2Se/EtOH, reflux

$\mathbf{L^{10}}$ $\mathbf{L^{11}}$

Scheme 4

their identities (Figures 11 and 12), and showed the Se atoms lying *exo*. Cyclic voltammetry experiments showed chemically reversible one electron oxidations shifted only marginally from that of ferrocene itself.[35]

Several small to medium ring macrocycles containing mixed Se/O donor sets have been described. Thus, reaction of Na_2Se with $Cl(CH_2)_2O(CH_2)_2O(CH_2)_2Cl$ in liquid ammonia–thf gave a mixture of [9]aneSeO_2 and [18]aneSe_2O_4, albeit both in low yield.[36] The same reagents gave moderate amounts of [18]aneSe_2O_4 under high-dilution conditions in EtOH, and the macrocycle has also been characterized as its selenonium salt, through reaction with CH_3I, and as its $PtCl_2$ complex—Section 3.8.

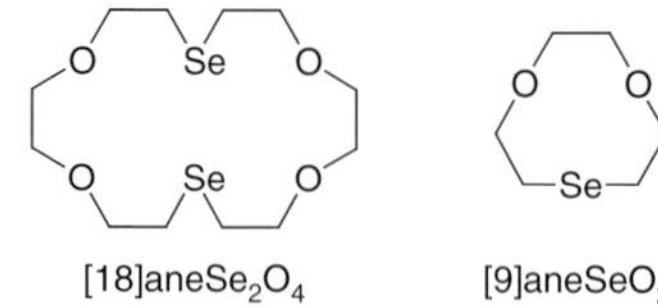

Selenium-rich rings have been prepared though high-dilution cyclization reactions as in Scheme 5..[37] Thus, reaction of $O(CH_2CH_2SeCN)_2$ with Na in liquid NH_3, followed by dropwise addition of a thf solution of $o\text{-}C_6H_4(CH_2Br)_2$ at $-40\,^{\circ}C$ produced all three mixed Se/O-donor macrocycles, $\mathbf{L^{12}}$, $\mathbf{L^{13}}$, and $\mathbf{L^{14}}$, formed by [1+1], [2+2], and [3+3] cyclizations, respectively. These were separated and purified by column chromatography. The [2+2] ring $\mathbf{L^{13}}$ was the dominant product under these conditions—it was suggested that this ring was promoted by the Na^+ acting as a template. In contrast, the same precursor compounds reacted under high-dilution conditions at room temperature with $NaBH_4$ in thf/EtOH to give only the [1+1] ring, $\mathbf{L^{12}}$.[37] The saturated Se/O-donor macrocycles, $\mathbf{L^{15}}$ and $\mathbf{L^{16}}$ were obtained similarly by simultaneous dropwise addition of

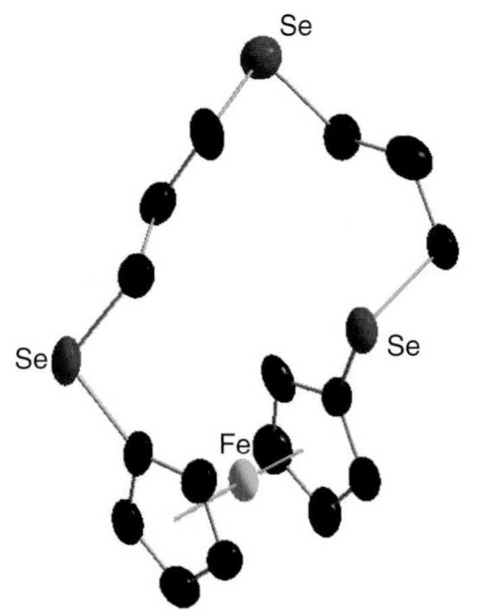

Figure 11 View of the structure of $\mathbf{L^{10}}$.[35]

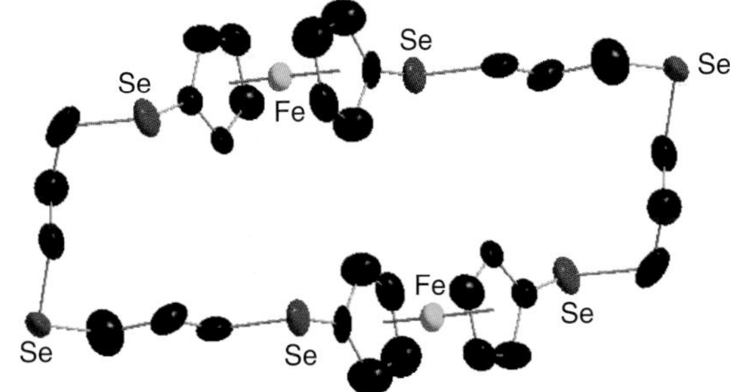

Figure 12 View of the structure of $\mathbf{L^{11}}$.[35]

solutions of $O(CH_2CH_2SeCN)_2$ and $Br(CH_2)_3Br$ to $NaBH_4$ suspended in thf/EtOH. The small tridentate Se_2O-donor ring, $\mathbf{L^{15}}$, was the major product under these conditions (71% yield), although the larger Se_4O_2-donor ring, $\mathbf{L^{16}}$,

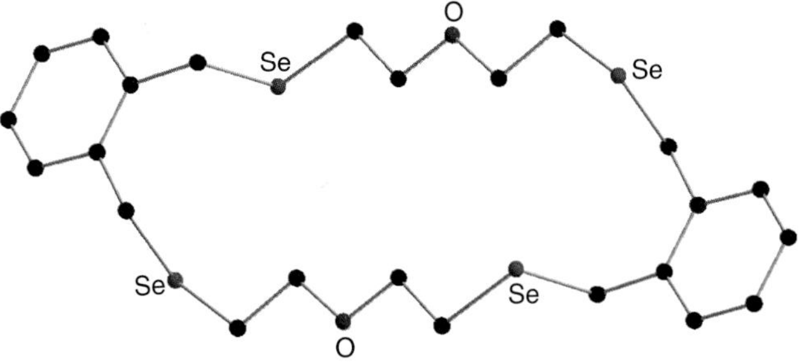

Figure 13 View of the structure of $\mathbf{L^{13}}$.[37]

was also obtained in low (8%) yield via [2+2] cyclization. The crystal structures of both $\mathbf{L^{12}}$ and $\mathbf{L^{13}}$ have been established (Figure 13), showing the Se-based lone pairs pointing *exo* to the ring and the C–Se–C bond angles more acute (~100°) compared to the C–O–C angles (~110°), consistent with increased p-orbital character in the Se–C bonds.[37]

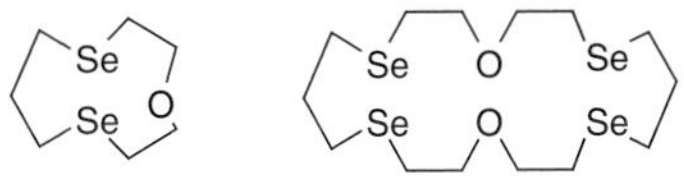

$\mathbf{L^{15}}$ $\mathbf{L^{16}}$

2.3 Telluroether macrocycles and mixed Te/O and Te/S crowns

Homoleptic telluroether macrocycles remain very elusive, and no new examples have been reported since earlier reviews. The mixed donor Te/O crown, $[18]aneTe_2O_4$ has been obtained in good yield (55%; $\delta(^{125}Te)$ 176) as a yellow, air-sensitive solid, by reaction of Na_2Te

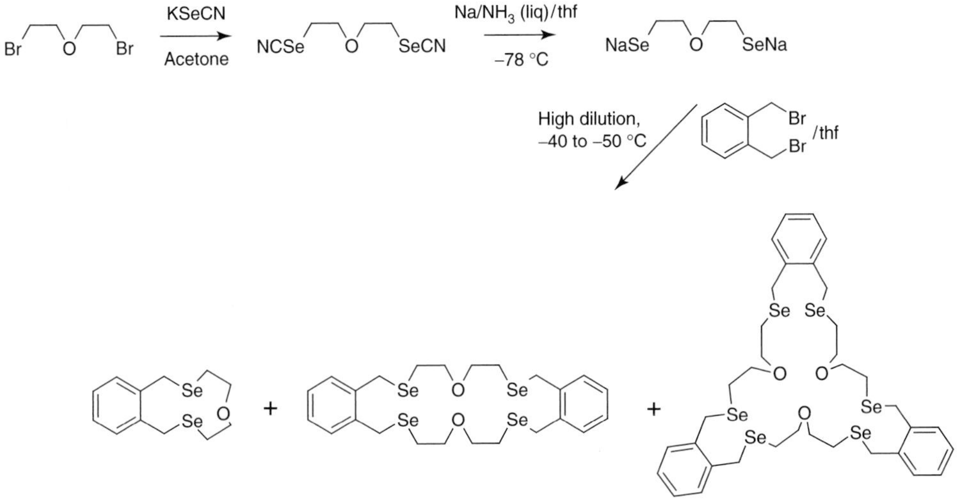

$\mathbf{L^{12}}$ $\mathbf{L^{13}}$ $\mathbf{L^{14}}$

Scheme 5

with $Cl(CH_2)_2O(CH_2)_2O(CH_2)_2Cl$ in liquid ammonia without the need for high-dilution conditions.[36] The yield of this compound was far superior to that of its Se analog (Section 2.2), attributed to the more nucleophilic Na_2Te leading to more complete substitution. There was supporting evidence that the Na^+ present in the reaction acts as a template favoring the [2+2] cyclization product; electrospray mass spectra obtained from solutions of [18]aneTe_2O_4 and NaCl in CH_2Cl_2–CH_3CN showed a strong multiplet as a result of $\{[18]aneTe_2O_4 + Na\}^+$ while other alkali metal cations did not give adducts.

The [1+1] ring, [9]aneTeO_2 was also obtained from this reaction as a minor by-product (4%; $\delta(^{125}Te) = 200$), separable by column chromatography.[36] The stability of the $-Te(CH_2)_2O-$ linkages in these compounds is noteworthy, since $-Te(CH_2)_2Te-$ readily eliminates $CH_2{=}CH_2$. Both [9]aneTeO_2 and [18]aneTe_2O_4 have been characterized as their stable telluronium derivatives through reaction with CH_3I, and the large ring as its Te(IV) tetrachloride (obtained through reaction with Cl_2).[36]

[18]aneTe_2O_4 [9]aneTeO_2

Several small ring mixed Te/S-donor macrocycles have been obtained from reaction of Na_2Te with various thioether reagents bearing terminal Cl groups—Scheme 6.[38,39] The structures of the 11- and 12-membered rings, [11]aneS_2Te and [12]aneS_2Te (Figure 14) showed *exocyclic* chalcogen atoms, with the C–Te–C angles (~94°) considerably smaller than the C–S–C angles (~100°), consistent with less s orbital character in the Te–C bonding. A crystal structure of the stable Te(IV) derivative [12]aneS_2TeI_2 (obtained from [12]aneS_2Te and diiodine) showed similar trends. The synthesis of [9]aneS_2Te was more sensitive to the precise reaction conditions than the larger rings and was accompanied by the formation of the ring-contraction products 1-thia-4-telluracyclohexane and 1,4-dithiane.[38,39]

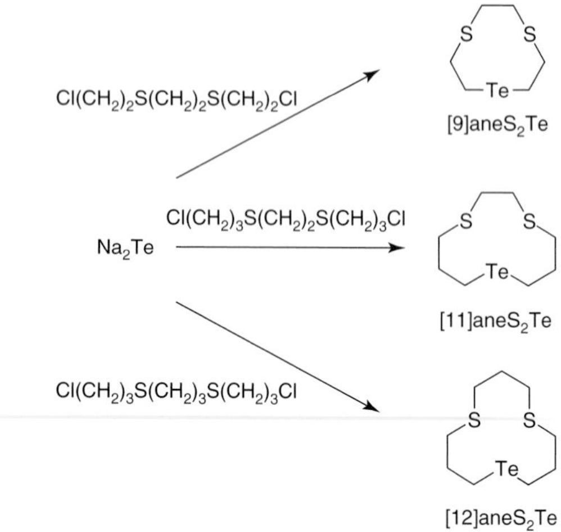

Scheme 6

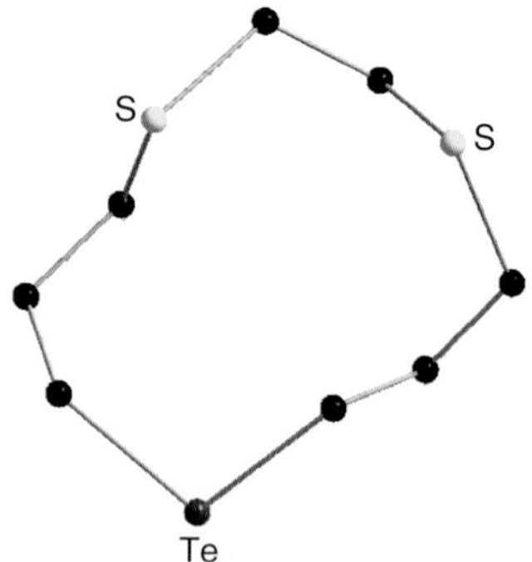

Figure 14 View of the structure of [11]aneS_2Te.[39]

3 HETERO-CROWN ETHER COMPLEXES WITH TRANSITION METALS

3.1 Group 3 and f-block complexes—Sc, Y, Ln, and An

The reaction of [9]aneS_3 with $[Sc(CH_2Si(CH_3)_3)_3(thf)_2]$ gave a high yield of *fac*-$[Sc(CH_2Si(CH_3)_3)_3([9]aneS_3)]$ (Figure 15), which is converted into the fluxional $[Sc(CH_2Si(CH_3)_3)_2([9]aneS_3)][BAr^F{}_4]$ ($Ar^F = C_6F_5$) by $[CPh_3][BAr^F{}_4]$. The latter adds thf to give *fac*-$[Sc(CH_2Si(CH_3)_3)_2([9]aneS_3)(thf)]^+$. These were the first examples of ethylene and α-olefin polymerization catalyst precursors based on a thia-macrocycle.[40,41] In contrast, attempts to isolate $[Y(CH_2Si(CH_3)_3)_3([9]aneS_3)]$ gave inseparable mixtures of the complex and starting materials. In anhydrous CH_3CN solution, [15]aneS_2O_3, $[ScCl_3(thf)_3]$, and $FeCl_3$ produce yellow $[ScCl_2([15]aneS_2O_3)][FeCl_4]$.[42] Although the structure was not determined, the ^1H NMR spectrum shows high-frequency shifts for *all* the methylene protons, suggesting that the macrocycle is pentadentate (S_2O_3), a conclusion supported by the ^{45}Sc NMR resonance which is some 70 ppm to high frequency of that in $[ScCl_2(15\text{-crown-}5)]^+$, consistent with sulfur coordination. The reactions of [9]aneS_3 with $[LaI_3(thf)_4]$ and $[UI_3(thf)_4]$ in CH_3CN gave, respectively, white $[LaI_3([9]aneS_3)(CH_3CN)_2]$ and green $[UI_3([9]aneS_3)(CH_3CN)_2]$; $[CeI_3([9]aneS_3)(CH_3CN)_2]$ was identified in solution.[43] The structures reveal distorted antiprismatic (8-coordinate) geometries and the bond lengths suggest stronger binding of the thia-macrocycle

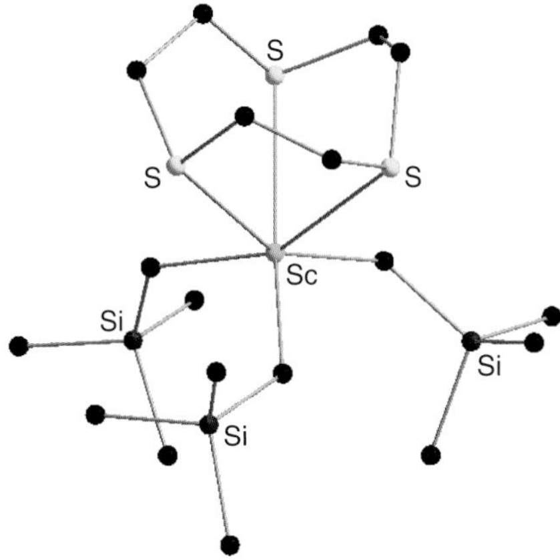

Figure 15 View of the structure of $[Sc(CH_2Si(CH_3)_3)_3([9]aneS_3)]$.[40]

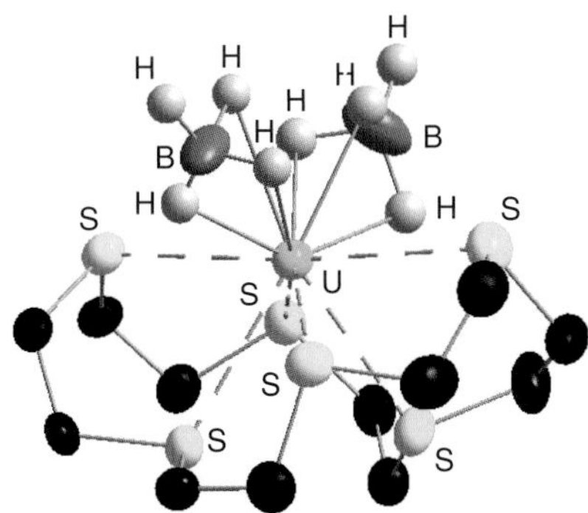

Figure 16 View of the structure of $[U([18]aneS_6)(BH_4)_2]^+$.[44]

to the U compared to La. Red $[U([18]aneS_6)(BH_4)_2]BPh_4$ was obtained from $[U(thf)_5(BH_4)_2]BPh_4$ and $[18]aneS_6$ in tetrahydrothiophene solution, and shown to contain cis κ^3-coordinated borohydride and hexadentate thia-crown (Figure 16).[44]

3.2 Group 4 complexes—Ti, Zr, and Hf

The reaction of TiX_4 (X = Cl, Br, or I) with $[n]aneS_3$ ($n = 9$ or 10) in anhydrous CH_2Cl_2 gave intensely colored, extremely moisture-sensitive powders, which were very poorly soluble in, or decomposed by, most solvents precluding growth of crystals for X-ray studies.[45] The spectroscopic data suggest a *fac*-$[TiX_3([n]aneS_3)]X$ constitution, which contrasts with the $[TiX_4\{k^2\text{-}CH_3C(CH_2SCH_3)_3\}]$ found with the tripodal trithioether. The reaction of $TiCl_4$, $[9]aneS_3$ and $SbCl_5$ gave $[TiCl_3([9]aneS_3)][SbCl_6]$.[45] The orange imido complex $[TiCl_2(N^tBu)([9]aneS_3)]$ contains 6-coordinate Ti with a markedly longer Ti–S bond trans to the imido group compared with $Ti\text{–}S_{transCl}$.[46] Both the UV–visible and the ^1H NMR spectra of the orange-yellow complex $[TiCl_4(\kappa^2\text{-}[15]aneS_2O_3)]$ showed that the macrocycle bonded via the thioether-S rather than the ether-O donor atoms to the hard oxophilic Ti(IV) center.[47] The distorted 7-coordinate $[ZrCl_4([9]aneS_3)]$ (Figure 17) is formed from $[ZrCl_4((CH_3)_2S)_2]$ and the ligand in anhydrous CH_2Cl_2.[48] The $[ZrCl_4([10]aneS_3)]$, $[HfCl_4([9]aneS_3)]$, and the yellow $[MI_4([9]aneS_3)]$ (M = Zr or Hf) were assigned similar structures based on spectroscopic data,[45,48,49] the 7-coordination contrasting with 6- or 8-coordination promoted by most acyclic polydentate ligands with group 4 tetrahalides.

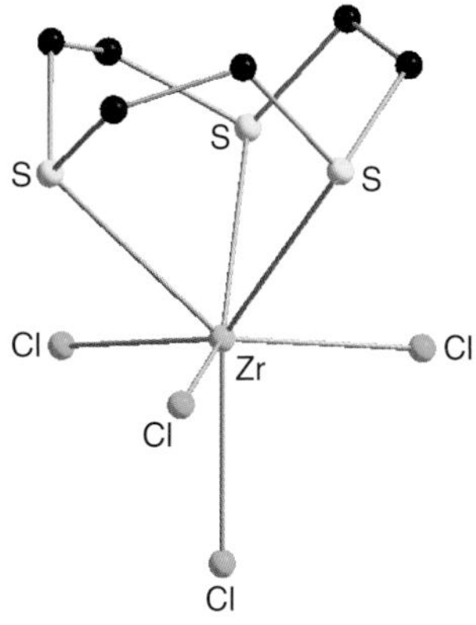

Figure 17 View of the structure of [ZrCl4([9]aneS3)].[48]

3.3 Group 5 complexes—V, Nb, and Ta

Dark red-brown $[VOCl_3([9]aneS_3)]$ and $[(VOCl_3)_2([18]aneS_6)]$ are formed from $VOCl_3$ and the ligands in anhydrous CH_2Cl_2.[50] The purple $[VOCl_2([9]aneS_3)][SbCl_6]$ was isolated in the presence of $SbCl_5$. Very poor solubility and solution instability has prevented growth of crystals for an X-ray study, although some structural data were obtained via vanadium K-edge extended X-ray absorption fine structure (EXAFS). The high-frequency shift in υ(VO) of $\sim 45\,cm^{-1}$ between $[VOCl_3([9]aneS_3)]$ and $[VOCl_2([9]aneS_3)][SbCl_6]$ suggests that the former is neutral with bidentate thia-macrocycle, while the latter is cationic, both containing 6-coordinate vanadium. Hydrolysis of $[VOCl_3([9]aneS_3)]$ produced the known $[VOCl_2([9]aneS_3]$ (blue).[50]

A comparative study of VCl_3 complexes with a range of crowns (12-crown-4, 15-crown-5 or 18-crown-6), thia-crown ($[12]aneS_4$) and thia-oxa-crowns ($[9]aneS_2O$, $[15]aneS_2O_3$, or $[18]aneS_3O_3$) revealed interesting differences in coordination behavior. While crown ethers gave $[VCl_3(\kappa^3\text{-crown})]$, $[VCl_3(\kappa^2\text{-crown})(H_2O)]$, and $[(VCl_3)_2(\kappa^3\kappa'^3\text{-18-crown-6})]$, the thia-crowns and thia-oxa-crowns gave only $[VCl_3(\kappa^3\text{-ligand})]$, all containing octahedral vanadium centers.[51] Most significantly, where a choice of O or S coordination was available (thia-oxa-crowns), softer S coordination was preferred. An analysis of X-ray structural data for complexes containing the chelate links $-OCH_2CH_2O-$ and $-SCH_2CH_2S-$ suggested that the short M–O distances result in considerable ring strain when macrocycles containing the former are κ^3-coordinated, whereas the longer M–S distances result in less acute chelate angles and hence less strain. The effect also rationalizes why $[VCl_3(\kappa^3\text{-crown})]$ are readily hydrolyzed to $[VCl_3(\kappa^2\text{-crown})(H_2O)]$, whereas the thia-crowns are much less sensitive to moisture.[51] The only new chemistry of Nb or Ta with thia-macrocycles is the synthesis and structural characterization of $[([9]aneS_3)H][NbF_6]$ described above (Section 2.1).[26]

3.4 Group 6 complexes—Cr, Mo, and W

The *fac*-tricarbonyls, $[M(CO)_3([n]aneS_3)]$ (M = Cr, Mo, or W, $n = 9$ or 10), are obtained from reaction of the thia-macrocycle with $[M(CO)_3(CH_3CN)_3]$ or $M(CO)_6$ and $NaBH_4$ in ethanol.[52] Detailed IR and multinuclear NMR studies comparing their properties to complexes of tripodal tridentates have been reported.[52] Chromium(III) chloride complexes of a range of crown, thia-crown, and thia-oxa-crown ligands have been reported[51]; the spectroscopic and structural data supporting the conclusions discussed in Section 3.3 for the vanadium analogs, about the preference for S- over O-coordination due to ring strain. Chromium trichloride complexes of triselena-crowns (those in Scheme 3) and Se_2N- and Se_2O-donor analogs have been prepared.[34] The Cr(V) complex $[Cr(N^tBu)Cl_2([9]aneS_3)][CF_3SO_3]$ has been prepared and its X-ray structure determined.[53]

Reaction of $[11]aneS_2Te$ with $[Mo(CO)_4(\text{norbornadiene})]$ produced *cis*-$[Mo(CO)_4([11]aneS_2Te)]$, the spectroscopic properties indicating the macrocycle was κ^2-STe coordinated, but the larger ring $[12]aneS_2Te$ decomposed on reaction with molybdenum carbonyl precursors.[39] The yellow $[MoO_2Cl_2([15]aneS_2O_3)]$ has cis-oxo, trans-chloro ligands and κ^2-S_2 coordinated macrocycle, and similar structures are probably present in $[MoO_2Cl_2([n]aneS_4)]$ ($n = 12$ or 14) although the poor solubility hindered spectroscopic studies.[54]

3.5 Group 7 complexes—Mn, Tc, and Re

Treatment of *fac*-$[Mn(CO)_3(CH_3CN)_3]^+$ with $[12]aneS_4$, $[14]aneS_4$, or $[15]aneS_5$ produced the corresponding *fac*-$[Mn(CO)_3(\kappa^3\text{-macrocycle})]^+$, which lose a carbonyl group on reaction with $(CH_3)_3NO$ to give rare *cis*-$[Mn(CO)_2(\kappa^4\text{-macrocycle})]^+$.[55] The spectroscopic data suggest high electron density on the manganese centers consistent with the thia-macrocycles bonding as σ-donors with minimal π interactions. *Fac*-$[Mn(CO)_3(\text{macrocycle})]^+$ (macrocycle = $[9]aneS_2Te$, $[11]aneS_2Te$, or $[12]aneS_2Te$) are formed from *fac*-$[Mn(CO)_3((CH_3)_2CO)_3]^+$ and the appropriate ligand.[39]

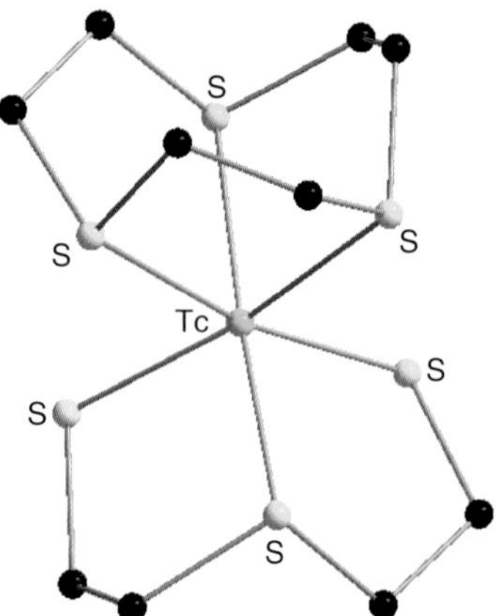

Figure 18 View of the structure of $[Te([9]aneS_3)\{S(CH_2)_2S(CH_2)_2S\}]^+$.[56]

$[9]aneS_3$ reacts with $[N^nBu_4][ReO_4]$ in acetic acid solution, followed by addition of $HBF_4{\cdot}Et_2O$ to give $[ReO_3([9]aneS_3)]BF_4$.[56] The reaction of $[N^nBu_4][MO_4]$ (M = Tc or Re) with $[9]aneS_3$, followed by $SnCl_2$ reduction, gives $[M([9]aneS_3)_2][BF_4]_2$, but attempted further reduction (e.g., with ascorbic acid or electrochemically) results in loss of ethene to give $[M([9]aneS_3)\{S(CH_2)_2S(CH_2)_2S\}][BF_4]$ (Figure 18).[56] This unusual cleavage of $[9]aneS_3$ which seems limited to Tc and Re has been investigated by density functional theory (DFT) calculations and seems to result from both $M(d\pi) \rightarrow C\text{–}S\ \sigma^*$ donation and the particularly suitable M t_{2g} orbital energies found in low-valent Tc and Re, but not in related metals.[57,58] Other $[9]aneS_3$ complexes of rhenium reported recently are $[ReBr(CO)_2([9]aneS_3)]$,[59] $[Re(CO)_3([9]aneS_3)][CF_3SO_3]$,[60] and $[Re(CO)_2(NO)([9]aneS_3)][CF_3SO_3]$.[61]

3.6 Group 8 complexes—Fe, Ru, and Os

Six-coordinate iron(II) complexes, $[Fe([11]aneS_3)_2][ClO_4]_2$ and $[Fe([n]aneS_6)][ClO_4]_2$, have been reported for two isomeric 11-membered ring thioethers, the 1,4,7- and 1,4,8-isomers of $[11]aneS_3$, and of the two hexathioethers, $[20]aneS_6$ and $[22]aneS_6$.[62,63] All the complexes are less stable to hydrolysis than the smaller ring analogs, and analysis of the UV/visible spectra shows that they exert weaker ligand fields than $[9]aneS_3$ or $[18]aneS_6$, although still produce low-spin complexes. Further X-ray crystal structures of the $[Fe([9]aneS_3)_2]^{2+}$ cation as ClO_4^-,[64] and BF_4^-[65] salts and in $[Fe([9]aneS_3)_2]Cl_2{\cdot}4H_2O$,[66] have been reported.

Comparison of the structure of $[Fe([9]aneS_3)_2][BF_4]_2$ with that of the PS_2-donor macrocycle analog $[Fe([9]anePS_2)_2][BF_4]_2$ ($[9]anePS_2$ = 1-phenyl-1-phospha-4,7-dithiacyclononane) reveal longer C–S bonds in the latter, consistent with the ready loss of ethene on heating.[65] The synthesis, structure, and detailed spectroscopic studies of $[Fe([9]aneS_3)_2][Fe(CN)_6]$ showed this to be a valence-trapped Turnbull's blue type Fe(II)–Fe(III) salt.[67]

Ring closure to form coordinated trithia-crowns $[n]aneS_3$ ($n = 8–12$) occurred on treatment of $[Ru(\eta^6\text{-}C_6(CH_3)_6)\{S(CH_2)_2S(CH_2)_2S\}]$ with $Br(CH_2)_yBr$ ($y = 1–5$), with the new ligands isolated as $[Ru(\eta^6\text{-}C_6(CH_3)_6)([n]aneS_3)][PF_6]_2$ (Figure 1).[17] X-ray structures for the complexes with $n = 8–11$ were reported and the ligand conformations compared. In contrast, treatment of $[Ru(\eta^5\text{-}C_5H_5)\{S(CH_2)_2S(CH_2)_2S\}]$ with $Br(CH_2)_2Br$ failed to achieve alkylation to form $[9]aneS_3$, although $[Ru(\eta^5\text{-}C_5H_5)([9]aneS_3)]PF_6$ was obtained from $[Ru(\eta^5\text{-}C_5H_5)Cl(PPh_3)_2]$ and $[9]aneS_3$.[68] Insertion of the diyne $(CH_3C{\equiv}CSCH_2CH_2)_2S$ into $[Ni(PEt_3)_2(benzyne)]$ resulted in the formation of the naphthalene-based trithiacyclononane, 2,3-$S(CH_2CH_2S)_2$-1,4-$(CH_3)_2C_{10}H_{14}$, which was characterized as its ruthenium complex $[Ru(PPh_3)Cl_2\{2,3\text{-}S(CH_2CH_2S)_2\text{-}1,4\text{-}(CH_3)_2C_{10}H_{14}\}]$ (Figure 19), which contains the *facially* coordinated trithia-macrocycle.[14] $[Ru([11]aneS_3)_2][PF_6]_2$,[62] $[RuL][ClO_4]_2$ (L = $[20]aneS_6$ and $[22]aneS_6$),[63] $[Ru([12]aneS_3)_2][CF_3SO_3]_2$,[69] and $[Ru((CH_3)_3\text{-}[9]aneN_3)([9]aneS_3)][CF_3SO_3]_2$[70] have been prepared. Ruthenium thia-crowns, including $[RuCl_2L([9]aneS_3)]$ and $[RuClL([n]aneS_4]PF_6$ (L = dmso (dimethylsulfoxide) or PPh_3), have been investigated as hydrogen-transfer reduction catalysts, although the activities were found to be generally rather poor.[71]

The past few years have seen a large amount of work using $Ru([9]aneS_3)$ or $Ru([n]aneS_4)$ as stable metal fragments from which to construct mixed ligand complexes, homo- and hetero-bimetallics, rings or larger units often via bridging polypyridyl type ligands. Mixed-valence dimers have been a particularly active area of research. Although the detailed structures, spectroscopic and redox properties vary with the thia-crown coligands, in many cases the interest focuses on the other ligands present, and thus detailed discussion of these complexes is not attempted here. The mononuclear complexes of $Ru([9]aneS_3)^{2+}$ include $[Ru([9]aneS_3)L_3][PF_6]_2$ (L = PhCN, pyrazine, pyridazine),[72] $[Ru([9]aneS_3)(L')(1,10\text{-phenanthroline})][ClO_4]_2$ (L′ = H_2O, CH_3CN, pyridine, Bu^tNC),[73–75] $[Ru([9]aneS_3)(1,10\text{-phenanthroline})(C{\equiv}CPh)]PF_6$,[76] $[RuCl([9]aneS_3)(dmso\text{-}S)_2]PF_6$, $[Ru([9]aneS_3)(dmso\text{-}S)_2(dmso\text{-}O)]^{2+}$, $[Ru([9]aneS_3)(dmso\text{-}S)(dmso\text{-}O)_2]^{2+}$,[77] $[RuCl([9]aneS_3)L_2]^+$ (L_2 = bidentate polypyridyl ligand),[78,79] $[RuCl_2([9]aneS_3)(PTA)]$, $[RuCl([9]aneS_3)(PTA)_2][CF_3SO_3]$, (PTA = 1,3,5-triaza-7-phosphaadamantane),[80] $[RuCl([9]aneS_3)(diamine)][CF_3SO_3]$,[80] and $[Ru([9]aneS_3)(dmso)(carboxylate)]$ (carboxylate = oxalate, malonate, etc.).[81] Dinuclear complexes include $[\{Ru([9]aneS_3)(dmso)\}_2(\mu\text{-oxalate})][CF_3SO_3]_2$, $[Ru([9]aneS_3)(dmso)(\mu\text{-oxalate})Ru(dmso)_3(CF_3SO_3)][CF_3SO_3]$,[81] and $[\{RuCl([9]aneS_3)\}_2(\mu\text{-}L)][PF_6]_2$ (L = various polypyridyl ligands) (Figure 20).[82] Cyclic trinuclear units are present in $[\{Ru([9]aneS_3)\}_3L_3][PF_6]_3$ (L = 9-methyladenine, adenosine) which can be electrochemically oxidized to four mixed-valence species.[83,84]

$[RuCl([14]aneS_4)\{NC(CH_2)_3Si(OEt)_3\}]PF_6$[85] has been prepared from the corresponding CH_3CN complex, and tethered to mesoporous silica via the silylnitrile group. 4-Ferrocenylpyridine and ferrocenyl-4-pyridylacetylene (L) complexes $[RuCl([14]aneS_4)(L)]PF_6$ have been prepared by displacement of dmso from $[RuCl([14]aneS_4)(dmso)]Cl$; electrochemical and spectroscopic studies suggest poor communication between the metal centers.[86] There are numerous examples of *cis*-$Ru([n]aneS_4)^{2+}$ units bridged by nitrogen heterocycles, including $[\{RuCl([12]aneS_4)\}_2(\mu\text{-}4,4'\text{-bipyridyl})][PF_6]_2$,[87] $[\{RuCl([12]aneS_4)\}_2(\mu\text{-pyrazine})][PF_6]_2$,[69] $[\{RuCl([12]aneS_4)\}_2(\mu\text{-bipyrimidine})][PF_6]_2$,[88] $[\{Ru([n]aneS_4)\}_2(\mu\text{-}2,$

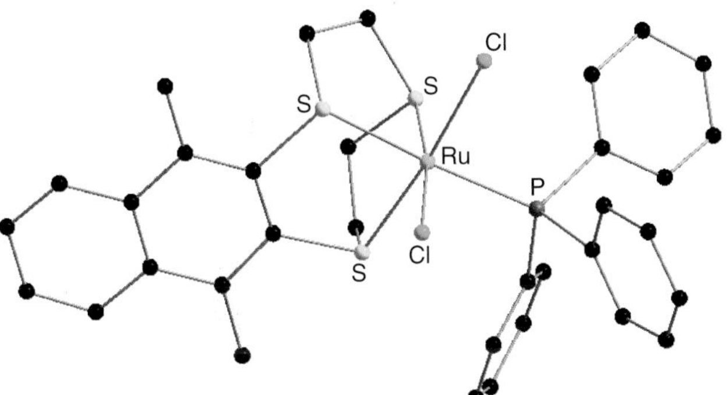

Figure 19 View of the structure of $[Ru(PPh_3)Cl_2\{2,3\text{-}S(CH_2CH_2S)_2\text{-}1,4\text{-}(CH_3)_2C_{10}H_{14}\}]$.[14]

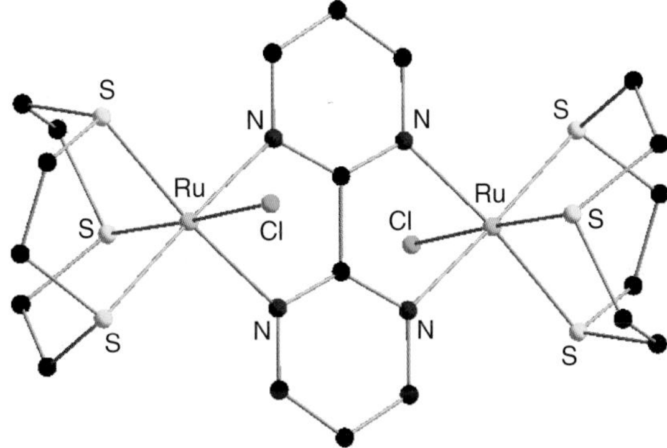

Figure 20 View of the structure of $[\{RuCl([9]aneS_3)\}_2(\mu\text{-}L)]^{2+}$ where L = 2,2′-bipyrimidine.[82]

3-bis(2-pyridyl)pyrazine)][PF_6]$_4$ ($n = 12$, 14, 16),[89] [{Ru([n]aneS_4)}$_2$(μ-3,6-bis(2-pyridyl)-1,2,4,5-tetrazine)][PF_6]$_4$ ($n = 12$, 14),[90] [{RuCl([n]aneS_4)}$_2${μ-2,3-bis(2-pyridyl)-pyrazine}][PF_6]$_2$ ($n = 12$, 14, 16),[89] [{RuCl([n]aneS_4)}$_2$ (μ-L)][PF_6]$_2$ ($n = 12$, 14, 16; L = pyrazine, 4,4′-bipyridyl, 3,6-bis(2-pyridyl)-1,2,4,5-tetrazine).[91] Detailed electrochemical, spectroscopic, and computational studies have been carried out on many of these complexes to establish the site(s) of electron loss/gain and to understand the subtle interplay of factors involved. Most of the mixed-valence systems belong to class II of the Robin and Day scheme. A series of complexes [RuCl([9]aneS_3)(L)$_2$] PF_6 (L = nicotinamide or isonicotinamide) and [Ru([n]andS_4)(L)][PF_6]$_2$ ($n = 12$, 14, 16), which are linked together by amide H-bonds into chains or tapes, have been structurally and spectroscopically characterized.[92]

Detailed structural, spectroscopic, and computational studies on [Ru([n]aneS_4)(2,2′-bipyridyl)][PF_6]$_2$ have concluded that the conformations of the macrocycles are determined by steric rather than electronic effects in these crowded cations and that the dynamic processes in solution previously attributed to reversible dissociation are in fact because of pyramidal sulfur inversion.[93]

Various series of mixed ligand [9]aneS_3 and [n]aneS_4 complexes have also been investigated for their biological activity as DNA intercalators, protein kinase inhibitors, and as anticancer agents.[78–80,94–96]

Little new chemistry of osmium thia-macrocycles has appeared, but both ruthenium and osmium [9]aneS_3 complexes with thiocarbonyl ligands and their migratory coupling reactions have been investigated.[97]

3.7 Group 9 complexes—Co, Rh, and Ir

Red-brown cobalt(II) complexes, [Co([11]aneS_3)$_2$][BF_4]$_2$, of both isomers of [11]aneS_3 have been prepared from nitromethane solutions of [Co(H_2O)$_6$][BF_4]$_2$ and the thiacrown using acetic anhydride as dehydrating agent.[62] The complexes are low-spin and very moisture sensitive. Oxidation with peroxodisulfate gives the orange Co(III) analogs.[62] Co(II) and Co(III) complexes of [22]aneS_6 have been prepared and, although the Co(II) complex is low spin, analysis of the UV/visible spectra of both suggest the ligand field exerted by this thia-crown is less than by most other hexathia-macrocycles.[63] The structure of the Co(III) polyiodide, [Co([9]aneS_3)$_2$]I_{11} has been determined; the cation is symmetrical octahedral as expected and the anions are three I_3^- groups linked by an I_2 molecule into 11-membered rings.[98] The complex [Co([9]aneS_3)$_2$][TCNQ]$_2$ is a one-dimensional antiferromagnet, while [Co([9]aneS_3)$_2$][TCNQ]$_3$ is a paramagnetic semiconductor.[99] The low-spin ($\mu = 1.77$ B.M.) purple [Co([9]aneS_2O)$_2$][BF_4]$_2$ is oxidized by $NOBF_4$ to the pink diamagnetic [Co([9]aneS_2O)$_2$]$^{3+}$.[100] In contrast, crimson-red [Co([18]aneS_4O_2)][BF_4]$_2$ is high spin ($\mu = 4.73$ B.M.) showing the larger ring produces a smaller ligand field. It too is oxidized by $NOBF_4$ to green diamagnetic [Co([18]aneS_4O_2)]$^{3+}$.[100]

Rhodium(III) and iridium(III) complexes with pentamethylcyclopentadienyl (Cp*) ligands [M(η^5-Cp*)([n]aneS_3)][PF_6]$_2$ ($n = 9$ or 10, M = Rh or Ir) have been prepared by reaction of the thiamcrocycle with [{M(η^5-Cp*)Cl_2}$_2$] in CH_3CN followed by metathesis with NH_4PF_6.[101] Crystal structures were reported for both Rh complexes and [Ir(η^5-Cp*)([9]aneS_3)][PF_6]$_2$, which reveal shorter M–S bonds than in the [M([n]aneS_3)$_2$]$^{3+}$, but curiously, only a single irreversible reduction wave is found in the cyclic voltammograms, attributed to M(III)/M(II) couples. Rhodium(III) complexes of [9]aneS_3 with polypyridyl coligands are active cytotoxic agents toward a variety of cancer cell lines.[102,103] The synthesis and crystal structures of *cis*-[Rh([12]aneS_4)Cl_2] PF_6 and *trans*-[Rh([16]aneS_4)(H_2O)Cl][CF_3SO_3]$_2$ have been reported.[104]

The reaction of [18]aneTe_2O_4 with $RhCl_3{\cdot}3H_2O$ in EtOH gave orange-yellow [$RhCl_2$([18]aneTe_2O_4)$_2$]Cl, which probably contains a Te_4Cl_2 donor set, although its poor solubility prevented NMR studies.[36] Distorted octahedral complexes [Rh(Cp*)([n]aneS_2Te)][PF_6]$_2$ ($n = 9$, 11, 12) were obtained in good yield from [{Rh(Cp*)Cl_2}$_2$], the ligands and $TlPF_6$.[39]

3.8 Group 10 complexes—Ni, Pd, and Pt

The X-ray structures of three [9]aneS_3 complexes of Ni(II) have been determined; [Ni([9]aneS_3)$_2$]$Br_2{\cdot}4H_2O$ in which the extensive H-bonding leads to double chain arrangements,[105] and two polyiodides, [Ni([9]aneS_3)$_2$][I_2]$_2$ and [Ni([9]aneS_3)$_2$][I_5]$_2$.[98] Like their Co(II) analogs (q.v.), [Ni([9]aneS_3)$_2$][TCNQ]$_2$ and [Ni([9]aneS_3)$_2$][TCNQ]$_3$ exhibit magnetic cooperativity effects.[99] Very moisture-sensitive crystals of [Ni([11]aneS_3)$_2$][BF_4]$_2$ (both 1,4,7- and 1,4,8-isomers) have been prepared, and shown to exhibit ligand field strengths intermediate between [9]aneS_3 and [12]aneS_3.[62] The large rings [20]aneS_6 and [22]aneS_6 form hydrolytically unstable [Ni([n]aneS_6)][ClO_4]$_2$ complexes and careful comparison of the UV/visible spectra show the ligand field strengths in NiS_6 chromophores fall: [18]aneS_6 > [20]aneS_6 > [22]aneS_6 > [24]aneS_6, and that the fields generated by the hexathia-crowns are smaller than that from two of the corresponding trithia-crowns (i.e., {[11]aneS_3}$_2$ > [22]aneS_6).[63] The thia-oxa-crowns [9]aneS_2O and [18]aneS_4O_2 have been complexed with

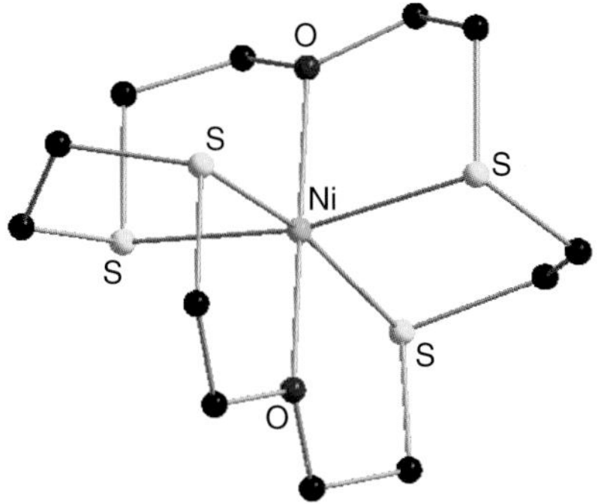

Figure 21 View of the structure of $[Ni([18]aneS_4O_2)]^{2+}$.[100]

nickel(II) fluoroborate under anhydrous conditions.[100] The structures of both octahedral nickel cations have been determined. That of $[Ni([18]aneS_4O_2)][BF_4]_2$ (Figure 21) shows trans-oxygen donors and rather short N–S bonds, although analysis of the UV/visible spectra show it exerts a weaker field than $[18]aneS_6$.[100] However, in $[Ni([9]aneS_2O)_2][BF_4]_2$, the ligand is disordered.

The chemistry of palladium and platinum with $[9]aneS_3$ continues to attract large research effort. Crystal structures have been reported for $[Pt([9]aneS_3)X_2]$ (X = Cl, Br, I),[106] which show an elongated square pyramidal geometry (Figure 22), with the long axial Pt–S interaction falling with halide: Cl > Br > I. In $[Pt([9]aneS_3)_2][Y]_2{\cdot}2CH_3NO_2$ ($Y = BF_4$, CF_3SO_3, PF_6),[107] the platinum shows four Pt–S in the plane (~2.3 Å) and two longer axial interactions (~3 Å) giving an overall $S_4 + S_2$ coordination, which contrasts with the $S_4 + S$ coordination in $[Pt([9]aneS_3)_2][PF_6]_2$,[108] showing the crystal packing has a significant effect on the structure. Under high pressure, the $[Pd([9]aneS_3)Cl_2]$ ($S_2Cl_2 + S$ coordination) converts into an intensely colored chain polymer with a distorted octahedral structure.[109] A very large number of heteroleptic complexes based on the $Pt([9]aneS_3)^{2+}$ core have been prepared to explore the effects of different donors on the axial Pt···S bond length and on the electrochemical and spectroscopic properties. Lack of space precludes detailed description of this chemistry, but $[Pt([9]aneS_3)L_2]^{2+}$ examples include: L_2 = 2,2′-bipyridyl, 4,4-$(CH_3)_2$-2,2′-bipyridyl, $(CH_3)_4$-1,10-phenanthroline,[110,111] substituted 2,2′-bipyridyls,[112] $2PPh_3$, $2AsPh_3$, $2SbPh_3$, Cl/PPh_3, $Cl/AsPh_3$, $Cl/SbPh_3$,[113] various diphosphines and diarsines,[113–117] cyclometallated (*NC* or *PC*) ligands,[118] 1,2-diaminoethane.[119] Di- and tetra-metallic complexes are formed using bridging N-donor bidentates.[104,120] An unusual Pt_2Ag complex $[\{Pt(2\text{-}C,N\text{-}phenylpyridine)([9]aneS_3)\}_2Ag(CH_3CN)_2][PF_6]_3$ with two Pt–Ag bonds was among the products of reacting $[\{Pt(2\text{-}C,N\text{-}phenylpyridine)Cl\}_2]$, $AgPF_6$, and $[9]aneS_3$ in CH_3CN solution.[121] The structures of $[Pt([9]aneS_3)L_2]^{2+}$ ($L = AsPh_3$, $SbPh_3$) are distorted trigonal bipyramids, contrasting with the square pyramidal geometry usually present.[113] The $[Pt([9]aneS_3)(SbPh_3)_2]^{2+}$ decomposes into $[Pt([9]aneS_3(SbPh_3)(\sigma\text{-}Ph)]^+$ in CH_3NO_2 solution.[122] Examples based on palladium are fewer,[104,113,114,117,118] but include the cyclometallated azobenzene complex $[Pd(C_6H_4N{=}NCC_6H_5)([9]aneS_3)][PF_6]$.[123]

The crystal structure and detailed spectroscopic data have been reported for the mononuclear octahedral Pt(III) complex $[Pt([9]aneS_3)_2][PF_6]_3$ (Figure 23).[124] The structure is consistent with a d^7 Pt(III) description and the EPR spectrum shows significant delocalization of the unpaired electron onto the thioether S-donor atoms.

Further studies on trithia-crown complexes with larger rings have been reported, including $[Pd([10]aneS_3)Cl_2]$,[106] $[Pt([10]aneS_3)(PPh_3)Cl]PF_6$,[125] and $[Pt([18]aneS_6)][BF_4]_2$.[107] Complexes of the larger rings $[Pd([11]aneS_3)_2][PF_6]_2$,[62] $[Pt([12]aneS_3)_2][PF_6]_2$, $[Pd([12]aneS_3)_2][BF_4]_2$,[126] and $[Pd([22]aneS_6)][PF_6]_2$,[63] which lack the two long axial M···S interactions found with smaller rings, do not exhibit M(II)/M(III) redox couples. Structures of the polyiodides $[Pd([12]aneS_4)]I_6$ and $[Pd([14]aneS_4)]I_{10}$ have been reported.[98] Liquid crystalline materials based on functionalized palladium(II)-$[n]aneS_4$ (n = 14 or 16) scaffolds have been synthesized.[127] Maleonitrile linkages have also been incorporated into tetrathia-crowns and their Pd(II) complexes obtained.[128]

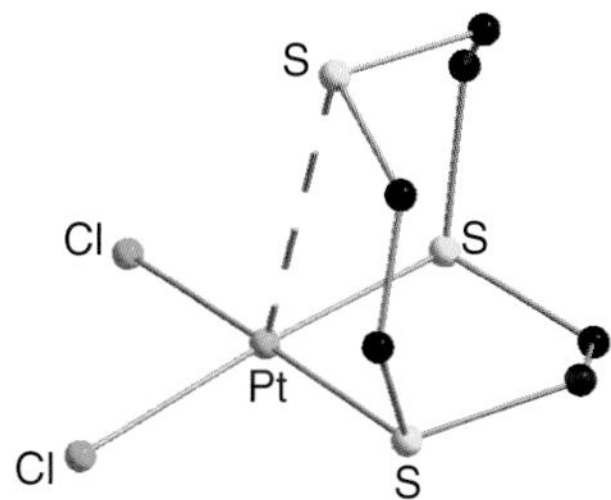

Figure 22 View of the structure of $[PtCl_2([9]aneS_3)]$.[107]

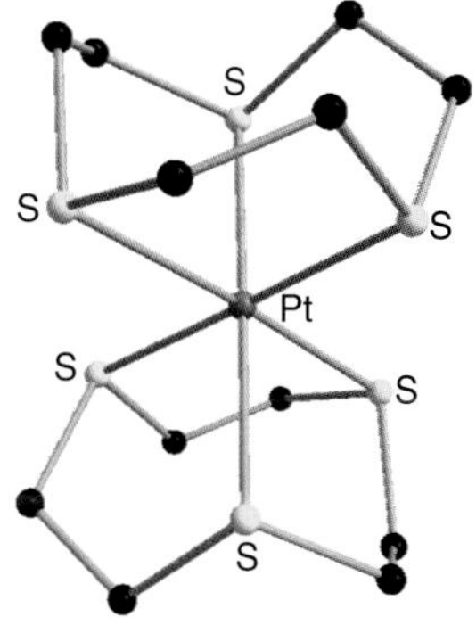

Figure 23 View of the structure of $[Pt([9]aneS_3)_2]^{3+}$.[124]

Considerable effort has been devoted to thia-oxa macrocyclic complexes. The small ring S_2O ligand *o*-$C_6H_4\{CH_2S(CH_2)_2\}_2O$ (L) is S,S′-coordinated in $[Pd(L)Y_2]$ (Y = SCN or NO_3).[129] A range of 14- and 15-membered ring S_2O_2 macrocycles have been complexed with Pd(II) and Pt(II)[130–135]—in all cases, only the sulfur donor atoms are coordinated. Examples of S_3O_2 and S_2O_4 oxa-thia-crowns are also known, again only the sulfur coordinates, but the ring conformations adopted are highly variable.[136–139] Maleonitrile thia-oxa-crowns with S_2O_x ($x = 2$–5) have been complexed with both Pd(II) and Pt(II); again coordination is via S.[140,141]

$[Pt(CH_3)_2(S(CH_3)_2)_2]$ reacts with [16]aneSe_4 to form $[Pt(CH_3)_2(\kappa^2$-[16]ane$Se_4)]$ and with $Pt(CH_3)_3I$ it gives $[Pt(CH_3)_3I(\kappa^2$-[16]ane$Se_4)]$; the latter can be converted into $[Pt(CH_3)_3(\kappa^3$-[16]ane$Se_4)]PF_6$ by treatment with $TlPF_6$.[142] The triselenoether macrocycles in Scheme 3 form poorly soluble $[PtCl_2(\kappa^2$-$Se_3)]$, but with $Pt(CH_3)_3I$ the products are the readily soluble $[Pt(CH_3)_3(\kappa^3$-$Se_3)]I$ (Figure 24).[34] The Se_2O-, Se_2O_4-, and Se_4O_2-donor selena-oxa-crowns (Scheme 5) coordinate via the selenium donors only in $[(MCl_2)_2(Se_4O_2)]$, $[MCl_2(Se_2O_4)]$ or $[MCl_2(Se_2O)]$ (M = Pd or Pt).[36,37] Similarly, the [9]aneTeO_2 and [18]aneTe_2O_4 ligands bind only via the tellurium in $[MCl_2([9]aneTeO_2)_2]$ and $[MCl_2([18]aneTe_2O_4)]$—confirmed both by the structures of $[MCl_2([18]aneTe_2O_4)]$ (Figure 25) and by multinuclear NMR studies.[36]

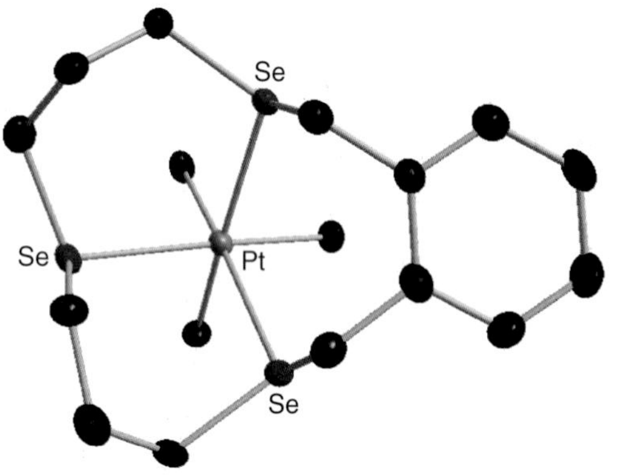

Figure 24 View of the structure of $[Pt(CH_3)_3(\kappa^3$-$Se_3)]^+$.[34]

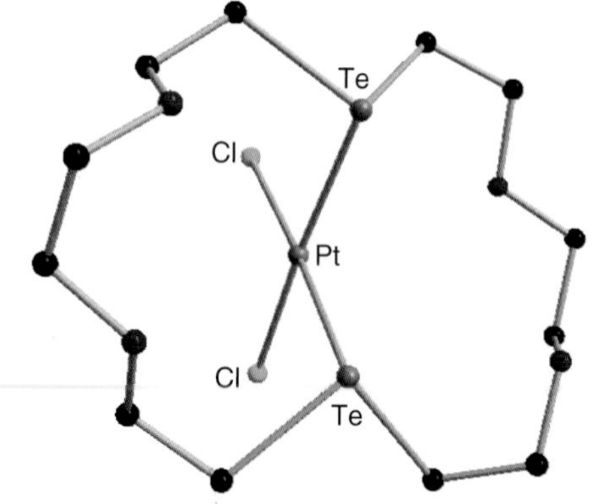

Figure 25 View of the structure of $[PtCl_2([18]aneTe_2O_4)]$ from.[36]

The presence of two soft donor types in [*n*]aneS_2Te (n = 11 or 12) could produce STe or S_2 coordinated isomers in $[MCl_2([n]aneS_2Te)]$, but the ^{195}Pt and ^{125}Te NMR spectra show only the STe forms are present.[39]

3.9 Group 11 complexes—Cu, Ag, and Au

$[Cu([9]aneS_3)L]PF_6$ (L = CO, C_2H_4) were prepared from $[Cu(CH_3CN)_4]PF_6$, [9]aneS_3, and L in acetone solution and characterized by vibrational spectroscopy and Cu K-edge EXAFS.[143] The electron-transfer kinetics and speciation in the Cu(II/I)-[9]aneS_3 systems have been reinvestigated.[144] The binuclear cyano-bridged complexes $[([9]aneS_3)Cu(CN)Cu([9]aneS_3)]BF_4$ and $[([9]aneS_3)Cu(CN)Cu([9]aneS_3)][TCNQ]_2$ have been prepared, the latter showing strong antiferromagnetic coupling.[145] Copper(I) halides form a variety of products with [12]aneS_4 or [16]aneS_4 depending on the Cu:ligand ratio used.[146] The structures of $[Cu([12]aneS_4)X]$ (X = Br or I) are square pyramidal (S_4X), while for X = Cl, the structure is tetrahedral (S_3Cl). Higher Cu:L ratios give $[Cu_4Br_4([12]aneS_4)_2]$, $[Cu_4I_4([12]aneS_4)]$, and $[Cu_2I_2([16]aneS_4)]$, all with different polymeric motifs. The synthesis, structures, and electron transfer kinetics have been reported for several series of copper complexes based on modified [14]aneS_4 ligands.[147–150]

The structure of $[CuCl_2([18]aneS_6)]$ shows a tetragonally elongated octahedral copper environment (S_4Cl_2) with bridging thia-crowns forming an infinite polymer chain.[151] In $[Cu([21]aneS_6)]ClO_4$ the Cu(I) is tetrahedrally coordinated, and electrochemical studies show this ligand strongly stabilizes Cu(I) compared to Cu(II).[152] The Cu(I) complex of an S_5 macrocycle involving a thiophene unit (4,7,10,13,19-pentathiabicyclo[14,2,1]-nonadeca-1(18),16-diene) has been structurally characterized, and shows a distorted tetrahedral (S_4) coordination with the thiophene group uncoordinated.[153]

The brown $[Cu([9]aneS_2O)_2][BF_4]_2$ contains a tetragonally distorted octahedral copper environment with equatorial S_4 coordination and long axial Cu–O bonds.[100] The structure of $[Cu([18]aneS_4O_2)][ClO_4]_2$ is similar, but $[Cu([18]aneS_4O_2)][ClO_4]$ is as expected tetrahedral (S_4).[100,150] S_2O_2 thia-oxa macrocycles with 14-, 17-, or 18-membered rings give a range of coordination polymers with CuX (X = Cl, Br, I, or CN), which contain S and X bridges.[100,154,155] Dark green [Cu(dibenzo-[17]aneS_3O_2)

$(H_2O)(ClO_4)]ClO_4$ contains a square pyramidal copper center coordinated to three sulfur donors and two oxygens from the water and the perchlorate ion.[156] However, the S_2O_3 analogs (and ligands with 16- and 18-membered rings) bring about reduction to form the tetrahedral $[Cu(S_2O_3)_2]ClO_4$ with CuS_4 centers.[156] Copper(I) halides produce mono-, di-, and polymeric complexes with various S_3O_2 crowns; the polymers have both X and crown ligand bridges.[157,158] $K[Cu_3I_4([18]aneS_3O_3)]$, containing 1,7,13-trithia-4,10,16-trioxacyclooctadecane, is built up of anionic sheets linked into a three-dimensional network by K^+ cations.[159]

New selena-crown complexes include the brown $[Cu([16]aneS_2Se_2)(O_3SCF_3)_2]$ with a disordered planar S_2Se_2 core and coordinated triflate ions completing 6-coordination.[33] The complex decomposes readily in solution to white tetrahedral $[Cu([16]aneS_2Se_2)][O_3SCF_3]$. The complexes $[Cu\{[16]aneSe_4(OH)_2\}(O_3SCF_3)_2]$ and $[Cu\{[16]aneSe_4(OH)_2\}][O_3SCF_3]$ were also fully characterized.[33] Copper(I) complexes of the selena-oxa-crowns $\mathbf{L^{12}}$ and $\mathbf{L^{13}}$ have been described.[37] Tellurium-containing examples are $[Cu([18]aneTe_2O_4)_2]BF_4$, $[Cu_2([18]aneTe_2O_4)][BF_4]_2$,[36] $[Cu([11]aneS_2Te)]BF_4$, and $[Cu([12]aneS_2Te)]BF_4$[39]; the last three are probably polymeric.

Both thia- and thia-oxa-crowns readily generate polymeric arrays with silver(I) centers. Crystallization of $[Ag([9]aneS_3)]O_3SCF_3$ from nitromethane gives the tetranuclear $[Ag_4([9]aneS_3)_4][O_3SCF_3]_4{\cdot}2CH_3NO_2$ (Figure 26) in which each silver is S_4-coordinated.[160] Thia-crowns also bridge between silver centers to produce polymeric structures in $[Ag([9]aneS_3)(hfpd)]$ and $[Ag_2([14]aneS_4)_2(hfpd)]$ (Hhfpd = hexafluoro-2,4-pentanedione), where the diketone blocks coordination sites, leading to one-dimensional polymers.[161] Changing the diketonate ligand R groups from $2 \times CH_3$ to ${}^tBu$ and C_3F_7 (Hfod) produces $[Ag([9]aneS_3)(fod)]$ which is a discrete monomer, although $[Ag_2([14]aneS_4)_2(fod)]$ is polymeric.[162] In $[Ag([16]aneS_4)(NO_3)]$, each Ag is coordinated to an S donor from three different thia-crowns and to the nitrate anion to give a three-dimensional network.[13] In contrast, mononuclear S_6 coordination is found in (1,4,7,11,14,17-hexathia-9,9,19,19-tetrahydroxymethylcycloicosane)silver(I) nitrate.[163]

Tetrathia-crowns with 12- and 13-membered rings containing a maleonitrile link form a range of Ag(I) complexes.[16,164] In $[Ag(\mathbf{L^4})_2]BF_4$, the silver is coordinated by eight S-donor atoms in a discrete cation with a distorted cube structure (Figure 27), whereas in $[Ag(\mathbf{L^5})_2]BF_4$, the silver is square pyramidally coordinated (Figure 28). In $[Ag_2(\mathbf{L^5})_3][PF_6]_2$, two square pyramidal AgS_5 units are linked by a bridging thia-crown. Silver complexes of conformationally restricted thia-crowns with unsaturated linkages have also been reported.[15,165] Macrobicyclic ligands with six thioether donors have been complexed to Ag(I), although the ligands usually adopt *exo*-coordination rather than encapsulation of the silver cation.[20]

The royal blue Ag(II) complex, $[Ag([18]aneS_6)][ClO_4]_2$, which contains an octahedral AgS_6 center (Figure 29), has been studied by multifrequency EPR spectroscopy

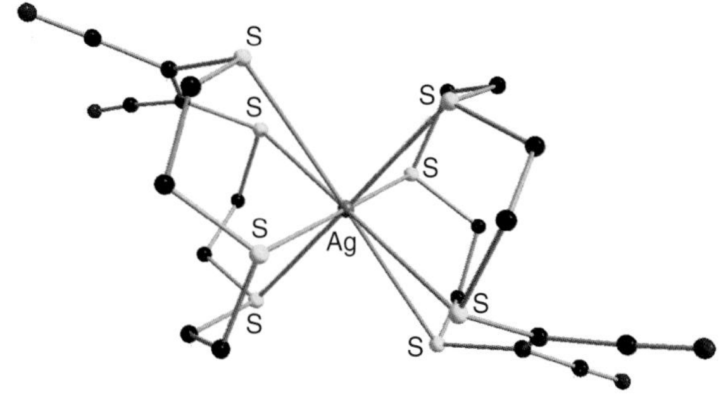

Figure 27 View of the structure of $[Ag(\mathbf{L^4})_2]^+$.[16]

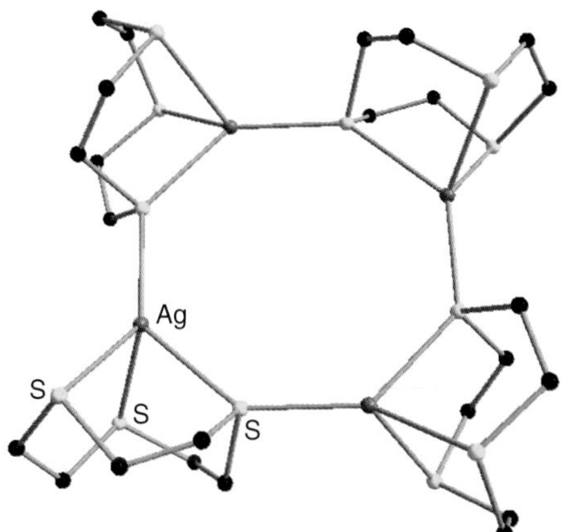

Figure 26 View of the structure of $[Ag_4([9]aneS_3)_4]^{4+}$.[160]

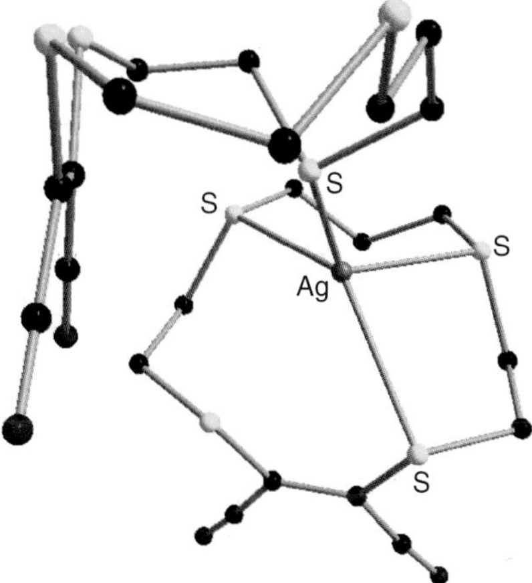

Figure 28 View of the structure of $[Ag(\mathbf{L^5})_2]^+$.[16]

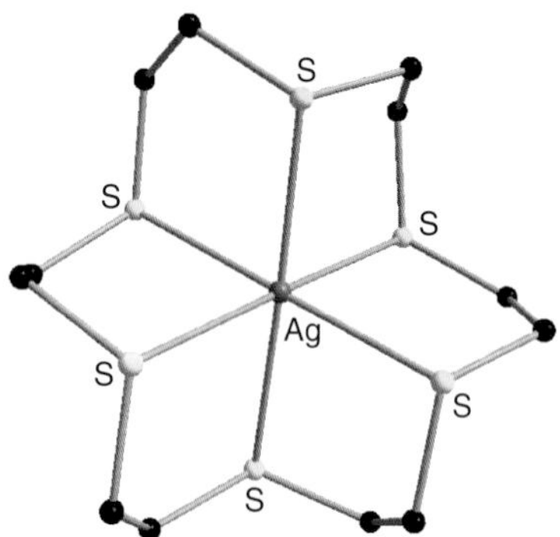

Figure 29 View of the structure of $[Ag([18]aneS_6]^{2+}$.[166]

and DFT calculations, which indicated extensive delocalization of the unpaired electron onto the S donors.[166] EPR studies of Ag(II) complexes of $[14]aneS_4$, $[16]aneS_4$, and $[27]aneS_9$ also showed extensive delocalization of the unpaired electron.[167]

A substantial amount of work on thia-oxa-crowns of Ag(I) has been reported; a bewildering array of coordination numbers and geometries have been identified, different motifs often resulting from small changes in ligand architecture, counter anions present or solvents used. For example, with three S_2O-donor ligands (L) with the sulfurs linked by *o*-, *m*-, or *p*-xylyl units, the first gives $[Ag(L)_2]PF_6$ which has a discrete AgS_4 tetrahedral core, and the third gives a chain polymer containing near linear 2-coordinate Ag centers.[168] The *m*-xylyl-linked ring gives complexes of Ag:L stoichiometry 1 : 1 (polymer chain), 2 : 3, and 2 : 4 (both discrete dimers). Further examples include ligands with S_2O_2-,[134,135,137,140,154,169–171] S_2O_3- and S_3O_2-,[21,22,165,172–175] and S_2O_4-donor sets.[174–176] The $[Ag([18]aneS_4O_2)]PF_6$ was electrochemically oxidized to blue $[Ag([18]aneS_4O_2)][PF_6]_2$ which has the unusual structure (Figure 30) with planar S_4 coordination, two crown oxygens cis and long contacts to a PF_6 group.[177]

Ag(I) complexes have been reported for **L**[12], **L**[13],[37] $[9]aneTeO_2$, and $[18]aneTe_2O_4$.[34] In $[Ag[11]aneS_2Te)]BF_4$,

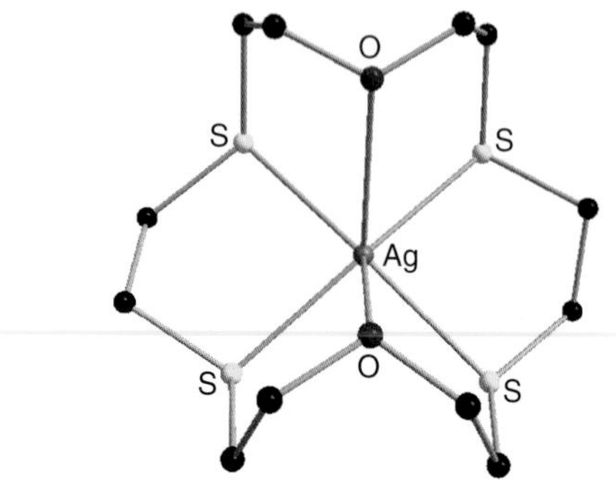

Figure 30 View of the structure of $[Ag([18]aneS_4O_2)]^{2+}$.[177]

the silver is trigonal planar, coordinated to three different ligands giving a chain polymer.[38] Complexes with $[9]aneS_2Te$ and $[12]aneS_2Te$ are also known.[39]

Detailed EPR studies of the Au(II) complex $[Au([9]aneS_3)_2][BF_4]_2$ showed significant delocalization of the unpaired electron onto the sulfur donor atoms.[166,178] Similar conclusions were reached for Au(II) complexes of other thia-crowns.[167,179] The three $[Au([9]aneS_2O)_2]^{n+}$ ($n = 1$, 2, or 3) containing, respectively, Au(I), Au(II), and Au(III) have been prepared, and spectroscopically and structurally characterized.[180] The Au(I) complex contains a tetrahedral AuS_4 cation, while the other two have planar AuS_4 cores with the oxygen atoms positioned approximately axial but with long Au···O distances. There are significant differences in the relative stabilities of the three oxidation states between the $[9]aneS_3$ and $[9]aneS_2O$ complexes.

3.10 Group 12 complexes—Zn, Cd, and Hg

Zinc chloride and nitrate complexes of a range of tetrathiacrowns have been prepared and are presumed to contain octahedral metal coordination, although structural authentication is lacking.[181] The structure of $[Zn([10]aneS_3)_2][ClO_4]_2$ shows a near regular octahedral cation geometry.[182]

Distorted octahedral cations are present in $[Cd([9]aneS_3)_2]Y_2$ ($Y = ClO_4$ or PF_6), and $[Cd([10]aneS_3)_2][ClO_4]_2$.[182–184] In contrast, $[Cd([14]aneS_4)_2][ClO_4]_2$ shows the first example of a CdS_8 cation which has a distorted square antiprismatic geometry,[184] while $[Cd([20]aneS_6)][ClO_4]_2$ seems likely to be octahedral.[63] Cadmium-113 NMR spectra have been recorded for a range of thia-crown complexes and appear to show systematic changes with ligand architecture.[184] Thia-oxa-crown complexes are represented by $[Cd([9]aneS_2O)_2][ClO_4]_2$ and $[Cd([18]aneS_4O_2)][ClO_4]_2$, both octahedral.[100,185]

Six-coordinated Hg(II) is present in $[Hg([10]aneS_3)_2]Y$ ($Y = ClO_4$, PF_6),[181,186] $[Hg([11]aneS_3)_2][ClO_4]_2$,[62] and $[Hg([18]aneS_6)][PF_6]_2$.[187] However, the 18-membered ring hexathia-crown I ($n = 4$) with unsaturated links, formed $[Hg(L)Cl_2]$, the structure of which revealed all six thioether donors coordinated in an approximately planar manner, with axial chlorines,[188] whereas $[Hg([18]aneS_6)Cl_2]$ is a chain structure with bridging thia-crown, giving 6-coordinate Hg centers (Figure 31). Mercury-199 NMR spectra have been recorded for a range of thia-crown complexes.[187] The methylmercury adduct $[CH_3Hg([9]aneS_3)]BF_4$ contains a 4-coordinate cation, a rare geometry for organomercury species, whereas in $[CH_3Hg([12]aneS_3)]BF_4$ and $[(CH_3Hg)_2([14]aneS_4)][BF_4]_2$ the mercury is essentially linearly C,S-coordinated.[189]

Thia-oxa-crowns mostly coordinate to Hg(II) only via the sulfur donors, usually producing distorted tetrahedral geometries,[185,190] although in some cases where the

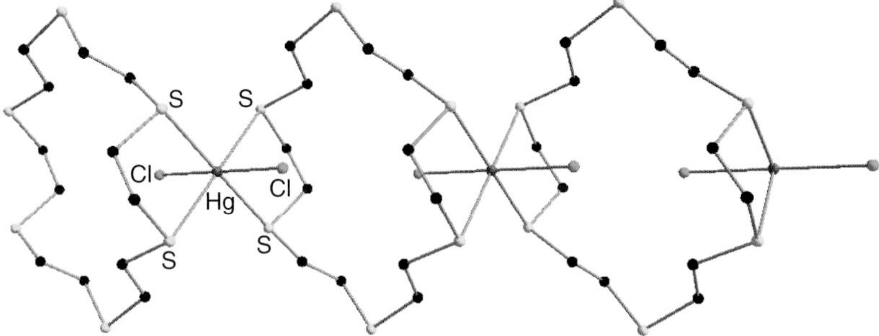

Figure 31 View if the structure of the $[Hg([18]aneS_6)Cl_2]$ polymer.[188]

crown geometry is appropriate, the ether donors are also involved.[159,191]

4 HETERO-CROWN ETHER COMPLEXES WITH P-BLOCK ACCEPTORS

4.1 Group 13 complexes—Al, Ga, In, and Tl

Indium(III) chloride complexes with tetrathia-crowns, $[InCl_2(\text{thia-macrocycle})][InCl_4]$ (thia-macrocycle = $[12]aneS_4$ or $[14]aneS_4$) were obtained from direct reaction of $InCl_3$ and the thia-crown in anhydrous CH_2Cl_2.[192] The crystal structure of $[InCl_2([14]aneS_4)][InCl_4]$ showed a cis-octahedral cation with the macrocycle in a folded conformation. The trans-influence Cl > S was evident with $In{-}S_{transCl}$ significantly longer (by ~0.1 Å) than $In{-}S_{transS}$. Under similar conditions, the poorly soluble *fac*-octahedral $[InCl_3([9]aneS_3)]$ was isolated. This compound was first prepared by Wieghardt and coworkers[193] from $InCl_3$ and the thia-crown in CH_3CN at 40 °C.

An electrode based on the thia-oxa-crown (**V**) showed very good sensitivity for selecting Tl^+ over Ag^+ ions. This was attributed to the combination of soft–soft donor–acceptor interactions and Tl-π-aromatic interactions. A crystal structure of [Tl(**V**)][PF_6] formed by reaction of (**V**) with $TlNO_3$ and NH_4PF_6, confirmed coordination via two S-donor atoms, two O-donor atoms, longer contacts to two frther S atoms and significant interactions with the π-aromatic ring subunits (Figure 32).[23]

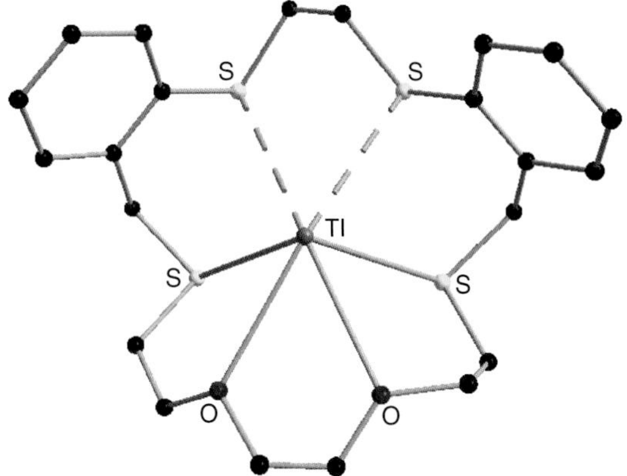

Figure 32 View of the structure of $[Tl(\mathbf{V})]^+$.[23]

4.2 Group 14 complexes—Ge, Sn, and Pb

There are no chalcogenoether crown complexes with silicon acceptor units. Within group 14, ^{19}F NMR spectroscopic evidence was reported for the formation of the tetravalent $[GeF_4(\kappa^2\text{-}[9]aneS_3)]$ in solution at low temperatures. Although this compound was not stable enough to be isolated, both of the related $[GeF_4\{RS(CH_2)2SR\}]$ (R = CH_3 or Et) were structurally authenticated.[194] Very recent work produced a series of divalent Ge halide complexes with thia-crowns, thia-oxa-crowns, and selena-crowns. The lability of the Ge(II) complexes in solution meant that NMR spectroscopic studies provided limited information on these systems. However, the structures of a range of representative examples have been established by X-ray crystallography. The thia-crowns $[9]aneS_3$ or $[14]aneS_4$ reacted with $[GeCl_2(\text{dioxane})]$ in anhydrous CH_2Cl_2 to produce the neutral compounds, $[GeCl_2([9]aneS_3)]$ and $[GeCl_2([14]aneS_4)]$.[195] Earlier work has demonstrated the strong preference of $[9]aneS_3$ in particular to behave as a tridentate ligand toward many metal ions across the periodic table. However, X-ray structural studies on $[GeCl_2([9]aneS_3)]$ showed a rare bridging coordination mode, forming an infinite zigzag chain structure based on *exocyclic* $[9]aneS_3$ coordination. The Ge(II) atom forms two Ge–Cl bonds of ~2.29 Å and two much longer Ge–S bonds of ~2.74 Å. The crystal packing revealed two additional long Ge···Cl contacts of ~3.7 Å between the chains, giving a supramolecular two-dimensional network with a highly distorted octahedral coordination environment at each Ge atom (Figure 33).[195]

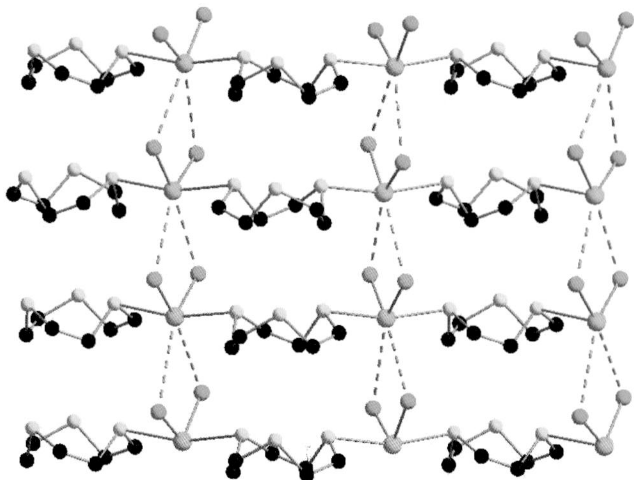

Figure 33 View of the structure of the polymeric [$GeCl_2$-([9]aneS_3)].[195] Turquoise = Ge, green = Cl, yellow = S, black = C.

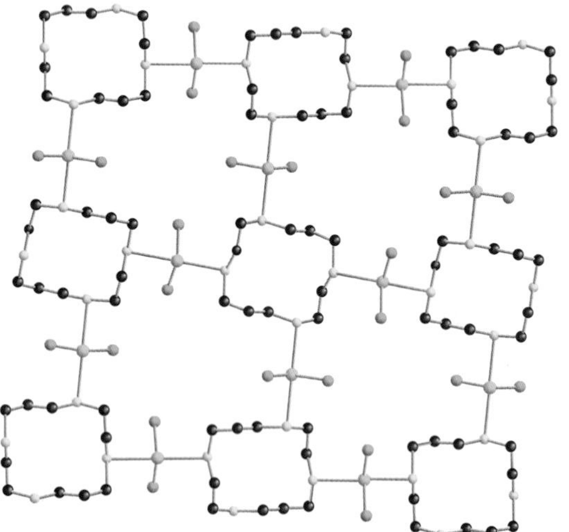

Figure 34 View of the polymeric [$(GeCl_2)_2$([14]aneS_4)].[195] Turquoise = Ge, green = Cl, yellow = S, black = C.

The 2 : 1 compound [$(GeCl_2)_2$([14]aneS_4)] was obtained unexpectedly from the filtrate of the preparation of [$GeCl_2$([14]aneS_4)]. This compound also adopted a two-dimensional network with similar trends in the bond lengths at Ge, although very different in detail from that described for [$GeCl_2$([9]aneS_3)]. In [$(GeCl_2)_2$([14]aneS_4)], a $GeCl_2$ unit was bonded weakly to each of the four macrocyclic S-donor atoms, in an *exocyclic* manner, and the μ_4-κ^1-coordinated tetrathia-crown units bridged $GeCl_2$ units to give the 2-D sheets (Figure 34).[195] [$GeCl_2$([18]aneS_6)] was obtained similarly and also formed chains through *exocyclic* coordination of the S atoms in the 1- and 10-positions to bridging $GeCl_2$ units (Figure 35).[196]

The Ge(II) bromide complexes [$GeBr_2$([14]aneS_4)] and [$GeBr_2$([16]aneS_4)] were isolated from the direct reaction of $GeBr_2$ with the thia-crown in anhydrous CH_3CN.[195] They showed infinite chain structures like those in [$GeCl_2$-([9]aneS_3)] and [$GeCl_2$([18]aneS_6)] above, through *exocyclic* μ_2-κ^1-coordination of the macrocycles to $GeBr_2$ units, with alternate S atom in the crowns coordinated to give linear chains (Ge–Br ~2.65 Å, Ge–S ~2.74 Å), weakly cross-linked through two weak Ge···Br contacts (~3.9 Å) between the chains.

Unexpectedly, using the mixed thia-oxa-crowns led to *endocyclic* coordination in [GeCl([18]aneS_3O_3)][$GeCl_3$] and [GeCl([15]aneS_2O_3)][$GeCl_3$]. Both complexes were comprised of discrete [GeCl(thia-oxa-crown)]$^+$ cations and [$GeCl_3$]$^-$ anions.[196] Within the [GeCl([18]aneS_3O_3)]$^+$ cation, the Ge was bonded to one axial Cl ligand, three weak Ge–O bonds (~2.5 Å) and two weak Ge–S contacts (~2.8 Å), giving a distorted pentagonal pyramid (Figure 36). Addition of excess $(CH_3)_3SiCF_3SO_3$ to the [$GeCl_2$(dioxane)]/[18]aneS_3O_3 reaction mixture gave [GeCl([18]aneS_3O_3)][CF_3SO_3], with the same cation structure. The crystal structure of the [GeCl([15]aneS_2O_3)]$^+$ cation (Figure 37) was also a distorted pentagonal pyramid, in this case with Ge coordinated to all of the macrocyclic donor atoms and the remaining Cl ligand, but the smaller ring occupied four equatorial coordination sites at Ge (O_3S donor set), with the Cl also equatorial, and the second S atom axial. The bond lengths around Ge were also quite different from those in [GeCl([18]aneS_3O_3)]$^+$; the Ge–S distances in [GeCl([15]aneS_2O_3)]$^+$ being ~0.1–0.2 Å shorter and the Ge–O distances 0.1–0.5 Å longer.[196]

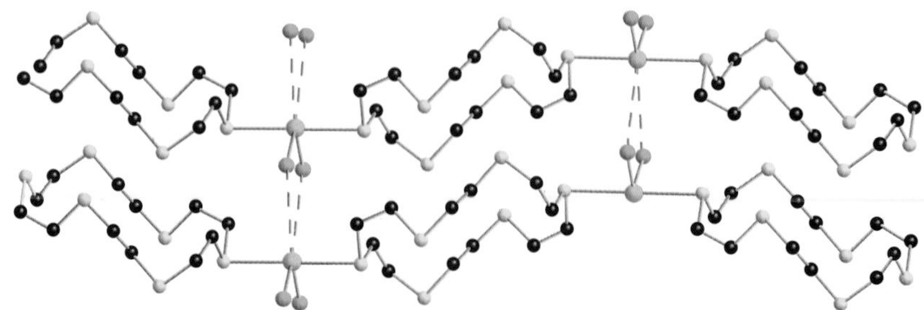

Figure 35 View of the structure of the polymeric [$GeCl_2$([18]aneS_6)].[196] Turquoise = Ge, green = Cl, yellow = S, black = C.

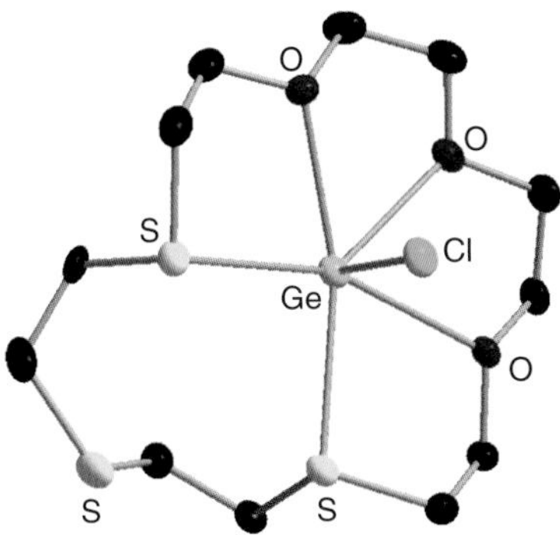

Figure 36 View of the structure of $[GeCl([18]aneS_3O_3)]^+$.[196]

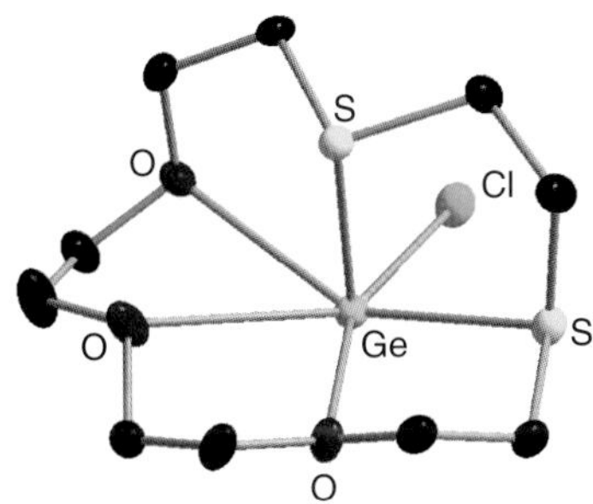

Figure 37 View of the structure of $[GeCl([15]aneS_2O_3)]^+$.[196]

The first selena-crown complexes of Ge(II) were $[GeCl_2([8]aneSe_2)]$, $[(GeCl_2)_2([16]aneSe_4)]$, $[GeBr_2([16]aneSe_4)]$, $[(GeI_2)_2([16]aneSe_4)]\cdot GeI_4$, and $[GeCl_2([24]aneSe_6)]$, all of which were isolated as very moisture-sensitive solids. The structure of $[GeCl_2([8]aneSe_2)]$ showed that the Ge atom coordinated to two cis Cl ligands (~2.3 Å), and two mutually trans-Se-donor atoms from $[8]aneSe_2$ molecules (Ge–Se = 2.8467(8) Å), giving a distorted saw-horse geometry for the $GeCl_2Se_2$ units. The selenoethers bridge to the next Ge atom to form infinite chains, while each Ge forms one further (weak) Ge···Cl interaction (3.589(2) Å) to a Cl on an adjacent chain, cross-linking them into a 2-D sheet (Figure 38).[196]

The 2 : 1 complex $[(GeCl_2)_2([16]aneSe_4)]$ resulted from both 1 : 1 and 2 : 1 Ge:selenoether ratios.[196] Its structure

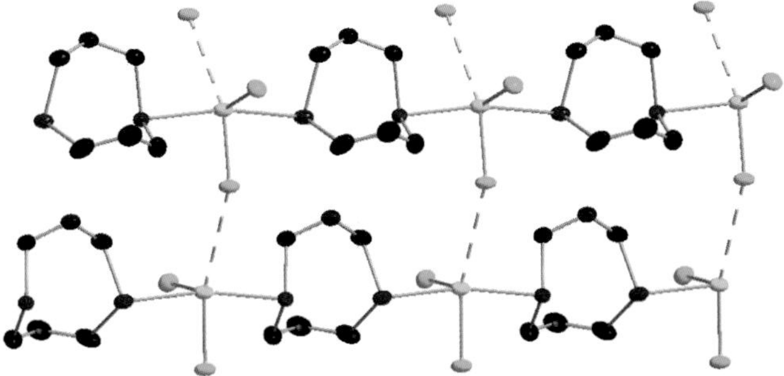

Figure 38 View of the structure of the polymeric $[GeCl_2([8]aneSe_2)]$.[196] Turquoise = Ge, green = Cl, red = Se, black = C.

showed that the macrocyclic ligand had fourfold symmetry and the complex was also a 2-D sheet polymer, with each Ge atom coordinated to two cis Cl atoms and two mutually trans-Se atoms, giving a *pseudo*-trigonal-bipyramidal geometry at Ge with the lone pair assumed to be in the equatorial void. The other Se atoms coordinate to other $GeCl_2$ units to give the sheet network, similar to that of $[(GeCl_2)_2([14]aneS_4)]$ above. $[GeBr_2([16]aneSe_4)]$ formed a chain polymer structure (Figure 39) with *exocyclic* selenoether coordination (Ge–Se ~2.8 Å) and weak Ge···Br contacts (~3.96 Å) between the chains to form a network similar to that for $[GeCl_2([18]aneS_6)]$, $[GeCl_2([9]aneS_3)]$, and $[GeBr_2([14]aneS_4)]$.

The compound $[(GeI_2)_2([16]aneSe_4)]\cdot GeI_4$ was isolated as a few crystals from reaction of GeI_2 with the macrocycle in anhydrous CH_3CN, and its structure was very similar to that in $[(GeCl_2)_2([16]aneSe_4)]$ above. The presence of molecular GeI_4 cocrystallized within the voids of the Ge(II) network was unexpected, and weak long-range I···I contacts between the network and the GeI_4 may contribute to stabilizing this arrangement (Figure 40). The GeI_4 may arise from oxidation or disproportionation of GeI_2 in the presence of the selenoether.[196]

In summary, it seems that *endocyclic* coordination to Ge(II) is favored for crown ethers and thia-oxa-crowns, whereas thia- and selena-crowns lead to *exocyclic* coordination and supramolecular assemblies, although the reasons are not yet understood.

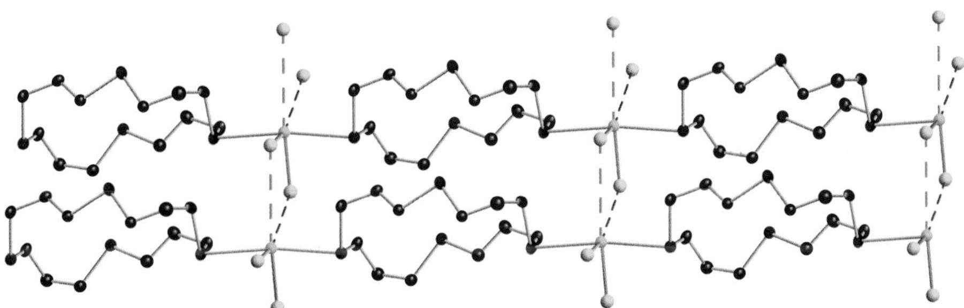

Figure 39 View of the structure of $[GeCl_2([16]aneSe_4)]$.[196] Turquoise = Ge, gold = Br, red = Se, black = C.

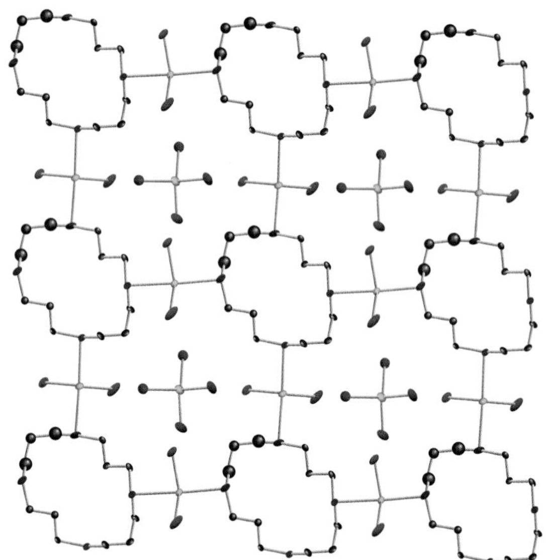

Figure 40 View of the structure of the $[(GeI_2)_2[16]aneSe_4)]$ polymer showing the cocrystallized GeI_4 molecules.[196] Turquoise = Ge, pink = I, red = Se, black = C.

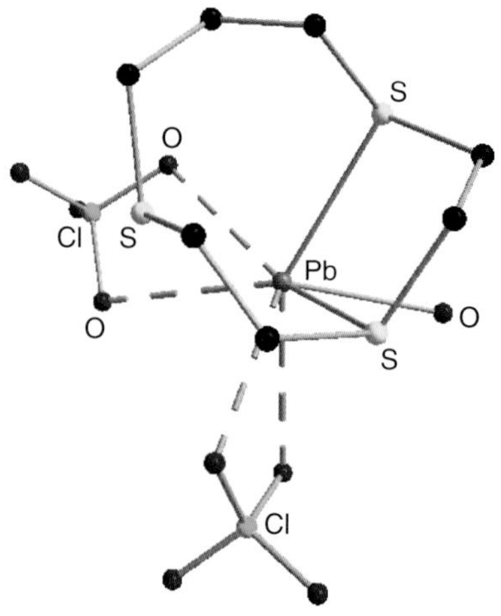

Figure 42 View of the structure of $[Pb([10]aneS_3)(H_2O)]^+$.[183]

The tetrathia-crown complexes $[(SnCl_4)_2([n]aneS_4)]$ ($n = 12$, 14, 16) and $[SnBr_4([n]aneS_4)]$ were obtained from reaction of 2 mol equivalents of the parent tin tetrahalide with 1 mol equivalent of macrocycle in anhydrous CH_2Cl_2. Far IR spectroscopic studies showed that the chloro complexes were cis-octahedral isomers, while crystal structures of the bromo complexes showed they were all chain polymers; $[SnBr_4([12]aneS_4)]$ adopts the cis-octahedral form, with the macrocycle bridging the Sn atoms through coordination at the 1-, and 7-positions, whereas $[SnBr_4([14]aneS_4)]$ and $[SnBr_4([16]aneS_4)]$ were trans isomers with the 14-membered ring coordinated to Sn atoms via the 1- and 8-positions (linear chain) and the 16-membered ring coordinated via the 1- and 5-positions (zigzag chain) (Figure 41).[197]

Reactions of $[16]aneSe_4$ with SnX_4 (X = Cl or Br) gave $[SnX_4([16]aneSe_4)]$, while similar reaction of $[8]aneSe_2$ gave $[SnCl_4([8]aneSe_2)]$, although structural data are lacking.[197]

A small number of Pb(II) complexes with thia-crowns have been reported,[7] More recent examples are $[Pb([10]aneS_3)(ClO_4)_2(H_2O)]$ whose structure showed $[Pb([10]aneS_3)(H_2O)]^+$ units with weakly coordinated bidentate perchlorate anions, giving a distorted 8-coordinate monomer structure, with a further weak interaction to an O atom from a $[ClO_4]^-$ anion in a neighboring molecule leading to a zigzag chain polymer, and 9-coordinate Pb(II) ions (Figure 42).[183]

4.3 Group 15 complexes—As, Sb, and Bi

Several series of thioether and selenoether macrocyclic complexes with MX_3 (M = Sb, Bi; X = Cl, Br, I) have been reported and were described in earlier review articles.[7] More recent work has included reports of a series of thia-crown complexes of $AsCl_3$, for example, $[AsX_3([9]aneS_3)]$ (X = Cl, Br, or I) and $[AsCl_3([14]aneS_4)]$. Although these compounds were extensively dissociated in solution, the crystal structure of $[AsCl_3([9]aneS_3)]$ showed that the tridentate macrocycle weakly coordinated to one face of the As atom, with three mutually *fac* Cl ligands giving a distorted octahedron. The As–S distances (~2.8 Å)

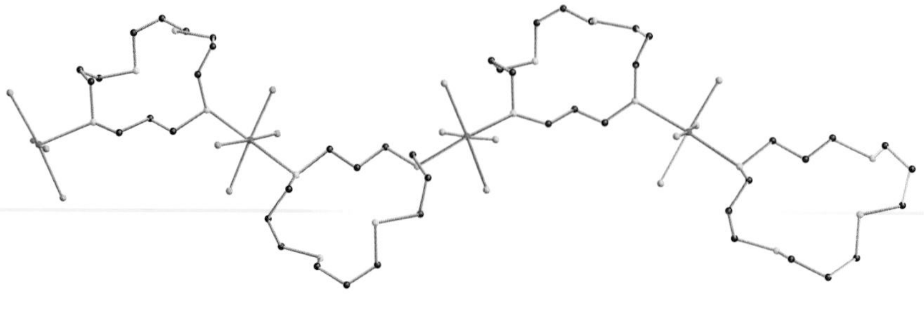

Figure 41 View of the structure of the $[SnBr_4([16]aneS_4)]$ polymer.[197] Turquoise = Sn, gold = Br, yellow = S, black = C.

Figure 43 View of the structure of $[AsCl_3([14]aneS_4)]$ polymer.[198] Turquoise = As, green = Cl, yellow = S, black = C.

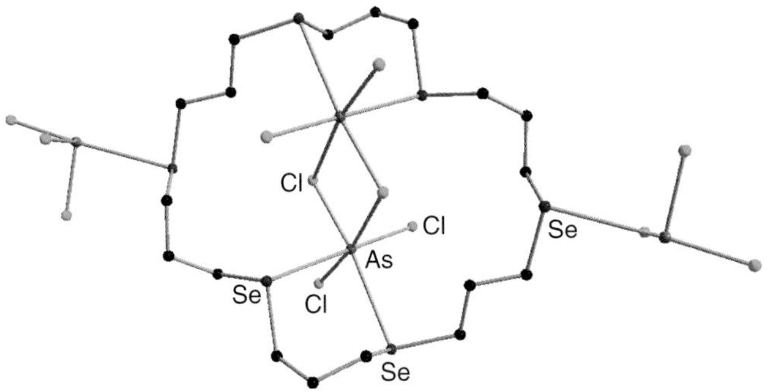

Figure 44 View of the structure of $[(AsCl_3)_4([24]aneSe_6)]$.[199]

are very much longer than the As–Cl bonds (~2.3 Å), suggesting weak (secondary) As–S interactions, with the As-based lone pair essentially stereochemically inactive.[198] The $[AsCl_3([14]aneS_4)]$ adopted a two-dimensional sheet polymer structure based on distorted 5-coordinate As, with bridging Cl ligands and bridging macrocycle (Figure 43).[198]

The first examples of selenoether macrocyclic complexes of $AsCl_3$ included the tetranuclear $[(AsCl_3)_4([24]aneSe_6)]$ the crystal structure of which showed a weakly coordinated *endocyclic* As_2Cl_6 unit with each As coordinated to two cis-Se-donor atoms, and two *exocyclic* $AsCl_3$ units each coordinated via one of the remaining Se atoms (Figure 44).[199] Other examples, also obtained from direct reaction of the cyclic selena-crown with AsX_3 in anhydrous CH_2Cl_2 (or thf for X = I), included $[AsX_3([8]aneSe_2)]$, $[(AsX_3)_2([16]aneSe_4)]$, and $[(AsBr_3)_2([24]aneSe_6)]$. The complexes $[(AsX_3)_2([16]aneSe_4)]$ (X = Cl or Br) both adopted similar three-dimensional polymer structures based on asymmetric weakly associated As_2X_6 dimers cross-linked by coordination to $[16]aneSe_4$ units.[198]

Structures of the homologous series, $[MCl_3([8]aneSe_2)]$ (M = As, Sb, or Bi) showed ladder structures formed from planar M_2Cl_6 units linked by bridging selenoether ligands, with the Se atoms occupying trans coordination sites at M, but unexpectedly with significant differences down group 15.[200] In $[AsCl_3([8]aneSe_2)]$, the 6-coordinate arsenic center was quite symmetrical, suggesting that the M-based lone pair is stereochemically inactive. In $[SbCl_3([8]aneSe_2)]$, the bond lengths and angles at Sb were much less regular, for example, the Sb–Se distances differed by ~0.5 Å, possibly due to a stereochemically active Sb lone pair. In contrast, $[BiCl_3([8]aneSe_2)]$ has been shown to adopt a regular 6-coordinate geometry, suggesting that the Bi-based lone pair occupies the 6s orbital, which is both stabilized and contracted because of relativistic effects.[201]

The effects on the conformations of complexes of Sb(III) and Bi(III) of the rigid maleonitrile units present in four related mixed thia-oxa macrocycles and of the ligands themselves has been investigated by variable temperature solution ^{1}H and $^{13}C\{^{1}H\}$ NMR studies together with molecular modeling. The Sb and Bi complexes were thought to show a preference for coordination via the ether-O atoms.[140]

5 CONCLUSIONS

The chemistry of hetero-crowns containing group 16 donor atoms is still an active area of research; while thia-crown chemistry is now a relatively mature field, macrocyclic ligands containing the heavier Se and Te donor atoms are relatively sparse, and correspondingly their coordination chemistry lags behind that of the sulfur-based systems. Whereas the vast majority of transition metal complexes of these ligands exhibit *endocyclic* coordination, complexes involving p-block acceptors often adopt *exocyclic* coordination modes. This is presumably because the energy cost of reorganizing the ligand conformations from the *exocyclic* form (typically seen in the uncomplexed ligands) to *endocyclic* is not recouped in forming the donor–acceptor bonds in these rather weakly coordinating systems. The fact that Ge(II) coordinates *exocyclic* to thioether and selenoether crowns, while substitution of some S atoms by O atoms switches the coordination to *endocyclic*, suggests that a very subtle balance exists and that there is much still to be understood about the bonding in these systems.

Projecting forward, much new early transition metal chemistry still remains to be explored and the properties imparted by the macrocyclic structures should enhance the relative stabilities of such species. The development of telluroether macrocycles remains a major synthetic challenge, but the lower electronegativity of the heavier group 16 element, combined with the additional stability provided by the rings should produce unprecedented coordination chemistry which is not possible with acyclic analogs. Finally, the transition metal chemistry of hetero-crowns based on the group 16 donor atoms is often directed by the geometric and electronic preferences of the metal center and the ability of the ligand to adapt its coordination properties to accommodate unusual oxidation states. In contrast, the weaker, secondary bonding in p-block complexes leads to the steric and architectural properties of the hetero-crown playing a more dominant role. Much new chemistry in this area is likely to be forthcoming and should provide new insight into the bonding in these systems.

ACKNOWLEDGMENTS

We thank Dr Wenjian Zhang and Sophie Benjamin for their help with assembling this chapter.

ABBREVIATIONS

The commonly accepted abbreviations used in this chapter are listed here. For less common ligands described, line drawings have been included for clarity.

dmso	dimethylsulfoxide,
LDA	lithium diisopropylamide,
NBS	*N*-bromosuccinimide,
Ts	*p*-toluenesulfonate,
TCNQ	tetracyanoquinodimethane,
$(CH_3)_3$-tacn	1,4,7-trimethyl-1,4,7-triazacyclonane,
15-crown-5	1,4,7,10,13-pentaoxacyclopentadecane,
18-crown-6	1,4,7,10,13,16-hexaoxacyclooctadecane,
cyclam	1,4,8,11-tetraazacyclotetradecane,
$(CH_3)_4$-cyclam	1,4,8,11-tetramethyl-1,4,8,11-tetraazacyclotetradecane,
$(CH_3)_3$-[9]aneN_3	1,4,7-trimethyl-1,4,7-triazacyclononane,
[9]aneS_3	1,4,7-trithiacyclononane,
[10]aneS_3	1,4,7-trithiacyclodecane,
[11]aneS_3	1,4,7-trithiacycloundecane or 1,4,8-trithiacycloundecane
[12]aneS_3	1,5,9-trithiacyclododecane,
[12]aneS_4	1,4,7,10-tetrathiacyclododecane,
[14]aneS_4	1,4,8,11-tetrathiacyclotetradecane,
[15]aneS_5	1,4,7,10,13-pentathiacyclopentadecane,
[18]aneS_6	1,4,7,10,13,16-hexathiacyclooctadecane,
[16]aneS_4	1,5,9,13-tetrathiacyclohexadecane,
[18]aneS_3O_3	1,4,7-trithia-10,13,16-trioxacyclooctadecane,
[15]aneS_2O_3	1,4-dithia-7,10,13-trioxacyclopentadecane,
[9]aneSO_2	1-thia-4,7-diaxaccylononane,
[9]aneS_2O	1,4-dithia-7-oxacyclononane,
[12]aneSO_3	1-thia-4,7,10-trioxacyclododecane,
[18]aneSO_5	1,thia-4,7,10,13,16-pentaoxacyclooctadecane,
[18]aneS_2O_4	1,10-dithia-4,7,13,16-tetraoxacyclooctadecane,
[18]aneS_4O_2	1,4,10,13-tetrathia-7,16-dioxacyclooctadecane,
[8]aneSe_2	1,5-diselenacyclooctane,
[16]aneSe_4	1,5,9,13-tetraselenacyclohexadecane,
[16]ane$Se_4(OH)_2$	3,11-dihydroxy-1,5,9,13-tetraselenacyclohexadecane,
[24]aneSe_6	1,5,9,13,17,21-hexaselenacyclotetracosane,
[12]aneSe_3	1,5,9-triselenacyclododecane,
[16]aneS_2Se_2	1,5-dithia-9,13-diselenacyclohexadecane,
[9]aneSeO_2	1-selena-4,7-dioxacyclononane,
[18]aneSe_2O_4	1,10-diselena-4,7,13,16-tetraoxacyclooctadecane,
[9]aneTeO_2	1-tellura-4,7-dioxacyclononane,
[18]aneTe_2O_4	1,10-ditellura-4,7,13,16-tetraoxacyclooctadecane.

REFERENCES

1. A. J. Blake and M. Schröder, *Adv. Inorg. Chem.*, 1990, **35**, 1.
2. S. R. Cooper and S. C. Rawle, *Struct. Bond.*, 1990, **72**, 1.
3. L. F. Lindoy, G. V. Meehan, I. M. Vasilescu, *et al.*, *Coord. Chem. Rev.*, 2010, **254**, 1713.
4. W. Levason and G. Reid, in *Comp. Coord. Chem. II*, eds. J. A. McCleverty and T. J. Meyer, Elsevier, Amsterdam, 2004, vol. 1, p. 399.
5. W. Levason, G. Reid, in *Handbook of Chalcogen Chemistry*, F. Davillanova, Royal Society of Chemistry, Cambridge, UK, 2006, p. 81.
6. W. Levason and G. Reid, *J. Chem. Res.*, 2002, **10**, 467.
7. W. Levason and G. Reid, *J. Chem. Soc., Dalton Trans.*, 2001, 2953.
8. E. G. Hope and W. Levason, *Coord. Chem. Rev.*, 1993, **122**, 109.
9. A. K. Singh and S. Sharma, *Coord. Chem. Rev.*, 2000, **209**, 49.
10. W. Levason, S. D. Orchard, and G. Reid, *Coord. Chem. Rev.*, 2002, **225**, 159.

11. A. Panda, *Coord. Chem. Rev.*, 2009, **253**, 1056.
12. A. Panda, *Coord. Chem. Rev.*, 2009, **253**, 1947.
13. M. Fainerman-Melnikova, L. F. Lindoy, S.-Y. Liou, *et al.*, *Aust J. Chem.*, 2004, **57**, 161.
14. A. J. Edwards, A. C. Willis, and E. Wenger, *Organometallics*, 2002, **21**, 1654.
15. T. Tsuchiya, T. Shimizu, and N. Kamigata, *J. Am. Chem. Soc.*, 2001, **123**, 11534.
16. H.-J. Holdt, H. Müller, M. Pötter, *et al.*, *Eur. J. Inorg. Chem.*, 2006, 2377.
17. R. Y. C. Shin, M. A. Bennett, L. Y. Goh, *et al.*, *Inorg. Chem.*, 2003, **42**, 96.
18. O. A. Fedorova, Y. V. Fedorov, A. I. Vedernikov, *et al.*, *J. Phys. Chem. A*, 2002, **106**, 6213.
19. A. I. Vedernikov, E. N. Ushakov, L. G. Kuz'mina, *et al.*, *J. Phys. Org. Chem.*, 2010, **23**, 195.
20. R. Alberto, D. Angst, K. Ortner, *et al.*, *New J. Chem.*, 2007, **31**, 409.
21. E. V. Tulyakova, E. V. Rakhmanov, E. V. Lukovskaya, *et al.*, *Chem. Heterocycl. Comp.*, 2006, **42**, 206.
22. L. G. Kuz'mina, A. I. Vedernikov, S. N. Dmitrieva, *et al.*, *Russ. Chem. Bull., Int. Ed.*, 2007, **56**, 1003.
23. K.-M. Park, Y. H. Lee, Y. Jin, *et al.*, *Supramol. Chem.*, 2004, **16**, 51.
24. S. M. Williams, J. S. Brodbelt, A. P. Marchand, *et al.*, *Anal. Chem.*, 2002, **74**, 4423.
25. A. Visnjevac, B. Kojic-Prodic, M. Vinkovic, and K. Mlinaric-Majerski, *Acta Crystallogr. C*, 2003, **59**, o314.
26. M. Jura, W. Levason, G. Reid, and M. Webster, *Dalton Trans.*, 2009, 7610.
27. M. R. J. Elsegood, L. M. Gilby, K. E. Holmes, and P. F. Kelly, *Can. J. Chem.*, 2002, **80**, 1410.
28. S. H. Dale, M. R. J. Elsegood, L. M. Gilby, *et al.*, *Acta Crystallogr. C*, 2005, **61**, o411.
29. S. M. Aucott, M. R. Bailey, M. R. J. Elsegood, *et al.*, *New J. Chem.*, 2004, **28**, 959.
30. M. R. J. Elsegood, P. F. Kelly, G. Reid, and P. M. Staniland, *Dalton Trans.*, 2007, 1665.
31. M. R. J. Elsegood, P. F. Kelly, G. Reid, *et al.*, *Dalton Trans.*, 2008, 5076.
32. I. Cordova-Reyes, E. VandenHoven, A. Mohammed, and B. M. Pinto, *Can. J. Chem.*, 1995, **73**, 1480.
33. R. J. Batchelor, F. W. B. Einstein, I. D. Gay, *et al.*, *Inorg. Chem.*, 2000, **39**, 2558.
34. W. Levason, J. M. Manning, G. Reid, *et al.*, *Dalton Trans.*, 2009, 4569.
35. S. Jing, C.-Y. Gu, W. Ji, and B. Yang, *Inorg. Chem. Commun.*, 2009, **12**, 846.
36. M. J. Hesford, W. Levason, M. L. Matthews, and G. Reid, *Dalton Trans.*, 2003, 2852.
37. W. Levason, J. M. Manning, M. Nirwan, *et al.*, *Dalton Trans.*, 2008, 3486.
38. W. Levason, S. D. Orchard, and G. Reid, *Chem. Commun.*, 2001, 427.
39. M. J. Hesford, W. Levason, M. L. Matthews, *et al.*, *Dalton Trans.*, 2003, 2434.
40. C. S. Tredget, F. Bonnet, A. R. Cowley, and P. Mountford, *Chem. Commun.*, 2005, 3301.
41. C. S. Tredget, E. Clot, and P. Mountford, *Organometallics*, 2008, **27**, 3458.
42. M. D. Brown, W. Levason, D. C. Murray, *et al.*, *Dalton Trans.*, 2003, 857.
43. L. Karmazin, M. Mazzanti, and J. Pecaut, *Chem. Commun.*, 2002, 654.
44. T. Arliguie, L. Belkhiri, S.-E. Bouaoud, *et al.*, *Inorg. Chem.*, 2009, **48**, 221.
45. W. Levason, B. Patel, and G. Reid, *Inorg. Chim. Acta*, 2004, **357**, 2115.
46. P. J. Wilson, A. J. Blake, P. Mountford, and M. Schroder, *Inorg. Chim. Acta*, 2003, **345**, 44.
47. W. Levason, M. C. Popham, G. Reid, and M. Webster, *Dalton Trans.*, 2003, 291.
48. R. Hart, W. Levason, B. Patel, and G. Reid, *J. Chem. Soc., Dalton Trans.*, 2002, 3153.
49. S. D. Reid, A. L. Hector, W. Levason, *et al.*, *Dalton Trans.*, 2007, 4769.
50. A. L. Hector, W. Levason, A. J. Middleton, *et al.*, *Eur. J. Inorg. Chem.*, 2007, 3655.
51. C. D. Beard, L. Carr, M. F. Davis, *et al.*, *Eur. J. Inorg. Chem.*, 2006, 4399.
52. A. J. Barton, J. Connolly, W. Levason, *et al.*, *Polyhedron*, 2000, **19**, 1373.
53. W.-H. Leung, M. C. Wu, T. K. T. Wong, and W.-T. Wong, *Inorg. Chim. Acta*, 2000, **304**, 134.
54. M. F. Davis, W. Levason, M. E. Light, *et al.*, *Eur. J. Inorg. Chem.*, 2007, 1903.
55. B. Patel and G. Reid, *J. Chem. Soc., Dalton Trans.*, 2000, 1303.
56. G. E. D. Mullen, P. J. Blower, D. J. Price, *et al.*, *Inorg. Chem.*, 2000, **39**, 4093.
57. A. Magistrato, P. Maurer, T. Fassler, and U. Rothlisberger, *J. Phys. Chem. A*, 2004, **108**, 2008.
58. P. Maurer, A. Magistrato, and U. Rothlisberger, *J. Phys. Chem. A*, 2004, **108**, 11494.
59. F. Zobi, L. Kromer, B. Spingler, and R. Alberto, *Inorg. Chem.*, 2009, **48**, 8965.
60. M. Casanova, E. Zangrando, F. Munini, *et al.*, *Dalton Trans.*, 2006, 5033.
61. P. Kurz, D. Rattat, D. Angst, *et al.*, *Dalton Trans.*, 2005, 804.
62. G. J. Grant, S. S. Shoup, C. L. Baucom, and W. N. Setzer, *Inorg. Chim. Acta*, 2001, **317**, 91.
63. G. J. Grant, C. G. Brandow, C. W. Bruce, *et al.*, *J. Heterocycl. Chem.*, 2001, **38**, 1281.
64. T. W. Green, J. A. Krause Bauer, and W. B. Connick, *Acta Crystallogr. E*, 2003, **E59**, m953.

65. F. E. Sowrey, P. J. Blower, J. C. Jeffery, *et al.*, *Inorg. Chem. Commun.*, 2002, **5**, 832.
66. J.-X. Dai, F.-H. Wu, W.-R. Yao, and Q.-F. Zhang, *Z. Naturforsch. B*, 2007, **62**, 491.
67. V. V. Pavlishchuk, I. A. Koval, E. Goreshnik, *et al.*, *Eur. J. Inorg. Chem.*, 2001, 297.
68. L. Y. Goh, M. E. Teo, S. B. Khoo, *et al.*, *J. Organometal. Chem.*, 2002, **664**, 161.
69. E. Zangrando, N. Kulisic, F. Ravalico, *et al.*, *Inorg. Chim. Acta*, 2009, **362**, 820.
70. M. Suedfeld and W. S. Sheldrick, *Z. Anorg. Allgem. Chem.*, 2002, **628**, 1366.
71. N. Shan, H. Adams, and J. Thomas, *Inorg. Chim. Acta*, 2006, **359**, 759.
72. S. Roche, S. E. Spey, H. Adams, and J. A. Thomas, *Inorg. Chim. Acta*, 2001, **323**, 157.
73. X. Sala, A. Poater, I. Romero, *et al.*, *Eur. J. Inorg. Chem.*, 2004, 612.
74. X. Sala, I. Romero, M. Rodriguez, *et al.*, *Inorg. Chem.*, 2004, **43**, 5403.
75. C.-Y. Wong, L.-M. Lai, H.-F. Leung, and S.-H. Wong, *Organometallics*, 2009, **28**, 3537.
76. C.-Y. Wong, L.-M. Lai, and P.-K. Pat, *Organometallics*, 2009, **28**, 5656.
77. E. Iengo, E. Zangrando, E. Baiutti, *et al.*, *Eur. J. Inorg. Chem.*, 2005, 1019.
78. J. Madureira, T. M. Santos, B. J. Goodfellow, *et al.*, *J. Chem. Soc., Dalton Trans.*, 2000, 4422.
79. I. Bratsos, S. Jedner, A. Bergamo, *et al.*, *J. Inorg. Biochem.*, 2008, **102**, 1120.
80. B. Serli, E. Zangrando, T. Gianferrara, *et al.*, *Eur. J. Inorg. Chem.*, 2005, 3423.
81. I. Bratsos, G. Birarda, S. Jedner, *et al.*, *Dalton Trans.*, 2007, 4048.
82. C. S. Araujo, M. G. B. Drew, V. Felix, *et al.*, *Inorg. Chem.*, 2002, **41**, 2250.
83. N. Shan, S. J. Vickers, H. Adams, *et al.*, *Angew. Chem. Int. Ed.*, 2004, **43**, 3938.
84. N. Shan, J. D. Ingram, T. L. Easun, *et al.*, *Dalton Trans.*, 2006, 2900.
85. M. Pillinger, I. S. Goncalves, A. D. Lopes, *et al.*, *J. Chem. Soc., Dalton Trans.*, 2001, 1628.
86. C. D. Nunes, T. M. Santos, H. M. Carapuca, *et al.*, *New. J. Chem.*, 2002, **26**, 1384.
87. D. E. Janzen, W. Chen, D. G. VanDerveer, *et al.*, *Inorg. Chem. Commun.*, 2006, **9**, 992.
88. C. D. Nunes, M. Pillinger, A. Hazell, *et al.*, *Polyhedron*, 2003, **22**, 2799.
89. M. Newell and J. A. Thomas, *Dalton Trans.*, 2006, 705.
90. M. Newell, J. D. Ingram, T. L. Easun, *et al.*, *Inorg. Chem.*, 2006, **45**, 821.
91. H. Adams, P. J. Costa, M. Newell, *et al.*, *Inorg. Chem.*, 2008, **47**, 11633.
92. N. Shan, S. M. Hauxwell, H. Adams, *et al.*, *Inorg. Chem.*, 2008, **47**, 11551.
93. H. Adams, A. M. Amado, V. Felix, *et al.*, *Chem. Eur. J.*, 2005, **11**, 2031.
94. T. Gianferrara, I. Bratsos, E. Iengo, *et al.*, *Dalton Trans.*, 2009, 10742.
95. J. Marques, T. M. Santos, M. P. Marques, and S. S. Braga, *Dalton Trans.*, 2009, 9812.
96. H. Bregman, P. J. Carroll, and E. Meggers, *J. Amer. Chem. Soc.*, 2006, **128**, 877.
97. J. C. Green, A. L. Hector, A. F. Hill, *et al.*, *Organometallics*, 2008, **27**, 5548.
98. A. J. Blake, W.-S. Li, V. Lippolis, *et al.*, *Acta Crystallogr. B*, 2007, **63**, 81.
99. J. Nishijo, A. Miyazaki, and T. Enoki, *Bull. Chem. Soc. Jpn.*, 2004, **77**, 715.
100. G. J. Grant, M. W. Jones, K. D. Loveday, *et al.*, *Inorg. Chim. Acta*, 2000, **300–302**, 250.
101. G. J. Grant, J. P. Lee, M. L. Helm, *et al.*, *J. Organomet. Chem.*, 2005, **690**, 629.
102. R. Bieda, I. Ott, M. Dobroschke, *et al.*, *J. Inorg. Biochem.*, 2009, **103**, 698.
103. R. Bieda, M. Dobroschke, A. Triller, *et al.*, *Chem. Med. Chem.*, 2010, **5**, 1123.
104. G. J. Grant, R. D. Naik, D. E. Janzen, *et al.*, *Supramol. Chem.*, 2010, **22**, 109.
105. A. J. Blake, N. R. Brooks, N. R. Champness, *et al.*, *Acta Crystallogr. E*, 2001, **57**, m376.
106. G. J. Grant, C. G. Brandow, D. F. Galas, *et al.*, *Polyhedron*, 2001, **20**, 3333.
107. G. J. Grant, W. Chen, A. M. Goforth, *et al.*, *Eur. J. Inorg. Chem.*, 2005, 479.
108. A. J. Blake, R. O. Gould, A. J. Holder, *et al.*, *Chem. Commun.*, 1987, 118.
109. D. R. Allan, A. J. Blake, D. Huang, *et al.*, *Chem. Commun.*, 2006, 4081.
110. G. J. Grant, K. N. Patel, M. L. Helm, *et al.*, *Polyhedron*, 2004, **23**, 1361.
111. D. E. Janzen, K. Patel, D. G. VanDerveer, and G. J. Grant, *J. Chem. Crystallogr.*, 2006, **36**, 83.
112. T. W. Green, R. Lieberman, N. Mitchell, *et al.*, *Inorg. Chem.*, 2005, **44**, 1995.
113. G. J. Grant, D. A. Benefield, and D. G. VanDerveer, *Dalton Trans.*, 2009, 8605.
114. G. J. Grant, S. M. Carter, A. L. Russell, *et al.*, *J. Organomet. Chem.*, 2001, **637–639**, 683.
115. G. J. Grant, D. F. Galas, I. M. Poullaos, *et al.*, *J. Chem. Soc., Dalton Trans.*, 2002, 2973.
116. G. J. Grant, J. A. Pool, and D. G. VanDerveer, *Dalton Trans.*, 2003, 3981.
117. G. J. Grant, I. M. Poullaos, D. F. Galas, *et al.*, *Inorg. Chem.*, 2001, **40**, 564.
118. D. E. Janzen, D. G. VanDerveer, L. F. Mehne, *et al.*, *Dalton Trans.*, 2008, 1872.

119. E. Pierce, E. Lanthier, C. Genre, *et al.*, *Inorg. Chem.*, 2010, **49**, 4901.
120. D. E. Janzen, K. N. Patel, D. G. VanDerveer, and G. J. Grant, *Chem. Commun.*, 2006, 3540.
121. D. E. Janzen, L. F. Mehne, D. G. VanDerveer, and G. J. Grant, *Inorg. Chem.*, 2005, **44**, 8182.
122. G. J. Grant, D. A. Benefield, and D. G. VanDerveer, *J. Organomet. Chem.*, 2010, **695**, 634.
123. R. Y. C. Shin, C. L. Goh, L. Y. Goh, *et al.*, *Eur. J. Inorg. Chem.*, 2009, 2282.
124. E. Stephen, A. J. Blake, E. S. Davis, *et al.*, *Chem. Commun.*, 2008, 5707.
125. G. J. Grant, D. F. Galas, and D. G. VanDerveer, *Polyhedron*, 2002, **21**, 879.
126. G. J. Grant, A. M. Goforth, D. G. VanDerveer, and W. T. Pennington, *Inorg. Chim. Acta*, 2004, **357**, 2107.
127. H. Richtzenhain, A. J. Blake, D. W. Bruce, *et al.*, *Chem. Commun.*, 2001, 2580.
128. H. Muller, A. Kelling, U. Schilde, and H. J. Holdt, *Z. Naturforsch. B*, 2009, **64**, 1003.
129. K.-M. Park, I. Soon, U.-H. Paek, *et al.*, *Anal. Sci.*, 2002, **18**, 1177.
130. I. Yoon, B. Soon, S. S. Lee, and B. G. Kim, *Acta Crystallogr. C*, 2000, **56**, 758.
131. S. Y. Lee, S. Park, and S. S. Lee, *Inorg. Chim. Acta*, 2009, **362**, 1047.
132. J. Seo, I. Yoon, J.-E. Lee, *et al.*, *Inorg. Chem. Commun.*, 2005, **8**, 916.
133. S. Y. Lee, K.-M. Park, and S. S. Lee, *Inorg. Chem. Commun.*, 2008, **11**, 307.
134. I. Yoon, K.-M. Park, J. H. Jung, *et al.*, *J. Incl. Phenom. Macrocyclic Chem.*, 2002, **42**, 45.
135. S. Y. Lee, S. Park, and S. S. Lee, *Inorg. Chem.*, 2009, **48**, 11335.
136. J. Seo, K.-M. Park, I. Yoon, *et al.*, *Inorg. Chem. Commun.*, 2008, **11**, 837.
137. A. Holzberger, H.-J. Holdt, and E. Kleinpeter, *Org. Biomol. Chem.*, 2004, **2**, 1691.
138. S. N. Dmitrieva, N. I. Sidorenko, A. I. Vedernikov, *et al.*, *Mendeleev Commun.*, 2009, **19**, 21.
139. I. Yoon, J. Seo, J.-E. Lee, *et al.*, *Inorg. Chem.*, 2006, **45**, 3487.
140. E. Kleinpeter, M. Grotjahn, K. D. Klika, *et al.*, *J. Chem. Soc., Perkin Trans. 2*, 2001, 988.
141. H.-J. Drexler, I. Starke, M. Grotjahn, *et al.*, *Inorg. Chim. Acta*, 2001, **317**, 133.
142. W. Levason, J. M. Manning, P. Pawelzyk, and G. Reid, *Eur. J. Inorg. Chem.*, 2006, 4380.
143. J. Hirsch, S. D. George, E. I. Soloman, *et al.*, *Inorg. Chem.*, 2001, **40**, 2439.
144. A. Kandegedara, K. Krylova, T. J. Nelson, *et al.*, *J. Chem. Soc., Dalton Trans*, 2002, 792.
145. T. Naito, K. Nishibe, and T. Inabe, *Z. Anorg. Allgem. Chem.*, 2004, **630**, 2725.
146. Q. Yu, C. A. Salhi, E. D. Ambundo, *et al.*, *J. Am. Chem. Soc.*, 2001, **123**, 5720.
147. P. Wijetunge, C. P. Kulatilleke, L. T. Dressel, *et al.*, *Inorg. Chem.*, 2000, **39**, 2897.
148. S. A. Kakos, L. T. Dressel, J. F. Bushendorf, *et al.*, *Inorg. Chem.*, 2006, **45**, 923.
149. S. Galijasevic, K. Krylova, M. J. Koenigbauer, *et al.*, *Dalton Trans.*, 2003, 1577.
150. G. Chaka, A. Kandegedara, M. J. Heeg, and D. B. Rorabacher, *Dalton Trans.*, 2007, 449.
151. A. J. Blake, V. Lippolis, S. Parsons, and M. Schroder, *Acta Crystallogr. C*, 2001, **57**, 36.
152. C. P. Kulatilleke, *Polyhedron*, 2007, **26**, 1166.
153. H. Meliani, C. Vinas, F. Teixidor, *et al.*, *Polyhedron*, 2001, **20**, 2517.
154. K.-M. Park, I. Soon, J. Seo, *et al.*, *Cryst. Growth Des.*, 2005, **5**, 1707.
155. H. J. Kim, M. R. Song, S. Y. Lee, *et al.*, *Eur. J. Inorg. Chem.*, 2008, 3532.
156. M. Jo, J. Seo, M. L. Seo, *et al.*, *Inorg. Chem.*, 2009, **48**, 8186.
157. E.-J. Kang, S. Y. Lee, H. Lee, and S. S. Lee, *Inorg. Chem.*, 2010, **49**, 7510.
158. M. Heller and W. S. Sheldrick, *Z. Anorg. Allgem. Chem.*, 2004, **630**, 1869.
159. M. Heller and W. S. Sheldrick, *Z. Anorg. Allgem. Chem.*, 2004, **630**, 1191.
160. T. Yamagushi, F. Yamasaki, and T. Ito, *Acta Crystallogr. C*, 2002, **58**, m213.
161. A. J. Blake, N. R. Champness, S. M. Howdle, and P. B. Webb, *Inorg. Chem.*, 2000, **39**, 1035.
162. A. J. Blake, N. R. Champness, S. M. Howdle, *et al.*, *CrystEngComm*, 2002, **4**, 88.
163. M. Fainerman-Melnikova, L. F. Lindoy, J. C. McMurtrie, and W. N. Setzer, *Acta Crystallogr. E*, 2003, **59**, m880.
164. A. Holzberger, H.-J. Holdt, and E. Kleinpeter, *J. Phys. Org. Chem.*, 2004, **17**, 257.
165. T. Tsuchiya, T. Shimizu, K. Hirabayashi, and N. Kamigata, *J. Org. Chem.*, 2002, **67**, 6632.
166. J. L. Shaw, J. Wolowska, D. Collison, *et al.*, *J. Am. Chem. Soc.*, 2006, **128**, 13827.
167. M. Kampf, J. Griebel, and R. Kirmse, *Z. Anorg. Allgem. Chem.*, 2004, **630**, 2669.
168. J. Seo, M. R. Song, J.-E. Lee, *et al.*, *Inorg. Chem.*, 2006, **45**, 952.
169. S. Y. Lee, J.-H. Jung, J. J. Vittal, and S. S. Lee, *Cryst. Growth Des.*, 2010, **10**, 1033.
170. K.-M. Park, I. Yoon, Y. H. Lee, and S. S. Lee, *Inorg. Chim. Acta*, 2003, **343**, 33.
171. S. Y. Lee, J. Seo, I. Yoon, *et al.*, *Eur. J. Inorg. Chem.*, 2006, 3525.
172. J. Seo, S. Y. Lee, and S. S. Lee, *Supramol. Chem.*, 2007, **19**, 333.
173. H. J. Kim, I. Yoon, S. Y. Lee, *et al.*, *New. J. Chem.*, 2008, **32**, 258.

174. M. R. Song, J.-E. Lee, S. Y. Lee, *et al.*, *Inorg. Chem. Commun.*, 2006, **9**, 75.
175. I. Yoon, J. Seo, J.-E. Lee, *et al.*, *Dalton Trans.*, 2005, 2352.
176. H. J. Kim, I. Yoon, S. Y. Lee, *et al.*, *Tet. Letts*, 2007, **48**, 8464.
177. D. Huang, A. J. Blake, E. J. L. McInnes, *et al.*, *Chem. Commun.*, 2008, 1305.
178. L. Ihlo, M. Kampf, R. Bottcher, and R. Kirmse, *Z. Naturforsch. B*, 2002, **57**, 171.
179. M. Kampf, R.-M. Olk, and R. Kirmse, *Z. Anorg. Allgem. Chem.*, 2002, **628**, 34.
180. D. Huang, X. Zhang, E. J. L. McInnes, *et al.*, *Inorg. Chem.*, 2008, **47**, 9919.
181. A. Kalam, S. Srivastava, Y. Pandey, *et al.*, *J. Ind. Chem. Soc.*, 2008, **85**, 203.
182. M. L. Helm, C. C. Combs, D. G. VanDerveer, and G. J. Grant, *Inorg. Chim. Acta*, 2002, **338**, 182.
183. M. L. Helm, K. D. Loveday, C. M. Combs, *et al.*, *J. Chem. Crystallogr.*, 2003, **33**, 447.
184. M. L. Helm, L. L. Hill, J. P. Lee, *et al.*, *Dalton Trans.*, 2006, 3534.
185. G. J. Grant, M. E. Botros, J. S. Hassler, *et al.*, *Polyhedron*, 2008, **27**, 3097.
186. M. L. Helm, D. G. VanDerveer, and G. J. Grant, *J. Chem. Crystallogr.*, 2003, **33**, 625.
187. M. L. Helm, G. P. Helton, and D. G. VanDerveer, *Inorg. Chem.*, 2005, **44**, 5696.
188. T. Tsuchiya, T. Shimizu, K. Hirabayashi, and N. Kamigata, *J. Org. Chem.*, 2003, **68**, 3480.
189. M. Wilhelm, S. Deeken, E. Berssen, *et al.*, *Eur. J. Inorg. Chem.*, 2004, 2301.
190. S. Y. Lee, S. Park, H. J. Lee, *et al.*, *Inorg. Chem.*, 2008, **47**, 1914.
191. H.-J. Holdt, H. Muller, A. Kelling, *et al.*, *Z. Anorg. Allgem. Chem.*, 2006, **632**, 114.
192. C. Gurnani, M. Jura, W. Levason, *et al.*, *Dalton Trans.*, 2009, 1611.
193. K. Wieghardt, M. Kleine-Boymann, B. Nuber, and J. Weiss, *Inorg. Chem.*, 1986, **25**, 1654.
194. M. F. Davis, W. Levason, G. Reid, *et al.*, *Dalton Trans.*, 2008, 533.
195. F. Cheng, A. L. Hector, W. Levason, *et al.*, *Chem. Commun.*, 2008, 5508.
196. A. L. Hector, W. Levason, G. Reid, *et al.*, *Dalton Trans.*, 2011,**40**, 694.
197. W. Levason, M. L. Matthews, R. Patel, *et al.*, *New J. Chem.*, 2003, **27**, 1784.
198. N. J. Hill, W. Levason, and G. Reid, *Inorg. Chem.*, 2002, **41**, 2070.
199. A. J. Barton, N. J. Hill, W. Levason, and G. Reid, *J. Am. Chem. Soc.*, 2001, **123**, 11801.
200. N. J. Hill, W. Levason, R. Patel, *et al.*, *Dalton Trans.*, 2004, 980.
201. A. J. Barton, A. R. J. Genge, W. Levason, and G. Reid, *J. Chem. Soc., Dalton Trans.*, 2000, 2163.

Cryptands and Spherands

Leonard F. Lindoy, Ki-Min Park, and Shim Sung Lee

Gyeongsang National University, Jinju, South Korea

1 INTRODUCTION

Naturally occurring antibiotic cyclic ionophores, such as valinomycin, have long been known to bind a range of alkali and alkaline earth metal ions, often wrapping around the metal cation such that its solvation shell is partially or completely replaced. A feature of particular systems of this type is that they are able to twist and adapt to the electronic and steric requirements of the metal ion involved and in one sense the bound ligand can be considered to act as a replacement solvent shell. The thermodynamics of complexation in these cases is thus necessarily complicated as it involves enthalpic and entropic terms associated with desolvation of the metal and, to a greater or lesser degree, also the ligand (resolvation of the bound ligand is a further contribution). Another contribution arises from the ligand conformational changes that occur in twisting around the metal center—normally an unfavorable contribution overall—and a largely favorable enthalpic term reflecting the electrostatic binding of the donor atoms to the metal cation. The cyclic antibiotics share many of their properties with the flexible synthetic crown ether macrocycles, first reported by Charles Pederson.[1] On the other hand, for ligand systems that are more rigid and have their donor atoms sets partially or fully preorganized for "ideal" interaction with a guest, enhanced binding is expected because unfavorable conformational ligand rearrangements are minimized. For this situation, there is less "loss of disorder" on complexation and an overall favorable (positive) contribution to the entropy of complexation is the result.

In this chapter, the synthesis and properties of two further categories of synthetic macrocyclic ligand types, exhibiting different degrees of restricted conformational flexibility, are discussed. These are the rigid spherands and the somewhat more flexible but still sterically restrained bi- or tricyclic cryptand ligands.

In 1987, the Nobel prize for Chemistry was shared by Charles Pedersen, Jean-Marie Lehn, and Donald Cram for their respective pioneering studies in the related fields of crown ether, cryptand, and spherand chemistry. Crown ether chemistry is presented elsewhere in this volume (see **Crown and Lariat Ethers**, Volume 3) and, in discussing the remaining two categories here, it is convenient to consider the less flexible category, the spherands, first.

2 SPHERANDS

2.1 What is a spherand?

Spherands, for which **1** may be considered prototypical, are a category of macrocyclic receptors with rigid cavities whose donor sites (normally oxygen) are fixed in space in relation to each other and directed inward for complexation with a range of complementary guests, which often have a spherical shape.[2–5] As a consequence of the rigidity, conformational changes on complex formation are minimal and the rigid electron-pair-lined cavities in uncomplexed

Supramolecular Chemistry: From Molecules to Nanomaterials.
Edited by Philip A. Gale and Jonathan W. Steed.
© 2012 John Wiley & Sons, Ltd. ISBN: 978-0-470-74640-0.

spherands have been postulated to be effectively nonsolvated.

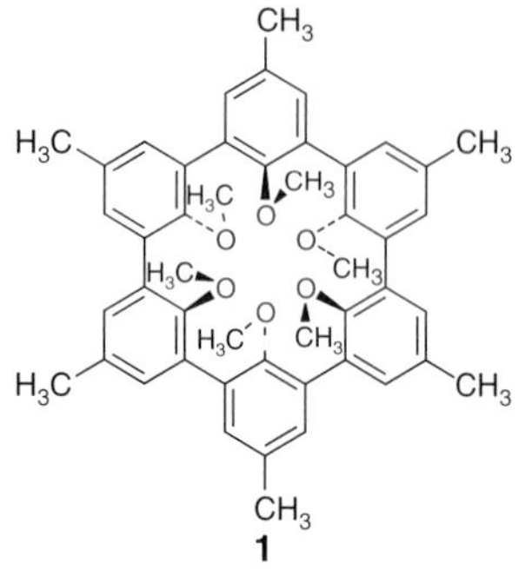

Preorganization is an important characteristic of this receptor type as spherands bind strongly to complementary guest molecules or ions in part due to their donor set having been positioned for complexation *during* spherand synthesis rather than this needing to occur during the complexation step. That is, in this receptor category, the donor atoms are orientated inward and so positioned that almost no conformational change is necessary for binding to the guest of interest. In particular, this will make a positive contribution to the entropy term for complexation as less "loss of disorder" is involved in the complexation step. This contrasts with the situation for simple flexible macrocyclic ligands, such as the cyclic antibiotics and crown ethers mentioned above, where some degree of conformational rearrangement is almost always necessary when binding to a suitable guest. Cram termed the enhanced stability arising from host–guest complexation in such cases the "*preorganization effect*." Since the term was first introduced by Cram, there have been a considerable number of spherand hosts reported in which the preorganization criterion is either fully or partially met.

Spherand **1** has a collar shape, D_{3d} symmetry, and is composed solely of *p*-methylanisole units. The six anisyl oxygens are arranged close to octahedrally being orientated alternatively below and above their mean plane. This rigid receptor was shown to be a highly selective and strongly binding complexing agent for Li^+ and Na^+ (while its cavity is too small to readily bind organic guests).[6,7] Both complexes are also very kinetically stable with respect to metal loss,[2] reflecting the operation of a significant macrocyclic effect.[8] Both the *p*-methyl and anisyl methyl groups of **1** provide lipophilicity, thus aiding solubility in organic solvents. This and other spherands, along with their complexes (the latter termed spheraplexes by Cram),[2] have hydrophobic exteriors and tend to be quite soluble in organic solvents while being virtually insoluble in water. Reflecting this, free spherands have been frequently employed as ionophores in two-phase (water:organic) solvent extraction processes for the selective uptake of particular metal ions from aqueous solutions (often as their picrate salts). For example, **1** selectively extracts Li^+ and Na^+ while ignoring K^+, Mg^{2+}, Ca^{2+}, and Sr^{2+}. The binding of Li^+ is stronger than that for Na^+ and, indeed, at the time it was first investigated, **1** was the strongest complexing agent for Li^+ thus far known.

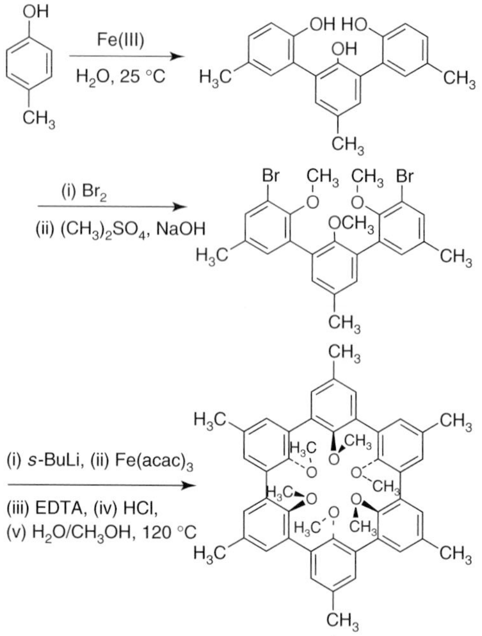

Scheme 1 Synthesis of the prototypical spherand **1**.[9]

2.2 A typical synthesis

The Cram group devised the multistep synthetic sequence shown in Scheme 1 to prepare **1** starting from *p*-cresol.[9] The latter precursor was oxidatively coupled using ferric chloride to give the corresponding triphenoxy derivative. This was then methylated with dimethyl sulfate and the product treated with butyl lithium followed by reaction with tris(acetonato)iron(III) to give (after workup) **1** as its lithium complex. Reaction of this product with EDTA (ethylenediaminetetraacetic acid) yielded metal-free **1** in 6.3% overall yield. A more general discussion of the synthesis of spherands and their derivatives is presented elsewhere.[10]

2.3 Spherand complexation

2.3.1 *Classical spherands*

The X-ray structure of **1** is illustrated in Figure 1 and confirms that the six oxygens are arranged octahedrally to produce a roughly spherical cavity (with a diameter

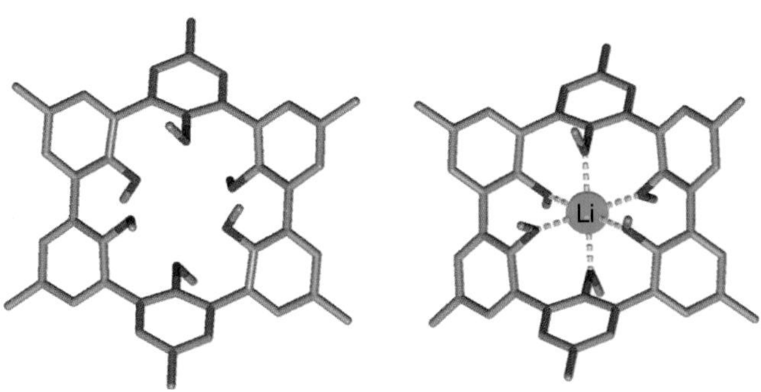

Figure 1 Schematic representation of the X-ray structure of spherand **1** and its Li^+ complex, $[Li(\mathbf{1})]^+$; hydrogen atoms are not shown (red, oxygen; gray, carbon).[11,12]

of 1.62 Å).[11] Clearly, there is significant repulsion present between the electronegative oxygen donors in the uncomplexed ligand; however, this is ameliorated to a greater or lesser degree on binding to a cationic guest. The cavity diameter of **1** lies between the diameter of Li^+ (1.48 Å) as observed in the X-ray structure of the lithium complex, $[Li(\mathbf{1})]^+$ (Figure 1), and Na^+ (1.75 Å), as observed in the X-ray structure of the corresponding 1 : 1 sodium complex; both the Li^+ and Na^+ structures are otherwise very similar.[12] Solution thermodynamic studies and subsequent gas phase density functional theory (DFT) calculations[13] have both confirmed the high affinity of **1** for each of these alkali metal cations. On the basis of $-\Delta G^\circ$ values obtained from picrate salt extraction experiments, Li^+/Na^+ selectivity was >600 in favor of Li^+ under the conditions employed. In this context, it needs to be noted that, because of their inherent rigidity, a strong correspondence is expected between the respective solid-state and solution structures.

The initial synthesis of the "bridged" spherand **2** involved seven steps from *p*-cresol and was obtained in 1.2% overall yield.[14] The rigid support framework in this species focuses the 12 pairs of electrons of the six attached oxygen atoms toward an enforced cavity. As might be anticipated from its similar cavity size to **1**, the system complexes Li^+ and Na^+ but rejects other alkali or alkaline earth ions.[15]

2

The results from molecular mechanics calculations confirmed that **2** (which can exist in either a *syn* or an *anti*

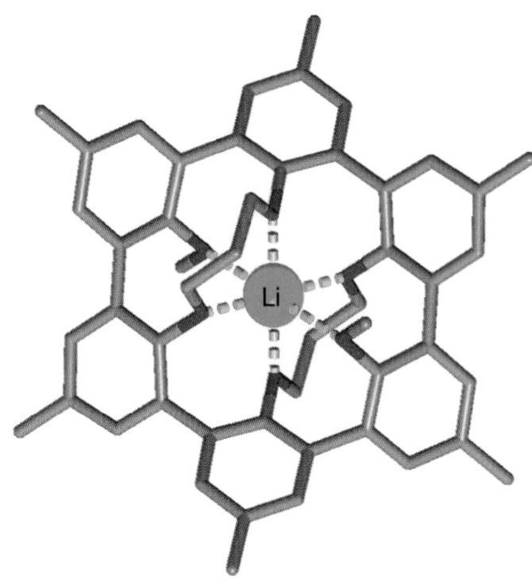

Figure 2 X-ray structure of the cationic Li^+ complex of the *anti*-isomer of **2** (red, oxygen; gray, carbon).[19]

form) would be an excellent binder of both Li^+ and Na^+[16] and that the *anti* form should bind the above metal ions more strongly than the *syn* form. Indeed, it was also predicted that *anti*-**2** would bind these cations even more strongly than **1**. These predictions were largely subsequently confirmed experimentally—for example, $[Li(anti\text{-}\mathbf{2})]^+$ does not release Li^+ when the complex is heated at ~100 °C in methanol/water, whereas, for $[Li(\mathbf{1})]^+$, the metal is lost under these conditions.[17] Further, it was suggested that the lithium complex, $[Li(anti\text{-}\mathbf{2})]^+$, which was synthesized by a Li^+ template procedure, appears to be unable to dissociate unless covalent bonds are broken.

The X-ray structures of the 1 : 1 complexes of Li^+ with both the *syn*[7,18] and *anti*[19] isomers of **2** have been reported. The coordination environment of the lithium is distorted octahedral in each complex. The structure of the Li^+/*anti*-**2** complex is shown in Figure 2; it displays four short [1.944(2)–1.998(2) Å] and two long [2.381(2) and 2.455(2) Å] Li–O bond lengths. The host *anti*-spherand is less strained in this complex than occurs in the structure of its *syn* analog.

3

Increasing the aryl framework to form the octaspherand **3** results in this larger cavity spherand showing its strongest

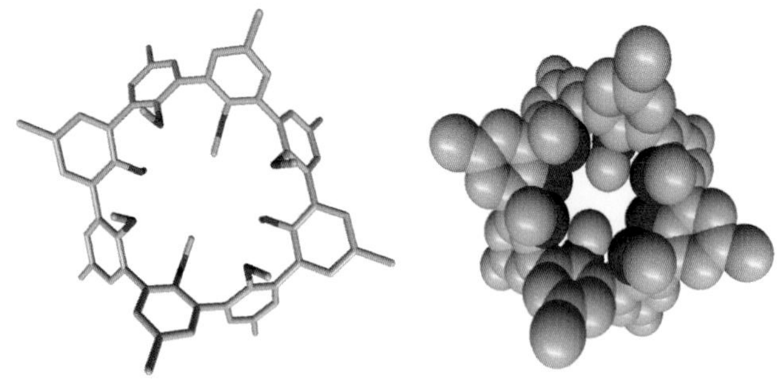

Figure 3 Schematic representations of the X-ray structure of the octaspherand **3**; hydrogen atoms are not shown (red, oxygen; gray, carbon).[20]

affinity for Cs^+ amongst the alkali metals.[20] Like **1** and **2**, this larger ring receptor was prepared by a multistep procedure; adding CsBr as a template during the synthesis resulted in an enhanced (but still low) yield of **3**, which in the latter case was isolated as its Cs^+ complex. The X-ray structure of guest-free **3** (Figure 3) shows that each of the eight oxygens is orientated *anti* to its neighbors in the ring such that an "up–down" conformational arrangement is present. However, it is noted that the free spherand is not perfectly preorganized for binding a spherical metal ion as two of the $-OCH_3$ groups are oriented such that they partially occupy the central cavity (see space-filling structure of **3** in Figure 3).[20] Nevertheless, the system is more flexible than **1** and hence the cavity can readily accommodate a wide range of guests, including ammonium and alkylammonium cations.

The crystal structure of the $CsClO_4$ complex of **3** (Figure 4) shows that the metal is bound by all eight oxygens of **3** located at the apices of a square antiprism, with two trans oxygens from two perchlorate anions completing the coordination shell.[20] Comparison of the crystal structure of this complex with that of its free spherand (**3**) indicates that complexation results in some host reorganization. Namely, the inward-turned methyl groups are reorientated outward and the square antiprism defined by the oxygens shrinks to better accommodate the diameter of the Cs^+ guest.

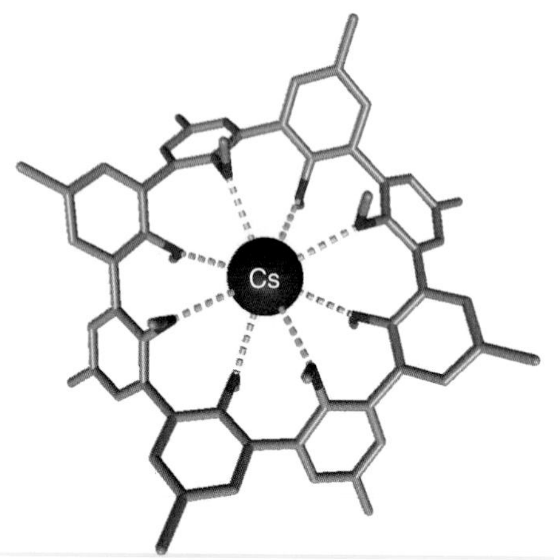

Figure 4 Schematic representations of the X-ray structure of $[Cs(\mathbf{3})]^+$ in the $CsClO_4$ complex of the octaspherand **3**, two oxygens from two perchlorate anions (not shown) complete the coordination shell; hydrogen atoms are not shown (red, oxygen; gray, carbon).[20]

4

Chiral spherand **4** incorporating four binaphthyl groups and eight oxygen donors has also been synthesized; with its larger cavity, it also binds strongly to Cs^+ and forms stable host–guest complexes with NH_4^+, $CH_3NH_3^+$, $(CH_3)_3CNH_3^+$, H_3NOH^+, $H_3NNH_3^{2+}$, $H_3N(CH_2)_2CH(CO_2H)NH_3^{2+}$, and 1,3-$H_3NC_6H_4NH_3^{2+}$.[20,21]

2.3.2 *Spherand derivatives*

In early studies, the Cram group produced examples of both hexa- and octafluoro analogs of the "classical" anisole-containing spherands **1** and **3** in which fluoro groups replaced the $-OCH_3$ groups. However, neither of these rings showed a tendency to complex alkali metal ions under the conditions employed, presumably reflecting the low donor capacity of the electronegative substituents.[22,23] Preparation of the corresponding octacyano derivative in which cyano groups replace the fluoro groups in the above octafluoro spherand did, however, lead to complexation with alkali metals as well as with ammonium and alkylammonium ions.[24]

5

The Cram group also developed syntheses for "hybrid" spherands that incorporated cyclic urea functions along with anisyl groups.[25,26] In part, the rationale for the synthesis of these derivatives was that it was anticipated that the urea oxygens would be significantly better donors than anisyl oxygens. An example of this receptor category is given by **5**. However, such hosts are more flexible than the related all $-OCH_3$ spherands and again, because of this, binding constants tend to be lower. Nevertheless, the additional flexibility is reflected by higher rates of complexation (and decomplexation). As the preorganization criterion is only partially met in **5**, it is perhaps better termed a "hemispherand" (see later). The binding of **5** peaks at K^+ and the urea oxygens on the "top" face are used for metal binding. The additional flexibility arising from the inclusion of urea functions enables the binding of larger cations than is possible for the prototype spherand **1**. Thus, **5** also binds ammonium and alkylammonium cations; in the case of the *t*-butylammonium cation, an X-ray diffraction structure shows that the three hydrogens of the ammonium function hydrogen bond to the three (similarly orientated) urea oxygens in a tripodal "perching" manner (Figure 5).[27] It is the lack of methyl substituents on the urea oxygens (and the resulting absence of steric hindrance) that makes such a perching arrangement possible in this case. The X-ray structures of the 1 : 1 complexes of **5** with Na^+ and Cs^+ have also been obtained.[25] In each structure, a water molecule also binds to the central metal from the side of the ring that is opposite to the methyl groups of the anisyl units. Surprisingly, one anisole oxygen does not coordinate to the central metal (even though molecular models suggest that binding does appear to be sterically possible).

A number of chromoionophore derivatives based on spherands (and hemispherands) have been reported.[28,29] For example, the chromoionophoric spherand **6** was synthesized by the Cram group.[30] In this, the chromophoric group corresponds to the 4-(2,4-dinitrophenylazo)phenol fragment. A dramatic shift in the pK_a of the (nitrophenylazo)phenol entity occurs from 13.0 for free **6** to 5.9 in the presence of Li^+ and to 6.3 in the presence of Na^+ (in a mixture v/v of 80% dioxane and 20% water). In contrast, the pK_a is only slightly influenced by the presence of K^+, Mg^{2+}, or Ca^{2+} ions. Hence, the strong bonding by Li^+ and Na^+ greatly promotes deprotonation of the phenol entity. This is accompanied by a color change from yellow in the free ionophore to deep blue-violet in the presence of these ions. By measuring the absorption increase at ~590 nm, both Li^+ and Na^+ can be detected at concentrations as low as 10^{-8} M.

6

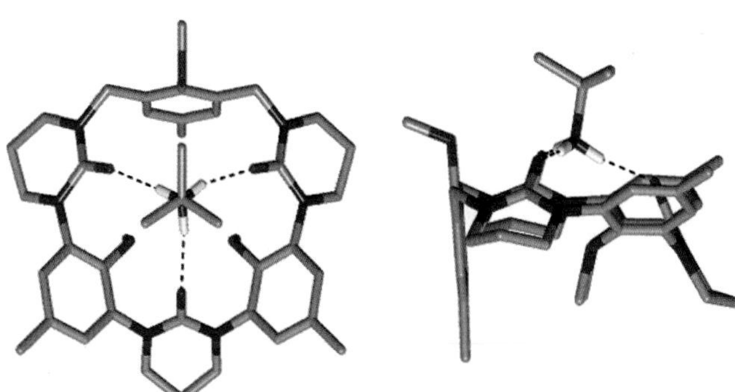

Figure 5 Two views of the X-ray structure of the perching *t*-butylammonium adduct of the cyclic urea-containing derivative **5**, hydrogen atoms are not shown (red, oxygen; blue, nitrogen; gray, carbon).[27]

Although they are not discussed in detail here, it is noted that a number of spherand-type calixarenes, combining aspects of the structural features of both calixarenes and spherands, have now been synthesized—with their structures and properties being the focus of a range of studies.[31–36] One example of this broad category is given by the calixspherand derivative **7**, which forms kinetically stable complexes with Na^+ and K^+.[37] As illustrated by the structure of its Na^+ complex (Figure 6),[38] the sodium ion is totally encapsulated on complexation, with some rearrangement of the free ligand structure having occurred.[39]

7

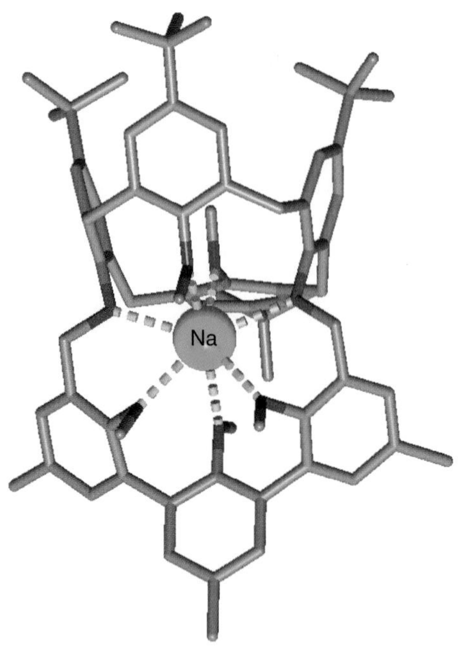

Figure 6 Schematic representation of the X-ray structure of the calixspherand derivative, $[Na(\mathbf{7})]^+$ (red, oxygen; gray, carbon).[38]

8

9

10 R = Methyl
11 R = 2,4,6-Trinitroanilino

Before leaving this section, it needs to be noted that a range of computational studies has been performed on spherands and related derivatives.[40–44] In particular, conformational and binding-free energy aspects have been probed using procedures that include molecular mechanics, molecular dynamics, and DFT. Overall, such calculations have proved to be quite successful in providing a fuller understanding of the nature and behavior of individual systems.

2.3.3 Hemispherands and cryptaspherands

In general terms and as already mentioned above, if some of the preorganization of a spherand is removed, then complexation to a wider range of ions becomes possible, usually with some loss of specificity. A hemispherand may be described as a ring system in which approximately half of the molecule is not rigidly preorganized. For example, the flexible (or semiflexible) part may be a polyether chain or a crown (or azacrown) ether unit (for example, see **8**–**10**). Clearly, a hemispherand such as **8** is less preorganized than the cryptahemispherands **9** and **10**. In the latter cases, the bicyclic cryptand nature limits both the conformational flexibility as well as solvent access to the central cavity; even so, both types are of course less preorganized than a "pure" spherand such as **1**.

The X-ray structures of the Na^+, K^+, and Cs^+ complexes of **10** have been determined and the tri-anisyl fragments in each case were found to possess the same preorganized configuration, with the unshared electron pairs of the three methyl groups turned inward and the methyl groups orientated outward.[45] Of the alkali metals, Cs^+ matches the size of the cavity well and an X-ray structure determination shows that this ion binds strongly to all nine donors of **10** (Figure 7). In contrast, a mismatch with the size of the cavity occurs for smaller Na^+ and K^+ ions. Indeed, for Na^+, only five of the potential donor sites of **10** bind to this ion. In accord with these observations, the 1 : 1 Cs^+ complex of **10** was shown to be more stable in solution than its Na^+ and K^+ analogs.

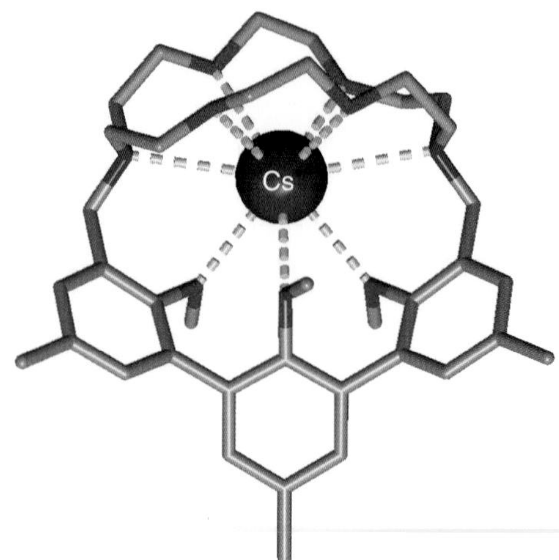

Figure 7 Schematic representation of the X-ray structure of the $[Cs(\mathbf{10})]^+$ cation illustrating the encapsulation of the Cs^+ ion by the cryptahemispherand **10** (red, oxygen; blue, nitrogen; gray, carbon).[45]

Functional derivatives of the present category of receptors have also been reported. For example, the chromogenic cryptahemispherand **11** was demonstrated to be a useful chromoionophore for K^+, with a detection limit of $\sim 4 \times 10^{-7}$ and selectivity over Na^+ of greater than 100 under the conditions employed.[46] For this system, the combination of the preorganized anisyl groups coupled with the flexibility of the diazacrown ether moiety (and the size of the corresponding 3D cavity) thus results in a reagent showing excellent selectivity for the spectrophotometric determination of K^+.

Finally, the cryptahemispherands as a class are seen to be stronger complexing agents than the cryptands (see next section) with, however, both classes exhibiting comparable ion selectivity behavior.[47]

3 CRYPTANDS

3.1 What is a cryptand?

Cryptands (from Greek: *cryptos* = cave) are cage-like ligands whose major categories are based on the generalized shapes given by **12** and the more spherical **13** where the bridgehead atoms (X) are typically nitrogen, carbon, or (less commonly) phosphorus and the "straps" contain donor (or in some cases acceptor) groups suitably spaced to interact and bind to a guest ion or molecule contained in the cavity. These cage systems thus differ from a classical spherand in that they commonly incorporate a more or less flexible framework while still maintaining a measure of preorganization with respect to the spatial positions of their functional sites.

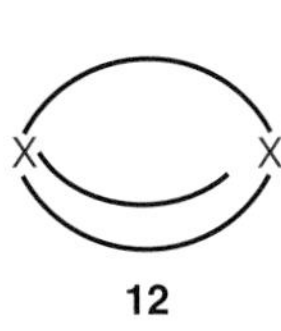

12

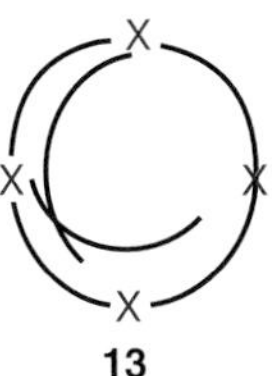

13

At the time of writing, the cryptand literature is enormous, with several thousand references mentioning cryptands having appeared. Hence, it is not possible to give a comprehensive coverage of the field. Instead a description of the characteristic features of this important cage category is presented on the basis of a selection of representative cryptand types. In particular, an attempt has been made to discuss the host–guest chemistry of representative cryptand types, with emphasis given to earlier studies during which the salient aspects of cryptand complexation behavior were established.

3.2 Early studies

3.2.1 An O,N-donor cryptand series

In a seminal paper appearing in 1969, Dietrich, Lehn, and Sauvage[48] described the first synthesis of a series of cryptands of the type **14**. In part, it was the known alkali metal binding (and membrane transport behavior) of the natural antibiotic ionophores that provided the inspiration for the design of these then novel cage systems.[49]

14

m	*n*	Abbreviation	Estimated cavity diameter (Å)
0	0	1.1.1	1.0
0	1	2.1.1	1.6
1	0	2.2.1	2.2
1	1	2.2.2	2.8
1	2	3.2.2	3.6
2	1	3.3.2	4.2
2	2	3.3.3	4.8

The preparation of the above systems allowed the two-dimensional diazacrown macrocyclic chemistry to be extended into three dimensions. These cage systems incorporate a moderately flexible framework that surrounds a central cavity, with oxygen and nitrogen atoms spaced throughout the framework such that stable five-membered chelate rings are able to form on binding a metal cation within the cavity. Along the series, the number of heteroatoms incorporated in the cage structure as well as the size of the cavity available for guest binding varies in a stepwise manner.[50,51]

3.2.2 Cryptand synthesis

A common synthesis of cryptands such as those of type **14** has involved condensation of an appropriate diacid dichloride with a diazapolyoxo macrocycle followed by reduction of the diamide product using lithium aluminum hydride or diborane. For example, one such synthesis for 2.2.2 is given in Scheme 2, where the macrocycle precursor is 1,10-diaza-18-crown-6. A procedure such as this is often carried out under high-dilution conditions in order to promote discrete cage formation over linear polymerization, which is promoted by higher concentrations. Apart from using a large volume of solvent and/or small amounts of reactants, high dilution during the condensation reaction has also often been achieved by the *very* slow addition of the reagents via the use of carefully controlled dropping funnels or with the aid of motorized syringes.

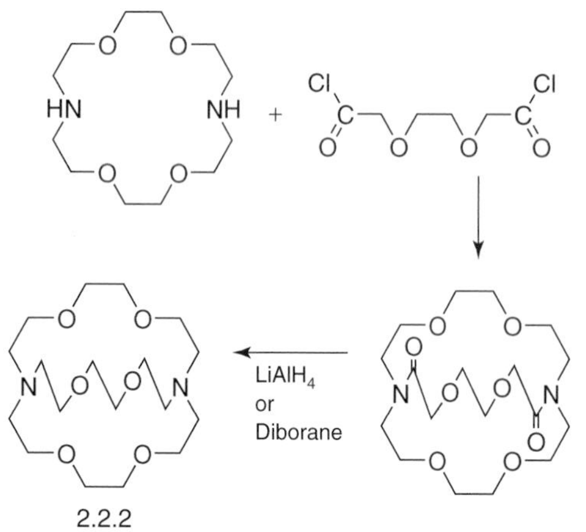

Scheme 2 A synthesis for cryptand 2.2.2 via its diamide precursor.

3.2.3 Host–guest complexation

Relative to monocyclic systems, the cavity in such bicyclic cryptands is more clearly delineated and, in principle, steric and electronic complementarity with a guest ion or molecule is able to be obtained more readily through the rational design of individual cages. This is an important advantage over single-ring systems in terms of designing for guest selectivity and, indeed, such selectivity, based on cation size match for the cryptand's cavity, has frequently been achieved. For example, the smaller members of the cryptand series given by **14** show such behavior—the increasing cavity size along the series 2.1.1, 2.2.1, and 2.2.2 is reflected by the preferential complexation of Li^+, Na^+, and K^+ respectively.[52] Nevertheless, the flexibility of polyether-containing bridges results in the ability of these cages to alter their conformation within defined limits in order to accommodate (at least partially) the electronic and steric requirements of different included guests. This is especially the case for the larger cryptands in the series. However, while a higher degree of rigidity would give rise to increased preorganization for complexation (and be expected to result in enhanced thermodynamics of complexation for individual complementary guests), as discussed earlier it will generally also contribute unfavorably to both entry of the guest into the cavity and its exit from the cavity—giving rise to slow kinetics of both formation and dissociation—with the effect being greater for dissociation.

"Peak" thermodynamic selectivity can occur for a given metal as the cavity size progressively increases across the series of cryptands of type **14**. Thus, the 1 : 1 stability constants (log values) for the interaction of K^+ with 2.1.1, 2.2.1, 2.2.2, and 3.2.3 are <2.0, 4.0, 5.4, and 2.2 respectively, reflecting that the match of the cavity size and the number of donor sites in 2.2.2 provide optimal complementarity for this cation.

Besides the alkali metals, cryptands of the above type also form stable complexes (termed *cryptates*) with the alkaline earth and lanthanide ions as well as with individual transition, posttransition, and main group ions.[8,49,51]

Aza-containing cryptands have been observed to display unusual protonation/deprotonation behavior. In principle, diaza-cryptands such as those given by **14** are able to exist in three isomeric forms, as the bridgehead nitrogen lone pairs are capable of being orientated both *endo* (that is, into the cavity), both *exo*, or one *endo* and one *exo*.[53,54] Such unusual protonation behavior has been observed for smaller cages of this type. For example, the 1.1.1 and 2.1.1 cryptands have been shown to be associated with slow proton exchange behavior and, in the case of the 1.1.1 cage, deprotonation remains very slow even in the presence of strong base—reflecting the presence of these protonated cages in their "in–in" configurations.

Complexation of the alkali and alkaline earth metal ions by "parent" cryptands of type **14** has commonly been followed in solution using ^1H and ^{13}C NMR and/or by spectrophotometric titration studies. In addition, many X-ray structures of metal cryptates of these ligands have now been reported. In these, the metal ion is frequently observed to occupy the central cavity of the cryptand. Typically, the cryptand is bound in its "in–in" configuration with the tertiary amines contributing to the binding of the metal and, at least in the case of the alkali metals, it is the electrostatic interaction with the ether oxygens that appears to dominate. The X-ray structure of the Na^+ complex of 2.2.2 is shown in Figure 8; as expected, the cryptand is present in its "in–in" configuration.[55,56]

A characteristic feature of cryptand complexes involving full encapsulation of a bound ion, such as frequently observed for cryptands of type **14**, is the increased stability of the product complex relative to related monocyclic ligand complexes. In this context, it is noted that a variety of methods have been employed to measure host–guest

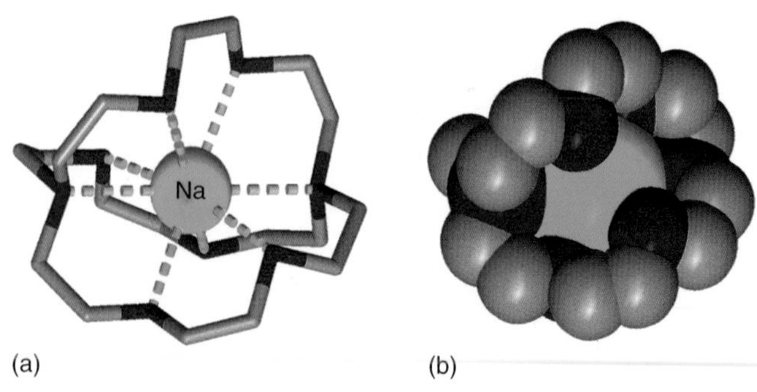

Figure 8 (a) Stick and (b) space-filling depictions of the X-ray structure of the $[Na(2.2.2)]^+$ cryptate showing (b) how the metal ion completely fills the cavity of this cryptand (red, oxygen; blue, nitrogen; gray, carbon).[56]

binding constants, including the routine use of pH, cation-selective electrode, UV–vis spectrophotometric, NMR, and calorimetric titrations.

Interestingly, the K^+ complex of 2.2.2 is approximately 10^5 more stable than the corresponding diazacrown complex. Such enhanced stability has been termed by Lehn the *cryptate effect* (which encompasses both enhanced kinetic as well as thermodynamic stability). The cryptate effect may be considered to be a special case of the well-known macrocyclic effect.[8] In this context, it needs to be noted, however, that a number of complexes exist in which a metal ion is coordinated *exo* to the cryptand's cavity (or which only partially occupy the cavity) and, of course, the cryptate effect will be absent or greatly reduced in such cases. Similarly, complexes of 2 : 1 (metal : cryptand) are known. For example, cryptand 2.1.1 forms a complex of type $[Pb_2(2.1.1)]^{4+}$ for which it was postulated that one or both metals are bound *exo* to the cavity as clearly both lead ions are unable to simultaneously occupy the cryptand's central cavity.[57]

Like the crown ethers, cryptands of type **14** have been demonstrated to solubilize a number of inorganic salts in both polar and nonpolar solvents. For example, 2.2.2 solubilizes $KMnO_4$ in organic solvents such as benzene allowing oxidations to be performed in such a nonaqueous medium. Further, 2.2.2 dramatically increases the solubility of barium sulfate in water. As cryptands are capable of completely encapsulating a metal ion and hence removing the ion from its solvent shell, they have also been favored reagents for phase-transfer applications, especially phase-transfer catalysis.

Complexation of the NH_4^+ cation by a range of cryptands has been well documented. For example, 2.2.2 forms a stable complex with NH_4^+ in water ($\log K = 4.5$); while this value is less than that for K^+ ($\log K = 5.3$) under similar conditions, it is still quite respectable.[52,58] The X-ray structure of this host–guest species (Figure 9) confirms that the NH_4^+ is contained in the cavity, being hydrogen bonded to one bridgehead nitrogen and three oxygen atoms.[59] Weaker electrostatic interactions are also present between the included NH_4^+ and the electronegative oxygen sites.[58]

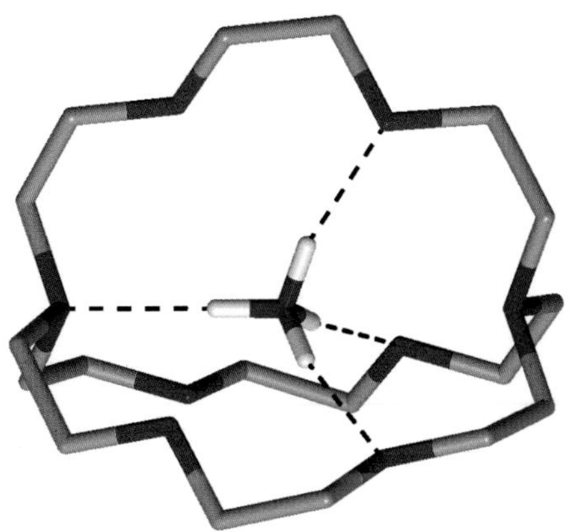

Figure 9 X-ray structure of the 1 : 1 complex between 2.2.2 and the NH_4^+ (red, oxygen; blue, nitrogen; gray, carbon).[59]

3.3 Further studies

3.3.1 Mixed heteroatom systems

Since these early studies, many cryptands have been reported—incorporating a wide variety of cavity sizes and potential donor atom types. Some further representative cryptands are given by **15**–**22**.[8,49]

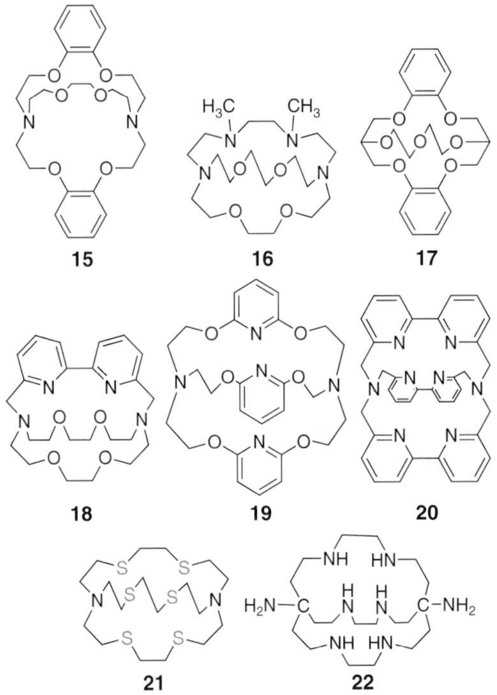

In the synthesis of many cryptands, a metal ion has been specifically employed as a template (often chosen to match the cavity size of the target cage) to boost the yield of the product. In other cases, a metal template effect is undoubtedly in operation through the use of an alkali metal reagent even though such a role had not always been anticipated when the synthetic procedure employed was chosen. However, it needs to be noted that apart from the above a number of other reaction types, employing a diverse range of precursors, have been employed for cryptand formation.[60]

As might be anticipated, the presence of substituents on the members of the "parent" cryptand series given by **14** may markedly affect their selectivity for individual metal ions. For example, incorporation of a benzo substituent on the backbone of 2.2.2 to yield **23** results in the corresponding Na^+ cryptate being more stable (in methanol) than occurs for 2.2.2, while the stability of the K^+ complex is decreased relative to that of 2.2.2.

23

24

Preliminary studies showed that the N_2O_6-donor cage **24** is a poor coordinating agent toward individual alkali metals.[61,62] On the basis of X-ray and molecular modeling (including DFT) studies, it was concluded that the systems preferentially adopt *exo* arrangements in which the lone pairs of the nitrogen bridgeheads are orientated outward away from the cavity. This results in the adjacent methylene groups being directed into the cavity, significantly reducing its available volume. In this configuration, the cavity is clearly too small to be occupied by an alkali metal ion. The X-ray structure of uncomplexed **24**, including a space-filling depiction, is shown in Figure 10.[61]

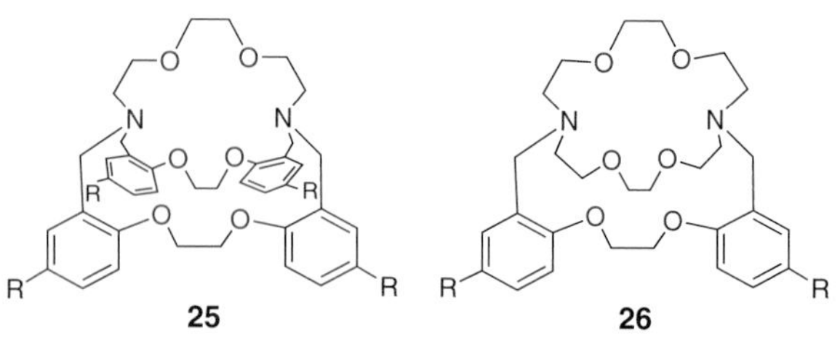

Contrasting with the above, when one or two of the aromatic ring-containing "straps" linking the bridgehead nitrogens were replaced with aliphatic ($-CH_2CH_2OCH_2CH_2OCH_2CH_2-$) groups to give **25** and **26** (R = *t*-Bu in each case), then Na^+ uptake was observed to occur in each case, confirmed for **26** by an X-ray structure determination of the Na^+ complex (Figure 11).[63]

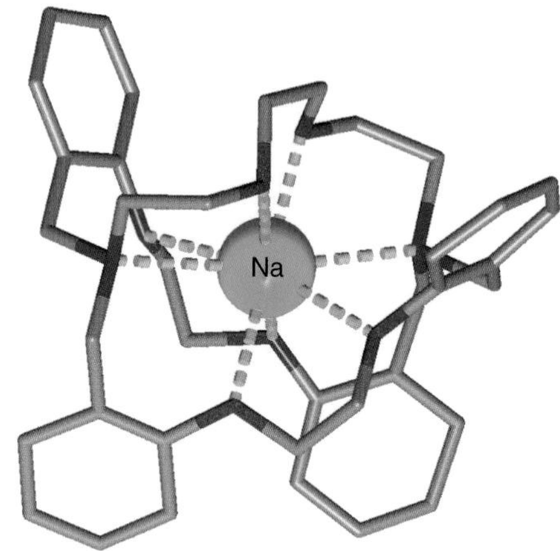

Figure 11 The X-ray structure of $[Na(\mathbf{26})]^+$ showing inclusion of the Na^+ ion in the cavity of **26** (red, oxygen; blue, nitrogen; gray, carbon).[63]

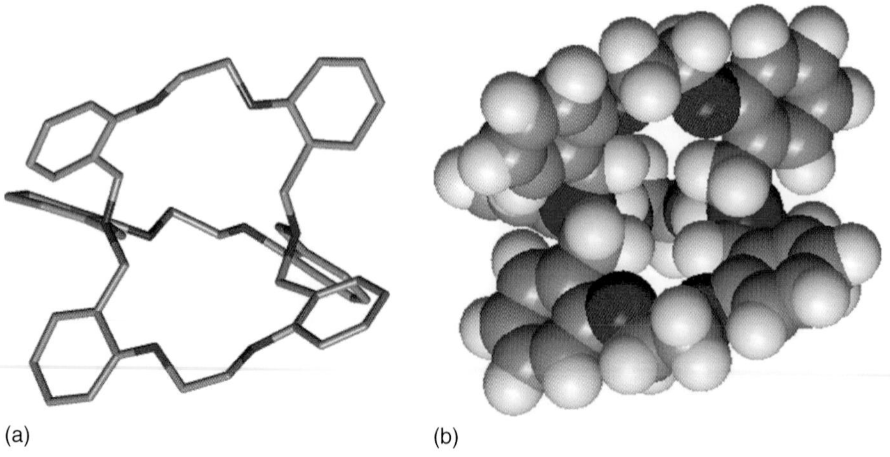

Figure 10 Two views of the X-ray structure of cryptand **24** illustrating (a) the *exo* orientation of the bridgehead nitrogen lone pairs and (b) how the volume of the cavity is substantially reduced due to steric crowding (red, oxygen; blue, nitrogen; gray, carbon).[61]

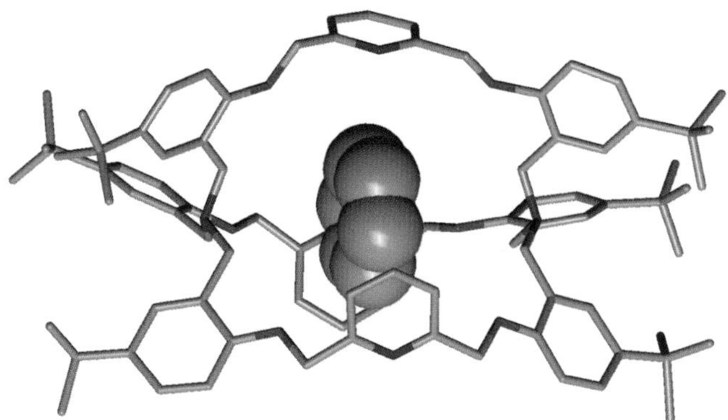

Figure 12 X-ray structure of the benzene inclusion complex of the tris-pyridyl cage **27** (red, oxygen; blue, nitrogen; gray, carbon).[64]

In view of the rigidity of the N_2O_6-cage **24**, an attempt was made to extend the size of the cavity that involved replacing each of the $-CH_2CH_2-$ bridges with pyridyl-containing bridges (see **27**). The rigidity associated with the tribenzylamine bridgeheads maintains a measure of preorganization into this cage. The X-ray structure of **27** recrystallized from benzene (Figure 12) shows that a benzene guest is centrally included in the slot-like cavity, held in position by edge-to-face π-stacking interactions.[64] Benzene uptake is also likely aided by the hydrophobic environment present within the cage. In deuterochloroform, this cage was also demonstrated to selectively bind the triphenolic derivative, phloroglucinol (with lesser affinity for the corresponding diphenolic derivative, resorcinol, also being event).[65] Attempts to isolate metal complexes of **27** were not successful.

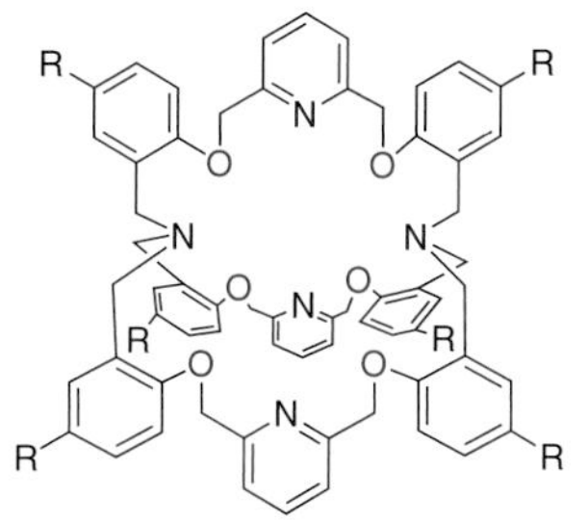

27

In an extension of the above studies, the corresponding cage systems **28** (R = H or *t*-Bu) incorporating a 2,2-bipyridyl rather than a 2,5-pyridyl fragment in each strap were synthesized to enlarge the cavity in order to enable metal ion binding.[66] Both cages were ultimately obtained in high yield employing a metal template procedure.[67] The design of this cage system was also directed at achieving helical coordination on binding to an octahedral metal ion. On the basis of molecular modeling, each triaryl-containing cap at either end of the cage structure was

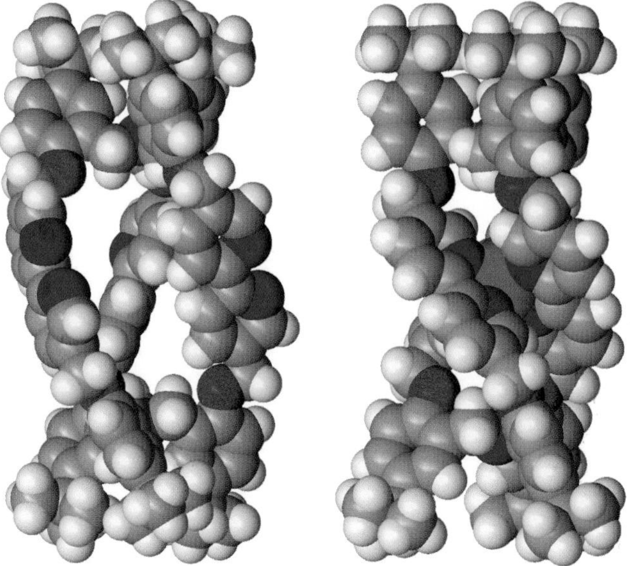

Figure 13 Comparison of the X-ray structure of (a) the metal-free cryptand **28** with that of its (b) triple-helical Cu(II) complex (red, oxygen; blue, nitrogen; gray, carbon).[67]

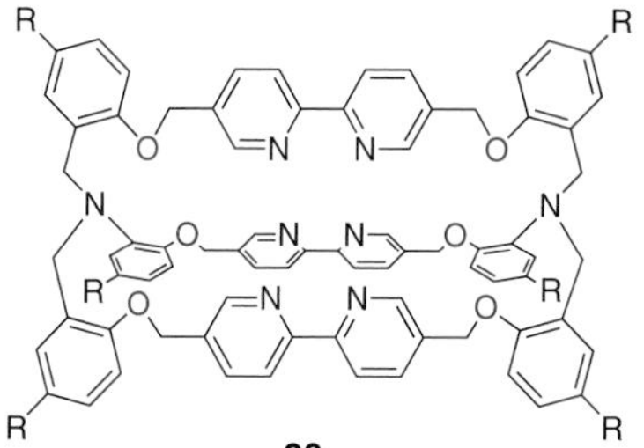

anticipated to adopt a three-bladed propeller configuration on octahedral metal coordination of the bipyridyl groups, provided the metal ion binds symmetrically in the cavity. This, in combination with the normal chiral twist about the octahedral metal ion site (around the C_3-axis of the coordination sphere), was predicted to promote the generation of a triple helix along the length of the cage's long axis. Mn(II), Fe(II), Ni(II), and Cu(II) each form 1 : 1 complexes with **28**. Intriguingly the X-ray structures of the above four metal ions fall into two groups: those incorporating Ni(II) and Cu(II) have their bridgehead nitrogen lone pairs orientated *exo*, while those of Mn(II) and Fe(II) have them orientated *endo*. The structure of the Cu(II) complex of **28** (R = *t*-Bu) is shown in Figure 13 alongside that of the free ligand; the Cu(II) complex is isostructural with its Ni(II) analog.

28

Despite the three straps in each **28** allowing some flexibility, the solid-state structure of the metal-free ligand (R = *t*-Bu) gives evidence for the desired propensity of this cage to adopt a helical arrangement (Figure 13a). This promise was met in the structure of the corresponding

Cu(II) complex (Figure 13b), which adopts a triple-helical configuration that extends approximately 22 Å along the long axis. Such behavior is unusual relative to other metal-containing helical structures reported previously in which more than one metal ion is employed to induce extended helicity of this type.

29 **30**

Macrotricyclic cage systems such as **29** have also been investigated as spherical receptors for cations, anions, as well as neutral molecules.[68–70] These symmetrical cages were synthesized using multistep, high-dilution procedures. Such cages form stable complexes with a range of alkali and alkaline earth metal ions as well as with the NH_4^+ ion. For example, spherical **29** readily includes an NH_4^+ ion, which is strongly held in place by a tetrahedral array of hydrogen bonds between the guest ammonium hydrogens and the four tetrahedrally orientated tertiary amine sites on the cryptand.

This ligand forms a more stable complex with NH_4^+ than with K^+ (log K values of 6.1 and 3.4, respectively). Interestingly, this is the reverse of the order observed for less symmetrical 2.2.2 with these ions.[71] Cage **29** also forms a moderately strong complex with Rb^+ in water (log $K = 4.2$).[71] The X-ray structure of the 1 : 1 complex of **29** with NH_4^+ shows that this cation occupies the cavity of the cage,[69] being bound via a tetrahedral array of four $N{-}H^+\cdots N$ hydrogen bonds involving each of the bridgehead nitrogens. The binding is presumably assisted by dipolar interactions between the positive charge on the included NH_4^+ ion and electronegative oxygen heteroatoms present in the cage's framework.

Although the area developed more slowly than that of cryptand–cation host–guest chemistry, examples of cryptand–anion binding have become increasingly common over the past two decades or so and examples of neutral molecule guest binding have also been documented, often involving the inclusion of solvent molecules in the cavity. An example of the latter type involves cage **29**, which in its diprotonated form also binds a water molecule.[69] In this host–guest species, the water is bound tetrahedrally by two $O{-}H\cdots N$ and two $NH^+\cdots O$ hydrogen bonds. ^1H NMR studies indicate that water exchange is slow on the NMR time scale. Subsequently, a limited number of reports of larger cage molecules including discrete water clusters have appeared, including a report of a tris 2,5-thiofuranyl-based octaazacryptand encapsulating a "hybrid" amine–water cyclic pentamer in its hydrophobic cavity.[72]

Protonated **29** also forms inclusion compounds with the spherical anions, Cl^- and Br^-, while I^- is too large to fit in the cavity. This cage is selective for Cl^- with the binding constant being $>10^4$ in water.[70] As might be expected, the tetramethylated analog **30** also gives rise to Cl^- and Br^- inclusion complexes but again I^- is too large to occupy the cavity.

31

A number of cryptand systems that include nonlinear anions have also been investigated. For example, the linear triatomic N_3^- ion forms an inclusion complex with the ellipsoidal-shaped hexaprotonated cryptand (see **31**) whose X-ray structure is given in Figure 14.[73] Not surprisingly, this N_8O_3 heteroatom receptor also binds spherical ions such as the halides. Anion selectivity by a range of such protonated "bis–tren" [where tren is tris(2-aminoethyl)amine] cryptand systems incorporating a range of linking groups between the tren caps has now been demonstrated.[74–76]

Of course, bis–tren species of the above type also readily bind a range of transition and posttransition metal ions to yield dinuclear species (see also Section 3.3.2).[74] In a number of cases, these dinuclear species have been shown to also bind small molecules or ions (including OH^-, N_3^-, and

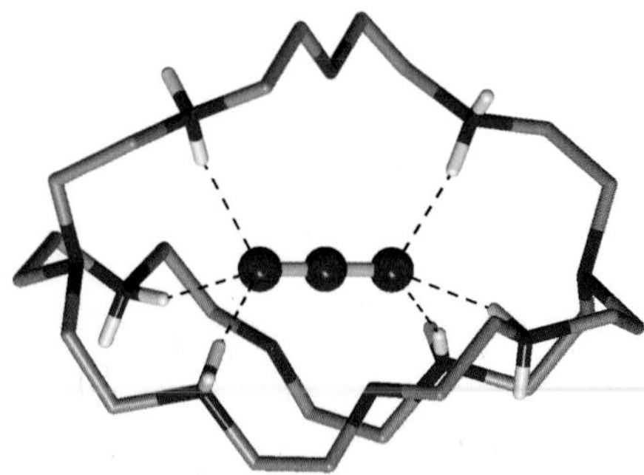

Figure 14 X-ray structure of **31** showing the inclusion of the linear triatomic N_3^- ion in the ellipsoidal cavity of the hexaprotonated cryptand (red, oxygen; blue, nitrogen; gray, carbon).[73]

NCO$^-$) to both metal centers,[77] thus occupying the remaining part of the cavity while simultaneously satisfying the outstanding coordination requirements of both metal centers. Such products have been termed "cascade" complexes.

3.3.2 All-nitrogen heteroatom systems

In a pioneering study, Schmidtchen prepared a series of symmetrical N_4 cages of type **32** ($n = 6$ or 8).[78,79] Interestingly, while these systems arguably may or may not be considered to be true cryptands, they readily bind halide ions, with the corresponding host–guest interactions being clearly electrostatic in nature as only positive quaternary nitrogen sites are present in their cage structures. As with the tricyclic cryptand **29** in its diprotonated form and the related tetramethylated derivative **30** discussed above, the Schmidtchen cages **32** display cavity size discrimination toward individual halide (Cl^-, Br^-, or I^-) ions.

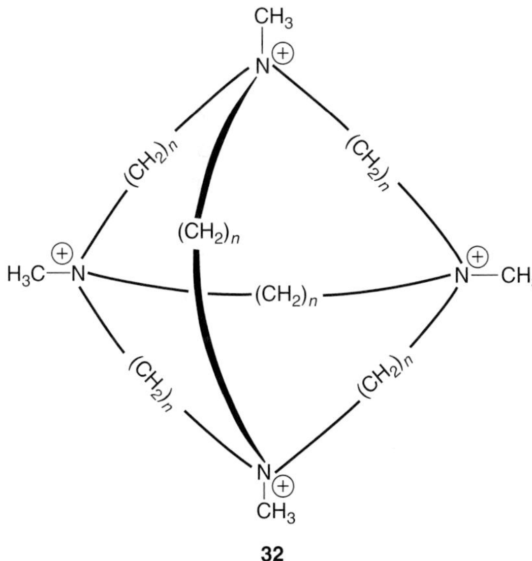

Along with systems of type **32**, a large number of all-nitrogen cryptands have now been reported. In a seminal study published in 1977, the Sargeson group reported the synthesis of the octa-azacryptate **33** ("sepulchrate") as its Co(III) complex.[80] This was followed by numerous other cage derivatives that included **34** ("sarcophagine"), the carbon-bridgehead analog of **33**.[81–83]

33 **34**

The initial synthesis of **33** involved the reaction of tris(ethylenediamine)cobalt(III) with formaldehyde and ammonia (Scheme 3) and led to formation of the Co(III) cage complex in greater than 95% yield. The role of the metal ion as a template is crucial to this reaction as the various precursor species need to be strategically positioned (as well as activated) in the coordination sphere for the required reaction to proceed. Interestingly, when one optical isomer of tris(ethylenediamine)cobalt(III) was employed in the synthesis, then the reaction proceeded with retention of chirality. As expected, the Co(III) cage product was found to be both exceedingly thermodynamically and kinetically stable—reflecting the steric complementarity occurring between this metal ion and the significantly preorganized cavity of **33** (with an additional contribution arising from the inherent kinetic stability of the Co(III) ion in its stable low-spin d^6 electronic configuration).

$+ 6CH_2O + 2NH_3 \xrightarrow{Li_2CO_3}$ $[Co(sepulchrate)]^{3+}$

Scheme 3 Preparation of $[Co(sepulchrate)]^{3+}$.[80]

Since the original synthesis of **33**, the chemistry of this and a wide range of related cages has been reported, much of which was performed by Sargeson and/or his collaborators.[81] As part of these studies, a variety of both all-nitrogen as well as mixed donor cages (including sulfur-nitrogen systems such as **21**) were investigated. Many of the resulting complexes were shown to exhibit unusual properties, including novel redox, optical activity, and bioactivity behavior.[84,85] The X-ray structure of $[Co(sepulchrate)]^{3+}$ (Figure 15) confirmed the sexadentate nature of **33** in this complex.[80]

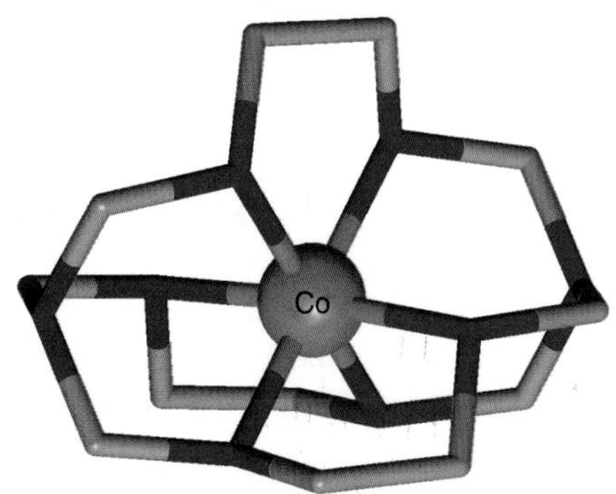

Figure 15 X-ray structure of the $[Co(sepulchrate)]^{3+}$ ion showing the octahedral coordination of the central Co(III) (blue, nitrogen; gray, carbon).[80]

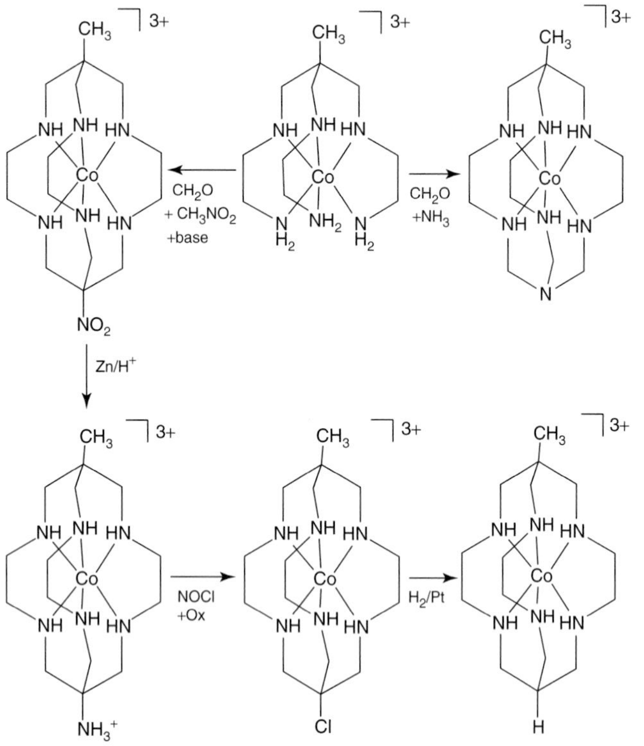

Scheme 4 Synthetic details for a selection of all-nitrogen donor cage systems bearing different substituents and incorporating one or two carbon bridgeheads.[8]

Metal-free cages may be obtained through demetallation of the corresponding metal-containing cages by prolonged heating in strong acid at elevated temperatures. A selection of substituted derivatives of such cage systems is shown in Scheme 4.[8] Much interesting metal ion coordination chemistry of such systems has now been recorded but these results are somewhat tangential to the scope of the present chapter and are not discussed here. A consequence of the availability of these cages is that many metal ions that normally give rise to labile complexes, when encapsulated, effectively become kinetically inert and, for example, yield optical isomers that do not rapidly racemize in solution.

In other studies, a number of related cryptand systems that can readily incorporate two metal ions have been reported. For one system of this type, reaction of Cu(I) with the saturated octaaza ligand **35**, synthesized by 2 : 3 condensation of tris(aminoethyl)amine and glyoxal followed by reduction of the resulting imine linkages, yielded a mixed Cu(I)/Cu(II) dinuclear species of type $[\mathrm{Cu(I)Cu(II)}(\mathbf{35})]^{3+}$. The assignment of the overall +1.5 oxidation state was based on EPR evidence, strongly suggesting that the saturated amine donor set in this case does not solely favor stabilization of the Cu(I) oxidation state.[86,87] In contrast, the corresponding hexa-imine cage **36** both includes and stabilizes two Cu(I) ions in its cavity.

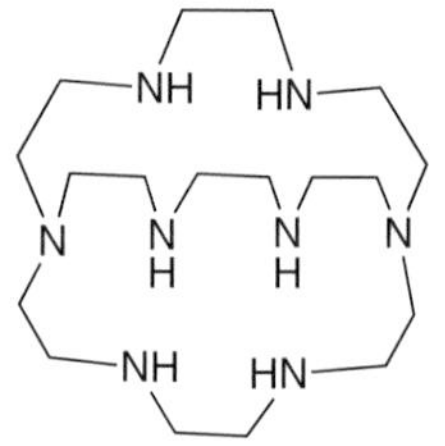

35 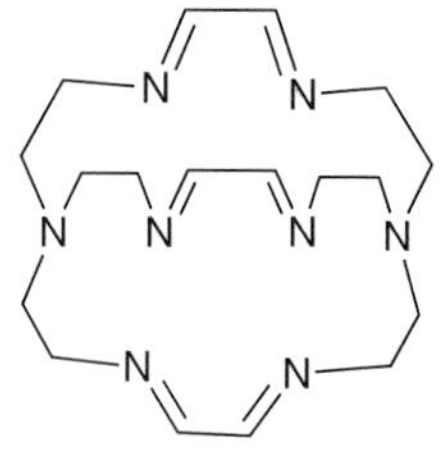 **36**

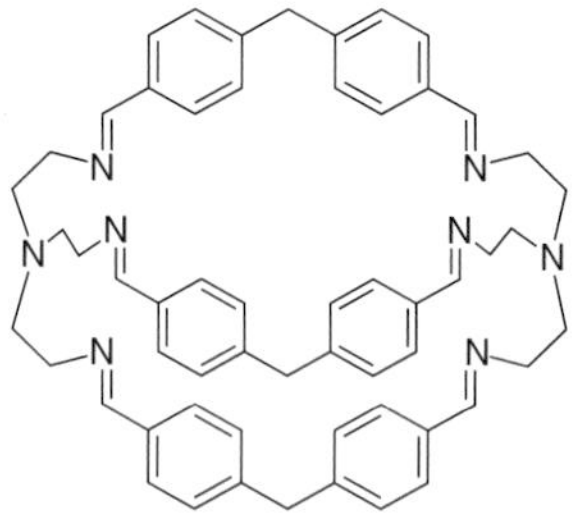

37

The expanded hexa-imine cryptand **37**,[88] incorporating three diphenylmethane bridges linking each tri-imine/amine capping group represents a further (and early) example of the series of both unsaturated (imine) and saturated (amine) bis–tren cryptands. On reaction with $[\mathrm{Cu(CH_3CN)_4}]\mathrm{ClO_4}$ in acetonitrile, **37** also takes up and stabilizes two Cu(I) ions. Each Cu(I) cation is bonded to three adjacent imine sites as well as to the corresponding bridgehead tertiary nitrogen site (Figure 16), with the coordination polyhedron exhibiting an unusual shape in which each copper cation is positioned 0.2 Å "outside" the approximate tetrahedron formed by the four nitrogen donors; there is a large separation between copper centers of 11.07 Å. The availability of "softer" imine linkages in both this and the smaller hexa-imine dinuclear species discussed above undoubtedly

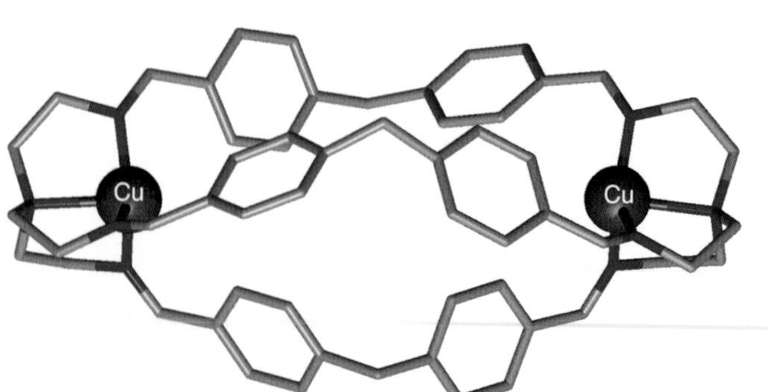

Figure 16 X-ray structure of the dinuclear Cu(I) complex of the expanded hexa-imine cage **37** (blue, nitrogen; gray, carbon).[88]

aids stabilization of the two (soft) Cu(I) guest ions in each complex.

4 CONCLUSION

In this chapter, an outline of the chemistry of two categories of receptor molecules (and their derivatives) showing different degrees of preorganization for guest binding are described. Both categories typically interact with suitable guests to yield products that often display unusual properties that include high selectivity for particular guests as well as enhanced thermodynamic and kinetic stabilities relative to their less preorganized analogs. It is these properties that have made, and undoubtedly will continue to make, such preorganized systems attractive candidates for use in a variety of applications—including, for example, in areas as diverse as separation science, analytical and environmental chemistry, and nuclear medicine.

REFERENCES

1. C. J. Pedersen, *J. Am. Chem. Soc.*, 1967, **89**, 7017.
2. D. J. Cram, *Angew. Chem. Int. Ed. Engl.*, 1986, **25**, 1039.
3. D. J. Cram, *J. Inclusion Phenom.*, 1988, **6**, 397.
4. E. Maverick and D. J. Cram, Spherands: hosts preorganized for binding cations, in *Comprehensive Supramolecular Chemistry*, ed. G. W. Gokel, Elsevier Science, Amsterdam/New York, 1996, vol. 1, pp. 213–244.
5. J. W. Steed and J. L. Atwood, *Supramolecular Chemistry*, John Wiley & Sons, Ltd, United Kingdom, 2009.
6. G. M. Lein and D. J. Cram, *J. Chem. Soc. Chem. Commun.*, 1982, 301.
7. D. J. Cram, G. M. Lein, T. Kaneda, *et al.*, *J. Am. Chem. Soc.*, 1981, **103**, 6228.
8. L. F. Lindoy, *The Chemistry of Macrocyclic Ligand Complexes*, Cambridge University Press, Cambridge, UK, 1989, pp. 1–269.
9. D. J. Cram, T. Kaneda, R. C. Helgeson, and G. M. Lein, *J. Am. Chem. Soc.*, 1979, **101**, 6752.
10. W. Verboom and D. N. Reinhoudt, Spherands, hemispherands and calixspherands, in *Macrocycle Synthesis*, ed. D. Parker, Oxford University Press, Oxford, 1996, pp. 175–206.
11. K. N. Trueblood, C. B. Knobler, E. Maverick, *et al.*, *J. Am. Chem. Soc.*, 1981, **103**, 5594.
12. K. N. Trueblood, E. F. Maverick, and C. B. Knobler, *Acta Crystallogr. Sect. B*, 1991, **47**, 389.
13. S. A. K. Elroby, K. H. Lee, S. J. Cho, and A. Hinchliffe, *Can. J. Chem.*, 2006, **84**, 1045, and references therein.
14. D. J. Cram, T. Kaneda, G. M. Lein, and R. C. Helgeson, *J. Chem. Soc., Chem. Commun.*, 1979, 948.
15. D. J. Cram, T. Kaneda, R. C. Helgeson, *et al.*, *J. Am. Chem. Soc.*, 1985, **107**, 3645.
16. P. A. Kollman, G. Wipff, and U. C. Singh, *J. Am. Chem. Soc.*, 1985, **107**, 2212.
17. E. Maverick and D. J. Cram, in *Comprehensive Supramolecular Chemistry*, ed. F. Vögtle, Elsevier Science Ltd, Oxford, 1996, vol. 2, p. 368.
18. C. B. Knobler, E. Maverick, and K. N. Trueblood, *J. Inclusion Phenom. Mol. Recognit. Chem.*, 1992, **12**, 341.
19. R. C. Helgeson, C. B. Knobler, E. F. Maverick, and K. N. Trueblood, *Acta Crystallogr. Sect. C: Cryst. Struct. Commun.*, 2000, **C56**, 795.
20. D. J. Cram, R. A. Carmack, M. P. deGrandpre, *et al.*, *J. Am. Chem. Soc.*, 1987, **109**, 7068.
21. R. C. Helgeson, J. P. Mazaleyrat, and D. J. Cram, *J. Am. Chem. Soc.*, 1981, **103**, 3929.
22. D. J. Cram, S. B. Brown, T. Taguchi, *et al.*, *J. Am. Chem. Soc.*, 1984, **106**, 695.
23. J. A Bryant, R. C. Helgeson, C. B. Knobler, *et al.*, *J. Org. Chem.*, 1990, **55**, 4622.
24. K. Paek, C. B. Knobler, E. F. Maverick, and D. J. Cram, *J. Am. Chem. Soc.*, 1989, **111**, 8662.
25. D. J. Cram, I. B. Dicker, C. B. Knobler, and K. N. Trueblood, *J. Am. Chem. Soc.*, 1982, **104**, 6828.
26. D. J. Cram, H. E. Katz, and I. B. Dicker, *J. Am. Chem. Soc.*, 1984, **106**, 4987.
27. E. F. Maverick, C. B. Knobler, S. Khan, *et al.*, *Helv. Chim. Acta*, 2003, **86**, 1309.
28. E. Chapoteau, M. S. Chowdhary, B. P. Czech, *et al.*, *J. Org. Chem.*, 1992, **57**, 2804.
29. T. Hayashita, N. Teramae, T. Kuboyama, *et al.*, *J. Inclusion Phenom. Mol. Recognit. Chem.*, 1998, **32**, 251.
30. D. J. Cram, R. A. Carmack, and R. C. Helgeson, *J. Am. Chem. Soc.*, 1988, **110**, 571.
31. V. Böhmer, D. Kraft, and M. Tabatabi, *J. Inclusion Phenom. Mol. Recognit. Chem.*, 1994, **19**, 17.
32. D. N. Reinhoudt, P. J. Dijkstra, P. J. A. In't Veld, *et al.*, *J. Am. Chem. Soc.*, 1987, **109**, 4761.
33. K. Agbaria, S. E. Biali, V. Böhmer, *et al.*, *J. Org. Chem.*, 2001, **66**, 2900.
34. K. Agbaria, O. Aleksiuk, S. E. Biali, *et al.*, *J. Org. Chem.*, 2001, **66**, 2891.
35. V. Böhmer, B. Costisella, J. Gloede, *et al.*, *Eur. J. Org. Chem.*, 2005, **2005**, 2788.
36. K. Ito and M. Nishiki, *Trends Org. Chem.*, 2008, **12**, 33.
37. P. J. Dijkstra, J. A. J. Brunink, K.-E. Bugge, *et al.*, *J. Am. Chem. Soc.*, 1989, **111**, 7567.
38. F. H. Allen, J. E. Davies, J. J. Galloy, *et al.*, *J. Chem. Inf. Comput. Sci.*, 1991, **31**, 187.
39. L. C. Groenen, J. A. J. Brunink, W. I. I. Bakker, *et al.*, *J. Chem. Soc., Perkin Trans. 2*, 1992, 1899.
40. P. A. Kollman, P. D. J. Grootenhuis, and M. A. Lopez, *Pure Appl. Chem.*, 1989, **61**, 593.

41. P. V. Maye and C. A. Venanzi, *J. Comput. Chem.*, 1991, **12**, 994.
42. I. Hataue, Y. Oishi, M. Kubota, and H. Fujimoto, *Tetrahedron*, 1991, **47**, 9317.
43. J. Vacek and P. A. Kollman, *J. Phys. Chem. A*, 1999, **103**, 10015 and references therein.
44. S. A. K. Elroby, *Int. J. Quantum Chem.*, 2009, **109**, 1515, and references therein.
45. D. J. Cram, S. P. Ho, C. B. Knobler, *et al.*, *J. Am. Chem. Soc.*, 1986, **108**, 2989.
46. R. C. Helgeson, B. P. Czech, E. Chapoteau, *et al.*, *J. Am. Chem. Soc.*, 1989, **111**, 6339.
47. D. J. Cram and S. P. Ho, *J. Am. Chem. Soc.*, 1986, **108**, 2998.
48. B. Dietrich, J.-M. Lehn, and J.-P. Sauvage, *Tetrahedron Lett.*, 1969, **34**, 2885.
49. B. Dietrich, Cryptands, in *Comprehensive Supramolecular Chemistry*, ed. G. W. Gokel, Elsevier Science, Amsterdam/New York, 1996, vol. 1, pp. 153–211.
50. B. Dietrich, J.-M. Lehn, and J.-P. Sauvage, *J. Chem. Soc., Chem. Commun.*, 1970, 1055.
51. B. Dietrich, J.-M. Lehn, J.-P. Sauvage, and J. Blanzat, *Tetrahedron*, 1973, **29**, 1629.
52. J.-M. Lehn and J.-P. Sauvage, *J. Am. Chem. Soc.*, 1975, **97**, 6700.
53. B. Metz, D. Moras, and R. Weiss, *J. Chem. Soc. Perkin Trans. 2*, 1976, **11**, 423.
54. J.-C. Chambron and M. Meyer, *Chem. Soc. Rev.*, 2009, **38**, 1663.
55. B. Mertz, D. Moras, and N. Weiss, *J. Chem. Soc., Chem. Commun.*, 1971, 444.
56. D. Moras and R. Weiss, *Acta Crystallogr.*, 1973, **B29**, 396.
57. F. Arnaud-Neu, B. Spiess, and M. J. Schwing-Weill, *J. Am. Chem. Soc.*, 1982, **104**, 5641.
58. B. Dietrich, J.-P. Kintzinger, J.-M. Lehn, *et al.*, *J. Phys. Chem.*, 1987, **91**, 6600.
59. A. N. Chekhlov, *Zh. Strukt. Khim.*, 2002, **43**, 949.
60. B. Dietrich, Cryptands, in *Macrocycle Synthesis*, ed. D. Parker, Oxford University Press, Oxford, 1996, pp. 93–118.
61. I. M. Atkinson, L. F. Lindoy, O. A. Matthews, *et al.*, *Aust. J. Chem.*, 1994, **47**, 1155.
62. I. M. Atkinson, A. M. Groth, L. F. Lindoy, *et al.*, *Pure Appl. Chem.*, 1996, **68**, 1231.
63. K. R. Adam, I. M. Atkinson, J. Kim, *et al.*, *J. Chem. Soc., Dalton Trans.*, 2001, 2388.
64. R. J. A. Janssen, L. F. Lindoy, O. A. Matthews, *et al.*, *J. Chem. Soc., Chem. Commun.*, 1994, 735.
65. I. M. Atkinson, A. R. Carroll, R. J. A. Janssen, *et al.*, *J. Chem. Soc., Perkin Trans. 1*, 1997, 295.
66. D. F. Perkins, L. F. Lindoy, G. V. Meehan, and P. Turner, *Chem. Commun.*, 2004, 152.
67. D. F. Perkins, L. F. Lindoy, A. McAuley, *et al.*, *Proc. Natl. Acad. Sci. U.S.A.*, 2006, **103**, 532.
68. E. Graf and J.-M. Lehn, *J. Am. Chem. Soc.*, 1975, **97**, 5022.
69. B. Metz, J. M. Rosalky, and R. Weiss, *J. Chem. Soc., Chem. Commun.*, 1976, 533.
70. E. Graf and J.-M. Lehn, *J. Am. Chem. Soc.*, 1976, **98**, 6403.
71. E. Graf, J.-P. Kintzinger, J.-M. Lehn, and J. Lemoigne, *J. Am. Chem. Soc.*, 1982, **104**, 1672 and references therein.
72. M. A. Saeed, B. M. Wong, F. R. Fronczek, *et al.*, *Cryst. Growth Des.*, 2010, **10**, 1486, and references therein.
73. B. Dietrich, J. Guilhem, J.-M. Lehn, *et al.*, *Helv. Chim. Acta*, 1984, **67**, 91.
74. V. Amendola, M. Bonizzoni, D. Esteban-Gómez, *et al.*, *Coord. Chem. Rev.*, 2006, **250**, 1451.
75. K. Wichmann, B. Antonioli, T. Söhnel, *et al.*, *Coord. Chem. Rev.*, 2006, **250**, 2987.
76. H.-J. Schneider and A. K. Yatsimirsky, *Chem. Soc. Rev.*, 2008, **37**, 263.
77. C. J. Harding, F. E. Mabbs, E. J. L. MacInnes, *et al.*, *J. Chem. Soc., Dalton Trans.*, 1996, 3227.
78. F. P. Schmidtchen, *Chem. Ber.*, 1980, **113**, 864.
79. F. P. Schmidtchen, *Chem. Rev.*, 1997, **97**, 1609, and reference therein.
80. I. I. Creaser, J. M. Harrowfield, A. J. Hertl, *et al.*, *J. Am. Chem. Soc.*, 1977, **99**, 3181.
81. A. M. Sargeson, *Pure Appl. Chem.*, 1984, **56**, 1603, and references therein.
82. A. M. T. Bygott, R. J. Geue, S. F. Ralph, *et al.*, *Dalton Trans.*, 2007, 4778, and references therein.
83. R. Barfod, J. Eriksen, B. T. Golding, *et al.*, *Dalton Trans.*, 2005, 491 and references therein.
84. N. Di Bartolo, S. V. Smith, E. Hetherington, and A. Sargeson, *Aust. J. Chem.*, 2009, **62**, 1261, and references therein.
85. A. M. Sargeson and P. A. Lay, *Aust. J. Chem.*, 2009, **62**, 1280, and references therein.
86. C. Harding, V. McKee, and J. Nelson, *J. Am. Chem. Soc.*, 1991, **113**, 9684.
87. J. Nelson, V. McKee, and G. Morgan, *Prog. Inorg. Chem.*, 1998, **47**, 167.
88. J. Jazwinski, J.-M. Lehn, D. Lilienbaum, *et al.*, *J. Chem. Soc., Chem. Commun.*, 1987, 1691.

Schiff Base and Reduced Schiff Base Ligands

Bellam Sreenivasulu
National University of Singapore, Singapore, Singapore

1 INTRODUCTION: A BRIEF GENERAL BACKGROUND

The Schiff base ligands are one of the most extensively studied and widely used class of ligands because of their easy and convenient methods of synthesis and their interesting coordination chemistry.[1] A Schiff base (or azomethine), named after Hugo Schiff, refers to a functional group that contains a carbon–nitrogen double bond with the nitrogen atom connected to an aryl or alkyl group but not hydrogen.[1] The Schiff base ligands are among the most fundamental chelating systems in bioinorganic and several relevant natural biological processes, and hence coordination chemistry of various forms of Schiff bases is enormously explored.[2–16] In general, these ligands, ranging from simple and compartmental, and from acyclic to macroacyclic type, usually contain N and O donors while other donor atoms such as S and P are also seen in several mixed donor ligands. Their mode of coordination to the metal ions, binding, and selectivity depend on the nature of the metal ion, nature of the donor atoms and their relative positions, the number and size of the chelate rings formed, the conformational flexibility of the ligand backbone, and their planar or three-dimensional architecture.[17] The steric match and electronic interaction around the metal core can be fine-tuned by an appropriate selection of electron-withdrawing or electron-donating substituents incorporated into the Schiff bases (Figure 1).

Synthesis of a Schiff base is quite straightforward, which involves the condensation reaction between a carbonyl precursor and primary amines and is usually carried out in alcoholic solution (sometimes under reflux conditions) or sometimes as azeotropic distillation with benzene to achieve high yields.[5] Catalysts are not generally required when aliphatic amines are involved though the reaction is acid catalyzed. Condensation of 2 equivalents of salicylaldehyde or salicylaldehyde derivatives with 1,2-diamines leads to the formation of an important class of chelating Schiff bases generally known as "Salens," which are generally characterized by four coordinating sites and two axial sites are open to any secondary or ancillary ligands. Although the term Salen is originally used only to describe the tetradentate Schiff bases derived from ethylenediamine, it has become a more general term in the relevant literature to describe the class of [O,N,N,O] tetradentate bis-Schiff bases as well, such as salophenes. Some examples of salen, salophens, and compartmental cyclic and acyclic Schiff bases are shown in Scheme 1.

Despite their convenient synthetic methods, one of the major concerns is the problems associated with the isolation of the free Schiff bases as they are not always stable. Their stability depends on many factors such as the polarity of

Supramolecular Chemistry: From Molecules to Nanomaterials.
Edited by Philip A. Gale and Jonathan W. Steed.
© 2012 John Wiley & Sons, Ltd. ISBN: 978-0-470-74640-0.

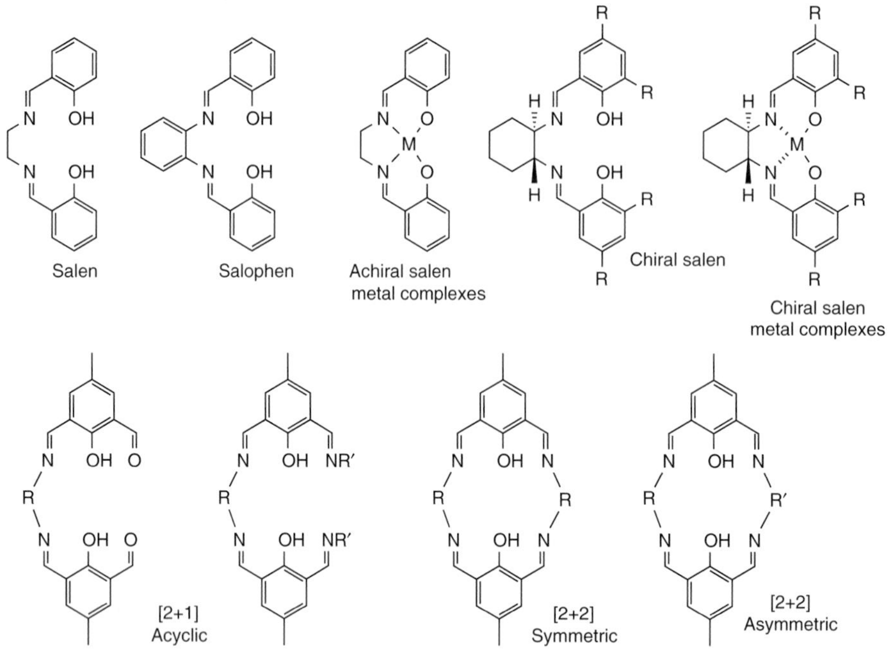

Scheme 1

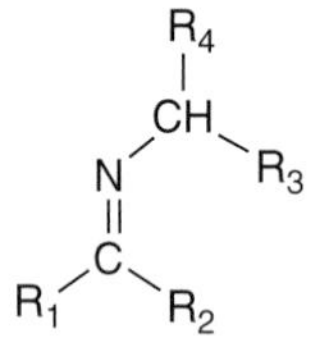

Figure 1 A general form of Schiff base ligands.

primary amine chain, pH, solvent, and temperature. Hence, in order to overcome the problems of isolating the Schiff base ligands, they can be stabilized by employing the suitable metal ions and obtained as *in situ* metal complexes.[5]

1.1 General synthetic and structural aspects

Under appropriate experimental conditions, it is much easier to synthesize acyclic systems directly from the reaction solution in high yields with a satisfactory purity.[18] Such condensation reactions are simple and generally do not lead to any formation of by-products (oligomers or polymers) in a considerable amount especially when the precursors bear only one formyl or primary amine group. Several forms of symmetric and asymmetric compartmental acyclic Schiff bases such as [1+1] acyclic, [1+1] asymmetric end-off, [1+2] or [2+1] symmetric end-off, [2+1] or [1+2] side-off asymmetric, and [3+1] or [1+3] systems have been classified according to their synthetic procedures, that is, the condensation ratio between the formyl (or keto) and amine precursors.[18]

On the contrary, synthesis of any designed cyclic (simple cyclic or macrocyclic) Schiff bases is often challenging as the use of di- or polyfunctional precursors can give rise to different cyclic or polymeric condensation products or, sometimes, by-products that make the purification of the designed macrocycle very difficult and thus drastically reduces the yield. To overcome this problem, in fact, the synthetic strategy that involves a metal ion to act as a template will be well suitable as a very common method for the synthesis and isolation of stable macrocyclic Schiff bases and their complexes.[3,4] In this direction, the chemistry of metal template cyclizations to form macrocyclic Schiff base ligands was first developed by Curtis,[19–21] Busch *et al.*,[22] and Jager,[23] and predates the first synthesis of crown ethers reported by Pederson.[24] Actually, this unique strategy aims at bringing the two metal centers into close proximity, with important implications for metal–metal interactions and magnetic exchange, and binuclear metal reactivity.[25] For this purpose, compartmental Schiff base ligands are more suitable class of ligands with two (or more) coordination chambers in close proximity, which can bind to accommodate either similar or dissimilar metal ions in the adjacent chambers in a well-defined stereochemistry and at an appropriate distance to each other.[26,27] Eventually, several acyclic and macrocyclic compartmental Schiff bases received a great deal of attention in the recent past owing to their relatively convenient synthesis, inherent molecular recognition ability for the selective binding of charged and/or neutral species at their adjacent chambers, and their versatility in the formation of stable supramolecular host–guest complexes as well as the traditional coordination compound.[1,18,25–28]

In terms of structural characterization, spectroscopic techniques such as ^{1}H NMR, UV–vis, and IR can be employed as powerful tools for the solution studies to elucidate the characteristic features of Schiff bases.[5] Actually, the phenomenon of tautomerism is one of the important structural aspects of Schiff bases that leads to the formation of phenol-imine and keto-amine forms depending on the intramolecular hydrogen bonding.[29–31] However, existence of a dominant tautomeric species depends on the nature of the carbonyl precursor—protic and aprotic nature of the solvents—irrespective of the stereochemistry of the molecule or substituent on the imine N atom. Protic and aprotic solvents with high dielectric constants shift the equilibrium toward the quinonoid tautomer. In such cases, compared to IR and UV–vis spectroscopies, NMR is the most suitable tool for the identification of existing tautomers.[10,31,32] Further, accurate information on tautomers can be derived on the basis of the single-crystal X-ray crystallographic study to confirm that the expected shortening in the C–O bond distance (from 1.279 to 2.263 Å) and the lengthening of the C=N distance (from 1.317 to 1.330 Å) is due to the dominance of the keto-amine form (quinoidal structure).[31]

Fundamentally, as all Schiff bases are characterized by the –C=N group, very well-defined IR stretching frequencies at a narrow range of 1600–1675 cm^{-1} allow the identification of this azomethine group.[5] On complexation, the stretching frequency of –C=N group is shifted about 10–15 cm^{-1} toward lower frequencies with respect to the free ligand. In case of dinuclear complexes, introduction of the second metal causes either a small or no further shifts. The C–O stretching frequency is also important that appears in the range of 1235–1288 cm^{-1} for the free base, which is also affected on coordination to a metal. The stretching frequency of the phenolic group (O–H), present in many Schiff bases such as those derived from salicylaldehydes, appears at 3400 cm^{-1}, and plays an important role for establishing whether or not deprotonation occurs on complexation resulting in the formation of a chelate complex or adduct.

The Schiff bases or their sodium or potassium derivatives can be reduced to the corresponding amines by reaction with slight excess of reducing agents such as $NaBH_4$. The reduced derivatives maintain the cyclic or acyclic nature of the corresponding Schiff bases that can be characterized by IR, NMR, and mass spectrometry. The IR spectra show absorption bands at 3385–3257 cm^{-1} due to ν(NH) while the strong bands at 1636–1630 cm^{-1}, characteristic of the azomethine group are completely absent because of the reduction of –C=N bond. Besides, the singlet at 8.50–8.15 ppm due to the imine protons (–CH=N) is expected to completely disappear while new peaks at 3.96–3.75 ppm corresponding to the $ArCH_2N$–H protons can be detected in ^{1}H NMR spectra of the reduced Schiff bases.[27,28]

1.2 Scope of the chapter

Research work on Schiff bases, in particular salen derivatives, has opened the possibility of producing a cavity, or lacuna, in the proximity of the coordination or metal-binding site that contains the metal atom.[33,34] Several receptor systems such as the crown ethers are unable to model metalloenzymes in which the enzyme function is conditioned by the metal atom despite their ability to generate interesting host–guest complexes. Hence, the traditional approach to metallohost–guest chemistry has been directed and shifted mainly to those and related macrocyclic Schiff bases and their reduced analogs. Further, the pendant arms bearing potential chelating groups that allow an opened bicyclic chelation or cryptand ability toward metal guests can be introduced to improve the binding ability both in terms of selectivity and strength.[35–39] Besides, the well-known synthetic and structural versatility of Schiff base complexes provides an alternative route to design such potential transition metal host systems in view of their guest-binding properties. Consequently, studies on several forms of Schiff bases and their reduced analogs are extended and directed toward exploring their potential to act as molecular hosts, and the resulting host–guest and supramolecular structures[33–39] are highly important to derive useful structure–reactivity relationships and molecular processes that are associated with bioinorganic/organic systems, metalla-supramolecular systems, magnetochemistry, material science, catalysis, transport and separation phenomena, and so on.

This chapter is intended to provide interesting supramolecular and host–guest phenomena associated with the Schiff base and reduced Schiff base ligands while focusing on some of the representative examples of the corresponding acyclic and macrocyclic hosts utilized for binding cations, anions, and neutral species. Compared to with the Schiff bases, the reduced Schiff bases are expected to give entirely different supramolecular structures because of conformational flexibility in the ligand that arises when imine is reduced to the corresponding amine groups. Thus, the utilization of the simple and polyamine systems further allows useful structure–reactivity comparison between Schiff bases and their reduced analogs about the different recognition processes. In view of this, several interesting examples of metalla-supramolecular structures and host–guest complexes, associated with

N–H···O, C=O···H–$O_{solvent}$, O–H···O, N–H···O=C hydrogen bonds and C=O···π, C–H···π, and $\pi \cdots \pi$ stacking interactions, derived from both Schiff base and reduced Schiff base ligands are discussed to provide the readers with some useful insights into the factors behind the observed structural diversity that direct the formation of supramolecular structures in the solution and/or solid state.

The structures of the compounds numbered **1**, **10**, **20–23**, **25**, **26**, **33**, **34**, **54–56**, **69**, **71**, **72**, and **85** have not been included in the artwork because that the structural details of the relevant and representative compounds have already been discussed wherever necessary. However these compounds are included deliberately and highlighted with the assigned structure numbers while suggesting the readers to refer to the relevant references cited for more details of their structures.

2 SCHIFF BASES AS HOSTS FOR CHARGED AND NEUTRAL SPECIES

The phenomenon of molecular recognition is pivotal to the development of supramolecular and host–guest chemistry as the recognition process is very fundamental and relevant to many biological processes such as enzyme catalysis and inhibition, membrane transport, and antibody–antigen interactions.[40–49] In the context of host–guest chemistry of different types of macrocyclic ligands,[50–52] cryptands,[53–56] cucurbit[*n*]urils,[57–60] calixarenes,[61–65] and cyclodextrins[66–69] are investigated as hosts for binding several types of charged or neutral guest species. Several classes of acyclic and cyclic Schiff bases are widely explored as hosts/receptors and the available information helps to gain more insights into the relevant molecular recognition processes occurring in bioinorganic/organic chemistry and metallasupramolecular and host–guest chemistry of binding cations or anions, as the phenomenon of molecular recognition is highly crucial in biological functions.[70–75] Among acyclic Schiff bases, particularly, salen-based Schiff base ligands can furnish novel metallohosts because of their multimetal complexation behavior and remarkable molecular recognition features that lead to interesting host–guest chemistry.[76] This section highlights some interesting examples of host–guest complexes derived from the metallohosts containing acyclic and cyclic/macrocyclic Schiff bases and their recognition and binding properties of charged and neutral guest species while emphasizing N_2O_2-based salens and related ligands.

2.1 Acyclic Schiff base receptors: cation binding

Salen-type Schiff base ligands, **L1–L4** with N_2O_2 donor set (Scheme 2), are suitable for the cooperative multimetal complexation systems that can form excellent metallohosts for various guest species when employed for the cooperative multimetal complexation.[76] A metallohost means it will consist of a series of host molecules having metal ion(s) that support and help to maintain the structural framework. A pseudomacrocycle and cyclic or 3D cage-like self-assembled metallacycles represent the examples of metallohosts. The metal ion(s) in the framework of a metallohost enhances its host property, and the guest binding by a metallohost is considered as a highly cooperative phenomenon. In addition, on the basis of the nature of the metal ion or the metal complex, a metallohost can provide useful information on chemical or physicochemical changes in the properties such as color, fluorescence, and redox potential on guest binding.

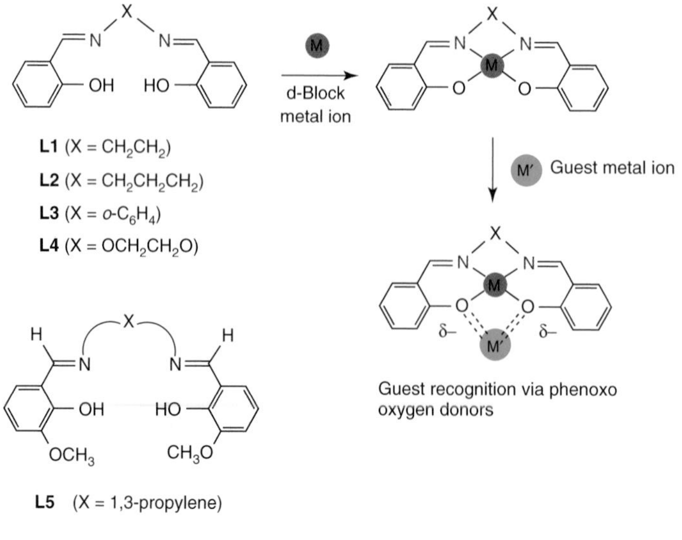

Scheme 2 Salen-type N_2O_2 ligands as metallohosts for binding a guest cation (M′).

On complexation of **L1**–**L4** as shown in Scheme 2, the two phenolic O–H groups in the resulting metallohost are converted into more polarized O–M (phenoxo–metal) bonds. The negatively charged phenoxo groups in metallohost can further coordinate to another metal and hence enhance the guest cation recognition to give the stable metallohost–guest complexes. These second metal ions can be regarded as "guest metals," and the phenoxo groups of the N_2O_2 metal complexes serve as a binding site for the guest cations. The way the oxygen atoms of the metallohost coordinating to the guest metal cations is similar to that of the ether oxygen atoms of crown ethers to form the metallohost–guest complexes. The preorganization of the metallohost is favored by the suitable orientation of the phenoxo groups for the cooperative metal complexation to furnish interesting metallocyclic or metallohelical structures.[76] Also, acyclic bis(N_2O_2) type of ligands binds guest cations such as d-block transition metal,[77] alkali metal,[78–80] alkaline earth metal,[81] and rare earth metal ions[82] in the central recognition site.

The complexes of the methoxy-substituted **L5** (Scheme 2) obtained with transition metal ions such as Cu, Zn, and Ni can bind Na^+ as in [Ni**L5**·Na(NO_3)(CH_3OH)], **1**, and other alkali metal ions to form different host–guest complexes. The mode of complexation, the stoichiometry, and structures of the resulting host–guest complexes depend on the nature of alkali metal salts and the number of carbon atoms of the groups bridging the imine nitrogen atoms. The sodium ions in **1** tend to reside within the planes of the salicylaldimine oxygens while potassium and cesium tend to locate between salicylaldimine ligands to form host–guest sandwich compounds.[83]

Incorporating additional N_2O_2 donor groups in a Schiff base results in a oligo(N_2O_2) host as in **L6**, so that multiple phenoxo groups are available for binding the guest ion. Hence, guest recognition can be enhanced in **L6** because an integration of the multiple phenoxo groups is the origin of strong guest binding (Scheme 3). Thus, compared to mono(N_2O_2) ligand such as **L5**, the bis(N_2O_2) chelating Schiff base **L6** can afford a C-shaped central recognition site with six oxygen donors when metalated with transition metals.[76,84]

The multiple phenoxo groups in **L6** can simultaneously interact with the guest ion in the central recognition site. The restricted conformation of the molecule leads to the inward orientation of the phenoxo oxygen atoms to form an O_6 cavity as a recognition site for binding any suitable third metal ion as shown in Scheme 3. For example, homotrinuclear d-block metal complexes of **L6** such as [**L6**·$Mn_3(OAc)_2(CH_3OH)_2$], **2**, can be easily obtained, and the structures can be determined by X-ray crystallography.[84] It is shown that the binding of guest metal ion in O_6 cavity is favored by the bridging of acetato and phenoxo groups of the metalated ligand. However, different divalent d-block transition metals such as Mn^{2+}, Co^{2+}, Ni^{2+}, Cu^{2+}, and Zn^{2+} ions can also be introduced through metal exchange by taking advantage of the different nature of the N_2O_2 and central O_6 cavity. Indeed, Mn^{2+} selectively occupies the central O_6 site as observed in the structure of heterotrinuclear [**L6**$Zn_2Mn(OAc)_2(CH_3OH)_2$], **3**, demonstrating the site-selective metalation of the N_2O_2 with Zn^{2+} ions and subsequent cooperative binding of guest Mn^{2+} ion in O_6 cavity.

The homotrinuclear d-block metal complexes such as [**L6**·Zn_3]$^{+2}$, **4**, derived from **L6** are further shown to bind several other cations such as alkaline earth and rare earth metal ions into the central acyclic C-shaped recognition site (Scheme 4) via transmetallation and guest exchange to form

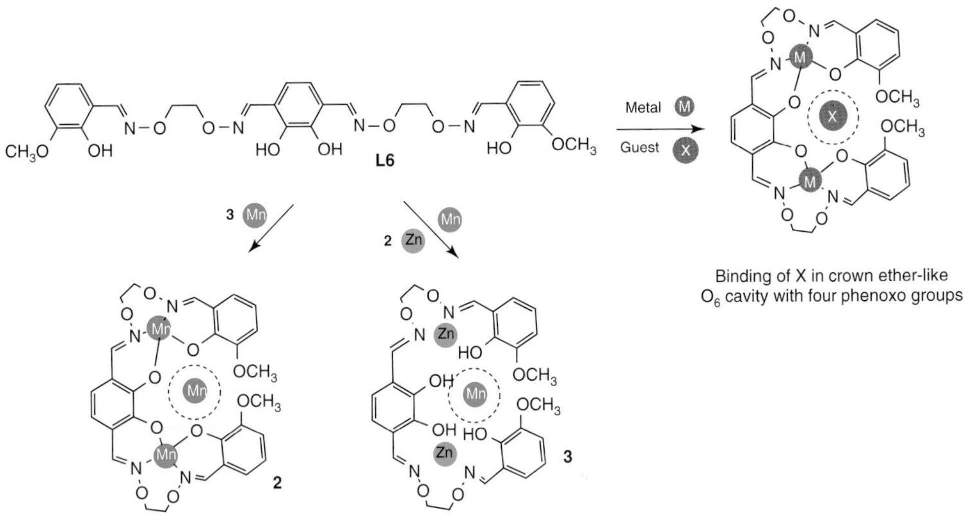

Binding of X in crown ether-like O_6 cavity with four phenoxo groups

Scheme 3 Principle of guest binding in **L6** with Mn^{2+} ion as a guest in homotrinuclear [**L6**$Mn_3(OAc)_2(CH_3OH)_2$], **2**, and heterotrinuclear [**L6**$Zn_2Mn(OAc)_2(CH_3OH)_2$], **3**.

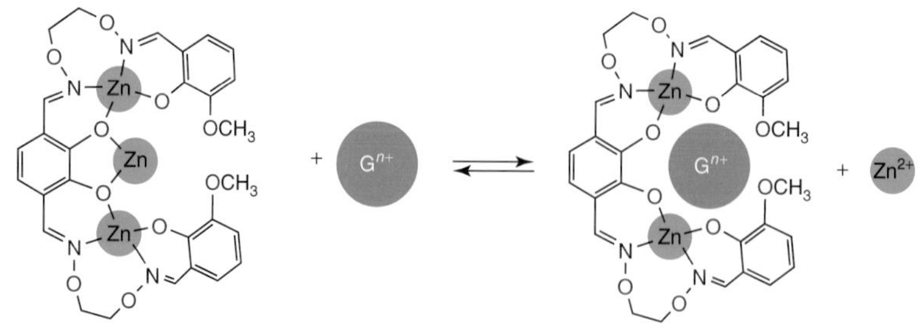

Scheme 4 Binding of G^+ (rare earth or alkali metal ion) in $[\mathbf{L6}\cdot Zn_3]^{+2}$, **4**, to form the host–guest complexes $[\mathbf{L6}Zn_2G]^{2+}$, **5**.

Table 1 Binding constants (K_G) of **4** for guest cation (G^{n+}) exchange.

Alkali cations		Alkaline earth cations		d or f metal ions	
Na^+	<0.001	Mg^{2+}	<0.001	Sc^{3+}	> 1000[a]
K^+	<0.001	Ca^{2+}	32 ± 3	Y^{3+}	> 1000[a]
Rb^+	<0.001	Sr^{2+}	3.9 ± 0.6	La^{3+}, Eu^{3+}, Lu^{3+}	> 1000[a]
Cs^+	<0.001	Ba^{2+}	0.16 ± 0.04		

[a] $[\mathbf{L6}Zn_3]^{2+}$ was completely (>97%) converted to $[\mathbf{L6}Zn_2G]^{n+}$ when 1 equivalent of G^{n+} was added.

$[\mathbf{L6}Zn_2G]^{2+}$, **5**, as metallohost–guest complexes.[85,86] The guest exchange in these metallohosts seems quantitative when rare earth metals are used as guest cations. For alkaline earth metals, selectivity followed the order of $Ca^{2+} > Sr^{2+} > Ba^{2+} \gg Mg^{2+}$.

The binding constants, K_G, of the metallohost, **4**, are summarized in Table 1, which indicate that **4** is highly selective for rare earth(III) metal ions ($K_G > 1000$) compared to the alkali metal ions ($K_G < 0.001$). Alkali metal guests have no ability to replace the zinc ion probably because monovalent alkali metals interact with the $[\mathbf{L6}Zn_2]$ moiety more weakly than the initially bound divalent zinc ion. Hence, it can be concluded that the $\mathbf{L6}Zn_2$ moiety of **4** is excellent in discriminating the charge of the guest ions.[85,86] Consequently, the equilibrium constants in the range up to 32 for alkaline earth metals suggest that the divalent alkaline earth metal ions interact sufficiently strongly to replace the divalent zinc ion at the central O_6 site from **4**. Further, based on the selectivity order of alkaline earth metal ions, it is obvious that Mg^{2+} is too small to interact effectively with all the six oxygen donors of the O_6 site. Highest binding constant for Ca^{2+} ($K_{Ca} = 32$) suggests that the O_6-binding site of the $[\mathbf{L6}Zn_2]$ moiety has an appropriate size to bind Ca^{2+} with little distortion. The binding constants for Sr^{2+} and Ba^{2+} are lower than that of Ca^{2+} probably because of the distortion of the $[\mathbf{L6}Zn_2]$ moiety caused by larger Sr^{2+} and Ba^{2+} that would destabilize $[\mathbf{L6}Zn_2G]^{2+}$, **5**.

Interestingly, the resulting metallohost–guest complexes, $[\mathbf{L6}Zn_2G]^{2+}$, are found to adopt helical structures in the form of molecular springs or coils in which the radius and winding angles are shown to be tunable, based on the size and exchange of the guest ions with a smaller guest metal producing a tighter helix.[84,85] If the metals in the complexes are paramagnetic, novel magnetic properties are expected due to magnetic exchange between 3d–3d and 3d–4f ions, and achieving such synergetic functions at the molecular level with more functionalized molecules using multimetal complexes is a highly challenging task.

Compared to the shorter chain ligands, when the chain length of the ligand with oligo(N_2O_2) donor set is increased as in **L7** (Scheme 5), the binding cavity of the resulting complex consists of more phenoxo groups, which might result in stronger guest binding. Complexation of the acyclic tris(salamo)-based **L7** with Zn(II) acetate produces a zinc(II) trinuclear complex $\mathbf{L7}\cdot Zn_3$ that can selectively bind Ba^{2+} to give a host–guest complex, $[\mathbf{L7}\cdot Zn_3Ba]$, **6** while no such binding is observed for other alkali metal ions (K^+ and Cs^+) and alkaline earth metal ions (Ca^{2+} and Mg^{2+}). Similarly, the host–guest complex, $[\mathbf{L7}\cdot Zn_3La]$, **7**, can also be obtained by complexation of **L7** with zinc(II) and lanthanum(III).[87] An X-ray crystallographic analysis revealed the single-stranded helical structure in **7**, in which three zinc(II) ions sit in the N_2O_2 sites and lanthanum(III) ion is in the central helical cavity. The helix inversion rates of complexes $[\mathbf{L7}\cdot Zn_3La]$, **7**, and $[\mathbf{L7}\cdot Zn_3Ba]$, **6**, determined by ^1H NMR spectroscopy are considerably different. The helix inversion of the complex $[\mathbf{L7}\cdot Zn_3La]$, **7**, is slow even at 353 K (in $CDCl_3/CD_3OD$, 1 : 1), whereas the complex $[\mathbf{L7}\cdot Zn_3Ba]$, **6**, undergoes inversion at room temperature. This fact indicates that the inversion rate of the helix can be easily modulated by a guest cation

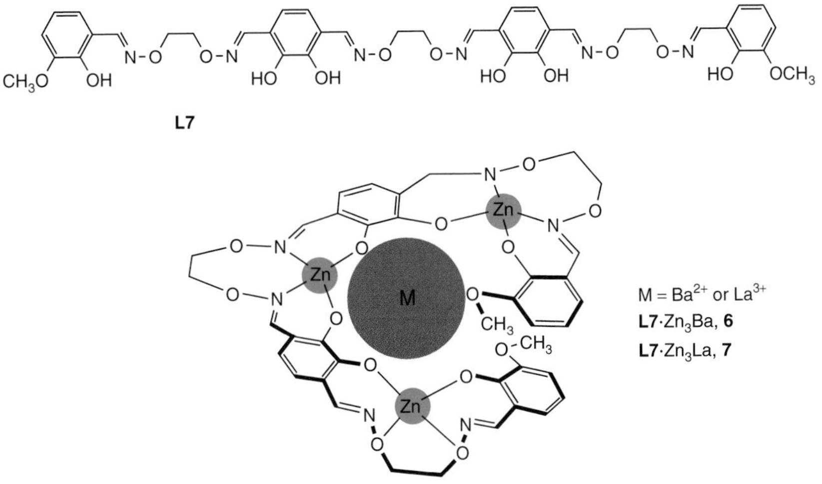

Scheme 5 Formation of [**L7**·Zn_3M] (M = Ba^{2+}, **6**; La^{3+}, **7**) after binding Ba^{2+} and La^{3+} in [**L7**·Zn_3].

without changing the organic ligand framework. Thus, the formation of host–guest complexes **5**–**7** suggests that by altering the number of N_2O_2 donors and hence the number of phenoxo oxygens and the chain length between the imine nitrogens of a Schiff base, it is possible to control the selective binding of cations and the resulting supramolecular structures of the host–guest complexes.

Cu(II) complexes of some of the chiral Schiff base ligands such as **L8** and their reduced forms are explored as catalysts for the asymmetric Henry reaction.[88] The ligand **L8** contains N_2O_2 donor set for coordinating to Cu(II) ion while the oxygen-rich domain formed by four oxygen donors is available for binding water through hydrogen bonds or for any other suitable cation (Scheme 6). The copper(II) complex, [**L8**Cu], derived from chiral **L8**, can selectively bind alkali metal and ammonium ions based on the size of the oxygen-rich O_4-binding site and the coordination number of the hosted metal ion.[88] When [**L8**Cu] obtained *in situ* is allowed to react with NH_4PF_6, the ammonium-bound host–guest complex, $[CuL8(NH_4)]^+$, **8**, is obtained (Scheme 6) in which the NH_4^+ guest is hydrogen bonded to the oxygen acceptors of O_4 recognition site as also suggested by the X-ray crystallographic structure.

Smaller cations (Na^+ and Ca^{2+}) reside in the plane of the O_4 donor set and axial sites are occupied by methanol molecules or coordinating anions as observed in [Cu**L8**Na(CH_3OH)($OClO_3$)]·CH_3OH, **9** (Scheme 6), which is obtained as reddish brown crystals on addition of $NaClO_4$ to [**L8**Cu] in CH_3OH–CH_2Cl_2. Larger cations (NH_4^+, K^+, Cs^+, and Ba^{2+}) are situated above the plane of the O_4 cavity that gives rise to sandwich-type complexes as in [(Cu**L8**)$_2$K]Br·CH_3OH, **10**, in which K^+ ion is sandwiched by two [**L8**Cu] units and is bound by eight oxygen donors in a *transoid* arrangement. These cation-binding studies of [**L8**Cu] containing a chiral Schiff base **L8** and the

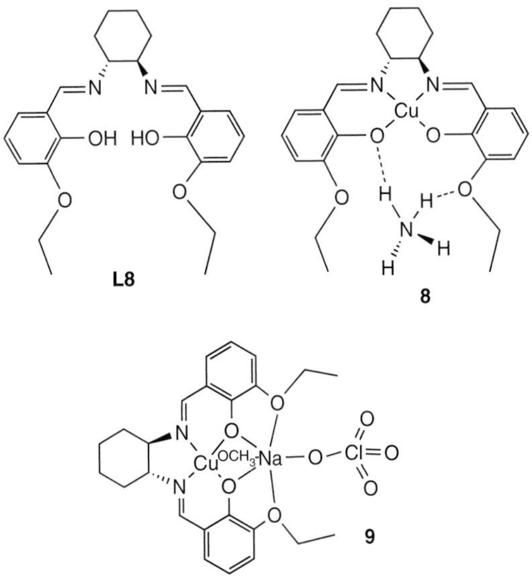

Scheme 6 Binding of NH_4^+ and Na^+ ions in [**L8**Cu] to form $[CuL8(NH_4)]^+$, **8**, and [Cu**L8**Na(CH_3OH)($OClO_3$)]·CH_3OH, **9**.

trend observed in the corresponding host–guest structures indicate the existence of a pseudoinversion center that is associated either with *transoid* sandwich complex **10** or with **9** containing the guest sodium ion. It also indicates that the tendency to form a centrosymmetric motif is dominant in the host–guest complexes of chiral [**L8**Cu], even though the presence of the chiral auxiliary renders a true inversion center impossible. In terms of the use of a chiral ligand, it is significant that a number of structures exhibit pseudoinversion centers, associated either with the formation of a *transoid* sandwich complex or, in the case of [Cu{(*R*,*R*)-**L8**}Na(CH_3OH)($OClO_3$)]·CH_3OH, **10**, with the formation of a hydrogen-bonded, dimeric unit. This observation indicates that the tendency to form a centrosymmetric motif is dominant in these systems, even though the presence of the chiral auxiliary renders a true inversion center impossible.[88]

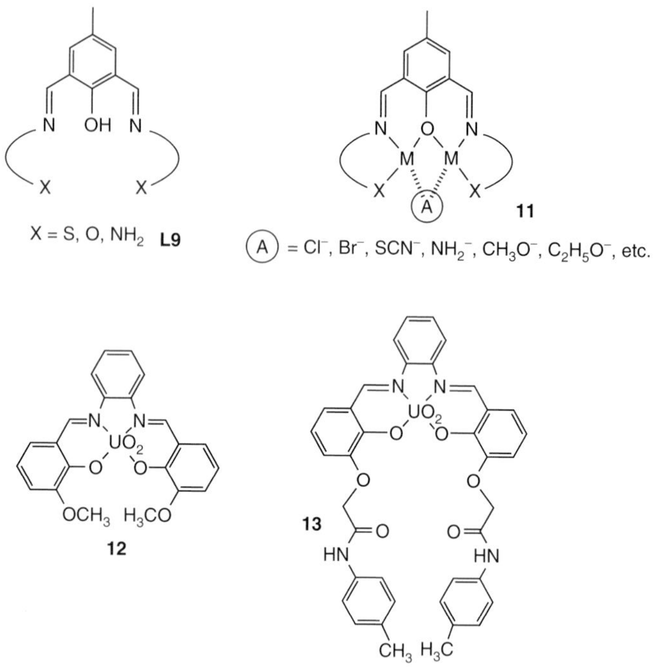

2.2 Acyclic Schiff base receptors: anion and ditopic binding

The phenomenon of anion recognition and binding is gaining increasing relevance and attention in supramolecular chemistry. Understanding the fundamental principles of host–guest chemistry associated with anion binding is a primary goal in the field of supramolecular chemistry because of the ubiquitous involvement of anions in biochemical regulation and control.[89–94] However, the chemistry of anion coordination/binding is still in its developing stage and is a highly unexplored area because of the intrinsic properties of anions that make them more difficult to bind than cations. Compared to cations, anions exist as larger species with different geometries such as spherical, linear, trigonal, tetrahedral, and octahedral. Anions possess higher free energies of solvation than cations, even when they have the same absolute charge and are of a comparable size. Most of the anions involve protonation equilibrium in aqueous solution and exist only within a narrow range of pH. Because of these reasons, the search for synthetic receptors that are able to recognize and bind anions is a real challenge. Several strategies have been adopted to reach the goal and different receptors reported can be grouped according to the noncovalent interactions used to host the anionic guest.[90]

Among the synthetic receptors developed for recognition of anionic species in organic solvents and in water, one can find numerous metal complexes derived from Schiff bases such as salen and salophen ligands as hosts for anions. According to the schematic classification of metal-containing anion receptors given by Steed in the recent review, salen- or salophen-based metal complexes as anion-binding hosts belong to the category in which the incorporation of the Lewis acidic metal ion into the host structure provides an electron-deficient acidic coordination site as the anion-binding site.[95] In this connection, dinuclear complexes such as $[M_2\mathbf{L9}(A)]^{n+}$, **11**, derived from acyclic [1+2] Schiff base ligand, **L9**, represent an example of the anion recognition host for a wide range of anions such as CH_3O^-, $C_2H_5O^-$, Cl^-, Br^-, SCN^-, NH_2^-, $C_6H_5CH_2O^-$, and $RCOO^-$ (Scheme 7). Such dinuclear complexes can also bind phosphate esters and increase their rate of hydrolytic cleavage,[96] which indicates their dual potential as anion host as well as catalytic properties.

The initial leading work carried out by Reinhoudt *et al.*[97,98] in the early 1990s demonstrated the potential of acyclic uranyl(VI) Schiff base complexes as receptors

X = S, O, NH_2 **L9**

11

(A) = Cl^-, Br^-, SCN^-, NH_2^-, CH_3O^-, $C_2H_5O^-$, etc.

12

13

Scheme 7

for anionic guests. Several salen- and salophene-based uranyl(VI) Schiff base complexes are shown to display potential anion recognition via coordination to UO_2^{2+} in its equatorial plane to complete the usual 5-coordination, whereas the peripheral functionalities of the ligand involve hydrogen bonding with the anions. For example, receptor **12** (Scheme 7) binds anions such as chloride with stability constants of 400 M^{-1} in acetonitrile/DMSO (dimethyl sulfoxide) (99 : 1). Crystallographic data suggested that the chloride ion to be bound to the Lewis acidic uranyl center. Further ^{1}H NMR titration experiments of **12** with tetrabutylammonium chloride in acetonitrile-d_3 (with 1% of DMSO-d_6) and the conductometry revealed high association constants of about $4.0 \times 10^2\ M^{-1}$. Thus, it can be suggested that, by virtue of the coordination geometry of uranyl moiety that forces the anion into the equatorial position, it is possible to control the strength of the binding through modification of the salophen skeleton.

Addition of hydrogen-bonding functionalities in combination with metal centers to produce a wide range of anion receptors is more promising and a successfully utilized strategy because it adds significant specificity to the recognition processes.[95] Hence, the idea of placing additional hydrogen-binding sites like –CONH fragments in a preorganized receptor molecule is to increase the selectivity and efficiency of anion complexation. In this connection, Reinhoudt *et al.*[98] reported uranyl–salophen complex **13** with flexible amido fragments in the 3,3′-positions and the secondary amide functionalities (Scheme 7), which significantly enhanced the binding of anions. X-ray crystallographic studies of **13** with bound dihydrogen phosphate indicated that the binding of anionic guest took place by a strong bond to the uranium atom (U···O–P distance 2.28(2) Å), although the two additional H bonds between the amido groups of the ligand and the complexed anion are present (N···O–P distance 2.79(2) Å), which clearly highlights the participation of the –CONH fragments in anion complexation and secondary amide hydrogen bonding.

Thus, at this point, it makes it clear that a supramolecular host/receptor is meant to display an ability to make use of the possible covalent and noncovalent interactions to achieve the enhancement of binding strength and/or of selectivity. Dihydrogen phosphate was found in **13** in polar organic solvents with stability constants as high as 10^5 M^{-1}. This result was remarkable for a neutral receptor system. It is suggested that the recognition of $H_2PO_4^-$ takes place highly selectively since binding of other anions such as Cl^-, HSO_4^-, SCN^-, and ClO_4^- is not observed under the same conditions. Thus, the combination of organometallic or transition metal Lewis acid units with amine- or amide-binding sites (and/or other functionalities) demonstrates their remarkable utility as metal-based hydrogen-bonding anion receptors.

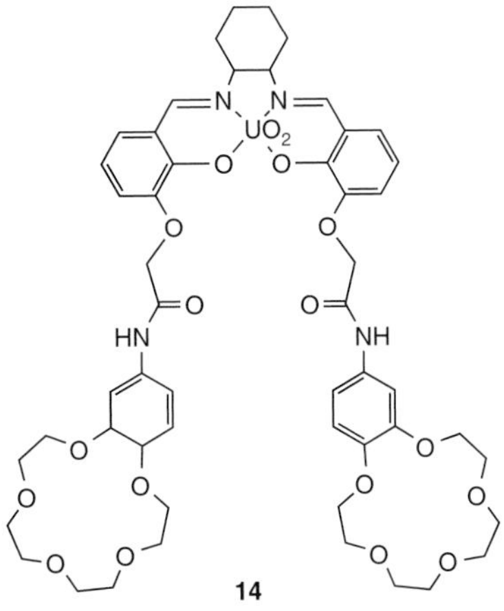

14

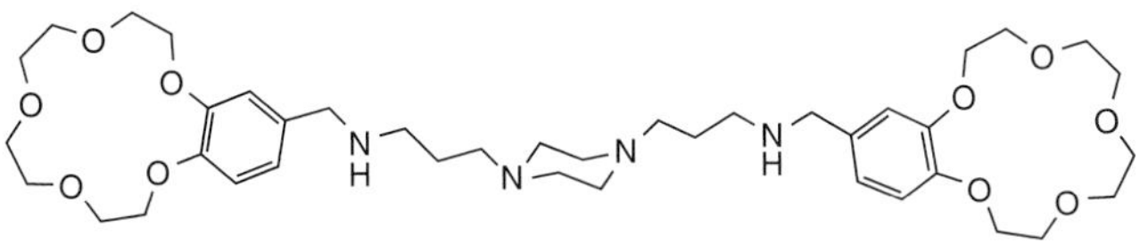

L10

A further modification of **13** by the introduction of crown ethers at the end of the side arms makes a ditopic receptor **14**[99,100] so that the receptor acquires an additional ability to bind both cations and anions simultaneously. Incorporation of cation-binding site makes it possible to introduce an additional electrostatic component to the anion-binding interaction through ditopic recognition. The receptor **14** is shown to exhibit significant transport of hydrophilic KH_2PO_4 through a supported liquid membrane. Combination of both electrostatic and hydrogen-bonding interactions is also possible when protonated form of a Schiff base receptor **L10** is used as it can simultaneously bind cations such as K^+ and Na^+ in the crown ether rings and anions such as Cl^- at the central protonated linker via N–H hydrogen bonding and electrostatic attraction. Binding of Cl^- is investigated using ^{35}Cl NMR techniques.

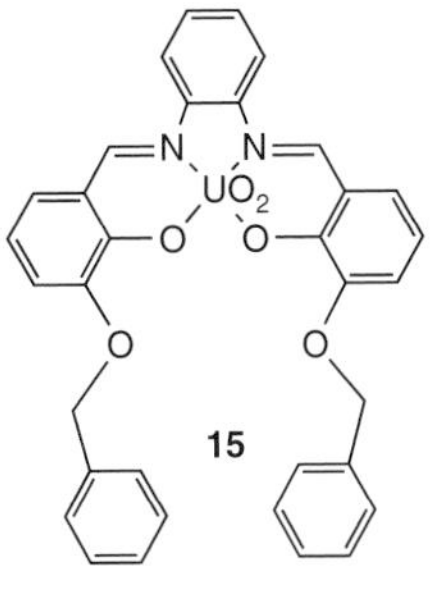

15 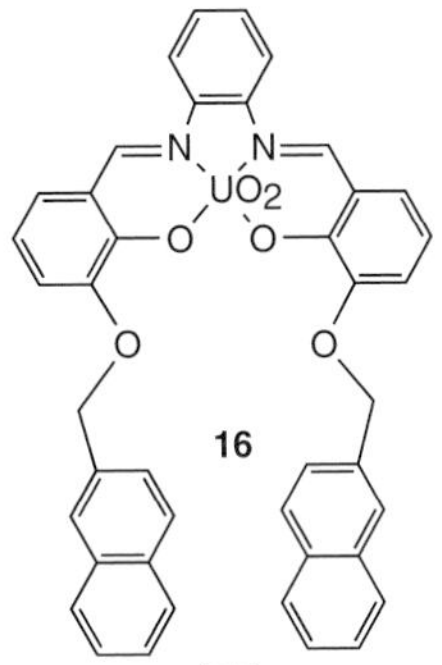 **16**

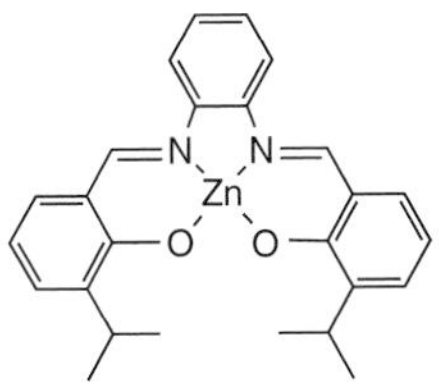

17 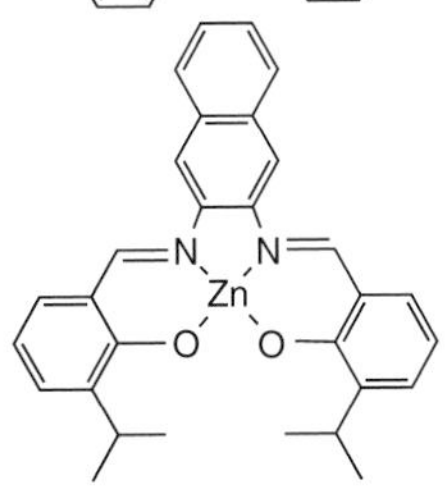 **18**

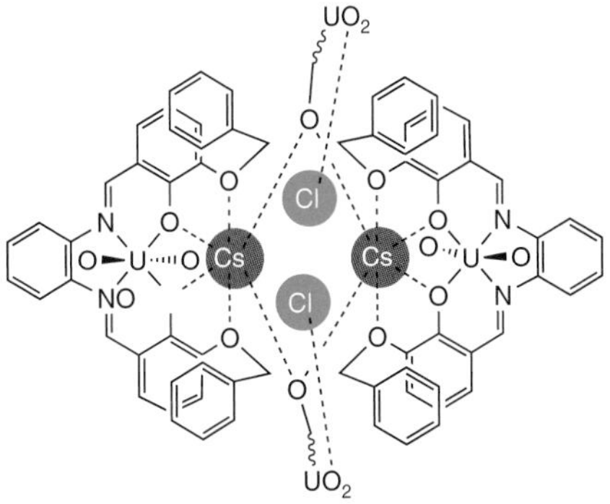

15·CsCl

Obviously, high binding affinity is expected via ditopic recognition when the salt is bound to the receptor as a contact ion pair.[101] For this purpose, besides having crown ether moieties, aromatic pendant arms can also be tagged to the receptor to achieve ditopic recognition of ion pairs or salt. Consequently, in this case, the recognition of cations involves cation···π interactions due to the aromatic pendant arms. For example, uranyl receptors, **15** and **16**, endowed with aromatic pendant arms, are used as ditopic receptors of contact ion pairs/salts such as alkali metal halides[102] and quaternary ammonium halides.[103,104]

Binding of hard anions occurs strongly at the hard Lewis acidic uranyl center, whereas cation$\cdots\pi$ interactions are established between the aromatic side arms and the cation counterpart of the ion pair. Thus, complexation of alkali metal salts (MX) such as CsCl and RbCl with **15** resulted in the formation of isomorphous supramolecular assemblies in the solid state. In the dimeric [**15**·CsCl], each cation is coordinated to six oxygens, three from each receptor, thus creating a pseudo-crown-ether-like environment for the cation. Additionally, each metal ion in the dimeric unit is coordinated to both halide ions and, most importantly, to two aromatic side arms, one from each of the receptors giving decacoordination for the cation. The closest metal ion–aromatic carbon distances of 3.44(1) Å for CsCl, 3.34–3.38(1) Å for RbCl, and 3.58(1) Å for CsF are observed in the respective alkali halide complexes [**15**·MX] indicating the conformational flexibility of the side arms and adaptability of the receptors **15** and **16** to form multiple cation$\cdots\pi$ interactions with the hosted cations.

Similar to the uranyl–salen-based anion receptors of Reinhoudt and coworkers,[97–100] several simple and inexpensive Zn^{2+}-salophen derivatives (**17** and **18**) are employed for the recognition of phosphates $PO_4{}^{3-}$, $P_2O_7{}^{4-}$, and $P_3O_{10}{}^{5-}$ and nucleotides AMP^{2-}, ADP^{3-}, and ATP^{4-}. ^{31}P NMR and mass spectrometry showed the binding of both inorganic phosphate and nucleotide anions.[105,106] Nevertheless, ^{1}H NMR, absorption, emission, and laser flash photolysis experiments demonstrated that the binding of only nucleotides produces relevant changes on the spectral properties of **17** and **18**. The observed trend, $ADP^{3-} > ATP^{4-} > AMP^{2-}$, has no correlation with the anion charge. The binding constants are tabulated in Table 2. The formation of 1 : 1 complexes is supported by the Job plots derived from the UV–vis titration data. There are no association constants observed for the inorganic phosphates due to no spectral variations. Interestingly, the binding is favored by two important interactions involved, namely, coordination of phosphate anions to zinc and $\pi\cdots\pi$ stacking between the salophen aromatic rings and the adenosine nucleobase of nucleotides. This simultaneous involvement of two clearly distinct recognition sites makes **17** and **18** effective supramolecular receptors for the detection of biologically relevant substrates like nucleotides.[106]

Table 2 Binding constants (log *K*) of **17** and **18** for nucleotides.

	17		18	
	Absorption	Emission	Absorption	Emission
AMP^{2-}	5.9	6.0	5.2	5.6
ADP^{3-}	6.7	6.8	7.0	7.1
ATP^{4-}	6.6	6.6	6.4	6.4

Another anion-binding strategy is to employ relatively simple [2+1] side-off Schiff bases **L11** and **L12** as ditopic zwitterionic extractants while complexation takes place through Lewis acid/base or electrostatic interactions that keep the anion close to an adequate signaling unit whose spectral properties are affected by coordination.[100,107,108]

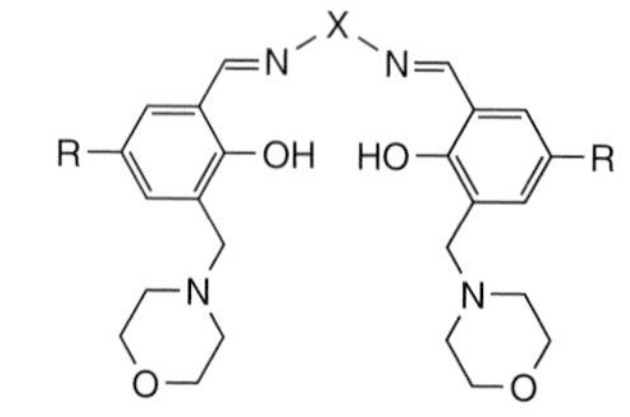

L11 X = $(CH_2)_2$, $(CH_2)_3$, C_6H_4, C_6H_{10}
R = CH_3, $C(CH_3)_3$

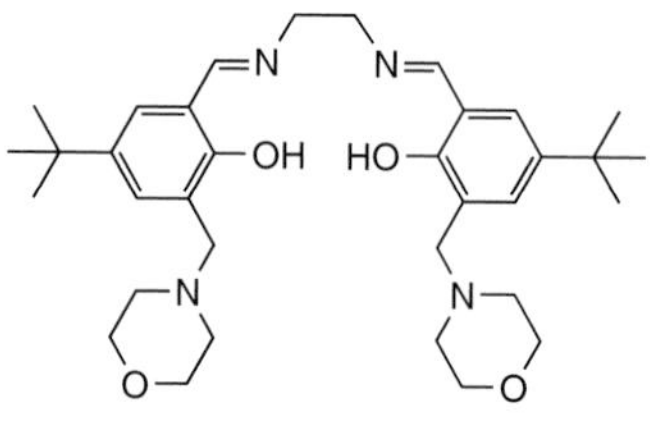

L12

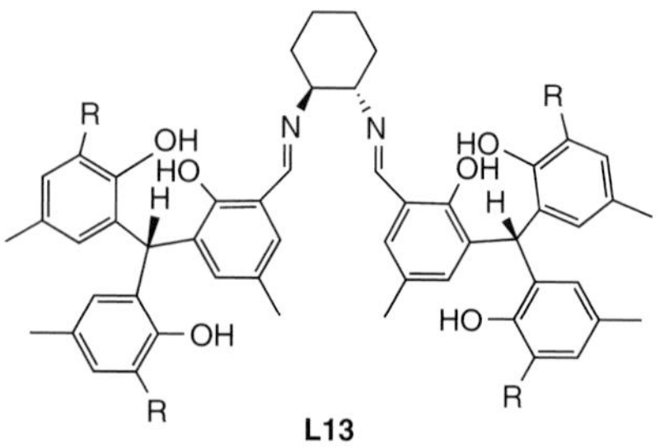

L13

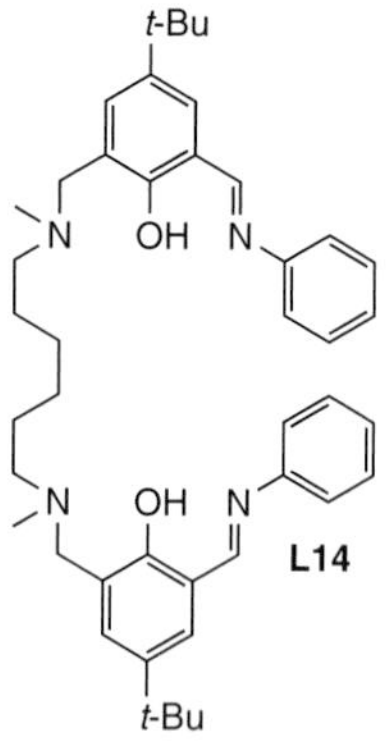

L14

A ligand with high efficiency for the solvent extraction of transition metal salts, [M^{2+} X^{2-}], can be achieved by incorporating a dianionic binding site for transition metal cations and a dicationic binding site for anions. For example, chloroform solution of **L12** can extract Cu^{2+} and sulfate ions from an aqueous $CuSO_4$ solution almost to 100%.[108] Further, X-ray crystallographic studies on [Ni**L12**SO_4)], **19**, obtained by the reaction of **L12** with $NiSO_4 \cdot 7H_2O$ in methanol demonstrated the ditopic binding of divalent Ni^{2+} and sulfate anions. The Ni^{2+} is coordinated in the planar N_2O_2 cavity of the Schiff base moiety. The coordination of the Ni^{2+} ion has two important effects on the structure of the ligand. The two protons liberated from the phenols of the metal-binding site have been transferred to the nitrogen atoms of the pendant morpholine groups, and the ligand has been organized to bring the two morpholine groups into close proximity to produce a dicationic binding site for the sulfate anion. This represents an example of cooperativity in which binding of the metal cation directly enhances the binding of the anion. The sulfate dianion is bound by the combination of both electrostatic interactions and two separate bifurcated hydrogen bonds, one to each quaternary amine.

19

Libra and Scott[109] demonstrated an efficient and size-selective anion binding by phenolic donors of **L13** functionalized with a chiral triphenoxymethane moieties to align four phenol groups into a tetrahedral orientation that can tightly bind F^- ion through OH···F hydrogen-bonding interactions. Anion-binding properties of the Ni^{2+}-containing metalloreceptor [Ni-**L13**], **20**, derived from **L13** are tested with n-Bu_4N^+ halides in different solvents, including chloroform, acetone, and DMSO. This represented an efficient anion-binding receptor system of its kind for halide binding while utilizing purely O–H donors. The receptor **20** can bind F^- selectively only at low concentrations to form $[NiL13F]^-$, **21**, when mixed with n-Bu_4N^+ halides. Addition of fluoride induced a redshift in the MLCT (metal to ligand charge transfer) transition in UV–vis spectra as shown in Figure 2 offering a visual colorimetric response for the binding event. The phenolic binding cavity is too small to host the larger anions, and even at higher concentrations of Cl^-, Br^-, I^-, NO_3^-, ClO_4^-, or HSO_4^- no changes could be detected in the

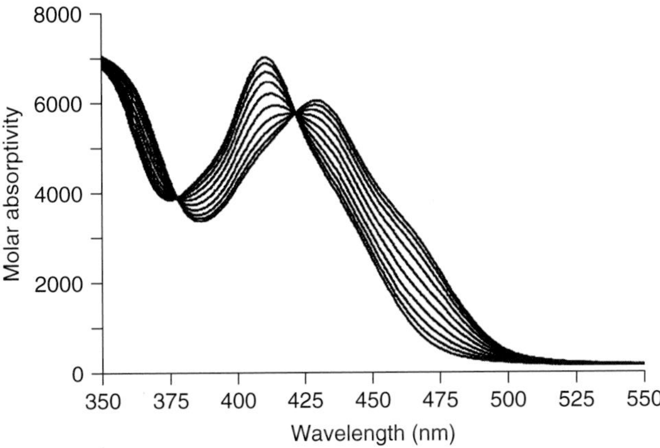

Figure 2 UV–vis titration of **20** with n-Bu_4N^+ F^- in acetone. Titration was complete after the addition of a single equivalent of fluoride.

^{1}H NMR or UV–vis spectra. Further, based on X-ray crystallographic studies, the receptors tightly bind fluoride ion by three short and one long O–H···F interactions (Figure 2) to represent a rare example of efficient anion binding by purely O–H hydrogen bonding in a well-defined sensor. In the case of **20**, the absorption shifts from 411 to 431 nm with a small decrease in the molar absorptivity, whereas the Pd(II)-based receptor exhibits a slightly larger shift from 407 to 440 nm. Job plots indicated that a single equivalent of fluoride binds in the phenolic cavity and from the titration data, the log K values (errors, 10%) were determined to be 5.6 for Ni-containing receptor **20** and 5.8 for Pd(II) receptor in acetone (log $K = 5.8$ for both complexes in DMSO).

The zwitterionic form of a copper-based metallohost, $[Cu_2(\mathbf{L14})_2]$, can be derived from a Schiff base ligand **L14** in which the salicylaldimine is modified by linking with 2,9-diazadecane units. This provides $[Cu_2(\mathbf{L14})_2]$ with an electropositive and hydrogen-bonding cavity into which anions such as SO_4^{2-} and BF_4^- can be encapsulated to form 2 : 2 : 1 host–guest complexes corresponding to Cu:ligand:anion as observed in $[Cu_2{\cdot}(\mathbf{L14})_2(SO_4)]^{2+}$, **22**, and $[Cu_2(\mathbf{L14})_2(BF_4)]^{3+}$, **23**, respectively. X-ray structure determination of the host–guest complex, $[Cu_2\mathbf{L14}(SO_4)]^{2+}$, **22**, shows this to be made up of two ligands in their zwitterionic forms (deprotonated phenols and protonated amines), two Cu(II) metal centers and two SO_4^{2-} anions, one of which is encapsulated inside the complex while the second sulfate acts as a counterion and lies outside the complex. The structure of tricationic tetrafluoroborate copper complex, $[Cu_2\mathbf{L14}(BF_4)]^{3+}$, **23**, shows similar ligand architecture, but the encapsulated BF_4^- is not coordinated to the tetrahedrally distorted copper atoms. The encapsulation of BF_4^- anion takes place via H bonds that are weak, as would be expected, compared to those involved in sulfate ion recognition and binding.

Anion-binding properties of $[Cu_2(\mathbf{L14})_2]$, which are further explored via solid-state and solution chemistry to define the effects of the strength of anion binding in the helical metallomacrocyclic cavity, have shown that these zwitterionic dicopper hosts can bind anions with an approximate volume of 0.09 nm^3 and smaller.[110] The dicopper host $[Cu_2(\mathbf{L14})_2]$ is capable of binding anions in 1 : 1 ratio on the order of $SO_4^{2-} > HPO_4^{2-} > ClO_4^- > BF_4^- > NO_3^-$. The schematic of the encapsulation of ClO_4^- and NO_3^- in $[Cu_2(\mathbf{L14})_2]$ is shown in Figure 3. X-ray crystallographic and spectroscopic analyses also indicated the encapsulation of ClO_4^-, NO_3^-, and I^- ions in cavity and the formation of host–guest complexes $[ClO_4 \subset Cu_2(\mathbf{L14})_2](ClO_4)_3$, **24**, $[NO_3 \subset Cu_2(\mathbf{L14})_2](NO_3)(PF_6)_2$, **25**, and $[I \subset Cu_2(\mathbf{L14})_2]_2{\cdot}I{\cdot}\ 5BF_4{\cdot}2H_2O{\cdot}C_4H_{10}O$, **26**.

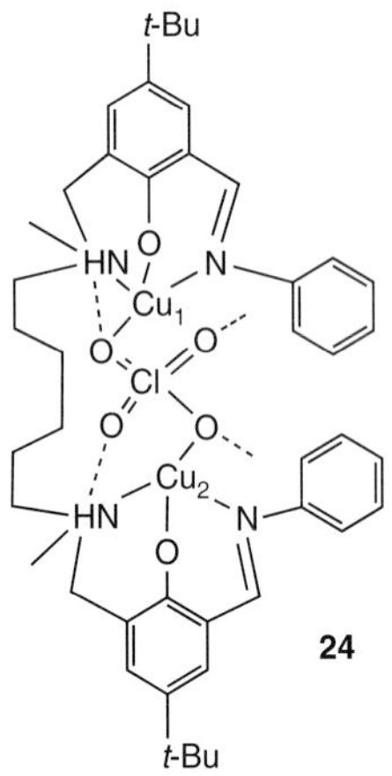
t-Bu
HN
O
N
Cu1
O
O
Cl
O
O
Cu2
HN
N
O
t-Bu

24

Figure 3 Binding of ClO_4^- ion in the copper complex, $[ClO_4 \subset Cu_2(\mathbf{L14})_2]^{3+}$, **24**, derived from **L14**.

Table 3 Binding constants for $[X^{n-} \subset Cu_2\mathbf{L14}]^{(4-n)+}$.

Anion or acid	log *K*
H_2SO_4	5.07 ± 0.24
H_3PO_4	4.30 ± 0.02
$[Bu_4N]H_2PO_4$	4.34 ± 0.13
$HClO_4$	3.38 ± 0.12
HBF_4	3.03 ± 0.29
HNO_3	2.95 ± 0.04

Based on spectrophotometric titrations of $[Cu_2(\mathbf{L14}\text{-}2H)_2]$ with H_2SO_4, H_3PO_4, $[Bu_4N]H_2PO_4$, $HClO_4$, HBF_4, and HNO_3 (in isopropanol at 294 K).

The binding constants given in Table 3 vary on the order nitrate < tetrafluoroborate < perchlorate < phosphate < sulfate. This order is consistent with the stability of the $[X^{n-} \subset Cu_2\mathbf{L14}]^{(4-n)+}$ (where X = nitrate, tetrafluoroborate, perchlorate, phosphate, and sulfate) complexes being dependent on a combination of the charge on the anion, its size, and its ability to form coordinate bonds to the copper atoms and arrays of hydrogen bonds to the N–H and C–H units in the alkylammonium groups. These factors, which favor the inclusion of sulfate and phosphate, appear to be more important than the hydration/solvation energies that are higher for these anions. For the weakly coordinating ClO_4^-, BF_4^-, and NO_3^- ions, the stronger binding of the perchlorate and tetrafluoroborate probably arises because their geometry allows them to affect weak coordinate interactions to the copper atoms while also forming four hydrogen-bonding interactions to the alkylammonium groups.

2.3 Cyclic Schiff base receptors: binding of cations

Macrocylic compartmental Schiff base ligands prepared from 2,6-diformylphenol and suitable diamines represent the N_2O_2-based ligands in which two N_2O_2 donor sites and phenoxo groups are utilized for deriving different homo- and heterodinuclear complexes. However, these complexes hardly display any vacant recognition site for binding guest species,[76, 109] and hence such compartmental ligands need to be functionalized appropriately to enhance their molecular recognition and host–guest properties. In this connection, an interesting series of compartmental ligands, **L15–L17** (Scheme 8), are functionalized to possess a cyclic polyether

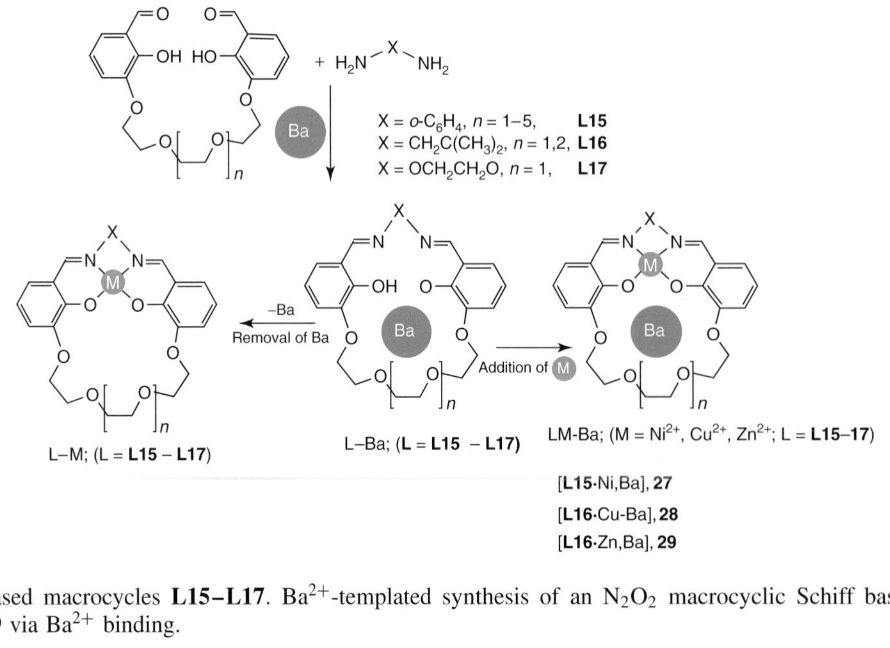
OH HO
+ H2N X NH2
Ba
X = o-C6H4, n = 1–5, L15
X = CH2C(CH3)2, n = 1,2, L16
X = OCH2CH2O, n = 1, L17
–Ba
Removal of Ba
Addition of M
L–M; (L = L15 – L17)
L–Ba; (L = L15 – L17)
LM-Ba; (M = Ni2+, Cu2+, Zn2+; L = L15–17)
[L15·Ni,Ba], 27
[L16·Cu-Ba], 28
[L16·Zn,Ba], 29

Scheme 8 Salen-based macrocycles **L15–L17**. Ba^{2+}-templated synthesis of an N_2O_2 macrocyclic Schiff base hosts of Ni^{2+}, **27**; Cu^{2+}, **28**; and Zn, **29** via Ba^{2+} binding.

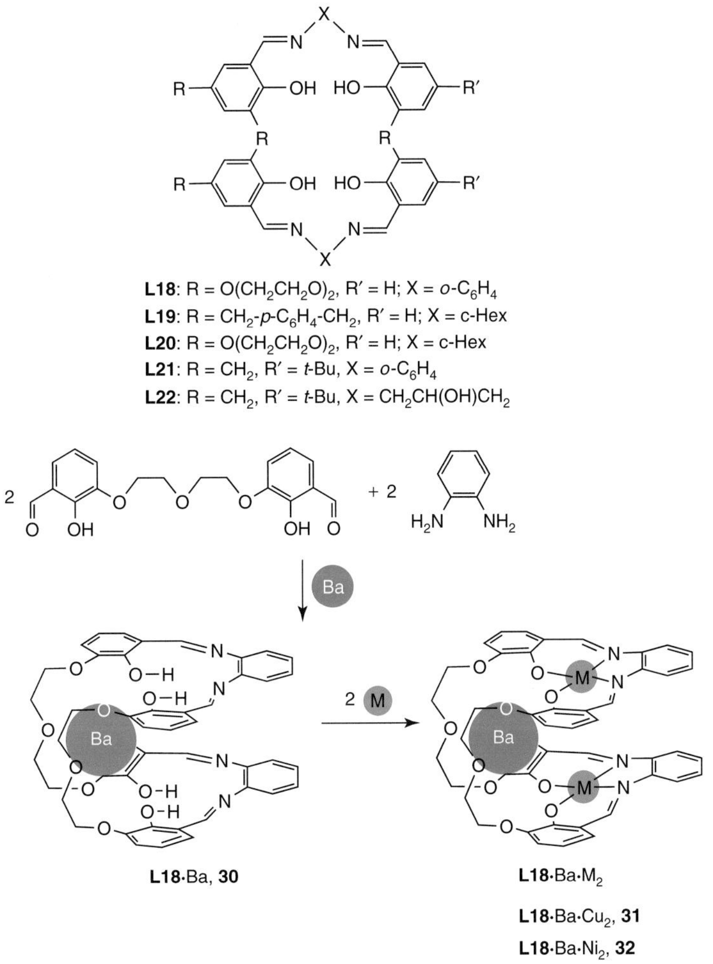

L18: R = $O(CH_2CH_2O)_2$, R′ = H; X = o-C_6H_4
L19: R = CH_2-p-C_6H_4-CH_2, R′ = H; X = c-Hex
L20: R = $O(CH_2CH_2O)_2$, R′ = H; X = c-Hex
L21: R = CH_2, R′ = t-Bu, X = o-C_6H_4
L22: R = CH_2, R′ = t-Bu, X = $CH_2CH(OH)CH_2$

L18·Ba, **30**

L18·Ba·M_2

L18·Ba·Cu_2, **31**

L18·Ba·Ni_2, **32**

Scheme 9 Template synthesis of heterotrinuclear host–guest complexes [**L18**·Cu_2Ba], **31**, and [**L18**·Ni_2Ba], **32**, via binding of Ba^{2+} having the macrocyclic **L18**.

compartment so as to recognize and accommodate guest species.[111–113]

In general, the size of the guest-binding cavity or recognition site of a macrocyclic Schiff base ligand depends on the nature of the metal cation employed during the template synthesis. In heterodinuclear complexes, a soft d-block transition metal (Ni^{2+}, Cu^{2+}, Zn^{2+}, etc.) and a hard cation, such as Ba^{2+} ion, can be bound to the N_2O_2 and crown ether sites, respectively.[76,109] For instance, synthesis of the N_2O_2 macrocyclic frameworks, **L15–L17**, is assisted by Ba^{2+} ion as a template as illustrated in Scheme 8. Accordingly, the barium-containing macrocyclic host [L-Ba] (where L = **L15–L17**) can have a vacant N_2O_2-binding site suitable for binding of other cations such as d-block transition metals to form host–guest heterodinuclear complexes (L–MBa) such as [**L15**·NiBa], **27**, [**L16**·CuBa], **28**, and [**L17**·ZnBa], **29**. On removal of Ba^{2+}, these host–guest complexes will form the macrocycle, L–M, with a vacant crown-ether-like cavity that can be further utilized for binding suitable cationic guests. With this promising strategy, different types of Schiff bases characterized by N_2O_2 donor sites are successfully functionalized with crown-ether-like cavity.[114–121] Interestingly, besides the cation-binding ability, the crown ether cavity associated with these metallomacrocyclic hosts also works as a binding site for neutral guests, such as urea derivatives, when the cavity is large enough.

A macrocyclic bis(N_2O_2) Schiff base ligand, **L18**, can be efficiently synthesized as barium complex [**L18**·Ba], **30**, using a template barium ion (Scheme 9) in which 10 oxygen donor atoms coordinate to the barium ion. Binding of the two vacant N_2O_2 sites with d-block metal ions (M = Cu(II) and Ni(II)) affords the heterotrinuclear complexes [**L18**·Cu_2Ba], **31**, and [**L18**·Ni_2Ba], **32**. In these complexes, the phenoxo groups coordinate to both the square planar d-block metal and the 10-coordinate barium ion in a μ_2 manner.[122,123] Further, the diuranyl complexes of **L18–L22** with bis(N_2O_2) moieties are also shown to efficiently recognize dicarboxylate anions[124] in which the two uranyl units work as a Lewis acid to bind anionic guests. The

Scheme 10 Binding of anionic guests to the pseudocalixarene-like dinuclear complex [**L21**·M_2G], **35**.

heterotrinuclear complex [**L20**·Zn_2Ba], **33**, derived from **L20** can recognize amine derivatives.[125]

When macrocyclic **L21** having 2-hydroxy-1,3-propylene bridges reacts with divalent d-block metals, pseudocalixarene type of dinuclear macrocyclic complexes [**L21**·M_2], **34** (M = Mn, Co, Ni, Cu, Zn), is obtained. Structures of these macrocyclic complexes are controlled by two strong O–H–O ($_{Ar}O \cdots H \cdots O_{Ar}$) interactions originating from the metal-ion-promoted monodeprotonation of the methylenediphenol units of **L21**. These hydrogen-bonding interactions keep the metal ions to be located in a cleft, within which either neutral or anionic guests can be accommodated to form [**L21**·M_2G], **35**[126] (Scheme 10).

The macrocylic **L23** is another bis(N_2O_2)-based Schiff base with 4,4′-methylene linkages which can furnish mono- and dinuclear Ni^{2+} complexes depending on the solvent used in the reaction. These complexes act as potential hosts for binding neutral molecules such as acetone and acetonitrile (Scheme 11). Accordingly, the mononuclear complex forms host–guest complex [**L23**·Ni·acetone], **36**, with acetone, whereas the dinuclear complex hosts acetonitrile to form [**L23**·Ni_2 CH_3CN], **37**. In the dinuclear nickel(II) complex, the two nickel atoms have a distorted square planar geometry and there is an oval-shaped cavity between the two Ni–N_2O_2 moieties, in which one acetonitrile molecule is hosted.[127]

Incorporation of rigid biaryl and binaphthyl units into the bis(N_2O_2) macrocyclic ligands as in ligands **L24**–**L26** significantly altered the complexation behavior when metalated using Ni and uranyl acetates, and hence their host–guest complexes. The reaction of the biphenol-based bis(N_2O_2) macrocyclic **L24** with U(acac)$_2$ affords a dinuclear complex [**L24**·U_2], **38**, in which only one of the two uranyl units resides in the N_2O_2 site and the other is situated in the central O4 site.[128] Similarly, the reaction of a binaphthyl-based bis(N_2O_2) macrocyclic **L26** with nickel(II) acetate results in the formation of a dinuclear complex [**L26**·Ni_2], **39**, in which only one of the nickel ions is in the N_2O_2 site.[129] (Scheme 12). It is thus obvious

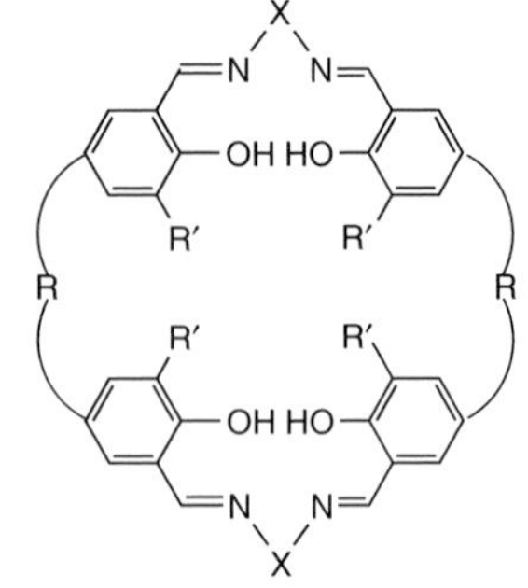

L23, R = CH_2; R′ = *i*-Pr or *t*-Bu; X = c-Hex

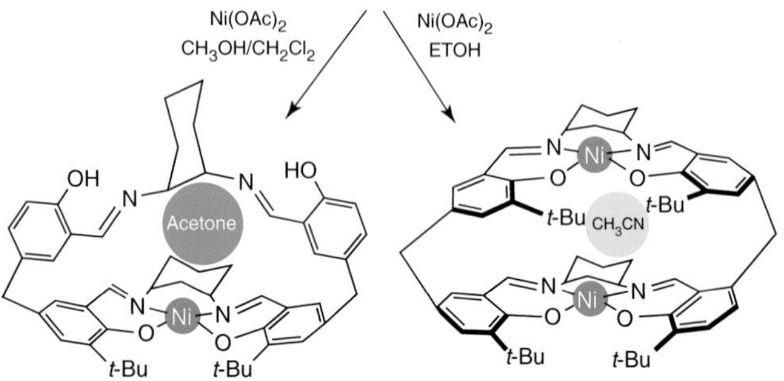

Scheme 11 Calixsalen-type macrocyclic metallohost derived from **L23** binding acetone as in **36** [**L23**·Ni·acetone] and acetonitrile as in **37** ([**L23**·Ni_2 CH_3CN]).

that the complexation at the first N_2O_2 site changes the second site conformation unfavorably for metal coordination in tetradentate N_2O_2 manner.

When three N_2O_2 units are joined in a cyclic manner in such a way that the neighboring N_2O_2 moieties also share the same benzene ring, the resulting six negatively charged phenoxo oxygen atoms can form a cavity similar to 18-crown-6 that can simultaneously interact with the guest ion and hence the binding of guest cations becomes much stronger (Scheme 13).

Indeed, several soluble derivatives of **L27** having peripheral alkoxy groups have been synthesized[70, 130–132] to study their complexation behavior and host–guest assembly. A trinuclear metallohost is expected to be formed on complexation of **L27** due to three N_2O_2 sites (Scheme13). However, the reaction with 3 equivalents of Zn^{2+} resulted in the formation of a mixture of complexes, while the addition of an excess amount of zinc acetate afforded a single product [**L27**·Zn_7], **40**. X-ray crystallography of **40** clearly demonstrated the heptanuclear structure with three zinc(II) in the N_2O_2 sites and a $Zn_4(\mu_4\text{-}O)$ core in the O_6 cavity.[133, 134] Complexation of **L27** with cadmium(II) also produced a similar heptanuclear complex that assembled into a dimeric capsule in the solid state.[135] Manganese(II) and nickel(II) gave a similar cluster, whereas copper(II) and cobalt(II) furnished trinuclear complexes as major products.[136] Thus, during the process of cluster formation, **L27** acts as a partial template to regulate the cluster size. The tris(N_2O_2)

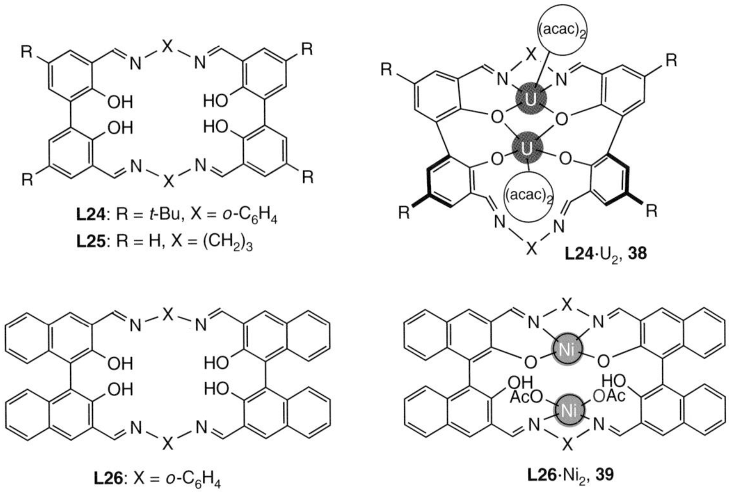

Scheme 12 Acetate ion binding in diuranyl **38** and dinickel **39** metallohosts derived from **L24–L26**.

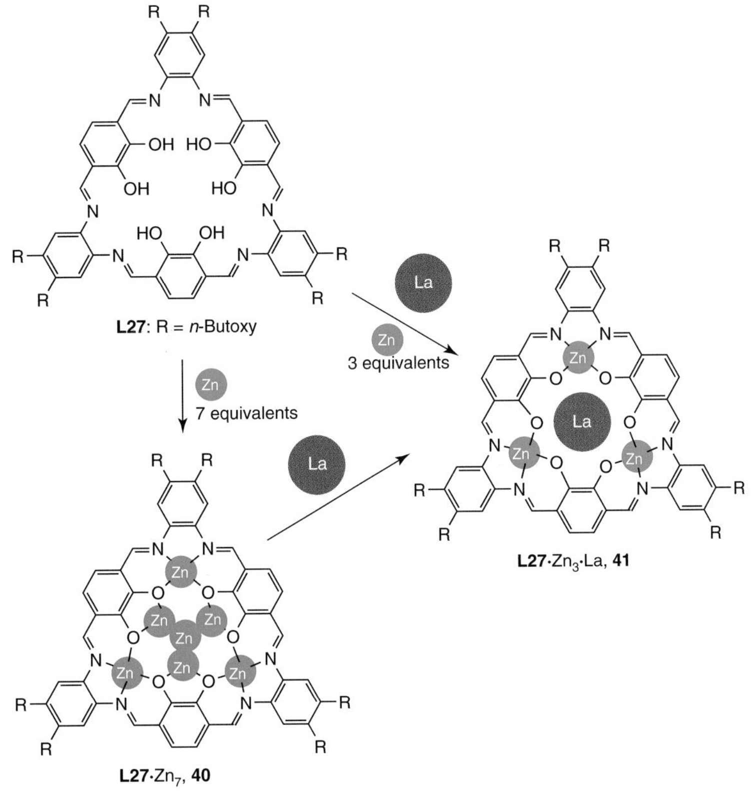

Scheme 13 Macrocyclic crown-ether-like **L27** employed to get [**L27**·Zn_7], **40**, and [**L27**·Zn_3La], **41**.

metallohost, **40**, has a higher affinity toward larger metal ions such as La(III) suitable to the O_6 cation-binding site. Accordingly, complexation of **L27** with both zinc(II) and lanthanum(III) quantitatively affords the heterotetranuclear host–guest complex [**L27**·Zn_3La], **41**, in which the peripheral N_2O_2 sites and central O_6 cation recognition site bind zinc(II) ions and the guest lanthanum(III) ion, respectively, in a site-selective manner. Interestingly, **41** can also be obtained by the direct reaction of the heptanuclear Zn_7 host, **40**, with the guest lanthanum(III) ion.[134]

Besides, macrocyclic Schiff base hosts for inorganic cationic guests (the macrocyclic **L28** and **L29**) with wide guest recognition cavities are suitable for binding different organic cations such as cetyltrimethylammonium, Bu_4N^+, and dimethylbipyridinium (or methyl viologen) when mixed with the corresponding halides in a solvent mixture of

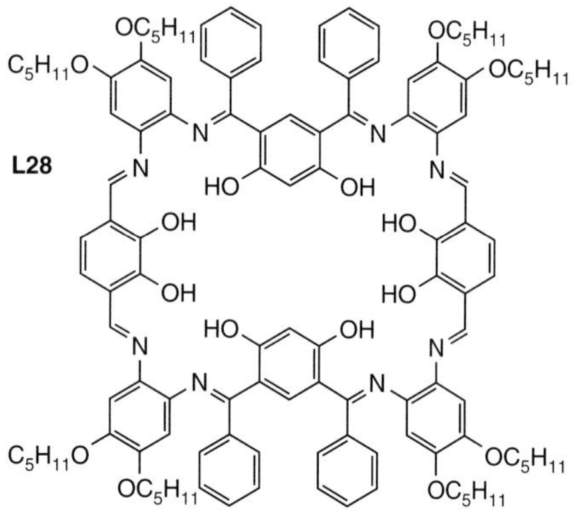

$CHCl_3$–CH_3OH. The cavities of these ligands are shown to be much wider than those of the most commonly used macrocyclic hosts.[137] Binding studies are quantified and verified by ^{1}H NMR and MALDI mass spectroscopic techniques, which suggested the formation of 1 : 1 host–guest complexes on cation binding. On the basis of these studies, methyl viologen exhibits the highest binding constant among the four cationic guests tested, which may be attributed to its +2 charge or enhanced π–π interactions. However, **L28** and **L29** are not sufficiently electron rich to bind larger organic cations such as cyclobis(paraquat-*p*-phenylene) ($CBPQT^{4+}$). Binding of $CBPQT^{4+}$ can be successfully achieved with another macrocyclic host **L30** with 10 hydroxyl groups to form an interesting macrocycle-in-macrocycle type of a host–guest complex.

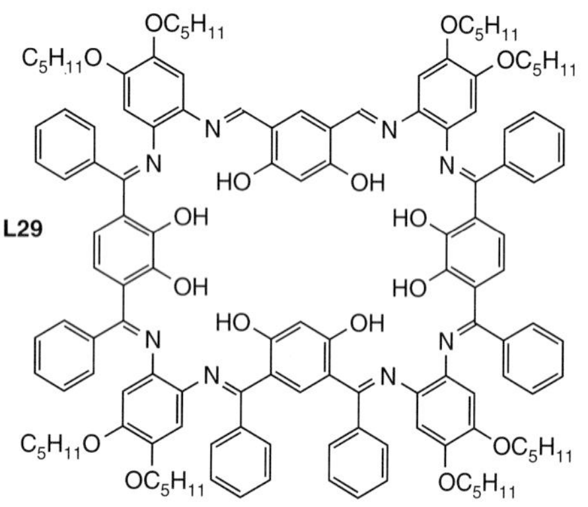

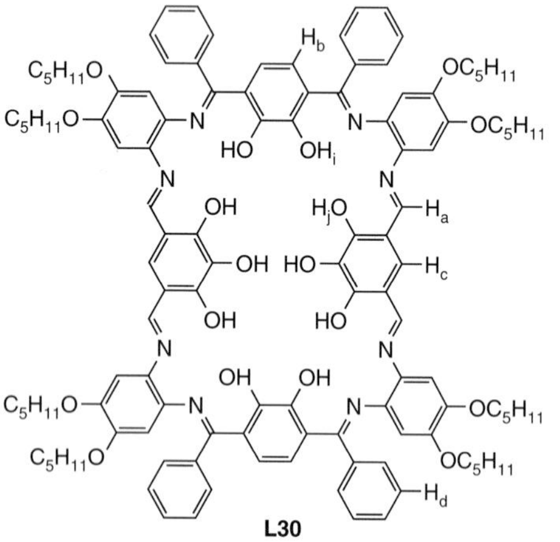

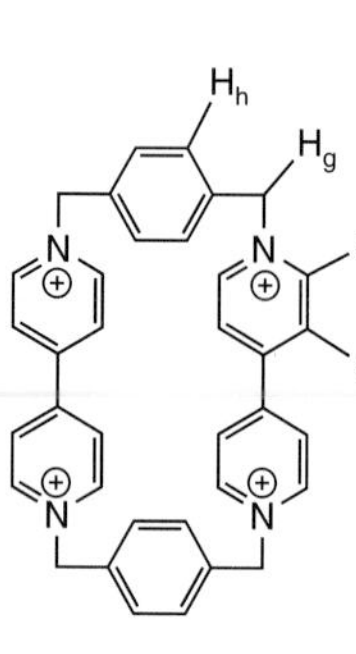

L28

L29

L30

$CBPQT^{4+}$

2.4 Cyclic Schiff base receptors: binding of anions

Besides many studies that have focused on exploring the molecular recognition and host–guest properties of different types of salen- and salophen-based acyclic and cyclic Schiff bases as discussed above, several pyrrole- and amide-based macrocyclic Schiff bases are also explored as synthetic receptors for anion complexation.[138] Pyrrole-based Schiff base macrocycles show anion and neutral guest recognition through hydrogen bonding.[139] These ligands that act as interesting receptors for tetrahedral oxoanions and spherical anions such as halides are studied in relation to anion complexation.[140,141] With this focus, anion complexation studies and binding properties of several macrocyclic Schiff base ligands containing polypyrroles and imine–amide-based moieties are all well established.

For example, 2,6-diamidopyridinedipyrromethane **L31** was derived from the condensation reaction of bis(2-aminophenyl)pyridine-2,6-dicarboxamide and diformyldimethyldipyrrolyl methane.[142] The host **L31** has a large cavity that favors the formation of well-oriented, directional N–H$\cdots$anion hydrogen bonds, favorable for binding hydrogen sulfate in a strong, 1 : 1 manner in acetonitrile ($K_a = 64\,000 \pm 2600\ M^{-1}$ from UV–vis spectroscopic titrations). Only weak binding interactions were seen in the case of cyanide, chloride, and bromide and no detectable affinity for nitrate anion. This feature makes it an attractive candidate for nuclear waste remediation applications, removing sulfate from nitrate-rich low-activity nuclear wastes.

Katayev and Sessler have further investigated the anion complexation behavior of the amido-imine-based hybrid-macrocycles **L32**, **L33**, and **L34** that are effective anion receptors (in the form of their tetrabutylammonium salts) for anions such as Cl^-, Br^-, CH_3COO^-, HSO_4^-, $H_2PO_4^-$, and ReO_4^- in acetonitrile solution with a maximal association constants on the order of $10^7\ M^{-1}$ as determined by UV–vis spectroscopic titration. The receptors **L32** and **L33** displayed selectivity toward the CH_3COO^-, HSO_4^-, and $H_2PO_4^-$, whereas the most structurally rigid **L34** showed preference for Cl^- ion. It has been shown that, for receptors containing pyrrolic hydrogen bonds, incorporation of dipyrromethanes into the receptor structure leads to the acetate ion selectivity,[143] whereas incorporation of 2,2-bipyrroles makes the receptor more selective for phosphate ion.[144] The ligand **L33** can be crystallized in both its free form and as complexed with sulfuric acid and DMSO as [**L33**·H_2SO_4], **42**, and [**L33**·DMSO], **43** (Scheme 14). In case of **L32** and **L34**, the solubility of the resulting complexes proved rather low.

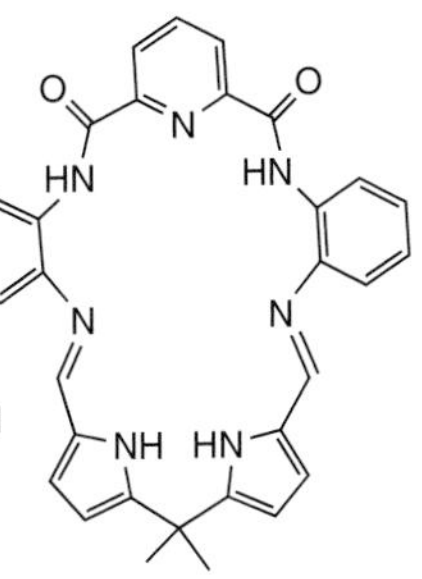

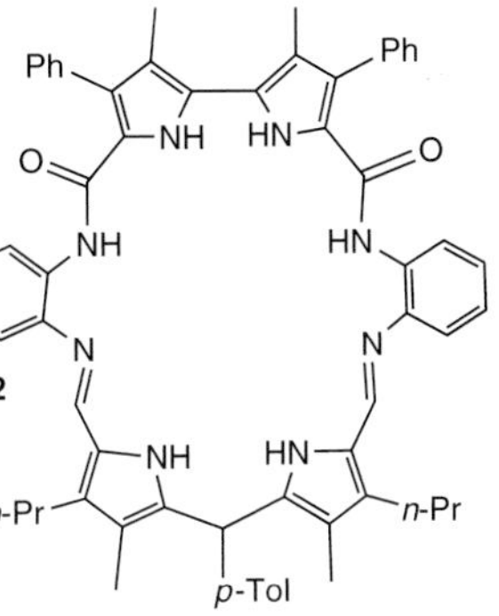

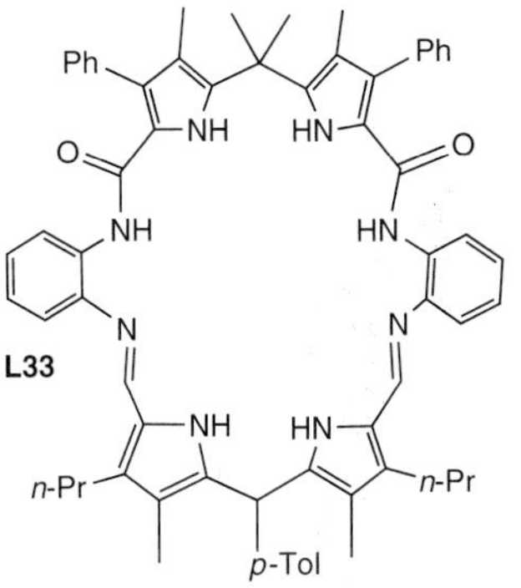

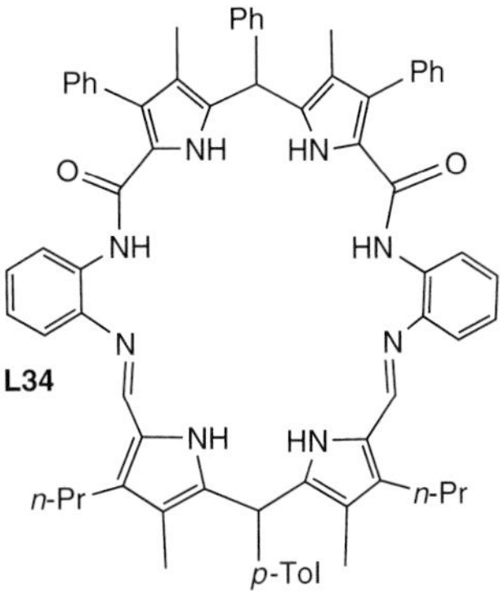

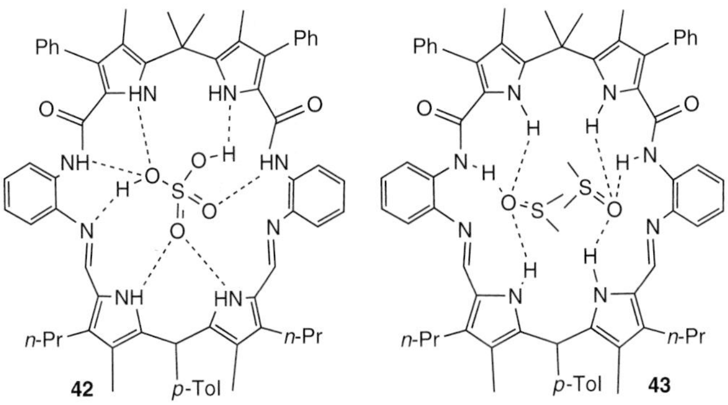

Scheme 14 Host–guest complexes [**L33**·H_2SO_4], **42**, and [**L33**·$DMSO_2$], **43**, derived from **L33**.

Table 4 Binding constants for **L35** and **L36**.

Anion	L35	L36	Anion volume (Å)^3
Cl^-	16600 ± 900	3 300 ± 300	24.8
Br^-	7100 ± 1000	7 100 ± 900	31.5
CH_3COO^-	3200 ± 600	3 600 ± 300	17.8
NO_3^-	15400 ± 2100	1 000 ± 300	24.0
HSO_4^-	18900 ± 1000	7 400 ± 800	28.7
$H_2PO_4^-$	9500 ± 400	Not fitted	33.5

Determined by UV–vis spectroscopic titrations in CH_4Cl_2 (anions are studied as their tetrabutylammonium salts with 1 : 1 binding stoichiometries).

When these studies are further extended to explore the impact of subtle variations in the structure of poypyrrolic macrocycles, described above, on anion-binding affinity/selectivity, the 2,5-diamidothiophene bipyrrole Schiff base macrocycles **L35** and **L36** having different linkages between the pyrrole rings in the structure[145] are investigated. The receptor **L35** has displayed a higher affinity for the anions of preferably smaller volumes (Cl_2, NO_3^-, and HSO_4^-), whereas the macrocycle **L36** interacts more strongly with the larger anions, and hence it can be inferred that the rigid and well-defined cavity of macrocycle **L36** prevents it from interacting with or accommodating anions of smaller dimensions and higher charge densities. Table 4 shows the anion-binding constants for bipyrrole receptors **L35** and **L36** obtained in CH_4Cl_2 and determined by UV–vis spectroscopic titrations.

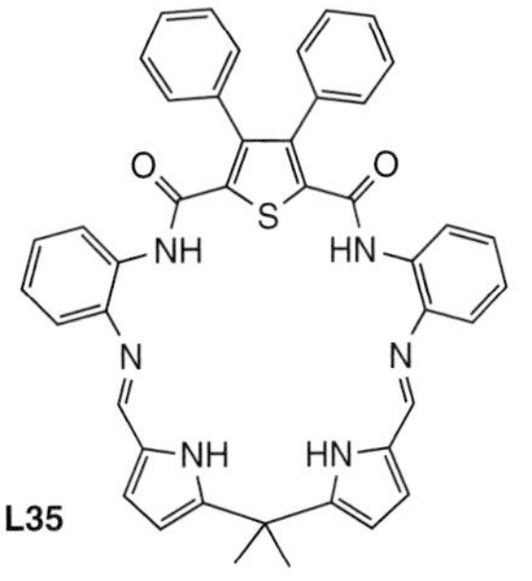

Interesting forms of macrocyclic N_2O_2 Schiff base analogs can be derived by inserting hexamethylene chain between the pendant amine groups as in **L37** to preorganize the sulfate-binding site if the hexamethylene strap would screen the complexed ion, increasing solubility in nonpolar- and water-immiscible solvents.[146]

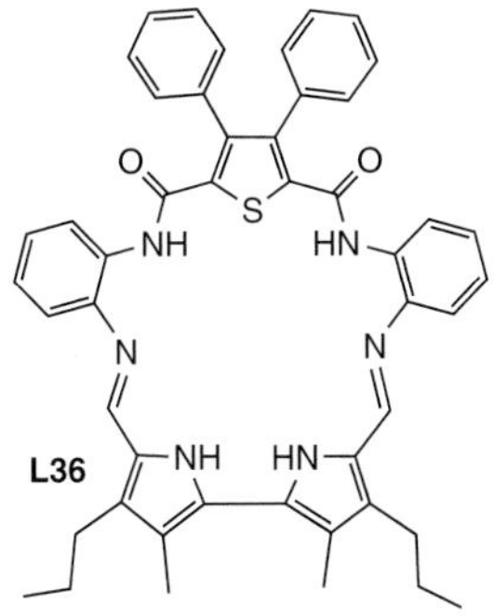

The macrocyclic Schiff base **L37** contains an hexyl linker between the two amine groups. In anion transport experiments, the corresponding metal complexes of **L37** are found to successfully extract sulfate anion from aqueous solution to organic solvents. In the case of the copper complex of receptor **L37**, it was found that more than

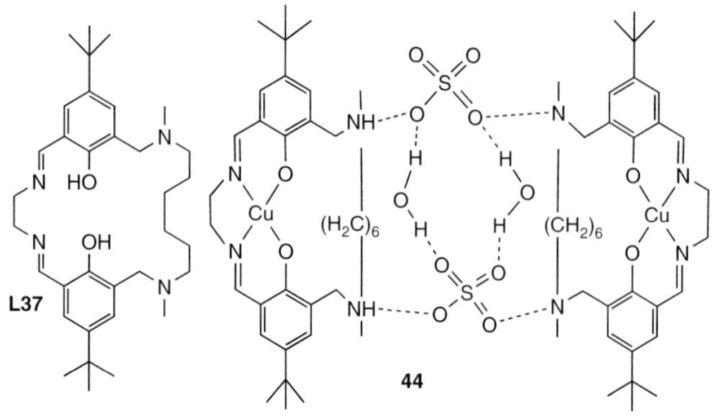

Scheme 15 Inclusion of sulfate inside the tetracationic macrocycle cavity to form a 2 : 2 : 2 : 2 host–guest assembly to form a dimeric $[CuL37(SO_4)(H_2O)]_2$, **44**, from the macrocyclic **L37**.

90% exists in the form of $[Cu(L37)(SO_4)]$, **44**, over a pH range of 3–5. However, less than 10% of the corresponding free ligand exists as sulfate complex in this pH range. These results were attributed to the cooperative effects of metal(II) and sulfate anion binding, which, in turn, provides an explanation for the efficient sulfate anion transport observed in the extraction experiments. X-ray structure determinations of the host–guest Cu(II) and Ni(II) sulfate complexes of **L37** suggest that the detuning of M^{2+} binding results from a distortion from planarity of the Schiff base $N_2O_2{}^{2-}$ donor set imposed by the incorporation of the hexamethylene moiety between imine nitrogens in the ligand and reveal that the sulfate is bound as a hydrate in a 2 : 2 : 2 : 2 ratio of ligand : M^{2+} : $SO_4{}^{2-}$: H_2O assembly (Scheme 15) as in $[Cu(L37)(SO_4)(H_2O)]_2$.[146]

3 REDUCED SCHIFF BASE LIGANDS AS HOST/RECEPTORS

Reduction of –C=N of a Schiff base to amine leads to the formation of the so-called reduced Schiff base ligands, and after reduction of a Schiff base, the corresponding reduced Schiff bases are expected to be more stable and adopt conformational flexibility to the ligand because the reduced $-CH_2-NH$ form will no longer be restricted to a rigid planar conformation. Supramolecular chemistry and molecular recognition properties of reduced Schiff base ligands remain largely unexplored compared with those of Schiff base ligands.[139, 147–150]

3.1 Recognition of anions, cations, and neutral species by reduced Schiff base ligands

3.1.1 Binding of anions

Reduced forms of Schiff bases containing amine/amide N–H groups form a wide range of receptors capable of coordinating anions. Macrocyclic polyamine/amide type of ligands can be readily synthesized by simple Schiff base condensations of either 2,2′-diaminodiethylamine (*dien*) or 2,2′,2″-triaminotriethylamine (*tren*) with aromatic or heterocyclic dialdehydes, followed by borohydride reduction. These ligands display potential binding properties toward different anions such as halides, nitrates, phosphates, and other biologically important carboxylates and nucleotides.[148–158] However, reduced forms of macrocyclic Schiff bases as receptors for perchlorate ions are not very common except a few examples.[159, 160] Very recently, reduced forms of hexaazamacrocycles, **L38** and **L39** derived from the cyclization of bis(2-aminoethyl)amine with 2,5-thiophenedicarboxaldehyde and teraphthalaldehyde, followed by the reduction with sodium borohydride,[160] are found suitable for binding perchlorate ion. Binding studies on these selective perchlorate ion receptors indicated that tetrahedral anionic species such as perchlorate is inert and interacts weakly with synthetic hosts.

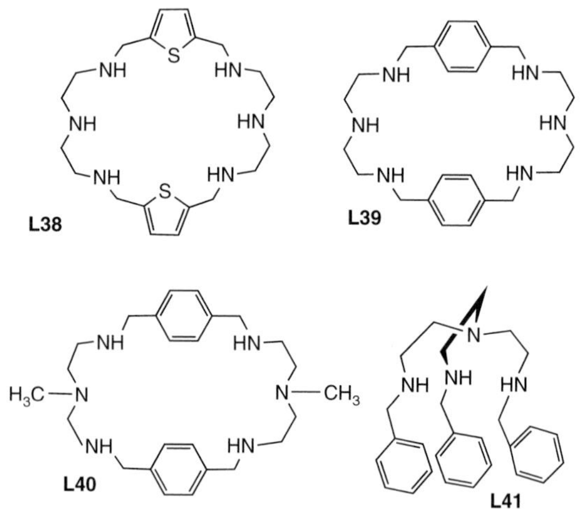

Ditopic binding of the perchlorate anions is observed in both **L38** and **L39** in hexa- and tetraprotonated forms, respectively, resulting into ditopic complexes in which the two perchlorate ions are coordinated by the hydrogen-bonding interactions within their protonated forms. The mode of ditopic encapsulation of perchlorate ions can be described as "face binding." In **L39**, all four protonated sites are utilized in binding anions; however, in **L38**, two protonated nitrogen atoms remain unbonded. X-ray crystal structure analysis of perchlorate complex of **L38** indicates the involvement of total six hydrogen bonds ($N\cdots O = 2.808–2.960$ Å) linked with the two perchlorate anions inside, where each anion binds to both of the central NH_2 and also one secondary $NH_2{}^+$ linked with aromatic unit (Scheme 16).

Besides "face-binding" mode of ditopic recognition, anion complexation by azamacrocycles involves other different modes of ditopic binding such as "bipyramidal

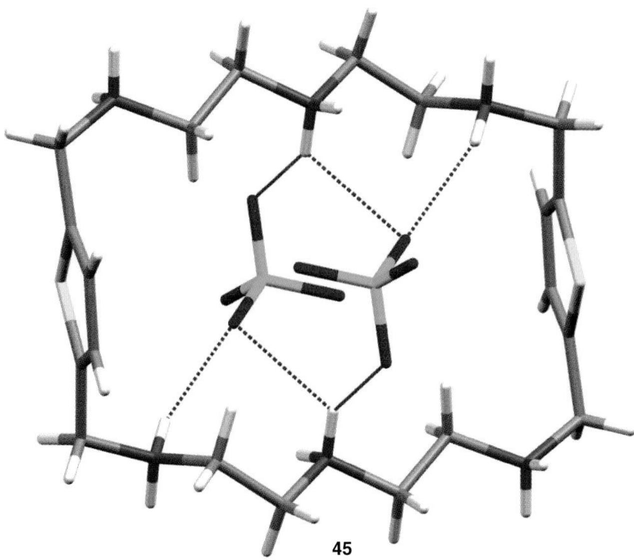

Scheme 16 Binding of the two perchlorate ions by **L38** in $[H_6\mathbf{L38}(ClO_4)_2]^{2+}$, **45**.

binding," "cross binding," and "complete encapsulation." Mode of binding of two anions as "complete encapsulation" is observed in a hexaprotonated azamacrocycle coupled with two methyl groups, **L40**, which is capable of encapsulating two chlorides inside its cavity.[161] All nitrogens in the anion complex are protonated, and the two chlorides remain within the cavity in close contact with binding units on opposite sides (Scheme 17). The six protons on the nitrogen centers point toward the cavity to form a ditopic recognition cavity for encapsulating two chlorides within the macrocyclic ring. Each of the remaining four chlorides binds outside the cavity with a single hydrogen bond from four secondary amines. In the complex, all six protons (four from secondary and two from tertiary nitrogens) on the charged nitrogen centers are effectively utilized in coordinating chloride anions. It can be concluded that the presence of two methyl groups coupled with fully charged nitrogens makes the ligand ideal for fitting two compatible chlorides inside the cavity through trigonal recognition with three hydrogen bonds.

L40·Cl_2, **46**

Scheme 17 Trigonal recognition and ditopic binding of Cl^- in **L40** to form **L40**·Cl_2, **46**, showing two chlorides bonded through N–H hydrogen bonds.

Bowman-James *et al.* described several monocyclic, bicyclic, and tricyclic hosts in terms of their affinities toward simple oxo anions, acetates, and halides.[162] The monocycles can form sandwich structures of host–guest complexes, whereas the bicycles tend to encapsulate their guests. On the other hand, multiple anions and water molecules are often found in the larger bicycles. There are several simple acyclic *tren*-based reduced Schiff base forms of receptors with tripodal amine-binding sites for effective binding of anions especially having C_3 axis of symmetry-like nitrate, phosphate, and sulfate.[163–167] For instance, a tripodal *tren*-based reduced Schiff base ligand, **L41**, in its triprotonated form can strongly bind $H_2PO_4^-$ and HSO_4^- and NO_3^- and halides as indicated by X-ray crystal structures. In the crystal structure of the phosphate and bromide complexes $[\mathbf{L41}][H_2PO_4]_3{\cdot}H_3PO_4$, **47**, and $[\mathbf{L41}]Br_3$, **48** (Scheme 18),[168] the ligand is triprotonated with the three arms pointing outward in a trigonal planar-like arrangement. In **47**, three $H_2PO_4^-$ counterions are located between each of the *tren* arms, whereas the neutral H_3PO_4 molecule lies above the quasi-planar *tren* moiety. The structure of the bromide complex **48** is slightly different, although again the *tren* receptor is triprotonated and quasi-planar, but in this case C_{2v}-like symmetry is observed with the two of the arms pointed in the same direction with a bromide ion in between. The other two bromides lie outside the *tren* arms. The binding constants (Table 5) derived based on the ^{1}H NMR titrations of $[H_3\mathbf{L41}]$[tosylate]$_3$ with $[n\text{-Bu}]_4N^+$ A^- (where $A^- = H_2PO_4^-$, HSO_4^-, NO_3^-, Cl^-, and Br^-) in $CDCl_3$. These studies indicate that the ligand **L41** binds $H_2PO_4^-$ and HSO_4^- more strongly than NO_3^- and halides.

Among macrocyclic *tren*-based reduced forms of Schiff bases, the polyamine-/amide-based cryptands represent an interesting class of ligands for anion binding as they can efficiently form ditopic and tritopic cascade complexes on binding of halide ions.[169] Octaaminocryptands, derived from Schiff base condensations between amines and aromatic or heterocyclic dialdehydes, followed by simple reductions of the resulting imines to amines, can also form

Table 5 Binding data[a] of $[H_3\mathbf{L41}]$ [tosylate]$_3$ in $CDCl_3$ for the anions.

Anions	log *K*
$H_2PO_4^-$	3.25
HSO_4^-	3.20
NO_3^-	1.55
Cl^-	1.80
Br^-	1.70

[a]Based on ^{1}H NMR titration curves of $[H_3L]$[tosylate]$_3$ with n-$[n\text{-Bu}]_4N^+$ salts of $H_2PO_4^-$, HSO_4^-, and Br^-.

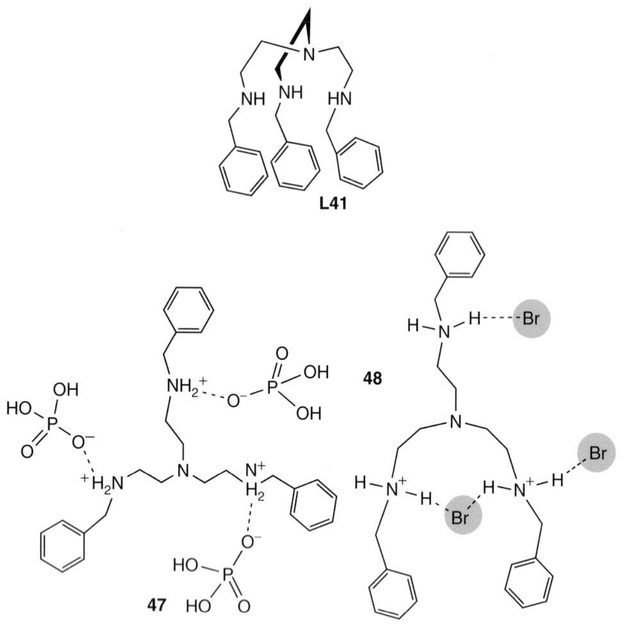

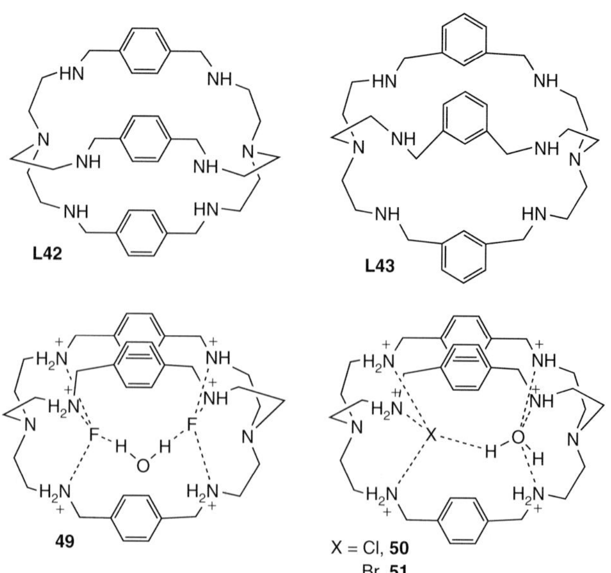

Scheme 19 Encapsulation of halide ions in the cavity of **L42** and **L43** forming **49, 50**, and **51**.

Scheme 18 Phosphate- and bromide-bound host–guest complexes $[H_3\mathbf{L41}][H_2PO_4]_3{\cdot}H_3PO_4$, **47**, and $[H_3\mathbf{L41}][Br]_3$, **48**, derived from **L41**.

similar types of anion-based cascade complexes where two spherical anions play the topological role similar to that of the two metal ions in traditional cascade complexes. Such octaaminocryptands with *p*-xylyl **L42** and *m*-xylyl **L43** display interesting anion-binding properties.[169] In this connection, the first example of anion-based cascade complex $[H_6\mathbf{L42}(F)_2(H_2O)][SiF_6]_2 12{\cdot}H_2O$, **49**, is obtained from the hexaprotonated form of **L42** and is surrounded by two external $SiF_6{}^{2-}$ counterions and 12 molecules of water. The two fluoride ions are encapsulated in the structure in the cavity of **L42** as shown in Scheme 19. The binding mode of F^- is described as tritopic recognition via bridging water. Each fluoride ion is coordinated in a much distorted tetrahedron manner through hydrogen-bonding interactions (F···HOH···F) to three protonated secondary amines of the *tren* unit and the bridging water molecule.[170]

However, with larger halide ions, Cl^- and Br^-, **L42** displayed ditopic recognition, unlike tritopic binding seen with smaller F^- ion. X-ray structural characterization of these chloro and bromo cascade complexes indicated the ditopic chloride/water and bromide/water binding without bridging water as shown in **50** and **51** (Scheme 19). Thus, the characterization of ditopic chloride/water and bromide/water complexes is obviously an indication of the effect of larger anionic radii for Cl^- and Br^- ions, which makes the fit more favorable for just one halide in the cavity.[171] Interestingly, on hexaprotonation of **L43** having slightly smaller *m*-xylyl spacer with HF, encapsulation of only F^- ion took place via similar ditopic binding mode as seen with Cl^- or Br^- as observed in **50** and **51**. X-ray crystallographic investigations on the encapsulation of higher homolog halides in **L43** in its different degree of protonated forms[169] indicated that hexaprotonated form of **L43** preferred monotopic spherical recognition and binding of larger I^- ion as in **52**, whereas the tetraprotonated form is suitable to encapsulate the bichloride to form $[HCl_2]^-$-bound host–guest complex, **53**, as shown in Scheme 20. The encapsulated iodide resides in distorted square pyramidal manner with four intramolecular $N{-}H^+\cdots I^-$ interactions in **52**. The ligand in **53** is tetraprotonated having one bridgehead protonated nitrogen center in one side of the *tren* cap and three protonated secondary nitrogen centers on the other side of the *tren* cap, encapsulating the linear bichloride via four coordinated distorted tetrahedral geometry.

Furthermore, higher degree of protonation (hexa and octa) and hence higher distribution of positive charge over the receptor **L42** (with *para*-xylyl spacer) makes the cavity more electrophilic toward binding of tetrahedral oxyanions, $ClO_4{}^-$ and $HSO_4{}^-$, as well as octahedral $HSiF_6{}^-$ ion.[169] The structures clearly illustrate that the overall conformation of the cavity is changed into ellipsoid and near spherical, due to the internal repulsive electrostatic forces, that is suitable for the geometry of anions allowing the encapsulation of anions like perchlorate, **54**, hydrogen sulfate, **55**, and hexafluorosilicate, **56**, inside the receptor.

3.1.2 *Binding of cations and neutral species*

Certain reduced Schiff base ligands, derived from salicylaldehyde and amino acids, are shown to form metallocrown type of hosts/receptors through cation-assisted self-assembly during complexation. It is observed that

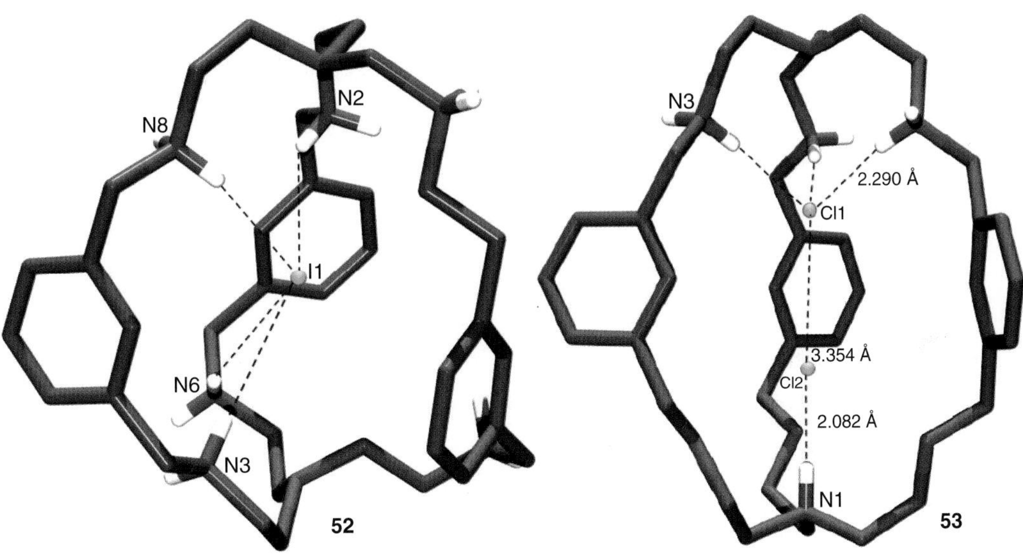

Scheme 20 Monotopic binding of I^- in **52** and **53** showing ditopic binding of Cl^- ions.

these complexation reactions are controlled by the solvents, pH, and the nature of the cations employed.[172,173] For example, Vittal *et al.* described the solvent-dependent formation of metallocrown-like copper complexes of the reduced Schiff base ligands **L44** and **L45** and encapsulation of alkali metal ions such as K^+ and Li^+ in the metallocrown cavity[172,173] as shown in Figure 4. Reaction of potassium salt of **L44** with copper perchlorate hexahydrate in CH_3CN/CH_3OH furnished a metallocrown complex $[K(ClO_4)_3\{Cu_3(\mathbf{L44})_3\}](ClO_4)$, **57**, in which the metallocrown cavity is occupied by potassium in the form of $\{K \subset [12\text{-MC-}3]\}$. When **57** is left in water, it resulted in the conversion of metallocrown complex into an infinite 1D coordination polymer $[Cu(\mathbf{L44})(H_2O)(ClO_4)]_n$, **58**. Interestingly, the direct complexation reaction of **L44** with Cu salt in aqueous solution can also furnish the same 1D coordination polymeric **58**. This solvent-dependent reactivity is attributed to the chiral center or the methyl substituent present in **L44**.

On the other hand, when $[Cu_3(\mathbf{L45})_3](ClO_4)_3$, **59**, obtained from the complexation of Na^+ or K^+ salt of **L45** with $Cu(ClO_4)_2{\cdot}6H_2O$ is further allowed to react with $LiClO_4$, a trinuclear copper complex, **59**, undergoes rearrangement to provide a metallocrown cavity [8-MC-3], suitable to accommodate Li^+ ion to form a host–guest complex $[(ClO_4)Li\{Cu_3(\mathbf{L45})_3\}]\text{-}(ClO_4)_3$, **60**. Interestingly, the same Lithium-bound **60** can also be prepared by the direct combination of $Cu(ClO_4)_2{\cdot}6H_2O$ with the Li^+ salt of **L45**. On binding Li^+ ion, $[Li\{Cu_3\text{-}(\mathbf{L45})_3\}]^{4+}$ cation in **60** acquired an idealized C_3 symmetry with three crystallographically independent Cu(II) centers, which is an indication of structural rearrangement in the $[Cu_3(\mathbf{L45})_3]^{3+}$ cation to favor the binding of lithium ion. Thus, the structural rearrangement of the $[Cu_3(\mathbf{L45})_3]^+$ cation of **59** resulting in the formation of $[(ClO_4)Li\{Cu_3(\mathbf{L45})_3\}](ClO_4)_3$, **60**, by selectively accommodating Li^+ ions illustrated the role of alkali metal ions on the reactivity.[172]

Ray *et al.*[174] in their recent study described metallocrown complexes $[M'\{M(\mathbf{L46})_2\}_3]^+$ (where M′ is either K^+ or Na^+ and M is either Ni^{2+} or Cu^{2+}) derived from the reduced Schiff base ligand, **L46** (Scheme 21), having L-leucine moiety. These Ni- and Cu-based metallocrown receptors are suitable for hosting alkali metal ions such as Na^+ and K^+ ions and furnish isostructural complexes and identical host–guest complexes. For example, in the X-ray crystallographic structure of $[K\{Ni(\mathbf{L46})_2\}_3]^+$, **61**, three $[Ni(\mathbf{L46})_2]$ units form a cavity for encapsulating K^+ ion as shown in Scheme 21. Interestingly, recognition properties of the metallocrown found independent of the nature of the metal ion employed as the similar self-assembly and

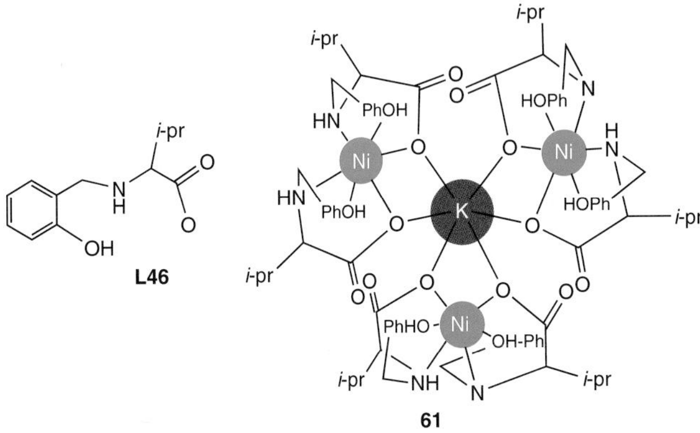

Scheme 21 Three $[Ni(\mathbf{L46})_2]$ units derived from **L46** forming a metallocrown ether-like cavity in **61** for binding alkali metal ions such as K^+.

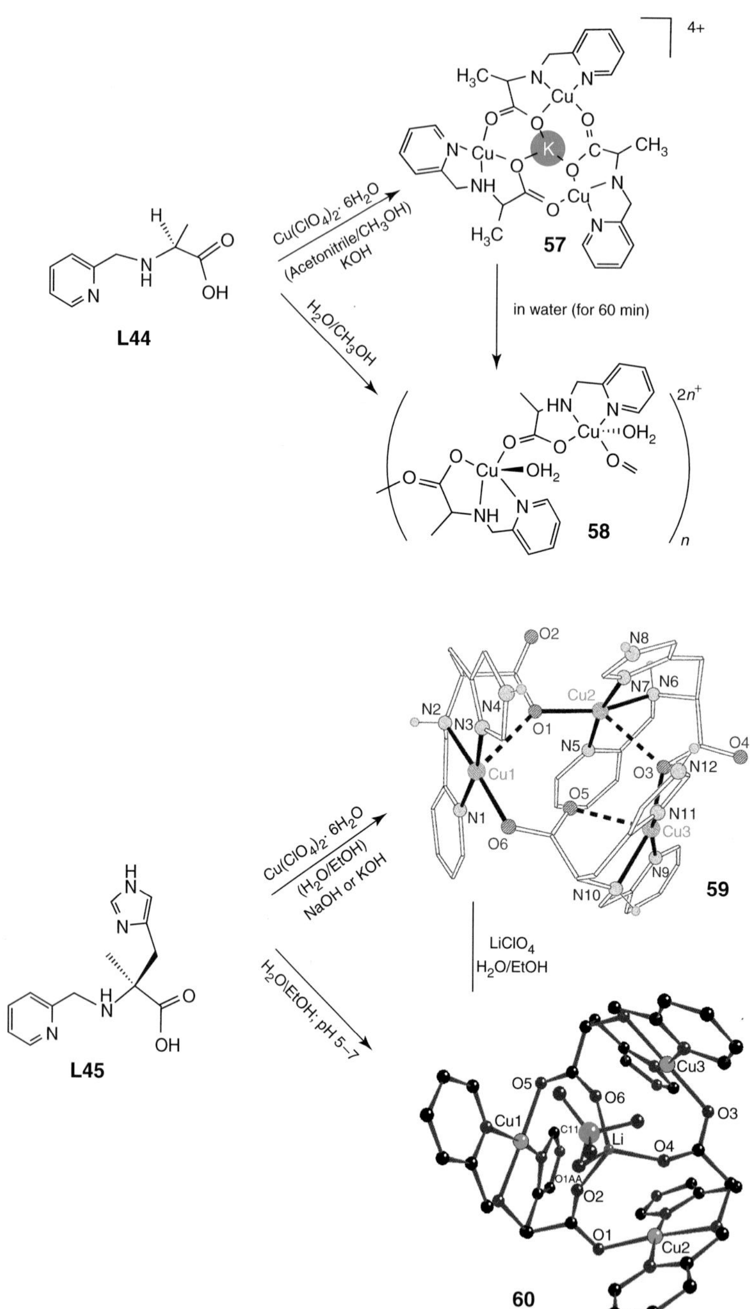

Figure 4 Solvent-dependent reactivity of **L44** and **L45** forming the host–guest complexes **57**–**60**.

subsequent binding process occurred even after replacing Ni with Cu. However, it is important that minor structural changes due to replacement of K^+ with Na^+ have shown to sufficiently shift the d–d transition of Ni(II) by ~70 nm, which provides an indirect way of distinguishing K^+ and Na^+ despite their nonspectroscopic identity in the visible range of wavelength.

The reduced Schiff base ligands have the potential for generating fascinating host–guest and supramolecular network structures while hosting neutral species such as pyridine and I_2. Ray *et al.*[175] have recently synthesized an octameric Cu(II) complex [Cu_8**L47**Py_{10}]·Py·3CH_3OH·$(C_2H_5)_2O$, **62**, with a novel capsule-like cavity capable of hosting neutral guests such as the pyridine molecules (Figure 5) by employing the tetradentate **L47**. The trapped pyridine molecules are held inside the cavity through interesting combination of hydrogen bonding and coordination to Cu(II). Such molecular cavities offer potential

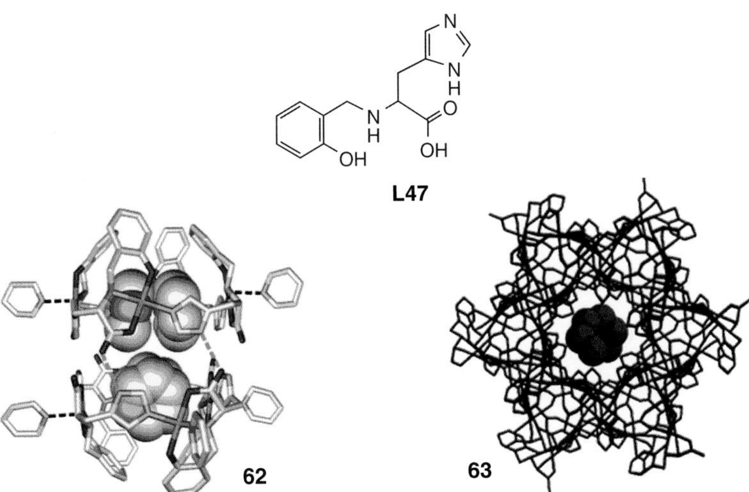

Figure 5 Molecular structure of [Cu_8**L47**Py_{10}] Py $3CH_3OH$ $(C_2H_5)_2O$, **62**, derived from **L47** showing the trapped pyridine molecules (the solvent molecules are excluded for clarity) and **63** filling I_2 inside the channels.

applications as selective hosts for anion sensing, catalysis, selective recognition, and separation.[176]

In another study dealing with **L47**, a microporous iron complex [Fe(III)$_2$(μ-OH)(μ-OAc)(**L47**)$_2$]·$4H_2O$, **63**, with an empty 1D helical hydrophilic channel is obtained. Thermal dehydration to remove loosely bound water molecules from the hydrophilic channel can create an inner lining of acidic protons without destroying the lattice.[177] Subsequently, resulting empty channels can be used to host the neutral iodine molecules (Figure 5). Thus, molecular recognition properties observed in **62** and **63** demonstrate that the enantiopure reduced Schiff base ligand with additional donor groups and multiple hydrogen-bonding functionality such as carboxylate, imidazole, amine, and phenolate can self-assemble differently with different metal ions giving rise to symmetrically ordered materials as diverse as capsules to helices.[174]

The ligand **L48** belongs to the class of reduced Schiff base ligands that contains ferrocenealdehyde, instead of salicylaldehyde, and amino acid side arm. Several copper complexes of **L48** are explored for their potential host–guest properties associated with the chiral cavity.[178] It is shown that the ferrocenylmethyl-substituted amino acid ligands can organize around the Cu center in C_2 symmetry in the chiral host [Cu(II)(**L48**)$_2$(CH_3CN)$_2$], **64**, which can form host–guest complexes while binding planar heterocyclic neutral guests with N-donors such as pyrazine [Cu(II)(**L48**)$_2$(pyz)] (**65**), pyridine, and bpy (4,4′-bipyridine) as shown in Scheme 22. The binding of these species into the host is found to stabilize the narrow cavity. Easily replaceable acetonitrile that is coordinated to Cu center is expected to facilitate the binding of these neutral guests.

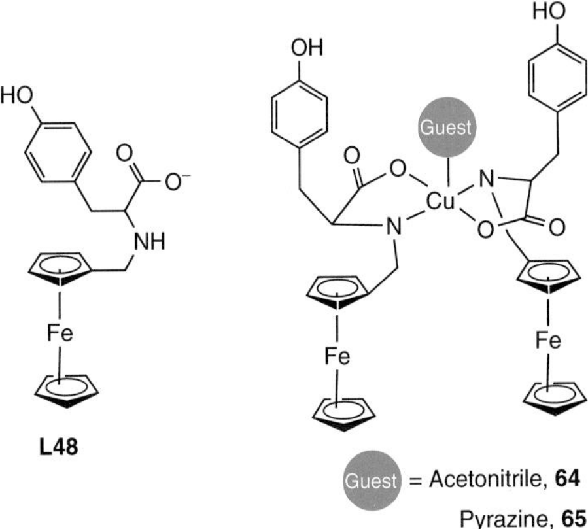

Scheme 22 Binding of neutral guest species: acetonitrile in **64** and pyrazine in **65** derived from **L48**.

3.2 Water as a guest: water aggregates hosted by reduced Schiff base complexes

Organic and metal–organic crystal lattices with a suitable environment for hydrogen bonding can serve as a potential solid-state media for hosting various water aggregates. With the help of available knowledge and clear understanding of the principles of supramolecular chemistry, many of such organic and inorganic lattices, as hosts for water aggregates, are well characterized for isolating and analyzing different types of hydrogen-bonded water topologies.[179–183] These guest water aggregates can assume different types of morphologies such as discrete chains, discrete rings, one-dimensional infinite chains without the formation of rings, and one-dimensional infinite tapes without rings, while the infinite chains of hydrogen-bonded water in one dimension can assume several conformations including zigzag or helix.[182,183]

Hydrogen-bonded 1D water chains, among several morphologies of water aggregates, deserve a special recognition in supramolecular chemistry of water particularly because of their pivotal role in fundamental biological processes such as transport of water, protons, and ions.[184] Formation and stabilization of 1D water chains are favored when there are suitable hydrophilic channels so that a strong hydrogen bonding between neighboring water molecules along the chain as well as between water molecules and donor–acceptor groups associated with the channels is facilitated.[185–190] In this connection, metal coordination compounds derived from reduced Schiff base ligands are shown to act as good hosts for various water aggregates because of their potential hydrogen bond donor and acceptor functionalities to facilitate the discrete water molecules to extend as 1D water chains or clusters via N–H···O and O–H···O hydrogen bonds.[191] For example, Cu(II) complexes [Cu(**L49**)Cl]·H_2O, **66**, [Cu(**L44**)Cl] H_2O, **67**, and [Cu(**L50**)(phen)] $2H_2O$, **68**, derived from **L49**, **L44**, and **L50** ligands are crystallized as hydrates in which the lattice water molecules are extended as infinite 1D chains via hydrogen bonding (Figure 6).[191] In the crystal structure of [Cu(**L49**)Cl] H_2O, **66**, and [Cu(**L44**)Cl] H_2O, **67**, bridging by asymmetrical chloride anion generated a zigzag 1D coordination polymeric structures sustained by N–H···O=C interactions. When these 1D coordination polymers are further connected by weak C–H···Cl and $\pi\cdots\pi$ interactions to form 2D sheets, the carbonyl oxygen atoms oriented along one side of the 2D sheets formed hydrophilic surfaces that are favorable for hosting hydrophilic guest molecules such as water. Consequently, water molecules are trapped in these channels (Figure 6). Each hydrogen atom of the water molecule interacts with the carbonyl oxygen atom and an oxygen atom of the neighboring water molecule in the chain forming C=O···H–O–H hydrogen bonds, which, in turn, can generate 1D zigzag water chains as shown in Figure 6. The water chains are sandwiched between the hydrophilic surfaces directed by the C=O···HOH interactions. Conformation of water chains in [Cu(**L44**)Cl]·H_2O, **67**, compared to that in [Cu(**L49**)Cl] H_2O, **66**, appears to have minor variation which can be attributed to the presence of bulkier methyl group in the chiral amino acid side chain of the **L44** on the hydrophilic surface.

When acetate ions involve bridging, in place of Cl^-, as in [Cu(**L44**)(CH_3COO)]·0.75H_2O, **69**, no such hosting of 1D water chains is observed which can be mainly due to the puckering caused by the acetate bridging

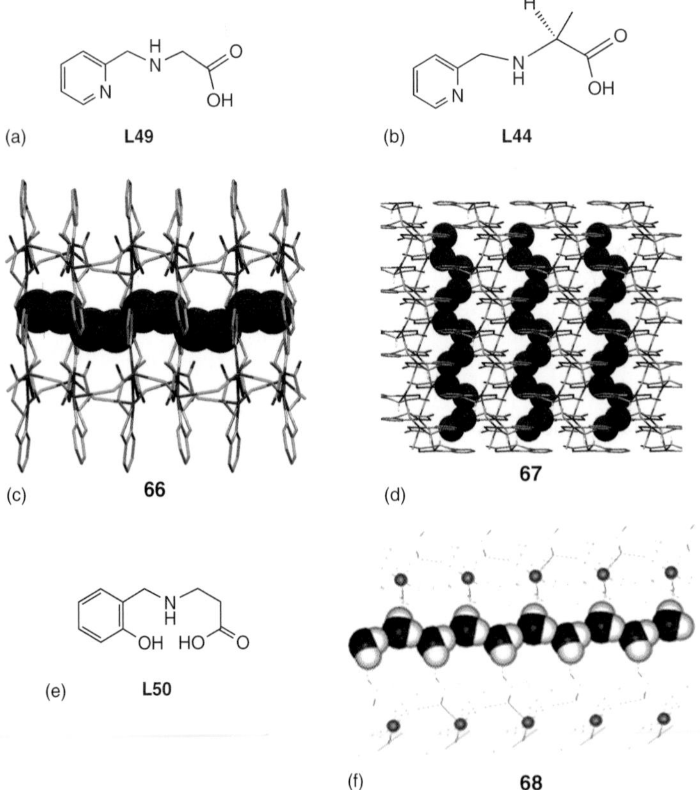

Figure 6 A view of showing the 1D water strand hosted by 1D coordination polymers in (c) [Cu(**L49**)Cl]·H_2O, **66**, and (d) [Cu(**L44**)Cl] H_2O, **67**. A perspective view of the single water chain hosted by hydrogen-bonded coordination complex [Cu(**L50**)(phen)] $2H_2O$, **68** (f).

group resulting in spiral conformation rather than zigzag polymeric arrangement, which, in turn, disfavored the formation of hydrophilic channels and hence water chains in the crystal packing. This makes clear the role of subtle effects of the substituents and conformational changes of coordination polymers on the formation of water chains in the lattice. Thus, the absence of water chains in **69** indicates the role of the nature of coordinating anion and the subsequent conformational changes on the formation of hydrophilic channels suitable for hosting 1D water chains.

In a ternary Cu(II) complex [Cu(**L50**)(phen)] $2H_2O$, **68**, derived from **L50** in the presence of auxiliary 1,10-phenanthroline, the two lattice water molecules are highly ordered, and generated interesting hydrogen-bonded network in the solid state. All the hydrogen atoms from the lattice water molecules and N–H protons involved hydrogen bonding. Hydrogen bonding by N–H proton to carboxylate oxygen (N–H···O) generated a water chain. The repeated hydrogen bonding between the lattice water molecules (O–H···O) resulted in the formation of 1D hydrogen-bonded water chain parallel to the N–H···O bonds (Figure 6).

Helical water chains are much more fascinating due to their intriguing structural and hydrogen-bonding features. Although 1D helical water chains are prevalent in biological systems, it is highly difficult to construct them in the synthetic hosts because the structural constraints required in stabilizing the 1D water chains are highly unpredictable and yet to be fully understood. In fact, a chiral ligand can often lead to the formation of helical structure,[192] and the presence of one or more nonchelating side arms in a chiral ligand may further facilitate the possibility for selective and complementary aggregation of the metal complexes into spiral or helical staircase-like network structure. One such promising possibility of deriving helical staircase coordination polymers is demonstrated by Sreenivasulu and Vittal[193] through their structural investigation based on a Ni(II) complex $[(H_2O)_2 \subset \{Ni(\mathbf{L51})(H_2O)_2\}]\cdot H_2O$, **70**, derived from the chiral ligand, **L51**. Intermolecular connectivity of octahedral Ni(II) centers via neighboring carboxylate oxygen atom generated a left-handed helical staircase-like coordination polymeric architecture in **70** that displayed an intriguing feature of hosting 1D helical chain of water molecules inside the chiral helical pores through hydrogen bonds.[193] Out of four lattice water molecules that are held inside the helical pore, two water molecules O(15) and O(16) are hydrogen bonded to produce 1D helical polymer with a pseudo-4_1 screw axis as shown in Figure 7.

Unlike the alternately arranged water molecules forming a helical water chain inside a hydrogen-bonded helical supramolecular host reported earlier by Chakravarty *et al.* and the left-handed 1D helical water chains trapped in a chiral 3D hydrogen-bonded supramolecular host reported

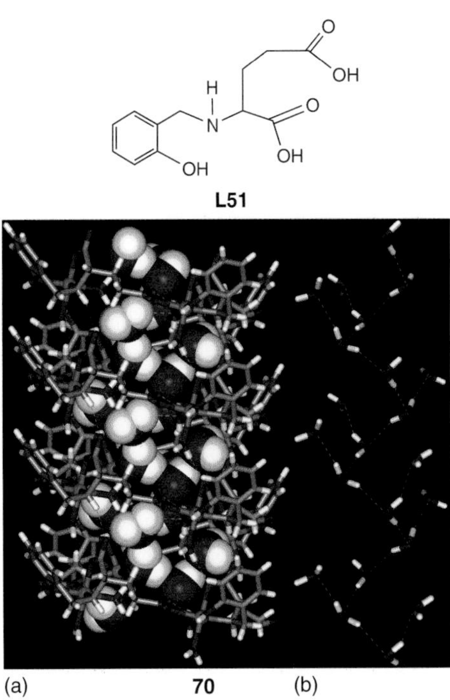

Figure 7 (a) Display of helical water chain encapsulated in staircase coordination polymer **70** derived from **L51**; (b) Hydrogen-bonded helical water chain with space-filling model.

by Hong *et al.*, the helix inside a helix type of host–guest structure of **70** exemplifies the effect of subtle changes in the constrained hydrogen-bonding environment in the lattice.

3.3 Metalla-supramolecular networks derived from reduced Schiff bases

Several forms of reduced Schiff base ligands, *N*-(2-hydroxybenzyl)-amino acids and *N*-(2-pyridylmethyl)-amino acids, *N*-(2-hydroxybenzyl)-aminocyclopentane/hexane carboxylic acids, and *N*-(2-hydroxybenzyl)-amino-methane/ethanesulfonic acids, derived from various aromatic aldehydes and different natural/unnatural amino acids are studied by Vittal and coworkers. Compared to the Schiff bases, the corresponding reduced Schiff bases are more advantageous in view of metalla-supramolecular chemistry because of their stability, conformationally flexible backbone (as they are not constrained to be planar due to reduction of the Schiff base, C=N group), and the resulting adaptability to form conformationally flexible five- or six-membered rings on complexation. Another important merit of these ligands is that the N–H fragment (from the reduced H_2C–NH–backbone) can generate intermolecular hydrogen-bonding interactions (N–H···O), apart from the hydrogen bonds from the carboxylate oxygen (–C=O···H–$O_{solvent}$, C=O···H–N) in the amino acid side

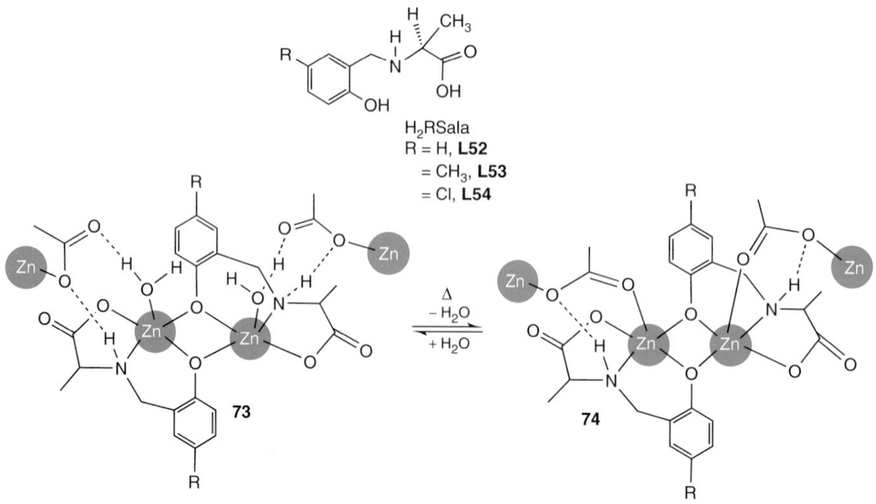

Scheme 23 Schematic illustration of thermally induced interconversion from 3D H-bonded network **73** to a coordination polymeric network **74** via dehydration.

arm. This is also one of the primary requirements from a ligand for the supramolecular self-assembly of the building blocks. In this connection, Vittal and coworkers described several solid-state metalla-supramolecular network structures ranging from hydrogen-bonded linear polymers and helical coordination polymers, and 2D sheets to 3D network architectures constructed via N–H···O, C=O···H–$O_{solvent}$, O–H···O, N–H···O=C hydrogen bonds and C=O···π, C–H···π, and $\pi\cdots\pi$ stacking interactions.[194] Some of the interesting examples are discussed here giving a brief account of different factors such as the role of different donors and acceptors, aqua ligands and solvents, nature of the ligands and metal ions, and coordination geometry around the metal ions that direct the formation of supramolecular structures.

It is possible to achieve supramolecular structural transformations when the structure is fine-tuned by modifying the components of the ligands employed and by controlling the noncovalent interactions. Such solid-state supramolecular structural transformations facilitated by thermal dehydration are observed in the Cu(II) and Zn(II) complexes derived from H_2Sala, **L52**, H_2CH_3Sala, **L53**, and H_2ClSala, **L54** ligands. On thermal dehydration, the hydrogen-bonded helical coordination polymeric structure of, $[Cu_2(\mathbf{L52})_2(H_2O)]_n$, **71** is irreversibly transformed into a 3D coordination network structure to form, $[Cu_2(\mathbf{L52})_2]_n$, **72**.[195] In an another similar irreversible conversion, the 3D hydrogen-bonded network structure in the hydrated complex, $[Zn_2(\mathbf{L52})_2(H_2O)]\cdot 2H_2O$, **73** is converted into a stable 3D coordination polymeric network architecture, $[Zn_2(\mathbf{L52})_2]$, **74**.[196] These structural transformations suggest that the conversion of one solid-state supramolecular structure into another involves breaking and forming of bonds in more than one direction. These Zn(II) complexes lose aqua ligands and lattice water molecules below 110 °C and convert to 3D network structure $[Zn_2(\mathbf{L52})_2]$, **74**. The close distance between the Zn(II) and oxygen atoms of the neighboring carboxylate group (3.74 Å) through a Zn–OH_2···O(carboxylate) interaction suggests that the removal of the aqua ligands by thermal dehydration in the solid state might lead to the formation of new Zn(II)–O(carboxylate) bonds and hence a 3D coordination network structure as confirmed by X-ray crystallography.

Supramolecular transformations that are irreversible are made reversible,[197] as shown in Scheme 23, when the *para-methyl-* and *chloro*-substituted ligands, **L53** and **L54**, are employed instead of **L52**. The reason for the dehydrated compounds to become hydrated during crystallization from aqueous solution is attributed to the repulsive interactions between the substituents and atoms in the other parts of the 3D coordination network, and the interconversion of structures is driven by these repulsive interactions.

The Cu(II) complex, $[Cu_2(\mathbf{L55})_2(H_2O)]\cdot H_2O$, **75** derived from **L55** crystallizes as a 1D helical coordination polymer assembled from the dimeric building blocks similar to **71**. But, the major difference is the presence of one more water molecule in the crystal lattice of **75**. Interestingly, despite the similar structures and crystal packing of helical coordination polymers in **71** and **75**, a stable dehydrated complex from **75** could not be obtained because the anhydrous species rapidly gets rehydrated in the presence of air. This indicates that the formation of new Cu–O bond by the carboxylate group is restricted in **75**. Further, in contrast to **71**, the reason why the reversible hydration takes place in the case of $[Cu_2(\mathbf{L55})_2(H_2O)]\cdot H_2O$, **75**, can be understood by the close examination of the crystal structure of **75**.[198]

The carboxylate group and the phenyl ring are almost parallel to each other and are separated by 3.352 Å in **75**, whereas in **71** the distance of separation is 3.650 Å. This C=O$\cdots\pi$ attractive interaction between these two groups, shown in Scheme 24, appears to hold back the carboxylate group to move closer to the aqua ligands thereby preventing the formation of a new Cu–O bond on thermal dehydration. Thus, it can be concluded that the topochemical transformation in **71** can be accounted for the absence of C=O$\cdots\pi$ attractive interaction. This was the first example of its kind as reported by Vittal *et al.*[198] for the effect of weak interactions like C=O$\cdots\pi$ on the hydrogen-bonding parameters and also on the thermal dehydration reactions.

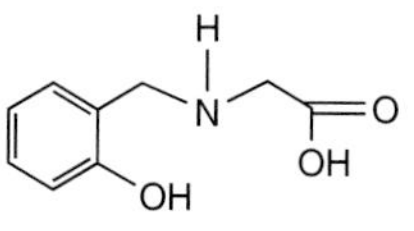

Scheme 24 A view of the intermolecular C=O$\cdots\pi$ interaction in $[Cu_2(\mathbf{L55})_2(H_2O)]\cdot H_2O$, **75**.

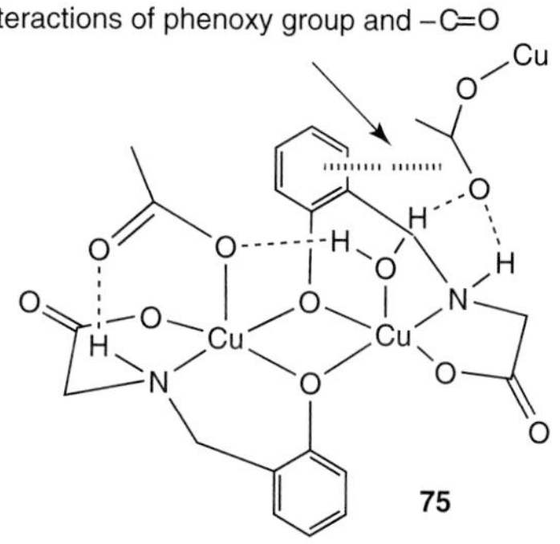

Besides the thermally induced solid-state structural transformations, the conversion of one structure to another is also possible by altering the crystallization conditions such as pH, counterions, solvent, and temperature as observed in the crystallization of $[Cu_2(\mathbf{L56})_2(H_2O)_2]$, **76**, containing a nonchiral ligand, **L56**, which is largely affected by the nature of the solvent system to furnish different structures.[199] Recrystallization of **76** from DMF (dimethylformamide)/acetone furnished rod-shaped bluish green crystals of $[Cu_2(\mathbf{L56})_2(H_2O)_2]\ 2(CH_3)_2CO$, **77**, along with greenish cubic crystals of $[Cu_2(\mathbf{L56})_2]$, **78**. Interestingly, while the solid-state structure of **77** is 1D hydrogen-bonded polymeric structure, the structure in **78** is an interesting three-dimensional coordination network with star-like channels[199] as shown in Figure 8. Actually, four dinuclear units of $[Cu_2(\mathbf{L56})_2]$ in **78** are fused to form a 16-membered square, and the carbonyl oxygen atoms at the periphery of this square are further connected to form a three-dimensional network with star-like channels. It is interesting to note that simple one-pot crystallization produced two different types of crystals that belong to two space groups at the extreme ends of the International Tables for Crystallography (No. 2 and No. 230).

When **L56** having –Cl, $-CH_3$, or –OH-substituted salicylaldehyde is employed for complexation, the resulting dicopper(II) complexes displayed interesting hydrogen-bonded polymeric structures depending on the nature of

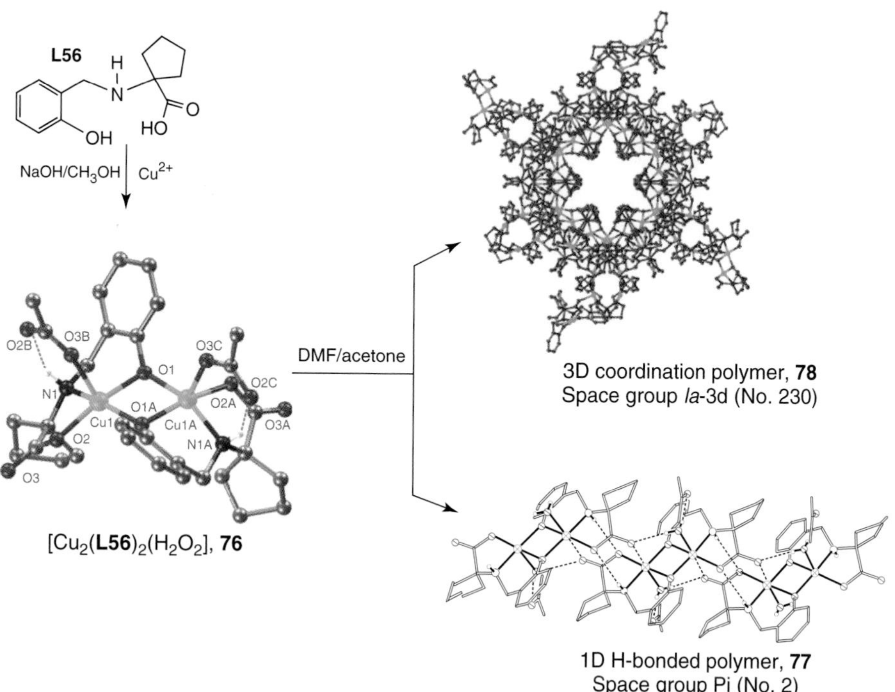

Figure 8 Formation of 1D hydrogen-bonded polymeric structure in $[Cu_2(\mathbf{L56})_2(H_2O)_2]\cdot 2(CH_3)_2CO$, **77**, and 3D coordination polymeric network architecture with star-shaped channels in $[Cu_2(\mathbf{L56})_2]$, **78**, in a single-pot crystallization of **76** indicating the effect of solvents.

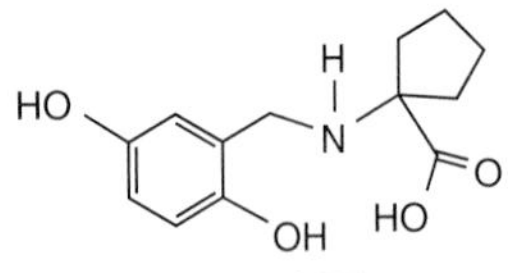

Figure 9 A segment of H-bonded 3D network in $[Cu_2(\mathbf{L57})_2(H_2O)_2]$, **79**.

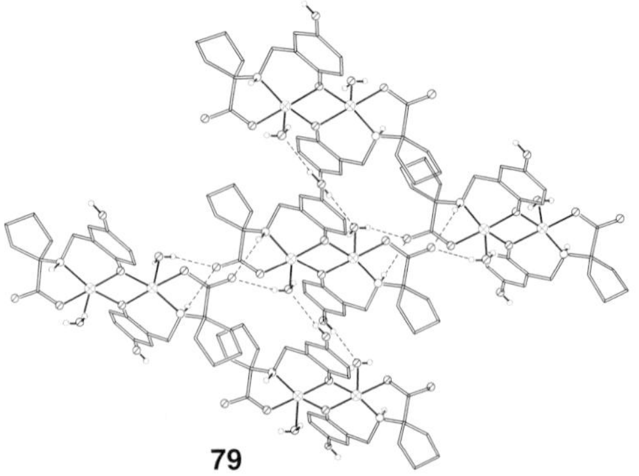

the solvents or the aqua ligands involved in the formation of intermolecular hydrogen bonds.[200] For example, in $[Cu_2(\mathbf{L57})_2(H_2O)_2]$, **79**, obtained with **L57** having –OH substitution at the phenyl ring, the intermolecular hydrogen bonds generated by the hydroxyl group on the phenolate moiety with the aqua ligands and carboxylate oxygen atoms and by the amine hydrogen with another carboxylate oxygen resulted in the formation of a 3D hydrogen-bonded network as shown in Figure 9.

Compared to aminocyclopentanecarboxylate group as in **L56**, the side arm in **L58** contains aminocyclohexanecarboxylate. Such alterations in the amino acid side arm is intended to induce some structural changes via conformational adjustments so that it is possible to derive different supramolecular networks. For instance, in $[\{Cu_2(\mathbf{L58})_2\}_2Cu_2(L58)_2(H_2O)_2]\cdot4H_2O$, **80**, derived from **L58**, the crystal structure shows three independent Cu(II)-**L58** units in the asymmetric unit. One of the Cu(II) centers has an aqua ligand in the apical position, and the oxygen atom of the carboxylato group from this unit is bound to another Cu(II) center.[200] The lattice water and the carbonyl oxygen participate in hydrogen bonding to produce a 2D coordination polymeric structure, and the interdimer connectivity leads to the formation of the well-known square grid network structure as shown in Figure 10.

At this point, it is interesting to compare the supramolecular structures derived from the Schiff base and the corresponding reduced Schiff base complexes to see the effect of inducing the conformational flexibility. Very drastic change in the mode of complexation and notable differences in supramolecular structures that may occur as expected by the reduction of C=N bond in Schiff base species can be illustrated by the two Cu(II) complexes, $[Cu_2(\mathbf{L59})_2(H_2O)_2]\ 2H_2O$, **81** and, $[Cu_2(\mathbf{L60})_2]\ 2H_2O$, **82** containing the Schiff base **L59** and the corresponding reduced Schiff base **L60**, respectively.[201] The copper centers in $[Cu_2(\mathbf{L59})_2(H_2O)_2]\ 2H_2O$ (**81**) assume a square pyramidal geometry with the apical positions occupied by a sulfonate oxygen atoms to form eight-membered sulfonato-bridged dinuclear complex as shown in Figure 11. The dimers are packed in the solid state to give a hydrogen-bonded 2D structure sustained by two types of complementary hydrogen bonds: the first one is between the phenoxo oxygen and one of the hydrogen atoms of the aqua ligand from the adjacent dimer and the next one is between the free oxygen atoms of the sulfonate group and another hydrogen

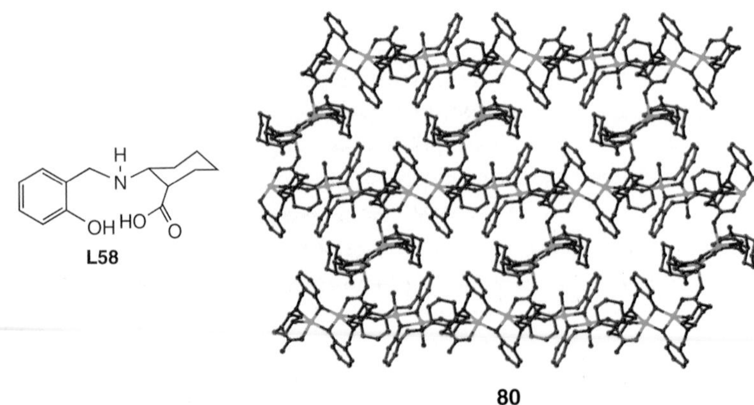

Figure 10 Portion of packing diagram showing the 2D square grid connectivity of dicopper units in $[\{Cu_2(\mathbf{L58})_2\}_2Cu_2(L58)_2(H_2O)_2]\cdot4H_2O$, **80**, derived from **L58**.

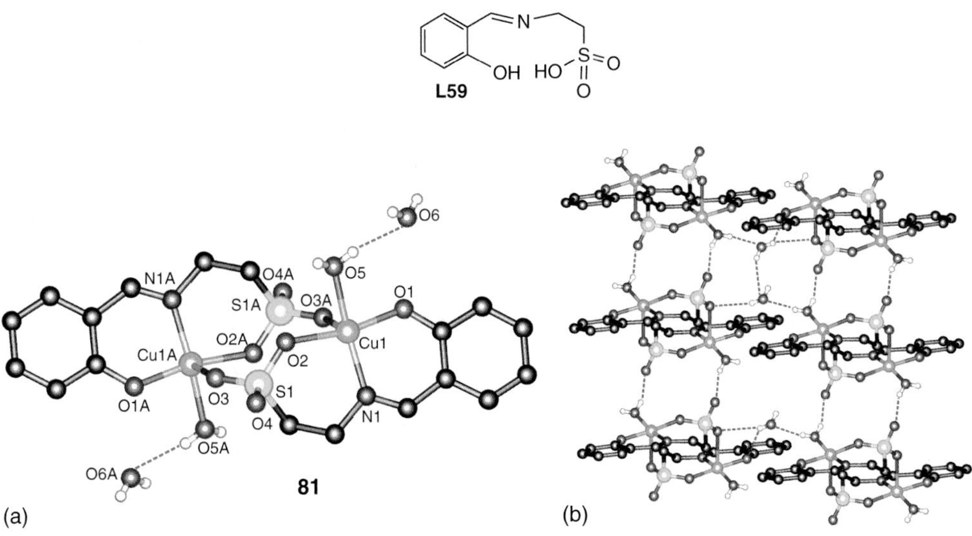

Figure 11 Structure of **81** showing the connectivity of Schiff base (**L59**) forming eight-membered sulfonate-bridged copper centers (a); hydrogen-bonded 2D sheet structure in **81**.

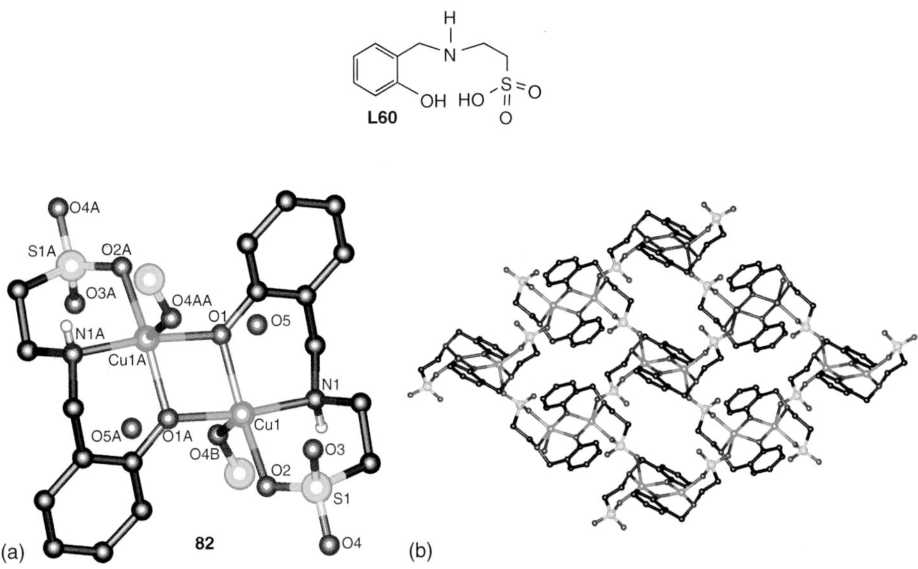

Figure 12 Phenoxo-bridged dicopper complex (a) and a portion of the 2D square grid structure (b) in **82**.

atom of the aqua ligand. In addition, the resulting 2D sheets are also supported by C–H· · ·O hydrogen-bonding interactions (Figure 11).

In $[Cu_2(\mathbf{L60})_2]\cdot 2H_2O$, **82**, the reduced Schiff base ligand **L60** with flexible backbone can show the similar mode of complexation as other reduced Schiff base ligands to form a phenolato-bridged Cu(II) dimer with Cu_2O_2 core (Figure 12). The apical site on each Cu(II) center is occupied by an oxygen atom of the sulfonate group from the neighboring dimer. This interdimer connectivity leads to the formation of an infinite 2D square grid network structure with (4,4) net structure as shown in Figure 12, which is the first of its kind observed in the Cu(II) and Zn(II) complexes containing reduced Schiff base ligands. Thus, the structural features of **81** and **82** exemplify an advantageous effect of reducing a Schiff base ligand to achieve the desired structural changes.

By incorporating the reactive functional groups in the side arm of the amino acids, the different molecular connectivity can be achieved resulting mainly in monomeric units to generate different hydrogen-bonded polymeric structures. The ligands **L51, L61**, and **L62** containing additional COO^- group in the amino acid side chain displayed a promising trend of forming 1D coordination polymeric structures in Cu(II) and Ni(II) complexes.[202] Hence, as shown by the staircase coordination polymeric structure in **70** encapsulating helical water chain,[193] it is essential to have additional donor groups such as

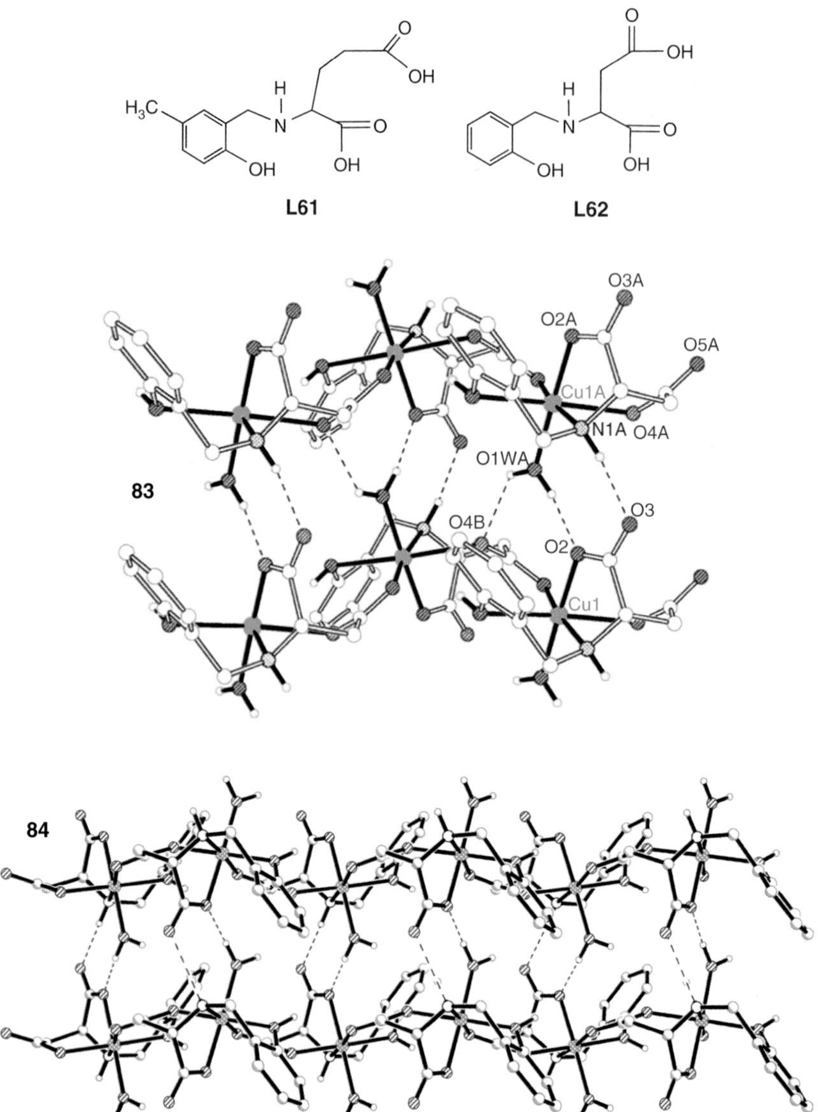

Figure 13 Packing diagrams of **83** and **84** showing hydrogen-bonded 2D sheets.

carboxylate in the ligand to achieve different intermolecular connectivity for getting interesting coordination polymers.

The compounds $[Cu(\mathbf{L62})(H_2O)]_n$, **83**, and $[Ni(\mathbf{L62})(H_2O)]_n$, **84**, are structurally similar 1D coordination polymers, and the repeating unit comprises mononuclear copper(II) and Ni(II) centers, respectively, with distorted octahedral geometry. Both these compounds furnished as shown in Figure 13. Interconnectivity of mononuclear repeating units via coordination of neighboring carboxylate oxygen generated 1D zigzag coordination polymeric structure, and further molecular packing resulted in the connectivity of all 1D zigzag chains to form 2D hydrogen-bonded network structures through intermolecular hydrogen bonding. These intermolecular hydrogen bonds are generated between the α-carboxylate oxygen atoms, aqua ligands, and amine hydrogen atoms. The α-carboxylate oxygen atoms are involved in syn–anti mode of hydrogen bonding with amine hydrogen atoms and aqua ligand.

Interestingly, complexation of **L62** with Zn(II) ions furnished a polynuclear aggregate $[Zn_6(\mathbf{L62})_4(H_2O)_8]\cdot 5H_2O$, **85**, made up of hydrated Zn(II) cations and $[Zn_5(\mathbf{L62})_4(H_2O)_4]_2$ anions.[203] Molecular packing of $[Zn(H_2O)_4]^{2+}$ cations and anions via bridging carboxylates furnished an interesting 1D coordination polymeric structure in **85**. The different coordination geometries of Cu(II) in **83** and Zn(II) in **85** significantly affect the overall solid-state polymeric structures and demonstrates that the overall topology depends on the nature of the metal ions. These differences in solid-state supramolecular assembly in **83** and **85** clearly indicate the uncertainty associated with the predictability of connectivity of the coordination polymer.

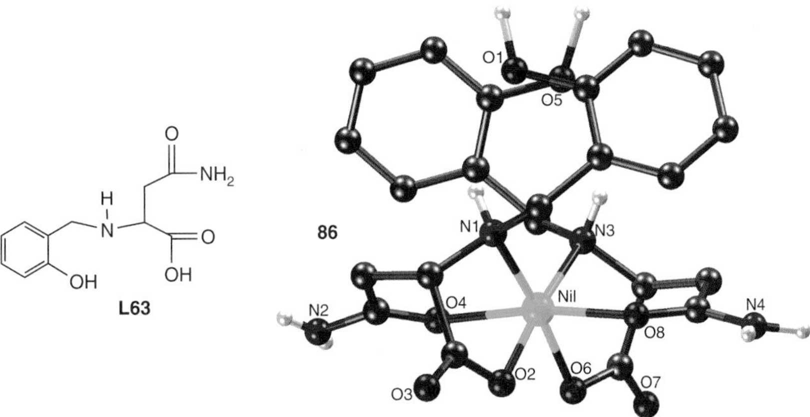

Figure 14 Perspective view of **86**.

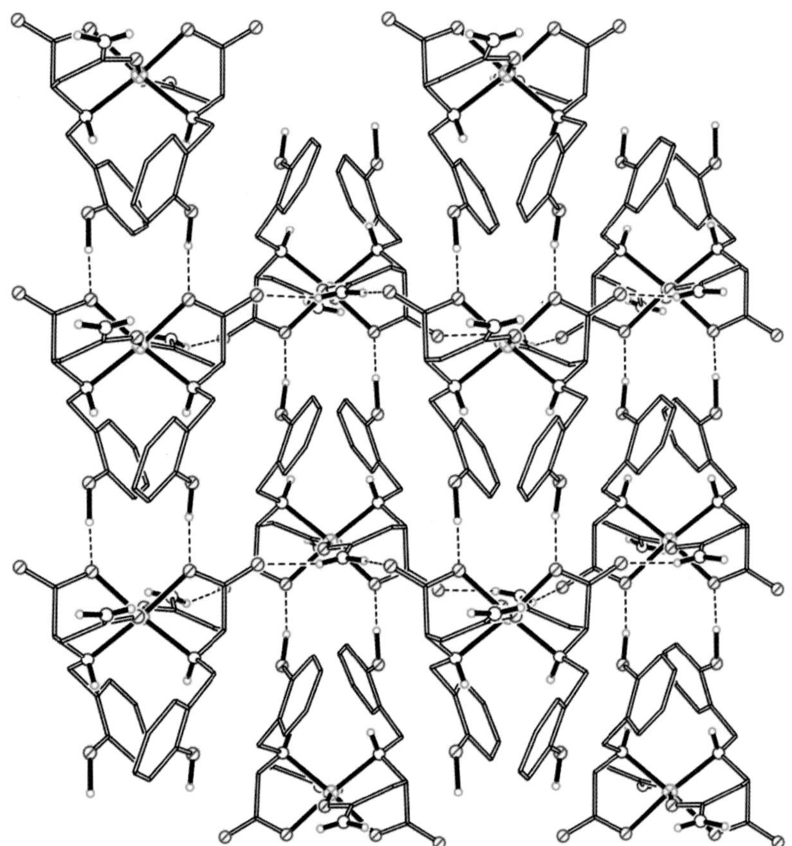

Figure 15 Hydrogen-bonded 3D network structure in **86** viewed from *b* axis.

Usually, the phenolate oxygen atom coordinates to the metal centers to give either dinuclear M_2O_2 cores via bridging mode or mononuclear metal complexes in a 1 : 1 stoichiometry with respect to the ligand to metal ratio. When the side arm of ligand is modified by incorporating the $-CONH_2$ donor group in place of COO^-, the mode of coordination of the ligand is different from the usual behavior observed for reduced Schiff base ligands as described above.

For example, in **L63**, the amino acid part of the ligand is L-asparagine that provides $-CONH_2$ donor group in the side arm. Thus, in the case of [Ni(**L63**)$_2$], **86**, which is a 2 : 1 compound phenolic group remained inert toward coordination, and this mode of formation of **86** (Figure 14) is probably due to the experimental conditions employed for synthesis.[202] Packing of molecules in **86** generated a hydrogen-bonded 3D network structure, as shown in Figure 15, in which the hydrogen atoms of each noncoordinating phenolate groups, the carboxylate oxygen atoms, and the hydrogen atoms from the $-CONH_2$ donor group are involved in generating three-dimensional intermolecular hydrogen bonds.

4 CONCLUSIONS

Recent developments in the field of Schiff base chemistry and several examples presented in this report suggest that the coordination chemistry of Schiff bases is constantly growing and transforming into metallasupramolecular and host–guest chemistry. Also, the research areas dealing with the Schiff base ligands have become highly broad and interdisciplinary. Owing to their simple synthetic procedures, flexible and adaptable coordination modes and applications, and potential molecular recognition features, several forms of Schiff bases and their reduced analogs are extensively employed for generating interesting metallasupramolecular structures derived from the corresponding homonuclear and/or heteronuclear complexes as building blocks.

This chapter attempted to provide an overview of several forms of acyclic and cyclic Schiff bases and their reduced homologs, with an emphasis on their potential molecular recognition properties to generate host–guest complexes as well as metallasupramolecular structures ranging from 1D to 3D networks via covalent and noncovalent interactions. Besides several forms of Schiff bases, this contribution also dealt with the reduced Schiff base ligands, containing natural as well as the unnatural amino acids, as a class of multidentate ligands and accounted for their mode of binding and coordination to the different transition metal ions such as Cu(II), Zn(II), and Ni(II). Several hydrogen-bonded and coordination polymeric types of metallasupramolecular structures of various metal complexes are highlighted, and these structures revealed various $C{=}O\cdots H{-}O_{solvent}$, $O{-}H\cdots O$, and hydrogen bonds and $C{=}O\cdots\pi$, $C{-}H\cdots\pi$, and $\pi\cdots\pi$ stacking interactions that are further sustained by hydrogen bonds via amine hydrogen atom. Further, the supramolecular aggregates are greatly influenced by the coordination geometries around the metal centers as defined by the ligands. Further, the observed structural diversity can be attributed to the role of different donors and acceptors, aqua ligands and solvents, nature of the ligands and metal ions, and coordination geometry around the metal ions and counterions besides the experimental conditions such as temperature and pH in directing the formation of supramolecular structures in the solid state. Many chiral molecules derived from reduced Schiff bases described here also exemplify an alternative approach that the chiral materials can be obtained from simple achiral molecular compounds.

Future directions relevant to this field of research may rely on, besides conventional procedures, the high-temperature hydrothermal or solvothermal synthesis and solid-state synthetic procedures, use of ionic liquids for green chemistry, and so on, which might open new avenues to make thermodynamically stable multidimensional coordination polymeric architectures. Finally, this review gives an insight into the understanding of the coordination chemistry of Schiff base and reduced Schiff base ligands in terms of their molecular recognition and supramolecular chemistry with various factors that should be considered for the design of coordination polymers with specific properties and functions.

At this point, as the supramolecular chemistry of Schiff base and reduced Schiff base ligands is getting highly established while producing vast literature, this chapter dealt with some of the representative classes of Schiff base and reduced forms as receptors for binding cations, anions, and neutral species and also as building blocks for generating metalla-supramolecular structures. However, a detailed comprehensive coverage of the whole available novel and interesting examples would be difficult and seems impossible due to space constraints. It is also possible, regrettably, that there may be any inadvertent omission and overlooking of excellent and novel structures and relevant reports. The readers and reviewers are welcomed to critically evaluate and suggest any update.

ACKNOWLEDGMENTS

Bellam Sreenivasulu thanks and acknowledges financial support from the Ministry of Education, Singapore, through National University of Singapore. It is a pleasure to thank Professor Vittal for all his support and also to thank other coworkers in the group who have made an excellent contribution to this area of research; their names are given in the references.

REFERENCES

1. R. Hernandez-Molina and A. Mederos, in *Comprehensive Coordination Chemistry II*, eds. J. A. McCleverty and T. J. Meyer, Elsevier, Amsterdam, 2004, vol. 1, pp. 411–446.
2. P. Guerriero, S. Tamburini, P. A. Vigato, *et al.*, *Inorg. Chim. Acta*, 1993, **213**, 279.
3. H. Okawa, H. Furutachi, and D. E. Fenton, *Coord. Chem. Rev.*, 1998, **174**, 51.
4. P. Guerriero, S. Tamburini, and V. A. Vigato, *Coord. Chem. Rev.*, 1995, **139**, 17.

5. M. Calligaris and L. Randaccio, in *Comprehensive Coordination Chemistry*, eds. G. Wilkinson, R. D. Guillard, and J. A McCleverty, Pergamon, Oxford, 1987, vol. 2, Chapter 20, pp. 715–738.
6. S. R. Collinson and D. E. Fenton, *Coord. Chem. Rev.*, 1996, **148**, 19.
7. D. E. Fenton and P. A. Vigato, *Chem. Soc. Rev.*, 1988, **17**, 69.
8. P. Zanello, S. Tamburini, P. A. Vigato, and G. A. Mazzocchin, *Coord. Chem. Rev.*, 1987, **77**, 165.
9. A. Mederos, S. Domínguez, R. Hernandez-Molina, *et al.*, *Coord. Chem. Rev.*, 1999, **193–195**, 913.
10. J. Costamagna, J. Vargas, R. Latorre, *et al.*, *Coord. Chem. Rev.*, 1992, **119**, 67.
11. M. M. Aly, *J. Coord. Chem.*, 1998, **98**, 89.
12. M. Calligaris, G. Nardin, and L. Randaccio, *Coord. Chem. Rev.*, 1972, **7**, 385.
13. S. Yamada and A. Takeuchi, *Coord. Chem. Rev.*, 1982, **43**, 187.
14. M. R. Maurya and A. Syamal, *Coord. Chem. Rev.*, 1989, **95**, 183.
15. J. S. Casas, M. S. García-Tasende, and J. Sordo, *Coord. Chem. Rev.*, 2000, **209**, 197.
16. S. Brooker, *Coord. Chem. Rev.*, 2001, **222**, 33.
17. K. Henrick, P. A. Tasker, and L. L. Lindoy, in *Progress in Inorganic Chemistry*, eds. S. J. Lippard, Wiley, New York, 1985, vol. 33, p. 1.
18. P. A. Vigato and S. Tamburini, *Coord. Chem. Rev.*, 2004, **248**, 1717.
19. N. F. Curtis, *J. Chem. Soc.*, 1960 4409.
20. M. M. Blight and N. F. Curtis, *J. Chem. Soc.*, 1962 1204.
21. N. F. Curtis, *Coord. Chem. Rev.*, 1968, **3**, 3.
22. D. H. Busch, K. Farmery, V. Goedken, *et al.*, *Adv. Chem. Ser.*, 1971, **44**, 100.
23. E.-G. Jager, *Z. Anorg. Allg. Chem.*, 1969, **364**, 177.
24. C. J. Pedersen, *J. Am. Chem. Soc.*, 1967, **89**, 2495.
25. A. J. Atkins, D. Black, A. J. Blake, *et al.*, *Chem. Commun.*, 1996 457.
26. N. H. Pilkington and R. Robson, *Aust. J. Chem.*, 1970, **23**, 2225.
27. U. Casellato, S. Tamburini, P. Tomasin, and P. A. Vigato, *Inorg. Chim. Acta*, 2004, **357**, 4191.
28. P. A. Vigato and S. Tamburini, *Coord. Chem. Rev.*, 2008, **252**, 1871.
29. M. Yildiz, Z. Kilic, and T. Hökelek, *J. Mol. Struct.*, 1998, **441**, 1.
30. T. Hökelek, Z. Kilic, M. Isiklan, and M. Toy, *J. Mol. Struct.*, 2000, **523**, 61.
31. H. Nazir, M. Yildiz, H. Yilmaz, *et al.*, *J. Mol. Struct.*, 2000, **524**, 241.
32. A. Mederos, S. Domínguez, R. Hernandez-Molina, *et al.*, *Coord. Chem. Rev.*, 1999, **193**, 913.
33. T. J. Maede and D. H. Busch, in *Progress in Inorganic Chemistry*, eds. S. J. Lippard, Wiley Interscience, New York, 1985, vol. 33.
34. S. Yamada, *Coord. Chem. Rev.*, 1966, **1**, 415.
35. A. Schepartz and J. P. McDevitt, *J. Am. Chem. Soc.*, 1989, **111**, 5976.
36. M. Boyce, B. Clarke, D. Cunningham, *et al.*, *J. Organomet. Chem.*, 1995, **498**, 241.
37. U. Casellato, S. Tamburini, P. Tomasin, and P. A. Vigato, *Inorg. Chim. Acta*, 1997, **262**, 117.
38. J. P. Costes, F. Dahan, and J. P. Laurent, *Inorg. Chem.*, 1994, **33**, 2738.
39. J. P. Costes, J. P. Laurent, P. Chabert, *et al.*, *Inorg. Chem.*, 1997, **36**, 656.
40. J.-M. Lehn, *Supramolecular Chemistry*, VCH Publishers, New York, 1995.
41. K. Gloe, ed., *Macrocyclic Chemistry: Current Trends and Future Perspectives*, Springer, Dordrecht, The Netherlands, 2005.
42. T. Schrader and A. D. Hamilton, *Functional Synthetic Receptors*, Wiley-VCH, Weinheim, Germany, 2005.
43. H. J. Schneider, *Angew. Chem. Int. Ed.*, 2009, **48**, 3924.
44. K. Ono, M. Yoshizawa, M. Akita, *et al.*, *J. Am. Chem. Soc.*, 2009, **131**, 2782.
45. A. Lledo, P. Restorp, and J. Rebek Jr, *J. Am. Chem. Soc.*, 2009, **131**, 2440.
46. T. B. Gasa, J. M. Spruell, W. R. Dichtel, *et al.*, *Chem.—Eur. J.*, 2009, **15**, 106.
47. B. H. Northrop, H. B. Yang, and P. J. Stang, *Chem. Commun.*, 2008 5896.
48. S. Gadde, E. K. Batchelor, J. P. Weiss, *et al.*, *J. Am. Chem. Soc.*, 2008, **130**, 17114.
49. C. Streb, T. McGlone, O. Brücher, *et al.*, *Chem.—Eur. J.*, 2008, **14**, 8861.
50. B. S. Kim, Y. H. Ko, Y. Kim, *et al.*, *Chem. Commun.*, 2008, 2756.
51. C. Caltagirone and P. A. Gale, *Chem. Soc. Rev.*, 2009, **38**, 520.
52. E. A. Katayev, G. V. Kolesnikov, and J. L. Sessler, *Chem. Soc. Rev.*, 2009, **38**, 1572.
53. M. X. Wang, *Chem. Commun.*, 2008, 4541.
54. J. C. Chambron and M. Meyer, *Chem. Soc. Rev.*, 2009, **38**, 1663.
55. T. Müller, *Angew. Chem. Int. Ed.*, 2009, **48**, 3740.
56. S. Li, M. Liu, J. Zhang, *et al.*, *Eur. J. Org. Chem.*, 2008, 6128.
57. P. A. Rupar, V. N. Staroverov, and K. M. Baines, *Science*, 2008, **322**, 1360.
58. L. Isaacs, *Chem. Commun.*, 2009, 619.
59. K. Kim, N. Selvapalam, Y. H. Ko, *et al.*, *Chem. Soc. Rev.*, 2007, **36**, 267.
60. I. Hwang, A. Y. Ziganshina, Y. H. Ko, *et al.*, *Chem. Commun.*, 2009, 416.

61. R. Wang and D. H. Macartney, *Org. Biomol. Chem.*, 2008, **6**, 1955.
62. D. M. Homden and C. Redshaw, *Chem. Rev.*, 2008, **108**, 5086.
63. A. Ikeda and S. Shinkai, *Chem. Rev.*, 1997, **97**, 1713.
64. W. Maes and W. Dehaen, *Chem. Soc. Rev.*, 2008, **37**, 2393.
65. N. Morohashi, F. Narumi, N. Iki, *et al.*, *Chem. Rev.*, 2006, **106**, 5291.
66. Y. Rudzevich, V. Rudzevich, F. Klautzsch, *et al.*, *Angew. Chem. Int. Ed.*, 2009, **48**, 3867.
67. A. Harada, Y. Takashima, and H. Yamaguchi, *Chem. Soc. Rev.*, 2009, **38**, 875.
68. R. Villalonga, R. Cao, and A. Fragoso, *Chem. Rev.*, 2007, **107**, 3088.
69. V. T. D'Souza and K. B. Lipkowitz, *Chem. Rev.*, 1998, **98**, 1741.
70. A. Ikeda, M. Matsumoto, M. Akiyama, *et al.*, *Chem. Commun.*, 2009, 1547.
71. M. J. MacLachlan, *Pure Appl. Chem.*, 2006, **78**, 873.
72. N. E. Borisova, M. D. Reshetova, and Y. A. Ustynyuk, *Chem. Rev.*, 2007, **107**, 46.
73. M. Botta, U. Casellato, C. Scalco, *et al.*, *Chem.—Eur. J.*, 2002, **8**, 3917.
74. D. A. Dougherty, *Chem. Rev.*, 2008, **108**, 1642.
75. J. C. Ma and D. A. Dougherty, *Chem. Rev.*, 1997, **97**, 1303.
76. S. Akine and T. Nabeshima, *Dalton Trans.*, 2009, 10395.
77. S. J. Gruber, C. M. Harris, and E. Sinn, *J. Inorg. Nucl. Chem.*, 1968, **30**, 1805.
78. L. G. Armstrong, H. C. Lip, L. F. Lindoy, *et al.*, *J. Chem. Soc., Dalton Trans.*, 1977, 1771.
79. A. Giacomelli, T. Rotunno, and L. Senatore, *Inorg. Chem.*, 1985, **24**, 1303.
80. A. Giacomelli, T. Rotunno, L. Senatore, and R. Settambolo, *Inorg. Chem.*, 1989, **28**, 3552.
81. L. Carbonaro, M. Isola, P. La Pegna, *et al.*, *Inorg. Chem.*, 1999, **38**, 5519.
82. G. Condorelli, I. Fragallà, S. Giuffrida, and A. Cassol, *Z. Anorg. Allg. Chem.*, 1975, **412**, 251.
83. D. Cunningham, P. McArdle, M. Mitchell, *et al.*, *Inorg. Chem.*, 2000, **39**, 1639.
84. S. Akine, T. Taniguchi, and T. Nabeshima, *Inorg. Chem.*, 2008, **47**, 3255.
85. S. Akine, T. Taniguchi, and T. Nabeshima, *J. Am. Chem. Soc.*, 2006, **128**, 15765.
86. S. Akine, T. Taniguchi, and T. Nabeshima, *Angew. Chem. Int. Ed.*, 2002, **41**, 4670.
87. S. Akine, T. Taniguchi, T. Matsumoto, and T. Nabeshima, *Chem. Commun.*, 2006, 4961.
88. E. C. Constable, G. Zhang, C. E. Housecroft, *et al.*, *CrystEngComm*, 2010, **12**, 1764.
89. C. Suksai and T. Tuntulani, *Chem. Soc. Rev.*, 2003, **32**, 192.
90. J.-L. Sessler, P. A. Gale, and W. S. Cho, *Anion Receptor Chemistry*, RCS Publishing, Baltimore, MD, 2006.
91. A. Bianchi, K. Bowman-James, and E. Garcýa-Espana, eds., *Supramolecular Chemistry of Anions*, Wiley-VCH, New York, 1997.
92. S. O. Kang, R. A. Begum, and K. Bowman-James, *Angew. Chem. Int. Ed.*, 2006, **45**, 7882.
93. C. A. Ilioudis, D. A. Tocher, and J. W. Steed, *J. Am. Chem. Soc.*, 2004, **126**, 12395.
94. E. A. Katayev, Y. A. Ustynyuk, and J. L. Sessler, *Coord. Chem. Rev.*, 2006, **250**, 3004.
95. J. W. Steed, *Chem. Soc. Rev.*, 2009, **38**, 506.
96. P. D. Beer, D. K. Smith, in *Progress in Inorganic Chemistry*, ed. K. D. Karlin, John Wiley & Sons, Inc., 1997, vol. 46, p. 1.
97. D. M. Rudkevich, W. P. R. V. Stauthamer, W. Verboom, *et al.*, *J. Am. Chem. Soc.*, 1992, **114**, 9671.
98. D. M. Rudkevich, W. Verboom, Z. Brzozka, *et al.*, *J. Am. Chem. Soc.*, 1994, **116**, 4341.
99. D. M. Rudkevich, Z. Brzozka, M. Palys, *et al.*, *Angew. Chem. Int. Ed. Engl.*, 1994, **33**, 467.
100. D. J. White, N. Laing, H. Muller, *et al.*, *Chem. Commun.*, 1999, 2077.
101. J. M. Mahoney, A. M. Beatty, and B. D. Smith, *J. Am. Chem. Soc.*, 2001, **123**, 5847.
102. M. Cametti, M. Nissinen, A. Dalla Cort, *et al.*, *J. Am. Chem. Soc.*, 2005, **127**, 3831.
103. M. Cametti, M. Nissinen, A. Dalla Cort, *et al.*, *Chem. Commun.*, 2003, 2420.
104. M. Cametti, M. Nissinen, A. Dalla Cort, *et al.*, *J. Am. Chem. Soc.*, 2007, **129**, 3641.
105. A. Dalla Cort, P. D. Bernardin, G. Forte, and F. Y. Mihan, *Chem. Soc. Rev.*, 2010, **39**, 3863.
106. M. Cano, L. Rodrıguez, J. C. Lima, *et al.*, *Inorg. Chem.*, 2009, **48**, 6229.
107. P. A. Tasker, C. C. Tong, and A. N. Westra, *Coord. Chem. Rev.*, 2007, **251**, 1868.
108. S. G. Galbraith, P. G. Plieger, and P. A. Tasker, *Chem. Commun.*, 2002, 2662–2663.
109. E. R. Libra and M. J. Scott, *Chem. Commun.*, 2006, 1485.
110. M. Wenzel, S. R. Bruere, Q. W. Knapp, *et al.*, *Dalton Trans.*, 2010, 2936.
111. C. J. M. van Veggel, W. Verboom, and D. N. Reinhoudt, *Chem. Rev.*, 1994, **94**, 279.
112. W. Verboom, D. M. Rudkevich, and D. N. Reinhoudt, *Pure Appl. Chem.*, 1994, **66**, 679.
113. M. M. G. Antonisse and D. N. Reinhoudt, *Chem. Commun.*, 1998, 443.
114. P. A. Vigato, S. Tamburini, and L. Bertolo, *Coord. Chem. Rev.*, 2007, **251**, 1311.
115. F. C. J. M. van Veggel, S. Harkema, M. Bos, *et al.*, *Inorg. Chem.*, 1989, **28**, 1133.
116. F. C. J. M. van Veggel, S. Harkema, M. Bos, *et al.*, *J. Org. Chem.*, 1989, **54**, 2351.
117. N. Brianese, U. Casellato, S. Tamburini, *et al.*, *Inorg. Chim. Acta*, 1999, **293**, 178.

118. V. B. Arion, J. P. Wignacourt, P. Conflant, *et al.*, *Inorg. Chim. Acta*, 2000, **303**, 228.
119. C. J. van Staveren, D. E. Fenton, D. N. Reinhoudt, *et al.*, *J. Am. Chem. Soc.*, 1987, **109**, 3456.
120. C. J. van Staveren, J. van Eerden, F. C. J. M. van Veggel, *et al.*, *J. Am. Chem. Soc.*, 1988, **110**, 4994.
121. W. F. Nijenhuis, A. R. van Doorn, A. M. Reichwein, *et al.*, *J. Am. Chem. Soc.*, 1991, **113**, 3607.
122. A. R. van Doorn, R. Schaafstra, M. Bos, *et al.*, *J. Org. Chem.*, 1991, **56**, 6083.
123. F. C. J. M. van Veggel, M. Bos, S. Harkema, *et al.*, *Angew. Chem. Int. Ed. Engl.*, 1989, **28**, 746.
124. F. C. J. M. van Veggel, M. Bos, S. Harkema, *et al.*, *J. Org. Chem.*, 1991, **56**, 225.
125. S. M. Lacy, D. M. Rudkevich, W. Verboom, and D. N. Reinhoudt, *J. Chem. Soc., Perkin Trans. 2*, 1995 135.
126. F. Gao, W.-J. Ruan, J.-M. Chen, *et al.*, *Spectrochim. Acta, Part A*, 2005, **62**, 886.
127. J. B. Fontecha, S. Goetz, and V. McKee, *Dalton Trans.*, 2005, 923.
128. Z. Li and C. Jablonski, *Inorg. Chem.*, 2000, **39**, 2456.
129. L. Salmon, P. Thuéry, E. Rivière, *et al.*, *New J. Chem.*, 2006, **30**, 1220.
130. H.-C. Zhang, W.-S. Huang, and L. Pu, *J. Org. Chem.*, 2001, **66**, 481.
131. A. J. Gallant and M. J. MacLachlan, *Angew. Chem. Int. Ed.*, 2003, **42**, 5307.
132. A. J. Gallant, B. O. Patrick, and M. J. MacLachlan, *J. Org. Chem.*, 2004, **69**, 8739.
133. A. J. Gallant, J. K.-H. Hui, F. E. Zahariev, *et al.*, *J. Org. Chem.*, 2005, **70**, 7936.
134. T. Nabeshima, H. Miyazaki, A. Iwasaki, *et al.*, *Chem. Lett.*, 2006, **35**, 1070.
135. A. J. Gallant, J. H. Chong, and M. J. MacLachlan, *Inorg. Chem.*, 2006, **45**, 5248.
136. P. D. Frischmann and M. J. MacLachlan, *Chem. Commun.*, 2007, 4480.
137. H. Nabeshima, A. Miyazaki, S. Iwasaki, *et al.*, *Tetrahedron*, 2007, **63**, 3328.
138. J. Jiang and M. MacLachlan, *Chem. Commun.*, 2009, 5695.
139. P. A. Gale and R. Quesada, *Coord. Chem. Rev.*, 2006, **250**, 3219.
140. E. A. Katayev, N. V. Boev, E. Myshkovskaya, *et al.*, *Chem.—Eur. J.*, 2008, **14**, 9065.
141. O. A. Okunola, P. V. Santacroce, and J. T. Davis, *Supramol. Chem.*, 2008, **20**, 169.
142. J. L. Sessler, E. Katayev, G. D. Pantos, and Y. A. Ustynyuk, *Chem. Commun.*, 2004 1276.
143. E. A. Katayev, N. V. Boev, V. N. Khrustalev, *et al.*, *J. Org. Chem.*, 2007, **72**, 2886.
144. E. A. Katayev, J. L. Sessler, V. N. Khrustalev, and Y. A. Ustynyuk, *J. Org. Chem.*, 2007, **72**, 7244.
145. J. L. Sessler, V. Roznyatovskiy, G. D. Pantos, *et al.*, *Org. Lett.*, 2005, **7**, 5277.
146. P. G. Plieger, P. A. Tasker, and S. G. Galbraith, *J. Chem. Soc., Dalton Trans.*, 2004, 313.
147. L. L. Koh, J. D. Ranford, W. T. Robinson, *et al.*, *Inorg. Chem.*, 1996, **35**, 6466.
148. C. R. Bondy and S. J. Loeb, *Coord. Chem. Rev.*, 2003, **240**, 77.
149. J. Bowmann, *Acc. Chem. Res.*, 2005, **38**, 671.
150. P. A. Gale, Amide and urea-based anion receptors, in *Encyclopedia of Supramolecular Chemistry*, eds. J.-L. Atwood and J. W. Steed, Marcel Dekker, New York, 2004, p. 31.
151. D. Chen and A. E. Martell, *Tetrahedron*, 1991, **47**, 6895.
152. R. J. Motekaitis, A. E. Martell, B. Dietrich, and J. M. Lehn, *Inorg. Chem.*, 1984, **23**, 1588.
153. P. D. Beer, J. W. Wheeler, and C. Moore, in *Supramolecular Chemistry*, eds. V. Balzani and L. De Cola, Kluwer Academic Publishers, Dordrecht, 1992, p. 105.
154. H. E. Katz, in *Inclusion Compounds*, eds. J. L. Atwood, J. E. D. Davies, and D. D. MacNichol, Oxford University Press, Oxford, 1991, p. 391.
155. C. Seel, A. Galan, and J. de Mendoza, *Top. Curr. Chem.*, 1995, **175**, 101.
156. G. Papoyan, K. Gu, J. Wiorkiewicz-Kuczera, *et al.*, *J. Am. Chem. Soc.*, 1996, **118**, 1354.
157. J. Wiorkiewicz-Kuczera, K. Kuczera, C. Bazzicalupi, *et al.*, *New J. Chem.*, 1999, **23**, 1007.
158. T. Clifford, A. Danby, J. M. Llinares, *et al.*, *Inorg. Chem.*, 2001, **40**, 4710.
159. C. Bazzicalupi, A. Bencini, A. Bianchi, *et al.*, *J. Chem. Soc., Perkin Trans. 2*, 1995, 275.
160. M. A. Saeed, J. J. Thompson, F. R. Fronczekb, and Md. Alamgir Hossain, *CrystEngComm*, 2010, **12**, 674.
161. M. A. Hossain, M. A. Saeed, F. R. Fronczek, *et al.*, *Cryst. Growth Des.*, 2010, **10**, 1478.
162. S. O. Kang, M. A. Hossain, and K. Bowman-James, *Coord. Chem. Rev.*, 2006, **250**, 3038.
163. O. A. Gerasimchuk, S. Mason, J. M. Llinares, *et al.*, *Inorg. Chem.*, 2000, **39**, 1371.
164. S. K. Valiyaveettil, J. F. J. Engbersen, W. Verboom, and D. N. Reinhoudt, *Angew. Chem. Int. Ed. Engl.*, 1993, **32**, 900.
165. P. D. Beer, Z. Chen, A. J. Goulden, *et al.*, *J. Chem. Soc., Chem. Commun.*, 1993, 1834.
166. P. D. Beer, P. K. Hopkins, and J. D. McKinney, *Chem. Commun.*, 1999, 1253.
167. A. Danby, L. Seib, N. W. Alcock, and K. Bowman-James, *Chem. Commun.*, 2000, 973.
168. M. A. Hossain, J. A. Liljegren, D. Powell, and K. Bowman-James, *Inorg. Chem.*, 2004, **43**, 3751.
169. I. Ravikumar, P. S. Lakshminarayanan, E. Suresh, and P. Ghosh, *Inorg. chem.*, 2008, **47**, 7992.
170. M. A. Hossain, J. M. Llinares, S. Mason, *et al.*, *Angew. Chem. Int. Ed.*, 2002, **41**, 2335.
171. M. A. Hossain, P. Morehouse, D. Powell, and K. Bowman-James, *Inorg. Chem.*, 2005, **44**, 2143.

172. X. Wang and J. J. Vital, *Inorg. Chem.*, 2003, **42**, 5135.

173. J. J. Vital, X. Wang, and J. D. Ranford, *Inorg. Chem.*, 2003, **42**, 3390.

174. M. Dubey, R. R. Koner, and M. Ray, *Inorg. Chem.*, 2009, **48**, 9294.

175. M. A. Alan, M. Nethaji, and M. Ray, *Angew. Chem. Int. Ed.*, 2003, **42**, 1940.

176. M. Fujita, K. Umemoto, M. Yoshizawa, *et al.*, *Chem. Commun.*, 2001 509.

177. M. A. Alam, M. Nethaji, and M. Ray, *Inorg. Chem.*, 2005, **44**, 1302.

178. S. C. Sahoo and M. Ray, *Dalton Trans.*, 2007, 5148.

179. R. Ludwig, *Angew. Chem. Int. Ed.*, 2001, **40**, 1808.

180. R. Ludwig, *ChemPhysChem*, 2000, **1**, 53.

181. J. L. Atwood and J. W. Steed, eds., *Encyclopedia of Supramolecular Chemistry*, Marcel Dekker, New York, 2004, vols 1–2.

182. L. Infantes and S. Motherwell, *CrystEngComm*, 2002, **4**(75), 454.

183. L. Infantes, J. Chisholm, and S. Motherwell, *CrystEngComm*, 2003, **5**, 480.

184. L. E. Cheruzel, M. S. Pometun, M. R. Cecil, *et al.*, *Angew. Chem. Int. Ed.*, 2003, **42**, 5452.

185. M. Mascal, L. Infantes, and J. Chisholm, *Angew. Chem. Int. Ed.*, 2006, **45**, 32.

186. Z. Fei, D. Zhao, T. J. Geldbach, *et al.*, *Angew. Chem. Int. Ed.*, 2005, **44**, 5720.

187. S. Banerjee and R. Murugavel, *Cryst. Growth Des.*, 2004, **4**, 545.

188. B. Q. Ma, H. L. Sun, and S. Gao, *Angew.Chem. Int. Ed.*, 2004, **43**, 1374.

189. A. Mukherjee, M. K. Saha, M. Nethaji, and A. R. Chakravarty, *Chem. Commun.*, 2004 716.

190. B. K. Saha and A. Nangia, *Chem. Commun.*, 2005, 3024.

191. B. Sreenivasulu and J. J. Vittal, *Synth. React. Inorg. Met.—Org. Nano Met. Chem.*, 2008, **38**, 1.

192. M. Albrecht, *Chem. Rev.*, 2001, **101**, 3457.

193. B. Sreenivasulu and J. J. Vittal, *Angew. Chem. Int. Ed.*, 2004, **43**, 5769.

194. R. Ganguly, B. Sreenivasulu, and J. J. Vittal, *Coord. Chem. Rev.*, 2008, **252**, 1027.

195. J. D. Ranford, J. J. Vittal, D. Wu, and X. D. Yang, *Angew. Chem. Int. Ed.*, 1999, **38**, 3498.

196. J. D. Ranford, J. J. Vittal, and D. Wu, *Angew. Chem. Int. Ed. Engl.*, 1998, **37**, 1114.

197. J. J. Vittal and X. D. Yang, *Cryst. Growth Des.*, 2002, **2**, 259.

198. X. Yang, D. Wu, J. D. Ranford, and J. J. Vittal, *Cryst. Growth Des.*, 2005, **5**, 41.

199. B. Sreenivasulu and J. J. Vittal, *Cryst. Growth Des.*, 2003, **3**, 635.

200. B. Sreenivasulu, F. Zhao, S. Gao, and J. J. Vittal, *Eur. J. Inorg. Chem.*, 2006, 2656.

201. B. Sreenivasulu, M. Vetrichelvan, F. Zhao, *et al.*, *Eur. J. Inorg. Chem.*, 2005, 4635.

202. B. Sreenivasulu and J. J. Vittal, *Inorg. Chim. Acta*, 2009, **362**, 2735.

203. L. Jia, N. Tang, and J. J. Vittal, *Inorg. Chim. Acta*, 2009, **362**, 2525.

Calixarenes in Molecular Recognition

Laura Baldini, Francesco Sansone, Alessandro Casnati, and Rocco Ungaro

Università di Parma, Parma, Italy

1 INTRODUCTION

The name calixarenes invented by David C. Gutsche in 1978 to indicate the $[1_n]$-metacyclophane (**I**) obtained by the condensation of phenols (usually para-substituted) and formaldehyde (Figure 1) has also been associated, during the years, to a variety of other macrocyclic compounds containing aromatic nuclei, which are known as resorcinarenes (or calixresorcarenes or resorcarenes), thiacalixarene, oxacalixarenes, azacalixarenes and, more recently, calixpyrroles and pillarcalixarenes.[1] Each class of such compounds developed so rapidly and so broadly that it occupies a distinct place in supramolecular chemistry. It will be impossible to cover, even in part, the supramolecular properties of these compounds, and, therefore, in this review we confine ourselves only to classical calixarenes. Several books[1–4] and few general review articles[5,6] have been published in the last 10 years highlighting the most recent aspects of the chemistry of calixarenes and their role in supramolecular science. The name of calixarenes was suggested to Gutsche by inspection of the CPK molecular model of the cyclic tetramers (Figure 2), which shows a cuplike structure resembling a Greek vase (*calix* in Latin and *χυλιξ* in Greek) when all four aryl groups are oriented in the same direction.

This name was also extended to larger cyclic oligomers. A bracketed number between the terms calix and arene specifies the number of phenolic units present in the macrocycle and therefore its size, whereas the name of the para-substituent is added to indicate from which phenol the calixarene is derived. The cyclic tetramer obtained from *p-tert*-butylphenol (**1**), for example, is named *p-tert*-butylcalix[4]arene. A more systematic nomenclature has evolved in which the term calixarene is referred only to the basic macrocyclic structure devoid of substituents. According to the numbering scheme of this nomenclature, *p-tert*-butylcalix[4]arene (**1**) is named 5,11,17,23,-tetra-*tert*-butylcalix[4]arene-25,26,27,28,-tetrol. Two regions can be distinguished in calixarenes (Figure 2), namely, the phenolic OH groups region and the para position of the aromatic rings, which are respectively called the *lower rim* and the *upper rim*, implying that the calixarene structure is drawn with the OH groups pointing downward (*endo*) and the para-substituents pointing upward (*exo*).

To render the designation independent from the orientation of the calix, Böhmer has suggested the alternative terms *narrow rim* and *wide rim*, which are also used in literature. In calix[4]arenes, adjacent nuclei have been named *vicinal* or *proximal* (or 1,2) and the opposite ones are indicated as *distal* or *diametral* (1,3).

Supramolecular Chemistry: From Molecules to Nanomaterials.
Edited by Philip A. Gale and Jonathan W. Steed.
© 2012 John Wiley & Sons, Ltd. ISBN: 978-0-470-74640-0.

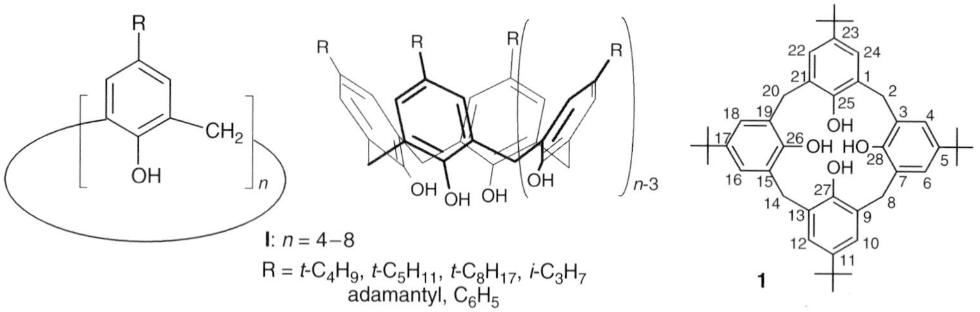

Figure 1 Different representations of calix[*n*]arenes (**I**) and numbering scheme for the *p-tert*-butylcalix[4]arene (**1**).

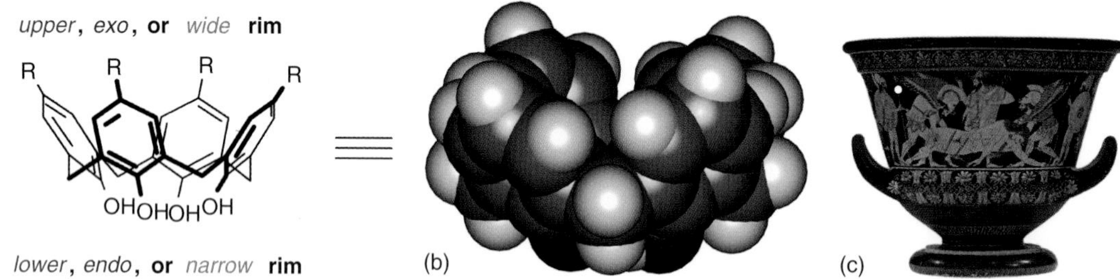

Figure 2 (a) The 3D representation of calix[4]arenes with an indication of the nomenclatures used to identify the two rims, (b) the CPK model, and (c) the *greek vase* that inspired their name.

1.1 Synthesis of calix[*n*]arenes

Although several methods for the synthesis of calixarenes have been developed during the years, the most general and useful is the one-step, base-induced condensation of para-substituted phenols and formaldehyde (Figure 3), which gives good yields of even-numbered ($n = 4$, 6, 8) cyclic products.[1] The odd-numbered calix[*n*]arenes ($n = 5$, 7, 9) can also be obtained by direct condensation but the yields are considerably lower. Very well established and reproducible procedures have been developed by Gutsche *et al.* for the synthesis of *p-tert*-butylcalix[*n*]arenes ($n = 4$, 6, 8).

The one-step synthesis produces calixarenes having the same substituent at the para-position. Calixarenes with different substituents can be obtained by the convergent stepwise synthesis (fragment condensation). This procedure (Figure 4) has been particularly useful for the synthesis of calix[5]arenes.[1,3]

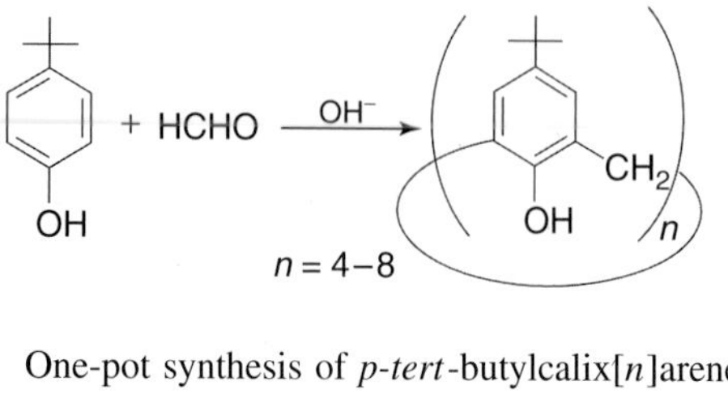

Figure 3 One-pot synthesis of *p-tert*-butylcalix[*n*]arenes.

1.2 Conformational properties of calixarenes

The water-insoluble *p*-H- or *p*-alkylcalix[*n*]arenes (**I**) having free phenolic OH groups at the *lower rim* (parent or native calix[n]arenes) are conformationally mobile in organic solution, and their preferred structures at different temperatures have been established by Dynamic ^1H NMR (nuclear magnetic resonance) and by theoretical calculations. In native calix[4]arenes and their simple ethers, conformational interconversion can only occur through the rotation of the OH or OR groups through the macrocyclic ring (*endo*-rim through the annulus pathway) and therefore the conformational behavior can be controlled by the size of the substituents at the *lower rim*. Tetramethoxy- (**2a**: $R = CH_3$) and tetraethoxy- (**2a**: $R = CH_3CH_2$) calix[4]arenes are conformationally mobile, but bulkier alkyl groups fix the macrocycle in one of the four possible conformations (Figure 5) differing for the orientation of each aryl group, which can project upward (u) or downward (d) with respect to an average plane defined by the methylene bridge carbon atoms. These conformations have been named by Gutsche as *cone* (u,u,u,u), *partial cone* (u,u,u,d), *1,3-alternate* (u,d,u,d), and *1,2-alternate* (u,u,d,d), and can be easily identified through their ^1H and ^{13}C NMR spectra.[1]

Starting from calix[5]arenes, the conformational interconversion can also take the *exo*-rim through the annulus pathway and the conformational control becomes more difficult, also considering that the number of possible

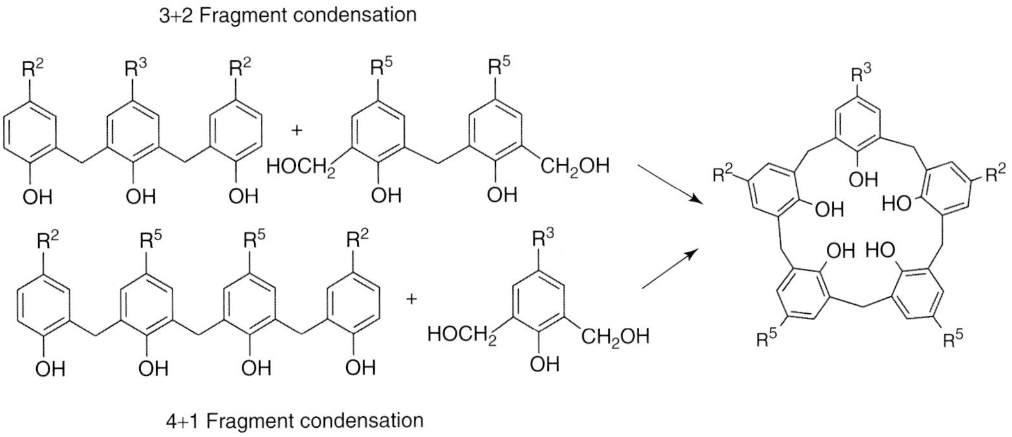

3+2 Fragment condensation

4+1 Fragment condensation

Figure 4 Fragment condensation applied to the synthesis of calix[5]arenes.

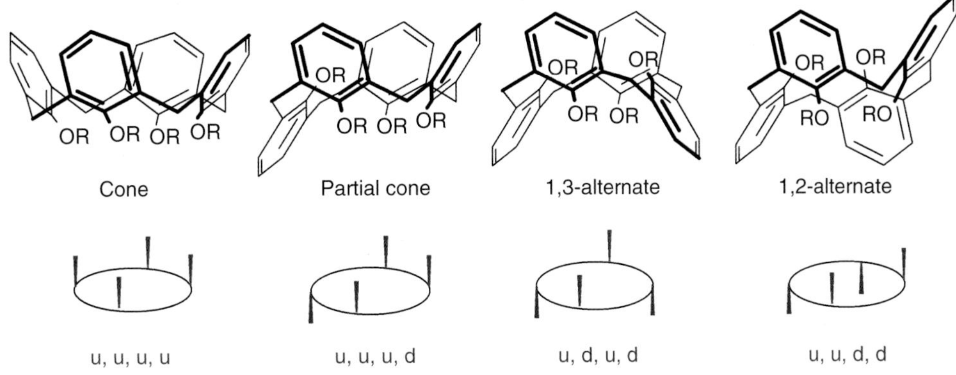

Figure 5 The four limiting conformations of calix[4]arene and their symbolic representations.

stereoisomers increases with the number of phenolic units. However, in *p-tert*-butylcalix[5]arene, the *exo*-rim through the annulus pathway is precluded for steric reasons, and again the conformational outcome can be controlled by the size of the substituents at the *lower rim*. *p-tert*-Butylcalix[5]arene derivatives blocked in C_{5v} or less symmetric structures have been reported, whereas extensive bridging at the *upper* or the *lower rim* is needed to fix the calix[6]- and calix[8]arenes in one particular conformation.[1]

1.3 Functionalization of calixarenes at the *lower rim*

The functionalization of the phenolic OH groups of calixarenes at the *lower rim* has been performed both for shaping the macrocycle through conformational control (especially for calix[4]- and calix[5]arenes) or for attaching recognition moieties for performing supramolecular functions. Obviously, direct acylation or alkylation (Williamson method), to achieve ester and ether derivatives, respectively, are the most extensively used functionalization methods at the *lower rim*. For certain purposes, it is desirable to have a complete functionalization, for others selective partial functionalization is pursued, and both goals have a scope in calixarene chemistry. The best method to achieve complete alkylation at the *lower rim* is to use a strong base (NaH) in DMF (dimethyl formamide) or DMSO (dimethyl sulfoxide) and a slight excess of alkylating agent, although milder procedures can be used with more reactive reagents. In the case of calix[4]arenes, this method also has the advantage of selectively producing tetraethers in the *cone* conformation (**2a**). If this reaction is performed using a reduced amount of alkylating agent (two equivalents instead of the four required to functionalize all the OH groups) and in a shorter reaction time, the proximal (1,2)-dialkylated products (**2b**) can be usually isolated in good yields. On the other hand, the use of a weaker base (K_2CO_3) and a solvent of lower donicity (acetone and acetonitrile) selectively gives the diametral (1,3)-dialkylated calix[4]arenes (**2c**) in very good yields. Using a similar method, it has also been possible to obtain, in reasonable yields, the 1,3,5-trimethoxy-*p-tert*-butyl-calix[6]arene (**3**), widely used for the synthesis of rotaxanes and other molecular receptors.[1,7,8]

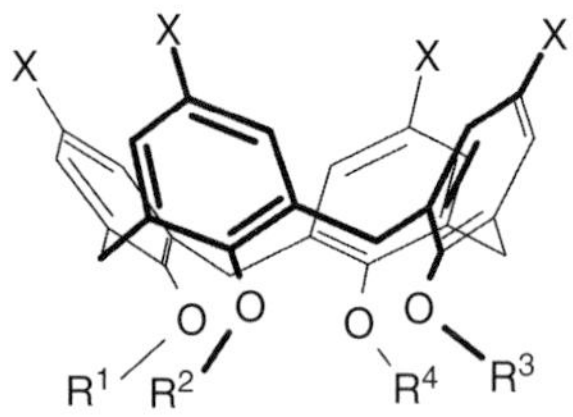

2a: R^1-R^4 = alkyl
2b: R^1, R^2 = alkyl, R^3, R^4 = OH
2c: R^1, R^3 = alkyl, R^2, R^4 = OH
2d: R^1 = alkyl, R^2–R^4 = OH
2e: R^1–R^3 = alkyl, R^4 = OH

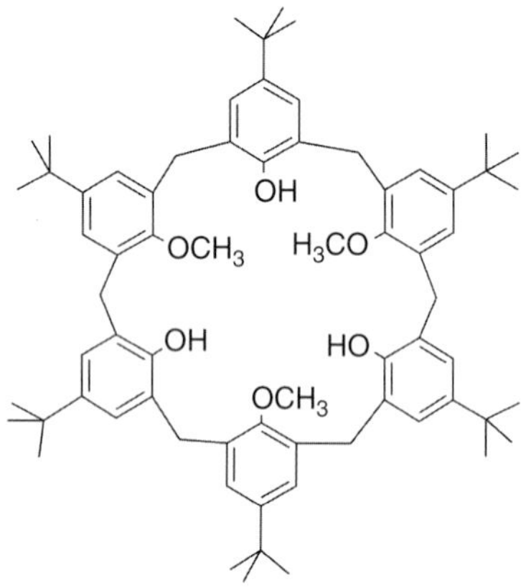

3

Selective trimethylation of *p-tert*-butylcalix[4]arene (**2e**: R^1–R^3 = CH_3) can be accomplished with $(CH_3)_2SO_4$ in DMF in the presence of BaO and $Ba(OH)_2$, whereas the monoalkyl ethers (**2d**) can be synthesized directly using CsF in DMF, or indirectly by submitting the easily available 1,3-dialkoxy and the tetralkoxy derivatives to selective dealkylation using, respectively, one or three equivalents of iodotrymethylsilane in $CHCl_3$.

1.4 Functionalization of calixarenes at the *upper rim*

It is a rather fortunate circumstance that the most easily synthesized calix[*n*]arenes are those derived from *p-tert*-butylphenol, as this group can be easily removed using Lewis or Brönsted acid catalysts, opening the way for further functionalization of the macrocycles at the para position (Figure 6).

Transfer of selectivity from the *lower* to the *upper rim* is the most useful method for the selective synthesis of partially functionalized calixarenes at the *upper rim*. Indeed, one can exploit the different reactivity of aryl ethers compared to phenols to introduce, regioselectively, additional functional groups at the *upper rim* of partially alkylated calixarenes. Moreover, if 1,3-dialkoxy-*p-tert*-butylcalix[4]arenes are submitted to the reverse Friedel–Crafts reaction, the *tert*-butyl groups are detached only from the para position of the phenolic nuclei, obtaining compounds where only two diametral aromatic rings are available for further functionalization.

2 MOLECULAR RECEPTORS

It is generally accepted that the great success of calixarenes in chemical research is mainly due to the favorable circumstances that they were rediscovered during the blossoming of supramolecular chemistry as an attractive building block for the synthesis of molecular receptors for a variety of guest species. It soon appeared evident that the presence of an apolar cavity in calixarenes and the possibility of attaching binding groups at the *lower* and the *upper rim*, of modulating their lipophilicity, and of controlling their conformational properties could be well exploited to control efficiency and selectivity in molecular recognition. The most interesting receptors were soon used in sensor devices, as such or after the introduction of suitable reporter groups of the molecular recognition event.[9] Subsequently, quite sophisticated molecular receptors and supramolecular architectures were built on calixarenes such as molecular capsules, catenanes, rotaxanes, molecular machines, supramolecular catalysts, supramolecular and nanomaterials, and so on, each probably deserving by themselves a chapter. Therefore, we do not treat these special topics and confine ourselves to the basic aspects of molecular recognition by calixarene receptors that could constitute the basis for a better understanding of these special topics, treated in other sections of the volume series.

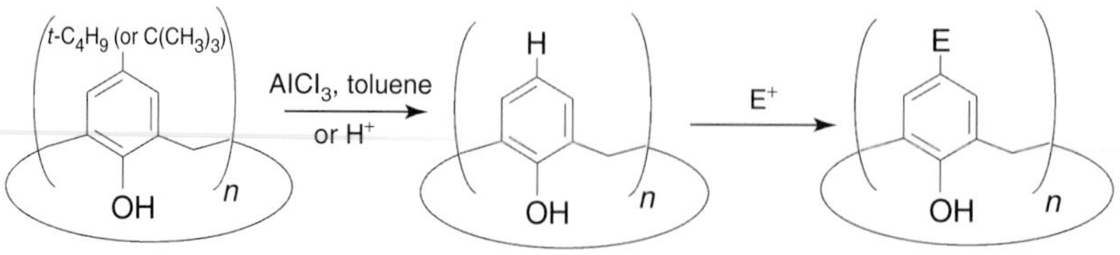

Figure 6 General scheme for the de-*tert*-butylation and complete functionalization of the *upper rim* (E = NO_2, Cl, Br, I, $CH_2CH{=}CH_2$, CHO, etc.).

2.1 Neutral molecules and Quats

2.1.1 Solid-state inclusion complexes of parent calixarenes with neutral molecules

The first evidence for the formation of an *endo*-cavity inclusion compound between a simple calixarene and a neutral molecule was reported by our group in 1979 and refers to the X-ray crystal structure of the 1 : 1 toluene complex (or clathrate, see below) of *p-tert*-butylcalix[4]arene **1** (Figure 7a). Since then, the X-ray crystal structures of many *endo*- and *exo*-cavity inclusion compounds involving calixarenes as hosts have been reported in literature.[1] The inclusion of guest molecules in *p*-alkyl calix[4]arenes having basic or acidic groups able to interact with the OH groups at the *lower rim* creates complex architectures in the solid state[10] but neutral molecules give mainly two types of structures, which were both observed in the early studies with aromatic guests. The first type is the already mentioned 1 : 1 complex (clathrate) between **1** and toluene (Figure 7a) and the second is the 2 : 1 cage compound formed by the same host with anisole (Figure 7b).

Whether these two type of compounds must be considered as true molecular complexes held together by specific CH–π interactions[11] between the CH bonds of the alkyl groups in para position of the host and the aromatic π cloud of the guests or, alternatively, simple clathrates, in which a variety of short-range interactions are involved in producing stable compounds, is still under debate. Indeed, for host **1** and its derivatives, the observed close contacts in the solid state between the methyls of *p-tert*-butyl groups at the *upper rim* and the aromatic nuclei of the guests ($CH_3 \cdots \pi \approx 3$ Å), the selectivity for aromatic guests found in crystallization experiments and molecular mechanics (MD) calculations[12] support the view of true molecular complexes for such compounds. On the other hand, Ripmeester *et al.* have isolated many 1 : 1 and 2 : 1 compounds between *p-tert*-butylcalix[4]arene **1** and aliphatic or alicyclic neutral molecules whose X-ray crystal structures resemble those shown in Figure 7. The extensive studies performed using a combination of several experimental techniques such as solid-state NMR (^{13}C, ^{2}H, and ^{1}H), X-Ray crystallography with different radiation sources, and DSC/TGA measurements, support the alternative view outlined by these authors of *p-tert-butylcalix[4]arene as a versatile inclusion host that in the solid state behaves more like a clathrate than a molecular receptor.*[10] Probably more work is needed, especially aimed at measuring the stability of the complexes in the solid state, obtaining selectivity data, and developing reliable theoretical models for calculations in order to clarify the relative importance of specific versus nonspecific interactions in these inclusion compounds.

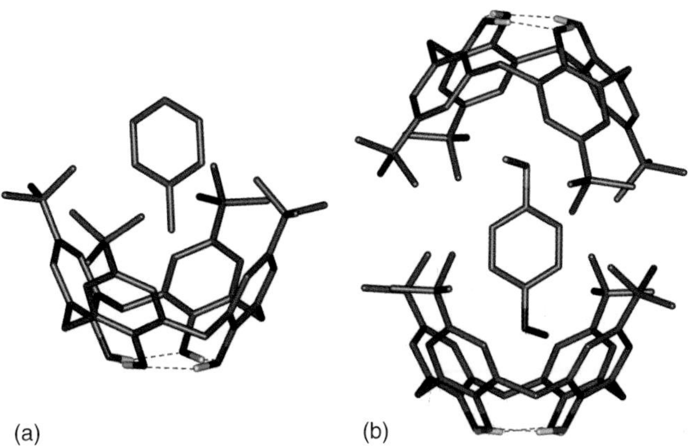

Figure 7 X-ray crystal structure of (a) 1 : 1 complex between **1** and toluene and (b) 2 : 1 complex between **1** and anisole (one of the two statistically equivalent OCH_3 groups is colored in orange).

2.1.2 Cavitands and their inclusion complexes in organic media

Owing to their conformational flexibility, parent calixarenes and resorcinarenes form very weak *endo*-cavity inclusion complexes with neutral molecules (with the exception of fullerenes, which are strongly bound) and quaternary ammonium cations (Quats) in organic media. Also, C_{4v} symmetric *cone* tetralkoxycalix[4]arenes form very weak complexes in organic solution with neutral molecules and charge diffused organic cations, due to residual conformational mobility between two C_{2v} *flattened cone* structures, as evidenced by Pochini *et al.*[3]

In order to observe such complexes in solution, calixarenes must be further rigidified by functionalization at the *lower rim*, thus obtaining receptors with enforced, open-ended cavities called *cavitands*. Rigidification of the flexible resorcinarenes to obtain cavitands (**4**) is usually achieved by intramolecular bridging of adjacent oxygen atoms on two different aromatic nuclei using a bifunctional reagent.[13] The apolar cavity can be further extended by functionalization at the *upper rim*, obtaining the *deep cavitands* (Figure 8).[3] Strictly speaking, the name of cavitands was coined for resorcinarene derivatives but it is also applicable to any host (thus also including calixarenes) possessing an enforced cavity large enough to accommodate guest molecules.[3]

The best way to rigidify calix[4]arenes is to bridge two adjacent (proximal) OH groups at the *lower rim* (1,2-functionalization) with two triethylene glycol units obtaining the *rigid cone* bis-crown-3 derivative **5**. Indeed, cavitand **5** is able to complex nitromethane in CCl_4 showing an association constant $K = 28\,M^{-1}$, which can be increased up to $230\,M^{-1}$ by introducing a *t*-butyl group at the *upper rim*.[3] Rigidification of the calix[4]arene at the *lower rim* and

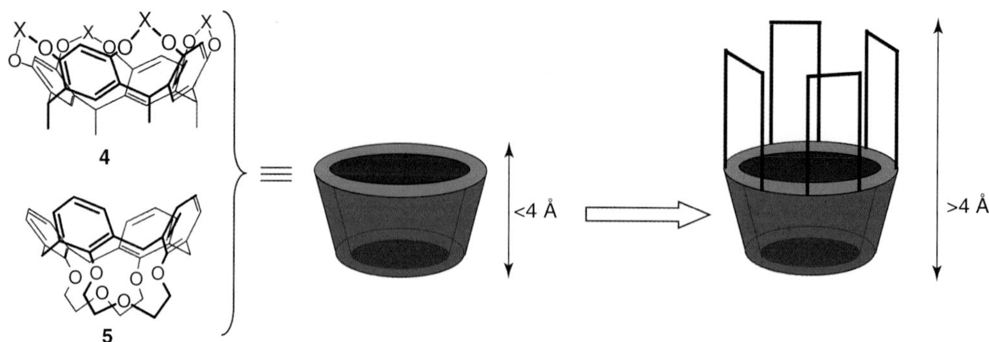

Figure 8 Normal (<4 Å) and deep (>4 Å) cavitands derived from the rigidification of the resorc[4]arene (**4**) and calix[4]arene (**5**) structures with the schematic representation of their cuplike structure.

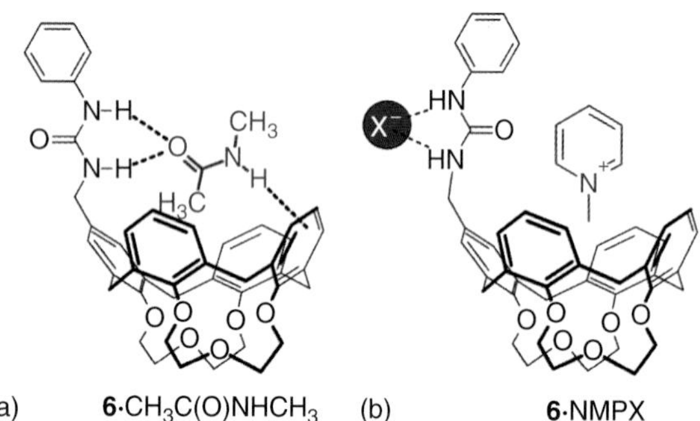

Figure 9 Schematic representations of (a) the **6**·*N*-methyl acetamide and (b) the **6**·NMPX inclusion complexes.

simultaneous introduction of a urea group at the *upper* gives cavitand **6**, which shows good efficiency and selectivity in the binding of primary and secondary amide derivatives owing to the cooperative action of hydrogen bonding and CH–π interactions (Figure 9). The same concept was used for the complexation of Quats in organic media.[14] The ditopic receptor **6** complexes *N*-methylpyridinium salts (NMPX) in chloroform solution more than 2 orders of magnitude stronger than the monotopic *rigid cone* cavitand **5**, and in both cases the association constants strongly depend on the counteranion X.

The general strategy of calixarene rigidification and introduction of additional binding groups is well illustrated by the design and synthesis of the *upper rim* bridged carbohydrate receptor **7**, which is able to complex several sugar derivatives in $CDCl_3$, showing selectivity for the β-anomer of octylglucoside (octyl-β-D-Glu), owing to the simultaneous action of hydrogen bonding between the OH groups of sugar with the charged phosphate and neutral amide groups of the receptor.[15]

2.1.3 *Water-soluble receptors*

Native calix[*n*]arenes are insoluble in water even at basic pH as only one OH group is acidic enough to be ionized,[1] and a single charge is not sufficient to make the macrocycles water soluble. In order to reach water solubility, more easily ionizable, permanently charged, or polar neutral groups must be appended either at the *upper* or at the *lower rim*.

7

7·Octyl-β-D-Glu

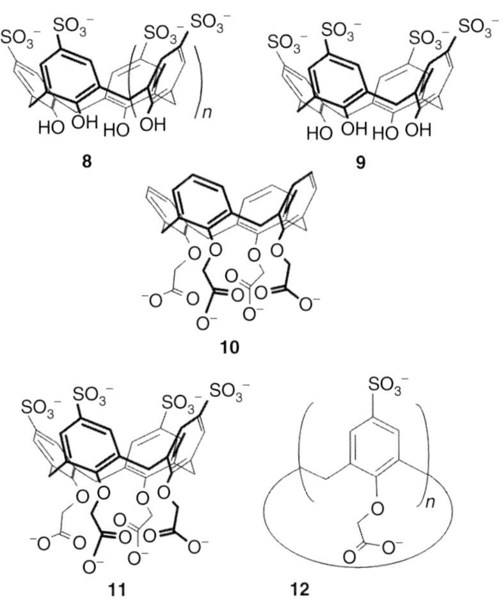

Figure 10 A collection of the most studied water-soluble calixarenes.

In Figure 10, a collection of the most studied water-soluble calix[*n*]arenes is reported. Depending on the nature of the substituents at the *upper* or at the *lower rim* and on the size and conformational flexibility of the macrocycle, water-soluble calixarenes can have an amphiphilic character that complicates the analysis of their host–guest properties. Therefore, these compounds are not discussed here. Conformationally mobile *p*-sulfonatocalix[*n*]arenes **8** are the most investigated water-soluble calixarenes,[16,17] and among them the best is tetramer **9**. At buffered neutral pH, **9** exists as a penta-anion because, besides the sulfonate groups, one rather acidic OH group at the *lower rim* ($pK_a = 3.34$) is also deprotonated, whereas at pH <2.5 it is a tetra-anion. Both these forms, however, adopt a *cone* conformation in solution. Symmetrical tetralkylammonium cations form stable *endo*-cavity 1 : 1 complexes with **9** in the log K range of 4.21–4.67 at pH = 2. Slightly higher values are obtained at neutral pH but the observed selectivity order is the same: $(CH_3CH_2)_4N^+ \geq (CH_3CH_2CH_2)_4N^+ \approx (CH_3)_4N^+ \geq (CH_3CH_2CH_2CH_2)_4N^+$, whereas $NH_4{}^+$ cation is not complexed at all by host **9**. Calorimetric studies reveal that the stability of the complexes is enthalpy driven with very small contribution coming from the entropy term, whereas the opposite is observed for the complexation of inorganic cations. This suggests a different mode of binding between the two classes of cations by **9**, and that in the case of Quats the alkyl chains of the cation enter the aromatic cavity of the calix[4]arene in *cone* conformation and the complexes are stabilized by van der Waals and hydrophobic forces, although the contribution of cation–π and CH–π interactions cannot be ruled out. A more complicated situation is encountered with nonsymmetric, ditopic Quats having one larger apolar tail (aromatic or aliphatic), as either the quaternary ammonium head group or the apolar tail can be included nonspecifically into the calix[4]arene cavity. The same ambiguity in the mode of binding exists in the complexation of toluene by **9** at neutral pH where either the methyl or the phenyl group of the guest unselectively enters the apolar cavity of the host. The association constants found for the inclusion of toluene and other monosubstituted benzene derivatives as well are 2 orders of magnitude lower than those observed with charged Quats.[18] Interestingly, the two water-soluble receptors **10** and **11**, in the *fixed cone* conformation, reported by Casnati *et al.*,[3] complex trimethylanilinium (TMA) chloride in a different but selective way (Figure 11). Compound **10**, devoid of the sulfonate groups at the *upper rim*, selectively complexes the trimethylammonium head group (log $K = 2.2$), whereas the sulfonate derivative **11**, blocked in *cone* conformation by four acetate groups at the *lower rim*, includes into the cavity only the aromatic nucleus of TMA (log $K = 3.3$).

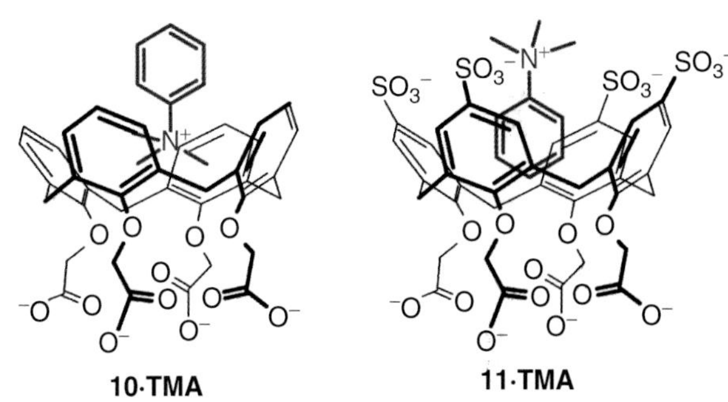

Figure 11 Schematic representation of the different ways of binding of water-soluble receptors **10** and **11**.

In both cases, the inclusion process is driven by the enthalpy, whose contribution is larger for the sulfonate derivative **11**. This explains the higher stability constant found for this host in the complexation of TMA compared with **10** and outlines the important role of the electrostatic interaction between the sulfonate groups at the *upper rim* and the alkylammonium group of TMA in stabilizing the complex and univocally orienting the ditopic guest inside the cavity. The important role of the conformational properties of the host and the sulfonate groups at the *upper rim* in determining the stability and structure of the complexes was also found by studying the inclusion of small aliphatic neutral molecules (alcohols, nitriles, and ketones) in water-soluble calix[4]arenes. In all the cases, the apolar part of the guest is included into the calixarene cavity and the stability constants of the complexes are in the $1.04 \leq \log K \leq 1.8$ range. Although, as already mentioned, the *p*-sulfonato calix[4]arene **9** does not complex NH_4^+ in water, it forms complexes both in the solid state and in water solution with aliphatic and aromatic organic cations, including those of amino acids and peptides, giving a biological perspective to these studies. However, in these cases it is always the apolar part and not the charged head group of the guest that should be included into the host cavity.[16,17]

Recently, Nau *et al.* reported[19] the inclusion properties of the positively charged and fluorescent bicyclic diazoalkane (DBO) which, as expected, forms a very stable complex in water with **9** ($K = 60\,000 \pm 16\,000\,M^{-1}$) owing to the complementarity of the spherical shape of the azoalkane and the calix[4]arene cavity and to the electrostatic interactions between the protruding ammonium head group and the sulfonate groups at the *upper rim*. Upon inclusion in **9**, the fluorescence of DBO is strongly quenched but it is switched on again when the azoalkane is displaced by a competitive guest and released into solution. This *indicator displacement* concept (see **Competition Experiments**, Volume 1) was exploited to set up a *supramolecular tandem assay* for detecting real-time enzyme activity as illustrated in Figure 12; this refers to the situation where the macrocycle binds the enzyme substrate weakly but the product of the enzymatic reaction strongly (*product selective assay*). This is the case, for example, of lysine and arginine decarboxylases studied at pH = 6; the substrates are only weakly bound to host **9** ($K = 870\,M^{-1}$ for lysine and $K = 2\,800\,M^{-1}$ for arginine), while decarboxylation products cadaverine ($K = 1.4 \times 10^7\,M^{-1}$) and agmatine ($K = 7.1 \times 10^6\,M^{-1}$) are bound more strongly and compete with DBO in occupying the calixarene cavity. As these competitive binding events are fast compared to the enzymatic reaction, the latter can be continuously followed by monitoring the fluorescence increase during the time. Very recently, Hof *et al.*[20] reported that **9** can efficiently distinguish the posttranslational methylation state of lysine on the side chain, which is important for the regulation of several genes and for other biological processes. Indeed, they confirmed the weak association of lysine with **9** ($K = 520\,M^{-1}$, in a more competitive 40 mM phosphate buffer at pH = 7.4) and found that the derivatives Lys (CH_3) ($K = 4000\,M^{-1}$), Lys($(CH_3)_2$) ($K = 16\,200\,M^{-1}$), and Lys($(CH_3)_3$) ($K = 37\,000\,M^{-1}$) formed, with the same host, 1 : 1 complexes whose stability increased with increasing the methylation degree. This is due to the insertion of the alkylammonium head group of the side chain into the cavity of **9** and to more effective cation–π and CH–π guest/host interactions as the methylation state of lysine proceeds.

Summarizing the available results, we can conclude that amino acid and peptide binding in solution is controlled by favorable enthalpy, mainly resulting from the tight inclusion of the apolar part of the guest into the hydrophobic cavity of the host through van der Waals (or CH–π and π–π stacking) interactions, although the electrostatic attraction between the charged groups at the

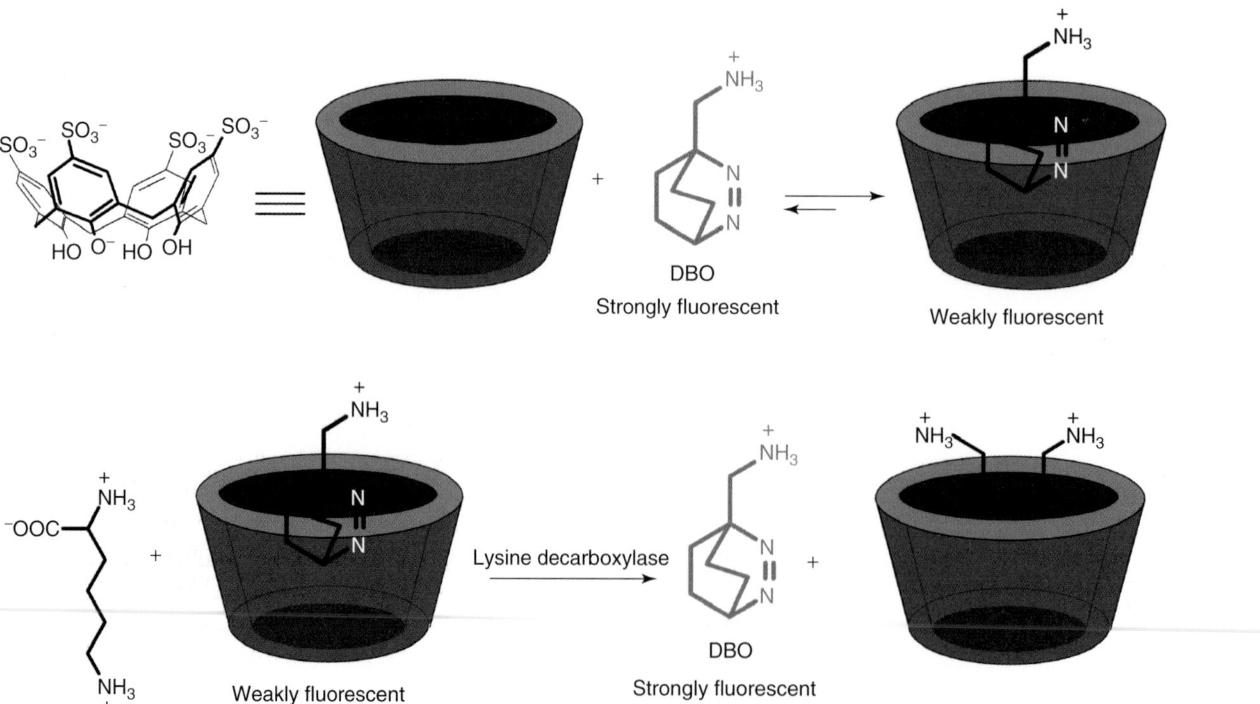

Figure 12 An indicator displacement assay based on calixarene **9** for the determination of lysine decarboxylase activity.

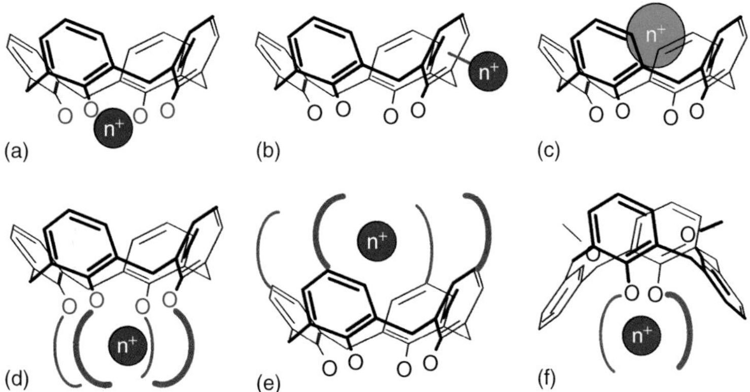

Figure 13 Different ways in which calixarenes interact with cations: (a) metallation at the *lower rim*; (b) *exo*-cavity (π-metallation); (c) *endo*-cavity; (d) at the *lower rim* with participation of additional binding sites; (e) at the *upper rim* by coordinating groups; (f) at the *lower rim* with participation of cation–π interactions. O atoms at the lower rim may be O^-, OH or OR.

upper rim of the host and additional polar or charged groups of the side chains of the guest (which also have a favorable entropic component due to desolvation) can also play a role in stabilizing the complexes. Pyridiniums and viologens form quite stable inclusion complexes with *p*-sulfonatocalix[4,6]arenes and *p*-sulfonatothiacalix[4]arene ($1.7 \leq \log K \leq 5.4$).[17] Calix[*n*]arenes having permanently charged positive groups or protonable amino groups at both rims are known[1,21] but they have not been extensively used for the inclusion of neutral molecules.

2.2 Cation complexation

Figure 13 illustrates all the known modes of cation binding by native and functionalized calixarenes exploiting cation–π, induced dipole, or electrostatic interactions.[1]

Strictly speaking, only for situations (c)–(f), we can talk of cation recognition by calixarene receptors, whereas in metallation with transition and f-element (situations (a) and (b)),[3] which are often exploited in anion complexation or catalysis, little control is experienced on selectivity.[22] Usually, the most important calixarene-based cation receptors are obtained by the introduction of chelating units at the *lower* rather than at the *upper rim*. This ensures a more convergent disposition of the chelating units and the direct involvement of the phenolic O atom in cation binding. In the following sections, we highlight the role of the macrocycle size and conformation and the nature of binding groups on efficiency and selectivity in cation recognition and sensing.

2.2.1 Alkali and alkaline-earth metal ions

Calixarenes fully functionalized at the *lower rim* with ether groups show a weak affinity for alkali metal ions. The case of the tetramethyl ether of *p*-*tert*-butylcalix[4]arene (**13**) is quite interesting, since it is conformationally mobile, and can adapt its conformation to different metal ions. The Na^+ ion is exclusively complexed by the *cone* isomer, while Cs^+ is preferentially bound by the *partial cone* isomer.

However, the binding properties of these simple ethers and even of the oligoethylene glycol derivatives **14** confirm that these podand-type polydentate compounds are rather poor binders ($\log K < 3$ in $CDCl_3$) of alkali metal ions and that they work only in low polarity media.[22] A burst in the understanding of the factors governing the recognition of alkali metal ions by calixarene receptors was obtained by Ungaro *et al.* through bridging of the distal O atoms at the *lower rim* of

13: R = CH_3
14: R = $CH_2CH_2OCH_2CH_3$
15: *n* = 4-6
16: *n* = 5
17: *n* = 6

Figure 14 Conformational interconversion upon caesium or sodium binding by calix-crown-6 (**17**).

calix[4]arenes with oligoethylene glycols. These compounds, known as calix[4]arene-crowns-*n* or simply calixcrowns (**15**),[3,22] exhibit extraordinary selectivity in binding Na^+, K^+, and Cs^+. The conformationally mobile 1,3-dimethoxycalix[4]arene-crown-5 (**16**) and -crown-6 (**17**) were used as probes to understand the preferred structure for binding. The larger 1,3-dimethoxycalixcrown-6 (**17**) assumes the *cone* structure in binding Na^+ but the *1,3-alternate* upon Cs^+ coordination (Figure 14).

Similarly, the smaller 1,3-dimethoxycalixcrown-5 (**16**) is present only in the *1,3-alternate* structure in the K^+ complex, while becomes *cone* with Na^+. Quite importantly, we found, both in solution and in the solid state (Figure 15), that the cations, besides interacting with the O atoms of the oligoethylene bridge, are also involved in relevant cation–π interactions with the aromatic nuclei of inverted anisole moieties.

The extensive studies on the binding-free energies of the alkali metal ion complexes for the different stereoisomers, *cone, partial cone,* and *1,3-alternate,* revealed a strong dependence of the binding efficiency upon calixarene conformation, although all the calix-crown-5 are selective for potassium and all the calixcrown-6 for caesium. The *cone* derivatives (**22** and **23**) are very poor receptors, even weaker than the conformationally mobile derivatives **16** and **17**, while the *partial cone* isomers **20** and **21** are rather efficient and selective. The preorganized calix-crown-5 (**18**) and -crown-6 (**19**) in *1,3-alternate* structure are, by far, the most efficient and selective receptors for K^+ and Cs^+ respectively. The 1,3-di-isopropoxycalixcrown-5 (**18**) is even more efficient and selective than natural valinomycin (see **Supramolecular Chemistry in *In Vitro* Biosensors**, Volume 4). The K^+/Na^+ selectivity is striking ($\Delta \log K > 6.4$ in CH_3OH) and was also confirmed in transport through supported liquid membranes (SLMs) and in detection studies on ion-selective field-effect transistors (ISFETs). The 1,3-di-isopropoxycalix[4]arene-crown-6 (**19**) in the *1,3-alternate* structure showed one of the highest Cs^+/Na^+ selectivity known ($\Delta \log K > 5$ in CH_3OH), which could be slightly improved by the introduction of benzo units in the crown-ether bridge (e.g., **24**). Owing

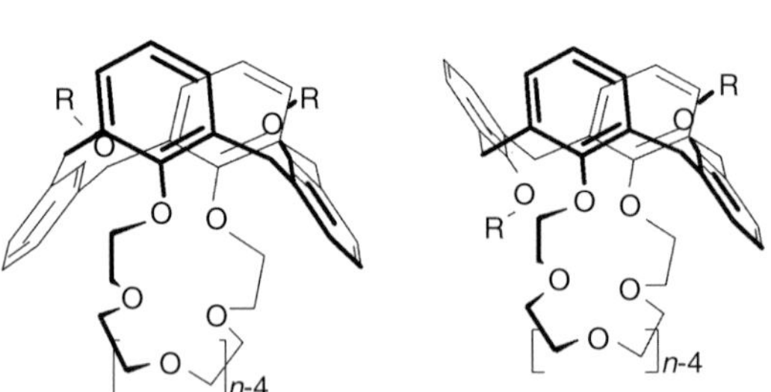

18: R = *i*-C_3H_7 (or $CH(CH_3)_2$), *n* = 5 **20**: R = *i*-C_3H_7(or $CH(CH_3)_2$), *n* = 5
19: R = *i*-C_3H_7(or $CH(CH_3)_2$), *n* = 6 **21**: R = *i*-C_3H_7(or $CH(CH_3)_2$), *n* = 6

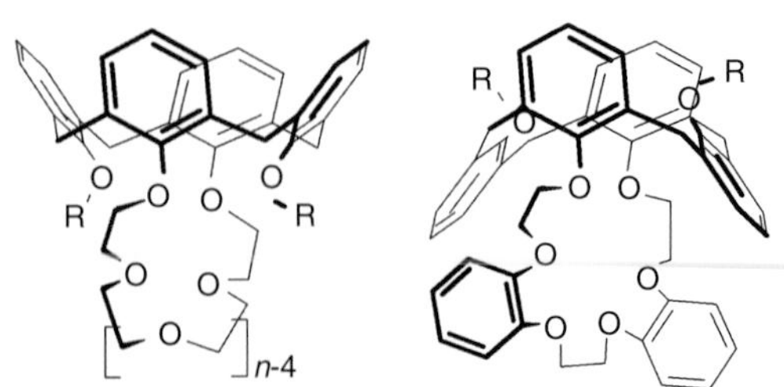

22: R = *i*-C_3H_7(or $CH(CH_3)_2$), *n* = 5 **24**
23: R = *i*-C_3H_7(or $CH(CH_3)_2$), *n* = 6

Figure 15 X-Ray crystal structure of the caesium picrate complex of **17**.

to this outstanding caesium selectivity, calixcrowns-6 were used in the separation of the ^{137}Cs radionuclide from radioactive wastes originated from the reprocessing of spent nuclear fuel. It was demonstrated even in hot tests that calixcrowns-6 are able to selectively and quantitatively extract caesium nitrate from a 10^{-3} M solution of this salt containing 1 M $NaNO_3$ and 4 M HNO_3.[23]

Shinkai *et al.* reported that reduction of the crown-ether bridge length gives a series of calix[4]arene-crown-4 derivatives among which the *partial cone* derivative **25** exhibits an exceptional Na^+/K^+ selectivity ($\log K^{pot}_{Na/K} = -5.3$), as determined by ion-selective electrodes (ISEs) (see review[22] for specific references).

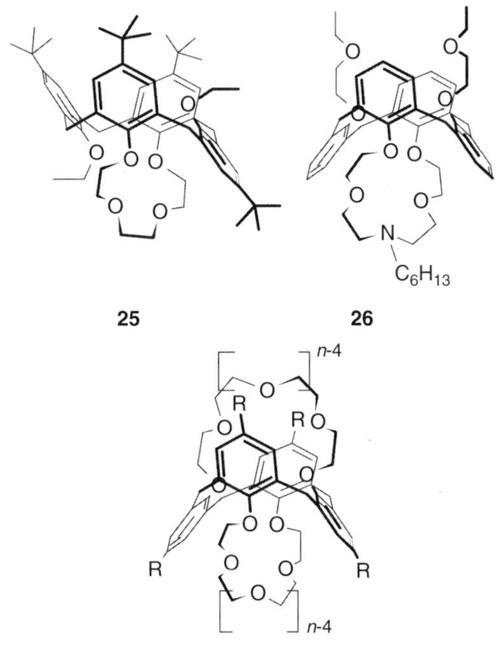

27: R = H, *t*-C_4H_9; *n* = 5, 6

A rather interesting phenomenon, named *molecular syringe,* was reported by the same authors using the aza-crown **26**.[22] Ag^+ is complexed in the crown ether side of the calixarene when the amine is free, while upon protonation the metal ion is pulled to the ethoxyethyl side of the calixarene through the π-basic tube. When the ammonium ion is again deprotonated with a hindered base or Li_2CO_3, the silver ion returns back to the bridged side. Calix[4]arene-1,3;2,4-biscrown ethers (distal biscrowns) **27**, reported by Vicens *et al*., on the other hand, show rather important complexing properties, quite similar to those of 1,3-dialkoxycalix[4]arene-crown-6 in the *1,3-alternate* structure: calix-bis-crowns-5 (**27**: $n = 5$) are also selective for K^+, while calix-biscrowns-6 (**27**: $n = 6$) for Cs^+, and evidence of the formation of cation–π interactions was also collected both in the solid state and in solution. The presence of *tert*-butyl groups in the para position hinders the cation uptake and therefore de-*tert*-butylated derivatives (**27**: R = H) are preferentially used as ionophores. Distal calix-biscrowns present two sites for cation binding. It was also demonstrated that the complexed cation can exchange from one site to the other both inter- and intramolecularly, the latter process taking place via tunneling of the cation through the π-basic tube of the calix. The head-to-head bridging of two calixarenes with polyether chains or ethylene moieties results in the so-called calix[4]tubes (**28** and **29**), reported by Beer and discussed in Ref. 22. In compound **28**, two distinct regions for the binding of the metal ions are present, and it was demonstrated that Na^+ or K^+ can oscillate between the two binding sites both intra- and intermolecularly, depending on the concentration and the temperature. The more rigid calixtube **29** is rather selective for potassium ion as it can transfer KI in solid–liquid extraction of alkali metal iodides.

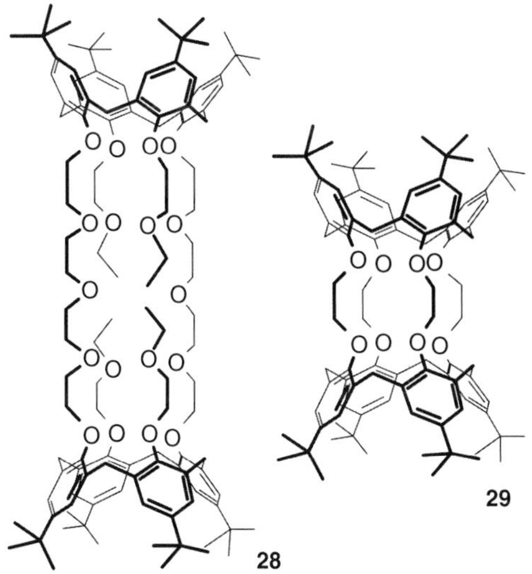

Among the most interesting calixarene-based ionophores for alkaline-earth metal ions are the podand-like derivatives (**30–40**) functionalized with $-CH_2C(=O)X$ groups (X = alkyl, aryl, OR, NR_1R_2, OH) at the *lower rim.* The $ArOCH_2C(=O)$ bidentate groups, once preorganized at the *lower rim* of the calixarene macrocycle, allow to recognize metal ions with selectivity and efficiency, depending on the donicity of the carbonyl group and the size and conformation of the macrocycle. Tetraesters (**31**) and tetraketones (**30**) of calix[4]arenes in the *cone* conformations show negligible log *K* values with alkaline-earth metal ions.[3,22]

Tetramides **32** are more efficient than tetraesters **31** and tetraketones **30,** and the association constants in CH_3OH of the *N*, *N*-diethylamides **33** and **34** show a marked preference for alkaline-earth over alkali metal ions with log *K* values comparable to those of the cryptands.[22] Compared to [2.2.1], however, the tetramides **33** and **34** show a higher

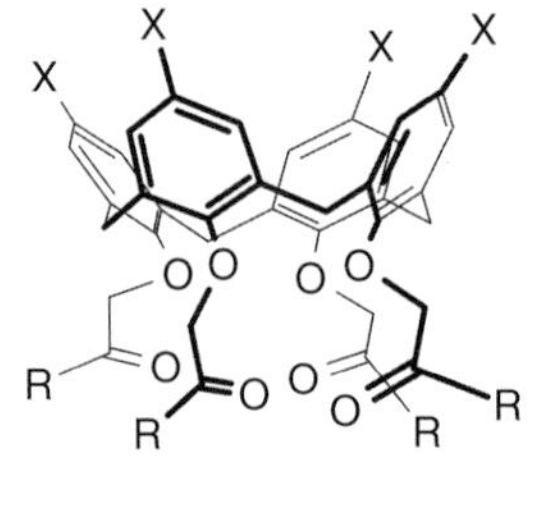

30: R = alkyl, aryl, X = H, t-C_4H_9
31: R = O-alkyl, X = H, t-C_4H_9

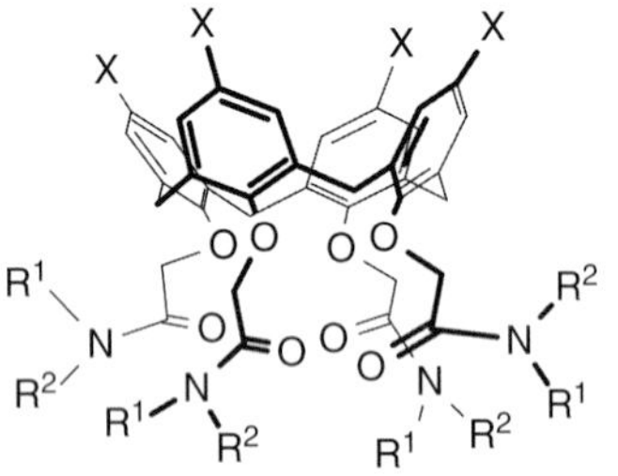

32: $R^1 = R^2$ = alkyl, X = H, t-C_4H_9
33: X = H, $R^1 = R^2 = CH_2CH_3$
34: X = t-C_4H_9, $R^1 = R^2 = CH_2CH_3$
35: $R^1 = R^2 = H$
36: $R^1 = H$, $R^2 = n$-C_4H_9

selectivity for Na^+ ion among alkali metal and for Ca^{2+} and Sr^{2+} among alkaline-earth metal ions. Primary (**35**) and secondary (**36**) amides are particularly inefficient in the extraction of metal ions into organic solvents, and they also show log K for the binding of both alkali and alkaline-earth metal ions at least 3 orders of magnitude lower than tertiary amides (**33** and **34**) in CH_3OH solution. The effect of the calixarene ring size is quite remarkable. Passing from tetramers (**33** and **34**) to hexamers (**37**) and octamers (**38**), the efficiency in the binding of small cations is strongly depressed, and selectivity is shifted toward larger metal ions, Rb^+ and Cs^+, among alkali and $Ba^{2+} > Sr^{2+} > Ca^{2+}$ among alkaline-earth metal ions. This results in a particularly high Ba^{2+}/Na^+ and especially Sr^{2+}/Na^+ selectivity, even higher than that of dicyclohexyl-18-crown-6, which could therefore be replaced by **37** and **38** for the removal of long-live ^{90}Sr radionuclides from radioactive waste.[23]

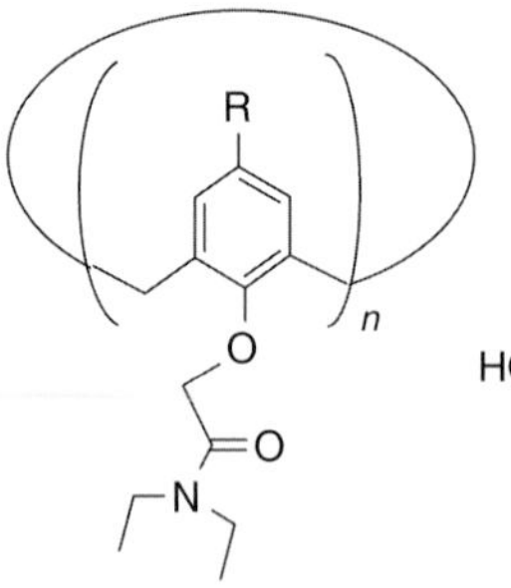

37: n = 6; R = H, t-C_4H_9, O-alkyl
38: n = 8; R = H, t-C_4H_9, O-alkyl

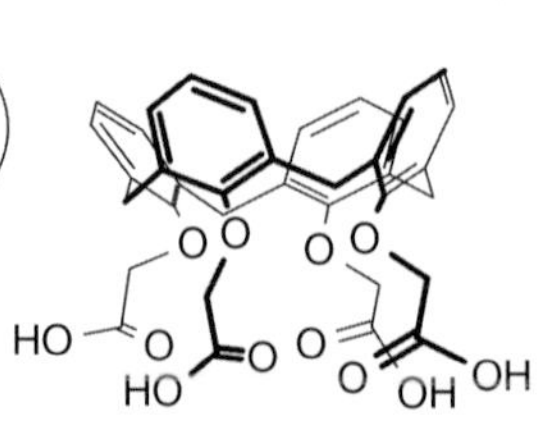

39

The introduction at the *lower rim* of calix[4]arenes of acetic acid moieties, easily ionizable to carboxylates upon metal ion binding, also resulted in receptors being quite selective for alkaline-earth and, particularly, calcium ions. One of the first receptors synthesized, which shows a high selectivity for alkaline-earth ion, was the tetracid **39**, but later on it was found that the diametrically 1,3-diamide-diacid derivatives **40** are even more efficient and selective ($Ca^{2+} \approx Sr^{2+} \gg$ alkali ions). The introduction of ionizable carboxylic acid groups close to the crown-ether moieties affords a receptor **41** showing high efficiency and selectivity in Ca^{2+} extraction, also as a consequence of the crown-ether size.

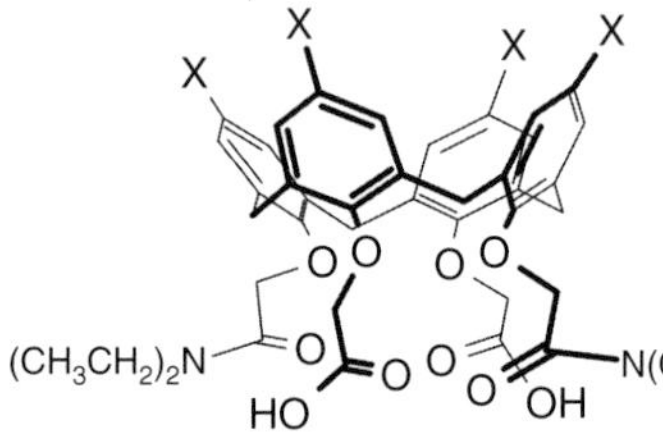

40: X = H, t-C_4H_9

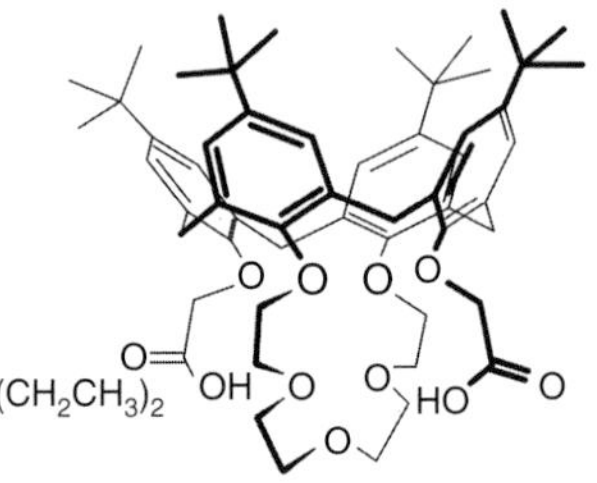

41

2.2.2 *Ammonium, alkylammonium, and guanidinium cations*

Early observations by Gutsche *et al.* indicated that parent calix[4]arenes interact strongly with primary alkyl amines in polar solvents, in consequence of a proton transfer. The formation of intracavity *endo*-complexes of the resulting alkylammonium ions was proposed and a selectivity for *tert*- over *sec*- and *n*-alkyl amines was observed.[1] Tripodal $NH^+ \cdots O$ interactions with the phenolic oxygen atoms and CH–π interactions were supposed to stabilize the complex. However, it is currently accepted that calix[4]arenes usually form *exo*-complexes with ammonium salts, while direct proofs of *endo*-complexes have been only collected for the larger calix[5]-, calix[6]-, and hexahomotrioxacalix[3]arenes. The insertion of cation-binding groups (OCH_2COX, X = OR, NR_2) at the *lower rim* of calix[6]arenes has the beneficial effect of increasing the extraction of NH_4^+ and guanidinium (Gua^+) salts from water to the organic solvents. A remarkable selectivity in the transport of Gua^+ over alkali salts through SLM was obtained by us with the highly complementary and preorganized C_{3v} symmetrical triamide **42**.[24] On the other hand, rigidification of the calixarene cavity by bridging or simply by insertion of bulk alkyl chains at the *lower rim* of calix[5]- and homooxacalix[3]arenes gives receptors with a preorganized cavity and high selectivity for alkylammonium salts.

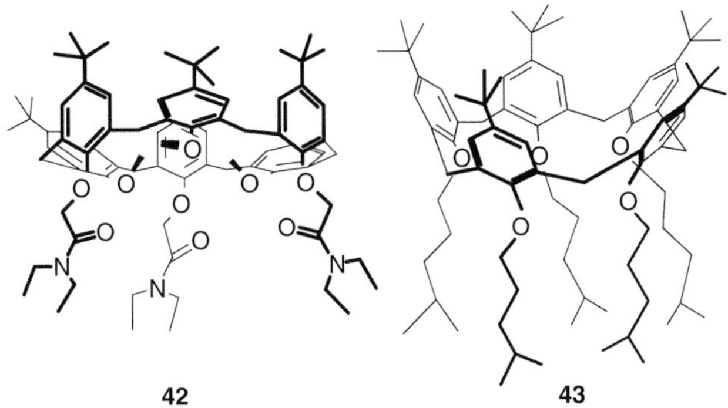

The calix[5]arene **43**, synthesized by Parisi *et al.*, for example, presents a marked shape and size selectivity even in ISEs for *n*-alkyl over *iso*-, *sec*-, and *tert*-alkylammonium salts. It was proven that linear *n*-alkyl ammonium salts form *endo*-complexes with the NH_3^+ head group interacting with the phenolic oxygen atoms and the alkyl residue with the calixarene cavity, while other branched alkyl ammonium salts generally form *exo*-complexes (see review[24] for specific reference).

2.2.3 *f-Element metal ions*

Complexation of f-element ions has been pursued in calixarene chemistry with ambitious aims, such as to separate lanthanides and actinides,[23,25] to obtain luminescent[3] and paramagnetic probes,[22] to study the reactivity of the coordinated metal ion at the metal-oxo surface, or to direct the self-assembly of calixarenes in nanostructured architectures in the solid state.[3] In this context, we mainly focus our attention on the use of calixarene receptors in solution for the recognition of trivalent f-element ions with the aim of obtaining molecular devices. *p*-Sulfonatocalixarenes **8** (especially the tetramer **9**) bearing free phenolic OH groups have been widely used to complex f-elements, especially lanthanide ions. Although it is known that some metal ions, such as transition metal ions and caesium, give polyhapto binding to the phenyl rings, there is no evidence that f-element metal ions can interact other than with O atoms. Several X-ray crystal structures[26] and isothermal titration calorimetry (ITC) measurements[17] indeed indicate that (partially) hydrated trivalent metal ions interact with the sulfonate groups. An important exception is offered by the complexes of the divalent uranyl ion with *p*-sulfonatocalix[5]- (**8**: $n = 5$) and -[6]arenes (**8**: $n = 6$) having free phenolic OH groups or acetic acid moieties (**12**: $n = 5$, 6) at the *lower rim* as reported by Shinkai *et al.* The exceptional efficiency in the binding of UO_2^{2+} ion (log K up to 19.2) and selectivity factors over divalent transition metal ions in the range $10^{10}-10^{17}$ granted the nickname *superuranophiles* to these receptors. The presence of at least five donor atoms in a plane is the reason of the observed selectivity for UO_2^{2+}, which requires a pentagonal–bipyramidal coordination.[1,17] Receptors for f-element ions have different structures and binding groups, depending on the final application they are designed for. Usually, as for alkaline-earth metal ions, amide and carboxylic acid binding groups are widely used. The role of the calixarene is, however, often not only limited to preorganize the donor atoms around a hydrophilic cavity but the macrocycle itself, in some cases, also plays an active role as illustrated by the Eu(III) and the Tb(III) complexes of the *p-tert*-butylcalix[4]arene tetramide (**34**). In particular, in the case of the Tb(III) complex of **34**, the calixarene backbone gives rise to an efficient *antenna effect* (Figure 16): a rather efficient energy transfer takes place between the excited states of the receptor and those of the Ln(III) ion, even though the phenols are poor chromophores ($\lambda_{max} = 270$ nm, $\varepsilon_{270} = 1500\,M^{-1}\,cm^{-1}$).

This, together with the efficient encapsulation of the metal ion even in water, where only one solvent molecule is still coordinated to the lanthanide ion in the $[\mathbf{34}\cdot Ln]^{3+}$ complex, originates a luminescent probe with a particularly high quantum yield ($\Phi = 0.20$).[3,22]

Different calix[4]arenes substituted with one to four 2,2′-bipyridine or phenanthroline nuclei were also synthesized. The Tb(III) complex of **44** ($\Phi = 0.12$) shows improved

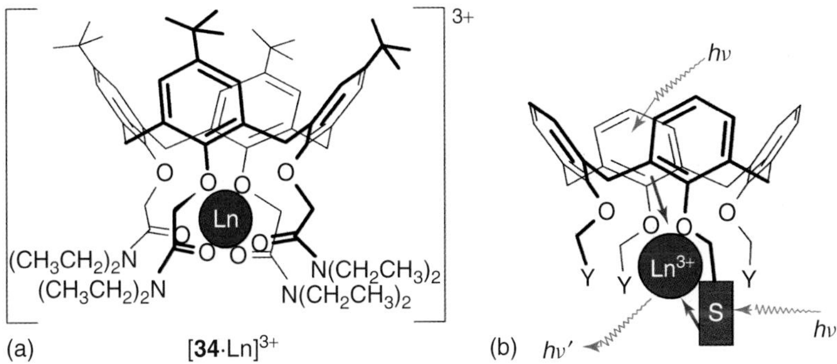

Figure 16 (a) Complex of receptor **34** with a lanthanide(III) ion and (b) *antenna effect* in calix[4]arene lanthanide complexes: S = sensitizer; Y = binding group.

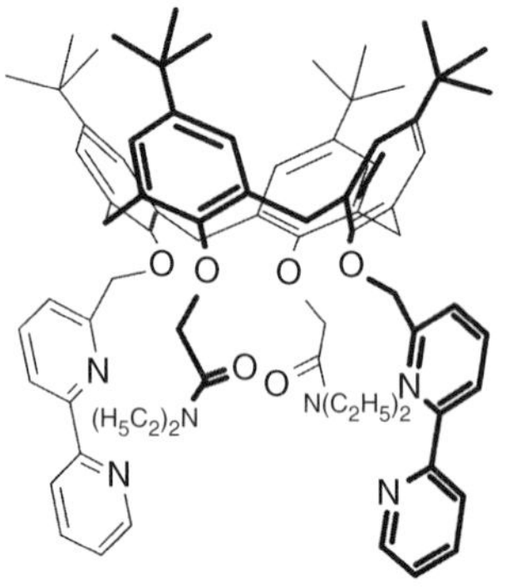

44

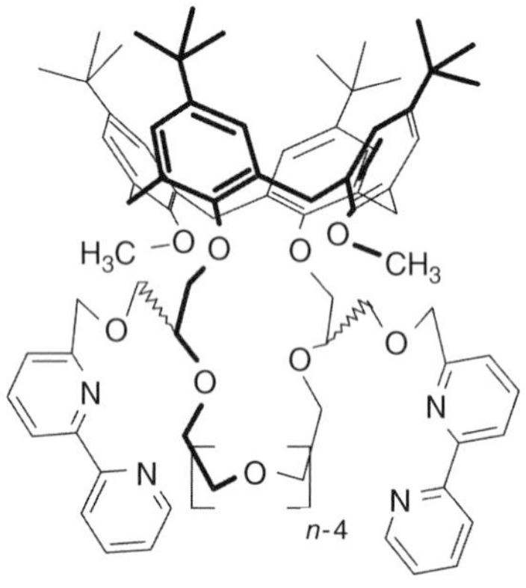

45

absorptivities ($\varepsilon_{306} = 29\,000\,\mathrm{M^{-1}cm^{-1}}$) and higher λ_{max} (306 nm) compared to tetramide **34**, owing to the presence of the heterocyclic aromatic sensitizers coordinated to the metal ion. Exceptional photophysical properties are shown by 2,2′-bipyridine lariat calixcrown-5 and -crown-4 (**45**: $n = 5$, 4) with quantum yields up to 0.39 for Tb(III) in CH_3CN. Quite interestingly, as the antenna effect mainly takes place through the bipyridine chromophores, the Eu(III) complexes are also highly luminescent (Φ up to 0.32).

The complexation of Gd(III) ion by synthetic receptors is attracting the interest of several research groups as it will potentially allow to prepare new contrast agents for magnetic resonance imaging (MRI) (see **Magnetic Resonance Imaging Contrast Agents**, Volume 5). A prerequisite for the use of Gd(III) complexes *in vivo* is an extremely high kinetic and thermodynamic stability with log β values in aqueous solution higher than 20. The bis-iminodiacetic acid **46** shows a relaxivity 2 times higher than that of [Gd-DOTA(H_2O)]$^-$; this is one of the best MRI contrast agents known so far, but with a stability constant log $\beta = 13$.[22] To ensure higher stability and higher relaxivity, especially at higher magnetic field strength (>1.5 T), where traditional contrast agents do not show good performances, two densely packed Gd-chelates (e.g., **47·Gd$_4$**) based on tetrapropoxycalix[4]arene were recently synthesized.[27]

Another important and rather urgent reason for complexation of f-element ions is the treatment of radioactive wastes originated from spent nuclear fuel. In this frame, it is of particular interest to selectively separate actinide from lanthanide ions in highly acidic media as this would allow the transmutation of long-lived minor actinides (Np, Am, and Cm) into short-lived radionuclides.[23] The introduction of multiple identical ligating functions onto molecular platforms gives polydentate receptors, which usually show extraction properties much higher than those of the corresponding simple monovalent compound. Following such design strategy, several calix[4]-, [6]-, and [8]arene receptors bearing CMPO (carbonylmethylphosphine oxides) functions at the *upper* (e.g., **48**) or at the *lower* (e.g., **49**) rim were synthesized, which show a large

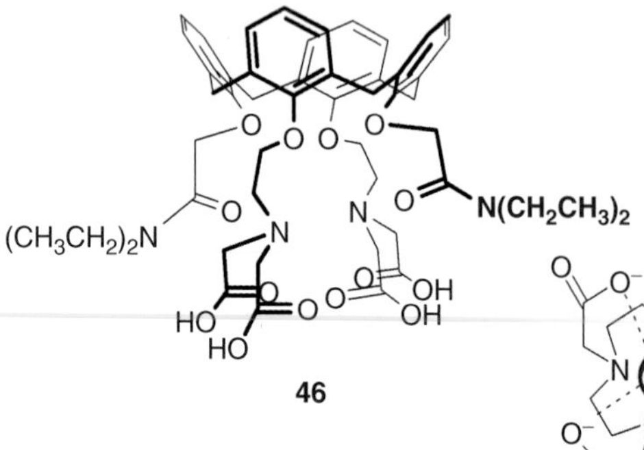

46

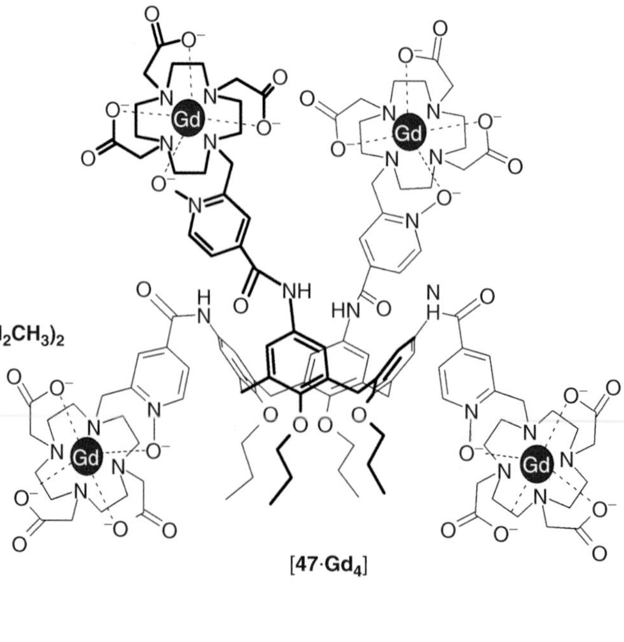

[47·Gd$_4$]

cooperative effect.[3,23,28] All these derivatives have high distribution coefficients (D_M is the ratio $[M]_{org}/[M]_{aq}$ at the equilibrium, with M = metal ion) of actinide and lanthanide ions from water to NPHE (*o*-nitrophenyl hexyl ether) even at high nitric acid concentration. Transport studies with compound **48** ($n = 4$) revealed high permeabilities of Pu and Am through SLMs having 1 M HNO_3/4 M $NaNO_3$ as a source phase.[22,23] Arduini *et al.* reported that the bis-crown-3 **50**, obtained by rigidification of the calixarene scaffold with two short bridges at the *lower rim*, shows a 10-fold increase in the extraction efficiency, as compared to **48** (see review[23] for specific reference).

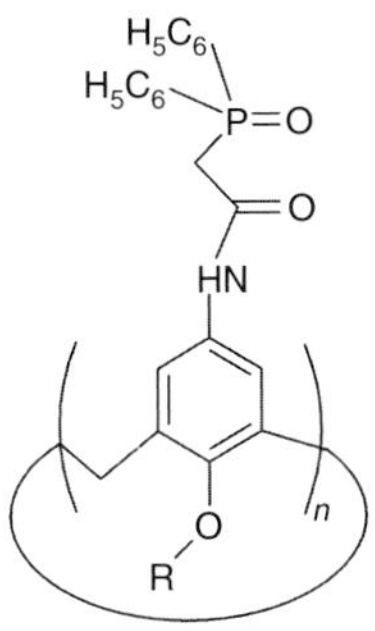

48: $n = 4, 6, 8$ 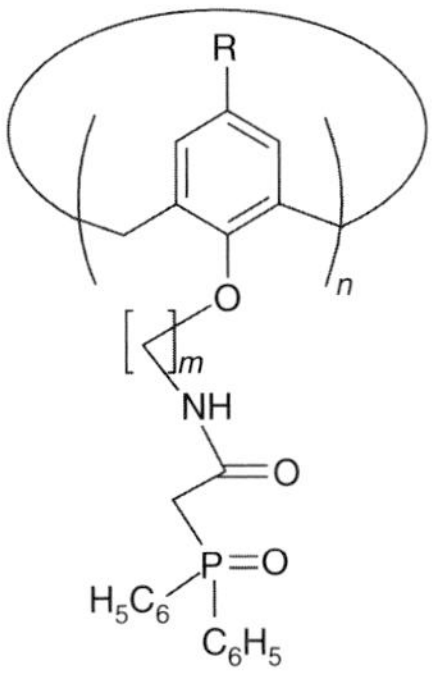 **49**: $m = 2, 3, 4$; $n = 4, 6, 8$

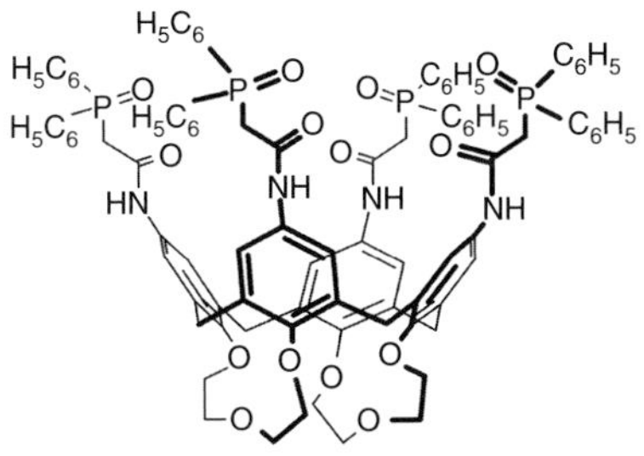

50

However, even if these phosphorylated receptors are very efficient in the extraction of trivalent f-element ions, they are not very selective. On the other hand, calixarenes bearing picolinamide binding groups at the *upper* or *lower* rim (e.g., **51**), synthesized by us, present an An/Ln selectivity ($SF_{Am/Eu}$ up to 13.8, where $SF_{Am/Eu} = D_{Am}/D_{Ln}$) suitable to be used in a reiterative extraction process. In the presence of a bromocosan synergizer, the efficiency of extraction is up to 3 orders of magnitude higher than that of a simple monovalent binding group. The introduction of electron-withdrawing substituents in the para positions of the pyridine rings (compound **52a**), especially the substitution of the pyridine with the less basic pyrazine nucleus (compound **52b**), consistently increases the efficiency of these receptors especially at high nitric acid concentration. This behavior suggests that, under strongly acidic conditions (highly competing $[H^+] \geq 0.01\,M^{-1}$), the protonation of the pyridine nuclei is a favored process compared to the metal ion coordination and that the latter process may only take place with less basic receptors such as **52**.

Calix[6]arene receptors **53a–d**, bearing additional carboxy groups in the 6-positions of the picolinamide nuclei, showed a remarkable efficiency of extraction of both Eu^{3+} and Am^{3+} ions even at very high nitric acid concentration. Indeed receptors **53c–d** are effective even at 3 M HNO_3 concentration and are more efficient than **51a**; for instance, at $[H^+] = 0.1$ M, D_M for **53c** are 6 order of magnitude higher than those for **51a**.[29]

2.2.4 *d-Block and heavy metal ions*

Complexation of d-block metal ions has been exploited in the supramolecular chemistry of calixarenes for several purposes including catalysis (see **Supramolecular Bioinorganic Chemistry**, Volume 4).[30] Owing to the vastness of

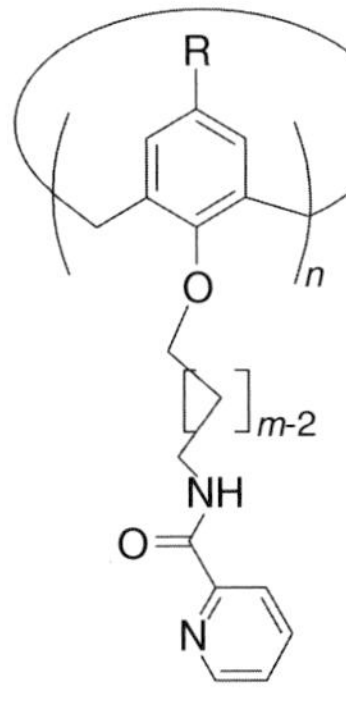

51a: $n = 6$, R = H, $m = 3$
51b: $n = 4$, R = H, $m = 3$
51c: $n = 4$, R = t-C_4H_9, $m = 4$

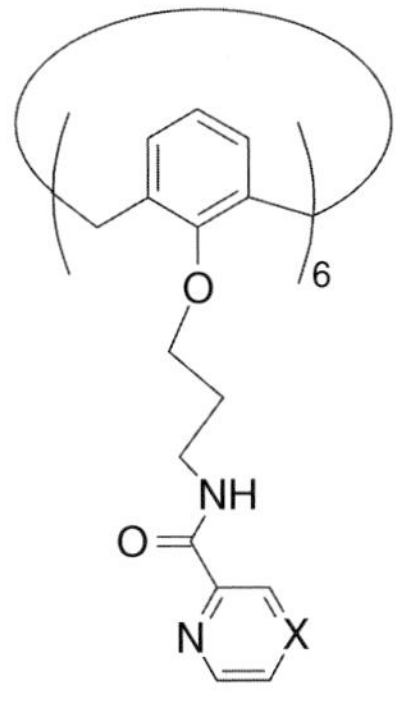

52a: X = C–Cl
52b: X = N

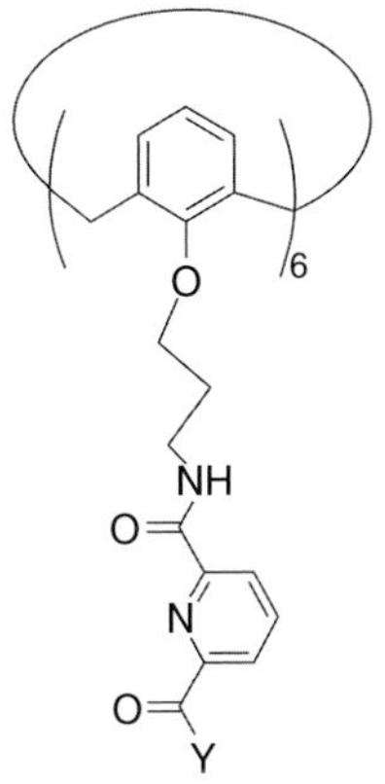

53a: Y = OCH_3
53b: Y = OBn
53c: Y = $N(CH_2CH_3)_2$
53d: Y = NH-n-C_4H_9

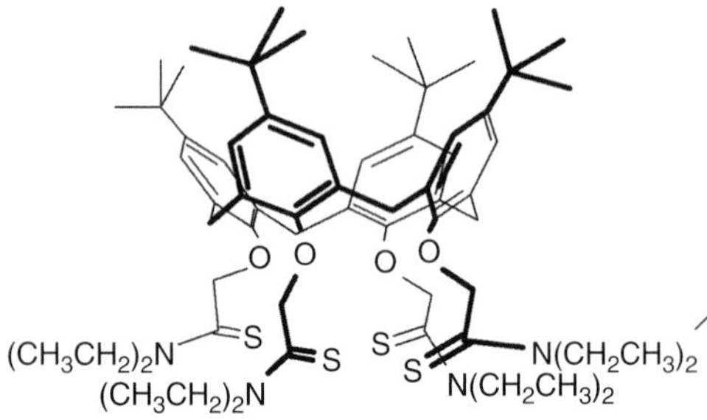

54 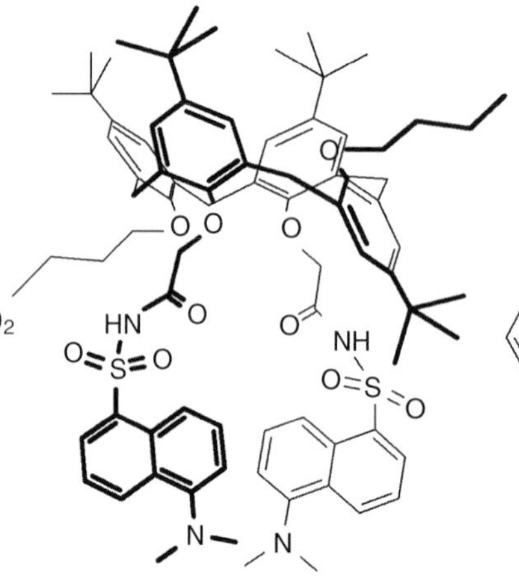 **55** 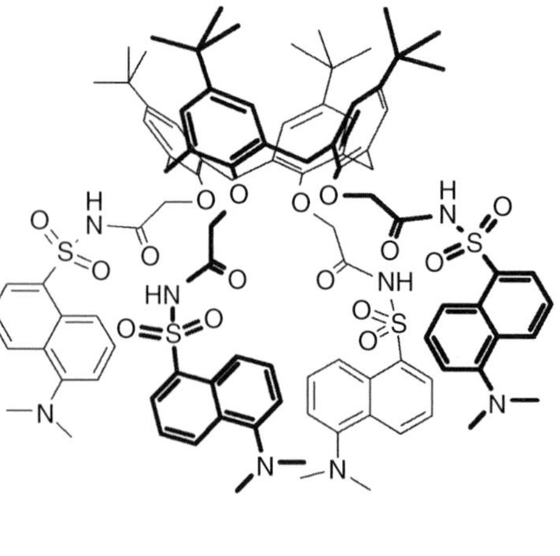 **56**

this field, here we briefly survey only the recognition properties of calixarene receptors for late transition and heavy metal ions designed for separation and detection. For this group of metal ions, in general, selectivity is primarily influenced by the number and type of donating atoms arranged around the cation and to a much lesser extent by the conformation of the calixarene. Moreover, most of the work has been carried out on calix[4]arenes with only very few examples concerning higher oligomers. The insertion of soft donor atoms into ligating groups appended to the calixarene scaffold shifts the selectivity of this class of macrocycles from spherical hard metal ions to soft transition and heavy metal ion recognition. A text-book case is reported by Reinhoudt *et al.* (see review[24] for specific reference), who showed that the substitution of the oxygen atoms of acetamide **34** to sulfur atoms to get thioacetamide **54** moves the selectivity measured in chemically modified field-effect transistor (CHEMFET) from alkali metal ions to Pb^{2+} with modest interference from Cu^{2+} ($\log K_{ij} = -3.4$), Ca^{2+}, and Cd^{2+} ($\log K_{ij}$ = about -4.2). Also, calix[4]arenes bearing $-CH_2CH_2SCH_3$ at the *lower rim* are remarkably selective for Ag(I), while those having thiocarbamoyl units ($CH_2CH_2SC(S)N(CH_2CH_3)_2$) are selective for Cu(II) over Pb(II) and Cd(II).

Since the first communication by Bartsch *et al.* that calix[4]arenes bearing proton-ionizable sulfonyl carboxamide ($-C(O)NHS(O)_2R$, $R = CF_3$, CH_3, $4\text{-}NO_2C_6H_4$) groups at the lower rim are quite effective and selective for lead extraction even in the presence of other mono- and divalent metal ions, a wide series of receptors bearing these chelating groups have been studied. When R is a dansyl residue, sensitive and selective fluorogenic turn-off sensors for lead could be prepared (see review[31] for specific references). Both the preorganized didansyl derivative **55** in the *partial cone* of Talanova *et al.* and the *cone* tetradansyl compound **56** of Leray *et al.* show lead detection limits at the parts-per-billion level, which are compatible with the World Health Organization on the limiting content of this hazardous pollutant in drinking water.[31] In order to attain silver complexes of high stabilities, the sole exploitation of cation–π interactions is usually not enough and (thio)ether groups or pyridine N atoms must be added to the calixarene.[32] The calix-crown-5 in the *1,3-alternate* structure (**18**) presents a high stability constant in methanol for silver ion ($\log K = 5.87$) owing to interactions with the crown-ether oxygen atoms and the inverted aromatic nuclei; however, potassium is bound better by more than 3 orders of magnitude. Calix-azacrowns proved to be selective for Hg^{2+}. The chromophoric 2,4-dinitrophenylazo derivative of calix-diazacrown-4 (O_2N_2) **57** undergoes deep color changes only in the presence of Hg^{2+} and to a minor extent of Cd^{2+}, while a similar compound (**58**) functionalized with pyrene units on the azacrown moiety turns off its fluorescence only upon addition of Hg^{2+} and not of Zn^{2+}, Cd^{2+}, Cu^{2+}, and Ni^{2+} (see review[24] for specific reference).

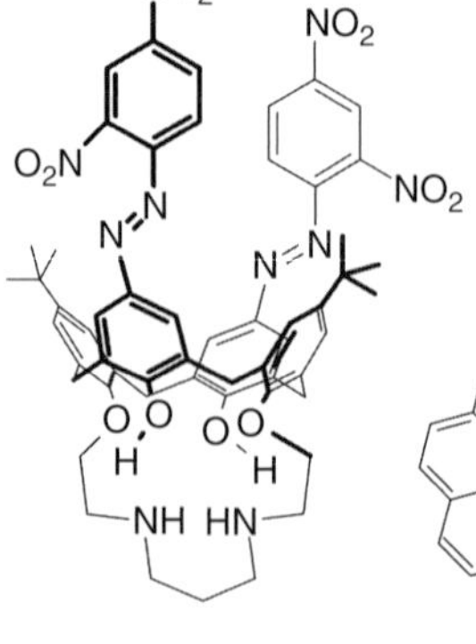

57 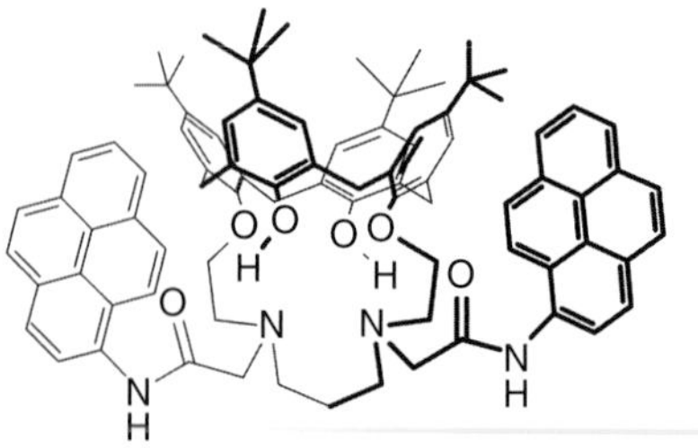 **58**

A calix-triazacrown-5 (O_2N_3) equipped with a dansylamide side arm (**59**) shows a pronounced quenching of the fluorescence only in presence of Hg^{2+}, and a binding

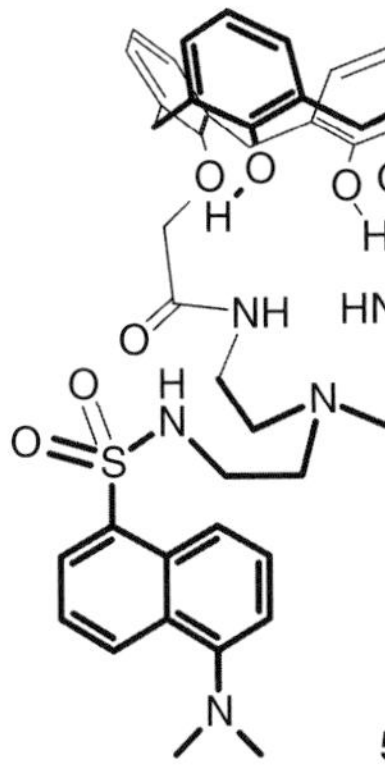

59 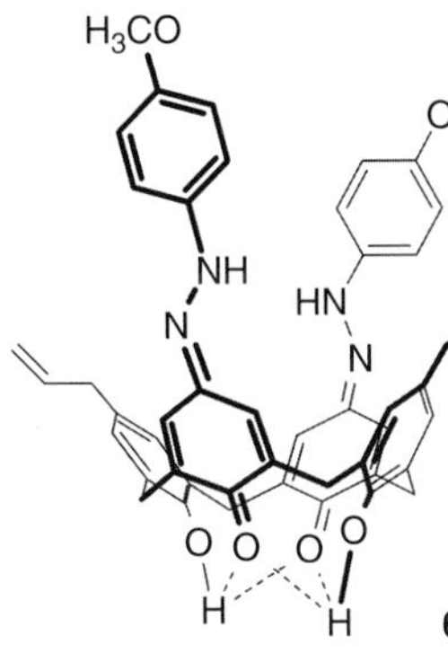 **60**

constant of roughly 3 orders of magnitude higher than that of Cu^{2+} was measured in $CH_3CN/H_2O = 4/1$ (see review[9] for specific reference). A series of calix[4]arenes functionalized at the *upper rim* with two or three 4-methoxyphenylazo units and two (**60**) or one allyl groups exhibit a substantial color change upon Hg^{2+} coordination by the azo and allyl moieties. The chromoionophore is insensitive to alkali and alkaline-earth metal, but undergoes a similar bathochromic change upon addition of other heavy metal salts.[33] The introduction of dansyl groups directly on the phenolic O atoms of calix[4]arenes was an interesting and easy way to obtain fluoroionophores for heavy metal ions with detection limits in the parts-per-billion region. Both the *p*-nitrocalix[4]arene bearing 4 dansyl groups and fixed in the *1,3-alternate* (**61**) structure and the *cone* bis-dansylated derivative (**62a**) are selective for Hg^{2+} but the former gives a 1 : 2 (M : L) complex,[34] while for the latter a 2 : 1 stoichiometry was determined.[35] On the other hand, bisdansyldiester of *p*-H and *p-tert*-butylcalix[4]arenes (e.g., **62b**) are selective for Cu^{2+} with no important interference from other metal ions[35]. This important inversion of selectivity originated by a change in the electron-donating power of the para-substituent; this was unexpected, and it is a clear indication of the importance of polar factors in determining selectivity among heavy metal ions.

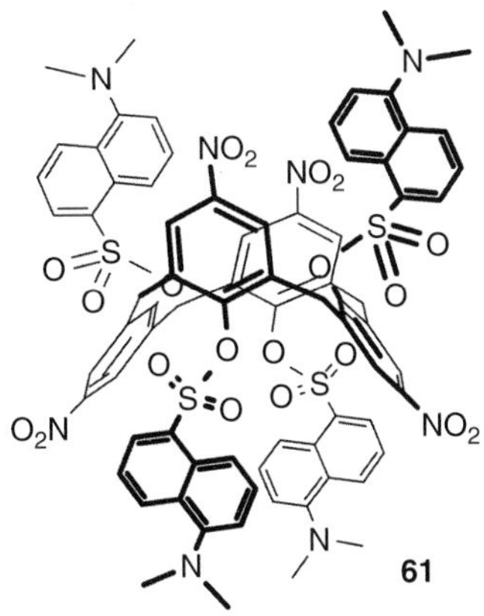

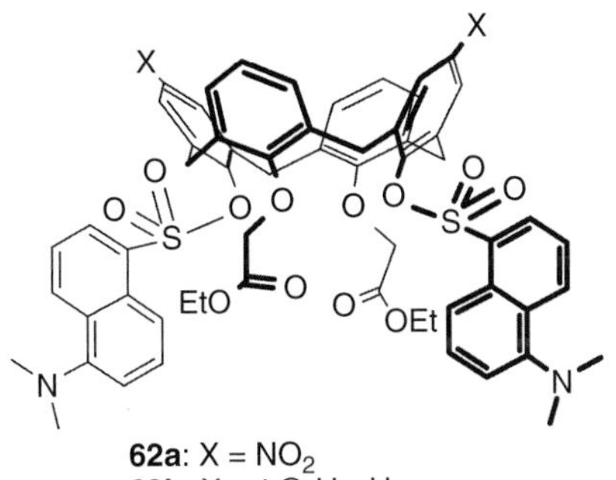

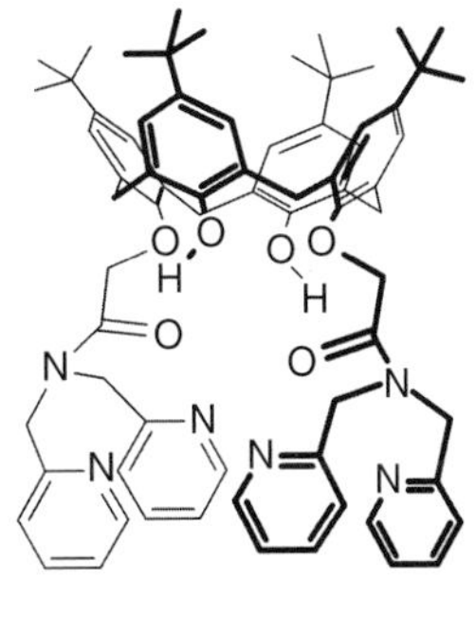

61

62a: X = NO_2
62b: X = t-C_4H_9, H

63 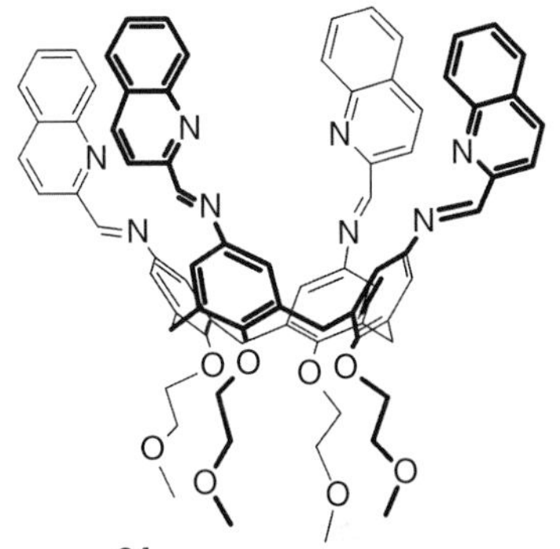 **64**

Among the copper-selective receptors, the derivatives of picolylamine (**63**) and iminoquinoline (**64**) should be mentioned because of their ability to switch on the fluorescence upon metal ion binding.[36] The tetraiminoquinoline derivative **64** is a highly efficient sensor for Cu^{2+}, exhibiting a remarkable fluorescence enhancement (1200-fold) and selectivity over a wide range of metal ions in acetonitrile.[37]

The derivatives of 1,3,5-trimethoxycalix[6]arene functionalized with three arms carrying nitrogen binding sites are quite interesting. The ditopic receptor **65** has a preference for hosting Cu(I) in the triazine site and Zn(II) in the imidazole site. Upon an electrochemical stimulus, a triggered translocation takes place (Figure 17). The oxidized Cu(II) moves to the imidazole site, while zinc goes to the triazole site and again the original situation can be restored upon reduction of Cu(II) to Cu(I).[38] The mechanism of this translocation has not yet been clarified, although it is supposed to take place intermolecularly.

2.3 Anion recognition

Calixarenes, resorcinarenes, and their simple ether derivatives have an electron-rich, π-basic cavity, which is able to host cations and neutral molecules but not anions, which, therefore, do not form *endo*-cavity inclusion complexes with these receptors. However, direct π-metallation of the exterior of the macrocycles provides a vacant cavity with altered electrostatic properties.[26] Indeed, the ability of certain transition metal ions

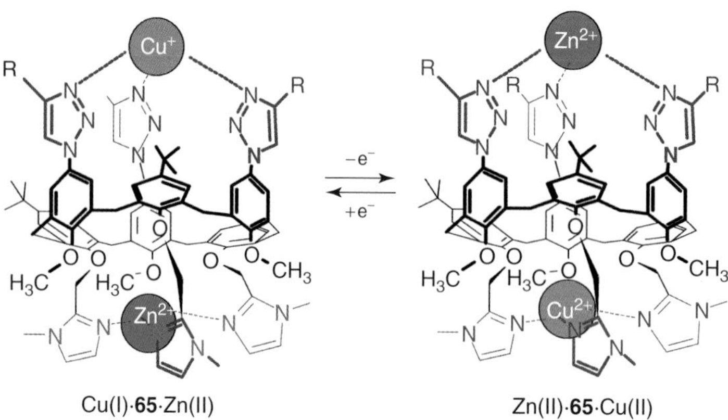

Figure 17 Electrochemically triggered translocation of zinc and copper in receptor **65**.

such as Ru(II), Ir(II), and Rh(II) to form π-metallated complexes with aromatic compounds was exploited to synthesize a new class of organometallic anion-binding calixarene hosts. The most interesting member of this class is the tetrametallic host **66**, which shows selectivity for the Cl^- anion, included in the calix[4]arene cavity ($K = 551\,M^{-1}$). On the other hand, the OH groups at the *lower rim* of calixarenes are poor H-bonding donors and, in addition, they are involved in strong intramolecular hydrogen bonds between themselves,[1] and both reasons explain why no example of H-bonded complexes of native calixarenes with anions has been reported, in spite of the fact that catechol and other simple phenolic compounds form complexes with chloride anion in organic media.[39]

In order to transform simple calixarenes into anion receptors, suitable anion-binding groups such as hydrogen-bonding donor moieties, positively charged groups, or metal ions must be introduced onto the calixarene scaffold.[3,22,39,40] As for cations, if a reporter group (redox active, fluorophoric, chromophoric, etc.) of the molecular recognition event is also incorporated, then we have a useful anion-sensing device.[3,9,22]

66

67a: R = $CH_2NHC(X)NHC_6H_5$, X = O
67b: R = $CH_2NHC(X)NHC_6H_5$, X = S
68a: R = H, X = O
68b: R = H, X = S

67a·CH_3COO^-

2.3.1 *Neutral, hydrogen-bonding receptors*

Hydrogen-bonding donor groups have been linked both at the *lower* and at the *upper rim* of calixarenes.[39] Among them, the most efficient are the (thio)urea groups and the activated amides, as they have NH protons of relatively high acidity and hydrogen-bonding ability. Many *lower rim* (thio)ureidocalixarenes have been reported in literature, which form 1 : 1 complexes with anionic species showing a general preference for inorganic anions, and some of them have also been used in ISEs and other sensor devices. However, the *upper rim* derivatives have, as additional attractive feature, the possibility for the apolar cavity of the calixarenes to cooperate in the anion binding, especially for calix[4]arenes and organic anions. A large variety of *upper rim* tetrafunctionalized calix[4]arene receptors in the *cone* conformation have been reported in literature, and, in general, they show selectivity for inorganic anions of different geometries but the participation of the cavity in the binding process has not been clearly evidenced. More interesting are the mono- and difunctionalized calix[4]arenes.[22,39,40] We introduced two (**67a** and **b**) or one (**68a** and **b**) (thio)urea units at the *upper rim* of the tetra-*n*-propoxycalix[4]arene blocked in the *cone* conformation. In DMSO-d_6, a highly competing solvent, an interesting selectivity of the bisurea **67a** was found for carboxylates, especially for acetate, over inorganic anions. The strong acetate binding ($K = 2200\,M^{-1}$) was explained assuming that each of the carboxylate oxygen atoms forms bifurcated hydrogen bonds with the urea NH groups, while the methyl group of acetate penetrates inside the calixarene cavity giving rise to CH_3/π interaction (**67a**·CH_3COO^-). Stibor *et al.* have recently confirmed, through a systematic study, the importance of the calix[4]arene conformation in anion binding by *upper rim* ureidocalix[4]arenes.[41] Calix[4]arenes (**69a–c**) functionalized at the *upper rim* with mono-, di-, and trichloroacetamide units show interesting recognition properties toward carboxylate anions. The dichloro compound **69b** shows the highest efficiency and a remarkable selectivity for the benzoate anion. The high efficiency of the dichloroacetamide binding group was explained to be due to a compromise between the hydrogen-bonding ability of the NH group and the steric hindrance to complexation, both increasing with halogen substitution.

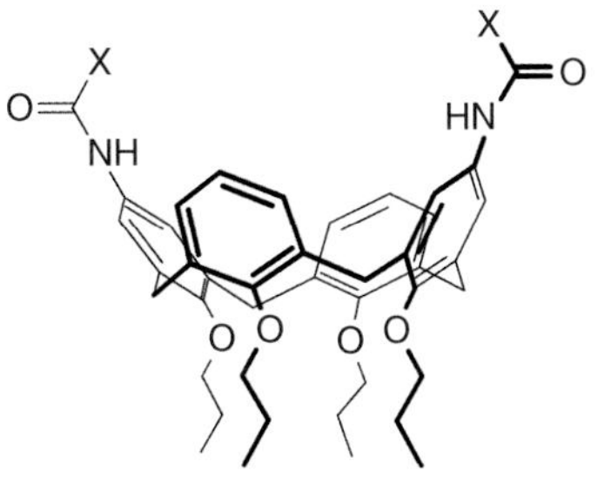

69a: X = CH_2Cl
69b: X = $CHCl_2$
69c: X = CCl_3

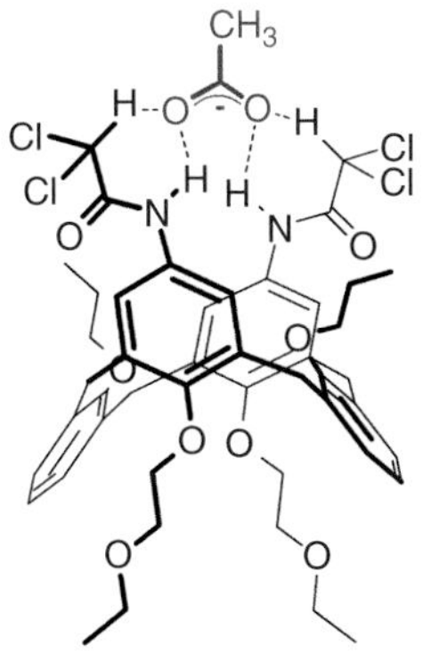

70·CH_3COO^-

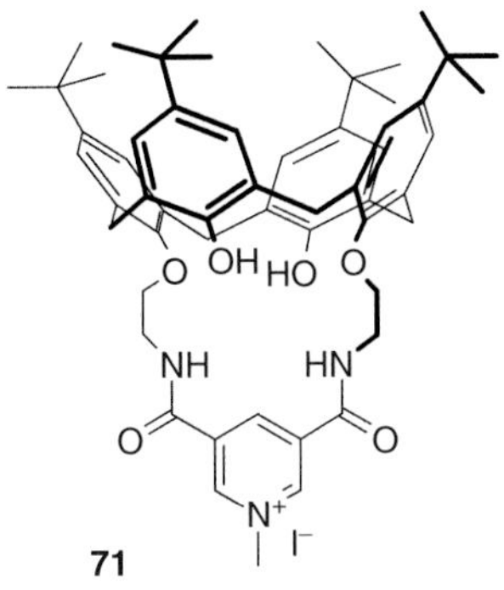

71

However, in studying the *1,3-alternate* calix[4]arene dichloroacetamide receptor **70**, which complexes the acetate anion in $CDCl_3$ ($K = 400\,M^{-1}$), spectroscopic evidence was obtained that the high anion-binding ability of the dichloroacetamide groups can be partly due to the presence of the acidic Cl_2CH–hydrogen atom, which can interact, together with the NH, with Y-shaped carboxylate anions in a bidentate fashion.[22] Several *upper rim* amidocalix[4]arenes have been incorporated in anion-sensing devices[9,39] or used as synthetic transporters for anions in liquid membranes[21,39]

2.3.2 *Charged anion receptors*

In addition to their general function of making the macrocycles water soluble (Section 2.1.3), positively charged groups have been introduced onto a calixarene scaffold to enhance the hydrogen-bonding ability of neutral binding groups or as direct interaction sites for anionic guests. An example of the first class of receptors is the macrobicyclic compound **71**, which forms 1 : 1 complexes in DMSO-d_6 with Cl^- ($K = 1015\,M^{-1}$) and Br^- ($K = 705\,M^{-1}$).[3]

The ammonium derivative **72** (Figure 18) is an example of the second type of receptors. It is able to interact electrostatically with phosphate anions of oligonucleotides in water (0.1 mM Hepes buffer) and the binding efficiency depends on the matching between the distance of the two positively charged macrocycles and the P–P distance of the terminal phosphate groups of the guest.[42] Among the examined guests, the highest binding to **72** ($K_d = 33\,\mu M$) was shown by the 5′-fluorescein-labeled adenine trinucleotide (FlAAA), whereas the cofactor flavine-adenine dinucleotide (FAD) is bound 1 order of magnitude less efficiently ($K_d = 333\,\mu M$). Thus, receptor **72** acts as a molecular ruler, which selectively binds well-separated diphosphates and oligonucleotides in buffered aqueous solution.

Amidinium and guanidinium cations are also very good binding groups for anionic species and are extensively used in supramolecular chemistry (see **Guanidinium-Based Receptors for Oxoanions**, Volume 3). However, they have been linked to calixarene scaffolds mainly to perform more complex supramolecular functions based on anion recognition (self-assembly, cell transfection, protein inhibition, see below), although examples of simple anion receptors have been reported.[39,40,43]

2.3.3 *Metal-containing anion receptors*

Metal ions, in the form of organometallic and coordination complexes, have been introduced in calixarenes mainly for the purpose of reporting the anion-binding event and transforming the anion receptors into anion sensors, although other functions of the metal ions can be envisaged.[44] A series of redox-active anion sensors (**73–75**) containing cobaltocenium units onto calix[4]arenes in the *cone* conformation have been reported. In general, the cobaltocenium derivatives show higher binding constants toward anions compared to neutral ferrocene analogs because the positive charge in the former receptors strengthens the anion-binding ability of the amide groups.

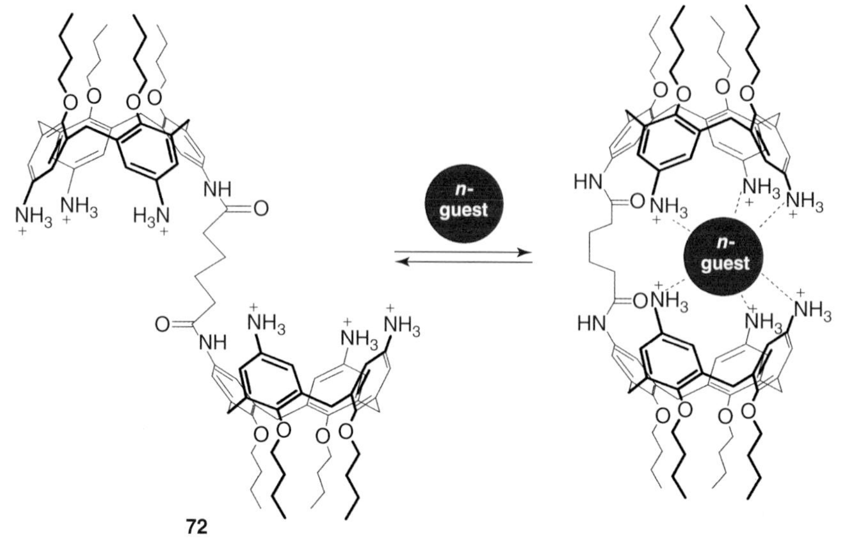

Figure 18 Anionic guest binding by host **72**.

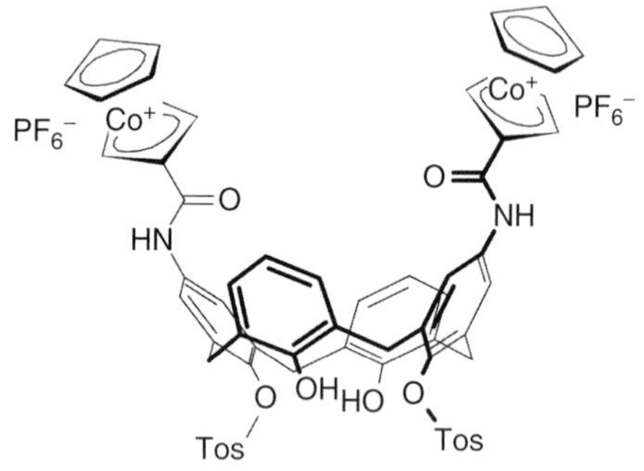

The anion-recognition properties of receptors **73–75** were found to be dependent on the structure of the *upper rim* of the calix[4]arene.[44] For example, in DMSO solution, the bis-cobaltocenium receptor **73** shows a greater affinity for acetate ($K = 21\,000\,\mathrm{M}^{-1}$) and other carboxylate anions over dihydrogen phosphate ($K = 3100\,\mathrm{M}^{-1}$), whereas for its isomer **74** the trend is reversed and the binding constants are lower for all anions. Interestingly, receptor **75**, in which the calix[4]arene is bridged by a single cobaltocenium moiety, displays significantly greater affinity for the above-mentioned anions ($K = 41\,520\,\mathrm{M}^{-1}$ for acetate) despite possessing only one positive charge. This selectivity preference was attributed to the *upper rim* bidentate amide hydrogen bond donor cavity of **75** being of complementary topology to bidentate anions such as carboxylates. The cobaltocene/cobaltocenium redox couple of **75** was found to undergo an anodic shift of 155 mV in the presence of acetate anion, which indicates that these compounds are very promising as amperometric sensors for anions. Other examples of metal-containing anion receptors can be found in Ref. 44.

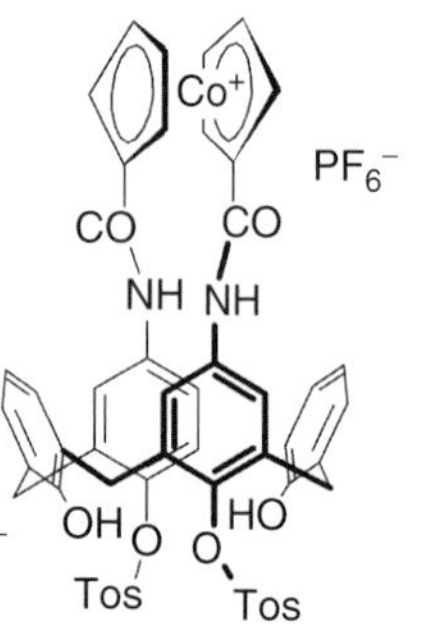

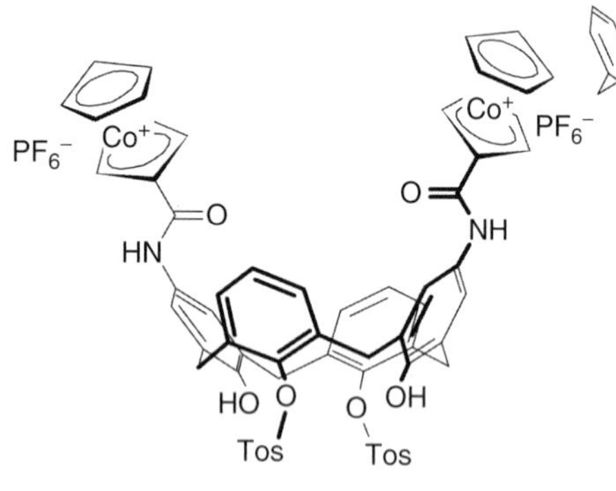

2.4 Heteroditopic receptors for simultaneous cation–anion complexation

In the previous sections, we have discussed the strategies for obtaining efficient and selective receptors for cations or anions of a salt, which were studied as single ions eliminating the effect of the counterion by using a noncoordinating anion (perchlorate, hexafluorophosphate, tetraphenylborate, etc.) or cation (usually tetra-alkylammonium), respectively. However, the simultaneous complexation of the cation and the anion of an ion pair by bifunctional receptors is quite attractive for several reasons. It can positively affect the extraction of salts in organic media or their transport through membranes (see **Membrane Transport**, Volume 4), influence the selectivity in complexation, or disclose interesting allosteric effects. For these reasons, it has become a distinct field of investigation in supramolecular chemistry (see also **Ion-Pair Receptors**, Volume 3).[45,46] The simultaneous complexation of cations and anions by an heteroditopic receptor may lead to three

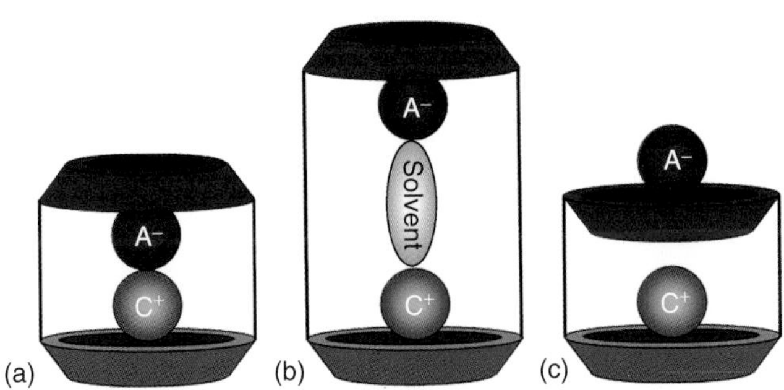

Figure 19 Three limiting situations for the complexation of ion pairs by a ditopic receptor: (a) contact, (b) solvent bridged, and (c) host separated.

limiting ion pair interactions[47] (Figure 19), each having its own advantages and disadvantages in relation with the particular objective which is pursued. As calixarenes are endowed with two rims, which could, in principle, be functionalized with two different types of binding units, one for cations and one for anions, they soon appeared to be quite attractive for the design and synthesis of heteroditopic receptors.[3,22,39]

Early contributions in this area came from Reinhoudt's group and refer to cases in which the ion-pair-binding mechanism is of the host-separated type (Figure 19c) (for original references see reviews[3,22]). They first exploited *cone* calix[4]arene tetraesters or *1,3-alternate* calix[4] crowns as binding units for alkali metal ions and uranyl-salophen complexes as anion-binding sites.

For instance, receptor **76b** shows selectivity toward NaH_2PO_4, as a consequence of the sodium-selective complexation at the *lower rim* by the *cone* calix[4]arene tetraester and the selective inclusion of $H_2PO_4^-$ at the *upper rim*, as previously found for the analogous receptor **76a**.[22] Compound **77**, acting as heteroditopic carrier, is able to transport the hydrophilic CsCl better than the more lipophilic $CsNO_3$ across a SLM, owing to the higher affinity of the anion-binding sites toward Cl^- rather than to NO_3^-. Interestingly, a very low flux is observed in the same conditions, using either a monotopic uranyl-salophen amide as anion carrier or a monofunctional *1,3-alternate* calix[4]crown-6 as a cesium-selective ionophore.

76a: R = n-C_3H_7
76b: R = $CH_2COOC_2H_5$

77

Addition of Cu(II) to the fluorescent heteroditopic receptor **78** in CH_3CN (Figure 20) causes a reduction of the metal to Cu(I) by the phenolic groups of the calix[4]arene and a strong enhancement in fluorescence, which was attributed to a rigidification of the receptor in the cationic complex to which acetate ($K = 159\,000\,M^{-1}$) and fluoride ($K = 59\,900\,M^{-1}$) anions can be further bound through H-bonding and electrostatic interactions, inducing a further increase in fluorescence. No anion binding occurs in the absence of copper cation. Receptor **78** probably

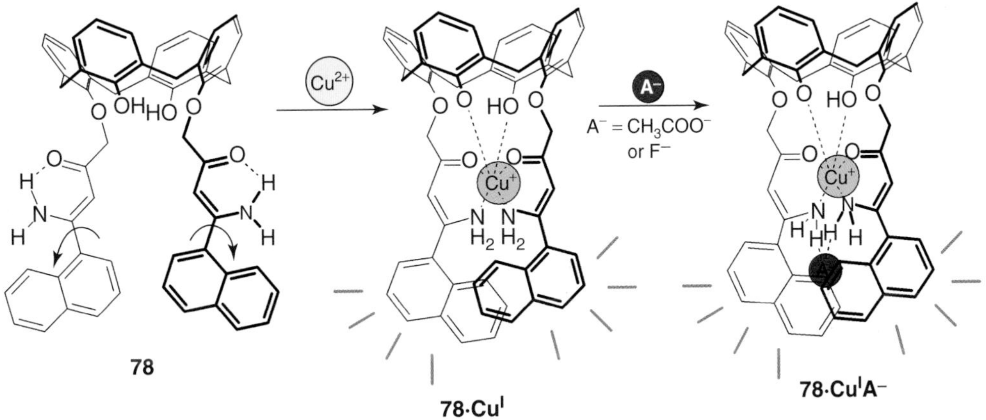

Figure 20 Complexation of copper salts by the heteroditopic receptor **78**.

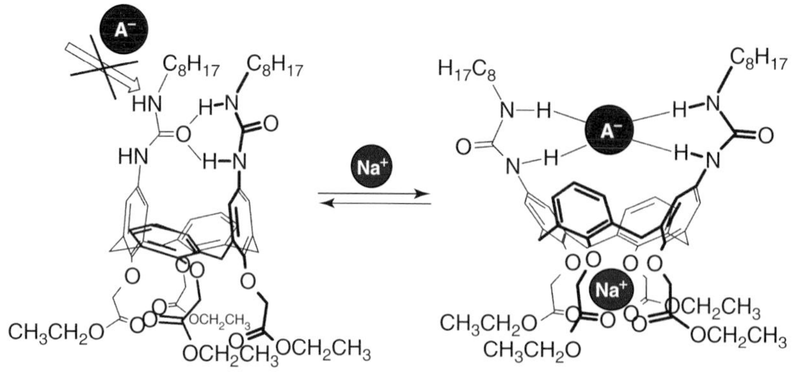

Figure 21 *Switching-on* anion binding by sodium complexation at the *lower rim* of heteroditopic receptor **79**.

represents the only reported example of a calixarene-based heteroditopic receptor in which Cu(I) is directly involved in ion pair complexation.[48] The *switching-on* anion binding by the cocomplexation of alkali metal cations is even more dramatically shown by receptor **79**, reported by Reinhoudt *et al.* and highlighted in several review articles,[3,22,40] which exploits urea groups at the calix[4]arene *upper rim* as anion-binding sites (Figure 21). The free receptor **79** adopts a *flattened cone* conformation in $CDCl_3$ due to the formation of strong intramolecular hydrogen bonding between the urea groups. In this situation, the *upper rim* cavity is closed, and no anion binding takes place when titration experiments are carried out using tetrabutylammonium salts. However, when sodium or potassium halides are used, the complexation is *switched on* and these salts are easily solubilized in $CDCl_3$. Evidently, the complexation of alkali metal ions to the ester groups at the *lower rim* causes the breaking of the intramolecular hydrogen bonds between the urea units and an opening of the calixarene cavity, which can now accommodate the anionic guests by hydrogen bonding with the NH groups.

A complementary *switching-on* cation binding promoted by addition of an anion was recently reported by Beer *et al.*, who synthesized several heteroditopic calixdiquinone receptors (e.g., **80a** and **80b**). These compounds show no affinity for separate NH_4^+ and Na^+ cations (paired with noncoordinating anions),[49] probably because the cation-binding sites (mainly the quinone groups) are intramolecularly hydrogen bonded to the amide N–H groups of the anionic binding sites. Addition of tetrabutyl ammonium (TBA) chloride induces a strong complexation of the NH_4Cl and NaCl ion pair, in this case through a contact-binding mechanism (Figure 19a), which was also supported by X-ray crystallography.

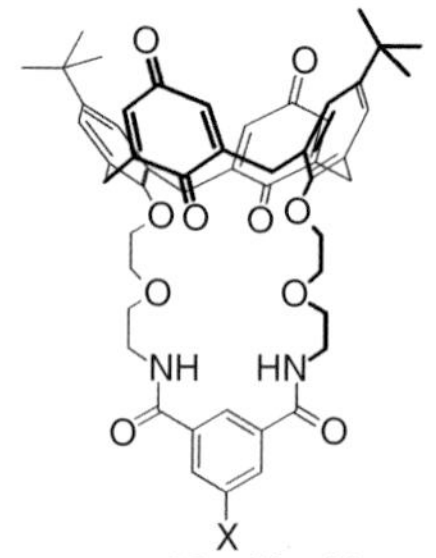

A contact ion pair mechanism was also observed by Jabin *et al.* in the complexation of alkylammonium chlorides by the heteroditopic receptor **81**, in which the calix[6]arene

80a: X = H
80b: X = NO_2

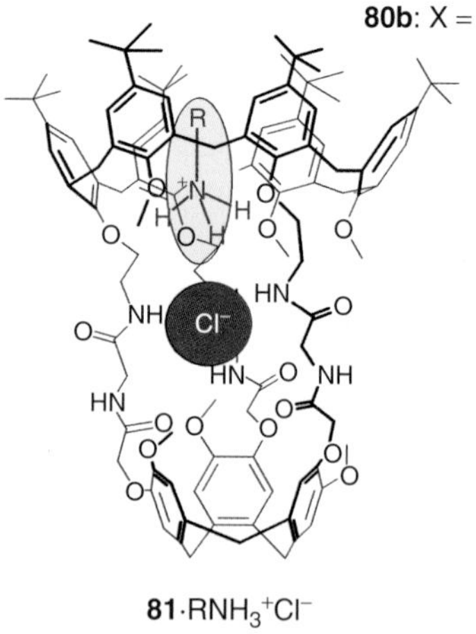

81·$RNH_3^+Cl^-$

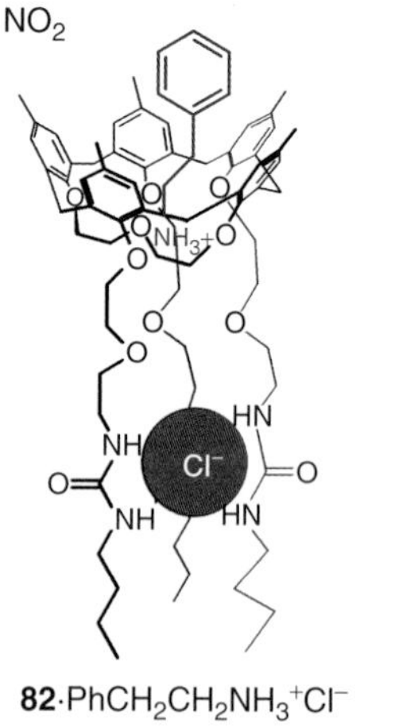

82·$PhCH_2CH_2NH_3^+Cl^-$

cavity acts as the alkylammonium binding site and a cryptamide unit binds the chloride anion. In addition, compound **81** is also able to form hydrogen-bonded inclusion compounds with polar neutral molecules.[7] On the contrary, the tris-ureido-calix[5]arene-crown-3 (**82**) efficiently binds 2-phenylethylamonium (PEA) chloride as a spatially separated ion pair, with the Cl^- hydrogen bonded to the urea groups and the PEA cation hosted in the calix[5]arene cavity.[50]

We reported that complexation of sodium ion by a crown[4]- moiety at the *lower rim* of an heteroditopic receptor **83** (Figure 22), bearing two 2,2,2-trifluoroethanol anion-binding groups at the *upper rim*, causes an increase

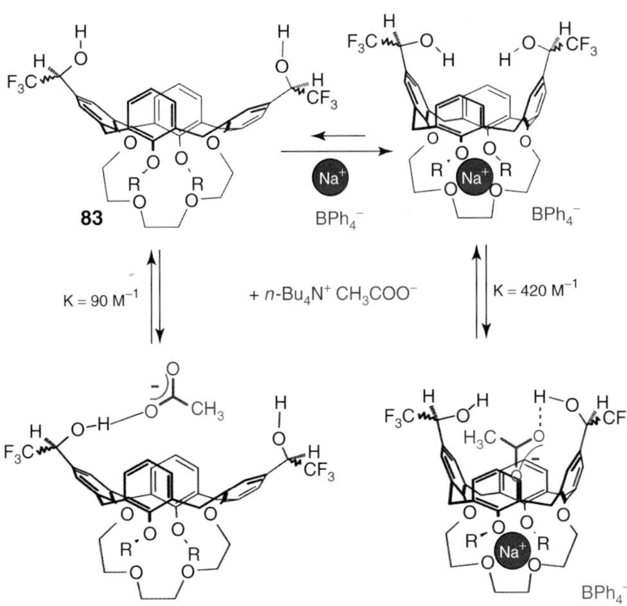

Figure 22 Allosteric effect in the binding of CH_3COO^- Na^+ by heteroditopic receptor **83**.

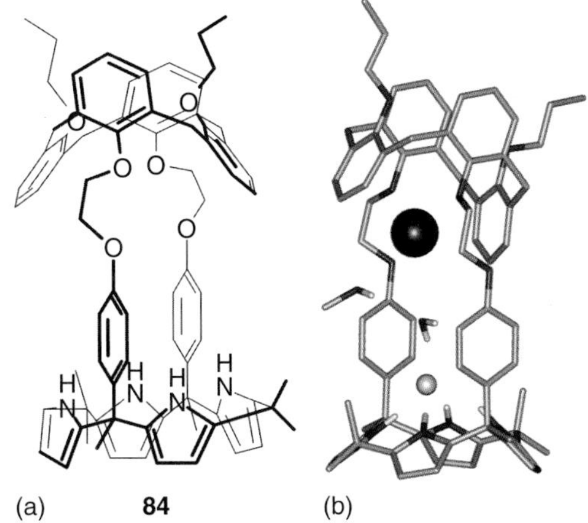

Figure 23 (a) The calix[4]arene strapped calix[4]pyrrole **84** and, (b) the X-ray crystal structure of its CsF complex.

(by a factor of 5) in acetate binding with respect to the Na^+-free receptor (positive allosteric effect).[39]

We close this section with a very recent example by Sessler *et al.* of an interesting *adjustable* heteroditopic receptor **84**, with which it was possible to stabilize all three limiting interaction modes shown in Figure 19 using a single molecular framework.[47] The calix[4]arene-strapped calix[4]pyrrole **84** possesses a strong anion-binding site (the calix[4]pyrrole) and a weak cation-binding site (the calix[4]arene in *1,3-alternate* conformation and the ethereal oxygen atoms). When caesium cation or fluoride anion, each coupled with a noncoordinating counterion, are separately added to receptor **84** dissolved in 10% CD_3OD in $CDCl_3$ (v/v), no complexation occurs due to strong competition of the polar solvent. On the contrary, receptor **84** forms a strong 1 : 1 complex with the CsF ion pair ($K = 1.3 \times 10^4\,M^{-1}$ by ITC) in the same conditions. The X-ray crystal structure (Figure 23b) of the **84**·CsF complex[47] shows the distance between the cation and the anion in the ion pair to be 5.62 Å, and a water molecule bridges the two constituent ions (interaction mode in Figure 19b). Compound **84** binds $CsNO_3$ in the form of contact ion pair and forms an unusual 2 : 2 complex with CsCl in which two different binding interactions are simultaneously present: a solvent loosened contact ion pair (with the solvent outside the binding cavity) and a host-separated ion pair. In this way, **84** reveals its ability to be an *adjustable* ion pair receptor sensitive, for a given cation, to the size and nature of the anion.

2.5 Chiral calixarenes

Chirality in calixarenes and resorcinarenes can be essentially of two different types, depending on the method used for generating it. The first type, called *chirality by attachment*,[51] consists of the introduction of a chiral, nonracemic, moiety onto the calixarene or resorcinarene framework. The second type, often referred to as *inherent chirality*,[51,52] has been particularly studied for calix[4]arenes, and it is usually observed when three (AABC or ABAC) or four (ABCD) different phenolic units are present in the macrocycles (Figure 24). As this type of chirality is linked to the nonplanar shape of calix[4]arenes, the well-known ring inversion of these compounds must be avoided through functionalization at the *lower rim* to obtain stable enantiomers.

The search for inherently chiral calixarenes has been steadily pursued since the very beginning of calixarene

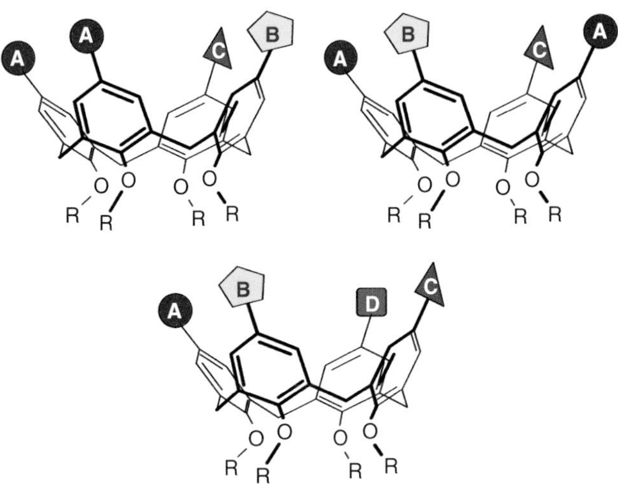

Figure 24 The three possible structures for inherently chiral *cone* calix[4]arenes.

chemistry,[1,2,51,52] but it has more or less remained a scientific curiosity of theoretical interest. Usually, researchers in this field were satisfied after showing, by spectroscopic or chromatographic methods, that two enantiomers were present in the reaction mixture and, only in a few cases, resolution of the racemic mixture was attempted. Moreover, the long synthetic procedures adopted and the low yields of the obtained products have hampered the use of these receptors for chiral recognition, sensing, and catalysis. The situation may change in the near future as better synthetic and resolution methods have been developed.[51,53] On the contrary, the synthesis of chiral calixarenes by attachment of chiral, nonracemic, moieties at the *lower* or at the *upper rim*, has a much wider scope. Depending on the particular enantiomeric guest to be recognized, the compounds available from the chiral pool or other sources have been attached to calixarenes and resorcinarenes to obtain receptors that are able to perform enantioselecive recognition and sensing. In general, it has been found that in calix[4]arenes the tetrafunctionalized receptors give poor enantioselective discrimination, perhaps because of their very high C_4 symmetry, while better results are obtained with C_2- and C_1-dissimmetric derivatives.[39,53]

2.5.1 *Enantioselective recognition of cations*

Very few examples of enantioselective, calixarene-based receptors for cations have been reported so far.[53,54] All of them target the ammonium head groups of chiral amines or of amino acid alkyl esters (Figure 25).

Among them, the chiral Schiff bases (**85**), rigidified at the *lower rim* by a tartaric acid bridge, show a weak preference in general ($K_D/K_L = 1.06-1.95$, when X = Phe, Cy) for D-amino acids methyl ester hydrochloride (D-AA-OCH_3·HCl) over the L-isomers, but the receptor derived from (R)-2-heptylamine (**85**, X = pentyl) has a remarkable preference for D-Phe-OCH_3·HCl over the L-Phe-OCH_3·HCl. Chiral calix[4]arene-aza-crowns-5 (**86**) preferentially bind L-amino acid methyl ester hydrochlorides over the D-enantiomers, but with a rather modest enantioselectivity ($K_L/K_D < 2.1$).[53] A smart, naked-eye sensor of chirality, based on a chiral calix[4]arene-binaphthol (**87**) bearing two indophenol chromophores, was developed by Kubo *et al.*[54] Upon proton transfer and ammonium ion complexation, the calix-crown can selectively recognize (R)-phenylglycinol, resulting in an immediate color change from red to blue–violet. The addition of up to 1000 equivalents of the (S)-enantiomer results in no significant color change. A three-point interaction mode involving the ammonium ion with the crown-ether bridge, the OH group of the guest with the phenoxide anion of the host, and the phenyl with the binaphthyl nuclei is proposed to take place.

2.5.2 *Enantioselective recognition of anions*

An interesting class of chiral hydrogen-bonding receptors for anions is represented by the peptido- (see **Synthetic Peptide-Based Receptors**, Volume 3) and glycocalixarenes obtained by the attachment of amino acids, peptides, and carbohydrates to the calixarene framework.[6,15,55] Our group has synthesized bridged N-linked peptidocalix[4]arenes and their open-chain analogs as well, with the aim of obtaining biologically active molecules based on peptide recognition (see **Supramolecular Chemistry in Medicine**, Volume 5). Indeed, some members of the receptor **88** family behave as Vancomycin mimics and bind the carboxylate terminus of dipeptide models of the cell wall mucopeptide precursors of Gram-positive bacteria.

We also synthesized bridged C-linked peptidocalix[4] arenes, containing an aromatic ring in the pseudopeptide loop. Solution studies performed in acetone-d_6 showed that

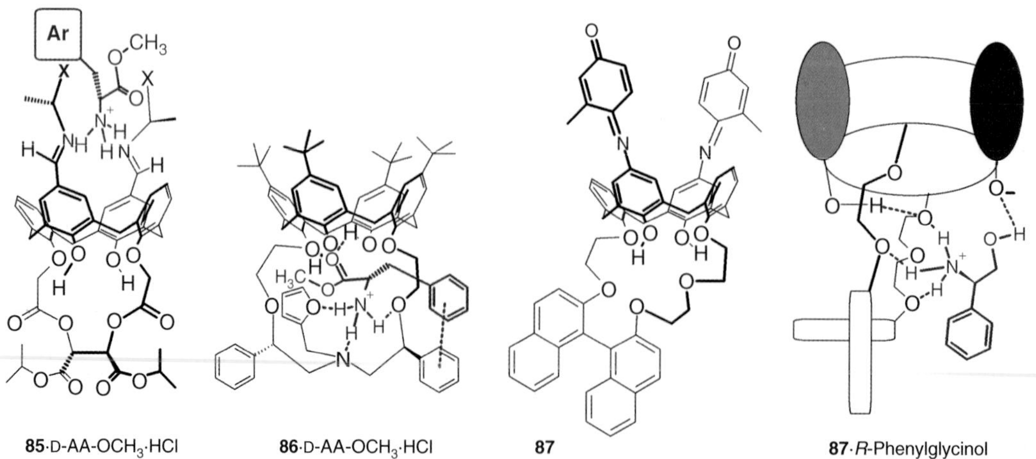

Figure 25 Schematic representations of the proposed structures of the most stable diasteromeric complexes of chiral receptors **85–87**.

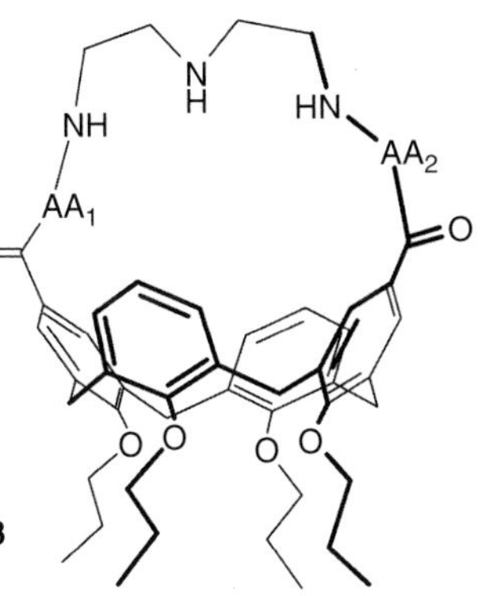

88

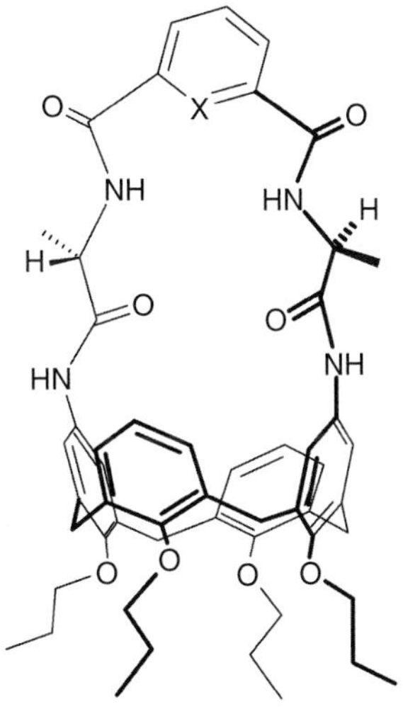

89a: X = CH
89b: X = N

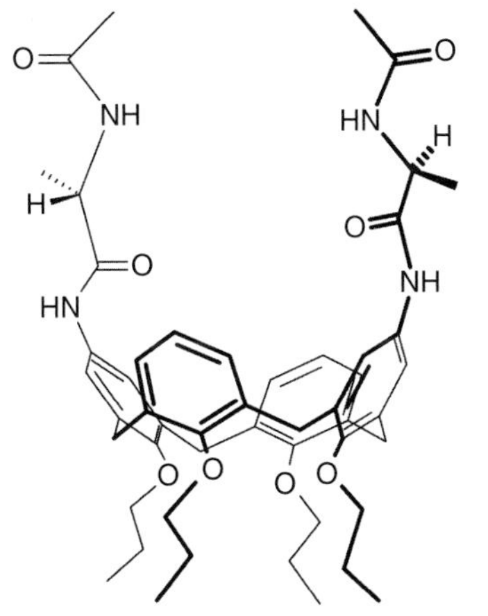

90

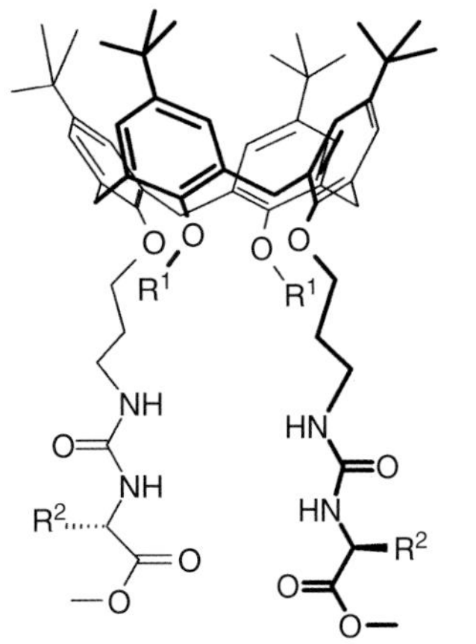

91a: R^1 = H, R^2 = H, CH_3, s-C_4H_9
91b: R^1 = n-C_3H_7, R^2 = H, CH_3, s-C_4H_9

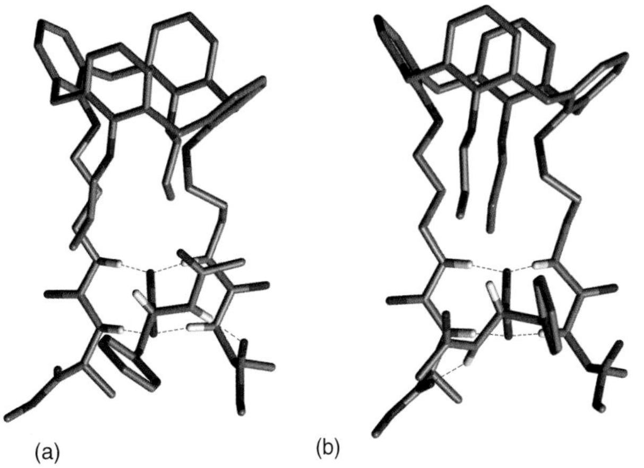

(a) (b)

Figure 26 Bisurea-peptidocalix[4]arenes **91** and minimized structures by semiempirical methods (*t*-Bu groups omitted) of the complexes of **91b** ($R^2 = CH_3$) with (a) D-*N*-Ac-Phe-COO⁻ and (b) L-*N*-Ac-Phe-COO⁻.

89a and **89b** strongly bind carboxylate anions (as tetrabutylammonium salts), including *N*-Ac-amino acids, showing association constants ($370 < K(M^{-1}) < 44\,000$) 2 orders of magnitude larger than those found for the cleft-like C-linked peptidocalix[4]arene analog **90**. However, these chiral bridged calix[4]arene derivatives as well as others reported in literature show moderate enantioselectivity in the recognition of chiral anions.[53] More recently, we have synthesized chiral bisurea-calix[4]arenes (**91**) bearing L-amino acid moieties at the *lower rim* and studied their anion recognition properties in the gas phase by ESI-MS and in $CDCl_3$ or acetone-d_6 by ^{1}H NMR, which evidenced the formation of 1 : 1 complexes with Cl^- and several chiral carboxylate anions (see reviews[53] and [55] for original references). Interestingly, the binding constants found in acetone-d_6 are substantially higher (up to 40 times) than those determined in $CDCl_3$ because, in this solvent, all receptors (but especially **91a**, which has two free phenolic OH groups) experience intramolecular hydrogen bonding between the donor and acceptor groups in the binding region at the *lower rim*. The highest efficiency is found for host **91b** ($R^2 = CH_3$), which also shows a remarkable enantioselectivity in the recognition of *N*-acetyl-D-phenylalaninate anion ($K_D/K_L = 4.14$), explained on the basis of a *three-point interaction* mode of binding (Figure 26). The analysis of the electrostatic potential indicates that the **91b**·L-*N*-Ac-Phe-COO⁻ complex (Figure 26b) is sterically more crowded than that with the D enantiomer (Figure 26a). In particular, in the L-Phe complex repulsion might originate between the phenyl moiety of the guest and the carboxymethyl group of one of the alanine of the host.

By linking two L-tryptophan units at the *lower rim* of calix[4]arenes through amide bonds, He *et al.* synthesized three chiral receptors (**92a–c**) that, together with other similar hosts reported by the same authors and discussed in recent review articles,[53,55] can be used as fluorescent chemosensors for the enantioselective recognition of amino acid anions and other chiral guest species. Host **92b** formed the most stable 1 : 1 complexes in DMSO-d_6, ranging from 62 M^{-1} for L-phenylglycinol to

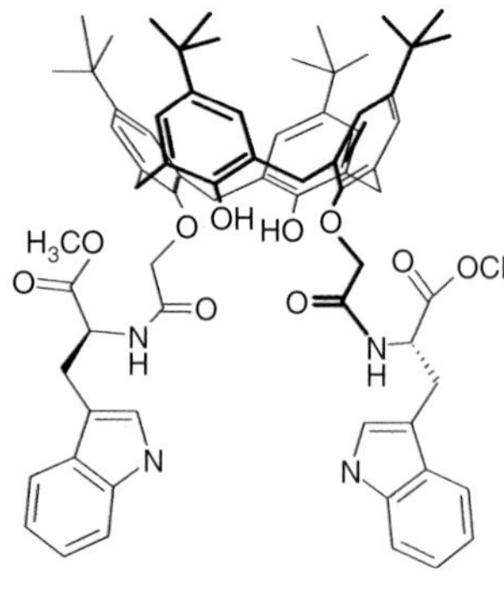

92a

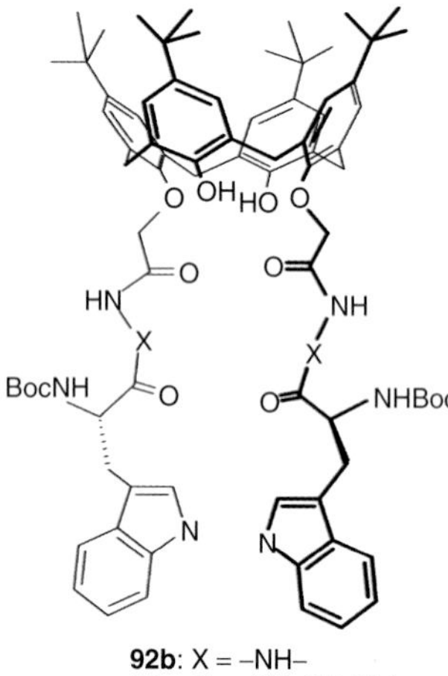

92b: X = –NH–
92c: X = –CH_2CH_2NH–

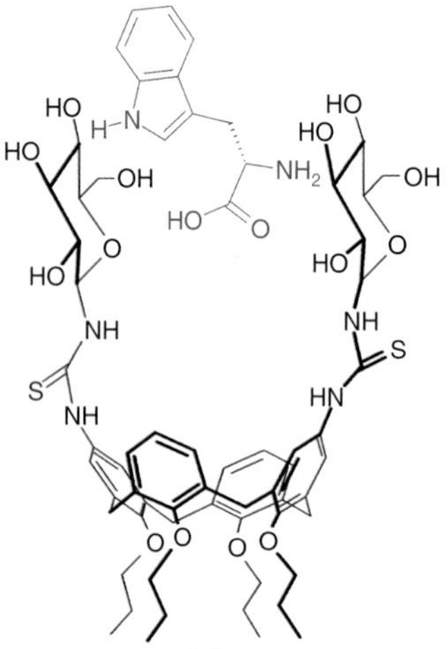

93·Trp

3.87 × 10^4 M^{-1} for D-malate, which was explained on the basis of the number of hydrogen-bonding donor groups available for complexation and of the rigidity of the host. The most interesting enantioselective discrimination was observed for host **92a** in the recognition of tartrate anion ($K_D/K_L = 5.70$) and for calix[4]arene **92b** in the recognition of *N*-Boc-alaninate ($K_D/K_L = 7.20$).

In collaboration with the group of Vainiotalo and Kalenius, we have recently reported the recognition of amino-acids by di- and tetra-glucosylthioureidocalix[4]arenes in the gas phase.[56] All receptors showed selectivity for aromatic amino acids and the most selective was the diglucosylthioureidocalix[4]arene **93,** which showed the following affinity order: Trp > Tyr > Phe > Ser. Compound **93** also showed the highest value of enantioselective discrimination in the recognition of tyrosine (D/L = 2.58). Very few examples of enantioselective recognition of zwitterionic amino acids by chiral calixarene receptors have been reported.[28,53]

3 MULTIVALENT LIGANDS

Calixarenes can play a relevant role in molecular recognition not only as receptors but also as ligands. In this latter case, they mainly work as scaffolds, which suitably organize several identical substituents in space linked to the *lower* or the *upper rim* and simultaneously orient these moieties toward the surface areas of large macromolecules or toward the recognition points of multiple binding site receptors. In other words, they act as core for the construction of multivalent ligands (see **Multivalency**, Volume 1).

3.1 Protein recognition, inhibition, and detection

Among the first examples of multivalent calixarene ligands are the tetraloop derivatives **94** designed by Hamilton

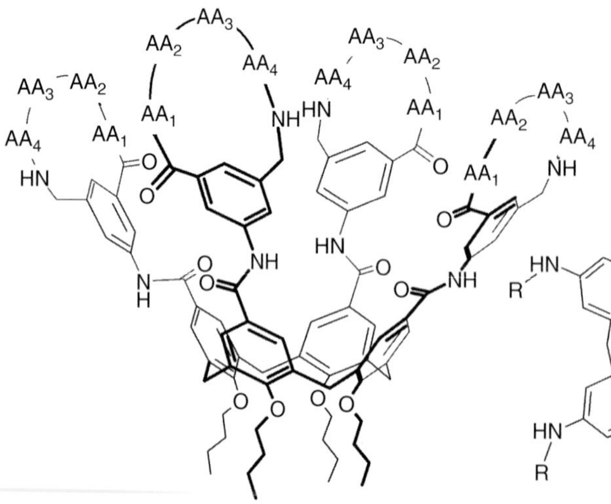

94: $AA_1 = AA_3$ = Gly; $AA_2 = AA_4$ = Asp
$AA_1 = AA_3$ = Gly; AA_2 = Asp; AA_4 = Tyr
$AA_1 = AA_3$ = Gly; $AA_2 = AA_4$ = Lys

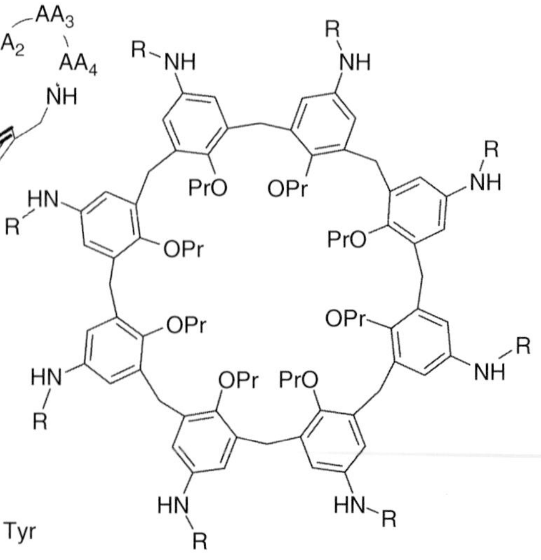

95: R = $CO(CH_2)_2NH_2$ or Lys

et al.[57] In these compounds, cyclic peptides anchored at the *upper rim* of a calix[4]arene constitute four contact points for a corresponding number of surface portions in selected proteins. By changing the amino acid composition of the loops, their charge and hydrophilic–lipophilic balance can be modulated, thus modifying the ligand selectivity. The derivative containing the sequence GlyAspGlyAsp and exposing two anionic aspartate residues per cyclopeptide shows, for example, inhibitory activity toward cytochrome *c*. This enzyme is, in fact, characterized by lysine-rich, positively charged zones around the heme edge, which represent its contact area for the interaction with the biological counterparts as the cytochrome *c* peroxidase. By replacing an aspartate with tyrosine in all the four loops, or both the aspartates with positively charged lysine residues, the peptidocalixarene ligand becomes a good antagonist for platelet-derived growth factor (PDGF) and for the vascular endothelial growth factor (VEGF), respectively, showing antiangiogenic and antitumoral activity in both cases. Through the complete substitution of the peptide loops with simpler isophthalic monoacid–monoester moieties, a ligand that features simultaneous inhibition of both the growth factors was obtained. In all the cases, the contacts between the four arms of the peptidocalix[4]arenes with complementary surface portions of the proteins determine the masking of the active site by the macrocycle cavity with the consequent disruption of the binding with their biological partners and suppression of their function. On the basis of the same approach, calix[4]arenes functionalized with linear peptides at the *upper rim* act as transglutaminase inhibitors. Despite a significantly lower preorganization degree, calix[8]arenes **95** functionalized with amino terminating chains, like a β-amino acid or lysine, was also shown to block the activity of a tetrameric recombinant tryptase having anionic aspartate side chains in the proximity of its active site. The macrocyclic ligands prevent access to the natural substrates affecting the enzyme activity. It is interesting to note that the same octamers were later studied as heparin binders, resulting as the only examples known up to now of calixarene derivatives able to interact with polysaccharides.[21]

The principle of charge complementarity between protein surfaces and calixarene arms is also the basis of the studies carried out on the anionic tetraphosphonate calix[4]arene **96** and the cationic tetra(ammonium-methylene)calix[4]arene **97**.[58] Owing to their lipophilic chains at the *lower rim*, these ligands were trapped in stearic acid monolayers and used in the recognition and detection of basic and acidic proteins, respectively, which interact with the charged groups of the ligands emerging from the lipophilic layer.

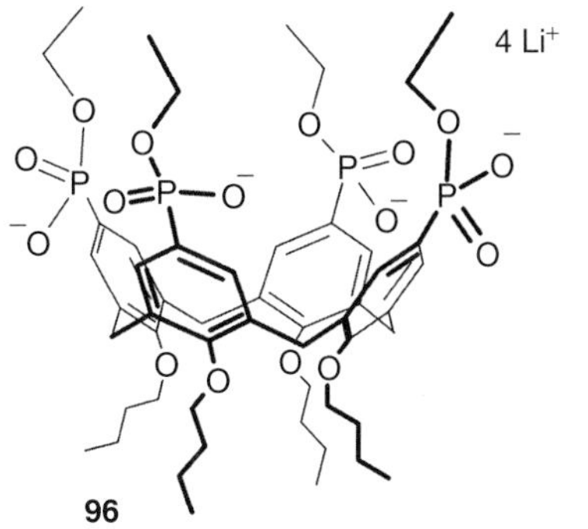

96 4 Li^+

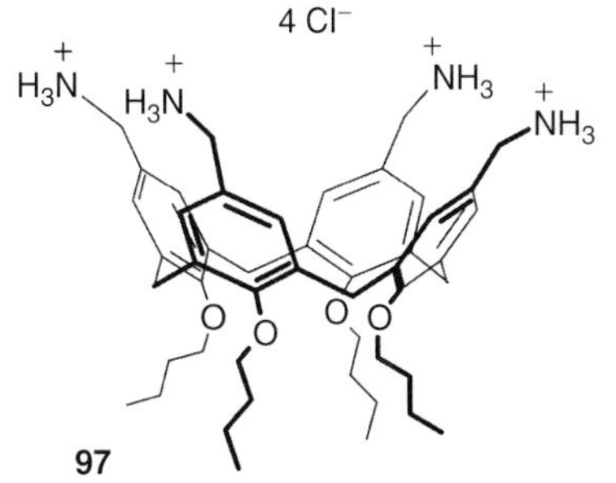

97 4 Cl^-

98 4 Cl^-

99 4 TFA^-

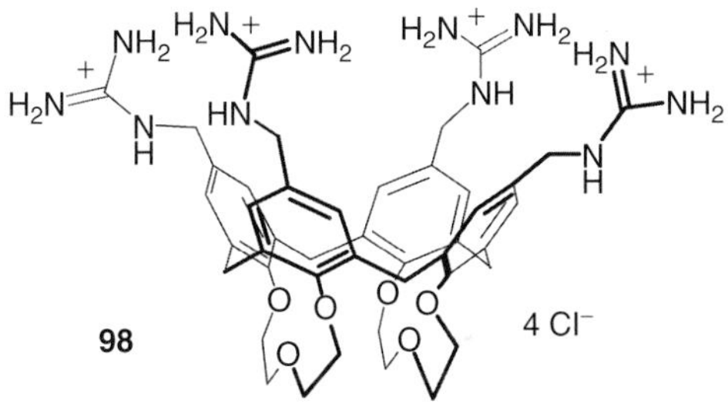

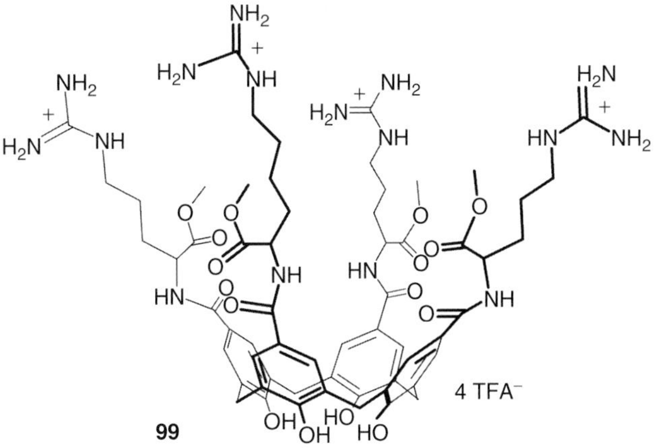

The synergy between charge–charge interactions, conical shape, C_4 symmetry, and conformational restrictions of guanidinocalix[4]arenes **98** and **99** was exploited by de Mendoza *et al.* to obtain ligands for the mutated p53 protein (p53-R337H) and the Kv1.2 potassium channel, two tetrameric proteins considered to be interesting and medically relevant targets. They present a squarelike distribution of carboxylate groups close to a pocket and a vestibule, respectively. The carboxylate residues are targeted by the guanidinium units, while in the concave zone the *cone* calix[4]arene is well accommodated like a stopper. Moreover, the preorganization of the macrocycles, due to short ethylene glycol bridges or hydrogen-bonding array between the free phenolic OH groups at the *lower rim*, adds to charge complementarity to bring about the remarkable biological activity of these protein ligands (see Ref. 21 for original references).

3.2 Lectin binding and inhibition

Proteins that act as carbohydrate receptors lacking enzymatic activity and not belonging to the immunoglobulin class are known as *lectins*. Their biological behavior is frequently characterized by a multivalent effect (see **Supramolecular Approaches to the Study of Glycobiology**, Volume 4) because, in order to increase the otherwise relatively low specificity and affinity, these proteins tend to establish a series of contacts through several identical recognition site–glycoside epitope pairs. For this reason, the use of multivalent ligands to interfere with processes where carbohydrates are involved is particularly interesting and the use of calixarenes as scaffolds for the attachment of a variable number of glycoside units, attractive. Different from what is observed in the interactions with the protein surfaces described above, where a 1 : 1 complex can be reasonably expected, often, in lectin binding, a large cross-linking phenomenon causing agglutination can be also observed. In several cases, the divergent orientation of the binding sites common to many lectins prevents their simultaneous interaction with a single multivalent ligand causing the formation of extended complex networks often evolving in the precipitation of an agglutinate. In the last 15 years, a large variety of glycoclusters based on calixarenes has been produced by anchoring the carbohydrate units on the macrocyclic skeletons through different linkages.[15,59] The calixarene-based glycoclusters show selectivity in the binding of lectins depending on the glycoside epitope present in their structure. Specificity in the interaction with several vegetal lectins like concanavalin A, peanut lectin, wheat germ agglutinin, *Vicia villosa* agglutinin, with agglutination due to crosslinking, was observed for polygluco-, polygalacto-, polysialo-, and poly-*N*-acetylgalactosamine-calixarenes, respectively, frequently accompanied by an amplified lectin affinity with respect to the corresponding single monosaccharides.[15] Even more remarkable are the results obtained with a family of galactosylthioureido- and lactosylthioureidocalixarenes (**100–103**), targeting human galectins of different topologies (Gal-1, -3, and -4), not only because of the medical relevance of this class of lectins but also for the high inhibition potency of these systems and the selectivity observed depending on their structural features. In fact, in bioassays with cells, the lactoclusters and, in particular, the conformationally mobile hexa- (**102d**) and octa- (**102f**) lactosyl derivatives exhibited the strongest inhibition activity ever found toward the Gal-4 binding to human pancreatic carcinoma cells, showing a large multivalent effect. Furthermore, the *1,3-alternate* tetralactocalix[4]arene (**101b**) resulted in the most efficient inhibitor of Gal-1 and the analog in *cone* conformation (**100b**) the worst, while the opposite was observed using Gal-3.

These results indicate the importance of the structural and conformational requirements of these polyglycosylated macrocycles for determining their biological activity toward structurally different and differently aggregated lectins.[6,21] As the thiourea-linked glycocalixarenes are able to interact with anions (Section 2.3.1), the potential development of site-directed drug delivery systems, which are able to exploit the carbohydrate recognition for organ and/or tissue specificity, can be envisaged.[21] Pathogens and their excreted products that use lectins to attack mammalian cells are also interesting targets for glycocalixarenes. Among these, cholera toxin (CT) and its homologous heat-labile toxin of *Escherichia coli* are classical and widely studied examples of toroid-like AB_5 proteins exploiting multivalency for establishing stable contacts with cells and starting the infection process. Each B subunit, in fact, simultaneously binds to the pentasaccharide unit (oGM1) of their natural ligand, the GM1 ganglioside, anchored onto the cell membrane. In collaboration with A. Bernardi, we designed and synthesized a divalent ligand (**104**) based on a *cone* calix[4]arene exposing two units of a oGM1 mimic (*pseudo*-GM1), where sialic acid and lactose are replaced by R-lactic acid and a 1,2-cis-cyclohexandiol, respectively. The synthetic divalent ligand was endowed with long spacers between epitopes and cavity, which, in principle, should allow the concomitant interaction with two nonadjacent binding sites. Indeed, **104** resulted a very strong antagonist of CT, showing a 50% saturation concentration of 48 nM, which is slightly better than that measured for the natural o-GM1. This result also corresponds to a remarkable affinity enhancement factor close to 4000-fold (2000-fold per sugar mimic) with respect to the single pseudo-GM1, which is one of the highest values ever observed for synthetic CT inhibitors. Saturation transfer difference (STD) and NOE-difference (nuclear overhauser effect) NMR experiments indicated that the stability of the toxin–calixarene complex can be attributed to a specific interaction of the two mimic units with the binding sites and to less specific interactions between the squaramide containing linkers and the protein surface.[6,21]

Targeting of the tetrameric PA-IL bacterial lectin from *Pseudomonas aeruginosa* was performed using different

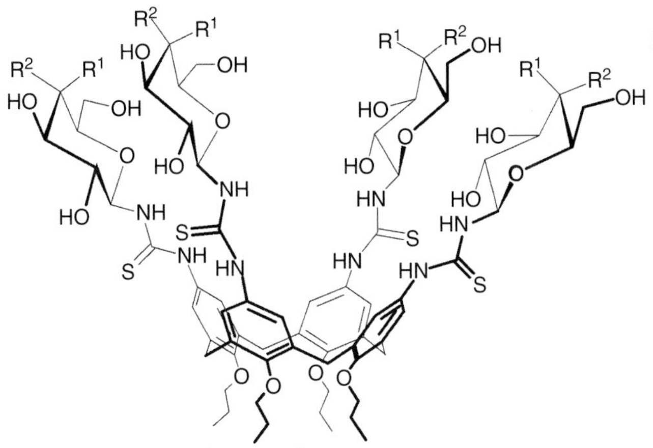

100a: $R^1 = OH$, $R^2 = H$
100b: $R^1 = H$, $R^2 = \beta$-galactose

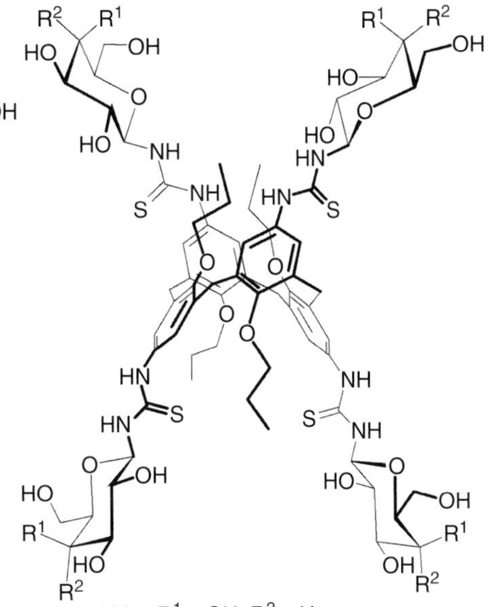

101a: $R^1 = OH$, $R^2 = H$
101b: $R^1 = H$, $R^2 = \beta$-galactose

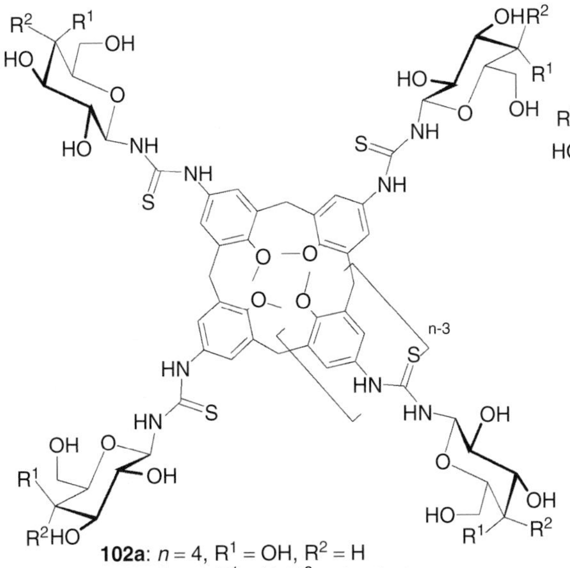

102a: $n = 4$, $R^1 = OH$, $R^2 = H$
102b: $n = 4$, $R^1 = H$, $R^2 = \beta$-galactose
102c: $n = 6$, $R^1 = OH$, $R^2 = H$
102d: $n = 6$, $R^1 = H$, $R^2 = \beta$-galactose
102e: $n = 8$, $R^1 = OH$, $R^2 = H$
102f: $n = 8$, $R^1 = H$, $R^2 = \beta$-galactose

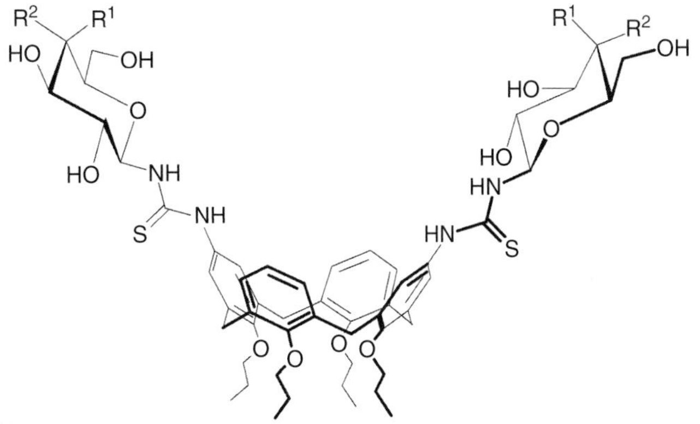

103a: $R^1 = R^3 = OH$, $R^2 = H$
103b: $R^1 = H$, $R^3 = OH$, $R^2 = \beta$-galactose

conformers of calix[4]arenes with variable valency by Cecioni *et al.*, while Marra *et al.* investigated polysialylated macrocycles as inhibitors of BK and influenza A virus (see review[21] for specific references).

3.3 DNA binding and cell transfection

Calixarenes (**105**–**108**) adorned with guanidinium groups at the *upper* or at the *lower rim* were recently synthesized for gene delivery (see **The Role of Supramolecular Chemistry in Responsive Vectors for Gene Delivery**, Volume 4), and some of them showed remarkable properties in cell transfection, representing the first examples of nonviral vectors based on this class of macrocycles. As observed by atomic force microscopy (AFM), *upper rim* guanidinocalix[4]arenes **105** bearing alkyl chains at the *lower rim* and blocked in the *cone* conformation efficiently bind to linear and plasmid DNA through charge–charge interactions with the phosphate groups and condense it in small particles owing to hydrophobic interactions between the aliphatic chains at the *lower rim*. DNA condensation is also observed in the presence of the *1,3-alternate* **106** and the conformationally mobile guanidinocalix[4]arenes **107a**, but in these cases both binding

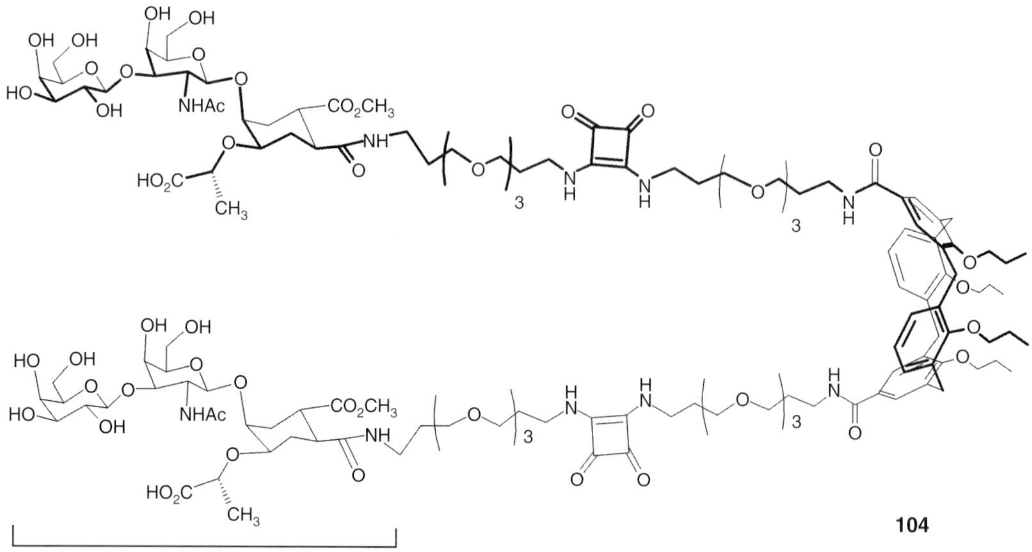

pseudo-GM1

104

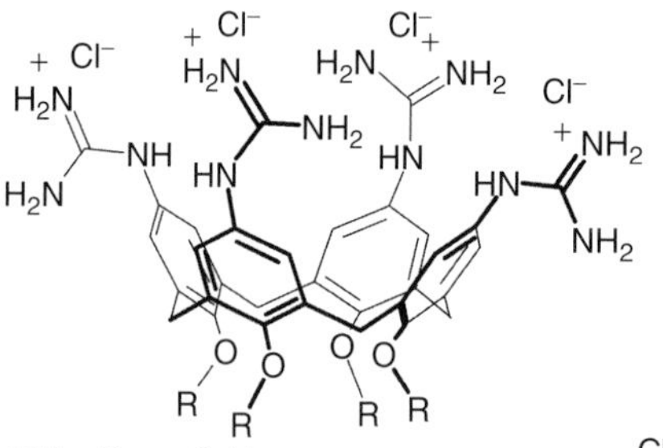

105a: R = n-C_3H_7
105b: R = n-C_6H_{13}
105c: R = n-C_8H_{17}

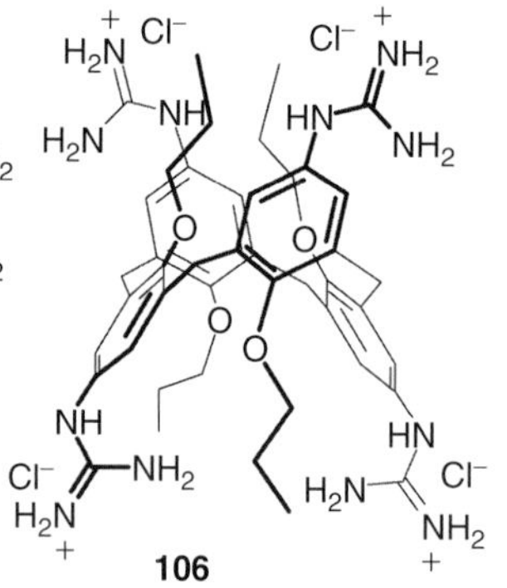

106

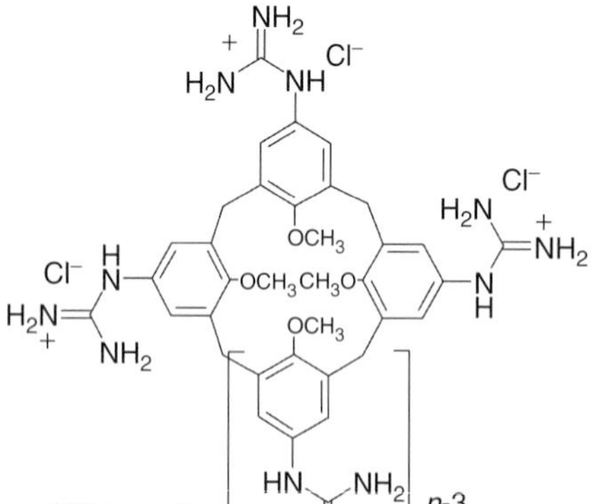

107a: n = 4:
107b: n = 6:
107c: n = 8:

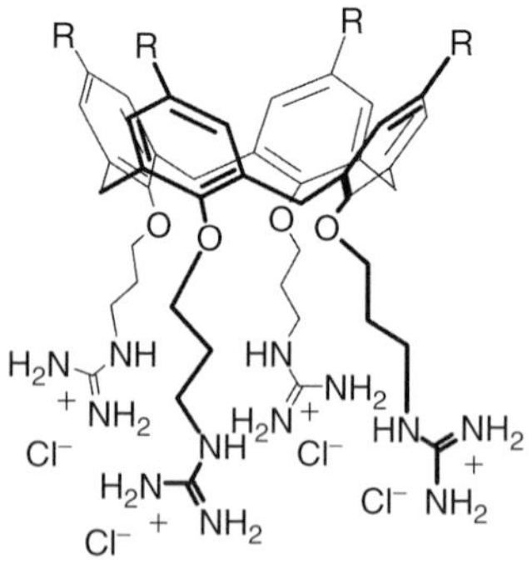

108a: R = t-C_4H_9
108b: R = H
108c: R = n-C_6H_{13}

and compaction are only due to charge–charge interactions. Also, the larger calix[6]- (**107b**) and calix[8]arene (**107c**) strongly interact with the two types of DNA but they are not able to give rise to nanometric condensates. According to this different condensation ability, the vectors based on the calix[4]arenes give transfection, as demonstrated through the delivery into human rhabdomiosarcoma cells of pEGFP-C1 DNA plasmid expressing for the green fluorescent protein, while those based on calix[6]- and calix[8]arene do not. However, with these compounds, citotoxicity is high and transfection efficiency low.[6] A very important improvement in these properties was achieved by bringing the guanidinium groups at the *lower rim* of *cone* calix[4]arenes (**108**). AFM studies revealed that these derivatives also compact DNA with a mechanism strongly dependent on the presence or absence of the lipophilic alkyl chains, in this case linked at the *upper rim*.

The most valuable result is that one compound of this second series of calixarene vectors (**108b**) is very efficient in transfection; it is not only better than the *upper rim* analogs but even more attractive than a commercially available lipofectamine (LTX), widely used for delivery protocols. Moreover, as a corresponding acyclic Gemini-type model shows a transfection efficiency that is definitely lower than that of **108b**, we conclude that, in the gene delivery processes studied, a *macrocyclic effect* related to the orientation and preorganization of the binding units and to the presence of a defined lipophilic cavity plays an important role.[21] The aminocalixarenes are markedly less effective in transfection, probably because only a partial protonation of the amino groups can be reached at physiological pH. These compounds, however, are able to interact with DNA.[21]

4 CONCLUSION

It is hard to find a topic in supramolecular chemistry that has not been touched, even briefly, by calixarenes or related macrocycles. This justifies the fact that we were compelled to give only a short summary of the properties of classical calixarenes as molecular receptors for ions and neutral molecules or as scaffolds for the construction of multivalent ligands able to interact with biomacromolecules. We believe, indeed, that these two aspects are the most peculiar and remarkable features of calixarenes, on which a variety of other complex supramolecular functions can be built. The future will certainly be bright for this class of synthetic macrocycles as their molecular recognition and scaffolding properties are likely to be exploited in the classical ways and also in novel forms within the framework of the third developing phase of supramolecular chemistry, namely, *constitutional dynamic chemistry* and *adaptive chemistry*.[60]

ACKNOWLEDGMENTS

The authors are grateful to the Ministero dell'Istruzione, dell'Università e della Ricerca (MIUR, PRIN 2008 projects n 2008HZJW2L and 200858SA98), to the Fondazione Cassa di Risparmio di Parma and to COST WG D34/0001/05 for the financial support.

REFERENCES

1. C. D. Gutsche, *Calixarenes. An Introduction*, ed. J. F. Stoddart, The Royal Society of Chemistry, Cambridge, 2008.
2. L. Mandolini and R. Ungaro, eds., *Calixarenes in Action*, Imperial College Press, London, 2000.
3. Z. Asfari, V. Böhmer, J. Harrowfield, and J. Vicens, eds., *Calixarenes 2001*, Kluwer Academic Publishers, Dordrecht, 2001.
4. J. Vicens and J. Harrowfield, eds *Calixarenes in the Nanoworld*, Springer, Dordrecht, 2007.
5. C. P. Rao and M. Dey, Calixarenes, in *Encyclopedia of Nanoscience and Nanotechnology*, ed. H. S. Nalwa, American Scientific Publishers, Stevenson Ranch, 2004, pp. 475–497.
6. L. Baldini, A. Casnati, F. Sansone, and R. Ungaro, *Chem. Soc. Rev.*, 2007, **36**, 254.
7. S. Le Gac and I. Jabin, *Chem. Eur. J.*, 2008, **14**, 548.
8. D. Coquiere, S. Le Gac, U. Darbost, *et al.*, *Org. Biomol. Chem.*, 2009, **7**, 2485.
9. J. S. Kim and D. T. Quang, *Chem. Rev.*, 2007, **107**, 3780.
10. J. A. Ripmeester, G. D. Enright, C. I. Ratcliffe, *et al.*, *Chem. Commun.*, 2006, 4986.
11. M. Nishio, Y. Umezawa, K. Honda, *et al.*, *CrystEngComm*, 2009, **11**, 1757.
12. R. Ungaro, A. Arduini, A. Casnati, *et al.*, Complexation of ions and neutral molecules by functionalized calixarenes, in *Computational Approaches in Supramolecular Chemistry*, ed. G. Wipff, Kluwer Academic Publishers, Dordrecht, 1994, pp. 277–300.
13. D. J. Cram and J. M. Cram, *Container Molecules and Their Guests*, ed J. F. Stoddart, Royal Society of Chemistry, Cambridge (UK), 1997.
14. L. Pescatori, A. Arduini, A. Pochini, *et al.*, *Org. Biomol. Chem.*, 2009, **7**, 3698.
15. A. Casnati, F. Sansone, and R. Ungaro, *Acc. Chem. Res.*, 2003, **36**, 246.
16. F. Perret, A. N. Lazar, and A. W. Coleman, *Chem. Commun.*, 2006, 2425.
17. D. S. Guo, K. Wang, and Y. Liu, *J. Inclusion Phenom. Macrocyclic Chem.*, 2008, **62**, 1.
18. M. Rehm, M. Frank, and J. Schatz, *Tetrahedron Lett.*, 2009, **50**, 93, and references therein.
19. W. M. Nau, G. Ghale, A. Hennig, *et al.*, *J. Am. Chem. Soc.*, 2009, **131**, 11558.
20. C. S. Beshara, C. E. Jones, K. D. Daze, *et al.*, *Chembiochem*, 2010, **11**, 63.
21. F. Sansone, L. Baldini, A. Casnati, and R. Ungaro, *New J. Chem.*, 2010, **34**, 2715.
22. A. Casnati, F. Sansone, and R. Ungaro, Calixarene receptors in ion recognition and sensing, in *Advances in Supramolecular Chemistry*, ed. G. W. Gokel, Cerberus Press Inc., South Miami, 2004, vol. 9, pp. 165–218.
23. J. F. Dozol and R. Ludwig, *Ion Exch. Solvent Extr.*, 2010, **19**, 195.

24. R. Ludwig and N. T. K. Dzung, *Sensors*, 2002, **2**, 397.

25. H. H. Dam, D. N. Reinhoudt, and W. Verboom, *Chem. Soc. Rev.*, 2007, **36**, 367.

26. J. T. Lenthall and J. W. Steed, *Coord. Chem. Rev.*, 2007, **251**, 1747.

27. D. T. Schuehle, M. Polasek, I. Lukes, *et al.*, *Dalton Trans.*, 2010, **39**, 185.

28. S. Cherenok and V. Kalchenko, *Top. Heterocycl. Chem.*, 2009, **20**, 229.

29. E. Macerata, F. Sansone, L. Baldini, *et al.*, *Eur. J. Org. Chem.*, 2010, 2675, and references therein.

30. D. M. Homden and C. Redshaw, *Chem. Rev.*, 2008, **108**, 5086, and references therein.

31. I. Leray and B. Valeur, *Eur. J. Inorg. Chem.*, 2009, 3525, and references therein.

32. B. S. Creaven, D. F. Donlon, and J. McGinley, *Coord. Chem. Rev.*, 2009, **253**, 893.

33. I.-T. Ho, G.-H. Lee, and W.-S. Chung, *J. Org. Chem.*, 2007, **72**, 2434.

34. S. Pandey, A. Azam, S. Pandey, and H. M. Chawla, *Org. Biomol. Chem.*, 2009, **7**, 269.

35. T. Gruber, C. Fischer, M. Felsmann, *et al.*, *Org. Biomol. Chem.*, 2009, **7**, 4904.

36. R. K. Pathak, S. Ibrahim, and C. P. Rao, *Tetrahedron Lett.*, 2009, **50**, 2730.

37. G. K. Li, Z. X. Xu, C. F. Chen, and Z. T. Huang, *Chem. Commun.*, 2008, 1774.

38. B. Colasson, N. Le Poul, Y. Le Mest, and O. Reinaud, *J. Am. Chem. Soc.*, 2010, **132**, 4393.

39. E. A. Shokova and V. V. Kovalev, *Russ. J. Org. Chem.*, 2009, **45**, 1275.

40. P. Lhotak, Anion receptors based on calixarenes, in *Topics in Current Chemistry 255: Anion Sensing*, ed. I. Stibor, Springer, Berlin, 2005, pp. 65–95.

41. P. Curinova, I. Stibor, J. Budka, *et al.*, *New J. Chem.*, 2009, **33**, 612.

42. R. Zadmard, S. Taghvaei-Ganjali, B. Gorji, and T. Schrader, *Chem. Asian J.*, 2009, **4**, 1458.

43. P. Blondeau, M. Segura, R. Perez-Fernandez, and J. de Mendoza, *Chem. Soc. Rev.*, 2007, **36**, 198.

44. S. R. Bayly and P. D. Beer, Metal-based anion receptor systems, in *Recognition of Anions*, ed. R. Vilar, Springer-Verlag, Berlin, 2008, pp. 45–94.

45. G. J. Kirkovits, J. A. Shriver, P. A. Gale, and J. L. Sessler, *J. Incl. Phenom. Macr. Chem.*, 2001, **41**, 69.

46. J. L. Sessler, P. A. Gale, and W. S. Cho, *Anion Receptor Chemistry*, ed. J. F. Stoddart, Royal Society of Chemistry, Cambridge, 2006.

47. S. K. Kim, J. L. Sessler, D. E. Gross, *et al.*, *J. Am. Chem. Soc.*, 2010, **132**, 5827.

48. A. Senthilvelan, I. T. Ho, K. C. Chang, *et al.*, *Chem. Eur. J.*, 2009, **15**, 6152, S6152.

49. M. D. Lankshear, I. M. Dudley, K. M. Chan, *et al.*, *Chem. Eur. J.*, 2008, **14**, 2248.

50. C. Gargiulli, G. Gattuso, C. Liotta, *et al.*, *J. Org. Chem.*, 2009, **74**, 4350.

51. M. J. McIldowie, M. Mocerino, and M. I. Ogden, *Supramol. Chem.*, 2010, **22**, 13.

52. V. Böhmer, D. Kraft, and M. Tabatabai, *J. Incl. Phenom. Mol. Recognit. Chem.*, 1994, **19**, 17.

53. A. Sirit and M. Yilmaz, *Turk. J. Chem.*, 2009, **33**, 159.

54. Y. Kubo, *J. Incl. Phenom. Mol. Recognit. Chem.*, 1998, **32**, 235.

55. S. Kubik, *Chem. Soc. Rev.*, 2009, **38**, 585.

56. M. Torvinen, R. Neitola, F. Sansone, *et al.*, *Org. Biomol. Chem.*, 2010, **8**, 906.

57. S. Fletcher and A. D. Hamilton, *J. R. Soc. Interface*, 2006, **3**, 215, and references therein.

58. S. Kolusheva, R. Zadmard, T. Schrader, and R. Jelinek, *J. Am. Chem. Soc.*, 2006, **128**, 13592, and references therein.

59. A. Dondoni and A. Marra, *Chem. Rev.*, 2010, **110**, 4949.

60. J. M. Lehn, *Chem. Soc. Rev.*, 2007, **36**, 151.

Cyclotriveratrylene and Cryptophanes

Michaele J. Hardie
University of Leeds, Leeds, UK

1 INTRODUCTION

1.1 Overview and scope

Cyclotriveratrylenes (CTVs) and cryptophanes have been developed as molecular hosts: that is, molecules with a molecular cavity that are capable of reversibly binding guest molecules in a noncovalent manner. Much of the pioneering work in the development of CTV and cryptophane chemistry came out of the laboratories of André Collet. Collet wrote extensive review chapters on both areas, which were published in *Comprehensive Supramolecular Chemistry* in 1996.[1,2] More recent reviews have been contributed by Brotin and Dutasta in the area of cryptophane chemistry[3] and by Hardie on recent applications of CTVs.[4] In this chapter, synthetic routes to CTVs and cryptophanes will be addressed, and also their host–guest chemistry. Further details of these, such as tables of yields and binding constants, can be found in the cited reviews.[1–3] Finally, the uses of CTVs and cryptophanes in other aspects of supramolecular chemistry will also be addressed.

Supramolecular Chemistry: From Molecules to Nanomaterials.
Edited by Philip A. Gale and Jonathan W. Steed.
© 2012 John Wiley & Sons, Ltd. ISBN: 978-0-470-74640-0.

1.2 Cyclotriveratrylene structure and conformations

Hexamethoxy-tribenzocyclononatriene, more commonly known as CTV, is a cyclic trimer of veratrole whose structure was first correctly elucidated in the 1960s.[1] The stable conformation of CTV is the rigid crown conformation with a distinctive pyramidal shape and shallow molecular cavity (Scheme 1). The flexible saddle conformation of CTV was isolated for the first time in 2004 by Luz and coworkers.[5] Pure samples of the saddle conformation can be isolated by rapid quenching of a melt or of a hot solution of CTV, and the saddle form then separated by chromatography. In the solid state, the saddle form is metastable but can be kept at room temperature for several months. In solution, an equilibrium is established between the crown and saddle forms with an isomerization half-life of approximately one day and activation energy of $\sim 97\,\mathrm{kJ\,mol^{-1}}$ in chloroform. Interestingly, the racemization equilibrium is solvent dependent, and polar solvents appear to disfavor the saddle conformation.

Most analogs of CTV share the tribenzo[a,d,g]cyclononatriene core but have different substitutions at the upper rim. Some common examples are shown in Scheme 1, including the fully demethylated cyclotricatechylene (CTC) and the C_3-symmetric cyclotriguaiacylene (CTG) and trimethoxy-triamino-tribenzo[a,d,g]cyclononatriene or triamino cyclotriguaiacylene (aCTG). As for CTV itself, the crown conformation is isolated on initial synthesis and is the most stable isomer. CTV analogs in which all six upper rim groups are identical are achiral. Different substitution patterns at the upper rim, however, produce chiral molecules. Both enantiomers of CTG are shown as an example in Scheme 1. Optically pure CTV analogs racemize in solution at rates and activation energies similar to those observed for the crown ⇔ saddle interconversion

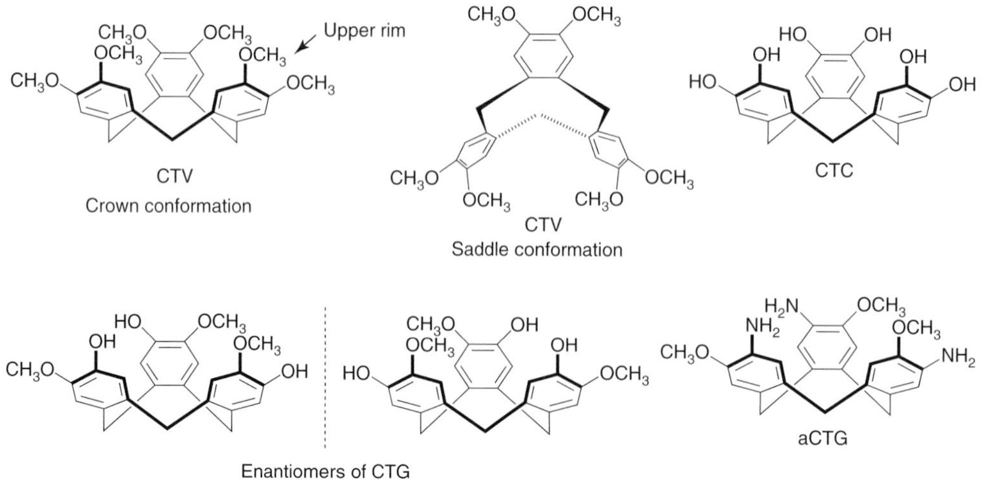

Scheme 1 Cyclotriveratrylene conformations and some common analogs of CTV.

of CTV itself. For example, the activation energy for the racemization of CTG is 112.5 kJ mol^{-1}.[6] Hence the racemization isomerization is likely to proceed via the saddle conformation.[1]

1.3 Cryptophane structure and conformations

Cryptophanes are cage-like molecules where two C_3-symmetric CTV fragments have been covalently joined together via three linker groups such that their upper rims face one another.[2] This creates a host with a well-structured internal molecular cavity. There are two geometric isomers, the D_3-symmetric anti-isomer and the C_{3h}-symmetric syn-isomer (Scheme 2). The anti-isomer cryptophanes are chiral as they are composed of only one CTV enantiomer. The C_{3h}-symmetric *syn*-crytophanes are achiral as they contain both CTV enantiomers. Early examples of cryptophanes were named according to both

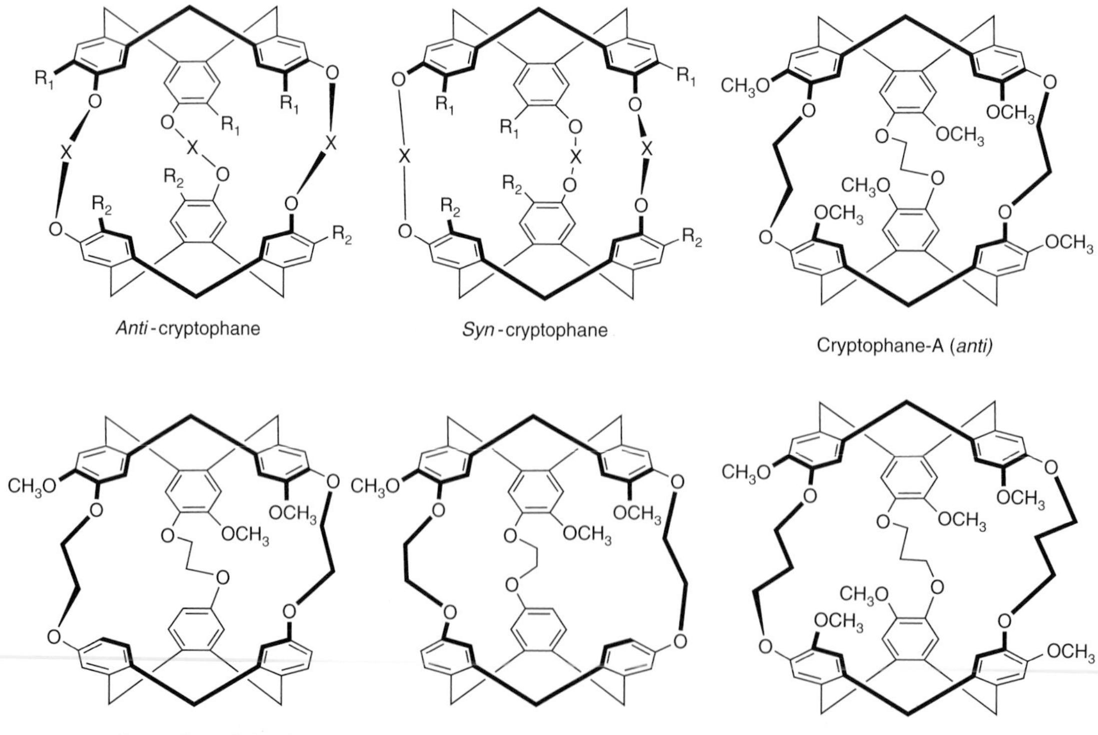

Scheme 2 Isomers of cryptophanes and early examples.

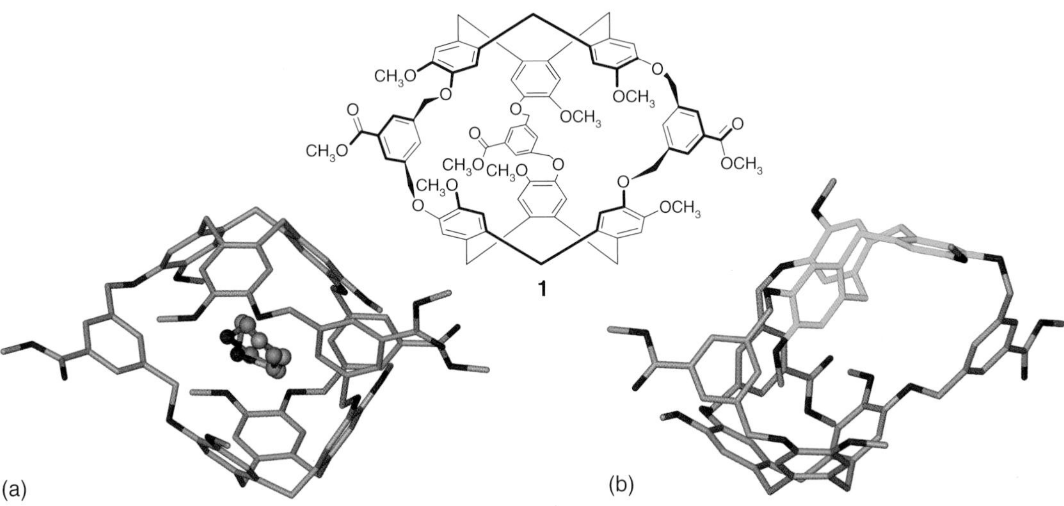

Figure 1 Crystal structures of cryptophane **1**. (a) Crown–crown conformation of cryptophane **1** with disordered tetrahydrofuran (THF) guest (shown as ball and stick); (b) Crown–saddle conformation of cryptophane **1**, with the saddle conformation tribenzocyclononatriene shown in light blue.

functional groups and the isomer. Cryptophane-A has (OCH_2CH_2O) linkers in the anti form; cryptophane-B is the as yet unreported syn-isomer of A; cryptophanes-C/D are the anti/syn isomers, respectively, of a similarly sized cryptophane but with $R_2 = H$ (Scheme 2); cryptophane-E is the anti-isomer with ($OCH_2CH_2CH_2O$) linkers.[2]

Cryptophanes are generally synthesized with both CTV fragments in their crown conformation; however, there have recently been reports of conformational isomerism between crown–crown and crown–saddle forms.[7,8] Holman and coworkers isolated and crystallized a crown–saddle conformation of the *m*-xylyl-linked cryptophane **1**. Initial synthesis gave the anti-isomer in crown–crown conformation with included tetrahydrofuran guest (Figure 1a). Evacuation of the solvent guest and recrystallization from chloroform allowed the crystal structure of a crown–saddle conformation to be elucidated (Figure 1b).[7]

2 SYNTHESIS AND TRANSFORMATIONS

2.1 Cyclotriveratrylenes

2.1.1 Condensation reactions to give CTV and analogs

CTV is synthesized from the self-condensation of the veratryl cation which is generated from veratrole alcohol or from a mixture of 1,2-dimethoxybenzene and formaldehyde (Scheme 3). Depending on the reaction conditions that are used, the tetrameric condensation product cyclotetraveratrylene (CTTV) may also be formed alongside CTV, along with small amounts of higher oligomers.[1] Reaction conditions are usually strongly dehydrating. Typical conditions and reported yields for condensation of veratrole alcohol to CTV include 60% perchloric acid (80% yield), trifluoroacetic acid in chloroform (19% yield), sulfuric acid in hot acetic acid (87% yield), BF_3 etherate in benzene

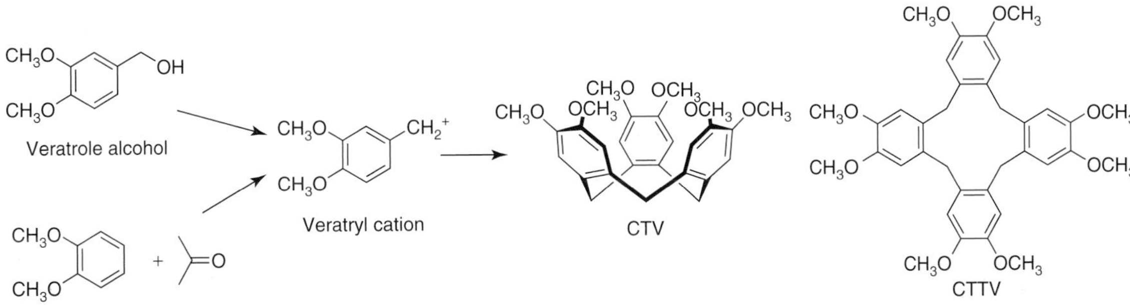

Veratrole alcohol
Veratryl cation
CTV
CTTV

Scheme 3 Synthesis of CTV and CTTV.

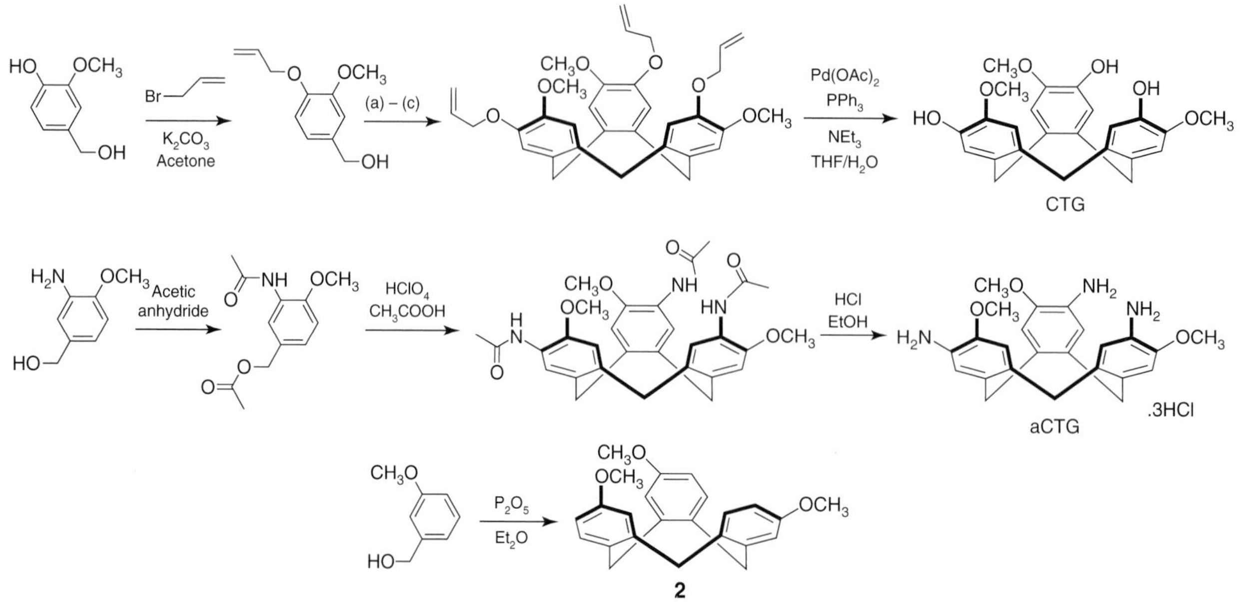

Scheme 4 Synthesis of the CTV analogs CTG, aCTG, and **2** from simple starting materials. Different conditions used for cyclization step of CTG synthesis include (a) cat. H_3PO_4, Δ; (b) cat. $Sc(OTf)_3$, CH_2Cl_2; (c) 60% $HClO_4$.

(45% yield),[1] and catalytic amounts of neat phosphoric acid in a solventless reaction at ~80 °C (41% yield).[9] Brotin *et al.* have recently demonstrated that much milder conditions involving catalytic levels of $Sc(OTf)_3$ in acetonitrile successfully gives CTV in 44% yield.[10]

Other 3,4-substituted benzyl alcohols can also be cyclized to give CTV analogs with extended arms on the upper rim. In general, these condensation cyclizations of the benzyl alcohols occur when the benzyl alcohol has functional groups in the meta- and para-positions that possess available electron pairs, such as OR, NR, and SR groups.[1] Most reported condensations of 3,4-substituted benzyl alcohols have used similarly strongly dehydrating conditions to those used for the synthesis of CTV, and the use of P_2O_5 is emerging as a useful strategy.[11] Milder conditions can be successful in some instances: for example, catalytic amounts of $Sc(OTf)_3$[10] or catalytic amounts of phosphoric acid in a solventless reaction or using ionic liquids as a medium.[9] The $Sc(OTf)_3$ approach is an interesting advance, as it allows for benzyl alcohols with acid-sensitive groups to be cyclized, and offers simpler work-up procedures. It should be noted that, to achieve cyclization of a particular benzyl alcohol, it may be necessary to try several different approaches. The condensation of substituted benzyl alcohols where the functional groups at the 3 and 4 positions are not the same invariably results in a C_3-symmetric product being formed. An alternative cyclization approach employing 1,2-substituted benzene and formaldehyde under acidic conditions gives a mixture of C_3- and C_2-symmetric products.[1]

As examples, the synthesis of CTG and the amine-functionalized aCTG from simple starting materials are shown in Scheme 4.[6,9,10,12] Both use acid condensation of a benzyl alcohol with a protected amine or an ether functional group that is subsequently deprotected. The synthesis of CTG is shown starting with 3-hydroxy-4-methoxy benzyl alcohol but the same procedure also works with 3-methoxy-4-hydroxy benzyl alcohol. Analogs with more or fewer functional groups at the upper rim can also be synthesized. The acid condensation of 3,4,5-trimethoxybenzyl alcohol gives the nonamethoxy analog nonamethoxy-tribenzo-cyclononatriene. This is initially synthesized in the crown conformation, but can be converted to a 1 : 1 mixture of crown and saddle forms by heating in a high-boiling solvent.[13] Cyclization of 3-methoxy benzyl alcohol in ether in the presence of an excess of phosphorous pentaoxide gives 2,7,12-trimethoxy-10,15-dihydro-5*H*-tribenzo[a,d,g]cyclononene **2** in 6% yield, and is a rare example of direct cyclization of a benzyl alcohol functionalized only in the meta-position.[11]

2.1.2 *Reactions of CTV and analogs*

Reactions at the upper rim

Cyclization of the appropriate benzyl alcohol is one route to the synthesis of extended-arm CTV analogs. Another, often more convenient route is by the functionalization of a preformed host framework. The most commonly used for this purpose are CTC and CTG. CTC is accessible from the complete demethylation of CTV using boron

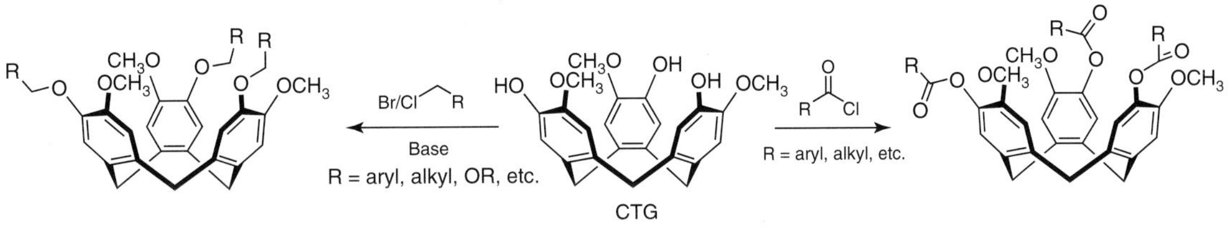

Scheme 5 Typical reactions at the upper rim of CTG.

tribromide. Partial demethylation of CTV may also be achieved using $TiCl_4$ or $AlCl_3$ as the demethylating agent.[14] The alcohol groups of CTC and CTG can undergo substitution reactions with alkyl halides in the presence of base to give ether-linked side arms, or with acid chlorides to give ester-linked side arms (Scheme 5). The amine-derived aCTG is another useful platform, as Schiff-base chemistry can be employed to synthesize imine or amine derivatives.

Reactions at the lower rim

Oxidation of CTV with sodium dichromate gives the monoketone, where one $-CH_2-$ group of the lower rim has been converted to a carbonyl.[1] Unlike CTV, the monoketone does not assume a crown conformation, but instead adopts the flexible saddle conformation.[15] Chemical derivatives of the CTV ketone, such as CTV oxime, likewise exist as a mixture of interconverting crown and flexible saddle forms. In the case of the oxime, and the ratios of the equilibrium mixtures is highly solvent dependent.[16] Derivation at the methylene site means that there is no negative steric crowding of the $-CH_2-$ group, which is thought to disfavor the saddle conformation in CTV.

2.2 Cryptophanes

2.2.1 Synthetic approaches to cryptophanes

An excellent and detailed account of cryptophane synthesis and characterization can be found in Brotin and Dutasta's review of 2009.[3] Therein, the authors outline in detail three main approaches to cryptophane synthesis: the direct method where both CTV fragments are formed in the one step; the template method where one preformed CTV fragment is used to template the cyclization of another; and the coupling approach where two preformed CTV fragments are linked together. These approaches will be summarized here. All three routes tend to give higher proportions of the anti-isomer over the syn-isomer, or only anti-isomer. Syn- and anti-isomers can usually be separated by column chromatography.

The direct method of cryptophane synthesis involves simultaneous formation of both CTV fragments from a

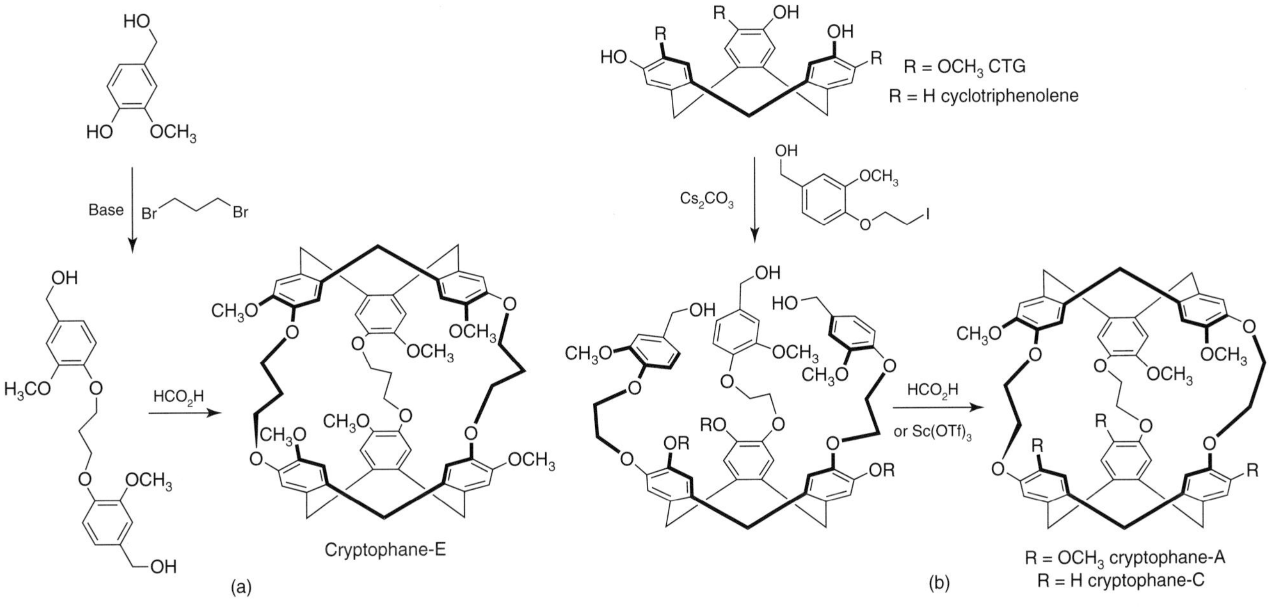

Scheme 6 Two approaches to the synthesis of cryptophanes: (a) direct synthesis of cryptophane-E; (b) template synthesis of cryptophane-A and cryptophane-C.

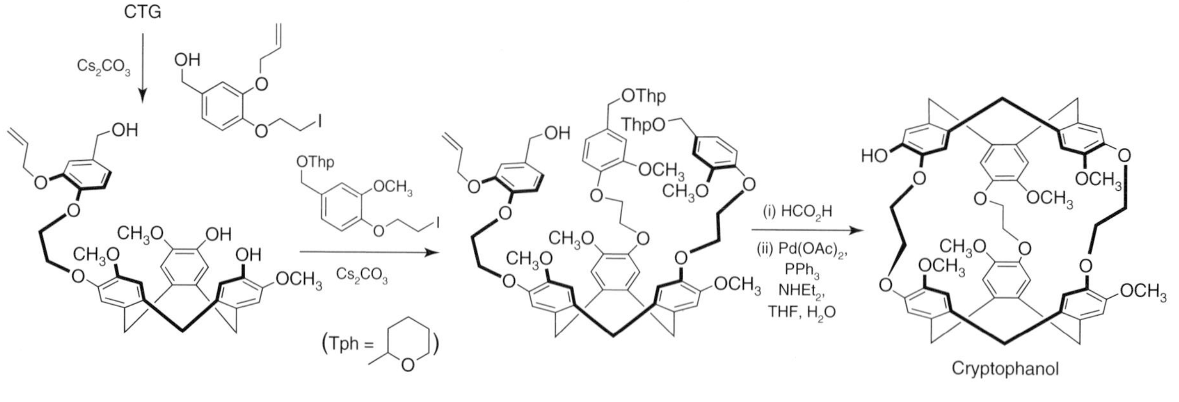

Scheme 7 The stepwise template synthesis of C_1-symmetric cryptophanol.

bis-(vanillyl alcohol) precursor. An example is shown in Scheme 6(a) for the formation of cryptophane-E. As for CTV synthesis, electron-donating groups are required at the meta- and para-positions, and formic acid is commonly used for the acid catalyzed cyclization. This approach has so far only been employed for symmetric cryptophanes of D_3 (anti) or C_{3h} (syn) symmetry, and yields tend to be low owing to the simultaneous formation of oligomeric products.

In the template method of cryptophane synthesis, CTG or another preformed CTV fragment is derivatized with further benzyl alcohol units and these subsequently cyclized to the cryptophane, as shown for cryptophane-A and -C in Scheme 6(b). As before, strongly electron-donating groups are required in the position meta to the benzylic alcohol. Reactions are performed at millimolar concentrations to avoid oligomerization, and the preorganization of the precursor means that overall yields are generally higher than for the direct approach. The ratio of syn/anti isomers obtained depends on the length of the linkers between the CTV fragments and may be different from those obtained from the direct method of synthesis.

The template method is particularly useful for the synthesis of cryptophanes of lower symmetry. Use of cyclotriphenolene as the preformed host, for example, allows for the formation of a C_3-symmetric cryptophane-C with different functionalizations at the upper rims of each CTV fragment (Scheme 6b). A stepwise variation of the template method can yield cryptophanes of C_2 or C_1 symmetry. For example, the synthesis of the C_1-symmetric cryptophanol is shown in Scheme 7. This involves initial monosubstitution of the CTG, then substitution at the final two positions with a different and protected benzyl alcohol. Treatment with formic acid both deprotects the alcohols and promotes the condensation reaction to cryptophane.[17]

The third approach to cryptophane synthesis involves coupling together two preformed and appropriately functionalized CTV fragments.[3] The smallest known cryptophane, cryptophane-1.1.1, was reported by Brotin and Dutasta using a coupling approach, where two cyclotriphenolene units were linked together through reaction with Cs_2CO_3 and $BrClCH_2$ (Scheme 8).[18] Higher yields for cryptophane-1.1.1 have been reported for an alternative coupling approach involving five steps from 3-hydroxy-4-iodobenzyl alcohol.[19] The coupling approach can also be used to effect resolution of CTV enantiomers (Section 2.3).

2.2.2 *Hemicryptophanes*

Hemicryptophanes are C_3-symmetric capsule-like hosts with only one CTV fragment.[1] They can be regarded as capped, extended CTVs. The capping group may be either an organic fragment or metal–ligand interactions can be used to cap an extended-arm CTV.[1] Organically

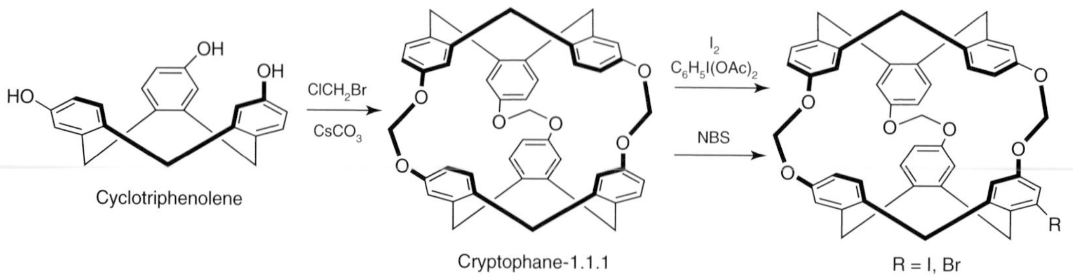

Scheme 8 The coupling synthesis of cryptophane-1.1.1 and its reactivity (Section 2.2.3).

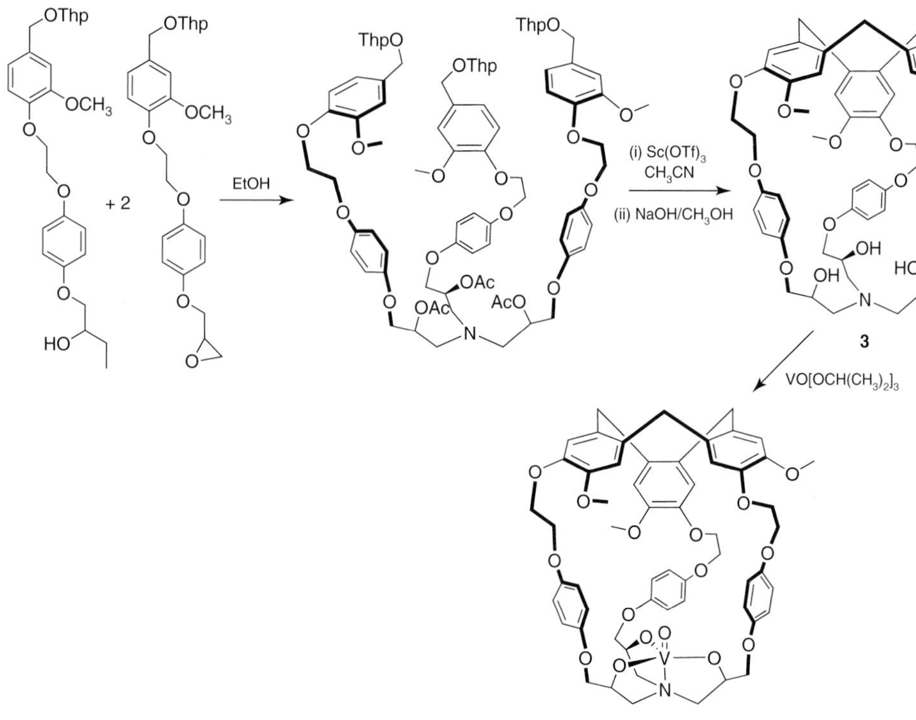

Scheme 9 Synthesis of hemicryptophanes **3** and vanadyl complex **4**.

capped hemicryptophanes have been synthesized through a variation on the template approach in which the C_3-symmetric capping group is preformed with three appended benzyl alcohol groups, which are then cyclized to form the CTV fragment. An example is given in Scheme 9 for the tris(alkanol)aminehemicryptophane **3**, which can be subsequently metalled to the oxovanadium complex **4**.[20] Use of an enantiopure chiral epoxide during the synthesis of **3** gives two diastereoisomeric hemicryptophanes, which can be separated after the cyclization stage using chiral chromatography.

2.2.3 *Reactions of cryptophanes*

Cryptophanes such as cryptophane-A can be demethylated using PPh_2Li to give hydroxyl derivatives.[2,3] These can then undergo substitution reactions akin to those described for CTC and CTG in Section 2.1.2, Reactions at the upper rim. Alternatively, C_3-symmetric cryptophanes with methoxy groups on one rim and hydroxyl groups on another can be synthesized by template strategies. These are convenient precursors for generating water-soluble carboxylated cryptophanes.[21]

One of the most important hydroxy-cryptophanes is the monosubstituted cryptophanol whose synthesis has been reported by a handful of stepwise template strategies including that of Scheme 7.[17,22] It can be used to generate bis-cryptophanes through reaction with dialkylbromides, and can be chirally resolved (Section 2.3).[3] The hydroxyl group is sterically hindered, leading to long reaction times and/or low yields for these reactions. Conversion of the alcohol to a carboxylic acid allows for the cryptophane to be appended onto various chemical moieties including peptides. This has been an important step for the development of cryptophane-based biosensors discussed in Section 3.2.1, Xe binding by cryptophanes and its applications.[22]

Direct functionalization of cryptophane in positions other than the upper rim of the CTV fragments is very rare. One example is the direct iodination or bromination of one arene ring of cryptophane-1.1.1 using $I_2/PhI(OAc)_2$ or *N*-bromosuccinamide, respectively (Scheme 9).[19] The monohalogenated cryptophanes have been used to generate deuterium, carboxylic acid, or allyl malonate derivatives.

2.3 Chiral resolution of CTVs and cryptophanes

Chiral stationary-phase HPLC (high-performance liquid chromatography) columns can be used to separate the enantiomers of *anti*-cryptophanes, or CTV derivatives[3]; however, it is not an ideal technique owing to its sensitivity to conditions, high expense, and inability to separate large quantities of material. A better approach for the resolution of chiral CTVs and cryptophanes is to attach a resolved chiral moiety onto the CTV or cryptophane framework. The

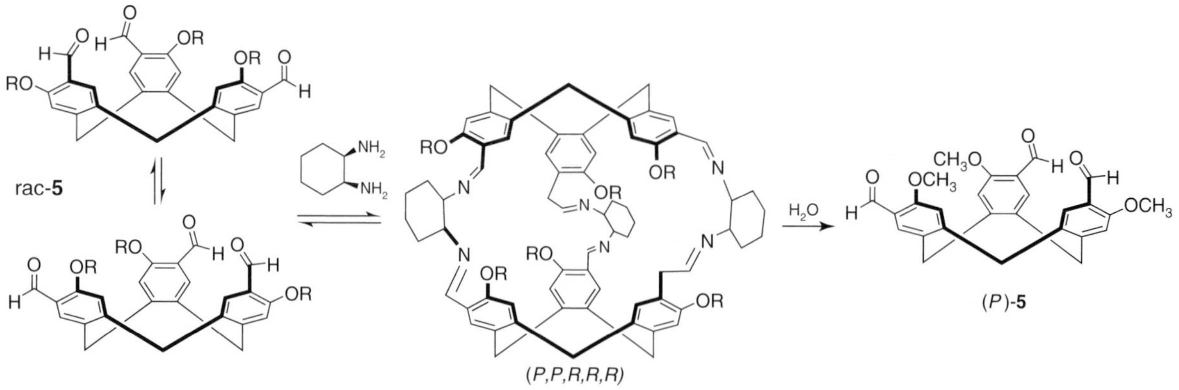

Scheme 10 Dynamic resolution of the tris-aldehyde CTV **5**.

subsequent diastereomers can then be separated by column chromatography or crystallization, and the appended chiral group removed to leave resolved CTV material.[3]

A recent example of this approach illustrates that chirally resolved cryptophanes may be accessed in a straightforward manner. Reaction of a racemic mixture of the monofunctionalized cryptophanol with (1*S*)-(−)camphanic acid gives two diastereoisomers. The diastereoisomers could be separated by fractional crystallization from toluene, although efficient separation occurs only at low concentrations. The resolved cryptophane camphanates are then converted to chirally pure cryptophanol by hydrolysis.[23]

The ability of chiral CTV derivatives to racemize can be manipulated to resolve the enantiomers by dynamic thermodynamic resolution as recently illustrated by Xu and Warmuth.[24] Their approach is outlined in Scheme 10. A racemic mixture of the tris-aldehyde **5** is reacted with (*R,R*)-diaminocyclohexane to give an enantiomerically pure *anti*-cryptophane. Only the (*P,P,R,R,R*) isomer of the cryptophane is formed from reaction of (*P*)-**5** with the amine; the (*M*)-**5** isomer is completely inverted. Hydrolysis of the cryptophane then gives (*P*)-**5** at >99% ee.

The chiroptic properties of resolved cryptophanes have recently garnered attention and can be observed through electronic circular dichroism (ECD) and vibrational circular dichroism (VCD).[3,25] The multiple chromophores of cryptophanes mean that their chiroptic properties can serve as useful models for more complicated supramolecular systems; VCD and *ab initio* studies can be combined to assign or study absolute configurations and conformations and host–guest interactions can be studied. Enantiopure cryptophanes should show specific CD (circular dichroism) responses on complexation of guests, reflecting conformational changes that occur in the bridges of the cryptophane on complexation.[3] The ECD spectra of aqueous basic solutions of the *M,M*-isomer of demethylated cryptophane-A, for example, shows different responses when different-sized chlorinated guests are encapsulated.[25]

3 HOST–GUEST CHEMISTRY

3.1 Inclusion and host–guest chemistry of cyclotriveratrylenes

3.1.1 Simple crystalline inclusion complexes

Unlike cryptophanes, the parent CTV does not exhibit extensive solution-phase host–guest chemistry. There is little by way of guest binding in solution, and the majority of crystalline inclusion complexes are α- or β-phase clathrate complexes. In these, the CTV molecules stack on top of one another in a bowl-in-bowl manner into misaligned columns. Solvent guests occupy channels created by packing of these columns. The phases can be identified by IR (infrared) spectra and by their differing *b* unit cell parameters, which in the α phase is about 9.5–9.7 Å while in the β phase it is much shorter at about 8.0–8.4 Å.[26] Unsolvated CTV adopts the α phase (Figure 2a), and the β phase is structurally distinguished by the rotation of one methoxy group.[26] Intracavity complexation of small organic guests by CTV in the solid state is rarer, and reports include CTV with acetone,[27] 1,1,1-trichloroethane (Figure 2b), or 1,1,2-trichloroethane[28] as the guest. When metallated, or embedded in hydrogen-bonded network structures or coordination polymers, intracavity guest complexation of solvent by CTV is more common. The so-called ball-and-socket complexes between CTV and large, spherical guests are also known, including fullerene C_{60},[29] *o*-carborane (Figure 2c),[30] and organometallic complexes,[31] among others.

Conversely, crystalline inclusion complexes of CTC feature intracavity complexation of the guest molecule.[14,31] The crystalline clathrates of CTG and their tris-substituted analogs are dominated by self-inclusion motifs. These include misaligned bowl-in-bowl self-stacking similar to that of α-phase CTV (Figure 2a),[32] but also the so-called "hand-shake" motif where the extended arm of one host occupies the molecular cavity of another and vice

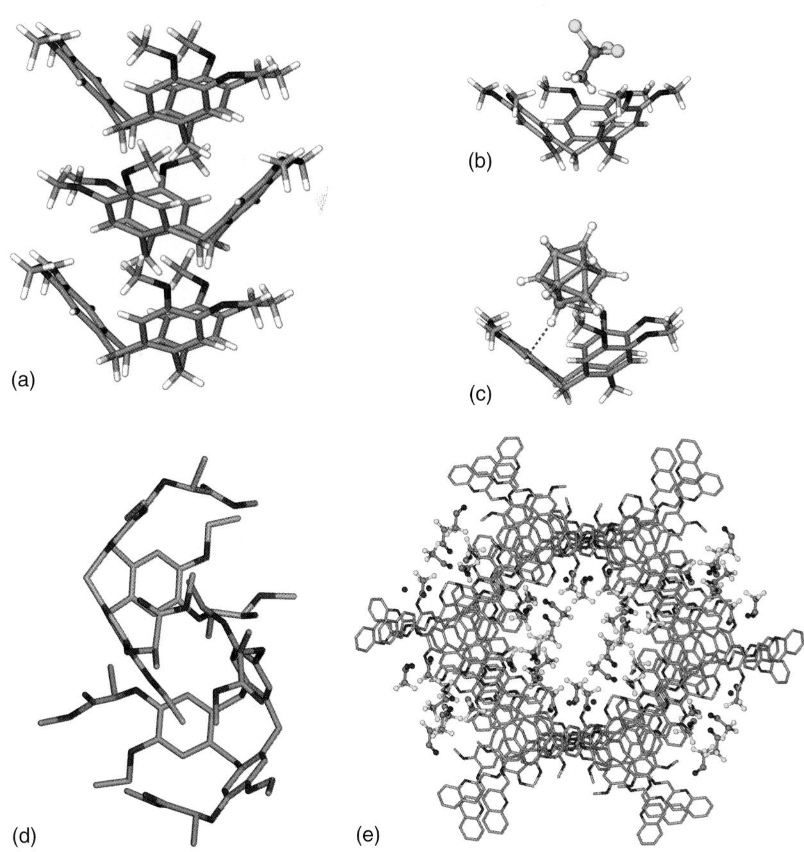

Figure 2 X-Ray structures of inclusion motifs in crystalline CTVs. (a) A misaligned stack of α-phase CTV; (b) intracavity complexation of 1,1,1-trichloroethane; (c) ball-and-socket structure from the complex (*o*-carborane)$(CTV)_2$: hydrogen bonds between the acidic carborane C–H and CTV are indicated as dashed lines; (d) "hand-shake" motif in **6**. (e) Solvent-filled channel in clathrate complex of **7**, where self-inclusion and $\pi-\pi$ stacking interactions result in an open structure.

versa.[33–35] This was first observed for the unsolvated structure of 2,7,12-tris-ethoxy-3,8,13-tris(1-methoxycarbonylethoxy)-10, 15-dihydro-5*H*-tribenzo[a, d, g]cyclononene **6**, shown in Figure 2(d).[33] Most other examples involve formation of C–H$\cdots\pi$ or face-to-face $\pi-\pi$ stacking interactions. Similar inclusion motifs can extend into chains,[34,35] or even 2D lattices, as shown in Figure 2(e) for the water/trifluoroethanol clathrate of tris-(2-quinolylmethyl)cyclotriguaiacylene **7**.[34] Crystalline inclusion complexes with intracavity guest complexation are known,[34] including those with capsule-like dimer formation around the guest.[36,37]

3.1.2 CTVs as hosts for fullerenes

The electron-rich nature of the CTV cavity and the electron-deficient nature of fullerenes, coupled with size and shape matching, mean that CTVs are good hosts for fullerenes. Several solid-state complexes have been characterized,[29,38–40] all forming 1 : 1 "ball-and-socket" superstructures where the fullerene perches above the host's molecular cavity. This is shown for the C_{60} complex with a close relative of CTVs, hexabromotribenzotriquinacene, in Figure 3(a).[40] In some cases, the overall complex is not 1 : 1 and additional C_{60} may be present within the crystal lattice.[29] C_{70} forms similar interactions with CTV within the quaternary crystalline complex [(CTV)(*o*-carborane)(C_{70})]·(1,2-dichlorobenzene).[38] In that complex, the *o*-carborane—itself a potential guest for CTV—acts as a hydrogen-bond donor, forming a chain structure through bifurcated hydrogen bonds to the methoxy groups of the CTV. Methoxy groups of CTV are also able to bind to alkali-metal cations to form coordination polymers. In the remarkable structure of complex $[Cs_2(DMF)_7(CTV)]\cdot\{(C_{70})_2\}\cdot$DMF·(toluene) where DMF = dimethylformamide, a 2D coordination polymer is formed between bridging CTV and $[Cs_2(DMF)_n]^{2+}$ units, and each CTV acts as a host for one anionic $(C_{70})_2{}^-$ dimer (Figure 3b).[39]

Unusually for CTV, it also binds C_{60} in solution, forming a 1 : 1 host/guest complex.[41] Fullerenes are highly colored in solution, and hence electronic spectroscopy is ideal for establishing solution binding constants with them. Functionalization of the CTV core can result in

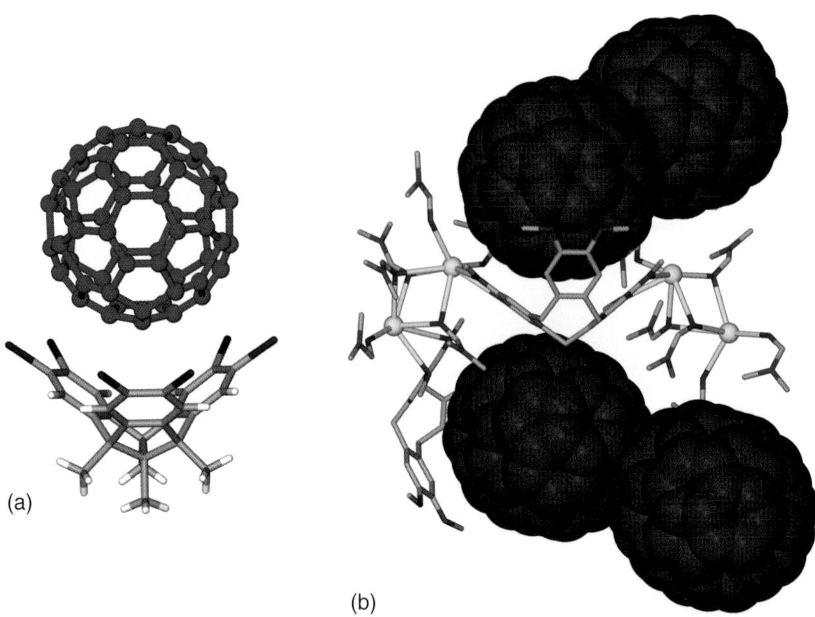

Figure 3 Crystal structure showing fullerene binding in the solid state by CTV or analogs. (a) C_{60} hexabromotribenzotriquinacene complex; (b) section of the $[Cs_2(DMF)_7(CTV)]^{2+}$ network binding $(C_{70})_2^-$ dimers, Cs^+ shown as orange spheres.

enhanced binding of C_{60} in solution. For example, the 1:1 binding constants for dendritic tris-CTV analogs with C_{60} increase on increasing the generation number of dendritic substituents of the CTV core.[42] Higher fullerenes can also be bound in solution by certain CTV derivatives, although not by CTV itself. For example, hosts **8–10** all bind C_{70} more strongly than they do C_{60},[43–45] with 1:1 binding strengths in the order **8** < **9** < **10**. These trends can be accounted for by considering the ability of the hosts to interact with C_{70}. The capsule-like **9** has two CTV fragments and is able to interact with two sides of the C_{70}. Density functional theory (DFT) calculations show that the highly electron-rich anthracene side arms of **10** are able to bend and wrap around the fullerenes to achieve maximal host–guest contact.[45]

The differences in binding abilities for different fullerenes of some hosts can be used to separate mixtures of fullerenes. Host **11** features three 4-ureidopyrimidinone side arms and is designed to form hydrogen-bonded dimeric capsules, as 4-ureidopyrimidinone has a self-complementary DDAA (D = donor, A = acceptor) hydrogen-bonding sequence. Fullerenes can be bound in solution inside the hydrogen-bonded capsules to form $(\mathbf{11})_2\cdot$F complexes where F is a fullerene.[46] The system is selective for higher fullerenes, and binding constants increase by an order of magnitude on binding fullerenes in the order of increasing size $C_{84} > C_{70} > C_{60}$.[46] The hydrogen-bonded capsules of **11** make the separation and recovery of fullerenes easy, as they are easily broken apart by changing the polarity of the solvent system (Scheme 11).

The fullerene-binding abilities of CTVs have also been exploited in order to solubilize fullerenes,[47] to immobilize them onto surfaces, and to incorporate them into liquid crystals. The hexa-substituted **12**, for example, forms a liquid-crystalline phase.[48] In the presence of C_{60}, a 2:1 host/guest complex is formed, which also displays liquid-crystalline behavior, with a fluid birefringent phase at temperatures below 70 °C.[48] The thio-derived CTV **13** forms self-assembled monolayers (SAMs) on gold surfaces.[49] SAMs of **13** on modified gold beads can bind C_{60} from solution when soaked in 1,2-dichlorobenzene solutions of C_{60}.

3.1.3 Binding and sensing of ions

Cation binding

While CTV is capable of complexing alkali-metal cations through its dimethoxy groups at the upper rim, examples of targeted cation binding have relied on appending cation-binding functional groups or through formation of capsule-like structures. The CTV molecular platform may serve to preorganize cation-binding groups. The azobenzene-derived host **14** forms a 1:1 complex with Hg(II), with significantly enhanced sensitivity and selectivity in comparison with 4-aminoazobenzene.[50] The presence of chromophores for a UV–visible response, selectivity over other heavy metal cations including Cd(II) and Pb(II), and high sensitivity make **14** a chemosensor for Hg(II) with a detection limit of 0.5 μM in aqueous environmental samples.[50] Host **15** is designed as a receptor for actinides, which can be bound by the hard oxygen donors on the side arms.[51] Use of

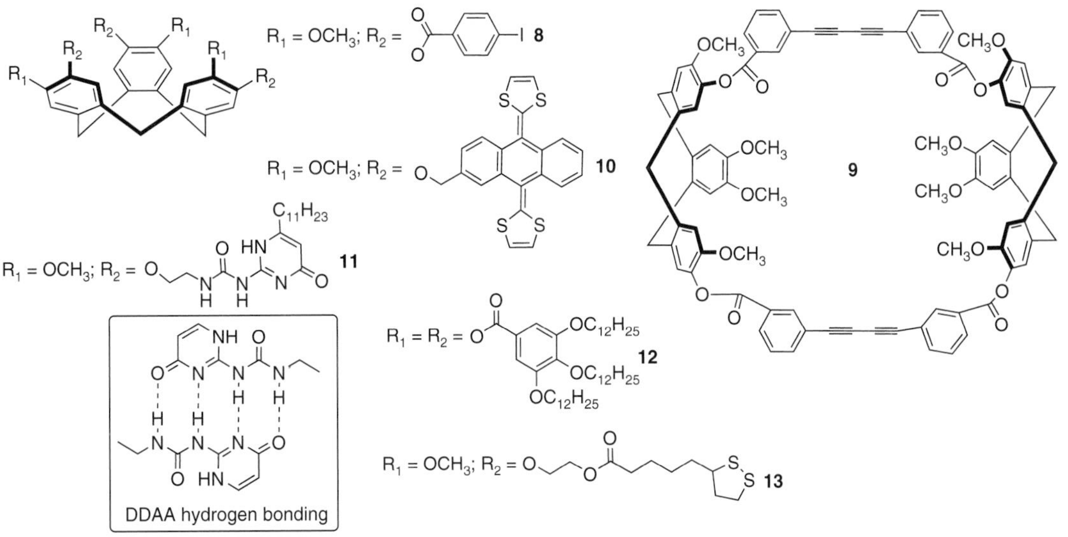

Scheme 11 Selected fullerene-binding CTV analogs.

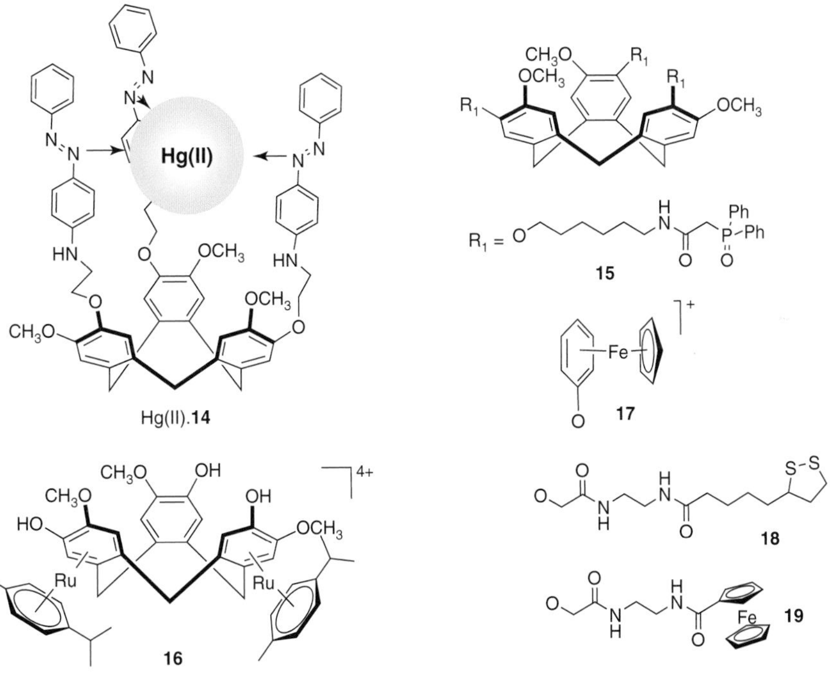

Scheme 12 Selected cation- and anion-binding CTV analogs.

the CTV platform gave better extraction distribution ratios over less preorganized platforms, but it did not lead to enhanced selectivity for extraction of Am(III) cf. Eu(III) (Scheme 12).[51]

Binding of cations inside the host molecular cavity is not common but has been demonstrated in the solid state for organometallic cations.[31] There are a small number of examples of cation capture within capsule-like assemblies. Deprotonated CTC forms "clam-like" capsule arrangement from hydrogen-bonding interactions between the CTC molecules. Encapsulated cationic guests include Rb^+ within a closed clam-shell arrangement and organic cations such as $NEt_4{}^+$ in an open clam-shell arrangement (Figure 4).[52] These complexes form from water-containing solutions, with the guest alkali-metal cations Rb^+ and Cs^+ preferring the aromatic environment inside the CTC capsule to coordination by hard donor O from the CTC or, indeed, from water. This is not the case with the smaller K^+ cation, where the K^+ sits inside the molecular cavity of only one monodeprotonated CTC and also is coordinated by the O sites on other monodeprotonated CTC hosts.

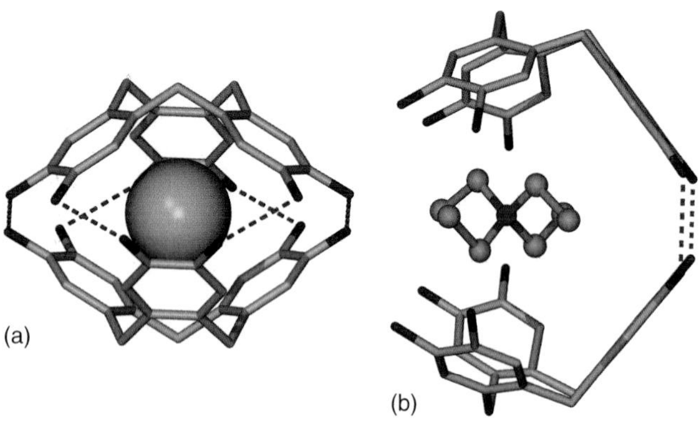

Figure 4 Crystal structures of "clam-like" CTC dimers encapsulating cations, with O···O consistent with hydrogen bonding shown as dashed lines. (a) Rb^+ complex; (b) NEt_4^+ complex with only one of two disordered positions of the cation shown.

Anion binding
One molecular approach to binding anions is metallation of the CTV. Bimetallated CTV hosts such as $[\{Ru(\eta^6\text{-}p\text{-cymene})\}_2(\eta^6:\eta^6\text{-CTV})]^{4+}$ **16**[53] exhibit solid-state binding of various anions including BF_4^-, $CF_3SO_3^-$, and ReO_4^-. In solution, **16** selectively binds ReO_4^- or $^{99}TcO_4^-$ in preference to Cl^-, nitrate, sulfate, triflate, and perchlorate. The ability of the CTV molecular cavity to bind anionic guests in these examples is attributed to the withdrawal of electron density from the arene rings on metallation. In contrast, $[\{Rh(\eta^4\text{-norbornadiene})\}_2(\eta^6:\eta^6\text{-CTV})]^{4+}$ does not bind anions in solution or in the solid state, due to back-bonding from the Rh(I) center rendering the arene rings relatively electron rich.[54] Appending a ferrocene-like moiety to the upper rim of CTV in host **17** gives a host with enhanced binding ability for anions in comparison with those of **16**, and is able to extract >95% of TcO_4^- from a saline solution containing a range of anions.[55]

CTV can be used as a platform for attaching functional groups with hydrogen-bonding capability for anion binding. For example, the tris-amide CTV derivative **18** strongly binds OAc^- and weakly binds $H_2PO_4^-$ in solution, but does not bind many other common anions.[56] SAMs of **18** have been shown to act as a sensor for OAc^-. Similar amide binding sites are utilized in **19**, where the ferrocene is an electrochemical reporter group for binding $H_2PO_4^-$ and ATP^- anions.[57]

3.2 Inclusion and host–guest chemistry of cryptophanes

3.2.1 Binding of neutral guests

Unlike CTVs, cryptophanes are able to bind small guest molecules in solution as well as within solid-state complexes.[3] In solution, cryptophanes show particularly strong affinity for small molecules including methane, chloroform, dichloromethane, and xenon.

Complexes with organic guests
Host–guest studies of cryptophanes with halogenated hydrocarbons demonstrate that a range of such guests can be bound, and that cryptophanes of different sizes are best matched with different guests.[3] Cryptophane-E, for example, binds chloroform most strongly while cryptophane-C favors dichloromethane. The crystal structure of the $\{CHCl_3 \cap \text{cryptophane-}E\}$ complex shows that the chlorine atoms of the trapped guest are in the equatorial plane of the cryptophane, while the C–H group points into a CTV cavity, albeit in a disordered manner (Figure 5a).[58] There is a slow exchange between guest and bulk chloroform in solution. Dynamics studies show that the entire complex behaves as a single molecule in solution, meaning there is no significant motion of the chloroform within the cryptophane.[59] This is in contrast to the $\{CH_2Cl_2 \cap \text{cryptophane-E}\}$ complex, which shows a lower binding constant and there is considerable rotation

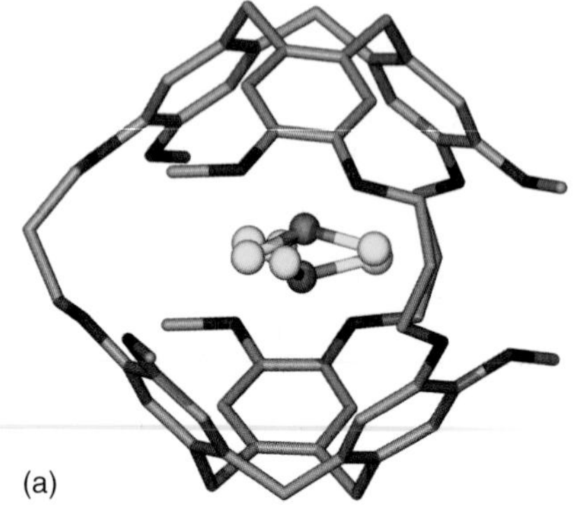

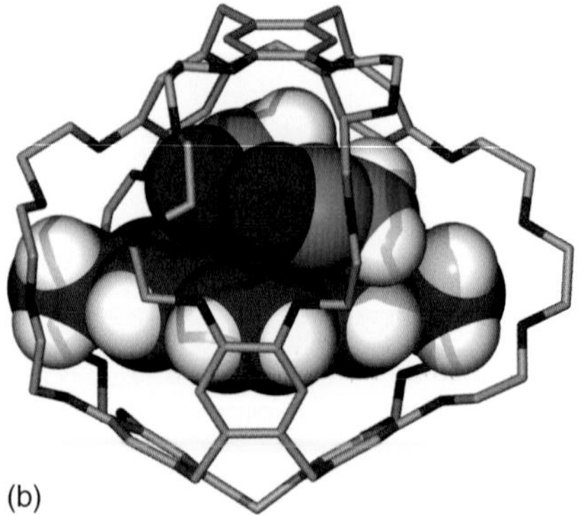

Figure 5 Crystal structures of guests within cryptophanes. (a) $\{CHCl_3 \cap \text{cryptophane-E}\}$ with disordered $CHCl_3$ shown in ball-and-stick mode (b) complex $\{(\text{dimethyldiazapyrenium})(CH_3CN)_2 \cap \mathbf{27}\}$. Guests shown in space-filling mode with dimethyldiazapyrenium in purple and CH_3CN in green.

of the guest in solution.[60] Analytical optical resolution of bromofluorochloromethane has been demonstrated through its complexation by cryptophane-C, with the diastereomeric host–guest complexes {(+)-CHBrClF ∩ (+)-cryptophane-C} and {(−)-CHBrClF ∩ (+)-cryptophane-C} distinguishable using NMR (nuclear magnetic resonance).[61]

The binding of methane by cryptophanes was first reported for cryptophane-A, which also binds a range of small chlorofluorocarbons.[62] This binding ability has been utilized for the design of new types of methane sensors.[63] Recently, Brotin and Dutasta and coworkers reported that cryptophane-1.1.1 exhibits slightly higher binding affinity for methane than does cryptophane-A.[64] They also reported weaker binding of C_2H_4 and C_2H_6 by cryptophane-1.1.1, but noted that the host shows no affinity for larger hydrocarbons or for small chlorinated hydrocarbons.[64] Akabori and coworkers report binding of various branched hydrocarbons by both the syn- and anti-isomers of the diethylenetrioxy-bridged cryptophane **20**, and by only the syn-isomer of similarly sized cryptophanes with arene linkers.[65]

The carboxylic acid-decorated and water-soluble cryptophane **21** is a versatile host, and solution binding has been reported for a diverse range of guests, namely organic radicals, ammonium cations (Section 3.2.3, Cation binding), and tetrahedral main group compounds.[3] Binding of piperidine aminoxyl radicals of type **22**, for example, was followed by EPR (electron paramagnetic resonance) spectroscopy, and binding constants K_a were estimated to be of the order 1000–2000 M^{-1}.[66] This is weaker binding than is seen with ammonium guests and $(CH_3)_3N(CH_2CH_2OH)^+$ (choline), for example, where $K_a \sim 5000\,M^{-1}$, can displace the aminoxyl radicals from within the cryptophane.

Both diastereomers of the hemicryptophane–oxidovanadium(V) complex **4** have also been shown to be effective supramolecular catalysts for sulfoxidation reactions.[67] The use of **4** as a catalyst gave significantly better yields than the use of a noncyclic vanatrane analog. Hence, binding of the reagents within the hemicryptophane strongly enhances the catalytic activity of the vanadium complexes. Each diastereomer of **4** exists as two diastereomeric conformers, owing to Δ or Λ configurations of the tricyclic atrane. Interconversion between the Δ and Λ forms has a high energy barrier and is strongly solvent dependent, and this effect has been used to demonstrate solvent-induced chirality switching in **4** (Scheme 13).[68]

Xe binding by cryptophanes and its applications

The Xe-binding abilities of cryptophanes were first reported by Bartik and coworkers in 1998. They used both ^1H and ^{129}Xe NMR studies to follow the reversible Xe encapsulation by cryptophane-A.[69] Since then, numerous analogs of cryptophane-A have been shown to bind Xe. An atom of Xe has a volume of 42 Å^3; hence cryptophanes with small internal volumes should be best suited to forming van der Waals host–guest complexes with it. Cryptophane-1.1.1 was designed as a cryptophane with a small internal cavity (~81–95 Å^3 for cryptophane-A) specifically for this purpose,[18] and is indeed the strongest known binder of Xe from an organic solvent with reported $K_a \sim 28\,000\,M^{-1}$.[11]

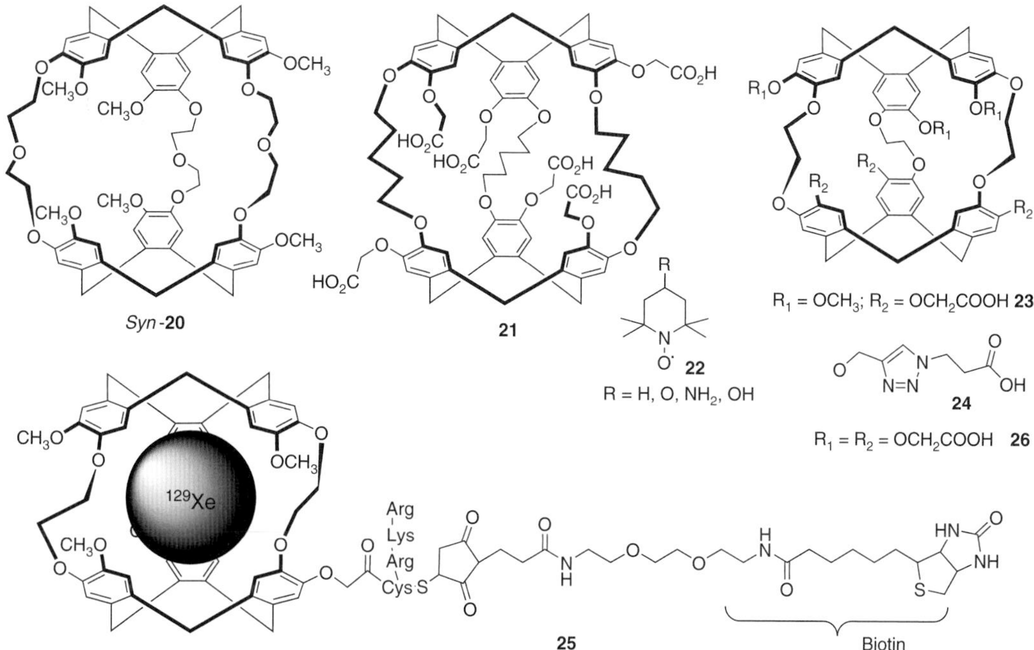

Scheme 13 Examples of cryptophanes that bind gaseous guests.

Even higher Xe-binding constants from aqueous solutions have been reported for the water-soluble, carboxylic acid-derived **23** and **24**.[21,70]

Considerable interest in Xe binding by cryptophanes has been prompted by Pines and coworkers' 2001 report demonstrating the feasibility of using a biotin-functionalized cryptophane-A **25** as a biosensor.[22] ^{129}Xe is a nontoxic, spin-1/2 isotope, which, on laser polarization, can have very high NMR sensitivity. Sensitivity coupled with large chemical shift differences between free and encapsulated Xe point to future applications in magnetic resonance imaging (MRI). Water-soluble carboxylic acid derivatives of cryptophane-A have been developed with such applications in mind.[8,21,70,71] Xe binding has been observed in human plasma (**24**),[70] and MRI of the encapsulated Xe has been demonstrated using the hexa-carboxylic acid-derived **26**.[71]

Bioactive groups can be attached to the cryptophane in order to generate targeted biosensors. For example, the Xe-containing cryptophane **25** with its biotin ligand group is designed to bind to proteins.[22] Binding of **25** to the protein avidin can be clearly monitored by ^{129}Xe NMR in D_2O solution. In another example, a DNA-modified Xe@cryptophane can recognize a complementary oligonucleotide strand, with binding again signaled by shifts in the ^{129}Xe NMR spectrum.[72] Viral capsids can be used as scaffold for cryptophane-based xenon hosts, which lowers the detection threshold of binding targets and improves solubility and biocompatibility.[73]

3.2.2 *Binding and sensing of ions*

Cation binding

Cryptophanes can bind ammonium cations, with the strongest binding occurring when there is a good size match between host and guest. Cryptophane-E, for example, shows moderate binding ability for $(CH_3)_3HN^+$, strong binding of $(CH_3)_4N^+$ from $(CDCl_2)_2$ solution, and a considerable drop off in binding abilities as the size of the ammonium cation increases until there is no detectable binding of $(CH_3)_3N^tBu^+$.[74] The larger water-soluble cryptophane **21** is a poor host for the small $(CH_3)_3HN^+$ from D_2O, but shows reasonable binding constants for larger cations such as acetylcholine and Et_4N^+.[74] The choice of isomer of the cryptophane employed also has an effect on the binding abilities for ammonium cations. The syn-isomer of the diethylenetrioxy-bridged cryptophane **20** binds Et_4N^+ and $Et_3CH_3N^+$ more strongly than does the anti-isomer.[75]

The unusual cryptophane analog **27** with two CTV fragments linked by six ethylene glycol chains is able to bind the larger organic cations dimethyldiazapyrenium or 4,4′-biphenylbisdiazonium. In both cases, crystal structures show the organic cation contained within the cryptophane along with two molecules of acetonitrile, shown for the dimethyldiazapyrenium complex in Figure 5(b) (Scheme 14).[76]

Kubo and coworkers report heterodimeric, covalently linked capsules that can be regarded as a type of hemicryptophane. These bind cations such as $NEt_4{}^+$ and Cs^+ in solution.[77] The heterocapsules are formed through a dynamic boronate esterification reaction that occurs in the presence of a base between CTC and a boronic acid-appended trioxacalix[3]arene with solvolysis, giving capsule **28**. Capsule formation is reversible and can be dynamically controlled using a pH switch. Neutral amines trigger the capsule formation, with their conjugate acids providing the cationic guest; for example, an NEt_3-triggered capsule formation results in encapsulation of NEt_3H^+. $N^nBu_3H^+$, however, is too large to be accommodated inside the cavity, so use of N^nBu_3 as the base gives the heterodimer with solvent as the guest, which can then be exchanged for Cs^+ guest cations.[77]

Anti- and syn-isomers of **20** can extract alkali-metal cations from aqueous solutions,[75] with binding of Cs^+ by the *syn*-cryptophane giving the best results. The *syn*-cryptophane **29** has internal carboxylic acid groups that

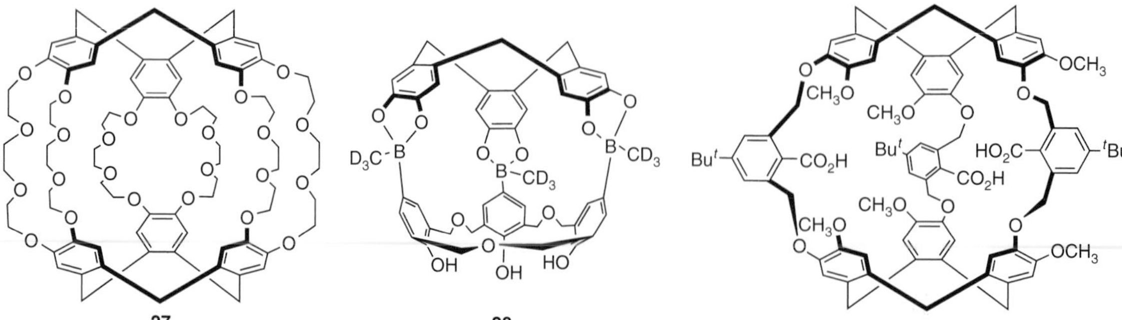

27 28 29

Scheme 14 Cation-binding cryptophanes.

are preorganized to bind hard metal cations. Weber and coworkers have shown that **29** can extract a range of alkali-metal and alkali-earth cations as well as Yb^{3+} and Eu^{3+} from water into chloroform solutions. The highest extractabilities were for Ca^{2+}, Sr^{2+}, and Yb^{3+}, in keeping with the likely interaction of these hard cations by the hard carboxylic acid groups.[78]

Anion binding

Holman and coworkers report the fully metalled cryptophane $[\{Ru(\eta^6\text{-}Cp^*)\}_6(\eta^6:\eta^6\eta^6\eta^6\eta^6:\eta^6\text{-cryptophane-E})]^{4+}$ where Cp^* = pentamethylcyclopentadiene. The metallated host binds anions inside the cryptophane cavity both in the solid state and in solution. It shows a preference for binding anions that fit snugly into the cavity including PF_6^- and $CF_3SO_3^-$, but does not bind smaller anions such as BF_4^- on size selectivity grounds.[79]

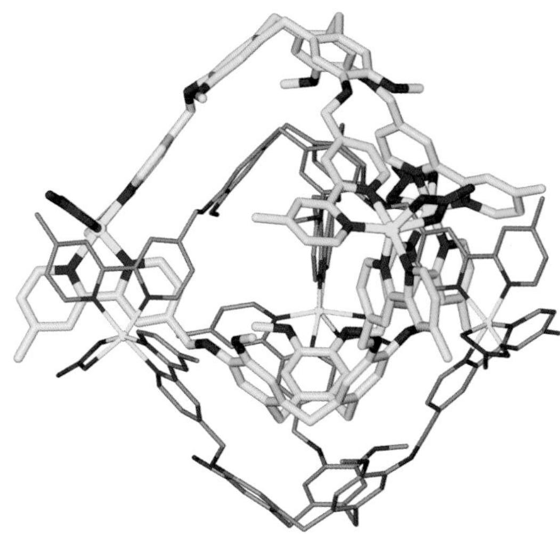

Figure 6 Crystal structures the [2]catenane structure of $[Zn_3(\mathbf{31})_2(NO_3)_3]_2^{6+}$ with two independent cages shown in different colors and line widths.

4 SELF-ASSEMBLED AND SELF-ORGANIZED SYSTEMS

Aside from their function in molecular recognition, CTV and analogs are interesting molecular platforms for more complex functions and assemblies. The inclusion properties of CTV manifest themselves through two mechanisms during these self-assembly processes. Firstly, typical inclusion motifs observed for the clathrate complexes of CTVs are also observed within the more complicated assemblies. This is particularly the case for the "hand-shake" motif discussed in Section 3.1.1. Secondly, binding of a guest molecule by the CTV may have a directing effect on the overall assembly.

4.1 Self-assembled cages

4.1.1 Metallo-cryptophanes and hydrogen-bonded capsules

Metallo-cryptophanes are analogous to cryptophanes, with both featuring two CTV fragments joined together in a head-to-head manner to create a capsule-like cage. For metallo-cryptophanes, CTV fragments are held together through labile metal–ligand bonding rather than the kinetically inert bonding of the purely organic cryptophanes. The first metallo-cryptophanes featured pyridyl-functionalized hosts that were linked together via cis-protected Pd(II) **30**.[80]

The 2,2′-bipyridine-derived ligand **31** forms unusual metallo-cryptophanes with Zn(II) or Co(II). Here, two capsules interlock to form the [2]catenane species $[M_3(\mathbf{31})_2(NO_3)_3]_2^{6+}$ (Figure 6),[81] which are detectable by mass spectrometry as well as crystallography. A [2]catenane is where two cyclic molecular components thread through one another so that they are chemically independent but mechanically interlocked. The triple interlocking motif of $[M_3(\mathbf{31})_2(NO_3)_3]_2^{6+}$ is relatively rare.

Homodimeric hydrogen-bonded capsules have been successfully employed to bind and separate fullerenes (Section 3.1.2). A heterodimeric capsule **32** forms through charge-assisted hydrogen bonding between a carboxylic acid and an amine-appended CTV and encapsulates a number of small neutral guests (Scheme 15).[82]

4.1.2 Larger cage-like assemblies

Three different types of M_4L_4 metallo-supramolecular cages with CTV-type ligands appended with N-donor groups have been reported by Hardie and coworkers. The tetrahedral cage $[Ag_4(\mathbf{33})_4]^{4+}$ crystallizes with a variety of counteranions, and is also present in solution.[83] The cage has Ag(I) centers at the vertices of a tetrahedron and the ligands on the faces of the tetrahedron. The internal cavity is filled with guest and coordinative acetonitrile molecules (Figure 7a). Each $[Ag_4(\mathbf{33})_4]^{4+}$ tetrahedron contains only one enantiomer of the ligand, although it crystallizes as a racemate. A different type of M_4L_4 tetrahedral assembly is isolated in the solid state from the self-assembly of the 2-quinoline-derived ligand **7** with $AgBF_4$.[34] As before, the four Ag(I) centers are arranged in a tetrahedron, but only two of the three quinoline groups of each ligand binds to an Ag(I) center, and the "hand-shake" inclusion motif occurs between pairs of ligand to give a dense tetrahedron with no internal space. The cationic tetranuclear assembly $[Pd_4(\mathbf{34})_4(NO_3)_2(H_2O)_2]^{6+}$ is not tetrahedral in nature but is more akin to a cube.[84]

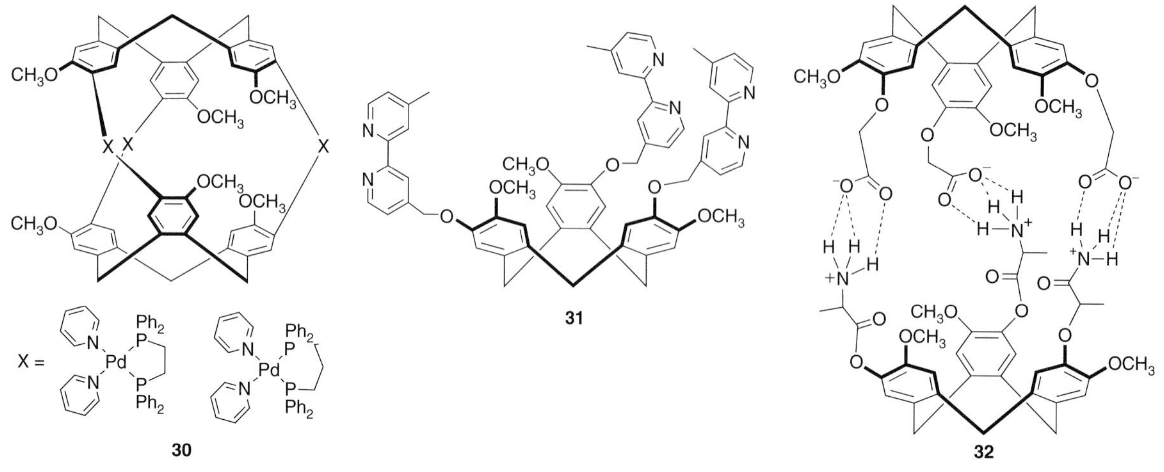

Scheme 15 Metallo-cryptophanes and hydrogen-bonded capsules.

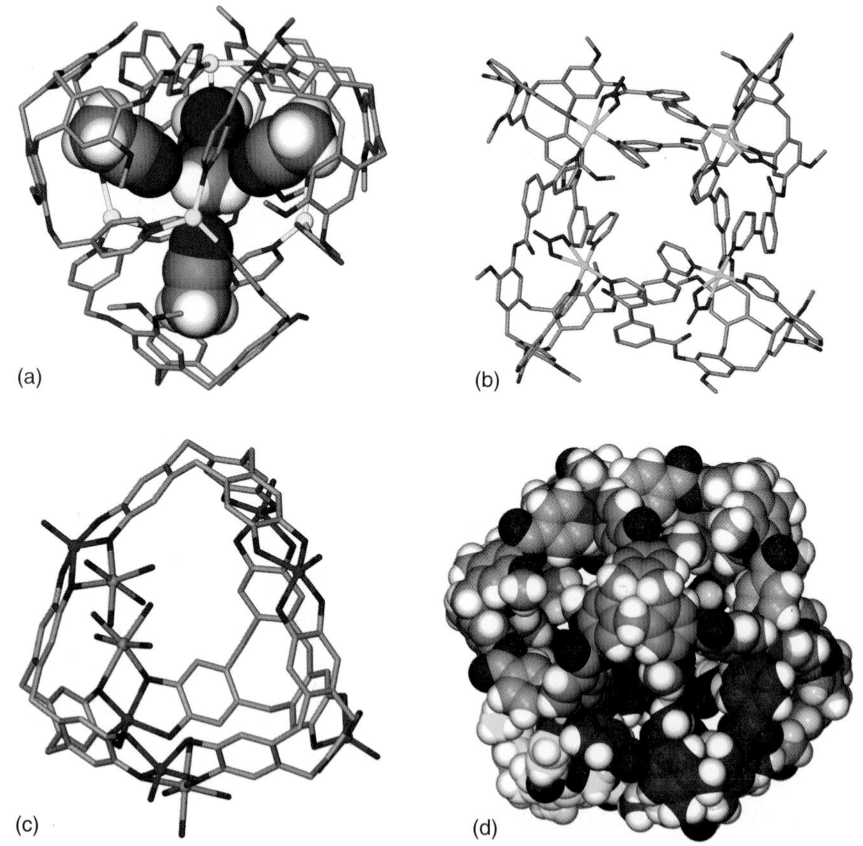

Figure 7 Crystal structures of prismatic metallo-supramolecular assemblies. (a) Tetrahedral $[Ag_4(\mathbf{33})_4]^{4+}$ with included CH_3CN shown in space-filling mode; (b) the topologically complicated complex $[Pd_4(\mathbf{34})_4(NO_3)_2(H_2O)_2]^{6+}$; (c) tetrahedral $[(VO)_6(CTC\text{-}6H)_4(Mg(H_2O)_4)_3]^{6-}$ anion with chelating deprotonated CTC ligands; Mg centers shown in green, V in purple; and (d) the stella octangula structure of $[Pd_6(\mathbf{35})_8]^{12+}$ shown in space-filling mode with each ligand in a different color.

The crystal structure reveals a square arrangement of both Pd(II) centers and ligands, and these are connected together in a unique topologically complex fashion, featuring interlocked ring structures (Figure 7b). Again the self-assembly process is self-selecting, with ligands of one hand within each $[Pd_4(\mathbf{34})_4(NO_3)_2(H_2O)_2]^{6+}$ assembly. Solution studies are consistent with the same species being present in solution. Tetrahedral assemblies with M_6L_4 stoichiometry have also been reported within crystalline complexes of deprotonated CTC. Vanadyl or manganese hydroxide

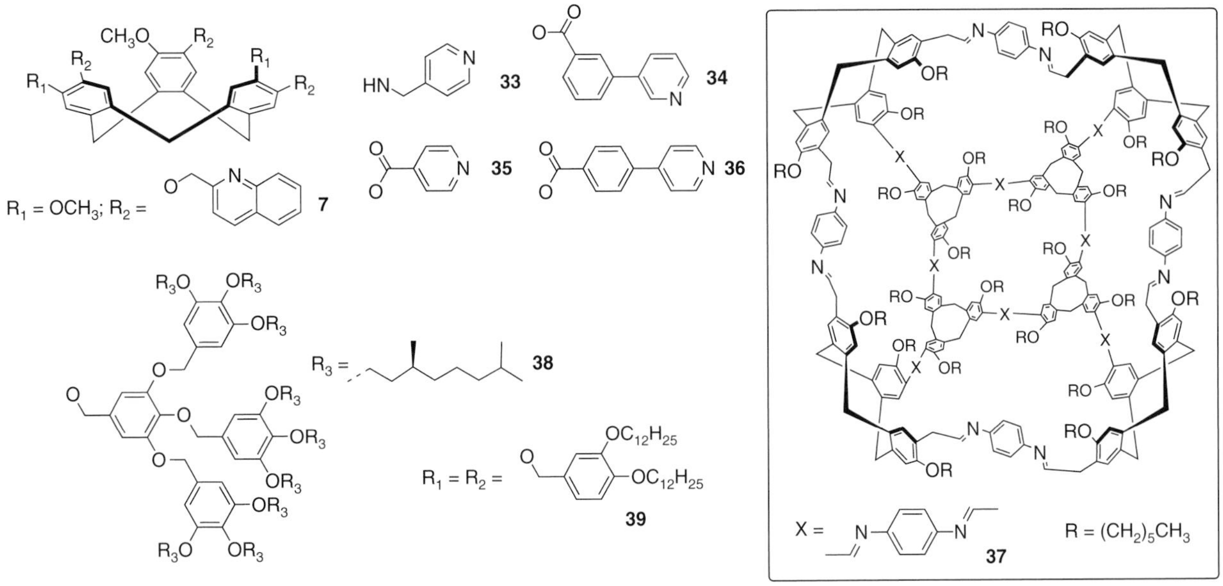

Scheme 16 Ligands involved in the formation of larger assemblies, and the organic homochiral cube **37**.

units bridge between deprotonated CTC units to form the tetrahedron.[85] These M_6L_4 units also interact with additional alkali-earth or alkali-metal cations, as shown for the $[(VO)_6(CTC\text{-}6H)_4(Mg(H_2O)_4)_3]^{6-}$ anion in Figure 7(c).

A series of $[Pd_6L_8]^{12+}$ stella octangula cage-like assemblies, where L includes **35** and **36** and has 4-pyridyl donor groups, self-assemble in solution. The smallest stella octangula is >3 nm in diameter. The crystal structure of $[Pd_6(\mathbf{35})_8]^{12+}$ shows that the Pd(II) centers are arranged in an octahedron, with the pyramidal ligands L occupying each face of the octahedron (Figure 7d).[86] Once again, the process is self-selecting, to give a racemic mixture of chiral cages.

Xu and Warmuth have described the synthesis of a covalently linked homochiral cube employing eight CTV fragments at the vertices of a cube **37** (Scheme 16).[24] As for the metallo-supramolecular assemblies described above, the self-assembly process relies on labile bond formation, in this case, imine bond formation from the reversible reaction of the aldehyde CTV derivative **5** and bridging 1,4-diaminobenzene. Resolution of **5** (Scheme 10) prior to the self-assembly generates a chirally pure nanocube.

Percec and coworkers recently reported a series of dendritic CTVs that self-assemble into even larger structures.[87] The dendrimers include both hexa- and tris-substituted CTV cores with either chiral or achiral alkyl chains. The dendrimers **38** and **39** are representative examples. The dendritic CTVs self-assemble into pyramidal columns that self-organize into hexagonal columnar lattices and/or into chiral supramolecular spheres.

4.2 Coordination polymers

4.2.1 Coordination polymers with CTV

The 1,2-dimethoxy groups of CTV are effective chelating ligand sites for group 1 metal cations. Coordination polymers have been reported for cations Na^+ to Cs^+ and form 1D chain or 2D polymeric structures, with the CTV acting as a bridging ligand.[36,39,88] For example, the 2D network of the complex $[Na(CTV)][Co(C_2B_9H_{11})_2](CF_3CH_2OH)$ is shown in Figure 8(a), where each CTV chelates to three Na^+ cations, which are in turn coordinated by three CTV ligands to give a 3-connected network.[88] One-dimensional chain structures usually form additional hydrogen-bonding interactions to give network structures of overall 2D topology.[36] Reported examples tend to involve large, globular counteranions such as carbaboranes[36,88] or the dimeric $(C_{70})_2^-$ anion,[39] and, unsurprisingly, the size and shape of the anion plays an important role in determining the network topology and overall crystal packing that is observed. In the majority of examples, the guest molecule residing within the CTV molecular cavity is an organic solvent molecule, with two notable exceptions. As discussed in Section 3.1.2, each CTV acts as a host for one $(C_{70})_2^-$ dimer within the 2D coordination polymer of $[Cs_2(DMF)_7(CTV)]\cdot\{(C_{70})_2\}\cdot$ DMF · (toluene) (Figure 3b). An extended diagram of the 2D network is shown in Figure 8(b).[39] In the complex $[K(CTV)(CB_{11}H_6Cl_6)(CF_3CH_2OH)_{0.5}]$ and related Na^+ complexes, the carbaborane anion is bound by the K^+ and is directed into the CTV bowl, where it forms a stabilizing $C\text{-}H_{carborane}\cdots\pi_{CTV}$ hydrogen-bonding interaction.[36]

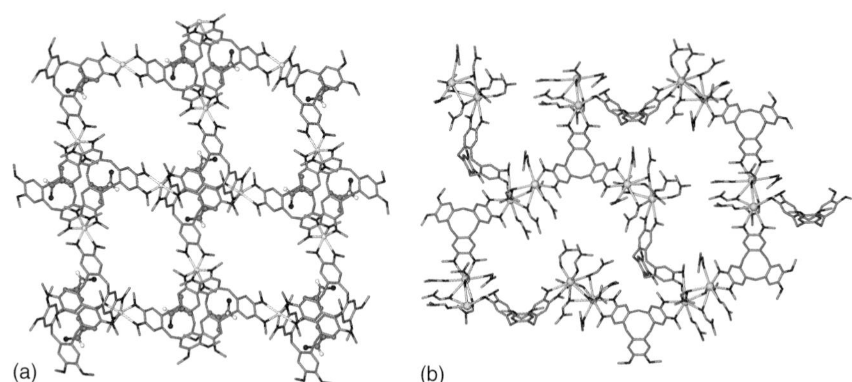

Figure 8 Examples of crystal structures 2D coordination networks with CTV. (a) $[Na(CTV)]^+$ network from the complex $[Na(CTV)][Co(C_2B_9H_{11})_2](CF_3CH_2OH)$, guest CF_3CH_2OH in ball-and-stick mode; (b) $[Cs_2(DMF)_7(CTV)]^{2+}$ network with bridging DMF from complex $[Cs_2(DMF)_7(CTV)] \cdot \{(C_{70})_2\} \cdot$ DMF · (toluene), guest excluded (Figure 4b).

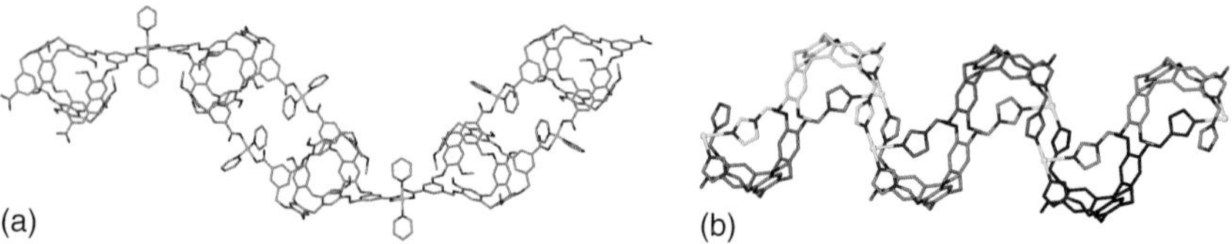

Figure 9 Examples of 1D coordination chains. (a) Section of the $[Cu_3(\mathbf{40}\text{-}3H)_2(py)_6]$ chain from Cu-linked cryptophanes; (b) section of the $[Ag(\mathbf{41})]^+$ chain from the structure of $[Ag(\mathbf{41})] \cdot ReO_4 \cdot CH_3CN$, illustrating the formation of "hand-shake" motif between different ligands within the chain.

4.2.2 Coordination polymers with CTV analogs and cryptophanes

The catecholate functionality of CTC should make it a more effective metal binder than is CTV, although there are few known examples. A discrete trinuclear chelate complex of deprotonated CTC with Pt(II)[89] has been reported. The manganese hydroxide-linked M_6L_4 tetrahedra of the type discussed in Section 4.1.2 link into a complicated 3D coordination polymer through bridging oxides and sandwich-type interactions to additional Cs^{2+} cations that occur between $\{(MnOH)_6L_4\}^{12-}$ tetrahedra.[85]

The carboxylic acid-decorated cryptophane **40** forms a 1D coordination polymer with Cu(II) in the complex $[Cu_3(\mathbf{40}\text{-}3H)_2(py)_6(CH_3OH)_2(DMF)]$, where py = pyridine, DMF = dimethylformamide (Figure 9a).[90] Each of the carboxylic acid groups of **40** is deprotonated and coordinates to a different Cu(II) center. In the crystal lattice, 1D chains pack together so as to create channels along one direction. The cryptophane encapsulates a DMF guest molecule, and additional DMF and CH_3OH solvent molecules occupy these channels (Scheme 17).

A number of transition-metal coordination polymers have also been reported that involve CTV-type ligands with N-donor functionality. One-dimensional chain structures and 2D networks predominate and a striking feature is that the host–guest behavior that is observed in the simple clathrate complexes of the host ligands can also be observed within coordination polymers. For example, the hand-shake motif is observed in 1D chain structures, for example, in

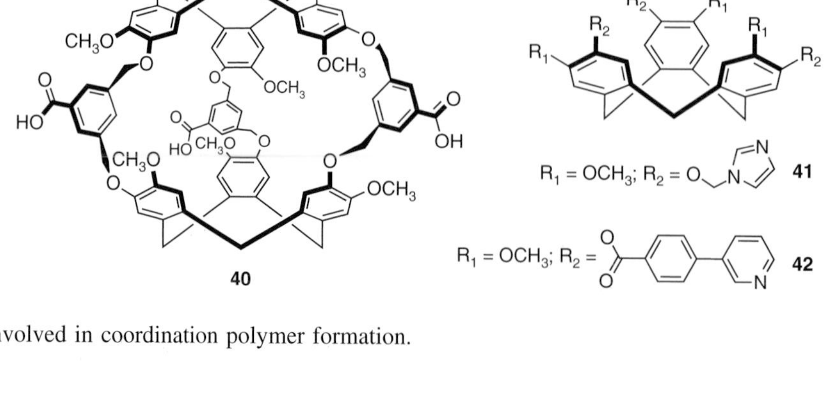

Scheme 17 Ligands involved in coordination polymer formation.

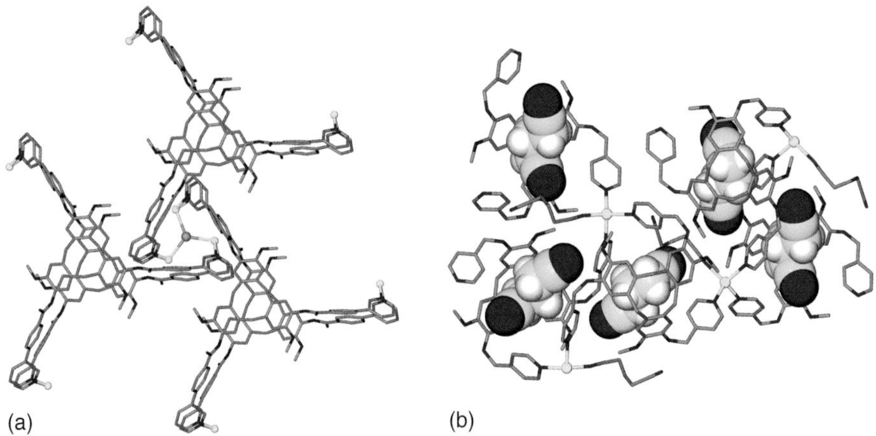

Figure 10 Examples of 2D coordination polymers of N-donor-functionalized CTV derivatives. (a) Section of the $[Ag_3(CH_3CN)_3(\mathbf{42})_2Cl]^{2+}$ polymer, central sphere shown in orange is an unusual μ_3-Cl; (b) section of the $\{[Ag(\mathbf{33})(NC(CH_2)_3CN)]^+\}$ coordination polymer with terminal glutaronitrile ligand in green and guest glutaronitrile in light blue and space-filling mode.

the complex $[Ag(\mathbf{41})]\cdot ReO_4\cdot CH_3CN$ (Figure 9b).[34] In the absence of any metal cations, ligands **35** and **42** both form head-to-head capsule-like arrangements around the guest solvent. Similar host–guest motifs occur between 1D chains of $[Ag(\mathbf{35})_2]^+$,[36] while $[Ag_3(CH_3CN)_3(\mathbf{42})_2Cl]^{2+}$ is a 2D polymer of linked metallo-cryptophanes, again showing the head-to-head ligand arrangement (Figure 10a).[37] A further trend is the influence of bulky guests to promote coordination polymer formation over formation of discrete metallo-supramolecular assemblies. For example, the discrete tetrahedral assembly $[Ag_4(\mathbf{33})_4]^{4+}$ is formed when acetonitrile is the guest molecule; however, if a larger glutaronitrile guest is bound in the molecular cavity, then the final self-assembly product is a 2D coordination polymer (Figure 10b).[83]

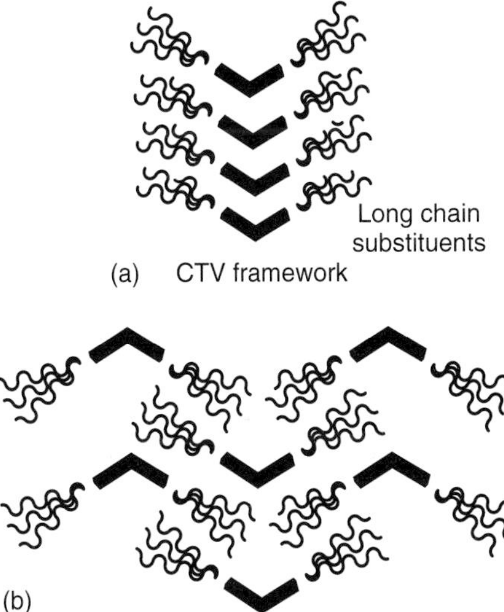

Figure 11 Cartoon showing proposed molecular arrays for the formation of columnar liquid-crystalline phases of hexa-substituted CTVs. (a) Bowl-in-bowl stacking; (b) ribbon-like assembly.

4.3 Soft materials

4.3.1 *Liquid crystals*

The ability of some CTV derivatives with long extended side arms to form liquid crystals is well established.[1,48] Such materials are known as *pyramidic liquid crystals* to distinguish them from more classical liquid crystals that are formed from flat-cored discotic molecules, which are also termed *mesogens*. The majority of known CTV-type mesogens are hexa-substituted with six side arms, usually involving long alkyl chains as well as other functional groups. Pyramidic liquid crystals show columnar[91–93] or, less commonly, smectic[93] mesophases. In the columnar phases, columns of CTV mesogens are proposed to form from either the bowl-in-bowl stacking of the CTV cores[92] or through a ribbon-like phase with stacking interactions of the side arms (Figure 11).[93] There is some evidence for ferroelectricity in the columnar mesophase where upper rim substituents are $O(CH_2)_7OCH(CH_3)COCH_3$, which indicates that the columns form with spontaneous polarization.[94] Pyramidic liquid crystals are also formed by nonasubstituted CTVs such as **43**.[95] Low-temperature esterification of the precursor nonahydroxy-CTV gives exclusively crown conformation mesogens; however, a high-temperature (200 °C) preparation gives a mixture of crown and saddle conformations, which can be separated using chromatographic techniques. Both conformations form liquid crystals, exhibiting columnar mesophases, but the saddle mesophase is not thermodynamically stable and gradually converts to the crown mesophase.[95] CTTV derivatives with long-chain substituents may also show liquid-crystalline behavior (Scheme 18).[93]

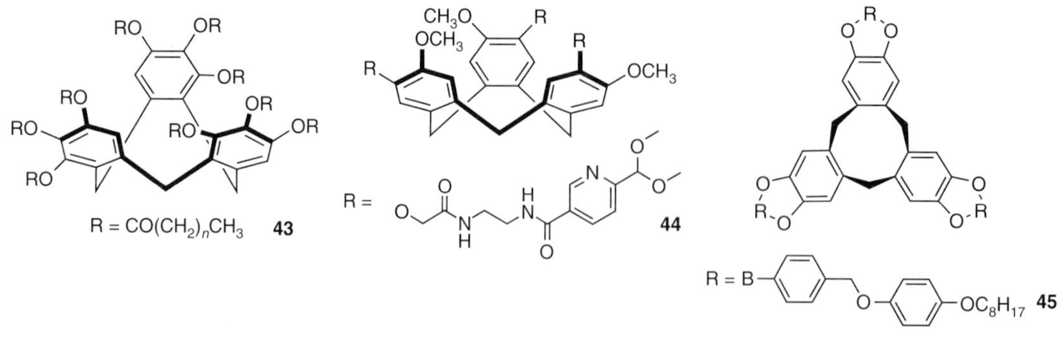

Scheme 18 Examples of CTV analogs known to form soft materials.

4.3.2 *Gels*

A gel is a material in which a liquid phase is trapped within a cross-linked network of gelator molecules and has properties between those of a solid and of a liquid. The 2-methylacetal-5-carbonylpyridine-derived **44**[96] and boronate ester **45**[97] both gelate a range of solvents and are proposed to form columnar aggregates in a hexagonal lattice, as revealed by X-ray diffraction (XRD) measurements. Morphological studies of gels with **44** show ribbon-like and fibrous structures, and gel formation is likely to be due to the self-assembly of **44** into 1D columns, which further aggregate to form ribbons and fibers. The columnar structure is supported by XRD studies, which are indicative of a columnar hexagonal 2D lattice, similar to the mesophases discussed above. Metallo-gels have also been reported for CTV-type ligands. Chelating ligand **31** is a gelator for DMF in the presence of 3 equivalents of $CuCl_2$, and for dimethylsulfoxide (DMSO) in the presence of $AgSbF_6$.[98] Both show fibrous morphologies. The $CuCl_2$/**31** gel is chemoresponsive, transforming to a sol on layering with ethyl acetate, which then re-gels. The isonicotinoyl-derived ligand **35** gives a fibrous gel with 3 equivalents of $CuBr_2$ and acetonitrile.[98] Both **31** and **35** form quite different crystalline complexes when reacted with different metal salts (Sections 4.1.1 and 4.1.2).

5 CONCLUSIONS

The chemistry and applications of CTVs and cryptophanes is currently an expanding field. Advances in synthetic approaches to CTVs and cryptophanes offer methods for both synthesizing them with a wider range of appended functional groups and their more selective functionalization. Resolution of chiral CTVs and cryptophanes can be achieved through HPLC or by functionalization with a chirally pure adjunct group. Examination and application of the chiroptic properties of cryptophanes is an expanding area of endeavor.

CTVs can be poor hosts in solution, with binding of guest molecules in solution often due to additional functional groups appended to the CTV framework rather than the CTV framework itself. An important exception to this is the affinity they display for fullerenes, and this has been utilized in a range of applications including purification of fullerenes and as a method to incorporate fullerene into SAMs and liquid crystals. Other exceptions exist: for example, metallation of CTVs converts them into anion-binding hosts. Most crystalline clathrates of CTVs show self-stacking structural motifs. Self-stacking motifs are also important in self-organized CTV systems, leading to long-established applications in liquid crystals, along with newer applications in giant dendritic assemblies and in soft materials such as gels. The pyramidal shape and convergent binding modes of CTV-based ligands make them excellent ligands for constructing organic or metallo-supramolecular prisms, and the internal binding sites of such prisms may show future nano-reaction-vessel applications.

The relatively enclosed binding space of cryptophanes, on the other hand, makes them very good hosts in solution. Cryptophanes can bind neutral organic guests or organic cations, and the majority of studies have been with small chlorinated guests. Smaller cryptophanes show particularly strong binding affinities for gases such as methane. This allows applications in methane detection, for example. One of the major current driving forces for the development of cryptophane chemistry and their host–guest studies is their outstanding Xe-binding abilities. The development of xenon-based biosensors for MRI applications using water-solubilized or otherwise functionalized cryptophanes is an exciting new area for cryptophane research.

Functionalizing the internal space of cryptophanes increases the range of applications that cryptophanes may have. For example, internal carboxylic acid groups allow the binding of metal cations. This, again, points to the development of more sophisticated hosts with nano-reactor capabilities, such as the hemicryptophane able to act as a supramolecular catalyst through an internal vanatrane unit.

REFERENCES

1. A. Collet, Cyclotriveratrylene and related hosts, in *Comprehensive Supramolecular Chemistry*, eds. J. L. Atwood, J. E. D. Davies, D. D. MacNichol, and F. Vögtle, Permangon, Oxford, 1996, vol. 6, pp. 281–303.
2. A. Collet, Cryptophanes, in *Comprehensive Supramolecular Chemistry*, eds J. L. Atwood, J. E. D. Davies, D. D. MacNichol, and F. Vögtle, Permangon, Oxford, 1996, vol. 2, pp. 325–365.
3. T. Brotin and J.-P. Dutasta, *Chem. Rev.*, 2009, **109**, 88.
4. M. J. Hardie, *Chem. Soc. Rev.*, 2010, **39**, 516.
5. H. Zimmermann, P. Tolstoy, H.-H. Limbach, *et al.*, *J. Phys. Chem.*, 2004, **108**, 18772.
6. J. Canceill, A. Collet, and G. Gottarelli, *J. Am. Chem. Soc.*, 1984, **106**, 5997.
7. S. T. Mough, J. C. Goeltz, and K. T. Holman, *Angew. Chem., Int. Ed.*, 2004, **43**, 5631.
8. G. Huber, T. Brotin, L. Dubois, *et al.*, *J. Am. Chem. Soc.*, 2006, **128**, 6239.
9. J. L. Scott, D. R. MacFarlane, C. L. Raston, and C. M. Teoh, *Green Chem.*, 2000, **2**, 123.
10. T. Brotin, V. Roy, and J.-P. Dutasta, *J. Org. Chem.*, 2005, **70**, 6187.
11. G. Huber, L. Beguin, H. Desvaux, *et al.*, *J. Phys. Chem. A*, 2008, **112**, 11363.
12. D. Bohle and D. J. Stasko, *Inorg. Chem.*, 2000, **39**, 5768.
13. O. Lafon, P. Lesot, H. Zimmermann, *et al.*, *J. Phys. Chem. B*, 2007, **111**, 9453.
14. A. Chakrabarti, H. M. Chawla, G. Hundal, and N. Pant, *Tetrahedron*, 2005,**61**, 12323.
15. M. N. Ponnuswamy and J. Trotter, *Acta Crystallogr. Sect. C*, 1984, **C40**, 1420.
16. D. C. French, M. R. Lutz Jr., C. Lu, *et al.*, *J. Phys. Chem. A*, 2009, **113**, 8258.
17. M. Darzac, T. Brotin, D. Bouchu, and J.-P. Dutasta, *Chem. Commun.*, 2002, 48.
18. H. A. Fogarty, P. Berthault, T. Brotin, *et al.*, *J. Am. Chem. Soc.*, 2007, **129**, 10332.
19. T. Traoré, L. Delacour, S. Garcia-Argote, *et al.*, *Org. Lett.*, 2010, **12**, 960.
20. A. Gautier, J.-C. Mulatier, J. Crassous, and J.-P. Dutasta, *Org. Lett.*, 2005, **7**, 1207.
21. P. A. Hill, Q. Wei, T. Troxler, and I. J. Dmochowski, *J. Am. Chem. Soc.*, 2009, **131**, 3069.
22. M. E. Spence, S. M. Rubin, I. E. Dimitrov, *et al.*, *Proc. Nat. Acad. Sci., U.S.A.*, 2001, **98**, 10654.
23. T. Brotin, R. Barbe, M. Darzac, and J.-P. Dutasta, *Chem. Eur. J.*, 2003, **9**, 5784.
24. X. Xu and R. Warmuth, *J. Am. Chem. Soc.*, 2008, **130**, 7520.
25. A. Bouchet, T. Brotin, D. Cavagnat, and T. Buffeteau, *Chem. Eur. J.*, 2010, **16**, 4507.
26. J. W. Steed, H. Zhang, and J. L. Atwood, *Supramol. Chem.*, 1996, **7**, 37.
27. B. T. Ibragimov, K. K. Makhkamov, and K. M. Beketov, *J. Incl. Phenom. Macro. Chem.*, 1999, **30**, 583.
28. M. R. Caira, A. Jacobs, and L. R. Nassimbeni, *Supramol. Chem.*, 2004, **16**, 337.
29. J. W. Steed, P. C. Junk. J. L. Atwood, *et al.*, *J. Am. Chem. Soc.*, 1994, **116**, 10346.
30. R. J. Blanch, M. Williams, G. D. Fallon, *et al.*, *Angew. Chem., Int. Ed. Engl.*, 1997, **36**, 504.
31. K. T. Holman, J. W. Steed, and J. L. Atwood, *Angew. Chem. Int. Ed. Engl.*, 1997, **36**, 1736.
32. S.-Q. Wang, G. Zeng, X.-F. Zheng, and K. Zhao, *Acta Cryst. E*, 2003, **59**, o1862.
33. A. Collet, J. Gabard, J. Jacques, *et al.*, *J. Chem. Soc., Perkin Trans. 1*, 1981, 1630.
34. C. Carruthers, T. K. Ronson, C. J. Sumby, *et al.*, *Chem. Eur. J.*, 2008, **14**, 10286.
35. Y.-Y. Shi, J. Sun, Z.-T. Huang, and Q.-Y. Zheng, *Cryst. Growth Des.*, 2010, **10**, 314.
36. M. J. Hardie, R. Ahmad, and C. J. Sumby, *New J. Chem.*, 2005, **29**, 1231.
37. T. K. Ronson and M. J. Hardie, *CrystEngComm*, 2008, **10**, 1731.
38. M. J. Hardie, P. D. Godfrey, and C. L. Raston, *Chem. Eur. J.*, 1999, **5**, 1828.
39. D. V. Konarev, S. S. Khasanov, I. I. Vorontsov, *et al.*, *Chem. Commun.*, 2002, 2548.
40. P. E. Georghiou, L. N. Dawe, H.-A. Tran, *et al.*, *J. Org. Chem.*, 2008, **73**, 9040.
41. J. L. Atwood, M. J. Barnes, M. G. Gardiner, and C. L. Raston, *Chem. Commun.*, 1996, 1449.
42. J.-F. Eckert, D. Byrne, J.-F. Nicoud, *et al.*, *New J. Chem.*, 2000, **24**, 749.
43. H. Matsubara, A. Hasegawa, K. Shiwaku, *et al.*, *Chem. Lett.*, 1998, 923.
44. H. Matsubara, S. Oguri, K. Asano, and Y. Yamamoto, *Chem. Lett.*, 1999, 431.
45. E. Huerta, H. Isla, E. M. Pérez, *et al.*, *J. Am. Chem. Soc.*, 2010, **132**, 5351.
46. E. Huerta, G. A. Metselaar, A. Fragoso, *et al.*, *Angew. Chem., Int. Ed.*, 2007, **46**, 202.
47. Y. Rio and J.-F. Nierengarten, *Tetrahedron Lett.*, 2002, **43**, 4321.
48. D. Felder, B. Heinrich, D. Guillon, *et al.*, *Chem. Eur. J.*, 2000, **16**, 3501.
49. S. Zhang, A. Palkar, A. Fragoso, *et al.*, *Chem. Mater.*, 2005, **17**, 2063.
50. Nuriman, B. Kuswandi, and W. Verboom, *Anal. Chim. Acta.*, 2009, **655**, 75.
51. H. H. Dam, D. N. Reinhoudt, and W. Verboom, *New J. Chem.*, 2007, **31**, 1620.
52. B. F. Abrahams, N. J. FitzGerald, T. A. Hudson, *et al.*, *Angew. Chem. Int. Ed.*, 2009, **48**, 3129.
53. K. T. Holman, M. M. Halihan, S. S. Jurisson, *et al.*, *J. Am. Chem. Soc.*, 1996, **118**, 9567.

54. K. S. B. Hancock and J. W. Steed, *Chem. Commun.*, 1998, 1409.
55. J. A. Gawenis, K. T. Holman, J. L. Atwood, and S. S. Jurisson, *Inorg. Chem.*, 2002, **41**, 6028.
56. S. Zheng and L. Echegoyen, *J. Am. Chem. Soc.*, 2006, **127**, 2006.
57. O. Reynes, F. Maillard, J.-C. Moutet, *et al.*, *J. Organometallic Chem.*, 2001, **637–639**, 356.
58. J. Canceill, M. Cesario, A. Collet, *et al.*, *Angew. Chem., Int. Ed. Engl.*, 1989, **28**, 1246.
59. J. Lang, J. J. Dechter, M. Effemey, and J. Kowalewski, *J. Am. Chem. Soc.*, 2001, **123**, 7852.
60. Z. Tošner, J. Lang, D. Sandstro1m, *et al.*, *J. Phys. Chem. A*, 2002, **106**, 8870.
61. J. Canceill, L. Lacombe, and A. Collet, *J. Am. Chem. Soc.*, 1985, **107**, 6993.
62. L. Garel, J.-P. Dutasta, and A. Collet, *Angew. Chem., Int. Ed. Engl.*, 1993, **32**, 1169.
63. S. Wu, Y. Zhang, Z. Li, *et al.*, *Anal. Chim. Acta*, 2009, **633**, 238.
64. K. E. Chaffee, H. A. Fogarty, T. Brotin, *et al.*, *J. Phys. Chem. A*, 2009, **113**, 13675.
65. S. Akabori, M. Takeda, and M. Miura, *Supramol. Chem.*, 1999, **10**, 253.
66. L. Garel, H. Vezin, J.-P. Dutasta, and A. Collet, *Chem. Commun.*, 1996, 719.
67. A. Martinez and J.-P. Dutasta, *J. Catal.*, 2009, **267**, 188.
68. A. Martinez, L. Guy, and J.-P. Dutasta, *J. Am. Chem. Soc.*, 2010, 10.1021/ja102873x.
69. K. Bartik, M. Luhmer, J.-P. Dutasta, *et al.*, *J. Am. Chem. Soc.*, 1998, **120**, 784.
70. P. Aru Hill, Q. Wei, R. G. Eckenhoff, and I. J. Dmochowski, *J. Am. Chem. Soc.*, 2007, **129**, 9262.
71. P. Berhault, A. Bogaert-Buchmann, H. Desvaux, *et al.*, *J. Am. Chem. Soc.*, 2008, **130**, 16456.
72. V. Roy, T. Brotin, J.-P. Dutasta, *et al.*, *ChemPhysChem*, 2007, **8**, 2082.
73. T. Meldrum, K. L. Seim, V. S. Bajaj, *et al.*, *J. Am. Chem. Soc.*, 2010, **132**, 5936.
74. L. Garel, B. Lozach, J.-P. Dutasta, and A. Collet, *J. Am. Chem. Soc.*, 1993, **115**, 11652.
75. S. Akabori, M. Miura, M. Takeda, *et al.*, *Supramol. Chem.*, 1996, **7**, 187.
76. M.-J. Li, C.-C. Lai, Y.-H. Liu, *et al.*, *Chem. Commun.*, 2009, 5814.
77. K. Kataoka, S. Okuyama, T. Minami, *et al.*, *Chem. Commun.*, 2009, 1682.
78. C. E. O. Roesky, E. Weber, T. Rambusch, *et al.*, *Chem. Eur. J.*, 2003, **9**, 1104.
79. R. M. Fairchild and K. T. Holman, *J. Am. Chem. Soc.*, 2005, **127**, 16364.
80. Z. Zhong, A. Ikeda, S. Shinkai, *et al.*, *Org. Lett.*, 2001, **3**, 1085.
81. A. Westcott, J. Fisher, L. P. Harding, *et al.*, *J. Am. Chem. Soc.*, 2008, **130**, 2950.
82. S. B. Lee and J.-I. Hong, *Tetrahedron Lett.*, 1996, **37**, 8501.
83. C. J. Sumby, J. Fisher, T. J. Prior, and M. J. Hardie, *Chem. Eur. J.*, 2006, **12**, 2945.
84. T. K. Ronson, J. Fisher, L. P. Harding, *et al.*, *Nat. Chem.*, 2009, **1**, 212.
85. B. F. Abrahams, N. J. FitzGerald, and R. Robson, *Angew. Chem., Int. Ed.*, 2010, **49**, 2896.
86. T. K. Ronson, J. Fisher, L. P. Harding, and M. J. Hardie, *Angew. Chem. Int. Ed.*, 2007, **46**, 9086.
87. V. Percec, M. R. Imam, M. Peterca, *et al.*, *J. Am. Chem. Soc.*, 2009, **131**, 1294.
88. M. J. Hardie and C. L. Raston, *Angew. Chem. Int. Ed.*, 2000, **39**, 3835.
89. D. S. Bohle and D. Stasko, *Chem. Commun.*, 1998, 567.
90. S. T. Mough and K. T. Holman, *Chem. Commun.*, 2008, 1407.
91. J. Malthete and A. Collet, *J. Am. Chem. Soc.*, 1987, **109**, 7544.
92. R. Poupko, Z. Luz, N. Spielberg, and H. Zimmerman, *J. Am. Chem. Soc.*, 1989, **111**, 6094.
93. R. Lunkwitz, C. Tschierske, and S. Diele, *J. Mater. Chem.*, 1997, **7**, 2001.
94. A. Jákli, A. Saupe, G. Scherowsky, and X. H. Chen, *Liq. Cryst.*, 1997, **22**, 309.
95. H. Zimmerman, V. Bader, R. Poupko, *et al.*, *J. Am. Chem. Soc.*, 2002, **124**, 15286.
96. D. Bardelang, F. Camerel, R. Ziessel, *et al.*, *J. Mater. Chem.*, 2008, **18**, 489.
97. Y. Kubo, W. Yoshizumi, and T. Minami, *Chem. Lett.*, 2008, **37**, 1238.
98. A. Westcott, C. J. Sumby, R. D. Walshaw, and M. J. Hardie, *New J. Chem.*, 2009, **33**, 902.

Carcerands and Hemicarcerands

Ralf Warmuth

Rutgers, The State University of New Jersey, Piscataway, NJ, USA

1 INTRODUCTION

Over the past 20 years, the field of molecular encapsulation has grown explosively and is beginning to make important contributions to molecular recognition, catalysis, medicinal chemistry, mechanistic organic chemistry, photochemistry and photophysics, molecular device fabrication, and materials science. Carcerands and hemicarcerands were the first container molecules with the ability to permanently encapsulate a single molecule.[1–6] Developed by Donald J. Cram in the mid-1980s, they are spherical, hollow hosts built up from two cup-shaped cavitands and enclose a smaller inner cavity, the inner phase, that is large enough to accommodate one or more guest molecules. Since the first report of a carcerand, a large number of container molecules have been designed and synthesized and their encapsulation properties studied. A very important and new direction in the field of molecular encapsulation has been the discovery of self-assembled molecular capsules and coordination cages.[7–12] These capsules form spontaneously under the correct conditions in multicomponent self-assembly processes involving hydrogen bonds, hydrophobic interactions, or metal-coordination bonds similar to assembly processes in nature. These systems have strongly contributed to what is perhaps the most exciting application of molecular containers, namely their use as molecular reaction flask or "nanoreactor,"[13–16] in which otherwise fleeting species, such as cyclobutadiene,[17] benzyne,[18] *anti*-Bredt olefins,[19] or labile tetrahedral intermediates,[20] can be generated and gain longevity at room temperature. In addition to being able to preserve highly labile species, molecular capsules may also serve as catalysts and accelerate, for example, Diels–Alder reactions,[21,22] ortho-formate ester hydrolysis, or Cope rearrangements inside their inner cavities by either concentrating reactants leading to higher effective concentrations or enthalpic transition-state (TS) stabilization or by preorganizing the reactant inside the capsule, thus lowering the activation entropy.[22–26] Finally, molecular capsules may create a microenvironment in which two encapsulated reactants are held together in an orientation that differs from their most reactive arrangement/approach in solution or in the gas phase. In some cases, this may prevent a reaction between both guests.[27] In others, it may lead to

Supramolecular Chemistry: From Molecules to Nanomaterials. Edited by Philip A. Gale and Jonathan W. Steed.
© 2012 John Wiley & Sons, Ltd. ISBN: 978-0-470-74640-0.

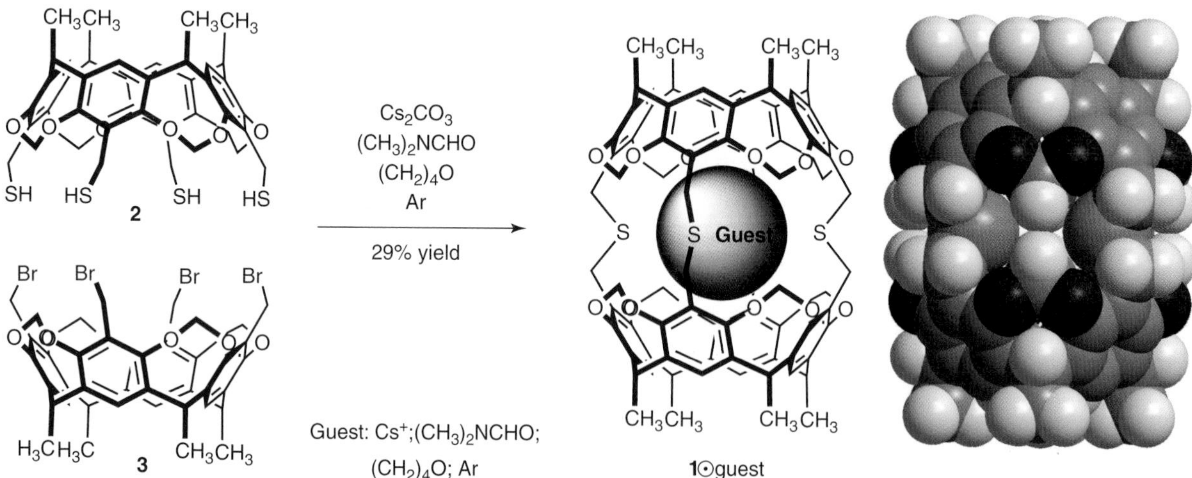

Figure 1 Templated synthesis of carceplex **1**⊙guest by fourfold connecting of cavitand **2** and **3** and space-filling model of **1**.

products that are disfavored in equivalent solution-phase reactions.[22,28,29]

In this chapter, we discuss aspects of the assembly of carcerands and hemicarcerands, which is a templated process, and discuss their molecular recognition properties. Hemicarcerands show high size and shape selectivity in equilibrium binding experiments as well as with respect to rates of complexation/decomplexation. Finally, we highlight some of the dynamic aspects of hemicarcerands and their guests and give examples of their use as molecular reactions flasks.

In 1985, Cram and coworkers synthesized the first carceplex **1**⊙guest by multiply linking cavitand **2** and **3** (Figure 1).[30,31] Because of the insolubility of these carceplexes, which prevented structural analysis by solution NMR techniques, their exact composition was determined via elemental analysis, IR, and fast atom bombardment mass spectrometry (FAB-MS) in combination with chemical tests. During its formation, **1** essentially incarcerated every component present in the reaction that was small enough to be accommodated in its inner cavity in Corey–Pauling–Koltun (CPK) space-filling models. The FAB-MS showed signals for **1**⊙guest, with the guests being argon, DMF, THF (tetrahydrofuran), Cs^+, and CsCl, as well as $ClCF_2CF_2Cl$ if the shell-closure reaction was performed in the presence of $ClCF_2CF_2Cl$ (Freon). If, in the latter case, the carceplex mixture was degraded with refluxing trifluoroacetic acid, free $ClCF_2CF_2Cl$ could be detected by gas chromatography–mass spectrometry (GC–MS).

With more soluble derivatives of **1**, ^1H NMR spectroscopic investigation became possible, which provided information about the structures and dynamics of molecular containers and their guests.[32] Encapsulated guests are subject to large shielding effects induced by the host's aryl units, which are the strongest in the polar region of the cavity and weaker in the central equatorial section. Typically, proton resonances of encapsulated guests are 1–5 ppm upfield-shifted, depending on guest orientation and dynamics, and are therefore, easily distinguished from those of the free guests.

The name carcerand is derived from the Latin word *carcer*, which means "prison." In carcerands, the encapsulated guest cannot leave the container even at high temperature. Complexes with permanently imprisoned guests are termed carceplexes. In space-filling models, carcerand **1** has an almost completely closed shell with two small openings in the center of each cavitand, through which only very small molecules, such as H_2, O_2, N_2, or water, may pass (Figure 1). In contrast, hemicarcerands form stable hemicarceplexes at ambient temperature, but release or exchange the guest at elevated temperature, whereby the guest passes through one of the size-restricted equatorial openings in the host shell.

2 HEMICARCERANDS

Since the first carceplex synthesis, a large variety of hemicarcerands have been prepared by connecting two cavitands with four appropriate linkers, and their binding properties have been studied (Figure 2).[33–42] These hemicarcerands have cavities suitable for the incarceration of guests of a wide range of sizes and shapes starting from molecular gases such as H_2, N_2, O_2, and Xe[43] to molecules as large as C_{60}.[40] Even larger multicavitand container molecules have recently been prepared by linking three and up to eight cavitands together (Figure 3).[44–49]

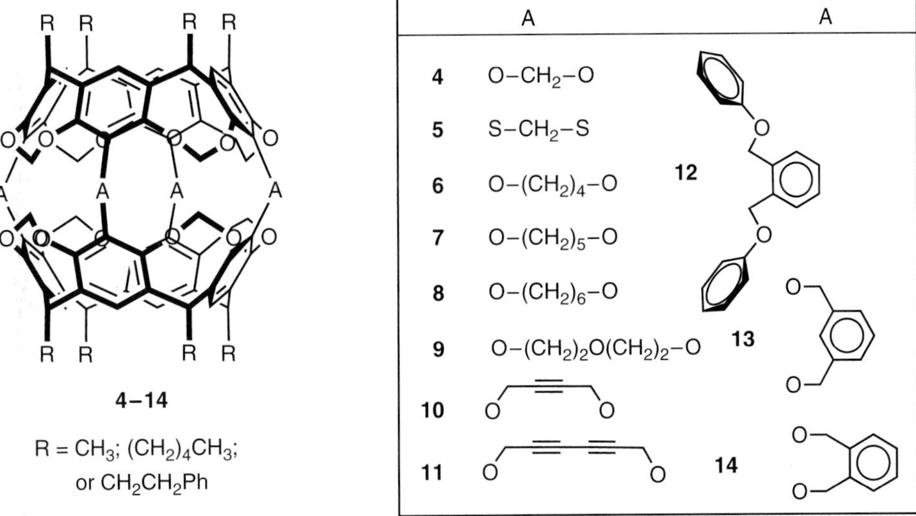

Figure 2 Selected examples of hemicarcerands.

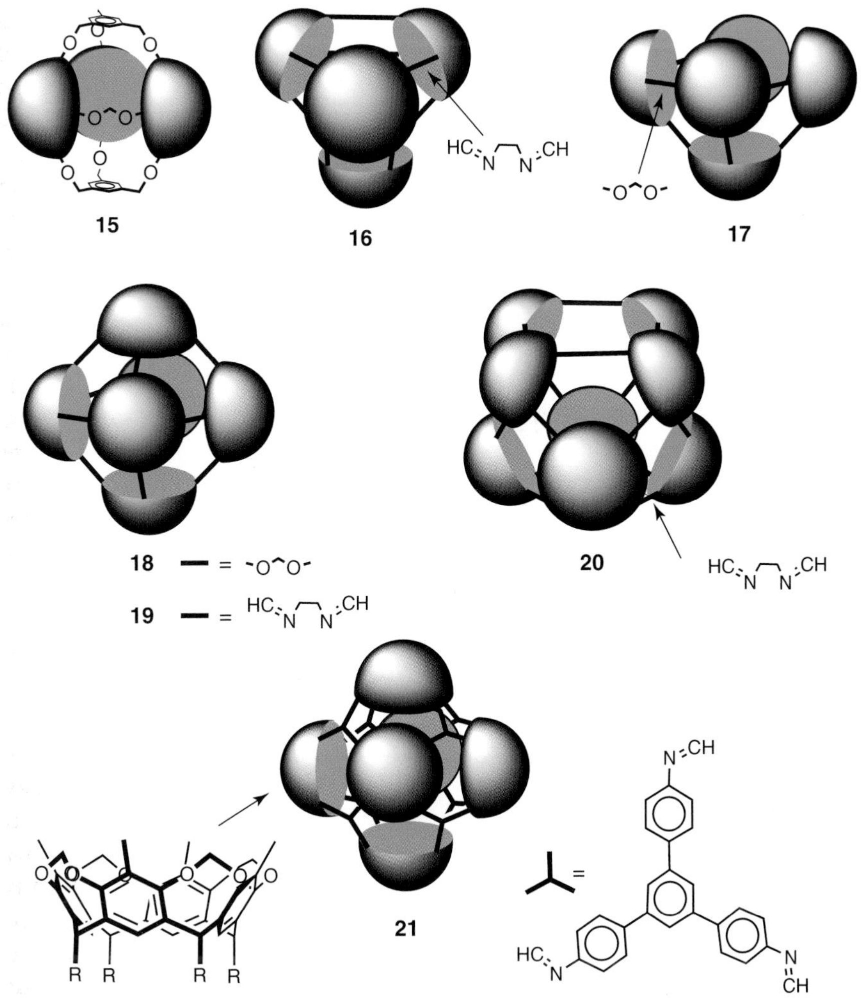

Figure 3 Multicavitand container molecules.

3 TEMPLATE EFFECTS AND MECHANISM OF CARCEPLEX AND HEMICARCEPLEX FORMATION

Carceplex and hemicarceplex formation is a templated process. This was noted by Cram *et al.* in the synthesis of the first carceplexes **1**⊙guest and **4**⊙guest.[32,33] The proper choice of the template molecule, which complements the inner phase of the target (hemi)carcerand, is vitally important. A good template adequately matches the size and shape of the inner cavity and roughly occupies 55% of its space,[50] which is the same space occupancy observed in many organic liquids. Formation of an empty host via an untemplated shell closure is entropically too unfavorable, as it would generate a larger "vacuum bubble." In addition to high structural recognition, electronic complementarity with the inner cavity in the TS of the guest-determining step, after which guest exchange under the reaction conditions is not possible anymore, is required. For example, if the synthesis of **1** was attempted in benzene with methanol, acetonitrile, or ethanol as cosolvent, **1** encapsulated only the polar cosolvent even though CPK models of **1**⊙benzene are easily assembled (Figure 4). Cram and coworkers explained this high structural recognition for the more polar component with the S_N2 TS of the product-determining step, which requires solvation of the sulfide anion by the templating polar solvent.[32]

Multiguest templation and preferential formation of carceplexes **1**⊙$(CH_3OH)_2$ and **1**⊙$(CH_3CN)_2$ with two encapsulated guests was observed for the smaller cosolvents CH_3OH and CH_3CN, which as a single molecule template poorly matches the size of the inner cavity. Interestingly, in **1**⊙$(CH_3CN)_2$, one of the acetonitrile guests could be expelled from the inner cavity under relatively mild conditions, whereas the resulting carceplex **1**⊙(CH_3CN) resisted further dissociation even at much harsher conditions. Cram rationalized this observation with a *billiard-ball effect* operating on **1**⊙$(CH_3CN)_2$. According to CPK models, both guests are roughly aligned along the longer polar axis of **1** such that high-energy collisions between both guests along that axis aid driving out one of the CH_3CN by the other. Calculations by Nakamura and Houk suggest that the likeliest trajectory for guest egress is through one of the equatorially located portals after a conformational change involving the cavitand's OCH_2O groups temporarily enlarges its size [*French Door* gating mechanism (see also Section 4 and Figure 9)].[51]

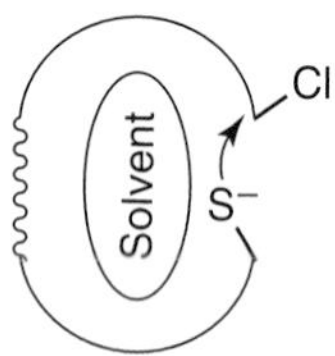

Figure 4 Transition-state model of the product-determining step in the assembly of carceplexes **1**⊙solvent.

A detailed investigation of template effects in carceplex formation was later launched by Sherman and coworkers, who provided a detailed mechanistic picture of templated carceplex formation.[52] They studied a large library of small molecules as templates for the formation of **4**⊙guest and of **6**⊙guest. For carceplex **4**⊙guest, yields varied from 0 to 87%, which is very impressive considering that six components react to form eight new bonds.[53–55] Yields correlate nicely with the structural complementarity of the inner phase and the guest, and are maximized when the cavitand bowls and the template guest can preassemble into a trimeric hydrogen-bonded intermediate (Figure 5b). NMR experiments and X-ray structures demonstrate the formation of the termolecular complex **22**⊙guest,[54,56] which shows the same structural recognition as **4**. Pyrazine, the best template, features a million-fold higher efficiency than the poorest template *N*-methylpyrrolidinone (Figure 5a). Pyrazine and other powerful templates not only fit snugly into the inner cavity of assembly **22**, where they occupy about 55%,[50] but they also perfectly match the electrostatic surface of the inner phase and allow van der Waals contacts in the polar regions of the carceplex and electrostatic interaction in the more central "equatorial" part.[54] The decrease in template ratio in the series pyrazine, pyridine, and benzene is a result of weakening the latter stabilizing host–guest interactions upon replacing the polar nitrogens with less polar CH groups. The poorest templates, which are also the weakest binders of **22**, are either too large and weaken the hydrogen bonds of **22** by distorting the complex, or they are too small and lack enough favorable interactions with the surrounding capsule.[57]

For the formation of the larger hemicarceplexes **6**⊙guest, termolecular complex **22**⊙guest is not important.[58] In this larger host, yields are lower and correlate neither with the template effect nor with the template's ability to stabilize **22**. Most likely, formation of the first linkage disrupts the remaining hydrogen bonds, which allows subsequent intermolecular and intramolecular reactions. Here, the template does not affect the course of the reaction until after the positioning of the second bridge has determined the fate of the product (Figure 5c). These mechanistic considerations may also apply to the formation of other hemicarcerands with longer linkers, whose yields rarely exceed 50%. Exceptions are dynamic covalent hemicarcerands, whose assembly is under thermodynamic control (Section 6).[48,59,60]

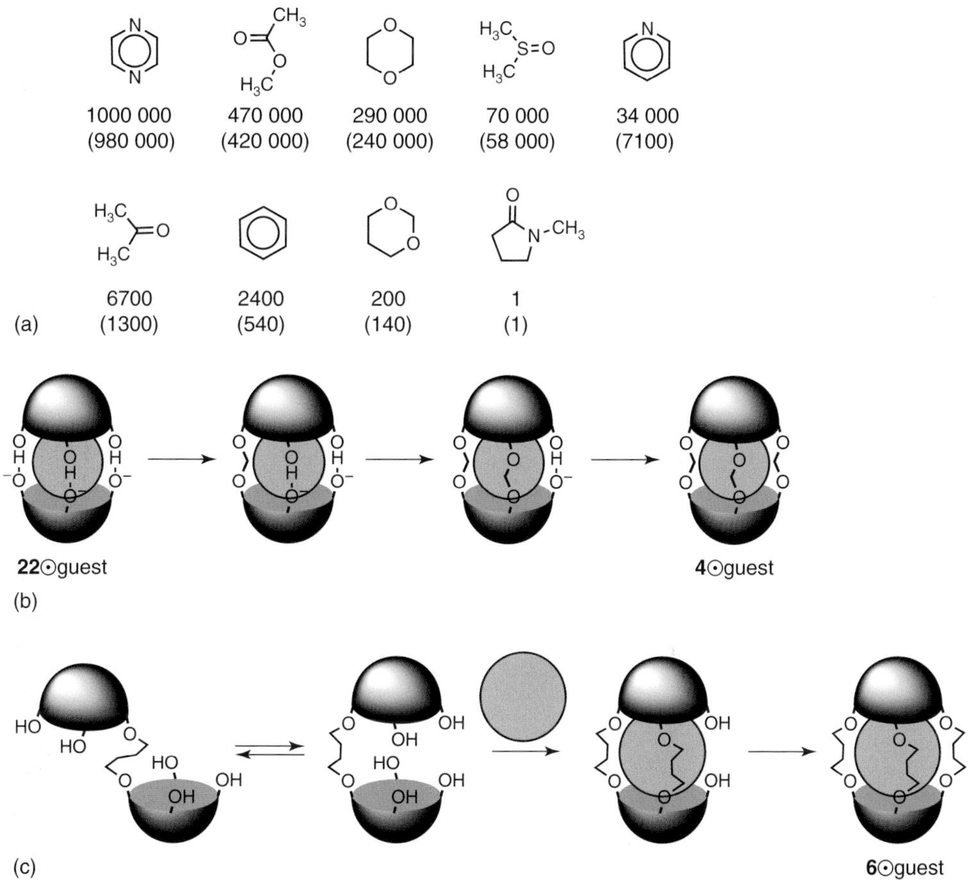

Figure 5 (a) Template ratio in the synthesis of **4**⊙guest and relative stability constant of complex **22**⊙guest (in parentheses). Mechanism of formation of (b) **4**⊙guest and (c) **6**⊙guest.

4 CONSTRICTIVE AND INTRINSIC BINDING IN HEMICARCEPLEXES

Investigation of the properties of hemicarceplexes and the kinetics of complexation and decomplexation led to the discovery of *constrictive binding*.[61,62] Hemicarceplexes are stabilized by intrinsic and constrictive binding (Figure 6a). Intrinisic binding, the free energy of complexation, depends on the magnitude of the noncovalent interactions between the guest and the host's inner surface. Constrictive binding, a physical barrier, is the activation energy required for a guest to enter the inner cavity of a hemicarcerand through a size-restricting portal in the host's skin. Related to constrictive binding are energies keeping mechanically interlocked systems, such as rotaxanes, assembled.[63] In rotaxanes, a macrocycle is slipped over a molecular dumbbell. The activation energy of rotaxane dissociation (slipping-off), which requires slipping of the macrocycle over a stopper, varies greatly and in some rotaxanes prevents slipping-off (Figure 6b and c).

Binding contributions for several hemicarceplexes **14**⊙guest have been investigated experimentally.[42] Intrinsic binding for **14** varied inversely with guest size, while constrictive binding correlated with the guest cross section. Calculations and a X-ray crystal study suggest that constrictive binding for **14**⊙guest involves the reorganization of host structure from a wrapped to an unwrapped state (Figure 7). Thus, **14**⊙guest and many other hemicarceplexes are stable at room temperature, yet slowly dissociate at temperatures above 100 °C. In solvents that fit into the inner phase, guest to solvent exchange is observed. Mechanistically, this guest exchange proceeds in two steps involving the empty hemicarcerand as intermediate[42]:

$$\mathbf{14}\odot\text{guest} \longrightarrow \mathbf{14} + \text{guest} \quad (1)$$

$$\mathbf{14} + \text{solvent} \longrightarrow \mathbf{14}\odot\text{solvent} \quad (2)$$

The effect of solvent in this process is substantial, and very large solvent effects on the rate of the first step (hemicarceplex dissociation) are observed. In combination with the large unfavorable entropic contributions to constrictive

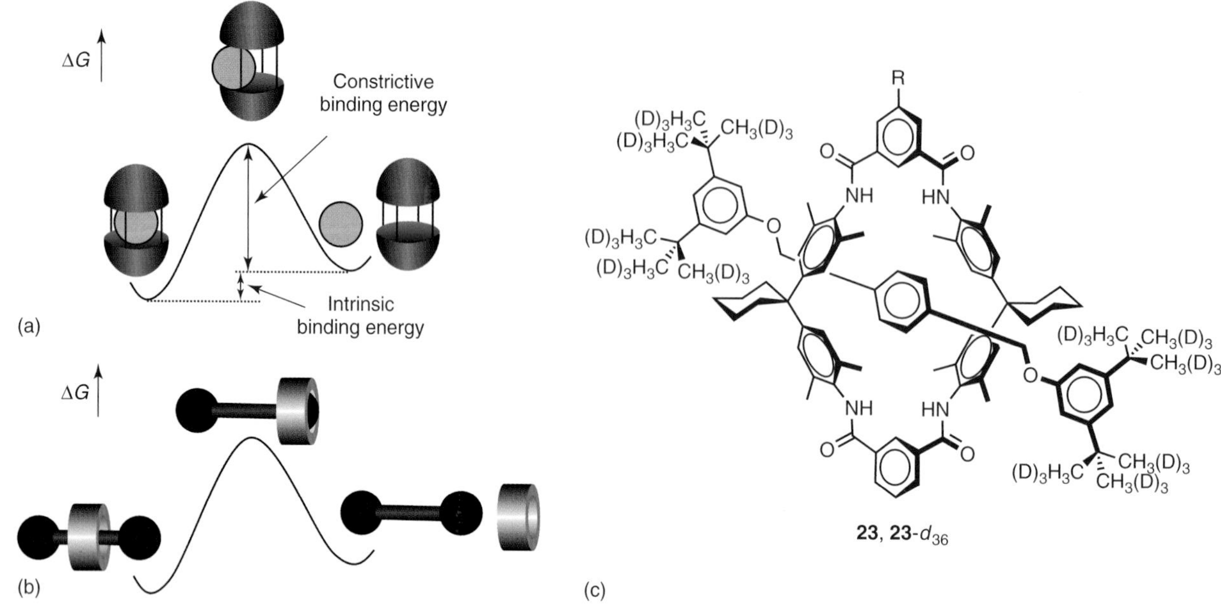

Figure 6 (a) Free-energy profile of hemicarceplex and (b) rotaxane dissociation. (c) Structure of rotaxane **23** and **23**-d_{36}. (Reproduced from Ref. 70. © Wiley-VCH, 2003.)

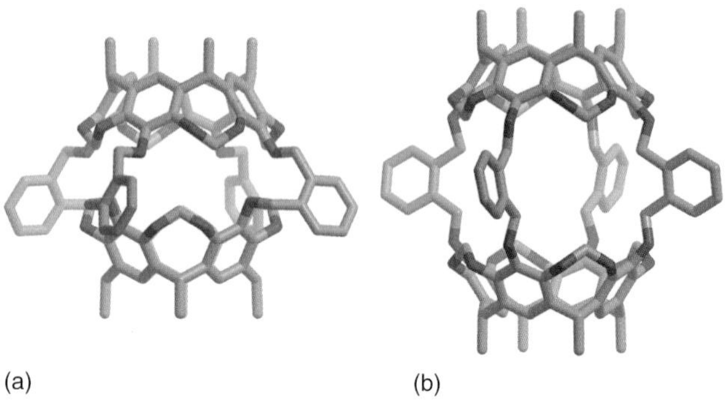

Figure 7 Stick models of wrapped (a) and unwrapped conformation (b) in hemicarceplex **14**⊙$(CH_3)_2NC(O)CH_3$.

binding, this suggests a late TS, in which the guest protrudes substantially from the portal and is stabilized by multiple solvent–guest contacts. For hemicarceplex **14**⊙$(CH_3)_2NC(O)CH_3$, dissociation rates decreased overall 45-fold at 100 °C in the following order of solvents: $C_6D_5Br > C_6D_5Cl > 1,2\text{-}(CD_3)C_6D_4 > 1,4\text{-}(CD_3)C_6D_4 > CDCl_2CDCl_2 > C_6D_5CD_3$. The different polarity and/or polarizability of these solvents are primarily responsible for this variation in rate constants. That the size and shape of the solvent are perhaps more important and may result in steric inhibition of solvation demonstrates the following observation. For hemicarceplex **24**⊙4-$CH_3OC_6H_4OCH_3$, the barrier of guest egress increased by $\Delta(\Delta G^{\ddagger}_{308}) = 4.1\,\text{kcal mol}^{-1}$ corresponding to an 800-fold longer decomplexation half-life at 35 °C upon changing the solvent from $CDCl_3$ to $CDCl_2CDCl_2$. Both have similar polarity but very different sizes and shapes.[64]

Hemicarceplexes are chiral in their wrapped state due to the twisting of the host's cavitands. Chapman and Sherman introduced the term *twistomers* for the two twisted enantiomers.[65] Recently, they observed the wrapping (twisting) and unwrapping (untwisting) of hemicarceplex **4**⊙guest ($R = CH_3$) by variable-temperature NMR spectroscopy and measured the energy barrier for this process. In hemicarceplex **4**⊙(*R*)-(−)-2-butanol, the twistomers are diastereomers, which interconvert with an activation barrier of $\Delta G^{\ddagger} = 12.6\,\text{kcal mol}^{-1}$ based on the coalescence temperature of guest and host NMR signals. The activation barriers of *twistomerism* and hence its detectability strongly depend on the nature of the bridges. Only a *half-twistomerism* was observed for hemicarceplexes **24**⊙guest with guest = *N*-methylpyrrolidinone (NMP), *N,N*-dimethylformamide (DMF), *N,N*-dimethylacetamide (DMA), owing to the different rates of twistomer isomerization in both hemispheres of **24**⊙guest, being slower in the thiahemisphere (Figure 8).[66]

According to molecular mechanics calculations by Houk and coworkers, "gates" control constrictive binding in some hemicarceplexes.[67] Two gating mechanisms were proposed (Figure 9a and b). In the *French door* mechanism, the methylene spanners (Figure 9c) open the portals by a chair-to-boat transition, which requires an activation energy of $17.5\,\text{kcal mol}^{-1}$ and raises the conformational energy by approximately $8.5\,\text{kcal mol}^{-1}$. Calculations for **14**⊙ethyl acetate predict that both French door and *sliding door* mechanisms lower the activation enthalpy for loss of the guest molecule from 26 to $20\,\text{kcal mol}^{-1}$, which is close to the experimentally measured activation barrier

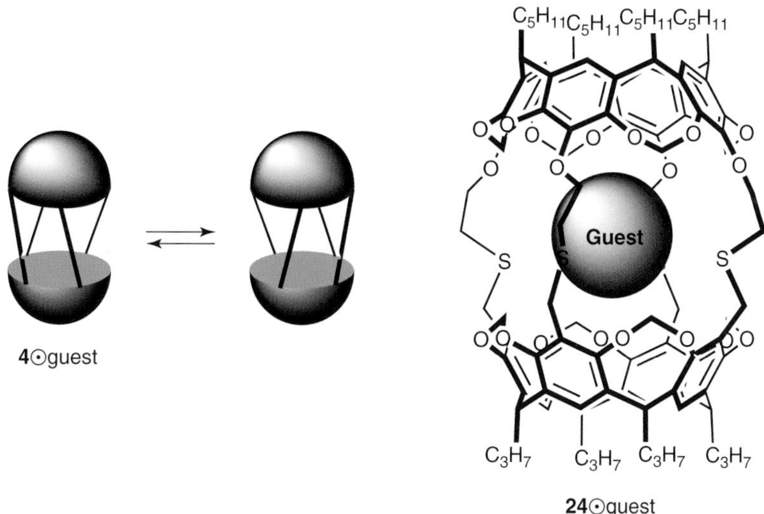

Figure 8 Twistomerism in hemicarceplexes **4**⊙guest and **24**⊙guest.

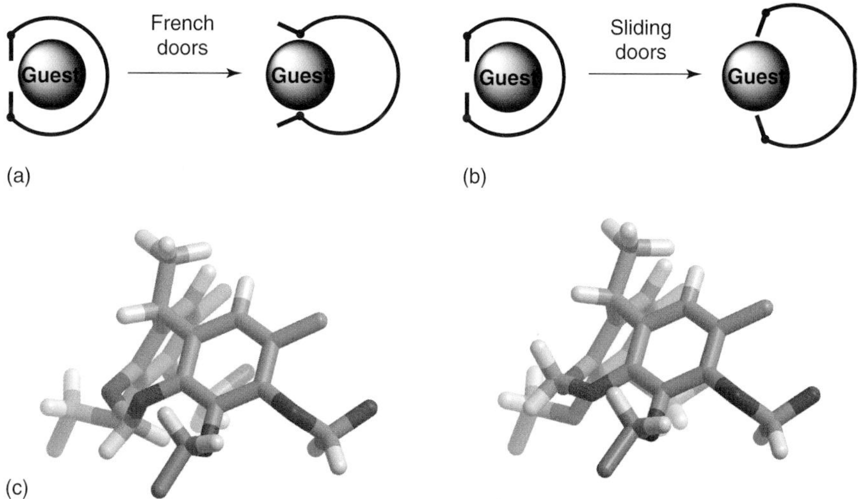

Figure 9 (a) French door and (b) sliding door gating mechanisms in hemicarceplex formation and dissociation. (c) Chair-to-boat transition of OCH_2O spanner during French door gating.

of 22.2 kcal mol^{-1}.[42] The extent to which French door gating affects constrictive binding depends strongly on the nature of the guest and of the host. Thus, French door gating lowered the barrier for the escape of toluene, which is flat, out of **14** only by 1 kcal mol^{-1}. On the other hand, for hemicarceplexes of **31** with $(CH_3)_2NCHO$ or $(CH_3)_2NCOCH_3$, a double French door mechanism is predicted to lower the barrier to guest escape by approximately 28 kcal mol^{-1}.[67]

An interesting way to analyze contributions that make up constrictive binding is to measure the steric kinetic isotope effects (SKIEs) by comparing decomplexation rates of hemicarceplexes with isotopically labeled guests. The physical origin of a steric isotope effect is the smaller zero-point energy (ZPE) of the C–D bond as compared to that of a C–H bond. In combination with the anharmonicity of the vibrational potential, this decreases the mean bond length and vibrational amplitude upon isotopic substitution, leading to a bond length difference of ~0.005 Å and a smaller molecular volume of the deuterated guest.[68] The absence of a secondary isotope effect for the isotopic hemicarceplex pairs **26**⊙naphthalene and **26**⊙naphthalene-d_8 ($k_H/k_D = 1$) and for **6**⊙1,4-$CH_3C_6H_4CH_3$ and **6**⊙1,4-$CH_3C_6D_4CH_3$ ($k_H/k_D = 1.01$) are consistent with French door and sliding door gating making up most of constrictive binding in these hemicarceplexes such that steric interactions in the TS for guest passage through an equatorial portal are minimized.[69] This is in vast contrast to constrictive binding in rotaxanes. Felder and Schalley measured a steric isotope effect of $k_H/k_D = 0.91$ for the deslipping

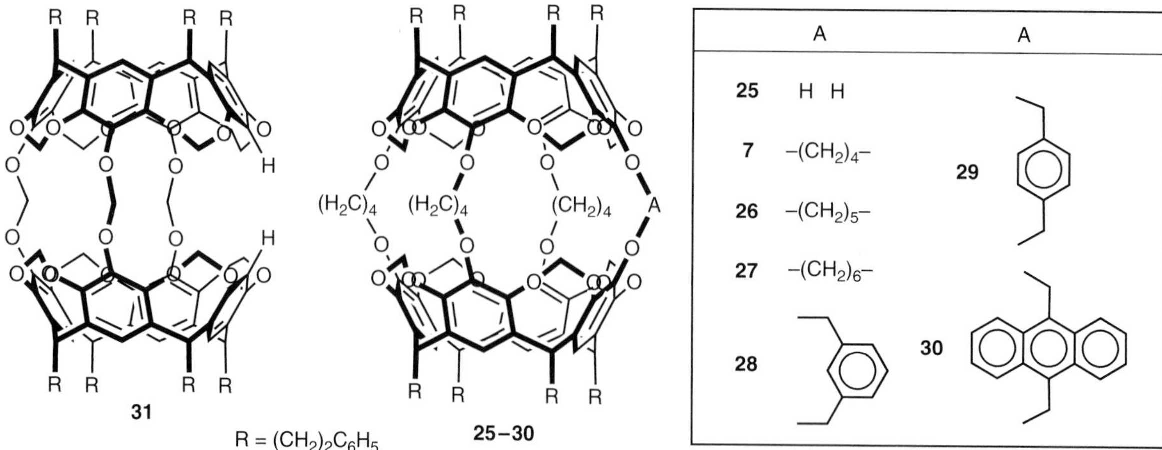

Figure 10 Unsymmetric hemicarcerands **25–31**.

reaction of **23** and **23**-d_{36} (Figure 6b and c).[70] Thus, host and guest reorganization make up most of constrictive binding in hemicarceplexes **6**⊙guest and likely also in other hemicarceplexes with longer flexible linkers,[71] whereas steric interactions seem to contribute substantially to constrictive binding in rotaxanes.

In addition to gating, the length of the linkers and the size of the guest control complex formation and stability in solution. A comparative study to probe the dependence of complexation properties on the portal size and shape was carried out for the unsymmetrical hemicarcerands **26–30**, which have one variable bridge that differs from the other three $O(CH_2)_4O$ linkers (Figure 10). These hemicarcerands are easily prepared from diol **25**, which is an intermediate in the synthesis of hemicarcerand **7** and can be isolated in high yield. The decomplexation rate constants for the 1,4-dimethoxybenzene complexes increase as the lengths of the fourth bridge increases in the order **7** ≪ **26** < **28** < **27**.[64] For **26**⊙4-dimethoxybenzene and **27**⊙4-dimethoxybenzene, the increase in portal size by one CH_2 group corresponds to an increase in the rate constant by a factor of 177. However, in **27**, **29**, and **30** the largest portals are nominally of the same size (28-membered rings). Increasing the *blocking power* as in **29** by substitution of 1,4-$(OCH_2)_2C_6H_4$ for the $O(CH_2)_6O$ bridge of **27** decreases the rate constant by a factor of 9.5.[64] This is consistent with the orientation of the 1,4-$(OCH_2)_2C_6H_4$ bridge in the crystal structure of **29**⊙$PhNO_2$.[72] When the 9,10-$(CH_2)_2$-anthracenyl bridge (**30**) is substituted for the $(CH_2)_6$ bridge of **27**, the rate constant decreases by a factor of at least 600.[64] The strong dependence of decomplexation rates on the nature of the unique bridges and the sizes and shapes of the guests is encouraging for a hemicarcerand approach to specificity of binding important organic compounds.

5 GATED HEMICARCEPLEXES

Because of considerable constrictive binding, many hemicarceplexes are very stable at ambient temperature allowing isolation and chromatographic separation, but slowly dissociate or exchange guests at high temperatures. However, for some applications of hemicarceplexes, such as delivery, spontaneous release of the guest(s) may be desirable. This is the idea behind "gated hemicarceplexes." An external stimulus or signal triggers a switch from a hemicarceplex with small portals and large constrictive binding (closed form) to a host with one or more enlarged openings, allowing spontaneous guest exchange at ambient temperature. The switching between closed and open state may be reversible or irreversible and may be accomplished via a conformational change of the host or the cleavage of a linker as a result of a photochemical or redox reaction, a pH jump, or the addition of a metal cation or nucleophile.

Piatnitski and Deshayes developed photosensitive hemicarcerands **32**, which has a photocleavable *ortho*-nitrobenzyl ether linkage (Figure 11a).[73] The DMA hemicarceplex of **32** is stable toward dissociation in the dark, but releases its guest upon UV irradiation, whereby the *ortho*-nitrobenzyl ether bond is broken in a one-photon process. Incorporation of three additional 3-nitro-*ortho*-xylyl linkers, as in **33**⊙DMA, which was prepared as a mixture of inseparable diastereomers, increased the rate of light-induced guest release 3.2-fold compared to **32**⊙DMA due to the higher probability of photon absorption.

The above photochemical reaction is irreversible. Hemicarcerands **32** and **33** only allow triggered release of encapsulated guests, but, once photolyzed, cannot be recycled. In a reversible gated hemicarcerand, switching between closed and open states is possible in both directions. Thus, target molecules can be released or trapped in response to

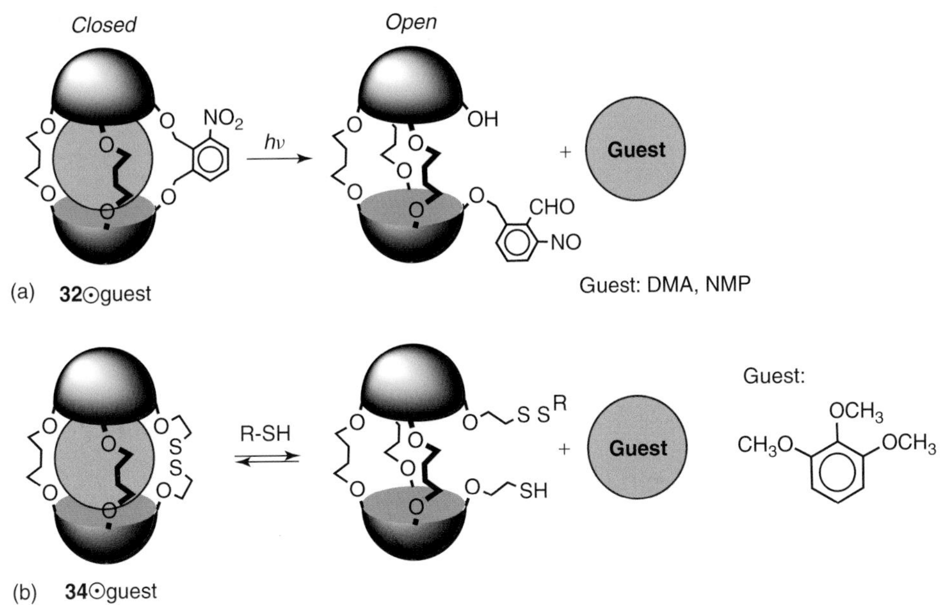

Figure 11 (a) Light-induced guest release in hemicarceplex **32**⊙DMA.[73] (b) Redox-active hemicarceplex **34**⊙guest and mechanism of thiol-triggered guest release.[74]

an external stimulus. An example of a potential reversible gated hemicarcerand is redox-active hemicarcerand **34**, which is structurally related to **32** but has one linker with a "dynamic" disulfide bond.[74] Hemicarcerand **34** forms a stable complex, for example, with 1,2,3-trimethoxybenzene, but slowly releases the guest upon addition of excess dithiothreitol or butane-1,4-dithiol and a base. In the presence of thiols or dithiols, disulfides undergo reversible disulfide–dithiol interchange reactions, leading to a temporarily enlarged portal in the host shell. Computationally, breaking the disulfide bond in **34** lowers constrictive binding and intrinsic binding of 1,2,3-trimethoxybenzene by 8 and 5 kcal mol^{-1}, respectively (Figure 11b).

Aside from breaking strategic bonds in a hemicarcerand, conformational changes that lead to a lowering of constrictive binding may be used to release a guest from a gated hemicarcerand. In the previous sections, we have seen that chair–boat transitions of intrahemispheral spanners (French door gating) reduced constrictive binding substantially. Diederich and coworkers made use of this type of gating in the development of reversibly switchable hemicarcerand **35**, which responds to changes in pH (Figure 12).[75] Hemicarcerand **35** is built up from a quinoxaline-bridged resorcin[4]arene cavitand, whose geometry can be interchanged between a concave "vase" form and a flat "kite" form by decreasing the temperature or by adding acid or metal

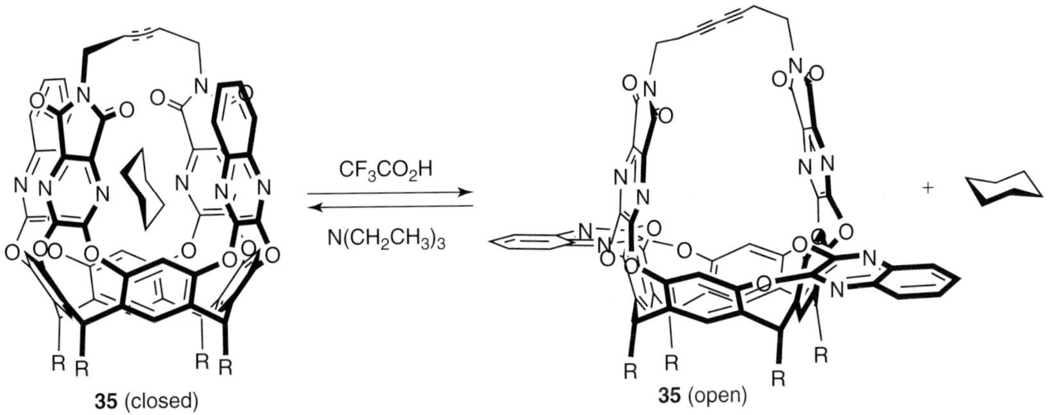

Figure 12 pH-sensitive reversible switching between closed (high affinity) and open (low affinity) hemicarcerand **35** releasing or capturing one cyclohexane molecule.[75]

ions. In mesitylene (1,3,5-trimethylbenzene), **35** resides exclusively in the closed concave form and encapsulates one cyclohexane guest with an affinity of $K = 3600\,M^{-1}$. Addition of excess CF_3COOH protonates the quinoxaline flaps and induces a vase-to-kite transition, which leads to spontaneous release of the guest as a consequence of the reduced binding affinity and constrictive binding of the open form. This process is fully reversible and neutralization with NEt_3 closes the gates and recaptures the cyclohexane guest.

6 DYNAMIC HEMICARCEPLEXES

Dynamic hemicarcerands are related to the reversible gated hemicarcerands discussed in the previous sections. In dynamic hemicarcerands, capsule building blocks are connected together through reversible (dynamic) covalent bonds. Among possible reversible bond forming and breaking processes, which include olefin metathesis, exchange of acetals, oximes, hydrazones, semicarbazones, imines, disulfide, or boronic esters, only the latter three have been utilized in hemicarcerand synthesis.[76] The reversibility of bond formation not only provides a gating mechanism for release and exchange of guests, but it also allows for continuous exchange of constituents leading to dynamic constitutional capsule libraries.[77–79] There are also important consequences for the assembly of dynamic hemicarcerands, which, unlike those discussed in previous sections, is controlled by thermodynamics. Reversibility and thermodynamic control provide error correction and proof-reading, leading, in the most favorable case, to quantitative formation of the desired capsule.

An example is octaimine hemicarcerand **36**, which forms in close to quantitative yield upon reacting two tetraformyl cavitands **37** with 4 equivalents of phenylene-1,3-diamine **38** in the presence of catalytic amounts of CF_3COOH.[60] The high efficiency of this assembly process expresses itself in effective molarities for the intramolecular imine bond formation that are close to the theoretical limit of 10^8, (J. Sun and R., Warmuth, 2011, unpublished results),[80] and is a result of the high degree of preorganization in the system. Opposite aryl planes of cavitand **37** form $\sim 60^\circ$ angles. In combination with the 120° diamine linker unit **38**, this leads to nearly strain-free imine linkages (Figure 13).

The dynamic features of hemicarcerand **36** allow construction of dynamic constitutional libraries. If a phenylene-1,3-diamine derivative is added to **36**, exchange of linkers takes place, leading eventually to a constitutional library composed of six octaimine hemicarceplexes affording approximately statistical ratio of products. Linker exchange is proposed to involve a sequence of two transiminations, in which the phenylene-1,3-diamine derivative first opens a bridge, followed by an intramolecular transimination that expels the original linker group. The same mechanism could also operate in the release of encapsulated guests through temporary opening of a diimino bridge. Hemicarceplex **36**⊙Fc (Fc = ferrocene) has a half-life of guest egress of $t_{1/2} > 4000\,h$ at room temperature in the absence of an acid catalyst. Addition of either CF_3COOH alone or a combination of excess free linker **38** and CF_3COOH decreases the half-life more than 20-fold in the latter case. Rate-limiting in all cases is the opening of the bridge. However, in addition to the linker exchange mechanism, opening of the hemicarcerand via hydrolysis of one imine in **36**⊙Fc with hydronium ions also contributes to the increased rate of Fc release.

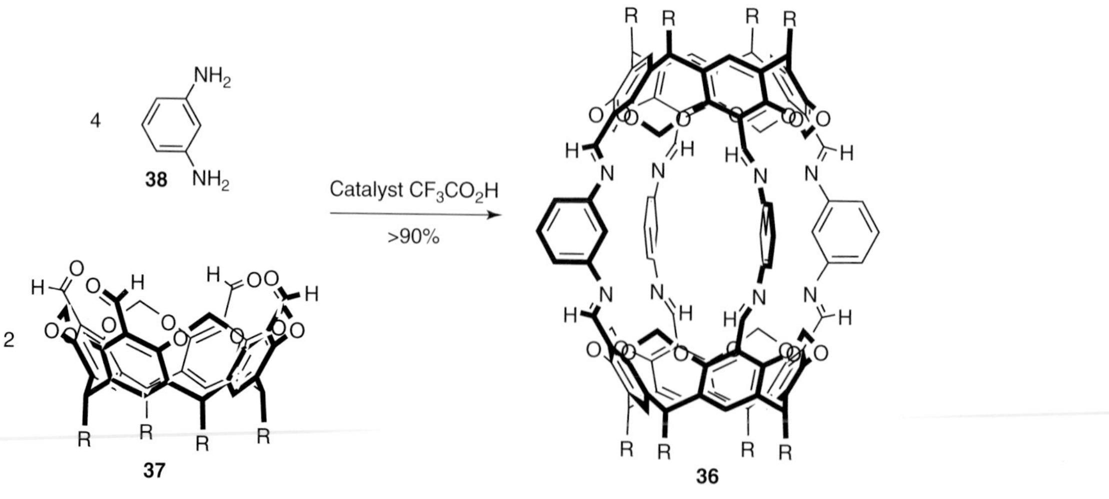

Figure 13 Thermodynamically controlled assembly of octaimine hemicarcerand **36** from 2 equivalents of cavitand **37** and four phenylene-1,3-diamines (**38**).

Further exploration of the thermodynamically controlled synthesis of octaimine hemicarcerands in our group uncovered that many other diamines, **39b–d**, that can adopt low-energy conformations with close to 120° angles between the C–NH_2 bonds also react with cavitand **37** to give related octaimine hemicarcerands. Linear diamines **39a,e,f** provide larger polycavitand nanocapsules with similar efficiency.[45,48,81] For example, ethylene-1,2-diamine **39a** undergoes an 18-component assembly to form octahedral nanocapsule **19** if mixed with **37** in 12 : 6 ratio in $CDCl_3$ containing catalytic TFA. Interestingly, changing the solvent to CH_2Cl_2 resulted in the even larger octa-cavitand nanocapsule **20** in 65% yield, which is impressive, considering that 24 molecules reacted to form 32 new imine bonds.[45] Under these conditions, **19** is only a kinetic product and slowly grows into **20** as the reaction reaches equilibrium. Yet, in THF, tetrahedral **16a** is the major condensation product, though only in 35% yield. However, the related tetrahedral nanocapsules **16e,f** are the sole products in the [4+8]-condensation reactions between **37** and **39e,f** (Figure 14).[81]

An alternative hexa-cavitand nanocapsule synthesis was discovered when we reacted 6 equivalents of **37** with 8 equivalents of trigonal planar triamine **41**, which gave nanocapsule **21** as the major product together with an octaimine hemicarcerand as the only other product.[49] Nanocapsule **21** may be viewed as a molecular rombicuboctahedron, in which the triamines and cavitands occupy the 8 triangular and 6 square faces leaving the remaining 12 squares unoccupied for possible guest entry or exit. Again, the geometry of cavitand **37** is nicely suited for a nearly strain-free nanocapsule (Figure 15).

The efficiency of these hemicarcerand and polycavitand nanocapsule syntheses approaches those of self-assembly processes using hydrogen bonding or metal–ligand coordination and results from selection of the least strained capsule through the thermodynamic control and reversibility of the system.

Boronic ester formation is an alternative process suitable for dynamic covalent chemistry. Nishimura and Kobayashi made use of the reversible condensation between an arylboronic acid and catechol in the development of dynamic hemicarcerand **42**.[82,83] Heating a 2 : 4 mixture of a tetraboronic acid cavitand and biscatechol **43** resulted in the quantitative formation of **42**. Again, the high efficiency is a result of the proper choice of the building blocks. Biscatechol **43**, which in its lowest energy conformation is a perfect 120° ditopic bis-1,2-diol unit, is complementary to the orientation of the boronic acids in the cavitand building block to yield **42** with little or no strain.

Hemicarcerand **42** forms 1 : 1-complexes with 4,4′-disubstituted biphenyls or 2,6-disubstituted anthracene derivatives, whose length matches that of the host's cavity. For example, **42** binds 4,4′-diacetoxybiphenyl **44a** and 4,4′-diethoxybiphenyl **45** in C_6D_6 with binding constants of $K = 12\,600$ and $13\,000\,M^{-1}$, respectively, but fails to form hemicarceplexes with 4-acetoxy-4′-methoxybiphenyl or 4-ethoxy-4′-propoxybiphenyl, which are too short or

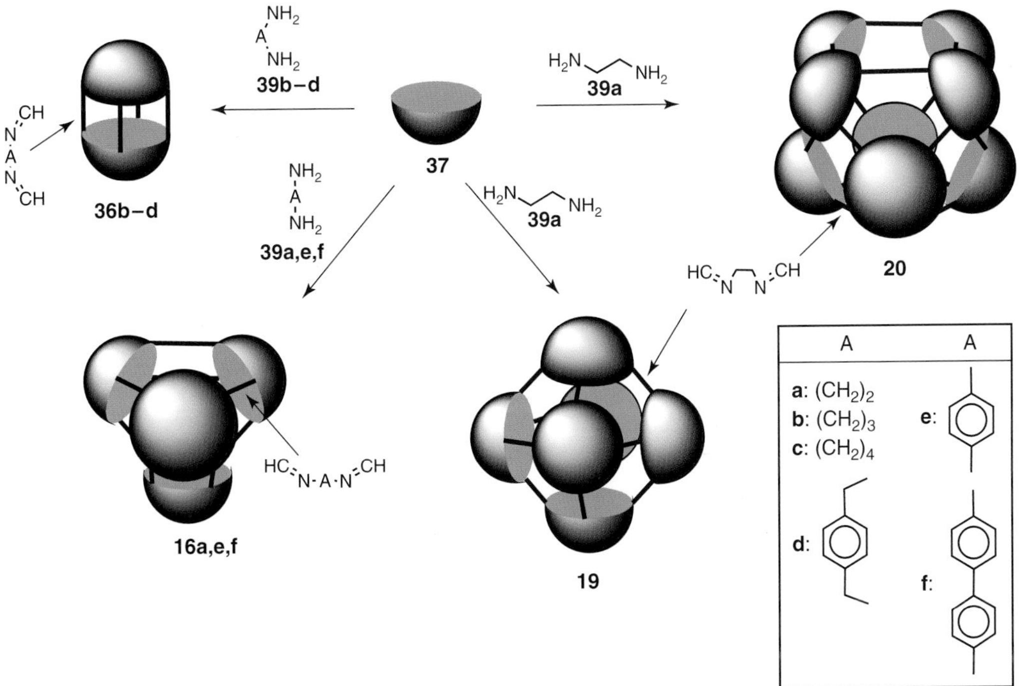

Figure 14 Thermodynamically controlled assembly of polycavitand nanocapsules from cavitand **37** and diamines **39a–f**.

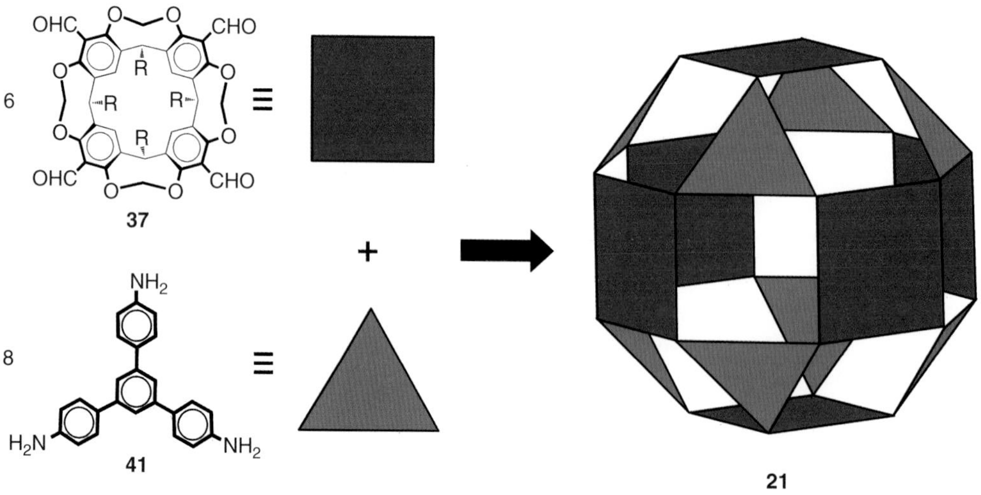

Figure 15 Thermodynamically controlled synthesis of a molecular rhombicuboctahedron **21** from six cavitands **37** and eight triamines **41**.

too long, respectively. The high affinity for these guests is remarkable and is a result of the low affinity of C_6D_6 for the host's cavity. In the more guest-competitive solvent $CDCl_3$, binding affinities decrease by 3–4 orders of magnitude. Nishimura and Kobayashi also probed the mechanism of guest exchange in and out of the dynamic hemicarceplex **42**⊙guest and proposed that hydrolysis of one or more boronic esters, by traces of water in the solvent, precedes guest egress through the enlarged hemicarcerand portal (Figure 16).[83]

In the previous sections, we have seen that disulfide–dithiol exchange is a dynamic process that was used in gated hemicarcerand **34** (Figure 11) to trigger guest release. Sherman and coworkers recently developed related dynamic disulfide-linked [4]carceplexes **46**⊙guest and [5]carceplexes **47**⊙guest$_2$, in which all interhemispheric bridges contain a disulfide bond.[84,85] Such systems are not only interesting for guest release applications, but they also allow a comparison between kinetic and thermodynamic template effects in the synthesis of carceplexes. In order to study kinetic template effects, tetrathiol cavitand **48** was air-oxidized in the presence of base and potential templating guests. Compared to the template effects on the synthesis of carceplex **4**⊙guest, for which template ability among suitable templates varied one million-fold, all suitable templates for formation of **46**⊙guest showed very similar template ability, which varied as little as 50-fold. Sun *et al.* explained the lower template ratio with an early TS of the guest-determining step—most likely involving formation of the second disulfide linkage—and with the

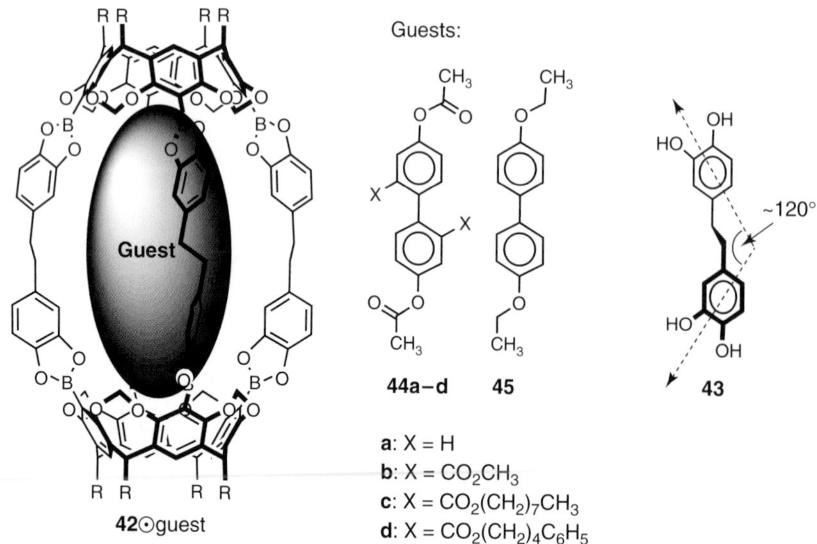

Figure 16 Dynamic boronic ester hemicarceplexes **42**⊙guest.

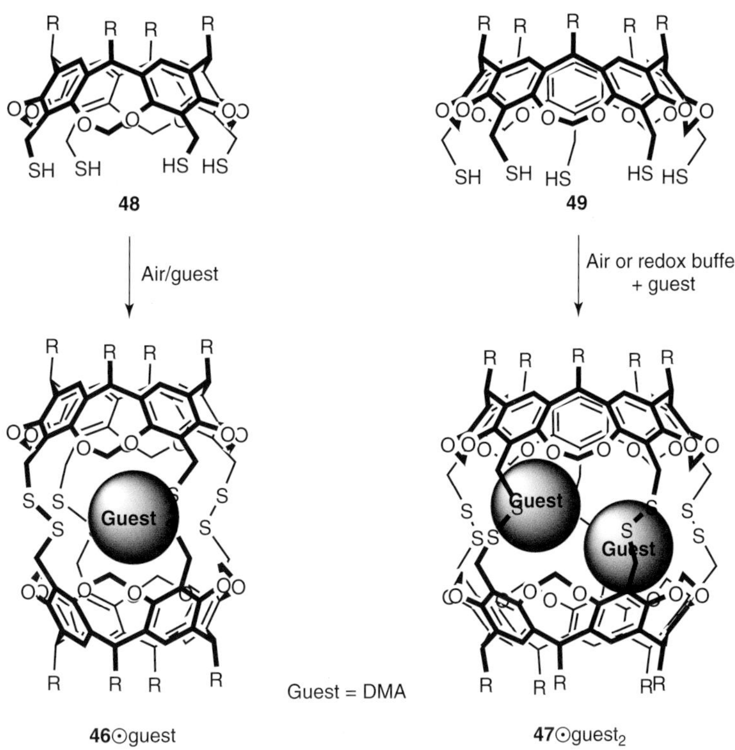

Figure 17 Kinetic and thermodynamic synthesis of disulfide-linked hemicarceplexes **46**⊙guest and **47**⊙guest$_2$ from cavitands **48** and **49**, respectively.

lack of strong inter-cavitand interactions at that stage,[85] which was important for the high template ratio observed for **4**⊙guest.[55]

Under similar conditions, **47**⊙DMF_2 was obtained in 25% yield in the air-oxidation of **49** in DMF.[84] However, when the shell closure was carried out under thermodynamic control in the presence of a redox buffer that allowed disulfide exchange, the yield of **47**⊙DMF_2 doubled. In the larger solvent DMA, **47**⊙DMA_2 was obtained in 42% yield under thermodynamic controlled conditions. Competition experiments between both solvents showed that DMF is the superior thermodynamic template, likely due to its smaller size and better complementarity to the size of [5]cavitand **49**. The thermodynamic capsule formation is not only reversible and addition of mercaptoethanol and base to **47**⊙DMF_2 quantitatively converted **47** back to **49**, but allows guest exchange, if, for example, **47**⊙DMF_2 is dissolved in DMA in the presence of a redox buffer (Figure 17).[85]

7 MOLECULAR RECOGNITION

The binding properties of several hemicarcerands have been studied to various extents. In general, hemicarcerands show high structural recognition in their binding abilities, which depend more on the size and shape of the guest rather than its electronic properties. Hemicarcerand **6**, which has four O–$(CH_2)_4$–O bridges, is an excellent example to highlight this point.[35] It is perhaps the most versatile binding host synthesized, and more than 100 hemicarceplexes **6**⊙guest are known with guests ranging in size from Xe and CH_3CH_2I to naphthalene, coumarin, and methoxyphthalide on the larger side (Figure 18).

Mercozzi and Rebek analyzed the space occupancy inside self-assembled and covalent container molecules using molecular mechanics calculations.[50] In self-assembled hydrogen-bonding capsules, high complex stability results if the encapsulated guest(s) occupy approximately 55% of the available space, which is the same space occupancy inside most weakly interacting organic solvents. Stability decreases at higher or lower space occupancy. The 55% rule also applies to other covalent containers such as carcerands and cryptophanes. The flexibility of its linkers allows **6** to adapt its inner phase to the space requirement of the guest through reorganization of its structure. For example, **6** nearly doubles the volume of its inner phase via stretching along its polar axis in order to form complexes with ideal space occupancy with benzene and 1,4-diiodobenzene (Figure 19). The elongation is accompanied by an untwisting of the host and the partial outward rotation of the linker ether oxygen lone pairs and

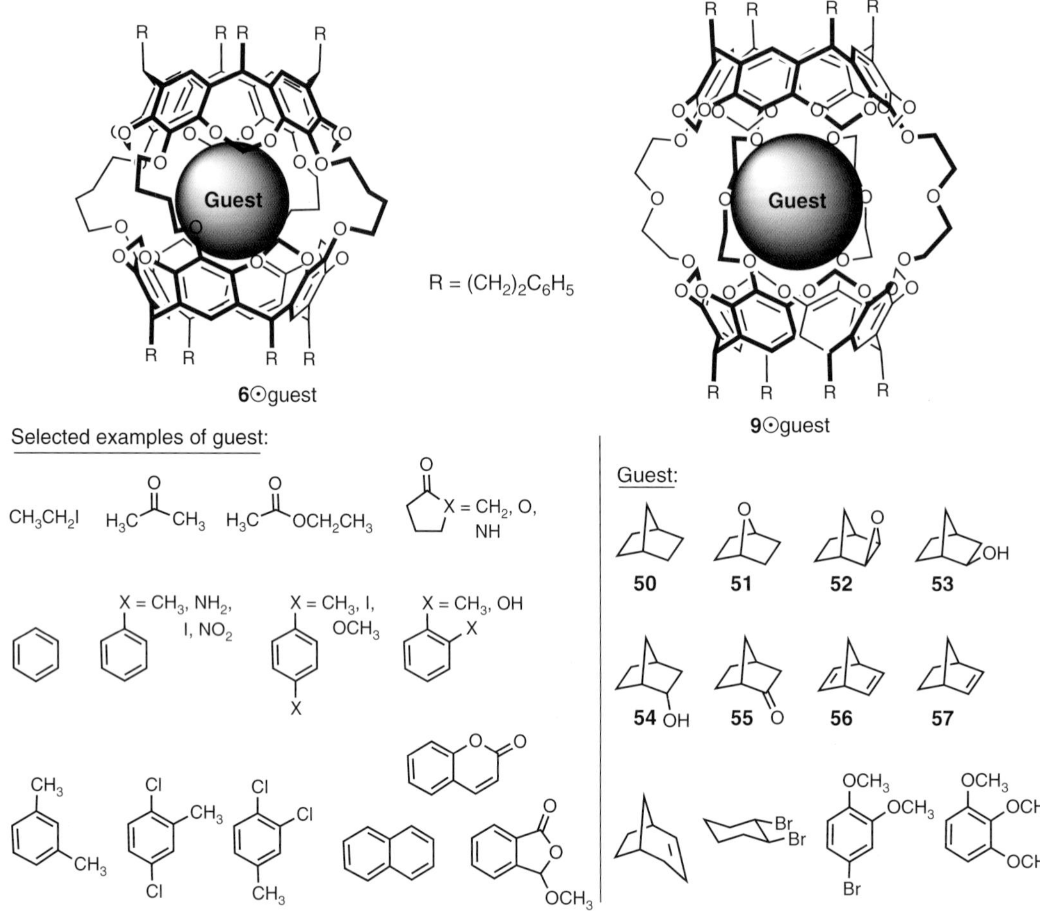

Figure 18 Molecular recognition properties of hemicarcerands **6** and **9**.

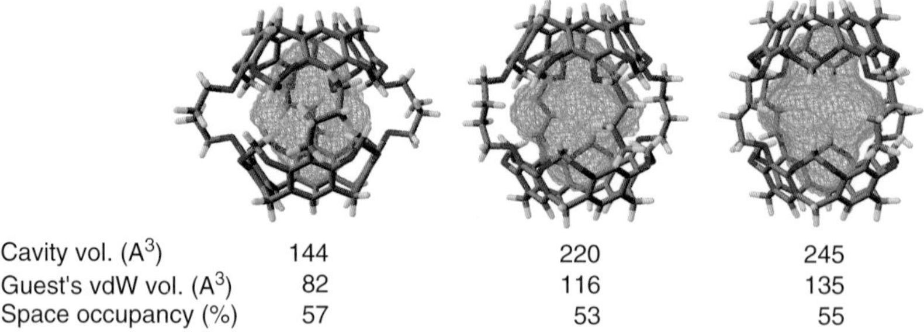

Cavity vol. (A^3)	144	220	245
Guest's vdW vol. (A^3)	82	116	135
Space occupancy (%)	57	53	55

Figure 19 Visualization of adaptability of the inner phase of hemicarcerand **6** to different guest dimensions through untwisting. From left to right: hemicarceplexes **6**⊙benzene, **6**⊙-xylene, and **6**⊙1,4-diiodobenzene. Hemicarcerands and cavities were modeled with the program Swiss-PdbViewer.[87] The inner phase and guest volumes as well as the space occupancy in the inner phase are given for each hemicarceplex.

results in increased hemicarceplex–surface interaction in silica gel chromatography. The effect is very sensitive and isotopomeric hemicarceplexes **6**⊙1,4-$CH_3C_6H_4CH_3$ and **6**⊙1,4-$CD_3C_6D_4CD_3$ are partially separable by normal phase recycle HPLC on the basis of the volume isotope effect of the guest.[86]

In complexation studies involving empty **6**, the ease of encapsulation decreased in the order: small acyclic guests > five-membered ring structures > monosubstituted benzenes > disubstituted benzenes > trisubstituted benzenes, which correlate with the guest's cross section (Figure 18). Among the disubstituted benzenes, the 1,4-substitution pattern is strongly preferred over the 1,3- and 1,2-patterns. The following example nicely highlights the high shape selectivity among isomeric guests. When empty **6** was heated for four days in 98% HPLC-grade *o*-xylene, a mixture of **6**⊙*o*-xylene, **6**⊙*m*-xylene, and **6**⊙*p*-xylene formed, of which the latter hemicarceplex dominated.

In >99% *m*-xylene, a 2 : 1-mixture of **6**⊙*m*-xylene and **6**⊙*p*-xylene was produced. Thus, the ability to complex isomeric xylenes increases in the order *o*-xylene < *m*-xylene ≪ *p*-xylene.

Increasing the interhemispheric bridges by one atom, as in hemicarcerand **7** and **9** (Figures 2 and 18), strongly increases structural recognition in complexation studies and substantially narrows the range of guests that form stable isolatable hemicarceplexes **7**⊙guest and **9**⊙guest.[36,38] For example, **9** formed stable hemicarceplexes with structurally related norbornane derivatives **50–57** and some trisubstituted benzenes, but, remarkably, failed to complex 2-chloro- or 2-bromonorbornane, which are only slightly larger than **50–57**, or with many mono- and disubstituted benzenes, even though they fitted comfortably into CPK models of **9**.[38] Molecular mechanical calculations suggest that constrictive binding is small and, more importantly, does not involve French door gating for the latter complexes (Figure 9). Thus, guests enter the inner phase, but also depart rapidly.[71] However, once French door gating starts contributing to the lowering of constrictive binding, **9**⊙guest can be isolated without decomplexation and constrictive binding becomes strongly guest-size-dependent to the extent that a small structural change, such as the substitution of the OH in **53** for a Cl, increases constrictive binding enough to prevent complexation.

In addition to the linker groups, the intrahemispheric spanners also influence the molecular recognition properties of a hemicarcerand. Recently, Cram *et al.* reported the syntheses and binding properties of **13** (**MM**), **58** (**EE**), **59** (**PP**), **60** (**EM**), **61** (**PM**), and **62** (**PE**) (Figure 20).[41,88] These hosts have either methylene (M), ethylene (E), or propylene (P) spanners in one cavitand.

In the crystal structure of cavitand **63** with propylene spanners, two spanners are outward and two are inward (Figure 20b).[41] As a consequence, the **P** bowl is more rectangular shaped and deviates substantially from C_4 symmetry. However, if bonded rim-to-rim with relatively rigid **E** or **M** bowls as in **61** or **62**, the **P** bowls possess perfect C_4 symmetry and assume a **bo-su** conformation, with the four **b**ridges **o**utward and the four **s**panners **u**pward (Figure 20c). Thus, **P** bowls reorganize substantially upon being incorporated into hemicarcerands. CPK models of hemicarcerands **13**, **58–62** provide the order **PP** > **PM** > **MM** > **PE** > **EM** > **EE** in maximum portal size. Because **P** bowls are flexible, the order of portal adaptability to guest shape for complexation–decomplexation is **PP** > **PM** > **PE** > **MM** > **EM** > **EE**. For these hosts, the inner cavity decreases in the order **PP** > **PE** > **EE** > **PM** > **EM** > **MM**. These hosts show high structural recognition in complexation. However, unlike hemicarcerand **6**, which preferentially binds 1,4-disubstituted benzenes,[35,58] they prefer 1,2-disubstituted benzene guests to 1,3- and 1,4-disubstituted isomers. For example, heating empty **EE** for four days in 3-$ClC_6H_4COCH_3$ gave a 1 : 1 mixture of **EE**⊙3-$ClC_6H_4COCH_3$ and empty **EE**. Under the same conditions, a 2 : 1 mixture of **EE**⊙2-$ClC_6H_4COCH_3$ and empty **EE** was formed in 4-$ClC_6H_4COCH_3$ as solvent. Thus, the relative rates of complexation of the three isomeric guests must be 1,2-isomer ≫ 1,3-isomer ≫ 1,4 isomer, which explains the host's ability to scavenge trace amounts of the 1,2-isomer in neat 4-$ClC_6H_4COCH_3$. Another example is the formation of a 2 : 1 mixture of **EE**⊙$(CH_3)_3CPh$ and **EE**⊙$PhCH(CH_3)CH_2CH_3$ after heating **EE** in $(CH_3)_3CPh$, which contained 2% $PhCH(CH_3)CH_2CH_3$ as impurity, indicating that **EE** complexes $PhCH(CH_3)CH_2CH_3$ around 25 times faster than its isomer $(CH_3)_3CPh$. The selectivity of hemicarcerand **EE** for 1,2-disubstituted benzenes contrasts that of **MM**, which encapsulates preferentially para-disubstituted benzenes. The cavity of **MM** is narrower with a longer polar axis than that of **EE** and therefore better suited for taller guests, whereas the shortest but most spherical of the three isomers prefers the more spherical

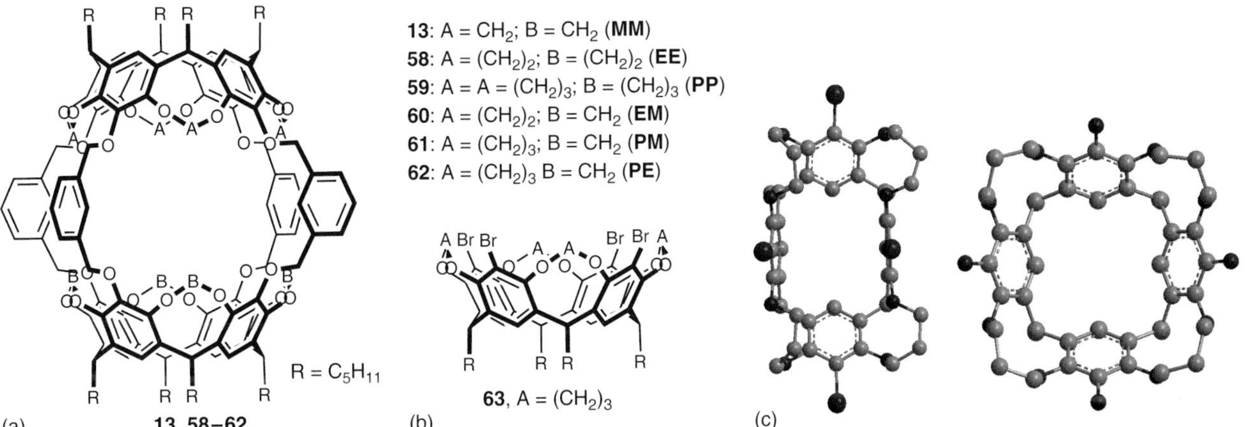

Figure 20 (a) Hemicarcerands **13**, **58–62**. Conformation of **P** bowl in X-ray structure of (b) **63** and (c) in hemicarcerands **61** and **62**.

cavity of **EE**. Such high structural recognition in combination with ease to tailor the dimension of the inner phase and the host's portals makes hemicarcerands ideal building blocks for hydrocarbon storage, separation, and purification applications.

8 WATER-SOLUBLE HEMICARCERANDS

In the previous examples, constrictive binding primarily controls selectivity in complexation studies. Water-soluble hemicarcerand **64** and **65** made possible a detailed analysis of differences in intrinsic binding among different hemicarceplexes (Figure 21).[89,90] In water, the hydrophobic effect, which is typically stronger than solvophobic effects in common organic solvents, drives complexation of nonpolar guests. Consequently, water-soluble hemicarceplexes are intrinsically more stable and their stabilities can be compared under equilibrium conditions.

Stable one-to-one hemicarceplexes of octaacid **64** with 14 guests were prepared in D_2O at pH 9.[89] Complexation was complete in a few minutes at room temperature, except for naphthalene, where dissolution of the lipophilic solid in D_2O was the rate-limiting step. Among the common guests 1,4-$(CH_3)_2C_6H_4$ and 1,4-$(CH_3O)_2C_6H_4$ that were studied for **64** and the structurally related nonpolar **13**, complexes of **64** are stable at room temperature in D_2O, whereas those of **13** decomplex rapidly at 25 °C in $CDCl_3$.

The four salts $(CH_3)_4N^+Br^-$, $Ph(CH_3)_3N^+Br^-$, $PhCH_2(CH_3)_3N^+Br^-$, and 3-$CH_3C_6H_4CO_2^-Na^+$ failed to complex **64** in D_2O buffer even though CPK models of hemicarceplexes can be assembled. Probably, D_2O solvates their charges better than it does the interior of **64**. It appears that the enthalpic solvation energies of the ions by water inhibit complexation, though the release of many inner-phase and guest-solvating water molecules would provide an entropic driving force for complexation.

Piatnitski *et al.* carried out a detailed thermodynamic analysis of the binding properties of the water-soluble hemicarcerand **65**.[90] This host lacks one of the linkers of **64**, which facilitates guest exchange. Thus, **65** displays thermodynamic selectivity in its binding properties, which differs from many other hemicarcerands, for which constrictive binding controls selectivity.

In water, **65** binds small hydrophobic guests with affinities that reach $K = 10^8\,M^{-1}$, which is higher than those measured for other receptors with hydrophobic cavities, such as cyclodextrins and cyclophanes.[91,92] Guest size, hydrophobicity, and charge are important factors in determining binding strength. An enthalpy–entropy compensation plot for binding of small hydrophobic guests provided a slope $\alpha = 0.75$ (Figure 22a). α varies from 0 to 1 and is a measure of to what extent the enthalpic gain is compensated by an entropic loss. Thus, it reflects the amount of host reorganization upon binding. Flexible enzymes reorganize substantially upon substrate binding and have $\alpha = 1$. The α value obtained for **67** is smaller than that for β-cyclodextrin ($\alpha = 0.9$) and comparable to that of cyclophanes ($\alpha = 0.78$) and indicates that **65** is relatively inflexible. Furthermore, the $T\Delta S$ intercept at $\Delta H = 0$, $T\Delta S_0 = 4.2\,\text{kcal mol}^{-1}$ is much higher than that for β-cyclodextrin and cyclophanes and has been taken as a measure for guest desolvation upon binding, which is an important driving force for complexation.[93]

Among aromatic guests with methyl or methoxy groups, *meta* and *para* substitution patterns are preferred over the ortho pattern, which can be rationalized with the ability of both methyl groups to undergo CH–π interactions if they are either para or meta (Figure 22b). Consistent with

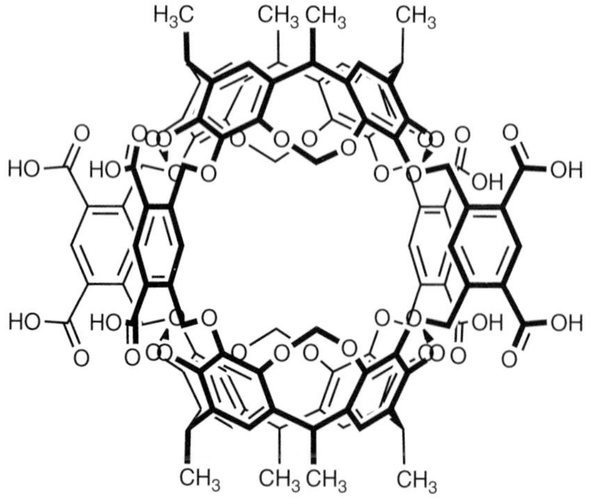

64

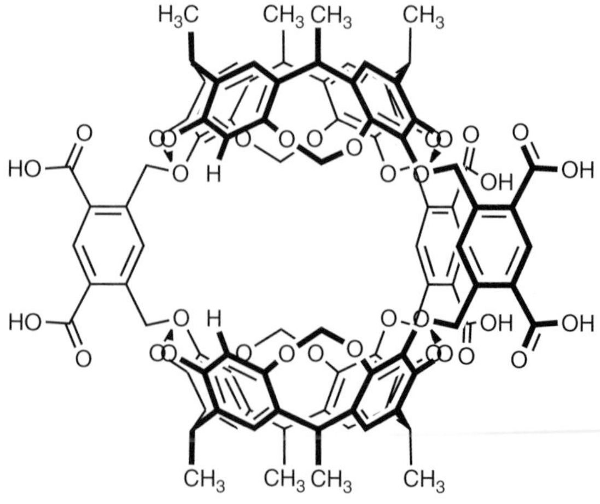

65

Figure 21 Water-soluble hemicarcerands **64** and **65**.

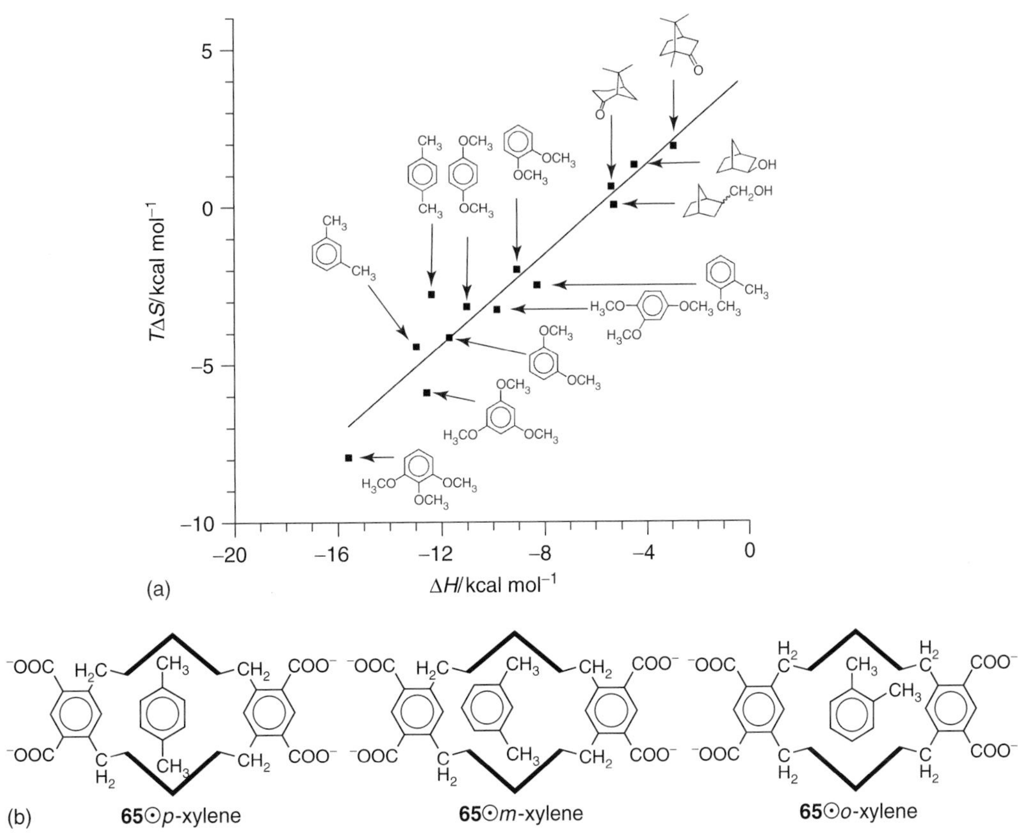

Figure 22 (a) Enthalpy–entropy plot for hemicarceplexes **65**⊙guest, and (b) illustration of interactions between xylene isomers and hemicarcerand **65**. (Reproduced from Ref. 90. © Wiley-VCH, 2000.)

this model are the measured binding enthalpies ($-\Delta H$), which decrease in the order *meta* > *para* ≫ *ortho* for xylenes and dimethoxybenzenes and the complexation-induced chemical shifts of the methyl protons, which are much larger for the meta and para isomers.

9 CHIRAL RECOGNITION PROPERTIES OF ASYMMETRIC HEMICARCERANDS

We have seen earlier that many hemicarcerands are chiral due to the twisting of the host's cavitands. However, interconversion between the two enantiomeric twistomers is typically fast at ambient conditions, thus preventing separation of twistomers. Introducing one or more chiral bridging units locks the hemicarcerand in one twisted conformation. Examples are asymmetric hemicarcerands $(S)_4$-**66**, $(SS)_4$-**67**, (S)-**68**, and (SS)-**69**, which have one and four bisoxymethylene-1,1′-binaphthyl or threonide bridges (Figure 23a).[94–96] In X-ray structures of $(SS)_4$-**67**⊙guest (guest = $CH_3C(O)N(CH_3)_2$ and $(CH_3)_2SO$), the host's cavitands are twisted along the polar axis by approximately 15°, which is slightly less than in the strongly twisted **14**⊙$CH_3C(O)N(CH_3)_2$ (24°).[95] These hosts typically display moderate thermodynamic chiral selectivity as a consequence of different interactions between enantiomers and the surface of the host's inner cavity (intrinsic binding energy). For example, heating $(SS)_4$-**67**⊙guest in Ph_2O containing excess racemic 2-butanol produces the two diastereomeric hemicarceplexes in a 2 : 1 ratio. This ratio amounts to an intrinsic binding energy difference of 300 cal mol^{-1}, which is comparable to the selectivity observed in chiral recognition studies of chiral cryptophanes and self-assembled molecular capsules.[6,97–99] Interestingly, both diastereomeric complexes have substantially different retention factors in thin-layer chromatography. The high sensitivity of the surface-adsorption properties of $(SS)_4$-**67** likely results from the host adapting its shape to the configuration of the guest. In complexation studies of (SS)-**69**, with only one threonide bridge, chiral recognition factors were smaller, for example, 1.4 for 2-butanol, and did not lead to changes in surface-adsorption properties of diastereomeric complexes. Under the same conditions, the chiral recognition factor for $PhS(O)CH_3$ was only 1.6.

Perhaps the most remarkable chiral selectivity is observed for binaphthyl bridged hosts $(S)_4$-**66** and (S)-**68**. Compared

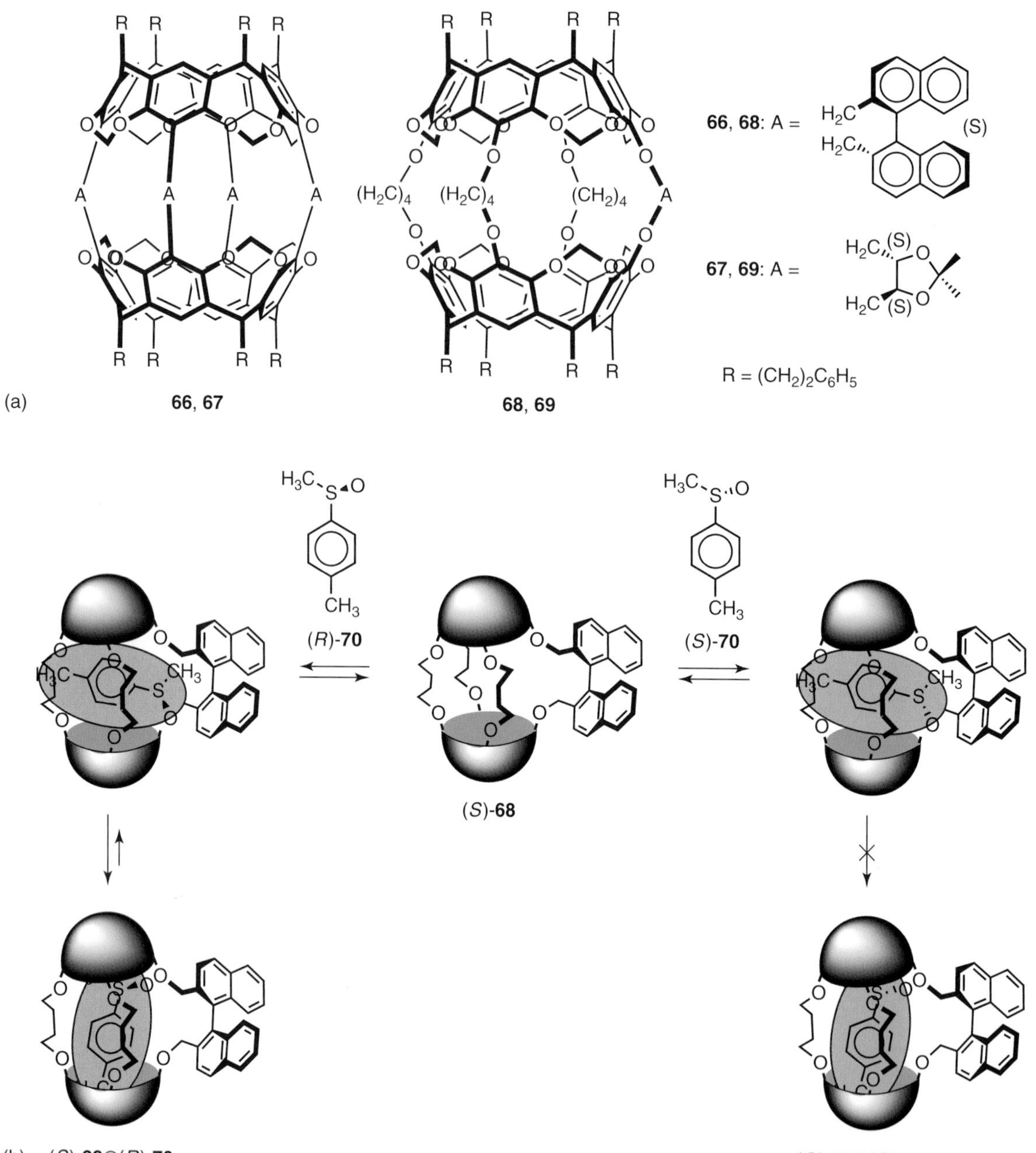

Figure 23 (a) Chiral hemicarcerands **66**–**69**. (b) Proposed mechanism for the high chiral selectivity in the complexation of *p*-$CH_3C_6H_4S(O)CH_3$ by host (*S*)-**68**.

to the fairly rigid threonide bridge, the bisoxymethylene-1, 1′-binaphthyl is more flexible and easily responds to differences in the degree of complementarity between host and guest by changing its naphthyl-to-naphthyl dihedral angle. When (*S*)-**68** was heated in the presence of excess racemic *p*-$CH_3C_6H_4S(O)CH_3$, only (*S*)-**68**⊙(*R*)-*p*-$CH_3C_6H_4S(O)CH_3$ formed. The chiral selectivity factor must be greater 20 : 1 and the free energy difference $\Delta G > 2.4\,\text{kcal mol}^{-1}$ for both diastereomeric complexes.[96] The fast decomplexation rate of (*S*)-**68**⊙(*R*)-*p*-$CH_3C_6H_4S(O)CH_3$ suggests that chiral recognition involves an equilibration between diastereomeric complexes rather than a kinetic resolution. Under the same conditions, racemic $C_6H_5S(O)CH_3$ gave diastereomeric complexes (*S*)-**68**⊙(*R*)-$C_6H_5S(O)CH_3$/(*S*)-**68**⊙(*S*)-$C_6H_5S(O)CH_3$ in 1.6 : 1 ratio. Yoon and Cram explained the high selectivity in the complexation of *p*-$CH_3C_6H_4S(O)CH_3$ with inability of (*S*)-*p*-$CH_3C_6H_4S(O)CH_3$ to adapt an orientation inside (*S*)-**68** that maximizes host–guest interactions. In CPK models, both guests can be easily pushed through one of the larger openings in the host shell. The resulting complexes have structures in which the guest lies roughly in the equatorial plane of the host. This orientation results in a minimal number of stabilizing host–guest contacts. However, only (R)-p-$CH_3C_6H_4S(O)CH_3$ is able to rotate about 90° around an equatorial axis of (S)-**68** to approximately align itself with the polar axis of the host such that both methyl groups can form favorable CH–π interactions with the host's cavitands (Figure 23b).

Apart from the cavity asymmetry, the chiral bridging units also create asymmetry in the shape of the host's portals, which gives rise to the kinetic chiral selectivity in complexation or decomplexation of these hemicarcerands or hemicarceplexes. Kinetic and thermodynamic selectivity may differ substantially as in the case of $(S)_4$-**66**.[94] For example, complexation studies with racemic $BrCH_2CH_2CHBrCH_3$ or $BrCH_2CHBrCH_2CH_3$ gave an equilibrated mixture of diastereomeric complexes in a ratio 2 : 1 for both guests. However, for $BrCH_2CH_2CHBrCH_3$, the thermodynamically less stable diastereomer dissociated five times faster than the more stable isomer. In the case of $BrCH_2CHBrCH_2CH_3$, the kinetic stability of the two diastereomeric complexes was reversed and the dissociation rate of the more stable isomer was ninefold larger than that of the less stable complex. Thus, whereas the difference in intrinsic binding in each diastereomeric pair is only $\Delta G_0 = 0.3\,\text{kcal mol}^{-1}$, the difference in constrictive binding, which is the $\Delta\Delta G^{\ddagger}$ value for the complexation diastereomeric TSs, is $\Delta\Delta G^{\ddagger} = 1.6\,\text{kcal mol}^{-1}$ for $BrCH_2CHBrCH_2CH_3$ and $\Delta\Delta G^{\ddagger} = 0.7\,\text{kcal mol}^{-1}$ for $BrCH_2CH_2CHBrCH_3$. Cram suggested that differences in steric repulsions in the diastereomeric TSs probably give rise to the observed chiral selectivities. Thus, the host is able to discriminate between the steric requirements of a CH_3 group versus Br atom or a CH_2CH_3 versus a CH_2Br group, whose volumes and surface areas differ by <10%. The high sensitivity of constrictive binding free energy to changes in the shape of the guest is particularly interesting for applications of asymmetric hemicarcerands in chiral separations, in asymmetric synthesis, or asymmetric catalysis.

10 DYNAMIC FEATURES OR HEMICARCEPLEXES

10.1 Guest rotation

Steric interactions also control rotational mobility of incarcerated guests. The inner phase of many hemicarcerands is shaped roughly like a rugby ball with the polar axis being longer than the equatorial diameter. Incarcerated guests align themselves such as to maximize favorable host–guest interactions and to minimize steric repulsions.

For example, in phenyldiazirine (**71**) and other monosubstituted aromatic guests, the guest's longer axis is aligned with the host's polar axis leading to large complexation-induced upfield shifts of the diazirine and the para aryl proton.[100] At ambient temperature, encapsulated **71** rapidly rotates around the longer C_4 axis of **6** and undergoes tumbling motions around equatorially located C_2 axes that are fast on the NMR time scale (Figure 24). Upon cooling below −15 °C, tumbling slows enough that different sets of proton signals for the northern and southern cap are observed. In the TS, the long axis of the guest passes through the smallest cross section of the inner cavity. Consequently, tumbling rates are very sensitive to the size of the substituent and to the presence of additional substituents especially in the para position.[35,58] For example, the free energy of rotation for acetophenone (**72**) inside hemicarcerand **6** is approximately $8\,\text{kcal mol}^{-1}$ higher than that of benzaldehyde (**73**).[101,102] An interesting situation arises for asymmetric hemicarcerands, in which the northern hemisphere differs from the southern hemisphere, such as in hybrid hemicarcerand **74**.[103] In this case, the restricted rotation of the guest inside the carcerand is referred to as *carcerisomerism* and the two stereoisomers that only differ in the orientation of the encapsulated guest as *carceromers* (Figure 25). In carceplex **74**⊙$(CH_3)_2NCOCH_3$, the two carceromers have different free energies ($\Delta G = 0.7\,\text{kcal mol}^{-1}$) and exchange fast on the NMR time scale at room temperature. However, at −30 °C, two different sets of signals are observed for each isomer.

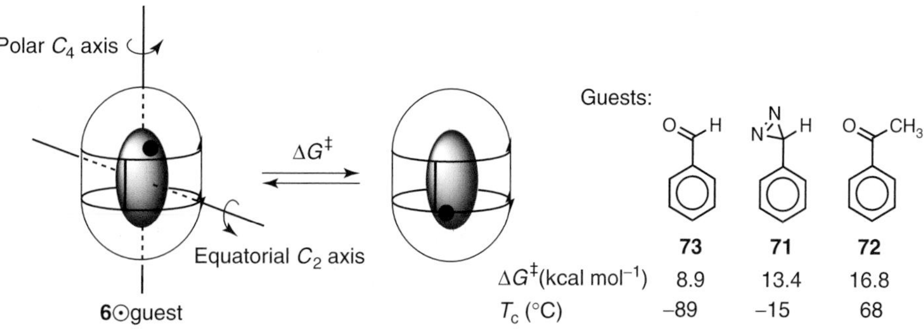

Figure 24 Orientation and mobility of incarcerated guest molecules.

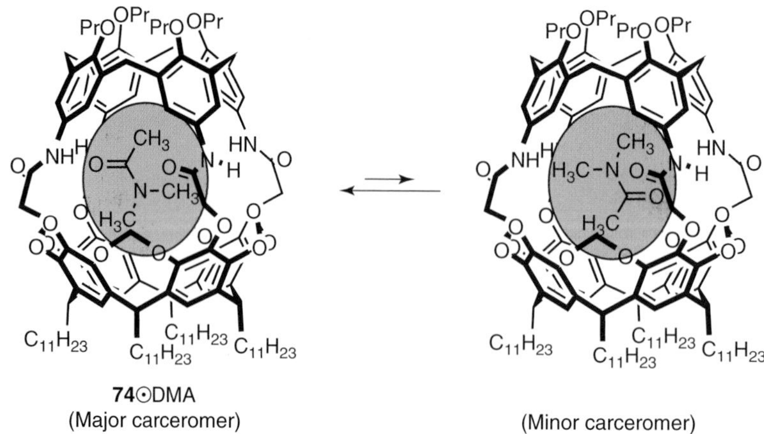

Figure 25 Carcerisomerism of **74**⊙$(CH_3)_2NCOCH_3$.

Guest rotation around the host's C_4 axis is typically very fast. In hemicarceplex **4**⊙C_6H_6, benzene spins around the host's polar axis with a rate constant of approximately 10^6–10^7 s^{-1} at room temperature.[104] To freeze out the spinning motion requires guests whose substituents protrude through equatorially located openings in the hemicarcerand shell. An interesting example was recently reported by Kobayashi and coworkers.[83] They studied the rate of rotation of 4,4′-diacetoxybiphenyl derivatives **44a–d** inside hemicarcerand **42** (Figure 16). These guests orient themselves inside the cavity such that one acetoxy methyl group is located inside a host's cavitand, where it forms stabilizing CH–π interactions. Guests **44a,b** spin rapidly around the long polar axis on the NMR time scale even at −60 °C. However, increasing the bulk of the substituents in 2 and 2′ positions slows the spinning motion in the order **44a, b** ≫ **44d** > **44c**, such that it is frozen at room temperature for **44c**. In the ground state, the equatorially located substituents of **44c** point toward an equatorial opening in the host shell, through which they can easily protrude in order to satisfy their space demand. Obviously, the guest experiences large steric interactions as it spins and the octyl groups have to pass by the rigid bridging units.

10.2 Bond rotation of amides and ring-flip of cyclohexanes

Confining a molecule inside a molecular container not only affects its rotational and vibrational degrees of freedom but also conformational changes of the guest, which are easily tractable by variable NMR spectroscopy and therefore ideal in order to study the effect of confinement on a TS. Cram and coworkers studied the effect of incarceration on the cis–trans isomerization of $(CH_3)_2NCHO$ and $(CH_3)_2NCOCH_3$ inside **4**.[33] For $(CH_3)_2NCHO$, the C–N rotational barrier decreased in the order liquid phase > inner phase > vacuum and was approximately 1 kcal mol^{-1} lower inside **4** than for the free amide in nitrobenzene. For $(CH_3)_2NCOCH_3$, the order was inner phase > liquid phase > vacuum and the barrier approximately 2 kcal mol^{-1} higher inside **4** as compared to the free amide. On the basis of examinations of molecular models of **4**⊙$(CH_3)_2NCHO$ and **4**⊙$(CH_3)_2NCOCH_3$, Cram explained these trends with the different ratio of free and occupied space in the inner phase. In **4**⊙$(CH_3)_2NCHO$, the guest is loosely held inside the container, whereas it is strongly compressed against the host walls in **4**⊙$(CH_3)_2NCOCH_3$ and even more so in the TS for bond rotation [**4**⊙$(CH_3)_2NCOCH_3$]‡. Thus, the rigid container resists being deformed more than the solvent cage resists being moved to accommodate the TS. Depending on the mix of free and occupied space, the inner phase may be more like vacuum, liquid, or even solid.

Large effects due to the rigidity of the container were also observed for the ring inversion of 1,4-thioxane and 1,4-dioxane inside carcerand **4** (Figure 26a). Chapman and Sherman measured 1.8 and 1.6 kcal mol^{-1} higher ring-flip barriers inside **4** as compared to the liquid phase.[105] The increase in barrier height was substantially more than inside the self-assembled hydrogen-bonding capsule $\mathbf{75}_2$⊙C_6H_{12}, in which the ring inversion barrier of cyclohexane-d_{11} increases by only 0.3 kcal mol^{-1} (Figure 26b).[106] In the latter case, Rebek and coworkers argued that steric interactions in the TS are unlikely the reason for the modest increase, since the TS is more planar than cyclohexane and should fit better into the "jelly doughnut"-shaped capsule. They suggested that the loss of favorable (C–H/D)–π contact stabilizes the ground state relative to the TS inside the container.

The origin for the increased barriers inside **4** is not fully clear and may result from similar ground state effects, such as stabilizing host–guest contacts that are lost in the TS, or from steric constraints in the TS. However,

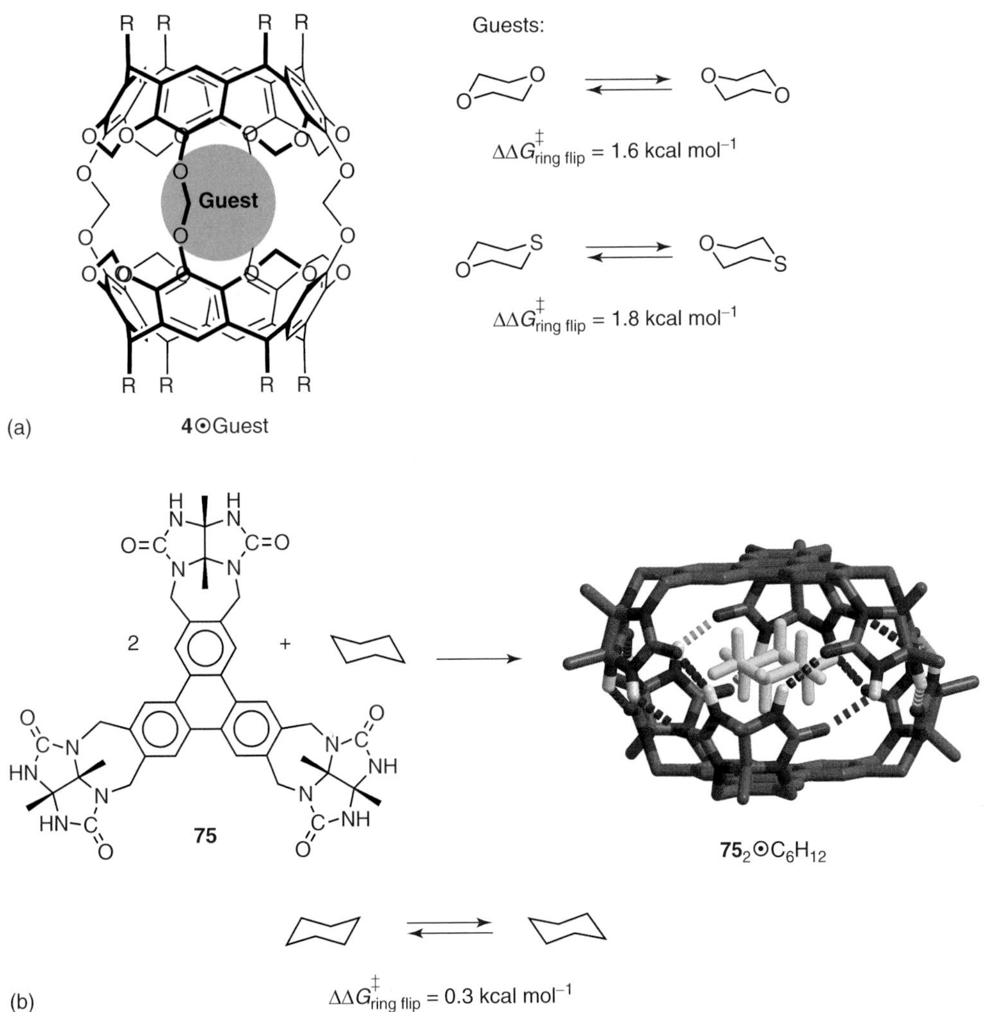

Figure 26 (a) Ring-flip dynamics of 1,4-thioxane and 1,4-dioxane inside carcerand **4**. (b) Structure and assembly of self-assembly capsule $\mathbf{75}_2\odot C_6H_{12}$ and ring-flip dynamics of encapsulated cyclohexane. $\Delta\Delta G^{\ddagger}_{\text{ring flip}} = \Delta G^{\ddagger}_{\text{ring flip}}(\text{encapsulated}) - \Delta G^{\ddagger}_{\text{ring flip}}(\text{free})$.

one can conclude that the effects are pronounced inside the carcerand, which is a result of its rigidity. Host rigidity translates into high sensitivity to small structural perturbations of the guest.

11 HEMICARCERANDS AS MOLECULAR REACTION FLASKS

The application of molecular containers as "molecular reaction flasks" has been a very exciting and rewarding venture in host–guest chemistry. In recent years, the exploration of reactions and reactivity inside covalent or self-assembled molecular capsules has produced very spectacular and unexpected discoveries.[15,16] For example, it has been demonstrated that molecular capsules may allow the taming of otherwise fleeting reactive intermediates,[107] alter the regiochemistry of reactions,[22,28,29] give rise to new forms of rate accelerations in pericyclic reactions,[21–23] and in some instances show enzyme-like behavior in ester and acetal hydrolysis reactions.[24]

Hemicarcerands were the first molecular containers in which chemical reactions involving encapsulated reactants have been investigated. In this section, some of the advances in inner-phase chemistry are reviewed. Inner-phase reactions may take place either entirely inside the carcerand, where they are influenced by the shape and size of the inner phase, or at the electrostatic inner surface of the hosts with its unusual high inner-phase polarizability.[108,109] Typically, these reactions involve one or two encapsulated reactants, in which case the host takes over the role of the solvent cage in equivalent condensed phase reactions. Proper solvation is particularly important in reactions involving zwitterionic intermediates or ion pairs. The absence of polar solvent molecules and the hydrophobicity and reduced deformability of the inner phase will be particularly felt in these types of reactions. A second kind of inner-phase reaction is best described as "through-shell" reactions. They involve both encapsulated and bulk-phase reactants, the latter being transferred through the host shell somewhere along the reaction path. Bond formation or breaking of through-shell reactions may take place inside one of the openings in the host shell. Thus, outcomes often depend on orientation and rotational mobility of the encapsulated reactant as well as the size, shape, and flexibility of the portals relative to those of the bulk-phase reactant.[102,110]

11.1 Through-shell reactions

11.1.1 Proton transfer reactions

Proton transfer between incarcerated bases and bulk-phase acids is a simple through-shell reaction and provides insight into the effect of incarceration on the guest's acidity or basicity. Consistent with observations in many enzyme-catalyzed reactions, the hydrophobicity of the inner phase should alter the pK_a of the incarcerated guest.[111] Cram and coworkers studied proton transfers between a strong bulk-phase acid and incarcerated amines **31**⊙pyridine, **31**⊙$(CH_3CH_2)_2NH$, and **31**⊙$CH_3(CH_2)_3NH_2$ (Figure 27).[33] Despite a large enough opening in the shell of **31**, attempts to protonate incarcerated pyridine with CF_3COOD in $CDCl_3$ failed. Cram suggested that the reduced basicity may result from ineffective solvation of the pyridinium ion by the rigid host, the inability to form a contact ion pair in the inner phase, and the larger size of the pyridinium ion compared to pyridine.

For **31**⊙$(CH_3CH_2)_2NH$, instantaneous decomplexation of **31**⊙$(CH_3CH_2)_2ND_2^+$ accompanied through-shell proton transfer. The ability to protonate **31**⊙$(CH_3CH_2)_2NH$ with CF_3COOD in $CDCl_3$ results from the location of the nitrogen of $(CH_3CH_2)_2NH$ in the equatorial region close to the portals. After protonation, the counterion pulled the guest out of the inner phase. Addition of excess CF_3COOD to **31**⊙$CH_3(CH_2)_3NH_2$ led to a 2 : 1 mixture of **31**⊙$CH_3(CH_2)_3ND_3^+$ and **31**⊙$CH_3(CH_2)_3NH_2$, which remained constant over time although slow decomplexation took place. Complete protonation of **31**⊙$CH_3(CH_2)_3NH_2$ required 100 equivalents of CF_3COOD. Excess CD_3COOD only H/D-exchanged the amine protons. These results show that the acidity of incarcerated $CH_3(CH_2)_3NH_3^+$ is comparable to that of CF_3COOH in $CDCl_3$. Furthermore, the strong upfield-shifted amine protons of **31**⊙$CH_3(CH_2)_3NH_2$ imply guest alignment along the polar axis of **31**. In this orientation, through-shell protonation most likely occurs through the holes in the polar caps of **31**.

11.1.2 Electron transfer reactions

Electron transfer reactions are well suited to be studied between an incarcerated guest and a bulk-phase reducing or oxidizing agent, since electron transfer processes do not require direct contact between donor and acceptor complexes but may take place over long distance through electron tunneling.[112] An oxidation–reduction cycle for different *ortho*- and *para*-hydroquinones could be carried out in the interior of **6** (Figure 28).[113] Oxidation with $Ce(NH_4)_2(NO_2)_6$–silica gel–CCl_4 or $Tl(O_2CCF_3)_3$–CCl_4 led to the parent incarcerated quinones in essentially quantitative yields.

Reduction back to the hydroquinones was possible with SmI_2/CH_3OH. The same reagent reduced nitrobenzene to *N*-hydroxyl aniline. Surprisingly, aniline, which is the product in the liquid phase, is not formed. The latter result—the high yields and the instability of free *o*-quinones—suggests that all reduction/oxidation took place inside **6** rather than

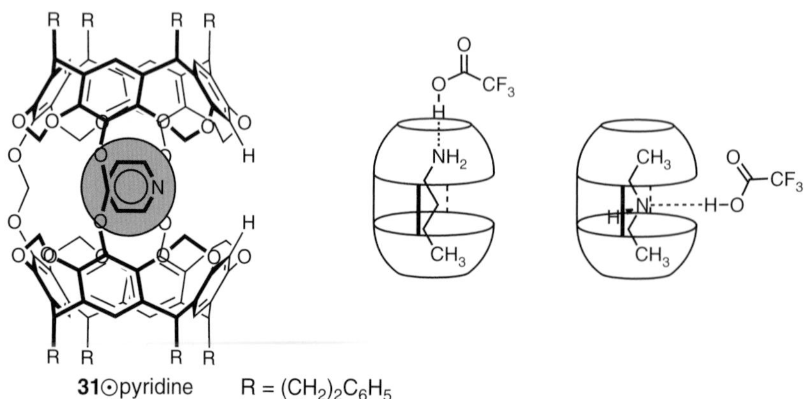

Figure 27 Structure of hemicarceplex **31**⊙pyridine and proposed proton transfer mechanism for **31**⊙$CH_3(CH_2)_3NH_2$ and **31**⊙$(CH_3CH_2)_2NH$.

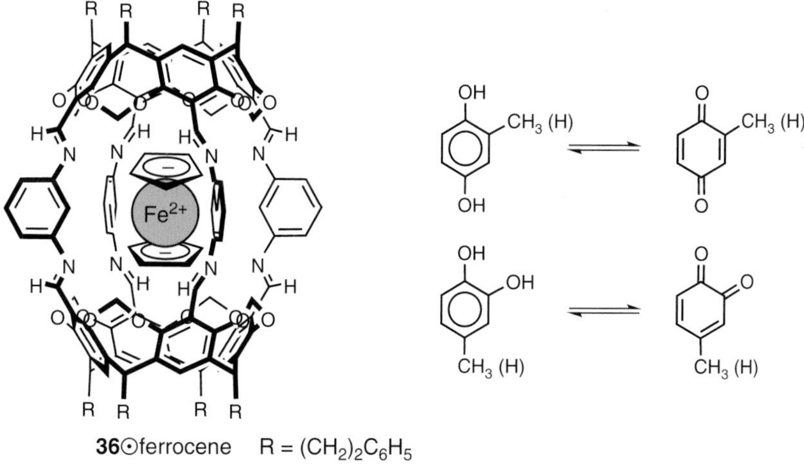

Figure 28 Through-shell oxidation–reduction cycles of *ortho*- and *para*-hydroquinones and nitrobenzene inside **6** and electrochemical oxidation of **36**⊙Fc.

by a dissociation (bulk-phase reaction)—association mechanism. It also demonstrates that electrons are transferred readily through the host shell in and out of the inner phase.

A second example in which through-shell electron transfer was examined quantitatively is the electrochemical oxidation of Fc incarcerated inside hemicarcerand **36**. Electron transfer was strongly hindered kinetically and thermodynamically compared to free Fc (Figure 28).[114] The half-way potential for the oxidation was more positive, due to the hydrophobicity of the inner phase, and the electron transfer rate was reduced 10-fold. The latter may result partially from the higher mass of **36**⊙Fc compared to Fc and also from a reduction of the electronic coupling between the Fc center and the electrode surface which is affected by the increase in distance from 3.5 to about 9 Å. Whether the hemicarcerand's aromatic structure mediates the electron coupling is not clear.

11.1.3 Nucleophilic substitutions and isotopic exchanges

The alkylation studies of Kurdistani *et al.* provide much insight into the interplay between guest reactivity, orientation, and bulk-phase reagent size.[110] Different phenols were alkylated in the inner phase of **6**. Two factors determined reactivity: (i) portal size and (ii) preferred guest orientation relative to the equatorially located portals. Alkylation with NaH/CH_3I in THF of 4-$HOC_6H_4CH_3$ (*p*-cresol) or 4-HOC_6H_4OH (*p*-hydroquinone) was impossible. Under the same conditions, 2-$HOC_6H_4CH_3$ (*o*-cresol), 3-$HOC_6H_4CH_3$ (*m*-cresol), and 3-HOC_6H_4OH (resorcinol) were quantitatively methylated. 2-HOC_6H_4OH (catechol) gave a mixture of mono- and dimethylated carceplexes. As discussed in Section 10.1, the preferred inner-phase orientation of 1,4-disubstituted benzene guests suggests that the OH group of 4-$HOC_6H_4CH_3$ is located in a protected polar cap of the host. In *ortho*- or *meta*-disubstituted benzenes, one substituent resides inside a shielded polar cap, whereas the second substituent is located near a portal. Therefore, these reactions must occur in the entryways through a linear TS, which is partially "solvated" by the alkoxy units that align the host's portals (Figure 29). Since this "pseudo solvent cage" has limited flexibility, larger alkylating agents failed to react.

Likewise, in D_2O-saturated $CDCl_3$ no H/D exchange of OH groups was possible when the guest was 4-$HOC_6H_4CH_3$, 2-HOC_6H_4OH, or 4-HOC_6H_4OH.[110] In the presence of diazobicyclo[5.4.0]undec-7-ene, 4-HOC_6H_4OH exchanged its hydroxyl protons, but not the rotationally more fixed 4-$HOC_6H_4CH_3$. In THF–NaH at 25 °C followed by D_2O quench, the hydroxyl protons of 2-HOC_6H_4OH, which are more exposed to the equatorial-located portals, exchange, but not the protected hydroxyl protons of 4-HOC_6H_4OH and 4-$HOC_6H_4CH_3$.

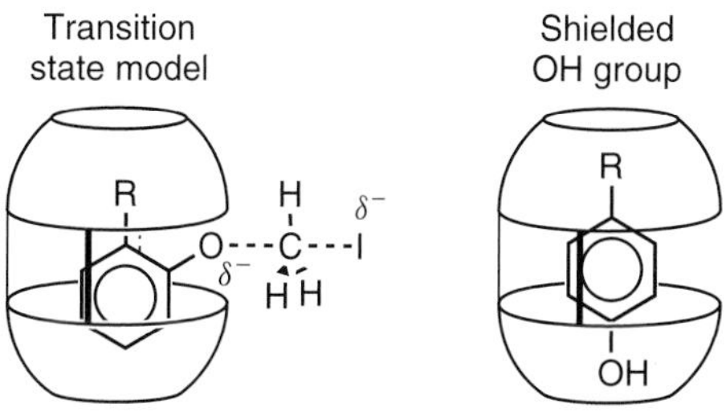

Figure 29 Transition-state model for alkylation of *ortho*-disubstituted phenolates and shielded OH group in *para*-disubstituted phenols.

11.1.4 Nucleophilic additions

Inner-phase guest orientation and mobility also control reactivity in through-shell borane and methyllithium additions to benzaldehyde **73**, benzocyclobutenone **76**, and benzocyclobutadione **77** inside hemicarcerand **6**.[102] BH_3·THF reduced all three incarcerated guests to benzyl alcohol, benzocyclobutanol, and 7-hydroxybenzocyclobutanone (**78**), respectively (Figure 30). Guest reactivity differed from that in the liquid phase and increased in the order **77** $\approx$ **76** $>$ **73**. Furthermore, incarcerated **77** added only 1 equivalent of BH_3·THF. An aqueous workup was required in order to reduce the second carbonyl group to **7**⊙**79**. Hydrolysis of **7**⊙**79** gave incarcerated *cis*-benzocyclobutenediol **80**.

Crystal structures helped to interpret the different reactivity of the guests. In hemicarceplexes **6**⊙**73** and **6**⊙**76**, the guest's carbonyl groups are located inside a host's cavitand and reduction requires reorientation of the guest. The additional conformational energy adds to the activation energy and is higher for **73** as compared to **76**. In hemicarceplex **6**⊙**77**, one carbonyl is shielded; the other is perfectly positioned for through-shell reaction inside an entryway. After addition to the exposed C=O, coordination of the boron of **81** to a host's ether oxygen hinders guest rotation and prevents exposure of the second C=O until **81** is hydrolyzed.

Guest orientation also explains outcomes of CH_3Li additions to incarcerated **73**, **76**, and **77** (Figure 31). Again, guest reactivity decreased in the order **77** $\gg$ **76** $>$ **73**. Compound **77** added 1 equivalent of CH_3Li already at $-78\,^\circ$C to yield **82** and Moore rearrangement product **83**. No double-addition took place. Hemicarceplex **6**⊙**76** required room temperature for reaction completion. Under the same conditions, **6**⊙**73** reacted sluggishly and incompletely.

Very interesting are the formation of host cleavage products **86** and **25** in these reactions. The former results from the cleavage of one of the host's spanners initiated by nucleophilic attack of lithium alcoholate **87** at the acetal carbon (Figure 32). On the other hand, at $0\,^\circ$C incarcerated lithium alcoholates **84** and/or **85** cleaved one of the O–$(CH_2)_4$–O linkers of **6** via β-eliminations.

Bulk-phase lithium alcoholates are not basic enough to induce this reaction. The incarcerated counterparts must be several orders of magnitude more reactive. Three factors contribute to these rate accelerations: (i) the absence of aggregation of R–OLi in the inner phase; (ii) the poor ability of **6** to "solvate" R–OLi, which increases its

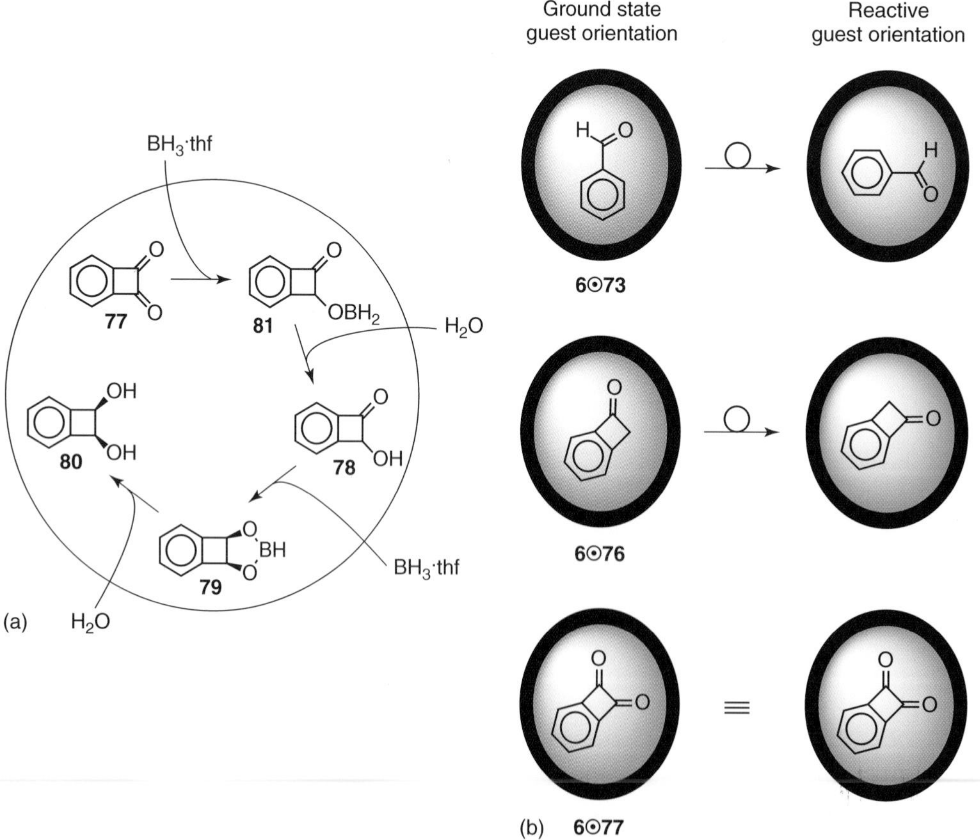

Figure 30 (a) Borane reductions of **77** inside **6**. (b) Ground state and "reactive" orientations of guests **73**, **76**, and **77** in inner-phase borane reductions and CH_3Li additions.

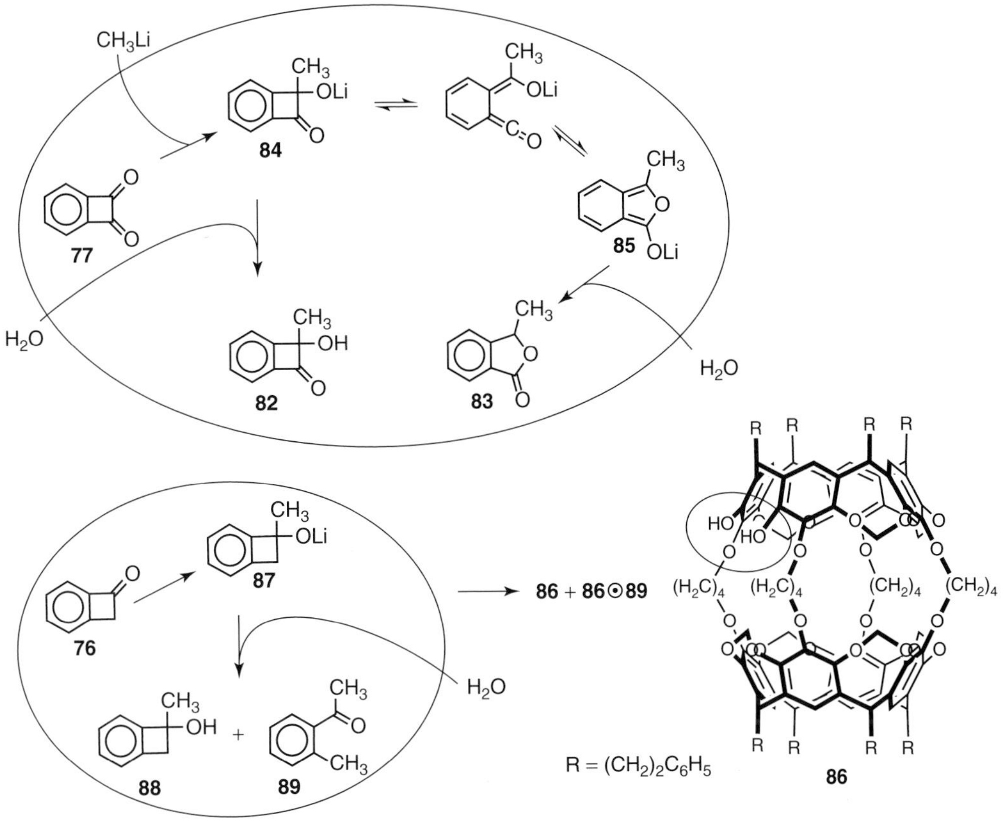

Figure 31 Through-shell CH_3Li additions to incarcerated **76** and **77**.

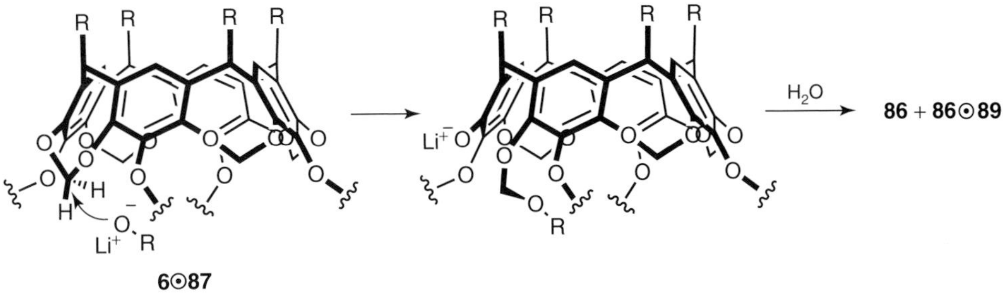

Figure 32 Proposed mechanism of formation of **86** and **86⊙89**.

basicity; (iii) lithium coordination to an oxygen lone pair of the cleaved C–O bond positions the alkoxide O in close proximity to the β-H of the bridge and provides charge compensation during the concerted *syn* elimination. These examples and those discussed in the previous sections show that through-shell and inner-phase chemistry clearly differs from "conventional" chemistry in the bulk phase with respect to reactivity and selectivity. Inner-phase and through-shell reactions show the following characteristic features:

1. Guest functional groups that reside inside a host's polar cap are less reactive than those exposed to an equatorial portal, which have the potential for high through-shell reactivity.
2. The reactivity of bulk-phase reactants is largely influenced by their size and shape relative to that of the host's equatorial portals.
3. If functional groups are protected in the guest's most favorable orientation, reactivity depends on the inner-phase rotational mobility of the guest.
4. The basicity and nucleophilicity of incarcerated lithium alcoholates exceed those of bulk-phase alcoholates by several orders of magnitude, resulting in efficient inner-molecular elimination or nucleophilic transacetalization and formation of hemicarcerands with one extended portal. In these inner-phase reactions, small structural changes of the guest have a sound effect on the reaction mode.

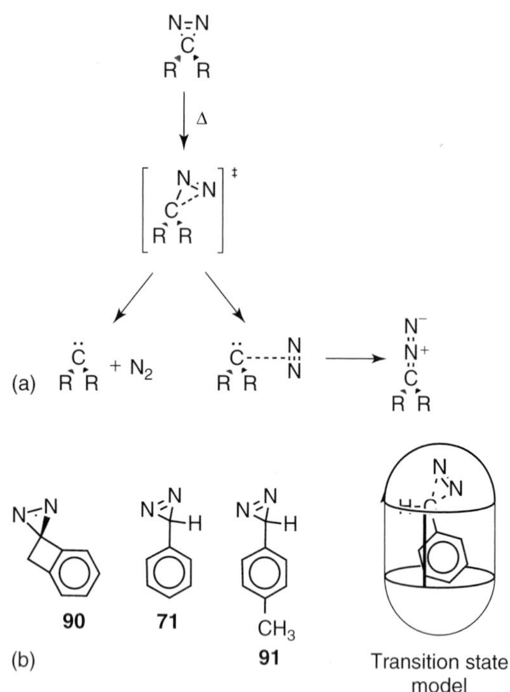

Figure 33 (a) Products and mechanism of diazirine thermolysis. (b) Aryldiazirines **71**, **90**, and **91** and transition-state model for the inner-phase phenyldiazirine fragmentation.

11.2 Intramolecular thermal reactions

Intramolecular reactions are well suited to compare the inner phase with other bulk phases in terms of reaction rate and to highlight special characteristics of the inner phase as a reaction environment. For example, the ring inversions of cyclic alkanes, which were discussed in Section 10.2, are slowed upon incarceration, likely due to steric interactions in the TS or selective stabilization of the ground state. In intramolecular reactions involving bond formation and or breaking, TS stabilization by the hemicarcerand is possible if bond breaking/formation takes place in close proximity of the host's aryl units. This was demonstrated in an investigation of the thermal fragmentation of aryldiazirines inside hemicarcerands.[109,115] The thermolysis of diazirines is a common method to produce carbenes and its mechanism has been studied in detail (Figure 33).[116,117] Compared to the bulk phase, inner-phase fragmentation of **90** is 15-fold accelerated, that of **71** slightly faster (1.2-fold), and that of **91** 2.4-fold slower.[115] Furthermore, all inner-phase TSs are stabilized enthalpically by 2–3 kcal mol^{-1}, which, in the case of **71** and **91**, is partially or fully compensated by unfavorable entropic contributions to $\Delta G^{\ddagger}$. The unfavorable $\Delta T \Delta S^{\ddagger}$ term likely results from loss of vibrational degrees of freedom as the guest expands upon reaching the TS, leading to a tighter hemicarceplex. The favorable enthalpic stabilization is interesting and was explained with the high polarizability of the inner phase.[108,109] The stretched C–N bonds of the TS are more polarizable than those of the ground state. Thus, the TS will be more strongly stabilized through dispersion interactions,[118] especially since bond breaking takes place in close proximity to the highly polarizable aryl units of a cavitand (Figure 33b).

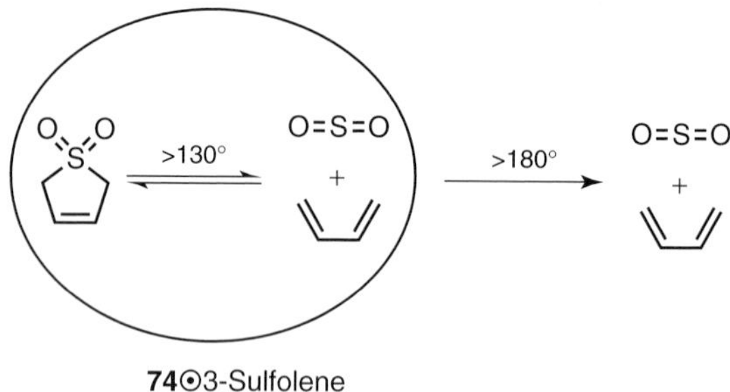

Figure 34 Thermal extrusion of SO_2 from **74**⊙3-sulfolene.

If the extrusion reaction is reversible, incarceration may strongly increase the thermal stability of the encapsulated reactant. For example, Reinhoudt and coworkers studied the extrusion of SO_2 and butadiene from 3-sulfolene incarcerated inside **74** by mass spectrometry (Figures 25 and 34).[119]

In the gas phase, the extrusion of SO_2 and butadiene from free 3-sulfolene readily takes place at 100–130 °C. Substantially higher temperatures were required for carceplex **74**⊙3-sulfolene. SO_2 was detected only above 170–180 °C. At lower temperature, only the intact carceplex was observed. Above 180 °C, also empty **74** was detected but not a SO_2 carceplex or a butadiene carceplex. Since **74** is stable at such high temperatures, guest escape due to the thermal destruction of **74** can be excluded. Hence, the detected SO_2 and butadiene must result from **74**⊙3-sulfolene and must escape from the inner phase through one of the larger side portals. Reinhoudt explained the unusually high thermal stability of incarcerated 3-sulfolene with a fast recombination in the inner phase (Figure 34). Below 180 °C, a thermal equilibrium among 3-sulfolene, SO_2, and butadiene is established. Above 180 °C, this equilibrium is pulled toward the extrusion products via their escape from the inner phase. This example shows how confinement changes the rates of bimolecular reactions by providing a very high local concentration of both reactants.[29,120]

11.3 Inner-phase stabilization of reactive intermediates

The possibility to photolyze incarcerated guest molecules presents a pathway to generate and protect highly strained and reactive molecules inside carcerands.[107] This allows NMR spectroscopic characterization of otherwise fleeting species, which complements matrix isolation spectroscopy, ultrafast spectroscopy, or flow techniques. The concept

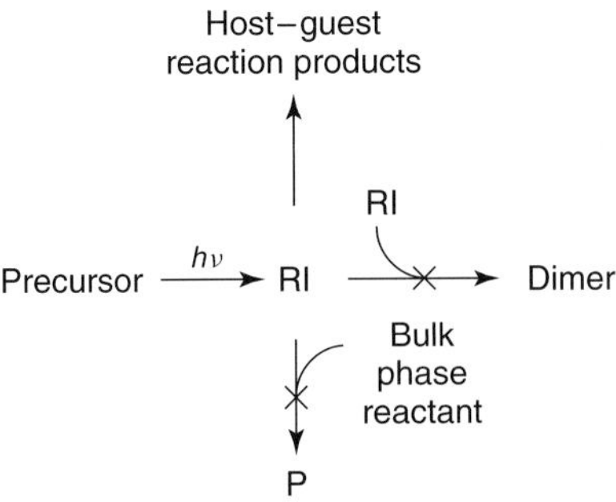

Figure 35 Reactive intermediate stabilization by incarceration.

of reactive intermediate stabilization by incarceration was introduced by Cram, Tanner, and Thomas with "the taming of cyclobutadiene" and is outlined in Figure 35.[17]

Photolysis of a suitable stable photochemical precursor yields the reactive intermediate in the inner phase. Once generated, the surrounding host prevents destructive reactions, such as dimerization or trapping, with bulk-phase reactants that are too large to pass through an opening in the host shell. Difficult to prevent are innermolecular reactions with the surrounding host, which may take place with incarcerated carbenes, nitrenes, radicals, and arynes, thus limiting their lifetime. In the following sections, several examples are discussed.

11.3.1 Cyclobutadiene

The "taming of cyclobutadiene" inside **31** is the first example of an inner-phase stabilization of a reactive intermediate and nicely demonstrates the power of this approach (Figure 36).[17] Cyclobutadiene **92** is the prototypical example to verify the theory of aromaticity.[121,122] It is severely angle-strained in addition to being antiaromatic.

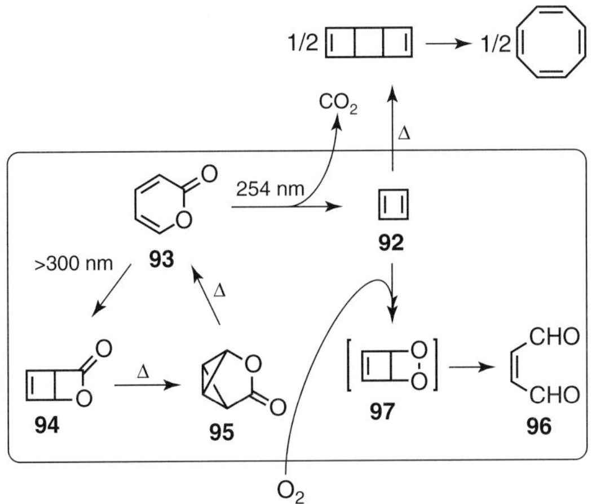

Figure 36 Photochemical generation of **92** reactions inside **31**.

Cram and coworkers generated **92** inside **31** by irradiating α-pyrone hemicarceplex **31⊙93**.

Irradiation above 300 nm converted **31⊙93** to photopyrone **31⊙94**, which, as a solid, rearranged to **31⊙95** at 90 °C. At higher temperature, **31⊙95** reverted quantitatively back to **31⊙93**. Controlled irradiation of **31⊙93** with unfiltered UV light produced cyclobutadiene nearly quantitatively. Prolonged photolysis gave acetylene. In the absence of oxygen, cyclobutadiene was stable up to 60 °C and could be characterized by ^{1}H NMR spectroscopy. Its lifetime is controlled by the barrier for passage through the larger opening inside **31**. If a solution of **77⊙92** was heated in a sealed tube at high temperatures, the guest escaped the protective shelter and dimerized. Also, oxygen, which easily passes through the host shell, trapped the guest as malealdehyde **96**, presumably via an intermediate dioxetane **97**.

11.3.2 Anti-Bredt bridgehead olefins

The approach of inner-phase stabilization of reactive intermediates works particularly well for species with highly strained multiple bonds, such as cyclobutadiene, which have a lower tendency to react with the surrounding host. Other examples are *anti*-Bredt bridgehead olefins,[123] which have a *trans*-cycloalkene and are unstable if their olefinic strain $OS \geq 21\,\text{kcal mol}^{-1}$.[124] For example, bicyclo[2.2.2]oct-1-ene **98** and (*Z*)-bicyclo[3.2.1]oct-1-ene **99** have $OS = 46.4$ and $21.9\,\text{kcal mol}^{-1}$, respectively. Both are fleeting in solution, due in part to their high tendency to dimerize or rearrange, but were stabilized recently at room temperature inside hemicarcerand **6**.[19] For the inner-phase synthesis of **98** and **99**, Jones' carbene route was chosen, in which carbene **100** rearranged to **98** (major) and **99** (minor).[125] Inner-phase photolysis of diazirine **101** gave a complex product mixture composed of hemicarceplexes **6⊙98**, **6⊙99**, and **6⊙103** and small amounts of carbene–hemicarcerand insertion products. Mechanistic studies suggest that photochemically excited **101*** directly rearranges to **98** and **99** without participation of carbene **100** (Figure 37). Both incarcerated *anti*-Bredt olefins are stable at room temperature in the absence of oxygen and oxidize to ketoaldehydes **104** and **105** in aerated solution. A thermal *retro*-Diels–Alder reaction of **98**, which in Jones' seminal pyrolysis studies had served as indirect proof for formation of **98**,[125] could also be induced inside **6**. At 62 °C, **98** slowly rearranged to triene **106**, which escaped the inner phase and was identified in the bulk by its characteristic ^{1}H NMR spectrum.

11.3.3 o-Benzyne

ortho-Benzyne, which has a highly strained triple bond, is another important reactive intermediate in some nucleophilic aromatic substitutions and was recently stabilized

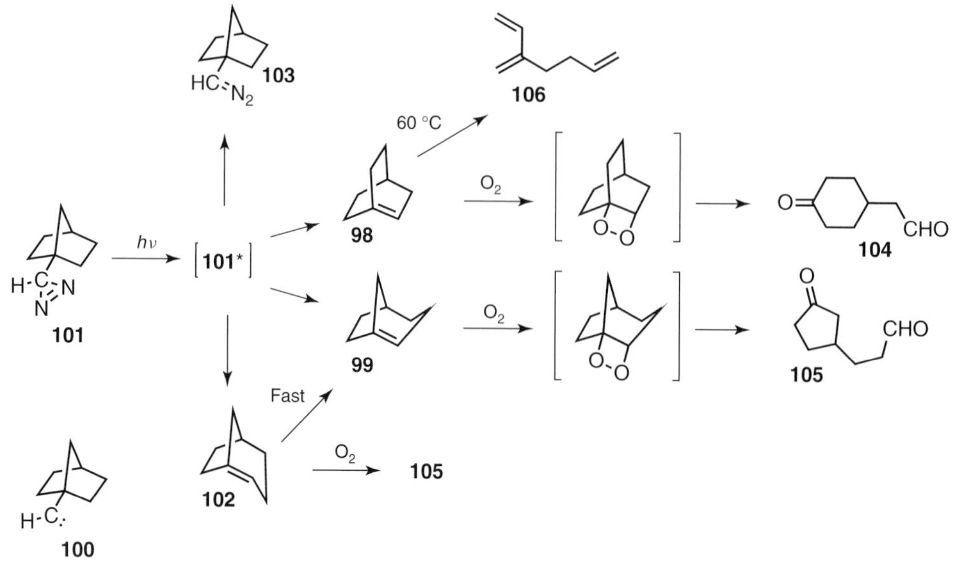

Figure 37 Inner-phase photochemistry of incarcerated **101**.

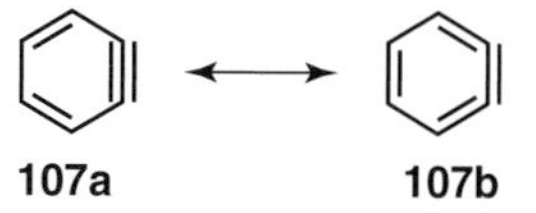

Figure 38 Resonance structures of *o*-benzyne.

in the inner phase of hemicarcerand **6** (Figure 38).[18,126] *o*-Benzyne is also of interest due to its unusual structural and electronic properties.[127,128] Chapman first matrix-isolated *o*-benzyne by photolyzing benzocyclobutenedione **77** at 8 K (Figure 39).[129] The same route led to the successful inner-phase synthesis of *o*-benzyne.[18]

Irradiation of hemicarceplex **6⊙77** at 400 nm gave hemicarceplex **6⊙109**, which upon further photolysis at 280 nm decarbonylated to yield **6⊙107**. Because of the high π-bond strain of 50 kcal mol^{-1},[130] *o*-benzyne underwent a Diels–Alder reaction with the surrounding **6**, which was fast above −75 °C (Figure 39b).[126] This reaction is very selective and **107** adds exclusively across the 1,4-position of an aryl unit of **6**. The MM3* minimum-energy conformer of **6⊙107** shows strong preorganization of the reactive triple bond for the observed Diels–Alder reaction with distances of 4.53 and 4.05 Å between the reacting carbons of the host and guest.[131] This high preorganization is reflected in the moderately negative activation entropy $\Delta S^{\ddagger}(298\,\text{K}) = -10.7\,\text{cal mol}^{-1}\,\text{K}^{-1}$.[126] Interestingly, the measured $\Delta H^{\ddagger}$ is slightly higher than the calculated $\Delta H^{\ddagger}_{\text{calc}}$ for the addition of **107** to benzene.[131] Thus, the increased reactivity of **6** must be compensated by steric interactions originating from a repulsion between H(1) and aryl unit A (Figure 39b). This suggests that an incarcerated 3,6-disubstituted *o*-benzyne may not be

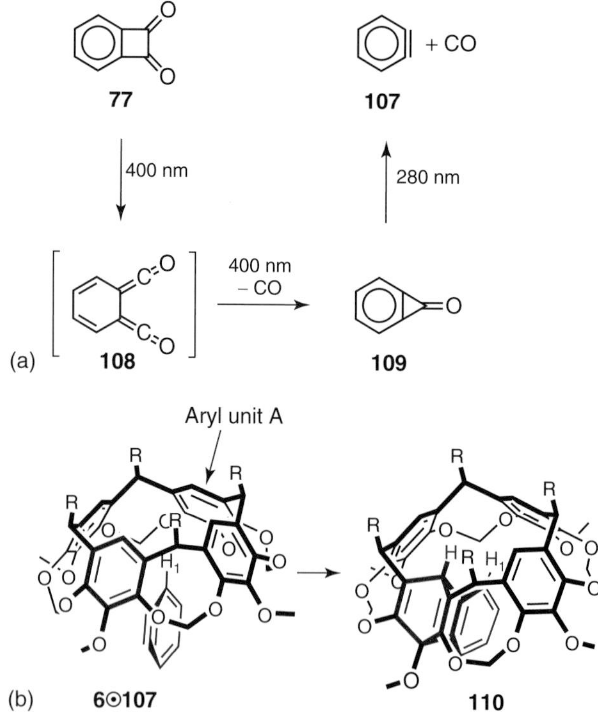

Figure 39 (a) Photochemistry of **77** in argon at 8 K and inside hemicarcerand **6** and (b) Diels–Alder addition of incarcerated *o*-benzyne to yield **110**.

able to react with the host and may be stable at higher temperatures.

At −75 °C, the lifetime of **6⊙107** was long enough to record a ^{1}H NMR spectrum. The protons of **107** resonated at δ 4.99 and δ 4.31. Under the assumption that they feel the same shielding by the surrounding host as the protons of benzene, the chemical shifts of "free" *o*-benzyne were estimated at δ 7.0 and δ 7.6, which are in excellent agreement with the calculated shifts.[128] Much less upfield shifted are the guest ^{13}C signals, which provide more insight into the electronic properties of *o*-benzyne. The measured chemical shift for the quaternary carbon of **107** at δ 181.33 is within the experimental error of the average of the three chemical shift tensor principal values $\delta 193 \pm 15$ of matrix-isolated, ^{13}C-enriched **107** at 20 K in argon.[132] The ^{13}C NMR spectrum of incarcerated *o*-benzyne also provided direct ^{13}C–^{13}C coupling constants. Comparison with the ^{13}C–^{13}C coupling constants of model compounds suggested a cumulenic character of *o*-benzyne (Figure 38), which, however, contradicts most recent results of *ab initio* calculations.[128,133] These calculations predict that *o*-benzyne is aromatic according to its geometric, energetic, and magnetic properties and that the in-plane π-bond induces a small amount of bond localization resulting in acetylenic character.

11.3.4 Phenylcarbene rearrangement

The phenylcarbene (PC) rearrangement was recently studied inside hemicarcerands **6** and **69**. In the gas phase, PC **111** ring-expands to cyclohepta-1,2,4,6-tetraene (CHTE) **113** involving bicyclo[4.1.0]hepta-1,3,5-triene **112** as intermediate (Figure 40).[134–136] CHTE, which is a bent and twisted allene with 40 kcal mol^{-1} strain energy,[136] is the local minimum on this part of the potential energy surface and enantiomerizes via the planar cyclohepta-1,3,5-trienylidene **114** as TS.[135,136]

Ring expansion can also be triggered photochemically via excitation of triplet PC ^{3}**111** generated from phenyldiazomethane or phenyldiazirine (Figure 41).[137–139]

However, photolysis of **6**⊙**71** at 77 K produced only insertion products **115** (85% yield) and **116** (4.5%). Both formed via insertion of transient **111** into an inward-pointing acetal C–H and linker α-C–H bonds of **6** respectively. The trick to rearrange transient ^{3}**111** to CHTE was the deuteration of **6**. In the partially deuterated **117**⊙**71**, a kinetic isotope effect of $k_H/k_D = 9.8$ slowed the carbene insertions and increased the lifetime of ^{3}**111**,[100] such that photochemical rearrangement to CHTE was possible in 17 and 30% yield at 77 and 15.5 K, respectively. Yields of 5-methyl-cyclohepta-1,2,4,6-tetraene (MeCHTE) **120** in the related inner phase *p*-tolylcarbene rearrangement were even higher. Photolysis of **6**⊙**91** and perdeuterated **118**⊙**91** at 77 K afforded **6**⊙**120** and **118**⊙**120** in 41 and 67% yield, respectively.[140] Likely steric interactions in the TSs for the *p*-tolylcarbene–host reactions increase the lifetime of incarcerated ^{3}**119** beyond that of ^{3}**111**, increasing its probability for photochemical ring expansion.

Both incarcerated allenes **113** and **120** persisted for months at room temperature in the absence of oxygen and could be characterized by ^{1}H NMR spectroscopy. Upon exposure of incarcerated CHTE to oxygen, rapid autoxidation to benzene took place. Under these conditions, oxygen diffused into the hemicarcerand, added to the central allene carbon and produced a spirocyclic dioxirane **121**, which subsequently rapidly decarboxylated to benzene.

The constrictively stabilized CHTE and MeCHTE allowed the measurement of several barriers of phenyl- and tolylcarbene rearrangements, which previously were only available from high-level calculations, and comparison between experiment and theory.[135,136,141] For example, in an attempt to measure the enantiomerization barrier of CHTE, **71** was photolyzed inside chiral hemicarcerand **69** and produced diastereomeric hemicarceplexes **69**⊙(+)-CHTE and **69**⊙(−)-CHTE in a 2 : 3 ratio.[139] In the asymmetric host environment, guest protons H2 experienced different host-induced shielding, allowing differentiation by ^{1}H NMR spectroscopy. The absence of coalescence at 100 °C gave a lower limit of 19.6 kcal mol^{-1} for the enantiomerization barrier, which agrees with all current calculations.[135,136] For the corresponding MeCHTE hemicarceplexes, exchange rate constants could be extracted from line-shape analysis of high-temperature NMR spectra.[142] Furthermore, photolysis of **69**⊙**91** produced**69**⊙(+)-MeCHTE and **69**⊙(−)-MeCHTE in the ratio = 1 : 1.15 (de = 7%), which slowly equilibrated into the thermodynamic ratio of 1 : 1.8 and allowed measurement of exchange rate constants. The experimental enantiomerization free energy, which was computed from these rate constants, agreed very well with the computed enantiomerization barrier.

111 112 113 114

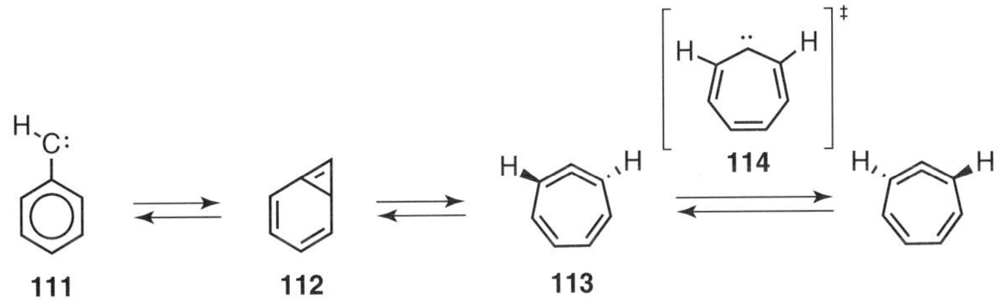

Figure 40 The Baron mechanism of the phenylcarbene rearrangement.

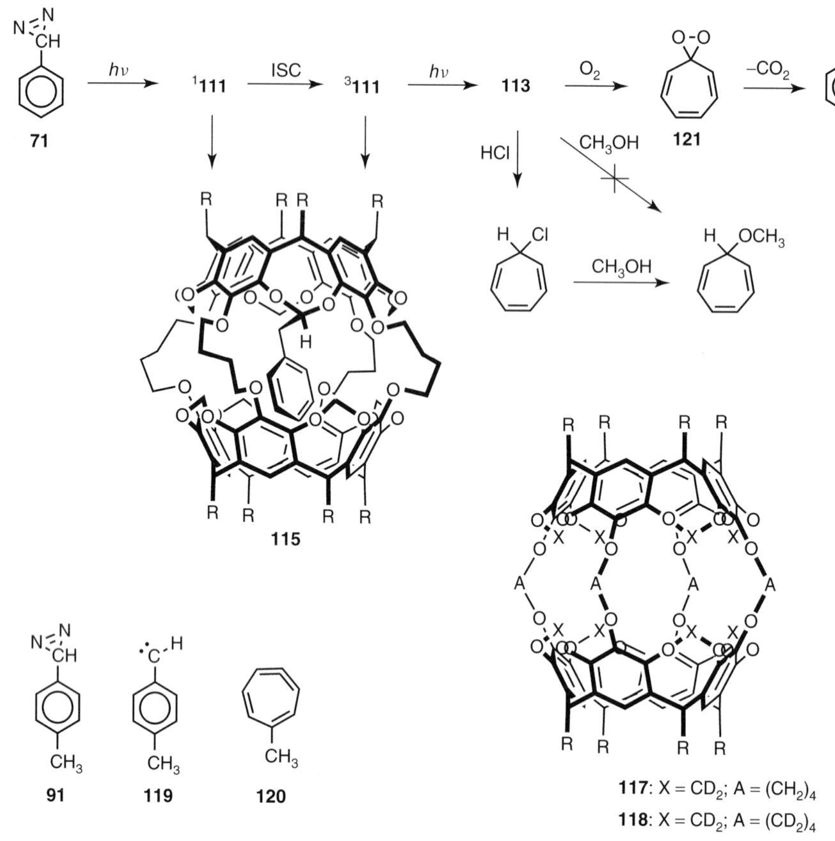

Figure 41 Chemistry of **71** and **91** inside **6** and **69** and structures of **117–118**.

This example illustrates nicely how kinetic experiments in confined space allow mapping of potential energy surfaces of important organic chemical processes involving highly reactive intermediates, which is difficult to achieve with other techniques such as laser flash photolysis, collision-induced dissociation (CID), or matrix isolation.

11.3.5 Carbenes

Most carbenes R–C–R′, in which R and R′ are H, alkyl, vinyl, or aryl, are too reactive to be observable inside a hemicarcerand.[143] For example, incarcerated arylcarbenes rapidly insert into C–H or C–O bonds of the hemicarcerand or add to one of the cavitand's aryl units even at very low temperature.[100,138,140] However, stability and reactivity of carbenes can be tailored especially with heteroatom substituents that stabilize the carbene's singlet state through electron donation (push effect).[144] In fact, many diaminocarbenes are stable and isolable at room temperature.[145,146] In cases where intrinsic stabilization (e-donation) is not sufficient, extrinsic effects (incarceration) may render an otherwise fleeting singlet carbene stable under normal conditions.

Fluorophenoxycarbene **122** is such a species and was recently room-temperature-stabilized by incarceration (Figure 42).[147] In **122**, the O- and F-substituents stabilize the singlet state by ~60 kcal mol^{-1} compared to singlet methylene.[148] This stabilization is, however, not large enough to render free **122** persistent at room temperature. On the contrary, if generated photochemically from

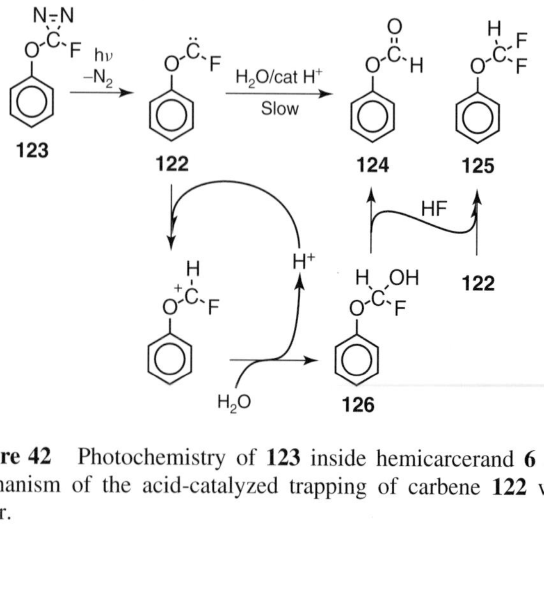

Figure 42 Photochemistry of **123** inside hemicarcerand **6** and mechanism of the acid-catalyzed trapping of carbene **122** with water.

diazirine **123, 122** dimerizes instantaneously, reacts with moisture, or is trapped in the presence of alkenes.[149]

Liu *et al.* generated incarcerated **122** by irradiation of fluorophenoxydiazirine hemicarceplex **6⊙123** at low temperature (Figure 42).[147] Incarcerated **122** persisted for weeks at room temperature. The ^{13}C and ^{19}F NMR spectra of **6⊙123** provided interesting insight into the electronic properties of **122**. The carbenic carbon resonated at δ 285.7 ppm, which compares well with chemical shifts of other persistent heteroatom-substituted carbenes.[146] The strongly downfield-shifted fluorine, the unusually large ^{19}F–^{13}C coupling constant, and the considerable upfield shift of the ipso carbon of **122** relative to that of **123** point toward strong participation of both O and F atoms in the carbene stabilization through push–push effects. In fact, the push effect of the O substituent is mostly responsible for the stability of **123** and its low tendency to react with the surrounding host. This was concluded from the high reactivity of fluorophenyl carbene **127**, which lacks the O substituent. Attempts to generate and observe through NMR spectroscopy **127** inside the same hemicarcerand via photolysis of **6⊙128** failed. Low-temperature UV–vis spectroscopy suggests that incarcerated Ph-C-F rapidly adds to one of the aryl units of **6** below −100 °C.[150]

127 **128**

In the presence of trace amounts of acid, incarcerated **122** slowly reacted with water in the bulk phase to yield phenylformate hemicarceplex **6⊙124** and phenyl difluoromethyl ether hemicarceplex **6⊙125** (Figure 42). The requirement of acid catalysis in the inner-phase water-trapping reaction is surprising since catalysis is not required for free **122**. This suggests that the water trapping of **122** is initiated by protonation and that water is not acidic enough in the inner phase to protonate **122**, contrary to residual water in an organic solvent. The hydrophobicity of the inner phase and lack of solvation of the hypothetical ion pair [**122**H]$^+$ [OH]$^-$ are likely reasons for the absence of this acid–base reaction similar to the examples discussed in Section 11.1.1.[151] This shows that incarceration not only prevents dimerization of **122** but also slows trapping reactions with water by many orders of magnitude.

11.3.6 Phenylnitrene

Very recently, phenylnitrene (PN) and its intramolecular rearrangement have been investigated inside hemicarcerand **6** (Figure 43).[152,153] PN is an important reactive intermediate for organic synthesis and photoaffinity labeling of biomacromolecules.[154] It is isoelectronic with PC and, above −100 °C, undergoes similar intramolecular rearrangements to the highly strained cyclic ketenimine **130** which can be trapped with amines or other nucleophiles.[155,156] Below −100 °C, ^{1}PN intersystem-crosses to triplet ^{3}PN, which is known to dimerize rapidly in solution (Figure 43).

Though at first glance PN and PC show many similarities in their chemistry, their reactivity differs dramatically, which has been subject of extensive investigations over the past decades, and reflects itself in the inner-phase chemistry of both species. For example, in solution, ^{1}PN ring-expands rapidly at room temperature to **130** in the subnanosecond time scale, whereas ring expansion of ^{1}PC can only be observed at elevated temperatures in the gas phase due to the substantially higher activation energy and the higher intermolecular reactivity of ^{1}PC.[135,157,158] Consequently, ^{1}PC does not rearrange to CHTE if generated inside **6**.[100,139] Ring expansion cannot compete with much faster insertions into hemicarcerand bonds. The situation is different for ^{1}PN, in which case intramolecular pathways (intersystem crossing and ring expansion) are much faster than reactions with the hemicarcerand. Thus, photolysis of incarcerated phenylazide at −86 °C, at which temperature ring expansion is faster than intersystem crossing, produced

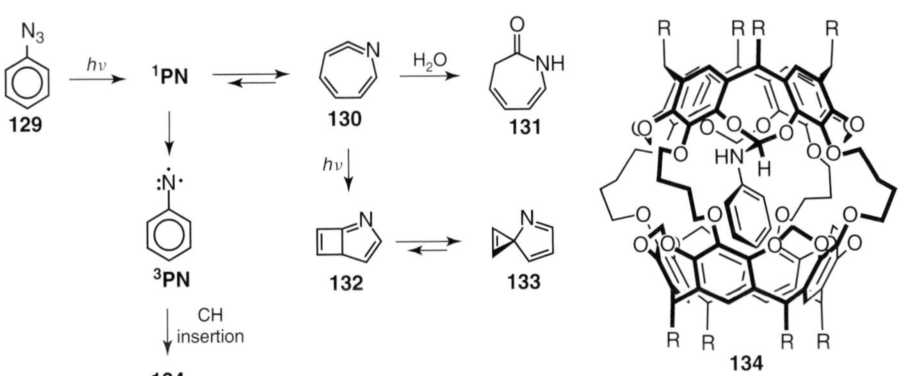

Figure 43 Inner-phase photochemistry of phenylazide **129** inside hemicarcerand **6**.

6⊙**130**, whose ^{13}C NMR spectrum could be recorded for the first time[1].[152] At this temperature, incarcerated **130** slowly decayed within 5 h to triplet ^{3}PN by ring contraction and intersystem crossing, allowing for a precise determination of the activation parameters of this process. If the photolysis was carried out in THF/water 8 : 1, water trapped **130** in the inner phase to produce lactam **131**.

The inner-phase photolysis studies of **6**⊙**129** uncovered a novel photochemical reaction of **130**.[153] Under the photolysis conditions, **130** underwent efficient photochemical electrocyclization to the *anti*-Bredt imine **132**, which at −5 °C thermally rearranged via a 1,5-shift to the slightly more stable 1-azaspiro[2.4]hepta-1,4,6-triene **133**.

The reactivity and lifetime of triplet ^{3}PN and triplet ^{3}PC differed remarkably in the inner phase of **6**. Whereas ^{3}PC reacts with **6** already at 15 K and persists probably only a few minutes at this temperature,[100,139] the lifetime of incarcerated ^{3}PN is 13.6 min at −3 °C.[153] Both ^{3}PC and ^{3}PN preferentially insert into inward-pointing acetal C–H bonds of **6** to produce **115** and **134**, respectively (Figures 41 and 43). The difference in reactivity toward C–H insertion between triplet carbenes and an isoelectronic triplet nitrenes is well known and is believed to proceed through a nitrogen rehybridization in the rate-limiting H-abstraction step of the nitrene. Rehybridization is not needed in the carbene reaction.[156]

C–H insertion reactions involving free PN are essentially impossible to study in solution by laser flash photolysis since C–H insertion cannot compete with dimerization, which is orders of magnitude faster.[160] Thus, confining ^{3}PN inside the molecular container[2], which eliminates dimerization, provided an elegant way to explore this important type of chemistry and allowed for the first time an accurate measurement of the activation parameters for a C–H insertion reaction involving ^{3}PN[3].

11.4 Photoelectron and triplet energy transfer

The concept of single-molecule incarceration inside a hemicarcerand, which provides an insulating multi-angstrom-thick wall around the guest, has also helped to better understand triplet excited state quenching by photoelectron transfer (PET) and energy transfer (ET). Both are important photophysical processes [163] and play a central role in biological photosynthesis,[164] visual transduction,[164] organic photochemistry,[165] semiconductor photocatalysis, and imaging.[166–169]

The idea behind through-shell PET and ET is to generate a triplet excited state inside a hemicarcerand and to measure rate constants and efficiencies of quenching the excited state by bulk-phase quenchers that are not covalently connected and are prevented from coming closer than approximately 7 Å to the incarcerated guest. Equations (3) and (4) schematically describe energy and electron transfer quenching of an incarcerated triplet state Host⊙G(T_1) with a bulk-phase quencher Q[170]:

$$\begin{aligned}&\text{Host}\odot\text{G}(\text{T}_1) + \text{Q}(\text{S}_0) \leftrightarrows \text{Host}\odot\text{G}(\text{T}_1)\cdots\text{Q}(\text{S}_0)\\&\quad\longrightarrow \text{Host}\odot\text{G}(\text{S}_0)\cdots\text{Q}(\text{T}_1) \leftrightarrows \text{Host}\odot\text{G}(\text{S}_0) + \text{Q}(\text{T}_1)\end{aligned} \tag{3}$$

$$\begin{aligned}&\text{Host}\odot\text{G}(\text{T}_1) + \text{Q} \leftrightarrows \text{Host}\odot\text{G}(\text{T}_1)\cdots\text{Q}\\&\quad\longrightarrow \text{Host}\odot\text{G}^-\cdots\text{Q}^+ \leftrightarrows \text{Host}\odot\text{G}^- + \text{Q}^+\end{aligned} \tag{4}$$

ET is a weakly coupled nonadiabatic process and proceeds by a Dexter electron exchange mechanism. Its rate constant k_{ET} can be approximated by the Golden Rule[171–173]:

$$k_{\text{ET}} = \left(\frac{2\pi}{h}\right) \times |\nu|^2 \times \text{FCWDS} \tag{5}$$

FCWDS is the Franck–Condon weighted density of states and ν the electronic coupling matrix element. In a semiclassical treatment, this equation can be separated into a pre-exponential factor A and an exponential term that relates k_{ET} to the driving force ΔG and nuclear reorganization energies of reactant λ_v and solvent λ_s:

$$k_{\text{ET}} = A \times \exp\left\{\frac{-(\lambda_s + \Delta G + \lambda_v)^2}{4\lambda_s k_B T}\right\} \tag{6}$$

The dependence of the rate constant for photoinduced electron transfer, k_{PET}, on the driving force and reorganization energies is similar. Equation 6 predicts a parabolic dependence of log k_{ET} on the driving force. At $-\Delta G = (\lambda_s + \lambda_v)$, k_{ET} is largest and decreases at smaller (normal region) and more exothermic driving force (inverted region). The experimental observation of the Marcus-inverted region for electron or energy transfer between noncovalently linked triplet excited state/quencher pairs has been very difficult mainly because ET and PET at high driving force are much faster than the rate of diffusional encounter. In the hemicarceplex/quencher encounter complexes (Host⊙G(T_1)···Q(S_0) and Host⊙G(T_1)···Q), the excited state and quencher are separated by about 7 Å. Because of their strong distance dependence,[172,174] ET and PET are substantially slower than diffusion, which has made observation of the inverted region possible in these systems.[170,175,176] In their seminal work on through-space triplet ET, Deshayes and coworkers studied acetophenone hemicarceplex **14**⊙Ac and probed through-shell triplet ET chemically via isomerization of *cis*-piperylene to *trans*-piperylene (Figure 44).[177]

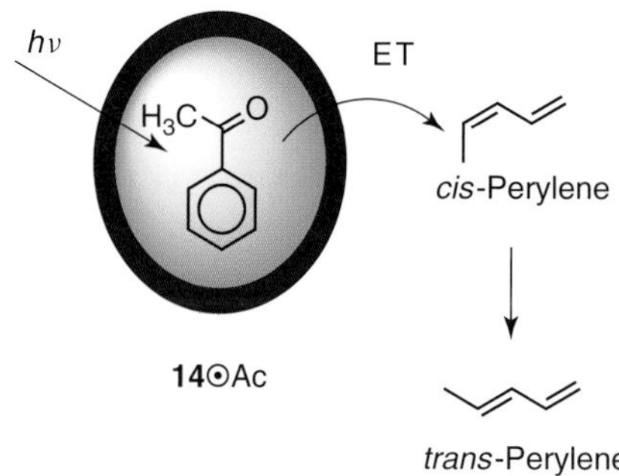

Figure 44 Photosensitized isomerization of *cis*-piperylene catalyzed by **14**⊙Ac.

Triplet ET was 2.7-fold slower for **14**⊙Ac compared to free acetophenone. This corresponds to an almost diffusion-controlled rate for **14**⊙Ac. Since triplet energy is transferred through an electron exchange mechanism, which necessitates a close contact between donor and acceptor, sufficient overlap of highest occupied molecular orbital (HOMO) and lowest unoccupied molecular orbital (LUMO) in the donor–acceptor pair must exist. Whether the intervening hemicarcerand plays a role in this through-space ET is uncertain. In a subsequent investigation, Farrán and Deshayes measured triplet ET rates between incarcerated biacetyl (**14**⊙biacetyl) and various bulk-phase acceptors.[175] Hemicarcerand **14** retarded triplet ET, which suggests a reduced electron coupling between donor and acceptor as a result of their larger separation. Also, log k_{ET} and ΔG showed a hyperbolic relationship as predicted by the Golden Rule.

Interesting is the extremely slow triplet ET rate to O_2. Oxygen is typically a very efficient quencher. Farrán and Deshayes concluded that the quenching rate drops off drastically if oxygen is prevented from making direct contact with the donor. Parola *et al.*, who independently measured triplet ET rates from **14**⊙biacetyl to quenchers used by Deshayes in addition to several others, agreed that the difference between k_{ET} of free and incarcerated biacetyl results from different electronic exchange matrix elements v.[170] However, they were careful in taking the observed parabolic-like relationship as firm evidence for inverted behavior especially since their data were strongly scattered. A parabolic-like correlation may simply reflect nonhomogeneity of the quenchers, as a consequence of their different sizes, which leads to different donor–acceptor distances and/or orientations and hence to different values for v. This may also be the reason for the failure to observe the inverted region in electron transfer quenching experiments between **14**⊙biacetyl and bulk-phase aromatic amine donors.

In a subsequent investigation, Deshayes and Piotrowiak provided clear support for the parabolic Marcus relationship and explained earlier data in a quantitative manner by

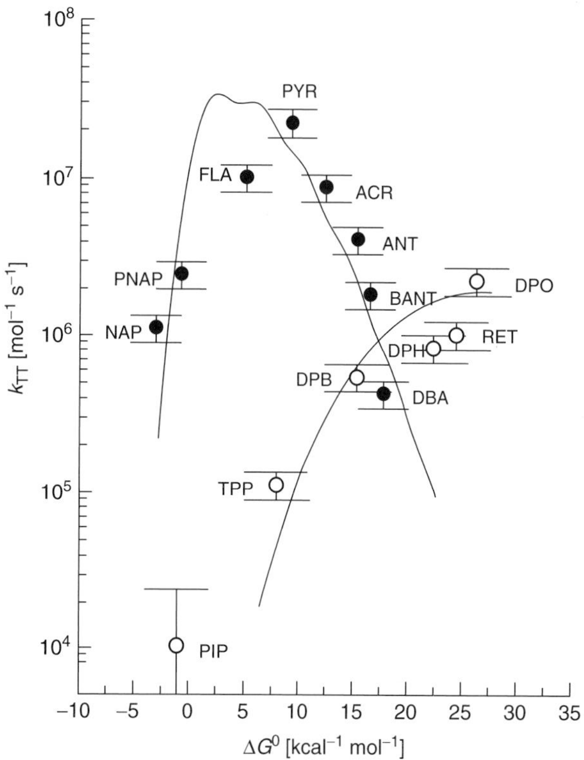

Figure 45 Rate constant versus driving force $-\Delta G$ of triplet ET from **14**⊙biacetyl to aryl (●) and alkene (○) acceptors and theoretical curves generated using the semiclassical Marcus–Jortner formalism of triplet energy transfer. (Reproduced from Ref. 176. © American Chemical Society, 1998.)

taking into account the different internal nuclear reorganization energies λ_v of the acceptors.[176] According to MO calculations, λ_v varies by more than $20\,\mathrm{kcal\,mol^{-1}}$ among the different acceptors. Thus, two acceptors with nearly identical driving forces and transfer rates may belong to different regions of the Marcus parabola. Examples are dibromoanthracene (DBA) and diphenylbutadiene (DPB) (Figure 45). Both were assigned to the correct region of a Marcus parabola based on their activation energy of transfer, which is negative for the former (typical for the inverted region) but positive for the latter (normal region behavior). Deshayes and Piotrowiak identified four groups of acceptors: (i) rigid aromatics that display small λ_v; (ii) acyclic olefins that twist around the double bond upon triplet excitation and therefore have large λ_v; (iii) cyclic olefins with even larger λ_v; and (iv) O_2, which has essentially no λ_v. Each acceptor group has its own log k_{ET} versus ΔG correlation (Figure 45). The remaining scattering in the experimental data may result from differences in size and shape of the acceptors, leading to different effective electronic couplings in the corresponding encounter complexes and possibly also to different encounter frequencies.

LIVERPOOL JOHN MOORES UNIVERSITY
LEARNING SERVICES

As pointed out earlier, electronic coupling between donor and acceptor is strongly reduced, leading to a reduced energy and electron transfer rate, if both are separated by an intervening hemicarcerand. Deshayes and Piotrowiak addressed the dependence of electronic coupling between incarcerated biacetyl and the bulk-phase quencher on the hemicarcerand size.[178] The electronic coupling between the incarcerated donor and bulk-phase acceptor can be described by a superexchange mechanism and viewed as a sequence of guest–hemicarcerand and hemicarcerand–solute interactions. The total electronic coupling matrix element ν_{total} is therefore the product of matrix elements for the guest–hemicarcerand ν_{GH} and the hemicarcerand–acceptor interaction ν_{HA}:

$$\nu_{\text{total}} \propto \nu_{\text{GH}} \times \nu_{\text{HA}} \qquad (7)$$

For hemicarceplexes **6**⊙biacetyl, **31**⊙biacetyl, and **14**⊙biacetyl, which vary in size and linker characteristics, k_{ET} increased with decreasing host size in the order $k_{\text{ET}}(\mathbf{6}) < k_{\text{ET}}(\mathbf{14}) \ll k_{\text{ET}}(\mathbf{31})$. The same trend was observed for the average electronic coupling matrix elements: $|\nu(\mathbf{6})| = 0.20\,\text{cm}^{-1}$, $|\nu(\mathbf{14})| = 0.26\,\text{cm}^{-1}$, and $|\nu(\mathbf{31})| = 0.66\,\text{cm}^{-1}$. Since ν_{HA} should be independent of the hemicarcerand size, the spread in ν_{total} reflects differences in the guest–hemicarcerand electronic coupling ν_{HA} among the hosts. These trends are consistent with predictions, according to which the time-averaged guest–hemicarcerand interaction should be cavity size dependent and should increase with decreasing cavity size. One can also conclude that the *o*-xylylene linkers in **14** do not cause special enhancement of electronic coupling.

These investigations not only unravel the role of the hemicarcerand in the mechanism of through-shell triplet ET, but also improve our understanding of solvent-mediated electron transfer,[179–185] in which a solvent molecule, separating donor and acceptor, provides the pathway for electronic coupling. Since the thickness of a hemicarcerand is comparable to that of common organic solvents, the measured electronic coupling matrix elements are good estimates for the magnitude of solvent-mediated contributions to electronic coupling in triplet excitation transfer.

12 CONCLUSIONS AND OUTLOOK

The conceptual idea and realization of molecular container compounds has opened a new and intellectually challenging research field: the chemistry of and within molecular container compounds and their complexes. The molecular architecture of hemicarcerands, which features a relatively rigid frame with smaller openings, through which guests have to pass in order to enter or leave the inner phase, leads to unique molecular recognition properties. Binding selectivity depends primarily on the size and shape complementarity between the guest's cross section and the dimension of the host's gates and inner phase. Such high structural recognition, in combination with the ease to tailor the dimension of the inner phase and the host's portals, makes hemicarcerands ideal building blocks for hydrocarbon storage, separation, and purification applications or as recognition sites in molecular sensors.[186]

Another future application of hemicarcerands that relies on their ability to fully embrace a guest molecule and to control guest egress is drug delivery.[187,188] The hydrophobic inner phase makes water-soluble hemicarcerands ideal for solubilizing and delivering highly water-insoluble drug molecules. The recent development of gated and dynamic hemicarcerands, which spontaneously release guests in response to photoirradiation or a change in the environment, shows great promise for such delivery applications. However, the functionality of these systems has to be demonstrated first in aqueous medium under physiological conditions.

Moreover, hemicarcerands have become interesting new tools for physical organic chemists to study reaction mechanisms and long-distance phenomena. Molecular containers made possible the investigation of highly strained and reactive molecules under normal working conditions by generating them in the protective inner phase. They also allowed the investigation of electronic interactions between encapsulated and bulk-phase molecules through the intervening hemicarcerand and have provided experimental support for theoretical models of long-distance spin–orbit coupling,[189] as well as electron and energy transfer. It is anticipated that this field of research will further grow and the recent development of multicavitand nanocapsules will make possible the investigation of chemical reactivity of macromolecular guests that are of interest to material and biological sciences.

NOTES

[1] An elegant alternative way to protect the strained cyclic keteneimine and arylnitrene is to incorporate the aryl unit of the nitrene into the host structure, such that the reactive nitrene group points into the inner cavity.[159]

[2] Recently, phenylnitrene has also been generated inside a deep cavitand.[161]

[3] An alternative method to suppress dimerization, is to generate the arylnitrene inside a polymeric matrix.[162]

ACKNOWLEDGMENTS

The author thanks the National Science Foundation (Grants CHE-0518351 & CHE-0957611) for financial support of his research.

REFERENCES

1. D. J. Cram and J. M. Cram, *Container Molecules and Their Guests*, The Royal Siciety of Chemistry, Cambridge, 1994.
2. D. J. Cram, *Nature*, 1992, **356**, 29.
3. A. Jasat and J. C. Sherman, *Chem. Rev. (Washington, DC)*, 1999, **99**, 931.
4. R. Warmuth and J. Yoon, *Acc. Chem. Res.*, 2001, **34**, 95.
5. E. Maverick and D. J. Cram, *Compr. Supramol. Chem.*, 1996, **2**, 367.
6. For cryptophane molecular containers, see T. Brotin and J.-P. Dutasta, *Chem. Rev. (Washington, DC)*, 2009, **109**, 88.
7. M. Fujita, M. Tominaga, A. Hori, and B. Therrien, *Acc. Chem. Res.*, 2005, **38**, 369.
8. J. L. Atwood, L. J. Barbour, and A. Jerga, *Perspect. Supramol. Chem.*, 2003, **7**, 153.
9. J. Rebek, *Acc. Chem. Res.*, 2009, **42**, 1660.
10. F. Hof, S. L. Craig, C. Nuckolls, and J. Rebek Jr., *Angew. Chem. Int. Ed.*, 2002, **41**, 1488.
11. S. J. Dalgarno, N. P. Power, and J. L. Atwood, *Coord. Chem. Rev.*, 2008, **252**, 825.
12. B. C. Gibb, *Org. Nanostruct.*, 2008, 291.
13. J. I. van der Vlugt, T. S. Koblenz, J. Wassenaarm, and J. N. H. Reek, in *Molecular Encapsulation*, eds. U. H. Brinker and J.-L. Mieusset, John Wiley & Sons, Ltd, Chichester, 2010, p. 145.
14. R. Warmuth, in *Molecular Encapsulation*, eds. U. H. Brinker and J.-L. Mieusset, John Wiley & Sons, Ltd, Chichester, 2010, 227.
15. M. D. Pluth, R. G. Bergman, and K. N. Raymond, *Acc. Chem. Res.*, 2009, **42**, 1650.
16. T. S. Koblenz, J. Wassenaar, and J. N. H. Reek, *Chem. Soc. Rev.*, 2008, **37**, 247.
17. D. J. Cram, M. E. Tanner, and R. Thomas, *Angew. Chem. Int. Ed.*, 1991, **30**, 1024.
18. R. Warmuth, *Angew. Chem. Int. Ed. Engl.*, 1997, **36**, 1347.
19. P. Roach and R. Warmuth, *Angew. Chem. Int. Ed.*, 2003, **42**, 3039.
20. M. Ziegler, J. Brumaghim, and K. N. Raymond, *Angew. Chem. Int. Ed.*, 2000, **39**, 4119.
21. J. Kang, J. Santamaria, G. Hilmersson, and J. Rebek Jr., *J. Am. Chem. Soc.*, 1998, **120**, 7389.
22. M. Yoshizawa, M. Tamura, and M. Fujita, *Science*, 2006, **312**, 251.
23. D. Fiedler, H. van Halbeek, R. G. Bergman, and K. N. Raymond, *J. Am. Chem. Soc.*, 2006, **128**, 10240.
24. M. D. Pluth, R. G. Bergman, and K. N. Raymond, *Science*, 2007, **316**, 85.
25. C. J. Hastings, M. D. Pluth, R. G. Bergman, and K. N. Raymond, *J. Am. Chem. Soc.*, 2010, **132**, 6938.
26. M. Yoshizawa, S. Miyagi, M. Kawano, *et al.*, *J. Am. Chem. Soc.*, 2004, **126**, 9172.
27. L. S. Kaanumalle, C. L. D. Gibb, B. C. Gibb, and V. Ramamurthy, *J. Am. Chem. Soc.*, 2005, **127**, 3674.
28. M. Yoshizawa, Y. Takeyama, T. Kusukawa, and M. Fujita, *Angew. Chem. Int. Ed.*, 2002, **41**, 1347.
29. J. Chen and J. Rebek Jr., *Org. Lett.*, 2002, **4**, 327.
30. D. J. Cram, S. Karbach, Y. H. Kim, *et al.*, *J. Am. Chem. Soc.*, 1985, **107**, 2575.
31. D. J. Cram, S. Karbach, Y. H. Kim, *et al.*, *J. Am. Chem. Soc.*, 1988, **110**, 2554.
32. J. A. Bryant, M. T. Blanda, M. Vincenti, and D. J. Cram, *J. Am. Chem. Soc.*, 1991, **113**, 2167.
33. J. C. Sherman, C. B. Knobler, and D. J. Cram, *J. Am. Chem. Soc.*, 1991, **113**, 2194.
34. J. Jung, H. Ihm, and K. Paek, *Bull. Korean Chem. Soc.*, 1996, **17**, 553.
35. T. A. Robbins, C. B. Knobler, D. R. Bellew, and D. J. Cram, *J. Am. Chem. Soc.*, 1994, **116**, 111.
36. Y.-S. Byun, T. A. Robbins, C. B. Knobler, and D. J. Cram, *Chem. Commun. (Cambridge, UK)*, 1995, 1947.
37. D. J. Cram, R. Jaeger, and K. Deshayes, *J. Am. Chem. Soc.*, 1993, **115**, 10111.
38. Y.-S. Byun, O. Vadhat, M. T. Blanda, *et al.*, *Chem. Commun. (Cambridge, UK)*, 1995, 1825.
39. C. N. Eid, C. B. Knobler, D. A. Gronbeck, and D. J. Cram Jr., *J. Am. Chem. Soc.*, 1994, **116**, 8506.
40. C. von dem Bussche-Heunnefeld, D. Buehring, C. B. Knobler, and D. J. Cram, *Chem. Commun. (Cambridge, UK)*, 1995, 1085.
41. R. C. Helgeson, K. Paek, C. B. Knobler, *et al.*, *J. Am. Chem. Soc.*, 1996, **118**, 5590.
42. D. J. Cram, M. T. Blanda, K. Paek, and C. B. Knobler, *J. Am. Chem. Soc.*, 1992, **114**, 7765.
43. M. E. Tanner, C. B. Knobler, and D. J. Cram, *J. Am. Chem. Soc.*, 1990, **112**, 1659.
44. D. A. Makeiff and J. C. Sherman, *Chem.—Eur. J.*, 2003, **9**, 3253.
45. X. J. Liu and R. Warmuth, *J. Am. Chem. Soc.*, 2006, **128**, 14120.
46. E. S. Barrett, J. L. Irwin, A. J. Edwards, and M. S. Sherburn, *J. Am. Chem. Soc.*, 2004, **126**, 16747.
47. D. A. Makeiff and J. C. Sherman, *J. Am. Chem. Soc.*, 2005, **127**, 12363.
48. X. J. Liu, Y. Liu, G. Li, and R. Warmuth, *Angew. Chem. Int. Ed.*, 2006, **45**, 901.
49. Y. Liu, X. Liu, and R. Warmuth, *Chem.—Eur. J.*, 2007, **13**, 8953.
50. S. Mecozzi and J. Rebek Jr., *Chem.—Eur. J.*, 1998, **4**, 1016.

51. K. Nakamura and K. N. Houk, *J. Am. Chem. Soc.*, 1995, **117**, 1853.
52. J. Sherman, *Chem. Commun. (Cambridge, UK)*, 2003, 1617.
53. R. G. Chapman, N. Chopra, E. D. Cochien, and J. C. Sherman, *J. Am. Chem. Soc.*, 1994, **116**, 369.
54. J. R. Fraser, B. Borecka, J. Trotter, and J. C. Sherman, *J. Org. Chem.*, 1995, **60**, 1207.
55. R. G. Chapman and J. C. Sherman, *J. Org. Chem.*, 1998, **63**, 4103.
56. R. G. Chapman, G. Olovsson, J. Trotter, and J. C. Sherman, *J. Am. Chem. Soc.*, 1998, **120**, 6252.
57. K. Nakamura, C. Sheu, A. E. Keating, *et al.*, *J. Am. Chem. Soc.*, 1997, **119**, 4321.
58. D. A. Makeiff, D. J. Pope, and J. C. Sherman, *J. Am. Chem. Soc.*, 2000, **122**, 1337.
59. M. L. C. Quan and D. J. Cram, *J. Am. Chem. Soc.*, 1991, **113**, 2754.
60. S. Ro, S. J. Rowan, A. R. Pease, *et al.*, *Org. Lett.*, 2000, **2**, 2411.
61. D. J. Cram, *Angew. Chem.*, 1988, **100**, 1041.
62. D. J. Cram, M. E. Tanner, and C. B. Knobler, *J. Am. Chem. Soc.*, 1991, **113**, 7717.
63. F. M. Raymo, K. N. Houk, and J. F. Stoddart, *J. Am. Chem. Soc.*, 1998, **120**, 9318.
64. J. Yoon and D. J. Cram, *Chem. Commun. (Cambridge, UK)*, 1997, 1505.
65. R. G. Chapman and J. C. Sherman, *J. Am. Chem. Soc.*, 1999, **121**, 1962.
66. K. Paek, H. Ihm, S. Yun, *et al.*, *J. Org. Chem.*, 2001, **66**, 5736.
67. K. N. Houk, K. Nakamura, C. Sheu, and A. E. Keating, *Science*, 1996, **273**, 627.
68. D. J. Lacks, *J. Chem. Phys.*, 1995, **103**, 5085.
69. Y. Liu and R. Warmuth, *Org. Lett.*, 2007, **9**, 2883.
70. T. Felder and C. A. Schalley, *Angew. Chem. Int. Ed.*, 2003, **42**, 2258.
71. C. Sheu and K. N. Houk, *J. Am. Chem. Soc.*, 1996, **118**, 8056.
72. J. Yoon, C. B. Knobler, E. F. Maverick, and D. J. Cram, *Chem. Commun. (Cambridge, UK)*, 1997, 1303.
73. E. L. Piatnitski and K. D. Deshayes, *Angew. Chem. Int. Ed.*, 1998, **37**, 970.
74. R. C. Helgeson, A. E. Hayden, and K. N. Houk, *J. Org. Chem.*, 2010, **75**, 570.
75. T. Gottschalk, B. Jaun, and F. Diederich, *Angew. Chem. Int. Ed.*, 2007, **46**, 260.
76. M. Mastalerz, *Angew. Chem. Int. Ed.*, 2010, **49**, 5042.
77. S. J. Rowan, S. J. Cantrill, G. R. L. Cousins, *et al.*, *Angew. Chem. Int. Ed.*, 2002, **41**, 899.
78. J.-M. Lehn, *Chem. Soc. Rev.*, 2007, **36**, 151.
79. P. T. Corbett, J. Leclaire, L. Vial, *et al.*, *Chem. Rev. (Washington, DC)*, 2006, **106**, 3652.
80. M. I. Page and W. P. Jencks, *Proc. Natl. Acad. Sci. U.S.A.*, 1971, **68**, 1678.
81. X. J. Liu, Y. Liu, and R. Warmuth, *Supramol. Chem.*, 2008, **20**, 41.
82. N. Nishimura and K. Kobayashi, *Angew. Chem. Int. Ed.*, 2008, **47**, 6255.
83. N. Nishimura, K. Yoza, and K. Kobayashi, *J. Am. Chem. Soc.*, 2010, **132**, 777.
84. C. Naumann, S. Place, and J. C. Sherman, *J. Am. Chem. Soc.*, 2002, **124**, 16.
85. J. Sun, B. O. Patrick, and J. C. Sherman, *Tetrahedron*, 2009, **65**, 7296.
86. Y. Liu and R. Warmuth, *Angew. Chem. Int. Ed.*, 2005, **44**, 7107.
87. N. Guex and M. C. Peitsch, *Electrophoresis*, 1997, **18**, 2714.
88. R. C. Helgeson, C. B. Knobler, and D. J. Cram, *J. Am. Chem. Soc.*, 1997, **119**, 3229.
89. J. Yoon and D. J. Cram, *Chem. Commun. (Cambridge, UK)*, 1997, 497.
90. E. L. Piatnitski, R. A. Flowers, and K. Deshayes II, *Chem.—Eur. J.*, 2000, **6**, 999.
91. M. V. Rekharsky and Y. Inoue, *Chem. Rev. (Washington, DC)*, 1998, **98**, 1875.
92. S. B. Ferguson, E. M. Seward, F. Diederich, *et al.*, *J. Org. Chem.*, 1988, **53**, 5593.
93. W. L. Jorgensen, T. B. Nguyen, E. M. Sanford, *et al.*, *J. Am. Chem. Soc.*, 1992, **114**, 4003.
94. J. K. Judice and D. J. Cram, *J. Am. Chem. Soc.*, 1991, **113**, 2790.
95. B. S. Park, C. B. Knobler, C. N. Eid Jr., *et al.*, *Chem. Commun. (Cambridge, UK)*, 1998, 55.
96. J. Yoon and D. J. Cram, *J. Am. Chem. Soc.*, 1997, **119**, 11796.
97. A. Scarso, A. Shivanyuk, O. Hayashida, and J. Rebek Jr., *J. Am. Chem. Soc.*, 2003, **125**, 6239.
98. C. Nuckolls, F. Hof, T. Martin, and J. Rebek Jr., *J. Am. Chem. Soc.*, 1999, **121**, 10281.
99. D. Fiedler, D. H. Leung, R. G. Bergman, and K. N. Raymond, *J. Am. Chem. Soc.*, 2004, **126**, 3674.
100. C. Kemmis and R. Warmuth, *J. Supramol. Chem.*, 2003, **1**, 253.
101. D. Place, J. Brown, and K. Deshayes, *Tetrahedron Lett.*, 1998, **39**, 5915.
102. R. Warmuth, E. F. Maverick, C. B. Knobler, and D. J. Cram, *J. Org. Chem.*, 2003, **68**, 2077.
103. P. Timmerman, W. Verboom, F. C. J. M. van Veggel, *et al.*, *Angew. Chem. Int. Ed. Engl.*, 1994, **33**, 2345.
104. N. Chopra, R. G. Chapman, Y.-F. Chuang, *et al.*, *J. Chem. Soc., Faraday Trans.*, 1995, **91**, 4127.
105. R. G. Chapman and J. C. Sherman, *J. Org. Chem.*, 2000, **65**, 513.
106. B. M. O'Leary, R. M. Grotzfeld, and J. Rebek Jr., *J. Am. Chem. Soc.*, 1997, **119**, 11701.

107. R. Warmuth, *Eur. J. Org. Chem.*, 2001, 423.

108. C. Marquez and W. M. Nau, *Angew. Chem. Int. Ed.*, 2001, **40**, 4387.

109. S. S. Carrera, J.-L. Kerdelhue, K. J. Langenwalter, *et al.*, *Eur. J. Org. Chem.*, 2005, 2239.

110. S. K. Kurdistani, R. C. Helgeson, and D. J. Cram, *J. Am. Chem. Soc.*, 1995, **117**, 1659.

111. A. Warshel, P. K. Sharma, M. Kato, *et al.*, *Chem. Rev. (Washington, DC)*, 2006, **106**, 3210.

112. H. B. Gray and J. R. Winkler, *Proc. Natl. Acad. Sci. U.S.A.*, 2005, **102**, 3534.

113. T. A. Robbins and D. J. Cram, *J. Am. Chem. Soc.*, 1993, **115**, 12199.

114. S. Mendoza, P. D. Davidov, and A. E. Kaifer, *Chem.—Eur. J.*, 1998, **4**, 864.

115. R. Warmuth, J.-L. Kerdelhue, S. S. Carrera, *et al.*, *Angew. Chem. Int. Ed.*, 2002, **41**, 96.

116. M. T. H. Liu, *Chem. Soc. Rev.*, 1982, **11**, 127.

117. M. T. H. Liu, Y. K. Choe, M. Kimura, *et al.*, *J. Org. Chem.*, 2003, **68**, 7471.

118. S. M. Ngola and D. A. Dougherty, *J. Org. Chem.*, 1996, **61**, 4355.

119. A. M. A. van Wageningen, P. Timmerman, J. P. M. van Duynhoven, *et al.*, *Chem.—Eur. J.*, 1997, **3**, 639.

120. J. Kang, G. Hilmersson, J. Santamaria, and J. Rebek Jr., *J. Am. Chem. Soc.*, 1998, **120**, 3650.

121. A. A. Deniz, K. S. Peters, and G. J. Snyder, *Science*, 1999, **286**, 1119.

122. G. Maier, *Angew. Chem. Int. Ed.*, 1988, **27**, 309.

123. H. Hopf, *Classics in Hydrocarbon Chemistry*, VCH, Weinheim, 2000.

124. W. F. Maier and P. V. R. Schleyer, *J. Am. Chem. Soc.*, 1981, **103**, 1891.

125. A. D. Wolf and M. Jones Jr., *J. Am. Chem. Soc.*, 1973, **95**, 8209.

126. R. Warmuth, *Chem. Commun. (Cambridge, UK)*, 1998, 59.

127. H. H. Wenk, M. Winkler, and W. Sander, *Angew. Chem. Int. Ed. Engl.*, 2003, **42**, 502.

128. H. J. Jiao, P. v. R. Schleyer, B. R. Beno, *et al.*, *Angew. Chem. Int. Ed. Engl.*, 1997, **36**, 2761.

129. O. L. Chapman, K. Mattes, C. L. McIntosh, *et al.*, *J. Am. Chem. Soc.*, 1973, **95**, 6134.

130. R. P. Johnson and K. J. Daoust, *J. Am. Chem. Soc.*, 1995, **117**, 362.

131. B. R. Beno, C. Sheu, K. N. Houk, *et al.*, *Chem. Commun. (Cambridge, UK)*, 1998, 301.

132. A. M. Orendt, J. C. Facelli, J. G. Radziszewski, *et al.*, *J. Am. Chem. Soc.*, 1996, **118**, 846.

133. S. G. Kukolich, M. C. McCarthy, and P. Thaddeus, *J. Phys. Chem. A*, 2004, **108**, 2645.

134. P. P. Gaspar, J. P. Hsu, S. Chari, and M. Jones Jr., *Tetrahedron*, 1985, **41**, 1479.

135. S. Matzinger, T. Bally, E. V. Patterson, and R. J. McMahon, *J. Am. Chem. Soc.*, 1996, **118**, 1535.

136. P. R. Schreiner, W. L. Karney, P. v. R. Schleyer, *et al.*, *J. Org. Chem.*, 1996, **61**, 7030.

137. R. J. McMahon, C. J. Abelt, O. L. Chapman, *et al.*, *J. Am. Chem. Soc.*, 1987, **109**, 2456.

138. R. Warmuth and M. A. Marvel, *Angew. Chem. Int. Ed.*, 2000, **39**, 1117.

139. R. Warmuth and M. A. Marvel, *Chem.—Eur. J.*, 2001, **7**, 1209.

140. J.-L. Kerdelhue, K. J. Langenwalter, and R. Warmuth, *J. Am. Chem. Soc.*, 2003, **125**, 973.

141. C. M. Geise and C. M. Hadad, *J. Org. Chem.*, 2002, **67**, 2532.

142. R. Warmuth, *J. Am. Chem. Soc.*, 2001, **123**, 6955.

143. M. Jones Jr. and R. A. Moss, in *Reactive Intermediate Chemistry*, eds. R. A. Moss, M. S. Platz, and M. J. Jones, Wiley-Interscience, Hoboken, NJ, 2004, p. 273.

144. R. A. Moss, *Acc. Chem. Res.*, 1989, **22**, 15.

145. R. A. Moss, in *Carbene Chemistry: From Fleeting Intermediates to Powerful Reagents*, ed. G. Bertrand, Fontis Media-Marcel Dekker, Lausanne, 2002, p. 57.

146. D. Bourissou, O. Guerret, F. P. Gabbai, and G. Bertrand, *Chem. Rev. (Washington, DC)*, 2000, **100**, 39.

147. X. Liu, G. Chu, R. A. Moss, *et al.*, *Angew. Chem. Int. Ed.*, 2005, **44**, 1994.

148. N. G. Rondan, K. N. Houk, and R. A. Moss, *J. Am. Chem. Soc.*, 1980, **102**, 1770.

149. R. A. Moss, G. Kmiecik-Lawrynowicz, and K. Krogh-Jespersen, *J. Org. Chem.*, 1986, **51**, 2168.

150. Z. Lu, R. A. Moss, R. R. Sauers, and R. Warmuth, *Org. Lett.*, 2009, **11**, 3866.

151. D. A. Makeiff, K. Vishnumurthy, and J. C. Sherman, *J. Am. Chem. Soc.*, 2003, **125**, 9558.

152. R. Warmuth and S. Makowiec, *J. Am. Chem. Soc.*, 2005, **127**, 1084.

153. R. Warmuth and S. Makowiec, *J. Am. Chem. Soc.*, 2007, **129**, 1233.

154. H. Bayley, *Photogenerated Reagents in Biochemistry and MolecularBiology*, Elsevier, Amsterdam, 1983.

155. N. P. Gritsan and M. S. Platz, *Chem. Rev. (Washington, DC)*, 2006, **106**, 3844.

156. W. T. Borden, N. P. Gritsan, C. M. Hadad, *et al.*, *Acc. Chem. Res.*, 2000, **33**, 765.

157. N. P. Gritsan, Z. Zhu, C. M. Hadad, and M. S. Platz, *J. Am. Chem. Soc.*, 1999, **121**, 1202.

158. M. S. Platz, *Acc. Chem. Res.*, 1995, **28**, 487.

159. G. Bucher, C. Toenshoff, and A. Nicolaides, *J. Am. Chem. Soc.*, 2005, **127**, 6883.

160. T. Y. Liang and G. B. Schuster, *J. Am. Chem. Soc.*, 1987, **109**, 7803.

161. G. Wagner, V. B. Arion, L. Brecker, *et al.*, *Org. Lett.*, 2009, **11**, 3056.

162. A. Reiser and L. Leyshon, *J. Am. Chem. Soc.*, 1970, **92**, 7487.

163. P. Piotrowiak, K. Deshayes, Z. S. Romanova, *et al.*, *Pure Appl. Chem.*, 2003, **75**, 1061.

164. E. Kohen, R. Santus, and J. G. Hirschberg, *Photobiology*, Academic Press, San Diego, CA, 1995.

165. G. J. Kavarnos, *Fundamentals of Photoinduced Electron Transfer*, VCH Publisher, New York, 1993.

166. M. Graetzel, *J. Photochem. Photobiol., A*, 2004, **164**, 3.

167. A. Hagfeldt and M. Graetzel, *Chem. Rev. (Washington, DC)*, 1995, **95**, 49.

168. B. O'Regan and M. Graetzel, *Nature*, 1991, **353**, 737.

169. D. Eaton, in *Photoinduced Electron Transfer I*, ed. J. Mattay, Springer, Berlin, Heidelberg, 1990, vol. 156, p. 199.

170. A. J. Parola, F. Pina, E. Ferreira, *et al.*, *J. Am. Chem. Soc.*, 1996, **118**, 11610.

171. G. L. Closs, M. D. Johnson, J. R. Miller, and P. Piotrowiak, *J. Am. Chem. Soc.*, 1989, **111**, 3751.

172. G. L. Closs, P. Piotrowiak, J. M. MacInnis, and G. R. Fleming, *J. Am. Chem. Soc.*, 1988, **110**, 2652.

173. J. Jortner, *J. Chem. Phys.*, 1976, **64**, 4860.

174. N. Koga, K. Sameshima, and K. Morokuma, *J. Phys. Chem.*, 1993, **97**, 13117.

175. A. Farran and K. D. Deshayes, *J. Phys. Chem.*, 1996, **100**, 3305.

176. I. Place, A. Farran, K. Deshayes, and P. Piotrowiak, *J. Am. Chem. Soc.*, 1998, **120**, 12626.

177. A. Farran, K. Deshayes, C. Matthews, and I. Balanescu, *J. Am. Chem. Soc.*, 1995, **117**, 9614.

178. Z. S. Romanova, K. Deshayes, and P. Piotrowiak, *J. Am. Chem. Soc.*, 2001, **123**, 11029.

179. N. E. Miller, M. C. Wander, and R. J. Cave, *J. Phys. Chem. A*, 1999, **103**, 1084.

180. E. W. Castner Jr., D. Kennedy, and R. J. Cave, *J. Phys. Chem. A*, 2000, **104**, 2869.

181. R. J. Cave, M. D. Newton, K. Kumar, and M. B. Zimmt, *J. Phys. Chem.*, 1995, **99**, 17501.

182. K. Kumar, Z. Lin, D. H. Waldeck, and M. B. Zimmt, *J. Am. Chem. Soc.*, 1996, **118**, 243.

183. R. W. Kaplan, A. M. Napper, D. H. Waldeck, and M. B. Zimmt, *J. Am. Chem. Soc.*, 2000, **122**, 12039.

184. H. Han and M. B. Zimmt, *J. Am. Chem. Soc.*, 1998, **120**, 8001.

185. A. M. Napper, I. Read, R. Kaplan, *et al.*, *J. Phys. Chem. A*, 2002, **106**, 5288.

186. B.-H. Huisman, D. M. Rudkevich, A. Farran, *et al.*, *Eur. J. Org. Chem.*, 2000, 269.

187. C. L. D. Gibb and B. C. Gibb, *J. Am. Chem. Soc.*, 2004, **126**, 11408.

188. T. V. Nguyen, H. Yoshida, and M. S. Sherburn, *Chem. Commun. (Cambridge, UK)*, 2010, **46**, 5921.

189. Z. S. Romanova, K. Deshayes, and P. Piotrowiak, *J. Am. Chem. Soc.*, 2001, **123**, 2444.

Cyclodextrins: From Nature to Nanotechnology

Stephen F. Lincoln and Duc-Truc Pham
University of Adelaide, Adelaide, South Australia, Australia

1 INTRODUCTION

1.1 Cyclodextrin characteristics

Cyclodextrins (CDs) are naturally occurring homochiral macrocycles composed of α-1,4-linked D-glucopyranose in the $^4C_1$ conformation, and as a class constitute one of the most widely used molecular components in supramolecular chemistry. The CDs were first reported by Villiers in 1891 and were produced from starch by the action of CD transferases produced by bacteria exemplified by *Bacillus macerans* and *Bacillus circulans*. Genetically engineered CD transferases of higher activity are now widely used in industrial CD production which amounts to thousands of tonnes annually. The smallest of these "native" CDs are α-, β-, and γ-CD, which consist of 6, 7, and 8 α-1,4-linked D-glucopyranose units, respectively, and are the most available and studied CDs (Figure 1).[1,2] Larger CDs have been characterized but are less studied.[3–5]

The D-glucopyranose units are labeled alphabetically from A in a clockwise direction when viewed from the narrower end of the truncated CD annulus delineated by a circle of primary hydroxyl groups on the C6 carbons. The wider end of the annulus is delineated by a circle of secondary hydroxyl groups on the C2 and C3 carbons. Together, these hydroxyl groups account for the solubilities of α-, β-, and γ-CDs in water, which are 145, 18.5, and 232 g dm^{-3} at 298.2 K, respectively.[1] The annular interiors are hydrophobic in nature and tend to complex the hydrophobic parts of guest species to form water-soluble host–guest or inclusion complexes. Such complexation may show substantial size discrimination and enantioselectivity between guests and these characteristics have led to a vast range of studies of both native and modified CD systems.

In this brief review, the intent is to build a basic understanding of native and modified CDs and their host–guest complexation chemistry and through this to reach the research frontiers where fascinating developments are occurring.

1.2 Cyclodextrin host–guest complexation

A very important property of CDs is their ability to partially or fully complex a wide range of guest species, X, within their annuli to form host–guest complexes in water as shown in Figure 2. There are a variety of interactions driving the complexation process, which include the relatively weak interactions between the hydrophobic guest and water and similarly weak interactions between water and the hydrophobic CD annular interior. By comparison, the interactions between water and the hydrophilic regions

Supramolecular Chemistry: From Molecules to Nanomaterials.
Edited by Philip A. Gale and Jonathan W. Steed.
© 2012 John Wiley & Sons, Ltd. ISBN: 978-0-470-74640-0.

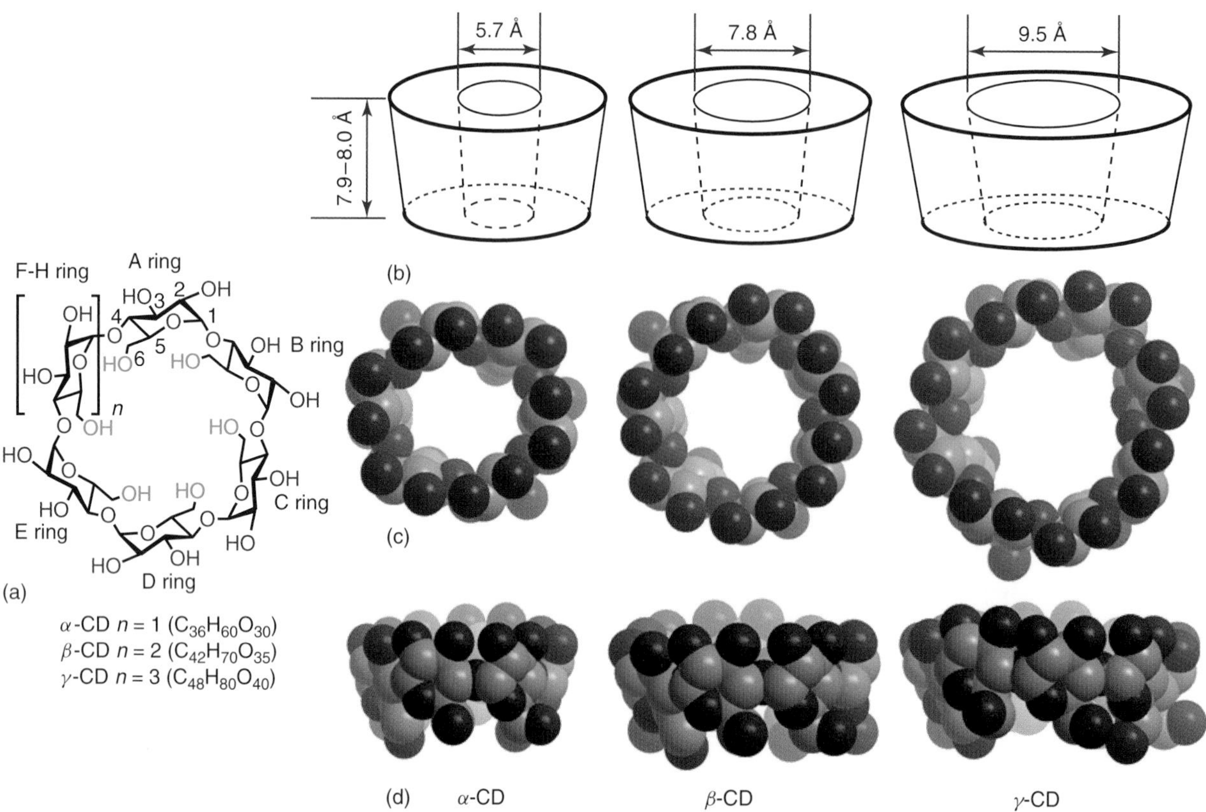

Figure 1 (a) The general formula for α-, β-, and γ-CDs. (b) The dimensions and simplified shapes of their annuli which have volumes of 174, 262, and 472 Å^3, respectively.[1] (c) Views from above the wider end of the CD annuli and (d) side views where carbon and oxygen atoms are shown in gray and red, respectively, and hydrogen atoms are omitted.

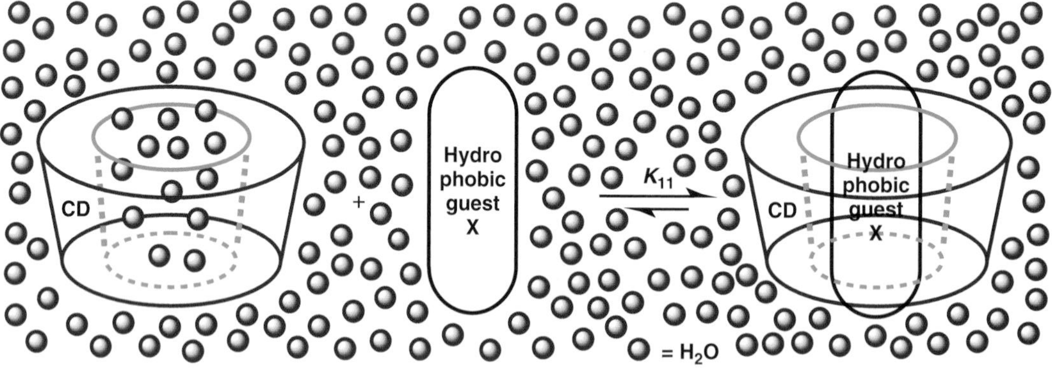

Figure 2 The CD complexation of an either completely or partially hydrophobic guest, X, at the left to form a 1 : 1 host–guest complex, CD·X, at the right characterized by a complexation constant $K_{11} = [\text{CD·X}]/([\text{CD}][\text{X}])$ in water.

of the CD and the guest and the secondary interactions between the hydrophobic guest and the annular interior are usually stronger. The overall stereochemical fit of the guest into the CD annulus also has an important effect on the complexation process. The balance of these interactions determines the stability of the host–guest complex which translates into substantial variations of ΔH° and ΔS° of complexation as the identities of the CD and the guest change.[6]

While there is no necessary relationship between ΔH° and ΔS° for CD complexation, a linear-free energy relationship between $T\Delta S^\circ$ and ΔH° according to (1) is observed for wide ranges of guests complexed by α-, β-, and γ-CDs. The slope of this linear relationship, α, indicates to what extent the enthalpic gain, $\Delta\Delta H^\circ$, caused by variations in the guest for a particular CD is counteracted by entropic loss, $\Delta\Delta S^\circ$, according to (2). When ΔH° is zero, the corresponding value of $T\Delta S^\circ = T\Delta S_0^\circ = \Delta G^\circ$ according to (3), which indicates that the CD host–guest complex is stable in the absence of enthalpic stabilization. Under these conditions, $T\Delta S_0^\circ = 8$ (0.79), 11 (0.80), and 15 (0.97) kJ mol^{-1} for α-, β-, and γ-CDs, respectively, where the numbers in brackets are the corresponding α values. Thus, without enthalpic stabilization, these complexes have significantly negative ΔG° values as a consequence of release of water from the CD annuli, dehydration of the peripheral CD hydroxy groups, and the guest upon complexation generating a substantial entropy gain.[6]

$$T\Delta S^\circ = \alpha \Delta H^\circ \quad (1)$$

$$T\Delta\Delta S^\circ = \alpha \Delta\Delta H^\circ \quad (2)$$

$$\Delta G^\circ = \Delta H^\circ - T\Delta S^\circ \quad (3)$$

In addition to the host–guest complexes composed of one CD and one guest as shown for the complex CD·X characterized by K_{11} in Figure 2, other CD to guest ratios arise in complexes as exemplified by CD·X_2, CD_2·X, and CD_2·X_2 characterized by the sequential complexation constants K_{12}, K_{21}, K_{22}, and $K_{22'}$, respectively, as shown in Figure 3. Hydrophobic guests are generally complexed more strongly than hydrophilic guests and some of the latter complex very weakly if at all. There is a tendency for those guests which best fit the CD annuli to form the more stable CD host–guest complexes, and because of this and their hydrophobic nature, aromatic guests are particularly widely employed in CD complexation studies. When CDs are modified through substitution of one or more hydroxyl groups, the range of possibilities for host–guest complexation is greatly extended and a variety of examples of the resulting complexes are considered in the following sections.

The structures of CD host–guest complexes determined by X-ray crystallography show the guest to reside either completely or partially within the CD annuli.[7] However, the structure in the solid state is not necessarily identical to that in solution in which most CD studies have been conducted using a range of spectroscopic and other techniques.[6] Nuclear magnetic resonance (NMR) provides the most direct evidence of complexation within the CD annulus

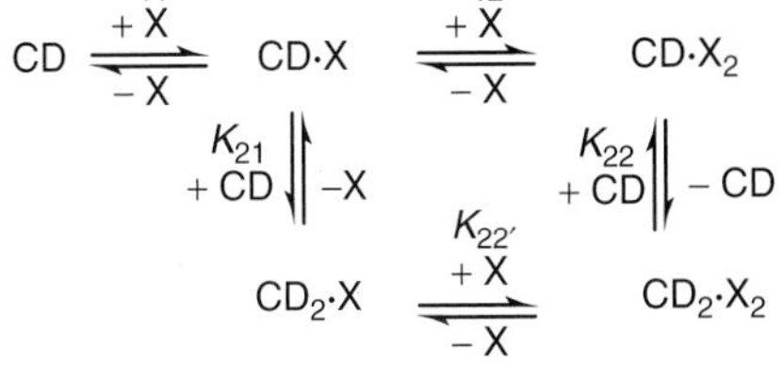

Figure 3 Multiple complexation equilibria for CD host–guest complexes of different stoichiometries.

in solution through 2D ^{1}H NMR rotating frame Overhauser enhancement spectroscopy, ROESY, and nuclear Overhauser enhancement spectroscopy, NOESY.[8] These techniques detect through-space NOE (nuclear Overhauser enhancement) interactions occurring between the H3, H5, and H6 protons lining the inside of the CD annulus and a proton of a guest species within a 4 Å or smaller distance which can only occur if the guest is wholly or partially within the annulus. Such NOE interactions are indicated by cross-peaks, the intensity of which is inversely proportional to the interaction distance raised to the power of six and directly to the number of protons in that environment.

2 CYCLODEXTRIN MODIFICATION

The arrays of primary hydroxyl groups on the C6 carbons and secondary hydroxyl groups on the C2 and C3 carbons at the narrow and wide ends of the native CDs, respectively, provide the opportunity for single substitution through to complete substitution.[2,9] Such modifications may be chosen to alter CD annular size, shape, charge, and polarity, and thereby better accommodate and orientate chosen guests within the CD annulus. In addition, CDs may be linked together[10] or attached to polymeric backbones[11,12] or surfaces[13,14] in a variety of ways to present a great array of supramolecular chemistry. In this section, only a very general picture of the multitudinous methods for modification of CDs is presented. This may be supplemented by referring to the methods used to prepare the CD systems considered in succeeding sections.

2.1 Monosubstituted cyclodextrins

Monosubstitution is the most studied of CD modifications and is generally carried out through well-established pathways; the most common of which appear in Figure 4.[2,9] Direct substitution may be achieved through alkylation, acylation, and sulfonation as shown in Figure 4(a). Alternatively, and because of their ease of preparation, CD sulfonates and CD halides may be used as convenient intermediates for a range of nucleophilic displacement reactions as shown in Figure 4(b). Amino groups may also be attached in this way and provide convenient routes to further modification as seen in Figure 4(c).

The modification of native CDs may be achieved through substitution of either a C2, C3, or C6 hydroxyl group and a wide range of methodologies generally employing the processes outlined in Figure 4, or similar ones, to achieve selectivity in substitution between these sites.[9] These methodologies frequently exploit the knowledge that the C6 hydroxyl groups are the most basic and usually the

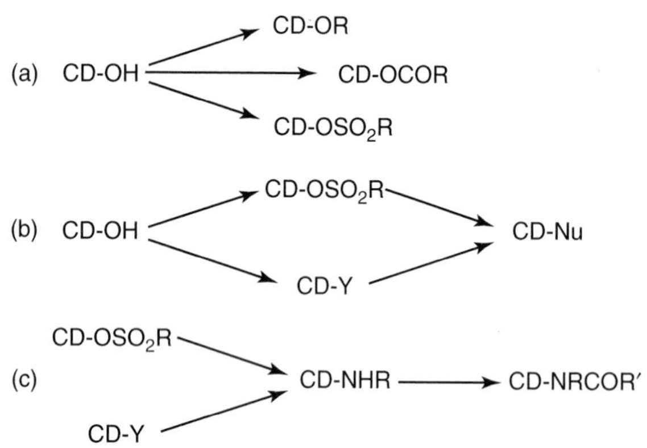

Figure 4 Some pathways to modified CDs through substitution at one or more primary or secondary hydroxyl groups where R is an alkyl or aryl group, Nu is a nucleophile, and Y is a halide.

most nucleophilic while the C2 and C3 hydroxyl groups are more acidic and the latter are sterically the most difficult to access. While direct substitution at C6^A retains the configuration of the D-glucopyranose ring substitution at C2^A and C3^A usually causes inversion at these carbons as shown for the preparation of (2^{A}S,3^{A}S)-3^A-amino-3^A-deoxy-β-CD in Figure 5.[15] The substitution of the A D-glucopyranose ring of β-CD, **1**, with a tosyl group at C2 is preceded by interaction with dibutyltin oxide, Bu_2SnO, which activates the C2 carbon toward reaction with tosylchloride, TsCl, to give the tosylated product 2^A-*O*-(4-methybenzenesulfonyl)-β-CD, **2**. Under basic conditions in the presence of ammonium bicarbonate in water, the manno-2^A, 3^A-epoxide-β-CD, **3**, is formed and with heating transforms into the (2^{A}S,3^{A}S)-3^A-amino-3^A-deoxy-β-CD, **4**, with inversion at C2 and C3.

Often, traces of di- and trisubstituted CD at either C2, C3, or C6 are synthesized along with the monosubstituted CD but purification by chromatography or recrystallization usually gives the pure major product. Such multiple substitutions are employed to give CDs with particularly desirable properties.[2,9] Selective substitution at either all C2, C3, or C6 or at all of these carbons simultaneously may also be achieved to give multiply substituted CDs in which the size of the annuli is greatly extended.[2,9]

2.2 Linked cyclodextrin dimers

Substitution with a bifunctional substituent may either form a bridge across a CD end[16] or link two CDs together in a dimer as shown in Figure 6.[17] The reaction of the succinate diester **5** with either 6^A-amino-(6^A-deoxy-β-CD) **6** or 3^A-amino-((2^{A}S,3^{A}S)-3^A-deoxy-β-CD) **7** or both produces three succinamide linked β-CD dimers *N,N′*-bis(6^A-deoxy-β-CD-6^A-yl)succinamide, **8**, *N,N′*-bis((2^{A}S,3^{A}S)-3^A-deoxy-β-CD-3^A-yl)succinamide, **9**, and *N*-((2^{A}S,3^{A}S)-3^A-deoxy-β-CD-3^A-yl)-*N′*-(6^A-deoxy-β-CD-6^A-yl) succinamide, **11**, through 6^A-(3-(4-nitrophenoxycarbonyl)-propionamido)-(6^A-deoxy-β-CD), **10**, in 70–85% yield. A substantial range of linked CD dimers has been studied.[10]

2.3 Cyclodextrin polymers

A wide range of polymers incorporating CDs has been prepared by adaptation of the methods employed for monosubstitution. A widely employed method is to attach a CD to a polymer with reactive groups as exemplified by the random substitution of polyacrylic acid by α-CD and β-CD in Figure 7(a).[18] The substitution of polyacrylate **12** with either α-CD or β-CD through reaction of their 6^A-(amino)-6^A-deoxy derivatives, **13**, in the presence of *N,N′*-dicyclohexyl carbodiimide (DCC) in water gives the randomly substituted polyacrylate **14** isolated from NaOH solution as the sodium salt. The extent of substitution may be varied at will over a substantial range by changing the reactant ratios. Similar substitutions of polyvinyl polymers have also been reported.[19]

An alternative approach is to build the polymer chain and incorporate the CD substituent as shown in Figure 7(b).[20] In this synthesis, β-CD substituted at the C6^A carbon with an aliphatic diamine, **15**, is reacted with glycidylmethacrylate, **16**, to form a monovinyl β-CD monomer, **17**. Subsequent copolymerization with *N*-isopropylacrylamide, **18**, yields the water-soluble β-CD substituted polyacrylamide copolymer **19** with molecular weights up to 10^4. An extensive

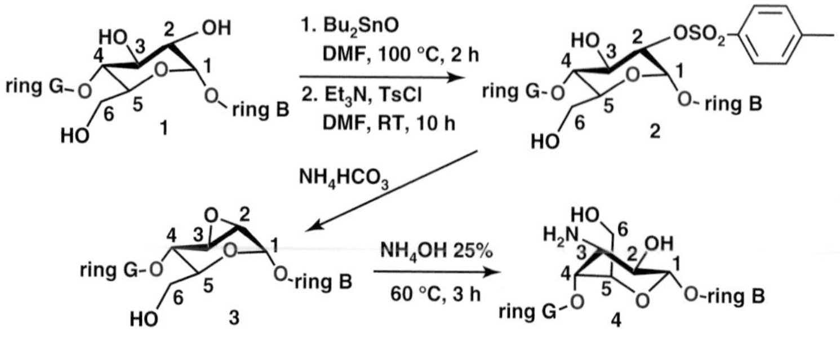

1. Bu_2SnO DMF, 100 °C, 2 h
2. Et_3N, TsCl DMF, RT, 10 h

NH_4HCO_3

NH_4OH 25% 60 °C, 3 h

Figure 5 The inversions at C2^A and C3^A of the A ring of β-CD through substitution.

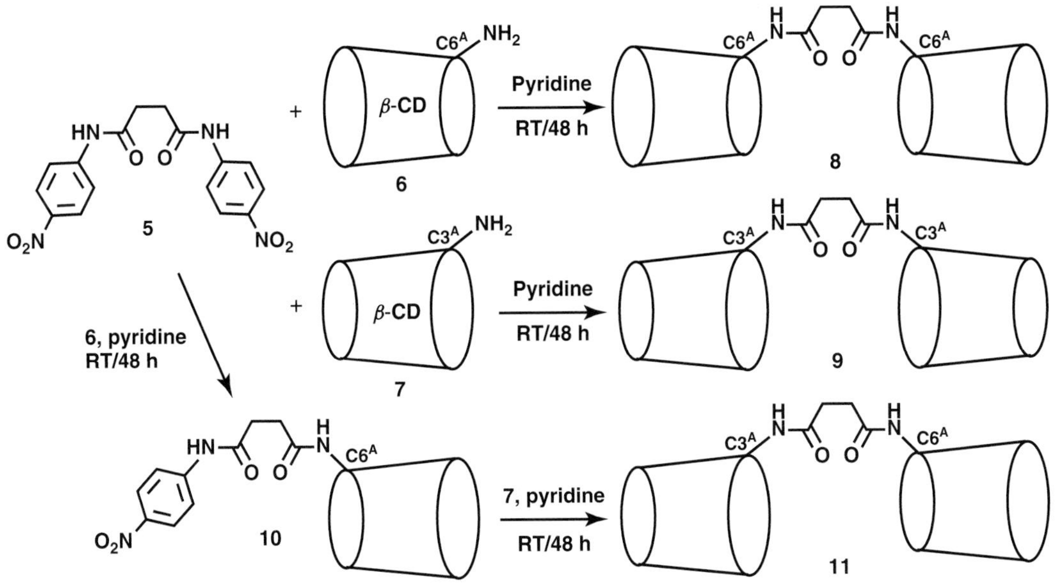

Figure 6 Reaction sequences for the syntheses of linked β-CD dimers **8**, **9**, and **11**.

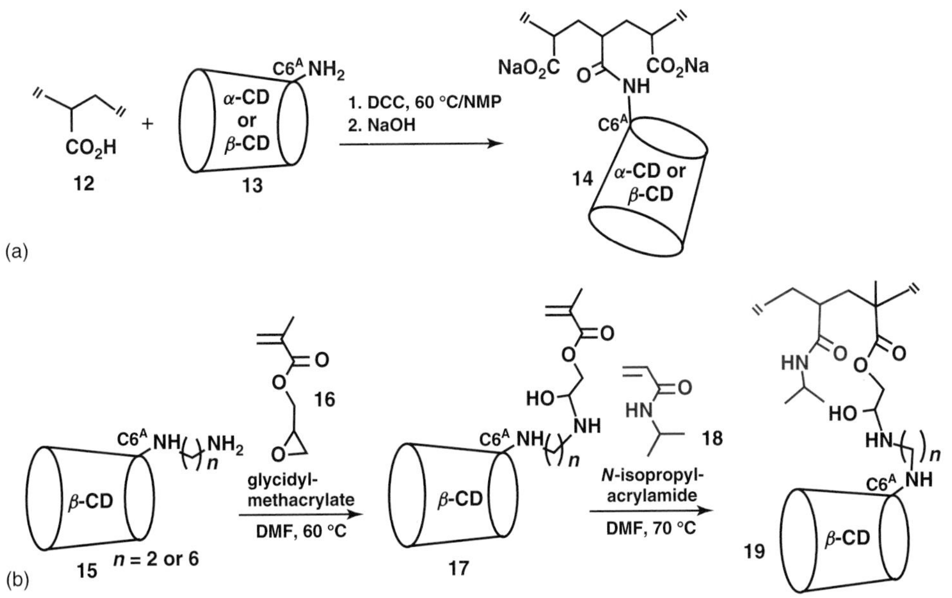

Figure 7 (a) The substitution of polyacrylic acid **12** with either α- or β-CDs to give the randomly substituted sodium polyacrylate **14**. (b) The formation of the β-CD substituted copolymer **19** through reaction of **15** with **16** to give **17** which is then reacted with **18**.

range of CD-substituted polymers and their properties and biomedical applications have been reported.[21]

3 CYCLODEXTRIN ENANTIOSELECTIVITY

As a consequence of the homochirality of native and modified CDs, there exists a possibility of chiral discrimination between enantiomeric guests, or enantioselectivity, in the formation of diastereomeric host–guest complexes.[22] This CD enantioselectivity is of major importance in a range of racemate resolution technologies as exemplified by thin layer, high-performance liquid, and gas–liquid phase chromatography and capillary electrophoresis.[23–26] In such usage, the CD may either be attached to a surface to form a stationary phase or may be part of a moving phase. Selective diastereomeric CD host–guest complex precipitation is also used in racemate resolution.[7,27] To gain a basic insight into the nature of the chiral interactions involved, several examples of enantioselectivity between guests by CDs and other aspects of the chiral nature of CDs are discussed below.

3.1 Chiral discrimination by native cyclodextrins

The liquid chromatographic separation of the *R*- and *S*-enantiomers of tryptophan and other aminoacids and several of their derivatives by α-CD bonded to silica gel[28,29] prompted a detailed study of enantioselectivity complexation of tryptophan by α-CD.[30,31] Both ^{1}H and ^{13}C NMR spectroscopies show that coupling constants, chemical shifts, and nuclear relaxation times change more for *R*-tryptophan upon complexation by α-CD than for *S*-tryptophan in D_2O at 298.2 K. For both enantiomers, ^{1}H NOESY NMR studies show the indole ring to be adjacent to the α-CD secondary hydroxyl groups with the benzene ring being deeper in the annulus in the host–guest complex (Figure 8a). Molecular modeling shows this orientation to be ~4 kJ mol^{-1} more stable than the reverse orientation. It also shows that the chirality of *R*-tryptophan allows it to form twice as many hydrogen bonds with α-CD as does *S*-tryptophan. Such hydrogen bonding stabilizes the α-CD·*R*-tryptophan complex by 12.7 kJ mol^{-1} more than it does the α-CD·*S*-tryptophan complex. It appears that for the α-CD diastereomeric host–guest complexes to show significant differences in stability the chiral guests must either fit snugly into the annulus or there should be a highly localized interaction with the annular interior, and that the guest stereogenic center should interact strongly with one of the α-CD C2 and C3 hydroxyl groups.[28–31] Thermodynamic studies of complexation of 43 enantiomer pairs by β-CD show that there is a fine balance between the orientation of the guests within the annulus and the strength of complexation in determining enantioselectivity.[32] Thus, guests with low symmetry nonpolar groups which complex in the β-CD annulus and those with larger distances between the stereogenic center and their most hydrophobic group are more likely to show enantioselectivity in their β-CD diastereomeric complexes. However, the extent of enantioselectivity tends to decrease when the guest enantiomers are modified to complex more strongly as the additional weak interactions induced diminish the chiral complementarity between the guest and the β-CD annulus.

For many chiral guests, the native CDs show small selectivities in complexing enantiomers and extensive thermodynamic studies show that for β-CD a range of stereochemical factors influence the mode of host–guest complexation.[32] Generally, the small extent of enantioselectivity is attributable to small differences in ΔH° for the formation of the diastereomeric host–guest complexes being offset by counteracting ΔS° differences such that differences in ΔG° are small. However, exceptions to this generalization do occur as exemplified by the strong enantioselectivity shown by β-CD in complexing helical 1,12-dimethylbenzo[c]phenanthrene-5,8-dicarboxylate for which the complexation constants, K_{11M} and K_{11P} for the left- and right-handed helical enantiomers, **22** and **23** in Figure 8(b), are 1.87×10^4 dm^3 mol^{-1} ($\Delta H^\circ = -51.1$ kJ mol^{-1}, $\Delta S = -90$ J mol^{-1} K^{-1}) and 2.2×10^3 dm^3 mol^{-1} ($\Delta H^\circ = -35.1$ kJ mol^{-1}, $\Delta S^\circ = -53.2$ J mol^{-1} K^{-1}), respectively, such that the difference in free energy for complexation between the enantiomers **22**(*M*−) and **23**(*P*−)$\Delta\Delta G^\circ = 5.2$ J mol^{-1} K^{-1} in aqueous solution at 298.2 K.[33]

Both 2D ^{1}H ROESY NMR and molecular modeling studies indicate that the carboxylate groups of **22** and **23** are close to the ring of secondary hydroxyl groups of β-CD in the host–guest complex and that **23** penetrates more deeply into the β-CD annulus. However, this deeper penetration into the cavity is enthalpically unfavorable but entropically favorable because it requires a greater dehydration of the carboxylate groups of **23** than is the case for the lesser penetration of **22**. Thus, the enantioselectivity of β-CD is dominated by the difference in enthalpy due to the deeper penetration of **23** into the β-CD annulus. Enantioselectivity between **22** and **23** is also shown by γ-CD for which $K_{11M} = 3.1 \times 10^3$ dm^3 mol^{-1} ($\Delta H^\circ = -30.2$ kJ mol^{-1}, $\Delta S^\circ = -34.4$ J mol^{-1} K^{-1}) and

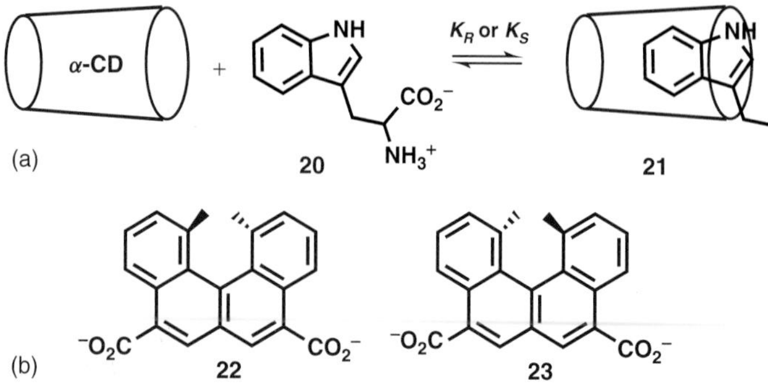

Figure 8 (a) Complexation of *R*- and *S*-tryptophan zwitterion, **20**, to form diastereomeric host–guest complexes **21** characterized by K_R and K_S, respectively. (b) The *M*-, **22**, and *P*-, **23**, enantiomers of helical 1,12-dimethylbenzo[c]phenanthrene-5,8-dicarboxylate.

$K_{11P} = 6.9 \times 10^2$ $dm^3 mol^{-1}$ ($\Delta H^\circ = -16.0\,kJ\,mol^{-1}$, $\Delta S^\circ = 0.45\,J\,mol^{-1}\,K^{-1}$) and $\Delta\Delta G^\circ = 3.7\,J\,mol^{-1}\,K^{-1}$. The lower stabilities of the γ-CD host–guest complexes are attributable to a deeper penetration of the carboxylate groups into the γ-CD annulus causing a more extensive and endothermic dehydration of **22** and **23**. It appears that the difference in stability of the diastereomeric host–guest complexes in the two systems is due to a chiral helical structure assumed by CDs in water.

Table 1 Host–guest complexation constants, K_{11}, for the hosts β-CD, D-**24** and L-**24** and the guests D-**26**, L-**26**, D-**27**, and L-**27** in the formation of **28** and **29** in water at pH 7.0 at 298.2 K.

Guest	K_{11} ($dm^3 mol^{-1}$)		
	Host = β-CD	D-24	L-24
D-**26**	179 ± 13	42 ± 13	113 ± 18
L-**26**	114 ± 13	54 ± 10	95 ± 17
D-**27**	197 ± 20	160 ± 36	139 ± 24
L-**27**	153 ± 14	83 ± 28	231 ± 45

3.2 Chiral discrimination by modified cyclodextrins

The introduction of molecular asymmetry into the CD structure by a single substitution renders all of the glucopyranose units inequivalent, which may affect the extent of enantioselectivity in the complexation of chiral guests. In addition, the substituent may interact directly with the guest, and thereby enhance enantioselectivity as appears to be the case with 6^A-amino-6^A-deoxy-β-CD in which the amino group is protonated in water at pH 6.9 and engenders a stronger interaction with negatively charged guests than is the case for β-CD.[34] Nevertheless, enantioselectivity remains quite small due to the enthalpy–entropy offset discussed above.

Sometimes there may be a competition between the substituent self-complexing inside the CD annulus and intermolecular host–guest complexation of chiral guests. This is exemplified by the 6^A-*N*-(*N'*-formyl-D-phenylalanyl)-6^A-deoxy-amino-β-CD and its L-analog, D-**24**, L-**24**, D-**25**, and L-**25**, which undergo self-complexation in competition with intermolecular host–guest complexation of the D- and L-enantiomers of *N*-dansylalanine, **26**, and *N*-dansylphenylalanine, **27**, to form the diastereomeric host–guest complexes, **28** and **29**, respectively, as shown in Figure 9.[35] From the host–guest complexation constants, K_{11}, in Table 1, it is apparent that K_{11} for the β-CD complexes is either larger than or similar to those for the D-**24** and L-**24** complexes with the exception of the L-**24**·D-**27** complex **29** which is more stable. Generally, enantioselectivity is small to moderate and tends to favor the complexation of D-**26** and D-**27** by β-CD, L-**26** and D-**27** by D-**24**, and D-**26** and L-**27** by L-**24**.

Hexa-coordinated tris-bidentate metal complexes exist as Δ- and Λ-enantiomers which may form diastereomeric host–guest complexes with native and modified CDs. This is exemplified by the Δ- and Λ-tris-(1,10-phenanthroline)ruthenium(II) complexes Δ- and Λ-$[Ru(phen)_3]^{2+}$, **30** and **31**, which complex with heptakis(6-carboxymethylthio-6-deoxy)-β-CD, **32**, to form host–guest complex, **33**, and its Δ-$[Ru(phen)_3]^{2+}$ analog as shown in Figure 10.[36] The complexation is largely dependent on the electrostatic attraction between dicationic **30** and **31** and heptaanionic **32**. The host–guest complexation constants $K_{11\Delta}$ and $K_{11\Lambda} = 1.25 \times 10^3$ and 5.90×10^2 $dm^3 mol^{-1}$ indicate an enantioselectivity of 2.12 by **32** in favor of **30** over **31** as determined by ^{1}H NMR spectroscopy in D_2O at pD 7 and 298.2 K. It appears from 2D ^{1}H ROESY NMR studies that the origin of the discrimination is that **30** penetrates more deeply into the annulus of **32** than does **31**. Under the same conditions, the enantioselectivity between Δ- and Λ-$[Rh(phen)_3]^{2+}$ is 1.43 in favor of the Δ-enantiomer. When **32** is replaced by its γ-CD analog, the enantioselectivity is 1.28 and 1.13 in favor of the Δ-enantiomer of the Ru(II) and Rh(II) complexes, respectively. This enantioselectivity is also detected by capillary zone electrophoresis where the two enantiomers are cleanly separated with the Δ-enantiomer showing the longer retention time. These data indicate that subtle

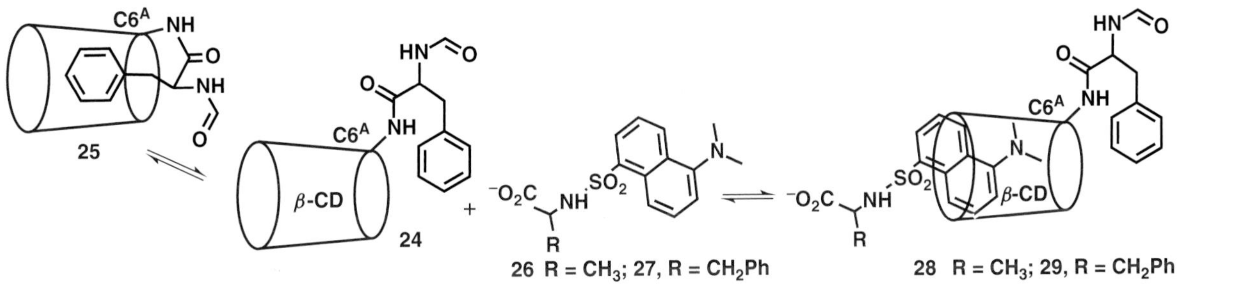

Figure 9 The competing intramolecular equilibrium between **24** and **25**, in which the substituent may be in either the D- or L-form, and the intermolecular equilibria between **24, 26, 27, 28**, and **29** where **26** and **27** may be in either the D- or L-form.

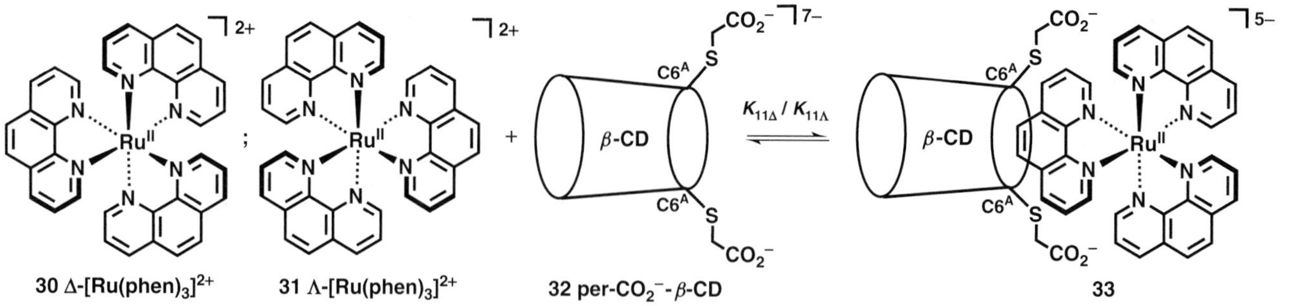

Figure 10 The complexation of Δ- and Λ-$[Ru(phen)_3]^{2+}$, **30** and **31** (viewed down the C_3 axes), by **32** for which only two of the seven—$SCH_2CO_2^-$ substituents are shown. Only the most stable diastereomeric host–guest complex, **33**, is shown.

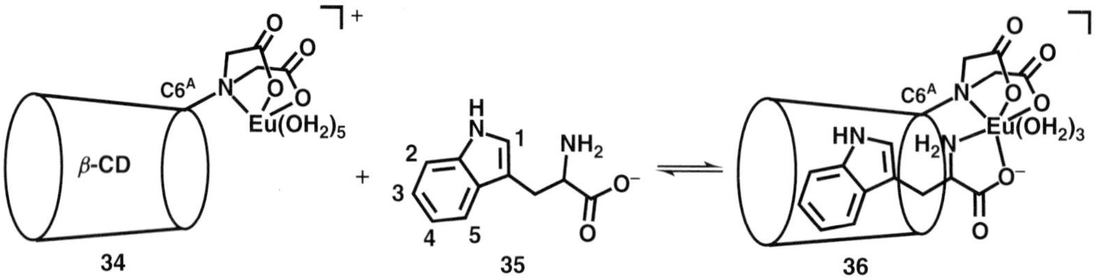

Figure 11 The complexation by the metallocyclodextrin **34** of either D- or L-tryptophan anion **35** to form the diastereomeric host–guest complexes **36**.

changes in enantioselectivity are induced by CD annular and metal complex size variation. In contrast, α-, β-, and γ-CDs show little tendency to complex either $[Ru(phen)_3]^{2+}$ or $[Rh(phen)_3]^{2+}$ largely because of the lack of electrostatic attraction between these CDs and the metal complexes.

Alternatively, the metal center may be coordinated by a CD substituent as shown for pentaaquo(6^A-[bis(carboxylatomethyl) amino]-6^A-deoxy-β-CD)europium(III), **34**, which enantioselectively complexes the D- and L-tryptophan anions, **35**, to form the diastereomeric host–guest complexes **36** in D_2O at pD 10 as shown in Figure 11.[37] The ^1H NMR doublet assigned to the tryptophan H2 proton resolves into two doublets with the upfield doublet being assigned to D-**35**, and a partial resolution of the H3 triplet into two triplets also occurs while the resolutions of the H1, H4, and H5 resonances are smaller. Separate resonances for free and complexed **35** are not observed consistent with exchange between these states being in the fast exchange limit of the 600 MHz ^1H NMR timescale with **34** acting as a chiral shift reagent through its complex **36**. Cross-peaks in the 2D ^1H ROESY NMR spectrum arising from dipolar interactions between the β-CD H3 and H5 annular protons of **36** and the H2 proton of complexed D-**35** show that they are in close proximity in **36**, whereas analogous cross-peaks show that the β-CD H3 and H5 annular protons of **36** and the H4 proton of complexed L-**35** are in close proximity and indicate that D-**35** and L-**35** are differently oriented in **36**.

Chiral discrimination also extends to the formation of the helical polymer **37** in water when the 4-*tert*-butoxyaminocinnamoylamino substituent at the $C3^A$ carbon of the modified α-CD monomer is complexed by a second monomer and so on to form a polymer composed of at least 15 such units in water as shown in Figure 12.[38] Negative and positive Cotton circular dichroic effects at 327 and 288 nm, respectively, are consistent with the polymer assuming a left-handed anticonfiguration and a slanted complexation of the substituents in the α-CD annuli.

4 CYCLODEXTRIN CATALYSTS

The complexation of hydrophobic substrates in the annuli of CDs where they may react catalytically with either the hydroxyl groups defining the CD ends or catalytic groups substituted on either end render them potential enzyme mimics.[39] A wide range of such CD-based catalysts has been studied. They generally exhibit kinetic characteristics similar to the Michaelis–Menten scheme typifying enzymes which include saturation, nonproductive substrate binding, and competitive inhibition although often at pH values distant from a physiological pH. While other CD catalysts also exhibit these characteristics, they do not have biochemical analogs. Four examples of CD-based catalysts are now discussed.

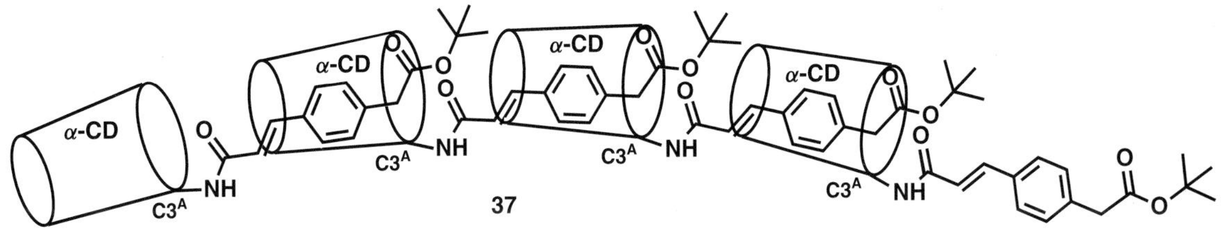

Figure 12 A left-hand helical rotating polymer, **37**, formed by 3^A-(4-*tert*-butoxyaminocinnamoylamino)-3^A-deoxy-α-CD.

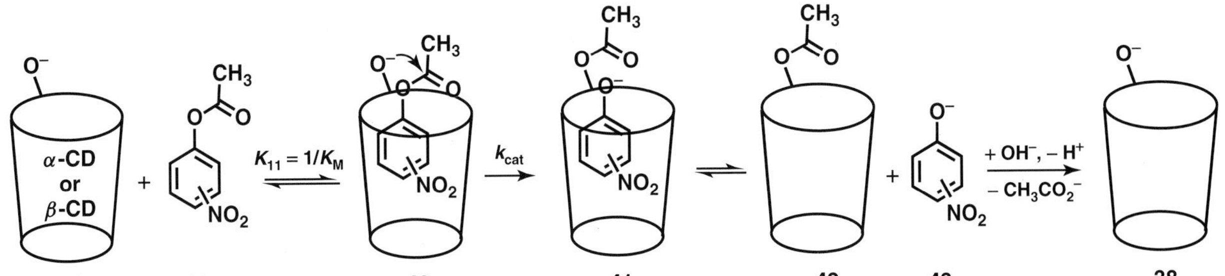

Figure 13 The sequence in which a secondary alkoxy group of **38** catalyzes the hydrolysis of **39**–**43** through the Michaelis-type complex **40**.

4.1 A cyclodextrin hydrolase mimic

An early example of hydrolase mimicry is the hydrolysis of 3-nitrophenyl acetate catalyzed by α-CD and β-CD at pH 10.6 in 5% acetonitrile water as shown in Figure 13.[40,41] Under these conditions, a small proportion of the CDs carry a deprotonated secondary hydroxyl, or alkoxy, group, **38** ($pK_a = 12.1$), which upon complexation of 3-nitrophenylacetate, **39**, makes a nucleophilic attack at the guest carboxylate carbon in **40** which resembles a Michaelis complex. This results in the hydrolysis of **39** to **43** and the attachment of an acetate group to the CD of **41** and **42**. For the system to be truly catalytic, this acetate group should subsequently hydrolyze to regenerate the catalyst **38** as shown in Figure 13. The K_{11} characterizing the formation of **40** is equivalent to the reciprocal of the Michaelis constant K_M and k_{cat} is the rate constant for the hydrolysis in **40** which is compared with that for the hydrolysis of **39**, k_{uncat}, under the same conditions but in the absence of the CD. It is seen from Table 2 that $k_{cat}/k_{uncat} = 300$ and 96, respectively, for 3-nitrophenyl acetate in the presence of α- and β-CDs. For 4-nitrophenyl acetate, the corresponding $k_{cat}/k_{uncat} = 3.4$ and 9.1 consistent with the magnitude of the catalysis being dependent on the nature of the guest stereochemistry and probably the positioning of the guest with respect to the catalytic center in the Michaelis complex. This is further emphasized by the data for 3- and 4-*tert*-butylphenylacetate for which $k_{cat}/k_{uncat} = 260$ and 1.1, respectively, in Table 2. It is also evident that the magnitude of K_M has little relationship to catalytic effectiveness. (A similar catalysis by a metallocyclodextrin (metalloCD) is discussed in Section 5.1).

4.2 A cyclodextrin enantioselective hydrolase mimic

The more effective CD-based catalysts are usually substituted CDs in which the substituent has an important role in the catalysis. Thus, the enantioselective aldol condensation of acetone and 4-nitrobenzaldehyde, **45**, to the corresponding aldol, **49**, by a β-CD substituted at the $C6^A$ carbon with a 1,2-diaminocyclohexane, **44**, appears to proceed through the sequence shown in Figure 14.[42] The reaction occurs in 5% v/v acetone/water at pH 4.80 and 298.2 K under which conditions the secondary amine is protonated ($pK_a \sim 5.7$). The $K_M = 6.31 \times 10^{-3}$ mol dm^{-3} and pertains to **46** prior to the formation of the enamine in **47** which is thought to be rate-determining and characterized by $k_{cat} = 1.05 \times 10^{-4}$ s^{-1}. The subsequent formation of the new carbon–carbon bond in **48** is followed by regeneration of the catalyst **44** and release of aldol **49** in the enantiomer ratio *R*-**49**/*S*-**49** = 97/3. The enantiomeric excess of *R*-**49** is attributed to the combined chiralities of the diamino substituent and the β-CD annulus positioning **45** in **47** such that the enamine attachment to the aldehyde carbon of **45** in **47** dominantly occurs from one side. Thus, while the rate of formation of **49** is only accelerated 6.45-fold by **44** over the reaction in the presence of the small molecule 1,2-diaminocyclohexane precursor to **44**, the enantioselectivity in the reaction is high, and although the reaction pH is low, it resembles the action of an aldolase.

Table 2 Constants for the hydrolysis of phenyl acetates in the presence of α- and β-CDs.[a]

Substrate	$10^4 k_{uncat}$ (s^{-1})[b]	$10^4 k_{cat}$ (s^{-1})	$10^2 K_M$ (mol dm^{-3})	k_{cat}/k_{uncat}
		Catalyst = α-CD		
3-Nitrophenylacetate	46.4	4250	1.9	300
4-Nitrophenylacetate	69.4	243	1.2	3.4
3-*tert*-Butylphenylacetate	4.90	1290	0.20	260
4-*tert*-Butylphenylacetate	6.07	6.7	0.65	1.1
		Catalyst = β-CD		
3-Nitrophenylacetate	46.4	4440	0.80	96
4-Nitrophenylacetate	69.4	634	0.61	9.1
3-*tert*-Butylphenylacetate	—	1220	0.013	250

[a]In 0.5% acetonitrile–water at pH 10.6 and 298.2 K.
[b]In the absence of CD.

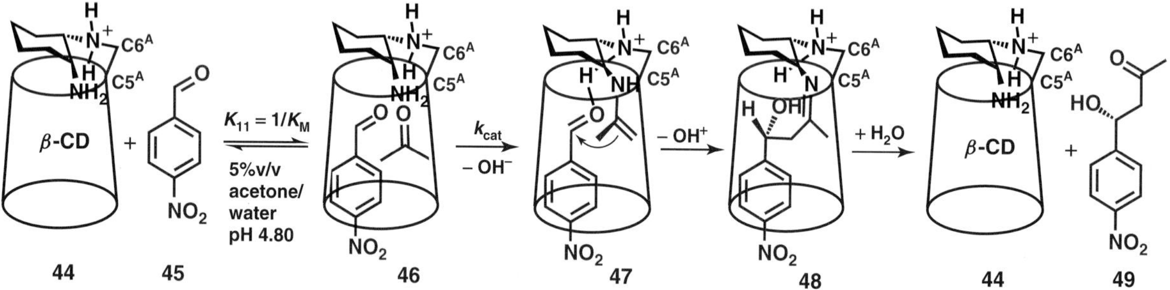

Figure 14 The enantioselective aldol condensation of 4-nitrobenzaldehyde, **45**, with acetone catalyzed by **44** to give the aldol *R*-**49** in 97% enantiomeric excess. The loss of hydroxide from **46** and the loss of a proton from **47** are likely to be highly synchronized.

4.3 A cyclodextrin oxidation catalyst

The oxidation of an alcohol catalyzed by a modified CD is illustrated in Figure 15, where the catalyst is β-CD substituted at two C6 carbons with dihydroxyacetone through ester bonds **50**.[43] In the presence of H_2O_2, which appears to add to the ketone function to form a hydroperoxide adduct as shown in **51**, benzylic alcohol, **52**, forms the Michaelis-type complex **53** with a $K_M = 2.0 \times 10^{-3}$ mol dm^{-3} at pH 7.25 and 298.2 K in aqueous solution. Subsequently, oxidation occurs in **53** in which **52** is oxidized to benzaldehyde, **54**, through an overall transfer of an electron pair with $k_{cat} = 2.69 \times 10^{-7}$ s^{-1} ($k_{cat}/k_{uncat} = 1690$) and regeneration of the catalyst. The α-CD analog of **50** is a similarly effective catalyst and both catalyze oxidations for a range of benzilic alcohols to aldehydes and anilines to nitrobenzenes.[44]

4.4 An organometallocyclodextrin hydrogenation catalyst

CDs may be used to change the characteristics of other catalysts as exemplified by the attachment of CDs to catalytic organometallic complexes to give organometallocyclodextrins. In such catalysts, the metal center has a low oxidation state and acts as a soft acid which interacts with soft base centers in organic molecules. Potentially, the CD component in such a catalyst may result in selectivity of complexation of reactant species and thereby selectivity in catalysis.

Figure 15 The catalyzed oxidation of benzylic alcohol, **52**, to benzaldehyde, **54**, in the presence of catalyst **50** and H_2O_2.

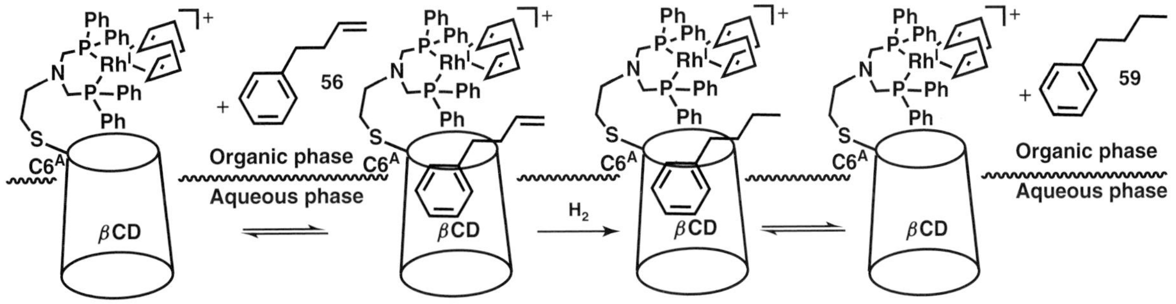

Figure 16 The hydrogenation of **56** at the organic/aqueous phase interface through complexation in the β-CD annulus of catalyst **55** to give **59**.

An interesting example of this is provided by the water-soluble Rh(I) metalloCD **55** catalyzing the hydrogenation of alkene **56** in a two-phase system in which the organic phase is *N,N*-dimethylformamide and the aqueous phase is 30% *N,N*-dimethylformamide and 70% water as shown in Figure 16.[45] It appears that the catalysis occurs at the phase interface where the organometallic Rh(I) component of **55** is also soluble in the organic phase together with **56**, and the β-CD component is soluble in the aqueous phase. An equilibrium exists between **55** and **56** and the host–guest complex **57** in which **56** resides largely in the β-CD annulus. This brings **56** into close proximity to the Rh(I) center to which it attaches through its alkene bond and hydrogenation occurs in a sequence of catalytic steps including oxidative addition and reductive elimination to give the hydrogenated product **59** in a second complex **58**. Subsequent release of **59** regenerates the catalyst **55**. When the phenyl group of **56** is replaced by n-C_6H_{13} to give alkene **60** it is found that **55** preferentially hydrogenates **56** over **60** in a ratio of 68/32 at 295.2 K probably because of preferential complexation of **56** in the β-CD annulus of **55**. In contrast, the catalyst in which a phenyl group replaces β-CD in **55** shows no discrimination in hydrogenation of the two alkenes. In the presence of carbon monoxide and hydrogen catalyst **55** also hydroformulates alkenes.

5 METALLOCYCLODEXTRINS

MetalloCDs are CDs bearing one or more groups which coordinate metal ions. By applying the principles of Lewis acid–base theory, the nature of the coordinating group may be varied to selectively coordinate a wide range of metal ions.[46,47] The simplest metalloCDs are those formed when either CD hydroxyl groups or deprotonated hydroxyl groups coordinate a metal ion, but they tend to be less stable in solution by comparison with metalloCDs incorporating multidentate coordinating groups and consequently are less studied. Here, we discuss six metalloCDs which are biological mimics, energy transfer systems, and organometallic complexes and which exemplify the great potential to design a range of fascinating systems.

5.1 Metallocyclodextrins as biological mimics

Apart from their intrinsic interest as metal complexes, a major interest in metalloCDs arises from the metal ion being coordinated adjacent to the hydrophobic CD annulus and thereby resembling the active site of a metalloenzyme. The formation of such a metalloCD in water is illustrated by the diamine in 6^A-(3-aminopropylamino)-6^A-deoxy-β-CD, **61**, coordinating Cu(II) from $[Cu(OH_2)_6]^{2+}$ with a $K_{coord} = 2.2 \times 10^7$ dm^3 mol^{-1} to give the metalloCD **62** in which only two molecules of water are shown coordinated to Cu(II) in Figure 17 although there could be up to four.[48,49] One coordinated water molecule deprotonates with a $pK_a = 7.84$ to give the hydroxo species **63**. While both **62** and **63** may complex the substrate 4-*tert*-butyl-2-nitrophenyl phosphate, **64**, in their β-CD annuli it is only **65** which is catalytically active (coordinated water usually shows little activity as a nucleophile). The formation of **65**, in which it is possible that one of the phosphate oxygens is coordinated by Cu(II), is characterized by a Michaelis constant $K_M = 1.2 \times 10^{-3}$ mol dm^{-3} at pH 7 and 298.2 K. Under these conditions, the hydroxo group of **65** makes a nucleophilic attack on the phosphorus of **64** with a rate constant $k_{cat} = 2.3 \times 10^{-2}$ s^{-1} at pH 7.0 and forms the intermediate **66** which dissociates to the hydrolysis products dimethylphosphate and 4-*tert*-butyl-2-nitrophenol and its conjugate base to regenerate the catalyst **61**. Thus, the rate of hydrolysis is accelerated in the presence of **64** as compared to that in its absence, $k_{uncat} = 3.2 \times 10^{-7}$ s^{-1}, such that k_{cat}/k_{uncat} is 7.2×10^4 at 298.2 K. The catalysis by **63** exhibits the Michaelis–Menton kinetic profile characterizing enzymatic catalysis and more than

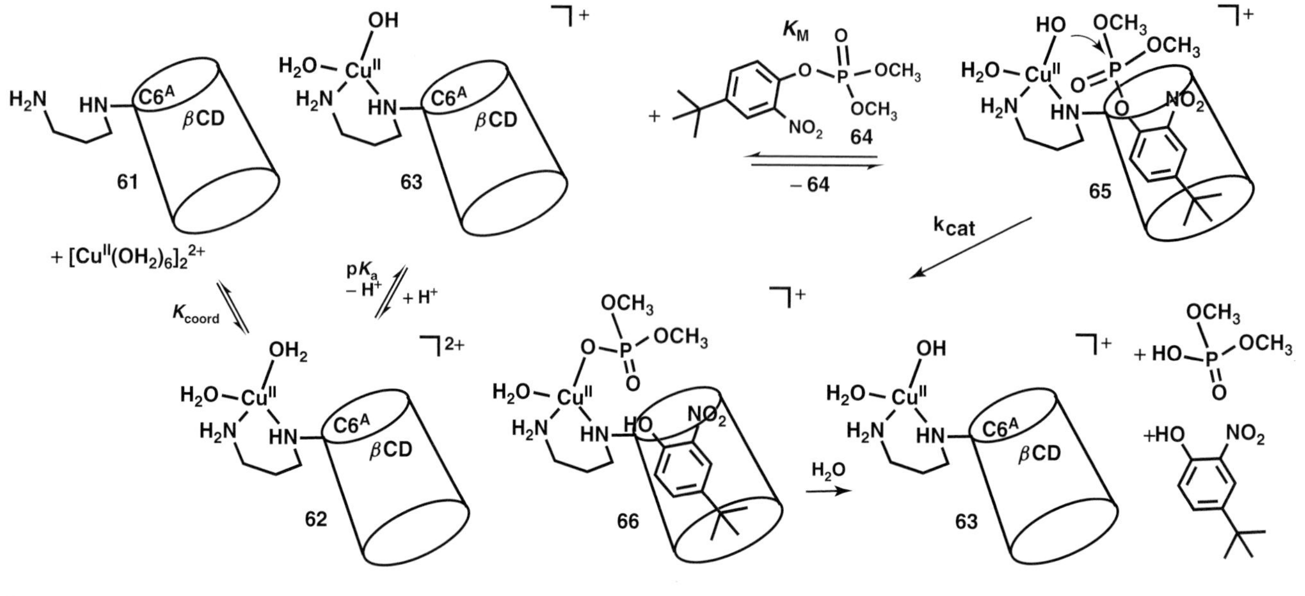

Figure 17 The catalysis of the hydrolysis of 4-*tert*-butyl-2-nitrophenyl phosphate **64** by **63**.

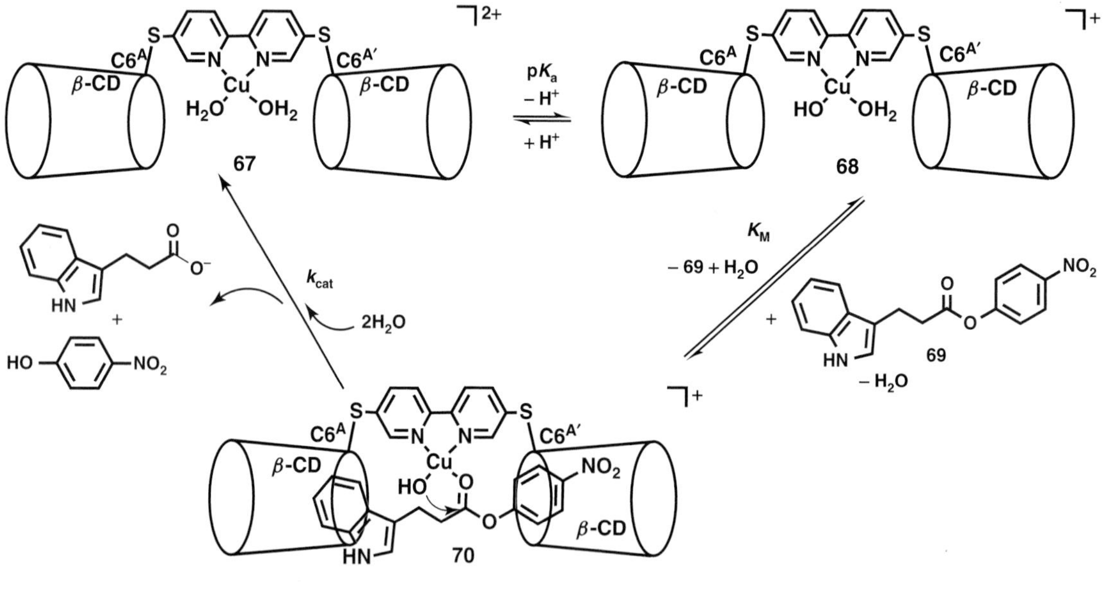

Figure 18 The catalyzed hydrolysis of 4-nitrophenyl indol-3-ylpropionate **69** by **68**.

ten **64** are catalytically hydrolyzed by each **63** representing a turnover rate >10.

Similar studies have been reported for the hydrolysis of carboxylate esters with Zn(II) as the metal ion in an imitation of carboxypeptidase.[50] Generally, the catalytic activity of such metalloCD enzyme mimics shows a substantial variation with change in the coordination site and metal ion.

MetalloCDs are also formed by linked CD dimers in which the linkers incorporate metal coordinating groups.[39,51] Such a metalloCD is represented by **67** in Figure 18 where two β-CDs are linked through sulfur substituted at the 6^A carbons and the Cu(II) coordinating group is the bidentate 2,2′-bipyridyl unit. One of the two waters molecules coordinated to Cu(II) in **67** has a $pK_a = 7.15$. The second water molecule is displaced as Cu(II) coordinates the carbonyl oxygen of 4-nitrophenyl indol-3-ylpropionate, **69**, as it is complexed in both β-CD annuli of **70** with a $K_M = 1.4 \times 10^{-5}$ mol dm^{-3}, such that the carbonyl group is adjacent to the catalytic center. The nucleophilic attack of the hydroxo ligand on the carbonyl carbon of **69** results in hydrolysis to indol-3-ylpropanoate and 4-nitrophenol with a high catalyst turnover characterized by $k_{cat} = 2.05 \times 10^{-4}$ s^{-1}. Thus, in water at pH 7 and 310.2 K, the observed rate constant, $k_{obs} = 5.5 \times 10^{-4}$ s^{-1}, compares with 3.0×10^{-8} s^{-1} observed in the absence of catalyst **68**; a 18 300-fold acceleration of the rate of hydrolysis of **69** when [**67**] = 10^{-4} mol dm^{-3}.

5.2 Metallocyclodextrins as myoglobin mimics

A particularly sophisticated example of a metalloCD is the myoglobin mimic **71**. Here, [5,10,15,20-tetrakis(*p*-sulfonatophenyl)porphinato]iron(II), Fe(II)TPPS, is bound within the linked CD dimer $2^A,2^{A'}$-*O*-[3,5-pyridinediyl-bis(methylene)]bis-per-*O*-methyl-*β*-CD in which all hydroxyl groups are methylated, except those on the $C2^A$ and $C2^{A'}$ carbons, to form the supramolecular complex **71** with a formation constant $K_{form} > 10^7$ mol dm^{-3} as shown in Figure 19.[52] This supramolecular complex binds dioxygen with a change in Fe(II) coordination from five to six and a $K = 3.4 \times 10^3$ dm^3 mol^{-1}, the ratio of the rate constants $k_{on} = 1.3 \times 10^7$ dm^3 mol^{-1} s^{-1} and $k_{off} = 3.8 \times 10^3$ s^{-1} at pH 7.0 and 298.2 K in complex **72**, and thereby mimics oxymyoglobin. Alternatively, **71** may bind carbon monoxide, but more strongly with $K = 5.0 \times 10^7$ dm^3 mol^{-1}, $k_{on} = 2.4 \times 10^6$ mol dm^{-1} s^{-1}, and $k_{off} = 4.8 \times 10^{-2}$ s^{-1}. In another reaction of **71**, nitric oxide is oxidized to nitrate and Fe(II) is oxidized to Fe(III).

An intriguing experiment has been reported for an analog of **71** in which the $-OCH_2PyCH_2O-$bridge joining the $C2^A$ and $C2^{A'}$ carbons of the linked permethylated *β*-CDs is replaced by a — SCH_2PyCH_2S — bridge joining the $C3^A$ and $C3^{A'}$ carbons of the two linked *β*-CDs in which all hydroxyl groups are methylated except those on the $C3^A$ and $C3^{A'}$ carbons.[53] When the oxy form of this analog was injected into a Wistar rat, it was found that the CO form was excreted in urine in accord with CO binding more strongly than O_2, and with no ill effects on the test animal. This promises an opportunity to both sequester and monitor CO in mammals.

5.3 Energy transfer in metallocyclodextrins

When several CDs bearing coordinating groups simultaneously form a metal complex, a variety of metalloCD structures may result. An example of this appears in Figure 20 where a 6-coordinate Ru(II) metalloCD, **73**, forms in aqueous solution as a result of bidentate coordination by the bipyridyl nitrogens of three 6^A-mono[4-methyl-(4′-methyl-2,2′-bypyridyl)]-per-*O*-methylated-*β*-CDs in which all of the C2, C3, and C6 (except $C6^A$) hydroxyl groups are methylated.[54] Each of the -per-*O*-methylated-*β*-CD, TM*β*-CD, annuli of **73** subsequently complex the adamantyl units of three 4′-((1-adamantyl)-2,2′ : 6′, 2″-terpyridyl)(2,2,2-terpyridyl)osmium(II) complexes, **74**, to form the assembly **75**. When **75** is irradiated at 324 nm, this energy is absorbed by the Ru(II) unit and is transferred with $k = 6.4 \times 10^{10}$ s^{-1} to the three complexed Os(II) complexes which then luminesce at 730 nm on a picoseconds timescale.

Energy transfer also occurs in the dimeric assembly **77** formed in dimethylformamide when the fullerene C_{60} is complexed by two Re(I) metalloCDs **76** shown in Figure 21.[55] When triscarbonyl(6^A-(4-pyridylmethyl) amino-6^A-deoxy-*β*-CD)(2,2′-pipyridyl)rhenium(I), **76**, is excited at 340 nm, it luminesces at 570 nm with a lifetime $\tau = 98$ ns and a quantum yield $\phi = 7 \times 10^{-3}$. However, when **76** is complexed in the dimer **77** $\tau = 12$ ns and $\phi = 1.4 \times 10^{-3}$; and reductions consistent with either energy or electron transfer between the Re(I) and C_{60} units of **77**.

5.4 Organometallocyclodextrins

Organometallic complexes have been less studied than conventional coordination complexes as substituents in metalloCDs.[56] Of these, ferrocene and its derivatives have been the most studied as both a guest in a range of CD host–guest complexes and as metalloCD substituents as exemplified by **78**–**80** in Figure 22.[57] In **78**, ferrocene

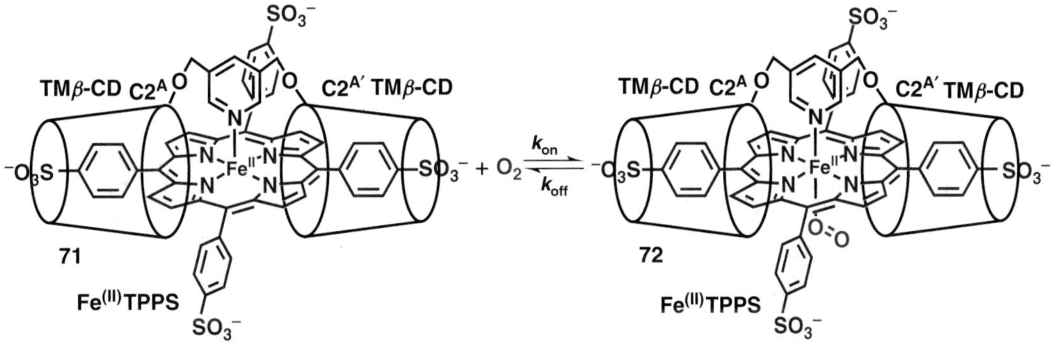

Figure 19 Dioxygen binding by the metalloCD myoglobin mimic **72**, where TM*β*-CD is the per-*O*-methylated CD.

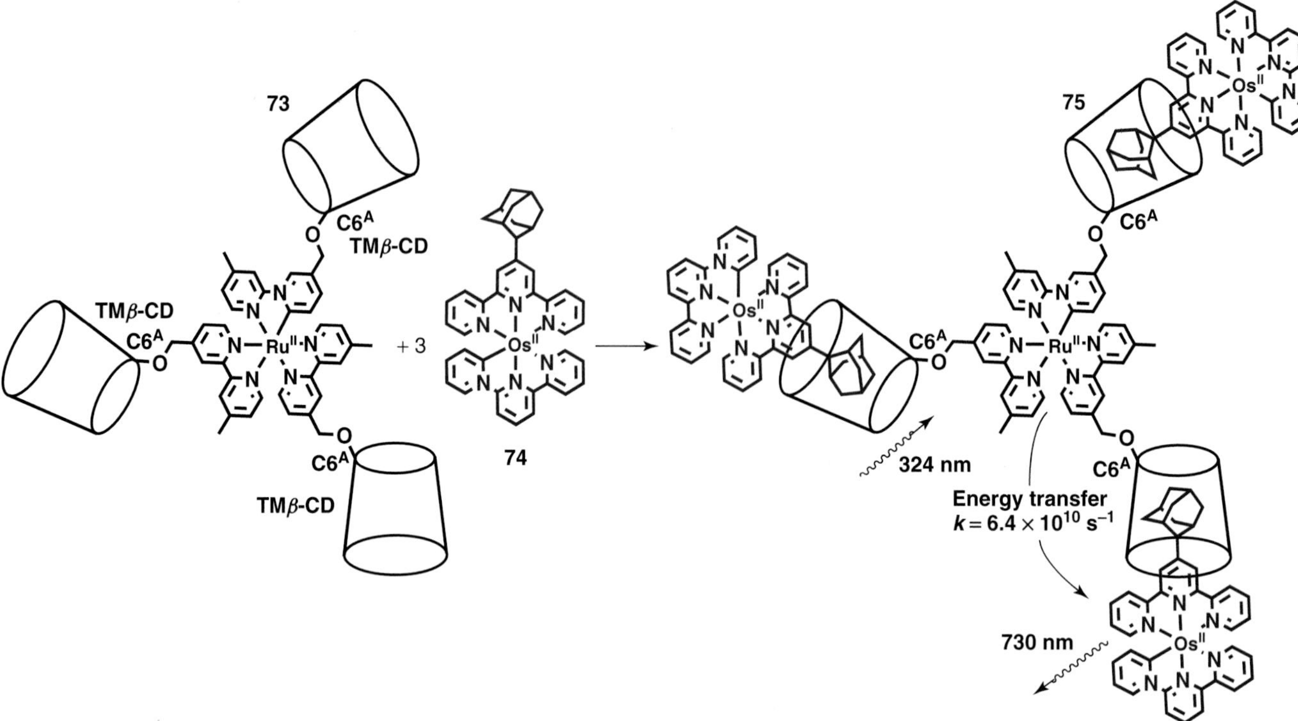

Figure 20 The complexation of the adamantyl groups of three Os(II) complexes, **74**, by the TMβ-CDs of **73** to form the multimetal centered assembly **75** in which rapid energy transfer from the Ru(II) center to the Os(II) centers occurs followed by luminescence at 730 nm.

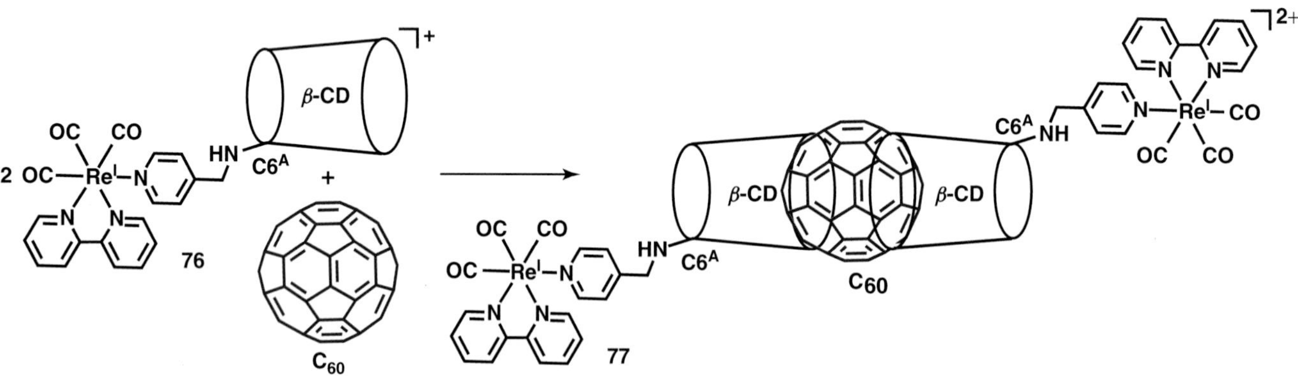

Figure 21 The complexation of the fullerene C_{60} by two Re(I) metalloCDs, **76**, to form the dimeric assembly **77** in which either photoinduced energy or electron transfer from the metal complex substituents of **77** to C_{60} occurs.

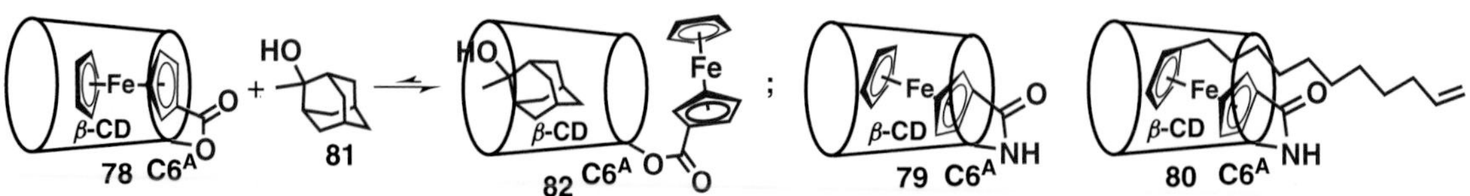

Figure 22 The self-complexation of the ferrocenyl substituents of **78**–**80** and the displacement of the ferrocenyl substituent from the β-CD annulus of **78** by 2-methyl-2-adamantol, **81**, to form the host–guest complex **82**.

is attached to β-CD at C6[A] through an ester link to a cyclopentadienyl ring while in **79** and **80** an amide link is used. In 20% ethylene glycol aqueous solution, circular dichroic and ^{1}H NMR studies show **78–80** to self-complex largely because of the hydrophobic nature of the ferrocenyl substituent. However, in the presence of 2-methyl-2-adamantanol, **81**, the ferrocenyl substituent is largely displaced from the β-CD annulus of **78** to form the host–guest complex **82** because of the hydrophobic nature of **81** and its good fit to the β-CD annulus. The ferrocenyl substituent of **79** is less readily displaced by **81,** which is attributed to the more rigid nature of its amide link to β-CD as compared with that of the ester link of **78**. There is no detectable displacement of the aliphatic substituted ferrocenyl substituent of **80** by **81** and this is attributed to the hydrophobic nature of the aliphatic substituent and the tighter fit of the ferrocenyl substituent of **80** to the β-CD annulus. The metalloCD **80** forms aggregates in solution and shows surfactant behavior as a consequence of its aliphatic substituent.

6 CYCLODEXTRIN ROTAXANES AND CATENANES

Rotaxanes and catenanes are unusual supramolecular assemblies held together by mechanical restraint.[58] A rotaxane consists of a macrocycle threaded onto a linear molecule in a similar manner to the mounting of a wheel on an axle. Accordingly, the name rotaxane is derived from Latin for wheel and axle, *rota* and *axis*, respectively. A catenane consists of macrocycles joined as links in a chain and the name catenane is derived from Latin for chain, *catena*. The macrocyclic nature and size of CDs render them ideal components for rotaxanes and catenanes; the examples of which are discussed below.

6.1 Cyclodextrin rotaxanes

The first CD rotaxanes were reported by Ogino in 1981 as exemplified in Figure 23.[59] The rotaxane is formed in dimethylsulfoxide by threading β-CD onto 1,12-diaminododecane **83** and two inert octahedral Co(III) complexes are then attached to either end to prevent dethreading as shown in Figure 23, which also illustrates the nomenclature for rotaxanes and the principle steps through which many CD rotaxanes are formed. A labile equilibrium exists between β-CD and 1,12-diaminododecane and the [2]-pseudorotaxane **84**, [2]-[1,12-diaminododecane]-[β-CD]-[pseudorotaxane]. Thus, the nomenclature has the form [number of entities]-[threading species]-[CD]-[type of rotaxane] where "pseudo" indicates that the CD may readily dethread. When *cis*-$[Co(en)_2Cl_2]^+$ is added, one end of **83** displaces a chloro ligand from *cis*-$[Co(en)_2Cl_2]^+$ to form *cis*-$[Co(en)_2(NH_2(CH_2)_{12}NH_2)Cl]^{2+}$, **85**, and the analogous [2]-pseudo rotaxane, **86**, which coexist in a labile equilibrium. Subsequently, a second *cis*-$[Co(en)_2Cl_2]^+$ adds to **85** and **86** to form **87** and **88**, respectively, where the latter is a [2]-rotaxane, [2]-[μ-(1,12-diaminododecane)bis(*cis*-chloro)(bis(1, 2-diaminoethane))cobalt(III)]-[β-CD]-[rotaxane] in which the Co(III) complex end groups are too large to allow β-CD to dethread. The Co(III) complexes at each end of the axle are of either Δ or Λ chirality such that the axles formed are a mixture of the combinations

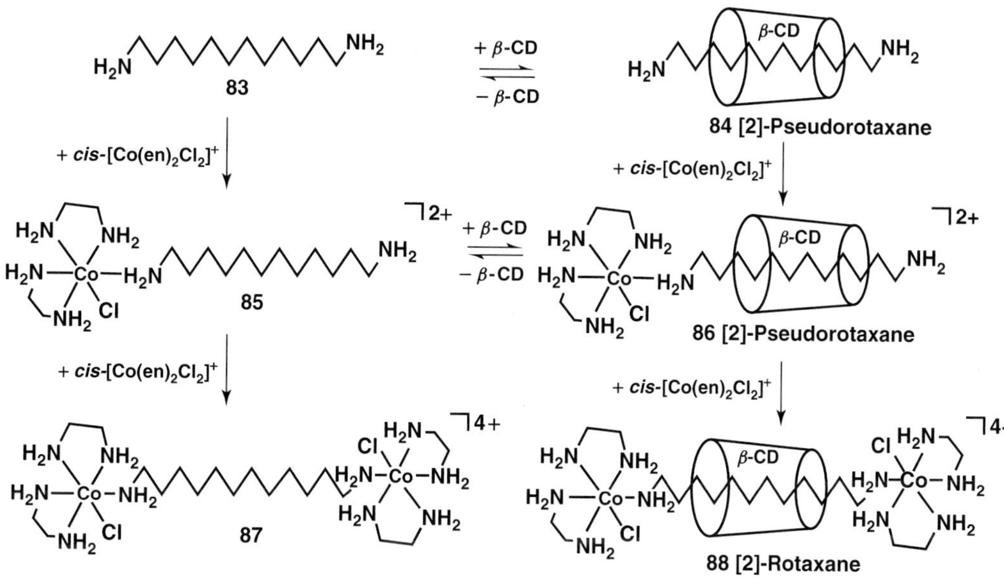

Figure 23 The sequence for the formation of the [2]-rotaxane **88**.

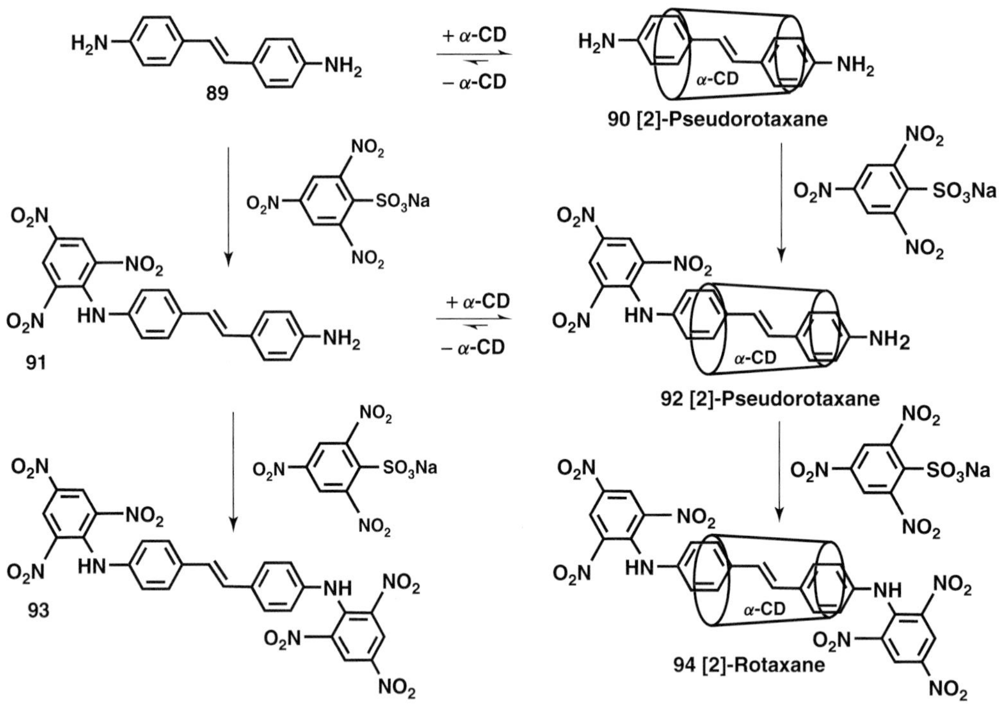

Figure 24 The sequence for the formation of the [2]-rotaxane **94**.

of ΔΔ, ΔΛ, ΛΔ, and ΛΛ chiralities and the [2]-rotaxane isolated in 7% yield exists as four diastereomers as the β-CD may either be oriented as shown in **87** or may possess the opposite orientation. The 1,10-diaminodecane, 1,14-diaminodotetradecane, and α-CD analogs of **88** have also been prepared.[60,61]

Organic end groups may also be employed as shown in Figure 24 for the formation of the [2]-rotaxane **94** in water.[62] Here, the α-CD and E-4,4′-diaminostilbene **89** are in equilibrium with the [2]-pseudorotaxane, **90**, and 2,4,6-trinitrobenzenesulfonate reacts with an axle amino group to form axle **91** which with α-CD is in equilibrium with the [2]-pseudorotaxane **92**. The substitution of the second amino group in **91** then forms the axle **93** whose trinitrophenyl end groups are too large to allow threading of α-CD. A similar substitution in **92** prevents dethreading of α-CD in the [2]-rotaxane, [(E)-4,4′-bis(2,4,6-trinitrophenylamino)stilbene]-[α-CD]-[rotaxane] **94** which was obtained in 10% yield in water:acetone 3:2 v/v solution.

While the threading of a CD onto an axle followed by attachment of blocking groups is a common method for preparing rotaxanes, alternative approaches have been developed. An innovative method is the use of photocyclodimerization of 2-anthracene carboxyl substituents on the 6^A carbons of α-CD in the presence of γ-CD in aqueous solution at 298.2 K as shown in Figure 25.[63] The sequential preassembly of the substituents of two 6^A-(2-anthracenecarbonyl)-6^A-deoxy-α-CDs, **95**, in a head-to-tail orientation in the annulus of γ-CD in the 1:2 host–guest complex **97** is favored by the steric hindrance arising from the α-CD entities. The sequential complexation constant for **96** is $K_{11} = 270\ \mathrm{dm^3\,mol^{-1}}$ and for **97** is $K_{12} = 21\,700\ \mathrm{dm^3\,mol^{-1}}$ where the latter species may be viewed as a [3]-pseudorotaxane with two axles. The large K_{12} magnitude as compared with that of K_{11} is attributable to a combination of π–π interactions between the anthracenyl substituents and the closer host–guest fit in **97**. Irradiation of **97** at >320 nm causes photocyclodimerization to give a bulky axle which is prevented from dethreading by the steric hindrance of the α-CD end groups in the *anti*- and *syn*-head-to-tail [2]-rotaxanes, **98** and **99**, in 60 and 35% yields, respectively. Only 3 and 2%, respectively, of the *anti*- and *syn*-head-to-head axle, which cannot form [2]-rotaxanes are produced.

Lengthening of the axle can lead to the threading of several CDs to form polypseudorotaxanes which become polyrotaxanes when bulky end groups are attached at either end of the axle.[11,12] An interesting example of such a polyrotaxane is shown in Figure 26 where the polyethyleneglycol-α-CD polyrotaxane **100** is reacted with epoxide **101** in 10% NaOH in water to form two or three links between the primary and secondary hydroxyls of the ends of adjacent α-CDs to make a molecular tube in the [2]-rotaxane **102**.[64] Treatment with 25% NaOH in water hydrolyzes the dinitrophenyl end groups from the axle to release the molecular tube **103**.

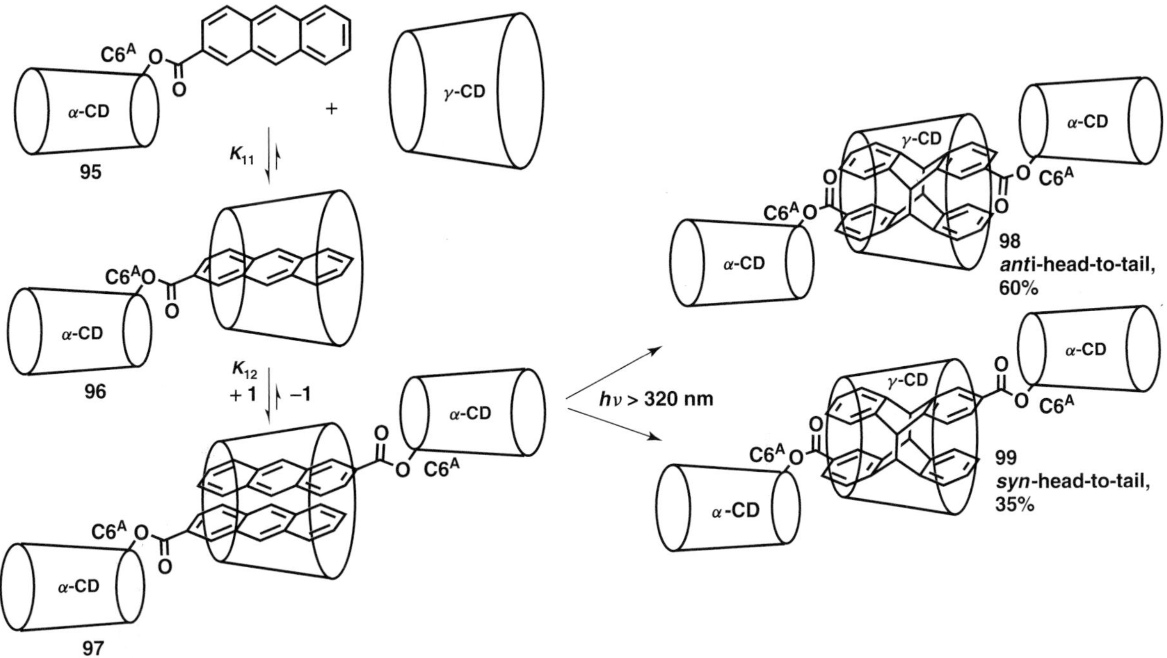

Figure 25 Sequential complexation of **95** to form **96** and **97** and the photocyclodimerization of **95** to the *anti*- and *syn*-head-to-tail axles in the [2]-rotaxanes **98** and **99**.

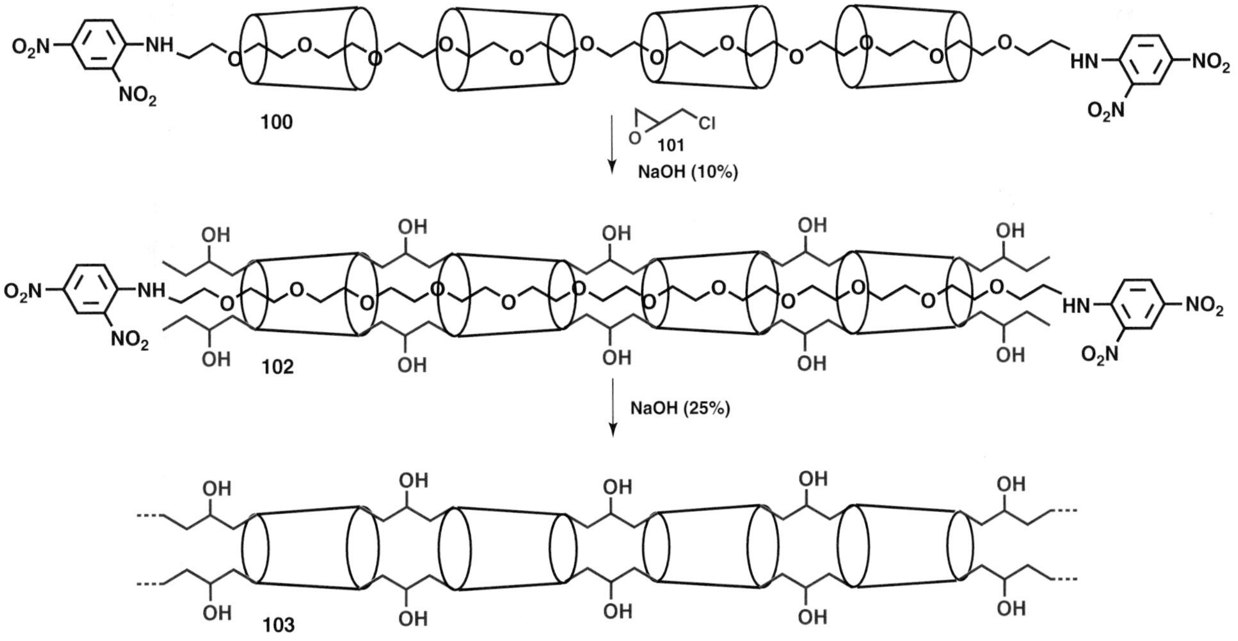

Figure 26 The reaction of the polyrotaxane **100** with epoxide **101** to form the molecular tube in the [2]-rotaxane **102** and release of the molecular tube **103**.

6.2 Cyclodextrin catenanes

The formation of catenanes by CDs is less explored than the formation of rotaxanes. Part of an extensive study of the formation of several catenanes in aqueous solution is shown in Figure 27.[65] When the bitolyl derivative diamine chain link precursor **104**, in which n is either 3 or 4, threads heptakis(2,6-di-*O*-methyl)-β-CD, DMβ-CD, to form the [2]-pseudorotaxane **105** subsequent reaction with teraphthaloyldichloride, **106**, gives the [2]- and [3]-catenanes **107** and **108** in yields of 3.0 and 0.8% when n is 3 and 2.4 and 0.3% when n is 4, respectively. The isomeric catenanes **109** and **110** are obtained as a 40 : 60 isomeric mixture in 1.1% yield when n is 3, and as a 50 : 50

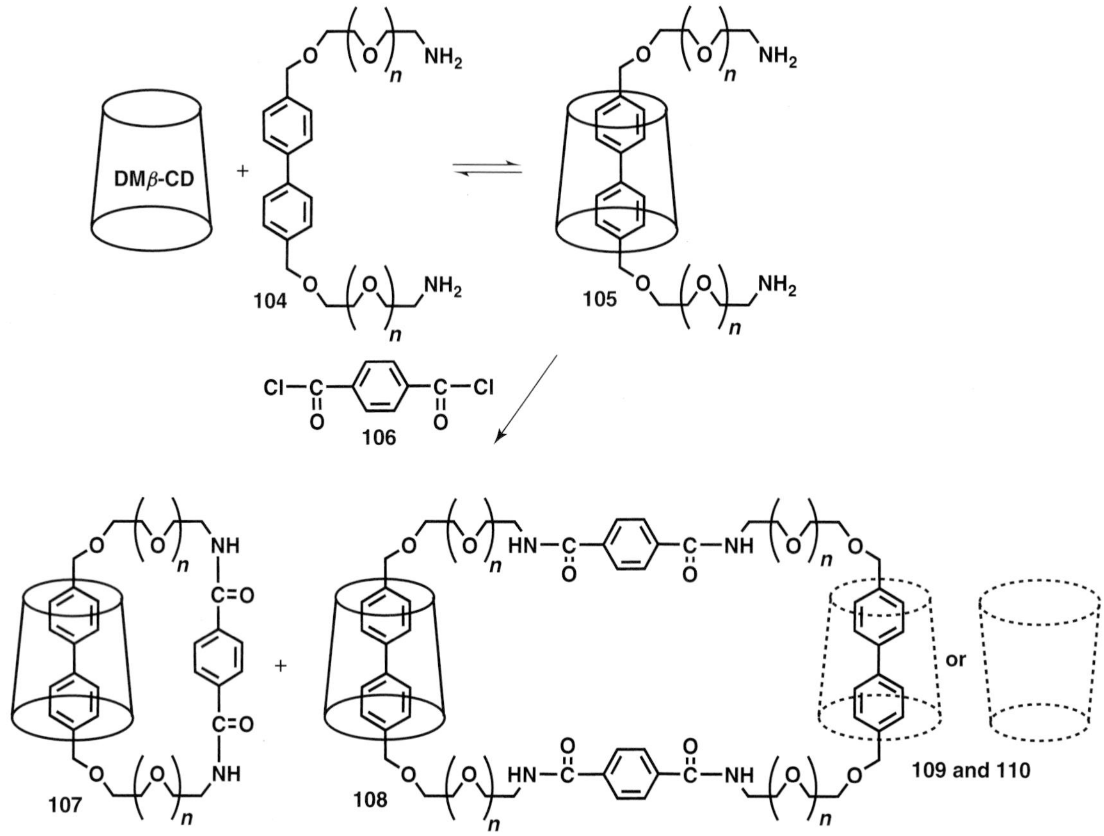

Figure 27 The formation of the catenanes **107** and **108** and of **109** and **110** in which two DMβ-CDs are present in either the head-to-head or the head-to-tail configurations shown by the dashed DMβ-CDs ($n = 3$ or 4).

isomeric mixture in 0.4% yield when n is 4. The separate macrocycles threading DMβ-CD in **107** and **108** are also obtained.

7 CYCLODEXTRIN MOLECULAR DEVICES AND NANOMACHINES

A molecular device is a supramolecular assembly which can be stimulated to perform operations resembling those of everyday macroscopic devices such as shuttles, switches, and hinges. These are explored herein in a sequence of increasing complexity leading to large supramolecular assemblies which combine several of these abilities and which are sometimes called nanomachines and potentially have a range of practical applications. There is increasing interest in constructing such nanomachines on surfaces as is illustrated in the later parts of this section.

7.1 Cyclodextrin rotaxane-based devices

When the length of the axle in a [2]-rotaxane is significantly longer than that of the CD, the possibility that the CD may be stimulated to shuttle along the axle arises. This is exemplified by the [2]-rotaxane shown in Figure 28 in which the redox properties of the tetrathiafulvalene unit allow control of the position of α-CD through a two-step electron transfer process.[66] In **111**, the α-CD complexes the hydrophobic and uncharged thiafulvalene unit in aqueous solution at 298.2 K. Upon addition of either D_2O_2 or Fe(II), the thiafulvalene is oxidized to either its radical cation or dication and loses its hydrophobicity such that the α-CD shuttles to complex the 1,2,3-triazole unit of the axle as shown in **112** and **113** through UV–vis, circular dichroic, and ^{1}H NMR studies. Electrochemical studies show that the process is reversible.

Shuttling is not always reversible as shown for the α-CD [2]-rotaxanes in Figure 29, where the tetracationic axle is composed of an azobenzene group linked at either side to a viologen group linked through a propylene group to a 2,4-dinitrobenzene end group.[67] In the initially formed [2]-rotaxanes, **114** and **115**, α-CD shuttles backward and forward on the azobenzene entity at temperatures up to 373.2 K in dimethylsulfoxide with a rate constant, $k(363.2\text{K}) = 94\,\text{s}^{-1}$ and $\Delta G^{\ddagger} = 80\,\text{kJ mol}^{-1}$. Upon heating above 373.2 K, α-CD shuttles irreversibly to positions over the propylene units as shown in **116**. Alternatively when

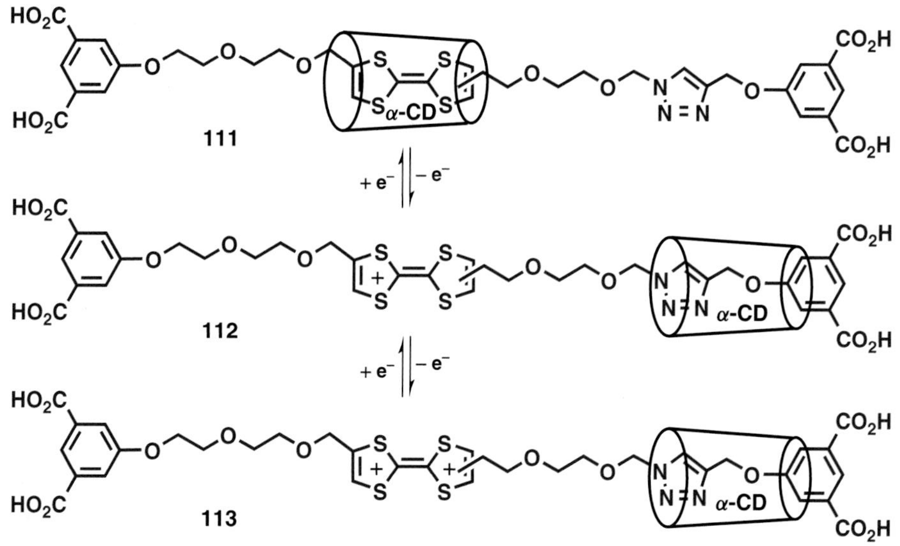

Figure 28 The sequence of reversible redox-driven shuttling of α-CD between tetrathiafulvalene and 1,2,3-triazole axle sites in response to changes in the axle oxidation state in the [2]-rotaxanes **111–113**.

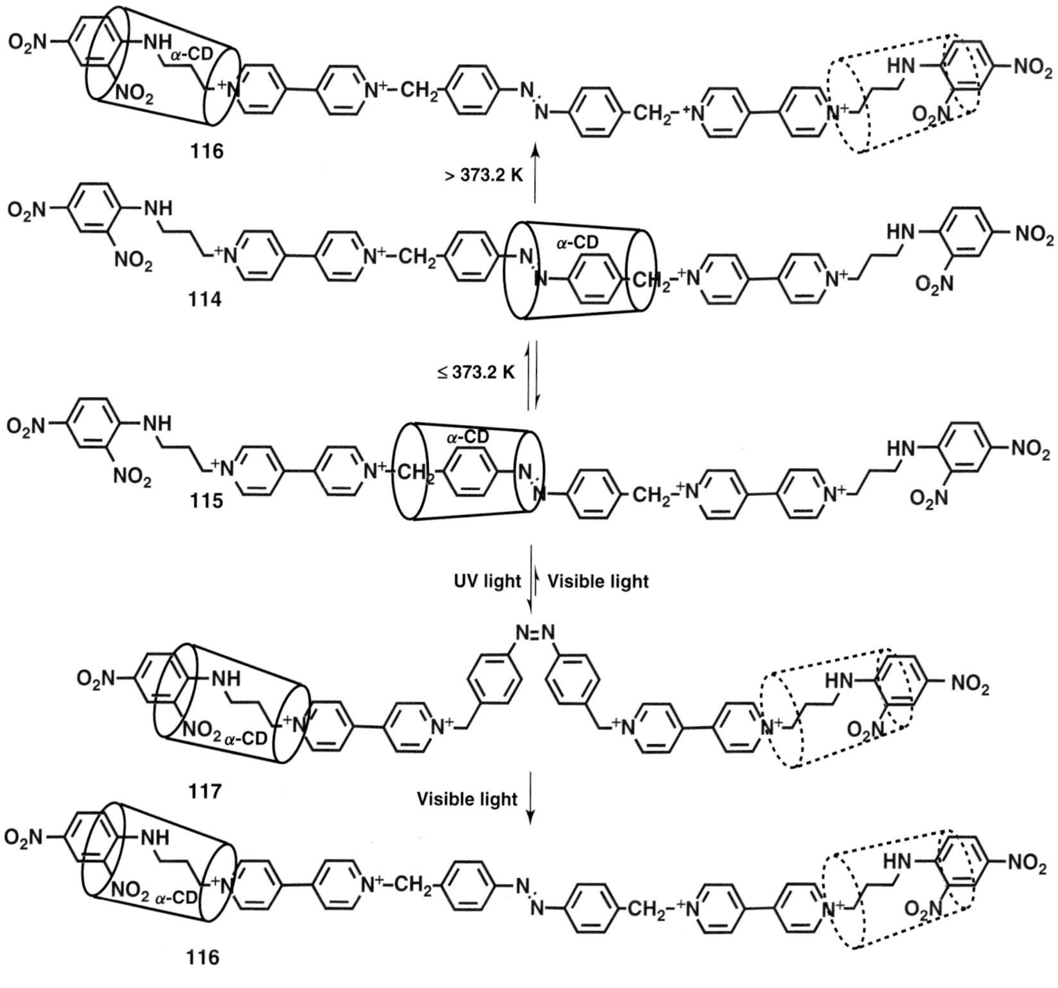

Figure 29 Multimode molecular shuttling of α-CD in the [2]-rotaxanes **114–117**. The alternative positions for α-CD in **116** and **117** are shown by the dashed outline of the α-CD annulus.

114 and **115** are irradiated with UV light, the azobenzene unit photoisomerizes from the *E* to the *Z* isomer and α-CD shuttles to the propylene unit as shown in **117**. Irradiation with visible light partially reverses the isomerization back to **114** and **115** but the dominant effect is the production of **116**. In contrast, no photoisomerization of **116** and **115** occurs in water and this is attributed to strong hydration of the cationic viologen units preventing passage of α-CD to

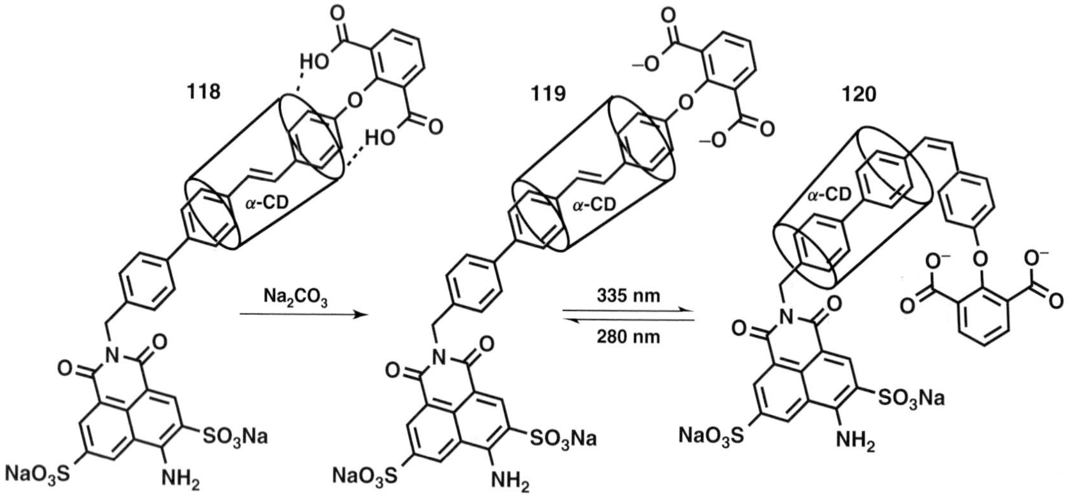

Figure 30 The pH-dependent photoisomerization and shuttling processes of the [2]-rotaxanes **118–120**.

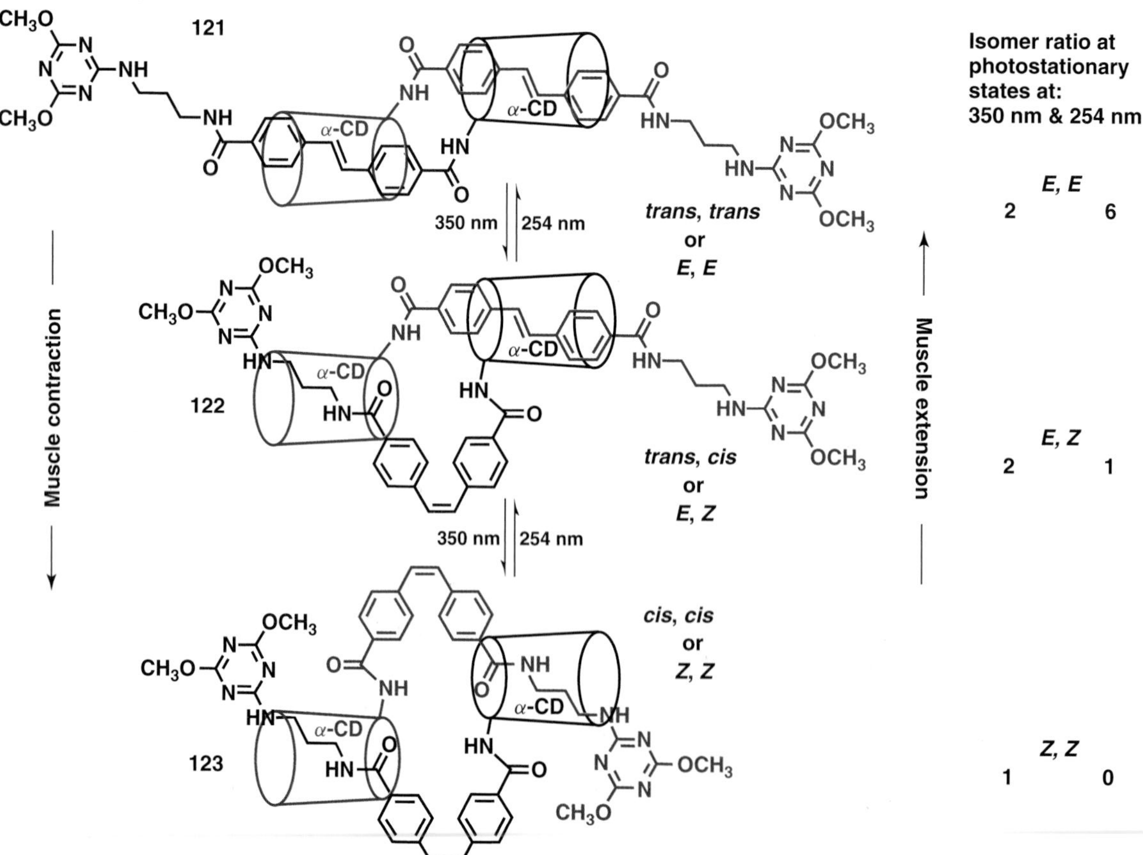

Figure 31 The extended, **121**, half contracted, **122**, and contracted, **123**, states of the photochemically controlled Janus [2]-rotaxane molecular muscle. The relative proportions of the three states at the 350 and 254 nm photostationary states in D_2O are shown on the left.

the alternative propylene unit and such that the photoisomerization is sterically hindered by α-CD remaining over the azobenzene entity. Similar shuttling processes and their solvent dependence have been reported.[68,69]

Another example of photoisomerization of a double bond and solvent conditions controlling CD shuttling is represented by the [2]-rotaxane **118** in Figure 30 in which the axle contains a stilbene unit.[70] UV–vis, fluorescence, and ^{1}H NMR studies show that in water no photoisomerization occurs consistent with hydrogen bonding between the α-CD hydroxy groups and the carboxylic acid substituents of the isophthalic acid unit of the axle increasing the barrier to photoisomerization. However, upon raising the pH to 10.5 with Na_2CO_3, a reversible photoisomerization of the stilbene unit between the *E* and *Z* forms in **119** and **120** occurs and drives the to-and-fro shuttling motion of α-CD. A similar use of stilbene photoisomerization to control shuttling in has been reported by Anderson.[71]

The incorporation of two isomerization centers increases the sophistication of the molecular devices as shown in the system resembling a molecular muscle depicted in Figure 31.[72] It consists of two interpenetrating [2]-rotaxanes where one provides the axle for the C6^A substituted α-CD of the other and vice versa in a hermaphroditic or Janus complex. Each [2]-rotaxane incorporates a stilbene unit such that when both are in the *E* isomeric form the molecular muscle is in its extended form, **121**, when one is in the *Z* form the muscle is partially contracted, **122**, and when both are in the *Z* form, **123**, the muscle is fully contracted. For the photostationary state under 350 nm radiation the ratio **121 : 122 : 123** is 2 : 2 : 1 in D_2O, whereas at the 254 nm stationary state the ratio **121 : 122 : 123** is 6 : 1 : 0. Thus, coupling of the two photoisomerizations increases the transformational control achievable in a molecular assembly.

7.2 Cyclodextrin intra- and intermolecular devices and nanomachines

The incorporation of both intra- and intermolecular interactions into molecular devices presents an increased range of design possibilities. Thus, both amide and alkene isomerizations and intermolecular host–guest complexation of 1-adamantol are exploited in the operation of the molecular device based on *E*-6^A-deoxy-6^A-(*N*-methylcinnamido)-β-CD shown in Figure 32.[73] The thermal *E* to *Z* isomerization about the amide bond controls the equilibrium between isomeric **124*E*** and **124*Z*** in which 1-adamantol,

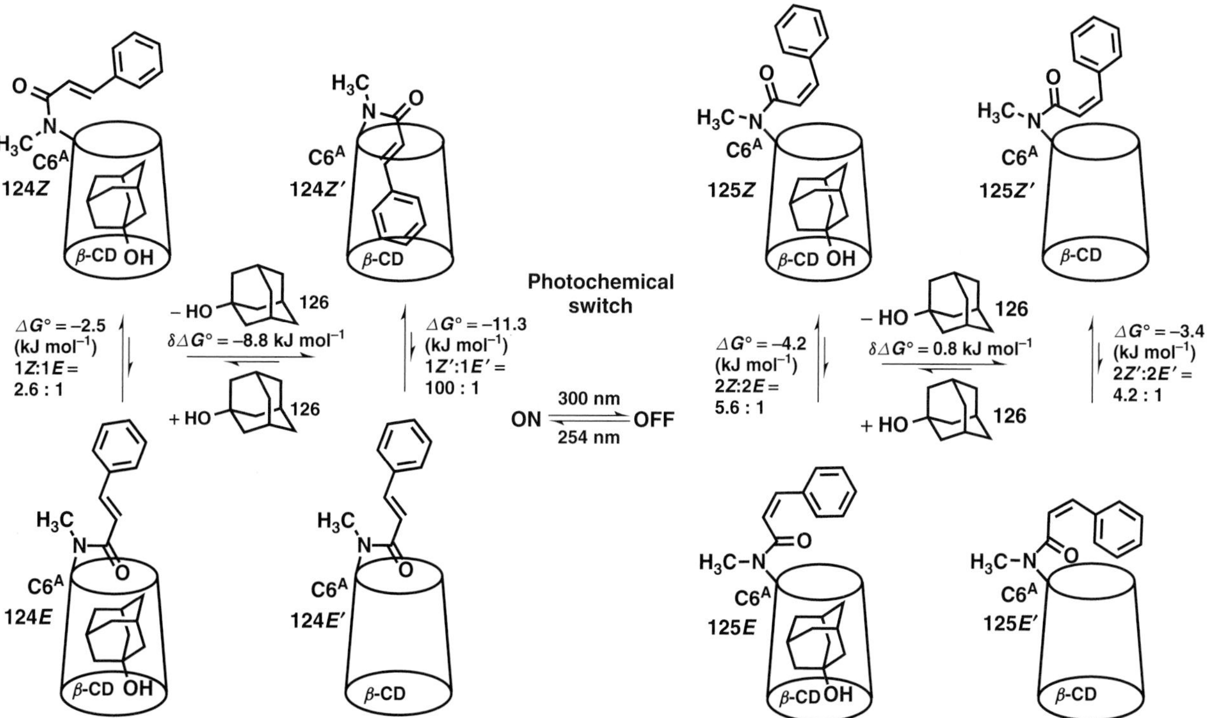

Figure 32 The **124*E***, **124*Z***, **124*E*′**, and **124*Z*′** equilibrium system in which intermolecular host–guest complexation of 1-adamantol competes with cinnamate self-complexation in **124*Z*′** and *E* to *Z* isomerization about the amide bond occurs while the *E* stereochemistry about the cinnamate alkene bond is retained. Photoisomerization by irradiating at 300 nm switches the stereochemistry about the cinnamate alkene bond to *Z* to produce the **125*E***, **125*Z***, **125*E*′**, and **125*Z*′** system in which no competitive cinnamate self-complexation occurs. The cinnamate alkene photoisomerization is reversed by irradiation at 254 nm to complete an on–off photoswitching process.

126, occupies the β-CD annuli of both in water. An equilibrium exists between **124*E*** and 1-adamantol and **124*E*′** in which the cinnamate group is unable to enter the vacant β-CD annulus. A similar equilibrium exists between **124*Z*** and 1-adamantol and **124*Z*′** in which the cinnamate group now self-complexes in the β-CD annulus because of the more favorable *Z* stereochemistry about the amide bond. Both of the last two equilibria may be driven to the right by extraction of 1-adamantol with hexane. Photoisomerization about the cinnamate alkene bond through irradiation at 300 nm converts **124*E***, **124*Z***, **124*E*′**, and **124*Z*′** to **125*E***, **125*Z***, **125*E*′**, and **125*Z*′**, respectively. Although both **125*E*** and **125*Z*** form intermolecular host–guest complexes with 1-adamantol, the cinnamate self-complexes in neither **125*E*′** nor **125*Z*′** because of the unfavorable *Z* stereochemistry about the cinnamate alkene bond. Irradiation at 254 nm photoisomerizes the stereochemistry about the alkene bond from *Z* to *E* and completes the photochemical switch between the two systems. The differences in the free energy changes, $\delta\Delta G^\circ = -11.3 + 2.5 = -8.8\,\text{kJ mol}^{-1}$ and $\delta\Delta G^\circ = -3.4 + 4.2 = 0.8\,\text{kJ mol}^{-1}$, associated with the host–guest complexation of 1-adamantol in the first and second systems, respectively, are directly related to the different constraining effects of the *E* and *Z* stereo chemistries about the cinnamide alkene bond between phenyl group and the amide function.

The formation of hydrogels through CDs attached to a polymer backbone complexing hydrophobic substituents on a second polymer backbone to form cross-links between polymer strands in water is attracting considerable attention.[74] Innovative examples of these macroscopic assemblies arise where the photoisomerization of an azobenzene-based substituent enables the viscosity of an aqueous polymer solution to be varied at will as shown in Figure 33.[75] Example Figure 33(a) is based on a polyacrylate with 2.2% of the carboxylic acid groups randomly substituted with β-CD attached through the C6^A carbon, **127**, which complexes the *E*-azobenzene substituent of a second polyacrylate with 2.7% of the carboxylic acid groups randomly substituted with azobenzene attached through a dodecyl tether, **128**, to form a viscous solution consistent with the complexation of **128** by **127** being characterized by $K = 1.2 \times 10^4$ $\text{dm}^3\,\text{mol}^{-1}$. On irradiation with UV light, the *E*-azobenzene photoisomerizes to *Z*-azobenzene, **129**, and the viscosity of the solution increases from $2.5 \times 10^2\,\text{Pa s}^{-1}$ two-fold consistent with the "locking" shown for the cross-link formed with **129**.

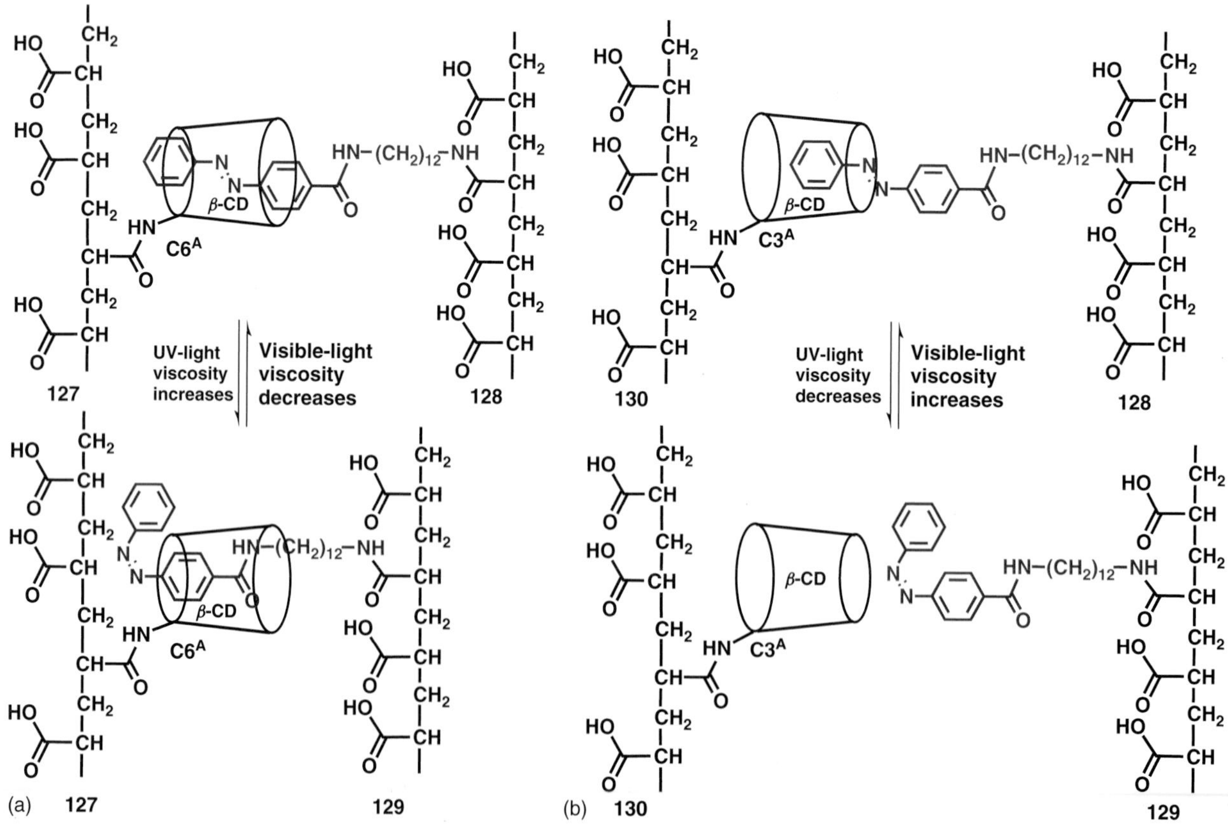

Figure 33 The opposite effects of irradiation with UV- and visible light in systems (a) and (b) are attributed to the *E* to *Z* and vice versa photoisomerization of the azobenzene substituents of polymers **128** and **129** impacting on the mode of complexation by the β-CD substituents of polymers **127** and **130**.

The reverse situation applies when the same experiment is carried out on polymer **130** where the 1.6% substituted polyacrylate has β-CD attached through the C3^A carbon as shown in Figure 33(b). Under the same conditions, the solution of polymers **130** and **128** is much less viscous at 6.5×10^{-1} Pa s^{-1} because of the weaker complexation of the *E*-azobenzene substituent by the β-CD substituent as indicated by a lower $K = 1.4 \times 10^2$ dm^3 mol^{-1}. Upon photoisomerization to form the *Z*-azobenzene substituent, the viscosity drops by an order of magnitude consistent with this substituent of **129** complexing even less effectively in the β-CD substituents of **130**. (It should be noted that the viscosity of **128** alone under the same conditions is 8.4×10^{-2} Pa s^{-1}). For both systems (Figure 33a and b) the photocontrolled viscosity switching is repetitively reversible. Thus, both systems may be viewed as nanomachines which demonstrate the impact of variations in CD complexation and azobenzene substituent photodimerization in controlling the characteristics of macroscopic systems.

Another type of macroscopic assembly formed by highly modified CDs is represented by CD substituted with long hydrophobic substituents.[76–78] This is exemplified in Figure 34 where the highly substituted β-CD **131** forms a spherical bilayer vesicle through hydrophobic interactions between the long hydrophobic substituents in water at pH 7.4.[78] This resembles the lipid bilayer membranes of biological cells. A further resemblance occurs through the complexation of the adamantyl group of the octapeptide **132** in the β-CD annuli on the vesicle exterior which mimics the selective docking sites on mammalian cell exteriors. The K_{11} for a single **131** on the exterior of the vesicle complexing **132** is 3.5×10^4 dm^3 mol^{-1} in water, and **132** adopts a random coil configuration when complexed. Upon changing the pH to 5.0, the vesicles change to nanotubes and **131** forms a β-sheet on the nanotube external surface. This transformation is reversed by changing the pH back to 7.4. When the vesicles are formed in the presence of tetrasodium 1,3,6,8-pyrenetetrasulfonate at pH 7.4 this dye is encapsulated and upon changing the pH to 5.0 it is released.

7.3 Cyclodextrin surface mounted nanomachines

Increasingly, CD-based molecular devices are being attached to surfaces to both increase their versatility and to extend their physical dimensions to the extent that they are often referred to as nanomachines. The first example shown in Figure 35 is based on the silica MCM-41 nanoparticle to which is attached an array of azobenzene derivatives in the *E* form, or "stalks," adjacent to the many pores which characterize the nanoparticle as represented for a single pore by **134**.[79] In water, the dye Rhodamine B, **133**, freely exchanges in and out of the pore. However, when β-CD subsequently complex the stalks egress of Rhodamine B

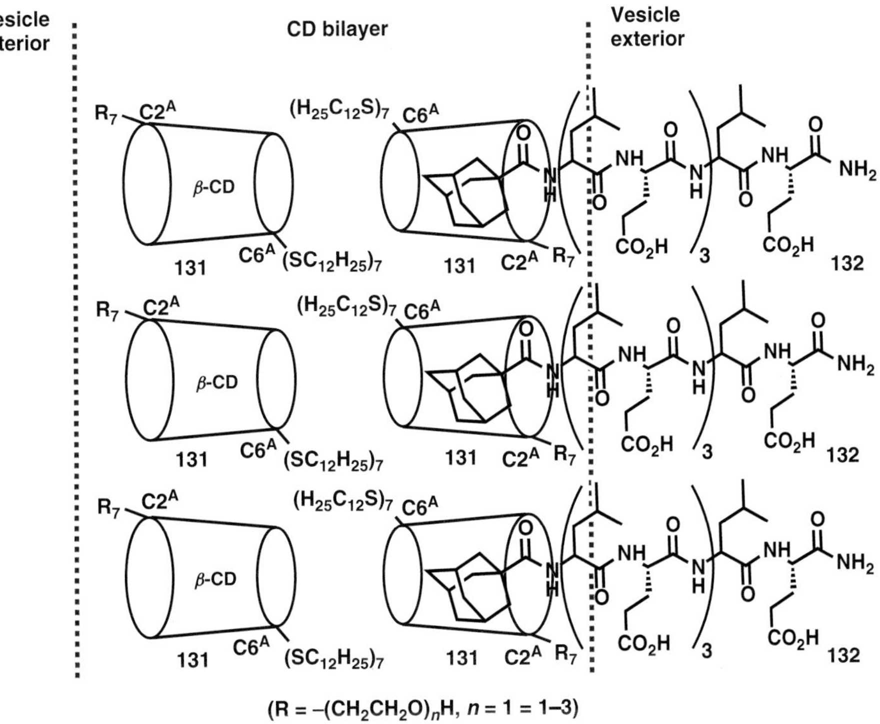

Figure 34 The vesicle bilayer formed by the highly modified β-CD **131** in water at pH 7.4 and the complexation of the adamantyl groups of the peptides, **132**, by **131** on the external surface of the vesicle.

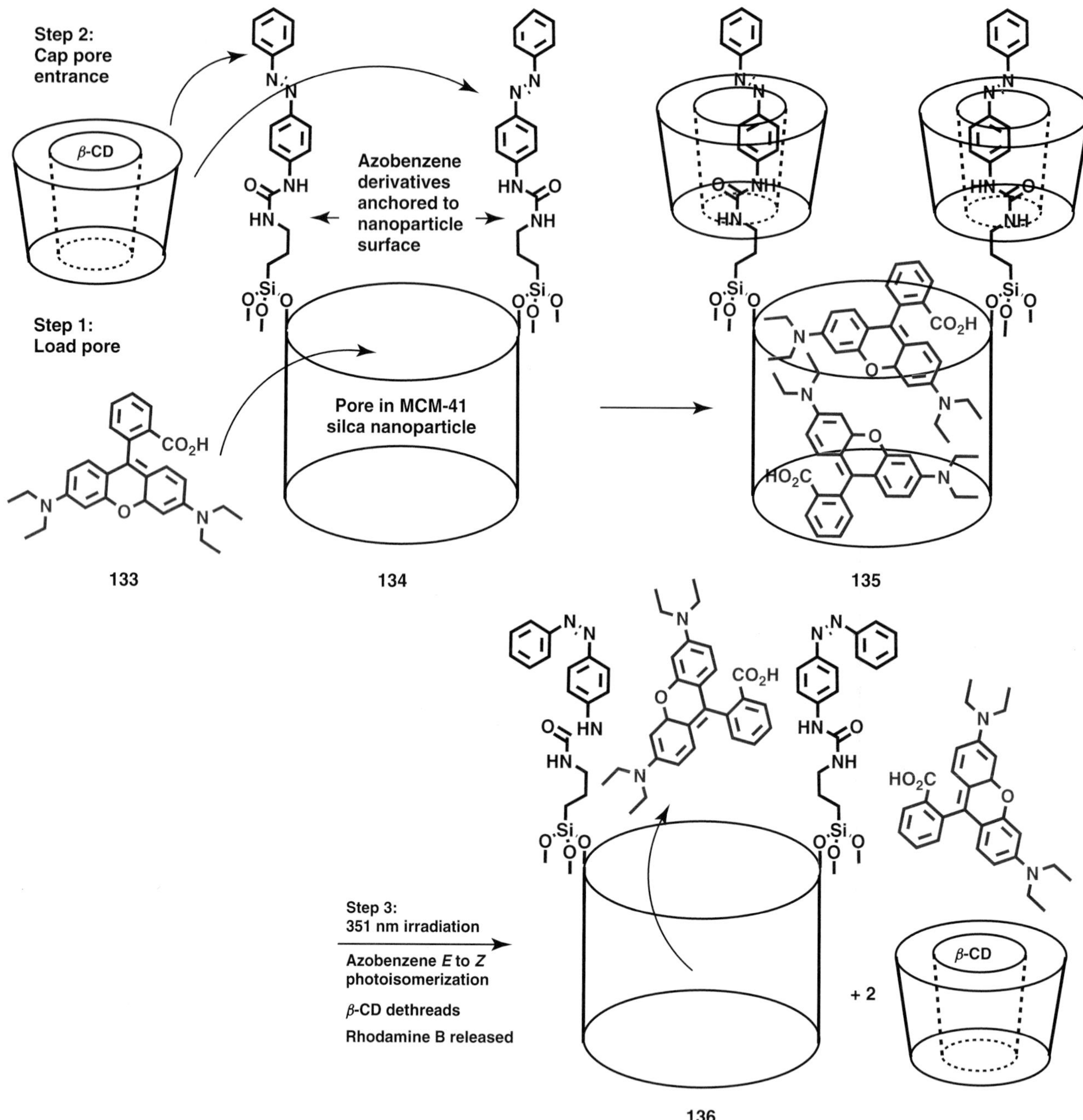

Figure 35 Scheme for the entrapment of Rhodamine B, **133**, in a pore of a MCM-41 nanoparticle, **134**, through capping the pore by complexation of β-CD on azobenzene derivatives attached to the nanoparticle surface, **135** (Steps 1 and 2). The Rhodamine B is subsequently released by uncapping the pore through photoisomerization of the azobenzene unit from the E to Z form and decomplexation of β-CD from **136** (Step 3).

from the pore of **135** is precluded and the system is effectively blocked. Irradiation at 351 nm causes the azobenzene derivative stalks to photoisomerize to the Z form and the β-CD to decomplex in response so that the pore is now opened and Rhodamine B exits from the pore **136** as shown in Figure 35. It is envisaged that such modified nanoparticles might act as photochemically controlled drug delivery systems when Rhodamine B is replaced by a drug molecule.

A second example of a nanomachine involves the attachment of multiple C6 sulfur substituent modified β-CD to a gold surface as shown in **138** in Figure 36.[80] When the gold surface is coated in this manner, it becomes a surface on which molecular printing is achieved by selectively

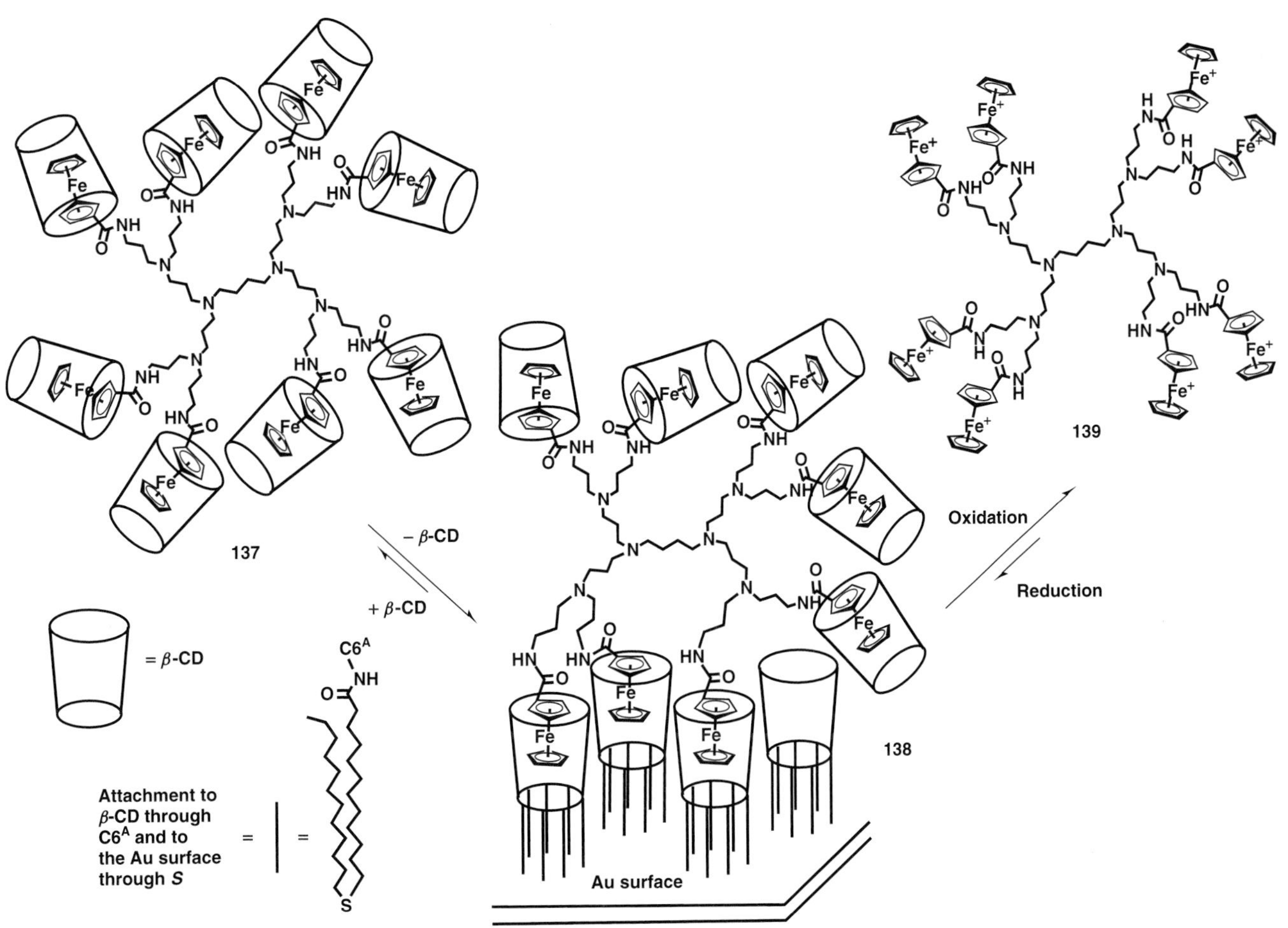

Figure 36 Scheme for attachment of a ferrocenyl dendrimer, **137**, to a gold surface functionalized with β-CD, **138**, and subsequent release upon oxidation of the ferrocenyl groups to ferrocinium groups, **139**, which do not complex significantly in β-CD.

complexing guest molecules in the β-CD annuli.[13] Thus, the water-soluble eight ferrocenyl terminated dendrimer/β-CD assembly **137** docks, or prints, on the modified gold surface to form a stable surface assembly **138**. However, when the ferrocenyl groups are electrochemically oxidized to ferrocium groups in **139** their positive charge greatly decreases the stability of their β-CD complexation such that the multicharged dendrimer is released from the surface. Recently, the versatility of the use of β-CD modified gold surfaces has been demonstrated by the assembly of a printboard for antibody recognition and lymphocyte cell counting.[81] Clearly, surface modifying nanomachines hold great promise for practical application.

8 CONCLUSION

This brief examination of CD chemistry commenced with the naturally occurring CDs and proceeded through an increasingly sophisticated sequence of supramolecular assemblies, some of which resemble metalloproteins, biological cells, and muscle components. Others have no counterpart in nature, but exhibit sophisticated characteristics in their own right. This is exemplified by rotaxanes, polymers, and light activated molecular devices and hydrogels. Increasingly, CDs are being attached to surfaces to generate versatile behaviors as shown by controlled guest release from silica nanospheres and molecular printboards. The great breadth of this supramolecular array is underpinned by synthetic chemistry which has been refined to the stage where almost any desired CD modification is achievable. In conjunction with the current powerful understanding of supramolecular chemistry, increases in the sophistication of CD supramolecular assemblies are likely to be limited only by imagination. Exactly where this will lead is not readily predictable, but it is likely that "smart" polymer, molecular device, and surface chemistry will be much to the fore. CDs are widely deployed in the agrochemical, cosmetic, food, and pharmaceutical industries and it seems inevitable that these and other uses will grow in both extent and sophistication as CD supramolecular chemistry progresses.

REFERENCES

1. J. Szejtli, *Chem. Rev.*, 1998, **98**, 1743.
2. C. J. Easton and S. F. Lincoln, *Modified Cyclodextrins: Scaffolds and Templates for Supramolecular Chemistry*, Imperial College Press, London, 1999.
3. W. Saenger, J. Jacob, K. Gessler, *et al.*, *Chem. Rev.*, 1998, **98**, 1787.
4. T. Endo and H. Ueda, *FABAD J. Pharm. Sci.*, 2004, **29**, 27.
5. H. Taira, H. Nagase, T. Endo, and H. Ueda, *J. Inclusion. Phenom. Macrocyclic. Chem.*, 2006, **56**, 23.
6. M. V. Rekharsky and Y. Inoue, *Chem. Rev.*, 1998, **98**, 1875.
7. K. Harata, *Chem. Rev.*, 1998, **98**, 1803.
8. H.-J. Schneider, F. Hacket, and V. Rüdiger, *Chem. Rev.*, 1998, **98**, 1755.
9. A. R. Khan, P. Forgo, K. J. Stine, and V. T. D'Souza, *Chem. Rev.*, 1998, **98**, 1977.
10. Y. Liu and Y. Chen, *Acc. Chem. Res.*, 2006, **39**, 681.
11. G. Wenz, B.-H. Han, and A. Müller, *Chem. Rev.*, 2006, **106**, 782.
12. A. Harada, A. Hashidzume, H. Yamaguchi, and Y. Takashima, *Chem. Rev.*, 2009, **109**, 5974.
13. O. Crespo-Biel, B. Jan Ravoo, J. Huskens, and D. N. Reinhoudt, *Dalton Trans.*, 2006, 2737.
14. J. Liu, W. Ong, E. Román, *et al.*, *Langmuir*, 2000, **16**, 3000.
15. T. Murakami, K. Harata, and S. Morimoto, *Tetrahedron Lett.*, 1987, **28**, 321.
16. C. Rousseau, B. Christensen, T. Petersen, and M. Bols, *Org. Biomol. Chem..*, 2004, **2**, 3476.
17. C. J. Easton, S. J. van Eyk, S. F. Lincoln, *et al.*, *Aust. J. Chem.*, 1997, **50**, 9.
18. X. Guo, A. A. Abdala, B. L. May, *et al.*, *Macromolecules*, 2005, **38**, 3037.
19. G. Crini, G. Torri, M. Guerrini, *et al.*, *Eur. Polym. J.*, 1997, **33**, 1143.
20. Y.-Y. Liu, X.-D. Fan, and L. Gao, *Macromol. Biosci.*, 2003, **3**, 715.
21. F. van de Manakker, T. Vermonden, C. F. van Nostrum, and W. E. Hennink, *Biomacromolecules*, 2009, **10**, 3157.
22. C. J. Easton and S. F. Lincoln, *Chem. Soc. Rev.*, 1996, **25**, 163.
23. R. Bhushan and S. Joshi, *Biomed. Chromatogr.*, 1993, **7**, 235.
24. S. Li and W. C. Purdy, *Chem. Rev.*, 1992, **92**, 1457.
25. V. Schurig and H.-P. Nowotny, *Angew. Chem. Int. Ed.*, 1990, **29**, 939.
26. H. Nishi and S. Terabe, *J. Chromatogr.*, 1995, **694**, 245.
27. J. A. Hamilton and L. Chen, *J. Am. Chem. Soc.*, 1988, **110**, 5833.
28. D. W. Armstrong, T. J. Ward, R. D. Armstrong, and T. E. Beesley, *Science*, 1986, **232**, 1132.
29. D. W. Armstrong, X. Yang, S. M. Han, and R. A. Menges, *Anal. Chem.*, 1987, **59**, 2594.
30. K. B. Lipkowitz, S. Raghothama, and J.-A. Yang, *J. Am. Chem. Soc.*; 1992, **114**, 1554.
31. K. B. Lipkowitz, *Chem. Rev.*, 1998, **98**, 1829.
32. M. V. Rekharsky and Y. Inoue, *J. Am. Chem. Soc.*, 2000, **122**, 4418.
33. K. Kano, H. Kamo, S. Negi, *et al.*, *J. Chem. Soc., Perkin Trans. 2.*, 1999, 15.
34. M. V. Rekharsky and Y. Inoue, *J. Am. Chem. Soc.*, 2002, **124**, 813.
35. K. Takahashi, H. Narita, M. Oh-Hashi, *et al.*, *J. Inclusion. Phenom. Macrocyclic Chem.*, 2004, **50**, 121.
36. K. Kano and H. Hasegawa, *J. Am. Chem. Soc.*, 2001, **123**, 10616.
37. D.-T. Pham, P. Clements, C. J. Easton, and S. F. Lincoln, *Tetrahedron Assym.*, 2008, **19**, 167.
38. M. Miyauchi, Y. Takashima, H. Yamaguchi, and A. Harada, *J. Am. Chem. Soc.*, 2005, **127**, 2984.
39. R. Breslow and S. D. Dong, *Chem. Rev.*, 1998, **98**, 1997.
40. R. L. van Etten, J. F. Sebastian, G. A. Clowes, and M. L. Bender, *J. Am. Chem. Soc.*, 1967, **89**, 3242.
41. R. L. van Etten, G. A. Clowes, J. F. Sebastian, and M. L. Bender, *J. Am. Chem. Soc.*, 1967, **89**, 3253.
42. S. Hu, J. Li, J. Xiang, *et al.*, *J. Am. Chem. Soc.*, 2010, **132**, 7216.
43. L. G. Marinescu and M. Bols, *Angew. Chem. Int. Ed.*, 2006, **45**, 4590.
44. L. Marinescu, M. Mølbach, C. Rousseau, and M. Bols, *J. Am. Chem. Soc.*, 2005, **127**, 17578.
45. M. T. Reetz and S. R. Waldvogel, *Angew. Chem. Int. Ed.*, 1997, **36**, 865.
46. J. M. Haider and Z. Pikramenou, *Chem. Soc. Rev.*, 2005, **34**, 120.
47. F. Bellia, D. La Mendola, C. Pedone, *et al.*, *Chem. Soc. Rev.*, 2009, **38**, 2756.
48. S. E. Brown, J. H. Coates, C. J. Easton, and S. F. Lincoln, *J. Chem. Soc., Faraday Trans.*, 1994, **90**, 739.
49. L. Barr, C. J. Easton, K. Lee, *et al.*, *Tetrahedron Lett.*, 2002, **43**, 7797.
50. D. H. Kim and S. S. Lee, *Biorg. Med. Chem.*, 2000, **8**, 647.
51. B. Zhang and R. Breslow, *J. Am. Chem. Soc.*, 1997, **119**, 1676; 1998, **120**, 5854.
52. K. Kano, Y. Itoh, H. Kitagishi, *et al.*, *J. Am. Chem. Soc.*, 2008, **130**, 8006.
53. H. Kitagishi, S. Negi, A. Kiriyama, *et al.*, *Angew. Chem. Int. Ed.*, 2010, **49**, 1312.
54. J. M. Haider, R. M. Williams, L. De Cola, and Z. Pikramenou, *Angew. Chem. Int. Ed.*, 2003, **42**, 1830.
55. A McNally, R. J. Forster, N. R. Russell, and T. E. Keyes, *Dalton Trans.*, 2006, 1729.
56. F. Hapiot, S. Tilloy, and E. Monflier, *Chem. Rev.*, 2006, **106**, 767.
57. Y. Han, K. Cheng, K. A. Simon, *et al.*, *J. Am. Chem. Soc.*, 2006, **128**, 13913.

58. S. A. Nepogodiev and J. F. Stoddart, *Chem. Rev.*, 1998, **98**, 1959.
59. H. Ogino, *J. Am. Chem. Soc.*, 1981, **103**, 1303.
60. H. Ogino and K. Ohata, *Inorg. Chem.*, 1984, **23**, 3312.
61. K. Takaizumi, T. Wakabayashi, and H. Ogino, *J. Sol. Chem.*, 1996, **25**, 947.
62. C. J. Easton, S. F. Lincoln, A. G. Meyer, and H. Onagi, *J. Chem. Soc., Perkins Trans. 1*, 1999, 2501.
63. C. Yang, T. Mori, Y. Origane, *et al.*, *J. Am. Chem. Soc.*, 2008, **130**, 8574.
64. A. Harada, J. Li, and M. Kamachi, *Nature*, 1993, **364**, 516.
65. D. Armspach, P. R. Ashton, R. Ballardini, *et al.*, *Chem. Eur. J.*, 1995, **1**, 33.
66. Y. L Zhao, W. R. Dichtel, A. Trabolsi, *et al.*, *J. Am. Chem. Soc.*, 2008, **130**, 11294.
67. H. Murakami, A. Kawabuchi, R. Matsumoto, *et al.*, *J. Am. Chem. Soc.*, 2005, **127**, 15891.
68. Y. Kawaguchi and A. Harada, *Org. Lett.*, 2000, **2**, 1353.
69. Y. Kawaguchi and A. Harada, *J. Am. Chem. Soc.*, 2000, **122**, 3797.
70. Q-C. Wang, D.-H. Qu, J. Ren, *et al.*, *Angew. Chem. Int. Ed.*, 2004, **43**, 2661.
71. C. A. Stanier, S. J. Alderman, T. D. Claridge, and H. L. Anderson, *Angew. Chem. Int. Ed.*, 2002, **41**, 1769.
72. R. E. Dawson, S. F. Lincoln, and C. J. Easton, *Chem. Commun.*, 2008, 3980.
73. R. J. Coulston, H. Onagi, S. F. Lincoln, and C. J. Easton, *J. Am. Chem. Soc.*, 2006, **128**, 14750.
74. L. Li, X. Guo, L. Fu, *et al.*, *Langmuir*, 2008, **24**, 8290.
75. I. Tomatsu, A. Hashidzume, and A. Harada, *J. Am. Chem. Soc.*, 2006, **128**, 2226.
76. B. J. Ravoo and R. Darcy, *Angew. Chem. Int. Ed.*, 2000, **39**, 4324.
77. R. Auzély-Veltry, F. Djedaïni-Pilard, S. Désert, *et al.*, *Langmuir*, 2000, **16**, 3727.
78. F. Versluis, I. Tomatsu, S. Kehr, *et al.*, *J. Am. Chem. Soc.*, 2009, **131**, 13186.
79. D. P. Ferris, Y.-L. Zhao, N. M. Khashab, *et al.*, *J. Am. Chem. Soc.*, 2009, **131**, 1686.
80. C. A. Nijhuis, J. Huskens, and D. N. Reinhoudt, *J. Am. Chem. Soc.*, 2004, **126**, 12267.
81. M. J. W. Ludden, X. Li, J. Greve, *et al.*, *J. Am. Chem. Soc.*, 2008, **130**, 6964.

Cucurbituril Receptors and Drug Delivery

Anthony Ivan Day and J. Grant Collins

UNSW@ADFA, Canberra, ACT, Australia

1 INTRODUCTION

Cucurbit[*n*]uril (Figure 1) is a relatively new family of macrocyclic cagelike molecules with a broad range of potential applications including drug delivery, the prime focus of the following discussion.[1] Cucurbit[*n*]uril is abbreviated in this chapter as Q[*n*] but is also often abbreviated as CB[*n*] in cited literature. The Q[*n*] possess a cavity and are shaped like open-ended empty barrels (Figure 2). Chemically, these macrocycles behave as molecular hosts and a variety of smaller molecules can be encapsulated within their cavities. Physically, the Q[*n*] are a macrocyclic framework of repeating five and eight membered rings of principally C–N bonds forming a relatively rigid and chemically robust structure. The portals are rimmed by carbonyl oxygens, which give them their electronegative property.

The factors that favor encapsulation are the hydrophobic nature of the cavity, van der Waals contact forces, cation attraction (ion–dipole interaction), and/or hydrogen bonding at the electronegative openings or portals to the cavity. The relative dimensions of guest molecules compared to the portal openings can be a determining factor (see Section 3), but a range of cavity sizes has been synthesized.[2–5]

The host–guest nature of Q[*n*] and its relatively rigid framework are two aspects of this macrocycle that facilitates its exploitation as a drug delivery vehicle.

The Q[*n*] are often compared to the cyclodextrins (CD) in terms of cavity sizes, shape, electropotential surfaces, and hydrophilic or hydrophobic characteristics.[1,6–8] CDs have had a history in drug formulations since 1953 and after a relatively slow beginning have eventually found reasonably broad application.[9–13] In comparison, the first connection for Q[*n*] to drug delivery was only made a decade ago in patents as a potential application.[14,15] The host–guest-binding characteristics of both CD and Q[*n*], which appear to be similar are in reality significantly different, especially in binding strength, driving forces to binding, compatibility to cations, and hydrophobic differences.[7,8,16] The Q[*n*] generally have higher binding constants, a much higher binding ratio, a high affinity for cations, and hydrophobic groups, and the cavity of the Q[*n*] is accessible from two equal sized portals.[1] In addition, the Q[*n*] are structurally much more robust molecules.

Since the discovery of the simplest Q[*n*] family (Figure 1a), some efforts have also been directed toward expanding their utility through the introduction of substitution (Figure 1b). Substitution of Q[*n*] is possible at the two carbons of the junction of the cis-fused imidazolone rings of the glycoluril moieties. There are two main approaches to this, either by direct reaction on these carbons for a specific Q[*n*] or by the introduction of a substituted glycoluril during the synthetic process. Direct reaction has been achieved through oxidation of the methine carbon to give perhydroxylated Q[*n*] ($R^1 = R^2 = OH$), which is most successful

Supramolecular Chemistry: From Molecules to Nanomaterials.
Edited by Philip A. Gale and Jonathan W. Steed.
© 2012 John Wiley & Sons, Ltd. ISBN: 978-0-470-74640-0.

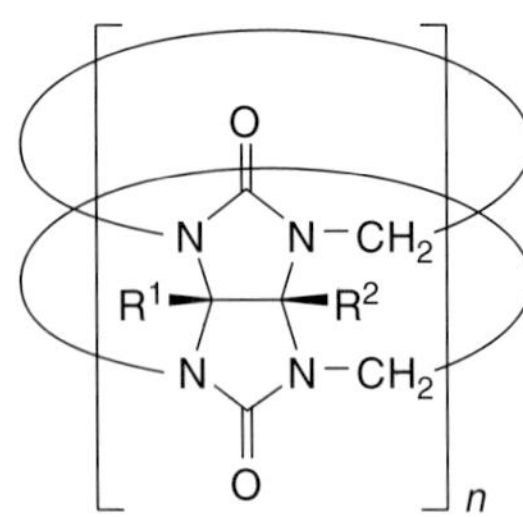

Figure 1 Structural representation of cucurbit[*n*]uril, Q[*n*] where $n = 5$–10, (a) $R^1 = R^2 = H$, (b) $R^1 = R^2 =$ substituent, or R^1 or R^2 are different substituent groups.

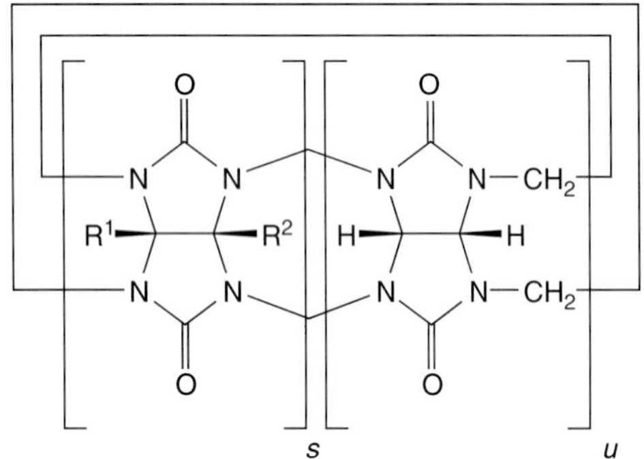

Figure 3 Structural representation of partially substituted Q[*n*], $s =$ substituted glycoluril moiety, $u =$ regular glycoluril moiety, and $n = s + u$. The number of *s* units in a substituted Q[*n*] is variable and their position relative to each other is variable.

where $n = 5$–6.[17,18] These are further elaborated through a variety of alkyl derivatives of the OH groups.[17] The synthetic approach of introducing a substituted glycoluril has to date, been most successfully developed for partially substituted Q[*n*] where only some of the introduced glycoluril moieties carry substitution (Figure 3).[18–25] This type of substitution has been possible for Q[5] and Q[6] and is likely to be extended to the higher homologs. Fully substituted Q[*n*] (Figure 1b) are limited to the smaller homolog Q[5] or very low yields of fully substituted Q[6].[21,27,28] Alternative methods to the introduction of substitution include cucurbituril analogs, and an example of a benzofuran substituent introduced as a linking group between two adjacent glycoluril N groups has been reported.[29,30]

2 CUCURBITURIL AS A DRUG DELIVERY VEHICLE

2.1 Toxicology and pharmacokinetics

As a drug delivery vehicle, the Q[*n*] need to be relatively harmless biologically or at least have low toxicity. Preliminary toxicology and pharmacokinetics studies have shown the Q[*n*] to fit the category of relatively low toxicity or nontoxic depending on the method of entry to the body. Toxicology has been evaluated for acute effects in mice orally and intravenously and in cell cultures of both animal and human cell lines.[31–33]

A single dose feed of Q[7] and Q[8] as a mixture in equal proportions has demonstrated no toxicity up to $600\,mg\,kg^{-1}$. The lack of toxicity is consistent with a separate study where only 3.6% of the total Q[*n*] was adsorb into the blood stream from the alimentary canal as measured using ^{14}C labeled Q[7] or Q[8].[31]

The evaluation of intravenous administration was limited to Q[7] as this Q[*n*] has sufficient aqueous solubility to reach a practical limit. Acute toxicity was found to be $>250\,mg\,kg^{-1}$, which is relatively low. A maximum tolerated dose was used as a measure of toxicity, where an animal's weight loss was observed to not fall below 10% following dosing. All animals begin to gain weight within five to eight days after dosing, indicating no extended effects. Intravenous administration studies with ^{14}C labeled Q[7] have shown that clearance into the urine has a mean half-life of clearance at 12.8 h. Q[7] does not cross the blood brain barrier, and accumulation in the liver and spleen is very low relative to the kidneys. Q[7]'s chemical and thermal stability and its relatively quick clearance into the urine following a higher activity in the kidneys suggest that Q[7] is excreted without modification.

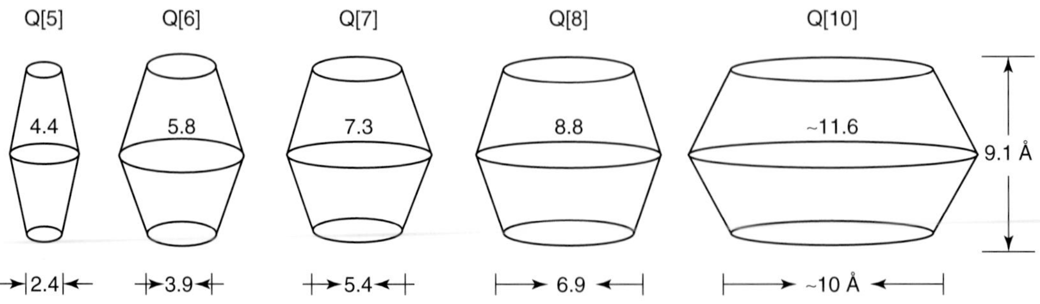

Figure 2 Barrel representation Q[5–10] with the dimensions of the portal, the cavity, and the depth as indicated.[1]

The relatively low toxicity of Q[7] in the intravenous trials has also been supported by *in vitro* cell culture studies. Q[7] and Q[8] have been shown to enter muscle mouse embryo cells,[34] and Q[7] is internalized by murine macrophage (RAW264.7).[33] Cells of human kidney (HEK293), human hepatocyte (HepG2), and murine macrophage (RAW264.7) were found to tolerate Q[5] and Q[7] up to 1 mM without significant effects,[33] while Q[7] resulted in an IC_{50} at 0.53 mM over a 48 h period (1 mM could be tolerated for 3 h without effect) for Chinese hamster ovary (CHO-K1) cells.[32] Metabolic function tests of CHO-K1 over 48 h established no cytotoxic activity up to 0.5 mM (~600 mg of Q[7] kg^{-1} of cells). The low aqueous solubility of Q[8] 20 μM precluded the possibility of a toxic limit being established.[32]

A number of Q[*n*] derivatives have been synthesized in order to modify Q[*n*] solubility, manipulate molecular guest-binding properties, and provide functionality for further structural developments.[5,6] Most of these derivatives have not been evaluated for biological compatibility, although some have been suggested as drug delivery vehicles.[6] There is an immerging potential for structural variants of Q[*n*] from small changes of substituents and functional groups to polymeric forms including nanocapsules.[19–37] As each of these find a place in drug delivery, their individual toxicology and pharmacokinetics will require evaluation.

2.2 Attributes of Q[*n*] as a delivery vehicle

The Q[*n*] in its simplest form or with substituents and functional groups have potential as an aid to drug delivery by providing one or more of the following benefits:

- increasing bioavailability,
- providing bioprotection,
- improve chemical stability,
- provide a method for slow release, and
- facilitate drug targeting.

The unique barrel-shaped cavity of Q[*n*] with its two openings that are slightly smaller in diameter than the internal cavity provides a site for drugs or parts of drugs to be bound or encapsulated without chemical modification. Supramolecular forces act on the drugs to hold them in place; the physical nature of the Q[*n*] molecular cage screens hydrophobic parts of drugs and protects sensitive functionality. The Q[*n*] are predicted to become another valuable tool in the box for the facilitation of drug efficiency and efficacy. A number of drugs have been evaluated for Q[*n*] encapsulation with a variety of structural variations (Figure 4). Significant binding features and biological findings are discussed throughout the following sections.

2.2.1 *Bioavailability*

One of the main factors governing bioavailability of a drug is its aqueous solubility and the ultimate requirement for the drug to be above required minimum plasma concentrations to be effective, whether the drug is administered orally or intravenously. The increasing challenges for drug applications today are the prevalence of drugs being developed with poor aqueous solubility.[38–40]

The Q[6–8] have been shown to facilitate the solubility of poorly soluble cytotoxic drugs such as camptothecin (CPT) and the benzimidazoles—albendazole (ABZ) and MEABZ.[41–44] The increase in solubility for ABZ and MEABZ was up to 2000- and 3000-fold, respectively.[41,44] It has also been found that the solubility of ABZ can be increased to an even greater extent (2400-fold) with the partially substituted Q[*n*], α,δ-$(CH_3)_4$Q[6] (Figure 3, $R^1 = R^2 = CH_3$, the two *s* units are separated by two *u* units).[44] This indicates that Q[*n*] assisted drug solubility has potential yet to be explored as new substituted Q[*n*] are developed. It should be noted that to increase the solubility of a drug using Q[*n*], it is not always necessary for the whole or a major part of the drug to be cavity encapsulation. The ABZ@Q[6] (1900-fold increase) and ABZ@α,δ-(CH_3)Q[6] only show encapsulation of the methyl carbamate group by ^{1}H NMR, but the close portal association of the remaining hydrophobic portion of the molecule is apparently sufficient to allow a suitable polarization of the overall complex or perhaps disrupting of π–π stacking and hence increased solubility. In general, solubility is achieved by encapsulating the hydrophobic parts of the drug molecule and creating a polar complex of drug@Q[*n*]. In some cases, the two individual components have considerably lower aqueous solubility when compared to the final complex.[44] The low aqueous solubility of Q[6][45] would appear to be an impediment to an intravenous administration especially once the drug has been released from Q[6], but Q[6] is highly soluble in physiological saline due to its affinity for Na^+ ions. Some of the known substituted Q[6] have significant natural water solubility.[46] However, Q[8] has very low aqueous solubility, which is not improved significantly by the presence of metal ions such as Na^+.[46] While Q[8] may not be suitable for intravenous use, there should be no impediment to oral applications. Examples of relatively high solubility of drug@Q[8] complexes are known; yet, the Q[8] alone has low solubility. Future derivatives of Q[8] are bound to change its solubility status.

2.2.2 *Bioprotection*

Bioprotection for drugs primarily refers to maintaining the maximum effectiveness of a drug by protecting it from unintended chemical reactions within the body. The

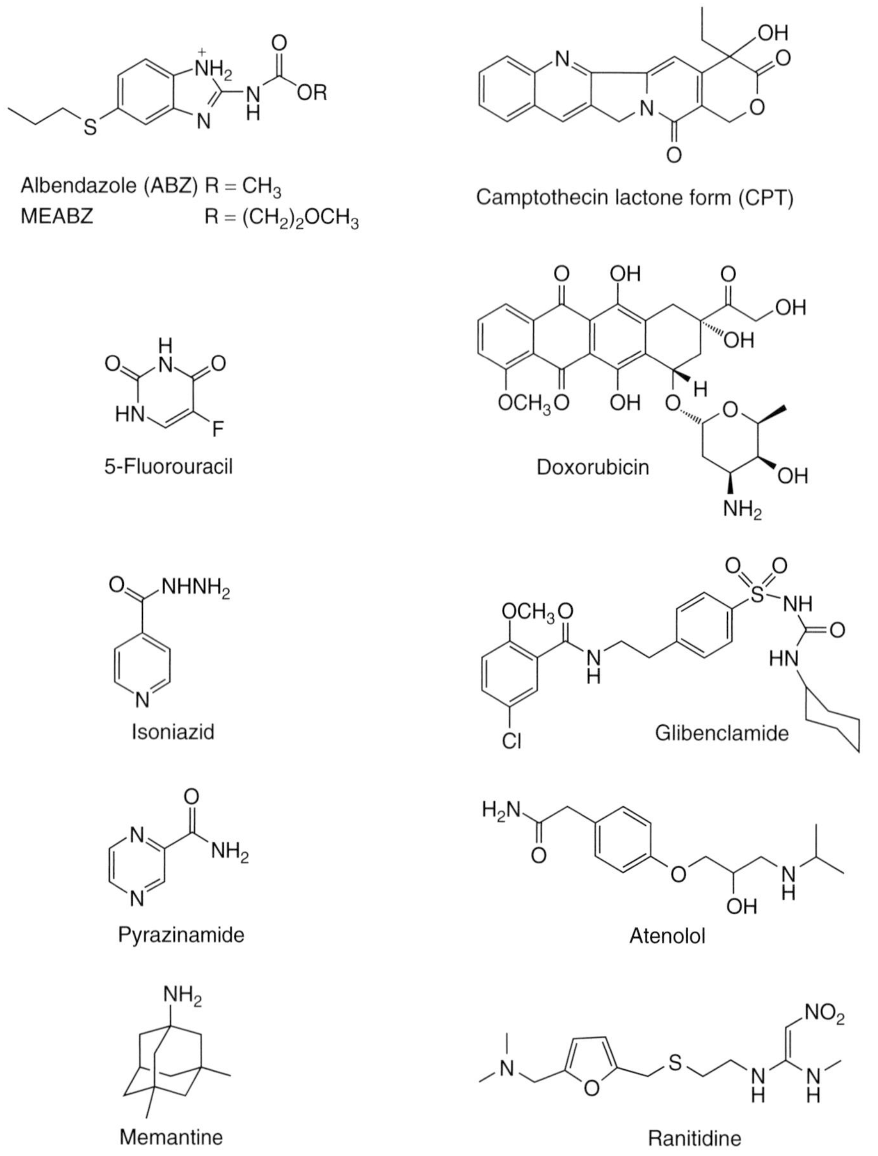

Figure 4 Drugs discussed in the text.

objective is to maximize the opportunity for the drug to complete its intended task with a sustained and desired concentration for the time required. In contrast, there is a natural detoxification by the body of drugs through direct excretion or chemical modification to aid excretion. As a consequence, a balance must be found between the pharmacokinetics and the pharmacological objective.

Drugs encapsulated in Q[n] have shown some promise in this regard especially in the area of cytotoxic dinuclear platinum drugs such as CT008, CT033, and CT233 derived from alkyl polyamines (Figure 4). Platinum-based cyctotoxic drugs are highly reactive to thiol-containing plasma proteins and are degraded to nonactive metabolites.[47,48] As a result, most of the platinum drugs are deactivated on administration. Studies have demonstrated protection of the platinum metal center of the three dinuclear platinum drugs CT008, CT033, and CT233 (Table 1).

Table 1 Half-life ($t_{1/2}$), in minutes, of the reaction of CT008, CT033, and CT233 with cysteine at 37 °C.

	$t_{1/2}$ CT033 (min)	$t_{1/2}$ CT008 (min)	$t_{1/2}$ CT233 (min)
Free	5	110	5
Encapsulated in Q[7]	15	1000	40
Encapsulated in Q[8]	10	1700	45

The $t_{1/2}$ is defined as the time taken for the free intact platinum complex to reduce in concentration to 50%.

Encapsulation in both Q[7] and Q[8] provides substantial protection. It should be noted that protection was achieved

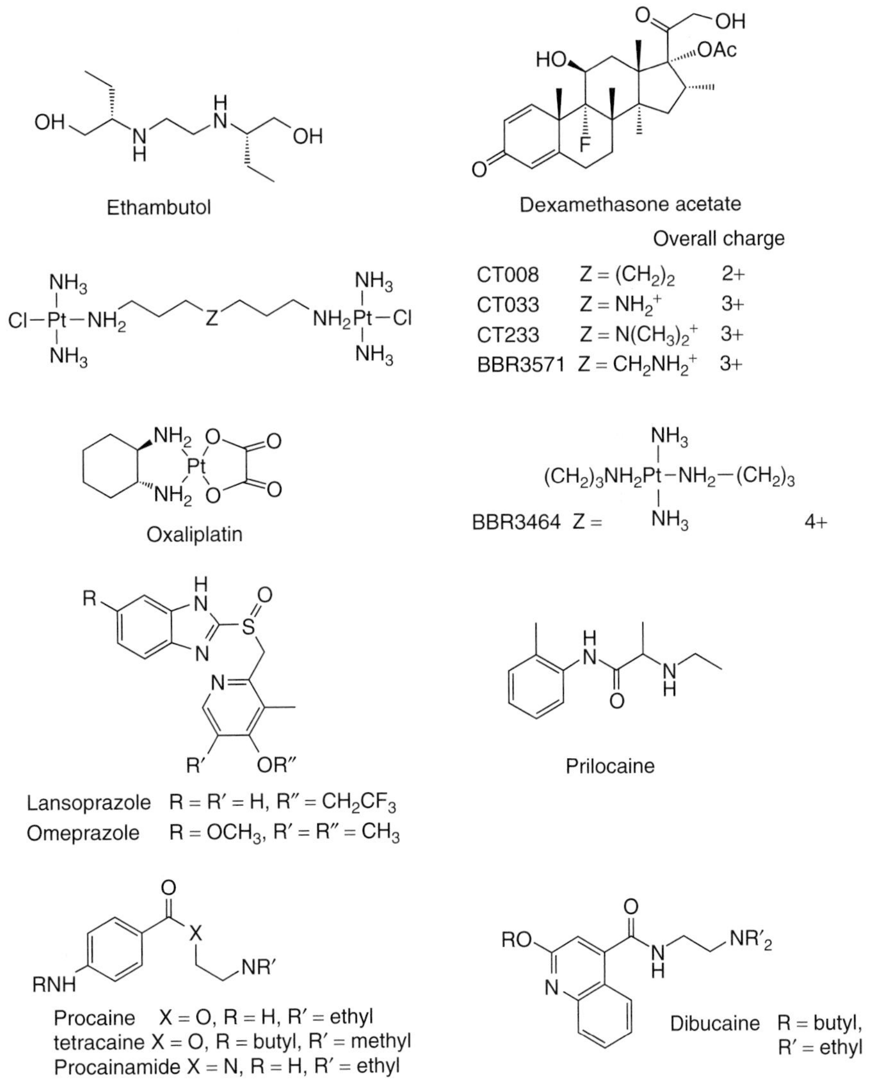

Ethambutol

Dexamethasone acetate

		Overall charge
CT008	Z = $(CH_2)_2$	2+
CT033	Z = NH_2^+	3+
CT233	Z = $N(CH_3)_2^+$	3+
BBR3571	Z = $CH_2NH_2^+$	3+
BBR3464	Z = $(CH_2)_3NH_2Pt(NH_3)_2-NH_2-(CH_2)_3$	4+

Oxaliplatin

Lansoprazole R = R′ = H, R″ = CH_2CF_3
Omeprazole R = OCH_3, R′ = R″ = CH_3

Prilocaine

Procaine X = O, R = H, R′ = ethyl
tetracaine X = O, R = butyl, R′ = methyl
Procainamide X = N, R = H, R′ = ethyl

Dibucaine R = butyl, R′ = ethyl

Figure 4 *Continued*.

even though the platinum metal centers are not completely inside the cavity of the Q[*n*] but rather reside just inside the portal (Figure 5). The degree of protection was found to be dependent on the ability of the metal center to enter the portal; hence, with a larger cavity as in Q[8], the linking ligand is able to fold drawing in the metal closer. This was clearly demonstrated with CT008 and CT233. The modest protection of CT033 was due to the limited ability of Q[*n*] to protect both platinum centers at the same time as a result of strong competing ion–dipole affinities between either Pt monocationic head or the protonated ligand amine and the portals. These competing affinities result in exposing at least one platinum center (Figure 6). Where maximum protection can be achieved, the degradation is significantly decreased. Compared to the free drug CT008 degrades 15 times slower, and the more reactive CT233 degrades nine times slower than its free state (Table 1).

In relation to the protective properties of Q[*n*], the inhibition of enzymatic hydrolysis of peptide-based drugs has been shown in isolated peptides. Therefore, a potential preservative effect is indicated for peptide substrates.[49]

2.2.3 *Improved chemical stability*

Improved chemical stability refers to a reduction or prevention of chemical change of a pharmaceutical guest following encapsulation. This could include, for example, aerial oxidation, hydrolysis of aqueous preparations, thermal degradation, and sensitivity to light. The stability of association complexes as a function of the ease or difficulty at which they dissociate is discussed in Section 5 "the mechanism of release." The importance of chemical stability is reflected in

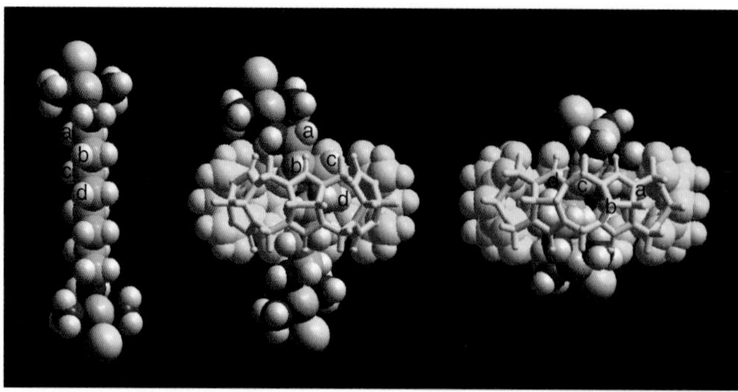

Figure 5 CT008 encapsulated in Q[7] and Q[8].

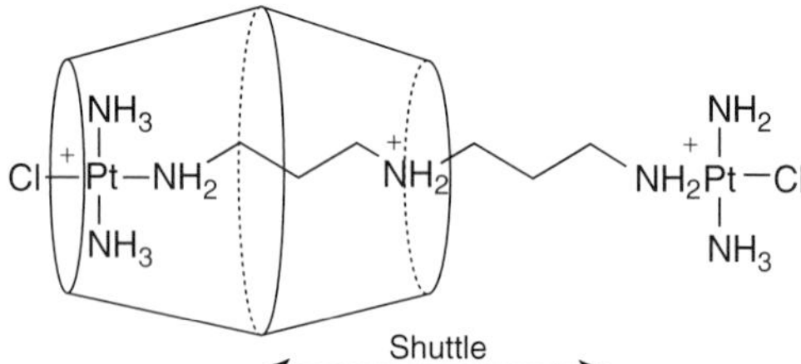

Figure 6 Q[*n*] shuttling over the CT033 ligand showing the competing affinities between Pt monocationic head and the protonated amine.

concerns for extended shelf lives of drugs, the ease of handling during the physical and perhaps chemical stresses of formulation such as compatibility with excipients and solvents including moisture. Amorphous materials and crystal polymorphs can also provide challenges to manufacture and medicinal availability.

Two forms of demonstrated improved stability through Q[*n*] drug encapsulation have been reported. These include thermal stability evaluation and a resistance to hydrolysis. Drugs such as atenolol, glibenclamide, memantine, paracetamol, pyrazinamide, and isoniazid (Figure 4) encapsulated in Q[7] as solid association complexes have all been shown to have higher thermal stability.[50,51] This has been consistently demonstrated using differential scanning calorimetry. However, a thorough evaluation of thermal stability using high-performance liquid chromatography (HPLC) analysis to determine minor degradation products has yet to be performed.

A dramatic improvement in stability to hydrolysis has been shown in the case of ranitidine (Figure 4). Formulations of ranitidine hydrochloride are known to have instability toward humidity and pH. Under test conditions of 50 °C at pH 1.5, the stability of ranitidine is extended to beyond two weeks in the presence of a slight excess of Q[7] with no decomposition compared to four days without Q[7] (50% decomposition).[52]

The stabilization of drug polymorphic states is an area of stability that has been proposed as being applicable to Q[*n*] encapsulation.[50] The separation of individual drug molecules by encapsulation could prevent conversion to alternative polymorphs in the solid state or aggregation in solution. The aggregation of laser dyes in solution can be avoided by Q[7] encapsulation, which suggests that this could also be applicable to hydrophobic drugs.[53]

2.2.4 Slow release or controlled release

Slow release is aimed at sustained plasma concentrations over time. The alternative of frequent administrations leads to concentration spikes or requires constant concentration maintenance, through continuous administration. The development of slow release and controlled release systems incorporating Q[*n*] has so far been focused primarily on the development of methods to active ingredient release and not necessarily specific to administrative application.

Biocompatible gels are recognized as one example where the potential exists for the controlled delivery of drugs.[54] 5-Fluorouracil has been shown to release slowly from Q[6]-mediated alginate hydrogel beads.[55] The alginate beads that formed in the absence of 5-fluorouracil are spherical with a diameter of 2.5 mm. Bead preparation with 5-fluorouracil (Figure 4) loading gave beads ranging from 3 to 4 mm in diameter with loading capacities of 3.87–6.13 wt%, respectively. It was found *in vitro* that slow release at pH 6.8, occurred with an optimal loading of 5.94 wt% (release of $t_{1/2} = 2.7$ h). The 5-fluorouracil is held in the network structure of the hydrogel and not encapsulated in the cavity of Q[6].

Encapsulation of drugs in the cavities of Q[*n*] is a more direct approach to slow release. However, the rate of release is dependent on a number of factors as discussed in mechanisms to release. Q[7] and Q[8] cavity bound drugs in solution are usually released in seconds, as exchange of a drug between the cavities or into a cell is facile. High binding constants are not necessarily a reflection of the ease of exchange as demonstrated by the saturation transfer experiment between Q[8] and Q[7] for the drug ABZ, which indicated a transfer of one molecule

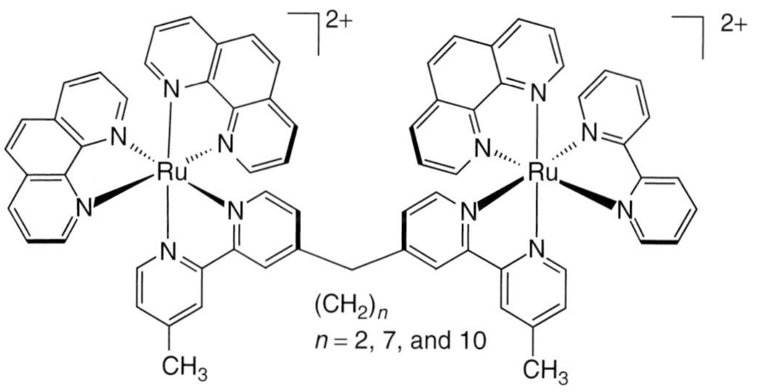

Figure 7 Dinuclear ruthenium drugs ($Rubb_n$).

per 3 s.[41] This shows that in the presence of a suitable receptor or through consumption, the drug would readily be released. However, slow release is possible through encapsulation when there are mechanical restrictions to the exit of a guest from a cavity. This is the case for the new drugs dinuclear ruthenium ($Rubb_n$) complexes (Figure 7). The $Rubb_n$ complexes are cytotoxic and have excellent antibiotic activity against bacteria.[56] As a preliminary study of encapsulation, the $Rubb_n$ complexes are found to be encapsulated slowly and are released slowly over several hours from the cavity of Q[10].[57]

There are three reported examples of controlled release using Q[6] as central components in nanostructures. Two of these require internal cellular triggers such as pH changes[58] or redox reactions[59] to release their drug "cargo," while the third responds to an external stimuli of an oscillating magnetic field that induces a local heating trigger.[60] The mechanism of controlled release for each of these nanostructures is discussed in Sections 2.2.5 and/or 6.

2.2.5 *Drug targeting*

Drug targeting is a desirable goal where ideally a drug can be delivered specifically to a set of diseased cells or invading cells (microorganisms) and the drug "off loaded" to only these cells. In the supramolecular context a molecular structure loaded with a drug acts as a vehicle that is carried in the blood stream and interstitial fluids, and attached to the vehicle is a cell specific (ideally) or selective, molecular tag. Certain cell types can be selectively targeted, such as cancer cells that over express a particular receptor or cell types with specific lectins. Drug targeting has the potential to limit unnecessary damage and reduce side effects.

A model of targeted drug delivery has been demonstrated using Q[6]-based carbohydrate clusters, which were synthesized from $(allyloxy)_{12}$Q[6] (Figure 1, where $R^1 = R^2 =$ allyloxy) to attach sugar units of choice to the "equator" of Q[6] (Scheme 1).[61] It was found by *in vitro* experiments with HepG2 hepatocellullar carcinoma cell with over expressed galactose receptors as a potential target that a Q[6]-based galactose cluster was translocated intracellularly. The experiment was extended to a drug-carrying model, where an encapsulated spermine (drug model) was also covalently bonded to a fluorescent dye molecule. The translocation of the dye indicated the validity of the model. It is suggested that the Q[6]-based galactose cluster carrying the drug model was a galactose receptor-mediated endocytosis.[61] Extending this type of targeting, where the cavity can be utilized for drug delivery, could find greater application for the higher homologs given the larger cavities.

An alternative targeted drug delivery system, also reported, incorporates derivatized Q[6] into a polymer nanocapsule or vesicle (Scheme 2).[59] Again contrary to the concept of encapsulating a drug in the cavity of a Q[n], the drug was encapsulated in the polymer nanocapsule's core and the cavities of the Q[6]s are used as a multifunctional platform for imaging probes and targeting groups. The structure of the nanocapsules, which are vesicle-like can be prepared at a uniform average diameter of 190 nm with a membrane depth of ~6 nm. It is suggested that the structure is a layering of 5-6 Q[6] with interdigitation of the disulfide "tails" forming a hollow core. These nanocapsules are robust structures that are not easily ruptured unlike conventional vesicles. However, the disulfide "tails" are sensitive to reduction, which provides a potential mechanism for release of the core contents into the cytoplasm of the cell (see Section 5). Targeted delivery of doxorubicin was demonstrated *in vitro* using a disulfide-tailed Q[6] nanocapsule with a surface elaboration of a folate–spermidine conjugate as the targeting group. The targeting group is attached by encapsulation of a spermidine moiety in the Q[6] cavity and the doxorubicin entrapped in the vesicle-like structure. Selective intracellular uptake was observed in the cancer cell line HeLa.[59] In addition, doxorubicin unloading was observed over a period of 2 h incubation at 37 °C (see Section 5).

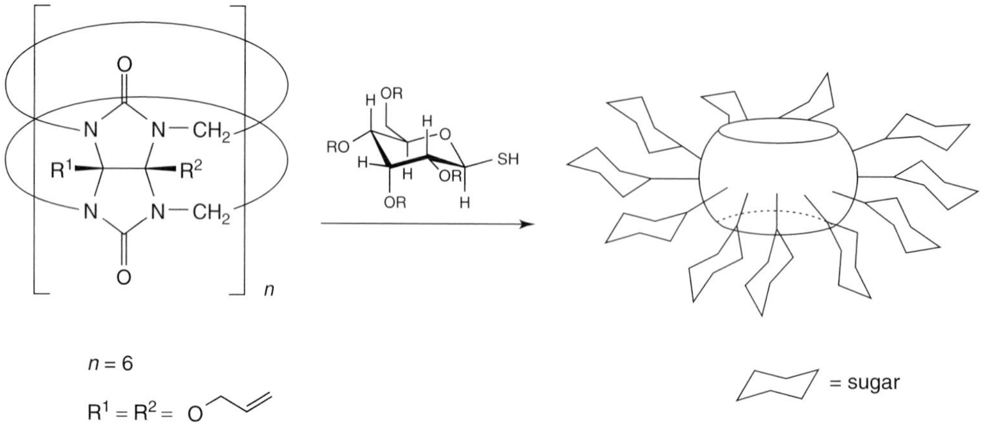

Scheme 1 An example of a Q[6]-based carbohydrate cluster and the general reaction scheme to its synthesis. The average number of sugar groups attached equals 11.2.

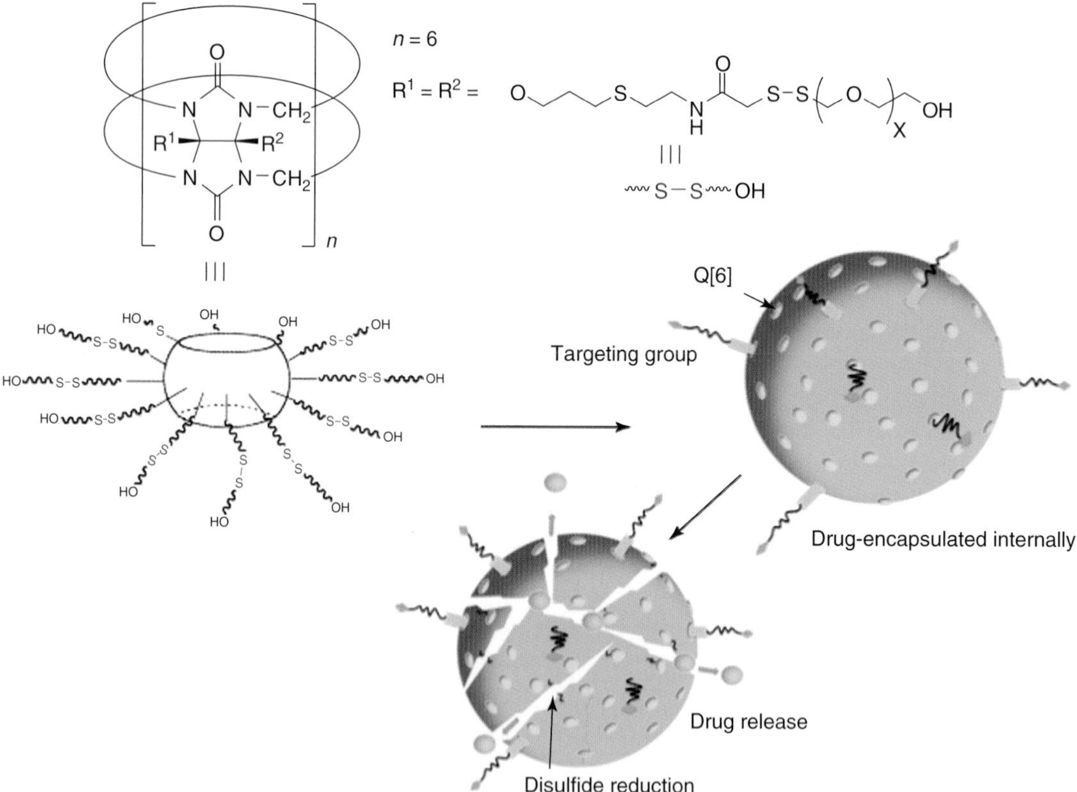

Scheme 2 The general scheme to the amphiphilic Q[6], the formation of nanocapsules and reductive rupture of the disulfide bonds, and collapse of the nanostructure to release the drug "cargo."

This was supported by doxorubicin accumulation in the cell nuclei and by a comparative cell viability study, where the folate-targeted nanocapsule performed ~1.9-fold better than equivalent concentrations (10 µM) of free doxorubicin. However, free doxorubicin—doxorubicin loaded disulfide-tailed Q[6] nanocapsules without targeting groups—and doxorubicin load regular Q[6] nanocapsule with targeting groups all have comparable activity. Perhaps, this indicates that disulfide-tailed Q[6] nanocapsules are translocated by endocytosis and easily spill their contents through cytoplasm reduction but are not as readily translocated as a targeted nanocapsule. The doxorubicin load regular Q[6] nanocapsule with targeting groups would appear to be translocated readily, and doxorubicin leakage is responsible for cytotoxicity. Therefore, the combination of targeting and reductive rupture provides the maximum cytotoxicity. The nanocapsules at 60 µM have been shown to be relatively nontoxic to the HeLa cells when incubated over a period of two days. Since the nanocapsules are robust self-assembled structures, their toxicity needs to be considered initially as a single entity, similarly to a polymer but as break down occurs further evaluation is required.

Robust nanocapsules of the type discussed present a promising potential future for highly active drugs as the loading capacity of a system of this type is relatively low and in the case discussed was found to be 1.5 wt%. Q[*n*] encapsulated drugs with targeting groups are likely to have a higher loading capacity but the controlled release of the drug requires further development.

2.2.6 *Encapsulation in Q[n] and drug activity*

It is important that the pharmacological activity of a drug not to be adversely affected when administered by molecular delivery vehicles such as Q[*n*] or any supramolecular structure utilizing Q[*n*]. Much of the pharmacological activity measured to date, where Q[*n*] are involved have been in the form of *in vitro* biological tests. In a number of examples, the importance of the delivery system is in increasing the aqueous solubility of a hydrophobic drug, and while most show no significant change in activity, there are some that have slightly lower activity. As the aqueous solubilities of the hydrophobic drugs are increased and plasma concentrations would also be higher, *in vivo* these drugs may prove equally active in spite of small differences found *in vitro*.

In vitro examination of ABZ@Q[7], on the cancer cells, human *T*-cell acute lymphoblastic leukemia cells (CEM), ovarian cancer cells (1A9), and human colorectal cancer cells (HT-29) showed no significant change in the cytotoxic activity of ABZ or MEABZ as a consequence of encapsulation.[41] This was also true for MEABZ@Q[7] and MEABZ@Q[8] for the cell lines HT-29 and PC-3 (human prostate cancer cells).[44] The encapsulated cytotoxic drug CPT also showed comparable activity to the free drug.[42] Although there are exceptions, as the activities of CPT encapsulated in Q[7] or Q[8] highlights, there are slight variations in activity for different cell lines, for example, human nonsmall lung cells (A549) are slightly better where as a moderate decrease in activity was found for the human leukemia cell line (K562) (Table 2).[42]

The antituberculosis drug ethambutol when encapsulated in Q[7] also showed no significant decrease in pharmacological activity.[33]

A series of platinum base cyctotoxic drugs encapsulated in Q[*n*] have also been examined for *in vitro* biological activity. The dinuclear polyamine platinum drugs, diPt and BBR3571, encapsulated in Q[7] showed no significant difference in cytotoxic activity between the free drug and the encapsulation complex for L1210 and L1210/DDP cell lines (murine leukemia cells and cisplatin-resistant

Table 2 Activity comparison (IC_{50} at μM) for camptothecin (CPT) free, encapsulated in Q[7] and Q[8] for the cell lines—human nonsmall lung cells (A549), human leukemia cells (K562), and murine macrophage cells (P388D1).

Cell Line	CPT Free	CPT@Q[7]	CPT@Q[8]
A549	7.76	6.36	6.78
K562	0.43	0.93	1.13
P388D1	2.47	2.38	2.98

cells, respectively).[62,63] In contrast, the oxaliplatin@Q[7] complex revealed a significant decrease in cytotoxic activity of varying degrees across five cell lines (A549 human nonsmall cell lung, SKOV-3 human ovarian, SKMEL-2 human melanoma, XF-498 human CNS, and HCT-15 human colon).[64] Similarly a series of platinum phenanthroline intercalating cytotoxic drugs with ancillary ligands of ethylene diamine or diaminocyclohexane as (1R,2R) and (1S,2S) led to mixed results for *in vitro* biological tests. Across the range of cavity sizes Q[6] to Q[8], the results were variable in activity, both positively and negatively. Perhaps, the surprising result was that the $[Pt(5\text{-Cl-phen})(S,S\text{-dach})]^{2+}$ @Q[6] complex had a 2.6-fold increased activity, $[Pt(5\text{-Cl-phen})(S,S\text{-dach})]^{2+}$ @Q[7] >400-fold decrease; and only a 4.4-fold decrease for the larger Q[*n*], $[Pt(5\text{-Cl-phen})(S,S\text{-dach})]^{2+}$ @Q[8].[65]

Where there have been decreases or even increases in activity *in vitro*, the exact reason for this change has not been established. It is not known whether this is simply a reflection of high binding affinities and therefore a decrease in available free drug and hence a decrease in activity or some other mechanism related to cellular uptake. The latter may also be related to whether the drug@Q[*n*] complex is entering the cell or the drug alone (see Section 2.1).

Clearly, a factor that can affect *in vitro* biological activity is the ease of release of the drug, as demonstrated by encapsulation of BBR3464 (Figure 4). This highly active cytotoxic drug is encapsulated in a ratio of 1 : 2 (BBR3464: Q[*n*]) for $n = 7$ and 8 and a ratio of 1 : 1 for Q[10]. The consequence is that the activity is moderated by the ease of release from the cavity. Almost all activities are lost through encapsulation with Q[7] with activity returning as the cavity size increases and/or the ratio decreases (Table 3).[62]

High binding affinity could also help to explain why oxaliplatin@Q[7] has decreased activity and the $[Pt(5\text{-Cl-phen})(S,S\text{-dach})]^{2+}$ @Q[*n*] complexes have variable results according to the Q[*n*] used. However, without a more detailed study to establish the mechanism of cellular membrane transport (drug release prior to cell uptake or drug @Q[*n*] and released after uptake), a proper analysis of variations in activity is not possible. Given the embryo cells and macrophage drug@Q[*n*] may occur at least in some cases.

Table 3 *In vitro* evaluation of BBR3464 with the murine leukemia cell line (L1210) and the cisplatin-resistant cell line (L1210/DDP) measured as IC_{50} the required concentration to induce 50% inhibition of growth.

Drug	Q[*n*]	IC_{50} (μM)	
		L1210	L1210/DDP
BBR3464	Nil	57 nM	24.5 nM
BBR3464	10	0.7	0.2
BBR3464	8	6.6	1.4
BBR3464	7	>37.5	>37.5

Molybdocene dichloride (Cp_2MoCl_2) is another cytotoxic metal complex that has been evaluated for its *in vitro* biological activity following Q[7] encapsulation. Cp_2MoCl_2@Q[7] was found to have improved activity when compared to free Cp_2MoCl_2 with the 2008 cell line and the MCF-7 cell line. It is not known whether the improvement is related to solubility or membrane permeability.[66]

The only *in vivo* drug@Q[*n*] study performed to date is that involving the dinuclear platinum drug BBR3571 (Figure 4), which is similar in structure to CT033 except that it has the unsymmetrical spermidine linking ligand. The complex BBR3571@Q[7] such as CT033@Q[7] has less than ideal protection when encapsulated in Q[*n*] (see Section 2.2.2). At physiological pH, the amine in the spermidine linking ligand is protonated and this drives one of the platinum centers away from the portal. In spite of the limited protection provided by Q[7], *in vivo* studies in mice determined that the maximum tolerated dose was 1.7 times less toxic (BBR3571@Q[7] compared to equivalent levels of free drug).[62] Whether this is a consequence of a reduction in toxic metabolites or not, this was not determined.

An examination of the cytotoxic activity *in vivo* on subcutaneous tumors in mice (2008 ovarian carcinoma cell line) at equivalent drug doses (BBR3571), comparing the free drug to the drug@Q[7] showed identical activity against the tumor.[62]

An alternative to drug activity maintenance is that of drug reversal. Q[7] has been proposed as drug moderator or reversing agent by the competitive binding of drugs that are administered for enzyme modulation. This relies on a balanced competitive binding of the drug relative to the enzyme.[16,67]

2.2.7 Conclusion

Considerable potential has been realized for Q[*n*] as a drug delivery vehicle through either cavity and/or portal encapsulation. There are a number of benefits that can be achieved that include bioavailability, bioprotection, improved chemical stability, slow release or controlled release, and drug targeting. In addition, the Q[*n*] are biocompatible and with careful evaluation can be used to improve or at least maintain the drugs pharmacological activity. A range of drugs have been evaluated for biological activity following encapsulation, and there is now sufficient evidence to suggest that drug activity can be maintained with the correct choice of Q[*n*] but each combination would require evaluation. The type and the structure of the drugs suitable for encapsulation are also becoming more predictable, although this process is yet to be finalized. The structural features required for encapsulation are discussed in Section 3.

2.3 Drug@Q[*n*] and administration

There is limited information to date with regard to which routes are applicable to drug@Q[*n*] administration or more specifically which Q[*n*] best serves each route. The solubility of Q[6] in saline solutions, reasonable solubility of Q[7] in water, and high solubility in saline suggest that these two Q[*n*] are applicable to intravenous administration especially as Q[7] is known to be readily excreted in urine.[31] The increased aqueous solubility of substituted Q[*n*] suggests that substitution would render all relevant Q[*n*] homologs suitable in the future to intravenous administration. The high thermal and chemical stability of all the Q[*n*] means that they would also be suitable for oral delivery of drugs as they would easily tolerate the pH changes of the alimentary canal.[68] The low solubility of Q[8] in nonacid conditions may favor its application in oral delivery. Formulation studies of Q[6] and excipients indicate that Q[*n*] can be readily compressed into stable and durable tablets for oral administration.[69]

Topical use for a water-dispersible ultraviolet (UV) shield (sunscreen agent) through supramolecular formulation with Q[*n*] by ball mill grinding has been patented. This is described as environment friendly with high UV absorbance, nontoxic, and a product removable with water.[70] Other medicinal applications or routes of entry to the body are yet to be reported.

3 DRUG TYPES THAT CAN BENEFIT FROM CUCURBITURIL ENCAPSULATION

3.1 Q[*n*] and drug matching

The whole of a drug or part of a drug can be bound to Q[*n*] through encapsulation within the cavity, or at either carbonyl portal, or can involve both sites of the Q[*n*]. There are a number of basic considerations required to determine the suitability of Q[*n*] as a vehicle for delivery.

These include:

- the molecular shape and size,
- the polarity or charge,
- the presence of hydrogen-bonding donors, and
- the presence of hydrophobic moieties.

Maximum association stability is achieved if a number of the criteria are satisfied.

3.1.1 *Size and shape*

Obviously, the size and shape of a drug molecule preclude it from being encapsulated in the cavity or to partially enter the portal if the molecule's physical dimensions are incompatible with a specific Q[*n*]. In some cases, only parts of a molecule need to be encapsulated to affect a desired outcome. The molecular size choices fall into a range for a molecular width or diameter of ~2.4–10 Å (Figure 2).[1] The proposed depth of a Q[*n*] is ~9.1 Å (including van der Waals radii), although theoretical calculations indicate that the extent of the electron influence is actually ~14 Å.[1,6] The length of the molecule is then only limited by how much of it is to be encapsulated. The equal access to two portal openings allows a molecule to thread to an undefined length. The width or diameter is dictated by the size of the portal opening, which can flex slightly to facilitate access to the larger diameter cavity. The Q[*n*] as relatively rigid structures can flex slightly into an ellipsoid shape, or the carbonyl portals can flex outward to accept a slightly larger width or diameter.[71]

The shape of a molecule is not only an important determinant to fitting a Q[*n*] cavity but also an added feature to increased binding strength. Maximum occupation of the space within the cavity increases the van der Waals contacts and therefore favors binding. Adamantyl ammonium ion or carborane are molecules that are roughly spherical in shape, and these molecules fit the cavity of Q[7] very well.[72,73] The binding affinity of the adamantyl ammonium ion in Q[7] is high ($K_a = 4 \times 10^{12}\,M^{-1}$) and only slightly lower in Q[8] ($K_a = 8 \times 10^{8}\,M^{-1}$).[7] While the shape and size in this example is important, there are other driving forces such as the charge and hydrophilicity.

3.1.2 *Polarity and charged functional groups*

An electropositive or cationic functional group is of particular importance in the binding of a molecule or ion to Q[*n*]. There is a high affinity for functional groups with positive character driven by the electronegative carbonyl oxygen atoms that form the rim of the portal. In contrast, anions are unfavorable to binding.[74,75] There are numerous examples of cationic assisted bindings. Dipole–ion interactions between a cationic functional group and the portal greatly increase the stability of a drug@Q[*n*] association complex. The most common cationic group for drug stability is an ammonium ion.[41–44,50,51,76–78] An ammonium ion-binding interaction is at its best when the charge carried is localized and not too diffuse.[79] A protonated amine has a higher affinity for the portal than a quaternary ammonium ion.[79,80] Diffuse cations of NR_4^+, SR_3^+, or PR_4^+ are often encapsulated within the cavity rather than at the portal.

Cationic groups of coordination metal complexes such as platinum or ruthenium cytotoxic drugs also form effective ion–dipole portal interactions.[57,62,65,80–82] Examples of affinity selection between cations that are carried on the same molecule are particularly prominent in the platinum-based drugs CT233, CT033, and BBR3571 (Sections 2.2.2 and 2.2.6). A protonated amine dominates the portal position over the Pt monocationic head in the latter two examples, whereas the Pt monocationic head is preferred over the NR_4^+, which can be accommodate within the cavity.[79,80] The strong portal association of a protonated amine results in a stabilization of the cation and an increase in the pK_a.[44,76,77]

Dipole–dipole interactions with the portal are primarily in the form of hydrogen bonding. This can be found in the neutral cytotoxic drug oxaliplatin, through the HN of the ligand, or the NH of amide groups of the organic drugs, atenolol or glibenclamide.[50,64] The nitrovinylamino group of ranitidine provides supporting binding strength, through hydrogen bonding at the opposite portal of Q[7] to a protonated amine, as this drug spans across the cavity (Figure 8).[52] Hydrogen bonding is also evident for the encapsulation of guests carrying carboxylic acid groups. In general, the carboxylic acid group sits at one portal and the remaining parts of the molecule thread through the cavity to the opposite portal.[75]

3.1.3 *Hydrophobic moieties*

Many drugs are difficult to administer because of their low water solubility. The Q[*n*] have polar portals that interact

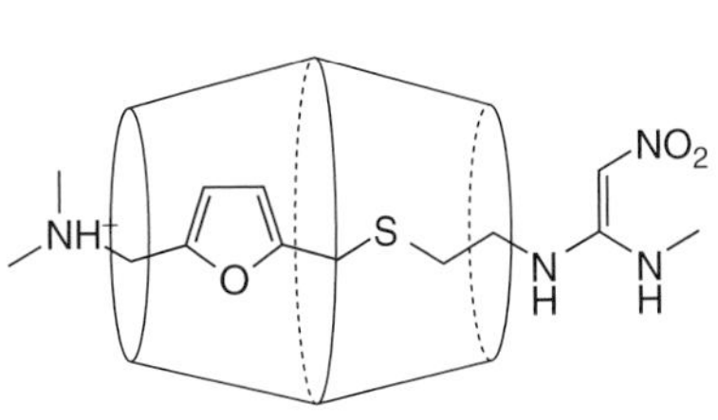

Figure 8 The ammonium salt of ranitidine encapsulated in Q[7] and stabilized by ion–dipole interaction at one portal and hydrogen bonding at the other.

with water and ions and possess a hydrophobic cavity compatible to hydrophobic moieties. A number of the drugs previously discussed (Sections 2.2.1 and 3) have both polar groups and hydrophobic moieties. It is this combination that favors a stable association complex with Q[*n*]. An obvious example is the improved solubility of the antiinflammatory steroid drug dexamethasone acetate (up to 17-fold increase) utilizing the α,δ-$(CH_3)_4$Q[6] (Figure 3; $R^1 = R^2 = CH_3$, $s = 2$).[83] This partially substituted Q[6] with aqueous solubility (1.4 mM) ~75 times higher than unsubstituted Q[6] has an ellipsoid cavity capable of accommodating slightly wider guests such as dexamethasone acetate. σ-Carborane, a potential therapeutic agent (boron neutron capture therapy), is normally water insoluble but as a molecular guest in Q[7], it is water soluble.[72,84]

3.2 Drug modification for improved Q[*n*] binding

Prodrugs are drugs that carry functional groups that are cleaved by a biological process such as an enzyme to release the active component. These functional groups are often applied to drugs to improve their water solubility.[85] The prodrug approach to improved bioavailability or toxicity reduction, foreseeably lends itself to an added potential when combined with the noncovalent process of encapsulation by Q[*n*]. Relatively minor changes to a drug could impart an advantage to encapsulation and stability by exploiting the favorable binding features required for better Q[*n*] encapsulation and therefore provide greater benefits to the activity of the drug component relatively simply and reversibly. Prodrugs via esters are relatively common, and an ideal ester functional group that could lead to a prodrug@Q[*n*] combination has, in principle, been demonstrated where choline or phosphonium analogs provide the cation as an ester, for the dipole–ion interacting function.[74,86] Adding cationic groups to a hydrophobic core has been shown to be very effective for increasing binding with Q[*n*].[7,8]

In some circumstances, minor changes to a drug to improve encapsulation may not even require that the drugs be prodrugs. The cytotoxic dinuclear platinum drug CT233 is a case in point. In order to achieve the best protection for the platinum metal centers, the central amine of CT033 was modified to a quaternary ammonium ion to give CT233 so that it could be accommodated within the cavity.[79,80] Through charge dispersion, the dominant ion–dipole drivers become the two Pt monocationic head centers, which act to fold the linking ligand within the cavity (Figure 4). Chain folding is favored because of improved van der Waals contacts and isolation of the relatively hydrophobic chain from the aqueous environment outside the cavity. This may also be facilitated by the small electronegative potential within the cavity.[6] The cytotoxic activity of CT233 relative to the parent platinum complex CT033 was found to be lower, but the resistance factor was excellent (Table 4).

Table 4 Cytotoxic IC_{50} in the murine leukemia line L1210 cancer cell line and its cisplatin-resistant cell line L1210/DDP.

Complex (μM)	L1210	L1210/DDP	RF
CT033	0.13	0.12	0.9
CT233	8.6	3.2	0.4

The resistance factor (RF) is defined as IC_{50} resistance/IC_{50} sensitive.

While some activity has been lost, the *in vitro* cysteine degradation model showed a threefold increase in protection of the platinum metal centers in CT233 (see Section 2.2.2). This protection may translate into a significant advantage *in vivo* given that platinum drug concentrations in plasma fall rapidly without protection.

As a second example of drug modification to gain advantage, the organic cytotoxic drug ABZ was modified by replacing the methyl carbamate with a methoxyethyl carbamate to give MEABZ that facilitates aqueous solubility of the carbamate group but has little effect on the overall solubility of MEABZ. In conjunction with this change and encapsulation of the remaining hydrophobic portion of the molecule in Q[8], the solubility was improved 1.3-fold compared to ABZ@Q[7] but 3.5-fold relative to ABZ@Q[8]. As a bonus, the cytotoxic activity of MEABZ was also found to be up to 10 times more active than ABZ.[44]

The potential for improvement through optimization of encapsulation as demonstrated for CT233 and MEABZ opens the possibilities of combining the concept of minor modification or prodrugs with Q[*n*] encapsulation.

4 PREPARATION OF ASSOCIATION COMPLEXES DRUG@CUCURBITURIL

4.1 General procedure

4.1.1 *Association complex—cavity or portal bound*

After evaluating the suitability of a drug for encapsulation according to size, shape, polarity, and hydrophilicity (see Section 3), one of the following methods can be used to prepare an association complex. Most examples involve the preparation of the association complex in water. Even though Q[6], Q[8], and Q[10] have low water solubility, a polar drug or a drug with a polar group can often increase the solubility of the association complex

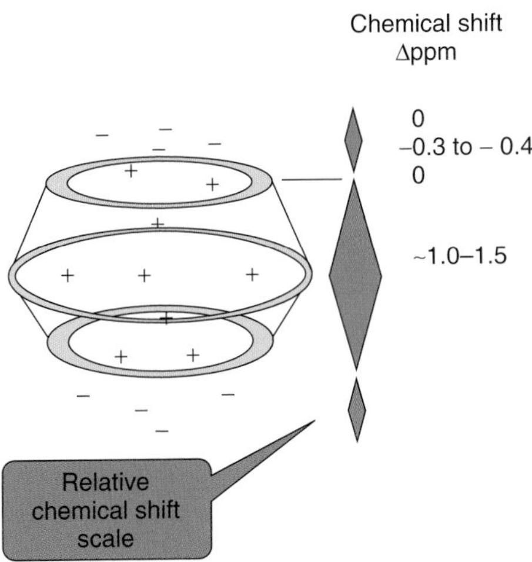

Figure 9 Proton resonance relative chemical shift differences compared to the free guest in the same solvent.

drug@Q[*n*] beyond the solubility of each individual component. The solubility of Q[6] can be assisted through the use of saline.

Given that most drug@Q[*n*] association complexes will be soluble to 1 mM, ^{1}H NMR spectroscopy provides a very powerful tool for analysis. Shifts in proton resonances for the free drug relative to the drug@Q[*n*] can provide information as to the location of each proton in the cavity or in the vicinity of the portal of a bound drug. Shifts upfield of 1–1.5 ppm generally indicates a location at the geometric center of the Q[*n*] cavity. A shift upfield <1 ppm generally indicates that a proton resides in the cavity but between the geometric center and the portal opening (smallest shift closest to the opening). A downfield shift <0.3–0.4 ppm indicates a position just outside the portal. These proton resonance shift indicators apply best to association complexes in slow kinetic exchange on the ^{1}H NMR timescale, although they also apply to combinations in fast or moderate exchange though the analysis requires careful consideration (Figure 9).[1] The binding ratio can also be determined using NMR by comparing the proton integrals of the guest relative to the Q[*n*]. Binding ratios for drug to Q[6] or Q[7] are generally no higher than 1 : 1, whereas the ratio for Q[8] or Q[10] can be 2 : 1. Drugs with more than one binding site can have a binding ratio, which also reflects this.

Soluble drugs and Q[n] complex preparation

In the cases where the drug already has aqueous solubility and encapsulation by Q[*n*] provides advantages other than increased solubility, then the following general procedure is applied.

The required mole ratio of drug is either added as a solid to a solution of Q[6], Q[7], or Q[10], or the drug is dissolved in water or saline and titrated into these solutions. The low solubility of Q[8] means that this is generally suspended in water and a solution of the drug added. Association complexes that are formed with high stability generally do not require filtration but this can be included. Solid products are obtained from aqueous solutions by lyophilization or evaporation.

- *Poorly Soluble Drugs and a Binding Ratio of 1 : 1.* A typical preparation involves mixing solid drug and solid Q[*n*] in H_2O. Mixtures are homogenized and sonicated. Following this process, mixtures are left to stand overnight. Filtration through a 0.5 μm PETE syringe filter gives clear solutions of drug@Q[*n*] complexes. Solid products are then obtained by lyophilization.
- *pH < 6.* Given that anions are unfavorable to encapsulation and that encapsulation is improved when cations can be formed such as amine protonation, the pH can be important in the formation of stable association complexes with Q[*n*]. The above preparative methods are also applicable to drugs that can benefit from acidic preparative media. As a consequence of the stabilization of cations and hydrogen-bonding groups, following encapsulation, the pH of medium can be increased without dissociation of the drug (see Section 5). The degree to which the pH can be increased is dependent on the degree of stabilization and is specific to each case.
- *Cosolvent Processing.* There are few examples of the use of cosolvents in the preparation of Q[*n*] association complexes, but the principle is sound. The problem exists for the unsubstituted Q[*n*] that there are few organic solvents that can dissolve both the Q[*n*] and an organic drug. While trifluoroacetic acid (TFA) has been used to encapsulate σ-carborane in Q[7] with water titrated into the mixture and eventual evaporation of TFA, it would not be broadly applicable.[72] With the development of substituted Q[*n*] that have a broader solvent range, this approach becomes more appropriate.
- *Ball Mill Grinding.* A relatively recent approach to improving bioavailability of insoluble drugs is ball mill grinding, which disrupts the crystal structure and creates ultrafine particles with better dissolution rates. Cogrinding of drugs in the presence of excipients or molecular hosts further improves bioavailability through nanoparticle formation.[87,88] The potential for ball mill cogrinding of Q[7] or Q[8] and poorly soluble or insoluble drugs has been reported with the highly water insoluble molecules such as [60]fullerene and a series of σ-phenylphenols.[69,89,90] The synthesis of Q[*n*] association complexes by ball milling is faster

and more efficient than heterogeneous slurries. It is also important to note that the Q[*n*] appear to be stable to the milling process.

4.1.2 Conclusion

There are a number of preparative approaches to drug@Q[*n*] complexes with most considerations centered on the solubility of the individual Q[*n*] and drug components. The pH of the solution is also an important consideration in achieving stable complexes, while temperature has little effect except in terms of increasing the rate of dissolution of one or both of the components and thereby increasing the rate of the formation. The exception is in the case of a guest that must squeeze through the portal, and here the rate of encapsulation is increased with temperature. The development of substituted Q[*n*] has broadened the potential for the preparation of association complexes of drugs and Q[*n*], and this is expected to expand.[5,7]

5 DRUG RELEASE MECHANISMS

The drug release mechanisms in the following discussion refer to two different relationships between drugs and Q[*n*]:

- The first relates to the release of drugs that are bound within the cavity or associated at the portal or both.
- The second refers to the release of a drug that is not encapsulated directly by the Q[*n*] but where the Q[*n*] is a functional part of a nanostructure and the drug is held within the nanostructure.

5.1 Cavity or portal release

5.1.1 Solution equilibrium and drug release

For a drug@Q[*n*] complex, whether the drug is associated at the portal or, in the cavity or, a combination of the two, there exists an equilibrium between the drug being free in the solution and associated in one of the manners described. The association constant provides a measure of the preference between the two states, free or bound. Predominantly, the solvent environment within the biological context is aqueous and the equilibrium constant is relevant to this. The highest association constant recorded to date for any guest@Q[*n*] combination is for a molecule with a hydrophobic core and two cationic arms. The guest bis(trimethyammoniomethyl)ferrocene encapsulated by Q[7] has been found to have an association constant of $3 \times 10^{15}\,M^{-1}$.[8] The association constant reflects the stability of the guest@Q[*n*] but not the ease with which the guest can exchange to a suitable alternative receptor or be released and removed through a consumptive process. Rates of exchange are usually in seconds or fractions of seconds (Section 2.2.4). The exception is when there is restricted egress due to mechanical constraints.[57] In the absence of a trigger mechanism, the release of a drug depends on available alternative receptors, contact with hydrophobic environments for hydrophobic drugs, or biochemical consumption.

Under some circumstances, a drug may be released through displacement by a cellular constituent, such as a functional group of a protein (e.g., an aromatic amino acid residue), but this only occurs if the binding is competitive with the drug.[91]

5.1.2 pH-controlled release

The observed stabilization of cationic centers through the stabilizing influence of the electronegative portals of the Q[*n*] also provides a mechanism to the immediate release of a guest through a pH increase or a slower release if the required pH for immediate release is not reached. This effect occurs as a consequence of shifts in pK_a to higher values for guest molecules that can be deprotonated.[52,76] It is particularly relevant in the physiological pH range 7–7.5 for blood plasma, the cytoplasm of cells (human, animal, and microbial), and the pH range 1.3–9 for the gastrointestinal tract.[68]

Notably, the gastrointestinal drugs, lansoprazole, omeprazole, and the histamine H_2-receptor antagonist ranitidine demonstrate Q[7] stabilization and pK_a shifts to high values of ~1.5–4. The extent of the shift is dependent on the degree of stabilization through hydrogen bonding and the ion–dipole interaction with portal oxygens.[52,76] The pK_a values of the anesthetics procaine, tetracaine, procainamide, dibucaine, and prilocaine are similarly increased on Q[7] encapsulation ($\Delta pK_a \sim 0.5$–1.9).[77] A pK_a shift of two units for the cytotoxic drug MEABZ@Q[7] has been found, which may provide a number of options for administration while maintaining protection for the drug prior to release.[44]

The pH range of the gastrointestinal tract of initially very acidic conditions progressing to a very alkaline state provides considerable scope for protection using Q[*n*], which are ideal for acidic media leading to release with increasing pH. With pH tuning of the drug@Q[*n*] complex, this could be used to target small intestine release (pH ~ 6–7) or at a later stage in the ileocecal region (pH up to 9).[68]

5.1.3 Drug action without release

[60]Fullerene (C_{60}) as a biologically active agent through light activation is prepared in aqueous media by association

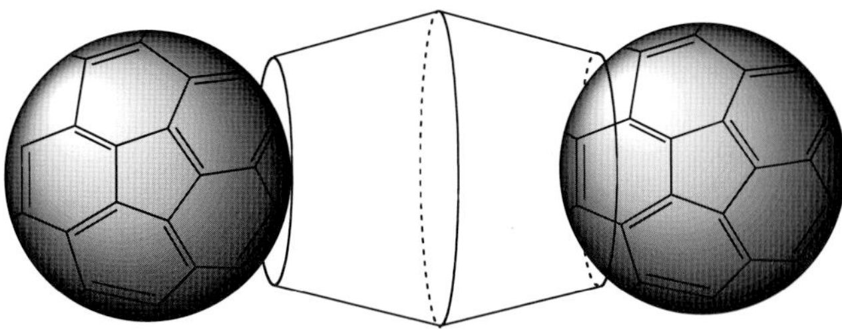

Figure 10 C_{60} association at the portals of Q[8].

with Q[8]. The C_{60} forms an association complex with Q[8] in a ratio of 2 : 1 as an aqueous soluble material (C_{60} is normally completely insoluble). An exclusion complex is proposed where each Q[8] portal is occupied by a C_{60} (Figure 10). Under these circumstances, the release of the active agent is unnecessary as an active surface would be available to cellular contact, which could allow the process of cytotoxicity to be initiated. This association complex was evaluated for cyctotoxic activity against the HeLa cancer cells.[90] Over a three-day period, a 30 min pulse of light, at one per day of 500 W (wavelength not specified, maximum absorption at λ 351 nm; light source distance 42 cm) resulted in an inhibition of growth. Inhibition of growth of 83% occurred at a concentration of 50 μg ml^{-1} and not more than 8% in the absence of the $(C_{60})_2$@Q[8] complex. This demonstrates external stimuli to drug activation without release. The precise mechanism has not been established nor is it known whether the complex enters the cell, but it was presumed by the authors to be as a result of phospholipid and protein damage within the membrane. A possible pathway is C_{60} excitation to a singlet excited state and intersystem crossing to the excited triplet state then quenching by O_2 to produce singlet oxygen the agent of damage.[92]

5.2 Release from nanostructures utilizing Q[*n*] components

5.2.1 Degradative release

One approach to controlled release of a drug is through degradative release. The disulfide-tailed Q[6] nanocapsule discussed in Section 2.2.5 was developed with the view to achieving cytoplasm release through reductive sensitive disulfides. Cellular glutathione is the reducing agent for the disulfide bonds of the disulfide-tailed Q[6] nanocapsule (Scheme 2). Reduction results in a significant shortening of the Q[6] tails, which leads to the disruption of the interlocking of disulfide tails followed by rupture. Glutathione at cellular concentrations of 5 μM was shown to affect capsule release of its contents using a fluorescent dye as a probe. This was further supported by *in vitro* cancer cell experiments where the nanostructure was loaded with doxorubicin and released (Section 2.2.5).[59]

5.2.2 pH-controlled release

Nanovalve technology for drug release: the nanovalve is constructed from mesoporous silica, which has a cavity, and on the rim of the cavity are attached polyamine strings. The polyamines provide a shaft to a rotaxane structure, where Q[6] is threaded. The rotaxane shaft is tunable to desired pK_a by the choice of amines employed. The shuttling of Q[6] up and down the shaft allows an opening and closing action to expose the cavity in the mesoporous nanostructure (Scheme 3b). This system relies on the shuttling movement of a Q[6] on a rotaxane axis with bistable states that are sensitive to subtle pH changes to be encountered inside a cell relative to the blood stream. These subtle pH changes are therefore autobiostimuli and cannot be as readily controlled except by careful tuning of the pK_a. The principle of the pH-driven Q[6]-based nanovalve system has been demonstrated through the pH-controlled release of a fluorescent dye but has yet to be demonstrated on cells for drug release.[58]

5.2.3 Hyperthermic release

Using a similar nanovalve structure described in the previous section, a second component is added to respond to local heating (Scheme 3a). The hyperthermic driven Q[6]-based nanovalve operates through a magnetic receiver structure of zinc-doped iron oxide nanocrystals (ZnNCs), which response to an external stimuli of an oscillating magnetic field that induces local heating. The nanostructure is referred to as the magnetic activating release system (MARS). It is the local heating that is the trigger, which increases the thermal induced molecular dynamics and opens the valve.[60] The mesoporous silica cavity in this case (MARS) has been loaded with the drug doxorubicin or the dye rhodamine B to a capacity of 4 wt% to demonstrate the effectiveness of controlled release. A single AC magnetic field pulse at 500 kHz and current amplitude of 37.4 kA m^{-1} gave a 40% release. The remaining cargo (dye) was released following successive pulses over ~150 min. As further support, the nanostructure was loaded with doxorubicin and then after being taken up by the breast cancer cell line MDA-MB-231, an AC field exposure resulted in a 37% cell death. In contrast, no AC field gave ~5% and an AC field without doxorubicin loading resulted in a 16% cell death. The latter result of 16% cell death indicated that there is some thermal effect on cell viability.

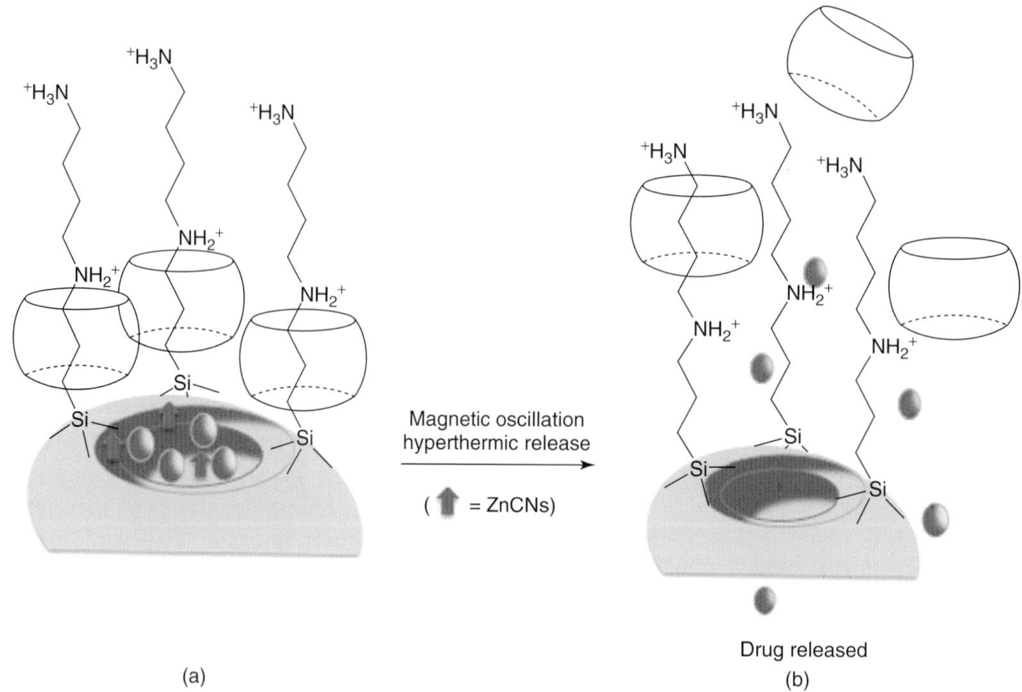

Scheme 3 (a) Q[6]-based nanovalve system showing the mechanisms of release by induced heating from an oscillating magnetic field. (b) An alternative mechanism using the same nanostructure but without the ZnCNs involves increasing the pH and deprotonation of the ammonium ions.

5.2.4 *Diffusion*

Hydrogels such as the Q[6]-mediated alginate hydrogel beads discussed in Section 2.2.4 release their drug loads through a process of diffusion and concentration gradients unless some other mechanisms are built into the gel to modify release.[55] This may be a direction for the future for certain drug applications.

6 CONCLUSION

Q[*n*] and drug delivery are emerging as an exciting area in the supramolecular context. Q[*n*] have the potential to contribute to the resolution of some drug delivery difficulties that include bioavailability, bio and chemical stability, improvements in efficacy and efficiency, limiting toxicity, and side effects. In relative terms, only a few examples of this potential have so far been realized. As we increase our understanding of the physical and chemical capabilities of Q[*n*] and develop new derivatives and supramolecular forms, Q[*n*] will be one of the drug delivery tools used in the clinic in future.

REFERENCES

1. J. Lagona, P. Mukhopadhyay, S. Chakrabarti, and L. Isaacs, *Angew. Chem. Int. Ed.*, 2005, **44**, 4844–4870.
2. W. A. Freeman, W. L. Mock, and N. Y. Shih, *J. Amer. Chem. Soc.*, 1981, **103**, 7367–7368.
3. J. Kim, I.-S. Jung, S.-Y. Kim, *et al.*, *J. Amer. Chem. Soc.*, 2000, **122**, 540–541.
4. A. Day, A. P. Arnold, R. J. Blanch, and B. Snushall, *J. Org. Chem.*, 2001, **66**, 8094–8100.
5. A. I. Day, R. J. Blanch, A. P. Arnold, *et al.*, *Angew. Chem., Int. Ed.*, 2002, **41**, 275–277.
6. R. V. Pinjari, J. K. Khedkar, and S. P. Gejji, *J. Incl. Phenom. Macrocycl. Chem.*, 2010, **66**, 371–380.
7. S. Mohaddam, Y. Inoue, and M. K. Gilson, *J. Amer. Chem. Soc.*, 2009, **131**, 4012–4021.
8. M. V. Rekharsky, T. Mori, C. Yang, *et al.*, *Proc. Natl. Acad. Sci. U.S.A.*, 2007, **104**, 20737–20742.
9. M. E. Davis and M. E. Brewster, *Nature Rev.*, 2004, **3**, 1023–1035.
10. R. Challa, A. Ahuja, J. Ali, and R. K. Khar, *AAPS Pharm. Sci. Tech.*, 2005, **6**, E329–E357.
11. J. Hu, Z. Tao, and S. Cao, *Recent Advances in Biomaterials Research*, in, ed. J. Hu, Research Signpost, Kerala, India, 2008, pp. 99–122.
12. M. Hamoudi, L. Trichard, J.-L. Grossiord, *et al.*, *Ann. Pharm. Fr.*, 2009, **67**, 391–398.
13. A. Rasheed, K. C. K. Ashok, and V. V. N. S. S. Sravanthi, *Sci. Pharm.*, 2008, **76**, 567–598.
14. A. I. Day, A. P. Arnold, and R. J. Blanch. 2000. PCT Int. Appl. WO 2000068232 A1 20001116.
15. K. Kim, Y. J. Jeon, S.-Y. Kim, and Y. H. Ko. 2003. PCT Int. Appl. WO 2003024978 A1 20030327.

16. I. W. Wyman and D. H. Macartney, *J. Org. Chem.*, 2009, **74**, 8031–8038.
17. A. I. Day, A. P. Arnold, and R. J. Blanch. 2005. PCT Int. Appl. WO 2005026168 A1 20050324.
18. A. I. Day, A. P. Arnold, and R. J. Blanch, *Molecules*, 2003, **8**, 74–84.
19. K. Kim, N. Selvapalam, Y. H. Ko, *et al.*, *Chem. Soc. Rev.*, 2007, **36**, 267–279.
20. S. Y. Jon, N. Selvapalam, D. H. Oh, *et al.*, *J. Amer. Chem. Soc.*, 2003, **125**, 10186–10187.
21. S. Sasmal, M. K. Sinha, and E. Keinan, *Org. Lett.*, 2004, **6**, 1225–1228.
22. H. Isobe, S. Sato, and E. Nakamura, *Org. Lett.*, 2002, **4**, 1287–1289.
23. Y. Zhao, S. Xue, Q. Zhu, *et al.*, *Chin. Sci. Bull.*, 2004, **49**, 1111–1116.
24. X.-L. Ni, J.-X. Lin, Y.-Y. Zheng, *et al.*, *Cryst. Growth Des.*, 2008, **8**, 3446–3450.
25. L. Zheng, J. Zhu, Y. Zhang, *et al.*, *Supramol. Chem.*, 2008, **20**, 709–716.
26. L.-H. Wu, X.-L. Ni, F. Wu, *et al.*, *J. Mol. Struct.*, 2009, **920**, 183–188.
27. A. Flinn, G. C. Hough, J. F. Stoddart, and D. J. Williams, *Angew. Chem., Int. Ed.*, 1992, **31**, 1475–1477.
28. J. Zhao, H.-J. Kim, J. Oh, *et al.*, *Angew. Chem., Int. Ed.*, 2001, **40**, 4233–4235.
29. J. Lagona, J. C. Fettinger, and L. Isaacs, *Org. Lett.*, 2003, **5**, 3745–3747.
30. W.-H. Huang, P. Y. Zavalij, and L. Isaacs, *Org. Lett.*, 2008, **10**, 2577–2580.
31. B. Wyse, *Assessing Pharmacokinetic and Tissue Distribution Characteristics of Cucurbit[7]uril and Cucurbit[8]uril as a single bolus dose in Sprague-Dawley rats Report*. TetraQ quality preclinical solutions, University of Queensland, 2009.
32. V. D. Uzunova, C. Cullinane, K. Brix, *et al.*, *Org. Biomol. Chem.*, 2010, **8**, 2037–2042.
33. G. Hettiarachchi, D. Nguyen, J. Wu, *et al.*, *PLOS One*, 2010, **5**, e10514.
34. P. Montes-Navajas, M. González-Béjar, J. C. Scaiano, and H. García, *Photochem. Photobiol. Sci.*, 2009, **8**, 1743–1747.
35. K.-M. Kim, D.-H. Oh, J.-Y. Kim. 2006. Repub. Korean Kongkae Taeho Kongbo. KR 2006110956 A 20061026.
36. K.-M. Kim, D.-W. Kim, E.-J. Kim, *et al.* 2006. PCT Int. Appl. WO 2006112673 A1 20061026.
37. K.-M. Kim, H.-K. Lee, K.-M. Park, *et al.*. 2005. PCT Int. Appl. WO 2005112890 A1 20051201.
38. R. K. Verma, D. M. Krishna, and S. Garg, *J. Control. Release*, 2002, **79**, 7–27.
39. H. L. van de Waterbeemd and P. Artursson, in *Drug Bioavailability: Estimation of Solubility, Permeability, Absorption and Bioavailability (Methods and Principles in Medicinal Chemistry)*, eds H. K. Raimund Mannhold, G. Folkers, John Wiley & Sons, Inc., Hoboken, 2003, p. 602.
40. C. A. S. Bergstrom, C. M. Wassvik, K. Johansson, and I. Hubatsch, *J. Med. Chem.*, 2007, **50**, 5858–5862.
41. Y. J. Zhao, D. P. Buck, D. L. Morris, *et al.*, *Org. Biomol. Chem.*, 2008, **6**, 4509–4515.
42. N. Dong, S. F. Xue, Q. J. Zhu, *et al.*, *Supramol. Chem.*, 2008, **20**, 659–665.
43. Y. Zhao, D. L. Morris, M. Pourgholami, *et al.*, *Org. Biomol. Chem.*, 2010, **8**, 3328–3337.
44. Y. Zhao, *Cucurbit[n]uril a delivery host for anti-cancer drugs*, PhD thesis, Australian Defence Force Academy, University of New South Wales, 2009.
45. H.-J. Buschmann, E. Cleve, K. Jansen, *et al.*, *Mater. Sci. Eng. C*, 2001, **14**, 35–39.
46. J. W. Lee, S. Samal, N. Selvapalam, *et al.*, *Acc. Chem. Res.*, 2003, **36**, 621–630.
47. K. J. Barnham, M. I. Djuran, P. del Socorro Murdoch, *et al.*, *Inorg. Chem.*, 1996, **35**, 1065–1072.
48. J. Holford, F. Raynaud, B. A. Murrer, *et al.*, *AntiCancer Drug Des.*, 1998, **13**, 1.
49. A. Hennig, G. Ghale, W. M. Nau, *Chem. Commun.* 2007, 1614–1616.
50. F. J. McInnes, N. G. Anthony, A. R. Kennedy, and N. J. Wheate, *Org. Biomol. Chem.*, 2010, **8**, 765–773.
51. N. J. Wheate, V. Vora, N. G. Anthony, and F. J. McInnes, *J. Incl. Phenom. Macrocycl. Chem.*, 2010, **68**, 359–367.
52. R. Wang and D. H. Macartney, *Org. Biomol. Chem.*, 2008, **6**, 1955–1960.
53. J. Mohanty, H. Haridas Pal, A. K. Ray, *et al.*, *Chem. Phys. Chem.*, 2007, **8**, 54–56.
54. N. A. Peppas, J. Z. Hilt, A. Khademhosseini, and R. Langer, *Adv. Mater.*, 2006, **18**, 1345–1360.
55. X. Huang, Y. Tan, Q. Zhou, and Y. Wang, *e-Polymer*, 2008, **95**, 1–11.
56. F. Li, Y. Mulyana, M. Feteri, *et al.*, *Dalton Trans.*, 2011. DOI: 10.1039/C1DT10250H.
57. M. J. Pisani, Y. Zhao, L. Wallace, *et al.*, *Dalton Trans.*, 2010, 2078–2086.
58. S. Sarah Angelos, N. M. Khashab, Y.-W. Yang, *et al.*, *J. Amer. Chem. Soc.*, 2009, **131**, 12912–12914.
59. K. M. Park, D. W. Lee, B. Sarkar, *et al.*, *Small*, 2010, **6**, 1430–1441.
60. C. R. Thomas, D. P. Ferris, J.-H. Lee, *et al.*, *J. Amer. Chem. Soc.*, 2010, **132**, 10623–10625.
61. J. Kim, Y. Ahn, K. M. Park, *et al.*, *Angew. Chem. Int.*, 2007, **119**, 7537–7539.
62. N. J. Wheate, A. I. Day, R. J. Blanch, and J. G. Collins. 2005. PCT Int. Appl. WO 2005068469 A1 20050728.
63. N. J. Wheate, A. I. Day, R. J. Blanch, *et al.*, *Chem. Commun.* 2004, 1424–1425.
64. Y. J. Jeon, S.-Y. Kim, Y. H. Ko, *et al.*, *Org. Biomol. Chem.*, 2005, **3**, 2122–2125.
65. S. Kemp, N. J. Wheate, S. Wang, *et al.*, *J. Biol. Inorg. Chem.*, 2007, **12**, 969–979.
66. D. P. Buck, P. M. Abeysinghe, C. Cullinane, *et al.*, *Dalton Trans.* 2008, 2328–2334.

67. S. Ghosh and L. Isaacs, *J. Amer. Chem. Soc.*, 2010, **132**, 4445–4454.
68. V. I. Dedlovskaya, *Bull. Exp. Biol. Med.*, 1968, **66**, 1292–1294.
69. S. Walker, R. Kaur, F. J. McInnes, and N. J. Wheate, *Mol. Pharm.*, 2010, **7**, 2166–2172.
70. K. E. Geckler, K. S. Ho, and D. Dipen. 2009. Repub. Korean Kongkae Taeho Kongbo. KR 2009050815 A 20090520.
71. V. Sindelar, M. A. Cejas, F. M. Raymo, and A. E. Kaifer, *New J. Chem.*, 2005, **29**, 280–282.
72. R. J. Blanch, A. J. Sleeman, T. J. White, *et al.*, *Nano Lett.*, 2002, **2**, 147–149.
73. S. Liu, C. Ruspic, P. Mukhopadhyay, *et al.*, *J. Amer. Chem. Soc.*, 2005, **127**, 15959–15967.
74. I. W. Wyman and D. H. Macartney, *Org. Biomol. Chem.*, 2010, **8**, 253–260.
75. V. Sindelar, S. Silvi, A. E. Kaifer, *Chem. Commun.* 2006, 2185–2187.
76. N. Saleh, A. L. Koner, and W. M. Nau, *Angew. Chem. Int. Ed.*, 2008, **47**, 5398–5401.
77. I. W. Wyman and D. H. Macartney, *Org. Biomol. Chem.*, 2010, **8**, 247–252.
78. N. Dong, S.-F. Xue, Z. Tao, *et al.*, *Acta Chim. Sin.*, 2008, **66**, 1117–1122.
79. A. D. St-Jacques, I. W. Wyman, D. H. Macartney, *Chem. Commun.* 2008, 4936–4938.
80. Y. J. Zhao, M. S. Bali, C. Cullinane, *et al.*, *Dalton Trans.* 2009, 5190–5198.
81. N. J. Wheate, D. P. Buck, A. I. Day, and J. G. Collins, *Dalton Trans.* 2006, 451–458.
82. M. S. Bali, D. P. Buck, A. J. Coe, *et al.*, *Dalton Trans.* 2006, 5337–5344.
83. G.-P. Li, F.-Q. Yu, J.-H. lin, and Y.-Z. Chen, *Yingyong Huaxue*, 2010, **27**, 563–566.
84. S. Morandia, S. Ristoria, D. Bertia, *et al.*, *Biochem. Biophys. Acta*, 2004, **1664**, 53–63.
85. C. E. Muller, *Chem. Biodiv.*, 2009, **6**, 2071–2083.
86. X.-Z. Fu, Y. Huang, Z. Tao, *et al.*, *Chin. J. Org. Chem.*, 2010, **30**, 675–683.
87. M. Jug, F. Maestrelli, M. Bragagni, and P. Mura, *J. Pharm. Biomed. Anal.*, 2010, **52**, 9–18.
88. P. N. Balani, W. K. Ng, R. B. H. Tan, and S. Y. Chan, *J. Pharm. Sci.*, 2010, **99**, 2462–2474.
89. F. Constabel and K. E. Geckeler, *Tetra. Lett.*, 2004, **45**, 2071–2073.
90. G. Jiang and G. Li, *J. Photochem. Photobiol. B: Biol.*, 2006, **85**, 223–227.
91. L. M. Heitmann, A. B. Taylor, P. J. Hart, and A. R. Urbach, *J. Amer. Chem. Soc.*, 2006, **128**, 12574–12581.
92. G. Jiang, Q. Zheng, and D. Yang, *J. Appl. Polym. Sci.*, 2006, **99**, 2874–2877.

Podands

Adam N. Swinburne and Jonathan W. Steed

Durham University, Durham, UK

1 CATION-BINDING PODANDS

1.1 Development and properties of podands

The name "podand" is derived from a combination of the combining form "pod" (having a foot) and lig*and*. The term was introduced by Vögtle and Weber in 1979[1] and as part of their systematic nomenclature of supramolecular hosts for metal cations. Vögtle and Weber envisaged an increasing progression of ligand preorganization from the very flexible podands, to macrocyclic "coronands" and macrobicyclic "cryptands," Figure 1. On binding a metal ion, podands form "podates" and coronands form "coronates." While the term coronand is little used, perhaps because of the dominance of the crown ether nomenclature, the word podand and, more importantly, the concept behind it are popular in modern supramolecular chemistry, and podand-type chemistry now encompasses hosts for anionic, cationic, and molecular guests. The key concept is one of the multiple-guest-binding sites linked to a common core within an acyclic host framework. The podands thus paved the way for a great deal of work on the generalized and important topic of *multivalency*.[2,3] Modern nanostructures such as dendrimers[4] also derive, at least conceptually, from podands. It is impossible in this brief, introductory chapter to present any comprehensive overview of the vast topic of acyclic receptors and instead we cover some of the core concepts and more interesting "classic" systems in this cation-binding section before moving on to some recent developments in anion-binding podands in Section 2. The burgeoning field of dendrimer chemistry is beyond the scope of this chapter and is covered in **Supramolecular Dendrimer Chemistry**, Volume 7.

Typical early podands were acyclic analogs of crown ethers such as pentaethylene glycol and its dimethyl ether (**1** and **2**) and we can distinguish mono-, di-, tri-, tetra-, or multipodands according to the number of guest-binding "feet" attached to the anchor group (A; absent in the case of monopodands such as **1**), Figure 2. Extensive reviews covering the diverse array of early polyether podands have appeared previously.[1,5] Compounds **1** and **2** are examples of monopodands since there is a single chain of donor atoms with no obvious anchor group. It is interesting to note that it was Pedersen's attempt to synthesize a podand-type diol ligand from a protected catechol that resulted in the isolation of the first crown ether—dibenzo[18]crown-6, Scheme 1. Pedersen proposed to use the podand as a ligand for vanadyl ion, VO^{2+}, in catalytic applications.[6]

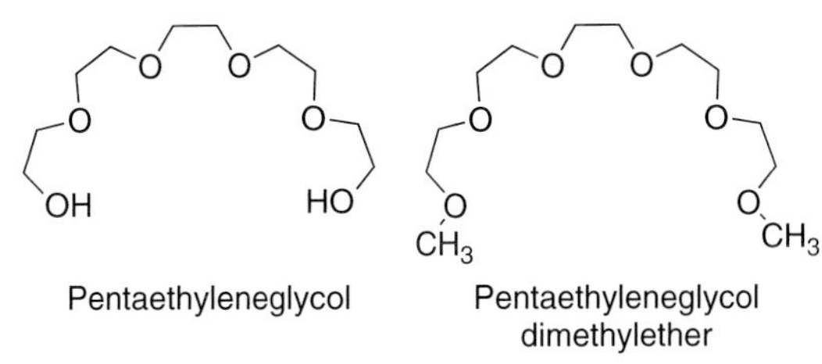

Supramolecular Chemistry: From Molecules to Nanomaterials.
Edited by Philip A. Gale and Jonathan W. Steed.
© 2012 John Wiley & Sons, Ltd. ISBN: 978-0-470-74640-0.

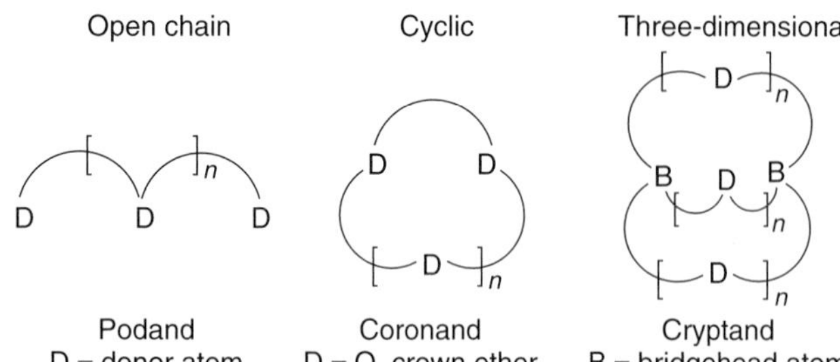

Figure 1 Classes of cyclic and acyclic ligands.

In the early years of supramolecular chemistry, a number of studies showed that in comparison to the cyclic crown ethers and bicyclic cryptands, alkali metal cation binding by podands of the oligoethylene glycol type is relatively weak in polar solvents such as water and methanol. Thus [18]crown-6 binds K^+ some four orders of magnitude more strongly in methanol solution compared to its open chain podand analog, **2**. Representative binding constants for cyclic and acyclic species are shown in Figure 3. The significantly lower cation affinity exhibited by podand **2** compared to [18]crown-6 and especially [2.2.2]cryptand was explained by the *macrocyclic* and *macrobicyclic effects*.[7] Both of these effects are manifestations of the principle of preorganization in supramolecular chemistry.[7–9] Rigidly preorganized cyclic receptors are in the correct conformation to bind their target guest and experience markedly less desolvation on guest complexation. The effect is particularly apparent in comparing extremely rigidly preorganized spherands such as **5** with their podand analog, **6**. The binding constant for the binding of Li^+ by the spherand is some 12 orders of magnitude higher!

The corollary to the rigid preorganization of the cryptands, and spherands in particular, however, is that guest complexation and decomplexation rates can be slow compared to those of the much more flexible podands.[10] While cation binding by oligoethylene-glycol-type podands occurs within the time of mixing, binding by spherands can take hours to reach equilibrium.[10] Broadly speaking, therefore, the characteristics of podands and macrocycles are mutually complementary with podands being more

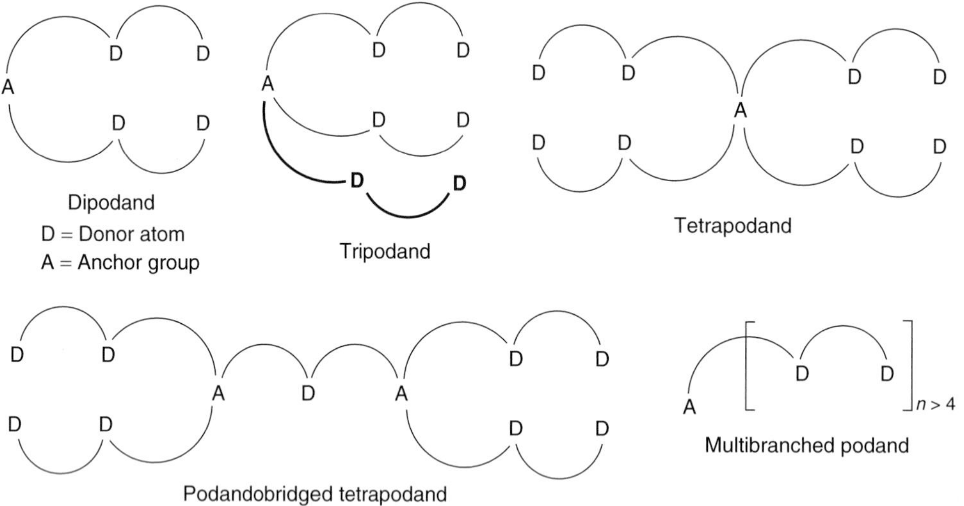

Figure 2 Types of podand according to the number of "feet."

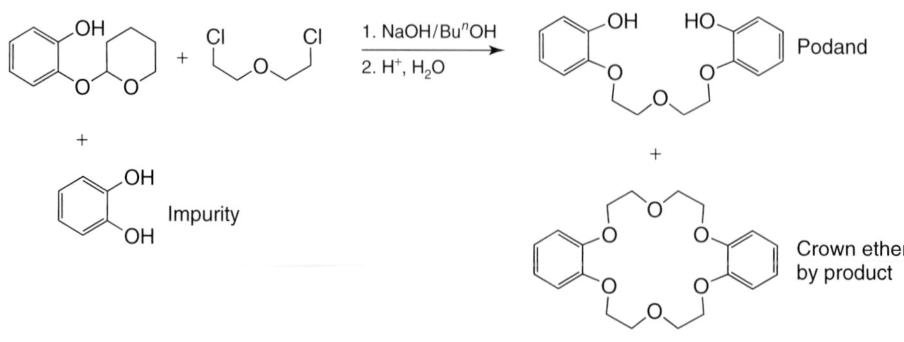

Scheme 1 Accidental synthesis of the first crown ether, dibenzo[18]crown-6, as part of the synthesis of a podand designed to act as a ligand for vanadyl ion.[6]

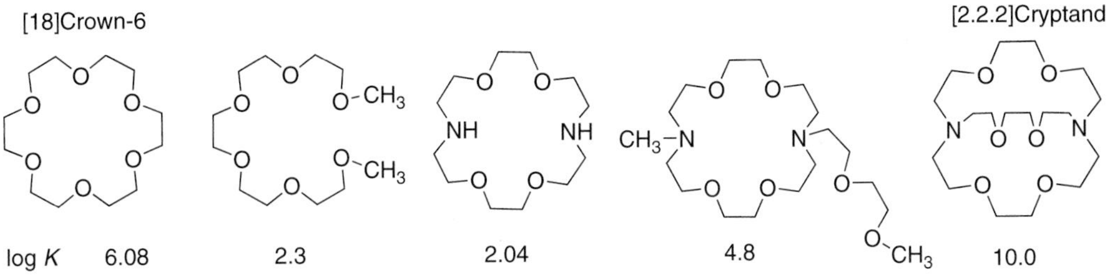

Figure 3 Binding constants (log K) for K^+ binding by simple podand, macrocyclic, and lariat ether compounds in methanol at 25 °C.

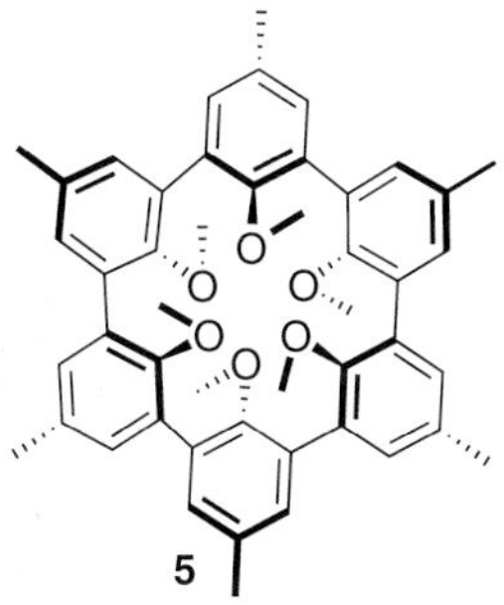

effective in situations where smaller binding constants but rapid complexation/decomplexation kinetics are required, whereas rigidly preorganized macrocycles are highly suited to sequestration-type applications. There is by no means a rigid divide, however. Macrocycles such as crown ethers are still relatively flexible and are highly effective in analytical applications, and the degree of donor group complementarity for the guest has a significant effect.

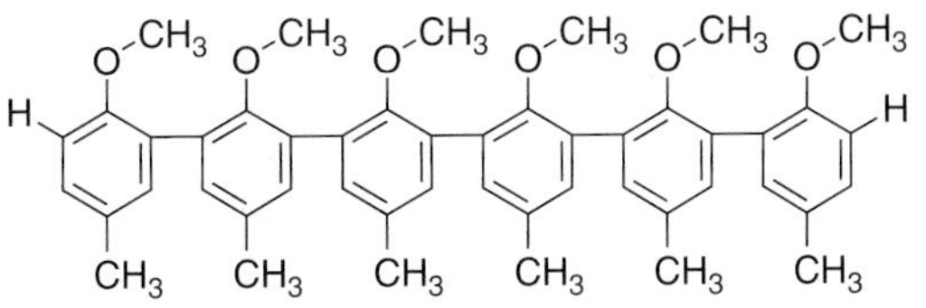

Compound **4** is an example of a lariat ether which represents a combination of the crown ether and podand-binding motifs. Lariat ethers are covered in detail in **Crown and Lariat Ethers**, Volume 3. Lariat ethers were designed to have both the strong binding imparted by the macrocyclic crown ethers and the flexibility offered by the podand side arm. The three-dimensional guest binding and additional oxygen donor atoms (complementary to K^+) offered by **4** compared to diaza[18]crown-6 (**3**) impart significant extra stability, but the receptor does not bind its guest cations nearly as effectively as the macrobicyclic cryptand analog.

To enhance the degree of preorganization in early podands, Vögtle and Weber[11] introduced the *terminal group concept*. By equipping cation-binding podands with rigid terminal end groups, the affinity for guest cations proved to be enhanced because the rigidity of the end groups imparts a certain degree of preorganization. Examples of these kinds of rigid terminal group podands are **7** and **8** (the latter is also called commercially *Kryptofix-5*; [2.2.2]cryptand is *Kryptofix-222*), in which the conjugated benzoic acid and hydroxyquinoline derived groups provide rigid, planar, and partially preorganized binding regions even though they are attached to a more flexible spacer chain. The compounds form crystalline complexes with metal salts such as alkali metal and transition metal thiocyanates, $AgNO_3$, and even uranyl nitrate. The X-ray structure of the RbI salt of **8** is shown in Figure 4. Applications of such compounds were envisaged in colorimetric ion detection as in the rigid-end-group chromoionophore **10**, terminated by quinone monoimine groups, which was studied as a possible means of sensing the presence of cations such as Ca^{2+} by changes in the UV–vis absorption spectrum of the host. The binding of divalent metal ions switches the podand into a zwitterionic form with highly basic phenolate ligating functionality and a delocalized chromophore (**10b**). In contrast, monovalent cations do not result in a shift to the zwitterionic form (**10a**), and hence the compound undergoes large spectral changes specifically on divalent metal ion binding.[5,12] The terminal group concept can also be extended to three dimensions as in tripodal receptors such as the "noncyclic cryptand" **9**,[13] giving more three-dimensional encapsulation and hence protection from the surrounding medium. Without the rigid end groups, such tripodal hosts are often highly flexible as a consequence of inversion at the bridgehead nitrogen atoms. In recent years, the ability of a podand to wrap around a central alkali metal cation has been used to make interesting ion pair receptors such as the system shown in Figure 5 which acts as a receptor for NaI, with the Na^+ cation bound by the ether oxygen atoms while the iodide anions interact with the iodotetrafluorophenyl substituents forming a bridge between two different receptors.[14]

The structural chemistry of polyether podands has been the subject of an extensive review which contrasts open and cyclic structures.[16] While podands often wrap around their guest metal cations in a similar way to crown ethers, forming, for example, five-membered chelate rings, their open structure leads to helical twisting as a result

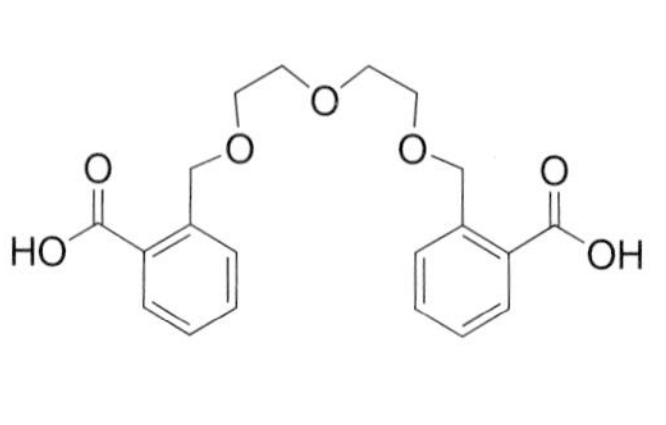

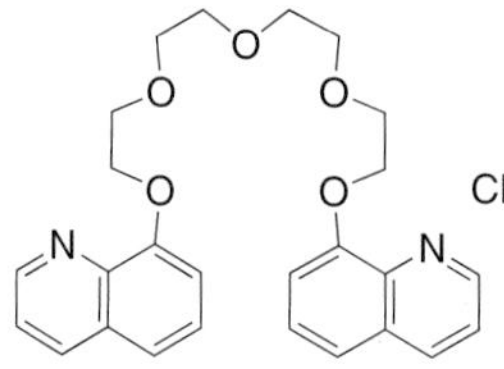

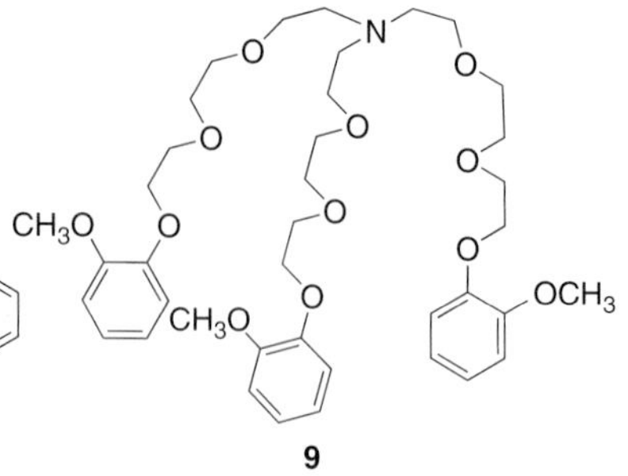

7 **8** **9**

Neutral form with monovalent cations

Zwitterionic form with divalent cations

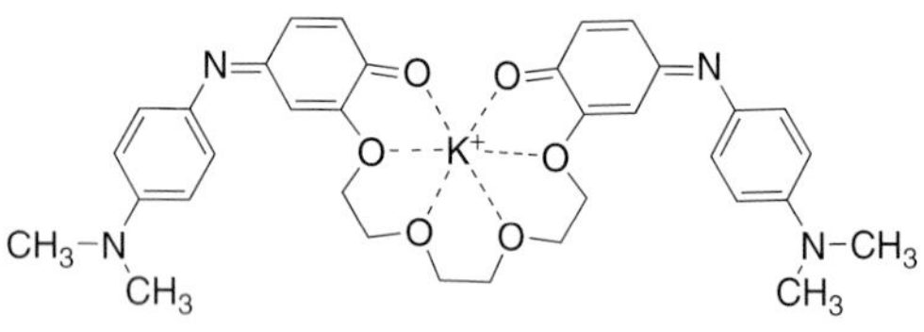

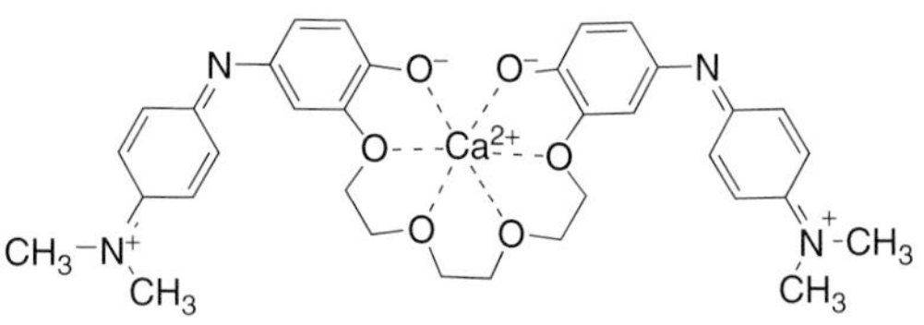

10a **10b**

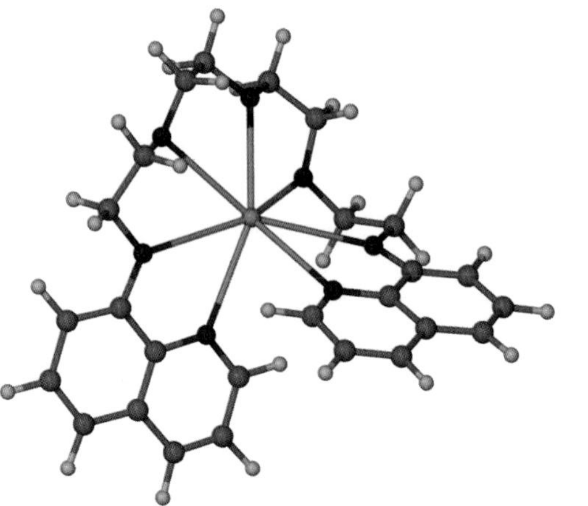

Figure 4 X-ray molecular structure of the cation in the RbI complex of **8**.[15]

of end-group steric interactions as in the neodymium complex of hexaethylene glycol (Figure 6a)[17] and much more frequent occurrence of bridged structures as in the cadmium and mercury complexes of triethylene glycol (Figure 6b and c).[18,19]

1.2 Factors affecting selectivity

The cation-binding selectivity of both cyclic and acyclic ligands including podands is assessed by the relative binding constants for the complexation of the guest cations in a given solvent at a particular temperature. The magnitude of the binding constant and hence thermodynamic stability of the resulting complex is influenced by a number of factors, of which the most important are listed below[7]:

- size complementarity between cation and host cavity;
- electronic complementarity between the cation and host binding sites;
- electrostatic charge;

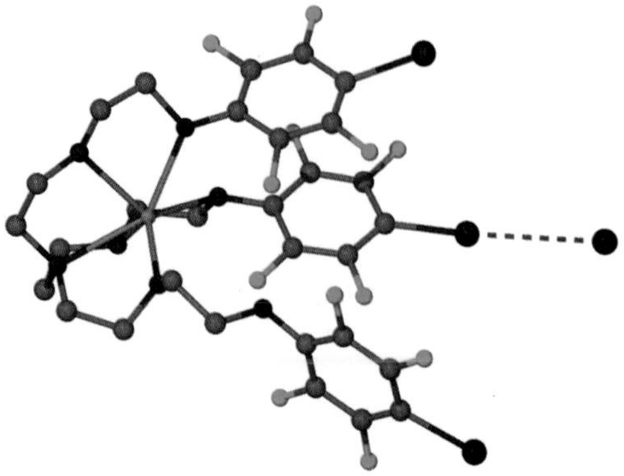

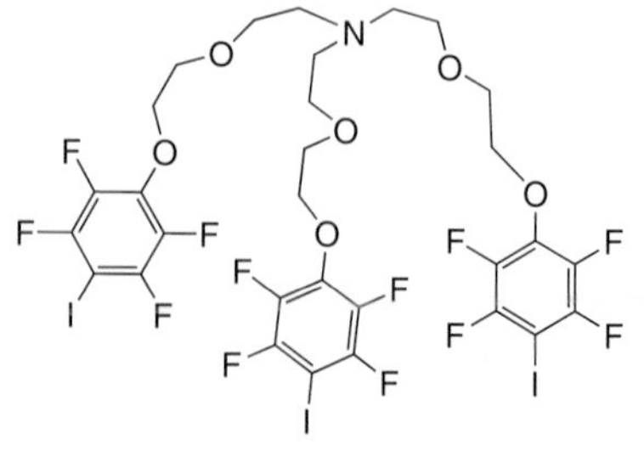

Figure 5 Line drawing and X-ray molecular structure of a podand host for NaI acting *via* halogen bonding interactions.[14]

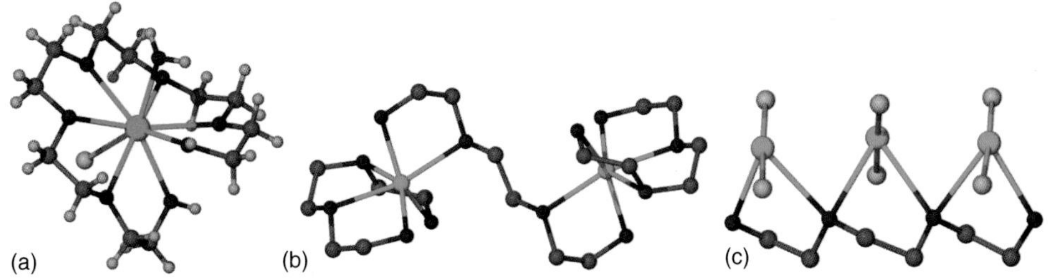

Figure 6 X-ray structure of podand complexes (a) $[NdCl(OH_2)(hexaethyleneglycol)]Cl_2$,[17] (b) $[Cd_2(triethyleneglycol)_2(\mu\text{-}triethyleneglycol)]Cl_4 \cdot 2H_2O$[18] (H atoms omitted), and (c) $[HgCl_2)_2(triethyleneglycol)]$.[19]

- solvent (polarity, hydrogen bonding, and coordinating ability);
- degree of host preorganization (terminal group concept);
- enthalpic and entropic contributions to the cation–host interaction;
- cation and host free energies of solvation;
- nature of the counter-anion and its interactions with solvent and the cation;
- cation-binding kinetics;
- chelate ring size and donor group orientation.

Among these factors the flexibility of podands means that criteria such as size complementarity and preorganization play less of a role than they do in macrocyclic hosts and podands do not benefit from the stabilization classically regarded as arising from the "macrocyclic effect." As a result, podands generally display "plateau selectivity" in their binding constants and extraction behavior of metal salts from water to organic media. In a recent study on the extraction of alkali metal picrates by the podands shown in Figure 7, for example, the ability of the podands to extract lithium picrate proved to be essentially independent

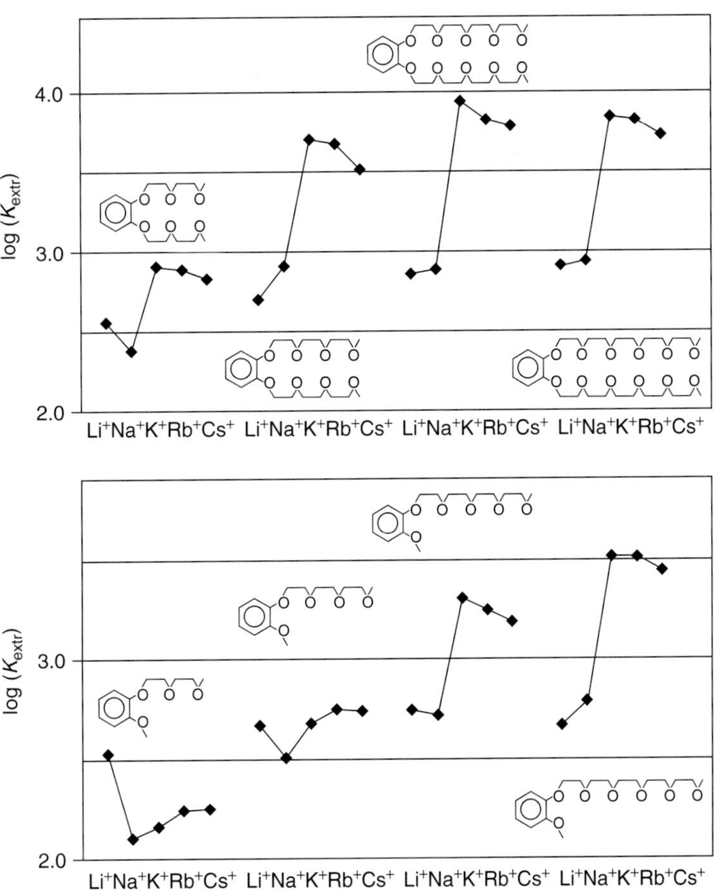

Figure 7 Values of extraction constant from water to dichloromethane at 25 °C (log K_{extr}) found for the podands shown for a range of alkali metal picrates. (Reproduced from Ref. 20. © Springer, 2010.)

of the number of ethyleneoxide units in the podand chain. Sodium extraction efficiency increases with the number of donor atoms up to ~7, which for potassium and cesium increases up to about eight oxygen atoms. These results were rationalized by the saturation of the cation's first coordination sphere once a critical chain length is reached. Podands with seven or more oxygen atoms were able to discriminate between sodium and potassium, but no discrimination was observed between potassium, rubidium, and cesium picrates.[20]

Even though they are relatively unselective, podands do chelate their guest species and hence chelate ring size and donor group orientation are of particular importance. The issue of optimal donor group orientation is crucial[21] and unlike macrocycles in which the orientation can be constrained by the host ring structure, podands can potentially adopt a more optimal donor group orientation and hence form enthalpically somewhat stronger individual metal–ligand bonds.

For polyamines such as ethylene diamine which involve sp^3 hybridized nitrogen atoms, optimal metal binding involves an M–N–X angle of 109.5° with a M–N–C–X torsion angle of ±60 or 180°. This arrangement results in the M–N axis being aligned with the orientation of the amine lone pair. As the complex deviates from this angle, the M–N interaction becomes weaker and so, for example, a 10° compression of the M–N–C angle results in a loss of ~2.9 kJ mol^{-1} to the bond strength. In the case of ethers, the ideal orientation involves alignment of the M–O bond with the ether dipole to give a trigonal planar oxygen atom; that is, an M–O–C angle of 123.5° with the metal in the plane of the C–O–C moiety and hence a M–O–C–X torsion angle of either 0 or ±120°. Again deviation weakens the bond; so, for example, a 30° displacement of Na^+ from the C–O–C plane weakens the bond by ~4 kJ mol^{-1}, Figure 8. This consideration appears strange given the commonly observed structures of crown

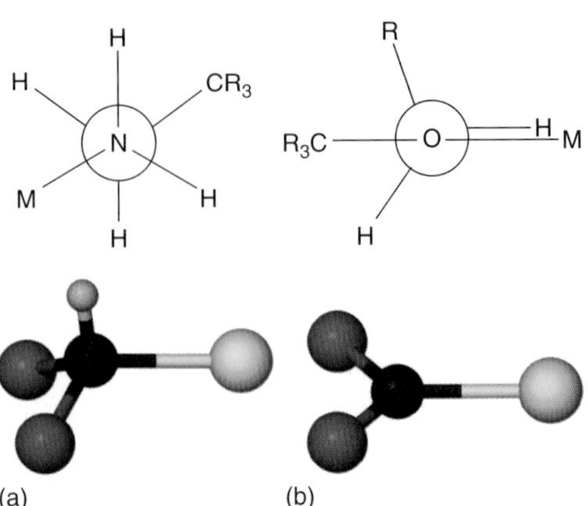

Figure 8 Optimal amine (a) and ether (b) donor group orientations. Note that metal–ether interactions are optimal with a planar C_2–O–M unit.

ether complexes which usually exhibit a pyramidal rather than planar C_2–O–M unit. The crown ethers induce strain into the bond because of their cyclic structure and are generally atypical of acyclic ligands.

Even in the case of chelating podands, however, the existence of multiple donor groups can mean that not all of them can simultaneously form optimal bond to a metal center because of ligand backbone orientational effects. The optimal metal–ligand "fit" in this regard can be visualized by drawing lines along the vectors of optimal metal ion approach. For a podand in an optimal conformation (either preorganized or able to rotate to adopt the most complementary conformation), these vectors will overlap at a single point at the optimal M–L bond length. If the vectors diverge from this situation, the binding is expected to be weaker and the conformation less optimal. Examples of the optimal donor group orientation for ethylene diamine, propylene diamine, and dimethoxy ethane are shown in Figure 9. There is a marked difference between ethylene diamine (five-membered chelate rings)

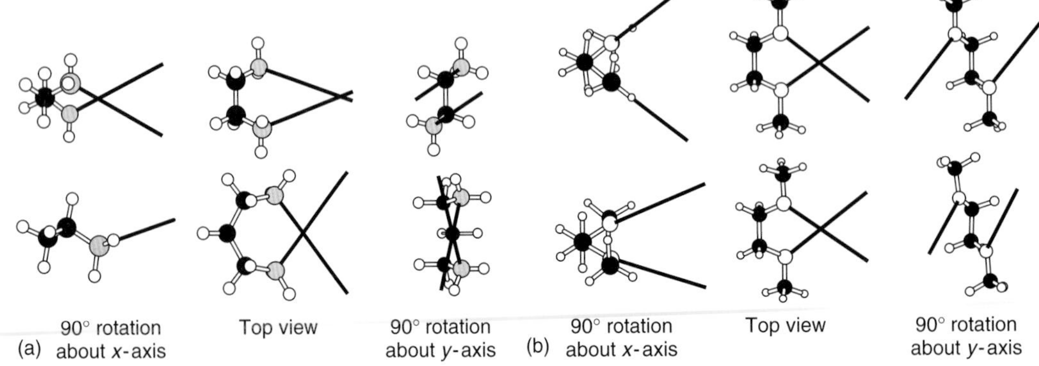

Figure 9 (a) Three views of ethylene diamine (top) and 1,3-propylene diamine (bottom) and (b) three views of 1,2-dimethoxyethane (dme) before (top) and after (bottom) Na^+ complexation. The 5-Å long vector attached to each oxygen illustrates the optimal line of approach for the metal ion. (Reproduced from Ref. 21. © Elsevier, 2001.)

Valinomycin

Nigericin, $R^1 = CH_2OH$, $R^2 = OH$
Grisorixin, $R^1 = OH$, $R^2 = CH_3$

Monesin

Figure 10 Naturally occurring cyclic and acyclic ionophores.

and propylene diamine (six-membered rings). For ethylene diamine (top), the optimal line of metal ion approach vectors crosses at a much longer M–L bond length than that in the six-membered ring analog. This observation is a general result, and hence ligands that form five-membered chelate rings form stronger interactions with larger metal ions, that is, those that form long M–L bonds, while six-membered rings favor smaller metal ions.

1.3 Natural ionophore podands

Natural evolution has produced a range of molecules that act as ionophores (cation carriers). Perhaps one of the best known is valinomycin, first isolated from the bacterium *Steptomyces fulvissimus* in 1955. This macrocyclic compound catalyzes the exchange of K^+ and H^+ across the membrane of mitochondria within cells, and is selective for K^+ over Na^+. This ion transport ability is linked to the function of this type of compound as an antibiotic. Chemically valinomycin is made up of a cyclic threefold repetition of four amino acid residues: L-valine (Val), D-hydroxyisovaleric acid (Hyi), D-valine, and L-lactic acid (Lac), and the cyclic conformation is supported by hydrogen bonding of type N–H···O=C. Interestingly, not all natural ionophores are macrocyclic like valinomycin, however, and a number of acyclic podands including the closely related nigericin, grisorixin, and monesin are also known (Figure 10). Monesin was the first such ionophore antibiotic to be characterized by X-ray crystallography, as its silver(I) salt.[22] The structure revealed that the podand adopts a *pseudo*cyclic structure by virtue of an intramolecular hydrogen-bonding interaction. The structure of the monohydrate (a complex with water as the guest) is also cyclic (Figure 11).[23] Selected alkali metal ion binding constants for some cyclic and podand ionophores are shown in Table 1. Clearly, the cyclic valinomycin is highly selective for K^+ over Na^+ because the macrocycle cannot contract down sufficiently to bind the smaller metal ion. However, Rb^+ and Cs^+ also bind strongly, but of course are not significant competing ions in biological media. In comparison, the acyclic nigericin is highly unselective. Monesin is unusual in that it is one of the few receptors that is Na^+ selective. The reasons for this selectivity are not straightforward since, while the fit for Na^+ is good, so is the fit for Ag^+ which is similar in size to K^+.

Monesin has proved to be a popular target for the design of artificial biomimetic ionophores. An example is **11** which possesses six ether oxygen atoms with the potential to bind cations assuming that the hydroxyl and carboxylate groups are involved in intramolecular hydrogen bonding.

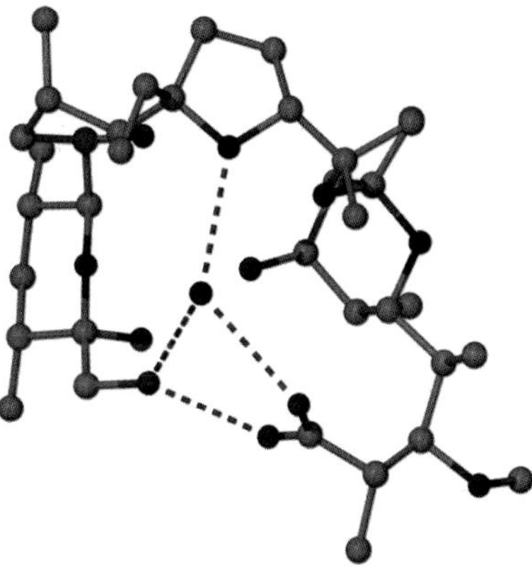

Figure 11 X-ray crystal structure of monesin hydrate showing the intramolecular hydrogen bonding (H atoms omitted).[23]

Table 1 Log K_{11} values for alkali metal ion binding by naturally occurring ionophores in methanol solvent at 25 °C.

Ligand	Li^+	Na^+	K^+	Rb^+	Cs^+	
Valinomycin	<0.7	0.67	4.90	5.26	4.42	Macrocycle
Monactin	<0.3	2.60	4.38	4.38	3.30	Macrocycle
Enniatin B	1.28	2.42	2.92	2.74	2.34	Macrocycle
Nigericin	—	4.7	5.6	5.0	—	Podand
Monensin	3.6	6.5	5.0	4.3	3.6	Podand

The ability of this compound and a number of analogs to transport metal ions was assessed by an H-cell apparatus with aqueous source and receiving phases at the bottom of each side of the H, separated by a bulk liquid membrane in the form of 1-hexanol. The transport rate follows the order $Li^+ > Rb^+ > Na^+ \approx K^+ > Cs^+$ which does not correlate particularly well with that observed for monesin.[24] A key design feature of **11** is its ability to mimic the monesin intramolecular hydrogen bond, a structural feature that is also integral to the design of **12** and related carboxylic acid derivatives.[25]

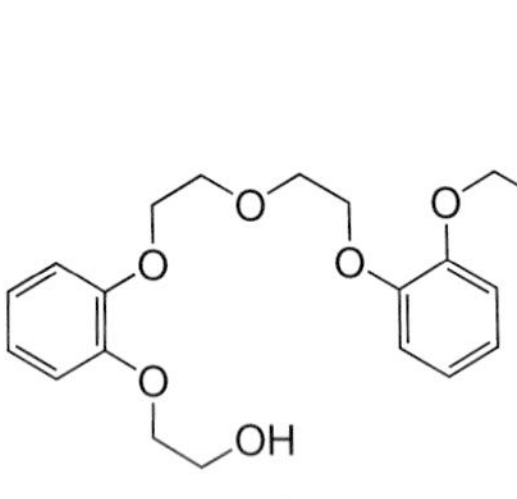

11 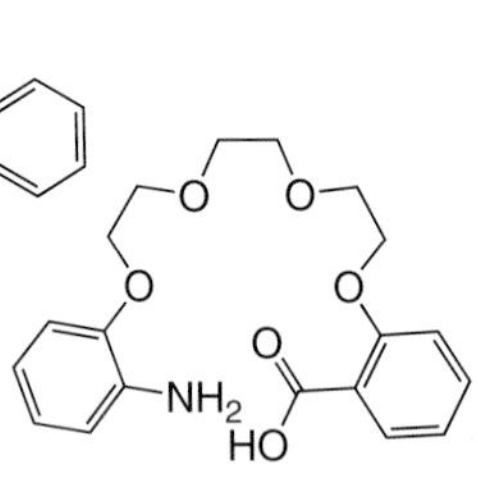 **12**

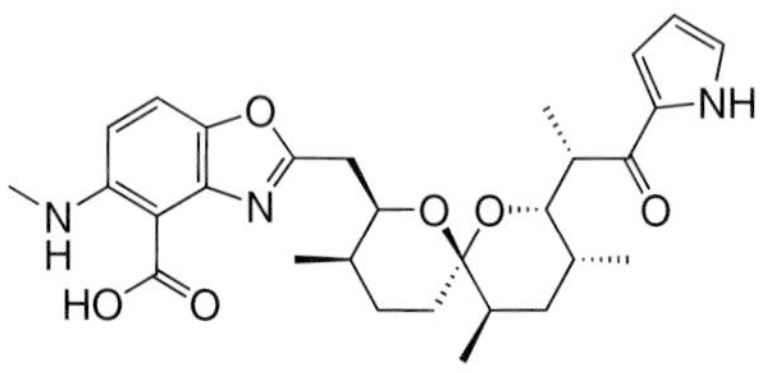

Calcimycin (**13**)

Monesin tends to bind most strongly to monovalent metal ions and hence is sometimes referred to as a "monovalent polyether antibiotic." There also exist a class of "divalent polyether antibiotics" such as lasalocid A and calcimycin (**13**) that binds to both monovalent and divalent metal ions. Lasalocid A exhibits a preference for the doubly charged species. The order of complexation strength $Ba^{2+} > Cs^+ > Rb^+ \approx K^+ > Na^+ \approx Ca^{2+} \approx Mg^{2+} > Li^+$ reflects a combination of ion size and charge density issues.[26] Calcimycin is produced by fermentation of *Streptomyces chartreusensis* and acts as an antibiotic against gram positive bacteria and fungi. It allows divalent cations to cross cell membranes with selectivity in the order $Mn^{2+} \gg Ca^{2+} > Mg^{2+} \gg Sr^{2+} > Ba^{2+}$.[27]

1.4 Nitrogen podands

In addition to the classic bidentate chelate ligand ethylene diamine, a tremendous variety of linear polyamine ligands such as spermine, spermidine, putrescine, and cadaverine are known, many of which have biochemical roles as their somewhat evocative names suggest (Figure 12). Both spermidine and spermine, which is formed from it, are involved in cellular metabolism in eukaryotic cells. Spermine is an essential growth factor in some bacteria and exists as a polycation at physiological pH because the propylene and butylene spacers reduce electrostatic repulsion on protonation, increasing basicity. As a result, protonated compounds of this type can also act as simple anion receptors as well as ligands for cations. For example, crystals of spermine phosphate were first described as early as 1678 by Anton van Leeuwenhoek, who obtained them from human semen. The proton binding ability of polyamines has been reviewed.[28] Spermine is associated with nucleic acids and is thought to stabilize helical structure, particularly in viruses. Putrescine and cadaverine are produced by the breakdown of amino acids, particularly in dead organisms, and both are toxic in large doses. They cause the foul smell of rotting flesh and are also implicated in the smell of bad breath.

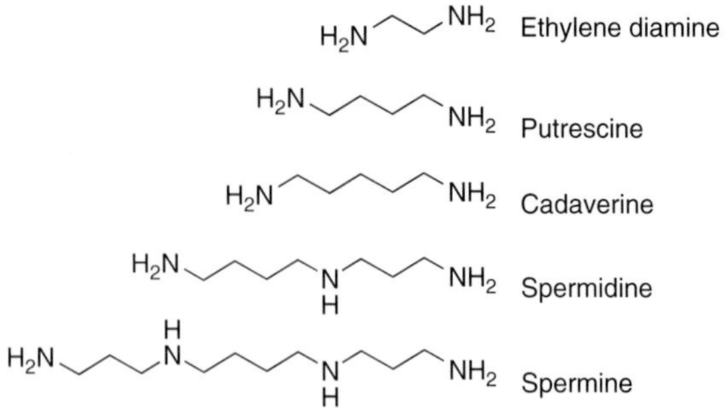

Ethylene diamine
Putrescine
Cadaverine
Spermidine
Spermine

Figure 12 Examples of polyamine podands.

Early work on complexes of these kinds of ligands clearly showed the additional stabilization imparted by the macrocyclic effect. The Zn(II) complexes **14** and **15**, both of which form complexes using four chelating donor atoms, are compared. However, the macrocyclic complex **14** is about 10^4 times more stable than the podand analog **15** as a consequence of the additional preorganization of the macrocyclic effect.[29] Despite this relative instability, however, polyamine podands bind reasonably effectively to a range of transition metals. Spermine, for example, forms a mononuclear complex with copper(II) sulfate containing both six- and seven-membered chelate rings, Figure 13(a).[30] In contrast, a binuclear bis(bidentate) complex is known with palladium(II) chloride in which the coordinating chloride ligands occupy two of the four sites on each square planar palladium(II) center to which the spermine is bound by six-membered chelate rings, Figure 13(b).[31] Interestingly, a related complex is known for lithium iodide even though alkali metal salts are not generally complementary to amine donor ligands in comparison to oxygen donors. The

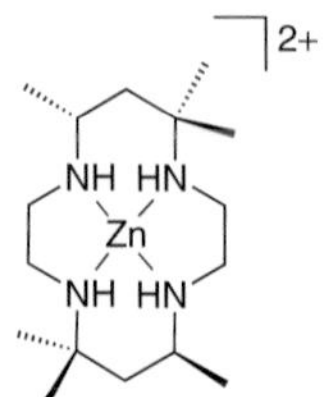

14 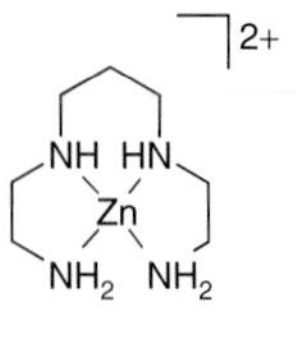 **15**

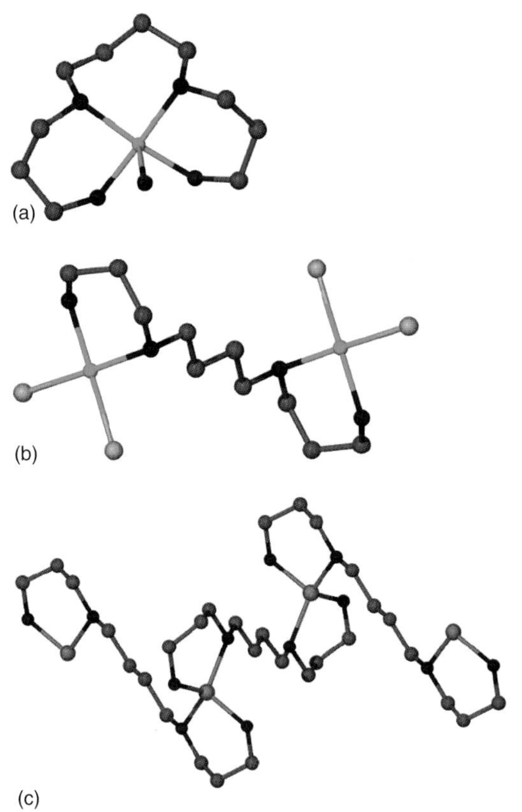

Figure 13 X-ray structures of metal complexes of spermine (H atoms omitted): (a) mononuclear copper(II) sulfate aquo complex,[30] (b) dinuclear palladium(II) chloride complex,[31] and (c) coordination polymeric lithium iodide complex.[32]

structure is a 1-D coordination polymer based on tetrahedral Li^+ ions, Figure 13(c).[32]

Perhaps one of the most important classes of N-donor podand-type ligand are the polypyridyls.[33] The combination of rigidity, excellent stability with a range of particularly lower oxidation state transition metals, and synthetic versatility has led to a cornucopia of metallosupramolecualr polypyridyl complexes. Broadly speaking, the three most important classes of compound are (i) helicates[34] (including circular helicates[35,36]), (ii) arrays such as grids, racks, and ladders,[37,38] and (iii) coordination polymers.[39] Work particularly by the groups of Constable,[40] and Lehn[37,41] among others in the late 1980s and early 1990s popularized helicates (helical metal complexes) as abiotic single, double, and triple helical supramolecular frameworks with interesting topology and as intermediates in topologically complex synthesis such as the preparation of a molecular trefoil knot.[42]

The simplest type of helicate is a mononuclear single helix exemplified by the silver(I) hexafluorophosphate complex of 2,2′:6′,2″:6″,2‴:6‴,2⁗-quinquepyridine which is near-planar but twisted into a shallow helical conformation as a result of unfavorable steric interactions between the pyridyl terminii of the podand, Figure 14(a).[43]

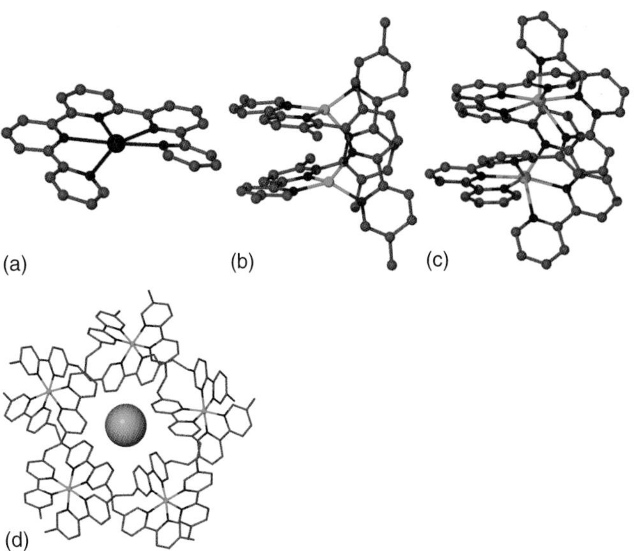

Figure 14 Helicates derived from oligopyridyl ligands: (a) mononuclear single helicate,[43] (b) binuclear double helicate based on four-coordinate metal ions,[44] (c) binuclear double helicate based on six-coordinate metal ions,[45] and (d) circular helicate templated by a chloride anion.[36]

Oligopyridyl ligands with greater terminal steric bulk or additional pyridyl rings as in 2,2′:6′,2″:6″,2‴:6‴,2⁗:6⁗,2⁗′-sexipyridine display a more pronounced helicity. In the case of metal ions with a tendency toward tetrahedral or octahedral coordination geometries, the [4 + 4] or [6 + 6] double helicates result, as in the dicopper(I) perchlorate complex of tetramethyl-2,2′:6′,2″:6″,2‴-quaterpyridine (**16**) in the case of 4-coordinate copper(I), Figure 14(b) or a range of complexes of 6-coordinate metals such as Cd^{2+}, Fe^{2+}, Co^{2+}, Ni^{2+}, and Cu^{2+} with 2,2′:6′,2″:6″,2‴:6‴,2⁗:6⁗,2⁗′-sexipyridine (**17**), Figure 14(c).[44,45] Interestingly, while the hexadenate sexipyridine can form a double helicate with the Jahn–Teller distorted copper(II), quaterpyridine does not because the preference for a distorted octahedral geometry of the metal ion is inconsistent with the helix-forming requirements of the ligand. As a result, reduction of the mononuclear copper(II) quaterpyridine complex results in redox-reversible helicate formation. One-electron oxidation of the compound gives a mixed-valence Cu(I)–Cu(II) helicate, which on further oxidation decomposes to give the mononuclear Cu(II) species (Scheme 2). The Cu(II)/Cu(I) redox interconversion in helicates with various podands has been extensively studied and different electrochemical behavior is observed according to the structural features such as denticity, rigidity, and steric hindrance of the helicand.[46] In one case, the mixed-valence Cu(II)/Cu(I) helicate complex is stabilized by specific metal–metal interactions and can be isolated and structurally characterized by X-ray crystallography.[47]

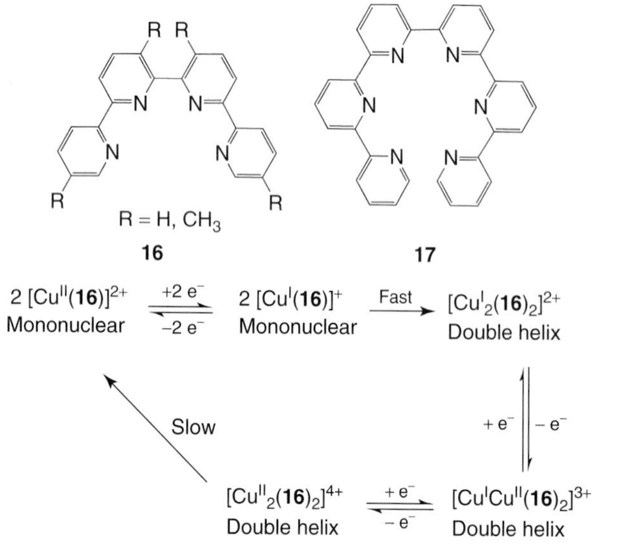

R = H, CH$_3$

16 **17**

2 [Cu^{II}(**16**)]$^{2+}$ Mononuclear ⇌ (+2 e$^-$ / −2 e$^-$) 2 [Cu^{I}(**16**)]$^+$ Mononuclear → (Fast) [Cu$^{I}_2$(**16**)$_2$]$^{2+}$ Double helix

[Cu$^{II}_2$(**16**)$_2$]$^{4+}$ Double helix ⇌ (+e$^-$ / −e$^-$) [Cu^{I}Cu^{II}(**16**)$_2$]$^{3+}$ Double helix; Slow

Scheme 2 Helix formation as a function of copper oxidation state.[48]

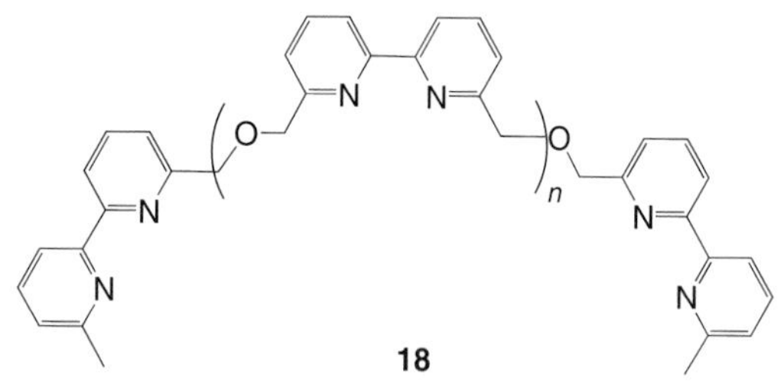

18

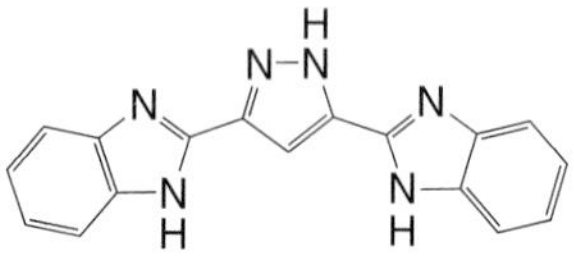

19

Dinuclear helicates are far from being the limit in terms of the number of metal centers, and the reaction of copper(I) salts with ligands of type **18** gives tri-, tetra-, and pentanuclear double helicates based on tetrahedral copper(I).[41] Helicates containing even more metal ions are known, often based on very simple ligands. For example, the pyrazole derivative **19** forms penta- and heptanuclear helicates with Zn(II) and Cd(II) salts.[49] A particularly unique form of helicate is the circular helicate discovered by Lehn's group in 1996, Figure 14(d). The complex is based on a podand with three bipyridyl binding domains and self-assembles around a chloride anion to give a pentanuclear complex with Fe(II).[50] The analogous hexafluorophosphate salt is hexanuclear while a variation in the ligand spacer gives a tetranuclear species.[35]

Pyridyl ligands are not the only podands to form helicates. Extensive work from Raymond's group in Berkley has been based on deprotonated 1,2-dihydroxy benzene-derived podands. Depending on the spacer unit between two of these binding domains, a large number of triple helices and hollow tetrahedral coordination shells have been prepared by self-assembly. The remarkable chemistry of the hollow shell compounds is covered in **Self-Assembly of Coordination Cages and Spheres**, Volume 5 and **Reactivity in Nanoscale Vessels**, Volume 4. For example, a mixture of three such ligands in a 3 : 2 ratio with Ga^{3+} gives the selective formation of three individual, homoleptic triple helices (Scheme 3).[51]

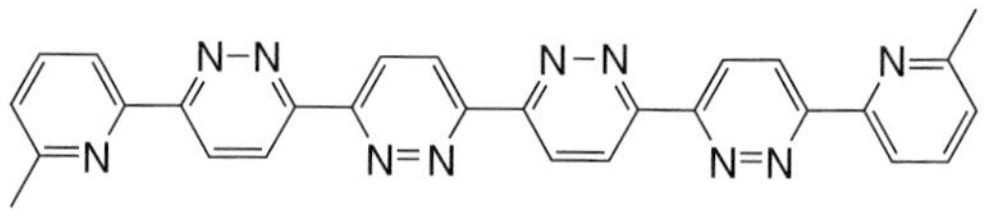

20

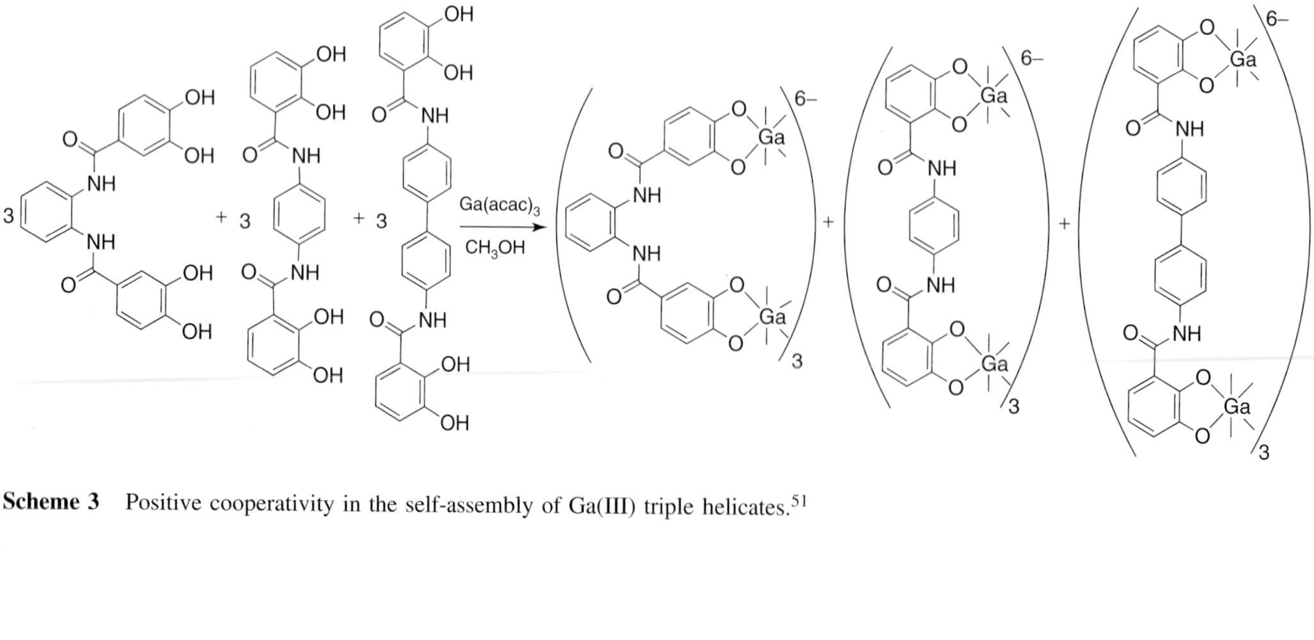

Ga(acac)$_3$, CH$_3$OH

Scheme 3 Positive cooperativity in the self-assembly of Ga(III) triple helicates.[51]

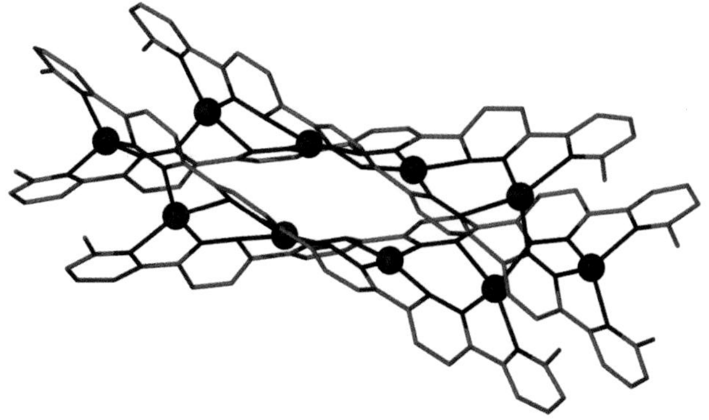

Figure 15 Decanuclear quadruple helicate.[52]

One of the most striking helicates is the decanuclear quadruple helicate formed from the reaction of silver(I) triflate and ligand **20**, Figure 15. The compound is formed at the same time as a 20 metal atom grid complex comprising two separated 2×5 arrays of Ag(I) ions.[52] Amid a growing range of spectacular structural studies, it is also noteworthy that helicates of lanthanoid metal cations have been used as model systems in a full thermodynamic analysis of the self-assembly process in the form of the extended site binding model.[53,54] Further details of helicates and related complexes are given in **Self-Assembly of Coordination Chains and Helices**, Volume 5.

1.5 Tripodal ligands

Among the most enduringly popular podand ligands are tripodal species based on tris(aminoethyl)amine (tren, **21**). A huge range of pyridyl-based tripodal ligands have also been realized. While they are relatively flexible molecules, metal complexation results in a rigid array of chelate rings which can be used, for example, in face capping an octahedral metal center. Ligand of type **21** finds application in the extraction of metal ions; for example, compounds **21** (R = benzyl or napthylmethyl) both extract perrhenate in protonated form and are significantly more effective than compound **22** and in the case of the benzyl derivative, even more effective than cryptand analogs.[55] Dipicolylamine (**23**) has been used very effectively in a range of tren-related tetrapodal architectures to produce dimetallic complexes with interesting sensing and catalytic functionality and the subject has been reviewed.[56] For example, compound **24** is an effective fluorescent sensor for dianionic phosphate derivatives, particularly peptides with phosphotyrosine residues.[57] The complex works by a three-component self-assembly mechanism with the phosphate derivative enhancing binding of the second zinc ion and preventing PET (photo-induced electron transfer) quenching from the uncoordinated amine nitrogen atom in the mono-zinc precursor. Compound **25** also acts as a sensor for phosphates, exhibiting a red-shift on addition of pyrophosphate. Unlike compound **24**, however, the compound functions as a single molecular unit.

Tripodal receptors based on an arene core have proved particularly popular, in part due to the degree of preorganization afforded by steric hindrance around the arene core. Early work by Lehn's group showed that the tripodal receptor **26** in conjunction with planar ligand **27** gives a discrete capped trinuclear complex on reaction with copper(I).[58]

R = H, CH_2Ph, $CH_2C_{10}H_7$

21 **22** **23**

4 NO_3^-

24 **25**

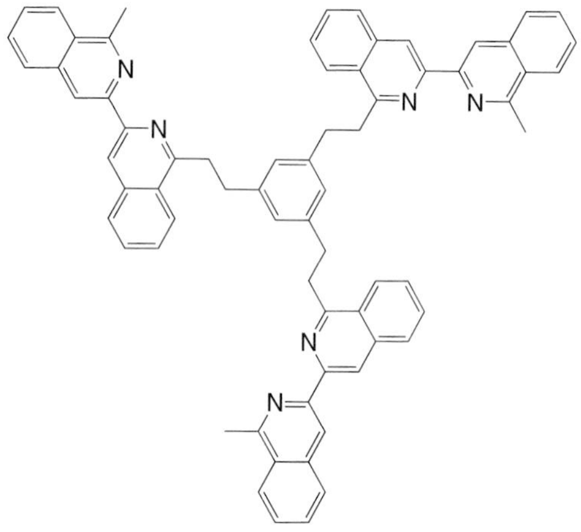

26

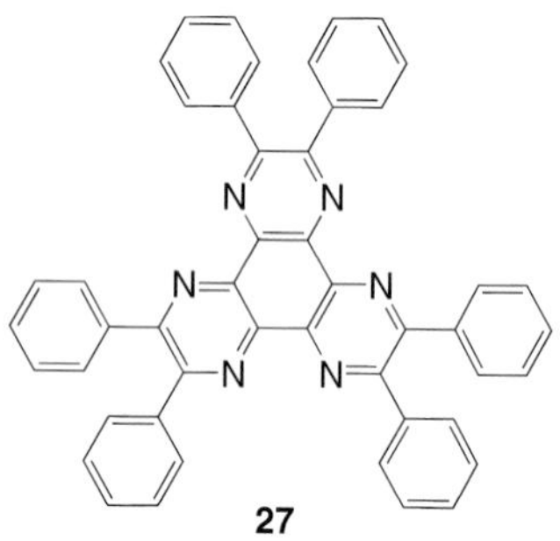

27

Discrete mononuclear complexes stable in solution over long periods are afforded by the novel "coelenterand" (hollow stomach) ligand **28**.[59] This coelenterand exhibits two different coordination modes, in which the metal ion is either inside or outside the aryl "stomach" (Figure 16). There is a relatively conventional tripodal chelating mode (analogous to face-capping ligands such as triphosphines, tris(pyrazolyl)borates, and tris(pyrazolyl)methanes). The more unusual encapsulating mode is exhibited by the complex produced by reaction of **28** with $[RuCl_2(DMSO)_4]$ which gives a fully encapsulating complex that can be crystallized in the presence of $[ZnCl_4]^-$. In the X-ray crystal structure, the Ru(II) is bound both to the pyrazole nitrogen atoms and the carbon atoms of the arene ring, exhibiting a short Ru–ring centroid distance of 1.58 Å (compared to normal values in organometallic complexes of about 1.67–1.70 Å), suggesting that the arene–π interaction is enhanced by the chelation of the ligand.

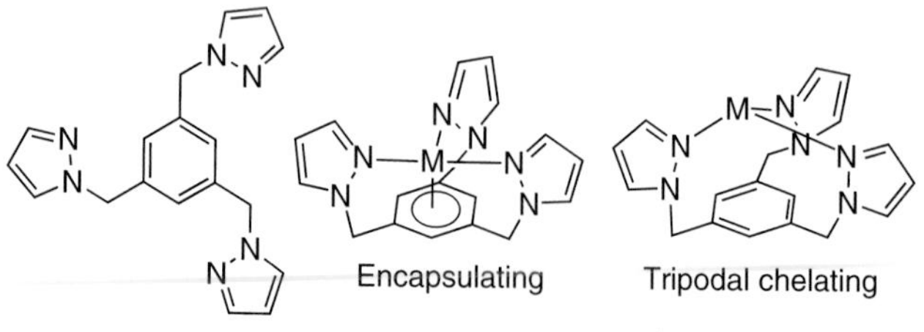

28 Encapsulating Tripodal chelating

Figure 16 Coordination modes of the coelenterands.[59]

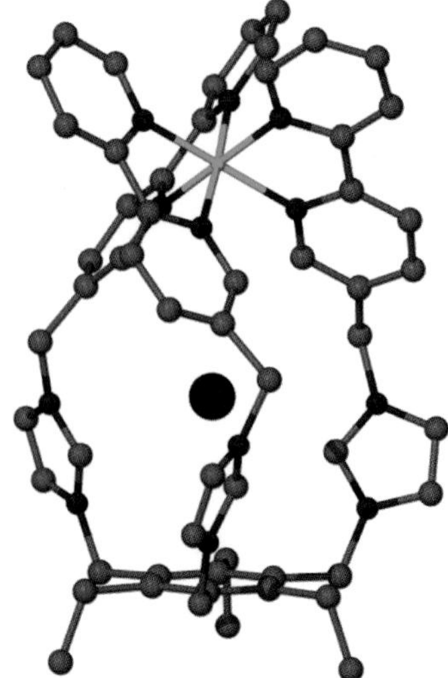

Figure 17 X-ray crystal structure of the Fe(II) complex of a 2,2′-bipyridyl derived tris(imidazolium) receptor including Br^-.[60]

This kind of concept has been extended to imidazolium-based ligands, in which three 2,2′-bipyridyl substituents bind an Fe(II) center resulting in the formation of an anion-binding pocket capable of complexing Cl^- and Br^- in acetonitrile solution with $\log K > 7$. The X-ray crystal structure of the metallocryptand shows that halides are bound solely by CH···anion interactions, Figure 17.[60]

Deprotonation of imidazolium ligands gives the highly topical *N*-heterocyclic carbene (NHC) class of ligands which exhibit strong σ-donor character and represent interesting alternatives to phosphines.[61] Carbenes are beginning to be used extensively in catalysis and coordination chemistry and a number of multidentate derivatives of the podand type have been developed. Ligands **29** and **30** represent dipodal carbenes, both of which form mononuclear palladium(II) complexes with catalytic activity in cross-coupling reactions.[62] The tren-based ligand **31** forms both a 2 : 3 complex with copper(I) in which the copper centers are linear, two-coordinate and a mononuclear tripodal tris-chelate.[63] In contrast, the dipodal ligand **32** is not geometrically disposed to chelate a single metal center and forms a 2 : 2 metallomacrocycle with palladium(II) chloride.[64]

Tripodal benzene-derived ligands have also been used as siderophores (strong complexants for Fe^{3+}) which are of significant biochemical and medical interest. The chemistry and biology of siderophores, which feature a number of interesting natural podands such as mycobactin, have recently been comprehensively reviewed.[65] One early example is the tricatecholate mesitylene derivative **33** which is structurally related to the naturally occurring enterobactin but without its ester linkages (which are sensitive to hydrolysis). This ligand is a remarkably strong binder of iron(III) with a binding constant of 10^{46}, although this is still some million times lower than that for the enterobactin natural analog. The mesityl spacer group in **33** is smaller than enterobactin's lactam ring perhaps resulting in a more strained complex geometry. Tests in *Escherichia coli* culture show that the complex is able to help the bacteria accumulate the very insoluble iron(III) and hence promote growth.[66] Unsurprisingly, the hexadeprotonated nature of the bound ligand makes iron complexation highly pH dependent, and under mildly acidic conditions (pH 5) there is significant competition for Fe^{3+} from $EDTA^{4-}$ (ethylenediaminetetracetic acid) which is a stronger acid in its protonated form. The chemistry of EDTA, one of the best known podands, is covered in the next section.

1.6 EDTA-type ligands

One of the best known podand-type ligands is EDTA which, in its tetraanionic form, is an extremely common chelate

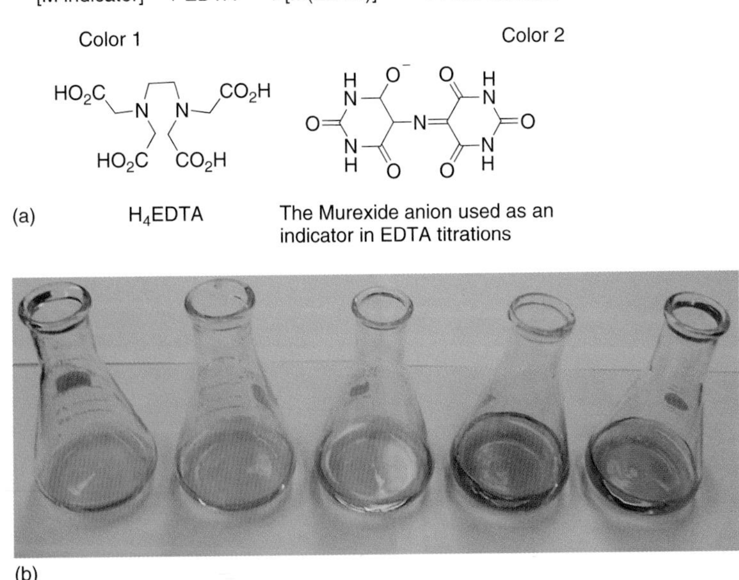

Scheme 4 (a) Metal ion analysis using indicator displacement by the strongly binding podand $EDTA^{4-}$. (b) Color changes during the titration of Ni^{2+} with $EDTA^{4-}$ and murexide under basic conditions.

ligand used as a strong complexant in analytical chemistry, for example, in the analysis of Ca^{2+} and Mg^{2+} in urine samples. Metal ion analysis is commonly carried out using the indicator displacement assay technique during complexometric titration as shown in Scheme 4, in which a colored indicator such as the murexide ion in the determination of calcium is displaced from the metal center by complexation with $EDTA^{4-}$. The flexible nature of $EDTA^{4-}$ means that it binds a wide variety of metal cations with selectivity depending mainly on the metal ion charge (highly charged ions are bound more strongly) and size. Hard metal ions are favored because of the carboxylate donors, and hence softer metals such as Ag(I) are bound relatively weakly, although notably the binding of the univalent alkali metal ions that cannot interact with the amine nitrogen atoms is very weak indeed, Table 2. While EDTA commonly binds octahedral metal ions as a hexadentate chelate as in the classic $NH_4[Co(EDTA)]\cdot 2H_2O$ published in 1959[67] (Figure 18a), larger metals such as Mn^{2+} can adopt a seven-coordinate structure involving coordinated water, as in $[Mg(H_2O)_6][Mn(EDTA)(H_2O)]\cdot H_2O$ (Figure 18b).[68]

The success of EDTA chelates has sparked extensive research on a wide range of analogs including the closely related trimethylenediamine tetraacetate (tmdta), which as its Fe(III) complex is used in the bleaching of photographic films and paper, to more unusual analogs such as **34**

Table 2 Stability constants (log K) in aqueous solution for metal complexes of $EDTA^{4-}$.[7]

Mg^{2+}	8.7	Zn^{2+}	16.7	La^{3+}	15.7
Ca^{2+}	10.7	Cd^{2+}	16.6	Lu^{3+}	20.0
Sr^{2+}	8.6	Hg^{2+}	21.9	Sc^{3+}	23.1
Ba^{2+}	7.8	Pb^{2+}	18.0	Ga^{3+}	20.5
Mn^{2+}	13.8	Al^{3+}	16.3	In^{3+}	24.9
Fe^{2+}	14.3	Fe^{3+}	25.1	Th^{4+}	23.2
Co^{2+}	16.3	Y^{3+}	18.2	Ag^{+}	7.3
Ni^{2+}	18.6	Cr^{3+}	24.0	Li^{+}	2.8
Cu^{2+}	18.8	Ce^{3+}	15.9	Na^{+}	1.7

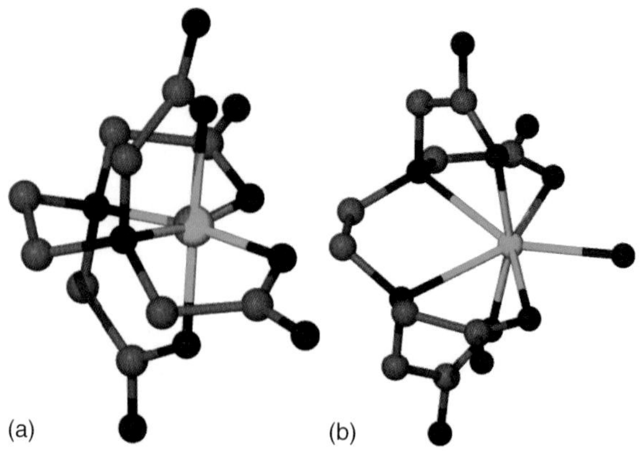

Figure 18 X-ray molecular structure of EDTA complexes (a) $NH_4[Co(EDTA)]\cdot 2H_2O$[67] and (b) $[Mg(H_2O)_6][Mn(EDTA)(H_2O)]\cdot H_2O$.[68]

which are designed to lower the overall ligand charge. This podand shows selectivity for Cd(II) and Pb(II) over Zn(II). The ligand forms significantly more stable complexes with Cd(II) and Pb(II) than EDTA and hence has potential applications in the extraction of these metals.[69] EDTA chemistry has also inspired a range of lariat-type ligands for lanthanoid metal ions that have found extensive applications in biomedical imaging, for example, as MRI (magnetic resonance imaging) contrast agents or as luminescent anion and pH sensors.[70] One such complex, **35**, has been shown to be an effective lactate and citrate sensor in diluted microliter samples of human serum, urine, or prostate fluids allowing a simple, fast method of detecting prostate cancer.[71] Similarly, paramagnetic lanthanoid complexes of the EDTA analog **36** find clinical applications as contrast agents in MRI.[72]

1.7 Octopus-type podands

Since podands benefit from multiple interactions with their guest cations, significant research has been carried

34 **35** **36**

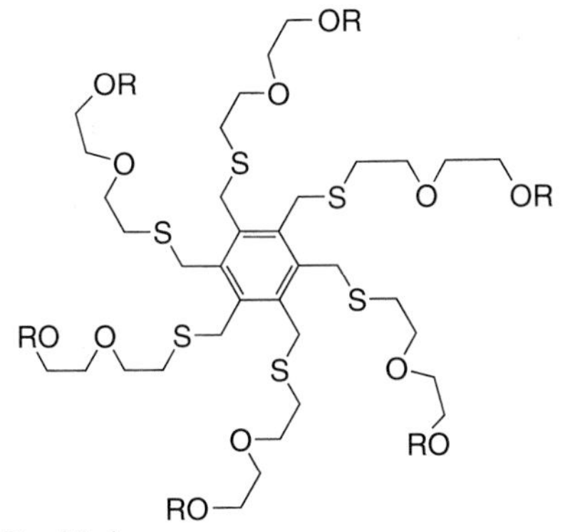

R = Alkyl
varying numbers of ethyleneoxide units

37

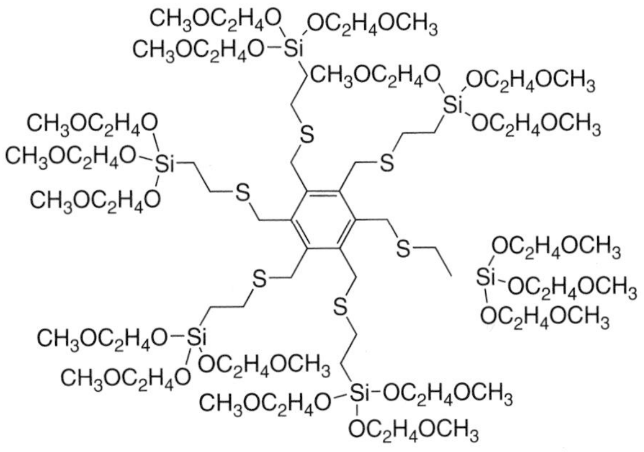

38

out on systems with multiple arms, of which the "octopus" podands are perhaps the best representative example. Named for their multiple (although not necessarily eight) tentacles, octopus podands such as **37** and **38** have six arms radiating out from a hexasubstituted aryl core.[73,74] Compounds of type **37** are extremely effective alkali metal ion extraction agents; even more so than the crown ethers despite their water insolubility. Their broad ^{1}H NMR spectra suggest restricted conformational motion about the hexasubstituted core. The siloxane analogs **38** can, interestingly, be polymerized into hybrid silicas.

1.8 Salt-binding ligands

A key application area in podand chemistry is in the binding and extraction of metal salts. Since metal ions necessarily come with an accompanying counter-anion, significant effort has been devoted to the design of systems capable of binding anions and metal cations simultaneously. By including both anion- and cation-specific recognition functionalities, the selectivity and, in some cases, catalytic specificity can be dramatically enhanced. This section looks at systems in which the metal ion is relatively labile and the focus is on binding both the metal and the anion. We resume the topic in Section 2.4 below which covers more inert metal complexes specifically designed to bind anions. Some of the earliest simultaneous receptors are "cascade complexes" which originally date back to the 1970s,[75] in which a podand such as a Schiff's base binds metal cations which in turn coordinate to the counter-anions. Cascade receptors have been extensively used as models for enzyme active sites.[76] Copper complexes are especially popular in this regard because of interest in dicopper hemocyanin respiratory proteins that are responsible for oxygen transport in molluscs and some arthropods, and studies on the copper-containing enzyme tyrosinase. For example, podands of type **39** exhibit modest antiferromagnetic coupling between the two copper(II) centers, which are ~3 Å apart. While this situation does not accurately reflect the properties of hemocyanins, the presence of different environments for the two copper atoms (one is distorted trigonal bipyramidal while the other is square pyramidal) may be related to the different modes of bonding proposed for the two copper atoms in tyrosinase.[77] Manganese(II) cascade complexes of another unsymmetrical ligand **40** represent functional models for manganese catalase, an enzyme that catalyzes the disproportionation of hydrogen peroxide into dioxygen and water. The complex is again asymmetrical as shown by the X-ray structure of $[Mn_2(\mathbf{40}-H)(CH_3CO_2)_2(NCS)]$ (**41**), in which one Mn(II) center is distorted square pyramidal while the other is distorted octahedral.[76] This bimetallic Schiff's base motif underlies a large number of bimetallic cascade podand complexes, as in sterically hindered divanadyl complex **42** which binds alkoxide ions in a cascade fashion, while the copper analog binds oxide in a 2 : 1 sandwich fashion, allowing the oxide ligand to simultaneously interact with four copper(II) centers in two separate complexes.[78]

Compounds **43** and **44** represent cascade complexes that bind hydrogen phosphate anion, HPO_4^{2-}, in water. The tetrahedral copper(II) center coordinates the phosphate anions which are further stabilized by charge-assisted hydrogen bonding to the tripodal peripheral groups. Complex **43** is protonated at neutral pH and binds HPO_4^{2-} with $K_a = 2.5 \times 10^4\ M^{-1}$ in water. The more preorganized **44** binds phosphate more weakly at $1.5 \times 10^4\ M^{-1}$ but is more selective for this anion. Thermodynamic studies show that phosphate binding by **43** is entropically driven, while complexation by **44** is enthalpy based as a result of decreased solvent organization around the guanidinium groups in **44** compared to the more exposed ammonium groups in **43**.[79]

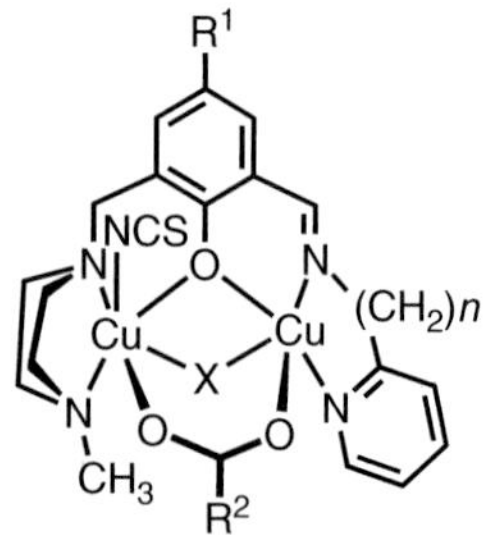

R^1 = Br, CH_3
R^2 = H, CH_3
X = Br, OAc
n = 1, 2

39

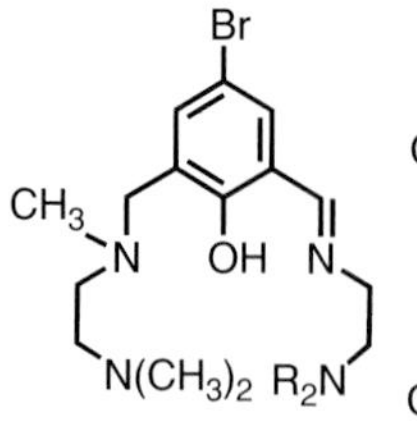

R = CH_3, Et

40

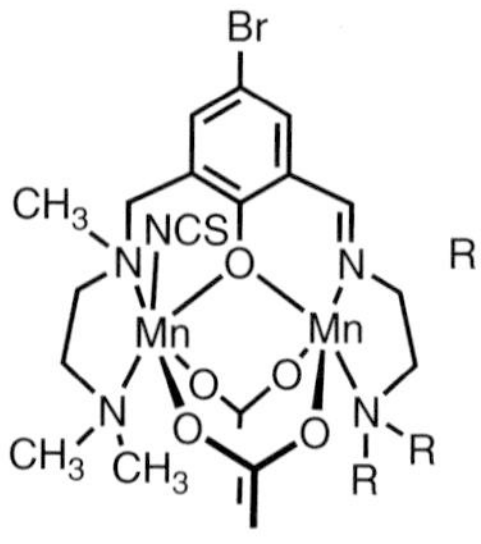

41

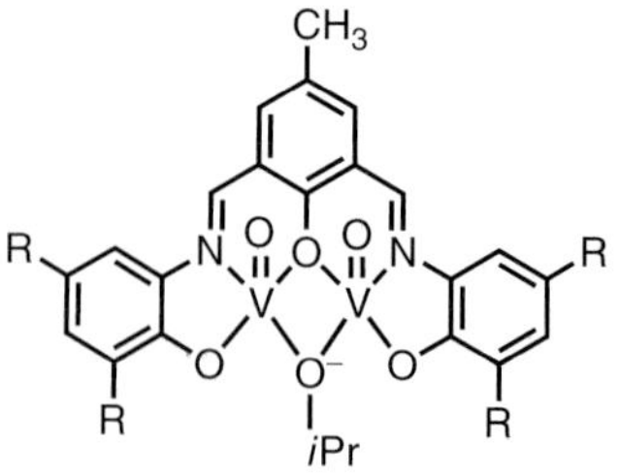

R = ^{t}Bu

42

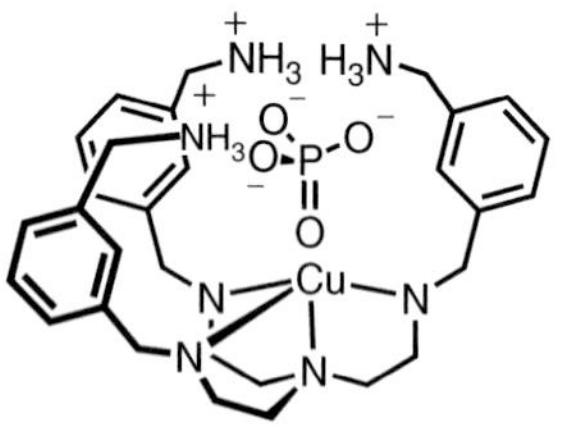

43

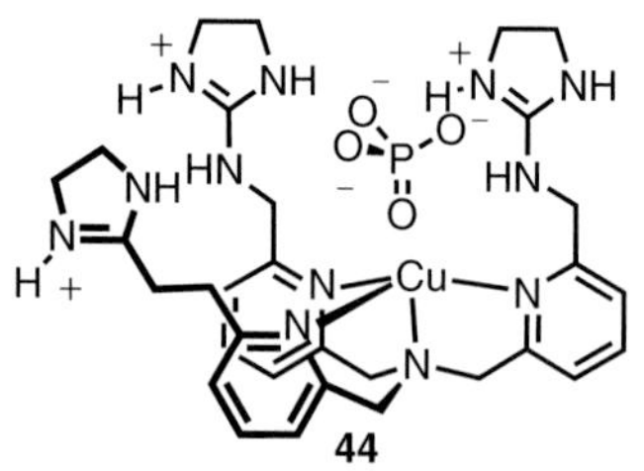

44

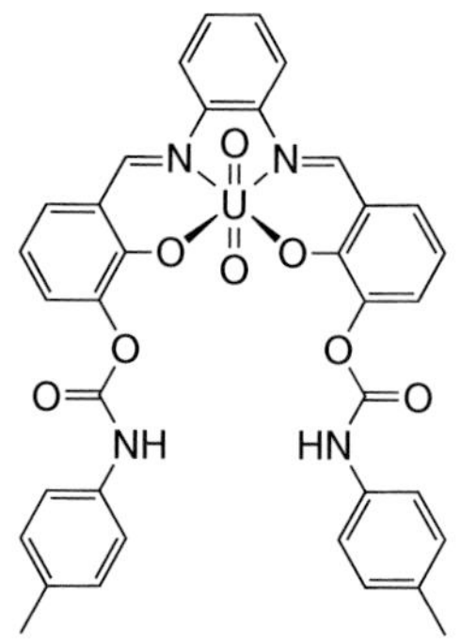

45

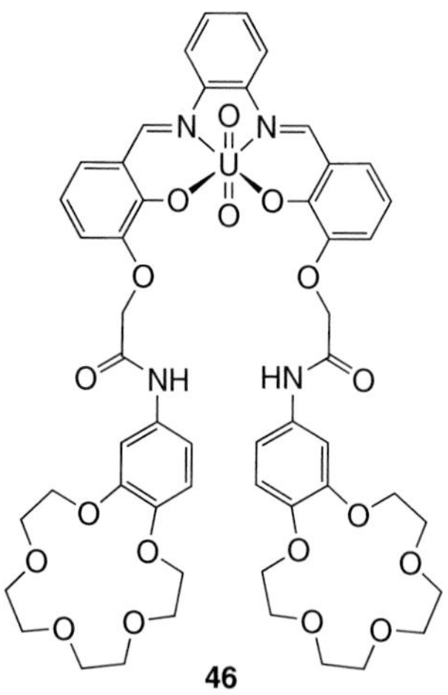

46

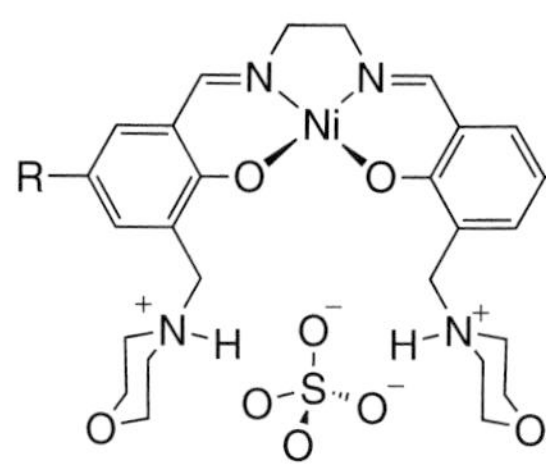

R = ^{t}Bu

47

A ditopic, Schiff's base podand that simultaneously binds uranyl ions while hydrogen bonding to anions such as dihydrogen phosphate via pendant arms is known (**45**).[80] The compound forms an interesting 1 : 2 host:guest complex with $H_2PO_4^-$, with one phosphate strongly bound to the metal center and the other hydrogen bonded to the first. Addition of crown ether moieties as in **46** results in binding of K^+ as well as phosphate.

The salen-based receptor **47** is an example of a selective extractant for metal sulfates. The ligand forms a complex involving deprotonated phenolic hydroxyl groups and protonated morpholine residues adapted for hydrogen bonding to the anion. The ligand readily extracts $CuSO_4$ into chloroform solution, for example, with essentially 100% efficiency.[81]

2 ANION-BINDING PODANDS

Anion binding has emerged over the last 20 years as a highly active research area, not least as exemplified by an entire issue of *Chemical Society Reviews* dedicated to the topic in 2010.[82] Anions are ubiquitous in biology with up to 75% of enzyme substrates and cofactors being anionic. They also play a role in disease with cystic fibrosis caused by a defective chloride transport protein.[7,83] Anions can also have a large environmental impact, for example, perchlorate pollution in the Colorado river and the radioactive anion $^{99}TcO_4^-$ which can leach from nuclear waste.[7] There is a wealth of literature on anion recognition with interest in both macrocyclic and podand systems.[7,83–85] Podands have a great deal of potential in anion binding and selected acyclic anion receptors have been the subject of a recent review.[86] As with cation binding, the preorganization of a podand for anion binding can be tuned quite readily by altering the rigidity or tailoring steric or other noncovalent interactions to finely tune the system.

The development of anion-binding podands over the past 20 years highlights the huge diversity and diversification of the podand concept and the majority of this chapter is devoted to anion-binding podands. Anion-binding podand receptors are dominated by NH hydrogen bond acceptors such as amide,[87,88] pyrrole,[89] indole,[90] and urea/thiourea[87] derivatives.[83] Charged systems containing guanidinium[91,92] and imidazolium[93] functional groups are also commonly used. Podands provide not only receptors which are often synthetically simpler than macrocyclic systems[87] but the flexibility in binding group choice and variable level of preorganization, which in turn allows a great deal of flexibility in receptor design to suit almost any application. Crucially, as in cation-binding podands, anion-binding systems offer flexibility generally leading to rapid binding and decomplexation kinetics[7,94] coupled with binding constants that are suitable for analytical and sensing applications. It is perhaps fair to say that anion binding is the major focus of work in the podand field at present and the following sections give an overview of some representative anion-binding podands that cover a range of structural and functional features, highlighting the wealth of variety that is possible by tuning the preorganization of a receptor, changing the binding moiety, and introducing reporter groups which can turn receptors into sensors.

2.1 Cholapods—preorganized anion-binding podands

2.1.1 *Concept and properties*

Podand systems are intrinsically more flexible than macrocyclic systems; however, this reduction in preorganization can also lead to lower binding affinities for anions. Rigid, aryl end groups can increase binding affinity[1]; however, to truly enhance binding strength, a rigid receptor design with a highly preorganized binding cavity is required. This concept is typified best in anion-binding receptors by the cholapods, pioneered by Anthony Davis of Bristol University, United Kingdom.[95] The cholapods are derivatives of the bile acid and cholic acid, and are based on a steroidal, fused alicyclic ring system. This provides a highly rigid scaffold on which binding functionality can be added. The cholic acid core allows for variation in design by both regio- and stereo-control.[96]

The conformation of the core cholic acid scaffold is curved with three hydroxyl groups on the α surface (Figure 19). The equatorial 3α-OH group is the least hindered, while the 12α-OH is less hindered than the 7α-OH because of unfavorable 1,3-diaxial interactions between the 7α-OH and CH_2 groups. The hydroxyl groups can themselves act as hydrogen-bond donors to bind anions; however, it is possible to convert all or individual hydroxyl functional groups to amine or amide moieties to further enhance the hydrogen-bonding capabilities or to provide a versatile receptor design. The large extended structure, with separated functionality of the cholapods, is an advantage when trying to bind often large anions.[96]

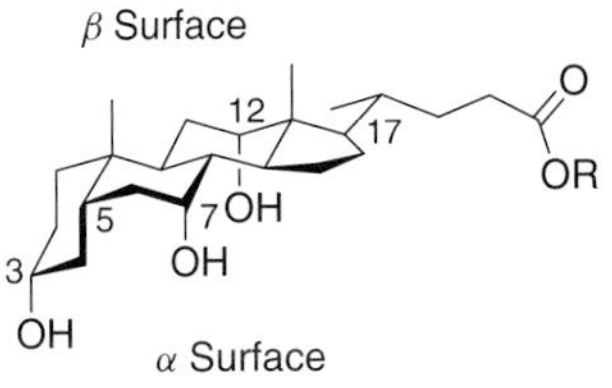

Figure 19 Basic structure of cholic acid derivatives.

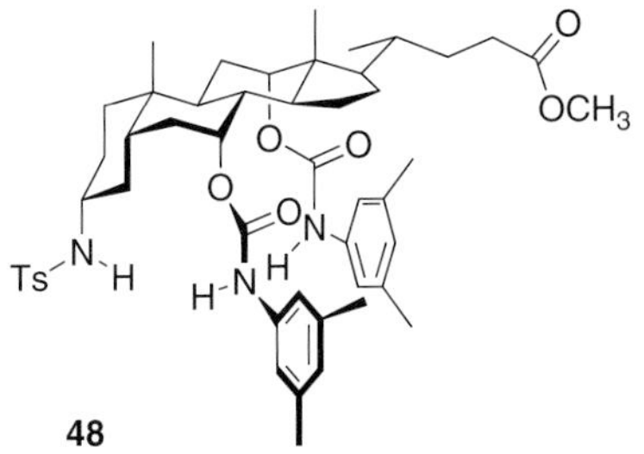

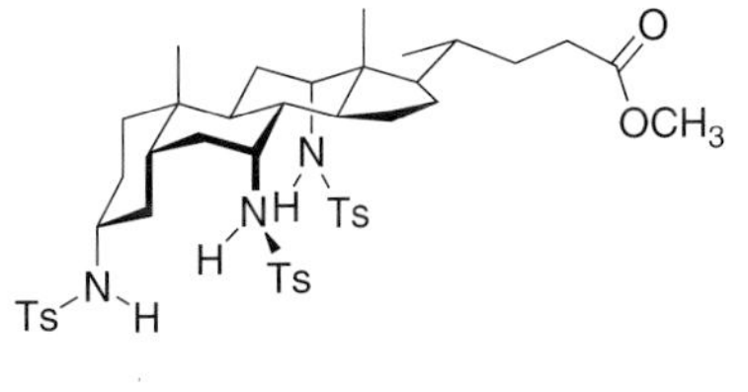

Neutral anion receptors have many potential advantages over charged systems. Primarily they may provide more selective binding by ensuring directionality, as anisotropic hydrogen bonding is the primary binding interaction rather than nondirectional electrostatic attraction. In addition, neutral systems do not have counterions which can compete to bind to the receptor. Binding constants measured are therefore absolute affinities and not relative to the counterion affinity. Neutral podand cholic acid derivatives have been studied extensively.[96,97] A range of neutral receptors have been produced, incorporating hydrogen-bond donor NH functionality to the α surface of the cholic acid in the form of amides, sulfonamides, carbamates, ureas, and thioureas. Examples of first generation receptors of this type are compounds **48** and **49**.[98] In **48**, free rotation is possible around the C_3–N bond; however, further preorganization (in addition to that provided by the core) is achieved by the restricted rotation of the carbamate-O/NH groups and the preferential *Z,Z*-conformation across the carbamate moieties. Binding affinity was measured by ^1H NMR spectroscopic titrations in $CDCl_3$, with **49** showing large affinity, for chloride (92 000 M^{-1} as the tetraethylammonium TEA salt), two orders of magnitude higher than its closest rival bromide. Binding constants are smaller for **48**, with fluoride binding being the strongest, $K_a = 15\,400\,M^{-1}$ (fluoride was not measured for **49**).

Further refinement of the receptor design has involved the addition of electron-withdrawing substituents on the binding arms, increasing the acidity of the hydrogen-bond donor and hence increasing the potential strength of the hydrogen bonds to guest anions. Receptors **50** and **51** exhibit a *p*-nitrophenylsulfonyl group, and the use of nitro or trifluoromethyl groups on the carbamates leads to a 10-fold increase in the binding constant ($3.4 \times 10^7\,M^{-1}$ for chloride (as the TEA salt) for **51** compared to $92\,000\,M^{-1}$ for **48**).[99]

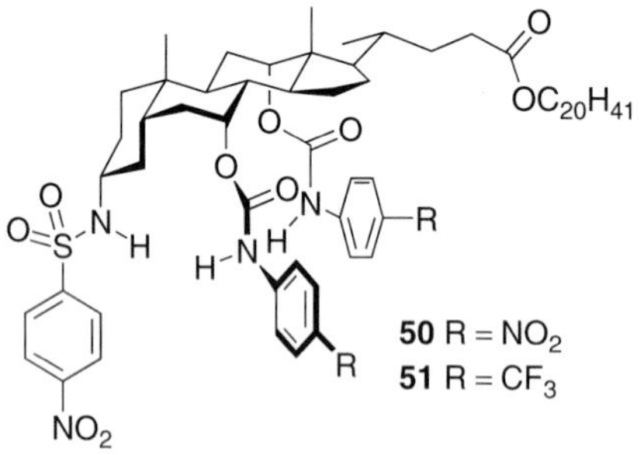

50 R = NO_2
51 R = CF_3

By increasing the number of hydrogen-bond donors to five using urea or thiourea moieties, exceptionally high binding constants can be achieved. For example, compound **52** shows a chloride (as the TEA salt) association constant of $1.03 \times 10^{11}\,M^{-1}$, the highest binding affinity measured for a neutral organic anion receptor, and highlights the level of binding strength that can be achieved through highly preorganized receptors, with convergent binding cavities and multiple, strong hydrogen-bond donors.[100]

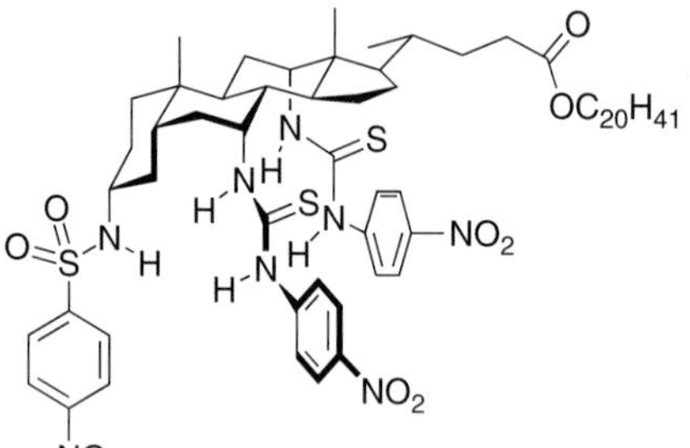

52

It is also possible to produce charged cholapods, typically using guanidinium or ammonium groups but also imidazolium[101] and triazolium[102] moieties. These charged systems have been employed as anion receptors, "smart transfer agents," and membrane transport anionophores.

2.1.2 *Membrane transport by cholapods*

Transmembrane transport of anions is a highly active area of research with many implications for biological and medicinal chemistry,[103] most notably in the case of the genetic disease cystic fibrosis which arises through mutations in the cystic fibrosis transmembrane conductance regulator protein, which acts both as a transmembrane chloride channel and as a regulator of other ion channels.[7]

Neutral cholapods have been shown to be effective anionophores, transporting nitrate, hydrogen carbonate, and notably chloride, particularly in the case of compound **52**. The anion transport was measured using a dye technique in which the fluorescence of lucigenin encapsulated in a vesicle is quenched by the inward flow of chloride ions.[104]

The charged guanidinium derivatives such as compound **53** proved capable of extracting *N*-acyl amino acids from an aqueous phosphate buffer into chloroform, often with good efficiencies (measured using NMR spectroscopy) and good enantioselectivies.[105, 106]

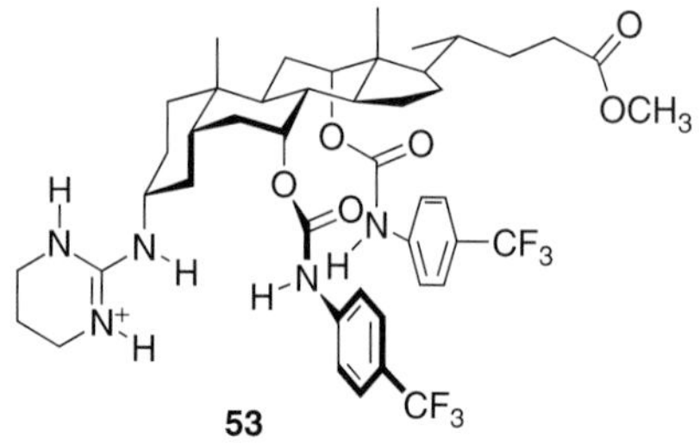

53

Membrane transport experiments using compound **53** and a U-tube apparatus showed that the compound can transport *N*-acetyl-DL-phenylaniline with 70% enantiomeric excess, with the anions bound by hydrogen bonds from the guanidinium and carbamate NH groups.

Ammonium-based cholapods have also been investigated as "smart transfer agents" as well as membrane transporters. The use of ammonium groups provides not only a positive charge but also hydrogen-bonding groups. Receptors of type **54** transfer anions from the aqueous to the organic phase.[107] The extent of anion transfer is dependent on the lipophilicity of the anion (given by the Hofmeister series, Table 3[7]). It was hoped that anion recognition would allow anti-Hofmeister behavior in which less lipophilic anions are extracted more readily. However, while the lipophilic anions were extracted less preferentially, the order of extraction in the Hofmeister series remained intact.

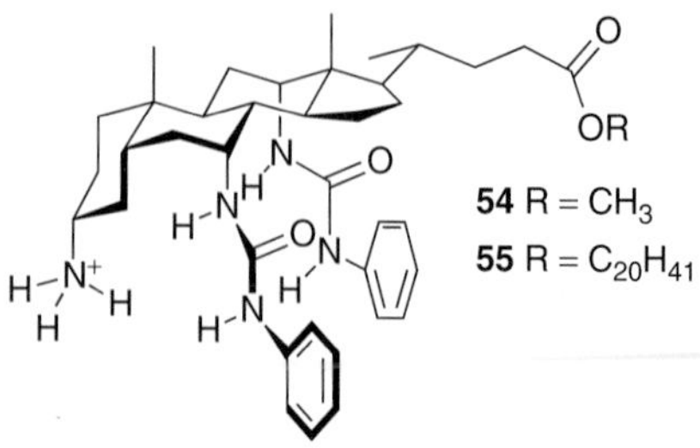

54 R = CH_3
55 R = $C_{20}H_{41}$

While compound **54** is ineffective at membrane transport of anions, the addition of a long alkyl chain as in **55** does result in the transport of anions through a chloroform

Table 3 The Hofmeister series.

Weakly hydrated (hydrophobic)	Strongly hydrated (hydrophilic)
Anions:	
Organic anions > ClO_4^- > I^- > SCN^- > NO_3^- > ClO_3^- > Br^- > Cl^- ≫ F^-, IO_3^- > $CH_3CO_2^-$, CO_3^{2-} > HPO_4^{2-}, SO_4^{2-} > citrate^{3-}	
Cations:	
$N(CH_3)_4^+$ > NH_4^+ > Cs^+ > Rb^+ > K^+ > Na^+ > H^+ > Ca^{2+} > Mg^{2+}, Al^{3+}	

liquid membrane (as part of a U-tube set-up) with a small selectivity for chloride.

A bicyclic guanidinium core has been utilized by de Mendoza and coworkers for transport of uronic acid salts. Receptors **56** and **57** contains a bicyclic guanidinium group connected to two modified deoxycholic acid motifs.[108] The binding of D-glucuronate, **58**, (as the tetrabutylammonium, TBA, salt) was evaluated using ^{1}H NMR spectroscopic titrations in acetonitrile. The receptors with the most flexible linker group showed the lowest binding constants, consistent with reduced preorganization. The glucuronate is bound to the guanidinium groups via the carboxylate moiety with the hydroxyl groups of the deoxycholic acid derivatives providing additional hydrogen bonds to the carbohydrate alcohol groups. The largest contribution to the binding strength is from the ion pairing of the carboxylate and the guanidinium groups.

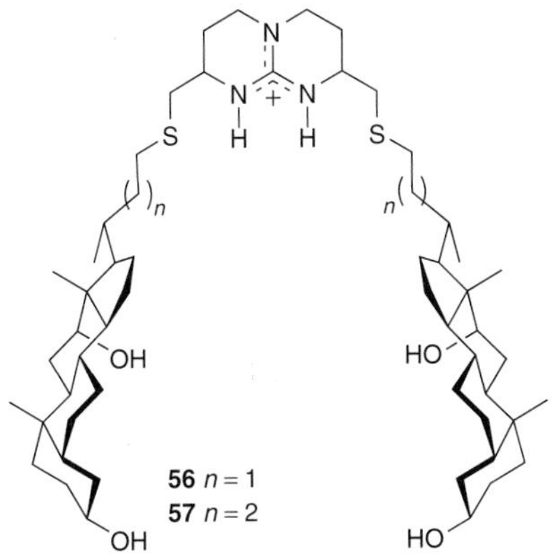

56 $n = 1$
57 $n = 2$

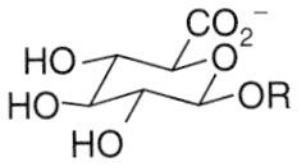

58

Regan and coworkers[109] have synthesized a range of charged anion transporters. An interesting example is that of **59**, which is known as a *molecular umbrella*. This dicationic species containing ammonium and guanidinium functionalities can bind adenosine triphosphate (ATP) through the interaction of the phosphate residues of the anion with the guanidinium moiety. The cholic acid groups can then encapsulate the anion by OH hydrogen bonds to the inner surface, presenting a lipophilic outer surface, allowing the complex to pass through the membrane.

2.1.3 *Anion sensing by cholapods*

The addition of fluorescent moieties can provide a reporter group which, in principle, can result in a selective receptor with high anion binding that can signal this binding through changes in the fluorescence emission. The advantage of a podand-type receptor is that binding and release are generally fast in comparison to rigidly preorganized macrocyclic systems.

Fang and coworkers[110,111] have developed cholapods functionalized at the C-24 position, which provides both additional hydrogen-bonding functionality and an anthracenyl reporter group, compounds **60** and **61**.

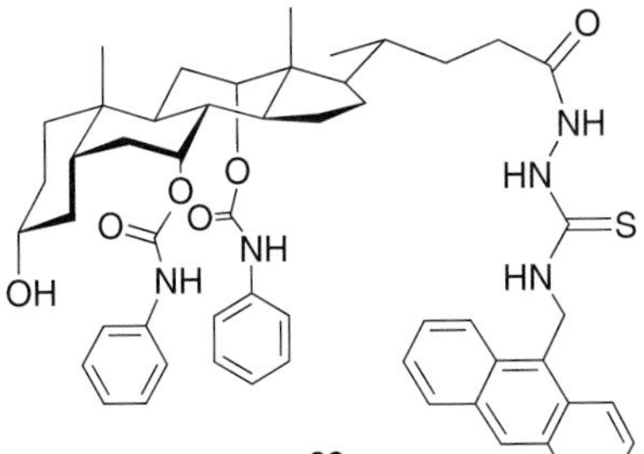

60

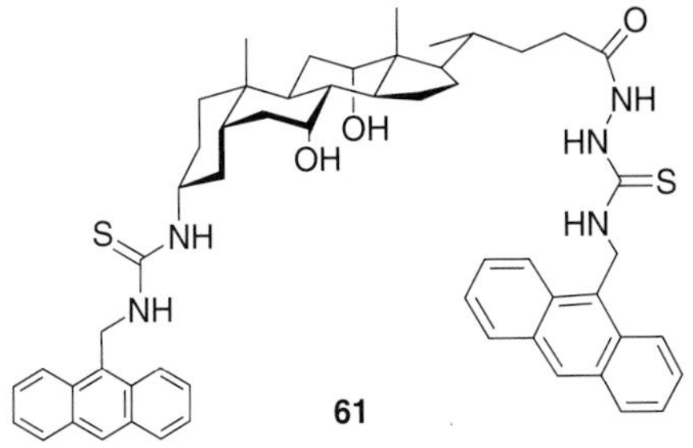

61

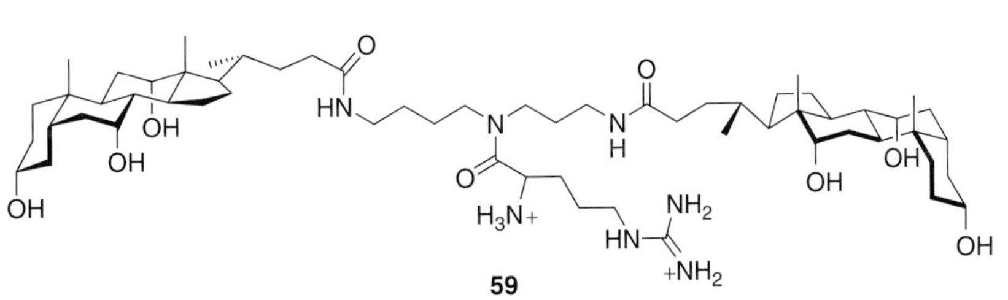

59

Compound **60** shows strong anion affinity in acetonitrile, particularly for carboxylates with an acetate binding constant of $7.69 \times 10^4\,M^{-1}$. The compound functions as a PET sensor, with fluorescence quenching observed on addition of anions. Compound **61** is also a PET sensor and shows remarkably strong anion binding to dicarboxylates in highly competitive solvents (1 : 1 v/v methanol:water); for example, L-glutamate is bound with a binding constant of $5.57 \times 10^6\,M^{-1}$. As with **60**, fluorescence quenching is observed on addition of anions; however, only a maximum of 20% reduction in intensity is achieved.

2.2 Induced-fit anion binding

2.2.1 *General considerations*

The principal advantage of highly preorganized receptors, such as the cholapods, is their specific and strong anion binding. However, this can also be a disadvantage in some circumstances, for example, in sensing. The small changes in conformation resulting from anion binding mean that the communication of the binding event to the reporter group, such as a fluorophore or chromophore, must be done via a change in the electronic distribution of the receptor. This may prove inefficient in comparison to processes requiring significant structural or conformational rearrangement. An alternative approach is to use conformationally flexible receptors which can change their shape or relative disposition of binding sites or chromophores on anion binding and in doing so, they lead to a change in the physical property of the molecule. Flexible systems also have faster complexation/decomplexation kinetics, which is also an advantage in sensing applications. Induced-fit binding can also lead to anion sensors capable of a significant degree of discrimination between anionic guests, because the induced conformation of the host is dependent on the size and geometry of the anion bound; therefore, each host/guest geometry is unique and can potentially affect a reporter group in a distinct way. The term discrimination in this context is distinct from binding selectivity (as measured by the magnitude of the binding constant) since it refers to the response of the receptor system to the guest-binding event. The distinction is particularly well exemplified in colorimetric sensor arrays in which each individual receptor is only very poorly selective, but the array as a whole can be highly discriminating in its ability to recognize particular guests based on the pattern of their response.[112,113]

A possible disadvantage of an induced-fit receptor is the intrinsically lower binding constants compared to those of the preorganized systems. Typically, an unfavorable entropic contribution is expected when binding an anion as the system becomes more ordered. In macrocyclic systems, with less conformational freedom, this entropy cost is paid during the synthesis of the molecule. However, the situation can be complicated by the restrictions on small conformation motion imposed by the binding.[114] For flexible systems this is not the case, and the reorganizational energy represents an unfavorable contribution to the overall binding free energy. The conformational flexibility can potentially mean that a large range of anions are bound with similar binding constants, resulting in reduced thermodynamic binding selectivity. However, as the sensing method is dependent on conformation rather than binding strength, induced-fit sensors can still discriminate for particular anions, even though that anion may not have the highest affinity for the receptor.

By using well-designed molecular architectures, it is also possible to increase the preorganization of a receptor (and hence increase the binding constant) while still allowing a degree of flexibility. The trialkylbenzene motif has been extensively used for this purpose because of its ability to balance these two competing attributes and is discussed in the following section.

2.2.2 *The trialkylbenzene motif*

The hexasubstituted benzene moiety, typically consisting of ethyl substituents in the 1, 3, and 5 positions and binding arms in the 2, 4, and 6 positions, has proved to be a highly versatile motif in both cation and anion binding. It has been combined with a wide range of binding groups to create both neutral and charged receptors and sensors.

The moiety provides a balance between flexibility and preorganization; steric interactions between adjacent substituents on the hexasubstituted benzene ring favor an alternating three-up, three-down arrangement of substituents by approximately $15\,kJ\,mol^{-1}$ (so-called "steric gearing").[115,116] This arrangement creates a convergent binding cavity, although the stability of the three-up, three-down conformation can be influenced by both electrostatic repulsion and steric interactions between binding arms. In cationic receptors, the C_3 symmetric "three-up" conformation is by no means the only conformation observed and can be in equilibrium with the "two-up, one-down" conformation also present in solution (Figure 20).[117] In some cases, additional conformational flexibility is also possible by involving the relative orientation of binding functional groups with respect to the cavity. Hence, *in* and *out* conformations of the binding groups are also possible; for example, in the hypothetical molecule in Figure 21, conformation (a) has all binding arms (R) in the *in* conformation, forming a convergent binding cavity, while (b) has all R groups in the *out* conformation and has a divergent cavity. Both *in* and *out* conformations have been shown to be present at room temperature in solution.[117]

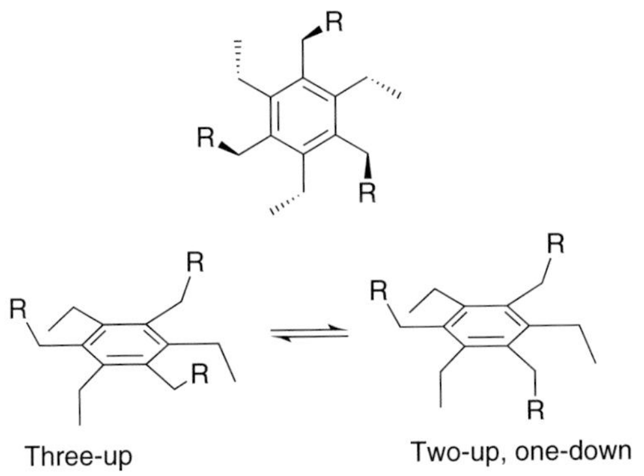

Figure 20 Trialkybenzene motif and its conformations.

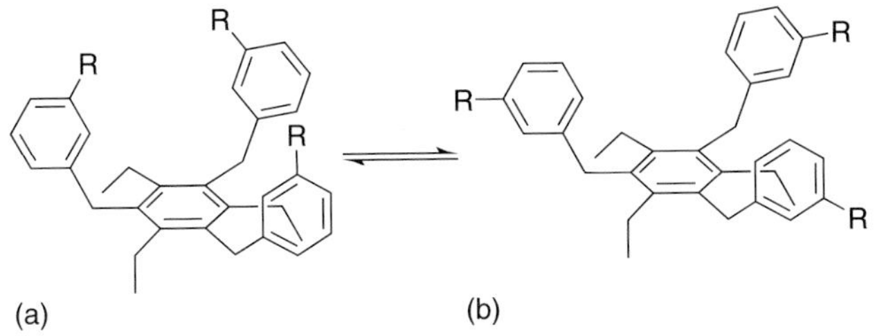

Figure 21 *In* (a) and *out* (b) conformations of the trialkylbenzene motif.

The trialkylbenzene motif in anion binding was pioneered by Ansyln and coworkers[118] and proved to be a highly successful platform for a range of anion sensors incorporated into an indicator displacement assay (IDA). An IDA system consists of a receptor such as **62** and a fluorophore or chromophore guest molecule such as the fluorescent indicator **63**, which weakly binds to the receptor in solution. When an anion that has a higher affinity than the guest dye is added to the system, it displaces the indicator from the binding cavity. The change in the microenvironment of the fluorophore causes a change in the UV–vis absorption and/or fluorescence emission of the indicator. In the case of **62**, the receptor proved to be highly selective for the tricarboxylate citrate, above even other carboxylates, as well as other salts and sugars.[115,119] The molecular structure of the **62**·citrate complex determined by X-ray crystallography can be seen in Figure 22. The assay is able to detect the concentration of citrate in water (a highly competitive solvent) through increases in the absorbance and emission intensities. The concentration of citrate was also determined in orange juice and other soft drinks.

Several other sensing arrays, for example, compound **64** with pyrocatechol violet (**65**), have been developed by the Ansyln group. The binding of gallate-like anions can occur through the reaction of the phenol group with the boronate ester and the binding of the carboxylate to the guanidinium group.[120] This results in the displacement of the indicator from the binding cavity, resulting in

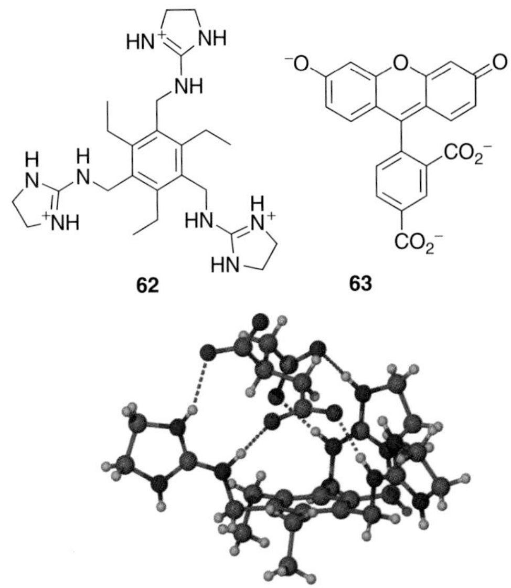

Figure 22 Molecular structure of **15**·citrate. Second complex in the asymmetric unit is omitted for clarity.[115]

an observable color change. This sensing system was used to accurately determine the age of Scotch whiskey, as gallate-like anions leach into the whiskey over time from the storage barrels. The same receptor with alizarin complexone (**66**) was able to determine the concentration of tartrate in wine and fruit juice.[121] Several other assays designed to sense the biologically important anions glucose-6-phosphate[122] and inositol triphosphate (IP_3) have also been developed.[123]

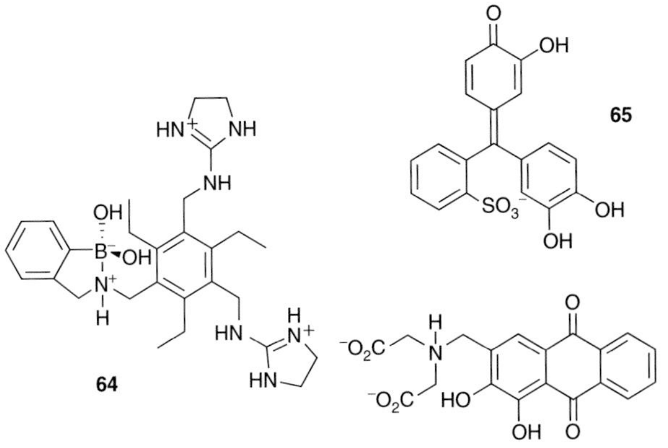

Combining several different anion receptors into a microarray can allow for the discrimination of anions present in a solution and the receptors are often known as *electronic tongues or noses*. Here, a series of receptors, each with often subtly different peak anion selectivity, can discriminate between various anions through their differential colorimetric, fluorescence, or electrochemical responses.

An example of this sensing array concept has been developed by Ansyln and coworkers who used a differential receptor system to discriminate between the nucleotides

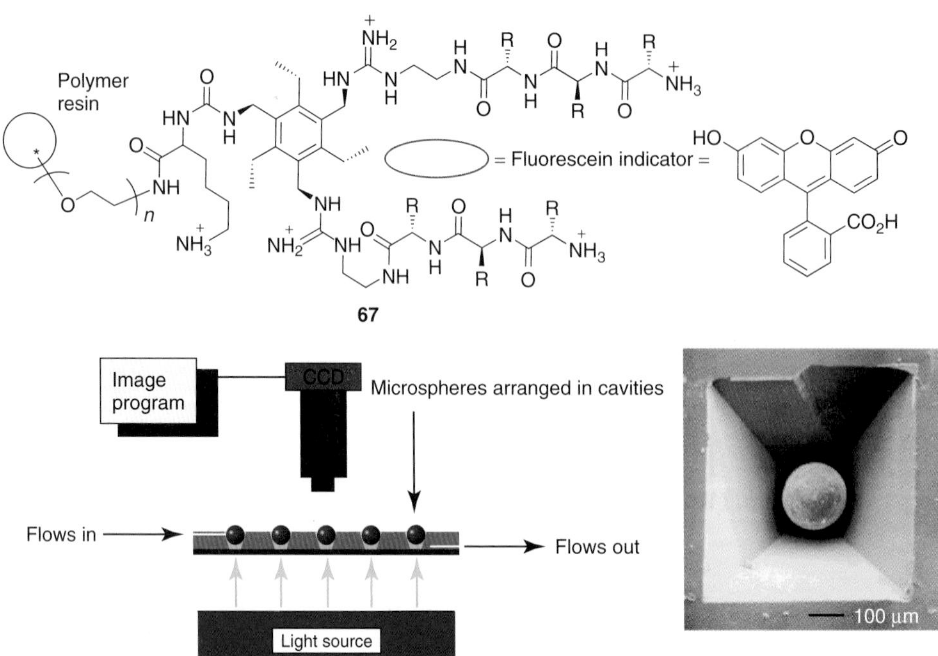

Figure 23 An array-sensing system with receptor with varying "R" groups and indicator is illustrated. (Reproduced from Ref. 112. © American Chemical Society, 2001.)

ATP, adenosine monophosphate (AMP), and guanidine triphosphate (GTP). The trialkylbenzene motif was used as the basis for the receptors, with each receptor immobilized onto a polymer resin by attachment to one of the binding arms (for example, **67** in Figure 23). The two remaining arms were functionalized with short peptides, synthesized using combinatorial synthesis. Thirty polymer beads were then used to generate the sensing array. The sensing method was based on an IDA, with fluorescein used as the chromophore. Analyte solutions were allowed to flow over the beads and the change in absorbance of the fluorescein as it was displaced from the receptor by the analyte was monitored using a CCD (charged coupled device) camera. Principal component analysis (PCA) was then successfully used to determine which analytes were present in the mixture.[112,113]

Steed and coworkers have used the trialkylbenzene motif in designing receptors such as **68** and **69** with pyridinium derivatives providing charge-assisted $CH\cdots X^-$ hydrogen bonds and/or $NH\cdots X^-$ hydrogen bonds, as can be seen in the molecular structure (determined by X-ray crystallography, Figure 24) of the **69**$\cdot Br^-$ complex. This shows the host in the three-up conformation with both *in* and *out* conformations of the binding arms present. ^1H NMR spectroscopic titrations show that compound **68** binds chloride strongly with a binding constant of $82\,000\,M^{-1}$ in CD_3CN/DMSO (v/v 95/5). The high affinity shows the complementarity between the binding cavity and the chloride.[117]

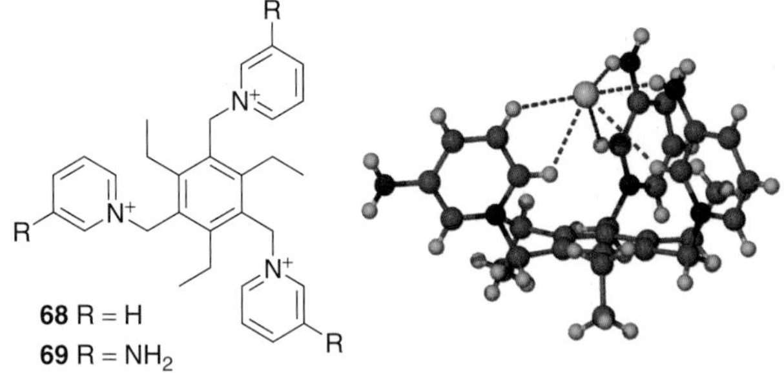

Figure 24 Receptor design for **68** and **69** along with the X-ray structure of **69**$\cdot Br^-$.[117]

Variable temperature (VT) NMR spectroscopy in acetone-d_6 revealed the presence of the two-up, one-down conformer, evidenced by the splitting and up-field shifting of the ethyl CH_3 resonance as the proton moves into the shielding area of the pyridinium groups for the "down" ethyl group. *In* and *out* binding arm fluctuations were also observed. VT NMR spectroscopic experiments with varying amounts of added chloride revealed that binding to 1 equivalent of Cl^- switches the system to the symmetric "three-up" conformation, while in the presence of substoichiometric amounts of Cl^- the Cl^-/PF_6^- exchange equilibrium can be observed on the NMR spectroscopic timescale. While such anion exchange equilibria are sometimes slow in macrocyclic systems,[124] they are generally fast in podands highlighting the degree of preorganization afforded by the steric crowding in **69**.

A similar receptor (**70**), which also utilizes NH and CH charge-assisted hydrogen bonding and shows strong binding to chloride, has been designed by Fabbrizzi and coworkers.[125]

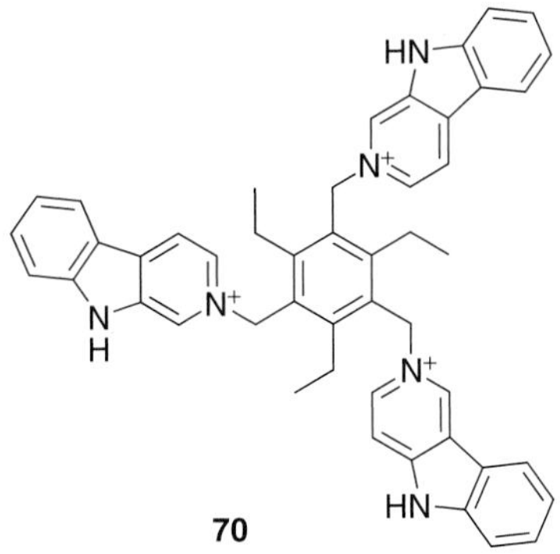

70

The incorporation of halogen atoms onto a pyridyl ring can also allow halogen bonding, for example, compound **71**.[126] The halogen bonding occurs between the positive region of electrostatic potential of the halogen, which is enhanced by the electron-withdrawing cationic pyridinium motif, and the negatively charged anion. When in the three-up conformation, compound **71** is capable of forming a convergent binding cavity with very strong binding to phosphate, PO_4^{3-} (as the sodium salt), with a binding constant of $\log K_a = 5.6$ in water (at pH 12.1). A related receptor, compound **72**, derived from a *para*-iodotetrafluorophenyl binding group, also binds anions strongly using halogen bonds.[127] Interestingly, receptor **72** displays a different anion selectivity with chloride ($K_a = 1.9 \times 10^4\ M^{-1}$ in acetone) having a larger binding constant than oxo-anions.

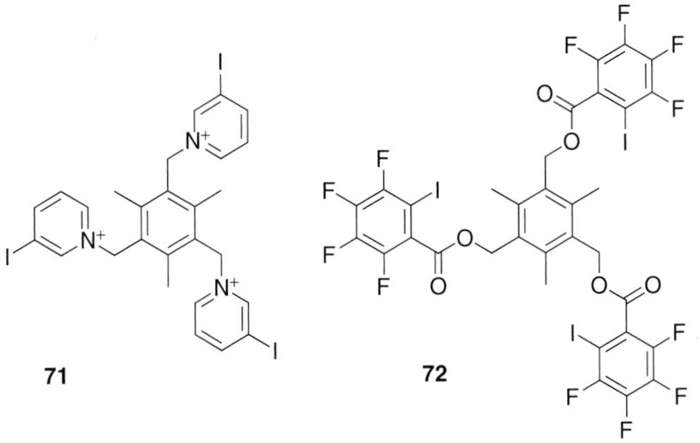

71 **72**

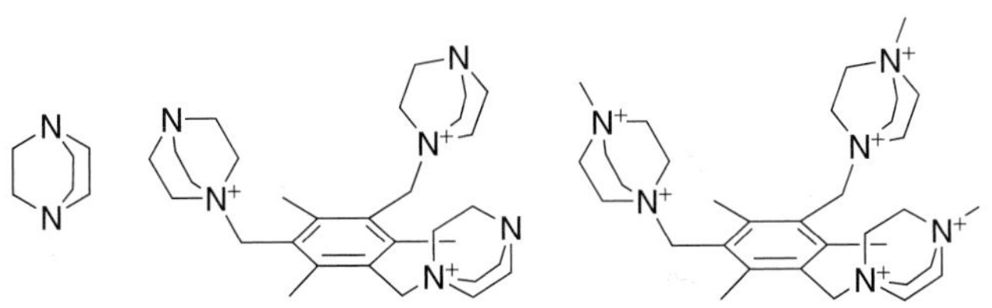

73 **74** **75**

It is likely that the smaller cavity size in **72** is a poor match to oxo-anions and is more complementary to chloride.

The binding arm 1,4-diazabicyclo[2.2.2]octane (DABCO, **73**) is useful as its derivatives can be both mono- and dicationic, leading to a tri- (**74**) and hexacationic (**75**) species and allows for the investigation of the effect of charge on the guest interaction. The strongest binding was found for complementary anions, for example, a tricationic host and a trianionic guest. An interesting example of this charge matching is in the selective binding and precipitation of ferricyanide, an Fe(III) species, over ferrocyanide, an Fe(II) species, with the tricationic host.[128]

Further development in the trialkylbenzene motif has led to receptor **76** by Schmuck and coworkers, which exhibits exceptionally high binding of tricarboxylates in water. The cationic guanidinium groups create a tricationic receptor which binds carboxylates in a 1 : 1 stoichiometry largely by electrostatic interactions. A binding constant of $>10^5\ M^{-1}$ was measured for citrate, by UV–vis and fluorescence spectroscopy, with excellent selectivity over monoanionic species.[129]

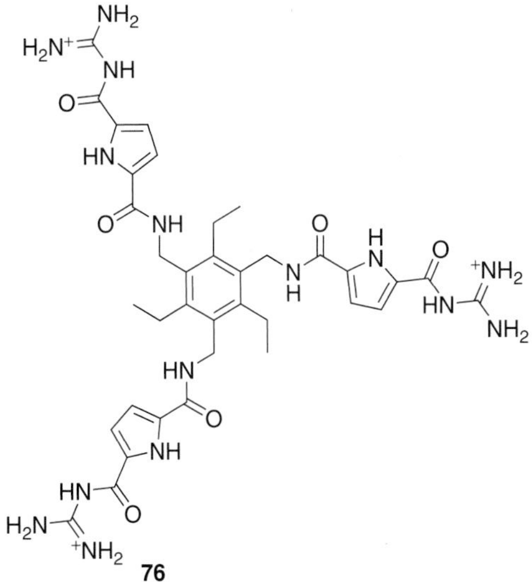

76

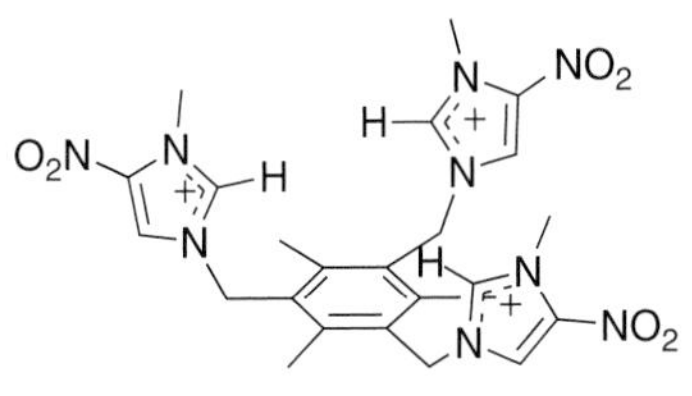

77

As an alternative to the more usual hydrogen-bonding groups such as urea, amides, and guanidinium, it is possible to use imidazolium groups to create $CH^+\cdots X^-$ hydrogen bonds as in receptor **77**.[130] The charged nature of this system means that the main interaction is electrostatic. To further enhance the strength of this form of hydrogen bond, an electron withdrawing nitro group can be attached to the imidazolium ring at the C-4 position, enhanced by the electron-withdrawing nature of the nitro group. *Ab initio* calculations and ^1H NMR spectroscopic titrations were used to determine the binding constant for halide anions, with both methods concurring to give a selectivity series of $Cl^- > Br^- > I^-$. The binding constant for Cl^- in DMSO-d_6 (dimethyl sulfoxide) is $4800\,M^{-1}$ with a 10-fold selectivity over Br^-. In addition, the compound also binds dihydrogen phosphate with a binding constant of $2500\,M^{-1}$ owing to the strong basicity of dihydrogen phosphate increasing its binding affinity.

The principle of using anion binding to shift the equilibrium of conformations of a host has been used by Duan and coworkers to synthesize the chloride-selective sensor **78**.[131] In this system, anions are bound by charge-assisted hydrogen bonding to the benzimidazolium group with a binding constant of $3.9 \times 10^3\,M^{-1}$ in DMSO. The **78**·$3BPh_4$ complex shows only monomer emission from the naphthyl reporter groups. However, on addition of chloride an excimer band is observed in the emission spectrum (Figure 25). The anion-induced conformational change brings two naphthyl groups into close contact and allows for excimer formation. Only chloride is able to induce a conformation where this can occur.

Building on previous work, Duan and coworkers have developed the ditopic and the tripodal ferrocenyl derivatives **79** and **80**.[132] Of the anions tested, both receptors bind chloride strongly with moderate selectivity. Differential pulse voltammograms (DPV) show a cathodic shift of 50 mV in the $E_{1/2}$ value of **79**. Smaller cathodic shifts were observed for other anions. No response was observed with the tripodal derivative **80**, despite anion ^1H NMR spectroscopic titrations confirming that the receptor does bind anions strongly, in the region of $10^3\,M^{-1}$.

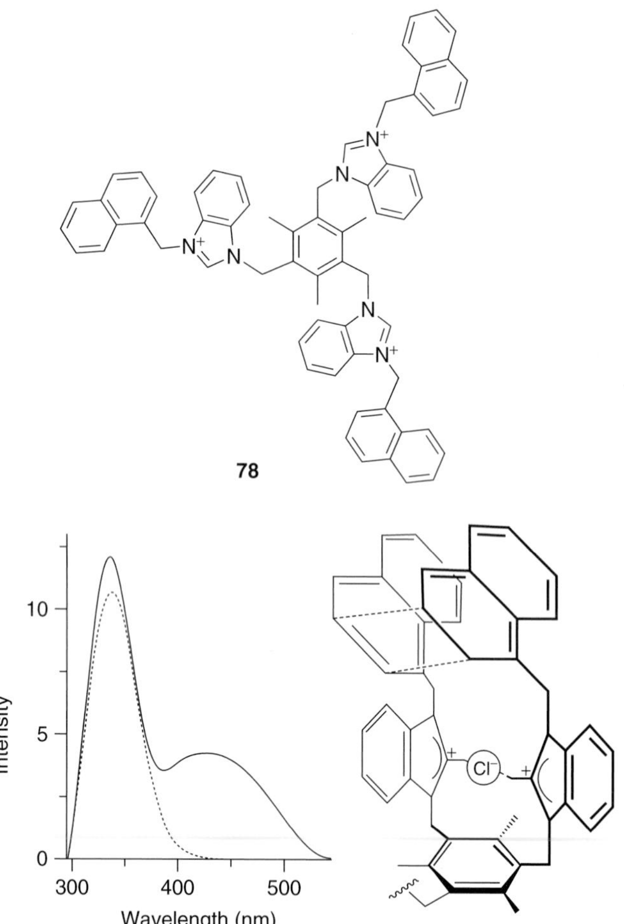

Figure 25 Emission spectrum of **78** in the absence (dotted line) and presence (solid line) of TBA-Cl. (Reproduced from Ref. 131. © Royal Society of Chemistry, 2005.)

2.2.3 *Anion-induced excimer formation in calixarene-based podands*

The design of induced-fit anion sensors in not limited to the trialkylbenzene motif. Calixarenes with pendant binding arms have also been explored with pyrene derivatives typically used as the fluorophore. Pyrene excimer emission has been used as a fluorescent reporter which requires close contact between adjacent pyrene molecules for its formation. They are an ideal candidate for induced-fit sensing as conformational changes can dramatically affect intramolecular pyrene–pyrene distances in both the ground and excited states.

Kim and coworkers[133] have synthesized receptor **81**, with two pendant pyrenyl binding arms. The system can shift from the dominance of a dynamic excimer (dimer formation in the excited state) in the free host to a static dimer (ground state dimer formation) on addition of fluoride. The binding of the fluoride in the receptor holds the pyrenyl groups in close proximity and allows a ground state dimer to form. This results in a 73-nm red-shift in the excitation spectrum and a 12-nm blue-shift in the excimer emission.

81

82

Steed and coworkers[134] have also synthesized a calix[4]arene-derived receptor **82**. The compound is locked into a 1,3-alternate conformation through steric interactions between the mesitylene rings. This creates a ditopic receptor with pyridinium binding groups and a pyrenyl reporter. The receptor binds 2 equivalents of dicarboxylates strongly ($K_{11} > 100\,000\,\mathrm{M}^{-1}$ in acetonitrile) with the dicarboxylate capable of spanning across the two binding arms. However, no significant change in the fluorescence emission was observed for this anion. The binding of chloride is an order of magnitude lower than that of dicarboxylates. However, chloride binding alters the conformation of the receptor in such a way as to allow ground state interactions between the pyrenyl groups and promote the formation of an excimer on excitation (Figure 26). Chloride binding also has an electronic effect in that it prevents fluorescence quenching

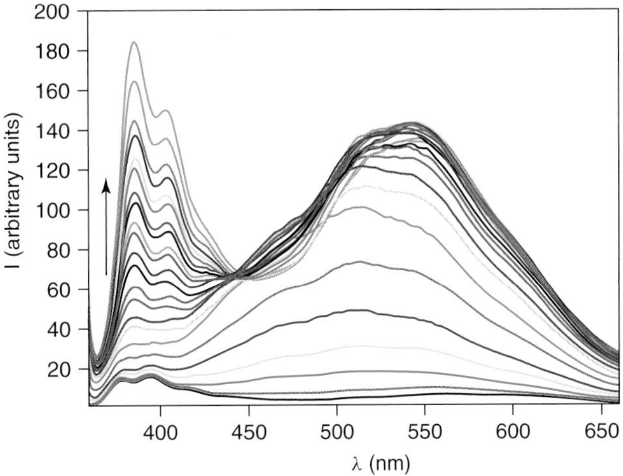

Figure 26 Emission spectrum of **82** on increasing amounts of TBA-Cl. (Reproduced from Ref. 134. © American Chemical Society, 2008.)

by intramolecular charge transfer (ICT) to the pyridinium moieties resulting in the observation of both pyrene excimer and monomer emission in the presence of excess chloride. This example shows the importance of conformation in this class of receptor. The sensor discriminates the anion which induces a conformation allowing excimer formation to occur and has the appropriate electronic effect on the system, rather than the one that is the most strongly bound.

2.2.4 *Induced-fit molecular clip sensors*

Molecular clips were first developed as hosts for molecular guests in the 1970s by Chen and Whitlock.[135] The concept has been further developed by Zimmerman,[136] Rebek,[137] and Harmata.[138] Typically, molecular clips are simple receptors consisting of two binding domains, which when binding molecular guests have generally consisted of aromatic groups tethered by a linker/spacer, which bind the guest through π–π and ion–dipole interactions (Section 3). Preorganization can depend on the rigidity of the receptor and its ability to maintain a convergent binding site.

The use of molecular clip receptors in induced-fit anion sensing has been relatively unexplored, even though the design of these receptors appears to be well suited to the task. In general, these systems consist of biaryl units with the addition of anion-binding groups. The relatively rigid nature of the coupled aromatic systems means that the conformational freedom is largely restricted to rotation about the inter-aryl bond. This provides a degree of preorganization to the system.

Receptor **83** shows a 2.4-fold increase in emission intensity after the addition of 2.5 equivalents of fluoride, which is attributed to a concept known as *conformational restriction*, in which one fluoride anion is bound by both urea moieties, rigidifying the receptor and reducing nonradiative decay. Further addition of F^- resulted in a decrease in the intensity. The binding of the second fluoride (one F^- bound to each urea group) allows more conformational flexibility and hence increases nonradiative decay.[139]

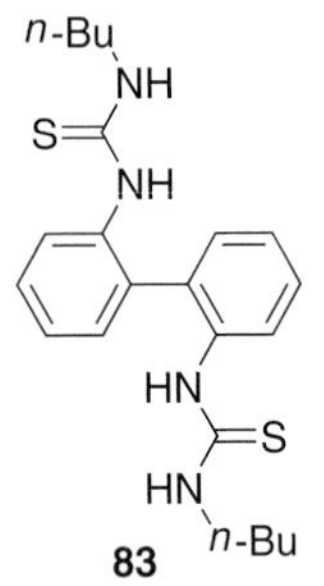

83 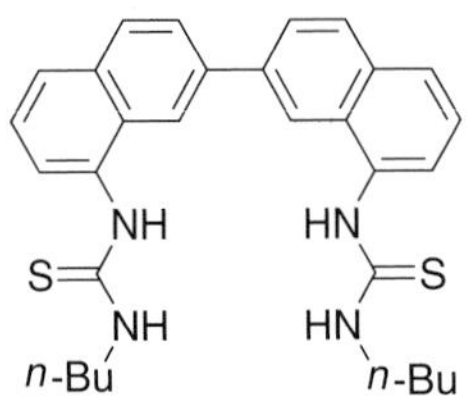 **84**

A similar receptor design based on a 2,2′-binaphthalene derivative (**84**) shows a red-shift in the UV–vis spectrum on addition of fluoride, consistent with planarizing of the binaphthyl groups and conformational restriction, although for this receptor, only small changes were observed in the fluorescence emission.[140]

Lin and coworkers have developed receptor **85** which consists of naphthyl groups at the end of urea-binding moieties.[141] On binding *ortho*-phthalate, there is a significant increase in emission at 460 nm, which is not observed with meta or para isomers of phthalate. The emission is from the excited state of the product of a photochemical reaction between the naphthyl groups, and it is only *ortho*-phthalate which induced the conformation necessary for this photochemical reaction to occur.

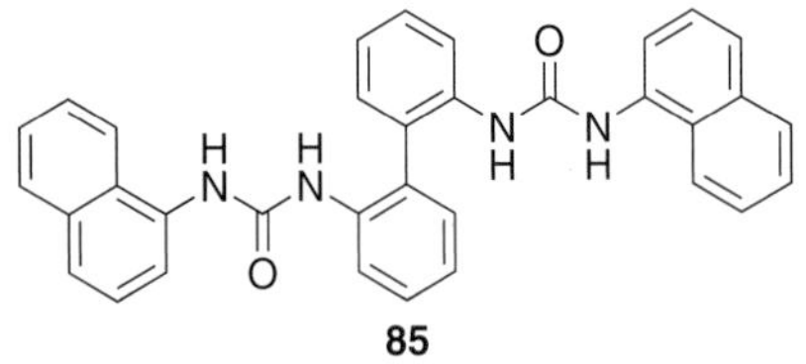

85

2.3 Small molecule anion receptors and sensors

2.3.1 *Prodigiosins and their analogs*

Prodigiosins are a class of compounds isolated from the microorganism of *Serratia* and *Streptomyces* genus. These naturally occurring pigments are dark red and colonies of the gram-negative bacteria often resemble droplets of blood and have been put forward as the scientific explanation to many apparent miracles.[142] More recently, it has been shown that prodigiosins have immunosuppressive and anticancer properties and the compounds are a promising lead in new drug therapies. The structure of prodigiosins consists of a tripyrrolic skeleton (**86**) as in prodigiosin 25-C (**87**), for example, which has been studied extensively for pharmaceutical use.[83]

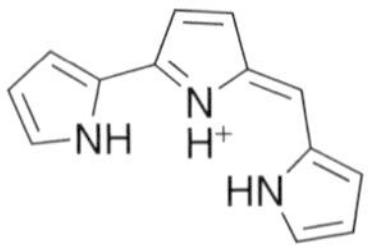

86 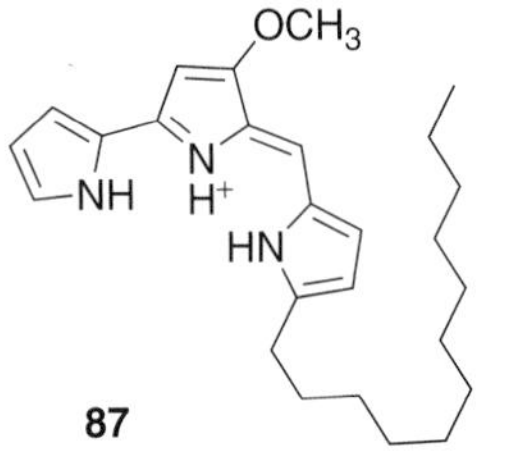 **87**

Two primary mechanisms have been suggested for the biological activity of prodigiosins. The first involves the ability of prodigiosins to mediate into-cell transport of HCl, while the alternative involves the coordination of copper and subsequent modification of DNA.[83,142] The transport of HCl into a cell is necessarily intimately linked with anion binding and transport, and so efforts have been made to understand the anion recognition behavior of prodigiosins.

Sessler and coworkers have investigated a series of prodigiosin analog compounds **88** and **89–92**.[83,143,144] Both types of compounds bind chloride strongly when protonated. Binding constants in the order of $10^6\ M^{-1}$ were observed for **88** and $10^5\ M^{-1}$ for **89** measured by isothermal titration calorimetry (ITC) in acetonitrile. These data show that even very simple acyclic pyrrole-derived sensors can bind anions strongly due to the relatively rigid nature of the skeleton, creating a preorganized binding cavity with NH hydrogen-bond donors and electrostatic attraction.

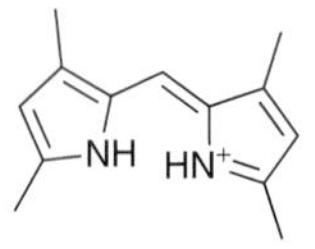

88

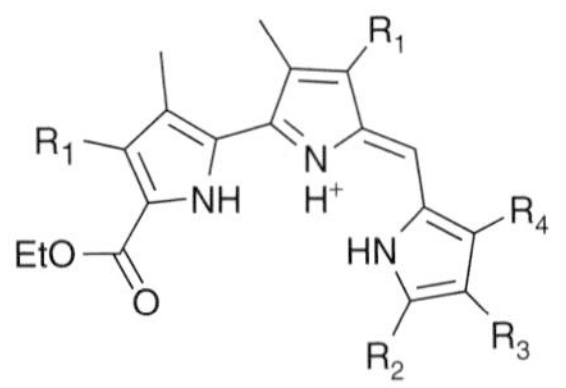

89 $R_1 = Et, R_2 = R_4 = CH_3, R_3 = H$
90 $R_1 = Et, R_2 = R_3 = R_4 = H$
91 $R_1 = Et, R_2 = H, = R_3 = R_4 = OCH_3$
92 $R_1 = R_2 = R_4 = CH_3, R_3 = H$

The molecular structure of compound **88** was determined by X-ray crystallography and confirms that chloride is bound by two NH hydrogen bonds and electrostatic interactions in an essentially planar conformation (Figure 27). The molecular structure of **89**·Cl^- determined by X-ray crystallography shows the chloride bound by three NH hydrogen bonds with the receptor in a slightly twisted conformation (Figure 27). Further modification to the prodigiosin analogs **90–92** also proved to be effective at binding chloride with binding constants in the range of 10^4–$10^5\ M^{-1}$. The synthesis of tetrapyrrolic receptors such as **93**, creating an additional hydrogen-bond donor site, also helps bind chloride effectively ($\sim 10^5\ M^{-1}$).[83,144]

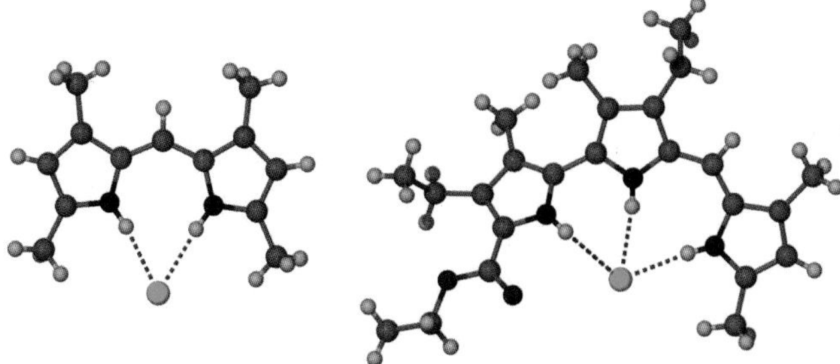

Figure 27 Molecular structure of **88**·Cl^- and **89**·Cl^- determined by X-ray diffraction.[143]

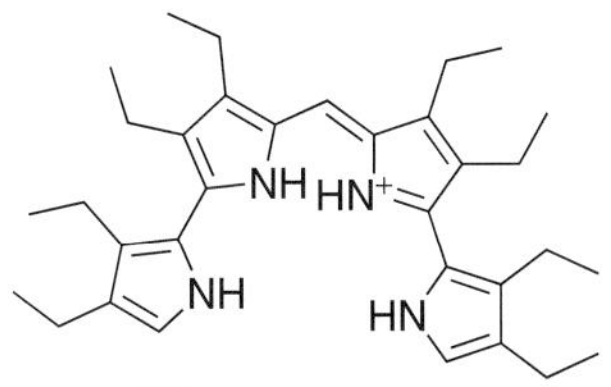

93

Membrane transport studies were also conducted on naturally occurring prodigiosins and their analogs, compounds **88** and **89**, and suggest that the anticancer properties of the receptors follow the trend for membrane transport efficiencies, rather than the anion-binding strengths. This result supports the theory of symport transport of H^+/Cl^- as the source of the anticancer properties of this class of compound and that large chloride binding constants impair the release of the chloride, that is, the kinetics of the systems are more important than the thermodynamics.[143]

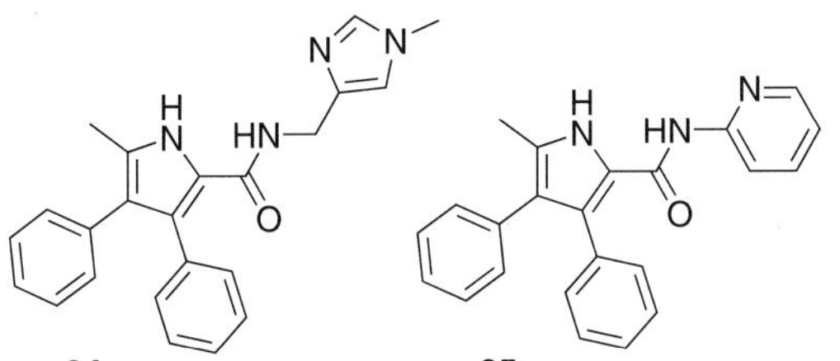

94 **95**

Prodigiosin mimics based on amidopyrroles synthesized by Gale and coworkers also showed membrane transport of HCl.[145] The amidopyyrole derivative **94** with a pendant imidazole group proved to be the most effective membrane transporter. Interestingly compound **95** showed relatively weak anion binding even when protonated, that is, 397 M^{-1} in acetonitrile, but proved effective at transporting HCl, highlighting the importance of kinetic lability in this process.

Prodigiosins have shown that small molecules consisting of a relatively rigid skeletal framework and convergent hydrogen-bond donors can be highly effective at anion binding. Research in this area is highly active and there are a wide variety of structural motifs which are discussed in the following section.

2.3.2 Small molecule podand receptors

Isophthalamide anion receptors, for example, compound **96**, were originally synthesized by Crabtree and coworkers as a binding motif with little rigidity or preorganization. However, reasonable binding affinities with halides were observed in the range of 10^3–10^4 M^{-1} in dichloromethane. The molecular structure determined by X-ray diffraction shows a distinctly nonplanar binding mode.[146,147]

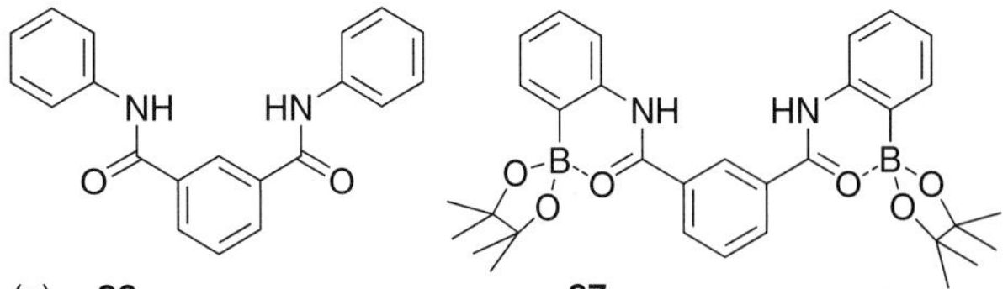

(a) **96** **97**

Smith and coworkers synthesized isophthalamide derivatives functionalized with boronate esters, **97**.[148] The carbonyl group is able to coordinate to the boron atom and increases the preorganization while also increasing the dipole moment, allowing stronger ion–dipole interactions. The NH residue has a greater positive charge and so has a stronger interaction with the acetate. Consequently, compound **97** has an order of magnitude higher binding constant than **96**.

Gale and coworkers have designed isophthalamide and 2,6-dicarboxyamidopyridine derivatives.[86] The addition of two indole motifs provides the molecule with a total of four NH hydrogen-bond donors.[149] Compound **98** is highly selective for fluoride (K_a of 1360 M^{-1} in DMSO) even compared to other basic anions such as acetate (K_a of 250 M^{-1} in DMSO). It was postulated that the twisted binding conformation observed in the molecular structure (measured by X-ray diffraction, Figure 28) is also found in solution and is more stable than the binding modes possible with other anions.

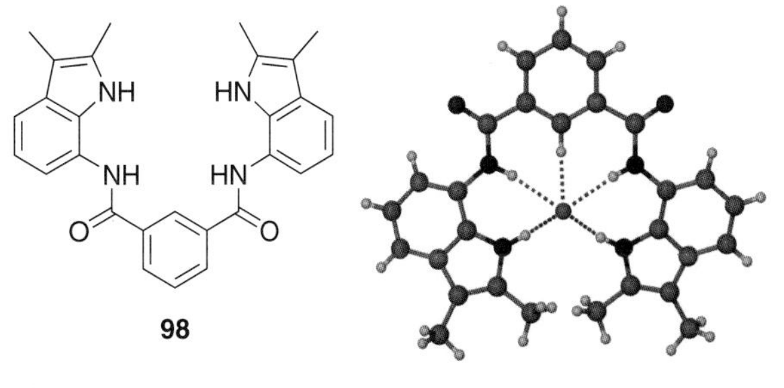

98

Figure 28 Compound **98** with the molecular structure of **52**·F^- determined by X-ray crystallography.[149]

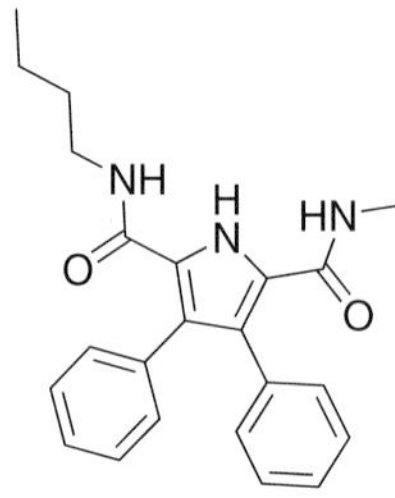

Figure 29 Compounds **102** and **103** with the molecular structure of **102**·Cl^- determined X-ray crystallography.[152]

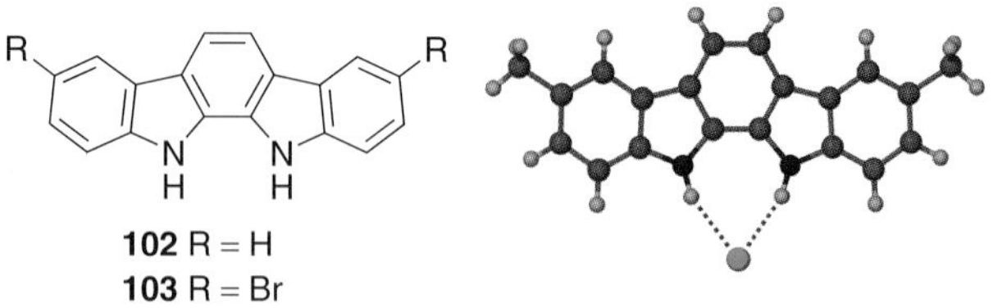

Modification of the core motif allows for 2,5-diamidopyrrole derivatives, **99**, having a additional NH hydrogen-bond donor compared with isophthalamide receptors.[150] A binding constant of $2500\,M^{-1}$ for benzoate in acetonitrile is observed with an asymmetric binding mode apparent in the molecular structure measured by X-ray crystallography.

ortho-Phenylaminediamine derivatives have also been investigated as anion receptors containing four NH hydrogen-bond donors in the form of urea groups or amide and pyrrole moieties.[151] Compounds **100** and **101** are selective for carboxylates with compound **100** showing significantly higher binding constants for carboxylates than **101**. It is suggested that the more open binding cleft in **100** is more structurally complementary to carboxylates than **101**. The presence of chloro groups in an analogous compound to **100** showed an increase in binding constant from $3210\,M^{-1}$ for acetate to $8079\,M^{-1}$. It is suggested that the increased acidity of the hydrogen atoms in the central ring allows for CH···O hydrogen bonding, effectively preorganizing the receptor into a planar conformation. Indeed, using thiourea derivatives reduces the acetate binding constant by an order of magnitude, as the large size of the sulfur atom results in steric hindrance between the sulfur and phenylene hydrogen atoms, distorting the binding cleft.

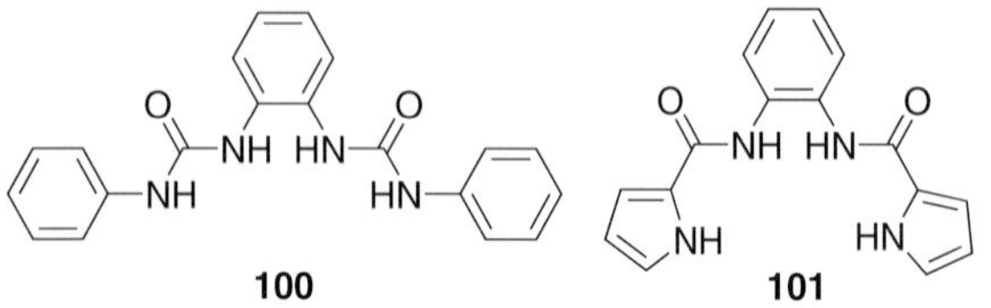

It has been shown above that intramolecular interactions can enhance the preorganization of small molecule acyclic receptors.[152] As an alternative, increasing structural rigidity using aryl rings can also provide preorganization. This is typified well by the indolocarbazoles. The rigidity of compound **102** is provided by five fused aromatic rings and it contains two convergent NH hydrogen-bond donors. Compound **102** is an effective anion receptor with benzoate and dihydrogen phosphate bound strongly ($\log K_a$ of 5.3 and 4.9, respectively, measured by UV–vis titration in acetone). The addition of bromine in the 3 and 8 positions (**103**) leads to a marked increase in binding strength through their electron-withdrawing effect. Figure 29 shows the molecular structure of the **102**·Cl^- complex determined using X-ray crystallography.

The indolocarbazole derivative **104** is able to bind anions in water in a helical conformation creating a tubular cavity in which multiple NH hydrogen bonds bind the anion. The binding conformation was confirmed using NOESY (nuclear Overhauser effect spectroscopy) NMR experiments with chloride bound strongest (K_a of $65\,M^{-1}$ in water). While this binding constant is small, it is measured in a highly competitive medium. The use of less competitive solvents on related compound shows strong anion binding.[153]

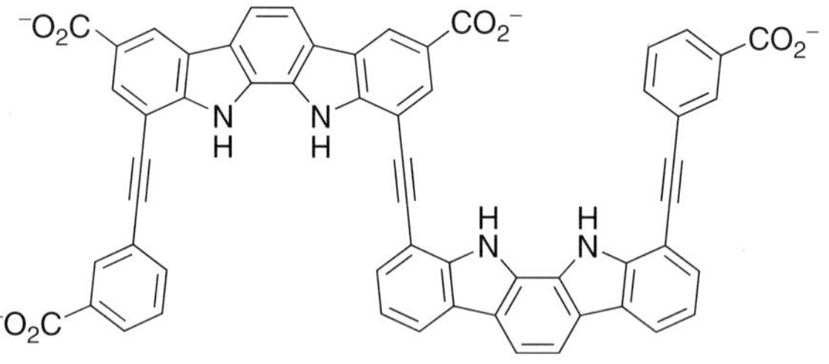

As final examples of acyclic receptors, compounds **105** and **106** are recently published examples of compounds with the potential to extract sulfate from solution through crystallization.[154] Sulfate is generally a difficult anion to bind due to its delocalized charge and low basicity. It has important environmental impact particularly as a component of nuclear waste. The receptors designed by Gale and coworkers are designed to provide NH hydrogen bonds from a variety of moieties such as ureas, amide, and pyrrolic groups. The molecular structure of **59**·SO_4^- determined by X-ray crystallography shows that each SO_4^- oxygen is bound by two NH hydrogen bonds, with a binding constant in DMSO/10% H_2O of $>10^4$ measured by ^{1}H NMR spectroscopic titration (Figure 30). Crystallization of **59**·SO_4^- from DMSO/10% H_2O occurs in as little as 20 min.

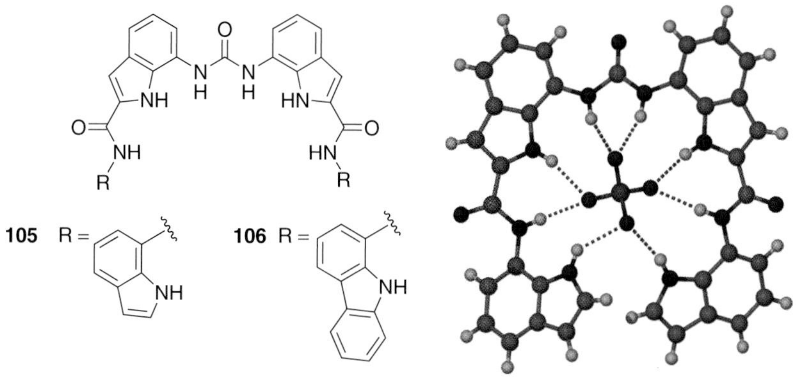

Figure 30 Compounds **105** and **106** and the molecular structure of **105**·SO_4^{2-} determined by X-ray diffraction.[154]

2.3.3 Small molecule podand anion sensors

Small molecule anion sensors have been an active field of research for the last 20 years. Fluorescent sensors have received significant attention and have been reviewed extensively.[155–157] They offer advantages such as high sensitivity and simple instrumentation. The incorporation of a fluorophore such as the molecular clip, indolocarbazoles, and indoles described previously not only provide a reporter group but can also provide a structural element to the receptor.

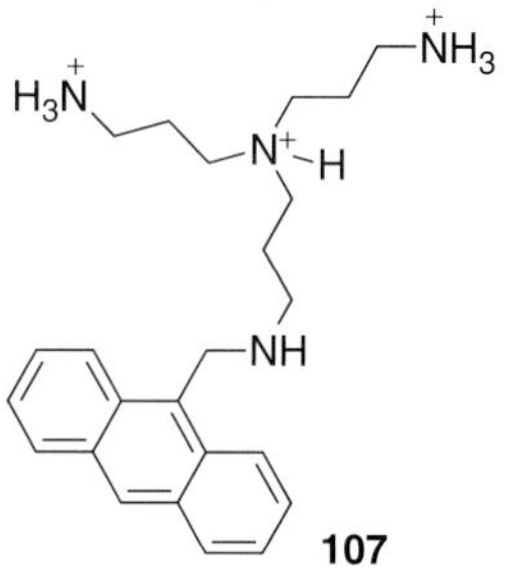

107

A luminescent sensor for anions (**107**) was synthesized by Czarnik and coworkers and is possibly the first example of its type.[158] In this system, an anthracenyl reporter and a tertiary amine receptor are linked by a methylene bridge. This sensor works via PET. Under the conditions of the experiment—an aqueous solution at pH 6—all amine groups except the benzylic amine (which is less basic) are protonated. The excitation of an electron to an excited state by a photon leaves an electron hole, into which the benzylic amine lone pair donates an electron. The excited electron cannot radiatively decay back down to the ground state, thereby giving fluorescence, and must relax via a nonradiative method. The fluorescence is therefore quenched.[157] Addition of HPO_4^{2-} to **107** leads to complexation, forming three NH···O^- hydrogen bonds by the primary amines and one N···HO hydrogen bond to the benzylic hydrogen with a log K_a of 0.82. It is likely that a partial or full proton transfer occurs leading to protonation of the nitrogen lone pair, preventing PET, and therefore luminescence is observed. Thermodynamically, this can also be rationalized as the nitrogen lone pair energy is greater than that of the anthracenyl HOMO (highest occupied molecular orbital); therefore, electron transfer is possible. The interaction with HPO_4^{2-} stabilizes the lone pair, so its energy is less than that of the anthracenyl HOMO; therefore, electron transfer is disfavored. This effect is also given the name chelation-enhanced fluorescence (CHEF).

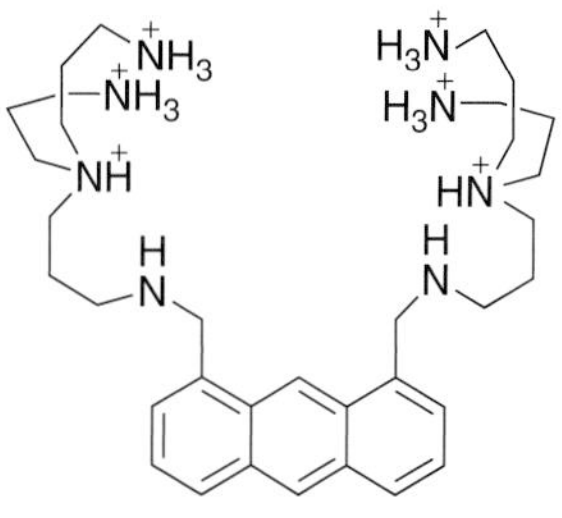

108

Citrate and sulfate can also give rise to the CHEF effect in **107**, although their binding constants are lower. Sulfate by itself does not have any acidic hydrogens, but leads to water dissociation, thereby leading to amine protonation. Similarly, the receptor **108** binds pyrophosphate in an analogous way to the above, with a K_a sufficient to allow micromolar fluorescent sensing. The mechanism of sensing is again a CHEF effect, found on pyrophosphate complexation.[159]

Gunnlaugsson and coworkers have synthesized a series of 4-amino-1,8-naphthalimide-based receptors **109** and **110**.[160] These receptors are quenched by acetate, $H_2PO_4^-$, and F^- through PET with the greatest quenching observed with F^- in DMSO. Usually for 4-amino-1,8-naphthalimide receptors, PET quenching is only observed when binding functionality is in the 4 position. However, in this instance the position of the binding groups does not affect the PET, with quenching found in both the 4-amino and the imide positions.

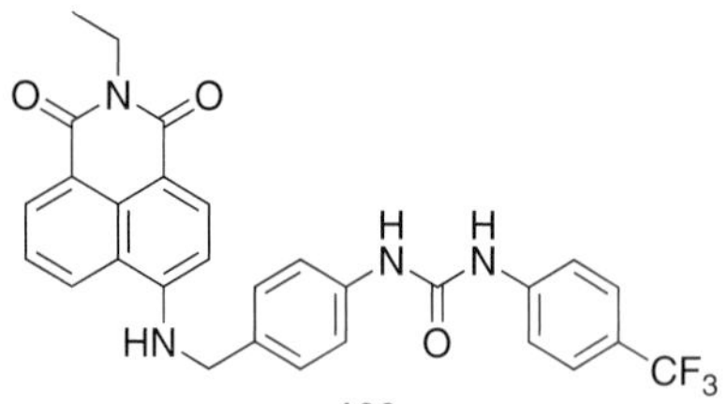

The Gunnlaugsson group has also synthesized the simple anion receptor **111** in which the addition of acetate, $H_2PO_4^-$, and F^- leads to fluorescence quenching by PET. Chloride, however, results in an increase in fluorescence emission. This was attributed to twisting of the molecule on binding chloride which reduces the efficiency of the pre-existing intramolecular PET process; hence, emission is increased.[161]

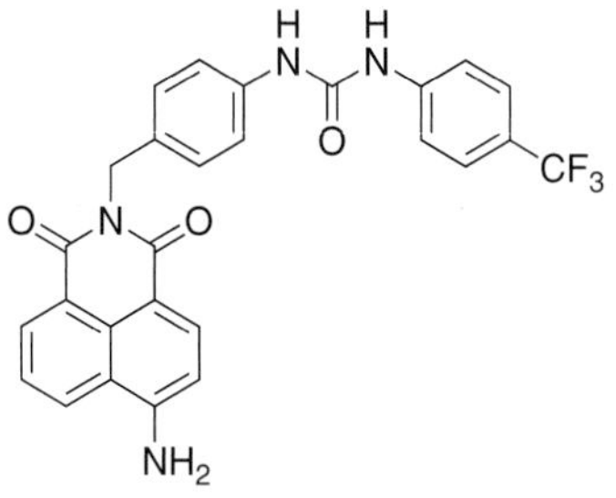

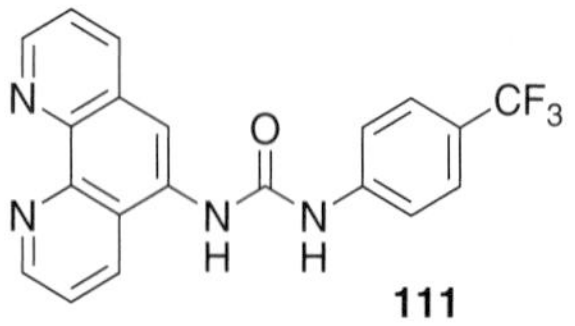

In addition to intramolecular excimer formation as a means of anion sensing, anion-induced self-assembly systems can also be used to good effect. The pyrene-functionalized guanidinium receptor **112** demonstrates a fluorescence emission at 376 nm in its monomeric state. On addition of pyrophosphate (PP_i), a 2 : 1 stoichiometry self-assembly system is formed between the receptor and pyrophosphate. A structureless emission band at 475 nm is then observed with quenching of the monomer emission. This is assigned to an excimer emission. Furthermore, significant excimer emission is only seen for pyrophosphate, showing good selectivity over other anions, for example, HPO_4^{2-}, $H_2PO_4^-$, and Cl^-.[162]

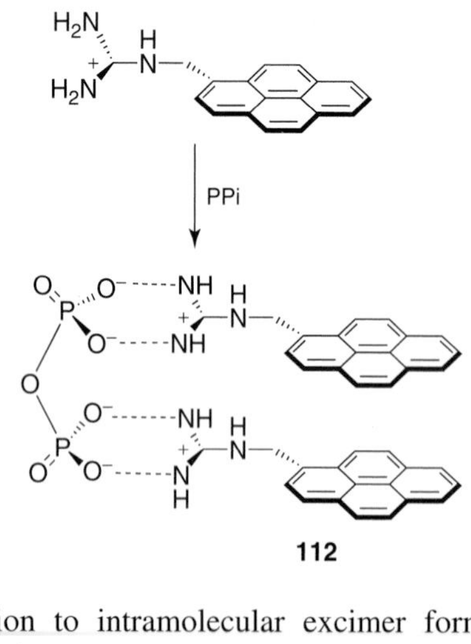

Colorimetric sensors are sensors in which binding of an anion results in a visible color change. They are a particularly attractive form of anion sensor as qualitative results can be achieved through "naked eye" sensing which does not require any equipment. There have been many examples of small molecule colorimetric sensors[156,163] with a selection discussed below to give a flavor of the concepts which can be used when designing receptors.

Hong and coworkers have combined two chormophores—azophenol and *p*-nitrophenyl into a receptor design, compound **113**.[164] Red-shifts were seen in the absorption bands of the receptor on addition of the highly basic anion dihydrogen phosphate to chloroform, resulting in a visible color change from light yellow to violet. Less basic anions such as Br^-, Cl^-, and HSO_4^- do not cause significant color changes; however significantly, anions of similar basicity to those of $H_2PO_4^-$, F^-, and acetate do not cause red-shifts to the same extent due to their different binding geometries.

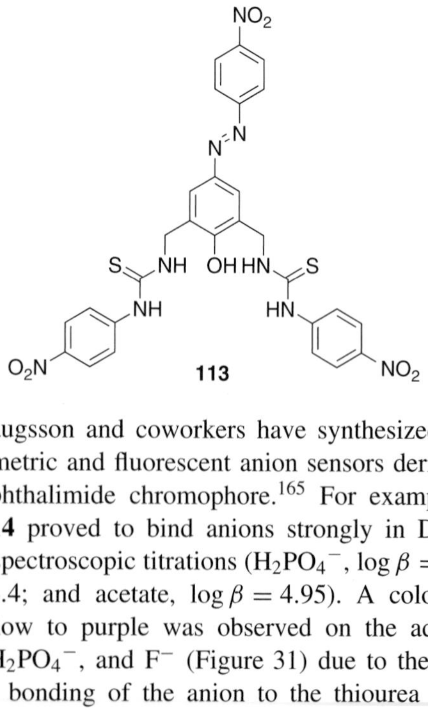

Gunnlaugsson and coworkers have synthesized a range of colorimetric and fluorescent anion sensors derived from a 1,8-naphthalimide chromophore.[165] For example, compound **114** proved to bind anions strongly in DMSO by UV–vis spectroscopic titrations ($H_2PO_4^-$, $\log\beta = 4.0$; F^-, $\log\beta = 4.4$; and acetate, $\log\beta = 4.95$). A color change from yellow to purple was observed on the addition of acetate, $H_2PO_4^-$, and F^- (Figure 31) due to the effect of hydrogen bonding of the anion to the thiourea group on the ICT. Interestingly, this receptor is effective in aqueous solvent mixtures as well as buffered aqueous systems with similar color changes observed.

Dipyrrolyl quinoxaline (DPQ)-derived receptors have been developed as colorimetric anion receptors.[166] The receptors proved effective at binding F^- ($K_a = 118\,000\,M^{-1}$

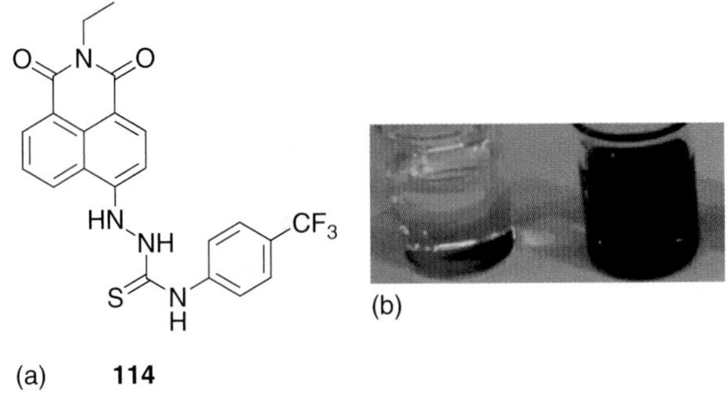

Figure 31 (a) Compound **114**. (b) Compound **114** without acetate (left) and on the addition of acetate (right). (Reproduced from Ref. 165. © American Chemical Society, 2005.)

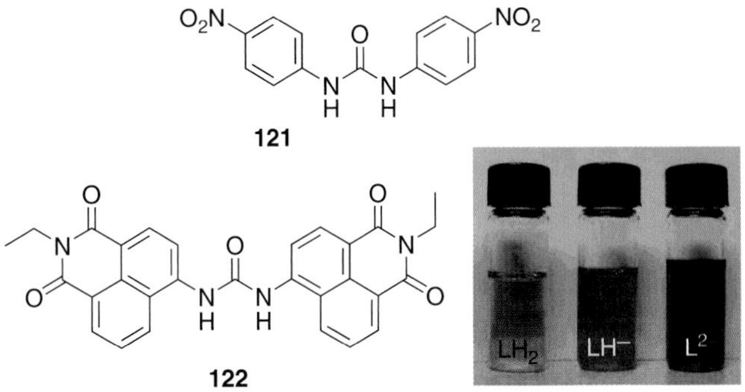

Figure 32 Compound **122** in the presence of TBA-F (L = **122**). (Reproduced from Ref. 169. © American Chemical Society, 2005.)

for F^- in $CHCl_3$ for **115**). A color change from yellow to blue is observed on addition of F^- to the receptors in $CHCl_3$ and DMSO. It is suggested that the color change is due to disruption of the conjugation of the molecule by puckering of the pyrrole moieties on binding F^-. DPQ derivatives also have a dual function as fluorescent anion sensors with quenching of fluorescence observed with receptors **115**–**117**.

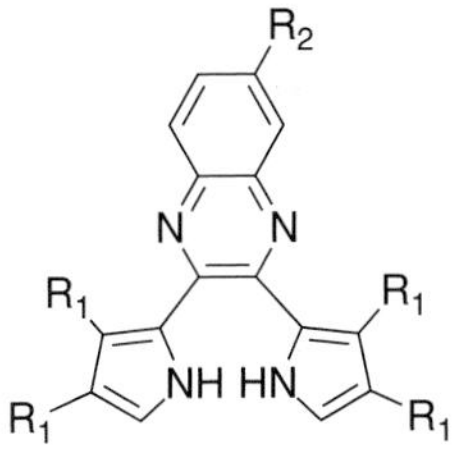

115 $R_1 = R_2 = H$
116 $R_1 = H$, $R_2 = NO_2$
117 $R_1 = F$, $R_2 = H$

Extended DPQ chromophores, compounds **118**–**120**, have been investigated by Anzenbacher and coworkers.[167] The effect of using 5,8-aryl substituents on the fluorescence was twofold with a red-shift in the emission maximum and an increase in the quantum yield. These modifications of the DPQ design also lead to an increase in binding affinity toward anions. Fluoride is bound strongly by all hosts (e.g., **118** $K_a = 51\,300\,M^{-1}$ in acetonitrile); however, pyrophosphate was bound very strongly ($K_a = 93\,700\,M^{-1}$ in acetonitrile). The addition of fluoride or pyrophosphate leads to the appearance of a new absorbance band at 500–550 nm and a decrease in the band at 400–450 nm. Besides a colorimetric response, fluorescence quenching is also observed.

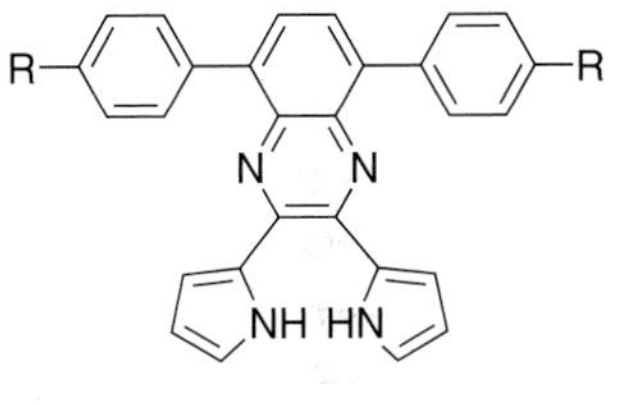

118 R = H
119 R = OCH_3
120 R = $N(CH_3)_2$

The phenomenon of urea deprotonation has been used extensively in the detection of fluoride and is one of the most common methods of producing a colorimetric sensor. The sensing of fluoride itself is a highly topical subject and has been recently reviewed.[168] The urea derivative **121** with two electron-withdrawing nitro-substituents deprotonates a single urea NH proton on the addition of fluoride leading to a color change.[169] This deprotonation was confirmed by ^1H NMR and crystallographic methods. The drive for the deprotonation is ascribed to the intrinsic acidity of the urea NH, enhanced by electron-withdrawing substituents and the high stability of the HF_2^- anion formed after deprotonation.

Fabbrizzi and coworkers have synthesized a naphthalimide-substituted urea capable of double deprotonation.[169,170] The addition of TBA-F to **122** in DMSO leads to a yellow to red color change after the addition of a few equivalents of anion, and on further addition, a second deprotonation step occurs leading to a blue coloration. This process can be monitored by UV–vis spectroscopy with the emergence of a new band at 540 nm for the single deprotonated species and a decrease in the free host band at 400 nm. With further addition of F^-, a new band at 600 nm forms corresponding to the doubly deprotonated species, with a decrease in the band intensity at 540 nm (Figure 32). Isosbestic points are observed for all new bands showing a clear transition between species. Carboxylates such as acetate also lead to a similar effect.

The use of electron-withdrawing substituents on pyrrole 2,5-diamides, for example, **123** synthesized by Gale and coworkers,[171] also leads to deprotonation of a urea NH with a concurrent color change from yellow to blue due to

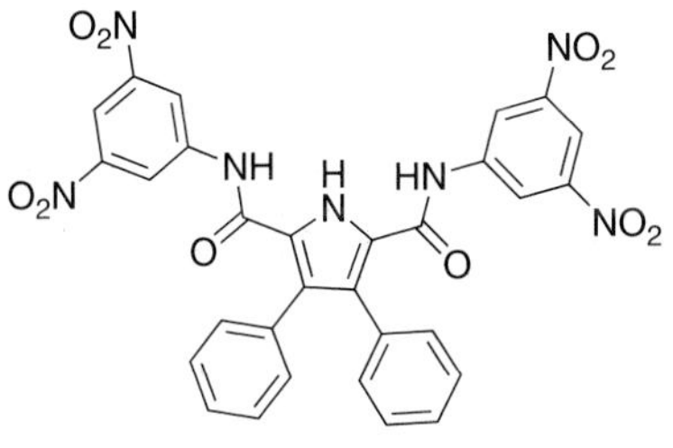

123

charge transfer from the deprotonated nitrogen atom to the nitrophenyl moiety.

2.4 Metal and Lewis acid-derived podand receptors

The use of metals as part of anion receptors is well established and the versatility of metal ion coordination chemistry has led to their varied use in receptor design, and we covered some examples in Section 1.7. In general, there are five major classes of metallic receptors:[172]

1. a substitutionally inert metal is used in a structural role;
2. a Lewis acidic metal ion forms part of the binding site;
3. self-assembled complexes of substitutionally labile metals involving thermodynamic anion templation;
4. anion-binding solid-state coordination polymers;
5. metal-based redox, colorimetric, or luminescence reporter groups.

There is often a large degree of overlap between classes; for example, it is possible for a metal to perform both a structural role and a reporter group role in the same molecule.

2.4.1 *Metals as structural elements*

By using substitutionally inert metals such a Ru(II), a low spin d^6 metal which when bonded to hard donors results in a large crystal field stabilization energy (CFSE), the metal can perform a structural role akin to that of the organic cholic acid or trialkylbenzene motifs.

The receptors **124–126** developed by Loeb and coworkers consist of a tetra-substituted Pt(II) complex.[173–175] Pyridyl amide derivatives are used as ligands to the metal and as anion-binding groups. This type of receptor can exist in many conformations, for example, a cone, partial cone, 1,2-alternate, and 1,3-alternate, analogous to that of calixarenes, with a 1,2-alternate conformation found in the solid state with PF_6^-.[173] The use of isoquinoline as a progression from pyridine resulted in an interesting anion selectivity. When bound to halides, complex **125** binds two anions strongly in DMSO, measured using ^{1}H NMR spectroscopic titrations. A 1,3-alternate conformation is found in the solid state, with the receptor behaving in a ditopic manner. The anion is bound with NH and CH hydrogen bonds as well as an electrostatic contribution from the Pt(II). In contrast, with $H_2PO_4^-$ and SO_4^{2-}, a cone conformation is observed in the solid state with a 1 : 1 host:guest stoichiometry (Figure 33).[174]

The use of pyrrolylpyridine ligands in complex **126** allows for competition between binding groups by allowing a choice between NH and CH hydrogen-bond donors.

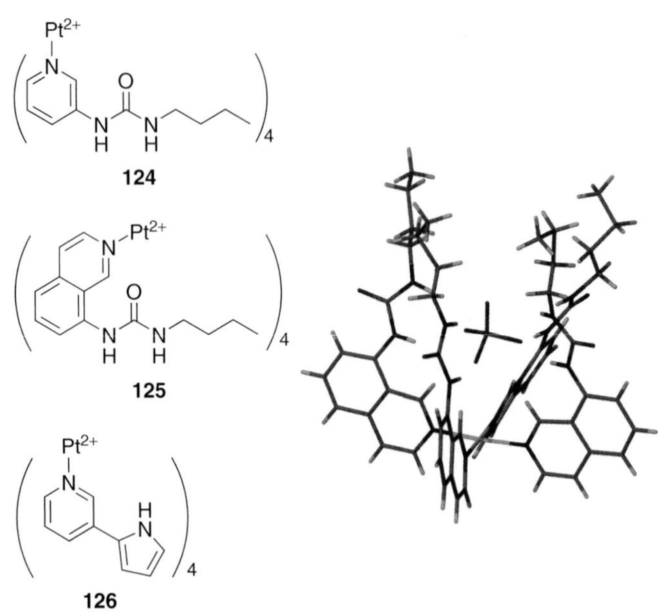

Figure 33 Compounds **124**, **125**, and **126** and the molecular structure of **125**·SO_4^+ determined by X-ray crystallography.[174]

For **126** with anions such as Cl^-, HSO_4^-, and NO_3^-, predominant downfield shifts were observed for the CH proton in DMSO, while for more basic anions such as acetate, NH downfield shifts were observed with acetate binding in a 1 : 2 host:guest stoichiometry. Binding in nitromethane shows significant downfield shifts in the NH proton only. It is suggested that the strong hydrogen-bond acceptor properties of the DMSO compete with the anion to bind with the NH, with only basic anions able to compete. The poor hydrogen-bond acceptor ability of nitromethane means that there is less competition and therefore the NH can interact with the anion.[175]

Steed and coworkers have synthesized a series of (arene)ruthenium(II) complexes such as complexes **127** and the analogous complex formed with the ligand **129**.[176] Receptors **127** and **128** have a low number of equivalent protons in the NMR spectrum suggestive of low symmetry in solution. Anions are bound strongly in acetonitrile with a

127 R = H
128 R = NO_2

129

1 : 1 and 2 : 1 host:guest stoichiometry observed. Interestingly, on addition of strongly bound anions such as chloride, the methylene protons collapse into a singlet suggestive of a more symmetric species and this is attributed to the loss of the ligated chloride to form a 16-electron species which would provide the required symmetry.

Inclusion of a carbazole fluorophore, as in the complex derived from ligand **129**, can create a fluorescent sensor, whereby the binding of chloride results in charge transfer from the chloride to a Ru-pyridyl-centered orbital which quenches the fluorescence emission.

Complexes **127**–**129** are on the border line of substitutionally stable and labile complexes with prolonged exposure to high equivalents of chloride leading to the displacement of a ligand and direct Ru–Cl complexation. This is on the timescale of hours to days, and the ligand exchange is slow on an experimental timescale.

Gale and coworkers have designed receptor **130** which is an interesting example of tuning preorganization, with Pt forming an integral structural role in preorganizing the receptor. Receptor **130** binds anions weakly in a DMSO-d_6—0.5% water solution with $H_2PO_4^-$ having an affinity constant of $90\,M^{-1}$ due to flexibility around the aryl–aryl bond of the bipyridine. However, when receptor **130** is reacted with $PtCl_2(DMSO)_2$, to form the Pt complex, **131**, the $H_2PO_4^-$ binding constant is increased to $3644\,M^{-1}$ as the binding groups are forced into a syn arrangement forming a preorganized, convergent binding cavity. The receptor also proved effective as a colorimetric sensor for F^-, through deprotonation of a urea NH group. A color change from yellow to purple is observed after deprotonation.[177]

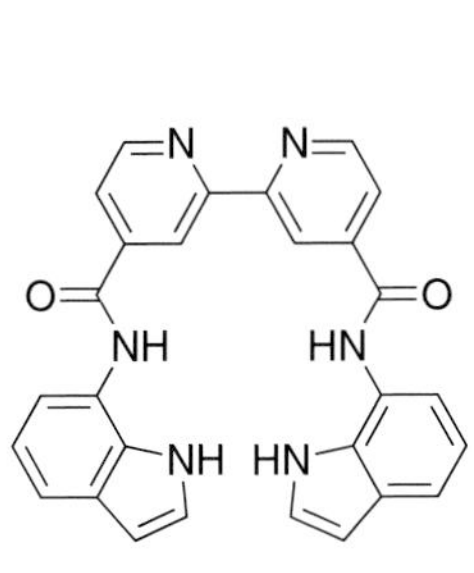

130 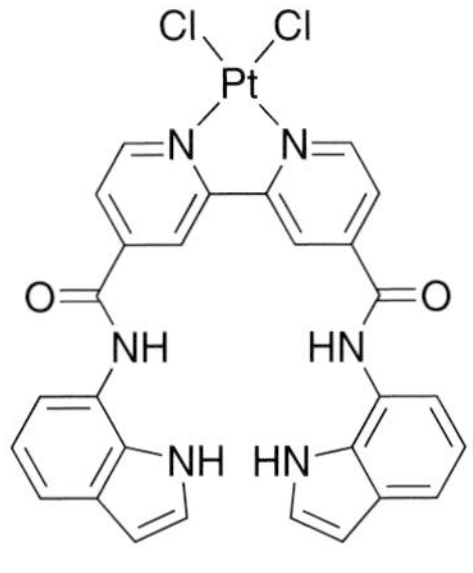 **131**

Beer and coworkers have developed $Ru(Bpy)_3^{2+}$ derivatives which are intrinsically chiral due to the helicity of the $Ru(Bpy)_3^{2+}$ but can also allow the incorporation of additional chiral functionality, for example, **132** and **133**.[178] Each receptor was isolated in an enantiomerically pure form and binds chiral anions such as *N*-Cbz-Glu and lactate in DMSO determined via ^1H NMR spectroscopic titrations. However, in all cases enantiomeric selectivity could not be achieved or was low.

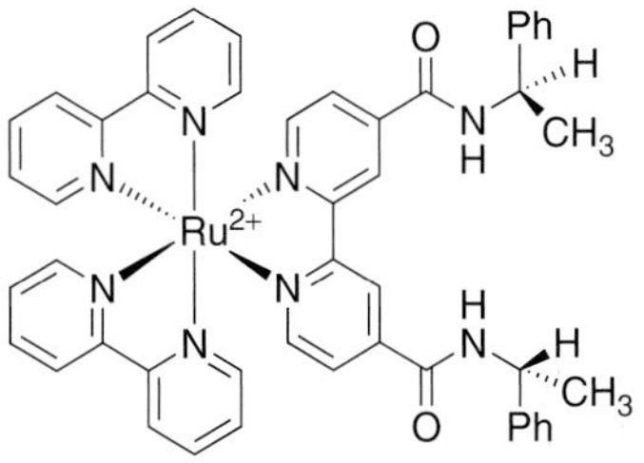

132

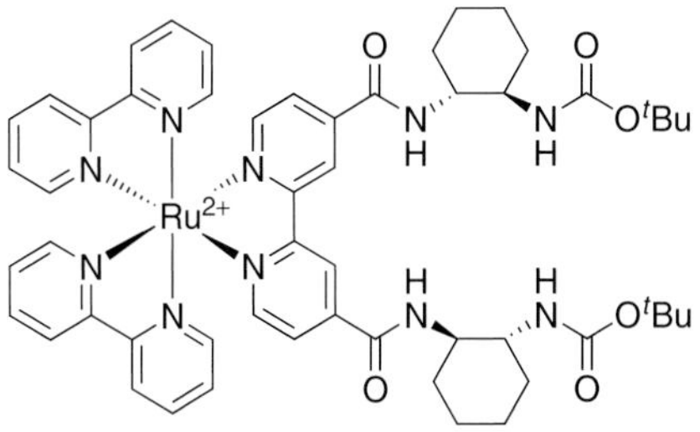

133

Stable transition metal pyrazole complexes using Mn (**134**), Re (**135**), and Mo (**136**) have been synthesized by Pérez and coworkers.[179–181] To create a convergent binding cavity, the metal fragment was functionalized with CO groups which prefer to adopt a fac arrangement of the CO groups to maximize back bonding. The rhenium complex proved to be substitutionally inert due to its d^6 configuration; however, a pyrazole ligand was displaced from the Mn and Mo complexes in the presence of anions. Hydrogen-bond formation between anions and the NH groups of the pyrazole is observed by ^1H NMR spectroscopic titrations as well as in the solid state. The solid-state structures reveal an unfavorable deformation in the N–Re–N bond angles of **135** when binding anions when compared to the tetrahedral B and Zn derivatives. Interestingly, HSO_4^- leads to protonation of the pyrazole unit and binding of a sulfate anion. Fluoride, on the other hand, deprotonates NH groups of a pyrazole.

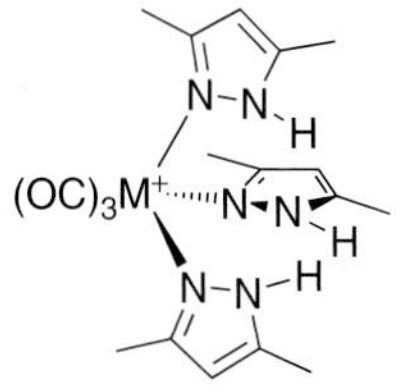

134 M = Mn **136**
135 M = Re

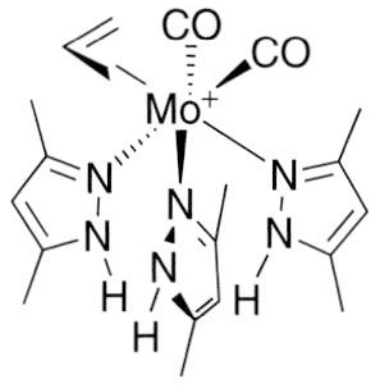

The Pérez group has also synthesized a range of d^6 transition metal-derived anion receptors based on rhenium and ruthenium, for example, complexes **137**[182] and **138**.[183] Compound **137** contains a bidentate pyrazolylamidino ligand which provides one NH hydrogen-bond donor and an additional pyrazolyl NH donor. The complex binds chloride particularly strongly because of its rigid nature, with a binding constant of $8725\,M^{-1}$ in CD_3CN measured by ^1H NMR spectroscopic titrations.

137 **138**

Receptor **138** utilized biimidazole as the anion-binding group. In this system, the metal center effectively preorganizes the biimidazole to bind anions by preventing rotation around the aryl–aryl bond and by also preventing self-association (Figure 34). The receptor binds anions strongly in CD_3CN, for example, $K_a = 5920\,M^{-1}$ for HSO_4^-.

2.4.2 *Labile metal-derived anion receptors*

The use of labile metal atoms, that is, those in which ligand association and disassociation are fast on an experimental timescale, has also been investigated, where the receptor–anion complex is essentially self-assembled in solution as the thermodynamic product.

Halcrow has designed a receptor involving the d^{10} Zn^{2+} metal and a pyrazole derivative (**139**).[184] This receptor is analogous to the systems Pérez described previously; however, the lability of these ligands with the Zn(II) metal means the complex, while being thermodynamically stable, is not kinetically stable in solution. The solid-state structure shows the assembled complex with a tetrahedral Zn^{2+} and three 3(5)*t*-butylpyrazole groups (Figure 35). The pyrazole derivative hydrogen bonds to the chloride of an adjacent complex in the crystal, forming a hydrogen-bonding polymer. Stable complexes of this type utilizing

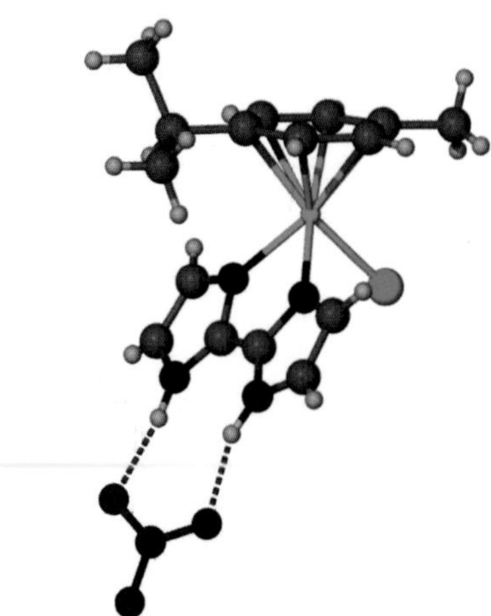

Figure 34 Molecular structure of $\mathbf{138}\cdot NO_3^-$ determined by X-ray diffraction.[183]

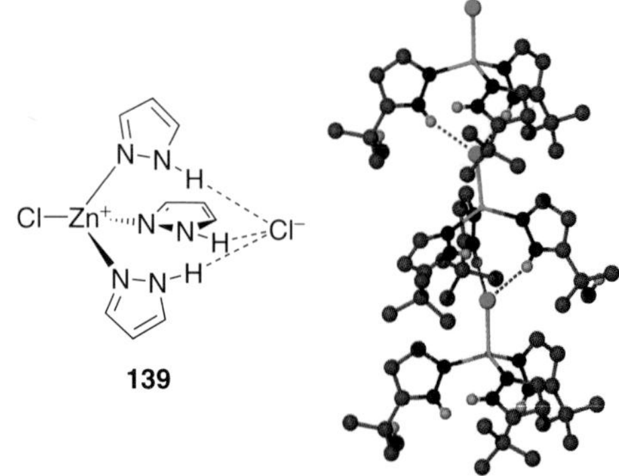

Figure 35 Compound **139** and the crystal structure of **139** determined by X-ray diffraction.[184]

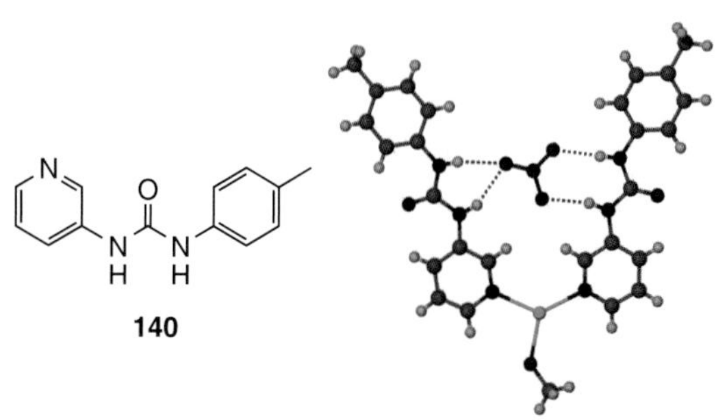

Figure 36 Compound **140** and molecular structure of $[Ag(\mathbf{140})_2(NO_3)(HOCH_3)]$ complex. Only one complex in the asymmetric unit is shown for clarity.[186]

covalent bonding can be synthesized using a triprotonated trispyrazolylborate dication and has been shown to bind chloride in the solid state.[185]

Metal complexes of pyridylurea ligands and silver salts have been investigated by Steed and coworkers.[186] In this work, the ligand self-assembles with a Ag^+ cation to form an $[Ag(\mathbf{140})_2]^+$ species. This is able to bind anions such as nitrate strongly in a 1 : 1 and 1 : 2 host:guest stoichiometry, measured by ^1H NMR spectroscopic titration ($K_{11} = 30\,200\,M^{-1}$, $K_{12} = 2900\,M^{-1}$ in CD_3CN), as well as a 1 : 3 stoichiometry due to ligation of nitrate to the silver center ($K_{13} = 550\,M^{-1}$). A solid-state structure of the $[Ag(\mathbf{140})_2(NO_3)]$ was determined by X-ray crystallography and is shown in Figure 36. The structure shows a convergent binding site with the nitrate bound by NH and CH hydrogen bonds.

2.4.3 *Lewis acid-metal-based receptors*

Metals that act as Lewis acids can increase the effectiveness of a receptor by directly interacting with an anion, polarizing a hydrogen-bond donor, hence increasing its acidity, providing a structural element, or more usually a combination of both. Direct coordination of anions to metal complexes is exemplified by cascade and related complexes as discussed in Section 1.6 and there is a history of metal centers such as tin[187] and mercury[188] being used in anion coordination chemistry. Rudkevich and coworkers have synthesized a range of cascade-type anion receptors containing a UO_2 fragment as a Lewis acid center. Compounds such as **141** bind anions strongly, particularly $H_2PO_4^-$, measured via a conductance method ($K_a > 10^5\,M^{-1}$ in CH_3CN:DMSO v/v 99 : 1) and a selectivity over Cl^- of 10^2. The molecular structure of **141**·$2H_2PO_4^-$ shows bond formation between the dihydrogen phosphate O and the UO_2 center, that is, a Lewis acid interaction. ^1H NMR spectroscopic titrations and the molecular structure also confirm hydrogen-bond formation between the amide functionality and the anion.[80]

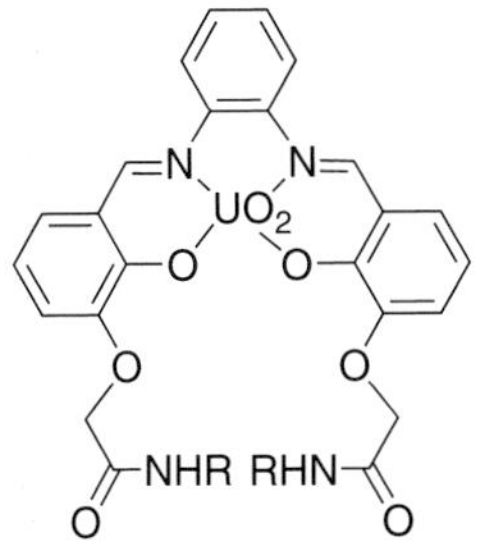

141 R = 4-$CH_3C_6H_4$

Receptor **142** is capable of self-organizing in the presence of NaX (where X is a halide) into dimeric structure held together by Lewis acid interactions between two crown ether moieties and the Na. This allows a binding cleft to be formed between the urea groups, which is able to bind anions, for example, chloride as shown in the crystal structure determined by X-ray crystallography (Figure 37).[189]

2.4.4 *Metals as sensing units*

Electrochemical-based anion receptors have been a versatile means of sensing since the pioneering work of Beer and coworkers in the late 1980s, principally using cyclic voltammetry. A range of redox active moieties, including cobaltocene, ferrocene, and $Ru(bpy)_3^{2+}$ derivatives, have been incorporated into these receptors.

Beer and coworkers have also described five mechanisms of sensing using redox groups[1]:

1. a through-space interaction with the anion-binding site in close proximity to the redox center;
2. direct coordination of the anion to the redox center;
3. a through-bond interaction through conjugation between binding site and redox center;
4. an anion-induced conformational change which leads to a perturbation in the redox center;
5. an interference mechanism whereby the interaction between several redox active centers is affected by anion binding.

Beer[190] has synthesized a large library of electrochemical sensors. The first cobaltocene sensor devised was compound **143** and its binding properties were investigated using Br^-.[191] Binding is via electrostatic interactions and the cobaltocene Cp_2Co^+/Cp_2Co redox couple displays a cathodic shift, that is, the anion increases the cobaltocene reduction potential.

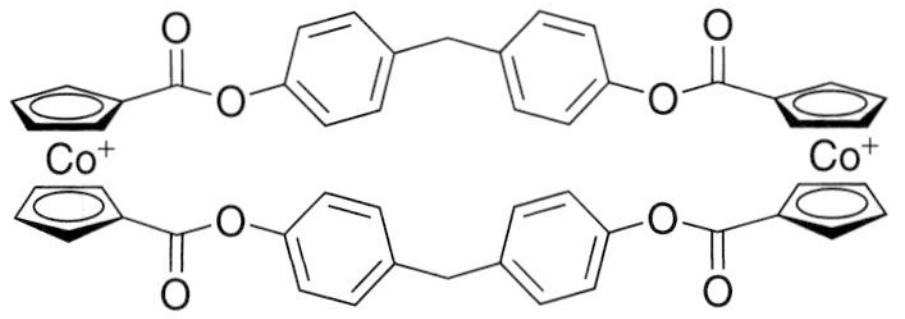

143

It is also possible to add hydrogen-bonding functionality to this class of sensors, for example, **144**[192] and **145**.[193,194] ^1H NMR spectroscopic titrations of compounds **144** and **145** reveal the highest binding affinity of $H_2PO_4^-$. As would be expected, the formation of macrocyclic systems

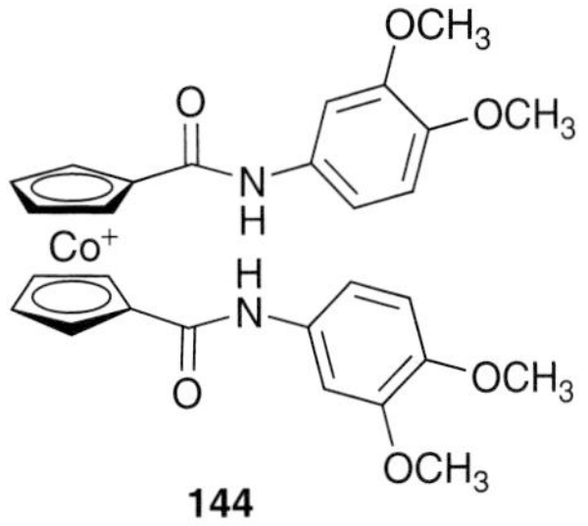

144 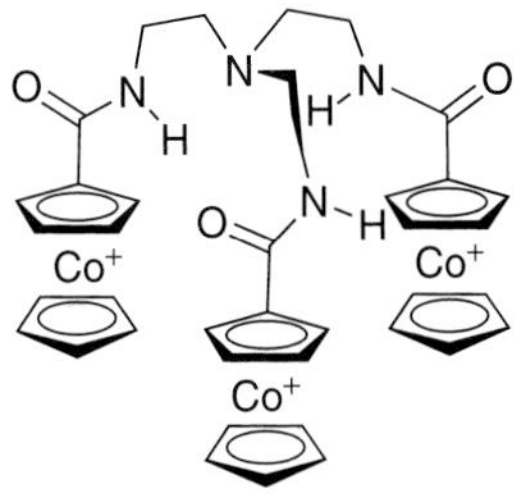 **145**

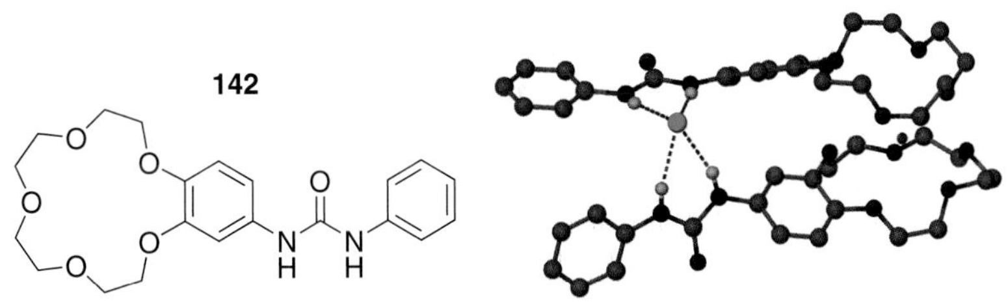

142

Figure 37 Compound **97** and the crystal structure of the **142**·NaCl complex, determined by X-ray crystallography.[189]

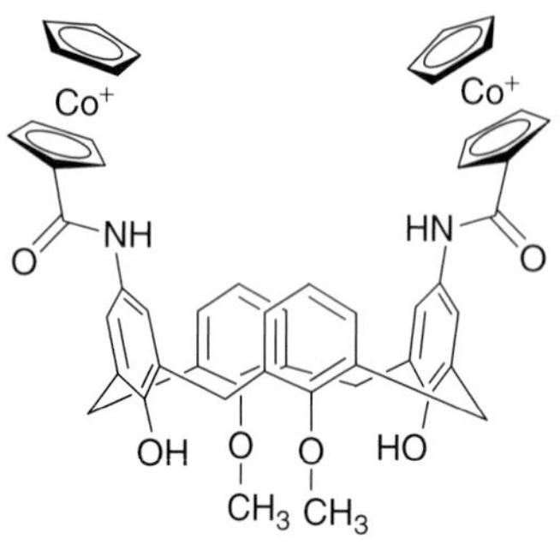

146

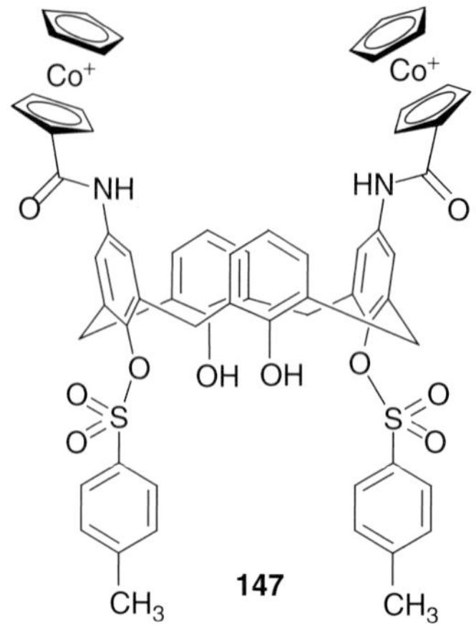

147

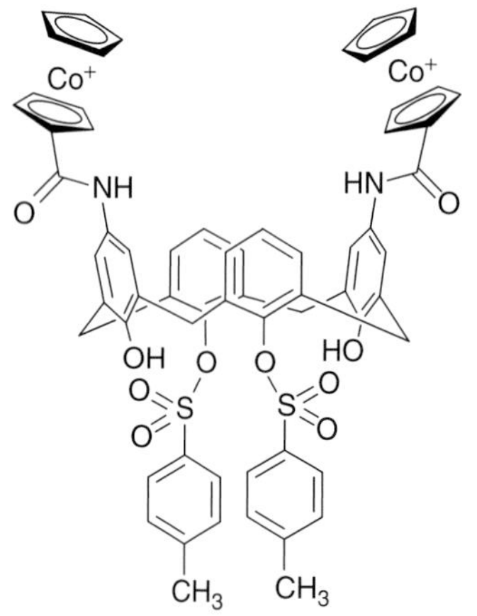

148

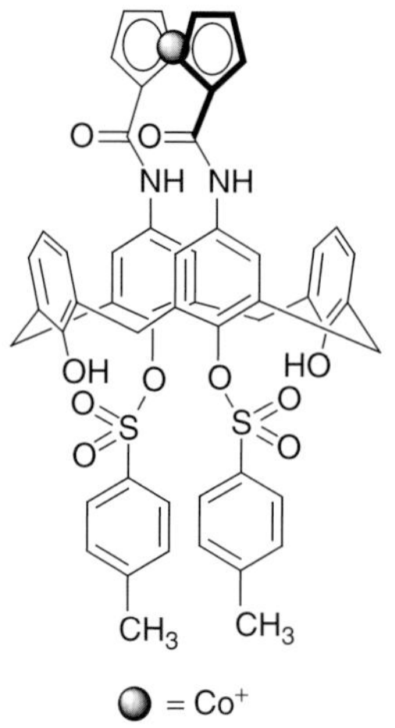

149

analogous to **101** increases the binding constants significantly, that is a 10-fold increase.

Calixarene-cobaltocene sensors, for example, compounds **146–149** have also been developed.[195,196] An interesting aspect of these sensors is that varying the topology of the receptor allows for the selective sensing of a specific anion. For example, the topology of **146** favors Cl^- determined via NMR spectroscopic titrations. Further functionalization of the lower rim with tosylate allows tuning of the binding properties. For example, when the tosyl groups were para to the cobaltocenium (**147**), $H_2PO_4^-$ binding was favored because of the tosyl groups forcing the cobaltocenium groups together. However, when ortho, **148**, the tosyl groups force the cobaltocenium groups slightly apart, favoring Cl^-. Finally, compound **149** is preorganized for carboxylate binding, because of the bridging cobaltocenium orientating the amide groups for carboxylate hydrogen-bond formation.

Ferrocene derivatives have also been studied for electrochemical sensing. However, as the sensor is neutral there is no intrinsic electrostatic interaction unless the ferrocene is oxidized to ferrocinium, in which case the receptor becomes cationic and binding is increased.[83]

Receptors **150** and **151** contain a mixture of hydrogen-bond acceptor and donor groups.[197] These receptors are able to selectively bind dihydrogen phosphate in acetonitrile because of its binding group complementarity and high basicity. Large cathodic shifts in the redox potential were observed in the presence of excess $H_2PO_4^-$ (**150**, 120 mV, **151**, 240 mV in acetonitrile), although **151** exhibits irreversible oxidation. Receptor **150** was able to sense $H_2PO_4^-$ even in the presence of competing anions such as Cl^-

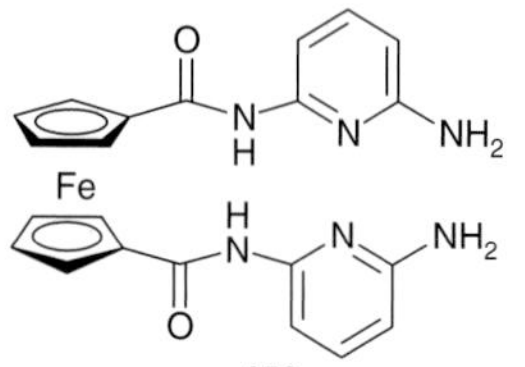

150 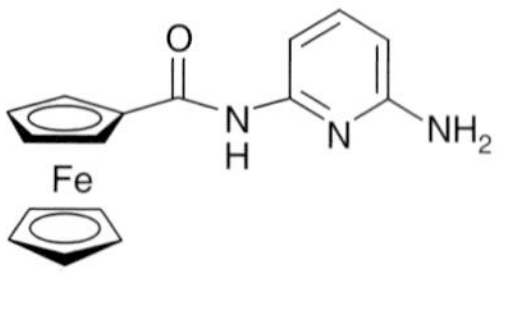 **151** 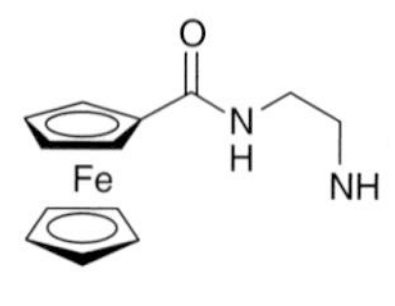 **152**

and HSO_4^-. Tripodal and calixarene derivatives have been synthesized. All display large cathodic shifts in the redox potentials.

The receptor **152** is interesting as it is difunctional.[197] Typically in neutral hydrogen-bonding receptors, $H_2PO_4^-$ binds more strongly than HSO_4^- as its higher basicity forms stronger hydrogen bonds. In this receptor, however, two binding modes are possible; the first operates for nonacidic guests and the receptors act as the hydrogen-bond donor from the amide. The second mode applies to acidic guests in which proton transfer occurs, allowing electrostatic and hydrogen-bonding interactions with the guest. In the case of $H_2PO_4^-$, it does not fit well into either category and therefore is not bound strongly. HSO_4^-, however, fits well into the second binding mode and is bound strongly, giving rise to this unusual selectivity. The electrochemical behavior of the receptor on addition of HSO_4^- showed a new oxidation peak, cathodically shifted by 220 mV from the free receptor, showing that anion binding greatly increases the ease of oxidation, although the receptor showed irreversible behavior.

Gale and coworkers[198] have appended the ferrocene moiety onto the 2,5-diamidopyrrole core described previously. Receptor **153** showed a large anodic shift in the redox potential of −130 mV with F^- in dichloromethane. This compares to only −75 mV for chloride, and this selectivity corresponds well with binding constants measured by NMR techniques ($K_a = 170\,M^{-1}$ for $H_2PO_4^-$ and <20 for Cl^- in dichloromethane).

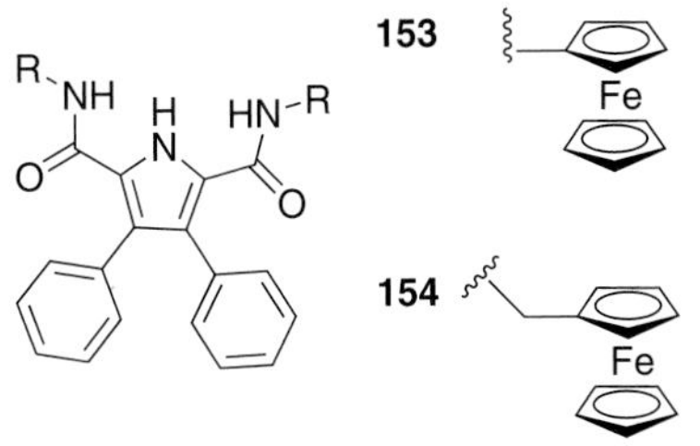

Receptor **154** also shows anodic shifts in the redox potential; however, on the addition of F^-, two redox waves are observed ($\Delta E = -125$ and −255 mV in dichloromethane) and are only observed with F^-. As with **153**, chloride shows a lower anodic shift of −55 mV, corresponding well with the binding affinity measured by NMR spectroscopy.

Besides iron- and cobalt-based receptors, ruthenium can also be used. This has the advantage of being both redox active and luminescent. A wide range of receptor designs, for example, compounds **155–157**, have been synthesized.[199] The molecular structure of **155** showing the bound Cl^- anion is shown in Figure 38 and shows two NH amide hydrogen bonds to the anion. All receptors show significant downfield shifts of NH resonance in the NMR spectrum on addition of anions, with chloride binding constants in the region of $10^2\,M^{-1}$ and those of $H_2PO_4^-$ in the region of $10^3\,M^{-1}$ in DMSO.

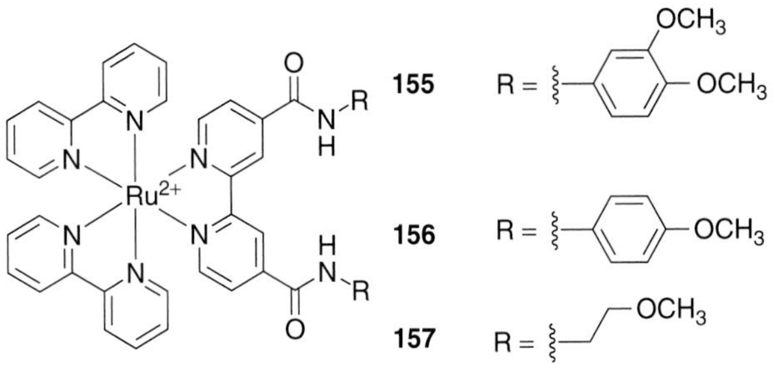

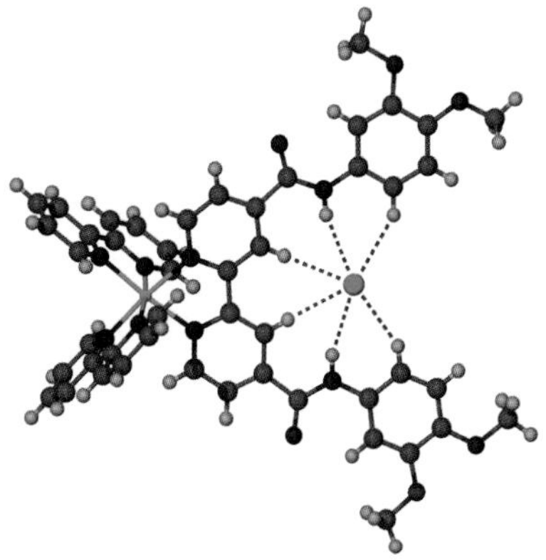

Figure 38 Molecular structure of **155**·Cl^- determined by X-ray diffraction.[199]

It is also possible to incorporate other binding motifs such as imidazole groups in receptor **158**.[200] Neutral and anionic phosphodiesters are bound by the receptor with an observed increase in emission intensity.

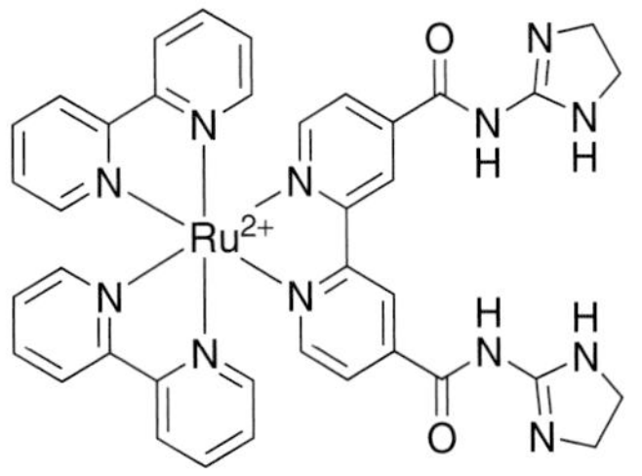

In addition to transition-metal-based receptors, lanthanoids have been used as a means to sense anions, by Parker and coworkers, as discussed in Section 1.6.[201] In these systems, the anion directly coordinates to the metal, displacing existing ligands, and changes the environment of the lanthanoid. This is modulated as a change in the emission spectrum of the receptor. These receptors can be ratiometric in that there is an increase in the intensity of one band at the expense of the intensity of another. This ratiometric approach is an advantage as it means that the sensor can work without the knowledge of the absolute concentration of the host in solution. Receptors **159**[202] and **160**[71] are able to bind and sense a range of biologically important anions such as citrate, lactate, and HCO_3^- under physiological conditions, with citrate being bound particularly strongly.

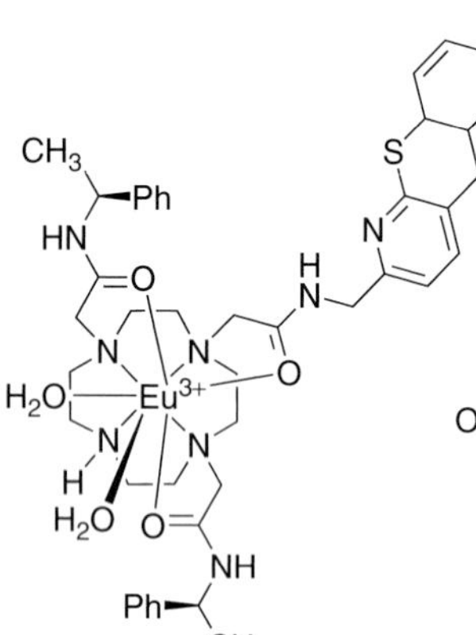

159 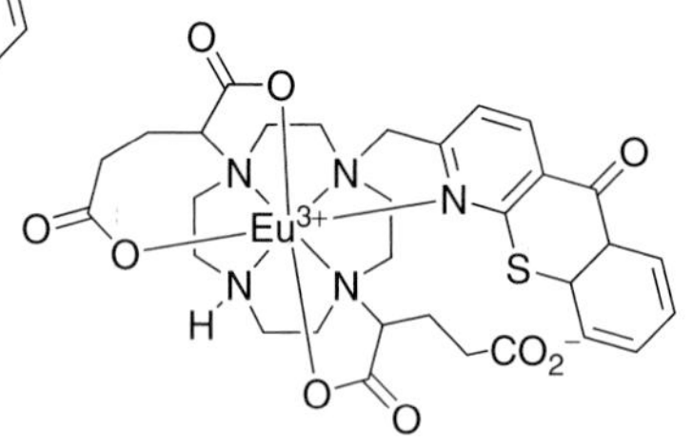 **160**

Receptor **159** has proved highly effective at binding citrate in biological fluids such as prostate fluid with high affinity ($\log K_a = 4.98$ in saline solution).[71] Citrate levels in prostate fluid have been shown to be a reliable marker for prostate cancer. The system was able to accurately determine the citrate concentration in prostate fluid samples compared with the currently used enzyme-based method, while proving to be experimentally simpler and quicker and may prove useful in the diagnosis of prostate cancer.

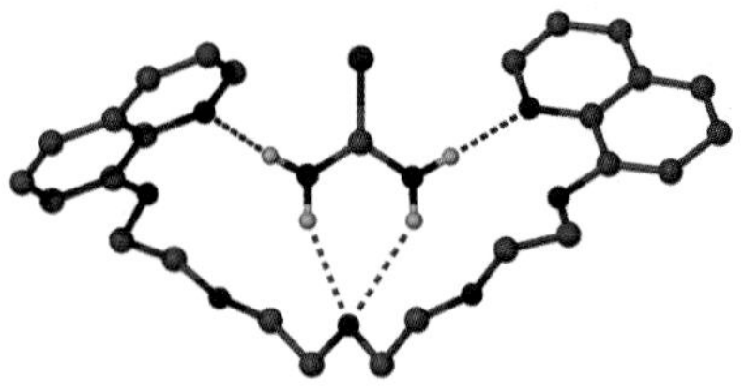

Figure 39 Thiourea complex of Kryptofix-5.[206]

3 COMPLEXATION OF MOLECULAR GUESTS

In comparison to anion and cation binding, the complexation of neutral guest molecules by podand-type receptors has received relatively little attention, possibly because the interactions with uncharged guests are relatively weak (chiefly hydrogen bonding and π-stacking) and, coupled with the lack of preorganization in podands, this factor makes podands generally not particularly effective at binding neutral molecules. Although stable, solution-phase neutral molecule complexes of simple oligoethyleneglycol derivatives are unknown, there have been some successes. For example, the rigid terminal group podand Kryptofix-5 (**8**) forms a complex with thiourea in which the guest binds to the receptor via NH···O and NH···N hydrogen bonds, Figure 39. Work by Rebek has resulted in a series of rigidly preorganized Kemp's triacid derivatives, Scheme 5. While the smallest member of the series engages in intramolecular hydrogen bonding and hence

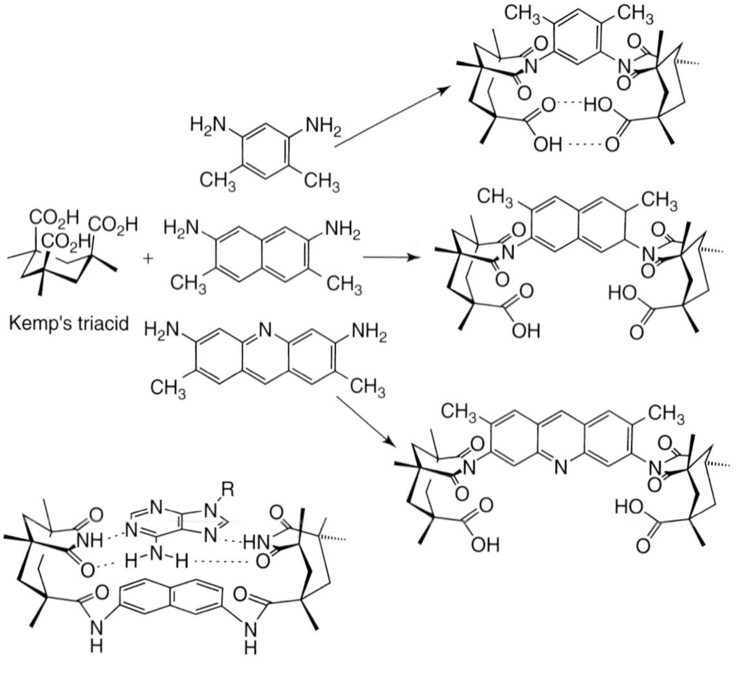

Kemp's triacid

Scheme 5 Molecular tweezers derived from Kemp's triacid and an examples of a complex with adenine.[137]

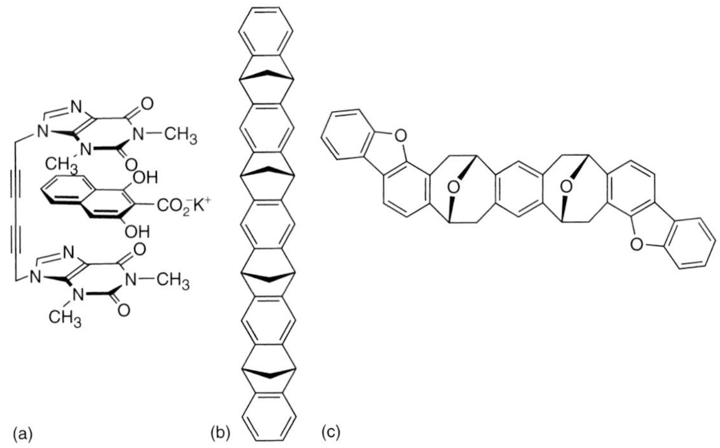

Figure 40 Molecular tweezers designed by the groups of (a) Whitlock, (b) Klärner, and (c) Harmata.

does not include guest molecules, the larger homologs can chelate molecular guests. Their ability to bind biologically relevant molecules has been studied in the context of biomimetic catalysis. A related compound, imide molecular tweezer, can extract nucleobases such as adenine, and nucleosides adenosine and deoxyadenosine from water and transport them across liquid organic membranes.[137] There have been a number of other studies on "molecular tweezer" type compounds, notably a number of early systems designed by Whitlock (Figure 40a),[135] rigid systems based on π-stacking designed by Klärner *et al.*,[203] (Figure 40b) and extensive work by Harmata based on Kagan's ether (Figure 40c) that may be considered podands. The field has been recently reviewed.[138] Finally, in terms of inspiration for the future, it may be that chemists can look toward the principle of multivalency to bring about complex multiple interactions between highly specific, bioinspired receptor substrate complexes. As a natural example, the drug–receptor complex formed between the new generation antibiotic vancomycin and proteins that are used in the synthesis of bacterial cell walls serves as a remarkable template, Figure 41.[204] Such biological inspiration is already resulting in the synthesis of remarkable

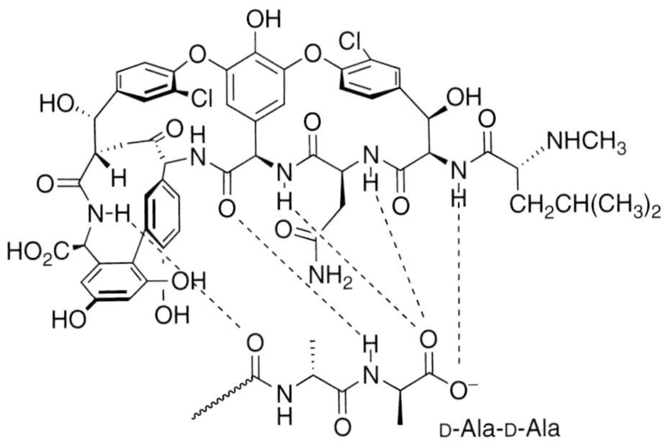

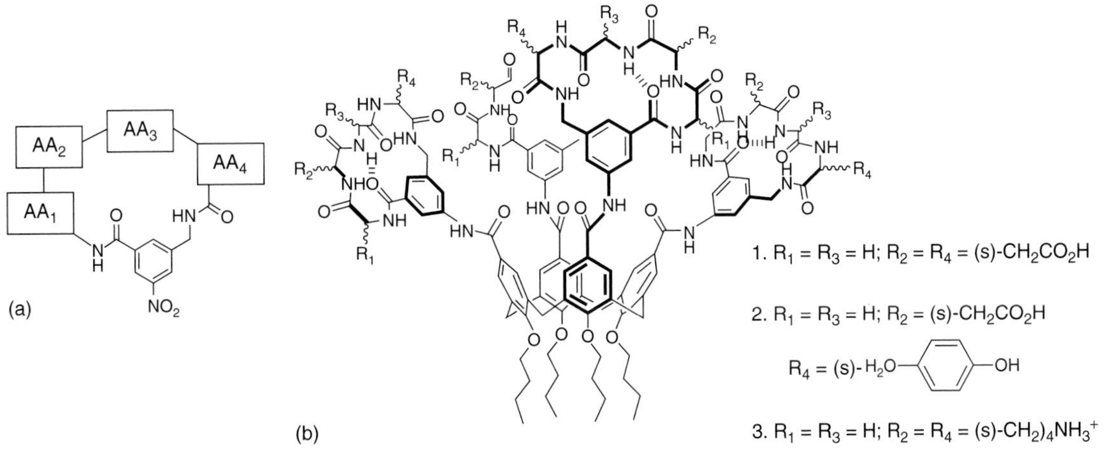

Figure 41 A drug–receptor complex formed by vancomycin.

Figure 42 Calixarene receptors (b) derived from tetrapeptide loops (a) designed to bind to the surface of chymotrypsin. (Reproduced from Ref. 205. © American Chemical Society, 1999.)

protein receptor podands such as the tetrapeptide loop calixarenes designed to bind the surface of chymotrypsin, Figure 42.[205] Similarly, water-soluble sugar-based multivalent calixarenes such as **161** bind strongly to lectin proteins such as Concanavalin A. They also exhibit interesting agglomeration behavior, forming micelles in solution.[3]

HO
O
OH
HN
HO
OH
HN
S
R = Alkyl
4
OR
161

4 CONCLUSION

The podand field has developed rapidly over the last 20 years, with the adaptability in design meaning that receptors can be rigidly preorganized or flexible allowing induced-fit binding or can be tuned for properties in between. Podands have also proved highly effective as molecular sensors, with fluorescent, colorimetric, and electrochemical sensors all being realized with many different sensing modes observed within each category, intimately linked to the receptor design. Many of these receptors also function in water in real-world applications, for example, in prostate cancer diagnosis. It is clear that the scope of podands has expanded greatly since the term was originally coined in the context of acyclic ligands, particularly for binding alkali metal cations, with the surge in interest in dendrimers representing a particular extensive new facet of podand chemistry.

ACKNOWLEDGMENTS

We thank Durham University for a Doctoral Fellowship (to ANS) that has facilitated our own work on anion-binding podands.

REFERENCES

1. F. Vögtle and E. Weber, *Angew. Chem. Int. Ed. Engl.*, 1979, **18**, 753.
2. J. D. Badjică, A. Nelson, S. J. Cantrill, *et al.*, *Acc. Chem. Res.*, 2005, **38**, 723.
3. L. Baldini, A. Casnati, F. Sansone, and R. Ungaro, *Chem. Soc. Rev.*, 2007, **36**, 254.
4. G. R. Newkome, C. N. Moorefield, and F. Vögtle, *Dendrimers and Dendrons: Concepts, Syntheses, Applications*, Wiley-VCH, Weinheim, 2001.
5. G. Gokel and O. Murillo, in *Comprehensive Supramolecular Chemistry*, 1st edn, eds. J. L. Atwood, J. E. D. Davies, D. D. Macnicol, and F. Vögtle, Pergamon, Oxford, 1996, vol. 1, p. 1.
6. C. J. Pedersen, *J. Am. Chem. Soc.*, 1967, **89**, 7017.
7. J. W. Steed and J. L. Atwood, *Supramolecular Chemistry*, 2nd edn, Wiley-Blackwell, Chichester, 2009.
8. R. W. Hoffmann, F. Hettche, and K. Harms, *Chem. Commun.*, 2002, 782.
9. D. J. Cram, *Angew. Chem. Int. Ed. Engl.*, 1986, **25**, 1039.
10. E. Maverick and D. J. Cram, in *Comprehensive Supramolecular Chemistry*, 1st edn, eds. J. L. Atwood, J. E. D. Davies, D. D. MacNicol, and F. Vögtle, Pergamon, Oxford, 1996, vol. 1, p. 213.
11. F. Vögtle and H. Sieger, *Angew. Chem. Int. Ed. Engl.*, 1977, **16**, 396.
12. M. Takagi, in *Cation Binding by Macrocycles*, 1st edn, eds. G. Gokel and Y. Inoue, Dekker, New York, 1991.
13. F. Vögtle, W. M. Muller, W. Wehner, and E. Buhleier, *Angew. Chem. Int. Ed. Engl.*, 1977, **16**, 548.
14. A. Mele, P. Metrangolo, H. Neukirch, *et al.*, *J. Am. Chem. Soc.*, 2005, **127**, 14972.
15. W. Saenger and H. Brand, *Acta Crystallogr., Sect. B*, 1979, **35**, 838.
16. R. D. Rogers and C. B. Bauer, in *Comprehensive Supramolecular Chemistry*, 1st edn, ed. G. W. Gokel, Pergamon, Oxford, 1996, vol. 1, p. 315.
17. R. D. Rogers, A. N. Rollins, R. F. Henry, *et al.*, *Inorg. Chem.*, 1991, **30**, 4946.
18. R. D. Rogers, A. H. Bond, S. Aguinaga, and A. Reyes, *Inorg. Chim. Acta*, 1993, **212**, 225.
19. R. D. Rogers, A. H. Bond, and J. L. Wolff, *J. Coord. Chem.*, 1993, **29**, 187.
20. M. Valente, S. Sousa, A. Lopes Magalhães, and C. Freire, *J. Solution Chem.*, 2010, **39**, 1230.
21. B. P. Hay and R. D. Hancock, *Coord. Chem. Rev.*, 2001, **212**, 61.
22. A. Agtarap, J. W. Chamberlin, M. Pinkerton, and L. Steinrauf, *J. Am. Chem. Soc.*, 1967, **89**, 5737.
23. W. K. Lutz, F. K. Winkler, and J. D. Dunitz, *Helv. Chim. Acta*, 1971, **54**, 1103.
24. N. Yamazaki, S. Nakahama, A. Hirao, and S. Negi, *Tetrahedron Lett.*, 1978, **19**, 2429.
25. H. Sieger and F. Vögtle, *Liebigs Ann. Chem.*, 1980, 425.
26. H. Degani and H. L. Friedman, *Biochemistry*, 1974, **13**, 5022.
27. B. J. Abbott, D. S. Fukuda, D. E. Dorman, *et al.*, *Antimicrob. Agents Chemother.*, 1979, **16**, 808.
28. A. Bencini, A. Bianchi, E. Garcia-España, *et al.*, *Coord. Chem. Rev.*, 1999, **188**, 97.
29. D. K. Cabbiness and D. W. Margerum, *J. Am. Chem. Soc.*, 1969, **91**, 6540.

30. H. Maluszyńska, A. Perkowska, and E. Skrzypczak-Jankun, *J. Chem. Crystallogr.*, 1995, **25**, 19.
31. G. Codina, A. Caubet, C. Lopez, *et al.*, *Helv. Chim. Acta*, 1999, **82**, 1025.
32. J. H. N. Buttery, N. C. Plackett, B. W. Skelton, *et al.*, *Z. Anorg. Allg. Chem.*, 2006, **632**, 1856.
33. E. C. Constable, in *Progress in Inorganic Chemistry*, ed. K. D. Karlin. John Wiley & Sons, Inc., New York, 1994, vol. 42, p. 67.
34. M. Albrecht, *Chem. Rev.*, 2001, **101**, 3457.
35. B. Hasenknopf, J. M. Lehn, N. Boumediene, *et al.*, *J. Am. Chem. Soc.*, 1997, **119**, 10956.
36. B. Hasenknopf, J. M. Lehn, B. O. Kneisel, *et al.*, *Angew. Chem. Int. Ed. Engl.*, 1996, **35**, 1838.
37. J.-M. Lehn, *Supramolecular Chemistry*, 1st edn, VCH, Weinheim, 1995.
38. M. Ruben, J. Rojo, F. J. Romero-Salguero, *et al.*, *Angew. Chem. Int. Ed.*, 2004, **43**, 3644.
39. S. R. Batten, S. M. Neville, and D. R. Turner, *Coordination Polymers: Design, Analysis and Application*, Royal Society of Chemistry, Cambridge, 2008.
40. E. C. Constable, *Tetrahedron*, 1992, **48**, 10013.
41. R. Kramer, J. M. Lehn, and A. Marquisrigault, *Proc. Natl. Acad. Sci. U.S.A.*, 1993, **90**, 5394.
42. C. O. Dietrich-Buchecker and J. P. Sauvage, *Angew. Chem. Int. Ed. Engl.*, 1989, **28**, 189.
43. E. C. Constable, M. G. B. Drew, G. Forsyth, and M. D. Ward, *J. Chem. Soc., Chem. Commun.*, 1988, 1450.
44. J. M. Lehn, J. P. Sauvage, J. Simon, *et al.*, *Nouv. J. Chem.*, 1983, **7**, 413.
45. E. C. Constable, M. D. Ward, and D. A. Tocher, *J. Am. Chem. Soc.*, 1990, **112**, 1256.
46. V. Amendola, M. Boiocchi, V. Brega, *et al.*, *Inorg. Chem.*, 2010, **49**, 997.
47. J. C. Jeffery, T. Riis-Johannessen, C. J. Anderson, *et al.*, *Inorg. Chem.*, 2007, **46**, 2417.
48. J. P. Gisselbrecht, M. Gross, J. M. Lehn, *et al.*, *Nouv. J. Chem.*, 1984, **8**, 659.
49. A.-X. Zhu, J.-P. Zhang, Y.-Y. Lin, and X.-M. Chen, *Inorg. Chem.*, 2008, **47**, 7389.
50. M. Dhaenens, L. Lacombe, J. M. Lehn, and J. P. Vigneron, *J. Chem. Soc., Chem. Commun.*, 1984, 1097.
51. D. L. Caulder and K. N. Raymond, *J. Chem. Soc., Dalton Trans.*, 1999, 1185.
52. P. N. W. Baxter, J. M. Lehn, G. Baum, and D. Fenske, *Chem.—Eur. J.*, 2000, **6**, 4510.
53. C. Piguet, M. Borkovec, J. Hamacek, and K. Zeckert, *Coord. Chem. Rev.*, 2005, **249**, 705.
54. J. Hamacek, M. Borkovec, and C. Piguet, *Dalton Trans.*, 2006, 1473.
55. D. Farrell, K. Gloe, K. Gloe, *et al.*, *Dalton Trans.*, 2003, 1961.
56. E. J. O'Neil and B. D. Smith, *Coord. Chem. Rev.*, 2006, **250**, 3068.
57. A. Ojida, Y. Mito-Oka, M. Inoue, and I. Hamachi, *J. Am. Chem. Soc.*, 2002, **124**, 6256.
58. E. Leize, A. Van Dorsselaer, R. Kramer, and J. M. Lehn, *J. Chem. Soc., Chem. Commun.*, 1993, 990.
59. C. M. Hartshorn and P. J. Steel, *Angew. Chem. Int. Ed. Engl.*, 1996, **35**, 2655.
60. V. Amendola, M. Boiocchi, B. Colasson, *et al.*, *Angew. Chem. Int. Ed.*, 2006, **45**, 6920.
61. F. E. Hahn and M. C. Jahnke, *Angew. Chem. Int. Ed.*, 2008, **47**, 3122.
62. B. P. Morgan, G. A. Galdamez, R. J. Gilliard Jr, and R. C. Smith, *Dalton Trans.*, 2009, 2020.
63. X. Hu, I. Castro-Rodriguez, and K. Meyer, *J. Am. Chem. Soc.*, 2003, **125**, 12237.
64. C. E. Willans, K. M. Anderson, M. J. Paterson, *et al.*, *Eur. J. Inorg. Chem.*, 2009, 2835.
65. R. C. Hider and X. L. Kong, *Nat. Prod. Rep.*, 2010, **27**, 637.
66. F. Vögtle, *Supramolecular Chemistry*, Wiley, New York, 1991.
67. H. A. Weakliem and J. L. Hoard, *J. Am. Chem. Soc.*, 1959, **81**, 549.
68. X. Solans, S. Gali, M. Font-Altaba, *et al.*, *Afinidad*, 1988, **45**, 243.
69. R. Ferreirós-Martínez, D. Esteban-Gómez, C. Platas-Iglesias, *et al.*, *Inorg. Chem.*, 2009, **48**, 10976.
70. D. Parker, *Chem. Soc. Rev.*, 2004, **33**, 156.
71. R. Pal, D. Parker, and L. C. Costello, *Org. Biomol. Chem.*, 2009, **7**, 1525.
72. D. Parker, in *Comprehensive Supramolecular Chemistry*, eds. J. L. Atwood, J. E. D. Davies, D. D. MacNicol, and F. Vögtle, Pergamon, Oxford, 1996, vol. 10, p. 487.
73. F. Vögtle and E. Weber, *Angew. Chem. Int. Ed. Engl.*, 1974, **13**, 814.
74. B. Gierczyk and G. Schroeder, *Mendeleev Commun.*, 2009, **19**, 75.
75. R. J. Motekaitis, A. E. Martell, and I. Murase, *Inorg. Chem.*, 1986, **25**, 938.
76. D. E. Fenton, *Chem. Soc. Rev.*, 1999, **28**, 159.
77. M. Lubben, R. Hage, A. Meetsma, *et al.*, *Inorg. Chem.*, 1995, **34**, 2217.
78. S. Mukherjee, T. Weyhermuller, E. Bothe, and P. Chaudhuri, *Eur. J. Inorg. Chem.*, 2003, 1956.
79. S. L. Tobey and E. V. Anslyn, *J. Am. Chem. Soc.*, 2003, **125**, 14807.
80. D. M. Rudkevich, W. Verboom, Z. Brzozka, *et al.*, *J. Am. Chem. Soc.*, 1994, **116**, 4341.
81. P. G. Plieger, P. A. Tasker, and S. G. Galbraith, *Dalton Trans.*, 2004, 313.
82. P. A. Gale and T. Gunnlaugsson, *Chem. Soc. Rev.*, 2010, **39**, 3595.
83. J. L. Sessler, P. A. Gale, and W.-S. Cho, *Anion Receptor Chemistry*, Royal Society of Chemistry, Cambridge, 2006.

84. P. A. Gale, *Coord. Chem. Rev.*, 2006, **250**, 2917.

85. P. A. Gale, *Coord. Chem. Rev.*, 2003, **240**, 1.

86. P. A. Gale, *Acc. Chem. Res.*, 2006, **39**, 465.

87. P. Gale, in *Encyclopedia of Supramolecular Chemistry*, eds. J. W. Steed and J. L. Atwood, Marcel Dekker, New York, 2004, p. 31.

88. S. O. Kang, R. A. Begum, and K. Bowman-James, *Angew. Chem. Int. Ed.*, 2006, **45**, 7882.

89. J. L. Sessler, S. Camiolo, and P. A. Gale, *Coord. Chem. Rev.*, 2003, **240**, 17.

90. P. A. Gale, *Chem. Commun.*, 2008, 4525.

91. M. D. Best, S. L. Tobey, and E. V. Anslyn, *Coord. Chem. Rev.*, 2003, **240**, 3.

92. P. Blondeau, M. Segura, R. Perez-Fernandez, and J. de Mendoza, *Chem. Soc. Rev.*, 2007, **36**, 198.

93. J. Yoon, S. K. Kim, N. J. Singh, and K. S. Kim, *Chem. Soc. Rev.*, 2006, **35**, 355.

94. H.-J. Schneider and A. K. Yatsimirsky, *Chem. Soc. Rev.*, 2008, **37**, 263.

95. P. R. Brotherhood and A. P. Davis, *Chem. Soc. Rev.*, 2010, **39**, 3633.

96. A. P. Davis and J. B. Joos, *Coord. Chem. Rev.*, 2003, **240**, 143.

97. A. P. Davis, *Coord. Chem. Rev.*, 2006, **250**, 2939.

98. A. P. Davis, J. J. Perry, and R. P. Williams, *J. Am. Chem. Soc.*, 1997, **119**, 1793.

99. A. J. Ayling, S. Broderick, J. P. Clare, *et al.*, *Chem.—Eur. J.*, 2002, **8**, 2197.

100. A. J. Ayling, M. N. Perez-Payan, and A. P. Davis, *J. Am. Chem. Soc.*, 2001, **123**, 12716.

101. M. Chahar, S. Upreti, and P. S. Pandey, *Tetrahedron*, 2007, **63**, 171.

102. A. Kumar and P. S. Pandey, *Org. Lett.*, 2008, **10**, 165.

103. A. P. Davis, D. N. Sheppard, and B. D. Smith, *Chem. Soc. Rev.*, 2007, **36**, 348.

104. B. A. McNally, A. V. Koulov, B. D. Smith, *et al.*, *Chem. Commun.*, 2005, 1087.

105. A. P. Davis and L. J. Lawless, *Chem. Commun.*, 1999, 9.

106. L. J. Lawless, A. G. Blackburn, A. J. Ayling, *et al.*, *J. Chem. Soc., Perkin Trans. 1*, 2001, 1329.

107. A. L. Sisson, J. P. Clare, L. H. Taylor, *et al.*, *Chem. Commun.*, 2003, 2246.

108. M. Segura, V. Alcazar, P. Prados, and J. de Mendoza, *Tetrahedron*, 1997, **53**, 13119.

109. V. Janout, B. W. Jing, I. V. Staina, and S. L. Regen, *J. Am. Chem. Soc.*, 2003, **125**, 4436.

110. L. Fang, W. H. Chan, Y. B. He, *et al.*, *J. Org. Chem.*, 2005, **70**, 7640.

111. S. Y. Liu, L. Fang, Y. B. He, *et al.*, *Org. Lett.*, 2005, **7**, 5825.

112. A. Goodey, J. J. Lavigne, S. M. Savoy, *et al.*, *J. Am. Chem. Soc.*, 2001, **123**, 2559.

113. S. C. McCleskey, M. J. Griffin, S. E. Schneider, *et al.*, *J. Am. Chem. Soc.*, 2003, **125**, 1114.

114. F. P. Schmidtchen, *Coord. Chem. Rev.*, 2006, **250**, 2918.

115. A. Metzger, V. M. Lynch, and E. V. Anslyn, *Angew. Chem. Int. Ed. Engl.*, 1997, **36**, 862.

116. D. J. Iverson, G. Hunter, J. F. Blount, *et al.*, *J. Am. Chem. Soc.*, 1981, **103**, 6073.

117. K. J. Wallace, W. J. Belcher, D. R. Turner, *et al.*, *J. Am. Chem. Soc.*, 2003, **125**, 9699.

118. S. L. Wiskur, H. Ait-Haddou, J. J. Lavigne, and E. V. Anslyn, *Acc. Chem. Res.*, 2001, **34**, 963.

119. A. Metzger and E. V. Anslyn, *Angew. Chem. Int. Ed. Engl.*, 1998, **37**, 649.

120. S. L. Wiskur and E. V. Anslyn, *J. Am. Chem. Soc.*, 2001, **123**, 10109.

121. J. J. Lavigne and E. V. Anslyn, *Angew. Chem. Int. Ed.*, 1999, **38**, 3666.

122. L. A. Cabell, M. K. Monahan, and E. V. Anslyn, *Tetrahedron Lett.*, 1999, **40**, 7753.

123. K. Niikura, A. Metzger, and E. V. Anslyn, *J. Am. Chem. Soc.*, 1998, **120**, 8533.

124. K. H. Choi and A. D. Hamilton, *J. Am. Chem. Soc.*, 2003, **125**, 10241.

125. V. Amendola, M. Boiocchi, L. Fabbrizzi, and A. Palchetti, *Chem.—Eur. J.*, 2005, **11**, 5648.

126. A. Abate, E. Cariati, A. Forni, *et al.*, *Private Communication*, 2010, in press.

127. M. G. Sarwar, B. Dragisic, S. Sagoo, and M. S. Taylor, *Angew. Chem. Int. Ed.*, 2010, **49**, 1674.

128. P. J. Garratt, A. J. Ibbett, J. E. Ladbury, *et al.*, *Tetrahedron*, 1998, **54**, 949.

129. C. Schmuck and M. Schwegmann, *J. Am. Chem. Soc.*, 2005, **127**, 3373.

130. H. Ihm, S. Yun, H. G. Kim, *et al.*, *Org. Lett.*, 2002, **4**, 2897.

131. Y. Bai, B. G. Zhang, J. Xu, *et al.*, *New J. Chem.*, 2005, **29**, 777.

132. Y. Bai, B.-G. Zhang, C. Y. Duan, *et al.*, *New J. Chem.*, 2006, **30**, 266.

133. S. K. Kim, J. H. Bok, R. A. Bartsch, *et al.*, *Org. Lett.*, 2005, **7**, 4839.

134. M. H. Filby, S. J. Dickson, N. Zaccheroni, *et al.*, *J. Am. Chem. Soc.*, 2008, **130**, 4105.

135. C. W. Chen and H. W. Whitlock, *J. Am. Chem. Soc.*, 1978, **100**, 4921.

136. S. C. Zimmerman, *Top. Curr. Chem.*, 1993, **165**, 71.

137. J. Rebek, *Angew. Chem., Int. Ed. Engl.*, 1990, **29**, 245.

138. M. Hamata, *Acc. Chem. Res.*, 2004, **37**, 862.

139. D. H. Lee, J. H. Im, J. H. Lee, and J. I. Hong, *Tetrahedron Lett.*, 2002, **43**, 9637.

140. S. Kondo and M. Sato, *Tetrahedron*, 2006, **62**, 4844.

141. Z. H. Lin, L. X. Xie, Y. G. Zhao, *et al.*, *Org. Biomol. Chem.*, 2007, **5**, 3535.

142. A. Furstner, *Angew. Chem. Int. Ed. Engl.*, 2003, **42**, 3582.

143. J. L. Sessler, L. R. Eller, W. S. Cho, *et al.*, *Angew. Chem. Int. Ed.*, 2005, **44**, 5989.

144. W. S. Cho, J. L. Sessler, L. R. Eller, *et al.*, *Abstr. Pap. Am. Chem. Soc.*, 2003, **225**, 486.

145. P. A. Gale, M. E. Light, B. McNally, *et al.*, *Chem. Commun.*, 2005, 3773.

146. K. Kavallieratos, S. R. de Gala, D. J. Austin, and R. H. Crabtree, *J. Am. Chem. Soc.*, 1997, **119**, 2325.

147. K. Kavallieratos, C. M. Bertao, and R. H. Crabtree, *J. Org. Chem.*, 1999, **64**, 1675.

148. M. P. Hughes and B. D. Smith, *J. Org. Chem.*, 1997, **62**, 4492.

149. G. W. Bates, P. A. Gale, and M. E. Light, *Chem. Commun.*, 2007, 2121.

150. S. Camiolo, P. A. Gale, M. B. Hursthouse, and M. E. Light, *Tetrahedron Lett.*, 2002, **43**, 6995.

151. S. J. Brooks, P. R. Edwards, P. A. Gale, and M. E. Light, *New J. Chem.*, 2006, **30**, 65.

152. D. Curiel, A. Cowley, and P. D. Beer, *Chem. Commun.*, 2005, 236.

153. J. M. Suk and K. S. Jeong, *J. Am. Chem. Soc.*, 2008, **130**, 11868.

154. P. A. Gale, J. R. Hiscock, C. Z. Jie, *et al.*, *Chem. Sci.*, 2010, **1**, 215.

155. A. P. de Silva, D. B. Fox, T. S. Moody, and S. M. Weir, *Pure Appl. Chem.*, 2001, **73**, 503.

156. R. Martinez-Manez and F. Sancenon, *Chem. Rev.*, 2003, **103**, 4419.

157. L. Fabbrizzi, M. Licchelli, G. Rabaioli, and A. Taglietti, *Coord. Chem. Rev.*, 2000, **205**, 85.

158. M. E. Huston, E. U. Akkaya, and A. W. Czarnik, *J. Am. Chem. Soc.*, 1989, **111**, 8735.

159. D. H. Vance and A. W. Czarnik, *J. Am. Chem. Soc.*, 1994, **116**, 9397.

160. E. B. Veale, G. M. Tocci, F. M. Pfeffer, *et al.*, *Org. Biomol. Chem.*, 2009, **7**, 3447.

161. C. M. G. Dos Santos, T. McCabe, and T. Gunnlaugsson, *Tetrahedron Lett.*, 2007, **48**, 3135.

162. S. Nishizawa, Y. Kato, and N. Teramae, *J. Am. Chem. Soc.*, 1999, **121**, 9463.

163. T. Gunnlaugsson, M. Glynn, G. M. Tocci, *et al.*, *Coord. Chem. Rev.*, 2006, **250**, 3094.

164. D. H. Lee, H. Y. Lee, K. H. Lee, and J.-I. Hong, *Chem. Commun.*, 2001, 1188.

165. T. Gunnlaugsson, P. E. Kruger, P. Jensen, *et al.*, *J. Org. Chem.*, 2005, **70**, 10875.

166. C. B. Black, B. Andrioletti, A. C. Try, *et al.*, *J. Am. Chem. Soc.*, 1999, **121**, 10438.

167. D. Aldakov and P. Anzenbacher, *Chem. Commun.*, 2003, 1394.

168. M. Cametti and K. Rissanen, *Chem. Commun.*, 2009, 2809.

169. D. Esteban-Gomez, L. Fabbrizzi, and M. Liechelli, *J. Org. Chem.*, 2005, **70**, 5717.

170. D. Esteban-Gomez, L. Fabbrizzi, M. Licchelli, and D. Sacchi, *J. Mater. Chem.*, 2005, **15**, 2670.

171. S. Camiolo, P. A. Gale, M. B. Hursthouse, and M. E. Light, *Org. Biomol. Chem.*, 2003, **1**, 741.

172. J. W. Steed, *Chem. Soc. Rev.*, 2009, **38**, 506.

173. C. R. Bondy, P. A. Gale, and S. J. Loeb, *Chem. Commun.*, 2001, 729.

174. C. R. Bondy, P. A. Gale, and S. J. Loeb, *J. Am. Chem. Soc.*, 2004, **126**, 5030.

175. I. E. Vega, P. A. Gale, M. E. Light, and S. J. Loeb, *Chem. Commun.*, 2005, 4913.

176. S. J. Dickson, M. J. Paterson, C. E. Willans, *et al.*, *Chem.—Eur. J.*, 2008, **14**, 7296.

177. C. Caltagirone, A. Mulas, F. Isaia, *et al.*, *Chem. Commun.*, 2009, 6279.

178. L. H. Uppadine, F. R. Keene, and P. D. Beer, *J. Chem. Soc., Dalton Trans.*, 2001, 2188.

179. S. Nieto, J. Pérez, V. Riera, *et al.*, *Chem. Commun.*, 2005, 546.

180. J. Pérez, D. Morales, S. Nieto, *et al.*, *Dalton Trans.*, 2005, 884.

181. S. Nieto, J. Perez, L. Riera, *et al.*, *Inorg. Chem.*, 2007, **46**, 3407.

182. M. Arroyo, D. Miguel, F. Villafane, *et al.*, *Inorg. Chem.*, 2006, **45**, 7018.

183. L. Ion, D. Morales, J. Pérez, *et al.*, *Chem. Commun.*, 2006, 91.

184. S. L. Renard, A. Franken, C. A. Kilner, *et al.*, *New J. Chem.*, 2002, **26**, 1634.

185. A. Looney, G. Parkin, and A. L. Rheingold, *Inorg. Chem.*, 1991, **30**, 3099.

186. D. R. Turner, B. Smith, E. C. Spencer, *et al.*, *New J. Chem.*, 2005, **29**, 90.

187. M. T. Blanda, J. H. Horner, and M. Newcomb, *J. Org. Chem.*, 1989, **54**, 4626.

188. X. G. Yang, C. B. Knobler, Z. P. Zheng, and M. F. Hawthorne, *J. Am. Chem. Soc.*, 1994, **116**, 7142.

189. M. Barboiu, G. Vaughan, and A. van der Lee, *Org. Lett.*, 2003, **5**, 3073.

190. P. D. Beer, *Acc. Chem. Res.*, 1998, **31**, 71.

191. P. D. Beer and A. D. Keefe, *J. Organomet. Chem.*, 1989, **375**, C40.

192. P. D. Beer, M. G. B. Drew, A. R. Graydon, *et al.*, *J. Chem. Soc., Dalton Trans.*, 1995, 403.

193. P. D. Beer, D. Hesek, J. Hodacova, and S. E. Stokes, *J. Chem. Soc., Chem. Commun.*, 1992, 270.

194. P. D. Beer, C. Hazlewood, D. Hesek, *et al.*, *J. Chem. Soc., Dalton Trans.*, 1993, 1327.

195. P. D. Beer, *Chem. Commun.*, 1996, 689.

196. P. D. Beer, D. Hesek, K. C. Nam, and M. G. B. Drew, *Organometallics*, 1999, **18**, 3933.

197. P. D. Beer, A. R. Graydon, A. O. M. Johnson, and D. K. Smith, *Inorg. Chem.*, 1997, **36**, 2112.

198. G. Denuault, P. A. Gale, M. B. Hursthouse, *et al.*, *New J. Chem.*, 2002, **26**, 811.

199. F. Szemes, D. Hesek, Z. Chen, *et al.*, *Inorg. Chem.*, 1996, **35**, 5868.

200. S. Watanabe, O. Onogawa, Y. Komatsu, and K. Yoshida, *J. Am. Chem. Soc.*, 1998, **120**, 229.

201. C. P. Montgomery, B. S. Murray, E. J. New, *et al.*, *Acc. Chem. Res.*, 2009, **42**, 925.

202. B. S. Murray, E. J. New, R. Pal, and D. Parker, *Org. Biomol. Chem.*, 2008, **6**, 2085.

203. F. G. Klärner, J. Panitzky, D. Blaser, and R. Boese, *Tetrahedron*, 2001, **57**, 3673.

204. D. H. Williams and M. S. Westwell, *Chem. Soc. Rev.*, 1998, **27**, 57.

205. H. S. Park, Q. Lin, and A. D. Hamilton, *J. Am. Chem. Soc.*, 1999, **121**, 8.

206. G. Weber and W. Saenger, *Acta Crystallogr., Sect. B*, 1980, **36**, 424.

Porphyrins and Expanded Porphyrins as Receptors

Jonathan L. Sessler[1,2], Elizabeth Karnas[1], and Elaine Sedenberg[1]

[1] *The University of Texas at Austin, Austin, TX, USA*
[2] *Yonsei University, Seoul, South Korea*

1 INTRODUCTION

Porphyrins and expanded porphyrins are attractive scaffolds from which receptors for a wide variety of entities, running the gamut from organic to inorganic and biological, can be constructed. Nature herself utilizes these versatile compounds for many necessary functions ranging from light harvesting in photosynthesis to respiration and electron transfer (ET). Porphyrins and related compounds have thus been objects of intense exploration and study for more than a century.[1,2] This large body of work has provided a foundation of knowledge which is facilitating current efforts to prepare porphyrinic frameworks with a variety of characteristics including variations in size, structure, solubility, and electronic properties. Further, many elements of porphyrins, including ease of synthetic modification, make them well suited for use as receptors. This chapter aims to familiarize the reader with key receptor systems and the concepts that underlie their design and construction. This is done by highlighting examples from the recent literature. Needless to say, this survey is illustrative, rather than comprehensive.

Prior to discussing porphyrinic receptors, however, a general introduction to porphyrins is appropriate. Porphyrins are intensely colored cyclic molecules, or macrocycles, which are composed of four smaller 5-membered heterocycles, called *pyrroles*, that contain one nitrogen and four carbon atoms. In porphyrins, one carbon, typically referred to as the *meso-carbon*, serves to connect each of the four pyrrole rings. One of the defining properties of porphyrins is that this connection permits electronic conjugation throughout the cyclic array of core nitrogen and carbon atoms. The associated conjugation path or the pattern of alternating double and single bonds, which is highlighted in Figure 1, allows for movement of π-electrons throughout what is formally an 18 π-electron periphery. As a result, the conditions for aromaticity are fulfilled. Consistent with the formulism, porphyrins are extremely stable. They also have distinct spectroscopic fingerprints, which facilitates their analysis via many techniques, including NMR (nuclear magnetic resonance) and UV–vis spectroscopy; these methods are discussed in more detail below, as they relate to the problems of receptor–substrate interactions. Other analogs of porphyrins, including larger systems known as *expanded porphyrins* (often with more than four pyrrole subunits in the macrocyclic ring), generally have their own spectroscopic features; they too have been exploited to study recognition phenomena.

While porphyrins can act as stand-alone receptors, functionalization with specific binding moieties is typically required in order to obtain selectivity for a chosen target. For example, in Seiji Shinkai's saccharide sensing systems, porphyrin scaffolds provide the chromogenic components

Supramolecular Chemistry: From Molecules to Nanomaterials.
Edited by Philip A. Gale and Jonathan W. Steed.
© 2012 John Wiley & Sons, Ltd. ISBN: 978-0-470-74640-0.

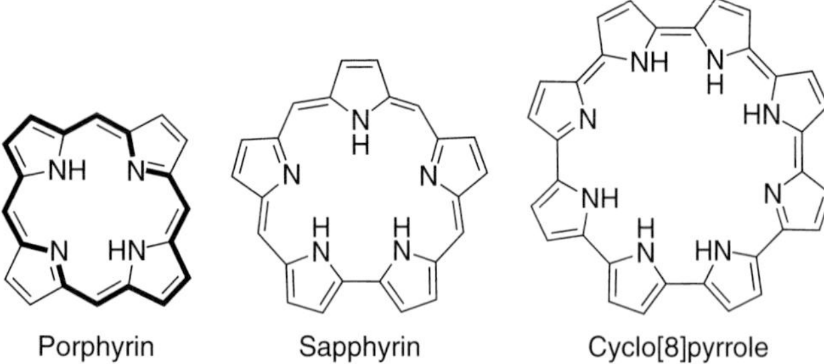

Figure 1 Core framework for porphyrin, sapphyrin, and cyclo[8]pyrrole, all of which are covered in this chapter. The cyclic path of conjugation for 18 π-electrons in the porphyrin is bolded.

and define a rigid backbone of a length commensurate with the guest. However, the diboronic acids appended on the periphery of the porphyrin are responsible for "attracting" the oligosaccharides. Another example, highlighted by a recent review, involves the use of crown-ether functionality to achieve cation recognition.[3] Nevertheless, it is important to appreciate that porphyrins are large molecules that are relatively electron rich; this makes them good stand-alone receptors for certain electron-deficient species. To illustrate this, one section in this chapter is dedicated to the fascinating and relatively new field of porphyrin–nanomaterial interactions, where the latter include C_{60} and single-walled carbon nanotubes (SWNTs). Here, the key binding events are mediated by donor–acceptor, charge-transfer-type interactions between the π-systems. As detailed in this chapter, expanded porphyrins can also act as stand-alone receptors. Here, much of the interest is in the protonated forms that can "capture" anionic species, as highlighted via a discussion of two prototypical species, namely, sapphyrin (with five pyrroles) and cyclo[8]pyrrole (which contains eight pyrroles) (Figure 1).

Because of limited space, this chapter covers porphyrins and expanded porphyrins as receptors through mostly noncovalent or supramolecular interactions. As a general rule, examples wherein the porphyrin acts as a receptor via metal coordination are excluded. For a further reading on this latter topic, the reader is referred to a recent *Chemistry Review* article entitled "Supramolecular Chemistry of Metalloporphyrins."[4] On the other hand, inspired by a recent review by Anslyn entitled "Supramolecular Analytical Chemistry," we expand our definition of receptor–substrate interactions to include certain boronic acid binding motifs and several representative examples of metalated porphyrins that are applied to sensing applications, wherein the proposed mode involves metal complexation.[5] Thus, included in this chapter is a brief summary of the seminal work carried out by Suslick and his group, where porphyrins are used to create the so-called artificial nose, that is, an array-based sensing device for various vapors.[6] Also included are examples wherein the metal serves to assist or increase the extent of association between the porphyrin and its guest. Finally, it is to be noted that the focus of this chapter is primarily on the material published since 2000. There are a number of reviews that cover the field up until that point; many are included in the very comprehensive *Handbook of Porphyrins and Phthalocyanines*, published in 2000.[7] Of course, in certain cases, materials published prior to 2000 are included, especially when it is needed to put current work in an appropriate context.

This chapter covers the following topics:

1. Properties of porphyrins that make them attractive for use in receptor design.
2. Porphyrins and expanded porphyrins as receptors for carbon nanostructures. Here, specific emphasis is placed on the design of porphyrin-based receptors that are optimized for interaction with carbon-rich surfaces and the tools that are used for the analysis of the resulting complexes. As is true throughout the chapter, key concepts are illustrated through the use of recently published, notable examples.
3. Porphyrins and expanded porphyrins as receptors for biological entities. Nature's prolific use of porphyrins and porphyrin analogs (e.g., chlorophylls) makes these kinds of macrocycles a "natural" choice for creating receptors that are targeted for use in biological settings. This utility is illustrated by summarizing recent advances involving receptors that are created for a specific subset of biologically important targets, including saccharides, peptides and proteins, and DNA.
4. Porphyrins and expanded porphyrins as receptors for anions. This section focuses on modified porphyrins that have been used for anion encapsulation, including receptors of the so-called "picket-fence" variety. Much of the discussion in this section involves expanded porphyrins, with a focus being placed on modified sapphyrins, and cyclo[8]pyrrole, as noted above. Included in this section is the use of anion recognition to create so-called ET dyads formed as a result of supramolecular anion-binding interactions.
5. Applications of porphyrin receptors as sensors and in the development of sensing arrays.
6. Conclusion and outlook for the future.

1.1 Properties of porphyrins: designing receptors

A salient feature of metalated porphyrins is that they contain three positions for further functionalization. Two are on the periphery and are known as the *meso*- and β-pyrrolic positions, respectively (Figure 2). The third point available for functionalization is the coordinated

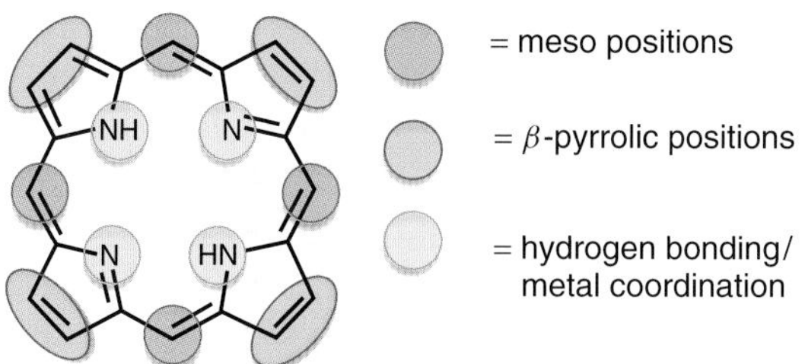

Figure 2 Features that make porphyrins an appealing scaffold for receptor design.

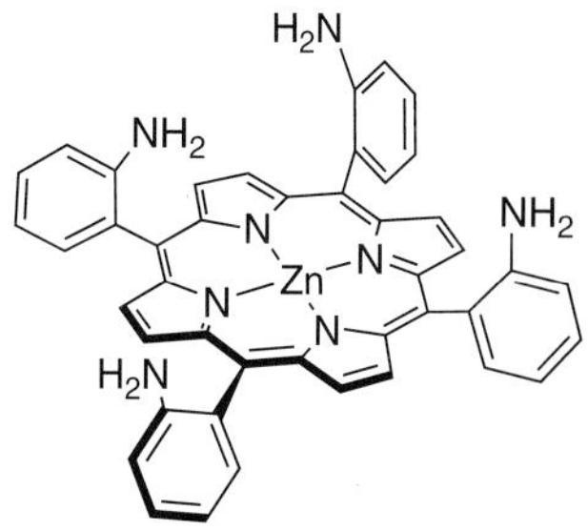

Figure 3 α,α,α,α-Atropisomer of 5,10,15,20-*meso*-tetrakis(*o*-aminophenyl)porphyrin. This is a widely used platform for "picket-fence" porphyrins that form cavities for binding-specific guests, depending on the substituents.

metal center; in many cases, the addition of apical ligands allows for the attachment of ancillary groups. There is an extensive literature involving the synthesis of porphyrins with substituents in the *meso*- or β-pyrrolic positions; metalation and complexation reactions involving porphyrins are also well documented.[2,8] But less is known about the expanded porphyrins. However, the structural diversity is greater, which makes these systems particularly attractive as scaffolds for creating new receptors.[9]

A survey of the literature involving porphyrin-based receptors, especially for biological entities and anions, shows a propensity for use of "picket-fence porphyrins." This latter is a term coined by Collman *et al.* in the 1970s when this basic structural motif was used to construct a functional model for oxymyoglobin.[10] In the case of receptors based on the picket-fence paradigm, the systems in question generally consist of a porphyrin with meso substitution of a phenyl ring that is functionalized in its ortho position. A popular starting point for some of the examples described in this chapter is the same α, α, α, α-atropisomer of 5,10,15,20-*meso*-tetrakis(*o*-aminophenyl)porphyrin, **1**, that was used by Collman to create his original picket-fence system (Figure 3). The free primary amine allows for easy modification and, as discussed below, results in structures wherein a cup or cavity is formed by the directed "pickets." Quite often, the pickets are the source of hydrogen bond donors, such as urea, for the binding of an anionic guest.[11–14]

The use of inward-pointing pyrrolic N-*H* protons is also a feature that has been exploited for recognition; this is especially true in the case of expanded porphyrins, where the cavity is larger than that in the case of porphyrins (something that allows for more facile protonation and recognition of larger guest species). For example, cyclo[8]pyrrole, **2**, is a conjugated, polypyrrolic macrocycle that is able to bind sulfate within the plane of the cavity as the result of N-*H*–anion hydrogen bonding interactions (Figure 4). Similar motifs are present in other expanded porphyrin anion complexes, as seen in several examples included in the anion-binding portion of this text.

The electronic features of porphyrins and expanded porphyrins also make them popular as receptor elements. Porphyrins and their conjugated analogs are generally highly

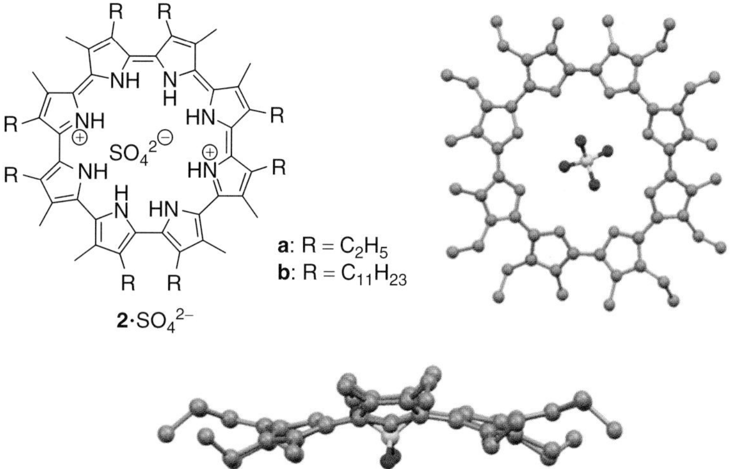

Figure 4 Line drawing and X-ray structure of cyclo[8]pyrrole with a bound sulfate (CCDC#: 176189).

colored. In the case of porphyrins, this color reflects an electronic spectrum that includes an intense band between 380 and 420 nm, known as a *Soret band*, that has its origins in a π, π^* transition from the ground state to the second excited singlet state.[15] For free-base porphyrins, four additional weaker absorptions in the visible region between 480 and 700 nm, called *Q-bands*, are typically seen; these are reflective of π, π^* transitions from the ground state to the first excited state.[15] When porphyrins are diprotonated or metalated, the molecule becomes more symmetrical and this leads the four Q-bands to collapse into two. Because of these unique spectral features, porphyrin-based receptor–guest interactions are typically studied in part by UV–vis spectroscopy; the rationale is that any electronic perturbations due to interaction with another species will result in a spectral change. Synthetic modifications can also result in predictable changes in the UV–vis spectrum.[1]

Free-base and most diamagnetic metalloporphyrins typically exhibit fluorescent behavior following photoexcitation. Further, it is not uncommon for metalated porphyrins to exhibit phosphorescence.[15] Not surprisingly, therefore, fluorescence spectroscopy is often used to monitor interactions of porphyrins with guest molecules. As a general rule, since photons are being detected directly, this method offers greater sensitivity than optical absorption-based methods. Furthermore, in cases involving donor–acceptor ensembles, quenching can indicate electron transfer (ET), and, subject to the proviso that this assumption is correct, fluorescence lifetimes can be used to determine the rate of ET.

The aromatic character of porphyrins and many expanded porphyrins has made analyses of their host–guest interactions by ^{1}H NMR spectroscopy exceedingly popular. Typically, the protons in the meso and β-positions are *deshielded* by the diamagnetic ring current, meaning that protons at either position will be shifted downfield. In general, the *meso*-protons are subject to a greater downfield shift since they are attached to particularly electron-deficient carbons. The same diamagnetic ring current serves to *shield* the inner or N-*H* protons in nonmetalated porphyrins. Therefore, these latter protons are shifted quite far upfield, even beyond TMS, the arbitrary zero point for NMR spectroscopy measurements. Analyzing receptor–guest chemistry by ^{1}H NMR spectroscopy is generally experimentally facile and relatively useful, since the movement of any of these protons as a function of guest addition can be used to construct a potential binding isotherm from which binding affinities can be calculated. For example, it is common to follow protons that are thought to be involved in hydrogen bonding, as ^{1}H NMR spectroscopy can reveal changes that are consistent with the formation of new species as the proton is shared between two entities.

2 CARBON NANOMATERIALS

Owing to their electronic properties *vide supra* there is much interest in the use of porphyrins and related macrocycles to create noncovalent assemblies with nanocarbons of all shapes and sizes, including the fullerenes, and SWNTs. This is a very active area and this section, therefore, only includes several representative examples with the goal of providing a general sense of research accomplishments and current challenges. There are sources that provide much more in-depth coverage of this topic, including some treatments that are very recent.[16,17]

2.1 Fullerenes

The relationship between porphyrins and fullerenes is of particular interest due to the potential applications of the resulting ensembles in the field of light harvesting or artificial photosynthesis.[18] This interest reflects the fact that porphyrins are typically good electron donors, and fullerenes are typically good electron acceptors.[17] In fact, there has been "official" identification of a new type of supramolecular interaction specifically related to that of flat porphyrins with round fullerenes.[19,20] While traditional π–π interactions were thought to be limited to donor–acceptor and solvatophobic interactions between two flat surfaces, it is now recognized that the interaction between the curved π-surface of a fullerene and the flat π-surface of a porphyrin is both real and important.[19,21] Even though this flat-to-round interaction is counterintuitive to the concepts of size- and shape-commensurate "fits," the distances between the porphyrin and fullerene in self-assembled ensembles are on the order of 2.7–3.0 Å, comparable to those seen in fullerene–fullerene or porphyrin–porphyrin adducts.

In a 1999 issue of *The Journal of the American Chemical Society*, two groups published crystal structures of noncovalently attached cocrystallate structures of porphyrins and fullerenes.[19,20] Figure 5 shows the packing of $H_2TPP{\cdot}C_{60}$ (with three omitted toluene molecules). In this case, the mean 24-atom porphyrin plane lies between 2.70 and 2.98 Å from the closest carbon atom in the double bond of a fullerene 6–6 ring juncture. Therefore, a "donor–acceptor" relationship between the electron-rich 6 : 6 ring juncture of the fullerene, C_{60}, and the electropositive center of a porphyrin was inferred.[21] Commonly, there is a bathochromic or red shift in the Soret band of the porphyrin on complexation with fullerene, providing further support for the notion of an electronic interaction. Numerous systems based on this interaction have been developed over the past decade. Typically, two to three porphyrins are employed for recognition, since this leads to higher binding affinities. The resulting systems can act as "molecular tweezers" that

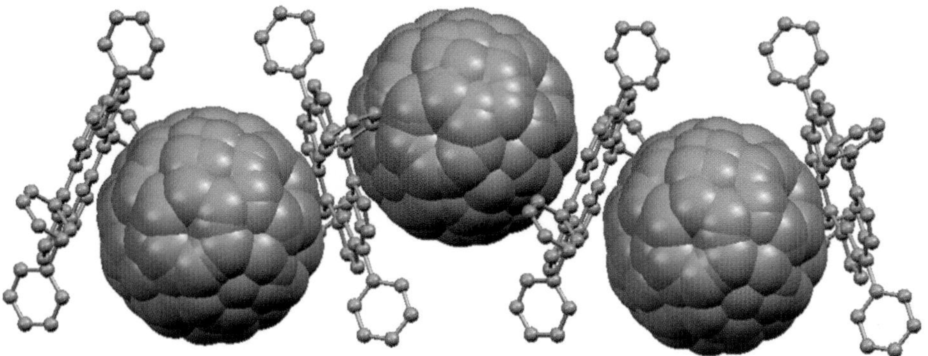

Figure 5 Cocrystallate of tetraphenylporphyrin (nonmetalated) with fullerene.

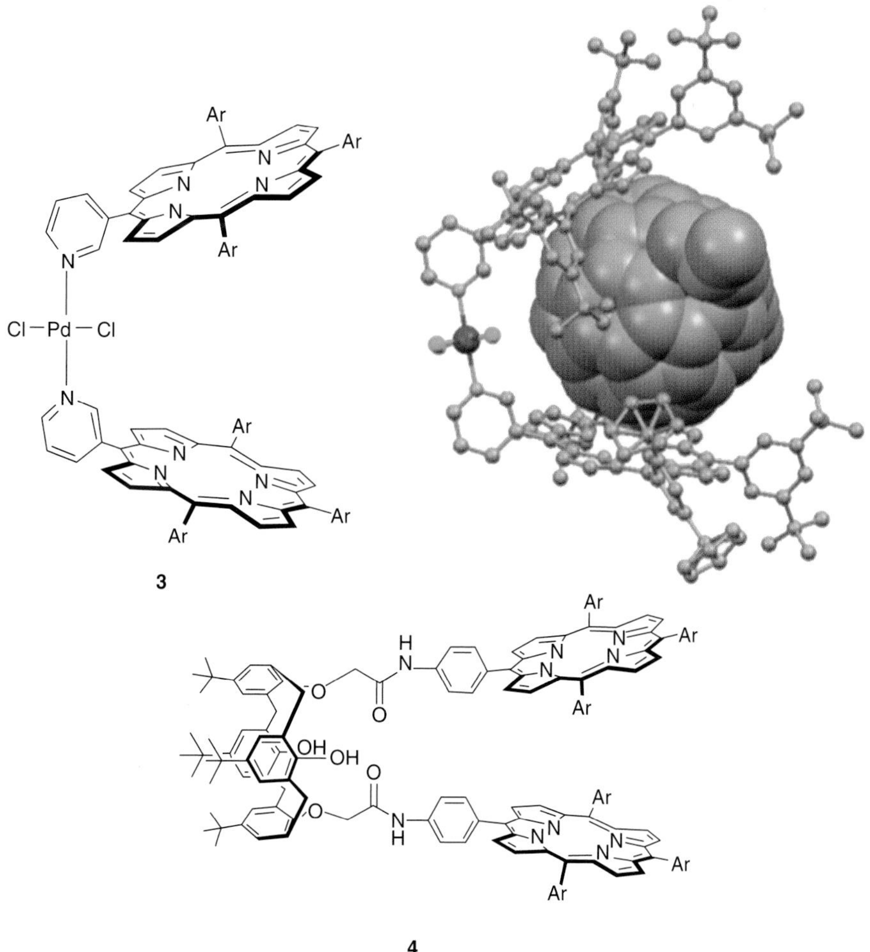

Figure 6 Jaw-like fullerene receptors developed by Boyd and coworkers. X-ray crystal structure of the complex of *N*-methylpyrrolidine-functionalized C_{60} (CCDC#: 205037). **3** is the schematic of the jaw structure shown in the crystal structure. Compound **4** is an example of a calixarene bisporphyrin host for fullerenes.

encapsulate fullerene. In fact, many motifs based on this paradigm have been published, with the major differences lying in the choice of linkers/scaffolds used to hold the constituent porphyrins in place.

Boyd and coworkers, who published the alternating porphyrin/fullerene aggregates shown in Figure 5 in 1999, followed up with a publication of a carefully designed "jaw" type system in 2000, wherein the covalently linked porphyrins were set an angle similar to that seen in polymer-like assemblies.[22] Figure 6 shows the X-ray crystal of **3**, bound to an *N*-methylpyrrolidine-functionalized C_{60}. More recently, calix[4]arenes have been used as a hinge for the "jaws" motif, **4**, as seen in Figure 6.[23] In this work, nonmetalated porphyrins were used with variations in either the substituents at the meso positions or the linker between the porphyrin and the calix[4]arene. It was found that affinities as high as 26 000 M^{-1} could be achieved in toluene; however, these relatively high affinities were ascribed to solvation effects rather than to specific electronic effects, such as donor–acceptor interactions.

Another interesting "tweezer" system utilizes hydrogen bonding moieties on the periphery of two zinc porphyrins to effect preorganization of the system into a U-shaped receptor, **5** (Figure 7).[24] Evidence for the proposed preorganization came from ^{1}NMR spectroscopic studies. Specifically, a downfield shift in the amide protons was taken as evidence of hydrogen bonding, with these interactions determined as being intramolecular in nature, as inferred from the fact that few spectral changes were seen on dilution. Addition of fullerene to these compounds in $CDCl_3$ led to substantial changes in the ^{1}NMR spectrum, presumably as the result of the strong putative intramolecular hydrogen bonds. The induced spectral changes were even greater in the case of C_{70}, a finding considered consistent with the presence of stronger interactions. Binding or association constants for the two-porphyrin tweezer systems and C_{60} were determined by UV–vis titrations in toluene. On this basis, the K_a values were found to range from 2.7×10^4 to $1.0 \times 10^5\,M^{-1}$ depending on the system in question. The same binding experiments with C_{70} yielded higher association constants, with values between 9.8×10^5 and $1.1 \times 10^6\,M^{-1}$ being recorded. This same report also includes a three-porphyrin variation on the foldamer theme, **6**. This elaborated system preorganizes itself to form an S-shaped receptor, which was found to bind fullerene in a 1 : 2 stoichiometry. Linked C_{60} "guests" were also synthesized in order to assess more completely the role that the presumed preorganization plays.

One of the first reports of porphyrin-based fullerene recognition that included crystallographic evidence for encapsulation in a cyclic structure came from the Aida group.[25,26] Here, the receptor, **7**, is a cyclic dimer of zinc porphyrins (Figure 8). At the time of publication in 2001, system **7** held the record for the largest host–C_{60} association constant ($K_a = 6.7 \times 10^5\,M^{-1}$ in benzene). The crystal structure also provided support for the presence of π interactions, as evidenced by separations of 2.765 and 2.918 Å between the Zn atom and fullerene carbon atoms.

This same report by Zheng *et al.* (Aida group) included evidence that variations in the coordinated metal within the porphyrin can have a great effect on the affinity for C_{60}.[26] The Zn metalloporphyrin showed affinities comparable to that of the protonated porphyrin. Coordination of group 9 metal ions, such as Co^{II} and Rh^{III}, served to increase greatly the affinity of the system, to as high as $10^8\,M^{-1}$. On the other hand, coordination of group 10 and 11 metals resulted in lower association constants. Another significant conclusion to be drawn from this work was the inference that across the board C_{70} is bound more strongly than C_{60}. This was ascribed to the more "oval" shape of C_{70} compared to C_{60}.

The original work was followed up with a paper that described the use of **7** to effect the selective extraction

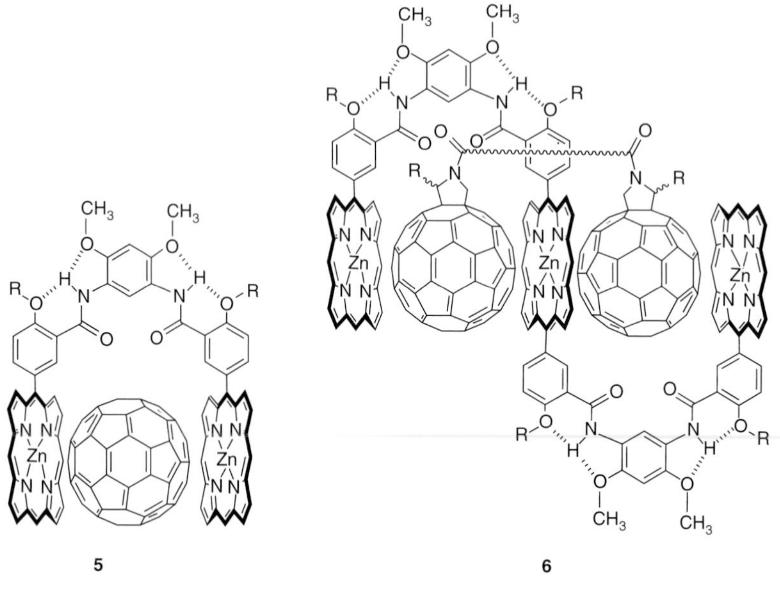

5

6

Figure 7 Hydrogen-bonding-driven preorganized fullerene receptors.

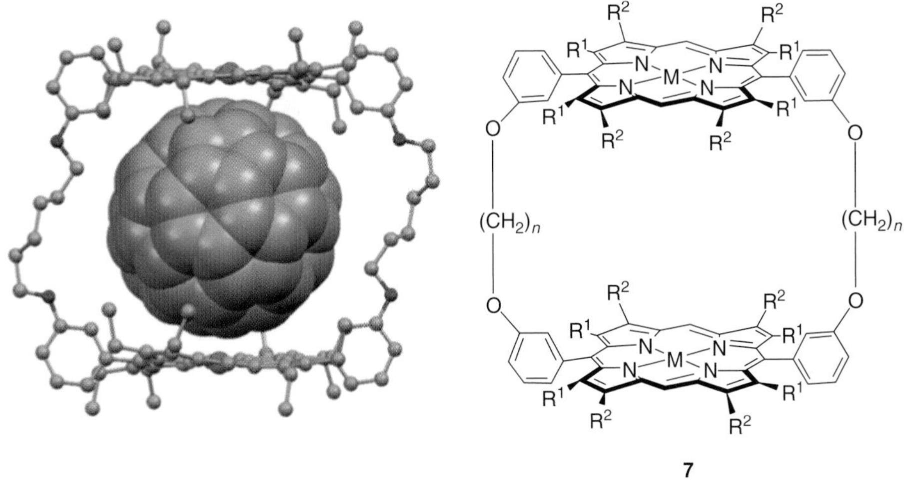

Figure 8 X-ray crystal structure of an inclusion complex of C_{60} within a cyclic dimer of zinc porphyrin where $n = 6$, and R^1 and R^2 = ethyl on structure **7**. (CCDC # 150074).

of higher fullerenes from a mixture of fullerenes. This was viewed as significant at the time because one of the then-current challenges was to isolate these products from the mixture obtained by commercial production of C_{60} and C_{70}.[27] Such separations are still difficult. With such objectives in mind, one goal is to incorporate these cyclic dimers onto solid supports, in accord with earlier work by Meyerhoff.[28] Another objective was to extract chiral C_{76}, a goal that was accomplished by using a "chiral" variation of the cyclic porphyrin dimer.[29]

Various tripod systems have been synthesized with the goal of fullerene complexation (Figure 9).[30] One such scaffold involves a tripod with rigid linkers; this system is proposed to make a supramolecular network wherein the net stoichiometry is expected to be 1 : 2 (tripod to fullerene). On the basis of an experimental analysis, it was concluded that each fullerene is always held by two porphyrins. These two porphyrins can be oriented such that a U-shaped arrangement is defined between two porphyrins contained within the same molecule, **8**. Or, the two porphyrins can from two separate tripods. In all cases, the distances between the surface of C_{60} and the porphyrin plane were found to range from 2.52 to 2.92 Å, as would be expected for a system in which π–π interactions persist.

A recently reported cyclic porphyrin trimer, **9**, shows an extremely high binding affinity for C_{60}, with a K_a on the order of $2.1 \times 10^7\,M^{-1}$ in cyclohexane and $1.6 \times 10^6\,M^{-1}$ in toluene.[31] For C_{70}, the affinity in toluene is even greater with a K_a of $2 \times 10^8\,M^{-1}$, and is one of the highest reported affinities for C_{70}. The binding constant is even larger for higher fullerenes, which may be due to the increased interaction between all three of the porphyrin walls.

Another tripod-type, porphyrin-based receptor is **10**. This system, characterized by flexible linkers, was readily synthesized through a one-step click reaction of an azide-functionalized porphyrin with a tertiary amine functionalized with three alkynyl groups. The resulting tripod system can capture a pyridine-functionalized fullerene with 1 : 1 stoichiometry. The corresponding association constant, K_a, was found to be $9.4 \times 10^4\,M^{-1}$ in *o*-dichlorobenzene as determined from UV–vis spectroscopic titrations. The supramolecular dyad formed via this capture event was then found capable of undergoing photoinduced ET from the zinc porphyrin to the fullerene with a charge-separated lifetime of 0.5 ns.[32]

An interesting system developed by Shinkai *et al.* combines two "tweezers" to produce the tetrameric porphyrin-containing receptor, **11**.[33] This system is capable of *cooperativity* with respect to fullerene binding, with the first and second binding constants being found to be 5800 and $2000\,M^{-1}$, respectively, in toluene. These values are consistent with weak positive homotropic allosterism since $K_2 > 1/4\,K_1$ – the definition for positive cooperativity in a 1 : 2 host–guest system (Figure 10). 'Monotweezers' were also synthesized and studied as controls; they displayed significantly lower affinities toward C_{60}.

These examples highlight the progress toward exploiting this porphyrin–fullerene host–guest interaction for purposes such as ET dyads and separation of mixtures of higher order fullerenes. However, a further application involves the construction of 3D structures, such as metal organic frameworks (MOFs).[34] In the context of these general efforts, Boyd and coworkers showed that sheets formed from tetrapyridyl porphyrins (H_2TpyP), which are connected in an infinite array via peripheral (i.e., pyridine-based) complexation of Pb^{2+} or Hg^{2+}, could then be intercalated with

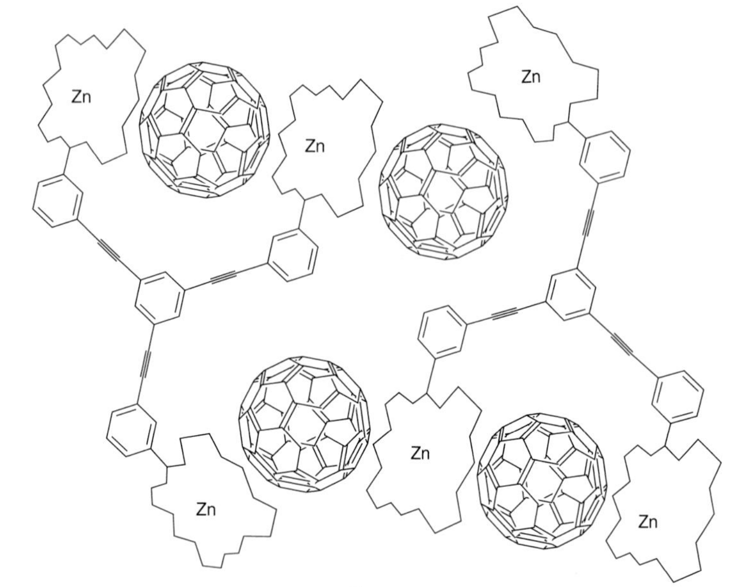

8

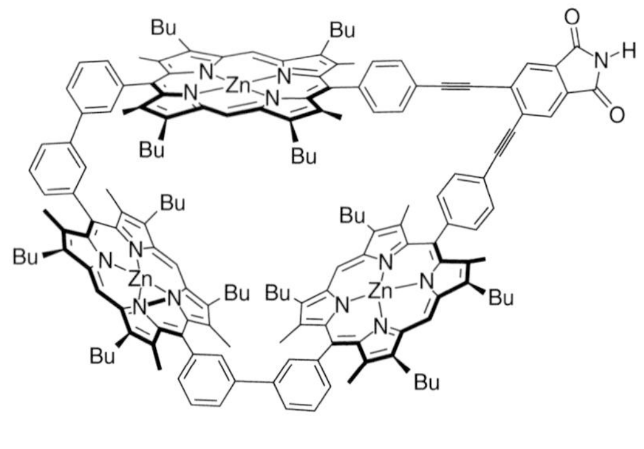

9

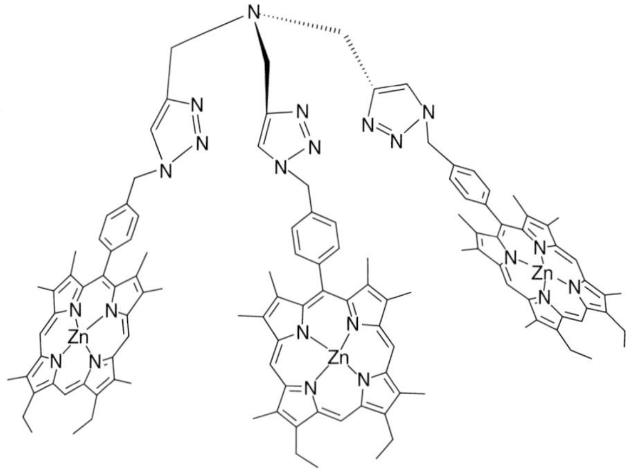

10

Figure 9 Recently developed tripod systems for fullerene encapsulation. **8** is a tripod on a rigid scaffold. **9** is a cyclic trimer. **10** is a trimer with flexible arms.

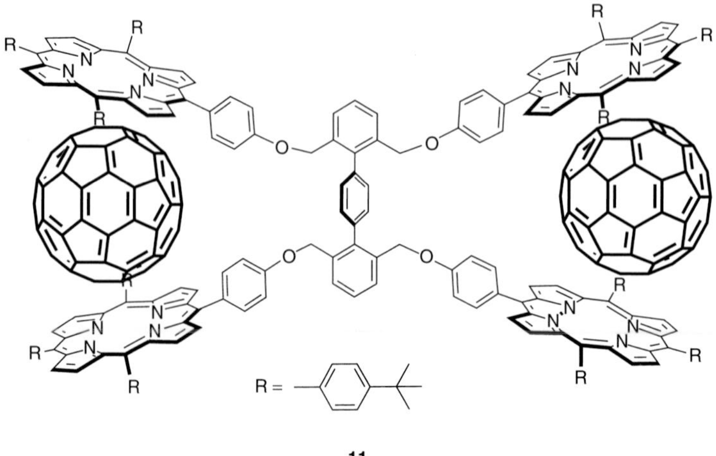

Figure 10 A tetramer capable of binding C_{60} in a cooperative manner.

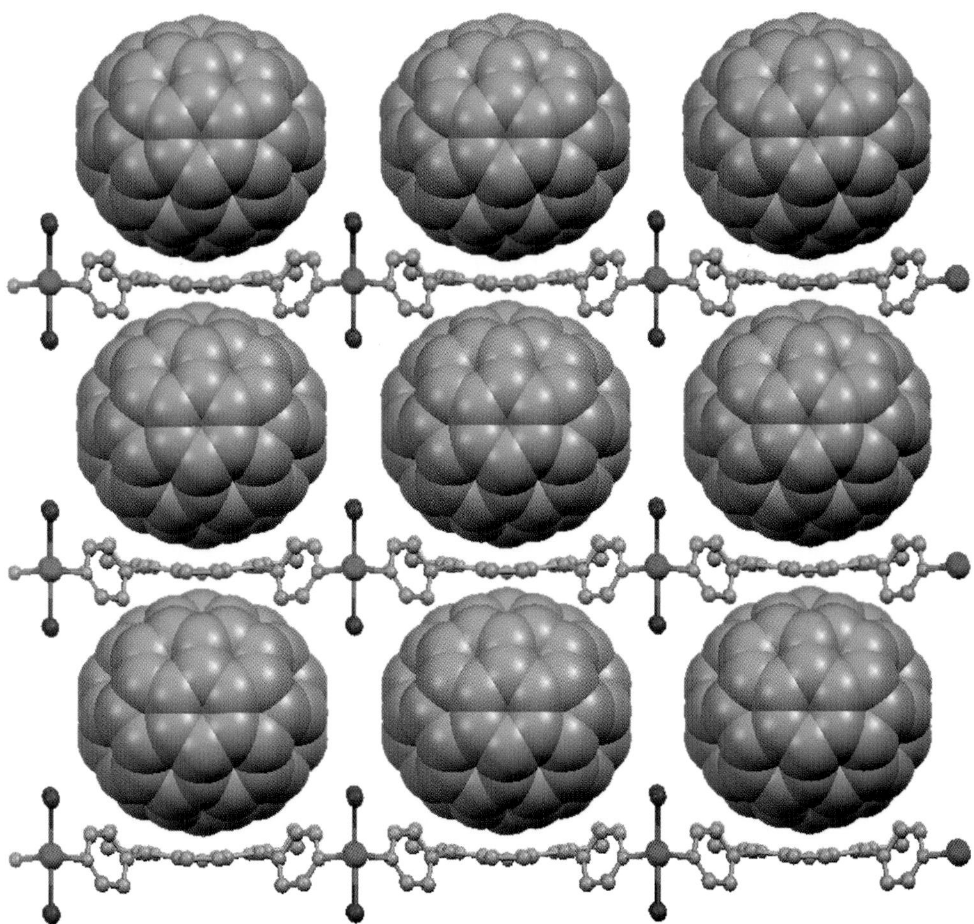

Figure 11 Cocrystallization of HgI_2, H_2TpyP, and C_{70}, illustrating the MOF-like nature of the assembly. (CCDC# 174254).

fullerenes. The result was the formation of fullerene pillars, with the C_{60} or C_{70} held above the porphyrin planes to give a structure that resembles a more classic MOF (Figure 11).

2.2 Nanotubes

A nonplanar donor–acceptor interaction also serves to underscore the binding between porphyrins and SWNTs. The basic interaction was first reported in 2003 by Murakami *et al.* who found that porphyrin molecules (zinc protoporphyrin IX) could solubilize SWNTs in organic solvents.[35] These researchers also reported a decrease in fluorescence intensity for the proposed ZnPP–SWNT ensembles in solution as compared to solutions of ZnPP alone. Because there is great potential for nanotubes in optoelectronic and photovoltaic applications, this discovery engendered tremendous interest in the study of the noncovalent interactions between porphyrins and SWNTs. Here, a key motivation is that the noncovalent functionalization of SWNTs via the use of porphyrins is expected to be benign and not destroy the SWNTs. In contrast, alternative covalent functionalization approaches result in the formal removal of two π-electrons from the conjugated π-system (thus altering the sp^2 hybridized walls[36] and degrading the desirable properties of SWNTs).[37]

In 2005, Guldi *et al.* reported the use of charged porphyrins (asymmetric 5-(4-hydroxyphenyl)-10,15,20-tri(4-sulfonatophenyl)porphyrin) incorporated into poly (methyl-methacrylate) (PMMA); the result is a modified polymer, **12**, that wraps around SWNTs on mixing and heating (Figure 12).[37] The resulting porphyrin–SWNT solution was cast on a copper grid and analyzed by transmission electron microscopy. This analysis revealed "bundles and ropes" – a finding interpreted in terms of the presence of PMMA–porphyrin polymers on the nanotubes. Absorption spectroscopic studies carried out in dimethylformamide (DMF) provided further support for the proposed interaction between the two constituent species, in that the porphyrin Soret bands are weaker, broader, and shifted; the Q-bands are also red-shifted by 1–2 nm.[37] Fluorescence quenching of the poly(H_2P) emission intensity also occurs in the presence of SWNTs in DMF, and the fluorescence

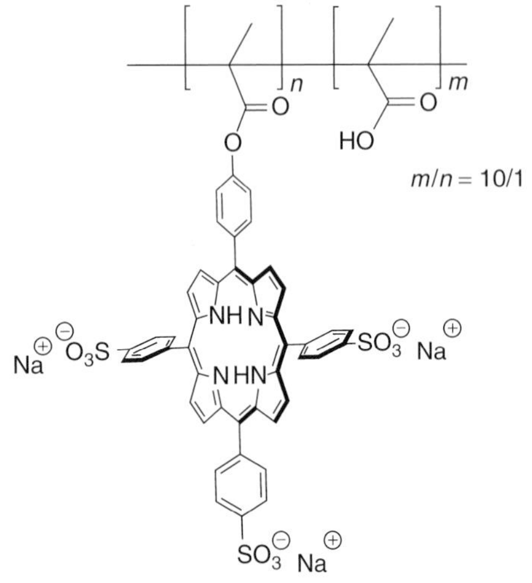

Figure 12 Structure of PMMA appended with sulfonate porphyrins for interaction with SWNTs.

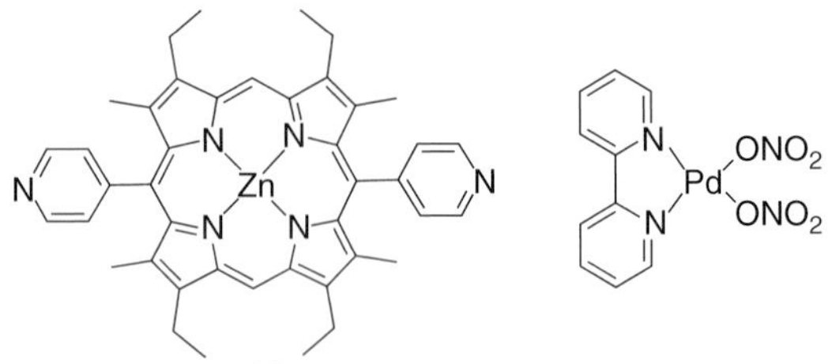

Figure 13 Components for the self-assembly of oligomers, which are found capable of binding single SWNTs by size.

lifetime is also altered. Specifically, a biexponential function with one component matching that of **12** alone was observed at approximately 10 ns, with the second, shorter component at 0.8 ns also being observed. The observation of the latter component implies an excited state quenching mechanism or deactivation of **12**. Transient absorption spectroscopic analyses carried out on photoirradiation at 532 nm revealed spectroscopic changes consistent with intracomplex ET from the porphyrin to the SWNT; for instance, features ascribed to the one-electron oxidized, or radical cation, porphyrin were explicitly seen.

Soon after Guldi *et al.* published on their system, Stoddart and coworkers published a report wherein "dynamic" chemistry was used to form self-assembled oligomers that were found to be capable of binding and, possibly even, sorting SWNTs by size.[38] Their system is composed of 5,15-*trans*-pyridyl metalloporphyrin, **13**, which can act as a linear divalent ligand and *cis*-protected Pd^{II} or Pt^{II} complexes (Figure 13). While neither the Pd/Pt ligand nor the porphyrin displayed an ability to solubilize SWNTs on its own, they were found to completely solubilize SWNTs in aqueous acetonitrile solutions when used in conjunction with one another.

A report in 2004 by the Sun group showed that 5,10,15,20-tetrakis(hexadecyloxyphenyl)-21*H*, 23*H*-porphine (THPP) could selectively solubilize SWNTs.[39] Thus, this readily prepared molecule allowed for efficient separation of mixtures of metallic and semiconducting SWNTs, which have different electrical conductivities among other properties. The mechanism was only speculated upon, but it is thought that the porphyrin is better able to interact with the semiconducting surface of the semiconducting SWNT than with the metallic versions. It was noted that coordination of Zn to the porphyrin hinders the interaction between the porphyrin and the SWNT.[39] This is in stark contrast to the case of fullerene, wherein it was found that metalation of the porphyrin with certain cations served to strengthen the interaction.[26]

Following the initial reports of porphyrin–nanotube interactions, sapphyrin–nanotube assemblies were produced by Sessler, Guldi, and coworkers. These ensembles were studied by several techniques, including UV/vis/NIR spectroscopy, fluorescence spectroscopy, and femtosecond laser flash photolysis. On the basis of this, the face-to-face binding model shown in Figure 14 was inferred.[40] Sapphyrin was chosen for this study because it may be excited at lower energy, or longer wavelengths than porphyrins. As such, it represents a better mimic of the natural photosynthetic system. Femtosecond laser flash photolysis studies provided support for the notion that this system undergoes intramolecular ET following photoexcitation, as evidenced by the observation of optical signatures for both the radical anion of sapphyrin and the radical cation of SWNTs. It was thus proposed that electrons from the valence band of the SWNT are excited into the conduction band, followed by transfer to the highest occupied molecular orbital (HOMO) of sapphyrin.[40]

Supramolecular interaction involving porphyrins also provides a means for solubilizing SWNTs in water. The first such example was reported in 2005 by Chen *et al.*,

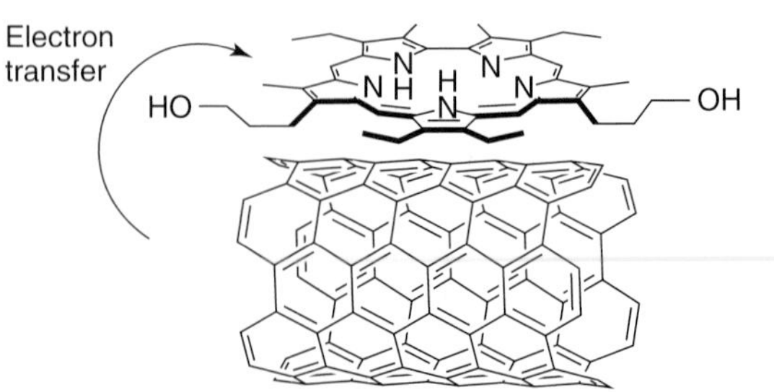

Figure 14 Graphical representation of the intramolecular electron transfer from the SWNT to sapphyrin, proposed to occur on photoirradiation at 775 nm.

who used the charged porphyrin [*meso*-(tetrakis-4-sulfonatophenyl) porphine dihydrochloride].[41] More recently, it has been found that the *uncharged*, water-soluble porphyrin, *meso*-tetrakis-(3,4,5-tri-[2-[2-(2-methoxy-ethoxy)-ethoxy]-ethoxy]-phenyl)-porphyrin, which contains multiple tri(oxyethylene)methyl ether chains, can also solubilize SWNTs in aqueous media, keeping them stable and in solution for weeks.[42]

3 BIOLOGICAL ENTITIES

As mentioned in the general introduction, the development of porphyrin chemistry as a chemical discipline was motivated in part by the fact that porphyrin and porphyrin-like compounds are so ubiquitous in Nature. Therefore, it is perhaps not surprising that porphyrins have been exploited to create systems that are capable of recognizing or sensing biological targets, including peptides, proteins, DNA, saccharides, and other functionally rich materials. There are also many reports of water-soluble porphyrins, both charged and neutral, that have been used as the starting point for the construction of more elaborated receptors, as detailed in the following section. The desire to create these systems is multifaceted, but includes an appreciation that the selective recognition of appropriate biological entities in aqueous environments could allow for a better definition of cellular processes, including those associated with cancerous transformations.[43,44] The simple identification of extracellular and blood-borne species, such as saccharides and certain critical proteins, would also be beneficial. While the rewards would be great, it is to be appreciated that in general terms biological substrate recognition is challenging. Biological milieus contain many constituents. So, a receptor not only has to bind a targeted substrate with an affinity high enough to permit sensing, extraction, transport, catalysis, and so on, but also has to do so in a way that is very selective. While much progress has been made recently, as highlighted below, there is still a long way to go before useful, application-ready systems are in hand.

3.1 Saccharides

There has been much attention devoted to the problem of saccharide recognition and sensing.[43] Carbohydrates and oligosaccharides are particularly important due to their specific implications in biological regulation. Previous molecular biology research has served to establish that oligosaccharides are involved in cell-to-cell recognition, target identification by pathogens during infection, immune response, intracellular protein distribution, and membrane transport. Oligosaccharides serve as biomarkers and provide the molecular signature of a cell phenotype. Unfortunately, saccharides, especially oligosaccharides, have proven to be one of the most difficult classes of biological compounds to recognize effectively and selectively. Complex carbohydrates relay vital signal information using distinct arrangements of structures composed of fairly simple linked monomeric subunits. This modular feature makes the creation of selective saccharide receptors particularly challenging. Nevertheless, since carbohydrates play such important biological regulatory and recognition roles, considerable work has been devoted to the challenge.[44,45]

Porphyrins are well suited for saccharide recognition because they can be substituted so as to create a hydrophobic pocket with binding sites endowed with attractive optical features (cf. discussion above).[45] To date, two main classes of porphyrin-based saccharide receptors have been developed. The first involves porphyrin scaffolds that are appended with boronic acids; these systems capture or sense saccharides via formation of a covalent linkage (usually in the form of boronic esters) between the receptor and the saccharide. The second main type of receptor relies on noncovalent interactions (e.g., ionic, hydrophobic, and hydrogen bonds) between the saccharide guest and the porphyrin receptor. As a general rule, the first approach relies on creating systems with well-defined geometries, while the second, noncovalent approach centers around the synthesis of receptors that provide for geometric flexibility during complexation.[46]

Research efforts at the beginning of the decade (2000–2002) reflected efforts to adapt well-characterized, water-soluble porphyrins for saccharide complexation. In early 2001, Král and coworkers reported progress toward developing a porphyrin receptor scaffold for sensing carbohydrates in competitive solvents.[47] This group developed a macrocyclic porphyrin "sandwich" system, **14**, that incorporates amido-azonia cryptands at the "hinges". The resulting combination of rigid, hydrophobic faces capped on two sides by groups that were expected to be well hydrated was designed to create a special polar environment into which saccharides would bind. According to molecular modeling studies, the cavity was appropriately sized to accommodate trisaccharides particularly well. As predicted, the sandwich system showed a preference for trisaccharides, most likely due to the complementary topography of the receptor and the creation of an effective microenvironment (Figure 15). Selectivity for maltotriose was inferred from UV–vis spectroscopic titrations, wherein the decrease in the porphyrin Soret band as a function of saccharide concentration was monitored. The resulting spectral changes could be fit to a 1 : 1 host–guest binding equation, and the stoichiometry was confirmed by Job plot analysis. Overall, an increase in the effective association constant (K_a) was seen as the saccharides were changed from mono to di to tri; however, the

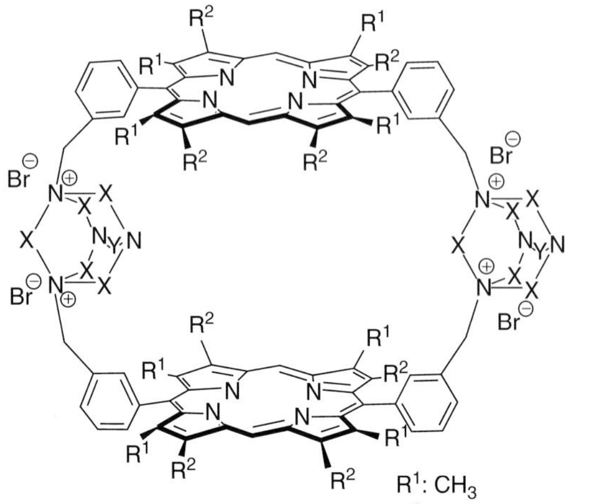

Figure 15 A porphyrin sandwich system that is designed to bind saccharides selectively via the use of a complementary hydrophobic/hydrophilic topology.

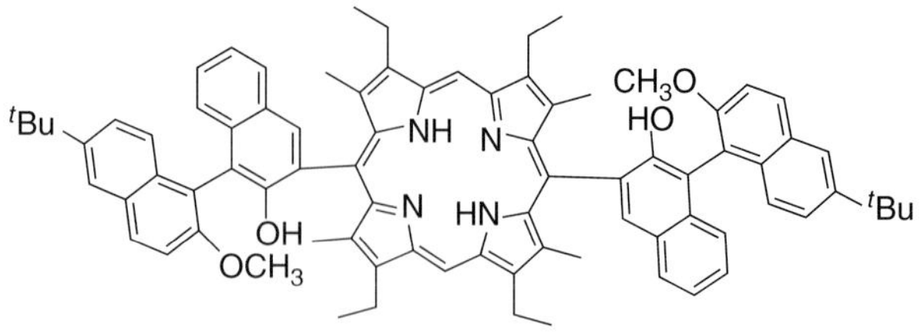

Figure 16 1,1′-Binaphthyl-substituted macrocycles as receptors for saccharide recognition based on the idea of a picket-fence porphyrin.

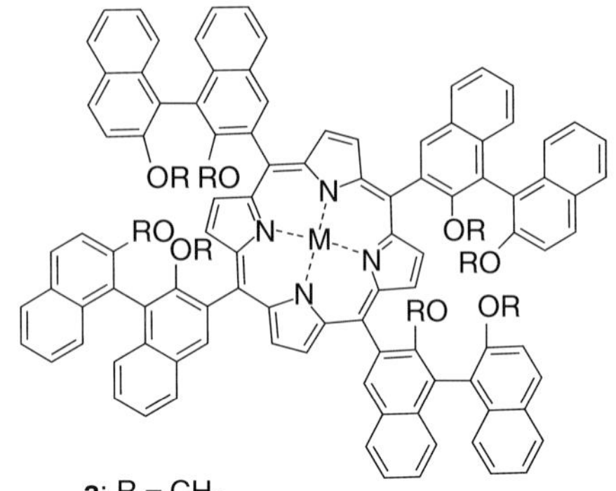

affinities were found to decrease on moving to higher glucose oligomers. A control, open chain porphyrin with the same functional groups, did not show the same selectivity for maltotriose, providing support for the contention that a well-defined cavity is important for effective recognition.

While hydrogen bonding is important for the recognition of sugars (substrates with multiple OH groups), the requisite interactions are often disrupted by water. Both self-aggregation and difficult-to-control solvation effects further militate against receptor-based saccharide recognition in aqueous media. However, Král and coworkers showed that these challenges could be overcome by using appropriately designed binaphthyl-substituted porphyrins and metalloporphyrins, such as **15** and **16**.[45] Again, the aim was to create a hydrophobic pocket through the use of bulky aryl groups while providing a source of hydrogen bond donors (Figure 16). Porphyrin **16b**, showed specificity for di- and trisaccharides based on a comparison of association constants obtained from UV–vis studies. This size preference was considered indicative of within-cavity binding. It was also inferred that in aqueous media, recognition likely occurs as a result of both CH–π interactions and hydrophobic effects, while in polar organic media binding is mediated by hydrogen binding. Metalation of porphyrin **16b** with Fe^{III} results in higher associations of the porphyrin receptor with saccharides, at least in aqueous solution.

The mechanism of binding of **16b** with D-glucose was assessed by ^{1}H NMR spectroscopy in DMSO-d_6 (DMSO was chosen to avoid problems with aggregation at NMR-relevant concentrations). It was found that the metal-free porphyrin complex, **16b**, binds D-glucose in a 1 : 1 stoichiometry, with a K_a of 30 M^{-1} in this medium. Saccharide-binding ability was also assessed by surface plasmon resonance (SPR). In SPR, the porphyrin receptor was physically adsorbed onto the surface of a chip, which was then incubated in the presence of aqueous D-glucose. The sample was then irradiated and the intensity of the reflected light beams was measured; the difference in the value of the refractive index obtained from the chip was then used to generate an SPR curve, which was compared to the corresponding curve obtained using the porphyrin in the absence of the saccharide substrate. In the case of the zinc porphyrin, **16d**, the curves were significantly different in the presence and absence of the sugar substrate. Control experiments, carried out with simple tetraphenylporphyrins (TPPs), gave SPR curves that were unchanged before and after incubation with D-glucose.

Recently, in 2010, an effort was made by Kalenius to create a supramolecular probe with improved selectivity and effectiveness for saccharide binding.[44] On the basis of the results from previous work by Král and coworkers, the Kalenius group used bile acid—a steroid derivative—as the porphyrin appendage with the goal of effecting saccharide-specific molecular recognition.[48] The attachment of two bile acid arms to a porphyrin core creates a host, **17**—a system expected to be capable of

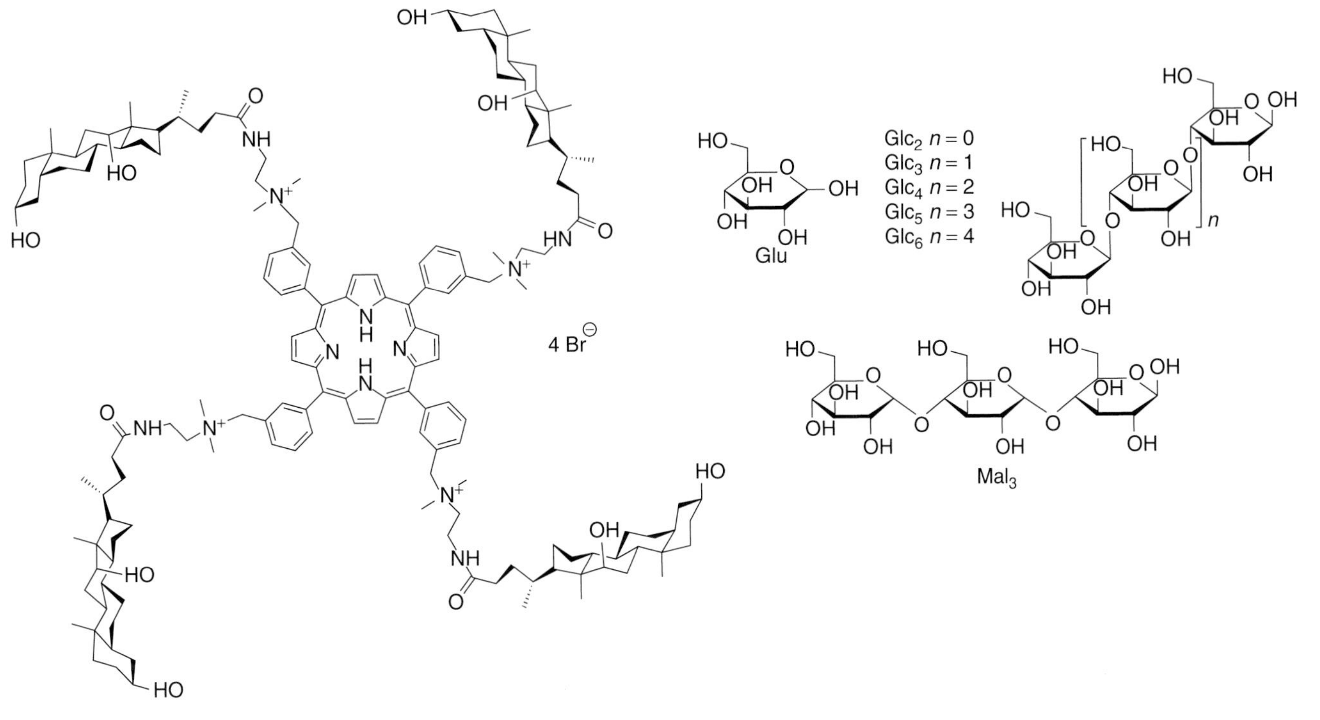

17

Figure 17 Tetrakis(bile acid)–porphyrin conjugate that is capable of selectively forming noncovalent bonds with saccharides, and representations of the sugars subject to study.

not only hydrogen bonding but also stabilizing hydrophobic and hydrophilic interactions (Figure 17). MS analyses revealed 1 : 1 complexation between **17** and **Glc$_4$** and provided support for this particular receptor being selective for oligosaccharides comprising *at least* three glucose residues. It was determined that the most stable complexes were formed with cellotetraose (four glucose residues) and cellotriose (three glucose residues). Electron capture dissociation (ECD) analyses revealed that a proper treatment of this receptor with a three-residue oligosaccharide makes it interact with the bile acid arms and provided support for the inference that the porphyrin center was essential for complexation and that at least one bile arm was necessary in order to capture an intact sugar. This very recent study highlights the potential of ECD for analyzing receptor–substrate interactions and leads to the suggestion that this technique may have a role to play in characterizing small, supramolecular assemblies.

Additional saccharide receptors based on the "picket-fence" porphyrin were developed by the Bonar-Law group, but these were found to bind saccharides strongly only in nonpolar media.[49] The starting material, the $\alpha,\alpha,\alpha,\alpha$-atropisomer of 5,10,15,20-*meso*-tetrakis(*o*-aminophenyl)porphyrin, **1** (cf. Figure 3), was functionalized with four different amino acid esters that were linked to the porphyrin scaffold by urea groups; this yielded receptors **18a–d** (Figure 18). The porphyrins were also metalated with Zn in order to produce stronger, substrate-induced changes in the porphyrin Soret band. However, in spite of expectations, there was no significant difference between the metalated and nonmetalated forms of the receptor. Binding affinities were analyzed by UV–vis titrations, and checked by

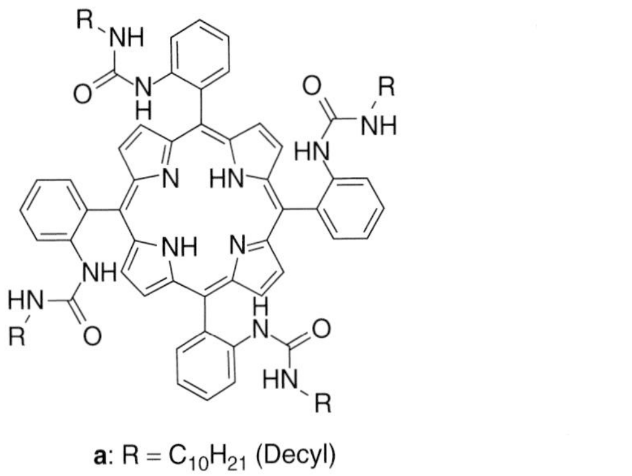

a: R = $C_{10}H_{21}$ (Decyl)
b: R = L-Valine methyl ester
c: R = L-Threonine methyl ester
d: R = L-Tryptophan methyl ester

18

Figure 18 A series of receptors made through the attachment of four identical amino acid esters to the base porphyrin through urea groups. The result is a set of "picket-fence" receptor for sugars.

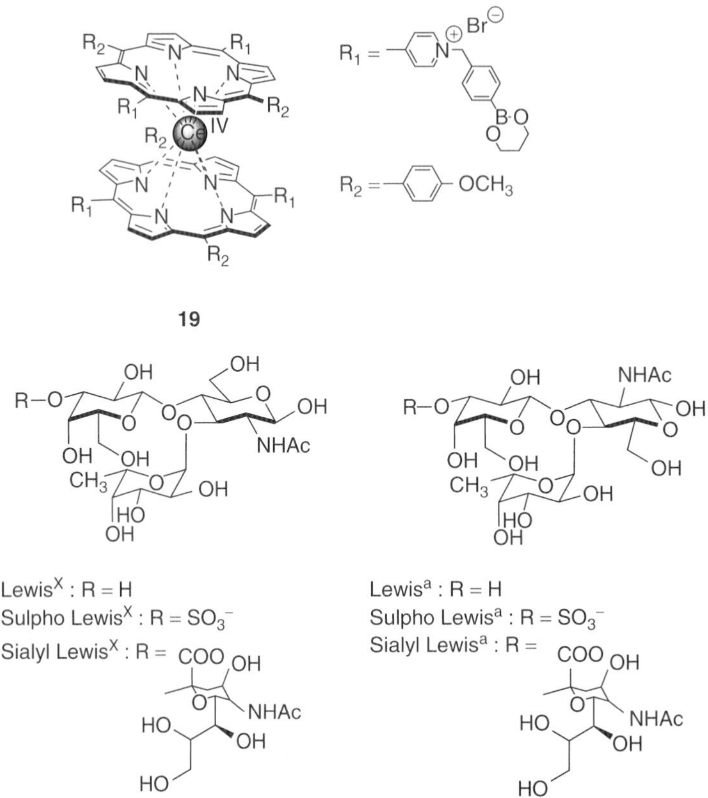

Figure 19 A phenylboronic-acid-group-appended cerium(IV) bis(porphyrinate) double decker receptor, **19**, which is capable of binding Lewis oligosaccharides in aqueous media. Lewis oligosaccharides are considered an indication of tumor progression/malignancy and are targets for molecular receptors.

fluorescence spectroscopy. Deviations from 1 : 1 binding modes were observed at high sugar concentrations; this was taken as an indication of possible second binding events. The receptors showed strong affinities for the sugars in dichloromethane, CH_2Cl_2, with K_a values ranging from 3×10^4 to $9 \times 10^5\ M^{-1}$. Some evidence of sugar binding was noted in DMSO, which leads to the suggestion that further development could yield systems that work in aqueous media.

In 2001, Shinkai and coworkers reported the use of positive homotropic allosterism for Lewis oligosaccharide recognition. Lewis oligosaccharides are important targets because they are involved in the adhesion of leukocytes and neutrophils to vesicular endothelial cells during both pathogenic and regular inflammatory response.[43] The appearance of Lewis oligosaccharides at high concentrations on the surface of cells or in the blood serum often indicates tumor progression and malignancy—specifically in gastrointestinal, pancreatic, and breast cancer. Development of an artificial receptor for these oligosaccharides would enable further study for the development of small-molecule antagonists.

The basic Shinkai approach is analogous to that described earlier for fullerene binding, with the exception that boronic acids were used as critical recognition elements. Boronic acids are well known in the saccharide sensing literature due to the ability of diboronic acid derivatives to react with four of the five hydroxyl groups on a saccharide to form intramolecular 1 : 1 complexes.[50] Two receptors were created by the Shinkai group: a phenylboronic-acid-appended cerium(IV) bis[porphyrinate] double decker system, **19**, and a porphyrin dimer that is linked through the meso positions (Figure 19).[43]

Addition of a Lewis oligosaccharide to solutions of **19** in buffered solution essentially resulted in no UV–vis or fluorescent changes. However, circular dichroism (CD) spectroscopic measurements revealed exciton-coupling-type CD bands. Such bands are observed only if a saccharide is bound to a boronic acid, while even stronger bands are expected should the saccharide be bound to two boronic acid groups in a macrocyclic form. Compound **19** was the only receptor that was shown to bind Lewis oligosaccharides. The existence of the positive homotropic allosterism was confirmed by a sigmoidal curve, obtained when CD intensity was plotted against the added concentration of oligosaccharide. Job plot analyses were also carried out using the CD intensity for the *y*-axis as a measure of complex formed. The maximum at 0.33, seen in the Job plot, is

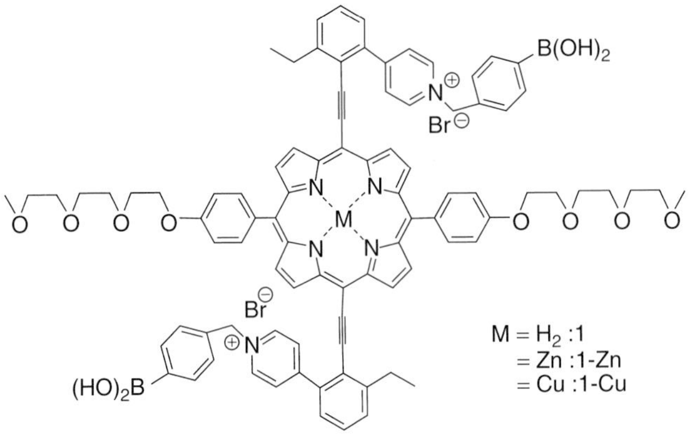

Figure 20 One of several sensors synthesized and tested for oligosaccharide sensing.

consistent with the formation of a 1 : 2 receptor for Lewis oligosaccharide complex. Finally, it is to be noted that this receptor permitted discrimination between different types of Lewis oligosaccharides, namely, Le^X and Le^a, as evidenced by the production of "opposite" CD spectra.

Prior work by Shinkai and coworkers showed that phenylboronic-acid-appended 5,15-bis(triarylethynyl)porphyrin, **20**, serves as an allosteric monosaccharide receptor in an aqueous environment, although it was proposed that allosterism may not be necessary for good association (Figure 20).[51] Favorable binding without the benefit of cooperative effects could occur under conditions wherein the receptor is sufficiently flexible that its phenyl boronic acid binding motifs can adjust to match the distance between the *cis*-1,2-diols and/or the 1,3-diols in the oligosaccharide. To test this latter hypothesis, the Shinkai group synthesized a diethynyl porphyrin containing an axis through the porphyrin. On rotation, this latter system can self-assemble to vary the distance between the two phenylboronic acids anywhere from 0.1 to 2.4 nm. These porphyrin receptors were metalated with Zn or Cu, in an effort to enhance the UV–vis signal (as previously discussed in the introduction) for optimal analysis of the interaction.

UV–vis and fluorescence binding studies were carried out in water/CH_3OH buffered solutions. Unfortunately, as above, the changes in UV–vis and fluorescence were too small to allow for a meaningful analysis; therefore, CD studies were performed. Overall, it was shown that the porphyrin receptor (**20**) performed better with Zn than with Cu. On the basis of these and other studies, it was concluded that **20** could bind both mono- and oligosaccharides. However, as true for other receptor systems, the lack of selectivity is still problematic.

3.2 Peptides and proteins

Another group of important biological targets is peptides and proteins. A variety of medical treatments and medical diagnostics could be developed using synthetic agents that are capable of specific protein or peptide recognition.

In 2003, the Hamilton group reported a family of tetrabiphenylporphyrin-based receptors for the protein surface recognition of cytochrome *c*.[52] Cytochrome *c* is a well-characterized protein that plays a key role in ET reactions; it is also involved in apoptosis. Cytochrome *c* has a surface that is well known for having patches of hydrophobicity, surrounded by charged residues, including aspartate and glutamate. Therefore, it was considered likely that an effective receptor would target both types of surfaces, namely, hydrophobic and charged. Porphyrins were considered attractive for this purpose by the Hamilton group due to the ease of creating a hydrophobic inner core, and a functionalized periphery with negative charge density. While previous work published by Hamilton and coworkers described the use of a TPP as the core scaffold (15.5 Å), later work described the expansion of the hydrophobic core to 24.0 Å using a tetrabiphenylporphyrin. The biphenyl groups were functionalized with multiple carboxyl groups to create an outer rim bearing negative charges (Figure 21). The binding affinities of a series of tetrabiphenylporphyrin receptors, differing only in the number of carboxyl groups, for horse heart cytochrome *c* were derived from fluorescence quench measurements, carried out in sodium phosphate buffer solution. A clear relationship was observed between binding affinity and the number of anionic groups present on the periphery of the receptor—receptors with four carboxyl groups were found to bind with a K_d of 12–17 μM, receptors with eight carboxyl groups were found to bind with a K_d of 1.3–1.7 μM, and finally, the receptor with 16 carboxyl groups, **21**, was found to bind with a K_d of 0.67 nM (K_a of $1.5 \times 10^9\ M^{-1}$)—which at the time was the strongest reported binding affinity for a synthetic receptor with a protein surface.

Further evaluation of **21** was carried out using CD spectroscopy, which traced the thermal denaturation of the

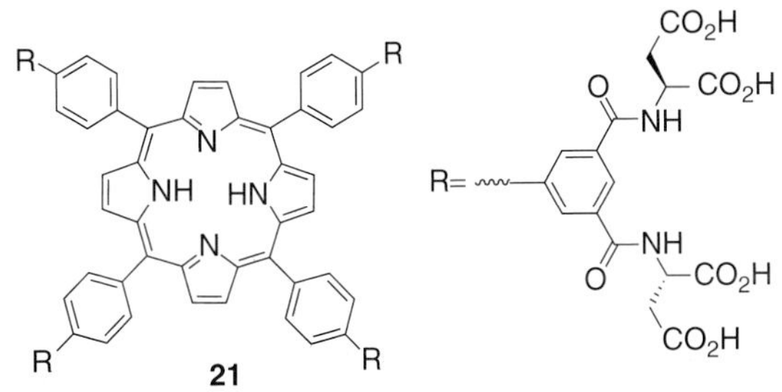

Figure 21 View of a receptor that demonstrated a subnanomolar affinity for binding to the surface of cytochrome *c*.

protein in the presence and absence of the receptor, and gave the melting temperature (T_m) of the protein. These analyses revealed a lowering of the T_m from 85 °C without the receptor to 35 °C with receptor. The selectivity of this receptor was also analyzed through binding studies with other proteins, including the closely related cytochrome c_{551} and ferredoxin, both of which were bound 270 and 25 000 times less well than cytochrome *c*. Overall, this work highlights the importance of charge and size matching, and suggests that porphyrins are a good scaffold in which to achieve this combination. An application of these types of porphyrins for pattern or array based sensing is described in a later section.[53]

Another interesting example of selective targeting of proteins by porphyrins was published in 2008 by Tanimoto *et al.* They reported what they believed was a similar type interaction to that described by Hamilton wherein hydrophobic and charged portions of the porphyrin receptor match a complementary portion of the targeted protein, which in this particular case is hER-α, or human estrogen receptor-α.[54] The goal of this study was to demonstrate that porphyrin derivatives could serve as protein photodegrading agents, as has been done for DNA via effects similar to those invoked in the case of photodynamic therapy (i.e., photoinduced singlet oxygen generation) (Figure 22). Initial studies involved bovine serum albumin (BSA) and the four porphyrin derivatives **22a–d**. All four derivatives showed light-induced degradation of this particular protein. However, when this same series of derivatives were tested with human estrogen receptor-α (hER-α), only one, **22a**, was able to show efficient degradation. This particular porphyrin is closest in structure to estrogens, which are known to bind hER. When two proteins, hER and Lyso (hen egg lysozyme) were mixed in the presence of **22a**, only hER was degraded. Additionally, it was found that catalytic amounts of the porphyrin were required to degrade hER, and when the same conditions were applied to BSA, no degradation occurred. Therefore, it was concluded that not only did the porphyrins provide a scaffold for recognition and binding of a particular protein target, but also provided a means for selectively degrading the target through irradiation of the porphyrin.

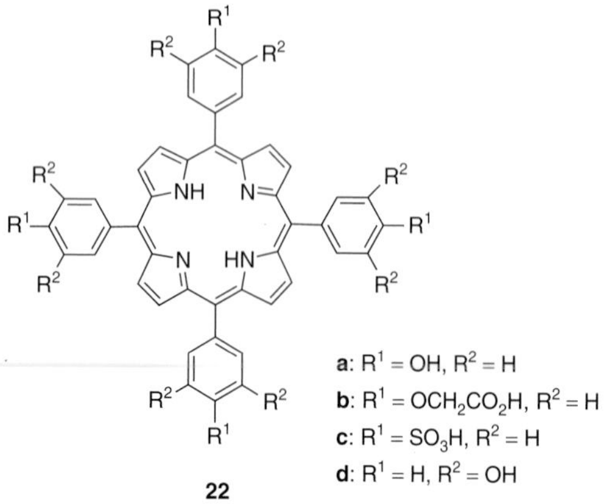

a: $R^1 = OH$, $R^2 = H$
b: $R^1 = OCH_2CO_2H$, $R^2 = H$
c: $R^1 = SO_3H$, $R^2 = H$
d: $R^1 = H$, $R^2 = OH$

22

Figure 22 Structures of porphyrin derivatives that selectively bind proteins targeted for photodegradation.

3.3 DNA

As noted above, an attractive feature of porphyrins is that they allow for the combination of different "binding" motifs within one ostensibly simple molecule. These can include both the hydrophobic core and the charged periphery. This makes such modified porphyrins attractive as receptors for DNA. In fact, water-soluble, cationic porphyrins (**23**) have an established history in the area of DNA recognition, with anionic porphyrins (**24**) also more recently being studied for this purpose (Figures 23 and 24).[1] While the mechanism for binding is not always known, or discernable, there are three main ways in which receptor binding could occur. The first one involves intercalation or insertion of the receptor between base pairs, something that generally occurs more easily between guanine and cytosine. The second is external binding, wherein the receptor sits in a "groove," while the third is aggregation on the surface.[1,7] These idealized limiting interactions have been highly investigated, and the

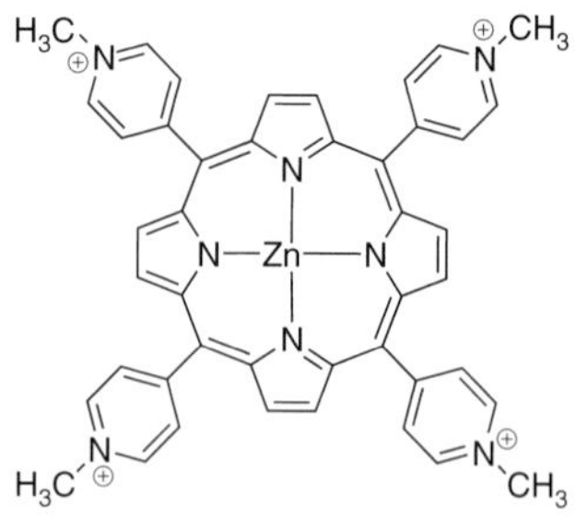

23

Figure 23 Tetramethylpyrdinium porphyrin: a water-soluble tetracationic porphyrin.

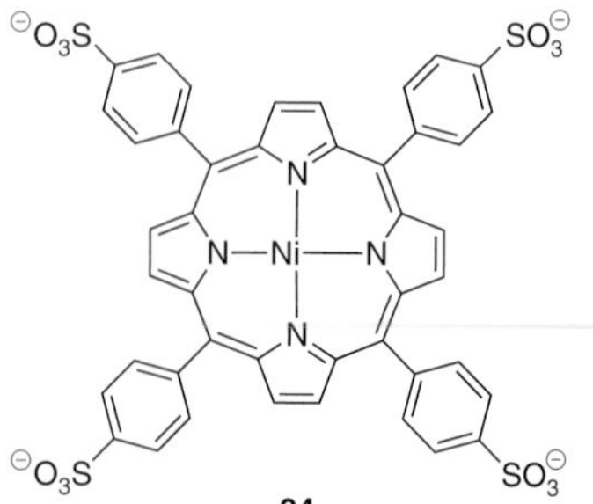

24

Figure 24 Tetrasulfonated porphyrin: a water-soluble tetraanionic porphyrin.

literature on the topic is extensive and detailed because of the potential therapeutic aspect. Porphyrins are easily photoexcited and produce singlet oxygen and other reactive oxygen species as a result. They can thus be used to initiate a reaction that results in cleavage or damage of the DNA, which leads to cell death in some cases. Some of the particular challenges that have been tackled over the past decade have centered on how to study or analyze the porphyrin–DNA interaction and how small changes in structure affect the activity and selectivity of certain forms or varieties of DNA.

One of the special features of DNA arises from the fact that it is essentially a polymer, and therefore can possess multiple secondary structures. Usually, DNA takes on a right-handed helical form, known as the *B conformation*. However, higher energy left-handed forms, the so-called Z conformations, can be accessed under certain conditions.[55] While the biological role of Z-DNA is still under scientific scrutiny, there is a need to identify it selectively. Methods for such identification include CD spectroscopy. Nevertheless, it is difficult to distinguish between these two forms of DNA in a true biological setting due to the presence of other analytes that interfere with accurate measurements. There are two recent reports wherein porphyrins are used to distinguish between these two types of DNA.[55,56]

In 2005, Purrello and coworkers reported the pentacoordinated cationic zinc-porphyrin-based chiroptical probe, **23**·H_2O, and demonstrated its ability to discriminate between B and Z structures. A desirable chiroptical probe is one that absorbs above 300 nm—an area free from interference of most biological species.[55] Porphyrins are attractive for this purpose due to their unique CD spectral properties (including absorbance above 300 nm); in fact, there is an extensive history of their use in detecting and amplifying the handedness of matrices and the application to other complex sterochemical problems.

Cationic zinc porphyrins are well-known DNA intercalators. Interestingly, however, they are not effective for Z-DNA since binding destabilizes the left-handed helix and causes it to revert to the B state. However, the simple modification of coordinating water in an axial position of a Zn porphyrin prevents intercalation, thereby obviating the reversion to the B-form. This was a feature that was exploited by Purrello and coworkers. They found that at a micromolar concentration of porphyrin, an intense bisignate induced circular dichroism (ICD) signal was seen in the case of Z-DNA; only a small ICD signal was seen in the case of B-DNA.

To study the interaction of the pentacoordinate zinc porphyrin with DNA further, the fluorescence emission intensity was monitored as a function of added DNA. B-DNA significantly quenches the emission, while Z-DNA results in a slightly blue-shifted emission and minimal quenching for one signal; the second signal increases in emission intensity. This was considered indicative of minimal $\pi-\pi$ interactions involving the porphyrins. Absorbance studies revealed a red-shift in the porphyrin Soret band upon interaction with either type of DNA. Finally, resonance light scattering (RLS) was used to assess the extent of aggregation. Aggregation is common when the DNA binding mode involves outside stacking. In the case of the Purrello system, an RLS spectrum of the porphyrin was taken. When B-DNA was added, another spectrum was taken, and this remained the same. When Z-DNA was added, the intensity of the RLS spectra increased. Thus, equal concentrations of **23**·H_2O and **23**·H_2O·Z-DNA were found to give two different RLS spectra supporting the notion that in the latter case, the porphyrins are coupled to each other. These findings were rationalized in terms of one open coordination site on the centrally coordinated Zn, which was bound to an open N atom in the Z-form of DNA, which is not accessible in the B-form. Competition with Ni^{II} cations showed restoration of free Z-DNA by CD, providing support for the conclusion that metal coordination assists in form-specific DNA recognition and binding.

In 2009, Purrello published another porphyrin that was able to bind Z-DNA selectively. In this case, an anionic porphyrin, nickel(II) *meso*-tetrakis(4-sulfonatophenyl)porphyrin, **24**·**Ni**, was used.[56] With this receptor, a "switching" effect was seen on the addition of spermine. Spermine is a positively charged species at biological pH, and this serves to stabilize the Z-form of DNA. CD spectroscopy was again used as the primary method of analysis. Receptor **24**·**Ni**, in the presence of B-DNA, produced no ICD signal in the visible region (~400 nm). On the other hand, in the presence of spermine, an induced exciton-coupled CD signal was seen in the 400 nm (Soret) region. When the pH is raised, the spermine becomes deprotonated, and the B-form of DNA dominates. These changes could be monitored by following the spermine-induced evolution of the porphyrin Soret band in the CD spectra.

Related to DNA duplex binding is G-quadruplex binding. A G-quadruplex is a tetrahelical derivative of DNA, formed by four guanines that are hydrogen bound to one another through what are called *Hoogsteen interactions* (Figure 25). This is most often formed from the G-rich human telomeric DNA, which has become a drug target because of its role in the proliferation of tumor cells.[57]

One of the challenges in the recognition and binding of G-quadruplexes is the competition with duplex (double-stranded) DNA. While it was reported that cationic tetra(*N*-methylpyridiniumyl) porphyrin, **23**·**Mn**, could target quadruplex DNA, there was no discrimination, since it was also found to bind duplex DNA well.[58] In 2007, Dixon *et al.* reported a pentacationic manganese(III)

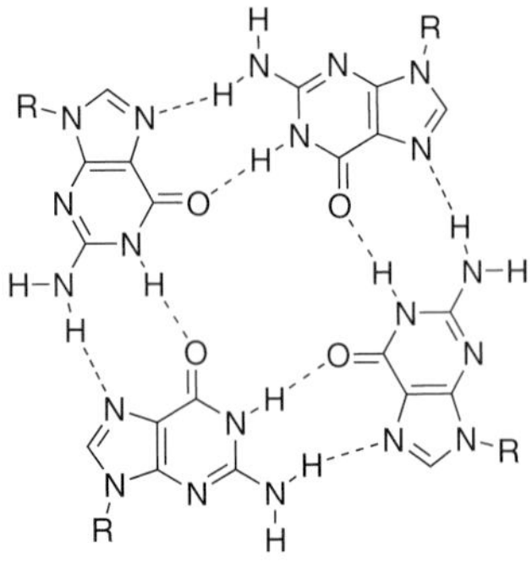

Figure 25 Structure of an individual G-tetrad showing the Hoogsteen interactions between the base pairs.

porphyrin, **25**, that displayed a 10 000-fold preference for G-quadruplexes, as opposed to duplex DNA (Figure 26).[59] SPR was used to measure the equilibrium constants for the porphyrin with both duplex and quadruplex DNAs; the resulting values were found to be 10^4 and $10^8\,M^{-1}$, respectively. In this case, it was proposed that the four bulky arms preclude binding to duplex DNA, while interactions with the porphyrinic core favor binding of the quartet form, with the flexible arms further inserting into grooves or loops. An assay, which is used to test inhibition of telomerase, namely, TRAP (telomeric repeat amplification protocol), revealed that micromolar concentrations of **25** could be used to inhibit telomerase; the IC_{50} was 580 nM.

4 ANIONS

The interaction between porphyrins and expanded porphyrins with anions has been well documented and thoroughly reviewed in the literature.[60] This field gained traction in the early 1990s with reports of the ability of sapphyrin to bind fluoride in its diprotonated form and also to bind and transport phosphates.[61,62] Sapphyrin is an expanded porphyrin that has a cavity equipped to host small anionic guests, especially in its protonated form. Porphyrins, however, because of their smaller cavity, are less basic. They are not easily protonated and usually need to be "elaborated" to allow for anion binding. A survey of the literature involving porphyrin-based anion receptors shows a propensity for use of "picket-fence porphyrins". To achieve anion recognition, the pickets are generally functionalized to provide a source of hydrogen bond donors that allow for anionic guest recognition. Urea subunits are particularly popular in this regard.

One report, coming from the group of Hong in 2001, details how the urea functionality may be combined with an azophenol group to produce a colorimetric anion sensor, **26** (Figure 27).[11] To gain insight into the operative binding mode, ^{1}H NMR spectroscopic experiments were carried out in DMSO-d_6. These studies revealed a shift for the urea NH signals, and a broadening of the phenolic OH resonances in the presence of both dihydrogenphosphate, $H_2PO_4^-$, and acetate, CH_3COO^-. UV–vis spectroscopic studies were also performed, confirming a binding interaction through

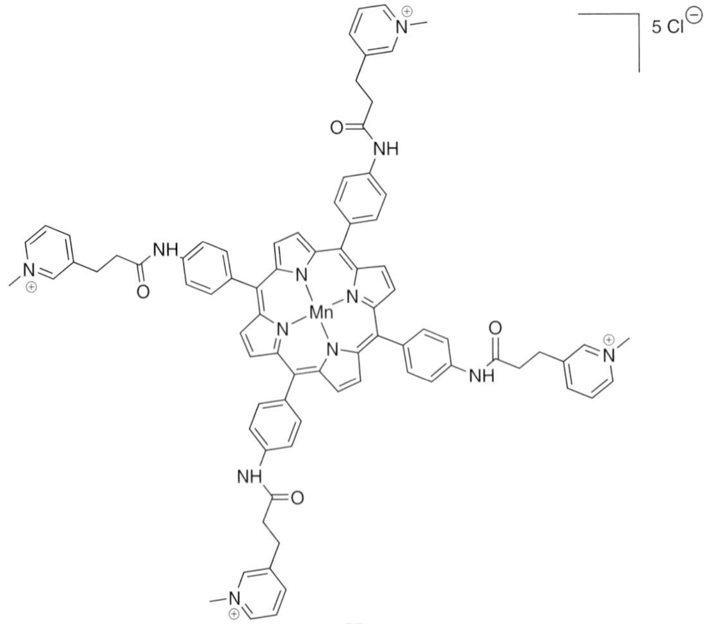

Figure 26 A pentacationic manganese (III) porphyrin that displayed a 10 000-fold preference for G-quadruplexes, as opposed to duplex DNA.

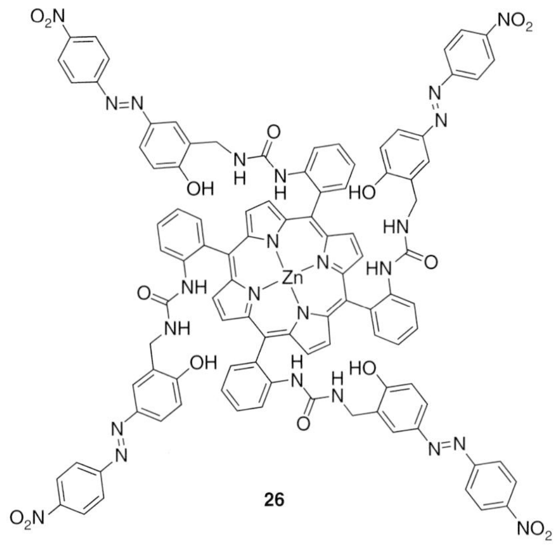

Figure 27 Picket-fence porphyrin with azophenol arms that are designed to act as a colorimetric anion sensor.

shift of the porphyrin Soret band. For two derivatives of this basic picket-fence porphyrin, selectivity was highest for CH_3COO^-, followed by $H_2PO_4^-$. Much less interaction was seen with all halides, except for fluoride, which is known to interact with the azophenol moiety. Therefore, the UV–vis studies provided evidence for a selectivity order, namely, $F^- > CH_3COO^- > H_2PO_4^-$, that follows the order of most to least basic anion.

In a similar vein, Beer and coworkers combined an anion-binding picket-fence porphyrin with a gold nanoparticle optical sensor.[12] Here, the tetra-amide-disulfide porphyrin **27** was synthesized and subsequently attached to a gold (Au) nanoparticle (Figure 28). UV–vis titrations were performed in CH_2Cl_2 and revealed high affinities for both Cl^- and $H_2PO_4^-$. However, there proved to be no significant

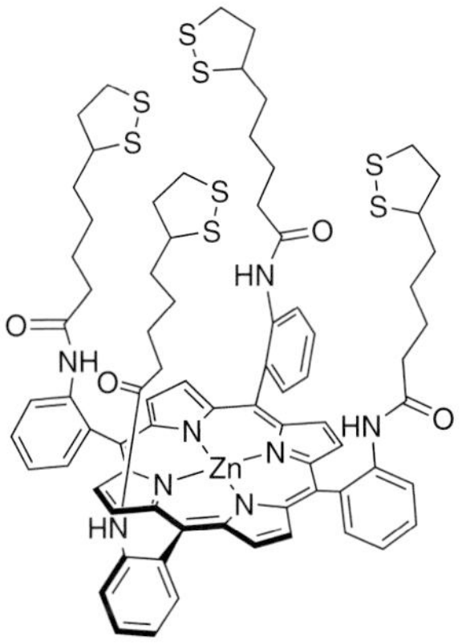

Figure 28 A tetra-amide-disulfide porphyrin synthesized by Beer and coworkers for attachment to gold nanoparticles through ligand substitution.

difference between the thiol-porphyrin (**27**) and the thiol-porphyrin nanoparticle (**27-NP**). On the other hand, when the solvent was changed to DMSO, a stark difference in anion affinity between **27** and **27-NP** was seen. For Cl^-, $\log K < 2$ in the case of **27** but about 4.3 in the case of **27-NP**. For $H_2PO_4^-$, the difference is similar, with $\log K = 2.5$ in the case of **27** and $\log K = 4.1$ in the case of **27-NP**. The reason for this difference could reflect preorganization on the surface of the nanoparticle that reduces the entropic cost on binding, thus increasing ΔS and making the interaction more favorable.

In another report, by the group of Beer, 5,10,15,20-*meso*-tetrakis(*o*-aminophenyl)porphyrin, **1**, was functionalized with imidazolium subunits to provide for anion recognition.[13] Imidazoliums are particularly useful for anion binding due to a combination of hydrogen bonding and electrostatic effects. The X-ray structure of the synthetic precursor **28**, shown in Figure 29, reveals the presence of three amide bonds at the center of the pocket. Unfortunately, the structure determined by crystallographic analysis shows only a pyridine coordinated to the Zn center of the porphyrin. On the other hand, ^{1}H NMR, ^{31}P NMR, and UV–vis spectroscopic analyses, as well as electrochemical and luminescence experiments, provided evidence for anion binding in the case of the final imidazolium picket-fence porphyrin, **29**. The luminescence experiments revealed dramatic changes during the course of spectral titrations carried out with $H_2PO_4^-$, Cl^-, and HSO_4^- (all used in the form of their tetrabutylammonium (TBA) salts) in DMSO. Accompanying electrochemical titrations revealed cathodic shifts in the oxidation potential of the porphyrin as a function of anion addition; however, the extent of the shift did not correlate with the binding affinities determined by UV–vis spectroscopy. These analyses gave the following log K values for **29:** 4.2 for Cl^-, 4.8 for $H_2PO_4^-$, and >6 for HSO_4^- in DMSO. In a DMSO–water mixture, **29** retained its affinity for HSO_4^-, with $\log K > 6$, ATP^{2-} ($\log K\,5.0$), and SO_4^{2-} ($\log K > 6$). In the control experiment, wherein the side arms lack imidazolium groups (**28**), the UV–vis changes proved too small to determine an association constant in the case of HSO_4^-.

Beer and coworkers have investigated other picket-fence porphyrin-type anion receptors (Figure 30). For instance, they reported on the effect of the metal center on the anion-binding ability of porphyrin **30**.[14] This study was motivated in part by the theory that adding or including a strong Lewis acid as part of the binding motif would increase the overall anion affinity. This indeed proved to be the case: in UV–vis spectroscopic studies, when Zn was replaced by the stronger Lewis acids Cd(II) and Hg(II), the affinity for Cl^-, Br^-, and I^- in CH_3CN increased. On the other hand, it stayed about the same for $H_2PO_4^-$. There was essentially no difference in the anion affinities for the two complexes

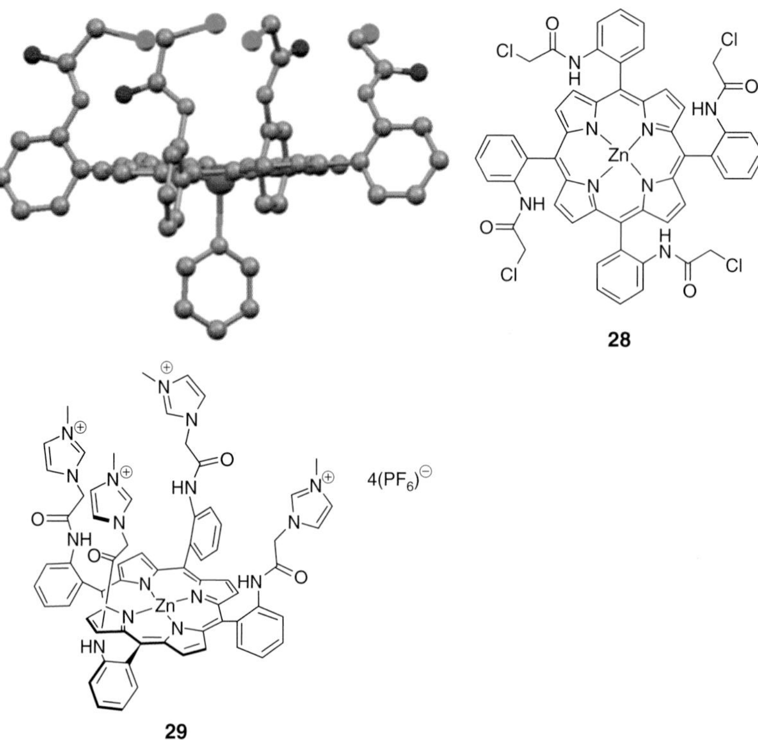

Figure 29 X-ray crystal structure (CCDC# 618393) of **28,** wherein hydrogens have been omitted for clarity. Also omitted is a pyridine coordinated to the Zn center. Receptor **29** was synthesized in two steps from **28**, and studied for anion-binding ability.

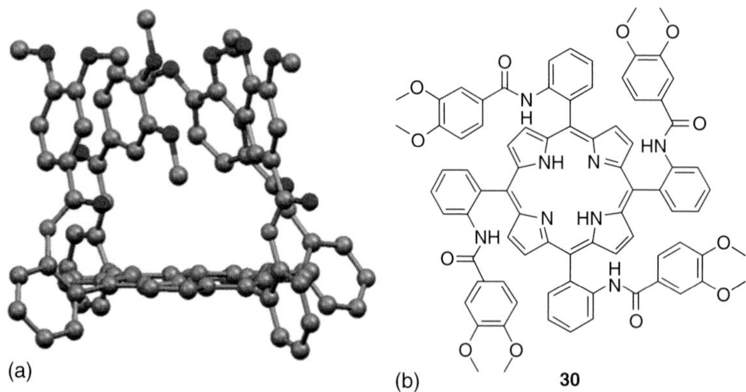

Figure 30 (a) X-ray crystal structure (CCDC# 686999) of the picket-fence porphyrin (b) wherein hydrogens have been omitted for clarity.

Cd·**30** and Hg·**30** in CH_3CN. In contrast, when the solvent was changed to DMSO, Zn·**30** was not found to bind anions at all, while Hg·**30** was found to bind Cl^-, Br^-, and I^- with a much higher affinity than Cd·**30**; the affinity for $H_2PO_4^-$ remained about the same.

The binding affinity of all studied anions for Hg·**30** could not be differentiated; this proved true even when the competitive nature of the solvent was increased via the addition of water to the initial DMSO solutions ($\log K > 6$ in all cases, except for $H_2PO_4^-$ for which $\log K$ is 4.2). The corresponding metal complexes of TPP were also made as controls. In most cases, M·**30** showed higher affinities than the corresponding M·TPP complexes for anions. The exceptions include similar $H_2PO_4^-$ affinity constants being recorded for both Hg·**30** and Hg·TPP, and a smaller binding constant for I^- in the case of Cd·**30**, as opposed to Cd·TPP. While further studies are warranted, on a very basic level, this study provides support for the appealing notion that the combination of amide groups and Lewis acidic metal centers can be used to enhance anion-binding effects.

The Burns group has exploited the picket-fence motif to investigate the influence of convergent solvent binding on

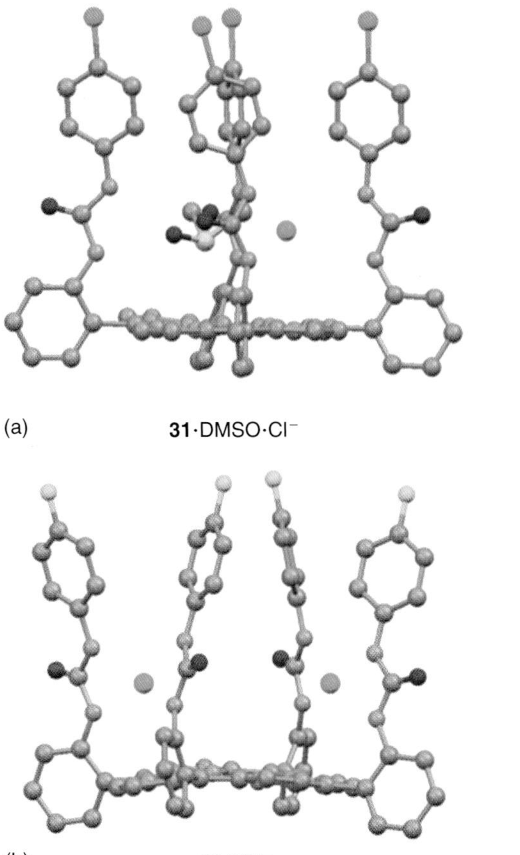

Figure 31 X-ray crystal structure of two picket-fence porphyrins: (a) complexation with one solvent molecule of DMSO and one chloride anion, (b) complexation with two chloride anions. (CCDC#s: 111804 and 673110).

the affinity, selectivity, and stoichiometry on anion recognition. These researchers appended four urea pickets onto the basic picket-fence porphyrin scaffold. This produced a receptor **31**, which allows for the simultaneous binding of a solvent molecule and an anion. This results in an anion-bound complex of greater stability than that produced by the corresponding porphyrin functionalized with just two urea pickets in the 5 and 10 positions (ortho).[63] On the basis of detailed studies of the thermodynamics of complexation, it was inferred that one buried solvent molecule adds the equivalent of an "additional" hydrogen bond in terms of anion affinity. These researchers also concluded that a buried solvent molecule can affect the receptor: anion-binding stoichiometry. Figure 31 shows a comparison between two crystal structures—one with a bound chloride and a bound DMSO molecule (grown in DMSO) and one with two chloride anions (grown in CH_2Cl_2). These solid state differences are consistent with the disparities in binding stoichiometry seen from Job plot analysis when anion-binding studies are carried out in different solvents.[63]

As noted above, sapphyrins were first appreciated for their anion-binding ability in the early 1990s. Several X-ray crystal structures have been elucidated and revealed the ability of the protonated macrocycle to bind anions; Figure 32 shows a sapphyrin derivative, **32**, in its diprotonated form, which is bound to two bromides. The resulting salt is **32**·2HBr.

Sapphyrins remain the subject of ongoing research since there are relatively few organic receptors that can bind phosphates selectively and efficiently, especially in aqueous media. In 2003, Sessler *et al.* published water-soluble versions of this macrocycle, **33a–f** (Figure 33). They also reported that it can act as a fluorescent sensor for phosphate.[64] This proposed sensing ability was based on a fluorescence quenching effect caused by the inherent "aggregation" of sapphyrins. Phosphate binding serves to deaggregate sapphyrin, resulting in a higher concentration of the monomeric form, which is appreciably fluorescent. The underlying aggregation effect was studied independently through dilution effects. At high concentrations of sapphyrin, the UV–vis spectra displayed

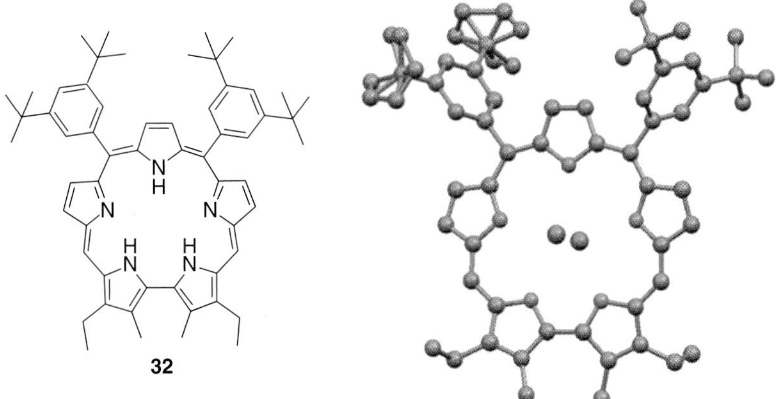

Figure 32 Single crystal X-ray structure of **32**·2HBr, viewed from the side. (CCDC#: 213468).

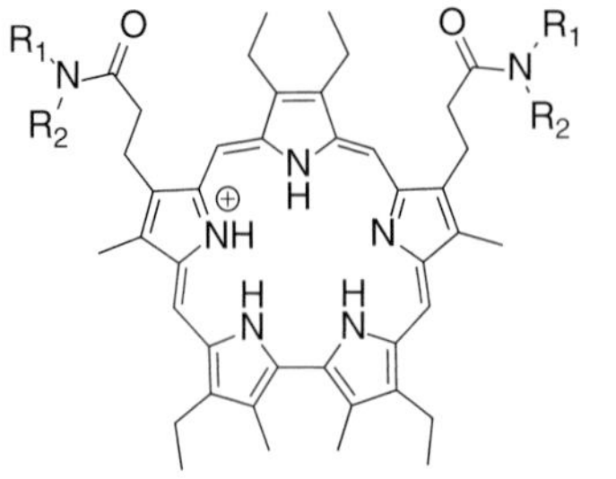

33

a $R_1 = R_2 = A$
b $R_1 = H, R_2 = B$
c $R_1 = H, R_2 = C$
d $R_1 = H, R_2 = D$
e $R_1 = H, R_2 = E$
f $R_1 = CH_3, R_2 = F$

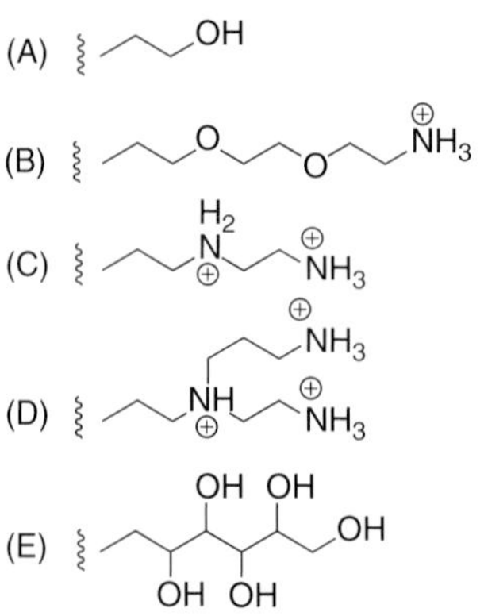

Figure 33 Various derivatives of sapphyrin, **33**, have been developed for the fluorescent sensing of phosphate in aqueous media.

a Soret band at around 410 nm, which was ascribed to higher order aggregates. On dilution, these higher aggregates break into dimers, which are characterized by a red-shifted Soret band at 420 nm. Further dilution results in a Soret band at 450 nm, characteristic of the monomeric species. The deaggregation phenomenon was also studied in methanol, and likewise explored as a function of sapphyrin structure.

Král *et al.* demonstrated that sapphyrin, an all-organic receptor for phosphate, could also be used as a catalyst for phosphate diester hydrolysis.[65] Because nucleophilic displacement plays an important role in this particular cleavage reaction, groups that are able to act as nucleophiles were appended to the periphery of the sapphyrin macrocycle (Figure 34). Binding of an appropriate phosphodiester, bis(4-nitrophenyl)phosphate (BNPP), was established in the solid state through X-ray diffraction analysis and in solution through ^{31}P NMR spectroscopic titrations. Rates of hydrolysis were measured by monitoring the amount of *p*-nitrophenolate (the hydrolysis product of BNPP and an anion characterized by a distinct UV–vis absorption at 400 nm) produced as a function of time. The analogous porphyrin complexes were also synthesized and studied as controls.

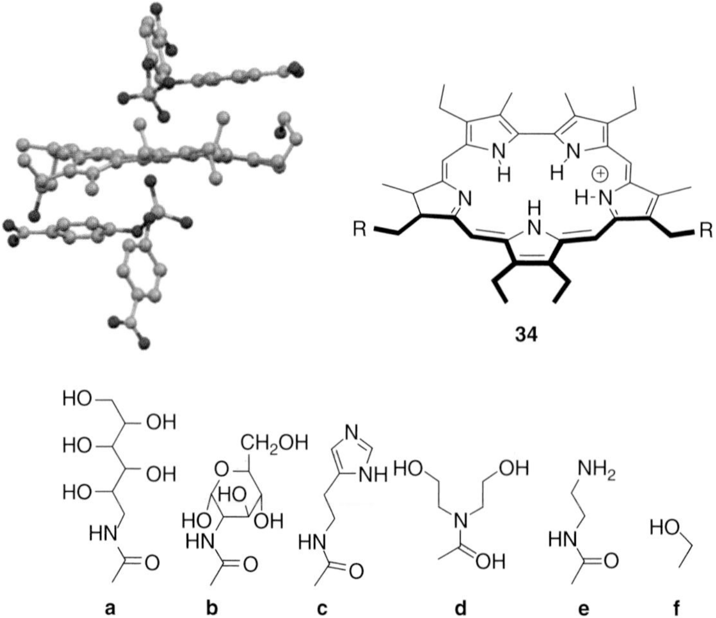

Figure 34 X-ray crystal structure of a sapphyrin (**34f**) bis(4-nitrophenyl)phosphate BNPP complex (CCDC# 296889). Hydrogens have been omitted for clarity.

A new sapphyrin derivative, dioxabenzosapphyrin, **36**, was synthesized by Sessler *et al.* with the goal of understanding the role that the pyrrolic protons play in anion binding.[66] In this system, two oxygens replace the nitrogens of the original bipyrrolic unit. An X-ray crystal structure was solved; it revealed a bound tosylate counteranion, which is held closest to the portion of the macrocycle via interaction with the three NH groups (Figure 35). When this species is subject to a titration of fluoride anion in the form of the TBA salt, the UV–vis spectrum undergoes changes consistent with a 1 : 1 binding stoichiometry. Such changes are similar to what is seen in the original pentaaza sapphyrin "parent." However, the calculated association constant, K_a, was found to be reduced by a factor of 170, reflecting presumably the difference between three NH binding sites as opposed to five. Other key findings were that chloride anion was still bound by this macrocycle, but in the presence of CH_3OH, competition with the solvent was appreciable. This was not the case when the all-nitrogen analog, **35**, was tested as a control; in this case, the chloride anion was bound with a clean 1 : 1 stoichiometry.[66]

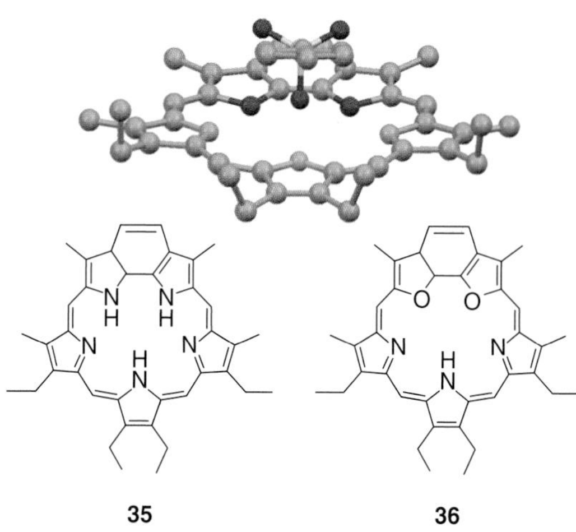

Figure 35 Rigidified sapphyrins that are prepared for the study of the anion-binding properties and designed to explore the question of what happens when two NHs are replaced by O. The X-ray crystal structure (CCDC# 703177) shows a bound tosylate that lies closer to the NH "bottom" of the macrocycle. Hydrogens have been omitted for clarity.

Cyclo[8]pyrrole, **2**, is a large aromatic expanded porphyrin reported by Sessler *et al.* in 2002. It is highly basic and is thus doubly protonated under a wide range of conditions. The result is a doubly positively charged system that shows great potential for anion binding.[67] It is speculated that anions play an important role as templates during synthesis, as the product distribution is highly dependent on the choice of acid in the ferric-ion-mediated oxidative couplings used to produce this system and its smaller congeners. The initial 2002 paper reports the use of sulfuric acid as the acid and the formation of the sulfate anion-complexed form of diprotonated cyclo[8]pyrrole as the major product (Figure 4). A follow-up paper in 2003 reported that, when hydrochloric acid was used instead of

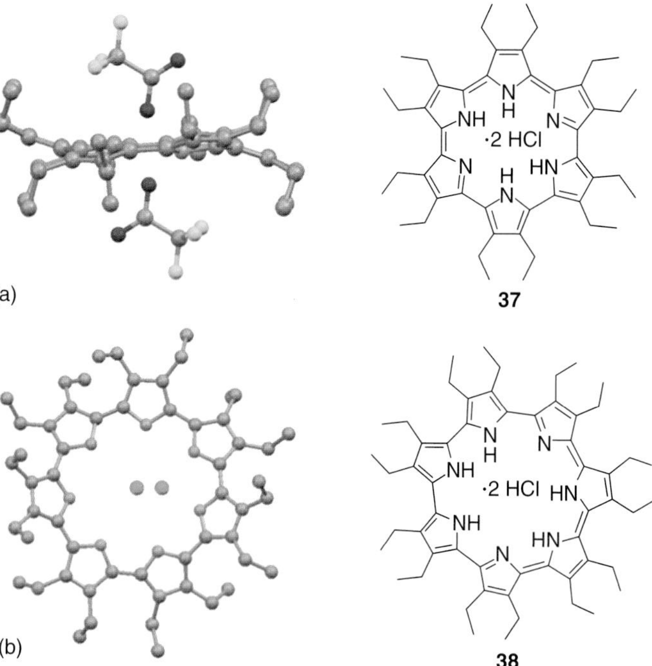

Figure 36 X-ray crystal structures of (a) cyclo[6]pyrrole·2TFA (CCDC# 213326) and (b) cyclo[7]pyrrole·2HCl (CCDC# 213327). Hydrogens have been omitted for clarity.

sulfuric acid, three products; namely, cyclo[6], cyclo[7], and cyclo[8]pyrrole, are formed in 15, 5 and 25% yields, respectively; these species were isolated as the respective bisHCl salts (Figure 36).[68]

Because of the anion complexation observed in the solid state, it was proposed that cyclo[8]pyrrole could function as an anion extractant, specifically for sulfate. Sulfate receptors that can act as extractants of this ion are highly desirable because sulfate is a problematic species in the vitrification process that is proposed for the disposal of certain radioactive wastes.[69] The original reported short-chained forms of cyclo[8]pyrrole presented solubility issues, but a newer derivative, octamethyl-octaundecylcyclo[8]pyrrole, **2b**, originally developed as a precursor for liquid crystals, proved to be amenable to extraction studies.[70] It was found that **2b** was able to selectively extract sulfate in the presence of high levels of nitrate. This cyclo[8]pyrrole was thus able to overcome the so-called Hofmeister bias or the inherent propensity for nitrate to partition before sulfate. While the kinetics are slow—reducing utility in the context of near-term applications—this is the first example where this level of selectivity is seen in a sulfate versus nitrate extraction experiment.[69]

4.1 Anion-binding electron transfer systems

Inspired by previous reports wherein both anion binding and an expanded porphyrin, sapphyrin, were used to successfully create an ET system, cyclo[8]pyrrole (**C8**), was investigated for potential utility as a component in the noncovalent assembly of an ET dyad.[71] It was found that the bis HCl salt of **C8** could bind carboxylate anions; this in turn allowed for the binding of a pyrene functionalized with a carboxylate to form the putative ET dyad shown in Figure 37.[72] On photoexcitation of the pyrene, ET occurs from **C8** to the pyrene. The lifetime of the charge-separated

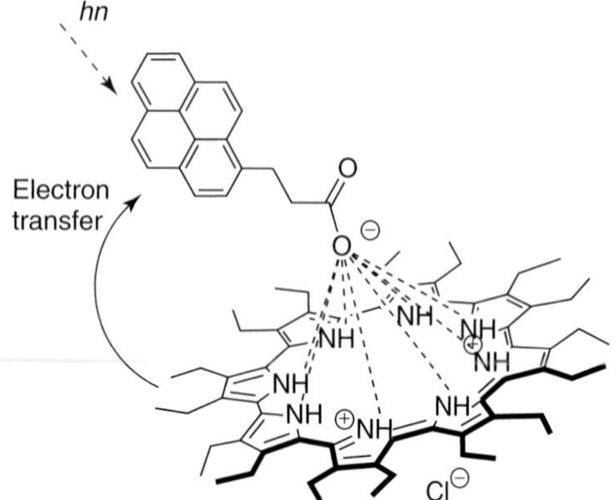

Figure 37 A carboxylate anion-tethered pyrene–cyclo[8]pyrrole ensemble.

state was determined by the following decay of the optical signature for the radical cation of **C8** via the use of nanosecond laser flash photolysis and transient absorption spectroscopy.[72] The rate of ET could also be determined by measuring the fluorescence lifetime of the pyrene subunit, both on its own and in the presence of cyclo[8]pyrrole.

A noncovalent ET system based on both protonated sapphyrins and fullerenes functionalized with dendrimers that were capped with multiple carboxylate groups was recently reported by Guldi, Sessler, and coworkers. The ability of the sapphyrin to bind carboxylates allowed for the formation of a donor–acceptor dyad with 1 : 1 (Sap/C_{60}–1) or 1 : 2 (Sap/C_{60}–2) (fullerene to macrocycle) stoichiometries (Figure 38). Under photoirradiation, intraensemble ET from the sapphyrin to the fullerene was observed.[73] It is worth noting that the lifetime of the charge-separated state is longer in the case of Sap/C_{60}–2 than that in the case of Sap/C_{60}–1; this was rationalized in terms of the two sapphyrin moieties being better able to delocalize the resulting charge than a single sapphyrin as in Sap/C_{60}–1.

Another recent noncovalent ET system, developed by D'Souza, is predicated on the anion-binding ability of oxoporphyrinogens (OPs), **39** and **40**. Although not porphyrins technically, the compounds have an extended π-system and are highly colored species (Figure 39). These systems also happen to be highly sensitive to anion coordination, especially by fluoride. Thus, even when two internal nitrogens were alkylated, as in **40**, there was still a strong chromogenic response to anions.[74] D'Souza and coworkers proposed incorporation of this OP into a supramolecular ET donor–acceptor ensemble with two distinct binding sites. One site like Zn-porphyrins is composed of "jaws" for the binding of bis(4-pyridyl)-substituted fullerene (Py_2C_{60}) and the second is the OP which can act as a receptor for anions (Figure 40).[75] It was found that complexation of anions by the OP resulted in a lower oxidation potential (by ~600 mV). This allows for stabilization of the purported Zn-porphyrin radical cation ($ZnP^{\bullet +}$) that would be produced following photoexcitation—essentially by sharing the positive charge. This effect should lower the driving force for charge recombination. The binding constant leading to the formation of ensemble **OP**(F^-)·Py_2C_{60} was measured through UV–vis and ^{1}H NMR spectroscopies; on this basis, a $K_a = 7.4 \times 10^4\ M^{-1}$ was derived in *o*-dichlorobenzene. Femtosecond and nanosecond laser flash photolysis experiments were carried out for **OP**·Py_2C_{60}, **OP**(F^-)·Py_2C_{60}, and **OP**(X^-)·Py_2C_{60} (X^- being anions other than fluoride) to determine lifetimes of the corresponding charge-separated states. It was found that, compared to other anions, fluoride resulted in a 90~fold enhancement in the lifetime of the charge-separated state produced from **OP**·Py_2C_{60}.

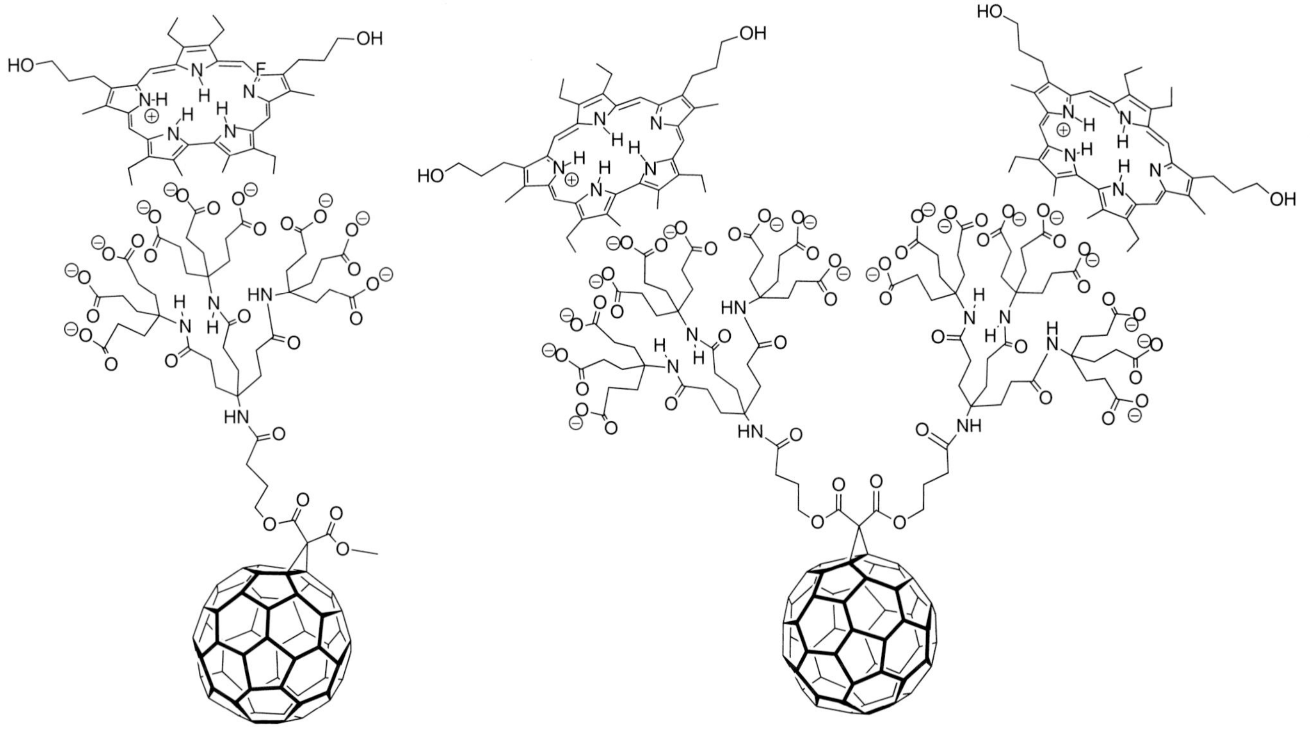

Sap / C_{60}-1

Sap / C_{60}-2

Figure 38 Fullerenes appended with dendrimers and capped with anionic carboxylate groups for the formation of complexes with sapphyrin and the construction of electron transfer dyads.

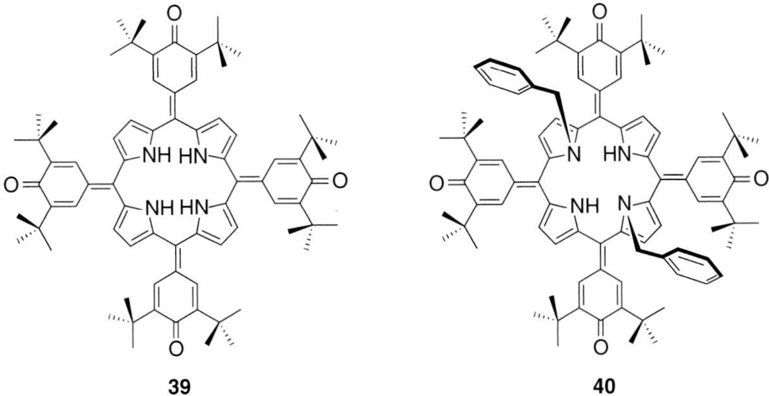

Figure 39 Structure of oxoporphyrinogens.

5 SENSING APPLICATIONS

While not a focus of this chapter *per se*, metalated porphyrins have a near unique ability to coordinate guests in the open axial positions, which produces what is often a ligand-specific spectral shift. This makes metalloporphyrins appealing as sensors. Recently, considerable effort has been made to exploit this trait, and some of the ground-breaking works in an application sense has been done by the Suslick group.[6] These researchers specialize in making "arrays," wherein a variety of chromogenic sensors, including metalloporphyrins, are immobilized onto a platform. The platform is then exposed to a "mixture" of analytes yielding a fingerprint. Typically, solutions of metalated TPPs are transferred onto reverse phase silica chromatography plates, which are then held fixed inside of a scanner. Initially, the plates are scanned, then exposed to vapors, and finally, scanned again. The color differential between the original and exposed plates generally gives a unique fingerprint for a particular analyte or mixture of analytes. This type of

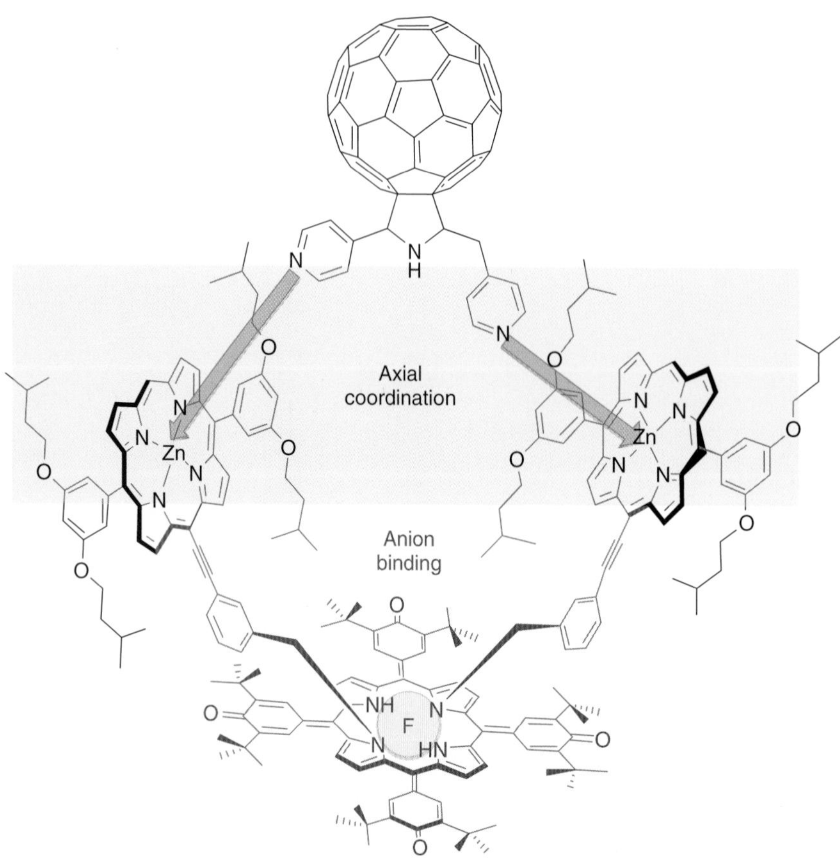

Figure 40 Oxoporphyrinogens that, although not technically expanded porphyrins, are characterized by an extended π-system. The result is highly colored species that are also highly sensitive to anion complexation.

approach to sensing is quite useful, as it circumnavigates the need to analyze a specific molecule, on an individual basis. The Suslick group colorimetric arrays were the first to be designed to permit the detection of volatile organic compounds, and, since the first *Nature* report by the group in 2000, have been used for unique analytical applications such as "fingerprinting" coffee aromas and the detection of toxic gases.[76,77]

Another sensing system in which porphyrins have found success involves protein recognition. Earlier in this chapter, Hamilton's use of porphyrins for cyctochrome *c* binding was described. Subsequently, Hamilton and his group have expanded on this discovery and have developed protein-detecting arrays based on functionalized TPPs as seen in **41**.[53,78] The protein binding event is easy to analyze as the TPPs are highly fluorescent. In analogy to what is expected of Suslick's systems, the Hamilton arrays may provide a fast and cheap means by which to quickly identify a mixture of highly complex protein components.

The Hamilton group first studied the viability of this concept by studying a small library of 35 different TPP derivatives, as shown in Figure 41.[53] This study was carried out by mixing two of the five possible combinations of amino acids with TPP, **41**, which are shown. An 8×5 array of quartz wells was then filled—eight rows were filled with eight different TPPs. All wells showed strong fluorescence. Next, four proteins, selected so as to cover a wide range of surface characteristics, were incubated with each of the eight porphyrins. This was done by filling one column per protein, leaving one column free as a control. In accord with expectations, the four columns yielded a distinctive response or fingerprint for a particular protein. Thus, by using only a relatively simple array of TPPs, modified only in the peripheral positions, unambiguous determination of proteins was achieved.

The Hamilton group has recently developed a more precise, more sensitive array that can be used to identify metal-containing proteins, metal-free proteins, and mixtures of these two classes.[78] In this case, pattern recognition techniques, namely, principal component analysis (PCA), which generates clusters in a 3D Euclidean space from the unique fingerprints of large arrays, were used. It was found that unambiguous identification could be achieved, and the concentration of the protein in question could be estimated in some cases. The resolving power also became much greater when the array of porphyrins was increased

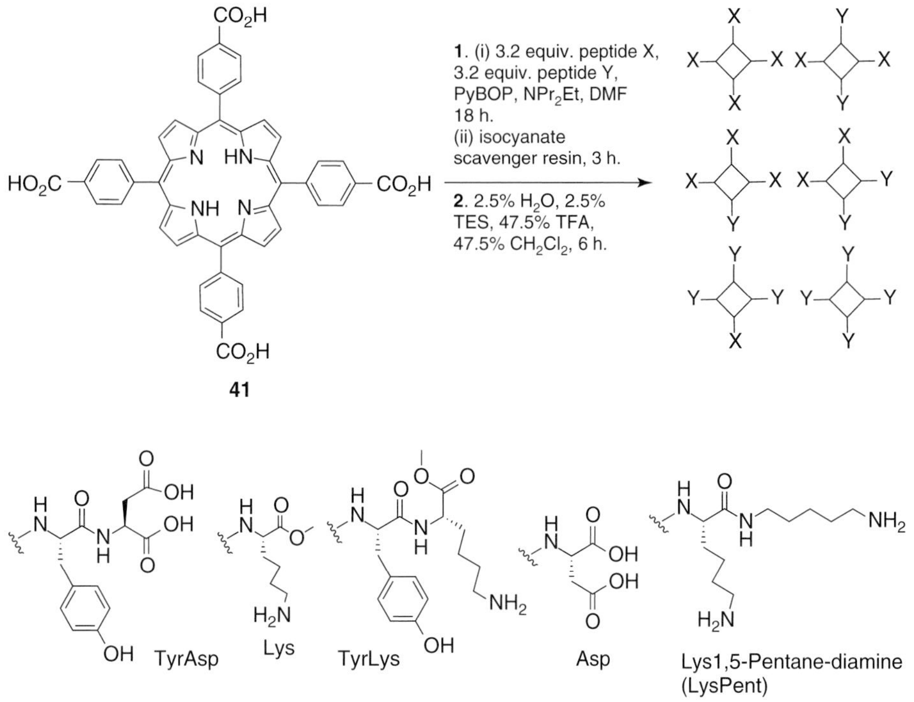

Figure 41 A synthetic tetra-*meso*-carboxylphenylporphyrin (TCPP) serves as the receptor core for a library of conjugates. The TCPP core was conjugated with amino acids/amino acid derivatives and was shown to exhibit differential binding properties to proteins with varying surface characteristics.

from 8 to 16. This especially helped with the resolution of nonmetalated proteins, which are generally hard to distinguish from one another. Metalated porphyrins such as cytochrome *c* and ferridoxin have paramagnetic iron, which serves to quench porphyrin fluorescence. Nonmetalated proteins can still affect fluorescence through electrostatics, or energy transfer between amino acid residues such as tryptophan and tyrosine, albeit to a much lesser extent. These recent developments are attractive and could lead to approaches to protein sensing that are far less laborious than current methods, which generally require the specific labeling of proteins with a chromophore.

6 CONCLUSIONS AND FUTURE OUTLOOK

An illustrative spectrum of porphyrins and expanded porphyrins, as well as examples of practical (or at least interesting!) uses of porphyrins in receptor design and host–guest chemistry, has been presented. The hope is that this survey has helped underscore a large number of opportunities that await those who take advantage of the porphyrin scaffold or who are keen to develop or exploit expanded porphyrins. It is expected that continued developments, perhaps sparked by readers of this chapter, will have an important impact on the timely fields of environmental chemistry, medical analysis, and alternative energy applications.

REFERENCES

1. L. R. Milgrom, *The Colours of Life: An Introduction to the Chemistry of Porphyrins and Related Compounds*, Oxford University Press, Oxford, 1997.
2. *The Porphyrin Handbook: Synthesis and Organic Chemistry*, Vol. 1, Academic Press, San Diego, 2000.
3. P. Even and B. Boitrel, *Coord. Chem. Rev.*, 2006, **250**, 519.
4. I. Beletskaya, V. S. Tyurin, A. Y. Tsivadze, *et al.*, *Chem. Rev.*, 2009, **109**, 1659.
5. E. V. Anslyn, *J. Org. Chem.*, 2007, **72**, 687.
6. N. A. Rakow and K. S. Suslick, *Nature*, 2000, **406**, 710.
7. H. Ogoshi, T. Mizutani, T. Hayashi, and Y. Kuroda, in *The Porphyrin Handbook*, eds K. M. Kadish, K. M. Smith, and R. Guilard, Academic Press, San Diego, 2000, Vol. 6, p. 279.
8. K. M. Kadish, K. M. Smith, and R. Guilard, *The Handbook of Porphyrin Science, Synthetic Methodology*, Vol. 3, World Scientific, Hackensack, 2010.
9. R. Misra and T. K. Chandrashekar, *Acc. Chem. Res.*, 2008, **41**, 265.
10. J. P. Collman, R. R. Gagne, T. R. Halbert, *et al.*, *J. Am. Chem. Soc.*, 1973, **95**, 7868.
11. C. Lee, D. H. Lee, and J.-I. Hong, *Tetrahedron Lett.*, 2001, **42**, 8665.

12. P. D. Beer, D. P. Cormode, and J. J. Davis, *Chem. Commun.*, 2004, 414.
13. D. P. Cormode, S. S. Murray, A. R. Cowley, and P. D. Beer, *Dalton Trans.*, 2006, 5135.
14. D. P. Cormode, M. G. B. Drew, R. Jagessar, and P. D. Beer, *Dalton Trans.*, 2008, 6732.
15. M. Gouterman, in *The Porphyrins: Physical Chemistry, Part A*, ed D. Dolphin, Academic Press, New York, 1978, Vol. 3, p. 1.
16. F. D'Souza and O. Ito, in *Handbook of Porphyrin Science*, eds K. M. Kadish, K. M. Smith, R. Guilard, World Scientific, Hackensack, 2010, Vol. 1, p. 307.
17. B. Grimm, A. Hausmann, A. Kahnt, *et al.*, in *Handbook of Porphyrin Science*, eds. K. M. Kadish, K. M. Smith, and R. Guilard, World Scientific, Hackensack, 2010, Vol. 1, p. 133.
18. D. M. Guldi, *Chem. Soc. Rev.*, 2002, **31**, 22.
19. P. D. W. Boyd, M. C. Hodgson, C. E. F. Rickard, *et al.*, *J. Am. Chem. Soc.*, 1999, **121**, 10487.
20. M. M. Olmstead, D. A. Costa, K. Maitra, *et al.*, *J. Am. Chem. Soc.*, 1999, **121**, 7090.
21. P. D. W. Boyd and C. A. Reed, *Acc. Chem. Res.*, 2005, **38**, 235.
22. D. Sun, F. S. Tham, C. A. Reed, *et al.*, *J. Am. Chem. Soc.*, 2000, **122**, 10704.
23. A. Hosseini, S. Taylor, G. Accorsi, *et al.*, *J. Am. Chem. Soc.*, 2006, **128**, 15903.
24. Z.-Q. Wu, X.-B. Shao, C. Li, *et al.*, *J. Am. Chem. Soc.*, 2005, **127**, 17460.
25. K. Tashiro, T. Aida, J.-Y. Zheng, *et al.*, *J. Am. Chem. Soc.*, 1999, **121**, 9477.
26. J.-Y. Zheng, K. Tashiro, Y. Hirabayashi, *et al.*, *Angew. Chem., Int. Ed*, 2001, **40**, 1857.
27. Y. Shoji, K. Tashiro, and T. Aida, *J. Am. Chem. Soc.*, 2004, **126**, 6570.
28. J. Xiao, M. R. Savina, G. B. Martin, *et al.*, *J. Am. Chem. Soc.*, 1994, **116**, 9341.
29. Y. Shoji, K. Tashiro, and T. Aida, *J. Am. Chem. Soc.*, 2010, **132**, 5928.
30. L. H. Tong, J.-L. Wietor, W. Clegg, *et al.*, *Chem. - Eur. J.*, 2008, **14**, 3035.
31. G. Gil-Ramirez, S. D. Karlen, A. Shundo, *et al.*, *Org. Lett.*, 2010, **12**, 3544.
32. A. Takai, M. Chkounda, A. Eggenspiller, *et al.*, *J. Am. Chem. Soc.*, 2010, **132**, 4477.
33. Y. Kubo, A. Sugasaki, M. Ikeda, *et al.*, *Org. Lett.*, 2002, **4**, 925.
34. D. Sun, F. S. Tham, C. A. Reed, and P. D. W. Boyd, *Proc. Natl. Acad. Sci.*, 2002, **99**, 5088.
35. H. Murakami, T. Nomura, and N. Nakashima, *Chem. Phys. Lett.*, 2003, **378**, 481.
36. S. I. Pascu, N. Kuganathan, L. H. Tong, *et al.*, *J. Mater. Chem.*, 2008, **18**, 2781.
37. D. M. Guldi, H. Taieb, G. M. A. Rahman, *et al.*, *Adv. Mater.*, 2005, **17**, 871.
38. K. S. Chichak, A. Star, M. V. P. Altoe, and J. F. Stoddart, *Small*, 2005, **1**, 452.
39. H. Li, B. Zhou, Y. Lin, *et al.*, *J. Am. Chem. Soc.*, 2004, **126**, 1014.
40. P. J. Boul, D.-G. Cho, G. M. A. Rahman, *et al.*, *J. Am. Chem. Soc.*, 2007, **129**, 5683.
41. J. Chen and C. P. Collier, *J. Phys. Chem. B*, 2005, **109**, 7605.
42. S. H. Jung and H.-J. Kim, *J. Porphyr. Phthalocyanines*, 2008, **12**, 109.
43. A. Sugasaki, K. Sugiyasu, M. Ikeda, *et al.*, *J. Am. Chem. Soc.*, 2001, **123**, 10239.
44. E. Kalenius, J. Koivukorpi, E. Kolehmainen, and P. Vainiotalo, *Eur. J. Org. Chem.*, 2010, 1052.
45. O. Rusin, K. Lang, and V. Kral, *Chem. Eur. J.*, 2002, **8**, 655.
46. W. B. Lu, L. H. Zhang, and X. S. Ye, *Sens. Actuators, B*, 2006, **B113**, 354.
47. V. Kral, F. P. Schmidtchen, K. Lang, and M. Berger, *Org. Lett.*, 2001, **4**, 51.
48. J. Kralova, J. Koivukorpi, Z. Kejik, *et al.*, *Org. Biomol. Chem.*, 2008, **6**, 1548.
49. K. Ladomenou and R. P. Bonar-Law, *Chem. Commun.*, 2002, 2108.
50. T. D. James, K. R. A. S. Sandanayaka, and S. Shinkai, *Angew. Chem., Int. Ed.*, 1996, **35**, 1910.
51. O. Hirata, Y. Kubo, M. Takeuchi, and S. Shinkai, *Tetrahedron*, 2004, **60**, 11211.
52. T. Aya and A. D. Hamilton, *Bioorg. Med. Chem. Lett.*, 2003, **13**, 2651.
53. L. Baldini, A. J. Wilson, J. Hong, and A. D. Hamilton, *J. Am. Chem. Soc.*, 2004, **126**, 5656.
54. S. Tanimoto, S. Matsumura, and K. Toshima, *Chem. Commun.*, 2008, 3678.
55. M. Balaz, N. M. De, A. E. Holmes, *et al.*, *Angew. Chem., Int. Ed.*, 2005, **44**, 4006.
56. A. D'Urso, A. Mammana, M. Balaz, *et al.*, *J. Am. Chem. Soc.*, 2009, **131**, 2046.
57. I. M. Dixon, F. Lopez, J. P. Estève, *et al.*, *ChemBioChem*, 2005, **6**, 123.
58. C. Vialas, G. Pratviel, and B. Meunier, *Biochemistry*, 2000, **39**, 9514.
59. I. M. Dixon, F. Lopez, A. M. Tejera, *et al.*, *J. Am. Chem. Soc.*, 2007, **129**, 1502.
60. (a) J. L. Sessler and J. M. Davis, *Acc. Chem. Res.*, 2001, **34**, 989; (b) J. L. Sessler, S. Camiolo, and P. A. Gale, *Coord. Chem. Rev.*, 2003, **240**, 17.
61. M. Shionoya, H. Furuta, V. Lynch, *et al.*, *J. Am. Chem. Soc.*, 1992, **114**, 5714.
62. H. Furuta, M. J. Cyr, and J. L. Sessler, *J. Am. Chem. Soc.*, 1991, **113**, 6677.
63. K. Calderon-Kawasaki, S. Kularatne, Y. H. Li, *et al.*, *J. Org. Chem.*, 2007, **72**, 9081.
64. J. L. Sessler, J. M. Davis, V. Kral, *et al.*, *Org. Biomol. Chem.*, 2003, **1**, 4113.

65. V. Kral, K. Lang, J. Kralova, *et al.*, *J. Am. Chem. Soc.*, 2006, **128**, 432.
66. D.-G. Cho, P. Plitt, S. K. Kim, *et al.*, *J. Am. Chem. Soc.*, 2008, **130**, 10502.
67. D. Seidel, V. Lynch, and J. L. Sessler, *Angew. Chem., Int. Ed.*, 2002, **41**, 1422.
68. T. Kohler, D. Seidel, V. Lynch, *et al.*, *J. Am. Chem. Soc.*, 2003, **125**, 6872.
69. L. R. Eller, M. Stepien, C. J. Fowler, *et al.*, *J. Am. Chem. Soc.*, 2007, **129**, 11020.
70. M. Stepien, B. Donnio, and J. L. Sessler, *Angew. Chem., Int. Ed.*, 2007, **46**, 1431.
71. S. L. Springs, D. Gosztola, M. R. Wasielewski, *et al.*, *J. Am. Chem. Soc.*, 1999, **121**, 2281.
72. J. L. Sessler, E. Karnas, S. K. Kim, *et al.*, *J. Am. Chem. Soc.*, 2008, **130**, 15256.
73. B. Grimm, E. Karnas, M. Brettreich, *et al.*, *J. Phys. Chem. B*, 2009, **114**, 14134.
74. J. P. Hill, A. L. Schumacher, F. D'Souza, *et al.*, *Inorg. Chem.*, 2006, **45**, 8288.
75. F. D'Souza, N. K. Subbaiyan, Y. Xie, *et al.*, *J. Am. Chem. Soc.*, 2009, **131**, 16138.
76. B. A. Suslick, L. Feng, and K. S. Suslick, *Anal. Chem.*, 2010, **82**, 2067.
77. S. H. Lim, L. Feng, J. W. Kemling, *et al.*, *Nat. Chem.*, 2009, **1**, 562.
78. H. Zhou, L. Baldini, J. Hong, *et al.*, *J. Am. Chem. Soc.*, 2006, **128**, 2421.

FURTHER READING

F. D'Souza and O. Ito, Tetrapyrrole-nanocarbon hybrids: self-assembly and photoinduced electron transfer, in *Handbook of Porphyrin Science*, eds. K. M. Kadish, K. M. Smith, R. Guilard, World Scientific, Hackensack, 2010, Vol. 1, pp. 307–437.

B. Grimm, A. Hausmann, A. Kahnt, *et al.*, Charge transfer between porphyrins and phthalocyanines and carbon nanostructures, in *Handbook of Porphyrin Science*, eds. K. M. Kadish, K. M. Smith, R. Guilard, World Scientific, Hackensack, 2010, Vol. 1, pp. 133–219.

Supramolecular Phthalocyanine-Based Systems

Christian G. Claessens[1], M. Victoria Martínez-Díaz[1], and Tomás Torres[1,2]

[1] *Universidad Autónoma de Madrid, Madrid, Spain*
[2] *IMDEA-Nanociencia, Madrid, Spain*

1 INTRODUCTION

In recent years, π-conjugated molecules have found many interesting applications in technological areas, including optical recording media and organic photovoltaics. In this regard, not only is the adequate choice of the π-conjugated molecular building block with particular physicochemical properties important, but also the control of the supramolecular organization of the molecular components, with the ultimate goal of extending or even improving the properties shown by individual molecules.[1] π-Conjugated molecules can be self-assembled as one-dimensional stacks by aggregation of their π-systems through $\pi-\pi$ interactions. In general, $\pi-\pi$ interactions have been described as a sum of noncovalent forces, namely, van der Waals interactions, electrostatic interactions, and solvophobic effect.[2]

Phthalocyanines (Pcs) are among the most charismatic functional π-conjugated molecules.[3] The propensity of Pcs to self-aggregate has been attributed to the strong $\pi-\pi$ interactions between the large, π-conjugated aromatic molecules and two-dimensional surface. Different studies have revealed the preferential association of the Pc rings to form dimers and the influence of the central metal ion on the aggregation properties of Pcs.[4] Moreover, Pc aggregation has been shown to be strongly solvent dependent.[5] Pcs exhibit a high propensity to dimerize in nonpolar solvents such as toluene, whereas they remain mainly monomeric in polar solvents such as dioxane or THF (tetrahydrofuran). Aggregation frequently causes insolubility of Pcs and can also dramatically affect the photochemical properties of these macrocycles.[6] For instance, aggregation produces a decrease in the absorption coefficient of Pcs and also induces a decrease in fluorescence quantum yield.[7] On the other hand, the formation of well-defined Pc aggregates is useful in material applications where the self-assembly of Pc cores in close proximity is beneficial.[8]

The process of Pc aggregation can be easily probed using electronic spectroscopy. Typically, monomeric Pcs exhibit strong absorption bands in the 300 and 700 nm spectral

Supramolecular Chemistry: From Molecules to Nanomaterials.
Edited by Philip A. Gale and Jonathan W. Steed.
© 2012 John Wiley & Sons, Ltd. ISBN: 978-0-470-74640-0.

regions, termed as B and Q bands, respectively. However, the presence of Pc aggregates can be observed in the Q band region of the electronic absorption spectrum: Frequently, Pc aggregation results in a decrease in intensity of the Q band corresponding to the monomeric species; meanwhile a new, broader, and blue-shifted band, around 630 nm, increases its intensity. This shift to lower wavelengths corresponds to the formation of H-type or face-to-face aggregates. More rarely, Pc aggregation causes red-shifting of the Q bands, attributed to the formation of J-type or head-to-tail aggregates.

2 FROM MOLECULES TO PHTHALOCYANINE $\pi-\pi$ AGGREGATES

One of the pioneering and more interesting applications of supramolecular assemblies made of π-conjugated molecules is the formation of well-aligned discotic liquid crystals made of functional molecules.

Pcs bearing long flexible hydrocarbon chains are known to form discotic mesophases, generally with a hexagonal columnar structure.[9] Very high charge carrier mobilities (about $0.5\,\mathrm{cm^2\,V^{-1}\,s^{-1}}$) have been found in the highly ordered liquid-crystalline phase of discotic molecules such as triphenylenes, hexabenzocoronenes, or Pcs, which have been used as anisotropic transport channels since the hopping of excitons and charge carriers is facilitated along the direction of the π-stacking.[10] This type of supramolecular organization was extensively studied in the 1990s and is out of the scope of this chapter.

Pcs have also been widely employed as electron donors in donor–acceptor (D–A) systems in which photoinduced electron transfer (PET) from the donor to the acceptor unit can efficiently take place as demonstrated by solution experiments.[11] Moreover, significant changes in the photophysical properties (e.g., lifetime of the charge-separated state) have been demonstrated in organized Pc-based D–A systems with respect to the monomeric species.

Wasielewski and coworkers demonstrated that ultrafast energy transfer can take place in ZnPc-perylenediimide (PDI)-based π-stacked aggregates.[12] Tetra-PDI-substituted ZnPc **1** self-assembles in solution forming stacked heptamers, as evidenced by SAXS/WAXS, and forms long fibrous structures in the solid state (Figure 1). Selective excitation of the PDI component results in ultrafast energy transfer (1.3 ps) from the stacked peripheral PDI chromophores to the central ZnPc stack. Subsequently, exciton hopping between the ZnPc units occurs in the femtosecond timescale.

Another example where photophysical properties were modulated by self-assembly was reported in 2006 by Torres, Echegoyen, and Guldi by employing D–A Pc–anthraquinone (AQ) triads.[13] Pc–AQ–Pc molecules **2** and **3** adopt an all-planar π-extended conformation, allowing for an efficient π-conjugation between the active units across the entire system and the formation of J-type oligomers

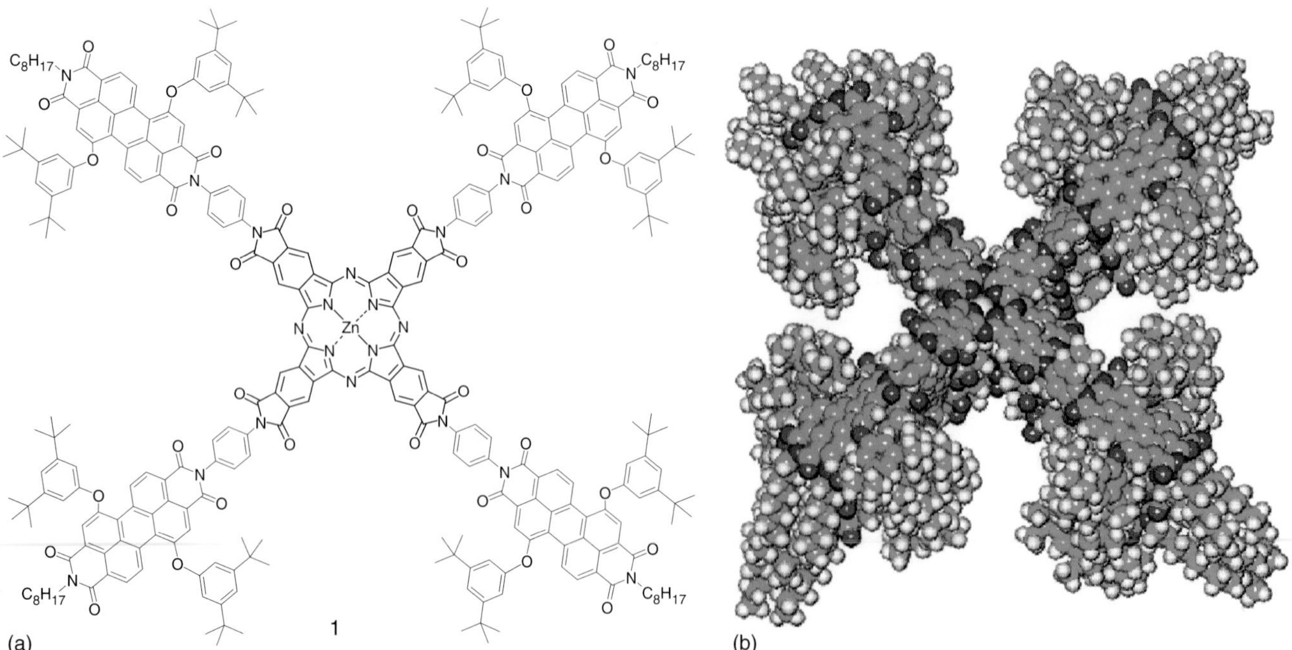

Figure 1 ZnPc-perylenediimide **1** (a) and its π-stacked aggregates (b). (Reproduced from Ref. 12. © American Chemical Society, 2004.)

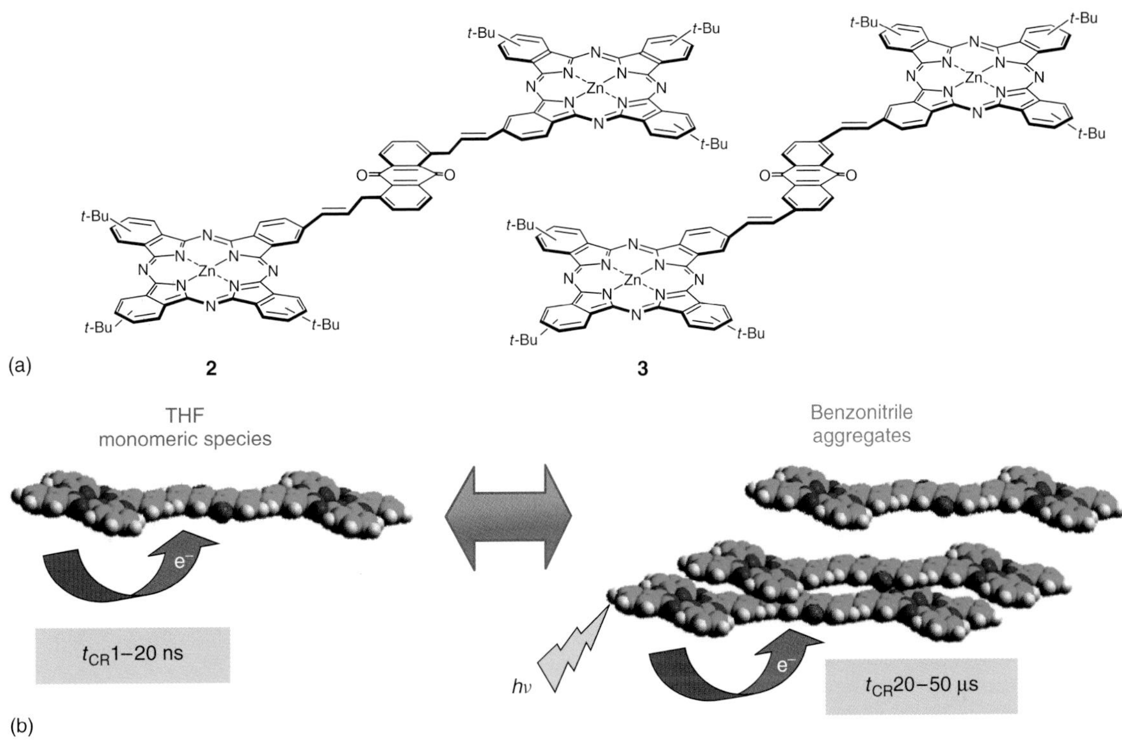

Figure 2 Donor–acceptor Pc–anthraquinone triads **2** and **3** (a). Effect of the aggregation on the charge recombination time (b). (Reproduced from Ref. 13. © American Chemical Society, 2006.)

by intermolecular $\pi-\pi$ stacking interaction in aromatic solvents such as benzonitrile (Figure 2). The formation of such aggregates was crucial for increasing the lifetime of the photoinduced charge-separated state by more than three orders of magnitude with respect to the values obtained for the same molecules in polar solvents such as THF, where they remain mainly monomeric as confirmed by UV–vis spectroscopy.

An even more pronounced effect in the stabilization of the photoinduced charge-separated state as a direct consequence of the supramolecular organization was reported for an amphiphilic Pc–C_{60} dyad salt **4**[14] (Figure 3). This dyad is able to form, because of a combination of solvophobic and $\pi-\pi$ stacking interactions, perfectly ordered 1D Pc–C_{60} nanotubules when dispersed in water, as demonstrated by transmission electron microscopy (TEM) analysis. A stabilization of about six orders of magnitude in the charge-separated lifetime was obtained for the supramolecular nanotubules with respect to the nonorganized Pc–C_{60} precursor of **4**.

More interestingly, the self-organization of Pc–C_{60} dyads on surfaces[15] (by drop-casting technique) and in liquid crystals[16] has also been reported. High electrical conductivity values were measured by conductive-AFM for some of these Pc-containing nanostructures, pointing to an evident relationship between the electrical properties and the supramolecular order of the Pc–C_{60} conjugates.

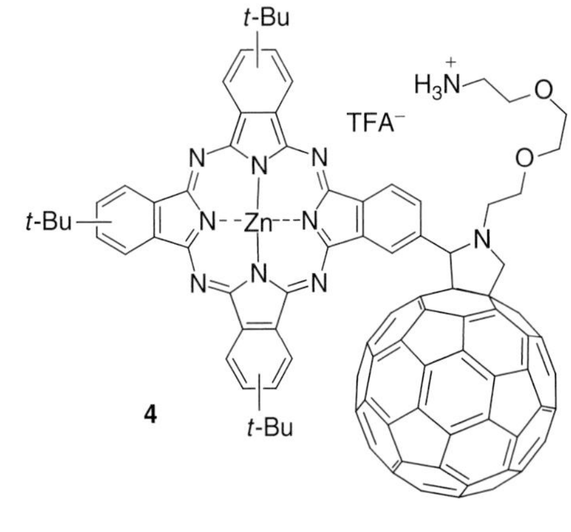

Figure 3 Pc–C_{60} dyad salt **4**.

3 DONOR–ACCEPTOR INTERACTIONS IN PHTHALOCYANINE AGGREGATES

D–A interactions between aromatic π-conjugated molecular components can be considered as a particular case of $\pi-\pi$ interactions, wherein the individual molecular components have different π-electron densities, with one of them being π-electron rich and the other being π-electron poor.[17] Therefore, it can be inferred that electrostatic forces would have a prominent contribution in D–A interactions.

The D–A recognition motif has been widely used for the construction of supramolecular ensembles: from the numerous and outstanding architectures published by the group of Stoddart based on the D–A interactions between dioxybenzene or dioxynaphthalene and bipyridinium complementary units,[18] to the extensive utilization of the tetrathiafulvalene (TTF) as π-electron donor building block,[19] or the incorporation of the PDI unit as π-electron acceptor unit in supramolecular systems. However, D–A π-interactions have not been explored in Pc systems until very recently.

In 2003, the groups of Torres and Nolte reported for the first time the self-assembly of a bis(phthalocyanine) molecule **5**, mainly driven by D–A interactions.[20] The molecular building block (Figure 4), which contained a π-electron donor component (namely, a Pc substituted with six electron donor alkoxy sustituents) and a complementary π-electron-poor counterpart, namely, a Pc substituted with six electron-withdrawing alkylsulfonyl substituents, underwent hetero-association in both solution and solid state. UV–vis dilution experiments of the bisPc **5** carried out from toluene solutions yielded an association constant of $3 \times 10^9\,M^{-1}$. Moreover, the formation of supramolecular aggregates was also confirmed by TEM studies, which showed the presence of 1D nanoaggregates, extremely monodisperse in both shape and size.

The same D–A recognition motif was further exploited by Torres and Guldi in the construction of photoactive supramolecular Pc–C_{60} triads.[21] Hetero-association between the complementary electron-deficient Pc **6** and electron-rich Pc **7**, bearing also a covalently linked C_{60} unit, was confirmed by different techniques, which provided evidence for the formation of a 1 : 1 D–A complex with a stability constant of about $10^5\,M^{-1}$ in chloroform (Figure 5). Moreover, the radical-ion-pair

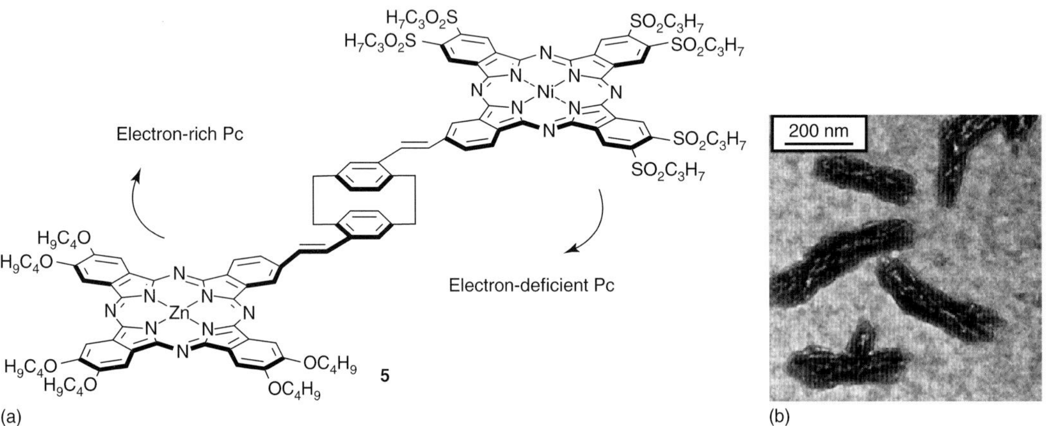

Figure 4 Bisphthalocyanine **5** (a) and transmission electron microscopy studies showing 1D nanoaggregates. (Reproduced from Ref. 20. © American Chemical Society, 2003.)

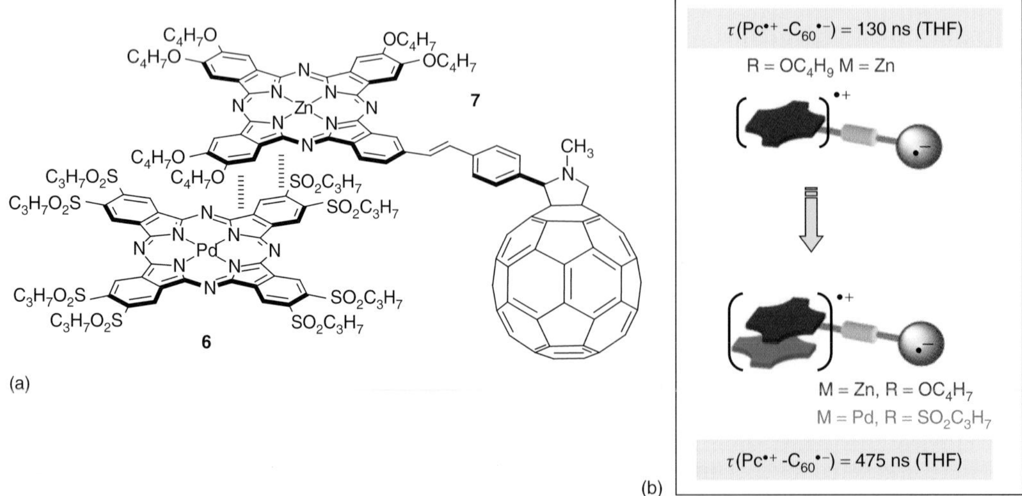

Figure 5 Heterosupramolecular triad formed by donor–acceptor interactions between Pc **6** and Pc–C_{60} dyad **7** (a). Schematic representation of the influence of the donor–acceptor interactions on the lifetime of the photoinduced charge-separated state.

formed on irradiating the supramolecular ensemble showed a lifetime that is appreciably longer (four times) than the one obtained for the corresponding ZnPc–C_{60} dyad **7**. The results were interpreted as a beneficial effect of the electronic coupling between the donor and the acceptor Pcs, which is expected to help in the delocalization of the radical cation over the two Pc moieties.

Intermolecular interactions between Pc moieties and electron-donor or electron-acceptor units other than Pcs have also been exploited in order to promote the organization in Pc-based systems. One of the examples has been reported by Amabilino, Rowan, and Nolte, who prepared a TTF-crown ether-substituted Pc **8** (Figure 6a), which was able to self-organize in solution mainly because of intermolecular interactions between the TTF and the Pc moieties as inferred from UV–vis studies.[22] TEM analysis of the gel formed by slow addition of dioxane to a chloroform solution of **8** revealed the formation with an equal distribution of both left and right helical fibers, approximately 20 nm long (Figure 6b).

D–A liquid-crystalline blends—made of liquid-crystalline Pc discotic mesogens such as **9**, as π-electron donors, and the liquid-crystalline perylenetetracarboxidiimide derivative **10**, as a π-electron acceptor (Figure 7)—have been reported by Zucchi and Lazzoroni.[23] The two components are fully miscible in blends containing at least 60 mol% of the Pc component **9**, whereas phase separation is systematically observed in blends containing less than 40% of this mesogenic Pc. The homogeneous D–A blends form a columnar hexagonal mesophase exhibiting enhanced thermal stability (over 100 °C) in comparison with the corresponding mesophase

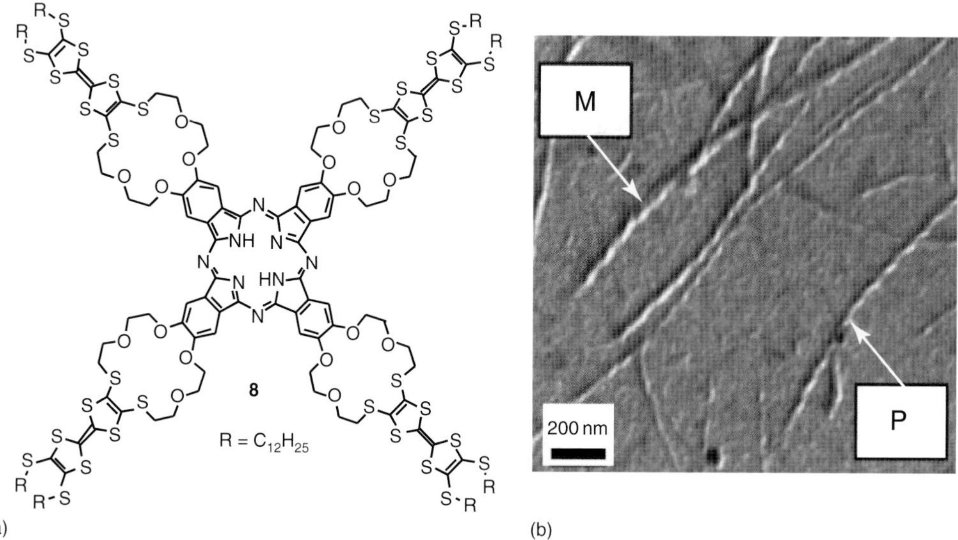

Figure 6 Tetrathiafulvalene (TTF)-crown ether-substituted Pc **8**. Transmission electron microscopy showing the formation of chiral helical fibers of P and M helicity. (Reproduced from Ref. 22. © Royal Society of Chemistry, 2005.)

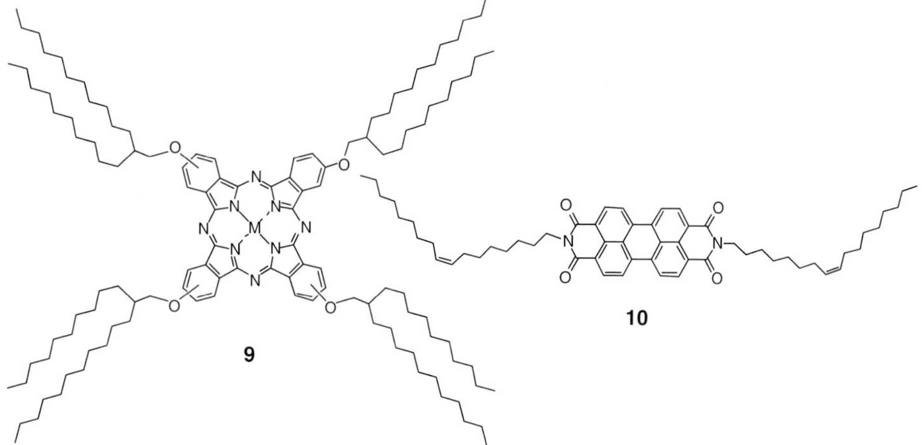

Figure 7 Phthalocyanine **9** and perylenediimide **10**.

of pure Pc. It has been proposed that the blended mesophase is mainly formed by the random insertion of acceptor PDI molecules within the Pc columns, thus allowing D–A interactions between the Pc and PDI components that can further stabilize the $\pi-\pi$ array. These blends present promising applications as active layers of photovoltaic devices.

4 Pc-BASED SUPRAMOLECULAR ASSEMBLIES HELD BY METAL–LIGAND INTERACTIONS

Metal–ligand interactions, which are (usually) strong and directional noncovalent interactions, are among the most popular molecular recognition tools for the construction of well-defined supramolecular architectures.[24] Thus, it is not surprising that over the years much effort has been devoted to the organization of metallophthalocyanines (MPcs), which possess extraordinary physical properties, by means of the metal coordination approach. Most of the transition metals have been incorporated within the central cavity of Pcs and many of them are able to coordinate extra ligands in the axial positions of the macrocycle. Owing to the limited extent of this chapter, which does not allow us to review all the possible modes of Pc–metal coordination, we focus on those metal–ligand interactions that have proved to be useful in the self-assembly of noncovalent superstructures. Thus, most of the chapter is dedicated to ZnPc and RuPc, which have, by far, been the most employed MPcs in metallosupramolecular chemistry.

4.1 Coordination of metal phthalocyanines to other electro- and photoactive units

4.1.1 Coordination to zinc phthalocyanines

The well-known coordination abilities of zinc porphyrins allowed the formation of many interesting metallosupramolecular structures, which have been reviewed elsewhere.[25] There has been a continuous effort toward the organization of zinc phthalocyanines in condensed phases owing to their outstanding photophysical properties, but these macrocycles have been shown to possess worse coordinating ability than zinc porphyrins. Thus, most of the work described so far involves the coordination of zinc phthalocyanine to other chromophores in order to study their photophysical properties.

The axial coordination of ZnPc to a pyridyl-functionalized porphyrin (Figure 8a) was first reported by Ng and Li in 2000.[26] In addition, a *meso*-tetra-4-pyridyl porphyrin was shown to form a supramolecular pentad in the presence of the same ZnPc (Figure 8b). Preliminary photophysical studies showed that there were no ground state $\pi-\pi$ interactions between the perpendicular moieties in the dyad or in the pentad.

SiPcs axially substituted with 3- or 4-pyridyloxy moieties have been used as building blocks in the formation of alternating coordination polymers when mixed with zinc(II) octathiobutoxy (R = *n*Bu) or octathiophenoxy phthalocyanine (R = Ph) (Figure 9).[27] The fairly low binding constant, about 300–400 M^{-1}, between the two Pc-based supramolecular monomers, SiPc and ZnPc, shows how weak the metal–ligand interaction between a pyridine

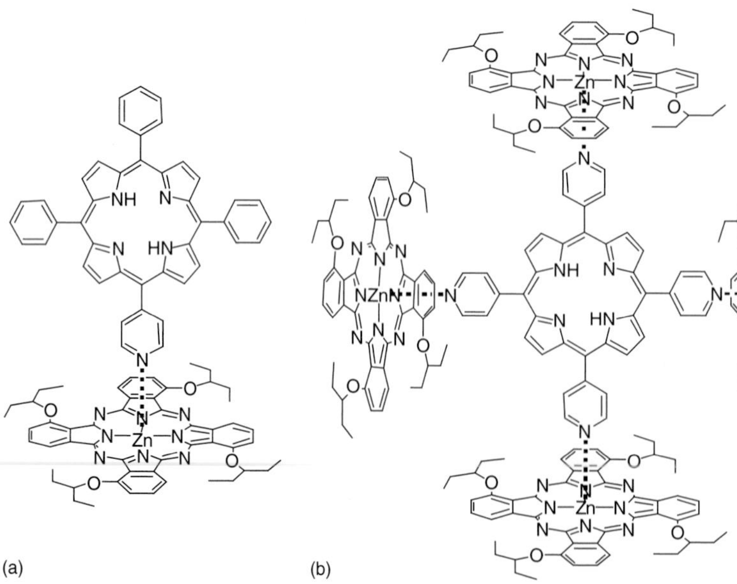

(a) (b)

Figure 8 Coordination of a tetrasubstituted zinc phthalocyanine to (a) a *meso*-4-pyridyl porphyrin and (b) a *meso*-tetra-4-pyridyl porphyrin.

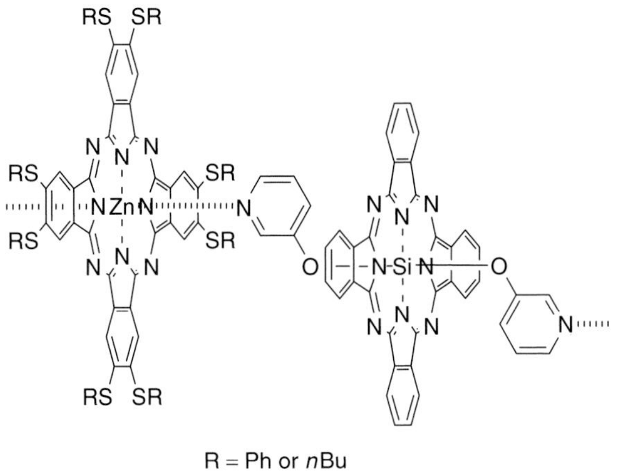

R = Ph or *n*Bu

Figure 9 Coordination polymer formed by a zinc phthalocyanine and bis-3-pyridyloxy silicon phthalocyanine.

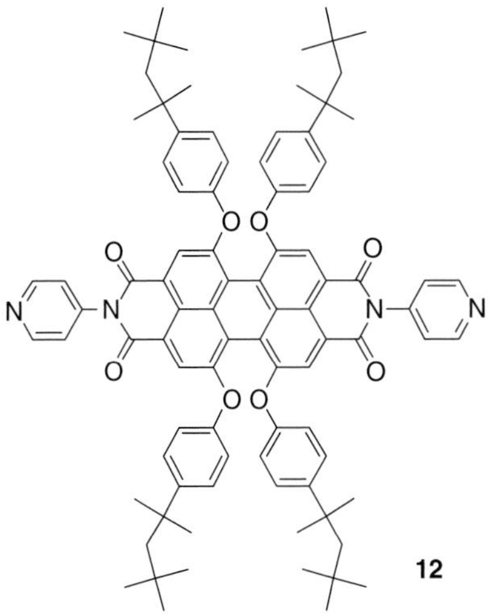

Figure 10 Dimer **11** formed by a zinc phthalocyanine and 3-pyridyloxysubphthalocyanine.

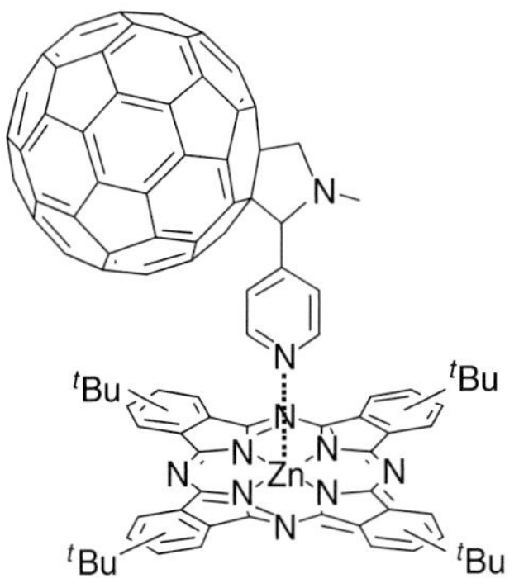

moiety and a donor-substituted ZnPc is when compared to that of pyridine and zinc porphyrins.

Axially substituted pyridyloxy-subphthalocyanines were shown to form coordination dimers **11** with zinc phthalocyanine with association constants (about $1-2 \times 10^4\,M^{-1}$) in the same range as those previously described for similar systems (Figure 10).[28] On addition of the pyridyl subphthalocyanine, the Q band of ZnPc was not changed in *N*, *N*-dimethylformamide (DMF), whereas in chloroform the shifts were extremely small. These observations suggested that the ground-state interactions between the two chromophores in these dyads are not significant, particularly in coordinating solvents. Preliminary photophysical studies revealed a strong quenching of the fluorescence, indicating at least good energy transfer between the subunits.

Dipyridyl PDI **12** (Figure 11) axially binds to zinc(II) phthalocyanine to form the corresponding 1 : 1 and 1 : 2

Figure 11 Dipyridyl perylenediimide **12**.

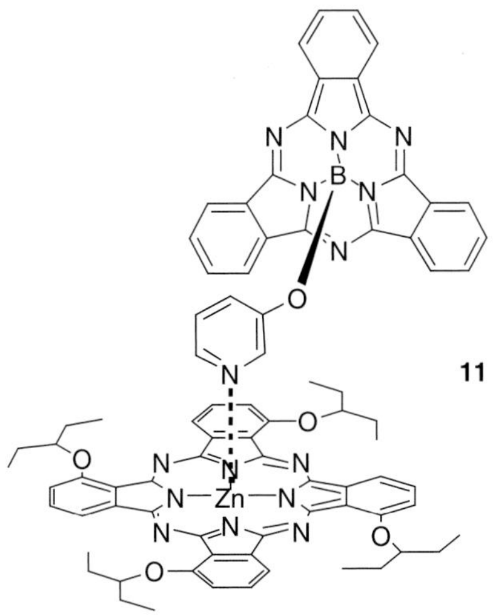

Figure 12 Coordination dimer **13** formed between a pyridyl-functionalized fullerene and tetra-*tert*-butyl zinc phthalocyanine.

supramolecular complexes.[29] It has been found that the coordination of pyridyl ligand to zinc phthalocyanine is relatively weak and labile with a binding constant of about $2 \times 10^3\,M^{-1}$ in $CDCl_3$. The two components mutually quench the fluorescence of their partner through an electron transfer process.

A huge amount of work has been dedicated to the coupling of the extraordinary photophysical properties of fullerenes with those of Pcs by means of coordination chemistry.[30]

The coordination of a pyridine-linked C_{60}-ligand to ZnPc was shown to form the 1 : 1 complex (**13**, Figure 12) with an association constant of $4.8 \times 10^3\,M^{-1}$, in the same range as other similar ZnPc–pyridyl complexes.[31] This weak equilibrium between dissociation and association of the zinc–pyridine coordination bond facilitates, after photoexcitation and rapid *intra*complex electron transfer, the crucial breakup of the radical-ion-pair into the free radical ions, $ZnPc^{\bullet+}$ and $C_{60}{}^{\bullet-}$. The lifetime of the latter

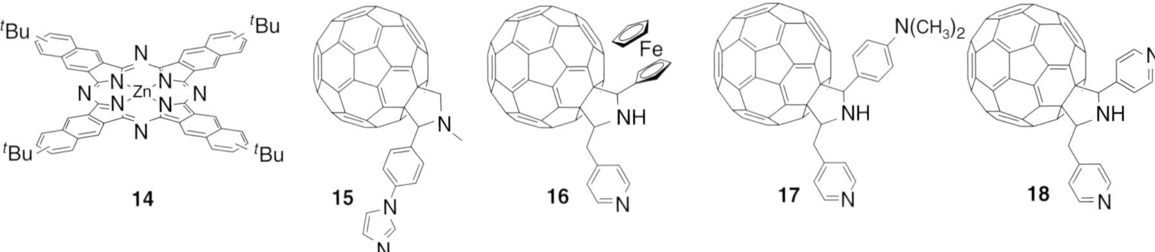

Figure 13 Tetra-*tert*-butyl zinc naphthalocyanine **14**. Imidazole and pyridine functionalized C_{60} fullerenes **15–18**.

was shown to be governed by a nearly diffusion-controlled, *inter*molecular back electron transfer characterized by a rate of about $10^9\, M^{-1}\, s^{-1}$.

The D–A dyad formed by axial coordination of zinc naphthalocyanine **14** (Figure 13) and fulleropyrrolidine **15** bearing an imidazole coordinating ligand (2-(4-imidazolylphenyl)-fulleropyrrolidine) was investigated in noncoordinating solvents such as toluene and *o*-dichlorobenzene.[32] The formation constant for the dyad was obtained from the absorption spectral data by using the Scatchard method. The K value thus calculated was found to be approximately $6.2 \times 10^4\, M^{-1}$ in toluene. This K value is an order of magnitude higher than the one previously reported for the same imidazole fulleropyrrolidine **15** and zinc tetraphenylporphyrin.[33] This observation suggests a better donor ability of ZnNc compared to that of the corresponding zinc porphyrin.

Direct evidence for the radical-ion-pair formation in the **15–14** dyad was obtained from picosecond transient-absorption spectral studies, which showed charge separation from the singlet-excited state of **14** to **15**. The calculated rates of charge separation and charge recombination were 1.4×10^{10} and $5.3 \times 10^7\, s^{-1}$ in toluene and 8.9×10^9 and $9.2 \times 10^7\, s^{-1}$ in *o*-dichlorobenzene, respectively. In benzonitrile, no charge transfer was observed.

Fullerene derivatives **16** and **17** (Figure 13) axially bind to ZnNPc(^{t}Bu)$_4$ **14** to form supramolecular triads in which the ZnNPc acts as an electron donor, the pyridyl fullerenes act as primary electron acceptors, and either the ferrocene (**16**) or the *N,N*-dimethylaminophenyl (**17**) unit acts as a secondary electron donor.[34] The binding constants of **16** and **17** with **14** are about $10^5\, M^{-1}$. These binding constants are comparable to that of **14** with **15** ($6.2 \times 10^4\, M^{-1}$), showing that the incorporation of a second electron donor on the pyrrolidine moiety increases the basicity of the pyridyl group—a fact that assists the binding to ZnNPc **14**. Photoexcitation of the naphthalocyanine unit **14** in the two supramolecular triads, **16–14** and **17–14**, leads to an efficient electron transfer to the fullerene ($k_{CS} = 2 \times 10^9\, s^{-1}$); the presence of the secondary electron donor was shown to extend the lifetime of the resulting ion pairs to about 10–15 ns.

The supramolecular triad **14**$_2$–**18** was self-assembled using fullerene derivative **18** bearing two pyridine ligands.[35] In this case, the charge separation rate of $5.7 \times 10^9\, s^{-1}$ was found to be slightly faster than those measured in the cases of **16–14** and **17–14**. This supramolecular entity was found to be extremely sensitive to coordinating solvents such as THF or benzonitrile in which the supramolecular structure is disassembled.

A series of fulleropyrrolidine containing pyridyl groups were evaporated on top of ZnPc thin films in order to produce efficient photovoltaic cells.[36] The idea was to improve the communication between the donor and acceptor layers by ligand–metal coordination. These bilayer solar cells exhibit efficiencies of up to 1.5%, about three times higher than those of reference cells based on the well-known C_{60}PCBM. The reference cells containing mostly C_{60}PCBM in the acceptor layer can be greatly improved (up to 4%) by mixing these chelating fulleropyrrolidine in small proportion with the C_{60}PCBM layer.

Single wall nanotubes (SWNTs) have been employed as building blocks for PET systems in combination with naphthalocyanine **19** (Figure 14).[37] In the first instance, SWNTs were solubilized by π–π stacking with a pyrene-functionalized imidazole moiety **20**. In the second step, zinc naphthalocyanine **19** was axially coordinated to the

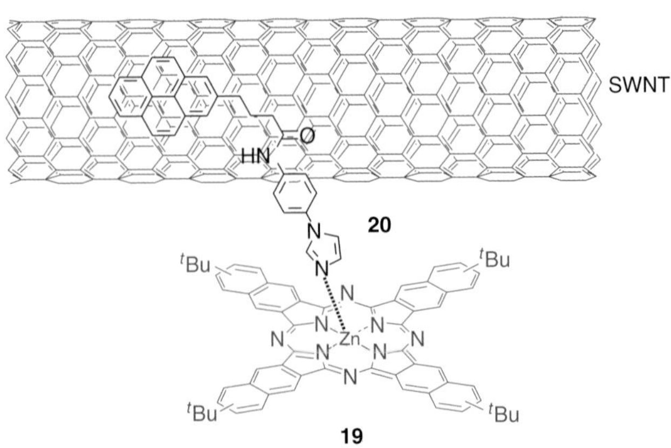

Figure 14 Supramolecular nanostructure formed between zinc naphthalocyanine **19**, imidazole-functionalized pyrene **20**, and a single wall carbon nanotube.

imidazole unit, thus forming the nanohybrid depicted in Figure 14. Photophysical studies revealed that the photoexcitation of the ZnNc moiety resulted in the one-electron oxidation of the donor unit with a simultaneous one-electron reduction of the SWNT. The presence of the charge-separated state **19**$^{\bullet+}$–**20**–SWNT$^{\bullet-}$ confirmed the importance of the present D–A nanohybrid in photogeneration of redox products.

4.1.2 Coordination to ruthenium phthalocyanines

The synthesis and properties of ruthenium phthalocyanines, RuPcs, have been well studied over the last 30 years, but only recently they have been rediscovered for their potential application in metallosupramolecular chemistry.[38] Basically, RuPcs are different from ZnPcs in their tendency to form stronger complexes with basic sp^2 nitrogen atoms (pyridine and imidazole) and in the possibility to form complexes on one side or on both sides of the macrocycle. Another interesting point of ruthenium phthalocyanines resides in the longer lifetime of their radical-ion-pair state when compared to that of zinc phthalocyanines. Thus, all things being equal, RuPcs display a richer potential for supramolecular chemistry than ZnPcs.

The synthesis of ruthenium-metallated Pc derivatives using $Ru_3(CO)_{12}$ in benzonitrile afforded ruthenium phthalocyanines either monocoordinated with CO or dicoordinated with benzonitrile. Cook and colleagues were the first to predict their utility in the preparation of further (pyridyl) ligated derivatives and showed that it was indeed straightforward.[39] This chemistry is sufficiently robust and efficient to permit elaborate supramolecular complexes to be prepared, as demonstrated by the synthesis of porphyrin–phthalocyanine multichromophores **21** and **22**, as illustrated in Figure 15. The absorption spectra of these arrays are essentially the sums of spectra of the starting materials. These observations indicate that there is little ground-state electronic interaction between the perpendicular macrocycles, in accordance with previous results published by Ng and Li (Section 4.1.1).[26]

Treatment of PDI, bearing two 4-pyridyl substituents at the imido positions, with ruthenium(II) phthalocyanine in chloroform affords the triad **23** in good yields (Figure 16).[40] This array shows remarkable stability in solution due to the robustness of the ruthenium–pyridyl linkage. Its electronic absorption spectrum is essentially the sum of the spectra of its molecular components, thus suggesting that

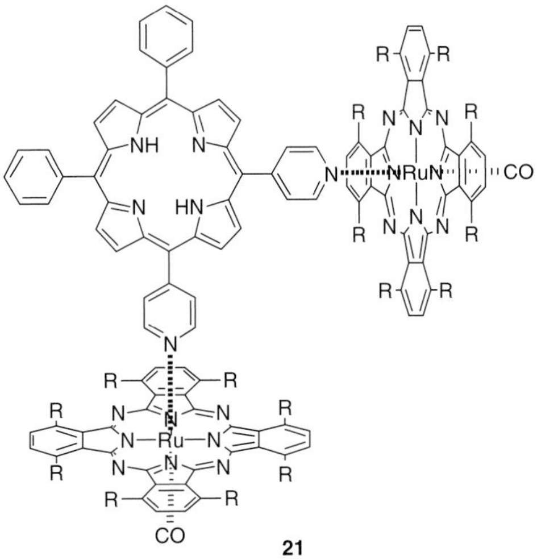

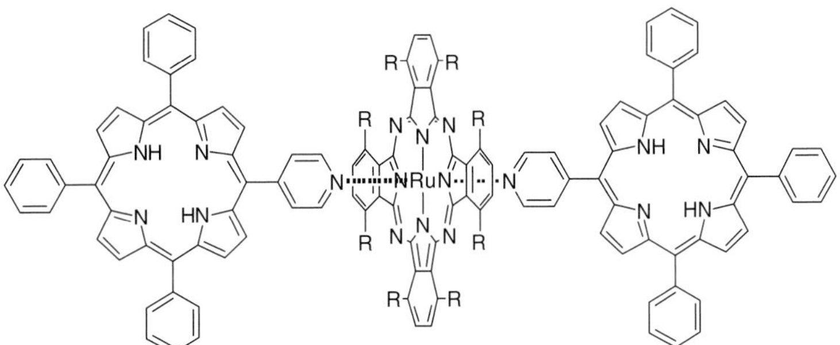

Figure 15 Supramolecular triads **21** and **22** formed between a ruthenium carbonyl phthalocyanine and *meso*-bis-4-pyridyl porphyrin (**21**), and ruthenium phthalocyanine and *meso*-4-pyridyl porphyrin (**22**).

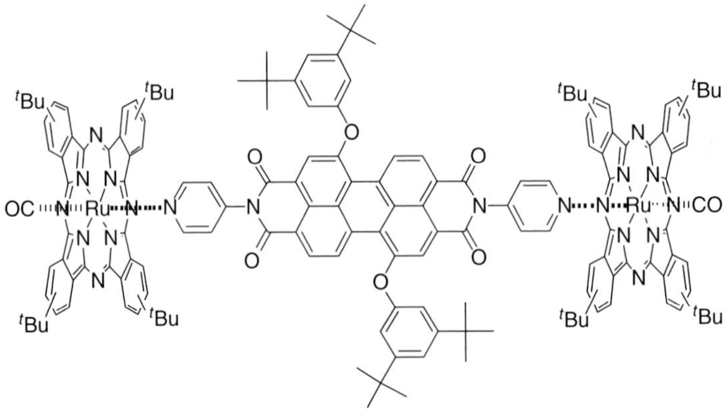

23

Figure 16 Supramolecular triad **23** formed by one dipyridyl perylenediimide and two ruthenium carbonyl phthalocyanines.

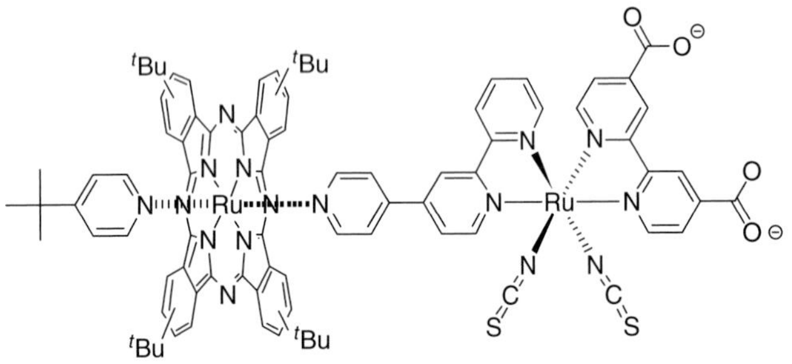

24

Figure 17 Supramolecular dyad **24** formed by one ruthenium 4-*tert*-butylpyridine phthalocyanine and a derivative of the N719 dye.

the subunits are not electronically coupled in the ground state as is usually the case with perpendicularly orientated chromophoric units. Photoexcitation of either RuPc or PDI leads to a rapid intraensemble charge separation, generating a long-lived radical-ion-pair which has a lifetime of about 115 ns.

Dyad **24** was self-assembled in a statistical manner by mixing the Rubis(bipyridyl) complex, 4-*tert*-butylpyridine, and the bisbenzonitrile ruthenium phthalocyanine in order to join two very good chromophores together in the same ensemble for dye-sensitized solar cells, that is, using Pc and N719 dyes (Figure 17).[41] The advantage of this system lies in the complementary absorption of the Rubis(bipyridyl) complex with respect to that of the RuPc. Unfortunately, this supramolecular dye gave rise to quite low efficiencies, mostly as a consequence of a very low open circuit voltage. Nevertheless, the photocurrent generation per molecule was found to be even better than that of the reference N719 dye molecule.

The coordination of dendritic oligothiophene-appended pyridines **25–27** to RuPc **28** gave rise to dyes possessing a panchromatic absorption in the visible range (Figure 18).[42] These compounds were incorporated into

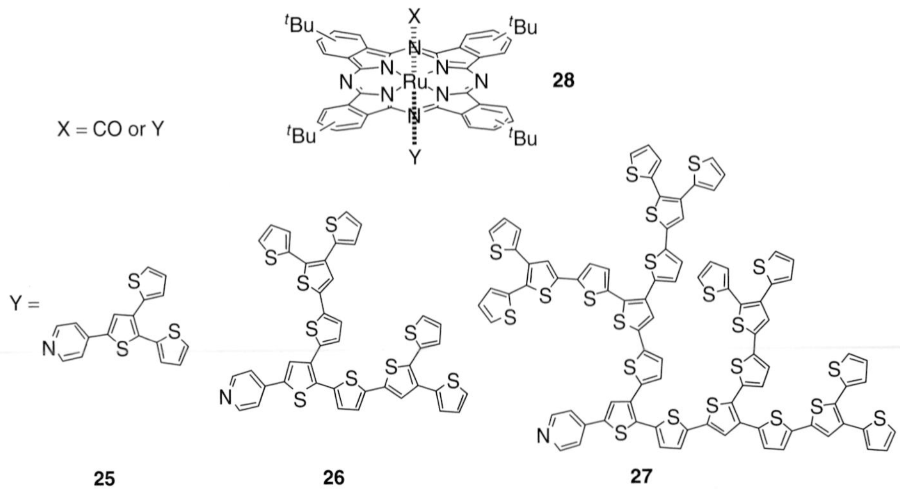

X = CO or Y

28

Y =

25 **26** **27**

Figure 18 Ruthenium phthalocyanine **28** and oligothiophene-functionalized pyridines **25–27**.

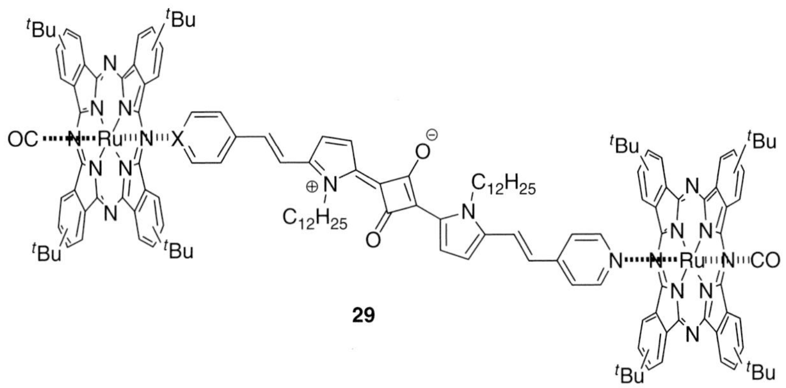

Figure 19 Supramolecular triad **29** formed between tetra-*tert*-butyl ruthenium carbonyl phthalocyanine and dipyridyl squaraine.

bulk-heterojunction solar cells mixed with conjugated donor polymers, giving rise to good efficiencies (about 1.6%), actually the best ones for phthalocyanines incorporated in this kind of devices.

The self-assembly of two RuCOPcs with one appropriate squaraine molecule bearing two 4-ethenylpyridyl moieties led to the triad **29** (Figure 19).[43] This ensemble was shown to absorb a large portion of the visible spectrum and also to be a good candidate for photovoltaic applications since, on excitation, it gives rise to a very long lifetime charge separation state (24 µs). Preliminary studies of incorporation of triad **29** into bulk-heterojunction solar cells gave rise only to modest results.

Pyridyloxysubphthalocyanine (Figure 10) was shown to bind to ruthenium phthalocyanine in the same way as with zinc phthalocyanine (Section 4.1.1).[28] The binding of pyridyloxysubphthalocyanine with Ru(II)Pc was shown to be stronger ($K \approx 2.5\text{–}4.7 \times 10^4\,M^{-1}$) than that with zinc(II) phthalocyanine ($K \approx 0.3\text{–}1.8 \times 10^4\,M^{-1}$).

Fullerene–phthalocyanine ensembles **30–32** (Figure 20) were self-assembled in good to excellent yields from the bisbenzonitrile Ru(II) Pc and the corresponding fullerene derivatives.[44] These dyads and triads gave rise to longer charge-separated lifetime than their zinc-based counterparts—see Section 4.1.1—the lifetime of radical-ion-pair state being several hundreds of nanoseconds.

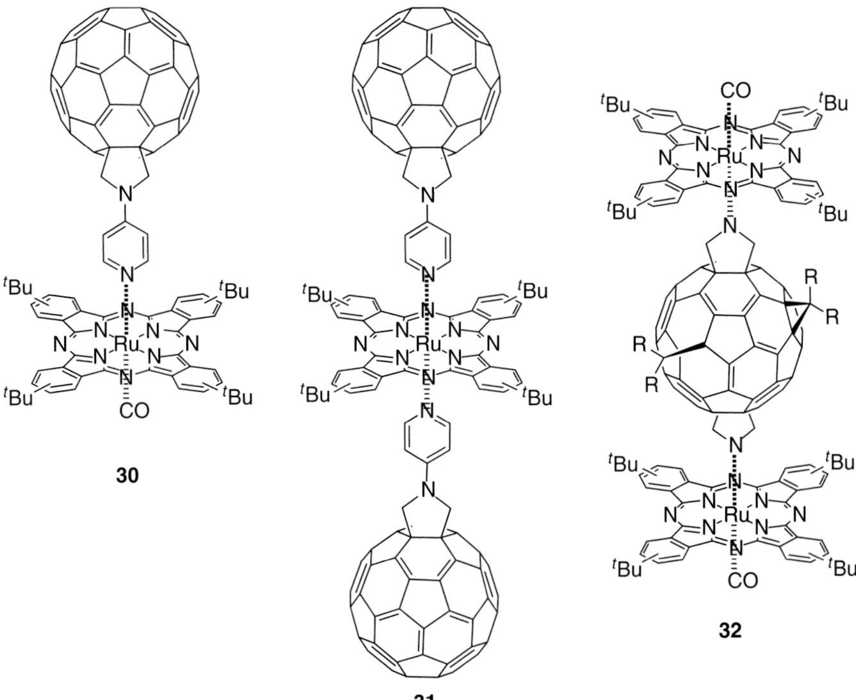

Figure 20 Supramolecular dyad **30** and triads **31** and **32** formed between ruthenium phthalocyanine and pyridyl-functionalized C_{60}.

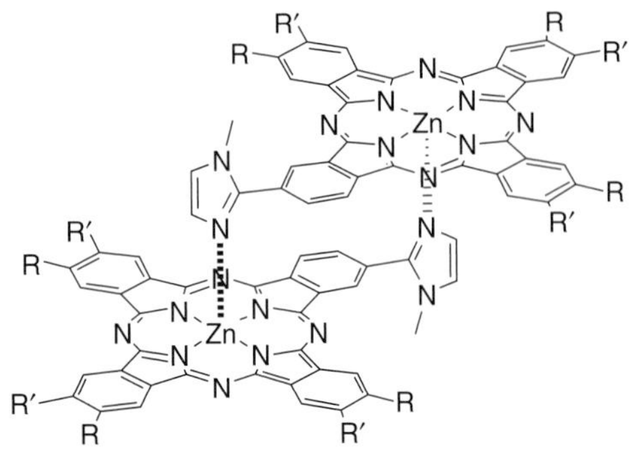

33

Figure 21 Self-coordinated dimer **33** formed between two imidazole-functionalized zinc phthalocyanines.

4.2 Phthalocyanine-based self-coordinated homodimers and polymers

4.2.1 Phthalocyanine ensembles

Kobuke demonstrated that imidazol-substituted porphyrins self-assemble into self-coordinated dimers with extremely high stability constant (about $10^{11}\,M^{-1}$).[45] Similarly, formation of phthalocyanine homodimers such as **33** in non-coordinating solvents such as toluene, dichloromethane, and chloroform was found to be extremely efficient with binding constants of about $1-10 \times 10^{11}\,M^{-1}$, depending on the nature of the peripheral substituents (Figure 21). The homodimers are so strong that they can break into the monomers only in the presence of a large excess of *N*-methylimidazole.[46]

When a phthalocyanine macrocycle is grafted onto the imidazole–porphyrin (**34**, Figure 22), no change is observed in the binding geometry, indicating that the imidazole–zinc porphyrin interaction is much stronger than the imidazol–zinc phthalocyanine one.[47]

In the dimeric species, the photoinduced energy transfer was shown to occur in the subpicosecond timescale, followed by charge separation and charge recombination with time constants of 47 and 510 ps, respectively.[48] In the case of the monomer, the charge-separated state was so short that it could not be observed. These results are consistent with a situation in which the slipped cofacial arrangement of the porphyrin rings induces a better charge separation in the supramolecular dimer than in the monomeric species, most probably because the formation of the dimer results in a decrease in the reorganization energy of the multicomponent ensemble.

Sonogashira cross-coupling methodology allowed the synthesis of a porphyrin–phthalocyanine–imidazole triad (**35**) that was found to form a supramolecular dimer both in solution and in the solid state (Figure 23).[49] In this case, since the zinc atom is no longer available within the porphyrin cavity, only the interaction between imidazole and ZnPc is responsible for the self-assembly process. Further, $\pi-\pi$ stacking between the large π surfaces of the complementary phthalocyanines and porphyrins (interplanar distance: 3.23 Å) accounts for the extremely high (about $10^{14}\,M^{-1}$) binding constant. The binding constant corresponding to the association of 1-methylimidazole and ZnPc

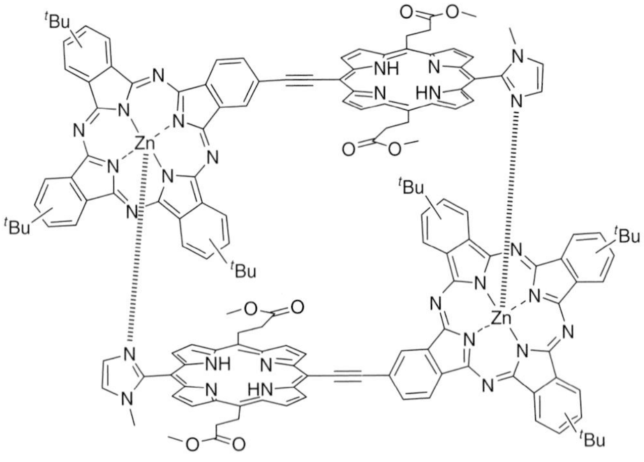

35

Figure 23 Self-coordinated dimer **35** formed between two imidazole-free-base porphyrin–zinc phthalocyanine ensembles.

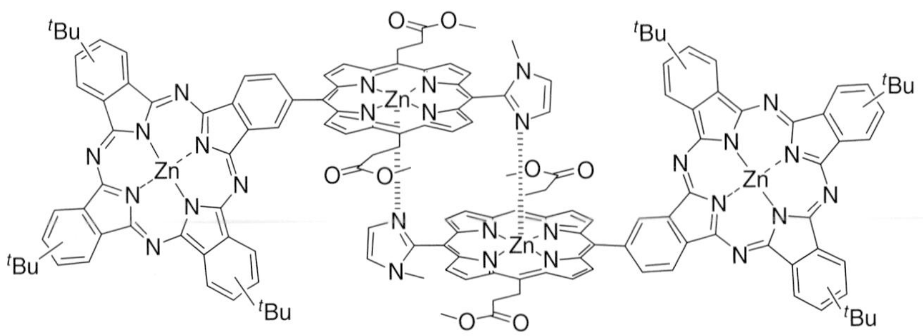

34

Figure 22 Self-coordinated dimer **34** formed between two imidazole–zinc porphyrin–zinc phthalocyanine ensembles.

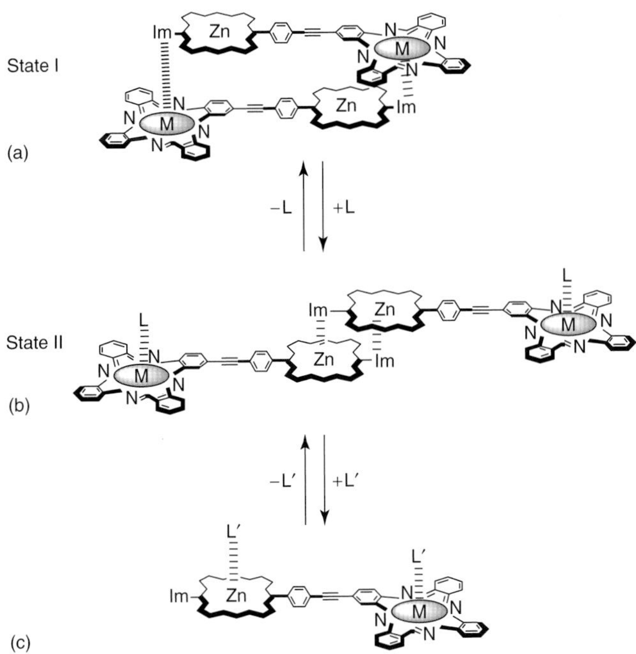

Figure 24 Self-assembly of stacked homodimers (state I, top), ligand-induced transformation into extended homodimers (state II, middle), and dissociation by coordination of ligand (bottom).[51] (Reproduced from Ref. 51. © World Scientific Publishing, 2009.)

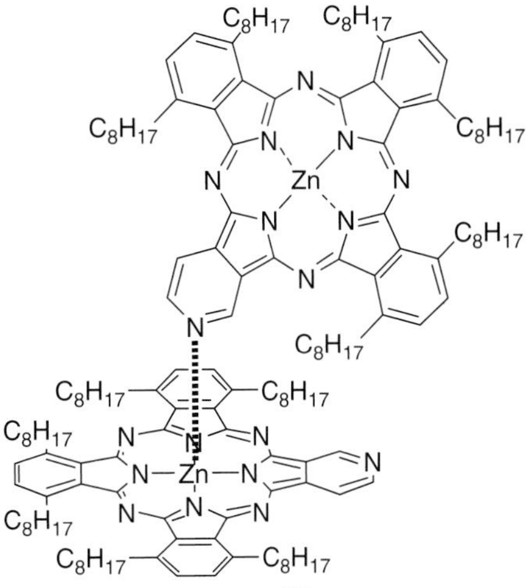

was found to be about $3 \times 10^5\ M^{-1}$,[50] a clearly lower value than that of the supramolecular dimer **35**, thus showing the relevance of the cooperative effects involved in the dimer formation.

Photophysical studies showed that, in the coordination dimer, the proximity between the two macrocycles induced significant π-electron interactions between the porphyrin and Pc planes. Tuning of the charge transfer extent between the porphyrin and Pc units was achieved by modulation of the donor ability of the ZnPc subunit (e.g., alkoxy instead of alkyl groups) or by changing the solvent polarity. The intermolecular charge transfer state in the supramolecular dimer was shown to be quenched by the addition of an axial ligand, thus switching to intramolecular energy transfer between H_2Por and ZnPc.

Kobuke and colleagues described a complementary zinc porphyrin–magnesium phthalocyanine dimer that can be considered as a supramolecular switch (Figure 24).[51] This switch is based on the variation of fluorescence induced by a change of ligands. More precisely, an imidazole-appended porphyrinatozinc–phthalocyaninatomagnesium complex was shown to spontaneously dimerize into a supramolecular short dimer (state I) in which the imidazole units are bound to the central magnesium atom of the Pcs. On addition of 2 equivalents (with respect to the dimer) of dimethyl sulfoxide (DMSO) or 1-methylimidazole to the dimer, a new extended dimer (state II) is formed in which

36

Figure 25 Coordinated homodimer **36** formed between two pyridino[3,4]tribenzoporphyrazines.

two DMSO molecules are bound to the magnesium atoms. Further addition of 1-methylimidazole to state II gives rise to the unbound monomeric dyad.

The metal-free and nickel derivatives of hexaoctyl-pyridino[3,4]tribenzoporphyrazine generate conventional face-to-face aggregates in solution; they exhibit columnar mesophases and form columnar assemblies on evaporation of solutions. By contrast, the zinc derivative forms alternative edge-to-face complexes such as **36** through axial coordination of one pyridine moiety to the zinc atom of another (Figure 25), a process which inhibits mesophase behavior.[52]

Kobayashi and colleagues went one step further in the study of a similar zinc tri-*tert*-butyl[3,4]tribenzoporphyrazine. This macrocycle has a natural tendency to form a homodimer such as **37** in noncoordinating solvents (Figure 26). In the presence of pyridine, the dimer dissociates into its components **38**, and in the presence of silver salts such as $AgPF_6$, it forms a peripherally Ag-coordinated homodimer **39**.[53]

1,4-Dibutoxy-2,3-di(4-pyridyl)-8,11,15,18,22,25-hexakis (hexyl)-phthalocyaninato zinc, **40**, was shown to form a coordinating polymer in the solid state (Figure 27).[54] Each zinc atom has square pyramidal coordination and the apical Zn–pyridine bonds link the molecules in an infinite polymer chain. Such a chain is reminiscent of that reported for zinc 5,10,15-triphenyl-20-pyridylporphyrin.[55] The behavior of **40** in solution was shown to be solvent dependent. In dichloromethane, it forms aggregates that break down into the monomeric Pcs in THF. Addition of pyridine to the solution evidently also breaks down the aggregates.

Diethylamino-substituted metal azaphthalocyanines **41** (M = Zn, Mg) were shown to self-assemble into *J*-dimers

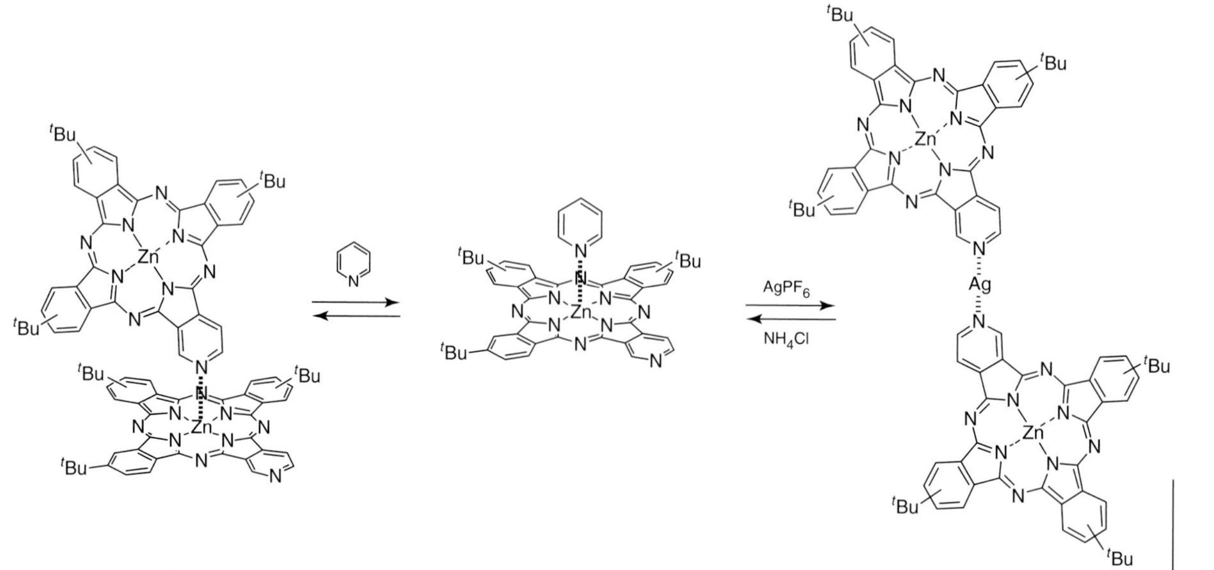

37 **38** **39**

Figure 26 Two modes of self-assembly of coordinated homodimers **37** and **39** from zinc pyridino[3,4]tribenzoporphyrazine **38**.

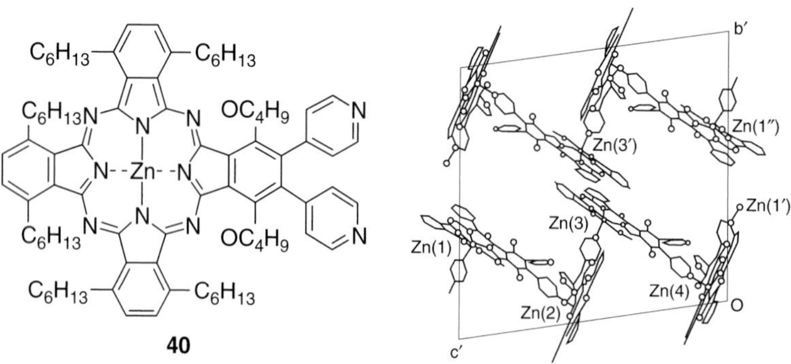

Figure 27 Bis-4-pyridyl zinc phthalocyanine **40**, and its coordinating polymer in the solid state.

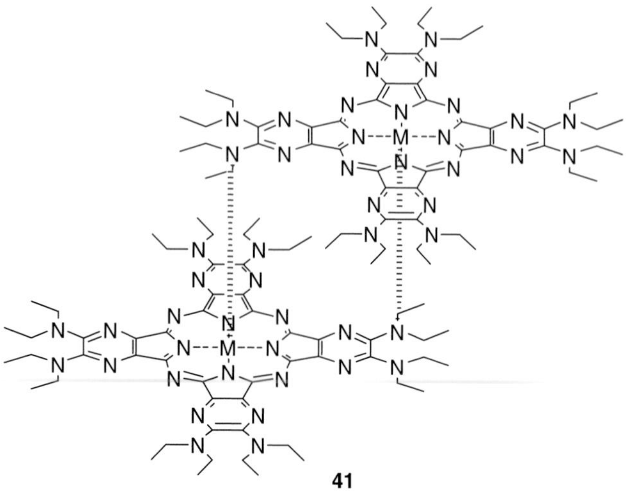

Figure 28 Coordinated homodimers **41** formed between two zinc or magnesium azaphthalocyanines.

in noncoordinating solvents (Figure 28).[56] The dimers are formed by the coordination of the free electron pair of one diethylamino group with the central metal of the adjacent molecule. As expected, the addition of pyridine leads to monomerization and considerable quenching of fluorescence and singlet oxygen formation as a result of intramolecular PET. PET is efficiently inhibited in dimers; therefore, dimers have higher fluorescence and singlet oxygen quantum yields than the corresponding monomers.

Zinc tetrakis(aryloxy)phthalocyanines such as **42** were shown to self-assemble into coordinating oligomers as demonstrated by UV–vis spectroscopy and mass spectrometry (MALDI-TOF) (Figure 29).[57] The compound was deposited onto carbon-coated grids and observed by TEM, which showed the formation of 50 nm diameter nanowires. As may be anticipated, similarly substituted free base and Ni(II) phthalocyanines do not form aggregates.

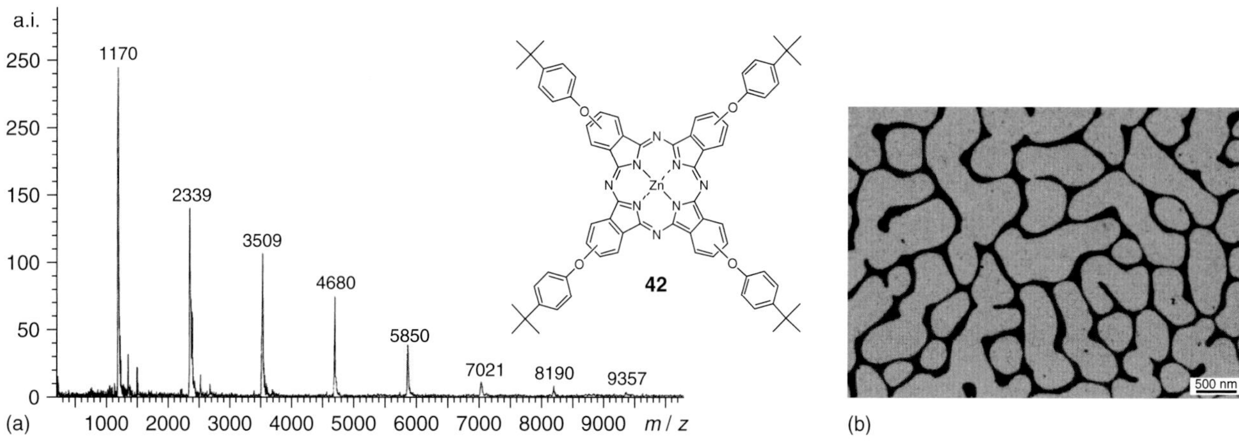

Figure 29 MALDI-TOF mass spectrum of zinc phthalocyanine **42** showing the formation of coordinating polymers (a). TEM images of samples of **42** prepared from a $CHCl_3$ solution (b).[57] (Reproduced from Ref. 57b. © American Chemical Society, 2007.)

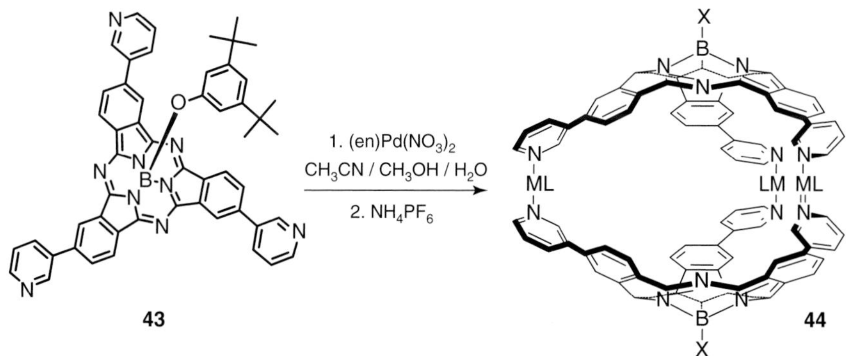

Figure 30 Self-assembly of M_3L_2 homodimeric capsule **44** from tris-3-pyridyl-subphthalocyanine **43**.

4.2.2 Subphthalocyanine ensembles

The hemispherical shape of subphthalocyanines (SubPcs) prompted Torres and Claessens to bring together two SubPcs in order to form a hollow capsule. This idea was made possible by synthesizing SubPc **43** bearing 3-pyridyl units in its periphery and by self-assembling **43** in the presence of a Pd(II) salt (Figure 30).[58]

The almost quantitative formation of the M_3L_2 capsule **44** was found to be accompanied by a self-discriminatory process. Since SubPc **43** was, in fact, a racemic mixture of two enantiomers [M] and [P], the self-assembly process was found to occur exclusively between two opposite enantiomers in a self-discriminatory manner.

Capsule **45**, made of two larger SubPcs **46** bearing an extra triple bond between the macrocycle and the 3-pyridyl units, was found to complex C_{60} fullerene in acetone (Figure 31), a solvent in which C_{60} is normally not soluble.[59] Thus, the appearance of a signal corresponding to C_{60} in the ^{13}C NMR spectrum of the cage in d_6

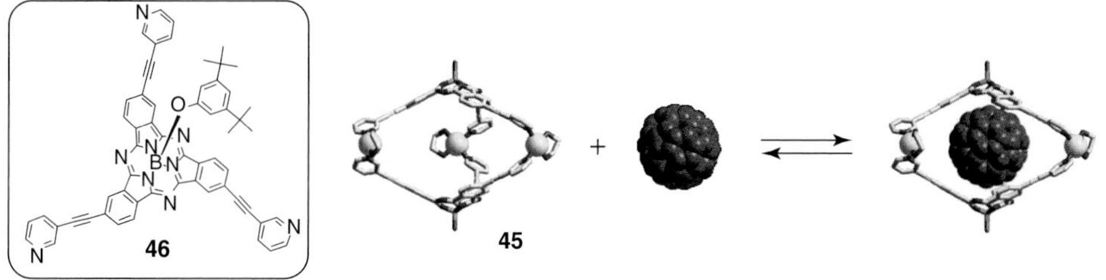

Figure 31 M_3L_2 homodimeric subphthalocyanine capsule **45**, formed by two tris-3-pyridylethynylsubphthalocyanines **46**, and its inclusion complex with C_{60} fullerene.

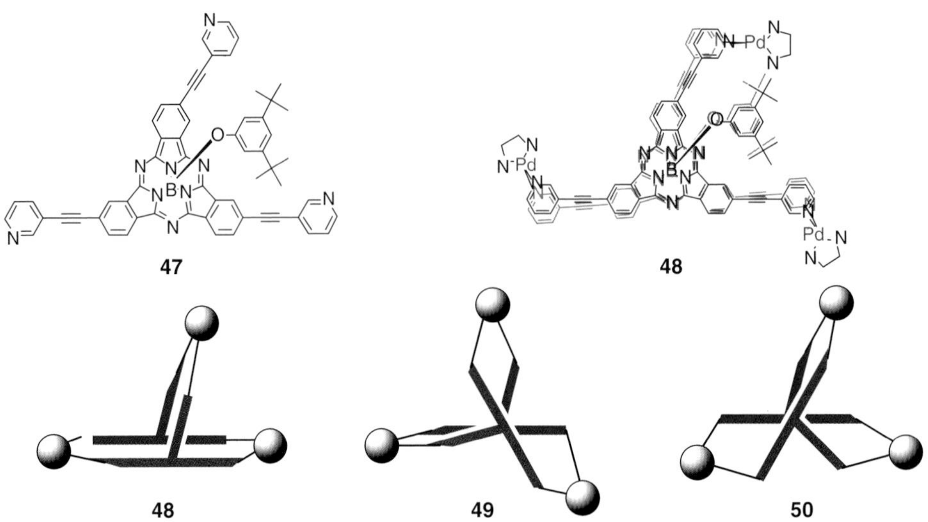

Figure 32 Self-sorting among M_3L_2 subphthalocyanine capsules **48–50** formed by two C_1 symmetrical subphthalocyanines **47** leading to the exclusive formation of cage **48**.

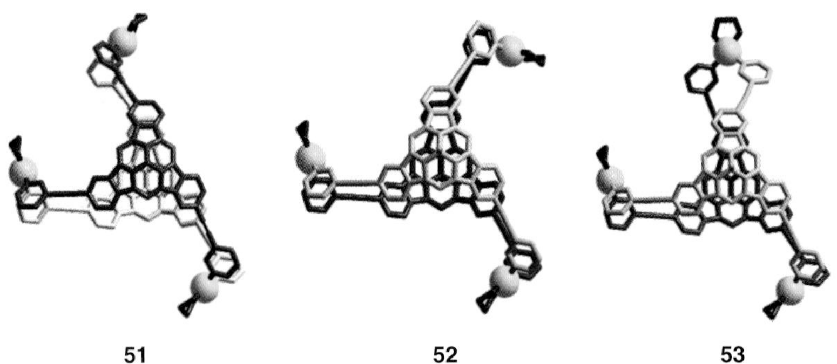

Figure 33 The three possible M_3L_2 subphthalocyanine capsules **51–53** formed from mixing C_3 and C_1 symmetrical subphthalocyanines, **46** and **47**, respectively, together with Pd(II).

acetone, accompanied by chemical shift changes of some relevant peaks and electrospray mass spectrometry, showed unambiguously the formation of the 1 : 1 adduct.

The formation of SubPc is a statistical process that gives rise to a mixture of C_1 and C_3 symmetrical regioisomers.[60] It was found that the C_1 regioisomer **47** was able to self-assemble as well as its C_3 counterpart into an M_3L_2 cage **48** in the presence of Pd(II) salts (Figure 32).[61] Interestingly, this self-assembly process was shown to form only one cage, that is, the most symmetrical one, **48**, among the three possible ones, **48, 49**, and **50**, as illustrated in Figure 32.

By mixing the two different regioisomers of C_3 and C_1 symmetries (**46** in Figure 31 and **47** in Figure 32, respectively), a statistical mixture of all the three possible cages **51, 52**, and **53** (Figure 33) was obtained in a first instance.[62] Remarkably, over a few hours, the mixture was found to equilibrate toward the two most symmetrical cages, **51** and **52**. This study was made possible by introducing different axial substituents in the axial positions of the C_1 and C_3 regioisomers of the starting SubPcs and by monitoring the self-assembly process by electrospray mass sprectrometry. In this way, the mixed C_3–C_1 cage was eliminated from the mixture by what could be called an error checking process.

5 PHTHALOCYANINE AGGREGATES HELD BY ELECTROSTATIC INTERACTIONS

Electrostatic interactions have been employed as a simple way to form Pc face-to-face hetero-ensembles by self-assembly of adequate components bearing oppositely charged substituents. Either tetraanionic Pcs such as those represented in Figure 34 (bearing SO_3Na groups) or tetracationic Pcs (bearing, e.g., $N(CH_3)_3^+$ groups) were shown to form stable cofacial complexes (equilibrium formation constants in the range of 10^8 M^{-1}) when mixed with complementary charged porphyrins. Interestingly, the stability

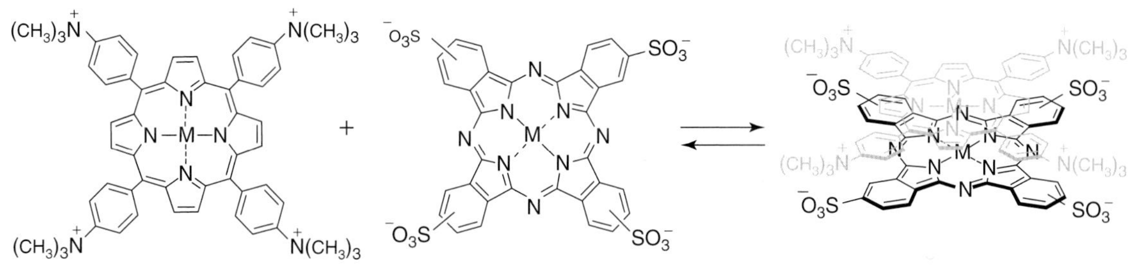

Figure 34 Porphyrin–phthalocyanine supramolecular dimer held by electrostatic interactions.

and stoichiometry of the assemblies mostly depend on the affinity of the Pc central metal for coordinating solvent molecules.[63]

6 CROWNED AND PSEUDOCROWNED PHTHALOCYANINES

The self-assembly of Pcs functionalized at the periphery with crown-ether moieties was intensively studied by Nolte, Bekaroglu, and Kobayashi in the late 1980s.[64–66] It was found that the aggregation properties of these types of Pcs are remarkably affected by the polarity of the solvent and/or the presence of alkali metal ions that can be hosted by the crown-ether units giving rise to Pc-based supramolecular complexes presenting different stoichiometries, depending on the size of the alkali metal ion used. Interestingly, when tetra-15-crown-5-substituted Pc **54** and alkali cations with diameters exceeding that of the crown-ether rings (such as K^+) were employed, cofacial dimeric supramolecular complexes were formed, in which four potassium ions were sandwiched between two crown-ether Pc units (Figure 35a).

Torres and coworkers reported structurally simpler octa(dialkylamido)Pcs **55** that showed a crown ether-like behavior. These Pcs in fact were able to aggregate in solution by the addition of methanolic solutions of alkali metal ions to the chloroform solution of the monomeric Pc **55** owing to complexation of the alkali metal cations by the dialkylaminocarbonylmethoxy units (Figure 35b).[67]

Rodgers *et al.* also elegantly employed tetrakis(18-crown-6)fused Pc/K^+ ensembles for indirect complexation of organic anions, such as tetracarboxy- or tetrasulfonated-substituted porphyrins **56**, by electrostatic interactions.[68] Cofacial complexes of **56** were formed (Figure 36), wherein the Pc and Por components showed strong electronic interactions in the ground state.

The complexation of primary alkyl ammonium cations ($R\text{-}NH_3^+$) presents an additional interest since it allows one to build functional supramolecular D–A conjugates where

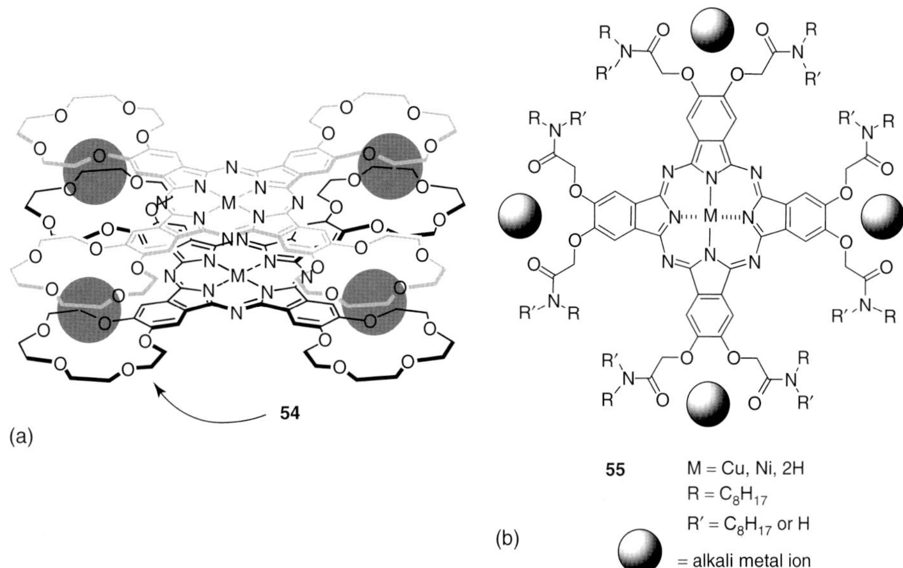

Figure 35 (a) Homodimeric supramolecular complex **54** formed by two crown-ether Pcs and four potassium ions. (b) Octa(dialkylamido)Pc (**55**) complexing four alkali metal cations.

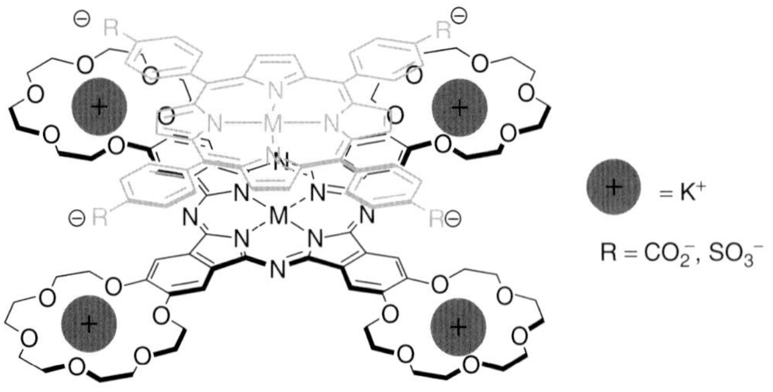

Figure 36 Porphyrin–phthalocyanine supramolecular dimer **56** held together by a combination of ion–dipole and electrostatic interactions.

PET can be studied. Despite the number of supramolecular complexes formed by ammonium-substituted fullerenes ($C_{60}-NH_3^+$) and 18-crown-8 appended porphyrins (Por-18C6), supramolecular ensembles held together exclusively by crown-ether Pcs/$C_{60}-NH_3^+$ interactions have not been prepared yet. The crown-ether/$R-NH_3^+$ recognition motif has been recently exploited in combination with other noncovalent interactions to build a sophisticated supramolecular Pc–fullerene system that is able to achieve a long-lived photoinduced charge-separated state.[69] The supramolecular system $(ZnPc)_2/(C_{60})_2$ **57** reported by D'Souza and coworkers consists of a cofacial Pc dimer held together by crown-ether/K^+ interactions. Then, metal–ligand interactions between the pyridine unit attached to the C_{60} component and the Zn(II) Pc central metal account for the supramolecular D–A Pc/C_{60} ensemble, which is further stabilized by alkylammonium cations/crown-ether interactions (Figure 37).

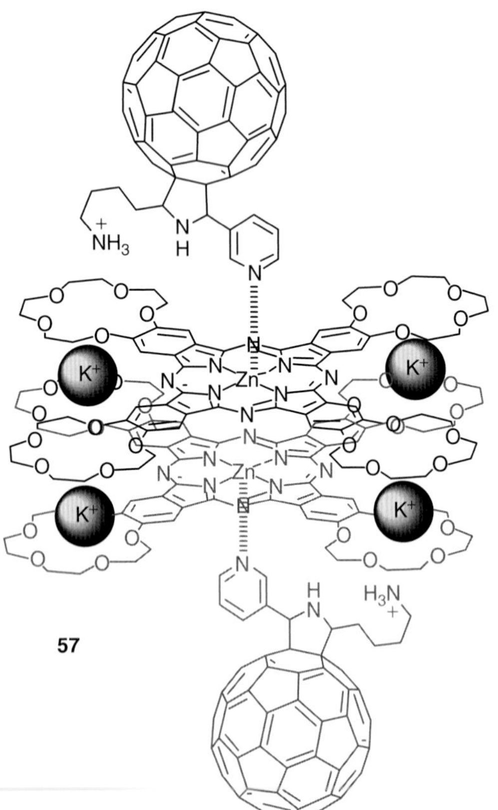

Figure 37 Pc–C_{60} ensemble **57** formed by a zinc crown-ether phthalocyanine supramolecular dimer similar to **54** (Figure 35a) coordinated to a pyridyl-functionalized C_{60} component bearing an alkylammonium cation that is able to further interact with one of the crown ether.

The robust molecular recognition discovered by Stoddart's group in the 1990s,[70] formed between medium-sized crown-ether unit (i.e., dibenzo-24-crown-8) and secondary dialkylammonium cation units ($RRNH_2^+$) (i.e., dibenzylammonium PF_6 salt), has been applied to Pcs.[71] Thus, unsymmetrically DB24C8-substituted crown-ether Pcs such as **58** have been assembled in solution with C_{60}-NH_2^+ CH_2Ph **59**, forming stable hydrogen-bonded complexes (K about $10^4\,M^{-1}$) with a pseudorotaxane-like geometry (Figure 38) by threading the dialkylammonium cation through the DB24C8 unit. These complexes represent simple models of the photosynthetic reaction center, achieving long-lived $Pc^{\bullet+}$ $C_{60}^{\bullet-}$ charge-separated states (1.5 µs) on exciting the Pc chromophore.

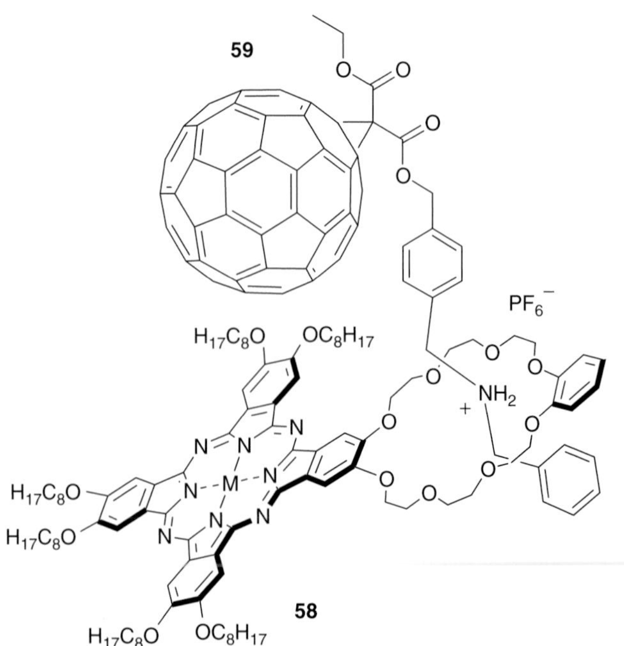

Figure 38 Phthalocyanine–C_{60} supramolecular dimer held by hydrogen-bonding interactions between crown-ether phthalocyanine **58** and dialalkylammonium-functionalized C_{60} **59**.

In addition, the stoichiometry of the resulting supramolecular complex can be modified using poly(dibenzylammonium) components. Thus, a supramolecular Pc dimer with a [3]pseudorotaxane geometry could also be formed by applying the DB24C8/$RRNH_2^+$ supramolecular interaction.[72] Furthermore, dethreading of the pseudorotaxane's components is possible by adding a base to the solution, so that the two interchangeable states of the pseudorotaxane can be switched simply by varying the solution's pH.

7 SUPRAMOLECULAR PHTHALOCYANINE ARRAYS HELD BY HYDROGEN BONDING

The influence of hydrogen-bonding interactions on the aggregation properties of Pcs has been well established.[73] For instance, it has been reported that octaalkylamide-substituted Pcs form discotic columnar hexagonal mesophases at higher temperatures than the structurally related dialkylamide derivatives because of further stabilization of the liquid crystal formed by NH amide hydrogen bonding.[74]

In the same way, unsymmetrically substituted Pc **60** bearing six chiral, long alkoxy chains and one terminal hydroxyl-substituted aliphatic substituent self-organizes, forming edge-to-edge dimers formed by lateral hydroxyl hydrogen bonding (Figure 39).[75] However, this hydrogen-bonded network is lost above 130 °C, causing the structure to change to a hexagonal columnar phase, in which the Pc molecules are arranged in a left-handed helix.

The Watson–Crick base-pairing interactions are well-established recognition motifs for the construction of robust hydrogen-bonding arrays.[76] Ng and Li first reported a series of Pc–nucleobase conjugates such as the tetra-adenine derivative **61** shown in Figure 40.[77] These authors pointed out the remarkable tendency of these Pcs to aggregate as a result of several noncovalent interactions, including $\pi-\pi$ interactions and metal–ligand coordination, which result in a poor solubility in organic solvents. However, in such nucleobase tethered systems, it is possible to take advantage of the possibility of forming Watson–Crick base pairs by adding a complementary nucleobase. Accordingly, the base-pairing capability of adenine-substituted Pcs with a complementary thymine-substituted 9,10-anthraquinone derivative **62** was investigated, suggesting the formation of a supramolecular complex through Watson–Crick base-pairing interactions.

Similary, complementary cytidine–guanosine nucleobases have been employed by Torres and Sessler as subunits to assemble a cytidine-appended Pc **63** to a guanosine-linked C_{60} **64**.[78] The UV-vis spectrum of ZnPc–cytidine indicated the formation of strong aggregates in organic solvents. However, the resulting aggregate(s) can be broken

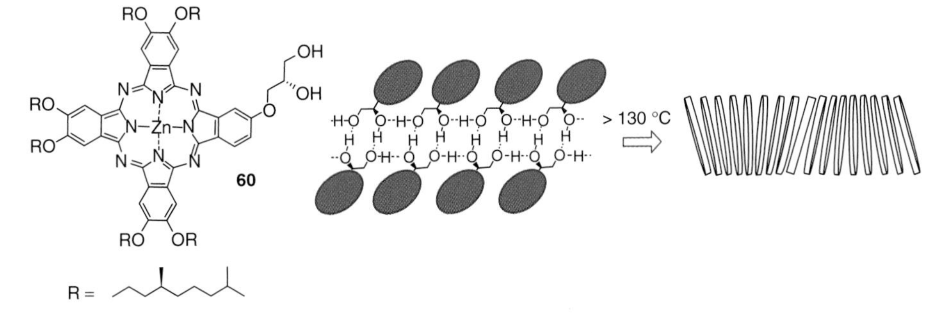

Figure 39 Supramolecular polymer of monomeric phthalocyanine **60** that is held by lateral OH-hydrogen bonding.

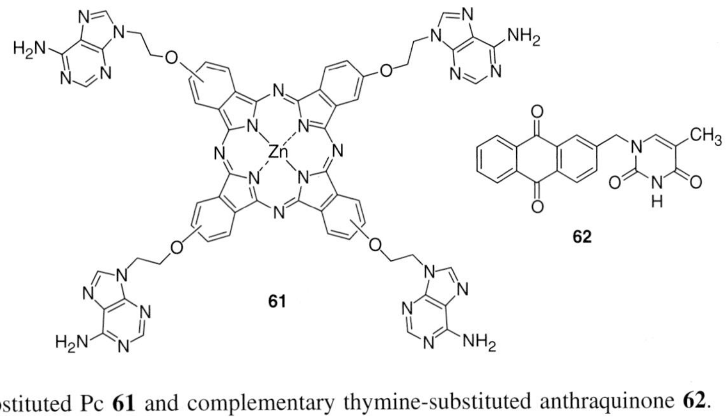

Figure 40 Tetraadenine-substituted Pc **61** and complementary thymine-substituted anthraquinone **62**.

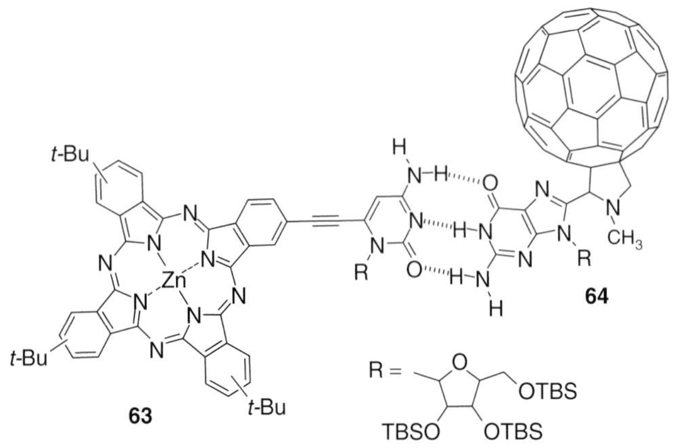

Figure 41 Cytidine-substituted Pc **63** hydrogen-bonded to complementary guanosine-appended C_{60} **64**.

up by the addition of guanosine, a nucleic acid base that can serve to tie up the cytidine functionality through base-pairing interactions. The addition of guanosine–C_{60} led to the formation of a supramolecular ZnPc–C : G–C_{60} ensemble depicted in Figure 41.

Recently, the melamine/PDI recognition motif has been employed as a new strategy to assemble Pcs in well-defined hydrogen-bonding supramolecular architectures. Thus, two Pc **65** molecules that are functionalized with a ditopic melamine moiety strongly bind to each of the complementary diimide subunits in PDI **66** forming a stable supramolecular assembly (Figure 42).[79] Photoexcitation of the PDI unit results in the transduction of singlet-excited state energy to the energetically lower lying Pc.

8 PHTHALOCYANINES AND DERIVATIVES ON SURFACES

8.1 Organization of phthalocyanines on surfaces

The study of the interaction between Pcs and all types of surfaces, of their bidimensional organization, and of their properties on conducting and semiconducting substrates is of paramount importance for the understanding of the way Pc-based devices work. All the physicochemistry involved in gas-sensing, photocurrent generation, and light emitting, among other processes, depends very much on the organization of the molecules at the nanoscale, more precisely on the way the molecules interact with each other and with the surface. This is why a huge amount of work has been devoted to the careful study of the behavior of Pcs on surfaces by utilizing the entire range of scanning probe microscopies.

8.1.1 Study of single phthalocyanines on surfaces

Pcs and, more precisely, CuPcs were among the first molecules to be observed individually by scanning tunneling microscopy (STM) in 1987.[80] Two years later, STM images under ultrahigh vacuum (UHV) of isolated vapor-deposited CuPc molecules on Cu(100) substrate exhibited subatomic-scale features that agree well with molecular-orbital calculations (Figure 43).[81] It was the first time that

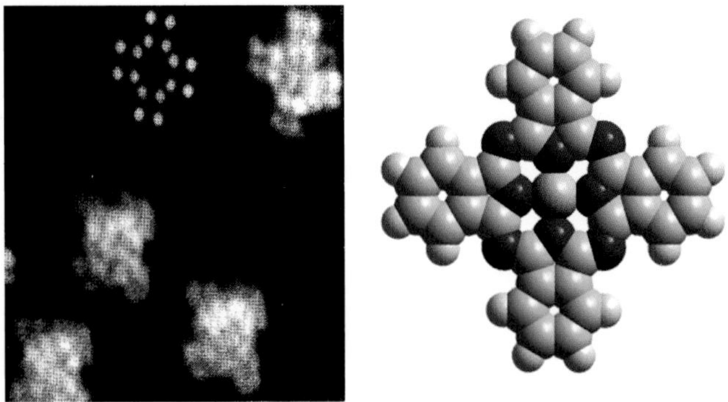

Figure 43 Observation of individual phthalocyanines by STM.[81] (Reproduced from Ref. 81. © American Physical Society, 1989.)

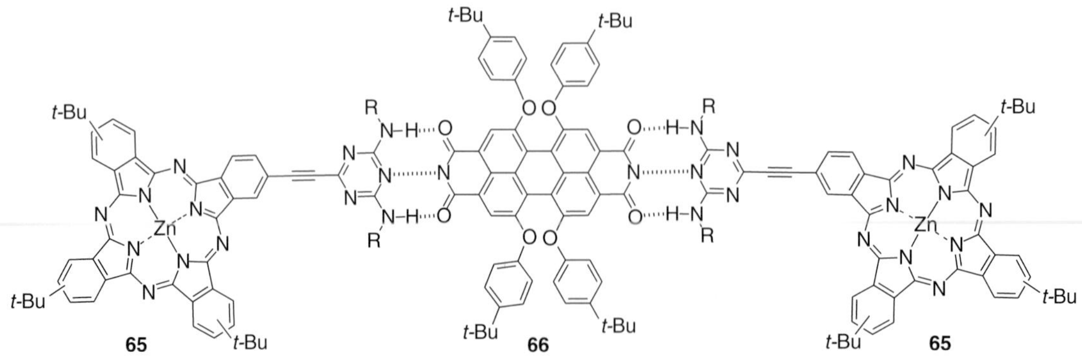

Figure 42 Melamine-substituted Pc **65** and complementary perylenediimide **66** forming a hydrogen-bonded supramolecular triad.

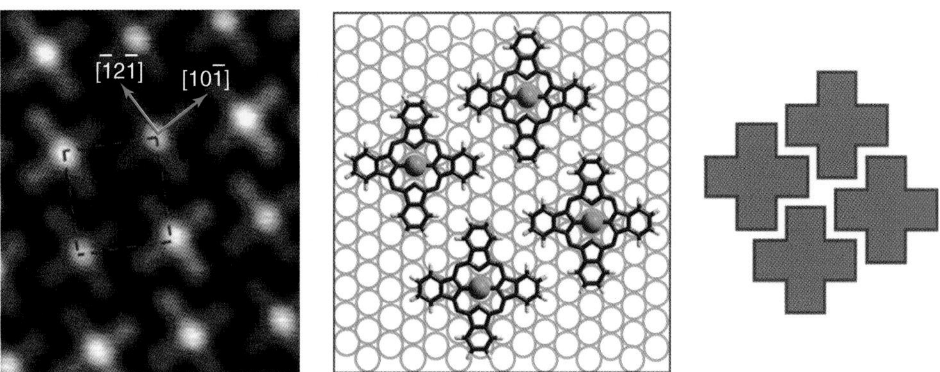

Figure 44 Close packing 2D organization of FePc on Au(111), as observed by STM.[83] (Reproduced from Ref. 83. © American Chemical Society, 1997.)

the internal structure of a molecule could be observed by STM.

Hipps and coworkers reported various MPcs (Cu,[82] Co,[82] Ni,[83] Fe,[83] and VO[84]) on reconstructed Au(111), and found that the brightness of the central spot of Pcs is dependent on the active central metal. The difference in contrast between the metal ions in STM images was explained in terms of the occupation of the d_{z^2} orbital. Similar conclusions could be drawn from the studies of metallated porphyrin derivatives.[85]

8.1.2 *Organization of phthalocyanines on surfaces*

To study the organization of the sole aromatic macrocyclic core of Pcs on surfaces, a series of unsubstituted and planar Pcs, that is, Pcs bearing 12 hydrogen atoms and a reasonably small central metal, were deposited onto metallic surfaces and studied by STM. CuPc,[82] CoPc,[82] NiPc,[83] and FePc[83] were all found to organize in the same close-packed ordered structure illustrated in Figure 44.

The initial stage of the growth of the first monolayer of FePc on Au(111) was carefully studied by Gao and colleagues with low-temperature STM.[86] The FePc molecules were shown to separately adsorb on the face-centered cubic and the hexagonal close-packed regions at the submonolayer regime, indicating that the molecular adsorption is greatly affected by the molecule–substrate interaction. At the initial adsorption stage, FePc molecules prefer to adsorb onto the terrace dispersedly as isolated adsorbates because of the molecule–substrate interaction that is stronger than the lateral intermolecular interaction. When increasing molecule coverage, the intermolecular interaction becomes more important. The FePc molecules self-assemble into dimers, trimers, short chains, and even cyclic hexamers (Figure 45). Interestingly, these hexamers are sustained by six intermolecular hydrogen bonds between

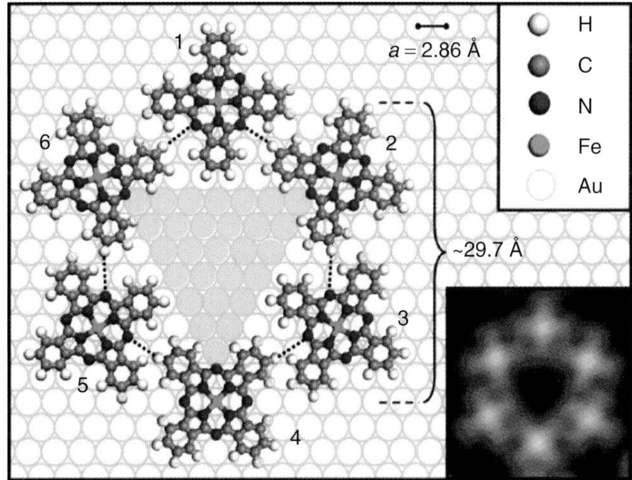

Figure 45 FePc hexamers on Au(111) at low coverage, as observed by STM.[86] (Reproduced from Ref. 86. © American Chemical Society, 2007.)

meso N atoms and nonperipheral H atoms in the vicinal macrocycle, as illustrated in Figure 45.

Interestingly, the second and third monolayers of some unsubstituted Pcs still show very good submolecular resolution when observed with low-temperature STM, suggesting a very good orbital overlap between aromatic macrocycles in adjacent layers. In the case of FePc, the unit cell of the formed molecular superstructure of the second layer shifts compared to the unit cell of the first layer. The Fe atoms in the top layer are shifted with respect to the ones underneath, and the plane of the Pc in the second layer was found to be tilted with respect to the plane of the substrate. Multilayers of CoPc (up to three layers) were also studied and it was shown that in this case the positions of both the unit cell and the metal atom are retained from one layer to the next.[87] Nevertheless, the plane of the Pc starts to tilt from the second layer on, being even more tilted in the third monolayer. To date, there is no explanation for this

Figure 46 Organization of nonplanar SnPc on Ag(111), as observed by STM.[90] (Reproduced from Ref. 90. © Wiley-VCH, 2009.)

increasing tilting of the Pc planes on going from one layer to the next.

There are several ways a Pc may be made nonplanar: by incorporating a central metal atom possessing a van der Waals radius greater than the size of the cavity, such as Pb, or by coordination of an axial ligand to the central metal atom. In the case of vanadium oxide phthalocyanine, the organization on Au(111) was found to be exactly identical to the one described earlier.[88] On the other hand, the molecular architecture of titanyl phthalocyanine (TiOPc) was found to be much more interesting.[89] Thus, three phases of TiOPc were found on Ag(111) under UHV, depending on the coverage.

Remarkably, tin phthalocyanine (SnPc), which is also a nonplanar Pc, was shown to organize in a different manner depending on the orientation of the Pc with respect to the surface. Thus, Pcs with tin pointing upward stay isolated from each other and from tin-down Pcs, while tin-down Pcs self-organize into 1D chains (Figure 46).[90]

In general, the adsorption geometry of nonplanar Pcs is not well understood and has not yet been definitely assigned.

Peripherally substituted Pcs were shown to possess a very different behavior on surfaces than their unsubstituted counterparts. The weak substrate–molecule and molecule–molecule interactions are replaced by relatively stronger intermolecular interactions that govern the self-organization macrocycle.

Alkylated and hexadecahalogenated Pcs are usually well packed, with their mean planes lying flat on the surface.[91] Their organization was shown to maximize the hydrophobic interactions between the lipophilic alkyl chains. Halogenated Pcs were found to be extremely difficult to observe on HOPG (highly oriented pyrolytic graphite) as a consequence of their extremely weak interactions with the substrate.[92] Jung and colleagues demonstrated that symmetrically substituted Pcs with eight peripheral di-(*tert*-butyl)phenoxy (DTPO) groups self-organize on both Ag(111) and Au(111) substrates into various assembly structures.[93] Mazur and colleagues described the organization of vanadyl oxide tetraphenoxy phthalocyanine on HOPG, showing that the adlayer formed three different stable architectures at HOPG–n-alkyl benzene interfaces.[94]

8.2 Organization of subphthalocyanines on surfaces

8.2.1 Study of single subphthalocyanines on surfaces

Single chlorosubphthalocyanine (ClSubPc) molecules were observed for the first time in 2000 under UHV on Si(111)-(7 × 7) substrates.[95] ClSubPc appears as three bright spots corresponding to the three outer aromatic rings when the chlorine atom points downward toward the surface (Figure 47) and as one bigger spot when the chlorine atom

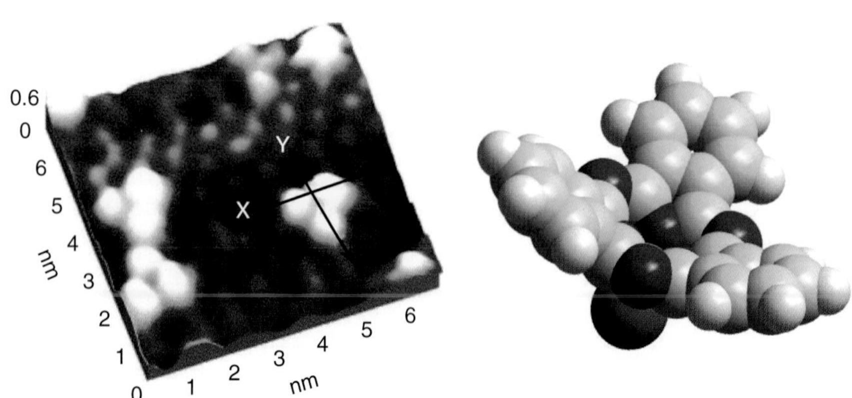

Figure 47 Single chlorosubphthalocyanine with its chlorine atom pointing toward the substrate on Si(111)-(7 × 7), as observed by STM.[95] (Reproduced from Ref. 95. © American Physical Society, 2000.)

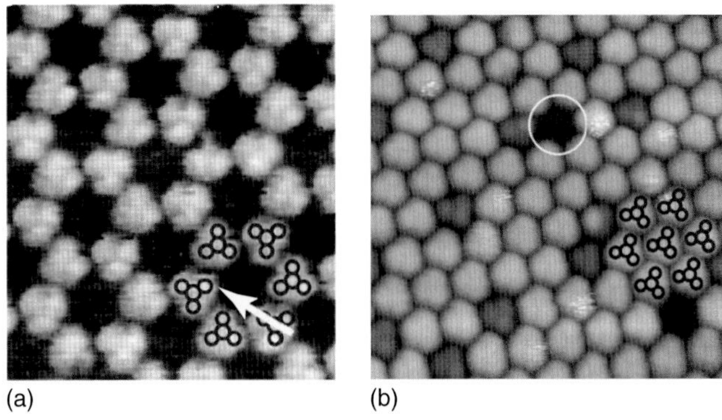

Figure 48 STM image of 2D honeycomb (a) and hexagonal compact (b) overlayers of SubPcs on Ag(111).[96] (Reproduced from Ref. 96c. © American Physical Society, 2003.)

8.2.2 *Organization of subphthalocyanines on surfaces*

Vacuum deposition of SubPcs onto silicon(111),[95] silver(111),[96] copper(111),[97] and gold(111)[98] substrates has been performed to yield highly ordered 2D arrays. From these studies, it can be concluded that the growth of SubPc monolayers is an epitaxial process that respects both the symmetry and the main geometric characteristics of the underlying substrate. For example, on Ag(111), ClSubPc self-organizes into a honeycomb pattern at low coverage (Figure 48). On increasing coverage, it starts to form a

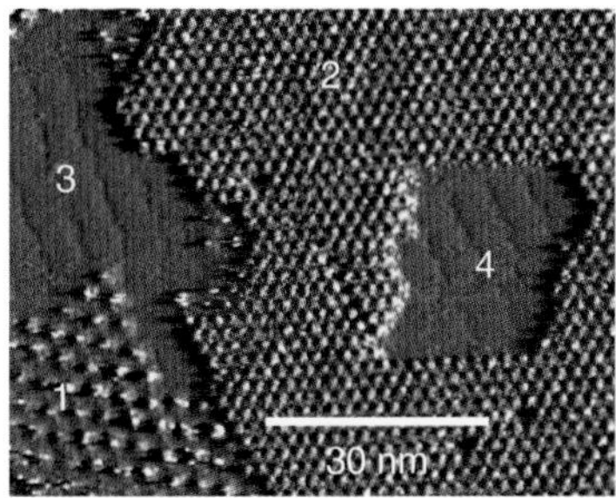

Figure 49 STM image of four coexisting Pc phases on HOPG.[99] (Reproduced from Ref. 99. © Wiley-VCH, 2001.)

points upward. ClSubPc with Cl pointing downward was observed to be shaped as a donut on Cu(100), most probably as a consequence of the fast rotation, with respect to the scanning rate of the microscope, of the macrocycle along its B–Cl axis. According to Jung and colleagues, who studied the 2D phase diagram of ClSubPcs on Ag(111), the Cl-down orientation is the most stable one. Nevertheless, the coexistence of 2D ClSubPc gas phase with more ordered phases demonstrates that ClSubPc–substrate interactions are very weak on Ag(111).

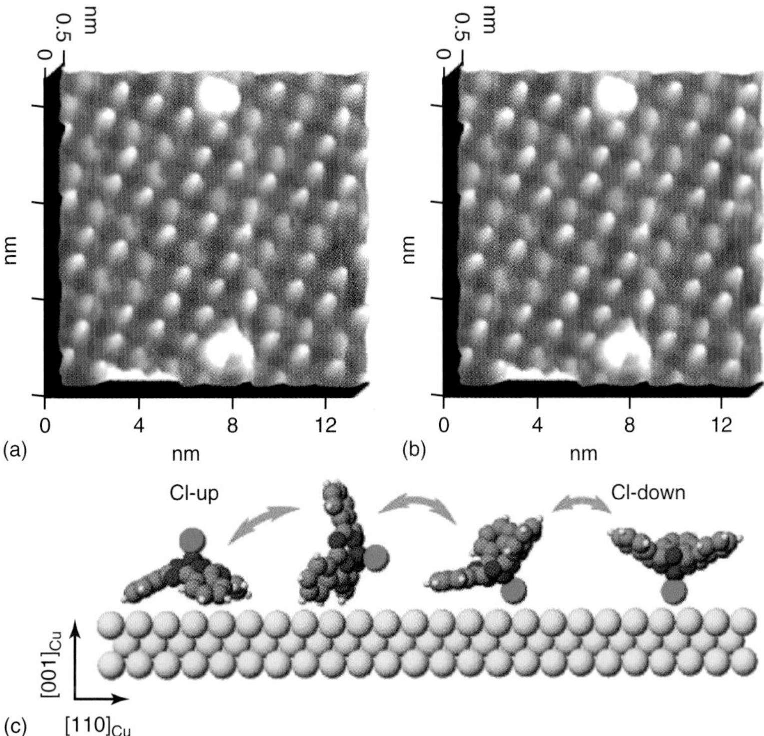

Figure 50 Organization of ClSubPc on Cu(100), as observed by STM. Depending on the voltage of the tip, the SubPc can be orientated with Cl up (a) or Cl down (b). (c) Schematic representation of the tip-induced flip-flop of the SubPc macrocycle.[97] (Reproduced from Ref. 97. © American Chemical Society, 2002.)

compact hexagonal pattern in which it is possible to distinguish between Cl-down and Cl-up molecules, with the latter representing the defects in the lattice.

8.3 Manipulation of phthalocyanines and subphthalocyanines on surfaces

Even before grasping a full understanding of the interactions between Pcs and surfaces, some scientists have undertaken the even more arduous task of manipulating and modifying their organization at a single molecule scale.

Rabe and colleagues proved that the interplay of intramolecular, intermolecular, and interfacial forces can be used to construct highly organized two- and three-dimensional architectures from Pc building blocks on the HOPG surface.[99] The switching between different nanostructures (Figure 49) coexisting at the surface can be stimulated with the tip of the STM with a resolution at the molecular scale.

Another example of nanostructure manipulation was recently given by Suzuki and colleagues in the case of a SubPc. When deposited on a Cu(100) surface,[97] SubPc molecules were found to align reversibly with Cl-up or Cl-down orientations (Figure 50), depending on the tunneling voltage applied between the STM tip and the metal surface. This phenomenon is a direct consequence of the quite high dipole moment of ClSubPcs, about 5 D. Control of these switching alignments at a single molecule scale would lead to an information density of around 60 Tb cm^{-2}.

9 CONCLUSION

Pcs have drawn considerable attention as molecular materials that give rise to outstanding electronic and optical properties. The supramolecular organization of Pcs is a fruitful concept for the design of new systems exhibiting specific macroscopic properties. New hierarchical structural ordering can be achieved by arranging the Pc moieties either with one another or with appropriate counterparters in different ways, using $\pi-\pi$ interactions, including D–A interactions, metal coordination, hydrogen bonding, and electrostatic interactions, or simply by placing them on surfaces. The development of these concepts, some old and many new, will allow the controlled organization of Pcs into highly ordered self-assembled arrays, which will pave the way for new potential applications of Pcs in different fields of materials science, particularly in nanotechnology.

ACKNOWLEDGMENTS

Financial support from the MICINN and MEC (Spain) (CTQ2008-00418/BQU, PLE2009-0070, and Consolider-Ingenio Nanociencia Molecular CSD2007-00010), Comunidad de Madrid (MADRISOLAR-2, S2009/PPQ/1533), and EU (MRTN-CT-2006-035533 Solar-N-type and Project ROBUST DSC FP7-Energy-2007-1-RTD, N° 212792) is gratefully acknowledged.

REFERENCES

1. (a) M. R. Wasielewski, *Acc. Chem. Res.*, 2009, **42**, 1910; (b) Z. Chen, A. Lohr, C. R. Saha-Möller, and F. Würthner, *Chem. Soc. Rev.*, 2009, **38**, 564; (c) Y. Kobuke, *Eur. J. Inorg. Chem.*, 2006, 2333.
2. C. A. Hunter and J. K. M. Sanders, *J. Am. Chem. Soc.*, 1990, **112**, 5525.
3. G. de la Torre, C. G. Claessens, and T. Torres, *Chem. Commun.*, 2007, 2000.
4. A. W. Snow, in *The Porphyrin Handbook*, eds. K. M. Kadish, K. M. Smith, and R. Guillard, Academic, San Diego, 2003, vol. 17, p. 129.
5. A. R. Monahan, J. A. Brado, and A. F. DeLuca, *J. Phys. Chem.*, 1972, **76**, 446.
6. R. D. George, A. W. Snow, J. S. Shirk, and W. R. Barger, *J. Porphyrins Phthalocyanines*, 1998, **2**, 1.
7. (a) P. C. Martin, M. Gouterman, B. V. Pepich, *et al.*, *Inorg. Chem.*, 1991, **30**, 3305; (b) S. Dhami, A. J. de Mello, G. Rumbles, *et al.*, *Photochem. Photobiol.*, 1995, **61**, 341.
8. A. Satake and Y. Kobuke, *Org. Biomol. Chem.*, 2007, **5**, 1679.
9. (a) C. Piechocki, J. Simon, A. Skoulios, *et al.*, *J. Am. Chem. Soc.*, 1982, **104**, 5245; (b) K. Ohta, L. Jacquemin, C. Sirlin, *et al.*, *New J. Chem.*, 1998, **12**, 751; (c) D. Masurel, C. Sirlin, and J. Simon, *New J. Chem.*, 1987, **11**, 455.
10. H. Eichhorn, *J. Porphyrins Phthalocyanines*, 2000, **4**, 88.
11. G. Bottari, G. de la Torre, D. M. Guldi, and T. Torres, *Chem. Rev.*, 2010, **110**, 6768–6816. DOI: 10.1021/cr900254z.
12. X. Y. Li, L. E. Sinks, B. Rybtchinski, and M. R. Wasielewski, *J. Am. Chem. Soc.*, 2004, **126**, 10810.
13. A. Gouloumis, D. González-Rodríguez, P. Vázquez, *et al.*, *J. Am. Chem. Soc.*, 2006, **128**, 12674.
14. D. M. Guldi, A. Gouloumis, P. Vázquez, *et al.*, *J. Am. Chem. Soc.*, 2005, **127**, 5811.
15. G. Bottari, D. Olea, C. Gómez-Navarro, *et al.*, *Angew. Chem. Int. Ed.*, 2008, **47**, 2026–2031.
16. (a) A. de la Escosura, M. V. Martínez-Díaz, J. Barberá, and T. Torres, *J. Org. Chem.*, 2008, **73**, 1475–1480; (b) Y. H. Geerts, O. Debever, C. Amato, and S Sergeyev, *Beilstein J. Org. Chem.*, 2009, **5**, No 49, DOI: 10.3762/bjoc.5.49.
17. F. J. M. Hoeben, P. Jonkheijm, E. W. Meijer, and A. P. H. J. Schenning, *Chem. Rev.*, 2005, **105**, 1491.

18. (a) C. G. Claessens and J. F. Stoddart, *J. Phys. Org. Chem.*, 1997, **10**, 254; (b) V. Balzani, A. Credi, F. M. Raymo, and J. F. Stoddart, *Angew. Chem. Int. Ed.*, 2000, **39**, 3348.
19. (a) M. R. Bryce, *J. Mater. Chem.*, 2000, **10**, 589; (b) J. L. Segura and N. Martín, *Angew. Chem. Int. Ed.*, 2001, **40**, 1372; (c) E. Toshiaki and M. Akira, *Chem. Rev.*, 2004, **104**, 5449.
20. A. de la Escosura, M. V. Martínez-Díaz, P. Thordarson, *et al.*, *J. Am. Chem. Soc.*, 2003, **125**, 12300.
21. A. de la Escosura, M. V. Martínez-Díaz, D. M. Guldi, and T. Torres, *J. Am. Chem. Soc.*, 2006, **128**, 4112.
22. J. Sly, P. Kasak, E. Gomar-Nadal, *et al.*, *Chem. Commun.*, 2005, 1255.
23. (a) G. Zucchi, B. Donnio, and Y. H. Geerts, *Chem. Mater.*, 2005, **17**, 4273; (b) G. Zucchi, P. Viville, B. Donnio, *et al.*, *J. Phys. Chem. B*, 2009, **113**, 5448–5457.
24. B. H. Northrop, Y.-R. Zheng, K.-W. Chi, and P. J. Stang, *Acc. Chem. Res.*, 2009, **42**, 1554–1563.
25. C. M. Drain, A. Varotto, and I. Radivojevic, *Chem. Rev.*, 2009, **109**, 1630–1658.
26. X. Y. Li and D. K. P. Ng, *Eur. J. Inorg. Chem.*, 2000, 1845.
27. M. T. M. Choi, C. F. Choi, and D. K. P. Ng, *Tetrahedron*, 2004, **60**, 6889.
28. H. Xu and D. K. P. Ng, *Inorg. Chem.*, 2008, **47**, 7921.
29. B. Gao, Y. Li, J. Su, and H. Tian, *Supramol. Chem.*, 2007, **19**, 207.
30. F. D'Souza and O. Ito, *Chem. Commun.*, 2009, 4913.
31. D. M. Guldi, J. Ramey, M. V. Martínez-Díaz, *et al.*, *Chem. Commun.*, 2002, 2774.
32. M. E. El-Khouly, L. M. Rogers, M. E. Zandler, *et al.*, *ChemPhysChem*, 2003, **4**, 474.
33. F. D'Souza, G. R. Deviprasad, M. E. Zandler, *et al.*, *J. Phys. Chem. A.*, 2002, **106**, 3243.
34. M. E. El-Khouly, Y. Araki, O. Ito, *et al.*, *J. Porphyrins Phthalocyanines*, 2006, **10**, 1156.
35. F. D'Souza, S. Gadde, M. E. El-Khouly, *et al.*, *J. Porphyrins Phthalocyanines*, 2005, **9**, 698.
36. P. A. Troshin, R. Koeppe, A. S. Peregudov, *et al.*, *Chem. Mater.*, 2007, **19**, 5363.
37. R. Chitta, A. S. D. Sandanayaka, A. L. Schumacher, *et al.*, *J. Phys. Chem. C*, 2007, **111**, 6947.
38. G. Berger, A. N. Cammidge, I. Chambrier, *et al.*, *Tetrahedron Lett.*, 2003, **44**, 5527.
39. A. N. Cammidge, G. Berger, I. Chambrier, *et al.*, *Tetrahedron*, 2005, **61**, 4067.
40. M. S. Rodríguez-Morgade, T. Torres, C. Atienza-Castellanos, and D. M. Guldi, *J. Am. Chem. Soc.*, 2006, **128**, 15145.
41. T. Rawling, C. Austin, F. Buchholz, *et al.*, *Inorg. Chem.*, 2009, **48**, 3215.
42. M. K. R. Fischer, I. López-Duarte, M. M. Wienk, *et al.*, *J. Am. Chem. Soc.*, 2009, **131**, 8669.
43. F. Silvestri, I. López-Duarte, W. Seitz, *et al.*, *Chem. Commun.*, 2009, 4500.
44. M. S. Rodríguez-Morgade, M. E. Plonska-Brzezinska, A. J. Athans, *et al.*, *J. Am. Chem. Soc.*, 2009, **131**, 10484.
45. Y. Kobuke, *Struct. Bond.*, 2006, **121**, 49.
46. K. Kameyama, M. Morisue, A. Satake, and Y. Kobuke, *Angew. Chem. Int. Ed.*, 2005, **44**, 4763.
47. K. Kameyama, A. Satake, and Y. Kobuke, *Tetrahedron Lett.*, 2004, **45**, 7617.
48. F. Ito, Y. Ishibashi, S. R. Khan, *et al.*, *J. Phys. Chem. A*, 2006, **110**, 12734.
49. M. Morisue and Y. Kobuke, *Chem. Eur. J.*, 2008, **14**, 4993.
50. K. Kameyama, M. Morisue, A. Satake, and Y. Kobuke, *Angew. Chem. Int. Ed.*, 2005, **44**, 4763.
51. A. Satake, T. Sugimura, and Y. Kobuke, *J. Porphyrins Phthalocyanines*, 2009, **13**, 326.
52. M. J. Cook and A. Jafari-Fini, *J. Mater. Chem.*, 1997, **7**, 2327.
53. (a) K. Ishii, S. Abiko, M. Fujitsuka, *et al.*, *J. Chem. Soc. Dalton Trans.*, 2002, 1735; (b) K. Ishii, M. Iwasaki, and N. Kobayashi, *Chem. Phys. Lett.*, 2007, **436**, 94; (c) K. Ishii, Y. Watanabe, S. Abiko, and N. Kobayashi, *Chem. Lett.*, 2002, 450.
54. S. Y. Al-Raqa, M. J. Cook, and D. L. Hughes, *Chem. Commun.*, 2003, 62.
55. E. B. Fleischer and A. M. Shachter, *Inorg. Chem.*, 1991, **30**, 3763.
56. V. Novakova, P. Zimcik, K. Kopecky, *et al.*, *Eur. J. Org. Chem.*, 2008, 3260.
57. (a) X. Huang, F. Zhao, Z. Li, *et al.*, *Chem. Lett.*, 2007, **36**, 108; (b) X. Huang, F. Zhao, Z. Li, *et al.*, *Langmuir*, 2007, **23**, 5167; (c) X. Huang, F. Q. Zhao, Z. Y. Li, *et al.*, *Chem. J. Chin. Univ.*, 2007, **28**, 487; (d) L. H. Niu, C. Zhong, Z. H. Chen, *et al.*, *Chin. Sci. Bull.*, 2009, **54**, 1169.
58. C. G. Claessens and T. Torres, *J. Am. Chem. Soc.*, 2002, **124**, 14522.
59. C. G. Claessens and T. Torres, *Chem. Commun.*, 2004, 1298.
60. C. G. Claessens and T. Torres, *Eur. J. Org. Chem.*, 2000, 1603–1607.
61. C. G. Claessens, I. Sánchez-Molina, and T. Torres, *Supramol. Chem.*, 2009, **21**, 44.
62. C. G. Claessens, M. J. Vicente-Arana, and T. Torres, *Chem. Commun.*, 2008, 6378.
63. (a) P. C. Lo, X. Leng, and D. K. P. Ng, *Coord. Chem. Rev.*, 2007, **251**, 2334; (b) T. Fournier, Z. Liu, T. H. Tran-Thi, *et al.*, *J. Phys. Chem. A.*, 1999, **103**, 1179.
64. (a) A. R. Koray, V. Ahsen, and O. Bekaroglu, *J. Chem. Soc. Chem. Commun.*, 1986, 932; (b) V. Ahsen, E. Yilmaser, M. Ertas, and O. Bekaroglu, *J. Chem. Soc. Dalton Trans.*, 1988, 401.
65. (a) N. Kobayashi and Y. Nishiyama, *J. Chem. Soc. Chem. Commun.*, 1986, 1462; (b) N. Kobayashi and A. B. P. Lever, *J. Am. Chem. Soc.*, 1987, **109**, 7433.
66. (a) O. E. Sielcken, L. A. Van de Kuil, W. Drenth, and R. J. M. Nolte, *J. Chem. Soc. Chem. Commun.*, 1986, 1464; (b) O. E. Sielcken, M. M. van Tilborg, M. F. M. Roks, *et al.*, *J. Am. Chem. Soc.*, 1987, **109**, 4261.

67. (a) J. A. Duro and T. Torres, *Chem. Ber.*, 1993, **126**, 269; (b) J. A. Duro, G. de la Torre, and T. Torres, *Tetrahedron Lett.*, 1995, **36**, 8079.

68. (a) A. V. Gusev and M. A. J. Rodgers, *J. Phys. Chem. A*, 2002, **106**, 1985; (b) A. V. Gusev, E. O. Danilov, and M. A. J. Rodgers, *J. Phys. Chem. A*, 2002, **106**, 1993.

69. F. D'Souza, E. Maligaspe, K. Ohkubo, *et al.*, *J. Am. Chem. Soc.*, 2009, **131**, 8787.

70. (a) P. R. Ashton, P. J. Campbell, E. J. T. Chrystal, *et al.*, *Angew. Chem. Int. Ed.*, 1995, **34**, 1865; (b) P. R. Ashton, E. J. T. Chrystal, P. T. Glink, *et al.*, *Chem. Eur. J.*, 1996, **2**, 709.

71. D. M. Guldi, J. Ramey, M. V. Martínez-Díaz, *et al.*, *Chem. Commun.*, 2002, 2774.

72. M. V. Martínez-Díaz, M. S. Rodríguez-Morgade, M. C. Feiters, *et al.*, *Org. Lett.*, 2000, **2**, 1057.

73. (a) A. Lützen, S. D. Starnes, D. M. Rudkevich, and J. Rebek Jr., *Tetrahedron Lett.*, 2000, **41**, 3777; (b) N. Kumaran, P. A. Veneman, B. A. Minch, *et al.*, *Chem. Mater.*, 2010, **22**, 2491–2501.

74. J. A. Duro, G. de la Torre, J. Barberá, *et al.*, *Chem. Mater.*, 1996, **8**, 1061.

75. M. Kimura, T. Kuroda, K. Ohta, *et al.*, *Langmuir*, 2003, **19**, 4825.

76. J. L. Sessler and J Jayawickramarajah, *Chem. Commun.*, 2005, 1939.

77. X.-Y. Li and D. K. P. Ng, *Tetrahedron Lett.*, 2001, **42**, 305.

78. (a) J. L. Sessler, J. Jayawickramarajah, A. Gouloumis, *et al.*, *Tetrahedron*, 2006, **62**, 2123; (b) T. Torres, A. Gouloumis, D. Sánchez-Garcia, *et al.*, *Chem. Commun.*, 2007, 292.

79. W. Seitz, A. J. Jiménez, E. Carbonell, *et al.*, *Chem. Commun.*, 2010, **46**, 127.

80. J. K. Gimzewski, E. Stoll, and R. R. Schlittler, *Surf. Sci.*, 1987, **181**, 267.

81. P. H. Lippel, R. J. Wilson, M. D. Miller, *et al.*, *Phys. Rev. Lett.*, 1989, **62**, 171.

82. (a) X. Lu, K. W. Hipps, X. D. Wang, and U. Mazur, *J. Am. Chem. Soc.*, 1996, **118**, 7197; (b) K. W. Hipps, X. Lu, X. D. Wang, and U. Mazur, *J. Phys. Chem.*, 1996, **100**, 11207.

83. X. Lu and K. W. Hipps, *J. Phys. Chem. B*, 1997, **101**, 5391.

84. D. E. Barlow and K. W. Hipps, *J. Phys. Chem. B*, 2000, **104**, 5993.

85. (a) L. Scudiero, D. E. Barlow, and K. W. Hipps, *J. Phys. Chem. B*, 2000, **104**, 11899; (b) L. Scudiero, D. E. Barlow, U. Mazur, and K. W. Hipps, *J. Am. Chem. Soc.*, 2001, **123**, 4073.

86. (a) Z. H. Cheng, L. Gao, Z. T. Deng, *et al.*, *J. Phys. Chem. C*, 2007, **111**, 2656; (b) Z. H. Cheng, L. Gao, Z. T. Deng, *et al.*, *J. Phys. Chem. C*, 2007, **111**, 9240.

87. M. Takada and H. Tada, *Chem. Phys. Lett.*, 2004, **392**, 265.

88. D. E. Barlow and K. W. Hipps, *J. Phys. Chem. B*, 2000, **104**, 5993.

89. Y. Wei, S. W. Robey, and J. E. Reutt-Robey, *J. Phys. Chem. C*, 2008, **112**, 18537.

90. Y. F. Wang, J. Kröger, R. Berndt, and W. Hofer, *Angew. Chem. Int. Ed.*, 2009, **48**, 1261.

91. S. Yoshimoto, A. Tada, K. Suto, and K. Itaya, *J. Phys. Chem. B*, 2003, **107**, 5836.

92. Z.-Y. Yang, S.-B. Lei, L.-H. Gan, *et al.*, *ChemPhysChem*, 2005, **6**, 65.

93. T. Samuely, S. X. Liu, N. Wintjes, *et al.*, *J. Phys. Chem. C*, 2008, **112**, 6139.

94. U. Mazur, K. W. Hipps, and S. L. Riechers, *J. Phys. Chem. C*, 2008, **112**, 20347.

95. H. Yanagi, D. Schlettwein, H. Nakayama, and T. Nishino, *Phys. Rev. B*, 2000, **61**, 1959.

96. (a) S. Berner, M. Brunner, L. Ramoino, *et al.*, *Chem. Phys. Lett.*, 2001, **348**, 175; (b) M. de Wild, S. Berner, H. Suzuki, *et al.*, *ChemPhysChem*, 2002, **3**, 881; (c) S. Berner, M. de Wild, L. Ramoino, *et al.*, *Phys. Rev. B*, 2003, **68**, 115410.

97. H. Yanagi, K. Ikuta, H. Mukai, and T. Shibutani, *Nano Lett.*, 2002, **2**, 951.

98. (a) H. Suzuki, H. Miki, S. Yokoyama, and S. Mashiko, *J. Phys. Chem. B*, 2003, **107**, 3659; (b) S. Mannsfeld, H. Reichhard, and T. Fritz, *Surf. Sci.*, 2003, **525**, 215.

99. P. Samorí, H. Engelkamp, P. de Witte, *et al.*, *Angew. Chem. Int. Ed.*, 2001, **40**, 2348.

Guanidinium-Based Receptors for Oxoanions

Matthew P. Conley, Julián Valero, and Javier de Mendoza

Institute of Chemical Research of Catalonia (ICIQ), Tarragona, Spain

1 INTRODUCTION

Nature frequently uses guanidinium moieties to coordinate different anions. The amino acid arginine contains a guanidinium group in the side chain which forms strong ion pairs with oxoanions and plays a crucial role in the active sites of enzymes. For example, the crystal structure of substrate-bound fumarate reductase shows that arginine-402 functions as a general acid and that arginine-544 helps to anchor one of the carboxylates of fumarate.[1] In the nucleosome, the arginine-rich histone wraps around DNA, and the amino acid side chains clearly show direct interactions with the anionic phosphodiester fragments of the DNA. Of the 39 arginine residues present in the four histone proteins forming the nucleosome core, 7 arginines enter the minor groove of DNA and are essential for histone–DNA binding. Methylation of arginine residues in the histone core leads to a conformational change, which in turn regulates DNA transcription.[2]

The capacity of the guanidinium group to bind oxoanions is due to its high pK_a value (about 12–13) [1], which ensures protonation over a wide pH range, and to its planar Y-shape, which directs hydrogen bonding. Four of the five NH bonds present in the guanidinium group of arginine can form hydrogen bonds with oxoanion acceptors along the two available edges (Figure 1). The positive charge is delocalized over the three nitrogen atoms, rendering the guanidinium a rather soft cation. From an energetic point of view, binding to oxoanions results from both ion pairing and hydrogen bonding. The favorable binding energy arises from the attractive electrostatic interaction (enthalpy) and removal of the solvation shell around the host–guest partners (entropy). In some cases, the entropic term becomes the dominant energetic driving force for the formation of the host–guest complex. As a result of these features, molecular recognition studies have extensively used artificial guanidinium-based compounds as receptors for anions.[3]

We should note that the oxoanion binding energetics is more complicated than the simple description above, because the formation of a guanidinium–oxoanion complex in highly polar or protic solvents remains a major challenge. In fact, the guanidinium–oxoanion interaction in proteins usually occurs within areas of low dielectric constant. Artificial systems that are designed to work in water or polar solvents are more exposed to solvation effects, which compete with the hydrogen-bond donor and acceptor sites, causing a substantial decrease of the binding constant. This is usually overcome by increasing the number of

Supramolecular Chemistry: From Molecules to Nanomaterials.
Edited by Philip A. Gale and Jonathan W. Steed.
© 2012 John Wiley & Sons, Ltd. ISBN: 978-0-470-74640-0.

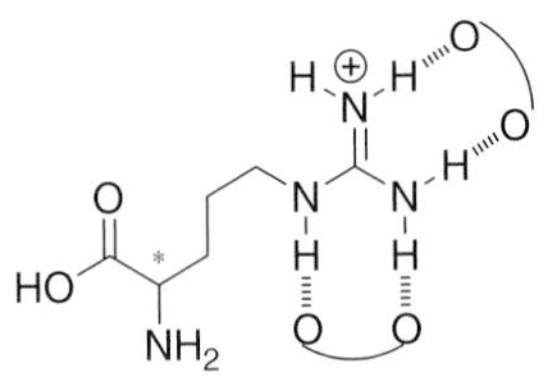

Figure 1 The guanidinium group of arginine, and the two possible oxoanion binding modes.

charges and hydrogen-bond donors, or designing receptors that limit access of the binding site to the solvent. In this chapter, a general overview and several recent examples of artificial guanidinium receptors as well as their oxoanion binding behavior are provided.

2 CARBOXYLATES

The NH hydrogen-bonding pairs of a substituted guanidinium can display either syn or anti conformations (Figure 2a), although only the syn arrangement is suitable to form hydrogen-bonding complexes with carboxylates. Insertion of the guanidinium into the bicyclic framework of decalin provides exclusive access to the syn isomer (Figure 2b). This conformational rigidity places the NH donors in an ideal location for the two syn lone pairs of oxoanions. Also, the lipophilic backbone of the bicyclic guanidinium increases its solubility in apolar solvents, where hydrogen bonds are stronger.

Two possible hydrogen-bonding modes could be considered depending on the position of the acidic proton (Figure 2b). The formally neutral AD–DA (acceptor-donor–donor-acceptor) complex results from hydrogen bonding between the bicyclic guanidine and a protonated carboxylic acid. This form is disfavored because of the large difference in pK_a between the carboxylic acid and the guanidinium (about 9 pK_a units in water), and contains a strong repulsive oxygen–nitrogen lone pair interaction (dashed line) [2].[4] Also, the guanidine is expected to be nonplanar, and would contain unsymmetrical C–N bond distances.[5] A more likely scenario is the ionic DD–AA (donor–donor–acceptor–acceptor) hydrogen-bonded complex. The NHs and oxygen lone pairs are geometrically well defined because both the carboxylate and guanidinium are planar. The complex gains stabilization from the ion pair, hydrogen bonding (hashed lines), and the favorable secondary interactions between the hydrogen-bond donors and the oxygen loan pairs (dashed lines).

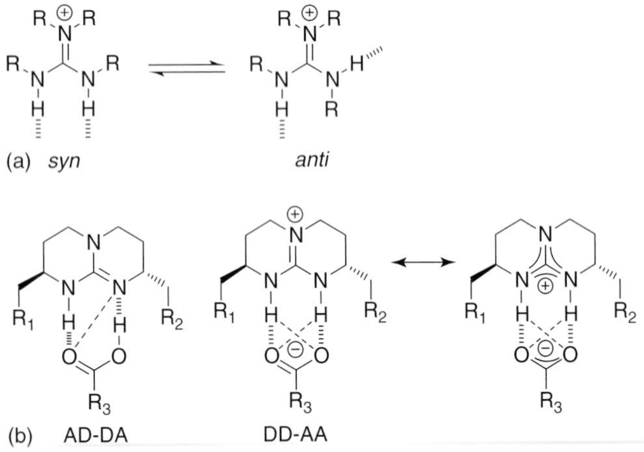

Figure 2 (a) Syn and anti conformations of the guanidinium group; (b) A chiral bicylclic guanidinium–carboxylate interaction, showing both the AD–DA (acceptor-donor–donor-acceptor) and DD–AA (donor-donor–acceptor-acceptor) hydrogen-bonded pairs. The delocalized resonance structure of the DD–AA adduct is also shown.

Initial studies by Schmidtchen and coworkers used the tetraphenylborate salt of bicyclic guanidinium **1**.[6] The acetate salt was characterized by X-ray crystallography. **1*OAc** contains two strong identical hydrogen bonds. The C–N bond distances in the receptor were identical, indicating that the DD–AA complex is formed in the solid state. UV titrations between **1** and tetrabutylammonium (TBA) *p*-nitrobenzoate in chloroform gave $K_a = 7 \times 10^6\ M^{-1}$. These results established that the structurally rigid bicyclic guanidinium **1**, possessing two hydrogen bonds in a linear array, forms stable complexes with carboxylates in solution and in the solid state.

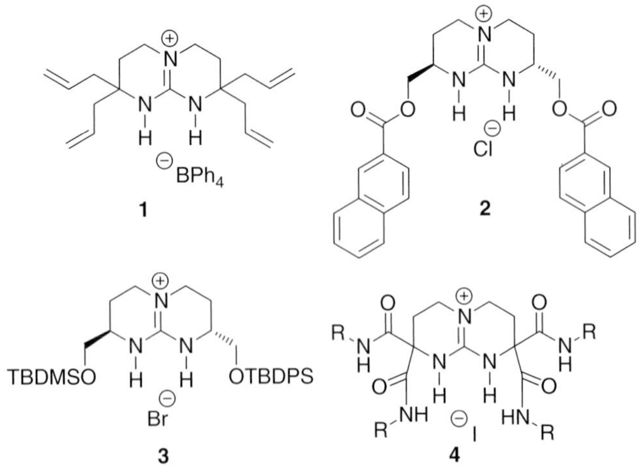

We developed the chiral guanidinium receptor **2**, which contains two flanking naphthoyl groups to promote π-stacking, for aromatic carboxylates binding.[7] The ^{1}H NMR (nuclear magnetic resonance) spectrum of **2** with tetraethylammonium *p*-nitrobenzoate in $CDCl_3$ displayed large downfield shifts of the guanidinium NH signals, which indicated that the hydrogen-bonded complex had formed. Stacking interactions between the naphthoyl side arms and the *p*-nitrobenzoate moieties were also present, based on chemical shift differences. However, a ^{1}H NMR titration study revealed that the binding constant for **2** ($K_a = 7 \times 10^3\ M^{-1}$) was much lower than the value obtained for **1**. This result indicates that the initial guanidinium–anion pair affects the binding strength. The tetraphenylborate anion in **1** is much less ion-paired to the guanidinium than the chloride in **2**. Therefore, weakly coordinating anions such as tetraphenylborate or hexafluorophosphate are necessary for high binding constants.[8]

Schmidtchen also studied carboxylate–guanidinium interactions by isothermal titration calorimetry (ITC).[9] The binding curve between tetraethylammonium acetate and **3** in acetonitrile ($K_a = 2.0 \times 10^5\ M^{-1}$) revealed that the process is both enthalpically and entropically favorable for a 1 : 1 complex. Experiments in CH_3OH produced too little heat to be measured. However, ITC experiments between **4**, which contains amide side arms as additional hydrogen-bond donors, and *p*-nitrobenzoate showed that stable 1 : 1 complexes are formed in CH_3OH ($K_a = 1.7 \times 10^4\ M^{-1}$).[10] Analysis of the ITC data for this system shows that the enthalpic term is highly endothermic, and the binding is driven by a very large entropic term. These results indicate that in polar aprotic solvents the guanidinium–carboxylate interaction is not only due to the electrostatic interaction (ΔH°) but also to the release of ordered solvent from the precomplexed species (ΔS°). In polar protic solvents, the complex is driven only by the desolvation of the precomplexed species (ΔS°). Though often overlooked, the importance of solvation in host–guest interactions must be considered.

Hamilton determined the thermodynamic parameters of glutarate (**5**) or 1,3-adamantane dicarboxylate (**6**) binding to four different receptors in competitive solvents (Figure 3).[11] The receptors were comprised of a rigid 1,4- or 1,3-substituted benzene spacer containing thiourea (**7**), urea (**8**), guanidinium (**9**), or acylguanidinium (**10**) units. The association constants, which increase as the hydrogen-bond donors become more acidic,[12] were studied by ITC experiments in DMSO (dimethyl sulfoxide). ITC studies of the binding between **7** and **8** with either dicarboxylate (**5** or **6**) gave good association constants ($10^3\ M^{-1} < K_a < 10^4\ M^{-1}$). Analysis of the binding curve showed that the association is enthalpically favored.

The association of **5** or **6** with receptors **9** or **10** (the guanidinium analogs of receptors **7** and **8**) in DMSO was too strong for accurate determination by ^{1}H NMR, and complicated by the formation of higher ordered structures in ITC experiments. Changing the solvent to CH_3OH resulted in clean 1 : 1 complexes with strong binding. In all cases, endothermic enthalpy and large positive entropic values were obtained. This result indicates that the hydrogen

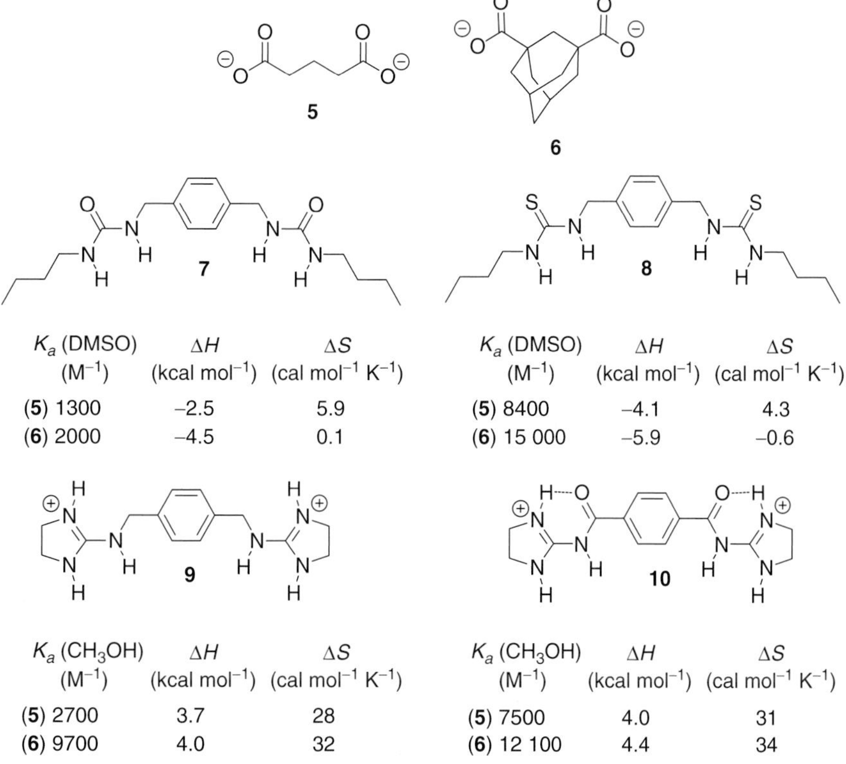

Figure 3 Receptors containing thiourea (**7**), urea (**8**), guanidinium (**9**), or acylguanidinium (**10**) units with their respective binding constants to **5** or **6**. The thermodynamic parameters to carboxylate binding are also given.

bonding between the substrate (carboxylate or guanidinium) and ordered solvent is stronger than the hydrogen bonds formed in the host–guest complex. The release of ordered solvent from both the solvated species is the likely source of the large entropy term. The desolvation process pushes the equilibrium toward the desired guanidinium–carboxylate complex.

Hof and coworkers have proposed that cation–π interactions can stabilize the guanidinium–carboxylate complex in polar solvents.[13] Molecular modeling suggested that ortho-substituted biaryl-benzylguanidinium **11** would adopt a geometry in which the guanidinium lies over one of the aromatic rings. ^{1}H NMR titration experiments of **11** with **5** in CD_3OD/D_2O give $K_a = 2680\,M^{-1}$. Control experiments with **12**, a receptor lacking the phenyl units necessary for the cation–π interaction, show a 4.7-fold decrease in the association constant for glutarate binding.

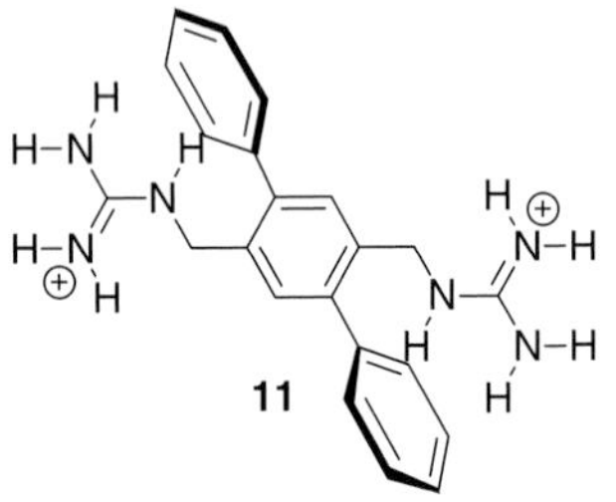

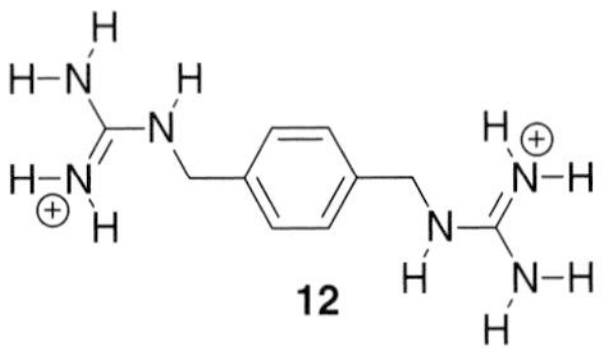

Anslyn and coworkers developed **13**, which contains three guanidinium units on a preorganized 2,4,6-triethylbenzene platform.[14] **13** selectively binds citrate (**14**) in water ($K_a = 6.9 \times 10^3\,M^{-1}$). The selectivity of citrate over other carboxylates was shown by extraction of only citrate by **13** from orange juice. Compound **13** can outcompete water for citrate, because the receptor contains three guanidinium units. Detailed work using a related metallo-receptor **15** to bind of di-, tri-, and tetracarboxylates revealed that a cooperative mechanism is operative for complexation.[15]

The examples described above show that complexation of carboxylates in competitive media (DMSO, CH_3OH, H_2O) requires multiple binding interactions (guanidinium ion pairing and/or additional hydrogen-bond donors) for reasonably high association constants. Schmuck and coworkers have investigated the more acidic, and readily tunable, acylguanidinium moiety to bind carboxylates in polar media. For example, the simple guanidinium cation **16** (pK_a = 13) does not complex carboxylates in aqueous DMSO. However, acylguanidinium **17** (pK_a = 7–8) weakly forms guanidinium–carboxylate complexes ($K_a = 50\,M^{-1}$) under identical conditions. Though the association constant is modest, the role of the carbonyl in the acyl guanidinium is worth mentioning. The carbonyl preorganizes the host by an intramolecular hydrogen bond between the acylcarbonyl and a guanidinium NH, and decreases the pK_a of the guanidinium NHs (but not to the extent as to provoke transprotonation to the weaker guanidine–carboxylic acid complex). The combination of these two factors allows stronger complexation for **17** in more polar solvents.

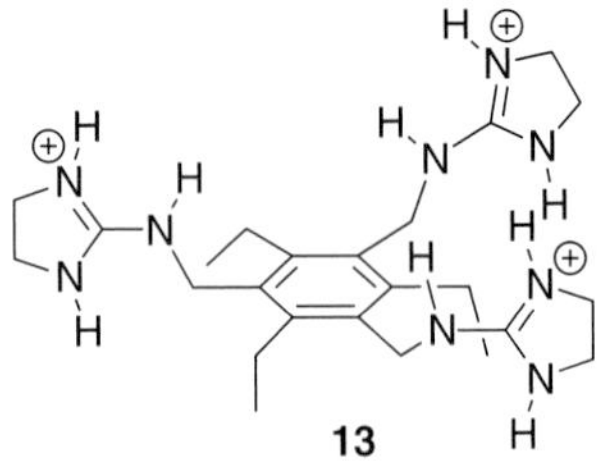

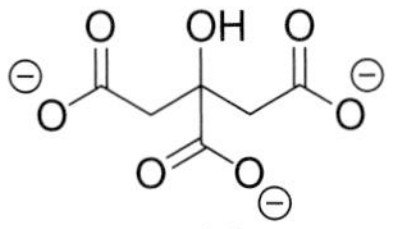

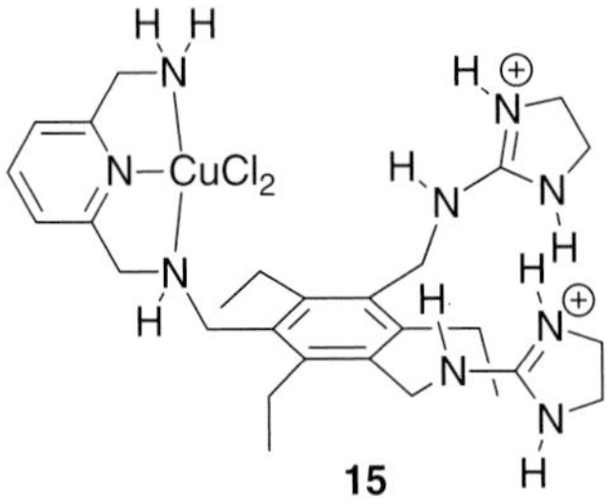

Direct modification of the simple acylguanidinium to include an additional hydrogen bond, from the pyrrole NH in **18**, increases the stabilization of the hydrogen-bonded

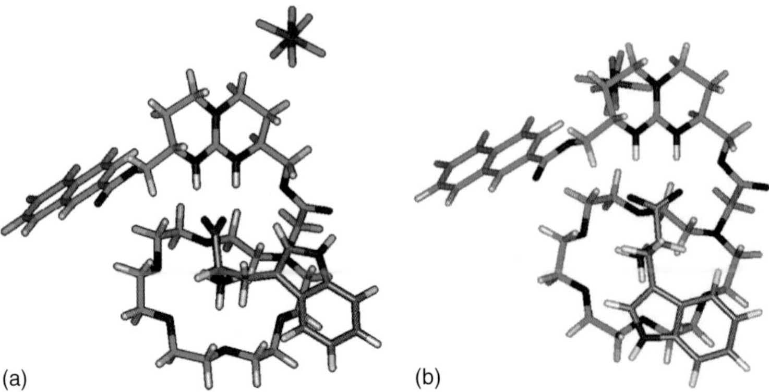

Figure 4 Molecular dynamics simulations for **23b*L-Trp** (a) and **23b*D-Trp** (b). Both structures were optimized and show no interactions between the π surfaces of the indole of tryptophan and the naphthoyl of the receptor.

adduct in aqueous DMSO by 2.2-fold ($K_a = 130\,M^{-1}$). A well-oriented NH hydrogen bond from the amide in **19** can reach the *anti* oxygen lone pair and provide even larger association constants ($K_a = 770\,M^{-1}$).[16] The size and electronic structure of the central aromatic ring are critically important to the binding behavior of **18** and **19**. Pyrrole systems are thus superior to the analogous benzene derivatives.[17] Lower carboxylate binding constants are obtained for systems containing pyridine or furan aromatic rings, likely due to the lone pair repulsion between the heteroatom and the bound carboxylate.[18]

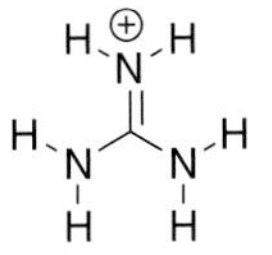

16 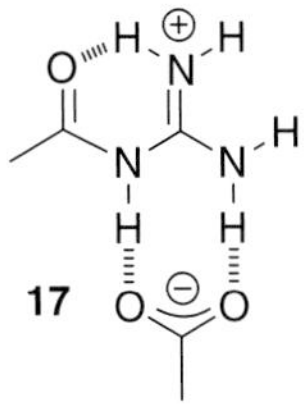 17 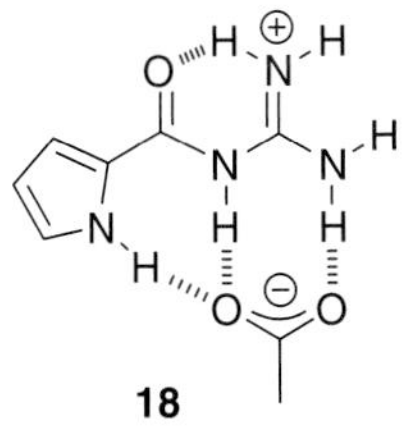 18 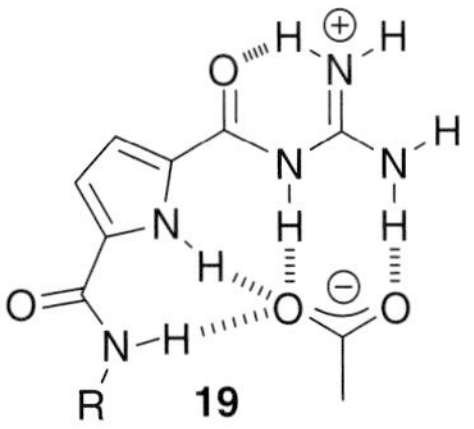 19

The much stronger binding of acylguanidinium groups has been shown by direct comparison of Anslyn's citrate receptor **13** and the triacylguanidinium pyrrole **20**.[19] The association constant between **20** and citrate in pure water was $1.6 \times 10^5\,M^{-1}$, nearly two orders of magnitude greater than the citrate binding constant for **13**. In aqueous solutions containing 200 equivalents of tris(hydroxymethyl)aminomethane (a buffering agent) and 1000 equivalents of NaCl relative to receptor **20**, the binding fell by only one order of magnitude. Under similar conditions, **13** showed a large decrease in citrate binding ($K_a < 10^2\,M^{-1}$) due to competition with the anions in solution. This result indicates that **20** is highly selective for citrate, even in the presence of large excesses of chloride and tris(hydroxymethyl)aminomethane.

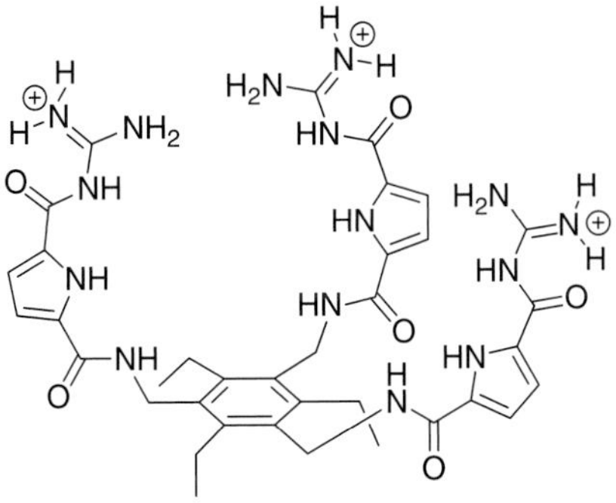

20

Schmuck has also shown that isostructural complexes that do not contain ion-pairing interactions form much weaker complexes than acylguanidinium–carboxylate adducts. Neutral **21** and zwitterionic **22** were characterized by X-ray crystallography.[20] The structures were shown to have nearly identical hydrogen-bonding distances and angles, suggesting that **21** and **22** can be directly compared in solutions as neutral and ionic congeners. Compound **21** forms only dimers in apolar solvents ($CDCl_3$, $K_a < 10^4\,M^{-1}$). The addition of 5% DMSO to $CDCl_3$ solutions of dimer shifts the equilibrium in such a way that only the monomer is present in solution. Zwitterionic **22** self-assembles in DMSO to form very stable dimers (^{1}H NMR, $K_a > 10^{10}\,M^{-1}$). Even in water the dimer forms ($K_a = 170\,M^{-1}$), which is remarkable since the assembly is formed by the electrostatic interaction between the acylguanidinium and carboxylate without additional secondary interactions (such as π-stacking, hydrophobic interactions, etc.). DFT (density functional theory) analysis suggests that the geometrical rigidity of **22** promotes formation of the dimer.[21] This result establishes that ion pairing and hydrogen bonding are necessary to form strong dimeric complexes in competitive media.

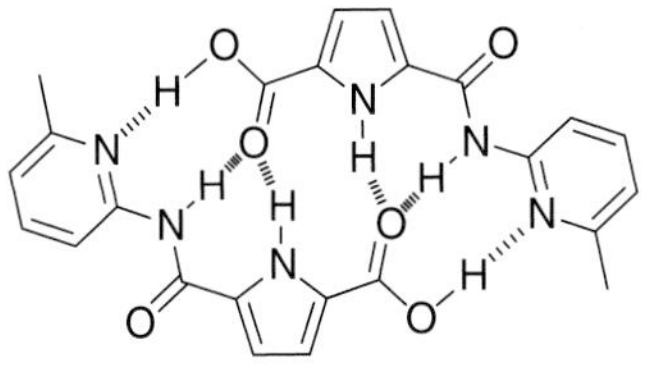

21

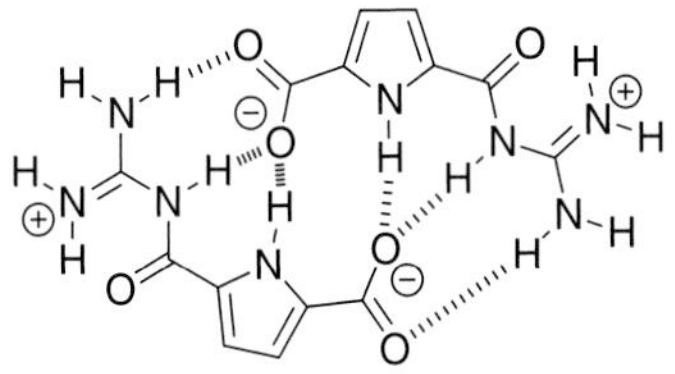

22

3 CHIRAL CARBOXYLATES

Enantiomerically pure compounds are usually obtained by asymmetric synthesis, crystallization of diastereomeric salts, kinetic resolution of racemic mixtures, or chiral chromatography. An interesting alternative to these methods is chiral discrimination of anions based on abiotic receptors. An ideal system could involve the translocation of a chiral guest between immiscible phases (chromatography, extraction, or membrane transport). For example, the chiral-lipophilic receptor would preferentially bind one enantiomer of a water-soluble substrate, and transport this molecule to another solution through a membrane. This concept has been applied to transport aqueous solutions of amino acids across bulk membranes using achiral ammonium salts.[22]

A useful concept for chiral separation using a transport or extraction reagent is the *three-point binding rule* developed for chiral chromatography.[23] The rule states that a minimum of three simultaneous interactions between the chiral stationary phase and one of the enantiomers are necessary to achieve enantioselection, with at least one of these interactions being stereochemically dependent. We describe chiral guanidiniums that apply this rule in the recognition, extraction, or transport separation of amino acids and peptides.

In 1989, de Mendoza reported the first example of chiral recognition of a carboxylate by a receptor based on the guanidinium moiety.[7] ^{1}H NMR titrations of triethylammonium salts of *N*-acetyltryptophan (*N*-Ac-Trp) with **2** in $CDCl_3$ gave $K_a = 1000$ and 500 M^{-1} for the L- and D-enantiomers, respectively. The difference in binding constants was exploited in extraction experiments. Racemic mixtures of *N*-Ac-Trp or *N*-Boc-Trp were extracted from an aqueous solution into $CHCl_3$ solutions containing **2** with modest enantioselectivity (up to 17% ee). The naturally occurring zwitterionic amino acids were not extracted from aqueous solution using **2**.

We designed the unsymetrically substituted receptor **23**, which contains an aza-crown ether to bind the ammonium of an amino acid, rendering the **23*AA** (AA = generic amino acid) complex more soluble in organic solvents.[24] **23** can be engineered to contain a variable functionality on the other arm of the guanidinium, to promote tertiary noncovalent interactions with the AA substrate. Experiments with 2-naphthoyl containing (*S*,*S*)-**23a** established that aromatic amino acids (phenylalanine, tryptophan) are extracted preferentially into the organic phase relative to alkyl-containing amino acids (ratio of extracted AAs is Phe : Trp : Leu : Val = 100 : 48 : 26 : 4). Chiral recognition was confirmed by extractions of aqueous solution containing racemic Phe or Trp with $CHCl_3$ solutions of (*S*,*S*)-**23a**. (*S*,*S*)-**23a** extracts L-Trp or L-Phe with high selectivity (about 80% ee). Complementary extractions with the (*R*,*R*)-**23a** isomer resulted in the isolation of the D-amino acids. The high affinity of **23a** for aromatic amino acids suggests that guanidinium–carboxylate, ammonium–crown ether, and π-stacking interactions are acting cooperatively during the enantioselective extractions.

23a X = O-2-naphthoyl
23b X = OSi^{t}BuPh$_2$
23c X = OSi^{i}Pr$_3$
23d X = OH

Systematic variation of the guanidinium side arm was investigated to determine the effect on amino acid extraction.[25] Efficient extraction of Trp was observed for all the receptors tested. Analysis of the extracted Trp revealed that the side arm has little effect on enantioselection. **23b** and **23c** gave very similar enantiomeric excess as **23a** (77 and 79% ee, respectively). Compound **23d**, which does not contain a lipophilic side arm, gives surprisingly good enantioselectivities in extraction experiments (51% ee). Apparently, the guanidinium–carboxylate and the ammonium–crown ether binding events have the greatest influence on the chiral discrimination of tryptophan.

Indeed, molecular dynamics simulations of **23a** with L- and D-Trp give very similar structures, neither of which contains π-stacking interactions between the indole of tryptophan and the naphthoyl of **23a**. The indole ring of **23a*L-Trp** is rigidly located over the aza-crown fragment, protecting this polar residue from apolar solvent interactions (Figure 4a). On the contrary, the indole ring of **23a*D-Trp** is dangling freely in the solvent (Figure 4b). The van der Waals and electrostatic energies were systematically more favorable for the more rigid **23a*L-Trp** than for **23a*D-Trp**. This result helps to explain the chiral discrimination of **23a** and suggests that the non-crown-ether arm is nearly entirely removed from enantioselection. These predictions are supported by the rather similar enantiomeric excess of the extracted tryptophan using **23a** or **23d**.

Compounds **23a**–**23d** were tested as tryptophan carriers across bulk model membranes. Transport was conducted with a U-tube that contained a dichloromethane phase between two water phases. At 1 mM concentration, **23a** transported 12% of available Trp with only 20% enantioselectivity after 2 h. Increasing transport times led to even less enantioselection. Compounds **23b**–**23d** behave similarly in Trp transport under these conditions. However, reducing the concentration of **23a** to 125 µM resulted in greatly enhanced enantioselection (73% ee), though with a reduced rate of transport. These results suggest that **23a** can select for the matched L-isomer of tryptophan, but competition with the D-isomer occurs when the concentration of L-Trp becomes depleted. Indeed, HPLC (high-performance liquid chromatography) experiments revealed that near the membrane interface the concentration of L-Trp is diminished relative to the top of the water phase. Gently stirring the aqueous source phase alleviates this problem, resulting in steady enantioselectivity over longer transport times.

König prepared guanidinium–crown ether receptors (**24**) for the recognition of zwitterionic substrates.[26] They investigated the effects of the spacer between the guanidinium and the crown ether units on the association of the guest. The 1,4-disubstituted benzyl spacer was optimal for binding peptides that contain carboxylates.

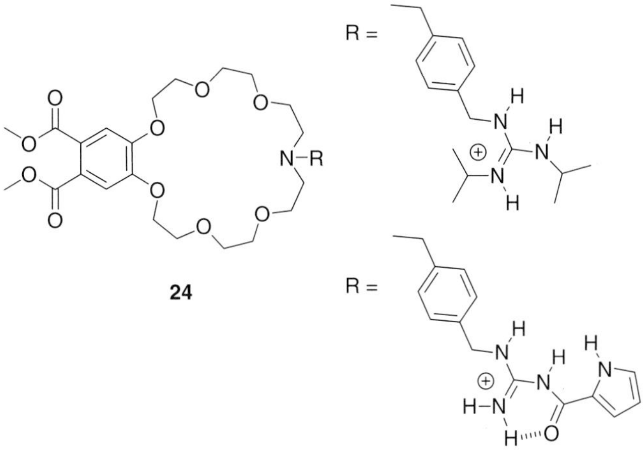

24

Davis reported cholic acid-derived **25a**–**25c**.[27,28] Compounds **25a**–**25c** contain both guanidinium and carbamate groups that allow the interaction with the derivatized amino acids. The host–amino acid interactions are stabilized by a guanidinium–carboxylate salt bridge and additional hydrogen bonding from the carbamate groups. The chirality is provided by the steroid backbone. Compound **25a** transports *N*-Ac-L-amino acids from aqueous phosphate buffered solutions (pH = 7.4) to chloroform with high efficiency and good selectivity for the L-isomer (7 : 1). The enantioselectivity of the transported amino acid substantially decreases in for the bulky *N*-Boc-L-amino acids. Chiral discrimination increased to 9 : 1 with **25b** and **25c**, probably due to the more acidic carbamoyl NHs.

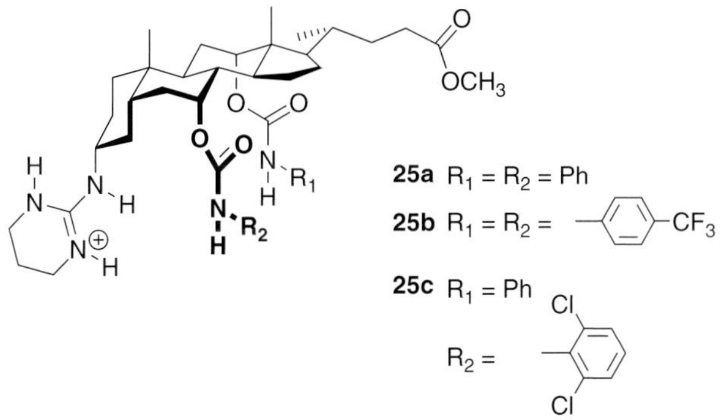

25a $R_1 = R_2 = Ph$
25b $R_1 = R_2 = C_6H_4\text{-}CF_3$
25c $R_1 = Ph$, R_2 = 2,6-dichlorophenyl

Schmuck reported the tris(dodecylbenzyl)-containing acylguanidinium **26** for amino acid transport.[29] The tris (dodecylbenzyl) unit ensures solubility of the receptor in organic solvents, and promotes cation–π interactions with aromatic amino acids allowing the transport of *N*-acetylated amino acids. *N*-Ac-L-Tyr and *N*-Ac-L-Trp are extracted into $CHCl_3$ in the presence of **26**. U-tube transport experiments established that release of the guest at the receiver phase is rate-determining. The lower pK_a of the acyl guanidinium gave active transport of the amino acid only when the source and receiver phases contained a pH gradient.

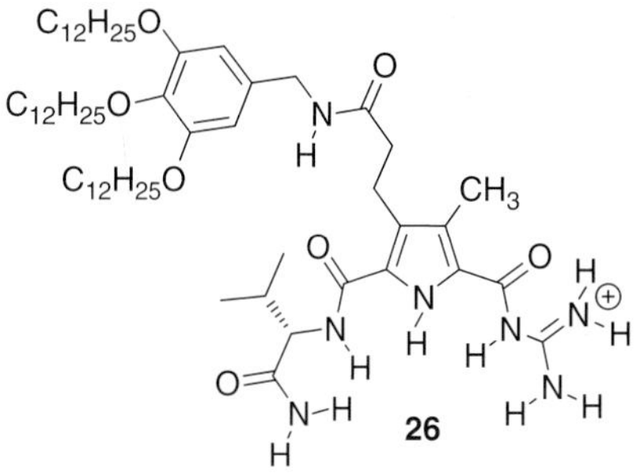

26

Schmuck and coworkers also developed acylguanidinium pyrrole receptors for the enantioselective binding of amino acids and short peptides. Receptor **27** contains a chiral center, and discriminates between the enantiomers of *N*-acetylalanine ($K_a = 1610$ and 910 M^{-1} for the L- and D-isomers, respectively).[16] The achiral receptor **28** forms strong complexes with dipeptides in aqueous DMSO ($1.5 \times 10^4\ M^{-1} < K_a < 5.4 \times 10^4\ M^{-1}$).[30]

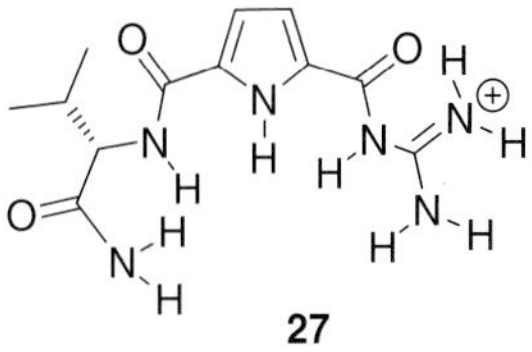

27

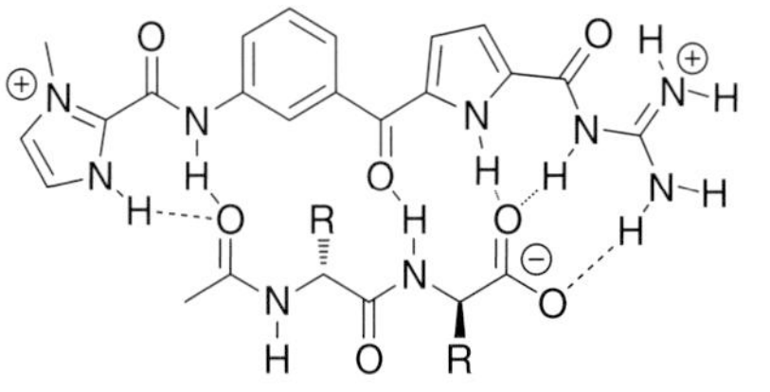

28

Receptor **29** contains a cyclotribenzylene (CTB) unit dangling from the chiral arm of a pyrrolo acylguanidinium moiety.[31] The CTB cavity provides a hydrophobic pocket which was designed to discriminate between the methyl groups of alanine and larger alkyl chains. They predicted that the chiral center in **29** would allow chiral recognition of D-Ala-containing peptided over L-Ala-containing ones. The Ac-D-Ala-D-Ala-OH dipeptide strongly binds to **29** in buffered water/DMSO solution ($K_a = 3.3 \times 10^5\,M^{-1}$). The similar dipeptides Ac-Gly-Gly-OH and Ac-Val-Val-OH were bound to **29** with much lower affinity ($K_a < 3000\,M^{-1}$). The D- and L-isomers of Ac-D-Ala-D-Ala-OH showed similar association constants with **29**, indicating that chiral discrimination does not occur with this receptor. Modeling studies suggest that both the *P*/D and *M*/D diastereoisomers of **29** contain nearly identical binding features. The C–C single bond linkage between the hydrophobic CTB cavity and the acylguanidinium is flexible, and allows the receptor to adopt a favorable conformation for either enantiomer of alanine.

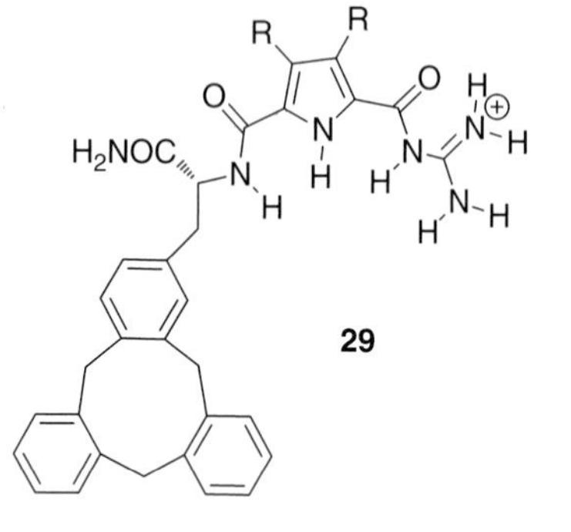

29

Kilburn and coworkers studied solid-phase-bound tweezer receptors that contain peptide arms linked through a guanidinium unit (**30**). The guanidinium binds the carboxylate terminus of a target peptide, and the peptide arms are varied through split-and-mix combinatorial libraries to optimize peptide–peptide interactions.[32] The guest peptide contains a tethered dye which reports binding to the host containing beads. A library containing 15 625 members was screened. The optimum binding receptor was identified and independently synthesized for solution-binding studies. However, the best candidates, once synthesized, were found to have poor binding in solution, likely due to the difference in solvation properties of a solution of the receptor relative to the resin-bound state.[33] Schmuck used a similar split-and-mix strategy for single-arm receptors that contain the pyrrolo acylguanidinium unit (**31**).[34] In this case, strong association constants were observed both on resin-bound beads and in solution.[35]

Macrocyclic receptors have been described for the recognition of chiral dicarboxylates. The use of a cyclic scaffold that incorporates two guanidinium groups could give extra rigidity and/or preorganization to the receptor. Kilburn reported the chiral bis-guanidinium macrocycle **32** that binds *N*-Boc-L-glutamic acid in 50% aqueous DMSO.[36] The 1 : 1 complex gives higher binding for *N*-Boc-L-Glu ($K_a = 3.8 \times 10^4\,M^{-1}$) than for *N*-Boc-D-Glu ($K_a = 2.9 \times 10^3\,M^{-1}$). Schmidtchen and coworkers reported the thermodynamic parameters of chiral dicarboxylates binding to diguanidinium macrocycle **33**.[37] Binding studies show that D-tartrate is selected over L-tartrate in a 3.5 : 1 ratio.

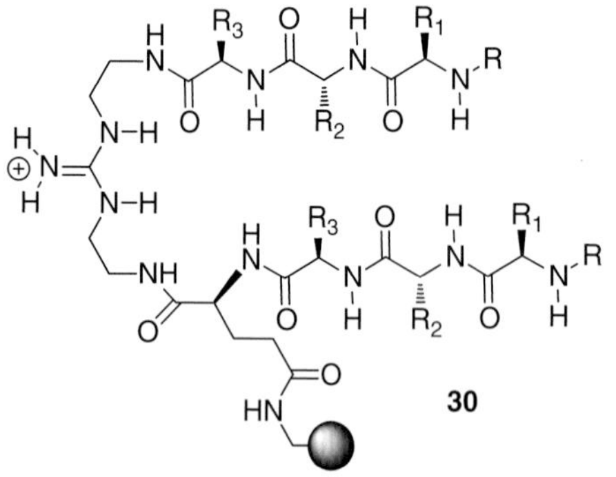

30

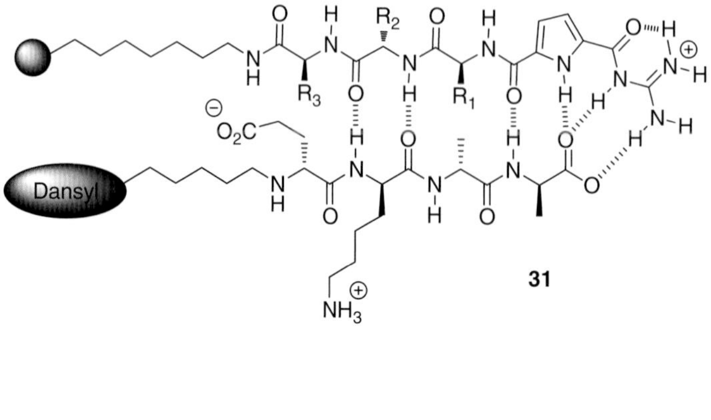

31

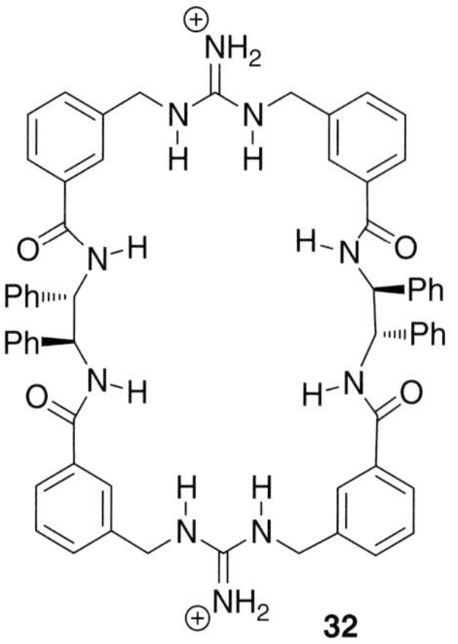

32

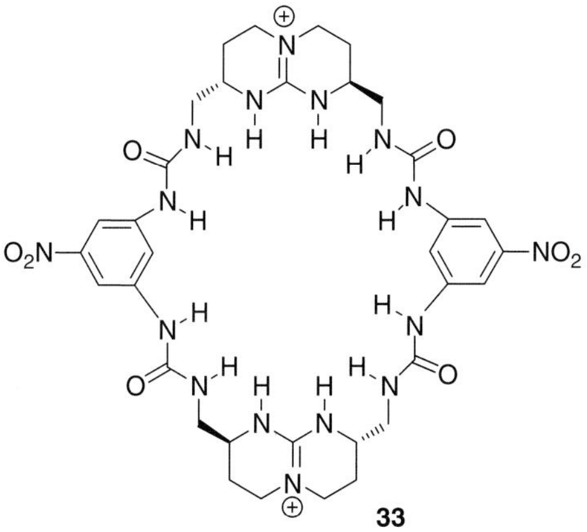

33

4 PEPTIDE AND PROTEIN RECOGNITION

Biological processes involving protein–protein recognition are of special interest since they are the basis of many relevant events in the cellular machinery. Enzyme–substrate binding, protein membrane signaling, immunologic response, and protein quaternary structural changes caused by cofactors or allosteric modulators are only a few of the examples that show the importance of protein–protein interactions at the molecular level.

From a supramolecular perspective, the design of small molecules able to selectively recognize the active, or allosteric, site of enzymes with higher affinities than the natural substrates or modulators represents a fundamental challenge in medicinal chemistry. This strategy uses molecules that complement the size, shape, and functional groups of the interacting active site. Most of these enzymatic sites are poorly solvated, enhancing the hydrophobic character of their interactions.

On the contrary, recognition processes occurring at the surface of proteins are more difficult to control due to the large featureless shape (about 750–1500 Å) and the highly solvated area of contact. Thus, recognition of protein surfaces requires multivalency and cooperative effects to achieve high binding constants and selectivity for specific proteins.[38]

In 1995, Hamilton reported the rigid, concave diguanidinium receptor **34**.[39] The two guanidinium groups are oriented to interact with the two aspartates of peptide **35** in the $i(i+3)$ relative positions. In an ideal α-helical conformation, the Asp residues of **35** are oriented in a favorable position to interact with the receptor. ^{1}H NMR titration studies established that a 1 : 1 complex was formed with $K_a = 2200\,M^{-1}$ in 10% water/methanol solution. Moving the aspartates to the $i(i+4)$ or $i(i+11)$ positions caused a large decrease in the binding constant ($K_a = 770$ and 390 M^{-1}, respectively). Circular dichroism (CD) studies revealed that the **34*35** complex was significantly more helical than **35** (8% helicity increase) in 10% water in CH_3OH. These results indicate that **34** stabilized the α-helical conformation in peptides containing properly located aspartate residues.

In a fruitful collaboration, the Hamilton, Giralt, and de Mendoza groups investigated peptide helix stabilization with tetraguanidinium **36** (Figure 5).[40] The tetra-anionic peptide **37** contains four aspartates in $i(i+3)$ positions and forms stable complexes with **36** ($K_a = 1.6 \times 10^5\,M^{-1}$ in 10% water/methanol, CD titrations).[40a,b] Most remarkably, **36** induces a 41% increase of helicity in the peptide. The large association constant, helix induction, and modeling studies suggested that the tetraguanidinium wraps around the peptide in a coiled-coil complementary helix (Figure 6).

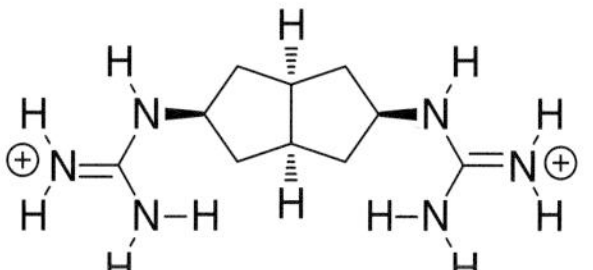

34

$i(i+3)$ Ac-Ala-Ala-Gln-**Asp**-Ala-Ala-**Asp**-Ala-Ala-Ala-Ala-Ala-Gln-Ala-Ala-Tyr-NH_2 **35**

$i(i+4)$ Ac-Ala-Ala-Gln-**Asp**-Ala-Ala-Ala-**Asp**-Ala-Ala-Ala-Ala-Gln-Ala-Ala-Tyr-NH_2

$i(i+11)$ Ac-Ala-Ala-Gln-**Asp**-Ala-Ala-Ala-Ala-Ala-Ala-Ala-Ala-Gln-Ala-**Asp**-Tyr-NH_2

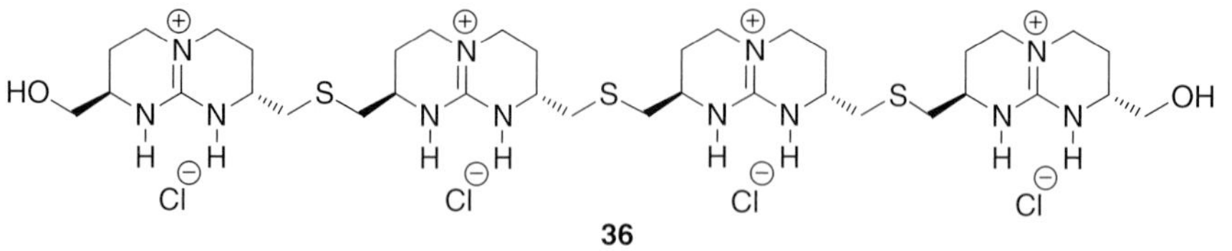

36

	Peptide	K_a (M^{-1})
37	Ac-Ala-Ala-Ala-**Asp**-Gln-Leu-**Asp**-Ala-Leu-**Asp**-Ala-Gln-**Asp**-Ala-Ala-Tyr-NH_2	1.6×10^5 (CH_3OH/H_2O)
		3.2×10^6 (TFE/H_2O)
38	Ac-Ala-Ala-Ala-**Asp**-Gln-Leu-**Asp**-Phe-Leu-**Asp**-Phe-Ala-**Asp**-Phe-Ala-Gln-Phe-Ala-Ala-Ala-NH_2	3.8×10^6 (TFE/H_2O)
39	Ac-Ala-Ala-Ala-**Asp**-Gln-Leu-**Asp**-Trp-Leu-**Asp**-Trp-Ala-**Asp**-Trp-Ala-Gln-Trp-Ala-Ala-Ala-NH_2	4.2×10^6 (TFE/H_2O)
40	Ac-Ala-Ala-Ala-Trp-Gln-Leu-Trp-**Asp**-Leu-Trp-**Asp**-Ala-Trp-**Asp**-Ala-Gln-**Asp**-Ala-Ala-Ala-NH_2	1.1×10^8 (TFE/H_2O)

Figure 5 Recognition of peptides with tetraguanidinium **36** and association constants in 10% aqueous CH_3OH or in 50% aqueous 1,1,1-trifluoroethanol (TFE).

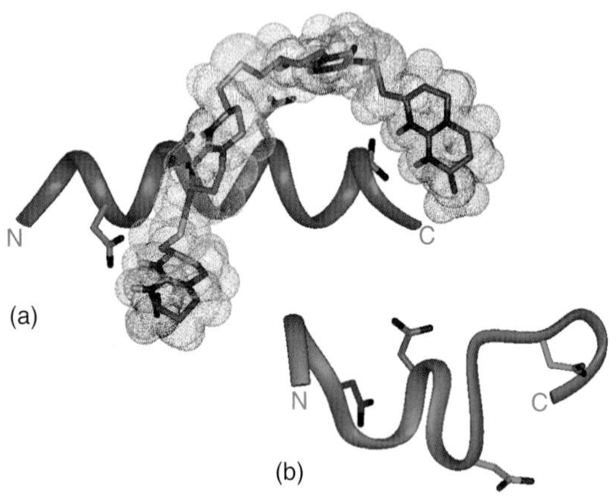

Figure 6 (a) Optimized model of an α-helical peptide backbone with four aspartates at $i(i+3)$ and tetraguanidinium receptor **36**. (b) Random coil structure of the free peptide.

Interpretation of the model in Figure 6 suggested that properly placed aromatic residues could form cation–π interactions. To test this hypothesis, peptide **38**, containing phenylalanine at complementary $i(i+3)$ positions, was synthesized.[40c] Little increase in binding was observed, relative to **37**, in 50% aqueous 1,1,1-trifluoroethanol (Figure 5). Similar results were obtained with **39**, which contains tryptophan residues at complementary $i(i+3)$ positions. However, exchanging the Asp and Trp positions produced a dramatic increase in the association constant between **40** and receptor **36**. The helicity of **40** increased by 38% over that of the free peptide as a consequence of the guanidinium–carboxylate and cation–π interactions. Modeling studies suggest that the indole rings of tryptophan in peptide **39** are staggered in either possible rotamer (Figure 7a and b) and have little overlap with the aspartate residues. In contrast, the indole rings of **40** can easily adopt the correct conformation for cation–π interactions with the guanidinium (Figure 7c).

We became interested in extending this methodology to more challenging and biologically relevant protein domains. The p53 protein is a natural tumor suppressor involved in the repairing of DNA damage or in apoptotic processes, and is commonly referred to as the guardian of the cell.[41] Tetramerization of p53 is essential for cellular

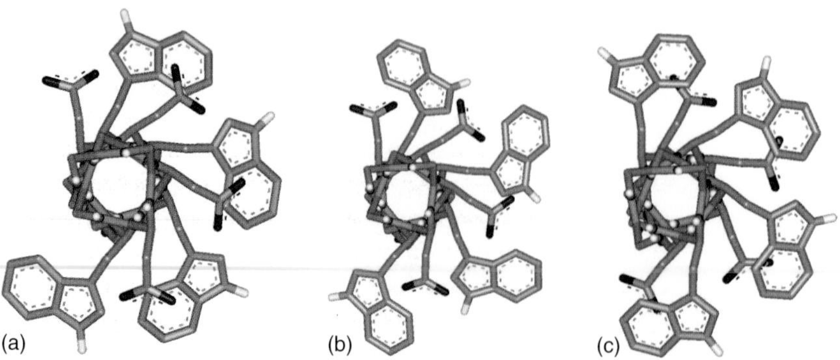

Figure 7 Comparison of the Trp and Asp side-chain overlap in (a) **39** with preferred side chain conformations, (b) **39** with the bond between the Trp and carbon atoms rotated 180°, and (c) **40** with the preferred side chain conformation. All views are down the helix axis from the N-terminus, and nonessential side chains have been removed for clarity.

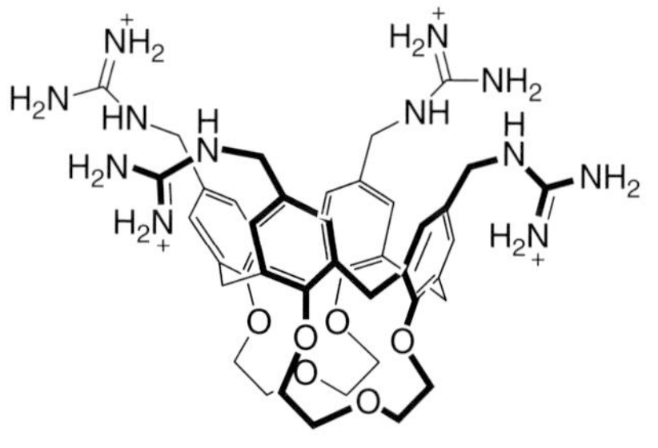

41

42

R^1 = COGuanid$^+$Cl$^-$, $R^2 = R^3$ = H
R^1 = CO-Arg-$OCH_3$$^+TFA^-$, $R^2 = R^3$ = H

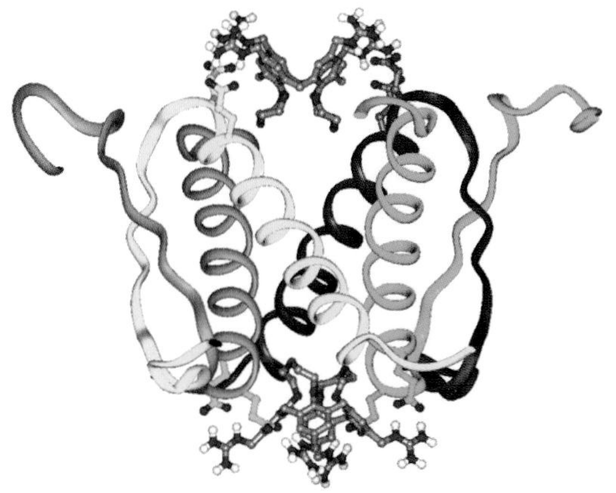

Figure 8 Molecular dynamics model of **41** docking to the Glu336 and Glu339 side chains of two different tetramerization domains of p53.

function. The structure of the tetramerized p53 domain (TDp53) is actually a dimer of dimers in which each domain contains an anionic patch formed by Glu and Asp residues (Glu336, Glu339, Glu343, Glu346, Glu349, and Asp352). The spacing between the anionic residues of the tetramerization domain is complementary to the distance between the guanidinium groups in **36**. Chemical shift perturbation and saturation transfer difference NMR (STD-NMR) techniques established that **36** binds the anionic sequence of amino acids present in the surface of the protein with high affinity and likely contributes to the stabilization of the protein.

Glu336 and Glu339 from two different strands of the tetramer are located in an almost perfect square just above a hydrophobic pocket in wild-type TDp53. The conformationally rigid calix[4]arene receptor **41** containing four guanidinium groups hanging at the upper rim was designed to form guanidinium–carboxylate interactions with these two separate tetramerization domains (Figure 8).[42] In addition, the hydrophobic ether loops on the lower rim were designed to fit inside the hydrophobic pocket of TDp53.^{1}H–^{15}N HSQC (heteronuclear single quantum resonance) and STD-NMR experiments established that **41** fits into the hydrophobic cavity of TDp53 with $K_d = 280\,\mu$M.

The tetramer of wild-type p53 is stabilized by a critical Arg337–Asp352 salt bridge. The mutant p53-R337H (mutation of arginine-337 to a histidine) results in destabilization of the tetramer, because histidine is not fully protonated at physiological pH. Differential scanning calorimetry shows that native TDp53 denatures at 85.5 °C, whereas mutated TDp53-R337H denatures at 62 °C. Introduction of 400 μM **41** to TDp53-R337H causes almost complete recovery of the thermal stability of the mutant.

A similar strategy was employed to block the entrance of the K_v1.x family of potassium ion channels.[43] In this study, a series of guanidinium-containing calix[4]arenes (**42**) were designed to complement the extracellular outer vestibule of the channel using guanidinium–carboxylate and hydrophobic interactions from the calixarene. Reversible inhibition of the channel was observed for acylguanidinium and tetra-arginine containing calix[4]arenes with –OH substitution at the lower rim. The dissociation constants of the guanidinium–calix[4]arenes were in the low micromolar range, and the Hill coefficient values close to 1 supported a single site of interaction.

Aida and coworkers have reported dendrimeric structures that contain guanidinium groups (**43**) as "molecular glue." These materials bind proteins without selectivity.

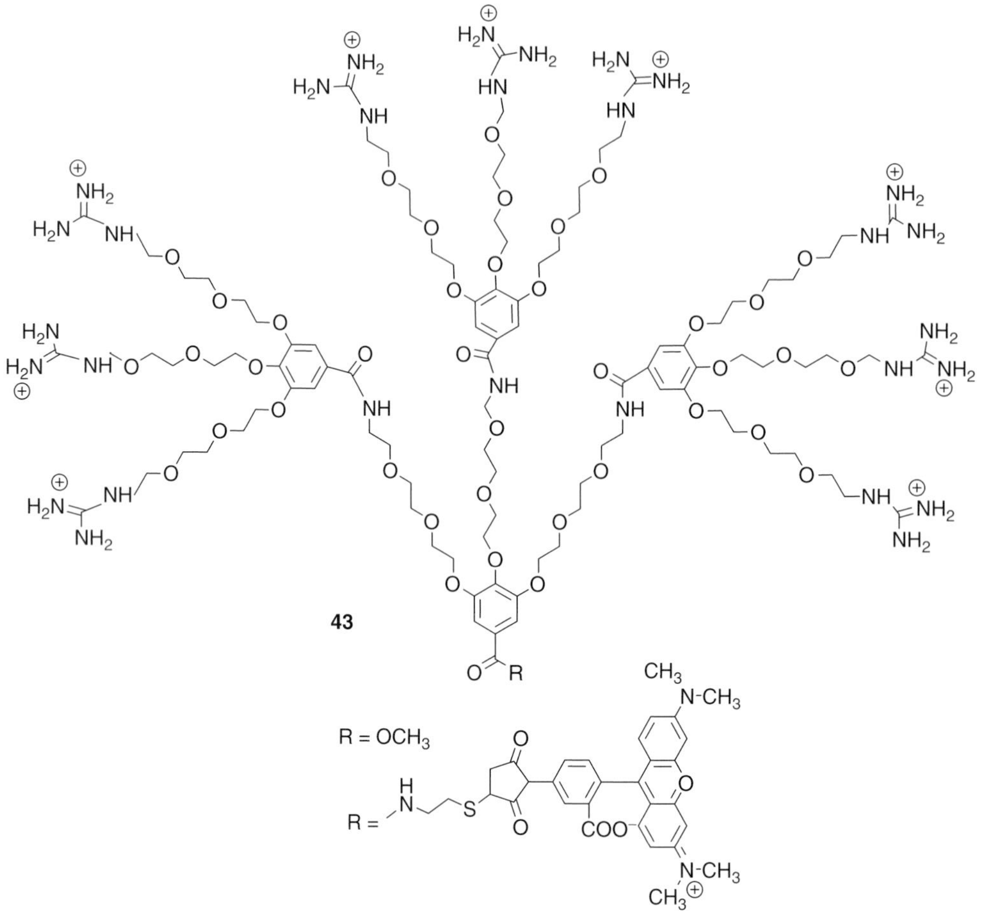

43

$R = OCH_3$

Compound **43** interacts with the oxoanionic groups of the α/β-tubulin heterodimer protein to stabilize their microtubule form against depolymerization.[44] The stabilization of the microtubules is comparable with paclitaxel, an anticancer drug that stabilizes microtubules by van der Waals interactions. The protein adhesion properties of **43** were also evaluated on the ATP-driven sliding motion of actomyosin.[45] The sliding motion of actomyosin is responsible for muscle contraction. Addition of **43** to actin–myosin filaments arrests the expected sliding motion. ITC and confocal laser scanning microscopy were used to probe the interaction of ligand **43** with myosin and actin filaments. Control experiments established that fluorescently labeled actin filaments bound to a myosin functionalized surface have an average velocity of 4.6 μm s^{-1}. Indeed, addition of ATP buffer solutions containing **43** stops the motion of the filament. Rinsing the surface with buffer, which removes **43** from the mixture, restores the filament motion. These results indicate that **43** interacts with myosin–actin filaments and can stop the motion necessary for muscle contraction *in vitro*.[46]

5 PHOSPHATES, SULFATES, AND NITRATES

Phosphate, sulfate, and nitrate are biologically relevant anions[47] and chemically challenging to recognize due to their weak basicity. At neutral pH, phosphate (as HPO_4^{2-}) and sulfate have tetrahedral binding modes with two negative charges. Nitrate has a trigonal planar binding motif with just one negative charge. Thus, two guanidinium moieties are required for phosphate and sulfate, but only one is needed for nitrate, to obtain neutral complexes. The design of suitable linkers between the guanidinium and additional hydrogen-bond donors with optimal geometry to accommodate the lone pairs of the guest constitutes a major issue in this field.

6 PHOSPHATES

Initial reports by Lehn and coworkers showed that guanidinium-containing macrocycles bind to phosphate (PO_3^{4-}) in water.[48] Weak association constants were obtained ($K_a = 50$ (**44**), 158 (**45**), 251 (**46**) M^{-1}, pH titrations). Despite the low association constants, this result is remarkable because the flexible macrocyclic framework allows free rotation of the guanidiniums groups to adopt the *syn* (productive for binding) and *anti* (nonproductive) conformations.

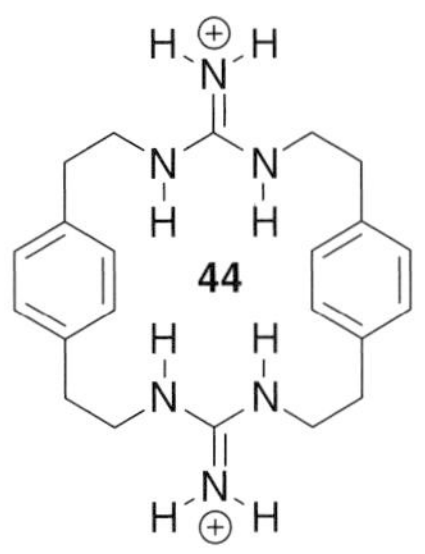

44 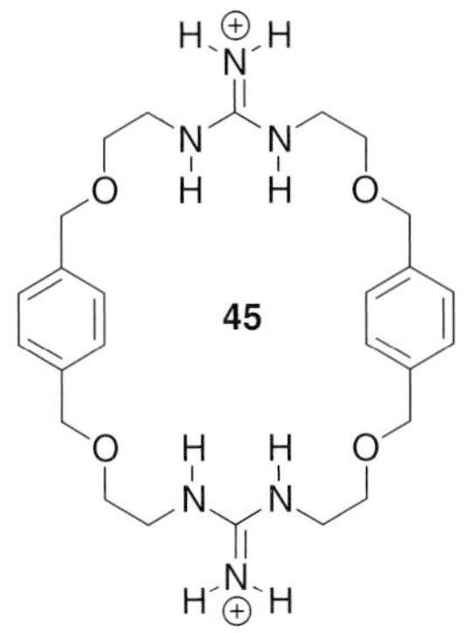 **45** 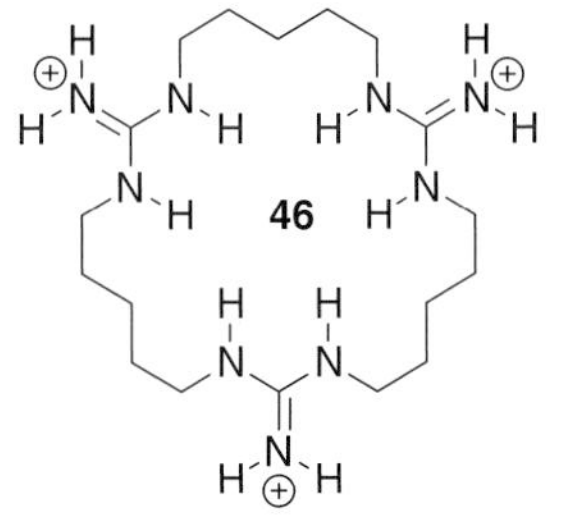 **46**

Hamilton and coworkers synthesized bis-acylguanidinium salt **47** as a receptor for phosphodiesters.[49] The binding constant with TBA diphenylphosphate was $K_a = 4.6 \times 10^4\,M^{-1}$, measured by UV–vis spectrometry in CH_3CN. The bicyclic guanidinium has also been shown to bind phosphate anions. Macrocycle **48** was designed to bind tetrahedral monoanions.[50] The cavity contains a total of six strong hydrogen-bond donors to complement the guanidinium–phosphate electrostatic interaction. Diphenylphosphate is extracted from water to form the stable guanidinium–phosphate complex. The binding constant from the chloride salt of **48** was $10^3\,M^{-1}$ in $CDCl_3$.

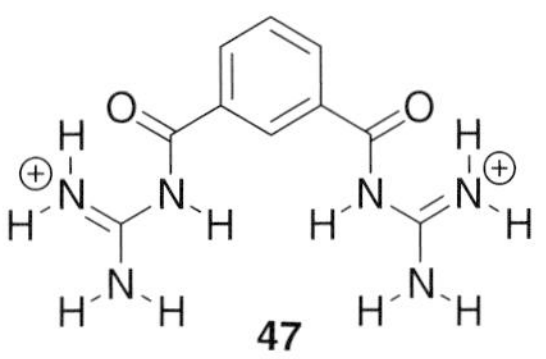

47 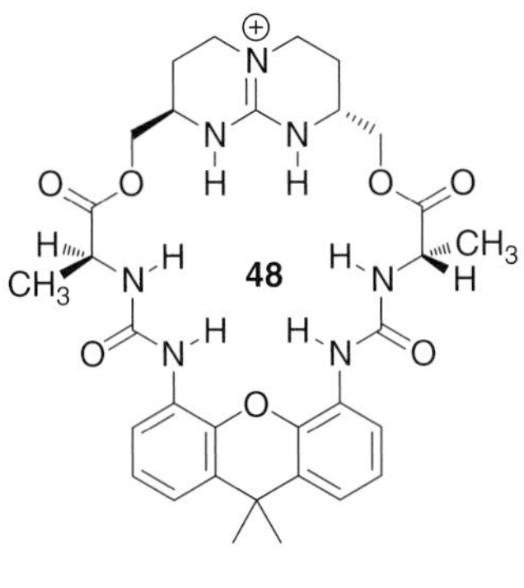 **48** 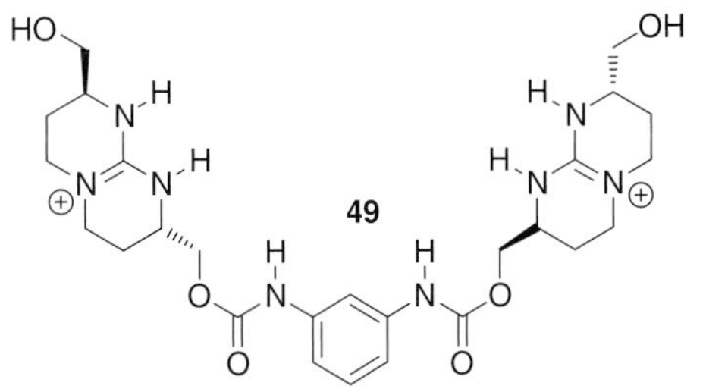 **49**

Schmidtchen synthesized a urethane-linked bis-guanidinium (**49**) to bind tetrahedral dianions.[51] Compound **49** binds *p*-nitrophenyl phosphate and cytidine-5′-phosphate in water with a $10^6\,M^{-1}$ binding constant. The very high binding constant in water was explained by the complexation of both guanidinium groups to the anionic guest.

Binding studies of **1** and **4** with phosphates of different sizes were measured by ^1H NMR and ITC.[52] The ^1H NMR data gave a 1 : 1 binding curve, but the ITC data predicted a 1 : 2 host–guest binding stoichiometry. In this case, calorimetry prevents misleading conclusions from the rapid interconversion of species in equilibrium on the NMR time scale. Thermodynamically, the binding event is driven by a strong entropy term. To understand the origin of this entropically driven binding, Schmidtchen and coworkers analyzed the thermodynamic data of phosphate binding to a small series of structurally different guanidinium compounds.[53] They concluded that the entropic driving force is not related to desolvation of the host and guest, but rather by direct host–guest binding which produces a conformationally flexible bound state.

The Anslyn group has investigated metallo-receptor **50** for the binding of phosphate (HPO_4^{2-}, $K_a = 10^4\,M^{-1}$, UV–vis and ITC titrations) and arsenate ($HAsO_4^{2-}$) in 98 : 2 water/CH_3OH at physiological pH.[54] Hosts that lack the guanidinium units, but contain the ligated Cu(II) center were two orders of magnitude less effective for phosphate association. This result shows that the guanidinium units in **50** participate in phosphate recognition. Thermodynamic data indicate that the association of HPO_4^{2-} with **50** is both enthalpically and entropically favorable. The ammonium metallo-receptor **51** binds HPO_4^{2-} with similar binding constants as **50**. However, ITC analysis reveals that the binding for **50** is driven only by the entropic term. These results were rationalized in terms of the different solvation energies of guanidinium (poor solvation) and ammonium (good solvation) groups.[55]

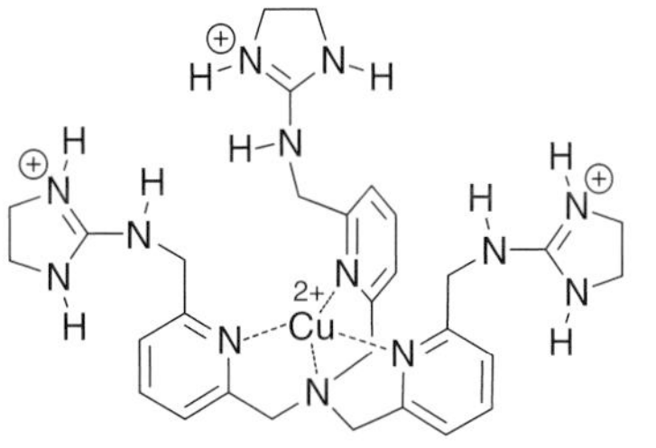

50 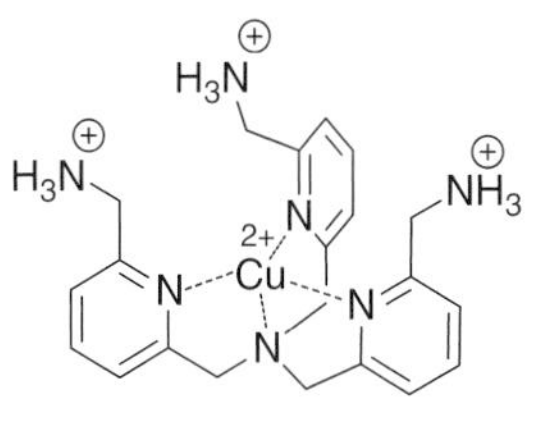 **51**

Compound **52** contains six guanidinium moieties on a tetrakis(2,4,6-triethylbenzene) core. As a result of steric gearing,[56] the guanidinium groups converge to form a cleft-like cavity that is ideal for the recognition of inositol–triphosphate.[57] The host and the guest do not contain a suitable chromophore for determination of the binding constant by UV–vis spectra. Therefore, the binding constant was measured by an indicator displacement assay. Competitive binding of **53** to **52***5-carboxyfluorescein liberates free 5-carboxyfluorescein. The binding constant was $K_a = 2.2 \times 10^4\,M^{-1}$ in 10 mM HEPES (4-(2-hydroxyethyl)-1-piperazineethanesulfonic acid) buffer, and $K_a = 1 \times 10^8\,M^{-1}$ in CH_3OH. The exceptionally high binding constant in CH_3OH allows detection of **53**-containing solutions with concentrations as low as 2 nM.

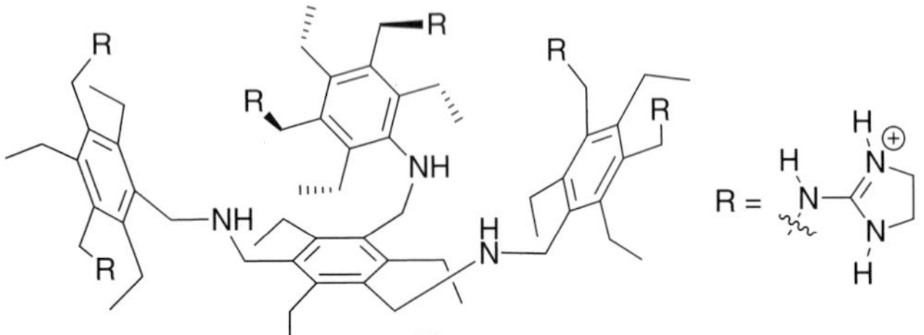

52

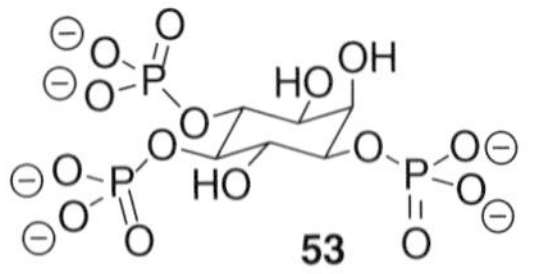

53

The combination of metal coordination and cleft-like preorganization has been exploited by Anslyn with receptor **54** to bind 2,3-bisphosphoglycerate (**55**).[58] The association constant for **55** binding was measured by an indicator displacement method in 1 : 1 water/CH_3OH, and determined to be $8 \times 10^8\,M^{-1}$. Structurally similar phosphoglycerates, in which of one phosphate is removed, gave binding constants 80–180 times lower than that of **55**. This result established that receptor **54** was very selective for **55**.

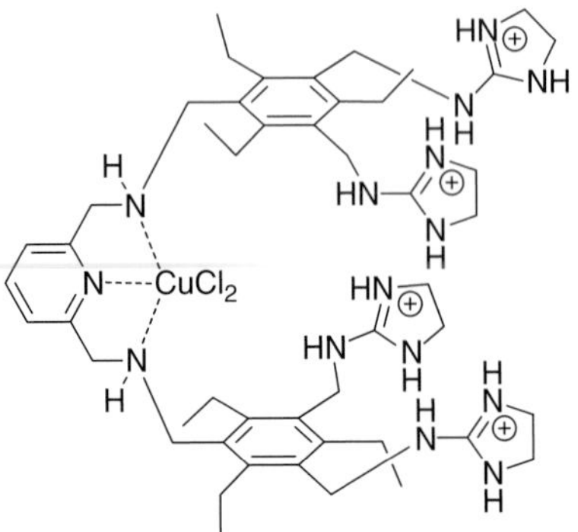

54

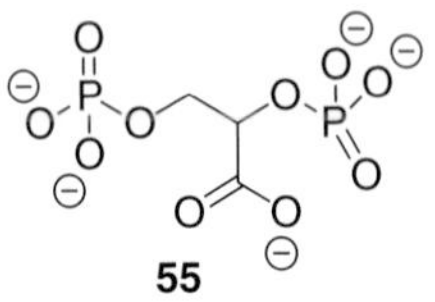

55

Compound **55** is an allosteric effector that regulates the oxygen release of hemoglobin, and is implicated in various inherited diseases.[59] The strong preference of **54** to selectively bind **55** suggested that similar behavior could occur in more complicated systems. The influence of receptor **54** on hemoglobin oxygenation was investigated in horse blood hemolyzate. The addition of **54** to the hemolyzate induces O_2 uptake by hemoglobin, indicating that **55** was bound to the receptor and not hemoglobin, causing a high oxygen affinity at hemoglobin.

The recognition of the biologically relevant pyrophosphate $P_2O_7^{4-}$ anion has been studied with various guanidinium-containing hosts. Ferrocenyl-based receptor **56** gives moderately strong 2 : 1 host–guest complexes with pyrophosphate in 50% methanol–water ($K_a = 4600\,M^{-2}$).[60] Receptor **57** also binds pyrophosphate in a 2 : 1 fashion ($K_a = 1.2 \times 10^8\,M^{-2}$), and displays selectivity for $P_2O_7^{4-}$ over a variety of anions.[61] ^1H NMR experiments indicate that the 2 : 1 host–guest complex is self-assembled through π-stacking interactions of the pyrene subunits. Acylguanidinium **58** binds pyrophosphate in a 1 : 1 fashion with low affinity ($K_a = 216\,M^{-1}$, 80% aqueous DMSO).[62] No significant binding was observed for a wide range of anions, including ATP, ADP, and AMP. This result indicates that **58** is highly selective for only pyrophosphate, though modification of the receptor is necessary to achieve a stronger binding.

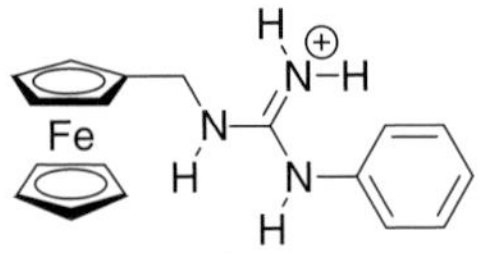

56

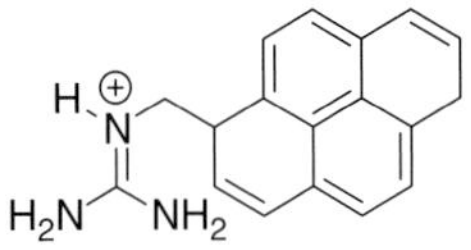

57

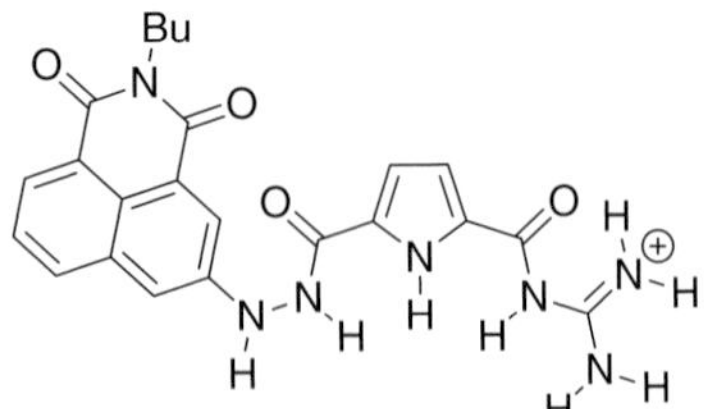

58

7 NUCLEOTIDES

Nucleotides, oligonucleotides, and related complex structures (such as DNA or RNA) contain phosphodiester anions. These structures have obvious relevance in a large variety of biological processes. The variable number of anionic groups in their structures makes this class of molecules interesting targets for recognition by guanidinium-containing receptors.

A family of bicyclic guanidinium receptors was studied for the complexation of nucleotides and dinucleotides.[63]

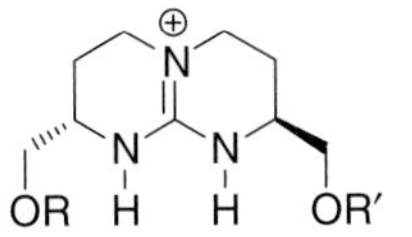

59a R = R′ = (**A**)
59b R = R′ = (**B**)
59c R = (**B**), R′ = (**C**)

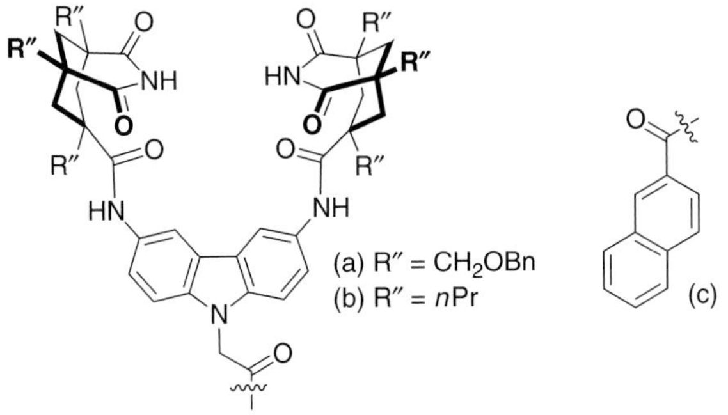

(a) R″ = CH_2OBn
(b) R″ = *n*Pr
(c)

The guanidinium units in **59a** and **59b** are flanked by two carbazole spacer units, each of which contains two Kemp triacid fragments. The receptor was designed to complex the anionic nucleotide through guanidinium–phosphate interactions, and form complementary hydrogen bonds to the nucleotide heterocycle with the Kemp triacid fragments. A full equivalent of diadenosine phosphate was extracted into organic solvents per equivalent of **59a**, indicating that the receptor has a very high affinity for the dinucleotide. ^{1}H NMR experiments established that the carbazole group π-stacks with the heterocyclic base of the nucleotide, and the Kemp triacid does indeed hydrogen-bond with the receptor, which is therefore an efficient nucleotide transporter, as determined by U-tube experiments. In these transport measurements, **59c**, which contains a 2-naphthoyl moiety, gave increased transport rates for adenosine monophosphate guests.

The diguanidinium tetrakis(β-cyclodextrin) tetrapodal **60** is a suitable host for different nucleotides.[64] **60** interacts with adenosine derivatives by combining the unspecific hydrophobic inclusion of the nucleobase with the electrostatic guanidinium–phosphate interaction. Nucleotide: receptor stoichiometries of 2 : 1 were obtained with binding constants approaching $10^6\,M^{-1}$ for the complexation of disodium 5′-adenosine triphosphate, monosodium 5′-adenosinediphosphate, and monosodium 5′-adenosinemonophosphte.

Schmuck and coworkers investigated the binding of DNA and RNA sequences to a series of guanidinium

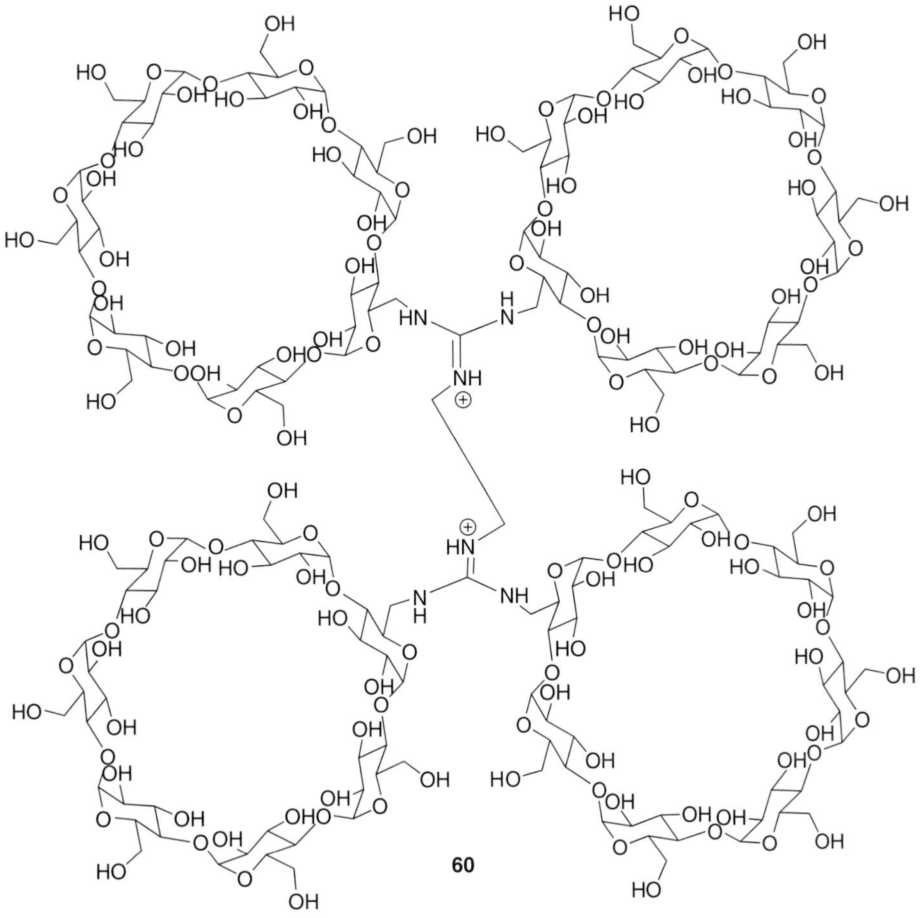

60

carbonylpyrrole-aryl derivatives.[65] The multipoint interaction of the receptors with DNA requires the pyrrolo acylguanidinium–phosphate interaction coupled to a flat aromatic surface for DNA/RNA intercalation. Compound **61**, which contains a pyrene moiety, binds double-stranded DNA with good affinity, and shows antiproliferative activity.

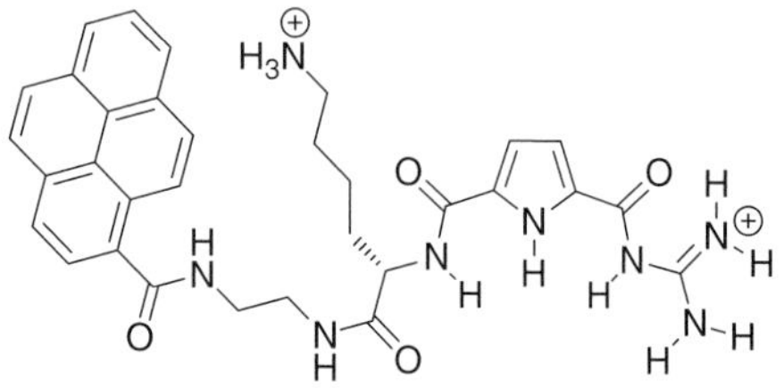

61

The anionic charge of DNA prevents it from crossing cellular barriers, because of repulsion with phosphate- and sulfate-containing lipids on the cell membrane. The complexation of DNA to lipophilic polycationic structures has been used to neutralize the anionic charge to effect efficient cellular uptake. Ungaro and coworkers studied a series of guanidinium calix[4–6]arenes that promoted transfection of DNA sequences into human rhabdomyosarcoma cells.[66] The calix[4]arenes containing hexyl or octyl substitution at the lower rim (**62b** and **62c**) are rigidly set in the cone conformation. These compounds efficiently transfect DNA into the cell. However, the members of the series that assume 1,3-alternate (**62a**) or mobile conformations (**62d**–**62f**) do not transfect DNA. AFM (atomic force microscopy) studies revealed that **62b** and **62c** form DNA condensates that are likely driven by intramolecular lipophilic interactions of the long alkyl chains. Compounds **62d**–**62f** form intermolecular aggregates, which inhibits cell transfection.

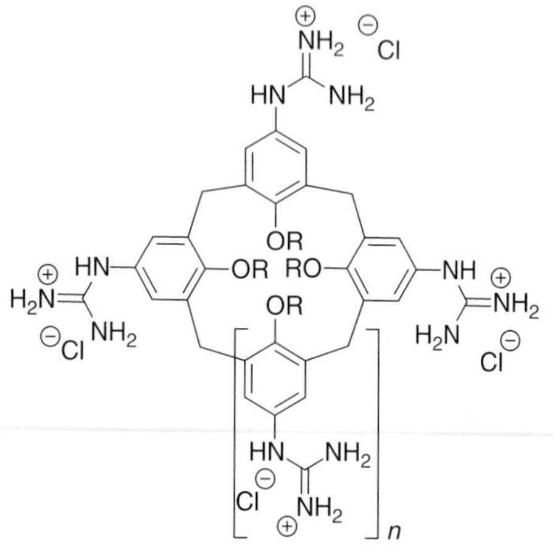

62a R = C_3H_7 $n = 1$; *1,3-alt* **62d** R = CH_3 $n = 1$; *mobile*
62b R = C_6H_{13} $n = 1$; *cone* **62e** R = CH_3 $n = 3$; *mobile*
62c R = C_8H_{17} $n = 1$; *cone* **62f** R = CH_3 $n = 5$; *mobile*

Tetraguanidinium-modified phthalocyanines recognize DNA containing G-quadruplex (G4) sequences. The phthalocyanine*DNA adducts efficiently transfect into cells.[67] The planar phthalocyanine structures are capable of intercalation into the DNA quadruplex loop and bind to the peripheral phosphates. Nonmetallated **63** stabilizes both the parallel and antiparallel forms of G4-DNA. The Zn-metallated derivative **64** selectivity stabilizes parallel G4-DNA.

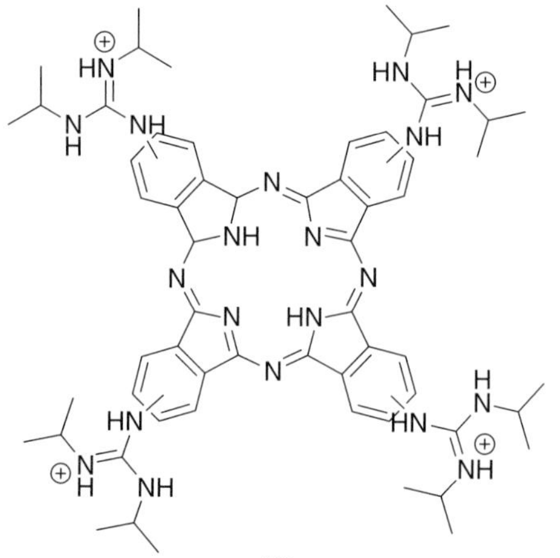

63

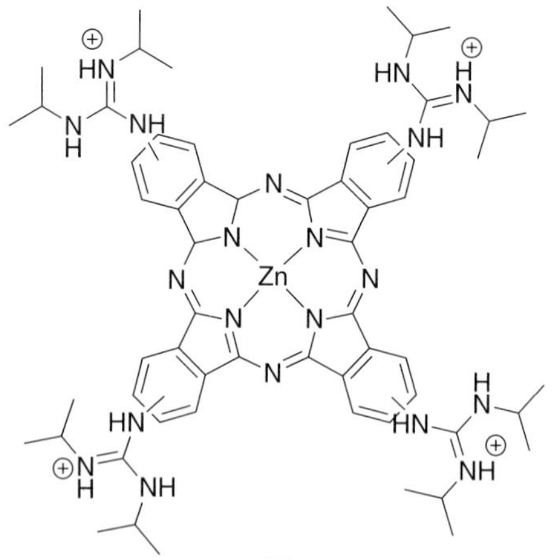

64

G4-DNA was targeted using a guanidinium-containing dynamic combinatorial library.[68] The planar aromatic scaffold **65** was modified from previous G4-DNA binding studies to include the thiol functionality necessary for dynamic combinatorial chemistry. Exposure of **65** to the library members **A**–**E** in the presence of G4-DNA caused amplification of **65*A** and **65*E**, relative to control experiments without G4-DNA. Additionally, **65*E** bound G4-DNA one order of magnitude tighter ($K_d = 6.6\,\mu M$) than **65** alone. These results suggest that hydrogen bonding is involved in the recognition of G4-DNA, though more studies are necessary to confirm this proposal.

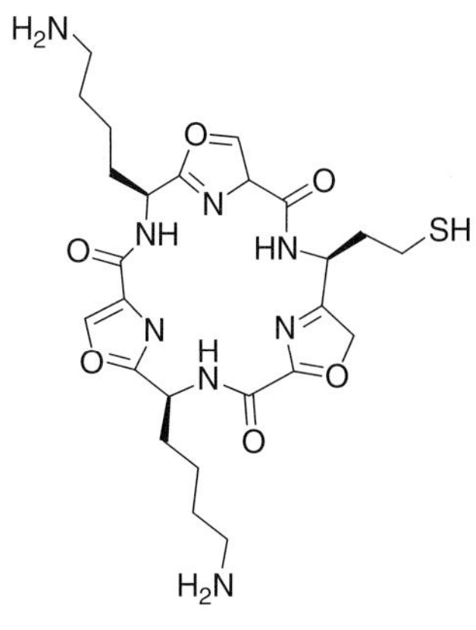

65

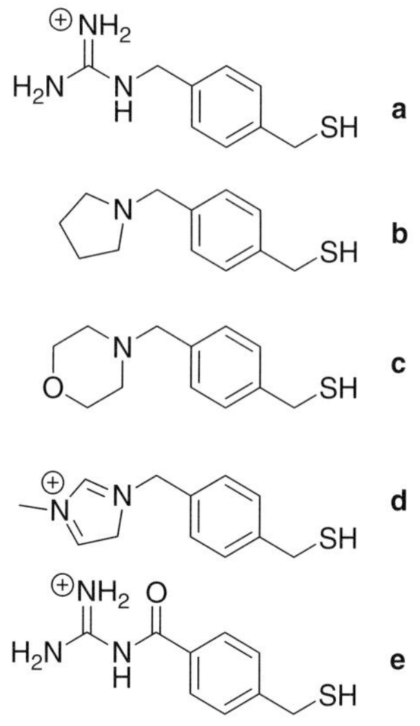

a **b** **c** **d** **e**

Overhauser effect spectroscopy) spectrum contained cross-peaks that could only be explained from the pro-helical conformation.

X-ray crystallography confirms that the closely related receptors **68** and **69** form 1 : 1 complexes with sulfate. Good docking of the anion into the cavity is observed for **68**. However, in preorganized **68** the sulfate is ion-paired but not bound in the guanidinium pocket. The pyridine nitrogen of **69** apparently causes repulsion of the anion due to the increased charge density in the pocket around the heteroatoms.

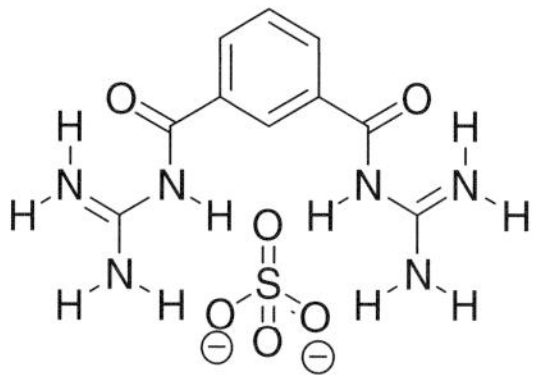

68 **69**

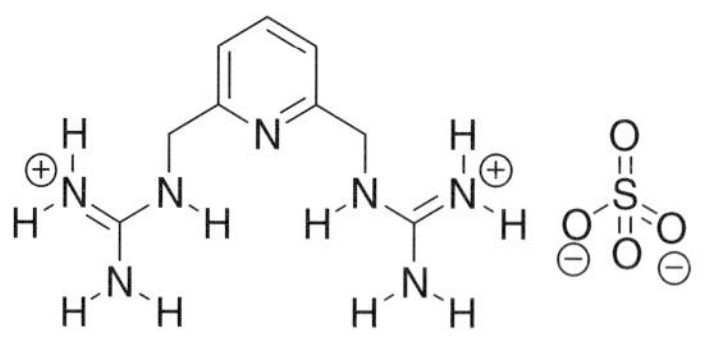

8 SULFATES

The sulfate anion is geometrically similar to phosphate, though less basic. In spite of the inherent difficulty for its recognition, there are a few examples in the field of guanidinium-based receptors for sulfate binding.

The structural and binding properties of di- and tetraguanidinium sulfate salts (**66** and **67**) were described in 1996.[69] Two guanidinium groups are necessary to satisfy the charge and hydrogen-bond acceptor requirements of sulfate. The CH_2SCH_2 linker is flexible, though too short to form an intramolecular arrangement. Two guanidinium receptors are required for binding. Each guanidinium orthogonally wraps around the tetrahedrical anion, forming a double-folded architecture. This structure was confirmed in solution by ^1H NMR analysis. Large downfield shifts of guanidinium protons indicated complexation of the guanidinium to the sulfate. The NOESY (nuclear

Theoretical studies suggest that simple guanidinium–sulfate interactions can access several energy minima.[70] Among the plausible structures are 1 : 1, 2 : 1, or various hydrogen-bonded complexes between charged species. This promiscuity has been used to recover denatured protein structure.[71] For instance, guanidinium chloride is a well-known protein denaturant. However, in the presence of sulfate salts this effect is reversed, thus the initial protein conformation is restored. The guanidinium–sulfate ionic complex likely affects the ability of guanidinium to disrupt intramolecular protein hydrogen bonds.

The energetics of the guanidinium–sulfate system have been analyzed by Schmidtchen using **70**.[72] Compound **70** contains two bicyclic guanidinium subunits linked through an isophtalamide spacer. ITC measurements reveal that sulfate complexation is strongly endothermic with an entropic

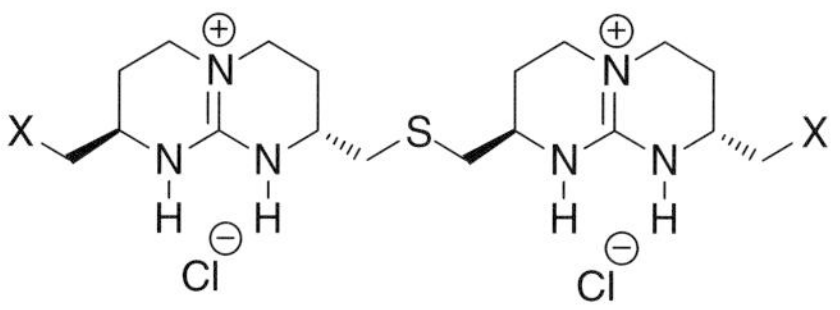

66a = OTBDPS

66b = $O(C{=}O)CH_2C_6H_{13}$

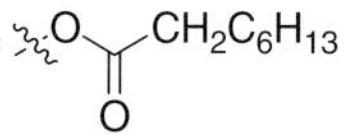

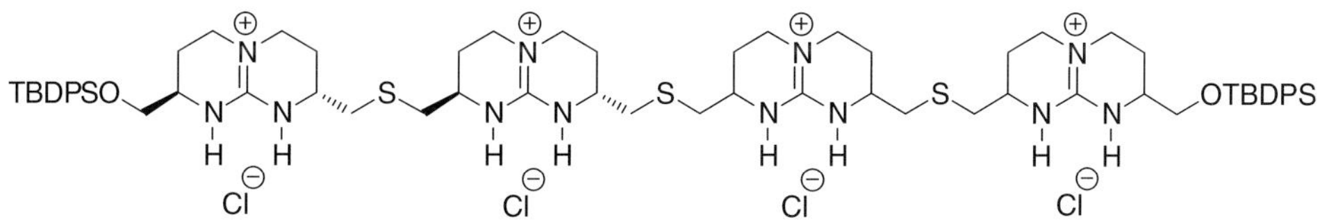

67

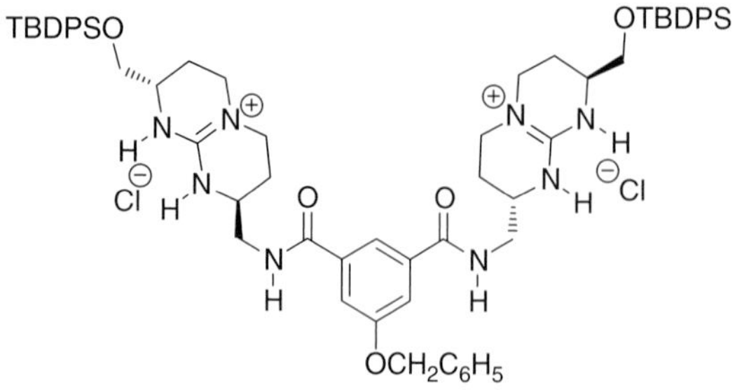

70

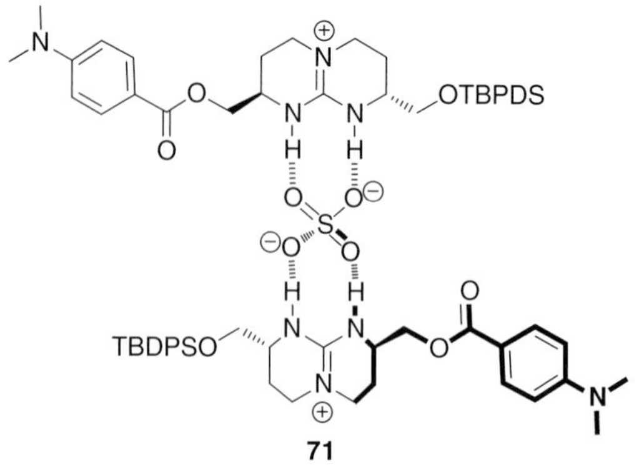

71

driving force. The association is strong enough in methanol ($K_a = 6.8 \times 10^6\ M^{-1}$) to overcome the positive enthalpy term. A comparison between receptor **70** and related monoguanidinium receptors, which show little or no interaction with sulfate in methanol, accounts for the importance of a bridging spacer between both cationic subunits.

Inoue *et al.* described bicyclic guanidinium **71** for sulfate binding in CH_3CN.[73] Compound **71** contains a chromophoric dimethylaminobenzoate group as reporter for sulfate complexation. ^1H NMR titrations indicate that stepwise binding occurs. Initial formation of a 1 : 1 complex ($K_a = 1.53 \times 10^6\ M^{-1}$) is followed by the formation of a 2 : 1 complex ($K_a = 4.84 \times 10^4\ M^{-1}$). CD spectroscopy reveals that the 2 : 1 guanidinium–sulfate complex is chiral, as expected from the chirality present in the receptor.

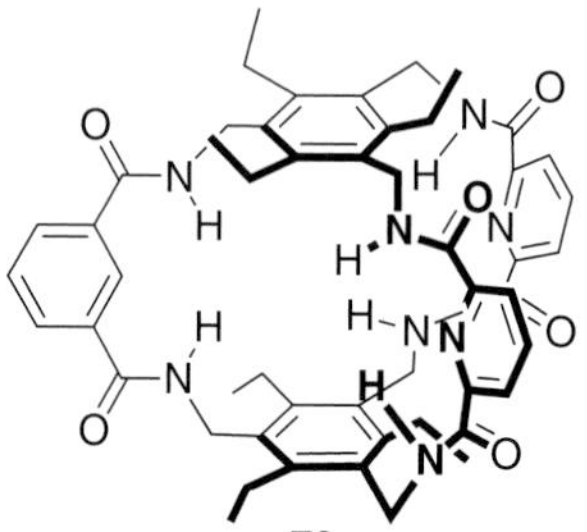

72

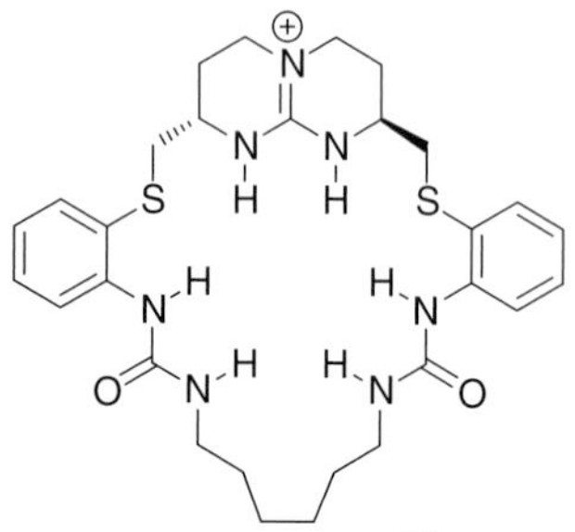

73

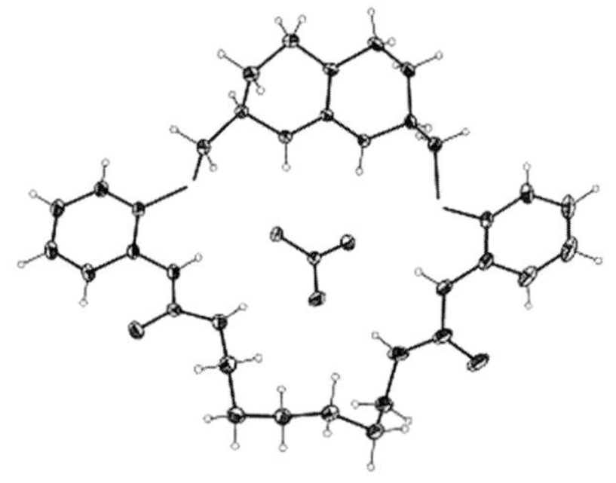

Figure 9 X-ray crystal structure of **73*NO₃**.

9 NITRATES

Nitrate is a weak, conjugated base with low hydrogen-bonding ability. Its trigonal planar geometry offers the possibility to establish up to six hydrogen bonds, one interaction per oxygen lone pair. The coordination features of nitrate with guanidinium cations have been investigated by theoretical calculations[74] and X-ray diffraction studies.[75] Owing to the relative difficulties to compete with other anions such as carboxylates, phosphates, or chloride, very few literature examples have been reported showing selectivity for this oxoanion.

An early example of nitrate binding is the cyclophane polyamide receptor **72**, which gives a modest association with nitrate ($K_a = 300\ M^{-1}$) relative to chloride (40 M^{-1}).[76] The bicyclic guanidinium-containing macrocycle **73** was designed to interact with all the possible oxygen nitrate lone pairs.[77] Indeed, **73** gave quite high binding constants ($K_a = 7.4 \times 10^4\ M^{-1}$, ITC measurements in CD_3CN). The thermodynamic parameters revealed that the interaction is favored both enthalpically and entropically. The crystal structure of **73*NO₃** (Figure 9) showed that the nitrate anion is hydrogen-bonded to all the available donors in the macrocycle, in good agreement with the predicted model.

10 CATALYSIS

Guanidiniums are finding increasing applications in organocatalysis. The scope of this review is not intended to provide a full description of guanidiniums in catalysis and this topic has been reviewed.[78] Therefore, only a few examples of guanidiniums participating in the stabilization of transition states are briefly disclosed.

Breit and coworkers have reported the Rh-catalyzed hydroformylation of carboxylic acids using ligand **74** (Scheme 1).[79] The catalyst displays high activity relative to PPh_3, and selectively reacts at the terminal position of

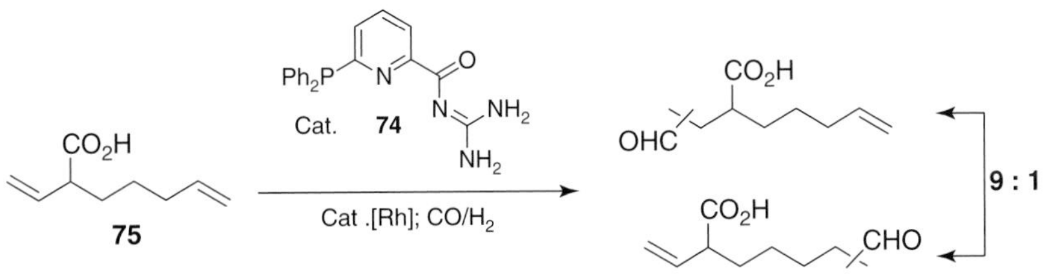

Scheme 1 Regioselective rhodium catalyzed hydroformylation of **75** in the presence of acylguanidinium containing ligand **76**.

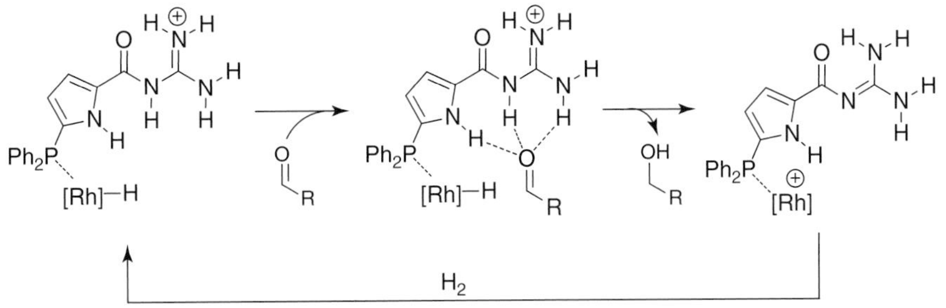

Scheme 2 Rhodium catalyzed hydrogenation of aldehydes using a pyrrolo-acylguanidinium containing a phosphine ligand.

double-bonds. Good regioselectivity was observed for the hydroformylation of diene **75** (9 : 1). Control experiments with PPh_3 show that the two olefins present in **75** react at similar rates, establishing that the olefins are essentially chemically equivalent. This is unambiguous evidence that the guanidinium–carboxylate ion pair directs the selectivity of the hydroformylation.

Modification of the central pyridine ring to a pyrrole, analogous to the design Schmuck has used for acylguanidinium receptors, gives lower reaction rates for the hydroformylation reaction of terminal olefins. However, this ligand promotes the hydrogenation of aldehydes under hydroformylation conditions.[80] Experiments conducted in CH_3OH or with the Boc-protected guanidine, which are conditions in which hydrogen bonding would be diminished, do not show hydrogenation reactivity. On the basis of these results, the authors propose the mechanism shown in Scheme 2. The guanidinium forms a hydrogen-bond complex with the aldehyde which lowers the LUMO (lowest unoccupied molecular orbital) of the carbonyl group. Transfer of the Rh-H to the carbonyl coupled with concerted proton transfer from the acylguanidinium gives the alcohol. Heterolytic H_2 cleavage regenerates the catalyst.

We reported an early example of anionic transition state stabilization with guanidinium derivatives. The addition of amines to α,β-unsaturated lactones is accelerated in the presence of catalytic amounts of guanidinium **76**.[81] The uncatalyzed reaction passes through an anionic enolate-type transition state, typical for Michael additions. The guanidinium offers charge stabilization of the negative charge and complementary hydrogen bonds analogous to the guanidinium–carboxylate interaction. In the presence of guanidinium, up to 10-fold rate enhancement over background was observed. Unfortunately, no enantioselectivity was observed using **76** or related derivatives, probably due to the large distance between the β-carbon of the lactone and the stereogenic centers of the guanidinium moiety.

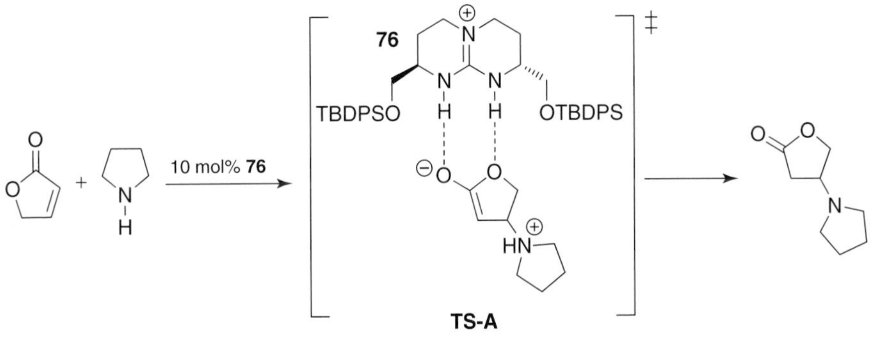

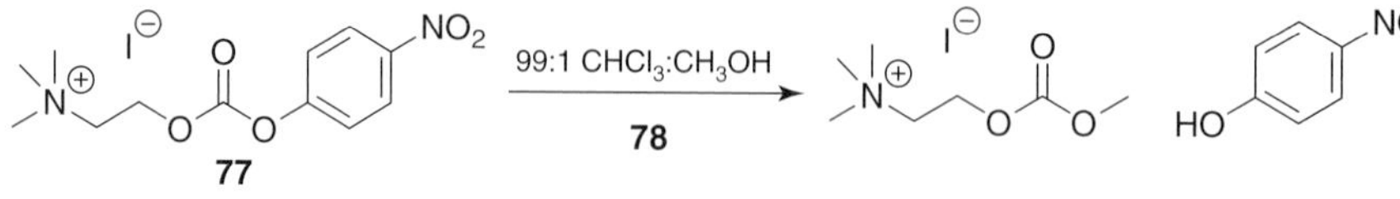

The transfer of acyl groups is commonly encountered in biology, and catalyzed by the acyltransferase family of enzymes. For example, acetylcholinesterase hydrolase hydrolyzes excess acetylcholine after nerve impulses. Many chemical models built to mimic the enzymatic function of acyltransferases have been developed, classically employing cyclodextrins which form hydrophobic complexation intermediates.[82]

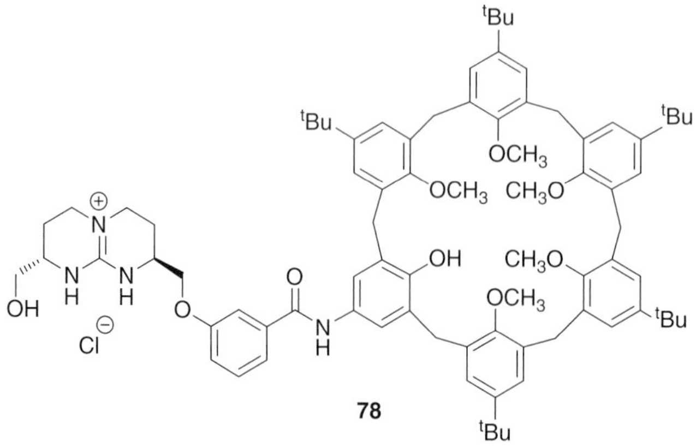

We also investigated the methanolysis of **77**, which is an acetylcholine model. The ditopic receptor **78** was designed to bind the trimethylammonium group of **77** by cation–π interactions and to stabilize the anionic tetrahedral transition state formed upon delivery of methoxide to the carbonyl through binding to the guanidinium unit. The overall reaction generates methyl carbonate; thus no carboxylates are produced that could inhibit catalysis by competitive binding to the guanidinium. The trimethylammonium group of acetylcholine was bound to **78** in the calixarene cavity as revealed by ^{1}H NMR experiments ($K_a = 730\,M^{-1}$). Independently, the calix[6]arene or the guanidinium do not provide large rate enhancements for the hydrolysis of **77** (two- and ninefold, respectively). However, **78** gave 149 times acceleration for their hydrolysis, suggesting that the calix[6]arene–guanidinium combination is able to stabilize the transition state cooperatively.

The formation and cleavage of phosphate esters in DNA and RNA are typically catalyzed by enzymes. For example, staphylococcyl nuclease is a metalloenzyme that contains a Ca(II) metal center and two arginines in the active site. Staphylococcyl nuclease hydrolyzes DNA 10^{16} faster than background hydrolysis.[83] However, site-directed mutagenesis to replace the arginine groups reduces the catalytic efficiency 35 000-fold. Replacement of the Ca(II) with Mn(II) maintains the tertiary structure of the enzyme, though reactivity is diminished on the same order of magnitude as the arginine mutagenesis.[84] Clearly, the Ca(II) center and the guanidinium groups are acting cooperatively to achieve such a high catalytic rate of hydrolysis. Intensive research efforts have focused on the cleavage of phosphate esters in an effort to mimic enzymatic behavior with a synthetic catalyst.[81,85]

Ansyln and coworkers designed **79** which contains a Zn(II) center within a cleft containing two guanidinium scaffolds.[86,87] The guanidinium groups are properly positioned to hydrogen-bond to the Zn(II) bound phosphate ester and stabilize the anionic transition state formed upon nucleophilic attach of hydroxide on the phosphoryl group.

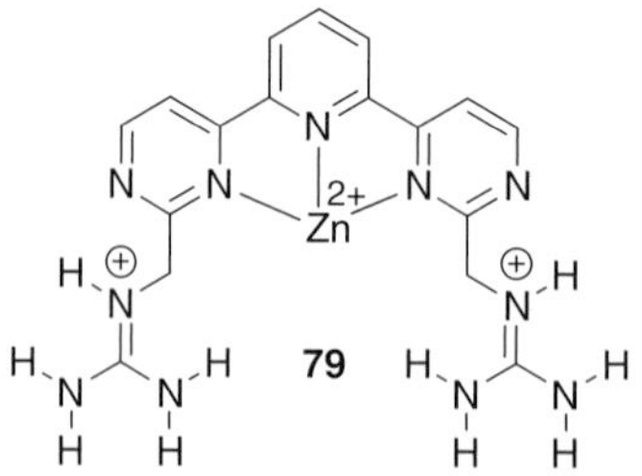

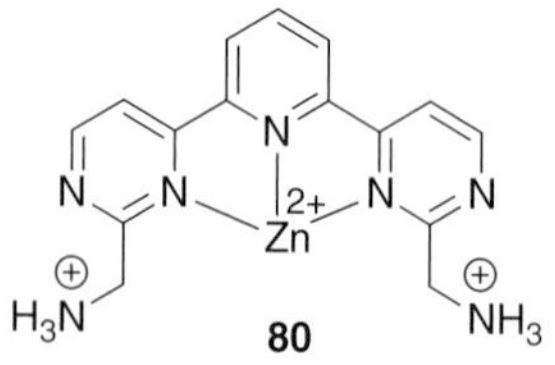

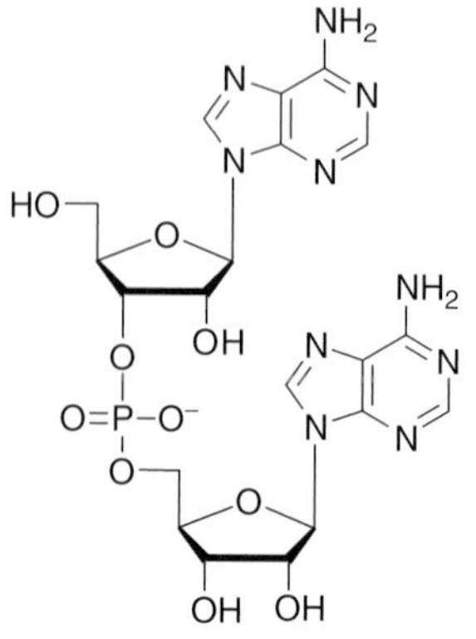

Adenylyl(3′->5′)phosphoadenine

The hydrolysis of the RNA dimer adenylyl(3′→5′) phosphoadenine displayed 10^6 rate enhancements over background hydrolysis. Compound **79** was 9 times faster than **80**, a related complex bearing ammonium groups, and 3300 times faster than the Zn(II) species lacking either functionality. Clearly, the catalyst operates by action of the guanidinium groups and the Zn(II) center, and approaches the cooperative behavior observed by staphylococcyl nuclease.

11 CONCLUSION

The guanidinium is a versatile functional group with unique properties. Well-structured hydrogen-bonded ion pairs with oxoanion substrates are possible with properly designed guanidinium receptors. When inserted into a bicyclic framework, the guanidinium function can be efficiently used for extractions or membrane transport. Inclusion of additional hydrogen-bond donors, or guanidinium functionalities, in the receptor molecule can stabilize the ion-pair interaction in polar protic solvents, allowing the detection of oxoanions near biologically relevant conditions. This approach has been used to recognize simple carboxylates and complex structures, such as proteins and DNA, in buffered water. Recent examples of catalysts containing guanidiniums show that this exceptional functional group can give both cooperative and regioselective reactivity to the system.

NOTES

[1] The guanidinium NHs become more acidic depending on the *N*-substitution. In general, the pK_a value increases from acyl > phenyl > alkyl (see ref. 3b).

[2] Though the difference in pK_a is reduced in apolar media, formation of an AD-DA complex remains unlikely. Neutral AD-DA complexes have been characterized and generally have small association constants.

REFERENCES

1. C. G. Mowat, C. S. Moysey, C. S. Miles, *et al.*, *Biochemistry*, 2001, **40**, 12292.
2. Y. Zhang and D. Reinberg, *Genes Dev.*, 2001, **15**, 2343.
3. For recent reviews see: (a) M. D. Best, S. L. Tobey, and E. V. Anslyn, *Coord. Chem. Rev.*, 2003, **240**, 3; (b) K. A. Schug and W. Lindner, *Chem. Rev.*, 2005, **105**, 67; (c) R. J. T. Houk, S. L. Tobey, and E. V. Anslyn, *Top. Curr. Chem.*, 2005, **255**, 199; (d) F. A Schmidtchen, *Coord. Chem. Rev.*, 2006, **250**, 2918; (e) C. Schmuck, *Coord. Chem. Rev.*, 2006, **250**, 3053; (f) P. Blondeau, M. Segura, R. Pérez-Fernández, and J. de Mendoza, *Chem. Soc. Rev.*, 2007, **37**, 198; (g) T. Rehm and C. Schmuck, *Chem. Comm.*, 2008, 801.
4. For a computational study of AD-DA complexes see: W. L. Jorgensen and J. Pranata, *J. Am. Chem. Soc.*, 1990, **112**, 2008.
5. For the X-Ray crystal structure of free-base guanidine see: T. Yamada, X. Liu, U. Englert, *et al.*, *Chem.—Eur. J.*, 2009, **15**, 5651.
6. G. Müller, J. Riede, and F. P. Schmidtchen, *Angew. Chem. Int. Ed. Engl.*, 1988, **27**, 1516.
7. A. M. Echavarren, A. Galán, J.-M. Lehn, and J. de Mendoza, *J. Am. Chem. Soc.*, 1989, **111**, 4994.
8. M. Haj-Zaroubi, N. W. Mitzel, and F. P. Schmidtchen, *Angew. Chem. Int. Ed.*, 2002, **41**, 104.
9. M. Berger and F. P. Schmidtchen, *J. Am. Chem. Soc.*, 1999, **121**, 9986.
10. V. D. Jadhav, E. Herdtweck, and F. P. Schmidtchen, *Chem.—Eur. J.*, 2008, **14**, 6098.
11. B. R. Linton, M. S. Goodman, E. Fan, *et al.*, *J. Org. Chem.*, 2001, **66**, 7313.
12. E. Fan, S. A. Van Arman, S. Kincaid, and A. D. Hamilton, *J. Am. Chem. Soc.*, 1993, **115**, 369.
13. X. Wang, O. V. Sarycheva, B. D. Koivisto, *et al.*, *Org. Lett.*, 2008, **10**, 297.
14. A. Metzger, V. M. Lynch, and E. V. Anslyn, *Angew. Chem. Int. Ed. Engl.*, 1997, **36**, 862.
15. A. D. Hughes and E. V. Anslyn, *Proc. Natl. Acad. Sci. U.S.A.*, 2007, **104**, 6538.
16. C. Schmuck, *Chem.—Eur. J.*, 2000, **6**, 709.
17. C. Schmuck and U. Machon, *Chem.—Eur. J.*, 2005, **11**, 1109.
18. For related systems: (a) K. Kavallieratos, C. M. Bertao, and R. H. Crabtree, *J. Org. Chem.*, 1999, **64**, 1675; (b) G. M. Kyne, M. E. Light, M. B. Hurthouse, *et al.*, *J. Chem. Soc., Perkin Trans. 1*, 2001, 1258; (c) S. Y. Chang, H. S. Kim, K. J. Chang, and K. S. Jeong, *Org. Lett.*, 2004, **6**, 181.
19. C. Schmuck and M. Schwegmann, *J. Am. Chem. Soc.*, 2005, **127**, 3373.
20. C. Schmuck and W. Wienand, *J. Am. Chem. Soc.*, 2003, **125**, 452.
21. S. Schlund, C. Schmuck, and B. Engels, *Chem.—Eur. J.*, 2007, **13**, 6644.
22. J. P. Behr and J. –M Lehn, *J. Am. Chem. Soc.*, 1973, **95**, 6108.
23. W. H. Pirkle and T. C. Pochapsky, *Chem. Ber.*, 1989, **89**, 347.
24. (a) A. Galán, D. Andreu, A. M. Echavarren, *et al.*, *J. Am. Chem. Soc.*, 1992, **114**, 1511; (b) J. de Mendoza and F. Gago, *NATO ASI Ser., Ser. C*, 1994, **426**, 79.
25. P. Breccia, M. Van Gool, R. Pérez-Fernández, *et al.*, *J. Am. Chem. Soc.*, 2003, **125**, 8270.
26. A. Späth and B. König, *Tetrahedron*, 2010, **66**, 1859.

27. L. J. Lawless, A. G. Blackburn, A. J. Ayling, *et al.*, *J. Chem. Soc., Perkin Trans. 1*, 2001, **11**, 1329.
28. B. Baragaña, A. G. Blackburn, P. Breccia, *et al.*, *Chem.—Eur. J.*, 2002, **8**, 2931.
29. C. Urban and C. Schmuck, *Chem.—Eur. J.*, 2010, **16**, 9502.
30. C. Schmuck and L. Geiger, *J. Am. Chem. Soc.*, 2004, **126**, 8898.
31. C. Schmuck, D. Rupprecht, and W. Wienand, *Chem.—Eur. J.*, 2006, **12**, 9186.
32. (a) K. B. Jensen, T. M. Braxmeier, M. Demarcus, *et al.*, *Chem.—Eur. J.*, 2002, **8**, 1300; (b) J. Shepherd, T. Gale, K. B. Jensen, and J. D. Kilburn, *Chem.—Eur. J.*, 2006, **12**, 713.
33. For a study of single peptide arm receptors using the bicyclic guanidinium see: J. Shepherd, G. J. Langley, J. M. Herniman, and J. D. Kilburn, *Eur. J. Org. Chem.*, 2007, **8**, 1345.
34. (a) C. Schmuck and P. Wich, *Angew. Chem. Int. Ed.*, 2006, **45**, 4277; (b) C. Schmuck and M. Heil, *Chem.—Eur. J.*, 2006, **12**, 1339.
35. (a) C. Schmuck and M. Heil, *Org. Biomol. Chem.*, 2003, **5**, 633; (b) C. Schmuck and M. Heil, *ChemBioChem*, 2003, **4**, 1232; (c) C. Schmuck, P. Frey, and M. Heil, *ChemBioChem*, 2005, **6**, 628.
36. S. Bartoli, T. Mahmood, A. Malik, *et al.*, *Org. Biomol. Chem.*, 2008, **6**, 2340.
37. (a) V. D. Jadhav and F. P. Schmidtchen, *Org. Lett.*, 2006, **8**, 2329; (b) V. D. Jadhav and F. P. Schmidtchen, *J. Org. Chem.*, 2008, **73**, 1077.
38. For recent reviews see: (a) V. Martos, P. Castreño, J. Valero, and J. de Mendoza, *Curr. Opin. Chem. Biol.*, 2008, **12**, 698; (b) H. Yin and A. D. Hamilton, *Angew. Chem. Int. Ed.*, 2005, **44**, 4130.
39. J. S. Albert, M. S. Goodman, and A. D. Hamilton, *J. Am. Chem. Soc.*, 1995, **117**, 1143.
40. (a) M. W. Peczuh, A. D. Hamilton, J. Sánchez-Quesada, *et al.*, *J. Am. Chem. Soc.*, 1997, **119**, 9327; (b) T. Haack, M. W. Peczuh, X. Salvatella, *et al.*, *J. Am. Chem. Soc.*, 1999, **121**, 11813; (c) B. P. Orner, X. Salvatella, J. Sánchez-Quesada, *et al.*, *Angew. Chem. Int. Ed.*, 2002, **41**, 117; (d) X. Salvatella, M. Martinell, M. Gairí, *et al.*, *Angew. Chem. Int. Ed.*, 2004, **43**, 196.
41. A. J. Levine, *Cell*, 1998, **88**, 323.
42. S. Gordo, V. Martos, E. Santos, *et al.*, *Proc. Natl. Acad. Sci. U.S.A.*, 2008, **105**, 16426.
43. V. Martos, S. C. Bell, E. Santos, *et al.*, *Proc. Natl. Acad. Sci. U.S.A.*, 2009, **106**, 10482.
44. K. Okuro, K. Kinbara, K. Tsumoto, *et al.*, *J. Am. Chem. Soc.*, 2009, **131**, 1626.
45. K. Okuro, K. Kinbara, K. Takeda, *et al.*, *Angew. Chem. Int. Ed.*, 2010, **49**, 3030.
46. For an application of these cationic dendrons related to hydrogels and molecular materials, see: Q. Wang, J. L. Mynar, M. Yoshida, *et al.*, *Nature*, 2010, **463**, 339.
47. (a) B. L. Jacobson and F. A. Quiocho, *J. Mol. Biol.*, 1988, **204**, 783; (b) H. Luecke and F. A. Quiocho, *Nature*, 1990, **347**, 402.
48. B. Dietrich, T. M. Fyles, J. – M. Lehn, *et al.*, *J. Chem. Soc., Chem. Commun.*, 1978, 934–936.
49. R. P. Dixon, S. J. Geib, and A. D. Hamilton, *J. Am. Chem. Soc.*, 1992, **114**, 365.
50. V. Alcázar, M. Segura, P. Prados, and J. de Mendoza, *Tetrahedron Lett.*, 1998, **39**, 1033.
51. (a) F. P. Schmidtchen, *Tetrahedron Lett.*, 1989, **30**, 4493; For a related study see: (b) P. Schiessl and F. P. Schmidtchen, *J. Org. Chem.*, 1994, **59**, 509.
52. V. D. Jadhav and F. P. Schmidtchen, *Org. Lett.*, 2005, **7**, 3311.
53. M. Haj-Zaroubi and F. P. Schmidtchen, *ChemPhysChem*, 2005, **6**, 1181.
54. S. L. Tobey and E. V. Anslyn, *J. Am. Chem. Soc.*, 2003, **125**, 14807.
55. For a study of the aqueous solvation sphere of guanidinium cations see: (a) P. E. Mason, G. W. Neilson, C. E. Dempsey, *et al.*, *Proc. Natl. Acad. Sci. U.S.A.*, 1990, **87**, 167.
56. (a) D. J. Iverson, G. Hunter, J. F. Blount, *et al.*, *J. Am. Chem. Soc.*, 1981, **103**, 6073; (b) K. V. Kilway and J. S. Siegel, *J. Am. Chem. Soc.*, 1992, **114**, 255; (c) T. D. P. Stack, Z. Hou, and K. N. Raymond, *J. Am. Chem. Soc.*, 1993, **115**, 6466; (d) H. – W. Marx, F. Moulines, T. Wagner, and D. Astruc, *Angew. Chem. Int. Ed. Engl.*, 1996, **35**, 1701.
57. K. Niikura, A. Metzger, and E. V. Anslyn, *J. Am. Chem. Soc.*, 1998, **120**, 8533.
58. Z. Zhong and E. V. Anslyn, *Angew. Chem. Int. Ed.*, 2003, **42**, 3005.
59. M. Delivoria-Papadopoulos, F. A. Oski, and A. J. Gottlieb, *Science*, 1969, **165**, 165.
60. P. D. Beer, M. G. B. Drew, and D. K. Smith, *J. Organomet. Chem.*, 1997, **543**, 259.
61. S. Nishizawa, Y. Kato, and N. Teramae, *J. Am. Chem. Soc.*, 1999, **121**, 9463.
62. Y. Sun, C. Zhong, R. Gong, and E. Fu, *Org. Biomol. Chem.*, 2008, **6**, 3044.
63. (a) A. Galán, J. de Mendoza, C. Toiron, *et al.*, *J. Am. Chem. Soc.*, 1991, **113**, 9424; (b) G. Deslongchamps, A. Galán, J. de Mendoza, and J. Rebek Jr. *Angew. Chem. Int. Ed. Engl.*, 1992, **31**, 61; (c) C. Andreu, A. Galán, K. Kobiro, *et al.*, *J. Am. Chem. Soc.*, 1994, **116**, 5501.
64. S. Menuel, R. E. Duval, D. Cuc, *et al.*, *New J. Chem.*, 2007, **31**, 995.
65. L. Hernandez-Folgado, D. Baretic, I. Piantanida, *et al.*, *Chem.—Eur. J.*, 2010, **16**, 3036.
66. (a) M. Dudic, A. Colombo, F. Sansone, *et al.*, *Tetrahedron*, 2004, **60**, 11613; (b) F. Sansone, M. Dudic, G. Donofrio, *et al.*, *J. Am. Chem. Soc.*, 2006, **128**, 14528.
67. A. Membrino, M. Paramasivam, S. Cogoi, *et al.*, *Chem. Commun.*, 2010, **46**, 625.
68. A. Bugaut, K. Jantos, J. – L. Wietor, *et al.*, *Angew. Chem. Int. Ed.*, 2008, **47**, 2677.
69. J. Sánchez-Quesada, C. Seel, P. Prados, and J. de Mendoza, *J. Am. Chem. Soc.*, 1996, **118**, 277.

70. I. Rozas and P. K. Kruger, *J. Chem. Theory Comput.*, 2005, **1**, 1055.
71. C. E. Dempsey, P. E. Mason, J. W. Brady, and G. W. Neilson, *J. Am. Chem. Soc.*, 2007, **129**, 15895.
72. M. Berger and F. P. Schmidtchen, *Angew. Chem. Int. Ed. Engl.*, 1998, **37**, 2694.
73. K. Kobiro and Y. Inoue, *J. Am. Chem. Soc.*, 2003, **125**, 421.
74. (a) B. P. Hay, M. Gutowski, D. A. Dixon, *et al.*, *J. Am. Chem. Soc.*, 2004, **126**, 7925; (b) B. P. Hay, T. K. Firman, and B. A. Moyer, *J. Am. Chem. Soc.*, 2005, **127**, 1810.
75. (a) A. Gleich, F. P. Schmidtchen, P. Mikulcik, and G. Müller, *J. Chem. Soc., Chem. Commun.*, 1990, **55**; (b) L. P. Lu, M. L. Zhu, and P. Yang, *Acta Crystallogr., Sect. C: Cryst. Struct. Commun.*, 2004, **C60**, m18.
76. A. P. Bisson, V. M. Lynch, M. K. C. Monahan, and E. V. Anslyn, *Angew. Chem. Int. Ed. Engl.*, 1997, **36**, 2340.
77. P. Blondeau and J. de Mendoza, *New J. Chem.*, 2007, **31**, 736.
78. (a) M. S. Taylor, E. N. Jacobsen, *Angew. Chem. Int. Ed.*, 2006, **45**, 1520; (b) A. G. Doyle and E. N. Jacobsen, *Chem. Rev.*, 2007, **107**, 5713; (c) D. Leow and C. H. Tan, *Chem. Asian J.*, 2009, **4**, 488.
79. (a) T. Šmejkal and B. Breit, *Angew. Chem. Int. Ed.*, 2008, **47**, 311; (b) T. Šmejkal, D. Gribkov, J. Geier, *et al.*, *Chem.—Eur. J.*, 2010, **16**, 2470; For the decarboxylative hydrofomylation of α,β-unsaturated acids see: (c) T. Šmejkal and B. Breit, *Angew. Chem. Int. Ed.*, 2008, **47**, 3946.
80. L. Diab, T. Šmejkal, J. Geier, and B, Breit, *Angew. Chem. Int. Ed.*, 2009, **48**, 8022.
81. (a) V. Alcázar, J. R. Morán, and J. de Mendoza, *Tetrahedron Lett.*, 1995, **36**, 3941; (b) M. Martín-Portugués, V. Alcázar, P. Prados, and J. de Mendoza, *Tetrahedron*, 2002, **58**, 2951.
82. R. Breslow and S. D. Dong, *Chem. Rev.*, 1998, **98**, 1997.
83. (a) D. Weber, A. K. Meeker, and A. S. Mildvan, *Biochemistry*, 1991, **30**, 610; (b) J. Aqvist and A. Warshel, *Biochemistry*, 1989, **28**, 4680.
84. E. H. Serpersu, D. Shortle, and A. S. Midvan, *Biochemistry*, 1987, **26**, 1289.
85. Reviews: (a) E. L. Hegg and J. N. Burstyn, *Coord. Chem. Rev.*, 1998, **173**, 133; (b) C. Liu, M. Wang, T. Zhang, and H. Sun, *Coord. Chem. Rev.*, 2004, **248**, 147; (c) N. H. Williams, B. Takasaki, M. Wall, and J. Chin, *Acc. Chem. Res.*, 1999, **32**, 485; For a few representative examples employing guanidinium derivatives to stabilize the anionic transition state for phosphate esters see: (d) V. Jubian, R. P. Dixton, and A. D. Hamilton, *J. Am. Chem. Soc.*, 1992, **114**, 1120; (e) J. Smith, K. Ariga, and E. V. Anslyn, *J. Am. Chem. Soc.*, 1993, **115**, 362; (f) V. Jubian, A. Veronese, R. P. Dixon, and A. D. Hamilton, *Angew. Chem. Int. Ed. Engl.*, 1995, **34**, 1237; (g) M. J. Belousolff, L. Tjioe, B. Graham, and L. Spiccia, *Inorg. Chem.*, 2008, **47**, 8641; (h) N. J. V. Lindgren, L. Geiger, J. Razkin, *et al.*, *Angew. Chem. Int. Ed.*, 2009, **48**, 6722.
86. H. Aït-Haddou, J. Sumaoka, S. L. Wiskur, *et al.*, *Angew. Chem. Int. Ed.*, 2002, **41**, 4014.
87. For related studies of catalysts for phosphate ester hydrolysis using ammonium subunits for stabilization see: (a) E. Koevari and R. Kräemer, *J. Am. Chem. Soc.*, 1996, **118**, 12704; (b) M. Wall, B. Linkletter, D. Williams, *et al.*, *J. Am. Chem. Soc.*, 1999, **121**, 4710.

Anion Receptors Containing Heterocyclic Rings

Philip A. Gale and Cally J. E. Haynes
University of Southampton, Southampton, UK

1 INTRODUCTION

Heterocycles play key roles in a wide variety of anion receptor systems, functioning as hydrogen bond donors, charged groups, and as organizational elements.[1] In this chapter, a variety of anion receptors containing heterocyclic components are reviewed and the roles the heterocycles play in forming stable complexes with anionic guests are highlighted. The chapter is organized according to the heterocycle employed, and pyrrole-based receptors, systems containing indole (and related biindole, carbazole, and indolocarbazole heterocycles), imidazolium-based receptors, pyridine (and related quinoline and *iso*-quinoline systems), and 1,2,3-triazole-based hosts are examined.

Supramolecular Chemistry: From Molecules to Nanomaterials.
Edited by Philip A. Gale and Jonathan W. Steed.
© 2012 John Wiley & Sons, Ltd. ISBN: 978-0-470-74640-0.

2 ANION RECEPTORS CONTAINING PYRROLE

2.1 Cyclic receptors

Pyrrole contains a single NH hydrogen bond donor group and forms the basis for a wide variety of charged and neutral anion receptor systems. In a pioneering early work, Sessler and coworkers at The University of Texas at Austin discovered that expanded porphyrins such as sapphyrin function as receptors for anions such as fluoride and chloride, binding the anionic guest via a combination of hydrogen bonding and electrostatic interactions (see **Porphyrins and Expanded Porphyrins as Receptors**, Volume 3).[2] Although these compounds are highly effective receptors, they are synthetically challenging to make, in some cases requiring over 20 synthetic steps. This led Sessler to consider alternative pyrrole-based macrocyclic systems as putative anion receptors. The calix[4]pyrroles are cyclic tetramers formed in one step by the acid-catalyzed condensation of pyrrole and a ketone. As the macrocycle contains two alkyl groups attached to each *meso*-carbon that links the pyrrole rings together, the system is reasonably stable toward oxidation, and thus is not a porphyrin precursor. The macrocycle was first synthesized by Baeyer in 1886[3] and has subsequently been studied as a ligand for transition metals[4] and lanthanides[5] (on deprotonation of the pyrrole NH groups). In 1996, Sessler and coworkers reported that *meso*-octamethylcalix[4]pyrrole **1** (Figure 1), formed by the acid-catalyzed condensation of acetone and pyrrole, complexes anions such as fluoride, chloride, and dihydrogen phosphate in deuterated dichloromethane (DCM) solution.[6] Solid-state studies showed that in the absence of an anionic guest, the macrocycle adopts the 1,3-alternate conformation in which the pyrrole rings are oriented alternatively up and down, whereas when binding an anion the four

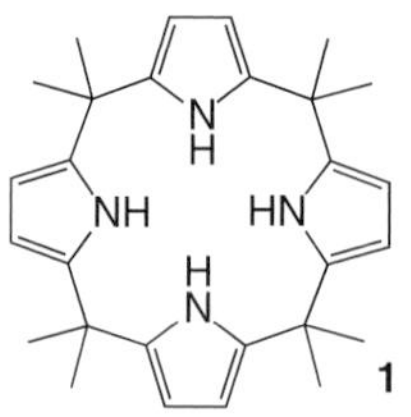

Figure 1 The chemical structure of *meso*-octamethylcalix[4] pyrrole **1**.

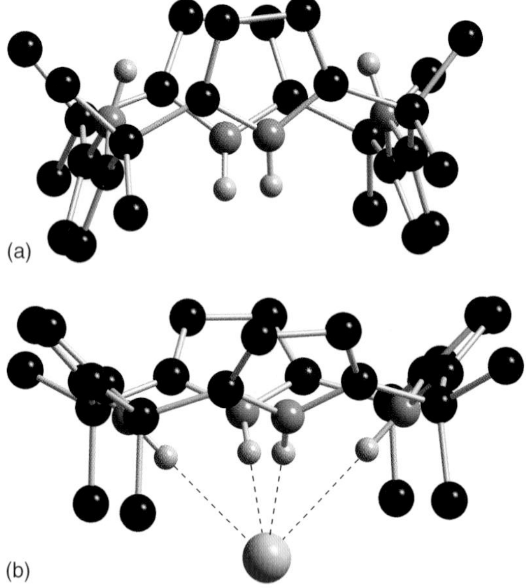

Figure 2 (a) The X-ray crystal structure of compound **1** and (b) the tetrabutylammonium chloride complex of compound **1** (countercation and solvent molecules have been omitted for clarity).

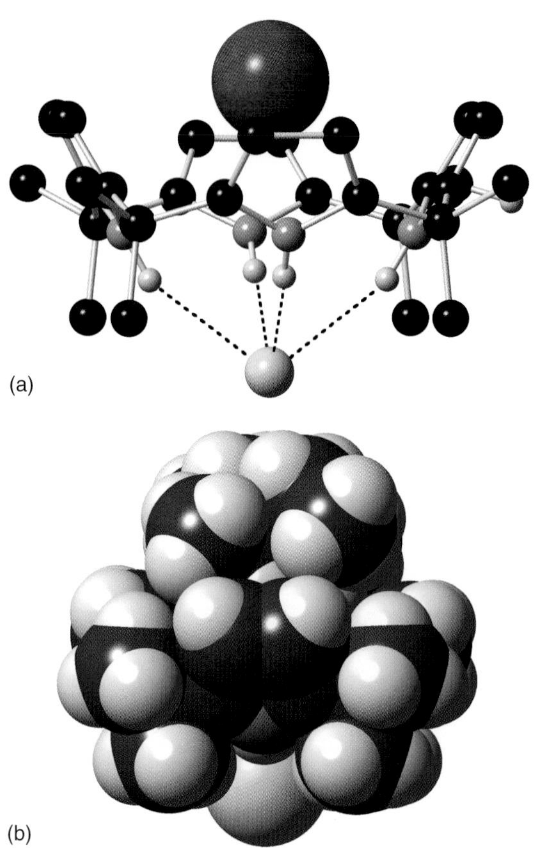

Figure 3 (a) The X-ray crystal structure of the cesium chloride complex of compound **1** and (b) a space-filling view of the tetraethylammonium chloride complex of compound **1** showing the cations occupying the cup-shaped cavity formed by the pyrrole rings (solvent molecules and nonacidic hydrogen atoms have been omitted for clarity).

pyrrole rings form a cone conformation in which all four pyrrole NH groups form hydrogen bonds to the anionic guest (Figure 2). In solution, in the absence of an anion, the macrocycle is flexible and the pyrrole rings can rotate through the annulus of the macrocycle.

In 2005, Moyer, Sessler, and Gale and coworkers reported that calix[4]pyrroles can function as receptors for ion pairs. On binding an anion, large charge-diffuse cations such as cesium, imidazolium, or pyridinium can occupy the cone-shaped cavity formed by the pyrrole rings (Figure 3a) in the solid state.[7–9] It was subsequently reported by Sessler, Schmidtchen, and Gale[10,11] that the stability constant of compound **1** with chloride varies with the nature of the countercation and solvent. The dependence of the stability constant for chloride binding on the countercation measured in DCM is shown in Table 1 using ITC (ionic thermocurrent) and NMR (nuclear magnetic resonance) techniques. The crystal structures of a series of tetraalkylammonium and tetraalkylphosphonium chloride complexes have been determined showing the association of the cations with the calixpyrrole cup-shaped cavity. For example, a space-filling view of the tetraethylammonium chloride complex of compound **1** is shown in Figure 3(b). These findings are evidence that anion complexation by *meso*-octamethylcalixpyrrole **1** in solution is not a simple 1 : 1 anion complexation process but involves countercations (as shown in Scheme 1) as well.

Sessler and coworkers found that higher order calix[n] pyrroles can be prepared by the acid-catalyzed condensation of 3,4-difluoropyrrole and acetone, as shown in Scheme 2.[12] Octafluoro-*meso*-octamethylcalix[4]pyrrole **2** shows enhanced binding affinities for a variety of anions in acetonitrile solution (as determined by ITC) compared to the parent macrocycle **1** because of the presence of the electron-withdrawing fluorine substituents increasing the acidity of the pyrrole NH groups (Table 2). Additionally, as the size of the fluorinated macrocycle increases, the ratio of the chloride and bromide stability constants decreases, demonstrating a preference for the larger macrocycles to complex larger anionic guests.

Table 1 Titrations of calixpyrrole **1** in dichloromethane with various chloride salts at 298 K (calorimetry) or 295 K (NMR).

	ΔH (kcal mol^{-1})	$T\Delta S$ (kcal mol^{-1})	ΔG (kcal mol^{-1})	K_a (M^{-1})	
				ITC	NMR
TEA–Cl	−9.91	−3.80	−6.11	3.1×10^4	3.7×10^4
TPA–Cl	^{a}	^{a}	^{a}	^{a}	6.6×10^2
TBA–Cl	^{a}	^{a}	^{a}	^{a}	4.3×10^2
TEP–Cl	−9.59	−4.68	−4.91	3.9×10^3	3.6×10^3
TBP–Cl	^{a}	^{a}	^{a}	^{a}	^{a}
TPhP–Cl	−6.8	−2.2	−4.6	2.8×10^3	ND

TEA, tetraethylammonium; TPA, tetrapropylammonium; TBA, tetrabutylammonium; TEP, tetraethylphosphonium; TBP, tetrabutylphosphonium; TPhP, tetraphenylphosphonium; ND, not determined.
^{a}No reliable fit to a 1 : 1 binding isotherm could be made.

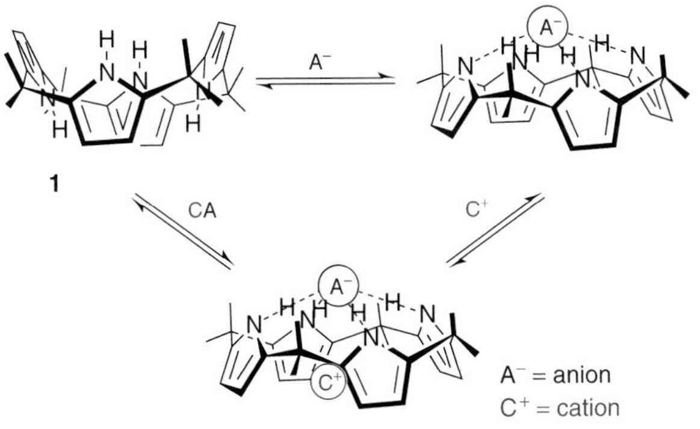

A$^-$ = anion
C$^+$ = cation

Scheme 1 Calix[4]pyrrole ion-pair complexation in solution.

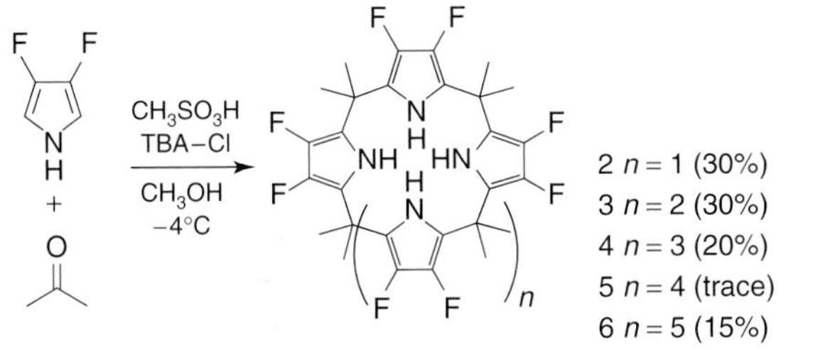

CH$_3$SO$_3$H, TBA–Cl, CH$_3$OH, −4°C

2 $n = 1$ (30%)
3 $n = 2$ (30%)
4 $n = 3$ (20%)
5 $n = 4$ (trace)
6 $n = 5$ (15%)

Scheme 2 The synthesis of fluorinated calix[n]pyrroles (n = 4–8).

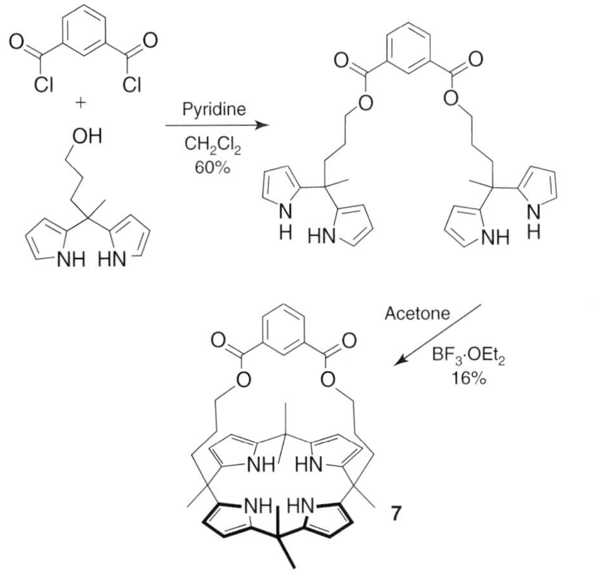

An alternative strategy to increase the affinity of the calixpyrrole framework for anions is to introduce a "strap" across the macrocycle containing additional hydrogen bond donor groups.[13]

For example, the synthesis of receptor **7**, a calixpyrrole with an isophthalate strap, is shown in Scheme 3.[14,15] This compound was synthesized by condensation of 5-hydroxy-2-pentanone with pyrrole in the presence of an acid catalyst to afford the dipyrromethane. Isophthaloyl dichloride was made to react with 2 equivalents of the dipyrromethane to afford the bis-dipyrromethane that was condensed with

Pyridine, CH$_2$Cl$_2$, 60%

Acetone, BF$_3$·OEt$_2$, 16%

Scheme 3 The synthesis of strapped calix[4]pyrrole **7**.

acetone in the presence of $BF_3 \cdot OEt_2$ to afford the ester strapped calixpyrrole in 16% yield. A single crystal X-ray structure of the chloride complex of **7** obtained shows that the chloride anion is bound to the four pyrrole rings of the calixpyrrole and that there is a further C–H hydrogen bond to the anion from the central Ar–H, as shown in Figure 4. The distance between the Ar–C and the bound chloride anion was 2.92 Å, a value consistent with a medium strength of CH–Cl hydrogen bond. Determination of the stability constant in acetonitrile by ITC reveals that the strapped system has one order of magnitude higher affinity for chloride than the parent receptor **1** (Table 3).

More recently reported systems include compound **8** that contains a fifth pyrrole hydrogen bond donor in the strap (Figure 5).[16] In the presence of substoichiometric quantities of chloride, this compound undergoes slow exchange on the

Table 2 Stability constants (K_a, M^{-1}) for anion binding by fluorinated calixpyrroles **2**–**4** and *meso*-octamethylcalix[4]pyrrole **1** in CH_3CN or DMSO as determined by ITC analysis at 30 °C using the corresponding tetrabutylammonium salts as the anion source.[a]

Anion	Solvent	**1**	**2**	**3**	**4**
Cl^-	CH_3CN	140 000	530 000	41 000	280 000
	DMSO	1 300	1 500		
Br^-	CH_3CN	3 400	8 500	4 500	110 000
I^-	CH_3CN	[b]	[b]	[b]	610
$CH_3CO_2^-$	CH_3CN	290 000	1 900 000	[c]	[c]
	CH_3CN[d]	350 000	2 400 000	520 000	1 000 000
	DMSO	6 100	48 000		
$C_6H_5CO_2^-$	CH_3CN	120 000	1 200 000	52 000	[c]
	CH_3CN[d]	170 000	1 400 000	83 000	580 000
$H_2PO_4^-$	DMSO	5 100	17 000	9 600	15 000
$K_{rel} = K_a(Cl^-)/K_a(Br^-)$		41	62	9	3

Errors < 15%.
[a]The host (macrocycle) solution was titrated with the guest (anion) solution unless otherwise indicated.
[b]Stability constant is too low to be determined by ITC.
[c]A good fit of the data to a 1 : 1 binding profile could not be made.
[d]The guest solution was titrated with the host solution (reverse titration).

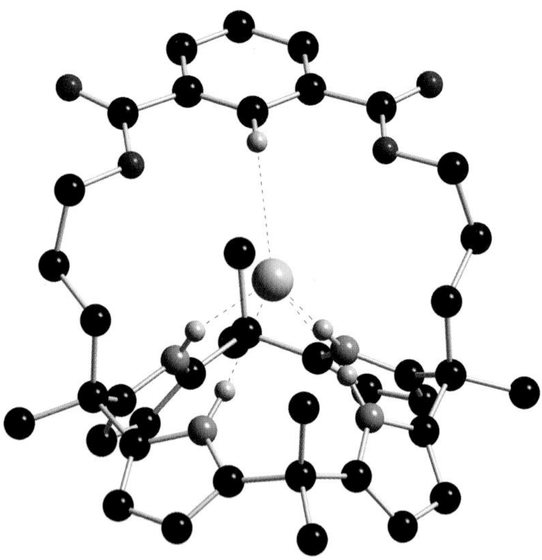

Figure 4 Single crystal X-ray structure of diester-strapped calix[4]pyrrole **7** binding chloride. Countercation and nonacidic hydrogen atoms have been omitted for clarity.

NMR timescale in DMSO (dimethyl sulfoxide)-d_6/water solution (4 : 1 v/v), that is, resonances corresponding to the free ligand and the chloride complex appear separately in the spectrum. This allowed the receptor to be used to determine the concentration of chloride in a sports drink by comparing the intensity of resonances from the free ligand and complex using simple ^1H NMR methods.[17]

Other macrocyclic pyrrole-based anion receptors include the cyclo[*n*]pyrroles that consist solely of an array of pyrrole groups linked directly to each other via the 2- and 5-positions of the rings. For example, cyclo[8]pyrrole can be prepared by direct coupling of bipyrroles in the presence of iron(III) chloride in 1 M sulfuric acid (Scheme 4).[18]

Table 3 Thermodynamic data for the interaction of calyx[4]pyrroles **1**, **7**, and **8** with chloride.[a]

Host	$T\Delta S$	ΔH	ΔG	$K_a(M^{-1})$
1	−2.91	−10.16	−7.29	2.2×10^5
7	−1.90	−10.54	−8.64	2.2×10^6
8	−1.44	−11.34	−9.90	1.8×10^7

[a]Units of $T\Delta S$, ΔH, and ΔG are kcal mol^{-1}; titrations were run at 25 °C in acetonitrile, and chloride was used in the form of its tetrabutylammonium salt.

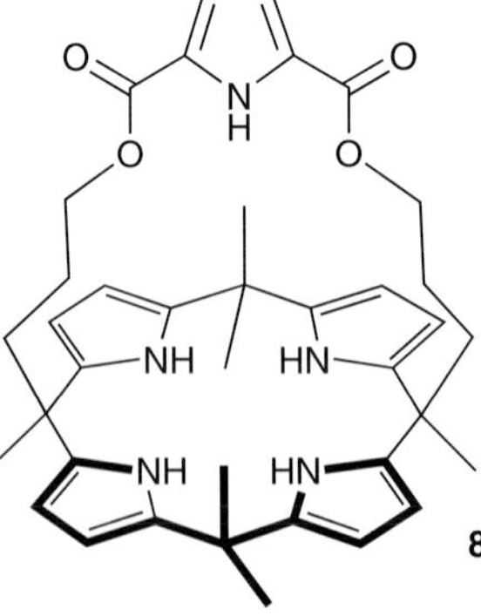

Figure 5 The chemical structure of pyrrole-strapped calix[4]pyrrole **8**.

Liquid crystals formed by functionalized cyclo[8]pyrrole have been employed as sensors for nitroaromatic explosive compounds such as TNT (trinitrotoluene) (Figure 6). The macrocycles stack with the aromatic compounds forming an ordered mesophase that can be detected by microscopy.[19]

Sessler and coworkers have used anion-induced synthesis to produce a new series of 2,6-diamidopyridine-bipyrrole

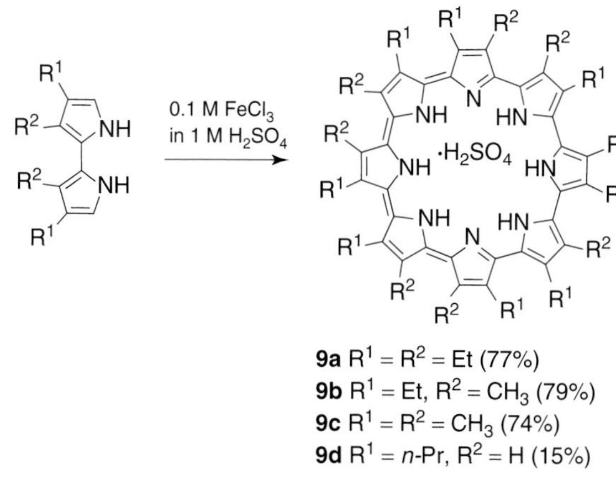

9a $R^1 = R^2 = Et$ (77%)
9b $R^1 = Et$, $R^2 = CH_3$ (79%)
9c $R^1 = R^2 = CH_3$ (74%)
9d $R^1 = n\text{-}Pr$, $R^2 = H$ (15%)

Scheme 4 The synthesis of compounds **9a–9d**.

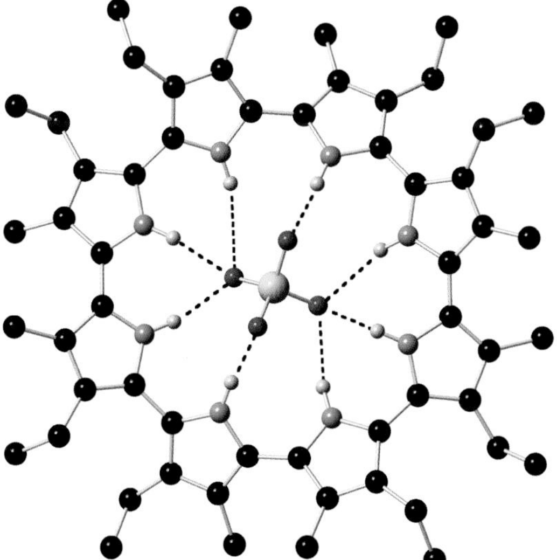

Figure 6 One of the two crystallographically independent complexes in the solid-state structure of **9b**. The sulfate anion is bound by eight hydrogen bonds from the diprotonated cyclo[8]pyrrole with NH···O distances in the range 1.91–2.49 Å.

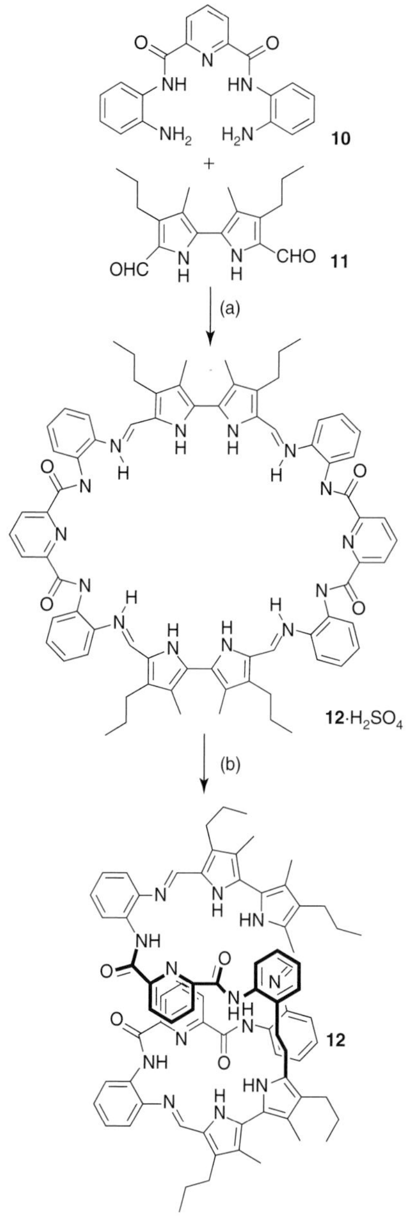

Scheme 5 The synthesis of macrocycle **12**. (a) Concentrated H_2SO_4 (2.2 equiv), 48 h, room temperature; (b) Et_3N (excess), CH_3OH/CH_2Cl_2.

macrocycles.[20] These compounds contain imine links, the formation of which are reversible. The macrocycles were obtained by the reaction of diamine **10** and diformylbipyrrole **11** in methanol under acidic conditions. The choice of the acid used (HCl, HBr, CH_3CO_2H, CF_3CO_2H, H_3PO_4, H_2SO_4, HNO_3) was shown to play a critical role in defining the distribution of products from this condensation reaction. The use of hydrochloric or hydrobromic acid gives rise to the formation of oligomers with high molecular weights ($m/z > 3000$) and a small amount of an unstable [1+1] macrocycle. However, the use of CH_3CO_2H, CF_3CO_2H, H_3PO_4 led to the formation of the [2+2] macrocycle **12** contaminated by uncharacterized oligomers, while the use of nitric acid affords only oligomeric species with high molecular weights. The use of sulfuric acid led to the formation the kinetic product $\mathbf{12} \cdot 2H_2SO_4$ under stirring, with the free macrocycle being liberated by suspension of the H_2SO_4 complex in dichloroethane and treatment with triethylamine (Scheme 5). Macrocycle **12** interacts strongly with tetrahedral anions such as hydrogen sulfate (1 : 1; $K_a = 63\,500 \pm 3000\,M^{-1}$) and dihydrogen phosphate (2 : 1; $K_{a1} = 191\,000 \pm 15\,400\,M^{-1}$; $K_{a2} = 60\,200 \pm 6000\,M^{-1}$), less strongly with acetate (1 : 1, $K_a = 26\,000 \pm 2400\,M^{-1}$), and not at all with chloride, bromide, or nitrate.

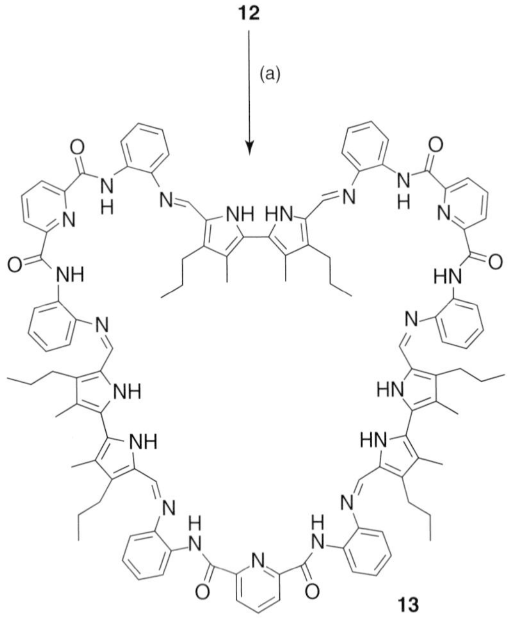

Scheme 6 The synthesis of macrocycle **13**. (a) $TBAHSO_4$, acetonitrile, five days, room temperature without stirring, followed by Et_3N; or $TBAH_2PO_4$, acetonitrile, five days, room temperature without stirring.

When compound **12** was dissolved in acetonitrile and allowed to stand for five days in the presence of tetrabutylammonium hydrogensulfate or dihydrogen phosphate, it was found to rearrange to give the [3+3] macrocycle **13** (Scheme 6). However, if the reaction mixture was stirred, the [3+3] macrocyclic product formed only in trace amounts and instead the predominant product was the [2+2] macrocycle **12** formed as the sulfuric acid complex. Thus, it seems that in this case the kinetics and thermodynamics of this reaction are in fine balance and depend on subtle changes in the reaction conditions. When stirred, the H_2SO_4 complex of **12** forms quickly and precipitates. In the absence of stirring the precipitation process is slow, allowing isolation of **13** as a thermodynamic product.

2.2 Acyclic receptors

Acyclic pyrrole-based receptors can also form stable and selective complexes with anionic guests. Perhaps, one of the most important groups of anion receptor from the perspective of biological activity is the prodigiosins. These compounds are naturally occurring tripyrrolic red pigments formed by microorganisms including *Serratia* and *Streptomyces*.[21–23] Naturally occurring prodigiosins, for example, compounds **14** and **15** (Figure 7), exhibit a range of potentially useful biological activities, including immunosuppression, induction of tumor cell apoptosis, and toxicity against

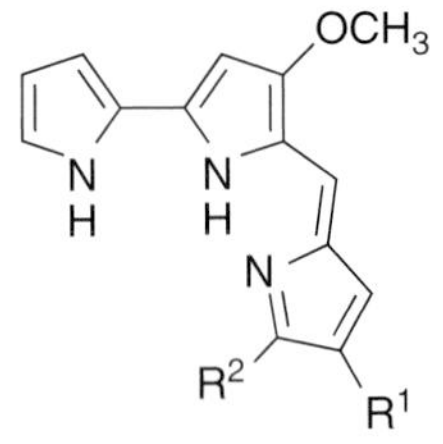

Figure 7 The structures of prodigiosin 25-C (**14**) and prodigiosin (**15**).

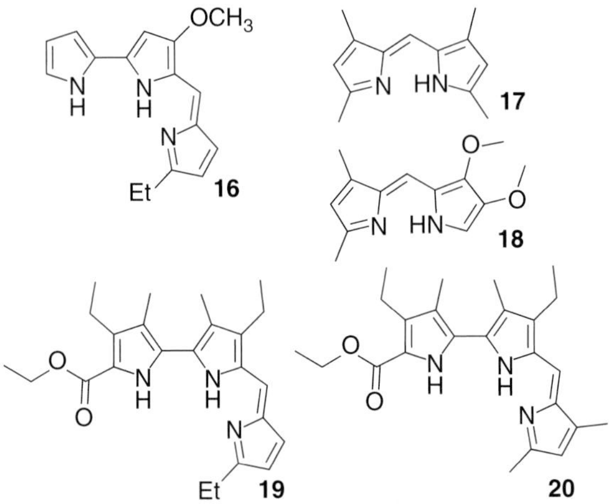

Figure 8 The structures of prodigiosin model compounds **16–20**.

bacteria, protozoa, fungi, and the malarial parasite.[24–27] A number of membrane-bound cellular compartments, such as the organelles of the biosynthetic and endocytic pathways, contain acidic interiors that are essential for organelle function and cell survival. There is evidence that connects abnormal changes in organelle steady-state pH with the pathology of several diseases.[28] The prodigiosins have been shown to promote the cotransport of H^+/Cl^- across bilayer membranes[29–33]; however, it is not known which, if any, of the biological activities are due to organelle deacidification.

Therefore, much effort has been directed toward the synthesis of compounds capable of the cotransport of HCl across lipid bilayer membranes. For example, Sessler and coworkers have studied the anion-binding and transport properties of prodigiosin model compounds **16–20** (Figure 8).[34] Binding studies conducted in acetonitrile solution by using ITC showed that the monoprotonated iodide salts of compounds **16–20** bound chloride with stability constants between 1.1×10^5 and 8.8×10^5 M^{-1} at 30 °C. The crystal structure of the HCl complex of compound **20** was elucidated, showing chloride bound to the protonated receptor by three hydrogen bonds in

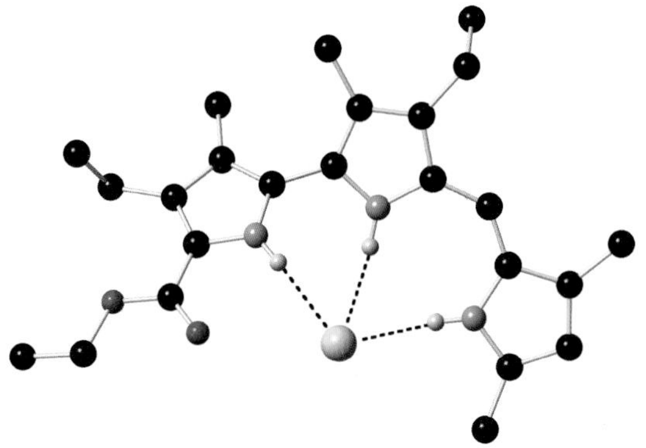

Figure 9 The crystal structure of the complex formed between monoprotonated prodigiosin **19** and chloride.

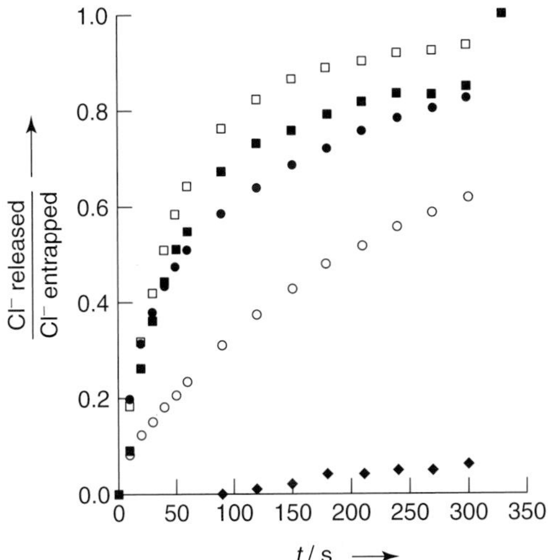

Figure 10 Time-dependent efflux of chloride ions from 200 nm vesicles loaded with a solution of NaCl (500 mM) and suspended in a solution of $NaNO_3$ (500 mM) and triethylsilane (5 mM), adjusted to pH 7.4. Prodigiosins are indicated by filled markers (**16** ●, **19** ■, **20** ♦); dipyrromethenes are indicated by open markers (**17** □, **18** ○).

the range 1.97–2.72 Å (Figure 9). Transport studies were conducted by monitoring the release of encapsulated chloride from 200 nm liposomes composed of a mixture of POPC (1-palmitoyl-2-oleoyl-*sn*-glycero-3-phosphocholine) and POPS (1-palmitoyl-2-oleoyl-*sn*-glycero-3-phospho-L-serine), using a chloride-selective electrode, as a function of time. The results (shown in Figure 10) reveal that the compound that is structurally most similar to the natural prodigiosins (i.e., compound **16**) released chloride at the fastest rate from the liposomes—this compound was consequently studied at a concentration four times lower than the other compounds. The transport efficiency was found to be **16** ≫ **19** ≈ **17** > **18** ≫ **20**. To investigate the mechanism of transport, the pH of the external solution was changed to 5.5. They found that under these conditions the chloride

21 R^1 = *n*-Bu; R^2 = Ph
22 R^1 = Ph; R^2 = Ph

23 R^1 = *n*-Bu; R^2 = Ph
24 R^1 = Ph; R^2 = Ph

25 R^1 = *n*-Bu; R^2 = Cl
26 R^1 = Ph; R^2 = Cl
27 R^1 = 4-nitrophenyl; R^2 = Ph
28 R^1 = 3,5-dinitrophenyl; R^2 = Ph

Figure 11 Structures of amidopyrroles **21–28**.

efflux was less efficient than when the external solution was neutral or basic—a finding consistent with a H^+/Cl^- cotransport mechanism. The anticancer ability of these compounds was studied using a cell proliferation assay with human lung cancer and PC3 human prostate cells. In both cases, the activity was found to be similar, namely, **16** > **19** > **17** > **18** ≈ **20**. Compounds **16–19** showed a significant cytotoxic activity, with 100% of the cancer cells killed at a concentration of 40 μM for both cell lines. Interestingly, the rate of chloride transport and the biological activity were found to correlate well—evidence that the proposed mode of action of prodigiosin via H^+/Cl^- cotransport from organelles is a reasonable hypothesis. More recent studies on prodigiosin have found that it is capable of exchanging both chloride and nitrate[35] and chloride and bicarbonate[36] across a POPC lipid bilayer in addition to functioning as an HCl cotransporter.

Gale and coworkers have combined pyrrole and amide groups to form a series of very simple anion receptors based on 2,5-disubstituted pyrroles (e.g., compounds **21** and **22**) (Figure 11). These compounds proved to be selective

Table 4 Stability constants of **21–23** with anionic guests.

Anion	Stability constants (K_a, M^{-1}) with anions[a,b]		
	Compound 21 in CD_3CN (0.03% water)	Compound 22 in DMSO-d_6/0.5% water	Compound 23 CD_3CN (0.03% water)
F^-	85[c]	74	134
Cl^-	138[c]	11	28
Br^-	<10	<10	<10
$H_2PO_4^-$	357	1450	89
$C_6H_5COO^-$	2500	560	202

[a] Anions added as tetrabutylammonium salts.
[b] Errors estimated to be <15%
[c] The amount of water present in the acetonitrile can have a dramatic effect on fluoride/chloride selectivity. In the presence of 0.5% water, fluoride is bound with a stability constant of 37.5 M^{-1} whereas chloride is bound more weakly ($K = 12.5\ M^{-1}$).

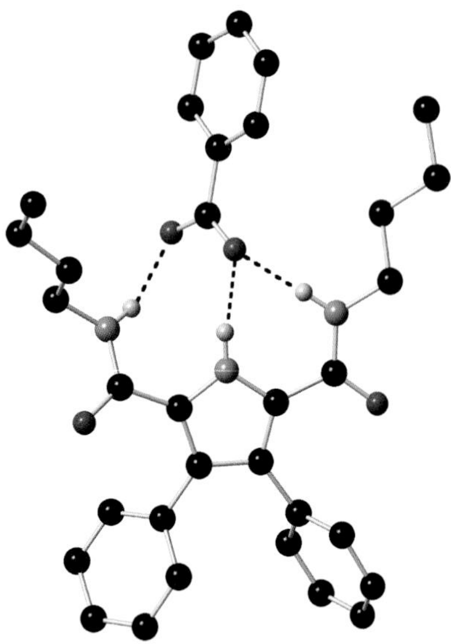

Figure 12 The crystal structure of the benzoate complex of receptor **21**. The tetrabutylammonium countercation and nonacidic hydrogen atoms have been omitted for clarity.

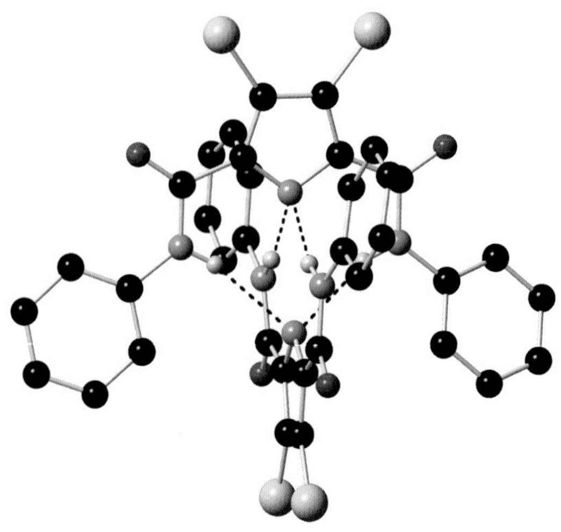

Figure 13 The X-ray crystal structure of the interlocked dimer $[(\mathbf{26}\text{-H}^+)_2]^{2-}$.

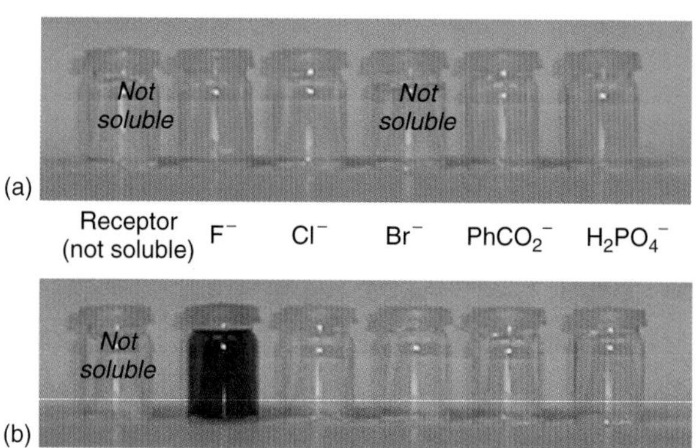

Figure 14 Solutions of (a) receptor **27** and (b) receptor **28** (2 mM) in acetonitrile with various anionic guests (added as their tetrabutylammonium salt at a concentration of 20 mM). In the absence of an anion, the receptors are not soluble in this solvent but are solubilized on addition of the anion (with the exception of receptor **27** and bromide).

receptors for oxo-anions (Table 4), and comparison of the anion-binding properties of these species with a mono-amidopyrrole (**23**), which has significantly lower affinity for anionic species than the bis-amidopyrroles, led the authors to suggest that all three of the hydrogen bond donor groups were involved in oxo-anion complexation in solution. This was confirmed in the solid state with the elucidation of the crystal structure of a benzoate complex (Figure 12).

Compounds **25** and **26** contain electron-withdrawing chlorine substituents in the 3- and 4-positions of the pyrrole ring and were synthesized in order to produce more acidic receptors that would form stronger hydrogen bonds to anionic guests. Addition of tetrabutylammonium chloride to solutions of compound **25** in acetonitrile-d_3 or DCM-d_2 or compound **26** in DCM-d_2 gave ^{1}H NMR titration curves indicative of 1 : 1 receptor : anion complex formation. For example, compound **25** binds chloride with a stability constant of $K_a = 2015\,\text{M}^{-1}$ in acetonitrile-d_3. The 3,4-diphenyl analog of compound **25**, compound **21**, is a much weaker chloride receptor, binding this anion in acetonitrile-d_3 with a stability constant of $138\,\text{M}^{-1}$. In contrast to these results, addition of excess fluoride to compound **26** in DCM-d_2 resulted in deprotonation of the receptor's pyrrole NH group. In the solid state, the deprotonated receptor formed an interlocked dimer as shown in Figure 13. This was the first indication that neutral hydrogen bond donor anion receptors could be deprotonated by basic anions such as fluoride. Similarly, when electron-withdrawing groups were appended to the amide groups (compounds **27** and **28**), addition of fluoride in acetonitrile resulted in deprotonation of the most acidic system (compound **28**) at the pyrrole NH group in a manner similar to compound **26** and an accompanying color change from colorless to blue in acetonitrile solution (Figure 14).

The amidopyrroles have also been employed as HCl membrane cotransport agents. Compounds **29** and **30** were synthesized as mimics for prodigiosin by Gale, Smith, and coworkers (Figure 15).[37] The compounds contain two hydrogen bond donors and a protonatable third site (a methylimidazole or a pyridine). They were designed to bind chloride when protonated (Figure 16) but to have a lower affinity for chloride when neutral, and hence they could potentially act as cotransporters for HCl across lipid bilayer membranes. It was found that compound **29** could facilitate the transport of Cl^- from POPC/cholesterol vesicles under a variety of different starting conditions (detected using an ion-selective electrode as described above). For example,

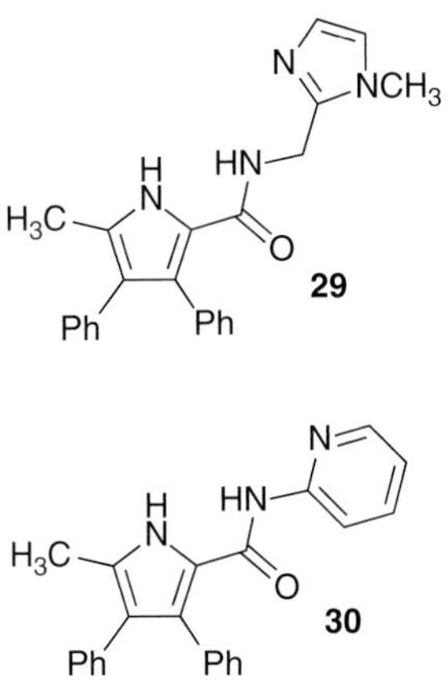

Figure 15 The structures of compounds **29** and **30**.

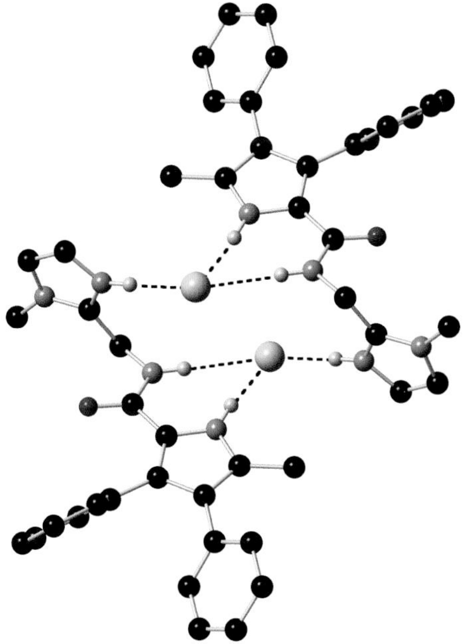

Figure 16 The X-ray crystal structure of the HCl complex of receptor **29**.

a moderate rate of Cl^- efflux was observed when the pH was 7.2 on both sides of the vesicle membrane, whereas no efflux was observed when both aqueous phases were acidic (pH 4.0). The highest efflux was observed when there was a pH gradient, specifically when the inside aqueous phase was acidic (pH 4.0) and the outside phase near neutral (pH 6.7). Receptor **30** was inactive as a chloride transporter under all of the above conditions. These findings, together with evidence gathered using a pH-sensitive fluorescent dye, showed that compound **29** could transport protons across a lipid bilayer but that compound **30** could not (presumably as it does not protonate under the conditions of the experiment), which indicates that compound **29** can cotransport HCl.

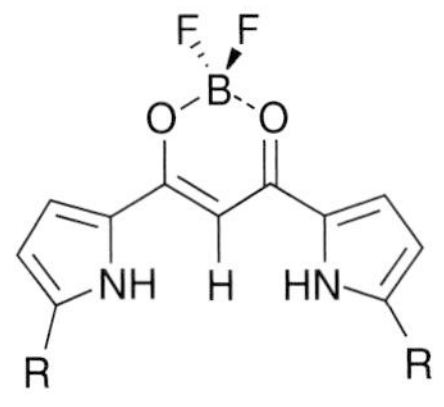

31 R = H
32 R = neopentyl

Figure 17 The structures of compounds **31** and **32**.

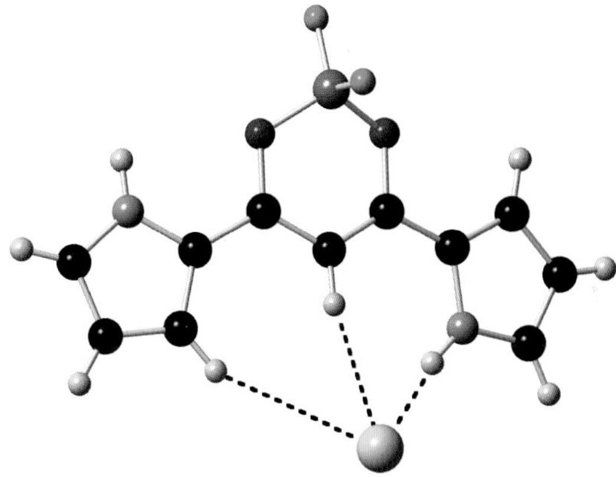

Figure 18 The X-ray crystal structure of the chloride complex of receptor **31**. Countercations and solvent have been omitted for clarity.

Maeda and workers have synthesized a variety of receptors based on the dipyrrolylketone difluoroboron complexes (e.g., compounds **31** and **32**; Figure 17).[38] These receptors employ both NH and CH hydrogen bond donors[39] to bind the anions—the difluroboron group acts as a Lewis acid that accepts electron density from the receptor's oxygen atoms and hence increases the acidity of the central CH group. Stability constants of compound **32** in DCM at room temperature determined by UV/vis titration techniques with tetrabutylammonium anion salts revealed a selectivity for acetate[40] with this anion bound with a stability constant of 110 000 versus 81 000 M^{-1} for fluoride, 13 000 M^{-1} for $H_2PO_4^-$ and 2000 M^{-1} for chloride. Interestingly, the crystal structure of the chloride complex of receptor **31** (shown in Figure 18) reveals that in the solid state the chloride anion is bound by one pyrrole NH group and one pyrrole CH group in addition to the central $CH\cdots Cl^-$ interaction.

3 INDOLE- AND CARBAZOLE-BASED ANION RECEPTORS

Indole and carbazole, like pyrrole, contain a single NH hydrogen bond donor group and are more acidic in

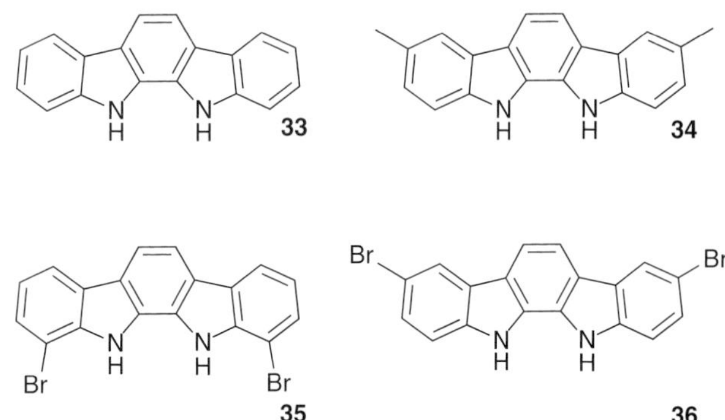

Figure 19 The structures of indolocarbazoles **33–36**.

DMSO solution (pK_a: pyrrole 23.0, indole 21.0, carbazole 19.9).[41] This has led to a wide variety of indole-, carbazole-, biindole-, and indolocarbazole-based receptors being reported in the literature in the last six years.[42]

Beer and coworkers first reported the anion complexation properties of indolocarbazoles in 2005.[43] Compounds **33–36** (Figure 19) were prepared using the Fischer indolization reaction and their anion-binding properties have been studied by UV/vis titration in acetone (Table 5). These results showed that the indolocarbazole skeleton is capable of forming strong complexes with anions under these conditions with a selectivity for carboxylates over the other anions studied. The crystal structure of receptor **34** with fluoride is shown in Figure 20(a), which reveals that four receptors assemble around two fluoride anions in the solid state. Beer and coworkers have investigated the use of indolocarbazoles in the formation of interlocked molecular structures.[44] The authors found that two indolocarbazoles can assemble around a single sulfate anion (Figure 20b). They then used a preexisting hybride crown ether isophthalamide anion-binding macrocycle together with sulfate and indolocarbazole **33** to form a pseudo-rotaxane structure (Figure 21).

Gale and coworkers have reported that receptors based on a diindolylurea skeleton form strong complexes with oxo-anions such as carboxylates and dihydrogen phosphate.

Table 5 Stability constants (log K_a)[a] in acetone solution.

	33	34	35	36
F^-	4.7	4.7	3.6	5.0
Cl^-	4.5	4.1	4.0	4.9
$PhCO_2^-$	5.3	5.4	4.4	5.9
$H_2PO_4^-$	4.9	5.2	4.1	5.3
HSO_4^-	—[b]	—[b]	—[b]	—[b]

[a]Determined by UV–vis spectroscopy; $T = 25\,^\circ$C; [Host] $= 3 \times 10{-}5$ M^{-1}.
[b]Very weak complexation. Stability constant could not be determined; error <10%.

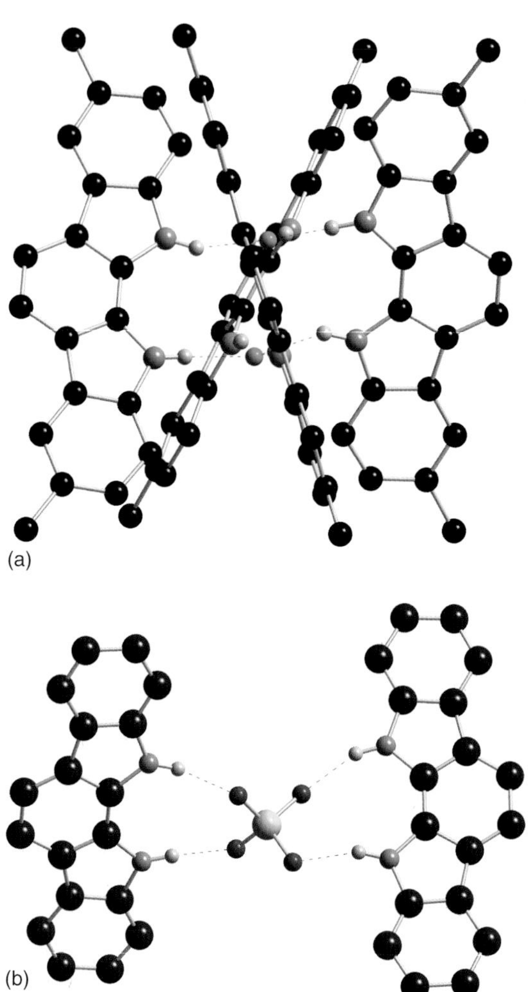

Figure 20 The X-ray crystal structures of (a) the fluoride complex of receptor **34** and (b) the sulfate complex of receptor **33**. Countercations and nonacidic hydrogen atoms have been omitted for clarity.

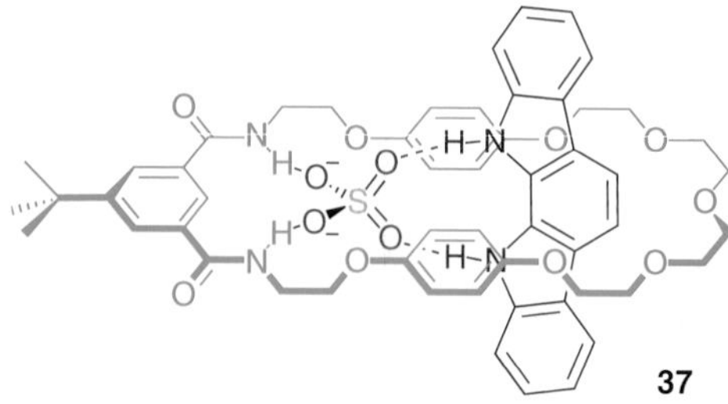

Figure 21 The structure of sulfate-templated pseudo-rotaxane **37**.

Receptors **38** and **39** (Figure 22) were formed by the reaction of triphosgene with 2,3-dimethyl-7-aminoindole and 7-aminoindole, respectively, in a two-phase mixture of DCM and saturated aqueous sodium bicarbonate.[45] This gave ureas **38** and **39** in 78 and 50% yields, respectively. ^{1}H NMR titration in DMSO-d_6/0.5% water demonstrated

38 R = CH_3
39 R = H

Figure 22 The structures of diindolylureas **38** and **39**.

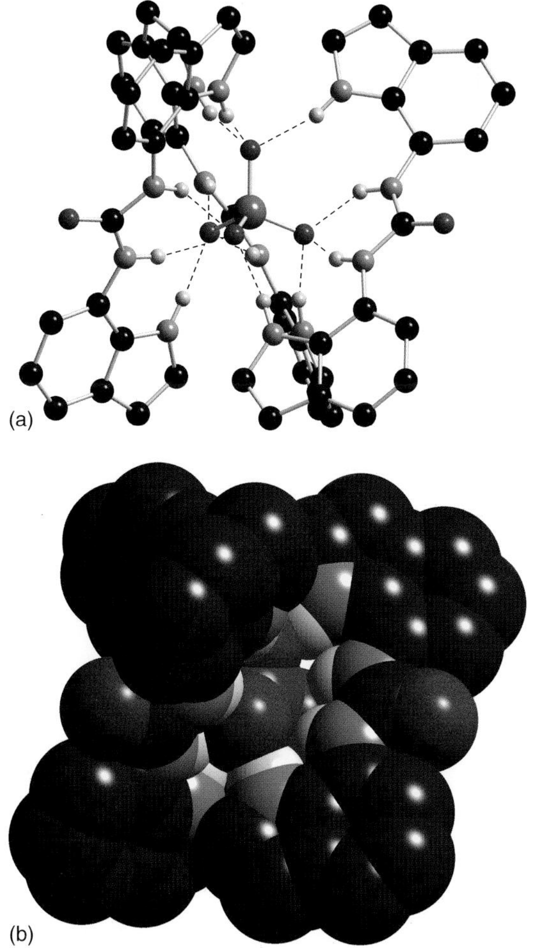

Figure 23 (a) Ball and stick and (b) space-filling views of the 3 : 1 complex formed between receptor **39** and PO_4^{3-}.

that compound **38** binds carboxylates and dihydrogen phosphate extremely strongly ($K_a > 10^4$ M^{-1}). Titration in the more competitive DMSO/10% water solvent system showed that compound **38** is selective for dihydrogen phosphate ($K_a = 4790$ M^{-1}) over acetate ($K_a = 567$ M^{-1}) and benzoate ($K_a = 736$ M^{-1}). Compound **39** exhibits a similar selectivity for dihydrogen phosphate in DMSO-d_6/10% water ($K_a = 5170$ M^{-1}) over acetate ($K_a = 774$ M^{-1}) and benzoate ($K_a = 521$ M^{-1}). Remarkably, even in the extremely competitive DMSO-d_6/25% water system, compound **39** complexes dihydrogen phosphate with $K_a = 160$ M^{-1}. Crystals of the tetrabutylammonium dihydrogen

40 R = *n*-Bu
41 R = Ph

Figure 24 The structures of diindolylureas **40** and **41**.

phosphate complex of **39** were grown from a DMSO-d_6/25% water solution of **39** with excess anion. The crystal structure shown in Figure 23(a) (ball and stick view) and 23(b) (space filling view) shows that the anion is bound by 3 equivalents of receptor **39**, providing a total of 12 hydrogen bonds to a single centrally bound anion, which has been completely deprotonated allowing each oxygen to accept 3 hydrogen bonds. However, there is no evidence for anion deprotonation on complexation in solution.

Subsequent work from Gale and coworkers describes the addition of further hydrogen bond donors to the diindolylurea skeleton in the conception of compounds **40** and **41** (Figure 24).[46] As with receptors **38** and **39**, strong binding was observed with oxo-anions in DMSO-d_6/0.5% water. Binding constants are shown in Table 6.

In the case of receptor **40** with dihydrogen phosphate in DMSO-d_6/0.5% water, examination of the ^{1}H NMR titration revealed that after the addition of 1 equivalent of anion, a second equilibrium was visible in slow exchange on the NMR timescale. This is shown in Figure 25. The addition of up to 1 equivalent of anion led to the downfield shift of the amide NH resonance; this was followed by the appearance of new peaks in the ^{1}H NMR spectrum as further aliquots of anion were added. These new peaks appeared significantly downfield compared to the resonance associated with the assumed 1 : 1 complex,

Table 6 Stability constants $(K_a)^a$ in DMSO-d_6/water solution.

Anion[b]	38 (0.5% water)	38 (10% water)	39 (0.5% water)	39 (10% water)
Cl^-	22	ND	<10	ND
BzO^-	$>10^4$	1100	[c]	481
AcO^-	$>10^4$	[d]	8460	1422
$H_2PO_4^-$	[e]	2310	[c]	[f]
HCO_3^-	2468	395	[c]	[c]

ND, not determined.
[a]Determined by ^{1}H NMR titration, calculated by following the change in chemical shift of the urea NH.
[b]Tetrabutylammonium or tetraethylammonium salt.
[c]Peak broadening prevented a stability constant being obtained.
[d]NMR spectrum indicates conformational changes.
[e]Fast and slow exchange.
[f]Isotherm could not be fitted to a 1 : 1 or 2 : 1 model.

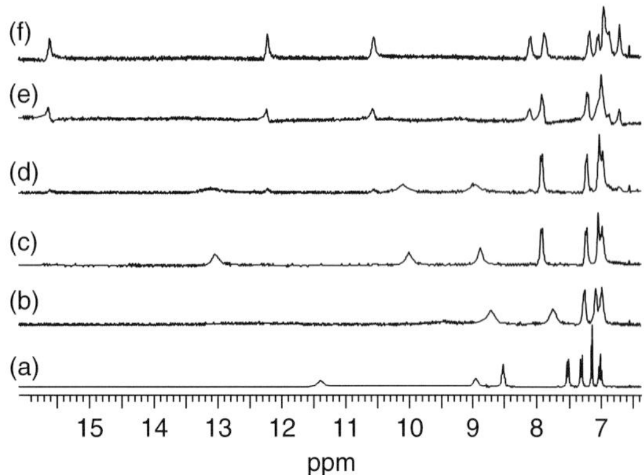

Figure 25 Proton NMR titration with compound **40** in DMSO-d_6/0.5% water: (a) free receptor; (b) 0.6 equivalent $TBAH_2PO_4$; (c) 1.0 equivalent $TBAH_2PO_4$; (d) 1.4 equivalents $TBAH_2PO_4$; (e) 1.4 equivalents $TBAH_2PO_4$ + 0.7 equivalent TBAOH; (f) 1.4 equivalents $TBAH_2PO_4$ + 1.4 equivalents TBAOH.

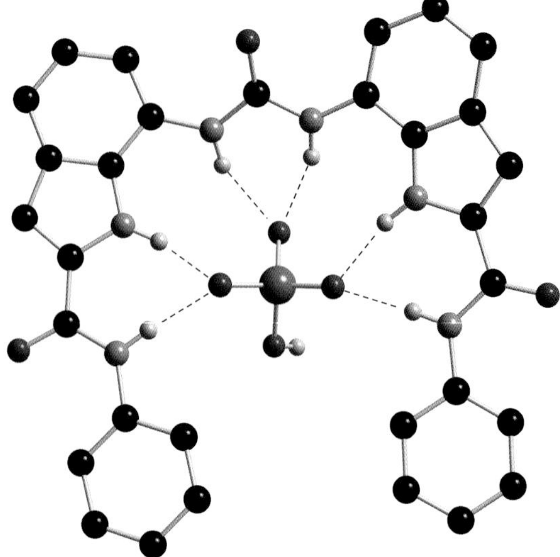

Figure 26 The X-ray crystal structure of the monohydrogen phosphate complex of receptor **41**. Tetrabutylammonium countercation and nonacidic hydrogen atoms have been omitted for clarity.

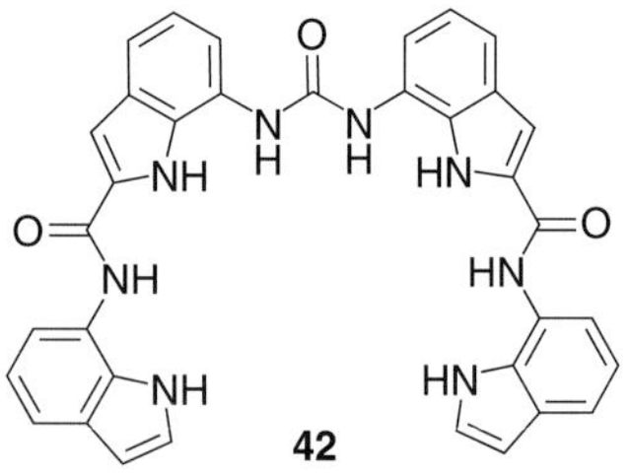

an observation not consistent with slow exchange with a 2 : 1 complex. It was hypothesized that the binding of dihydrogen phosphate by the six hydrogen bond donors available in compound **40** could lead to the lowering of the pK_a of the bound anion. This could allow deprotonation of the bound anion by the more basic free dihydrogen phosphate in solution, resulting in a new complex of receptor **40** with monohydrogen phosphate. The double negative charge on this complex would explain the large downfield shift of the NH resonances. This theory was investigated by the titration of compound **40** with up to 1.4 equivalents of $TBAH_2PO_4$ followed by the addition of aliquots of TBA hydroxide. This led to an increase in the intensity of the new peaks, an observation consistent with the formation of more monohydrogen phosphate complex in solution.

The crystals of receptor **41** with TBA dihydrogen phosphate were grown by slow evaporation of a DMSO solution of **41** with excess anion. The crystals formed showed a 1 : 1 receptor : anion stoichiometry and the dihydrogen phosphate was found to be deprotonated, leaving monohydrogen phosphate bound in the cleft (Figure 26). Three of the phosphate oxygen atoms are bonded to six NH groups. This is further evidence for proton transfer from anions bound by these receptors.

The diindolylurea skeleton was further extended in a recent publication by the same authors.[47] Compound **42** (Figure 27) contains four indole groups and a total of eight potential hydrogen bond donors. ^{1}H NMR titration of compound **42** in DMSO-d_6/0.5% water showed that in all cases, complex binding equilibria were present and as such no binding constants were obtained. However, by examining the changes in chemical shift of the protons associated with each type of NH hydrogen bond donor, different binding modes were demonstrated for Y-shaped carboxylate anions acetate and benzoate, and trigonal planar or tetrahedral anions such as bicarbonate and dihydrogen phosphate. In the case of carboxylate anions, addition of up to 1 equivalent of anion led to the downfield shift of the urea and top indole NH resonances. The addition of further equivalents of anion then resulted in the downfield shifting of the amide and pendant indole NH resonances. This implies that only four NH donors are utilized in binding the first equivalent of anion. However, in the case of dihydrogen phosphate and bicarbonate, the resonances of all four types of NH donor shift downfield from the start of the titration, implying that all NHs are involved in binding the first equivalent of anion. As with compound **40**, titration with dihydrogen phosphate led to deprotonation of the bound species. A similar effect was observed on titration with bicarbonate, resulting in deprotonation of the HCO_3^- anion to CO_3^{2-}.

Figure 27 The structure of diindolylurea **42**.

Crystals of the tetrabutylammonium benzoate complex of receptor **42** were obtained by slow evaporation from a DMSO solution containing excess tetrabutylammonium benzoate. The structure shows that each receptor binds three

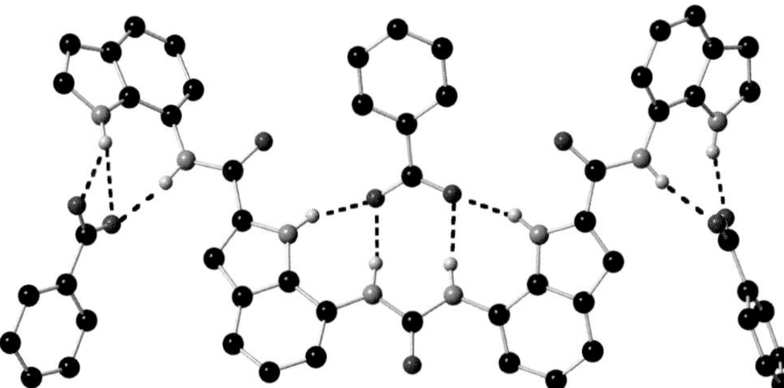

Figure 28 The X-ray crystal structure of the tetrabutylammonium benzolate complex of receptor **42**. Countercations and nonacidic hydrogen atoms have been omitted for clarity.

benzoate anions in the solid state (Figure 28). The central diindolylurea group binds 1 equivalent of benzoate by four hydrogen bonds, while the pendant amidoindole groups are orientated away from the central cavity and each binds a further equivalent of benzoate by two or three hydrogen bonds.

Crystals of the tetrabutylammonium sulfate complex of receptor **28** were obtained by slow evaporation of a DMSO solution of the receptor in the presence of excess tetrabutylammonium sulfate. The crystal structure reveals that the sulfate is centrally bound by all eight hydrogen bond donors from the receptor (Figure 29).

Sessler and coworkers expanded the calix[4]pyrrole binding cavity by employing bipyrrole in place of pyrrole, yielding new calixpyrrole analogs such as calix[n]bipyrroles[48] ($n = 3, 4$), a calix[2]bipyrrole[2]furan, and a calix[2]bipyrrole[2]thiophene.[49] The carbazole moiety was also an attractive linker to include as the fluorescent nature of

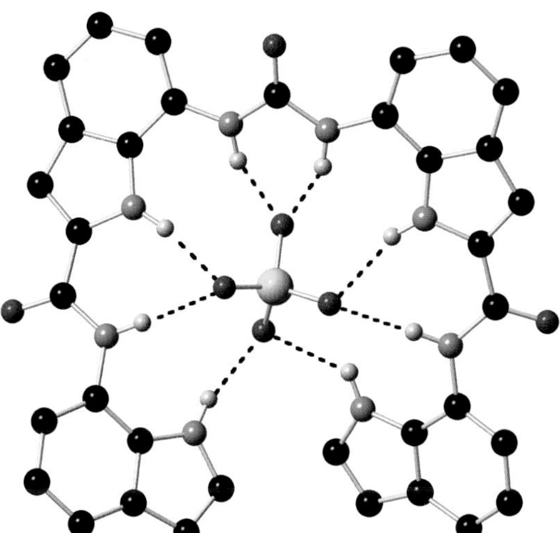

Figure 29 The X-ray crystal structure of the tetrabutylammonium complex of receptor **42**. Countercations and nonacidic hydrogen atoms have been omitted for clarity.

Br, HN, Br; Sn(CH$_3$)$_3$, N, Boc; Toluene, K$_3$PO$_4$ aq, 95 °C, 48 h; NR, HN, NR; 180 °C, vacuum; R=Boc → R=H; Acetone, TFA, r.t

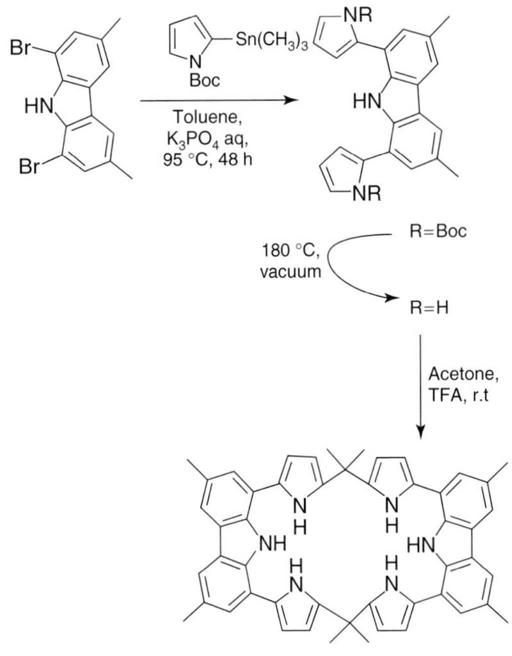

43

Scheme 7 Synthesis of calix[4]pyrrole[2]carbazole **43**.

this group can lead to potential applications of carbazole-containing receptors in anion sensing. Calix[4]pyrrole[2] carbazole, **43**, was synthesized according to Scheme 7. 1-*tert*-Butoxycarbonyl-2-(trimethylstannyl)pyrrole was coupled with 1,8-dibromo-3,6-dimethyl-9H-carbazole in the presence of a Pd catalyst. Removal of the *Boc* protecting groups was followed by condensation with acetone in the presence of trifluoroacetic acid to give **43** in 40% yield.

Crystals of **43,** suitable for X-ray diffraction analysis, were grown by slow diffusion of pentane into a solution of **43** in THF. The structure was elucidated by single crystal X-ray diffraction, which reveals that compound **43** adopts a "wing-like" conformation (Figure 30). Crystals of the tetrabutylammonium benzoate complex of **43** were

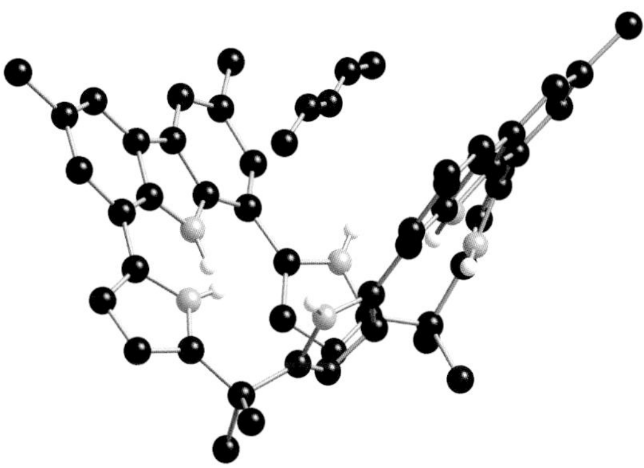

Figure 30 The X-ray crystal structure of calix[4]pyrrole[2] carbazole **43**.

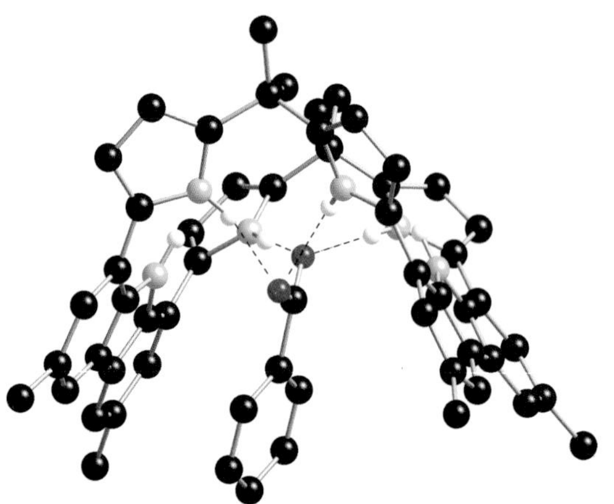

Figure 31 The X-ray crystal structure of the benzoate complex of **43**.

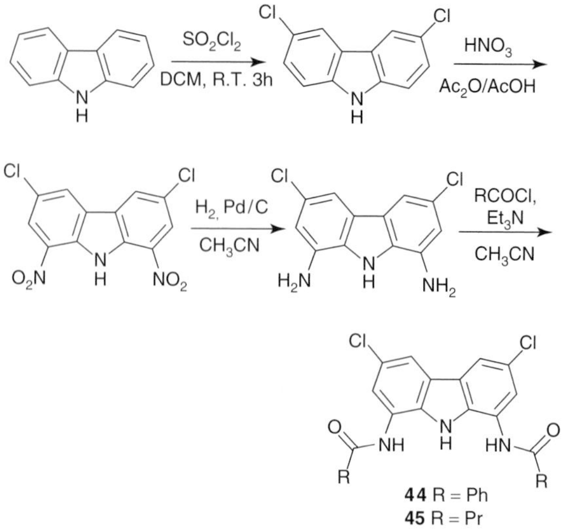

obtained by slow evaporation of a DCM solution of **43** and tetrabutylammonium benzoate. The crystal structure of the 1 : 1 receptor : benzoate complex shows that the carboxylate group is bound by four NH groups, while the anion is found to be within the "saddle" formed by the wings of compound **43** (Figure 31). Solid-state analysis of the benzoate complex implied that the solution-phase binding of benzoate would be expected to take place from only one side of the saddle-like cavity as the other side is blocked by the *meso*-methyl groups.

Proton NMR titrations provided evidence of complexation of tetrabutylammonium acetate, benzoate, and chloride. Peak broadening of the pyrrolic NH resonances and subsequently the carbazole NH resonances at higher anion concentrations was observed, followed by the reappearance of these peaks after the addition of a large excess of anion. This behavior was attributed to the strong complexation of the anions by compound **43**, but binding constants could not be obtained due to the peak broadening. The binding constants were therefore determined by fluorescence titration methods in DCM. Titrations with the tetrabutylammonium anion salts at a receptor concentration of 0.5 μM showed that this receptor is selective for acetate ($K_a = 229\,000\ M^{-1}$) over other anions, including chloride ($K_a = 35\,000\ M^{-1}$) and dihydrogen phosphate ($K_a = 72\,000\ M^{-1}$). Lower binding constants were observed at higher concentrations of compound **43**, which was attributed to receptor aggregation at higher concentrations.

The carbazole moiety has also been utilized as a scaffold in anion receptor design. Jurczak and coworkers have reported the synthesis of receptors **44** and **45**, which combine amide NH hydrogen bond donors in a cleft converging with the carbazole NH.[50] Compounds **44** and **45** were synthesized as shown in Scheme 8. Carbazole was selectively chlorinated in 60% yield using sulfuryl chloride in DCM. The resulting 3,6-dichlorocarbazole was nitrated

SO_2Cl_2, DCM, R.T. 3h; HNO_3, Ac_2O/AcOH; H_2, Pd/C, CH_3CN; RCOCl, Et_3N, CH_3CN

44 R = Ph
45 R = Pr

Scheme 8 Synthesis of compounds **44** and **45**.

using nitric acid in AcOH/$(AcO)_2$ to give 1,8-dinitro-3,6-dichlorocarbazole in 73% yield. The nitro compound was reduced under an atmosphere of H_2 with catalytic Pd/C in acetonitrile. Finally, the resulting diamine was coupled with the necessary acid chloride to give receptors **44** and **45** in 69 and 60% yields, respectively.

Compounds **44** and **45** were found to be insoluble in many common organic solvents; however, compound **44** dissolves in DCM and 1,2-dichloroethane on the addition of tetrabutylammonium chloride or tetrabutylammonium acetate—an early indication of the anion-binding capacity of this receptor. Diffusion of ether into a DCM solution of **44** in the presence of tetrabutylammonium chloride yielded crystals of the chloride complex of **44** suitable for X-ray diffraction. The crystal structure revealed that two distinct complexes were present in the solid state (Figure 32). In both cases, the strongest hydrogen bonds to the chloride anion are from the carbazole NH (NH···Cl 2.22 and 2.33 Å). In the first complex, the chloride anion is asymmetrically positioned within the cavity.

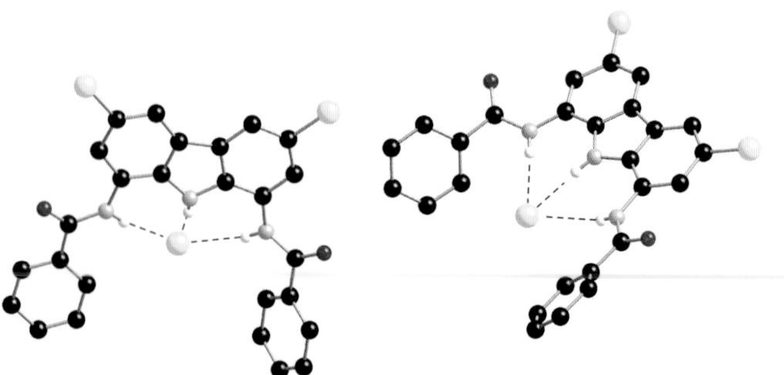

Figure 32 The X-ray crystal structure of the chloride complex of receptor **44**. Two different complexes are present in the solid state.

One amide–chloride distance is shorter than the other ($NH \cdots Cl^-$ 2.60 vs 2.94 Å). This finding implies that the cavity is too large for the chloride anion. This observation is further evidenced by the second complex, in which the chloride is found to be almost centrally located above the plane of the cavity with longer $N \cdots Cl^-$ distances (3.57 and 3.58 Å). This confirms the observation about the poor geometric fit of the anion with the cavity.

Solution-state anion binding was investigated by ^1H NMR titration with various tetrabutylammonium anion salts in DMSO-d_6/water (0.5%). Both receptors were found to be selective for dihydrogen phosphate and benzoate over chloride (binding constants for **44** were found to be 1230 and 1910 M^{-1} with benzoate and dihydrogen phosphate, respectively, compared to 13 M^{-1} with chloride). This trend would be expected on the basis of anion basicity, but may be enhanced by the better fit of these larger anions within the cavity. Small shifts of the resonance associated with the amide NH protons when both receptors were titrated with chloride (0.20 and–0.18 ppm) indicated that the interaction of these hydrogen bond donors with chloride is minimal. Conversely, the carbazole NH was found to undergo a much larger shift (2.60 and 2.92 ppm), indicating that interaction with this donor is much stronger. Interestingly, higher binding constants were found for each anion with **45** than with **44**. This is the opposite of what would be expected when considering the greater electron-donating capacity of the alkyl groups to the amide moieties.

Jeong and coworkers have also incorporated the biindole scaffold into receptor design.[51] Compound **48** contains two NH hydrogen bond donors and the biindole scaffold is appended with two pyridyl units through rigid ethynyl linkers, which can function as hydrogen bond acceptors. Compounds **47** and **48** were synthesized as shown in Scheme 9. Compound **46** and 2 equivalents of 2-trimethylsilanylpyridine were coupled through a Pd-catalyzed Sonogashira reaction. This yielded compound **46** in 33% yield, whereas compound **47** was synthesized by coupling with 1 equivalent 2-trimethylsilanylpyridine in 34% yield.

The anion-binding properties of compounds **46**–**48** were investigated by UV/vis titration with tetrabutylammonium anion salts in CH_3CN. Compound **48** was found to bind dihydrogen phosphate extremely strongly with $K_a = 1.1 \times 10^5\ M^{-1}$. Compound **48** was shown to be selective for dihydrogen phosphate over the other anions tested, including acetate ($K_a = 22\,000\ M^{-1}$), chloride ($K_a = 5000\ M^{-1}$), and cyanide ($K_a = 2100\ M^{-1}$). In general, the more basic anions were bound the most strongly. Compounds **47** and **46** were shown to have a much lower affinity for dihydrogen phosphate under these conditions, with $K_a = 6800\ M^{-1}$ and 500 M^{-1}, respectively. This indicates that the pyridyl functionality is important in binding dihydrogen phosphate, probably because of hydrogen bonding between the pyridine nitrogen and the hydroxyl groups of the anion.

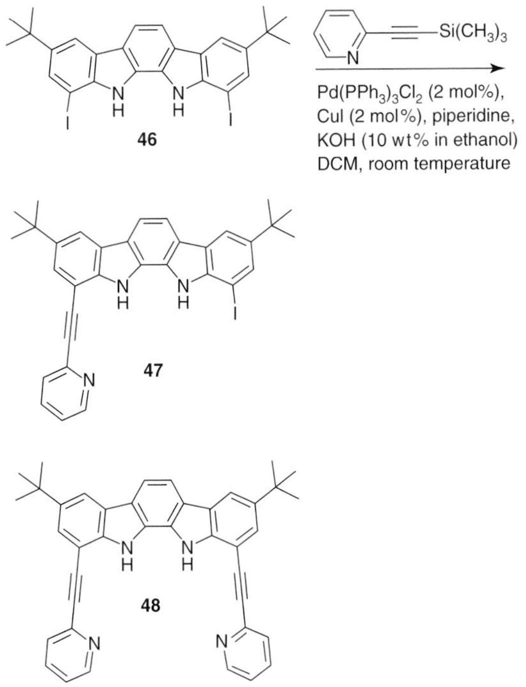

Following on from this and earlier work on helical oligoindole-based foldamers,[52] Jeong *et al.* also reported the synthesis of receptors **49**–**51** (Figure 33).[53] UV/vis

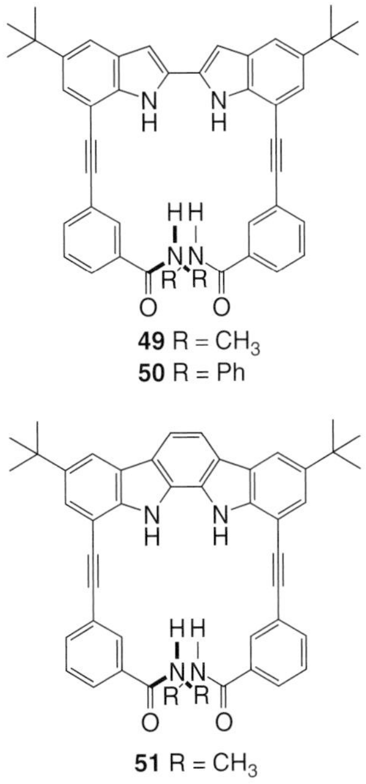

46

$Si(CH_3)_3$

$Pd(PPh_3)_3Cl_2$ (2 mol%), CuI (2 mol%), piperidine, KOH (10 wt% in ethanol) DCM, room temperature

47

48

Scheme 9 The synthesis of compounds **47** and **48**.

49 R = CH_3
50 R = Ph

51 R = CH_3

Figure 33 The structure of compounds **49**–**51**.

titration of compound **49** with tetrabutylammonium chloride in acetonitrile gave $K_a = 5.1 \times 10^3\ M^{-1}$. Computer modeling revealed that the biindolyl scaffold exists in an *s-trans* conformation in which the two indole NH groups are orientated in opposite directions to minimize dipole–dipole repulsions. On complexation of chloride, the receptor adopts an *s-cis* conformation with both indole NH groups converging to hydrogen bond to the anion. This reorganization decreases the entropic and enthalpic favorability of binding. Thus, compound **51**, in which the two indoles are bridged by an ethyno group at the 3,3′-position, was synthesized. This more rigid receptor is preorganized in a conformation in which the indole hydrogen bond donors are convergent. Correspondingly, compound **51** was found to complex chloride more strongly ($K_a = 1.1 \times 10^5\ M^{-1}$) than compound **49** under the same conditions. Further titrations revealed that compound **51** complexes a range of anions more strongly than compound **49**, and that both receptors show anion selectivity in the order acetate > chloride > bromide > hydrogen sulfate > iodide as anticipated from the electrostatic nature of the binding interaction. The relative ratios of association constants are in the range of 20–40, which indicates that the binding cleft preorganization of **51** results in an additional stabilization energy ($-\Delta\Delta G$) of $2.0 \pm 0.2\ \text{kcal mol}^{-1}$.

The propensity of indoles to bind anions has also been exploited by Jeong and coworkers in the design of oligoindole-based foldamers **54**–**56**, which form helicates in the presence of chloride.[52] Compounds **54**–**56** were synthesized by the repeated Sonogashira coupling of monomers **52** and **53** as shown in Scheme 10. Computer modeling studies indicate that tetramer **54** exists in the *s-trans* conformation in solution but in the presence of an anion will fold into an *s-cis* conformation with all indole NHs converging to bind the anion. The folded structure generates an internal cavity of complimentary size for the chloride anion. Hexame **55** and octamer **56** fold in the presence of an anion to generate a helical conformation with one turn comprising four indole units; **55** and **56** yield 1.5 and 2 turns, respectively.

The behavior of compounds **54**–**56** in the presence of chloride were investigated by observing changes in the ^{1}H NMR spectrum on stepwise addition of TBACl in CD_2Cl_2. All receptors showed a significant response to chloride. As might be expected, all of the indole NH resonances were shifted downfield. In addition to this, the aromatic CH resonances of compounds **55** and **56** were shifted upfield as a result of π–π stacking induced by the helical folding. All spectral changes were saturated after addition of 1 equivalent of chloride, implying that the binding stoichiometry is 1 : 1. Further evidence of stacking was provided by two-dimensional NMR techniques such as TOCSY (*to*tal *c*orrelation *s*pectroscop*y*), NOESY (*n*uclear *o*verhauser *e*ffect *s*pectroscop*y*), and ROESY (*r*otational nuclear *o*verhauser *e*ffect *s*pectroscop*y*). Enhanced NOE (*n*uclear *o*verhauser *e*ffect) cross peak signals between protons not in close proximity to each other in the unfolded structure were observed on addition of chloride.

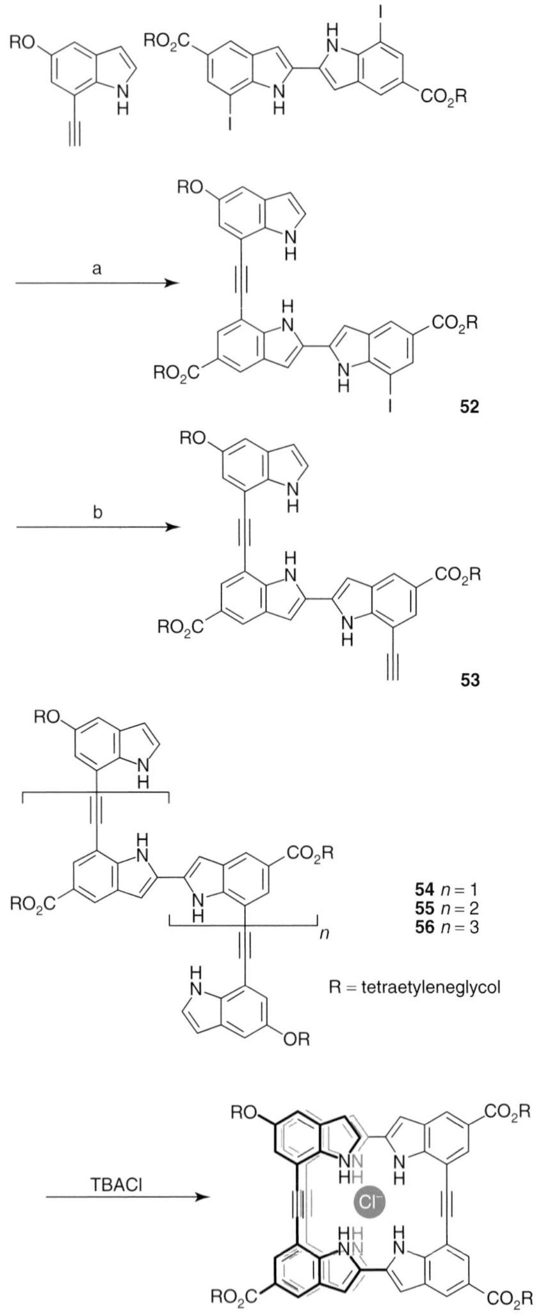

The binding of chloride was also investigated by UV/vis titration of **54**–**56** with tetrabutylammonium chloride in

RO ... RO2C ... CO2R
a → **52**
b → **53**
54 $n = 1$
55 $n = 2$
56 $n = 3$
R = tetraetyleneglycol
TBACl → Cl^-

Scheme 10 The synthesis of compounds **52**–**54**. Reagents and conditions: (a) $PdCl_2(PPh_3)_2$, CuI, Et_3N, 52–54 °C; (b) $Pd(dba)_2$, CuI, Ph_3P, trimethylsilylethyne, Et_3N/THF, 52–54 °C, and then TBAf–AcOH, rt.

Figure 34 Compound **57**.

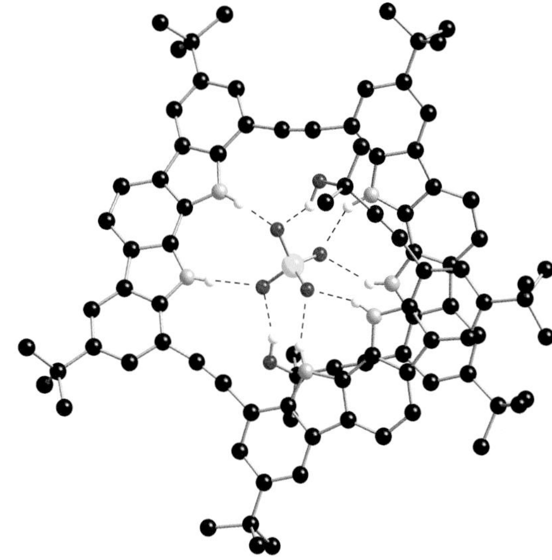

Figure 35 The X-ray crystal structure of the sulfate complex of **57**.

acetonitrile, with Job plot analysis confirming a 1 : 1 binding stoichiometry. Chloride was bound extremely strongly by all of these receptors ($K_a = 1.3 \times 10^5\ M^{-1}$, $1.2 \times 10^6\ M^{-1}$, and $>10^7\ M^{-1}$ for receptors **54**, **55**, and **56**, respectively). Strong binding of chloride was also observed in the more competitive acetonitrile/water (10%) mixture (compound **56** binds chloride under these conditions with $K_a = 2.3 \times 10^4\ M^{-1}$). In both solvent mixtures, compound **56** was found to bind chloride the most strongly. The binding constant was found to increase with the number of hydrogen bond donors, which may not always be expected in an aqueous medium.

More recently, work from the same authors describes the hybrid indolocarbazole oligomer **57** which, on binding sulfate, adopts a helical conformation (Figure 34).[54] Folding of **57** generates a helical cavity containing eight convergent hydrogen bond donors. The helical structure is proved unambiguously in the solid state by the crystal structures of **57** $\cdot$ $(TBA)_2SO_4$, as shown in Figure 35. Crystals of this complex were obtained by slow diffusion of hexanes into a 10% v : v DCM–ethyl acetate solution of compound **57** and 1 equivalent of $(TBA)_2SO_4$. The structure clearly shows the helical nature of the complex. Each sulfate anion is bound by eight hydrogen bond donors; six from the indole NHs and two from the terminal OH groups. Interestingly, individual helices of the complex aggregate into a higher order structure in which M- and P-helices are alternately stacked; in this way, the crystal is racemic.

The solution-phase binding of anions was investigated by fluorescence titration with excitation at 320 nm in 10% v : v methanol/acetonitrile with anions added as their tetrabutylammonium salts. Addition of sulfate leads to hypochromic and bathochromic shifts of a high intensity band originally found at 413 nm. This is characteristic of π–π stacking due to helical folding of aromatic strands. Smaller bathochromic shifts were observed on addition of dihydrogen phosphate and chloride. The binding constants obtained are shown in Table 7. Although a preference for sulfate is observed, other anions tested show similar lower binding constants.

Investigation of the binding by ^1H NMR spectroscopy in 1 : 1 : 8 $CD_3OH/CD_2Cl_2/CD_3CN$ revealed that as expected, the addition of anions as $(TBA)^+$ salts leads to a downfield shift of the NH signals, which is indicative of hydrogen bond formation. However, only the addition of sulfate caused a downfield shift of the terminal OH signals, which implies that these protons are involved only in the complexation of sulfate. This may explain the selectivity of this receptor for sulfate.

Table 7 Stability constants $(K_a)^a$ for compound **57** in 10% v/v CH_3OH/CH_3CN solution.

Anion^b	K_a (M^{-1})
SO_4^{2-}	640 000
$H_2PO_4^-$	3600
Cl^-	8800
Br^-	2800
I^-	<100
$CH_3CO_2^-$	5700
CN^-	1600
N_3^-	790

^{a}Determined by fluorescence spectroscopy at 298 K.
^{b}Tetrabutylammonium salt.

4 PYRIDINE- AND PYRIDINIUM-BASED ANION RECEPTORS

The neutral pyridine group has been widely utilized in the synthesis of receptors for anions. This group contributes no classical hydrogen bond donors for anion complexation. However, early work from Hamilton and coworkers in the field of nucleotide base and barbituate complexation highlighted that receptors containing 2,6-diamidopyridines form preorganized binding clefts.[55–57] In every case where an X-ray crystal structure was available, both substrate-free and substrate-bound forms of the receptors were found to take up the parallel orientation of hydrogen bonding sites. This was attributed to intramolecular hydrogen bonding between the pyridine nitrogen and amide NH groups, as shown in Figure 36.

The application of the 2,6-diamidopyridine unit to anion complexation was pioneered by Crabtree, who reported the anion-binding studies of receptors **58–60** (Figure 37).[58] These compounds were readily prepared by the reaction of commercially available acid dichlorides with the corresponding amine followed by recrystallization from benzyl alcohol, methanol, or ethanol. The binding constants were determined by ^{1}H NMR titration in deuterated DCM. The results are shown in Table 8. Binding constants for **58** could not be determined because of the poor solubility of this compound. Although pyridyl compound **59** showed lower binding constants than **60** for all of the anions tested, it was found to be selective for F^- while **58** was selective for Cl^-. This observation was attributed to enhanced rigidity of the binding cleft due to the intramolecular hydrogen bonding as described above and the electrostatic repulsion of the pyridine-N lone pair with larger anions.

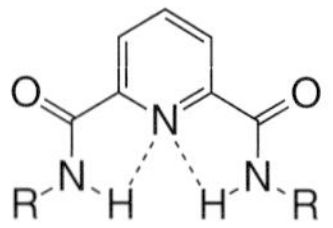

Figure 36 Preorganization of 2,6-diamidopyridines by intramolecular hydrogen bonding.

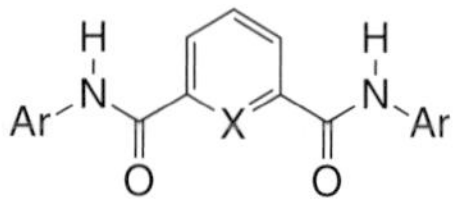

58 Ar = Ph, X = CH
59 Ar = Ph, X = N
60 Ar = *p*-(*n*-Bu)C_6H_4, X = CH

Figure 37 Receptors **58–60**.

Table 8 Stability constants $(K_a)^a$ in CD_2Cl_2 solution.

Anion[b]	**58**	**60**
F^-	24 000	30 000
Cl^-	1500	61 000
Br^-	57	7100
I^-	<20	460
$CH_3CO_2^-$	525	19 800

[a]Determined by ^{1}H NMR titration.
[b]Tetrabutylammonium salts.

More recently there have been examples of this preorganizational effect leading to a stronger binding by 2,6-diamidopyridine-containing receptors than their isopthalamide-containing analogs. Jurczak and coworkers have reported the synthesis of macrocycles **61** and **62**,[59–61] and it has been shown by ^{1}H NMR titration in DMSO-d_6 that the binding constants with all of the anions studied were higher for receptor **61** than for **62**. In the absence of anions, compound **62** adopts a conformation in which two carbonyl oxygen atoms are directed toward the center of the cavity, which is stabilized by hydrogen bonds (Scheme 11). However, this does not occur in the case of compound **61**, presumably because of the preorganizing effect of the pyridine moiety. It was found that hybrid compound **63** displayed higher binding constants for almost every anion tested under the same conditions and in the

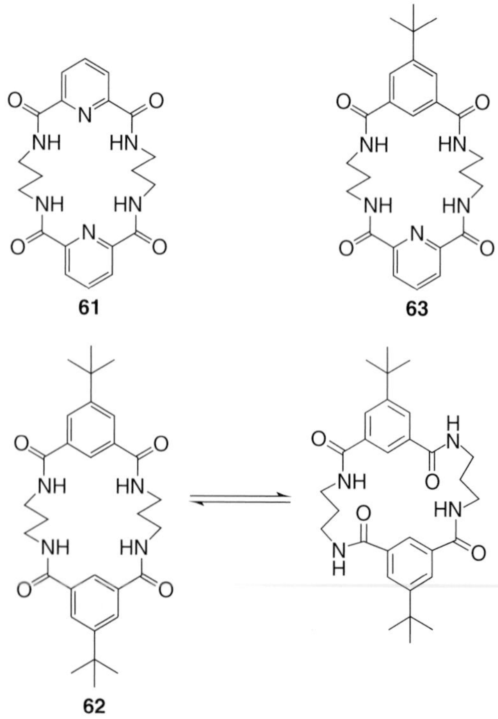

Scheme 11 Receptors **61–63**.

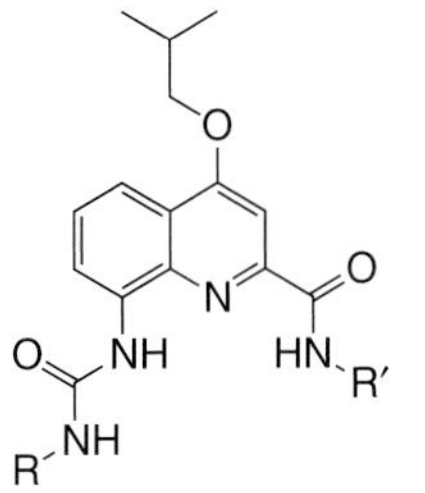

Figure 38 The structures of compounds **64–68**.

more competitive DMSO-d_6/5% water mixture than both **61** and **62**. Macrocycle **63** is preorganized for anion binding in a manner similar to **61** because of the presence of one 2,6-diamidopyridine moiety; however, the higher anion affinity of the isopthalamide unit as demonstrated by Crabtree resulted in receptor **63** showing the highest affinity for most anions.

In a related work, Albrecht and coworkers have reported the synthesis of anion receptors **64–68** based on a quinoline backbone (Figure 38).[62,63] Previous work from the same authors had demonstrated that similar quinoline derivatives such as 5,7-dibromo-8-hydroxyquinoline-2-carboxylic acid adopt a conformation in the solid such that the two acidic hydrogens are orientated in the same direction.[64,65] This conformation is maintained by hydrogen bonding interactions between these protons and the quinoline nitrogen atom. The binding affinities of these receptors with various tetrabutylammonium anion salts were investigated by ^1H NMR and fluorescence spectroscopic titration in $CDCl_3$ and $CHCl_3$, respectively. Job's plot analysis indicated that **64–68** form 1 : 1 complexes with all anions tested. The results of the NMR analysis showed that all receptors display a preference for smaller anions over larger ones. Decreasing binding constants may be observed for the series chloride, bromide, nitrate; for example, compound **64** has binding constants of $K_a = 1000$, 500, and 420 M^{-1} for these anions, respectively. The binding constants were found to be highly substituent dependant. Compound **68** forms the strongest complexes with all anions tested, with $K_a = 7700\ M^{-1}$ for chloride. This is due to the greater electron-withdrawing power of the phenyl substituents compared to the alkyl chain substituents of the other receptors. Thus, the hydrogen bond donors of **68** are the most acidic, leading to higher binding constants. The results of the fluorescence titrations demonstrate a similar preference of all the receptors for smaller anions, with the highest binding constants found for fluoride (the binding constants for fluoride could not be determined by NMR methods). For example, compound **67** has $K_a = 150\,000\ M^{-1}$ for fluoride and $K_a = 10\,380\ M^{-1}$ for chloride.

Another potential application of the pyridine group to anion receptor design is to coordinate a metal via the Lewis

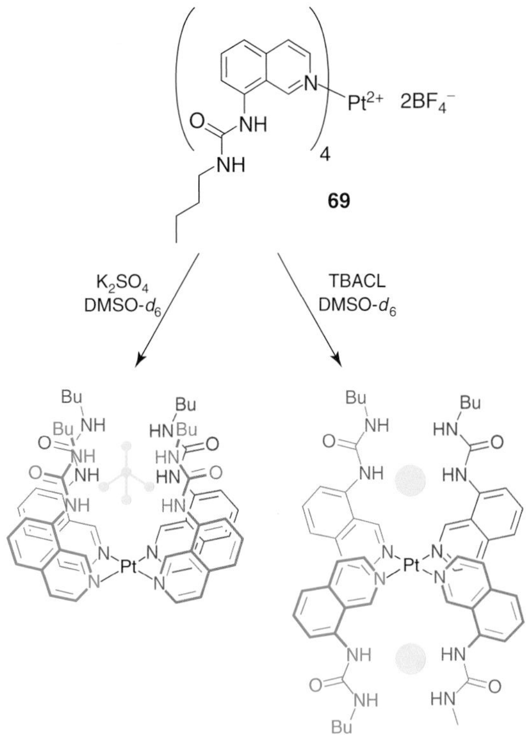

Scheme 12 The conformational changes suggested by the results of titration of receptor **69** with sulfate and chloride.

basic pyridyl nitrogen, thereby rendering the receptor more electron deficient and increasing the anion affinity. This has been utilized by Gale, Loeb, and coworkers in the design of receptors **69–72**.[66,67] By appending the pyridine moiety with various hydrogen bond donors they created a series of ligands that can coordinate a Pt(II) metal center to create metal–organic anion receptors, as shown in Scheme 12(a) and (b).

The anion-binding properties of complex **69** were investigated by ^1H NMR titration in DMSO-d_6. The binding constants are shown in Table 9. The data collected from titration with halide anions was best fitted to a 1 : 2 receptor : anion binding model. This implies that on binding

Table 9 Binding constants $(K_a)^a$ in DMSO-d_6 solution.

Anion[b]	K_a (M^{-1})
Cl^-	$K_1 = 11693$, $K_2 = 2223$
Br^-	$K_1 = 1364$, $K_2 = 450$
I^-	$K_1 = 1431$, $K_2 = 52$
$H_2PO_4^-$	$> 10^5$
SO_4^{2-}	$> 10^5$

[a]Determined by NMR titration.
[b]Anions added as tetrabutylammonium salts except SO_4^{2-}, which was added as K_2SO_4 (the effects of ion pairing have been ignored). In the case of the K^+ salt, this effect would likely decrease the observed affinity if the receptor for an anion is relative to the tetrabutylammonium salt.

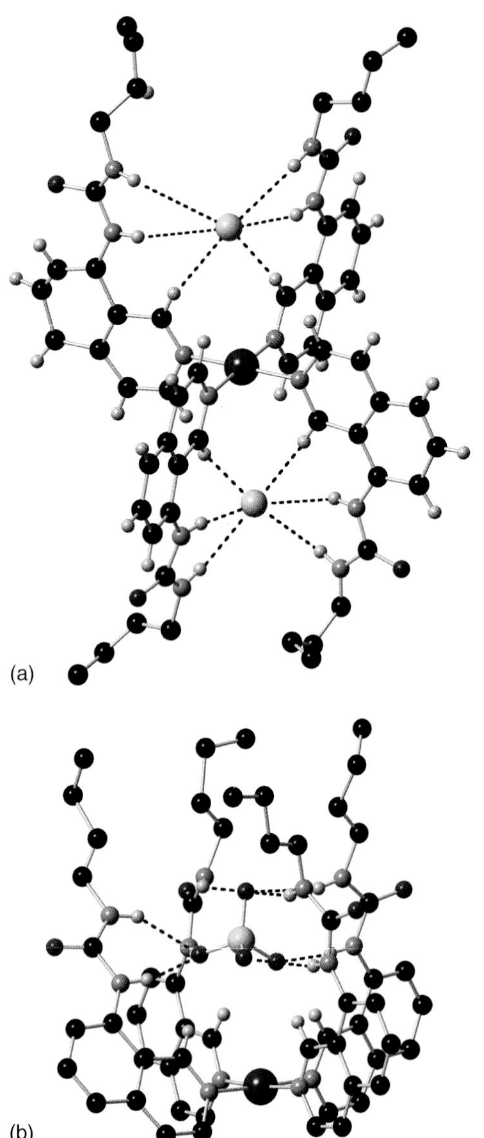

Figure 39 (a) The X-ray crystal structure of the chloride complex of receptor **69** and (b) the X-ray crystal structure of the sulfate complex of receptor **69**.

halide anions, the complex must adopt a 1,2- or 1,3-alternate conformation with two ligands binding each anion, for example, as depicted in Scheme 12(a). The structure of at least the chloride complex of **69** in the solid state was elucidated by X-ray crystallography (Figure 39a). It was found that in the solid state at least does indeed adopt a 1,2-alternate conformation in which two pairs of adjacent urea groups form hydrogen bonds to two separate chloride anions. The urea groups are orientated such that the NH···Cl^- interactions are almost linear. There is also a CH···Cl^- interaction in the solid state from the quinoline proton, which is directed toward the anion.

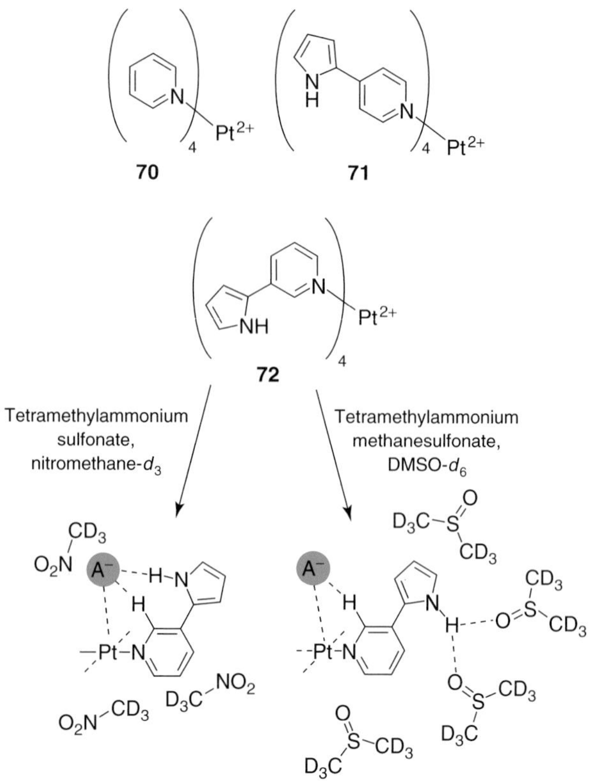

Tetramethylammonium sulfonate, nitromethane-d_3

Tetramethylammonium methanesulfonate, DMSO-d_6

Scheme 13 Receptors **70**–**72** and the solvent-dependant conformation of receptor **72**.

The titration data indicated that sulfate and dihydrogen phosphate were bound in a 1 : 1 receptor : anion stoichiometry. This implies that all four urea-containing ligands are orientated in the same direction to complex a single anion in a cone-like conformation as depicted in Figure 39(a). The structure of the sulfate complex of **69** in the solid state was elucidated by X-ray crystallography, which revealed a cone-like receptor conformation with all eight urea hydrogen bond donors converging to bind a single anion (Figure 39b).

The affinity of receptors **70**–**72** for a variety of anions was investigated by ^{1}H NMR titration in DMSO-d_6 with the tetrabutylammonium anion salts (Scheme 13). In most cases, data could be fitted to a 1 : 1 binding model. Receptor **72** was found to have the highest affinity for all anions tested (for example, with chloride $K_a = 195$, 216, and 960 M^{-1} for receptors **70**, **71**, and **72**, respectively). Shifts of the pyridine CH resonance in the 2-position were observed on addition of anions to all receptors, indicating that this proton is important in the binding process. Indeed, receptors **70** and **71** do not contain a convergent array of NH-based hydrogen bond donors. Interestingly, on titration of receptor **72** with $CH_3SO_3^-$, no downfield shift of the pyrrole NH proton was observed. This indicates that this proton is not involved in binding the anion. However, when titration was repeated in nitromethane-d_3 more usual behavior was

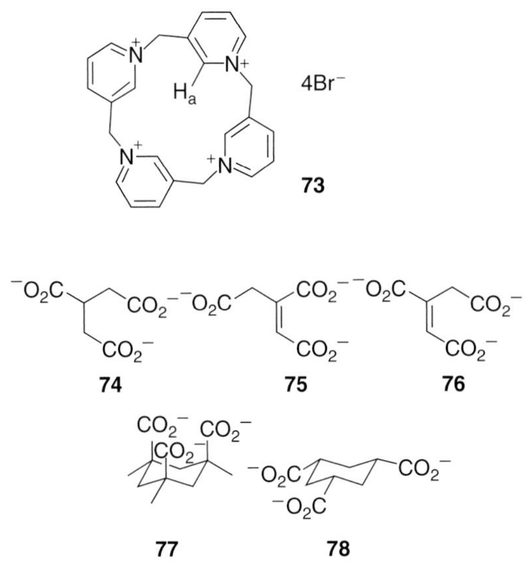

Figure 40 The structure of macrocycle **73** and tricarboxylates **74–78**.

observed, with the pyrrole NH resonance shifting downfield by 1.77 ppm (after 2 equivalents of anion). This evidence highlights a strong solvent dependence of the binding mode. DMSO-d_6 is a good hydrogen bond acceptor. It is possible that when the less basic anions (such as $CH_3SO_3^-$) are complexed, it is more favorable for the pyrrole NH to hydrogen bond to the surrounding solvent molecules, as shown in Scheme 12(b). However, nitromethane-d_3 does not form such strong hydrogen bonds, and as such this effect is not observed.

The pyridinium moiety may also be utilized in the construction of cationic receptors. For example, Shinoda *et al.* have reported the quaternary tetrapyridinium macrocycle **73** which may be prepared directly from an aqueous solution of bromomethylpyridinium bromide (Figure 40).[68] This solution was neutralized with sodium bicarbonate and the bromomethylpyridinium bromide extracted into DCM. When the solvent was removed from the organic layer at room temperature, N-alkylation occurred, giving a mixture of quaternary pyridinium salts. Crystalline **73** · $2H_2O$ was obtained in 2% yield by recrystallization from water.

X-ray diffraction analysis of **73** · $2H_2O$ indicates that the macrocycle is almost planar (Figure 41). Of the four counter–bromide anions, two are hydrogen bonded inside the macrocycle cavity and two are located outside the macrocycle. The distances between these bromides and the H_a protons of the pyridinium rings are in the range 2.65–2.79 Å. The binding of a series of tricarboxylate anions **74–79** by macrocycle **73** were investigated by ^1H NMR titration in D_2O. To prepare solutions of the anions, the corresponding tricarboxylic acids were dissolved in D_2O and the initial pH value was adjusted to be 7–8 with sodium bicarbonate to ensure all guests were present

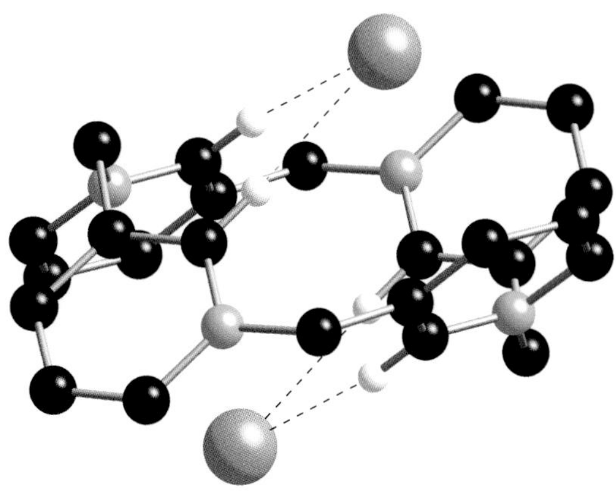

Figure 41 The X-ray crystal structure of **73** · $2H_2O$.

in trivalent form. Titration data was fitted to a 1 : 1 complexation model, except with anion **79**, which showed little interaction. The calculated binding constants showed the following selectivity trend; **75** ($\log K_a = 5.1$) > **76** (4.5) ≈ **74** (4.4) > **78** (4.1). Of the acyclic anions, the strongest complex was formed with macrocycle **73** and anion **76** with two carboxylates at the *cis*-1,2 positions, indicating that the geometry of the guest is an important factor.

Beer and coworkers have used pyridinium groups in the synthesis of anion-templated interlocked molecular species such as rotaxane **80**.[69] They observed that the yield of pyridine-containing macrocycle **81** was significantly improved in the presence of 3,5-bis-hexylamide-substituted pyridinium derivative **82** · Cl, as shown in Scheme 14. This was attributed to favorable π–π stacking interactions between the electron-rich hydroquinone groups of the precursor and the electron-poor pyridinium of **82** · Cl. This inspired the analogous synthesis of **80** as depicted below.

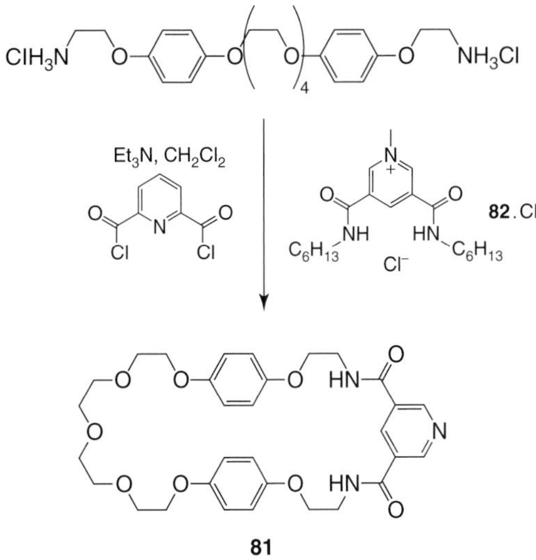

Scheme 14 The synthesis of macrocycle **81**.

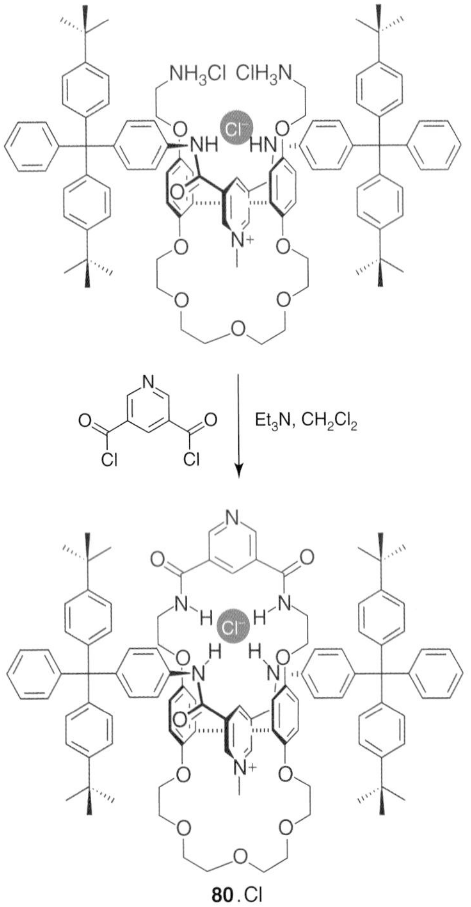

Scheme 15 The chloride-templated synthesis of rotaxane **80** · Cl.

Rotaxane **80** · Cl was prepared as shown in Scheme 15 and isolated in 32% yield after purification by preparative thin layer chromatography. Evidence for rotaxane formation was initially obtained by ^{1}H NMR spectroscopy. On comparison with macrocycle **81**, a downfield shift of the amide protons is observed, consistent with hydrogen bonding to the templating chloride anion. The inward-pointing pyridyl proton is also deshielded by the negative charge of the encapsulated anion. The resonances associated with the protons of the hydroquinone moieties are split and shifted upfield. This is representative of $\pi-\pi$ donor–acceptor interactions with the pyridinium group of the axle component. Resonances of the axle component of the pyridyl aromatic protons are shifted upfield because of the binding of the chloride by additional hydrogen bond donors from the macrocycle. The methyl protons of the axle component are also shifted downfield owing to hydrogen bonding with the polyether oxygens from macrocycle **81**.

The chloride anion of **80** · Cl may be replaced with PF_6^- by repeatedly washing the chloroform solution of **80** · Cl with aqueous NH_4PF_6. The hexafluorophosphate anion is much larger and charge diffused, and correspondingly does not coordinate within the cavity of the rotaxane, as evidenced by the upfield shift of the hydrogen bond donor resonances in the ^{1}H NMR spectrum. The anion-binding affinity of **80** · PF_6 was then investigated by ^{1}H NMR spectroscopy. Titration with TBA salts in $CDCl_3 : CD_3OD$ (1 : 1) mixture demonstrated that compound **80** binds chloride extremely strongly ($K_a > 10^4$ M^{-1}). Even in the highly competitive solvent mixture $CDCl_3 : CD_3OD : D_2O$ (45 : 45 : 10), chloride was bound with $K_a = 1500$ M^{-1}. Titration with other anions under these conditions revealed that they were bound far less strongly (K_a (Br^-) = 780 M^{-1}, K_a ($H_2PO_4^-$) = 60 M^{-1}, K_a (AcO^-) = 110 M^{-1}). This suggests that the interlocked cavity of **80** · PF_6 is of complimentary size and shape for the small spherical chloride anion. **80** · PF_6 is therefore highly selective for chloride in aqueous solvent media.

5 1,2,3-TRIAZOLE-BASED RECEPTORS

1,2,3-Triazoles may be readily prepared by the so-called "click chemistry," the Cu(I)-catalyzed 1,3-dipolar cycloaddition of an azide with an alkyne.[70] Various reports indicate that the hydrogen bond donor ability of C5 CH bond is comparable to that of an amide NH bond[71]; this has made the 1,2,3-triazole unit a useful building block in anion receptor design. Flood and coworkers have prepared a series of rigid triazolophanes such as compounds **83–86**, as shown in Figure 42.[72] These compounds provide a well-defined cavity for anion encapsulation and potential CH hydrogen bond donors from the triazole and phenylene units. Compounds

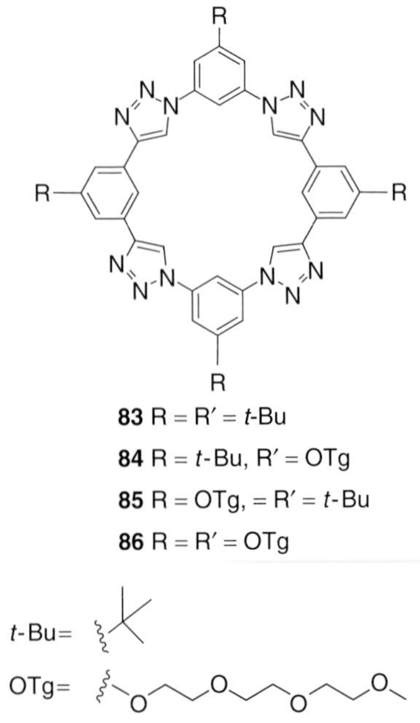

83 R = R′ = *t*-Bu
84 R = *t*-Bu, R′ = OTg
85 R = OTg, = R′ = *t*-Bu
86 R = R′ = OTg

Figure 42 The structure of macrocycles **83–86**.

84 and **85** differ in their substitution of N-linked phenylene units and C-linked phenylene units. The *tert*-butyl group is less electron donating than the methoxy-terminated triethyleneglycol (OTg) group; thus, OTg-substituted receptors were expected to have a lower anion affinity than *t*-butyl-substituted receptors as the inward-pointing phenylene CH protons are correspondingly less acidic.

The affinity of these compounds for halide anions was investigated by UV titration with tetrabutylammonium halide salts in DCM. Aggregation was found to occur at higher concentrations, as evidenced by diffusion NMR studies. The propensity to aggregate was found to increase with OTg substitution. UV titrations were therefore performed at receptor concentrations below 10 μM. Job's plot analysis confirmed a 1 : 1 binding model for all anions tested. An extremely strong binding affinity ($K_a = 1.1 \times 10^7\ M^{-1}$) was found for compound **83** with chloride. The same receptor was also shown to bind bromide strongly ($K_a = 7.5 \times 10^6$), whereas fluoride ($K_a = 2.8 \times 10^5\ M^{-1}$) and iodide ($K_a = 1.7 \times 10^4\ M^{-1}$) were bound less strongly. Molecular modeling indicated that iodide was too large to be bound within the cavity, while fluoride was small enough to be located in one corner of the cavity between two triazole CHs and a single N-linked phenylene CH. As predicted, replacement of the *tert*-Bu groups with OTg groups led to reduced halide affinity. Interestingly, it was found that substitution of the *tert*-Bu groups on N-linked phenylenes had a greater effect on binding constant than substituting on the C-linked phenylenes. Binding constants for chloride were determined to be $K_a = 5.1 \times 10^6$, 3.7×10^6, and $2.9 \times 10^6\ M^{-1}$ for **84**, **85**, and **86**, respectively. However, it was clear that the major component of the binding was due to the triazole CHs as all of these receptors show a high affinity for these anions. Proton NMR titrations performed in deuterated DCM produced the results in agreement with the findings of the UV titration experiments. In the case of compound **85**, fluoride was observed to associate almost exclusively with the CHs of the triazoles and the N-linked phenylenes and to hardly interact with the C-linked phenylenes. The results suggest the following order of hydrogen bond donor strengths: triazole CH> N-linked phenylene CH> C-linked phenylene CH. An acyclic analog of compound **83** was synthesized and shown to have a reduced affinity for halides, thus highlighting the importance of preorganization in receptor design.

Hecht and coworkers have included the 1,2,3-triazole motif in a series of oligomeric molecules **87** and **88** as shown in Figure 43.[73] The 2,6-bis(1,2,3-triazol-4-yl) pyridine units of the backbone have a strong preference for the *anti, anti* conformation, with the building blocks linked via *meta*-phenylene "hinges." It was thought that rotation around these hinges results in a helical conformation, stabilized by $\pi-\pi$ stacking interactions. UV spectra recorded in acetonitrile and chloroform solution showed no evidence for this $\pi-\pi$ stacking. However, by increasing the water content of acetonitrile solutions, hypochromic and bathochromic shifts of bands at 210 and 306 nm were observed, which is characteristic of chromophore stacking. This indicated that the solvophobic effect could provide the necessary driving force for the formation of the helical conformation. Further evidence for the formation of these $\pi-\pi$ stacking interactions was provided by fluorescence spectroscopy in acetonitrile : water (>60%) mixture, in which a transition from a sharp, monomer-like emission at 373 nm to a broader, excimer-like emission at 409 nm was observed. This is also characteristic of a system of stacked chromophores.

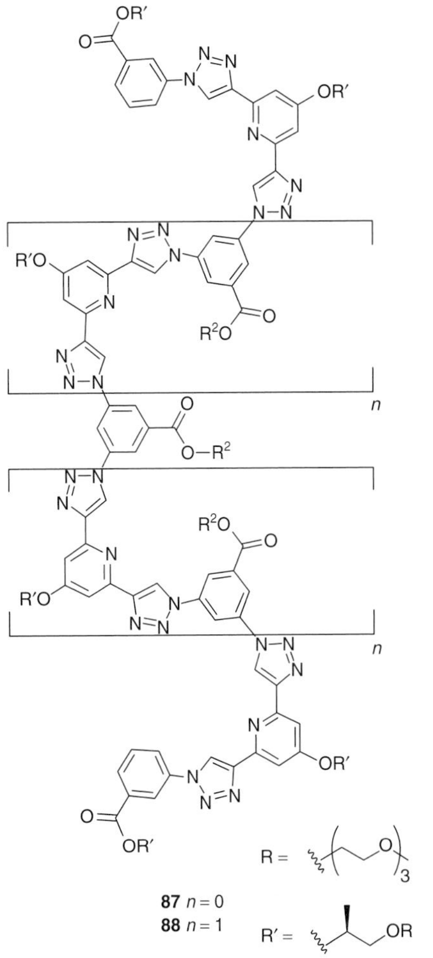

Figure 43 The structure of oligomers **87** and **88**.

Isolated 2,6-bis(1-aryl-1,2,3-triazol-4-yl)pyridines undergo a change in conformation from *anti, anti* to *syn, syn* on complexation of the metal cations or protons. Therefore, the response of clickamer **88** to HCl was investigated by CD (circular dichroism) spectroscopy in CH_3CN : water

(75%). Surprisingly, the addition of HCl did not result in the loss of the helical conformation but complete inversion of the CD signal. The effect of the counteranions on the clickamer conformation was investigated by the addition of potassium halide salts. Addition of KF simply resulted in a slight increase in the intensity of the CD signal associated with the free clickamer host. Addition of KCl resulted in an inverted signature of low intensity, whereas addition of KBr yielded an inverted spectrum of high intensity. This implies that the coordination of the larger halide anions caused an inversion of the helical twist but no structural changes. This represents a change in the chirality of a substrate induced by the binding of an achiral guest.

Concurrent with this work, Craig *et al.* independently reported the structurally related triazole-containing oligomers **89**–**91**.[74] Molecular modeling indicated that **89** had no preference for a particular conformation, a prediction in agreement with the NOESY spectrum of **89** in acetone-d_6. However, the modeling predicted that on addition of chloride compound **89** would form a helical structure with all of the triazole CHs pointing into the cavity to hydrogen bond to the centrally bound anion, as depicted in Figure 44.

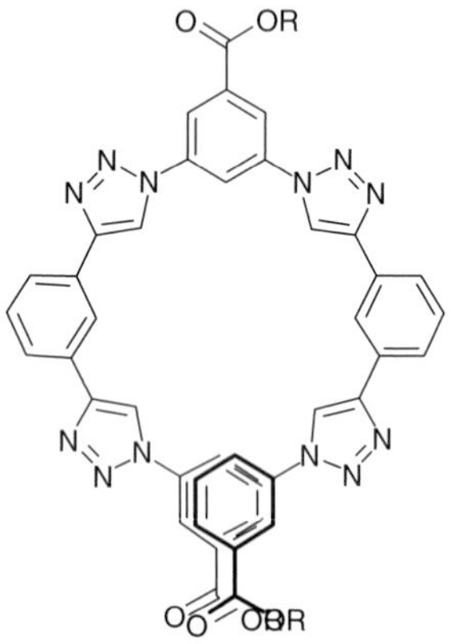

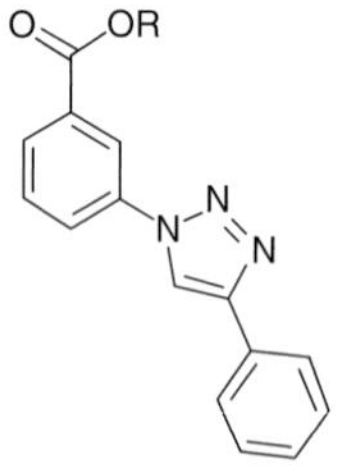

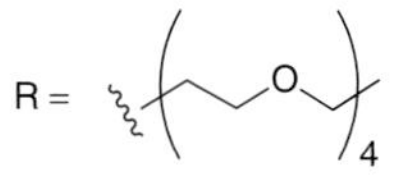

Figure 44 Triazole receptors **89**–**91**.

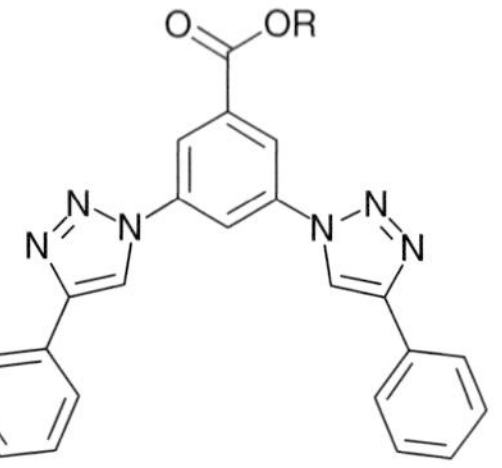

Table 10 Stability constants $(K_a)^a$ in acetone-d_6 solution.

Receptor	Anion^b	K_a (M^{-1})
89	Cl^-	1.7×10^4
89	Br^-	1.2×10^4
89	I^-	1.3×10^2
90	Cl^-	1.2×10^1
91	Cl^-	1.3×10^3

^{a}Determined by ^{1}H NMR titration.
^{b}Anions were added as the tetrabutylammonium salts.

The proton NMR titrations of compound **89** with tetrabutylammonium anion salts in acetone-d_6 revealed that as expected, the resonances of the triazole CHs were shifted downfield by 1.4 and 0.9 ppm on addition of chloride, indicative of hydrogen bonding to the anion. Similar but less significant downfield shifts are observed in the aryl protons which are orientated into the interior of the cavity. The chloride-induced folding was confirmed by 2D NOESY experiments. In the absence of chloride, the various rotamers of compound **89** are equally populated, but the addition of chloride results in a preference for the helical conformation. Job's plot analysis confirmed that the halides were bound with a 1 : 1 stoichiometry. The binding constants are shown in Table 10. Compound **89** was found to be selective for chloride and bromide over iodide. The importance of individual triazole units in chloride binding was assessed by titration of the simpler receptors **90** and **91** under the same conditions. As might be expected, increasing the number of triazole units leads to a higher chloride affinity. Compound **90** has only a single triazole CH hydrogen bond donor. The binding constant was found to be 12 M^{-1}, approximately 10 times less than the conventional hydrogen bond donor of 1,4-diphenylpyrrole.[75,76]

6 IMIDAZOLIUM-BASED ANION RECEPTORS

Another heterocycle that provides an acidic CH proton that can be utilized for hydrogen bonding interactions is the imidazolium moiety.[77,78] The imidazolium group also has the advantage of being cationic, that is, incorporating favorable electrostatic interactions for anion binding in addition to hydrogen bonds. Alcalde and coworkers reported the first example of such work. Dicationic macrocycles

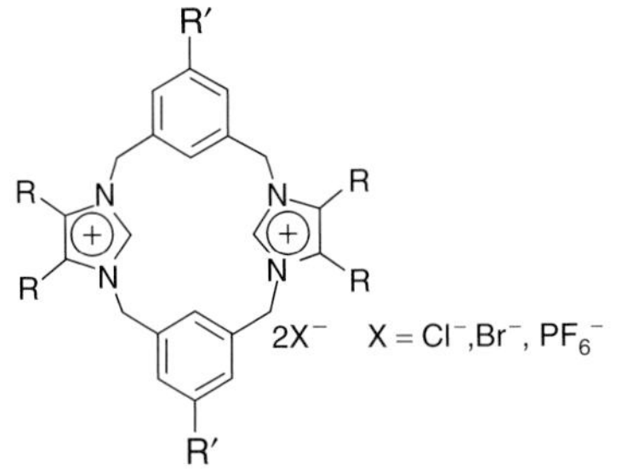

R′ R N R N R N N R R′ 2X⁻ X = Cl^-, Br^-, PF_6^-

92 R = H, R′ = H
93 R = H, R′ = *t*-Bu
94 R = *t*-Bu, R′ = *t*-Bu

Figure 45 The structures of imidazolium-based macrocycles **92–94**.

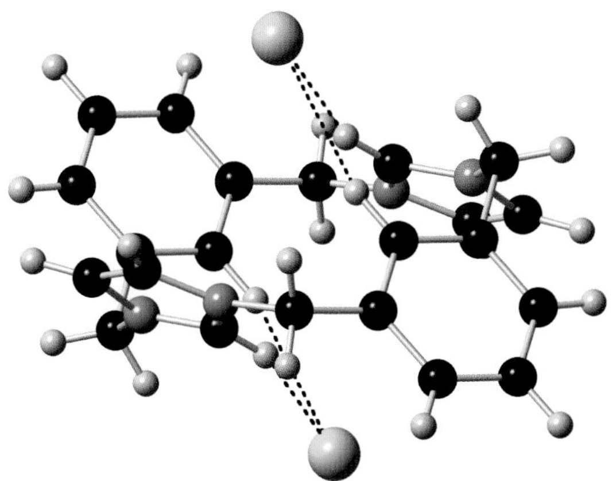

Figure 46 The X-ray crystal structure of the chloride complex of receptor **92**.

92–94 contain two imidazolium units in a 1,3-alternate conformation (Figure 45).[79]

The structures of **92** · 2Cl · $2H_2O$ and **93** · 2Cl · $3.5H_2O$ were elucidated by X-ray crystallography (Figures 46 and 47). It was found that macrocycle **92** · 2Cl · $2H_2O$ adopts a chair-like conformation with each chloride counteranion hydrogen bonded to an inward-pointing imidazolium and *m*-xylyl proton. The cavity takes the form of a square with each edge approximately 5 Å in length. Meanwhile macrocycle **93** · 2Cl · $3.5H_2O$ was found to adopt a cone-like conformation. The crystal structure contained two complexes; in the first, one of the chloride anions is hydrogen bonded within the cavity; the second complex consists of the receptor bound to a molecule of water.

The addition of various tetrabutylammonium salts to a solution of **94** · $2PF_6$ in DMSO-d_6 caused deshielding of the ^1H NMR resonances of the inward-pointing imidazolium protons due to binding of the added anion. The magnitude of the downfield shift may be correlated with the strength of the hydrogen bonding interaction formed. In this case, the

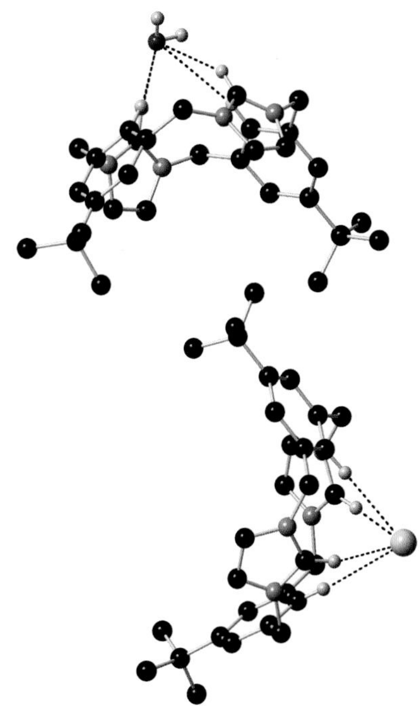

Figure 47 The X-ray crystal structure of the chloride/water complex of receptor **93**. Nonacidic hydrogens and nonbound chloride anions and water have been omitted for clarity.

observed deshielding followed the order $H_2PO_4^- > F^- > CH_3CO_2^- > CN^- > Cl^-$.

7 CONCLUSIONS

The discovery by Sessler and coworkers that expanded porphyrins could bind anions (see **Porphyrins and Expanded Porphyrins as Receptors**, Volume 3) has led to the synthesis of a range of both cyclic and acyclic, and charged and neutral molecules containing a variety of different heterocyclic rings that function as NH or CH hydrogen bond donors. These compounds have high affinities for anionic guests and have begun to find applications in roles such as mediating the transport of anions across lipid bilayer membranes. Heterocycles will continue to be an important part of the anion receptor design toolbox.

ACKNOWLEDGMENTS

We would like to thank the EPSRC for funding (CJEH).

REFERENCES

1. J. L. Sessler, P. A. Gale, and W.-S. Cho, *Anion Receptor Chemistry*, The Royal Society of Chemistry, Cambridge, 2006.

2. J. L. Sessler and J. M. Davis, *Acc. Chem. Res.*, 2001, **34**, 989–997.
3. A. Baeyer, *Ber. Dtsch. Chem. Ges.*, 1886, **19**, 2184–2185.
4. L. Bonomo, E. Solari, G. Martin, *et al.*, *Chem. Commun.*, 1999, 2319–2320.
5. E. Campazzi, E. Solari, C. Floriani, and R. Scopelliti, *Chem. Commun.*, 1998, 2603–2604.
6. P. A. Gale, J. L. Sessler, V. Kral, and V. Lynch, *J. Am. Chem. Soc.*, 1996, **118**, 5140–5141.
7. R. Custelcean, L. H. Delmau, B. A. Moyer, *et al.*, *Angew. Chem. Int. Ed.*, 2005, **44**, 2537–2542.
8. G. W. Bates, P. A. Gale, and M. E. Light, *Crystengcomm*, 2006, **8**, 300–302.
9. G. W. Bates, P. A. Gale, and M. E. Light, *Supramolecular Chem.*, 2008, **20**, 23–28.
10. J. L. Sessler, D. E. Gross, W. S. Cho, *et al.*, *J. Am. Chem. Soc.*, 2006, **128**, 12281–12288.
11. D. E. Gross, F. P. Schmidtchen, W. Antonius, *et al.*, *Chem.-Eur. J.*, 2008, **14**, 7822–7827.
12. J. L. Sessler, W. S. Cho, D. E. Gross, *et al.*, *J. Org. Chem.*, 2005, **70**, 5982–5986.
13. C. H. Lee, H. Miyaji, D. W. Yoon, and J. L. Sessler, *Chem. Commun.*, 2008, 24–34.
14. D. W. Yoon, H. Hwang, and C. H. Lee, *Angew. Chem. Int. Ed.*, 2002, **41**, 1757–1759.
15. C. H. Lee, H. K. Na, D. W. Yoon, *et al.*, *J. Am. Chem. Soc.*, 2003, **125**, 7301–7306.
16. D. W. Yoon, D. E. Gross, V. M. Lynch, *et al.*, *Angew. Chem. Int. Ed.*, 2008, **47**, 5038–5042.
17. D. W. Yoon, D. E. Gross, V. M. Lynch, *et al.*, *Chem. Commun.*, 2009, 1109–1111.
18. D. Seidel, V. Lynch, and J. L. Sessler, *Angew. Chem. Int. Ed.*, 2002, **41**, 1422–1425.
19. M. Stępien, B. Donnio, and J. L. Sessler, *Angew. Chem. Int. Ed.*, 2007, **46**, 1431–1435.
20. E. A. Katayev, G. D. Pantos, M. D. Reshetova, *et al.*, *Angew. Chem. Int. Ed.*, 2005, **44**, 7386–7390.
21. N. N. Gerber, *Crit. Rev. Microbiol.*, 1974, **3**, 469–485.
22. J. W. Bennett and R. Bentley, *Adv. Appl. Microbiol.*, 2000, **41**, 1–32.
23. A. Fürstner, *Angew. Chem. Int. Ed.*, 2003, **42**, 3582–3603.
24. R. D'Alessio, A. Bargiotti, O. Carlini, *et al.*, *J. Med. Chem.*, 2000, **43**, 2557–2565.
25. M. S. Melvin, J. T. Tomlinson, G. Park, *et al.*, *Chem. Res. Toxicol.*, 2002, **15**, 734–741.
26. A. J. Castro, *Nature*, 1967, **213**, 903–904.
27. J. E. H. Lazaro, J. Nitcheu, R. Z. Predicala, *et al.*, *J. Nat. Toxins*, 2002, **11**, 367–377.
28. O. A. Weisz, *Traffic*, 2003, **4**, 57–64.
29. C. Yamamoto, H. Takemoto, H. Kuno, *et al.*, *Hepatology*, 1999, 894–902.
30. K. Tanigaki, T. Sato, Y. Tanaka, *et al.*, *FEBS Lett.*, 2002, **524**, 37–42.
31. T. Sato, H. Konno, Y. Tanaka, *et al.*, *J. Biol. Chem.*, 1998, **273**, 21455–21462.
32. R. A. Gottlieb, J. Nordberg, E. Showronski, and B. M. Babior, *Proc. Nat. Acad. Sci.*, 1996, **93**, 654–658.
33. S. Ohkuma, T. Sato, M. Okamoto, *et al.*, *Biochem. J.*, 1998, **334**, 731–741.
34. J. L. Sessler, L. R. Eller, W.-S. Cho, *et al.*, *Angew. Chem. Int. Ed.*, 2005, **44**, 5989–5992.
35. J. L. Seganish and J. T. Davis, *Chem. Commun.*, 2005, 5781–5783.
36. J. T. Davis, P. A. Gale, O. A. Okunola, *et al.*, *Nature Chem.*, 2009, **1**, 138–144.
37. P. A. Gale, M. E. Light, B. McNally, K. Navakhun, K. E. Sliwinski and B. D. Smith, *Chem. Commun.*, 2005, 3773–3775.
38. H. Maeda and Y. Kusunose, *Chem. Eur. J.*, 2005, **11**, 5661–5666.
39. C. Fujimoto, Y. Kusunose, and H. Maeda, *J. Org. Chem.*, 2006, **71**, 2389–2394.
40. H. Maeda and Y. Ito, *Inorg. Chem.*, 2006, **45**, 8205–8210.
41. F. G. Bordwell, G. E. Drucker, and H. E. Fried, *J. Org. Chem.*, 1981, **46**, 632–635.
42. P. A. Gale, *Chem. Commun.*, 2008, 4525–4540.
43. D. Curiel, A. Cowley, and P. D. Beer, *Chem. Commun.*, 2005, 236–238.
44. M. J. Chmielewski, L. Zhao, A. Brown, *et al.*, *Chem. Commun.*, 2008, 3154–3156.
45. C. Caltagirone, J. R. Hiscock, M. B. Hursthouse, *et al.*, *Chem.-Eur. J.*, 2008, **14**, 10236–10243.
46. P. A. Gale, J. R. Hiscock, S. J. Moore, *et al.*, *Chem.-Eur. J.*, 2010, **5**, 555–561.
47. P. A. Gale, J. R. Hiscock, C. Z. Jie, *et al.*, *Chem. Sci.*, 2010, **1**, 215–220.
48. J. L. Sessler, D. Q. An, W. S. Cho, and V. Lynch, *Angew. Chem. Int. Ed.*, 2003, **42**, 2278–2281.
49. J. L. Sessler, D. Q. An, W. S. Cho, and V. Lynch, *J. Am. Chem. Soc.*, 2003, **125**, 13646–13647.
50. M. J. Chmielewski, M. Charon, and J. Jurczak, *Org. Lett.*, 2004, **6**, 3501–3504.
51. T. H. Kwon and K. S. Jeong, *Tetrahedron Lett.*, 2006, **47**, 8539–8541.
52. K. J. Chang, B. N. Kang, M. H. Lee, and K. S. Jeong, *J. Am. Chem. Soc.*, 2005, **127**, 12214–12215.
53. K. J. Chang, M. K. Chae, C. Lee, *et al.*, *Tetrahedron Lett.*, 2006, **47**, 6385–6388.
54. J. I. Kim, H. Juwarker, X. Liu, *et al.*, *Chem. Commun.*, 2010, **46**, 764–766.
55. F. Garciatellado, S. Goswami, S. K. Chang, *et al.*, *J. Am. Chem. Soc.*, 1990, **112**, 7393–7394.
56. S. J. Geib, S. C. Hirst, C. Vicent, and A. D. Hamilton, *J. Chem. Soc. Chem. Commun.*, 1991, 1283–1285.

57. A. D. Hamilton and D. van Engen, *J. Am. Chem. Soc.*, 1987, **109**, 5035–5036.
58. K. *Ka*vallieratos, C. M. Bertao, and R. H. Crabtree, *J. Org. Chem.*, 1999, **64**, 1675–1683.
59. M. J. Chmielewski and J. Jurczak, *Tetrahedron Lett.*, 2005, **46**, 3085–3088.
60. M. Chmielewski and J. Jurczak, *Tetrahedron Lett.*, 2004, **45**, 6007–6010.
61. M. J. Chmielewski, A. Szumna, and J. Jurczak, *Tetrahedron Lett.*, 2004, **45**, 8699–8703.
62. M. Albrecht, Triyanti, M. de Groot, *et al.*, *Synlett*, 2005, 2095–2097.
63. M. Albrecht, Triyanti, S. Schiffers, *et al.*, *Eur. J. Org. Chem.*, 2007, 2850–2858.
64. P. Roychowdhury, B. N. Das, and B. S. Basak, *Acta Cryst.*, 1978, **B34**, 1047–1048.
65. M. Albrecht, K. Witt, R. Frohlich, and O. Kataeva, *Tetrahedron*, 2002, **58**, 561–567.
66. C. R. Bondy, P. A. Gale, and S. J. Loeb, *J. Am. Chem. Soc.*, 2004, **126**, 5030–5031.
67. I. E. D. Vega, P. A. Gale, M. E. Light, and S. J. Loeb, *Chem. Commun.*, 2005, 4913–4915.
68. S. Shinoda, M. Tadokoro, H. Tsukube, and R. Arakawa, *Chem. Commun.*, 1998, 181–182.
69. L. M. Hancock and P. D. Beer, *Chem. Eur. J.*, 2009, **15**, 42–44.
70. M. G. Finn and V. V. Fokin, *Chem. Soc. Rev.*, 2010, **39**, 1231–1232.
71. W. S. Horne, M. K. Yadav, C. D. Stout, and M. R. Ghadiri, *J. Am. Chem. Soc.*, 2004, **126**, 15366–15367.
72. Y. L. Li and A. H. Flood, *J. Am. Chem. Soc.*, 2008, **130**, 12111–12122.
73. R. M. Meudtner and S. Hecht, *Angew. Chem. Int. Ed.*, 2008, **47**, 4926–4930.
74. H. Juwarker, J. M. Lenhardt, D. M. Pham, and S. L. Craig, *Angew. Chem. Int. Ed.*, 2008, **47**, 3740–3743.
75. J. L. Sessler, N. M. Barkey, G. D. Pantos, and V. M. Lynch, *New. J. Chem.*, 2007, **31**, 646–654.
76. J. L. Sessler, D. E. Gross, W. S. Cho, *et al.*, *J. Am. Chem. Soc.*, 2006, **128**, 12281–12288.
77. J. Yoon, S. K. Kim, N. J. Singh, and K. S. Kim, *Chem. Soc. Rev.*, 2006, **35**, 355–360.
78. Z. Xu, S. K. Kim, and J. Yoon, *Chem. Soc. Rev.*, 2010, **39**, 1457–1466.
79. E. Alcalde, C. Alvarez-Rua, S. Garcia-Granda, *et al.*, *Chem. Commun.*, 1999, 295–296.

FURTHER READING

Anion Receptor Chemistry by J. L. Sessler, P. A. Gale, and W.-S. Cho, *Monographs in Supramolecular Chemistry*, ed. J. F. Stoddart, Royal Society of Chemistry, Cambridge, 2006.

There are special issues of Chemical Society Reviews (2010, Vol. 39, Issue 10) and Coordination Chemistry Reviews (2003, Vol. 240, Issue 1–2 and 2006, Vol. 250, Issue 23–24) devoted to the area of anion complexation to which the reader is directed for further information.

Amide and Urea-Based Receptors

Md. Alamgir Hossain[1], Rowshan Ara Begum[2], Victor W. Day[2], and Kristin Bowman-James[2]

[1]*Jackson State University, Jackson, MS, USA*
[2]*University of Kansas, Lawrence, KS, USA*

1 EARLY HISTORY

In 1968, Park and Simmons discovered that diazabicyclic compounds, named *katapinands*, formed inclusion complexes with halide anions in water by H-bonding interactions.[1] Encapsulation was not confirmed, however, until the partial crystal structure of the chloride complex ($n = 9$) was reported by Bell *et al.* in 1975.[2] Indeed the interesting aspect of anion binding was not the primary objective of this 1975 *Science* article, rather it was the "short, symmetric hydrogen bond" in an unanticipated $H_{13}O_6^+$ cation. These rather overlooked but landmark contributions led to the field of anion recognition becoming a major area of research at the interface of chemistry and biology.[3–7]

Supramolecular Chemistry: From Molecules to Nanomaterials.
Edited by Philip A. Gale and Jonathan W. Steed.

© 2012 John Wiley & Sons, Ltd. ISBN: 978-0-470-74640-0.

$(CH_2)_n$ $(CH_2)_n$ N N $(CH_2)_n$

Katapinands

Although polyammonium hosts dominated the area initially, in the 1980s researchers started to focus on other H-bond donor groups. Some of the early forays into other H-bonding units included amide/thioamide, urea/thiourea, and sulfonamides. For example, in 1986, Pascal and coworkers reported an amide-based cryptand, **1**, that was found to bind fluoride ion in DMSO-d_6.[8] Seven years later, Reinhoudt and coworkers published a seminal paper that explored the tren-based amide [tren = $N,N'N''$-tris(2-aminoethyl)amine], **2**, and sulfonamide corollary and their complexes with a variety of anions.[9] In 1992, another landmark paper, this time involving urea hosts for anions, was published by Wilcox, who observed that an acycle containing a single urea binding site, **3**, was able to complex ion pairs containing phosphonates, sulfates, and carboxylates in $CHCl_3$.[10] This was followed by a report by Hamilton a year later, who used both a very simple monotopic dimethyl urea host, **4**, and a ditopic host based on a *p*-xylyl framework with urea and thiourea arms, **5**, to explore anion binding.[11] The monotopic dimethyl urea host was found to bind acetate in DMSO, $K = 45$, and the ditopic urea host **5** was found to bind glutarate in DMSO, $K = 6.4 \times 10^2$. Thioamides were not explored much until the early 2000s, when the Yamamoto and Bowman-James groups almost simultaneously published the first of the thioamide-based anion hosts **6** and **7**, respectively.[12,13] Even now, this class of anion hosts is not as frequently studied. The main thrust of this chapter is on amide and urea hosts, and, because of their structural similarities, thioamides and thioureas.

1

2 $R = CH_2Cl$
$R = (CH_2)_4CH_3$
$R = C_6H_5$
$R = 4\text{-}CH_3OC_6H_4$

3

4

$X = O, S$

5

6

7

2 BINDING CONCEPTS

The basic functional groups of the four types of receptors discussed in this chapter are shown in Scheme 1. Indeed, interest in hosts involving amide functionalities has been rapidly increasing because of the structural analogy with the amides in proteins, C(O)NHR, that assist in the formation of α helices and β sheets through the interactions of amide NH and carbonyl C=O groups. While amine hosts function as H-bonding donors primarily when in their protonated forms, amides readily form H-bonds when neutral. Fundamentally, the electronegative oxygen atom, with its heightened electron-withdrawing property, weakens the NH bond, making it more available as a H-bond donor for an anion. Going one step further by replacing the oxygen atom in amides with the more polarizable sulfur atom weakens the NH bond by about 10 kcal mol^{-1}.[14]

By adding a second NH moiety to a carbonyl group, the urea functionality is obtained. Ureas and thioureas have also been the subject of significant interest as ligands in anion coordination chemistry, especially because of the directional and potentially chelating properties of both of the two nearby NH groups in a given urea unit, resulting in a "mini" chelate. Even a very simple monourea host such as **4** can chelate with an anionic guest, forming a complex that is crystallographically characterizable. Like

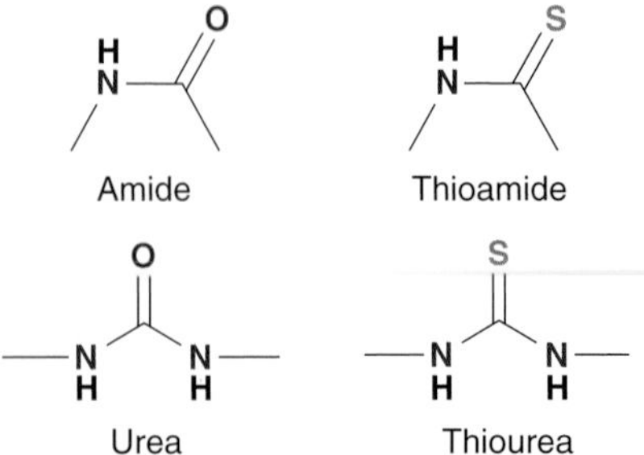

Scheme 1 Basic functional groups involved in the four neutral receptors: amides, thioamides, ureas, and thioureas.

in the progression from amide to thioamide, the acidity of a urea is generally increased by replacing the oxygen with a sulfur atom.[15–18] For example, in DMSO, the pK_a of thiourea is 21.1, while that of urea is 26.9.[15] Therefore, the sulfur analogs, the thioamides and thioureas, present possible advantages for anion binding over their oxo counterparts.

In the following two sections amide/thioamide and urea/thiourea hosts are described, respectively. Discussions within each section follow in order of increasing host dimensionality, that is, acyclic, monocyclic, and bicyclic/polycyclic.

3 AMIDES AND THIOAMIDES

3.1 Acycles

A seminal paper that illustrated the efficacy of isophthalamide spacers was that of Crabtree and coworkers in 1997.[19] From the reaction of the commercially available diacid chloride and aniline in DMF, this group synthesized the simple host **8** (Scheme 2). The authors found that **8** could bind bromide ion, holding it by H-bonds from the two amide nitrogen atoms (N···Br = 3.634 and 3.437 Å) (Figure 1), providing bidentate chelation. The *m*-xylyl group CH is even closer, but is probably held in position because of the NH···Br H-bonds. The bromide ion lies slightly above the plane of the isophthalamide, possibly to escape the nearness of the phenyl CH group. The solution-binding behavior of **8** was not determined because of its poor solubility in CD_2Cl_2. By replacing the isophthalamide unit with a 2,6-diamidopyridine, the authors obtained **9**, which was selective for the fluoride anion in CD_2Cl_2 ($K = 2.40 \times 10^4$ M^{-1}).[20] However, **9** exhibited negligible affinity for larger halides, which the authors attributed to the electrostatic repulsion between the negatively charged anions and the lone pair of the pyridine nitrogen (Table 1).

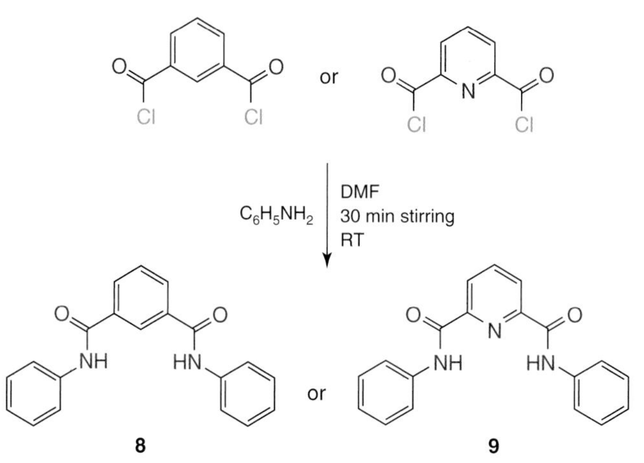

Scheme 2 Acylation of primary amines by acid chlorides.

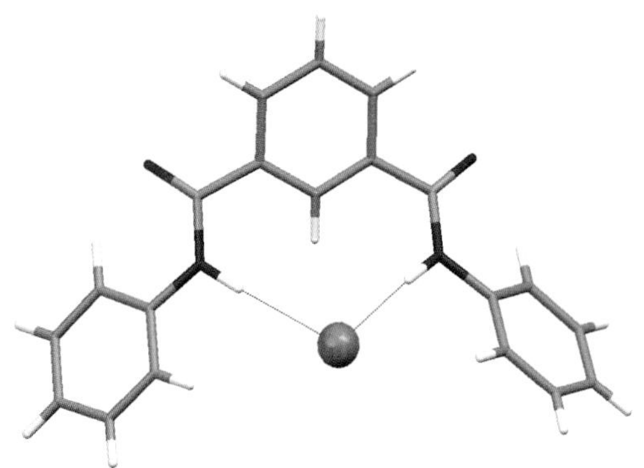

Figure 1 Overhead view of the bromide complex of **8**. The $[Ph_4P]^+$ counterion is not shown for clarity.

Table 1 Association constants (K) of acyclic amides **9**, **10** and **12** for various anions determined by ^{1}H NMR titrations.

Host	Anions	K (M^{-1})	Solvent	Ref.
9	F^-	2.40×10^4	CD_2Cl_2	20
	Cl^-	1.50×10^3		
	Br^-	57.0		
	I^-	<20.0		
	CH_3COO^-	525		
10	Cl^-	5.23×10^3	CD_3CN	21
	Br^-	716		
	I^-	152		
12	Cl^-	45	DMSO-d_6	22
	CH_3COO^-	163		
	$H_2PO_4^-$	346		

Davis, Gale, Quesada, and coworkers designed a related system in which they functionalized the central phenyl group to provide planned preorganization. They proposed that, by appending hydroxy or methoxy groups to obtain **10** and **11**, the H-bonding tendencies would direct the organization.[21] The hydroxyl group would potentially yield the *syn–syn* conformation with intramolecular –O–H···O=C H-bonds between the hydroxy and carbonyl groups. This directed H-bond arrangement would presumably allow the ligand to bind a chloride ion in the cleft formed by the inwardly oriented diamides, and indeed K is high (5230 M^{-1}) in CD_3CN. Crystallographic evidence indicates the assumed orientation to be the case only in part, since the *anti–anti* form is also present in the crystal lattice. On the other hand, in the solid state, **11** exists only in the *anti–anti* conformation, which may also be the predominant isomer in solution, with –NH···O–CH_3 H-bonds between the amide and methoxy groups. The host **11** does not bind chloride, bromide, or iodide, as expected if the amide NH groups are tied up in H-bonding with the adjacent methoxy substituent.

10 **11**

Jurczak and coworkers synthesized an expanded amide, **12**, in a study aimed at evaluating the macrocyclic effect.[22] The crystal structure of the free ligand shows a twisted conformation with intramolecular H-bonds between the carbonyl group on one side of the molecule and the two amide groups associated with the other 2,6-diamido pyridine group. This fourfold H-bonding network results in a diamond-like H-bonding pattern that includes the carbonyl oxygen, amide NH group, pyridine lone pair, and the amide NH group. The carbonyl thus possibly occupies the anion complexation site in solution as well, necessitating a conformational change before complexation takes place. In support of this theory, **12** only binds anions weakly in solution. This tendency of amides to undergo intramolecular H-bonding has been cited (especially by urea enthusiasts) as a drawback to amides as H-bond donor groups. Indeed, this type of intramolecular bonding is responsible for the success of amides in forming secondary structural patterns such as α helices and β sheets in proteins. Structural characterization of the chloride complex with **12** shows that the chloride ion is held by four H-bonds from the surrounding amide hydrogen atoms, ranging from 3.246 to 3.307 Å (Figure 2).[22]

12

Johnson and coworkers designed an acyclic host made more rigid by acetylene groups inserted adjacent to a central pyridine unit. The result was the highly conjugated, H-bonding host **13**.[23] Upon protonation of the pyridine, **13** undergoes a conformational change to form a helical assembly with chloride (Figure 3), in which the anion is held by three NH···Cl and two CH···Cl H bonds (N···Cl distances range from 3.004 to 3.271, and C···Cl distances are 3.631 and 3.644 Å). In the crystal lattice, the chloride is H-bonded to an adjacent molecule via a CH···Cl bond (3.639 Å) to form a six-coordinate complex.

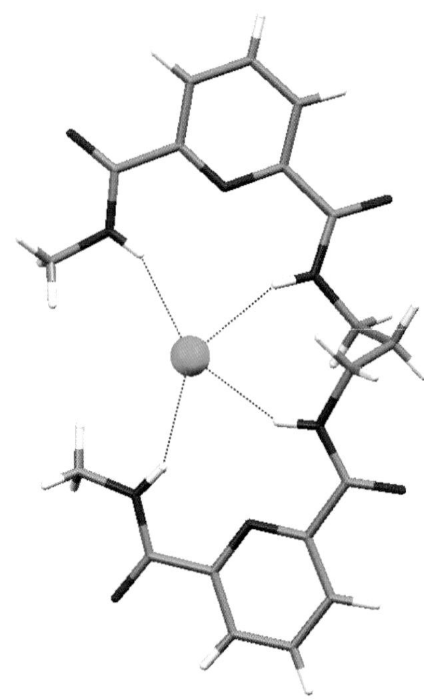

Figure 2 Overhead view of the X-ray structure of the chloride complex of **12**. The $[n\text{Bu}_4\text{N}]^+$ counterion is not shown for clarity.

13

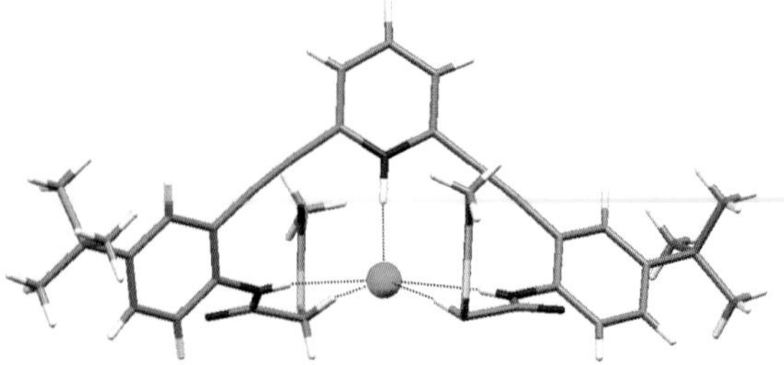

Figure 3 A view of the chloride complex of H[**13**]$^+$ that leads to a helical assembly.

Arunachalam and Ghosh found that a 1,3,5-phenyl-substituted host with *p*-nitrobenzoyl amido chains, **14**, can form an interlinked capsule that captures two nitrate guests.[24] One of the nitrate ions is held more strongly to the tripodal host by the amide NH groups, with N···O distances of 2.984, 3.040, and 3.132 Å. The second nitrate is held loosely by two N···O separations of 3.108 and 3.181 Å. The two nitrates are essentially parallel with phenyl caps at distances of 3.225 and 3.229 Å to the phenyl centroids, and the N···N distance between the two nitrates is 3.562 Å (Figure 4).[24]

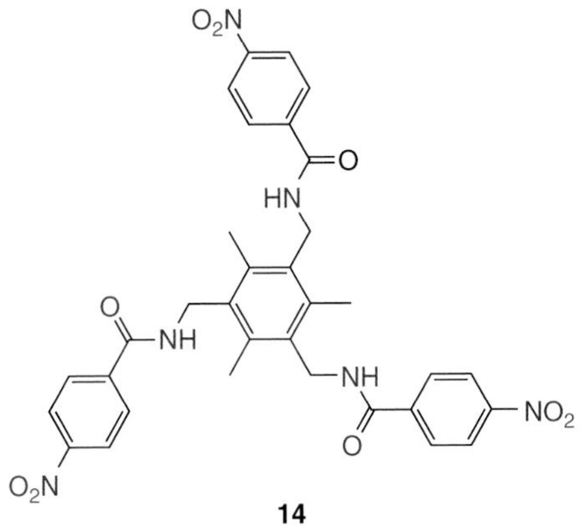

14

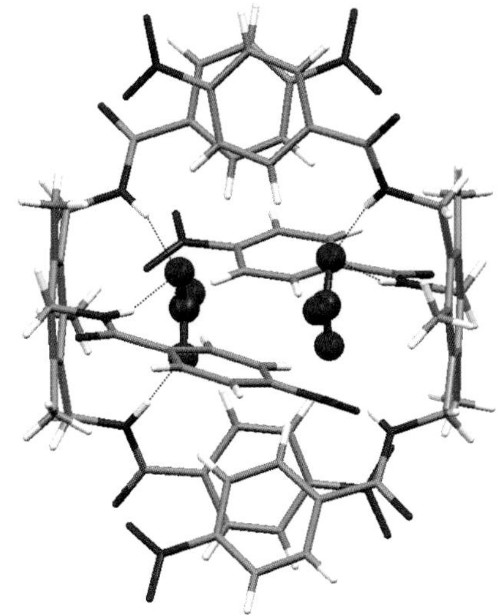

Figure 4 The X-ray structure showing two nitrates inside the capsule formed by two molecules of **14**. The $[n\mathrm{Bu_4N}]^+$ counterions are not shown for clarity.

While amides have been utilized extensively as anion hosts, their thio counterparts, thioamides, are less frequently explored, possibly due to the unappealing smells involved in the syntheses. In order to compare the relative binding affinity of amides and thioamides in anion binding, Zielinski and Jurczak synthesized a thioamide **16** from the corresponding amide **15** by refluxing with Lawesson's reagent (LR) in THF (Scheme 3).[25] Association constants were determined to be $K = 3.8$ and $11.1\ \mathrm{M^{-1}}$ for chloride ion for **15** and **16**, respectively, by titrations in DMSO-d_6 with 0.5% H_2O. The higher binding of **16** by a factor of 2.9 was attributed to the enhanced acidity of the thioamide NH group. In the solid state, a single chloride ion sits in the pincer cleft, held via two amide NH···Cl and one pyrrole NH···Cl H-bonds (Figure 5a). The thioamide, **16**, forms

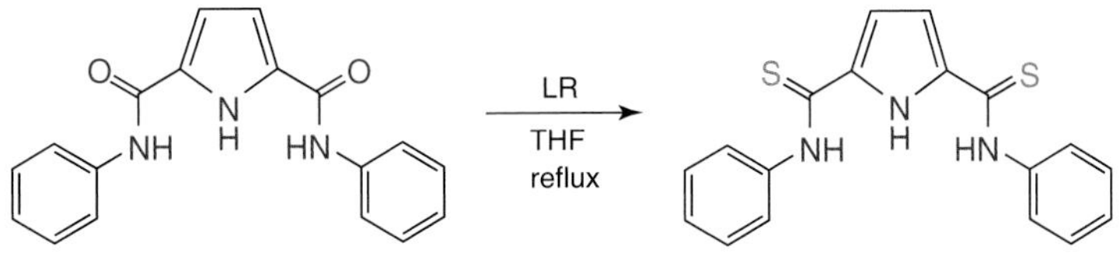

15 → (LR, THF, reflux) → **16**

Scheme 3 Simple one-step conversion of amides to thioamides by Lawesson's reagent (LR).

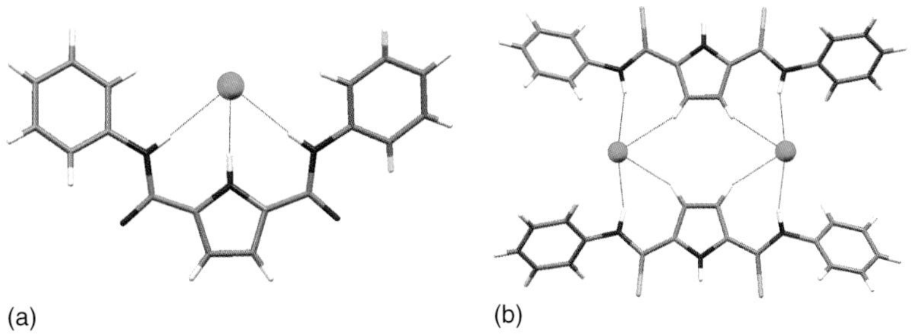

(a) (b)

Figure 5 (a) Crystal structures of the chloride complexes of **15** and (b) of **16**. The accompanying $[n\mathrm{Bu_4N}]^+$ counterions are not shown for clarity.

a dimer via two bound chlorides, each being held by two amide NH and two aromatic CH H-bonds (Figure 5b).

Because of the potential C_3 symmetry that can be imposed on the binding site, a number of researchers have taken advantage of the rather simple tren-based ligand framework first introduced by Reinhoudt.[9] The Bowman–James group examined the influence of chain length on binding and isolated an interesting lipid-bilayer-like complex of the host **17** with nitrate. The long chains of the lipid-like **17** orient in a parallel manner throughout the crystal lattice. The result is a channel-like opening between the opposing protonated ammonium "heads" where the nitrate counterions lie (Figure 6).[26] The nitrate–nitrate distance along the channels is 5.75 Å. One of the amide carbonyl groups is intramolecularly H-bonded to an adjacent amide NH group, which, as noted previously, is a drawback to amide-based receptors.[22]

17 **18**

19

Ghosh and coworkers have examined a simple tren-based amide host with nitro substituents (**18**).[27] Again, intramolecular N· · ·O H-bonds prevent the encapsulation of bromide in the cavity. As seen in Figure 7, one bromide links two neighboring acyclic units by two NH· · ·Br and one CH· · ·Br H-bonds.

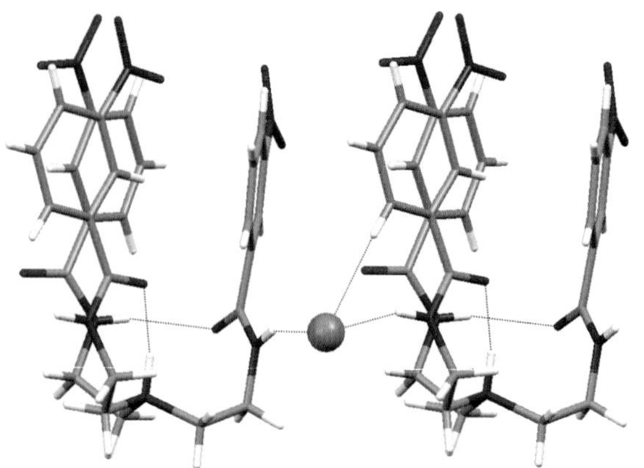

Figure 7 A view of the H-bonded bromide between two $[H\mathbf{18}]^+$ hosts showing the internal H-bonds preventing the anion from entering the cavity.

A tren framework is again utilized in a ferrocene-containing host reported by Beer and coworkers as potential redox sensors for selected anions, **19**.[28] Competition experiments were performed between $H_2PO_4^-$, HSO_4^-, and Cl^-, in which the redox couple shifts were the same irrespective of whether HSO_4^- or Cl^- were present, and depended only on the concentration of the $H_2PO_4^-$. Two related ferrocenium-based hosts also displayed similar properties, a simple ferrocene with a pendent amide and pyridine-bearing arm, and a calixarene-based host with two long arms, each attached to one of the cyclopentadienide units of a single ferrocene below.

3.2 Macrocycles

The Szumna and Jurczak group reported a pyridine-based tetraamide macrocycle, **20**. NMR titration studies in DMSO-d_6 suggested that this receptor should be suitable for fluoride ($K = 830\ M^{-1}$) (Table 2). The crystal structure of the fluoride complex (Figure 8a), indicates that the anion is coordinated by four amide H-bonds with an average N· · ·F distance of 2.78 Å.[29] The fluoride sits almost in the plane

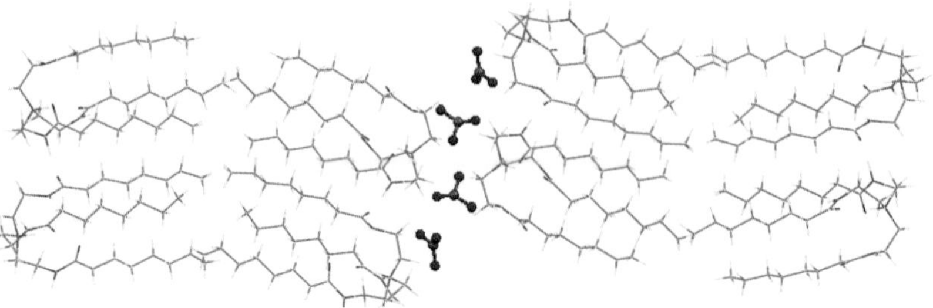

Figure 6 Formation of a lipid-like bilayer by $[H\mathbf{17}]^+$ with the nitrate counterions lying along the channels.

Table 2 Association constants (K) of **6, 20, 21a, 21b, 22a, 22b, 23, 24a**, and **24b** for anions determined by ^{1}H NMR titrations in DMSO-d_6.

Host	Anions	K (M^{-1})	Ref.
6	F^-	9.60×10^3	12
	Cl^-	1.10×10^3	
	Br^-	1.50×10^2	
	I^-	<2.00	
	$H_2PO_4^-$	3.90×10^3	
20	F^-	830	29
	Cl^-	65.0	
	CH_3COO^-	2.64×10^3	
	$H_2PO_4^-$	1.68×10^3	
	HSO_4^-	<5.00	
	p-$O_2NC_6H_4O^-$	67.0	
21a	Cl^-	25.0	13
	Br^-	20.0	
	I^-	<10.0	
	$H_2PO_4^-$	830	
	HSO_4^-	780	
	NO_3^-	<10	
	ClO_4^-	<10	
21b	F^-	407	13
	Cl^-	490	
	Br^-	513	
	I^-	<10.0	
	$H_2PO_4^-$	1.12×10^4	
	HSO_4^-	107	
	NO_3^-	<10.0	
	ClO_4^-	<10.0	
22a	F^-	707	13
	Cl^-	105	
	Br^-	100	
	I^-	28.0	
	$H_2PO_4^-$	9.33×10^4	
	HSO_4^-	1.41×10^3	
22b	F^-	1.29×10^{3a}	13
	Cl^-	398	
	Br^-	25.0	
	I^-	<10.0	
	$H_2PO_4^-$	4.27×10^{4a}	
	HSO_4^-	9.77×10^{3a}	
23	F^-	^{b}	30
	Cl^-	33.0	
	Br^-	<10.0	
	$H_2PO_4^-$	73.0	
	HSO_4^-	500	
24a	F^-	479	31
	Cl^-	1.70×10^3	
	Br^-	138	
	I^-	100	
	$H_2PO_4^-$	1.15×10^4	
	HSO_4^-	^{b}	
	NO_3^-	45.0	
	ClO_4^-	40.0	
24b	F^-	110	31
	Cl^-	5.62×10^4	
	Br^-	2.40×10^4	

Table 2 (*continued*).

Host	Anions	K (M^{-1})	Ref.
24b	I^-	162	31
	$H_2PO_4^-$	2.09×10^5	
	HSO_4^-	7.94×10^3	
	NO_3^-	209	
	ClO_4^-	251	

[a]Slow exchange.
[b]Value was not reported due to irregular chemical shifts.

of the macrocycle. However, in the crystal structure of the chloride complex, the larger anion hovers above the plane of the macrocycle, although it is also associated with four H-bonds (Figure 8b). The counterion in these two cases is $[Ph_4P]^+$.

Yamamoto and coworkers reported a macrocyclic thioamide, **6**, that was synthesized from its tetraamide precursor **20** using Lawesson's reagent.[12] This ligand was found to complex F^-, Cl^-, and I^-, with binding constants of 9.60×10^3, 1.10×10^3 and 1.50×10^2 M^{-1}, respectively, in DMSO-d_6 (Table 2). The high binding constant for the fluoride is perhaps due to its size complementarity with the macrocyclic cavity. The authors also isolated crystals of the chloride complex in which one chloride is H-bonded with the four thioamide groups (Figure 8c). The macrocycle folds slightly, and the chloride sits above the plane formed by four amides, similar to what is seen for the chloride structure of **20**.

In one of the crystal forms of the acetate structure of **20**, however, a sandwich-like, 2 : 1 complex is formed (Figure 8d) when $[nBu_4N]^+$ is the counterion, while a simple 1 : 1 complex is formed in the presence of $[(CH_3)_4N]^+$. In the sandwich complex, each oxygen atom of the carboxylate group is H-bonded to the host via four NH···O bonds (Figure 8d). In the 1 : 1 complex of **20** with acetate, the acetate hovers above the macrocyclic cavity, much like the chloride structure, with four H-bonds to the macrocyclic amides (Figure 8e). The anion is also coordinated by an H-bond to an axial water molecule, an occurrence that is frequently seen in complexes with anions.

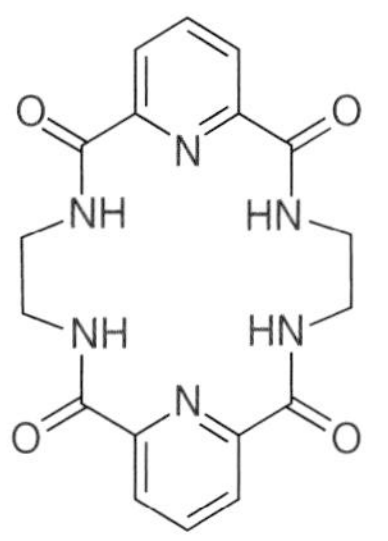

20

LIVERPOOL JOHN MOORES UNIVERSITY
LEARNING SERVICES

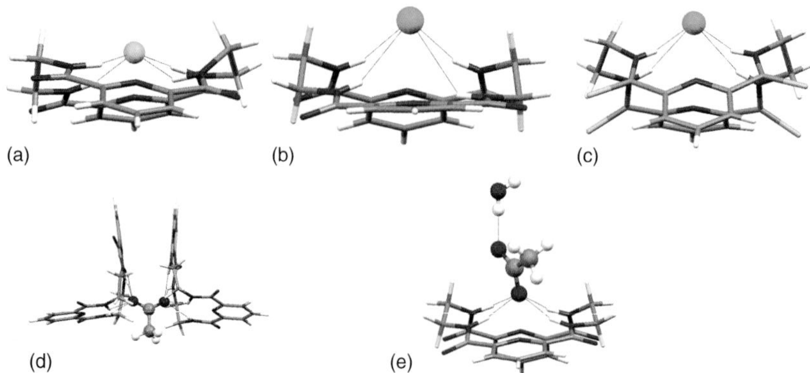

Figure 8 Views of the crystal structures of complexes of **20** with (a) fluoride and (b) chloride; of **6** with chloride (c); and the 2 : 1 (d) and 1 : 1 (e) complexes of **20** with acetate. $[R_4N]^+$ counterions are not shown for clarity.

21a (X = CH)
21b (X = N)

22a (X = CH)
22b (X = N)

23

Another sandwich structure with a slightly expanded ring system containing a mixed amide/amine framework, **21a** (X=CH), was reported by the Bowman–James group (Figure 9a).[32] The complex is neutral and the SO_4^{2-} is held by eight H-bonds, four to each of the two macrocycles. The corresponding thioamide macrocycles, **22**, were also synthesized. Although no crystal structures of anion complexes of the thioamides were isolated, these macrocycles displayed enhanced affinities for anions over the amide, **21**.[13] Both the amides and thioamides exhibited strong affinities for the oxo acids, $H_2PO_4^-$ and HSO_4^- (Table 2). The trend in the higher affinities for oxoacids may be related to a synergistic luring effect for the protonated anions by initial proton capture by the tertiary amines of the hosts.

When the *m*-xylyl group is replaced by pyridine in these mixed amine/amide macrocycles, **21b**, folded structures normally ensue as seen in the perrhenate structure (Figure 9b).[33] In this case, the two secondary amines of the macrocycle are protonated, resulting in a host with charge complementarity for a dianionic guest. A similar structure is seen for dichromate as well. Protonation of the tertiary amines appears to be common in the pyridine series, potentially assisting in the

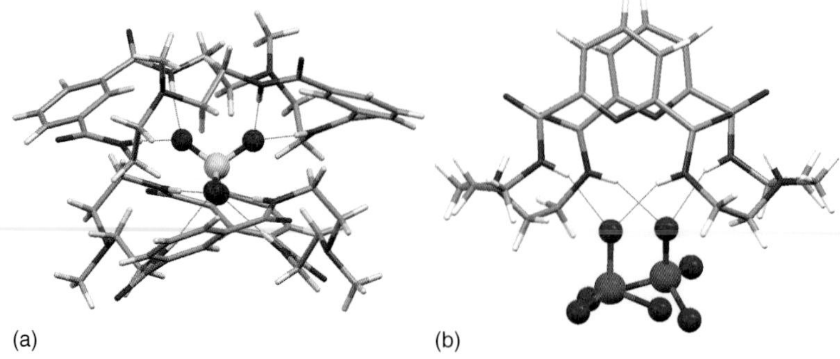

Figure 9 (a) Views of the sulfate sandwich complex of **21a** and (b) the perrhenate complex of $H_2[\mathbf{21b}]^{2+}$. The $[n\mathrm{Bu}_4\mathrm{N}]^+$ counterions are not shown in (a) for clarity. Complex (b) is neutral due to the protonated amines.

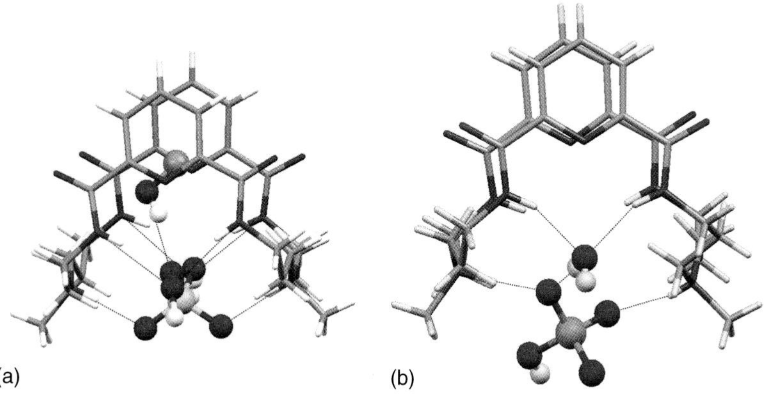

Figure 10 Views of the crystal structures of (a) $H_2[\mathbf{23}]^{2+}$ with SO_4^{2-} and methanol (hydrogen atoms not shown on the carbon) and (b) HPO_4^{2-} with water inside the cavity, both structures showing the folded macrocycle.

generally enhanced binding over the *m*-xylyl analogs (Table 2).[33] However, similar higher affinities for the pyridine analogs have been reported by Jurczak and coworkers (without the presence of amines),[34] which has been attributed to the tendency of the pyridine to preorganize the amide hydrogen atoms inward for chelation with an anion.

The slightly larger macrocycle, **23**, with propyl linkers, crystallizes in a similar folded conformation with both SO_4^{2-} and HPO_4^{2-}.[30] As in the perrhenate complex with **21b**, the two tertiary amines are protonated; thus the net charge of each complex is zero. As shown in Figure 10, each macrocycle folds along the line between the two tertiary amines. The "fit" appears to be better for the SO_4^{2-} complex, where the sulfur atom lies almost on the $N_{amine}–N_{amine}$ axis. A methanol molecule of crystallization also lies in the cavity above the SO_4^{2-} ion. In the HPO_4^{2-} structure, the anion lies somewhat below the cavity, and a water molecule is also chelated to the anion. However, in both cases the macrocycle is large enough so that solvent can also be included.

While amine-containing, amide-based hosts can protonate to satisfy charge needs, quaternization of the amines provides a sustainable charge to the host. In a systematic study of the influence of charge on anion binding, **24a** and **b** were synthesized by the reaction of **21a** and **b** respectively with CH_3I.[31] Binding data comparing the neutral and quaternized ligands (Table 2) show enhanced binding for the quaternized hosts compared to the unquaternized **21** (Scheme 4).

There are several possible orientations of the amide carbonyl groups for the dicationic **24**. As seen in the complex of chloride with the isophthalamide ligand **24a**, the macrocycle elongates, with the two charged ends at the farthest distances of the elliptical host (Figure 11a) and the anions outside of the cavity. Each chloride ion is held by an amide NH group with an average distance of 3.24 Å.

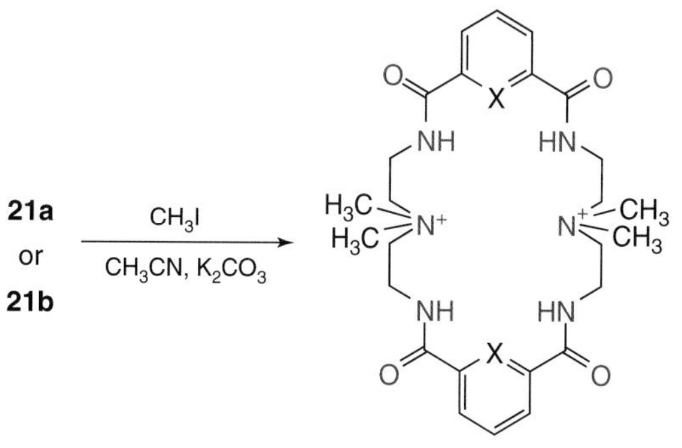

21a or **21b** → (CH_3I; CH_3CN, K_2CO_3)

24a (X = CH)
24b (X = N)

Scheme 4 Simple, one-step quaternization of the tertiary amine in **21** by CH_3I.

The crystal structure of the iodide complex of the pyridine-containing ligand **24b** adopts a different conformation. As in the case with the pyridine-containing macrocycles, **24b** folds (Figure 11b). The two iodides are H-bonded to two of the macrocyclic amides with an average N···I distance of 3.68 Å, and the complexes form chains of ···I···**24b**···I···**24b**··· throughout the lattice. The similarity in the conformations between the charged and neutral pyridine hosts indicates the strength of the preorganization influence of the pyridine groups, which seem to prefer stacking in a parallel manner.

An unanticipated structure was obtained for the bisulfate complex with **24a**.[35] As in the other structures of **24a**, the bisulfates were found to reside outside of the macrocyclic cavity. However, a third bisulfate was found in the channels between the macrocycles, and the third positive charge necessary for charge balance was identified in the center of the macrocycle, a single proton, bridging the two amide carbonyl groups (Figure 11c and d). This very short O···H···O separation (2.45 Å) falls in the category of a low barrier hydrogen bond (LBHB).

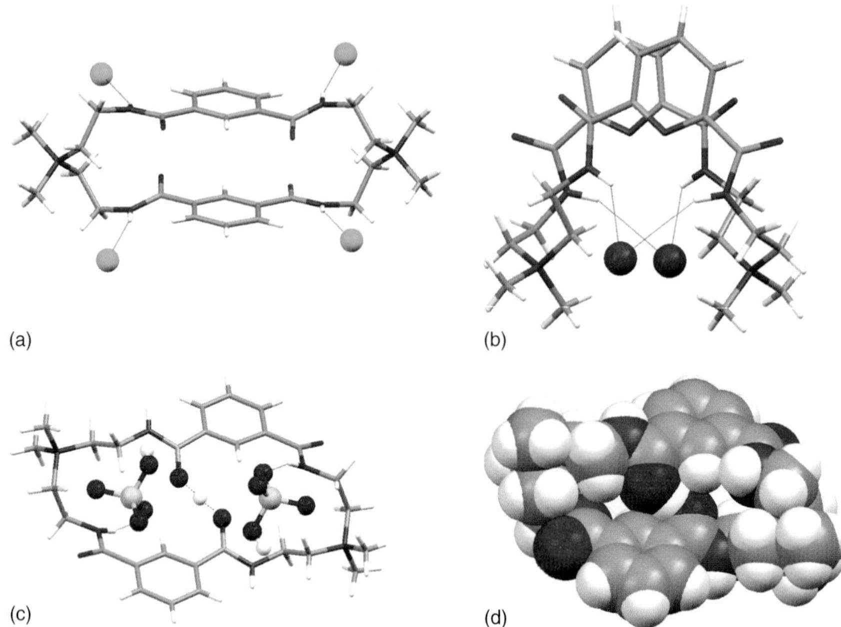

Figure 11 Crystal structures for complexes of **24a** with chloride (a), iodide (b), and bisulfate (c). View (d) is a space-filling version of (c) without the bisulfate anions showing the encapsulated proton, H-bonded to the two carbonyl oxygen atoms.

While the positive charge needed for host–guest complementarity can be designed into the ligand, an alternative is a dual host system, with components for both anion and cation binding. Smith and coworkers have been among the pioneers in this area with their dual ether/amide hosts. In addition to facilitating anion binding by a dual host effect (Table 3), the unsymmetrical mixed ether/amide/amine macrocycle **25** provided rather unanticipated solution chemistry.[36] The lone pair of the tertiary amine nucleophilically attacks the CH_2Cl_2 solvent, resulting in a quaternized ammonium group and a chloride guest (Figure 12a).

In an extension of the dual host concept, a bicyclic analog, **26**, containing a crown ether fitted with a diamide-containing bridge, was designed and synthesized.[37] The ether portion readily binds an alkali metal, while the upper diamide half holds the anion, as shown for the NaCl complex (Figure12b). The binding enhancement provided in the presence of different alkali metal ions clearly shows that the crown ether fit is best with a potassium ion (Table 3).

Table 3 Association constants (K) of **26** for chloride determined by ^{1}H NMR titrations in DMSO-d_6 (from Ref. 37).

Host	Anions	K (M^{-1})
26	Cl^-	35.0
26 + Na^{+a}	Cl^-	50.0
26 + K^{+b}	Cl^-	460

[a]Determined after the addition of one equivalent of Na[BPh_4].
[b]Determined after the addition of one equivalent of K[BPh_4].

Kubik and coworkers synthesized a cyclopeptide-based amide receptor **27** from the reaction of Boc-L-proline and 6-aminopicolinic acid benzyl ester.[38] They isolated crystals of the iodide complex showing a single iodide ion sandwiched between the two macrocycles. As the authors note, the peptide-based host acts somewhat like its biological counterparts in that it serves to shield the bound ion from its solvent shell. In the complex, the iodide is held by three NH bonds from each macrocycle (Figure 13), resulting in a six-coordinate, nearly trigonal prismatic geometry. The $N\cdots I^-$ distances are in the range

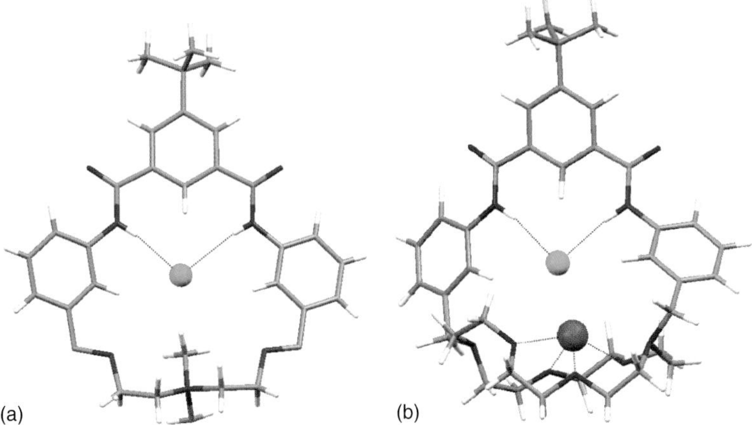

Figure 12 Crystal structures of (a) [CH_3**25**(Cl)] and (b) [**26**(NaCl)] showing similarly bound chloride ions in the macrocyclic cavities.

of 3.724–3.935 Å. ^1H NMR titration experiments also indicated the formation of a 2 : 1 complex in solution.

27

3.3 Bicycles (cryptands) and tricycles

Some of the first amide-based cryptands synthesized in order to target anion binding were prepared in the Bowman–James group.[39] These amide analogs of the much-studied azacryptands readily showed similar encapsulation properties without the need for extra charge on the host. In complexes with fluoride, the cryptand remains neutral, although it has the potential to protonate at the bridgehead amines. Fluoride complexes of the prototype hosts were especially interesting in that not only was encapsulation verified by X-ray techniques but also ^{19}F NMR spectroscopy indicated the complex to remain intact in solution.[39–41] The fluoride structures of both **28a** and **28b** show a centralized fluoride ion, with H-bonding contacts to six amide hydrogen atoms for **28b** and an additional three H-bonds to the three *m*-xylyl hydrogen atoms that stick into the cavity in **28a** (Figure 14). The solution ^{19}F NMR spectra also display patterns indicative of fluoride coupling with nine and six hydrogen atoms for **28a** and **28b**, respectively.

28a (X = CH)
28b (X = N)

29

30

It is of interest to compare and contrast the influence of size and potentially charge/protonation on binding in a series of three amide-based cryptands, **28b**, **29**, and **30**.[42,43] The ligands **28b** and **29** differ in size, ethyl linkers in the former versus propyl in the latter. By quaternizing the larger propyl-based cryptand, **29**, with CH_3I, **30** was isolated, with methylated bridgehead ammonium groups. The quaternized form of the smaller **28b** could not be obtained, possibly due to steric restrictions.

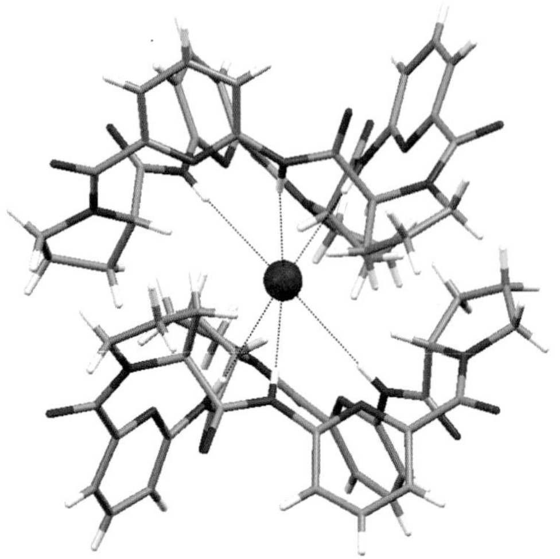

Figure 13 Side view of the sandwiched iodide ion in **27**.

For the nonquaternized **28b** and **29**, the bridgehead amines are protonated upon complexation with oxo anions such as sulfate, probably due to the presence of extra protons from the H_2SO_4 or [nBu$_4$N][HSO$_4$] reagents used in obtaining the crystals. In the smaller host (**28b**), a single sulfate is bound internally and is held via eight H-bonds in a bicapped pseudo-trigonal prismatic geometry. All six amide hydrogen atoms and the protonated bridgehead amines (Figure 15a) are utilized in H-bonding with the sulfate ion.

Increasing the size of the host increases its capacity for encapsulation of multiple guests as seen in the sulfate complex with **29**. Furthermore, the sulfate complex crystallizes as a dimer pair, formed when the pyridine groups of two macrocycles intercalate (Figure 15b). Two molecules of water are found between two of the intercalated rings, and water molecules also reside alongside the sulfate ions, "resting" on the arms of the cryptands. Each sulfate is held by seven H-bonds that include the neighboring water molecules.

Upon quaternization with CH_3I, the conformation of the cryptands changes drastically. As seen in the chloride structure with **30**, all three arms of the cryptand fold in the same direction and the "guests" lie on top with a chain of water molecules threading down into the host

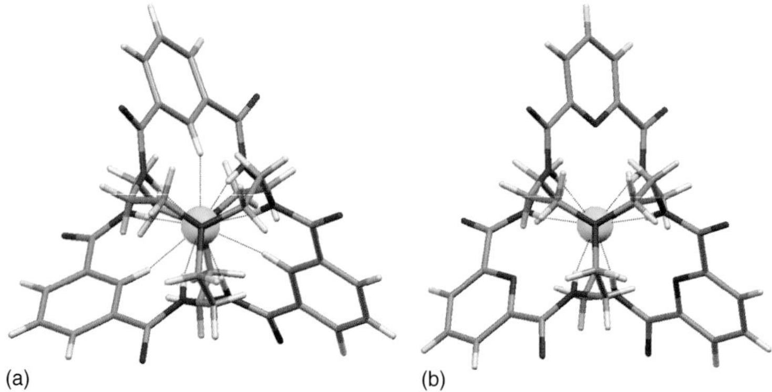

Figure 14 View down the bridgehead amines of the structure of the fluoride complexes of (a) **28a** and (b) **28b**.

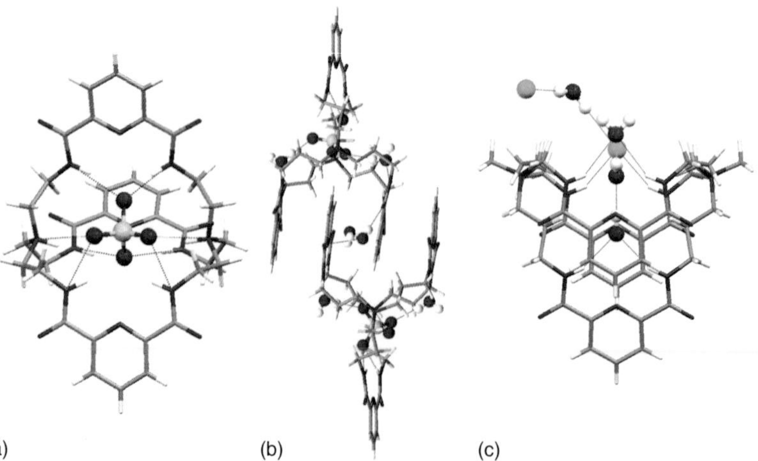

Figure 15 Crystal structure for (a) the sulfate complex of H_2[**28b**]$^{2+}$, (b) the sulfate complex of H_2[**29**]$^{2+}$, and (c) the chloride complex of the quaternized **30**.

Table 4 Association constants (K) of **28a**, **28b**, **29**, **30**, **32**, and **34** for selected anions determined by ^{1}H NMR titrations.

Host	Anions	K (M^{-1})	Solvent	Ref.
28a	F^-	3.30×10^4	DMSO-d_6	43
28b	F^-	$>1.00 \times 10^5$	DMSO-d_6	39
	Cl^-	2.95×10^3		
	Br^-	40.0		
	$H_2PO_4^-$	2.00×10^3		
	HSO_4^-	68.0		
	HSO_4^-	9.12×10^4	$CDCl_3$	
	HSO_4^-	5.50×10^4	CD_3CN	
29	F^-	$>1 \times 10^{5b}$	DMSO-d_6	41
	Cl^-	180		
	Br^-	7.00		
	$H_2PO_4^-$	170		
	HSO_4^-	2.70×10^3		
30	F^-	^{a}	DMSO-d_6	41
	Cl^-	3.10×10^3		
	Br^-	1.30×10^3		
	$H_2PO_4^-$	1.20×10^4		
	HSO_4^-	340		
32	F^-	2.10×10^4	DMSO-d_6	44
	FHF^-	110		
	$H_2PO_4^-$	3.00×10^3		
	$HP_2O_7^{3-}$	9.80×10^3		
34	FHF^-	5.50×10^3	DMSO-d_6	45
	$H_2PO_4^-$	740		
	N_3^-	340		
	CH_3COO^-	100		

[a]Not reported because of the peak broadening and irregular shift change.
[b]Slow equilibrium.

(Figure 15c). The binding data for **28, 29**, and **30** are provided in Table 4 and indicate that **28b** in general binds ions with higher affinity than **29** with the exception of the relatively high affinity of **29** for HSO_4^-. There appears to be less selectivity in the binding of anions observed for **30**, which binds most ions at K value about 10^3, with the exception of the fairly weak binding constant of HSO_4^-.

While tren is a handy bridgehead for cryptands, cyclophanes provide an alternative for bicyclic and tricylic hosts. Anslyn and coworkers synthesized a cryptand containing 1,3,5-trisubstituted cyclophane bridgeheads connected by rigid 2,6-diamidopyridine chains, **31**.[44] This host was found to be selective for acetate and nitrate, which are assumed to be encapsulated. Indeed, the acetate structure shows the guest neatly nestled between the two phenyl bridgeheads (Figure 16a). The chloride structure shows incorporation of two chloride ions bridged by a "cascading" molecule of water. This is similar to the chloride structure of **29** (Figure 16b and c), and an increasingly common occurrence for some of the larger cryptand-like hosts.

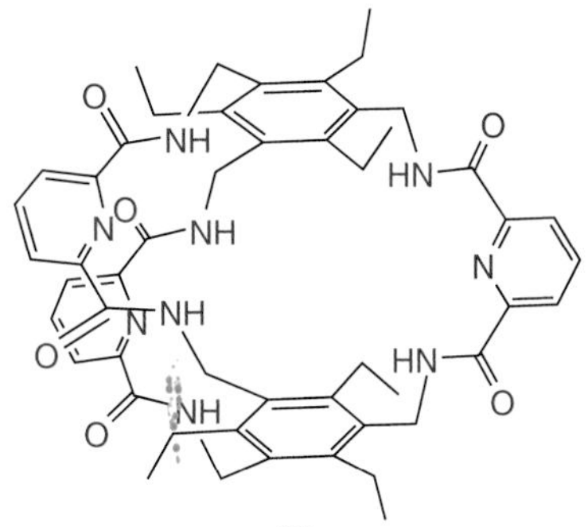

31

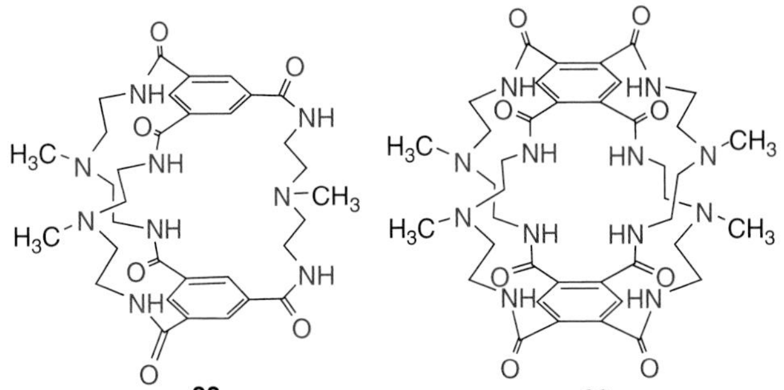

32 **33**

In an attempt to examine hosts with expanded cyclophane bridgeheads, the Bowman–James group synthesized a series of complexes with three (triad), **32**, and four (tetrad), **33**, chains linking the two phenyl caps.[46] The linkers chosen contained flexible diethylamine chains. Encapsulation of anionic guests was not observed; however, in crystal structures anions lined the clefts formed by adjacent arms. For the triad, there was only a single pocket for anions, which was formed by two of the chains lying in the same direction, with the third arm dangling in the opposite direction. The single cleft in **32** appeared to be well suited for binding bifluoride ion, which was held by just two H-bonds to the host, one to each of the chains, with two H-bonds to $CHCl_3$ molecules above and below (Figure 17a). The F· · ·F distance was a short 2.18 Å.

In order to offset the dangling arm in the triad **32** and to increase the binding capability of the host, a host containing 1,2,4,5-tetrasubstituted phenyl caps, **33**, was synthesized. The plan for incorporation of twice the number of anions worked, and the complex displays an affinity for phosphate ions, including dihydrogen phosphate, pyrophosphate, and triphosphate. An overhead view of the complex with $H_3P_3O_{10}^{2-}$ is shown in Figure 17(b). Interestingly, in the bifluoride structure, the host is not protonated, but it is in the phosphate structures, resulting in neutral complexes in the latter without the need for counterions. Of the two hosts, however, the triad, **32**, exhibited the strongest anion binding, with special affinity for bifluoride and phosphates

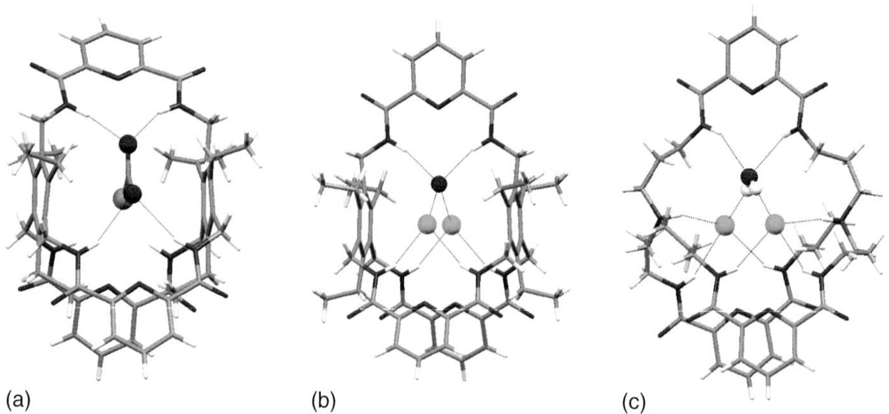

Figure 16 Crystal structures of (a) [**31**(CH_3COO)]$^-$, (b) [**31**($Cl)_2(H_2O$)]$^{2-}$, (c) and [**29**($Cl)_2(H_2O$)]. The [$n$$Bu_4N$]$^+$ counterions are not shown in (a) and (b) for clarity.

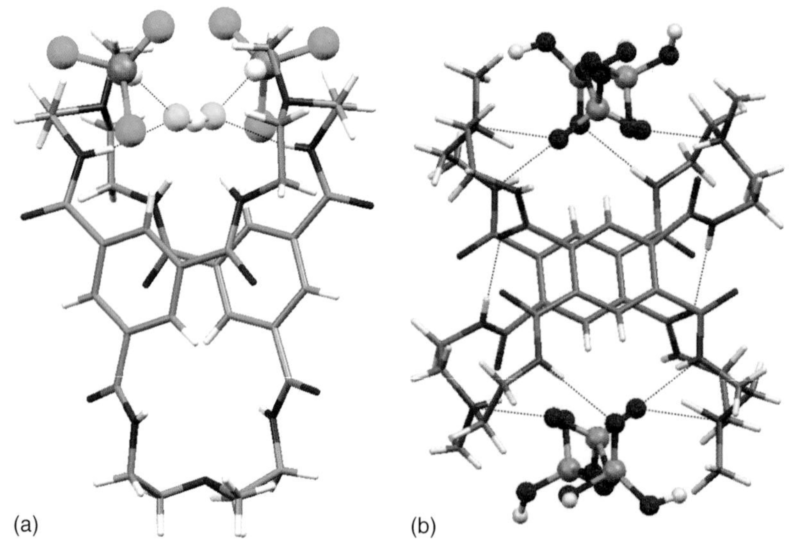

Figure 17 Crystal structures of (a) [**32**(FHF)]$^-$·2$CHCl_3$ and (b) [H_4(**33**)($H_3P_3O_{10})_2$]. The [($CH_3)_4N$]$^+$ counterion is not shown for clarity in (a).

(Table 4). The diminished binding of the tetrad **33** for anions was attributed to intramolecular H-bonds between an amide and a carbonyl group on an adjacent arm, thereby reducing the binding capabilities.

A tricyclic host was synthesized in the Bowman–James group by joining two amide-based macrocycles through their two secondary amine bridgeheads with an ethylene spacer, **34**.[45,47] While studying the affinity of the host for fluoride ion, crystals were isolated that revealed an encapsulated bifluoride complex, the first example of an encapsulated bifluoride in an organic host (Figure 18). The linear anion is held by four H-bonds from the amide hydrogen atoms. The multiple H-bonds serve to lengthen the F···F distance considerably to 2.475 Å (compared to the 2.18 Å in **32**). The suitability of the tricyclic host for linear triatomic molecules was further supported by the isolation of the essentially isomorphous azide structure (Figure 18b). Although one might surmise that **34** would have a rather flexible conformation, an extensive $N_{amine}\cdots HN_{amide}\cdots N_{py}$ H-bonding network tends to rigidify the large molecule so that the same pseudo-S_4 symmetry is observed in several of the structures. In a second preferred conformation, **34** is also held together by a H-bonding network, but a D_{2h} symmetry is observed. Here, the sulfate guests lie outside the cavity, lining the tricycle (Figure 18c). Like in other oxo acid complexes, all four tertiary amines of the tricyclic host are protonated, and the resulting disulfate complex is neutral. This tricycle is exceedingly selective for bifluoride, as seen in Table 4, and less so for other anions.

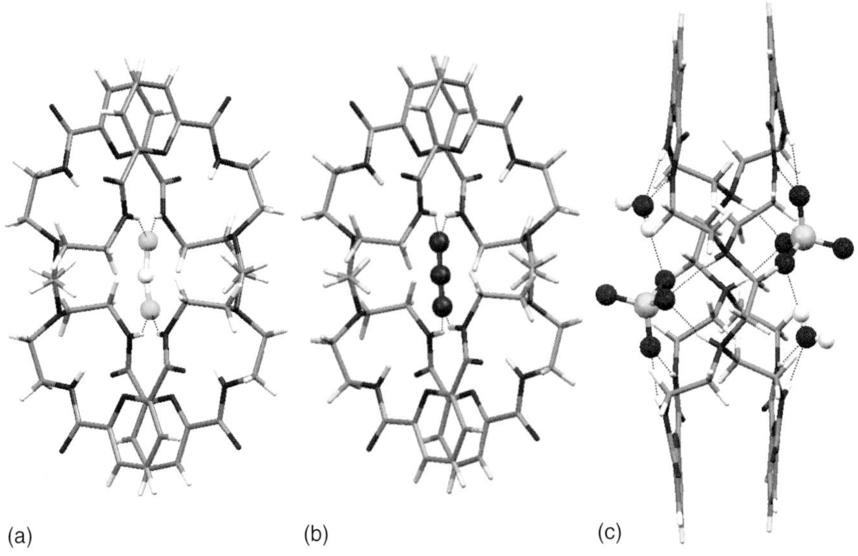

Figure 18 Crystal structures of (a) bifluoride, (b) azide, and (c) sulfate complexes with **34**. The $[n\text{Bu}_4\text{N}]^+$ counterions in (a) and (b) are not shown for clarity.

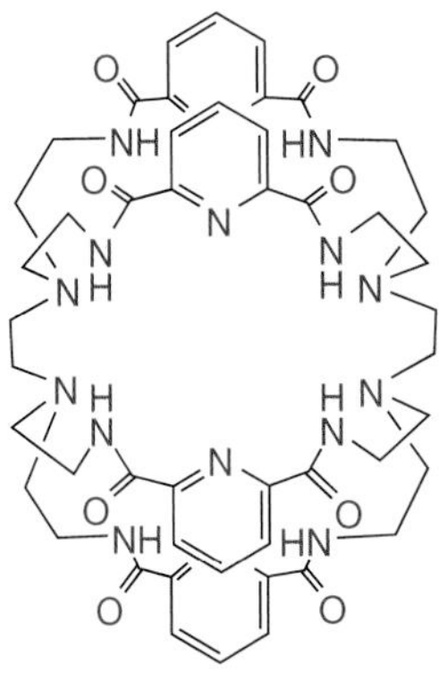

34

4 UREA AND THIOUREA

4.1 Acycles

The very simple functionalized urea, **35**, of Fabbrizzi and coworkers serves to illustrate the binding capabilities of the minichelates with adjacent H-bond donors. The host is fabricated from the reaction of an equimolar ratio of 4-nitrophenylisocyanate and 4-nitroaniline in dry dioxane (Scheme 5).[48] With acetate, a 1 : 1 complex is formed with each of the acetate oxygen atoms bound to a urea NH group (Figure 19a). However, while bicarbonate binds in a similar mode, it forms a dimer, which is made possible because of the OH groups that link the two ions together (Figure 19b). This urea host shows high binding for both fluoride and acetate, and slightly less for $H_2PO_4^-$ (Table 5).

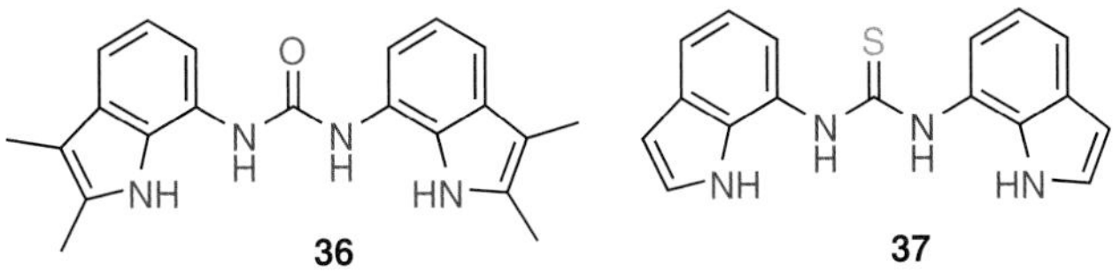

36 **37**

Gale and coworkers reported simple urea, **36**, and thiourea, **37**, ligands with attached indole groups.[49,50] A carbonate complex of **36** was isolated as the $[\text{Et}_4\text{N}]^+$ salt with the anion surrounded by two ligands in a 2 : 1 complex. The N···O H-bond distances ranged from 2.739 to 2.938 Å. Each ligand was multiply coordinated to the carbonate, resulting in an eight-coordinate vice holding the anion (Figure 20a). The thiourea bicarbonate analog with **37** was isolated as the $[\text{Et}_4\text{N}]^+$ salt of a 2 : 2 complex. The H-bond range was slightly smaller than that in **36**

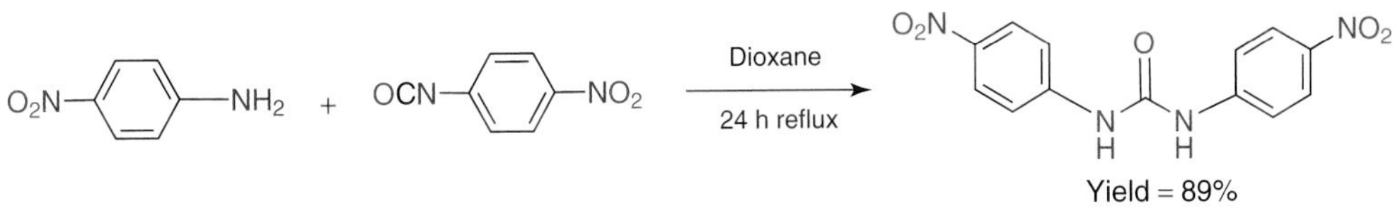

Dioxane, 24 h reflux; Yield = 89%; **35**

Scheme 5 Simple, one-step synthesis of **35**.

Figure 19 Crystal structures of (a) $[35(CH_3COO)]^-$ and (b) $[35(HCO_3)]_2^{2-}$. The $[nBu_4N]^+$ counterions are not shown for clarity.

Table 5 Association constants (K) of **35**, **36**, and **37** with anions.

Host	Anions	K (M^{-1})	Solvent	Ref.
35[a]	F^-	2.40×10^7	CH_3CN	47
	Cl^-	3.55×10^4		
	NO_3^-	4.47×10^3		
	$H_2PO_4^-$	2.34×10^5		
	CH_3COO^-	4.04×10^6		
36[b]	Cl^-	128	DMSO-d_6^c	49
	CH_3COO^-	$>10^4$		
	$C_6H_5CO_2^-$	$>10^4$		
	$H_2PO_4^-$	$>10^4$		
37[b]	Cl^-	74		49
	CH_3COO^-	1620	DMSO-d_6^c	
	$C_6H_5CO_2^-$	477		
	$H_2PO_4^-$	1630		

[a]Determined by UV–vis titrations.
[b]Determined by ^1H NMR titrations.
[c]Mixed with 0.5% H_2O.

(N···O = 2.798 to 2.887 Å), and the anions formed chains of H-bonded bicarbonate dimers passing through a channel between chains of the tetradentate ligands (Figure 20b). Hydrogen atoms were not located due to disorder and therefore do not appear in the figure. Binding constants for **36** and **37** are summarized in Table 5 and show significant binding for oxo anions, especially for **36**.

38 **39** **40** **41**

Colorimetric sensors containing amide, urea, thiourea, and carbamate groups appended to a dye, 4-nitroazobenzene, **38**–**41**, were reported by Martinez-Máñez, Rurack, and coworkers.[51] Each of the hosts exhibited slightly different binding characteristics. Two types of spectral shifts occurred when the hosts bound anions, which were dependent on the actual process taking place. Initially all of the receptors were bright yellow, owing to charge transfer bands around 375–400 nm. When simple anion binding (i.e., via H-bonds) occurred, a small bathochromic shift was observed with a change in color to a light orange. However, when deprotonation of the NH group occurred (in the presence of fluoride ion), a

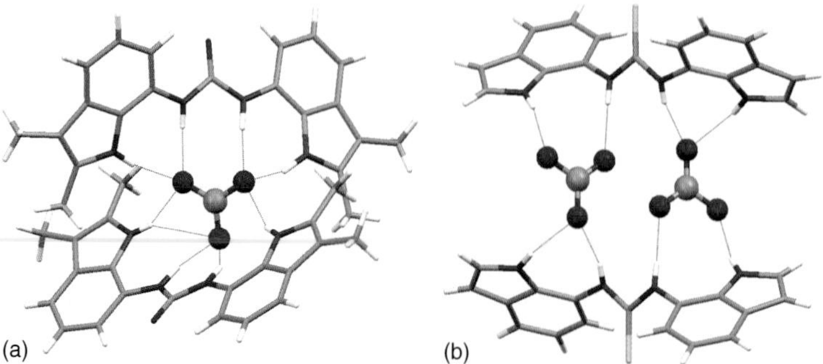

Figure 20 Crystal structures for (a) $[36_2(CO_3)]^{2-}$ and (b) $[37_2{\cdot}2HCO_3]^{2-}$. The $[Et_4N]^+$ cations are not shown for clarity, and the protons were not located for the bicarbonate complex.

Table 6 Association constants (M^{-1}) for **38–41** with anions measured by UV–vis titrations (taken from Ref. 51) in CH_3CN.

Host	Anions	K (M^{-1})
38	CH_3COO^-	6.00×10^4
	BzO^-	4.10×10^4
	$H_2PO_4^-$	7.40×10^3
	Cl^-	1.25×10
39	CH_3COO^-	2.71×10^4
	BzO^-	1.32×10^4
	$H_2PO_4^-$	3.90×10^3
	Cl^-	370
40	CH_3COO^-	194
	BzO^-	149
	$H_2PO_4^-$	258
	Cl^-	—
41	CH_3COO^-	142
	BzO^-	108
	$H_2PO_4^-$	134
	Cl^-	—

large absorption band shift to about 580 nm occurred, resulting in a deep blue color. Only the urea and thiourea hosts bound oxo anions such as acetate, benzoate, and $H_2PO_4^-$ significantly as determined by UV–vis titrations. The thiourea analog was the best in this regard (Table 6). In the deprotonated form (i.e., in the presence of fluoride ion), the sensor was found to react with atmospheric CO_2, forming the carbonate (or bicarbonate) complex in the presence of water.

During the last few years, a number of research groups have taken advantage of the fluorescent properties of anthracene attached to urea/thiourea groups. For example, Gunnlaugsson and coworkers synthesized several fluorescent photoinduced electron transfer (PET) receptors with monofunctional and difunctional urea/thiourea units to explore how the functional groups influence the anion binding.[52,53] Using titration techniques and fluorescence spectroscopy, they observed that **42b**, which contains a single thiourea and a phenyl group with an attached trifluoromethane, binds F^-, CH_3COO^-, and $H_2PO_4^-$ in DMSO with log $K = 3.35$, 2.55, and 2.05, respectively. These affinities are higher than when there is no electron-withdrawing group (**42a**).[52] The same binding trends are also observed in **43a** and **43b** (Table 7), with generally

Table 7 Association constants (K) of **42b** and **43a–c** for anions determined by fluorescence emission titrations in DMSO.

Host	Anions	K (M^{-1})	Ref.
42b	F^-	2.24×10^3	52
	CH_3COO^-	178	
	$H_2PO_4^-$	112	
43a	F^-	1.07×10^3	53
	Pyrophosphate	1.18×10^3	
	Malonate	1.41×10^3	
	Glutarate	1.86×10^3	
43b	F^-	1.35×10^4	53
	Pyrophosphate	2.51×10^3	
	Malonate	219	
	Glutarate	5.50×10^3	
43c	F^-	2.00×10^3	53
	Pyrophosphate	525	
	Malonate	457	
	Glutarate	5.89×10^3	

higher binding of the thiourea over the urea, and the trifluoromethane-substituted host over the nonsubstituted analog, with the exception of malonate.

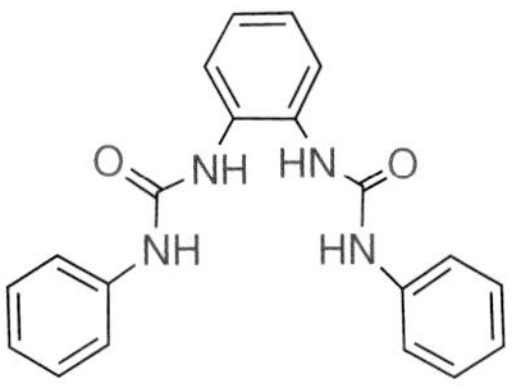

44

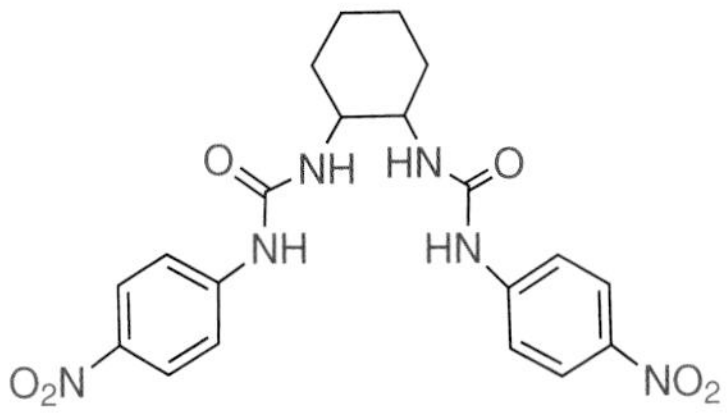

45

A host with a phenyl spacer between two ureas, **44**, binds benzoate, with each urea chelating just a single carboxylate oxygen atom.[54] The four H-bonds in the complex range from 2.740 to 2.939 Å (Figure 21a). However, in an analog that involves a saturated, more flexible cyclohexane spacer, **45**, two $H_2PO_4^-$ guests reside in the binding area. As seen in Figure 21(b), each anion is bound with one urea subunit through two NH bonds. In addition, the two anions are involved in intramolecular O–H···O H-bonding interactions that extend throughout the crystal lattice, providing additional stability to the 1 : 2 complex. Chemically, the binding occurs in sequential steps, and log K_2 (3.46) is higher than log K_1 (2.96) in DMSO (Table 8), making this

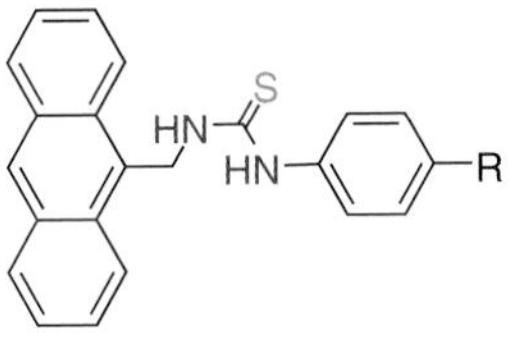

42a (R = H)
42b (R = CF_3)

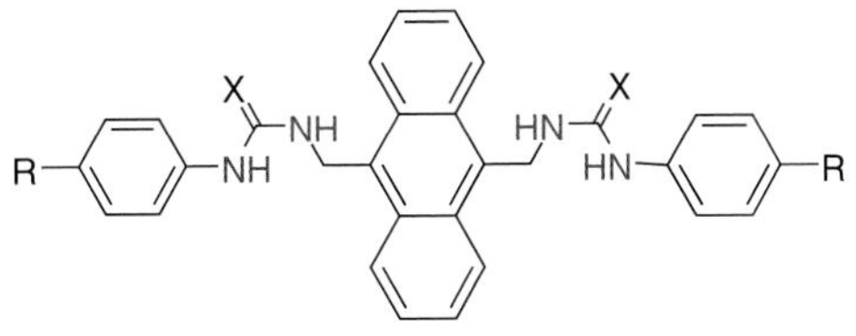

43a (R = H, X = S)
43b (R = CF_3, X = S)
43c (R = CF_3, X = O)

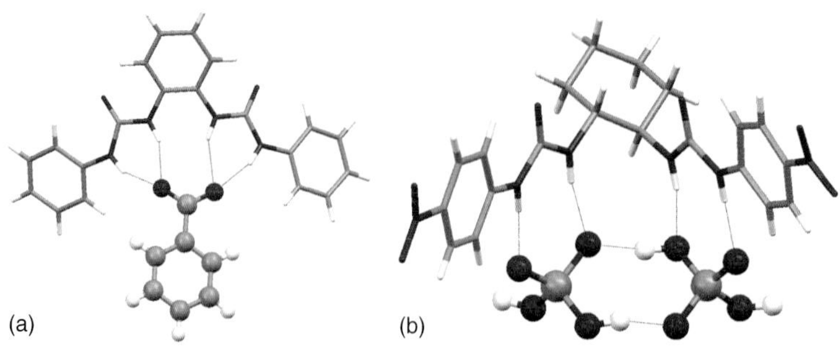

Figure 21 Overhead views of the crystal structures of (a) $[\mathbf{44}(C_6H_5COO)]^-$ and (b) $[\mathbf{45}(H_2PO_4)_2]^{2-}$ showing chelation of the anionic guests. The $[n\text{Bu}_4\text{N}]^+$ counterions are not shown for clarity.

Table 8 Association constants (K) of **44, 45, 47**, and **48** with anions.

44[a]	Cl^-	43.0	DMSO-d_6[c]	54
	Br^-	<10.0		
	HSO_4^-	10.0		
	$H_2PO_4^-$	732		
	CH_3COO^-	3.21×10^3		
45[b]	$H_2PO_4^-$	912 (2.88×10^3)[d]	DMSO	55
	$HP_2O_7^{3-}$	4.27×10^4		
	CH_3COO^-	724		
47[a]	Cl^-	2.10×10^3	$CDCl_3$	57
	Br^-	400		
H[**47**]$^+$[b]	Cl^-	4.27×10^4	CH_3CN	57
	Br^-	6.12×10^4		
48[a]	F^-	1.15×10^4	DMSO-d_6	58
	Cl^-	2.63×10^3		
	Br^-	18.0		
	SO_4^{2-}	5.37×10^4		
	$H_2PO_4^-$	$>10^5$		
	CH_3COO^-	2.82×10^4		

[a]Determined by ^{1}H NMR titrations.
[b]Determined by UV–vis titrations.
[c]Mixed with 0.5% H_2O.
[d]K_2.

an example of a cooperative effect, where the second event is facilitated by the first.[55]

46

Gale and coworkers synthesized an expanded ditopic "compartmental" receptor, using a biphenyl linker, **46**.[56] The structure shows a single acetate bound in each cleft by three H-bonds ranging from N···O = 2.769 to 2.918 Å. Evidently the orientation of the urea groups is not quite right for each carboxylate to be held by two H-bonds. By mixing the ditopic compartmental ligand with a ditopic guest, a linear polymer network is obtained as shown for the terephthalate complex (Figure 22b).

47

Johnson and coworkers isolated and characterized **47**, a urea corollary of the amide-containing diacetylene framework described earlier (**13**). When protonated, **47** is an even more effective anion-binding agent.[57] The ligand binds chloride ions in a pentadentate manner, resulting in a five-coordinate chloride complex (Figure 23). The H-bond distance to the pyridinium NH group (N···Cl = 3.03 Å) is quite short, while NH···Cl distances range up to 3.64 Å. In fact, the protonated ligand, **47**, shows significant binding for both chloride and bromide (Table 8).

48

In order to enhance the acidity of NH protons, Ghosh and coworkers introduced electron-withdrawing groups in

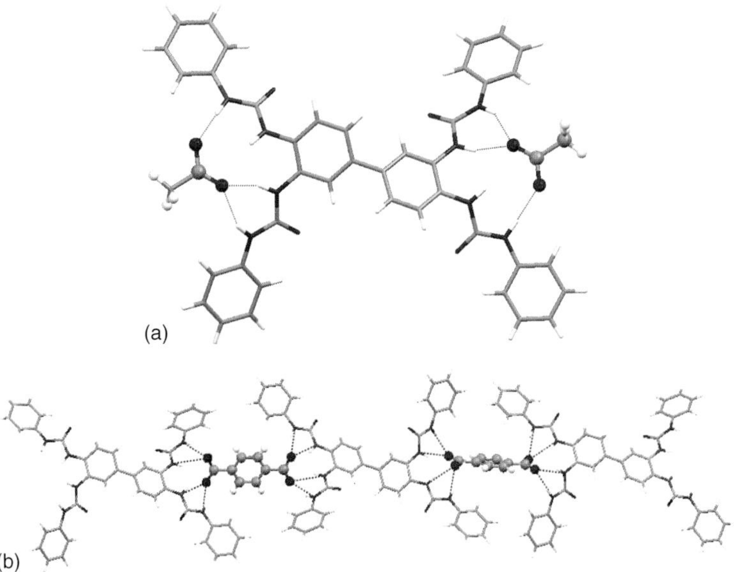

Figure 22 Overhead view of (a) the crystal structure of the acetate complex of **46** and (b) the polymeric chain obtained for **46** with the ditopic terephthalate ion. The $[n\text{Bu}_4\text{N}]^+$ counterions are not shown for clarity.

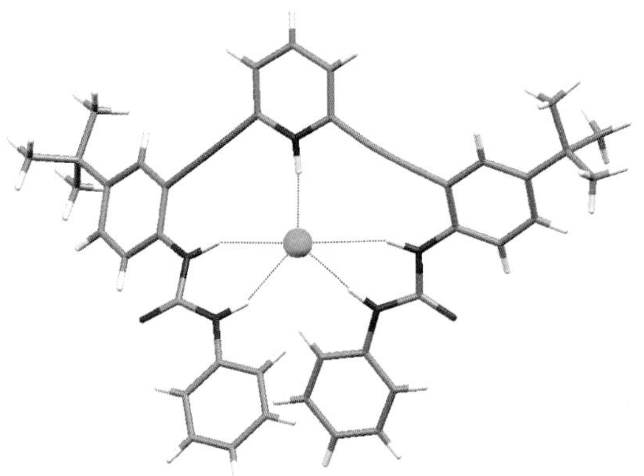

Figure 23 Overhead view of the crystal structure of the chloride complex of the pyridinium ligand [H**47**(Cl)].

the aromatic units of a tren-based host to obtain the pentafluorophenyl-substituted tripodal urea receptor **48**.[58] The host showed a high affinity for fluoride over other halides, with an association constant of log $K = 4.06$ in DMSO-d_6. Interactions of fluoride with the ligand were also observed in the ^{19}F NMR spectrum, and were indicated by a downfield shift of 5 ppm from the free fluoride ion in DMSO-d_6. In the solid state, **48** surrounds the fluoride ion with six H-bonds, N···F ranging from 2.700 to 2.884 Å (Figure 24a) in a nearly perfect trigonal prismatic geometry. The tripodal ligand binds oxo anions quite strongly (Table 8). Crystals of the sulfate complex were also isolated. In this structure, two symmetry-related molecules form a capsule-like cavity that is occupied by a sulfate anion, held by 11 H-bonds to the urea NH groups ranging from N···O = 2.777 to 3.062 Å (Figure 24b). This tren-based host shows a proclivity for forming interlocked capsules, and also complexes a phosphate dimer.[59]

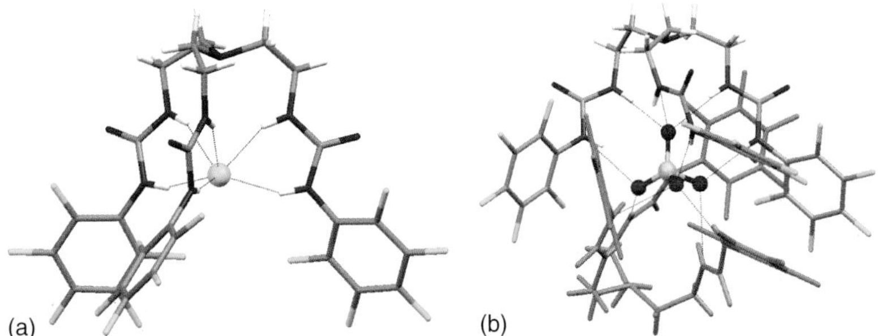

Figure 24 Crystal structures for complexes of **48** with (a) fluoride and (b) sulfate. The second host in (b) is shown in orange to highlight the interlocking encapsulation. The $[n\text{Bu}_4\text{N}]^+$ counterions are not shown for clarity.

49

Recently, Gale and coworkers isolated a carbonate complex from a mixture of the thiourea tripodal host **49**, with $[Et_4N][HCO_3]$.[60] In the process, the bicarbonate becomes deprotonated and is held between the two symmetry-related molecules by 12 N···O H-bonds in the range of 2.824–3.070 Å (Figure 25). The simple tripod is also capable of transporting bicarbonate across lipid membranes, and is superior in that regard to its urea analog.

50

Anion binding with the assistance of metal ions has become increasingly popular in the last decade. Barboiu *et al.* reported a dual host heteroditopic (ureido) crown ether

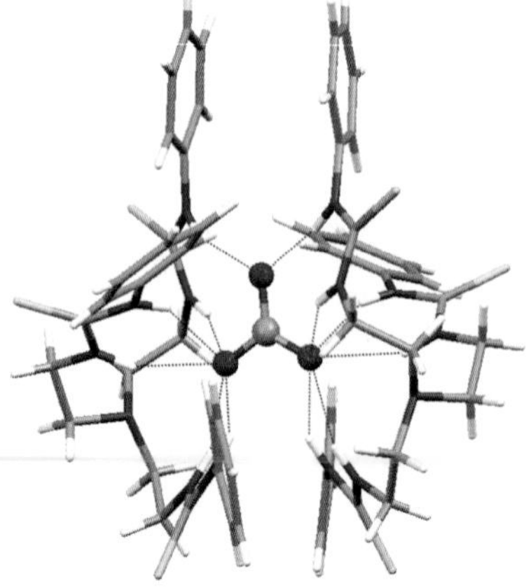

Figure 25 View of the crystal structure of the carbonate complex of **49**, showing how the dimer sandwiches the carbonate guest. The $[Et_4N]^+$ counterions are not shown for clarity.

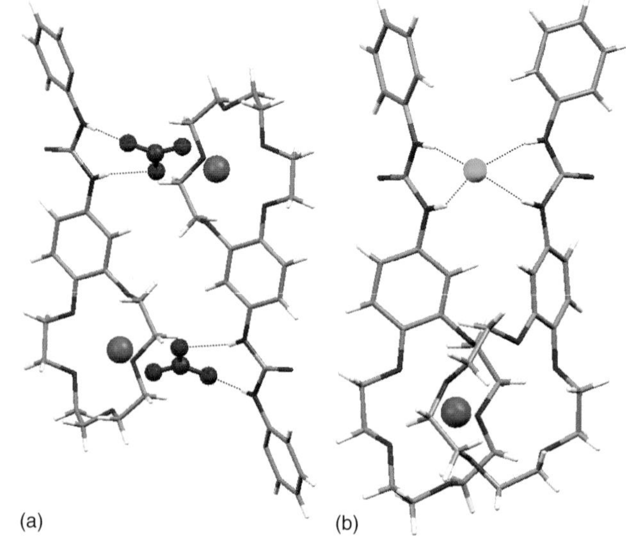

Figure 26 Views of the crystal structures of the dual host crown ether complexes of **50** with (a) $NaNO_3$ and (b) NaCl, illustrating the antiparallel and parallel coordination modes observed for the two guests, respectively.

receptor **50**.[61] The coordination modes differ depending on the anion, as illustrated by two different complexes, one with NaCl and the other with $NaNO_3$. In the nitrate structure, two hosts work in tandem to bind the salt by lining in an antiparallel manner. The crown portion, containing the sodium ion, lies adjacent to a neighboring urea podand chain that binds the nitrate by two rather long H-bonds (N···O = 3.063 and 3.073 Å) (Figure 26a). Two of the nitrate oxygen atoms are associated via an electrostatic interaction with the sodium ion at 2.666 and 2.516 Å. In the NaCl structure, the dimer pair again works in tandem, but lies in a parallel orientation, with the sodium ion serving to hold the two crown ethers in a sandwich structure. The appended urea chains then have four donor hydrogen atoms available to hold the chloride ion, but with very long urea N···O distances ranging from 3.333 to 3.430 Å (Figure 26b).

51

As seen for the NaCl structure above (Figure 26b), the sodium ion can act as a template holding the two anion-binding chains in proximity. Transition metal complexes potentially provide a more inert environment for

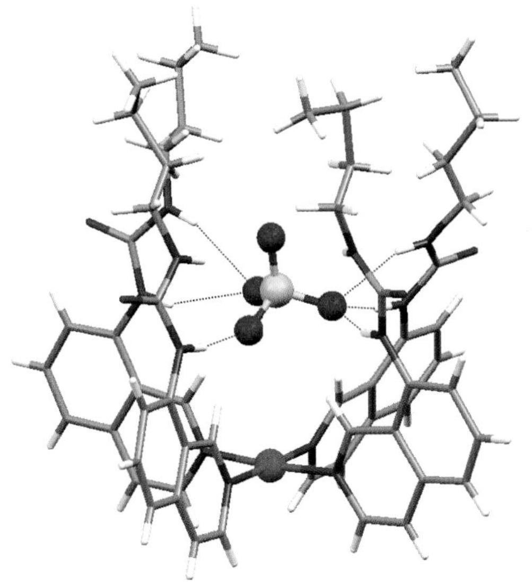

Figure 27 Side view of the crystal structure of the complex formed with sulfate and the platinum-based scaffold **52**.

accomplishing a similar goal by providing the base for scaffolds that include multiple binding sites for anions. For example, Loeb and coworkers formed a square planar platinum(II) complex **52** by binding it with four equivalents of the monodentate isoquinoline ligand **51**, functionalized with a pendant urea group. The anticipation was that the four scaffolds would point in the same direction, and would be oriented in such a way as to bind anionic guests within the four walls.[62] Indeed, all four walls were oriented in the same direction, providing maximum H-bonding interactions with the sulfate anion (Figure 27). The sulfate forms a total of six H-bonds with urea NH groups that range from 2.890 to 3.061 Å. The other three interactions with the upper oxygen atom and urea NH groups are weak to very weak with N···O distances of 3.136, 3.258, and 3.301 Å. The Pt-O distance is 3.719 Å.

52

The potential of using transition metals as anchors for a variety of scaffolds is seemingly unlimited as seen for an modified version of the platinum complex **52**. Loeb, Gale, and coworkers reported a platinum-based scaffold, **53**, with a 3 : 2 host:anion stoichiometry.[63] The groups modified the complex slightly so that they bound only two of the quinoline ligands, in a trans configuration, while using simple pyridine groups for the other two sites. In this case, rather than the platinum acting as a template for binding the anion, the two sulfate guests became the template. By binding with urea chains of neighboring complexes, they brought three platinum complexes together to form a 3 : 2 Pt^{2+}:SO_4^{2-} medley. The trimer preferentially binds sulfate even though another counterion, BF_4^-, is present in the unit cell. The BF_4^- ions balance the remaining charge of the dicationic complex. Each sulfate ion is surrounded by a sevenfold array of H-bonds with N···O distances ranging from 2.771 to 3.108 Å. It is pertinent to mention that sulfate is also bound via seven H-bonds in the sulfate binding protein,[64] but seven coordination for sulfate ion is rather rare. However, the authors note that there could be as many as 14 interactions around each of the sulfate ions in the 3 : 2 Pt-scaffold species if one includes all of the aromatic interactions from the α-carbon atoms in the pyridine and isoquinoline moieties (Figure 28).[63]

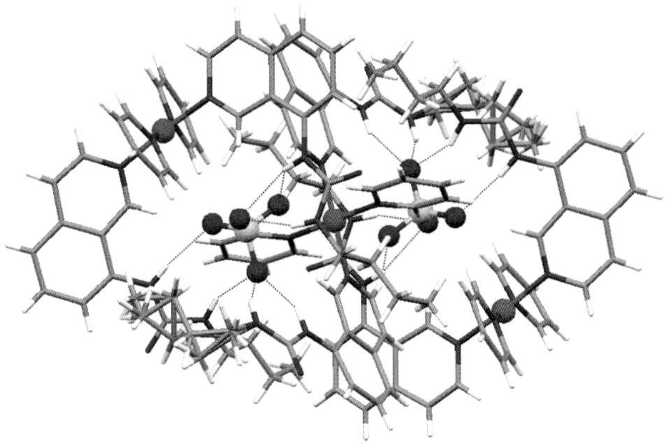

Figure 28 Structure of the 3 : 2 host:guest complex $[\mathbf{53}_3(SO_4)_2]^{2+}$. The BF_4^- counterions are not shown for clarity.

53

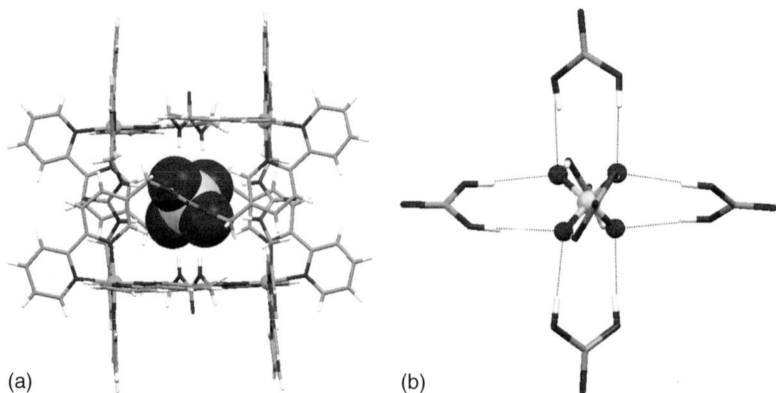

Figure 29 (a) Self-assembly of **54** and $NiSO_4$: crystal structure of the encapsulated sulfate as a space-filling model in a cage formed by four molecules (sticks) and four nickel ions (green balls) and (b) coordination sphere of the sulfate ion of 12 H-bonds formed with six urea groups (eight clearly shown with four obscured, two in front and two behind).

Metal organic frameworks (MOFs), extended organic polymeric networks held together by metal ions, have exploded into mainline chemical research in recent years. In most of the cases, these systems comprise extended polymeric networks as described below. There is significant potential for these systems to bind targeted guests, and to serve as catalysts for important reactions, to say the least. An elegant example of the use of a discrete (nonpolymeric) MOF as a sequestering agent is provided by the square arrangement of nickel(II) ions at the vertices of a urea-based cage. The precursor acyclic monourea host contains two bipyridine units at each end for binding the metal ion, **54**.[65] When $NiSO_4$ is used as the template, the square assembles. Each of the urea chelates binds two of the sulfate oxygen atoms (Figure 29), resulting in a total of 12 H-bonds for the sulfate. The immediate coordination sphere is also provided (Figure 29b), and can be described as a cube, with each urea "tong" in the middle of the faces, with the tetrahedral sulfate oxygen atoms pointed toward opposing corners of the cube. As noted above, the tetrameric nickel complex is a discrete unit, unlike the silver MOF described next. N···O distances range from 2.904 to 2.942 Å, showing only a small variation in H-bond distance for the 12 interactions.

54

In another example of, this time a polymeric, MOF system, Custelcean, Moyer, and Hay reported an extended silver-based MOF that binds sulfate. They used the labile silver(I) d^{10} salt, Ag_2SO_4, as a template in the assembly of two tripodal tren-based urea hosts functionalized with a *m*-cyano-substituted phenyl group for binding with the silver ion, **55**.[66] Interestingly, the use of other salts did not result in MOFs; therefore, clearly the sulfate is playing a major role in the template process. Once again, the sulfate exhibits a full 12-coordinate H-bond sphere, which the Hay group has proposed to be maximally suited for sulfate[67] (Figure 30). H-bonding interactions vary over a slightly wide range with N···O distances ranging from 2.852 to 3.174 Å, compared to a smaller range as seen for the nickel complex in **54**.

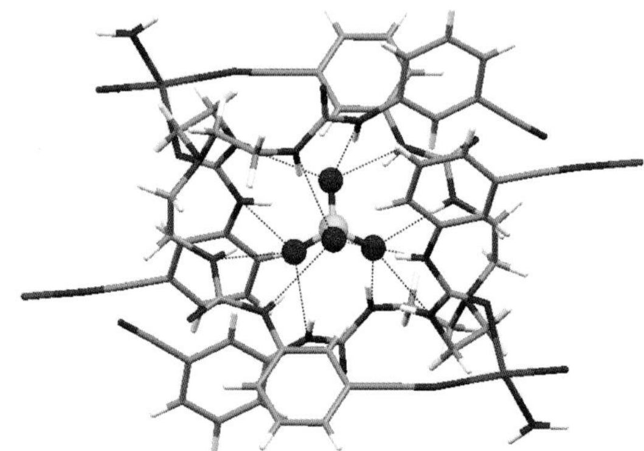

Figure 30 Encapsulation of sulfate inside the crystalline MOF formed by **55** and silver ions.

55

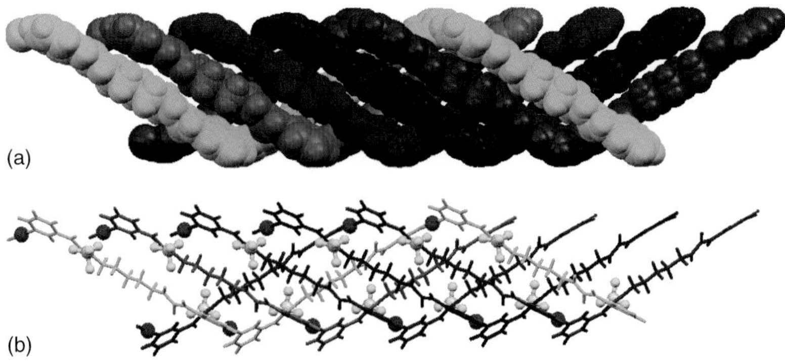

Figure 31 Anion-assisted helical assembly of **56** and $AgBF_4$ in (a) space filling and (b) ball and stick views. In (b) Ag^+ and BF_4^-, atoms are shown as balls and urea receptors are shown as sticks.

The crucial role anions can play in assembling building blocks for polymeric networks is not restricted to **55**. Steed and coworkers synthesized a bifunctional bis(pyridylurea) ligand **56** with a long pentamethylene spacer that forms a coordination polymer using pyridyl groups and silver ions. Interestingly, polymer formation can be influenced by H-bonding interactions of the urea functionality with counter anions (NO_3^- or BF_4^-).[68] Each of the urea units in **56** can act as an H-bond donor for anions through the NH groups. The pyridine groups serve to bind the silver ions that form the backbone of the weave. Figure 31 shows the polymeric species resulting from the reaction of **56** with $AgBF_4$. The bis(pyridylurea) ligands are linked with silver ions through pyridine nitrogens to make a helical assembly, and the BF_4^- anions interact orthogonally with the urea units to assist in the formation of a quintuple helical molecular braid.

56

57

An alternative way to construct an anion-controlled, self-assembled polymer is with **57**, which has a shortened ethylene chain instead of the pentamethylene spacer used in **56**.[69] In this case, the molecules are assembled through trigonal Ag(I) nodes to form a hexanuclear ring with a diameter of about 31.5 Å. As shown in Figure 32, three urea groups are directed toward the cavity and involved in coordinating two nitrates via a total of 12 H-bonds ($N\cdots O = 2.960$ to 2.944 Å) to form a Borromean weave. Each nitrate ion is surrounded by three urea groups from three different molecules with six H-bonds, whereas the other nitrate (interanion distance of 3.21 Å) is linked with

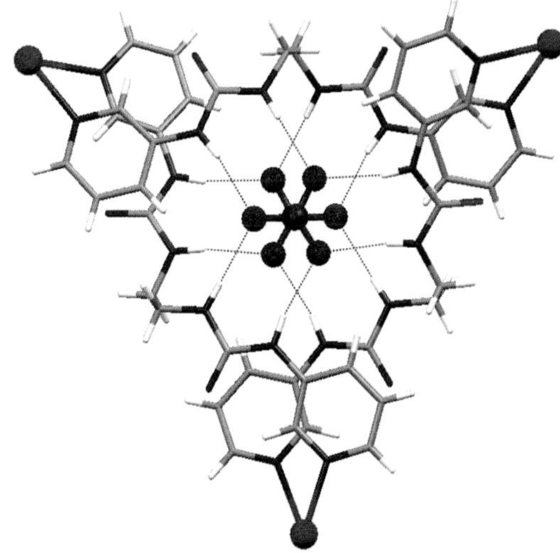

Figure 32 Borromean weave consisting of portions of three of the ligand(s) **57** held together by silver(I) ions, and surrounding two nitrate ions.

the remaining three urea groups to form a 2D H-bonding network. According to the authors, this is the first example of anion binding as an integral part of other coordination interactions used to achieve Borromean weaves.

4.2 Monocycles

Possibly because of the more interesting host:guest possibilities of ureas without the need for cyclization, there are fewer macrocyclic examples of these H-bond donors. An early report of a macrocyclic urea is **58**, synthesized by the Reinhoudt group. This host contains a flexible geometry with four urea groups[70] and was found to bind the tetrahedral $H_2PO_4^-$ anion with an association constant log $K = 3.0$ in DMSO. Another example of a macrocyclic urea is **59**, which incorporates both urea and amide linkages.[71] Since there are two isophthalamide linkages, one might consider this host under the amide section. However, due

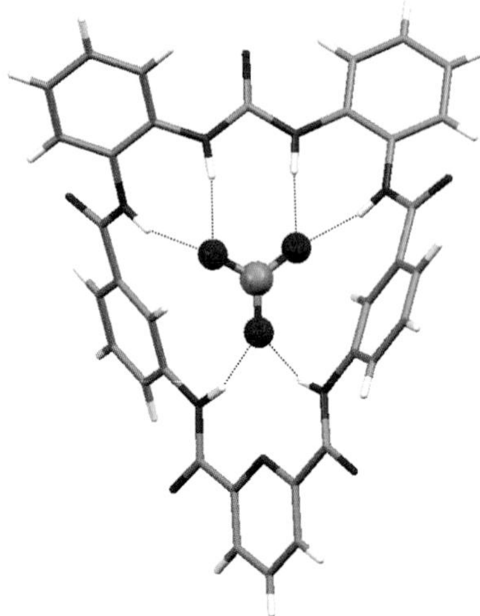

Figure 33 Crystal structure of the encapsulated carbonate in **59**$[CO_3]^{2-}$ showing six H-bonds. The $[n\mathrm{Bu_4N}]^+$ counterions are not shown for clarity.

to the rarity of urea macrocycles, it is included here. In the example shown, each carbonate oxygen atom is held by two H-bonds (Figure 33).

58

59

60

Böhmer and coworkers synthesized a cyclic urea **60** by incorporating ether linkages in positions ortho to the urea groups.[72] As shown in Figure 34, this molecule encapsulates one chloride in its cavity with six NH···O H-bonds. Because of the directional characteristics of urea groups, where all six NH groups are pointed toward the center, the host is perfectly aligned for this spherical anion. Such orientation of NH groups could be particularly useful for a nitrate as predicted by Hay from the theoretical calculations.[67]

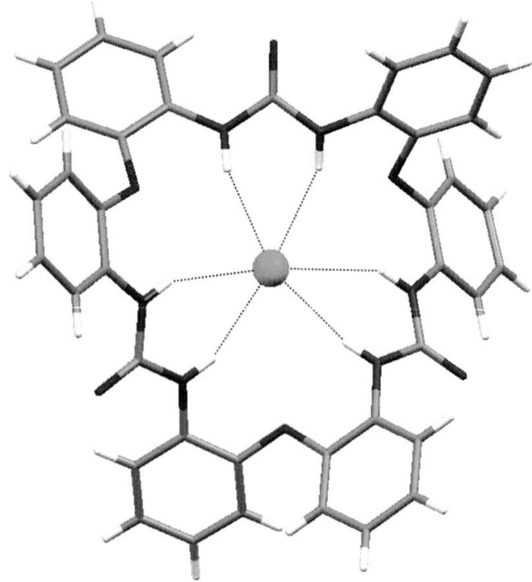

Figure 34 Crystal structure of the encapsulated chloride in **60**$[Cl]^-$ showing six H-bonds. The $[n\mathrm{Bu_4N}]^+$ counterion is not shown for clarity.

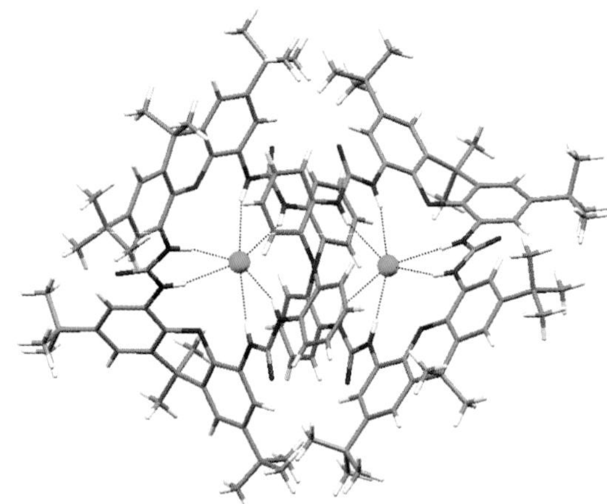

Figure 35 The figure-eight conformation observed in the crystal structure the chloride complex of **61**. Each "pocket" of the figure eight contains a chloride template held by six H-bonds.

An interesting example of a hexa-urea xanthene-derived macrocycle, **61**, was isolated as a 2 × 2 side product from a condensation that was designed to provide a tris-urea host.[73] This figure-eight complex was obtained using two chloride ions as templates (Figure 35). The N···Cl distances range from 3.18 to 3.34 Å, and the Cl···Cl separation is 6.029 Å. In this gigantic figure-eight-like host, the presence of multiple binding sites allows for a higher coordination number, in this case 12, for the mega macrocycle.

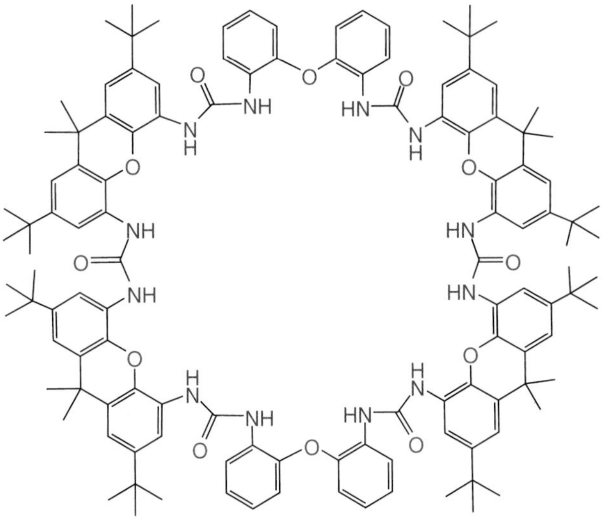

61

5 SUMMARY AND CONCLUSIONS

Considerable research has focused on both enhancing the binding of anions by taking advantage of the various properties of diverse H-bond donor groups, and imparting selectivity and sensor behavior to the hosts. As examples, the presence of adjacent donor sites in urea/thiourea hosts allows for the mini-chelate effect to take place, providing an added binding clout to the unit. Furthermore, the replacement of oxygen by sulfur in carbonyl groups increases the H-bonding ability of NH groups, thereby in many instances enhancing the binding affinity. Additional binding enhancement can be achieved by incorporating adjacent C=O/S groups and/or appropriately placed electron-withdrawing groups. Thus, while this chapter covers only some key examples of each class of hosts, the above discussion clearly suggests that the ability to form superior H-bonding ligands also depends on other secondary groups linked to these functional units as well. Ultimately, the placement of "working" groups within the hosts, that is, colorimetric or redox sensors, is leading to new and exciting areas of exploration for the field of anion coordination chemistry that began slowly back in the late 1960s.

ACKNOWLEDGEMENTS

K.B.-J. gratefully acknowledges support from the National Science Foundation CHE-0809736.

REFERENCES

1. C. H. Park and H. E. Simmons, *J. Am. Chem. Soc.*, 1968, **90**, 2431–2433.
2. R. A. Bell, G. G. Christoph, F. R. Fronczek, and R. E. Marsh, *Science*, 1975, **190**, 151–152.
3. A. Bianchi, K. Bowman-James, and E. García-España, eds., *Supramolecular Chemistry of Anions*, Wiley-VCH, New York 1997.
4. I. Stibor, ed., *Anion Sensing. Topics in Current Chemistry*, Springer, Berlin 2005.
5. J. L. Sessler, P. A. Gale, and W.-S. Cho, eds., *Anion Receptor Chemistry*, RSC Publishing, Cambridge, UK 2006.
6. P. A. Gale, guest ed., *Coord. Chem. Rev.*, 2003, **240**(1 and 2).
7. P. A. Gale and T. Gunnlaugsson, guest eds., *Chem. Soc. Rev.*, 2010, **39**(10).
8. R. A. Pascal, J. Spergel, and D. Van Eggen, Jr., *Tetrahedron Lett.*, 1986, **27**, 4099–4102.
9. S. Valiyaveettil, J. F. J. Engbersen, W. Verboom, and D. N. Reinhoudt, *Angew. Chem. Int. Ed. Engl.*, 1993, **32**, 900–901.
10. P. J. Smith, M. V. Reddington, and C. S. Wilcox, *Tetrahedron Lett.*, 1992, **33**, 6085–6088.
11. E. Fan, S. A. Van Arman, S. Kincaid, and A. D. Hamilton, *J. Am. Chem. Soc.*, 1993, **115**, 369–370.
12. Y. Inoue, T. Kanbara, and T. Yamamoto, *Tetrahedron Lett.*, 2003, **44**, 5167–5169.
13. M. A. Hossain, J. M. Llinares, D. Powell, and K. Bowman-James, *Inorg. Chem.*, 2003, **42**, 5043–5045.
14. F. G. Bordwell, D. J. Algrim, and J. A. Harrelson, Jr., *J. Am. Chem.Soc.*, 1988, **110**, 5903–5904.
15. F. G. Bordwell, *Acc. Chem. Res.*, 1988, **21**, 456–463.
16. A.-F. Li, J.-H. Wang, F. Wang, and Y.-B. Jiang, *Chem. Soc. Rev.*, 2010, **39**, 3729–3745.
17. Z. Zhang and P. R. Schreiner, *Chem. Soc. Rev.*, 2009, **38**, 1187–1198.
18. R. Custelcean, *Chem. Commun.*, 2008, 295–307.
19. K. Kavallieratos, S. R. Gala, D. J. Austin, and R. H. Crabtree, *J. Am. Chem. Soc.*, 1997, **119**, 2325–2326.
20. K. Kavallieratos, C. M. Bertao, and R. H. Crabtree, *J. Org. Chem.*, 1999, **64**, 1675–1683.
21. P. V. Santacroce, J. T. Davis, M. E. Light, *et al.*, *J. Am. Chem. Soc.*, 2007, **129**, 1886–1887.
22. M. J. Chmielewski, T. Zieliski, and J. Jurczak, *Chem. Eur. J.*, 2005, **11**, 6080–6094.
23. C. A. Johnson, O. B. Berryman, A. C. Sather, *et al.*, *Cryst. Growth Des.*, 2009, **9**, 4247–4249.
24. M. Arunachalam and P. Ghosh, *Inorg. Chem.*, 2010, **49**, 943–951.
25. T. Zielinski and J. Jurczak, *Tetrahedron*, 2005, **61**, 4081–4089.
26. A. Danby, L. Seib, N. W. Alcock, and K. Bowman-James, *Chem. Commun.*, 2000, 973–974.

27. P. S. Lakshminarayanan, E. Suresh, and P. Ghosh, *Inorg. Chem.*, 2006, **45**, 4372–4380.

28. P. D. Beer, Z. Chen, A. J. Goulden, *et al.*, *J. Chem. Soc. Chem. Commun.*, 1993, 1834–1836.

29. B. A. Szumna and J. Jurczak, *Eur. J. Org. Chem.*, 2001, 4031–4039.

30. S. O. Kang, V. W. Day, and K. Bowman-James, *Org. Lett.*, 2009, **11**, 3654–3657.

31. M. A. Hossain, S. O. Kang, D. Powell, and K. Bowman-James, *Inorg. Chem.*, 2003, **42**, 1397–1399.

32. M. A. Hossain, J. M. Llinares, D. Powell, and K. Bowman-James, *Inorg. Chem.*, 2001, **40**, 2936–2937.

33. S. Ghosh, B. Roehm, R. A. Begum, *et al.*, *Inorg. Chem.*, 2007, **46**, 9519–9521.

34. M. J. Chmielewski, T. Zieliski, and J. Jurczak, *Pure Appl. Chem.*, 2007, **79**, 1087–1096.

35. V. W. Day, M. A. Hossain, S. O. Kang, *et al.*, *J. Am. Chem. Soc.*, 2007, **129**, 8692–8693.

36. J.-J. Lee, K. J. Stanger, B. C. Noll, *et al.*, *J. Am. Chem. Soc.*, 2005, **127**, 4184–4185.

37. J. M. Mahoney, A. M. Beatty, and B. D. Smith, *J. Am. Chem. Soc.*, 2001, **123**, 5847–5848.

38. S. Kubik, R. Goddard, R. Kirchner, *et al.*, *Angew. Chem. Int. Ed. Engl.*, 2001, **40**, 2648–2651.

39. S. O. Kang, J. M. Llinares, D. Powell, *et al.*, *J. Am. Chem. Soc.*, 2003, **125**, 10152–10153.

40. S. O. Kang, D. VanderVelde, D. Powell, and K. Bowman-James, *J. Am. Chem. Soc.*, 2004, **126**, 12272–12273.

41. S. O. Kang, V. W. Day, and K. Bowman-James, *J. Org. Chem.*, 2010, **75**, 277–283.

42. S. O. Kang, M. A. Hossain, D. Powell, and K. Bowman-James, *Chem. Commun.*, 2005, 328–330.

43. S. O. Kang, D. Powell, and K. Bowman-James, *J. Am. Chem. Soc.*, 2005, **127**, 13478–13479.

44. A. P. Bisson, V. M. Lynch, M. K. C. Monahan, and E. V. Anslyn, *Angew. Chem. Int. Ed.*, 1997, **36**, 2340–2342.

45. S. O. Kang, D. Powell, V. W. Day, and K. Bowman-James, *Angew. Chem. Int. Ed.*, 2006, **45**, 1921–1925.

46. S. O. Kang, V. W. Day, and K. Bowman-James, *Org. Lett.*, 2008, **10**, 2677–2680.

47. S. O. Kang, V. W. Day, and K. Bowman-James, *Inorg. Chem.*, 2010, **49**, 8629–8636.

48. M. Boiocchi, L. Del Boca, D. Esteban-Gómez, *et al.*, *J. Am. Chem. Soc.*, 2004, **126**, 16507–16514.

49. C. Caltagirone, P. A. Gale, J. R. Hiscock, *et al.*, *Chem. Commun.*, 2008, 3007–3009.

50. C. Caltagirone, J. R. Hiscock, M. B. Hursthouse, *et al.*, *Chem.Eur. J.*, 2008, **14**, 10236–10243.

51. J. V. Ros-Lis, R. Martinez-Máñez, F. Sancenon, *et al.*, *Eur. J. Org. Chem.*, 2007, **15**, 2449–2458.

52. T. Gunnlaugsson, A. P. Davis, and M. Glynn, *Chem. Commun.*, 2001, 2556–2557.

53. T. Gunnlaugsson, A. P. Davis, J. E. O'Brien, and M. Glynn, *Org. Biomol. Chem.*, 2005, **3**, 48–56.

54. S. J. Brooks, P. A. Gale, and M. E. Light, *Chem. Commun.*, 2005, 4696–4698.

55. V. Amendola, M. Boiocchi, D. Esteban-Gómez, *et al.*, *Org. Biomol. Chem.*, 2005, **3**, 2632–2639.

56. S. J. Brooks, P. A. Gale, and M. E. Light, *Cryst. Eng. Comm.*, 2005, **7**, 586–591.

57. C. N. Carroll, O. B. Berryman, D. A. Johnson II, *et al.*, *Chem. Commun.*, 2009, 2520–2522.

58. I. Ravikumar, P. S. Lakshminarayanan, M. Arunachalam, *et al.*, *Dalton Trans.*, 2009, 4160–4168.

59. P. S. Lakshminarayanan, I. Ravikumar, E. Suresh, and P. Ghosh, *Chem. Commun.*, 2007, 5214–5216.

60. N. Busschaert, P. A. Gale, C. J. E. Haynes, *et al.*, *Chem. Commun.*, 2010, **46**, 6252–6254.

61. M. Barboiu, G. Vaughan, and A. van der Lee, *Org. Lett.*, 2003, **5**, 3073–3076.

62. C. R. Bondy, P. A. Gale, and S. J. Loeb, *J. Am. Chem. Soc.*, 2004, **126**, 5030–5031.

63. M. J. Fisher, P. A. Gale, M. E. Light, and S. J. Loeb, *Chem. Commun.*, 2008, 5695–5697.

64. J. W. Pflugrath and F. A. Quiocho, *Nature*, 1985, **314**, 257–260.

65. R. Custelcean, J. Bosano, P. V. Bonnesen, *et al.*, *Angew. Chem. Int. Ed.*, 2009, **48**, 4025–4029.

66. R. Custelcean, B. A. Moyer, and B. P. Hay, *Chem. Commun.*, 2005, 5971–5973.

67. B. P. Hay, T. K. Firman, and B. A. Moyer, *J. Am. Chem. Soc.*, 2005, **127**, 1810–1819.

68. P. Byrne, G. O. Lloyd, K. M. Anderson, *et al.*, *Chem. Commun.*, 2008, 3720–3722.

69. P. Byrne, G. O. Lloyd, N. Clarke, and J. W. Steed, *Angew. Chem. Int. Ed.*, 2008, **47**, 5761–5764.

70. B. H. M. Snellink-Ruël, M. M. G. Antonisse, J. F. J. Engbersen, *et al.*, *Eur. J. Org. Chem.*, 2000, 165–170.

71. S. J. Brooks, P. A. Gale, and M. E. Light, *Chem. Commun.*, 2006, 4344–4346.

72. D. Meshcheryakov, F. Arnaud-Neu, V. Böhmer, *et al.*, *Org. Biomol. Chem.*, 2008, **6**, 1004–1014.

73. D. Meshcheryakov, V. Böhmer, M. Bolte, *et al.*, *Angew. Chem. Int. Ed.*, 2006, **45**, 1648–1652.

Synthetic Peptide-Based Receptors

Stefan Kubik
Technische Universität Kaiserslautern, Kaiserslautern, Germany

1 INTRODUCTION

Nature efficiently responds to the huge demand of ligands for the numerous substrates involved in biochemical processes by recruiting biomolecules, mostly proteins, whose structure can be varied in a wide range while being assembled from only a limited number of building blocks, namely, the 20 proteinogenic amino acids. Variation of the sequence of these building blocks along the protein backbone (primary structure), of folding pattern (secondary and tertiary structures), and of chain length, combined with the possibility to stabilize certain protein conformations by cross-linking or to assemble two or more protein chains (quaternary structure), gives rise to an almost unlimited number of unique structures that can be optimized in an evolutionary manner to fit a given task. Nonpeptidic cofactors sometimes play an additional role, but there are many examples of proteins accomplishing their task without the help of a cofactor.

Supramolecular Chemistry: From Molecules to Nanomaterials.
Edited by Philip A. Gale and Jonathan W. Steed.
© 2012 John Wiley & Sons, Ltd. ISBN: 978-0-470-74640-0.

Because of the complexity of such systems, which makes understanding and prediction of their molecular recognition properties difficult, synthetic receptors in supramolecular chemistry are generally based on structurally simpler abiotic molecular frameworks. Some of these receptors have been shown to rival natural ligands in terms of binding affinity and/or selectivity, showing that efficient molecular recognition can be realized without natural building blocks. However, the construction of a host molecule from the building blocks used in natural systems, for example, from amino acids, or the combination of abiotic and natural building blocks in a synthetic receptor can have a number of advantages:

- The chirality of a synthetic receptor containing enantiomerically pure amino acids could give rise to enantioselective recognition.
- Receptors containing a linear sequence of amino acids can be prepared by using combinatorial methods, thus allowing rapid screening of receptor properties.
- Depending on the structure and sequence of the subunits, linear peptides containing natural and/or nonnatural amino acids can fold into defined conformations enclosing a cavity, in which substrate binding can take place.

One also has to consider that not all amino acid-based natural ligands are structurally so complex as a protein. There are a number of relatively low-molecular-weight natural peptides, for example, cyclopeptides, that very efficiently and selectively interact with a complementary substrate. Notable examples are the antibiotics valinomycin, **1**, and vancomycin, **2**, as well as the natural macrocyclic ligand patellamide A, **3** (Scheme 1).

Valinomycin, **1**, is obtained from *Streptomyces fulvissimus* and several other Streptomyces strains and is made up of D- and L-valine (Val), D-hydroxyisovaleric acid (Hyv),

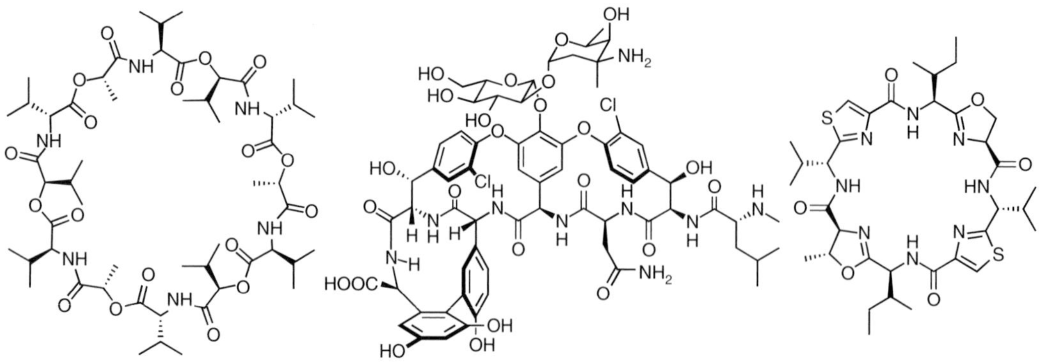

1

cyclo(L-Lac-Val-D-Lac-D-Hyv)$_3$

2

3

Scheme 1

and L-lactic acid (Lac), with ester and amide bonds alternating along the ring.[1] Antibiotic activity of **1** is due to the ability of this cyclodepsipeptide to selectively transport potassium ions across lipid membranes. The associated disturbance of the membrane potential ultimately leads to cell death. The glycopeptide antibiotic vancomycin, **2**, is produced by the bacterium *Amycolatopsis orientalis* and possesses high activity against Gram-positive bacteria.[2] The aglycon of **2** consists of a linear heptapeptide with several nonproteinogenic subunits. Five aromatic subunits are involved in the formation of three macrocycles, one containing a biaryl and two containing a diphenylether moiety. The antibiotic activity of vancomycin and related glycopeptides is due to their inhibitory effect on the mechanical stabilization of the bacterial cell wall during cell wall biosynthesis. Specifically, vancomycin binds to a pendant peptide target containing a terminal D-Ala–D-Ala unit, which is involved in the transpeptidase-catalyzed cross-linking of the peptidoglycan cell wall precursors. Vancomycin thus prevents the cell wall from being cross-linked efficiently, which eventually leads to the death of the bacterium. The sea squirt *Lissoclinum patella* produces several families of structurally closely related cyclic peptides, of which patellamide A, **3**, is one example.[3] Besides L-isoleucine and D-valine, this cyclic octapeptide contains two oxazoline and two thiazole rings, which derive from, respectively, threonine and cysteine. Patellamides bind transition metals such as Cu^{2+} and Zn^{2+} within their cavity—an ability that is supposedly the reason why *L. patella*, the organism from which **3** has been isolated, can contain several metals, including copper, up to 10 000 times their concentration found in the local marine environment. Several conclusions can be drawn from these few examples:

- There are no principal differences between natural and abiotic receptors. Macrocyclic compounds, independent of whether they are constructed from natural or nonnatural building blocks, can efficiently interact with a complementary substrate as long as they possess a cavity that can accommodate the substrate and suitable binding sites that can engage in specific interactions with it. As a consequence, one decisive factor that controls substrate binding is the ability of the receptor to adopt a conformation prior to or at least during complex formation, which allows tight interactions with the guest.
- Natural macrocyclic receptors are not solely constructed from α-amino acids. To stabilize a certain receptor conformation, that is, preorganize the receptor, or introduce binding sites beyond those available in the side chains of the proteinogenic amino acids, other building blocks can also be incorporated.
- The building blocks that constitute a natural receptor can generally be assembled in a sequential manner to give the final product or a precursor of the product.

These general guidelines are easily transferable to the development of synthetic receptors. Compounds prepared by using this strategy usually possess a characteristic relationship with their natural analogs, both in terms of structure and properties. Synthetic peptide-based receptors also have the advantage of allowing facile systematic structural variation by the exchange and/or modification of building blocks. Structure/property relationships that provide detailed insight into how receptor structure influences binding behavior can therefore be established. Finally, characterization of biomimetic receptors may provide information about the behavior of more complex proteins. Because of these reasons, several groups have focused their attention on the development and investigation of peptide-based receptors, and selected examples of successful work in this area are summarized in this chapter.

Peptide-based receptors can be distinguished according to the type of substrate recognized (cationic, anionic, and neutral), or according to structural criteria. Since the focus of this chapter strongly lies on structural aspects of this class of receptors, the latter classification was chosen, and receptors were therefore divided into the following four categories:

- cyclic peptides and cyclic pseudopeptides,
- conformationally constrained cyclic peptides containing rigid nonnatural subunits,
- linear peptides with natural and/or nonnatural subunits, and
- conjugates of abiotic receptors and peptide-based subunits.

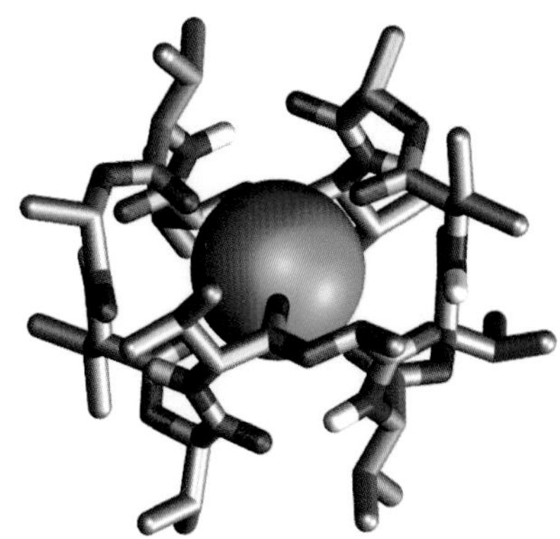

Figure 1 Crystal structure of the K^+ complex of **1**.

2 MACROCYCLIC PEPTIDES AND PSEUDOPEPTIDES

The elucidation of the mode of action of valinomycin, **1**, and the almost simultaneous discovery of the cation binding properties of cyclic oligo(ethylene glycols) (crown ethers) triggered intensive research activities among several groups to develop synthetic receptors on the basis of cyclopeptides or cyclic pseudopeptides. One appeal of such receptors lies in the fact that even the simplest cyclopeptide that has a cavity large enough for the inclusion of a guest contains an alternating arrangement of hydrogen-bond donors (NH groups) and hydrogen-bond acceptors (C=O groups) along the ring, both of which can contribute to binding. In addition, the relative ease with which individual subunits along the ring of a cyclopeptide can be deliberately exchanged or structurally varied should allow fine-tuning of receptor properties. Structural variation in a cyclopeptide usually has a pronounced effect on its conformational flexibility and preferred solution conformations. Prediction of binding properties, which strongly depend on receptor conformation, is therefore much more difficult for a cyclopeptide than for a macrocyclic abiotic receptor, whose arrangement of subunits along the ring is usually varied to a much lesser extent. In addition, one has to consider that cyclopeptides, particularly larger ones, have a tendency to adopt conformations in solution that are stabilized by intramolecular hydrogen bonds between NH and C=O groups, which are unsuitable for the inclusion of a guest molecule. To avoid this adverse effect on receptor properties, rigid subunits such as proline residues (or nonnatural amino acids, see Section 2) are therefore often incorporated into cyclopeptide-derived synthetic receptors. Because of the pronounced influence of the conformational behavior of a cyclopeptide on binding properties, characterization of cyclopeptide-derived receptors usually involves extensive conformational analyses.

First work in the area of cyclopeptide-derived synthetic receptors took inspiration from the binding mechanism of valinomycin. This mode of action involves conformational stabilization of the otherwise rather flexible valinomycin ring on potassium complexation to form a bracelet-like structure containing six hydrogen bonds between each NH group and the C=O of the preceding valine moiety (Figure 1).[4] In this structure, the six remaining ester carbonyl groups point to the center of the cavity, where they can interact with the included metal ion in an almost perfect octahedral fashion. Cavity dimensions are optimal for K^+, but also allow the inclusion of larger metal ions such as Rb^+. The valinomycin backbone is, however, too rigid to allow tight binding of smaller ions such as Na^+ or even Li^+.

To probe how sensitive coordination of the metal reacts to structural changes along the macrocycle, various valinomycin analogs were synthesized. Examples are compounds **4a–d**, all containing D-proline at the positions of the D-hydroxyisovaleric acid residues in **1** whose introduction should not affect the intramolecular hydrogen bonding pattern.[5] Compound **4a** (Scheme 2), in which the lactic acid residues of **1** were additionally replaced by L-proline, was shown in a two-phase extraction assay to exhibit about 3 orders of magnitude larger affinity for alkali picrates than **1** while the selectivity sequence $K^+ \approx Rb^+ > Cs^+ > Na^+ > Li^+$ is retained. Similar results were obtained for **4b**, showing that the side chains in the residues, which are responsible for intramolecular hydrogen bond formation, have no large effect on receptor properties. The improved binding affinities of **4a** and **4b** with respect to **1** were attributed to the interaction of the amide carbonyl groups, which have higher electron-donating properties than the ester carbonyl groups in **1**, with the metal. In addition, the proline subunits in **4a** and **4b** render these peptides to be more rigid. In line with this interpretation is that reduction in the nucleophilicity of the cation-coordinating carbonyl groups by replacement of the tertiary L-proline amides in **4a** and **4b**

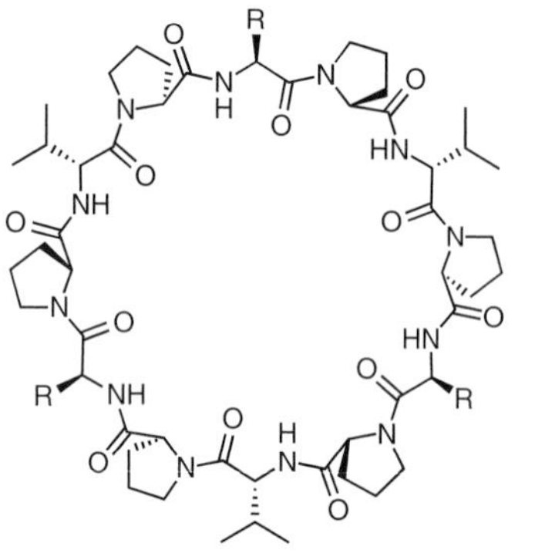

4a R = $CH(CH_3)_2$ cyclo(Pro-Val-D-Pro-D-Val)$_3$
4a R = CH_3 cyclo(Pro-Ala-D-Pro-D-Val)$_3$

4c
cyclo(Ala-Val-D-Pro-D-Val)$_3$

4d
cyclo(Pro-Val-D-Pro-D-Val)$_2$

Scheme 2

5
cyclo(Pro-Val-Gly-Gly)$_3$

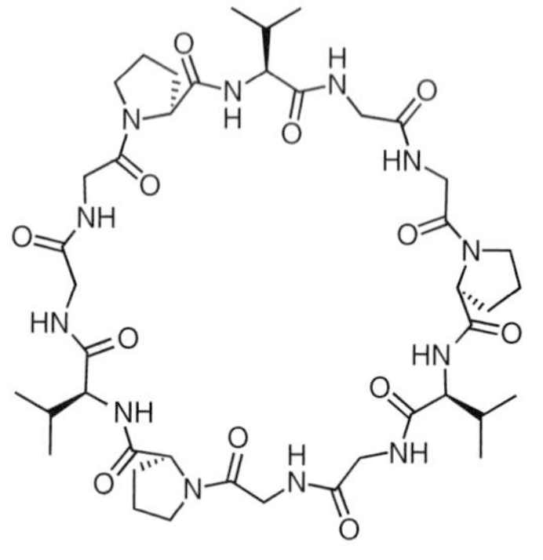

Scheme 3

Table 1 Stability constants K_a (in M^{-1}) of the 1 : 1 complexes of valinomycin **1** or receptor **5** with the perchlorate salts of K^+, Ca^{2+}, and Ba^{2+} in acetonitrile.

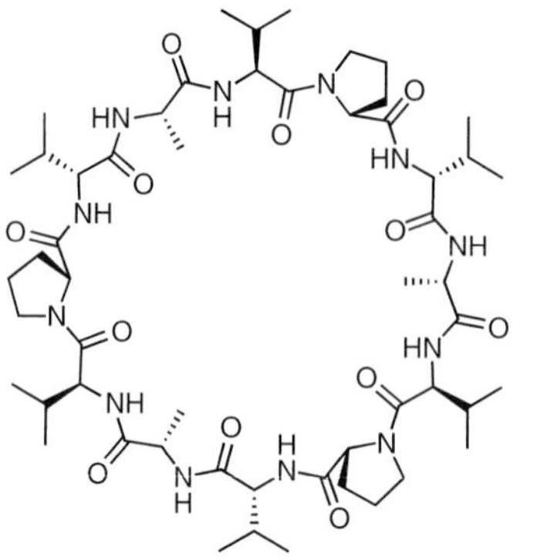

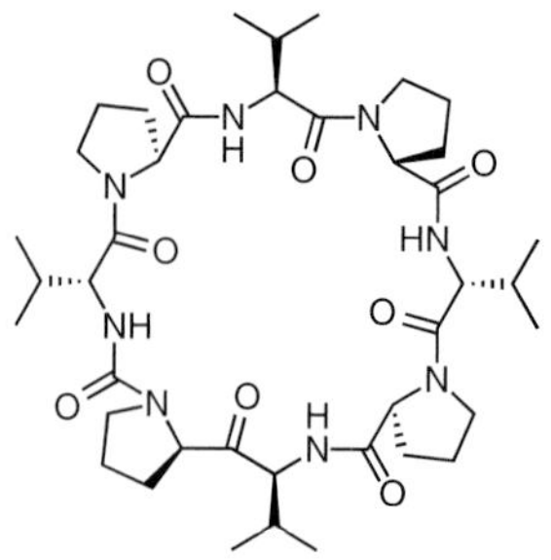

Receptor	Cation	K_a
1	K^+	3.0×10^5
5	K^+	1.7×10^5
5	Ca^{2+}	3.0×10^8
5	Ba^{2+}	2.0×10^{10}

by secondary amides in **4c** causes a drop in cation affinity. Decrease in ring size not only diminishes cation affinity but also reverses cation selectivity. Sodium affinity of **4d** is, for example, slightly larger than the affinity of this receptor for potassium.

Similar investigations were performed with the significantly more flexible valinomycin analog **5** (Scheme 3). This compound adopts a conformation in CD_2Cl_2 and $CDCl_3$ that lacks a defined cavity because all six glycine NH groups engage in intramolecular hydrogen bonds.[6] No complexation of metal ions was therefore observed in these solvents. In acetonitrile, **5** is conformationally more mobile and able to form complexes with metal ions such as K^+, Ba^{2+}, and Ca^{2+}, which are structurally related to the complex between **1** and K^+. Intramolecular hydrogen bonding involves the NH groups of the valine residues and the glycine residues that are directly adjacent to the proline moieties. The stability of the potassium complex of **5** in acetonitrile is similar to that of valinomycin (Table 1). However, earth-alkaline metal ions are bound much stronger. Interestingly, also complexes of higher stoichiometry were observed, in which **5** binds to two ions or two molecules of **5** bind a single ion. The latter type of complexes, the so-called peptide sandwich complexes, is formed especially with smaller ions such as Li^+, Na^+, and Mg^{2+}.

Synthetic analogs have also been prepared for other naturally occurring metal binding cyclopeptides. The relatively recent analysis of the solution conformation and metal ion binding properties of cyclopeptide **6**, containing the same postulated active tripeptide sequence Pro-Phe-Phe of cyclolinopeptide A **7** (Scheme 4), shows that this area of research is still a matter of considerable interest.[7]

These investigations provided detailed insight into how structural variation of natural ionophores affects metal binding. They did not demonstrate, however, the relevance of cyclopeptides as synthetic receptors in general. Metal binding cyclopeptides were therefore subsequently developed, for which there are no natural analogs.[8] Cyclohexapeptide **8** containing alternating glycine and L-proline residues is a good example of such a receptor, which also nicely demonstrates the close interplay between solution conformation and binding properties. A conformational analysis based on circular dichroism (CD) and NMR spectroscopy revealed that **8** (Scheme 5) adopts a C_3 symmetric conformation, termed S, in dioxane and chloroform, which has all peptide bonds trans, and is stabilized by three intramolecular hydrogen bonds. Addition of metal ions induces a conformational reorganization, leading to

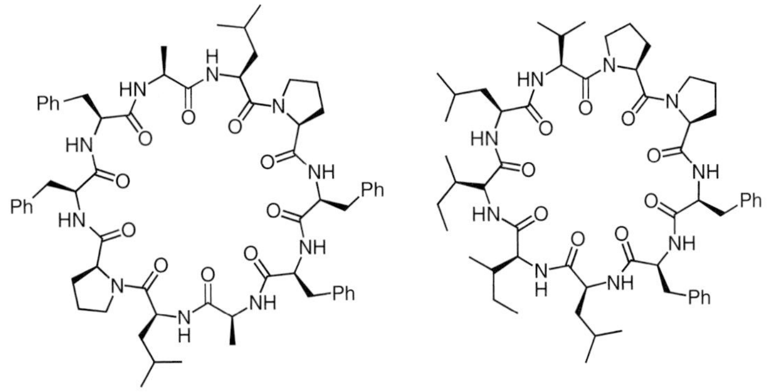

6
cyclo(Pro-Phe-Phe-Ala-Leu)$_2$

7 Cyclolinopeptide A
cyclo(Pro-Pro-Phe-Phe-Leu-Ile-Ile-Leu-Val)

Scheme 4

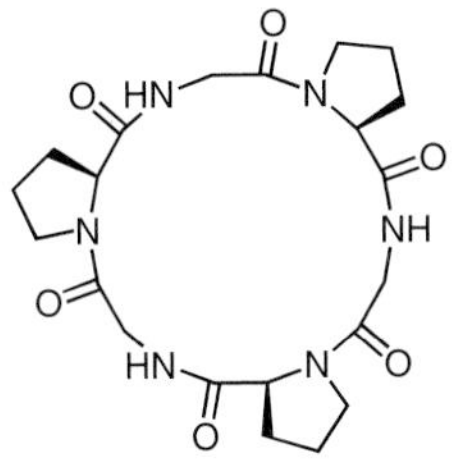

8
cyclo(Pro-Gly)$_3$

Scheme 5

Table 2 Stability constants K_a (in M^{-1}) of the complexes of receptor **8** with the perchlorate salts of various alkali and earth-alkaline metal ions in 80% methanol/water.

Cation	K_a
Li^+	180
Na^+	110
K^+	29
Ca^{2+}	1400
Ba^{2+}	420

conformation S_1* with no intramolecular hydrogen bonds. Instead, all six carbonyl groups of the amino acids converge toward the center of the cavity where cation coordination takes place. To illustrate this conformational reorganization, conformations S and S_1*, which were calculated on the basis of the reported results, are depicted in Figure 2. Among the alkali metal ions, **8** shows selectivity for Li^+ and Na^+ over K^+ and larger ions in 80% methanol/water (Table 2). Among the divalent earth-alkaline ions **8** binds Ca^{2+} in preference to Ba^{2+}. Similar to **5**, **8** also forms complexes of higher stoichiometry, especially with Mg^{2+} in acetonitrile. Formation of the complex in which one molecule of **8** binds to two magnesium ions requires the cyclopeptide to adopt conformation S_2*, which slightly differs from S_1*. In polar solvents such as water and in the absence of salts, an asymmetrical conformation of **8** is preferred. Similar studies are described for a variety of other synthetic cyclopeptides.

Cyclopeptide **8** has also been shown to interact with ammonium ions—a property it shares with crown ethers. Complex formation is believed to be due to hydrogen bond formation between the acidic protons of the guests and

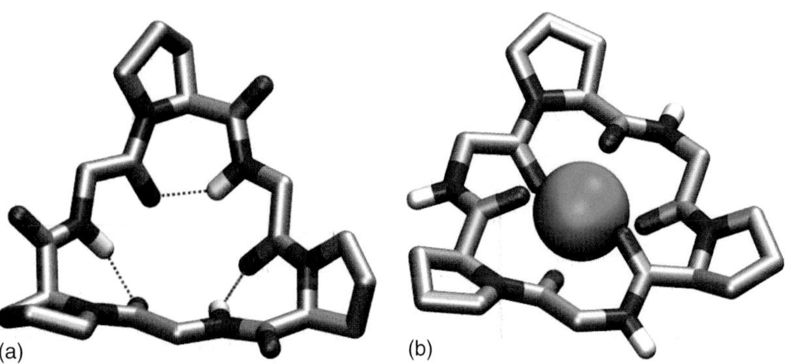

Figure 2 Calculated structures of conformations S (a) and S1* with included Mg^{2+} (b) of cyclopeptide **8**. The calculations were performed on the basis of the reported results by using MacSpartan 04 (Wavefunction, Inc.) and the MMFF force-field.

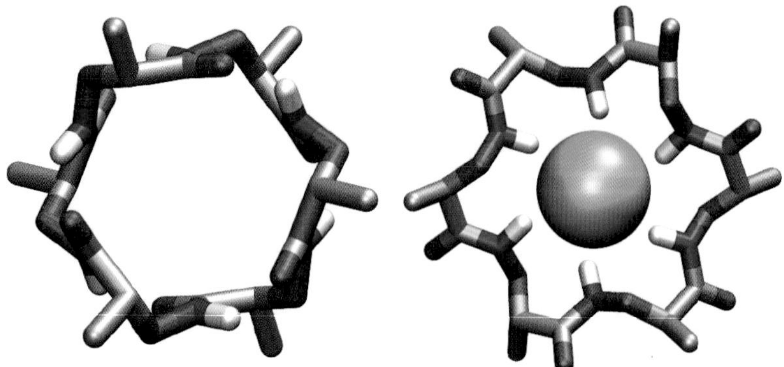

Figure 3 Calculated structures of the preferred conformation of **7** in the absence of anions and in the chloride complex. Nonacidic hydrogen atoms are omitted and amino acid side chains are replaced by methyl groups for clarity.

carbonyl groups of the cyclopeptides. Interestingly, while the hydrochloride of L-proline benzyl ester forms discrete 1 : 1 complexes with **8**, a complex of higher stoichiometry with two amino acids bound to one cyclopeptide was observed with the hydrochloride salt of L-valine methyl ester. A stepwise binding mechanism is proposed for the formation of this complex, in which interaction of three carbonyl groups of **8** with the ammonium group of the first valine molecule induces a cyclopeptide conformation that allows binding of a second guest to the remaining three carbonyl groups on the opposite side of the macrocycle. Protonated primary amines are required for this to occur, as shown by the inability of proline to induce binding of a second guest to **8**.

While metal ions can interact with the C=O groups along the ring of a cyclopeptide, the NH groups can serve to bind anionic guests. Most cyclopeptide-derived anion receptors, however contain rigid nonnatural amino acid subunits, and they are therefore treated in the next section. An anion receptor that fits well within the topic of this section is cyclic hexapseudopeptide **9**, which is composed of D,L-α-aminoxy acids (Scheme 6), since its conformation in the absence of guests is similar to that of **1**.

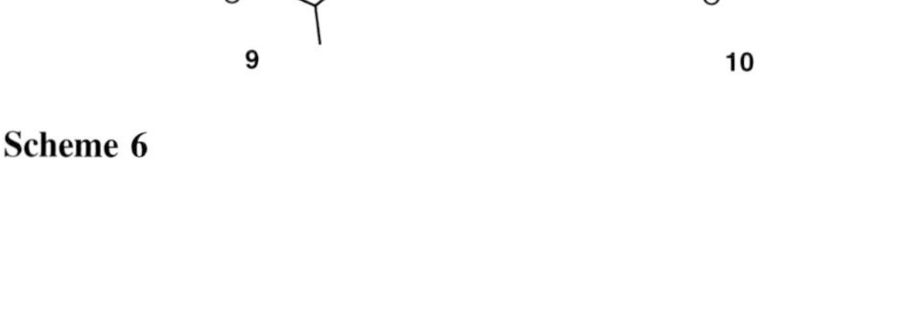

Scheme 6

Because of their well-defined secondary structures, pseudopeptides consisting of α-, β-, or γ-aminoxy acids give rise to a whole family of foldamers.[9] In the case of **9**, intramolecular C=O $\cdots$H–N hydrogen bonds cause this macrocycle to adopt a C_3 symmetric bracelet-like conformation in $CDCl_3$, as shown in Figure 3. All α-protons of the aminoxy acid residues point inward and the side chains point outward, with those of the D- and the L-aminoxy acid residues residing on opposite sides of the ring plane. Since all C=O and NH groups are involved in hydrogen bonding, prediction of whether **7** binds ions at all and if it does, which type of ion is preferred was not straightforward. Binding studies using cations and anions were therefore carried out, which revealed that **9** selectively interacts with anions in $CDCl_3$. The association constant for the chloride complex was determined to be $11\,880\,M^{-1}$. Structural investigations showed that for binding to occur, a conformational reorganization leads to a structure of **9**, in which all NH groups converge toward the center of the macrocycle. To illustrate this behavior, the calculated structure of the chloride complex of **9** is also depicted in Figure 3.

These investigations were subsequently extended to cyclic hexapeptide **10** comprising alternating D-α-aminoxy acids and D-α-amino acids. Conformational analysis revealed that **10** adopts a C_3 symmetric conformation with alternating seven-membered rings (γ-turns) and eight-membered rings (N—O turns), which are stabilized by intramolecular hydrogen bonds. Quantitative assessments of anion affinity in CD_2Cl_2 showed that chloride affinity of **10** is slightly larger than that of **9** ($Cl^- \subset$ **10** $K_a = 15\,000\,M^{-1}$). In addition, **10** possesses excellent anion selectivity; the chloride complex of **10** is at least 16 times more stable than all other anion complexes investigated (Br^-, I^-, NO_3^-).

Besides the peptide C=O and NH groups along the ring, many cyclopeptides contain functional groups in the amino acid side chains that are also available for substrate binding. Imanishi *et al.* showed, for example, that

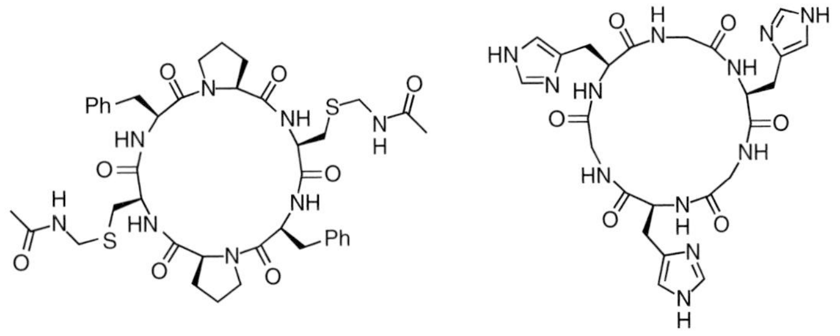

11
cyclo[Pro-Cys(Acm)-Phe]$_2$

12
cyclo(His-Gly)$_3$

Scheme 7

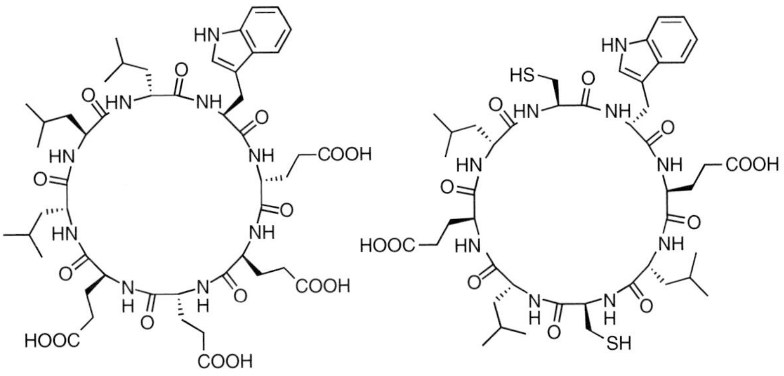

13 cyclo[Trp-(D-Glu-Glu)$_2$-D-Leu-Leu-D-Leu]

14 cyclo[D-Trp-Glu-D-Leu-Cys-D-Leu-Glu-D-Leu-Cys]

Scheme 8

the carbonyl groups in the side chains of **11** (Scheme 7) cooperatively contribute to the Ba^{2+} affinity of this peptide.[10] Another example is cyclic hexapeptide **12**, in which the imidazole residues of the histidine moieties act as ligands for copper(II). Interestingly, the metal ion coordinates to three nitrogen atoms of the imidazole residues at neutral pH, while at basic pH, copper coordination involves three deprotonated amide groups along the cyclopeptide ring.[11]

The side chains in cyclopeptides **13** and **14** that contain alternating D- and L-amino acids (Scheme 8) serve to induce affinity for transition metal ions. Isothermal titration microcalorimetry showed that peptide **13** forms complexes with the divalent transition metal ions Cu^{2+}, Zn^{2+}, Cd^{2+}, Hg^{2+}, and Pb^{2+} and with the trivalent group III metal Al^{3+}, but not with alkali or earth-alkaline metal ions in water.[12] Highest affinity, amounting to a K_a of $2.2 \times 10^6\ M^{-1}$, was observed for Hg^{2+}. The interaction with Pb^{2+}, Hg^{2+}, and Cu^{2+} modulates the fluorescence emission properties of the tryptophan chromophore in **13**, even in the presence of an excess of other metal ions (Li^+, Na^+, K^+, Mg^{2+}, Ca^{2+}, Ba^{2+}, Zn^{2+}, and Cd^{2+}), allowing this cyclopeptide to be used as an optical sensor for such analytes in water. Affinity and fluorescence response of mercury ions in water could be improved by the introduction of cysteine residues in cyclopeptide **14** ($Hg^{2+} \subset$ **14**: $K_a = 7.6 \times 10^7\ M^{-1}$).

A recent investigation targeted the development of a cyclopeptide with affinity for vitamin B_{12}.[13] In this context, libraries of cyclic decapeptides were prepared on solid support and screened with vitamin B_{12} derivatives. Of the 15 625 unique cyclopeptides thus tested, those that contained histidine and cysteine as coordinating residues showed the most promising binding properties. Two hits, namely, *cyclo*(His-Asp-Glu-Pro-Gly-Ile-Ala-Thr-Pro-D-Gln) and *cyclo*(Val-Asp-Glu-Pro-Gly-Glu-Asp-Cys-Pro-D-Gln), were resynthesized and binding to vitamin B_{12} was investigated in solution (20 mM HEPES buffer, pH 7.0). The peptides bind aquocobalamin with coordination of the His or Cys side chains to the cobalt with high affinities (K_a about $10^5\ M^{-1}$). Additional interactions between the peptide side chains and the corrin moiety of vitamin B_{12} were observed by ^1H NMR spectroscopy. Moreover, the cyclopeptide–cobalamin complex of histidine-containing cyclopeptides showed enhanced stability toward cyanide exchange, demonstrating the shielding effect of the ligand on the metal center.

Cyclopeptides have also been used for the construction of chemosensors. In a collaboration between the Jung and the Gauglitz group, for example, cyclopeptides were immobilized on a solid support. Depending on the side chains in the periphery of the cyclopeptide ring, these materials allowed for the selective sensing of volatile organic compounds such as tetrachloroethene, anisole, ethylacetate, and *n*-octane in the gaseous phase or of amino acid derivatives in aqueous solution.[14,15]

The examples presented in this section clearly show that cyclopeptides containing only natural amino acid subunits (or structurally closely related analogs) can serve as receptors for cations, anions, or neutral substrates if the subunits and their sequence are appropriately chosen. Receptor/substrate interactions can involve the C=O or NH groups along the ring and/or side chain functional groups. Allowing also nonnatural building blocks or nonpeptidic linkages in the ring substantially increases the structural space available for macrocyclic peptide-derived receptors. In turn, receptor conformation can be controlled in a more subtle way as can the affinity toward a certain type of substrate. Relevant work in this area is summarized in the following section.

3 CONFORMATIONALLY CONSTRAINED CYCLIC PEPTIDES

An important strategy to control the properties of macrocyclic peptide-based receptors is the introduction of more or less rigid amino acids, which restrict and/or control conformational mobility, along the ring. Rigid subunits can, for example, stabilize certain bioactive conformations, increasing the affinity of a cyclopeptide to a natural receptor. Conformationally constrained cyclopeptides therefore comprise attractive lead structures for pharmaceutical applications. Rigid subunits can, however, also be useful to induce cyclopeptide conformations that are well suited for interactions with a substrate of interest. Conformational constraints can derive from natural amino acids, but nonnatural amino acids are used more frequently. In addition, some synthetic cyclopeptides have been developed that solely comprise rigid nonnatural building blocks, giving rise to synthetic receptors with similarly well-defined macrocyclic cavities like calixarenes.

The most important motif to introduce conformational constraints into a cyclopeptide containing only natural building blocks involves disulfide formation between the side chains of two cysteine residues. The advantages of the incorporation of disulfide bonds into a cyclopeptide are (i) the tendency of the disulfide bond to adopt a *gauche* conformation with a potential energy minimum at a dihedral angle of around $|90^\circ|$ and a rotational barrier intermediate between those of a typical carbon–carbon single bond and an amide bond, which allows conformational control, (ii) the dimeric symmetrical character of cystine that facilitates cyclopeptide synthesis, and (iii) the possibility of controlling the relative direction of the two chains attached to a cystine residue. One of the fist examples of a cyclopeptide containing disulfide bonds whose receptor properties were studied is cyclopeptide **15** (Scheme 9). This peptide, designed to mimic the cation binding of valinomycin, indeed possesses good cation affinity and selectivity in acetonitrile.[16] The highest affinity amounting to a log K_a of 9.09 was observed for Ba^{2+}. While smaller ions (Li^+, Na^+, and Mg^+) are bound with 1 : 1 stoichiometry, peptide–sandwich complexes were observed upon binding of **15** to larger ions (Ca^{2+}, Sr^{2+}, and Ba^{2+}).

Systematic work on cyclopeptides containing cysteine residues combined with nonnatural aromatic amino acids was more recently carried out in the Ranganathan group.[17] This family of receptors, an example of which is **16** (Scheme 9), was termed cystinophanes. Compound **16** was shown to bind to tetrabutylammonium salts of 1,ω-dicarboxylic acids with 1 : 1 stoichiometry in $CDCl_3$. The highest binding affinity was observed for glutaric acid ($K_a = 369\ M^{-1}$).

In a somewhat similar approach, two linear peptides, $HSCH_2CO$-Tyr-(D-Ala-Lys)$_4$-D-Ala-Gln-$NHCH_2CH_2SH$ and $HSCH_2CO$-Tyr-(D-Ala-Lys)$_6$-D-Ala-Gln-$NHCH_2CH_2$SH, were recently described that are able to cyclize on

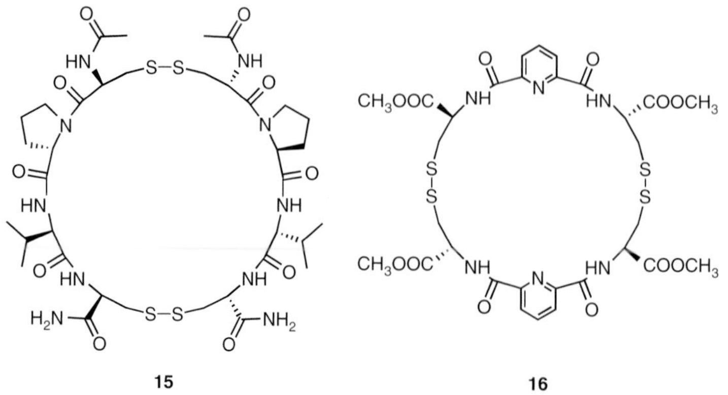

15

16

Scheme 9

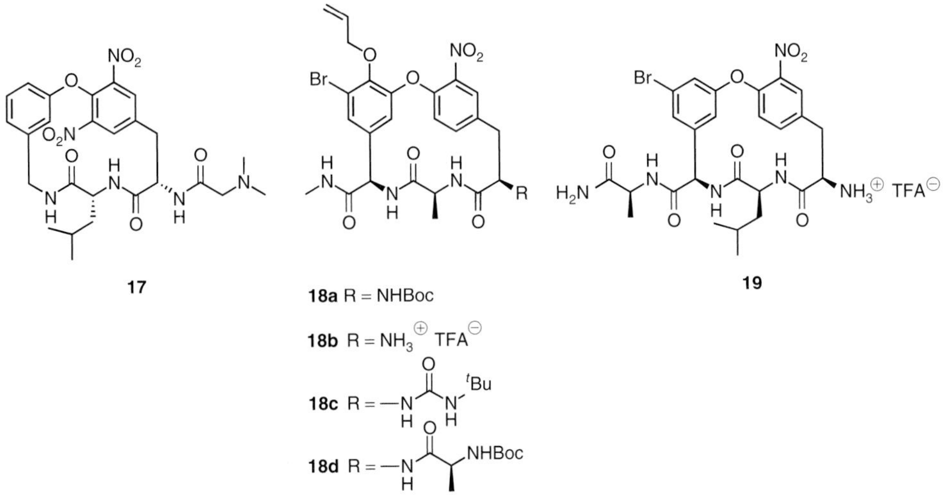

17

18a R = NHBoc

18b R = $NH_3^{\oplus}$ $TFA^{\ominus}$

18c R = —NH–C(O)–NH–^tBu

18d R = —NH–C(O)–CH(CH_3)–NHBoc

19

Scheme 10

oxidation of the two terminal thiol moieties.[18] Reversible cyclization of the peptides in the presence of single-walled carbon nanotubes causes the peptides to wrap around the tubes, improving their solubility in aqueous solution and leading to a noncovalent surface functionalization. The correlation between the length of the peptide and the diameter of the preferentially recognized nanotube is a strong indication of the specific interactions. Cyclopeptide rings encircling the carbon nanotubes are unable to slide off unless the disulfide bonds are cleaved by reduction.

While many cyclopeptides described above were designed to mimic the cation binding properties of valinomycin **1**, there has also been considerable research activity in the design of vancomycin analogs. A number of these compounds were inspired by the intricate structure of **2** comprising a chain of amino acids with appropriate side chains linked via aromatic moieties. For example, in receptor **17** developed by the Hamilton group, the AB and the CD rings of the vancomycin skeleton were deleted, leaving only the DE ring framework, that is, the carboxylate binding pocket, intact. (Scheme 10).[19] Binding of carboxylates by **17** involves a combination of hydrogen bonding from the amide and ammonium functionalities, thus resembling the binding mode of vancomycin. A series of related receptors **18a–d** were synthesized by Pieters *et al.* and investigated with respect to binding of Ac-D-Ala-O^- in $CDCl_3$ (Scheme 10). Complex stability is almost independent of the nature of the residue R (Table 3). The leucine-containing receptor **19**, however, proved to possess about four times weaker affinity to Ac-D-Ala-O^- than **18b**, showing that variation of the amino acid in the center of the carboxylate binding pocket has an influence on complex stability.[20]

Table 3 Stability constants K_a (in M^{-1}) of the complexes of receptors **18a–d** and **19** with the tetrabutylammonium salt of Ac-D-Ala-O^- in $CDCl_3$.

Receptor	K_a
18a	9.7×10^3
18b	3.2×10^4
18c	3.7×10^4
18d	2.3×10^4
19	7.3×10^3

Another means of controlling the conformational mobility and in turn binding properties of a macrocyclic peptide is the introduction of rigid aromatic subunits along the ring (and not as linking units between the side chains). An example of the successful use of this strategy is cyclic hexapeptide **20**, in which 3-aminobenzoic acid (Aba) subunits alternate with L-alanine (Scheme 11). The ability of **20** to strongly interact with phosphate esters was demonstrated by Ishida and coworkers.[21] A 1 : 1 complex is formed with *p*-nitrophenyl phosphate, for example, whose association constant amounts to $K_a = 1.2 \times 10^6\ M^{-1}$ in DMSO (dimethyl sulfoxide). Replacement of alanine in **20** with serine causes a slight reduction in complex stability. A much larger drop was, however, observed for acyclic and larger analogs of **20**, for example, the corresponding octapeptide, demonstrating that the hexapeptide has the correct size and shape for phosphate ester recognition. The downfield shift of the NH signals in the ^1H NMR spectrum of **20**, observed on complex formation, indicates that all NH groups of this cyclopeptide form hydrogen bonds to the substrate. The same group subsequently used these peptides as serine protease mimics and for the construction of ion channels.[22]

The Kubik group is also interested in the development of synthetic receptors based on conformationally constrained cyclopeptides, and their approach mainly involves

20 cyclo(Ala-Aba)$_3$

21 cyclo(Pro-Aba)$_3$

22a R = OCH_3
22b R = $COOCH_3$

Scheme 11

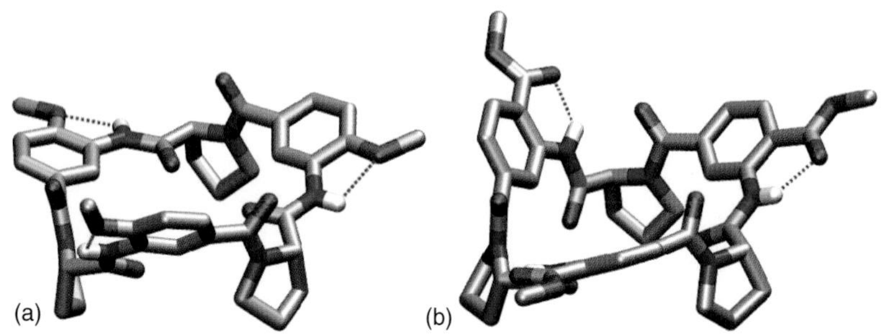

Figure 4 Molecular structures of cyclopeptides **22a** (a) and **22b** (b). Nonacidic protons are omitted for clarity.

combining Aba-derived amino acids and L-proline subunits in an alternating fashion along the ring. The cyclic proline residues in these cyclopeptide were expected to cause a further reduction in conformational mobility in comparison to **20**. The first cyclopeptide synthesized in this context, compound **21**, was shown to interact with quaternary ammonium ions such as *n*-butyltrimethylammonium ($BTMA^+$) picrate in 0.2% d_6-DMSO/$CDCl_3$. A conformational analysis indicated that **21** preferentially adopts conformations in chloroform with a tilted arrangement of the aromatic subunits. These subunits line the wall of a dish-shaped cavity in which binding is believed to occur. Preorganization of **21** for cation binding is, however, not optimal because rotation of the secondary amide groups along the peptide ring causes a variation in cavity dimensions.

An improvement in cation affinity was achieved by the introduction of substituents in the 4-position of the aromatic peptide subunits, which restrict the rotation of the neighboring amide groups. For example, in peptides **22a** containing methoxy groups and **22b** containing methoxycarbonyl groups (Scheme 11), intramolecular hydrogen bonds between the aromatic amide groups and the aromatic substituents stabilize a C_3 symmetric conformation with divergent NH groups and a well-defined cavity, as nicely demonstrated by the crystal structures of both cyclopeptides (Figure 4). As a consequence, both

Table 4 Stability constants K_a (in M^{-1}) of the complexes of cyclopeptides **21**, **22a**, and **22b** with *n*-butyltrimethylammonium ($BTMA^+$) picrate in 0.2% d_6-DMSO/$CDCl_3$.

Cyclopeptide	K_a
21	1260
22a	2700
22b	10 800

compounds bind $BTMA^+$ picrate significantly stronger in 0.2% d_6-DMSO/$CDCl_3$ than **21** (Table 4). This work is a good example how the deliberate control of the conformation of a peptide-derived receptor can lead to improved binding properties.

Interestingly, **21** is also able to bind both components of an ion pair simultaneously if ammonium salts are used as substrates, whose anion is a better hydrogen-bond acceptor than picrate. Complexation of, for example, $BTMA^+$ iodide or tosylate causes stabilization of a cyclopeptide conformation, with all NH groups converging toward the center of the cavity to allow interactions with the anion. The cation then enters the bowl-shaped cavity thus produced, where it interacts with the aromatic cyclopeptide subunits. Strong Coulomb-attraction between the oppositely

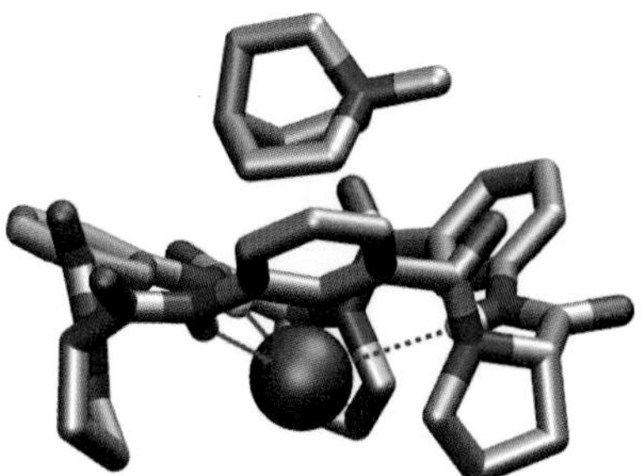

Figure 5 Molecular structure of the *N*-methylquinuclidinium iodide complex of **21**. Nonacidic hydrogen atoms are omitted for clarity.

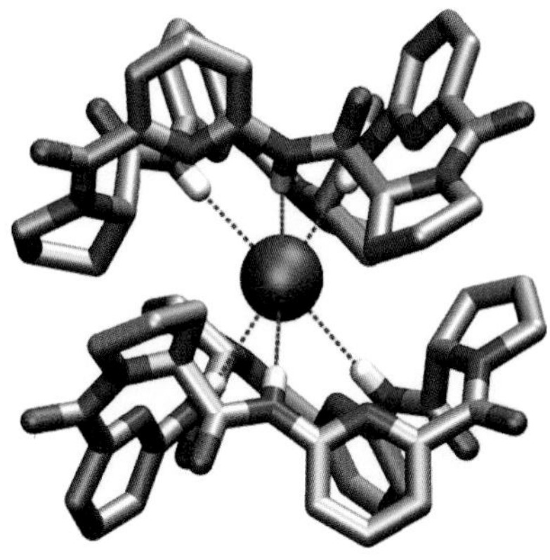

Figure 6 Molecular structure of the iodide complex of **23**. Nonacidic hydrogen atoms are omitted for clarity.

charged ions causes a further stabilization of the complex. To illustrate this arrangement, the X-ray crystallographically determined structure of the complex between **21** and *N*-methylquinuclidinium iodide is depicted in Figure 5.

No coordination of the NH groups of peptides **22a** and **22b** to anions was detected because the aromatic substituents in these peptides stabilize a conformation with divergent NH groups that are unsuitable for interactions with anionic guests. Conversely, conformational control also allowed the improvement of the anion binding of such peptides by stabilization of a conformation with converging NH groups. This was achieved by replacing the Aba subunits in **21** by 6-aminopicolinic acid (Apa) subunits.[23] The corresponding peptide **23** (Scheme 12) preferentially adopts conformations in polar protic or aprotic solvents, with all NH groups pointing toward the cavity center. As a consequence, **23** exhibits a slightly larger affinity than the more flexible **21** for tosylate in DMSO-d_6. More interesting is, however, the fact that anion binding of **23** was also observed in strongly competitive aqueous solvent mixtures such as 80% D_2O/CD_3OD. In the case of inorganic anions, binding in aqueous solution involves the formation of 2 : 1 complexes, in which a completely desolvated anion is bound by six hydrogen bonds in a cavity between two perfectly interdigitating peptide rings well shielded from the surrounding solvent. As an example, the structure of the iodide complex of **23** is depicted in Figure 6. Larger organic anions such as tosylate cannot be fully encapsulated because of steric reasons, explaining why they are bound in the form of 1 : 1 complexes also in the aqueous environment.

The 2 : 1 complexes formed by **23** were subsequently converted into 1 : 1 complexes by covalently linking two peptide units together. Examples of bis(cyclopeptides) prepared in this context are compounds **24a,b** and **25a,b**, all of which possess substantially improved anion affinity with respect to **23**. While **24a,b** resulted from molecular modeling studies,[24] bis(cyclopeptides) **25a,b** were identified by using dynamic combinatorial chemistry.[25] With a log K_a of almost 7 in 33% water/CH_3CN, the latter two compounds

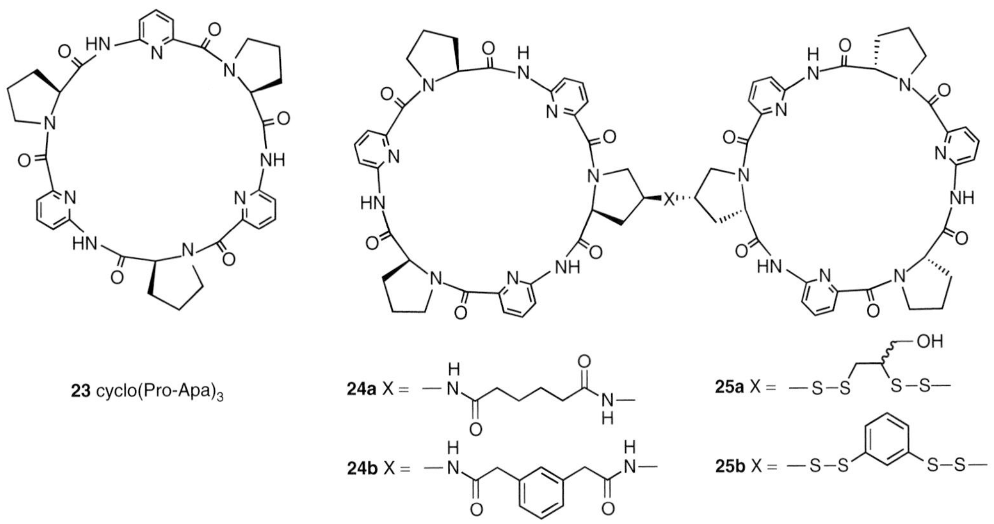

23 cyclo(Pro-Apa)$_3$

24a X =

24b X =

25a X =

25b X =

Scheme 12

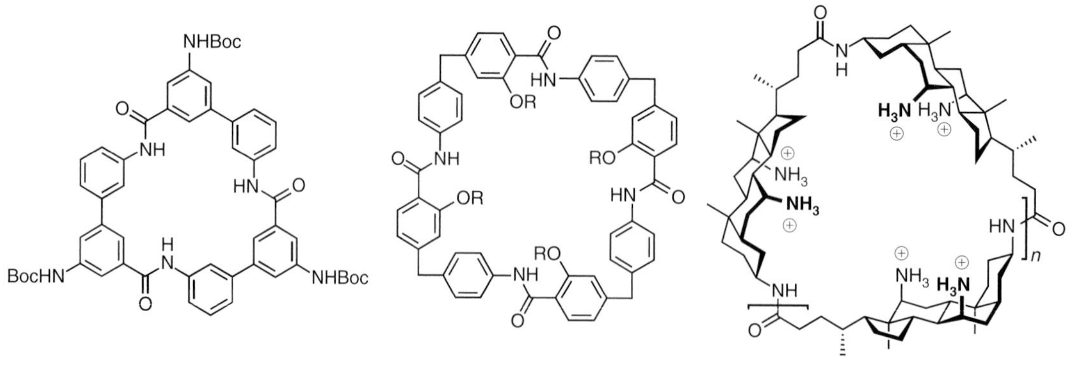

26

27 R = $(CH_2)_3NH_2 \cdot TFA$

28a $n = 1$
28b $n = 2$

Scheme 13

represent, to the best of my knowledge, so far the most efficient *neutral* sulfate receptors active in aqueous media. A detailed characterization of the thermodynamics of complex formation provided valuable information about the causes that are responsible for the unique binding properties of such bis(cyclopeptides) in aqueous solution.[26]

Compounds **26**, **27**, and **28** show that synthetic peptide-derived receptors can also be assembled solely from rigid nonnatural amino acid subunits. The macrocyclic triamide **26** (Scheme 13) was developed by the Hamilton group.[27] This rigid macrocycle contains three NH groups pointing toward the center of the about 5 Å wide cavity, thus inducing affinity for tetrahedral anions. Interaction of **26** with Bu_4NOTs in 2% d_6-DMSO/$CDCl_3$ leads to a 1 : 1 complex whose K_a amounts to 2.1×10^5 M^{-1}. Inorganic anions such as halides, nitrate, hydrogensulfate, and dihydrogenphosphate anions give rise to more complex binding equilibria in the same solvent mixture due to the formation of complexes of higher stoichiometry containing two receptor molecules and one anion. In this respect, **26** behaves similarly to **23**.

In the more competitive solvent d_6-DMSO, formation of 2 : 1 complexes by **26** was not observed. While the stabilities of the halide and nitrate complexes of this receptor are significantly smaller in this solvent, high affinity for tetrahedral anions is retained with the stability constant of the dihydrogenphosphate complex amounting to 1.5×10^4 M^{-1}, for example. Binding selectivity of **26** was correlated with the size and shape of the receptor cavity as well as with the arrangement of the NH groups. Thus, iodide binds more strongly to **26** than chloride despite the fact that the charge density of a chloride ion is higher because the fit of iodide inside the receptor cavity better allows simultaneous interactions with all converging NH groups.

The group of Nowick introduced cyclopeptide **27** (Scheme 13), which is built up from four identical synthetic ι-amino acids (ι = iota).[28] A slow equilibrium was observed between two conformations of **27** in water, a square one with all amide bonds adopting the *trans*

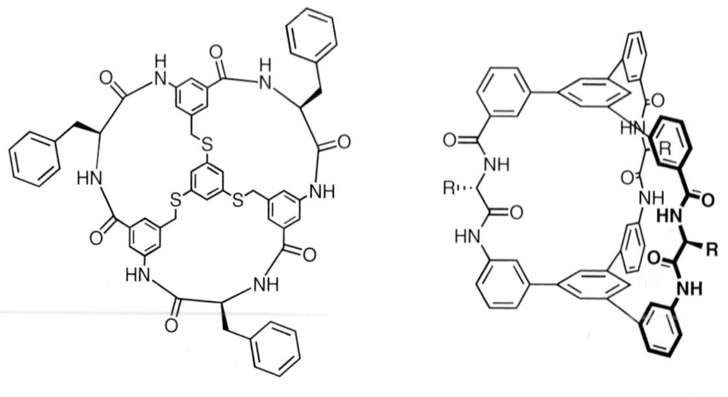

29

30a R = $CH_2CH(CH_3)_2$
30b R = CH_2Ph

Scheme 14

conformation and a rectangular one with alternating *cis*- and *trans*-amides. Binding of sodium cholate in water shifts this equilibrium toward the square conformer and leads to a complex whose stability amounts to $10^4\,M^{-1}$ in water. The main driving forces behind complex formation are most probably hydrophobic effects with additional contributions coming from electrostatic interactions between the anionic substrate and the cationic host.

The macrocyclic cyclocholamides **28a,b** were developed by the Davis group.[29] Calculations suggested that both macrocycles possess flexibility due to the steroidal side chain. In both cases, however, toroidal conformations with inward directed ammonium groups are among the energetically most favorable ones. The cavity of the trimer has a diameter of about 9 Å while the average internal diameter of the larger analog amounts to about 14 Å. Compound **28b** was shown to mediate the transport of chloride anions across vesicle bilayer membranes, most probably via a carrier mechanism.

The conformationally very rigid polycyclic receptors **29** and **30a,b** also contain amino acids as important structural elements (Scheme 14). These amino acids serve to introduce chirality as well as to provide hydrogen-bond donors and acceptors with which substrate can specifically interact. The remaining aromatic building blocks define the overall receptor structure and induce preorganization for substrate recognition. Receptor **29** developed by the Still group has been shown to enantioselectively recognize Boc-protected amino acid *N*-methylamides, *N*-methoxycarbonyl amino acid esters, or Boc-protected amino acid esters in chloroform with a strong preference for the L-enantiomer over the D-enantiomer. Also interactions of this receptor with 1-*O*-octyl glycosides were demonstrated.[30] Binding occurs inside the bowl-shaped cavity created by the aromatic subunits along the ring. Subsequently, analogs of **29** were prepared containing, for example, naphthyl units along the wall of the cavity. In addition, this receptor was immobilized on solid support and shown to allow chromatographic separation of amino acid enantiomers.

Receptors **30a,b** possess affinity for amino acid derivatives. ^{1}H NMR titrations in $C_2D_2Cl_4$ revealed that these receptors bind N-Cbz-protected L- and D-glutamate enantioselectively, with differences in the free energy of complex formation between the diastereomeric complexes of up to 4.6 kJ mol^{-1}.[31] These receptors failed, however, to resolve mixtures of amino acid enantiomers when anchored to silica because of incompatibility of the HPLC (high performance liquid chromatography) conditions, which require the use of methanol as cosolvent for elution of the polar solutes, with efficient host–guest association *via* hydrogen bonding.

Affinity for transition metals, specifically Cu^{2+} and Zn^{2+}, is a feature shared by cyclopeptides containing oxazoline, thiazole, or thiazoline subunits, most of which are natural products isolated from the sea squirt *L. patella*. One example for this structurally diverse family of cyclopeptides is patellamide A **3** (Scheme 1) whose metal binding properties have been reviewed in detail.[32] To illustrate a binding mode in transition metal binding, the crystal structure of the dicopper(II) complex of ascidiacyclamide **31** is depicted in Figure 7. In this complex, two copper ions separated by a bridging carbonate anion are included into the cavity of **31**, each metal ion coordinating to two nitrogen atoms of adjacent heterocyclic subunits. A third coordination site is provided by a deprotonated amide NH that is situated between the azole units involved in metal binding. To accommodate the two copper ions, the cyclopeptide adopts a saddle-shaped conformation.

These natural cyclopeptides also served as starting points for several groups to devise synthetic receptors. In this respect, cyclopeptides **32a–d** (Scheme 15) were prepared as analogs of patellamide A **3** by the Comba group to probe the importance of the oxazoline moieties in metal coordination.[33]

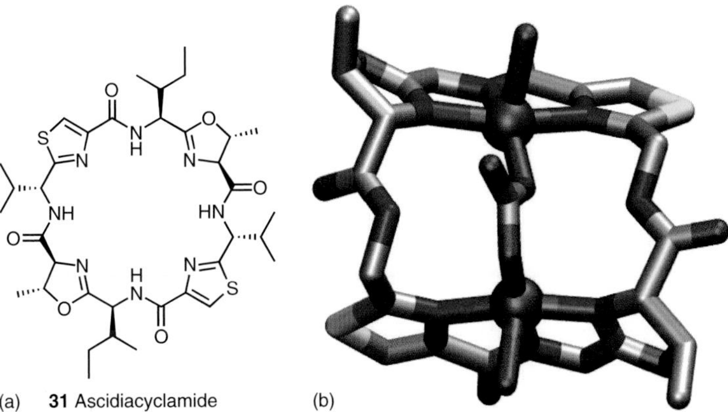

Figure 7 Structure of **31** (a) and molecular structure of the dicopper(II) complex of this cyclopeptide (b). Hydrogen atoms and valine and isoleucine side chains are omitted for clarity.

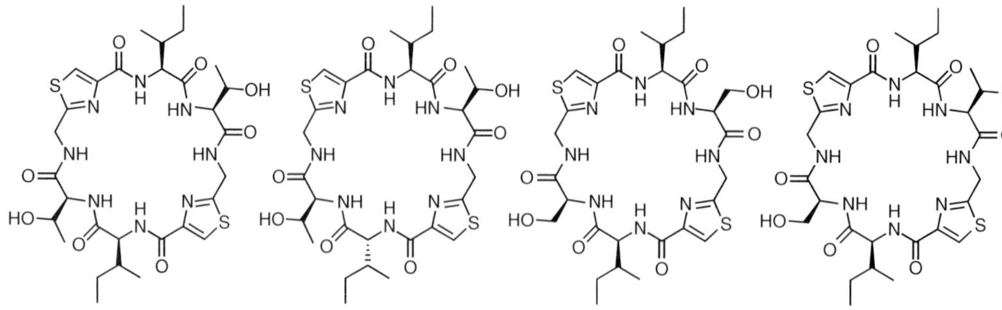

32a PatJ[1]
cyclo[Ile-Thr-(Gly)Thz]$_2$

32b PatJ[2]
cyclo[Ile-Thr-(Gly)Thz-D-Ile-Thr-(Gly)Thz]

32c PatL
cyclo[Ile-Ser-(Gly)Thz-Ile-Ser-(Gly)Thz]

32d PatN
cyclo[Ile-Thr-(Gly)Thz-Ile-Ser-(Gly)Thz]

Scheme 15

In the crystal, PatJ[1] **32a** adopts a conformation with two almost coparallel thiazole rings. The thiazole-N and amide-N atoms of **32a** are well preorganized for transition metal ion binding. In solution, Cu^{2+} coordination was monitored by UV/vis spectroscopy, revealing several mono- and dinuclear Cu^{2+} species whose stoichiometries were confirmed by mass spectrometry, with dinuclear complexes being the more stable ones. Two types of dicopper complexes, $[Cu_2(L-2H)(OH_2)_n]^{2+}$ (n = 6, 8) and $[Cu_2(L-4H)(OH_2)_n]$ (n = 4, 6; L = **32a–d**) were structurally analyzed by EPR (electron paramagnetic resonance) spectroscopy and a combination of spectra simulations and molecular mechanics calculations. These investigations showed that the cyclopeptides in these complexes adopt a conformation similar to the one of the oxazoline-containing analogs. Thus, metal complexation is not significantly affected by the absence of the oxazoline residues, although slight but significant structural differences were found in the dinuclear Cu^{2+} complexes depending on cyclopeptide structure.

Most work on the coordination chemistry of azole-based cyclopeptides concentrated on transition metals, in particular, Zn^{2+} and Cu^{2+}, as substrates, but in some cases complexation of earth-alkaline metal ions such as Ca^{2+} was also studied. To gain insight into structural factors that influence Ca^{2+} affinity, derivatives **33a–c** (Scheme 16), in which subunit structure and number of heterocyclic subunits were varied, were synthesized. All of these peptides interact with Ca^{2+} significantly more strongly than their natural counterparts, for example, **31** (Table 5). This result shows that oxazoline rings are detrimental for calcium binding.[34] That calcium affinity correlates with the conformational mobility of these cyclopeptides and not with the number of coordination sites (four carbonyl groups in **31**, six in **33a** and **33b**, and eight in **33c**) can be deduced from the similar calcium affinity of **33b** and **33c**.

The group of Haberhauer has used analogs of patellamide A **3**, and also ones having a smaller ring size, as basis for the construction of synthetic receptors. The relatively straightforward synthesis of such cyclopeptides renders them versatile starting materials for further structural elaboration. By systematically varying the substituents around the ring, binding properties can be deliberately controlled, for example. In addition, ring size as well as the type of heterocyclic subunits may be varied. Scheme 17

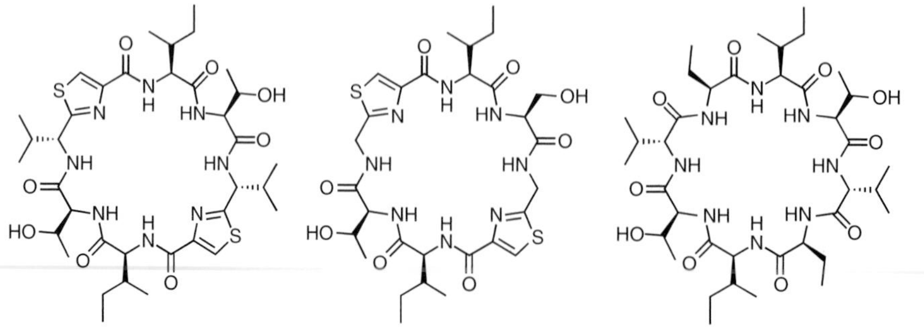

33a
cyclo[Ile-Thr-(D-Val)Thz]$_2$

33b
cyclo[Ile-Ser-(Gly)Thz-Ile-Thr-(Gly)Thz]

33c
cyclo[Ile-Thr-D-Val-αAbu]$_2$

Scheme 16

Table 5 Stability constants, log K_a, of the complexes of cyclopeptides **31**, and **33a–c** with Ca^{2+} (as perchlorate salt) in acetonitrile.

Cyclopeptide	log K_a
31	2.9
33a	4.0
33b	5.7
33c	5.5

Table 6 Stability constants K_a (in M^{-1}) of the complexes of receptors **36a–c** with $H_2PO_4^-$ or AcO^- (as tetrabutylammonium salt) in 5% $CDCl_3/d_6$-DMSO.

Receptor	$H_2PO_4^-$ K_a	AcO^- K_a
36a	2640	270
36b	24 700	7120
36c	30 000	23 400

shows a selection of cyclopeptide derivatives synthesized by the Haberhauer group.

The X-ray structure of **34a** revealed that all the substituents on the imidazole moieties point to one direction opposite to the isopropyl groups of the adjacent α-C-atoms. Moreover, these substituents do not point downward perpendicularly but are arranged in a triple helix-like fashion. This suggests that coordination of all three 2,2′-bipyridyl groups to a metal ion would give one of the two diastereomeric octahedral complexes in excess, possibly exclusively. Indeed, on reaction of **34a** with ruthenium(II) or osmium(II) salts only one diastereomer of the corresponding complexes [**34a**·Ru]$^{2+}$ and [**34a**·Os]$^{2+}$ could be isolated.[35] A crystal structure showed that the octahedral metal complexes possess the Λ-configuration in these compounds.

Also compounds **34b**, **34c**, **35a**, and **35b** contain appropriate ligands in the side chains that allow for transition metal binding. As a consequence, coordination of metal ions was observed in all cases, leading to the highly diastereoselective formation of the corresponding complexes. Receptor **34d** enantioselectively interacts with quaternary ammonium ions in $CDCl_3$, presumably via cation–π interactions between the substrates and the pendant aromatic subunits.[36] For example, affinity toward *R*-1-phenylethylammonium perchlorate in $CDCl_3$ amounts to 30 000 M^{-1}, while the enantiomer of this cation is bound with a K_a of only 4500 M^{-1}. Anion binding was investigated by using cyclopeptides **36a–c**.[37] These peptides interact with various inorganic anions in 5% $CDCl_3/d_6$-DMSO, with stability constants of the complexes correlating with the basicity of the anions (more basic anions are more strongly bound) and with the fit of the anions into the available cavity. The highest affinity was observed for acetate and dihydrogenphosphate. In addition, the nature of the heterocyclic subunits plays a large role in complex stability. Thus, independent of the anion, receptor **36c** forms the most stable anion complexes, followed by **36b** and **36a** (Table 6). The lowest anion affinity of **36a** was rationalized by unfavorable interactions of the anions with the lone pair of the nitrogen atoms in the imidazole subunits. These lone pairs can, however, serve as a hydrogen-bond acceptor

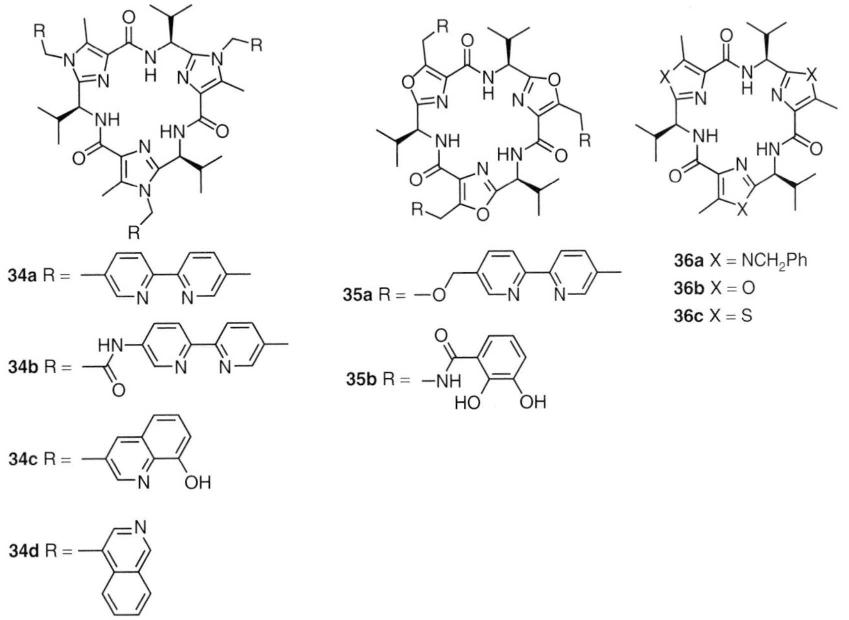

34a R =
34b R =
34c R =
34d R =
35a R =
35b R =
36a X = NCH_2Ph
36b X = O
36c X = S

Scheme 17

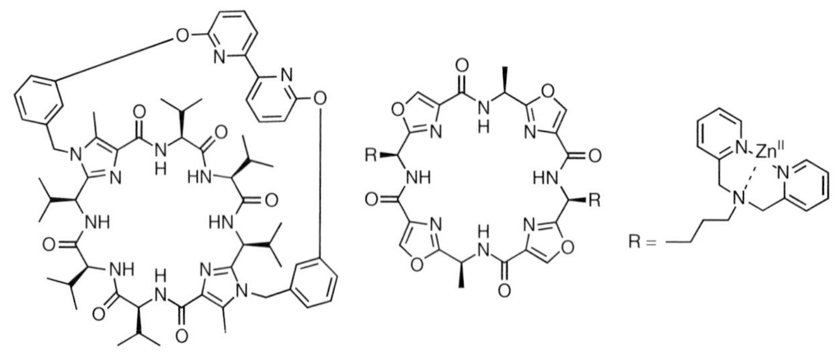

Scheme 18

for protonated anions, which is presumably the reason why the dihydrogenphosphate complex of **36a** is about 10 times more stable than the complex with acetate.

Another interesting system introduced by Haberhauer is the molecular hinge **37** (Scheme 18). The uncomplexed form of **37** represents the open state of the hinge and exhibits planar chirality. The 2,2′-bipyridine moiety in this form can only adopt the *M* conformation because the *P* conformation is strongly destabilized by the chiral peptidic scaffold. Copper(II) binding induces rotation around the central bond in the bipyridine moiety and stabilizes a ligand conformation with *syn*-oriented nitrogen atoms. This closing of the hinge destroys its planar chirality, which can nicely be followed by CD spectroscopy. The hinge can again be opened by the addition of cyclam which causes decomplexation of [**37**$\cdot Cu^{2+}$], but since only the *M* conformation can be reached from the closed state the resulting molecular motion is unidirectional, a characteristic feature of molecular machines. A recent investigation showed that it is possible to modify the amplitude and height of the hinge motion by structural variation of the cyclopeptide scaffold or by changing the position at which the bipyridine moiety is attached to it.[38] A similar cyclopeptide scaffold containing an azobenzene moiety in place of the bipyridine unit allows control over the reversible switching of the azo chromophore between the achiral *trans* configuration and one of the chiral *cis*-configurations.[39]

In a somewhat related approach, the Jolliffe group synthesized the C_2 symmetric cyclic octapeptide **38** (Scheme 18), in which two pendant dipicolylamino groups complexed to Zn(II) served as binding sites for phosphate anions.[40]

4 LINEAR PEPTIDES

Proteins are linear molecules whose molecular recognition properties are crucially dependent on the correct folding of the chain. In recent years, various groups demonstrated that oligomers or polymers can be designed solely composed of synthetic building blocks which also fold into well-ordered secondary structures in solution or the solid state and that folding pattern is controlled by the structure of the subunits.[41,42] So far, there are only few examples of such foldamers, however, whose folded conformations also exhibit molecular recognition properties. Examples are the "molecular apple-peels" introduced by Huc.[43] Although these compounds contain amide groups along the chain, they are not peptides in the strict sense and they are therefore not considered here.

Synthetic receptors derived from linear peptides were developed in several groups by using surprisingly similar approaches. Most receptors were, for example, synthesized in a combinatorial fashion to facilitate identification of the most potent receptor from a large pool of related structures. In addition, all receptors contain at least one synthetic building block, which either controls the spatial arrangement of pendant peptide chains or serves as an additional binding site with known receptor properties. The receptors developed in the Schmuck group, for example, contain a guanidiniocarbonyl pyrrole moiety at the end of a peptide chain to induce affinity for substrates with a free carboxylate group. Guanidinium moieties located between two peptide chains serve the same purpose in the receptors described by Kilburn *et al.* The diketopiperazine subunits used by the Wennemers group allow two peptide chains to be arranged in a parallel manner. These examples also illustrate that most receptors developed in this context were used to sequence-selectively recognize small peptides.

The contribution of the Schmuck group to the development of peptide-derived receptors is based on the finding that the guanidiniocarbonyl pyrrole moiety can serve as a highly efficient binding motif for carboxylates.[44] This residue efficiently interacts with carboxylates by a combination of hydrogen-bonding and Coulomb attraction, allowing complex formation to proceed in competitive polar solvents such as DMSO, DMSO/water mixtures, and

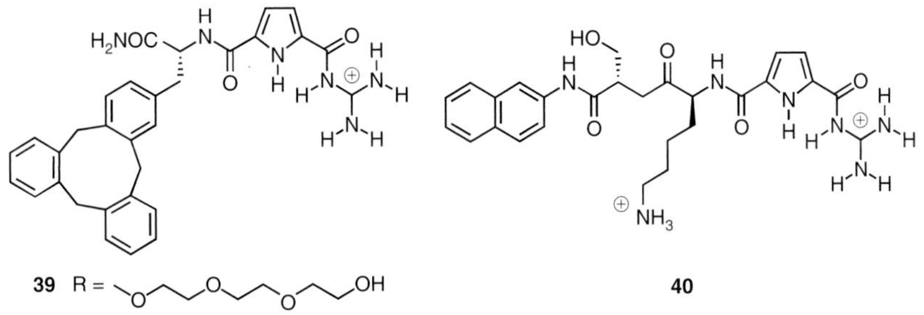

Scheme 19

water. In addition, receptor properties can be deliberately controlled by varying the substituents in 5-position of the pyrrole ring. In a rational approach, receptors **39** and **40** (Scheme 19) were devised, for example, comprising short peptide residues (composed of natural and/or nonnatural building blocks) and a terminal guanidiniocarbonyl pyrrole group. Receptor **39** was shown to efficiently recognize the deprotonated dipeptide Ac-D-Ala-D-Ala-O^- in buffered water with a K_a of 33 100 M^{-1}.[45] The cyclotribenzylene-substituted alanine derivative at the N-terminus of this compound serves as a binding site for side chains of the substrate. Thus, substrates lacking side chains (Ac-Gly-Gly-O^-) or ones with sides chains too large to be included into the bowl of the cyclotribenzylene group (Ac-D-Val-D-Val-O^-) are bound significantly less efficiently by factors of more than 10. Complex formation is not stereoselective as Ac-D-Ala-D-Ala-O^- and the corresponding enantiomer are bound with almost the same affinity.

Bis-cationic host **40** contains a lysine residue for improved anion affinity, serine for water solubility, and a naphthyl group to allow for favorable hydrophobic contacts with the substrate. Binding studies showed that **40** possesses millimolar affinities ($K_a = 2–5 \times 10^5\,M^{-1}$) for dipeptide carboxylates in 20% DMSO/water (pH 6.0) and exhibits some preference for alanine in the C-terminal position.[46]

Alternatively, optimization of peptide affinity and selectivity of guanidiniocarbonyl pyrrole-derived peptide receptors has been achieved by using combinatorial methods.[47] In this context, a library of hosts of the general structure **41** (Scheme 20) was prepared on solid support, and binding affinity was screened toward the N-protected tetrapeptide Ac-Val-Val-Ile-Ala-O^-, the C-terminal sequence of the amyloid-β-peptide. The first investigation involved a library of 125 receptors of which about 7% showed selective binding to the target substrate. After selection of the most efficient receptors, an on-bead binding assay was used to determine relative affinities for Val-Val-Ile-Ala-O^- in methanol. Substrate affinities of the two best receptors **41a** and **41b** amounted to 9800 M^{-1} and 9300 M^{-1}, respectively. Subsequently, a larger library containing 512 potential hosts was synthesized and screened in water (5 mM bis-TRIS, pH 6). In this solvent, binding affinities of the most potent receptors toward the target are less than half of those observed in methanol. More important is, however, that the best receptors in water are structurally quite different from the ones in methanol; the largest affinity for the target peptide on solid support ($K_a = 8800\,M^{-1}$) and in bis-TRIS buffer at pH 6.1 ($K_a = 6025\,M^{-1}$) was detected, respectively, for receptor **41c** and the analog Gua-Lys-Leu-Lys-NH_2. This result clearly demonstrates the important influence of the solvent on supramolecular complex formation, both in terms of binding affinity and selectivity. Notably, the best receptors that have thus been identified inhibit the formation of amyloid plaques *in vitro*, but only for amyloid-β(1–42) and not amyloid-β(1–40) which has the wrong C-terminal sequence.

In related studies, Schmuck and coworkers also identified an efficient receptor for the tetrapeptide Ac-D-Glu-Lys-D-Ala-D-Ala-O^-. A receptor library containing 512 members of the general structure **41** was prepared, and affinity toward the target peptide was evaluated using an on-bead binding assay. Affinities of the receptors within the library varied between <20 M^{-1} and 17 100 M^{-1}. To validate these results, analogs of the two best receptors, containing a substituent with a fluorescent dye at the N-terminus, were resynthesized and their binding properties in solution

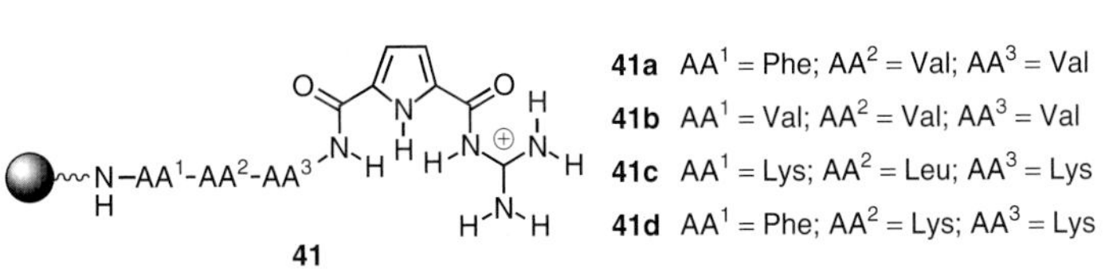

Scheme 20

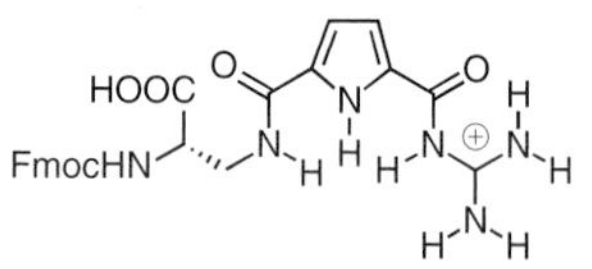

42

Scheme 21

were characterized. These experiments confirmed that both receptors form stable complexes also in solution. For example, stability of the complex between Ac-D-Glu-Lys-D-Ala-D-Ala-O$^-$ and the best receptor with the same amino acid sequence as **41d** (Scheme 20) amounts to 15 400 M^{-1}. Subsequent binding studies then showed that altering the sequence of amino acids in the target peptide, or the configuration of a single subunit, causes a considerable drop in affinity. Such one-armed guanidiniocarbonyl pyrrole-containing short peptides thus represent a remarkably potent family of selective receptors despite the fact that complex formation is mainly based on a single charge interaction and both substrates and receptors are rather flexible.

Schmuck *et al.* have also developed the arginine analog **42** (Scheme 21) containing a 2-guanidiniocarbonyl pyrrole moiety in the side chain. Incorporation of **42** into a library of octapeptides containing 625 individual members has furnished peptides that accelerate phosphoester hydrolysis in water by a factor of 175 over the uncatalyzed background reaction.[48] Since the most active peptides all contain serine or histidine in addition to **42**, it is postulated that the guanidiniocarbonyl pyrrole moiety is responsible for substrate binding by interacting with the negatively charged subunit of the substrate while the nucleophilic amino acids histidine and serine induce the actual bond cleavage.

This approach was recently extended to target the cleavage of RNA model compounds.[49] A polypeptide that adopts a characteristic helix-loop-helix conformation and accelerates the hydrolysis of the RNA model substrate 2-hydroxypropyl 4-nitrophenyl phosphate (HPNP) by about 2 orders of magnitude was used in this investigation. Two histidine and two arginine units in this peptide were believed to be responsible for catalytic activity. Analogs of this peptide, in which the two histidine, the two arginine, or all four amino acid residues were replaced by **42,** were prepared. Interestingly, the latter peptide proved to possess the highest activity, accelerating the hydrolysis of HPNP by a factor of 150 over the rate of the reaction catalyzed by the parent peptide and by a factor of 1500 over the rate of the imidazole-catalyzed reaction. It was thus concluded that **42** cannot only serve as a binding site for the substrate but that is also provides general-base catalysis in the reaction investigated.

The tweezer-type receptor **43** that was developed by Kilburn and coworkers contains a disubstituted guanidinium group in the middle of the chain which was expected to bind to the terminal carboxylate group of peptidic guests (Scheme 22).[50] The two peptide arms that are arranged in a parallel fashion serve to induce substrate selectivity. To test this idea, **43** was incubated in aqueous sodium borate buffer (pH 9.2, 16.7% DMSO) with a 1000-member library of tripeptides attached to a TentaGel resin via the amino terminus. Mainly hydrophobic amino acid residues were incorporated into these tripeptides to ensure that receptor substrate interactions are largely due to hydrophobic interactions. Receptor **43** was found to bind to about 3% of the library members and showed 95% selectivity for Val at the carboxylate terminus of the tripeptides and 40% selectivity for Glu(O*t*Bu) at the amino terminus. A stability constant of $4 \times 10^5\ M^{-1}$ was determined for the complex between **43** and *Z*-Glu(O*t*Bu)-Ser(O*t*Bu)-Val-O$^-$ in 16.7% DMSO/water (1 mM sodium borate buffer, pH 9.2) by means of isothermal titration microcalorimetry.

The reverse experiment involved a resin-bound library of symmetrical tweezer-type receptors **44a**, with identical peptide fragments appended to both sides of the guanidinium scaffold (Scheme 22). Screening with a dye-labeled tripeptide as substrate allowed for the identification of a host that was shown to interact with this tripeptide in 15% DMSO/water with appreciable selectivity over the corresponding enantiomer. This approach was subsequently

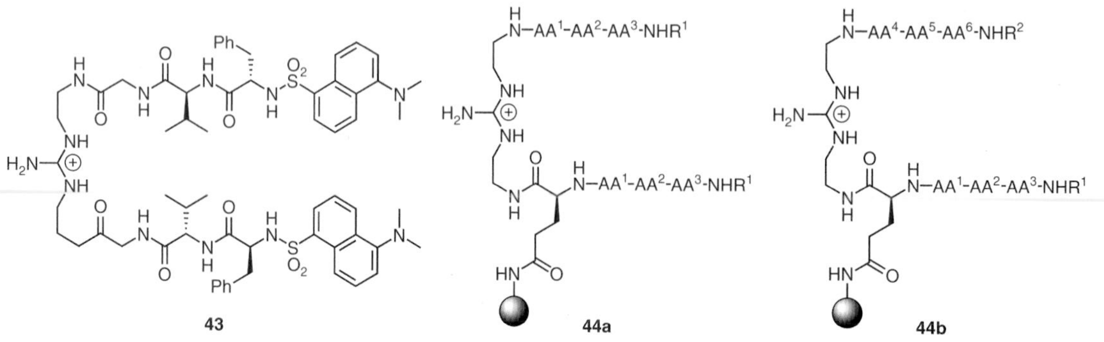

43 44a 44b

Scheme 22

extended to "unsymmetrical" receptors **44b**, whose arms were synthesized independently.[51] Libraries of **44b** were screened to identify receptors for the Ac-Lys-D-Ala-D-Ala-O$^-$ tripeptide sequence. A receptor thus identified (AA^1 = Gly; AA^2 = Val; AA^3 = Val; AA^4 = Met; AA^5 = His; AA^6 = Ser; R^1 = Ac; and R^2 = H) was resynthesized and used to determine the association constant with Ac-Lys-D-Ala-D-Ala-O$^-$. In free solution, although a UV and NMR binding study provided evidence for an association between the receptor and the tripeptide, the data did not allow the estimation of complex stability using a 1 : 1 binding model. When attached to the solid support, however, the receptor binds to the tripeptide in an aqueous buffered solution with a millimolar association constant. The failure of binding in the solid phase to translate to equivalent binding in free solution was ascribed to the environment created by the resin that could affect binding affinity by excluding or organizing solvent molecules within the resin matrix or by suppressing aggregation of the tweezer receptor molecules, which occurs readily in free solution.

The peptide receptors developed by Wennemers *et al.* contain no extra binding site to target certain functional groups in the substrate. These two-armed receptors are composed of two identical tripeptide units linked via their terminal carboxylate groups to the di(*trans*-4-aminoproline)-diketopiperazine (systematic name: (2*R*, 5a*S*, 7*R*,10a*S*)-2,7-diaminooctahydrodipyrrolo[1,2-a:1′, 2′-d]pyrazine-5,10-dione) moiety (Scheme 23). This central subunit serves to induce a parallel arrangement of the pendant substituents, allowing them to simultaneously engage in substrate recognition. To probe the ability of these receptors to sequence-selectively recognize short peptides, the dye-labeled derivatives **45a–e** were equilibrated with a library of resin-bound tripeptides containing 24 389 different members.[52] Notably, each of these five receptors was shown to select for different tripeptides within the library, despite their close structural relationship. For example, while receptor **45a** exclusively selected peptides containing a D-His following two hydrophobic D-amino acids, receptor **45e** solely chose peptides with an Asn following a hydrophobic L- and a hydrophobic D-amino acid. Receptor **45b**, differing from **45a** in the configuration of the tyrosine residue, selected for sequences with an Asn in the middle or N-terminal position being flanked by a combination of a D- and L-hydrophobic amino acids. These results clearly demonstrated that slight structural changes in the peptide residues of these receptors can cause pronounced differences in binding selectivity. Quantitative evaluation of peptide affinity showed that **45a** binds to the tripeptide Ac-D-Val-D-Val-D-His with a K_a of 1430 M^{-1} in chloroform. Affinity of **45d** to the same tripeptide is significantly lower (K_a = 260 M^{-1}), and small modifications of the substrate such as inversion of the absolute configuration at the central D-Val residue cause a further substantial drop in affinity (K_a = 15 M^{-1}). Thus, such diketopiperazine-derived receptors are not only highly selective for certain tripeptides but also show a significantly higher binding affinity toward the selected peptides in comparison to nonselected peptides.

The same series of receptors was also shown to very selectively interact with resin-bound side chain-protected tripeptides. Receptor **45f** containing four aspartic acid residues with free carboxylate groups in the side chains was subsequently shown to allow peptide binding in water (100 mM TRIS buffer, pH 7.2).[53] Not unexpectedly,

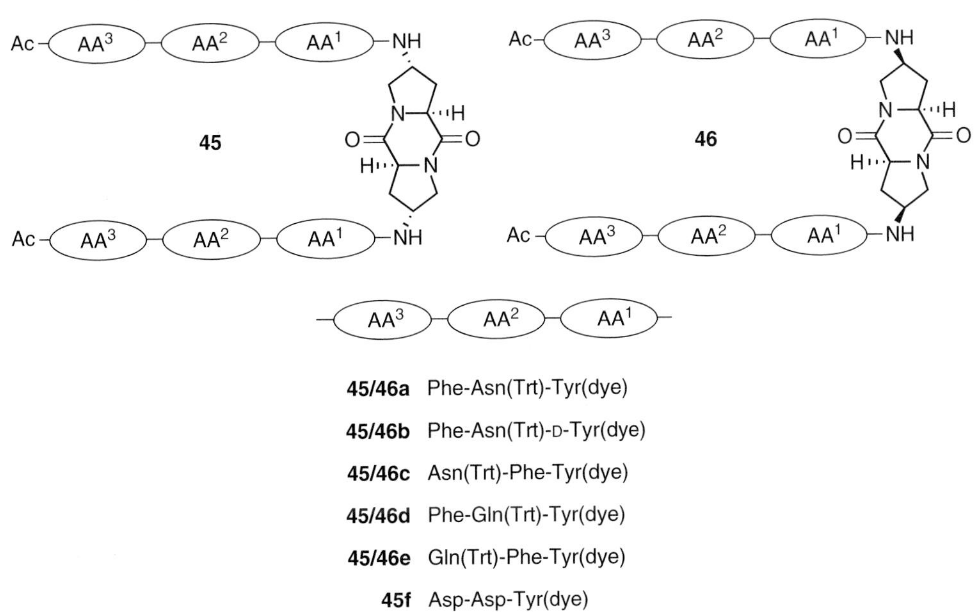

Scheme 23

this receptor preferentially binds arginine-rich tripeptides. Comparison of the properties of receptors of the general structure **45** containing the di(*trans*-4-aminoproline)-diketopiperazine unit with those of analogs derived from di(*cis*-4-aminoproline)-diketopiperazine showed that the configuration at the positions in these receptors carrying the peptide side arms has a pronounced effect on binding properties.[54] Although the di(*cis*-4-aminoproline)-diketopiperazine unit in receptors **46a–e** can be expected to orient the two substituents in a similar fashion as the corresponding diastereomeric diketopiperazine derivative, peptide affinities of receptors **45a–e** are overall superior. No staining of beads was observed, for example, when solutions of derivatives **46b**, **46c**, or **46e** were mixed with the resin-bound tripeptide library. Receptors **46a** and **46d** did show interactions, both preferentially interact with tripeptides containing a D-His followed by two hydrophobic D-amino acids.

The different properties of receptors **45** and **46** were attributed to differences in their overall structures imposed by the diketopiperazine subunits. While the di(*trans*-4-aminoproline)-diketopiperazine preferentially adopts a turn conformation arranging the two peptide subunits in close spatial proximity, the di(*cis*-4-aminoproline)-diketopiperazine has an almost linear structure. As a result, the distance of the substituents is significantly larger, which makes cooperative action in substrate binding more difficult. Connecting the two peptide side arms of receptor **45d** with appropriate linkers leads to macrocyclic receptors.[55] Although macrocyclization causes an overall decrease in binding affinity, subtle effects of the structure and the length of the linker on binding selectivity were noted.

Several examples of receptors mentioned above show that the proteins that are believed to be responsible for the deposition of insoluble plaques in the brains of patients suffering from neurodegenerative diseases such as Alzheimer's, Parkinson's, and Creutzfeld–Jacob disease are important targets in peptide recognition. For widely unknown reasons, these native proteins change conformation and adopt a β-sheet rich structure, which acts as a seed to nucleate misfolding of other proteins and subsequently aggregation. Insights into the mechanism of aggregation and the development of possible β-sheet binders, which slow down or even prevent this pathological process, have huge relevance from both a mechanistic and a therapeutic view. Low-molecular-weight templates that force a peptide strand into the β-sheet conformation by hydrogen bonding to the NH and C=O groups along the backbone have been developed, of which the 3-aminopyrazole derivative **47** (Scheme 24) described by Schrader *et al.* is one example. This simple heterocyclic building block contains a donor–acceptor–donor (DAD) hydrogen-bond pattern perfectly complementary to that of a β-sheet. Acylated aminopyrazoles were shown to lock even glycine-containing dipeptides in the β-sheet conformation on complexation by forming three hydrogen bonds with the top face of the peptidic guest.

The Schrader group subsequently devised oligomers containing 3-aminopyrazole subunits, for example, compounds **48a–c** (Scheme 24). Moderate stabilities were observed in chloroform for the complexes between receptors **48a–c** and peptides of similar length, with stability increasing with the number of possible hydrogen bonding interactions (Table 7).[56]

Interestingly, complex stability decreases only slightly when changing the solvent to phosphate buffer (pH 5.2), indicating that the decrease in complex stability that can be expected as a result of the weaker hydrogen bonding interaction in the aqueous environment is compensated to a large extent by a favorable effect such as hydrophobic

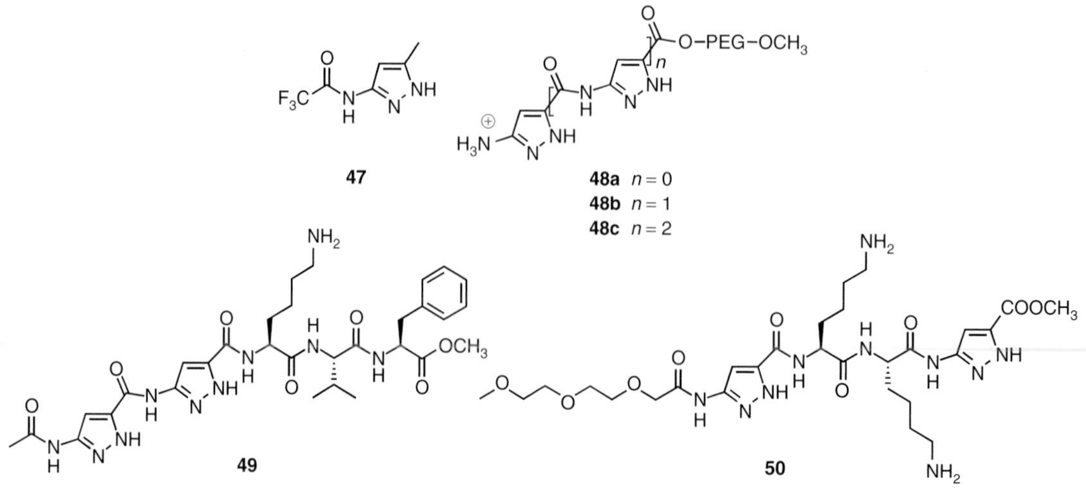

Scheme 24

Table 7 Stability constants K_a (in M^{-1}) of the complexes of receptors **48a–c** with peptides of varying lengths in $CDCl_3$.

Receptor	Substrate	K_a
48a	Ac-Val-Val-OCH_3	25
48b	Boc-Phe-Ala-Val-Leu-OCH_3	100
48c	Ac-Lys-Leu-Lys-Leu-Lys-Leu-OEt	790

interactions between the binding partners. A disadvantage of oligomers solely composed of 3-aminopyrazole moieties is their tendency to strongly self-aggregate in aqueous solution because of aromatic interactions between the planar heterocyclic subunits. To circumvent this problem, hybrid receptors containing heterocyclic amino acid and natural amino acid building blocks were devised. Examples of these receptors are compounds **49** and **50**, in which two adjacent pyrazole units in the chain are followed by a tripeptide fragment and in which a dipeptide is flanked by pyrazole moieties at both ends, respectively.[57] Since the nucleation site for the pathogenic aggregation of the Alzheimer's peptide has the Lys-Leu-Val-Phe-Phe sequence, the di- and tripeptide were taken from this key fragment. Binding studies revealed that of the two receptors **49** and **50**, **50** significantly and more efficiently interacts with Lys-Leu-Val-Phe-Phe in water, forming a 1 : 1 complex whose stability amounts to $1700\,M^{-1}$. A detailed conformational analysis provided experimental evidence of an increased β-sheet content induced in the peptidic substrate on complex formation. To optimize binding selectivity of these receptors toward the target peptide, a combinatorial receptor synthesis combined with an on-bead binding assay has recently been established.

Almost all of the oligomeric peptide-derived receptors presented so far contain at least one natural amino acid so that the term peptide is warranted. Only compounds **48**, which can be regarded as precursors in the development of amino acid-containing receptors **49** and **50**, are, in the strictest sense, not peptides but oligoamides. Structurally related to **48** are the synthetic receptors developed by the Dervan group, comprising a linear sequence of heterocyclic pyrrole or imidazole-derived amino acid building blocks. These oligomers sequence-selectively bind within the minor groove of DNA, recognizing the hydrogen-bond donor and acceptor groups of the nucleobase protruding into this part of the DNA molecule. They thus mimic the binding mode of natural DNA binders such as the antibiotic netropsin. These highly versatile receptors have been shown to exhibit a wide range of biological activities. They also possess important hallmarks of peptide-derived receptors. They are, for example, accessible via sequential solution-phase or solid-phase synthesis, they can adopt certain secondary structure motifs such as hairpin structures, or they can be transferred into cyclic derivatives, which also exhibit characteristic DNA binding properties. One could therefore argue that these compounds are members of the family of linear peptide-derived receptors, only ones composed of only nonnatural and nonchiral amino acid subunits. Although this argument is not unreasonable, these receptors are not treated in more detail within this chapter and the interested reader is referred to relevant reviews.[58,59]

5 ABIOTIC RECEPTORS WITH PEPTIDE-DERIVED SUBSTITUENTS

There are numerous receptors containing one or more amino acid residues attached to, for example, a calixarene, a crown ether, or a cyclodextrin. These residues serve as binding sites complementing the properties of the underlying scaffold, or, often in the case or tryptophan residues, allowing the binding event to be followed by fluorescence spectroscopy. An overview of this huge and structurally very diverse class of receptors is outside the scope of this chapter. Instead, only selected examples of receptors are presented, which exhibit unusual binding properties or contain not only monomeric amino acid residues but also oligomeric chains of amino acid sequences arranged around suitable core molecules.

One popular receptor platform that was decorated with amino acid residues in many different ways is that of calixarenes. Substantial contributions in this context came from the Ungaro and co-workers, who developed two types of so-called peptidocalix[4]arene derivatives, namely, C-linked peptidocalixarenes containing amino acids linked to the calixarene moiety via their carboxylate function, and N-linked derivates, in which the amino group of the amino acid serves as the anchor group.[60,61] Particularly interesting receptor properties were found for receptors with an additional strap between the amino acid residues, for example, compound **51** (Scheme 25). This receptor was shown to form a 1 : 1 complex with the dipeptide Ac-D-Ala-D-Ala, that has a log K_a of 3.4 in 3% DMSO-d_6/$CDCl_3$. Because of its affinity for the same dipeptide that is also targeted by the antibiotic vancomycin, biological activity of **51** and derivatives thereof has been investigated. Interestingly, activity of **51** against several Gram-positive bacterial strains is close to that of vancomycin. Increasing the length of the bridge in **51** by using the Ala–Ala dipeptide chain segment instead of a simple Ala causes a pronounced drop in activity, while protecting the central NH group with a Boc group or substituting it with a methylene group completely inhibits biological activity. The active compounds show a behavior very close to that of vancomycin: no activity is observed against Gram-negative bacteria (*Escherichia*

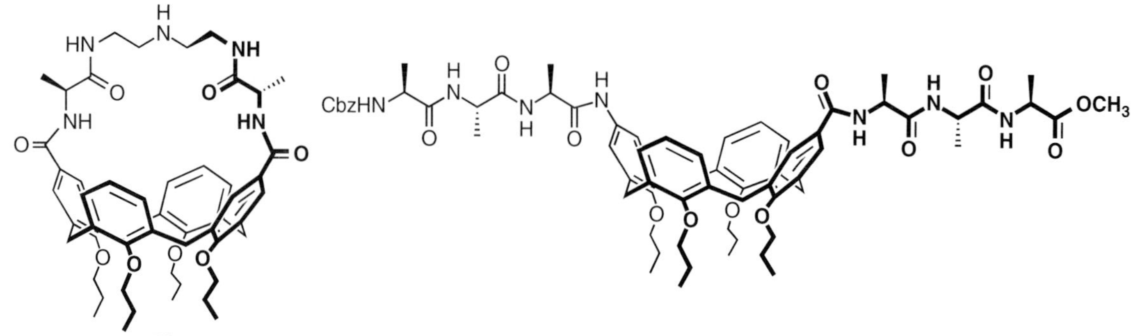

51 52

Scheme 25

coli), yeast (*Saccharomyces cerevisiae*) or cell-wall-lacking bacteria (*Acholeplasma laidlawii*), indicating that the biological target of this class of calixarene-based antimicrobials is, as for vancomycin, the terminal D-Ala–D-Ala part of the peptidoglycan that, after cross-linking, constitutes the bacterial cell wall.

The Ungaro group recently also described the synthesis of a calixarene derivative containing an amino group and a carboxylate group on opposite aromatic subunits. The attachment of two amino acids or small peptides in opposite orientation (one C-linked and the other N-linked) to this scaffold afforded self-complementary linear peptides containing one abiotic subunit such as the formal heptapeptide **52** (Scheme 25). These compounds were shown to self-assemble in apolar media into dimeric structures held together by a hydrogen bonding pattern between the C=O and NH groups analogous to that of an antiparallel β-sheet. In the dimer formed, two calixarene moieties are arranged in a capsule-like fashion. The dimerization constants of these aggregates depend on the chain length of the peptide residues, increasing with the number of hydrogen-bond donors and acceptors.

Peptidocalixarenes anchored to the solid support were also used as scaffolds to attach *four* tetrapeptide chains in a combinatorial fashion.[62] The ability of these products to serve as peptide receptors was investigated in a similar manner to the peptide-derived receptors described in section 4. Also trisubstituted cyclotriveratrylenes[63] or the 1,3,5-trisubstituted 2,4,6-triethybenzene core has been shown to give rise to synthetic receptors on attachment of suitable peptide residues. Important contributions concerning the properties of the latter type of receptors came from the Anslyn group.[64] Receptors of the general structure **53** (Scheme 26), containing a hexasubstituted 1,3,5-triethylbenzene core as scaffold besides two guanidinium groups for anion binding, two identical peptide side chains for selectivity, and two fluorophores F^1 and F^2 for optical sensing, were, for example, synthesized in a combinatorial approach to achieve ATP sensing in aqueous media.

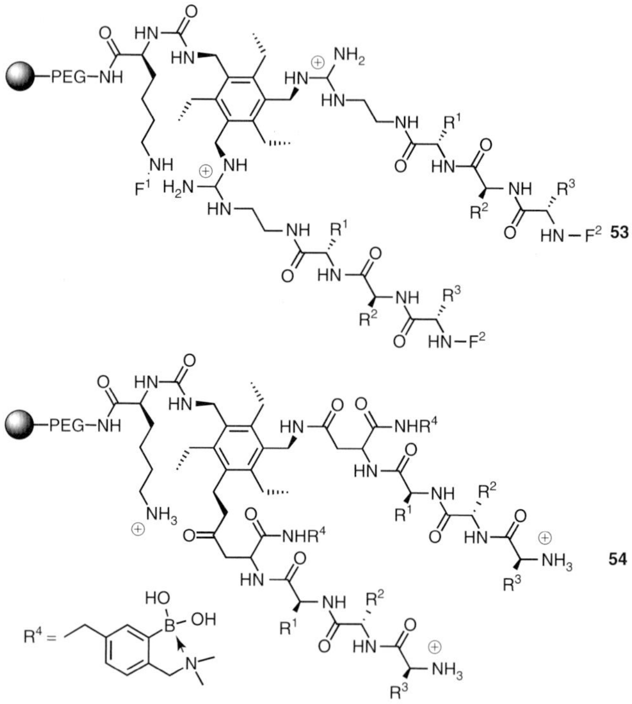

53

54

Scheme 26

By screening a host library containing 4913 members and differing in the peptide sequence along the side chains, a receptor was identified with a remarkably high affinity ($K_a = 3.4 \times 10^3\ M^{-1}$ in 200 mM HEPES, pH 7.4) and high selectivity for ATP over GTP and AMP. In a slightly different approach, a selection of 12 beads of the host library (before attachment of the fluorophores) was chosen to construct a chip-based array for the optical differentiation between structurally similar anions such as AMP, GTP, and ATP using an indicator displacement assay.

To differentiate proteins and glycoproteins, a 6859-member library containing receptors of the general structure **54** (Scheme 26) was synthesized, which incorporates one of 19 natural amino acids (cysteine was excluded) at each of

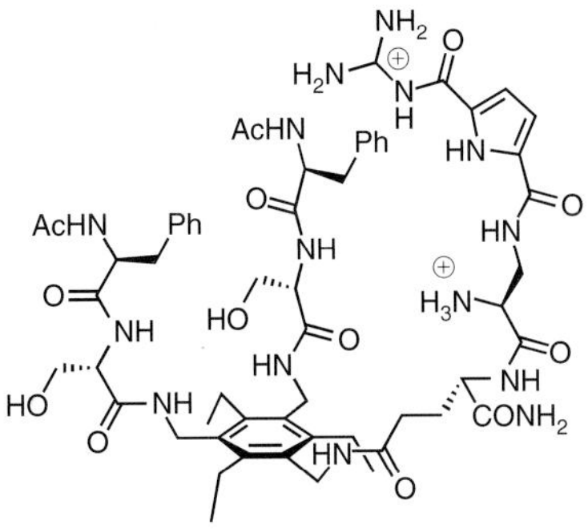

Scheme 27

three sites on two different binding arms. The peptide arms were expected to provide binding sites for proteins while the boronic acids were introduced because of their ability to reversibly form esters with diols in aqueous media. This reaction was expected to permit the differentiation of glycoproteins from conventional proteins. Interaction of an array of 29 randomly selected resin-bound receptors with a series of representative protein types was analyzed spectroscopically by following the rate of uptake of a dye in the presence of a protein solution. Principal component analysis of the results demonstrated that the array of receptors adequately separated proteins from glycoproteins and, to a lesser extent, even separated proteins within the classes.

The Schmuck group demonstrated that attachment of dipeptide fragments in combination with a guanidiocarbonyl pyrrole group to the 1,3,5-trisubstituted 2,4,6-triethylbenzene platform gives rise to a receptor with affinity for anionic carbohydrates in 20% buffered water in DMSO.[65] For example, the stability constants of the complexes of the corresponding tripodal receptor **55** (Scheme 27) with glucose-1-phosphate and cAMP amount to, respectively, $3000\,M^{-1}$ and $2400\,M^{-1}$. Binding selectivity is influenced by a complex interplay of steric effects, solvation effects, and ion pairing of the individual binding partners.

Conjugates **56a–c** containing a calix[4]arene core with four identical conformationally constrained cyclopeptides, arranged along the upper rim, were developed by the Hamilton group to achieve selective recognition of protein surfaces.[66,67] Receptor design was inspired by the realization that many protein surfaces have the characteristic feature to contain hydrophobic patches surrounded by a ring of polar and/or charged areas. These areas can be specifically addressed by the appropriate choice of amino acid residues in the cyclopeptide moieties of **56**. The calixarene core, on the other hand, not only preorganizes the cyclopeptide units in a circular fashion but also features a sufficiently large hydrophobic surface to cover the hydrophobic region on the protein. If interaction of these receptors with protein surfaces alters or inhibits protein activity, for example, because binding of **56** to the protein surface prevents the substrate from entering the active center, these systems would represent promising lead structures for pharmaceutical applications, whose mode of action differs profoundly from that of substances that directly bind to the active site (Scheme 28).

Receptor **56a** was shown to interact with the surface of cytochrome *c*, where a hydrophobic region is surrounded by positively charged lysine residues. This interaction significantly reduces the rate at which Fe(III)-cytochrome *c* is reduced by ascorbate, indicating that the ligand blocks access to the heme center of the protein. A similar inhibitory effect of **56a** is observed for the cytochrome *c*-cytochrome *c* peroxidase complex. Receptor **56a** also acts as a competitive inhibitor of α-chymotrypsin by binding to a patch of several cationic residues, which is found near the active site of the enzyme. Ligand **56b** was shown to bind to the platelet-derived growth factor (PDGF), whose receptor-binding region is composed of cationic and

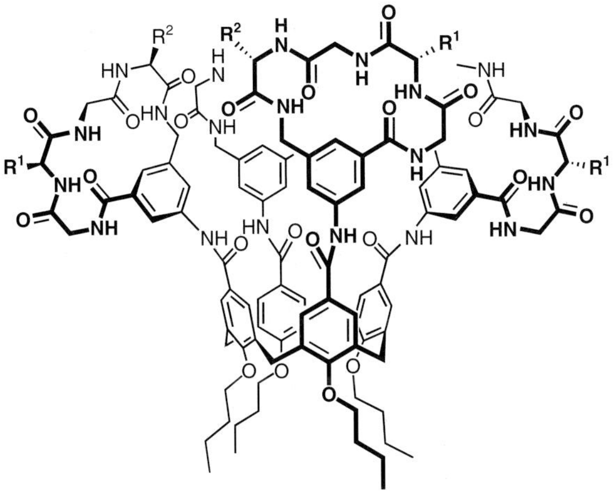

56a $R^1 = R^2 = -CH_2COOH$
56b $R^1 = -CH_2COOH$; $R^2 = -CH_2\text{-}p(C_6H_4OH)$
56c $R^1 = R^2 = -(CH_2)_4NH_2$

Scheme 28

hydrophobic residues. *In vivo* studies revealed that tumor growth and angiogenesis can be inhibited in nude mice bearing human tumors on treatment with **56b**. This effect is presumably due to binding of the ligand to PDGF, which in turn becomes unable to interact with its membrane-bound receptor. As a consequence, cell-signaling pathways, which would be triggered by PDGF–receptor interactions and would eventually lead to cell growth and angiogenesis, are shut off. In a similar way, compound **56c** selectively disrupts binding of the vascular endothelial growth factor (VEGF) to its receptor, thus also inhibiting angiogenesis both *in vitro* and *in vivo*, and tumorigenesis and metastasis *in vivo*. In a related approach, Neri and coworkers screened a small library of tetrasubstituted calix[4]arenes bearing tetrapeptide substituents along the upper rim for potential inhibitors of tissue and microbial transglutaminase.[68]

6 CONCLUSION

First work on the development of peptide-derived synthetic receptors dates back to the early years of supramolecular chemistry. While the initial studies were mainly concerned with the systematic structural variation of ionophores to better understand or to improve their binding properties, the field more and more moved toward the conception of innovative new systems only remotely related to natural analogs. Because many different approaches were pursued in this context involving, for example, cyclic or linear receptors, or the use of nonnatural subunits of considerable structural diversity, the family of peptide-derived receptors turned out to be structurally much less homogeneous than, for example, crown ether-, cyclodextrin-, or calixarene-based systems. A systematic classification is therefore difficult. Still, the examples presented in this chapter clearly demonstrate that peptide-derived receptors came full circle from mimics for natural systems over receptors that may help to address basic issues in molecular recognition to completely abiotic receptors that possess many characteristic features of natural systems such as activity in the aqueous environment and high affinity to biological targets. Thus, practical application of such receptors in sensing or in medicinal chemistry can now be foreseen.

The remarkable development of this field owes to the highly versatile synthetic approach with which peptide-derived receptors are accessible. This approach principally differs from the syntheses of many other receptor types in that peptide-derived receptors are obtained by the sequential assembly of suitable building blocks. Thus, every subunit along the chain or the ring can be exchanged easily, which makes fine-tuning of receptor properties relatively straightforward. In contrast, a defined sequence of different subunits around the cavity of a preformed macrocyclic receptor like a calixarene is much more difficult to achieve. Moreover, with the advent of combinatorial chemistry, the synthesis of libraries of peptide-derived receptors has also become possible, which significantly facilitated receptor screening and the identification of potent receptors. Use of combinatorial methods also provided a solution to one of the main problems of peptide-derived receptors: their conformation (secondary structure), which is crucial for selective substrate recognition, is usually difficult to predict, making a rational design particularly of acyclic peptide-derived receptors difficult. This problem is less pronounced in the case of cyclic systems whose conformational mobility is generally lower and can be further reduced by the incorporation of rigid (nonnatural) amino acids along the ring. Finally, the possibility to combine peptides with almost all other receptor types relevant in molecular recognition studies significantly extends the structural space accessible with peptide-derived receptors.

Intensive research efforts during recent decades have established peptide-derived receptors as an important class of synthetic receptors in supramolecular chemistry. Remarkable advances, many of which have been made only recently, indicate that this family of receptors will continue to play an important role in the years to come.

REFERENCES

1. M. Dobler, *Ionophores and Their Structures*, John Wiley & Sons, Inc., New York, 1981.
2. D. H. Williams and B. Bardsley, *Angew. Chem. Int. Ed.*, 1999, **38**, 1173–1193.
3. B. S. Davidson, *Chem. Rev.*, 1993, **93**, 1771–1791.
4. Y. A. Ovchinnikov and V. T. Ivanov, *Tetrahedron*, 1975, **31**, 2177–2209.
5. B. F. Gisin, H. P. Ting-Beall, D. G. Davis, *et al.*, *Biochim. Biophys. Acta*, 1978, **509**, 201–217.
6. K. R. K. Easwaran, L. G. Pease, and E. R. Blout, *Biochemistry*, 1979, **18**, 61–67.
7. G. Saviano, F. Rossi, E. Benedetti, *et al.*, *Chem. Eur. J.*, 2001, **7**, 1176–1183.
8. E. R. Blout, *Biopolymers*, 1981, **20**, 1901–1912.
9. X. Li, Y.-D. Wu, and D. Yang, *Acc. Chem. Res.*, 2008, **41**, 1428–1438.
10. E. Ozeki, S. Kimura, and Y. Imanishi, *Int. J. Pept. Protein Res.*, 1989, **34**, 111–117.
11. J. P. Laussac, A. Robert, R. Haran, and B. Sarkar, *Inorg. Chem.*, 1986, **25**, 2760–2765.
12. M. Ngu-Schwemlein, P. Butko, B. Cook, and T. Whigham, *J. Pept. Res.*, 2006, **66**(Suppl. 1), 72–81.
13. V. Duléry, N. A. Uhlich, N. Maillard, *et al.*, *Org. Biomol. Chem.*, 2008, **6**, 4134–4141.
14. D. Leipert, F. Rathgeb, M. Herold, *et al.*, *Anal. Chim. Acta*, 1999, **392**, 213–221.

15. D. Leipert, D. Nopper, M. Bauser, *et al.*, *Angew. Chem. Int. Ed.*, 1998, **37**, 3308–3311.
16. C. García-Echeverría, F. Albericio, E. Giralt, and M. Pons, *J. Am. Chem. Soc.*, 1993, **115**, 11663–11670.
17. D. Ranganathan, *Acc. Chem. Res.*, 2001, **34**, 919–930.
18. A. Ortiz-Acevedo, H. Xie, V. Zorbas, *et al.*, *J. Am. Chem. Soc.*, 2005, **127**, 9512–9517.
19. N. Pant and A. D. Hamilton, *J. Am. Chem. Soc.*, 1988, **110**, 2002–2003.
20. C. J. Arnusch and R. J. Pieters, *Eur. J. Org. Chem.*, 2003, 3131–3138.
21. H. Ishida, M. Suga, K. Donowaki, and K. Ohkubo, *J. Org. Chem.*, 1995, **60**, 5374–5375.
22. H. Ishida and Y. Inoue, *Rev. Heteroat. Chem.*, 1999, **19**, 79–142.
23. S. Kubik, *Chem. Soc. Rev.*, 2009, **38**, 585–605.
24. C. Reyheller, B. P. Hay, and S. Kubik, *New J. Chem.*, 2007, **31**, 2095–2102.
25. S. Otto and S. Kubik, *J. Am. Chem. Soc.*, 2003, **125**, 7804–7805.
26. Z. Rodriguez-Docampo, S. I. Pascu, S. Kubik, and S. Otto, *J. Am. Chem. Soc.*, 2006, **128**, 11206–11210.
27. K. Choi and A. D. Hamilton, *J. Am. Chem. Soc.*, 2003, **125**, 10241–10249.
28. S.-W. Kang, C. M. Gothard, S. Maitra, *et al.*, *J. Am. Chem. Soc.*, 2007, **129**, 1486–1487.
29. S. D. Whitmarsh, A. P. Redmond, V. Sgarlata, and A. P. Davis, *Chem. Commun.*, 2008, 3669–3671.
30. R. Liu and W. C. Still, *Tetrahedron Lett.*, 1993, **34**, 2573–2576.
31. R. J. Pieters, J. Cuntze, M. Bonnet, and F. Diederich, *J. Chem. Soc., Perkin Trans. 2*, 1997, 1891–1900.
32. A. Bertram and G. Pattenden, *Nat. Prod. Rep.*, 2007, **24**, 18–30.
33. P. V. Bernhardt, P. Comba, D. P. Fairlie, *et al.*, *Chem. Eur. J.*, 2002, **8**, 1527–1536.
34. R. M. Cusack, L. Grøndahl, D. P. Fairlie, *et al.*, *J. Chem. Soc., Perkin Trans. 2*, 2002, 556–563.
35. G. Haberhauer, T. Oeser, and F. Rominger, *Chem. Commun.*, 2005, 2799–2801.
36. M. Schnopp and G. Haberhauer, *Eur. J. Org. Chem.*, 2009, 4458–4467.
37. M. Schnopp, S. Ernst, and G. Haberhauer, *Eur. J. Org. Chem.*, 2009, 213–222.
38. S. Ernst and G. Haberhauer, *Chem. Eur. J.*, 2009, **15**, 13406–13416.
39. G. Haberhauer and C. Kallweit, *Angew. Chem. Int. Ed.*, 2010, **49**, 2418–2421.
40. M. J. McDonough, A. J. Reynolds, W. Y. G. Lee, and K. A. Jolliffe, *Chem. Commun.*, 2006, 2971–2973.
41. S. Hecht and I. Huc, *Foldamers–Structure, Properties, and Applications*, Wiley-VCH, Weinheim, 2007.
42. D. J. Hill, M. J. Mio, R. B. Prince, *et al.*, *Chem. Rev.*, 2001, **101**, 3893–4011.
43. J. Garric, J.-M. Léger, and I. Huc, *Chem. Eur. J.*, 2007, **13**, 8454–8462.
44. C. Schmuck, *Coord. Chem. Rev.*, 2006, **250**, 3053–3067.
45. C. Schmuck, D. Rupprecht, and W. Wienand, *Chem. Eur. J.*, 2006, **12**, 9186–9195.
46. C. Schmuck and L. Hernandez-Folgado, *Org. Biomol. Chem.*, 2007, **5**, 2390–2394.
47. C. Schmuck and P. Wich, *Top. Curr. Chem.*, 2007, **277**, 3–30.
48. C. Schmuck and J. Dudaczek, *Org. Lett.*, 2007, **9**, 5389–5392.
49. N. J. V. Lindgren, L. Geiger, J. Razkin, *et al.*, *Angew. Chem. Int. Ed.*, 2009, **48**, 6722–6725.
50. N. Srinivasan and J. D. Kilburn, *Curr. Opin. Chem. Biol.*, 2004, **8**, 305–310.
51. J. Shepherd, T. Gale, K. B. Jensen, and J. D. Kilburn, *Chem. Eur. J.*, 2006, **12**, 713–720.
52. H. Wennemers, M. Conza, M. Nold, and P. Krattiger, *Chem. Eur. J.*, 2001, **7**, 3342–3347.
53. P. Krattiger and H. Wennemers, *Synlett*, 2005, 706–708.
54. H. Wennemers, M. C. Nold, M. M. Conza, *et al.*, *Chem. Eur. J.*, 2003, **9**, 442–448.
55. J. Bernard and H. Wennemers, *Org. Lett.*, 2007, **9**, 4283–4286.
56. K. Černovská, M. Kempter, H.-C. Gallmeier, *et al.*, *Org. Biomol. Chem.*, 2004, **2**, 1603–1611.
57. P. Rzepecki and T. Schrader, *J. Am. Chem. Soc.*, 2005, **127**, 3016–3025.
58. P. B. Dervan, *Bioorg. Med. Chem.*, 2001, **9**, 2215–2235.
59. P. B. Dervan and B. S. Edelson, *Curr. Opin. Struct. Biol.*, 2003, **13**, 284–299.
60. A. Casnati, F. Sansone, and R. Ungaro, *Acc. Chem. Res.*, 2003, **36**, 246–254.
61. L. Baldini, A. Casnati, F. Sansone, and R. Ungaro, *Chem. Soc. Rev.*, 2007, **36**, 254–266.
62. M. Kubo, R. Nishimoto, M. Doi, *et al.*, *Chem. Commun.*, 2006, 3390–3392.
63. C. Chamorro and R. M. J. Liskamp, *J. Comb. Chem.*, 2003, **5**, 794–801.
64. A. T. Wright and E. V. Anslyn, *Chem. Soc. Rev.*, 2006, **35**, 14–28.
65. C. Schmuck and M. Heller, *Org. Biomol. Chem.*, 2007, **5**, 787–791.
66. M. W. Peczuh and A. D. Hamilton, *Chem. Rev.*, 2000, **100**, 2479–2494.
67. H. Yin and A. D. Hamilton, *Angew. Chem. Int. Ed.*, 2005, **44**, 4130–4163.
68. S. Francese, A. Cozzolino, I. Caputo, *et al.*, *Tetrahedron Lett.*, 2005, **46**, 1611–1615.

Biological Small Molecules as Receptors

Carolina Godoy-Alcántar[1] and Anatoly K. Yatsimirsky[2]

[1] *Universidad Autónoma del Estado de Morelos, Cuernavaca, México*
[2] *Universidad Nacional Autónoma de México, México D.F., México*

1 INTRODUCTION

Many natural low-molecular weight compounds that function as antibiotics or have other biological activities, for example, as receptor antagonists, possess structural elements such as specific arrangements of charged or donor–acceptor groups, cavities or clefts, or chiral centers, making them promising host molecules for recognition of both ionic and neutral guests. Among these compounds, biological ionophores really use the classical host–guest complexation mechanism for their action, but in other cases the ability of natural compounds to bind guests of different types is not directly related to their biological functions; nevertheless, it can be fairly significant. Since many of these compounds are commercially available and inexpensive, they find practical applications as host molecules in analysis and separation.

Usually small biomolecules have one or more functional groups which are not essential for their recognition properties but can be used for their modification or incorporation into more sophisticated hosts. Such semisynthetic receptors with improved recognition or sensor properties are more easily accessible than purely synthetic receptors, and some of them also are discussed in this review.

Particularly popular biological small molecules with receptor properties are cyclodextrins, which are covered in a separate chapter in this book (see **Cyclodextrins: From Nature to Nanotechnology**, Volume 3). This review is focused on structures and physicochemical properties of other small biological molecules, mostly antibiotics, relevant to their use as receptors. Reported data on their recognition properties, as well as analytical applications, are discussed. Natural ionophors, which are extensively covered in previous literature, are touched upon only briefly.

2 BIOLOGICAL IONOPHORES

Ionophores, which are lipophilic molecules capable of transporting ions across the lipid bilayer of the cell membrane, are synthesized by microorganisms and have antibiotic properties by disrupting transmembrane ion concentration gradients. Ionophores are not used in human medicine because of their potent cardiovascular effects, but they are used in farming for the prevention of coccidiodomycosis in poultry and are fed to cattle to improve growth. They may be relatively large molecules, such as the polypeptide gramicidin which forms a transmembrane channel for ion transport, or smaller molecules acting as mobile

Supramolecular Chemistry: From Molecules to Nanomaterials.
Edited by Philip A. Gale and Jonathan W. Steed.
© 2012 John Wiley & Sons, Ltd. ISBN: 978-0-470-74640-0.

ion carriers. Ionophores of this last type were the subject of intensive studies during 1960–1980 and have served in many aspects as reference compounds for synthetic ionophores. They also were used for the development of the first neutral carrier-based selective electrodes for alkali-metal ions.[1] Thermodynamic parameters of cation binding to natural ionophores, measured mostly in methanol and other organic solvents, can be found in several reviews.[2–4] The properties of siderophores, an important group of ionophores specifically designed for transport of iron ions, are covered in several recent reviews[5–7] and are not discussed in this chapter.

In biological systems, ionophores selectively transport alkali- and alkaline-earth metal ions: for example, valinomycin is a selective carrier for K^+ and monensin for Na^+, but they also can bind other cations for which they have not been designed by nature[8] and this is an area of active current research with natural ionophores. One such nonnatural function of natural ionophores is the complexation of organic ammonium cations: for example, protonated amino acid esters. Since natural ionophores are chiral compounds, the process can be enantioselective. Binding of ammonium cations by such receptors as monensin or lasalocid (Figure 1) is efficient but lack the expected enantioselectivity.[9] However, cyclic or podand-type derivatives of monensin like **1** or **2** (Figure 1) show significant enantioselectivity: the ratios of binding constants of the *R* and *S* enantiomers of protonated methyl esters of phenylglycine, phenylalanine, and leucine to **2** are $K_R/K_S = 5.1$, 6.2, and 7.6, respectively.[9] Crown-ether type derivatives like **1** show lower enantioselectivities.

Nonnatural guests of another type are transition- and heavy-metal ions. Thus, monensin and nigericin (Figure 1) act as selective carriers for toxic Pb^{2+} cations, as illustrated in Figure 2.[10]

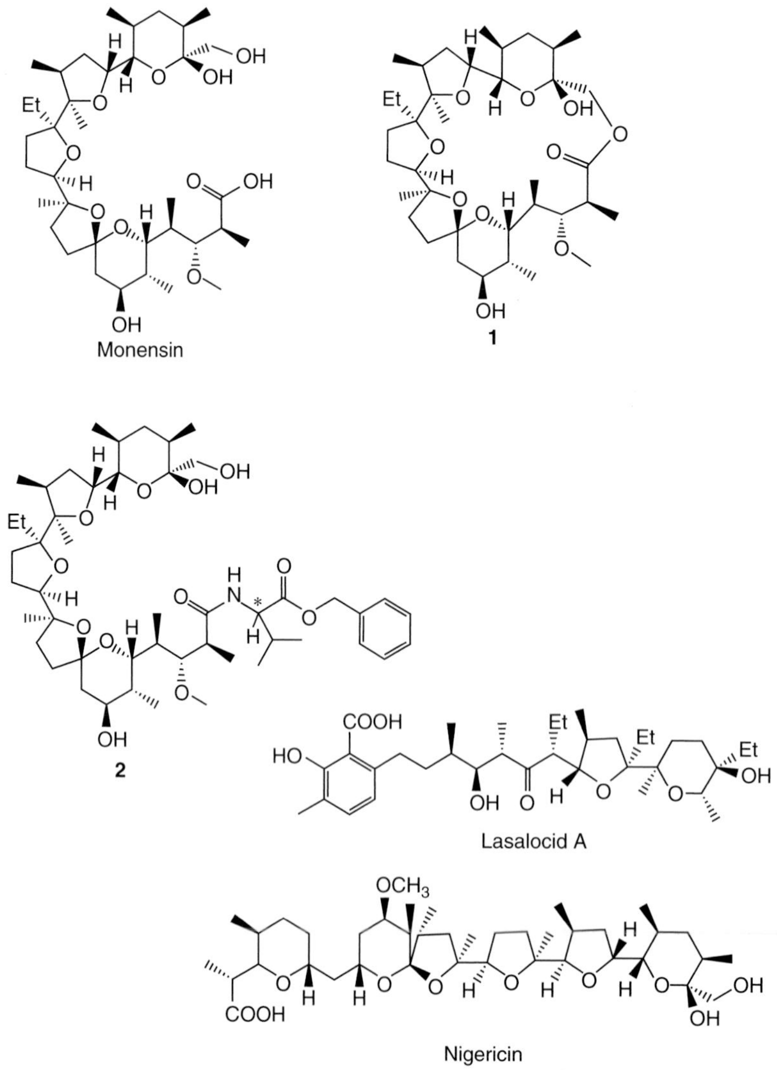

Monensin

1

2

Lasalocid A

Nigericin

Figure 1 Some biological ionophores and their derivatives with improved recognition properties.

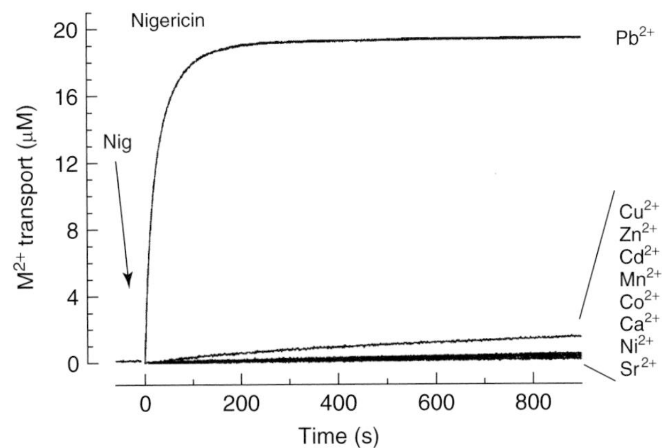

Figure 2 Transport of Pb^{2+} and other divalent cations through 1-palmitoyl-2-oleoyl-*sn*-glycerophosphatidylcholine vesicles by nigericin. (Reproduced from Ref. 10. © American Chemical Society, 2004.)

The binding constants for Pb^{2+} to nigericin and monensin anions ($\log K = 7.57$ and 7.25, respectively, in 80% methanol/water) by far surpass those for biological cations ($\log K$ for Mg^{2+}, Ca^{2+}, Na^{+}, and K^{+} are in the range 2.5–3.8), which enables these antibiotics to be used for treatment of lead poisoning.

Extensive studies including both inner-sphere and outer-sphere complexation of cations were performed with lasalocid A, which is a small natural ionophore containing a salicylic acid fragment (Figure 1).[11] The ability of lasalocid to form neutral outer-sphere complexes with species like $Co(NH_3)_6{}^{3+}$, $Cr(bpy)_3{}^{3+}$, $Pt(bpy)(NH_3)_2{}^{2+}$ allows one to use it as an ionophore for the membrane transport (including chiroselective transport) of such species. The lasalocid ionophore also was shown to be an efficient carrier for toxic water-soluble metal cations such as Pb^{2+} and Cd^{2+} across artificial flat-sheet-supported liquid membranes, which represent a potential system for separation of these cations.[12]

3 MACROCYCLIC GLYCOPEPTIDE ANTIBIOTICS

There are over a hundred glycopeptide antibiotics that typically are cross-linked polymacrocyclic heptapeptides having sugar substituents in different positions.[13] Structures of several frequently employed commercially available antibiotics are shown in Figure 3. The first discovered member of the group and the only one that is widely used as a therapeutic agent is vancomycin. Glycopeptide antibiotics act by inhibiting the biosynthesis of the bacterial cell wall. They specifically bind to polypeptide intermediates terminating with the sequence -D-Ala-D-Ala-COOH and prevent the transpeptidation reaction necessary for the synthesis of peptidoglycan. The same reaction is inhibited by β-lactam antibiotics, which block the transpeptidation enzymes. Therefore, vancomycin is used in patients allergic to β-lactam antibiotics and it has become the drug of last resort for the treatment of methicillin-resistant *Staphylococcus aureus*.

Some of these compounds are mixtures of several derivatives, usually with similar binding properties. Thus, teicoplanin is a mixture of five derivatives with different groups R in the acyl side chain of β-D-glucosamine, among which a derivative with $R = (CH_3)_2CH(CH_2)_7$ is the major component (tiecoplanin A_2-5).

Owing to the presence of several ionogenic groups, binding of guest molecules to glycopeptide antibiotics is pH-dependent with a wide optimum between pH values of about 4 and 7.5 (in case of vancomycin). The pK_a values for vancomycin and teicoplanin can be found in the literature.[14–17] Binding constants of model peptides and some low-affinity guests to most important antibiotics are summarized in Table 1 (data from Refs 18–23).

The highest affinity is observed with Ac_2-L-Lys-D-Ala-D-Ala and it is not improved even for a natural target undecaprenyl-*N*-acetylmuramyl-L-Ala-D-Glu-L-Dap-D-Ala-D-Ala.[13] The binding is remarkably enantioselective: substitution of the terminal D-Ala with L-Ala eliminates the binding completely. Shorter peptides show progressively smaller affinities with similar trends for all antibiotics. The interactions involved in peptide binding were identified mostly on the basis of detailed NMR (nuclear magnetic resonance) studies of associations with vancomycin[24] and later supported by determinations of crystal structures of vancomycin complexes.[25] Figure 4 illustrates the hydrogen-bonding interactions between the carboxylate group of the guest and peptide groups of both host and guest molecules. Additional binding contributions come from hydrophobic interactions between methyl groups of the guest molecule and the aromatic rings of the antibiotic.

With more frequent use of vancomycin, bacteria resistant to this antibiotic have emerged. The mechanism of resistance involves the reprogramming of the intermediate peptide synthesis to one that contains -D-Ala-D-Lac terminus instead of -D-Ala-D-Ala terminus and this simple substitution of O for NH (Figure 5) does not prevent the biosynthesis of peptidoglican but strongly reduces affinity to vancomycin.[26]

The reason for the decrease in the binding constant by three orders of magnitude (Table 1, row 9) is the lack of one hydrogen-bonding interaction and a repulsion between lone pairs of lactate oxygen and the carbonyl oxygen of a vancomycin peptide group as illustrated in Figure 6. This model was recently supported by analysis of the crystal structure of the vancomycin complex with Ac_2-L-Lys-D-Ala-D-Lac.[25] In order to separate

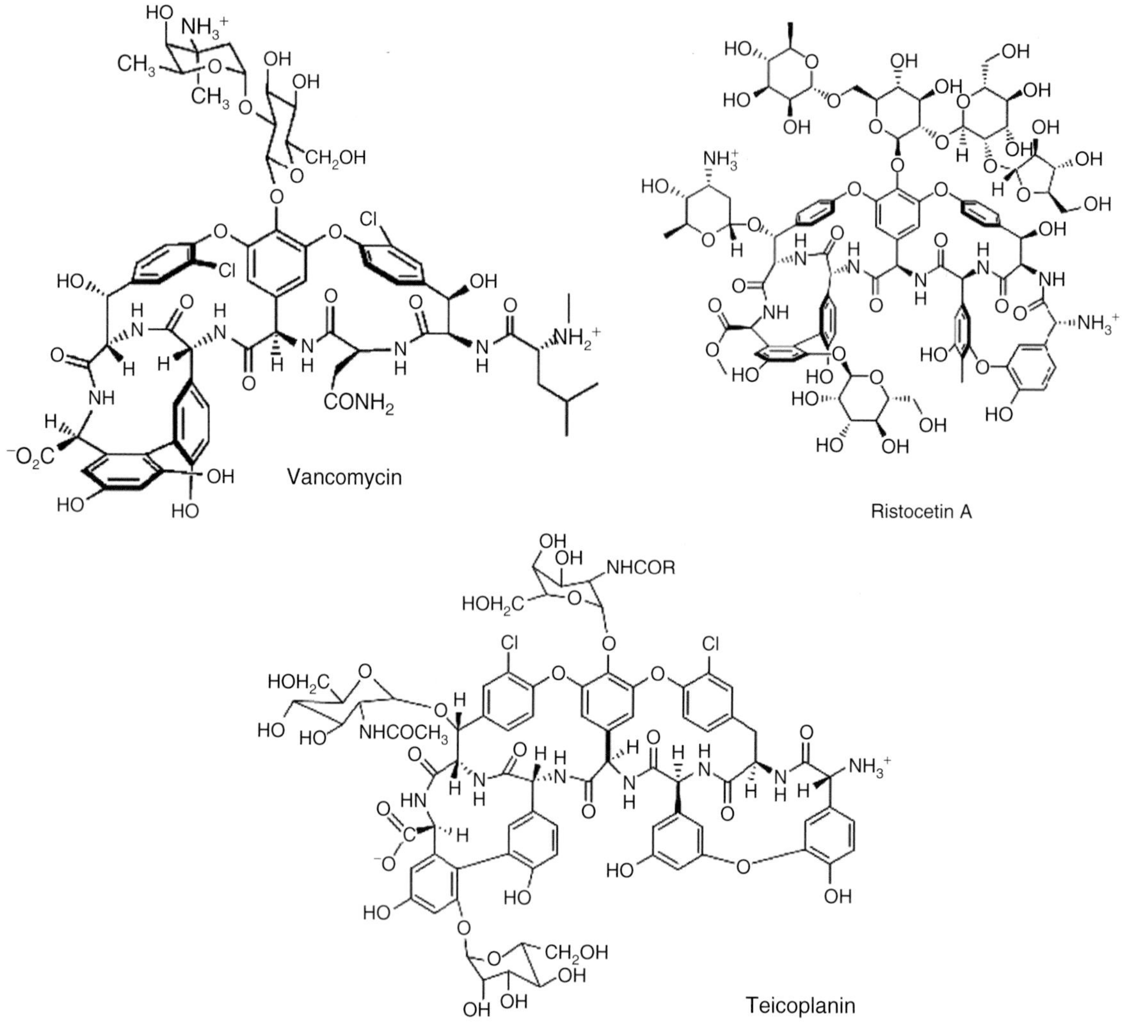

Figure 3 Glycopeptide antibiotics.

Table 1 Logarithms of association constants of peptides and some low-affinity substrates with glycopeptide antibiotics.

	Ligand	Vancomycin	Ristocetin[a]	Teicoplanin	Teicoplanin aglicone
1	Ac_2-L-Lys-D-Ala-D-Ala	5.8	5.2	6.5	6.1
2	Ac_2-L-Lys-D-Ala-L-Ala	No binding	No binding		
3	Ac-D-Ala-D-Ala	4.3	4.9	5.5	4.2
4	Ac-Gly-D-Ala	4.0	4.7	5.0	4.9
5	Ac-D-Ala-Gly	3.7	3.3	4.1	3.1
6	Ac-D-Ala	2.5	3.1	3.2	3.0
7	Ac-Gly	1.9			
8	Acetate	1.5	1.1		
9	Ac_2-L-Lys-D-Ala-D-Lac	2.5			

[a]Data for ristocetins A and B.

these effects, a substrate **3** (Figure 5) was prepared, which contains lacking lone pairs CH_2 group instead of O.[27] The binding constant of **3** to vancomycin (log $K = 4.5$) is roughly one order of magnitude less than that for Ac_2-L-Lys-D-Ala-D-Ala, but still 100-fold greater than that for Ac_2-L-Lys-D-Ala-D-Lac, indicating that the lone pair repulsion is the principal contribution to the low affinity of the lactate derivative.

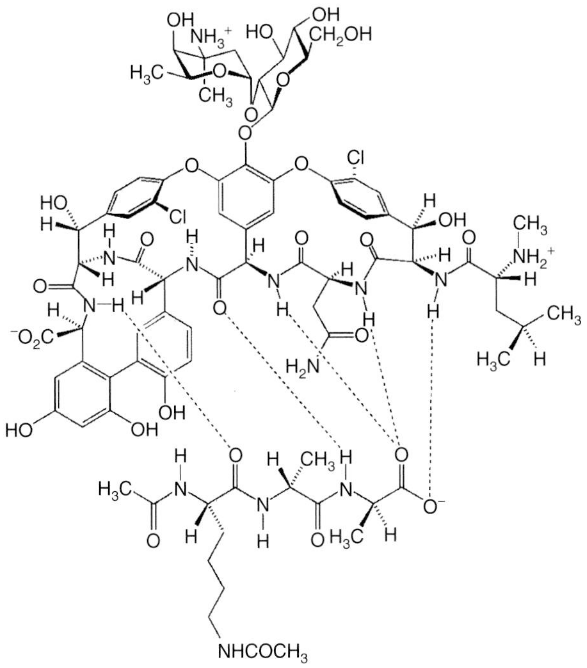

Figure 4 The binding interaction between vancomycin and the peptide ligand Ac_2-L-Lys-D-Ala-D-Ala. (Reproduced from Ref. 24. © Wiley-VCH, 1999.)

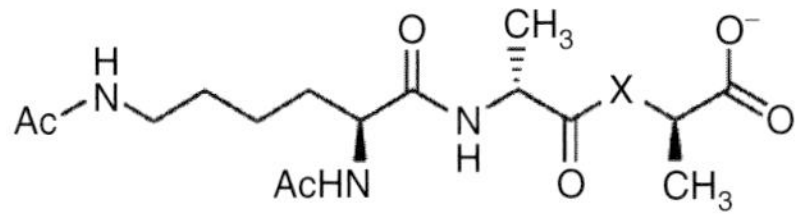

Figure 5 Structures of mutant peptides employed as ligands for vancomycin.

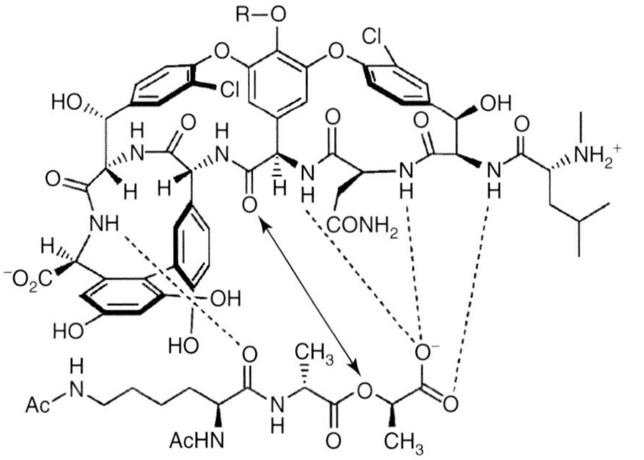

Figure 6 The interaction between vancomycin and the mutant peptide ligand Ac_2-L-Lys-D-Ala-D-Lac. (Reproduced from Ref. 27. © American Chemical Society, 2003.)

Glycopeptide antibiotics undergo dimerization in solution, which in some cases significantly enhances the affinity to peptides.[28] Recent results on the crystal structures of vancomycin complexes emphasize the importance of self-association for complex stability.[29] The relationship between the dimerization and affinity is not simple, however. Eremomycin, an antibiotic structurally similar to vancomycin, has two orders of magnitude larger dimerization constant,[28] but it binds the target peptide Ac_2-L-Lys-D-Ala-D-Ala with a 23-fold smaller stability constant.[30] On the other hand, eremomycin is a significantly more potent antibiotic than vancomycin. It seems, therefore, that self-association may be a more important factor for interactions with surface-bound ligands.[13]

Since first demonstration in 1994 of the potential use of macrocyclic antibiotics as chiral selectors in analysis,[31] glycopeptide antibiotics have been successfully applied for enantiomer separations by liquid chromatography, as recognition components of chiral stationary phases,[32–34] and by capillary electrophoresis (CE) as soluble chiral selectors.[35–38] Four chiral stationary phases for chromatography with the selectors vancomycin, ristocetin, teicoplanin, and the teicoplanin aglycone are commercialized under the trade name Chirobiotic by Astec and Supelco.[33] Various aspects of analytical applications of glycopeptide antibiotics have been extensively covered in the recent reviews cited above. As an example, Table 2 shows some representative results for CE enantioseparations with vancomycin, ristocetin A, and teicoplanin, which were taken from Ref. 39.

Note that complete baseline resolution has $R_s > 1.5$, and this level can be achieved for each analyte with at least one selector. In general, the best results are observed with ristocetin A, and teicoplanin is the most distinct of the three. This last antibiotic has particular aggregation properties. It is surface active due to the presence of a long hydrophobic chain in the radical R in the acyl group of β-D-glucosamine (see above) and forms micellar aggregates above the critical micelle concentration of 0.18 mM. Apparently, this may affect its recognition properties.

The structures of analytes that can be successfully resolved with antibiotics have very little in common with structures of specifically bound peptides. They all have a carboxylate group, which apparently binds to the antibiotic site occupied by the terminal carboxylate of specifically bound peptides, but since also a phosphate can be resolved (Table 2, line 4) the presence of carboxylate group seems to be generally unnecessary. Indeed, chiral antibiotic-based stationary phases resolve also uncharged compounds by liquid chromatography.[31] An inspection of Table 2 shows that, generally, compounds bearing large aromatic fragments are better resolved. Consequently, the principal interactions responsible for enantioseparation may involve the stacking and hydrophobic binding of analytes with

Table 2 CE enantioseparations with glycopeptide antibiotics as chiral selectors (0.1 M phosphate buffer, pH 6.0, containing 2 mM antibiotic).[39]

	Compound	Resolution (R_s)		
		Vancomycin	Ristocetin A	Teicoplanin
1	Ketoprofen	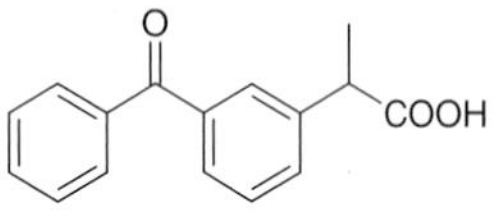 6.2	5.7	1.1
2	Fenoprofen	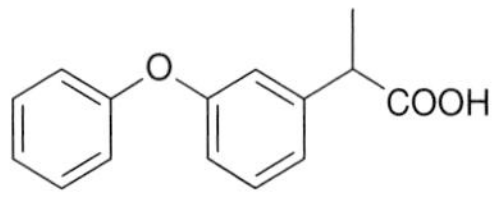 3.0	0.9	1.1
3	Mandelic acid	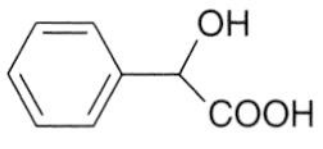 0.0	2.0	3.4
4	1,1′-Binaphthyl-2,2′-diyl hydrogen phosphate	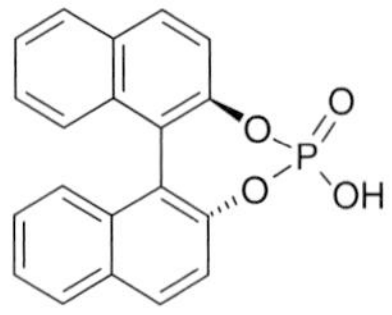 0.6	4.1	0.0
5	2-Phenoxypropionic acid	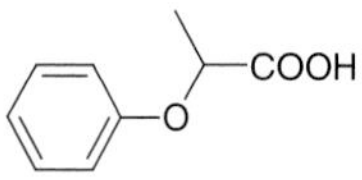 1.2	0.8	0.0
6	2-(3-Chlorophenoxy)propionic acid	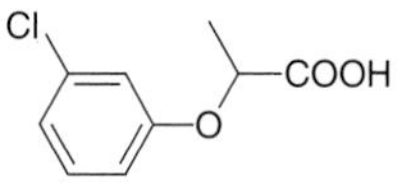 1.0	8.9	20
7	Methotrexate	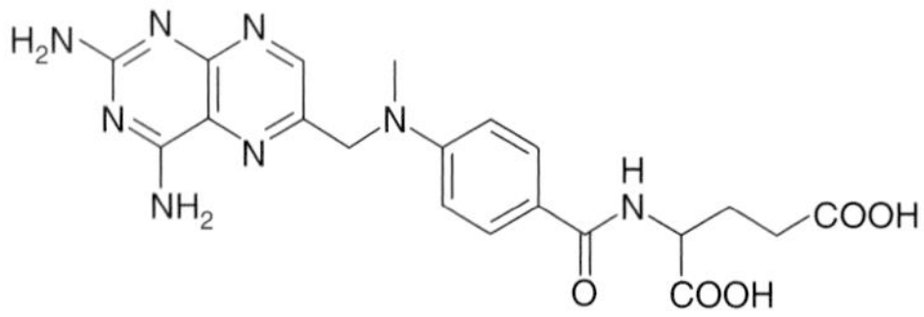 2.3	11.4	11.0
8	AQC-DL-homoserine	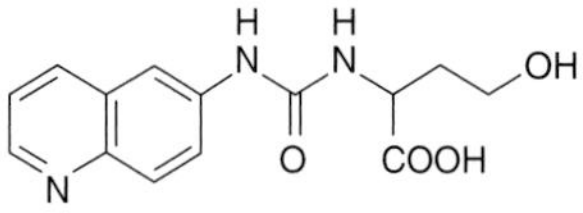 7.9	11.1	6.8
9	*N*-Formyl-DL-phenylalanine	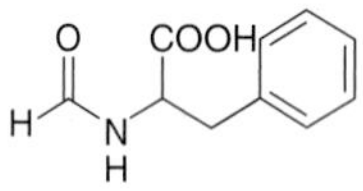 2.3	0.7	1.1
10	*N*-Acetyl-DL-phenylalanine	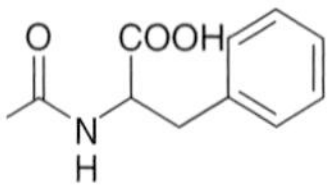 4.8	0.6	1.5
11	Dansyl-DL-phenylalanine	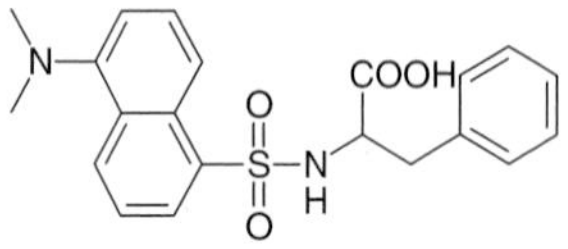 2.6	9.9	0.0
12	*N*-benzoyl-DL-valine	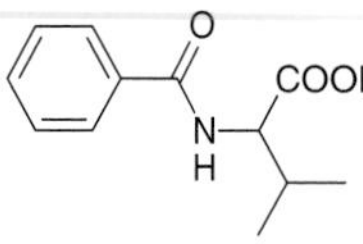 1.7	5.5	1.4

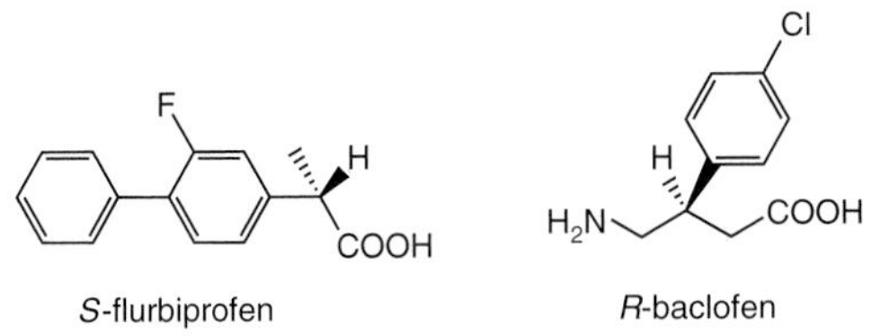

S-flurbiprofen *R*-baclofen

Figure 7 Structures of chiral drugs determined with electrodes based on glycopeptide antibiotics.

the aromatic rings of antibiotics. Unfortunately, no studies with nonpeptide substrates have been reported yet that could clarify the nature of enantioselectivity for practically important analytes.

Further analytical applications of glycopeptide antibiotics involve the development of enantioselective potentiometric membrane electrodes employed as sensors for chiral drugs. A carbon paste electrode impregnated with vancomycin or teicoplanin was used for detection of *S*-flurbiprofen or *R*-baclofen (Figure 7) with high sensitivity and selectivity.[40,41]

4 BILE ACIDS

Two most important classes of biological amphiphiles are lipids and salts of bile acids (bile salts). Lipids serve as the building blocks of biological membranes, while the bile salts are soluble compounds that play an important role in digestion and other biological processes. They are produced by the liver and stored in the gall bladder, and solubilize apolar compounds, in particular, cholesterol and fat-soluble vitamins. Structures of several most frequently used bile salts are shown in Figure 8.

Bile salts are chiral rigid molecules. They serve as important building blocks in the synthesis of both cyclic and acyclic hosts.[42,43] Natural bile acids are employed for enantioseparation of racemates of various classes of organic compounds by enantioselective inclusion complexation in the solid state.[44,45] Crystals of bile acids contain

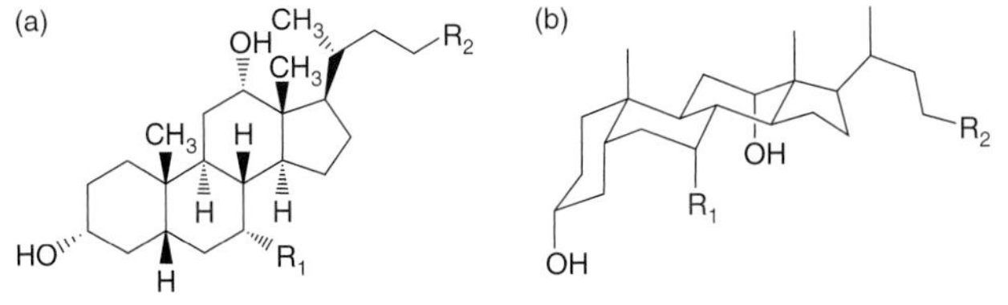

R_1 = OH, R_2 = COONa — Sodium cholate
R_1 = H, R_2 = COONa — Sodium deoxycholate
R_1 = OH, R_2 = $CONHCH_2CH_2SO_3Na$ — Sodium taurocholate
R_1 = H, R_2 = $CONHCH_2CH_2SO_3Na$ — Sodium taurodeoxycholate

Figure 8 (a) Structural formula and (b) three-dimentional structure of the most common bile salts.

chiral cavities in which one of enantiomers is included, predominantly leaving the other enantiomer in the liquid phase. The separation is generally highly efficient with enantiomeric excess often reaching 99%. The host–guest interaction within the chiral cavity is so strong that it can force the flexible guest to adopt a chiral conformation. An interesting example of induced chirality is the isolation of chiral nitrosamines by inclusion complexation with cholic or deoxycholic acids.[46] Owing to the hindered rotation about the N–N bond ($\Delta G^{\neq} \approx 25\,\mathrm{kcal\ mol^{-1}}$), in compounds like **4** (Figure 9) the induced chirality does not disappear immediately after removal of the guest from the inclusion solid complex and can be observed for a certain period in solution. Figure 9 shows the crystal packing of the complex between cholic acid and **4**, with guest molecules accommodated in channels typical of the bile acid bilayered structure. Only one enantiomer of **4** is selected by complexation. After dissolution of the crystals in methanol, circular dichroism measurements confirmed that the guest still remained chiral for a period of about 1 h.

Another important aspect is the size selectivity of inclusion, which can be regulated by the cavity size. This can be achieved by the crystallization of bile salts with cations of different sizes, which partially fill the cavity leaving variable space for guest molecules.[47] Thus, crystals of deoxycholate salts of alkylammonium cations $CH_3(CH_2)_nNH_3^+$,

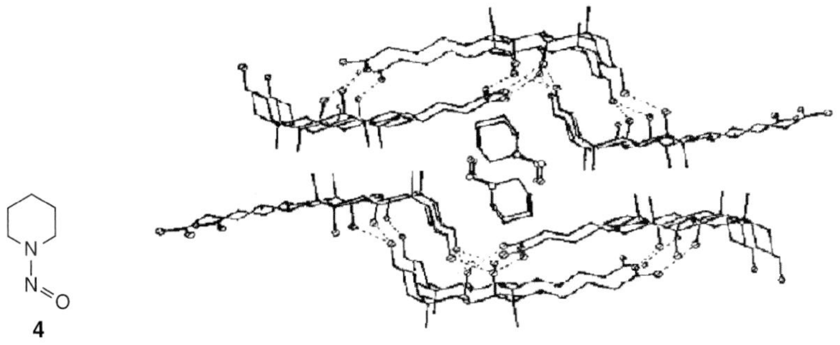

Figure 9 Crystal packing of the inclusion complex between nitrosamine **4** and cholic acid. (Reproduced from Ref. 46. © Wiley-VCH, 1999.)

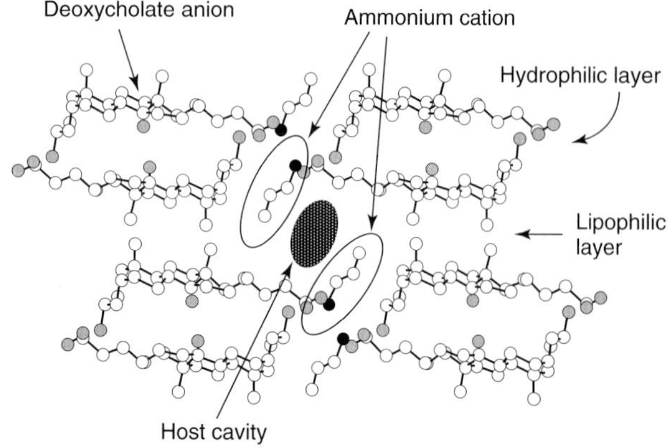

Figure 10 Crystal structure of an inclusion complex of *n*-propylammonium deoxycholate with 2-propanol (1 : 1). Hydrogen atoms are omitted. Empty, filled, and gray circles represent carbon, nitrogen, and oxygen atoms, respectively. (Reproduced from Ref. 47. © American Chemical Society, 1998.)

where $n = 0$–4, contain cavities of progressively reduced size on increase in number n of methylene groups in the ammonium cation. To test the size selectivity, a series of aliphatic alcohols from methanol to isomeric butanols were employed as guests. Figure 10 shows the structure of *n*-propylammonium salt with 2-propanol as the guest. The host cavity indeed is partially occupied by the cation hydrocarbon chain, which, however, leaves enough space to accommodate alcohols of different sizes from methanol to *tert*-butanol. Crystals with smaller methylammonium and ethylammonium cations possessing larger cavities do not form inclusion complexes with small alcohols (methanol and ethanol), but form 1 : 1 complexes with butanols, which better fit larger cavities. Crystals with the larger *n*-butylammonium cation still bind isomeric butanols, but with stoichiometry less than 1 : 1, and crystals with *n*-pentylammonium cation do not form inclusion compounds at all since the cavity is completely filled with the long *n*-pentyl group.

Natural bile salts are employed for enantiodiscrimination in micellar electrokinetic chromatography (MEKC).[37] The principle of separation is the enantioselective inclusion of analytes in micellar aggregates of bile salts, which are formed in the concentration range 1–10 mM. Several studies of host properties of bile salt micelles by using spectroscopic probes were reported, which allowed the characterization of sites of guest localizations inside micelles.[48–50] Owing to the rigidity of the hydrophobic moieties of bile salts, their micellar properties are significantly different from those of common surfactants possessing long, flexible hydrocarbon chains.[51] Several discrete types of premicellar and micellar aggregates are formed in solution at increased bile salt concentration, and the aggregation is driven not only by hydrophobic interactions but also by hydrogen bonding.

5 RIFAMYCINS

A large group of natural macrocylic compounds potentially useful as receptors constitutes the clinically important antibiotics rifamycins. Rifamycins are antibiotics belonging to the group of naphthalenic ansamycins which exert their activity by specific inhibition of bacterial DNA-dependent RNA polymerase. They have activity against a large variety of organisms, such as bacteria, eukaryotes, and viruses; in particular, they are used to treat tuberculosis and leprosy. The molecule is made up of a naphthoquinonic system condensed to a furanone ring (chromophore) spanned by a 17-membered ansa chain connecting two sides of the chromophore.[52] Rifamycins are biosynthesized by fermentation. The biosynthesis proceeds through the assembly of a polyketide by addition of acetate and propionate chain extension units to a unique start unit 3-amino-5-hydroxybenzoic acid. The chemical structures of the most important derivatives are shown in Figure 11 and some of their physicochemical properties are collected in Table 3 (data from Refs 53–55).

Rifamycins form yellow to orange solutions depending in part on pH and organic cosolvents. As the compounds begin to degrade in solution, they turn dark red and eventually begin to precipitate as degradation continues. The ^{13}C and ^{1}H NMR spectra of rifamicins were assigned using 2D homo- and heteronuclear correlation NMR spectroscopy.[56] Twenty-six crystal structures of rifamycins were grouped into two classes (active and nonactive).[57]

Recently, the association of amino acids and some other low-molecular weight compounds with rifamycin SV in water was studied by ^{1}H NMR titrations.[58] Rifamycin binds aromatic amino acids with pronounced enantioselectivity in favor of L-enantiomers and forms complexes with heterocyclic compounds but does not interact with aliphatic amino acids and with simple benzene derivatives. The binding constants for several guests are given in Figure 12.

Binding constants correlate with the lowest unoccupied molecular orbital (LUMO) energies and hydrophobicities

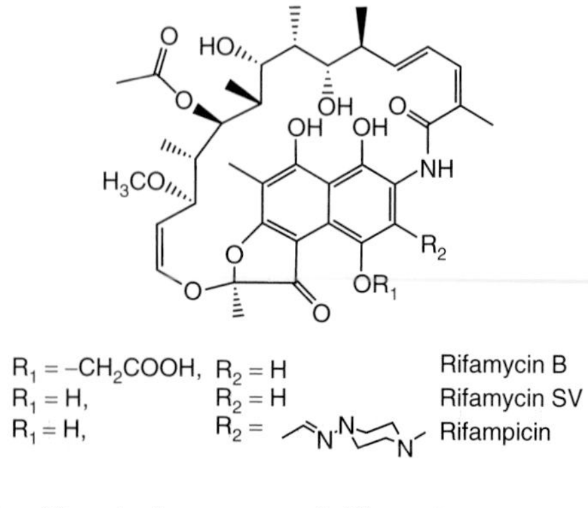

$R_1 = -CH_2COOH$, $R_2 = H$ Rifamycin B
$R_1 = H$, $R_2 = H$ Rifamycin SV
$R_1 = H$, $R_2 =$ Rifampicin

Figure 11 Chemical structure of rifamycins.

Table 3 Physicochemical properties of some rifamycins.

Characteristics	Rifamycin B	Rifamycin SV	Rifampicin
Chemical formula	$C_{38}H_{49}O_{14}N$	$C_{37}H_{46}NO_{12}Na$ as sodium salt	$C_{43}H_{58}N_4O_{12}$
Formula weight (Da)	755.8	719.75	822.94
Solubility	Light alcohols and acetone; slightly soluble in water	Ether; water in basic medium	$CHCl_3$, water in acid or basic media
Stereogenic centers	9	9	9
Hydroxyl groups	4	5	5
Aromatic rings	2	2	2
Ionogenic groups and pK_a's values	pK_{a1} = 2.8 (COOH), pK_{a2} = 6.7 (Ph-OH)	pK_a = 1.8	pK_{a1} = 1.7 (Ph-OH), pK_{a2} = 7.9 (piperazine)
UV–vis spectrum	Maxima at approximately 220, 304, 425 nm	Maxima at approximately 220, 304, 425 nm	Maxima at 237, 255, 334, 475 nm

$K_L = 34$, $K_D = 18$ 74 M^{-1} 10 M^{-1} 7 M^{-1} 150 M^{-1}

Figure 12 The binding constants (M^{-1}) for several guests to rifamycin SV in water at pH 9.0.

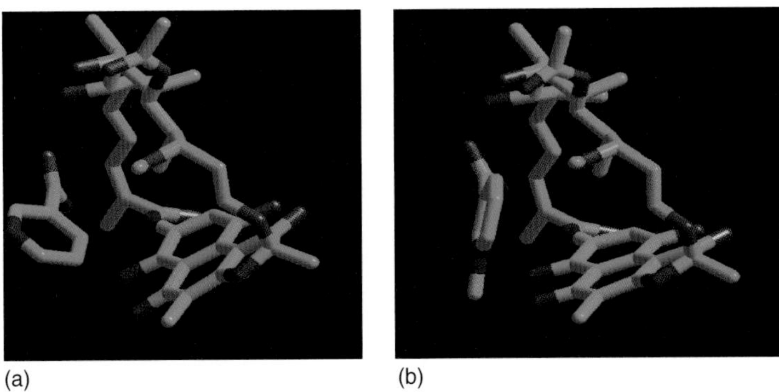

Figure 13 Simulated structures of the rifamycin SV complexes with (a) nicotinamide and (b) 1-methylnicotinamide. (Reproduced from Ref. 58. © Springer, 2009.)

(expressed as log P values) of guest molecules, indicating contributions to the binding free energy from charge-transfer interactions with the naphtohydroquinone fragment of rifamycin SV acting as an electron donor as well as from hydrophobic interactions. The proposed mode of binding is supported by semiempirical calculations of structures of host–guest complexes. The situation with nicotinamide and its N-methylated derivative, which show large difference in the binding constants (Figure 12), is illustrated in Figure 13. Nicotinamide forms an edge-to-face contact with the naphthohydroquinone fragment observed in several cyclophane inclusion complexes with aromatic guests,[59] but 1-methylnicotinamide turns into a more coplanar orientation favorable for stacking interaction. The reason for this is the strongly reduced LUMO energy of the methylated compound, which makes more favorable charge-transfer interaction with the naphtohydroquinone fragment.

Like glycopeptide antibiotics, rifamycins were tested as chiral selectors in liquid chromatography and CE,[31,60] but appeared to be less efficient. They are not anymore used for the preparation of stationary phases, but rifamycins B and SV are used as selectors in CE.[36]

Rifamycin SV was found to have ionophoric properties and was successfully employed for the development of a potentiometric membrane sensor for potassium ion.[61]

The potentiometric selectivity to alkali and alkaline-earth cations is in the order $K^+ > Rb^+ > Cs^+ > Na^+ > NH_4^+ > Ba^{2+} > Mg^{2+} > Ca^{2+} > Sr^{2+} > Li^+$ and the interference from transition-metal ions is negligible. No binding constants were reported.

6 AMINOGLYCOSIDES

The aminoglycosides constitute a large and diverse class of clinically important antibiotics which contain two or more aminosugars linked by glycoside bonds to 2-deoxystreptamine with exception of streptomycin which is based on a streptidine fragment (Figure 14). Natural aminoglycosides are obtained from actinomycetes of either genus *Streptomyces* (names terminated with "-mycin") or *Micromonospora* (names terminated with "-micin"). Chemically modified aminoglycosides, for example, amikacin, also are used clinically.

Aminoglycoside antibiotics bind to the A-site decoding region of the bacterial 16S ribosomal RNA and disrupt the protein biosynthesis, leading to the death of bacteria. Recently, RNA has become an important target for the development of new pharmaceuticals and consequently interactions of aminoglycosides with RNA have acquired particular significance as a model for understanding of small molecule–RNA recognition.[62] In fact, aminoglycosides are very promiscuous ligands, which interact not only with different types of RNA but also with DNA and proteins. In these interactions, aminoglycosides behave actually as guests for macromolecular biological hosts, but analysis of these interactions is important for understanding of possible receptor properties of aminoglycosides since the same intermolecular forces are involved in both cases.

To a large extent, the above-mentioned promiscuity results from the fact that aminoglycosides in neutral aqueous solutions are polycations and the principal contribution to their binding free energy is the long-range nondirectional Coulombic attraction to the negatively charged sites of their biological targets. Detailed analysis of structural and thermodynamic data revealed, however, large, additional non-electrostatic contributions, with electrostatic interactions contributing about 50% to the total binding free energy.[63] Figure 15(a) shows the interactions observed in the crystal structure of the paramomycin complex with the 30S subunit, which is a portion of the 16S ribosomal RNA.[64] There is only one direct salt bridge together with large number of hydrogen-bonding interactions and a stacking interaction between a nucleobase and an aminosugar ring probably of hydrophobic character. Besides these types of interactions, there are also pseudo-base pair interactions, characteristic of aminoglycosides, as illustrated in Figure 15(b) for a pair between ring I of tobramycin and adenine of RNA observed in the crystal structure of the corresponding complex.[65]

The presence of vicinal OH groups in aminosugars reduces the basicity of amino groups as compared to that in aliphatic polyamines. Usually, the ammonium group in position 3 of the 2-deoxystreptamine moiety has the

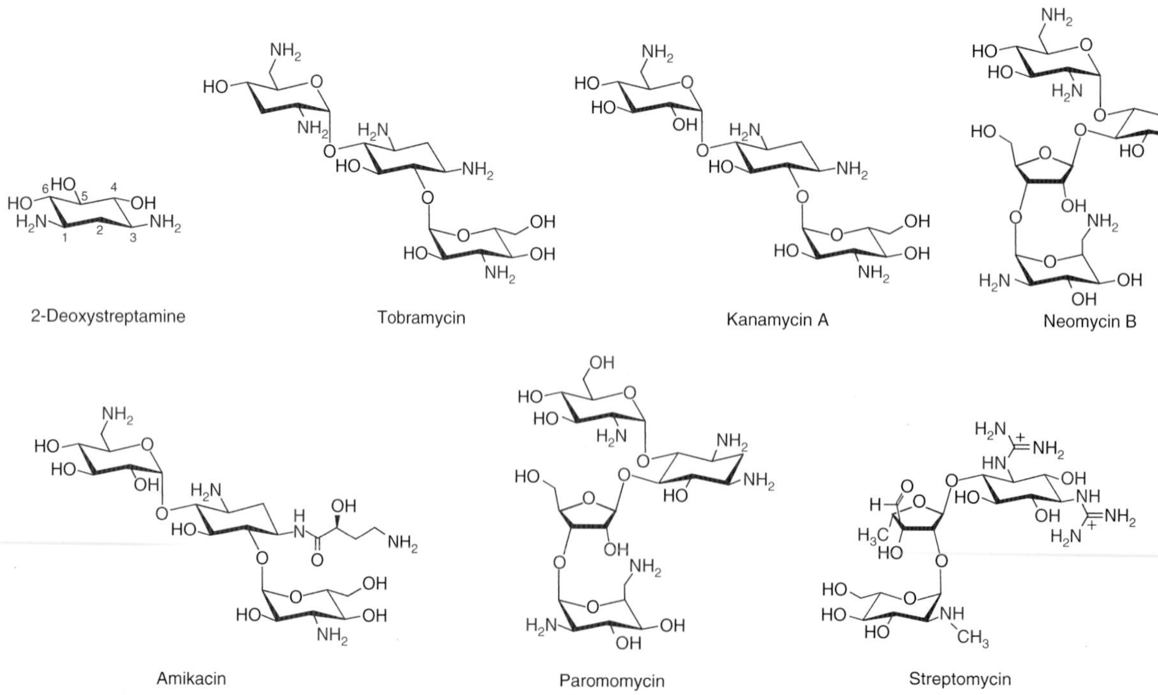

Figure 14 Structures of 2-deoxystreptamine and some aminoglycosides.

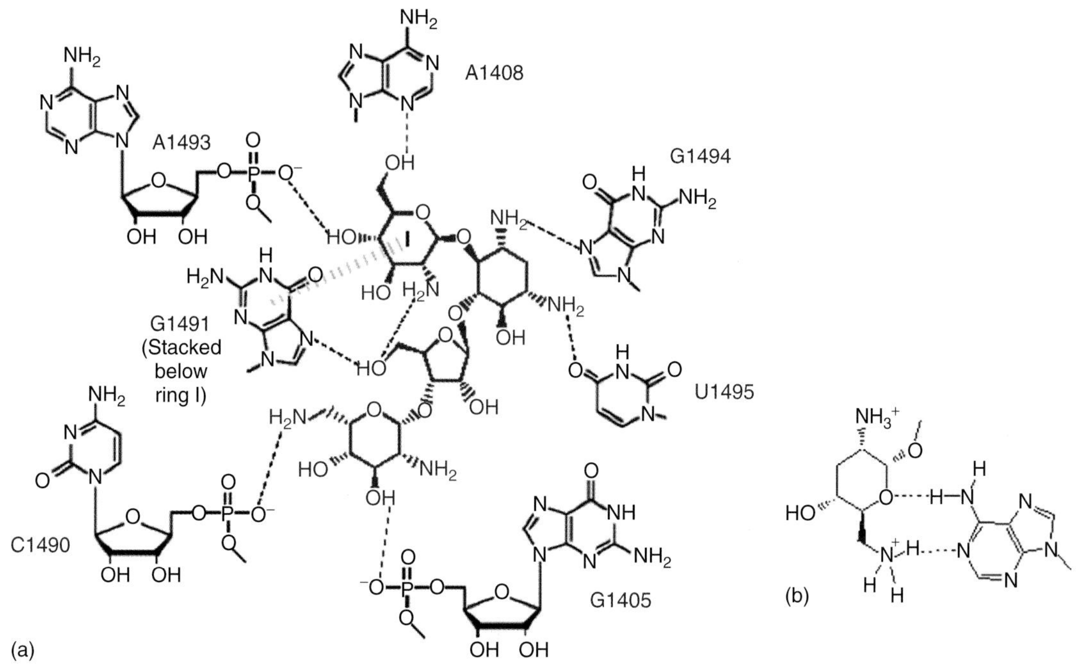

Figure 15 (a) Binding interactions in the crystal structure of paramomycin complex with the 30S subunit, a portion of the 16S ribosomal RNA.[64] (b) Pseudo-base pair interaction between ring I of tobramycin and adenine of RNA.[65]

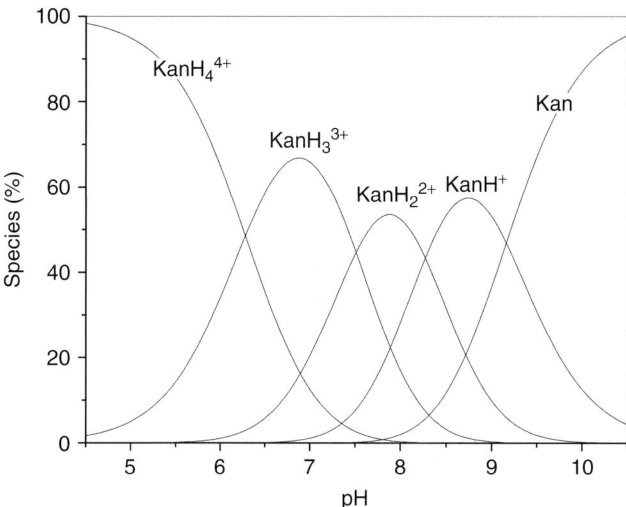

Figure 16 Distribution of protonated species for kanamycin A (Kan); pK_a values are 6.28, 7.52, 8.27, and 9.16.

lowest pK_a in the range of 6.2–6.5, while other ammonium groups have pK_a values above 7.5. A typical distribution of protonated species is illustrated for kanamycin A containing four amino groups in Figure 16, from which one can see that at pH 7 the predominant form is a trication. Protonation of aminoglycosides is exothermic like that of aliphatic polyamines, but there is no correlation between the protonation enthalpy and pK_a,[66,67] usually observed for the latter.[68]

The first report on the use of aminoglycosides as receptors was related to their chirality. Sulfates of fradiomycin, kanamycin, and streptomycin were used as CE chiral resolving agents for compounds like 1,1′-binaphthyl-2,2′-diyl hydrogen phosphate, or 1,1′-antibiotbinaphthyl-2,2′-dicarboxylic acid.[69] Although the resolution was good, the absence of significant advantage over other chiral selectors and the strong adsorption of aminoglycosides to the capillary wall precluded further use of these compounds. Interestingly, simple neutral disaccharides and trisaccharides like maltose or maltotriose[70] and even monosaccharides like glucose or mannose[71] are effective for the separation of the racemic mixture of 1,1′-binaphthyl-2,2′-diyl hydrogen phosphate in CE. This observation is in line with the significant nonelectrostatic contribution mentioned above to the binding of aminoglycosides to target molecules, which occurs via contacts of guests with their sugar fragments.

The use of protonated aminoglycosides for anion recognition was first demonstrated for interaction of neomycin B with anions of different types studied by isothermal titration calorimetry at pH 5.5 ensuring complete protonation of the receptor.[72] For inorganic anions, the binding constants correlated well with the anion charge increasing from $3.3 \times 10^2\ M^{-1}$ for CrO_4^{2-} to $1.3 \times 10^5\ M^{-1}$ for $Fe(CN)_6^{4-}$. For a series of nucleotides ($ADPH^{2-}$, $ATPH^{3-}$, $AtetraPH^{4-}$), the charge dependence was less significant, with binding constants varying from 5.4×10^3 to $7.3 \times 10^4\ M^{-1}$. It seems therefore that for nucleotide guests there is a significant nonelectrostatic contribution probably due to interaction

Table 4 Logarithms of association constants of nucleotides and nucleosides with differently protonated forms of kanamycin A (Kan) and tetrandrine derivatives at 25 °C.

Receptor	AMP^{2-}	ADP^{3-}	ATP^{4-}	CTP^{4-}	GTP^{4-}	Ado	Guo	References
Kan					2.45		3.12	74
Kan H^+	1.75	2.35	1.90	1.86	2.60	2.28	3.04	
Kan $H_2{}^{2+}$	1.86	2.51	2.44	2.34	3.16	2.37	3.41	
Kan $H_3{}^{3+}$	2.20	3.02	3.20	3.32	3.76	2.46	3.49	
Kan $H_4{}^{4+}$	2.47	3.40	4.15	4.26	5.05	2.29	3.35	
10	1.68	1.60	2.04					75
11	2.69	2.93	3.73	About 3	About 3			76
12	1.89	2.04	3.86					

with the nucleobase. In all cases, the binding is enthalpy driven, indicating that the interaction is not a simple ion pairing which normally is entropy-driven.[73]

A more systematic study of nucleotide recognition by protonated forms of kanamycin A, streptomycin, and amikacin demonstrated that guest binding constants are similar to those observed with aliphatic macrocyclic polyamines of similar charge although aminoglycosides are linear polyamines and that, more importantly, there is a noticeable selectivity to the type of the nucleobase (Table 4).[74] As follows from Table 4, kanamycin binds with increased affinity AMP^{2-}, ADP^{2-}, and ATP^{4-} due to the charge effect and in a series of nucleotide triphosphates binds GTP^{4-} approximately one order of magnitude stronger than ATP^{4-} and CTP^{4-}. In line with this, kanamycin binds neutral guanosine one order of magnitude stronger than adenosine. Binding of nucleosides is independent of the receptor charge, which is quite significant, indicating large nonelectrostatic contribution to nucleotide recognition. The binding of ATP^{4-} to all protonated forms of kanamycin is exothermic.

Aminoglycosides are expected to act as ligands for metal coordination, but only Cu(II) complexes were studied in detail.[77] With kanamycin A as a lignad, a set of monuclear complexes with differently protonated/deprotonated forms ranging from CuH_2L to $CuH_{-2}L$ were found with stability constant $\log \beta_{11} = 8.83$ for the neutral form.[78] Copper complexes of aminoglycosides are active catalysts for several hydrolytic and oxidative reactions.[77]

7 CYCLIC POLYPEPTIDE ANTIBIOTICS

Peptide antibiotics operate through numerous different mechanisms.[79] They are large molecules approaching proteins in their structural complexity and often have cyclic structures (Figure 17), making them promising compounds as receptors for small guests. Nevertheless, the potential of these compounds as receptors remains practically unexplored.

Thiostrepton, a thiopeptide antibiotic (a group of naturally occurring, sulfur-containing, highly modified, macrocyclic peptides),[80] was applied for the preparation of chiral stationary phases for liquid chromatography.[31] Its performance was later improved by chemical modification of the antibiotic.[81] Thiostrepton is insoluble in water, but it was studied as a hydrogen-bonding receptor for anions in organic solvents.[82] It forms both 1 : 1 and 1 : 2 complexes with anions with $K_2 > K_1$ for F^- and $K_2 \ll K_1$ for all other anions studied. Stability constants of 1 : 1 complexes range from 10^2 to $5 \times 10^3\,M^{-1}$. In DMSO (dimethyl sulfoxide), they follow the order $AcO^- \approx F^- \gg Cl^-$, Br^-, $HSO_4{}^-$, $H_2PO_4{}^-$, which roughly correlates with basicities of anions, but in $CHCl_3$ they follow a different order: $Cl^- \approx HSO_4{}^- > F^- \approx AcO^- > Br^- > H_2PO_4{}^-$. A reason for this change is the stronger solvation of more basic anions by chloroform, which acts as a weak proton donor solvent.

Bacitracin A_1 is a lariat-type macrocycle (Figure 17) well "designed" for complexation of metal ions. Indeed, it needs the presence of a metal ion for the antibiotic activity, with maximum activity observed with Zn^{2+}.[83] Table 5 shows the pK_a values of ionogenic groups of bacitracin and logarithms of binding constants of some divalent metal ions to differently protonated forms of the peptide.[84] Affinity of the neutral form, predominating at pH 7, is not very large and surprisingly does not follow the Irwing–Williams series ($Ni^{2+} < Cu^{2+} > Zn^{2+}$).

The polymyxins, polymyxin B (Figure 17), and colistin (contains D-Leu instead of D-Phe in the macrocycle), are used as last-line antibiotics to treat infections caused by Gram-negative bacteria that are resistant to essentially all other currently available antibiotics.[85] They bind to the lipid A (a β-1′-6-linked D-glucosamine disaccharide phosphorylated at the 1- and 4′-positions containing six long fatty acyl chains attached to OH and NH_2 groups of the disaccharide) through a combination of electrostatic and hydrophobic interactions disrupting the bacterial outer membrane. Owing to the high total positive charge (5+)

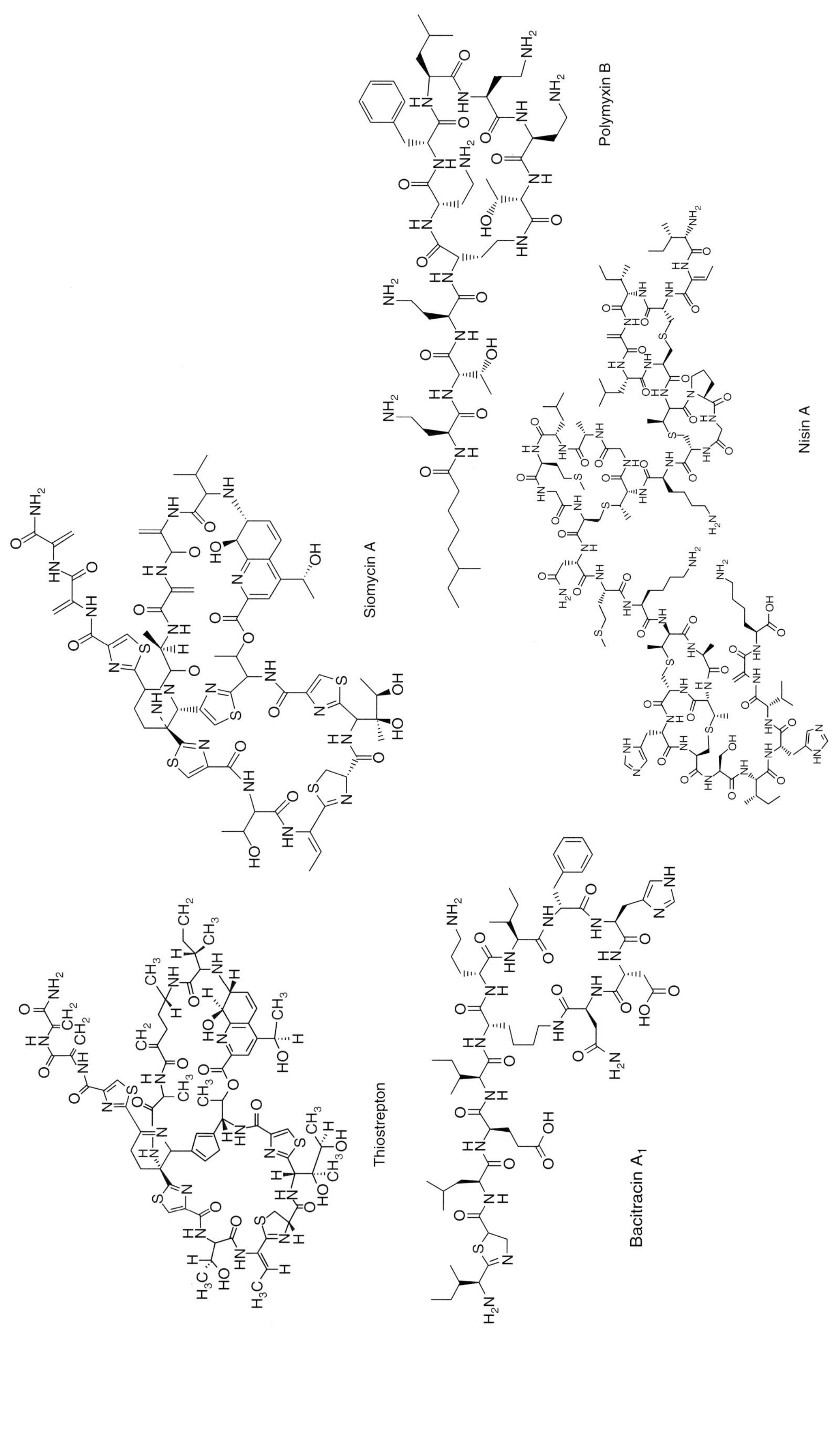

Thiostrepton

Siomycin A

Polymyxin B

Bacitracin A_1

Nisin A

Figure 17 Structures of several commerially avaliable polypeptide antibiotics.

Table 5 Proton dissociation (pK_a) and metal binding ($\log K$) constants for bacitracin A_1.[84]

Group	pK_a	Peptide species	$\log K$		
			Ni^{2+}	Cu^{2+}	Zn^{2+}
D-Asp (macrocycle)	3.6	P^0	4.36	3.69	3.18
D-Glu (lateral chain)	4.4	P^{1-}	7.08	6.30	4.98
His (macrocycle)	6.4	P^{2-}	9.66	9.08	<6.9
Ile (N-terminal)	7.6				
D-Orn (macrocycle)	9.7				

and specific arrangement of ammonium groups, it may be expected to act as a receptor for various anionic species.

A lantibiotic nisin (Figure 17) widely used for food preservation targets the cell wall biosynthetic intermediate lipid II.[86] The structure of the complex of nisin A and a lipid II analog solved by NMR studies in DMSO[87] demonstrates trapping of the pyrophosphate moiety of the lipid between two rings (first and second rings from the Ile terminus of the molecule) through five H-bonds to amide NH groups. It seems probable that nisin may act as a hydrogen-bonding receptor for free pyrophosphate as well as for some other anions.

8 ALKALOIDS

Alkaloids have been used for chiral separations from a long time ago. In 1853, Pasteur achieved the resolution of isomers of tartaric acid by precipitation of their salts with cinchona alkaloids (Figure 18), which are still perhaps the most popular chiral building blocks for preparation of chiral catalysts and reactants.[88] However, the receptor properties of these compounds and some of their simple derivatives in solution were studied only recently.

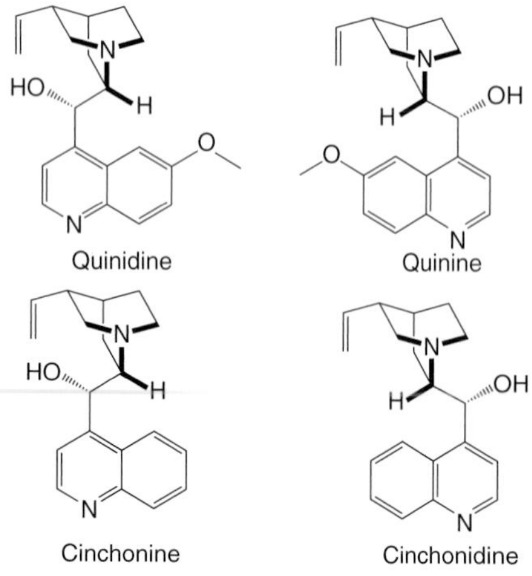

Quinidine

Quinine

Cinchonine

Cinchonidine

Figure 18 Cinchona alkaloids.

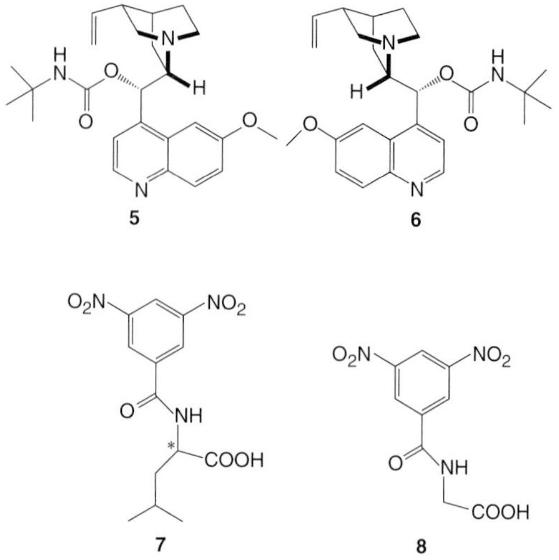

5

6

7

8

Figure 19 Carbamate derivatives of cinchona alkaloids and amino acid derivatives studied as guests.

Binding of model chiral compounds *R*- and *S*-**5** and achiral guest **6** to quinidine, quinine, and their carbamates **7** and **8** (Figure 19) was studied by spectroscopic and calorimetric titrations in CH_3OH (methanol).[89] The results are summarized in Table 6. Perhaps, the most unexpected result is that unsubstituted alkaloids do not discriminate between enantiomers of **5**, but their carbamates bind the enantiomers with large 10-fold difference in the association constants. The crystal structure of the solid-state complex between *S*-**5** and **8** (Figure 20)[90,91] indicates a strong steric interaction between the *tert*-butyl fragment of the carbamate group of the receptor and the side isobutyl group of leucine, as well as hydrogen bonding between leucine NH donor and carbamate carbonyl group. Two other binding interactions involve ionic hydrogen bond inside the salt bridge between the carboxylate group of the amino acid and the ammonium group of **8** obtained by proton transfer to the quinuclidine nitrogen and $\pi-\pi$ stacking interaction between dinitrobenzoyl group of **5** and quinoline group of **8**. Complexes with unsubstituted alkaloids are more stable than those with carbamates for all guests studied, including the glycine derivative **6** lacking the isobutyl group (Table 6). Obviously, the presence of the carbamate group is critical for the overall affinity and enantioselectivity, but there is no clear explanation of its role. The binding of all guests to all receptors is enthalpy driven, indicating predominantly nonelectrostatic binding contributions.

Bisbenzylisoquinoline alkaloids form a large group (>150) of naturally occurring compounds in which two benzylisoquinoline moieties are joined by up to three ether links.[92] Many of them have macrocyclic cyclophane-type structures, which makes them potentially capable of forming inclusion compounds with small guests. In addition, they have several stereogenic centers and may be useful

Table 6 Logarithms of association constants and thermodynamic parameters (kJ mol^{-1}) for binding of (*R*)-**5**, (*S*)-**5**, and **6** to the chiral selectors **7, 8**, quinine, and quinidine in methanol at 25 °C.[89]

	S-**5**			*R*-**5**			**6**
	log *K*	ΔH	$T\Delta S$	log *K*	ΔH	$T\Delta S$	log *K*
8	3.57	−33	−13	2.34	−27	−14	3.34
7	3.00	−30	−13	3.95	−37	−14	3.36
Quinine	4.04	−38	−15	4.08	−36	−13	4.20
Quinidine	4.00	−30	−7	4.11	−29	−6	4.28

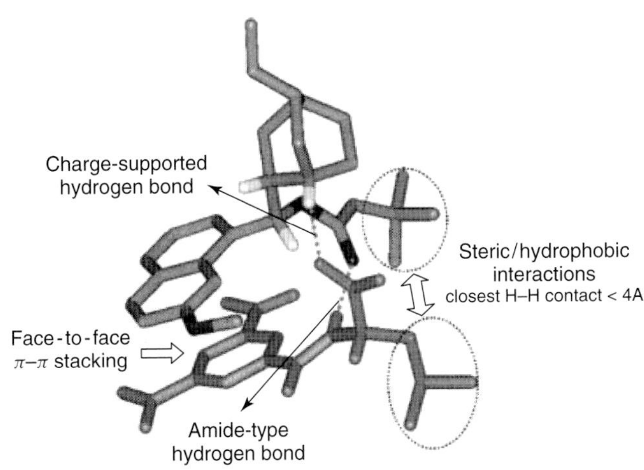

Figure 20 Crystal structure of the complex between cinchona alkaloid carbamate **8** and *S*-**5**. (Reproduced from Ref. 91. © Wiley-Liss, Inc, 2003.)

for chiral recognition. Two alkaloids of this type, *d*-(+)-tubocurarine and tetrandrine, the latter in chemically modified forms, have been studied as anion receptors.

d-Tubocurarine (Figure 21) is obtained from the bark of the South American plant *Chondodendron tomentosum*. For the first time, it was isolated from an old specimen of tube curare stored in the British Museum in 1935; hence the name tubocurarine.[93] It is a competitive antagonist of nicotinic neuromuscular acetylcholine receptors and works as a neuromuscular blocking drug or skeletal muscle relaxant. Since 1942, it has been used as an anesthetic adjunct but later it was substituted with safer drugs.

d-Tubocurarine has three ionogenic groups with pK_a values 7.6, 8.1–8.65, and 9.1–9.65. The protonation sites were not strictly assigned, but since tetrahydroisoquinoline has pK_a 9.41, it seems that two lower pK_a values belong to phenolic groups and the last pK_a to the protonated nitrogen of the tetrahydroisoquinoline moiety. Thus at pH below 7, tubocurarine is a dication and at pH about 9 it is a neutral zwitterion (Figure 21, structure **9**). In its dicationic form, *d*-tubocurarine efficiently binds dianions of dicarboxylic acids (log $K = 2.8$ for terephthalate and 2.4 for malonate dianions) and shows significant enantioselectivity in the binding of anions of *N*-acylated amino acids (Table 5).[94]

Tubocurarine was used as a chiral selector for the separation of optical isomers of a series of organic carboxylates (amethopterin, ketoprofen, *N*-protected amino acids) using CE in the pH range 5–7.[95] In several cases, R_S values of about 2 were observed, but there were no clear correlations between structures of analytes and efficiency of chiral resolution.

Affinity of anions to the neutral zwitterionic form **9** is expectedly much lower, and for *N*-Ac-Phe, which still shows a noticeable affinity, the enantioselectivity is inverted (Table 7).[96] The zwitterion is an efficient catalyst for the hydrolysis of *p*-nitrophenyl carboxylates, which involves the nucleophilic attack of the deprotonated phenol hydroxyls on the ester group of the substrate, with a small kinetic enantioselectivity in the cleavage of esters of *N*-protected phenylalanine.

Tetrandrine (Figure 21) isolated from the root of the plant *Stephania tetrandra* has a number of medicinal properties. These include blockage of Ca^{2+} channels, antiinflammatory and anticancer activities, and stimulation of prostaglandin production by macrophages. It also induces apoptosis in many cell types including human leukemic (U937), human lung carcinoma (A549), and human hepatoblastoma (HEPG2).

The protonated form of tetrandrine has pK_a values 6.33 and 9.6, which means that it exists as a dication only in acid medium; however, the molecule can be easily converted

Table 7 Logarithms of binding constants of enantiomers of *N*-acetyl and free amino acids as anions to bisbenzylisoquinoline macrocyclic alkaloids.[96]

	d-Tubocurarine dication	*d*-Tubo-curarine zwitterion (**9**)	**10**	**11**
N-AcGly			6.3	
N-Ac-L-Ala	<20	<5	72	
N-Ac-D-Ala	85	<5	—[a]	
N-Ac-L-Phe	<20	56	17.4	
N-Ac-D-Phe	270	32	16.4	
L-Phe			8.8	167
D-Phe			10.6	23

[a]No binding.

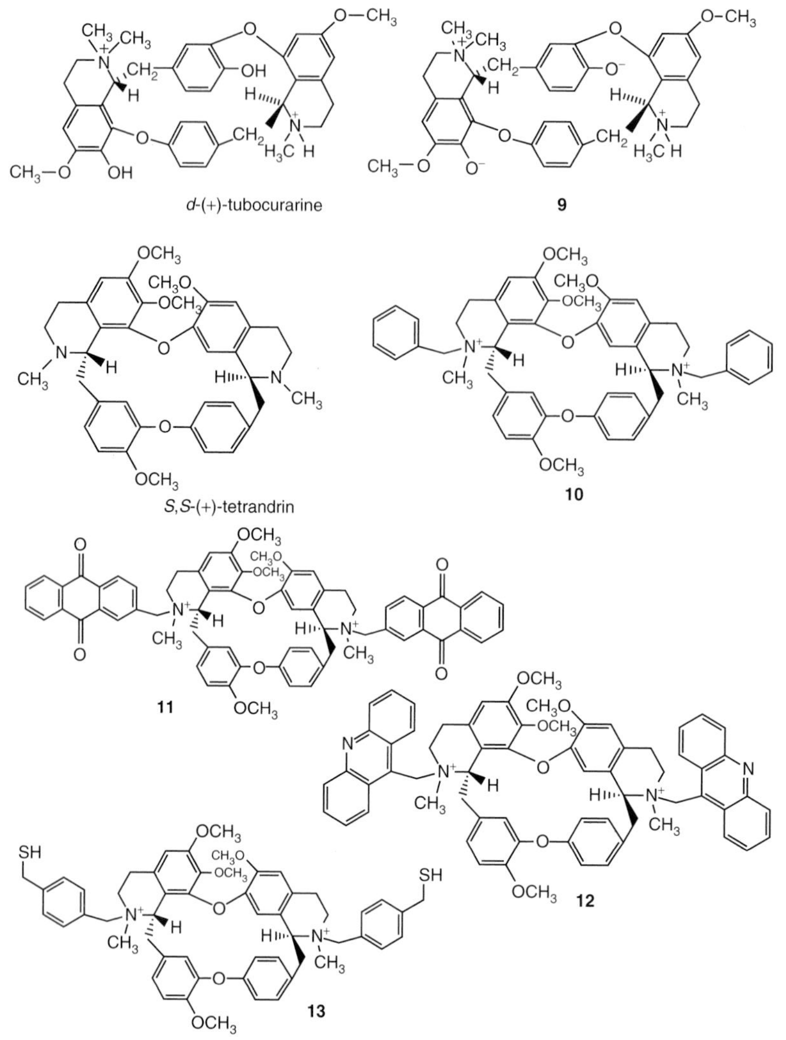

Figure 21 Chemical structures of macrocyclic bisbenzylisoquinoline alkaloids and some semisynthetic derivatives.

into a dication by quaternization of both nitrogen atoms. Bisbenzylated tetrandrine **10** (Figure 21) was employed as a receptor for carboxylate and phosphate anions.[75,97] The dication **10** undergoes dimerization with the association constant $K = 34\,\mathrm{M}^{-1}$ in 0.1 M NaCl. In the complexation of α,ω-dicarboxylates, a peak selectivity for succinate ($K = 58\,\mathrm{M}^{-1}$) is observed. Binding of *N*-acetyl amino acid anions and free amino acids as anions at high pH values shows generally modest enantioselectivity with exception of *N*-acetyl-alanine (Table 5). On the basis of detailed NMR studies, it was demonstrated that these guests interact with **10** preferably from one side of the macrocycle, as shown in Figure 22. Molecular mechanics simulations showed that methyl group of *N*-Ac-D-Ala fits to the small cavity of **10**, providing a significant hydrophobic binding contribution, while in the complex with *N*-Ac-L-Ala this group is directed outside the macrocycle cavity. More voluminous amino acid side groups cannot fit into a small receptor cavity and the respective guests are bound weaker than *N*-Ac-D-Ala and without any enantioselectivity. Interestingly, there is a certain analogy in the recognition properties of **10** and vancomycin. This involves the binding specificity to alanine, although with inverted enantioselectivity, contribution of ion pairing between guest terminal carboxylate and the host cationic site, similar complexation-induced shifts of proton NMR signals, and similar hydrophobic contributions of aliphatic guest moieties.

Further development of this type of anion receptors was the synthesis of antraquinone and acridine derivatives **11** and **12** (Figure 21) as possible electro-active and fluorescent receptors, respectively.[76] Both receptors have significantly improved, compared to **10**, binding properties toward nucleotides (Table 5) and **11** also binds much stronger and with larger enantioselectivity the anionic form of phenylalanine (Table 7). These effects can be attributed to stronger stacking and/or hydrophobic interactions of guests

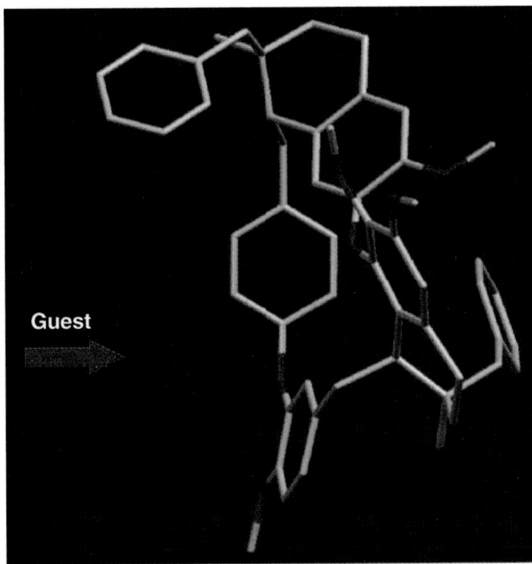

Figure 22 Conformation of receptor **10**. The arrow shows the side of the receptor used for the guest complexation. (Reproduced from Ref. 97. © Royal Society of Chemistry, 2004.)

with antraquinone and acridine groups as compared to the smaller benzyl groups of **10**. The mercapto derivative **13** was employed for enantioselective cleavage of *para*-nitrophenyl esters of amino acids.[98]

Unsubstitute tatrandrine binds alkaline-earth cations in CH_3CN and tetrahydrofuran (THF) with log K values in the range 3.5–5.5 depending on the solvent and counterion.[99] No interaction with alkali cations was detected.

9 CONCLUSIONS

Small biological molecules are capable of molecular recognition of essentially all types of guests—anionic, cationic, or neutral. The most significant advantage of natural compounds as receptors is their chirality. Chemical synthesis of chiral receptors in an optically pure form is often difficult and laborious, whereas natural compounds typically are isolated as optically pure compounds. Usually, they can be easily attached to a carrier for applications in different analytical techniques or further modified for improvement of their recognition and/or sensing properties with conservation of their chirality. There are already several successful practical applications of small biological molecules as receptors, for example, for the preparation of chiral stationary phases for HPLC (high-performance liquid chromatography) or as soluble selectors in CEs. However, systematic studies on molecular recognition of analytically important compounds by small biological molecules are scarce and their applications are developed mostly on purely empirical basis. This gap should be filled in the future. Also, there are still a large number of small biological molecules, for example, among polypeptide antibiotics, which possess suitable structural features for using as receptors but never were tested in this capacity.

REFERENCES

1. P. Bühlmann, E. Pretsch, and E. Bakker, *Chem. Rev.*, 1998, **98**, 1593.
2. R. M. Izatt, J. S. Bradshaw, S. A. Nielsen, *et al.*, *Chem. Rev.*, 1985, **85**, 271.
3. R. M. Izatt, K. Pawlak, J. S. Bradshaw, and R. L. Bruening, *Chem. Rev.*, 1991, **91**, 1721.
4. H. Tsukube, Cation binding by natural and modified ionophores: from natural ionophore to synthetic ionophore, in *Cation Binding by Macrocycles*, eds. Y. Inoue and G. W. Gokel, Marcel Dekker, New York, 1990.
5. A. Butler and R. M. Theisen, *Coord. Chem. Rev.*, 2010, **254**, 288.
6. R. C. Hider and X. Kong, *Nat. Prod. Rep.*, 2010, **27**, 637.
7. J. B. Neilands, *J. Biol. Chem.*, 1995, **270**, 26723.
8. H. Tsukube, K. Takagi, T. Higashiyama, *et al.*, *Inorg. Chem.*, 1994, **33**, 2984.
9. K. Marauyama, H. Sohmiya, and H. Tsukube, *Tetrahedron*, 1992, **48**, 805.
10. S. A. Hamidinia, B. Tan, W. L. Erdahl, *et al.*, *Biochemistry*, 2004, **43**, 15956.
11. L. F. Lindoy, *Coord. Chem. Rev.*, 1996, **148**, 349.
12. L. Canet and P. Seta, *Pure Appl. Chem.*, 2001, **73**, 2039.
13. P. J. Loll and P. H. Axelsen, *Annu. Rev. Biophys. Biomol. Struct.*, 2000, **29**, 265.
14. M. Nieto and H. R. Perkins, *Biochem. J.*, 1971, **123**, 773.
15. M. Swiatek, D. Valensin, C. Migliorini, *et al.*, Jeżowska-Bojczuk, *Dalton Trans.*, 2005, 3808.
16. K. Takács-Novák, B. Noszál, M. Tókés-Kövesdi and G. Szász, *Int. J. Pharm.*, 1993, **89**, 261.
17. M. Brzezowska, M. Kucharczyk-Klaminska, F. Bernardi, *et al.*, *J. Inorg. Biochem.*, 2010, **104**, 193.
18. M. Nieto and H. R. Perkins, *Biochem. J.*, 1971, **123**, 789.
19. M. Nieto and H. R. Perkins, *Biochem. J.*, 1971, **124**, 845.
20. P. J. Loll, J. Kaplan, B. S. Selinsky, and P. H. Axelsen, *J. Med. Chem.*, 1999, **42**, 4714.
21. P. Arriaga, J. Laynez, M. Menendez, *et al.*, *Biochem. J.*, 1990, **265**, 69.
22. D. H. Williams, M. S. Searle, J. P. Mackay, *et al.*, *Proc. Natl. Acad. Sci. U.S.A.*, 1993, **90**, 1172.
23. A. Rodríguez-Tebar, D. Vázquez, J. L. P. Velázquez, *et al.*, *J. Antibiot.*, 1986, **39**, 1578.
24. D. H. Williams and B. Bardsley, *Angew. Chem. Int. Ed.*, 1999, **38**, 1172.
25. Y. Nitanai, T. Kikuchi, K. Kakoi, *et al.*, *J. Mol. Biol.*, 2009, **385**, 1422.

26. C. T. Walsh, S. L. Fisher, I.-S. Park, *et al.*, *Chem. Biol.*, 1996, **3**, 21.
27. C. C. McComas, B. M. Crowley, and D. L. Boger, *J. Am. Chem. Soc.*, 2003, **125**, 9314.
28. D. McPhail and A Cooper, *J. Chem. Soc., Faraday Trans.*, 1997, **93**, 2283.
29. P. J. Loll, A. Derhovanessian, M. V. Shapovalov, *et al.*, *J. Mol. Biol.*, 2009, **385**, 200.
30. V. M. Good, M. N. Gwynn, and D. J. C. Knowles, *J. Antibiot.*, 1990, **43**, 550.
31. D. W. Armstrong, Y. Tang, S. Chen, *et al.*, *Anal. Chem.*, 1994, **66**, 1473.
32. I. D'Acquarica, F. Gasparrini, D. Misiti, *et al.*, *Adv. Chromatogr.*, 2008, **46**, 109.
33. A. Berthod, *Chirality*, 2009, **21**, 167.
34. M. Lämmerhofer, *J. Chromatogr. A*, 2010, **1217**, 814–856.
35. T. J. Ward, *Anal. Chem.*, 2006, **78**, 3947.
36. T. J. Ward and A. B. Farris III, *J. Chromatogr. A*, 2001, **906**, 73.
37. M. Blanco and I. Valverde, *Trends Anal. Chem.*, 2003, **22**, 428.
38. F. Hui and M. Caude, *Analusis*, 1999, **27**, 131.
39. M. P. Gasper, A. Berthod, U. B. Nair, and D. W. Armstrong, *Anal. Chem.*, 1996, **68**, 2501.
40. A. A. Rat'ko and R.-I. Stefan, *Analyt. Lett.*, 2004, **37**, 3161.
41. R.-I. Stefan-van Staden, J. F. van Staden, and H. Y. Aboul-Enein, *Anal. Bioanal. Chem.*, 2009, **394**, 821.
42. J. Tamminen and E. Kolehmainen, *Molecules*, 2001, **6**, 21.
43. A. P. Davis, *Molecules*, 2007, **12**, 2106.
44. M. Miyata, N. Tohnai, and I. Hisaki, *Molecules*, 2007, **12**, 1973.
45. O. Bortolini, G. Fantin, and M. Fogagnolo, *Chirality*, 2005, **17**, 121.
46. M. Gdaniec, M. J. Milewska, and T. Połonski, *Angew. Chem. Int. Ed.*, 1999, **38**, 392.
47. K. Sada, N. Shiomi, and M. Miyata, *J. Am. Chem. Soc.*, 1998, **120**, 10543.
48. S.-Z. Zhang, J.-W. Xie, and C.-S. Liu, *Anal. Chem.*, 2003, **75**, 91.
49. C. M. Hebling, L. E. Thompson, K. W. Eckenroad, *et al.*, *Langmuir*, 2008, **24**, 13866.
50. L. L. Amundson, R. Li, and C. Bohne, *Langmuir*, 2008, **24**, 8491.
51. D. Madenci and S. U. Egelhaaf, *Curr. Opin. Colloid In.*, 2010, **15**, 109.
52. A. Bacchi, G. Pelizzi, M. Nebuloni, and P. Ferrari, *J. Med. Chem.*, 1998, **41**, 2319.
53. P. Sensi, A. M. Greco, and R. Ballota, *Antibiotics Annual (1959–1960)*, Antibiotics, Inc, New York, 1959.
54. G. G. Gallo, C. R. Pasqualucci, and P. Radaelli, *Farmaco (Pavia) Ed. Prat.*, 1963, **18**, 78.
55. C. R. Pasqualucci, A. Vigevani, and P. Radaelli, *Farmaco (Pavia) Ed. Prat.*, 1969, **24**, 46.
56. L. Santos, M. A. Medeiros, S. Santos, *et al.*, *J. Mol. Struct.*, 2001, **563–564**, 61.
57. A. Bacchi and G. Pelizzi, *J. Comput. Aided Mol. Des.*, 1999, **13**, 385.
58. C. Godoy-Alcántar, F. Medrano, and A. K. Yatsimirsky, *J. Incl. Phenom. Macrocycl. Chem.*, 2009, **63**, 347.
59. E. A. Meyer, R. K. Castellano, and F. Diederich, *Angew. Chem. Int. Ed.*, 2003, **42**, 1210.
60. D. W. Armstrong, K. Rundlett, and G. L. Reid III, *Anal. Chem.*, 1994, **66**, 1690.
61. S. S. M. Hassan, W. H. Mahmound, and A. H. M. Othman, *Talanta*, 1997, **44**, 1087.
62. M. Chittapragada, S. Roberts, and Y. W. Ham, *Perspect. Med. Chem.*, 2009, 21.
63. J. R. Thomas and P. J. Hergenrother, *Chem. Rev.*, 2008, **108**, 1171.
64. A. P. Carter, W. M. Clemons, D. E. Brodersen, *et al.*, *Nature*, 2000, **407**, 340.
65. Q. Vicens and E. Westhof, *Biopolymers*, 2003, **70**, 42.
66. C. M. Barbieri and D. S. Pilch, *Biophys. J.*, 2006, **90**, 1338.
67. Y. Fuentes-Martínez, C. Godoy-Alcántar, F. Medrano, *et al.*, *Bioorg. Chem.*, 2010, **38**, 173.
68. A. Bencini, A. Bianchi, E. Garcia-España, *et al.*, *Coord. Chem. Rev.*, 1999, **188**, 97.
69. H. Nishi, K. Nakamura, H. Nakai, and T. Sato, *Chromatographia*, 1996, **43**, 426.
70. B. Chankvetadze, M. Saito, E. Yashima, and Y. Okamoto, *Chirality*, 1998, **10**, 134.
71. H. Nakamura, A. Sano, and H. Sumii, *Anal. Sci.*, 1998, **14**, 375.
72. T. Ohyama, D. Wang, and J. A. Cowan, *Chem. Commun.*, 1998, 467.
73. C. Bazzicalupi, A. Bencini, A. Bianchi, *et al.*, *J. Am. Chem. Soc.*, 1999, **121**, 6807.
74. Y. Fuentes-Martínez, C. Godoy-Alcántar, F. Medrano, *et al.*, *Supramol. Chem.*, 2010, **22**, 212.
75. K. O. Lara, C. Godoy-Alcántar, I. L. Rivera, *et al.*, *J. Phys. Org. Chem.*, 2001, **14**, 453.
76. R. Moreno-Corral and K. Ochoa Lara, *Supramol. Chem.*, 2008, **20**, 427.
77. N. D'Amelio, E. Gaggelli, N. Gaggelli, *et al.*, *Dalton Trans.*, 2004, 363 and references therein.
78. W. Szczepanik, P. Kaczmarek, J. Sobczak, *et al.*, *New J. Chem.*, 2002, **26**, 1507.
79. C. J. Dutton, M. A. Haxell, H. A. I. McArthur, and R. G. Wax, eds. *Peptide Antibiotics: Discovery, Modes of Action and Applications*, Marcel Dekker, 2002.
80. M. C. Bagley, J. W. Dale, E. A. Merritt, and X. Xiong, *Chem. Rev.*, 2005, **105**, 685.
81. Y. L. Hsiao and S. S. Chen, *Chromatographia*, 2009, **70**, 1031.
82. C. Godoy-Alcántar, I. León-Rivera, and A. K. Yatsimirsky, *Bioorg. Med. Chem. Lett.*, 2001, **11**, 651.

83. L.-J. Ming and J. D. Epperson, *J. Inorg. Biochem.*, 2002, **91**, 46.

84. M. Castagnola, D. V. Rossetti, R. Inzitari, *et al.*, *Electrophoresis*, 2004, **25**, 846.

85. T. Velkov, P. E. Thompson, R. L. Nation, and J. Li, *J. Med. Chem.*, 2010, **53**, 1898.

86. C. Chatterjee, M. Paul, L. Xie, and W. A. van der Donk, *Chem. Rev.*, 2005, **105**, 633.

87. S. T. Hsu, E. Breukink, E. Tischenko, *et al.*, *Nat. Struct. Mol. Biol.*, 2004, **11**, 963.

88. C. E. Song, ed., *Cinchona Alkaloids in Synthesis and Catalysis*, Wiley-VCH, 2009.

89. J. Lah, N. M. Maier, W. Lindner, and G. Vesnaver, *J. Phys. Chem. B*, 2001, **105**, 1670.

90. N. M. Maier, L. Nicoletti, M. Lämmerhofer, and W. Lindner, *Chirality*, 1999, **11**, 522.

91. K. H. Krawinkler, N. M. Maier, R. Ungaro, *et al.*, *Chirality*, 2003, **15**, S17.

92. M. Shamma, *The Isoquinoline Alkaloids*, Academic Press, New York, 1972.

93. H. King, *J. Chem. Soc.*, 1935, 1381.

94. C. Godoy-Alcántar, A. V. Eliseev, and A. K. Yatsimirsky, *J. Mol. Recognit.*, 1996, **9**, 54.

95. U. B. Nair, D. W. Armstrong, and W. L. Hinze, *Anal. Chem.*, 1998, **70**, 1059.

96. C. Godoy-Alcántar, M. I. Nelen, A. V. Eliseev, and A. K. Yatsimirsky, *J. Chem. Soc. Perkin Trans. 2*, 1999, 353.

97. K. Ochoa Lara, C. Godoy-Alcántar, A. V. Eliseev, and A. K. Yatsimirsky, *Org. Biomol. Chem.*, 2004, **2**, 1712.

98. K. Ochoa Lara, C. Godoy-Alcántar, A. V. Eliseev, and A. K. Yatsimirsky, *Arkivoc*, 2005 (vi), 293.

99. I. Stanculescu, C. Mandravel, F. Delattre, *et al.*, *J. Photochem. Photobiol. A*, 2003, **161**, 79.

Receptors for Nucleotides

Enrique García-España[1], Raquel Belda[1], Jorge González[1], Javier Pitarch[1], and Antonio Bianchi[2]

[1]*Universidad de Valencia, Valencia, Spain*
[2]*University of Florence, Florence, Italy*

1 INTRODUCTION

Supramolecular chemistry, chemistry beyond the molecule, was established as an interdisciplinary research field following the pioneering work of C. Pedersen, J.-M. Lehn, and D. J. Cram in the 1960s and 1970s[1–8] on the recognition of alkaline and alkaline earth cations by crown ethers, cryptands, and spherands. After the initial studies dealing with the coordination chemistry and transport properties of spherical metal ions by synthetic receptors, the target substrates in supramolecular chemistry became much larger, covering neutral and anionic species. Anions have great relevance from the environmental and biological points of view. Over 70% of all cofactors and substrates involved in biology are of anionic nature. Proteins and nucleic acids themselves are negatively charged, water-soluble polymeric species.

Interestingly enough, the birth of the first recognized synthetic halide receptors occurred practically at the same time as the discovery by Charles Pedersen of the crown ethers. While C. Pedersen submitted his first paper on crown ethers entitled "Cyclic Polyamines and their Complexes with Metal Salts" on April 1967 to JACS.[1] Park and Simmons, who were working in the same company as Pedersen, submitted their paper on the complexes formed by bicyclic diammonium receptors with chloride entitled "*Macrobicyclic Amines. III. Encapsulation of Halide ions by in, in-1, (k+2)-diazabicyclo[k.l.m]alkaneammonium ions*" to JACS in November of the same year.[9] These cage-type receptors were called *katapinands* taking after the Greek term describing the swallowing up of the anionic species toward the interior of the cavity. However, while the investigations of crown ethers rapidly evolved and many of these compounds were prepared and their chemistry widely explored, the studies on anion coordination chemistry remained at this initial stage and were not further developed until J.-M. Lehn and his group focused on this area in the late 1970s and beginning of the 1980s.[10–16] An initial reference book on this topic was published in 1997,[17] which has been followed by two more recent volumes[10,18] and a good number of review articles, many of them appearing in special journal issues dedicated to anion coordination. Some of these review articles are included in Refs 11–16, 19–53. Very recently, a whole issue of the journal *Chemical Society Reviews* has been devoted to the supramolecular chemistry of anionic species.[54] Also, several chapters of this multivolume set deal with anion coordination chemistry.

Supramolecular Chemistry: From Molecules to Nanomaterials.
Edited by Philip A. Gale and Jonathan W. Steed.
© 2012 John Wiley & Sons, Ltd. ISBN: 978-0-470-74640-0.

1.1 Nucleotides

Nucleotides were one of the first classes of anionic substrates investigated by the pioneers in the field of supramolecular chemistry.[55,56] This is not surprising taking into account the many biological implications of these molecules. As is well known, nucleotides are the building blocks of DNA- and RNA-biopolymers, which store the genetic information of organisms and afford the blueprint of protein construction.[57] The binding by proteins of 5′-adenosine triphosphate (ATP) [1], the biological energy currency, is one of the most important recognition events in nature.[58] ATP plays important roles in energy transduction in organisms and controls several metabolic processes including the synthesis of cyclic adenosine monophosphate.[59] 5′-Guanosine triphosphate (GTP) is required for many biological activities such as synthesis of DNA, RNA, and proteins; nutrient metabolism; and cell signaling.[60]

GTP binding proteins play a diversity of roles as switches in cell growth, receptor activation, exocytosis, ion channel conductivity, and change in cell shape.[61] 5′-Uridine triphosphate (UTP) and 5′-uridine diphosphate (UDP), which are key building blocks in the synthesis of RNA and in glycotransfer pathways, play pivotal roles in various biological events. UTP serves as a donor in energy transduction in organisms and as a control element in metabolic processes by its participation in enzymatic reactions. Besides, UTP and UDP are involved in many glycosylation processes that are catalyzed by glycotransferases. Thymidine nucleotides, including 5′-thymidine monophosphate (TMP), 5′-thymidine diphosphate (TDP), and 5′-thymidine triphosphate (TTP) are essential building blocks in DNA replication and cell division.[62] Imbalances in the intracellular level of TTP have been related to the development of diseases characterized by defects in the repair of replication of mitochondrial DNA (mtDNA). These include Alper syndrome, progressive external ophtalmoplegia, mtDNA depletion syndrome, and recessive mitochondrial neurogastrointestinal encephallomyopathy.[63]

In this chapter, we describe, in a comprehensive way, examples of the recognition, activation, and detection of this important class of biologically relevant anionic substrates.

2 NUCLEOTIDE BINDING

2.1 Characteristics of the mononucleotide species

From the recognition point of view, nucleotides offer distinctive chemical features that make these molecules appealing targets for multipoint binding (Figure 1). Apart from their inherent negative charge, which is typically the main property to address when constructing anion receptors, nucleotides can either accept or donate hydrogen bonds through their three constitutional moieties namely, the

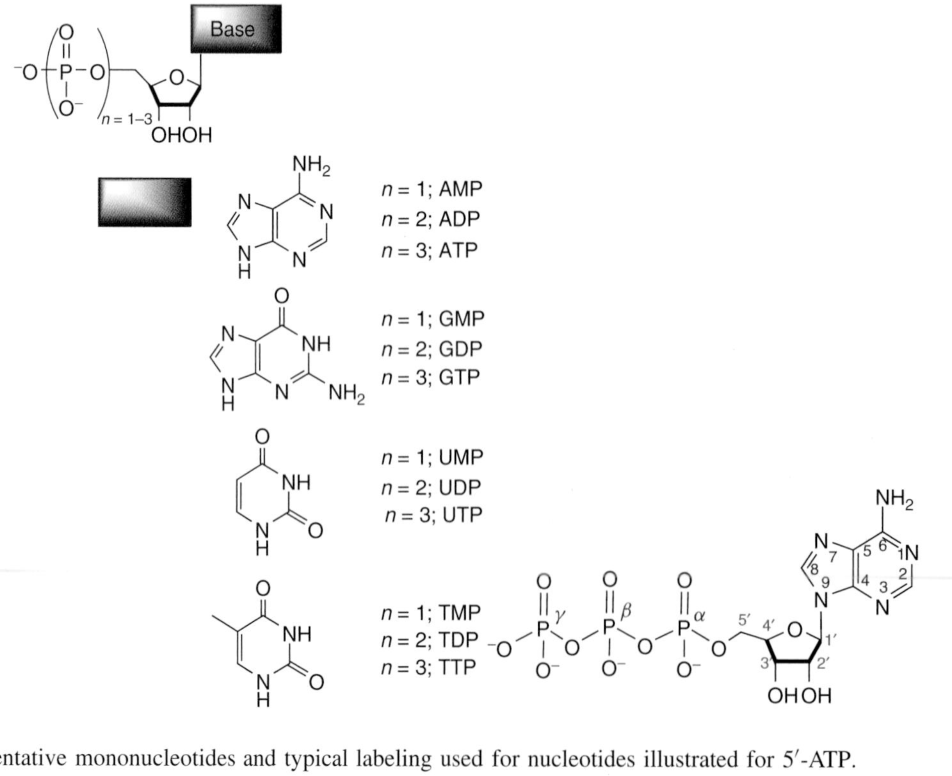

Figure 1 Representative mononucleotides and typical labeling used for nucleotides illustrated for 5′-ATP.

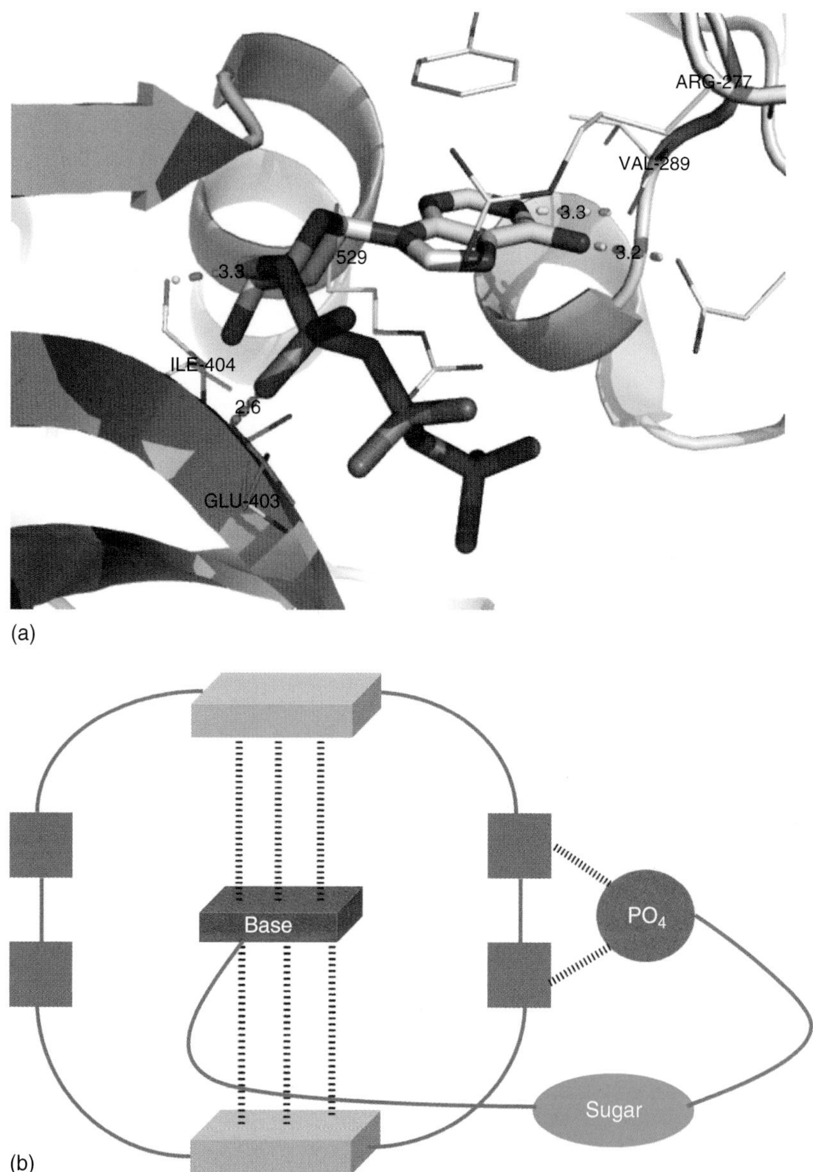

Figure 2 (a) Representation of ATP nucleotide binding site in the human glycyl–tRNA synthetase (GlyRS) in complex with glycine and ATP.[64] (b) Schematic representation of the different binding motifs in a nucleotide.

phosphate chain, the sugar unit, and the nucleobase, at the time that they can give stacking interactions through their nucleobases with electronically matching aromatic functionalities of the receptor species. For instance, Figure 2(a) shows how ATP is bound in human glycyl–tRNA synthetase (GlyRS) through a set of seven hydrogen bonds involving the three components of ATP. Additionally, π-stacking interactions between adenine and a phenylalanine residue and a π-cation interaction between adenine and an arginine residue occur.[64]

Therefore, nucleotide receptors have been built trying to simultaneously satisfy several of these characteristics: complementary charged functions, hydrogen-bonding capacity to any of the nucleotide functions, and stacking interactions with the nucleobase (Figure 2b). Also, the obvious point that the net charge in the nucleotide depends on pH should be stressed; at a physiological pH of 7.4, mononucleotide triphosphates, mononucleotide diphosphates, and mononucleotide monophosphates possess 4-, 3- and 2- charge respectively.[65]

2.2 Receptors for nucleotides

2.2.1 *Receptors without aromatic components*

From a historical point of view, protonated polyamines were the first class of receptors used for nucleotide binding.

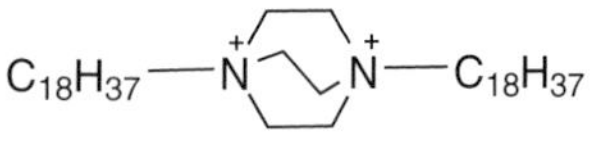

1

Scheme 1

One of the first reports that can be found in the literature is a study performed by Tabushi *et al.*[66] in which it was described that the quaternized DABCO with stearyl chains (**1**, Scheme 1) was able to transfer 5′-adenosine diphosphate (ADP) very efficiently from aqueous solutions at pHs of 3 and 5 into a chloroform phase. The selectivity found for the transfer of ADP over AMP was explained by the formation of a double salt bridge between the phosphate groups and the quaternized receptor.

The Lehn and Kimura groups provided initial evidences of the strong interaction occurring between polyammonium receptors and the nucleotides ATP, ADP, and AMP.[55,56] Lehn's group in a communication to JACS[55] reported the interaction of the hexaprotonated [24]aneN_6 ($H_6(\mathbf{2})^{6+}$), octaprotonated [32]aneN_8 ($H_8(\mathbf{3})^{8+}$), and hexaprotonated [27]aneN_6O_3 ($H_6(\mathbf{4})^{6+}$) with AMP^{2-}, ADP^{3-}, and ATP^{4-} (Scheme 2). The values of the binding constants obtained by computer analysis of the pH-metric titration curves varied from 3.4 to 9.1 logarithmic units, following the sequence $AMP^{2-} < ADP^{3-} < ATP^{4-}$.

Kimura *et al.* reported in a full paper in JACS[56] the interaction with nucleotides of a series of polyazamacrocycles including [18]aneN_4 (**5**), [15]aneN_5 (**6**), [18]aneN_6 (**7**), the mixed amino-amide receptor (**8**), [16]aneN_5 (**9**),

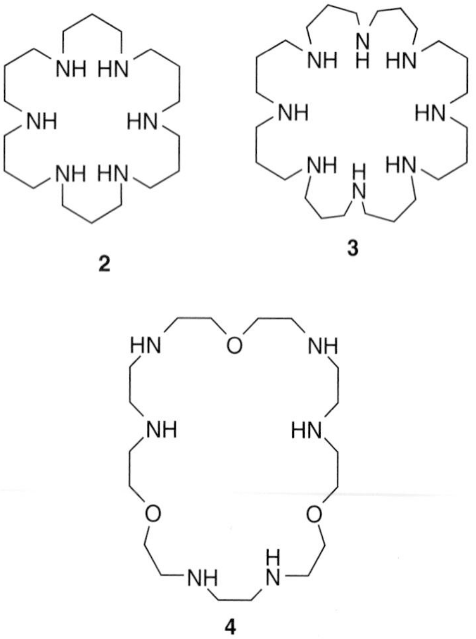

2 3 4

Scheme 2

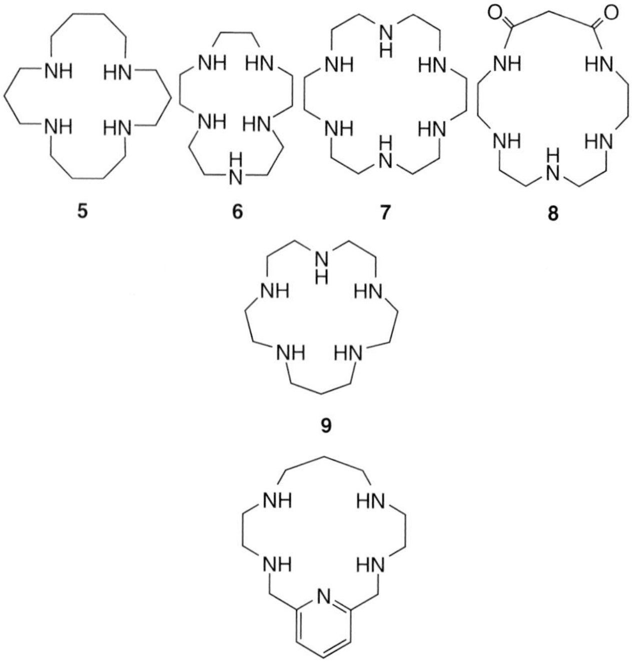

5 6 7 8 9 10

Scheme 3

and a tetraazapyridinophane (**10**) by means of polarographic techniques (Scheme 3). The polarographically determined association constants in general followed the expected order with higher values for the more highly charged ATP anion. However, a reversed sequence seemed to be found for the interaction of diprotonated pyridinophane receptor **10** with the three nucleotides.

The relevance of charge factors in nucleotide binding is evidenced in the interaction with ATP of the series of "so-called" large [3*k*]aneN_K macrocycles (Scheme 4).[67,68]

To have a reliable comparison of the relative stabilities of the different members of the series and a clear picture of the selectivity trend, the cumulative and stepwise constants obtained from the analysis of the pH-metric titrations were translated into conditional stability constants. Such constants are calculated, for every pH value, as the quotient between the sum of the complexed species and the product

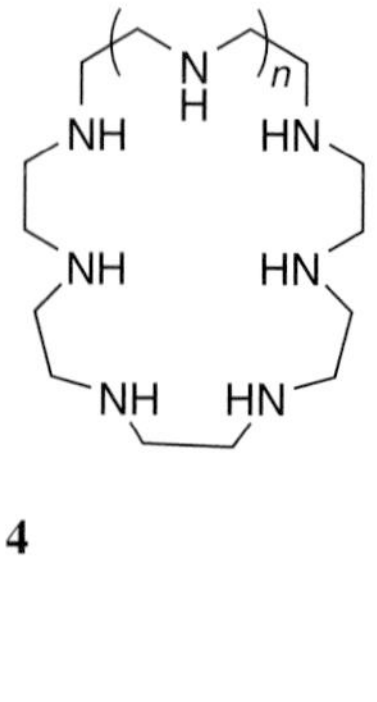

$n = 1$ [21]ane N_7 (**11**)
$n = 2$ [24]ane N_8 (**12**)
$n = 3$ [27]ane N_9 (**13**)
$n = 4$ [30]ane N_{10} (**14**)
$n = 5$ [33]ane N_{11} (**15**)
$n = 6$ [36]ane N_{12} (**16**)

Scheme 4

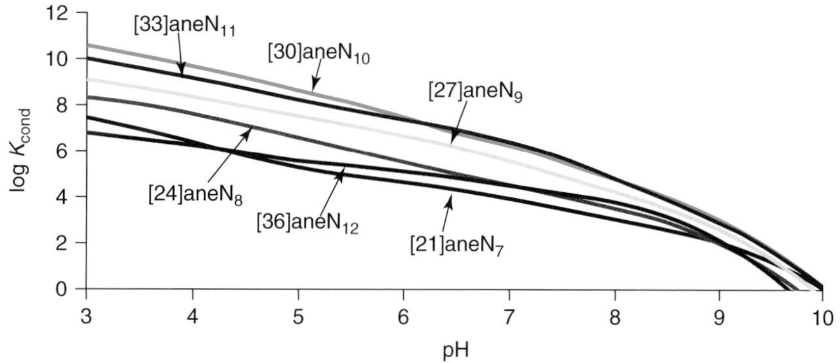

Figure 3 Plot of the logarithms of the conditional stability constant versus pH for the system [3k]aneN$_k$-ATP (k = 7–11 (**11**–**16**)).

of the sums of the free reagents.[69]

$$K_{\text{cond}} = \Sigma[\text{H}_{j+i}\text{LA}]/\Sigma[\text{H}_j\text{L}] \times \Sigma[\text{H}_i\text{A}]$$

This method of treating the data provides, for all the pH ranges of study, equilibrium constants comparable to those obtained by other techniques at fixed pH values. Inspection of Figure 3 shows that the values of the conditional constants increase progressively from [21]aneN$_7$ (**11**) to [30]aneN$_{10}$ (**14**) decreasing then for the two largest members of the series [33]aneN$_{11}$ (**15**) and [36]aneN$_{12}$ (**16**). Therefore, it seems that for a certain macrocyclic size there is an optimum matching between the charges of the polyammonium receptor and the polyphosphate chain. To have a more complete picture of the situation, the interaction of these receptors with AMP and ADP was also studied, and the stability order that was found indicated that these macrocycles interact more strongly with the more charged ATP nucleotide.

It is well established that the length of the hydrocarbon chains between the nitrogen atoms in a monocyclic receptor is a key factor in determining its basicity and consequently its affinity for the anionic species.[70] In this respect, Bianchi, Micheloni, and Paoletti reported that the tetraaza monocycle [20]aneN$_4$ (**17**), which has butylenic chains between the amino groups and is fully protonated at the physiological pH of 7.4, has a significant affinity for ATP^{4-} in water (H$_4$(**17**)$^{4+}$ + ATP^{4-} = H$_4$(**17**)(ATP), log K = 3.81(3).[71] Also, related to this point, C. Burrows prepared a chiral [18]aneN$_4$ derivative including an alcohol C-linked side chain, which was proved to interact with ATP by ^{31}P NMR.[72] Bis-macrocyclic polyamines **18** and **19** (Scheme 5) also have interesting complexing properties toward nucleotide species.[73]

Another strategy in nucleotide binding is based on using tetraazacycloalkanes with pendant arms containing additional ammonium groups such as in receptor **20** (TAEC) (Scheme 6).[74]

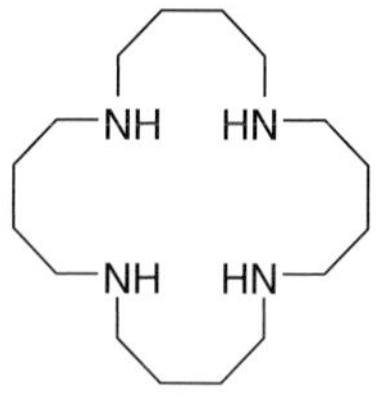

17

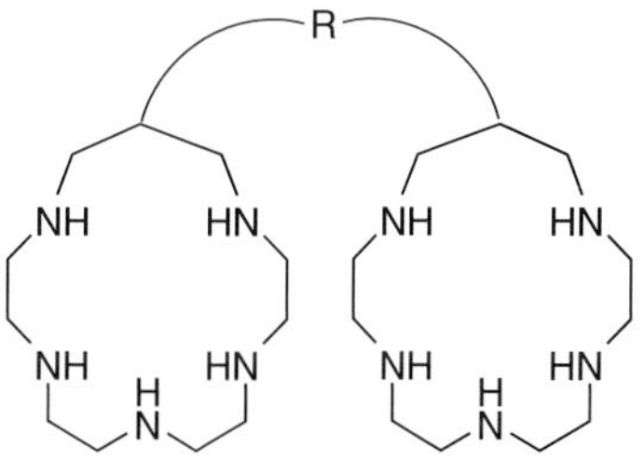

18 R = –CH$_2$CH$_2$OCH$_2$CH$_2$OCH$_2$CH$_2$–
19 R = –CH$_2$CH$_2$CH$_2$–

Scheme 5

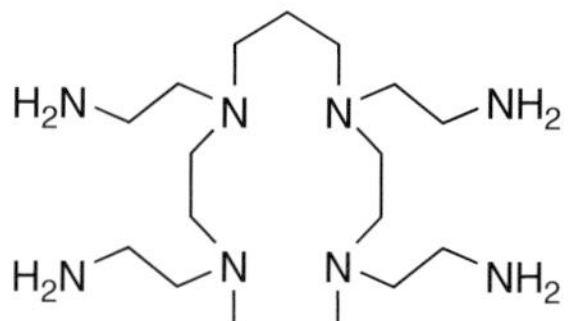

20

Scheme 6

However, a comparison of the interaction of ATP with **20** (TAEC) and with macrocycle **12** ([24]aneN$_8$), containing the same number of amino groups, by means of the conditional constant clearly shows higher conditional constants for [24]aneN$_8$ (Figure 4). The same perspective is offered by an analysis of the amount of complexed ATP in a system containing equimolar amounts of ATP, **12**, and **20**, assuming that mixed ATP-**12**-**20** species are not formed, in

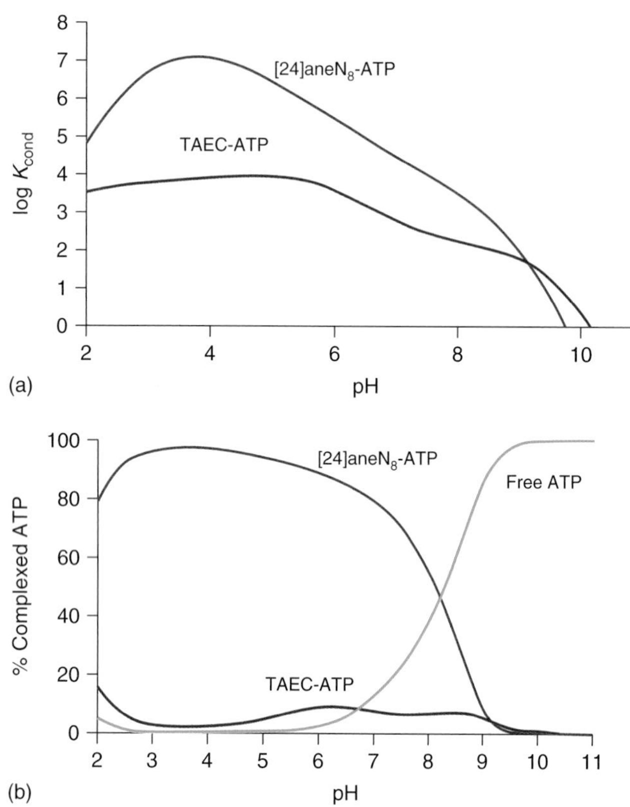

Figure 4 (a) Plot of log K_{cond} versus pH for the systems TAEC (**20**)-ATP and [24]aneN_8(**12**)-ATP. (b) Plot of the overall amount of complexed ATP in a mixed system containing ATP, TAEC (**20**), and [24]aneN_8 (**12**) in equimolar amounts.

which it can be seen that the ATP is, at a larger extent, bound to [24]aneN_8. The orientation and higher solvation energy of the charged primary amino groups have to be among the factors responsible for this tendency.

Comparison of the relative affinities of nucleotides of acyclic and cyclic receptors is also interesting. Although not much data regarding this point can be found in the literature, several years ago a study was published that compared the relative affinities of the cyclic receptors [18]aneN_6 (**7**), $(CH_3)_4$[18]aneN_6 (**21**) and of the acyclic receptor $(CH_3)_2$pentaen (**22**) (Scheme 7) for ATP.[75]

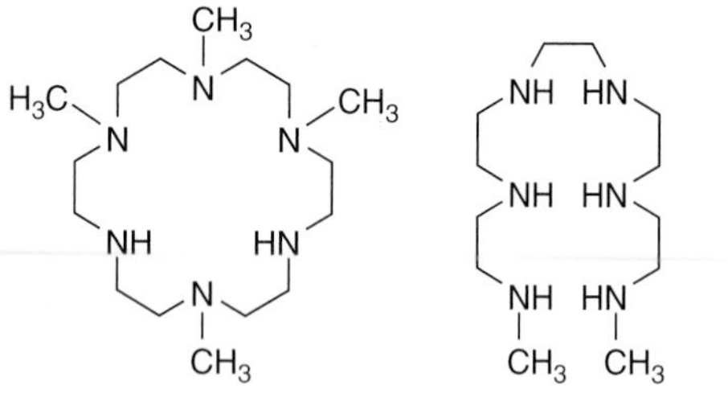

$(CH_3)_4$[18]aneN_6 (**21**) $(CH_3)_2$pentaen (**22**)

Scheme 7

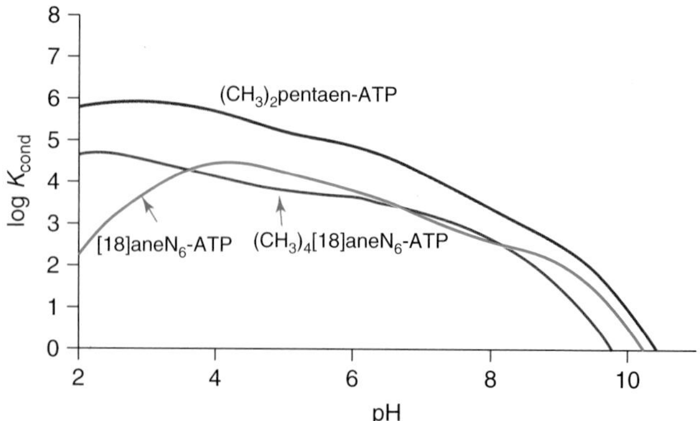

Figure 5 Plot of log K_{cond} versus pH for the systems [18]aneN_6 (**7**)-ATP, $(CH_3)_4$[18]aneN_6(**21**)-ATP and $(CH_3)_2$pentaen (**22**)-ATP.

The analysis of the conditional stability constants for the three receptors (Figure 5) shows that, in this case, the acyclic ligand interacts more strongly with ATP than the cyclic ligands throughout the pH range studied.

However, in cases where the complementarity between the nucleotide and the cyclic guest species is optimum, higher interaction should be expected for the cyclic receptor.

Stoichiometry of the formed adducts

A point that deserves to be discussed is the stoichiometry of the adducts formed by the nucleotide and receptors. This can be addressed by analyzing the variations produced in the ^{31}P NMR spectra of the nucleotides upon addition of the receptors.

For instance, Figure 6 plots the ^{31}P NMR spectra of solutions in which increasing amounts of TAEC (**20**) are added to a solution of ATP at pH = 6. Typically, the doublet signal of P_γ is the one experiencing the largest downfield chemical shift upon addition of the nucleotide at pH values where protonation of the phosphate chain occurs. The displacement of the chemical shift of P_β is much reduced, while the triplet signal of P_α practically does not move. The interaction of ATP with a receptor leads to a reduction in the actual basicity of ATP and an increase in the basicity of the receptor. The maximum displacement that can generally be observed in the ^{31}P NMR shifts corresponding to the difference between the ^{31}P NMR chemical shifts of the ATP^{4-} and $HATP^{3-}$ species, that is, the pH range in which the jump between both limiting values is produced changes as a function of the binding strength. This is exemplified in Figure 7 in which the displacements of the P_α, P_β and P_γ signals are plotted versus the pH for ATP alone and for ATP in the presence of two macrocyclic polyamines with different binding affinities.

The larger the interaction, the more acidic is the ATP and the more basic is the receptor. However, it has to be

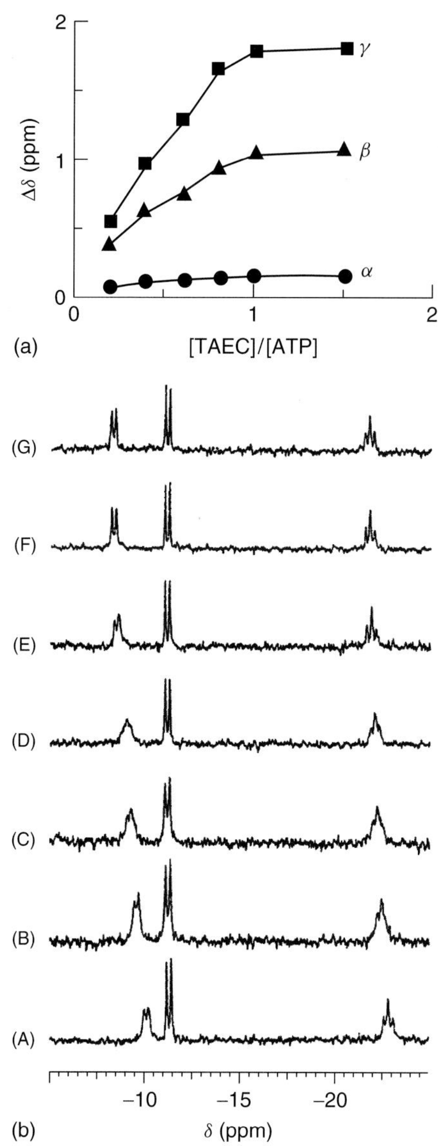

Figure 6 ^{31}P NMR spectra of the system TAEC (**20**): ATP as a function of $R = [\text{TAEC}]/[\text{ATP}]$, (a) $R = 0$ (A), $R = 0.2$ (B), $R = 0.4$ (C), $R = 0.6$ (D), $R = 0.8$ (E), $R = 1{,}0$ (F), and $R = 1.5$ (G). (b) ^{31}P NMR shifts ($|\Delta\delta|$) in ppm for the α, β, and γ signals of ATP as a function of increasing R at pH 6. (Adapted from Ref. 74 © Royal Society of Chemistry, 1991.)

taken into account that conformational factors can affect the variation of these signals and a clear-cut interpretation cannot be offered.

Although stoichiometric methods should be applied to define the stoichiometry of the nucleotide:receptor adducts formed, a representation of the displacement of the ^{31}P NMR chemical shift versus the receptor:ATP mole ratio gives a good indication about this point (Figure 6b). In this respect, although most of the mononucleotide:receptor adducts described in the literature have 1 : 1 stoichiometries, there are a few cases in which other stoichiometries have

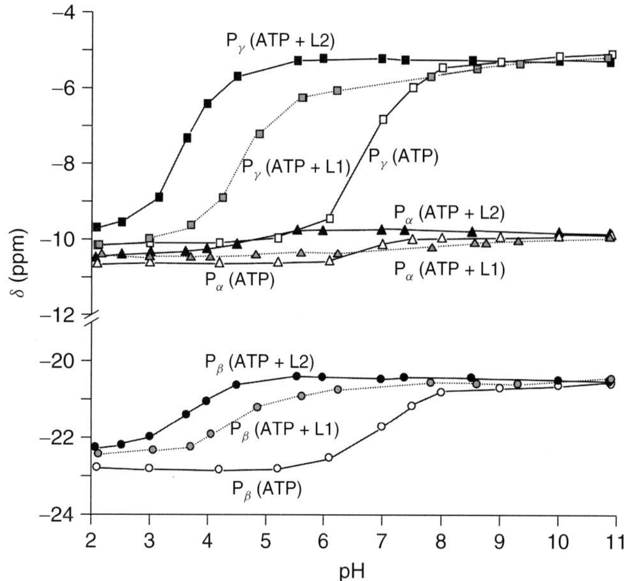

Figure 7 Variations with pH of the P_γ, P_β, and P_α ^{31}P NMR signals of free 5′-ATP and in the presence of saturation amounts of the phenathrolinophane receptors L1 and L2. The larger the interaction, the larger are the observed upfield shifts. (Adapted from Ref. 86. © American Chemical Society, 2009.)

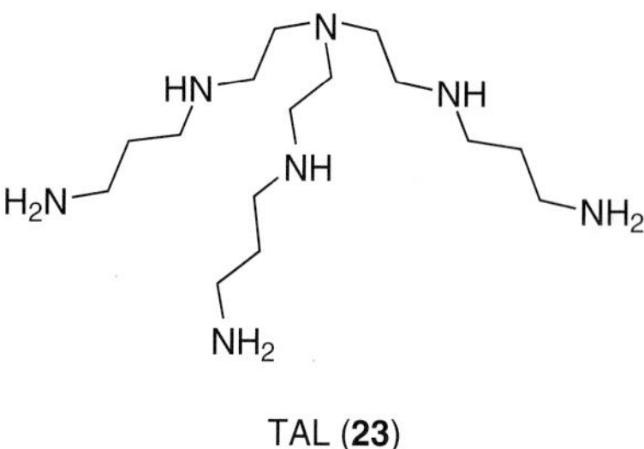

Scheme 8

been reported. This is the case of the system formed by the protonated forms of the tripodal polyamine TAL (**23**, Scheme 8) and AMP.[76]

Both the analysis of the ^{31}P NMR data as well as the mass spectra permitted the formation of complexes with AMP:**23** stoichiometry 1 : 1, 2 : 1, and 3 : 1 to be inferred. Comparison of the constants for the equilibrium $\text{AMP}^{2-} + \text{H}_6(\mathbf{23})^{6+} = [\text{H}_6(\mathbf{23})(\text{AMP})]^{4+}$, $\log K_1 = 5.36$, $[\text{H}_6(\mathbf{23})(\text{AMP})]^{4+} + \text{AMP}^{2-} = [\text{H}_6(\mathbf{23})(\text{AMP})_2]^{2+}$, $\log K_2 = 2.89$, and $[\text{H}_6(\mathbf{23})(\text{AMP})_2]^{2+} + \text{AMP}^{2-} = [\text{H}_6(\mathbf{23})(\text{AMP})_3]$, $\log K_3 = 3.85$, shows the trend $K_1 > K_2 < K_3$. This reversed trend was explained assuming that all three arms of the receptors are involved in the binding of the first AMP anion, with the entrance of the second AMP producing an opening of the receptor due to electrostatic repulsion between the anionic guests. The third AMP finds an adequate receptor arrangement for binding, and the constant again increases.

2.2.2 *Receptors containing aromatic fragments*

As indicated in the introduction of this chapter, the aromatic rings in the nucleotide structure provide a very important motif for their binding through stacking interactions. With this purpose, Lehn and coworkers[77] prepared receptors **25** and **26** (Scheme 9) in which acridine moieties were added to the classical receptor O-BISDIEN (**24**) to take advantage from electrostatic attractions, hydrogen bonding, and π-stacking interactions. O-BISDIEN[78] is one of the receptors that has been most broadly studied in Lehn's group by virtue of its catalytic activity (*vide infra*).

^{1}H NMR experiments showed that **25** and **26** form stronger complexes with ATP than with **24** due to the additional involvement of π-stacking interactions between the acridine rings and the nucleic base.

Lehn and coworkers have also prepared a series of bis-intercaland compounds and their reference mono-intercaland partners to see the affinity enhancement provided by this double binding motif (**27**–**36**, Scheme 10).[79] Evidence of intercalation was obtained from the hypochromism observed in the UV–vis spectra of the nucleotide-ligand systems. While differences in stability were obtained for the different nucleic bases, the constants for the nucleotides containing the same base pair and different

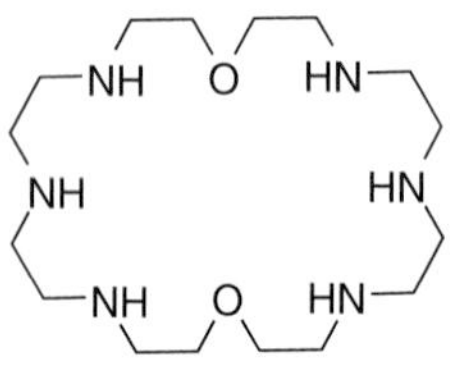

24

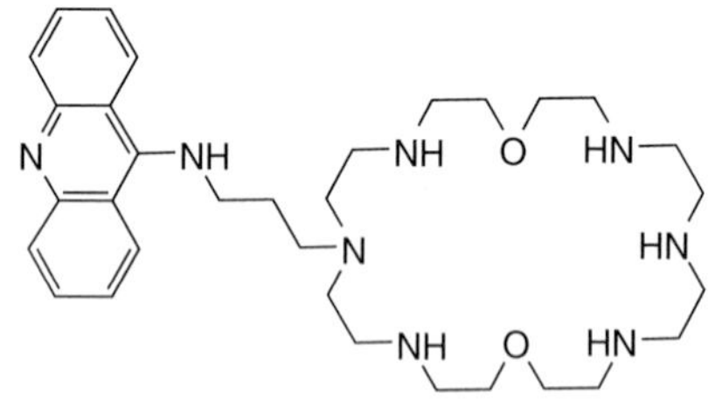

25

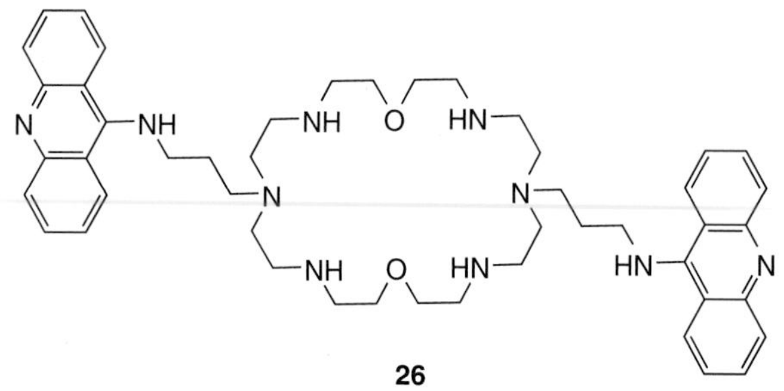

26

Scheme 9

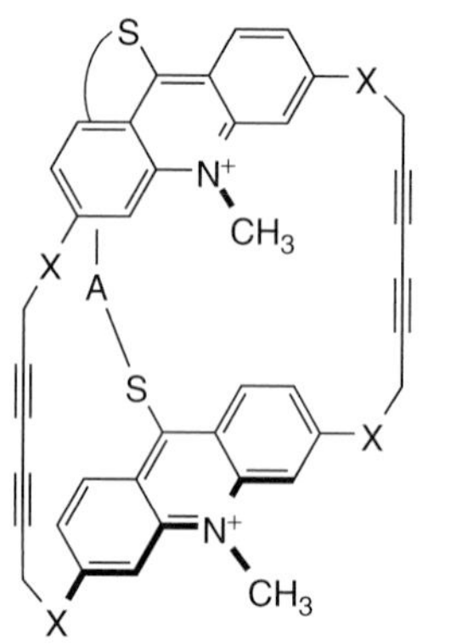

27 X= O, A = $-(CH_2)_6-$
28 X = O, A = $-(CH_2)_2-O-(CH_2)_2-$
29 X = O, A = $-(CH_2)_2-O-(CH_2)_2-O-(CH_2)_2-$
30 X = NH, A = $-(CH_2)_2-O-(CH_2)_2-$

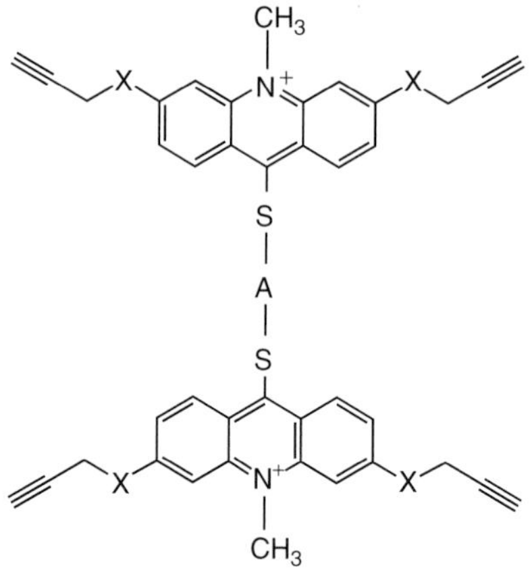

31 X = O, A = $-(CH_2)_6-$
32 X = O, A = $-(CH_2)_2-O-(CH_2)_2-$
33 X = O, A = $-(CH_2)_2-O-(CH_2)_2-O-(CH_2)_2-$
34 X = NH, A = $-(CH_2)_2-O-(CH_2)_2-$

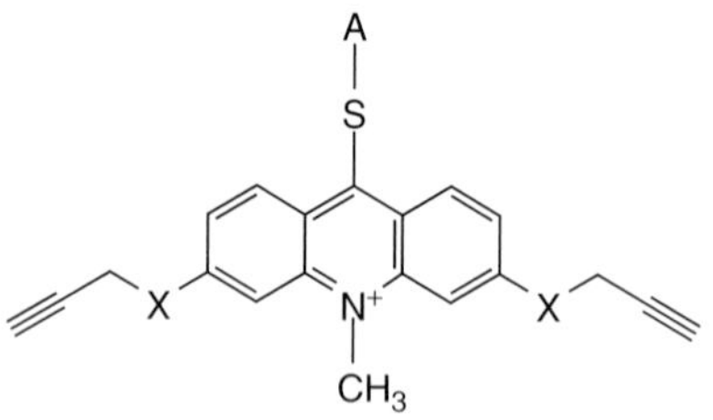

35 X = O, A = $-(CH_2)_2-CH_3$
36 X = NH, A = $-(CH_2)_2-CH_3$

Scheme 10

phosphate chains were practically the same. For instance, in spite of their different net charges, the constants for the interaction of AMP^{2-}, ADP^{3-}, and ATP^{4-} with **28** were 3.79, 3.84, and 3.91 logarithmic units respectively.

The importance of π-stacking in nucleotide binding is even more evident in the case of receptors **37**–**40** (Scheme 11). For example, the equilibrium constants for the binding of charged ATP^{4-}, ADP^{3-}, and AMP^{2-} are $\log K_s = 5.80$, 5.65, and 5.38 respectively.[80]

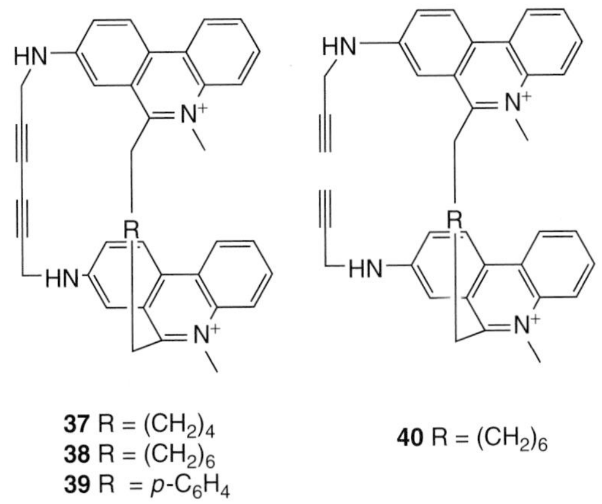

37 R = $(CH_2)_4$
38 R = $(CH_2)_6$
39 R = $p\text{-}C_6H_4$

40 R = $(CH_2)_6$

Scheme 11

Monocyclic azacyclophanes with appropriate polyamine chains can behave as multipoint binders of nucleotides,[81,82] as was evidenced for **41** (Scheme 12) by García-España, Luis, and coworkers using ^{1}H NMR and potentiometric techniques. While the phosphate chain is a good electrostatic binding point and a hydrogen bond acceptor, the nucleoside part operates as an adequate site for stacking with the aromatic part of the macrocycle. $\pi-\pi$ interactions are in this case evidenced in the ^{1}H NMR spectra by upfield shifts of the signals of the anomeric proton of the nucleotides and of both the aromatic protons of nucleotides and **41**, as exemplified for the system ADP-**41** in Figure 8. Such shifts are observed over the entire pH range where interaction occurs.

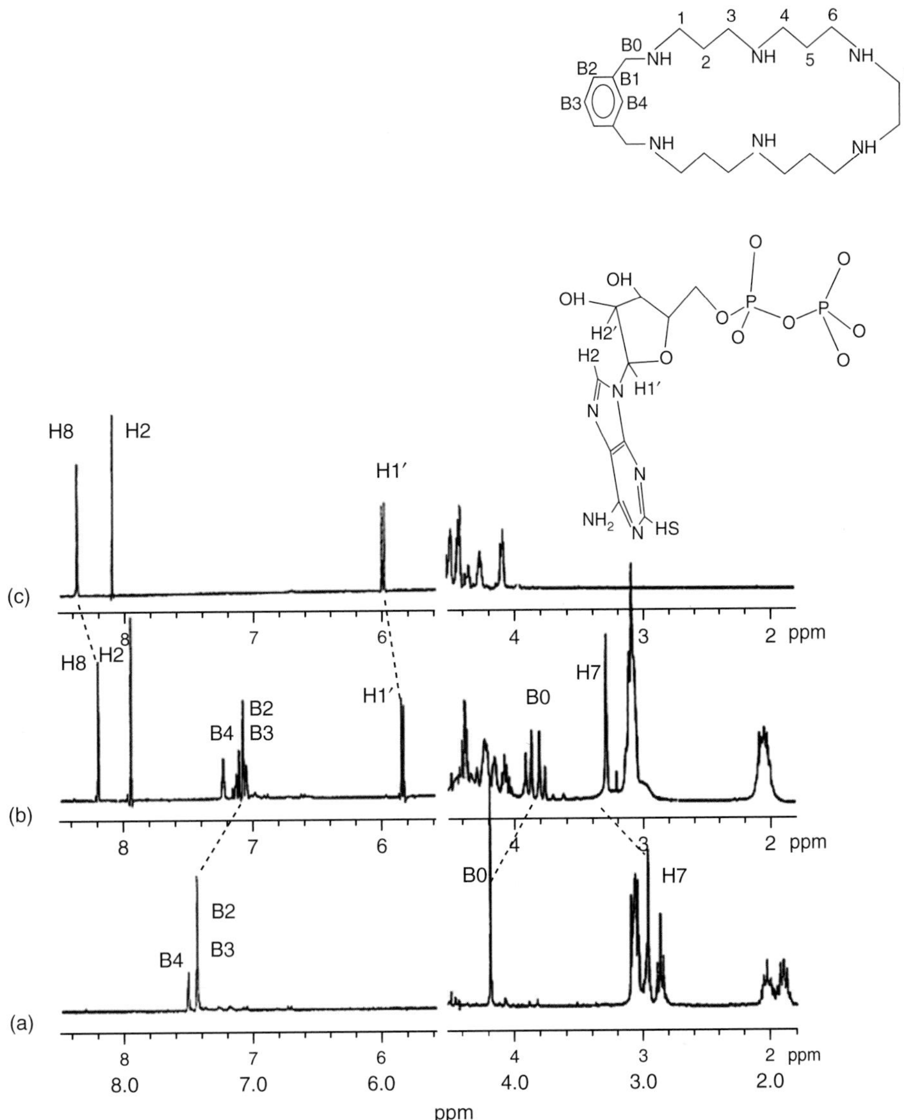

Figure 8 ^{1}H NMR spectra recorded in D_2O of (a) **41**, (c) ADP and (b) an equimolar mixture of **41** and ADP at pD = 7. (Adapted from Ref. 82 © Elsevier, 1996.)

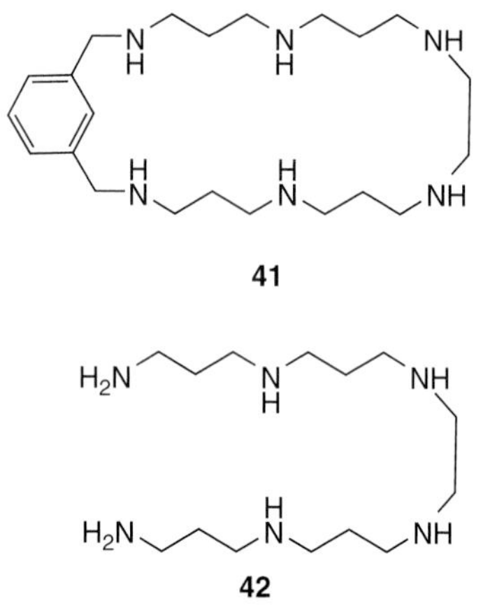

41

42

Scheme 12

The stacking interaction might be an important component in the difference in conditional constants found for the systems ATP-**41** and ATP-**42**, in the pH range where the interaction occurs, pH 3–8 (Figure 9a). This difference in stability is also consistent with the higher percentage of ATP complexed by **41** than by **42**, observed in the same pH range (Figure 9b). Nevertheless, it needs to be remarked that **41** and **42** are not strictly comparable due to the presence of primary amino groups in **42,** which have higher hydration energies.

Several acyclic receptors (Scheme 13), in which either one or two naphthalene or anthracene aromatic fragments have been appended to polyamine chains, have been constructed to take advantage of $\pi-\pi$ stacking binding mode in water solutions.[69,83]

Among the receptors sharing the same polyamine chain, those containing one anthrylmethyl fragment interact stronger with ATP than those with just one naphthalene. However, they present comparable stability constant values to those containing naphthalene at both ends. NMR studies show that π-stacking is present throughout the pH range where interaction occurs. NOE experiments confirm the proximity of the aromatic rings of the receptors and ATP and have allowed for proposing preliminary models for the interaction (Figure 10).

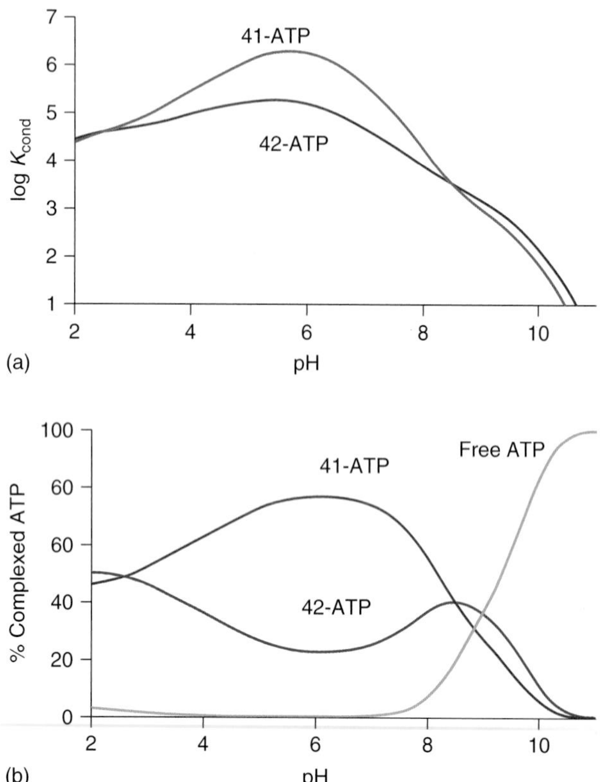

Figure 9 (a) Plot of log K_{cond} versus pH for the systems **41**-ATP and **42**-ATP. (b) Plot of the overall amount of complexed ATP in a mixed system containing ATP, **41**, and **42** in equimolar amounts. Concentration in all reagents is 1×10^{-3} M.

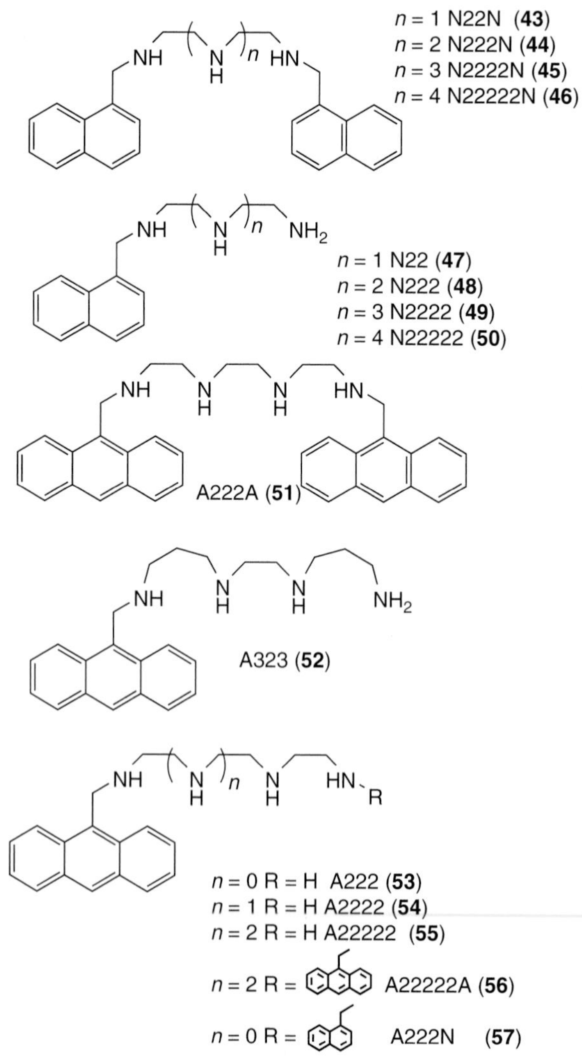

n = 1 N22N (**43**)
n = 2 N222N (**44**)
n = 3 N2222N (**45**)
n = 4 N22222N (**46**)

n = 1 N22 (**47**)
n = 2 N222 (**48**)
n = 3 N2222 (**49**)
n = 4 N22222 (**50**)

A222A (**51**)

A323 (**52**)

n = 0 R = H A222 (**53**)
n = 1 R = H A2222 (**54**)
n = 2 R = H A22222 (**55**)
n = 2 R = A22222A (**56**)
n = 0 R = A222N (**57**)

Scheme 13

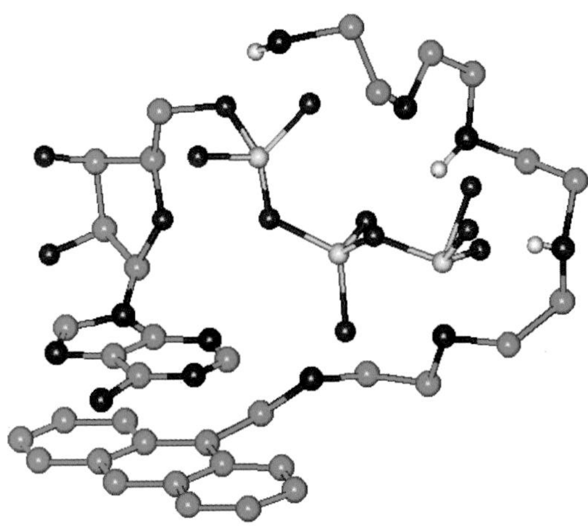

Figure 10 Molecular Model for the interaction of A22222 (**55**) with ATP showing the stacking between the aromatic fragments. (Adapted from Ref. 83 © Wiley-VCH, 2003.)

Recently, a couple of acyclic receptors in which two tetra-amine fragments have been interconnected by *meta*-xylene or 2,6-bis(methyl)pyridine aromatic units (**58** and **59**, Scheme 14) have been shown to be capable of forming stable complexes with ATP at neutral pH values.[84] Although stacking interactions are evidenced from the NMR spectra, the percentages of ATP and tripolyphosphate (TPP) complexed by the receptors do not differ much throughout the pH range in which complexation occurs. The receptors are flexible enough to organize around the guest species maximizing different binding schemes in both substrates. For example, for ATP, stacking interactions were identified.

Several macromonocycles including phenathroline units have shown an interesting multipoint binding ability for nucleotides.[85,86] In particular, receptor **60** (Scheme 15) shows a very good discrimination for ATP over the other triphosphate nucleotides TTP, CTP, and GTP. Molecular dynamic calculations indicated that ATP can assume a conformation that permits the simultaneous involvement of strong π-stacking, charge–charge, and hydrogen bond contacts, which reinforce the overall substrate–receptor interaction. Furthermore, the complete quenching of the fluorescence of **60** upon addition of ATP allows for competitively signaling this guest in aqueous solution.

Tripier, Bencini, and coworkers have recently prepared a series of receptors in which 1,4,7,10-tetrazadecane units

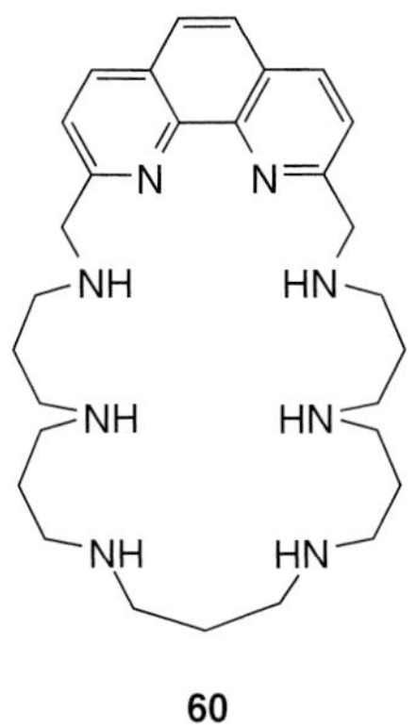

Scheme 15

have either been reinforced or have been connected through benzene or pyridine spacers (Scheme 16).[87–89]

Related receptors and their Zn^{2+} complexes had previously shown interesting properties within a range of applications, including a remarkable antiviral activity.[90]

For compounds **61** and **66,** there are clear spectroscopic evidences for the participation of π-stacking interactions in the binding to AMP, ADP, and ATP. Additionally, the protonation of pyridine in **62** at around pH 2 permits a more efficient interaction with this compound at acidic pH. Tris-macrocycle **63**[88] behaves as a double proton sponge yielding a $H_2(\mathbf{63})^{2+}$ species, which is protonated at the peripheral secondary amines and cannot be deprotonated even at strongly alkaline pH values. Addition of two further protons results in the protonation of each one of the lateral macrocycles. The protonated species of **63** interact quite effectively with ATP and ADP forming 1 : 1 adduct species. However, the stability of the formed adducts relies exclusively on charge–charge interactions and hydrogen bonding. The particular calixlike shape adopted by **62** in aqueous solution precludes the involvement of the central macrocycle and of the aromatic fragments in the binding; only the lateral macrocycles seem to be participating in the interaction. ^{31}P NMR and molecular dynamics suggest that the phosphate terminal group γ and the central group β of ATP mainly interact with the protonated lateral macrocycles.

Receptors **64–66** (Scheme 16) offer interesting perspectives about the relative contribution of charge, hydrogen bonding, and rigidity in the binding of nucleotide species.

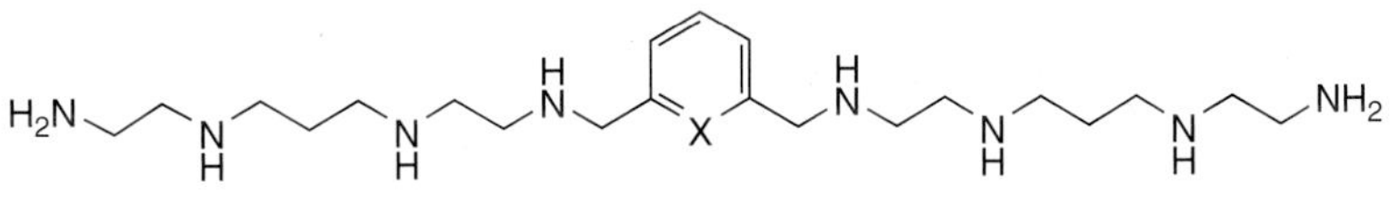

H_2N N H N H N H X N H N H N H NH_2

X = H **58**
X = N **59**

Scheme 14

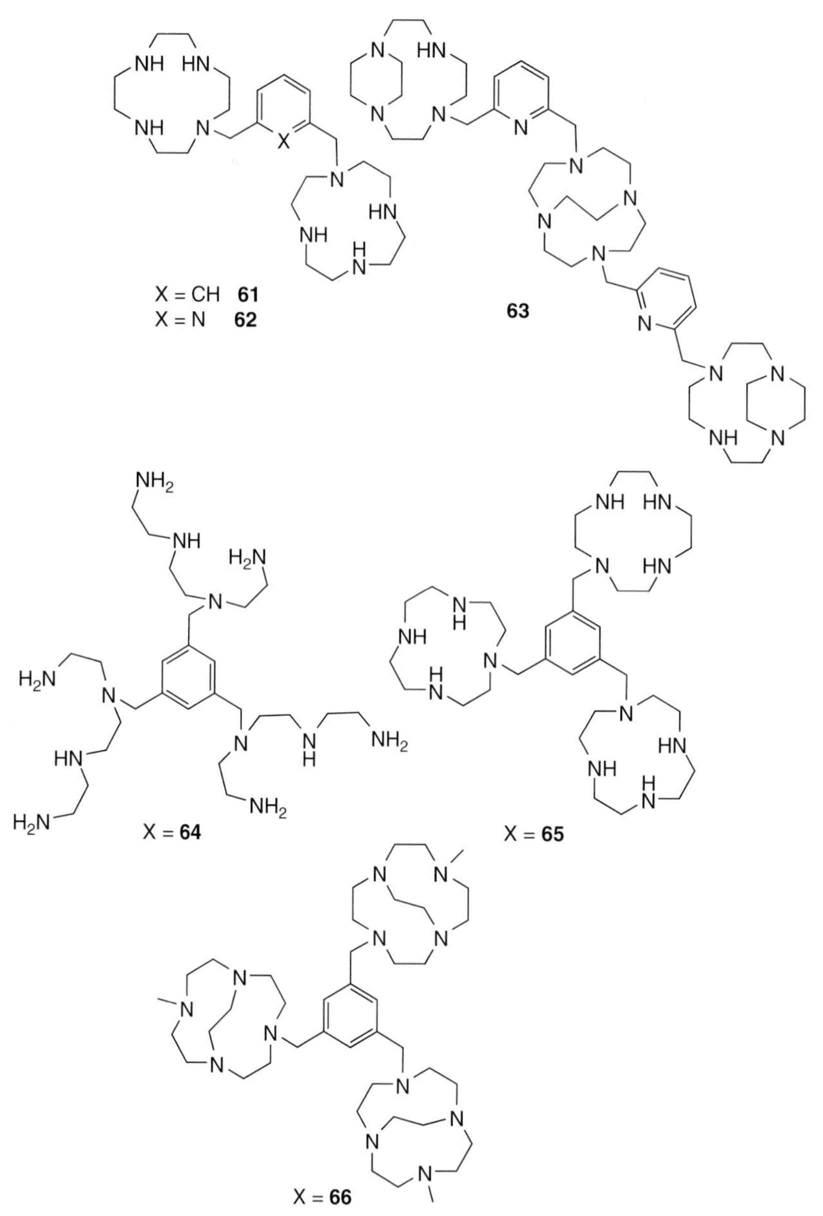

X = CH **61**
X = N **62**

63

X = **64**

X = **65**

X = **66**

Scheme 16

The interaction of the three ligands with ATP provides comparable stabilities but in different pH domains. Spectroscopic studies clearly highlight that the formation of adducts is significantly different for each receptor, the lack of protonated sites in **66** being compensated by the rigidity of the structure. On the other hand, while the cyclic nature of **65** and **66** facilitates their interaction at acidic pH, the extensive protonation of acyclic **64** at these pH values implies strong repulsion, which results in weak binding.[89]

Receptor **67** (Scheme 17), which can be classified as a "double-aza-scorpiand", is built by linking together two "scorpiand" receptors by a 2,9-dimethylphenanthroline fragment.[91] Apart from the amino groups of the lateral macrocycles and the secondary amino groups of the central chains, **67** displays two different aromatic fragments that can interact with nucleotides. ^{1}H NMR and molecular dynamic studies show that the macrocycle in its free non-complexed form adopts a close conformation with the phenanthroline ring stacked with the pyridine rings. This situation seems to be preserved until the protonation of phenanthroline ring occurs (protonation degree 7), which produces an opening of the receptor conformation due to electrostatic repulsion between the positive charges of its different components. As proved by ^{1}H NMR spectra and molecular dynamics calculations, interaction of $H_4(\mathbf{67})^{4+}$ or $H_6(\mathbf{67})^{6+}$ with ATP leads to a disruption in the phenanthroline–pyridine stacking with the consequent formation of phenanthroline–adenine stacked pairs (Figure 11). While

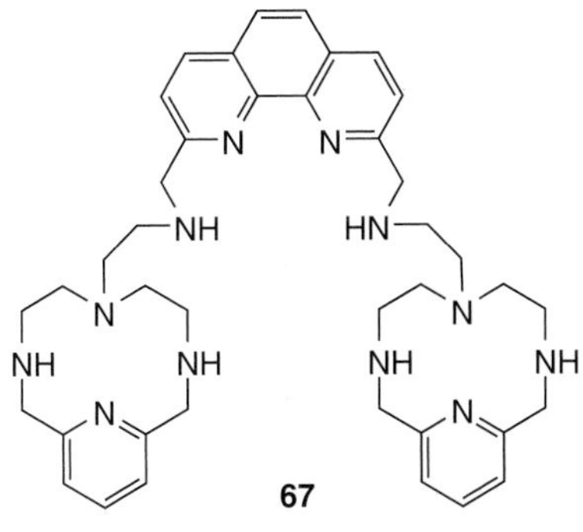

67

Scheme 17

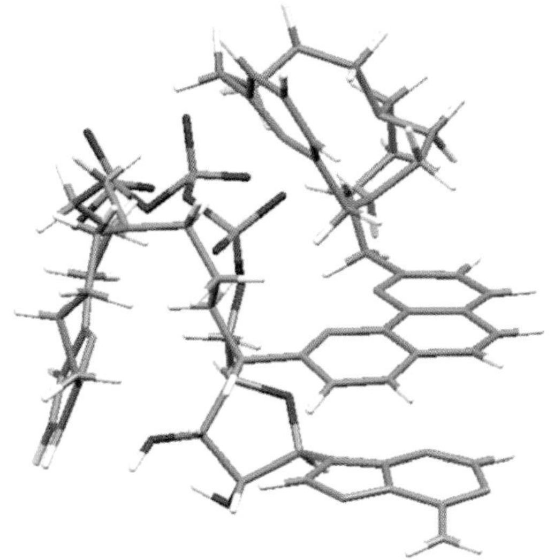

Figure 11 Minimum energy conformer for the adduct formed between $H_6(\mathbf{67})^{6+}$ and ATP^{4-}. (Adapted from Ref. 91 © Royal Society of Chemistry, 2010.)

upon addition of ATP, the ^{1}H signals of the pyridine protons of $H_4(\mathbf{67})^{4+}$ or $H_6(\mathbf{67})^{6+}$ shift downfield, those of the phenanthroline ring and of the anomeric and aromatic protons of ATP move upfield.

The relative importance of electrostatic attraction, hydrogen bonding, and π-stacking is clearly highlighted by the interaction of receptors $Ph_2(CH_3)_6N_6$ (**68**) and $Ph_2Pip_2(CH_3)_4N_8$ (Scheme 18) (**69**) with ADP and ATP.[92]

These two receptors present a similar molecular architecture composed of two polyamine subunits linked by aromatic spacers. The presence of aromatic moieties, the short ethylenic chains connecting the amino groups, nitrogen methylation, and, in $Ph_2Pip_2(CH_3)_4N_8$ (**69**), the piperazine rings, gives rise to rigid macrocyclic frameworks. The protonation patterns of these ligands, determined by means of potentiometric and NMR studies, showed that, despite the different number of nitrogen atoms in the molecules, similar arrangements of positive charges are achieved in the polyprotonated species.[92,93] For instance, both in $H_4(Ph_2(CH_3)_6N_6)^{4+}$ ($[H_4(\mathbf{68})]^{4+}$) and in

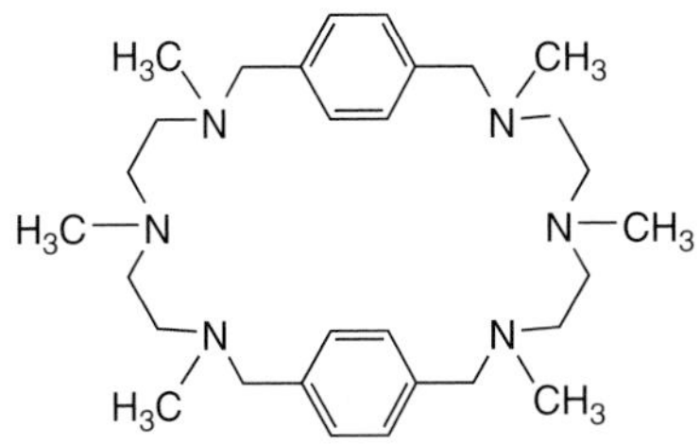

$Ph_2(CH_3)_6N_6$ (**68**)

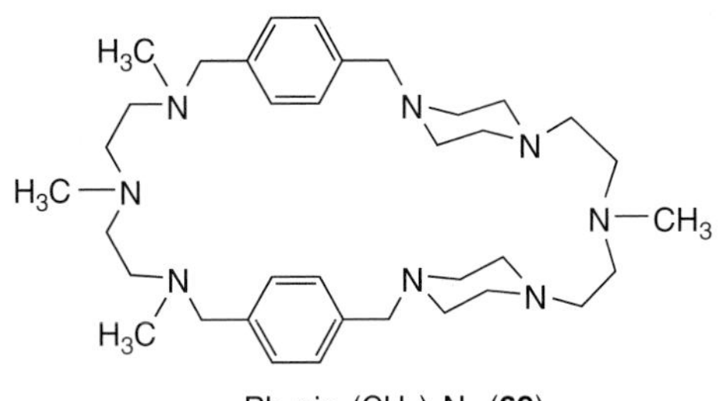

$Ph_2pip_2(CH_3)_4N_8$ (**69**)

Scheme 18

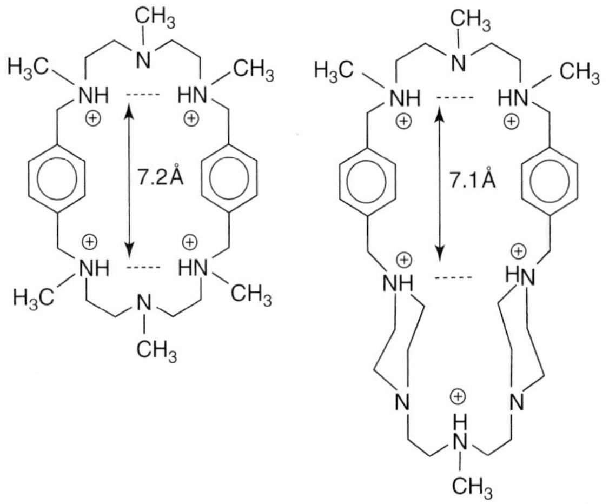

Figure 12 Charge disposition in $H_4(Ph_2(CH_3)_6N_6)^{4+}$ ($[H_4(\mathbf{68})]^{4+}$) and in $H_5(Ph_2Pip_2(CH_3)_4N_8)^{5+}$ ($[H_5(\mathbf{69})]^{5+}$) macrocycles. (Adapted from Ref. 92 © Royal Society of Chemistry, 1997.)

$H_5(Ph_2Pip_2(CH_3)_4N_8)^{5+}$ ($[H_5(\mathbf{69})]^{5+}$) four acidic protons are located on the benzylic nitrogens forming rectangular arrangements of positive charges of very similar dimensions (Figure 12). Of course, an additional positive charge is present in $H_5(Ph_2Pip_2(CH_3)_4N_8)^{5+}$.

Such similarity in the spatial distribution of charge in the two polyammonium cations leads one to assume that a similar charge–charge matching between the phosphate chains of ATP or ADP and the receptors can be achieved. Moreover, it seems reasonable to expect that $Ph_2Pip_2(CH_3)_4N_8$ (**68**) is a better receptor than $Ph_2(CH_3)_6N_6$ (**67**) since the former, owing to its higher

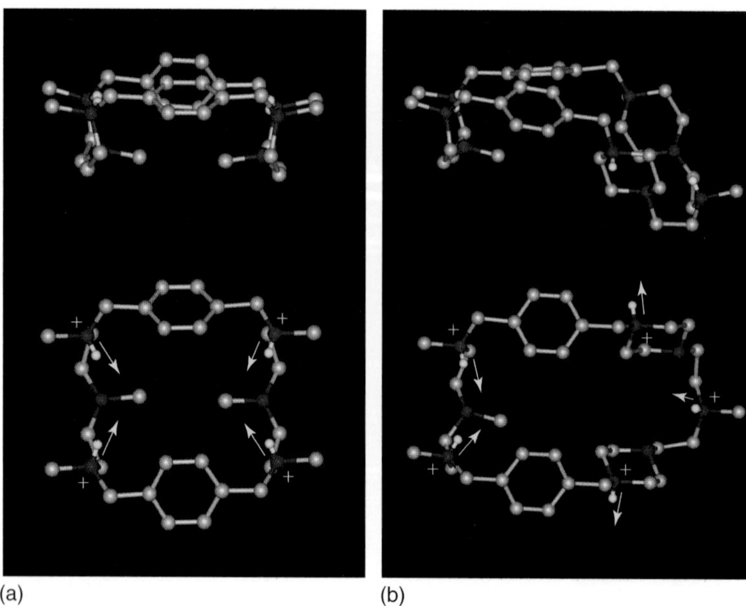

Figure 13 Lateral and top views of the crystal structure of $H_4(Ph_2(CH_3)_6N_6)^{4+}$, $([H_4(\mathbf{68})]^{4+})$ (a) and in $H_5(Ph_2Pip_2(CH_3)_4N_8)^{5+}$, $([H_5(\mathbf{69})]^{5+})$. (b) The orientation of $N{-}H^+$ groups is indicated by the arrows. (Adapted from Ref. 93 © Royal Society of Chemistry, 1995.)

number of amino groups, forms more charged species than the latter under similar conditions. Rather surprisingly, however, while the two nucleotides formed stable complexes with various protonated forms of $Ph_2(CH_3)_6N_6$ (**67**), no interaction with $Ph_2Pip_2(CH_3)_4N_8$ (**68**) was detected by means of potentiometric and NMR measurements over the entire pH range. A reasonable interpretation of such behavior can be drawn by considering the crystal structures of the $H_4(Ph_2(CH_3)_6N_6)^{4+}$ and $H_5(Ph_2Pip_2(CH_3)_4N_8)^{5+}$ cations (Figure 13). In $H_4(Ph_2(CH_3)_6N_6)^{4+}$, the four $N{-}H^+$ groups are convergent, allowing the receptor to form four hydrogen-bonded salt bridges with the phosphate chain of the nucleotides. On the other hand, in $H_5(Ph_2Pip_2(CH_3)_4N_8)^{5+}$, the five $N{-}H^+$ groups are divergent, preventing the simultaneous interaction with the nucleotides and the formation of complexes. Hence, although the charge–charge attraction is the driving force for most ion-pairing processes in water, it does not always produce enough energy to stabilize anion complexes with polyammonium receptors. As shown in this case, as well as in previous cases, the formation of salt bridges reinforced by hydrogen bonds can be the key to achieve efficient anion binding. Complexation-induced ^1H chemical shifts of the adenine protons of ATP and ADP evidenced that a modest π-stacking interaction between the nucleobase and the aromatic groups of the receptors furnishes some stabilization to the nucleotide complexes of $Ph_2(CH_3)_6N_6$.

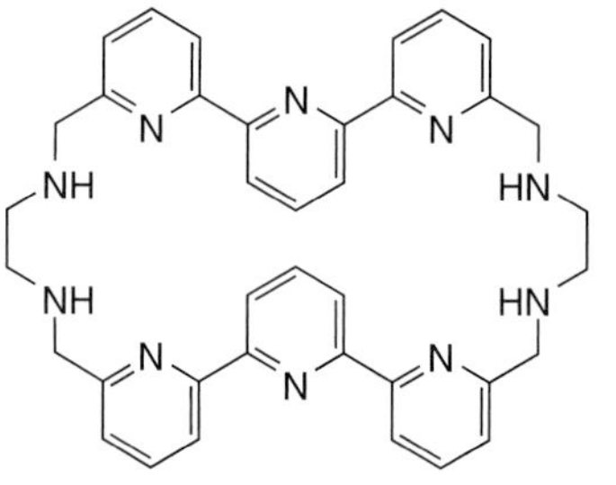

Recently Bianchi *et al.*[94] have reported a crystal structure for the interaction of the terpyridinophane macrocycle **70** (Scheme 19) with 5′-TTP that includes many of the binding

70

Scheme 19

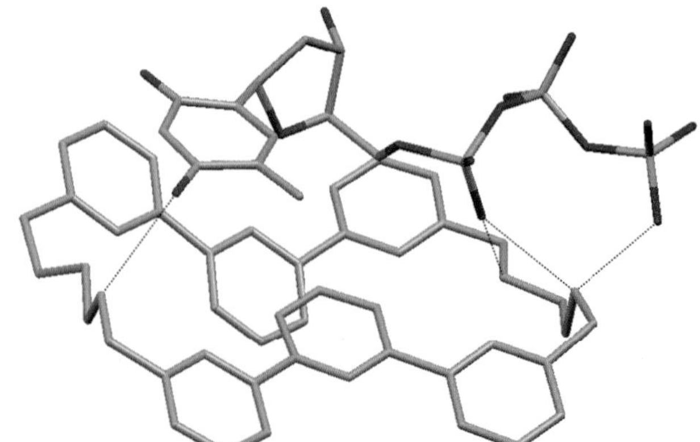

Figure 14 View of the crystal structure of complex $[(H_4(\mathbf{70}))HTTP]$ showing some of the intermolecular hydrogen bonds between the partners.

mechanisms[95] of nucleotides by polyazamacrocycles containing aromatic functions (Figure 14). Three hydrogen bonds between the polyphosphate chain of the nucleotide and protonated nitrogen atoms of **70**, one hydrogen bond between a carbonyl oxygen of TTP and one ammonium group of **70**, one CH$\cdots\pi$ interaction involving TTP carbon atoms and ligand pyridine units, and one O$\cdots\pi$ interaction between the TTP carbonyl oxygen O(14) and the N(10) pyridine ring of **70** determine the overall solid-state structure of the [H_4(**70**)(TTP)] complex, whose stability in water ($K = 4.57 \times 10^4\,M^{-1}$) is significantly higher than the stability of nucleotide complexes with polyazamacrocycles, bearing the same positive charge and having comparable size with H_4(**70**)$^{4+}$, but not including aromatic groups.

2.2.3 *Receptors other than polyamines*

Although this chapter is mainly devoted to polyamine receptors, there are other classes of receptors that are very relevant in nucleotide recognition. A few examples are discussed in this section.

Apart from polyamine-based receptors that make use of charge–charge and hydrogen bonding as principal binding motifs for nucleotide recognition, guanidinium and amide groups also constitute interesting anchorage points for nucleotides. Indeed, a recent critical revision of 3003 crystal structures of proteins with bound phosphates from the RSCB Protein Data Bank has shown that, for instance, glycine residues are very recurrent motifs in these sites. This has been explained because glycine (Gly) residues allow the folding and wrapping of the amino acidic residues around the phosphate with a nearly macrocyclic organization. Also, strong dependence of the nature of the binding site with its location was found. Binding sites located at the surface of the protein have higher contents in cationic amino acidic residues (arginine, lysine) than those located in buried hydrated sites where neutral amino acids predominate.[34]

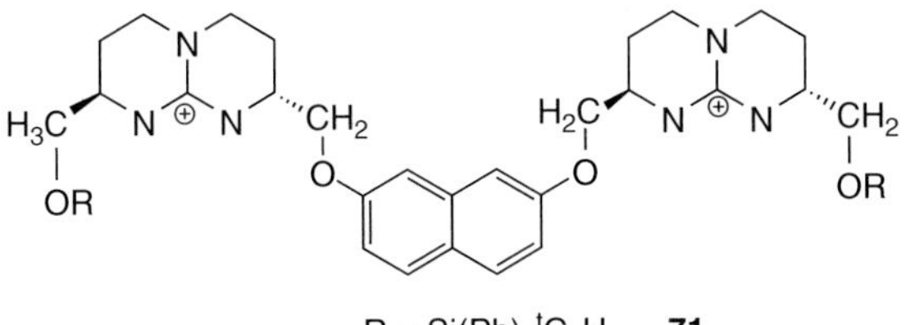

R = Si(Ph)$_2$^{t}C$_4$H$_9$ **71**
R = H **72**

Scheme 20

Inspired by the widespread occurrence of guanidinium anchor groups for oxoanions in natural receptors, numerous groups have introduced such functions in abiotic receptors. Regarding this point, Schmidtchen and Schiessl have reported the preparation and binding studies with organic phosphates of receptors **71** and **72** (Scheme 20).[96] Compound **71** forms in methanol complexes of relatively high

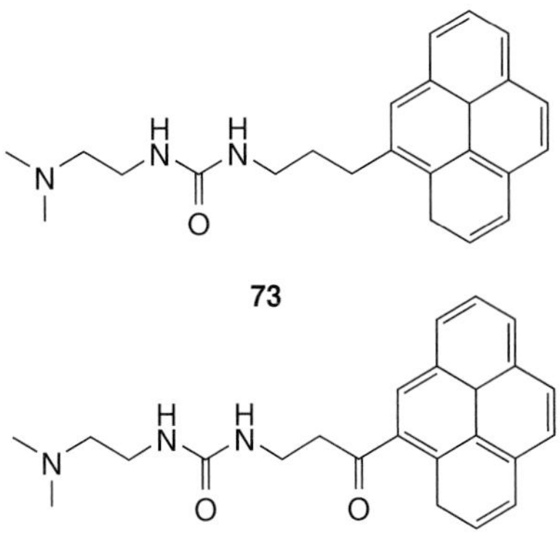

73

74

Scheme 21

stability with AMP ($K = 38\,000\,M^{-1}$) as determined by ^{1}H NMR and ^{31}P NMR experiments. The authors of this work indicated that they could not obtain meaningful binding constants for the more hydrophilic host **72** because the ^{1}H NMR and ^{31}P experiments showed an odd behavior indicative of complexes of higher stoichiometry. In the case of AMP, a binding constant of $9330\,M^{-1}$ was inferred. However, in pure water, **72** forms complexes of only 1 : 1 host:guest stoichiometry. The analysis of the data provided values of association constants of $204\,M^{-1}$ for AMP (reduction factor of over 40 with respect to methanol), $140\,M^{-1}$ for NAD, and $840\,M^{-1}$ for ATP. The low value obtained for ATP indicates an inefficient matching of the receptor with the different binding sites of this nucleotide. Nevertheless, the results show that low charge yet well-designed artificial receptors such as **71** can form, even in pure water, adducts of reasonable stability with biologically relevant phosphates of matching characteristics.

Receptors **73** and **74** combine urea groups and pyrene aromatic fragments as potential hydrogen-bonding donors or acceptors and stacking units, respectively (Scheme 21). Compounds **73** and **74** show fairly good complexation with nucleotide monophosphates in D_2O with log K values, determined by fluorescence techniques, of around three logarithmic units showing an slight increase in the order A > C > G > T. However, comparison with unsubstituted pyrene just provides ΔG° values 2–5 kJ mol^{-1} smaller. The moderate contribution of the urea sites to the binding was interpreted by a lack of geometrical complementary between the pyrene unit and the urea with the nucleotides. Studies performed in DMSO gave lower constants, which was attributed to the diminished efficiency of π-stacking in this solvent.

In 1994, Eliseev and Schneider published a pioneering study about the recognition of nucleotides by aminocyclodextrins.[97] β-Aminocyclodextrines **75** and **76** (Scheme 22), having two or seven aminomethyl groups

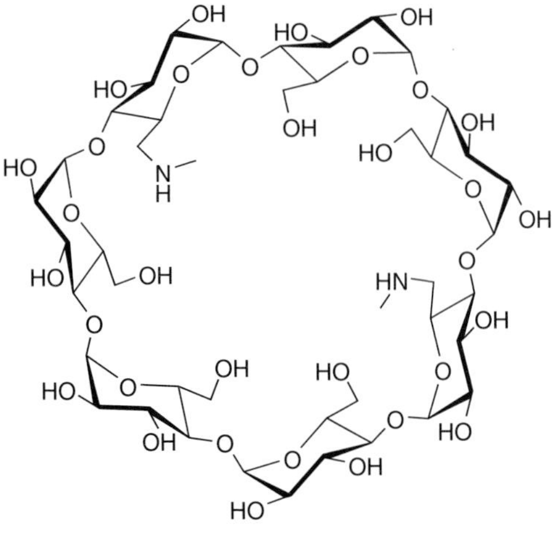

75

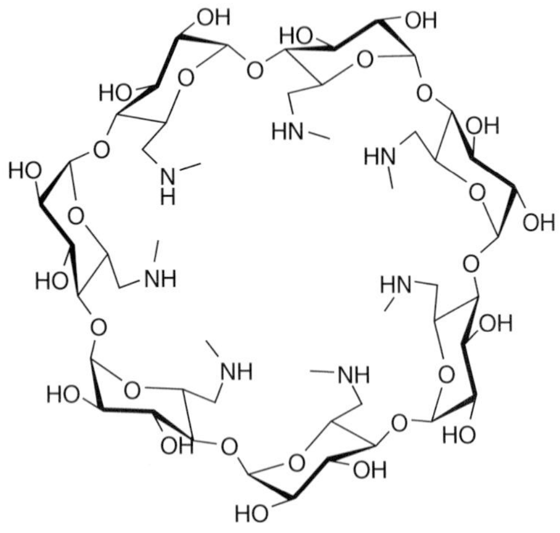

76

Scheme 22

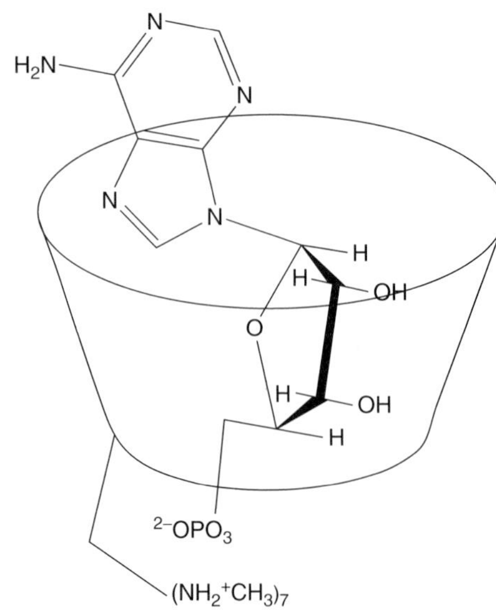

Figure 15 Schematic representation of the binding of a nucleotide by a polyamine modified cyclodextrin.

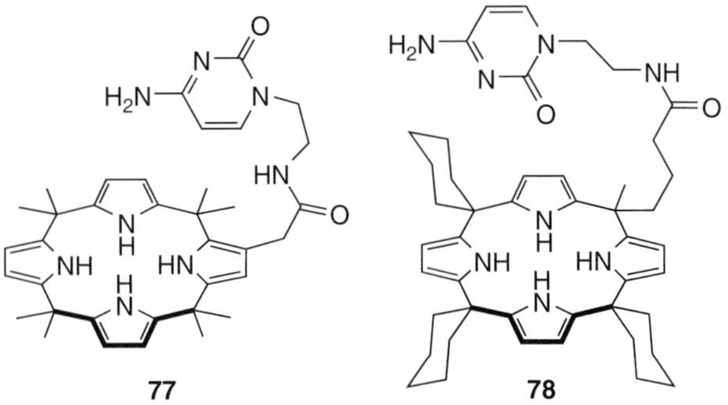

77

78

Scheme 23

respectively, behave as polytopic receptors for nucleotides. Compound **75** discriminates the guest nucleotides making use of the nucleobase nature, sugar type (oxy- or deoxyribose), and position of the phosphate group (3′ or 5′) (Figure 15). For instance, while the constant determined for the interaction of 5′-AMP with **75** was of 14 100 M^{-1}, that obtained with 3′-AMP, was about 1 order of magnitude lower, 1510 M^{-1}. Among mononucleotides, the constants for 5′-GMP, 5′-UMP, and 5′-CMP were 6160, 830, and 830 M^{-1}. Therefore, a moderate selectivity for 5′-GMP and an interesting selectivity for 5′-AMP were achieved. Although **76** shows much stronger interactions with the nucleotides, it displays poorer selectivity.

Porphyrins and calix[4]pyrrole receptors (Scheme 23) appended with chains containing selective recognition motifs have been revealed as an interesting class of binders and/or carriers of nucleotidic species.[98,99]

Neutral calix[4]pyrrole macrocycles **77** and **78** containing appended nucleobases can mediate the selective transport through lipophilic membranes of the complementary Watson–Crick nucleotides as exemplified in Figure 16 for the pair cytosine–guanidine.[98]

Aminoglycosides are natural polyamines, which in their protonated forms may have rather unique properties for anion recognition since they have charged groups disposed in a fixed manner over a relatively rigid scaffold. Kanamicyn-A (**79**), which can take up four protons at the amino groups, discriminates triphosphate nucleotides with different nucleobases, in particular GTP over ATP, as proved by NMR and pH-metric analysis (Scheme 24).[100] The constants displayed by the tetraprotonated form of this receptor compare well with those reported recently for other receptors especially designed for nucleotide recognition, making these kinds of natural molecules appealing hosts for this purpose.

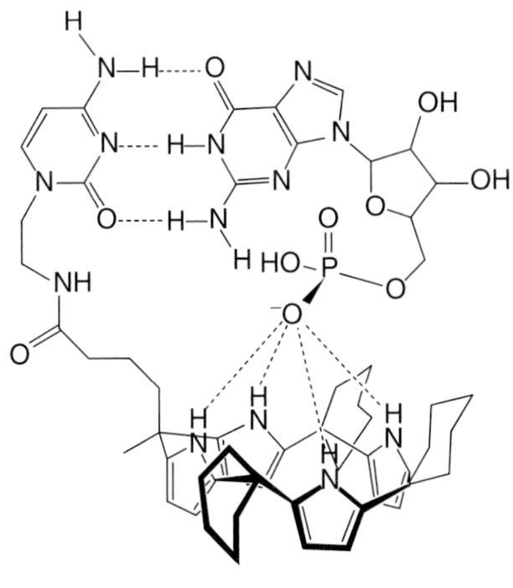

Figure 16 Schematic representation of the binding of a nucleotide by a calix[4] pyrrole macrocycle.

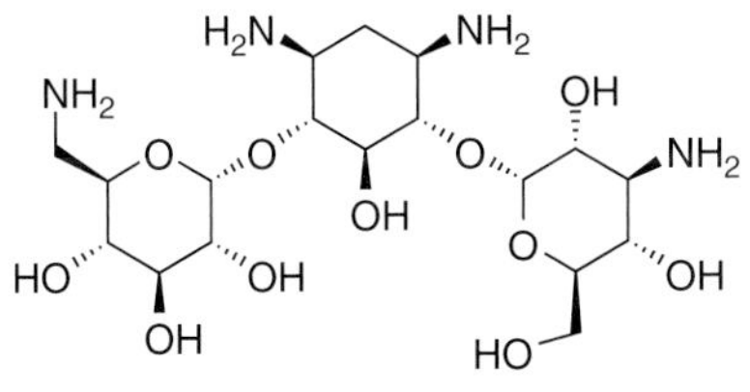

Kanamicyn A (**79**)

Scheme 24

Other relevant examples of receptors that are not essentially based on polyammonium sites, which combine binding motifs with chromogenic or fluorogenic groups for nucleotide signaling, are presented in the last section of this chapter.

2.2.4 Metal complexes as nucleotide receptors

Complexed metal ions can express their Lewis acid characteristics if they are coordinatively unsaturated or if they have coordination positions occupied by labile ligands that can be easily replaced by the incoming guests. If this occurs, metal complexes are well suited for interacting with additional Lewis bases, which very often are of anionic nature. Metal ions are involved in the coordination of ATP in a number of metalloenzymes. Figure 17 shows Zn^{2+} participating in the binding of ATP in pyridoxal kinase (DOI:10.2210/pdb1lhr/pdb).[101]

In this respect, Kimura and coworkers provided initial evidence for nucleobases incorporated as exogen ligands either in the coordination sphere of the monomer Zn^{2+} complex of the tetraazamacrocycle 1,4,7,10-tetraazacyclododecane ([12]aneN_4) or inserted as bridging ligands in dinuclear or trinuclear complexes containing two or three [12]aneN_4 subunits linked by different aryl groups (Scheme 25).[102]

In their initial work, these authors reported a potentiometric study on the ability of the Zn^{2+} complex of cyclen (**80**) to interact with the deoxyribonucleotides 2′-deoxyadenosine (dA), 2′-deoxyguanosine (dG), 2′-deoxycytidine (dC), 2′-deoxythymidine (dT), uridine (U), and 3′-azido-3′- deoxythymidine (AZT). [Zn(**80**)]$^{2+}$ had a good selectivity for dT, U, AZT, and the related derivatives Ff (ftorafur, 5-fluoro-1-(tetrahydro-2-furyl)uracil), and

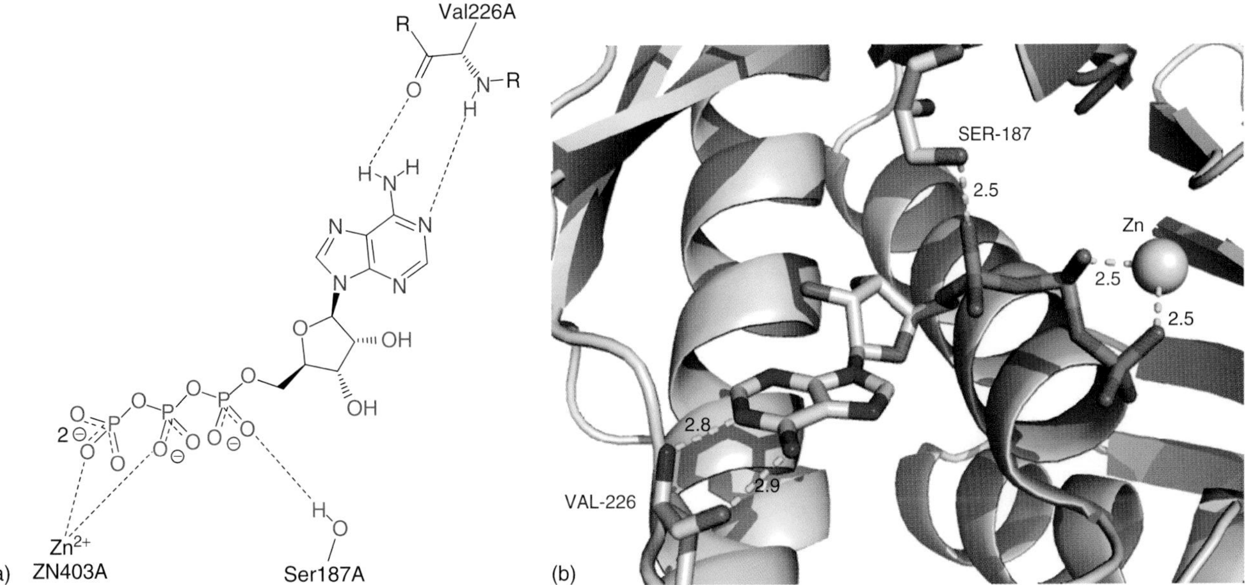

Figure 17 (a) Schematic representation of interaction of ATP with active center of pyridoxal kinase. (b) Binding center of pyridoxal kinase with ATP.

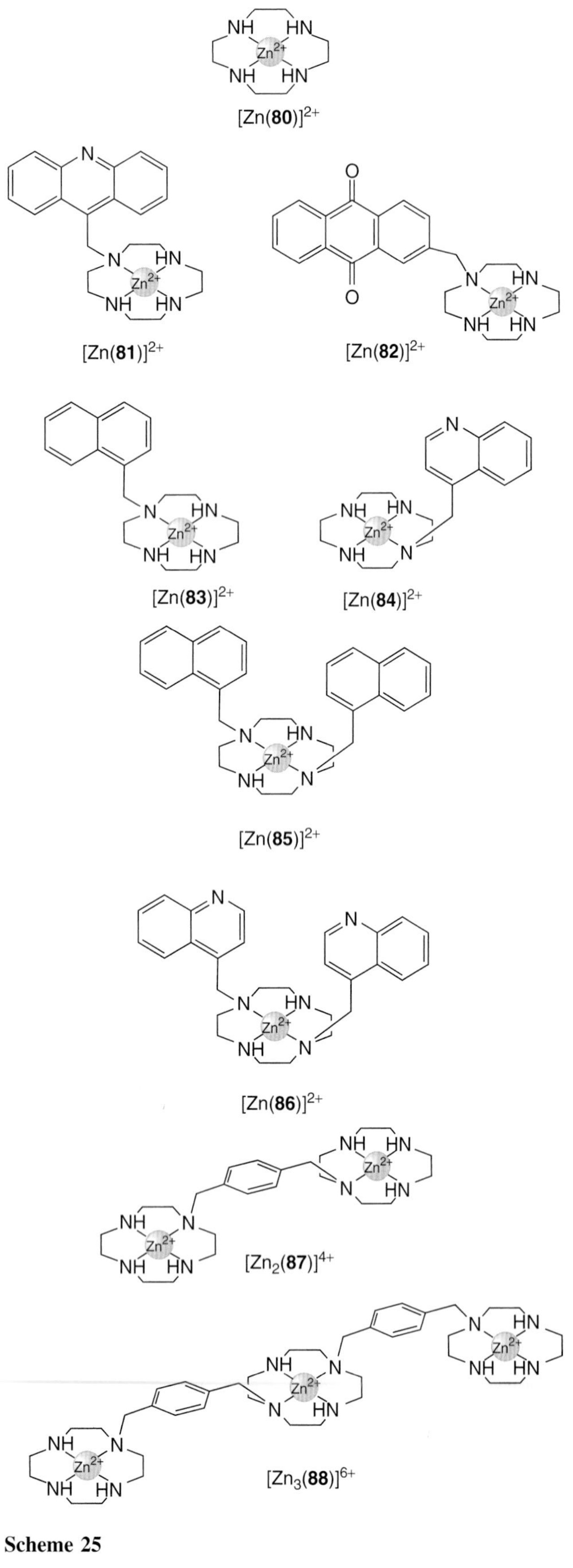

$[Zn(\mathbf{80})]^{2+}$

$[Zn(\mathbf{81})]^{2+}$ $[Zn(\mathbf{82})]^{2+}$

$[Zn(\mathbf{83})]^{2+}$ $[Zn(\mathbf{84})]^{2+}$

$[Zn(\mathbf{85})]^{2+}$

$[Zn(\mathbf{86})]^{2+}$

$[Zn_2(\mathbf{87})]^{4+}$

$[Zn_3(\mathbf{88})]^{6+}$

Scheme 25

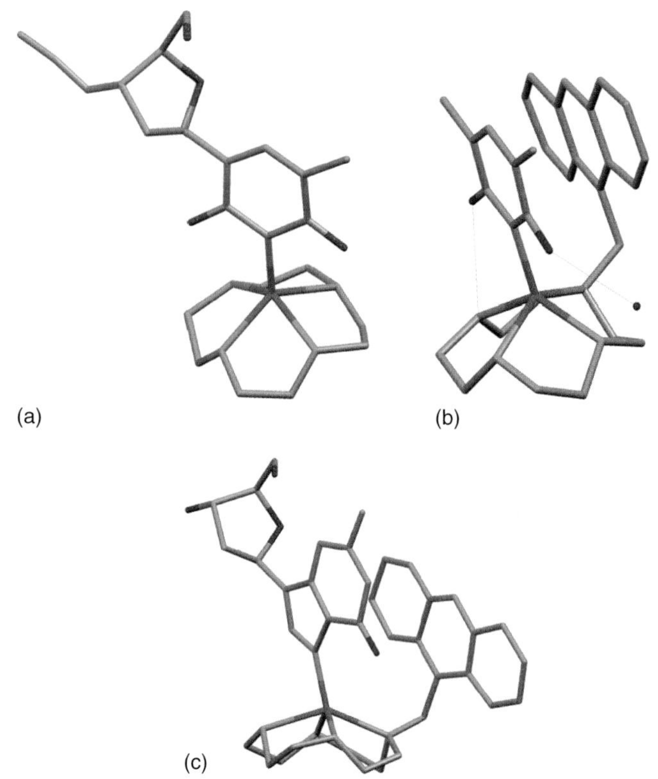

Figure 18 Crystal structure showing the mixed complexes (a) $[Zn(\mathbf{80})]^{2+}$-AZT, (b) $[Zn(\mathbf{81})]^{2+}$-methylthymine. (c) $[Zn(\mathbf{81})]^{2+}$-2′-deoxyguanidine.

riboflavin. UV absorption spectra proved that coordination occurred through the deprotonated N(3) imide. Other nucleosides containing an amino group in place of the carbonyl oxygen of dT, such as dG, or without imide NH groups, did not bind to $[Zn(\mathbf{80})]^{2+}$, which was ascribed, respectively, either to steric hindrance in the case of dG or to the lack of N^- formation. This behavior was confirmed by an X-ray crystal structure of the mixed complex of $[Zn(\mathbf{80})]^{2+}$ with AZT, which showed the metal ion coordinated in a distorted square pyramidal geometry to the four nitrogen atoms of the macrocycle and to the deprotonated imide group (Figure 18a). Although quite long in the solid state, the position of the two pyrimidine carbonyl groups suggests possible direct or indirect hydrogen-bonding formation with the amine groups of cyclen.

Moving on with these studies and trying to improve the binding ability of the Zn^{2+} complexes, the same research group prepared cyclen derivatives appended with the intercalator acridine (**81**). Potentiometric, UV, and NMR studies revealed that the acridine moiety yielded enhanced binding with dT and dT derivatives due to π-stacking interactions between acridine and the nucleobase in addition to the Zn^{2+}-deprotonated imide group coordinative bond and to likely hydrogen bonding between the carbonyl pyrimidine groups and alternated amine groups of cyclen.

The crystal structure of the mixed complex $[Zn(\mathbf{81})]^{2+}$ with methylthymine (Figure 18b) revealed square pyramidal N_5 coordination geometry with a strong interaction between the metal and the deprotonated imide nitrogen (Zn – N(3″) = 1.987 Å). One of the carbonyl oxygen atoms forms a direct hydrogen bond with a cyclen NH, while the other one binds to a diagonal NH amino group indirectly through a water molecule.

Contrary to $[Zn(\mathbf{80})]^{2+}$, $[Zn(\mathbf{81})]^{2+}$ is also able to interact with dG. Crystals of $[Zn(\mathbf{81})]^{2+}$ with the free form of dG grown in CH_3CN-H_2O display a different binding pattern in which Zn^{2+} is coordinated in a distorted square pyramidal manner involving the four nitrogen atoms of cyclen and N(7″) of the purine ring (Figure 18c). Stacking interactions and hydrogen bonding between oxygen O(6″) of the purine ring and an NH of cyclen completes the binding. Solution studies show, however, that $[Zn(\mathbf{81})]^{2+}$ is still selective for thymidine.

Motivated by these and related findings, these authors have developed several related Zn^{2+} complexes (Scheme 25) for monophosphate, diphosphate, and triphosphate nucleotide recognition. For example, highly efficient recognition of thymidine and uridine nucleotides such as 3′-dTMP, 5′-dTMP, 2′-UMP, 3′-dUMP, 5′-dUMP, 5′-dTTP, 3′-azido-3′- deoxythymidene 5′-monophosphate (AZTMP), or azido-3′-deoxythymidene 5′-diphosphate (AZTDP) was achieved with complex $[Zn_2(\mathbf{87})]^{4+}$ and the analogous complex containing *meta*-xylene junctions between the cyclen units. Moreover, the trinucleotide d(TpTpT) was found to be selectively bound by $[Zn_3(\mathbf{88})]^{6+}$.

The coordination mode described for thymine and uracil containing nucleobases depicted in the structures of Figure 18 has been advantageously used for the selective detection of nucleotides and polynucleotides containing these functions (Section 4).

Lin *et al.* have developed receptors consisting of phenanthroline units with appended amine or pycolylamine chains (Scheme 26).[103] Binding of Zn^{2+} by these ligands has been shown to enhance nucleotide complexation with respect to the free receptors in aqueous solution, which is a typical behavior in this type of chemistry.

An example of recognition of nucleotides through formation of mixed complexes with polyamine ligands is provided by the system Cu^{2+}-4,7,10,13-tetrazahexadecane-1,16-diamine (**100**)- AMP (Scheme 27).[104]

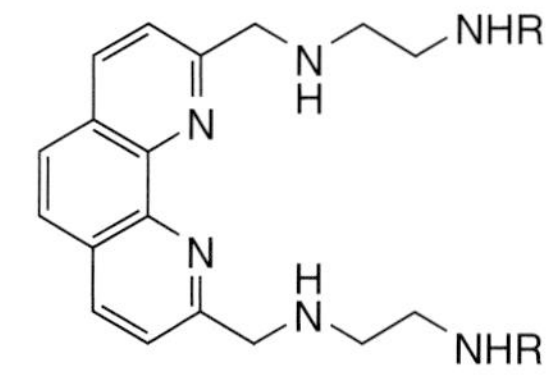

In the binary systems, **100**-AMP formation of $[H_q(\mathbf{100})(AMP)]^{(q-2)+}$ species with protonation degrees varying from $q = 3$ to $q = 8$ were observed. On the other hand, **100** forms Cu^{2+} complexes of 1 : 1 and 2 : 1 stoichiometry. The insertion of all these data, together with the protonation constants of polyamine and AMP and the formation constants of Cu^{2+}-AMP as known parameters in the titrations

R = CH_3 **89**, R = Et **90**, R = *n*-Pr **91**
R = *n*-Bu **92**, R = *i*Pr **93**

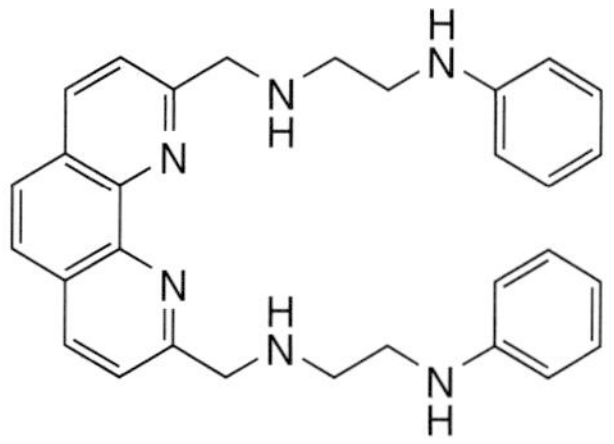

94

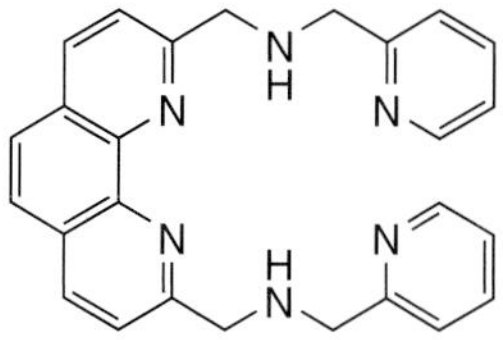

95

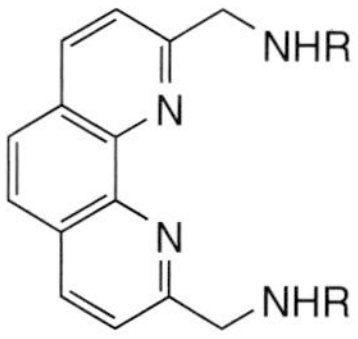

R = CH_3 **96**, R = Et **97**, R = *n*-Pr **98**
R = *n*-Bu **99**

Scheme 26

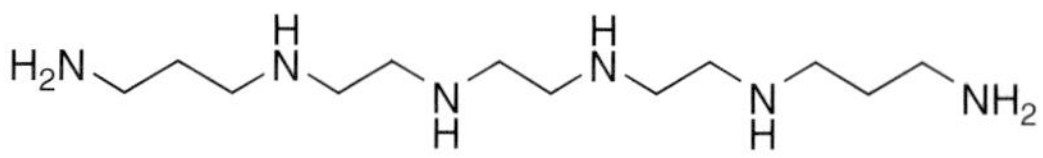

100

Scheme 27

performed for the ternary systems, in which different Cu^{2+}-**100**-AMP molar ratios were used, as a result formed mixed species of stoichiometries $[CuH_rL(AMP)]^{r+}$ with r varying from 4 to 0 and binuclear ones $[Cu_2H_rL(AMP)]^{(2+r)+}$ with r between +1 and −1. Distribution diagrams for this system show the prevalence of the ternary complexes throughout the pH range of the study. Another example regarding this kind of ternary system is included in the next section devoted to catalytic aspects.

3 CATALYTIC ASPECTS: A BRIEF NOTE OF ATPASE AND KINASE ACTIVITY

One of the most interesting points of the supramolecular chemistry of nucleotides is the capacity that some receptors have to induce catalytic processes mimicking the behavior of ATP-ases or kinases.

The 24-membered macrocyclic receptor **24** ranks among the abiotic receptors producing highest enhancements of ATP cleavage into ADP and the so-called inorganic phosphate.[78,105–110] Contrary to other polyammonium receptors, ATP-cleavage activation with **24** was not only observed at acidic pH but also at neutral pH values. In neutral conditions, the hydrolysis of ATP by this receptor was enhanced by a factor of 100. The ^{31}P NMR was used to monitor the course of the reaction and revealed the formation of a transient phosphorylated species, which appeared in the ^{31}P NMR spectrum as a singlet downfield shifted 10 ppm with respect to an external H_3PO_4 reference. The hydrolysis of ATP in the presence of protonated **24** was proved to be a true catalytic process since, for 10-fold excess of ATP with respect to **24**, the change in the ATP concentration was linear with time in the early period. The products coming from the reaction of ADP and inorganic phosphate did not interfere with the initial course of the reaction due to their reduced affinity for the protonated receptor.

The authors of this work proposed the classical mechanism depicted in Figure 19 for ATP hydrolysis catalyzed by **24**.

One of the central amines of the bridges is not protonated at neutral pH and performs the nucleophilic attack to the terminal phosphorous atom (P_γ) of ATP to form the phosphoramidate transient species (steps A and B in Figure 19). The electrostatic and hydrogen-bonding interactions of the phosphate chain of the nucleotide with the protonated macrocycle places the γ-phosphorous atom and the nonprotonated amine of the partners in an appropriate position for facilitating the nucleophilic attack. The final steps of the catalytic cycle are the dissociation of the ADP complex and the hydrolysis of the phosphoramidate, not necessarily in this order (steps C and D).

Studies of the ATP cleavage conducted with the series of azacycloalkanes [3k]aneN$_k$ ($k = 6–12$) (**7, 11–16**) revealed the key role played by the macrocycle size.[111] The rates of dephosphorylation were dependent on ATP concentration and pH. At pH 3, the rate of ATP hydrolysis in the presence of [21]aneN$_7$ (**11**) was $k = 0.029\,\text{min}^{-1}$ at 20 °C for an initial ATP concentration 10^{-5} M. At higher temperatures and/or concentrations, the rates for this macrocycle were too fast to be monitored using HPLC and NMR techniques. All the other macrocycles showed considerably lower rates, [24]aneN$_8$ (**12**) being the next fastest one with $k = 0.0045\,\text{min}^{-1}$ ($T = 40\,°\text{C}$, $[\text{ATP}]_0 = 10^{-5}$ M). Either the four larger macrocycles [27]aneN$_9$ (**13**), [30]aneN$_{10}$ (**14**), [33]aneN$_{11}$ (**15**), and [36]aneN$_{12}$ (**16**) or the smaller [18]aneN$_6$ (**7**) were clearly slower. At pH 7, all macrocycles showed slower rates than at pH 3. This can be attributed to a lessened participation of general acid catalysis.

The highest efficiency of [21]aneN$_7$ (**11**), particularly at lower pH values, can be attributed to several factors. First, the magnitude of the interaction does not seem to be a key point since the larger macrocycles forming stronger complexes show less efficiency. Second, if the

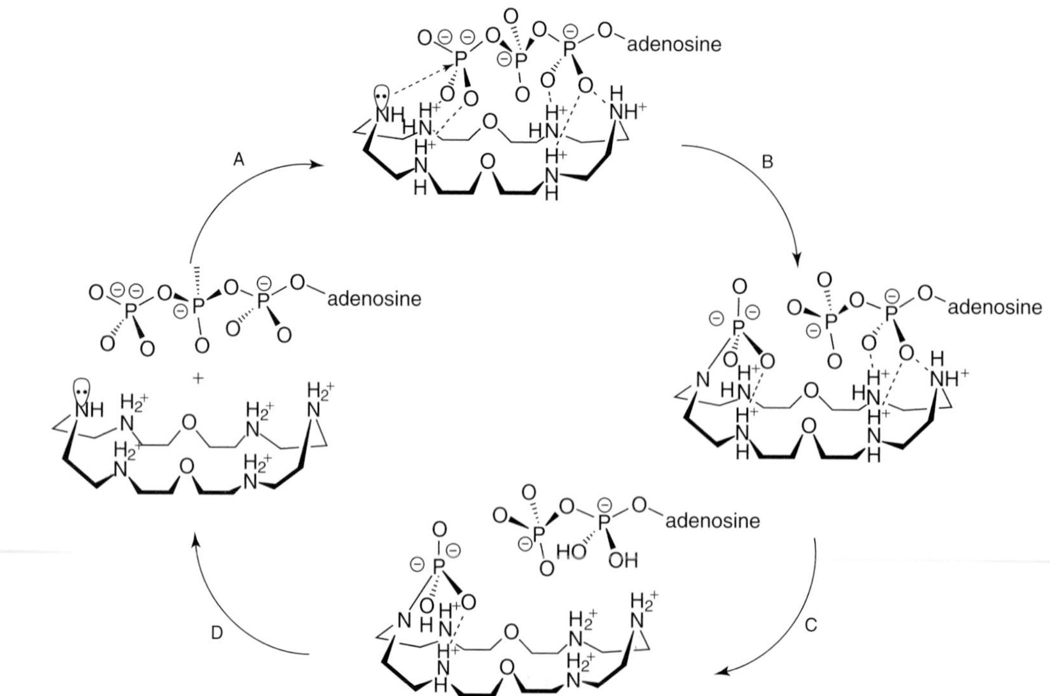

Figure 19 Catalytic cycle for ATP in the presence of **24** at neutral pH.

mechanism proceeded through formation of a covalent intermediate phosphoramidate species, the nucleophilicity of the macrocyclic amines should be critical. In the case of the fastest [21]aneN_7 macrocycle (**11**), evidence of formation of such covalent intermediate was provided by a ^{31}P NMR signal appearing at about 10 ppm, which was coincident with that displayed by a synthetically prepared phoshoramidate. Factors that favor nucleophilicity are as follows: (i) reduced overall positive charge at a given protonation, (ii) hydrophobicity of the environment where the reaction occurs, and (iii) electron donor substituents. As at a given pH, [21]aneN_7 (**11**) is less charged than the larger ligands; this should favor a nucleophilic attack. However, the same reasoning should drive to a higher efficiency of the smaller [18]aneN_6 (**7**); however, this was not the case.

Therefore, other factors affecting the electronic and stereochemical matching between the receptor and the substrate have to play a key role in this activity. [24]aneN_6O_2 (**24**) (O-BISDIEN), which also produces a very significant rate enhancement that is persistent from acidic to neutral pH, shares the feature of having the same number of atoms in the macrocyclic ring with [24]aneN_8 (**12**).

Table 1 collects the lengths of the minor and major axes, defining the elliptical shapes of various macrocycles whose structures have been determined by X-ray diffraction analysis.

As seen in Table 1, the macrocycles showing highest efficiencies **11** and **24** (forms $H_4([21]aneN_7)^{4+}$ and $H_6([24]aneN_6O_2)^{4+}$ in Table 1) have very close lengths of the minor axis. Modeling studies indicated that ATP would be located along this common minor axis.

The relevance of the macrocyclic size and charge density on rate catalysis was later checked by introducing structural modifications such as N-methylation of some of the nitrogen atoms in the macrocycles [18]aneN_6 (**7**) and [21]aneN_7 (**11**)[112,113] to yield macrocycles (**21**, **101**, **102**) (Scheme 28).

While dimethylated **101** leads to a decrease in the rate of the hydrolytic cleavage of ATP with respect to unmethylated **7**, the tetramethylated **21** produces a significant rate enhancement. On the other hand, trimethylation

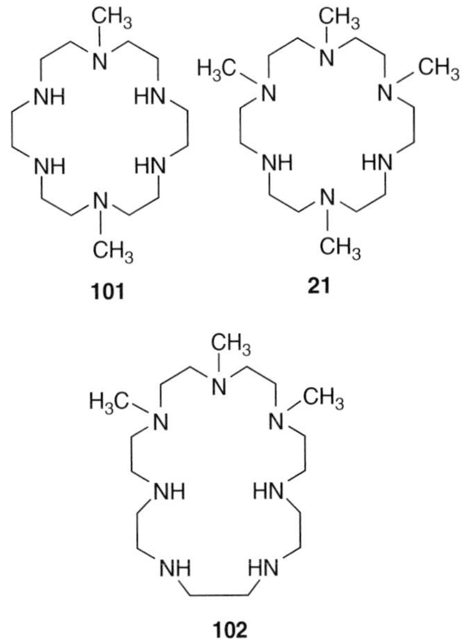

Table 1 Lengths of the major and minor axes for the average ellipsoid of polyammonium macrocycles.

Macrocycle	Major axis (Å)	Minor axis (Å)
$H_4([18]aneN_6)^{4+}$	7.682	6.212
$H_6([18]aneN_6)^{6+}$	7.647	6.236
$H_4([21]aneN_7)^{4+}$	7.653	6.725
$H_6([24]aneN_6O_2)^{4+}$	10.005	6.714
$H_8([30]aneN_{10})][Co(CN)_6]_2Cl_2\cdot10H_2O$	15.486	6.873

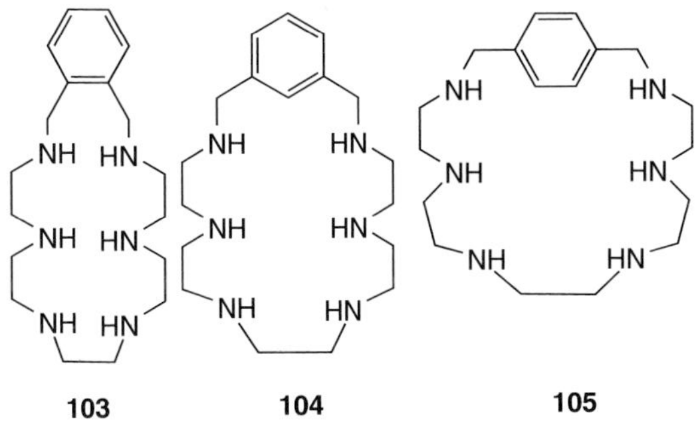

Scheme 28

of [21]aneN_7 (**11**) to give $(CH_3)_3$[21]aneN_7 (**102**) produces a diminution in the rate of the process. Taking into account that methyl groups donate electron density to electronegative atoms and should increase the nucleophilicity of the nitrogens, the slower rate of cleavage must be closely related to the alteration of the optimal cavity size brought about by the functionalization. However, solvation effects need to be always taken into account when one makes these kinds of considerations.

Additional evidence supporting the key role played by the size cavity in ATP hydrolysis was provided by García-España and coworkers for a series of cyclophane receptors **103–105** (Scheme 29).[114]

The critical size required is manifested in the fact that either the ortho-(**103**) or the para-isomers (**105**), containing 20- and 22-membered cavities, yield much poorer ATP cleavage rate enhancements than the meta-isomer **104** with a 21-membered cavity. This lack of activity is particularly noticeable in the case of the ortho-derivative.

Scheme 29

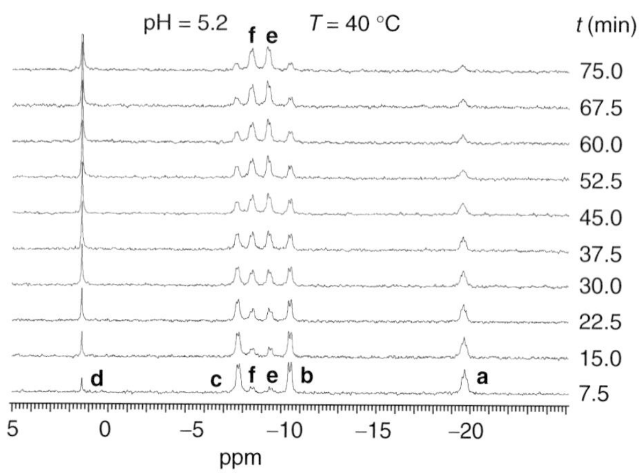

Figure 20 Time evolution for ^{31}P NMR spectra of solutions containing ATP and **104** in 10^{-2} M at 40 °C and pH = 5.2 (**a**) P_β (ATP); (**b**) P_α ATP); (**c**) P_γ (ATP); (**d**) π; (**e**) P_α (ADP); (**f**) P_β (ADP).

^{31}P NMR studies revealed that **104** produced rate enhancements comparable to those of O-BISDIEN (**24**) and slightly lower than those of [21]aneN_7 (**11**), confirming the critical role of size in this catalytic processes (Figure 20). Interestingly enough, the reaction using **103** as a catalyst is not only efficient but also very specific, since it stops at the formation of ADP and does not proceed significantly further.

Molecular dynamics studies suggest that the optimal size is that for which just γ-phosphate of ATP perfectly resides at the macrocyclic cavity. Figure 21 shows the perfect fit between the phosphate group and the macrocyclic cavity of **104**.

Another macrocycle that presented a 21-membered size was *m*Ph22222 (**106**, Scheme 30), in which the *meta*-benzene spacer had been substituted by a *meta*-phenolic one. Although not as high as in the *m*B22222-ATP system, relevant rate accelerations were also found for *m*Ph22222

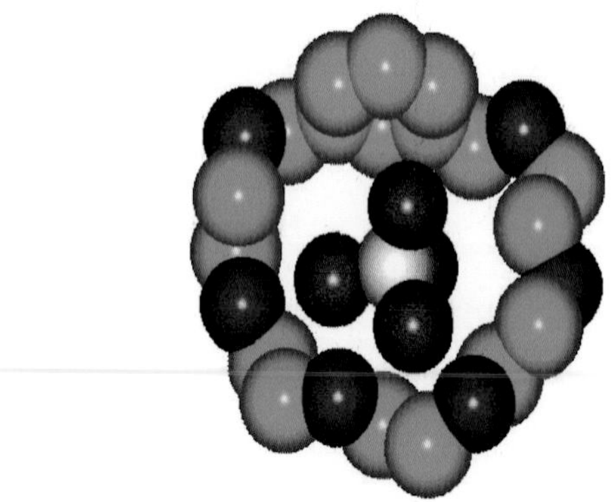

Figure 21 Comparison between the size of the macrocyclic cavity of **105** and the γ-phosphate of ATP. The rest of the ATP has been obscured for clarity.

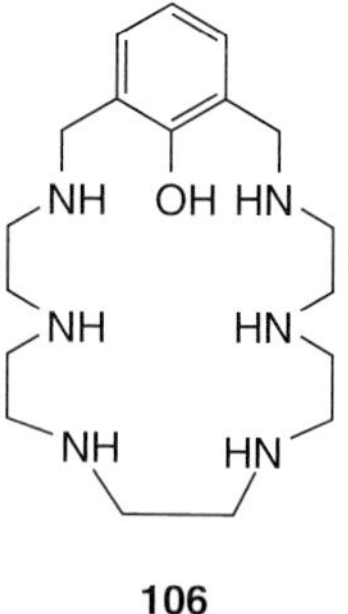

Scheme 30

from acidic pH to pH values about 7.5. At these pH values, the reaction slows down in concordance with the deprotonation of the phenolic group, which corresponds to the passage from the $H_4(Ph22222)^{4+}$ species to the $H_3(Ph22222)^{3+}$ species. Deprotonation of the phenolic group, apart from introducing a negative charge that somewhat repels the phosphate chain of ATP, yields intramolecular hydrogen bonding between the phenolate group and the protonated benzylic ammonium groups that at this pH are still protonated. This hydrogen bonding leads to a significant size reduction as illustrated in Figure 22, which prevents a good fit with the phosphate group.

O-BISDIEN (**24**) exhibited other characteristics of naturally occurring enzymes, such as the ability to modify the course of reaction and to transfer phosphoryl to other substrates. The protonated forms of **24** were shown to catalyze acetylphosphate hydrolysis and pyrophosphate formation through formation of a phosphorylated macrocyclic intermediate, which transfers a phosphoryl group to a phosphate substrate.[115] ATP phosphotransferases, hydrolases, and synthetases often require mono- or divalent anions in order to carry out their functions. Therefore, to generate systems that resembled their biological counterparts even more closely, protonated **24** was coupled to metal ions.[116,117] The investigations of the effect of the biologically significant metal ions Ca^{2+}, Mg^{2+}, and Zn^{2+}, added to **24**, revealed striking influences of the metal ions on ATP hydrolysis and on the formation of the phosphoramidate intermediates and pyrophosphate. While addition of both Ca^{2+} and Mg^{2+} increased the observed percentage of phosphoramidate, only Ca^{2+} provided a significant acceleration in ATP hydrolysis, almost doubling the first-order rate constant found for free **24** in the same experimental conditions. Ln^{3+} also led to a considerable increase in the rate of ATP dephosphorylation. The presence of Mg^{2+} had no apparent effect on the catalytic rate of ATP cleavage, while addition of Zn^{2+} or Cd^{2+} to macrocycle-ATP solutions decreased the rate of hydrolysis. The most striking finding in this respect was the ready appearance of pyrophosphate in the presence of **24** and either Mg^{2+}, Ca^{2+}, or Ln^{3+} at pH 4.5.

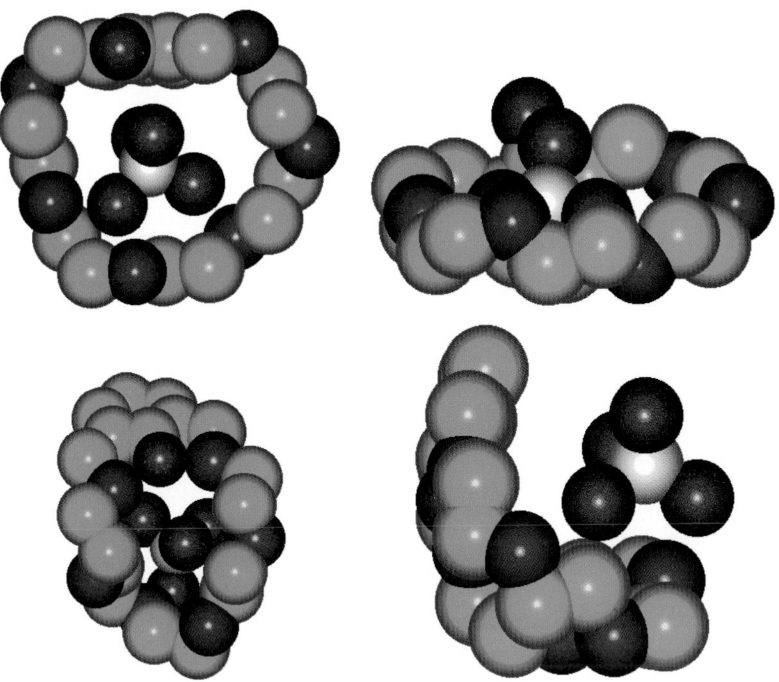

Figure 22 Model showing the dimensions of $H_4(m\mathrm{Ph}2222)^{4+}$ and $H_3(m\mathrm{Ph}2222)^{3+}$ in which phenolate is deprotonated in comparison to the γ-phosphate of ATP. The remaining portion of the ATP molecule has been omitted for clarity.

No pyrophosphate formation was observed in the absence of metals under these experimental conditions.

Mertes *et al.* indicated that macrocycle **24** was able to activate formate in the presence of ATP and Ca^{2+} or Mg^{2+} ions, yielding, as a final product, the macrocycle formylated at the central nitrogen of the chain.[118] Hosseini and Lehn proved that **24** in the presence of Mg^{2+} as promoter, catalyzed the generation of ATP from acetyl phosphate and ADP in dilute aqueous solution at neutral pH.[119]

The presence of a not-coordinated metal ion can play a decisive role in the activation of ATP cleavage by ternary metal complex formation.[120] This is the case of the protonated mononuclear Zn^{2+} complexes with the ligand Terpy2222 (**107**, Scheme 31). This ligand forms both mono- and binuclear Zn^{2+} complexes depending on the metal:ligand molar ratio and the solution pH. The first Zn^{2+} ion interacting with the ligand binds to the terpyridine moiety, while the second one occupies the polyamine chain. Both mono- and binuclear complexes form stable ternary complexes with ATP but do not show any significant ability in ATP hydrolysis. Only the tetraprotonated $[ZnH_4(\mathrm{Terpy2222})(\mathrm{ATP})]^{2+}$ species in the presence of not-coordinated Zn^{2+} ions gives rise to fast ATP cleavage to produce ADP and phosphate. The analysis of the time dependence of the ^{31}P NMR spectra recorded at different pHs clearly shows that only the tetraprotonated species is able to activate ATP hydrolysis, while both tri- and penta-protonated ones are completely inactive. The hydrolytic process proceeds through the formation

Terpy2222 (**107**)

Scheme 31

of a phosphoramidate (PN) intermediate, which is then rapidly hydrolyzed to hydrogen phosphate. The hydrolysis rate ($k_{\mathrm{OBS}} = 3.2 \times 10^{-2}\,\mathrm{min}^{-1}$ at pH 4) is among the highest observed for ATP dephosphorylation promoted by polyammonium receptors.

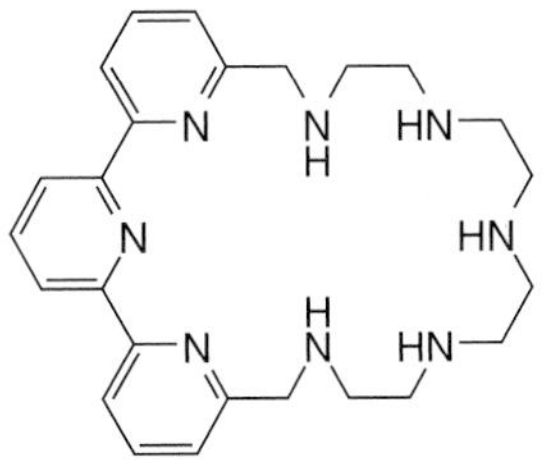

The fact that ATP cleavage takes place only in the presence of a "second" Zn^{2+} ion with second-order kinetic suggests that the transition state could be stabilized by this metal ion, probably through coordination of the metal to unprotonated amine groups of the macrocycle and to the γ-phosphate of ATP, leading to a higher activation of the γ-phosphorus to the nucleophilic attack. At the same time, the PN intermediate could be stabilized via coordination to the metal, accounting for the observed relatively high percentage of PN accumulating during the cleavage process.

The present system, therefore, represents a unique case of ATP dephosphorylation promoted by the simultaneous action of a metal complex, which is used essentially to anchor the anionic substrate, and of a second metal, which acts as a cofactor, assisting the phosphoryl transfer from ATP to an amine group of the receptor.

4 RECEPTORS FOR NUCLEOTIDE SENSING

Although anion sensing is a topic that has been broadly discussed in other chapters of this multivolume set encyclopaedia, owing to the great relevance that this topic has in nucleotide chemistry, we devote the last section of this chapter to advance several recent examples of selective nucleotide receptors including signaling units.[121]

Regarding this point, acyclic receptors **43–57** were shown to display significant fluorescence quenching at acidic pH upon interaction with 5′-ATP.[69,83] Comparison of the steady-state fluorescence titration curves with the distribution of the species formed as a function of pH, derived from the pH-metric studies, demonstrated that quenching occurred following protonation of the adenine N(3) atom. This protonation should render the adenine ring more electron deficiency, permitting a more efficient stacking with the naphthalene or anthracene rings of the receptor. Moreover, steady-state fluorescence and time-correlated single-photon counting analysis of a system made by ATP and the bischromophoric receptor **56** containing naphthalene and anthracene fluorophores at both ends of the polyamine showed that the energy transfer between the naphthalene and anthracene chromophores was not prevented by the presence of the nucleotide guest.

Recently, monochromophoric amines of this type have been grafted onto the surface of small-sized boehmite nanoparticles to generate recyclable devices for ATP sensing to be able to operate in an aqueous solution (Figure 23).[122] In another strategy for ATP detection, organic fluorophores were attached to the surface of mesoporous silica materials.[123]

5′-GTP sensing was achieved using a water-soluble imidazolium receptor including an anthracene group as the fluorophore (**108**, Scheme 32) that was synthesized by reaction of 1,6-bis(imidazolylmethyl)anthracene with

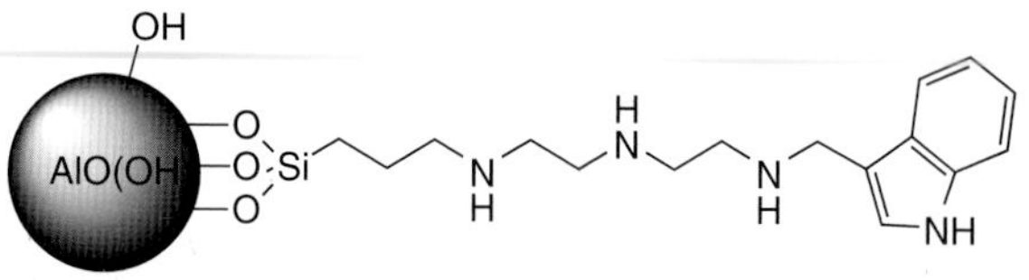

Figure 23 Scheme of a monochromophoric amine attached to the surface of boehmite nanoparticles.

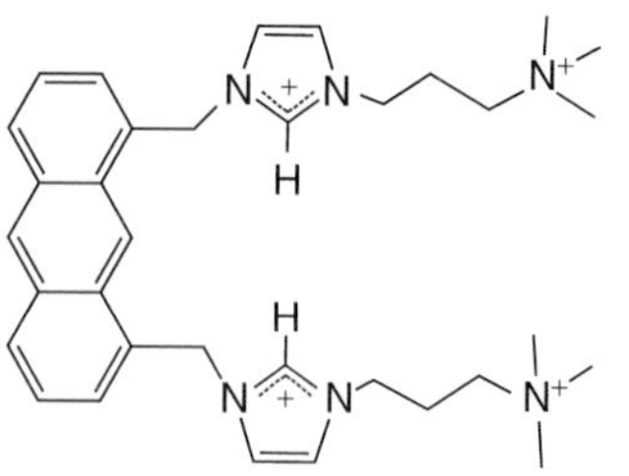

108

Scheme 32

(3-bromopropyl)-trimethylammonium bromide in acetonitrile.[124] It has to be remarked that receptors containing imidazolium or pyridinium groups can be strong hydrogen bond donors to anions through the $(C-H)^+$ fragment.[125] Host **108** displays a 6-fold selectivity for GTP over ATP and 100-fold over ADP, AMP, pyrophosphate, phosphate anions, and halide anions. Moreover, while ATP interaction with **108** at the physiological pH of 7.4 leads to an enhanced fluorescence (CHEF effect), GTP produces a marked quenching of the fluorescence of **108** (CHEQ effect).

An interesting cavitand-type receptor functionalized with four imidazolylmethyl fragments (**109**, Scheme 33) was reported in 2005 by J. Yoon *et al.*[126] Receptor **109** shows fluorescence behavior similar to **108**, with a moderate selectivity for GTP over ATP and CTP.

Zn^{2+} complexes with coordinatively unsaturated binding sites have been used for nucleoside and nucleotide sensing in a number of cases.[121] One example of this chemistry is provided by the dinuclear Zn^{2+} chemosensor **110** included in Figure 24,[127] which comprises two

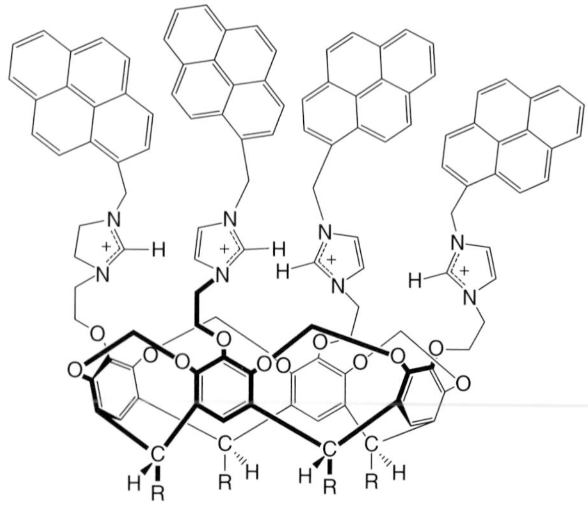

109

Scheme 33

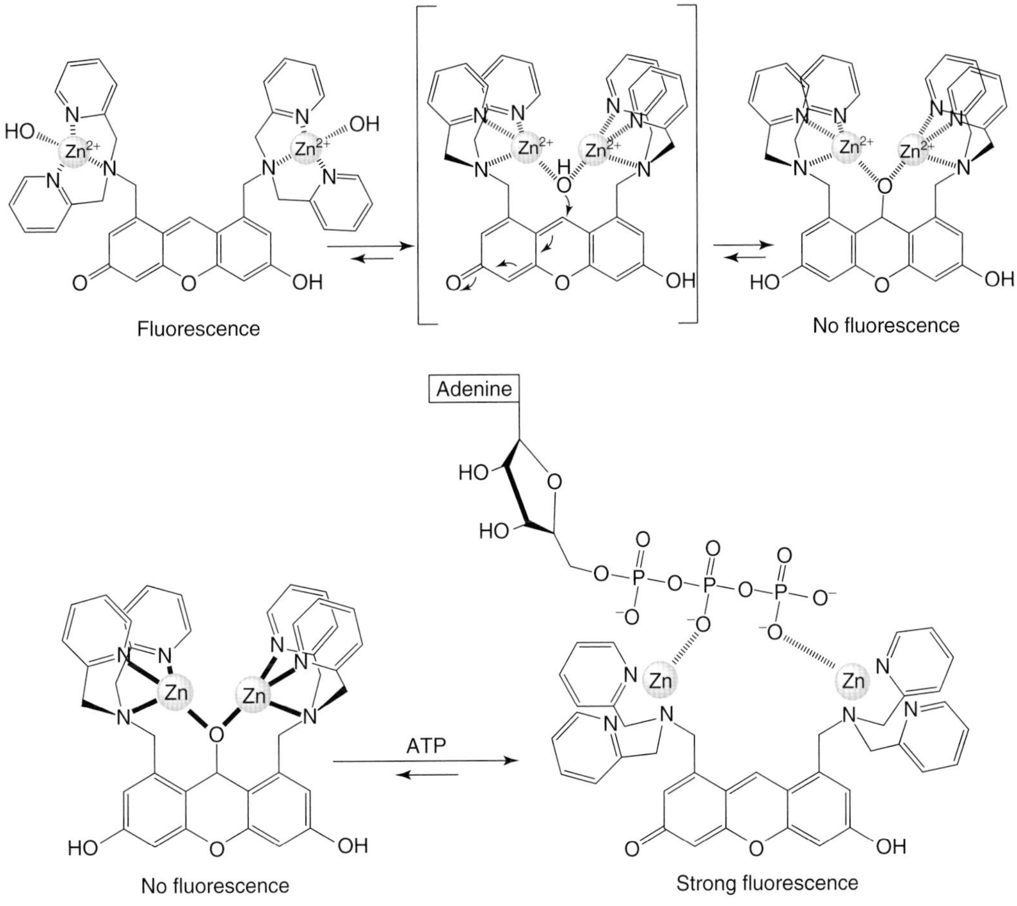

Figure 24 Schematic illustration of the turn-on fluorescence-sensing mechanism of ATP by $[Zn_2(\mathbf{110})]^{4+}$.

2,2′-dipicolylamine Zn^{2+} binding sites connected by a xanthene chromophore. The interaction of this binuclear Zn^{2+} complex with nucleotide polyphosphates ($K_{cond} \sim 1 \times 10^{-6}\,M^{-1}$) was accompanied by a large turn-on in the fluorescence emission, while no changes were observed in fluorescence upon addition of nucleotide monophosphates. The enhanced fluorescence of the complex $[Zn_2(\mathbf{110})]^{4+}$ in the presence of nucleotide di- and triphosphates is based on a mechanism involving the binding-induced recovery of the conjugated form of the xanthene ring from nonfluorescent deconjugated state, which is formed by a nucleophilic attack of a zinc-bound water molecule.

An alternative mechanism for sensing nucleotides relies on the formation or dissociation of highly emitting excimer species. This is the case of the pinzerlike benzene-bridged imidazolium receptor **111** depicted in Figure 25.[128]

The pyrene rings in **111** are π-stacked, forming an intramolecular excimer, which leads to an appearance in the fluorescence spectra, apart from the classical emission band at 375 nm of the monomeric pyrene, of a red-shifted band at 475 nm. Upon addition of different nucleotides, a unique switch of the excimer versus monomer fluorescence is observed for ATP due to a sandwich pyrene–adenine–pyrene π-stacking (Figure 25). Other nucleotide triphosphates, such as GTP, CTP, UTP, and TTP, can interact only from the outside with the already stabilized stacked pyrene–pyrene dimer of **111**. This results in that, although all the nucleotides quench the emission of the excimer with a trend ATP $\sim$ GTP $>$ TTP $\sim$ UTP $>$ CTP, only ATP induces a large enhancement in the monomeric fluorescent peak of **111**. Although the other nucleotides containing adenine AMP and ADP should respond in the same manner, the observed changes are practically negligible for AMP and smaller for ADP. The I_{375}/I_{487} ratio for ATP is much larger, permitting its selective sensing. Sensor **111** has been successfully applied either to ATP staining experiments or to monitor ATP and ADP hydrolysis by apyrase.

An opposite strategy to the alteration of preformed pyrene dimers following nucleotide interaction, discussed in the previous example, would be that stacking between the pyrene rings occurs as a consequence of nucleotide

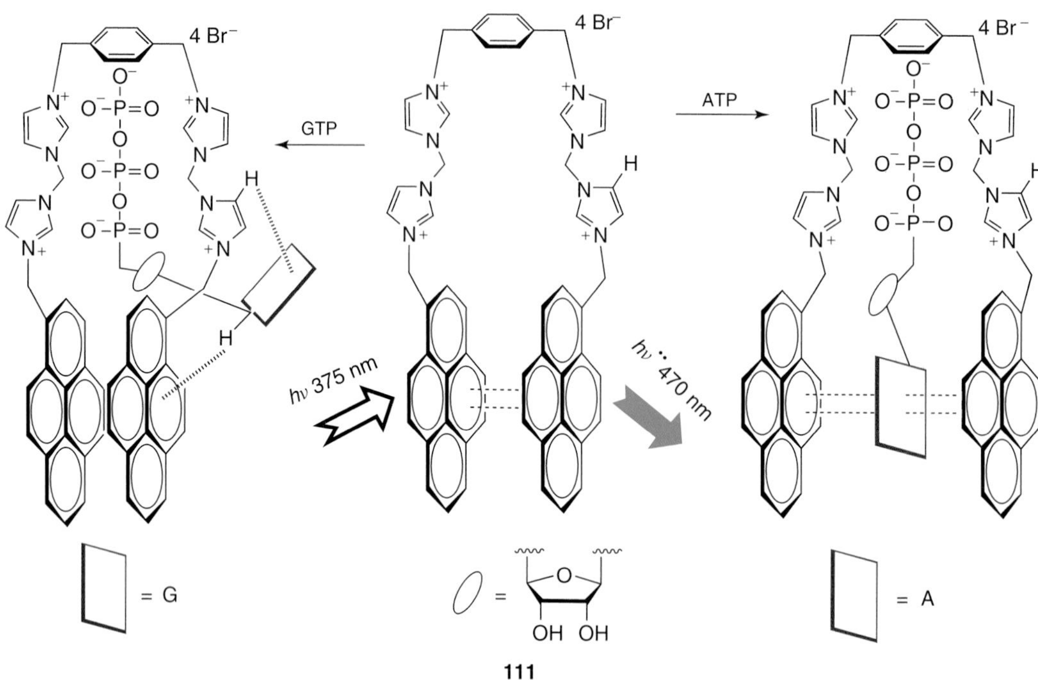

Figure 25 Proposed binding mode of receptor **111** with ATP and GTP.

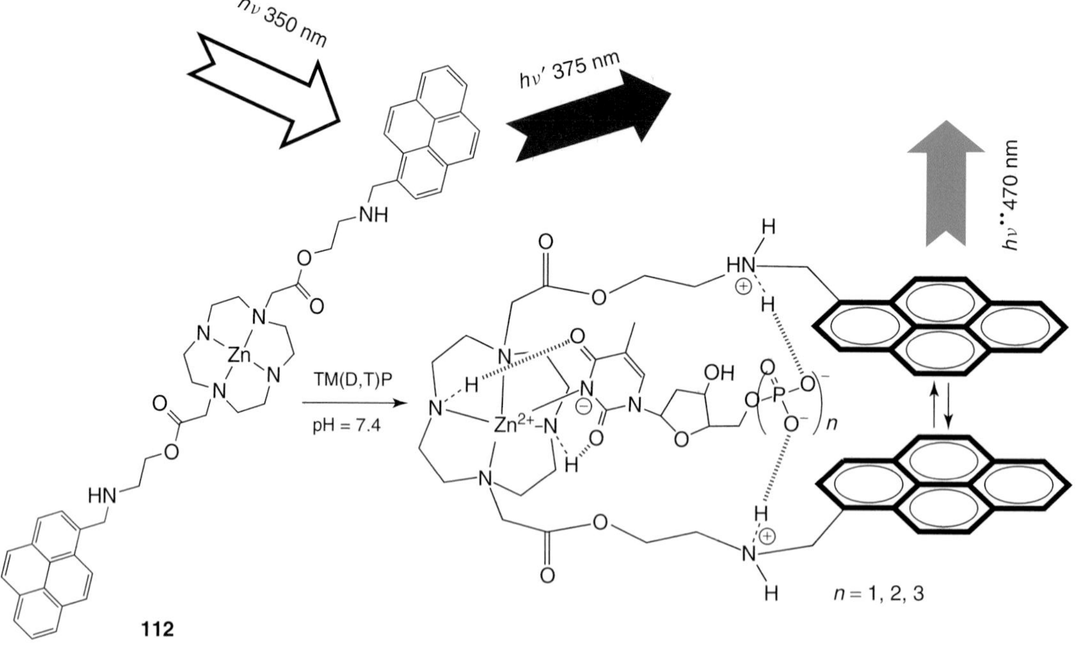

Figure 26 Proposed binding mode of thymidine phosphate to $[Zn(\mathbf{112})]^{2+}$.

binding. In this respect, pyrene fluorophores were implemented in the well-known tetra- or triazacycloalkane scaffolds $[12]aneN_4$ and $[9]aneN_3$ (receptors **112** and **113** in Figure 26 and Scheme 34).[129]

Receptor **112** also has the ability, of the Zn^{2+} complex of $[12]aneN_4$, to bind thymidine nucleotides previously discussed[102] in such a way that they can stack upon nucleotide binding, resulting in excimer formation between rings attached to the macrocyclic core. While the thymine nucleobase binds to Zn^{2+} through a deprotonated imide group, the phosphate groups of the chain are hydrogen bonded to the secondary pendant amine groups, linking the macrocycle with the fluorophores that, at pH 7.4, are protonated. This leads to excimer emission upon thymidine

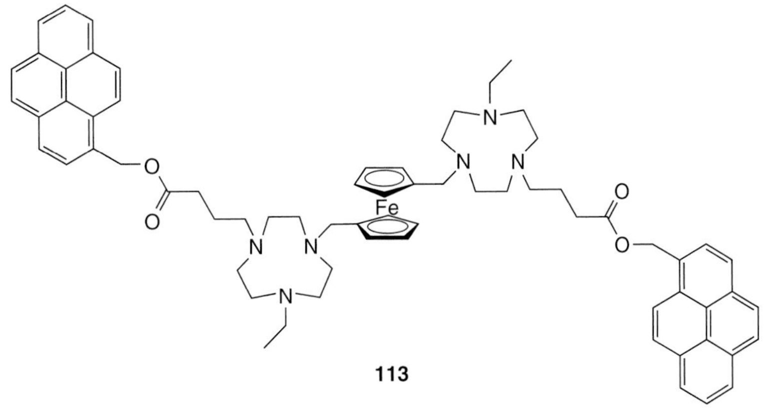

113

Scheme 34

Table 2 Summary of the binding constants K (M^{-1}) for the interaction between Zn(**113**) and various nucleotide anions measured in a 1 : 9 mixture of CH_3CN/TRIS HCl buffer (10 mM; pH = 7.4).

Anion/Nucleotide	K (M^{-1})	Nucleotide	K (M^{-1})
PPi	$(4.45 \pm 0.41) \times 10^6$	CDP	$(1.55 \pm 0.14) \times 10^4$
ATP	$(9.31 \pm 0.84) \times 10^4$	TDP	$(2.03 \pm 0.18) \times 10^4$
CTP	$(2.32 \pm 0.19) \times 10^5$	AMP	ND
GTP	$(2.13 \pm 0.29) \times 10^5$	CMP	ND
GDP	$(1.41 \pm 0.12) \times 10^4$	GMP	ND
TTP	$(5.05 \pm 0.46) \times 10^5$	TMP	ND
ADP	$(9.63 \pm 0.93) \times 10^3$	Π	ND

ND, not determined.

nucleotide binding, allowing for the selective sensing of this type of nucleotides.

Receptor **113** consists of two triazacyclononane [9]aneN_3 units with appended pyrene fragments, linked by a covalently attached ferrocene subunit. Analogous with the previous example, interaction with pyrophosphate anions (PPi), AMP, ADP, and ATP nucleotides promotes excimer formation between the pyrene units. Fluorescence titrations indicate that the largest binding constant is obtained for PPi, followed by a group formed by TTP, CTP, and GTP (Table 2). The fluorescence changes observed for nucleotide monophosphates and inorganic phosphate were too small for allowing an accurate measurement of the stability constants.

Following a somewhat similar strategy, interaction of UTP, UDP, UMP, and TTP or PPi to the Zn^{2+} complexes Zn(**114**)$^{2+}$ and Zn_2(**115**)$^{4+}$ containing pyrene labels (Scheme 35) were shown to induce naked-eye pyrene excimer emission.[130] Such an emission was explained in the case of the monomer complexes by the binding of pyrophosphate or nucleotides as bridging ligands between two Zn(**114**)$^{2+}$ monomer complexes, as depicted in Figure 27.

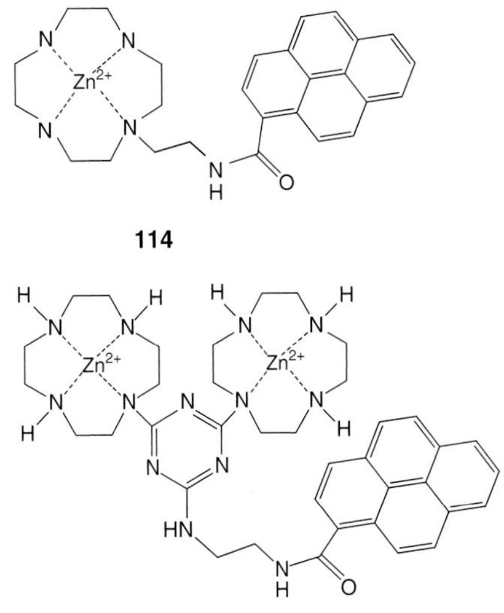

114

115

Scheme 35

A different class of receptors for nucleotides are the bipyridinium open-chain and cyclic ligands included in Scheme 36. Compounds **116** and **118** are selective for

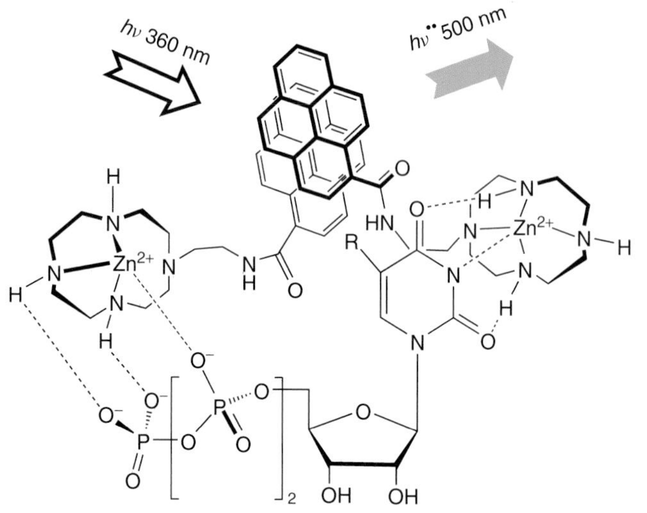

Figure 27 Intermolecular pyrene excimer formation in $[Zn(\mathbf{114})]^{2+}$ mediated by the interaction with UTP.

ATP and GTP as compared to other nucleotides, whereas negligible selectivity is found for **117** and **119**.[131]

Compounds **116**, **117**, and **118** exhibited significant binding interactions with the fluorescent, water-soluble dye 8-hydroxyl-1,3,6-pyrene trisulfonate (HPTS) yielding a quenching of the fluorescence. Titration of the adducts with the nucleotides produces the displacement of the dye and the revival of the fluorescence.

An elegant system for the recognition of phosphate anions from other inorganic anions is provided by a glycosylated amino acetate hydrogel having nanosized fibers with well-developed hydrophobic domains and microsized cavities filled with water (Scheme 37).[132] When the fluorescent dinuclear Zn^{2+} complex of **120** is entrapped in the hydrogel matrix, three different fluorescence patterns are identified: (i) increase in the emission intensity with blue-shift of the dansyl band for the hydrophobic phenyl phosphate anion, (ii) decrease in intensity and red-shift for the hydrophilic phosphate, phosphate–tyrosine, and 5′-ATP anions, and (iii) no change in nonphosphate anionic guests. Additionally, the extent of the change observed for phosphate, phosphate–tyrosine, and ATP was different, which permitted discrimination from each other. It is interesting to note that such a clear-cut discrimination between hydrophilic and hydrophobic anions was not achieved either in aqueous solution or using the nonhydrophobic agarose gel as a support. Therefore, the hydrophobic domains of the hydrogel are fundamental for discrimination; the Zn^{2+} complex with **120** is sufficiently mobile in the supramolecular gel matrix to move toward hydrophobic fibers when capturing the more hydrophobic anionic guests such as phenyl phosphate or toward the aqueous cavities when interacting with the more hydrophilic anions such as phosphate or ATP. This dynamic behavior was advantageously used to make FRET mechanisms of recognition operative by introducing the coumarin-appended receptor **121** and the hydrophobic styryl dye as a fluorescence resonance energy

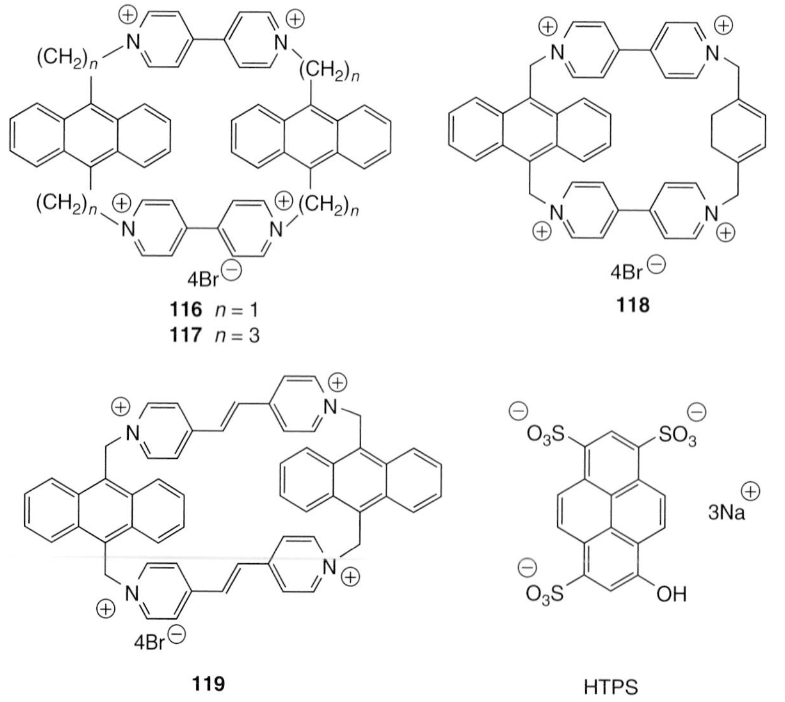

116 $n = 1$
117 $n = 3$

118

119

HTPS

Scheme 36

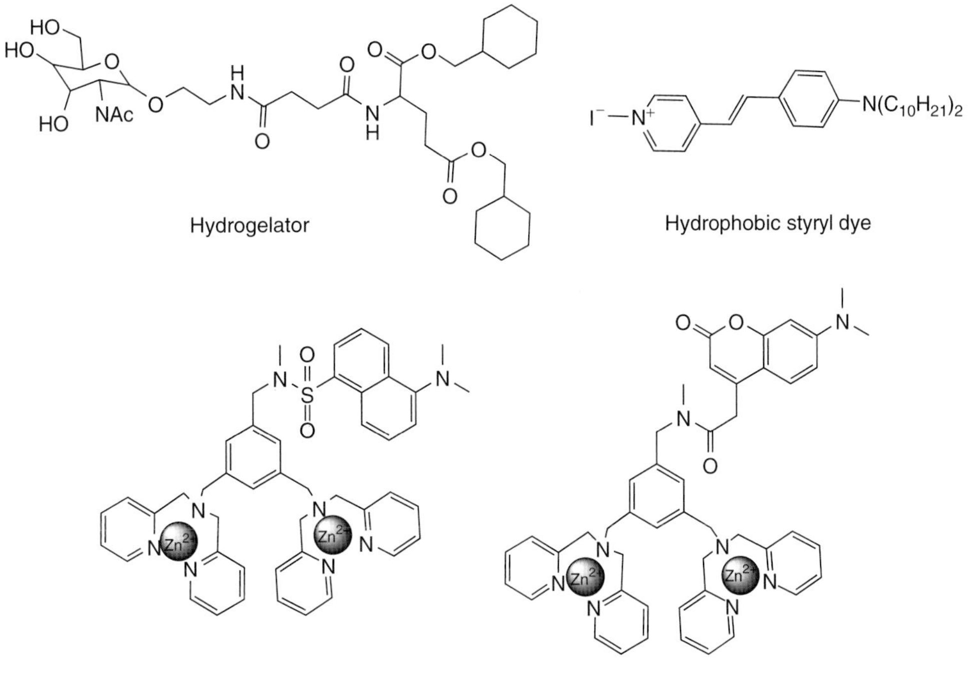

Hydrogelator

Hydrophobic styryl dye

120

121

Scheme 37

transfer (FRET) donor and acceptor (Scheme 37). The magnitude of the FRET depends on the donor–acceptor distance and therefore the effect is greater in the presence of the hydrophobic phosphates, which are bound by the Zn^{2+} receptor closer to the fibers where the styryl acceptor dye resides.

5 POLYMERS AS NUCLEOTIDE CHEMOSENSORS

To conclude this chapter, we refer to a couple of examples regarding the use of polymers for molecular recognition and sensing of nucleotide anions, a topic that has received increasing attention in recent years.[133]

The cationic polythiophene (PT) derivative PMTPA depicted in Scheme 38 allows the differentiation of 15 nucleotides with a confidence limit of 100% using absorption spectroscopy in combination with linear dynamic analysis.[134]

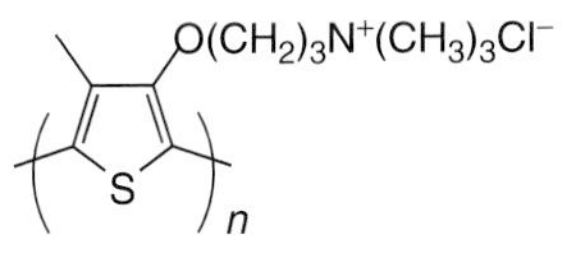

Poly(3-alkoxy-4-methylthiophene)

Scheme 38

Interaction of PMTPA and mononucleotides carried out in a HEPES buffer (10 mM, pH 7.4) showed unique responses to the different nucleotides with absorption maxima in the visible spectra red shifted from 405 nm for PMTPA to 416 nm for AMP, 455 nm for ADP, 542 nm for ATP, 499 nm for UDP, 460 nm for TTP, 540 nm got GTP, and 469 nm for CTP. This unique response toward different nucleotides was assigned to changes in the conformation and aggregation mode of the PT backbone, and the formation of an ordered phase of PTs driven by ionic self-assembly between the quaternary ammonium groups of the conjugated polymer and the phosphate groups of the polynucleotides. Multiple negative charges (triphosphates as compared to di- and monophosphates) and more hydrophobic nucleobases (purine relative to pyrimidine) facilitate the formation of PMTPA aggregates. These differences in the polymer ordering lead to specific signal transductions of the nucleotide binding, generating an array of wavelengths that has been advantageously used for 100% selective discrimination of the target nucleotides.

The binding of guest species to supramolecular recognition sites incorporated into elastic polymers can lead to dimensional changes in macroscopic dimensions that can be used for different applications such as in actuators, sensors, process control, and, in particular, drug release systems. Schneider and coworkers have carried out extensive work in this field. A few years ago, they synthesized a suitable precursor by reaction of polymethyl methacrylate with diethylenetriamine and various amounts of long-chain alkylamines.[135] Transparent thin films of the elastic polymers were preconditioned in a buffer solution and cut into small pieces. Changes in the dimensions were observed upon addition of several anions. Changes as large as 126% were observed when adding AMP. UV–vis spectroscopy permitted to monitor the depletion of the AMP in the supernatant solution and its absorption by the polymer. The changes in dimensions proved to be completely reversible.

6 CONCLUSIONS

Mononucleotides were one of the first substrates investigated under the umbrella of supramolecular chemistry. Recognition and catalysis of nucleotides were investigated by some of the pioneer researchers in the field. The first investigated receptors used for nucleotide recognition were saturated polyammonium receptors that use attraction between opposite charges as the main binding force. The cyclic or acyclic nature of the receptor, size, and distribution of charge density are important parameters governing the interaction. Hydrogen bonding between the amino groups and matching donor or acceptor groups of the phosphate, the sugar moiety, and the nucleobase moieties of the nucleotide is also a key contribution to the overall interaction. Receptors with appropriately located aromatic fragments can also interact with the nucleobases through π-stacking interactions. Receptors with given characteristics can, apart from interacting with the nucleotides, display either an ATPase and/or a kinase activity. The dioxatetraaza receptor O-BISDIEN is one of most interesting receptors used for this purpose. Nucleotide chemosensors have been constructed containing different binding sites and chromophoric groups. Excimer formation or disruption following nucleotide interaction has been the strategy of choice in a number of cases. Polyconjugated polymers displaying differentiated optical signals or modified elastic polymers experiencing macroscopic dimensional changes represent new strategies in nucleotide recognition or sensing addressed to find applications in environmental or biomedical fields.

NOTES

[1] 5′-ATP will be named throughout the text as ATP. The same type of abbreviation will be used for the other 5′-nucleotides.

ACKNOWLEDGEMENTS

We would like to thank MICINN (Spain) for financial support projects CONSOLIDER INGENIO CSD-2010-00065 and CTQ2009-14288-CO4-01. JGG, RBV and JPJ would like to thank MICINN for their fellowships.

REFERENCES

1. C. J. Pedersen, *J. Am. Chem. Soc.*, 1967, **89**, 7017–7036.
2. B. Dietrich, J.-M. Lehn, and J.-P. Sauvage, *Tetrahedron Lett.*, 1969, **10**, 2889–2992.
3. B. Dietrich, J.-M. Lehn, J.-P. Sauvage, and J. Blanzat, *Tetrahedron*, 1973, **29**, 1629–1645.
4. D. J. Cram and J. M. Cram, *Science*, 1974, **183**, 803–809.
5. D. J. Cram and J. M. Cram, *Acc. Chem. Res.*, 1978, **11**, 8–14.
6. J.-M. Lehn, *Acc. Chem. Res.*, 1978, **11**, 49–57.
7. J.-M. Lehn, *Angew. Chem. Int. Ed. Engl.*, 1988, **27**, 89–112.
8. J.-M. Lehn, *Supramolecular Chemistry. Concepts and Perspectives*. VCH, Weinheim, 1995.
9. C. H. Park and H. E. Simmons, *J. Am. Chem. Soc.* 1968, **90**, 2431–2432.
10. I. Stibor, ed., *Anion Sensing. Topics in Current Chemistry*, Springer, Berlin, Heidelberg, New York, 2005.
11. A. Bianchi, M. Micheloni, and P. Paoletti, *Coord. Chem. Rev.*, 1990, **110**, 17–113.
12. B. Dietrich, *Pure Appl. Chem.*, 1993, **65**, 1457–1464.
13. R. M. Izatt, K. Pawlak, and J. S. Bradshaw, *Chem. Rev.*, 1995, **95**, 2529–2586.
14. P. D. Beer, P. A. Gale, and G. Z. Chen, *Coord. Chem. Rev.*, 1999, **185–186**, 3–36.
15. P. D. Beer, and J. Cadman, *Coord. Chem. Rev.*, 2000, **205**, 131–155.
16. P. A. Gale, *Coord. Chem. Rev.*, 2000, **199**, 181–233.
17. A. Bianchi, K. Bowman-James, and E. García-España, eds., *Supramolecular Chemistry of Anions*, Wiley-VCH, New York, 1997.
18. J. L. Sessler, P. A. Gale, and V. S. Cho, *Anion Receptor Chemistry*, Royal Society of Chemistry, Cambridge, UK, 2006.
19. P. D. Beer and P. A. Gale, *Angew. Chem. Int. Ed.*, 2001, **40**, 486–516.
20. V. Amendola, L. Fabbrizzi, C. Mangano, *et al.*, *Coord. Chem. Rev.*, 2001, **219–221**, 821–837.
21. J. M. Llinares, D. Powell, and K. Bowman-James, *Coord. Chem. Rev.*, 2003, **240**, 57–75.
22. V. McKee, J. Nelson, and R. M. Town, *Chem. Soc. Rev.*, 2003, **32**, 309–325.
23. C. R. Bondy and S. J. Loeb, *Coord. Chem. Rev.*, 2003, **240**, 77–99.

24. K. Choi and A. D. Hamilton, *Coord. Chem. Rev.*, 2003, **240**, 101–110.

25. K. Bowman-James, *Acc. Chem. Res.*, 2005, **38**, 671–678.

26. S. Kubik, C. Reyheller, and S. Stüwe, *J. Inclusion Phenom. Macrocyclic Chem.*, 2005, **52**, 137–187.

27. M. A. Hossain, S. O. Kang, and K. Bowman-James, *Current Trends and Future Perspectives*, Springer, Dordrecht, The Netherlands, 2005, pp. 178–183.

28. P. A. Gale, *Acc. Chem. Res.*, 2006, **39**, 465–475.

29. E. García-España, P. Díaz, J. M. LLinares, and A. Bianchi, *Coord. Chem. Rev.*, 2006, **259**, 2952–2986.

30. S. O. Kang, R. A. Begum, and K. Bowman-James, *Angew. Chem. Int. Ed.*, 2006, **45**, 7882–7894.

31. S. O. Kang, M. A. Hossain, and K. Bowman-James, *Coord. Chem. Rev.*, 2006, **250**, 3038–3052.

32. V. Amendola, M. Bonizzoni, D. Esteban-Gómez, *et al.*, *Coord. Chem. Rev.*, 2006, **250**, 1451–1470.

33. K. Wichmann, B. Antonioli, T. Söhnel, *et al.*, *Coord.Chem. Rev.*, 2006, **250**, 2987–3003.

34. A. K. H. Hirsch, F. R. Fischer, and F. Diederich, *Angew. Chem. Int. Ed.*, 2007, **46**, 338–352.

35. G. W. Gokel and I. A. Carasel, *Chem. Soc. Rev.*, **36**, 2007, 378–389.

36. M. J. Chmielewski, T. Zieliński, and J. Jurczak, *Pure Appl. Chem.*, 2007, **79**, 1087–1096.

37. R. Custelceann and B. A. Moyer, *Eur. J. Inorg. Chem.*, 2007, 1321–1340.

38. P. Blondeau, M. Segura, R. Pérez-Fernández, and J. de Mendoza, *Chem. Soc. Rev.*, 2007, **36**, 198–210.

39. P. A. Gale, S. E. García-Garrido, and J. Garric, *Chem. Soc. Rev.*, 2008, **37**, 151–190.

40. P. A. Gale, *Chem. Commun.*, 2008, 4525–4540.

41. I. Alfonso, *Mini Rev. Org. Chem.*, 2008, **5**, 33–46.

42. C.-H. Lee, H. Miyaji, D.-W. Yoon, and J. L. Sessler, *Chem. Commun.*, 2008, 24–34.

43. N. A. Itsikson, Y. Y. Morzherin, A. I. Matern, and O. N. Chupakhin, *Russ. Chem. Rev.*, 2008, **77**, 751–764.

44. P. Prados and R. Quesada, *Supramol. Chem.*, 2008, **20**, 201–216.

45. Y. Z. Voloshin and A. S. Belov, *Russ. Chem. Rev.*, 2008, **77**, 161–175.

46. T. Nabeshima and S. Akine, *Chem. Record.*, 2008, **8**, 240–251.

47. J. W. Steed, *Chem. Soc. Rev.*, 2009, **38**, 506–519.

48. V. L. Amendola and L. Fabbrizzi, *Chem. Commun.*, 2009, 513–531.

49. S. Kubik, *Chem. Soc. Rev.*, 2009, **38**, 585–605.

50. E. A. Katayev, G. V. Kolesnikov, and J. L. Sessler, *Chem. Soc. Rev.*, 2009, **38**, 1572–1586.

51. C. Caltagirone and P. A. Gale, *Chem. Soc. Rev.*, 2009, **38**, 520–563.

52. Y. Hua and A. Flood, *Chem. Soc. Rev.*, 2010, **39**, 1262–1271.

53. P. Mateus, N. Bernier, and R. Delgado, *Coord. Chem. Rev.*, 2010, **254**, 1726–1747.

54. P. A. Gale and T. Gunnlaugson, *Chem. Soc. Rev.*, 2010, **39**(10), 3581–4008.

55. B. Dietrich, M. W. Hosseini, and J.-M. Lehn, *J. Am. Chem. Soc.*, 1981, **103**, 1282–1283.

56. E. Kimura, M. Kodama, and T. Yatsunami, *J. Am. Chem. Soc.*, 1982, **104**, 3182–3187.

57. J. M. Berg, J. L. Tymozco, and L. Stryer, *Biochemistry*, W. F. Freeman & Company, 2002.

58. W. N. Lipscomb and N. Strater, *Chem. Rev.*, 1996, **96**, 2375–2433.

59. D. H. Le, S. Y. Kim, and J.-I. Hong, *Angew. Chem. Int. Ed.*, 2004, **43**, 4777–4780.

60. B. Alberts, A. Johnson, J. Lewis, *et al.*, *Molecular Biology of the Cell*, Garlnad Science, New York, 2002.

61. D. Communiamd and J.-M. Boeynaems, *Trends Pharmacol. Sci.*, 1997, **18**, 83–86.

62. S. Pullarkat, J. Stoehlmacher, V. Ghaderi, *et al.*, *Pharmacogenesis. J.*, 2001, **1**, 65–70.

63. (a) G. Pontarin, P. Ferraro, M. L. Valentino, *et al.*, *J. Biol. Chem.*, 2006, **281**, 22720–22728; (b) C. Rampazzo, S. Fabris, E. Franzolin, *et al.*, *J. Biol. Chem.*, 2007, **282**, 34758–34769; (c) N. Ashley, S. Adams, *et al.*, *Hum. Mol. Gen.*, 2007, **16**, 1400–1411.

64. R. T. Guo, Y. E. Chong, M. Guo, and X. L. Yang, *J. Biol. Chem.*, 2009, **284**(42), 28968–28976.

65. H. Sigel and R. Griesser, *Chem. Soc. Rev.*, 2005, **34**, 875–900.

66. I. Tabushi, Y. Kobuke, and J.-I. Imuta, *J. Am. Chem. Soc.*, 1981, **103**, 6152–6157.

67. A. Bianchi, S. Mangani, M. Micheloni, *et al.*, *Inorg. Chem.* 1985. **24**, 1182.

68. A. Bencini, A. Bianchi, E. García-España, *et al.*, *Bioorg. Chem.*, 1992, **20**, 8–29.

69. M. T. Albelda, M. A. Bernardo, E. García-España, *et al.*, *J. Chem. Soc. Perkin Trans. 2*, 1999, 2545–2549.

70. A. Bencini, A. Bianchi, E. García-España, *et al.*, *Coord. Chem. Rev.*, 1999, **188**, 97–156.

71. A. Bianchi, M. Micheloni, and P. Paoletti, *Inorg. Chim. Acta*, 1988, **151**, 269–272.

72. J. F. Marececk and C. J. Burrows, *Tetrahedron Lett.*, 1986, **27**, 5943–5946.

73. E. Kimura, Y. Kuramoto, T. Koike, *et al.*, *J. Org. Chem.*, 1990, **55**, 42–46.

74. A. Bencini, A. Bianchi, M. I. Burguete, *et al.*, *J. Chem. Soc. Perkin Trans. 2*, 1991, 1445–1451.

75. A. Andrés, J. Aragó, A. Bencini, *et al.*, *Inorg. Chem.*, 1993, **32**, 3418–3424.

76. M. T. Albelda, E. García-España, H. R. Jimenez, *et al.*, *Dalton Trans.*, 2006, 4474–4482.

77. M. W. Hosseini, A. J. Blacker, and J.-M. Lehn. *J. Am. Chem. Soc.*, 1990, **112**, 3896–3904.

78. M. W. Hosseini, J.-M. Lehn, S. R. Duff, *et al.*, *J. Org. Chem.*, 1987, **52**, 1662–1666.

79. S. Claude, J.-M. Lehn, F. Schmidt, and J.-P. Vigneron, *J. Chem. Soc. Chem. Commun.*, 1991, 1182–1185.

80. P. Cudic, M. Zinic, V. Tomisic, *et al.*, *J. Chem. Soc. Chem. Commun.*, 1995, 1182–1185.

81. J. Aguilar, E. García-Espana, J. A. Guerrero, *et al.*, *J. Chem. Soc. Chem. Commun.*, 1995, 2237–2239.

82. J. Aguilar, E. García-España, J. A. Guerrero, *et al.*, *Inorg. Chim. Acta*, 1996, **246**, 287–294.

83. M. T. Albelda, J. Aguilar, S. Alves, *et al.*, *Helv. Chim. Acta*, 2003, **86**, 3118–3135.

84. A.-S. Deléphine, R. Tripier, N. Le Bris, *et al.*, *Inorg. Chim. Acta*, 2009, **362**, 3829–3834.

85. C. Bazzicalupi, S. Biagini, A. Bencini, *et al.*, *Chem. Commun.*, 2006, 4087–4089.

86. C. Bazzicalupi, A. Bencini, S. Biagini, *et al.*, *J. Org. Chem.*, 2009, **74**, 7349–7363.

87. A.-S. Deléphine, R. Tripier, and H. Handel, *Org. Biomol. Chem.*, 2008, **6**, 1743–1750.

88. A. Bencini, S. Biagini, C. Giorgi, *et al.*, *Eur. J. Org. Chem.*, 2009, 5610–56121.

89. A.-S. Deléphine, R. Tripier, M. Le Baccon, and H. Handel, *Eur. J. Org. Chem.*, 2010, 5380–5390.

90. (a) G. J. Bridger, R. T. Skerlj, S. Padmanabhan, *et al.*, *J. Med. Chem.*, 1996, **39**(1), 109–119; (b) Z. Guo, and P. J. Sadler, *Adv. Inorg. Chem.*, 2000, **48**, 183–306; (c) E. Kikuta, S. Aoki, and E. Kimura, *J. Am. Chem. Soc.*, 2001, **123**, 7911–7912.

91. J. González. J. M. Llinares, R. Belda, *et al.*, *Org. Biomol. Chem.*, 2010, **8**, 2367–2376.

92. C. Bazzicalupi, A. Bencini, A. Bianchi, *et al.*, *J. Chem. Soc. Perkin Trans. 2*, 1997, **4**, 775–782.

93. C. Bazzicalupi, A. Bencini, A. Bianchi, *et al.*, *J. Chem. Soc. Perkin Trans. 2*, 1995, **2**, 275–280.

94. C. Bazzicalupi, A. Bencini, A. Bianchi, *et al.*, *J. Am. Chem. Soc.*, 2008, **130**, 2440–2441.

95. H.-J. Schneider, *Angew. Chem. Int. Ed.*, 2009, **48**, 3924–3977.

96. (a) P. Schiessl and F. P. Schmidtchen, *Tetrahedron Lett.*, 1993, **34**, 2449–2452; (b) P. Schiessl and F. P. Schmidtchen, *J. Org. Chem.* 1994, **59**, 509–511; (c) M. Haj-Zaroubi and F. P. Schmidtchen, *ChemPhysChem*, 2005, **6**, 1181–1186.

97. A. V. Eliseev and H.-J. Schneider, *J. Am. Chem. Soc.*, 1994, **116**, 6081–6088.

98. J. L. Sessler, V. Kral, T. V. Shishkanova, and P. A. Gale, *Proc. Natl. Acad. Sci. U. S. A.*, 2002, **99**, 4848–4853.

99. V. Malinowski, L. Tumir, I. Piantanida, *et al.*, *Eur. J. Org. Chem.*, 2002, 3785–3795.

100. Y. Fuentes-Martínez, C. Godoy-Alcántar, F. Medrano, *et al.*, *Supramol. Chem.*, 2010, **22**, 212–220.

101. M.-H. Lui, F. Kowk, W.-R. Chang, *et al.*, *J. Biol. Chem.*, 2002, **29**, 46385–46390.

102. (a) M. Shionoya, E. Kimura, and M. Shiro, *J. Am. Chem. Soc.*, 1993, **115**, 6730–6737; (b) M. Shionoya, T. Ikeda, E. Kimura, and M. Shiro, *J. Am. Chem. Soc.*, 1994, **116**, 3848–3859; (c) S. Aoki, Y. Honda, and E. Kimura, *J. Am. Chem. Soc.*, 1998, **120**, 1018–1026; (d) E. Kikuta, M. Murata, N. Katsube, *et al.*, *J. Am. Chem. Soc.*, 1999, **121**, 5426–5436; (e) S. Aoki and E. Kimura, *J. Am. Chem. Soc.*, 2000, **122**, 4542–4548; (f) S. Aoki and E. Kimura, *Chem. Rev.*, 2004, **104**, 769–787.

103. (a) Y. Guo, Q. Ge, H. Lin, *et al.*, *J. Mol. Recognit.*, 2003, **16**, 102–111; (b) Y. Guo, Q. Ge, H. Lin, *et al.*, *Inorg. Chem. Commun.*, 2003, **6**, 308–312; (c) Y. Guo, Q. Ge, H. Lin, *et al.*, *Biophys. Chem.*, 2003, **105**, 119–131.

104. J. Aguilar, P. Díaz, F. Escartí, *et al.*, *Inorg. Chim. Acta*, 2002, **339**, 307–316.

105. M. W. Hosseini and J.-M. Lehn, *Helv. Chim. Acta*, 1987, **70**, 1312–1319.

106. M. W. Hosseini, J.-M. Lehn, and M. P. Mertes, *Helv. Chim. Acta*, 1983, **66**, 2454–2466.

107. M. W. Hosseini, J.-M. Lehn, and M. P. Mertes, *Helv. Chim. Acta*, 1985, **68**, 818.

108. M. W. Hosseini, J.-M. Lehn, L. Maggiora, *et al.*, *J. Am. Chem. Soc.*. 1987, **109**, 537–544.

109. G. M. Blackburn, G. R. J. Thatcher, M. W. Hosseini, and J.-M. Lehn, *Tetrahedron Lett.*, 1987, **28**, 2779–2782.

110. R. C. Bethell, G. Lowe, M. W. Hosseini, and J.-M. Lehn, *Bioorg. Chem.*, 1988, **16**, 418–428.

111. A. Bencini, A. Bianchi, E. García-España, *et al.*, *Bioorg. Chem.*, 1992, **20**, 8–29.

112. A. Andrés, C. Bazzicalupi, A. Bencini, *et al.*, *J. Chem. Soc. Perkin Trans. 2*, 1994, 2367–2373.

113. A. Bencini, A. Bianchi, C. Giorgi, *et al.*, *Inorg. Chem.*, 1996, **35**, 1114–1120.

114. J. A. Aguilar, A. B. Descalzo, P. Díaz, *et al.*, *J. Chem. Soc. Perkin Trans. 2*, 2000, 1187–1192.

115. M. W. Hosseini and J.-M. Lehn, *J. Chem. Soc. Chem. Commun.*, 1985, 1155–1157.

116. P. G. Yohannes, M. P. Mertes, and K. B. Mertes, *J. Am. Chem. Soc.*, 1985, **107**, 8288–8289.

117. P. G. Yohannes, K. E. Plute, M. P. Mertes, and K. B. Mertes, *Inorg. Chem.*, 1987, **26**, 1751–1758.

118. K. Jahansouz, Z. Jiang, R. H. Himes, *et al.*, *J. Am. Chem. Soc.*, 1989, **111**, 1409–1412.

119. M. W. Hosseini and J.-M. Lehn, *J. Chem. Soc. Chem. Commun.*, 1991, 451–453.

120. C. Bazzicalupi, A. Bencini, A. Bianchi, *et al.*, *Chem. Commun.*, 2005, 2630–2632.

121. Y. Zhou, Z. Xu and J. Yoon, *Chem. Soc. Rev.*, 2011, **40**, 2222–2235.

122. (a) R. Aucejo, J. Alarcón, C. Soriano, *et al.*, *J. Mater. Chem.*, 2005, **15**, 2920–2927; (b) R. Aucejo, P. Díaz, E. García-España, *et al.*, *New J. Chem.*, 2007, **31**, 44–51.

123. A. B. Descalzo, M. D. Marcos, R. Martínez-Máñez, *et al.*, *J. Mater. Chem.*, 2005, **15**, 2721–2731.

124. J. Y. Kwon, N. J. Singh, H. N. Kim, *et al.*, *J. Am. Chem. Soc.*, 2004, **126**, 8892–8893.

125. W. J. Belcher, M. Fabre, T. Farhan, and J. W. Steed, *Org. Biomol. Chem.* 2006, **4**, 781–786.

126. S. K. Kim, B.-S. Moon, J. H. Park, *et al.*, *Tetrahedron Lett.*, 2005, **46**, 6617–6620.

127. A. Ojida, I. Takashima, T. Kohira, *et al.*, *J. Am. Chem. Soc.*, 2008, **130**, 12095–12101.

128. Z. Xu, N. J. Singh, J. Lim, *et al.*, *J. Am. Chem. Soc.* 2009, **131**, 15528–15533.

129. (a) Z. Zeng and L. Spiccia, *Chem. Eur. J.*, 2009, **15**, 12941–12944; (b) Z. Zeng, A. J. Torriero, A. M. Bond, and L. Spiccia, *Chem. Eur. J.*, 2010, **16**, 9154–9163.

130. F. Schmidtch, S. Stadlbauer, and B. König, *Dalton Trans.*, 2010, **39**, 7250–7261.

131. P. Neelakandan, P. C. Nandajan, B. Subymol, and D. Ramaiah, *Org. Biomol. Chem.*, 2001, **9**, 1021–1029.

132. S. Yamaguchi, I. Yoshimura, T. Kohira, *et al.*, *J. Am. Chem. Soc.*, 2005, **127**, 11835–11847.

133. S. W. Thomas, G. D. Joly, and T. M. Swager, *Chem. Rev.*, 2007, **107**, 1339.

134. Z. Yao, X. Feng, W, Cheng, *et al.*, *Chem. Commun.*, 2009, 4696–4698.

135. (a) H.-J. Schneider, L. Tianjun, and N. Lomadze, *Angew. Chem. Int. Ed.*, 2003, **42**, 3544–3546; (b) H.-J. Schneider, L. Tianjun, and N. Lomadze, *Eur. J. Org. Chem.*, 2006, 677–692.

Receptors for Zwitterionic Species

Santiago V. Luis[1], Ignacio Alfonso[2], and Francisco Galindo[1]
[1] *University Jaume I, Castellón, Spain*
[2] *IQAC-CSIC, Barcelona, Spain*

1 INTRODUCTION

The appropriate design of receptors to selectively interact with a given zwitterionic species requires a careful balance between a complex set of factors, some of which seem to act in opposite directions. A reasonable structure developed for this purpose should be a ditopic receptor containing two well-defined sites, one being able to interact with cations and the other being able to interact with anions. Each of the sites needs to be optimized for its specific function, but, at the same time, they need to be properly located within the corresponding molecule, providing relative orientations, distances, and shapes matching those of the zwitterionic species to be recognized. Taking this generic design into consideration, it can be expected that both recognition sites in such a receptor would display complementary features. Thus, an important factor to be solved is that the intramolecular interaction between those complementary fragments should never compete with the interaction with the corresponding betainic guest. Of course, the same applies to the zwitterionic guest itself. In the same way, a second concern comes from the possibility of dimerization or oligomerization of either the receptor or the betainic substrate competing with the recognition process. Those alternative interaction possibilities are illustrated by the general scheme depicted in Figure 1. Thus, a proper design effort has also to be allocated to the fragment or fragments connecting the two complementary sites in the receptor, in order to disfavor all those competing pathways.

The occurrence of those alternative interaction pathways has been studied in detail by Schmuck and coworkers, using relatively simple heterocycle-based guanidinium-carboxylate zwitterions. Their results not only illustrate the key importance of those processes when considering ditopic systems having complementary sites, but also clearly highlight the critical role that structural elements different from the primary recognition sites can have in determining the final outcome. Intramolecular interactions leading to self-folding molecules were studied for pyrrole-derived structures such as **1** and **2** (Figure 2).[1] Molecular folding associated with the strong interaction between the two complementary carboxylate and guanidinium sites was observed for **1b** ($n = 4$) in DMSO (dimethyl sulfoxide) and for **2** in methanol. No intermolecular interactions leading to the formation of dimers or polymers were detected. Although a complex set of additional intramolecular interactions also seems to be present, ion pairing is characterized as the main factor determining folding. Thus, when the

Supramolecular Chemistry: From Molecules to Nanomaterials.
Edited by Philip A. Gale and Jonathan W. Steed.
© 2012 John Wiley & Sons, Ltd. ISBN: 978-0-470-74640-0.

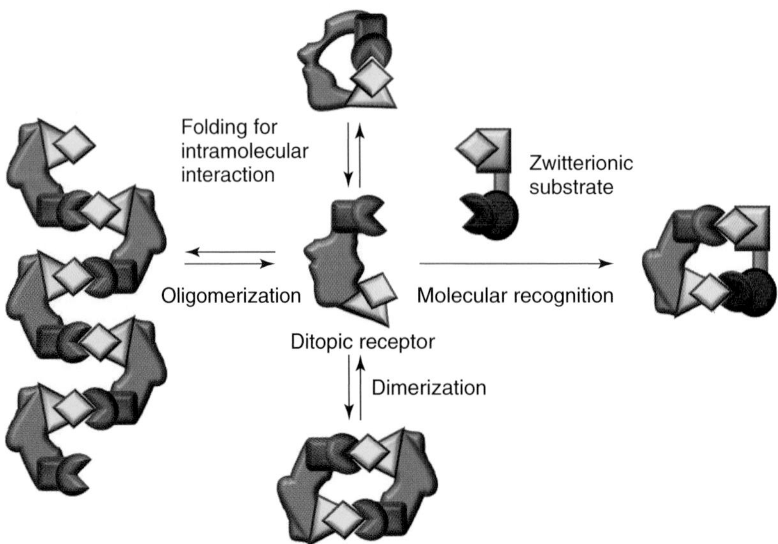

Figure 1 Receptor–guest interaction versus intramolecular site–site or intermolecular receptor–receptor interactions.

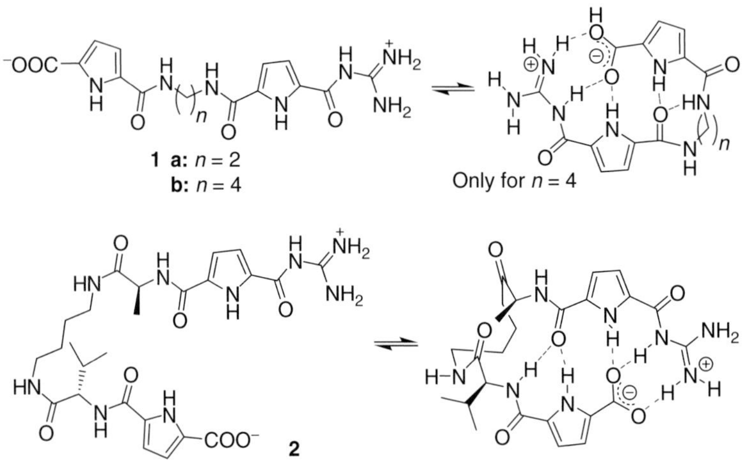

Figure 2 Examples of flexible pseudopeptidic structures that fold to favor the ion pairing between the chain ends with complementary charges.

precursor of **2** containing the protected carboxyl (Bn) and guanidinium (Boc) groups was studied, no folding but only a weak intermolecular self-association, most likely leading to a dimer, was detected.[2] It is important to note that the folding process requires the presence of a flexible enough spacer and accordingly folding cannot be detected for **1a**, for which the aliphatic spacer has been reduced from four to two methylene groups.[1]

The self-assembly of zwitterions to form stable dimers could be observed with the simple pyrrole-derived structure **3** and with the indole-derived zwitterionic analog **4** (Figure 3). The design of **3** involves the presence of a triethyleneglycol chain, whose only significant role is to improve the solubility of the resulting compound.[3] In both cases, the dimerization process was studied using NMR (nuclear magnetic resonance) and/or MS (mass spectrometry) techniques and the corresponding solid-state structures for the dimers could be determined by X-ray diffraction.[3,4] Dimerization constants determined by NMR for **3** in polar solvents were surprisingly high, ranging from $10^{10}\,M^{-1}$ (estimated) in DMSO to $170\,M^{-1}$ in water. This strong interaction is associated with the ion pairing between the two complementary bonding groups. Thus, substituting the guanidinium fragment by a pyridine group leads to the formation of a dimer containing a similar H-bonding pattern but lacking the ion-pair interaction.[3] In this case, the dimer is stable only in the low-polarity solvents with $K < 10\,M^{-1}$ in 5% DMSO in $CDCl_3$. Zwitterion **4** also forms less stable dimers than **3** ($K > 10^5$ in DMSO), and this has been associated with the lower acidity of its NH groups and with the lack of planarity in the structure of its dimer. This last factor involves an unfavorable steric interaction between

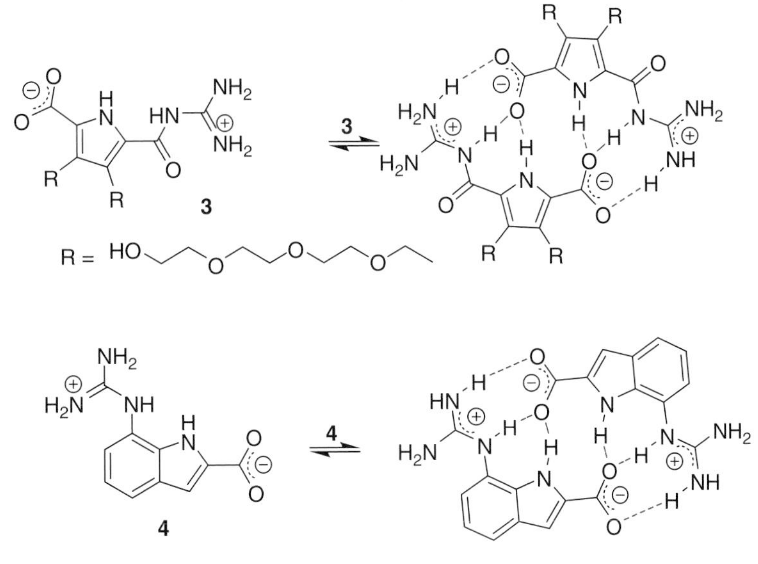

Figure 3 Zwitterionic structures that self-associate to form dimers.

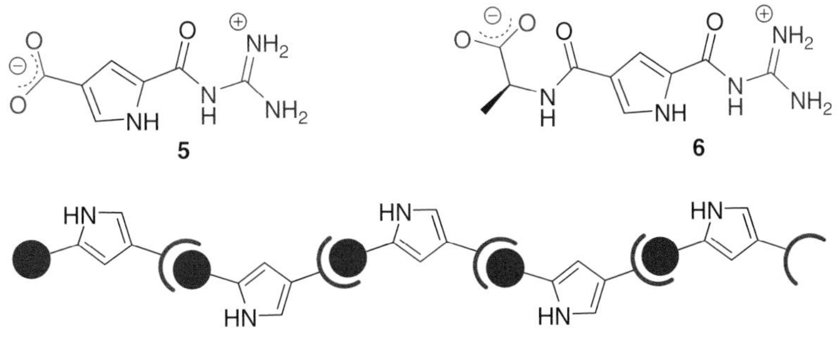

Figure 4 Oligomerization of zwitterion-containing self-complementary sites.

the guanidinium group and the vicinal C–H fragment of the indole group.[4]

Finally, oligomerization to form extended linear-like chains has been observed with the pyrrole derivative **5** (Figure 4).[5] This compound is related to **3**, but the change in the position of the carboxylate hinders the self-assembly to form dimers and only the oligomerization process is allowed as the rigid spacer between the two self-complementary sites does not permit the folding. NMR determination of K_{ass} at different temperatures (values range from 22 M^{-1} at 303 K to 71 M^{-1} at 363 K in DMSO) reveals that the association process is endothermic and, accordingly, entropically driven ($\Delta H = 17.6\,kJ\,M^{-1}$; $\Delta S = 83.4\,J\,K^{-1}\,M^{-1}$ in DMSO). The insertion of an L-alanine fragment between the pyrrole and the carboxylate end in **6** also hinders the dimerization and the intermolecular oligomerization is again favored.[6] Here, however, the self-association process seems to be more complex. At concentrations above 10 mM, compound **6** self-associates to form soft vesicles in DMSO in spite of the fact that it lacks the amphiphilic character usually associated with this behavior.

The same variety of structures that is found in the construction of receptors for cations or anions can be found in the case of the receptors for zwitterionic species. Nevertheless, this is a field that has been much less studied, not only because of the above-mentioned design difficulties but, in particular, also because of the relatively reduced number of zwitterionic species that have been explored. The most obvious examples of such species can be easily found in the biological world. Most likely, amino acids along with some related species including some peptides represent a very significant percentage of the total species considered up to now in this area.

According to the former, the organization of this chapter has not considered a classification based on the nature of the zwitterionic species. On the contrary, the following discussion is mainly based on the structures of the receptors being used. It is obvious that, as mentioned above, an optimized structure of this class should be ditopic, containing a site for the interaction with the cationic part of the zwitterions and a site for the interaction with the anionic part. In most cases, however, we can find a predominant structure, optimized for the interaction of either the cationic or the anionic part, to which an additional functionality has been added, connected through a given spacer, which is able to interact with the other charged moiety of the betaine.

2 RECEPTORS CONTAINING CROWN ETHERS AND RELATED STRUCTURES

The discovery of the interaction of crown ethers with alkali and alkali-earth cations by Pedersen[7] is at the origin of the field to which the name of Supramolecular Chemistry, as suggested by Lehn,[8] has been assigned. Very soon, after this seminal work, many researchers found that crown ethers were also able to interact efficiently, and even selectively depending on different structural parameters, with ammonium groups, in particular, those formed through protonation of primary amines. Taking their biological relevance into consideration, it is not surprising that the interaction of crown ethers with amino acids and some derivatives was also studied at the earlier stages of the development of the chemistry of crown ethers. Many, if not most, of those initial studies were carried out in organic solvents and involved the use of the ammonium salts of the corresponding amino acid esters. Examples, however, can be found, in which zwitterionic nonderivatized amino acids were directly studied and even efficiently extracted to organic phases through the formation of the appropriate supramolecular complex. Those receptors had not been specifically designed for the interaction with the amino acids as zwitterions and were essentially monotopic systems, lacking the essential design considerations briefly described in the introduction. According to this we do not describe those systems in more detail.

Here we concentrate on the receptors containing a crown-ether fragment, to which an additional functional moiety, connected through the appropriate spacer, has been specifically added for the interaction with the anionic part of the zwitterions. Podands and aza or polyaza crown ethers have also been used and should be considered here, when appropriate. In particular, the use of aza or polyaza crown ethers is very common as the presence of the nitrogen atoms greatly facilitates the synthetic effort for the introduction of the spacer or the complementary functionality.

2.1 Ammonium group as anion receptor

Many different structural fragments have been described for the interaction with anions, but one of the simplest ones is an ammonium group. Figure 5 contains some examples of receptors for zwitterionic species based on a molecule having a crown-ether fragment and an ammonium group.

Compound **7**, developed by Barboiu as a receptor for zwitterionic α-amino acids,[9] contains a dibenzo 18-crown-6 unit as the receptor site for the ammonium group, while a simple primary ammonium group, separated from the crown subunit by a long aliphatic spacer, is the site for the interaction with the carboxylate group of the amino acid. According to the protonated nature of the ammonium group, the behavior of this receptor is pH sensitive and care needs to be taken to guarantee its full protonation at

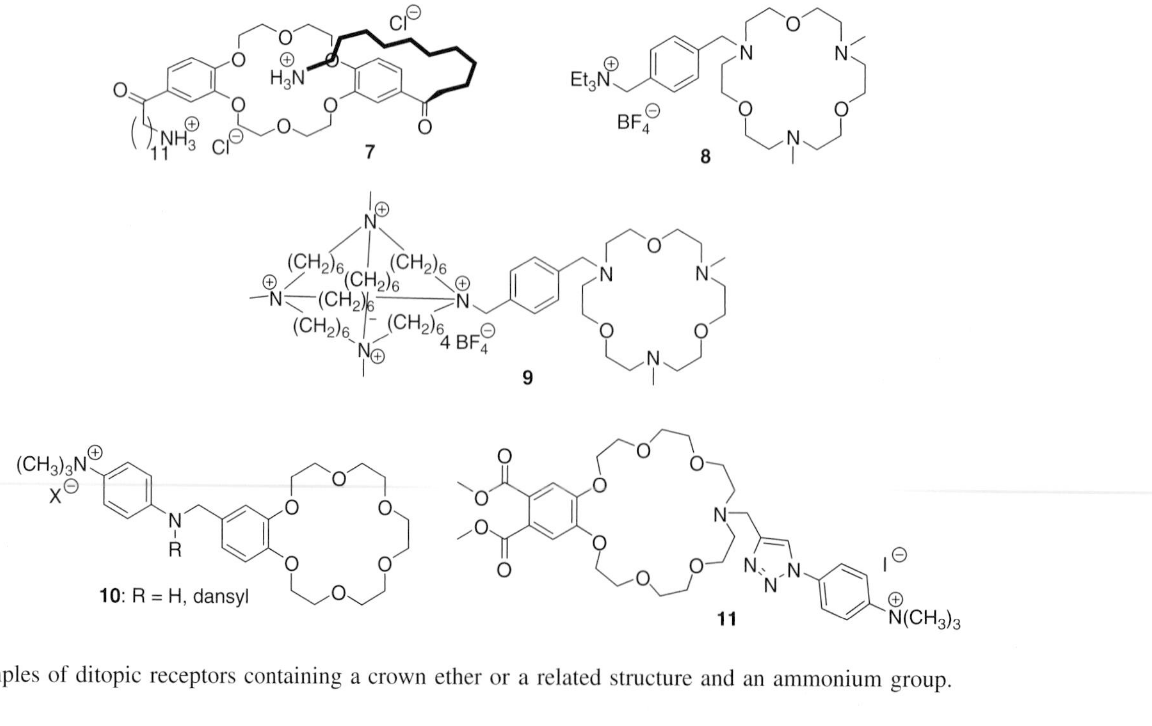

Figure 5 Examples of ditopic receptors containing a crown ether or a related structure and an ammonium group.

the isoelectric point of the amino acid. On the other hand, because of the full complementarity of the two recognition sites in the receptor and the lack of any steric or spatial constraint, the interaction of **7** with the guests needs to compete with the intramolecular process of ammonium–crown interaction, as discussed above. Taking into consideration that quaternary ammonium salts do not significantly interact with crown ethers, a logic alternative to avoid this shortcoming is to substitute the primary ammonium group by a quaternary one. This has been achieved in receptors **8–10**. Compound **8**, containing a triaza-18-crown-6, was developed by Schmidtchen also as a receptor for ω-amino acids.[10] Although the strength of the interaction, in 90% v/v aqueous methanol, showed not to be sensitive to the length of the spacer connecting both interaction sites in the ω-amino acids (selectivity γ-amino butyric acid (GABA)/ω-amino pentenoic acid ≈ 1.2) owing to the large flexibility of the receptor, it selectively recognized those over some ω-amino alcohols or phenols (selectivities in the order of 2). The related receptor **9**, containing a much more complex ammonium group, was shown, by the same authors, to have a slightly lower affinity for zwitterionic amino acids ($K_{ass} \approx 250\,M^{-1}$ for GABA vs 360 M^{-1} in the case of **8**) although displaying some increased selectivity over the corresponding amino alcohols. Receptor **10** was designed by Schneider for the selective recognition of zwitterionic peptides in water.[11] Some modularity is present in **10** through the modification of the nature of the pendant group R. By changing from R = H to R = dansyl, they could follow the recognition process not only by NMR studies but also by fluorescence techniques. The authors found that, according to their design parameters, both receptors were selective for tripeptides, with the interaction with similar tetrapeptides being one order of magnitude weaker ($K = 200\,M^{-1}$ for Gly-Gly-Gly, 50 M^{-1} for Gly-Gly, and 45 M^{-1} for Gly-Gly-Gly-Gly). The importance of ion pairing in a bidentate binding was demonstrated with the use of the methyl triglycylester (Gly-Gly-Gly-OCH_3), for which the value of the association constant was reduced to $K = 30\,M^{-1}$. The introduction of the dansyl (DNS) group affords a receptor that is able to show a clear side chain selectivity for different tripeptides. Thus, higher association constants were found for tripeptides containing amino acids with large lipophilic side chains in the central position. This suggests a participation of the DNS group through lipophobic/solvophobic interactions. As could be expected, for the possibility of aromatic–aromatic interactions, this effect is maximized for tripeptides containing Phe or Trp in the central position. In this case, the corresponding constants are 1 order of magnitude bigger than those for other tripeptides ($K = 2150\,M^{-1}$ for Gly-Trp-Gly). The conceptually related system **11**, prepared by König in a broader context (see below), containing a crown and a quaternary ammonium group linked through an aromatic spacer was also shown to selectively interact with tripeptides over dipeptides or ω-amino acids.[12] The association constant, determined by UV titrations in methanol/water 9/1 for Gly-Gly-Gly, was only moderate (600 M^{-1}) but significantly higher than the ones obtained for Gly-Gly (200 M^{-1}), ε-aminohexanoic acid (AHX) (200), or GABA (<100 M^{-1}).

In some instances, the substitution of the amino/ammonium group by a protonated nitrogen heterocycle has been used to study the interaction with zwitterionic species. Changes in the UV–vis or fluorescent signals derived from the heterocyclic subunit have been used to study the corresponding recognition process. Some examples are given in Figure 6. Thus, receptor **12** contains an acridinium subunit that along with the thiourea fragments can participate in the binding to the carboxylate group of zwitterionic amino acids.[13] Experimental studies by NMR, UV–vis, and fluorescence spectrophotometry in DMSO revealed that **12** selectively interacts with aromatic α-amino acids, and in particular with Trp (K_{ass} for Trp 3157 M^{-1}, 900 M^{-1} for Ala). On the basis of the gathered data, the authors suggest a structure for the complexes in which electrostatic interactions of the carboxylate moiety with the acridinium ring and hydrogen bonding with the acridine and thiourea protons exist, but the ammonium group does not interact with the oligoethylene chain. Instead, $RNH_3^+ - \pi$ interactions are observed, and the oligoethylene chain merely acts as a hydrophobic cavity to accommodate the aromatic side chain. This explains that when receptor **13**, lacking the oligoethylene fragment, is studied, very similar results are obtained (K_{ass} for Trp 2873 M^{-1}), although the values of the association constants for the aromatic amino acids are slightly smaller,

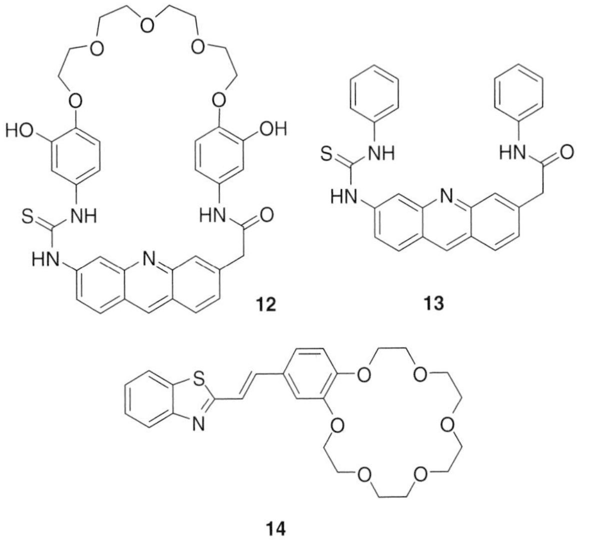

Figure 6 Receptors containing a crown ether or a related structure and a protonable heterocycle.

being relatively higher than those of the aliphatic amino acids, giving place to a reduced selectivity. Compound **14** was prepared as a potential receptor for ω-amino acids, considering that under conditions providing protonated amino acids, the H-bonding interaction of the heterocyclic nitrogen with the proton of the carboxylic group would provide a hydrogen transfer to produce a protonated receptor and a zwitterionic amino acid.[14] Nevertheless, the results obtained through fluorescence experiments in CH_3CN revealed that this expectation is not completely fulfilled. No selectivity was observed for the amino acids studied ($^-OOC-(CH_2)_n-NH_3^+$, $n = 2$, 10) and the values of the constants were similar to those determined for simple ammonium cations, which indicates that a monotopic interaction of the ammonium group with the B18C6 cavity is the main process involved.

2.2 Guanidinium fragments for anion interaction

Guanidinium salts are ideally suited for the interaction with carboxylates. On the one hand, they are easily protonated even under conditions that do not allow the protonation of an amine, thus guaranteeing the ion-pair interaction over a wide pH range. On the other hand, the geometry of the guanidinium groups fits nicely for the simultaneous formation of two N–H···O hydrogen bonds with a carboxylate group. This explains that a large variety of receptors, for amino acids and related species, that contain a crown ether and a guanidinium group have been designed. Some of those structures are depicted in Figure 7.

Compound **15a** was initially designed by de Mendoza as a three-point binding receptor for the enantioselective recognition of α-amino acids.[15] The basic elements for the design involved the presence of a guanidinium function as a carboxylate receptor, an aza-18C6 as the ammonium receptor, and a planar aromatic structure (the naphthaloyl group) for the interaction with the side chains of the amino acids. The chirality was provided by the bicyclic guanidinium fragment. Because of its low solubility in water, initial experiments were carried out by liquid–liquid single extraction involving an aqueous phase containing a single amino acid or a mixture of amino acids and a CH_2Cl_2 phase containing receptor **15a**. Those experiments revealed that this compound was able to selectively interact with amino acids having aromatic side chains, in particular, Phe and Trp. On the other hand, only the L-enantiomers were extracted when the configuration of the guanidinium

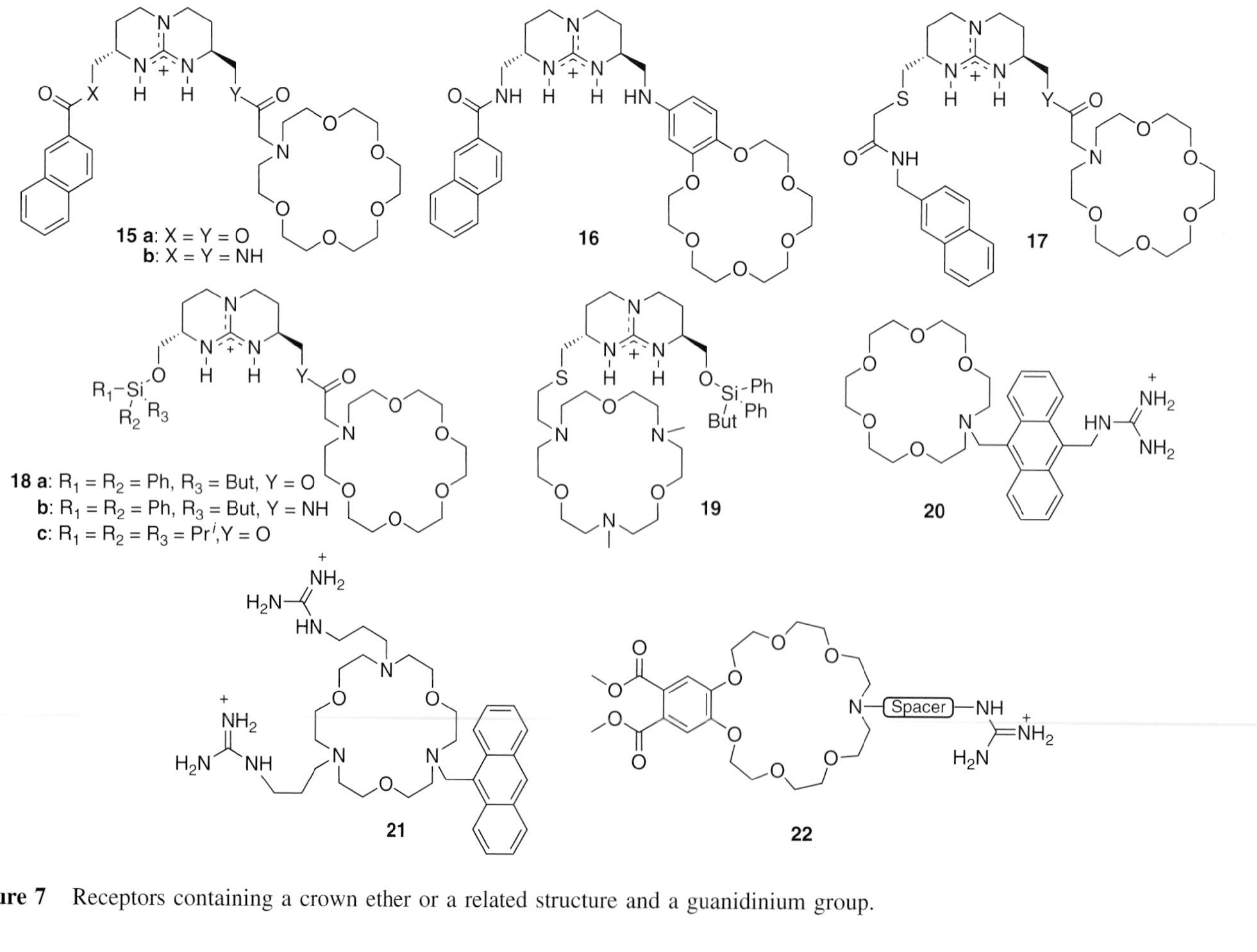

15 a: X = Y = O
b: X = Y = NH

16

17

18 a: $R_1 = R_2 = Ph$, $R_3 = Bu^t$, Y = O
b: $R_1 = R_2 = Ph$, $R_3 = Bu^t$, Y = NH
c: $R_1 = R_2 = R_3 = Pr^i$, Y = O

19

20

21

22

Figure 7 Receptors containing a crown ether or a related structure and a guanidinium group.

fragment was the one shown in the Figure. In a further attempt to optimize the structure of this receptor and to fully understand the selective recognition process, the same group prepared a large variety of related structures and studied all of them in transport experiments across a synthetic organic membrane (U-tube cell) of 1,2-dichloroethane or dichloromethane.[16] The structural modifications considered involved the use of amide instead of ester groups (**15a** and **b**), the use of other crowns such as B18C6 (**16**), the introduction of longer spacers between the guanidinium fragment and the crown or the aromatic moiety as in **17,** and, finally, the substitution of the naphthaloyl fragment by a silyloxy group (**18**). The general trends observed were similar to those found in extraction experiments. Amides were shown to be better carriers than esters but, on the contrary, they were less enantioselective. Best results were obtained with the original structure **15a** and with **18a**, allowing the efficient transport of Trp and Phe with enantioselectivities close to 80% *ee* under optimized conditions. It must be noted that those studies, including receptors lacking the fragment for the interaction with the side chain of the amino acid, as well as detailed calculation revealed that the three-point interaction model was unnecessary to explain the high enantioselectivity achieved. The formation of the two ion-pairs seems to be the major contribution for the enantioselectivity, while the other substituent (naphthaloyl or silyloxy) plays a role in increasing transport rates and improving selectivity. The related receptor **19**, prepared by Schmidtchen, was reported to extract even small hydrophilic amino acids such as Ser or Gly much more efficiently than **15a**.[17] Nevertheless, enantioselectivity was only moderately achieved in the case of Phe (40% *ee*).

The introduction of an anthracene fragment in receptors **20** and **21** was considered in order to associate a fluorescence response with the recognition event. In both cases, the fluorophore is bound to a nitrogen atom of an aza or polyaza crown through a methylene group and accordingly the nitrogen atom can participate in a PET (photoinduced electron transfer) process leading to the deactivation of the fluorescence. Thus, the interaction of a guest with the aza crown can produce a partial recovery of this fluorescence and the measurement of the corresponding FE (fluorescence enhancement) factors can be used to assess the selectivity of the interaction. In the case of the receptor **20** reported by de Silva, the studies carried out in methanol–water (3 : 2 v/v) showed a clear length selectivity for ω-amino acids ($^{-}OOC{-}(CH_2)_n{-}NH_3^{+}$).[18] Although the response to GABA ($n = 3$) was not the higher one, it deserves to be mentioned that this biologically important compound produces a significant fluorescent enhancement (FE = 2.2) while glutamic acid, its physiological precursor, or Gly provides essentially no response (FE 1.1 and 1.2). The compound **21** contains two guanidinium fragments attached to a triaza-18C6 cavity, but showed trends similar to those of **20**.[19] From the different guests studied, the selectivity order was GABA > Lys > n-$BuNH_3^{+}$ > Gly (FE values 2.2, 2.1, 1.6, and 1.3), while other amino acids examined did not show any fluorescence response. In both cases, the studies were carried out at basic pH values (pH = 9.4–9.5) to avoid that protonation of the nitrogen atoms of the aza crown could compete in the generation of a FE, but maintaining, at the same time, the zwitterionic character of the amino acids.

Very recently, König has prepared and studied a series of ditopic receptors with the general structure **22**, containing an aza-B21C7 as the site for binding ammonium and a substituted or unsubstituted guanidium as the site for binding carboxylate. Using these compounds, the studies could be carried out at pH values closer to the physiological ones. Although the selectivities found were not high enough for practical applications, some of the compounds prepared showed a moderate selectivity for GABA over other related species.

2.3 Miscellaneous groups for the interaction with the anionic fragment

Finally, Figure 8 gathers some additional structures containing ditopic receptors containing crown or aza crown fragments. Thus, receptor **23** combines a 15C5 structure and a phenyl boronic fragment for the recognition of the carboxylate functionality and was shown to increase the rate of transport of Phe through a $CHCl_3$ membrane by a factor of 352.[20] In compound **24**, the carboxylate binding site is a urea moiety associated to a B18C6 macrocycle.[21] Extraction and transport studies with **24** and α-amino acids revealed that this receptor is only moderately selective. Amino acids with aromatic or large aliphatic side chains (lipophilic amino acids) are those that are more efficiently transported. Pyrilium, thiapyrilium, and pyridinium cations have also been used as the subunit for the interaction with the negative part of a zwitterionic ion, with the added advantage that their fluorescent properties can be modified through this interaction. Structures **25** and **26** represent illustrative examples of this family of receptors.[22,23] Fluorescence studies of those compounds in methanol with different amino acids showed that a good selectivity could be achieved in some instances. Thus, **25** was selective toward lysine in the presence of glutamine and asparagine, while **26** was selective for asparagine. Large polyaza macrocycles act, in many cases, as ditopic or even polytopic receptors. This is the case of macrobicyclic (or tricyclic) polyamines containing pyrazole subunits related to structure **27** and studied by Navarro and García-España. In this regard, the receptors of this family (i.e., **27**, R = H, Bn) were shown

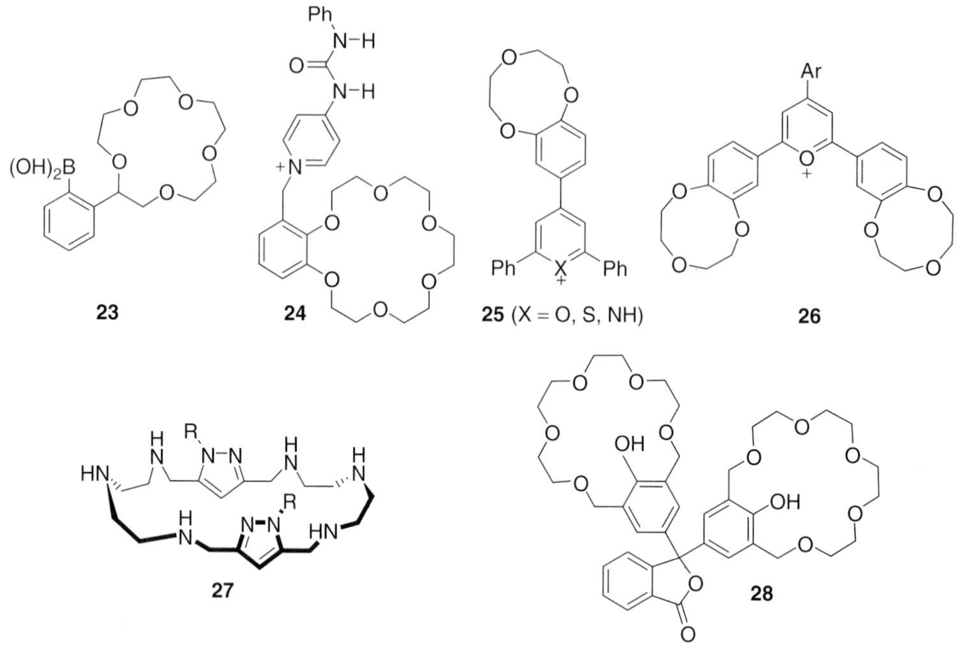

Figure 8 Miscellaneous receptors for zwitterionic species containing a crown-ether fragment.

to be able to interact in water with the zwitterionic form of dopamine, with stability constants (log K) ranging from 3 to 6.[24] Obviously, the interaction is stronger with the Cu_2L complex species, but, as we see below, this is a common strategy to develop sites for the efficient binding of the anionic part of zwitterions. The same family of macrocycles has been shown to efficiently act as receptors for L-glutamate in water at physiological pH values.[25] The receptor displaying the largest interaction at pH 7.4, according to potentiometric titrations, is a derivative of **27** (R = H), N-benzylated at the two nitrogen atoms that are located at a higher distance of the pyrazole rings ($K_{eff} = 2.04 \times 10^4$). A very particular case is that of compound **28**, containing two crown-ether subunits and a phenolphthalein fragment, which is able to selectively develop a brilliant purple color in the presence of dipeptides containing a C-terminal Lys.[26]

3 RECEPTORS RELATED TO CYCLODEXTRINS

For zwitterionic species, such as peptides and amino acids, a number of reports have been described, both with natural CDs (cyclodextrins) and with synthetically modified hosts. In the former case, association constants are found to be very low, but can be substantially increased by a judicious functionalization of the CD. Thus, β-CD binds D-Trp in water forming a weak 1 : 1 complex ($K_{ass} = 13\,M^{-1}$, $\Delta G = -1.5\,kcal\ mol^{-1}$) and the binding process is mainly driven by the entropy change ($\Delta S = +3.0\,cal\ K^{-1}\,mol^{-1}$ vs $\Delta H = +0.3\,kcal\ mol^{-1}$).[27] Functionalization with ammonium and carboxylate groups, as shown in **29a** (Figure 9), leads to a negligible increase in the binding strength ($K = 15\,M^{-1}$, $\Delta G = -1.6\,kcal\ mol^{-1}$), but when the introduction of polar groups is accompanied by the presence of an apolar environment surrounding such groups (like in **29b**, Figure 9), the association constant rises to 54 M^{-1} ($\Delta G = -2.63\,kcal\ mol^{-1}$), which is still a very moderate value. A thermodynamic analysis of the polar interaction energies leads to the conclusion that even in the most favorable case ($\Delta H = \sim -5\,kcal\ mol^{-1}$), ionic interactions for CDs in water are far less important than the London's dispersion forces (accounting for about $\Delta H = -10\,kcal\ mol^{-1}$).

This early result by Tabushi is illustrative of the trends followed by research in the following years, introducing additional binding points of coordinative nature (i.e., metallic complexes) or new CD units (leading to bis-CDs) to enhance the binding ability of CDs toward zwitterionic species. Substructures introduced into the CD rims (not forming organometallic complexes) include sulfonylamines,[28] sulfonic acids,[29] esters,[29] and organoselenium pendant arms.[30]

As anticipated, a great enhancement of the interaction was achieved by the introduction of Cu(II) complexes as one of the binding sites of the host, as reported by Marchelli and Rizzarelli.[31] CD **30**, bearing an arm containing a Cu(II) complex with histamine, forms a ternary complex with free amino acids in water, with high stability constants and a good degree of chiral recognition. Aromatic D-amino

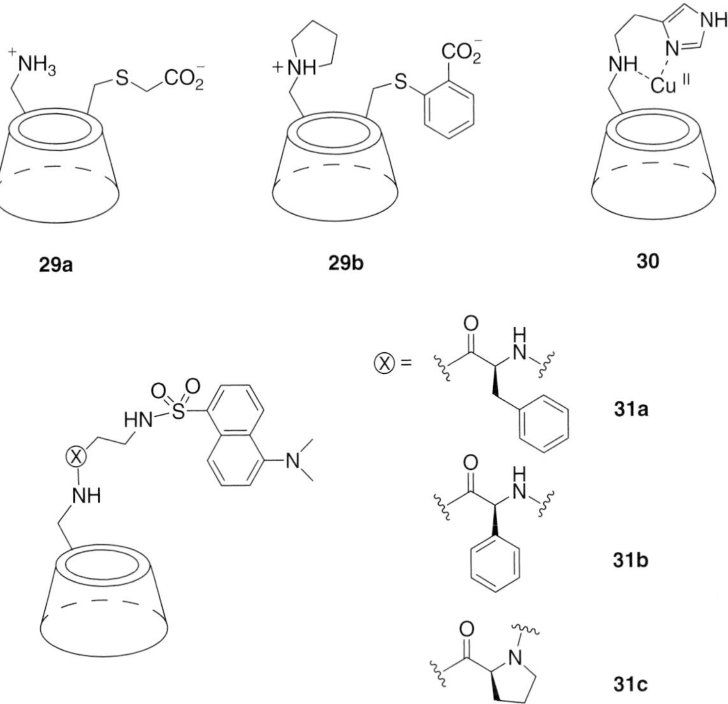

Figure 9 Selected examples of cyclodextrins used as receptors for amino acids.

acids are better recognized than the L counterparts by **30**, leading to the hypothesis of an inclusion of the side chain of the D substrates into the CD cavity. The enthalpy gain displayed by such D-isomers on inclusion into the cavity (−1 to −2 kcal mol^{-1} in comparison to the L-isomer) is accompanied by a loss of conformational freedom, inducing an entropic penalty which is lower for the L-isomers ($\Delta\Delta S_{D,L} = -5$ cal K^{-1} mol^{-1} for Trp and −2 cal K^{-1} mol^{-1} for Phe). Overall, the process is enthalpically driven. Chiral separation of Trp, Tyr, and Phe was accomplished by adding **30** to solutions of the racemates and using an achiral C_{18} column.

One interesting evolution of this type of organometallic CD has been the receptors **31**, bearing a fluorescent reporter such as DNS as a part of the ligand coordinating the metallic center.[32,33] According to NMR, ESI-MS (electrospray ionization-mass spectrometry), and time-resolved fluorescence measurements, the DNS chromophore should be buried into the CD cavity when the host is synthesized with a L-Phe in the side arm, whereas this is not favorable when the building block is a D-Phe. This difference makes the corresponding Cu(II) complexes good enantioselective receptors for several free amino acids, especially for the rigid proline. Thus, whereas the receptor from D-Phe shows an enantioselectivity factor of 0.77 for Pro, the most compact receptor from D-Phe discriminates Pro with a selectivity of 3.93. This has allowed the development of an analytical protocol for the enantiomeric analysis of mixtures, using a microplate reader with fluorescence detection.[34]

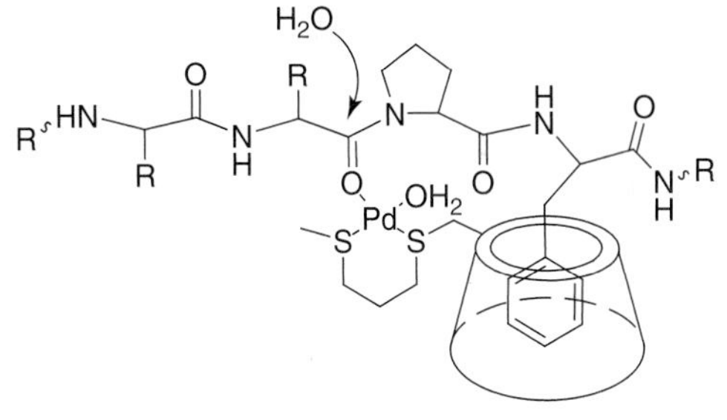

Figure 10 Catalytic function of supramolecular receptor **32**.

A related system has been developed, by Kostic, with the development of a conjugate of a Pd(II) complex and β-CD capable of acting as a biomimetic peptidase (**32** in Figure 10).[35] The ability of $[Pd(H_2O)_4]^{2+}$ to cleave internal X-Pro bonds in peptides has been utilized for the design of receptor **32**, in which the hydrophobic cavity of the CD accommodates the aromatic residue of an aromatic amino acid and the contiguous Pd(II) complex binds to the carbonyl of the neighboring His. This combination makes **32** an excellent host for X-Pro-Ar peptides and catalyzes the hydrolysis of the X-Pro bond with outstanding specificity for cleaving at a specific sequence. Thus, treatment of bradikynin (Arg-Pro-Pro-Gly-Phe-Ser-Pro-Phe-Arg) with **32** resulted in a specific cleavage of the Ser-Pro bond.

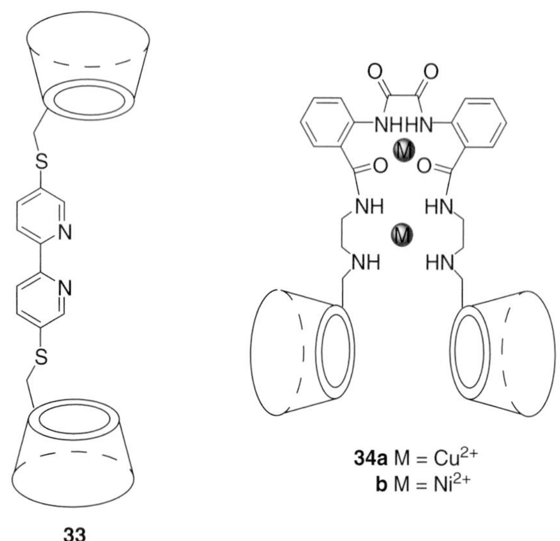

Figure 11 Bis-cyclodextrins developed as supramolecular receptors.

An antecedent of this peptidase mimic can be found in the bis-CD **33** developed earlier by Breslow (Figure 11).[36] The two cavities are involved in the recognition of a dipeptide, such as Trp-Trp, with two hydrophobic groups, with an association constant of $2.5 \times 10^6\,M^{-1}$, which is increased by one order of magnitude when Zn(II) is also bound to the bipyridyl central bridge. Although the Cu(II) complex of **33** catalyzes the hydrolysis of ester bonds in hydrophobic substrates fitting into the cavities of the CDs, the hydrolysis of the amide bond in Trp-Trp was not reported.

Other bis-CD receptors have been described for the recognition or sensing of amino acids and oligopeptides. One remarkable example is the case of receptor **34** (Figure 11).[37] This polyfunctional host moderately binds aliphatic di- and tripeptides, but the introduction of a metallic center such as Cu(II) or Ni(II) dramatically enhances the strength of the binding process. Thus, the free bis-CD binds to Glu-Glu with a log $K = 2.13$ ($\Delta G = -12.16\,\text{kJ mol}^{-1}$) but the complex **34a** recognizes Glu-Glu with an association constant that is two orders of magnitude higher (log $K = 4.25$, $\Delta G = -24.27\,\text{kJ mol}^{-1}$). An even stronger complex is formed with the host with two Ni(II) cations **34b** (log $K = 4.83$, $\Delta G = -27.59\,\text{kJ mol}^{-1}$). A systematic study by means of UV–vis, circular dichroism, fluorescence, and 2D NMR spectroscopy allowed rationalization of the results by proposing a cooperative binding of two CDs and an additional chelation interaction.

The development of a bis-CD with a functional tether capable of discriminating nonaromatic dipeptides without the help of any metallic center is also worth mentioning.[38] The complexation of Glu-Glu takes place with a $K = 590\,M^{-1}$ ($\Delta G = -15.8\,\text{kJ mol}^{-1}$) but Gly-Leu binds with $K = 16\,600\,M^{-1}$ ($\Delta G = -24.1\,\text{kJ mol}^{-1}$). Notably, the host is also sequence selective, since the related dipeptide Leu-Gly forms a much weaker complex ($K = 3300\,M^{-1}$, $\Delta G = -20.1\,\text{kJ mol}^{-1}$).

Finally, fluorescent Cu(II) metallobridged bis-CDs have recently been used for the development of analytical protocols that are aimed at the sensing of tripeptide glutathione, with good sensitivity (limit of detection 63.8 nM) and selectivity (no interference from the plasma constituents).[39]

4 CALIXARENE/RESORCINARENE-LIKE RECEPTORS

The calixarene/resorcinarene-like molecular structure is usually characterized by a medium- to large-sized macrocycle formed by aromatic rings joined through some of their carbon atoms (Figure 12). The most relevant structural peculiarity of these compounds is their conical-like shape with a large aromatic internal cavity (able to interact with hydrophobic and/or cationic moieties) and with one wider and one narrower rim (upper and lower rims, respectively). The functionalities on the rims (R) can modulate the binding abilities of the aromatic moieties and also interact with additional functional groups of the guests.

Considering these structural peculiarities, one can guess that most of the examples of molecular recognition of zwitterionic species using calixarene-type compounds are based on the main interaction between the inner cavity formed by the aromatic groups (π-basic surface) and the cationic part of the zwitterions, leading to an inclusion complex. Regarding the anionic moiety of the zwitterionic guests, this can interact by H-bonding with the substituents in any of the rims, or can be used for obtaining some kind of selectivity in the binding process. One of the seminal examples of these kinds of complexes was reported by Rebek and coworkers.[40] In that paper, the authors described the binding of carnitine to a deep cavitand (**35**, Figure 13), derived from a simple resorcinarene frame. The binding constant (determined by ^{1}H NMR in DMSO-d_6 at 298 K) was measured to be $15\,000\,M^{-1}$, corresponding to a $\Delta G^\circ = -5.6\,\text{kcal mol}^{-1}$. The NMR data, with the help of modeling studies, suggested the inclusion of the cationic

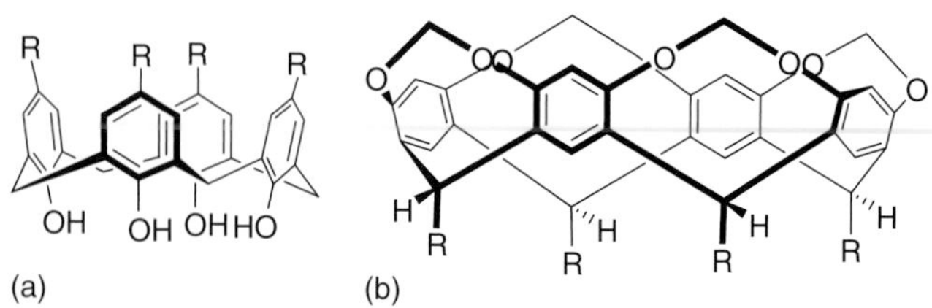

Figure 12 Chemical structures for calix- and resorcinarenes ((a) and (b), respectively).

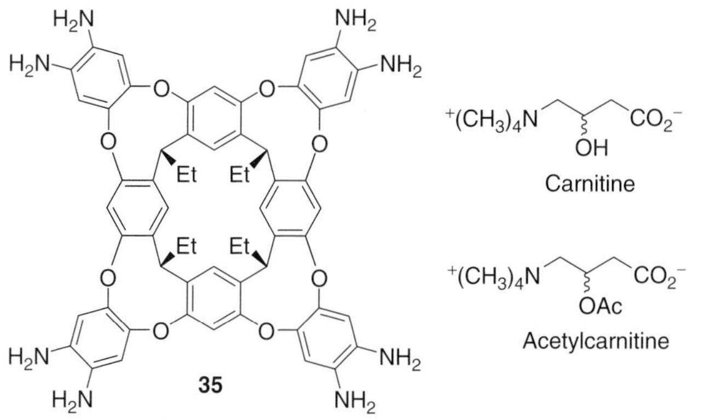

Figure 13 A deep cavitand host for carnitine guests.

tetraalkylammonium moiety within the inner space of the cavitand, which showed a vaselike conformation. Interestingly, protonation of the carboxylic moiety of carnitine reduced the binding interaction in 1.0 kcal mol^{-1}, which supports the binding of the anionic group of the carnitine guest with the pendant amino groups in the upper rim of the host. This is a solid proof for the recognition of the substrate in its zwitterionic form. Moreover, acetylation of the hydroxyl group of carnitine further decreased the binding (in 3.6 kcal mol^{-1}), which was ascribed by the authors to steric hindrance inside the cavitand space. Later on, a subsequent study from the same group deepened on the binding abilities of a wider family of these receptors toward different kinds of cationic guests, with carnitine being among them.[41] From this study, the authors concluded that apart from the complementary size, shape, and charge distribution, some other factors such as hydrophilicity and solvent polarity play a fundamental role in the binding process.

Probably, the calixarene-type compounds that are more deeply studied in the field of molecular recognition of zwitterions are the family of *p*-sulfonatocalix[*n*]arenes ($n = 4$, 6, and 8). In a seminal study, the binding constants between those compounds and Lys or Arg, as well as their corresponding di- and tripeptides, in aqueous buffer at pH = 8.0 were studied by NMR and titration microcalorimetry.[42] At this pH value, the carboxylic end should be in the anionic form, while the amino groups are partially protonated. Despite the fact that this work was not claimed as molecular recognition of zwitterions, some of the general conclusions are very interesting and must be commented in this context. The binding interactions are within the range of $K = 1000–40\,000\,M^{-1}$ and generally increase when the number of residues of the guest (and thus, the number of hydrophobic-cationic binding sites) also increases. For calix[4]arene sulfonate ($n = 4$), only complexes of 1 : 1 stoichiometry were detected in solution. Interestingly, this system was characterized in the solid state by single crystal X-ray diffraction as a 2 : 4 host : guest complex.[43] The binding was controlled by a favorable enthalpic contribution, mainly dictated by the inclusion of the nonpolar moiety of the guest into the host hydrophobic cavity, through van der Waals contacts. Besides, the favorable entropic contribution due to the desolvation of ionic groups of the guest on ionic interactions seems to be important in the recognition event. The presence of additional lysine residues (di- and tripeptides) did not afford significant changes. With $n = 4$, mixtures of complexes with 1 : 1 and 1 : 2 stoichiometries were detected and an even more complex behavior was observed for the higher order derivative ($n = 8$). The data showed the classical enthalpic–entropic compensation, which suggested that desolvation and folding of the guests must be playing important roles in the recognition process. More recently, the binding between *p*-sulfonatocalix[4]arene (**36**, Figure 14) and cationic moieties of zwitterionic guests has been used for the development of a fluorescent supramolecular sensor.[44] The system made use of the displacement method with 2,3-diazabicyclo[2.2.2]oct-2ene (DBO, Figure 14) as the fluorescent probe. The binding to L-carnitine was determined to have an association constant of 6300 M^{-1} at p$D = 2.4$, and it decreased to 1700 M^{-1} at p$D = 7.4$. Also in this case, the O-acetylation of the guest reduced the interaction ($K = 2900$ and 1400 M^{-1} at p$D = 2.4$ and 7.4, respectively). Although the system did not show very impressive results in terms of selectivity, the simplicity of the molecules used and the possibility of working in pure water, without added organic cosolvents, represented an interesting starting point for further development of similar fluorescent sensors for zwitterions under practical conditions.

Pyrogallol[4]arenes are also interesting potential guests for zwitterions, since they have a concave π-basic surface for the binding of cations and a polyhydroxylated upper rim for H-bonding to basic anionic functions. The binding processes of one of these guests (**37**) toward betaine and L-carnitine have been recently studied by isothermal titration calorimetry (ITC) in ethanolic solutions at 298 K (Figure 15).[45] Betaine rendered a binding constant of 3200 M^{-1}, while for L-carnitine the interaction was found to be much stronger ($K = 18\,000\,M^{-1}$). This

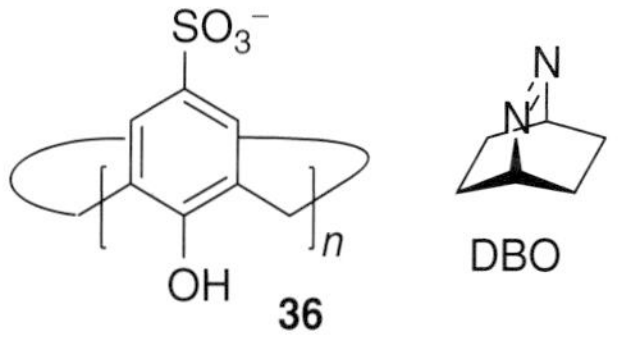

Figure 14 General formula of *p*-sulfonatocalix[n]arene receptors and DBO.

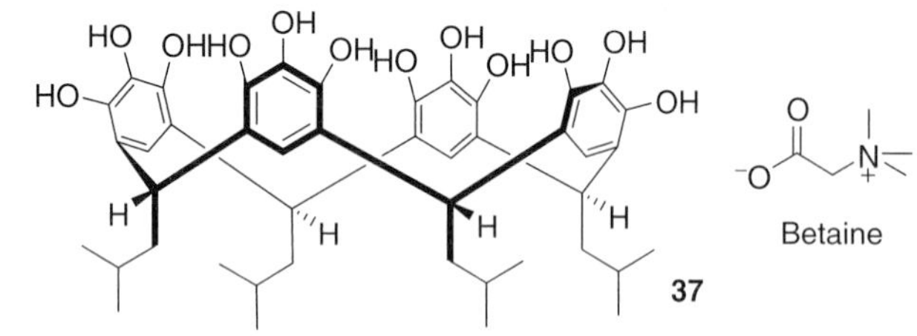

Figure 15 Chemical structure of a pyrogallol[4]arene host and the betaine guest.

result suggested that all the functional groups (trimethylammonium, carboxylate, and hydroxyl) as well as the additional methylenes contribute to the enhanced affinity by ion–dipole, cation–π, H-bonding, and van der Waals interactions, as observed by molecular modeling. The protonation of the carboxylate group in both guests reduced the binding constants, and the thermodynamic functions showed the typical enthalpy–entropy compensation. These data again suggested that the inclusion process in these systems is strongly affected by conformational and solvation effects.

5 CUCURBIT[*N*]URIL RECEPTORS

Cucurbit[*n*]urils ($n = 7$ and 8) (**38**) have also been used for the molecular recognition of zwitterionic species in aqueous medium (Figure 16). These systems show a hydrophobic cavity with the urea-type carbonyls pointing to the corresponding entrances. This peculiarity implies binding properties with some resemblance to other cavitands (CDs, calixarenes, or resorcinarenes). The size of the cucurbituril receptor can be controlled by the number of monomers present in the structure, and they are usually named as CB*n* (with *n* being the number of repeating units).

The stability of the corresponding supramolecular complexes is strongly dependent on the size of the CB (cucurbituril) host. Thus, the CB7 macrocycle formed strong complexes with carnitine and phosphorylated choline in water at neutral pH, where both guests are zwitterions.[46] The NMR data suggest the main inclusion of the trimethylammonium cation within the host cavity. The same host (CB7) also recognizes different amino acids, showing a good selectivity for the aromatic nonpolar ones.[47] This fact has been used to design supramolecular tandem enzyme assays for multiparameter sensor arrays and for enantiomeric excess determination of amino acids, taking advantage of the fluorescence properties of dapoxyl in a displacement assay approach.

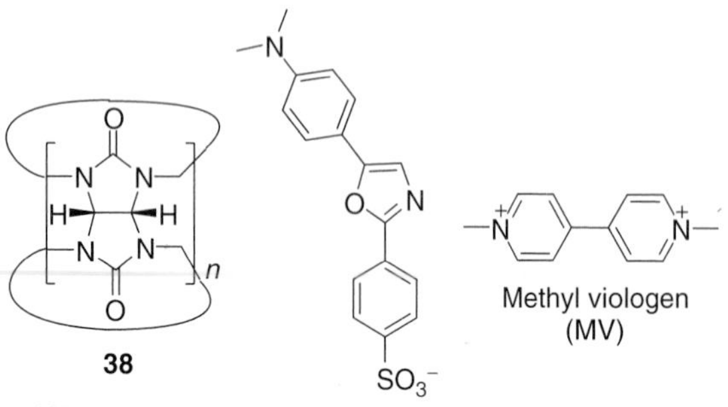

Figure 16 Chemical structures of cucurbit[*n*]uril hosts and guests used as probes for the formation of ternary complexes.

The larger host (CB8) has also been studied in detail in several reports. For instance, Urbach and coworkers determined the scope of amino acid recognition by CB8 using ITC.[48] They found that CB8 exclusively binds three of the natural amino acids: Trp, Phe, and, to a lower extent, Tyr. The observed complexes showed a 1 : 2 host : guest stoichiometry, because of the expanded inner cavity. These data suggest that the interaction mainly occurs by inclusion of the aromatic side chain into the flat hydrophobic inner surface of the receptor. The ability of CB8 to cooperatively bind two aromatic rings inside its cavity has been used to prepare ternary complexes by coinclusion of methyl viologen (MV) and the intended substrate into CB8.[49] These ternary complexes proved to exert some selectivity for Trp over other aromatic amino acids. Even more interestingly, when the authors studied different short oligopeptides containing Trp, they observed that the position of this amino acid is not innocent. Thus, they found that CB8·MV binds Trp-Gly-Gly with sixfold selectivity over Gly-Trp-Gly, and with 40-fold selectivity over Gly-Gly-Trp. The unambiguous existence of the corresponding ternary complexes was thoroughly studied using ITC, mass spectrometry, UV–vis, fluorescence, and ^{1}H NMR spectroscopy. Finally, the authors concluded that the selectivity observed for the binding of N-terminal Trp is due to what they called charge-mediated peptide recognition. In this model, the Trp side chain would be included inside the CB8·MV inner hydrophobic space, while the N-terminal ammonium cation could be H-bound to the carbonyl oxygen atoms of the portal of CB8. This group also studied the sequence-specific recognition and cooperative dimerization of N-terminal aromatic peptides,[50] using CB8 as the host, which is able to simultaneously bind two aromatic guests. They prepared a series of peptides of the types X-Gly-Gly, Gly-X-Gly and Gly-Gly-X (X = Trp, Phe, Tyr, and His), and studied their interaction with CB8 by ITC, NMR, and X-ray crystallography. CB8 selectively recognized Trp-Gly-Gly and Phe-Gly-Gly with high affinity with the binding constants for the other 10 peptides being too small to be measured by ITC. Both peptides formed 1 : 2 host : guest complexes in a stepwise manner, and Phe-Gly-Gly was bound with positive cooperativity. All these data also supported the charge-mediated peptide recognition, which was unambiguously demonstrated by the solid-state structures of [CB8·Trp-Gly-Gly] and [CB8(Phe-Gly-Gly)$_2$] cocrystals.

6 METAL COMPLEXES

Free amino acids are able to coordinate transition metal cations via the carboxylate and the amino groups. This is a well-established fact, which has been utilized for the design of synthetic receptors since a long time ago. In this regard, a number of examples can be found in which metal complexes are at the core of the design of synthetic receptors for amino acids or other zwitterionic guests

Nitrogenated ligands are typically employed, forming metallic complexes that are used as receptors (see selected examples in Figure 17). One early example of amino acid recognition with high stereospecificity is the pyridine ligand forming the Co(III) complex **39**.[51] This dichlorocobalt complex binds alanine with high and predictable regio- and stereospecificity. Apart from pyridine, numerous other examples, including terpyridine, pyridinebisoxazoline, phenanthroline, and linear polyamines, have been described

Bencini and Bianchi have thoughtfully studied the binding behavior of the Zn(II) complex of the phenanthroline-containing macrocycle **40** as a receptor for free amino acids and peptides by NMR and potentiometry.[52] The unsaturated environment of the Zn(II) complex allows for the coordination of the aforementioned substrates. In their zwitterionic form, the binding to Zn(II) occurs through the carboxylate group, whereas in their anionic form, it takes place via the amine functionality, and probably with additional participation of the amide carbonyl of the substrates. It is remarkable that amino acids containing aromatic side chains form the most stable complexes, likely due to hydrophobic and/or π-stacking interactions between the aromatic units of the receptor and those of Phe and Trp. Moreover, complex **40** shows hydrolytic activity toward the amide bond of L-leucine-p-nitroanilide (LNA).

Preferential recognition of aromatic amino acids has also been described by Fabbrizzi using the tripodal ligand **41**, which forms a series of Zn(II) complexes.[53] The recognition process is described to occur through (i) metal–ligand interaction between the Zn(II) ion and the carboxylate group of the amino acid, and (ii) a π-stacking interaction involving the aromatic moieties of the receptor and substrates. This stacking interaction has been estimated at about 7–8 kJ mol^{-1}. The interaction with Trp is accompanied by a strong quenching of the fluorescence of the receptor.

Other metallic complexes showing certain preference for aromatic versus aliphatic amino acids in the recognition process include lanthanide(III) tris(β-diketonate) complexes,[54] which extract Phe, Trp, Phg, and other aromatic guests from an aqueous phase to dichloromethane, with some degree of enantioselectivity, while Ala and Gly were rarely extracted. Nevertheless, the more hydrophobic aliphatic amino acid Leu was also extracted. Another recent example of enantioselective recognition is the case of the Zn-salophen receptor **42** described by Dalla Cort,[55] which binds L- and D-Phe with a notable selectivity of $K_L/K_D = 9.6$.

As has been seen, the Zn(II) cation is one of the most commonly used metallic centers in organometallic receptors. García-España and Abarca have designed a triazolopyridine ligand that on complexation with Zn(II) affords chemosensor **43** (Figure 18), which is capable of recognizing two similar amino acids such as L-Asp and L-Glu with moderate specificity ($K_{Asp} = 10^4\,M^{-1}$; $K_{Glu} = 5.5 \times 10^3\,M^{-1}$).[56] More recently, Kwong *et al.* have described

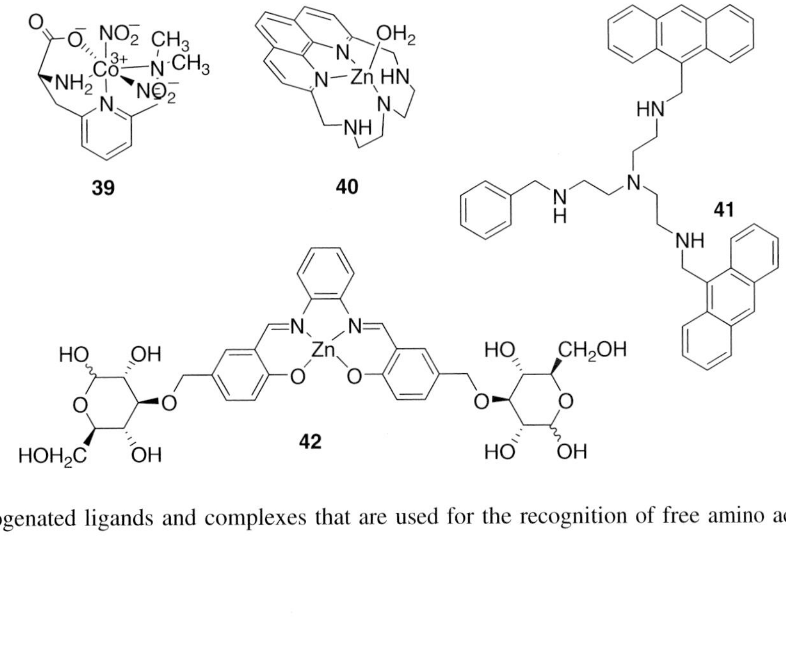

Figure 17 Selected nitrogenated ligands and complexes that are used for the recognition of free amino acids and peptides.

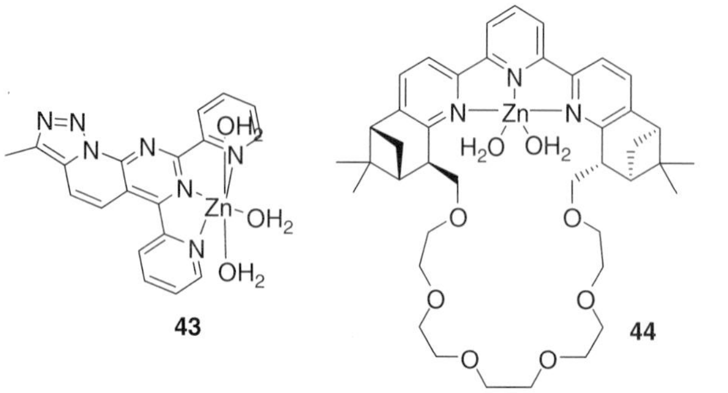

Figure 18 Selected terpyridine complexes that are used for the recognition of free amino acids and peptides.

the Zn(II) complex of a terpyridine-crown macrocycle (**44**) which binds L-Asp with $K = 4.5 \times 10^4\,M^{-1}$ in water–DMF (1 : 3 v/v).[57] In a similar way, Hong described a pyridine bis(oxazoline)–copper(II) complex as a receptor for seven amino acids (Ala, Val, Phe, Asn, Gln, Asp, and Glu) in CH_3OH–water (1 : 1, v/v), with moderate binding constants (typically log $K = 4$–5, except for Asp for which a log $K = 6$ is described).[58] Isothermal microcalorimetric analysis (ITC) showed that the complexation is accompanied by a small entropy loss, which is compensated by a large enthalpy gain. Chiral discrimination of amino acids by this pybox–copper complex was not large, although enough to resolve the corresponding enantiomers by capillary electrophoresis.

As has been shown, binding of amino acids or peptides in the free form to monometallic complexes usually takes place via the carboxylate group or by simultaneous chelation of COO^- and $-NH_2$ at the metal center. Interactions with side chains of substrates are typically secondary and weak in comparison to the chelation with the metallic ion, and they do not involve the coordination to the metal center of the organometallic complex. One approach followed by several groups to increase the importance of interactions with the side chains is to design receptors with two metallic positions precisely engineered to bind such pending residues. Typically, His and Asp/Glu have been the targets due to the coordinative nature of their side chains.

Fabbrizzi developed receptor **45** that is able to bind L-His, with log $K = 5.5$, in aqueous solution, and rationalized this behavior by the simultaneous coordination of the imidazole of L-His to both Cu(II) centers of the receptor.[59] A further development was reported by the same group two years later with the synthesis of a binuclear Zn complex of an anthracenic ligand (**46**) for the recognition of L-His (log $K = 2.92$) (Figure 19).[60] This receptor is also a fluorescent chemosensor for this amino acid, provided the imidazole ring acts both as a coordinating bridge between two zinc centers and as a quencher of the anthracenic fluorescence (via PET).

Other ditopic complexes for amino acid recognition include the binuclear Cu(II) pyridinophane **47**, developed by García-España, forming ternary complexes with L-Asp and L-Glu.[61] In this case, the receptor forms much more stable complexes with L-Glu, which allowed its electrochemical sensing using adsorbates of the bimetallic complex over glassy carbon electrodes.

A receptor for tripeptides, having the sequence L-xxx-L-Lys-L-Lys (with xxx = His, Cys, and Met) and operating through cooperative metal-coordination and ion pairing, has been reported by Anslyn (Cu(II) complex **48**, Figure 20).[62] Thus, **48** binds L-His-L-Lys-L-Lys with a $K = 10^6\,M^{-1}$, in contrast to related tripeptides such as L-His-Gly-Gly ($K = 1.5 \times 10^4\,M^{-1}$) and L-Cys-L-Lys-L-Lys ($K = 3.0 \times 10^5\,M^{-1}$).

A competitive assay (indicator displacement assay, IDA) was reported by Anslyn using the Zn(II) complex **49** shown in Figure 21.[63] This 2,2′ : 6, 2″-terpyridine complex forms a chemosensing ensemble with pyrocatechol violet because of the ability of the dye to chelate the Zn(II) cation. On addition of a series of unprotected amino acids to a solution of the chemosensing ensemble, the release of the dye was visually verified and spectroscopically measured (1 : 1

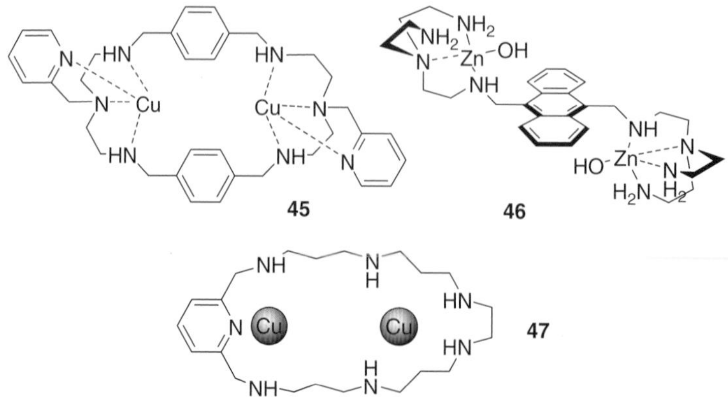

Figure 19 Selected bimetallic complexes that are used as receptors for amino acids.

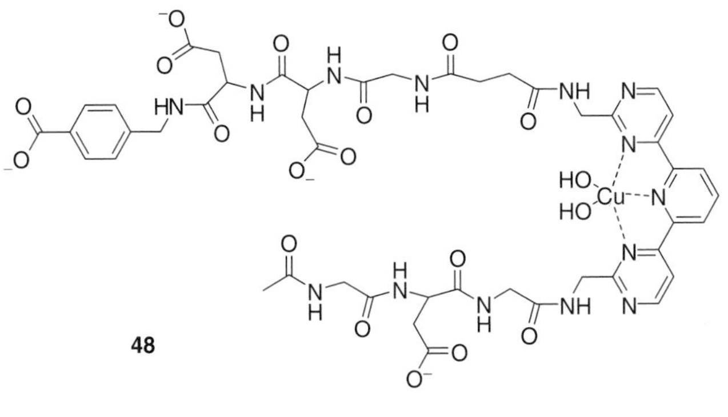

Figure 20 Cu(II) complex as a receptor for L-xxx-L-Lys-L-Lys (with xxx = His, Cys, and Met).

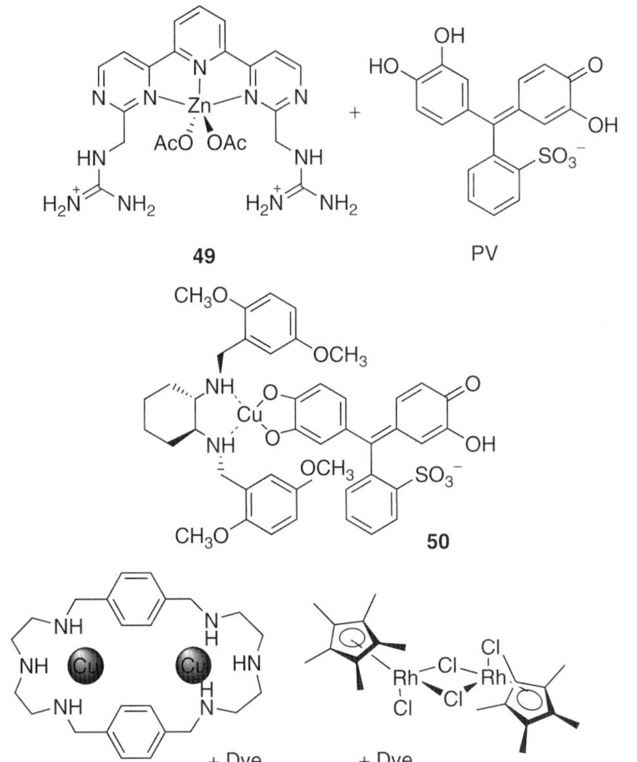

Figure 21 Selected organometallic systems that are utilized for the development of indicator displacement assays.

methanol : water, HEPES buffer, pH 7.4). Fitting of the data to the appropriate models afforded binding constants with hydrophobic amino acids such as Gly, Val, and Phe in the range of $10^4\,M^{-1}$. Notably, the highest affinity was found for L-Asp ($1.5 \times 10^5\,M^{-1}$) in contrast to the homologous L-Glu ($2.2 \times 10^4\,M^{-1}$). Overall, the binding process for all the amino acids tested is dominated by the chelation of Zn(II) and the specificity for Asp can be explained by taking into account the existence of an extra H-bond between the pending COO^- in Asp and one of the guanidinium groups of the receptor.

A further development from the same group was the chemosensing ensemble **50**, in which pyrocatechol violet is again the signaling unit that is bound to a Cu(II) cation coordinated to a *trans*-diaminocyclohexane derivative.[64] A significative degree of enantiomeric discrimination for amino acids in 1 : 1 CH_3OH : water mixtures (buffered at pH 7.0) was observed. Thus, the interaction was more favorable in all the cases for the D-enantiomers of Val ($K_D/K_L = 2.6$), Leu ($K_D/K_L = 1.7$), Phe ($K_D/K_L = 2.1$), and Trp ($K_D/K_L = 2.2$). Very similar aliphatic amino acids such as Leu, Val, and Ile could be clearly differentiated using the two-dimensional plots generated from an array of 21 chemosensing ensembles based on this general approach. Moreover, this allowed the unambiguous establishment of the enantiomeric nature of each target, in close analogy to the human taste response. The viability of this idea for the development of a high-throughput screening protocol was also demonstrated by adapting the experimental conditions to the 96-well plate analysis format.[65]

Fabbrizzi has also reported an assay (IDA) for the recognition of His, in this case based on a dicopper complex and polyamine macrocycle **51**, in combination with three different dyes.[66] A Cu(II) complex was also described by Anslyn for the sensing of His, although, in this case, the mechanism is the metal stripping from a ligand–Cu–fluorescein complex by His to form a 2 : 1 His:Cu(II) complex, with the concomitant release of free unquenched fluorescein.[67]

Following the same principle, Severin has developed assays for His- and Met-containing peptides in water at neutral pH, for concentrations of analytes as low as 0.3 μM.[68] The organometallic host was a Cp*Rh(III) complex (**52**) combined with the indicator azophloxine. With the same Rh

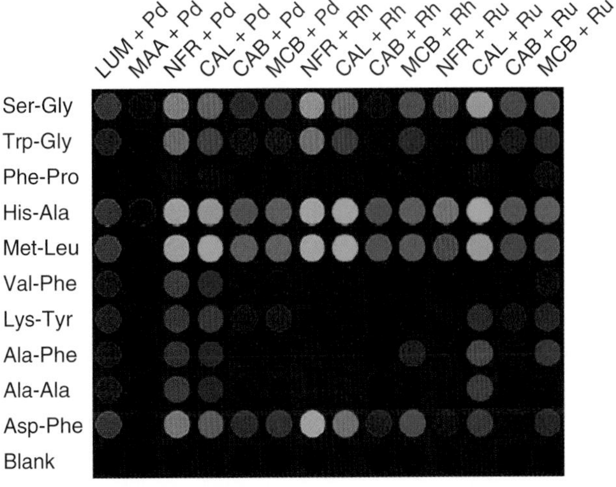

Figure 22 An example of sensing array for the analysis of dipeptides, which was described by Severin,[70] by employing different metallic complexes of know dyes.

complex and three different dyes, the same group was able to identify 20 natural amino acids colorimetrically, making use of a multivariate analysis.[69] Other assays involving Cu, Ni, Pd, and Ru as metallic cations, in combination with the concept of combinatorial libraries of metal–dye complexes, have appeared recently, for the analysis of a large variety of peptides (Figure 22).[70]

7 PORPHYRINS

Metal porphyrins represent a very particular case of metal complexes and they have been frequently studied as receptors for amino acids and peptides. Not only have Zn(II) porphyrins been used as hosts for this purpose, but also the corresponding Cu, Co, Er, Gd, and Yb porphyins. Those cations have taken advantage of their Lewis acid character to act as the recognition site for the amine end of amino acids and peptides. Nevertheless, this sole interaction has proven to be insufficient for an efficient binding and for the (enantio)discrimination of the substrates. Positively charged Zn(II) porphyrins were synthesized and studied by Imai (compounds **53a** and **b** in Figure 23).[71] Such receptors are decorated with trimethylammonium groups that are capable of binding the carboxylate end of several amino acids and dipeptides in water. It is noteworthy that the racemic mixture resulting from the synthetic process was optically resolved by means of a chiral HPLC (high performance liquid chromatography) column, leading to optically pure receptors **53a** and **b**, which showed important enantiorecognition toward the pure L- or D-amino acids or dipeptides. Besides the coordination of the amine end to the Zn(II) metal and the coulombic interactions between carboxylates and ammonium groups, the existence of relevant hydrophobic forces was noticed in the presence of aromatic side chains (Trp or Phe).

Uncharged neutral porphyrins have also been the subject of study for the recognition of amino acids. Thus, Villari reported that a series of noncharged Co(II) porphyrins were able to bind amino acids through N-coordination to the Co(II) along with $\pi-\pi$ interactions without the need of extra contribution from any coulombic attractive forces.[72]

Another strategy for the recognition of peptides and amino acids by means of porphyrins has been developed by Schneider with the attachment of crown-ether rings to the skeleton of the porphyrin ring, leading to a bismacrocycle capable of binding those substrates even in the absence of the metallic center (Figure 24).[73] In the system **54e**, the crown-ether moiety would act as the binding site for the ammonium group, thus replacing the function of the Lewis acid center. A variety of di-, tri-, and tetrapeptides can bind to **54e** with log K values ranging from 2 to 5. Computer calculations and NMR measurements of the complex with Gly-Gly-Phe revealed three recognition events: (i) the ammonium-crown ether ring, (ii) the carboxylate–pyridinium unit, and (iii) stacking between the terminal phenyl ring and the porphyrin.

Similar crown-ether-containing porphyrins have been described by Tsukube.[74] In the porphyrins **55a–c**, the metal was an erbium cation, serving as the binding site for the free carboxylate of amino acids. Good extraction efficiencies from water to dichloromethane were described with L-Trp, through the formation of a 1 : 1 complex. Notably, the conjugates with several L-amino acids and dipeptides gave

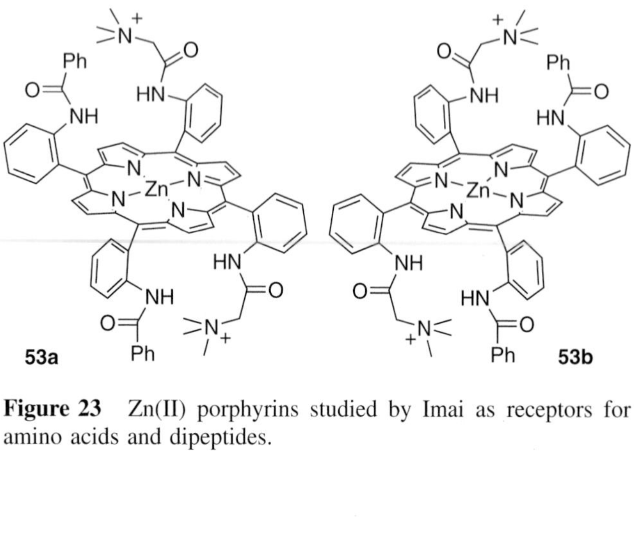

Figure 23 Zn(II) porphyrins studied by Imai as receptors for amino acids and dipeptides.

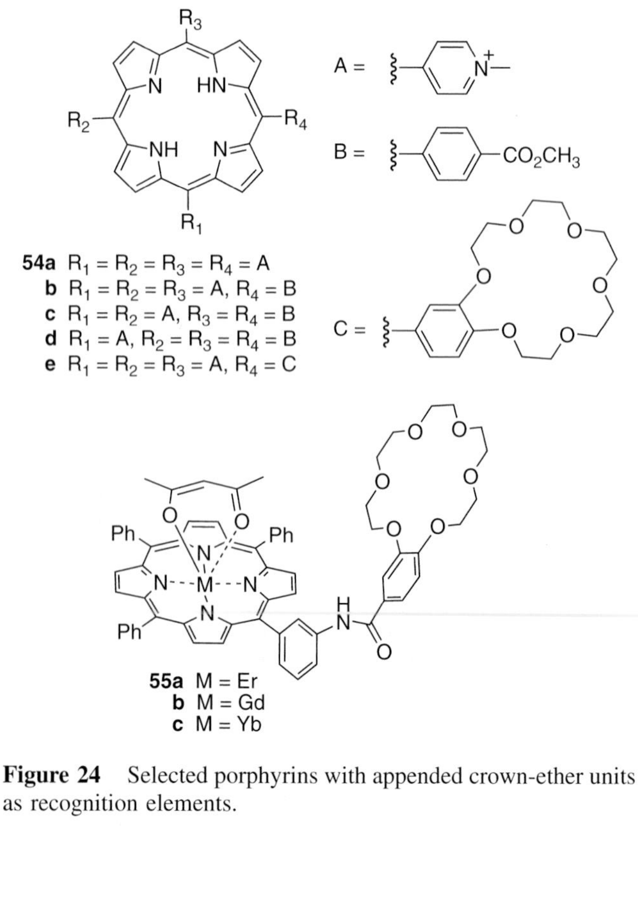

Figure 24 Selected porphyrins with appended crown-ether units as recognition elements.

specific circular dichroism signals around the Soret band, with the opposite shape being recorded with the D-isomers.

8 PEPTIDE-LIKE RECEPTORS

Some zwitterions form a part of important sequences for biomolecular processes, such as the specific recognition of peptides, proteins, or protein fragments. In most cases, this specific recognition takes place through electrostatic and H-bonding contacts, despite the high solvation energies of charged functional groups in polar water. The full understanding of these processes could be of great importance for the development of sensors or drug candidates. Learning from nature, chemists have also used peptide-like molecules for the development of artificial receptors for zwitterionic species.

A combinatorial approach to this problem has been developed by Schmuck and coworkers,[75] since the preparation of large libraries of peptide-like molecules is relatively simple. They focus on optimizing an artificial receptor for a highly polar tetrapeptide, D-Glu-Lys-D-Ala-D-Ala (Figure 25). For the appropriate screening toward the receptor library, the corresponding fluorescent labeled (*N*-dansyl substituted) substrate was prepared. Regarding the general formula of the artificial receptors (Gua-Xxx1-Xxx2-Xxx3), they were designed as follows. They possessed a carboxylate binding site, using the guanidinocarbonyl pyrrole unit (abbreviated Gua), which had been found to be an efficient motif for the binding of the C-terminus of oligopetides.[76] In addition, the molecular diversity was implemented in the three adjacent positions. In each of the three coupling steps, the same eight amino acids were used (Lys, Tyr, Ser, Glu, Phe, Val, Leu, and Trp), rendering the corresponding library of 512 different members. The amino acids were chosen from the 20 proteinogenic amino acids to provide a representative range of varying polar, charged, and hydrophobic residues. A quantitative on-bead binding assay in water was performed with the soluble fluorescent peptide substrate and the solid-supported library of peptide-like receptors. The binding affinities vary from $K = 17\,100\,\mathrm{M}^{-1}$ (the best receptor) to $K < 20\,\mathrm{M}^{-1}$ (worst receptor), which means a difference in activity of more than two orders of magnitude. Receptor Gua-Lys-Lys-Phe was the most efficient in the on-bead screening, whereas Gua-Lys-Tyr-Lys showed a medium affinity. UV-titration experiments of the nonsupported corresponding receptors in solution also confirmed the trends observed in the screening experiments, rendering binding interactions of $K = 15\,400\,\mathrm{M}^{-1}$ for Gua-Lys-Lys-Phe and $K = 6200\,\mathrm{M}^{-1}$ for Gua-Lys-Tyr-Lys, which is a solid proof for the validity of the on-bead binding screening approach. An inspection of the binding affinities reveals that complex formation is strongly correlated with the positive charge of the receptors. Those receptors bearing two Lys residues next to the Gua cationic site form the strongest complexes with binding affinities higher than $10^4\,\mathrm{M}^{-1}$. Introduction of anionic residues in the receptor structure reduced the affinity. Besides, hydrophobic interactions are not important for the recognition event.

Additional work from the same research group demonstrated that the same combinatorial approach can be efficiently used for checking the substrate selectivity.[77] They assayed the original library on a new peptidic substrate with an inverse sequence: D-Ala-D-Ala-Lys-D-Glu. The on-bead quantitative binding screening showed that the binding constant is around 2–3 times smaller for the inverse substrate. Moreover, binding studies in solution for selected receptors again correlated with the on-bead binding assay. Molecular modeling studies on the substrates alone revealed that the inverse peptide substrate forms an intramolecular ion pair between the Lys residue and the two Glu carboxylates (Figure 26). However, in the initial peptide, only the Lys and Glu side chains interact, leaving the C-terminal carboxylate free for complexation, thus explaining the differences in the binding properties with the receptor. Once again in this case, avoiding the self-interaction of the

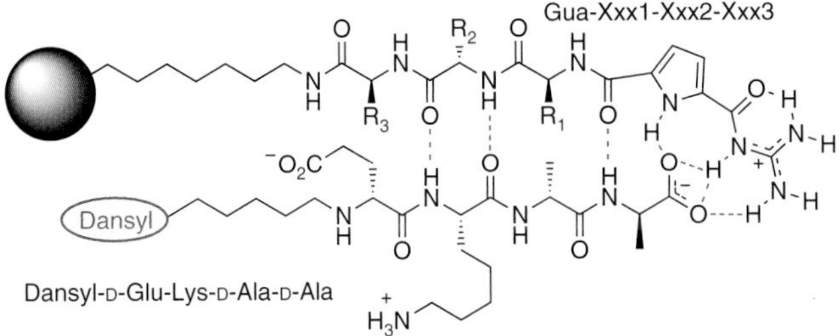

Figure 25 On-bead supported library for the selective interaction with the sequence D-Glu-Lys-D-Ala-D-Ala.

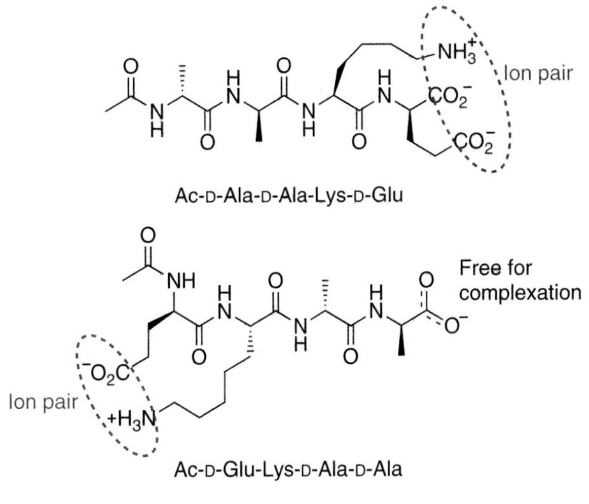

Figure 26 Comparison of the structures of D-Ala-D-Ala-Lys-D-Glu and the reverse sequence.

zwitterionic species is fundamental for obtaining efficient molecular recognition properties. The molecular modeling studies with the supramolecular complex formed by the Gua-Lys-Lys-Phe receptor and the D-Ala-D-Ala-Lys-D-Glu substrate revealed the main interactions responsible for the efficient molecular recognition. First of all, the C-terminal carboxylate of the substrate is electrostatically H-bound to both the Gua and the adjacent Lys moieties of the receptor. In addition, the second Lys residue of the receptor forms an ion pair with the Glu side chain of the substrate, which is also stabilized by the adjacent Lys of the tetrapeptide.

The receptor Gua-Lys-Lys-Phe also showed a remarkable stereoselectivity toward D/L-isomers on some of the sites of the tetrapeptides.[78] More interestingly, the stereoselectivity is highly dependent on the position where the stereochange is introduced, leading to a sequence-dependent stereoselectivity binding. The stereoselectivity is only observed when the D-Ala/L-Ala exchange takes place at the position 3 of the tetrapeptide, but not at the N-terminus. Thus, the two substrates, D/L-Ala-D-Glu-D-Glu-D-Glu, are bound with similar affinities to the receptor ($K = 5100\,M^{-1}$ and $4800\ M^{-1}$, respectively). However, from the two diastereomers, D-Glu-D-Glu-D/L-Ala-D-Glu, the one with D-Ala is preferred ($K = 4500\,M^{-1}$ vs $1400\ M^{-1}$, respectively). Most likely, stereoselectivity required a rather well-defined structure of the complex, which is only possible when the D-Ala/L-Ala exchange is fixated at both sides by strong charge interactions between the substrate and the receptor. In good agreement with this multipoint-binding model is the observation that the exchange of D-Ala/D-Lac at the C-terminal position of the substrate also decreased the interaction ($K = 2100\,M^{-1}$ vs $\approx 300\,M^{-1}$, for the peptide and the depsipeptide, respectively).

Regarding a rational approach to this problem, similar peptide-like receptors for zwitterions have also been studied.[79] The authors decided to target the Arg-Gly-Asp peptide sequence, which is biologically important for cell–cell and cell–matrix adhesion processes, as the substrate. To mimic a more realistic situation, where this sequence is an internal fragment of a longer peptide, the N- and C-capped derivatives (Ac-Arg-Gly-Asp-NH_2) were also studied (Figure 27). The chemical structures of the synthetic receptors (**56–59**) were constructed by connecting two well-known binding sites for cationic guanidinium and carboxylate anions. Thus, the already commented guanidiniocarbonyl pyrrole served as a carboxylate binding site, while the *m*-xylylene biphosphonate bisanion was selected as the guanidinium binder. To modulate the distance between the binding sites, as well as the overall flexibility of the receptors, different spacers were used.

Since the corresponding binding sites of the hosts are mutually complementary, the authors started to study the

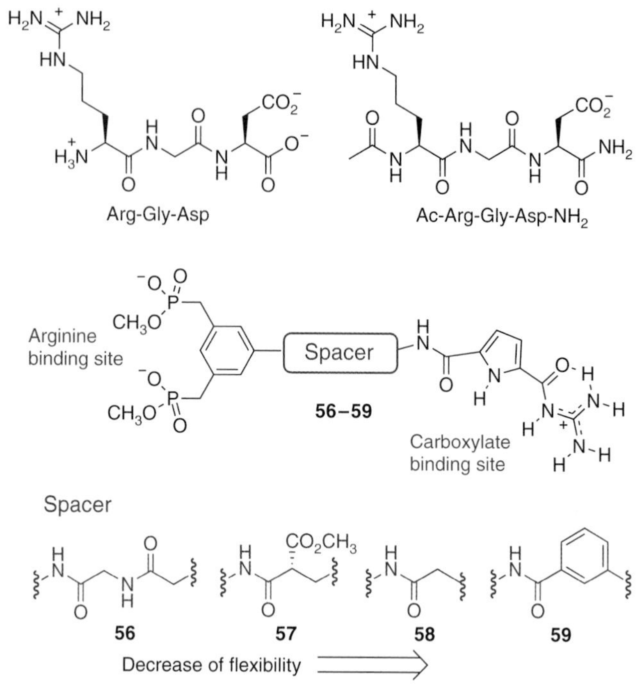

Figure 27 Chemical structures of Arg-Gly-Asp mimic sequences (substrates) and those for the synthetic ditopic receptors (**56–59**).

self-complexation of the receptors, since this would be competitive with the binding of the substrates. Interestingly, **56** and **57**, having the most flexible spacers, did not show any concentration dependence of their spectroscopic properties (NMR and UV–vis spectroscopy), suggesting the total absence of aggregation in the micromolar concentration range. However, molecular modeling calculations revealed that these receptors are flexible enough to adopt a loop conformation where the two oppositely charged binding motifs of the same molecule can interact with each other (Figure 28). This situation should be entropically favored versus the self-aggregation process. For the shorter receptor **58**, the same intramolecular interactions led to a very strained and a less favorable geometry in the modeling studies. In this case, the NMR dilution studies showed a clear aggregation behavior, which could be attributed to a dimerization process with an apparent $K_{dimer} \approx 400\,M^{-1}$. Molecular modeling studies showed that two molecules of **58** can stack in an antiparallel fashion so that each cationic Gua interacts with a negatively charged biphosphonate anion (Figure 28). For the receptor with the most rigid spacer (**59**), the folded intramolecular self-complexation is prevented, as shown by modeling studies, while the dilution experiments by UV–vis spectroscopy served to discard the formation of dimers in the micromolar concentration range. Thus, this receptor should be the most suitable for the binding of the target peptides.

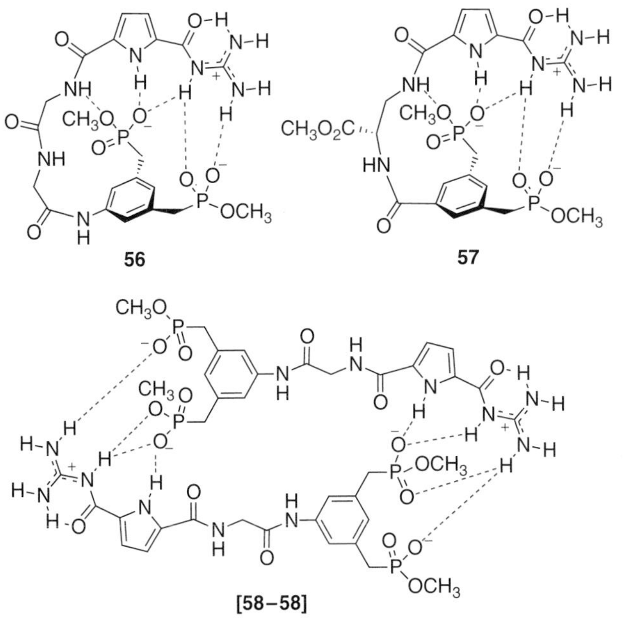

Figure 28 Intramolecular interaction versus dimerization in receptors **56**–**58**.

The binding properties of the different receptors with the two target peptides were studied by NMR and/or UV–vis spectroscopy (depending on the solubility of the corresponding host) in aqueous solutions (either in pure water or in the presence of different buffers). The authors did not observe any interaction between **58** and Arg-Gly-Asp or Ac-Arg-Gly-Asp-NH_2, suggesting that the dimerization of the host was completely inhibiting the molecular recognition process. For the more flexible compounds (**56** and **57**), NMR titration studies determined that the binding to both substrates was weak ($K \approx 100\,M^{-1}$), most likely due to the efficient intramolecular self-association. Therefore, the most rigid receptor, **59**, should be more efficient in the binding of the target peptides. Interestingly, **59** was not able to interact with the densely charged and highly solvated Arg-Gly-Asp, but efficiently recognized its caped version Ac-Arg-Gly-Asp-NH_2, rendering a binding constant of $K \approx 3000\,M^{-1}$ (varying from 2400 to 3200 M^{-1}, depending on the technique used for its measurement) (Figure 29).

The selectivity of the proposed binding interactions was also studied. Accordingly, no complexation with closely related peptides (Ac-Arg-Gly-Gly-NH_2, Ac-Gly-Gly-Asp-NH_2, or Ac-Gly-Gly-Gly-NH_2) was detected by UV, supporting the high selectivity of **59** toward the Ac-Arg-Gly-Asp-NH_2 sequence. To evaluate the effect of the bisphosphonate binding site, they carried out titrations with the corresponding bis(methyl) ester, which was a synthetic precursor of **59**. Surprisingly, they found a higher binding constant of $K = 4700\,M^{-1}$, in spite of the absence of the negative charges in the binding site for the guanidinium group. Since this receptor is more hydrophobic than the original one, the authors concluded that the solvation properties of the host could be playing a fundamental role in the molecular recognition.

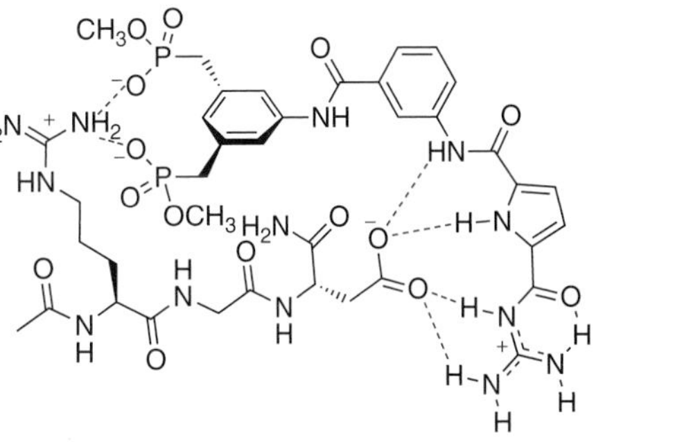

Figure 29 Schematic representation of the complex [**59**·Ac-Arg-Gly-Asp-NH_2].

9 MISCELLANEOUS RECEPTORS

Even though the former categories for the classification of the receptors are quite flexible and many structures considered could be easily classified in more than one of those categories, there are still quite a few receptors that do not fit well on those classifications. Figure 30 gathers some of those receptors including clefts, tweezers, cyclophanes, and so on. Thus, for instance, in this seminal work in this area, Rebek reported the convergent diacid **60** as an efficient receptor for the selective recognition and transport, across a bulk liquid membrane, of neutral amino acids bearing aromatic side chains.[80] In this regard, the transport of Trp across a $CHCl_3$ liquid membrane was about 5 times faster than that of Phe, while the transport of Leu was negligible. NMR data suggested the formation of a 2 : 1 (receptor : substrate) complex in which the carboxylate fragment interacts with one carboxyl moiety of each receptor and the same occurs with the ammonium group, while π–π stacking takes place between the aromatic side chain and one of the acridine rings. Schrader and Klämer, on the other hand, prepared the molecular tweezer **61** in an attempt to develop a selective receptor for Lys and Arg and for peptides containing those amino acids.[81] The design of this receptor considered two main components: the phosphonate groups in the periphery of the cavity for the interaction with positively charged fragments, such as ammonium and guanidinium, and a cleftlike cavity whose walls are formed by electron-rich fragments that are essential for the interaction with the side chain of the amino acid. It is worth mentioning that binding studies were carried out in D_2O and in a neutral phosphate buffer using NMR measurements as well as ITC titrations. Although, in general, the association constants observed for N/C protected amino acids were higher than those for zwitterionic

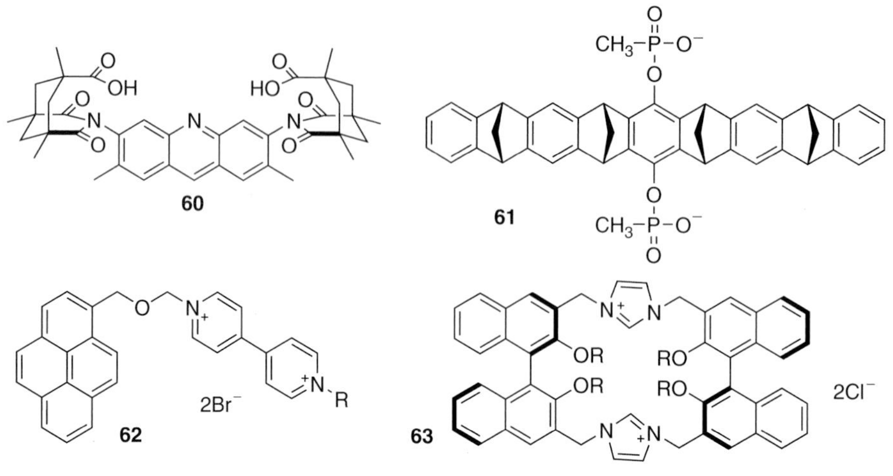

Figure 30 Some examples of miscellaneous receptors for zwitterionic species.

amino acids, the values obtained for basic amino acids were remarkable (i.e., $K_{ass} = 1400\,M^{-1}$ for Lys). The same selectivity trends were found for short peptides including those amino acids at different positions. A partly related design is present in the structure **62**, prepared and studied by Ramaiah.[82] This compound contains two aromatic subunits, one of which is strongly electron rich (alkoxy pyrene) while the other is electron poor (viologen). Thus, these two subunits are able to adopt a folded conformation of the molecule affording a fluorescent intramolecular charge-transfer (CT) complex involving the two complementary subunits. The cleft formed on folding is also able to easily adapt to accommodate different amino acids, although Trp is selectively recognized over the rest of amino acids, efficiently inhibiting the fluorescence of the initial CT complex. Electrostatic, π–π stacking, and donor–acceptor interactions seem to act cooperatively to produce this selective recognition. The last receptor (**63**) is based on a BINOL structure, and the resulting bis-imidazolium macrocycle is also able to act as a chemosensor for the highly selective recognition of Trp in aqueous solutions. Not only do the imidazolium subunits contribute with electrostatic interaction, but it is also reported that the C2 hydrogen of this ring is able to act as an efficient hydrogen bond donor.[83] On the other hand, some of the derivatives that were studied are shown to display a remarkable enantio-differentiation for the two enantiomers of Trp (up to $K_D/K_L = 6.2$).

10 CONCLUSIONS AND OUTLOOK

Although many different kinds of structures have been assayed for the selective recognition of zwitterionic species, the progress in this field has been, up to now, relatively limited, and a significant effort in this area can be envisaged for the next years. As we have seen in this chapter, the number of zwitterionic species that has been studied is essentially limited to amino acids (mainly focusing on the reduced number of natural α-amino acids) and short peptidic sequences. Studies directed toward the selective recognition of other zwitterionic species are rather scarce. In this regard, broadening of the range of guests under consideration should be expected for the coming years.

A second important limitation is associated with the fact that most zwitterionic compounds are water soluble. On the contrary, most designed receptors continue being soluble in organic solvents and having a much reduced solubility in water or aqueous solutions. Thus, it is not surprising that the analysis of the receptor–substrate interaction is carried out by means of extraction and/or transport experiments. Besides the interest of selective recognition for these kinds of processes, this fact also reflects, in many cases, the very different solubilities of both components, the zwitterionic substrate and the receptor.

Finally, a current third limitation in this area is the rather restricted development of practical applications. Even though many potential applications, including the development of specific sensors for amino acids or for peptides containing specific sequences rich in one given amino acid (electrochemical, UV–vis, or fluorescent), the purification by extraction, transport or chromatography of amino acids or short peptides, and, in particular, enantioseparation of racemic mixtures, are considered for studies in this field, it is not clear until now if they have achieved the practical implementations for being of true practical use. In this regard, much effort needs to be devoted in the near future to transform the basic knowledge that is generated up to now into efficient realistic systems for the *in vivo* or *in vitro* monitoring of zwitterionic species, for their separation and purification, and, whenever possible, for the resolution of enantiomeric mixtures.

REFERENCES

1. C. Schmuck, *J. Org. Chem.*, 2000, **65**, 2432.
2. C. Schmuck and J. Dudaczek, *Eur. J. Org. Chem.*, 2007, 3326.
3. C. Schmuck and W. Wienand, *J. Am. Chem. Soc.*, 2003, **125**, 452.
4. C. Rether, W. Sicking, R. Boese, and C. Schmuck, *Beilstein J. Org. Chem.*, 2010. DOI: 10.3762/bjoc.6.3.
5. C. Schmuck, *Tetrahedron*, 2001, **57**, 3063.
6. T. Rehm, V. Stepanenko, S. Zhang, *et al.*, *Org. Lett.*, 2008, **10**, 1469.
7. C. J. Pedersen, *Angew. Chem. Int. Ed. Engl.*, 1988, **27**, 1021.
8. J.-M. Lehn, *Angew. Chem. Int. Ed. Engl.*, 1988, **27**, 89.
9. M. Barboiu, C. Luca, C. T. Supuran, *et al.*, *Liebigs Ann.*, 1997, 1853.
10. F. P. Schmidtchen, *J. Org. Chem.*, 1986, **51**, 5161.
11. M. L. Hossain and H.-J. Schneider, *J. Am. Chem. Soc.*, 1998, **120**, 11208.
12. A. Späth and B. König, *Tetrahedron*, 2010, **66**, 1859.
13. A. Sirikulkajorn, P. Duanglaor, V. Ruangpornvisuti, *et al.*, *Supramol. Chem.*, 2009, **21**, 486.
14. O. A. Fedorova, E. N. Andryukhina, Y. V. Fedorov, *et al.*, *Org. Biol. Chem.*, 2006, **4**, 1007.
15. A. Galán, D. Andreu, A. M. Echavarren, *et al.*, *J. Am. Chem. Soc.*, 1992, **114**, 1511.
16. P. Breccia, M. Van Gool, R. Pérez-Fernández, *et al.*, *J. Am. Chem. Soc.*, 2003, **125**, 8270.
17. A. Metzger, K. Gloe, H. Stephan, and F. P. Schmidtchem, *J. Org. Chem.*, 1996, **61**, 2051.
18. A. P. de Silva, H. Q. N. Gunaratne, C. McVeigh, *et al.*, *Chem. Commun.*, 1996, 2191.
19. S. Sasaki, A. Hashizume, D. Citterio, *et al.*, *Tetrahedron Lett.*, 2002, **43**, 7243.
20. M. T. Reetz, J. Huff, J. Rudolph, *et al.*, *J. Am. Chem. Soc.*, 1994, **116**, 11588.
21. K.-S. Jeong and T.-Y. Park, *Bull. Kor. Chem. Soc.*, 1999, **20**, 129.
22. A. Moghini, M. F. Rastegar, M. Ghandi, *et al.*, *J. Org. Chem.*, 2002, **67**, 2065.
23. A. Moghini, B. Maddah, A. Yari, *et al.*, *J. Mol. Struct.*, 2005, **752**, 68.
24. L. Lamarque, P. Navarro, C. Miranda, *et al.*, *J. Am. Chem. Soc.*, 2001, **123**, 10560.
25. C. Miranda, F. Escartí, L. Lamarque, *et al.*, *J. Am. Chem. Soc.*, 2004, **126**, 823.
26. K. Tsubaki, T. Kusumoto, N. Hayashi, *et al.*, *Org. Lett.*, 2002, **4**, 2313.
27. I. Tabushi, Y. Kuroda, and T. Mizutani, *J. Am. Chem. Soc.*, 1986, **108**, 4514.
28. V. P. Kumar, I. Suryanarayana, Y. V. D. Nageswar, and K. R. Rao, *Helv. Chim. Acta*, 2008, **91**, 753.
29. C. Kahle, R. Deubner, C. Schollmayer, *et al.*, *Eur. J. Org. Chem.*, 2005, 1578.
30. Y. Liu, C.-C. You, H.-Y. Zhang, and Y.-L. Zhao, *Eur. J. Org. Chem.*, 2003, 1415.
31. R. Corradini, A. Dossena, G. Impellizzeri, *et al.*, *J. Am. Chem. Soc.*, 1994, **116**, 10267.
32. S. Pagliari, R. Corradini, G. Galaverna, *et al.*, *Tetrahedron Lett.*, 2000, **41**, 3691.
33. S. Pagliari, R. Corradini, G. Galaverna, *et al.*, *Chem. Eur. J.*, 2004, **10**, 2749.
34. R. Corradini, C. Paganuzzi, R. Marchelli, *et al.*, *J. Mater. Chem.*, 2005, **15**, 2741.
35. N. M. Milovic, J. D. Badjic, and N. M. Kostic, *J. Am. Chem. Soc.*, 2004, **126**, 696.
36. R. Breslow and B. Zhang, *J. Am. Chem. Soc.*, 1992, **114**, 5882.
37. Y. Liu, Y.-L. Zhao, Y. Chen, *et al.*, *Bioconjug. Chem.*, 2004, **15**, 1236.
38. Y. Liu, Y.-W. Yang, Y. Chen, and F. Ding, *Bioorg. Med. Chem.*, 2005, **13**, 963.
39. B. Tang, F. Liu, K. Xu, and L. Tong, *FEBS J.*, 2008, **275**, 1510.
40. P. Ballester, A. Shivanyuk, A. R. Far, and J. Rebek Jr, *J. Am. Chem. Soc.*, 2002, **124**, 14014.
41. S. M. Biros, E. C. Ulrich, F. Hof, *et al.*, *J. Am. Chem. Soc.*, 2004, **126**, 2870.
42. N. Douteau-Guével, F. Pret, A. W. Coleman, *et al.*, *J. Chem. Soc., Perkin Trans. 2*, 2002, 524.
43. M. Selkti, A. W. Coleman, I. Nicolis, *et al.*, *Chem. Commun.*, 2000, 161.
44. H. Bakirci and W. M. Nau, *Adv. Funct. Mater.*, 2006, **16**, 237.
45. B. Scnatwinkel, M. V. Rekharsky, V. V. Borovkov, *et al.*, *Tetrahedron Lett.*, 2009, **50**, 1374.
46. I. W. Wyman and D. H. Macartney, *Org. Biomol. Chem.*, 2010, **8**, 253.
47. D. M. Bailey, A. Henning, V. D. Uzunova, and W. M. Nau, *Chem.-A Eur. J.*, 2008, **14**, 6069.
48. P. Rajgariah and A. R. Urbach, *J. Incl. Phenom. Macrocycl. Chem.*, 2008, **62**, 251.
49. M. E. Bush, N. D. Bouley, and A. R. Urbach, *J. Am. Chem. Soc.*, 2005, **127**, 14511.
50. L. M. Heitmann, A. B. Taylor, P. J. Hart, and A. R. Urbach, *J. Am. Chem. Soc.*, 2006, **128**, 12574.
51. J. Chin, S. S. Lee, K. J. Lee, *et al.*, *Nature*, 1991, **401**, 254.
52. C. Bazzicalupi, A. Bencini, E. Berni, *et al.*, *Eur. J. Inorg. Chem.*, 2003, 1974.
53. L. Fabbrizzi, M. Licchelli, A. Perotti, *et al.*, *J. Chem. Soc., Perkin Trans. 2*, 2001, 2108.
54. H. Tsukube, S. Shinoda, J. Uenishi, *et al.*, *Inorg. Chem.*, 1998, **37**, 1585.
55. A. Dalla Cort, P. De Bernardin, and L. Schiaffino, *Chirality*, 2009, **21**, 104.

56. M. Chadlaoui, B. Abarca, R. Ballesteros, *et al.*, *J. Org. Chem.*, 2006, **71**, 9030.

57. H.-L. Kwong, W.-L. Wong, C.-T. Yeung, and P.-F. Teng, *Inorg. Chem. Commun.*, 2009, **12**, 815.

58. H.-J. Kim, R. Asif, D. S. Chung, and J.-I. Hong, *Tetrahedron Lett.*, 2003, **44**, 4335.

59. L. Fabbrizzi, P. Pallavicini, L. Parodi, *et al.*, *Chem. Commun.*, 1995, 2439.

60. L. Fabbrizzi, G. Francese, M. Licchelli, *et al.*, *Chem. Commun.*, 1997, 581.

61. B. Verdejo, J. Aguilar, A. Doménech, *et al.*, *Chem. Commun.*, 2005, 3086.

62. B. E. Collins and E. V. Anslyn, *Chem. Eur. J.*, 2007, **13**, 4700.

63. H. Aït-Haddou, S. L. Wiskur, V. M. Lynch, and E. V. Anslyn, *J. Am. Chem. Soc.*, 2001, **123**, 11296.

64. J. F. Folmer-Andersen, V. M. Lynch, and E. V. Anslyn, *J. Am. Chem. Soc.*, 2005, **127**, 7986.

65. D. Leung and E. V. Anslyn, *J. Am. Chem. Soc.*, 2008, **130**, 12328.

66. M. A. Hortalá, L. Fabbrizzi, N. Marcotte, *et al.*, *J. Am. Chem. Soc.*, 2003, **125**, 20.

67. J. F. Folmer-Andersen, V. M. Lynch, and E. V. Anslyn, *Chem. Eur. J.*, 2005, **11**, 5319.

68. A. Buryak and K. Severin, *Angew. Chem. Int. Ed.*, 2004, **43**, 4771.

69. A. Buryak and K. Severin, *J. Am. Chem. Soc.*, 2005, **127**, 3700.

70. S. Rochat, J. Gao, X. Qian, *et al.*, *Chem. Eur. J.*, 2010, **16**, 104.

71. H. Imai, H. Munakata, Y. Uemori, and N. Sakura, *Inorg. Chem.*, 2004, **43**, 1211.

72. V. Villari, P. Mineo, N. Micali, *et al.*, *Nanotechnology*, 2007, **18**, 1.

73. M. Sirish, V. A. Chertkov, and H.-J. Schneider, *Chem. Eur. J.*, 2002, **8**, 1181.

74. H. Tsukube, M. Wada, S. Shinoda, and H. Tamiaki, *Chem. Commun.*, 1999, 1007.

75. C. Schmuck, M. Heil, J. Scheiber, and K. Baumann, *Angew. Chem. Int. Ed.*, 2005, **44**, 7208.

76. C. Schmuck and L. Geiger, *Curr. Org. Chem.*, 2003, **7**, 1485.

77. C. Schmuck and M. Heil, *Chem. Eur. J.*, 2006, **12**, 1339.

78. C. Schmuck and P. Wich, *Angew. Chem. Int. Ed.*, 2006, **45**, 4277.

79. C. Schmuck, D. Rupprecht, M. Junkers, and T. Schrader, *Chem. Eur. J.*, 2007, **13**, 6864.

80. J. Rebek, B. Askew, D. Nemeth, and K. Parris, *J. Am. Chem. Soc.*, 1987, **109**, 2432.

81. M. Fokkens, T. Schrader, and F.-G. Klärner, *J. Am. Chem. Soc.*, 2005, **127**, 14415.

82. M. Hariharan, S. C. Karunakaran, and D. Ramaiah, *Org. Lett.*, 2007, **9**, 417.

83. L. Yang, S. Qin, X. Su, *et al.*, *Org. Biomol. Chem.*, 2010, **8**, 339.

Ion-Pair Receptors

Antonella Dalla Cort

Universitá "La Sapienza," Roma, Italy

1 INTRODUCTION

The development of synthetic receptors able to bind charged species in efficient and selective modes is one of the most challenging research topics in supramolecular chemistry. Historically, the evolution of this discipline has been, and still is, deeply related to the search of hosts able to mimic recognition events occurring in Nature. Crucial is indeed the role played by ions in biological processes. Many enzyme substrates and cofactors[1] are anionic, DNA itself is a polyanion, whereas several neurotransmitters (i.e., dopamine, serotonin, or acetylcholine) are ammonium ions, and cations such as sodium and calcium are involved in fundamental biological mechanism critical to the maintenance of life, for example, the conductance process in human red blood cells.[2] Nevertheless, the greatest interest of researchers in this area, at least up to now, has been mainly devoted to the development of artificial receptors that selectively bind either a cation or an anion. In the last decades, several receptors, acyclic and macrocyclic, with high affinities for cations have been reported.[3] Past years have also witnessed a fast increase, after a slower start, in the number of papers devoted to the synthesis and properties of receptors for anions.[4,5] All these efforts have resulted in less attention being payed to the role of counterions in modulating the strength and the selectivity of the recognition process.[6] We must consider that the two partners in the ion pair are hardly present as separate ions unless the medium is strongly solvating. Salts generally exist as solvent-separated ion pairs, contact ion pairs, or aggregated contact ion pairs (Figure 1).[7] Obviously, competition for the charged guest species between the host and the counterion is almost inevitable when using uncharged abiotic hosts in organic solvents. A way to circumvent the problem consists of using noncompetitive counterions, low demanding in terms of the formation of associated (contact) ion pairs, such as tetrabutylammonium (TBA) cation or tetraarylborate and picrate anions. But, of course, in the real world of the biological processes and of the potential applications of these receptors to practical problems such as salt extraction or transport across membranes (symport processes), use of cations such as those that were just mentioned is not implicated. These considerations lead to the conclusion that counterion competition cannot be neglected since it might hinder the single-ion recognition process leading to an ambiguous interpretation of the phenomena; therefore, other ways of tackling this subject should be found.

A strategy that has become quite popular in recent years implies the design of receptors capable of simultaneously forming complexes with both a cation and an anion.[8,9] These heteroditopic hosts endowed with specific cation- and anion-binding sites present a number of advantages, first being the fact that the complex formed by the receptor and the ion pair is charge neutral and can help in salt solubilization, extraction, detection, and symport processes. Moreover, as described later in this chapter, heteroditopic hosts may display affinities for ion pairs or for the single

Supramolecular Chemistry: From Molecules to Nanomaterials.
Edited by Philip A. Gale and Jonathan W. Steed.
© 2012 John Wiley & Sons, Ltd. ISBN: 978-0-470-74640-0.

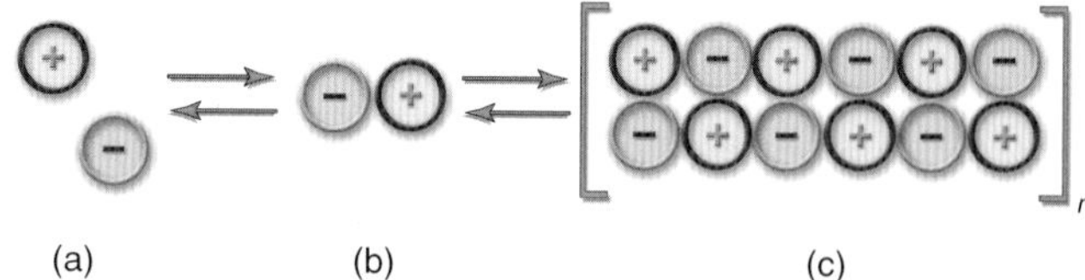

Figure 1 Possible states of ions in solution: (a) solvated ions, (b) contact ion pairs, and (c) aggregated contact pairs.[7]

component of the pair, which are increased with respect to simple ion receptors. Often this is a direct consequence of allosteric effects and enhanced electrostatic interactions between the cobound ions.

1.1 Classifications of ion-pair receptors

Usually, receptors designed for ion-pair complexation are molecules comprising well-known anion-binding motifs, for example, Lewis-acidic centers, and familiar cation-binding motifs, such as crown ethers, calixarenes. The distance occurring between these sites classifies receptors as follows:

1. *Cascade receptors* are those in which the distance between the two binding functionalities is so short as to allow cooperative contact ion-pair binding. These receptors first bind one of the ions, often it happens to be the cation, which becomes the new binding site for the counterion, showing a higher affinity because of maximized coulombic attraction.
2. *Heteroditopic receptors* are those in which the distance between the two binding sites allows the accomodation of solvent-bridged ion pairs, in which solvent molecules help link the anion to the cobound cation.
3. Finally, receptors that possess *two compartmentalized* and independent binding sites for which the complexation is that of host-separated ion pairs (Figure 2).

Ion-pair receptors can also be classified by considering their possibility of binding ion pairs in either a sequential or a simultaneous fashion, as recently examined by Sessler in his review.[8] Of course, this classification includes the previous one, since the result of a sequential binding is a cascade receptor, and simultaneous binding can be performed by a heteroditopic receptor with proximal or compartmentalized binding sites. In the case of sequential binding, the receptor can interact with one ion of the pair. Consequently, the affinity toward the counterion may enhance through allosteric effect. For example, the association with the first ion can induce the receptor conformational changes that influence the specificity and the affinity of the second binding site, thus resembling the regulation of enzyme activities.[10]

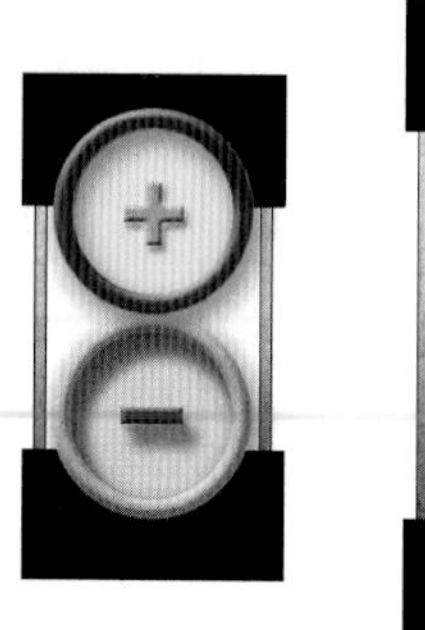

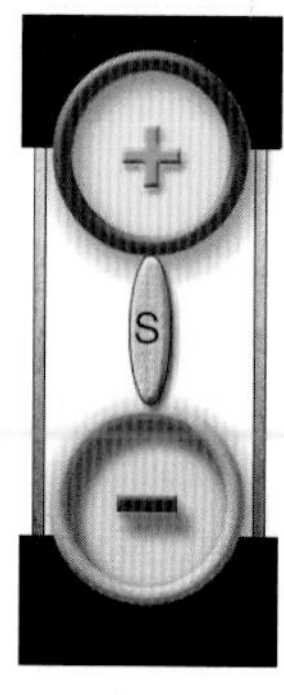

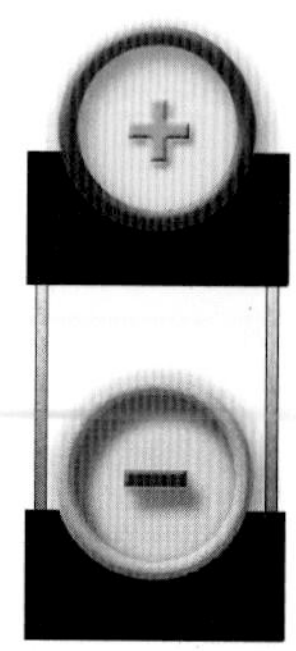

Figure 2 Schematic representation of ion-pair receptors (S, solvent).

Positive cooperativity can be thought as an optimal mode of binding in which the various discreet binding events that collectively drive a host–guest interaction are used to their full ability, and indeed, go over the binding strength that could be expected from the sum of the individual Gibbs free energies of binding. Achieving positive cooperativity in host–guest systems would signify a comprehensive understanding of binding phenomena, a major goal of the supramolecular chemistry community.

Schröder *et al.*[11] reported the example of a functional heteroditopic molecule, **1**, which contains both an azathioether macrocyclic unit and an appended acylurea component for metal and anion coordination, respectively. In the absence of any potential guest, an internal hydrogen bond is observed between the macrocyclic nitrogen atom and the urea pendant group, whereas an additional intramolecular hydrogen bond is formed between the outermost urea hydrogen and the acyl oxygen, thus holding the urea in a twisted conformation, that is, hydrogens pointing in opposite directions. The crystal structure of the compound (Figure 3) clearly shows the presence of both types of hydrogen bonds.

The addition of 1 equivalent of silver nitrate ($AgNO_3$) induces the reorganization of the macrocycle to the endo conformation, which is necessary to accommodate the metal cation. The direct consequence is disruption of the internal hydrogen bonding and subsequent counteranion binding by the "free" pendant unit. Thus, the complexation of the cation acts as a switch for selective anion coordination showing cooperative ion-pair binding.

Another example is reported by Kubik[12] and includes cyclopeptides composed of natural amino acids and 3-aminobenzoic acid in an alternating sequence. It was found that the cation affinity of receptor **2** depends on the anion present in solution. With anions, such as phosphonates and sulfonates, that bind to the NH groups of the cyclopeptides, allosteric effects have been observed. NMR and FT-IR (Fourier transform infrared) spectroscopic investigations show that these anions stabilize a particular conformation of the originally relative flexible host, which is ideally suited

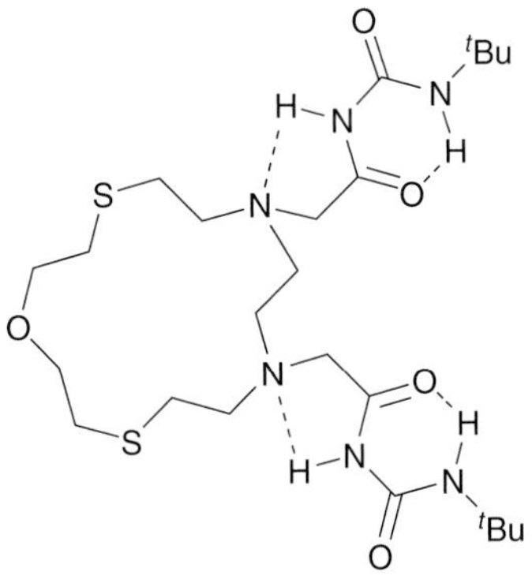

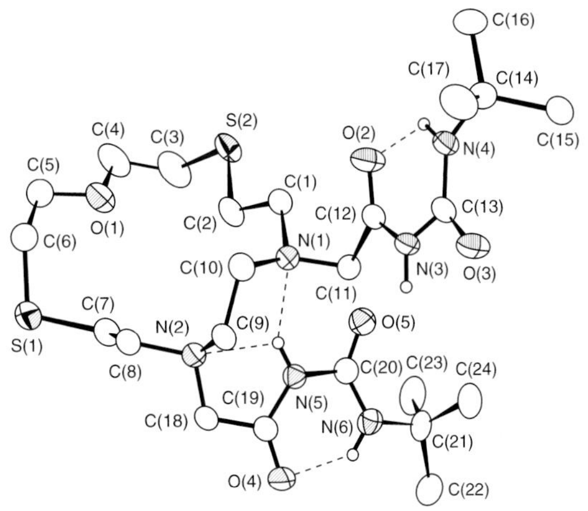

Figure 3 Crystal structure of **1** showing the twisted conformation of the pendant arms, intramolecular hydrogen bonding, and exo arrangement of the N and S donor atoms. Displacement ellipsoids are drawn at the 50% probability level. (Reproduced from Ref. 11. © Royal Society of Chemistry, 2003.)

for an interaction with cations. Binding of the positively charged guests with the cyclopeptide–anion complexes is due to cation–π interactions with the aromatic subunits of the peptide as well as electrostatic interactions with the anion. Thus, the positive cooperativity of the counteranions is an effect of the conformational reorganization of the receptor, and, in this case, it is also due to the proximity between the cation and anion in the final complex. The combination of both binding mechanisms leads to an increase in complex stability by a factor of 10^3–10^4 in comparison to that of cation complexes without the influence of counter anions.

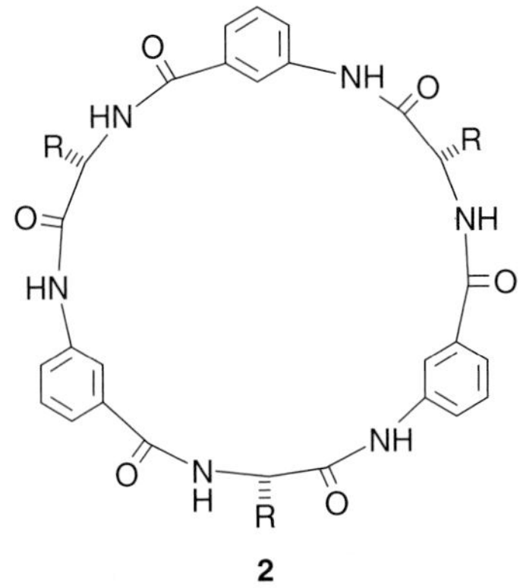

2

R = $CH_2CH_2COOCH(CH_3)_2$

The case of calix[4]pyrrole, **3**, described by Sessler *et al.*[13] as a stoichiometric ion-pair receptor is quite similar. The origin of the cation dependence on the anion-binding affinities (i.e., the observed cooperative ion-pair effect) clearly has a structural basis. Calix[4]pyrrole, **3**, in its uncomplexed form is extremely flexible and interconverts rapidly between all possible conformations while on average favoring the 1,3-alternate arrangement. On interaction with a tightly bound anion, such as benzoate or chloride anion, it becomes frozen into its so-called cone conformation. This structural arrangement creates a cavity distal to the pyrrole NH donor atoms that can complex a tetraalkylammonium cation of suitable size and shape. The increase in the resulting anion-binding association constant values up to a 10^3-fold variation in the case of benzoate or chloride in dichloromethane is highly dependent on the choice of countercation (Scheme 1, Figure 4).

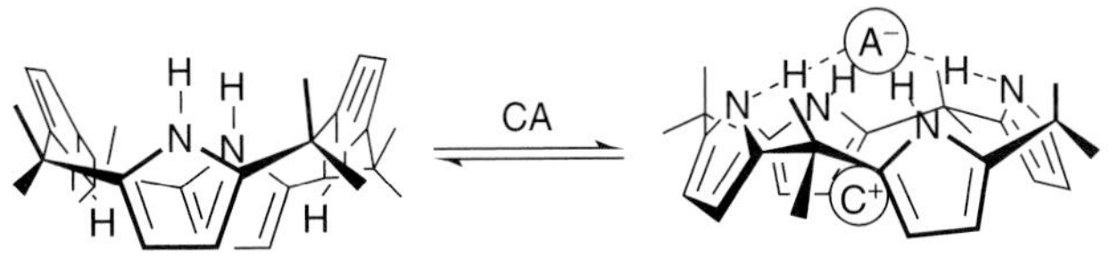

Scheme 1 Proposed calix[4]pyrrole, **3**, binding motif (C, cation; A, anion).

Cooperativity in binding potassium cyanide is shown in the system reported by Miyaji *et al.*,[14] in which the receptor, a ferrocenyl derivative decorated with a crown ether and a trifluoroacetylcarboxanilide group, **4**, recognizes anions by the formation of a reversible covalent adduct.

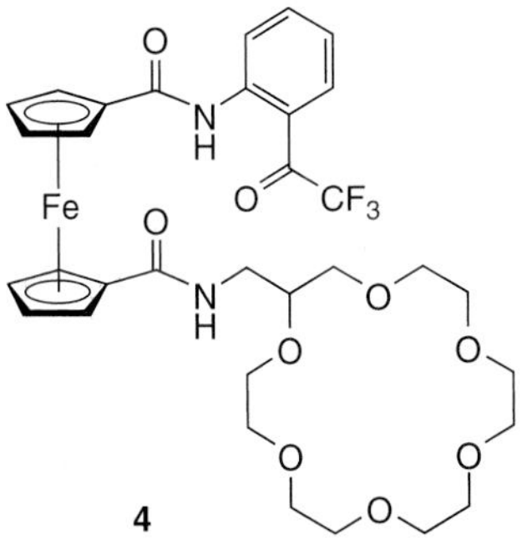

4

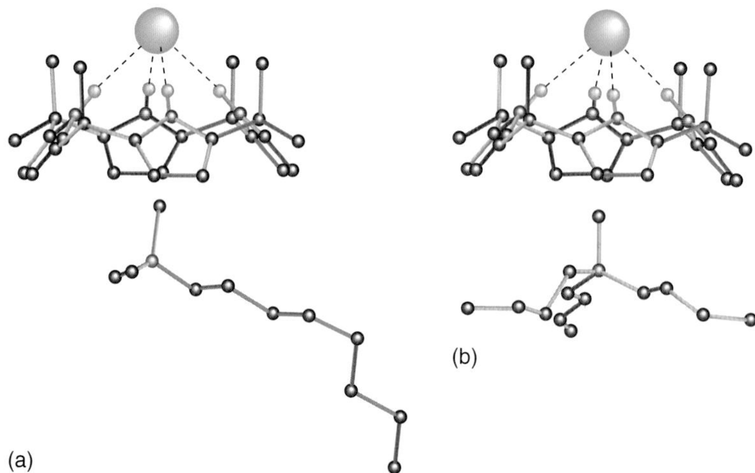

Figure 4 X-ray crystal structure of (a) **3**, octyltrimethylammonium chloride; and (b) **1**, methyltributylammonium chloride. Both structures have dichloromethane. Solvent and CH hydrogens removed for clarity. (Reproduced from Ref. 13. © Wiley-VCH, 2008.)

The binding motif for anion recognition is *o*-trifluoroacetylcarboxanilide (TFACA) that is known to efficiently recognize and sense anions such as cyanide and carboxylates. The idea is that the alkoxide adduct resulting from the addition of cyanide to the carbonyl group may interact with a cation in close proximity by means of electrostatic interactions. Ferrocene was chosen as a spacer and as an electroactive label in which the cyclopentadienyl (Cp) units can rotate like a ball bearing, readily adjusting the two binding sites to a specific substrate in such a way to favor the ion-pair complex formation. Also, in this case, a conformationally flexible receptor can favor the formation of the ternary complex, showing a large positive cooperative effect ($\sim 10^2$) (Scheme 2).

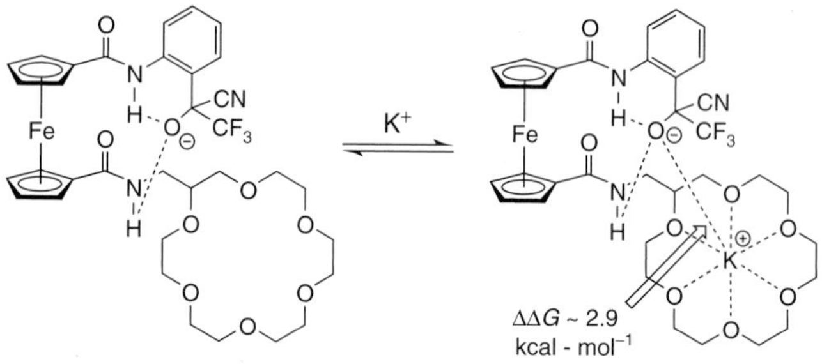

Scheme 2 Suggested binding mode for receptor **4**

Another interesting example of a heteroditopic receptor capable of cooperative recognition of ion pairs through a contact binding mechanism was recently reported by Beer and coworkers.[15] Receptors **5** and **6** were designed to provide a good complementarity to contact ion pairs.

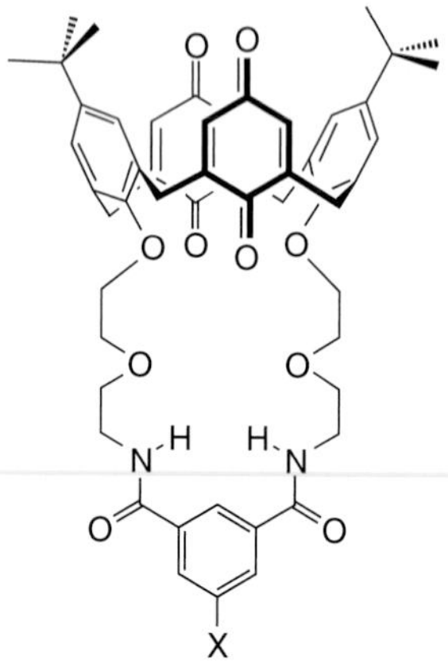

5, X=H
6, X=NO_2

The addition of increasing amounts of tetrabutylammonium chloride (TBACl) was found to induce only negligible downfield shift in the signals corresponding to the amide and isophthalyl protons (^{1}H NMR in d_3-acetonitrile), thus indicating that no anion binding was occurring. On the other hand, the addition of 1 equivalent of TBACl to a 1 : 1 mixture of the receptor and a group 1 metal or ammonium salt results in the broadening of the above-mentioned signals and a marked downfield shift. This interaction was estimated to be higher than $10^4\ M^{-1}$. Thus the receptors show absolutely no affinity for chloride in the absence of a suitable cationic guest, but display a strong affinity toward the halide anion when a suitable coordinating cationic species (sodium, potassium, ammonium, etc.) is present. This is a phenomenon of ion-pair cooperativity, in which the receptor binds an ion pair really strongly such that no affinity for either of the "free" ions is observed. The recognition is postulated to arise from the self-inhibition of the receptor by intramolecular hydrogen bonds, which may only be disrupted by the presence of both a suitable cation and anion with change of conformation (Figure 5).

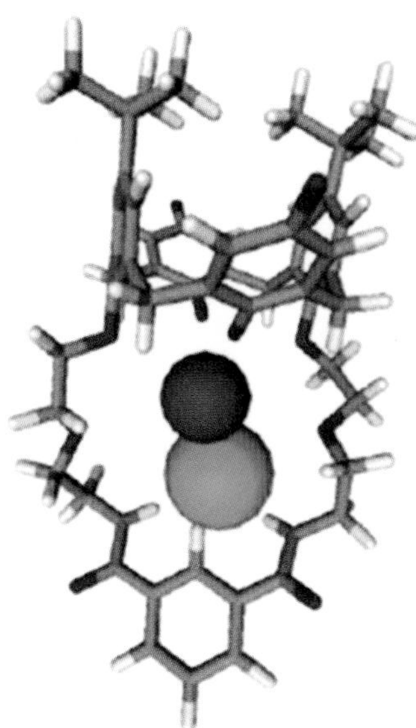

Figure 5 Optimized CAChe model of **5**·KCl recognition. Potassium is in purple, chloride in green. (Reproduced from Ref. 15. © Wiley-VCH, 2008.)

A well-known example of the cooperative binding of ion pairs was from the pioneering work of Smith *et al.*[16] A ditopic, macrobicyclic receptor with adjacent anion- and cation-binding sites, **7**, is able to distinguish between various monoalkylammonium salts by binding them as contact ion pairs. In the case of *i*-$PrNH_3Cl$ and *n*-$PrNH_3Cl$, the association constants are 2.0×10^2 and $2.0 \times 10^4\ M^{-1}$, respectively. The strong steric selectivity exhibited by the receptor for *n*-$PrNH_3Cl$ over *i*-$PrNH_3Cl$ is easily understood if we imagine that selectivity is due to the deep penetration of the ammonium cation into the receptor cavity, driven by the electrostatic attraction to the simultaneously bound Cl^-. Furthermore, receptor **7** has affinity for *n*-propylammonium cation, which is counteranion dependent. The affinity for linear *n*-propylammonium chloride is at least two orders of magnitude greater than that for *n*-propylammonium acetate and *n*-propylammonium *p*-toluenesulfonate. So in this case, the identity of the counteranion can be used to modulate receptor/cation affinity.

7

In 2005, Nabeshima *et al.*[17] from Japan reported the synthesis and the binding behavior of a receptor, in which three different types of ion-binding sites are arranged on a calix[4]arene skeleton. The result was a perfect biomimetic model capable of the effective and efficient multistep regulation of anion recognition by utilizing two different cationic guests. On the lower rim of the conical calix[4]arene framework, compound **8** possesses two ester substituents and two polyether units containing a urea group linked to a bipyridine moiety. The mechanism of the stepwise regulation of anion recognition is illustrated in Figure 6.

The ^1H NMR titration in $CDCl_3/CD_3CN$ (9 : 1) showed quantitative 1 : 1 complexation of **8** with Na^+ or Ag^+. As the concentrations of Na^+ increased, the signals of free **8** decreased and new signals of **8**·Na^+ appeared. For Ag^+, instead, signals with averaged chemical shifts between free **8** and **8**·Ag^+ were observed. Hence, the capture and release of Na^+ and Ag^+ are slow and fast on the NMR timescale, respectively. The experiments clearly indicate the quantitative formation of the ternary complex **8**·Ag^+·Na^+. The soft and hard cationic guests are bound by the bipyridine and ester units, respectively. In contrast, the anion affinity of receptor **8** is considerably weaker. But this ability remarkably increases in the presence of the two bound cations Na^+ and Ag^+. On complexation of the bipyridine moieties with a cationic guest, a conformational change occurs in **8** that brings the two urea moieties in close proximity, and this change favors anion binding. The enhancement of the binding constants in the end reaches factors of 1500 and 2000 for NO_3^- and $CF_3SO_3^-$, respectively, when compared to the free **8**. This large stepwise enhancement is reasonably ascribed to the electrostatic interactions between the anions and cations. As far as we know, this has been the first description of a multiresponsive anion receptor that is regulated stepwise by different cationic effectors.

The few selected examples reported till now illustrate situations in which the binding of the ion pair is most probably sequential and the initial binding of one of the two ions significantly alters the affinity toward the counterion.

In the case of receptors that can accommodate solvent-bridged ion pairs or that are provided with well-separated binding sites for the two partners of the ion pair, the inherent binding ability toward one of the ion is not expected to be significantly altered by the presence of the other ion. But some exceptions can be found, thus confirming the well-established limits of any classification.

The ion-pair receptor **9** was designed to bring together an anion-binding core and a cation-recognizing subunit in such a way that a large separation between the constituent ions

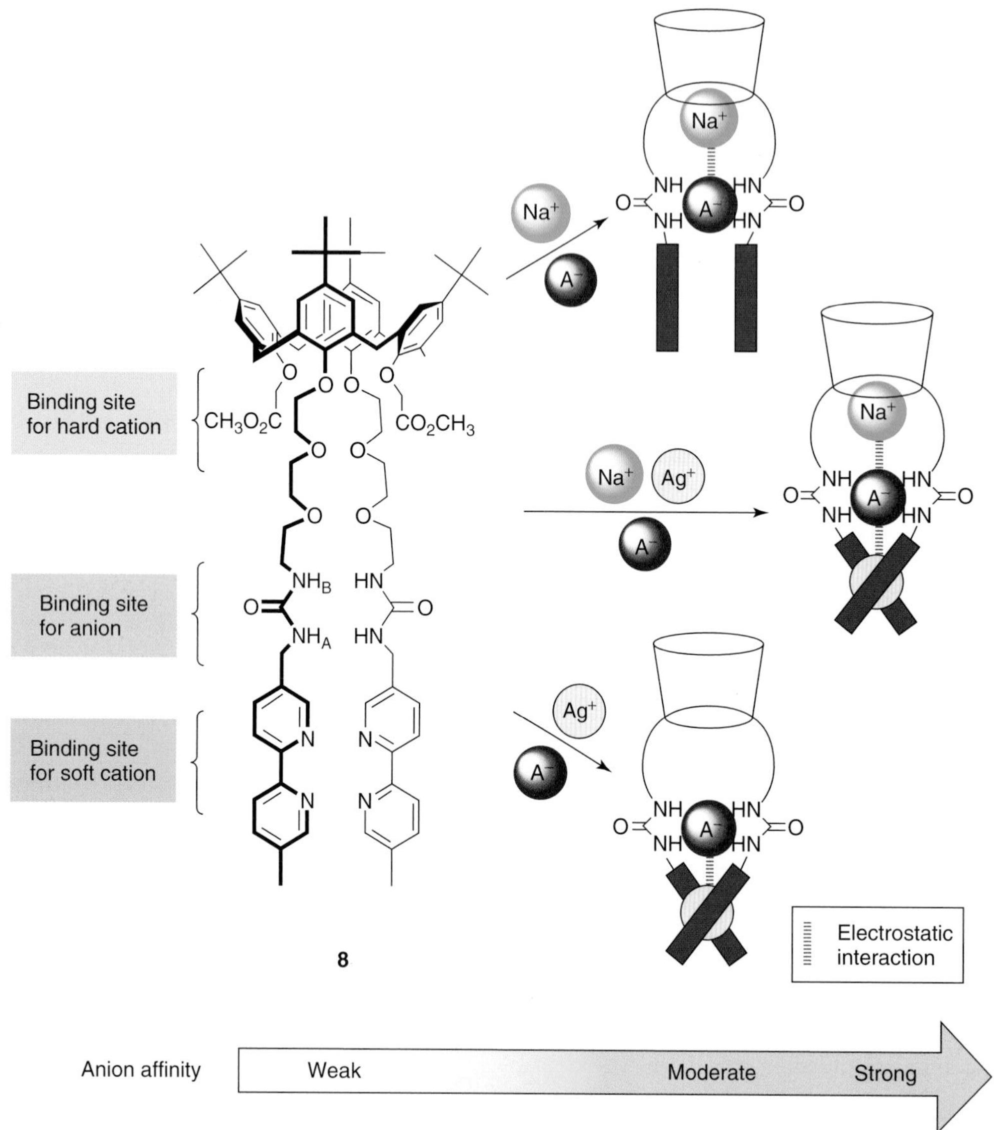

Figure 6 Anion recognition by host **8** is regulated by two different cationic effectors. (Reproduced from Ref. 17. © American Chemical Society, 2005.)

of a bound ion pair would be enforced.[18] Calix[4]pyrrole and calix[4]arene crown-6 were chosen as the anion- and cation-binding units, respectively. The X-ray crystal structure provides support to the formation of a stable 1 : 1 complex with CsF in spite of the large separation imposed between the anion and the cation. ^{1}H NMR studies carried out in 10% (v/v) CD_3OD in $CDCl_3$ in the presence of tetrabutylammonium fluoride (TBAF) provide no evidence of fluoride anion binding, even after two days when **9** was treated with 5 equivalents of salt. In contrast to what was seen with TBAF, addition of 5 equivalents of cesium perchlorate induces remarkable changes in the signals for both the aromatic protons of the calix[4]arene core and the aliphatic protons of the crown-6 ring, consistent with the expectation that the addition of $CsClO_4$ leads to the formation of a cation-bound complex wherein the cesium cation is encapsulated in the calix[4]arene crown-6 ring and the perchlorate anion is bound either weakly or not at all by the calix[4]pyrrole core. The addition of 5 equivalents of CsF to a solution of receptor **9** in the same solvent mixture leads to spectral variations indicating that the receptor binds both Cs^+ and F^- ions behaving as an ion-pair complex **9**·CsF in a manner analogous to what is seen in the solid state. It was thus concluded that the binding of the cesium cation to the crown ether ring plays a very important role in inducing binding of the fluoride anion to the calix[4]pyrrole moiety of receptor **9**, which is otherwise not observed in the absence of Cs^+ in this solvent system. None of the other cations tested (specifically, Li^+, Na^+, and K^+) were found to produce such an effect (Figure 7).

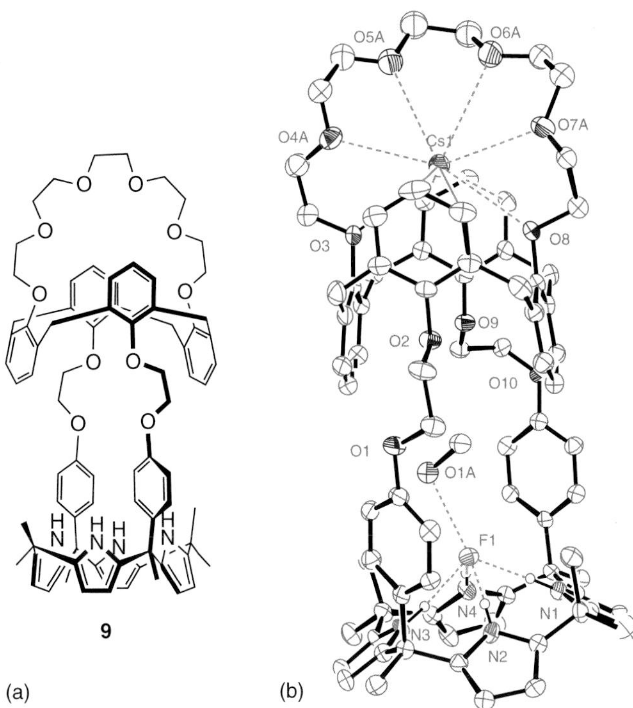

Figure 7 (a) Crown-6-calix[4]arene-capped calix[4]pyrrole **9**. (b) The **9**·CsF complex showing a partial atom-labeling scheme. Displacement ellipsoids are scaled to the 30% probability level. (Reproduced from Ref. 18. © American Chemical Society, 2008.)

A receptor for solvent-separated ion pairs is receptor **10**,[19] a combination of a dibenzo-18-crown-6 and a bridging 1,3-phenyldicarboxamide. In the presence of 1 mol equivalent of metal cation, chloride affinities are enhanced in the following order: K^+ (ninefold enhancement), Na^+ (eightfold enhancement), and Cs^+ (no enhancement). An X-ray crystal structure shows that the receptor binds sodium chloride as a solvent-shared ion pair (Figure 8).

The central cavity of the macrocycle contains either a chloroform molecule (62% relative occupancy) or two water molecules (38% relative occupancy) that are not shown. The molecular dipole of the chloroform is aligned with the dipole generated by the **10a**·NaCl complex. Na^+ is complexed within the dibenzo-18-crown-6 with average Na–O distances of 2.67 Å and is also coordinated by an axial water molecule. A Cl^- ion is hydrogen bonded to the two receptor NH residues (Cl–N distances are 3.34 and 3.31 Å; N–H–Cl angles are 152 and 177°).

2 STRUCTURAL KEY-MOTIFS IN ION-PAIR RECEPTORS

The following part focuses on a number of heterotopic ion-pair receptors reported in the literature that feature simultaneous complexation of both cationic and anionic species. The combination of structural motifs used to coordinate the ionic species is the interpretative key that is principle to this session.

2.1 Receptors employing Lewis-acidic centers for anion recognition

2.1.1 *With donor units for cation recognition*

Reetz and coworkers in 1991[20] provided an early example of a receptor in which a crown ether is covalently linked to a Lewis-acidic boron center, **11**. This compound is able to complex potassium and fluoride ions simultaneously. The F^- ion binds to the Lewis-acidic boron center, whereas the potassium ion is captured by the crown ether moiety. It is established that ether oxygens are good donor groups for alkali metals. One perceptible consequence of the binding is that this receptor allows the otherwise insoluble salt, KF, to dissolve in dichloromethane within 4 h.

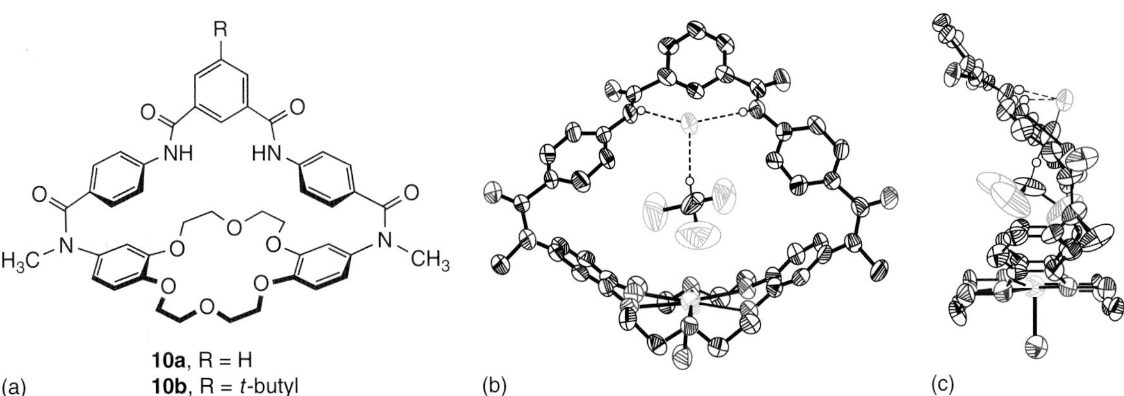

Figure 8 (a) The structure of compounds **10**. (b, c) Front and side views of the X-ray crystal structure of [**10a**·Na^+·$CHCl_3$·Cl^-] showing 50% probability ellipsoids. Absent are disordered solvent molecules found in the lattice voids away from the macrocyclic cavity. (Reproduced from Ref. 19. © American Chemical Society, 2000.)

11

The Lewis-acidic uranyl dication, UO_2^{2+}, center has been used by Reinhoudt and his group to build up receptors based on uranyl–salen complexes and crown ether units, **12**. In particular, the two benzo[15]crown-5 units form a sandwich-like complex with potassium while the metal binds to the dihydrogen phosphate counterion.[21]

12

A similar strategy can be used, employing a zinc metal center that acts as an efficient binding site for anions. A couple of ditopic porphyrin receptors, **13** and **14**, incorporating a diaza-15-crown-5 unit and a diaza-18-crown-6 unit, respectively, were synthesized by Liu and Chen.[22] The Zn-porphyrin moiety is well known to coordinate nitrogen ligands. The two artificial receptors, **13** and **14**, are indeed able to selectively recognize sodium cyanide and potassium cyanide in a polar solvent, such as methanol. Binding affinities were measured by UV–vis and ^{1}H NMR titrations that pointed out that sodium and potassium cyanide were bound in a ditopic fashion (Figure 9). In the case of NaCN, its addition to a solution containing **13** caused the solution color to change from purple to pale green.

2.1.2 *With π-systems for cation recognition*

Crown ethers are quite ineffective for the recognition of quaternary ammonium salts. The recognition of these salts can still imply coordination of the anion to a Lewis acid center, but the presence of these salts in the structure of aromatic subunits able to interact with the counterion through cation–π interactions can provide the second binding site. Cation–π interaction can be defined as the noncovalent molecular interaction occurring between the face of an electron-rich π system (e.g., benzene, ethylene) and an adjacent cation. It plays an important role in molecular recognition and is used in nature. For example, the side chains of the aromatic amino acids, phenylalanine, tyrosine, and tryptophan, provide a surface of negative electrostatic potential that binds to a wide range of cations by a predominantly electrostatic interaction.[23] Such an interaction has been used to build up ditopic receptors for ion pairs.

It is known that complex **15** formed by the salophen ligand and the uranyl dication binds strongly to hard anions in organic solvents.[24,25] The introduction to the ligand skeleton of appended aromatic units gives the possibility to establish stabilizing cation–π interactions with tetraalkyl cations. It has been established that uranyl–salophen complexes decorated with aromatic sidearms behave as effective ion-pair receptors for tetraalkylammonium halides and iminium salts in solution and in the solid state. Consistent with the hard Lewis acid character of the uranyl, the higher the binding affinities for complexation of quaternary salts to uranyl–salophen receptors in solution, the harder the anion, confirming the idea that a major driving force for

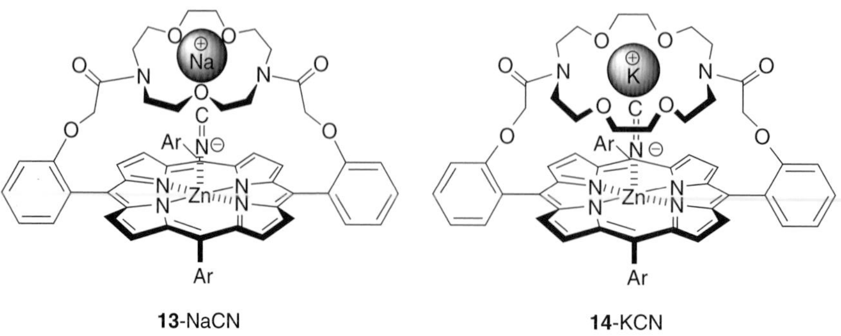

Figure 9 Proposed ditopic binding modes of receptors **13** and **14**. (Reproduced from Ref. 22. © Elsevier, 2005.)

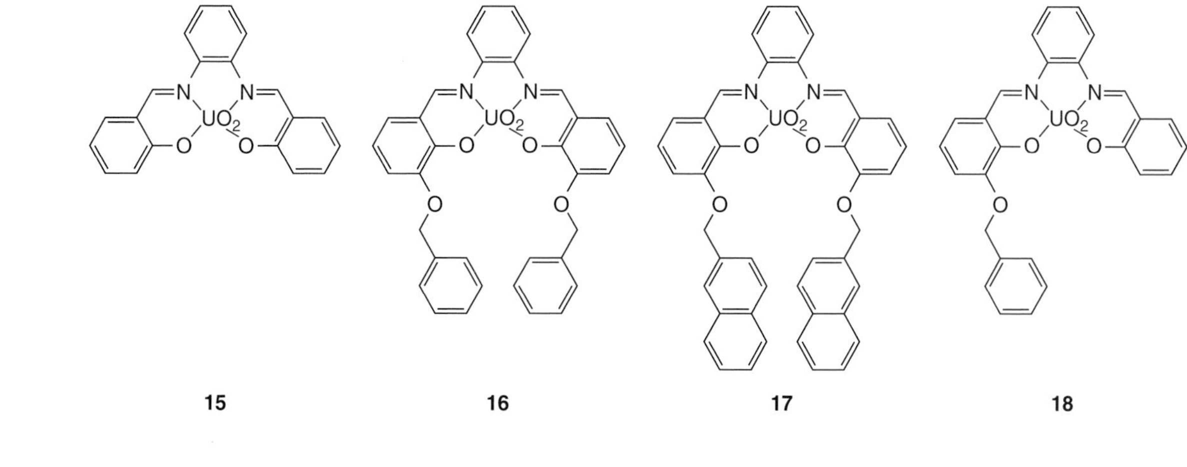

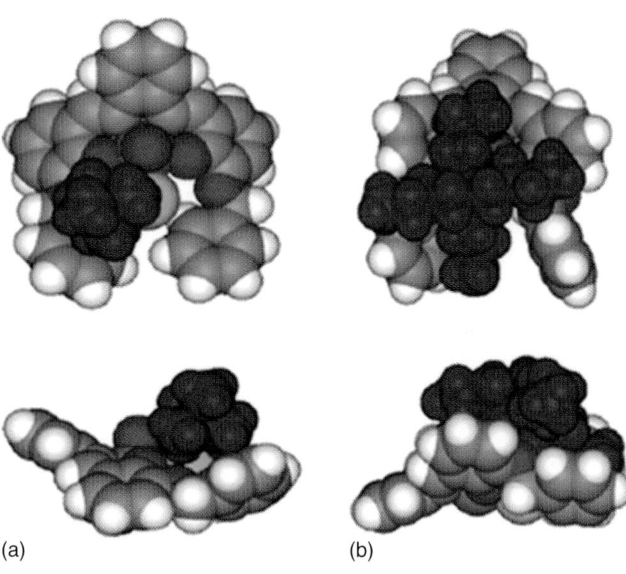

Figure 10 VDW (Van der Waals) presentations of the crystal structure of **16**·TMACl (a) and **16**·TBACl (b), top and side views. TMA and TBA cations are dark red, and chloride is green. For the sake of clarity, only one of the two ligand–TBACl pairs in the asymmetric unit is shown. (Reproduced from Ref. 26. © Royal Society of Chemistry, 2003.)

complexation arises from anion coordination to the metal center. Less important, yet significant, contributions to complex stability in solution arise from cation–π/CH–π interactions of the quaternary ions with the aromatic pendants. Compared with the parent receptor, the one without the aromatic arms, receptors **16** and **17** were found to bind tetramethylammonium chloride (TMACl) and TBACl with higher efficiency in $CHCl_3$. The existence of stabilizing cation–π interactions and CH···O/Cl^- hydrogen bonds to oxygens of the ligand and to the chloride anion coordinated to uranium (closest donor–acceptor distances are 3.16–3.38 Å for C···O and 3.68–3.97 Å for C···Cl interactions) is also revealed by solid-state structures of complexes of receptor **16** with TMACl and TBACl (Figure 10).[26]

Receptors **16** and **17** also form complexes with alkaline halide salts in the solid state.[27] In particular, the solid-state structures of complexes formed by the two-armed receptor **16** with CsF and with the chlorides of K^+, Rb^+, and Cs^+ reveal the existence of dimeric supramolecular assemblies in which two receptor units assemble into capsules fully enclosing $(MX)_2$ ion quartets (Figure 11). In addition to the strong coordinative binding of the anion to the uranyl center and to electrostatic cation–anion interactions, stabilizing interactions arise from coordination of each cation with six oxygens, three from each receptor, and most importantly to two aromatic sidearms belonging to different receptors. ^1H NMR data provide strong evidence that cation–π(arene) interactions with the sidearms participate in binding also in solution. Receptor **18**, having only one appended benzylic arm, was found to bind the CsCl ion pair, but in this case, four uranyl–salophen units instead of two are assembled in a capsule-like arrangement housing a $(CsCl)_2$ ion quartet.

A series of organotin-substituted crown ethers, **19**, were prepared by Jurkschat *et al.*[28] In these derivatives, the

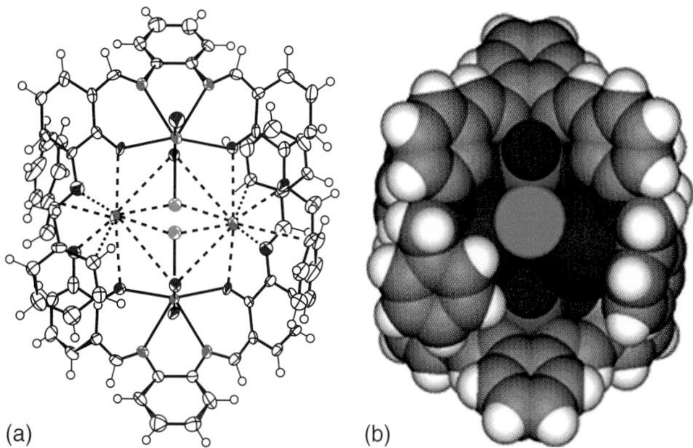

Figure 11 Dimeric assembly of the CsCl complex of **16** presented as (a) Ortep plot (50% probability level) and (b) VDW presentation (chloride, green). (Reproduced from Ref. 27. © American Chemical Society, 2005.)

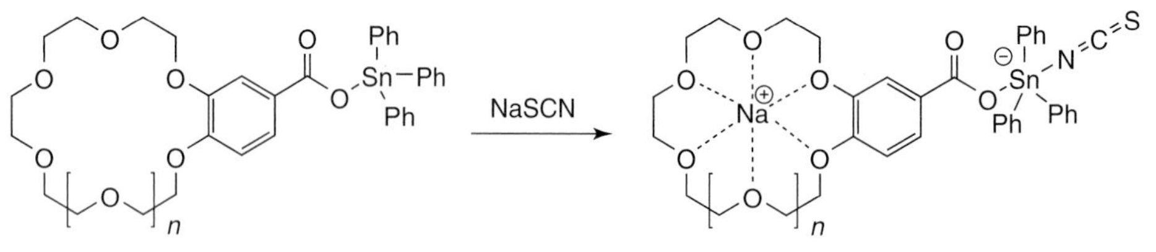

19, $n = 0, 1$

Figure 12 Binding of NaSCN by compound **19**.

O–Sn intramolecular coordination is broken by addition of a chloride anion (as its $[(Ph_3P)_2N]^+$ salt) or a sodium cation (as its Ph_4B^- salt), as well as by addition of sodium chloride or sodium thiocyanate, thus forming the corresponding ditopic complex (Figure 12).

By linking a bicentric organotin-based Lewis acid with a crown ether moiety, the same group obtained a robust ditopic host $Ph_2ISnCH_2Sn(I)(Ph)$-CH_2-[16]crown-5, **20**, that, in contrast to its monotin-substituted analog $Ph_2ClSnCH_2$-[16]crown-5, is capable of overcoming the high lattice energy of sodium fluoride and solubilizing the latter in organic solvents, such as acetonitrile.[29]

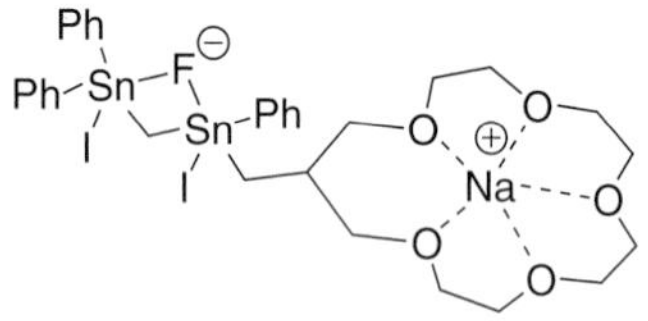

20 · NaF

2.2 Receptors employing hydrogen bonds for anion recognition

This chapter reports some examples of ditopic receptors that use hydrogen-bonding interactions to recognize anions. Lately, many binding motifs have been employed for this purpose (amide, indole and pyrrole units, urea and thiourea moieties, hydroxyl units, etc.) since the area of anion complexation continues excitingly to grow fast. Thus, the following examples should not be regarded as being comprehensive but just descriptive of the synthetic design chosen to reach the goal.

2.2.1 With donor units for cation recognition

In 2000, Kubo and coworkers[30] synthesized an ion-pair receptor by attaching two thiourea groups to macrocyclic dibenzo-diaza-30-crown-10, **21**. The use of the highly flexible cycle was suggested by the large conformational change that this compound can undergo on complexation. The incorporation of thiourea units as an efficient anion-binding site into the building block creates a multisite receptor for the unique cooperative recognition of cationic and anionic species. Quantitative assessment of the cooperative ion-pair binding properties was made by ^1H NMR titration using CD_3CN solution. The K^+-assisted organization of the receptor that leads the dibenzo-30-crown-10 portion of it to wrap around the bound cation also brings the two thiourea moieties into close proximity, providing a preorganized binding site for the phosphate anion. Compared to the cation-free form of the receptor

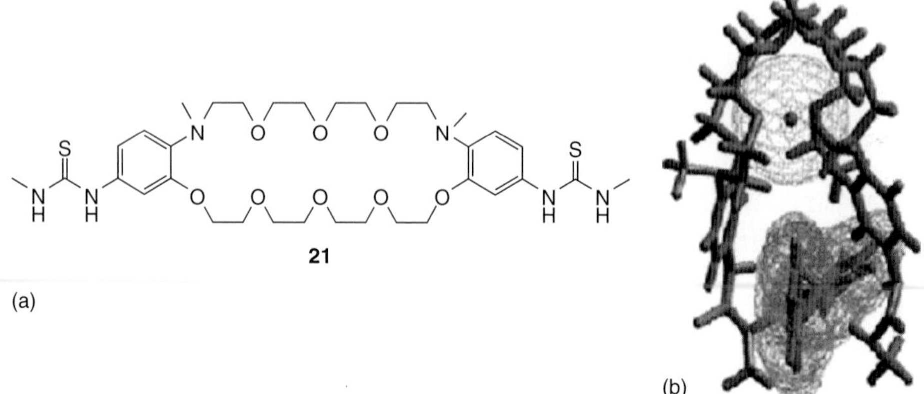

Figure 13 (a) Compound **21**. (b) An energy-minimized complex structure $[\mathbf{21}\cdot K^+\cdot(PhO)_2P(O)O^-]$. (Reproduced from Ref. 30. © Elsevier, 2000.)

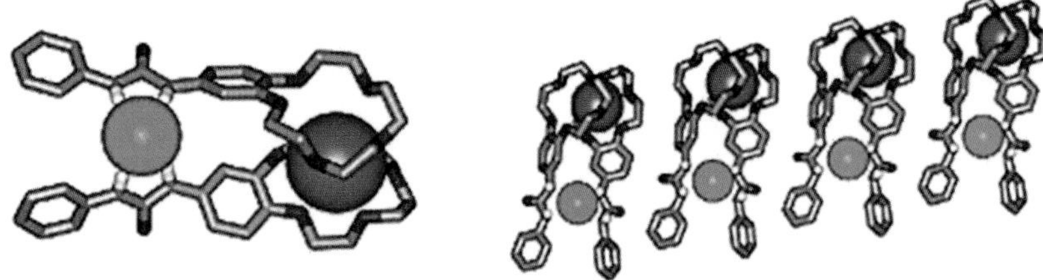

Figure 14 Crystal structure of the complex **22**·NaCl. (a) Stick representation. (b) Side view of the crystal packing. (Reproduced from Ref. 31. © American Chemical Society, 2003.)

21 ($K_a = 490\,M^{-1}$), the association constant values are significantly enhanced for the phosphate anion in the presence of K^+ ($K_a = 9200\,M^{-1}$). Similar effect, although less pronounced, is obtained with Cs^+. Thus, the establishment of cooperative interactions requires a prebound Na^+ cation (Figure 13).

An interesting example reported by Barboiu[31] is based on the neutral heteroditopic (ureido)crown ether, 4-phenylurea-benzo-15-crown-5 receptor **22**, complexing both anions and cations and being able to give self-organized structures in solution and in solid state.

22

Heteroditopic receptor **22** extracts solid NaX (X = F^-, Cl^-, NO_3^-, $CF_3SO_3^-$) salts into $CDCl_3$ as worked out through the observation of changes in host NMR spectra, consistent with the formation of the exchanging receptor–salt complexes. The largest change is the downfield shift of 0.54–1.45 ppm for the NH protons, which indicates hydrogen bonding of the urea moiety with the anion. Solid-state structures of **22**, **22**·NaCl, and **22**·$NaNO_3$, reveal that the urea moiety and the phenyl ring are disposed in an orthogonal position (87.5°) with respect to the plane of the macrocycle (Figure 14a) and this structure may be propagated in one direction; two adjacent macrocycles are present as a tight contact. The crystal structure of **22**·NaCl (Figure 14b) shows the $\mathbf{22}_2$·NaCl parallel dimers. Na^+ is sandwiched between two macrocycles.

A series of diamide-based cryptands derived from bis(*m*-phenylene)-[32]crown-10 have been synthesized to complex divalent salts such as paraquat (*N*,*N*-dimethyl-4,4′-bipyridinium) dichloride. The synthetic project started from the observation that readily prepared bicyclic crown ether **23** containing two 1,3,5-phenylene units linked by three tetra(ethyleneoxy) units forms a "pseudorotaxane-like" inclusion complex with *N*,*N*-dimethyl-4,4-bipyridinium bis(hexafluorophosphate), with an association constant $K_a = 6.1 \times 10^4\,M^{-1}$ that is 100-fold greater than that of the analogous simple crown ether.[32] On this basis, additional anion-binding moieties, such as isophthalamide or 2,6-pyridinedicarboxamide unit, were introduced in the third chain of the cryptand receptor.[33] The anion-assisted complexation of bypyridinium dications was analyzed by a combination of different techniques (^{1}H NMR,

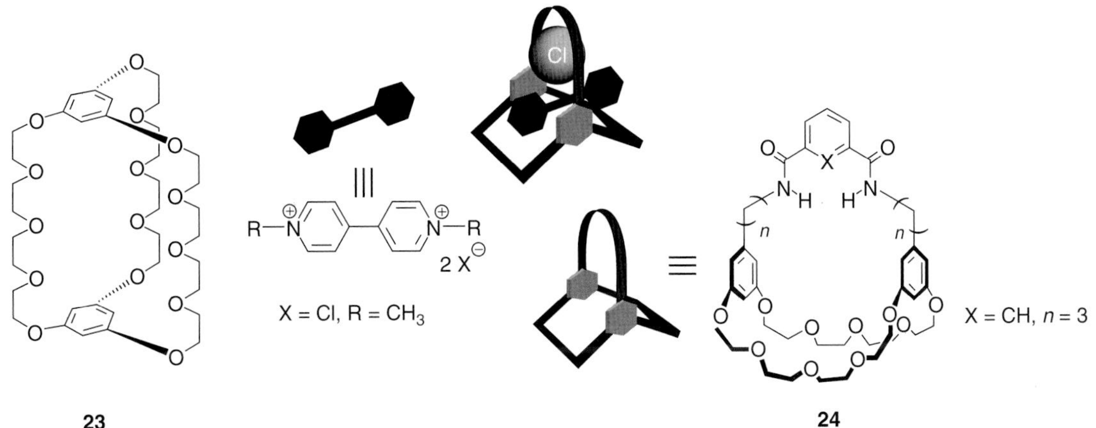

Figure 15 (Left) Bicyclic crown ether containing two 1,3,5-phenylene units linked by three tetra(ethyleneoxy) units. (Right) Derivative in which one chain contains anion-binding moieties. (Reproduced from Ref. 33. © Wiley-VCH, 2010.)

UV–vis, electrospray ionization (ESI), single-crystal X-ray analysis). Derivative **24**, with 13 atoms and an isophthalamide moiety in the third chain, showed the best binding affinity for tightly ion-paired paraquat dichloride in CH_3CN/CH_3OH mixtures. This was attributed to the combination of its spatial complementarity and to the presence of the additional anion-binding site (Figure 15).

An ion-pair receptor in which dramatic increases in halide-binding abilities when alkali metal cations are coordinated to the adjacent crown ether unit is that formed by a rigid biindole scaffold with two indole NHs as hydrogen-bond donors, **25**.[34] Diaza-18-crown-6 has been selected in this case as the cation-binding site. The cation effect on the binding strengths becomes pronounced in the order of $Li^+ < K^+ < Na^+$, regardless of the kind of halides.

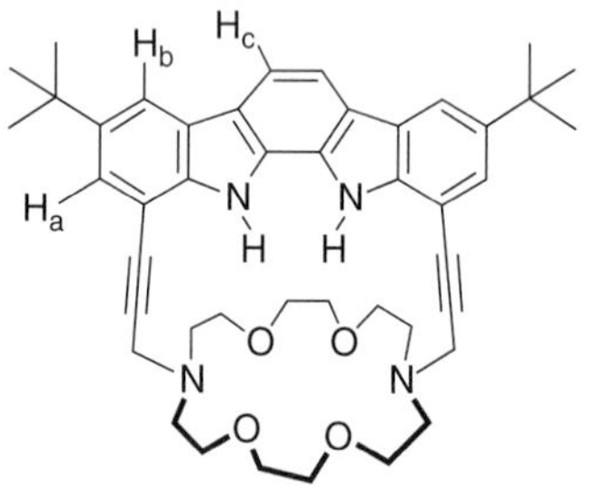

25

The common procedure to detect cooperativity in ion-pair binding is to titrate the host molecule by adding increasing amounts of the cation salt in which the counteranion is not involved in binding. In this case, the innocent anions were perchlorate for Li^+ and Na^+ and PF_6^- for K^+. A second set of titrations are then performed using generally TBA halide salts to determine the host binding affinities toward halide in the absence of any alkali metal salt. Next, the titrations are carried out in the presence of alkali metal salts to reveal the cation effect on the binding affinity of halides to the receptor. Such a protocol when applied to **25** showed that in the presence of alkali metal salts, the binding affinities of halides greatly increase up to three orders of magnitudes, see Table 1.

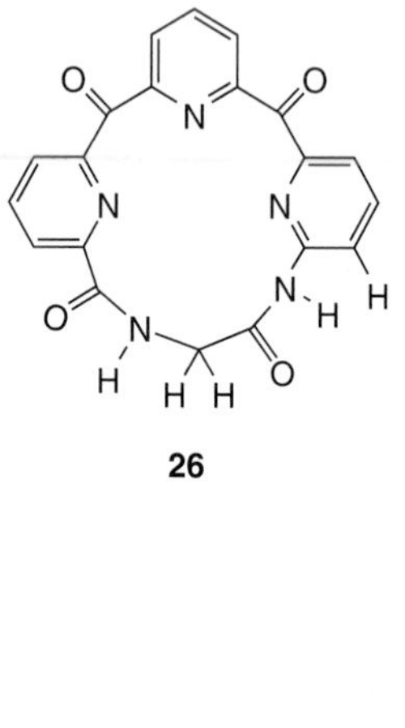

26

Table 1 Association constants (K_a, M^{-1}) between receptor **25** and halides in the presence of alkali metal salts (1 equivalent) in 10% (v/v) DMSO-d_6/CD_3CN at $24 \pm 1\,°C$.

Halide[a]	Cation additive	Binding constant (M^{-1})
Cl^-	None	7
	$LiClO_4$	120
	$NaClO_4$	14 000
	KPF_6	6 200
Br^-	None	—[b]
	$LiClO_4$	24
	$NaClO_4$	600
	KPF_6	200
I^-	None	—[b]
	$LiClO_4$	9
	$NaClO_4$	61
	KPF_6	45

[a]Halides were used as tetrabutylammonium salts.
[b]No interaction could be detected.

Macrocycle **26** consists of three basic pyridines, four hydrogen-bond-accepting carbonyls, and two hydrogen-bond-donating amide groups and binds monoalkyl ammonium salts in a manner that is dependent on the counterion of the ammonium guest.

The monoalkyl ammonium salt binding properties of **26** were examined with salts of L-phenylalanine methyl ester.[35] In general, all the amino acid salts bind more strongly than their TBA counterparts. For example, the binding constant for the association of the receptor with the nitrate salt of the amino acid was $18\,400\ M^{-1}$, whereas the corresponding association constant for the TBA salt was only $70\ M^{-1}$. The binding differences between monoalkyl ammonium and TBA salt confirm the notion that when both components of the guest can bind, the association is stronger (Table 2).

To sum up, overall binding is enhanced when the anionic component is a better guest. Accordingly, strongly binding nitrate ion leads to a cooperativity factor of 257, whereas

Table 2 Binding constants K_a (M^{-1}) between **26**, 1.0 mM, and various ammonium salts at 25 °C in $CDCl_3$.

TBA salt	K_a	$^+H_3NCH(Bn)CO_2CH_3$ salt	$K_a \times 10^3$	Cooperativity factor[a]
F^-	110	F^-	—[b]	—[b]
Cl^-	63	Cl^-	12.1	190
Br^-	40	Br^-	1.9	48
I^-	32	I^-	0.4	13
PF_6^-	—[c]	PF_6^-	—[b]	—[b]
NO_3^-	70	NO_3^-	18.4	257
$CF_3CO_2^-$	52	$CF_3CO_2^-$	6.3	121
TsO^-	42	TsO^-	1.5	36

[a] $K_{a(\text{amino acid})}/K_{a(\text{TBA})}$.
[b]Guest insoluble in $CDCl_3$.
[c]No interaction could be detected.

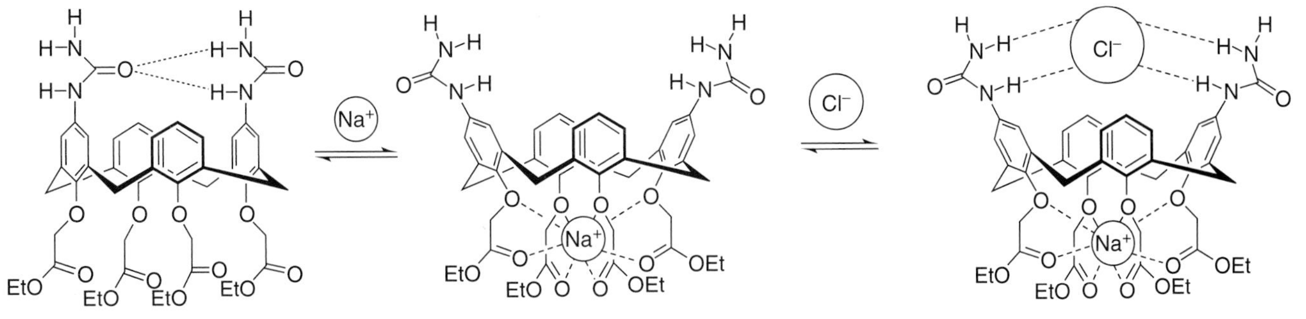

Figure 16 Proposed sequential binding of Na^+ and Cl^- by receptor **27**.

weakly binding iodide results in a cooperativity factor of 13.

Calixarene scaffold has been intensively used as the building block for ion-pair receptors. An early example of such hosts comes from the study by Reinhoudt group,[36] who prepared a calix[4]arene scaffold with cation-binding ester groups on the lower rim and anion-binding urea groups on the upper rim, **27**. The calixarene adopts a pinched-cone conformation in $CDCl_3$ solution at room temperature because of the intramolecular hydrogen bonding between the opposite urea moieties. On addition of Bu_4NCl or Bu_4NBr to a solution of **27** in $CDCl_3$, no complexation of halide anions was observed, and this was ascribed to the intramolecular hydrogen bonding that blocks the anion-binding site. However, complexation of Na^+ ions at the lower rim of **27** converts the pinched-cone conformation to a symmetrical cone conformation. Intramolecular hydrogen bonding at the upper rim is not possible in this conformation because of the rigidification of the calix[4]arene skeleton, and consequently, ions, such as Cl^- and Br^-, are now bound to the urea groups on the upper rim of the calixarene, as evidenced by downfield shift of the urea protons observed in the ^1H NMR spectra. No evidence of Cs^+ cation binding within the cavity was seen. Likely, this reflects the fact that the Cs^+ ion is too big to fit into the cavity (Figure 16)

A calix[4]arene receptor featuring both a cation-binding site in the form of ether/amide functionality and an anion-binding site in the form of a bisthiourea together on the lower (narrow) rim of the calixarene scaffold was reported by Kilburn in 2003.[37]

The crystal structure of **28** showed that the compound adopts a pinched-cone conformation with the ethylene glycol substituents properly disposed to fold in and bind a metal cation at the base of the calixarene and that the two thiourea units are effectively oriented to provide an anion-binding site through hydrogen bonding. ^1H NMR titrations in CD_3CN of **28** with the TBA salts of acetate, phenylphosphinate, and diphenyl phosphate pointed out significant downfield shifts of the thiourea NH signals (>1 ppm) and amide NH signals (0.3–0.7 ppm), indicating hydrogen-bond formation and suggesting that the amide proton is appreciably involved in anion binding. The resulting association constants show that **28** binds the tetrahedral anion phenylphosphinate ($K_a = 24\,000\,M^{-1}$) more strongly than the Y-shaped acetate anion ($K_a = 11\,000\,M^{-1}$). Diphenyl phosphate, however, binds significantly less strongly ($K_a = 1800\,M^{-1}$). Instead, in the presence of the Na^+ cation, the selectivity is reversed. A possible explanation is that while the unexpected selectivity for acetate anion over diphenyl phosphate in the absence of sodium cation might be attributed to electrostatic repulsion between the phenoxy groups of the diphenylphosphate and the ether oxygens of the calixarene, in the presence of Na^+ the phenoxy groups can interact favorably with the bound cation, leading to the observed strong binding of the ion pair.

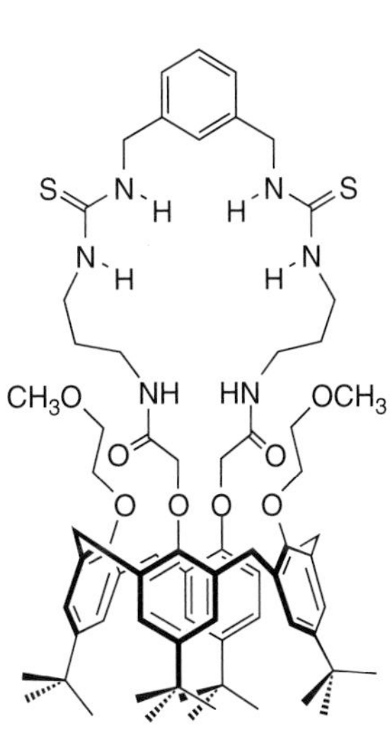

28

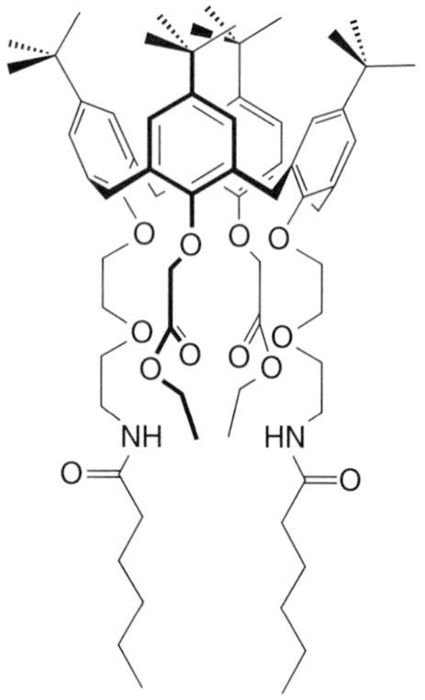

29

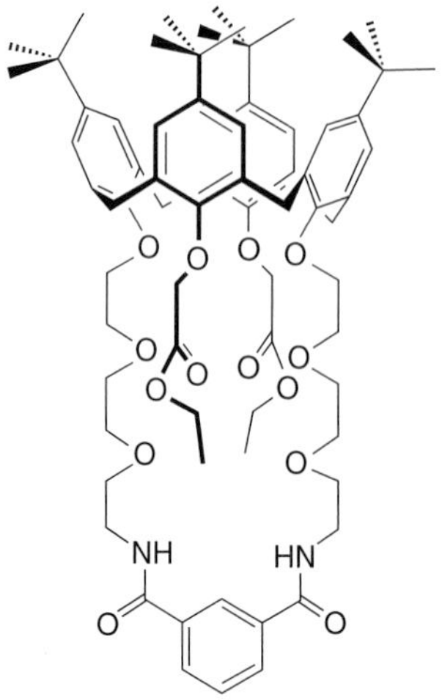

30

Beer and his group in 2007 reported the synthesis and binding properties of new heteroditopic calix[4]arene bis-ester compounds **29** and **30**, which are shown to have biased selectivity toward sodium- and lithium-containing ion pairs, as well as a macrocyclic effect in ion-pair recognition.[38] By ^{1}H NMR titrations, the addition of sodium or lithium perchlorate salts to a solution of receptor **29** in acetone-d_6 induced significant changes in the aliphatic region of the spectrum, indicating interaction of the components. The results are that in this solvent, **29** binds sodium strongly, lithium less strongly, and displays no affinity for the larger alkali metal cation potassium, or ammonium. When sodium or lithium perchlorate was added to a 1 : 1 solution of **29** and TBACl or TBA bromide salts, precipitation was observed, thus preventing estimation of the association binding constant. By changing the solvent, it was found that in acetonitrile, the addition of TBA halide salts to compound **30** induces small downfield shifts in the signals arising from the amide protons of the receptor. Significantly, the chemical shift change induced on addition of 1 equivalent of halide anion was larger for **29**·$NaClO_4$ than for **29** alone, suggesting that the bound cation enhances anion recognition. From Table 1 it can be seen that although the interaction of **29** with halide anions in this solvent is weak, a clear cooperative enhancement of anion recognition is affected by the presence of sodium cations. Same behavior was observed for **30** in acetone-d_6. The addition of lithium and sodium perchlorate induces perturbations in the aliphatic region of the ^{1}H NMR spectrum of the receptor, whereas the addition of potassium, rubidium, or ammonium produces no effect. From data elaboration, it was possible to establish that **30** is selective for sodium and lithium cations and, like in the previous case, when lithium and sodium perchlorate are added to a 1 : 1 mixture of **30** and TBA bromide, the cation-binding behavior is influenced. In particular, when lithium binds, the induced chemical shift and the resulting association constant increase, indicating that bromide enhances the affinity of **30** for this cation. An anticooperative effect on cation binding was instead observed for sodium in the presence of bromide. It is also interesting to report that addition of sodium or lithium cations to a 1 : 1 solution of **30** and TBACl did not reveal any cation-binding interaction. Investigation of the the anion-binding properties of **30** in acetone provided the expected selectivity order of $Cl^- > Br^- > I^-$. So the amount of anion added to derivative **30** strongly influences the association. The addition of chloride to **30**·M^+ does not initially lead to any significant changes in the ^{1}H NMR spectrum, but after the addition of 1 equivalent of anion, downfield shifts in the amide and isophthalyl protons are observed, consistent with the binding of chloride. This was attributed to the sequestration of cation by the first equivalent of added anion, followed by unassisted binding of the rest. Although not accompanied by precipitation in this case, the behavior is similar to that observed for receptor **29** in this solvent. Same behavior is observed when TBA bromide or iodide was added to 1 : 1 mixtures of **30** and potassium, rubidium, or ammonium salts. This is due to the ion pairing rather than sequestration, as these cations are not bound by the receptor in the first place. On the other hand, on addition of TBA bromide and iodide salts to a 1 : 1 mixture of **30** and lithium or sodium cations, significant downfield shifts in the amide and isophthalyl protons are observed. These are larger than the corresponding changes in chemical shift induced on addition of anion to the free receptors, suggesting an increase in the association strength. Estimation of the binding constants points out a cooperative binding effect for lithium and sodium bromide and iodide. Indeed, anion affinity is increased by nearly an order of magnitude (Table 3). This means that both the receptors bind ion pairs in a cooperative fashion, if the correct conditions are provided. For **29**, a modest cooperative interaction can be inferred in acetonitrile-d_3 solution from the halide-binding properties in the presence and absence of sodium cations, but the association constants observed are very small, and studies in the less-polar solvent acetone-d_6

Table 3 Anion-binding affinities (K_a, M^{-1}) of receptor **29** in CD_3CN, at 25 °C.

	29	**29**·$NaClO_4$
TBACl	<5	20
TBABr	10	20
TBAI	—[a]	—[a]

TBABr, TBA bromide; TBAI, TBA iodide.
[a]No interaction could be detected.

Table 4 Cation-binding affinities (K_a, M^{-1}) of receptor **30** in acetone-d_6, at 25 °C.

	30	**30**·TBABr
$LiClO_4$	2840	$>10^4$
$NaClO_4$	3350	740
KPF_6	—[a]	—[a]
NH_4PF_6	—[a]	—[a]

TBABr, TBA bromide.
[a]No interaction could be detected.

were not possible because of precipitation problems. Receptor **30** provides an example of a system in which the ion-pair cooperativity is highly selective for certain salts, that is, bromide and iodide salts of sodium and lithium, while displaying little or no affinity for other ion pairs. Even if it is impossible to compare the binding properties of the two receptors, it is felt that **30**, where the anion and cation-binding sites are within the same macrocycle, behaves as a better binder for ion pairs, being less prone to solvent ion pairing or precipitation. Maybe a macrocyclic effect for ion-pair recognition can be envisioned (Table 4).

Not only calix[4] but also calix[5]arene and calix[6]arene have been used as scaffolds for the construction of ditopic receptors for ion pairs.

An example of a heteroditopic receptor for organic salts based on calix[5]arene is the one reported by Gattuso *et al.* in 2009.[39] In this case, the synthetic approach led to a spatially separated ion-pair receptor, the tris-ureido-calix[5]arene-crown-3, **31**.

This derivative forms stable complexes with the cations of alkali metal picrates ($K_a > 10^6\,M^{-1}$ for Na^+, K^+, Rb^+, and Cs^+) in 4 : 1 $CDCl_3/CD_3OD$. Moreover, it can also solubilize and bind alkali metal chloride salts ($K_a > 10^6 M^{-1}$)

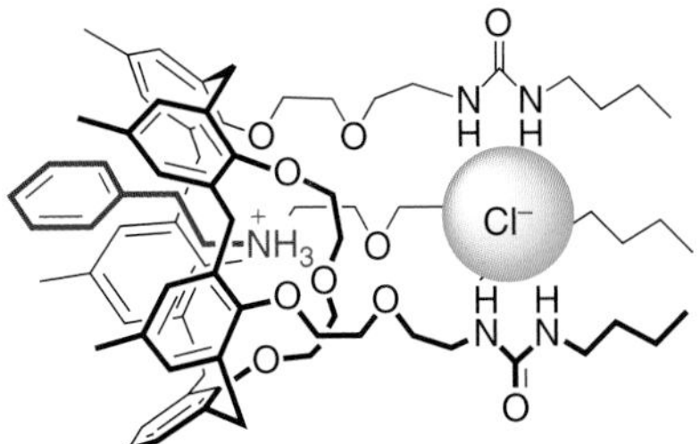

31

very effectively in the same solvent mixture, suggesting the idea that the three ureido moieties, present at the lower rim of **31**, are involved in the uptake of the chloride counterion. Receptor **31** is also able to bind the biologically relevant 2-phenylethylamine hydrochloride (PEA·HCl) salt with an association constant ($K_a = 1.89 \times 105\,M^{-1}$) for both cation and anion. Semiempirical calculations performed at the PM3 level on the Cl^-·**31**·PEA·H^+ complex showed that the cation is, as expected, embedded in the calix[5]arene cavity, whereas the chloride anion is encircled by the ureido-containing chains, which effectively encapsulate the ion. The two ions are physically separated by the crown-3 bridging loop, and the $^+N\cdots Cl^-$ distance is 7.17 Å. The ortho hydrogens and the α,β-CH_2 groups of PEA·HCl are placed in the shielding cone of the π-rich aromatic cavity of **31**, and at the same time, the ammonium group is placed in an optimal position to form tripodal hydrogen bonding with three of the phenolic oxygen atoms.

In 2008, Jabin and Le Gac reported the synthesis of a calix[6]cryptamide that exhibits two favorably positioned binding sites for the complexation of organic-associated ion pairs in close proximity: a well-defined calix[6]arene cavity suitable for the inclusion of ammonium ions and a cryptamide unit for the coordination of anions, **32**.[40]

Addition of an excess (up to 15 equivalents) of ammonium salts $RNH_3^+Cl^-$ (R = Et or Pr) to a solution of **32** in $CDCl_3$ produced the corresponding endo-complexes (Figure 17) with, in both cases, high-field signals characteristic of the presence of the alkyl chain of the ammonium ions in the heart of the calixarene cavity. The *in* and *out* guest exchange appears to be slower than that of the NMR timescale and involves an induced-fit process with the expulsion of methoxy groups from the calixarene cavity. In addition to the endo-complexation of the ammonium ion RNH_3^+, the simultaneous binding of the chloride counterion by the amide groups of the host was clearly evidenced by a significant downfield shift of the resonances of the NHCO protons close to the calixarene subunit, indicating a strong hydrogen-bonding interaction of these protons with the counter anion occurring through a convergent arrangement of hydrogen-bonding NH groups. Most probably, the ammonium ion, in addition to the electrostatic interaction with its counterion, interacts through hydrogen bond with the ethereal oxygen atoms and is further stabilized through CH–π interactions with the aromatic rings of the calixarene. The use of low-coordinating counterions (picrate and TBA cations) demonstrated the significant cooperative binding of the associated ion pair. When a large excess (about 30 equivalents) of n-$Bu_4N^+Cl^-$ was added to a solution of **32** in $CDCl_3$, the NMR spectrum of the receptor remained almost unchanged. On the other hand, when the bulkier and less densely charged picrate anion was used in place of the chloride anion, no inclusion of the propylammonium ion was detected even after a long time. The conclusion is that the complexation of the chloride can take place only when an ammonium ion is present in the calixarene cavity, and conversely, in the absence of Cl^-, **32** is ineffective to binding the ammonium ion. This significant positive cooperativity is probably due to both structural and electronic modifications of the receptor: each partner of the associated ion pair polarizes the receptor, and conversely,

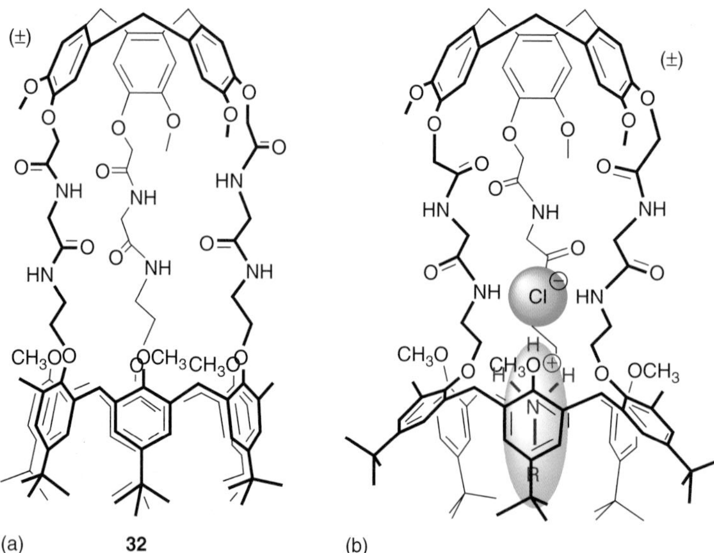

Figure 17 (a) Calix[6]cryptamide **32**. (b) Complex with $RNH_3^+Cl^-$.

the anion coordination contributes to the preorganization of the binding site of its counterion by an induced-fit process. The expulsion of the OCH_3 groups from the cavity caused by the chloride binding brings the oxygen atoms of the ammonium binding site closer.

A similar system is that consisting of the calix[6]cryptamide, **33**, bearing a *tren*-based cap.[41] In this case, fluoride anion can be encapsulated into the tris-amido cap of the host and acts as an allosteric activator by favoring the inclusion of ammonium ions into the calixarene cavity. As in the previous case, a strong electrostatic interaction occurs at the level of the ion pair, while the ammonium ion is further stabilized by H-bonding interaction with a phenoxy oxygen and by a combination of CH–π interactions with the aromatic walls of the host. In this particular case, the binding of contact ion pairs can exclusively proceed with F^- as the anionic partner, and the remarkable selectivity is mostly because of the tininess of the binding site provided by the convergent NH groups of the cryptamide cap. Only the small fluoride anion can be encapsulated inside the cap and located at a suitable distance for the crucial electrostatic interaction with the ammonium ion. A final consideration is that receptor **33** possesses a proton-sensitive site at the level of the cap that permits acid–base control of the binding properties of the receptor. The protonation of the aza cap results in a positively charged receptor in which an intramolecular hydrogen-bonding interaction at the level of the tris-amido cap "turns off" the host binding properties. Hence, the receptor displays an allosteric regulation of the recognition processes (the fluoride anion and the NH^+ proton can be considered, respectively, as an allosteric activator and allosteric inhibitor) (Figure 18).

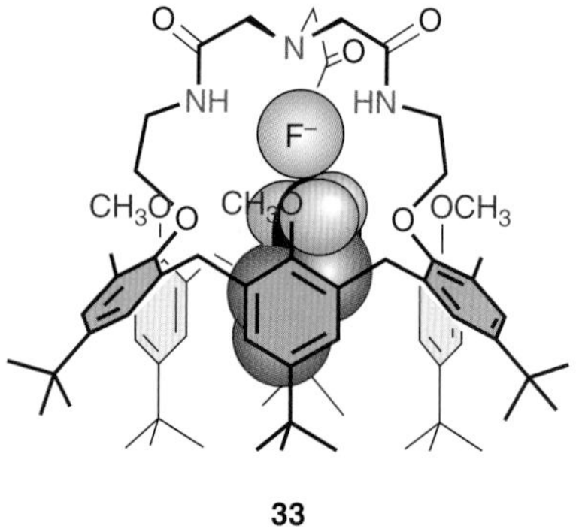

Figure 18 Representation of the ternary complex of calix[6]cryptamide **33** and $PrNH_3^+F^-$. (Reproduced from Ref. 41. © Royal Society of Chemistry, 2010.)

34

A very recent example of an ion-pair receptor based on cavitands was reported by Dalcanale and Diederich.[42] Receptor **34** has been proved to be capable of selective ion-pair complexation by a synergistic combination of CH–anion and hydrogen-bonding interactions to P=O groups.

2.2.2 *With π-systems for cation recognition*

In 2009, the group of Arduini and Secchi[43] reported the synthesis of the ion-pair receptors **35** and **36**, where an anion-recognizing urea group is linked through a methylene unit to the upper rim of a calix[4]arene. Binding studies employing *N*-methylpyridinium (NMP) and *N*,*N*′-dimethylviologen salts in $CDCl_3$ were carried out by means of ^{1}H NMR and UV–vis titrations.

^{1}H NMR spectroscopic titrations showed that on complexation, an extensive upfield shift and broadening of the resonances of the NMP cation. Similarly, the resonances of the phenylurea NH protons of the calixarene receptor underwent a large downfield shift. Such broadenings prevented the determination of reliable binding constants, but they clearly suggested that, as anticipated in the solid state[44] (Figure 19), the ion-pair binding mainly takes place through a "soft" cation–π interaction and a "hard" H-bond interaction. The former interaction occurs between the NMP cation and the π-rich aromatic cavity of the calixarene scaffold, whereas the latter stronger interaction operates between the phenylurea group and the anion.

The comparison of the binding constants calculated for **35** and **36** from UV–vis titrations with those calculated for the corresponding monotopic derivatives, **37** and **38**, shows that the introduction of the ancillary anion-binding site on the calix[4]arene cavity increases the affinity for the NMPX salts up to two orders of magnitude. Owing to the very low solubility of paraquat salts in halogenated solvents, preliminary binding studies were accomplished in the gas phase using ESI mass spectrometry. These studies together with ^{1}H NMR titrations pointed out complexation of the bipyridinium dication within an extended aromatic cavity created by the cooperation of two self-assembled cavities of receptor **36**.

In a previous work, the same group synthesized an adaptable heteroditopic cavitand by using a flexible cone conformer of calix[4]arene, bearing four 4-hydroxybenzyl groups, **39**.[45] The derivative was able to bind cooperatively to a series of tetramethylammonium (TMA) salts (tosylate, chloride, acetate, trifluoroacetate, and picrate).

The complexing abilities of the host **39** toward a set of tetraalkylammonium (TMA) salts having different counteranions, chloride, tosylate (TsO), acetate (Ac), trifluoroacetate (TFA), and picrate (Pic), were studied by ^{1}H NMR in $CDCl_3$ solutions at 300 K. The extent of cation complexation is dependent on the type of counteranion. In addition,

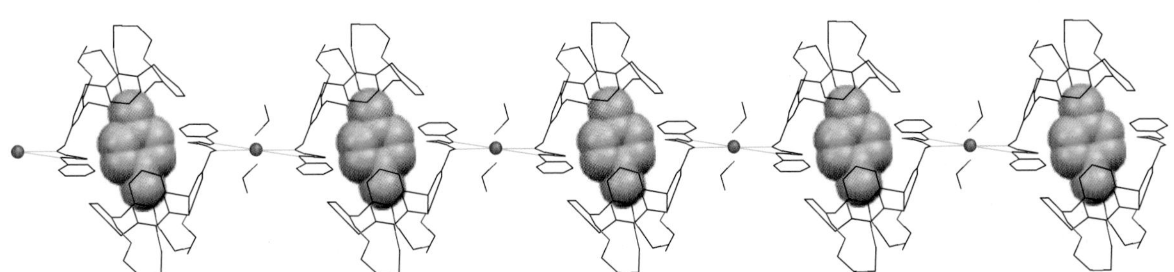

Figure 19 Partial view of an infinite hydrogen-bonded one-dimensional polymeric chain of self-assembled 2 : 1 complexes between heteroditopic calix[4]arene **35** and *N*-methylpyridinium chloride ion pairs (Reproduced from Ref. 43. © Royal Society of Chemistry, 2009.)

during titration experiments, the host OH groups experience extensive broadening and downfield shift ($\Delta\delta \sim 5$ ppm), whereas the aromatic signals of the sidearms tend to coalesce to $\delta = 6.80$ ppm. These data can be explained by hypothesizing a pivoting role of the anion that, depending on its ability to act as hydrogen-bond acceptor, interacts with the OH groups, thus preorganizing the extended cavity of the host for cation binding. This is reflected by the efficiency order (Ac > TsO > Cl > TFA > Cl > Pic) that supports the hypothesis of a positive anion allosteric effect. This hypothesis is further confirmed by the lack of efficiency showed by the analogous compound in which the OH groups have been transformed in the corresponding methyl ethers thus preventing any influence of the anion on the binding event.

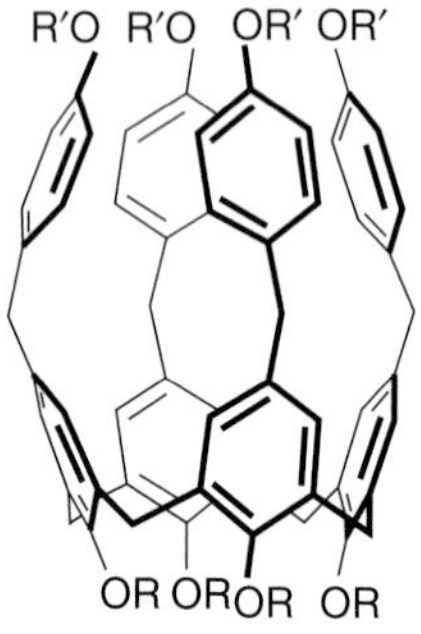

R'O R'O OR' OR'

OR OROR OR

39

R = n-C_3H_7
R' = H

A ditopic receptor consisting of two calix[4]arene units linked by two ethylene chains at the lower rim for cation recognition and urea-based hydrogen-bonding recognition sites for anions, **40**, has been developed by the group of Beer in the United Kingdom.[46]

^{1}H NMR investigations with alkali metal cations revealed that the host displayed a remarkable selectivity and fast kinetics of complexation for potassium cation over all other group 1 metal cations. Compound **40** only binds halide and acetate anions very weakly in 2 : 1 $CDCl_3$:CD_3CN, but its sodium and potassium ion complexes form much more stable complexes with these anions, with anion-binding enhancements of over 30-fold in the case of bromide anion.

The increase in stability constant can be ascribed to (i) an increase in acidity of the urea NH protons because of the electron withdrawing effect of the bound cation; (ii) the rigidification of the calixarene units by cation complexation; and (iii) an electrostatic attraction between the bound cation and anion. Thus, this receptor cooperatively binds a range of sodium and potassium halides with the bound group 1 metal cation effectively "switching on" anion complexation. Extraction experiments demonstrated that the host is capable of solubilizing sodium and potassium salts in chloroform.

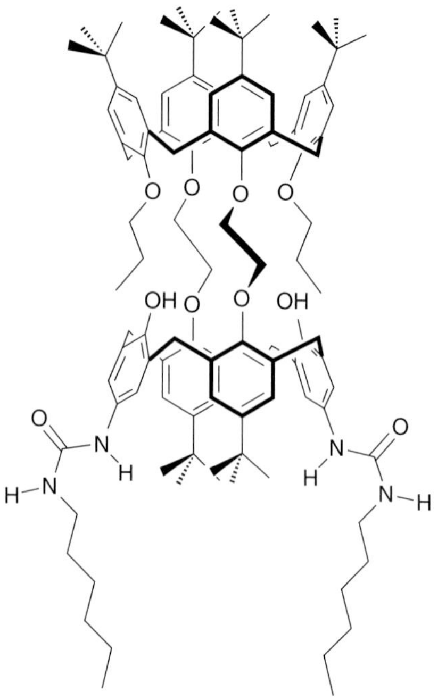

40

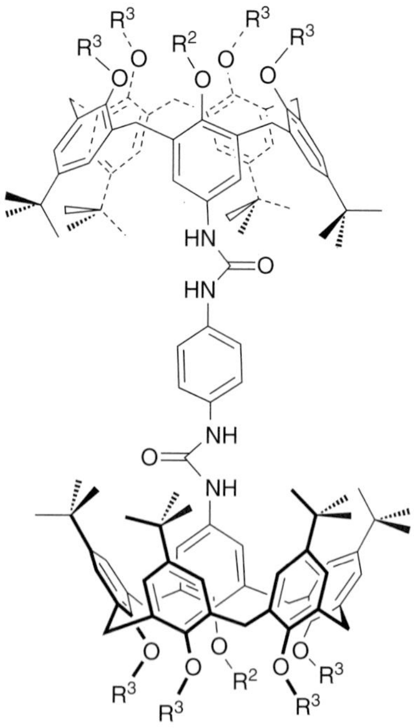

$R^2 = R^3 = (CH_2)_3CH(CH_3)_2$

41

In 2005, the group of Parisi and Pappalardo[47] reported on the synthesis and unique molecular recognition abilities of the heterotetratopic receptor, **41**, which consists of two convergent, conformationally fixed, cone calix[5]arene units (cation-binding sites) that are covalently linked at their upper rims by means of a 1,4-bis(ureido)phenylene spacer (anion-binding sites). Host **41** readily forms overall charge-neutral (ligand-separated ion-pair complexes) unimolecular capsules of nanoscale dimensions by tight encapsulation of linear α,ω-alkanediyldiammonium ions, which range from 1,12-dodecane- to 1,16-hexadecanediammonium, inside the inner space defined by the two converging calix[5]arene cavities by simultaneously binding each of the two counteranions to the peripheral ureido functions through hydrogen bonds. The binding affinity of receptor **41** for linear, long-chain α,ω-alkanediyldiammonium (C8–C16) dichloride salts, as well as the host–guest architectures they form, was investigated by a combination of ^{1}H NMR spectroscopy and ESI mass spectrometry. The affinity of **41** for α,ω-alkanediyldiammonium dichloride salts was investigated by ^{1}H NMR titration experiments in $(CDCl_2)_2/CD_3OD$ (2 : 1 v/v). Addition of the guest salt (up to 4 equivalents) to a solution of **41** (1.0×10^{-3} M) led to the formation of very strong inclusion complexes, whose host–guest stoichiometries (1 : 1 and/or 2 : 1) and geometries depend on the length of the diammonium ion and the host/guest ratio (Figure 20).

The formation of capsular complexes between **41** and long-chain α,ω-alkanediyldiammmonium dichloride salts is further corroborated by the positive ESI mass spectra of equimolar host–guest mixtures, which show prominent doubly charged ions that correspond to $[\mathbf{41}{\cdot}C_n]^{2+}$ (base peak) and low-intensity ion peaks for $[\mathbf{41}{\cdot}C_nCl]^+$.

2.3 Receptors employing charged components

Compounds incorporating positively charged group have been employed in the construction of ion-pair receptors that also rely on charge complementarity for recognition. A neutral multidentate ligand present in the molecule is often deputed to provide the recognition site for the cation.

In 1986, Schmidtchen[48] reported the synthesis of a ditopic receptor containing a triaza[18]crown-6 subunit covalently linked to a tetraammonium macrocyclic anion recognition motif through a p-xylene moiety (Figure 21). Binding studies by ^{1}H NMR showed that receptor **42** is able to recognize γ-aminobutyric acid (GABA) and 6-aminohexanoic acid in 90% aqueous methanol. The selectivity toward the resultant zwitterionic amino acids increased by a factor of 2.5 as compared to the control compound composed of a triaza[18]crown-6 subunit and a single tetraalkylammonium unit.

In 1992, de Mendoza[49] and his group reported the synthesis of a receptor for amino acids, **43**, in zwitterionic form featuring a guanidinium function to bind carboxylate, and a crown ether to recognize ammonium unit. An aromatic planar surface, the naphthalene ring, for an

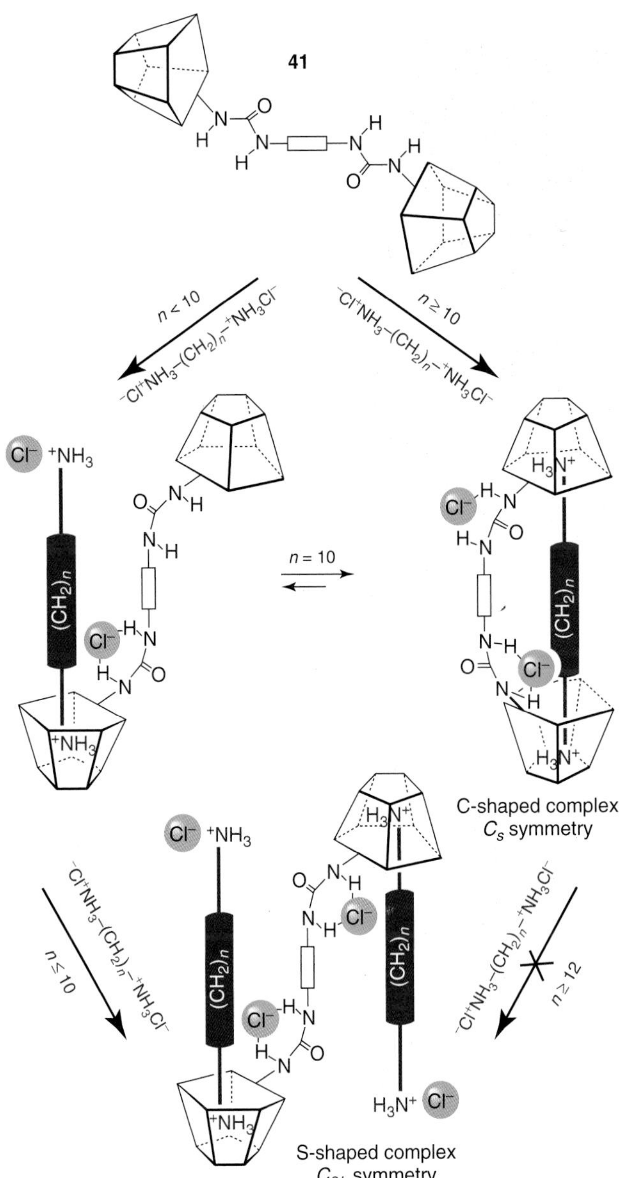

Figure 20 Schematic representation of the influence of the spanning of the guest α,ω-alkanediyldiammonium ion on the complexation pathways leading to C-shaped and/or S-shaped inclusion complexes with host **41**. (Reproduced from Ref. 47. © Wiley-VCH, 2005.)

additional selective stacking interaction with the side chain of aromatic amino acids, and a chiral structure for enantioselective recognition were also present. The affinity of the receptor toward amino acids was established by liquid–liquid single-extraction experiments, in which 0.5 ml of an aqueous solution of L-Trp, L-Phe, or L-Val (0.2 M) was extracted into 2 ml of a CH_2Cl_2 solution of **43**. The extraction efficiencies (i.e., fraction of receptor molecules occupied by substrate) in the organic phase, determined by ^{1}H NMR integration, were ~40% for both L-Trp and L-Phe, whereas L-Val, without any aromatic side chain,

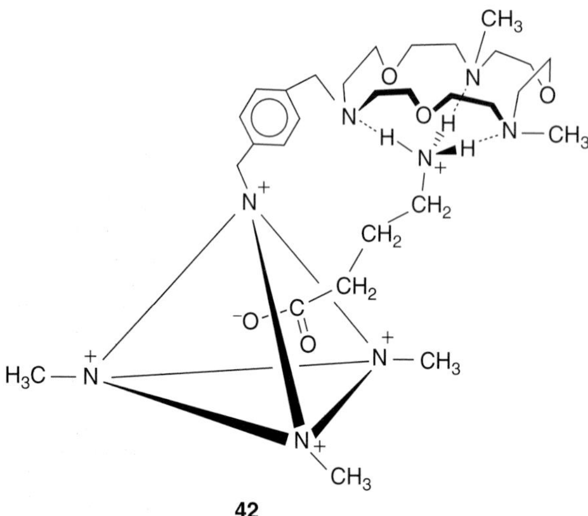

Figure 21 Binding mode of a zwitterionic ω-amino carboxylate to the ditopic host **42**. (Reproduced from Ref. 48. © American Chemical Society, 1986.)

was not detected. This supports the idea that the aromatic naphthalene ring is strongly involved in the recognition of amino acids endowed with aromatic units.

43

2.4 Receptors based on G-quartets

As shown, in the majority of cases ion-pair recognition calls for receptors with separate cation- and anion-binding sites. On the other hand, an alternative approach could be to use noncovalent interactions to build the ion-pair receptor from multiple components and employ molecular self-assembly for ion-pair binding.

The G-quartet is a macrocycle formed by hydrogen-bonded guanosine units (Scheme 3). Alkali metal cations template its formation from guanosine nucleotides in water, and these cation-filled G-quartets stack to give octamers, dodecamers, hexadecamers, and higher aggregates. In 2001, Davis reported an example of how self-assembly provides a supramolecular complex with Lewis basic and Lewis acidic sites for the simultaneous binding of cations and anions.[50] Hydrogen-bonding, ion–dipole, and base-stacking interactions provide a tubular complex with a cation- loaded interior, while an array of hydrogen-bond donors on the receptor's surface favors anion coordination.

Davis and coworkers demonstrated that the lipophilic **G1**, a guanosine nucleotide, forms G-quadruplex-based hexadecamers in the presence of barium picrate, $BaPic_2$, and strontium picrate, $SrPic_2$, both in the crystalline state and in CD_2Cl_2 solution (Figure 22). The divalent cations, each interacting with eight nucleosides, are well separated from their picrate counterions (>8.5 Å) that, on the other hand, are not innocent spectators. In solid state, the four picrate groups join the $G_8{\cdot}M^{2+}$ octamers, thanks to hydrogen bonding with the NH_b amino protons that point outward the two inner G-quartets as shown in Scheme 3. So, 16 nucleosides, 2 cations, and 4 anions form a complex that has dimensions of $25 \times 25 \times 30$ Å and a molecular weight greater than 7600 Da. Picrate anions that clamp together the two inner G-quartets in the hexadecamer's crystal structure also associate with the G-quadruplex in solution as it was clearly pointed out by proton NMR investigations. Crossover experiments in CD_2Cl_2 proved the impressive kinetic stability of these guanosine hexadecamers. Instead, the use of thiocyanate, an anion that does not bridge $(\mathbf{G1})_8{\cdot}M^{2+}$ octamers as effectively as the picrate anion, lowers the kinetic stability of the hexadecamer, thus demonstrating that the anion's identity plays a dramatic role in modulating the stability of aggregates.

A couple of years later, Davis described the synthesis of a calix[4]arene–guanosine conjugate that forms water-stabilized dimers via the formation of G-quartets in water-saturated $CDCl_3$, **G2**.[51] The resulting dimer extracts alkali halide salts from water into organic solution (Figure 23). Two-dimensional NMR and ion-induced NMR shifts indicated that the cation-binding site is formed by an intermolecular G-quartet and the anion-binding site is provided by the amide NH groups. Extractions showed selectivity for K^+ over Na^+ and Br^- over Cl^-. As in the previous example, the anion's identity influences the association process. Changing the anion from a halide to the noncoordinating BPh_4^- switches the assembly from a discrete dimer to a noncovalent polymer.

Lippert computations demonstrated that two stacked DNA-base quartets, namely, a G_4 quartet and an A_4 quartet (A, adenine), can bind NaCl in a cooperative fashion.[52] In this case, the stable G_4 quartet is a template that assists, through stacking, the formation of the A_4 quartets. The latter can, in principle, bind an anion, such as Cl^-, through N-H$\cdots$$Cl^-$ hydrogen bonds (Figure 24). This assembly constitutes a stack of two purine quartets G_4A_4 with the potential to act as a NaCl ion pair.

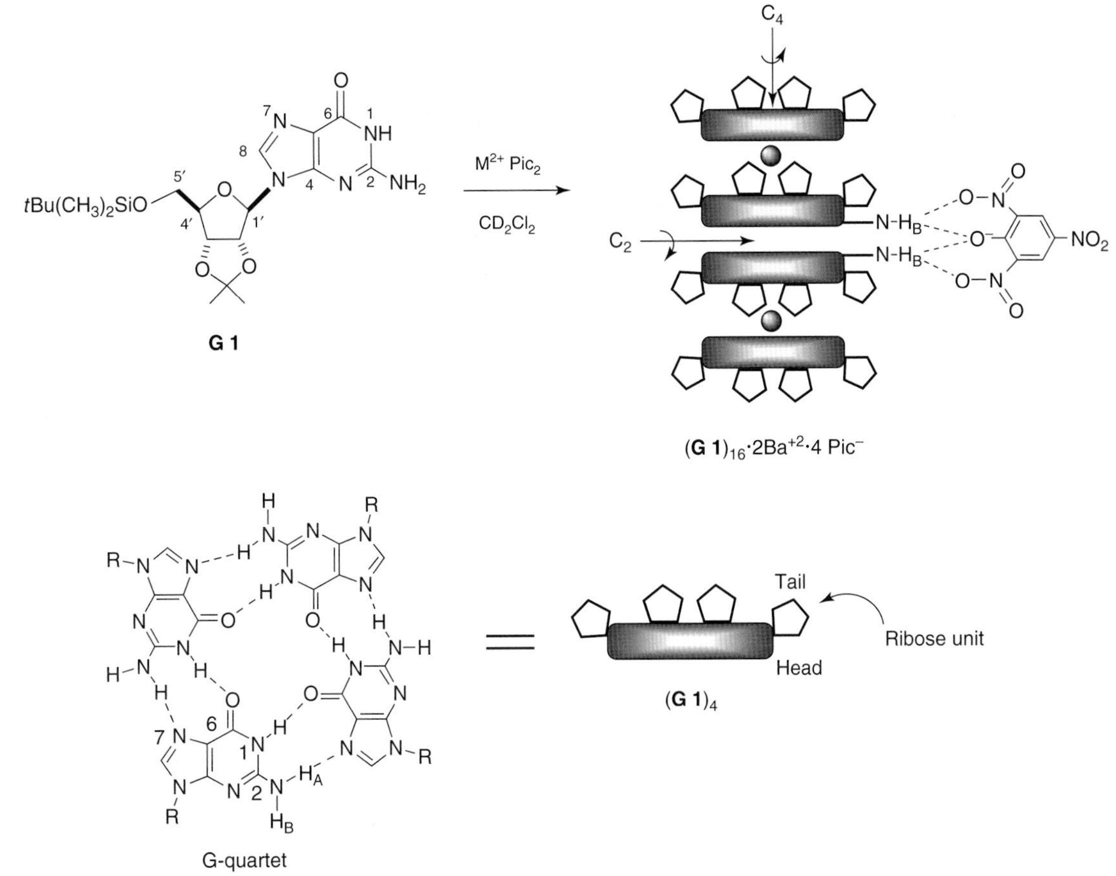

Figure 22 A lipophilic G-quadruplex that binds ion pairs. Pic, picrate; R, 5′-silyl-2′,3′-isopropylidene-D-ribose. (Reproduced from Ref. 50. © Wiley-VCH, 2001.)

2.5 Receptors based on "dual-host" combination

An easier approach to the construction of systems to recognize ion pairs is the use of mixtures of synthetically more accessible anion and cation receptors, so avoiding laborious multistep syntheses of ditopic receptors. The versatility of this method was demonstrated, for example, by the complexation of paraquat dichloride by the use of calix[6]pyrrole, **44**, in conjunction with β-cyclodextrin **45** in CD_2Cl_2. Receptor **45** is known to complex paraquat (preferably as the hexafluorophosphate salt) in polar solvents, such as CH_3CN, acetone. However, this salt, which is scarcely soluble in a solution of CD_2Cl_2 containing 1 equivalent of cyclodextrin, is mostly dissolved (80%) when 2 equivalents of **44** are also present. The ^1H NMR spectrum is consistent with the formation of the expected

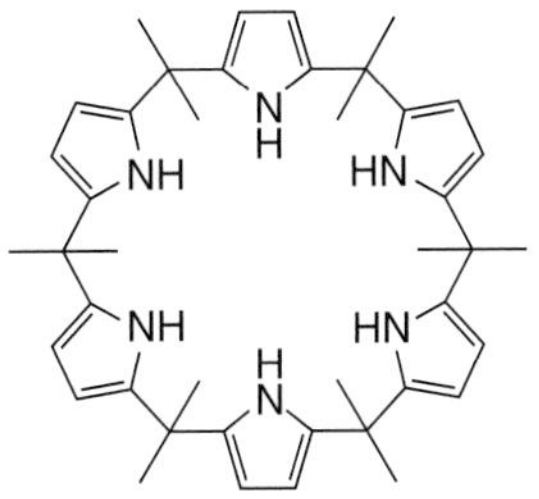

44

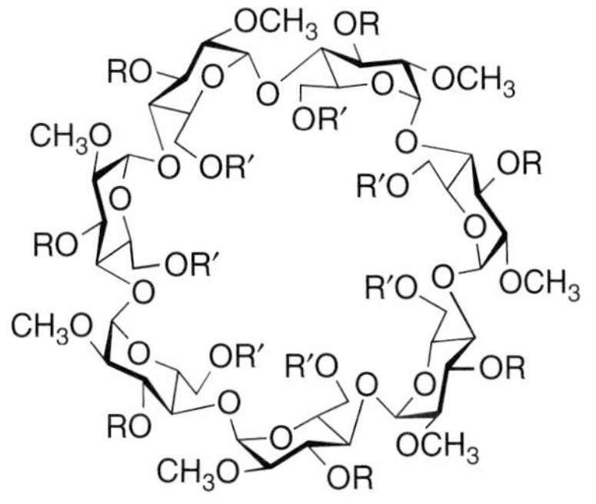

45

R = Ac R′ = CH_3
R = Ac R′ = Ac
R = Bz R′ = CH_3

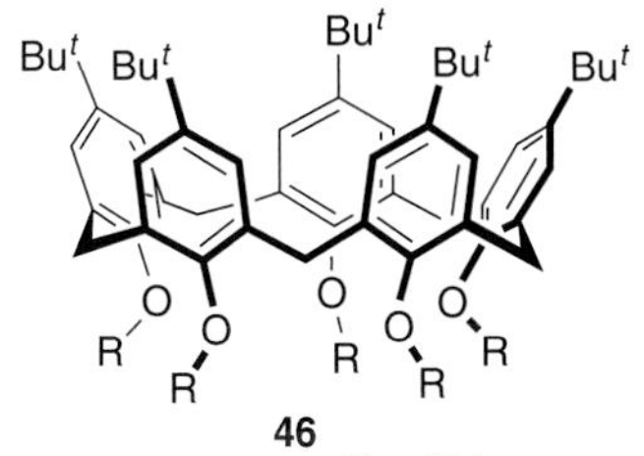

46

R = CH_3

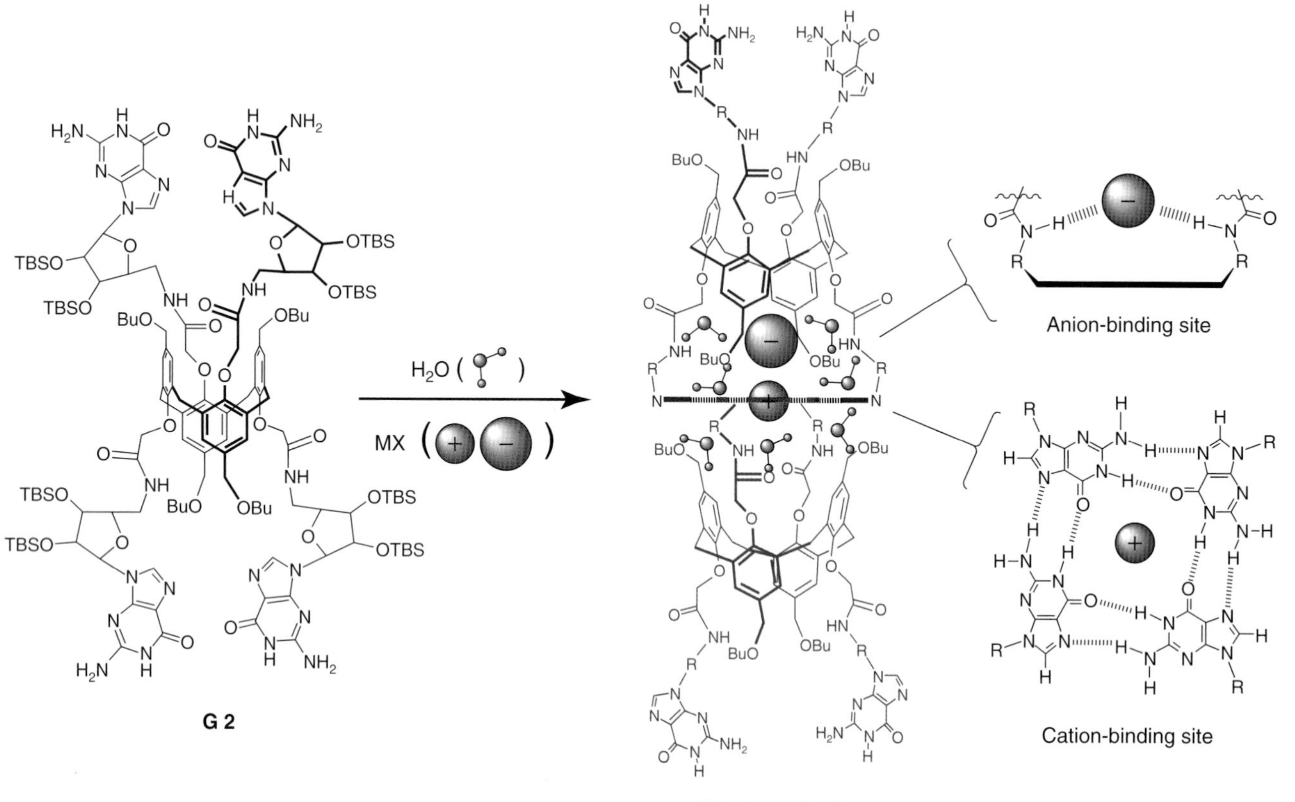

Figure 23 Schematic representation of $(\mathbf{G2})_2 \cdot MX \cdot (H_2O)_n$ formation from **G2**, indicating the anion and self-assembled cation-binding sites. (Reproduced from Ref. 51. © American Chemical Society, 2003.)

supramolecular system (paraquat·cyclodextrin·supercation and Cl^-·calix[6]pyrrole·superanion). Of course, in this approach, pairs of receptors must be carefully chosen for optimum binding of a target ion-paired guest, and it should be pointed out that the increased binding efficiency of these binary host systems does not in any way affect the intrinsic selectivity of each of the two host components. However, a weak anion or cation receptor can be transformed into a powerful one via the synergic action of a complementary receptor. This was clearly demonstrated by this group considering the complexation of *n*-butylammonium chloride, $n\text{BuNH}_3{}^+\text{Cl}^-$, with calyx[6]pyrrole, **44**, as the anion receptor, and *p-tert*-butylcalix[5]-arene **46** as the cation receptor. When a solution of the calixpyrrole (5×10^{-3} M) in CD_2Cl_2 is exposed to *n*-butylammonium chloride (0.5 equivalent), only a weak interaction occurs, as evidenced by the small downfield shift experienced by the pyrrole NH resonance. This is consistent with a fast complexation process of Cl^- ions on the NMR timescale. On the other hand, the addition of $n\text{BuNH}_3{}^+\text{Cl}^-$ (1 equivalent) to a solution of the highly flexible **46** in CD_2Cl_2 leads to the doubling of the signals for the free and complexed host and guest species, from which a 2% host–guest complex formation can be evaluated. Although **46** alone shows a very poor affinity for the $n\text{BuNH}_3{}^+$ ion, after the addition of the calixpyrrol (1 equivalent) the signals of the free guest disappear, whereas those of the included ammonium ion show up in the characteristic high-field region of the spectrum ($\delta = -0.4$ to -2.0 ppm). At the same time, chloride is completely complexed by **44**, as pointed out by the shift of the pyrrole NH resonance from $\delta = 7.74$ (free host) to $\delta = 10.91$ ppm (complex). So, a weak non-preorganized receptor such as **46** is transformed into an efficient one by the synergic action of a proper "complementary" receptor.

3 ION-PAIR RECOGNITION "AT WORK"

Mimicking biological systems and biological ion transport is one of the main research activity of supramolecular chemistry. The transport of both anions and cations is a vital part of biochemistry. There are two kinds of primary transport processes: the ion exchange or *antiport* that couples the transport of two components of like charge across a membrane in opposite directions and synport that occurs when two ions of different charges are transported at the same time in the same direction.[53] Receptors for ion

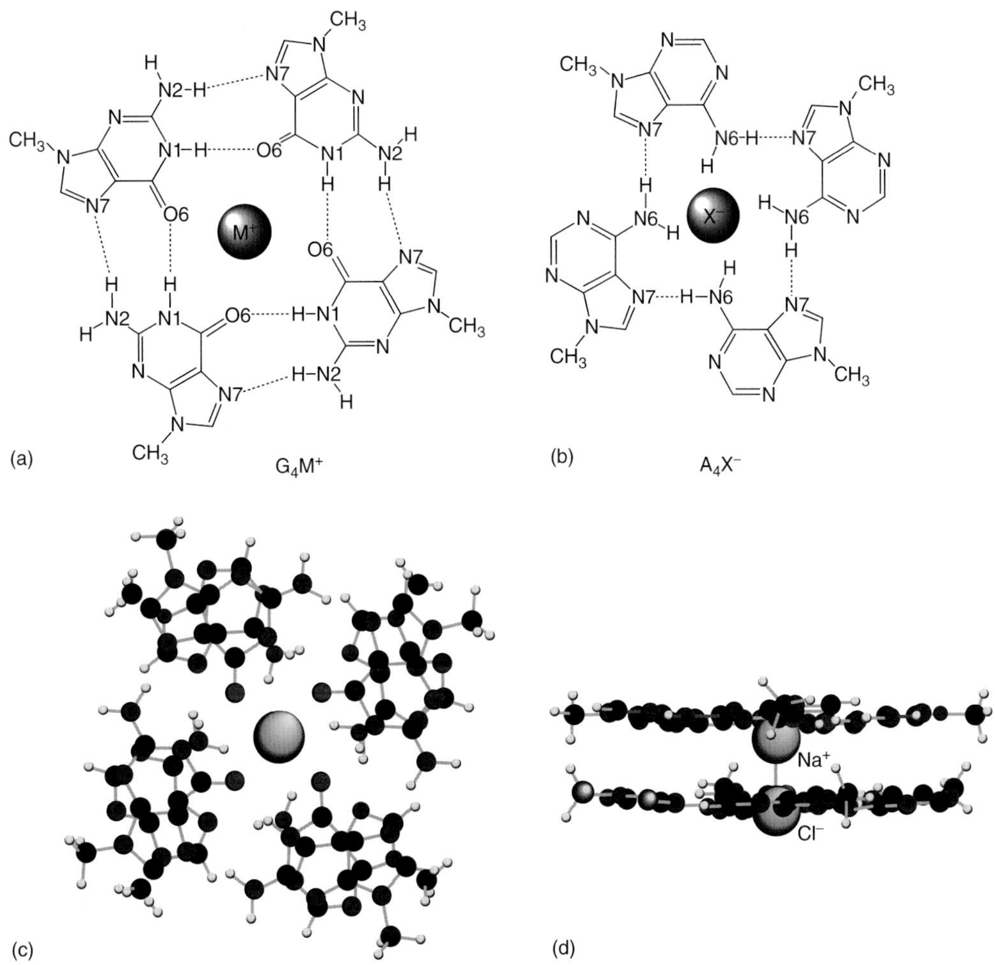

Figure 24 (a) G_4 cation receptor. (b) A_4 anion receptor. (c) Ditopic receptor G_4NaClA_4 in aqueous solution, top view; (d) side view. Na^+ and Cl^-, yellow; O, red; N, blue; C, black; and H, white. (Reproduced from Ref. 52. © Wiley-VCH, 2009.)

pairs might perform such kind of transport and be useful in symport precesses.

The very simple macrocycle, the easy-to-prepare system meso-octamethylcalix[4]pyrrole **3**, which has been already mentioned in this chapter, can act as both a cation (Cs^+) and an anion (Cl^- or Br^-) receptor forming a complex of 1 : 1 : 1 cesium:calix[4]pyrrole:halide stoichiometry, under conditions of aqueous nitrobenzene solvent extraction (Figure 25).[54] Control studies involving nitrate revealed no evidence of ion pairing for $CsNO_3$ under conditions identical to those where it was observed for CsCl and CsBr, thus enlightening the high selectivity of the process.

Ungaro and Reinhoudt with their group developed ion-pair receptors by employing the "classical" combination of calix[4]arenes and crown ether bridges for the problem of selective removal of $^{137}Cs^+$ from medium level radioactive waste, an important environmental and technological problem.[55] The objective was to selectively extract long-lived ^{137}Cs isotopes from aqueous solution containing 1 M HNO_3 and 4 M $NaNO_3$. They demonstrated that 1,3-dialkoxycalix[4]arene-crown-6 derivatives, **47**, fixed in the 1,3-alternate conformation, are exceptionally selective ionophores for cesium cations. A number of modifications to this basic framework have been reported, including both calix[4]arene biscrown ethers and dialkoxy calix[4]arene monocrown ethers. One of the best receptors for the cesium cation in solvent extraction is the crown-strapped calixarene known as BOBcalixC6, **48**.[56]

47 **48**

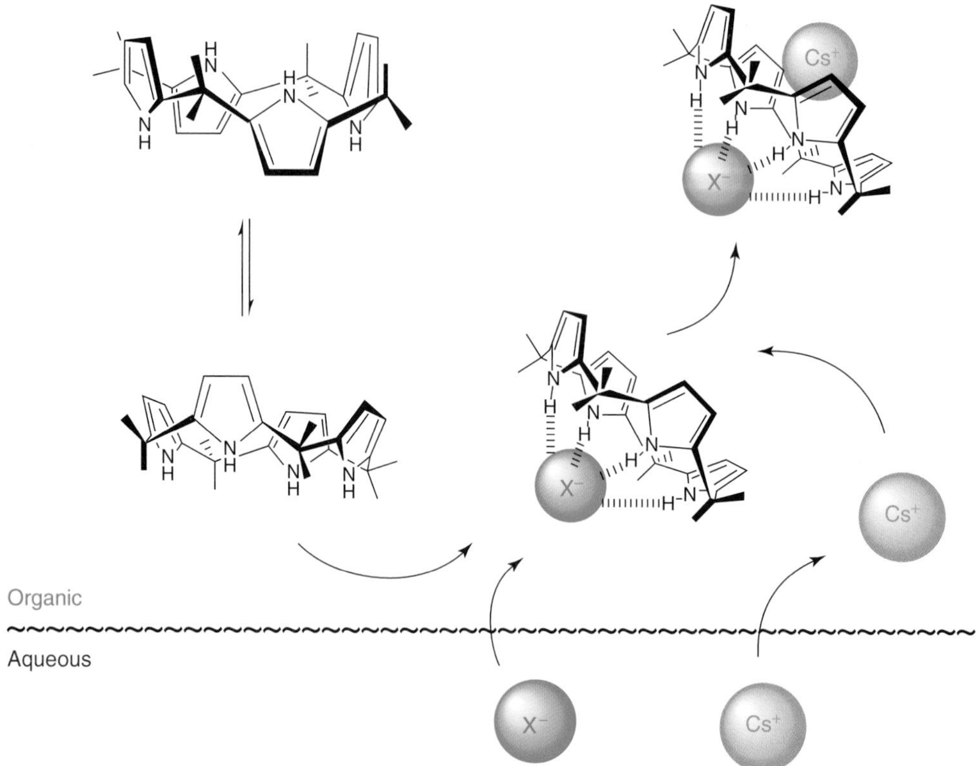

Figure 25 Proposed steps in cesium salt extraction. (Reproduced from Ref. 54. © American Chemical Society, 2008.)

The selective extraction of lithium salts is quite important because of the potential applications in high technology and medicine. So many efforts had focused on the transfer of lithium salts from an aqueous liquid phase into an organic liquid phase, to develop ionophores that can overcome the Hofmeister bias[57] against lithium cation. The small lithium cation is indeed highly solvated in water, and an organic soluble lithium ionophore has to have an affinity that is high enough to overcome the extremely unfavorable Gibbs free energy for transfer from aqueous to organic phase. Using a neutral ditopic receptor, the efficiency of the extraction process is also strongly dependent on the nature of the extracted anion.

Transport experiments that involve the use of ion-pair receptor **7** as a carrier for alkali halide salt have been carried out by Smith and coworkers.[58] Such a receptor can transport alkali metal halide salts out of an aqueous phase through a phospholipid bilayer or a liquid organic membrane. The receptor can transport these salts from an aqueous phase through a liquid organic membrane with a cation selectivity of $K^+ > Na^+ > Li^+$ for a given anion. Nevertheless, the selectivity order is strongly reversed when receptor **7** extracts solid alkali metal chlorides and bromides into organic solution; that is, the process is highly lithium selective. For a three-component mixture of solid LiCl, NaCl, and KCl, the ratio of salts extracted and complexed to the receptor in $CDCl_3$ was 94 : 4 : 2, respectively. The same strong lithium selectivity was also observed in the case of a three-component mixture of solid LiBr, NaBr, and KBr, where the ratio of extracted salts was 92 : 5 : 3. This behavior is ascribed to the unusually high solubility of lithium salts in organic solvents. Thus, receptor **7** may have application as an extractant in lithium salt purification processes.

The already mentioned heteroditopic ureido derivative **22** designed by Barboiu is another example of an ion-pair receptor able to eliminate the anion Hofmeister bias selectivity and to show discrimination in transporting NaX salts. U-tube transport experiments showed a direct relation between the synergetic recognition properties of the heteroditopic macrocyclic receptor **22** and the transport rate of the NaX salts by liquid chloroformic membranes. Generally, small hydrophilic anions (i.e., F^-, CH_3COO^-, $SO_4{}^{2-}$) go into water in low percentages because of the highly positive Gibbs energy. In this case, the cooperativity between anion and cation recognition favors the transport of such highly hydrophilic anions. The presence of the cation-binding site on the same host increases the sodium partitioning into the membrane phase, synergizing the anion extraction by hydrogen bonding.

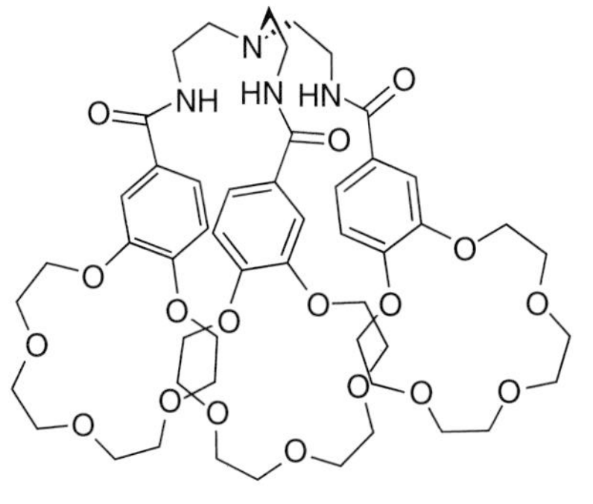

49

Ion-pair cooperativity has been also employed by Beer's group[59] to extract the radioactive pertechnetate anion, TcO_4^-, as its sodium salt, whose presence in water is a consequence of discharges from nuclear fuel reprocessing plants. The ability of tripodal tris(amidobenzo[15]crown-5) **49** to extract and transport sodium halide salts and sodium pertechnetate has been proved. A preliminary investigation was carried out on sodium iodide and sodium perrhenate, $NaReO_4$, in $CDCl_3$ solution. The reason for the choice of these two anions is that I^- has approximately the same size and charge density as TcO_4^-, and ReO_4^- is an isostructural analog. In the presence of 1 equivalent of sodium cation, in the case of ReO_4^-, the increase is more than 20-fold. This positive cooperativity has been attributed to the increase of electrostatic attraction between the positively charged complex **49**$\cdot Na^+$ complex and the guest anion. The complexed metal cation is supposed also to influence the spatial arrangement of the ligand and boost the relative acidity of the amide protons, leading to stronger hydrogen bonding. The anion selectivity trend displayed by the host is changed in the presence of sodium. In the absence of such cation, the receptor binds preferentially Cl^- over ReO_4^-, while the addition of sodium makes sure that ReO_4^- is more strongly bound. On the basis of these results, pertechnetate extraction experiments were carried out using conditions that simulated nuclear waste streams. It was found that the ditopic receptor **49** is able to extract sodium pertechnetate from water and transport it more efficiently than the monotopic tripodal ligand and benzo-15-crown-5.

Another example is related to the so-called dual-host approach (Section 2.4) applied to the problem of transporting a selected salt versus targeting only a single ion. It was reported that benzene-1,3,5-tricarboxamide, **50**, enhances $CsNO_3$ extraction by tetrabenzo-24-crown-8 via hydrogen bonding to the nitrate anion.[60] The chosen benzene-1,3,5-tricarboxamide skeleton combines the desired C_3 symmetry environment for recognition of the nitrate anion with synthetic ease and flexibility. Tetrabenzo-24-crown-8 was used as the cation host because it effectively complexes the cesium cation in 1,2-dichloroethane, while keeping the encapsulated Cs^+ cation and the NO_3^{2-} anion fully dissociated. Indeed, distribution results show that amides **50** and **51** increase the extraction of $CsNO_3$ by tetrabenzo-24-crown-8. At 10 mM, tripodal amide **50** enhances the extraction by a factor of 1.9, whereas the same concentration of bipodal amide **51** enhances Cs^+ extraction by a factor of 1.3. Of course, the increase is dependent on the amide concentration while no extraction was detected in the presence of the amides alone.

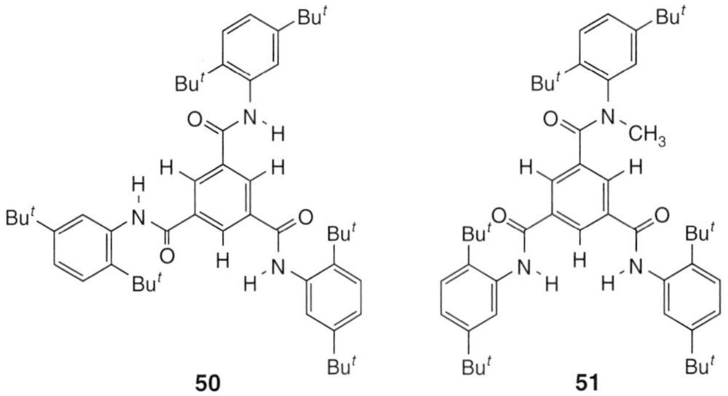

50 **51**

Polymeric systems are particularly attractive materials for the selective separation of alkaline salts from aqueous media. In 2008, Sessler and coworkers[61] reported that copolymers of a polymerizable derivative **52** (Scheme 3) of octamethylcalix[4]pyrrole and methyl methacrylate (MMA) are effective at extracting TBACl or TBAF from aqueous media. But unfortunately, these polymeric materials display relatively low affinities for the corresponding salts containing cations less soluble in organic media (e.g., Na and K). Thus, they developed mixed MMA copolymers containing pendant calix[4]pyrrole subunits known to bind halide anions in a 1 : 1 ratio in organic media and benzo-[15]crown-5 subunits capable of forming 2 : 1 sandwich complexes with potassium cations.

Initial qualitative evidence that copolymer **54**, which contains both calix[4]pyrrole and crown ether subunits, could extract chloride salts into organic media came from a visual test involving a water-soluble dye, **57**, that contains a chloride counteranion. Treatment of an aqueous solution of the dye (25.5 mM) with a CH_2Cl_2 solution of copolymer **54** (effective concentration of the calix[4]pyrrole and crown ether repeat units was 1.56 and 1.22 mm, respectively) resulted in a colored organic phase.

Control experiments showed that the mixture in CH_2Cl_2 of calyx[4]pyrrole and crown ether, as well as the solution in the same solvent of derivatives in which the two units were linked together covalently, when exposed to an aqueous solution of the dye did not show color transfer at all. On the basis of these observations, the problem of extracting "hard" potassium salts, such as KF or KCl, from

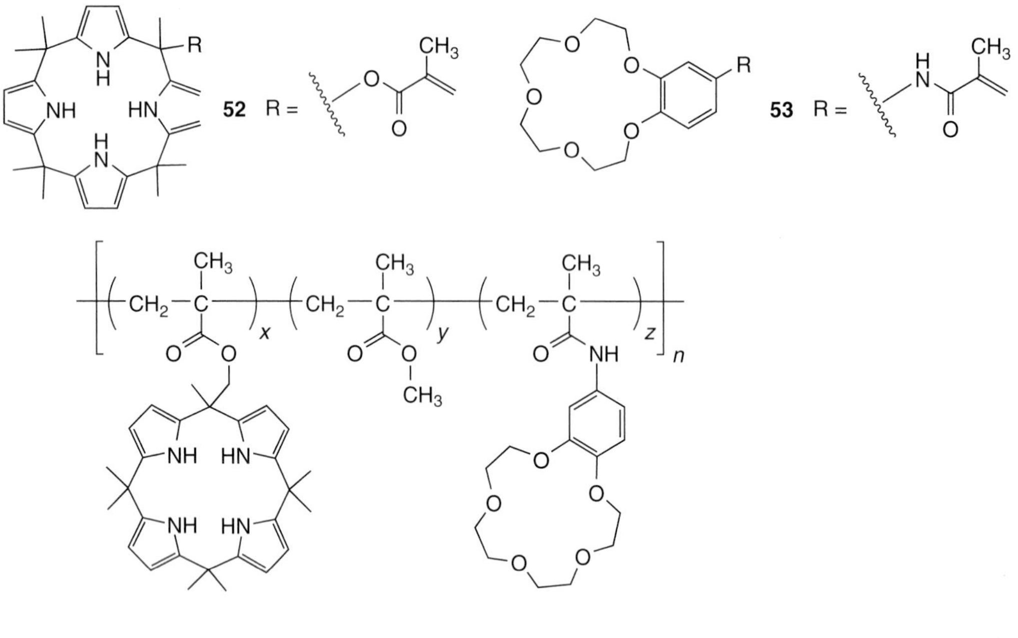

54: $x = 1.0$, $y = 14$, $z = 0.8$; $M_n = 57$ kDa
55: $x = 0.0$, $y = 12$, $z = 1.0$; $M_n = 33$ kDa
56: $x = 1.0$, $y = 14$, $z = 0.0$; $M_n = 90$ kDa

Scheme 3

aqueous media was tackled. NMR studies demonstrated that the homogeneous polymeric systems allow to overcome the relatively high hydration energies of KF and KCl, thus enabling their extraction from aqueous media with efficiencies that exceed those expected on the basis of the effective concentration of the individual receptors (crown ether and calixpyrrole).

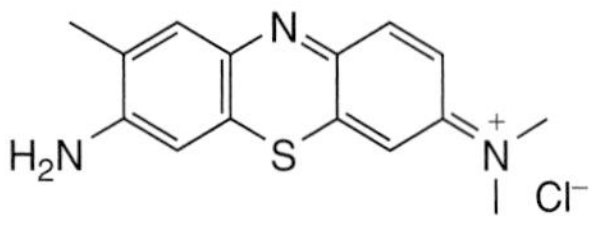

57

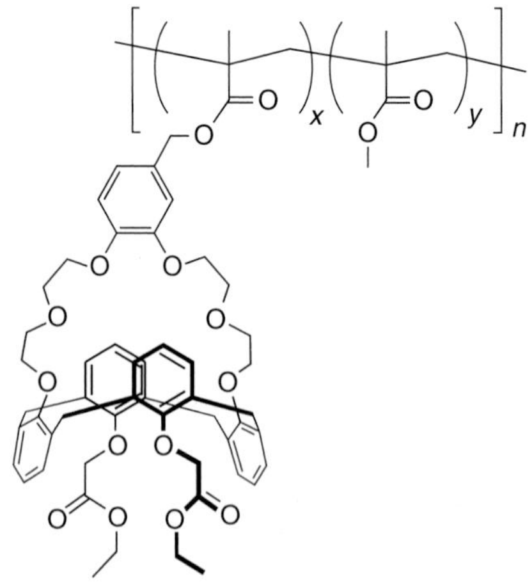

58

Another copolymer composed of a 10 : 1 molar ratio of MMA and a methacrylate functionalized benzocrown-6-calix[4]arene was reported very recently by the same group, **58**.[62]

This system is capable of extracting the Cs^+ cation from aqueous media with a selectivity over K^+ and Na^+ that exceeds that displayed by the constituent (monomeric) calixcrown receptor. It was further found that precipitation of the salt-bound form of the polymer, followed by polymer "digestion" under oxidizing acidic conditions, can be used to isolate the extracted cesium species.

4 CONCLUSIONS AND OUTLOOK

This chapter attempted to show that receptors designed to bind specifically ion pairs might present numerous advantages in terms of affinity and selectivity. Often ion-pair receptors containing binding sites for both cations and anions display affinities for ion pairs that are enhanced relative to simple ion receptors. Unfortunately, in spite of their potential applications in various fields, such as salt solubilization, extraction, and membrane transport, the number of ion-pair receptors reported in the literature is still quite limited, if compared with that of receptors for cations and anions. The synthetic challenge implied in the development of such kind of compounds is one of the reason for the slower growth; also, experimental complexities associated with the treatment of data[63] for quantitative estimation of binding affinities and assessment of possible cooperative effects make these studies quite demanding.[64] The achievement of positive cooperativity in these host–guest systems would signify a comprehensive understanding of binding phenomena that still represent one of the main goals of supramolecular chemistry.

REFERENCES

1. (a) A. Bianchi, K. Bowman-James, and E. García-España, eds. *Supramolecular Chemistry of Anions*, Wiley, New York, 1997; (b) P. D. Beer and P. A. Gale, *Angew. Chem., Int. Ed.*, 2001, **40**, 486; (c) J. M. Llinares, D. Powell, and K. Bowman-James, *Coord. Chem. Rev.*, 2003, **240**, 57.
2. C. Duranton, S. M. Huber, and F. J. Lang, *J. Physiol.*, 2002, **539**, 847.
3. G. W. Gokel, Molecular Recognition Receptors for Cationic Guests, in *Comprehensive Supramolecular Chemistry*, eds. J.-M. Lehn, J. L. Atwood, J. E. D. Davies, *et al.*, Pergamon, Oxford, 1996, vol. 1.
4. V. Amendola and L. Fabbrizzi, *Chem. Commun.*, 2009, 513.
5. C. Caltagirone and P. A. Gale, *Chem. Soc. Rev.*, 2009, **38**, 520.
6. V. Boehmer, A. Dalla Cort, and L. Mandolini, *J. Org. Chem.*, 2001, **66**, 1900.
7. P. A. Gale, Simultaneous cation and anion receptors, in *Core concepts in Supramolecular Chemistry and Nanochemistry*, eds J. W. Steed, D. R. Turner, and K. J. Wallace, John Wiley & Sons, Inc., UK, 2007, 73.
8. S. K. Kim and J. L. Sessler, *Chem. Soc. Rev.*, 2010, **39**, 3784.
9. B. D. Smith, Ion Pair Recognition by Ditopic Receptors, in *Macrocyclic Chemistry: Current Trends and Future*, eds. K. Gloe and B. Antonioli, Kluwer, London, 2005, 137.
10. Koshland. D. E. Jr. In *The Enzymes*, ed. P. Boyer, Academic Press, New York, 1970, vol. 1, 341.
11. M. W. Glenny, A. J. Blake, C. Wilson, and M. Schröder, *Dalton Trans.*, 2003, **21**, 1941.
12. S. Kubik, *J. Am. Chem. Soc.*, 1999, **121**, 5846.
13. D. E. Gross, F. P. Schmidtchen, W. Antonius, *et al.*, *Chem. Eur. J.*, 2008, **14**, 7822.
14. H. Miyaji, D.-S. Kim, B.-Y. Chang, *et al.*, *Chem. Commun.*, 2008, 753.
15. M. D. Lankshear, I. M. Dudley, K.-M. Chan, *et al.*, *Chem. Eur. J.*, 2008, **14**, 2248.
16. J. M. Mahoney, J. P. Davis, A. M. Beatty, and B. D. Smith, *J. Org. Chem.*, 2003, **68**, 9819.
17. T. Nabeshima, T. Saiki, J. Iwabuchi, and S. Akine, *J. Am. Chem. Soc.*, 2005, **127**, 5507.
18. J. L. Sessler, S. K. Kim, D. E. Gross, *et al.*, *J. Am. Chem. Soc.*, 2008, **130**, 13162.
19. M. J. Deetz, M. Shang, and B. D. Smith, *J. Am. Chem. Soc.*, 2000, **122**, 6201.
20. M. T. Reetz, C. M. Niemeyer, and K. Harms, *Angew. Chem. Int. Ed. Engl.*, 1991, **30**, 1472.
21. D. M. Rudkevich, Z. Brzozka, M. Palys, *et al.*, *Angew. Chem. Int. Ed. Engl.*, 1994, **33**, 467.
22. H. Liu, X.-B. Shao, M.-X. Jia, *et al.*, *Tetrahedron*, 2005, **61**, 8095.
23. J. C. Ma and D. A. Dougherty, *Chem Rev.*, 1997, **97**, 1303.
24. M. Cametti, M. Nissinen, A. Dalla Cort, *et al.*, *J. Am. Chem. Soc.*, 2007, **129**, 3641.
25. A. Dalla Cort, P. De Bernardin, G. Forte, and F. Yafteh Mihan, *Chem. Soc. Rev.*, 2010, **39**, 3863.
26. M. Cametti, M. Nissinen, A. Dalla Cort, *et al.*, *Chem. Commun.*, 2003, 2420.
27. M. Cametti, M. Nissinen, A. Dalla Cort, *et al.*, *J. Am. Chem. Soc.*, 2005, **127**, 3831.
28. G. Reeske, M. Schürmann, B. Costisella, and K. Jurkschat, *Organometallics*, 2007, **26**, 4170.
29. G. Reeske, G. Bradtmoller, M. Schürmann, and K. Jurkschat, *Chem. Eur. J.*, 2007, **13**, 10239.
30. T. Tozawa, Y. Misawa, S. Tokita, and Y. Kubo, *Tetrahedron Lett.*, 2000, **41**, 5219.
31. M. Barboiu, G. Vaughan, and A. van der Lee, *Org. Lett.*, 2003, **5**, 3073.
32. W. S. Bryant, J. W. Jones, P. E. Mason, *et al.*, *Org. Lett.*, 1999, **1**, 1001.
33. K. Zhu, L. Wu, X. Yan, *et al.*, *Chem. Eur. J.*, 2010, **16**, 6088.
34. M. K. Chae, J.-I. Lee, N.-K. Kim, and K.-S. Jeong, *Tetrahedron Lett.*, 2007, **48**, 6624.
35. J. Gong, B. C. Gibb, *Chem. Commun.*, 2005, 1393.
36. J. Scheerder, J. P. M. van Duynhoven, J. F. J. Engbersen, and D. N. Reinhoudt, *Angew. Chem. Int. Ed*, 1996, **35**, 1090.
37. G. Tumcharern, T. Tuntulani, S. J. Coles, *et al.*, *Org. Lett.*, 2003, **5**, 4971.
38. M. D. Lankshear, I. M. Dudley, K.-M. Chan, and P. D. Beer, *New J. Chem.*, 2007, **31**, 684.
39. C. Gargiulli, G. Gattuso, C. Liotta, *et al.*, *J. Org. Chem.*, 2009, **74**, 4350.
40. S. Le Gac and I. Jabin, *Chem. Eur. J.*, 2008, **14**, 548.
41. A. Lascaux, S. Le Gac, J. Wouters, *et al.*, *Org. Biomol. Chem.*, 2010, **8**, 4607.
42. F. Tancini, T. Gottschalk, W. B. Schweizer, *et al.*, *Chem. Eur. J.*, 2010, **16**, 7813.
43. L. Pescatori, A. Arduini, A. Pochini, *et al.*, *Org. Biomol. Chem.*, 2009, **7**, 3698.
44. L. Pescatori, A. Arduini, A. Pochini, *et al.*, *CrystEngComm*, 2009, **11**, 239.
45. A. Arduini, G. Giorgi, A. Pochini, *et al.*, *J. Org. Chem.*, 2001, **66**, 8302.
46. P. R. A. Webber and P. D. Beer, *Dalton Trans.*, 2003, **21**, 2249.

47. D. Garozzo, G. Gattuso, A. Notti, *et al.*, *Angew. Chem. Int. Ed.*, 2005, **44**, 4892.
48. F. P. Schmidtchen, *J. Org. Chem.*, 1986, **51**, 5161.
49. A. Galán, D. Andreu, A. M. Echavarren, *et al.*, *J. Am. Chem. Soc.*, 1992, **114**, 1511.
50. X. Shi, J. C. Fettinger, and J. T. Davis, *Angew. Chem. Int. Ed.*, 2001, **40**, 2827.
51. F. W. Kotch, V. Sidorov, Y.-F. Lam, *et al.*, *J. Am. Chem. Soc.*, 2003, **125**, 15140.
52. T. van der Wijst, C. Fonseca Guerra, M. Swart, *et al.*, *Angew. Chem. Int. Ed.*, 2009, **48**, 3285.
53. J. W. Steed and J. L. Atwood, *Supramolecular Chemistry*, John Wiley & Sons, Inc., U.K, 2009.
54. M. P. Wintergerst, T. G. Levitskaia, B. A. Moyer, *et al.*, *J. Am. Chem. Soc.*, 2008, **130**, 4129.
55. A. Casnati, A. Pochini, R. Ungaro, *et al.*, *J. Am. Chem. Soc.*, 1995, **117**, 2767.
56. R. A. Sachleben, P. V. Bonnesen, T. Descazeaud, *et al.*, *Solvent Extr. Ion Exch.*, 2010, **17**, 1445.
57. F. Hofmeister, *Arch. Exp. Pathol. Pharmakol.*, 1888, **24**, 247.
58. J. M. Mahoney, G. U. Nawaratna, A. M. Beatty, *et al.*, *Inorg. Chem.*, 2004, **43**, 5902.
59. P. D. Beer, P. K. Hopkins, and J. D. McKinney, *Chem. Commun.*, 1999, 1253.
60. K. Kavallieratos, R. A. Sachleben, G. J. Van Berkelb, and B. A. Moyer, *Chem. Commun.*, 2000, 187.
61. A. Aydogan, D. J. Coady, S. K. Kim, *et al.*, *Angew. Chem. Int. Ed.*, 2008, **47**, 9648.
62. B. M. Rambo, S. K. Kim, J. S. Kim, *et al.*, *Chem. Sci.*, 2010, **1**, 716.
63. J. W. Jones and H. W. Gibson, *J. Am. Chem. Soc.*, 2003, **125**, 7001.
64. S. Roelens, A. Vacca, O. Francesconi, and C. Venturi, *Chem. Eur. J.*, 2009, **15**, 8296.

Metal Complexes as Receptors

Stephen J. Loeb

University of Windsor, Windsor, Ontario, Canada

1 INTRODUCTION

1.1 Metal complexes as receptors: limitations

A metal complex is the product of the interaction of a Lewis-acidic, transition metal ion and a set of Lewis-basic ligands. In order for a metal complex to be studied as a receptor (host) for a target substrate (guest), it must have sufficient stability under the experimental conditions to allow measurement of the receptor–substrate (host–guest) interaction. For example, it is not possible to study the host–guest chemistry of a metal complex if interaction with the guest of interest results in ligand substitution by the guest. This necessarily limits the type of metal complexes that can be studied as receptors to those containing robust metal–ligand interactions. This is achieved by utilizing an inert metal center, a chelating ligand, or more often both. Complexes formed from labile metal–ligand interactions via self-assembly often contain interesting structural features but are very difficult to evaluate in terms of a measurable receptor–substrate (host–guest) interaction.

1.2 First- and second-sphere coordination

The concept that a transition metal complex can interact in an orderly manner with neutral molecules or ions to give an outer sphere complex dates back over 100 years to Alfred Werner's original description of coordination chemistry (Figure 1). Werner found the idea, we now call second-sphere coordination, essential to explain such simple phenomena as (i) the dependence of optical rotation on the nature of the anion and solvent, (ii) the formation of adducts between amines and saturated complexes, and (iii) solvents of crystallization.[1] Indeed, aspects of the second-sphere coordination are known to be important in such diverse areas as the biological activity of siderophores[2] and the function of MRI contrast agents.[3]

1.3 Role of the metal ion in a metal complex receptor

The five basic roles of a metal ion in a metal complex receptor or host are as follows:

1. Direct involvement in the binding event via metal–ligand coordination to the guest in the first-sphere of coordination.
2. Indirect involvement in the binding event via noncovalent interactions with the guest in the second-sphere of coordination.
3. Promotion of a significant conformational change that effects binding of the guest; positive and negative allosteric effects.

Supramolecular Chemistry: From Molecules to Nanomaterials.
Edited by Philip A. Gale and Jonathan W. Steed.
© 2012 John Wiley & Sons, Ltd. ISBN: 978-0-470-74640-0.

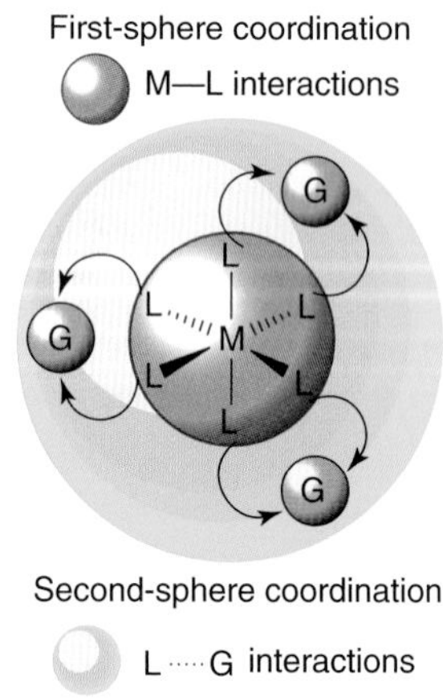

Figure 1 Schematic representation of the binding of guest molecules or ions using a metal complex as a receptor; first-sphere = red and second-sphere = yellow.

4. A building block for the creation of a structural framework that can encapsulate a guest as in container molecules and metal–organic framework (MOF) materials.
5. A reporter group that signals the binding of a guest by producing some measurable outputs (electrochemical, optical, etc.).

1.4 Scope

This chapter is focused solely on robust metal complexes in which the metal ion contributes significantly to the host–guest interaction (1 and 2 above). Since the simple interaction of a ligand (guest or substrate) with an open coordination site of a metal complex is ubiquitous in coordination chemistry, this first-sphere interaction must be accompanied by some other noncovalent interaction with the guest for the metal complex to be classified as a receptor (i.e., 1 plus 2); simultaneous first- and second-sphere coordination. On the other hand, a variety of metal complexes can act as receptors utilizing only noncovalent interactions between the complex and a guest (i.e., 2 only); second-sphere coordination. This is true for neutral guests to a certain extent and is prominent in the design of receptors for anions.

This chapter does not cover the use of metal ions as allosteric effectors (3), skeletal components (4) (see **Crystal Engineering**, Volume 6), or sensing elements (5) (see **Molecular Redox Sensors**, **Colorimetric Sensors** and **Luminescent Sensing**, Volume 5). It also does not address the interaction of metal complexes with biological systems (see **Synthetic Peptide-Based Receptors**, **Biological Small Molecules as Receptors**, Volume 3 and **Supramolecular Bioinorganic Chemistry**, Volume 4) or the use of metal complexes as drugs (see **Supramolecular Chemistry in Medicine**, Volume 5).

2 METAL COMPLEXES AS RECEPTORS FOR NEUTRAL GUESTS

2.1 Water and ammonia—the simplest of guests

A water molecule (H_2O) can interact directly with a transition metal ion by donation of a lone pair of electrons from oxygen. This leaves the two H atoms available for hydrogen bonding with an appropriate hydrogen-bond acceptor atom on the receptor. A molecule of ammonia (NH_3) can interact directly with a transition metal ion by donation of the lone pair of electrons from nitrogen. This leaves the three H atoms available for hydrogen bonding with an appropriate hydrogen-bond acceptor atom on the receptor. Figure 2 illustrates the multipoint binding of these simplest of guests using what is often referred to as simultaneous first- and second-sphere coordination.

The first example of this type of host–guest interaction verified by X-ray crystallography was the complex $[Rh(CO)(\mathbf{1})(H_2O)][PF_6]$.[4] The cavity formed by trans coordination of the chelating diphosphino-polyether ligand **1** (Figure 3) is ideal for binding a water molecule. The H_2O molecule acts as a first-sphere ligand by forming a σ-bond directly to the rhodium(I) center, while simultaneously being involved in second-sphere coordination through hydrogen bonding to the ether oxygen atoms. The coordination of H_2O to the metal center enhances the hydrogen bonding by increasing the acidity of the water H atoms. The overall effectiveness of this multipoint binding was demonstrated by the fact that the water molecule could not be removed by dissolution in polar solvents or evacuation at 10^{-3} Torr.[5]

In a similar manner, the rhodium(III) complex of macrocycle **2**, $[RhCl_2(H_2O)(\mathbf{2})][PF_6]$,[6] and the palladium(II) complex of the macrocycle **3**, $[Pd(\mathbf{3})(H_2O)][BF_4]$,[7] were found to have a molecule of water bound to the metal center,

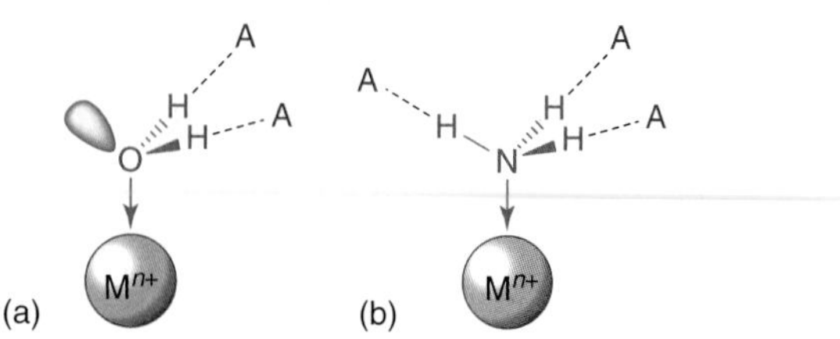

Figure 2 Schematic representations showing the first- and second-sphere coordination of (a) a water molecule and (b) a molecule of ammonia.

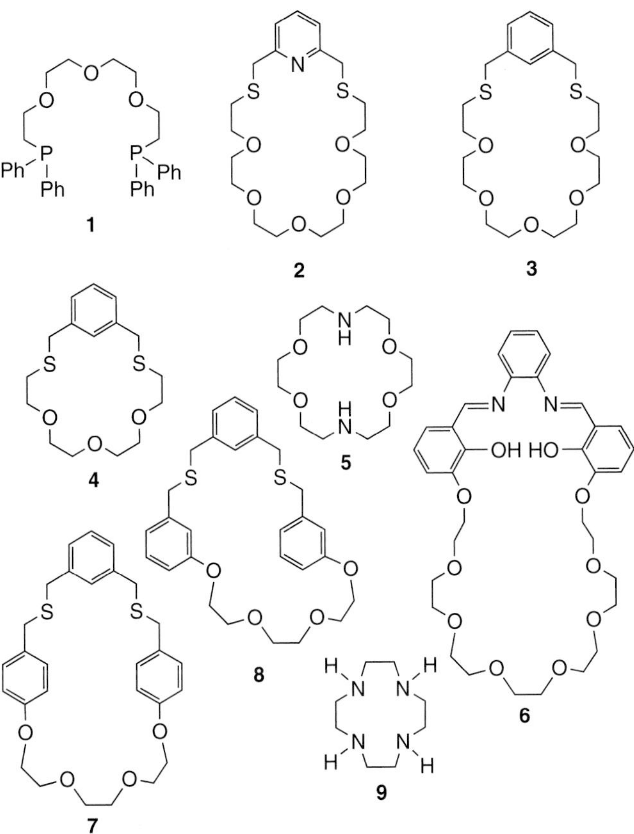

Figure 3 Ligands **1**–**9** used to form metal complex receptors that bind guests through the first- and second-sphere coordination.

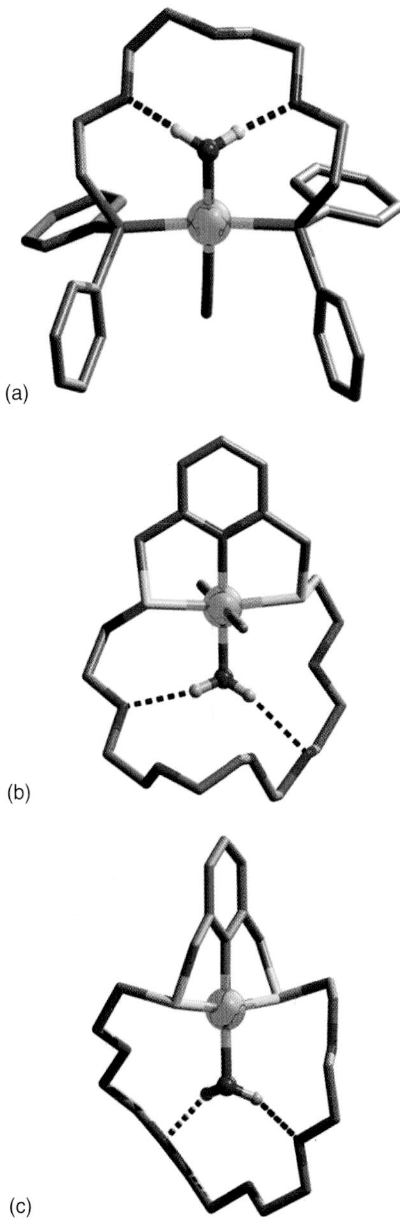

Figure 4 Ball-and-stick representations of the single-crystal X-ray crystal structures of (a) $[Rh(CO)(\mathbf{1})(H_2O)]^+$, (b) $[RhCl_2(\mathbf{2})(H_2O)]^+$, and (c) $[Pd(\mathbf{3})(H_2O)]^+$. Noncoordinating anions and H atoms not involved in hydrogen bonding are omitted for clarity.

which is augmented by hydrogen bonding to ether oxygen atoms inside the macrocyclic ligand (Figure 4).[8]

None of the three complexes depicted in Figure 4 were originally created as receptors for H_2O. Their isolation was more a product of serendipity than design. In contrast, the complex $[Pd(\mathbf{4})(CH_3CN)][BF_4]$ was designed specifically to allow the substitution of a labile CH_3CN ligand by NH_3.[7] The ^1H NMR spectrum of $[Pd(\mathbf{4})(NH_3)][BF_4]$ at 300 K in CD_3CN showed a single broad resonance at 2.49 ppm for the three hydrogen atoms of the bound NH_3 molecule due to rapid rotation about the Pd–N bond. However, at 213 K, this rotation is slowed and separate signals were observed at 2.60 and 2.49 ppm in a 2 : 1 ratio consistent with two hydrogen-bonded NH protons and one free NH proton. This conformation was also observed in the X-ray structure in which the palladium(II)-bound NH_3 guest is involved in two NH···O hydrogen bonds to ether oxygen atoms (Figure 5). In the complex $[Rh_2(cod)_2(NH_3)_4(\mathbf{5})][PF_6]_2$ (cod = 1,5-cyclooctadiene), the diazacrown ether **5** binds 2 equivalents of $[Rh(cod)(NH_3)_2)]^+$ through the receptor NH groups and the rhodium(I)-bound NH_3 guests enter into weak hydrogen-bonding interactions with the ether oxygen atoms.[9]

2.2 Other guests combining metal coordination and hydrogen bonding

The Schiff base macrocycle **6** binds the $[UO_2]^{2+}$ ion through nitrogen and phenolate oxygen atoms. The complex $[UO_2(\mathbf{6})]$ then acts as a receptor binding urea ($(NH_2)C{=}O$) via direct interaction of the carbonyl oxygen with the Lewis-acidic, uranium(IV) center in concert with a series of NH···O hydrogen bonds to peripheral ether oxygen atoms (Figure 6).[10] Similar $[UO_2]$ complexes have been designed,

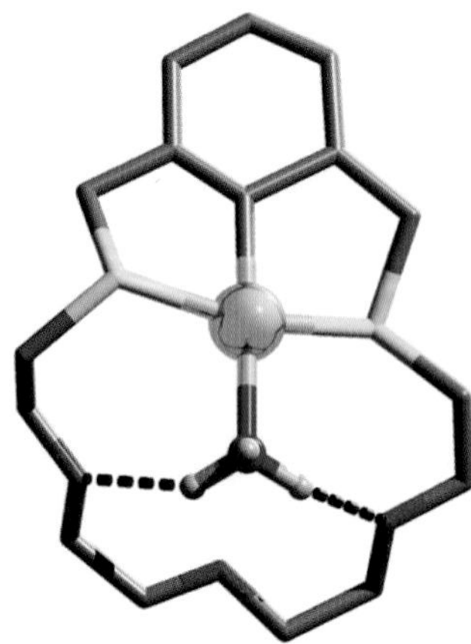

Figure 5 Ball-and-stick representation of the single-crystal X-ray crystal structure of $[Pd(\mathbf{4})(NH_3)]^+$. Noncoordinating anions and H atoms not involved in hydrogen bonding are omitted for clarity.

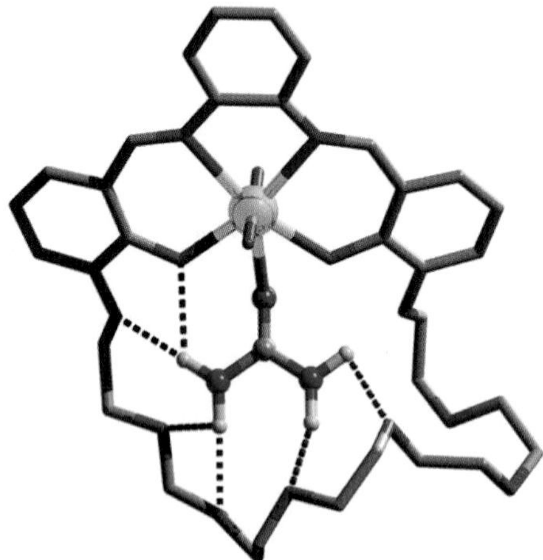

Figure 6 Ball-and-stick representation of the single-crystal X-ray crystal structure of $[UO_2(\mathbf{6})(urea)]^{2+}$. Noncoordinating anions and H atoms not involved in hydrogen bonding are omitted for clarity.

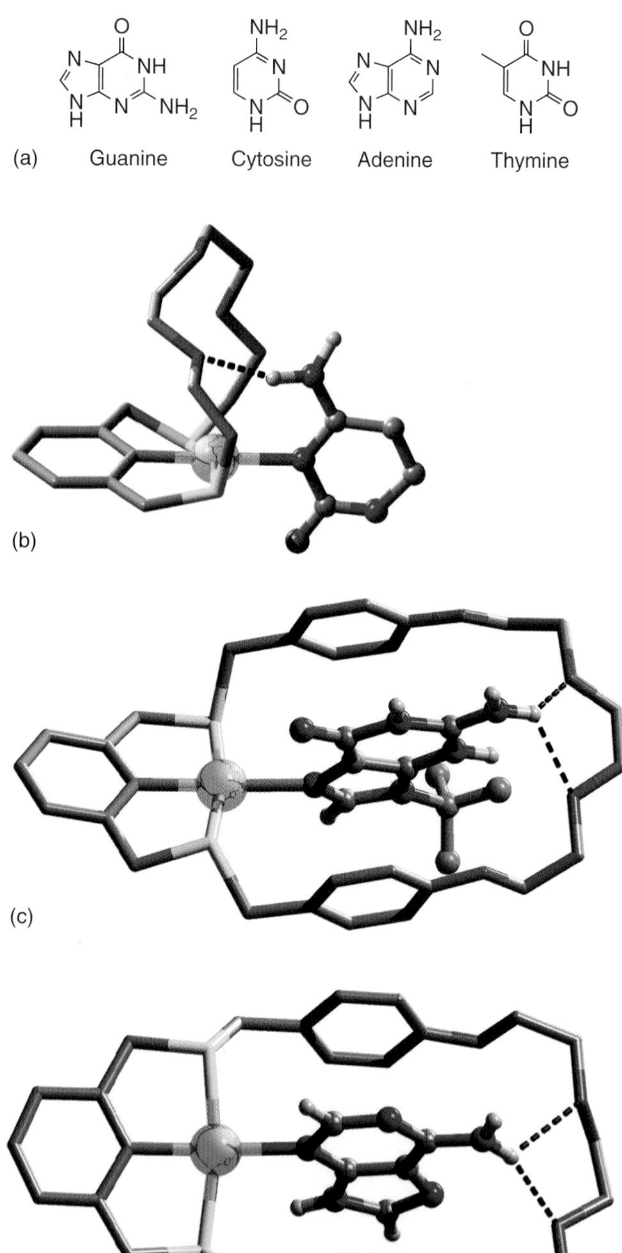

Figure 7 Complexes of nucleobases. (a) The four complementary nucleobases of DNA. Ball-and-stick representations of the single-crystal X-ray crystal structures of (b) $[Pd(\mathbf{4})(cytosine)]^+$, (c) $[Pd(\mathbf{7})(guanineBF_3)]^+$, and (d) $[Pd(\mathbf{7})(adenine)]^+$. Noncoordinating anions and H atoms not involved in hydrogen bonding are omitted for clarity.

which interact with barbiturates[11] in an analogous manner and have been used to perform chiral recognition of a number of neutral and charged guests.[10a,12]

In an interesting application of first- and second-sphere coordination, the complexes $[Pd(\mathbf{4})(CH_3CN)][BF_4]$, $[Pd(\mathbf{7})(CH_3CN)][BF_4]$, and $[Pd(\mathbf{8})(CH_3CN)][BF_4]$ were investigated for the ability to selectively recognize the DNA nucleobases cytosine, guanine, adenine, and thymine.[13] $[Pd(\mathbf{4})(CH_3CN)][BF_4]$ was found to be selective for cytosine,[14] whereas $[Pd(\mathbf{7})(CH_3CN)][BF_4]$ and $[Pd(\mathbf{8})(CH_3CN)][BF_4]$ showed a very strong affinity for adenine and guanine.[15] The X-ray structures of $[Pd(\mathbf{4})(cytosine)][BF_4]$, $[Pd(\mathbf{7})(guanineBF_3)][BF_4]$, and $[Pd(\mathbf{7})(adenine)][BF_4]$ are shown in Figure 7. The guanine derivative is bound through palladium(II) at N7 while simultaneously interacting in the second sphere via π-stacking of the purine ring between the parallel aromatic rings of the receptor and hydrogen bonding of the amino group to two aliphatic ether oxygen atoms. The adenine guest of $[Pd(\mathbf{7})(adenine)][BF_4]$ is bound in a very similar manner using this same trio of noncovalent interactions: metal coordination, π-stacking, and hydrogen-bonding interactions.[15] Most interestingly, the coordination of adenine through N7 is not the normal mode of transition metal coordination to this nucleobase[16]; the great majority of adenine complexes involve coordination to a metal through the more basic N3 donor.[17] It was rationalized that the observed preference for N7 in [Pd(**7**)(adenine)][BF_4] was a result of the strength and directionality of the complementary hydrogen-bonding and π-stacking interactions in the second-sphere of coordination.

A metal complex receptor for thymine derivatives was devised based on the zinc(II) complex of macrocycle **9**.[18] This receptor interacts with AZT (3′-azido-3′-deoxythymidine) such that the AZT pyrimidine ring diagonally bisects the symmetrical N4 plane of the [Zn(**9**)]$^+$ unit with NH···O hydrogen bonds formed between the two pyrimidine carbonyls and the NH groups of **9**. This type of molecular recognition, employing metal complexes, has also been successful using rhodium(I) and zinc(II) porphyrin complexes to form adducts with neutral amino acids.[19] The strong binding and transport of biologically important molecules such as nucleic acid constituents and the activation of selective guests is reminiscent of enzymatic catalysis. An example is the rhodium(I) complex of **10** (Figure 8) that was shown to selectively hydrogenate propene-substituted 1,3-dihydroxybenzenes.[20]

The cavity of a calix[4]arene has also been utilized as the source of second-sphere bonding and selectivity in the molecular recognition of appropriately shaped guests. The palladium(II) complex [Pd(**11**)(CH_3CN)][BF_4] of the strapped calix[4]arene ligand **11** was shown to be capable of distinguishing between the *ortho, meta*, and *para* isomers of phenylpyridine.[21] On coordination to the palladium(II) center, only *para*-phenylpyridine was the correct shape to project the substituent phenyl ring of the guest into the bowl-shaped cavity of the calix[4]arene (Figure 9).

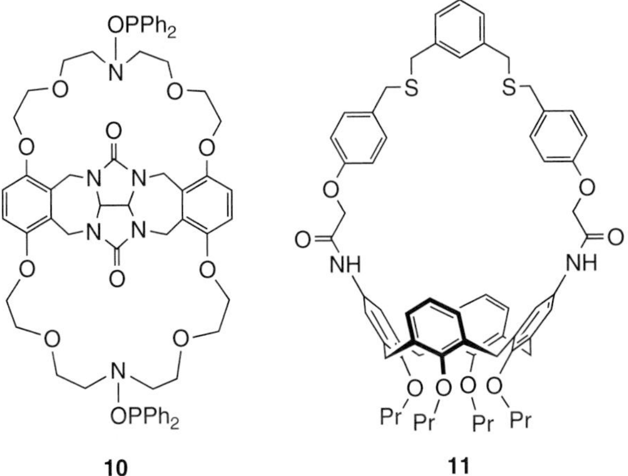

Figure 8 Ligands **10** and **11** used to form metal complex receptors that bind guests through the first- and second-sphere coordination.

Figure 9 Space-filling representation of the MM3-derived structure of [Pd(**11**)(4-phenylpyridine)]$^+$. Noncoordinating anions are omitted for clarity.

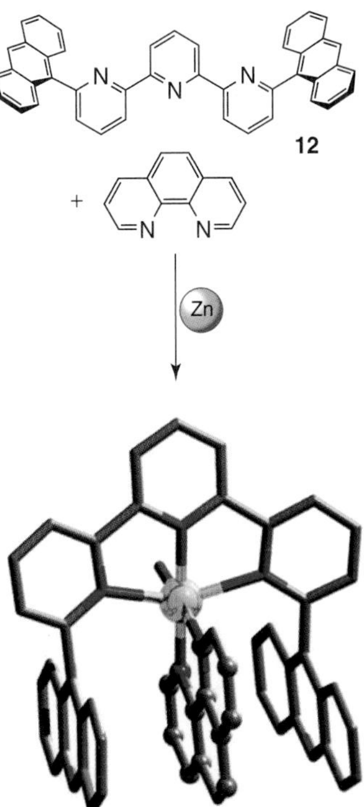

Figure 10 Schematic showing how W-shaped ligand **12** adopts a U-shape to bind 1,10-phenanthroline in the presence of zinc(II) ions. The single-crystal X-ray crystal structure of the product [Zn(**12**)(1,10-phen)]$^+$ is shown as a ball-and-stick representation. Noncoordinating anions and H atoms not involved in hydrogen bonding are omitted for clarity.

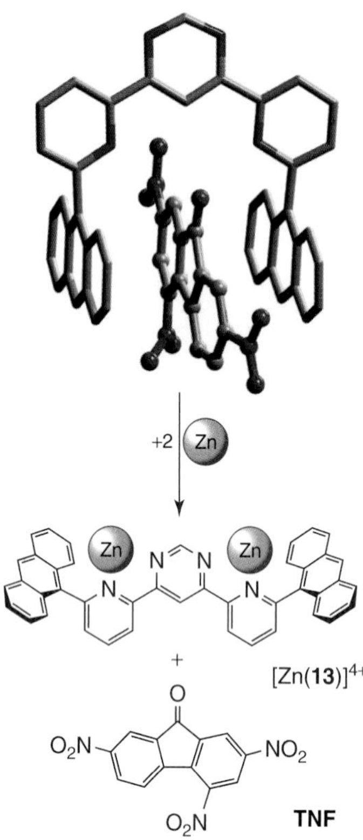

Figure 11 Schematic showing how host **13** releases a guest (**TNF**) when it converts from a U-shaped receptor to a W-shaped complex on coordination to zinc(II). The single-crystal X-ray crystal structure of the host–guest adduct [(**13**)(**TNF**)] is shown as a ball-and-stick representation. H atoms not involved in hydrogen bonding are omitted for clarity.

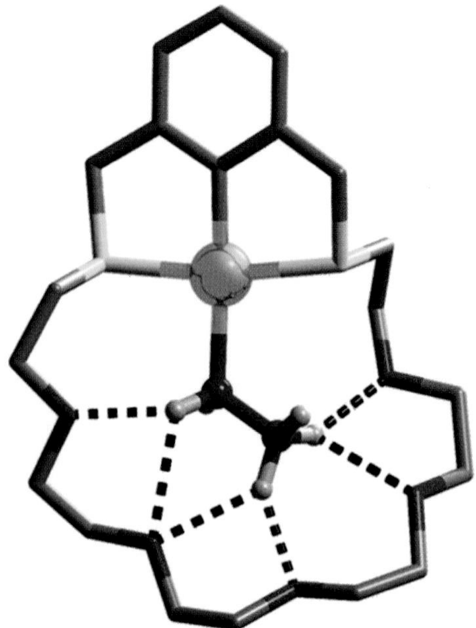

Figure 12 Ball-and-stick representation of the single-crystal X-ray crystal structure of $[Pd(\mathbf{4})(NH_2NH_3)]^{2+}$. Noncoordinating anions and H atoms not involved in hydrogen bonding are omitted for clarity.

The bistable receptor **12** converts from a W- to a U-shape on complexation with zinc(II)[22] (Figure 10). The U-shaped conformation was shown to be capable of binding an aromatic guest such as 1,10-phenanthroline by insertion between the pyrene branches of the tweezers with the participation of the metal cation.[22] Conversely, the coordination of copper(I) cations to **13** provokes the release of an intercalated substrate such as trinitrofluorene (**TNF**), by inducing a conformational change from a U- to a W-shape[22] (Figure 11). This latter sequence was described by the authors as an example of a multistate supramolecular switching device and a nanomechanical chemical system.[22,23]

3 METAL COMPLEXES AS RECEPTORS FOR CATIONIC GUESTS

3.1 An example of first- and second-sphere coordination to a cationic guest

Since most metal complexes involve a positively charged Lewis-acidic center, the number of metal complexes used to bind positively charged guests is very small. There are, however, some very interesting examples which demonstrate that the counterintuitive idea of using a cationic metal complex to bind a cationic guest is not impossible and should not be dismissed. For example, the complexes $[Pd(\mathbf{3})(CH_3CN)][BF_4]$ and $[Pd(\mathbf{4})(CH_3CN)][BF_4]$ have been shown to bind the positively charged hydrazinium ion, $[NH_2NH_3]^+$, and the structure of $[Pd(\mathbf{3})(NH_2NH_3)]^{2+}$ is shown in Figure 12.[7,24] The hydrazinium ion bonds directly to the palladium(II) center through Pd–NH_2 coordination while forming an array of NH···O hydrogen bonds to the five ether oxygen atoms of the macrocycle.

4 METAL COMPLEXES AS RECEPTORS FOR ANIONIC GUESTS

4.1 Design principles

On the basis of simple electrostatic arguments, a cationic transition metal complex will always undergo some degree of ion-pairing interaction with a guest anion. In the solid state, this electrostatic interaction is essential since there is the need to balance charge while optimizing crystal-packing forces, but in solution, the degree of cation–anion interaction depends on the nature of the solvent. With respect to designing a metal complex receptor, relying on only the electrostatic interaction between a cation and an anion has rarely resulted in an effective receptor. Thus, the best strategy for designing metal complex receptors for anions combines both the electrostatic attraction of the metal center and the directional influence of appended H-donor functional groups on the ligands.[25] The metal ion in these receptors brings together functional groups on separate ligands into a new binding site that takes advantage of the flexibility and directionality inherent in a hydrogen-bonding scheme; that is, through second-sphere coordination.[1]

4.2 Preorganization via metal–ligand coordination

A well-established tenant of supramolecular chemistry is that the preorganization of the binding site leads to enhanced recognition. When metal complexes act as anion receptors, it is coordination of the ligands to the metal center that acts to preorganize the hydrogen bonding or other functional groups for anion binding. It can then be argued that metal–ligand bonding reduces the entropic cost to attain the conformation required to bind the guest and any entropic loss associated with rigidifying the system is outweighed by the net enthalpic gain from formation of a set of strong M–L bonds.

4.3 Organizing hydrogen-bond donor groups on ligands

Mono-, bi-, and tridentate, aromatic amine ligands containing appended functional groups for hydrogen bonding to anions have been coordinated to a single transition metal ion. Formation of this type of complex results in the organizing of previously independent hydrogen-bonding groups into pockets that can then operate in unison to efficiently bind anions. Figure 13 shows the ligands of this type based on the classical frameworks of pyridine, bipyridine, and terpyridine in combination with well-established anion hydrogen-bonding groups such as amide, urea, and pyrrole.

The first example of a metal complex designed to act as an anion-binding receptor was the ruthenium(II) complex $[Ru(\mathbf{14})_2][PF_6]_2$ reported by Hamilton in 1995.[26] Hamilton showed that coordinating 2 equivalents of ligand **14** to ruthenium(II) and employing the noncoordinating PF_6^- anion created a complex that demonstrated effective binding for dicarboxylate ions through hydrogen bonding to the thiourea groups attached to **14** (Figure 14). Thus, for the first time, it was shown that two sets of two NH hydrogen-bond donors on *different* ligands could be organized into a single binding site by virtue of their coordination to a metal center. One of the key design features was the use of an inert ruthenium(II) metal center. This has two distinct

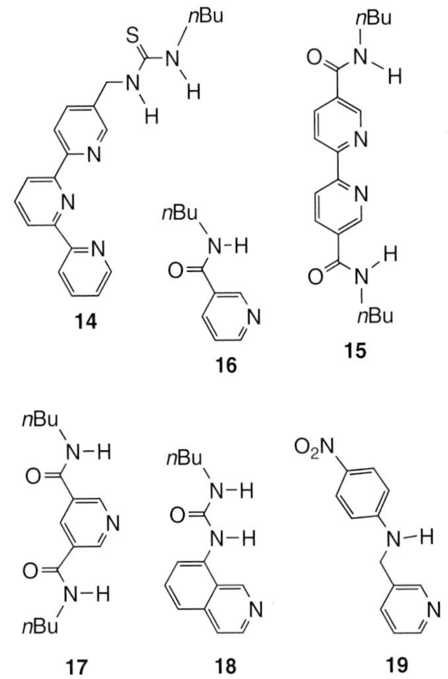

Figure 13 Ligands **14–19** used to form metal complex receptors that bind anions through second-sphere coordination.

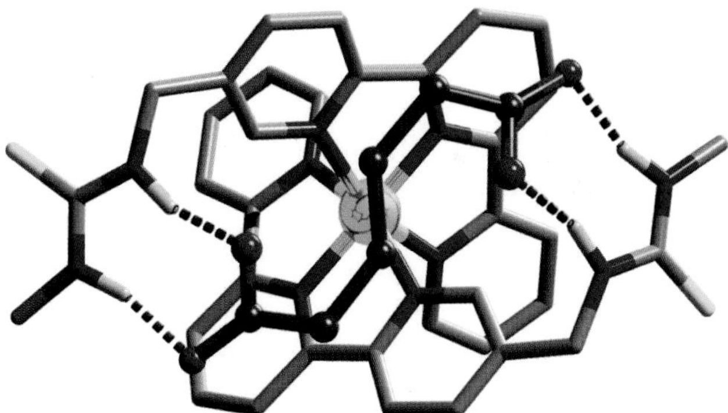

Figure 14 Ball-and-stick representation of the MM3-derived structure of [Ru(**14**)(adipate)]. The adipate anion carbon atoms are shown in black, and the terminal *n*-butyl groups have been shortened to one carbon to aid in visualization. H atoms not involved in hydrogen bonding are omitted for clarity.

advantages. It allows for isolation and characterization of a stable receptor prior to any interaction with a target anion, and it eliminates the problem of ligand exchange resulting from the competition between anionic guests and ligands for coordination sites. Indeed, Hamilton showed that the Fe(II) analog of this ruthenium(II) receptor was substitution labile as carboxylate anions easily displaced the terpy ligands. Although there was no solid-state structural data available for directly probing the thiourea–anion interaction, the strong binding of this receptor with a variety of dicarboxylate anions such as glutarate, adipate, pimelate, *iso*-phthalate, and 1,3-phenylenediacetate led to the conclusion that the binding cavity was relatively flexible.

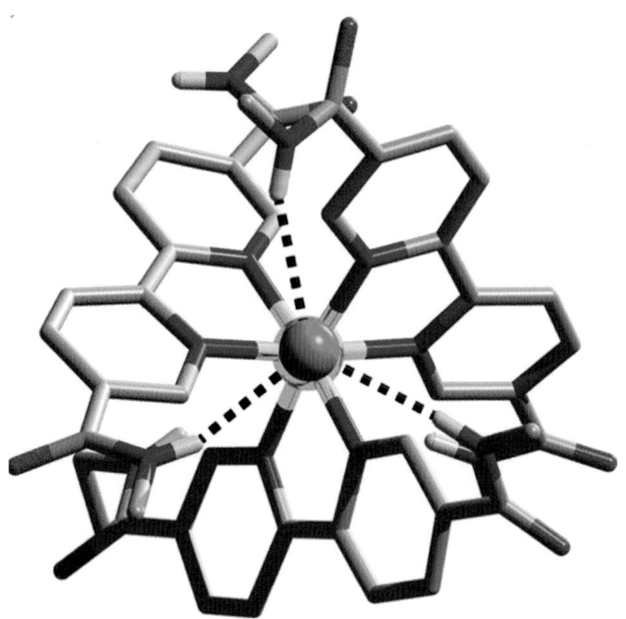

Figure 15 Ball-and-stick representation of the single-crystal X-ray crystal structure of $[Ru(\mathbf{15})(Cl)]^+$ viewed along the Cl···Ru vector. The terminal n-butyl groups have been shortened to one carbon to aid in visualization, and H atoms not involved in hydrogen bonding are omitted for clarity.

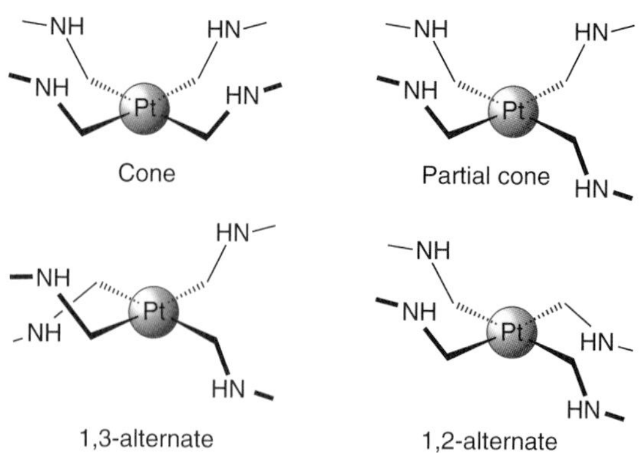

Figure 16 Schematic illustrating the analogy between the four conformations of a calix[4]arene and the four conformations of a receptor with formulation $[Pt\mathbf{L}_4]^{2+}$ where **L** is a mondentate ligand-containing hydrogen-bond donors such as **16**.

Indeed, the presence of a connecting methylene group between the terpyridine ligand and the thiourea groups would certainly provide any needed flexibility. A molecular mechanics (MM3)-derived model[27] of this receptor clearly shows the possibility of a significant binding site between the two terpy ligands (Figure 15). In this case, the adipate ion is a good fit and nicely spans the gap between the hydrogen-bond acceptor thiourea groups.

Coordination of 3 equivalents of the bidentate ligand **15** to ruthenium(II) yielded the metal complex receptor $[Ru(\mathbf{15})_3]^{2+}$.[28] This complex contains two identical binding sites resulting from the convergence of two sets of three hydrogen-bond donor, amide NH groups. Since octahedral, tris-chelated complexes are chiral, Beer isolated both enantiomers and investigated the ability of Δ-$[Ru(\mathbf{15})_3]^{2+}$ to bind chloride, nitrate, and acetate ions (dimethyl sulfoxide (DMSO)-d_6).[29] The X-ray structure of $[Ru(\mathbf{15})_3]Cl_2$ is shown in Figure 15. It is clear that the threefold symmetry of the complex is superimposed on the binding site, which provides for three NH···Cl^- interactions; each from a different molecule of **15**. The amide N-atoms define an equilateral triangle that is approximately 5.7 Å on a side. Given the threefold symmetry of the molecule, it is not surprising that this receptor also binds well to the trigonal planar anion NO_3^-.

The monodentate ligand **16** was designed so that the pyridine to metal bond and the potential amide to anion hydrogen bond would be parallel. The idea was that coordination of four pyridine atoms to a square-planar metal center such as platinum(II) would act to converge all of the hydrogen-bond donors toward a single anion within the sphere of electrostatic influence of the metal ion.[30] Since, in the absence of significant steric demands, rotation about the Pt–N bonds should be facile, there are four geometric isomers possible for $[Pt(\mathbf{16})_4]^{2+}$ based on the relative positioning of the amide groups above or below the square coordination plane. In an analogy to the well-studied chemistry of calix[4]arenes, these isomers were labeled, 1,2-alternate, 1,3-alternate, partial cone, and cone for ease of identification (Figure 16).

$[Pt(\mathbf{16})_4][PF_6]_2$ was shown to function as an effective host for the oxoanions $CF_3SO_3^-$, ReO_4^-, NO_3^-, HSO_4^-, $H_2PO_4^-$, and $CH_3CO_2^-$ but showed particular selectivity toward the bidentate anions NO_3^- and $CH_3CO_2^-$ (CD_3CN)[30] These planar anions were bound in a 1 : 2 receptor:anion ratio, and K_2 was found to be greater than K_1. It was concluded that the first cation–anion interaction acted to preorganize two pairs of *cis* amide groups (the 1,2-alternate conformation) due to the close structural match between these hydrogen-bond donor groups and a bidentate anion. This resulted in enhanced binding of the second anion demonstrating a positive allosteric effect. Figure 17 shows the X-ray structures of the receptor $[Pt(\mathbf{16})_4][PF_6]_2$ (a) and the interaction of $[Pt(\mathbf{16})_4]^{2+}$ with the ReO_4^- anions (b). It should be noted that in the structure of $[Pt(\mathbf{16})_4][PF_6]_2$ no hydrogen-bonding interactions were observed between the amide NH groups of the receptor and the PF_6^- anions. The amide carbonyl oxygen atoms are actually involved in hydrogen bonding to CH_2Cl_2 solvent molecules. This was offered as evidence that competition by PF_6^- anions for the recognition site in these complexes was negligible.[30]

These initial results with receptor $[Pt(\mathbf{16})_4][PF_6]_2$ encouraged the authors to design a metal complex receptor

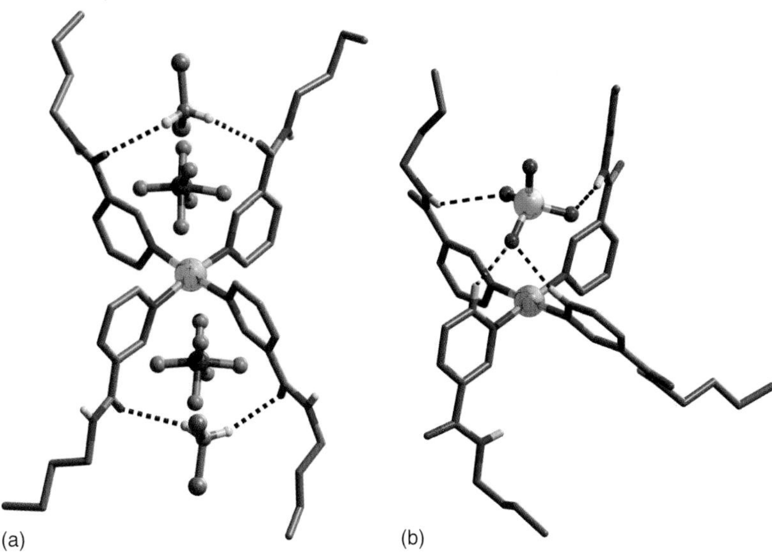

Figure 17 Ball-and-stick representations of the single-crystal X-ray crystal structures of: receptor [Pt(**16**)$_4$][PF_6]$_2$·2CH_2Cl_2 (a) and complex: [Pt(**16**)$_4$][ReO_4]$^+$ (b). Noncoordinating anions and H atoms not involved in hydrogen bonding are omitted for clarity.

for planar anions utilizing a ligand already preorganized into the 1,2-alternate conformation. The complex [Pt(4,4′-*tert*-Bu(2,2′-bipyridine))(**17**)$_2$][PF_6]$_2$ utilizes the cis coordination of two 3,5-di-(*n*-butylamide)pyridine ligands to mimic the two NH hydrogen-bonding sites found in the 1,2-alternate conformation of [Pt(**16**)$_4$][PF_6]$_2$.[31] An X-ray structure showing the interaction of complex cation [Pt(4,4′-*tert*-Bu(2,2′-bipyridine))(**17**)$_2$]$^{2+}$ with NO_3^- anions is shown in Figure 18. Note that one of the NO_3^- anions is bound parallel to the platinum(II) coordination plane as envisaged in the initial design, but the other adopts a position that is perpendicular to the plane.

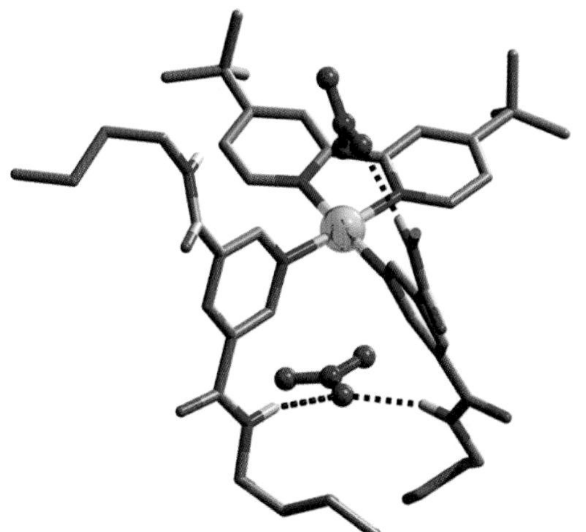

Figure 18 Ball-and-stick representation of the single-crystal X-ray crystal structure of [Pt(**17**)$_2$][NO_3]$_2$. H atoms not involved in hydrogen bonding are omitted for clarity.

It was rationalized that since two amide groups on one side of the coordination plane could not act independent of the other set and binding the first nitrate anion caused pinching of the ligands toward each other, this simultaneously opened up the binding site on the opposite side of the plane. This ultimately prevented the other amide groups from being able to bind a small NO_3^- anion in a two-point interaction. This inhibition or negative allostery is shown schematically in Figure 19.[31]

Utilizing the better donating ability of an *iso*-quinoline ligand and the increased hydrogen-bonding capabilities of a urea group, it was found that [Pt(**18**)$_4$][BF_4]$_2$ was more resistant to metal–ligand exchange and a superior anion receptor to [Pt(**16**)$_4$][PF_6]$_2$.[32] Relatively strong 1 : 1 interactions were observed in solution for halides and symmetrical oxyanions, which involved surrounding the anion with the cone conformation of the receptor. In particular, addition of $H_2PO_4^-$ or SO_4^{2-} to a DMSO-d_6 solution of the receptor resulted in very strong binding and the observance of only a host–guest complex with a 1 : 1 ratio of receptor and anion. The ^{1}H NMR titration of [Pt(**18**)$_4$][BF_4]$_2$ with 0.0, 0.5, and 1.0 equivalents of SO_4^{2-} ion is shown in Figure 20. It can be seen that receptor–anion exchange is slow on the NMR timescale and that at 1.0 equivalent of added SO_4^{2-} ion only bound receptor is observed.[32]

The 1 : 1 interaction of [Pt(**18**)$_4$][BF_4]$_2$ with SO_4^{2-} in the cone conformation was demonstrated unambiguously by X-ray crystallography (Figure 21). All eight NH donors from the four urea functional groups as well as the four CH atoms adjacent to the N-donors (10-position) are organized so that they are directed at the anion that is nestled into the binding pocket.[32] Although the threefold symmetry of the

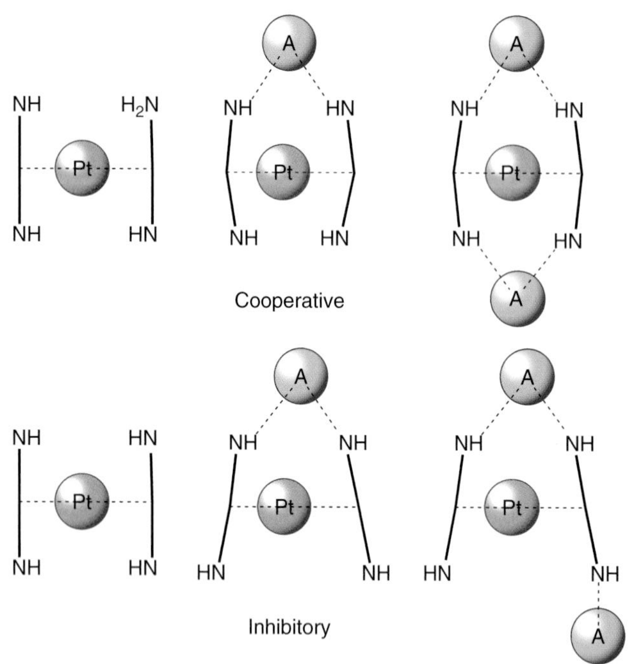

Figure 19 Schematic illustrating the difference between (a) the cooperative effects observed for $[Pt(\mathbf{16})_4]^{2+}$ when binding CH_3COO^- or NO_3^- ion and (b) the inhibitory effects observed for $[Pt(\mathbf{17})_2]^{2+}$ when binding NO_3^- ion.

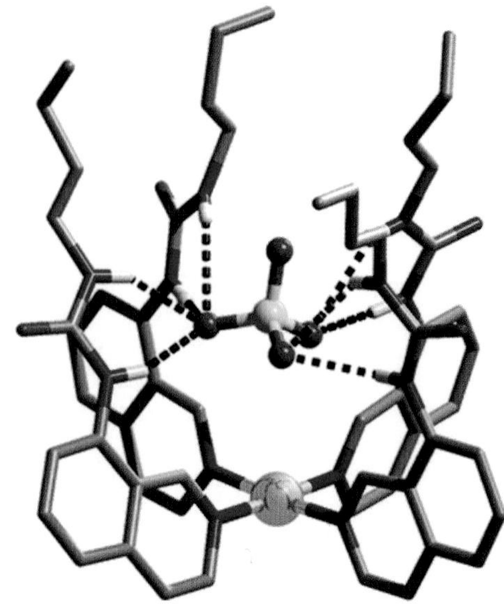

Figure 21 Ball-and-stick representation of the single-crystal X-ray crystal structure of $[Pt(\mathbf{18})_4][SO_4]$. H atoms not involved in hydrogen bonding are omitted for clarity.

anion does not perfectly match the fourfold symmetry of the receptor, the anion is completely engulfed in hydrogen-bonding interactions which is charge assisted by a strong electrostatic interaction between the dication and dianion.

The same authors also explored complexes in which the ligand **18** was mixed with similar ligands that lacked the urea functionality (**Q** = *iso*-quinoline). Although the metal complex *trans*-$[Pt(\mathbf{18})_2(\mathbf{Q})_2]^{2+}$ only contains two urea functions, it was rationalized that three of these receptors could provide six urea groups to bind two sulfate ions.[33] This turned out to be the case, and the X-ray structure of the resulting cluster $\{[Pt(\mathbf{18})_2(\mathbf{Q})_2]_3[SO_4]_2\}^{2+}$ is shown

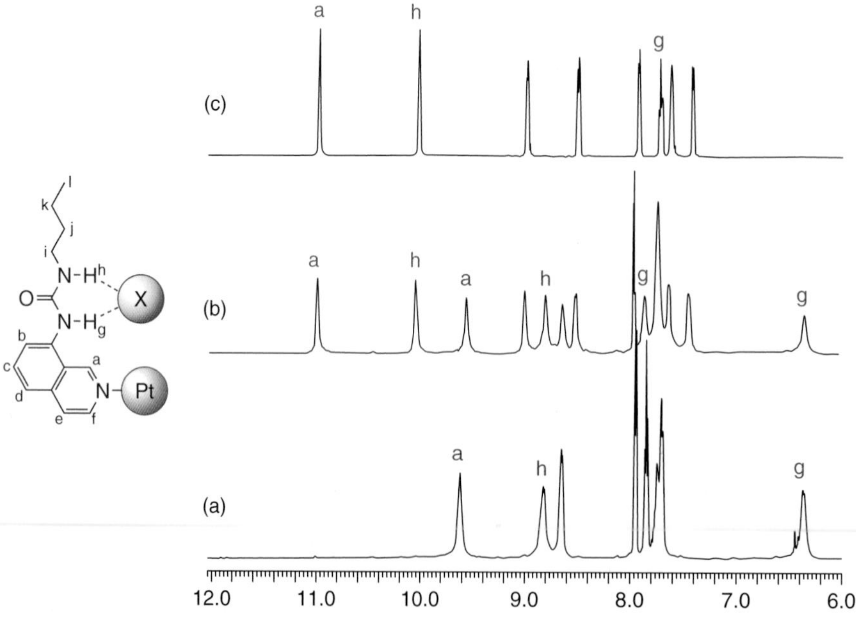

Figure 20 ^1H NMR spectra (5.0×10^{-3} M, DMSO-d_6, 298 K) for $[Pt(\mathbf{18})_4]^{2+}$ and (a) 0.0 equivalent of SO_4^{2-} added, (b) 0.5 equivalent of SO_4^{2-} added, and (c) 1.0 equivalent of SO_4^{2-} added. Labels represent the H atoms of the ligand with red indicating free receptor and blue representing SO_4^{2-}-bound receptor.

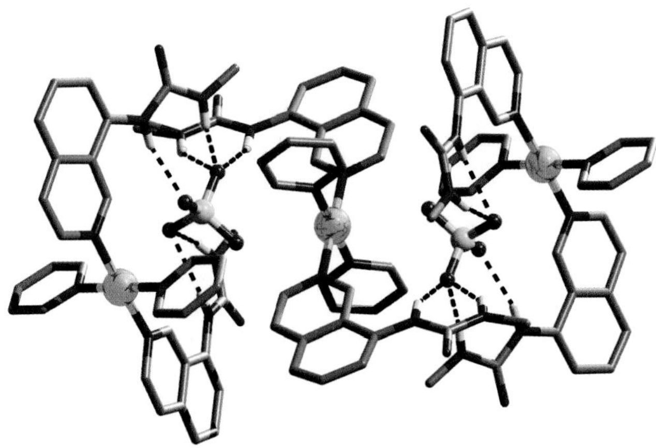

Figure 22 Ball-and-stick representation of the single-crystal X-ray crystal structure of $\{[Pt(\mathbf{18})_2(\mathbf{Q})_2]_3[SO_4]_2\}^{2+}$. Noncoordinating anions and H atoms not involved in hydrogen bonding are omitted for clarity.

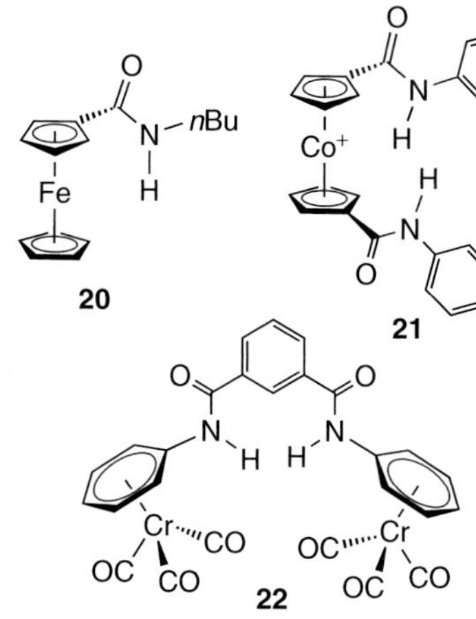

Figure 23 Organometallic compounds **20**–**22** that act as metal complex receptors binding anions through second-sphere coordination.

in Figure 22. It was noted that the CH hydrogen bonds observed with $[Pt(\mathbf{18})]^{2+}$ were also present in the structure of $\{[Pt(\mathbf{18})_2(\mathbf{Q})_2]_3[SO_4]_2\}^{2+}$ when the much simpler *iso*-quinoline (**Q**) was used as the ligand.

Simple organometallic species such as **20**–**22** have been shown to act as metal complex receptors for anions (Figure 23).[28b] The direct attachment of an organometallic fragment, such as $[Cp_2Fe]$, $[Cp_2Co]^+$ or $[Cr(CO)_3]$, to a ligand-containing hydrogen-bond donor groups is a good strategy for creating metal complex receptors for anions as these complexes are particularly robust with respect to metal–ligand substitution in the presence of anions. The disadvantage of this class of metal complex receptor is that it usually requires an increase in synthetic work to prepare and these species are often air- and water-sensitive.

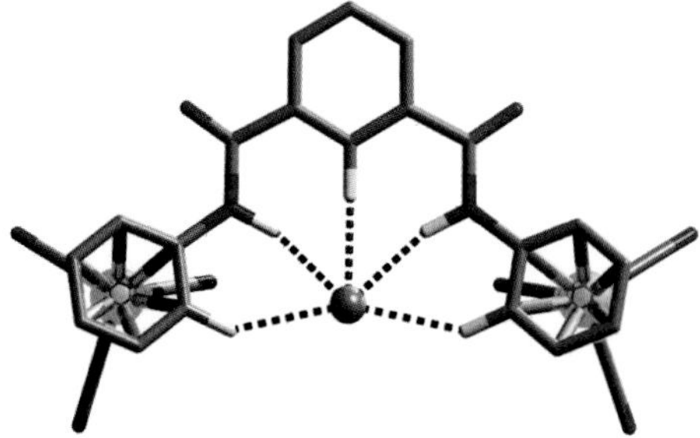

Figure 24 Ball-and-stick representation of the MM3-derived structure of $\{[(Cr(CO_3))_2(\mathbf{22})][Cl]\}^-$. H atoms not involved in hydrogen bonding are omitted for clarity.

Complex **20** was shown to act as an anion receptor by utilizing both the amide NH hydrogen and one of the acidic CH hydrogens of the Cp ligand.[34] Variations in the well-behaved electrochemistry of the ferrocene derivatives were used to signal anion binding in these early organometallic sensors reported by Beer. In a very similar manner, cobaltocene derivative **21** binds anions via NH and CH hydrogen atoms on both the Cp rings.[35] The positive charge on cobalt also contributes an extra electrostatic component to the binding of anions when compared to analogous ferrocene derivatives. Organometallic complex **22** contains strongly withdrawing $[Cr(CO)_3]$ fragments[36] and is a derivative of the well-studied *iso*-phthaloyldiamide cleft-shaped anion receptor first introduced by Crabtree.[37] An MM3-derived structure[27] depicting the interaction of a chloride anion with the organometallic hydrogen-bonding cleft is shown in Figure 24.

It would appear that multidentate chelating ligands such as the terpyridine derivative **14** provide a degree of complex stability and a more rigid platform but limit the number of hydrogen-bond donors that can be incorporated. Conversely, monodentate ligands, such as **16**, are more flexible and allow more functional groups to be incorporated but are more likely to suffer from exchange of the ligand by anionic guests. The success of $[Pt(\mathbf{18})_4]^{2+}$ was attributed to its combination of robust metal–ligand interactions, incorporation of good hydrogen-bond donors, and conformational flexibility.

4.4 Inducing conformational changes that enhance anion binding

An alternative strategy for creating a metal complex receptor involves using metal–ligand coordination to influence the conformation of ligands such that hydrogen-bond donor sites on the same ligand can work together to bind a single anion. The conformational change could be anything from single-bond rotation to a substantial rearrangement that acts to *template* the formation of an anion-binding site.

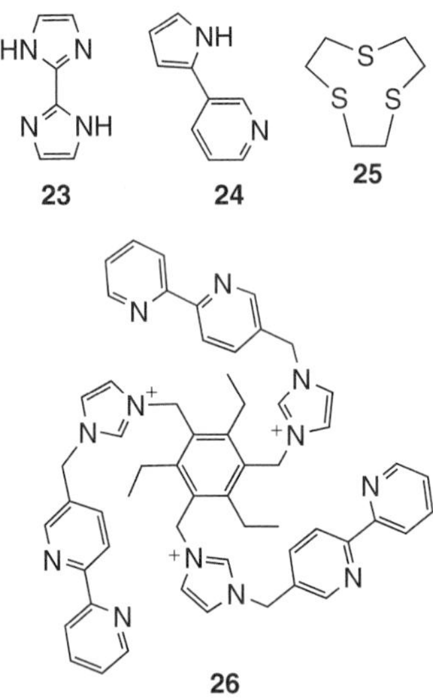

Figure 25 Ligands **23–26** used to form metal complex receptors that bind anions through second-sphere coordination.

The ligands **23–26** shown in Figure 25 form metal complex receptors in which a conformational change occurs on binding that aids anion recognition.

The free ligand 2,2′-biimidazole, **23**, prefers an *anti* conformation as shown in Figure 25. Chelation to a metal center requires rotation about the single bond linking the rings and adoption of the syn conformation, which orients both the imidazole NH groups outward so that they can participate in a second-sphere interaction with an anion.[38] This is clearly evident from the X-ray crystal structure depicting the interaction of [Mo(CO)$_2$(*tert*-BuNC)(η^3-allyl)(**23**)] with a Cl^- ion as shown in Figure 26.[39] In this example, the metal ion organized multiple hydrogen-bond donors on a single ligand via beneficial conformational changes that occurred on coordination.

Since anion receptors are designed to operate in solution and binding studies conducted in solution are used to validate receptor designs, an example of the importance of solvation seems prudent. One of the major advantages of the pyrrole group is that it has only a single hydrogen-bond donor. Unlike ligands with amide and urea groups, pyrrole ligands cannot self-associate owing to the presence of hydrogen-bond-accepting C=O functionalities. When the binding of anions to the complex [Pt(**24**)$_4$]$^{2+}$ was studied,[40] a comparison of the solution-binding experiments and the X-ray crystal structure of the complex with the $CH_3SO_3^-$ anion served to demonstrate that solid-state data does not always allow prediction of solution structure. The X-ray structure of [Pt(**24**)$_4$][CH_3SO_3]$_2$ shown in Figure 27 clearly indicates that the pyrrole NH group interacts with the O atoms of the $CH_3SO_3^-$ anion.[40]

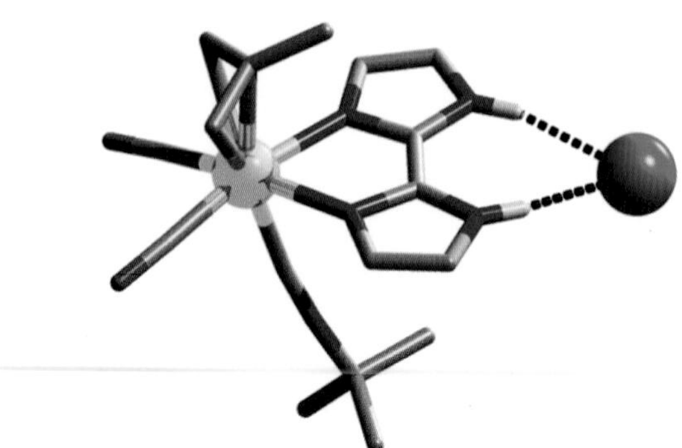

Figure 26 Ball-and-stick representation of the single-crystal X-ray crystal structure of {[Mo(CO)$_2$(*tert*-BuNC)(η^3-allyl)(**23**)][Cl]}$^-$. H atoms not involved in hydrogen bonding are omitted for clarity.

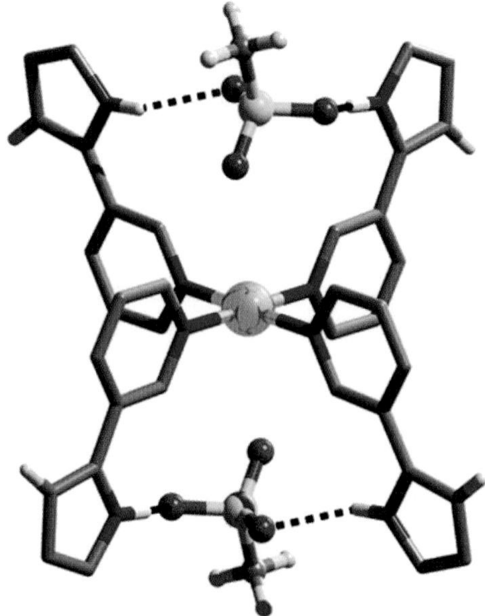

Figure 27 Ball-and-stick representation of the single-crystal X-ray crystal structure of [Pt(**24**)$_4$][CH_3SO_3]$_2$. H atoms not involved in hydrogen bonding are omitted for clarity.

In CH_3NO_2 solution, both the pyrrole NH peak and the α-pyridine CH peak shift downfield in the presence of $CH_3SO_3^-$ anions (1.77 and 1.24 ppm, respectively). However, in DMSO solution, there is no evidence that the pyrrole NH groups interact with the $CH_3SO_3^-$ anion (0.00 ppm) but rather the pyrrole CH group (0.17 ppm) appears to be involved in a CH···O hydrogen bond with the anion (Figure 28). Since CH_3NO_2 is a relatively poor hydrogen-bond acceptor, when $CH_3SO_3^-$ was added to the receptor, it bound to the most acidic hydrogen-bond donor; the pyrrole NH. However, in DMSO, which is a good hydrogen-bond acceptor, the anion only bound to the pyrrole NH if it was basic enough (e.g., chloride and acetate). For less basic anions, the pyrrole NH groups preferred to orientate into the solvent allowing the formation of hydrogen bonds to the DMSO, leaving the anion involved in

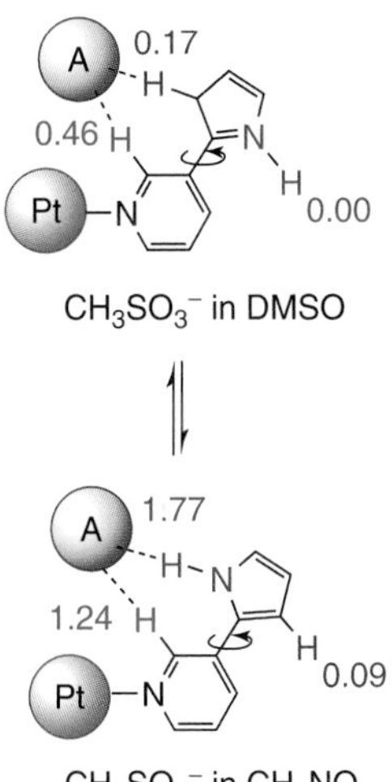

Figure 28 Schematic illustrating the difference between chemical shift data and conformation of **24** for the interaction of $[Pt(\mathbf{24})_4]^{2+}$ with $CH_3SO_3^-$ anions in DMSO and CH_3NO_2 solutions.

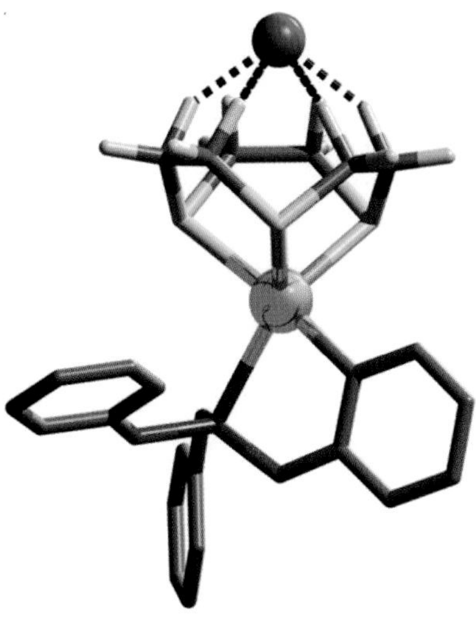

Figure 29 Ball-and-stick representation of the molecular mechanics (MM3)-derived model of the complex $[Pd(\mathbf{25})(P(OAr)_3)]^+$ with Cl^- ion. The six CH···Cl interactions in the second-sphere coordination are shown as dashed lines. The 2,4-*tert*-Bu groups on the aromatic rings have been removed for clarity.

weaker hydrogen bonding with the pyrrole CH group.[40] This conformational change occurred because it led to a greater overall stability for the system. The profound effect of solvent on the binding mode of this simple anion receptor is an important phenomenon and should be taken into account when analyzing receptors containing NH hydrogen-bond donor groups.

Coordination of the macrocyclic ligand, 1,4,7-trithiacyclononane, (9S3), **25**, to palladium required an endo-conformation for the S-donors that resulted in *facial* coordination to the palladium(II) center.[41] This simultaneously organized the backbone ethylene groups so that a set of six *endo* CH bonds could act as hydrogen-bond donors to an anion in the second-sphere of coordination. Although, strongly coordinating anions such as chloride were shown to be competitive with **25**, good recognition was observed for weaker coordinating, spherical ions such as iodide.[42] No solid-state structure was available for unambiguous verification of this arrangement, but MM3 calculations yielded the structure shown in Figure 29, which matches that predicted from the solution data.

The large and flexible ligand **26** was designed as a tris(bipyridyl) style ligand.[43] The 1,3,5-triethylbenzene platform is well known to orient alternating groups on the same side of the aromatic ring and complexation of **26** to iron(II) organized the three arms of the ligand by forming a classic $[Fe(bipy)_3]$ type complex. The cage formed was the appropriate size for encapsulating halide ions and the azide ion. ^1H NMR titration experiments showed transfer of electron density on the imidazolium ring, suggesting that the anion occupies a position inside the cage and is hydrogen bonding to three CH bonds from the imidazolium group and three CH bonds from pyridyl rings which were activated by coordination to the iron(II) center. The X-ray structure of the Br^- adduct, shown in Figure 30, verified the interactions observed solution; encapsulation of the Br^- ion with six hydrogen bonds formed from two types of CH groups. Since a fairly labile iron(II) center was used to organize the

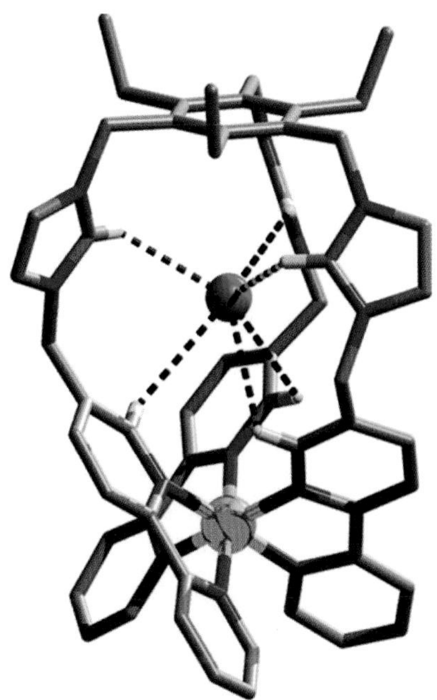

Figure 30 Ball-and-stick representation of the single-crystal X-ray crystal structure of $[Fe(\mathbf{26})]^{2+}$ with a Br^- ion. The six CH···O interactions in the second-sphere coordination are shown as dashed lines.

three bipy groups and create the anion-binding cavity, it is important to distinguish this from formal anion templation. It is important to note that in this case, the anion does not influence the nature of the final assembly. The metal–ligand assembly was stabilized by the chelate effect and the formation of multiple Fe–N bonds made it a robust complex that functioned as a metal complex receptor for anions.

4.5 Building a cage to converge on an anion

The sulfate ion has a relatively large hydration energy making the design of receptors for this ion difficult. Pflugrath and Quiocho's crystal structure of the SO_4^{2-} ion bound within the sulfate-binding protein showed that nature employs seven hydrogen bonds from neutral donor groups to selectively complex the anion within the protein.[44] With an association constant of $\sim 8 \times 10^6\,M^{-1}$ for binding in water, the protein is by far the most effective receptor known for this anion.[45] Theoretical calculations on water clusters had indicated that sulfate could accept up to 12 hydrogen bonds in its first hydration shell,[46] and electronic structure calculations confirmed this pattern could also be achieved with six urea groups.[47] On the basis of this knowledge, Hay and coworkers focused on identifying urea-containing molecular components that could assemble to form architectures in which urea groups would line the inside of a cavity to bind the SO_4^{2-} ion.[47a,48] For host construction, a tetrahedron was seen as particularly attractive as it had the correct symmetry elements to construct the target $[SO_4(urea)_6]^{2-}$ unit. Since in a cage, the four metal ions serve as C_3-symmetric vertices and the six C_2-symmetric ligands serve as the edges of the tetrahedron, it was rationalized that appropriately functionalizing the edge molecules with urea groups would lead to an ideal arrangement for sulfate binding. As cages typically contain tris-chelate metal complexes, $[Ni(bipy)_3]^{2+}$ was chosen for the vertices (Figure 31).

Compound **27** was identified as the best candidate ligand and the complex assembled using $NiSO_4$ in a water–methanol mixture.[49] Electrospray ionization mass spectrometry verified that the capsule remained intact in solution showing a peak at $m/z = 451.2$ corresponding to $[Ni_4(\mathbf{27})_6(SO_4)]^{6+}$ (expected $m/z = 451.3$) at concentrations as low as 1 µM. As a test of complex stability, 4 equivalents of $Sr(NO_3)_2$ were added to a solution containing 1 equivalent of $[Ni_4(\mathbf{27})_6(SO_4)][SO_4]_3$ in water. The resulting precipitate contained only 3 equivalents of $SrSO_4$ as the encapsulated sulfate was completely isolated from interaction with the reagent. The association constant was estimated to be $\sim 6 \times 10^6\,M^{-1}$; the X-ray structure of $[Ni_4(\mathbf{27})_6(SO_4)]^{6+}$ is shown in Figure 32.[49]

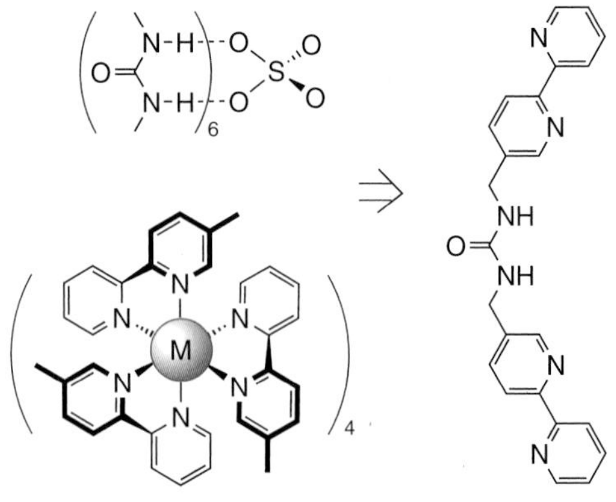

Figure 31 The ligand **27** was designed by optimizing the required molecular components for SO_4^{2-} binding using computational chemistry.

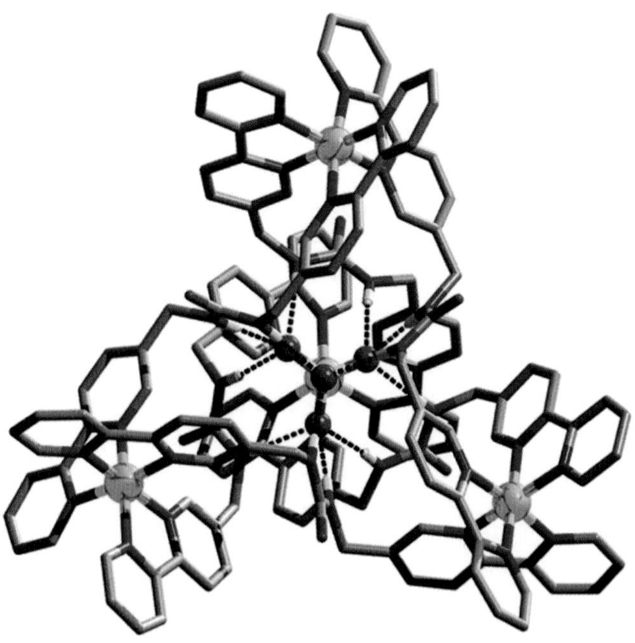

Figure 32 Ball-and-stick representation of the single-crystal X-ray crystal structure of the metal–ligand assembly $[Ni_4(\mathbf{27})_6]^{8+}$ wrapped around a single SO_4^{2-} ion. The 12 NH···O interactions are shown as dashed lines. Noncoordinating anions and H atoms not involved in hydrogen bonding are omitted for clarity.

5 CONCLUSIONS

The incorporation of hydrogen-bond donor or acceptor groups into ligands followed by their coordination to inert transition metals has proven to be a successful methodology for creating unique receptors for anions, cations, and neutral molecules. The metal ion can participate indirectly via second-sphere coordination or directly via simultaneous first- and second-sphere coordination. The major advantage of metal complex receptors designed using this strategy is the ease of synthesis. Although many of the resulting complex–substrate assemblies have complicated structures, the ligands themselves do not require tedious synthetic routes and coordination to the metal center is often facile. The combination of coordination chemistry and noncovalent interactions such as hydrogen bonding and π-stacking has proven to be a powerful tool for molecular and ion recognition. The state of the art is sufficiently advanced enough to allow us to target specific guests by tailoring both the metal–ligand and host–guest interactions. Compounds such as $[Pd(\mathbf{7})(adenine)]^{+}$ and $[Ni_4(\mathbf{27})_6(SO_4)]^{6+}$ described in this chapter are particularly good examples of that art.

ACKNOWLEDGMENTS

I would like to acknowledge the aid of my assistant Ginny and the encouragement of my friend and collaborator Phil. Without their motivation and constant encouragement, this chapter would never have made it to press.

REFERENCES

1. S. J. Loeb, *Compr. Supramol. Chem.*, 1996, **1**, 733.
2. A. L. Crumbliss and J. M. Harrington, *Adv. Inorg. Chem.*, 2009, **61**, 179.
3. (a) E. L. Que and C. J. Chang, *Chem. Soc. Rev.*, 2010, **39**, 51; (b) V. Kubicek and E. Toth, *Adv. Inorg. Chem.*, 2009, **61**, 63.
4. N. W. Alcock, J. M. Brown, and J. C. Jeffery, *J. Chem. Soc., Chem. Commun.*, 1974, 829.
5. N. W. Alcock, J. M. Brown, and J. C. Jeffery, *J. Chem. Soc., Dalton Trans.*, 1976, 583.
6. G. Ferguson, K. E. Matthes, and D. Parker, *Angew. Chem.*, 1987, **99**, 1195.
7. J. E. Kickham and S. J. Loeb, *Inorg. Chem.*, 1995, **34**, 5656.
8. K. Brandenburg, Diamond, Diamond 3.2, Crystal Impact, Postfach 1251, D-53002, Bonn, Germany, 2009.
9. (a) H. M. Colquhoun, S. M. Doughty, J. M. Maud, *et al.*, *Isr. J. Chem.*, 1985, **25**, 15; (b) H. M. Colquhoun, J. F. Stoddart, and D. J. Williams, *Angew. Chem.*, 1986, **98**, 483.
10. (a) V. Van Axel Castelli, A. Dalla Cort, L. Mandolini, *et al.*, *Supramol. Chem.*, 2002, **14**, 211; (b) C. J. Van Staveren, D. E. Fenton, D. N. Reinhoudt, *et al.*, *J. Am. Chem. Soc.*, 1987, **109**, 3456; (c) C. J. Van Staveren, J. Van Eerden, F. C. J. M. Van Veggel, *et al.*, *J. Am. Chem. Soc.*, 1988, **110**, 4994.
11. A. R. Van Doorn, D. J. Rushton, W. F. Van Straaten-Nijenhuis, *et al.*, *Recl. Trav. Chim. Pays-Bas*, 1992, **111**, 421.
12. (a) A. Dalla Cort, C. Pasquini, and L. Schiaffino, *Supramol. Chem.*, 2007, **19**, 79; (b) A. Dalla Cort, L. Mandolini, C. Pasquini, and L. Schiaffino, *J. Org. Chem.*, 2005, **70**, 9814; (c) A. Dalla Cort, J. I. M. Murua, C. Pasquini, *et al.*, *Chem. Eur. J.*, 2004, **10**, 3301; (d) A. Dalla Cort, L. Mandolini, G. Palmieri, *et al.*, *Chem. Commun.*, 2003, 2178.
13. J. E. Kickham, S. J. Loeb, and S. L. Murphy, *J. Am. Chem. Soc.*, 1993, **115**, 7031.
14. J. E. Kickham and S. J. Loeb, *Inorg. Chem.*, 1994, **33**, 4351.
15. J. E. Kickham, S. J. Loeb, and S. L. Murphy, *Chem. Eur. J.*, 1997, **3**, 1203.
16. (a) H. Sigel, *Pure Appl. Chem.*, 2004, **76**, 375; (b) D. Choquesillo-Lazarte, M. d. P. Brandi-Blanco, I. Garcia-Santos, *et al.*, *Coord. Chem. Rev.*, 2008, **252**, 1241.
17. (a) C. Price, M. R. J. Elsegood, W. Clegg, and A. Houlton, *J. Chem. Soc., Chem. Commun.*, 1995, 2285; (b) C. Price, M. A. Shipman, N. H. Rees, *et al.*, *Chem. Eur. J.*, 2001, **7**, 1194; (c) A. Houlton, *Adv. Inorg. Chem.*, 2002, **53**, 87; (d) M. A. Galindo, D. Amantia, A. Martinez Martinez, *et al.*, *Inorg. Chem.*, 2009, **48**, 10295.
18. (a) M. Shionoya, E. Kimura, and M. Shiro, *J. Am. Chem. Soc.*, 1993, **115**, 6730; (b) M. Shionoya, T. Ikeda, E. Kimura, and M. Shiro, *J. Am. Chem. Soc.*, 1994, **116**, 3848; (c) E. Kimura, T. Ikeda, and M. Shionoya, *Pure Appl. Chem.*, 1997, **69**, 2187.
19. (a) Y. Aoyama, A. Yamagishi, M. Asagawa, *et al.*, *J. Am. Chem. Soc.*, 1988, **110**, 4076; (b) T. Hayashi, T. Miyahara, N. Hashizume, and H. Ogoshi, *J. Am. Chem. Soc.*, 1993, **115**, 2049.
20. H. K. A. C. Coolen, P. W. N. M. Van Leeuwen, and R. J. M. Nolte, *Angew. Chem.*, 1992, **104**, 906.
21. (a) B. R. Cameron and S. J. Loeb, *Chem. Commun.*, 1996, 2003; (b) B. R. Cameron, S. J. Loeb, and G. P. A. Yap, *Inorg. Chem.*, 1997, **36**, 5498; (c) S. J. Loeb and B. R. Cameron, *ACS Symp. Ser.*, 2000, **757**, 283.
22. A. Petitjean, R. G. Khoury, N. Kyritsakas, and J.-M. Lehn, *J. Am. Chem. Soc.*, 2004, **126**, 6637.
23. S. Ulrich, A. Petitjean, and J.-M. Lehn, *Eur. J. Inorg. Chem.*, 2010, 1913.
24. J. E. Kickham and S. J. Loeb, *J. Chem. Soc., Chem. Commun.*, 1993, 1848.
25. (a) J. Perez and L. Riera, *Chem. Soc. Rev.*, 2008, **37**, 2658; (b) J. W. Steed, *Chem. Soc. Rev.*, 2009, **38**, 506; (c) V. Amendola and L. Fabbrizzi, *Chem. Commun.*, 2009, 513; (d) V. Amendola, M. Bonizzoni, D. Esteban-Gómez, *et al.*, *Coord. Chem. Rev.*, 2006, **250**, 1451; (e) D. J. Mercer and S. J. Loeb, *Chem. Soc. Rev.*, 2010, **39**, 3612.
26. (a) M. S. Goodman, V. Jubian, and A. D. Hamilton, *Tetrahedron Lett.*, 1995, **36**, 2551; (b) M. S. Goodman, V. Jubian, B. Linton, and A. D. Hamilton, *J. Am. Chem. Soc.*, 1995, **117**, 11610.
27. *Scigress Explorer Software*, Biosciences Group, Fujitsu Computer Systems Corp., Westwood, MA 02052, 2009.
28. (a) L. H. Uppadine, M. G. B. Drew, and P. D. Beer, *Chem. Commun.*, 2001, 291; (b) P. D. Beer and E. J. Hayes, *Coord. Chem. Rev.*, 2003, **240**, 167.
29. L. H. Uppadine, F. R. Keene, and P. D. Beer, *J. Chem. Soc., Dalton Trans.*, 2001, 2188.

30. C. R. Bondy, S. J. Loeb, and P. A. Gale, *Chem. Commun.*, 2001, 729.

31. C. R. Bondy, P. A. Gale, and S. J. Loeb, *J. Supramol. Chem.*, 2003, **2**, 93.

32. C. R. Bondy, P. A. Gale, and S. J. Loeb, *J. Am. Chem. Soc.*, 2004, **126**, 5030.

33. M. G. Fisher, P. A. Gale, M. E. Light, and S. J. Loeb, *Chem. Commun.*, 2008, 5695.

34. P. D. Beer, C. A. P. Dickson, N. Fletcher, *et al.*, *J. Chem. Soc., Chem. Commun.*, 1993, 828.

35. P. D. Beer and S. R. Bayly, *Top. Curr. Chem.*, 2005, **255**, 125.

36. S. Camiolo, S. J. Coles, P. A. Gale, *et al.*, *Chem. Commun.*, 2000, 275.

37. K. Kavallieratos, S. R. d. Gala, D. J. Austin, and R. H. Crabtree, *J. Am. Chem. Soc.*, 1997, **119**, 2325.

38. L. Ion, D. Morales, J. Perez, *et al.*, *Chem. Commun.*, 2006, 91.

39. L. Ion, D. Morales, S. Nieto, *et al.*, *Inorg. Chem.*, 2007, **46**, 2846.

40. I. E. D. Vega, P. A. Gale, M. E. Light, and S. J. Loeb, *Chem. Commun.*, 2005, 4913.

41. S. R. Cooper, *Acc. Chem. Res.*, 1988, **21**, 141.

42. R. B. Bedford, M. Betham, C. P. Butts, *et al.*, *Chem. Commun.*, 2008, 2429.

43. V. Amendola, M. Boiocchi, B. Colasson, *et al.*, *Angew. Chem. Int. Ed.*, 2006, **45**, 6920.

44. J. W. Pflugrath and F. A. Quiocho, *Nature*, 1985, **314**, 257.

45. B. L. Jacobson and F. A. Quiocho, *J. Mol. Biol.*, 1988, **204**, 783.

46. B. Gao and Z.-F. Liu, *J. Chem. Phys.*, 2004, **121**, 8299.

47. (a) B. P. Hay, T. K. Firman, and B. A. Moyer, *J. Am. Chem. Soc.*, 2005, **127**, 1810; (b) R. Custelcean, B. A. Moyer, and B. P. Hay, *Chem. Commun.*, 2005, 5971.

48. B. P. Hay, A. A. Oliferenko, J. Uddin, *et al.*, *J. Am. Chem. Soc.*, 2005, **127**, 17043.

49. R. Custelcean, J. Bosano, P. V. Bonnesen, *et al.*, *Angew. Chem. Int. Ed.*, 2009, **48**, 4025.

Hydrogen-Bonding Receptors for Molecular Guests

Andrew J. Wilson
University of Leeds, Leeds, UK

1 INTRODUCTION

Hydrogen bonding[1,2] because of its strength and directionality represents a powerful tool within the toolbox of noncovalent interactions for the elaboration of receptors. The theory and properties of hydrogen bonds have been and continue to be discussed at length[1]; however, a few considerations will be summarized here. Hydrogen-bonding interactions have been defined as strong, for example, those of HF; moderate, for example, those of amide NH and carbonyl groups; or weak, for example, those of CH groups—this text focuses largely on the latter two interactions. Similar to all noncovalent interactions, electrostatic, dipole, repulsive, and dispersive forces contribute to the hydrogen-bonding interaction. The predominant contribution to the strength of a hydrogen bond is the electrostatic component, and Hunter has elegantly described how such interactions in solution can be quantified purely on the basis of electrostatics[3]; however, the directionality of the hydrogen bond stems from the fact that the electrostatic component is primarily a dipole–dipole interaction. The geometry of a hydrogen bond can vary significantly in solution or the solid state and is determined by shape and the influence of other forces: while strong hydrogen bonds involving one acceptor are close to being linear, moderate hydrogen bonds can involve two or more acceptors and are referred to as three-centered or bifurcated.

Much of the early research in the area of hydrogen-bonding receptors focused on the development of hosts for neutral molecules in organic solvents. The general principles for the elaboration of any receptor are the following: (i) the receptor should maximize noncovalent interactions with its target guest, (ii) the receptor and target guest should have complementary shapes, (iii) the receptor should be preorganized and/or macrocyclic. One such example is receptor **1** for barbiturate derivatives **2**[4] shown in Figure 1; (i) the receptor makes six hydrogen bonds with the target satisfying its hydrogen-bonding requirements, which engenders selectivity over, for example, urea which can make only four contacts, (ii) the 3D projection of hydrogen-bonding donor and acceptor motifs on the guest is perfectly matched to the host, and (iii) the macrocyclic nature of the host confers affinity to the interaction versus an acyclic and consequently less preorganized host. Other early appealing targets for receptor development were the DNA bases, and this simple approach for generating receptors has been applied extensively to such targets. For instance, Rebek and coworkers developed a "cleft" type receptor **3** employing functionalized Kemp's triacid motifs for recognition of adenine (Figure 2a).[5] Similarly, in combination with other interactions, a "molecular hinge" receptor **4** for adenine was reported by Hamilton,[6] which exploits complementary hydrogen-bonding and $\pi-\pi$ stacking interactions (Figure 2b).

Supramolecular Chemistry: From Molecules to Nanomaterials.
Edited by Philip A. Gale and Jonathan W. Steed.
© 2012 John Wiley & Sons, Ltd. ISBN: 978-0-470-74640-0.

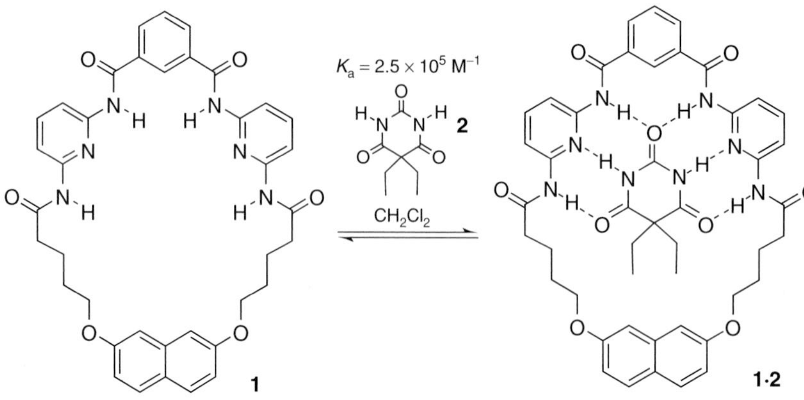

Figure 1 Hamilton's barbiturate receptor **1** in equilibrium with its target guest diethyl barbital **2**.

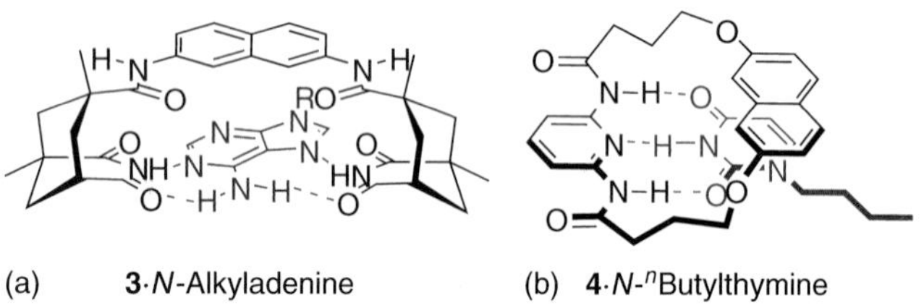

Figure 2 Examples of hydrogen-bonding receptors based on (a) a cleft **3** and (b) a hinge **4**.

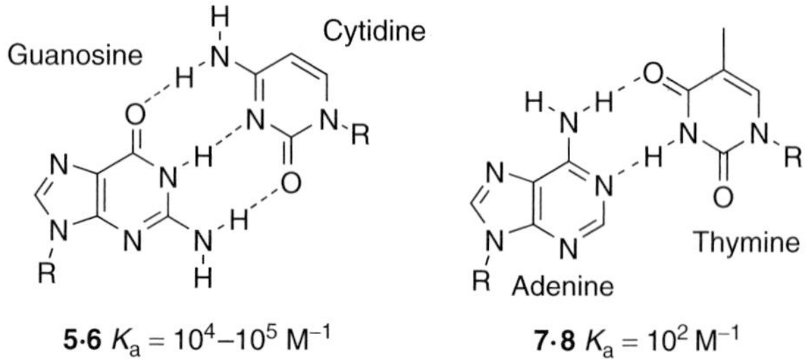

Figure 3 The DNA base pairs guanosine–cytidine **5·6** and adenine–thymine **7·8** (association constant in chloroform).

2 HOST–GUEST COMPLEXES

2.1 Linear arrays: model systems for understanding hydrogen-bonding interactions

The early focus on DNA bases has generated significant interest in the study of synthetic linear arrays[7] and such studies have furnished significant fundamental understanding of hydrogen bonding for receptor design. Although more complicated architectures may assemble, arrays of hydrogen bonds such as the nucleobases fall into this family. At the simplest level, homodimers result when a linear arrangement of interacting groups is complementary and heterodimers where two different linear motifs with complementary arrangements of donor and acceptor functionality are used. The stability of the resultant complex is strongly influenced by the number of hydrogen bonds—in general, two hydrogen bonds will be stronger than one hydrogen bond, three hydrogen bonds will be stronger than two, and so on. This cooperative effect can be seen for the base pairs where the GC complex **5·6** is two to three orders of magnitude more stable than the AT complex **7·8** (Figure 3). Research within the last 10–15 years has moved toward the development of quadruple hydrogen-bonding arrays such as the self-complementary ureidopyrimidinone or (UPy) motif **9**[8] introduced by Meijer and Sijbesma, the self-complementary deazapterin or (DeAp) array **10** introduced by Zimmerman[9] and the 2,7-diamido-1,8-naphthyridine (DAN) motif **11**[10] introduced by Li which forms heterocomplexes with both UPy[10] and DeAp[9] (Figure 4).

Key factors that affect the affinity of interaction within such systems are acidity/basicity of hydrogen-bond donor and acceptors, preorganization, the arrangement of interacting groups, and tautomerization. For instance, intramolecular hydrogen bonding can be advantageous; for example, exchanging the ureido group for an acyl group in self-complementary ureidotriazine derivatives **12** and **13**[11] results in diminished association constants owing to the loss of preorganization (Figure 5a). Intramolecular H-bonding can also be problematic with the classic example being 2-pyridyl ureas **14**; which can present a DDA arrangement, but because of intramolecular H-bonding, prefer to present only two sites for intermolecular contact (Figure 5b).[12] The use of a conformer independent array **15** can negate this problem as described by the Wilson group (Figure 5c).[13]

An important observation made by Jorgenson is that adjacent donor and acceptor functionalities, which are significantly close in space, have an effect on complex stability—the so-called secondary electrostatic effect.[14] Thus, a DAD-ADA array would have a lower stability than an AAD-DDA array possessing similar donor and acceptor functionality (Figure 6). Schneider has developed a method

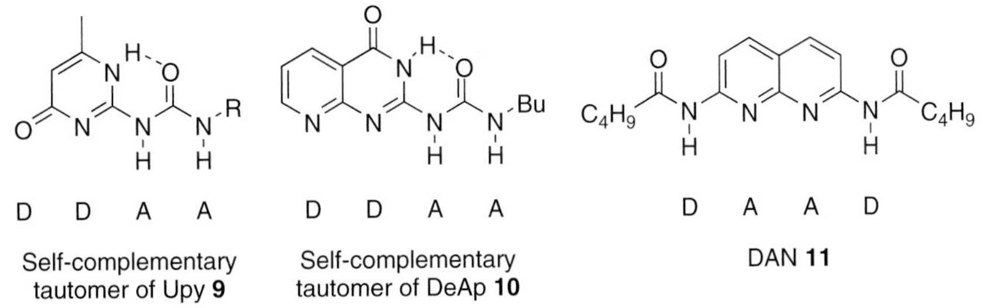

Figure 4 Linear arrays **9–11** capable of quadruple hydrogen bonding.

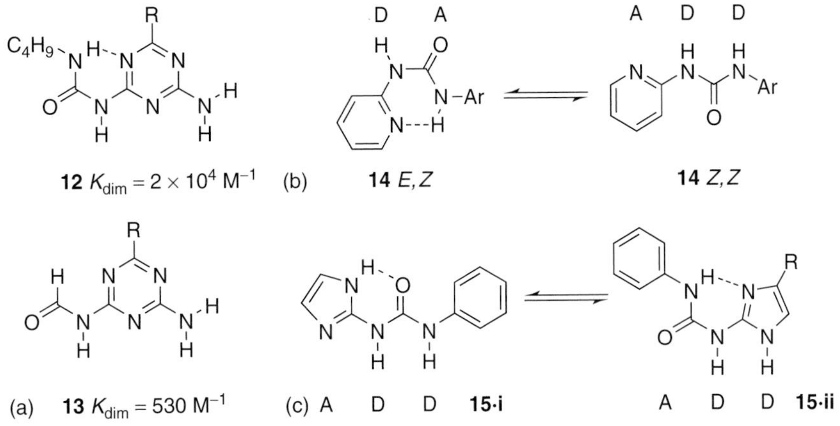

Figure 5 (a) Positive effect of preorganizing hydrogen bonds on self-dimerization of triazine derivatives **12** and **13** (K_{dim} given in chloroform), (b) negative effect of intramolecular hydrogen bonds on desired presentation of an ADD array **14**, and (c) conformer independent linear array **15**.

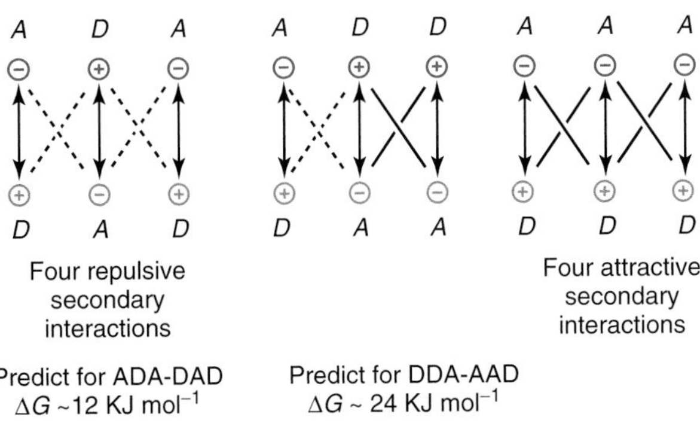

Figure 6 Cartoon representation illustrating secondary electrostatic effects on the dimerization affinity of linear arrays of hydrogen bonds.

that accurately accounts for this effect for a range of complexes.[15] Recently, Leigh and coworkers have reported on the development of a series of AAA-DDD arrays[16]; heterodimers **16**·**17** with extremely high association constants $K_a \sim 10^7\ \text{M}^{-1}$ in fully support this concept (Figure 7a). An approach to circumvent this problem has been described by Gong and coworkers, who reported on the synthesis of hydrogen-bonded molecular duplexes from 5-alkoxy-3-aminobenzoic acid monomers or 2,4-bisalkoxyisophthalate and 1,3 diaminobenzene monomers. Similar earlier work on duplexes from the Hunter group, for example, **20**·**21** (Figure 7c) has been exploited in a wide range of fundamental studies directed toward understanding individual contributions to noncovalent interactions such as $\pi-\pi$ interactions through chemical double mutant cycles.[17] Because the individual hydrogen-bond donor and acceptor atoms are separated, the precise arrangement has little effect on the dimerization constant, for example, the DDAA motif **18** (with dimerization constant $>4.4 \times 10^4\ \text{M}^{-1}$ in $CDCl_3$) and DADA motif **19** (with dimerization constant $\sim 6.5 \times 10^4\ \text{M}^{-1}$ in $CDCl_3$) have similar association constants (Figure 7b).[18]

The use of heteroaromatic motifs can create problems with tautomerization or prototropy. For instance, the UPy motif adopts three different tautomers with two different conformers adopted for two of these (Figure 8). This is advantageous in that it permits the UPy motif to form heterodimers with DAN-type units; however, it is disadvantageous in that it diminishes the strength of self-association via the preferred DDAA presentation of functional groups.

It is also possible to control hydrogen-bonding affinity using external stimuli,[19] particularly light and electrochemistry, given their potential use as environmental

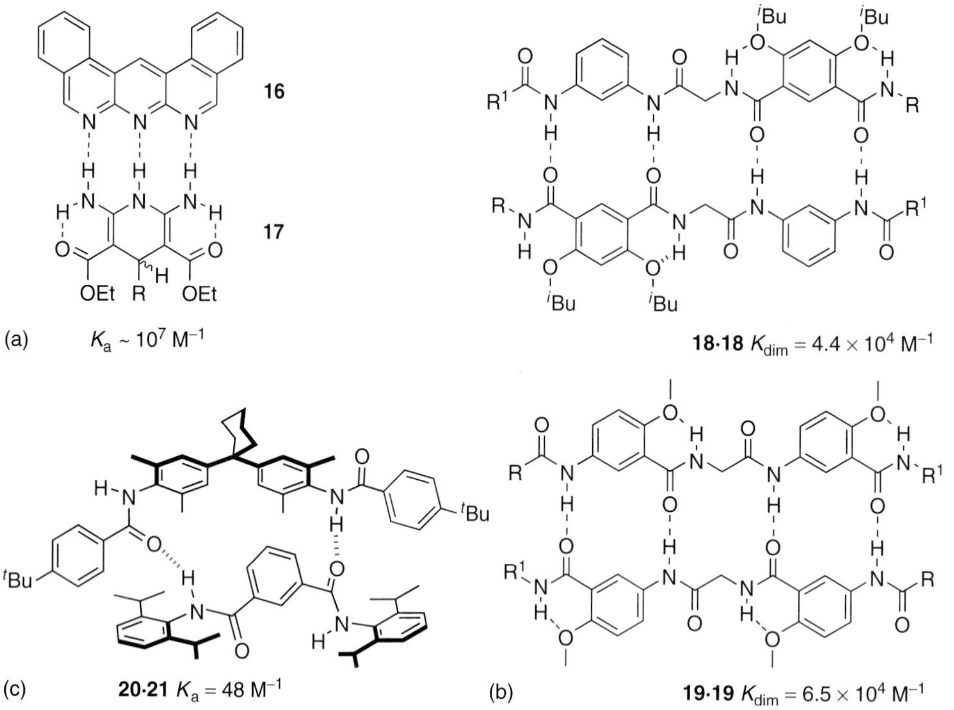

Figure 7 (a) DDD-AAA triply hydrogen-bonded array **16**·**17**, (b) homoduplexes **18**·**18** and **19**·**19**, and (c) heteroduplex **20**·**21**.

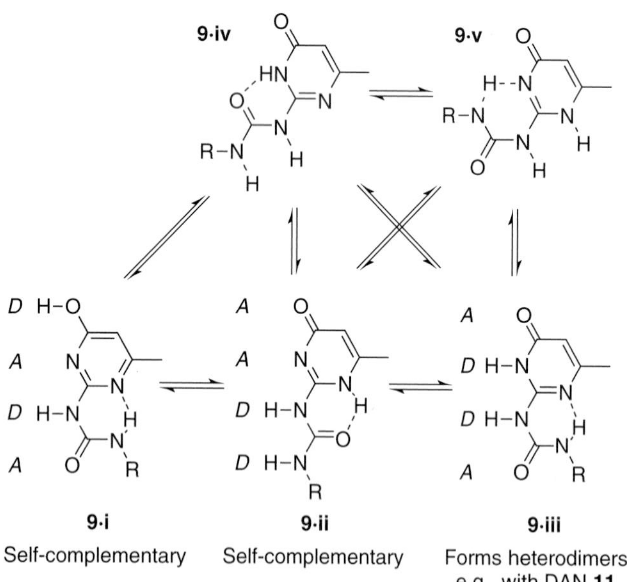

Figure 8 Different tautomers/conformer adopted by ureidopyrimidinone derivatives, for example, **9**.

switches for stimuli responsive materials. Several important motifs have been reported including flavins **22**[20] or 1,8-naphthalimides **23**[21]; these motifs possess the complementary hydrogen-bond donor and acceptor motifs typical of linear arrays, but have different binding affinity toward 2,6-dialkylamidopyridine **24** derivatives depending on redox state (Figure 9a and b). Motifs that change their behavior in response to light are less well studied; however, the strength of interaction between two linear arrays **25**·**26** has been modulated via the photoinduced cis–trans isomerization of an azobenzene motif:—the cis form reinforcing the hydrogen-bonding interaction through aromatic $\pi-\pi$ stacking (Figure 9c).[22] In addition, caged systems that can be activated once to unmask a hydrogen-bonding motif have been described—one such example has been described by Folmer where a UPy motif **28** is unmasked by UV-mediated photocleavage of an *o*-nitrobenzyl ether **27**.[23]

2.2 Receptors exploiting hydrogen bonding

As model systems, linear arrays have served as perfect vehicles to understand, manipulate, and teach the general concepts of cooperative hydrogen-bonding interactions. It is in essence these principles that have been exploited in the development of more elaborate receptors that exploit hydrogen-bonding interactions. Initially, studies on neutral targets such as those of the Hamilton[4] and Rebek[5] groups arose out of the desire to recapitulate the exquisite molecular recognition properties observed in natural systems where hydrogen-bonding interactions are used extensively to impart selectivity and directionality. Examples include the high-affinity interaction between the antibiotic vancomycin **29** and D-Ala-D-Ala **30** (Figure 10a)[24] and the

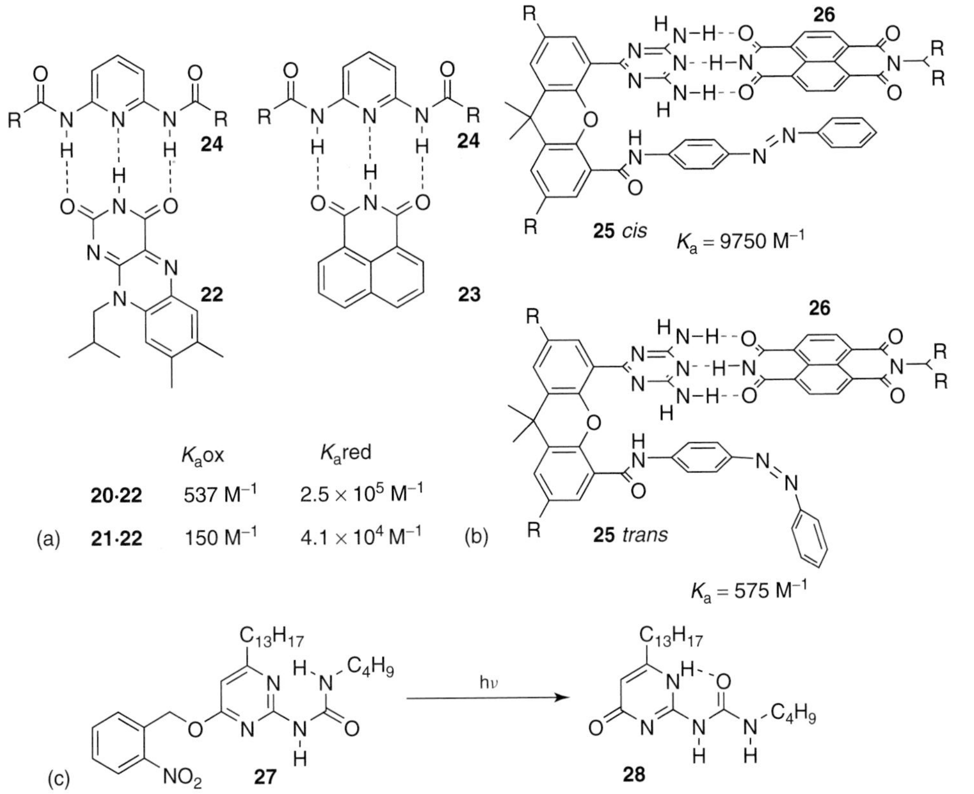

Figure 9 (a) Electrochemically active hydrogen-bonded motifs **22·24** and **23·24**, (b) light responsive hydrogen-bonded heterodimers **25·26**, (c) light activated transformation of masked UPy derivative **27** to homodimerizing UPy **28**.

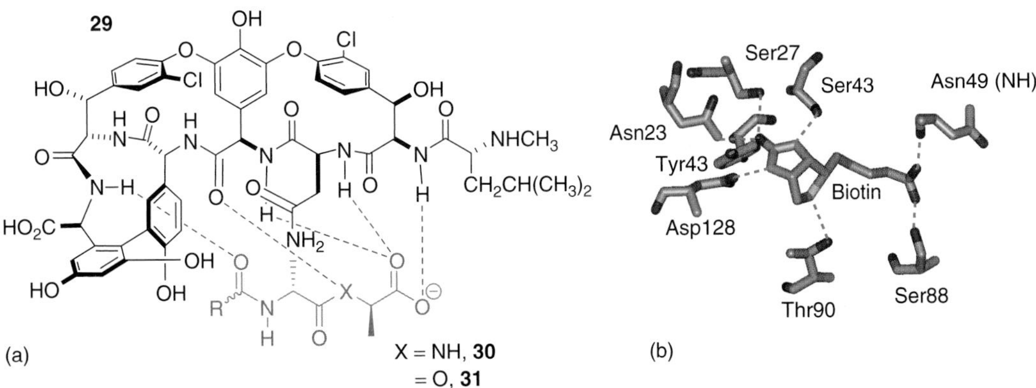

Figure 10 (a) Structure of vancomycin **29** bound through H-bonding to dipeptides **30** and **31** and (b) X-ray structure of biotin–streptavidin (PDB ID: 1STP) illustrating the biotin binding site.

interaction between biotin and streptavidin (Figure 10b).[25] For the former, the role of hydrogen bonding is illustrated beautifully through the mutation to D-Ala-D-Lac **31** that confers antibiotic resistance (Figure 10a).[24] As the field has developed beyond the studies of the late 1980s and early 1990s, other targets have been identified, in acknowledgment that neutral-hydrogen-bonding donor and acceptor groups can be used to recognize ionic guests (including those of biological relevance). An additional driver for these studies has been that such targets have served as a useful means to make the transition from organic solvents to more challenging systems. The following section highlights some of the pertinent examples and will illustrate how the key elements of receptor design (i.e., complementarity of shape and noncovalent interactions, maximal site occupancy, cooperativity, and preorganization) alongside the special selectivity that can be imparted through use of directional noncovalent interactions can be exploited to recognize a range of targets.

2.2.1 *Neutral guests*

A number of studies in the late 1980s and early 1990s represented key developments in this area. In addition to the examples highlighted in the introduction (Figures 1 and 2), a particularly elegant example is the work of Hunter, who described a macrocyclic host **32** for *p*-benzoquinone **33**,[26] cooperative bifurcated H-bonds between either end of the tetralactam and the carbonyl of the preorganized cyclic host result in high-affinity recognition in chloroform (Figure 11). Higher affinity interactions with diketopiperazine derivatives **35** could be observed as a consequence of additional CH/NH–π interactions.[27]

One of the simplest early acyclic examples of neutral H-bond-mediated recognition is exemplified through the recognition of dicarboxylic acids described by Hamilton.[28] By exploiting cooperative two point hydrogen bonding between the amidopyridine motifs and the carboxylic acid, receptors, for example, **37** with high affinity for targets such as glutaric acid **38** could be identified as illustrated in Figure 12. This type of receptor might be considered to fall within the "cleft" class of receptor and a number of these were used for recognition of uric-acid-type molecules **39·40**,[29] ureas **41·42** and **45·46**,[30,31] and polyols **43·44**[32] as illustrated in Figure 12, alongside "tweezer" based receptors such as **47** described by Zimmerman, Wu, and Zeng[33] which supplement two point hydrogen bonding with additional π–π stacking interactions to grip the target DNA base **48** (Figure 12). In all cases, a common feature of these receptors is the use of multiple geometrically matched hydrogen-bonding interactions alongside shape complementarity.

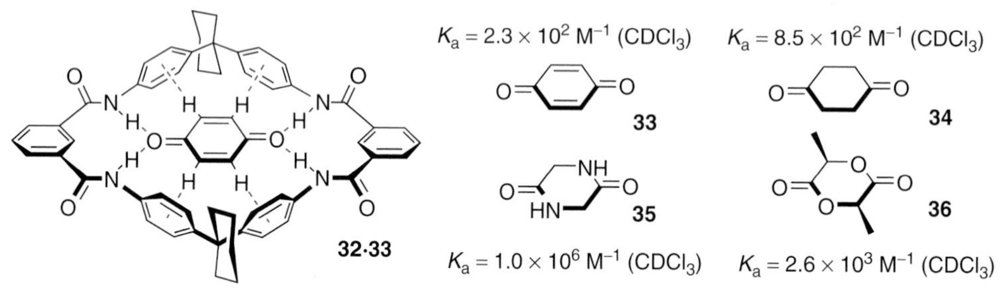

Figure 11 Structure of a tetralactam host **32** bound to *p*-benzoquinone **33** and association constants for binding with guests **33–36**.

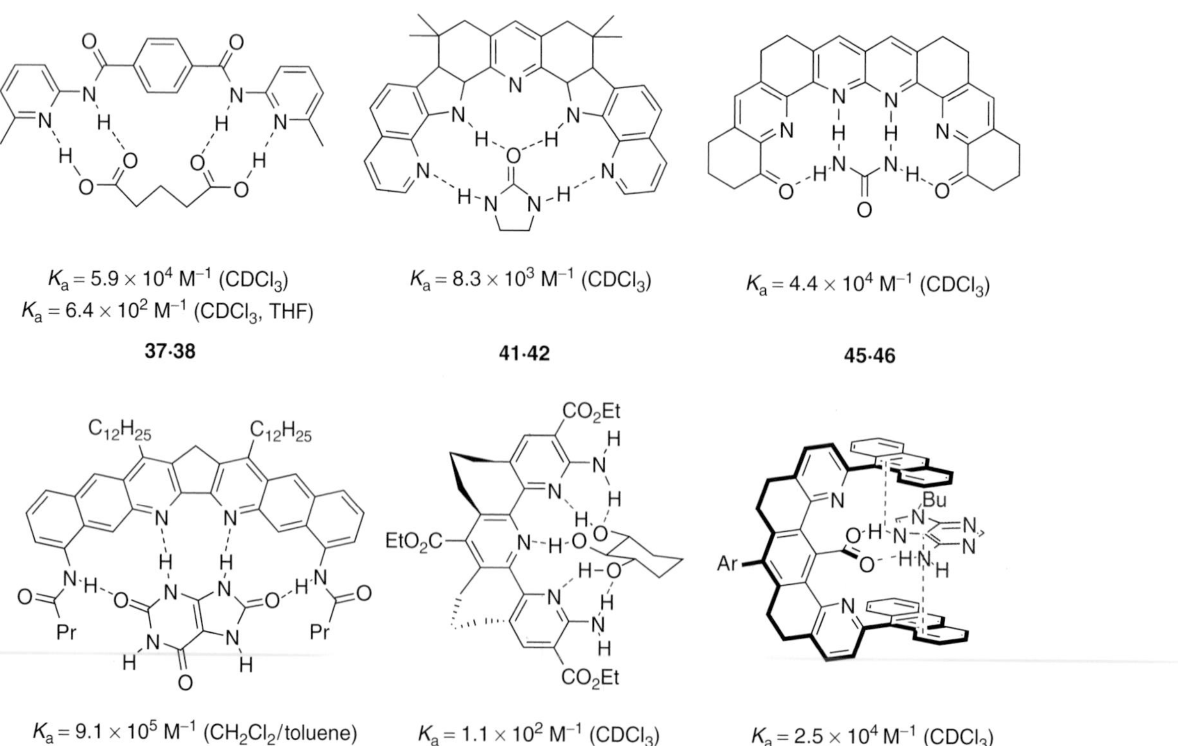

Figure 12 Structures and association constants for host–guest complexes **37–48**.

2.2.2 Cations

Cations that posses hydrogen-bond donor atoms (i.e., ammonium and guanidinium ions) were some of the earliest examples targeted for recognition using synthetic hosts. The crown ethers originally described by Pedersen[34] for recognition of alkali metal cations were also studied extensively for recognition of ammonium ions by Cram during the 1970s.[35] A key feature of the recognition is the directional hydrogen-bonding interactions between the ammonium NH groups and the crown ether oxygen atoms. Although binding of ammonium groups by crown ethers is weaker than for alkali metal ions such as potassium, 18-crown-6 **49** can bind to ammonium ions such as *t*-butyl ammonium **50** and extract them into chloroform.[36] The importance of both preorganization and matching of donor and acceptor vectors is illustrated by the weaker binding exhibited by acyclic **51** and smaller analogs such as the 15-crown-5 analog **52**.[36] Guanidinium ions **54** can similarly be recognized through hydrogen-bonding interactions with the benzo-27-crown-9 receptor **53**.[35] This area has become a huge area of focus with recognition of chiral ammonium derivatives, a major theme and the reader is directed to review articles on the topic.[37] In extension of these studies, a number of groups have exploited crown ethers for recognition of dialkylammonium motifs; elegant examples have been reported by the Stoddart group, who have illustrated that by using a larger crown ether such as dibenzo-24-crown-8 **55**, the dialkyl ammonium motif **56** is able to thread through the annulus of the ring and bind through hydrogen-bonding interactions between both the NHs of the ammonium and adjacent CHs to the crown ether O interaction; these adjacent CHs are capable of enhanced H bonding due to the polarized nature of the C–N bond.[38] The importance of this work is evidenced by the extensive use of such recognition motifs in self-assembly and template-directed syntheses described elsewhere in this chapter and text (Figure 13) (see **Crown and Lariat Ethers**, Volume 3, **Self-Assembly of Macromolecular Threaded Systems**, **Self-Assembled Links: Catenanes**, **Rotaxanes—Self-Assembled Links**, Volume 5).

There are a number of additional more recent contributions that are worthy of note concerning cation recognition; Gong and coworkers have elegantly illustrated that both aromatic oligoamide macrocycles **57**[39] and more recently the foldamers **58** from which they are derived are capable of recognizing guanidinium cations[40] as illustrated in Figure 14. The foldamers are particularly interesting; they are highly preorganized as a consequence of rigidifying backbone intramolecular hydrogen bonding and project carbonyl acceptor groups into the cavity but the gap between the two ends of the foldamer permits an alkyl group to be appended to the guest head group.

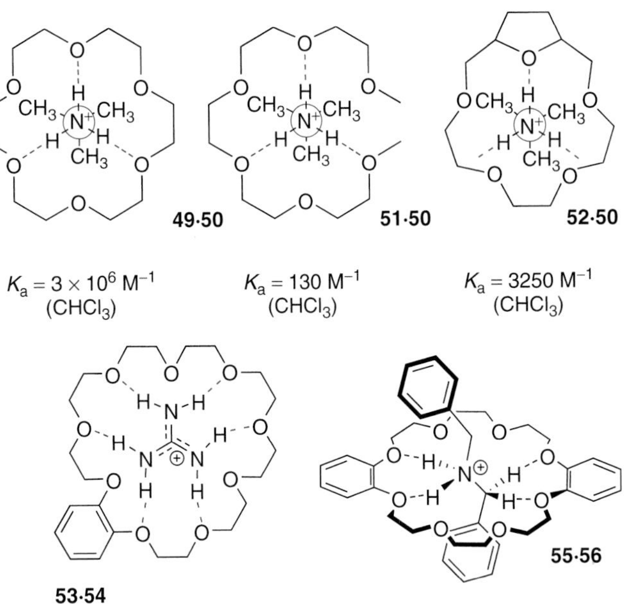

Figure 13 Structures and association constants for host–guest complexes **49**–**56**.

2.2.3 Anions

Until the 1990s, most anion receptors exploited charged or lewis acidic binding functionality; a key development in the area was the report by Sessler and coworkers in 1996 on calix[4]pyrrole (porphyrinogen) derivatives such as **60**. This initial report illustrated that directional pyrrole donor atoms could be exploited within the framework of a macrocycle for recognition of halide anions in halogenated solvents with fluoride being the most tightly bound anion.[41] The convergent projection of H-bond donor atoms is ideal for recognition of spherical anions. Since this report an enormous effort in the area of calix[4]pyrrole anion recognition[42] and more generally the use of pyrrole containing macrocycles and receptors has been described and the reader is directed to relevant review articles (see **Porphyrins and Expanded Porphyrins as Receptors**, Volume 3).[42,43] A similarly influential report by Crabtree and coworkers in 1997 described the use of an isophthalamide derivative **61a** as a receptor for halide anions in dichloromethane with the highest affinities for chloride anion ($K_a = 6.1 \times 10^4$ M^{-1}).[44] More recently, it has been shown that more preorganized isopthalamide derivatives, for example, **62** which exhibits an order of magnitude higher binding affinity for chloride than the corresponding Crabtree-based receptor **61b** can be used to effect chloride transport through membranes[45] while Yang and coworkers have shown non-preorganized α-aminoxyisophthalamide derivatives **63** possessing stronger H-bond donors exhibit high affinity toward anions and can also effect chloride transport across membranes.[46] Isophthalamide and related motifs have subsequently found extensive use elsewhere,

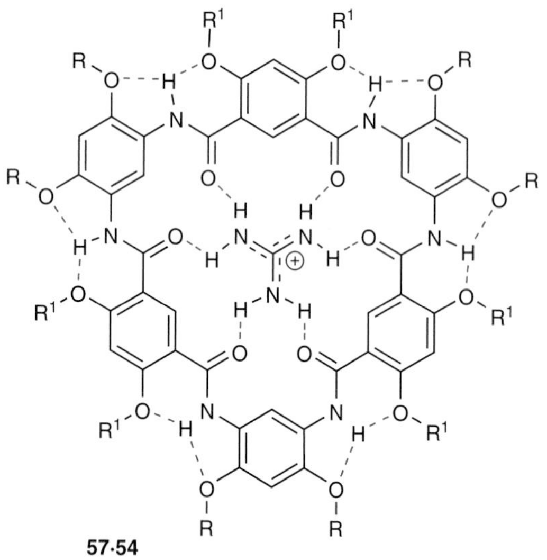

57·54

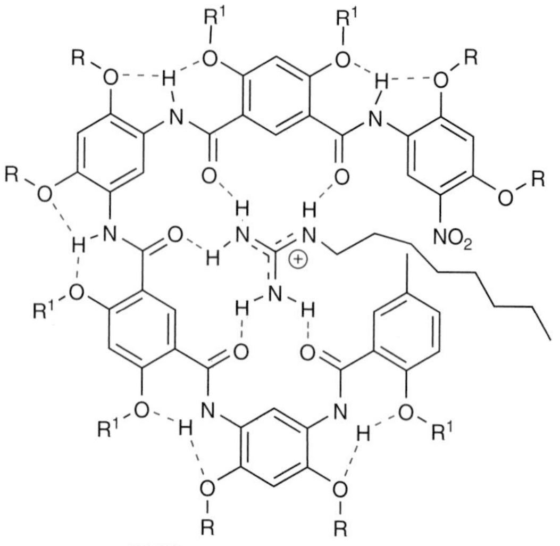

58·59

Figure 14 Structures for host–guest complexes **54**, **57–59**.

a particularly impressive example was reported by Kristin Bowman-James and coworkers, who illustrated macrobicyclic derivatives incorporating 2,6-pyridinedicarboxamide motifs such as **64** could bind with high affinity to anions even in DMSO-d_6 with fluoride exhibiting the highest affinity log $K > 5.00$.[47] An elegant fusion of the pyrrole and dicarboxamide concepts has been elaborated by the Gale group, who have shown that 2,5-diamidopyrroles, for example, **65** act as simple receptors for a range of anions, in particular, benzoate and phosphate in relatively competitive solvents such as acetonitrile and even aqueous DMSO.[48] A further example of a "cleft" type receptor has been described by Hamilton and coworkers, which turned to anions in the search for H-bond-mediated recognition in challenging solvents. By extension of their earlier studies on receptors for dicarboxylic acids, replacement of the amidopyridine motif with (thio)urea afforded receptors **66** with high affinity for dicarboxylates such as glutarate **67**.[49] An alternative type of acyclic anion recognition motif has been described by Davis and coworkers; Cholapod receptors such as **68** have been elaborated that project H-bond donor groups into the cleft formed on the underside of the steroidal framework resulting in receptors with remarkable affinity for anions (Figure 15).[50]

2.2.4 *Zwitterions and ion pairs*

The preceding two sections have focused on cation and anion recognition using neutral hydrogen-bond donor atoms; however, a particularly important interaction that has not been considered thus far is the interaction between a cation and anion—the classic salt bridge between an ammonium or guanidinium motif is found extensively in biomolecular recognition events and has been a key interaction in making the transition from recognition in low dielectric media to competitive solvents such as water. Similarly, posttranslational modifications to proteins such as phosphorylation and anions such as sulfate are recognized in biological systems using ammonium groups. These interactions are attractive for high-affinity binding because they offer both the special directionality of hydrogen bonds and enhanced affinity (see **Guanidinium-Based Receptors for Oxoanions**, Volume 3). As a starting point in addition to (thio)ureas, Hamilton and coworkers' studies on dicarboxylate recognition also exploited imidazole motifs. Receptor **69** exhibited fivefold greater affinity for glutarate **67** than the corresponding receptor **66** and could even function in aqueous DMSO (Figure 16).[49] Similarly, the Schmidtchen group[51] have elaborated receptors such as **70** for sulfate that exploit two rigidified bicyclic guanidinium groups for recognition of the tetrahedral anion in methanol while Schrader's group have used phosphonate moieties within cleft-type receptors such as **71** for recognition of guanidinium cations, for example, **72** (Figure 16).[52,53] Additional noncovalent interactions can be used to enhance binding affinity based on salt bridges; for instance, appending flat naphthyl groups to a rigid bicyclic guanidinium as in **73** motif helps binding of benzoates (e.g., **74**) as a result of flanking π–π stacking interactions made by the tweezer motif.[54] Amino acids as zwitterions possess both an anion and a cation binding site—in a classic example **75**, aromatic amino acids, for example, phenyl alanine **76** can be recognized by a receptor possessing a rigid bicyclic guanidinium motif for interaction with the carboxylate, a crown for recognition of the primary ammonium and a flanking naphthyl group for π–π stacking with the aromatic side chain—a preference for phenylalanine was observed in liquid–liquid extraction experiments.[55] A more recent

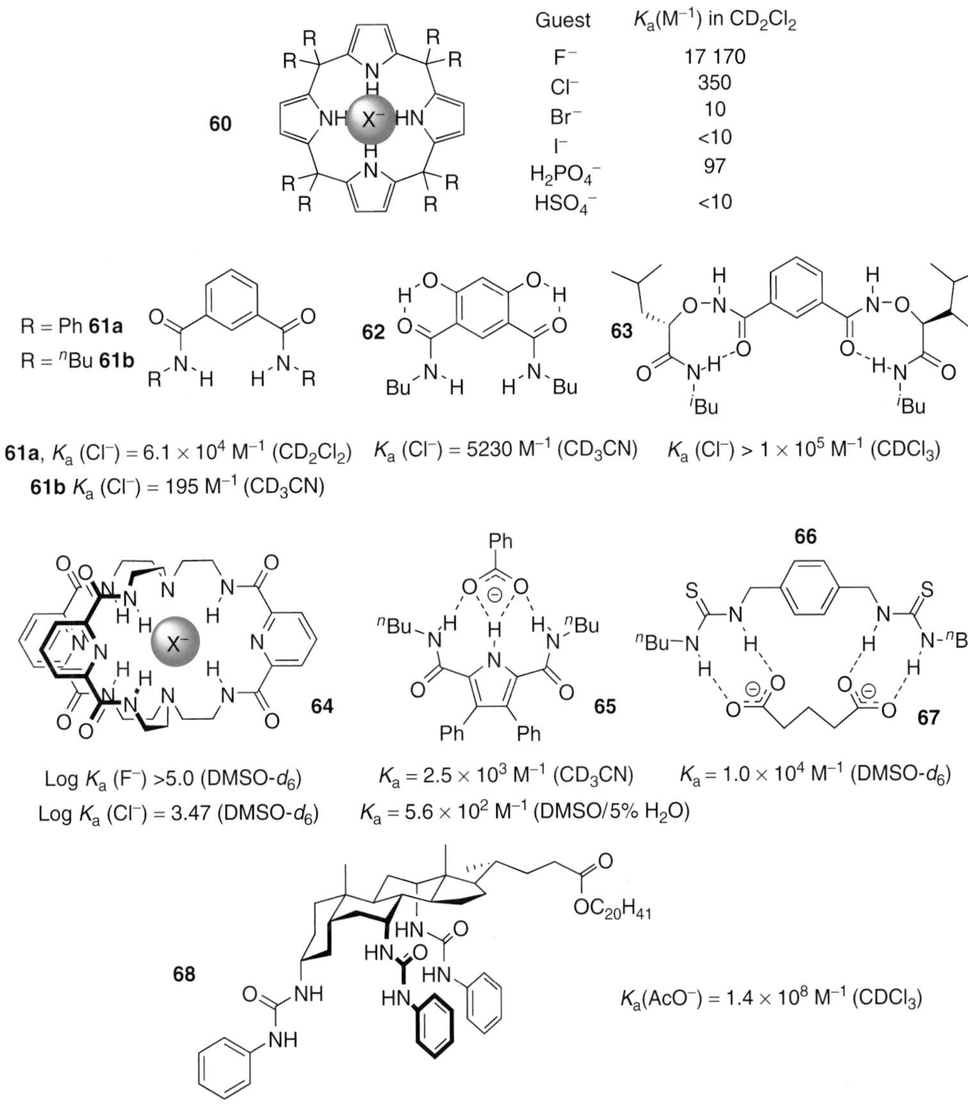

Guest	$K_a(M^{-1})$ in CD_2Cl_2
F^-	17 170
Cl^-	350
Br^-	10
I^-	<10
$H_2PO_4^-$	97
HSO_4^-	<10

Figure 15 Structures and association constants for host–guest complexes **60–68**.

example is the tweezer molecule **77**, which recognizes noradrenaline **78** using a head group capable of binding the catechol head group through hydrogen-bonding, biphenyl arms for interaction with the hydrophobic aromatic ring and phosphonates for recognition of the ammonium group.[56]

2.3 Modern challenges: recognition in competitive media

More recently, the focus has turned toward recognition in water which remains a challenge, not only because this is the solvent for recognition of biologically important guests, but also because of the competitive nature of the solvent. Some of the early attempts to do this built upon receptor designs employed in less demanding solvents in tandem with the hydrophobic effect. For instance, a water-soluble version **79** of Hunters *p*-benzoquinone tetralactam host **29** was shown to exhibit some affinity for diketopiperazine **32** in water, albeit vastly reduced in comparison to the corresponding interaction in chloroform—some of this affinity could be recovered by increasing the hydrophobic surface of the diketopiperazine as in **80** (Figure 17).[57]

Successful recognition of a much more complicated target in water was addressed in work reported by Davis.[58] Sugars are challenging targets for selective molecular recognition because they possess only weak hydrogen-bond donor and acceptor moieties that do not differ from those of water and because they are only subtly variable. Even in nature, binding is usually weak, however, Nature tends to recognize oligosaccharides, where more noncovalent contacts can lead to higher affinities. The Davis receptor **81** (Figure 18) was designed to take advantage of this and was targeted toward disaccharides possessing

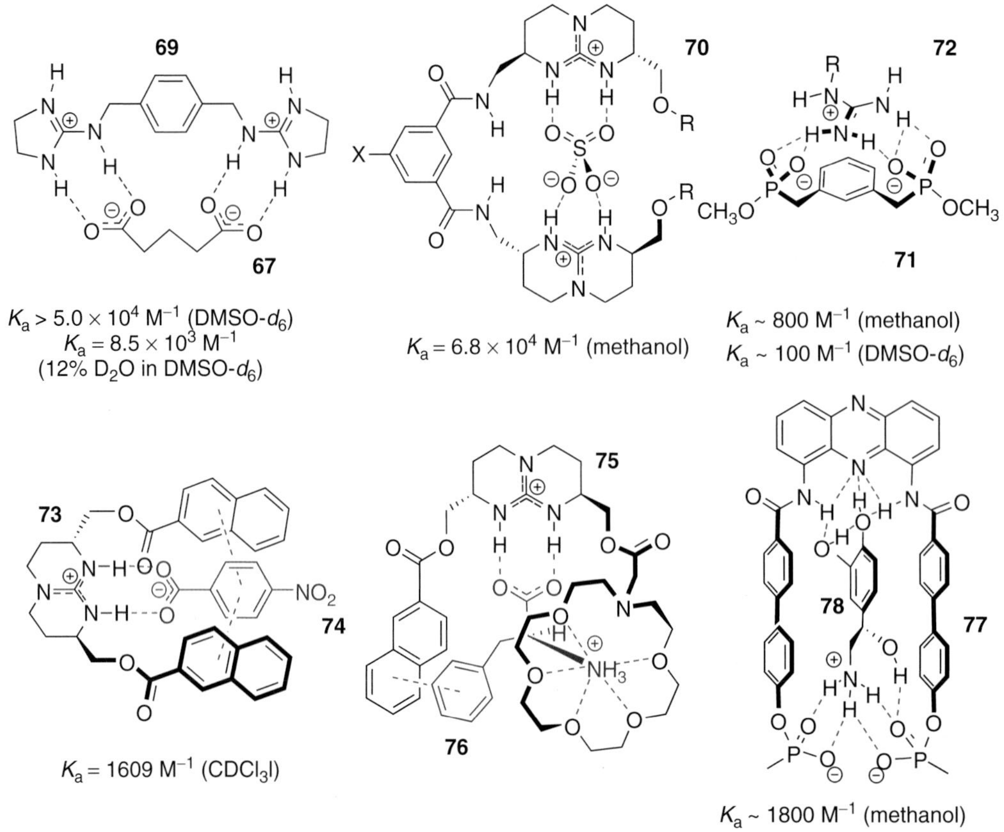

$K_a > 5.0 \times 10^4\ M^{-1}$ (DMSO-d_6)
$K_a = 8.5 \times 10^3\ M^{-1}$
(12% D_2O in DMSO-d_6)

$K_a = 6.8 \times 10^4\ M^{-1}$ (methanol)

$K_a \sim 800\ M^{-1}$ (methanol)
$K_a \sim 100\ M^{-1}$ (DMSO-d_6)

$K_a = 1609\ M^{-1}$ ($CDCl_3$l)

$K_a \sim 1800\ M^{-1}$ (methanol)

Figure 16 Structures and association constants for host–guest complexes **69**–**78**.

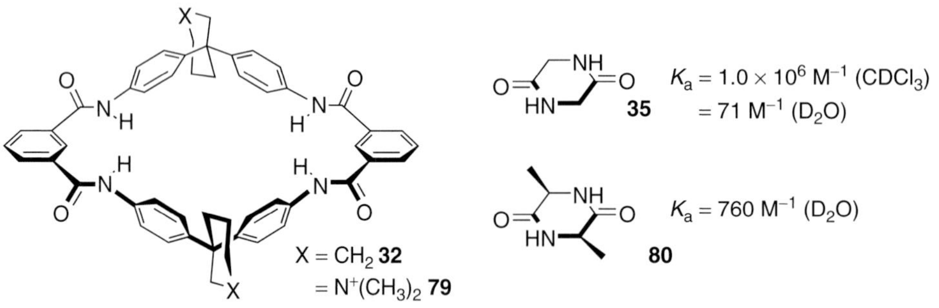

$K_a = 1.0 \times 10^6\ M^{-1}$ ($CDCl_3$)
$= 71\ M^{-1}$ (D_2O)

$K_a = 760\ M^{-1}$ (D_2O)

X = CH_2 **32**
= $N^+(CH_3)_2$ **79**

Figure 17 Structure of a tetralactam host **32** and **79** bound to guests **35** and **80**.

all axial substituents, that is, cellobiose. This equatorial positioning of hydroxyls positions CH groups axially forming apolar patches at the top and bottom. Receptor **81**, with an aromatic hydrocarbon roof and bottom bridged by isophthalamides capable of hydrogen-bond formation was functionalized with acidic solubilizing groups at the periphery recognizes, cellobiose with $K_a \sim 600\,M^{-1}$. NOE (nuclear Overhauser effect) studies suggested that the sugar was indeed bound within the cavity of the macrotetracycle while binding studies with a series of sugars (comprising 10 disaccharides and 3 monosaccharides) demonstrated a degree of selectivity in favor of disaccharides possessing all equatorial substituents.

One of the classic examples of receptors that operate through neutral hydrogen-bonding interactions is the Dervan polyamides, for example, **82** used for recognition of DNA. These aromatic oligoamide ligands employ different sequences of monomer building blocks to recognize the Hoogsteen edge along the major groove of DNA—shape complementarity and the hydrophobic effect confer affinity.[59,60] Using a series of five-membered heterocycles, it is possible to recognize pairwise each of the four base pairs as illustrated in Figure 19.

Within the context of charged hydrogen-bonding motifs, there has also been considerable progress. Many of the charged motifs described in the preceding section,

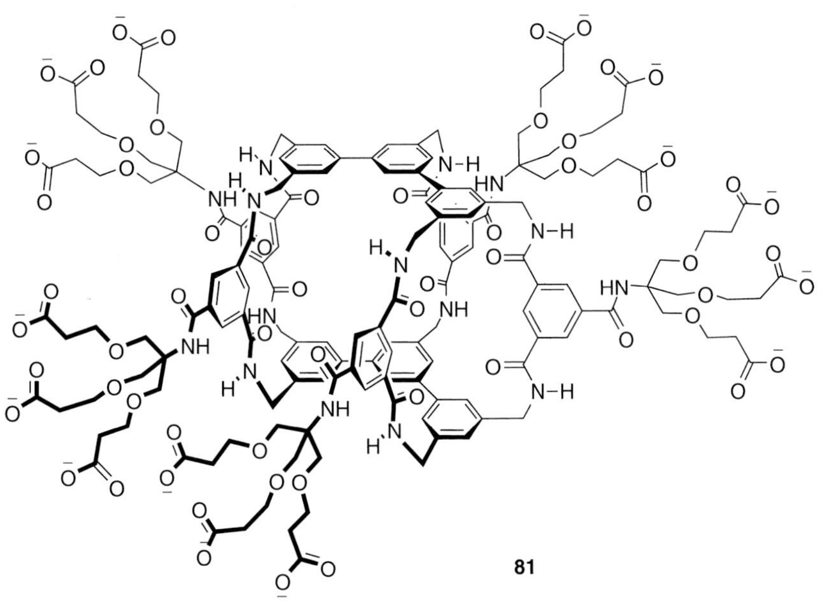

Figure 18 Structure of disaccharide binding macrotetracycle **81**.

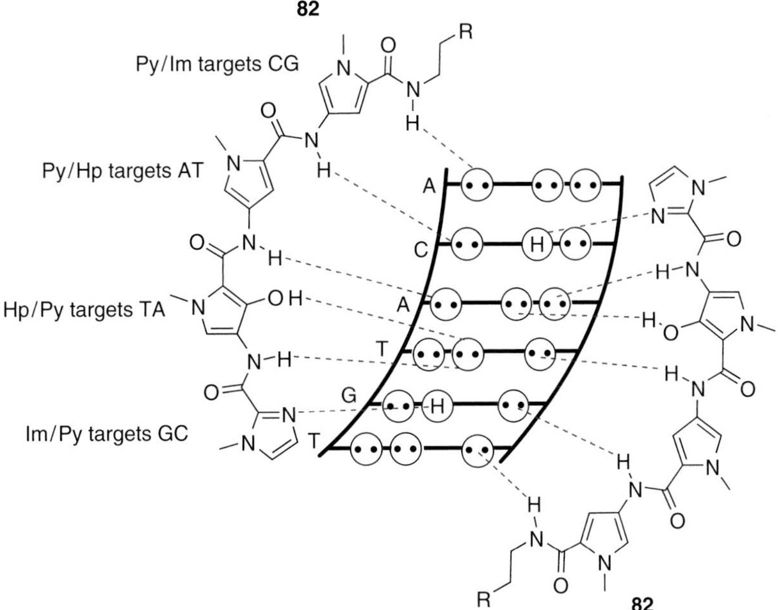

Figure 19 Cartoon illustrating the binding of a representative Dervan polyamide **82** bound to double-stranded DNA (lone pairs denote acceptors and H denotes donors in the major groove).

for example, the bis-imidazolium dicarboxylate receptors, the rigidified bicyclic guanidinium sulfate receptors, and the bis-phosphonate motifs were also tested in aqueous or mixed organic/aqueous solvents; however, as for the tetralactam host **77** described above, affinities are vastly reduced compared to wholly organic solvent systems and additional noncovalent interactions (often the hydrophobic effect) usually have to be employed to confer high affinity. It is worth noting in this context that one class of host which has received immense attention is the cucurbituril family. These toroidal macrocycles possess a hydrophobic core with carbonyl acceptor groups at the periphery—these acceptor groups are typically employed to anchor target guests through ion dipole hydrogen-bonding interactions with ammonium groups as is illustrated for the host–guest interaction between bis-ammonium hexane **83** and cucur[6]bituril **84** in Figure 20.[61]

Figure 20 Binding constant and cartoon representation of the binding between cucur[6]bituril **83** binding to a diammonium motif **84**.

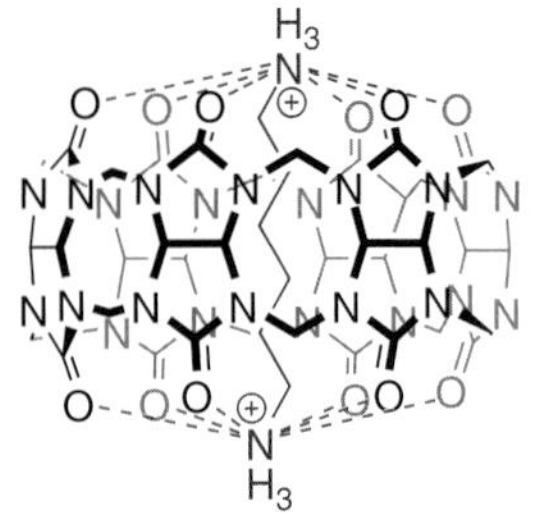

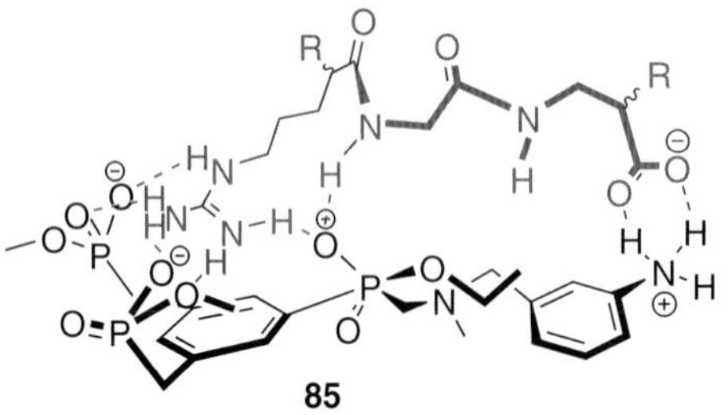

Figure 21 Binding constant and cartoon representation of the recognition of the RGD peptide sequence using receptor **85**.

Further examples of this approach included the elaboration of the bisphosphonate moiety for recognition of RGD peptides.[52] Extension of the bisphosphonate moiety with an ammonium group for recognition of the C-terminus of the peptide allows recognition of the RGD tripeptide sequence using receptor **85** as illustrated in Figure 21.

A more recent contribution that exploits similar concepts has been elaborated by Schmuck and coworkers.[62,63] They have developed a guanidiniocarbonylpyrrole motif that has high affinity for carboxylates—the interaction is typical of an ion-pairing interaction in that a cation and anion are involved; however, it exploits multiple directional hydrogen-bonding interactions. Initially this has been used to recognize dipeptides,[62] but the approach has been extended to the recognition of amino acid carboxylates.[63] Receptor **86** exhibiting strong affinity for alanine in 90% water/DMSO was observed ($K_a = 2100\,M^{-1}$) as suggested in Figure 22(a). In looking toward recognition of more challenging targets, Schmuck and coworkers have used a combinatorial approach to recognize tetrapeptides. They used the guanidinocarbonylpyrrole as a tetracationic anchor point for recognition of negatively charged tetrapeptides. The group had previously observed that similar structures bound to D-Glu-L-Lys-D-Ala-D-Ala-OH (a tetrapeptide bearing similarity to an important peptidoglycan, which is crucial to cell wall synthesis) with $K_a > 10^4\,M^{-1}$ in buffered water.[64] The receptor **87** was constructed with an N-terminal dansyl fluorophore to facilitate on-bead screening and to allow a full structure binding relationship to be established (Figure 22b).[65] From a library of 320 different tetrapeptides, derived from D-Glu, L-Lys, and D/L-Ala, they were able to observe that the receptor **87** binds most efficiently to the D-Glu-D-Glu-D-Glu-D-Glu sequence (K_a, 26 500 M^{-1}) and interestingly, the authors were able to show that the receptor exhibited sequence-dependent stereoselective recognition specifically for D-Ala over its enantiomer in positions 2 and −4 of the tetrapeptide.

A final set of examples that are appropriate to include at this point are the receptors described by the Anslyn group—these typically employ ammonium or guanidinium functional groups and in some instances reversible covalent linkages, for example, boronate ester formation between diol and boronic acids. The representative examples **88** and **89** in Figure 23 have been used to sense citrate in beverages[66] and "gallate-like" compounds[67] in scotch using indicator displacement assays.

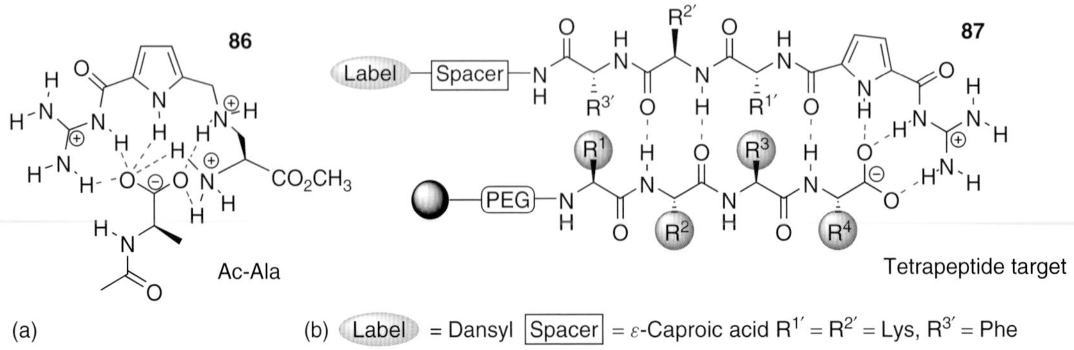

Figure 22 Structural representation of (a) the interaction between guanidinopyrrole receptor **86** and acetyl-alanine and (b) the interaction between guanidinopyrrole tripeptide **87** and tetrapeptides.

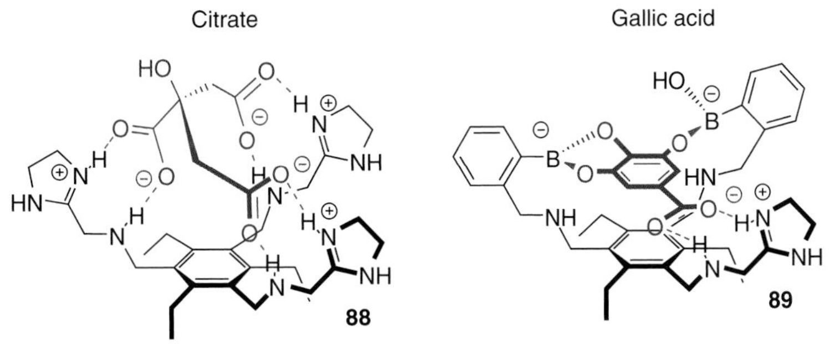

Figure 23 Structural representation of the interaction between receptor **88** and citrate and the interaction between receptor **89** and gallic acid.

3 FUNCTIONAL HOST–GUEST COMPLEXES

3.1 Template-directed synthesis

The simplest role that a host–guest complex can play is within the context of a template-directed synthesis. For instance, benzo-27-crown-9 receptor **53** can be synthesized using its guest (guanidine) as a template.[35] Similarly, the macrocyclic dibenzo-24-crown-8 **55** serves as a receptor for the formation of a host–guest complex with a dialkyl ammonium motif **90**—because the alkyl chains must "thread" either side of the annulus of the ring, this has been termed a *pseudo*-rotaxane. This can be converted into a rotaxane by reacting the ends of the dialkyl ammonium with an appropriate stopper resulting in a mechanically interlocked species **91** (Figure 24).[68]

Similarly, macrocyclic host molecules can serve as templates for the synthesis of catenanes. In catenane **93**, the first ring formed acts as a template for the formation of the second ring through multipoint recognition of the amide as illustrated in Figure 25(a).[69] Such a process of interlocking seemingly "magic-rings" can be achieved under thermodynamic control (Figure 25b); incorporating

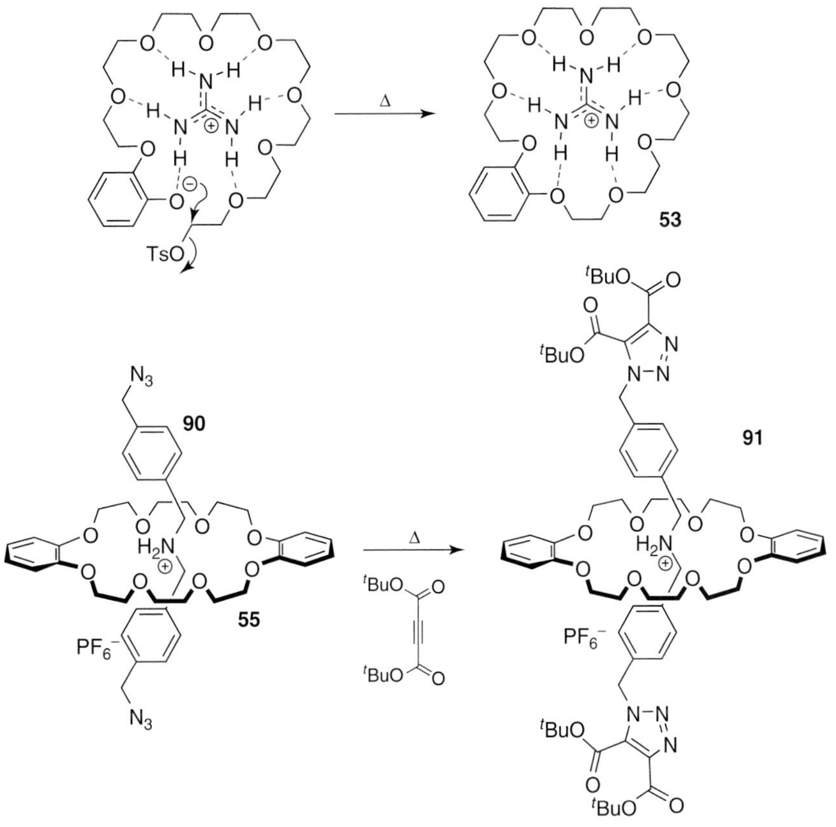

Figure 24 Syntheses of receptor **53** and rotaxane **91** using template-directed synthesis.

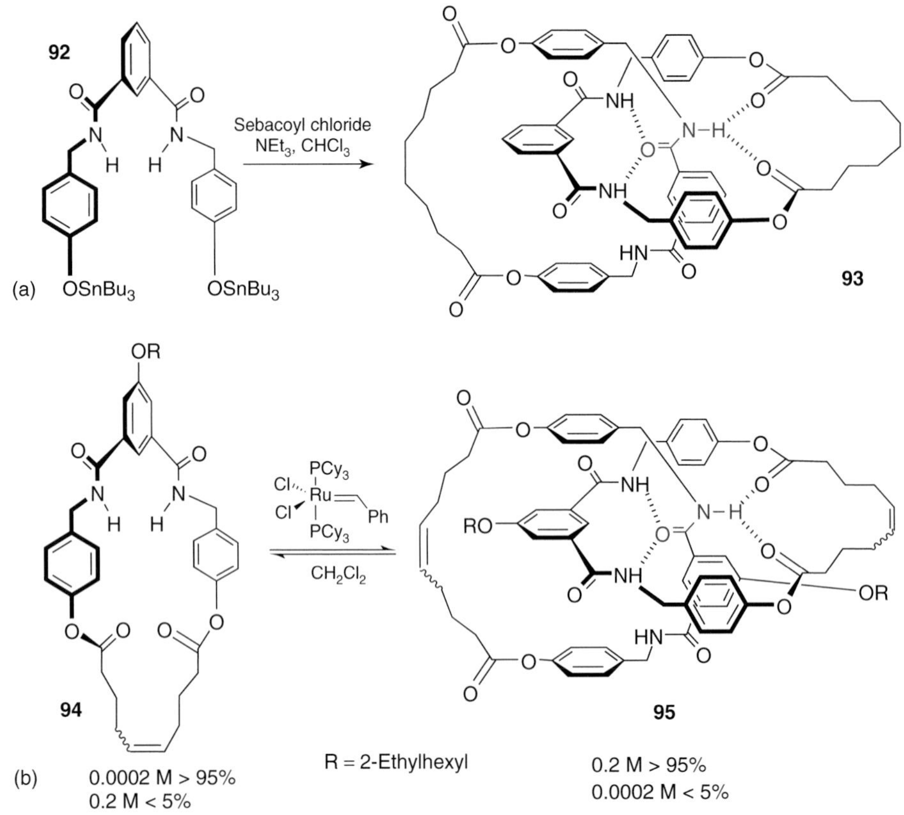

(a)

92 Sebacoyl chloride NEt_3, $CHCl_3$ 93

(b)

94 0.0002 M > 95% 0.2 M < 5%

CH_2Cl_2 R = 2-Ethylhexyl

95 0.2 M > 95% 0.0002 M < 5%

Figure 25 (a) Synthesis of mixed benzylic amide ester catenane **93** by template-directed synthesis and (b) "magic-ring catenane" **95** using reversible ring-opening ring-closing metathesis.

an alkene into the macrocycle **94** permits interconversion between the macrocyclic **94** and interlocked architectures **95** using reversible olefin metathesis—product distribution can be controlled simply through concentration.[70]

3.2 Supramolecular catalysis

A major area of focus for supramolecular chemistry has been catalysis and a number of examples exploiting hydrogen-binding receptors have been elaborated. For instance, Breit and coworkers have exploited the hydrogen-bonding complementarity of the adenine–thymine base pair analogs to combinatorially screen bidentate ligands for catalytic activity.[71] The approach is proposed to be well suited to the discovery of structurally diverse and meaningful ligand libraries and to the discovery of ligands with nonsymmetric bidentate binding sites. The aminopyridine **96–98** and isoquinoline ligands **99** and **100** in Figure 26 were anticipated to have significant impact on ligand bite angle and coordination geometry at the metal center on self-assembly, thus affecting catalytic activity. In addition to a study of the coordination properties of each ligand toward platinum, each ligand combination was screened for the ability to mediate the rhodium catalyzed hydroformylation reaction in toluene. Regioselectivities of up to 99 : 1 for the linear over branched product were observed. Thus, while the hydrogen-bonding ability is weak, it is sufficient to direct assembly of a structurally defined complex when reinforced by the chelating metal ligand interactions—however, polar solvents were able to reduce the observed regioselectivities, and the ligands behave as simple monodentate ligands under these conditions.

An alternative approach for development of functional receptors has focused on using hydrogen bonding to bias a reaction pathway. Specifically, Thorsten Bach and coworkers[72] have shown that two point hydrogen bonding can be used to (a) mediate the stereochemical outcome of a radical cyclization reaction and (b) achieve catalysis through substrate activation as a consequence of proximity to the receptor. Two point hydrogen bonding between receptor **101** and substrate **102** blocks one face of the substrate and permits radical cyclization to occur only from the unhindered side permitting the formation of only **103**. Activation of the host molecule is achieved via UV excitation and results in PET (photoinduced electron transfer) to the amine generating a radical cation. Subsequent proton loss presumably leads to an α-aminoalkyl radical, which after intramolecular addition undergoes back electron transfer and protonation of the enolate generating the product (Figure 27).

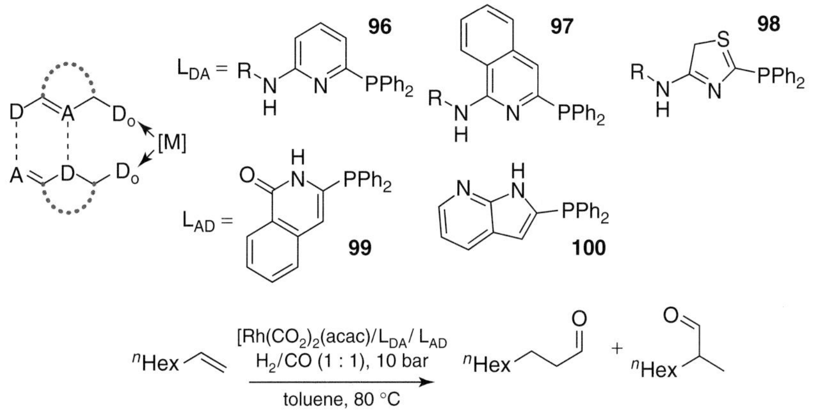

Figure 26 Structures of monodentate phosphine ligands **96–100**, which self-assemble via hydrogen-bonding interactions to generate bidentate ligands that mediate hydroformylation reactions.

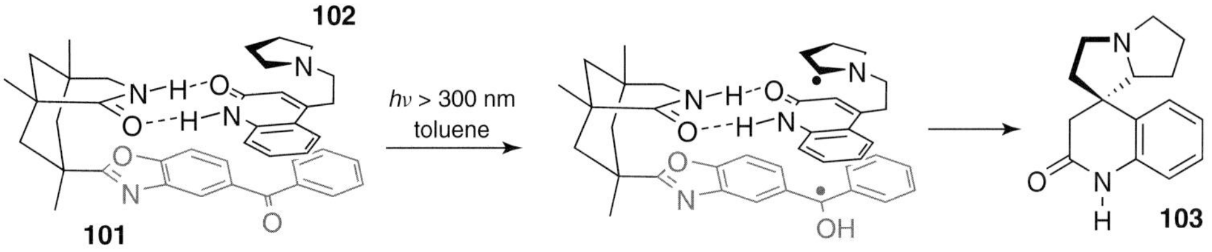

Figure 27 Proposed structure via which receptor **101** mediates cyclization of **102** to generate **103**.

A different type of catalysis is represented in the form of self-replication—in such systems, the product acts as a template for (and therefore accelerates) its own formation. In order for a self-replicating system to work effectively, the following criteria must be met: (i) product inhibition resulting from formation of catalytically inactive dimerization of products must be minimized, (ii) intramolecular product formation leading to catalytically inactive species must be minimized, and (iii) the catalytic efficiency within the ternary complex must be maximized. In an elegant study by Philp (Figure 28),[73] the cycloaddition reaction between an *N*-aryl nitrone and a maleimide was employed together with computational screening to identify systems with the potential to undergo self-replication. The system makes use of the two point interaction between carboxylic acids and amidopyridines also utilized by Hamilton and described earlier in this chapter.[28] *Endo*-**107** with an extended structure was proposed as a template for its own formation from **104** and **105** while the *exo* isomer was proposed to yield the catalytically inactive intramolecular product. In the absence of molecular recognition, that is, using **106** which is incapable of hydrogen bonding to the amidopyridine no control over the *endo:exo* is observed for this reaction. In contrast, when the carboxylic acid **105** is used, molecular recognition takes place and an *endo:exo* ratio of 33 : 1 resulted. Further kinetic studies revealed that the formation of the *endo*- is faster for **105** than it is for **106** and that it is faster again when substoichiometric quantities of *endo*-**107** are added at the start of the reaction, proving that the system operates as a self-replicator.

3.3 Self-assembly of polymers using hydrogen bonds

Supramolecular polymers have emerged over the last two decades as an area of immense scientific interest.[74] At the simplest level—the potential to construct stimuli responsive materials from small, easy to synthesize and purify molecular components is enormously attractive and offers opportunities for error-checking synthesis of polymers under thermodynamic control, development of "green" easy to process materials, construction of modular bioactive and biodegradable materials, and elaboration of self-healing materials. Hydrogen-bonding receptors have played a major role in the development of this field, and a number of methods for forming main chain polymers and cross-linked polymers held together by hydrogen bonds have been developed. The first use of arrays of hydrogen bonds in the assembly of supramolecular polymers was reported by Lehn and coworkers in 1993, who described the assembly of polymeric, liquid crystalline, and lyotrophic mesophases using 2,6-diamidopyridine **108** and thymine functionalized **109** building blocks (Figure 29).[75,76]

It was not until 1997, when the UPy unit was discovered, that it became possible to assemble supramolecular polymers in dilute solution.[77] This work highlighted

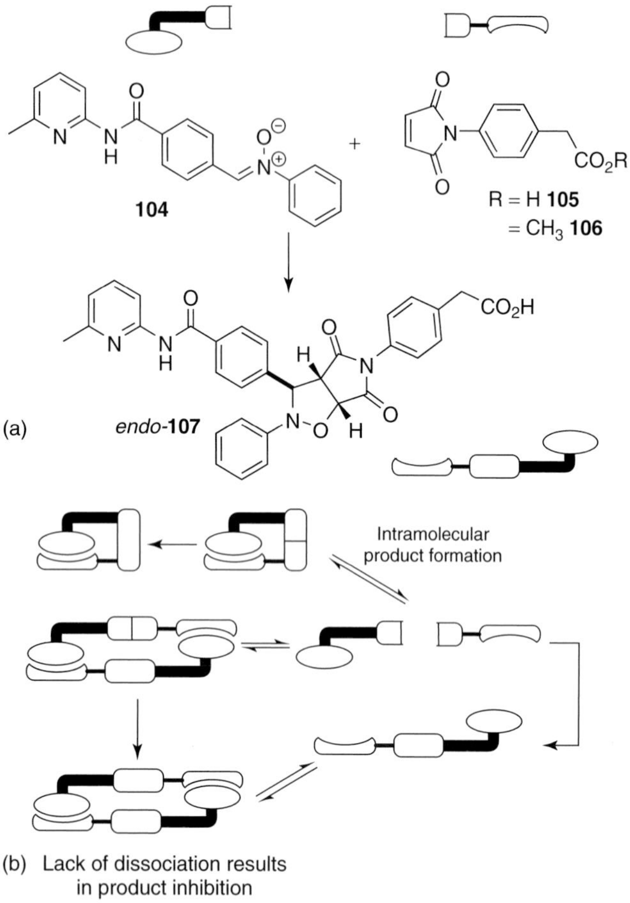

Figure 28 (a) Molecular structures **104–107** used to create (b) a minimal self-replicating cycle.

how linear polymers could be constructed using hydrogen bonds in the same way and obeying the same principles as traditional condensation polymers. In this seminal work, a series of bifunctional UPy motifs **110** were shown to exhibit concentration and temperature-dependent viscosity changes—indicative of polymer self-assembly in solution (Figure 30).

In terms of forming more complex architectures, for example, supramolecular block copolymers, heterodimerizing motifs are more desirable. Lehn and coworkers have published a series of papers on the use of barbiturate/cyanurate—isophthaloyldiamidopyridine as a connecting unit for supramolecular polymers (Figure 31) **111** and **112**.[78,79] The use of this supramolecular motif is attractive because although the interaction is of lower affinity (e.g., compared to the interaction of UPy), it is well defined. In apolar solvents, the two bifunctional monomers were shown to form AA-BB-type supramolecular polymers; rigid fibers were observed by electron microscopy in toluene, while small-angle neutron scattering in decane demonstrated gel-like systems resulted at low concentration from formation of rigid fibers.

Binder and coworkers have also used this motif to induce microphase separation in strongly phase separating poly(etherketone) PEK and poly(isobutylene) PIB (Figure 32).[80] Importantly, they compared the properties of the resulting mixtures against those formed with triazine functionalized PEK and thymine functionalized PIB. For the weakly hydrogen-bonding triazine–thymine linked polymers, irreversible phase separation occurred at the glass transition temperature of the PEK (145 °C), whereas for the barbiturate/cyanurate—isophthaloyldiamidopyridine, (compounds **113** and **114**) only minor and reversible changes in the structure were observed up to 230 °C. This study illustrates beautifully the importance of well-defined high-affinity hydrogen bonding in defining material's properties (Figure 32).

3.4 The use of hydrogen-bonding receptors in biological settings

A significant proportion of the reported receptors that recognize their guests through hydrogen bonding have been targeted toward molecules of biological interest; however, this is a little different from the development of functional molecules. This section provides additional examples of receptors that function in a real setting to achieve sensing—in part, the difficulty relates to the challenge of recognition in water described earlier where hydrogen bonding alone is often insufficient to achieve the necessary affinities to be of practical interest. One area where considerable success has been achieved is in the development of artificial ion channels and transporters: here recognition takes place within the more favorable environment of the membrane or at the membrane–water interface; examples include the anion receptors discussed earlier (Figure 15). Pertinent examples of sensors where recognition in water is achieved that have already been discussed include the tetrapeptide sensors developed by Schmuck (Figure 22) and the citrate and gallate receptors from the Anslyn group (Figure 23)—in these cases, a sensing element is either covalently built into the receptor or used as a noncovalently bound indicator as the signaling element. Two further examples are discussed here—one of the simplest examples is in the application of linear arrays to the detection of mismatches in DNA. Nakatani has employed a simple bivalent DDA array **115** (Figure 33b) for the recognition of G–G mismatches in duplex DNA[81]—the approach works because the receptor can fill the bulge created by two mismatch bases and hydrogen bonding is augmented by hydrophobic shielding and π–π stacking interactions (Figure 33a). Binding can be detected by the change in melting temperature of the target oligonucleotide sequence or through a sensor element attached to the central amino

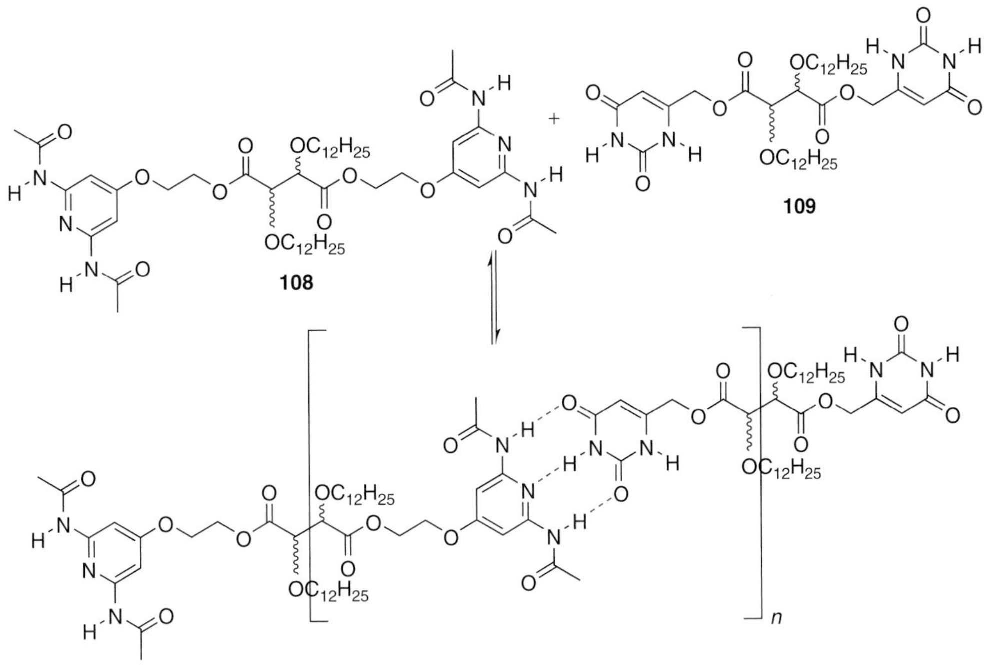
108
109
n

Figure 29 Self-assembly of polymeric systems using **108** and **109**.

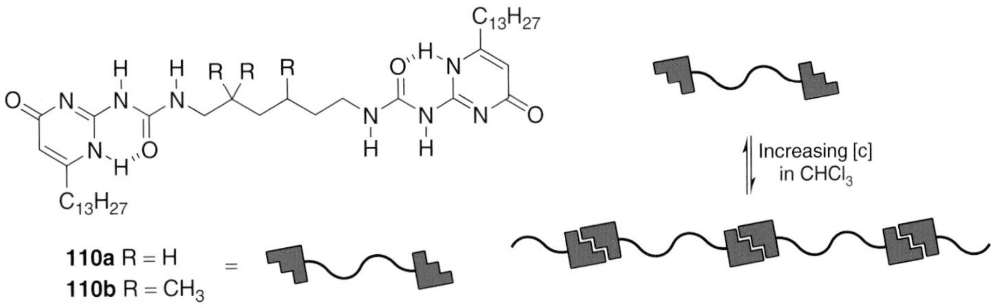
110a R = H
110b R = CH3
Increasing [c]
in CHCl3

Figure 30 Structure and cartoon illustrating the assembly of a supramolecular polymer using self-complementary ditopic motif **110**.

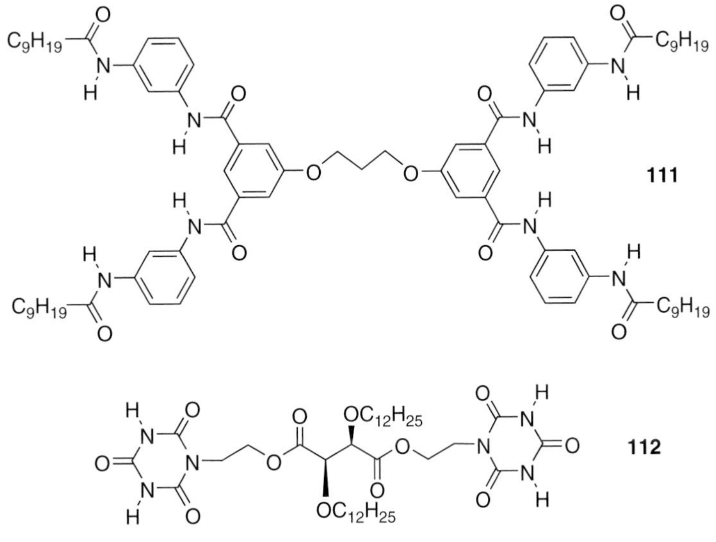
111
112

Figure 31 Self-assembly of polymeric systems using **111** and **112**.

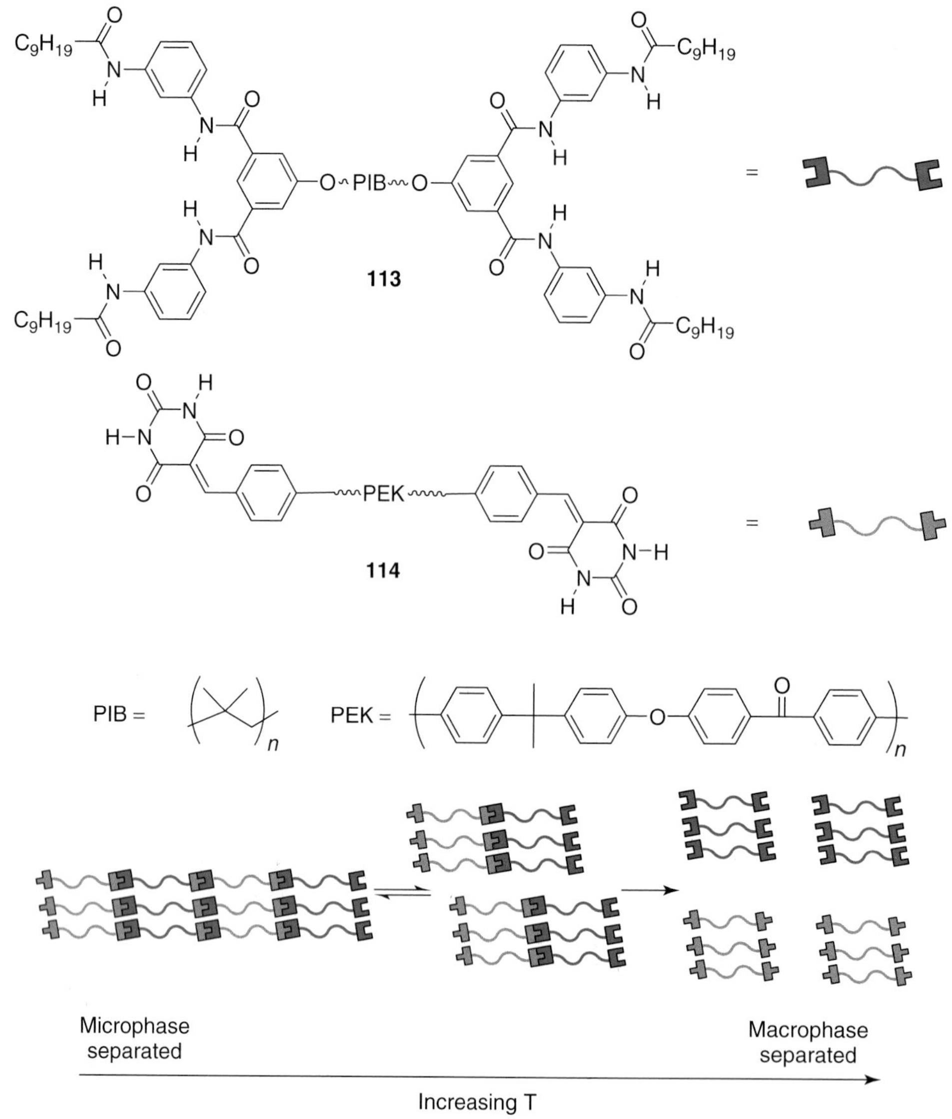

Figure 32 Self-assembly of supramolecular polymers **113** and **114** together with cartoon depicting temperature-dependent micro/macrophase separation behavior.

group of the receptors. In the last example, Zimmerman and coworkers have also recently exploited a fluorescently labeled Janus-type linear array for the detection of T-T or U-U mismatches (Figure 33c).[82] The approach is similar in that complementary hydrogen bonding along both faces of the Janus wedge to the linear array on the target nucleobase is augmented by hydrophobic shielding and $\pi-\pi$ stacking interactions. Binding can be detected by the change in melting temperature of the target oligonucleotide sequence. Importantly, binding of the small molecule was shown to prevent binding of a protein to the mismatch site—this DNA protein interaction had previously been shown to play a key regulatory role in RNA splicing and the development of myotonic dystrophy type I.

4 CONCLUSION

The last 20–30 years have seen enormous progress in our understanding and use of hydrogen bonding within the context of supramolecular chemistry, in particular host–guest chemistry. The breadth of appropriate published research worthy of inclusion far exceeds that which can be covered here. Instead, the author has within this chapter attempted to illustrate the key concepts and step changes that have occurred within this area to place the community at the point where it can exploit hydrogen bonding in a functional setting. In tandem with our understanding of other noncovalent interactions, it is now possible to design receptors that can target a range of guests in competitive

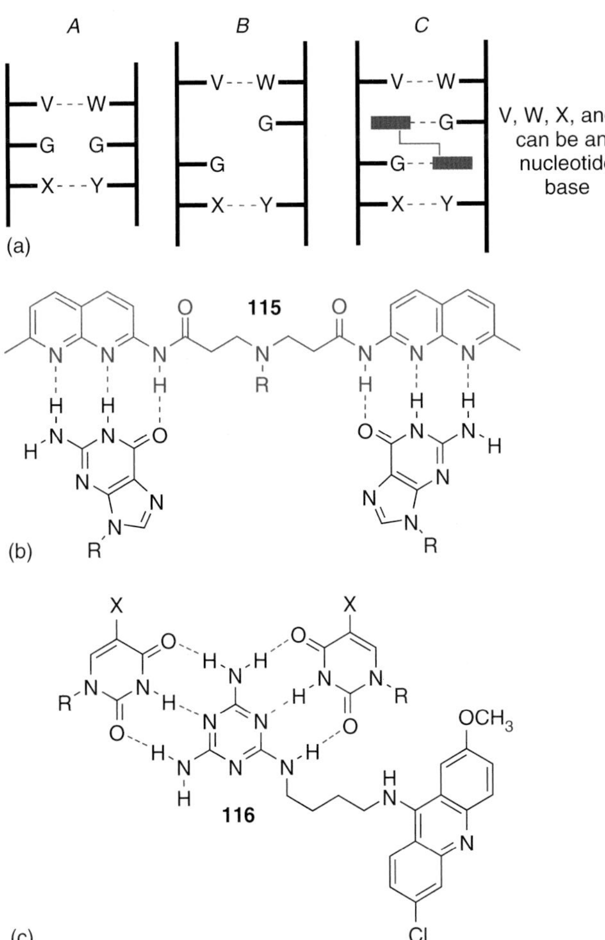

Figure 33 (a) Illustration of a duplex containing a G-G mismatch A, represented as two consecutive guanine bulges B, intercalated with a dimeric synthetic base pair mimic C. (b) Dimeric synthetic DNA base pair mimic **115** and in its proposed hydrogen-bonding interaction with guanine, and (c) Janus wedge molecule **116** in its proposed hydrogen-bonding interaction with thymine/cytosine.

media and exploit such receptors for simple applications including assembly of complex architectures, sensing, and catalysis. Aside from an ongoing requirement to progress in these areas, enormous challenges remain including the requirement to develop hydrogen-bonding receptors capable of self-sorting[83] within a "systems" setting where high fidelity[84] orthogonal and promiscuous receptors are required.

ACKNOWLEDGMENTS

The author wishes to thank current and former members of his research group for their outstanding contributions and ongoing commitment to his research program.

REFERENCES

1. G. A. Jeffrey, *An Introduction to Hydrogen-Bonding*, Oxford University Press, New York, 1997.
2. L. J. Prins, D. N. Reinhoudt, and P. Timmerman, *Angew. Chem. Int. Ed.*, 2001, **40**, 2382.
3. C. A. Hunter, *Angew. Chem. Int. Ed.*, 2004, **43**, 5310.
4. S. K. Chang, D. Van Engen, E. Fan, and A. D. Hamilton, *J. Am. Chem. Soc.*, 1991, **113**, 7640.
5. J. Rebek Jr, *Acc. Chem. Res.*, 1990, **23**, 399.
6. A. D. Hamilton and D. Van Engen, *J. Am. Chem. Soc.*, 1987, **109**, 5035.
7. A. J. Wilson, *Soft Matter*, 2007, **3**, 409.
8. F. H. Beijer, R. P. Sijbesma, H. Kooijman, *et al.*, *J. Am. Chem. Soc.*, 1998, **120**, 6761.
9. P. S. Corbin and S. C. Zimmerman, *J. Am. Chem. Soc.*, 1998, **120**, 9710.
10. X.-Z. Wang, X.-Q. Li, X.-B. Shao, *et al.*, *Chem. Eur. J.*, 2003, **9**, 2904.
11. F. H. Beijer, H. Kooijman, A. L. Spek, *et al.*, *Angew. Chem. Int. Ed.*, 1998, **37**, 75.
12. C.-H. Chien, M.-K. Leung, J.-K. Su, *et al.*, *J. Org. Chem.*, 2004, **69**, 1866.
13. A. M. McGhee, C. Kilner, and A. J. Wilson, *Chem. Commun.*, 2008, 344.
14. W. L. Jorgensen and J. Pranata, *J. Am. Chem. Soc.*, 1990, **112**, 2008.
15. J. Sartorius and H.-J. Schneider, *Chem. Eur. J.*, 1996, **2**, 1446.
16. S. Djurdjevic, D. A. Leigh, H. McNab, *et al.*, *J. Am. Chem. Soc.*, 2007, **129**, 476.
17. H. Adams, F. J. Carver, C. A. Hunter, *et al.*, *Angew. Chem. Int. Ed. Engl.*, 1996, **35**, 1542.
18. B. Gong, Y. Yan, H. Zeng, *et al.*, *J. Am. Chem. Soc.*, 1999, **121**, 5607.
19. G. Cooke and V. M. Rotello, *Chem. Soc. Rev.*, 2002, **31**, 275.
20. Y.-M. Legrand, M. Gray, G. Cooke, and V. M. Rotello, *J. Am. Chem. Soc.*, 2003, **125**, 15789.
21. Y. Ge, R. Lilienthal, and D. K. Smith, *J. Am. Chem. Soc.*, 1996, **118**, 3976.
22. A. Goodman, E. C. Breinlinger, M. Ober, and V. M. Rotello, *J. Am. Chem. Soc.*, 2001, **123**, 6213.
23. B. J. B. Folmer, E. Cavini, R. P. Sijbesma, and E. W. Meijer, *Chem. Commun.*, 1998, 1847.
24. D. H. Williams and M. S. Westwell, *Chem. Soc. Rev.*, 1998, **27**, 57.
25. P. C. Weber, D. H. Ohlendorf, J. J. Wendoloski, and F. R. Salemme, *Science*, 1989, **243**, 85.
26. C. A. Hunter, *Chem. Commun.*, 1991, 749.
27. H. Adams, F. J. Carver, C. A. Hunter, and N. J. Osborne, *Chem. Commun.*, 1996, 2529.
28. F. Garcia-Tellado, S. Goswami, S. K. Chang, *et al.*, *J. Am. Chem. Soc.*, 1990, **112**, 7393.

29. T. R. Kelly and M. P. Maguire, *J. Am. Chem. Soc.*, 1987, **109**, 6549.
30. T. W. Bell and J. Liu, *J. Am. Chem. Soc.*, 1988, **110**, 3673.
31. C. Y. Hung, T. Hopfner, and R. P. Thummel, *J. Am. Chem. Soc.*, 1993, **115**, 12601.
32. C.-Y. Huang, L. A. Cabell, and E. V. Anslyn, *J. Am. Chem. Soc.*, 1994, **116**, 2778.
33. S. C. Zimmerman, W. Wu, and Z. Zeng, *J. Am. Chem. Soc.*, 1991, **113**, 196.
34. C. J. Pedersen, *Angew. Chem. Int. Ed. Engl.*, 1988, **27**, 1021.
35. D. J. Cram and J. M. Cram, *Acc. Chem. Res.*, 1978, **11**, 8.
36. J. M. Timko, S. S. Moore, D. M. Walba, *et al.*, *J. Am. Chem. Soc.*, 1977, **99**, 4207.
37. X. X. Zhang, J. S. Bradshaw, and R. M. Izatt, *Chem. Rev.*, 1997, **97**, 3313.
38. P. R. Ashton, E. J. T. Chrystal, P. T. Glink, *et al.*, *Chem. Eur. J.*, 1996, **2**, 709.
39. A. R. Sanford, L. Yuan, W. Feng, *et al.*, *Chem. Commun.*, 2005, 4720.
40. K. Yamato, L. Yuan, W. Feng, *et al.*, *Org. Biomol. Chem.*, 2009, **7**, 3643.
41. P. A. Gale, J. L. Sessler, V. Kral, and V. Lynch, *J. Am. Chem. Soc.*, 1996, **118**, 5140.
42. W.-S. Cho and J. L. Sessler, in *Functional Synthetic Receptors*, eds. T. Schrader and A. D. Hamilton, Wiley, Weinheim, 2005, 165.
43. J. L. Sessler and D. Seidel, *Angew. Chem. Int. Ed.*, 2003, **42**, 5134.
44. K. Kavallieratos, S. R. de-Gala, D. J. Austin, and R. H. Crabtree, *J. Am. Chem. Soc.*, 1997, **119**, 2323.
45. P. V. Santacroce, J. T. Davis, M. E. Light, *et al.*, *J. Am. Chem. Soc.*, 2007, **129**, 1886.
46. X. Li, B. Shen, X.-Q. Yao, and D. Yang, *J. Am. Chem. Soc.*, 2007, **129**, 7264.
47. S. O. Kang, J. M. Llinares, D. Powell, *et al.*, *J. Am. Chem. Soc.*, 2003, **125**, 10152.
48. S. Camiolo, P. A. Gale, G. J. Tizzard, *et al.*, *J. Org. Chem.*, 2001, **66**, 7849.
49. E. Fan, S. A. Van Arman, S. Kincaid, and A. D. Hamilton, *J. Am. Chem. Soc.*, 1993, **115**, 369.
50. J. P. Clare, A. J. Ayling, J.-B. Joos, *et al.*, *J. Am. Chem. Soc.*, 2005, **127**, 10739.
51. M. Berger and F. P. Schmidtchen, *Angew. Chem. Int. Ed.*, 1998, **37**, 2694.
52. S. Rensing and T. Schrader, *Org. Lett.*, 2002, **4**, 2161.
53. T. Schrader, *Tetrahedron Lett.*, 1998, **39**, 571.
54. A. Echavarren, A. Gálan, J.-M. Lehn, and J. de Mendoza, *J. Am. Chem. Soc.*, 1989, **111**, 4994.
55. A. Gálan, D. Andreu, A. M. Echavarren, *et al.*, *J. Am. Chem. Soc.*, 1992, **114**, 1511.
56. O. Molt, D. Rübeling, and T. Schrader, *J. Am. Chem. Soc.*, 2003, **125**, 12086.
57. C. Allott, H. Adams, P. L. Bernad, *et al.*, *Chem. Commun.*, 1998, 2449.
58. Y. Ferrand, M. P. Crump, and A. P. Davis, *Science*, 2007, **318**, 619.
59. C. L. Kielkopf, S. White, J. W. Szewczyk, *et al.*, *Science*, 1998, **282**, 111.
60. S. White, J. W. Szewczyk, J. M. Turner, *et al.*, *Nature*, 1998, **391**, 468.
61. J. Lagona, P. Mukhopadhyay, S. Chakrabarti, and L. Isaacs, *Angew. Chem. Int. Ed.*, 2005, **44**, 4844.
62. C. Schmuck and L. Geiger, *J. Am. Chem. Soc.*, 2004, **126**, 8898.
63. C. Schmuck and L. Gieger, *J. Am. Chem. Soc.*, 2005, **127**, 10486.
64. C. Schmuck, M. Heil, J. Scheiber, and K. Baumann, *Angew. Chem. Int. Ed.*, 2005, **44**, 7208.
65. C. Schmuck and P. Wich, *Angew. Chem. Int. Ed.*, 2006, **45**, 4277.
66. A. Metzger and E. V. Anslyn, *Angew. Chem. Int. Ed.*, 1998, **37**, 649.
67. S. L. Wiskur and E. V. Anslyn, *J. Am. Chem. Soc.*, 2001, **123**, 10109.
68. P. R. Ashton, P. T. Glink, J. F. Stoddart, *et al.*, *Chem. Eur. J.*, 1996, **2**, 729.
69. D. A. Leigh, K. Moody, J. P. Smart, *et al.*, *Angew. Chem. Int. Ed. Engl.*, 1996, **35**, 306.
70. T. J. Kidd, D. A. Leigh, and A. J. Wilson, *J. Am. Chem. Soc.*, 1999, **121**, 1599.
71. C. Waloch, J. Wieland, M. Keller, and B. Breit, *Angew. Chem. Int. Ed.*, 2007, **46**, 3037.
72. A. Bauer, F. Westkämper, S. Grimme, and T. Bach, *Nature*, 2005, **436**, 1139.
73. E. Kassianidis and D. Philp, *Angew. Chem. Int. Ed.*, 2006, **45**, 6344.
74. L. Brunsveld, B. J. B. Folmer, E. W. Meijer, and R. P. Sijbesma, *Chem. Rev.*, 2001, **101**, 4071.
75. M. Kotera, J.-M. Lehn, and J. P. Vigneron, *J. Chem. Soc., Chem. Commun.*, 1994 197.
76. T. Gulick-Krymicki, A. M. Fouquey, and J.-M. Lehn, *Proc. Natl. Acad. Sci. U. S. A.*, 1993, **90**, 163.
77. R. P. Sijbesma, F. H. Beijer, L. Brunsveld, *et al.*, *Science*, 1997, **278**, 1601.
78. E. Kolomiets, E. Buhler, S. J. Candau, and J.-M. Lehn, *Macromolecules*, 2006, **39**, 1173.
79. V. Berl, M. Schmutz, M. J. Krische, *et al.*, *Chem. Eur. J.*, 2002, **8**, 1227.
80. W. H. Binder, S. Bernstorff, C. Kluger, *et al.*, *Adv. Mater.*, 2005, **17**, 2824.
81. K. Nakatani, S. Sando, and I. Saito, *Nat. Biotechnol.*, 2001, **19**, 51.
82. J. F. Arambula, S. R. Ramisetty, A. M. Baranger, and S. C. Zimmerman, *Proc. Natl. Acad. Sci. U. S. A.*, 2009, **106**, 16068.
83. A. Wu and L. Isaacs, *J. Am. Chem. Soc.*, 2003, **125**, 4831.
84. T. Park and S. C. Zimmerman, *J. Am. Chem. Soc.*, 2006, **128**, 14236.

Boronic Acid-Based Receptors

John S. Fossey[1] and Tony D. James[2]

[1] *University of Birmingham at Birmingham, Birmingham, UK*
[2] *University of Bath, Bath, UK*

1 INTRODUCTION

The challenge of recognizing a molecular analyte with a synthetically prepared receptor has inspired many supramolecular chemists. Research in this area has been instrumental in elucidating the mechanisms of many biological events encompassing recognition and catalysis. The significance of this work was underlined with the awarding of the Nobel Prize in Chemistry to Cram, Lehn, and Pederson in 1987 "for their development and use of molecules with structure-specific interactions of high selectivity." Since then, the diversity of compounds studied within the remit of supramolecular chemistry has significantly grown. Of particular interest are molecular sensors (chemosensors), single molecules that can both recognize and signal analyte presence in real time.

Supramolecular Chemistry: From Molecules to Nanomaterials.
Edited by Philip A. Gale and Jonathan W. Steed.
© 2012 John Wiley & Sons, Ltd. ISBN: 978-0-470-74640-0.

Molecular recognition underpins all sensor chemistry; the process involves the interaction between two substances. Importantly, recognition is not just defined as a binding event but requires an element of selectivity between the guest and the host. Optimal selectivity occurs between compounds with carefully matched electronic, geometric, and polar elements. For synthetic receptors, the potential exists to engineer receptors for a chosen analyte through selective structural design and complementary functional groups. The power of this concept is illustrated within nature, where biological systems have evolved with exquisite binding sites, sequestering guest molecules with high selectivity.

A receptor becomes a sensor when a channel of communication is established between the receptor and the "outside world." For a sensor to function, it must allow selective binding between the host and the guest and report these binding events by generating an output signal. Sensors have the potential to relay information concerning the presence and location of important species in a potentially quantifiable manner, acting as a conduit for information about events occurring at the molecular level.

Chemical sensors may be categorized as either biosensors or synthetic chemosensors. Biosensors make use of existing biological units for recognition. Many physiologically important analytes already have corresponding biological receptors with intrinsically selectivity; if these receptors can be coupled to a signal transducer, a biosensor can be developed.

Synthetic chemosensors incorporate a synthetically manufactured element for recognition. While biomimetic receptors have been prepared, with synthetic receptors mimicking the active sites of naturally occurring biological molecules, synthetic receptors can be designed entirely from first principles.

The development of coherent strategies for the selective binding of analyte molecules, by rational design of synthetic receptors, remains one of chemistry's most challenging goals. Research conducted to this end is driven by a fundamental curiosity and the need to monitor compounds of industrial, environmental, and biological importance.

2 BORONIC ACIDS

Boronic acids, both alky and aryl, have been available for around 150 years. Currently, the most common method used to prepare boronic acids is the reaction of trialkyl borates with Grignard reagents. While boronic acids are widely used in synthetic chemistry, the aim of this chapter is to introduce the use of boronic acids in molecular recognition and supramolecular chemistry.

The interaction of boronic acids with saccharides[1–11] and anions[12,13] has been extensively investigated, and boronic acids have also been exploited in a range of applications as diverse as sensing and separation,[14] NMR (nuclear magnetic resonance) shift reagents,[15–18] functional polymers for electrophoresis,[19,20] and molecular self-assembled capsules and materials.[21–25]

2.1 Boron–diol interaction

Both boric and boronic acids have been used to determine the configuration of saccharides for around 50 years. These early investigations determined that cyclic boric and boronic esters are formed when boronic acids are mixed with polyhydroxylated compounds such as sugars. A seminal paper by Lorand and Edwards contains the first quantitative investigation of the interactions between boronic acids and polyols.[26] A key observation of Lorand and Edwards was that the conjugate base of phenylboronic acid has a tetrahedral, rather than trigonal structure (Scheme 1). For phenylboronic acid, the acidity arises from the dissociation of a proton from an associated water molecule. When phenylboronic acid and water react to generate a tetrahedral boronate, a hydrated proton is liberated, thereby defining the acidity constant K_a'. Typically, the pK_as of phenylboronic acid ranges between ~8.7 and 8.9, and is 8.70 in water at 25 °C.[27]

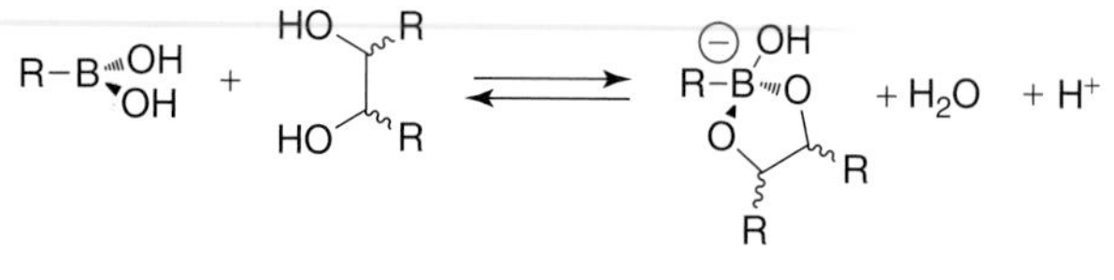

Scheme 1 The rapid and reversible formation of a cyclic boronate ester between a boronic acid and 1.2-diol.

2.2 Boron–nitrogen interaction

The recognition of saccharides using the esterification with boronic acids is facilitated by the interaction between *o*-methylphenylboronic acid and a proximal tertiary amine. The precise nature of the Lewis acid–base interaction (N···B) has been the subject of some controversy. However, the simple fact that the proximal amine has a positive effect on the binding efficiency of boronic acids is not in debate. What, however, is in debate is the nature of the (N···B) interaction.[8,9,27–29] So what are the positive effects of the proximal amine? (i) The interaction between a boronic acid and proximal amine reduces the pK_a of boronic acid,[30] allowing binding to occur at neutral pH. (ii) The contraction of the O–B–O bond angle upon complex formation with a saccharide and the concomitant increase in acidity at the boron atom. The increase in acidity of the already Lewis acidic boron enhances the N···B interaction, which in turn influences the fluorescence. A reduction in pK_a at boron on saccharide binding has the overall effect of modulating fluorescence intensity.

An extensive study of compounds with coordinative N–B bonds concluded that steric interactions along with ring strain (in the case of cyclic diesters) weakens and elongates the N–B bond, which is accompanied by a reduction in the tetrahedral geometry of the boron atom.[31] The *N*-methyl-*o*-(phenylboronic acid)-*N*-benzylamine system has been used as a model system by a number of groups to investigate the nature of the (N···B) interaction.[27–29,32–34] Through these investigations, it was shown that the upper and lower limits of the N–B interaction lie between 15 and 25 kJ mol^{-1}.[27] These numbers agree well with computational analyses, which estimate the N–B interaction to be 13 kJ mol^{-1} or less in the absence of solvent.[32] To help in the characterization of the nature of these N–B interactions using a more familiar bonding regime, the N–B interaction, in these systems, is kept at approximately the same energy as that in a hydrogen bond. This needs to be contrasted with the experimentally and computationally derived dative N–B bond energy of 58–152 kJ mol^{-1}.

So what is the structure of the (N···B) interaction? The question was answered by Anslyn, who performed detailed structural investigations of the N–B interaction in model *o*-(*N*,*N*-dialkyl aminomethyl) arylboronate systems.[29] Using ^{11}B NMR measurements (and X-ray crystallographic data), it was shown that, in aprotic media, the dative N–B bond is usually present. However, in protic solvent, insertion of the solvent into the N–B *bond* occurs to afford a hydrogen-bonded zwitterionic species owing to the investigations of Anslyn[29] and other groups.[27–29,32,33] The N–B interaction can in many cases now be ascribed to a hydrogen-bonding interaction mediated through a bound solvent molecule (Figure 1).

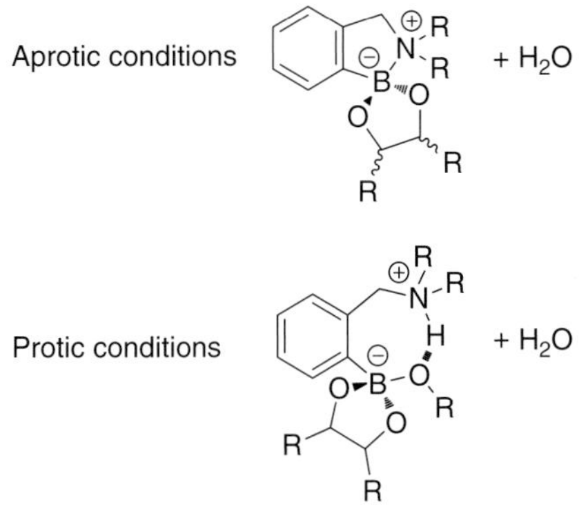

Figure 1 The N–B interaction in aprotic and protic conditions.

2.3 Boron–anion interactions

Anions are involved in many fundamental processes in all living things. Recognition, transport, and concentration control of anions such as chloride, phosphate, and sulfate is carried out by biological systems on a continual cycle. Fluoride, nitrate, and pertechnetate are potentially dangerous contaminants that can gain access to our water systems by various means.

The boron atom of trisubstituted boron species possesses an sp^2 trigonal planar geometry with an empty p orbital perpendicular to the plane of the molecule. This feature dominates both the synthetic and receptor chemistry of boron compounds. Nucleophiles are able to interact with or donate into this vacant site, causing a subsequent change in geometry and hybridization. The tetrahedral nature of the phenylboronate anion was confirmed by Lorand and Edwards[26] (Scheme 2).

A significant contribution to anion recognition chemistry—particularly pertinent to the chemistry disclosed in this chapter—came in 1967 when Shriver and Biallas identified the complex formed between the bidentate Lewis acid (Figure 2) and the methoxide anion.[35]

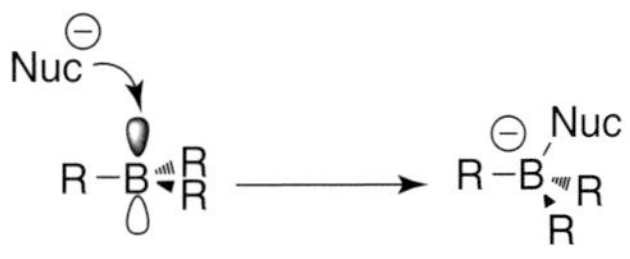

Scheme 2 Diagram showing the change in geometry undergone at the boron center when the vacant p orbital is filled by an attacking nucleophile.

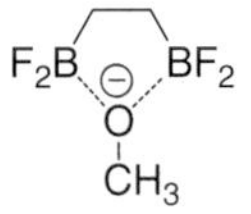

Figure 2 First-known example of a bisboron compound binding an anion.

This was a seminal piece of work given that, over 40 years later, molecules with single, dual, or multiple Lewis acidic sites (including those with mercury, tin, germanium, and silicon centers in addition to the more commonly employed boron) remain attractive recognition units.

The relatively weak Lewis acidity of the boron center creates a wealth of synthetic chemistry but also allows boron to act as a receptor for hard anions, particularly fluoride, hydroxide, and cyanide. The use of boron as a Lewis acid extends to formation of coordinate bonds with a wide variety of heteroatoms including oxygen, sulfur, phosphorus, and nitrogen.

3 SACCHARIDE RECEPTORS

The primary interaction of a boronic acid with a diol is covalent and involves the reversible and rapid formation of a cyclic boronate ester. An array of hydroxyl groups presented by saccharides provides an ideal architecture for these interactions and has led to the development of boronic acid-based sensors for saccharides (Scheme 1).

Many synthetic receptors developed for neutral guests have relied on noncovalent interactions, such as hydrogen bonding, for recognition, yet in aqueous systems neutral guests may become heavily solvated. Biological systems are able to expel water from their binding pockets to sequester analytes, using noncovalent interactions, until recently synthetic monomeric receptors using hydrogen bonding were not able to compete with water (solvent) for low concentrations of monosaccharides.[36] Major progress has been made in this area by Davis, who reported a hydrogen-bonding receptor capable of binding D-glucose in water with a weak but significant stability constant.[37] Continuing these excellent development, Davis has recently been able to develop synthetic lectin analogs for biomimetic disaccharide,[38] *O*-linked *β*-*N*-acetylglucosamine,[39] and *β*-glucosyl (also D-glucose)[40] recognition, all of which work very well in water.[41]

The most popular class of the boronic acid-based sensors utilize an amine group proximal to boron coupled to a fluorescence output. The Lewis acid–Lewis base interaction between the boronic acid and the tertiary amine has a dual role. Firstly, it enables molecular recognition at neutral pH. Secondly, it can be used to signal binding by modulating the intensity of fluorescence emissions.

3.1 Fluorescent sensors

Optical signals convey information through space and fluorescent sensors can be used in dynamic systems, such as living tissue and relay information remotely. Submillisecond response times are usual, allowing information to be communicated essentially in real time. If targeted correctly, fluorophores can be located with subnanometer accuracy, in effect permitting real-space monitoring. Fluorescence also demonstrates exceptionally high sensitivity; under controlled conditions, detection of responses from single fluorescent molecules and, in the case of fluorescent sensors, from single guest molecules.

As fluorescent sensors are capable of reporting a wealth of physical information at low concentrations (micromolar concentrations are typical), they can operate with the minimum disruption to the system being investigated. From an analytical perspective, these characteristics are attractive, and commercially the tiny quantities of compound required can offset synthetic costs.

Fluorescent sensors can be found in many recent analytical advances, such as the continuous monitoring systems developed by immobilizing fluorescent sensors onto fiber-optic sensing arrays, or the live imaging of analytes within cells through confocal microscopy. Commercially available fluorescent sensors include clinical tools such as the blood gas analyzers that are now commonplace within hospital high-dependency wards and ambulances allowing point-of-care diagnostic monitoring, or the glucose-responsive contact lenses currently being pioneered by the Lakowicz research group.[42] These examples underline the robust and adaptable nature of fluorescent sensors, which in turn permit rapid and accurate analyte detection by portable devices.

The use of boronic acids in the development of fluorescent sensors for saccharides is a comparatively new field (Scheme 3). Following the first report by Yoon and Czarnik[43] D-glucose selectivity was achieved in 1994 by James *et al.*[44,45] A year later, this was followed up by enantioselective saccharide recognition.[46] The intervening years have seen the field grow to the point where hundreds of publications now report on boronic acid–saccharide recognition.[1–11]

As mentioned above, the first fluorescent sensor for saccharides was reported by Yoon and Czarnik.[43] The internal charge transfer (ICT) sensor **1** consisted of a boronic acid fragment directly attached to anthracene. On addition of saccharide, it was noted that the intensity of the fluorescence emission for the 2-anthrylboronic acid **1** was reduced by ~30%. This change in fluorescence emission intensity is ascribed to the change in electronics that accompanies rehybridization at boron. For boronic acid **1** (below its pK_a), the neutral sp^2 hybridized boronic acid displayed a strong fluorescence emission (above its pK_a) and the anionic sp^3 boronate displayed a reduction in the intensity of fluorescence emission.

1 **2**

The formation of a boronic acid–saccharide complex acidifies the boron atom, making the resultant boronic ester more acidic than the initial uncomplexed boronic acid. In this instance, a pK_a of 8.8 was reported for the neutral 2-anthrylboronic acid and a pK_a' of 5.9 was reported for the 2-anthrylboronic acid complex formed in saturated fructose solution. Exploiting this phenomenon, the system was buffered to a pH of 7.4, a value between the corresponding pK_a and pK_a' values reported. With this constraint in place, a high-fluorescence emission intensity was observed from the uncomplexed boronic acid ($pH < pK_a$). However, under these buffered conditions, addition of a saccharide to the solution formed the boronic ester, lowering the acidity of the boronic species below the pH of the solution ($pK_a' < pH$). As a direct result, the boronate anion was generated inducing a decrease in fluorescence observed on addition of saccharide. The corresponding isomer 9-anthrylboronic **2** was also examined but displayed smaller changes in fluorescence emission, a feature attributed to the unfavorable *peri*-interactions that would be expected at the 9-position.[47]

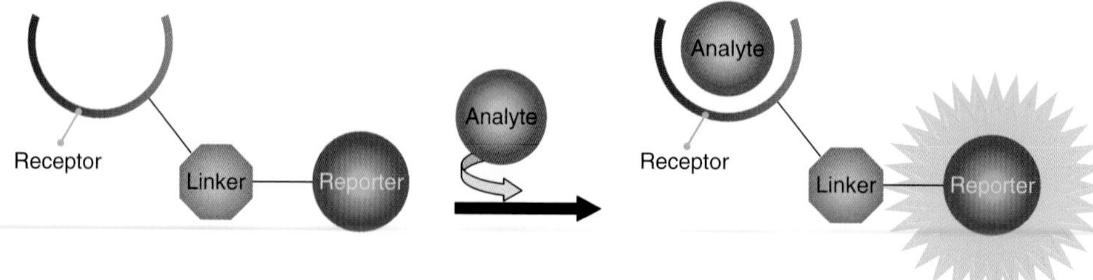

Scheme 3 The complementary interaction between a guest analyte and a host binding pocket, illustrated here by a red–orange guest analyte and blue host, allows selective binding to occur between two elements. Linking a unit capable of reporting this binding event converts the receptor into a sensor. In this cartoon, an optical "off–on" response is depicted from an appended fluorophore, illustrated in green.

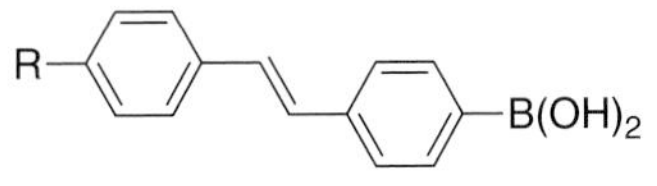

One of the best uses of ICT in boronic acid-based saccharide receptor designs comes from the Lakowicz group who have examined a range of stilbene boronic acid derivatives.[48,49] The neutral sp^2 hybridized boronic acid displays acceptor group properties and the anionic sp^3 hybridized boronic acid acts as a donor group. With 4′-(dimethylamino)stilbene-4-boronic acid **3**, the dimethylamino moiety is the donor group. When boron is sp^2 hybridized, and therefore an acceptor, excited-state ICT can occur between the amino donor and boron acceptor, red shifting the emission wavelength of the sp^2 species. On rehybridization of boron to sp^3, its acceptor properties are lost. This leads to a loss of the ICT effect in the excited state of the sp^3 species and shifts the emission wavelength of the fluorophore to higher energy. The inability of the sp^3 hybridized species to lower the energy of its excited state by a mechanism available to the sp^2 hybridized species causes a dramatic change in the properties of the emission band. On sp^2 to sp^3 interconversion, a 45-nm blue shift is induced in the emission wavelength coupled with an increase in the emission intensity. Conversely, with 4′-cyanostilbene-4-boronic acid **4**, the electron-withdrawing cyano moiety is the acceptor group. When boron is sp^2 hybridized and therefore is also an acceptor, no excited state ICT is feasible. On rehybridization of boron to an sp^3 form, the boron becomes a donor group, allowing ICT to occur between the boron donor and cyano acceptor, red shifting the emission wavelength of the sp^3 species.

Reversing the roles of the donor and acceptor groups for the boronic acid and 4′ moiety yielded values for the changes in emission wavelength and intensity for sensor **4** that were similar in magnitude to sensor **3** but occurred toward opposite ends of the electromagnetic spectrum. On interconversion of sp^2 to sp^3, a 40-nm red shift was induced in the emission wavelength coupled with a decrease in the emission intensity.

These ICT sensors have one drawback for potential real-world sensor development, which is pH sensitivity. In the case of fluorescent photo-induced electron transfer (PET) sensors, the interaction between *o*-methylphenylboronic acids (Lewis acids) and proximal tertiary amines (Lewis bases) has been exploited, while the precise nature of the amino base–boronic acid (N···B) interaction is debated.[8,9,27,29] The interaction between a boronic acid and proximal amine lowers the pK_a of the boronic and allows pH-insensitive binding over a wide range.

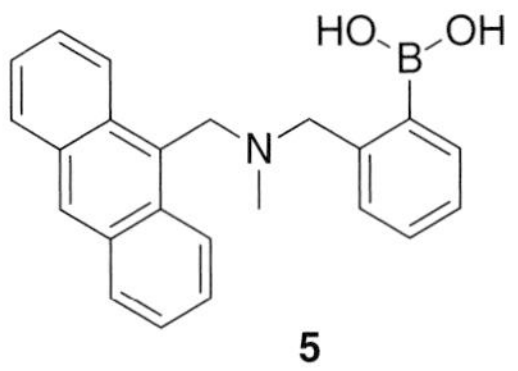

3: R = $N(CH_3)_2$
4: R = CN

5

The first fluorescent PET sensor **5** for saccharides to employ the *N*-methyl-*o*-(aminomethyl)phenylboronic acid fragment was reported by James *et al.*[44,50] Owing to the amino base–boronic acid (N···B) interaction sensor **5** behaves as "off–on" sensor, producing a large increase in fluorescence enhancement on addition of saccharide, as well as functioning over a broad pH range.

The naphthalimide fluorophore is of particular interest because of its relatively long excitation and emission wavelengths. Heagy was the first to work with the naphthalimide fluorophore.[51,52] Mohr prepared naphthalimide **6** producing fluorescence changes at long wavelength on addition of saccharides.[53] Wang has produced several long-wavelength sensor systems **7a–c** using the naphthalimide fluorophore. Sensor **7c** was the most water soluble and produced twofold fluorescence enhancement on saccharide binding.[54,55]

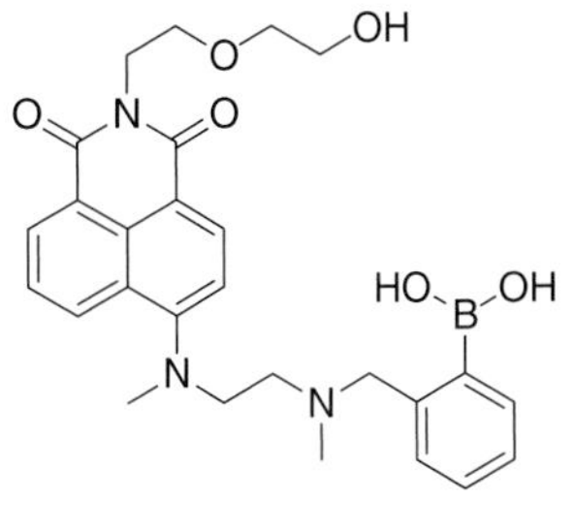

6

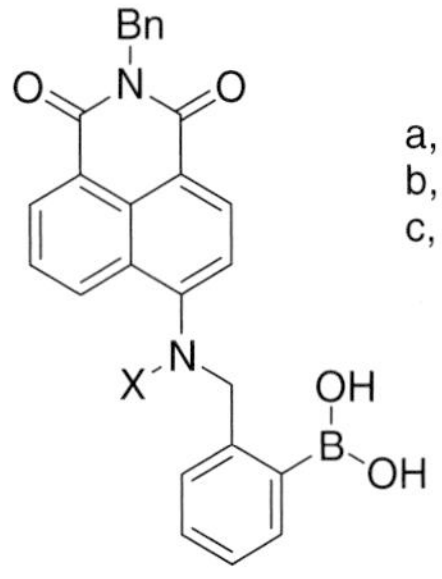

7 a, X = Bn; b, X = CH_3; c, X = H

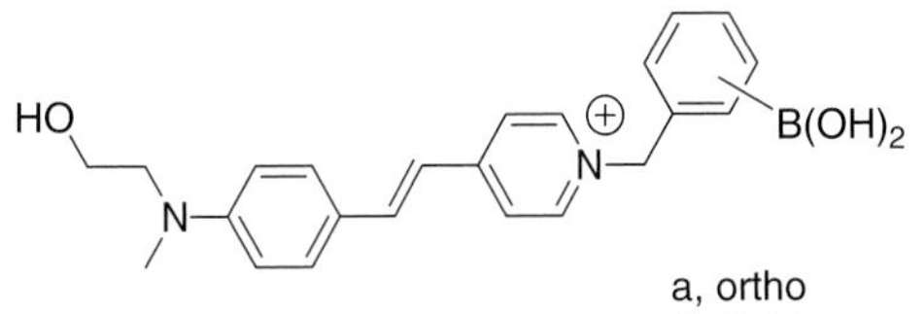

8 a, ortho; b, meta; c, para

In the search for a long-wavelength fluorescent boronic acid receptor, Mohr has prepared a series of fluorescent hemicyanine dyes receptors **8a–c**. Hemicyanine **8a** was the best receptor with a binding constant of 280 M^{-1}, while **8b** was 40 M^{-1} and **8c** was 200 M^{-1} with D-fructose.[56]

Zhao and James have developed some interesting reverse PET (d-PET) sensors, where excited-state electron transfer occurs from the fluorophore to the protonated amine/phenylboronic acid unit. Such sensors have advantages over standard PET systems (a-PET) in that they function better under acidic pH conditions.[57–59]

The D-fructose selective monoboronic acid-based sensor **5** was enhanced by James in 1995 with the introduction of a

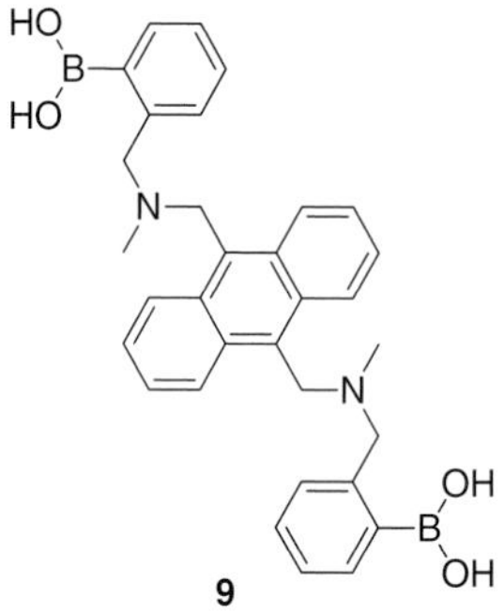

9

Figure 3 The first rationally designed boronic acid-based fluorescent PET sensor to display selectivity for D-glucose. The receptor–spacer–fluorophore–spacer–receptor assembly requires binding to occur at both receptors in order to restore fluorescence.

second boronic acid group to form the diboronic acid sensor **9** (Figure 3).[44,45] This receptor–spacer–fluorophore–spacer–receptor system retained the advantage of utilizing PET to modulate an "off–on" response to saccharides while introducing an advanced recognition site. The cooperative action of two boronic acid receptors permitted a number of possible binding modes to occur with saccharides. However, for fluorescence to be restored, both boronic acid moieties must be complexed, which requires formation of either an acyclic 2 : 1 or a cyclic 1 : 1 (saccharide/sensor) complex.

The modification proved successful and fortuitously the spacing of the two boronic acid groups provided an effective binding pocket for D-glucose. D-Glucose complexation occurred with a 1 : 1 stoichiometry with the saccharide binding to form a macrocyclic ring. While the inherent selectivity of monoboronic acids is for D-fructose, in this compound, the stabilization derived from the rigid macrocyclic ring produces a D-glucose selective system.

In refining the selectivity of boronic acid sensors for saccharides, the structure of the guest species must be addressed. Although there is a long history of research into the structural character of boronic acid–saccharide complexes, the rapid isomerization of monosaccharides in water precludes the description of a simple generic binding motif. The hemiacetal ring of a monosaccharide is readily cleaved in water, often reforming rings of different sizes and anomeric configurations. The equilibrium between linear, pyranose, and furanose configurations as well as the α and β anomers of the pyranose and furanose rings substantially increases the number of possible structures that may be formed on complexation.

Sensor **9** provided the first structural elucidation of a diboronic acid sensor with D-glucose complexed within the binding site. The ^{1}H NMR spectrum of this complex indicated that in deuterated methanol the D-glucose was bound in the α-pyranose form at the 1,2 and 4,6 positions, as in Figure 4.[44]

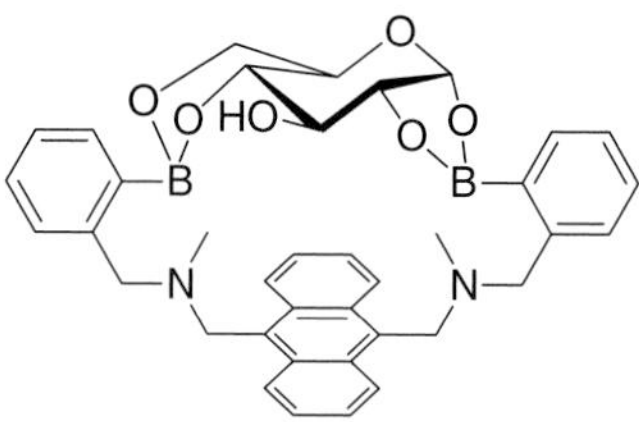

Figure 4 The initial 1,2 : 4,6 complex formed between sensor **9** and D-glucose in CH_3OD.

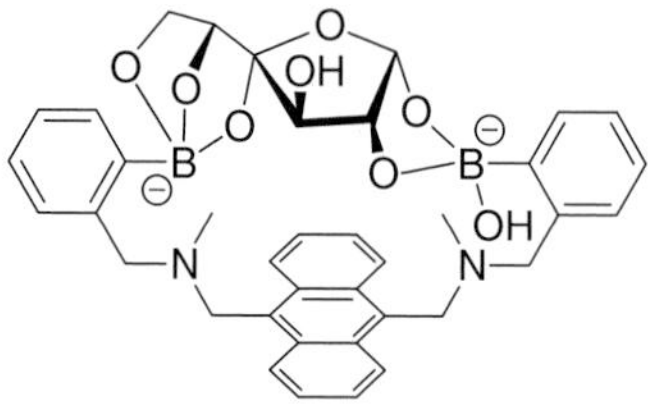

Figure 5 The thermodynamically stable 1,2 : 3,5,6 complex observed to form between sensor **9** and D-glucose in basic aqueous media (given the current understanding of N–B interactions, the boron bound at the 1,2 position is illustrated as a tetrahedral boronate).

A reexamination of this work was conducted by Norrild and Eggert. Employing ^{13}C-1- and ^{13}C-6-labeled D-glucose, the $^1J_{C1-C2}$ and $^1J_{C5-C6}$ coupling constants were monitored.[60] Exploiting the observed reduction in the $^1J_{CC}$ value upon formation of a five-membered boronic ester,[61] the analysis determined that the previous ^{1}H NMR assignment was correct, but that the interpretation was only valid as the initial complex formed under anhydrous conditions. With time, the α-D-glucopyranose isomerized to the α-D-glucofuranose form. In deuterated methanol, this process was slow; 20 h elapsed before the emergence of new NMR peaks became clear, with a complete disappearance of the original α-D-glucopyranose signals occurring after eight days. However, if water was added to the system, isomerization was accelerated dramatically, and after 10 min in a 1 : 2 water/methanol solution isomerization was complete.

In the case of sensor **9**, it was concluded that once formed the complex in Figure 4 rearranges to the thermodynamically more stable 1,2 : 3,5,6 bound α-D-glucofuranose complex (Figure 5) as a function of time and the water content of the medium.

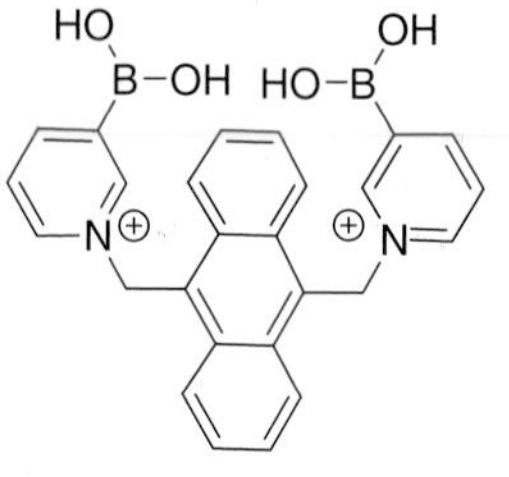

10

Norrild has developed the diboronic acid system **10**, which works by reducing the quenching ability of the pyridine groups of **10** on saccharide binding.[60] The structure of the complex was determined to be a 1,2 : 3,5 bound α-D-glucofuranose. Evidence of the furanose structure was obtained from ^{1}H and ^{13}C NMR data with emphasis on the information from $^1J_{C-C}$ coupling constants.

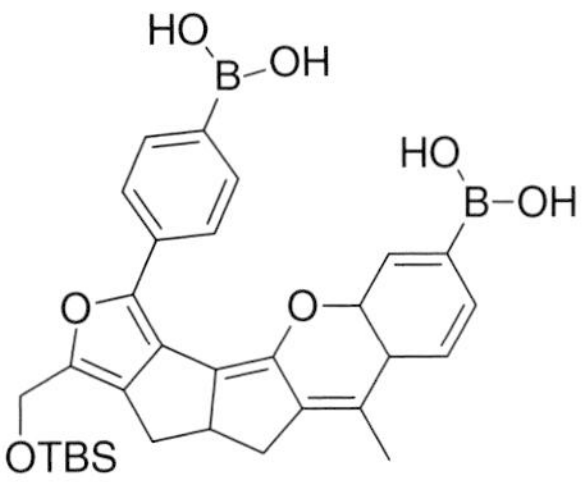

11

Drueckhammer used a computer-guided approach to engineer sensor **11**, a receptor specifically designed to complex α-D-glucopyranose at the 1,2 and 4,6 positions.[62] The computational approach produced a rigid molecular scaffold anchoring the two boronic acid groups precisely in space. Defined spatial architecture led the receptor to exhibit a 400-fold greater affinity for D-glucose over any of the other saccharides the receptor was evaluated against. Significantly, ^{1}H NMR confirmed that D-glucose was formed and retained as a stable complex in its pyranose form. Sensor **11** indicates that where two-point binding is achieved between boronic acids of fixed distances and enforced geometries, specific isomeric forms of a saccharide guest may be observed within the binding cleft.

In refining the selectivity of boronic acid-appended sensors for saccharides, it therefore seems that preempting the structure of the guest species or the thermodynamic complex that it will form is nontrivial. It is known that only saccharides with the ability to interconvert between pyranose and furanose forms with an available anomeric hydroxyl pair have so far been reported to interact strongly with boronic acids. It is also generally the case that in aqueous solutions the furanose form of the saccharide will be thermodynamically favored. However, as illustrated by Drueckhammer and others, in the case of recognition sites with two linked boronic acid fragments, this is a function of substrate structure and geometry.

Extending the design parameters of sensor **9**, James developed the chirally selective sensors *R*- and *S*-**12**.[46,63] On the basis of the intramolecular fluorescence quenching of 1-1′-binaphthyls, sensors *R*- and *S*-**12** demonstrated that selectivity of a diboronic acid-appended sensor could not only be tuned toward specific ligands, such as monosaccharides, but also toward the specific enantiomers of these ligands.

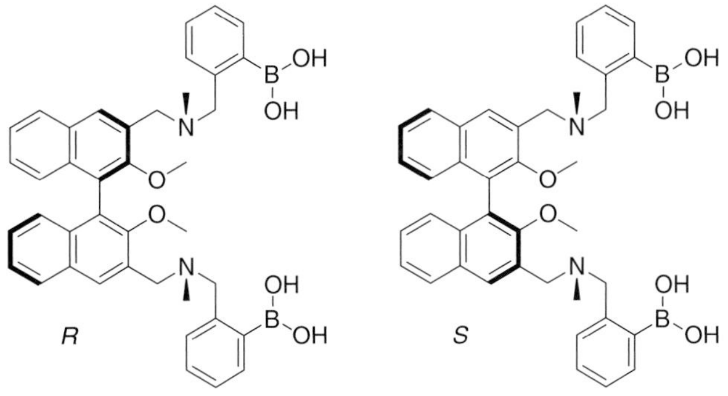

R *S*

12

Following observations by Houston and Gray[64] that the racemate of sensor **12**-bound sugar acids such as tartaric acid had comparable strength to the monosaccharides, the chiral complexation of these ligands was investigated by Zhao *et al.*[63] It was found that the recognition of D- and L-tartaric acid by sensors *R*- and *S*-**12** was strongly pH dependent. At pH 8.3, the fluorescence enhancements behaved as expected. For sensor *R*-**12** with D-tartaric acid and sensor *S*-**12** with L-tartaric acid, the fluorescence enhancements were large; conversely, for sensor *S*-**12** with D-tartaric acid and sensor *R*-**12** with L-tartaric acid, the fluorescence enhancements were small.

Accordingly, when the acidity was adjusted to pH 5.6, the fluorescence of sensor *R*-**12** with D-tartaric acid displayed an increase in fluorescence intensity. Quite astonishingly, however, the use of tartaric acid's L-enantiomer with sensor *R*-**12** produced a decrease in the fluorescence intensity. The same relationship was observed with the mirror-image host–guest complexes. Sensor *S*-**12** and L-tartaric acid displayed a fluorescence increase, while sensor *S*-**12** and D-tartaric acid displayed a fluorescence decrease. These results signify that sensors *R*- and *S*-**12** have the unusual property of allowing the fluorescence intensity of the reporting signal to be enantioselectively diminished or enhanced, relative to the fluorescence emission of the unbound species.

Evaluation of sensors *R*- and *S*-**12** with D-gluconic acid produced a similar enantioselective diminution or enhancement in the fluorescence intensity; curiously this was not the case with D-glucaric acid, D-glucuronic acid, or D-sorbitol. It is understood that on complexation of an asymmetric guest to sensors *R*- and *S*-**12**, the orientations of the two receptor units relative to the fluorescent BINOL core are locked in an asymmetrically distorted conformation. These results imply that in all probability the normal PET quenching that mediates fluorescence intensity must be susceptible to the induced geometrical changes that occur between receptors and fluorophores within individual host–guest complexes; the degree of geometric strain varies on a case-by-case basis.

Considering the effect of geometry on both the stability of the complexes formed and the fluorescence enhancements generated, the two properties appear to be independent. For changes in geometry, the observed stability constants (K_{obs}) increased up to ~25-fold, while the fluorescence intensity increased only up to ~3-fold. The geometry of the receptor therefore has a far greater influence on substrate recognition than it does on fluorescence.

More recently, Zhao and James have prepared 6,6′-*cis*-substituted BINOL boronic acids, which display improved enantioselectivity toward the sugar alcohols D-sorbitol and D-manitol.[65]

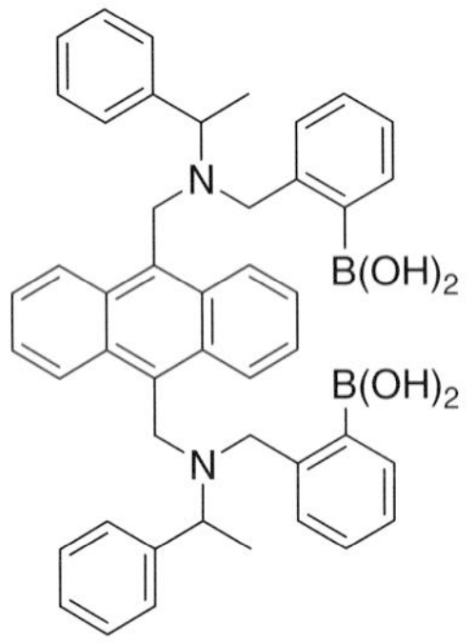

R,R or *S,S*

13

James prepared chiral boronic acid-based sensors *R,R*- and *S,S*-**13**, which are direct structural derivatives of the original D-glucose-selective sensor **9**.[66,67] The pH titrations demonstrated that sensors *R,R*- and *S,S*-**13** were highly selective for sugar acids such as tartaric, glucaric, gluconic, and glucuronic acid, with chemoselectivity of up to 11 000 : 1 (pH 5.6) being reported between species and enantioselectivity between the D- and L-forms of the ligands of up to 500 : 1 (pH 7.0) being reported with sensors *R,R*- and *S,S*-**13**. The same titrations were performed with monoboronic acid analogs; these analogs displayed no enantioselective discrimination between D- and L-tartaric acid indicating that for enantioselectivity a 1 : 1 cyclic complex must be formed.

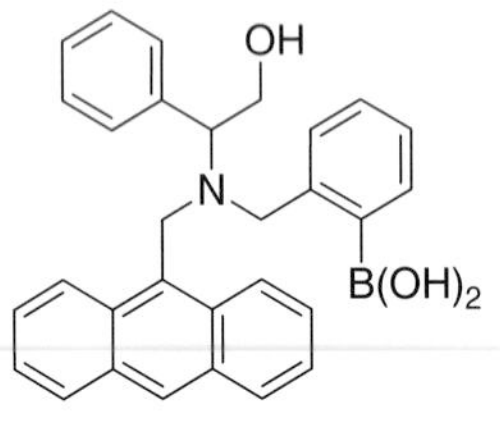

R or *S*

14

Zhao and James have investigated a related monoboronic acid system *R*-**14** and *S*-**14**, which was found to be an enantioselective fluorescence sensor for mono α-hydroxyl carboxylic acids such as mandelic acid and lactic acid.[68]

The bottleneck in accessing new receptors is typically the synthetic chemistry. Therefore, James and Fossey have developed a convenient fluorescent boronic acid unit, that is, a "click-fluor" assembled via Huisgen [3+2] cycloadditions,[69] which are ideally suited to a modular synthetic approach toward sensors. The so-called "*click reaction*" forms an aromatic 1,2,3-triazole ring following the addition of an azide to a terminal alkyne and has the potential to create a fluorescent sensor from nonfluorescent constituent parts. Fluorescent boronic acids have also been prepared using Huisgen [3+2] cycloadditions by Smietana and Vasseur, who employed the Seyferth–Gilbert procedure in a one-pot process to generate alkynes from boronic acid aldehydes, which were then coupled in a Huisgen [3+2] cycloaddition with an azide fluorophore. Wang has prepared acetylene-substituted boronic acids as potential [3+2] cycloaddition units[70] and suggested their use as units for *click reactions*. Wang has also recently used [3+2] cycloaddition reactions to prepare a boronic-acid-labeled thymidine triphosphate for incorporation into DNA.[71,72]

Fluorescent sensor **15** was prepared by employing a copper (I) catalyzed azide-alkyne [3+2] cycloaddition developed by Ham, enabling the synthesis of 1,2,3-triazole ring as predominantly the 1,4-regioisomer. At this point during the synthesis, X-ray quality crystals of the pinacol-protected intermediate **16** were obtained. The target sensor **15** was then obtained via a two-step deprotection of the pinacol ester.[73]

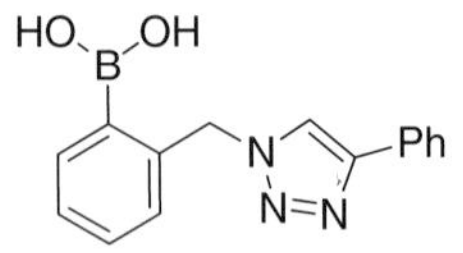

15 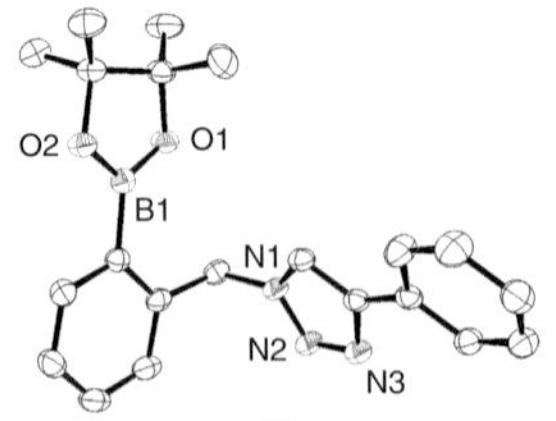 **16**

The fluorescence enhancements (I/I_0) obtained for **15** on the addition of D-fructose, D-galactose, and D-mannose are 27-, 20-, and 16-fold respectively. We believe that these large fluorescence enhancements can be attributed to fluorescence recovery of the 1,2,3-triazole fluorophore. In the absence of saccharides, the normal fluorescence of the locally excited (LE) state of the 1,2,3-triazole donor of sensor **15** is quenched by energy transfer to the neutral phenylboronic acid acceptor, weakly Lewis acidic boron center. When saccharides are added, a negatively charged boronate anion is formed due to the enhancement of the Lewis acidity of the boron center on saccharide binding. Under these conditions, energy transfer from the

1,2,3-triazole donor becomes unfavorable and fluorescence is recovered.

We coined the phrase "*click-fluor*" to describe the generation of a fluorophore from nonfluorescent constituent parts via a so-called "*click reaction.*" In this case, the triazole ring forms an integral part of the fluorophore; that is, a new property is imparted on a molecule conceived by a "*click reaction.*" Using our simple "*click-fluor*" unit, we believe that it will become possible to develop fluorescent modular sensor arrays for rapid screening of target saccharides. The two most attractive aspects of "*click-fluor*" are that a fluorophore is generated upon triazole formation and the wide availability of acetylene units.

The utility of the "*click-fluor*" as a scaffold for probing reactions of boronic acids has been investigated by Frost; however, the sensing ability of the constructs was not investigated.[74] The *in situ* generation of fluorophores, somewhat reminiscent of the "*click-fluor*" approach, was used by Jiang, who has elegantly used a Suzuki reaction of a boronic acid to enhance the sensing of guests.[75]

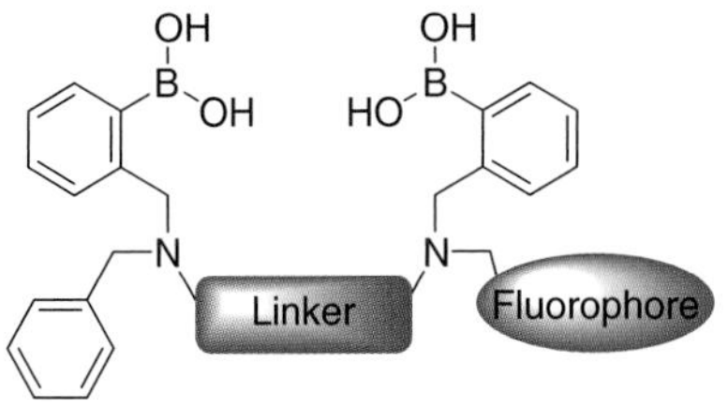

OH OH B OH HO B N N Linker Fluorophore

17

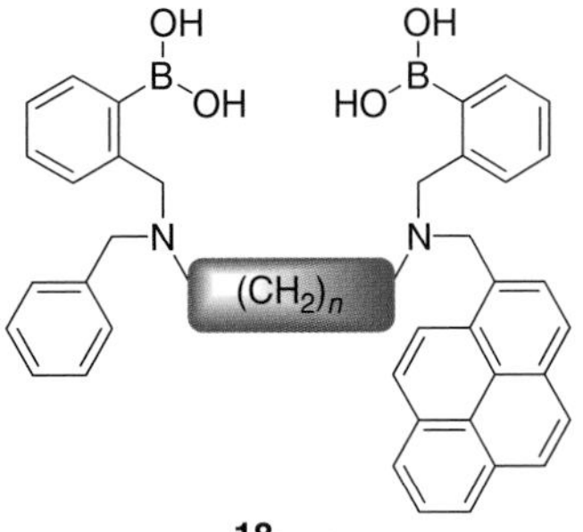

OH OH OH B OH HO B HO B N N N $(CH_2)_n$

$\mathbf{18}_{(n=3)}$
$\mathbf{19}_{(n=4)}$
$\mathbf{20}_{(n=5)}$
$\mathbf{21}_{(n=6)}$
$\mathbf{22}_{(n=7)}$
$\mathbf{23}_{(n=8)}$

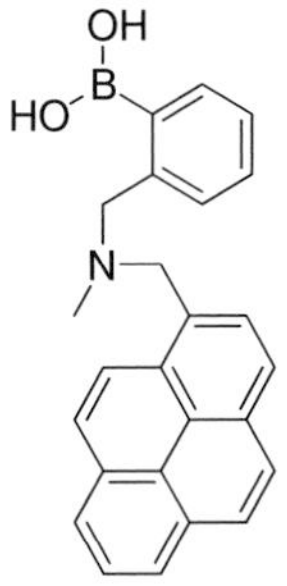

$\mathbf{24}_{(\text{pyrene})}$

3.2 Modular fluorescent sensors

In the design of boronic acid-based sensor systems, it has been established that two receptor units are required if saccharide selectivity is to be achieved.[44] Retaining the same dual boronic acid recognition units throughout, a modular system in which the linker and fluorophore units of these sensors could be modified independently needed to be developed. The use of modular or core scaffolds for saccharide recognition has been championed by James,[76–81] Wang,[11,82–85] Hall,[86,87] and Singaram.[88–91]

While sensors developed around an anthracene core **9** unit have proved to be selective for saccharides such as D-glucose,[44,45] the rigid core unit that acts triply as scaffold, linker, and fluorophore limits the modifications that can be made to any one part of the system without influencing the sensor as a whole. A generic template **17** has been designed by James on which saccharide-selective sensors can be developed. The design includes two boronic acid groups required for selectivity but allows the separation between them to be varied by altering the linker. It also permits the fluorophore to be varied independently and, by using only one fluorophore, overcomes the problems that may arise from excimer emission, insolubility, excessive hydrophobicity, and steric crowding at the binding pocket.

James prepared modular PET sensors $\mathbf{18}_{(n=3)}$–$\mathbf{23}_{(n=8)}$ containing two phenylboronic acid groups, a pyrene fluorophore and a variable linker. The linker was varied from trimethylene $\mathbf{18}_{(n=3)}$ to octamethylene $\mathbf{23}_{(n=8)}$.[76,77] In most cases, the observed stability constants (K_{obs}) with diboronic acid sensors $\mathbf{18}_{(n=3)}$–$\mathbf{23}_{(n=8)}$ are higher than those for the monoboronic acid sensor $\mathbf{24}_{(\text{pyrene})}$. D-Glucose and D-galactose bind to diboronic acids readily using two sets of diols, thus forming stable, cyclic 1 : 1 complexes. The allosteric binding of the two boronic acid groups is clearly illustrated by the relative difference between the observed stability constants (K_{obs}) of the equivalent di- and monoboronic acid compounds. The observed stability constants (K_{obs}) for the diboronic acid sensors with D-fructose and D-mannose are up to twice as strong as with the monoboronic acid sensor $\mathbf{24}_{(\text{pyrene})}$. Each D-fructose and D-mannose molecule will only bind to one boronic acid unit through one diol. This allows complexes to form with an overall 2 : 1 (saccharide/sensor) stoichiometry. The relative binding values of about 2 are indicative of two independent saccharide binding events on each sensor, with no concomitant increase in stability derived from cooperative binding. The highest observed stability constants (K_{obs}) for D-glucose ($962 \pm 70\,M^{-1}$) within these systems was obtained by sensor $\mathbf{21}_{(n=6)}$. The flexible six-carbon linker provided the optimal selectivity for D-glucose over other saccharides.

Curiously, there is an inversion in the selectivity displayed by these systems upon moving from a six- to a seven-carbon linker. The trimethylene linked $\mathbf{18}_{(n=3)}$ shows little specificity between D-glucose and D-galactose. Increasing the size of the binding pocket, tetramethylene $\mathbf{4}_{(n=4)}$ through to hexamethylene $\mathbf{21}_{(n=6)}$, induces a clear selectivity for D-glucose, with $\mathbf{21}_{(n=6)}$ providing the strongest binding. However, there is an inversion in this selectivity on increasing the linker length to heptamethylene $\mathbf{22}_{(n=7)}$ and octamethylene $\mathbf{23}_{(n=8)}$, with the enlarged binding pocket being D-galactose selective.

James has developed a molecular tweezer **25** that selectively opens for certain saccharides.[81] The fluorescence intensity at 377 nm of the tweezer **25** increases with increasing concentration of D-glucose, D-fructose, D-galactose, and D-mannose, while fluorescence intensity changes at 470 nm differ among the four carbohydrates. The 470-nm band decreases with increasing D-glucose and D-mannose concentrations. The intensity of the 470 band is invariant with added D-fructose. Finally, D-galactose shows an initial quenching of the 470-nm band at low concentrations followed by fluorescence recovery as the concentration increases. In the case of D-glucose, D-galactose, and D-mannose, the complex formed is a cyclic 1 : 1 structure (Scheme 4). This explains the quenching of the intramolecular excimer emission at 470 nm as the binding of the saccharide separates the pyrene units. The more complex behavior of D-galactose indicates formation of a 1 : 1 complex at low concentration and subsequent formation of a 1 : 2 complex at higher concentration. For molecular tweezer **25** with D-fructose, only the noncyclic 1 : 2 complex forms even at low saccharide concentration (Scheme 4).

Wang and coworkers have documented a range of diboronic acid sensors (Figure 6a–z). It can be seen by examining the generic template used that the sensors are designed around the known core of sensor **9**, the first diboronic acid sensor to display selectivity for D-glucose. In this construct, the number of carbon atoms from one N-methyl-o-(aminomethyl)phenylboronic acid nitrogen atom to the other is increased substantially. Six carbon atoms separate each of the adjacent amine–amide nitrogen atoms with anthracene cores rigidifying this section of the molecule and introducing possible interactions through either π–π stacking or steric encumbrance. The variable linkers examined further augment the length of these rigid linkers.

Sensor (**f**) with the *para*-benzene linker (Figure 6) was found to be selective for sialyl Lewis X,[82,83] while sensor (**g**) with the *ortho*-xylene linker (Figure 6) was found to be selective for D-glucose.[84] In replacing the *ortho*-xylene linker of sensor (**g**) with the flexible butyl linker of sensor (**b**), the number of carbon atoms in the linker remained the same but the structural rigidity of the linker was lost; this led to halving of the observed stability constant (K_{obs}). In reintroducing the rigidity but changing the geometry and spacing of the core unit to the *ortho*-benzene linker sensor (**h**) the observed stability constant (K_{obs}) was seen to decrease further.

sialyl Lewis X

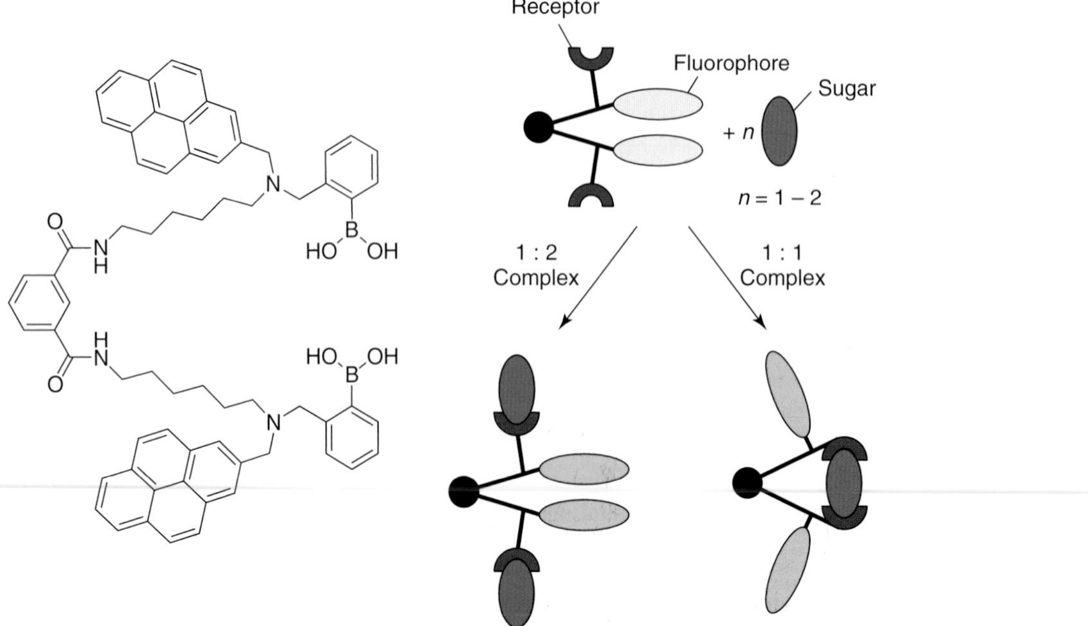

Scheme 4 Schematic representation showing the two possible ways of binding for the molecular tweezer **25** with sugars.

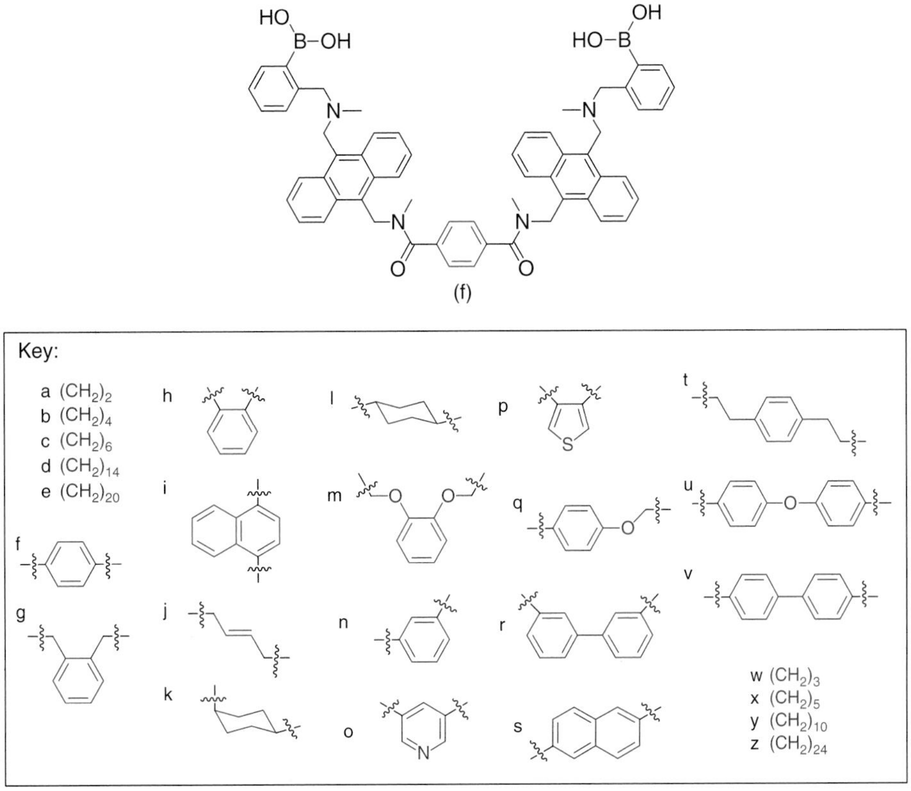

Figure 6 Range of diboronic acid sensors (a)–(z) with the interchangeable linker fragments highlighted in red, (f) displayed selectivity for sialyl Lewis X, and (g) displayed selectivity for D-glucose.

In an article published in 2004, Hall and coworkers documented the first parallel, solid-phase synthesis of modular boronic acid-based sensors.[86] The series was developed from a range of common components allowing the rapid assembly of a library of compounds, with the use of semipreparative, high-performance liquid chromatography (HPLC) to ensure sensors of satisfactory purity. This approach allowed the structures of the interamine linkers to be altered (selectivity was once again found for D-glucose with a linker six-carbon atoms in length); however, the investigation went on to assess the potential role of a third boronic acid receptor moiety in the recognition of disaccharides (Figure 7) and the effect of introducing unencumbering electron-withdrawing and electron-donating groups para to the arylboronic acid (Figure 8).

The triboronic acid sensors were titrated against four disaccharides: lactulose, melibiose, turanose, and trehalose. Across the range of sensors and guests examined, no benefit was found from three (vs two) boronic acid receptor units. For example, the observed stability constant (K_{obs}) of triboronic acid sensor (**1c, 2a**) with lactulose was $200\,M^{-1}$ (Figure 6). This value can be contrasted with the value obtained for the analogous diboronic acid (**2a** derivative) that displayed an observed stability constant (K_{obs}) of $220\,M^{-1}$ with lactulose in $0.010\,mol\,dm^{-3}$ phosphate buffer at pH 7.8 in a 1 : 1 water/methanol mixture.

Building on these observations, Hall and coworkers examined the dependence of complexation on the electronic characteristics of the arylboronic acid receptors (Figure 8). By altering the Lewis acidity at boron, it was believed that two main features in the molecular recognition event could be altered: the strength of the binding interaction with the saccharides and the strength of the N–B interaction controlling the fluorescence intensity. This hypothesis was tested by introducing electron-withdrawing and electron-donating groups at the para position of the arylboronic acid ring.

Five substituent groups were considered in all: methoxy, fluoro, methoxycarbonyl, cyano, and nitro, with para-substituent parameters, σ_p: −0.12, 0.15, 0.44, 0.70, and 0.81 respectively. If Figure 8 is considered and lactulose is used as the model disaccharide, the binding measurements display a qualitative trend that electron-poor phenylboronic acids are preferable for binding. This observation was rationalized on the basis that on increasing the Lewis acidity at boron the N–B interaction becomes stronger, developing more of a tetrahedral character at boron. This in turn reduces the ring strain in the developing boronic ester.

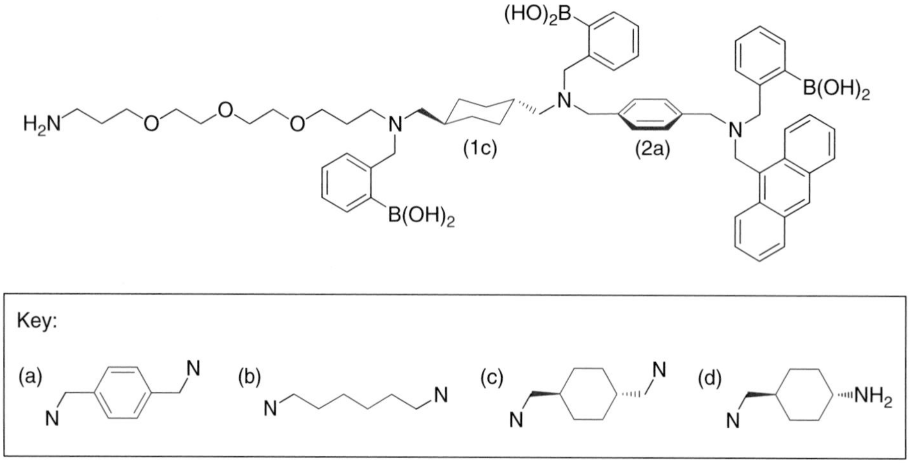

Key:

(a) (b) (c) (d)

Figure 7 Triboronic acid sensors assembled via combinatorial synthesis to evaluate the potential of increased allosteric binding effects through three-point binding (the interchangeable linker units are highlighted in red).

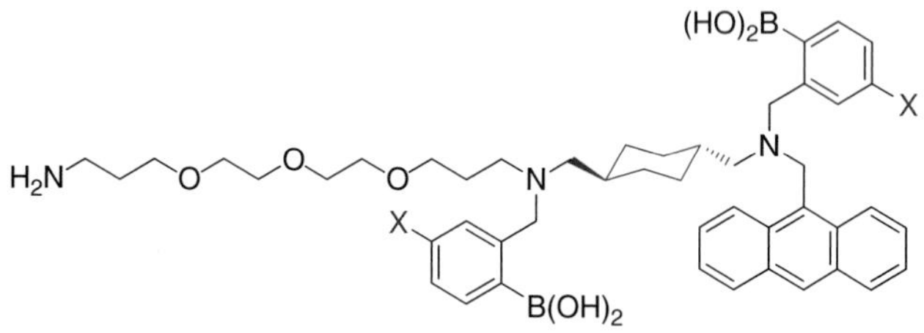

Figure 8 Diboronic acid modular systems with an interchangeable linker section (highlighted in red) and interchangeable electron-withdrawing and electron-donating groups para to the arylboronic acid (highlighted in blue).

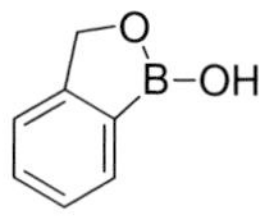

In addition to this, acidifying boron and reducing its pK_a also provides the substantial benefit of allowing the sensor to function at a lower pH. In enhancing the observed stability constants (K_{obs}), the use of methoxycarbonyl, cyano, and nitro groups appeared to be particularly effective. Overall, the largest observed stability constant (K_{obs}) reported with lactulose was with the diboronic acid illustrated in Figure 8 with a hexamethylene linker (**b**) and a *para*-cyano electron-withdrawing group.[86]

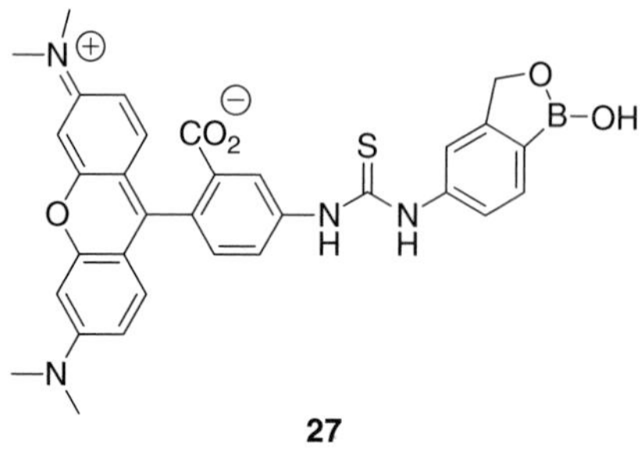

Hall has recently taken a different approach to improve the binding efficacy of simple boronic acid units. The design takes its inspiration from the enhanced binding of *o*-benzoboroxoles **26** with 4,6- or *cis*-3,4-diols.[92] The binding constants of *o*-benzoboroxoles with glycopyranosides were obtained by using an ARS UV assay in neutral water. The K_a with methyl α-D-glucopyranoside was 22 M^{-1}, while the binding constant of phenyl boronic acid (PBA) with glucose was about 5 M^{-1} under similar conditions. The importance of this binding motif was quickly recognized by Hindsgaul[93] who used the receptor unit connected to a tetramethylrhodamine dye **27** for the visual analysis of the terminal glycosylation of glycoproteins.

26 **27**

Hall also employed the receptor unit and produced a library of synthetic receptors targeted against tumor-associated carbohydrate antigen, the Thomsen–Friedenreich (TF) disaccharide (Gal-β-1,3-GalNAc).[87] The group made two changes in the design of the library from the earlier modular receptors. The linkers employed were peptides to make the synthesis easier while also adding hydrogen-bonding units to the receptors. They also employed benzoboroxoles rather than boronic acids because this unit binds particularly well with 4,6- or *cis*-3,4-diols found in the TF antigen out of a library of 400 receptors **28** (varying R_1 and R_2), one particularly potent system **29** was discovered with an IC_{50} of 20 μm.

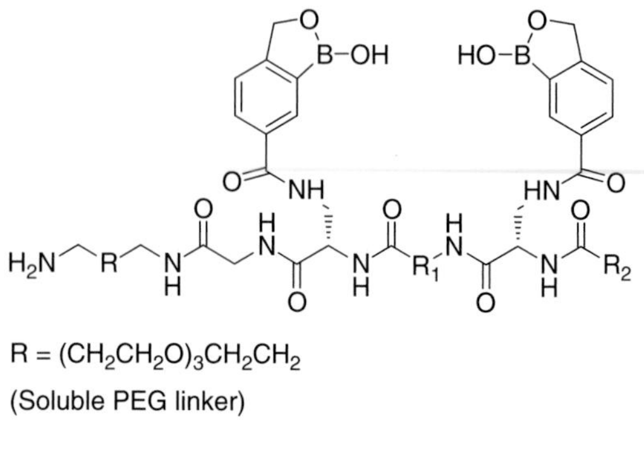

28

R = $(CH_2CH_2O)_3CH_2CH_2$

(Soluble PEG linker)

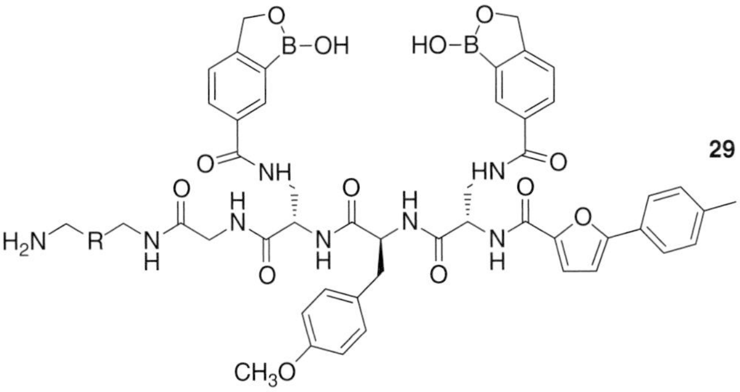

29

Anslyn has pioneered peptide-based boronic acid receptors for pattern-based carbohydrate sensing. Using an array system, they were able to identify sucralose from a complex beverage sample.[94] Peptide-based boronic acid receptors have also been studied by Lavigne, who prepared novel peptide boronolectins on beads and used the receptors to probe glycoproteins and oligosaccharides with fluorescent labels.[95] Dugan has also investigated the potential of solid-supported receptors derived from 4-borono-L-phenylalanine.[96]

3.3 Colorimetric sensors

Colorimetric sensors for saccharides are of particular interest in a practical sense. If a system with a large color change can be developed, it could be incorporated into a diagnostic test paper for saccharides, similar to universal indicator paper for pH. Such a system would make it possible to measure D-glucose concentrations without the need of specialist instrumentation. This would be of particular benefit to diabetic patients in developing countries.

Boronic acid azo dyes have been known for over 40 years as they have been used for investigations in the treatment of cancer by a technique called boron neutron capture therapy (BNCT). However, it was not until the 1990s that related dyes and their interaction with saccharides were studied. A patent by Russell concerns a boronic acid azo dye derived from *m*-aminophenylboronic acid.[97] Nagasaki investigated two diazo chromophores (which aggregate in water) containing boronic acid moieties, changed color, and deaggregated upon complexation with saccharides.[98] The behavior was rationalized by the boronic acid–saccharide complexation, increasing the hydrophilicity of the bound species. Takeuchi has prepared a boronic acid diazo dye, which undergoes an absorption spectral change on addition of nucleosides.[99]

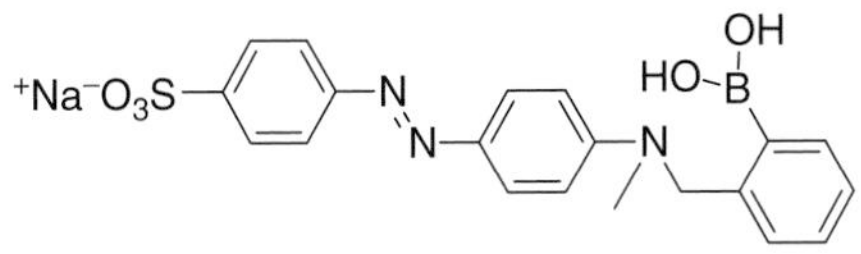

30

The ICT sensor **30** prepared by Sandanayake employs an intramolecular interaction between the tertiary amine and the boronic acid group to promote color changes on addition of saccharides.[100] The electron-rich amine creates a basic environment around the electron-deficient boron center, which has the effect of inducing the boronic acid–saccharide interaction and reducing the working pH of the sensor. Electronic changes associated with this decrease in the pK_a of the boronic acid moiety on saccharide complexation are transmitted to the neighboring amine. This creates a spectral change in the connected ICT chromophore, which can be detected as a change in color. The observed stability constant (K_{obs}) for **30** was 138 M^{-1} for D-fructose in water at pH 7.6. The stability constant with D-glucose could not be determined due to small spectral changes.

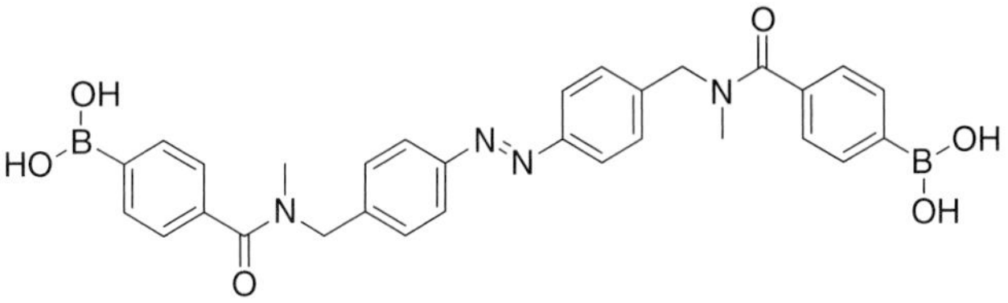

31

A light-gated saccharide sensor **31** has been studied by Shinmori *et al.*[101] When the azobenzene unit is switched from the more stable transconformation to the thermodynamically unfavorable *cis*-isomer by photoirradiation, the system displays high D-glucose and D-allose selectivity. A cyclic 1 : 1 complex forms between the saccharide and the dye in its *cis*-geometry and explains the enhanced binding.

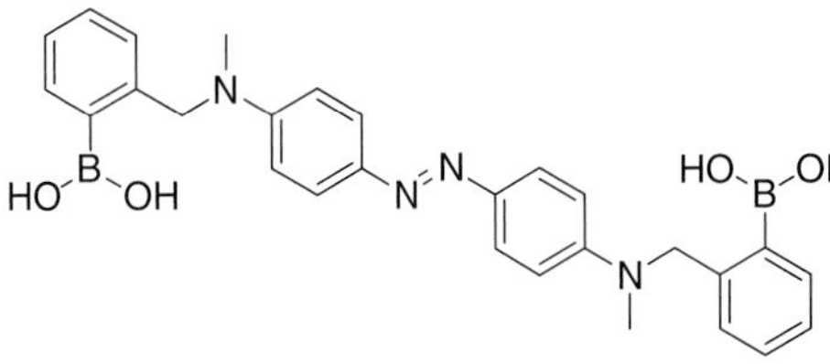

32

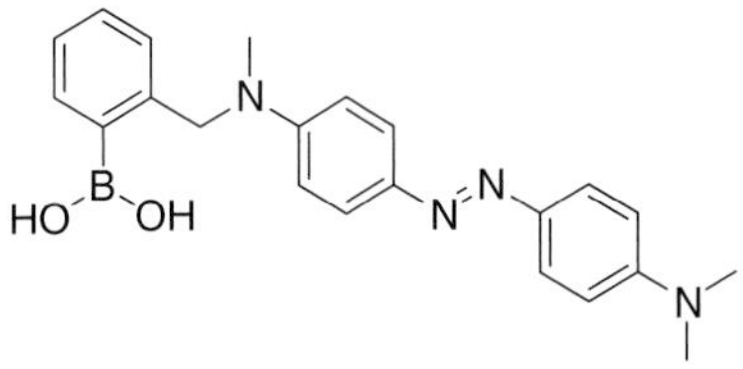

33

Koumoto demonstrated that azobenzene derivatives bearing one or two aminomethylphenylboronic acid groups **32** and **33** can be used for practical colorimetric saccharide sensing in "neutral" aqueous media.[102] The observed stability constants (K_{obs}) for **33** were 433 M^{-1} for D-fructose and 13.0 M^{-1} for D-glucose in 1 : 1 (v/v) methanol/water at pH 7.5 (phosphate buffer).

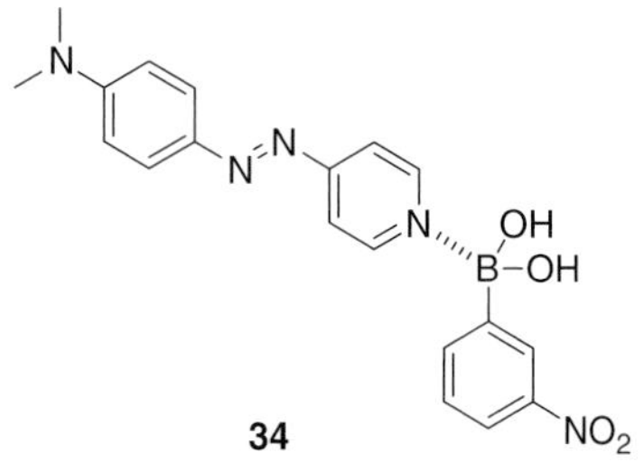

34

The boronic acid–amine interaction has been cleverly used by Koumoto in the molecular design of a sensing system for saccharides. 3-Nitrophenylboronic acid interacts with the pyridine nitrogen of 4-(4-dimethylamino-phenylazo)-pyridine **34** in methanol and changes its color from yellow to orange.[103] Saccharides form complexes with the boronic acid and enhance the acidity of the boronic acid group. A stronger boron–nitrogen interaction intensifies the intramolecular charge-transfer band and changes the solution to red in color.

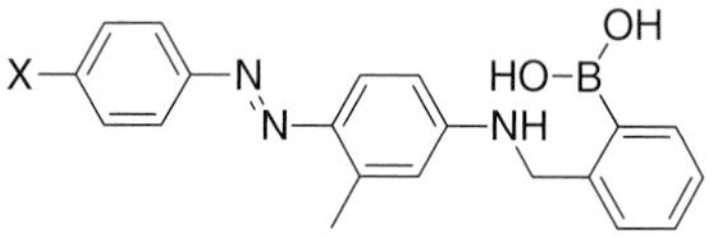

35: X = *p*-NO_2
36: X = *p*-SO_3H
37: X = *p*-CO_2H
38: X = *p*-OCH_3
39: X = *m*-CO_2H

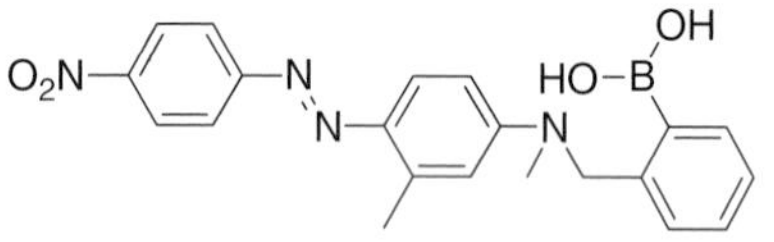

40

James has prepared a diazo dye system **35**, which shows a large visible color change from purple to red on saccharide binding.[104, 105] With azo dye **35**, the wavelength maximum shifts by ~55 nm to a shorter wavelength upon saccharide complexation.

With dye molecule **30**, it was proposed that at intermediate pH, a boron–nitrogen interaction exists, whereas at high and low pH this interaction is broken. What makes the equilibria of dye molecule **35** more interesting is the presence of the *anilinic hydrogen*, which can give rise to different species at high pH. In the absence of saccharide, at pH 11.32, the observed color of **35** is purple and in the presence of saccharide the color is red. In the presence of saccharide, the N–B interaction becomes stronger. The increased N–B interaction causes the N–H proton to become more acidic. Therefore, at pH 11.32, the saccharide–boronate complex dehydrates (loss of H^+ from aniline and OH^- from boronate) to produce a red species with a covalent N–B bond. These equilibrium species explain why dye molecule **30** did not give a visible spectral shift on saccharide binding. With **30**, there is no possibility of dehydration, so a strong boron–nitrogen bond cannot be formed, hence no spectral shift is observed. This hypothesis has been confirmed by evaluating **40**, which does not have an anilinic hydrogen. No color change was observed for **40** on addition of saccharides at pH 11.32.[105] A detailed investigation of a series of azo dyes with both electron-donating and withdrawing groups **35–39** indicated that a strong electron-withdrawing group is required to produce a color change.[105]

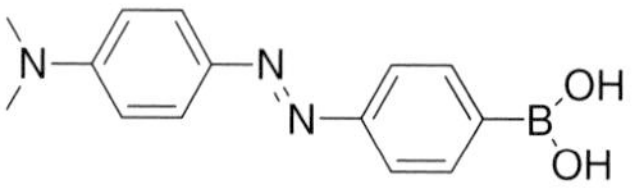

41

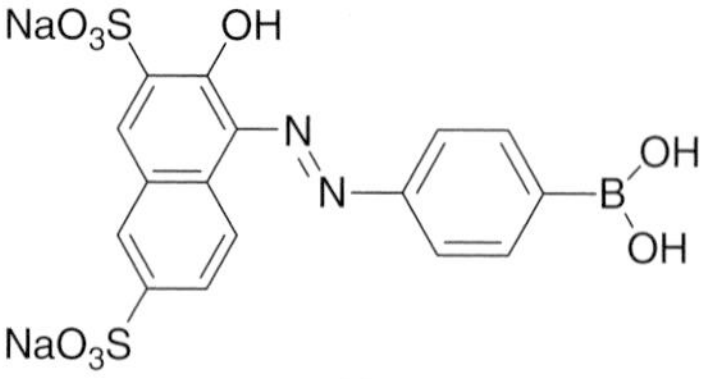

42

Lakowicz has also prepared boronic acid azo dye molecules **41** and **42** in which direct conjugation with the boron center is possible.[106] In particular, the azo dye **42** produces a visible color change from yellow to orange at pH 7.

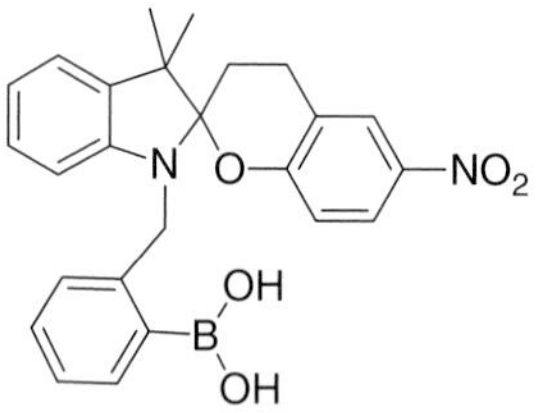

43

Shinmori has shown that a boronic acid-appended spirobenzopyran **43** undergoes changes in the absorption spectra on the addition of saccharides.[107] Added saccharides change the position of the merocyanine (MC) to spiropyran (SP) equilibrium and hence change the color of the system. With added saccharide, the SP structure is favored due to a stronger N–B interaction in the saccharide complex (Scheme 5).

Strongin has prepared a tetraboronic acid resorcinarene system for the visual sensing of saccharides.[108, 109] Characteristic color changes were observed for specific carbohydrates, D-glucose phosphates, and amino sugars on gentle heating in DMSO (dimethyl sulfoxide). Further work by Strongin with another resorcinol derivative has shown that oxygen promotes the color changes and that the resorcinol hydroxyl groups play a key role in the color formation of the solutions.[110] The mechanism of color change points to xanthenes as *in situ* chromophores are formed by heating resorcinols in DMSO. Nonboronic acid receptors

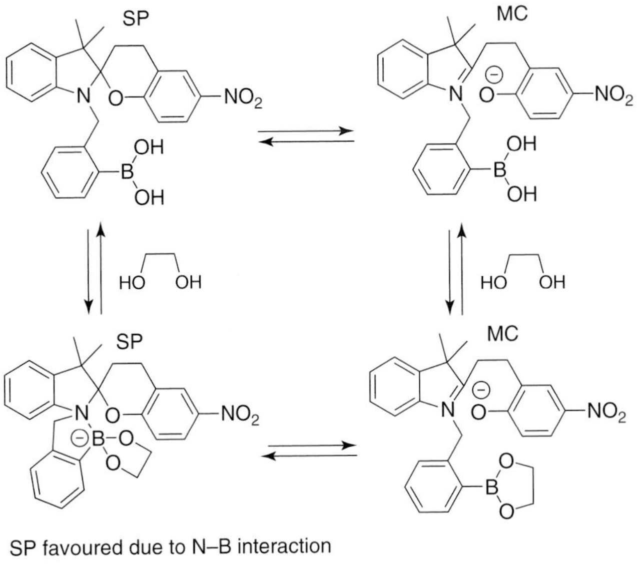

SP favoured due to N–B interaction

Scheme 5 The effect of saccharides on the spiropyran versus merocyanine equilibrium.

also produce colored solutions but to a lesser extent. In these cases, the color is due to hydrogen bonding between aldonic acids (heating sugars in DMSO produces aldonic acid derivatives) and the hydroxyls of the *in situ* xanthene chromophore.[109, 110]

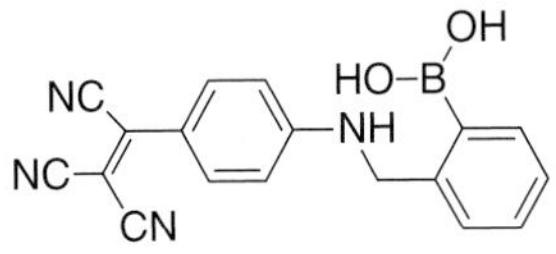

44

The results obtained with **35–39** led James and coworkers to prepare the strongly electron-withdrawing tricyanovinyl dye **44**. The pK_a of **44** (7.81) was much less than **35** (10.2), resulting in a visible color change on addition of saccharides at a much lower pH (8.21).[111]

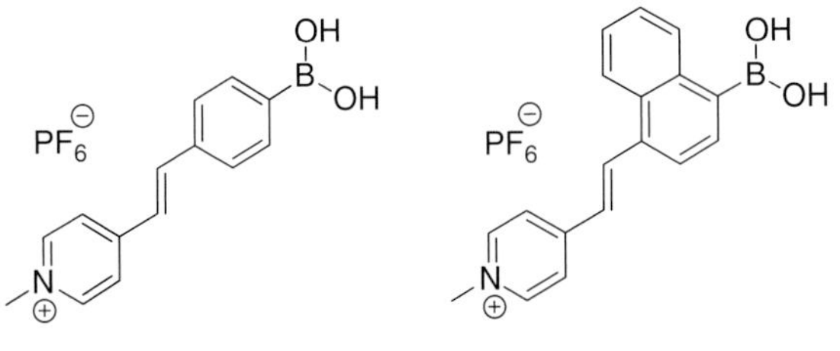

45 **46**

Sato has prepared stilbazolium boronic acids **45** and **46** and demonstrated the suitability of this unit for the optical sensing of saccharides.[112] Wang has prepared nitrophenol boronic acids, which show large shifts in the UV on addition of saccharides.[113] The changes have been attributed to a change in the balance of the phenolate to boronate equilibria in the presence of saccharides.

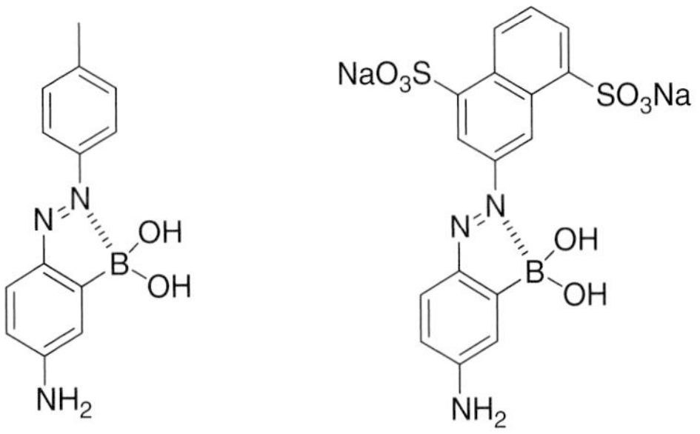

47 **48**

Egawa has taken a novel approach to use the N–B interaction to influence the color of azo dyes **47** and **48** on saccharide binding.[114] The ortho boronic acid systems show substantial changes in the UV–vis absorption spectra on saccharide binding. The same system has also been attached to a poly(ethyleneimine) polymer.[115]

3.4 Displacement systems

So far we have discussed the development of integrated molecular sensors using boronic acids. The systems contain a receptor and reporter (fluorophore or chromophore) as part of a discrete molecular unit. However, another approach toward boronic acid-based sensors is also possible where the receptor and a reporter unit are separate as in a competitive assay. A competitive assay requires that the receptor and reporter (typically a commercial dye) to associate under the measurement conditions. The receptor–reporter complex is then selectively dissociated by the addition of the appropriate guests. When the reporter dissociates from the receptor, a measurable response is produced (Scheme 6).

A dye displacement assay is primarily a colorimetric competitive binding sensor system where an analyte displaces a dye from a receptor; this displacement results in some color change that can be related to the amount of analyte present. Pioneering reports in this arena include the seminal protocols of Anslyn[94, 116–123] and Buryak and Severin.[124]

Anslyn reported two very elegant systems based on boronic acid receptors. The C_3 symmetric tri-podal boronic acid **49** is a selective receptor for D-glucose-6-phosphate.[125] The binding of D-glucose-6-phosphate is measured through the competitive displacement of 5-carboxyfluorescein. Addition of the D-glucose-6-phosphate caused a decrease in the absorption of light at 494 nm allowing the concentration of the guest to be monitored directly within the visible spectrum. Anslyn has also prepared more elaborate C_3 symmetric tri-podal boronic acid receptors. The binding of heparin and **50**, is monitored through displacement of pyrocatechol violet.[126]

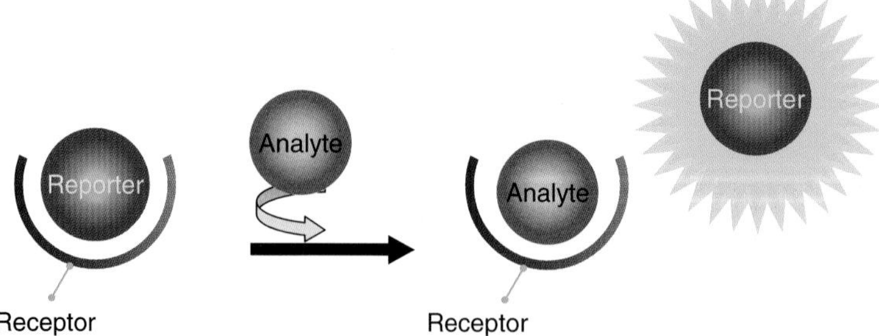

Scheme 6 Cartoon depicting the function of an assay system.

49

5-Carboxyfluorescein

50

Pyrocatechol violet

With sensor **51**, the binding of the tartrate or malate anions can be detected through the competitive displacement of alizarin complexone. The same sensor system was used for the analysis of malate in Pinot noir grapes.[127] When **52** was paired with pyrocatechol violet, an assay suitable for the detection of gallic acid in Scotch whiskies was developed. An increase in the concentration of gallic acid correlated with the age of the whiskies.[128] A combination of **51** and **52** and two indicators pyrocatechol violet and bromopyrogallol red can be used to detect the concentrations of tartrate and malate in mixtures.[129] Using **52** and pyrocatechol violet, the reaction kinetics for the formation of tartaric acid by the dihydroxylation of malic acid could be followed.[130] Anslyn has also elegantly paired chiral boronic acids with a variety of indicators to develop enantioselective assays for α-hydroxyl carboxylates and diols.[131,132]

Lakowicz has also used competitive interactions between a ruthenium metal–ligand complex, a boronic acid derivative, and D-glucose.[133] The metal–ligand complex forms a reversible complex with 2-tolylboronic acid or 2-methoxyphenyl boronic acid. Complexation is accompanied by a several-fold increase in the luminescent intensity of the ruthenium complex. Addition of D-glucose results in decreased luminescent intensity, which appears to be the result of decreased binding between the metal–ligand complex and the boronic acid. Ruthenium metal–ligand complexes are convenient for optical sensing because their long luminescent decay times allow lifetime-based sensing with simple instrumentation.

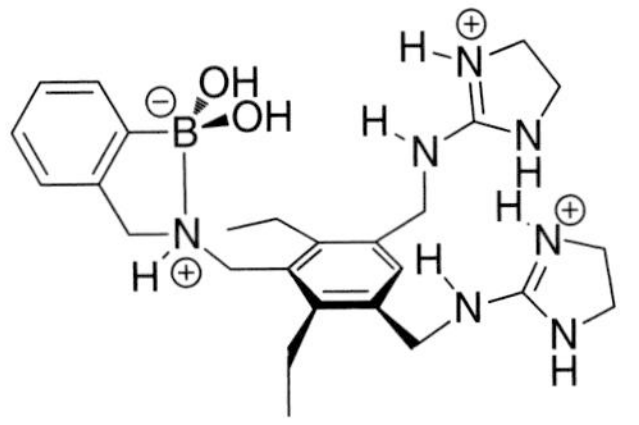

51

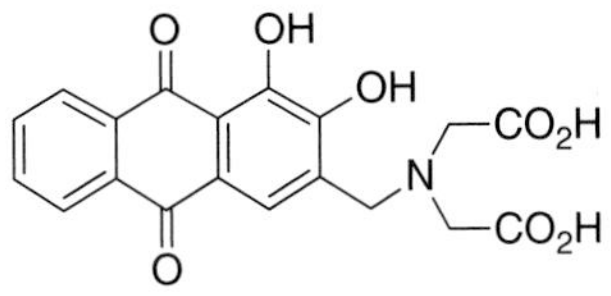

Alizarin complexone

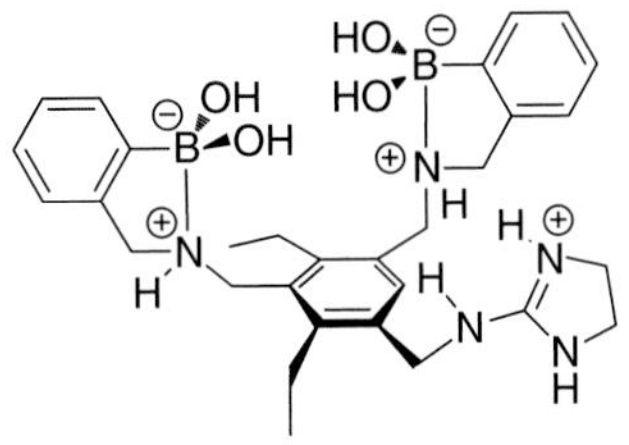

52

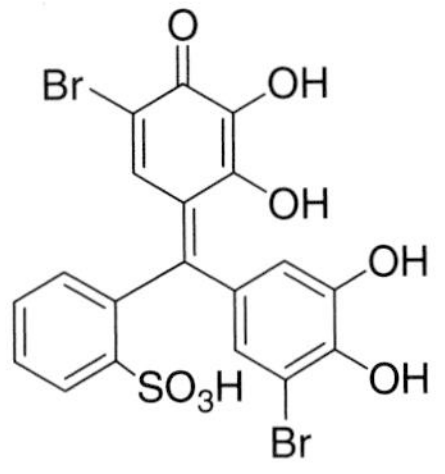

Bromopyrogallol red

An interesting multicomponent system has been devised by Singaram where quenching of an anionic pyranine dye **54** by bisboronic acid viologen units **53, 55**, and **57** is modulated by added saccharide.[134–136] Viologen **53** binds well with D-fructose ($K_{obs} = 2600\,M^{-1}$) and weakly with D-glucose ($K_{obs} = 43\,M^{-1}$) in pH 7.4 phosphate buffer.[134] Viologen **55** also binds weakly with D-glucose[136] but, more importantly, both systems only produce a 4–6% fluorescence recovery, whereas viologen **57** binds well with both D-fructose ($K_{obs} = 3300\,M^{-1}$) and D-glucose ($K_{obs} = 1800\,M^{-1}$) in pH 7.4 phosphate buffer. Together with the enhanced selectivity for D-glucose, this system also produces a 45% fluorescence recovery on addition of saccharides.[135] When viologen **55** and tetrakis(4-sulfophenyl)porphine fluorophore were used in combination, the system produced a 33% fluorescence recovery on saccharide addition along with a D-glucose binding constant of $14\,M^{-1}$.[137] Viologen **53** has also been

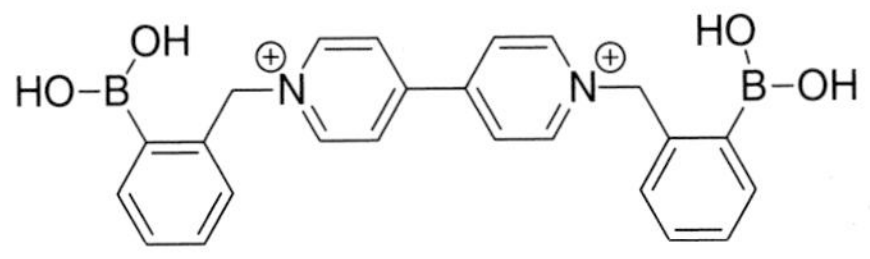

53

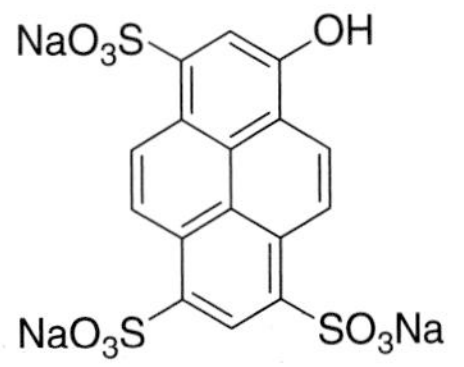

8-Hydroxypyrene-1,3,6-trisulfonic acid

54

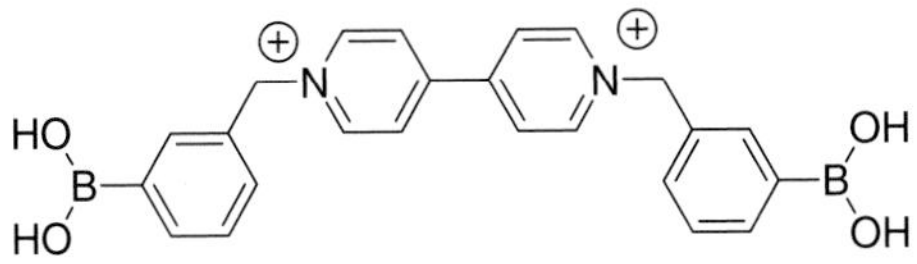

55

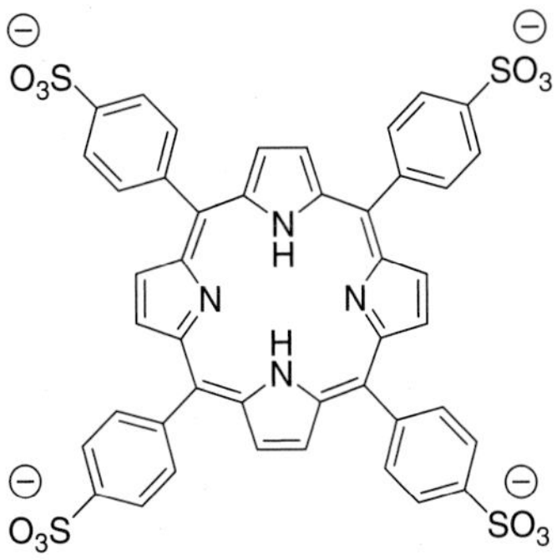

Tetrakis(4-sulfophenyl)porphine

56

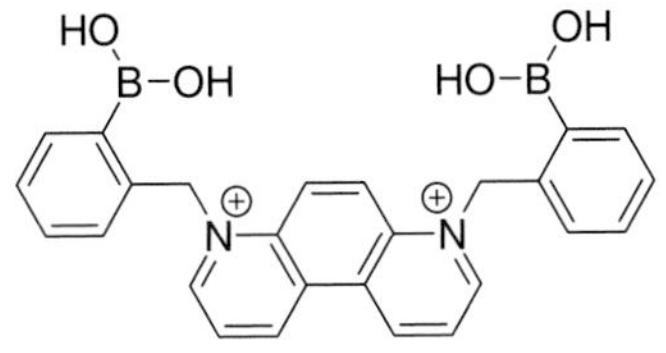

57

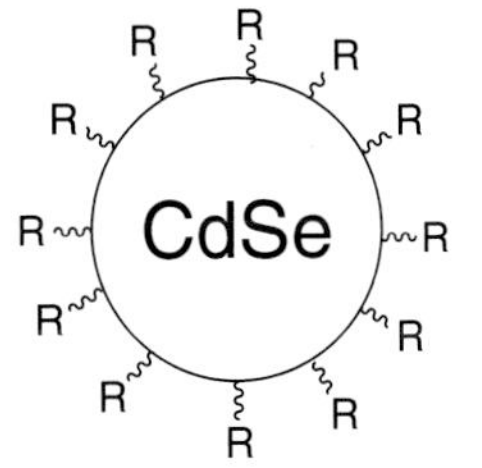

R = CO_2H or NH_2

58

used in combination with fluorescent quantum dots **58**, which possesses broad absorption, narrow emission, are bright, and have good photostability.[90] The quantum dot viologen system displayed a reasonable fluorescent recovery on D-glucose binding.

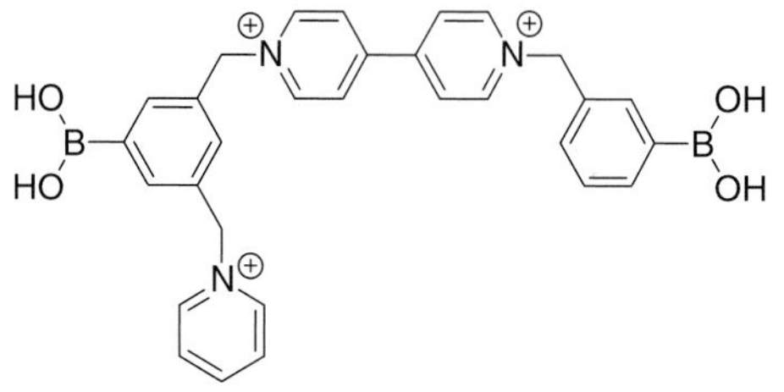

59

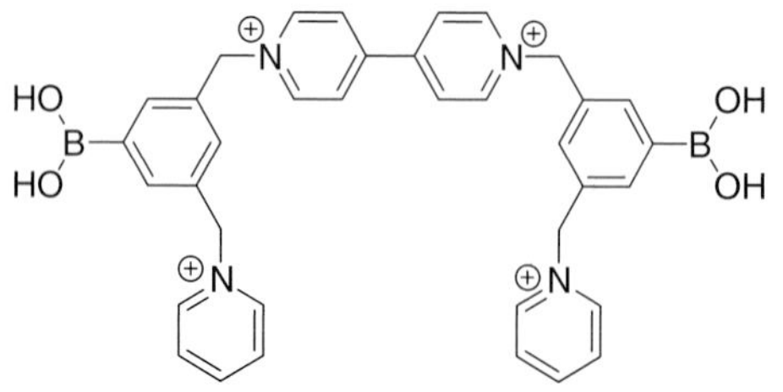

60

Singaram used viologens **59** and **60** to determine the effect of increasing the charge of viologen unit **55**. The higher the charge on the viologen unit, the greater is the quenching of the anionic dye, that is, interaction of the anionic dyes with higher charged species is greater. However, the glucose-sensing ability is reduced because dissociation of the viologen–dye duplex is made more difficult.[138] The effect of varying the anionic dye component used with viologen **60** was also investigated. Using 11 different anionic dye molecules, Singaram determined that anionic dyes with higher charge worked better in saccharide sensory systems.[139]

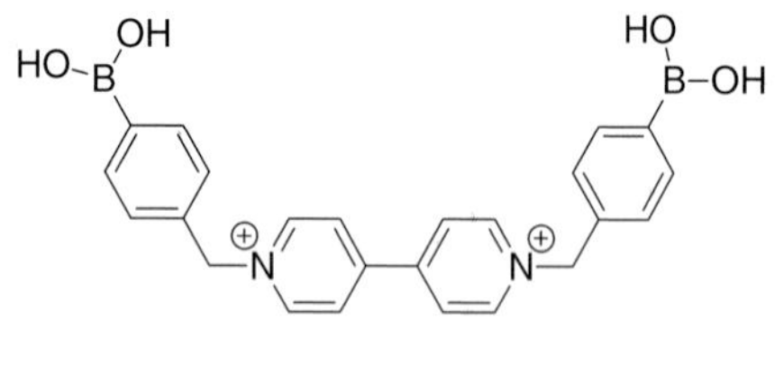

61

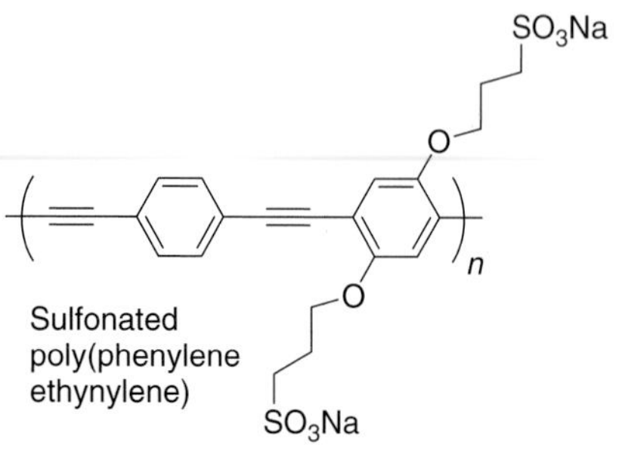

Sulfonated poly(phenylene ethynylene)

Lakowicz has also examined the quenching and recovery of a sulfonated poly(phenylene ethynylene) by a bisboronic acid viologen **61** on addition of saccharides.[140] The system is D-fructose selective and produces up to 70-fold fluorescence enhancement on addition of saccharides.

Wang has shown that alizarin red S and PBA could be used in competitive assays for saccharides.[141–143] The system is D-fructose selective, which is the expected selectivity for a monoboronic acid system.[26] This system takes advantage of the known interaction of alizarin red S with boronic acids.[144] The observed stability constants (K_{obs}) for the PBA alizarin red S assay were $160\,M^{-1}$ for D-fructose and $4.6\,M^{-1}$ for D-glucose in water at pH 7.4 (phosphate buffer). Developing this elegant system, Hu employed 3-pyridinylboronic acid and pyrocatechol violet in a competitive assay for D-glucose.[145] The observed stability constants (K_{obs}) for assay were $272\,M^{-1}$ for D-glucose in water at pH 7.4 (phosphate buffer). Therefore, this very simple system can be used to detect millimolar D-glucose.

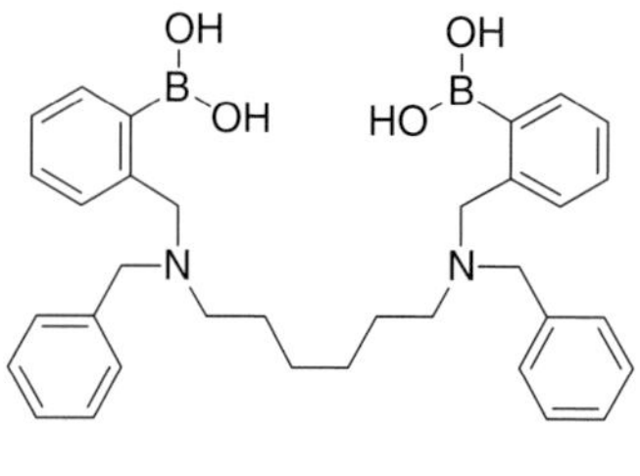

62

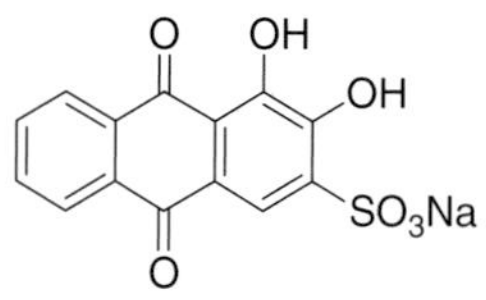

Alizarin red S

James has used alizarin red S in the design of a D-glucose-selective fluorescent assay.[146] Sensor **62** and alizarin red S show a sixfold enhancement over PBA for D-glucose. Sensor **62** can also be used at a concentration 10 times lower than PBA. The observed stability constants (K_{obs}) for **62** were $140\,M^{-1}$ for D-fructose and $66\,M^{-1}$ for D-glucose in 52.1 wt% methanol/water at pH 8.21 (phosphate buffer). Alizarin red S has also been used by Basu with a number of commercial monoboronic acids; it was found that 3-methoxycarbonyl-5-nitrophenyl boronic acid was more efficient than PBA in competitive assays.[147]

Elmas had detailed a temperature-sensitive copolymer, which contained boronic acid units; its interaction with ARS and loss of fluorescence upon exposure to a series of analytes was discussed.[148] However, boronic acid containing hydrogels that had been developed for electrophoresis applications (discussed later)[19,73] were reasoned to be suitable in the stationary phase for a similar supported system. Thus, this platform allows a dye displacement assay to be performed.[149]

Cox, Fossey, and James have prepared hydrogel spheres, 5 mm in diameter, incorporating a phenylboronic acid functionality that was exposed to an ARS dye. Exposure to an analyte diol (e.g., fructose) proportionally released the dye

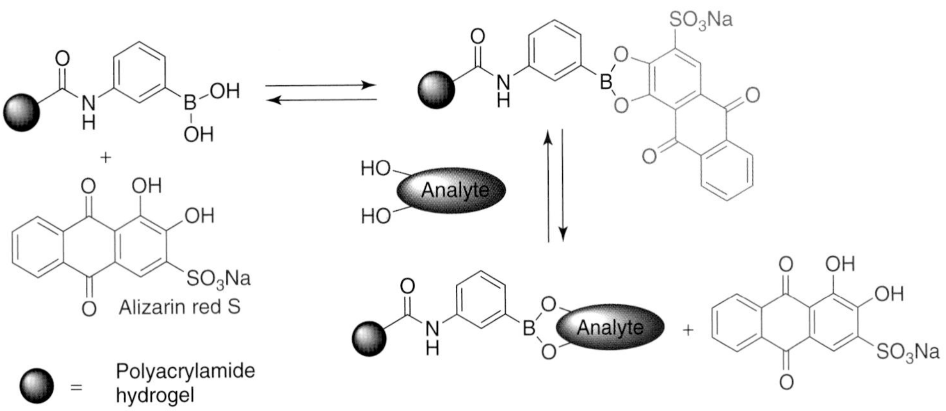

Scheme 7 Binding and analyte-mediated release of alizarin red-S with hydrogel-bound boronic acid.[149]

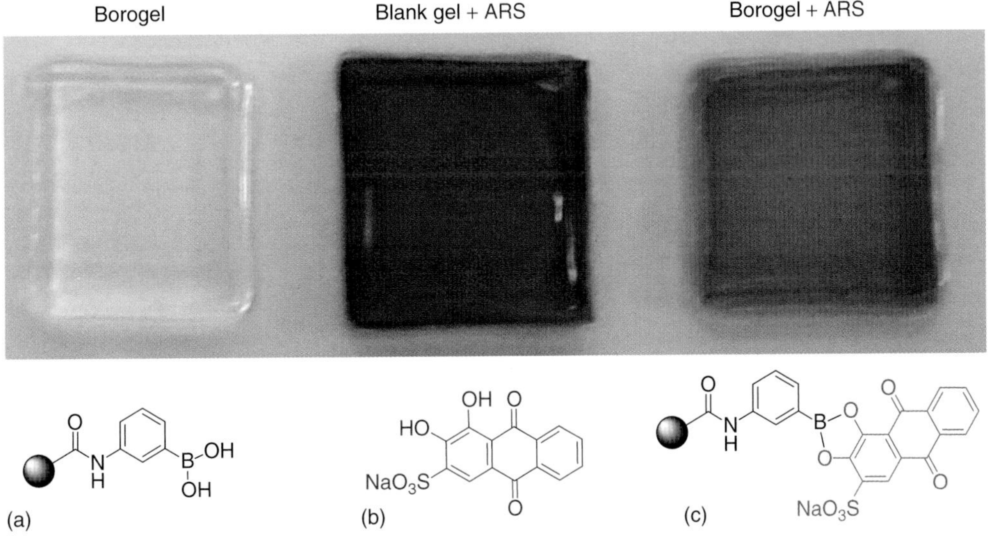

Figure 9 Gel slabs: (a) Borogel, (b) blank gel plus alizarin red-S, and (c) borogel plus alizarin red-S.[149]

into solution (represented in Scheme 7). To demonstrate the hydrogel displacement assay, the relative amounts of saccharides in samples of fruit juices were determined.

It is noteworthy that a hypsochromic shift is indicative of ARS binding to boron; that ARS develops an orange color upon incorporation into a boronic ester motif provided further visual corroboration of ARS binding to boron,[150] exemplified by comparing $10 \times 10 \times 1.5\,mm^3$ gel slabs that did and did not contain boron, as shown in Figure 9(b) and (c), respectively (a boron-containing gel prior to exposure to ARS is shown for comparison—Figure 9a).

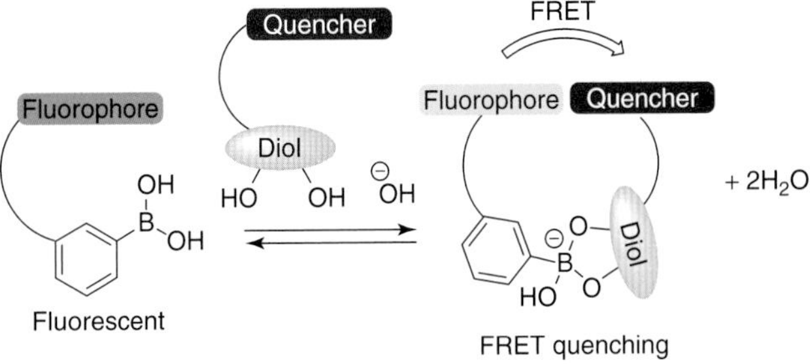

Figure 10 A fluorophore-appended boronic acid interacting with a diol-appended quencher.

3.5 Fluorophore quencher systems

The "molecular beacon" fluorophore—quencher pairs methodology used in quantitative PCR (polymerase chain reaction) assays such as Taqman™—may also be a useful signaling regime in the study of boron diol interactions. In order to probe this hypothesis, Fossey and James conceived a new signaling regime where a fluorescent boronic acid was quenched when exposed to an analyte diol, appended with a quencher, thus signaling the presence of the diol-appended quencher (Figure 10).

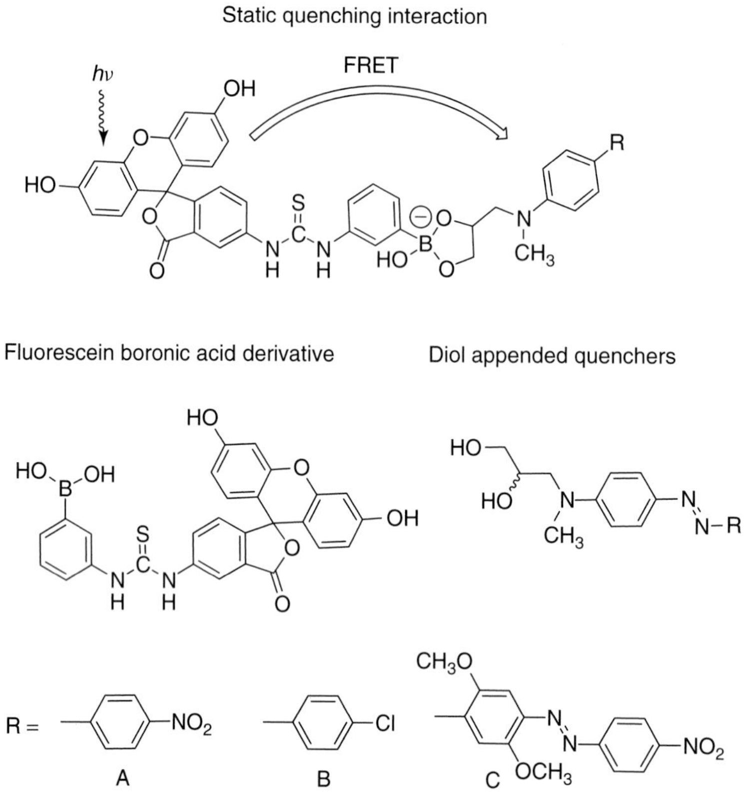

Static quenching interaction

Figure 11 A fluorescein boronic acid derivative, three diol appended quenchers (**a**–**c**), and a representation of the FRET quenching interaction.

A fluorescein boronic acid derivative was prepared to function as the fluorescent partner and a series of methyl-red-inspired diols were synthesized as quencher partners to probe the Förster resonance energy transfer (FRET) quenching sensing regime based on boronate ester formation (Figure 11).[151]

A detailed study of the combination of fluorescein boronic acid with diol-appended quenchers **a**–**c** and comparison with the fluorescence outputs of nonboron or nondiol-containing systems (i.e., fluorescein or methyl red were employed directly) revealed that the boronate ester formation results in enhanced quenching in each case, and that compound **c** is the best overall quencher.[151] Nucleosides were also shown to bind to the same fluorescein boronic acid derivative. While the quenching ability of each nucleoside tested was different, the same ratiometric quenching enhancement was observed in each case, suggesting similar binding affinities.

Fossey and James have also developed a self-assembled boronic acid hybrid system for surface-plasmon-enhanced fluorescence detection of quencher-labeled diols.[152] Surface plasmon excitation of the read-out fluorophore has dual advantages: firstly, surface plasmon resonance (SPR) may be concomitantly conducted, and no incident light enters the sample chamber, implying that any observed photons are only due to excitation and emission from the surface-appended fluorophores. Utilizing a boronic acid receptor to catch quencher analytes is a generic sensing format that is schematically illustrated in Figure 12 with an underlying mechanism closely related to the schematic of Figure 11.

In order to assemble a sensor construct at a gold-streptavidin surface, the molecule fluorophorelinker boronic acid biotin (FLAB) was prepared. The design incorporated a terminal biotin for attachment to surface-bound streptavidin, a boronic acid receptor, and a fluorophore (Alexafluor 647, Invitrogen, ex_{max} 647 nm). A quencher–diol conjugate was prepared utilizing a quencher for Alexa-Fluor 647, BHQ-3 (Biosearch Tech) (Figure 13).[152]

Attachement of FLAB to a streptavidin-appended gold surface was confirmed by both SPR and concomitant fluorescence surface plasmon resonance (f-SPR). Exposure of the surface thus prepared to BHQ-diol gave rise to both fluorescence quenching and an SPR response demonstrating the potential for the dual techniques for SPR and fluorescence to work in unison in a sensor regime under the guise of f-SPR.

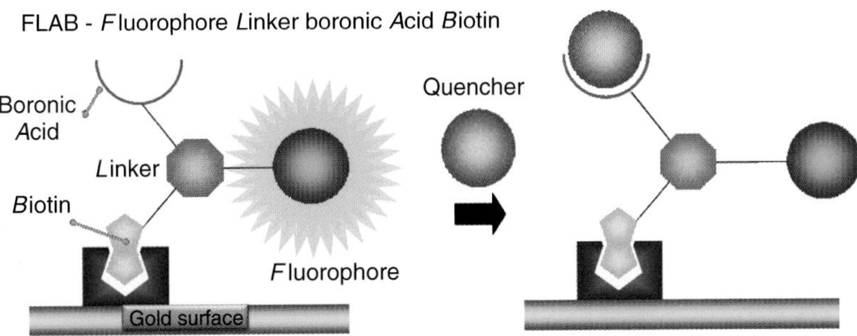

Figure 12 Surface-appended fluorophore linker boronic acid biotin (FLAB) for use in a fluorescence surface plasmon resonance (f-SPR).

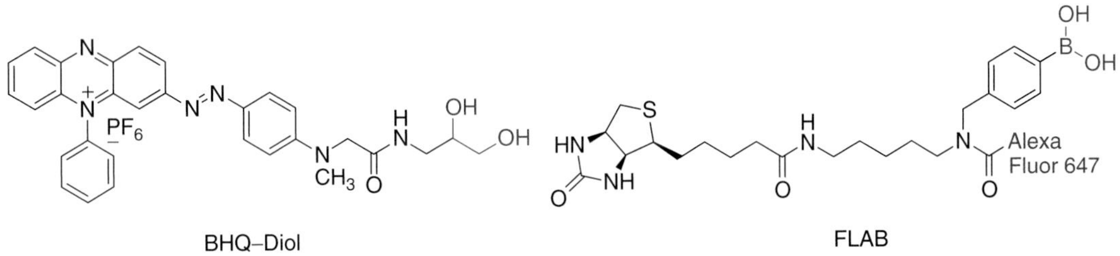

BHQ–Diol

FLAB

Figure 13 BHQ–diol and FLAB units for f-SPR assay.

4 ELECTROCHEMICAL SENSORS

Electrochemical detection of saccharides by enzymatic decomposition of saccharides is the basis of most current commercial D-glucose biosensors.[153] The development of boronic acid-based electroactive saccharide receptors for D-glucose is also possible. However, the main value of the boronic acid-based synthetic systems is that they can provide selectivity for a range of saccharides other than D-glucose.

Chiral ferroceneboronic acid derivatives have been synthesized by Ori and tested for chiral electrochemical detection of monosaccharides.[154] The best discrimination was observed for L-sorbitol and L-iditol at pH 7.0 in 0.1 mol dm^{-3} phosphate buffer solution. Moore and Wayner have explored the redox switching of carbohydrate binding with commercial ferrocene boronic acid.[155] From their detailed investigations, they have determined that binding constants of saccharides with the ferrocenium form are about 2 orders of magnitude greater than those for the ferrocene form. The increased stability is ascribed to the lower pK_a of the ferrocenium (5.8) than ferrocene (10.8) boronic acid.

Niwa has employed ferrocenylboronic acid and an enzyme-modified electrode for the electrochemically detection of lipopolysaccharides (LPS).[156] The electrode produced a rapid response to LPS, and the detection limit for the LPS from *Escherichia coli* O127 : B38 was 50 ng ml^{-1}.

Fabre has investigated the electrochemical sensing properties of the boronic acid-substituted bipyridine iron(II)

63

complex **63**. On addition of 10 mM D-fructose, the oxidation peak was shifted by 50 mV toward more positive values.[157]

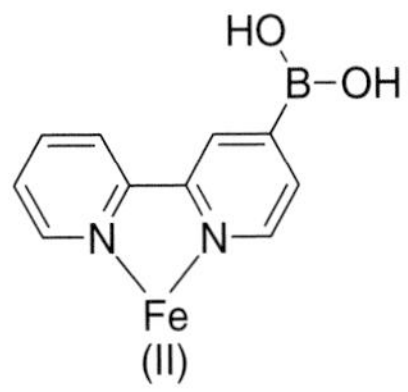

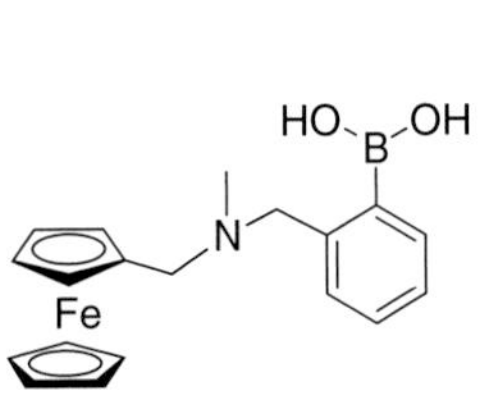

64 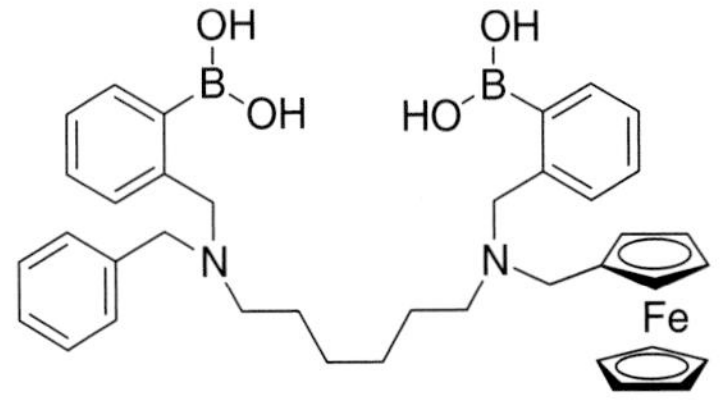 **65**

James has prepared a ferrocene monoboronic acid **64** and diboronic acid **65** as electrochemical saccharide sensors.[158] The monoboronic acid system **64** has also been prepared and proposed as an electrochemical sensor for saccharides by Norrild and Sotofte.[159] The electrochemical saccharide sensor **65** contains two boronic acid units (saccharide selectivity), one ferrocene unit (electrochemical read out), and a hexamethylene linker unit (for D-glucose selectivity). The electrochemical sensor **65** displays enhanced D-glucose (40 times) and D-galactose (17 times) selectivity when compared to the monoboronic acid **64**.

Marken has studied the transfer of the α-hydroxy-carboxylates of glycolic, lactic, mandelic, and gluconic acid from the aqueous electrolyte phase into an organic 4-(3-phenylpropyl)-pyridine (PPP) phase using a triple-phase boundary electrode system.[160] On addition of naphthalene-2-boronic acid, transfer of the α-hydroxy-carboxylates is facilitated and a shift of the reversible transfer potential to more negative values is observed, indicating fast and reversible binding. Marken has also prepared *N*-hexadecyl-pyridinium-4-boronic acid hexafluorophosphate monolayer films for the electrochemical sensing of catechol and L-dopa, dopamine, caffeic acid, and alizarin red S.[161] The same group has also used alizarin red S to probe nanofibrillar cellulose membranes modified by addition of boronic acid dendrimers.[162]

Tucker has recently employed chiral ferrocene boronic acid **66** to electrochemically determine the enantiomeric excess of the chiral diol (binol),[163] which took its inspiration from the three component systems used to determine enantiomeric excess of amines and diols by NMR developed by Bull and James.[15–18, 164–167]

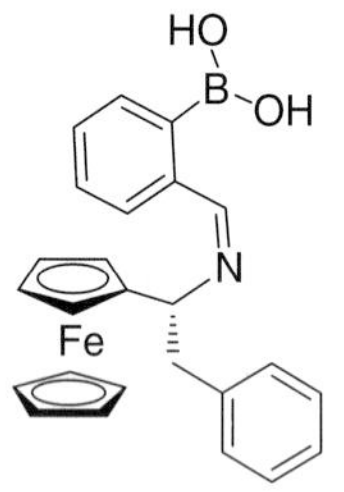

66

5 ANION RECEPTORS

5.1 Fluoride receptors

The first detailed investigation of Lewis acidic boron binding to fluoride ion was published in 1985 by Katz (compound **67** pictured in Scheme 8). The adduct **67**·F^- was characterized successfully by $^{19}F-^{1}H$ and $^{19}F-^{13}C$ coupling in its ^{1}H and ^{13}C NMR spectra, with large $^{19}F-{}^{1}H_{CH_3}$ coupling verified on a 60-MHz spectrometer. For compound **67** on fluoride binding, the B–B distance was found to be significantly shorter and sp^3 character at the boron centers was indicated by ^{11}B NMR.[168] Katz also prepared an analogous system where one boron is replaced by a trimethyl silyl group and using X-ray analysis found that the fluoride bridges the boron and silicon.[169]

In 1991, Reetz *et al.* observed that a crown ether, the archetypal cation-binding skeleton, appended with a Lewis acidic boron center, served to solubilize a stoichiometric amount of a suspension of KF in dichloromethane.[170] The proposed dual host–guest system **68** was examined by ^{11}B and ^{13}C NMR spectroscopies. The crown ether displayed low-field shifts in the ether carbon atoms expected due to cationic binding. More noteworthy, the complex showed an upfield shift in ^{11}B NMR from 30 ppm characteristic of sp^2 to 10 ppm, indicating a pseudo tetrahedral sp^3 environment at the boron center. KCl and KBr did not bind monotopically or heterotopically even after two weeks. The system was also very selective because, in the presence of the potassium salts of fluoride, chloride, bromide, and iodide, only the fluoride adduct was observed (Scheme 9).

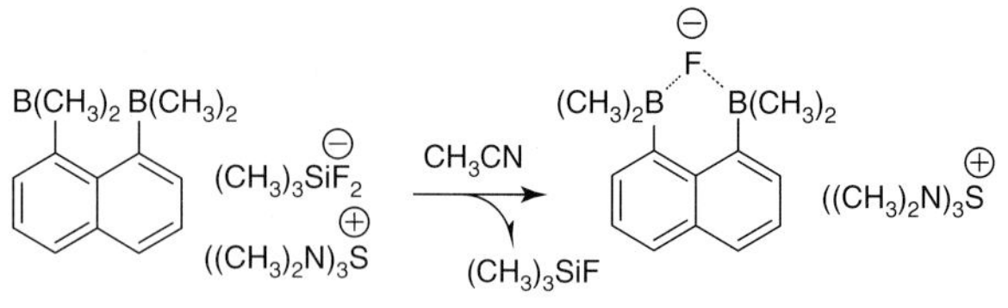

$B(CH_3)_2$ $B(CH_3)_2$ $(CH_3)_3SiF_2^{\ominus}$ $((CH_3)_2N)_3S^{\oplus}$ $\xrightarrow{CH_3CN}$ $(CH_3)_3SiF$ $F^{\ominus}$ $(CH_3)_2B$ $B(CH_3)_2$ $((CH_3)_2N)_3S^{\oplus}$ **67**

Scheme 8 The first Lewis acid-based bidentate fluoride receptor **67**.

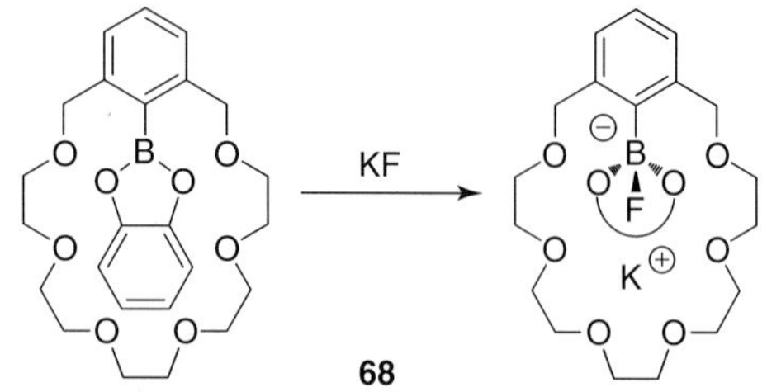

KF **68**

Scheme 9 A ditopic receptor for potassium fluoride based on a crown ether.

Jacobsen *et al.* performed a theoretical investigation of organoboron macrocycles and determined that it was possible to design different Lewis acid macrocycle hosts for optimum binding with specific anions.[171] Molecular orbital (AM1) calculations were performed on trimethylboron, two classes of boron-containing macrocycles, and their anion complexes with H^-, F, Cl^-, and O^{2-}. These calculations revealed that ion binding occurs with a change in boron hybridization from sp^2 to sp^3. The nature of the anion–boron interaction was found to be more like a covalent bond than a purely ionic interaction. A significant amount of charge is transferred from the anion to the host upon complexation. An important consequence of this is that anions can fit into cavities that are smaller than their ionic diameters. Size exclusion, μ-anionic bridge formation, and cage flexibility are all key factors for anion complex formation. Formation of a μ-anion bridge was found to improve ion binding; one key outcome was the tendency of fluoride to bind exclusively in a μ-F bridge, bound to only two boron atoms even in the presence of four Lewis acidic sites. However, chloride recognition appears to be favored by the inclusion of four boronic acid moieties. Another finding was that host flexibility facilitates μ-F type binding at the expense of selectivity, a particularly pertinent point when considering molecular design.

5.2 Fluoride sensors

In 1995, Shinkai showed that the interaction between the Lewis acidic boron and strongly basic fluoride could be exploited to create a means of determining the concentration of fluoride present in aqueous solution, even in the presence of other anions including halides.[172] This landmark paper was the first example of transforming a receptor unit for fluoride anion into a chemosensor. The electron-withdrawing nature of the ferrocenyl group in compound **69** increases the affinity of the boron for fluoride while simultaneously providing the redox active center. Increasing the concentration of fluoride ions incurs a decreasing polarographic half-wave potential, which is linear over a range of 200 mM. In 9 : 1 H_2O–CH_3OH, selectivity for fluoride over chloride was 500-fold, and over sulfate it was 50-fold. Hydroxide interacts with the boron center in a similar fashion to that with fluoride, but this only occurs at high pH when the concentration of hydroxide is high.

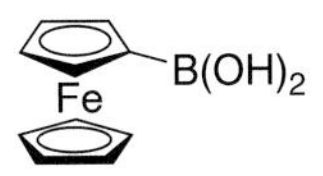

B(OH)2 Fe Ph O B O Ph Fe B O O Ph Ph

69 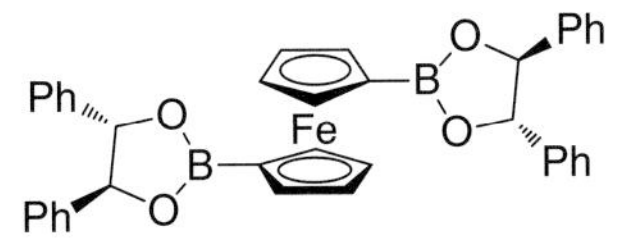 **70**

N(CH3)2 B(OH)2 Fe

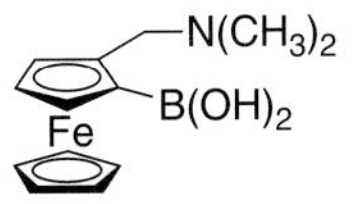

71

A related system **70** containing two boronic esters has been developed by Aldridge *et al.*, who observed that the bis-fluoride adduct undergoes a color change from orange to green as the ferrocene is aerobically oxidized to ferrocenium.[173] The same group has also developed a boronic acid with dimethyl amine **71** to act as a ditopic receptor for HF.[174]

Shinkai futher elaborated his ferrocenyl system and created a colorimetric system capable of visually detecting fluoride binding, thus devising an optical sensor for fluoride.[175] Taking advantage of a redox reaction between the dye molecule methylene blue and ferrocenyl boronic acid, it was possible to visually determine fluoride concentrations. Decolorization of the dye occurs over a 4×10^{-3}–3×10^{-2} mM fluoride range (monitored by UV–vis spectroscopy as a decreasing absorbance at 665 nm).

In 1998, James published the first fluorescent sensors with a selectivity for fluoride.[176] Fluorescence quenching of a series of simple aromatic boronic acids (**72**–**74**)

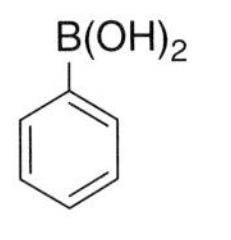

B(OH)2 B(OH)2

72 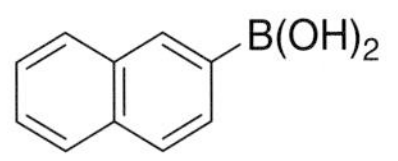 **73**

B(OH)2 N HO F H HO B N⊕ ⊖ Ph

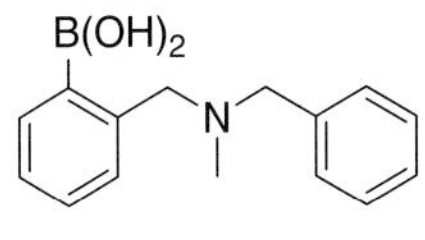

74 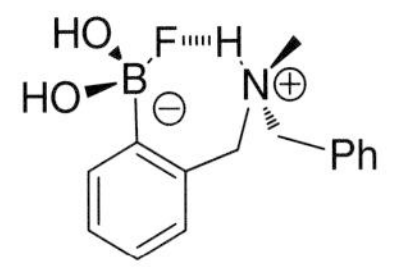 **75**

was observed in buffered aqueous methanol solution at pH 5.5 upon addition of KF. Tetrahedral boronate anions had already been shown to quench the fluorescence of directly attached fluorophores in research toward saccharide sensors and this ICT mechanism was shown to proceed upon fluoride binding (in the chapter, we have suggested that the observed fluorescence changes were caused by PET, it has, however, become clear that ICT is the more reasonable explanation of the observed fluorescence changes). The ^{11}B NMR spectra of **72** and **73** displayed shifts consistent with a change from sp^2 to sp^3 boron center as the concentration of fluoride was increased from 1 to 5 equivalents. Compounds **72**–**73** allow fluoride concentrations to be determined over a 50–70 mM range. With **74**, a tertiary amine component was introduced to provide an additional hydrogen-bonding site. The amine proton of **74** has a pK_a of 5.5, so under the measurement conditions the nitrogen is partially protonated, allowing a hydrogen-bonding interaction with fluoride (**75**). The two binding sites of **75** significantly enhance the binding, permitting determination of fluoride at lower concentrations (5–30 mM). This family of compounds serves to introduce the concept of "tunability" by making simple alterations to the molecule; it is also possible to change the binding strength without interfering with the mode of action. This is an essential feature of sensors required for commercial applications, as different environments require monitoring across varying concentration ranges. The fluorescence sensor **76**, closely related to **74**, was prepared by Yoon *et al.* where the fluorophore was a fluorescein; this system produced large fluorescence changes and a visible response to added fluoride.[177]

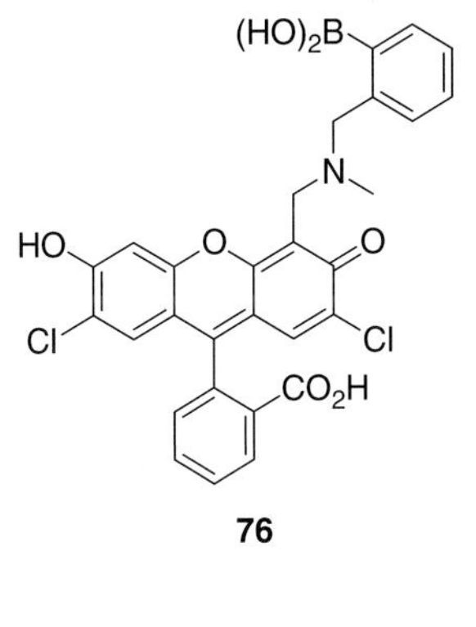

(HO)2B N HO O O Cl Cl CO2H

76

In order to develop receptor systems with enhanced fluoride binding, we investigated the use of rigid frameworks as scaffolds for boronic acid-based anion sensors. We found that the bis(bora)calix[4]arene **77** acts as a sensor for tetra-*n*-butylammonium fluoride (Bu_4NF) in chloroform.[178] Subsequently, in order to probe the factors affecting fluoride binding, we prepared the related boronates **80** and **81**.[179] On addition of an excess of Bu_4NF to solutions of **80** or **81** in chloroform, dramatic color changes were observed from colorless to yellow (**80**) or purple (**81**). In both the cases, ^1H NMR spectroscopic analysis clearly indicates Bu_4NF-mediated cleavage of the boron–aryloxide bond, an observation in agreement with a recent report by Bresner *et al.*[180] These observations led us to investigate the addition of fluoride to the parent phenols **78** and **79**, and they also exhibited similar color changes to those observed with **80** and **81**. In light of the similar behavior of both phenols and arylboronates in the presence of fluoride, a feasible mechanism for this colorimetric response involves a common phenolate anion intermediate obtained via either fluoride-mediated deboronation or deprotonation, followed by ambient oxidation to a colored radical that is stabilized by the 2,6-dialkyl substitution. The importance of radical formation in the generation of colored species was established using electrochemical techniques.[179] While fluoride caused deboronation of compound **81** and dramatic color changes, Bu_4NCl and Bu_4NBr produced no color change. Bu_4NCl caused fluorescence quenching of compound **81** but did not quench **80**, or alcohols **78** and **79**. Bu_4NBr did not cause a significant change in the fluorescence spectra of compounds **78–81**. We attributed the fluorescence quenching by chloride to a conformational change in the fluorophore caused by bidentate binding of chloride through two BOH hydrogen bonds.

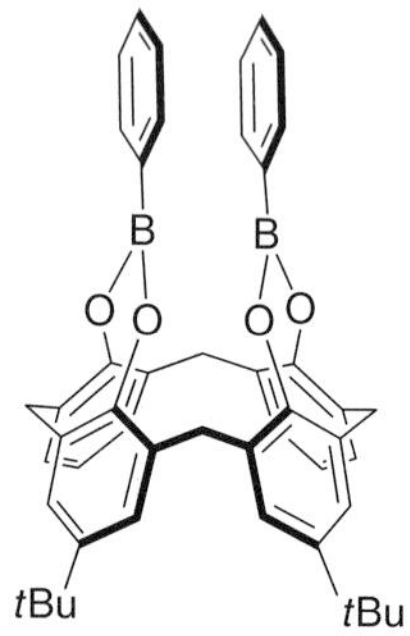

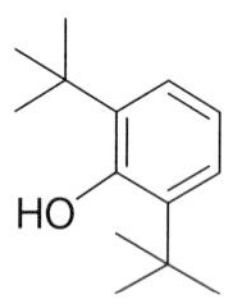

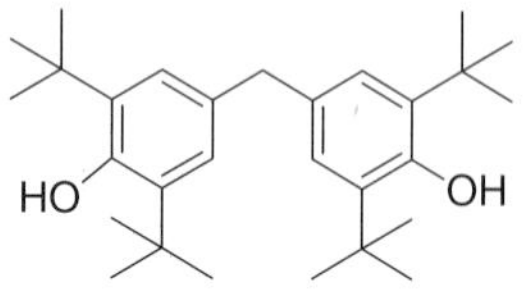

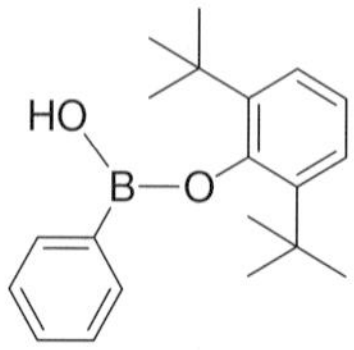

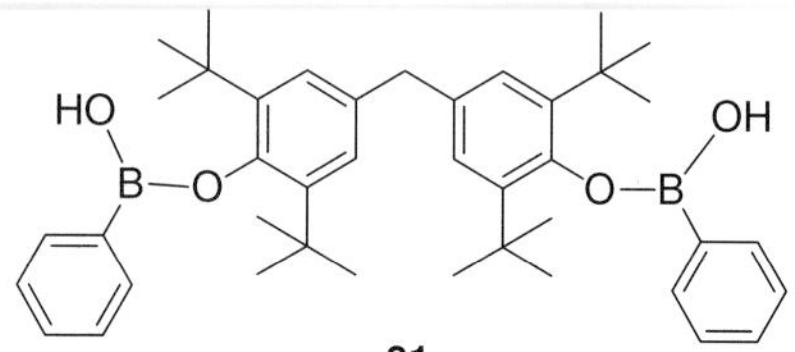

tBu tBu

77

HO

78

HO OH

79

HO B—O

80

HO OH B—O O—B

81

In 2001, Yamaguchi *et al.* reported a range of boron-containing species **82a–c** and **83** that showed a visible color change upon fluoride binding in THF media.[181] The highly conjugated system was interrupted by the boron–fluoride interaction and the concurrent change from sp^2 to sp^3 boron.

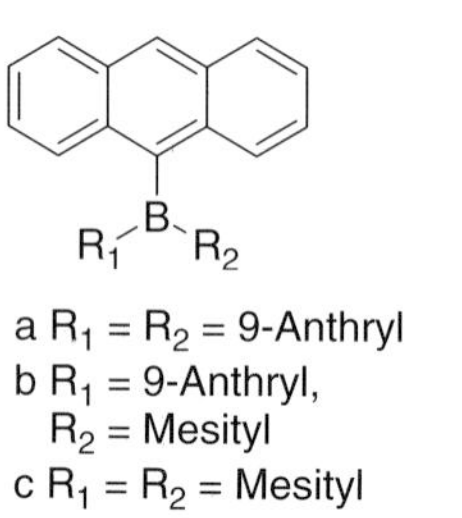

R_1 B R_2

a $R_1 = R_2$ = 9-Anthryl
b R_1 = 9-Anthryl, R_2 = Mesityl
c $R_1 = R_2$ = Mesityl

82

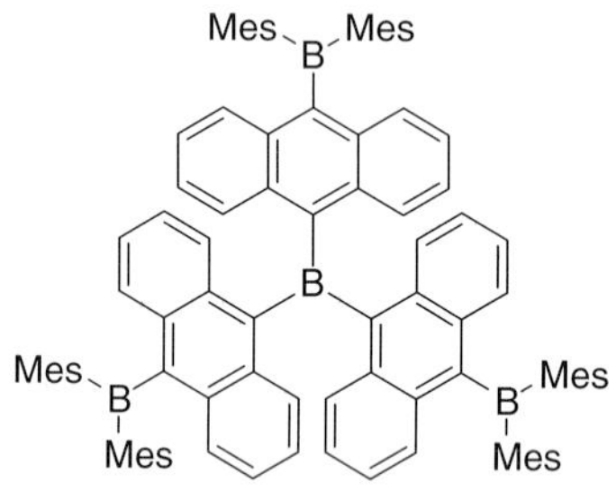

Mes B Mes

Mes B Mes B Mes B Mes

83

In 2003, Shinkai developed a colorimetric and ratiomeric fluorescence chemosensor **84** that displayed three emission responses to fluoride ions at 356, 670, and 692 nm.[182] The system comprised a porphyrin and a triarylborane center connected via a conjugated linker. Changing the conjugation of a laterally expanded porphyrin results in a significant hypsochromic shift of the Soret band and a bathochromic shift of the Q band. Fluoride coordination to the boron center generates an anionic sp^3 boron, thus disrupting the linker conjugation and changing the energy pathway. Evidence for the latter was provided by measuring the fluorescence decay of the emission band. The fluorescence lifetime at 515 nm of free **84** corresponds to less than 100 ps, which lengthened to 0.53 ns in the fluoride-bound species. Compound **85** does not contain a porphyrin-linked component, and has a longer lifetime than that of **84** at 4.52 ns. The shorter lifetime of **84** can be rationalized by Dexter-type energy transfer from the triaryl borane to the porphyrin ring system. However, only limited energy transfer can occur across the host–guest complex **18**$\cdot F^-$. What is particularly noteworthy about this publication is that the authors have developed a ratiomeric sensor. For the majority of reported fluorescent sensors for fluoride, the binding of the anion causes a quenching of the fluorescence emission. Only a few sensors in which the binding of a fluoride ion causes an increase in the fluorescence have been reported.[183] However, in most practical applications, changes in fluorescence intensity (fluorescence quenching or enhancement) can also be caused by many other poorly quantified or variable factors such as photobleaching, sensor concentration, the environment around

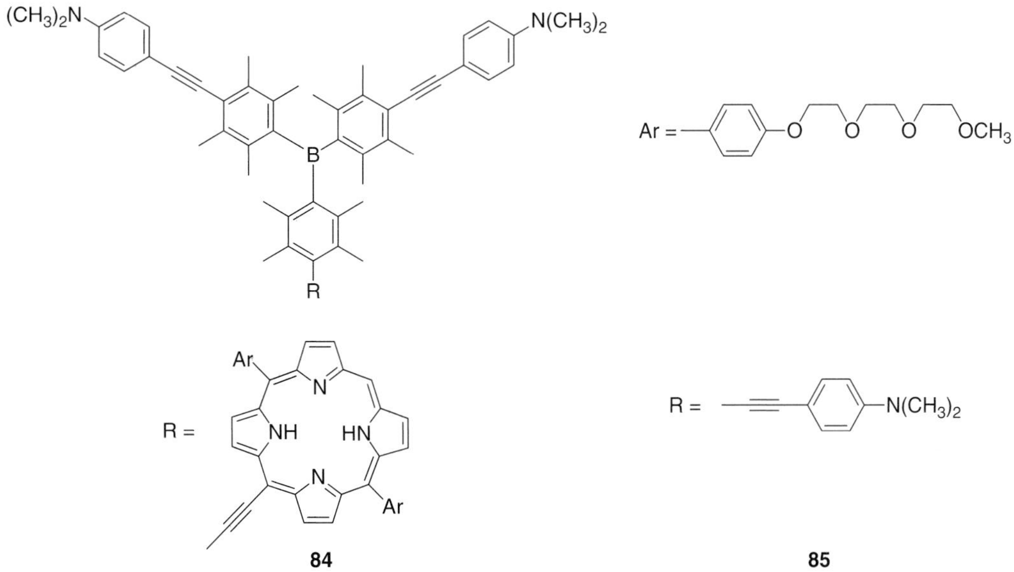

84 **85**

the sensor molecule (polarity, temperature, and so forth), and the stability of the sensory system under illumination. To increase the selectivity and sensitivity, ratiometric measurements are utilized, which involve the observation of changes in the ratio of the intensities of the absorption or the emission at two wavelengths. Ratiometric fluorescent probes have an important feature in that they permit signal rationing and thus increase the dynamic range and provide built-in correction for environmental effects.

Gabbaï is at the forefront of the challenging prospect of anion recognition in aqueous media. One of his first forays into this area came in 2004 when his group used a neutral bidentate diborane species to bind fluoride, creating a colorimetric sensor **86** that is not affected by the presence of water. The cornerstone of the group's research is the use of a rigid 1,8-naphthalene backbone with two proximal Lewis acidic sites promoting fluoride anion chelation. The association constant of **86** (Scheme 10) with fluoride was $5 \times 10^9\,M^{-1}$ in THF, which was higher than that observed for any previously documented monofunctional borane receptor. Addition of $B(C_6H_5)_3$ permits conversion back to the unbound sensor, confirming reversible binding.[184] The visual cue to the binding event is the dissipation of the vivid yellow color, recordable as a decreasing absorption at 340–390 nm by UV spectroscopy. The bridging fluoride species $[\mathbf{86}\mu_2\text{-F}]^-$ was successfully isolated and characterized, confirming the bidentate binding by shifts in the ^{11}B NMR spectra; the expected -188 ppm ^{19}F NMR signal and X-ray analysis showed B–F bonds and pyramidal B centers.

Kubo and James have reported a novel sensor system in which anions induce self-organization of 3-nitrophenylboronic acid and alizarin.[150,185,186] In the presence of fluoride or acetate ions, the two components bind, "switching on" the fluorescence. In the absence of a strongly basic anion, the reporter and receptor barely interact. Once the boronic acid is converted into the phenylfluoroboronate, the change from sp^2 to sp^3 and the increased electron density on the boron stabilizes the ensemble (Figure 14). The system is not completely selective but successfully demonstrates an interesting way to bring unbound components of a sensor together in the presence of a target analyte.

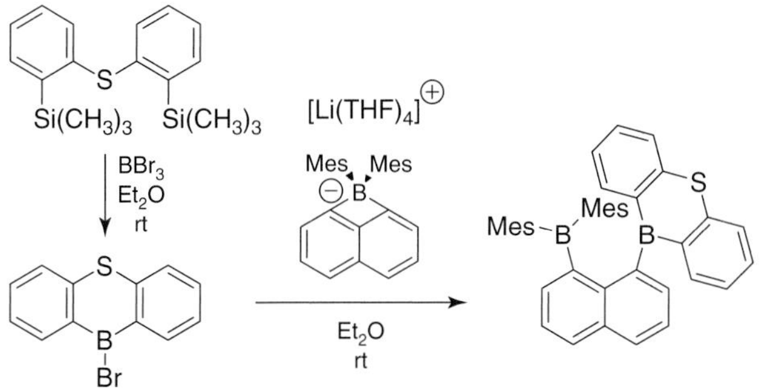

Scheme 10 Facile synthesis of Gabbaï's first bidentate system for fluoride **86**.

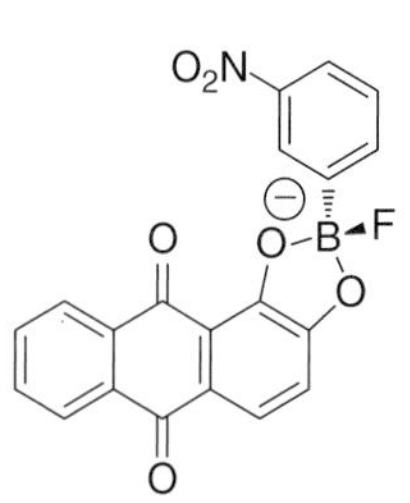

Figure 14 3-Nitrophenylboronic acid and alizarin self-organize in the presence of various anions.

Kubo and James have shown that it is possible to use anion-directed self-assembly to control the formation of molecular capsules between boronic acid and diol trimeric units.[22,23] The system consists of cyclotricatechylene and a boronic acid-appended hexahomotrioxacalix[3]arene. The two components do not interact with each other until Et_4NOAc is added to the solution. On addition of Et_4NOAc, quantitative formation of a capsule by boronate esterification is observed. The self-assembly process is a result of anion-directed boronate ester formation around the Et_4N^+ template. Reversible boronate esterification also allowed for selective control of capsule formation as a function of pH.

A group headed by Liu has reported a highly sensitive and selective sensor for fluoride based on an organic borane **87**.[187] A color change of bright green to colorless accompanies the fluorescence quenching on fluoride binding. The dimesityl tri-coordinated boron species showed interesting two-photon excited fluorescence (TPEF) and single-photon excited fluorescence (SPEF) activity on binding with fluoride. The trigonal planar boron is shielded by the dimesityl groups, increasing the selectivity toward fluoride ion due to its small size. TPEF chemosensors have been widely used in conjunction with laser scanning microscopy in the imaging of ions in cellular processes due to the greater 3D spatial selectivity than SPEF techniques.

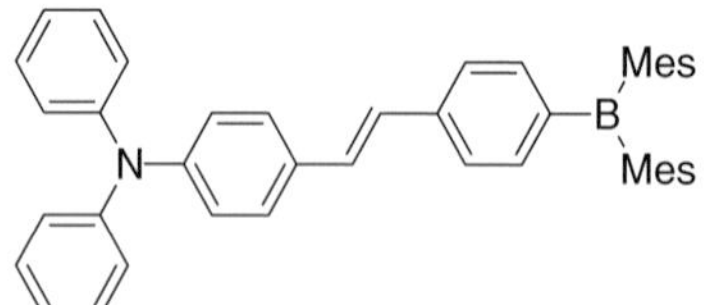

87

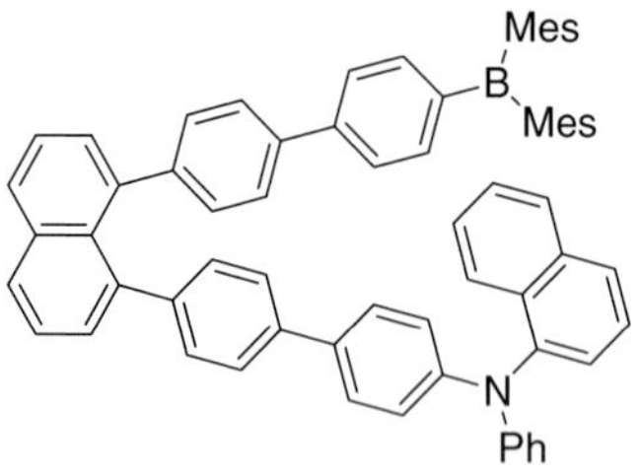

88

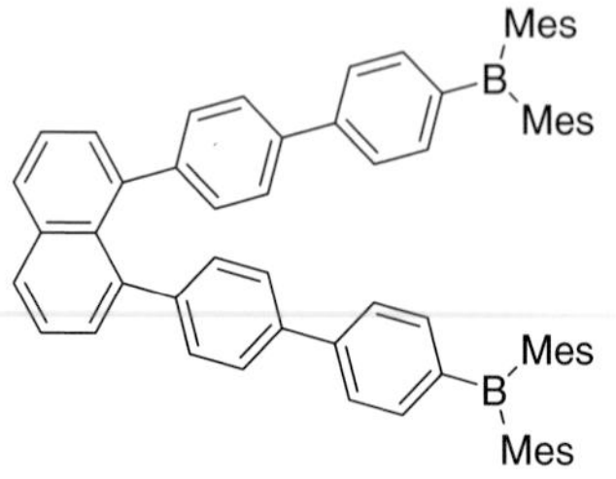

89

Most monodentate boron-based chemosensors display fluorescence quenching on anion binding, termed "switch-off" sensors. However, practical applications would greatly benefit from a "switch-on" response, particularly if accompanied by a visible color change. Liu reported the synthesis of fluorescent sensors **88** and **89** and observed selective fluorescent enhancement for fluoride over other halides with **88** in dichloromethane.[188] Molecular modeling predicts that the two biphenyl "arms" of compound **88** sit orthogonally to the naphthalene ring system, implying charge transfer through space from the "donor" nitrogen to the dimesitylboron "acceptor." Compounds 4,4′-bis((1-naphthyl) (phenyl)amino) biphenyl and 1,8-bis (4-[(1-naphthyl)(phenyl)amino]biphenyl-4′-ylnaphthalene), molecules closely related to the donor motif, exhibit a strong emission in the blue region of the spectrum. In solution, compound **88** emits in the green but on binding to fluoride the charge transfer is interrupted and a change to the characteristic blue color is observed. Compound **89** displayed a typical fluorescence quenching behavior, indicating that the amine functionality of **88** and its interaction with the neighboring boron center was critical to the desired enhancement effect.

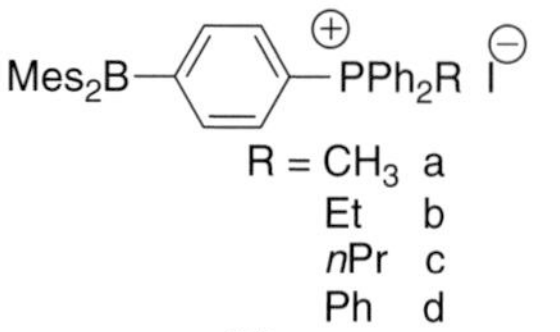

R = CH_3 a, Et b, *n*Pr c, Ph d

90

In perhaps the most compelling example of a fluoride sensor system to date, Gabbaï *et al.* synthesized an impressively simple series of sensors that incorporated a charged phosphonium unit and a triarylborane **90**. Binding constants were calculated at optimum pH (pH 4.6–4.9) in 9 : 1 H_2O–CH_3OH giving a maximum value of $10.5 \times 10^3\ M^{-1}$ for species **90d**.[189] The increased Lewis acidity required to overcome the large hydration enthalpy ($504\ kJ\ mol^{-1}$) of the fluoride anion in water has been proposed by the authors to be provided by the Coulombic effect of the cationic substituent. Across the series **90a–d**, it was found that as the hydrophobic character increases there is a corresponding increase in the Lewis acidity at the boron centers. To highlight the success of this body of work, compound **90d** is capable of binding fluoride in pure water at pH 4.9, at a concentration level of 4 ppm below the US Environmental Protection Agency's recommended maximum level for drinking water.[189]

5.3 Ditopic anion receptors

Sensors containing boronic acids (or derivatives thereof) as binding sites in combination with tandem binding sites have also been developed. Several groups have introduced additional functionality next to the boron center to either complement the binding or to introduce additional signaling units.

In parallel to their work on neutral bisboron bidentate frameworks for anion sensing, the Gabbaï group has looked at mixed Lewis acid centers and charged receptors for fluoride, aiming to develop systems capable of working in aqueous media. In 2005, Gabbaï developed a highly selective phosphorescent sensor **91** (Scheme 11) for the fluoride anion similar to sensor **86** but containing a mercury center.[190] Again, a μ_2-F^- bridged species was isolated and characterized fully by X-ray diffraction analysis. The distinct green solid-state phosphorescence changes to red upon binding. With a binding constant higher than that measurable by direct titration in THF ($K_a > 10^8\,M^{-1}$), this system also showed, albeit reduced, binding in aqueous media ($K_a = 3.3 \times 10^5\,M^{-1}$). Even the highly competitive acetate anion did not bind, possibly due to the fact that acetate does not have a μ_2-bridging binding mode. Phosphorescence is rarely utilized as an optical property for anion sensors in comparison to fluorescence. However, in this example, the authors took advantage of mercury's ability to induce phosphorescence of hydrocarbon chromophores via spin–orbit coupling at room temperature. These values are much higher in comparison to the boron only species, adding to the cooperative binding argument. In a related communication, the group outlined the synthesis of two very closely related cationic bidentate Lewis acids. This was done in an attempt to assess whether Coulombic attractions improved the binding constant. Stronger binding was indeed observed with the cationic version in partially aqueous—THF/H_2O (9 : 1)—media.[191] However, this system suffers from a weak response to acetate ions, so some degree of selectivity is lost.

Continuing their exploration into Coulombic factors, Gabbaï made a significant receptor development when they discovered that cationic borane **92** (Scheme 12) could capture fluoride ion across a phase barrier, thus extracting fluoride from water in a chloroform/water biphasic system.[192] X-ray diffraction studies confirmed the presence of a C–H···F–B hydrogen bond arrangement that was contributing to the binding motif. A striking piece of evidence for this behavior persisting in solution came from ^1H and ^{19}F NMR spectral analysis, paying particular attention

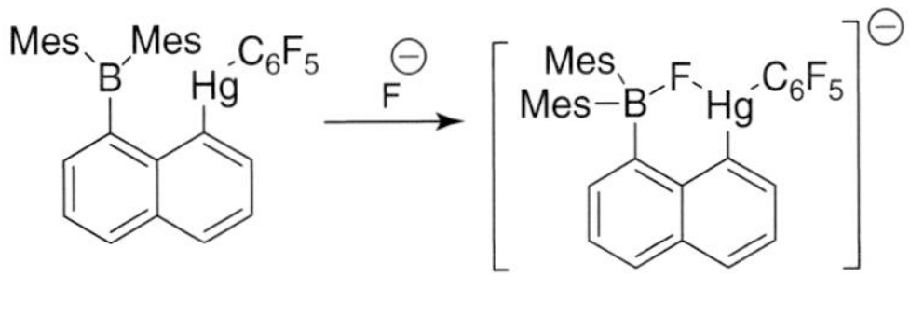

Scheme 11 Fluoride capture by a mixed Lewis acid system **91**.

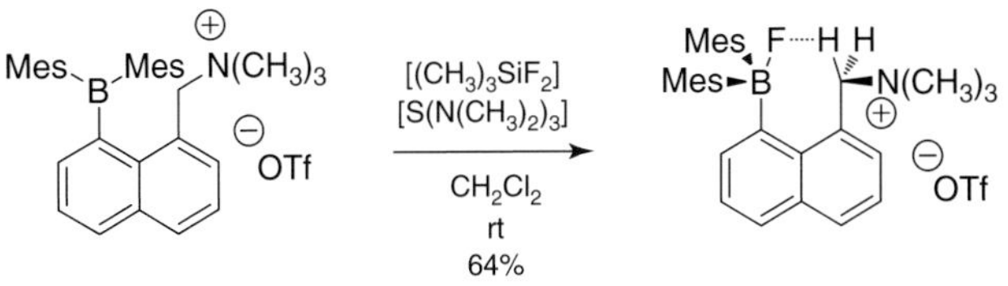

Scheme 12 Compound **92** combines Lewis acidic boron interaction with a neighboring hydrogen-bonding interaction to selectively extract fluoride across a biphasic barrier.

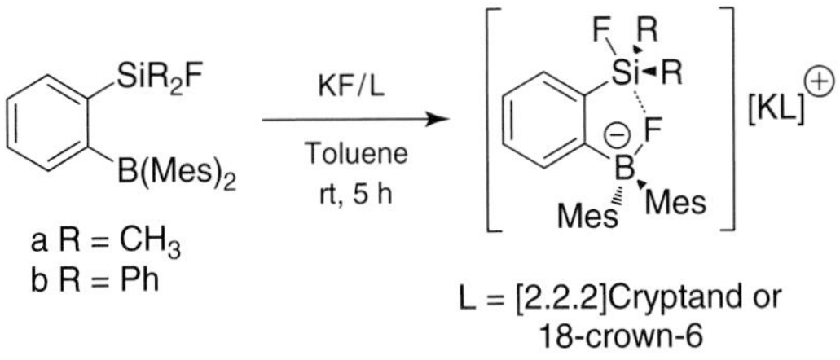

Scheme 13 A mixed silicon–boron bidentate receptor for fluoride **93**.

to the diastereotopic methylene protons on the nitrogen-containing substituents. Regarding species **92**·F^-, one of the resonances showed coupling to the fluorine nucleus ($^1J_{H-F} = 9.2\,Hz$) and was resolved as a doublet of doublets ($^2J_{H-H} = 12.9\,Hz$).

One recent mixed Lewis acid system, looking at an area that was surprisingly neglected considering the obvious relationship to Katz's system, containing both boron and silicon centers in a geometry set to encourage fluoride bridging has been synthesized.[193] Katz showed the 1,8-naphthyl arrangement to be relatively weakly binding but **93a–b** (Scheme 13), based on an *o*-phylene backbone, have shown stronger binding than the monodentate boron analogs. Fluoride-bound species **93a–b** that were analyzed using X-ray crystallography and Si–F coupling in the ^{19}F NMR spectra provided clear evidence of the silicon adopting a pseudo-pentacoordinate geometry and those with ^{11}B NMR analysis indicated a strong tetrahedral character. For this system, at present the high affinity of fluoride has not been utilized to develop a useful chemosensor.

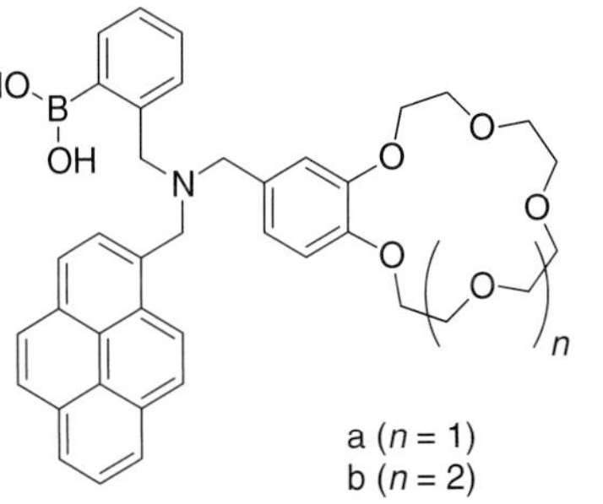

94

James has developed a ditopic receptor reminiscent of Reetz's crown ether boronic acid system **68**. The ditopic fluorescent sensors **94a–b** were developed as reversible AND logic gates, with selectivity for potassium fluoride.[194] The sp^2 hybridized boronic acid, which is a hard Lewis acid, interacts strongly with the fluoride anion, which is a hard Lewis base, to become sp^3 hybridized. The potassium cation is held *in situ* partly by the crown ether and partly by the electrostatic interaction with the fluoride anion. This cooperative complexation allows the cationic and anionic guests to be bound to the host as an ion pair while allowing the host to discriminate between potassium fluoride and other similar ion pairs such as potassium chloride and potassium bromide. The dynamic characteristics of this sensor are particularly attractive. Not only are both guests required at the binding sites to generate a fluorescence response but it is also possible to add and remove individual guests producing a controlled and reversible read-out signal derived from the AND logic functionality inherent to the sensor. This example provides an elegant representation of the dynamic and versatile disposition of these recognition systems, permitting selective and reversible binding to be designed into sensors and switches alike. The observed stability constants (K_{obs}) for sensor **94a** were $10 \times 10^3\ M^{-1}$ with potassium and $320\ M^{-1}$ with fluoride. For sensor **94b**, the constants were $50 \times 10^3\ M^{-1}$ with potassium and $250\ M^{-1}$ with fluoride in methanol.

Yoon has developed a bidentate receptor **95** for fluoride anions that employs a boronic acid site and an imidazolium group.[195] In a competitive aqueous solvent system (95 : 5 CH_3CN-HEPES), selectivity was achieved for fluoride over challenging acetate and phosphate anions, and a ratiomeric fluorescent response was observed. The orientation of the boronic acid group was critical, with only the ortho derivative (shown) displayed selectivity; the para and meta derivatives failed to display binding selectivity. The C–H hydrogen bond donor was anticipated to stabilize the binding, allowing recognition to occur in the competitive media. Binding as depicted in Scheme 14 was confirmed by ^{19}F NMR spectroscopic analysis, which indicated that one fluoride was hydrogen bonded.

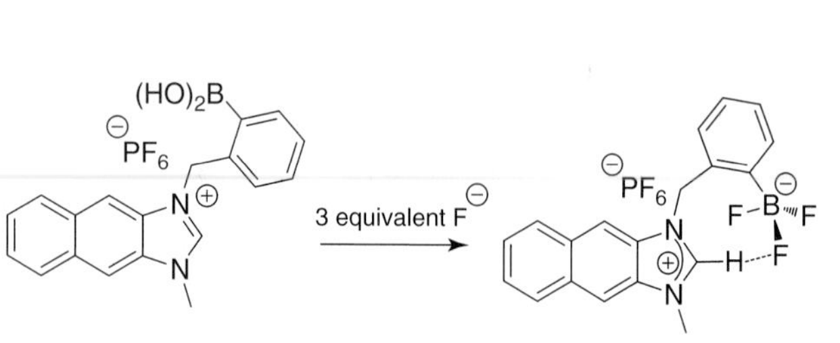

Scheme 14 A sensor that combines a boronic acid receptor with CH–anion interaction **95**.

5.4 Sensors for other anions

So far we have concentrated on the use of boronic acids and as sensors for fluoride ions. One noticeable exception is compound **81**, which also displayed a fluorescence response to chloride anions. Surprisingly, given that cyanide (pK_a 9.1) is more basic than fluoride (pK_a 3.2), the development of cyanide anion sensors based on boron centered Lewis acids has been relatively limited. Perhaps the most elegant and practical demonstration of how to tip the balance in selectivity between fluoride and cyanide comes from the Gabbaï group,[196] who have shown that cationic boranes **96** and **97** are selective receptors for cyanide and fluoride respectively in water at neutral pH (Scheme 15). The cyanide binding ability of **96** was attributed to favorable Coulombic effects, which increase the Lewis acidity of the boron center and strengthen the receptor–cyanide interaction. The selectivity could then be switched to fluoride by positioning the trimethylammonium functionality ortho to the boron center in **97**. Steric crowding of the boron center prevents coordination of the larger cyanide anion.

Kubo has developed a new strategy for the fluorescent detection of multiphosphates in aqueous solution using Zn(II)-DPA (DPA, dipicolylamine)-appended phenylboronic acid. **98**·Zn forms an assembly with alizarin-dye in CH_3OH–10 mM HEPES (1 : 1 v/v) containing 10 mM NaCl at pH 7.4 at 25 °C, in which the dye binds favorably to the coordinated zinc(II) in the DPA moiety. Addition of pyrophosphate (PPi) causes reorganization of the complex to produce an alternative boronate ester assembly and produces an increase in fluorescence and visible response. It is interesting to note that the system exhibited PPi-selectivity over other phosphates such as ATP (adenosine 5′-triphosphate), ADP (adenosine 5′-diphosphate), AMP (adenosine 5′-monophosphate), and Pi (inorganic phosphate); the competitive assay used to determine the apparent

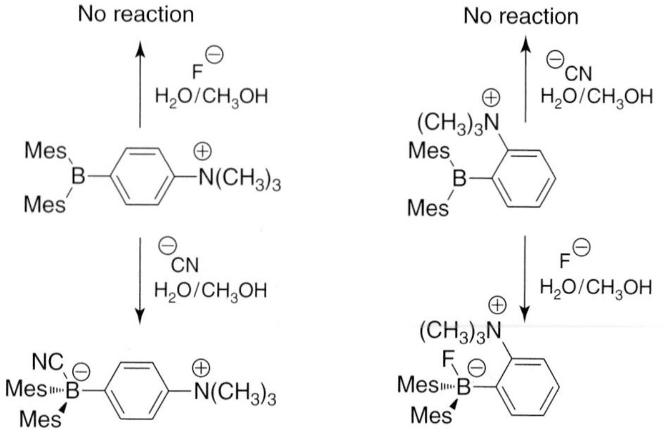

Scheme 15 Cyanide and fluoride selectivity of **96** and **97**.

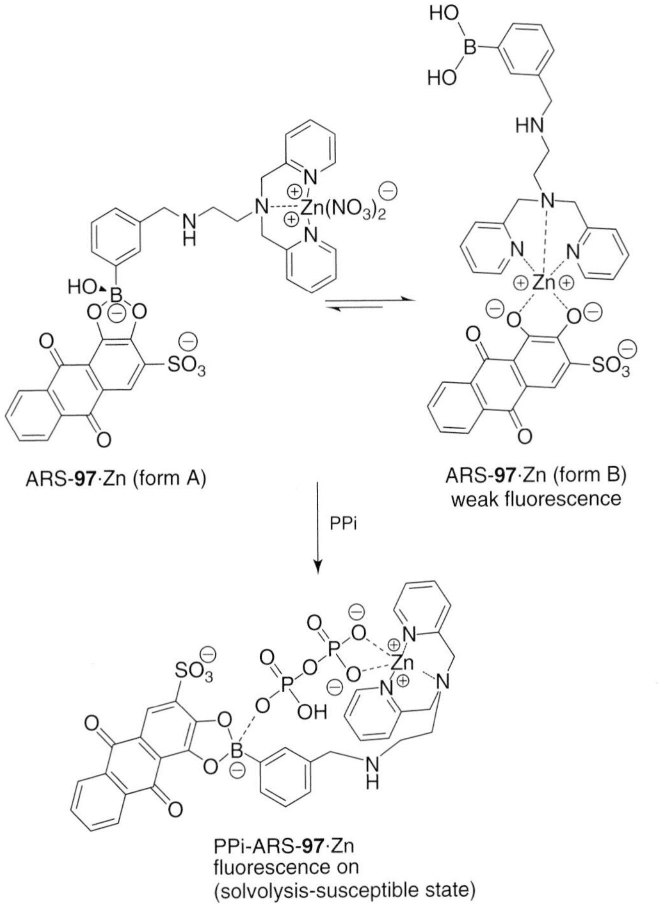

Scheme 16 Mechanism for PPi-induced fluorescent enhancement of **98**·Zn.

association constants of **98**·Zn with anions allowed us to estimate that the binding with PPi $(1.6 \pm 0.04) \times 10^6\ M^{-1}$ is 10-fold and 84-fold higher than that with ATP and ADP, respectively (Scheme 16).[197]

6 CHROMATOGRAPHY, ANALYSIS, AND SEPARATION

The reversible interaction of boronic acids with diol motifs has been exploited in separation science to great effect. Incorporation of boronic acids into the various stationary phases employed in chromatographic techniques has allowed for saccharide-selective or specific separation protocols to be developed. A particularly noteworthy area that lies out with the remit of this current chapter is the development of boron affinity columns used in HPLC, a collection of pertinent references are, however, provided.

Polyacrylamide gel electrophoresis (PAGE) exploits hydrogel polymers to separate molecules on a size and charge basis. Among the useful separations of biomolecules that electrophoresis is commonly employed for a technique for the separation of carbohydrates called fluorophore assisted carbohydrate electrophoresis (FACE) is especially notable for its ability to separate and sort oligosaccharides on a size and charge basis. However, the FACE, which requires fluorophore labeling of analytes, does not separate saccharides of similar size and charge and is limited to reducing sugars, so is not applicable to the analysis of many glycated proteins and other sugar-appended glycoconjugates.

In order to address the need to provide a separation tool for similar mass saccharides, Jackson *et al.* have developed boron affinity saccharide electrophoresis (BASE).[19] Through exploitation of the different affinities of saccharides for boronic acid, it was assumed that it would be possible to arrest gel electrophoresis mobilities to varying extents if boron could be incorporated into the stationary phase of an electrophoretic FACE experiment. Thus, a protocol for incorporation of boronic acid motifs into hydrogel domains was sought. Previous examples of boronic acid saccharide sensors served as an inspiration, and the team synthesized a range of acrylamide boronic acids, which were readily incorporated into a polyacrylamide hydrogel matrix of the kind employed in electrophoresis.[19] For comparison of the existing FACE technique with the new BASE technique, an optimal formulation was found to consist of 60 wt% water, 0.5 wt% boron-containing monomer, 1 wt% methylene bisacrylamide (cross-linker), and 38.5% acrylamide (Scheme 17). Blank (nonboron-containing) gels were prepared with 39 wt% acrylamide, all other conditions and reagents were unchanged.

Hydrogels could be prepared that contained *ortho*, *meta*, and *para* boronic acids, although solubility allowed up to about 3% reliable incorporation, but this was sufficient to obtain excellent results in electrophoresis. For the majority of investigations, the meta derivative was preferred due to its overall higher synthetic yield, in comparison to the para derivative, in which no differences were observed in terms of electrophoresis applications; however, the *ortho*

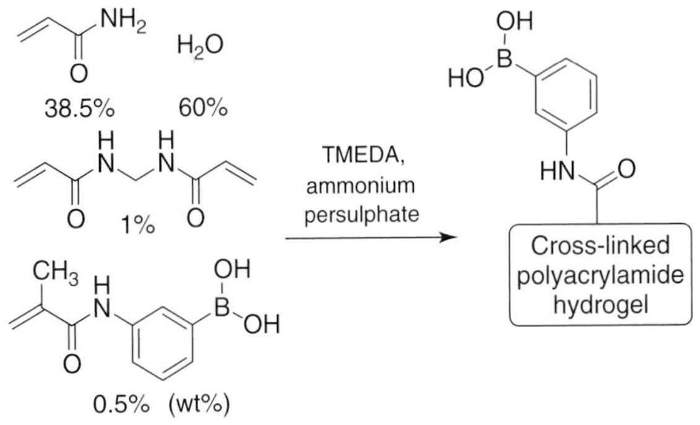

Scheme 17 Gel formulation protocol.

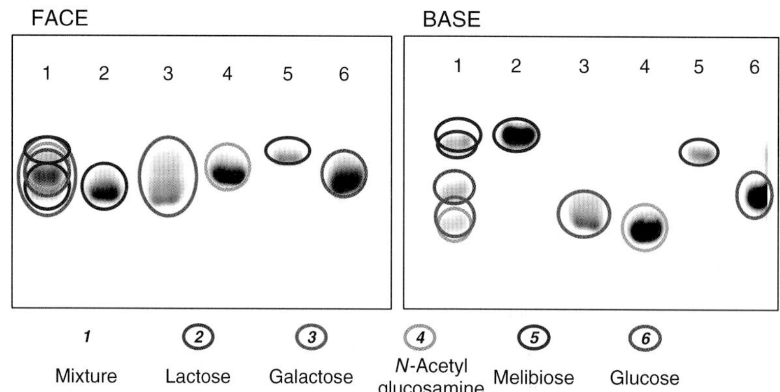

Figure 15 Comparison of electrophoretic separation of AMAC-labeled saccharides utilizing FACE and BASE (with and without boron, respectively).

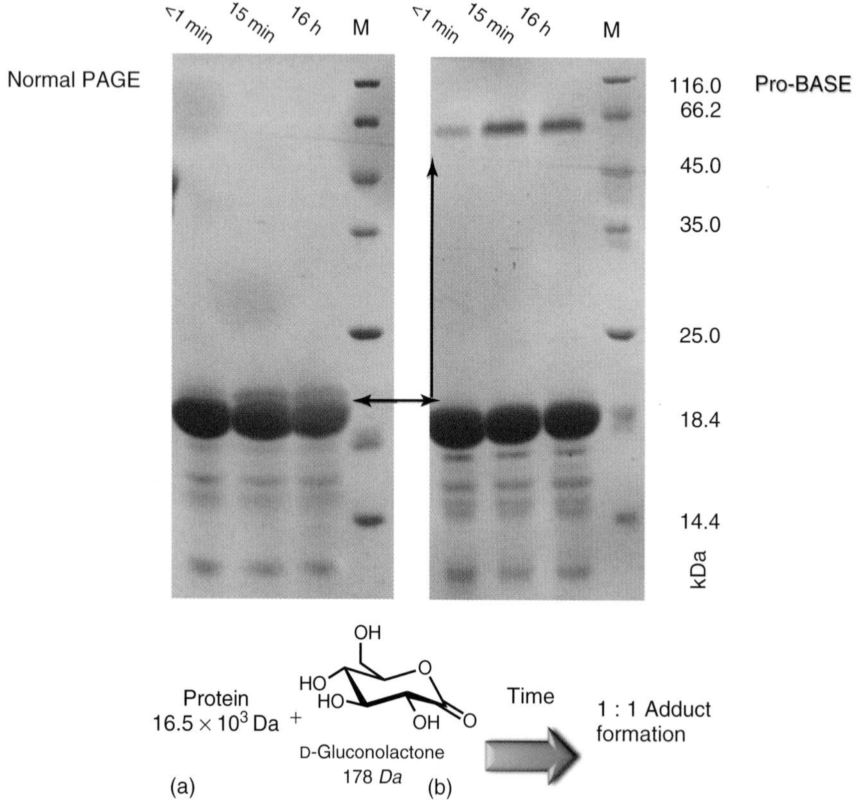

Figure 16 Separation by protein–boron assisted saccharide electrophoresis (pro-BASE) of a glyconoylated protein utilizing stationary phases that do not (a) and do (b) contain boronic acids.

derivative gave less effective separation (analyte mobility was more facile) and was prone to degradation (*ortho*-heteroatom-assisted increase in liability of the boronic ester part and self-polymerization are speculated to be the origin of less effective separation and degradation, respectively).

Gels for the FACE and BASE techniques were compared for the electrophoretic separation of fluorophore-labeled saccharides. Exemplified in Figure 15 where FACE performs poorly in separating a series of 2-AMAC labeled saccharides, the BASE system induces dramatic mobility differences among the saccharides employed. Using a BASE gel, previously inseparable saccharides are now clearly resolved; even though the resolution is not perfect in some cases, the modulated mobility achieved as a function of reversible boron diol interactions within a hydrogel domain resulted in a range of new applications, not only in electrophoresis[20] but also for applications such as hydrogel sensors mentioned earlier.[149]

BASE was next employed in the detection of D-gluconolactone modification of a protein that had been shown to inhibit the innate immune system and is under development as a therapy for complement-mediated acute inflammatory diseases. The protein contains a 25-residue *N*-terminal tag (MSYHHHHHHDYDIPTTENLYFQGAM); mass spectrometry analysis of similar constructs containing the same tag have shown it to be especially prone to 6-phosphogluconoylation (6PGL). Upto this point, *N*-terminal adducts had only been detected by mass spectrometry of protein (an increase in mass of 258 Da, representing 6PGL, and/or an increase in mass of 178 Da corresponding to the D-gluconolactone adduct, from dephosphorylation).

Purified protein was exposed to gluconolactone, and its electrophoretic analysis was performed at various time intervals by standard PAGE and protein-boron assisted saccharide electrophoresis (coined Pro-BASE) (Figure 16).

Even after <1 min of protein incubation with gluconolactone, Pro-BASE reveals a new band, which is almost indistinguishable by a normal electrophoresis experiment. The band grew more intense over 16 h, as shown in the center of Figure 16(a) and (b) lanes in the Pro-BASE gel. What appears as a small shadow to the main 16.5-kDa band in a normal experiment has an increased apparent or *virtual* molecular weight in the Pro-BASE system.

The term *virtual molecular weight* is used to describe the apparent molecular weight against the molecular weight standard ladder (right-hand lane) and the term *relative virtual molecular weight* is the relative apparent mass increase due to boron incorporation, in this case the virtual molecular weight is $\sim$60 kDa corresponding to an almost fourfold increase in the apparent molecular weight. Mass spectroscopic analysis indicates that the new band is indeed the monogluconoylated adduct and further experiments over a range of boron monomer inclusion percentages confirmed that its retention (or *virtual molecular weight*) is proportional to the boron content in the gel, confirming that the new band is directly modulated by boron. The team then went on to describe how glycated and glycosylated proteins (nonenzymatic and enzymatic addition of sugars to proteins, respectively) could also be distinguished by this technique as the molecular construction of the linkage to the protein is different, the sugar part's interaction with the boron in the gel is also different and hence separation is possible.[20]

7 CONCLUSIONS

This chapter highlights some of the most interesting and important developments in boronic acid-based sensor design. The chapter is by no means a comprehensive treatise but does contain reference to a number of excellent reviews in the area should the reader wish to delve further into the molecular recognition of boronic acids.

> *"There is no harm in doubt and skepticism, for it is through these that new discoveries are made."*
> Richard Feynman (1918–1988).

REFERENCES

1. T. D. James, K. R. A. S. Sandanayake, and S. Shinkai, *Angew. Chem. Int. Ed. Engl.*, 1996, **35**, 1911–1922.
2. T. D. James, P. Linnane, and S. Shinkai, *Chem. Commun.*, 1996, 281–288.
3. T. D. James and S. Shinkai, *Top. Curr. Chem.*, 2002, **218**, 159–200.
4. S. Shinkai and M. Takeuchi, *Bull. Chem. Soc. Jpn.*, 2005, **78**, 40–51.
5. J. Yan, H. Fang, and B. Wang, *Med. Res. Rev.*, 2005, **25**, 490–520.
6. A. P. Davis and T. D. James, in *Functional Synthetic Receptors*, eds. T. Schrader and A. Hamilton, Wiley-VCH, Weinheim, 2005, pp. 45–109.
7. T. D. James, in *Boronic Acids*, ed. D. G. Hall, Wiley-VCH, Weinheim, 2005, pp. 441–480.
8. T. D. James, M. D. Phillips, and S. Shinkai, *Boronic Acids in Saccharide Recognition*, RSC, Cambridge, 2006.
9. T. D. James, *Top. Curr. Chem.*, 2007, **277**, 107–152.
10. H. S. Mader and O. S. Wolfbeis, *Microchim. Acta*, 2008, **162**, 1–34.
11. S. Jin, Y. F. Cheng, S. Reid, *et al.*, *Med. Res. Rev.*, 2010, **30**, 171–257.
12. T. W. Hudnall, C.-W. Chiu, and F. P. Gabbai, *Acc. Chem. Res.*, 2009, **42**, 388–397.
13. E. Galbraith and T. D. James, *Chem. Soc. Rev.*, 2010, **39**, 3831–3842.
14. R. Nishiyabu, Y. Kubo, T. D. James, and J. S. Fossey, *Chem. Commun.*, 2011, **47**, 1106–1123.
15. Y. Pérez-Fuertes, A. M. Kelly, J. S. Fossey, *et al.*, *Nat. Protoc.*, 2008, **3**, 210–214.
16. A. M. Kelly, Y. Pérez-Fuertes, J. S. Fossey, *et al.*, *Nat. Protoc.*, 2008, **3**, 215–219.
17. Y. Perez-Fuertes, A. M. Kelly, A. L. Johnson, *et al.*, *Org. Lett.*, 2006, **8**, 609–612.
18. A. M. Kelly, Y. Perez-Fuertes, S. Arimori, *et al.*, *Org. Lett.*, 2006, **8**, 1971–1974.
19. T. R. Jackson, J. S. Springall, D. Rogalle, *et al.*, *Electrophoresis*, 2008, **29**, 4185–4191.
20. M. P. Pereira Morais, J. D. Mackay, S. K. Bhamra, *et al.*, *Proteomics*, 2010, **10**, 48–58.
21. K. Severin, *Dalton Trans.*, 2009, 5254–5264.
22. K. Kataoka, S. Okuyama, T. Minami, *et al.*, *Chem. Commun.*, 2009, 1682–1684.

23. K. Kataoka, T. D. James, and Y. Kubo, *J. Am. Chem. Soc.*, 2007, **129**, 15126–15127.
24. N. Fujita, S. Shinkai, and T. D. James, *Chem. Asian J.*, 2008, **3**, 1076–1091.
25. R. Nishiyabu, Y. Kubo, T. D. James, and J. S. Fossey, *Chem. Commun.*, 2011, **47**, 1124–1150.
26. J. P. Lorand and J. O. Edwards, *J. Org. Chem.*, 1959, **24**, 769–774.
27. L. I. Bosch, T. M. Fyles, and T. D. James, *Tetrahedron*, 2004, **60**, 11175–11190.
28. J. D. Larkin, J. S. Fossey, T. D. James, *et al.*, *J. Phys. Chem. A*, 2010, **114**, 12531.
29. L. Zhu, S. H. Shabbir, M. Gray, *et al.*, *J. Am. Chem. Soc.*, 2006, **128**, 1222–1232.
30. G. Wulff, *Pure Appl. Chem.*, 1982, **54**, 2093–2102.
31. H. Höpfl, *J. Organomet. Chem.*, 1999, **581**, 129–149.
32. S. Franzen, W. Ni, and B. Wang, *J. Phys. Chem. B*, 2003, **107**, 12942–12948.
33. W. J. Ni, G. Kaur, G. Springsteen, *et al.*, *Bioorg. Chem.*, 2004, **32**, 571–581.
34. S. L. Wiskur, J. J. Lavigne, H. Ait-Haddou, *et al.*, *Org. Lett.*, 2001, **3**, 1311–1314.
35. D. F. Shriver and M. J. Biallas, *J. Am. Chem. Soc.*, 1967, **89**, 1078–1081.
36. A. P. Davis and R. S. Wareham, *Angew. Chem. Int. Ed.*, 1999, **38**, 2978–2996.
37. E. Klein, M. P. Crump, and A. P. Davis, *Angew. Chem. Int. Ed.*, 2005, **44**, 298–302.
38. Y. Ferrand, M. P. Crump, and A. P. Davis, *Science*, 2007, **318**, 619–622.
39. Y. Ferrand, E. Klein, N. P. Barwell, *et al.*, *Angew. Chem. Int. Ed.*, 2009, **48**, 1775–1779.
40. N. P. Barwell, M. P. Crump, and A. P. Davis, *Angew. Chem. Int. Ed.*, 2009, **48**, 7673–7676.
41. A. P. Davis, *Org. Biomol. Chem.*, 2009, **7**, 3629–3638.
42. R. Badugu, J. R. Lakowicz, and C. D. Geddes, *Bioorg. Med. Chem.*, 2004, **13**, 113–119.
43. J. Yoon and A. W. Czarnik, *J. Am. Chem. Soc.*, 1992, **114**, 5874–5875.
44. T. D. James, K. R. A. S. Sandanayake, R. Iguchi, and S. Shinkai, *J. Am. Chem. Soc.*, 1995, **117**, 8982–8987.
45. T. D. James, K. R. A. S. Sandanayake, and S. Shinkai, *Angew. Chem. Int. Ed. Engl.*, 1994, **33**, 2207–2209.
46. T. D. James, K. R. A. S. Sandanayake, and S. Shinkai, *Nature*, 1995, **374**, 345–347.
47. J. Yoon and A. W. Czarnik, *Bioorg. Med. Chem.*, 1993, **1**, 267–271.
48. N. DiCesare and J. R. Lakowicz, *J. Phys. Chem. A*, 2001, **105**, 6834–6840.
49. N. DiCesare and J. R. Lakowicz, *J. Photochem. Photobiol., A*, 2001, **143**, 39–47.
50. T. D. James, K. R. A. S. Sandanayake and S. Shinkai, *J. Chem. Soc., Chem. Commun.*, 1994, 477–478.
51. H. Cao, D. I. Diaz, N. DiCesare, *et al.*, *Org. Lett.*, 2002, **4**, 1503–1505.
52. Z. Cao, P. Nandhikonda, and M. D. Heagy, *J. Org. Chem.*, 2009, **74**, 3544–3546.
53. S. Trupp, A. Schweitzer, and G. J. Mohr, *Org. Biomol. Chem.*, 2006, **4**, 2965–2968.
54. J. F. Wang, S. Jin, S. Akay, and B. H. Wang, *Eur. J. Org. Chem.*, 2007, **13**, 2091–2099.
55. S. Jin, J. F. Wang, M. Y. Li, and B. H. Wang, *Chem.—Eur. J.*, 2008, **14**, 2795–2804.
56. S. Trupp, A. Schweitzer, and G. J. Mohr, *Microchim. Acta*, 2006, **153**, 127–131.
57. F. Han, L. N. Chi, X. F. Liang, *et al.*, *J. Org. Chem.*, 2009, **74**, 1333–1336.
58. X. Zhang, L. N. Chi, S. M. Ji, *et al.*, *J. Am. Chem. Soc.*, 2009, **131**, 17452–17463.
59. X. Zhang, Y. B. Wu, S. M. Ji, *et al.*, *J. Org. Chem.*, 2010, **75**, 2578–2588.
60. H. Eggert, J. Frederiksen, C. Morin, and J. C. Norrild, *J. Org. Chem.*, 1999, **64**, 3846–3852.
61. J. C. Norrild and H. Eggert, *J. Am. Chem. Soc.*, 1995, **117**, 1479–1484.
62. W. Yang, H. He, and D. G. Drueckhammer, *Angew. Chem. Int. Ed.*, 2001, **40**, 1714–1718.
63. J. Z. Zhao, T. M. Fyles, and T. D. James, *Angew. Chem. Int. Ed.*, 2004, **43**, 3461–3464.
64. C. W. Gray and T. A. Houston, *J. Org. Chem.*, 2002, **67**, 5426–5428.
65. X. F. Liang, T. D. James, and J. Z. Zhao, *Tetrahedron*, 2008, **64**, 1309–1315.
66. J. Zhao, M. G. Davidson, M. F. Mahon, *et al.*, *J. Am. Chem. Soc.*, 2004, **126**, 16179–16186.
67. J. Z. Zhao and T. D. James, *J. Mater. Chem.*, 2005, **15**, 2896–2901.
68. L. Chi, J. Z. Zhao, and T. D. James, *J. Org. Chem.*, 2008, **73**, 4684–4687.
69. D. K. Scrafton, J. E. Taylor, M. F. Mahon, *et al.*, *J. Org. Chem.*, 2008, **73**, 2871–2874.
70. S.-L. Zheng, S. Reid, N. Lin, and B. Wang, *Tetrahedron Lett.*, 2006, **47**, 2331–2335.
71. N. Lin, J. Yan, Z. Huang, *et al.*, *Nucleic Acids Res.*, 2007, **35**, 1222–1229.
72. C. Dai, Y. Cheng, J. Cui, and B. Wang, *Molecules*, 2010, **15**, 5768–5781.
73. F. D'Hooge, D. Rogalle, M. J. Thatcher, *et al.*, *Polymer*, 2008, **49**, 3362–3365.
74. J. R. White, G. J. Price, S. Schiffers, *et al.*, *Tetrahedron Lett.*, 2010, **51**, 3913–3917.
75. S. Y. Xu, Y. B. Ruan, X. X. Luo, *et al.*, *Chem. Commun.*, 2010, **46**, 5864–5866.
76. S. Arimori, M. L. Bell, C. S. Oh, *et al.*, *Chem. Commun.*, 2001, 1836–1837.
77. S. Arimori, M. L. Bell, C. S. Oh, *et al.*, *J. Chem. Soc., Perkin Trans. 1*, 2002, 803–808.

78. S. Arimori, M. L. Bell, C. S. Oh, and T. D. James, *Org. Lett.*, 2002, **4**, 4249–4251.
79. S. Arimori, M. D. Phillips, and T. D. James, *Tetrahedron Lett.*, 2004, **45**, 1539–1542.
80. M. D. Phillips and T. D. James, *J. Fluoresc.*, 2004, **14**, 549–559.
81. M. D. Phillips, T. M. Fyles, N. P. Barwell and T. D. James, *Chem. Commun.*, 2009, 6557–6559.
82. W. Yang, H. Fan, X. Gao, *et al.*, *Chem. Biol.*, 2004, **11**, 439–448.
83. W. Yang, S. Gao, X. Gao, *et al.*, *Bioorg. Med. Chem. Lett.*, 2002, **12**, 2175–2177.
84. V. V. Karnati, X. Gao, S. Gao, *et al.*, *Bioorg. Med. Chem. Lett.*, 2002, **12**, 3373–3377.
85. G. Kaur, H. Fang, X. M. Gao, *et al.*, *Tetrahedron*, 2006, **62**, 2583–2589.
86. D. Stones, S. Manku, X. Lu, and D. G. Hall, *Chem.—Eur. J.*, 2004, **10**, 92–100.
87. A. Pal, M. Berube, and D. G. Hall, *Angew. Chem. Int. Ed.*, 2010, **49**, 1492–1495.
88. Z. Sharrett, S. Gamsey, L. Hirayama, *et al.*, *Org. Biomol. Chem.*, 2009, **7**, 1461–1470.
89. S. Gamsey, A. Miller, M. M. Olmstead, *et al.*, *J. Am. Chem. Soc.*, 2007, **129**, 1278–1286.
90. D. B. Cordes, S. Gamsey, and B. Singaram, *Angew. Chem. Int. Ed.*, 2006, **45**, 3829–3832.
91. J. T. Suri, D. B. Cordes, F. E. Cappuccio, *et al.*, *Angew. Chem. Int. Ed.*, 2003, **42**, 5857–5859.
92. M. Berube, M. Dowlut, and D. G. Hall, *J. Org. Chem.*, 2008, **73**, 6471–6479.
93. M. D. Sørensen, R. Martins, and O. Hindsgaul, *Angew. Chem. Int. Ed.*, 2007, **46**, 2403–2407.
94. N. Y. Edwards, T. W. Sager, J. T. McDevitt, and E. V. Anslyn, *J. Am. Chem. Soc.*, 2007, **129**, 13575–13583.
95. Y. Zou, D. L. Broughton, K. L. Bicker, *et al.*, *Chembiochem*, 2007, **8**, 2048–2051.
96. P. J. Duggan and D. A. Offermann, *Aust. J. Chem.*, 2007, **60**, 829–834.
97. A. P. Russell, Photometric method and means involving dyes for detecting vicinal polyhydroxyl compounds. WO 91/04488, 1991.
98. T. Nagasaki, H. Shinmori, and S. Shinkai, *Tetrahedron Lett.*, 1994, **35**, 2201–2204.
99. M. Takeuchi, M. Taguchi, H. Shinmori, and S. Shinkai, *Bull. Chem. Soc. Jpn.*, 1996, **69**, 2613–2618.
100. K. R. A. S. Sandanayake and S. Shinkai, *J. Chem. Soc., Chem. Commun.*, 1994, 1083–1084.
101. H. Shinmori, M. Takeuchi, and S. Shinkai, *J. Chem. Soc., Perkin Trans. 2*, 1998, 847–852.
102. K. Koumoto and S. Shinkai, *Chem. Lett.*, 2000, 856–857.
103. K. Koumoto, M. Takeuchi, and S. Shinkai, *Supramol. Chem.*, 1998, **9**, 203.
104. C. J. Ward, P. Patel, P. R. Ashton, and T. D. James, *Chem. Commun.*, 2000, 229–230.
105. C. J. Ward, P. Patel, and T. D. James, *J. Chem. Soc., Perkin Trans. 1*, 2002, 462–470.
106. N. DiCesare and J. R. Lakowicz, *Org. Lett.*, 2001, **3**, 3891–3893.
107. H. Shinmori, M. Takeuchi, and S. Shinkai, *J. Chem. Soc., Perkin Trans. 2*, 1996, 1–3.
108. C. J. Davis, P. T. Lewis, M. E. McCarroll, *et al.*, *Org. Lett.*, 1999, **1**, 331–334.
109. O. Rusin, O. Alpturk, M. He, *et al.*, *J. Fluoresc.*, 2004, **14**, 611–615.
110. P. T. Lewis, C. J. Davis, L. A. Cabell, *et al.*, *Org. Lett.*, 2000, **2**, 589–592.
111. C. J. Ward, P. Patel, and T. D. James, *Org. Lett.*, 2002, **4**, 477–479.
112. K. Sato, A. Sone, S. Arai, and T. Yamagishi, *Heterocycles*, 2003, **61**, 31–38.
113. W. Ni, H. Fang, G. Springsteen, and B. Wang, *J. Org. Chem.*, 2004, **69**, 1999–2007.
114. Y. Egawa, R. Gotoh, S. Niina, and J. Anzai, *Bioorg. Med. Chem. Lett.*, 2007, **17**, 3789–3792.
115. Y. Egawa, R. Gotoh, T. Seki, and J. Anzai, *Mater. Sci. Eng., C*, 2009, **29**, 115–118.
116. S. L. Wiskur, H. Ait-Haddou, J. J. Lavigne, and E. V. Anslyn, *Acc. Chem. Res.*, 2001, **34**, 963–972.
117. S. C. McCleskey, P. N. Floriano, S. L. Wiskur, *et al.*, *Tetrahedron*, 2003, **59**, 10089–10092.
118. A. Goodey, J. J. Lavigne, S. M. Savoy, *et al.*, *J. Am. Chem. Soc.*, 2001, **123**, 2559–2570.
119. Y. S. Sohn, A. Goodey, E. V. Anslyn, *et al.*, *Biosens. Bioelectron.*, 2005, **21**, 303–312.
120. A. P. Umali, E. V. Anslyn, A. T. Wright, *et al.*, *J. Chem. Educ.*, 2010, **87**, 832–835.
121. B. T. Nguyen and E. V. Anslyn, *Coord. Chem. Rev.*, 2006, **250**, 3118–3127.
122. T. Z. Zhang and E. V. Anslyn, *Org. Lett.*, 2007, **9**, 1627–1629.
123. T. Z. Zhang and E. V. Anslyn, *Org. Lett.*, 2006, **8**, 1649–1652.
124. A. Buryak and K. Severin, *Angew. Chem. Int. Ed.*, 2004, **43**, 4771–4774.
125. Z. Zhong and E. V. Anslyn, *J. Am. Chem. Soc.*, 2002, **124**, 9014–9015.
126. L. A. Cabell, M. K. Monahan, and E. V. Anslyn, *Tetrahedron Lett.*, 1999, **40**, 7753–7756.
127. A. M. Piatek, Y. J. Bomble, S. L. Wiskur, and E. V. Anslyn, *J. Am. Chem. Soc.*, 2004, **126**, 6072–6077.
128. S. L. Wiskur and E. V. Anslyn, *J. Am. Chem. Soc.*, 2001, **123**, 10109–10110.
129. S. L. Wiskur, P. N. Floriano, E. V. Anslyn, and J. T. McDevitt, *Angew. Chem. Int. Ed.*, 2003, **42**, 2070–2072.
130. B. T. Nguyen, S. L. Wiskur, and E. V. Anslyn, *Org. Lett.*, 2004, **6**, 2499–2501.
131. L. Zhu and E. V. Anslyn, *J. Am. Chem. Soc.*, 2004, **126**, 3676–3677.

132. L. Zhu, Z. Zhong, and E. V. Anslyn, *J. Am. Chem. Soc.*, 2005, **127**, 4260–4269.
133. Z. Murtaza, L. Tolosa, P. Harms, and J. R. Lakowicz, *J. Fluoresc.*, 2002, **12**, 187–192.
134. J. N. Camara, J. T. Suri, F. E. Cappuccio, *et al.*, *Tetrahedron Lett.*, 2002, **43**, 1139–1141.
135. J. T. Suri, D. B. Cordes, F. E. Cappuccio, *et al.*, *Langmuir*, 2003, **19**, 5145–5152.
136. F. E. Cappuccio, J. T. Suri, D. B. Cordes, *et al.*, *J. Fluoresc.*, 2004, **14**, 521–533.
137. D. B. Cordes, A. Miller, S. Gamsey, and B. Singaram, *Anal. Bioanal. Chem.*, 2007, **387**, 2767–2773.
138. D. B. Cordes, S. Gamsey, Z. Sharrett, *et al.*, *Langmuir*, 2005, **21**, 6540–6547.
139. D. B. Cordes, A. Miller, S. Gamsey, *et al.*, *Org. Biomol. Chem.*, 2005, **3**, 1708–1713.
140. N. DiCesare, M. R. Pinto, K. S. Schanze, and J. R. Lakowicz, *Langmuir*, 2002, **18**, 7785–7787.
141. G. Springsteen and B. Wang, *Chem. Commun.*, 2001, 1608–1609.
142. G. Springsteen and B. Wang, *Tetrahedron*, 2002, **58**, 5291–5300.
143. J. Yan, G. Springsteen, S. Deeter, and B. Wang, *Tetrahedron*, 2004, **60**, 11205–11209.
144. S. Arimori, C. J. Ward, and T. D. James, *Chem. Commun.*, 2001, 2018–2019.
145. S. Boduroglu, J. M. El Khoury, D. Venkat Reddy, *et al.*, *Bioorg. Med. Chem. Lett.*, 2005, **15**, 3974–3977.
146. S. Arimori, C. J. Ward, and T. D. James, *Tetrahedron Lett.*, 2002, **43**, 303–305.
147. H. R. Mulla, N. J. Agard, and A. Basu, *Bioorg. Med. Chem. Lett.*, 2004, **14**, 25–27.
148. B. Elmas, S. Senel, and A. Tuncel, *React. Funct. Polym.*, 2007, **67**, 87–96.
149. W. M. J. Ma, M. P. Pereira Morais, F. D'Hooge, *et al.*, *Chem. Commun.*, 2009, 532–534.
150. Y. Kubo, A. Kobayashi, T. Ishida, *et al.*, *Chem. Commun.*, 2005, 2846–2848.
151. S. A. Elfeky, S. E. Flower, N. Masumoto, *et al.*, *Chem. Asian J.*, 2010, **5**, 581–588.
152. S. A. Elfeky, F. D'Hooge, L. Poncel, *et al.*, *New J. Chem.*, 2009, **33**, 1466–1469.
153. G. S. Wilson and H. Yibai, *Chem. Rev.*, 2000, **100**, 2693–2704.
154. A. Ori and S. Shinkai, *J. Chem. Soc., Chem. Commun.*, 1995, 1771–1772.
155. A. N. J. Moore and D. D. M. Wayner, *Can. J. Chem.*, 1999, **77**, 681–686.
156. D. Kato, S. Iijima, R. Kurita, *et al.*, *Biosens. Bioelectron.*, 2007, **22**, 1527–1531.
157. M. Nicolas, B. Fabre, and J. Simonet, *Electrochim. Acta*, 2001, **46**, 1179–1190.
158. S. Arimori, S. Ushiroda, L. M. Peter, *et al.*, *Chem. Commun.*, 2002, 2368–2369.
159. J. C. Norrild and I. Sotofte, *J. Chem. Soc., Perkin Trans. 2*, 2002, 303–311.
160. N. Katif, R. A. Harries, A. M. Kelly, *et al.*, *J. Solid State Electrochem.*, 2009, **13**, 1475–1482.
161. Y. J. Huang, Y. B. Jiang, J. S. Fossey, *et al.*, *J. Mater. Chem.*, 2010, **20**, 8305–8310.
162. M. J. Bonne, E. Galbraith, T. D. James, *et al.*, *J. Mater. Chem.*, 2010, **20**, 588–594.
163. G. Mirri, S. D. Bull, P. N. Horton, *et al.*, *J. Am. Chem. Soc.*, 2010, **132**, 8903–8905.
164. A. M. Kelly, S. D. Bull, and T. D. James, *Tetrahedron: Asymmetry*, 2008, **19**, 489–494.
165. S. L. Yeste, M. E. Powell, S. D. Bull, and T. D. James, *J. Org. Chem.*, 2009, **74**, 427–430.
166. M. E. Powell, A. M. Kelly, S. D. Bull, and T. D. James, *Tetrahedron Lett.*, 2009, **50**, 876–879.
167. E. Galbraith, A. M. Kelly, J. S. Fossey, *et al.*, *New J. Chem.*, 2009, **33**, 181–185.
168. H. E. Katz, *J. Org. Chem.*, 1985, **50**, 5027–5032.
169. H. E. Katz, *J. Am. Chem. Soc.*, 1986, **108**, 7640–7645.
170. M. T. Reetz, C. M. Niemeyer, and K. Harms, *Angew. Chem. Int. Ed. Engl.*, 1991, **30**, 1472–1474.
171. S. Jacobson and R. Pizer, *J. Am. Chem. Soc.*, 1993, **115**, 11216–11221.
172. C. Dusemund, K. R. A. S. Sandanayake, and S. Shinkai, *J. Chem. Soc., Chem. Commun.*, 1995, 333–334.
173. S. Aldridge, C. Bresner, I. A. Fallis, *et al.*, *Chem. Commun.*, 2002, 740–741.
174. C. Bresner, S. Aldridge, I. Fallis, *et al.*, *Angew. Chem. Int. Ed.*, 2005, **44**, 3606–3609.
175. H. Yamamoto, A. Ori, K. Ueda, *et al.*, *Chem. Commun.*, 1996, 407–408.
176. C. R. Cooper, N. Spencer, and T. D. James, *Chem. Commun.*, 1998, 1365–1366.
177. K. M. K. Swamy, Y. J. Lee, H. N. Lee, *et al.*, *J. Org. Chem.*, 2006, **71**, 8626–8628.
178. S. Arimori, M. G. Davidson, T. M. Fyles, *et al.*, *Chem. Commun.*, 2004, 1640–1641.
179. E. Galbraith, T. M. Fyles, F. Marken, *et al.*, *Inorg. Chem.*, 2008, **47**, 6236–6244.
180. C. Bresner, J. K. Day, N. D. Coombs, *et al.*, *Dalton Trans.*, 2006, 3660–3667.
181. S. Yamaguchi, S. Akiyama, and K. Tamao, *J. Am. Chem. Soc.*, 2001, **123**, 11372–11375.
182. Y. Kubo, M. Yamamoto, M. Ikeda, *et al.*, *Angew. Chem. Int. Ed.*, 2003, **42**, 2036–2040.
183. G. Xu and M. A. Tarr, *Chem. Commun.*, 2004, 1050–1051.
184. S. Sole and F. P. Gabbai, *Chem. Commun.*, 2004, 1284–1285.
185. Y. Kubo, T. Ishida, A. Kobayashi, and T. D. James, *J. Mater. Chem.*, 2005, **15**, 2889–2895.
186. Y. Kubo, T. Ishida, T. Minami, and T. D. James, *Chem. Lett.*, 2006, **35**, 996–997.

187. Z. Q. Liu, M. Shi, F. Y. Li, *et al.*, *Org. Lett.*, 2005, **7**, 5481–5484.

188. X. Y. Liu, D. R. Bai, and S. N. Wang, *Angew. Chem. Int. Ed.*, 2006, **45**, 5475–5478.

189. Y. Kim and F. P. Gabbai, *J. Am. Chem. Soc.*, 2009, **131**, 3363–3369.

190. M. Melaimi and F. P. Gabbai, *J. Am. Chem. Soc.*, 2005, **127**, 9680–9681.

191. M. H. Lee and F. P. Gabbai, *Inorg. Chem.*, 2007, **46**, 8132–8138.

192. C. W. Chiu and F. P. Gabbai, *J. Am. Chem. Soc.*, 2006, **128**, 14248–14249.

193. A. Kawachi, A. Tani, J. P. Shimada, and Y. Yamamoto, *J. Am. Chem. Soc.*, 2008, **130**, 4222–4223.

194. S. J. M. Koskela, T. M. Fyles, and T. D. James, *Chem. Commun.*, 2005, 945–947.

195. Z. C. Xu, S. K. Kim, S. J. Han, *et al.*, *Eur. J. Org. Chem.*, 2009, **18**, 3058–3065.

196. T. W. Hudnall and F. P. Gabbai, *J. Am. Chem. Soc.*, 2007, **129**, 11978–11986.

197. A. Nonaka, S. Horie, T. D. James, and Y. Kubo, *Org. Biomol. Chem.*, 2008, **6**, 3621–3625.